ENCYCLOPEDIA *of* NATURAL HAZARDS

Encyclopedia of Earth Sciences Series

ENCYCLOPEDIA OF NATURAL HAZARDS

Volume Editor
Peter Bobrowsky is an Adjunct Professor at the Center for Natural Hazards Research, Department of Earth Sciences, Simon Fraser University, Burnaby, BC, Canada (ptbobrow@sfu.ca). With over 30 years of professional experience as an environmental and engineering geologist he has worked in Africa, China, India, North America, Middle East and South America. He has published extensively on a variety of subjects and has served/serves on a number of bodies and organizations: Secretary General of IUGS, President of the Geological Association of Canada, President of the Canadian Quaternary Association, Vice President of the International Consortium on Landslides, editorial board for Landslides Quaternary International and several others. The project to compile and publish this volume was completed during his tenure as SG of IUGS.

Editorial Board

Pedro Basabe
UN/ISDR Africa Programme
United Nations Complex, Gigiri
PO Box 47074
Nairobi, Kenya

Tom Beer
Australian Commonwealth Scientific and Industrial Research Organisation (CSIRO)
PB1, Aspendale
Victoria, Australia

Norm Catto
Department of Geography
Memorial University
St. John's, Newfoundland, Canada

Viacheslav K. Gusiakov
Siberian Division, Russian Academy of Sciences
Pr.Lavrentieva, 6
Novosibirsk, Russia

Michael K. Lindell
Hazard Reduction & Recovery Center
Texas A&M University
College Station, Texas, USA

Bill McGuire
Department of Earth Sciences
University College London
136 Gower Street (Lewis Building)
London, UK

Jay Melosh
Department of Earth and Atmospheric Sciences
Civil Engineering Building, Room 3237
550 Stadium Mall Drive
Purdue University
West Lafayette, Indiana, USA

Farrokh Nadim
Norwegian Geotechnical Institute (NGI)
PO Box 3930 Ullevall Stadion
Oslo, Norway

Paul Slovic
Decision Research
1201 Oak Street, Suite 200
Eugene, Oregon, USA

Philipp Schmidt-Thomé
Geological Survey of Finland
PO Box 96
02151 Espoo, Finland

Aims of the Series
The *Encyclopedia of Earth Sciences Series* provides comprehensive and authoritative coverage of all the main areas in the Earth Sciences. Each volume comprises a focused and carefully chosen collection of contributions from leading names in the subject, with copious illustrations and reference lists.

These books represent one of the world's leading resources for the Earth Sciences community. Previous volumes are being updated and new works published so that the volumes will continue to be essential reading for all professional earth scientists, geologists, geophysicists, climatologists, and oceanographers as well as for teachers and students. See the back of this volume for a current list of titles in the *Encyclopedia of Earth Sciences Series*. Go to http://www.springerlink.com/reference-works/ to visit the "Earth Sciences Series" online.

About the Series Editor
Professor Charles W. Finkl has edited and/or contributed to more than eight volumes in the Encyclopedia of Earth Sciences Series. For the past 28 years he has been the Executive Director of the Coastal Education & Research Foundation and Editor-in-Chief of the international *Journal of Coastal Research*. In addition to these duties, he is Professor Emeritus at Florida Atlantic University in Boca Raton, Florida, USA. He is a graduate of the University of Western Australia (Perth) and previously worked for a wholly owned Australian subsidiary of the International Nickel Company of Canada (INCO). During his career, he acquired field experience in Australia; the Caribbean; South America; SW Pacific islands; southern Africa; Western Europe; and the Pacific Northwest, Midwest, and Southeast USA.

Founding Series Editor
Professor Rhodes W. Fairbridge (deceased) has edited more than 24 Encyclopedias in the Earth Sciences Series. During his career he has worked as a petroleum geologist in the Middle East, been a WW II intelligence officer in the SW Pacific and led expeditions to the Sahara, Arctic Canada, Arctic Scandinavia, Brazil and New Guinea. He was Emeritus Professor of Geology at Columbia University and was affiliated with the Goddard Institute for Space Studies.

ENCYCLOPEDIA OF EARTH SCIENCES SERIES

ENCYCLOPEDIA *of* NATURAL HAZARDS

edited by

PETER T. BOBROWSKY
Simon Fraser University
Canada

Springer Reference

Library of Congress Control Number: 2012944445

ISBN: 978-90-481-8699-0
This publication is available also as:
Electronic publication under ISBN 978-1-4020-4399-4 and
Print and electronic bundle under ISBN 978-94-007-0263-9

Springer Dordrecht, Heidelberg, New York, London

Printed on acid-free paper

Cover photo: Volcanic Lava, Hawaii. Photo reproduced courtesy of Paul Souders.

Every effort has been made to contact the copyright holders of the figures and tables which have been reproduced from other sources. Anyone who has not been properly credited is requested to contact the publishers, so that due acknowledgement may be made in subsequent editions.

All rights reserved for the contributions *Communicating Emergency Information*; *Dose Rate*; *Harmonic Tremor*; *Natural Radioactivity*; *Pore-Water Pressure*; *Radon Hazards*

© Springer Science+Business Media Dordrecht 2013

No part of this work may be reproduced, stored in a retrieval system, or transmitted in any form or by any means, electronic, mechanical, photocopying, microfilming, recording or otherwise, without written permission from the Publisher, with the exception of any material supplied specifically for the purpose of being entered and executed on a computer system, for exclusive use by the purchaser of the work.

Contents

Contributors	xv	Asteroid Impact Predictions *Brian G. Marsden (Deceased)*	29
Preface	xxxix		
Acknowledgments	xli	Automated Local Evaluation in Real Time (ALERT) *Lev I. Dorman*	31
AA-LAVA *Robert Buchwaldt*	1		
Accelerometer *Zhengwen Zeng and Lin Fa*	2	Avalanches *Chris Stethem*	31
Acid Rain *Mary J. Thornbush*	2	Aviation (Hazards to) *Thomas Gerz and Ulrich Schumann*	34
Adaptation *Philipp Schmidt-Thomé and Sirkku Juhola*	3	Avulsion *Joann Mossa*	40
Airphoto and Satellite Imagery *J. D. Mollard*	5	Base Surge *Catherine J. Hickson*	41
Albedo *Alan W. Harris*	9	Beach Nourishment (Replenishment) *Charles W. Finkl*	42
Antecedent Conditions *Michael James Crozier, Nick Preston and Thomas Glade*	10	Beaufort Wind Scale *Tom Beer*	42
		Biblical Events *Jerry T. Mitchell*	45
Arsenic in Groundwater *Arindam Basu*	13	Body Wave *Zhengwen Zeng and Lin Fa*	47
Asteroid *Alan W. Harris*	14	Breakwaters *Giovanni Cuomo*	48
Asteroid Impact *Christian Koeberl*	18	Building Code *Rohit Jigyasu*	49
Asteroid Impact Mitigation *Clark R. Chapman*	28	Building Failure *Tiziana Rossetto*	50

Buildings, Structures, and Public Safety *John M. Logan*	51	Cost-Benefit Analysis of Natural Hazard Mitigation *Sven Fuchs*	121
Calderas *James W. Cole*	57	Costs (Economic) of Natural Hazards and Disasters *Howard Kunreuther and Erwann Michel-Kerjan*	125
Casualties Following Natural Hazards *Kerrianne Watt and Philip Weinstein*	59	Creep *Piotr Migoń*	129
Challenges to Agriculture *Julie A. March*	65	Critical Incident Stress Syndrome *Ann M. Mitchell and Kirstyn Kameg*	130
Civil Protection and Crisis Management *Scira Menoni and Antonio Pugliano*	69	Critical Infrastructure *Susanne Krings*	132
Classification of Natural Disasters *Thomas Glade and David E. Alexander*	78	Cryological Engineering *Lukas U. Arenson and Sarah M. Springman*	132
Climate Change *Jasper Knight*	82	Cultural Heritage and Natural Hazards *Piotr Migoń*	135
Cloud Seeding *Steven T. Siems*	92	Damage and the Built Environment *Adriana Galderisi and Andrea Ceudech*	141
Coal Fire (Underground) *Glenn B. Stracher*	92	Debris Avalanche *Marco Giardino*	145
Coastal Erosion *Wayne Stephenson*	94	Debris Flow *Oldrich Hungr*	149
Coastal Zone Risk Management *Norm Catto*	97	Deep-seated Gravitational Slope Deformation *Mauro Soldati*	151
Cognitive Dissonance *Jaroslaw Dzialek*	98	Desertification *Nicholas Lancaster*	155
Collapsing Soil Hazards *Andrew J. Stumpf*	99	Disaster Diplomacy *Ilan Kelman*	158
Comet *Paul R. Weissman*	105	Disaster Relief *Jane Carter Ingram*	159
Communicating Emergency Information *John H. Sorensen*	110	Disaster Research and Policy, History *J. C. Gaillard and Ilan Kelman*	160
Community Management of Natural Hazards *William T. Hartwell*	112	Disaster Risk Management *N. Nirupama*	164
Complexity Theory *William H. K. Lee*	117	Disaster Risk Reduction *Walter J. Ammann*	170
Concrete Structures *Murat Saatcioglu*	118	Disasters *Ian Stewart*	175
Convergence *Ilan Kelman*	118	Dispersive Soil Hazards *Andrew J. Stumpf*	186
Coping Capacity *Virginia R. Burkett*	119		

Doppler Weather Radar *Rodger A. Brown*	188	Emergency Management *Michael K. Lindell*	263
Dose Rate *Cathy Scheib*	188	Emergency Mapping *Frank Fiedrich and Sisi Zlatanova*	272
Drought *Suzanne Hollins and John Dodson*	189	Emergency Planning *Scira Menoni*	276

CASE STUDY

Dust Bowl *Richard Seager and Benjamin I. Cook*	197	Emergency Shelter *Camillo Boano and William Hunter*	280
Dust Devil *Nilton O. Rennó*	201	Epicenter *Valerio Comerci*	284
Dust Storm *Nilton O. Rennó*	202	Epidemiology of Disease in Natural Disasters *Gilbert M. Burnham*	285
Dvorak Classification of Hurricanes *Raymond Zehr*	203	Erosion *Matija Zorn and Blaž Komac*	288
Early Warning Systems *Graham S. Leonard, Chris E. Gregg and David M. Johnston*	207	Erosivity *Matija Zorn and Blaž Komac*	289
Earthquake *John F. Cassidy*	208	Eruption Types (Volcanic Eruptions) *Catherine J. Hickson, T. C. Spurgeon and R. I. Tilling*	290
Earthquake Damage *Nicolas Desramaut, Hormoz Modaressi and Gonéri Le Cozannet*	223	Evacuation *Graham A. Tobin, Burrell E. Montz and Linda M. Whiteford*	293
Earthquake Prediction and Forecasting *Alik T. Ismail-Zadeh*	225	Expansive Soils and Clays *Ghulappa S. Dasog and Ahmet R. Mermut*	297
Earthquake Resistant Design *Tiziana Rossetto and Philippe Duffour*	231	Expert (Knowledge-Based) Systems for Disaster Management *Jean-Marc Tacnet and Corinne Curt*	300
Economic Valuation of Life *Mohammed H. I. Dore and Rajiv G. Singh*	240	Exposure to Natural Hazards *Jörn Birkmann*	305
Economics of Disasters *Pierre-Alain Schieb*	242	Extensometers *Erik Eberhardt*	306
Education and Training for Emergency Preparedness *Kevin R. Ronan*	247	Extinction *Ross D. E. MacPhee*	307
Elastic Rebound Theory *John Ristau*	249	Extreme Value Theory *Gianfausto Salvadori*	310

CASE STUDY

Electromagnetic Radiation (EMR) *Norman Kerle*	250	Eyjafjallajökull Eruptions 2010 *Freysteinn Sigmundsson*	311
El Niño/Southern Oscillation *Michael Ghil and Ilya Zaliapin*	250	Fault *William A. Bryant*	317

Federal Emergency Management Agency (FEMA) *Vincent R. Parisi*	321	**CASE STUDY**	
		Galeras Volcano, Colombia *Barry Voight and Marta L. Calvache*	369
Fetch *Norm Catto*	322	Gas-Hydrates *Harsh K. Gupta and Kalachand Sain*	377
Fire and Firestorms *John Radke*	323	Geographic Information Systems (GIS) and Natural Hazards *Paolo Tarolli and Marco Cavalli*	378
Flash Flood *Yang Hong, Pradeep Adhikari and Jonathan J. Gourley*	324	Geographic Information Technology *Brigitte Leblon*	385
Flood Deposits *János Kovács*	325	Geohazards *Blaž Komac and Matija Zorn*	387
Flood Hazard and Disaster *Yang Hong, Pradeep Adhikari and Jonathan J. Gourley*	326	Geological/Geophysical Disasters *Richard Guthrie*	387
		Glacier Hazards *John J. Clague*	400
Flood Protection *Fernando Nardi*	336	Global Change and Its Implications for Natural Disasters *Gonéri Le Cozannet, Hormoz Modaressi and Nicolas Desramaut*	405
Flood Stage *Fernando Nardi*	336		
Floodplain *Klement Tockner*	337	Global Dust *Edward Derbyshire*	409
Floodway *Armand LaRocque*	338	Global Network of Civil Society Organizations for Disaster Reduction *Terry Gibson*	416
Fog Hazard Mitigation *Steve LaDochy and Michael R. Witiw*	338	Global Positioning Systems (GPS) and Natural Hazards *Norman Kerle*	416
Fog Hazards *Paul J. Croft*	342		
Föhn *Anita Bokwa*	346	Global Seismograph Network (GSN) *Allison Bent*	417
Forest and Range Fires *George Eftychidis*	346	**CASE STUDY**	
		Haiti Earthquake 2010: Psychosocial Impacts *James M. Shultz, Louis Herns Marcelin, Zelde Espinel, Sharon B. Madanes, Andrea Allen and Yuval Neria*	419
Frequency and Magnitude of Events *Lionel E. Jackson, Jr.*	359		
Frost Hazard *Leanne Webb and Richard L. Snyder*	363	Harmonic Tremor *Melanie Kelman*	425
Fujita Tornado Scale *Thomas W. Schmidlin*	366	Hazard *Farrokh Nadim*	425
Fumarole *Travis W. Heggie*	367	Hazard and Risk Mapping *Brian R. Marker*	426

Hazardousness of a Place *Netra Raj Regmi, John Rick Giardino and John D. Vitek*	435	Impact Firestorms *Tamara Goldin*	525
Heat Waves *Gerd Tetzlaff*	447	Impact Tsunamis *Galen Gisler*	525
High-Rise Buildings in Natural Disaster *Murat Saatcioglu*	451	Impact Winter *Owen Brian Toon*	528
Historical Events *Suzanne A. G. Leroy and Raisa Gracheva*	452	Inclinometers *Erik Eberhardt*	529

CASE STUDY

Hospitals in Disaster *Jeffrey N. Rubin*	471	Indian Ocean Tsunami, 2004 *Franck Lavigne, Raphaël Paris, Frédéric Leone, J. C. Gaillard and Julie Morin*	529
Human Impacts of Hazards *Douglas Paton, David Johnston and Sarb Johal*	474	Induced Seismicity *Maurice Lamontagne*	535
Humanity as an Agent of Natural Disasters *Thomas Glade and Andreas Dix*	478	Information and Communication Technology *Peter S. Anderson*	536
Hurricane (Typhoon, Cyclone) *Robert Korty*	481	Insect Hazards *Philip Weinstein*	540

CASE STUDY

Hurricane Katrina *Joann Mossa*	494	Insurance *Jaroslaw Dzialek*	542
Hydrocompaction Subsidence *Andrew J. Stumpf*	496	Integrated Emergency Management System *Frank Fiedrich*	544
Hydrograph, Flood *Fernando Nardi*	497	Intensity Scales *David Giles*	544
Hydrometeorological Hazards *Gordon McBean*	497	International Strategies for Disaster Reduction (IDNDR and ISDR) *Karl-Otto Zentel and Thomas Glade*	552
Hyogo Framework for Action 2005–2015 *Pedro Basabe*	508	Internet, World Wide Web and Natural Hazards *Lucy Stanbrough*	563
Hypocenter *Maurice Lamontagne*	516	Isoseismal *Valerio Comerci*	565
Ice and Icebergs *Norm Catto*	519	Jökulhlaups *Marten Geertsema*	567
Ice Storms *Ronald E. Stewart*	520	Karst Hazards *Viacheslav Andreychouk and Andrzej Tyc*	571
Impact Airblast *Natalia Artemieva*	522		

CASE STUDY

Impact Ejecta *Christian Koeberl*	523	Krakatoa (Krakatau) *Bill McGuire*	576
Impact Fireball *Peter Brown*	524	Lahar *Richard B. Waitt*	579

Land Degradation *Matija Zorn and Blaž Komac*	580	Magma *Catherine J. Hickson, T. C. Spurgeon and R. I. Tilling*	639
Land Subsidence *Brian R. Marker*	583	Magnitude Measures *David Giles*	640
Land Use, Urbanization, and Natural Hazards *Brian Marker*	590	Marginality *Ben Wisner*	651
Landsat Satellite *María Asunción Soriano*	594	Marine Hazards *Tore Jan Kvalstad*	652
Landslide *John J. Clague*	594	Mass Media and Natural Disasters *Wojciech Biernacki*	655
Landslide Dam *Reginald L. Hermanns*	602	Mass Movement *Roy C. Sidle*	657
Landslide Impacts *Michael James Crozier, Nick Preston and Thomas Glade*	606	Megacities and Natural Hazards *Norman Kerle and Annemarie Müller*	660
Landslide Inventory *Javier Hervás*	610	**CASE STUDY**	
Landslide Triggered Tsunami, Displacement Wave *Reginald L. Hermanns, Jean-Sébastien L'Heureux and Lars H. Blikra*	611	Mega-Fires in Greece (2007) *George Eftychidis*	664
		Mercalli, Giuseppe (1850–1914) *Valerio Comerci*	671
Landslide Types *David Cruden*	615	Meteorite *Jay Melosh*	672
Land-Use Planning *Stefan Greiving and Philipp Schmidt-Thomé*	618	Methane Release from Hydrate *Graham Westbrook*	672
Lateral Spreading *Steven L. Kramer*	623	Mining Subsidence Induced Fault Reactivation *Laurance Donnelly*	673
Lava *Robert Buchwaldt*	623	Misconceptions About Natural Disasters *Timothy R. H. Davies*	678
Levee *Joann Mossa*	624	Mitigation *Farrokh Nadim*	682
Lightning *Leopoldo C. Cancio*	625	Modified Mercalli (MM) Scale *Valerio Comerci*	683
Liquefaction *Steven L. Kramer*	629	Monitoring Natural Hazards *Michel Jaboyedoff, Pascal Horton, Marc-Henri Derron, Céline Longchamp and Clément Michoud*	686
Livelihoods and Disasters *J. C. Gaillard*	633		
Loess *János Kovács and György Varga*	637		
Macroseismic Survey *Roger M. W. Musson*	639	Monsoons *Song Yang, Viviane Silva and Wayne Higgins*	696

CASE STUDY

Montserrat Eruptions — Katherine Donovan — 697

Mortality and Injury in Natural Disasters — Shannon Doocy — 699

CASE STUDY

Mt Pinatubo — Katherine Donovan — 703

Mud Volcanoes — Behruz M. Panahi — 705

Mudflow — Christophe Ancey — 706

Myths and Misconceptions in Disasters — Alejandro López Carresi — 706

Natural Hazard — Anita Bokwa — 711

Natural Hazards in Developing Countries — Paolo Paron — 718

Natural Radioactivity — Cathy Scheib — 726

Neotectonics — James P. McCalpin — 730

CASE STUDY

Nevado del Ruiz Volcano, Colombia 1985 — Barry Voight, Marta L. Calvache, Minard L. Hall and Maria Luisa Monsalve — 732

North Anatolian Fault — Thomas Rockwell — 738

Nuée Ardente — Catherine J. Hickson, T. C. Spurgeon and R. I. Tilling — 740

Overgrazing — Norm Catto — 741

Ozone — Tom Beer — 741

Ozone Loss — Mary J. Thornbush — 743

Pacific Tsunami Warning and Mitigation System (PTWS) — Laura S. L. Kong — 747

Pahoehoe Lava — Robert Buchwaldt — 748

Paleoflood Hydrology — Gerardo Benito — 748

Paleoseismology — Alan R. Nelson — 749

Paraglacial — Jasper Knight — 750

Perception of Natural Hazards and Disasters — Jaroslaw Dzialek — 756

Permafrost — Julian B. Murton — 759

Piezometer — Sylvi Haldorsen — 764

Piping Hazard — Michael James Crozier, Nick Preston and Thomas Glade — 764

Planning Measures and Political Aspects — Brian R. Marker — 765

Plate Tectonics — John Ristau — 769

Pore-Water Pressure — Mark E. Reid — 772

Post Disaster Mass Care Needs — Frank Fiedrich, John R. Harrald and Theresa Jefferson — 773

Posttraumatic Stress Disorder (PTSD) — Fran H. Norris — 776

Primary Wave (P-Wave) — Allison Bent — 777

Probable Maximum Flood (PMF) — Armand LaRocque — 777

Probable Maximum Precipitation (PMP) — Gerd Tetzlaff and Janek Zimmer — 778

Psychological Impacts of Natural Disasters — James M. Shultz, Yuval Neria, Andrea Allen and Zelde Espinel — 779

Pyroclastic Flow — Robert Buchwaldt — 791

CASE STUDY

Queensland Floods (2010–2011) and "Tweeting" 797
France Cheong and Christopher Cheong

Quick Clay 803
Marten Geertsema

Quick Sand 804
János Kovács

Radiation Hazards 807
Lev I. Dorman

Radon Hazards 808
James D. Appleton

Recovery and Reconstruction After Disaster 812
Michael K. Lindell

Recurrence Interval 824
Glenn Biasi

Red Cross and Red Crescent 825
Donald J. Shropshire

Red Tides 826
Philip Weinstein

Reflections on Modeling Disaster 827
David A. Etkin

Release Rates 835
Pat E. Rasmussen

Religion and Hazards 836
Heather Sangster, Angus M. Duncan and David K. Chester

Remote Sensing of Natural Hazards and Disasters 837
Norman Kerle

Reservoir, Dams, and Seismicity 847
Maurice Lamontagne

Resilience 849
Adriana Galderisi and Floriana F. Ferrara

Richter, Charles Francis (1900–1985) 850
Susan Hough

Rights and Obligations in International Humanitarian Assistance 851
George Kent

Rip Current 855
Wayne Stephenson

Risk 856
Jörn Birkmann

Risk Assessment 862
Suzanne Lacasse

Risk Governance 863
Stefan Greiving and Thomas Glade

Risk Perception and Communication 870
Michael K. Lindell

Rock Avalanche (Sturzstrom) 875
Reginald L. Hermanns

Rockfall 875
Fausto Guzzetti

Rogue Wave 877
Norm Catto

Rotational Seismology 877
William H. K. Lee

Sackung 881
Michael J. Bovis

Saffir–Simpson Hurricane Intensity Scale 882
Ilan Kelman

San Andreas Fault 883
William A. Bryant

CASE STUDY

Santorini, Eruption 884
Yuri Gorokhovich

Sea Level Change 895
Peter J. Hawkes

Secondary Wave (S-Wave) 901
Allison Bent

Sedimentation of Reservoirs 901
Anton J. Schleiss

Seiche 905
Giovanni Cuomo

Seismic Gap 906
John F. Cassidy

Seismograph/Seismometer 907
Allison Bent

Seismology Alik T. Ismail-Zadeh	907	Sunspots David H. Boteler	986
Shear Murat Saatcioglu	908	Supernova Lev I. Dorman	986
Shield Volcano Raphaël Paris	910	Surge Giovanni Cuomo	987
Sinkhole María Asunción Soriano	911	Susceptibility María José Domínguez-Cuesta	988
Slide and Slump Lionel E. Jackson, Jr.	913	**CASE STUDY**	
Slope Stability Kaare Høeg	919	Tangshan, China (1976 Earthquake) Zhengwen Zeng and Chenghu Wang	989
Snowstorm and Blizzard Thomas W. Schmidlin	924	Tectonic and Tectono-Seismic Hazards James P. McCalpin	994
Social–Ecological Systems Fabrice G. Renaud	926	Tectonic Tremor David Shelly	1004
Sociology of Disaster Alison Herring	926	Thunderstorms Colin Price	1006
Solar Flares David H. Boteler	936	Tidal Bores Hubert Chanson	1007
Solifluction Piotr Migoń	936	Tiltmeters Erik Eberhardt	1009
Space Weather David H. Boteler	937	Time and Space in Disaster Thomas Glade, Michael James Crozier and Nick Preston	1009
Storm Surges Gonéri Le Cozannet, Hormoz Modaressi, Rodrigo Pedreros, Manuel Garcin, Yann Krien and Nicolas Desramaut	940	**CASE STUDY** Tohoku, Japan (2011 Earthquake and Tsunami) Kenji Satake	1015
Storms Norm Catto	941	Torino Scale Norm Catto	1019
Stratovolcanoes Shane J. Cronin	941	Tornadoes Matthew R. Clark and R. P. Knightley	1019
Structural Damage Caused by Earthquakes Murat Saatcioglu	947	Triggered Earthquakes Harsh K. Gupta	1031
Structural Mitigation Murat Saatcioglu	959	Tsunami William Power and Graham S. Leonard	1036
Subduction Alik T. Ismail-Zadeh	979	Tsunami Loads on Infrastructure Dan Palermo, Ioan Nistor and Murat Saatcioglu	1046
Subsidence Induced by Underground Extraction Devin L. Galloway	979	Uncertainty Philipp Schmidt-Thomé	1055

United Nations Organizations and Natural Disasters 1056
Badaoui Rouhban

Universal Soil Loss Equation (USLE) 1062
Armand LaRocque

Unreinforced Masonry Buildings 1062
Fabio Taucer

Urban Environments and Natural Hazards 1063
Pat E. Rasmussen

CASE STUDY
Usoi Landslide and Lake Sarez 1065
Alexander Strom

CASE STUDY
Vaiont Landslide, Italy 1069
Monica Ghirotti and Doug Stead

CASE STUDY
Vesuvius 1073
Bill McGuire

Volcanic Ash 1074
Thomas Wilson and Carol Stewart

Volcanic Gas 1076
Travis W. Heggie

Volcanoes and Volcanic Eruptions 1077
Sue C. Loughlin

Vulnerability 1088
Susan L. Cutter

Warning Systems 1091
Graham S. Leonard, David M. Johnston and Chris E. Gregg

Waterspout 1096
Miquel Gayà

CASE STUDY
Wenchuan, China (2008 Earthquake) 1097
Zhengwen Zeng and Chenghu Wang

Wildfire 1102
Brigitte Leblon and Laura Bourgeau-Chavez

World Economy, Impact of Disasters 1107
Ilan Noy

Worldwide Trends in Natural Disasters 1111
Margreth Keiler

Zoning 1115
Philipp Schmidt-Thomé and Stefan Greiving

Author Index 1117

Subject Index 1119

Contributors

Pradeep Adhikari
School of Civil Engineering and Environmental Science
Atmospheric Radar Research Center, University of Oklahoma
120 David L. Boren Blvd
Norman, OK 73072
USA
and
Department of Geography and Environmental Sustainability, University of Oklahoma
Norman, OK 73072
USA
pradhikari@ou.edu

David E. Alexander
Global Risk Forum
Promenade 35
7270 Davos Platz
Switzerland
david.alexander@grforum.org
d.alexander@alice.it

Andrea Allen
School of Adult and Continuing Education
Barry University
11300 NE 2nd Avenue
Miami Shores, FL 33161
USA
aallen@mail.barry.edu
aallen70@aol.com

Walter J. Ammann
Global Risk Forum GRF Davos
Promenade 35
7270 Davos Platz
Switzerland
walter.ammann@grforum.org

Christophe Ancey
Ecole Polytechnique Fédérale de Lausanne
Laboratoire Hydraulique Environnementale ENAC/ICARE/LHE
station 18 Ecublens
1015 Lausanne
Switzerland
christophe.ancey@epfl.ch

Peter S. Anderson
School of Communication
Simon Fraser University
8888 University Drive
Burnaby, BC V5A 1S6
Canada
anderson@sfu.ca

Viacheslav Andreychouk
Regional Geography and Geotourism Unit
Faculty of Earth's Sciences
University of Silesia
Będzińska 60
41-200 Sosnowiec
Poland
geo@wnoz.us.edu.pl

James D. Appleton
British Geological Survey
Keyworth
Nottingham NG12 5GG
UK
jda@bgs.ac.uk

Lukas U. Arenson
BGC Engineering Inc.
Suite 500-1045 Howe Street
Vancouver, BC V6Z 2A9
Canada
larenson@bgcengineeirng.ca
lukas.arenson@gmail.com

Natalia Artemieva
Planetary Science Institute
1700 E. Ft. Lowell, suite 106
Tucson, AZ 85719
USA
artemeva@psi.edu

Pedro Basabe
United Nations Complex
Gigiri Block T, Room T-121
P.O. Box 47074
Nairobi
Kenya
pedro.basabe@unep.org
basabe@un.org

Arindam Basu
Health Sciences Centre & Health Services
Assessment Collaboration
University of Canterbury
Christchurch
New Zealand
arindam.basu@canterbury.ac.nz

Tom Beer
CSIRO Marine and Atmospheric Research
Centre for Australian Weather and Climate Research
Energy Transformed Flagship
P.B1, Aspendale
VIC 3195
Australia
tom.beer@csiro.au

Gerardo Benito
CSIC-Centro de Ciencias Medioambientales
Serrano 115 bis
28006 Madrid
Spain
benito@ccma.csic.es

Allison Bent
Geological Survey of Canada
Natural Resources Canada
7 Observatory Crescent
Ottawa, ON K1A 0Y3
Canada
abent@nrcan.gc.ca

Glenn Biasi
Nevada Seismological Laboratory
University of Nevada Reno
MS-174
Reno, NV 89557
USA
glenn@seismo.unr.edu

Wojciech Biernacki
University School of Physical Education in Cracow
Al. Jana Pawla II 78
31-571 Krakow
Poland
wojciech.biernacki@interia.pl
wojciech.biernacki@awf.krakow.pl

Jörn Birkmann
Institute for Environment and Human Security
United Nations University
Hermann-Ehlers-Str. 10
53113 Bonn
Germany
birkmann@ehs.unu.edu

Lars H. Blikra
Åknes/Tafjord Early-Warning Centre
Ødegårdavegen 176
6200 Stranda
Norway
lhb@aknes.no

Camillo Boano
Development Planning Unit
University College London
34 Tavistock Square
London WC1H 9EZ
UK
camillo.boano@gmail.com
c.boano@ucl.ac.uk

Anita Bokwa
Institute of Geography and Spatial Management
Jagiellonian University
7 Gronostajowa St
30-387 Kraków
Poland
anita.bokwa@uj.edu.pl

David H. Boteler
Geomagnetic Laboratory
Earth Science Sector, Natural Resources Canada
7 Observatory Crescent
Ottawa, ON K1A 0Y3
Canada
dboteler@nrcan.gc.ca

Laura Bourgeau-Chavez
Forestry and Environmental Management
University of New Brunswick
P.O. Box 4400
Fredericton, NB E3B 5A3
Canada

Michael J. Bovis
Department of Geography
University of British Columbia
Vancouver, BC V6T 1Z2
Canada
mbovis@geog.ubc.ca

Peter Brown
Department of Physics and Astronomy
Centre for Planetary Science and Exploration (CPSX)
University of Western Ontario
London, ON N6A 3K7
Canada
pbrown@uwo.ca

Rodger A. Brown
NOAA/National Severe Storms Laboratory
120 David L. Boren Blvd
Norman, OK 73072
USA
Rodger.Brown@noaa.gov

William A. Bryant
Senior Engineering Geologist
California Geological Survey
801 K Street, MS 12-31
Sacramento, CA 95814
USA
William.Bryant@conservation.ca.gov
bbryant@consrv.ca.gov

Robert Buchwaldt
Department of Earth, Atmospheric, and
Planetary Sciences
Massachusetts Institute of Technology
Building 54-1117
Cambridge, MA 02139-4307
USA
buch_1@mit.edu

Virginia R. Burkett
United States Geological Survey, Climate and
Land use Change Mission Area
540 North Courthouse Street
Many, LA 71449
USA
virginia_burkett@usgs.gov

Gilbert M. Burnham
The Johns Hopkins Bloomberg School of Public Health
Center for Refugee and Disaster Response
615 N Wolfe St., Suite E8132
Baltimore, MD 21205
USA
gburnham@jhsph.edu

Marta L. Calvache
INGEOMINAS
Bogota
Colombia

Leopoldo C. Cancio
US Army Institute of Surgical Research
Colonel, Medical Corps, U.S. Army
Fort Sam Houston, TX 78234-6315
USA
lee.cancio@us.army.mil

Alejandro López Carresi
Centre of Studies on Disasters and Emergencies
C/Hileras 4
28013 Madrid
Spain
info@cedemformacion.com

John F. Cassidy
Geological Survey of Canada
Natural Resources Canada
9860 West Saanich Road
Sidney, BC V8L 4B2
Canada
jcassidy@nrcan.gc.ca

Norm Catto
Department of Geography
Memorial University of Newfoundland
St. John's, NL A1B 3X9
Canada
ncatto@mun.ca

Marco Cavalli
National Research Council – Research Institute for
Geo-Hydrological Protection
Corso Stati Uniti 4
35127 Padova
Italy
marco.cavalli@irpi.cnr.it

Andrea Ceudech
Dipartimento di Pianificazione e Scienza del Territorio
University of Naples "Federico II"
Naples, P.le V. Tecchio, 80
80125 Naples
Italy
ceudech@unina.it

Hubert Chanson
School of Civil Engineering
The University of Queensland
Brisbane, QLD 4072
Australia
h.chanson@uq.edu.au

Clark R. Chapman
Department of Space Studies
Southwest Research Institute
1050 Walnut, Suite 300
Boulder, CO 80302
USA
cchapman@boulder.swri.edu

Christopher Cheong
School of Business IT and Logistics
RMIT University
Melbourne, VIC 3001
Australia

France Cheong
School of Business IT and Logistics
RMIT University
Melbourne, VIC 3001
Australia
france.cheong@rmit.edu.au

David K. Chester
Department of Geography
University of Liverpool
Liverpool L69 3BX
UK
jg54@liv.ac.uk

John J. Clague
Centre for Natural Hazard Research
Department of Earth Sciences
Simon Fraser University
8888 University Drive
Burnaby, BC V5A 1S6
Canada
jclague@sfu.ca

Matthew R. Clark
11B Lower North Street
Exeter EX4 3ET
Devon, UK
and
TORRO
P.O. Box 972
Thelwall WA4 9DP
Warrington, UK
matthew.clark@metoffice.gov.uk

James W. Cole
Department of Geological Sciences
University of Canterbury
Private Bag 4800
Christchurch 8140
New Zealand
jim.cole@canterbury.ac.nz

Valerio Comerci
Geological Survey of Italy
ISPRA – Institute for Environmental Protection
and Research
Via Vitaliano Brancati 48
00144 Roma
Italy
valerio.comerci@isprambiente.it

Benjamin I. Cook
NASA Goddard Institute for Space Studies
2880 Broadway
New York, NY 10025
USA
bc9z@ldeo.columbia.edu
bcook@giss.nasa.gov

Paul J. Croft
School of Environmental &Life Sciences
College of Natural, Applied, & Health Sciences
Kean University
1000 Morris Avenue
Union, NJ 07083
USA
pcroft@kean.edu
croftemj@aol.com

Shane J. Cronin
Institute of Natural Resources
Massey University
Private Bag 11 222
Palmerston North 4442
New Zealand
s.j.cronin@massey.ac.nz

Michael James Crozier
Institute of Geography, School of Geography
Environment and Earth Sciences
Victoria University of Wellington
PO Box 600 Wellington 6140
New Zealand
michael.crozier@vuw.ac.nz

David Cruden
Department of Civil and Environmental Engineering
University of Alberta
3-064 Markin/CNRL Natural
Edmonton, AB T6G 2W2
Canada
dcruden@ualberta.ca

Giovanni Cuomo
Hydraulics Applied Research & Engineering Consulting (HAREC) s.r.l
Via Gregorio VII, 80
00165 Rome
Italy
gcuomo@harec.net

Corinne Curt
Irstea
UR OHAX, Hydraulic Engineering and Hydrology Research Unit
3275 route de Cézanne CS 40061
13182 Aix-en-Provence, Cedex 5
France
corinne.curt@irstea.fr

Susan L. Cutter
Department of Geography, Hazards & Vulnerability Research Institute
University of South Carolina
Columbia, SC 29208
USA
scutter@sc.edu

Ghulappa S. Dasog
University of Agricultural Sciences
580005 Dharwad
India
gdasog@gmail.com

Timothy R. H. Davies
Department of Geological Sciences
University of Canterbury
Private Bag 4800
Christchurch 8140
New Zealand
tim.davies@canterbury.ac.nz

Edward Derbyshire
3, 104 Evesham Road
Cheltenham, Gloucestershire GL52 2AL
UK
and
Centre for Quaternary Research
Royal Holloway, University of London
Egham, Surrey TW20 0EX
UK
ed4_gsl@sky.com

Marc-Henri Derron
Institute of Geomatics and Analysis of Risk
AMPHIPOLE – 338, Faculté des géosciences et de l'environnement
University of Lausanne
1015 Lausanne
Switzerland
Marc-Henri.Derron@unil.ch

Nicolas Desramaut
French Geological Survey, Natural Risks and CO_2 Safety Storage Division
Landslide Risk Unit
BRGM – French Geological Survey
3 avenue Claude Guillemin, BP 36009
45060 Orléans, Cedex 2
France
n.desraumaut@brgm.fr

Andreas Dix
Bamberg Institut for Geography, Historical Geography
Department of Geography
Otto-Friedrich-University Bamberg
Am Kranen 12
96045 Bamberg
Germany
andreas.dix@uni-bamberg.de

John Dodson
Institute for Environmental Research
Australian Nuclear Science and Technology Organisation
New Illawarra Road
Lucas Heights, NSW 2234
Australia
John.Dodson@ansto.gov.au
John.Dodson@bigpond.com

María José Domínguez-Cuesta
Departamento de Geología
Universidad de Oviedo
C/Arias de Velasco s/n
33005 Oviedo
Spain
mjdominguez@geol.uniovi.es

Laurance Donnelly
Engineering & Exploration Geologist, Forensic Geologist & Police Search Adviser
Wardell Armstrong LLP
2 The Avenue
Greater Manchester WN7 1ES
UK
ldonnelly@wardell-armstrong.com

Katherine Donovan
Department of Earth Sciences
University of Oxford
South Parks Road
Oxford, Oxfordshire OX1 3AN
UK
katherine.donovan@earth.ox.ac.uk

Shannon Doocy
Department of International Health, Center for Refugee and Disaster Response
Johns Hopkins Bloomberg School of Public Health
615 N Wolfe Street, Suite E8132
Baltimore, MD 21205
USA
sdoocy@jhsph.edu

Mohammed H. I. Dore
Department of Economics
Brock University
4th Floor, Plaza Building
St Catharines, ON L2S 3A1
Canada
dore@brocku.ca

Lev I. Dorman
Israel Cosmic Ray and Space Weather Center and Emilio Segre' Observatory, affiliated to Tel Aviv University, TECHNION, and Israel Space Agency
P.O. Box 2217
Qazrin 12900
Israel
and
Cosmic Ray Department of N.V. Pushkov IZMIRAN
IZMIRAN of Russian Academy of Sciences
142190 Troitsk, Moscow Region
Russia
lid@physics.technion.ac.il
lid010529@gmail.com

Philippe Duffour
Department of Civil, Environmental and Geomatic Engineering
University College London
Gower Street
London WC1E 6BT
UK
p.duffour@ucl.ac.uk

Angus M. Duncan
Institute for Research in Applied Natural Sciences
University of Bedfordshire
Park Square
Luton LU1 3JU
UK
angus.duncan@beds.ac.uk

Jaroslaw Dzialek
Institute of Geography and Spatial Management
Jagiellonian University
7 Gronostajowa St
30-387 Krakow
Poland
jarek.dzialek@uj.edu.pl

Erik Eberhardt
Geological Engineering
Department of Earth and Ocean Sciences
University of British Columbia
6339 Stores Rd
Vancouver, BC V6T 1Z4
Canada
erik@eos.ubc.ca

George Eftychidis
Algosystems S.A.
206, Leoforos Syggrou
Kallithea 17672
Greece
and
Pangaiasys Ltd.
28, Marathonos Avenue
Pikermi 19009
Greece
g.eftychidis@gmail.com

Zelde Espinel
Center for Disaster & Extreme Event Preparedness (DEEP Center)
University of Miami Miller School of Medicine
Clinical Research Building Suite 1512
1120 NW 14 St.
Miami, FL 33136
USA
z.espinel@gmail.com

David A. Etkin
York University
4700 Keele Street
Toronto, ON M3J 1P3
Canada
etkin@yorku.ca

Lin Fa
Department of Geology and Geological Engineering
University of North Dakota
Grand Forks, ND 58202
USA
and
Department of Electronics and Information Engineering
Xi'an Institute of Post and Telecommunications
Xi'an, Shaanxi 710121
China
lin.fa@und.edu

Floriana F. Ferrara
Environmental Engineering Consultant
Via Morghen, 72
80129 Naples
Italy
floriana.ferrara@gmail.com

Frank Fiedrich
Public Safety and Emergency Management
Faculty D "Safety Engineering"
Wuppertal University
Gaussstr. 20
42119 Wuppertal
Germany
fiedrich@uni-wuppertal.de

Charles W. Finkl
Professor Emeritus
Department of Geosciences
Florida Atlantic University
Boca Raton 33431 FL
USA
cfinkl@cerf-jrc.com

Sven Fuchs
Institute of Mountain Risk Engineering
University of Natural Resources and Life Sciences
Peter Jordan Strasse 82
1190 Vienna
Austria
sven.fuchs@boku.ac.at

J. C. Gaillard
School of Environment
The University of Auckland
Private Bag 92019
Auckland 1142
New Zealand
jc.gaillard@auckland.ac.nz

Adriana Galderisi
Dipartimento di Pianificazione e Scienza del Territorio
University of Naples "Federico II"
P.le V. Tecchio, 80
80125 Naples
Italy
galderis@unina.it

Devin L. Galloway
Water Mission Area Modoc Hall, Ste. 3005, CSUS
US Geological Survey
3020 State University Dr. E.
Sacramento, CA 95819-2632
USA
dlgallow@usgs.gov

Manuel Garcin
Natural Risks and CO2 Storage Security Division/Coastal Risk Unit
BRGM-French Geological Survey – Natural Risks and CO_2 Storage Division
3 avenue Claude Guillemin
45060 Orléans
France
m.garcin@brgm.fr

Miquel Gayà
Delegación Territorial en Illes Balears
Agencia Estatal de Meteorología
Muelle de Ponent, s/n
P.O. Box 07015
Palma
Spain
fiblo.miquel@gmail.com

Marten Geertsema
British Columbia Forest Service
1011 4th ave
Prince George V2L 3H9 BC
Canada
Marten.Geertsema@gov.bc.ca

Thomas Gerz
Deutsches Zentrum für Luft- und Raumfahrt e.V. DLR
Institut für Physik der Atmosphäre
82234 Oberpfaffenhofen
Germany
thomas.gerz@dlr.de

Michael Ghil
Geosciences Department and Laboratoire de
Météorologie Dynamique (CNRS and IPSL)
Ecole Normale Supérieure
24, rue Lhomond
75231 Paris
France
and
Department of Atmospheric and Oceanic Sciences and
Institute of Geophysics and Planetary Physics
University of California
Los Angeles, CA
USA
ghil@lmd.ens.fr

Monica Ghirotti
Department of Earth and Geo-Environmental Sciences
Alma Mater-University of Bologna
Via Zamboni, 67
40127 Bologna
Italy
monica.ghirotti@unibo.it

John Rick Giardino
Department of Geology and Geophysics
Texas A&M University
College Station, Texas 77843
USA
rickg@tamu.edu

Marco Giardino
Department of Earth Sciences
University of Torino
Via Valperga Caluso, 35
10125 Torino
Italy
marco.giardino@unito.it

Terry Gibson
Global Network of Civil Society Organisations for
Disaster Reduction (GNDR)
100 Church Road
Teddington, Middlesex TW11 8QE
UK
terry.gibson@globalnetwork-dr.org

David Giles
Centre for Applied Geosciences, School of Earth and
Environmental Sciences
University of Portsmouth
Burnaby Building, Burnaby Road
Portsmouth PO1 3QL
UK
dave.giles@port.ac.uk

Galen Gisler
Physics of Geological Process
University of Oslo
Sem Sælandsvei 24
0316 Oslo
Norway
galen.gisler@fys.uio.no

Thomas Glade
Geomorphic Systems and Risk Research
Department of Geography and Regional Science
University of Vienna
Universitaetsstr. 7
1010 Vienna
Austria
thomas.glade@univie.ac.at

Tamara Goldin
Department of Lithospheric Research
Center for Earth Sciences, University of Vienna
Althanstrasse 14
1090 Vienna
Austria
tamara.goldin@univie.ac.at
tgoldin@lpl.arizona.edu

Yuri Gorokhovich
Department of Environmental, Geographic and
Geological Sciences
Lehman College, City University of New York
250 Bedford Park Blvd. West
Bronx, NY 10468
USA
yuri.gorokhovich@lehman.cuny.edu

Jonathan J. Gourley
NOAA National Severe Storms Laboratory
120 David L. Boren Blvd
Norman, OK 73072
USA
jj.Gourley@noaa.gov

Raisa Gracheva
Institute of Geography of RAS
Staromonetny per. 29
119017 Moscow
Russia
gracheva04@list.ru

Chris E. Gregg
Department of Geosciences
East Tennessee State University
Yoakley Hall, Rm 204
P.O. Box 70357
Johnson City, TN 37614
USA
gregg@etsu.edu

Stefan Greiving
Faculty of Spatial Planning, IRPUD Institute of Spatial Planning
TU Dortmund University
August-Schmidt-Str. 10
44227 Dortmund
Germany
stefan.greiving@uni-dortmund.de

Harsh K. Gupta
National Disaster Management Authority
Government of India, NDMA Bhawan
A–1, Safdarjung Enclave
New Delhi 110029
India
and
National Geophysical Research Institute
Uppal Road
Hyderabad 500 606
India
harshndma@gmail.com
harshg123@gmail.com

Richard Guthrie
Director of Geohazards and Geomorphology
MDH Engineered Solutions, SNC Lavalin Group
909 5th Ave SW
Calgary, AB T2P 3G5
Canada
rguthrie@mdhsolutions.com
rhguthrie@gmail.com

Fausto Guzzetti
CNR – IRPI
via della Madonna Alta 126
06128 Perugia
Italy
F.Guzzetti@irpi.cnr.it

Sylvi Haldorsen
Department of Plants and Environmental Science
Norwegian University of Life Sciences
P.O. Box 5003
1432 Aas
Norway
sylvi.haldorsen@umb.no

Minard L. Hall
Escuela Politecnica
Quito
Ecuador

John R. Harrald
The Center for Community Security and Resilience
Virginia Tech
900 N. Glebe Road
Arlington, VA 22203
USA
jharrald@vt.edu

Alan W. Harris
Institute of Planetary Research
German Aerospace Center (DLR)
Rutherfordstr. 2
12489 Berlin
Germany
alan.harris@dlr.de

William T. Hartwell
Division of Earth and Ecosystem Sciences
Desert Research Institute, Nevada System of Higher Education
755 E. Flamingo Rd
Las Vegas, NV 89119
USA
Ted.Hartwell@dri.edu

Peter J. Hawkes
Hydrodynamics and Metocean Group
HR Wallingford Limited
Howbery Park
Wallingford, Oxfordshire OX10 8BA
England
p.hawkes@hrwallingford.com

Travis W. Heggie
Recreation & Tourism Studies Program
University of North Dakota
225 Centennial Drive, Mail Stop 7116
Grand Forks, ND 58202-7116
USA
travis.heggie@und.edu

Reginald L. Hermanns
Head of Landslides Department
Geological Survey of Norway, International Centre for Geohazards
Leiv Eirikssons vei 39
7491 Trondheim
Norway
Reginald.Hermanns@ngu.no

Alison Herring
Department of Sociology
University of North Texas
1155 Union Circle #311157
Denton, TX 76203
USA
Alison.Herring@unt.edu

Javier Hervás
Institute for Environment and Sustainability
Joint Research Centre, European Commission
21027 Ispra (Va)
Italy
javier.hervas@jrc.ec.europa.eu

Catherine J. Hickson
Magma Energy Corp.
410-625 Howe Street
Vancouver, BC V6C 2T6
Canada
and
Alterra Power Corp.
600 - 888 Dunsmuir Street
Vancouver, BC V6C 3K4
Canada
chickson@magmaenergycorp.com
chickson@alterrapower.ca

Wayne Higgins
Climate Prediction Center, NCEP/NWS/NOAA
5200 Auth Rd. Room 800
Camp Springs, MD 20746
USA
Wayne.Higgins@noaa.gov

Kaare Høeg
Norwegian Geotechnical Institute
Postboks 3930 Ullevål Stadion
0806 Oslo
Norway
Kaare.Hoeg@ngi.no

Suzanne Hollins
Institute for Environmental Research
Australian Nuclear Science and Technology Organisation
New Illawarra Road
Lucas Heights, NSW 2234
Australia
Suzanne.Hollins@ansto.gov.au

Yang Hong
School of Civil Engineering and Environmental Science
Center for Natural Hazards and Disaster Research,
National Weather Center, University of Oklahoma
120 David L. Boren Blvd
Norman, OK 73072
USA
yanghong@ou.edu

Pascal Horton
Institute of Geomatics and Analysis of Risk
AMPHIPOLE – 338, Faculté des géosciences et de l'environnement
University of Lausanne
1015 Lausanne
Switzerland
Pascal.Horton@unil.ch

Susan Hough
Southern California Earthquake Center
525 South Wilson Avenue
Pasadena, CA 91106
USA
hough@gps.caltech.edu

Oldrich Hungr
Earth and Ocean Sciences
University of British Columbia
6339 Stores Rd
Vancouver, BC V6T 1Z4
Canada
ohungr@eos.ubc.ca

William Hunter
Development Planning Unit
University College London
34 Tavistock Square
London WC1H 9EZ
UK
william.hunter.81@gmail.com

Jane Carter Ingram
Department of Ecology, Evolution and Environmental Biology
Wildlife Conservation Society
2300 Southern Boulevard
Bronx, NY 10460-1099
USA
cingram@wcs.org

Alik T. Ismail-Zadeh
Institut für Angewandte Geowissenschaften
Karlsruher Institut für Technologie
Bldg. 50.40, Adenauerring 20b
76131 Karlsruhe
Germany

and
Institut de Physique du Globe de Paris
1, rue Jussieu
75238 Paris
France
and
Institute of Earthquake Prediction Theory and
Mathematical Geophysics
Russian Academy of Sciences
Profsoyuznaya 84/32
117997 Moscow
Russia
alik.ismail-zadeh@gpi.uka.de
alik.ismail-zadeh@kit.edu
aiz@ipgp.fr
aismail@mitp.ru

Michel Jaboyedoff
Institute of Geomatics and Analysis of Risk
AMPHIPOLE – 338, Faculté des géosciences et de
l'environnement
University of Lausanne
1015 Lausanne
Switzerland
Michel.Jaboyedoff@unil.ch

Lionel E. Jackson, Jr.
Pacific Division
Geological Survey of Canada
15th floor, 605 Robson Street
Vancouver, BC V6B 5J3
Canada
lijackso@nrcan.gc.ca

Theresa Jefferson
Information Systems and Operations Management
Department
Loyola University Maryland
4501 North Charles Street
Baltimore, MD 21210
USA
tljefferson@loyola.edu

Rohit Jigyasu
Research Center for Disaster Mitigation of Urban
Cultural Heritage
Ritsumeikan University
Kita-Cho 58
603-8341 Komatsubara, Kita-ku, Kyoto
Japan
rohit.jigyasu@gmail.com

Sarb Johal
Joint Centre for Disaster Research
GNS Science/Massey University
P.O. Box 30 368
Wellington 5040
New Zealand

David M. Johnston
Joint Centre for Disaster Research
GNS Science, Massey University
1 Fairway Drive
Lower Hutt 5010
New Zealand
David.Johnston@gns.cri.nz

Sirkku Juhola
Centre for Urban and Regional Studies (YTK)
Aalto University
P.O. Box 12200
00076 Aalto
Finland
sirkku.jukola@tkk.fi

Kirstyn Kameg
School of Nursing and Health Sciences
Robert Morris University
6001 University Boulevard
Moon Township, PA 12108
USA
kameg@rmu.edu

Margreth Keiler
Division of Earth & Ocean Sciences
Duke University
203 Old Chemistry Building
Durham, NC 27708
USA
margreth.keiler@duke.edu

Ilan Kelman
Center for International Climate and Environmental
Research – Oslo (CICERO)
P.O. Box 1129
Blindern, 0318 Oslo
Norway
ilan_kelman@hotmail.com

Melanie Kelman
Geological Survey of Canada
Natural Resources Canada, Geological Survey of Canada
625 Robson Street, 16th floor
Vancouver, BC V6B 5J3
Canada
mkelman@nrcan.gc.ca

George Kent
Department of Political Science
University of Hawai'i
2424 Maile Way, Saunders 610
Honolulu, HI 96822
USA
kent@hawaii.edu

Norman Kerle
Department of Earth System Analysis (ESA)
Faculty of Geo-Information Science and Earth
Observation (ITC), University of Twente
Hengelosestraat 99
P.O. Box 6
7500 AA Enschede
The Netherlands
Kerle@itc.nl

Jasper Knight
School of Geography, Archaeology and Environmental
Studies
University of the Witwatersrand
Private Bag 3
Johannesburg WITS 2050
South Africa
jasper.knight@wits.ac.za

R. Paul Knightley
TORRO
P.O. Box 972
Thelwall WA4 9DP
Warrington, UK
Paul.Knightley@meteogroup.com

Christian Koeberl
Department of Lithospheric Research
Center for Earth Sciences, University of Vienna
Althanstrasse 14
1090 Vienna
Austria
and
Natural History Museum
Burgring 7
1010 Vienna
Austria
christian.koeberl@univie.ac.at

Blaž Komac
Scientific Research Centre of the Slovenian Academy of
Sciences and Arts
Anton Melik Geographical Institute
Novi trg 2
1000 Ljubljana
Slovenia
blaz.komac@zrc-sazu.si

Laura S. L. Kong
UNESCO/IOC-NOAA International Tsunami
Information Center
737 Bishop St., Ste. 2200
Honolulu, HI 96813
USA
laura.kong@noaa.gov

Robert Korty
Department of Atmospheric Sciences
Texas A&M University
1204 Eller O and M Bldg., 3150 TAMU
College Station, TX 77843-3150
USA
korty@tamu.edu

János Kovács
Department of Geology
University of Pécs
Ifjúság u. 6
7624 Pécs
Hungary
jones@gamma.ttk.pte.hu

Steven L. Kramer
Department of Civil and Environmental Engineering
University of Washington
132E More Hall
Seattle, WA 98195-2700
USA
kramer@u.washington.edu

Yann Krien
Natural Risks and CO2 Storage Security Division/Coastal
Risk Unit
BRGM-French Geological Survey – Natural Risks and
CO2 Storage Division
3 avenue Claude Guillemin
45060 Orléans
France
y.krien@brgm.fr

Susanne Krings
Institute for Environment and Human Security
(UNU-EHS)
United Nations University
UN Campus, Hermann Ehlers Straße 10
53113 Bonn
Germany
s.u.krings@googlemail.com

Howard Kunreuther
Risk Management and Decision Processes Center
The Wharton School, University of Pennsylvania
3730 Walnut Street, Jon Huntsman Hall, Suite 500
Philadelphia, PA 19104-6340
USA
Kunreuther@wharton.upenn.edu
hellerc@wharton.upenn.edu

Tore Jan Kvalstad
Department of Offshore Geotechnics
Norwegian Geotechnical Institute
Sognsveien 72, 3930 Ullevaal Stadion
0806 Oslo
Norway
Tore.Jan.Kvalstad@ngi.no

Suzanne Lacasse
Norwegian Geotechnical Institute
Ullevaal Stadion
P.O. Box 3960
0806 Oslo
Norway
suzanne.lacasse@ngi.no

Steve LaDochy
Department of Geosciences & Environment
California State University, Los Angeles
5151 State University Dr.
Los Angeles, CA 90032
USA
sladoch@calstatela.edu
sladoch@exchange.calstatela.edu

Maurice Lamontagne
Geological Survey of Canada
Natural Resources Canada
615 Booth St Room 216
Ottawa, ON K1A 0E9
Canada
maurice.lamontagne@nrcan.gc.ca

Nicholas Lancaster
Division of Earth and Ecosystem Sciences
Desert Research Institute
2215Raggio Parkway
Reno, NV 89512-1095
USA
nick.lancaster@dri.edu
nick@dri.edu

Armand LaRocque
Forestry Remote Sensing Laboratory
Faculty of Forestry and Environmental Management
University of New Brunswick
28 Dineen Drive
Fredericton, NB E3B 6C2
Canada
larocque@unb.ca

Franck Lavigne
Laboratoire de Géographie Physique
UMR 8591 CNRS, Paris 1 Panthéon-Sorbonne University
1 Place A. Briand
92190 Meudon
France
franck.lavigne@univ-paris1.fr

Gonéri Le Cozannet
Natural Risks and CO2 Storage Security Department/
Risks Division/Coastal Risks unit
BRGM-French Geological Survey – Natural Risks and
CO2 Storage Division
3 avenue Claude Guillemin
45060 Orléans
France
g.lecozannet@brgm.fr

Brigitte Leblon
Faculty of Forestry and Environmental Management
University of New Brunswick
P.O. Box 4400
Fredericton, NB E3B 5A3
Canada
bleblon@unb.ca

William H. K. Lee
U.S. Geological Survey
MS 977, 345 Middlefield Road
Menlo Park, CA 94025
USA
Lee@usgs.gov

Graham S. Leonard
GNS Science
Massey University
1 Fairway Drive
Lower Hutt 5010
New Zealand
G.Leonard@gns.cri.nz

Frédéric Leone
Department of Geography
University of Montpellier and GESTER Laboratory
Route de Mende
34199 Montpellier
France
Frederic.Leone@univ-montp3.fr

Suzanne A. G. Leroy
Institute for the Environment
Brunel University
Kingston Lane
Uxbridge (London) UB8 3PH
UK
suzanne.leroy@brunel.ac.uk

Jean-Sébastien L'Heureux
Norwegian Geotechnical Institute (NGI)
P.O. Box 1230
Pirsenteret
7462 Trondheim
Norway
jean-sebastien.lheureux@ngi.no

Michael K. Lindell
Hazard Reduction and Recovery Center
Texas A&M University
3137 TAMU
College Station, TX 77843-3137
USA
mlindell@tamu.edu

John M. Logan
University of Oregon
P.O. Box 1776
Bandon, OR 97411
USA
jmllogan@aol.com

Céline Longchamp
Institute of Geomatics and Analysis of Risk
AMPHIPOLE – 338, Faculté des géosciences et de l'environnement
University of Lausanne
1015 Lausanne
Switzerland
Celine.Longchamp@unil.ch

Sue C. Loughlin
British Geological Survey
West Mains Road
Edinburgh EH9 3LA
UK
sclou@bgs.ac.uk

Ross D. E. MacPhee
Division of Vertebrate Zoology
American Museum of Natural History
Central Park West @ 79th St
New York, NY 10024
USA
macphee@amnh.org

Sharon B. Madanes
Department of Psychiatry
Columbia University
New York, NY
USA

Louis Herns Marcelin
Department of Anthropology
Interuniversity Institute for Research and Development
(INURED), Port-au-Prince, Haiti, and
University of Miami
5202 University Drive (Merrick Building, Room 102)
Coral Gables, FL 33146
USA
lmarcel2@med.miami.edu

Julie A. March
Agriculture & Food Security Advisor, Office of US Foreign Disaster Assistance
United States Agency for International Development (USAID)
1300 Pennsylvania Ave, NW, RRB 8.7.92
Washington, DC 20523, 8602
USA
jmarch@usaid.gov

Brian R. Marker
Independent consultant
40 Kingsdown Avenue
London W13 9PT
UK
brian@amarker.freeserve.co.uk

Brian G. Marsden (Deceased)

Gordon McBean
Institute for Catastrophic Loss Reduction
The University of Western Ontario
London, ON N6A 5B2
Canada
gmcbean@uwo.ca.

James P. McCalpin
GEO-HAZ Consulting Inc.
600 E. Galena Ave
Crestone, CO 81131
USA
mccalpin@geohaz.com

Bill McGuire
Department of Earth Sciences
Aon Benfield UCL Hazard Research Centre
University College London
136 Gower Street, Lewis Building
London WC1E 6BT
UK
w.mcguire@ucl.ac.uk

Jay Melosh
Department of Earth and Atmospheric Sciences
Civil Engineering Building, Room 3237, 550 Stadium Mall Drive
Purdue University
West Lafayette, IN 47907
USA
jmelosh@purdue.edu

Scira Menoni
Dipartimento di Architettura e Pianificazione
DIAP-Politecnico di Milano
Via Bonardi 3
20133 Milan
Italy
menoni@polimi.it
menoni@mail.polimi.it

Ahmet R. Mermut
Department of Soil Science
Harran University
63200 Şanlıurfa
Turkey
a.mermut@usask.ca
ar.mermut@gmail.com

Erwann Michel-Kerjan
Risk Management and Decision Processes Center
The Wharton School, University of Pennsylvania
3730 Walnut Street, Jon Huntsman Hall, Suite 500
Philadelphia, PA 19104-6340
USA
ErwannMK@wharton.upenn.edu

Clément Michoud
Institute of Geomatics and Analysis of Risk
AMPHIPOLE – 338, Faculté des géosciences et de l'environnement
University of Lausanne
1015 Lausanne
Switzerland
Clement.Michoud@unil.ch

Piotr Migoń
Department of Geography and Regional Development
University of Wrocław
pl. Uniwersytecki 1
50-137 Wrocław
Poland
piotr.migon@uni.wroc.pl

Ann M. Mitchell
School of Nursing and School of Medicine
University of Pittsburgh
3500 Victoria Street, 415 Victoria Building
Pittsburgh, PA 15261
USA
ammi@pitt.edu

Jerry T. Mitchell
Hazards and Vulnerability Research Institute
University of South Carolina
Columbia, SC 29208
USA
mitchell@sc.edu

Hormoz Modaressi
Development Planning and Natural Risks
French Geological Survey, Natural Risks and CO2 Safety Storage Division
BRGM – French Geological Survey – RNSC/D
3 avenue Claude Guillemin, BP 36009
45060 Orléans, Cedex 2
France
h.modaressi@brgm.fr

J. D. Mollard
J.D. Mollard and Associates Ltd.
810 Avord Tower, 2002 Victoria Ave.
Regina, SK S4P 0R7
Canada
mollard@jdmollard.com

Maria Luisa Monsalve
INGEOMINAS
Bogota
Colombia

Burrell E. Montz
Department of Geography
East Carolina University
A-228 Brewster
Greenville, NC 27858
USA
montzb@ecu.edu

Julie Morin
Planet Risk Association, Montgeron, France and
GeoSciences Reunion
UMR 7154 CNRS – IPGP
La Reunion University
Saint Denis
France
julieapi@yahoo.fr

Joann Mossa
Department of Geography
University of Florida
P.O. Box 117315
Gainesville, FL 32611-7315
USA
mossa@ufl.edu

Annemarie Müller
Department Urban and Environmental Sociology
(SUSOZ)
Helmholtz-Centre for Environmental Research (UFZ)
Permoserstr. 15
04318 Leipzig
Germany
annemarie.mueller@ufz.de

Julian B. Murton
Permafrost Laboratory, Department of Geography
University of Sussex
Brighton BN1 9QJ
UK
j.b.murton@sussex.ac.uk

Roger M. W. Musson
British Geological Survey
Murchison House, West Mains Road
Edinburgh EH9 3LA
UK
rmwm@bgs.ac.uk

Farrokh Nadim
International Centre for Geohazards
Norwegian Geotechnical Institute
Ullevall Stadion
P.O. Box 3930
0806 Oslo
Norway
Farrokh.nadim@ngi.no

Fernando Nardi
GEMINI Department
University of Tuscia
Via Camillo De Lellis snc
01100 Viterbo (VT)
Italy
and
Hydraulics Applied Research & Engineering
Consulting S.r.l.
Corso Trieste 142
00198 Rome
Italy
fernando.nardi@unitus.it
fnardi@harec.net

Alan R. Nelson
Geologic Hazards Science Center – Golden
U.S. Geological Survey
MS 966
P.O. Box 25046
Golden, CO 80225
USA
anelson@usgs.gov

Yuval Neria
Department of Psychiatry
Columbia University, The New York State Psychiatric
Institute
1051 Riverside Drive, Unit #69
New York, NY 10032
USA
ny126@columbia.edu

N. Nirupama
Disaster and Emergency Management Program
York University
4700 Keele Street
Toronto, ON M3J 1P3
Canada
nirupama@yorku.ca

Ioan Nistor
Department of Civil Engineering
University of Ottawa
161 Louis Pasteur Street
Ottawa, ON K1N 6N5
Canada
inistor@uOttawa.ca

Fran H. Norris
Department of Psychiatry, Department of Veterans Affairs
Dartmouth Medical School, National Center for PTSD
116 North Main Street
White River Junction, VT 05009
USA
fran.norris@dartmouth.edu

Ilan Noy
Department of Economics
University of Hawaii, Manoa
2424 Maile Way, Saunders Hall 542
Honolulu, HI 96822
USA
and
School of Economics and Finance
Victoria Business School
Wellington 6140
New Zealand
noy@hawaii.edu

Dan Palermo
Department of Civil Engineering
University of Ottawa
161 Louis Pasteur Street
Ottawa, ON K1N 6N5
Canada
Dan.Palermo@uOttawa.ca

Behruz M. Panahi
Geology Institute
Azerbaijan National Academy of Sciences
29-a, H.Javid Avenue
370143 Baku
Azerbaijan
bpanahi@rambler.ru

Raphaël Paris
Laboratoire Magmas et Volcans, UMR 6524
CNRS - Université Blaise Pascal
Clermont University
5 rue Kessler
63038 Clermont-Ferrand
France
raparis@univ-bpclermont.fr

Vincent R. Parisi
740 Florence Drive
Park Ridge, IL 60068
USA
vincent.parisi@dhs.gov

Paolo Paron
Department of Water Science & Engineering
UNESCO–IHE
Westvest 7
2611 AX Delft
The Netherlands
and
St Cross College
University of Oxford

St. Giles
Oxford OX1 3LZ
UK
paoloparon@gmail.com
P.Paron@unesco-ihe.org

Douglas Paton
School of Psychology
University of Tasmania
Launceston, TAS 7250
Australia
Douglas.Paton@utas.edu.au

Rodrigo Pedreros
Natural Risks and CO2 Storage Security
Division/Coastal Risk Unit
BRGM-French Geological Survey – Natural Risks and
CO_2 Storage Division
3 avenue Claude Guillemin
45060 Orléans
France
r.pedreros@brgm.fr

William Power
Earthquakes, Volcanoes and Tectonics
GNS Science
1 Fairway Drive
P.O. Box 30-368
Lower Hutt 5040
New Zealand
W.Power@gns.cri.nz

Nick Preston
Institute of Geography, School of Geography
Environment and Earth Sciences
Victoria University of Wellington
PO Box 600, Wellington 6140
New Zealand
Nick.Preston@vuw.ac.nz

Colin Price
Department of Geophysics and Planetary Sciences
Tel Aviv University
Levanon Road
Ramat Aviv 69978
Israel
cprice@flash.tau.ac.il

Antonio Pugliano
Lombardia Firemen Regional Headquarters
Milan
Italy

John Radke
Department of Landscape Architecture and
Environmental Planning, Department of City and
Regional Planning
University of California, Center for Catastrophic Risk
Management
202 Wurster Hall #2000
Berkeley, CA 94720-2000
USA
ratt@berkeley.edu

Paris Raphaël
Maison des Sciences de l'Homme
CNRS – GEOLAB UMR 6042 CNRS – UBP
Clermont University
4 rue Ledru
63057 Clermont-Ferrand
France
raparis@univ-bpclermont.fr

Pat E. Rasmussen
Earth Sciences Department
Health Canada, University of Ottawa
50 Columbine Driveway, Tunney's Pasture 0803C
Ottawa, ON K1A 0K9
Canada
Pat_Rasmussen@hc-sc.gc.ca

Netra Raj Regmi
Division of Earth and Ecosystem Sciences
Desert Research Institute
2215 Raggio Parkway
Reno, NV 89512
USA
netra.regmi@dri.edu

Mark E. Reid
U.S. Geological Survey
345 Middlefield Road, MS 910
Menlo Park, CA 94025
USA
mreid@usgs.gov

Fabrice G. Renaud
Institute for Environment and Human Security
United Nations University, UN Campus
Hermann-Ehlers-Strasse 10
53113 Bonn
Germany
renaud@ehs.unu.edu

Nilton O. Rennó
College of Engineering
University of Michigan
1531 Space Research Building, 2455 Hayward Ave
Ann Arbor, MI 48109
USA
renno@alum.mit.edu

John Ristau
GNS Science
1 Fairway Drive
Avalon, Lower Hutt 5040
New Zealand
J.Ristau@gns.cri.nz

Thomas Rockwell
Department of Geological Sciences
San Diego State University
5500 Campanile Drive
San Diego, California, CA 92182
USA
trockwell@geology.sdsu.edu

Kevin R. Ronan
Department of Health and Human Services
CQUniversity Australia
Bruce Highway
North Rockhampton, QLD 4702
Australia
k.Ronan@cqu.edu.au

Tiziana Rossetto
Department of Civil, Environmental and Geomatic
Engineering
Earthquake and People Interaction Centre (EPICentre)
University College London
Gower Street
London WC1E 6BT
UK
t.rossetto@ucl.ac.uk

Badaoui Rouhban
Section for Disaster Reduction
UNESCO
1, rue Miollis
P.O. Box 15
75732 Paris
France
b.rouhban@unesco.org

Jeffrey N. Rubin
Tualatin Valley Fire & Rescue
11945 SW 70th Ave.
Tigard, OR 97223
USA
jeff.rubin@tvfr.com
jnrubin@aya.yale.edu

Murat Saatcioglu
Department of Civil Engineering
University of Ottawa
161 Louis Pasteur Street
Ottawa, ON K1N 6N5
Canada
Murat.Saatcioglu@uOttawa.ca

Kalachand Sain
National Geophysical Research Institute
Uppal Road
Hyderabad 500 606
India
kalachandsain@yahoo.com

Gianfausto Salvadori
Dipartimento di Matematica
Università del Salento
Provinciale Lecce-Arnesano
P.O. Box 193
73100 Lecce
Italy
gianfausto.salvadori@unisalento.it

Heather Sangster
Department of Geography
University of Liverpool
Liverpool L69 3BX
UK
H.Sangster@liv.ac.uk

Kenji Satake
Earthquake Research Institute, University of Tokyo
1-1-1 Yayoi
Bunkyo-ku, Tokyo 113-0032
Japan
satake@eri.u-tokyo.ac.jp

Cathy Scheib
British Geological Survey
Kingsley Dunham Centre, Keyworth
Nottingham NG12 5GG
UK
cemery@bgs.ac.uk

Pierre-Alain Schieb
OECD SGE/AU IFP
2, rue André Pascal
75775 Paris, Cedex 16
France
pierre-alain.schieb@oecd.org

Anton J. Schleiss
Laboratory of Hydraulic Constructions (LCH)
Ecole polytechnique fédérale de Lausanne (EPFL)
Station 18
1015 Lausanne
Switzerland
anton.schleiss@epfl.ch

Thomas W. Schmidlin
Department of Geography
Kent State University
P.O. Box 5190
Kent, OH 44242-0001
USA
tschmidl@kent.edu

Philipp Schmidt-Thomé
Geological Survey of Finland (GTK)
Land use and environment
P.O. Box 96
Espoo 02151
Finland
Philipp.schmidt-thome@gtk.fi

Ulrich Schumann
Deutsches Zentrum für Luft- und Raumfahrt e.V. DLR
Institut für Physik der Atmosphäre
82234 Oberpfaffenhofen
Germany
ulrich.schumann@dlr.de

Richard Seager
Lamont Doherty Earth Observatory of Columbia
University, Palisades Geophysical Institute
61 Rt 9W
Palisades, NY 10964
USA
rich@maatkare.ldgo.columbia.edu
seager@ldeo.columbia.edu

David Shelly
U.S. Geological Survey
MS 977, 345 Middlefield Rd
Menlo Park, California, CA 94025
USA
dshelly@usgs.gov

Donald J. Shropshire
Community Development
Municipality of Chatham-Kent
312 Wellington Street W
Chatham, ON N7M 1K1
Canada
donshropshire@sympatico.ca

James M. Shultz
Center for Disaster & Extreme Event Preparedness
(DEEP Center)
University of Miami Miller School of Medicine, Clinical Research Building Suite 1512
1120 NW 14 St.
Miami, FL 33136
USA
jamesmichaelshultz@gmail.com
jshultz1@med.miami.edu

Roy C. Sidle
Department of Geology
Appalachian State University
P.O. Box 32067
Boone, NC 28608
USA
and
US EPA, ORD–NERL, Ecosystems Research Division
Athens, GA
USA
roysidle@gmail.com

Steven T. Siems
School of Mathematical Sciences
Monash University
Monash, VIC 3800
Australia
steven.siems@sci.monash.edu.au

Freysteinn Sigmundsson
Institute of Earth Sciences, University of Iceland
Building of Natural Sciences, Askja
Room 322, Sturlugata 7
101 Reykjavík
Reykjavík, Iceland
fs@hi.is

Viviane Silva
Climate Prediction Center, NCEP/NWS/NOAA
5200 Auth Rd. Room 800
Camp Springs, MD 20746
USA
Viviane.Silva@noaa.gov

Rajiv G. Singh
Department of Economics
Brock University
4th Floor, Plaza Building
St Catharines, ON L2S 3A1
Canada
rajivgsingh@hotmail.com

Richard L. Snyder
Extension Biometeorologist, Department of Land Air and Water Resources
University of California
243 Hoagland Hall, One Shields Avenue
Davis, CA 95616-8627
USA
rlsnyder@ucdavis.edu

Mauro Soldati
Dipartimento di Scienze della Terra
Università degli Studi di Modena e Reggio Emilia
Largo S. Eufemia 19
41121 Modena
Italy
soldati@unimore.it

John H. Sorensen
Oak Ridge National Laboratory
1 Bethel Valley Road
Oak Ridge, TN 37831-6422
USA
jhs@ornl.gov

María Asunción Soriano
Departamento de Ciencias de la Tierra
Facultad de Ciencias
Universidad de Zaragoza
50009 Zaragoza
Spain
asuncion@unizar.es

Sarah M. Springman
Eidgenössische Technische Hochschule Zürich
Institut für Geotechnik/Institute for Geotechnical Engineering
Wolfgang Pauli Strasse 15, Hönggerberg, HIL C13.1
8093 Zurich
Switzerland
sarah.springman@igt.baug.ethz.ch

T. C. Spurgeon
Alterra Power Corp.
600 - 888 Dunsmuir Street
Vancouver, BC V6C 3K4
Canada

Lucy Stanbrough
Department of Earth Sciences, Aon Benfield UCL Hazard Centre
University College London
Gower Street
London WC1 6BT
UK
l.stanbrough@ucl.ac.uk
lucy@lucystanbrough.com
lucy.stanbrough@ucl.ac.uk

Doug Stead
FRBC Chair in Resource Geoscience and Geotechnics
Department of Earth Sciences
Simon Fraser University
8888 University Drive
Burnaby, BC V5A 1S6
Canada
dstead@sfu.ca

Wayne Stephenson
Department of Geography
University of Otago
P.O. Box 56
Dunedin 9016
New Zealand
wjs@geography.otago.ac.nz

Chris Stethem
Chris Stethem & Associates Ltd.
409 8th Avenue
Canmore, AB T1W 2E6
Canada
cstethem@snowsafety.ca

Carol Stewart
Joint Centre for Disaster Research
GNS Science/Massey University
Lower Hutt 5010
P.O. Box 756
New Zealand
stewart.carol@xtra.co.nz

Ian Stewart
School of Geography, Earth, & Environmental Sciences
University of Plymouth
Plymouth PL4 8AA
UK
istewart@plymouth.ac.uk

Ronald E. Stewart
Department of Environment and Geography 213 Isbister Building
University of Manitoba
183 Dafoe Road
Winnipeg, MB R3T 2N2
Canada
ronald.e.stewart@gmail.com
ronald_stewart@umanitoba.ca

Glenn B. Stracher
Division of Science and Mathematics, East Georgia College
University System of Georgia
131 College Circle
Swainsboro, GA 30401
USA
stracher@ega.edu

Alexander Strom
Geodynamic Research Center – Branch of JSC "Hydroproject Institute"
Volokolamsk Highway, 2
119225 Moscow
Russia
a_strom2002@yahoo.co.uk
a.strom@g23.relcom.ru

Andrew J. Stumpf
Illinois State Geological Survey, Institute of Natural Resource Sustainability Prairie Research Institute
University of Illinois at Urbana-Champaign
615 East Peabody Drive
Champaign, IL 61820
USA
stumpf@isgs.illinois.edu

Jean-Marc Tacnet
Irstea
UR ETGR, Unité Erosion Torrentielle Neige et Avalanches (Snow Avalanche Engineering and Torrent Control Research Unit)
2, rue de la papèterie BP 76
38402 Saint Martin d'Hères, Cedex
France
jean-marc.tacnet@irstea.fr

Paolo Tarolli
Department of Land, Environment, Agriculture and Forestry
University of Padova
viale dell'Università 16
35020 Legnaro, Padova
Italy
paolo.tarolli@unipd.it

Fabio Taucer
ELSA Unit, Institute for the Security and
Safety of the Citizen
European Commission – Joint Research Centre
Via E. Fermi, 2749
21027 Ispra (VA)
Italy
fabio.taucer@jrc.ec.europa.eu

Gerd Tetzlaff
Institut für Meteorologie, Universität Leipzig
Stephanstr 3
04103 Leipzig
Germany
tetzlaff@uni-leipzig.de

Mary J. Thornbush
School of Geography, Earth and Environmental Sciences
University of Birmingham
Edgbaston, Birmingham B15 2TT
UK
m.thornbush@bham.ac.uk

R. I. Tilling
Volcano Science Center
U.S. Geological Survey
Menlo Park, CA 94025
USA
and
Alterra Power Corp.
Vancouver, BC
Canada

Graham A. Tobin
Office of the Provost; Department of Geography
University of South Florida
4202 East Fowler Avenue, ADM 226
Tampa, FL 33620
USA
gtobin@acad.usf.edu
gtobin@cas.usf.edu
gtobin@usf.edu

Klement Tockner
Leibniz Institute of Freshwater Ecology and Inland
Fisheries, IGB
Mueggelseedamm 310
12587 Berlin
Germany
tockner@igb-berlin.de

Owen Brian Toon
Department of Atmospheric and Oceanic Sciences and
Laboratory for Atmospheric and Space Physics
University of Colorado
Campus Box 392
Boulder, CO 80309-0392
USA
btoon@lasp.colorado.edu

Andrzej Tyc
Department of Geomorphology
Faculty of Earth's Sciences
University of Silesia
Będzińska 60
41-200 Sosnowiec
Poland
andrzej.tyc@us.edu.pl

György Varga
Department of Geology
University of Pécs
Ifjúság u. 6
7624 Pécs
Hungary
gyoker@gamma.ttk.pte.hu

John D. Vitek
Department of Geology and Geophysics
Texas A&M University
College Station, Texas 77843
USA
jvitek@neo.tamu.edu
jvitek@tamu.edu

Barry Voight
Department of Geosciences
Penn State University
334A Deike Building
University Park, PA 16802
USA
and
U.S. Geological Survey
Cascades Volcano Observatory
Vancouver, WA
USA
voight@ems.psu.edu

Richard B. Waitt
Cascades Volcano Observatory
U.S. Geological Survey
1300 SE Cardinal Court, Ste. 100
Vancouver, WA 98683
USA
waitt@usgs.gov

Chenghu Wang
Department of Geology and Geological Engineering
University of North Dakota
81 Cornell Street MS 8358
Grand Forks, ND 58202–8358
USA
and
Institute of Crustal Dynamics
China Earthquake Administration
Beijing 100085
China
chenghu.wang@und.edu

Kerrianne Watt
School of Public Health
Tropical Medicine & Rehabilitation Sciences; James Cook University
Townsville, QLD 4811
Australia
kerrianne.watt@jcu.edu.au

Leanne Webb
University of Melbourne, Institute of Land and Food Resources, CSIRO Division of Marine and Atmospheric Research
PMB 1
Aspendale, Vic 3195
Australia
leanne.webb@csiro.au

Philip Weinstein
Professor of Ecosystem Health
University of South Australia
GPO Box 2471
Adelaide, SA 5001
Australia
philip.weinstein@unisa.edu.au

Paul R. Weissman
Jet Propulsion Laboratory/California Institute of Technology
Mail stop 183-301, 4800 Oak Grove Drive
Pasadena, CA 91109
USA
paul.r.weissman@jpl.nasa.gov

Graham Westbrook
School of Geography, Earth & Environmental Sciences
University of Birmingham
Edgbaston, Birmingham B15 2TT
UK
g.k.westbrook@bham.ac.uk

Linda M. Whiteford
Office of the Provost; Department of Anthropology
University of South Florida
4202 East Fowler Avenue, ADM 226
Tampa, FL 33620
USA
lwhiteford@acad.usf.edu
lwhiteford@usf.edu

Thomas Wilson
Department of Geological Sciences
University of Canterbury
Private Bag 4800
Christchurch 8140
New Zealand
thomas.wilson@canterbury.ac.nz

Ben Wisner
Oberlin College
373 Edgemeer Place
Oberlin, OH 44074
USA
and
Aon Benfield UCL Hazard Research Centre
University College London
UK
bwisner@igc.org

Michael R. Witiw
Embry-Riddle Aeronautical University Worldwide
Everett Campus
Everett, WA 98203
USA
MichaelWitiw@aol.com

Song Yang
Climate Prediction Center, NCEP/NWS/NOAA
5200 Auth Rd. Room 800
Camp Springs, MD 20746
USA
Song.Yang@noaa.gov

Ilya Zaliapin
Department of Mathematics and Statistics
University of Nevada Reno
Ansari Business Bldg
Reno, NV 89557
USA
zal@unr.edu

Raymond Zehr
NOAA/NESDIS Regional and Mesoscale Meteorology Branch
CIRA/CSU
Fort Collins, CO 80521
USA
Zehr@cira.colostate.edu

Zhengwen Zeng
Department of Geology and Geological Engineering
University of North Dakota
81 Cornell Street MS 8358
Grand Forks, ND 58202-8358
USA
zeng@und.edu

Karl-Otto Zentel
Deutsches Komitee Katastrophenvorsorge
Friedrich-Ebert-Allee 38
53113 Bonn
Germany
Zentel@dkkv.org

Janek Zimmer
Helmholtz Center Potsdam
GFZ German Research Centre for Geosciences
Haus C4
14473 Potsdam
Germany
jzimmer@gfz-potsdam.de

Sisi Zlatanova
GIS Technology / Research Institute for the Built Environment
Delft University of Technology
Jaffalaan 9
2628 BX Delft
The Netherlands
S.Zlatanova@tudelft.nl

Matija Zorn
Scientific Research Centre of the Slovenian Academy of Sciences and Arts
Anton Melik Geographical Institute
Novi trg 2
1000 Ljubljana
Slovenia
matija.zorn@zrc-sazu.si

Preface

Few subjects have caught the attention of the entire world as much as those dealing with natural hazards. The first decade of this new millennium provides a litany of tragic examples of various hazards that turned into disasters affecting millions of individuals around the globe. The human losses (some 225,000 people) associated with the 2004 Indian Ocean earthquake and tsunami, the economic costs (approximately 200 billion USD) of the 2011 Tohoku Japan earthquake, tsunami, and reactor event, and the collective social impacts of human tragedies experienced during Hurricane Katrina in 2005 all provide repetitive reminders that we humans are temporary guests occupying a very active, angry, and ancient planet. Any examples here to stress the point that natural events on Earth may, and often do, lead to disasters and catastrophes when humans place themselves into situations of high risk.

Few subjects share the true interdisciplinary dependency that characterizes the field of natural hazards. From geology and geophysics to engineering and emergency response to social psychology and economics, the study of natural hazards draws input from an impressive suite of unique and previously independent specializations. Natural hazards provide a common platform to reduce disciplinary barriers or boundaries and facilitate a beneficial synergy in the provision of timely and useful information and action on the critical subject matter.

As social norms change regarding the concept of acceptable risk and human migration leads to an explosion in the number of megacities, costal overcrowding and unmanaged habitation in precarious environments such as mountainous slopes, the vulnerability of people and their susceptibility to natural hazards increase dramatically. Coupled with the concerns of changing climates, escalating recovery costs, and a growing divergence between more developed and less developed countries, the subject of natural hazards remains on the forefront of issues that affect all people, nations, and environments all the time.

At the start of a new decade, in the first few months of 2010 alone, a magnitude 7 earthquake near Port-au-Prince, Haiti, killed an estimated 230,000 people, exposed the inadequacies of their infrastructure and emergency response capacity, and virtually crippled an entire nation. In contrast, a month later, a significantly larger magnitude 8.8 earthquake off the coast of Chile provided a sober lesson that those areas with a long history of exposure to natural hazards are indeed much more capable to cope with the consequences of unexpected events. Shortly thereafter, the eruptive events from Eyjafjallajökull volcano in Iceland virtually paralyzed air traffic in the United Kingdom and Western Europe for days. Travelers from around the world were impacted and inconvenienced. The economic repercussions were significant and all nations quickly realized how unprepared they are to natural hazards occurring outside of their borders.

This treatise provides a compendium of critical, timely, and very detailed information and essential facts regarding the basic attributes of natural hazards and concomitant disasters. For instance, swelling soils cost some $2.3 billion in damage annually in the United States alone, some 3 billion people were affected and about 7 million people died from flooding in the twentieth century, there are over 45,000 dams around the world and 80% of the useful storage capacity for hydropower in Asia will be lost by 2035, wildfires release an estimated 3,431 million tons of CO_2 into the atmosphere annually, lightning strikes cost the insurance industry more than $5 billion annually, an individual's social perception of the environment has radically changed in recent decades with the advent of near-real-time media reporting (CNN effect), in 1800 only 3% of the global population was urbanized as compared to over 50% by 2008, and resilience is frequently the most common outcome among survivors exposed to natural disasters. This volume clarifies and defines many fundamental concepts and terms, for instance, P waves vs. S waves, comets vs. asteroids, debris flow vs. debris avalanche, dispersive soil vs. expansive soil, or shield volcano vs. stratovolcano. Under a single cover, a diverse suite of topics ranging from solar flares, droughts, and blizzards to tiltmeters, seismographs, and GIS to cryological engineering and structural mitigation to post-traumatic stress disorder, coping capacity and religion, and hazards are addressed.

The *Encyclopedia of Natural Hazards* effectively captures and integrates contributions from an international portfolio of almost 300 specialists whose range of expertise addresses over 330 topics pertinent to the field of natural hazards. Disciplinary barriers are overcome in this comprehensive treatment of the subject matter. Clear illustrations and numerous color images enhance the primary aim to communicate and educate. The inclusion of a series of "classic case study" events interspersed throughout this volume provides tangible examples linking concepts, issues, outcomes, and solutions. These case studies illustrate different but notable recent, historic, and prehistoric events that have shaped the world as we now know it. They provide excellent focal points linking the remaining terms to the primary field of study. This *Encyclopedia of Natural Hazards* will remain a standard reference of choice for many years.

January 2013

Peter Bobrowsky

Acknowledgments

The editorial team is extremely grateful to the ca. 300 contributing authors who often willingly, but sometimes reluctantly, agreed to participate in this important endeavor. The success, utility, and widespread adoption of this encyclopedia owes considerable gratitude to all of these authors for sharing their knowledge and expertise toward a common cause of educating a broader population on the topic of natural hazards. All of the contributions were first read by the editor in chief; similarly, all of the contributions were also read by at least one of the ten editorial board members. Each contribution was critically reviewed by at least one external specialist. We are also grateful to the following individuals who each reviewed multiple papers to ensure the scientific integrity of the submissions was of the highest caliber: Orhan Altan, Mohsen Ashtiany, Tom Baldock, Martin Batterson, Allison Bent, Bruce Broster, Robert Buchwaldt, Tom Casadevall, Hubert Chason, Paul Croft, Heather Crow, Johannes Dahl, Miranda Dandoulaki, Tim Davies, A. Deganutti, Fabio Dell'Acqua, Angus Duncan, Frank Fiedrich, Charles Finkl, Duane Froese, Sven Fuchs, J.C. Gaillard, Matthias Garschagen, David Gauthier, Marten Geertsema, Thomas Gerz, Yuri Gorokhovich, Raisa Gracheva, Rick Guthrie, Fausto Guzetti, Reginald Hermanns, Lynn Highland, David Huntley, Kovacs Janos, Rohit Jigyasu, Kate Jones, Margareth Keiler, Julia Kloss, Jasper Knight, Robert Korty, Migala Krzystof, Steve LaDochy, Maurice Lamontagne, Craig Landry, Julie Lantos, Jean Francois Lenat, Suzanne Leroy, Jean-Sebastian L'Heureux, Colin Mackay, Brian Marker, Bruce Masse, James McCalpin, Scira Menoni, John Miles, Joann Mossa, Ilian Noy, Douglas Paton, Toon Pronk, Rajmund Przybylak, Fabrice Renaud, Tiziana Rosetto, Jeff Rubin, Zita Sebesvari, Vladimir Sokolov, Mauro Soldati, Anders Solheim, Ian Spooner, Sasha Strom, Bert Struik, Andy Stumpf, Peter Suhadolc, Brian Tang, Jakob Tendel, Vincent Parisi, Gonghui Wang, Brent Ward, Dennis Wenger, Tom Wilson, and Karl Otto Zendel. Following revisions of the contributions, the entries were read by the editorial board. The editorial board very much appreciates the guidance, perseverance, and patience of the managing staff at Springer Publishing, in particular Petra van Steenbergen, Sylvia Blago, and Simone Giesler. They were instrumental in the delivery of this volume. Peter Bobrowsky expresses his personal and sincere thanks to his wife Theresa for her unselfish support, appreciation, and understanding as to why so many precious evenings and weekends were sacrificed trying to influence and get along with 300 personalities during the past few years. Additional thanks go to the smaller members of the family, Michiko and Toba, for providing regular and amusing distractions during the entire process. Collectively we all hope that the information in this book serves its intended purpose to better inform, educate, and prepare all of those people who are directly or indirectly affected by natural hazards. Our aim is to help reduce the risk and losses associated with "things that go bump in the night."

A

AA-LAVA

Robert Buchwaldt
Massachusetts Institute of Technology, Cambridge, MA, USA

Synonyms

a'a; 'a'a; a-aa

Definition

Aa (derived from the Hawaiian word a'a' for rough lava) lavas is the term for cooling textures of a highly viscous lava flow and was introduced as a technical term by Clarance Dutton in the year 1883. Aa is characterized by a rough rubbly surface composed of broken lava blocks called clinkers. When the molten rock cools, the lava flow increases its viscosity due to degassing and crystallization of minerals. At some point, the lava flows so slowly that it allows a thick crust to form. As flow continues to move, the cooler and brittle skin surface breaks into rough jagged blocks, or clinkers. The clinkers are carried along at the surface. At the leading edge of an aa flow, however, these cooled fragments often tumble down the steep front and are subsequently buried by the advancing flow. This process produces a layer of lava fragments both at the bottom and top of an aa flow. Accretionary lava balls as large as 3 m are common on aa flows (Figure 1).

Aa flows are emitted from vents at high rates ranging up to 50 km/h, often with much lava fountaining. They are characteristic of viscous magmas. Aa flows are animated with sporadic bursts of energy. Such flows may overturn houses, walls, and forests.

AA-LAVA, Figure 1 An aa flow front on the coastal plain of Isabella, one of the islands of the Galapagos archipelago, Ecuador (Robert Buchwaldt).

Bibliography

Dutton, C. E., 1883. 4th Annual Report U.S. Geological Survey, 95.

Cross-references

Eruption Types (Volcanic)
Lahar
Shield Volcano
Stratovolcano
Volcanoes and Volcanic Eruptions

ACCELEROMETER

Zhengwen Zeng[1], Lin Fa[1,2]
[1]University of North Dakota, Grand Forks, ND, USA
[2]Xi'an Institute of Post and Telecommunications, Xi'an, Shaanxi, China

Synonyms
Geophone; Gravimeter

Definition
An accelerometer is defined as a transducer whose output is proportional to acceleration.

An accelerometer measures the proper acceleration it experiences relative to freefall. This is equivalent to inertial acceleration minus the local gravitational acceleration, where inertial acceleration is understood in the Newtonian sense of acceleration with respect to a fixed reference frame, which the Earth is often considered to approximate.

An accelerometer can be used for detecting magnitude and direction of the acceleration as a vector quantity.

An accelerometer consists of a proof mass, a reference frame, a sensor (an induction coil or a potentiometer and so on), a spring, a damper, and a casing. Because of the constraint of the reference frame, the proof mass can only move along an axis-line. According to Newton's law, when the casing experiences an acceleration in the axis, the proof mass with a certain inertia resists a change in its state of motion. The motion of the proof mass relative to the casing takes place. Therefore, the string creates deformation, and the proof mass begins to experience acceleration motion under the effect of the string force. Until the inertia force created by the moving proof mass equals the string force, the motion of the proof mass relative to the casing stops. The deformation of the string at this time reflects the magnitude of the measured acceleration. The induction coil or the potentiometer then converts the acceleration signal into the output electric-signal. The damper is used to improve the dynamic quality of the accelerometer.

A geophone is a type of accelerometer, and its output is proportional to the acceleration of earth particles. It is used to measure the seismic signals created by an earthquake or a seismic energy source.

A gravimeter also is a type of accelerometer, which is used for measuring the local gravitational field.

Bibliography
Sheriff, R. E., 1984. *Encyclopedic Dictionary of Exploration Geophysics*. Tulsa: Society of Exploration Geophysicists.
Wikipedia Online, 2009. Accelerometer. Wikimedia Foundation.

Cross-references
Body Wave
Earthquake
Primary Wave (P Wave)
Secondary Wave (S Wave)

ACID RAIN

Mary J. Thornbush
University of Birmingham, Edgbaston, Birmingham, UK

Synonyms
Acid deposition; Acid precipitation

Definition
The US Environmental Protection Agency (EPA) refers to acid rain as a broad term that encompasses acidic particles and vapors deposited through dry and wet deposition. Wet deposition occurs in wet precipitation (such as rain, sleet, hail, snow, fog, or mist), with a pH normally less than 5.6. Half of atmospheric acidity falls to the ground as dry deposition (such as fly ash, sulfates, nitrates, and gases), where acidity is incorporated into dust and smoke, and then falls dry onto natural and human-made surfaces, including buildings. When gases come into contact with water, they become acids (such as sulfuric and nitric acids).

Sources: Natural (as from volcanoes and decaying vegetation) and anthropogenic (as from emissions into the atmosphere from human activities, including combustion). Primary anthropogenic emissions are of sulfur dioxide (SO_2) and nitrogen oxides (NO_x) from the burning of fossil fuels. Energy production through the combustion of coal remains a problem even in the developed world, where many power plants still rely on coal-burning. Coal-fired power stations account for more than 50% of global SO_2 emissions (Mohanty et al., 2009). Asian countries are now experiencing severe environmental degradation due to acidic precipitation, for example, China (cf. Wang et al., 2008).

Effects: Acidic pollution affects lakes, rivers, forests, soils, fish and wildlife populations, and buildings. Areas that comprise acidic bedrock (such as granite), as on the Precambrian Shield in Eastern Canada (Scanlon, 2001), have a low natural buffering capacity and support "acid-sensitive" ponds and lakes with an alkalinity less than 8 mg $CaCO_3$ per liter (Hagar et al., 2000). For instance, Sudbury, Ontario in Canada was artificially acidified by nickel mining and processing in the 1950s, which thinned soils and led to nutrient depletion, killed trees from soil acidity and chlorosis, and blackened surfaces by acid corrosion. Lakes in that area and in windward areas also suffered acidification. However, a sample of 44 acidic Sudbury lakes monitored since 1981 have shown a reduction in the number of highly acidic lakes (pH < 5.0) from 28 to 6 by 2004, indicating gradual recovery (Keller et al., 2007).

Extent: Acid rain is a trans-boundary environmental issue because wind-blown atmospheric pollutants can be carried great distances across political boundaries. For example, half of the acid deposits in Eastern Canada come from the USA (in the Muskoka-Haliburton area north of Toronto, this rises to three-quarters); similarly, acid rain in Japan comes from

Korea and China; whereas, acid rain in Norway is derived from England and Scotland (Scanlon, 2001).

Control: Acid rain monitoring programs were initiated in the 1970s in North America (for example, the USA Federal Clean Air Act of 1970), and policy has since changed to control environmental degradation in urban areas as well as in forests, rivers, and lakes associated with acid precipitation. The 1985 Eastern Canada Acid Rain Control Program and 1990 USA Clean Air Act Amendments both called for reduced annual SO_2 emissions by 40% from 1980 levels by 1994 and 2010, respectively (Venkatesh et al., 2000).

Bibliography

Hagar, W. G., Crosby, B. A., and Stallsmith, B. W., 2000. Comparing and assessing acid rain-sensitive ponds. *Journal of Hazardous Materials*, **74**, 125–131.
Keller, W., Yan, N. D., Gunn, J. M., and Heneberry, J., 2007. Recovery of acidified lakes: lessons from Sudbury, Ontario, Canada. *Water, Air and Soil Pollution*, **7**, 317–322.
Mohanty, C. R., Adapala, S., and Meikap, B. C., 2009. Removal of hazardous gaseous pollutants from industrial flue gases by a novel multi-stage fluidized bed desulfurizer. *Journal of Hazardous Materials*, **165**, 427–434.
Scanlon, J., 2001. Increasingly intolerable boundaries: future control of environmental pollution. *Journal of Hazardous Materials*, **86**, 121–133.
Venkatesh, S., Gong, W., Kallaur, A., Makar, P. A., Moran, M. D., Pabla, B., Ro, C., Vet, R., Burrows, W. R., and Montpetit, R., 2000. Regional air quality modelling in Canada – applications for policy and real-time prediction. *Natural Hazards*, **21**, 101–129.
Wang, W., Qin, Y., Song, D., and Wang, K., 2008. Column leaching of coal and its combustion residues, Shizuishan, China. *International Journal of Coal Geology*, **75**, 81–87.

Cross-references

Challenges to Agriculture
Erosivity
Global Dust
Land Degradation
Release Rate
Volcanic Ash
Volcanic Gas

ADAPTATION

Philipp Schmidt-Thomé[1], Sirkku Juhola[2]
[1]Geological Survey of Finland (GTK), Espoo, Finland
[2]Aalto University, Aalto, Finland

Synonyms

Adjustment; Modification

Definition

In a geo-scientific context, the term adaptation is the ability of a species to adjust to (varying) environmental conditions. In association with a changing climate, adaptation refers to the processes, practices, or structures to moderate or offset potential damages or to take advantage of opportunities (Smit and Pilifosova, 2001). Adaptation of a society is dependent on the adaptive capacity of that particular society, irrespective of whether adaptation is automatic or planned.

Not all life forms are able to survive in all climate zones on the planet earth; human beings have permanently settled on all but one continent and in basically all climate zones, thus representing one of the most adaptive mammal species. By manufacturing clothing, tools, and other devices that enabled life in all kinds of natural conditions, humans have increased their adaptive capacity.

In the context of natural hazards, adaptation refers to the potential of humans to understand and respect natural extreme events that may affect life and human assets. In a strict sense, adaptation means the ability to respect extreme events in the design and shape of the living environment, for example, the structure of houses or the location of fields and pastures. In this sense, adaptation does not incorporate protective structures, as it refers to the possibilities to cope with the living conditions without restricting the total extent and exposure of an extreme event.

In practice (e.g., see entry *Land-Use Planning*), adaptation also encompasses protection and retreat. In the example of flood-prone areas, adaptation to life in regularly flooded areas is often met with the construction of houses on pillars or elevated mounds. In this case, protection means the construction of sea walls or dams to protect an area from being flooded whereas retreat refers to permanent abandonment of flood-prone areas. Historical examples of adaptation to flood-prone areas can be found in many places around the world, for example, on the island of Chiloe (Chile) to respect local tidal changes and on Lake Tong Le Sap (Cambodia) to maintain settlements in the yearly reoccurring flooding. One of the most flood-protected countries in the world is the Netherlands, where the entire coastline and the hinterland are integrated into flood protection measures and plans. Retreat can be found in areas that experienced sea level rise, such as historic settlements in the Black Sea Basin or on the Mediterranean shorelines.

With the introduction of protective measurements, humans have further increased their adaptive capacity, such as living closer to areas threatened by natural hazards. On the other hand, technical solutions to protect from natural hazards have also increased the exposure of life and assets to hazards and have thus increased the overall vulnerability (see entry *Vulnerability*). For instance, the building of dams along riverbeds has enabled the use of agriculture for crop production in flood plains, which were formerly used only as pastureland. The construction of dams has also often increased the river flow velocity and thus modified flood-prone areas, also because simultaneously natural flood retention basins have been removed. The benefits of increasing agricultural productivity have been acquired at the cost of increasing the vulnerability to floods.

Several disasters that occurred in the early twenty-first century have been attributed to climate change (see entry *Climate Change*), meanwhile the real causes for the increasing extent of damage is, at least to a great extent, caused by poorly-adapted location of settlement and inappropriate land use. The continuous development and change of vulnerability patterns therefore imply that adaptation and adaptive capacity are not static, but continuously evolving concepts.

Within the context of climate change, adaptation is seen as a response strategy involving adjustments to reduce vulnerability of communities, regions, or activities. Adaptive capacity, then, is defined as the ability or potential of a system to respond successfully to climate variability and change, and it includes adjustments in both behavior and in resources and technologies (IPCC, 2007). It has been argued that adaptive capacity, first and foremost, is context specific and varies from country to country and region to region and within social groups and individuals. It also varies over time, responding to society's changing economic, institutional, political, and social conditions (Smit and Wandel, 2006). In spatial terms, adaptive capacity is a nested concept: Capacities of regions are tied to the capacity of countries in terms of enabling or constraining environments for adaptation. The adaptive capacities at a national level may not always correspond to those at local level.

Irrespective of the complex nature of the concept, identifying the determinants of adaptive capacity is of importance to both scientists and policy-makers. Table 1 summarizes the most often used categories.

The list is not exhaustive, and these determinants are not independent of each other, nor are they mutually exclusive; rather, it should be recognized that the combination of these determinants varies between regions and countries. Several studies exist that have focused on identifying determinants of adaptive capacity – at the national level (Haddad, 2005; Yohe and Tol, 2002; Adger et al., 2005; Moss et al., 2001); at the local level (Posey, 2009; Engle and Lemos, in press); and across all levels of governance (Westerhoff et al., submitted).

The adaptive capacity within communities is extremely heterogeneous by locality, and also scale-dependent. There are two main possibilities on defining adaptation strategies: anticipatory and reactive (IPCC, 2007), with the former focusing on measures that can be taken as a result of expected changes, whilst the latter is in response and in reaction to changes that have already taken place. In the historical past, when scientific knowledge forecasts and planning were limited, basically all adaptation was taken in reaction to perceived climatic conditions. As the knowledge and information of climate change is steadily increasing, societies have an opportunity to engage in anticipatory adaptation.

Anticipatory adaptation mainly takes place through planned adaptations, although some automatic adaptation can take place (IPCC, 2007). Planned adaptation can be directed toward reducing the potential impacts of climate change or to reduce the general vulnerability of a society or toward building adaptive capacity. Planned adaptations

Adaptation, Table 1 Determinants of adaptive capacity (Adapted from Smit and Pilifosova, 2001)

Economic resources	Economic assets, capital resources, financial means, and wealth
Technology	Technological resources enable adaptation options
Information and skills	Skilled, informed, and trained personnel enhance adaptive capacity, and access to information is likely to lead to timely and appropriate adaptation
Infrastructure	Greater variety of infrastructure enhances adaptive capacity
Institutions	Existing and well-functioning institutions enable adaptation and help to reduce the impacts of climate-related risks
Equity	Equitable distribution of resources contributes to adaptive capacity

can, for example, demand changes in current land-use and planning practices, highlighting the local nature of decisions related to adaptation. These are difficult and challenging and decisions often need to be taken on case-by-case basis, involving cost-benefit calculations of the measures involved. Thus, planned adaptation measures demand not only scientific knowledge and a balanced analysis of uncertainties of potential future developments, but also societal acceptance in order to be implemented and sustainable. The issuance and implementation of adaptation guidelines can be steered at a national level. The assessment of both the adaptive capacity and adaptation potential is to be assessed and planned for at regional or local levels that is, the level of detailed land-use planning.

Summary

Adaptation has reentered spatial development concepts in the wake of the discussion on climate change since the end of the twentieth century. In historic times, people generally were more used to adapting to extreme events, such as avoiding settlements in flood-prone areas or using crop types adjusted to existing water resources. A general attitude to use technical solutions to shape the living environment for human benefit and the belief one can control natural extreme events has not only led to straightening of river beds but also to inappropriate agriculture processes and overuse of water resources. These phenomena are accompanied by population increase, mainly in developing countries and, as an overall phenomenon, in coastal and flood-prone areas. The rise in costs caused by damages of natural hazards since the end of the twentieth century can so far not be attributed to a change in weather extremes (e.g., Barredo, 2009, 2010), but it shows an increase in overall vulnerability and thus the importance to adapt to natural hazards, as well as potential impacts of the changing climate. Traditional adaptation measures were non-structured and often reactive. The current debate on climate change impacts leads to the need to incorporate adaptation into land-use planning and have a stronger focus on cost-benefit analysis of anticipatory adaptation measures.

Bibliography

Adger, W. N., Arnell, N. W., and Tompkins, E. L., 2005. Successful adaptation to climate change across scales. *Global Environmental Change Part A*, **15**(2), 77–86.

Barredo, J. I., 2009. Normalised flood losses in Europe: 1970–2006. *Natural Hazards Earth System Sciences*, **9**, 97–104.

Barredo, J. I., 2010. No upward trend in normalised windstorm losses in Euope: 1970–2008. *Natural Hazards Earth System Sciences*, **10**, 97–104.

Engle, N. L., and Lemos, M. C., (In Press, Corrected Proof). Unpacking governance: Building adaptive capacity to climate change of river basins in Brazil. *Global Environmental Change*.

Haddad, B. M., 2005. Ranking the adaptive capacity of nations to climate change when socio-political goals are explicit. *Global Environmental Change Part A*, **15**(2), 165–176.

IPCC, 2001. *Climate Change 2001: Impacts, Adaptation and Vulnerability*. Cambridge: Cambridge University Press.

IPCC, 2007. *Assessment of Adaptation Practices, Options, Constraints and Capacity*. Cambridge: Cambridge University Press.

Moss, R. H., Brenkert, A., and Malone, E. L., 2001. *Vulnerability to Climate Change: A Quantitative Approach*. Richland Washington: Pacific Northwest National Library.

Posey, J., 2009. The determinants of vulnerability and adaptive capacity at the municipal level: Evidence from floodplain management programs in the United States. *Global Environmental Change*, **19**(4), 482–493.

Smit, B., and Pilifosova, O., 2001. Adaptation to climate change in the context of sustainable development and equity. In *Climate Change 2001: Impacts, adaptation, and vulnerability- Contribution of the Working Group II to the Third Assessment report of the Intergovernmental Panel on Climate Change*. Cambridge: Cambridge University Press.

Smit, B., and Wandel, J., 2006. Adaptation, adaptive capacity and vulnerability. *Global Environmental Change*, **16**(3), 282–292.

Westerhoff, L., Keskitalo, E. C. H., and Juhola, S., (Submitted). Capacities across scales: enabling local to national adaptation policy in four European countries. *Climate Change*.

Yohe, G., and Tol, R. S. J., 2002. Indicators for social and economic coping capacity—moving toward a working definition of adaptive capacity. *Global Environmental Change*, **12**(1), 25–40.

Barredo, J. I. 2010 No upward trend in normalised windstorm losses in Euope: 19070–2008. Natural Hazards Earth System Sciences, 10, 97–104.

Cross-references

Climate Change
Coastal Zone, Risk Management
Land-Use Planning
Uncertainty
Vulnerability
Zoning

AIRPHOTO AND SATELLITE IMAGERY

J. D. Mollard
J.D. Mollard and Associates Ltd., Regina, SK, Canada

Synonyms

3D airphoto; Multispectral satellite imagery; Remote sensing

Definition

Stereoscopic (3D) and multispectral analysis of aerial photography and satellite imagery is used by skilled terrain analysts and interpreters to identify (remote sense) Earth processes, environments, landforms, and materials, and to use the information to map and evaluate the physical and cultural terrain characteristics and conditions that adversely affect people and property. These remotely sensed data are used with supporting literature and networked information to investigate, assess, and manage (monitor, mitigate, avoid, eliminate, treat) natural hazards, such as earthquake and volcanic activities, excessive *Erosion* and landslides, and climate effects that precipitate drought and wildfires among other hazards.

Discussion

Aerial photographs and satellite imagery are useful tools for viewing, identifying, and assessing natural hazards. Used in conjunction with easily accessible aids such as maps (topography, geology, climate), pocket stereoscopes, and computer programs, aerial and space imagery allow terrain analysts and interpreters to acquire information regarding the Earth's landscapes and environments at a distance without direct contact, a science commonly termed "remote sensing."

Some remote sensing specialists see differences between airphoto terrain analysis and airphoto terrain interpretation. Airphoto terrain analysis involves identifying physical environments and their geomorphic features (landforms) *on* the Earth's surface and inferring earth materials (soils and rocks) in them, and their material properties (e.g., permeability, compressibility, strength) and conditions (e.g., wet or dry, frozen or unfrozen) *under* the earth's surface. Airphoto terrain interpretation involves assessing what is important about the physical environment – the earth materials, properties, and conditions in it – for a specific use or objective. Both techniques are used in identifying, mapping, and assessing natural hazards that adversely affect human lives (loss of life, safety, property) in addition to damage to infrastructure, resources, and the natural environment.

Remote sensing requires an understanding of Earth materials, processes, and environments as well as their identifying features and characteristics in airphoto and satellite images. Today, airphoto and satellite image remote sensing technologies are widely used in locating, mapping, planning, exploring, and developing social and industrial infrastructure, energy, and natural resources. In terms of natural hazard assessment, these studies tend to emphasize the more densely populated places on Earth, where natural hazards and people come together.

Airphotos and global natural hazards

Remote sensing technologies involve information acquired through electromagnetic radiation (EMR) reflected and emitted from the Earth's surface. Useful information can be obtained from computer analysis of the captured EMR signals along with the analysis and interpretation of conventional panchromatic, true color and false color airphotos, thermal infrared imagery, and radar images. Several wavebands (visible, near and thermal infrared, radar) are used to sense natural hazards from the air and space. Airphoto terrain analysis and interpretation applied to natural hazard appraisal are used by geologists, geographers, engineers, meteorologists, and environmental scientists with training and experience in understanding and recognizing natural hazard phenomena.

Both airphoto and satellite images display a wide field of view of the Earth's surface, revealing natural and cultural landscape features and patterns that may be completely obscured to viewers on the ground. This reorientation of perspective permits terrain analysts and interpreters to make sense of the Earth's surface and its hazardous features and conditions. Airphotos can also be used to map and evaluate terrain hazard features and conditions in two, three, and four dimensions – the fourth being time from different ages of imagery spanning several decades. Stereoscopic airphotos (pairs of photos taken of the same area from two different positions) allow the airphoto terrain analyst to study landscape relief in three dimensions (3D) using a simple handheld stereoscope. Two or more images taken years or decades apart reveal changes in the natural physical environment (topography, drainage, geology and geomorphology, and climate effects) as well as changes in human cultural activities that can contribute significantly to natural hazards. The airphoto and satellite image study of hazard-indicator clues over time is part of the technique of "change detection."

Small-scale airphotos (e.g., 1:60,000) and satellite images show landscapes in a regional context, allowing remote sensing interpreters to view diagnostic identification features of natural hazards in plan view and 3D. Large-scale airphotos on the other hand show subtle soil and rock tones, minor relief variations, and small drainage features in greater detail. Even though smaller and less obvious, these special characteristics are among a group of significant geoindicator clues used to assess different kinds of past, present, and future natural hazards.

Because most natural hazards involve ground movements – thus change on and below the Earth's surface – aerial and space imagery remote sensing specialists take advantage of three useful attributes of airphoto and satellite imagery: *multitemporal*, *multispatial*, and *multispectral* capabilities. Separately and together, these attributes are used to infer natural hazard *exposure*, *sensitivity*, and *vulnerability*. Exposure is the degree to which a particular hazard or phenomena is likely to occur and is largely related to topography, geology, and climate in the natural environment. Sensitivity is the potential degree to which a community and infrastructure could be affected by a natural hazard. Vulnerability is the degree to which a person or a system is likely to be adversely affected by a natural hazard. Vulnerability depends on both exposure and sensitivity. Although the hazard exposure may be high, the hazard sensitivity may be low because people have learned how to cope, having "adapted" to the hazard exposure conditions. (Adaptation is the response to natural hazards, environmental conditions, or stresses, especially those associated with climate change and variation.) Conversely, if the sensitivity is high and exposure is low, the vulnerability will be low. If both exposure and sensitivity are high, then the vulnerability could be very high.

Many terrain analysts and interpreters prefer conventional stereoscopic (3D) panchromatic airphotos because they are available in most areas of the world, because their coverage dates back to the 1930s and sometimes earlier, and because their applications are familiar to most users. Using good quality 3D airphotos, experienced terrain analysts can detect many shades of gray tone that aid inferring subtle changes in topography, soil moisture, and vegetation in addition to compositional, structural, and strength properties of soils and rocks in which earthquake and volcanic activity create landslides and other natural hazards.

In addition to conventional airphotos, specialists use true color and false color airphotos in their terrain analysis studies. Using special filters during photography, it is possible to isolate a "slice" of specific wavelengths of light, which may allow interpreters to pick out potential hazard targets for more precise, in-depth target classification and assessment. By combining black-and-white film with a filter that cuts out visible wavelengths, it is possible to produce airphotos in the reflected near infrared (light beyond the range of the human eye), which expresses vegetation differences more clearly. Near-infrared airphotos also show water clearly because water absorbs the infrared wavelengths and appears black, or nearly so, making it easier to assess the critical roles that surface water and groundwater play in causing landslides, snow avalanches (see entry *Avalanches*), major storms, flooding, drought, and other climate-variation effects.

Thermal infrared images in the far infrared wavelength capture heat emitted from the Earth's surface features. These images can be taken in both the daytime and nighttime to display differences in the way terrain materials absorb, hold, and release heat. Highly absorbent surfaces (e.g., sandstone) are strong nighttime emitters. Among the more dramatic thermal images are those taken of volcanoes, particularly ones with active vents that eject hot lava flows (lahars) and airborne rock fragments, ash, and gases, which can show up conspicuously in thermal imagery. Infrared anomalies in thermal images may reveal microearthquakes in terrains that are hydrothermally active. Thermal infrared images can also be used to detect warm and cold groundwater circulating in bedrock fracture systems and detect massive ground ice at shallow depths in permafrost terrain as well as the effects of global warming on permafrost and Arctic Ocean ice melting.

While visible, reflected near-infrared, and emitted thermal images are passive systems, radar is an active system that supplies its own energy source (the acronym *radar* comes from "radio detecting and ranging"). Radar uses radio and microwave pulses to "see" great distances despite fog, rain, snow, clouds, and darkness. It is also used to highlight natural hazard features like steep slopes on mountain and valley sides, and different types of landslide and ground subsidence collapse and heave, as well as geological structures like faults, solution-widened joint systems and differentially eroded strata in sedimentary rock folds. There are several radar systems in use today: SAR (synthetic-aperture radar), SLAR (side-looking airborne radar), and InSAR (interferometric synthetic aperture radar). Some terrain attributes, such as topography, display dramatic three-dimensional relief in radar images, emphasizing long lineament trenches and troughs, a few of which might be structural in origin and therefore the location of past and future earthquake activity.

LiDAR (from "light detecting and ranging") uses laser pulses instead of microwaves, and is a highly useful remote sensing tool in transportation and communication system route-selection studies because it penetrates tree cover and other heavy vegetation, and records minor topographic features in remarkable detail, especially those distinguishing characteristics of different landslide types (see entry *Landslide Types*) and their change and evolution over time, important in determining monitoring strategies. Large-area mosaics assembled from them are a useful resource in the study of global natural hazards and disasters (i.e., catastrophic natural hazards).

Satellite images and global natural hazards

Satellite imagery has the timing, location, and larger area advantage of providing immediate information about Earth, its surface features and conditions, for those investigating the vulnerability of natural hazards in particular physical, geologic, and geomorphic settings and environments. Viewed digitally, Landsat images are used to identify landforms and infer the kinds of earth materials in them in addition to important soil and rock physical hazard properties, like strength. These kinds of information are used to variously help predict, prevent, monitor, mitigate, avoid, eliminate, and treat natural hazards. The challenge is which one to select and why and how. A list of hazards could include several landslide types and their distinguishing features in airphotos, snow avalanche tracks, faults and other geologic structural lineaments, active volcanic craters and calderas, lava flows and ejected pyroclastic materials (tephra), actively eroding coastlines and lake shorelines, flash-flooding river valleys, and melting sea ice and permafrost. Infrared and radar satellite images are also used to chart wave and current patterns, study pack ice on large lakes and the oceans, and track oil spills and thermal pollution in lakes and the oceans.

Meteorological satellites using visible and infrared wavebands are used to forecast weather and track major offshore and onshore storms. They are also used to detect, map, and monitor climate change effects and map different sizes and ages of wildfire burn areas and vegetation disease, land cover and land use change, and population growth. Remote sensing satellite images are also used to assess global natural hazard damage affecting people and property caused by such atmospheric and hydrometeorological events as hurricanes, typhoons, cyclones and tornadoes, major snow, rain and hail storms, flash floods, and droughts that cause desertification.

Earthquake hazards

Existing regional and local geological maps and data on regional and local seismicity (e.g., seismicity and earthquake zone maps) are collected and reviewed to provide a database for airphoto and satellite imagery studies of earthquake hazards and their damage. Maps showing the clustering, recurrence, and magnitude of past earthquakes are used to focus the analysis and interpretation of their visible effects from remotely sensed images. Visible effects include fault (see entry *Fault*) and rift lineaments and scarps, offset physical features like ridges and stream channels, tsunami effects, various slides (snow, ice and rock avalanches, and soil liquefaction), open fissures, ground deformations, and blocked groundwater. Most of these are recognizable and their significance inferred from suitably scaled aerial photography and space images, including visible, thermal infrared and radar. The technology can be used to map and evaluate earthquake damage (see entry *Earthquake Damage*) to buildings and critical infrastructure: bridges and tunnels, pipelines and power plants, and dams and reservoirs.

One can apply remote sensing technology to a major fault system like the *San Andreas Fault* – a 430-km long strike-slip fault that passes through San Francisco, the site of an April 18, 1906, catastrophic earthquake. Estimated at 8.3 magnitude, the earthquake caused ground displacements up to 6 m, resulted in the death of 450 people, and caused enormous destruction. It should be noted that the San Andreas is not a single fault; rather, it is a series of interconnected shorter, subparallel and interwoven fault ruptures that make up the width zone of seismic activity, where the tears can be viewed and measured from airphotos and passive and active satellite imageries. Longer lengths of main faults and wider zones of seismic activity and branching fault ruptures – as viewed along the San Andreas Fault system – can be measured from remotely sensed satellite images and correlated broadly with higher earthquake magnitudes, longer duration of ground shaking, and greater amounts and degrees of caused damage. Other relationships include fault displacement versus earthquake magnitude, fault length versus earthquake magnitude, and distance from a fault versus acceleration due to gravity – all useful in the study of earthquake features in aerial photographs and satellite images.

Volcanic hazards

Large numbers of volcanoes (as well as earthquakes) are largely concentrated along the edge of tectonic plates in the Earth's crust. The plates subduct, collide, pull apart, and slide past one another creating earthquakes and volcanic activity. The most famous occurrence of tectonic plate action is located along the margins of continents encircling the Pacific Ocean, called the "Ring of Fire." Along this "ring of fire," airphotos and satellite images can be used to map, monitor, and track the behavior of active and potentially active volcanic vents as they heat up, stir, swell, and vibrate before they erupt, ejecting hot lavas (lahars) and hurling rock fragments and clouds of ash, dust, and poisonous gases (CO_2, H_2S, SO_2, Cl) into the air. Remote sensing specialists have been able to track clouds of ash across an entire continent using satellite radar imagery (AVHRR bands 4 and 5) and InSAR measurements can be made to assess the bulging side of composite volcanoes, such as Mount St. Helens.

Volcanic explosions under snow and glacier-ice covers can create tremendous devastation. Many of their frozen features that are visible on the ground are also identifiable in aerial and space imageries. They include rock avalanches and lahars (consisting of mixtures of water, rock fragments, and mud), terrain effects from acid rain, flooding, fires, surface water and groundwater contamination, and forest destruction that denudes volcanic slopes and the surrounding outlying area. Precursor signs that aerial and space analysts look for and track when hot magma begins to rise in volcanoes are swelling, cracks and fractures, and microearthquakes– warning signs that can be used with seismic meter and tilt meter data to assess the level of volcanic hazard activity.

Slope failure (landslide) hazards

There are many recognizable landslide types. Virtually all are identifiable from their associated distinguishing characteristics in 3D airphotos and satellite images. The slope failures include creep (see entry *Creep*) (a slow earthflow), soil and rock topples and falls, planar translational rockslides and rock avalanches; rotational slides and slumps in soils and rocks; debris avalanches (see entry *Debris Avalanche*), slides, flows and torrents; lahars (volcanic mudflows); rapid earthflows (also called spreads, flowslides, and mudflows), shallow active-layer detachment (skin) flows and bimodal flows involving retrogressive ice-rich headscarp melting and long gently sloping tongue-like flows in horseshoe-shaped cavities in permafrost terrain.

Airphotos and satellite imagery terrain analysts and interpreters familiar with different soil and rock landslides are aware that certain types tend to be concentrated in regionally defined physical environments. Three common landslide-prone natural environments are tectonic plate boundaries, steep slopes in high mountain terrains, and deep river valleys where the toe of valley sides is being actively stream-eroded. Natural hazard investigators of landslides are familiar with geological materials that are landslide-prone. Three examples are Cretaceous marine clay shales, commonly with bentonite seams on steep, high valley sides; stacked rows of columnar basalt flows with weak weathered clayey surfaces between some of the basalt rows on canyon cliffs; and weathered volcaniclastic and serpentine rocks in steep, high slopes – mainly in tropical terrains. Steeply dipping interbedded sedimentary rock strata having weak clay shale layers that daylight on steeper slopes are also prone to translational slide failures. Eroded slopes on silty and clayey glaciolacustrine and sensitive glaciomarine sediments are also landslide-prone. Several geological rock structures– faults, joints and other rock fractures, bedding plane surfaces, and foliation– are rarely visible yet locally inferable from airphotos and satellite images, and they all can be significant factors in creating natural hazards.

As well as precursor topographic and geologic geoindicators of landslides in airphotos and satellite images, there can be adverse human indicators too. They include deep human-made excavations on lower slopes and loading placed on upper slopes, irrigation, mining activity, tree clearing, and broken water and sewer pipelines, among others. Any one of these can affect or trigger landslide instability. Landslides are usually triggered by specific natural events, such as earthquake shaking, volcanic explosions, and weather and climatic effects like intense prolonged rainfall and ice melting that causes soil and rock saturation and elevated porewater pressures on failure slip surfaces.

By evaluating the roles that topography, geology (especially geomorphology), human activities, and triggering factors play in creating landslides, aerial and space imagery interpreters may (depending on the situation on the ground) recommend avoiding, eliminating, monitoring, mitigating, or treating (e.g., stabilizing) the hazard conditions.

Ground subsidence, collapse, and heave hazards

Hazardous vertical ground subsidence is caused by seismic activity and oil, gas, and groundwater extraction, resulting in faulting and ground cracking, flooding, and destruction of human structures –visible, mappable, and assessable effects detectable from multidate, multiscale, and multispectral aerial and space images. Ground collapse can be caused by subsurface coal and metal mining, underground carbonate and evaporite dissolution forming collapse sinkholes in karstic topography, where near-surface cavities and caverns grow large enough for the roof above to collapse. Near-vertical piping erosion of cavities can develop in thick, loose loessial and glaciolacustrine silty and fine sandy soils. The resulting, commonly roundish hole-like depressions are usually recognizable from their multiple dot pattern in 3D airphotos. And shield volcanoes may develop collapse caldera (see entry *Calderas*), where surface rocks drop into the magma chamber. All these collapse features are identifiable on 3D aerial photographs. Ground swelling (heave) and shrinkage are significant economic hazards affecting buildings in large population centers constructed on high plasticity volcanic and lacustrine soils, where significant changes in moisture content occur in the upper

6 m or so. Three well-known North American cities where these natural hazard effects are major concerns are Mexico City, Long Beach, California, and Houston, Texas.

While other natural hazard examples could be given (e.g., river flooding, tsunamis, and climate variation causing permafrost melting, wildfires, drought, and desertification, etc.), the primary focus here are a few illustrative applications of airphoto and satellite remote sensing tools and technologies in natural hazard studies, rather than a comprehensive list of natural hazards and their identifying characteristics on the ground, in 3D airphotos and satellite images.

Summary

Geologists, geographers, engineers, and environmental scientists with training and experience in interpreting terrain conditions use 3D airphotos and satellite images to identify, map, and assess hazardous landforms and earth materials in natural and cultural landscapes. Such studies involve assessing potentially unstable and active hazard features and conditions, many of which are associated with adverse topography and geology along with severe weather and climate variation. In some cases, harmful human activities may also play a part. Aided by topographic, geologic, and other maps, remotely sensed information from airphoto and satellite images can be used to predict and evaluate hazards as well as make recommendations that may involve avoiding, eliminating, monitoring, mitigating, or treating (e.g., stabilizing) hazards – for example, in locating, planning, and developing new urban developments, social and industrial infrastructure, and energy and natural resources.

Bibliography

Brooks, G. R. (ed.), 2001. *A Synthesis of Geological Hazards in Canada*. Ottawa, Canada: Natural Resources Canada, 281 pp.
Evans, S. G., and DeGraff, J. V. (eds.), 2002. *Catastrophic Landslides: Effects, Occurrences and Mechanisms: Reviews in Engineering Geology*. Boulder, CO: The Geological Society of America, Vol. XV, p. 411.
Hunt, R. E., 2007. *Geologic Hazards: A Field Guide for Geotechnical Engineers*. Boca Raton, FL: CRC Press/Taylor & Francis Group, 323 pp.
Highland, L. M., and Bobrowsky, P., 2008. *The Landslide Handbook – A Guide to Understanding Landslides*. Reston, VA: U.S. Geological Survey Circular 1325, 129 pp.
Hyndman, D., Hyndman, D., and Catto, N., 2009. *Natural Hazards and Disasters*, First Canadian Edition. Scarborough, ON, Canada: Nelson Education Ltd., 526 pp.
Liang, Ta., 1952. *Landslides: An Aerial Photographic Study*. PhD thesis, New York, Cornell University Engineering Library, 274 pp.
Mollard, J. D., 1977. Regional landslide types in Canada. In Donald R. C. (ed.), *Reviews in Engineering Geology*. Boulder, CO: Geological Society of America, Vol. III, pp. 29–56.
Mollard, J. D., and Robert Janes, J., 1984. *Airphoto Interpretation and the Canadian Landscape*. Ottawa, Canada: Energy Mines and Resources Canada, Canadian Government Publishing Centre, Supply and Services, p. 415.
National Geographic Society, 1997. *Restless Earth: Disasters of Nature*. Washington, D.C.: The Book Division, p. 338.
Philipson, W. R., editor in chief. 1997. *Manual of Photographic Interpretation*, 2nd edn. Falls Church, VA: American Society of Photogrammetry and Remote Sensing, Vol. III, 689 pp.
Rivard, L. A., 2009. *Geohazards – Associated Geounits: Atlas and Glossary*. Berlin/Heidelberg: Springer Environmental and Engineering, 1052 pp.

Cross-references

Avalanches
Avulsion
Caldera
Challenges to Agriculture
Coal Fire (Underground)
Coastal Erosion
Creep
Debris Avalanche
Debris Flow
Earthquake Damage
Emergency Mapping
Erosion
Fault
Floodplain
Forest and Range Fires (wildfire)
Glacier hazards
Hazard and Risk Mapping
Ice and Icebergs
Karst Hazards
Lahar
Land-Use Planning
Landslide Inventory
Landslide Types
Monitoring Natural Hazards
North Anatolian Fault
Remote Sensing and Natural Hazards and Disasters
Rockfall
Sackung
San Andreas Fault
Sinkhole
Subsidence Induced by Underground Extraction

ALBEDO

Alan W. Harris
German Aerospace Center (DLR), Berlin, Germany

Definition

The albedo of a surface is a measure of its reflectivity. The albedos of planetary surfaces depend on their compositions and physical properties. Two types of albedo are often used in connection with planetary surfaces. One is the *geometric albedo* in a particular photometric band (normally denoted by p, with a subscript for the photometric band, e.g., p_V for visual geometric albedo), which is the ratio of the body's brightness at zero solar phase angle (i.e., as seen from the direction of the Sun) to the brightness of a perfectly diffusing (Lambertian) disk with the same apparent size and at the same position as the body. The visual geometric albedo is useful for most observational work because it is relevant to the visual brightness of the object. Values of p_V for

Albedo, Table 1 Values of visual geometric albedo for a selection of solar system bodies

Body	Visual geometric albedo, p_V
Moon	0.12
Venus	0.46
1 Ceres (dwarf planet in main asteroid belt)	0.09
4 Vesta (main-belt asteroid)	0.3
433 Eros (near-Earth asteroid)	0.23
C-type asteroid	0.06 (typical value)
S-type asteroid	0.20 (typical value)
E-type asteroid	0.50 (typical value)
Comet nucleus	0.04 (typical value)

a selection of bodies are given in Table 1. The second is the *bolometric Bond albedo*, which is the ratio of the total reflected radiation, summed over all wavelengths and directions, to the incident solar radiation. The bolometric Bond albedo, A, is relevant to the energy balance of a planetary body. For example, if the effects of thermal inertia and surface roughness are negligible, the thermal emission from any point on an atmosphereless body's surface can be considered to be in instantaneous equilibrium with the solar radiation absorbed at that point, thus:

$$\text{Total emitted thermal radiation} = (1 - A).S.a_p, \quad (1)$$

where S is the total incident solar radiation per unit area and a_p is the instantaneous sunward projected area of the body.

Overview

Albedo and size are two of the most fundamental physical properties of asteroids, especially for considerations of the impact hazard (e.g., Stuart and Binzel, 2004). Since the visual brightness of an asteroid depends on the *product* of p_V and projected area, an asteroid's size cannot be determined from optical observations alone without knowledge of the geometric albedo, p_V. Whereas (1) is an oversimplification because effects such as rotation and thermal inertia are ignored, a combination of infrared observations of an asteroid's thermal emission and observations of its visual brightness, together with an appropriate thermal model, can be used to derive both size and p_V (e.g., Delbo' et al., 2003). In the absence of infrared data, or other means of size determination, the sizes of asteroids are often estimated on the basis of measurements of the visual brightness and an assumed geometric albedo by means of the simple relationship:

$$D(\text{km}) = 1329.p_V^{-0.5}.10^{-H/5}, \quad (2)$$

where H is the absolute magnitude, defined as the brightness in the V band referred to zero solar phase angle and unit heliocentric and geocentric distances (in astronomical units, AU). However, in the case of near-Earth asteroids, which display a large range of visual geometric albedo, for example, 0.03–0.6, use of this relationship with an assumed value of p_V can result in significant errors. For example, if a "typical" albedo, for example $p_V = 0.16$, is assumed, the resulting diameter could be in error by a factor of 2 and the mass estimate by a factor of 8. Due to the present lack of albedo measurements, the size distribution of the near-Earth asteroid population is still largely based on estimates made with the help of (2) and an assumed value of p_V. As a consequence, only crude estimates of the destructive impact potential within the population of near-Earth asteroids can be made at present.

Summary

The reflectivities of planetary surfaces are governed by their compositions and physical characteristics. A measure of reflectivity is the albedo, which is often used in the sense of backscattered sunlight in a given spectral region, normally the visual (visual geometric albedo), or the total amount of reflected solar radiation in all directions and spectral bands (bolometric Bond albedo). In the case of asteroids, albedos are an important indicator of mineralogy and are often derived by means of observations in the thermal-infrared and visual spectral regions. The distribution of albedos in the population of near-Earth asteroids is of crucial importance for the determination of the size distribution of potential impactors on the Earth.

Bibliography

Bottke, W. F., Cellino, A., Paolicchi, P., and Binzel, R. P. (eds.), 2002. *Asteroids III*. Tucson: University of Arizona Press.

Delbo', M., Harris, A. W., Binzel, R. P., Pravec, P., and Davies, J. K., 2003. Keck observations of near-Earth asteroids in the thermal infrared. *Icarus*, **166**, 116–130.

Harris, A. W., 2006. The surface properties of small asteroids from thermal-infrared observations. In Lazzaro, D., Ferraz-Mello, S., and Fernandez, J. A. (eds.), *Proceedings IAU Symposium 229*. Cambridge: Cambridge University Press, pp. 449–463.

Stuart, J. S., and Binzel, R. P., 2004. Bias-corrected population, size distribution, and impact hazard for near-Earth objects. *Icarus*, **170**, 295–311.

Cross-references

Asteroid
Asteroid Impact
Torino Scale

ANTECEDENT CONDITIONS

Michael James Crozier[1], Nick Preston[1], Thomas Glade[2]
[1]Victoria University of Wellington, Wellington, New Zealand
[2]University of Vienna, Vienna, Austria

Synonyms

Preceding event

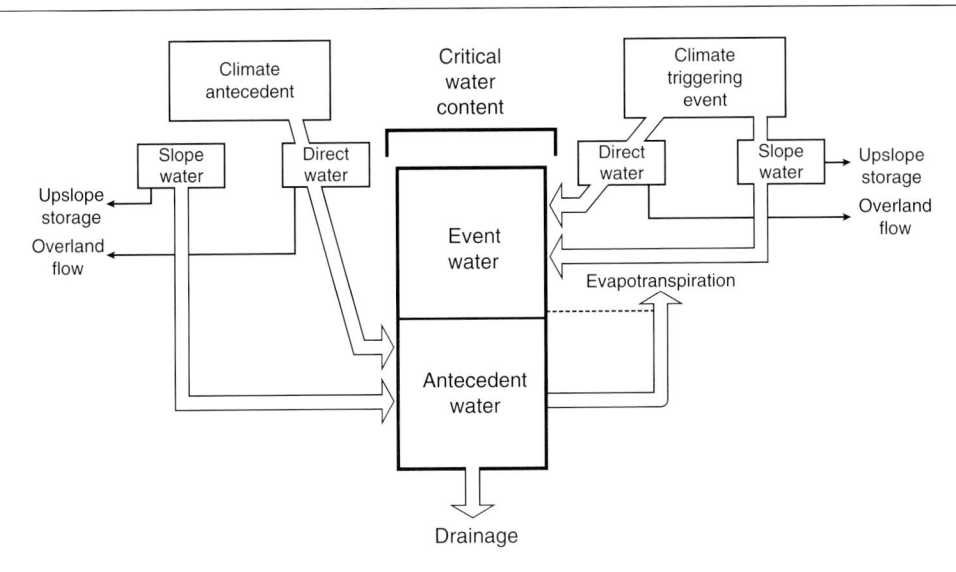

Antecedent Conditions, Figure 1 Conceptual climatic-hydrological model for shallow rainfall-triggered landslides.

Definition

Antecedent conditions represent a temporary state within dynamic natural and social systems that precedes and influences the onset and magnitude of a hazard and its consequences. They are distinct from, but influenced by, what are commonly referred to as preconditions (preexisting conditions). Preconditions are generally static or slow changing and influence the inherent (as opposed to temporary) susceptibility of an area. For example, in natural systems, rock type, soil structure, and topographic geometry are common preconditions that affect susceptibility to landslide occurrence, whereas groundwater level, soil moisture content, and under certain circumstances, vegetation cover are dynamic factors representing influential antecedent conditions for landsliding. In social systems, coping capacities such as the presence of emergency response organizations or availability of insurance schemes are preconditions whereas time of occurrence (e.g., day/night; workday/weekend; holiday, etc.) is a dynamic factor strongly influencing the consequences of a triggering event.

Examples of antecedent conditions for specific hazards include tidal phase (tsunami and storm surge), vegetation moisture levels (forest fire), humidity (heat waves), groundwater level (liquefaction and flooding), wind direction and strength (volcanic eruption), temperature and freeze/thaw history of snow packs (snow avalanching), and amount of debris accumulated in source areas (debris flow). Antecedent conditions can also be represented by hazard history. For instance, forest fires can induce hydrophobic conditions in soils that favor the development of debris flows during heavy rainfall, and foreshocks may weaken natural and man-made structures causing amplified damage in subsequent earthquakes.

An example of rainfall-triggered landslides illustrates the rationale and methods for assessing antecedent conditions. Antecedent conditions are represented in this case by the *antecedent soil water* (water accumulated in the slope over a period preceding landslide occurrence) which, along with *event water* (water accumulated on the day of landslide occurrence), forms the *critical water content* (CWC) within a slope, that is, the amount of water required to initiate landslide movement (Figure 1). Rainfall thresholds for landslide initiation established by historical observation essentially represent an approximation of the CWC (Crozier, 1989; Glade et al., 2000; Guzzetti et al., 2008). In a location where CWC for landsliding is known, real-time monitoring of the antecedent soil water, allows a continuous estimate of the amount of event water required to reach critical conditions. This in turn has been used to derive a probability of landslide occurrence based on frequency–magnitude distributions of rainfall conditions derived from the historical climate record (Crozier, 1999).

Water within the slope reduces stability by either decreasing cohesion or increasing buoyancy through the development of positive porewater pressures, to the point where strength of slope material is lowered below the prevailing shear stress and consequently failure occurs.

In practical terms, the *event water* component of the CWC can be represented by some parameter of rainfall at the time of landslide occurrence whereas *antecedent soil water* is represented by the *antecedent soil water status*. The value of the *antecedent soil water status* can be determined at any one point in time given knowledge of past precipitation and evapotranspiration rates, the regolith storage capacity (porosity and depth), and the drainage rate of *excess precipitation* (i.e., rainfall in excess of storage and evapotranspiration requirements).

In short, the *antecedent soil water status* is an index of the water content of the soil based on the climatic water balance, and provides a scale with negative values

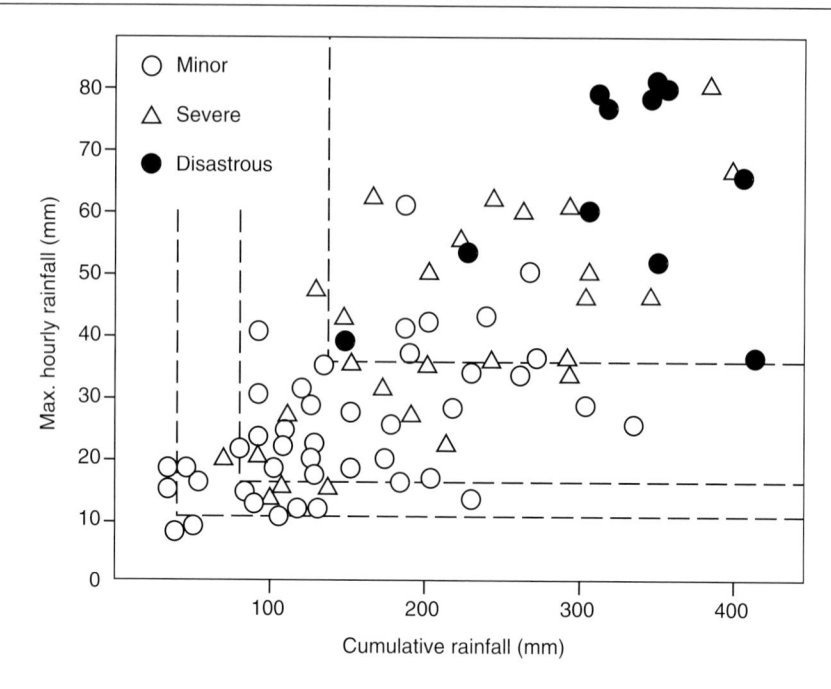

Antecedent Conditions, Figure 2 Landslide event thresholds defined by intensity and 2-day antecedent rainfall, Korea (Source Kim et al., 1992).

representing soil storage below field capacity, held in the form of capillary or hygroscopic water, and positive values representing gravitational water that accumulates as groundwater in certain slope locations (Crozier, 1999).

Landsliding is usually (but not always) associated with positive values of the *antecedent soil water status index* calculated from *excess rainfall,* that is, rainfall exceeding potential evapotranspiration and soil storage requirements.

In the calculation of the *antecedent soil water status index* (Crozier, 1999), *excess rainfall* is decayed on a daily basis and accumulated over a given period (often about 10 days) to represent antecedent excess rainfall values. These constitute the positive values of the *antecedent soil water status index*.

$$EPa_0 = kEP_1 + k^2EP_2 + \cdots + k^n EP^n$$

where EPa_0 = antecedent excess rainfall on day 0 (mm), EP^n = excess rainfall on the nth day before day 0 (mm), and k = constant decay factor.

The decay factor represents the rate of drainage from the soil and in some cases can be determined from the exponential decline of the recessional limb of flood hydrographs for streams within the locality under study (Glade et al., 2000). The use of stream hydrographs to indicate antecedent water status assumes that flow regimes have not been artificially modified by structures such as reservoirs or other drainage works.

Kim et al. (1992) (Figure 2), Glade (2000), and Garland and Olivier (1993) have demonstrated that, in certain regions, antecedent conditions have a major influence on the initiation of landslides, whereas in other regions, storm event characteristics appear to dominate (Caine, 1980; Wilson and Wieczorek, 1995). Similar findings for debris flows have been summarized by Wieczorek and Glade (2005).

Although critical preconditions and triggering factors can be established for a number of different hazards, the temporal variability of antecedent conditions provides a significant level of uncertainty to hazard assessments. The investigation of antecedent conditions is a critical component for both the prediction and explanation of hazard occurrence.

Bibliography

Caine, N., 1980. The rainfall intensity-duration control of shallow landslides and debris flows. *Geografiska Annaler,* **62A**(1–2), 23–27.

Crozier, M. J., 1989. *Landslides: Causes, Consequences and Environment*. London: Routledge.

Crozier, M. J., 1999. Prediction of climatically-triggered landslides – a test of the antecedent water status model. *Earth Surface Processes and Landforms,* **24**(9), 825–833.

Garland, G. G., and Olivier, M. J., 1993. Predicting landslides from rainfall in a humid sub-tropical region. *Geomorphology,* **8**(2–3), 165–174.

Glade, T., 2000. Modelling landslide-triggering rainfalls in different regions in New Zealand – the soil water status model. *Zeitschrift für Geomorphologie,* **122**, 63–84.

Glade, T., Crozier, M. J., and Smith, P., 2000. Establishing landslide-triggering rainfall thresholds using an empirical antecedent daily rainfall model. *Journal of Pure and Applied Geophysics,* **157**, 1059–1079.

Guzzetti, F., Peruccacci, S., Rossi, M., and Stark, C. P., 2008. The rainfall intensity-duration control of shallow landslides and debris flows: an update. *Landslides*, **5**(1), 3–18.

Kim, S. K., Hong, W. P., and Kim, Y. M., 1992. Prediction of rainfall-triggered landslides in Korea. In Bell, D. H. (ed.), *Landslides*. Rotterdam: Balkema, Vol. 2, pp. 989–994.

Wieczorek, G. F., and Glade, T., 2005. Climatic factors influencing triggering of debris flows. In Jakob, M., and Hungr, O. (eds.), *Debris flow hazards and related phenomena*. Heidelberg: Springer, pp. 325–362.

Wilson, R. C., and Wieczorek, G. F., 1995. Rainfall thresholds for the initiation of debris flows at La Honda, California. *Environmental and Engineering Geoscience*, **1**(1), 11–27.

Cross-references

Early Warning Systems
Exposure to Natural Disasters
Hazards
Uncertainty
Vulnerability

ARSENIC IN GROUNDWATER

Arindam Basu
University of Canterbury, Christchurch, New Zealand

Synonyms

Arsenate; Arsenite; As

Definition

Inorganic arsenic (abbreviation: As) is a trace element found in the earth's crust. Arsenic is present in groundwater as inorganic arsenate (As(V)) and arsenite (As(III)).

Origins of arsenic in groundwater

Arsenic enters groundwater from the surrounding substrate (soil and bedrock). It is generally believed that weathering reactions, microbial and human activities result in mobilization of inorganic arsenic from the substrate to groundwater (Appelo and Heederik, 2006).

In the substrate, As is commonly concentrated in sulfide bearing minerals, has strong affinity for pyrites, and is concentrated in hydrous iron oxides. Sulfide minerals usually contain high concentrations of As; consequently, oxidation of sulfide minerals releases inorganic As in groundwater. For instance, in the aquifers of alluvial river basins of the Ganges delta (parts of West Bengal, an Indian state, and Bangladesh), presence of organic matter in soil leads to complex redox reactions and loss of dissolved oxygen from groundwater; this, in turn, leads to dissolution of iron oxides and release of As in groundwater (Bhattacharyya et al., 2003; Nickson, 2000). Geothermal activities contain high concentration of sulfides, and are other sources of As in groundwater in those regions that are characterized by hot springs.

Anthropogenic sources of As in groundwater include mining, use of arsenic-based pesticides, and groundwater abstraction. Mineralized zones often contain arsenopyrites. Heavy and long-term mining is often associated with dissolution of these minerals, and release of As in the groundwater. Mine wastes and tailing ponds may have high concentrations of As in the water. It is believed that in the Ganges delta region of West Bengal and Bangladesh, overpumping of groundwater for irrigation has led to dewatering of sediments, anoxic soil conditions, and mobilization of high As into the groundwater. Alternatively, high As concentration in groundwater in this region may be due to the activity of naturally occurring organic carbons in a naturally occurring process.

Acceptable levels of as concentration in groundwater

In 1993, the World Health Organization (WHO) reduced the provisional guideline limit from 50 to 10 μg/L (WHO, 2010). This was based on the observation that at concentrations of 10 μg/L, there is a risk of 1 extra case of cancer per 100, 000 individuals. Whereas the developed countries follow this standard, the developing countries still limit the groundwater As concentration at 50 μg/L.

Geographical regions that have high concentrations of As in groundwater include young, unconsolidated sediments usually of Quarternary (Holocene) age, inland closed basins in arid or semiarid settings (Argentina, Mexico, Nevada, California), or alluvial or deltaic plains (Ganges delta in West Bengal and Bangladesh, Yellow River basin, Mekong Valley, Red River delta, Irrawaddy delta). The Ganges delta region has about 36 million individuals that are potentially exposed to arsenic over 50 μg/L. The Chaco-Pampean plain of Argentina is the largest high-As groundwater province in South America (1 million square kilometers).

Consequences of high concentrations of inorganic as in groundwater

Arsenic primarily enters the human body through consumption of groundwater since it is tasteless and odor-free. Small mass transfer of As to water is required for As toxicity to be manifested in humans.

Exposure to inorganic As in groundwater results in noncancerous health effects and cancers. The International Agency for Research on Cancer (IARC) classifies As as a class I carcinogen. The cancers associated with As in groundwater include those of skin, lungs, urinary bladder, kidney, and liver. Arsenic is also related to noncancer effects, including skin lesions, chronic cough, and possibly diabetes.

Bibliography

Appelo, T., and Heederik, J., 2006. Arsenic in groundwater – a world problem. In Appelo, T. (ed.), *Arsenic in Groundwater - A World Problem*. Utrecht: International Association of Hydrogeologists, pp. 1–142.

Nickson, R., 2000. Mechanism of arsenic release to groundwater, Bangladesh and West Bengal. *Applied Geochemistry*, **15**(4), 403–413, doi:10.1016/S0883-2927(99)00086-4.

Nordstrom, D., 2002. Worldwide occurrences of arsenic in ground water. *Science*, **296**(5576), 2143–2145.

WHO, 2010. Arsenic in drinking water. Retrieved from http://www.physics.harvard.edu/~wilson/arsenic/arsenic_project_introduction.html.

Cross-references

Dose Rate
Erosion
Land Degradation
Release Rates

ASTEROID

Alan W. Harris
German Aerospace Center (DLR), Berlin, Germany

Synonyms

Minor planet; Planetoid; Small solar system body

Definition

An asteroid is an irregularly shaped, rocky body orbiting the Sun. In contrast to planets and dwarf planets, asteroids do not have sufficient mass for their self-gravity to form a spheroid. In contrast to comets, asteroids display neither a coma nor a tail.

Overview

Asteroids and comets are considered to be remnant bodies from the epoch of planet formation. Planet embryos formed in the protoplanetary disk about 4.5 billion years ago via the accretion of dust grains and collisions with smaller bodies (planetesimals). A number of planet embryos succeeded in developing into the planets we observe today; the growth of other planet embryos and planetesimals was terminated by catastrophic collisions or a lack of material in their orbital zones to accrete. Most asteroids are thought to be the fragments of bodies that formed in the inner Solar System and were subsequently broken up in collisions. Comets and related icy bodies are thought to have accreted in the cold, outer regions of the protoplanetary disk where volatile material, such as water and carbon dioxide, were abundant as ices.

The main asteroid belt between the orbits of Mars and Jupiter contains most of the known asteroids. An asteroid is assigned a permanent designation, that is, a sequential number, once its orbit has become accurately established through a sufficient number of astrometric observations. There are over 300,000 numbered asteroids in the main belt, consisting largely of silicates and metals and having sizes up to about 1,000 km in diameter.

Asteroids are classified dynamically according to their orbital elements (semimajor axis, period, inclination, eccentricity, etc.). Within the main belt, between about 2.0 and 3.5 astronomical units (AU, the mean Sun-Earth distance) from the Sun, there exist a number of "families" of asteroids, the members of which have very similar dynamical characteristics and may be fragments from relatively recent (relative to the history of the Solar System) collisions.

As a result of subtle thermal effects and the very strong gravitational field of Jupiter, small main-belt asteroids can drift into certain orbital zones from which they may be ejected under the influence of Jupiter into the inner Solar System. As a result there exists a population of near-Earth asteroids (NEAs), with orbits that can cross that of the Earth. NEAs are further categorized dynamically as Amors, Apollos, Atens, or Inner-Earth objects (IEOs) according to the semimajor axes, aphelion and perihelion distances of their orbits (Table 1). Interest in the population of NEAs is focused mainly on the associated impact hazard (see below), but close approaches of NEAs to the Earth facilitate detailed telescope observations, including radar investigations, which provide insight into the characteristics of asteroids in general.

The taxonomic classification and mineralogy of asteroids

Sunlight incident on the surface of an asteroid is absorbed in particular wavelength bands depending on the minerals present; reflected light, therefore, carries a spectral signature of the mineralogical composition of the asteroid's surface. Attempts have been made to classify asteroids according to details of the absorption features in their optical reflection spectra observed with astronomical telescopes. A number of classification schemes have been devised based on letters of the alphabet. Table 2 lists some of the main taxonomic types in use today and the likely mineralogical compositions associated with them.

Asteroid physical properties

The shapes of asteroids and their observed size distribution are consistent with a scenario in which collisions over billions of years have led to a grinding down of objects in the main belt and an ever increasing number of small collision fragments. Some insight into the internal structure of asteroids can be gained by considering their bulk densities. Reliable estimates of density are difficult to obtain, since techniques for obtaining accurate masses and sizes of asteroids are complex and subject to large uncertainties. There are various methods of determining asteroid masses (see Britt et al., 2002, for a review), which all require measurement of the asteroid's gravity field by means of a spacecraft, observations of the perturbations of the orbits of other asteroids or Mars (applicable to large asteroids only), or observations of a satellite or companion asteroid by means of precision optical or radar observations. Asteroid sizes can be determined from spacecraft, thermal-infrared measurements, radar observations, polarimetry, and occultation observations.

Density estimates for some 40 asteroids are available to date. An important finding is that asteroid bulk densities

Asteroid, Table 1 Dynamical groupings and numbers of near-Earth asteroids

Dynamical category	Semimajor axis [AU]	Perihelion/aphelion	Approximate number known
Amors	>1	$1.017 <$ perihelion ≤ 1.3 AU	2,800
Apollos	≥ 1	Perihelion ≤ 1.017 AU	3,300
Atens	<1	Aphelion ≥ 0.983 AU	540
Inner Earth objects	<1	Aphelion < 0.983 AU	10

The orbital characteristics of the different classes of NEAs are defined with respect to the Earth's perihelion (0.983 AU) and aphelion (1.017 AU) distances. The listed approximate numbers of known objects are valid as of December 2009; these numbers increase rapidly with time as new objects are discovered. For more details and updates see the Minor Planet Center web site: http://www.cfa.harvard.edu/iau/mpc.html

Asteroid, Table 2 Important asteroid taxonomic (spectral) classes

Class	Probable mineralogy	Associated geometric albedo[a] (p_V) range	Approximate % of all classified asteroids[b]
D, P	Carbon, organic-rich silicates	0.03–0.06	2
C, B	Carbon, organics, hydrated silicates	0.03–0.1	30
M	Fe, Ni, enstatite	0.1–0.2	2
S	Olivine, pyroxene, metals	0.1–0.3	40
Q	Olivine, pyroxene, metals	0.2–0.5	1
V	Pyroxene, feldspar	0.2–0.5	3
E	Enstatite, other Fe-poor silicates	0.3–0.6	1
X	Unknown (signifies otherwise unclassifiable featureless spectrum)	0.03–0.6	15

[a]The geometric albedo p_v is the ratio of a body's (V-band) brightness at zero phase angle to the brightness of a perfectly diffusing (Lambertian) disk with the same apparent size and at the same position as the body
[b]Estimated from the data of Bus and Binzel (2002), and Tholen and Barucci (1989). Note that discovery and observation bias against objects with dark surfaces implies that the distribution of taxonomic classes in the known population may not be representative of the entire asteroid population

tend to be significantly lower in general than expected from measurements of meteorites and terrestrial analogues of asteroid material (2–8 g cm^{-3}). In fact the bulk densities of some asteroids appear to be similar to that of water (1 g cm^{-3}), implying that these bodies must be highly porous. Such results have increased speculation that some asteroids, for example 25143 Itokawa, may be aggregates of components of various sizes weakly bound by gravity, or "rubble piles." An asteroid that has been shattered by collisions with other objects may survive under the collective weak gravitational attraction of the resulting fragments as a cohesionless, consolidated rubble pile; this idea is supported by the images of Itokawa returned by the Hayabusa spacecraft (Figure 1), revealing a highly irregular object apparently consisting of separate component blocks of various shapes and sizes.

The known NEA population contains a confusing variety of objects. Some NEAs are thought to be largely metallic, indicative of material of high density and strength, whereas others are carbonaceous and probably of lower density and less robust. A number of NEAs may be evolved cometary nuclei that are presumably porous and of low density but otherwise with essentially unknown physical characteristics. In terms of large-scale structure NEAs range from monolithic slabs to rubble piles and binary systems. Some 40 NEAs in the currently known population are thought to be binary systems and many more are probably awaiting discovery.

Even fundamental parameters such as size and mass are very difficult to determine accurately, especially in the case of small asteroids, such as NEAs, which generally have irregular shapes and widely differing surface properties. Since the brightness of an asteroid is proportional to its surface albedo and cross-sectional area, a very rough size estimate can be obtained from the observed brightness and knowledge of the asteroid's orbit. Telescope observations in the thermal infrared combined with knowledge of an asteroid's optical brightness offer a means of obtaining more accurate information on size and albedo (e.g., Delbo' et al., 2003; Harris, 2006). Darker, low-albedo asteroids are less reflective in the visible spectral region and absorb more solar radiation; they

Asteroid, Figure 1 Near-Earth asteroid 25143 Itokawa was the target of the Japanese Hayabusa spacecraft in 2005–2006. Itokawa is 535 m in length and of type S. Itokawa may be an aggregate of components weakly bound by gravity, or "rubble pile" (Courtesy of JAXA).

are, therefore, warmer and brighter in the thermal-infrared spectral range. The opposite is true for high-albedo asteroids. The two different physical relationships governing visible and thermal-infrared brightness enable simultaneous solutions for size and albedo to be obtained. While the number of asteroids observed in the thermal infrared is still only a very small fraction of the total known, this technique has provided the vast majority of size and albedo determinations to date (e.g., Tedesco et al., 2002), and on-going and future infrared surveys promise further results, especially for NEAs, using this technique.

An important result from thermal-infrared observations is the apparent size-dependence of the albedos of some types of NEA in the 0.1–10 km size range, such that the mean albedo appears to increase with decreasing size (Delbo' et al., 2003; Harris, 2006). As explained above, the mean albedo of NEAs is important for current estimates of their size distribution. If the mean albedo is higher, the diameter derived from a particular observed brightness is smaller and the overall impact hazard from NEAs is reduced (Stuart and Binzel, 2004).

The rate of discovery of NEAs has increased dramatically in recent years and is now seriously outstripping the rate at which the population can be physically characterized. The NEA population is still largely unexplored.

Near-earth asteroids and the impact hazard

The phenomenon of collisions in the history of our Solar System is very fundamental, having played the major role in forming the planets we observe today. Asteroids may have contributed to the delivery of water and organic materials to the early Earth necessary for the development of life, but later impacts of asteroids probably played a role in mass extinctions and they currently pose a small but significant threat to the future of our civilization. Collisions of asteroids with the Earth have taken place frequently over geological history and it is an irrefutable scientific fact that major collisions of asteroids and comets with the Earth will continue to occur at irregular, unpredictable intervals in the future. The risk of a comet impact is thought to be much lower than that of an NEA impact, although, given the potentially high relative velocities, the effects in the case of a comet impact could be much more devastating.

Near-earth object search programs

A number of observatories are operated specifically to discover near-Earth objects and establish their orbits. The main currently active asteroid search programs, such as the Catalina Sky Surveys, Lincoln Near-Earth Asteroid Research (LINEAR), and Spacewatch (e.g., see Yeomans and Baalke, 2009) are funded primarily by NASA and the US Air Force. The telescopes employed for asteroid searching have modest mirror diameters of around 1 m, large fields of view, and state-of-the art CCD detectors. Asteroids appear as star-like point sources in the focal planes of such telescopes and are discovered by virtue of their motion against the fixed stellar background (Figure 2). Sophisticated software enables the detection of asteroids to be carried out automatically. Observations of an asteroid over periods of months or years are required to enable its orbit to be calculated accurately. Many observers, including amateur astronomers, around the world contribute to the tracking of asteroids and submit their astrometric data to the Minor Planet Center in Cambridge, Massachusetts, which acts as a clearing house for asteroid position data and maintains databases of orbits.

As of December 2009 the number of known NEAs was about 6,500 (Figure 3), of which some 800 have diameters of 1 km or more; this number is close to the predicted final tally of NEAs with diameters of 1 km or more, including as yet undiscovered objects. A special term, "potentially hazardous asteroids" (PHAs), is reserved for those with diameters above 120 m and orbits that bring them within 0.05 AU of the Earth's orbit. PHAs are large enough to survive passage through the Earth's atmosphere and cause extensive damage on impact. As of December 2009 the number of known PHAs stands at about 1,100.

Summary

Planet embryos formed some 4.5 billion years ago from dust accretion and the collisional coalescence of primordial building blocks called planetesimals. Most asteroids are thought to be remnant collisional fragments of

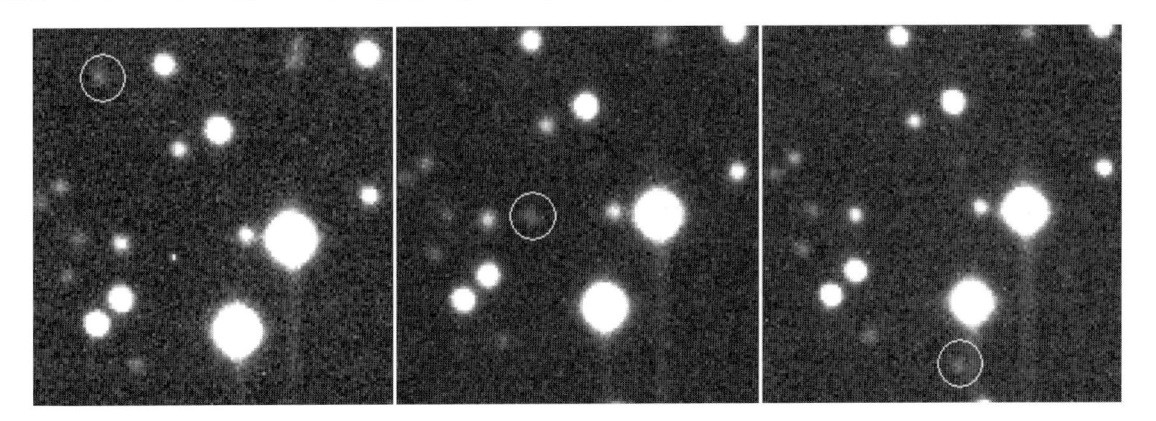

Asteroid, Figure 2 Asteroids are discovered by virtue of their apparent motion on the sky relative to distant stars. A candidate asteroid (*encircled*) is seen to move between successive exposures of this field of stars taken some minutes apart.

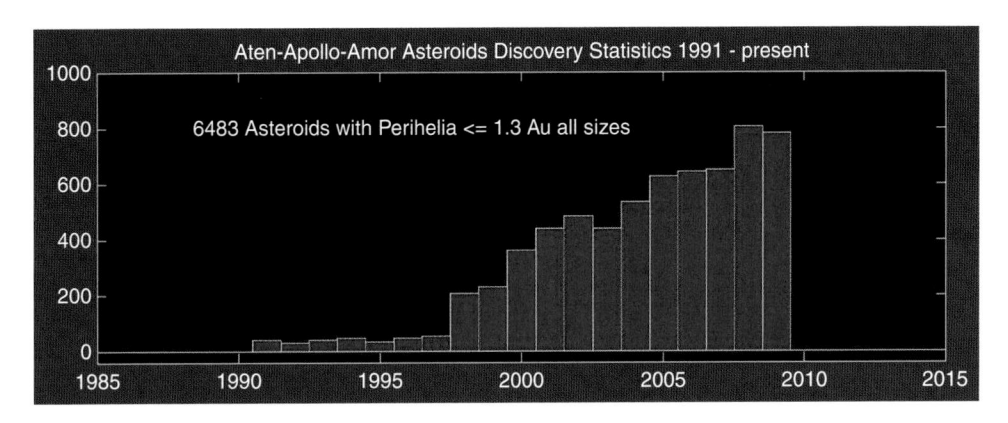

Asteroid, Figure 3 Recent annual discovery rates of near-Earth asteroids (Courtesy of G. Hahn (DLR) and the European Asteroid Research Node, EARN: http://earn.dlr.de/).

planetesimals and planet embryos from the inner Solar System that were not incorporated into planets; comets, on the other hand, are remnants from the cold, outer regions of the Solar System. Subsequent impacts of asteroids and comets have played a major role in the development of planets, especially the Earth. Asteroids and comets probably contributed significantly to the early Earth's inventory of water and organic materials, facilitating the development of life. In contrast, impacts of asteroids and comets present a small but scientifically well-founded threat to the future of our civilization: There is no scientific reason why asteroids and comets should not continue to hit the Earth at irregular and unpredictable intervals in the future. As a result of modern observing techniques and directed efforts thousands of near-Earth asteroids have been discovered over the past 20 years and the reality of the impact hazard has been laid bare. Future observation programs and space missions will be crucial for a better understanding of the processes and conditions governing the development of planets and life, the composition and physical nature of asteroids, and the techniques that would be most effective in preventing a collision of an asteroid with the Earth.

Bibliography

Britt, D. T., Yeomans, D., Housen, K., and Consolmagno, G., 2002. Asteroid density, porosity, and structure. In Bottke, W. F., Cellino, A., Paolicchi, P., and Binzel, R. P. (eds.), *Asteroids III*. Tucson: University of Arizona Press, pp. 485–500.

Bus, S. J., and Binzel, R. P., 2002. Phase II of the small main-belt asteroid spectroscopic survey. A feature-based taxonomy. *Icarus*, **158**, 146–177.

Delbo', M., Harris, A. W., Binzel, R. P., Pravec, P., and Davies, J. K., 2003. Keck observations of near-Earth asteroids in the thermal infrared. *Icarus*, **166**, 116–130.

Harris, A. W., 2006. The surface properties of small asteroids from thermal-infrared observations. In Lazzaro, D., Ferraz-Mello, S., and Fernandez, J. A. (eds.), *Proceedings IAU Symposium 229*. Cambridge: University Press, pp. 449–463.

Stuart, J. S., and Binzel, R. P., 2004. Bias-corrected population, size distribution, and impact hazard for near-Earth objects. *Icarus*, **170**, 295–311.

Tedesco, E. F., Noah, P. V., Noah, M., and Price, S. D., 2002. The supplemental *IRAS* minor planet survey. *The Astronomical Journal*, **123**, 1056–1085.

Tholen, D. J., and Barucci, M. A., 1989. Asteroid taxonomy. In Binzel, R. P., Gehrels, T., and Matthews, M. S. (eds.), *Asteroids II*. Tucson: University of Arizona Press, pp. 298–315.

Yeomans, D., and Baalke, R., 2009. *Near Earth Object Program*. Available from World Wide Web: http://neo.jpl.nasa.gov/programs.

Cross-references

Albedo
Asteroid Impact
Asteroid Impact Mitigation
Asteroid Impact Prediction
Torino Scale

ASTEROID IMPACT

Christian Koeberl
Natural History Museum, Vienna, Austria
Center for Earth Sciences, University of Vienna, Vienna, Austria

Definition and introduction

Craters are a fundamental and common topographic form on the surfaces of planets, satellites, and asteroids. On large planetary bodies, of the size of the Moon and larger, craters can form in a variety of processes, including volcanism, impact, subsidence, secondary impact, and collapse. On smaller bodies (e.g., of the size of minor planets), impact may be the only process that can form craters. In the explanation of terrestrial crater-like structures, the interpretation as volcanic features and related structures (such as calderas, maars, cinder cones) has traditionally dominated over impact-related interpretations. The importance of impact cratering on terrestrial planets (Mercury, Venus, Mars), our Moon, and the satellites of the outer planets is obvious from the abundance of craters on their surfaces. On most bodies of the Solar System that have a solid surface, impact cratering is the most important surface-modifying process even today. On Earth, active geological processes rapidly obliterate the cratering record. To date only about 180 impact structures have been recognized on the Earth's surface. They come in various forms, shapes, and sizes, from 300 km to less than 100 m in diameter, and from recent to 2 billion years in age.

On the Moon and other planetary bodies that lack an appreciable atmosphere, it is usually easy to recognize impact craters on the basis of morphological characteristics. On the Earth, complications arise as a consequence of the obliteration, deformation, or burial of impact craters. Thus, it is ironic that despite the fact that impact craters on Earth can be studied directly in the field, they may be much more difficult to recognize than on other planets. Thus, the following diagnostic criteria for the identification and confirmation of impact structures on Earth were developed: (a) crater morphology, (b) geophysical anomalies, (c) evidence for shock metamorphism, and (d) the presence of meteorites or geochemical evidence for traces of the meteoritic projectile – of which only (c) and (d) can provide confirming evidence. Remote sensing, including morphological observations, as well as geophysical studies, cannot provide confirming evidence – which requires the study of actual rock samples.

Impacts influenced the geological and biological evolution of our own planet; the best known example is the link between the 200-km-diameter Chicxulub impact structure in Mexico and the Cretaceous-Tertiary boundary. Understanding impact structures, their formation processes, and their consequences should be of interest not only to Earth and planetary scientists, but also to society in general.

History of impact cratering studies

In the geological sciences, it has only recently been recognized how important the process of impact cratering is on a planetary scale. During the last few decades, planetary scientists and astronomers have demonstrated that our Moon, Mercury, Venus, Mars, the asteroids, and the moons of the outer gas planets are all covered (some surfaces to saturation) with meteorite impact craters (Figure 1). However, it is fairly recent that this observation has become accepted among astronomers and geologists, because up to the first third of the twentieth century, it was commonly accepted that all lunar craters are of volcanic origin (and at that time the presence of craters on planetary bodies other than the moon had not yet been established).

The origin of craters on the Moon was discussed since 1610, when Galileo Galilei first discovered them. Geologists paid no interest to the Moon for the following centuries, so that the discussion of lunar craters was left to the astronomers. One of the earliest researchers to speculate about the origin of lunar craters was Robert Hooke in 1665, who proposed two alternative hypotheses. First, he dropped solid objects into a mixture of clay and water and found that these experiments resulted in crater-like features. However, he rejected the possibility that the lunar craters could have formed by such "impact" processes, because it was not clear from "whence those bodies should come," as the interplanetary space was, at that time, considered to be empty. After all, Hooke made his experiments 135 years before the first asteroid, Ceres, was discovered by Piazzi in Palermo. Thus, he preferred a second hypothesis, in which, from experiments with "boiled alabaster," he concluded that the lunar craters formed by some kind of gas – rejecting a perfectly correct explanation because the "boundary conditions" were missing.

The eighteenth and nineteenth centuries were dominated by the volcanic theory, and only a few researchers at the end of the nineteenth century expressed the idea that

Asteroid Impact, Figure 1 The full moon, showing the large circular impact basins that are the remnants of large impacts during the late heavy bombardment (about 4 billion years ago); photo by the Galileo spacecraft (NASA).

impacts might form craters (on the moon and elsewhere). For example, at the end of the nineteenth century, in 1892, Grove Karl Gilbert, chief geologist of the US Geological Survey, concluded – partly based on experiments made in his hotel room during a lecture tour – that the formation of lunar craters can be best explained by the impact theory. In contrast, he rejected the hypothesis that Meteor Crater in Arizona was formed by impact. This was odd because fragments of iron meteorites were actually found around this crater.

Real progress was only made in the first decades of the twentieth century, when the mining engineer Daniel Moreau Barringer (1860–1929) studied the "Coon Butte" or "Crater Mountain" structure (as the "Meteor Crater" was then called) in Central Arizona. Despite the opinion of several leading geologists (including Gilbert) that this structure was of volcanic origin, and that the presence of the meteorite fragments was only a coincidence, Barringer was convinced that this was an impact crater, and his work laid the foundations for a wider acceptance of the existence of impact craters on Earth.

A well-known case in point is the Cretaceous-Tertiary (K-T) boundary, where the discovery of an extraterrestrial signature (Alvarez et al., 1980), together with the presence of shocked minerals (Bohor et al., 1984), led not only to the identification of an impact event as the cause of the end-Cretaceous mass extinction (Smit, 1999), but also to the discovery of a large buried impact structure about 200 km in diameter, the Chicxulub structure. Earlier, the idea that an extraterrestrial object would have influenced the geological and biological evolution on the Earth was not even seriously considered. This might explain the mixture of disbelief, rejection, and ridicule with which the suggestion was greeted that an asteroid or comet impact wiped out the dinosaurs and other species at the end of the Cretaceous. It was the debate that followed this suggestion, which, over the past 20–30 years, finally led to a more general realization that impact cratering is an important process on the Earth as well (Figure 2), and not only on the other planetary bodies of the Solar System.

Formation of impact craters

Impact cratering is a high-energy event that occurs at more or less irregular intervals (although over long periods of time, an average cratering rate can be established). Part of the problem regarding recognition of the remnants of impact events is the fact that terrestrial processes (weathering, plate tectonics, etc.) either cover or erase the surface expression of impact structures on Earth. Many impact structures are covered by younger (i.e., post-impact) sediments and are not visible on the surface. Others were mostly destroyed by erosion. In some cases, the ejecta have been found far from any possible impact structure. The study of these ejecta led, in turn, to the discovery of some impact craters.

Before discussing the detailed mineralogical, petrographic, and geochemical characteristics of impact craters, it is important to give a short overview of their formation, criteria of recognition, and their general geology. From the morphological point of view, it is necessary to distinguish between an impact *crater*, that is, the feature that results from the impact, and an impact *structure*, which is what is observed today, that is, long after formation and modification of the crater. On Earth, two distinctly different morphological forms are known: *simple* craters (small bowl-shaped craters) with diameters of up to ≤ 2–4 km, and *complex* craters, which are larger and have diameters of ≥ 2–4 km (the exact changeover diameter between simple and complex crater depends on the composition of the target). Complex craters are characterized by a peak or peak ring that consists of rocks that are uplifted from greater depth and would not normally be exposed on the surface. The stratigraphic uplift amounts to about 0.1 of the crater diameter (e.g., Melosh, 1989). Craters of both types have an outer rim and are filled by a mixture of fallback ejecta and material slumped in from the walls and crater rim during the early phases of formation. Such crater infill may include brecciated and/or fractured rocks, and impact melt rocks. Fresh simple craters have an apparent depth (measured from the crater rim to present-day crater floor) that is about one third of the crater diameter, whereas that value for complex craters is closer to one sixth. On Earth basically all small craters are relatively young, because erosional processes obliterate small (0.5–10 km diameter) craters after a few million years, causing a severe deficit of such small craters.

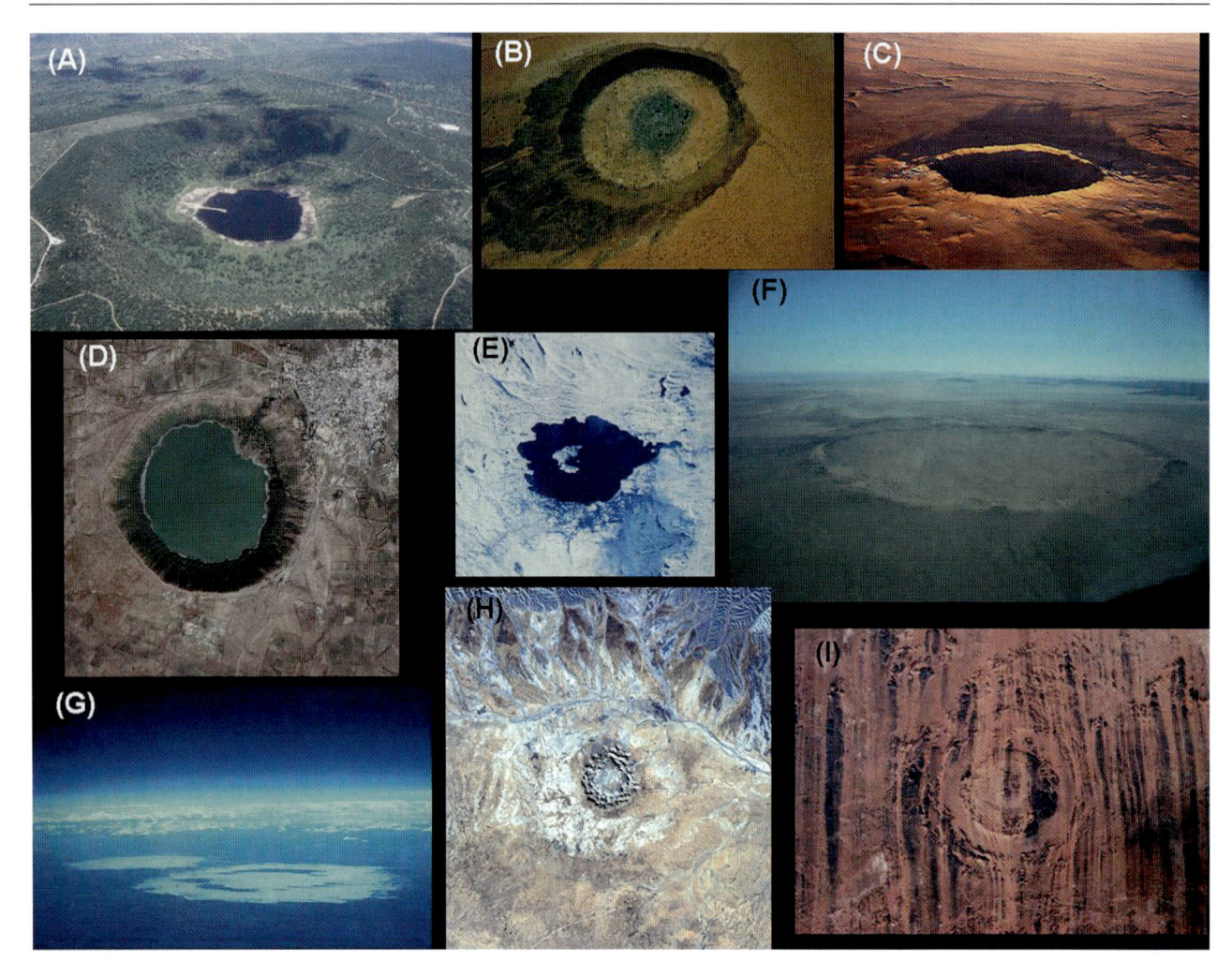

Asteroid Impact, Figure 2 Examples of simple and complex impact craters on Earth. Craters in the upper row, and the one in the center row on the right, are simple craters, and the others are complex craters. Upper row: (**a**) Tswaing (Saltpan)-crater in South Africa (1.2 km diameter, 250,000 years old); (**b**) Wolfe Creek crater in Australia (1 km diameter, 1 Ma old); (**c**) Meteor Crater in Arizona, USA (1.2 km diameter, 50,000 years old); Center row: (**d**) Lonar crater, India (1.8 km diameter, age ca. 50,000 years); (**e**) Mistastin crater in Canada (28 km diameter, age ca. 38 Ma); (**f**) Roter Kamm crater in Namibia (2.5 km diameter, age ca. 4 Ma); bottom row: (**g**) Clearwater-double crater in Canada (24 and 32 km diameter, age ca. 250 Ma); (**h**) Gosses Bluff crater in Australia (24 km diameter, age 143 Ma); and (**i**) Aorounga crater in Chad (18 km diameter, age unknown but younger than ca. 300 Ma).

The formation of a crater by hypervelocity impact is – not only in geological terms – a very rapid process that is customarily divided into three stages: (1) contact/compression stage, (2) excavation stage, and (3) post-impact crater modification stage. For more detailed discussions of the physical principles of impact crater formation, the interested reader is referred to the literature (e.g., Melosh, 1989, and references therein). The most important aspect of impact cratering concerns the release of large amounts of kinetic energy (equal to $\frac{1}{2}mv^2$, m = mass, v = velocity) when an extraterrestrial body hits the surface of the Earth with cosmic velocities (ranging from about 11–72 km/s). The physical processes that govern the formation of an impact crater are the result of the extremely high amounts of energy that are liberated almost instantaneously when the projectile hits the ground. For example, a meteorite with a diameter of 250 m has a kinetic energy that is roughly equivalent to about 1,000 mt of TNT, which would lead to the formation of a crater about 5 km in diameter. There is a difference between the behavior of a stony impactor and an iron one. Due to the difference in mechanical strength, smaller iron meteorites can reach the ground intact, in contrast to stony meteorites, which may undergo catastrophic disintegration in the atmosphere. The impact energy can be compared to that of "normal" terrestrial processes, such as volcanic eruptions or earthquakes. During a small impact event, which may lead to craters of 5–10 km in diameter, about 10^{24-25} ergs (10^{17-18} J) are released, comparable to the about 6.10^{23} ergs (6.10^{16} J) that were released over several months during the 1980 eruption of Mount St. Helens (see, e.g., French, 1998). In the case of an impact, the kinetic energy is concentrated more or less at a point on the Earth's surface, leading to an enormous local energy increase.

Schematically, the formation of an impact crater can be summarized as follows: First, a relatively small extraterrestrial body, traveling at a velocity of several tens of kilometers per second, hits the surface; this marks the beginning of the contact and compression stage. Almost immediately, a small amount of material is ejected from the impact site during a process called jetting with velocities that can approach about one half of the impact velocity. The jetted material is strongly contaminated with projectile material. When the projectile hits the surface, a shock wave is propagated hemispherically into the ground. Because the pressures in the shock waves are so high, the release of the pressure (decompression) results in almost instantaneous melting and vaporization of the projectile – and of large amounts of target rocks. Results of the interaction of the shock wave with matter can be observed in various forms of shocked minerals and rocks, all of which originate during the contact (or compression) stage, which only lasts up to a few seconds even for large impacts. After the passage of the shock wave, the high pressure is released by a so-called rarefaction wave (also called release wave), which follows the shock front. The rarefaction wave is a pressure wave, not a shock wave, and travels at the speed of sound in the shocked material. The rarefaction wave leads to the creation of a mass flow that opens up the crater, marking the beginning of the excavation phase. Important changes in the rocks and minerals occur upon decompression, when the material follows a release adiabat in a pressure versus specific volume diagram. Excess heat appears in the decompressed material, which may result in phase changes (e.g., melting or vaporization).

The actual crater is excavated during this stage. Complex interactions between the shock wave(s) and the target, as well as the release wave(s), lead to an excavation flow. In the upper layers of the target, material moves mainly upward and out, whereas in lower levels material moves mainly down and outward, which results in a bowl-shaped depression, the *transient cavity*. This cavity grows in size as long as the shock and release waves are energetic enough to excavate material from the impact location. At this point a note of caution is necessary. For a crater about 200 km in diameter, the depth of the transient cavity can easily reach 60 km. However, only about one third of this is excavation, the rest is simply material that is pushed down. Thus, even the largest craters known on Earth have not resulted in excavation of mantle material, and impact-induced volcanism is implausible and has never been found. Afterward, gravity and rock-mechanical effects lead to a collapse of the steep and unstable rims of the transient cavity, and widening and filling of the crater. Compared to the contact stage, the excavation stage takes longer, but still only up to a minute or 2 even in large craters of more than 200 km diameter.

Details on the physics and mechanics of the formation of impact craters can be found in the publication by Melosh (1989).

Recognition criteria for an impact crater

The affected rocks are important witnesses for the characteristics of the impact process. As mentioned above, crater structures are filled with melted, shocked, and brecciated rocks (Figure 3). Some of these are in situ, and others have been transported, in some cases to considerable distances from the source crater. The latter are called ejecta. Some of that material can fall back directly into the crater, and most of the ejecta end up close to the crater (<5 crater radii; these are called *proximal* ejecta), but a small fraction may travel much greater distances and are then called *distal* ejecta. The book by Montanari and Koeberl (2000) contains more detailed information on impact ejecta (see also entry *Impact Ejecta*).

How to recognize an impact crater is an important topic. On the Moon and other planetary bodies that lack an appreciable atmosphere, impact craters can commonly be recognized from morphological characteristics, but on Earth complications arise as a consequence of the obliteration, deformation, or burial of impact craters. This problem made it necessary to develop diagnostic criteria for the identification and confirmation of impact structures on Earth (see French, 1998; French and Koeberl, 2010). The most important of these characteristics are as follows: (a) crater morphology, (b) geophysical anomalies, (c) evidence for shock metamorphism, and (d) the presence of meteorites or geochemical evidence for traces of the meteoritic projectile. Morphological and geophysical observations are important in providing supplementary (or initial) information. Geological structures with a circular outline that are located in places with no other obvious mechanism for producing near-circular features may be of impact origin and at least deserve further attention. Geophysical methods are also useful in identifying promising structures for further studies, especially in the case of subsurface features. In complex craters, the central uplift usually consists of dense basement rocks and usually contains severely shocked material. This uplift is often more resistant to erosion than the rest of the crater, and thus, in old eroded structures the central uplift may be the only remnant of the crater that can be identified. Geophysical characteristics of impact craters include gravity, magnetic properties, reflection and refraction seismics, electrical resistivity, and others (see Grieve and Pilkington, 1996, for a review).

Of the criteria mentioned above, only the presence of diagnostic shock metamorphic effects and, in some cases, the discovery of meteorites, or traces thereof, are generally accepted to provide unambiguous evidence for an impact origin. Shock deformation can be expressed in macroscopic form (shatter cones) or in microscopic form. The same two criteria apply to distal impact ejecta layers and allow to confirm that material found in such layers

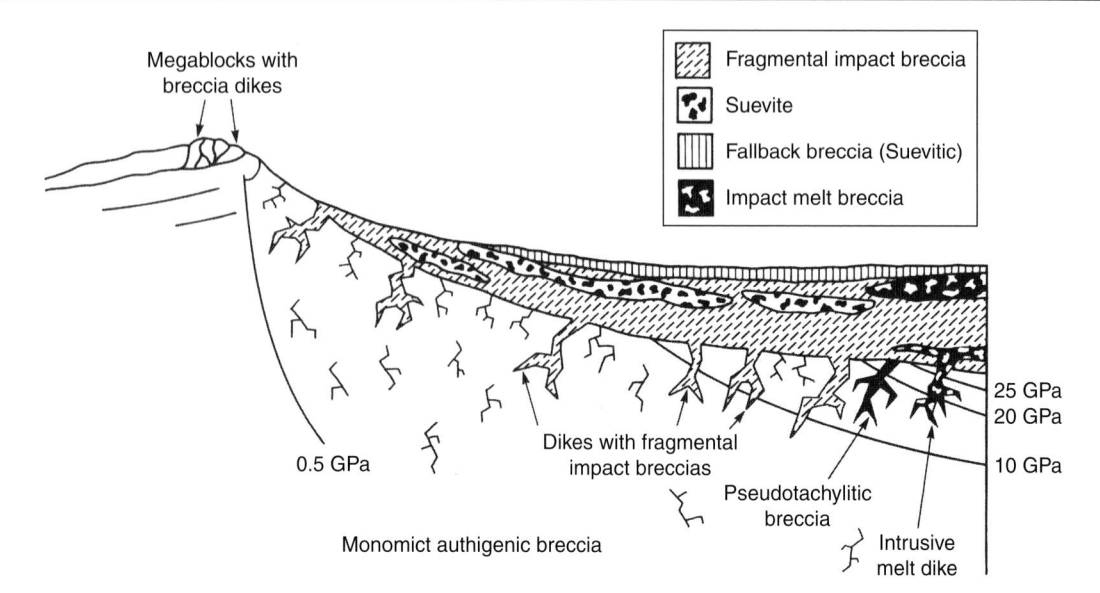

Asteroid Impact, Figure 3 Cross-section through a hypothetical simple impact crater, showing the distribution of possible impact-derived materials and rock formations (After Koeberl, 2002).

originated in an impact event at a possibly still unknown location. As of 2010 about 180 impact structures have been identified on Earth based on these criteria.

Shock metamorphism

In nature, shock metamorphic effects are uniquely characteristic of shock levels associated with hypervelocity impact. The response of materials to shock has been the subject of study over much of the second half of the twentieth century, in part stimulated by military research. Using various techniques, controlled shock wave experiments, which allow the collection of shocked samples for further studies, have led to a good understanding of the conditions for the formation of shock metamorphic products and a pressure-temperature calibration of the effects of shock pressures up to about 100 GPa (see, e.g., French and Short, 1968; Stöffler and Langenhorst, 1994; Grieve et al., 1996; French and Koeberl, 2010; and references therein).

For the identification of meteorite impact structures, suevites and impact melt breccias (or impact melt rocks) are the most commonly studied units. It is easy to distinguish between the two impact formations, as suevites are polymict breccias that contain inclusions of melt rock (or impact glass) in a clastic groundmass, and impact melt breccias have a melt matrix with a variable amount of (often shocked) rock fragments as clasts (they are also referred to in the literature as "melt-matrix breccias"). Whether or not these various breccia types are present and/or preserved in a crater depends on factors including the size of the crater, the target composition (e.g., crystalline or sedimentary rocks), the degree of porosity of the target, and the level of erosion for an impact structure. In cases of very deeply eroded structures, only remnants of injected impact breccias in the form of veins or dikes may remain. Besides injections of suevite and impact melt rock, and local (in situ) formations of monomict or polymict clastic impact breccia, this may involve veins and pods of so-called "pseudotachylitic breccia" that are recorded from a number of impact structures. This material may closely resemble what is known as "pseudotachylite," the term for "friction melt." However, it has become clear in recent years that not all of the formations of such appearance actually represent friction melt, but may also include impact melt rock and even tectonically produced fault breccias (friction melt, mylonite, or cataclasite).

The rocks in the crater rim zone are usually only subjected to relatively low shock pressures (commonly <2 GPa), leading mostly to fracturing and brecciation, and often do not show shock-characteristic deformation. Even at craters of several kilometers in diameter, crater rim rocks that are in situ rarely show evidence for shock deformation. However, there may be injections of impact breccias that may contain shock metamorphosed mineral and rock fragments. In well-preserved impact structures, the area directly outside the crater rim is covered by a (vertical) sequence of different impactite deposits, which often allow the identification of these structures as being of impact origin.

The presence of shock metamorphic effects constitutes confirming evidence for impact processes. In nature, shock metamorphic effects are uniquely characteristic of shock levels associated with hypervelocity impact. Shock metamorphic effects are best studied in the various breccia types that are found within and around a crater structure, as well as in the formations exhumed in the central uplift area. During the impact, shock pressures of about

≥100 GPa and temperatures ≥3,000°C are produced in large volumes of target rock. These conditions are significantly different from conditions for endogenic metamorphism of crustal rocks, with maximum pressures of usually <2 GPa and temperatures <1,200°C. Shock compression is not a thermodynamically reversible process, and most of the structural and phase changes in minerals and rocks are uniquely characteristic of the high pressures (diagnostic shock effects are known for the range from 8 to >50 GPa) and extreme strain rates (10^6–10^8 s^{-1}) associated with impact. The products of static compression, as well as those of volcanic or tectonic processes, differ from those of shock metamorphism, because of lower peak pressures and strain rates that are different by many orders of magnitude.

A wide variety of microscopic shock metamorphic effects have been identified (see Table 1). The most common ones include planar microdeformation features; optical mosaicism; changes in refractive index, birefringence, and optical axis angle; isotropization (e.g., formation of diaplectic glasses); and phase changes (high-pressure phases; melting). Kink bands (mainly in micas) have also been described as a result of shock metamorphism, but can also be the result of normal tectonic deformation (for reviews and images of examples, refer to Stöffler and Langenhorst, 1994; Grieve et al., 1996; French, 1998; French and Koeberl, 2010).

Planar microstructures are the most characteristic expressions of shock metamorphism and occur as planar fractures (PFs) and planar deformation features (PDFs). The presence of PDFs in rock-forming minerals (e.g., quartz, feldspar, or olivine) provides diagnostic evidence for shock deformation, and thus, for the impact origin of a geological structure or ejecta layer (see, e.g., Stöffler and Langenhorst, 1994; Montanari and Koeberl, 2000; French and Koeberl, 2010, and references therein). Good examples are shown in Figure 4. PFs, in contrast to irregular, non-planar fractures, are thin fissures, spaced about 20 μm or more apart. While they are not considered shock diagnostic per se, should they be observed in significant abundance and particularly in densely spaced sets of multiple orientations, they can provide a strong indication of shock pressures around 5–10 GPa. To an inexperienced observer, it is not always easy to distinguish "true" PDFs from other lamellar features (fractures, fluid inclusion trails, tectonic deformation bands).

The most important characteristics of PDFs are that they are extremely narrow, closely and regularly spaced, completely straight, parallel, extend often (though not always) through a whole crystal, and, at shock pressures above about 15 GPa, occur in more than one set of specific crystallographic orientation per grain. This way, they can be distinguished from features that are produced at lower strain rates, such as the tectonically formed Böhm lamellae, which are not completely straight, occur only in one set, usually consist of bands that are >10 μm wide, and are spaced at distances of >10 μm. Transmission electron microscopy (TEM) studies demonstrate that PDFs consist of amorphous silica, that is, they are planes of amorphous quartz that extend through the quartz crystal. This allows them to be preferentially etched by, for example, hydrofluoric acid, thus accentuating the planar deformation features. PDFs occur in planes that correspond to specific rational crystallographic orientations (for details, see, e.g., Stöffler and Langenhorst, 1994). With increasing shock pressure, the distances

Asteroid Impact, Table 1 Characteristics of shock deformation features in rocks and minerals

Pressure (GPa)	Features	Target characteristics	Feature characteristics
2–45	Shatter cones	Best developed in homogeneous fine-grained, massive rocks.	Conical fracture surfaces with subordinate striations radiating from a focal point.
5–45	Planar fractures and Planar deformation features (PDFs)	Highest abundance in crystalline rocks; found in many rock-forming minerals (e.g., quartz, feldspar, olivine, and zircon).	PDFs: Sets of extremely straight, sharply defined parallel lamellae; may occur in multiple sets with specific crystallographic orientations.
30–40	Diaplectic glass	Most important in quartz and feldspar (e.g., maskelynite from plagioclase)	Isotropization through solid-state transformation under preservation of crystal habit as well as primary defects and sometimes planar features. Index of refraction lower than in corresponding crystal but higher than in fusion glass.
15–50	High-pressure polymorphs	Quartz polymorphs most common: coesite, stishovite; but also ringwoodite from olivine, and others	Recognizable by crystal parameters, confirmed usually with XRD or NMR; abundance influenced by post-shock temperature and shock duration; stishovite is temperature-labile.
>15	Impact diamonds	From carbon (graphite) present in target rocks; rare	Cubic (hexagonal?) form; usually very small but occasionally up to mm-size; inherits graphite crystal shape.
45–>70	Mineral melts	Rock-forming minerals (e.g., lechatelierite from quartz)	Impact melts are either glassy (fusion glasses) or crystalline; of macroscopically homogeneous, but microscopically often heterogeneous composition.

XRD X-ray diffraction, *NMR* nuclear magnetic resonance, *PDF* planar deformation features
Table after Montanari and Koeberl (2000)

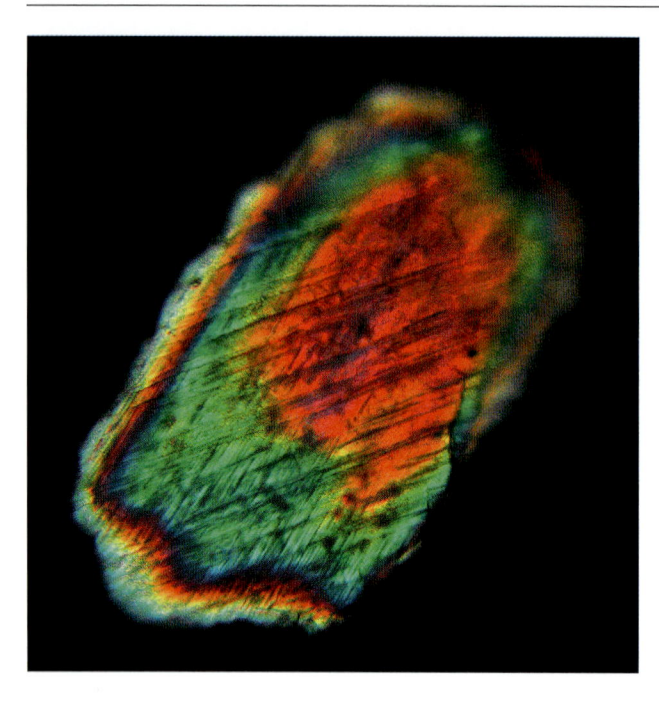

Asteroid Impact, Figure 4 Shock-characteristic planar deformation features (PDFs) in a quartz grain (in distal ejecta from the Manson impact crater, found in South Dakota). Width of the grain ca. 100 μm. Multiple intersecting sets of PDFs are clearly visible.

between the planes decrease, and the PDFs become more closely spaced and more homogeneously distributed through the grain, until at about 30–35 GPa the grains show complete isotropization. Depending on the peak pressure, PDFs are observed in about 2–10 orientations per grain. To confirm the presence of PDFs, it is necessary to measure their crystallographic orientations by using either a universal stage or a spindle stage with an optical microscope, or to characterize them by TEM. Because PDFs are well developed in quartz (Stöffler and Langenhorst, 1994), a very widely observed rock-forming mineral, and because their crystallographic orientations are easy to measure in this mineral, most studies report only shock features in quartz. However, other rock-forming minerals, as well as accessory minerals, also develop PDFs.

Higher shock pressures than those recorded in PDFs in quartz and other rock-forming minerals lead to shock-induced amorphization (without melting) of the minerals (producing "diaplectic" minerals, such as diaplectic quartz or feldspar), thermal decomposition or melting of selected minerals (e.g., the monomineralic melt of quartz is lechatelierite), and whole-rock melting. Impact glasses can form directly at a crater, or be ejected to great distances (e.g., tektites). These melt rocks and glasses are often the objects of geochemical investigations. For example, the detection of small amounts of meteoritic matter in breccias and melt rocks can also provide confirming evidence of impact (Koeberl, 2007).

Tektites and microtektites

Tektites are chemically homogeneous, often spherically symmetric natural glasses, with most being a few centimeters in size (Figure 5). Mainly due to chemical studies, it is now commonly accepted that tektites are the product of melting and quenching of terrestrial rocks during hypervelocity impact on the Earth. The chemistry of tektites is in many respects identical to the composition of upper crustal material. Tektites are currently known to occur in four strewn fields of Cenozoic age on the surface of the Earth. Strewn fields can be defined as geographically extended areas over which tektite material is found. The four strewn fields are the North American, Central European (moldavite), Ivory Coast, and Australasian strewn fields. Tektites found within each strewn field have the same age and similar petrological, physical, and chemical properties. Relatively reliable links between craters and tektite strewn fields have been established between the Bosumtwi (Ghana), the Ries (Germany), and the Chesapeake Bay (USA) craters and the Ivory Coast, Central European, and North American fields, respectively. The source crater of the Australasian strewn field has not yet been identified. Tektites have been the subject of much study, but their discussion is beyond the scope of the present review. For details on tektites see the reviews by Koeberl (1994) and Montanari and Koeberl (2000).

In addition to the "classical" tektites on land, microtektites (<1 mm in diameter) from three of the four strewn fields have been found in deep-sea cores. Microtektites have been very important for defining the extent of the strewn fields, as well as for constraining the stratigraphic age of tektites, and to provide evidence regarding the location of possible source craters. Microtektites have been found together with melt fragments, high-pressure phases, and shocked minerals and, therefore, provide confirming evidence for the association of tektites with an impact event. The variation of the microtektite concentrations in deep-sea sediments with location increases toward the assumed or known impact location.

There has been some discussion about how to define a tektite, but the following characteristics should probably be included (see Koeberl, 1994; Montanari and Koeberl, 2000): (1) they are glassy (amorphous); (2) they are fairly homogeneous rock (not mineral) melts; (3) they contain abundant lechatelierite; (4) they occur in geographically extended strewn fields (not just at one or two closely related locations); (5) they are distal ejecta and do not occur directly in or around a source crater, or within typical impact lithologies (e.g., suevitic breccias, impact melt breccias); (6) they generally have low water contents and a very small extraterrestrial component; and (7) they seem to have formed from the uppermost layer of the target surface (see below).

An interesting group of tektites are the Muong Nong-type tektites, which, compared to "normal" (or splash-form) tektites are larger, more heterogeneous in

Asteroid Impact, Figure 5 Several Australasian tektites (from Thailand), showing the variety in shapes and forms. Tektites are distal impact ejecta, which formed by total melting of continental crustal target rocks.

composition, of irregular shape, have a layered structure, and show a much more restricted geographical distribution. They are also important because they contain relict mineral grains that indicate the nature of the parent material and contain shock-produced phases that indicate the conditions of formation. The occurrence of relict minerals in some tektites points to sedimentary source rocks. Muong Nong-type tektites contain unmelted relict inclusions, including zircon, chromite, quartz, rutile, and monazite, all showing evidence of various degrees of shock metamorphism. Coesite, stishovite, and shocked minerals were found in the North American and Australasian microtektite layers.

Despite knowing the source craters of three of the four tektite strewn fields, we still do not know exactly when and how during the impact process tektites form. Besides the characteristics mentioned above, tektites contain a relatively high amount of the cosmogenic isotope beryllium-10 (^{10}Be), which cannot have originated from direct irradiation with cosmic rays in space or on Earth, but can only have been introduced from sediments that have absorbed ^{10}Be that was produced in the terrestrial atmosphere. Tektites might be produced in the earliest stages of impact, which are poorly understood. It is clear, however, that tektites formed from the uppermost layers of terrestrial target material (otherwise they would not contain any ^{10}Be). However, the question of which process was responsible for tektite production and distribution remains the subject of further research.

The hazards of asteroid impact

In July 1994, the fragments of comet Shoemaker-Levy 9 that crashed into the atmosphere of the planet Jupiter, producing impact plumes the size of our own planet (Figure 6), brought home to billions of people all over the world that impact catastrophes are not only a matter of the distant past but that they could happen any time, any place in the Solar System – also on Earth. And just about 100 years ago the Tunguska explosion of June 30, 1908, devastated some 2,000 km^2 of Siberian forests and caused environmental effects as far as London, and an air pressure signal around the world. Having been debated as the result of impact as much as of the explosion of a nuclear UFO, it is now widely accepted that this explosion was caused by the explosive disruption of a small (maybe 30–50 m wide) asteroid within the atmosphere, with an explosive energy of about 5–10 Mton TNT-equivalent.

Even more recently, a small cosmic projectile impacted in Peru on September 15, 2007. The Carancas meteorite created a ca. 14-m-wide impact crater, in a rather remote area of that country, and besides several people sustaining a significant scare, no casualties are known. However, even though this impact event was of insignificant size on a planetary scale (the largest impact structure known in the Solar System, the South Pole-Aitken basin in the south polar region of the Moon, measures some 2,500 km in diameter), there would have been hundreds of victims if this event had occurred in densely populated areas of our Planet. Clearly the understanding of impact processes is an important issue for mankind – especially if one considers what happened to the dinosaurs and their contemporaries 65 Ma ago. The impact of a 5–15-km asteroid at Chicxulub on the Yucatán peninsula generated then a 180 km diameter impact structure with global catastrophic effects (e.g., papers in Ryder et al. 1996).

The 1994 impact of fragments of comet Shoemaker-Levy 9 into the atmosphere of Jupiter demonstrated to humankind that impact is not a threat of the past. Hardly a year goes by without a call of danger from a "rogue asteroid" to Earth – although so far, all of them seem to have remained without cause. However, the terrestrial cratering record predicts a Tunguska event – impact of a roughly 50-m-sized projectile – for every 1,500 years or so, and if this event had occurred in a highly populated region, this relatively minor event comparable to ca. 1,000 Hiroshima atom bombs could have caused a loss of life potentially going into the millions (Figure 7). A large event of Chicxulub magnitude might occur every 100 Ma, but this statistical approach does not predict when it might happen next. Depending on the actual velocity of the Chicxulub impactor, that projectile might have measured between 5 and 15 km in diameter, although scaling from the amount of extraterrestrial material around the world indicates a value of about 10 km. However, current predictions of impact consequences suggest that even a 1-km-sized bolide might be sufficient to cause potential harm to mankind through global environmental catastrophe.

Projectiles enough are out there – comets in the Kuiper-Edgeworth belt and in the Oort cloud, and near-Earth objects (those asteroids that approach the Sun closer than 1.3 AU) are not particularly rare in the asteroid belt, largely a result of the NASA-sponsored international Spaceguard surveying project. Data from these studies indicate that "the risk of any person's death by impact prior to Spaceguard was estimated at about the same as dying in an airplane crash" – something like 1 in 30,000, but since the successful work of Spaceguard, this estimate has changed to 1 in 600,000 as the danger from likely still undetected asteroids.

And still, much remains to be learned about the catastrophic impact effect. Space search programs for possible still unknown asteroids continue, and the danger from suddenly appearing comets remains anyway. The latter case provides a challenge for impact workers, as the behavior of large, low-density projectiles, and the effects of their impacts onto various planetary surfaces are far from being understood. The terrestrial impact record has to be improved to further constrain the past impact flux onto Earth, and the energy threshold for truly global devastation by impact still needs to be identified. Does it take an impact event of the magnitude of Chicxulub? Or was that impact particularly lethal because of the sulfate- and carbonate-rich target area at Yucatán? Can our highly evolved and, through its intricate civilization, highly vulnerable race survive a much smaller impact event, perhaps only creating a 40 km impact structure? And that will only take a 2–3-km-sized impactor. …

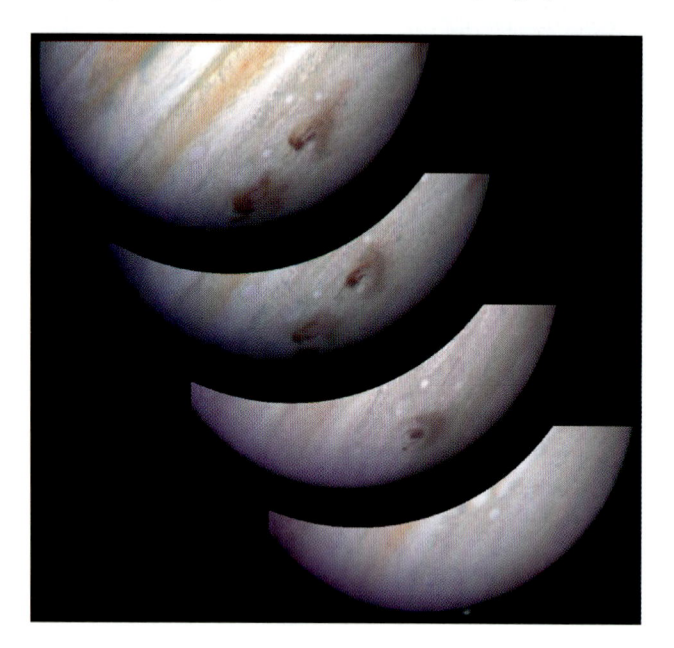

Asteroid Impact, Figure 6 Hubble space telescope image sequence of Comet Shoemaker-Levy 9 impacting Jupiter in 1994. Shown is a time-lapse sequence of the result of two fragments striking Jupiter. The fragments hit the atmosphere at about 60 km/s and left Earth-sized traces that could be observed for many weeks. Jupiter has a diameter that is about 12 times that of the Earth.

Conclusions

Mineralogical, petrographic, and geochemical methods have been used for many decades in the study of the effects of asteroid impact in the formation of terrestrial meteorite impact craters. Currently about 180 impact structures are known on Earth (Figure 8). A clear hiatus in the history of impact-related studies was the realization

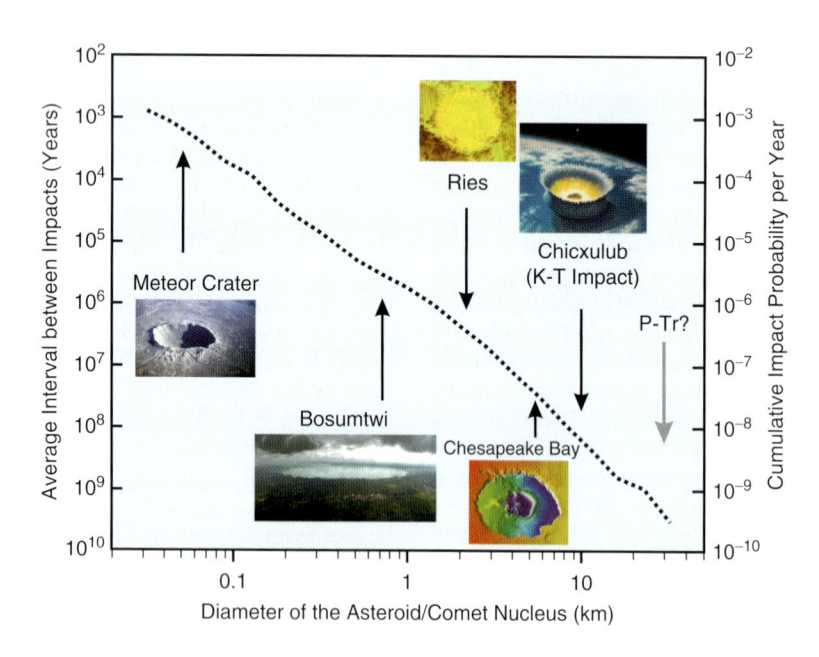

Asteroid Impact, Figure 7 Relation between impact probability and crater (as well as impactor) diameter (Updated from Montanari and Koeberl, 2000).

Diameters (km)
- ○ 0 - 10
- ● 11 - 20
- ● 21 - 50
- ● 51 - 100
- ● 101 - 300

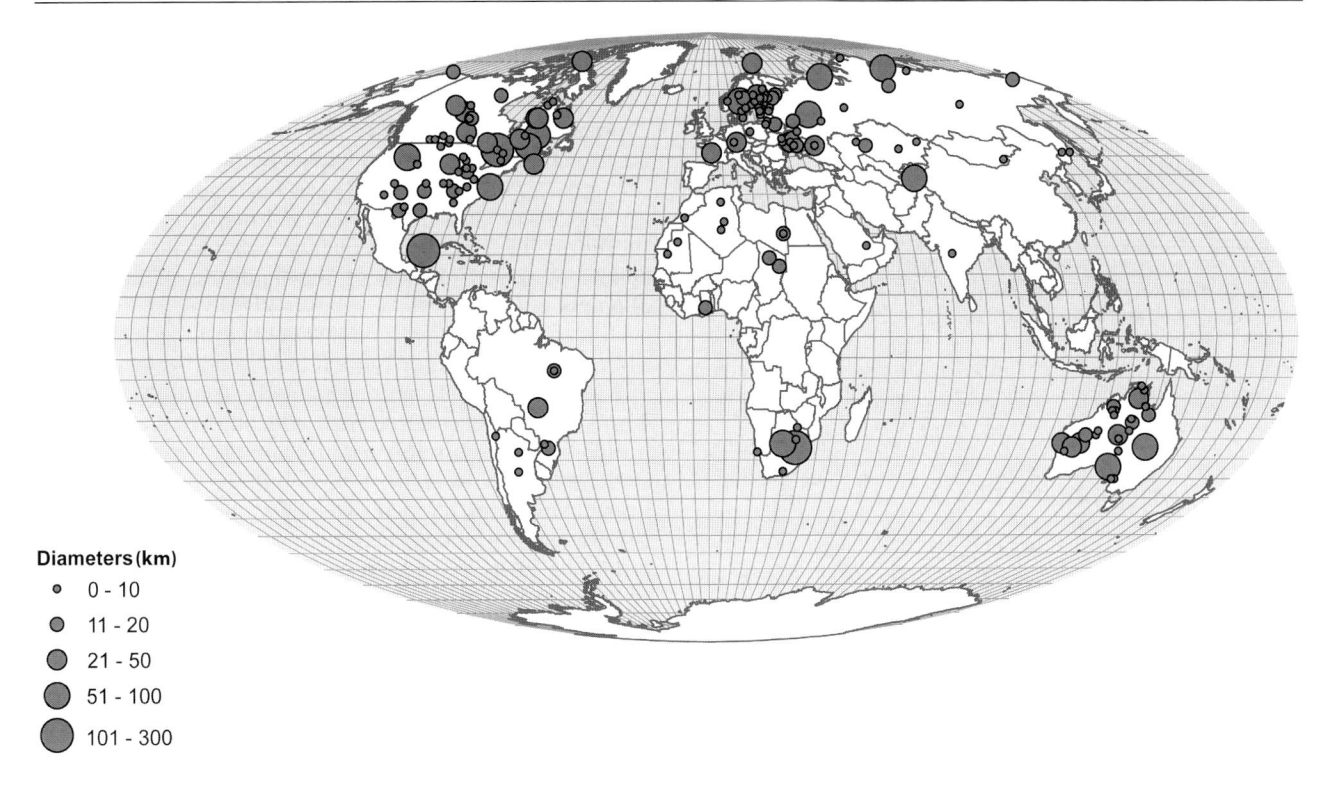

Asteroid Impact, Figure 8 Map of Earth with the currently known impact craters.

that K-T boundary bears unambiguous evidence for a large-scale catastrophic impact event (related to the formation of the 200-km-diameter Chicxulub impact structure, Mexico). Analyses of the K-T ejecta layers led to improved detection sensitivities for impact markers, allowing identification of smaller events and the study of their effects. Distal impact ejecta layers can be used to study a possible relationship between biotic changes and impact events, because it is possible to study such a relationship in the same outcrops, whereas correlation with radiometric ages of a distant impact structure is always associated with larger errors. Investigations of impact markers yield important information regarding the physical and chemical conditions of their formation, such as temperature, pressure, oxygen fugacity, and composition of the atmosphere. These data are necessary to understand the mechanisms of interaction of impact events with the environment and should ultimately lead to a better appreciation of the importance of impact events in the geological and biological evolution of the Earth. New geochemical techniques, such as the use of the Cr and Os isotopic systems, or analyses of comet-dust-derived ^{3}He in sediments (e.g., Farley et al., 1998), have helped to confirm or better explain several important impact events. Recent improvements in analytical methods and techniques will certainly continue to influence our understanding of the interaction between cosmic bodies and the Earth.

Bibliography

Alvarez, L. W., Alvarez, W., Asaro, F., and Michel, H. V., 1980. Extraterrestrial cause for the Cretaceous-Tertiary extinction. *Science*, **208**, 1095–1108.

Bohor, B. F., Foord, E. E., Modreski, P. J., and Triplehorn, D. M., 1984. Mineralogical evidence for an impact event at the Cretaceous/Tertiary boundary. *Science*, **224**, 867–869.

Farley, K. A., Montanari, A., Shoemaker, E. M., and Shoemaker, C. S., 1998. Geochemical evidence for a comet shower in the late Eocene. *Science*, **280**, 1250–1253.

French, B. M., 1998. *Traces of Catastrophe: A Handbook of Shock-Metamorphic Effects in Terrestrial Meteorite Impact Structures*. Houston: Lunar and Planetary Institute. LPI Contribution 954. 120 pp.

French, B. M., and Koeberl, C., 2010. The convincing identification of terrestrial meteorite impact structures: what works, what doesn't, and why. *Earth-Science Reviews*, **98**, 123–170.

French, B. M., and Short, N. M. (eds.), 1968. *Shock Metamorphism of Natural Materials*. Baltimore: Mono Book Corp. 644 pp.

Grieve, R. A. F., and Pilkington, M., 1996. The signature of terrestrial impacts. *AGSO Journal Australian Geology and Geophysics*, **16**, 399–420.

Grieve, R. A. F., Langenhorst, F., and Stöffler, D., 1996. Shock metamorphism in nature and experiment: II. Significance in geoscience. *Meteoritics and Planetary Science*, **31**, 6–35.

Koeberl, C., 1994. Tektite origin by hypervelocity asteroidal or cometary impact: target rocks, source craters, and mechanisms. In Dressler, B. O., Grieve, R. A. F., and Sharpton, V. L. (eds.), *Large Meteorite Impacts and Planetary Evolution*. New York: Geological Society of America. GSA, Special Paper 293, pp. 133–152.

Koeberl, C., 2002. Mineralogical and geochemical aspects of impact craters. *Mineralogical Magazine*, **66**, 745–768.

Koeberl, C., 2007. The geochemistry and cosmochemistry of impacts. In Davis, A. (ed.), *Treatise of Geochemistry*. Oxford: Elsevier-Pergamon, Vol. 1, pp. 1.28.1–1.28.52, doi:10.1016/B978-008043751-4/00228-5. Online edition.

Melosh, H. J., 1989. *Impact Cratering – A Geologic Process*. New York: Oxford University Press. 245 pp.

Montanari, A., and Koeberl, C., 2000. *Impact Stratigraphy – The Italian Record*. Heidelberg-Berlin: Springer Verlag. 364 pp.

Ryder, G., Fastovsky, D., and Gartner, S. (eds.), 1996. *The Cretaceous-Tertiary Event and other Catastrophes in Earth History*. New York: Geological Society of America. GSA, Special Paper 307. 576 pp.

Smit, J., 1999. The global stratigraphy of the Cretaceous-Tertiary boundary impact ejecta. *Annual Reviews of Earth and Planetary Science*, **27**, 75–113.

Stöffler, D., and Langenhorst, F., 1994. Shock metamorphism of quartz in nature and experiment: I. Basic observations and theory. *Meteoritics*, **29**, 155–181.

Cross-references

Asteroid
Asteroid Impact Mitigation
Asteroid Impact Predictions
Comet
Extinction
Impact Airblast
Impact Ejecta
Impact Fireball
Impact Firestorms
Impact Tsunami
Impact Winter
Meteorite
Torino Scale

ASTEROID IMPACT MITIGATION

Clark R. Chapman
Southwest Research Institute, Boulder, CO, USA

Synonyms

NEA preparation; NEA response

Definition

The deflection of Near-Earth Asteroids and/or avoidance of the consequences of an impact.

Discussion

Near-Earth Asteroids (NEAs) present a potential hazard. Meteorites falling from the sky are a minimal threat. An NEA larger than ~20 m diameter could explode close enough to the Earth's surface to cause local damage whereas one larger than ~150 m could reach the ground or ocean with most of its cosmic velocity, causing an explosion far larger than any nuclear weapon tested. An NEA larger than ~2 km diameter could damage the global ecosphere sufficiently to threaten civilization.

Rescue and recovery would resemble that following most other natural disasters, should an impact not be forecast in advance. With warning, there are two basic approaches to NEA impact mitigation: (1) deflecting the NEA well before predicted impact by one of several technological approaches using spacecraft, and/or (2) evacuating regions around ground zero and preparing in advance for more widespread consequences, should there be inadequate warning to deflect the body or if deflection attempts fail.

Telescopic searches, and thus long-term warnings, are directed toward larger NEAs, that is, those larger than 100 m in diameter, and especially those larger than 1 km. If one is discovered with a decade or more advance notice, a space mission could probably be deployed in sufficient time to deflect it, and thus ensure that the impact does not happen. The best approach would be to launch one or more massive (e.g., 1 t or greater) spacecrafts, as early as possible before the expected impact, with an aim to collide with the NEA, triggering a small velocity change in the optimal direction such that it eventually arrives either early or late to the point in space where it would otherwise strike the Earth. Such "kinetic impactors" might be insufficient to deflect an NEA larger than ~1 km, in which case it might be necessary to detonate a nuclear device near the NEA in order to deflect it.

An NEA's response to a kinetic impactor or nuclear explosion cannot be perfectly predicted. Thus, it is possible that the NEA could be deflected into a so-called keyhole, resulting in it returning to impact Earth in a subsequent year. For this reason, it is highly desirable to have an observer spacecraft in the vicinity of the NEA to characterize its properties in advance of deflection and, afterward, to determine precisely how much deflection was achieved. If the observer spacecraft is equipped with thrusters so as to act as a "gravity tractor", the mutual gravity between the spacecraft and the NEA could be used to precisely change the NEA's velocity sufficiently so as to preclude any further danger.

NEAs smaller than ~100 m diameter are much more likely to impact Earth than larger NEAs, but telescopes are less likely to detect them years or decades prior to impact. There is a good chance (perhaps as high as 50%) that one would be found during the last days or weeks before impact. Such short notice precludes deflection by spacecraft (exploding an NEA on its way in is an especially bad idea), but there would likely be opportunity for warning, evacuation, and other approaches to mitigating the expected damage.

Bibliography

Committee to Review Near-Earth Object Surveys and Hazard Mitigation Strategies, National Research Council, 2010. Defending planet earth: near-earth object surveys and hazard mitigation strategies: final report, 136 pp (Available from the National Academies Press.).

Cross-references

Asteroid
Asteroid Impact
Asteroid Impact Predictions
Comet
Impact Ejecta

ASTEROID IMPACT PREDICTIONS

Brian G. Marsden (Deceased)

Synonyms

Asteroid; Minor planet

Definition

AU. Astronomical unit – approximately the mean distance between the earth and the sun, about 149 597 871 km.
Aphelion. Point in an orbit that is farthest from the sun.
Apophis. Asteroid (99942) = 2004 MN$_4$.
Minor planet center. The International Astronomical Union's central clearinghouse for positional observations and orbital data for asteroids, located at the Smithsonian Astrophysical Observatory, Cambridge, Mass.
MOID. Minimum orbit intersection distance – the closest separation of two orbits. A PHA (potentially hazardous asteroid) is an asteroid with a MOID relative to the earth of less than 0.05 AU.
Near-earth asteroid. An asteroid whose distance from the sun at perihelion is less than 1.30 AU.
Palermo scale. A more quantified version of the Torino scale, including consideration of the time remaining to potential impact.
Perihelion. Point in an orbit that is closest to the sun.
Purgatorio ratio. PR – the ratio of the time interval covered by the observations of an asteroid to the time remaining until its next impact possibility.
Resonant return. A situation where the actual configuration of the earth and an asteroid approximately repeats after an integral number of years.
Torino scale. An integer from 0 to 10 equating impact probability and energy, events with scale 0 having no consequence and events with scale 10 the certainty of worldwide destruction.
Väisälä orbit. Orbit in which the asteroid is assumed to be at perihelion or aphelion.

Introduction

There are two main aspects to the problem of asteroid impact prediction. The first involves the situation in which a newly discovered object might impact the earth within a matter of days or weeks. The second considers the possibility that an object that is already well known will hit the earth many years or decades in the future.

Newly discovered objects

In principle, any object that has just been discovered has some nonzero chance of striking the earth. Initially, there are measurements of its direction from the discovery site as it tracks essentially uniformly along a small arc of the sky over the course of an hour or so. There is no direct information about the distance from the observer, although if the object is not substantially brighter than and its angular motion is not greatly different from those of known main-belt asteroids in the same region of the sky, it is very likely to be just another member of this enormous population of harmless asteroids.

To quantify the situation, one can therefore examine the consequences of *assuming* various values of the object's distance from the observer at the two instants, t_0 and t_1. A more enlightened procedure (Marsden, 1999a), routinely used for unusual new discoveries reported to the Minor Planet Center, considers values of the distance ρ_0 at t_0 but rotates the coordinate system so that the xy plane passes through the two observations and the xz plane through the observation at t_1. This immediately provides values for the heliocentric vector $\mathbf{r}_0(x_0, y_0, z_0)$ at t_0, and, with the help of the first two terms of the well-known "f and g" expansions, the components $\dot{y}_0$ and $\dot{z}_0$ of the heliocentric velocity immediately follow.

For each selected value of ρ_0, it therefore remains only to assume a variety of values of $\dot{x}_0$. Use of the energy integral leads to $1/a$, the reciprocal of the orbital semimajor axis, and taking $\dot{x}_0 = 0$ will yield $1/a_M$, its largest possible value for the chosen ρ_0. If $1/a_M$ is positive (i.e., the $\dot{x}_0 = 0$ solution is elliptical), it follows that the choice of $\dot{x}_0 = \pm\sqrt{1/a_M}$ will produce a pair of parabolic solutions. Since $1/a_M$ shows a general decrease as ρ_0 increases from the value zero, all the orbits eventually become hyperbolic (and hence of no interest), although the decrease is not necessarily monotonic.

For a given ρ_0, there is readily found a particular value of $\dot{x}_0$ that ensures that the heliocentric radial velocity $\dot{r}_0 = 0$, i.e., the object is at either perihelion or aphelion, in what is often termed a "Väisälä" orbit (Väisälä, 1939). The transition between perihelic and aphelic solutions implies the existence of a ρ_0 value that yields an exactly circular orbit, and if this solution has a small-to-moderate direct inclination and is in the distance range corresponding to the main asteroid belt or the Jupiter Trojans, this can support the conclusion from the apparent motion and brightness that the object is not a threat.

On the other hand, the existence of a circular solution that is similar to the orbit of the earth is obviously cause for worry. Such was the situation with the initial observations of 2008 TC$_3$, an object discovered in October 2008 and routinely reported to the Minor Planet Center, which automatically solicited follow-up observations. The availability of several such observations over the course of a couple of hours was sufficient to show that the orbital eccentricity was as high as 0.3, but they confirmed that this object was at roughly the distance of the moon and approaching in such a way that it would collide with the earth only 20 h after discovery. The observed faintness made it clear that 2008 TC$_3$ was at most a few meters across and thus of no danger, but this unique example provided a valuable exercise for studying the first of the situations posed in the first paragraph of this article.

More established objects

To study the second of those situations, a useful point to consider is that the minimum orbital intersection distance (MOID) (Bowell and Muinonen, 1994) between the orbits of an asteroid and the earth is less than 0.05 AU. Of course, the MOID will change over time due to the actual gravitational influence of the earth and other planets, so it is also helpful to be aware of the asteroid's MOIDs with respect to the other planets. Furthermore, if an asteroid does actually approach the earth within the MOID, a further approach around the same date some integral number of years later is to be expected, the uncertainty in our knowledge of the orbit being magnified during the initial encounter and thereby increasing the likelihood of a subsequent collision.

The significance of such "resonant returns" for impact prediction was first recognized for (35396) 1997 XF$_{11}$ and (137108) 1999 AN$_{10}$ (Marsden, 1999b; Milani et al., 1999; Milani et al., 2000), with impact probabilities in the range 1 in 100,000–1,000,000 computed for several dates a decade or two after the earth approaches in 2028 and 2027, respectively, a time when the observations spanned just a few months. With the recognition now of pre- and post-discovery observations covering decades, it can be unequivocally stated that we are safe from these objects for the foreseeable future. With a 6-month observed arc, the more extreme case of (99942) Apophis = 2004 MN$_4$ actually had a 1-in-37 chance of earth impact in 2029 (Chesley, 2006), and as the arc (including radar data) increased to some four years, the fact that this object approaches in 2029 to within some 35,000 km of the earth means that there are still impact probabilities in the 1 in 100,000–1,000,000 range in 2036 and during 2056–2076.

As of early 2010, there were 200–300 cases of near-earth asteroids with some nonzero (i.e., more than 1 in a billion) probability of earth impact during the next century. Computed independently by groups in Pasadena and Pisa, these are routinely tabulated at http://neo.jpl.nasa.gov/risk/ and http://newton.dm.unipi.it, respectively. In addition to specifying the impact probabilities for each object on the relevant dates (with hundreds of possibilities in some cases), these tabulations also estimate the impact energy and attempt to assess the relative danger of the potential impacts. For public consumption, this assessment is done in terms of the Torino Scale (Binzel, 2000), simply an integer from 0 to 10 that equates probability and energy, with no consideration of when the impact might occur. The time to impact is included in the logarithmic Palermo Scale (Chesley et al., 2002), largely with regard to the likely statistical background hazard due to impacts by unknown objects during the intervening time.

An alternative way of assessing impact threats is to consider the "Purgatorio Ratio" (PR), which is simply the ratio of the time interval covered by the observations to the time remaining until the date of the next impact possibility (Marsden, 2007). As of early 2010, the PR for Apophis was 15%. If the PR is less than 1%, it is reasonable to ignore the threat, in the general expectation that, as further observations are obtained, computations will show the threat to vanish. Obviously, this does not always happen, and as a potential impact date approaches, the PR will then not only rise well above 1%, but it will eventually become infinite. If, indeed, this continues to be the case with the acquisition of current observations, there is reason for concern, particularly if the impact probability is more than, say, 10%. But the PR eventually becomes infinite even if there are no further observations. Given that the vast majority of the entries on the Pasadena and Pisa lists are "lost," having had observed arcs of only a few days (even only 3 h!) several years ago, such threats can almost certainly be dismissed, although if this causes unease, one can endeavor to take the opportunity of checking beforehand to see if the objects are located at the points in the sky they would have to occupy if impacts really were to take place.

Summary

The procedures currently in place for the prediction of potential impacts of asteroids on the earth are described. Firstly, there is consideration of the possibility that an object may be near the earth and have a distinct chance of impacting it within a matter of days, as in the case of 2008 TC$_3$. Secondly, when an object has been under more extensive observation, like Apophis, there is some meaningful possibility of establishing whether there is even a remote chance of impact up to a century hence. Various ways of assessing the impact hazard are discussed.

Bibliography

Binzel, R. P., 2000. The Torino impact hazard scale. *Planetary and Space Science*, **48**, 297–303.

Bowell, E., and Muinonen, K., 1994. Earth-crossing asteroids and comets: ground-based search strategies. In Gehrels, T., Matthews, M. S., and Schumann, A. M. (eds.), *Hazards Due to Comets and Asteroids*. Tucson and London: University of Arizona Press, pp. 149–197.

Chesley, S. R., 2006. Potential impact detection of Near-Earth asteroids: the case of 99942 Apophis (2004 MN$_4$). In Lazarro, D., Ferraz Mello, S., and Fernandez, J. A. (eds.), *Asteroids, Comets, Meteors–ACM 2005, International Astronomical Union Symposium, No. 229*. Cambridge/New York/Melbourne/Madrid/Cape Town: Cambridge University Press, pp. 215–228.

Chesley, S. R., Chodas, P. W., Milani, A., Valsecchi, G. B., and Yeomans, D. K., 2002. Quantifying the risk posed by potential earth impacts. *Icarus*, **159**, 423–432.

Marsden, B. G., 1999a. Ephemerides of small bodies in the solar system. In Fiala, A. D., and Dick, S. J. (eds.), *Proceedings,*

Nautical Almanac Office Sesquicentennial Colloquium. Washington DC: US Naval Observatory, pp. 333–351.

Marsden, B. G., 1999b. A discourse on 1997 XF$_{11}$. *Journal of the British Interplanetary Society*, **52**, 195–202.

Marsden, B. G., 2007. Impact risk communication management (1998–2004): has it improved? In Bobrowsky, P., and Rickman, H. (eds.), *Comet Asteroid Impacts and Human Society*. Berlin/Heidelberg/New York: Springer-Verlag, pp. 505–519.

Milani, A., Chesley, S. R., and Valsecchi, G. B., 1999. Close approaches of asteroid 1999 AN$_{10}$. *Astronomy and Astrophysics*, **346**, L65–L68.

Milani, A., Chesley, S. R., and Valsecchi, G. B., 2000. Asteroid close encounters: risk assessment. *Planetary and Space Science*, **48**, 945–954.

Väisälä, Y., 1939. Eine Einfache Methode der Bahnbestimmung. *Annales Academiæ Scientiarum Fennicæ Series A*, **52**, 5–32.

Cross-references

Asteroid
Asteroid Impact
Asteroid Impact Mitigation
Comet
Meteorite
Torino Scale
Tsunami

AUTOMATED LOCAL EVALUATION IN REAL TIME (ALERT)

Lev I. Dorman
Tel Aviv University, TECHNION, Qazrin, Israel
Russian Academy of Science, Moscow, Russia

Definition

ALERT (Automated Local Evaluation in Real Time) is a system for the automated production of a special signal forecasting different types of natural phenomena: earthquakes, tsunami, magnetic storms, changes in weather, solar cosmic ray events, radiation hazard, and so on. The forecasting is based on real-time observation data and theoretical and/or correlation models.

Importance for practice

The production of ALERT is especially important if the phenomenon may be dangerous to people, infrastructure, and/or technology. In many cases, if developed and deployed in time, ALERT can help to completely avoid or sufficiently minimize the negative consequences which can result from different natural and dangerous phenomena.

The probabilities T and F that ALERT is true or may be false

Any ALERT system that is adopted is done so with the expectation that some probability T is to be true (adjust accurately, fit exactly) and another probability F that it can be false (not true; counterfeit; deceitful; wrong), so T + F = 1.

The probability M that ALERT will be missing

Some natural phenomena may occur without producing a corresponding ALERT. The missing of ALERT may be the result of poor quality data or an insufficient number of observations as well as by not applying a fully adequate model. The probability M, that ALERT will be missing, decreases considerably on increasing the quality and number of observations used in the real-time data. It is important that the value M decreases with an increasing amplitude of the natural hazard phenomenon.

The forecasting time Δt between ALERT and expected natural hazardous phenomenon

This forecasting time Δt is a very important characteristic of any ALERT system. It may concern millions or thousands of years of forecasting of certain natural phenomena such as crossings of our solar system through galactic arms and galactic dust-molecular clouds (which may lead to long-term changes of the Earth's climate – decreasing of planetary temperature on several degrees; these natural hazard phenomena have occurred in the past many times), or impacts of asteroids with the Earth (which can lead to significant disasters in the biosphere), or nearby supernova explosions (accompanied with an increase of ultraviolet and neutrino radiation as well as cosmic ray fluxes, see entry *Supernova*). The forecasting time Δt can be hundreds or tens of years for estimation of expected global climate change and expected changes in seismic activity in some local regions. The value Δt can be years or months for forecasting phenomena connected with variations of solar and geomagnetic activity, and other similar processes in interplanetary space.

It is especially important in practice that now and in the near future that the forecasting time Δt be reduced, as days and hours for many natural phenomena such as magnetic storms, radiation hazards from large fluxes of solar cosmic rays, local meteorological conditions, volcanic eruptions, tsunami, and hurricanes. The forecasting of especially large natural hazards is usually recurrent, and with a subsequent reduction in forecasting time Δt the quality of the forecasts may increase.

Cross-references

Asteroids
Radiation Hazards
Supernova

AVALANCHES

Chris Stethem
Chris Stethem & Associates Ltd., Canmore, AB, Canada

Definition

A snow avalanche ensues when a pent up snow mass loses its hold and is discharged from the mountainside (Seligman, 1936, 292).

Introduction

Human interaction with snow avalanches undoubtedly began before recorded history when travelers first encountered steep slopes and deep snow. Written accounts and studies on snow avalanches began in the European Alps in the mid-nineteenth century (Seligman, 1936). Whereas at one time the majority of loss of life and property damage was associated with natural avalanches affecting those living and working in the mountains, today the majority of victims are recreationists who trigger such avalanches themselves. Over the 5-year period from 2004–2005 to 2008–2009, between 120 and 195 avalanche fatalities per year were reported to the International Commission on Alpine Rescue (www.ikar-cisa.org).

Avalanche phenomena

Avalanche types include *loose snow* avalanches which start from a point and spread downhill (Figure 1) and *slab* avalanches which have a distinctive fracture line or crown at their top (Figure 2). Dry slab avalanches initiate by a rapidly propagating shear fracture in a thin weak layer that underlies a more cohesive planar slab (McClung and Schaerer, 2006, 80). *Avalanche size* is dependent on *terrain* and *snow supply*. Slab avalanches are usually larger in size than loose snow avalanches and hence more dangerous.

Avalanches may initiate in either *dry* or *wet* snow. Dry snow avalanches have been observed to reach speeds exceeding 60 ms^{-1} (McClung, 1990) and consequently will tend to overrun terrain features in their path. Wet snow avalanches will move more slowly and are more easily diverted or channeled by terrain features such as gullies. A moving avalanche can exhibit a variety of flow characteristics depending on speed, terrain, and moisture content. A dry avalanche is often observed to have a *dense flow* at the base and a *powder* component above, termed a *mixed motion* avalanche (McClung and Schaerer, 2006). A fluidized saltation layer is also described to surround the dense core (European Commission, 2009). It is also possible that a powder avalanche will be observed without the dense flowing core, especially on very steep terrain, or that the powder component may separate from the dense flow in the course of its descent. Wet snow avalanches move with a flowing motion without the powder characteristic.

The *impact pressure* of avalanches is proportional to the avalanche speed squared, v^2, and the flowing density, ρ, of the moving material (McClung and Schaerer, 2006). Impact pressure measurements show that the highest impact pressures result from the initial impact of the dense flowing core of large, dry avalanches. Measured impact pressures range from <50 KPa to as much as 700 KPa (McClung and Schaerer, 1985; Sovilla et al., 2008). The impact pressures of wet avalanches are also significant given the higher density of flow and speeds which can exceed 30 ms^{-1}.

Avalanches, Figure 1 A loose snow avalanche starting from a point below the rocks at the top of the slope.

Avalanches, Figure 2 A dry slab avalanche fracture line at the convex feature near the top of a steep slope.

Avalanche terrain

An avalanche path (Figure 3) consists of a *starting zone*, a *track*, and a *runout zone*. Within the starting zone, the avalanche releases and begins to accelerate. The track connects the starting zone and runout and is the section of the avalanche path where maximum speeds are usually observed. Significant entrainment of additional snow or debris by the moving avalanche may also be observed in the track (Sovilla et al., 2004). The runout zone is the location where the avalanche decelerates and stops.

The primary terrain characteristic for avalanche formation is slope *incline*. The lower limit of incline in the starting zone for dry snow avalanches is >25° (McClung and Schaerer, 2006). Large slab avalanches are usually observed on slopes with inclines in the range of 30–45°,

Avalanches, Figure 3 A group of avalanche paths with adjacent starting zones (*top*), tracks (*middle*), and runout zones (*bottom*). Wet avalanche deposits are evident in the runout zones.

with a mean incline of 38° (Perla, 1977). More frequent and smaller avalanches are observed on steeper terrain. The lower limit of incline can be reduced in wet snow as liquid water content rises, the extreme case of which is a slush flow where average starting zone inclines have been reported in the range of 5–20° (Hestenes, 1985, 1998).

Slope *aspect* or orientation in regard to the wind and sun is generally regarded as the second most important terrain factor. Exposure to wind (Schaerer, 1977) will determine the extent to which wind transport increases or decreases snow load in potential starting zones. Certain terrain features such as bowls and gullies act as natural catchments for blowing snow. The orientation in regard to the sun and time of year will influence the radiation balance at the snow surface and resultant changes in snow temperatures and properties.

Other important terrain characteristics include slope *size and configuration*, *ground surface roughness*, *vegetation*, *and elevation*. Starting zone area and avalanche path length are important determinants of potential avalanche size. Certain terrain configurations such as gullies will tend to channel moving avalanches. Where the ground is rough, more snow is required to reach the threshold depth for avalanche formation. Slab avalanches are rare where the forest cover density reaches 500–1,000 stems per hectare (McClung and Schaerer, 2006), depending on slope incline and forest type (Schneebeli and Myer-Grass, 1992). Different snow, weather, ground conditions and vegetation can be expected at different elevations.

Avalanche *runout distance* is the farthest point reached by an avalanche in the runout zone. Determination of maximum runout distance is an important task in engineering and planning which combines physical evidence gathering with modeling of the runout distance of design avalanches. Physical evidence includes the historic record of avalanche occurrence, anecdotal history, topographic mapping, vegetation evident in aerial photography, and field observation. These are analyzed to identify key terrain features and indications of past damage or avalanche deposition. A variety of statistical and dynamic methods can then be applied to model the runout of maximum events with runout intervals of 100 years or more. The various dynamic models and their development are described by Salm (2004). Statistical models are described by Lied and Bakkehøi (1980) and McClung and Mears (1991). Different preferences in method are apparent in different parts of the world, and a combined approach may yield the best results (Jamieson et al., 2008). Experience in application of the methods is required for selection of appropriate data sets for statistical models or input parameters for dynamic models and expression of uncertainty.

Predicting avalanches

The data elements for interpreting *snow instability* can be grouped into direct evidence of instability (Class I), snowpack factors (Class II), and meteorological factors (Class III) (LaChapelle, 1980; McClung and Schaerer, 2006).

The most direct information on snow instability (Class 1) is gained through observation of avalanche occurrences, cracking or sudden collapse during slope tests, and quality of shear failure gained in instability tests.

The potential for slab avalanche formation and especially the identification of deep-seated instability is dependent on snowpack structure (Class II) or the layering of the snow cover. Where a weak layer or poor bond between layers exists, there is a greater potential for shear failure and slab avalanche release. Rule-based approaches to interpretation of snowpack data can be applied (Schweizer et al., 2004).

The most important snow and weather factors (Class III) in avalanche formation include amount and rate of loading by precipitation and wind, temperature (air and snow), and temperature trend (Perla, 1970; McClung, 2002; Exner and Jamieson, 2009). Other snow and weather variables which are monitored include precipitation type, new snow density, total snow depth, radiation, humidity, resistance to penetration, and surface condition.

The interaction of these factors with mountain terrain is a complex and uncertain process, and targeted education and experience are required to accurately identify existing snow instability or forecast future trends in instability and avalanche hazard.

Bibliography

European Commission, 2009. The design of avalanche protection dams. In Jóhannesson, T., Gauer, P., Issler, P., and Lied, K. (eds.), *Project Report EUR 23339*. Climate Change and Natural Hazards Research, Series 2.

Exner, T., and Jamieson, B., 2009. The effects of daytime warming on snowpack creep. In *Proceedings International Snow Science Workshop 2009*, Davos, Switzerland, pp. 271–275.

Hestenes, E., 1985. A contribution to the prediction of slush flows. *Annals of Glaciology*, **6**, 1–4.

Hestenes, E., 1998. Slushflow hazard-where, why and when? 25 years of experience with slushflow consulting and research. *Annals of Glaciology*, **26**, 370–376.

Jamieson, B., Margreth, S., and Jones, A., 2008. Applications and limitations of dynamic models for snow avalanche mapping. In *Proceedings International Snow Science Workshop 2008*, Whistler, BC, pp. 730–739.

LaChapelle, R. E., 1980. The fundamental processes in conventional avalanche forecasting. *Journal of Glaciology*, **26**, 75–84.

Lied, K., and Bakkehøi, S., 1980. Empirical calculation of snow-avalanche runout-distance based on topographic parameters. *Journal of Glaciology*, **26**(94), 165–177.

McClung, D. M., 1990. A model for scaling avalanche speeds. *Journal of Glaciology*, **36**(123), 188–198.

McClung, D. M., 2002. The elements of applied avalanche forecasting Part II: the physical issues and rules of applied avalanche forecasting. *Natural Hazards*, **26**, 131–146.

McClung, D. M., and Mears, A. I., 1991. Extreme value prediction of snow avalanche runout. *Cold Regions Science and Technology*, **19**, 163–175.

McClung, D. M., and Schaerer, P. A., 1985. Characteristics of flowing snow and avalanche impact pressures. *Annals of Glaciology*, **6**, 9–14.

McClung, D. M., and Schaerer, P. A., 2006. *The Avalanche Handbook*. Seattle, WA: The Mountaineers.

Perla, R. I., 1970. On contributory factors in avalanche hazard evaluation. *Canadian Geotechnical Journal*, **7**(4), 414–419.

Perla, R. I., 1977. Slab avalanche measurements. *Canadian Geotechnical Journal*, **14**(2), 206–213.

Salm, B., 2004. A short and personal history of avalanche dynamics. *Cold Regions Science and Technology*, **39**, 83–92.

Schaerer, P. A., 1977. Analysis of snow avalanche terrain. *Canadian Geotechnical Journal*, **14**(3), 281–287.

Schneebeli, M., and Meyer-Grass, M., 1992. Avalanche starting zones below the timberline – structure of forest. In *Proceedings International Snow Science Workshop*, October 1992, Breckenridge, CO, pp. 176-181.

Schweizer, J., Fierz, C., and Jamieson, J.B., 2004. Assessing the probability of skier triggering from snow layer properties. In *Proceedings International Snow Science Workshop*, September 2004, Jackson Hole, WY, pp. 192–198.

Seligman, G., 1936. *Snow Structure and Ski Fields*. London: MacMillan and Co. Reprint 1980 Cambridge: International Glaciological Society.

Sovilla, B., Bartelt P., and Margreth, S., 2004. The importance of snow entrainment in avalanche dynamics calculations. In *Proceedings International Snow Science Workshop 2004*, Jackson Hole, WY, pp. 65–73.

Sovilla, B., Schaer, M., and Ramer, L., 2008. Measurements and analysis of full scale impact pressures at the Vallée de Sion test site. *Cold Regions Science and Technology*, **51**(2–3), 122–137.

Cross-references

Emergency Planning
Geohazards
Glacier Hazards
Hazard
Hazard and Risk Mapping
Land-Use Planning
Magnitude Measures
Prediction of Hazards
Probable Maximum Precipitation
Snowstorm and Blizzard
Uncertainty

AVIATION (HAZARDS TO)

Thomas Gerz, Ulrich Schumann
Institut für Physik der Atmosphäre, Oberpfaffenhofen, Germany

Definition

Aviation. Worldwide commercial and noncommercial flying aircraft.

Hazard. Weather condition that impacts the safe flight of an aircraft. Such conditions include phenomena as strong wind, turbulence, clear-air turbulence, wind shear, downbursts, strong convection, storms, heavy precipitation, hail, lightning, icing, snow, fog, sandstorm, volcanic ash, and aircraft wake vortices.

Introduction

Aircraft by definition move through airspace. Therefore, the natural hazards to aviation described in this section are atmospheric events, collectively called weather. The safety (and efficiency) of flight in general is most influenced by weather events. "No other industry is more sensitive to weather than the aeronautical industry" (Sprinkle and Macleod, 1991). Although weather can be observed and forecast with considerable accuracy, adverse weather still has a large impact to operational safety and efficiency. The bulk of the fatal accidents in aviation are associated with visual flight-rules general-aviation pilots in single engine fixed gear aircraft attempting to operate under instrument meteorological conditions (IMC). As the capability of the pilot (e.g., IMC qualified) and plane (e.g., multiengine with retractable wheels) increases, thunderstorms become a much more important factor in accidents (Evans, 2006). Accident and incident analyses show that there is always more than one cause. Weather very often contributes to an accident or incident even if it is not the primary cause. Nevertheless, 13% of all hull losses between 1995 and 2004 were primarily caused by weather (Boeing, 2005), and in 33% of all aviation accidents, weather was a primary or secondary factor (NTSB, 2007). Weather impacts aviation in terms of safety, efficiency, and capacity. Risks (safety) and efficiency (delays) are under certain conditions positively correlated in aviation when their common source is insufficient weather hazard awareness by the flight crew.

Phenomena such as volcanic ash and aircraft wake vortices, themselves not atmospheric hazards, are also considered here since their hazardous attributes are controlled by weather conditions, in particular wind.

When flying an aircraft from one place to another, certain atmospheric and weather conditions have to be favorable. The atmospheric state parameters have to stay within certain ranges to allow for safe aircraft attitudes, and the weather conditions have to be suitable to allow the crew to maneuver the aircraft.

Most of the weather-related accidents occur during takeoff or landing, because several weather hazards such as turbulence, down-draughts, heavy rain, and shear winds are found predominantly in the lower atmospheric levels. Moreover, due to the necessity of having to land at a certain location and on a runway in a given direction, weather hazards in that last flight phase are not able to be circumvented. Visibility and a minimum cloud ceiling are still required for a safe landing and/or takeoff. During these flight phases, the margin to stall the aircraft is at its lowest since the aircraft is flying relatively slow and at a relatively high angle of attack.

Atmospheric state parameters, flight envelope, and hazardous weather

Atmospheric state parameters include wind, temperature, pressure, and density, as well as the chemical composition of the atmosphere, dust and aerosol concentration and liquid water content. The safe operation of a specific aircraft requires that the actual state parameters, together with flight state parameters such as true airspeed, Mach number, pressure altitude, and angle of attack, lay within a certain range, usually referred to as the flight envelope of the given aircraft. For example, a certain range of air density values is required to provide the necessary lift for a given aircraft in order to fly. A typical hazard relating to this is the incorrect calculation of the cruising or maximum flight altitude of an aircraft when flying over high topography or at high temperatures. As another example, the liquid water content along the flight track may be too high, resulting in an engine flameout. Further, parts of the aircraft such as the engine blades or air inlets may be degraded, leading to substantial loss of performance of the engine up to its flameout when flying through air with a high density of sand particles (e.g., in sandstorms) or through clouds of volcanic ash or smoke from fires.

If we consider weather as the variable part of the atmospheric state parameters, suitable weather conditions mean a range of weather conditions in which aircraft are certified to fly and in which the pilot in command can safely maneuver the aircraft. One may then call "hazardous weather" situations or conditions during flight which put an aircraft outside its flight envelope or where the crew may lose control over the aircraft.

The critical "aviation-weather" events which generally reduce safety margins and facilitate the occurrence of incidents and possibly accidents can be listed as:

- Severe storms
- Fog, low clouds (causing low visibility)
- Thunderstorms and tornadoes
- Supercooled clouds (causing in-flight icing)
- Clear-air turbulence
- Sandstorms and volcanic ash
- Aircraft wake vortices

The physical features characterizing these phenomena and causing the hazard include:

- Turbulence (in clouds or in clear air)
- Wind shear, gusts, downbursts
- Lightning
- Heavy precipitation (rain, snow), hail
- Supercooled liquid water in air (with temperatures below $0°C$)

Typically, wind shear and gusts, fog and heavy snow or rainfall affect the takeoff and landing phases, whereas clear-air turbulence is a characteristic en route hazard. Basic characteristics of an airport such as orientation of runways, precision landing equipment, topography, and local weather phenomena make it susceptible to certain weather events. An airport or an airspace also become more vulnerable to adverse weather when they operate at their capacity limits, in particular when the demand exceeds the capacity.

In the following sections, the text focuses on the weather hazards to aviation and identifies the related impacts. It closely follows a report by Theusner and Röhner (2006) who investigated aviation weather hazards, aviation weather impact areas, and evaluation methods in the framework of the European Integrated Project FLYSAFE (http://www.eu-flysafe.org/). The reader is also referred to the weather section of EUROCONTROL's "SKYbrary" (http://www.skybrary.aero/index.php/Weather).

Severe storms

Severe storms, such as hurricanes or typhoons in the tropics or severe storms in the mid-latitudes and polar regions, are large-scale, long-lasting, and highly energetic weather patterns. They are a general threat and, as such, can cause considerable damage also to airports, the air traffic control infrastructure, and aircraft. Today, the development and movement of large storms can be forecast with an accuracy that is high enough to allow for appropriate actions, like closing of airports, diversion of traffic to other airports, alternate flight routing, adaptation of traffic density, etc. These mitigation measures may not prevent the destruction of infrastructure but will likely save human lives.

Low visibility

Today, reduced or low visibility due to fog, clouds, rain, snow, a moonless night, a smoke plume, or a sandstorm is still the main hazardous weather condition for aviation. Controlled flight into terrain resulting from a loss of orientation due to poor visibility is one of the most frequent primary causes of accidents in aviation (Boeing, 2005). Fog, a cloud sitting on the surface, also has a significant impact on flight operations particularly during takeoff, landing, and taxiing. The safety situation in poor visibility will improve in the future due to fully automated navigation and guidance as well as enhanced vision systems.

Thunderstorms and tornadoes

A thunderstorm (cumulonimbus with lightning) describes a cloud of strong and deep convection (vertical motion) of warm and moist air. Within thunderstorms, areas of downbursts, strong wind shear, turbulence, icing, heavy rain, hail, and lightning strikes exist simultaneously in the same volume of air or close by. All of these phenomena are a possible threat to aircraft as they can alter the aerodynamic state of flight, damage the hull or engines of the aircraft, or cause malfunctions of the onboard equipment.

A *tornado* is a violently spinning vortex of air extending down from the base of a cumulonimbus cloud and in contact with the surface, often visible as a funnel cloud. The central core of the funnel exhibits an extremely low pressure and winds surrounding the tornado typically reach 50 m/s (180 km/h). A tornado can cause considerable damage to airport and air traffic control infrastructure if it touches down on an aerodrome and, clearly, any aircraft caught on the ground in the path of a tornado will most likely be damaged beyond repair.

Lightning

Lightning is a transient, high-current electric discharge over a path length of several kilometers in the atmosphere (Betz et al., 2009). The majority of lightning in Earth's atmosphere is associated with convective thunderstorms. Lightning is the result of a breakdown of electrical charge separation in thunderstorms. Charge separation is strongest in powerful updrafts containing supercooled liquid water, ice crystals, and hail or graupel. The lightning discharge in its totality is called a flash. One distinguishes between cloud-to-ground lightning and various other lightning types, including intracloud, intercloud, and cloud-to-air lightning.

Most lightning strikes to aircraft (90% of events) are triggered by the intrusion of the aircraft into a region of the thunderstorm where the electrostatic field is sufficiently strong. The other case may be the interception by the aircraft of a branch of natural lightning. An aircraft passing close to an area of charge can initiate a discharge, and this may also occur some distance from a thunderstorm. Lightning also occurs within clouds of volcanic ash, again because of the vertical movement and collision between particles within the cloud which generates static charge.

Physical damage by a lightning strike to an aircraft is rare and not likely to threaten the safety of the aircraft. Damage is usually confined to parts of the avionics system, aerials and compasses, and the burning of small holes in the fuselage. There is, however, a potential for engine shutdown caused by the transient airflow interruption associated with lightning.

Lightning strikes from cloud to ground are dangerous for humans on the ground and in the area. Therefore, airport ground operations including baggage handling or refueling of the aircraft have to be suspended when a lightning strike has been observed within a radius of 5 km of an airport.

Downburst or microburst and wind shear

A downburst or microburst is created by an area of significantly rain-cooled, descending air (downdraught) that, after hitting ground level, spreads out in all directions producing strong and rapidly changing winds (wind shear). Surrounding dry air is mixed into the rain fall region which causes the rain to evaporate. The process of evaporation cools the air which then becomes heavier than the surrounding air, and thus, descends even faster. Microbursts normally cover an area of approximately 4 km in diameter, although the area of resulting high winds, created when the microburst hits the ground, can be much larger. Winds of up to 65 m/s (240 km/h) may result when a microburst is formed. Most microbursts last for a few minutes, and generally less than 10 min. Microbursts can occur also in rain showers or virga. (Virga is rain that evaporates before it reaches the ground and is associated with a "dry" microburst.)

Severe wind shear is a rapid change in wind direction and/or velocity and causes horizontal velocity changes of at least 15 m/s over distances of 1–4 km, or vertical speed changes greater than 2.5 m/s. Close to the ground, where the impact upon safety is largest, the phenomenon is called low-level wind shear. The terms "microburst" and "wind shear" are often used interchangeably because the vast majority of dangerous wind shears result from microbursts.

Downbursts can be an extreme hazard to aircraft during the landing and takeoff phases of flight, since an aircraft flying through a microburst may experience extremely hazardous fluctuations in airspeed and lift. Roughly 50% of all microbursts are truly hazardous to aircraft (Chambers, 2003). Although the downdraught region from a microburst can be dangerous to an aircraft by itself, the greatest threat comes from the change in wind direction – or wind shear – near the center of the microburst which may result in a corresponding loss of airspeed: An aircraft approaching a downburst will first encounter a strong headwind often combined with a turbulent upward motion which leads to an increase in indicated airspeed and lift. The temptation to reduce the power of the engines would be very dangerous because soon the aircraft enters the downdraught area, forcing the aircraft downward, and finally, as the aircraft has passed through it, the wind becomes a tailwind where the indicated airspeed decays and the lift drops even further. Once caught in a downburst, escape is only possible by flying straight ahead; whichever way the aircraft turns, it will encounter the tail winds and the associated performance impact.

Hail and heavy rain

Hail develops in thunderstorm clouds in several steps and cycles. When rising moist warm air becomes saturated and liquid water particles form, these particles continue to be carried upward by the updraught. When they rise beyond the freezing level, they start to turn to ice. These small ice particles, due to their shape and weight, rise slower or even descend where they meet and melt with water droplets (coalescence). They may again be carried upward by strong updraughts in the cloud, freezing again to become an ice pellet or hail. The cycle repeats and the ice pellets grow each time until the cumulonimbus becomes mature. Then the updraughts weaken and are insufficient to carry the weight of the ice pellets which fall out of the cloud, start to melt, and reach the ground as large rain drops (heavy precipitation) or still in solid form as hail.

Hail pellets may grow to a size larger than a golf ball and can cause considerable damage to aircraft whether on ground or in the air. Other precipitation, such as snow, sleet, or rain, may contaminate airfield and runway surfaces creating a hazard to aircraft attempting to take off or land. Heavy rain can result in the flameout of an engine.

In-flight icing

In-flight icing is the accretion of ice on an aircraft structure, its engines or the air inlets as a result of flight in atmospheric icing conditions. Airframe icing can lead to reduced performance, loss of lift, altered controllability and ultimately stall and subsequent loss of control of the aircraft. Longitudinal stability may also be affected by a degradation of the tail control surfaces. The modified airflow pattern may significantly alter the pressure distribution around flight control surfaces such as ailerons and elevators which can severely affect flight. Blockage of the air inlet to any part of a pitot static system can produce errors in pressure instruments such as altimeters, airspeed indicators, and vertical speed indicators.

Water in the air continues to exist in liquid form well below $0°C$ and is typically observed in stratus clouds, convective clouds embedded in stratus clouds, or in cumulonimbus clouds. This considerable quantity of "supercooled" water steadily decreases with lower temperature until most has frozen at about $-40°C$. Ice accretion can be defined as (1) the freezing of supercooled liquid cloud droplets after impacting the cold surface of an aircraft, (2) liquid drops freezing on the supercooled aircraft, or (3) the accumulation of solid particles like snowflakes and ice crystals by melting and refreezing to warm or wet parts of the aircraft.

Icing is observed in the atmosphere in the temperature range between $-40°C$ and $+6°C$. (Carburettor icing on piston engines can occur at positive air temperatures due to expansion and subsequent cooling of air.) The size of supercooled water droplets and the nature of the airflow around the aircraft surface determine how many of the droplets will impact the surface. The size of a droplet also determines what happens after impact, for example, larger droplets tend to splash and break up into smaller ones. Finally, the size is related to the mass of water in the droplet, which determines the time required for the physical state change. Larger droplets which do not break up into smaller ones will take longer to freeze and can form a surface layer of liquid water before freezing takes place.

Small cloud droplets with diameters of approximately 20 μm impact the aircraft and freeze immediately to an opaque coating of the impact area (rime ice). Larger drizzle drops (50–500 μm), also called supercooled large drops (SLD), take longer to freeze and form clear ice. The larger the droplets and the slower the freezing process, the more transparent is the ice. If the freezing process is sufficiently slow to allow the water to spread more evenly before freezing, the resultant transparent sheet of ice may be difficult to detect. Furthermore, SLD do not freeze completely when impacting the leading edge of the wing (like the 20 μm sized cloud drops do), but they remain liquid long enough to flow back across the wing chord for up to 1 m and freeze there, potentially outside the area covered by de-icing equipment (runback ice). Freezing rain drops (>500 μm) finally originate from rain drops which fall through a supercooled layer of air before impacting the aircraft. If such a layer of air reaches the ground, it results in freezing rain at the surface and all runways and airport buildings and structures are affected.

In-flight icing is observed more in winter than in summer because of the generally lower air temperatures in winter. Icing mainly affects turboprop and small aircraft, since they usually cannot climb high enough to fly above the clouds and, therefore, spend a large fraction of flight time inside the clouds. Also, small aircraft can remove ice only pneumatically by so-called rubber boots, since such systems consume less energy than thermal ones that are used by larger aircraft to melt away the ice using hot exhaust gases.

The special risk emanating from freezing drizzle is that aircraft are only tested for average drop sizes of up to 50 μm (FAA, 1999). Though aircraft comply with international certification regulations, they are not necessarily qualified for all types of icing conditions. SLD therefore pose a special risk also because the atmospheric conditions leading to their occurrence are not well known and even more difficult to forecast.

Turbulence

The irregular fluctuation of the wind is called turbulence. It originates from thermal or mechanical forces and may occur either within a cloud or in clear air (clear-air turbulence, CAT). Turbulence is patchy and intermittent. The absolute severity of turbulence depends directly upon the rate at which the wind vector is changing; however, the perception of the severity of the turbulence is also affected by the inertia and hence the mass of the encountering aircraft. Significant mechanical turbulence, most often as CAT, results from the airflow over irregular terrain (combined with mountain waves) or obstacles and also occurs in certain regions of the jet stream at cruising level. Thermally induced turbulence arises from air movements associated with convective activity, especially in or near deep convection like a thunderstorm where it may occur inside or outside the cloud. Also zones of downbursts and wind shear are always accompanied by turbulence.

Turbulence is graded on a relative scale, according to its perceived or potential effect on a "typical" aircraft, as "light," "moderate," "severe," and "extreme." Light turbulence results in slight, erratic changes in attitude and/or altitude of the aircraft. Moderate turbulence has similar effects as light turbulence, but of greater intensity – variations in speed as well as altitude and attitude may occur, but the aircraft remains in control all the time. Severe turbulence is characterized by large, abrupt changes in attitude and altitude with large variations in airspeed. There may be brief periods where effective control of the aircraft is impossible. Loose objects may move around the cabin and damage to aircraft structures may occur. Extreme turbulence is capable of causing structural damage and a loss of control of the aircraft.

Even though turbulence is seldom damaging to modern aircraft, which are designed to withstand associated stresses, it is the leading cause of in-flight nonfatal accidents of passengers and flight attendants. Whereas turbulence in the context of clouds can be "seen" and respective measures can be taken, there is currently no effective warning system for clear-air turbulence. In June 1995, the FAA issued a public advisory to airlines urging the use of seat belts at all times when passengers are seated as a precaution against unexpected turbulence (FAA, 1997). Nevertheless, the number of turbulence accidents per million departures (USA carriers) has increased markedly between 1982 and 2003 (FAA, 2005) and cannot be explained by the increase in air traffic volume alone. Instead, load factors might be the controlling factor, too.

Convective turbulence

Convective turbulence is a phenomenon that can be found throughout the troposphere in the vicinity of convection of varying strength. It can be found in the planetary boundary layer (the lower few hundred meters of the Earth's atmosphere) where it is especially dangerous as the aircraft usually are in the stages directly prior to landing or directly after takeoff at that altitude and are flying at low speeds (risk of stalling). Any turbulence enforced change in altitude therefore bears the risk of collision with terrain.

In mid-level shower clouds and shallow convection, turbulence usually does not pose a serious risk to the aircraft itself. It does, however, reduce the flight comfort of the passengers and may even lead to serious injuries, if no seat belt is worn, which is usually the case for cabin attendants. The typical up- and downdraught speeds are of the order of up to ± 5 m/s.

The real threat to aircraft is convective turbulence found in thunderstorms. There, updraughts of up to $+65$ m/s and downdraughts of -25 m/s can exist in close proximity. The strong wind shear and turbulence induced by these draughts is a serious safety hazard for passengers and also the structural integrity of aircraft. Turbulence can not only be found within the convective systems, but convection can induce turbulence far away from the originating systems through the excitation of gravity waves.

Clear-air turbulence, CAT

Patches or regions of turbulence without the presence of clouds at high altitudes (usually above 4,500 m) are referred to as clear-air turbulence (CAT). Its sudden appearance in an otherwise calm and clear atmosphere is mostly a safety and convenience problem for passengers. Nevertheless, CAT can also become a safety problem for the structural integrity of the aircraft, for example, if inappropriate movements of the elevators or rudders are performed. Turbulence also leads to frequent load alternation and, consequently, to structural loads with resulting material fatigue.

CAT develops mostly in the vicinity of jet streams, above and below the wind maximum, where the wind shear is strongest. The three ingredients of jet stream curvature, upward vertical moist convection (from, e.g., thunderstorms), and cold air advection together dramatically increase the risk of severe, possibly accident-producing CAT (Kaplan et al., 2005). Another important source for CAT is by breaking gravity waves above mountain ranges and above convective weather systems.

Dust storms and volcanic ash

A dust storm is a cloud of blowing very fine sand or dust. The sediment is lifted by the wind when it exceeds the threshold speed necessary to lift and transport the surface particles. Dust storms can be created by an advancing gust front ahead of a cumulonimbus cloud (typical in the Sahara desert) or along a cold front where cool air, passing over hot ground, creates instability in the air above (typical in the midwest of the USA). The height of the dust "wall" can be roughly 2 km, but with the support of updraughts in an unstable atmosphere, the dust can reach heights of 6 km. The sediment significantly reduces visibility and its ingestion into engines, pitot static system, conditioning packs causes blockage and erosion.

Volcanic ash clouds are clouds of very small particles of rock ejected from a volcano during an eruption. Large quantities of material can be ejected into the atmosphere during an eruption; the smaller particles reach great heights where they are transported horizontally by the wind. Since moist deposition processes and sedimentation are slow, the small particles can remain for a long time at aircraft cruising levels and, thus, a threat to aviation. The ash does not show up on an aircraft's weather radar because of the small size of the particles. These particles conduct electricity, and the electrical charges within a cloud of volcanic ash can give rise to thunder and lightning. St. Elmo's fire, created when the charged particles hit the aircraft, may be the first indication to a crew that they are flying into volcanic ash clouds. Other indications are the smell of sulfur and dust in the cabin.

The rock particles cause significant damage to the surface of the aircraft skin and engine components. Rock particles entering a jet engine can, if the inlet temperature is hot enough, melt and/or burn leaving deposits which clog the engine and may result in engine surges and ultimately flameout.

Aircraft wake vortices

Wake vortices shed by an aircraft are a natural consequence of its lift. The vortices can be described as two coherent counterrotating swirling flows, like horizontal tornadoes, of about equal strength. Each vortex has a strong circulation, which is proportional to the weight and inversely proportional to the speed of the aircraft, and a significant size in the order of the wing span. The vortices are transported by the wind and sink downward due to mutual velocity induction. The vortices generally persist for between one and three minutes, sometimes much longer. They decay due to turbulent mixing processes in the atmosphere. The calmer the atmosphere the longer the life time of the vortices (Gerz et al., 2002). Decay is faster in the atmospheric boundary layer in turbulent conditions and slower in the free atmosphere at cruising level.

Mature aircraft wake vortices can produce a significant rotational or structural influence on an aircraft encountering them for several minutes after they have been generated. Especially, during an aircraft's critical landing phase, the vortices can endanger any aircraft following close behind. Today, wake vortices are a capacity issue rather than a safety problem because a following aircraft must maintain a safe distance from a landing aircraft up ahead of it to avoid wake-vortex encounters. But in the light of increasing air traffic, airspace becomes a limiting factor, en route but even more in the vicinity of airports. Hence, in the future, wake vortices may become a safety issue, too.

Conclusions

Hazards to aviation are weather hazards. An abrupt or vigorous change of the atmospheric state parameters (wind, temperature, pressure, density, liquid water content, dust, and aerosols) may lead to a hazard when it results in an unfavorable attitude of the aircraft, or in a lack of awareness of the crew maneuvering the aircraft, or both.

Safety, efficiency, and capacity of aviation are influenced by weather. Risks and delays have a common source: insufficient weather hazard awareness. Our skills to observe and forecast weather with high accuracy and our ability to communicate worldwide within seconds from ground to air, air to air, and air to ground will be combined to build new expert systems for aviation weather. The situational awareness of pilots and controllers regarding weather will be enhanced when all actors perceive the coming threat by consistent, tailored, dedicated, unambiguous, and timely weather information. Flights will become safer when all (re-)act according to that information and avoid the hazard.

Bibliography

Betz, H. D., Schumann, U., and Laroche, P. (eds.), 2009. *Lightning: Principles, Instruments and Applications.* Heidelberg: Springer, 641 pp. ISBN 978-1-4020-9078-3.

Boeing, 2005. *Statistical summary of commercial jet airplane accidents – worldwide operations 1959–2004.* Available from Internet: http://www.boeing.com/news/techissues/.

Chambers, R., 2003. *Concept to Reality: Contributions of the NASA Langley Research Center to United States Civil Aircraft of the 1990s.* Washington, DC: National Aeronautics and Space Administration, 301 pp.

Evans, J. E., 2006. The weather impact on aviation. In *Proceedings of the American Meteorological Society Conferences on Aviation, Range, and Aerospace Meteorology.* MIT Lincoln Laboratory.

FAA, 1997. *CNN – A primer on air turbulence* – December 29, 1997. Available from World Wide Web: http://cgi.cnn.com/US/9712/29/turbulence.explainer/.

FAA, 1999. *Appendix C, Airworthiness standard: transport category airplanes, Part 25, Aeronautics and space.* U.S. Code of Regulations, Office of the Federal Register, National Archives and Records Administration (Available from Superintendent of Documents). U.S. Government Printing Office, Washington, DC, Vol. 14.

FAA, 2005. *Preventing injuries caused by turbulence.* U.S. Department of Transportation – Federal Aviation Administration, Advisory Circular No. 120-88, 28 pp. Available from Internet: http://www2.airweb.faa.gov/Regulatory_and_Guidance_Library/rgAdvisoryCircular.nsf/0/47baa2ba78366a0a862570b4007670f2/$FILE/AC120-88.pdf.

Gerz, T., Holzäpfel, F., and Darracq, D., 2002. Commercial aircraft wake vortices. *Progress in Aerospace Sciences*, **38**, 181–208.

Kaplan, M. L., Huffman, A. W., Lux, K. M., Charney, J. J., Riordan, A. J., and Lin, Y.-L., 2005. Characterizing the severe turbulence environments associated with commercial aviation accidents. Part 1: A 44-case study synoptic observational analyses. *Meteorology and Atmospheric Physics*, **88**, 129–152.

NTSB, 2007. Query for weather related accidents between 2004 and 2007 (Online). Available from World Wide Web: http://www.ntsb.gov/ntsb/query.asp.

Sprinkle, C. H., and Macleod, K. J., 1991. Impact of weather on aviation: a global view. In *Preprints of 4th International Conference on Aviation Weather Systems*, June 24–28, Paris, France. AMS, Boston, MA, pp. 191–196.

Theusner, M., and Röhner, P., 2006. *Report on weather hazards, weather impact areas and evaluation methods.* Report FLY_210UNI_DEL_D2.1-1 of Project FLYSAFE, 98 pp.

Cross-references

Doppler Weather Radar
Early Warning Systems
Expert (Knowledge-Based) Systems for Disaster Management
Exposure to Natural Hazards
Fog Hazards
Hazard
Impact Winter
Lightning
Mitigation
Monitoring Natural Hazards
Natural Hazard
Prediction of Hazards
Risk
Shear
Snowstorm and Blizzard
Storms
Thunderstorms
Volcanic Ash
Warning Systems

AVULSION

Joann Mossa
University of Florida, Gainesville, FL, USA

Synonyms
Channel abandonment; Channel shifting; Diversion

Definition
Avulsion is recognized as an abrupt change in the course of a stream.

Background
An avulsion is when a river channel switches location, often abruptly, along part of its course. Avulsions are characteristic of fluvial and deltaic environments, including alluvial fans and rivers with multiple channels. Partial or complete abandonment of the active channel results in an inactive channel or a channel with substantially less flow. Avulsions usually occur during floods, either through levee breaching via lowlands or occupying a preexisting channel with which it is connected. Frequency, spatial scale, and causes vary widely depending on the type of river and environmental setting (Jones and Schumm, 1999; Slingerland and Smith, 2004; Stouthamer and Berendsen, 2007). Human activities may cause or promote avulsions where levees have been destroyed during floods, intentionally harming some areas with the aim to protect others, and during wartime, with the intent to create havoc. Excavated pits on mined floodplains create a landscape vulnerable to avulsions during overbank flooding (Mossa and Marks, 2011). Also, avulsion frequency may increase inadvertently from dam-building (Ethridge et al., 1999).

Avulsion hazards
The redistribution of discharge and sediment, and associated channel geometry change, can be hazardous in a number of ways. Where the channel is abandoned, hazards include loss of water supply, navigational benefits, increased coastal land loss, and loss of connections to irrigation canals. A potentially catastrophic avulsion of the Mississippi River into the Atchafalaya River in south Louisiana was averted in the mid-1900s. As the Atchafalaya received increased amounts of flow, it threatened New Orleans' water supply and major ports there and in Baton Rouge. Flow control structures were built near the juncture of these rivers to control diversion and prevent an avulsion from occurring; one of the first structures was partially undermined in the 1973 flood and subsequently two other diversions were built.

The new path that the channel now occupies experiences other hazards. Sudden flooding is a considerable danger as has happened repeatedly on the Yellow River in China (Slingerland and Smith, 2004) and in lowland Bangladesh. Economic costs of avulsions include property loss and subsequent homelessness, damage to bridges and infrastructure, ecosystem losses, and engineering mitigation such as artificial levee building to protect new floodplains.

Bibliography
Ethridge, F. G., Skelly, R. L., and Bristow, C. S., 1999. Avulsion and crevassing in the sandy, braided Niobrara River: complex response to base-level rise and aggradation. In Smith, N. D., and Rogers, J. (eds.), *Fluvial Sedimentology VI. International Association of Sedimentologists Special Publication 28*. Oxford: Blackwell Science, pp. 179–191.

Jones, L. S., and Schumm, S. A., 1999. Causes of avulsion: an overview. In Smith, N. D., and Rogers, J. (eds.), *Fluvial Sedimentology VI. International Association of Sedimentologists Special Publication 28*. Oxford: Blackwell Science, pp. 171–178.

Mossa, J., and Marks, S. R., 2011. Pit avulsions and planform change on a mined river floodplain: Tangipahoa River, Louisiana, USA. *Physical Geography*, **32**.

Slingerland, R., and Smith, N. D., 2004. River avulsions and their deposits. *Annual Review of Earth and Planetary Sciences*, **32**, 257–285.

Stouthamer, E., and Berendsen, H. J. A., 2007. Avulsion: the relative roles of autogenic and allogenic processes. *Sedimentary Geology*, **198**, 309–325.

Cross-references
Flood Deposits
Flood Hazard and Disaster
Flood Protection
Floodplain
Floodway
Levee

B

BASE SURGE

Catherine J. Hickson
Magma Energy Corp., Vancouver, BC, Canada
Alterra Power Corp., Vancouver, BC, Canada

Synonyms

Blast; Ground-based surge; Ground surge; Pyroclastic density flow; Pyroclastic surge; Surtseyan eruption

Definition

A destructive, dilute, fast-moving (30 m/s) turbulent density current (flow) of particles and gas and/or liquid that is the result of an explosion.

Discussion

Base surges are highly destructive and dangerous (Nakada, 2000). The term has been used to describe high-velocity (up to 30 m/s) flows of material emanating from explosions. The term was first used to describe the ground-hugging clouds observed after underwater and underground nuclear explosions (e.g., Trinity Atomic Web Site). In photographs of nuclear explosions, there is a characteristic ring-shaped cloud that moves outward close to the ground – the base surge (Trinity Atomic Web Site Figure 2.97). It was first documented at the eruption of Capelinhos volcano, Azores October 10th, 1957 and then adopted by the volcanological community based on the work of James G. Moore (1966a, b) and his observations of the eruption of Taal volcano in the Philippines. Moore adopted the term "base surge" to describe phreatic to phreatomagmatic explosions and their fine grained (pulverized, broken clasts), bedded deposits. The term has become synonymous with "Surtseyan eruptions" which are phreatomagmatic (hydromagmatic). From these early observations, the term has been used in a number of ways making a precise definition in volcanological literature problematic. The term has also been used to describe the turbulent, dilute flow fronts visible in pyroclastic flows. Thus, the usage of the term now spans cold, wet, phreatic explosions, to moderate temperature wet phreatomagmatic (or hydromagmatic) explosions (usually now referred to as "pyroclastic surges") to the basal portions of hot, dry pyroclastic flows. Work by Sulpizio et al. (2008) and others using large-scale experiments show the continuum and the use of the term "pyroclastic density current" (PDC) has now become more common. In all cases (wet or dry), the explosive discharges can be extremely vigorous and will propel eruption plumes of particles and gases many kilometers into the air. The resultant surges can sculpt the landscape by being highly erosive near source, stripping and scouring underlying soils and vegetation and more distally, depositing material as pyroclastic surge (base surge) deposits.

Bibliography

Moore, J. G., Nakamura, K., and Alcaraz, A., 1966a. The September 28–30, 1965 Eruption of Taal Volcano, Philippines. *Bulletin of Volcanology*, **29-1**, 75–76.
Moore, J. G., Nakamura, K., and Alcaraz, A., 1966b. The 1965 eruption of Taal Volcano. *Science*, **151–371**, 955–960.
Nakada, S., 2000. Hazards from pyroclastic flows and surges. In Sigurdsson, H., et al. (eds.), *Encyclopedia of Volcanoes*. San Diego: Academic, pp. 945–955.
Sulpizio, R., Dellino, P., Mele, D., and La Volpe, L., 2008. Generation of pyroclastic density currents from pyroclastic fountaining or transient explosions: insights from large scale experiments, IOP Conference Series: Earth and Environmental Science 3 (2008) 012020 doi:10.1088/1755-1307/3/1/012020.
Trinity Atomic Web Site, Nuclear weapons: History, Technology, Consequence in Historic Documents, Photos and Video. http://www.cddc.vt.edu/host/atomic/nukeffct/undrwtr.html; http://www.cddc.vt.edu/host/atomic/nukeffct/enw77b2.html.
Valentine, G. A., and Fisher, F. V., 2000. Pyroclastic surges and blasts. In Sigurdsson, H., et al. (eds.), *Encyclopedia of Volcanoes*. San Diego: Academic, pp. 571–580.

Cross-references

Galeras Volcano
Mt. Pinatubo
Nuee Ardente
Pyroclastic Flow
Vesuvius

BEACH NOURISHMENT (REPLENISHMENT)

Charles W. Finkl
Florida Atlantic University, Boca Raton, FL, USA

Synonyms

Renourishment; Replenishment; Restoration

Definition

Beach nourishment is the artificial placement of sand on an eroded shore to stabilize the shoreline position and provide protection from shoreline retreat and coastal flooding. This multipurpose procedure provides coastal protection, habitat, and recreation sites in an attempt to mitigate coastal erosion hazards. Replacement of sand lost to erosion can take place naturally (e.g., by alongshore littoral transport or onshore sand movement) or artificially (e.g., by deposition of sediment dredged from offshore or mined inland).

History and procedures

The term *beach nourishment* came into general use after the first renourishment project in the United States at Coney Island in 1922 (Dornhelm, 1995). In engineering parlance, the terms *beach (re)nourishment*, *beach replenishment*, and *beach restoration* are often used more or less interchangeably in reference to the artificial (mechanical) placement of sand along an eroded stretch of coast where only a small beach, or no beach, previously existed. There are, however, subtle connotations in the application of each term (Finkl and Walker, 2002, 2005), as described below.

Sediments that accumulate along the shore in the form of beaches are naturally derived from a variety of sources including fluvial transport in rivers to deposition of sediments in deltas, from preexisting sediments on the offshore seabed, from chemical precipitates (e.g., oolites on carbonate banks in tropical and subtropical environments), or from organisms living along the shore (e.g., shells and exoskeletons from marine organisms). When the natural sediment supply is interrupted, beaches become sediment-starved and the shoreline retreats landward due to volume loss. Efforts to artificially maintain beaches that are deprived of natural sediment supply thus attempt to proxy nature and (re)nourish the beach by mechanical placement of sand. The beach sediment is thus *replenished* by artificial means. *Beach restoration* implies an attempt to restore the beach to some desired previous condition. A nourished or *constructed beach* could be placed along a previously beach-less shore, whereas a restored beach is revitalized by the mechanical placement of sediment. Beach nourishment projects involve placement of sand on beaches to form a *designed structure* so that an appropriate level of protection from storms is achieved. The placement of sand is commonly by methods involving dredging sand from borrow areas on the seafloor (e.g., Finkl et al., 1997), bypassing sand around deepwater inlets or other obstructions (e.g., groins) along the coast that interrupt the littoral drift, or trucking sand from inland quarries to the coast. Pumping sand from offshore is the most widespread method of application due to the large volumes of sediment that are required to combat erosion hazards.

Along undeveloped shores, beaches provide natural habitat (e.g., nesting grounds for sea turtles and shorebirds) but on developed shores, beaches additionally protect coastal infrastructure from storms and are important recreational sites for a globally expanding tourist industry. When beaches are degraded by decreased width and lowering of berms, many communities choose to replace lost sediment by pumping beach quality sand from offshore to eroded sites onshore. Sand artificially placed on an eroded shore is sacrificial in the sense that it must be periodically replaced as erosion progresses in order to provide coastal defense. The beach is thus said to be renourished, replenished, or restored.

Bibliography

Dornhelm, R. B., 1995. The Coney Island public beach and boardwalk improvement of 1923. *Shore and Beach*, **63**(1), 7–11.
Finkl, C. W., and Walker, H. J., 2002. Beach nourishment. In Chen, J., Hotta, K., Eisma, D., and Walker, J. (eds.), *Engineered Coasts*. Dordrecht: Kluwer Academic, pp. 1–22.
Finkl, C. W., and Walker, H. J., 2005. Beach nourishment. In Schwartz, M. (ed.), *The Encyclopedia of Coastal Science*. Dordrecht: Kluwer Academic, pp. 37–54.
Finkl, C. W., Khalil, S. M., and Andrews, J. L., 1997. Offshore sand sources for beach replenishment: Potential borrows on the continental shelf of the eastern Gulf of Mexico. *Marine Georesources and Geotechnology*, **15**, 155–173.

Cross-references

Breakwater
Coastal Erosion
Coastal Zone Risk Management
Erosion
Hurricane
Storm Surge

BEAUFORT WIND SCALE

Tom Beer
Centre for Australian Weather and Climate Research,
Energy Transformed Flagship, Aspendale, VIC, Australia

Synonyms

Beaufort wind force; Wind intensity

Beaufort Wind Scale, Table 1 Beaufort scale of wind

Beaufort Number	Descriptive term	Velocity equivalent at a standard height of 10 m above open flat ground					Specifications			Probable wave height* in metres	Probable wave height* in feet
		Mean velocity in knots	ms^{-1}	km h^{-1}	m.p.h.	Land	Sea	Coast		
0	Calm	<1	0–0.2	<1	<1	Calm; smoke rises vertically	Sea like a mirror	Calm	—	—
1	Light air	1–3	0.3–1.5	1–5	1–3	Direction of wind shown by smoke drift but not by wind vanes	Ripples with the appearance of scales are formed, but without foam crests	Fishing smack just has steerage way	0.1 (0.1)	¼ (¼)
2	Light breeze	4–6	1.6–3.3	6–11	4–7	Wind felt on face; leaves rustle; ordinary vanes moved by wind	Small wavelets, still short but more pronounced; crests have a glassy appearance and do not break	Wind fills the sails of smacks which then travel at about 1–2 knots	0.2 (0.3)	½ (1)
3	Gentle breeze	7–10	3.4–5.4	12–19	8–12	Leaves and small twigs in constant motion; wind extends light flag	Large wavelets; crests begin to break; foam of glassy appearance; perhaps scattered white horses	Smacks begin to careen and travel about 3–4 knots	0.6 (1)	2 (3)
4	Moderate breeze	11–16	5.5–7.9	20–28	13–18	Raises dust and loose paper; small branches are moved	Small waves, becoming longer; fairly frequent white horses	Good working breeze, smacks carry all canvas with good list	1 (1.5)	3½ (5)
5	Fresh breeze	17–21	8.0–10.7	29–38	19–24	Small trees in leaf begin to sway; crested wavelets form on inland waters	Moderate waves, taking a more pronounced long form; many white horses are formed (chance of some spray)	Smacks shorten sail	2 (2.5)	6 (8½)
6	Strong breeze	22–27	10.8–13.8	39–49	25–31	Large branches in motion; whistling heard in telegraph wires; umbrellas used with difficulty	Large waves begin to form; the white foam crests are more extensive everywhere (probably some spray)	Smacks have double reef in mainsail; care required when fishing	3 (4)	9½ (13)
7	Near gale	28–33	13.9–17.1	50–61	32–38	Whole trees in motion; inconvenience felt when walking against wind	Sea heaps up and white foam from breaking waves begins to be blown in streaks along the direction of the wind	Smacks remain in harbour and those at sea lie to	4 (5.5)	13½ (19)
8	Gale	34–40	17.2–20.7	62–74	39–46	Breaks twigs off trees; generally impedes progress	Moderately high waves of greater length; edges of crests begin to break into the spindrift; the foam is blown in well-marked streaks along the direction of the wind	All smacks make for harbor, if near	5.5 (7.5)	18 (25)

Beaufort Wind Scale, Table 1 (Continued)

Beaufort Number	Descriptive term	Mean velocity in knots	ms⁻¹	km h⁻¹	m.p.h.	Specifications — Land	Sea	Coast	Probable wave height* in metres	Probable wave height* in feet
9	Strong gale	41–47	20.8–24.4	75–88	47–54	Slight structural damage occurs (chimney pots and slates removed)	High waves; dense streaks of foam along the direction of the wind; crests of waves begin to topple, tumble and roll over; spray may affect visibility	—	7 (10)	23 (32)
10	Storm	48–55	24.5–28.4	89–102	55–63	Seldom experienced inland; trees uprooted; considerable structural damage occurs	Very high waves with long overhanging crests; the resulting foam, in great patches, is blown in dense white streaks along the direction of the wind; on the whole, the surface of the sea takes on a white appearance; the tumbling of the sea becomes heavy and shocklike; visibility affected	—	9 (12.5)	29 (41)
11	Violent storm	56–63	28.5–32.6	103–117	64–72	Very rarely experienced; accompanied by widespread damage	Exceptionally high waves (small and medium-sized ships might be for a time lost to view behind the waves); the sea is completely covered with long white patches of foam lying along the direction of the wind; everywhere the edges of the wave crests are blown into froth; visibility affected	—	11.5 (16)	37 (52)
12	Hurricane	64 and over	32.7 and over	118 and over	73 and over	—	The air is filled with foam and spray; sea completely white with driving spray; visibility very seriously affected	—	14 (—)	45 (—)

*This table is only intended as a guide to show roughly what may be expected in the open sea, remote from land. It should never be used in the reverse way; i.e., for logging or reporting the state of the sea. In enclosed waters, or when near land, with an off-shore wind, wave heights will be smaller and the waves steeper. Figures in brackets indicate the probable maximum height of waves.

Definition

An empirical measure (0–12) for describing the force (speed) of the wind.

Discussion

The Beaufort scale of wind speed, which is displayed in the accompanying table, provides an ordinal scale for wind speed that ranges from 0 to 12 and is based on visual estimates (Table 1). It is named after Admiral Sir Francis Beaufort who developed the measure as a young naval officer and later introduced it for use by the British navy. In 1853, it was adopted for international use. Kinsman (1969), List (1951: p. 119), and WMO (1970) document the history.

Experienced observers will regularly estimate the same Beaufort number for the same wind. There is less agreement on the actual wind speed to which this corresponds. The wind speed conversions presently in use (WMO, 1995) are based on measurements taken 10 m above the ground of the Scilly Isles by Dr. G.C. Simpson in 1906. His formula relates wind speed V (in m/s) and Beaufort number B by $V = 0.836 * B^{1.5}$.

As noted from the formula, the Beaufort scale relates wind speed and the Beaufort number. The scale officially ceases at Beaufort number 12, which equates to hurricane force winds (above 33 m/s). The WMO manual of codes provides a guide to conditions that may be expected from a particular Beaufort number.

There have been attempts to extend the Beaufort scale beyond 12 up to Beaufort number 16, so as to describe tornadoes and hurricanes. It is, however, more usual to measure tornadoes using the Fujita scale, and hurricanes with the Saffir–Simpson scale.

Bibliography

Kinsman, B., 1969. Historical notes on the original Beaufort scale. *Marine Observer*, **39**, 116–124.

List, R. J., 1951. *Smithsonian Meteorological Tables, 6th revised edition*. Washington: Smithsonian Institution.

World Meteorological Organization, 1970. The Beaufort Scale of Wind Force. Marine Science Affairs. Geneva, WMO: (Note: The recommendations in this report advocating new wind speed equivalents were not adopted).

World Meteorological Organization, 1995. Manual on codes, 1995 edition. http://www.wmo.int/pages/prog/www/WMOCodes.html#ManualCodes.

Cross-references

Fujita Tornado Scale
Saffir–Simpson Hurricane Intensity Scale

BIBLICAL EVENTS

Jerry T. Mitchell
University of South Carolina, Columbia, SC, USA

Synonyms

Biblical disaster; Biblical hazard

Definition

Biblical disaster. As described in the Christian Bible, but also in other non-Christian, influential texts, a Biblical disaster is an extraordinary disruption resulting in massive human suffering. Usually represented by a natural occurrence such as a flood, earthquake or plague, the event symbolizes God's displeasure with humanity, demonstrates God's supremacy and timelessness, or reveals aspects of the End Time (Christian eschatology). Each of these aspects, perhaps simplified as *punishment*, *power*, and *prophecy*, varies in interpretation (allegorical versus literal) among Christian followers.

Introduction

The need to explain natural hazards and their impact upon humans extends across societies, time, and geographic location (Taylor, 1978; Hoffman and Oliver-Smith, 1999). Differing explanations have been identified and include the naturalistic, fatalistic (chance), and supernaturalistic (Dynes and Yutzey, 1965). The latter holds a supernatural power, typically a Deity, as responsible for the event. The persistent reference to disaster as an *act of God* or *bolt from the blue* confirms this link with religion (Mitchell, 2000; Steinburg, 2000; Harbaugh, 2001) and has been used in the legal arena to excuse human culpability. Although disaster events appear within the many important texts of other faith traditions (Schmuck, 2000), the Christian Bible makes use of "Biblical events" or "Biblical disasters" in a variety of ways, often as a show of God's power or his plan for Earth. Most references are far from subtle; in fact, describing an event today as one of "Biblical proportions" conjures a notion of something quite large, direct, and significant.

Regarding the topic of disaster, as well as many others, controversy and debate exist within the Christian church over the timing and order of certain events and whether a literal or figurative interpretation of Scripture is appropriate. An additional confound is the numerous translations of the Bible, with a dozen plus available in English alone. For ease, here, all scripture cited is from the *New International Version*. The Apocrypha (e.g., Gospel of Thomas), the Septuagint (e.g., Sirach), and other non-Christian texts were not considered for these examples, but also are worthy of inspection (see, e.g., Sirach 41:9).

Punishment

For some Christians the God of the Old Testament differs from his depiction in the New Testament. The former is vengeful, the latter is forgiving. Whereas Christians acknowledge that the God of all Biblical books is the same, the oft portrayal of God as cause of disaster in the Old Testament lends itself to this dualistic view. Several passages of the Bible call attention to God as punishing humanity for its disobedience or acts of evil. The fact that humans face natural hazards at all can be initially ascribed to the Fall (Genesis 3:17–19) where the disobedience of Adam and Eve brought about a difficult life for them and their descendants.

Perhaps the best known story is of the Flood, a global catastrophe brought about by wickedness and evil. As written in Genesis 6:13, 17, "so God said to Noah, "I am going to put an end to all people, for the earth is filled with violence because of them. I am surely going to destroy both them and the earth . . . I am going to bring floodwaters on the earth to destroy all life under the heavens, every creature that has the breath of life in it. Everything on earth will perish." The Bible is far from alone, here, as other ancient texts, such as *The Epic of Gilgamesh*, also recount large flood events. Proof of a great deluge as a historic event remains of interest to both religionists and scientists (Ryan and Pitman, 2000).

In another example from Genesis (19:1–29), the cities of Sodom and Gomorrah are destroyed when ". . .the Lord rained down burning sulfur on Sodom and Gomorrah – from the Lord out of the heavens" (v. 24). But this "angry" God also rescues his people by bringing catastrophe upon their enemies. Exodus 7–10 recounts the plagues of frogs, flies, and hail, among others that befall the Egyptian captors of the Israelites. The Red Sea is parted (a reverse flood?) to allow the chosen to escape and then set back in its regular course to drown the pursuers. God is shown as capable of enacting disaster *and* controlling the events set in motion.

Supernatural interpretations of disaster intersperse human history beyond these Biblical accounts. The Lisbon earthquake, tsunami, and fire of November 1 (All Saints Day), 1755, is an excellent example (Dynes and Yutzey, 1965; Shrady, 2008). John Wesley, founder of Methodism, published the tract *Serious Thoughts occasioned by the late Earthquake at Lisbon*. Within he faulted human misbehavior and the Portuguese inquisition (a telling stab at Catholicism) for the earthquake, even noting that "men of fortune" without faith could find no comfort (de Boer and Sanders, 2004). Although proclamations of this type have generally lessened over time, some religious commentators today continue to explain disaster in terms of human "decadence," for example, making connections between Hurricane Katrina (2005) and homosexual behavior and the Haitian earthquake (2010) and voodoo. The Biblical allegory is this: a lack of faith and/or right behavior leads to a poor foundation unable to withstand rains, winds, or flood (Matthew 7:24–27).

Power

Another representation is the natural event designed as a show of God's power. The ultimate example is at the very beginning with the Creation story where chaos surrenders to order. Light is separated from darkness, mountains are hewn, waters are gathered, and Earth is populated. The theme of God in control of all nature extends through his Old Testament prophets up to the work of his Son (Jesus) in the New Testament. A passage from Luke 8 (also Mark 4:35–41 and Matthew 8:23–27) is illustrative:

> One day Jesus said to his disciples, "Let's go over to the other side of the lake." So they got into a boat and set out. As they sailed, he fell asleep. A squall came down on the lake, so that the boat was being swamped, and they were in great danger. The disciples went and woke him, saying, "Master, Master, we're going to drown!" He got up and rebuked the wind and the raging waters; the storm subsided, and all was calm. "Where is your faith?" he asked his disciples. In fear and amazement they asked one another, "Who is this? He commands even the winds and the water, and they obey him." (v. 22–25)

As this text illustrates, Jesus' power is above that of the "natural," physical world. The storms of the natural world and the storms that rock the spiritual boats of humanity both have the same master.

Later, an earthquake is felt at the death of Jesus (Matthew 27:51: "At that moment the curtain of the temple was torn in two from top to bottom. The earth shook and the rocks split"). In neither this nor the previous example does the event rise to the level of disaster; no one is hurt and property remains intact. But the supremacy of God over all things, including a tumultuous natural world, is unquestionable. This, however, opens up an enduring debate among Christian followers: how to reconcile an all-knowing, powerful, and *loving* God with human suffering (Chester and Duncan, 2009).

Prophecy and prediction

Biblical disasters also emerge within both Old and New Testament discourses about the End Time (see the Books of Daniel or Revelation). Christian eschatology is the study of the beliefs that surround final events, including the Second Coming of Christ and Armageddon. Several historic examples exist where theologians, other thinkers, and even the laity have attempted to draw parallels between disaster events and Biblical prophecy, but Biblical passages also are used to explain the occurrence of contemporary disaster events. For example, in Revelation 8:11 the text reads ". . .the name of the star is Wormwood. A third of the waters turned bitter, and many people died from the waters that had become bitter." An arguably loose translation of *Wormwood* in Ukrainian is "*polyn* . . . a very close botanical cousin to *chornobyl*" (Mycio, 2005). Suddenly, the nuclear accident at Chernobyl (1986) took on prophetic meaning for some Christian believers. Prophetic visions are not the sole purview of Christians as in this case the seer Nostradamus is also credited with a Chernobyl prediction.

Other present challenges on Earth have been interpreted Biblically. From Jeremiah 47:2: "See how the waters are rising in the north; they will become an overflowing torrent. They will overflow the land and everything in it, the towns and those who live in them. The people will cry out; all who dwell in the land will wail." For a few the rising waters from the north refer to glacial ice melt as a result of global climate change, specifically from human-induced warming. Irrespective of prophecy, this reading is also compatible with creation-care, a growing environmental stewardship movement within Christianity where believers focus on how "God so loved the *world*" (John 3:16; author emphasis), not just the human part of it.

Another common reference to the End Time specifically uses disaster as a sign. From Matthew 24:7 (and its Gospel parallels Luke 21:11 and Mark 13:8), we read that "nation will rise against nation, and kingdom against kingdom. There will be famines and earthquakes in various places." How to read contemporary events as signs within this prophecy is challenged by a later passage in the same entry: "No one knows about that day or hour, not even the angels in heaven, nor the Son, but only the Father." (Matthew 24:36).

Biblical disasters are part of an ongoing story in Christianity, reflected not only in Biblical text but also in the present and the future. Punishment, power, and prophecy are among the most recognizable themes. But for many Christians, especially aid societies, Biblical events also provide direction to help people mitigate, prepare, respond, and rebuild (i.e., the emergency management cycle). Davis and Wall (1992) provide several examples: Noah exhibited preparedness before the Flood (Genesis 6:13–22); relief for famine is detailed in Acts 11:27–30; Jerusalem is rehabilitated and reconstructed following disaster (Nehemiah 6:15); and starvation is mitigated in Egypt with the storage of food (Genesis 41:34–36). As the perception of disaster varies between Christian faith groups (Mitchell, 2000), among other religions, and among the unchurched, a delicate balance is needed. Within the scientific realm the notion of an *act of God* has been replaced by a demoralized nature, yet "religion is an essential element of culture and must be carefully considered in the [disaster] planning process, and not simply dismissed as a symptom of ignorance, superstition and backwardness" (Chester, 2005). In other words, hazard protection efforts will be more effective when respect is given to cultural beliefs. As such, the historic Biblical disaster event does bear on the present.

Bibliography

Chester, D. K., 2005. Theology and disaster studies: the need for dialogue. *Journal of Volcanology and Geothermal Research*, **146**, 319–328.

Chester, D. K., and Duncan, A. M., 2009. The Bible, theodicy and Christian responses to historic and contemporary earthquakes and volcanic eruptions. *Environmental Hazards*, **8**, 304–332.

Davis, I., and Wall, M., 1992. *Christian Perspectives on Disaster Management: A Training Manual*. London: Interchurch Relief and Development Alliance.

de Boer, J. Z., and Sanders, D. T., 2004. *Earthquakes in Human History: The Far-Reaching Effects of Seismic Disruptions*. Princeton: Princeton University Press.

Dynes, R. R., and Yutzey, D., 1965. The religious interpretation of disaster. *Topic 10: A Journal of the Liberal Arts*, 34–48.

Harbaugh, G., 2001. *Act of God/Active God: Recovering from Natural Disasters*. Minneapolis: Fortress Press.

Hoffman, S., and Oliver-Smith, A., 1999. Anthropology and the angry earth: an overview. In Oliver-Smith, A., and Hoffman, S. (eds.), *The Angry Earth*. New York: Routledge, pp. 1–16.

Mitchell, J. T., 2000. The hazards of one's faith: hazard perceptions of South Carolina Christian clergy. *Environmental Hazards*, **2**, 25–41.

Mycio, M., 2005. *Wormwood Forest: A Natural History of Chernobyl*. Washington, DC: Joseph Henry Press.

Ryan, W., and Pitman, W., 2000. *Noah's Flood: The New Scientific Discoveries About the Event that Changed History*. New York: Simon and Schuster.

Schmuck, H., 2000. An act of Allah: religious explanations for floods in Bangladesh as survival strategy. *International Journal of Mass Emergencies and Disasters*, **18**, 85–95.

Shrady, N., 2008. *The Last Day: Wrath, Ruin, and Reason in the Great Lisbon Earthquake of 1755*. New York: Penguin.

Steinburg, T., 2000. *Acts of God: The Unnatural History of Natural Disaster in America*. Oxford: Oxford University Press.

Taylor, V. A., 1978. Future directions for study. In Quarantelli, E. L. (ed.), *Disasters: Theory and Research*. London: Sage, pp. 251–280.

Cross-references

Coping Capacity
Disaster Relief
Perception of Natural Hazards and Disasters
Psychological Impacts of Natural Disasters
Recovery and Reconstruction After Disaster
Risk Perception and Communication

BODY WAVE

Zhengwen Zeng[1], Lin Fa[1,2]
[1]University of North Dakota, Grand Forks, ND, USA
[2]Xi'an Institute of Post and Telecommunications, Xi'an, Shaanxi, China

Synonyms

Seismic waves

Definition

Seismic waves that travel through the entire earth.

Discussion

Body waves are defined as waves that travel through the body of a medium, as opposed to surface waves that travel near the surface of the medium. They follow ray paths bent by the varying density and modulus (stiffness) of the medium. The density and modulus, in turn, vary according to temperature, composition, and phase. This effect is similar to the refraction of light waves. Seismic waves are body waves, and they are the primary means of transferring information regarding the Earth's interior. Main kinds of body waves are the *P*-wave and the *S*-wave.

The mediums of Earth's interior are generally considered as isotropic or transversely isotropic. A transversely isotropic medium with a vertical axis of symmetry is called a VTI medium.

P-wave (primary wave) is a longitudinal or compressional wave. It can travel in both solids and fluids.

S-wave (secondary waves) is a transverse or shear wave. It can only propagate through a solid.

Generally, the ray or phase velocity of the *P*-wave is greater than the ray or phase velocity of the *S*-wave.

For an elastic isotropic medium, the particle motion of the *P*-wave is in its propagation direction whereas that of the *S*-wave is perpendicular to its propagation direction. For either the *P*-wave or *S*-wave, the direction and magnitude of its ray (travel) velocity are the same as those of its phase velocity.

For an elastic VTI medium, generally the *P*-wave is not polarized in either the phase velocity or energy (ray) velocity direction. There are two types of shear waves: (1) the *SV*-wave which is polarized in a plane constructed by the measurement line (a source and a line array receiver) and its ray path, and generally its particle motion is not perpendicular to either the slowness or ray directions and (2) the *SH*-wave which is polarized perpendicularly to a plane constructed by the measurement line and its ray path. The magnitude and direction of phase velocity are different from those of the ray velocity for each of the *P*-, *SV*-, and *SH*-waves, respectively. The magnitude of the phase (or ray) velocity for either the *P*-wave or *SV*-wave is not related to the modulus of medium, but also to its propagation direction. The magnitude of the phase (or ray) velocity for the *SH*-wave only depends on the modulus of the medium.

Bibliography

Sheriff, R. E., 1984. *Encyclopedic Dictionary of Exploration Geophysics*. Tulsa: Society of Exploration Geophysicists.
Wikipedia Online, 2009. Seismic wave. Wikimedia Foundation, Inc.

Cross-references

Accelerometer
Earthquake
Earthquake Damage
Primary Wave (P Wave)
Secondary Wave (S Wave)

BREAKWATERS

Giovanni Cuomo
Hydraulics Applied Research & Engineering Consulting (HAREC) s.r.l, Rome, Italy

Synonyms

Barrier; Caisson; Dam; Rubble mound; Screen

Definition

Breakwaters are maritime structures constructed to provide shelter from waves.

Discussion

There are two basic types of breakwaters: rubble mound and vertical breakwaters. Rubble mound breakwaters are generally used in shallow water, whereas vertical breakwaters become more convenient in deeper water. A typical rubble mound breakwater consists of an armor layer of heavy rocks or concrete units usually protecting a less permeable filter beneath and a core comprising smaller sized rocks; the mound can be submerged or extend above the sea level. When above sea level, the crest of the breakwater can be completed by a concrete crownwall, usually cast in situ. A vertical breakwater consists of a vertical structure of various designs resting on a rubble foundation. Composite breakwaters consist of a combination (either vertical or horizontal) of a rubble mound and a vertical superstructure.

Due to the enormous power of waves, the construction of breakwaters can be very difficult and the history of constructed breakwaters includes a long list of damages and failures. The oldest known existing rubble-mound breakwater is located in Civitavecchia (Italy) and was constructed by the Roman Emperor Trajanus (AD 53–117). The age of modern breakwaters nevertheless began only in the eighteenth century with the introduction of concrete armor units and composite breakwaters. Armor units were made possible by the invention of the Portland cement in 1824, allowing the production of large, geometrically complex, un-reinforced concrete armor units with improved interlocking capability (e.g., Cubes, Tetrapods, Antifers, Dolos, Accropodes, X-blocks). The possibility of casting very large units allowed the construction of breakwaters in very exposed locations such as the Atlantic coast of Spain and Portugal. The use of slender armor units was nevertheless suddenly interrupted when their structural fragility became evident during the disastrous failure of the Sines breakwater in 1978.

Composite breakwaters originally consisted of high mounds and block work superstructures; over large tidal excursions, the mound could nevertheless induce wave breaking onto the structure causing severe damage and destruction of many breakwaters. The stability of vertically composite breakwaters was later substantially improved with the reduction of the mound and the introduction of taller and heavier superstructures, which was made possible by the use of superimposed Cyclopean blocks, again later replaced by concrete caissons. The introduction of float over caissons made possible the construction of Cyclopean structures such as the Tsunami breakwater at Kamaishi (Japan) in a water depth of 60 m.

Recent development of vertical caisson breakwaters include sloping top, perforated and recurve-wall caissons, respectively, to increase their stability, reduce wave reflection, and transmission. Caissons have also been equipped with OWC wave energy devices.

Berm (or reshaping) breakwaters have a large berm along their seaward face. These are dynamically stable structures made of medium-sized, high-quality rocks and their front reshapes to adapt to wave action during severe storm. Other "special type breakwaters" include piled structures (including wave screens), horizontal plate breakwater, floating breakwater (supported on piles or chained to the seabed), and pneumatic breakwaters, which reduce wave agitation by forcing an airflow.

Bibliography

Goda, Y., 2000. *Random Seas and Design of Maritime Structures*. Singapore: World Scientific. Advanced Series on Ocean Engineering, Vol. 15, p. 443.

Oumeraci, H., Kortenhaus, A., Allsop, N. W. H., de Groot, M. B., Crouch, R. S., Vrijling, J. K., and Voortman, H. G., 2001. *Probabilistic design tools for Vertical Breakwaters*. Rotterdam: Balkema, p. 392.

Tsinker, G., 1996. *Handbook of Port and Harbor Engineering: Geotechnical and Structural aspects*. Springer, 1054 pp.

Cross-references

Beach Nourishment
Coastal Erosion
Erosion
Fetch
Hurricane
Marine Hazards
Rogue Wave
Storm Surges
Tsunami

BUILDING CODE

Rohit Jigyasu
Ritsumeikan University, Komatsubara, Kita-ku, Kyoto, Japan

Synonyms

Building control; Building law

Definition

A building code is a set of rules or standards that specify the minimum acceptable level of safety for buildings to secure life and property from all hazards such as earthquakes, hurricanes, flood, fires, etc. to which the building maybe exposed.

Introduction

Natural hazards have potential to cause immense loss of life and property if buildings are not designed according to the standards specified in the code. This is demonstrated by recent disasters such as the 2010 Haiti Earthquake and 2001 Gujarat Earthquake in India. Moreover building codes ensure safe evacuation in the event of any emergency such as fire. The standards may apply not only to individual buildings but also to the planning of neighborhoods, towns, and cities.

Criteria for designing codes

Since various types of buildings have different vulnerabilities and capacities with respect to different hazards, such codes are designed according to the types of buildings. Building codes may therefore be different for professionally engineered urban structures in contrast to non-engineered buildings that are commonly found in rural areas. Also the codes vary according to the nature of materials; for example, random rubble stone masonry structures, brick, and reinforced cement concrete constructions would all require different codes. These may also differ according to the performance requirement based on usage; for example, schools and office buildings will have different requirements compared to housing. Significantly, codes for repair and retrofitting of historic buildings would need to take into account not only the special nature of the building but also its heritage values.

The codes are formulated by deciding the thresholds of acceptable risks depending on the nature of hazard, building, and the local context. However building codes go beyond hazard mitigation and also provide standards to regulate the structural integrity, adequate ventilation, natural light, sanitation, means of egress, density, topography as well as minimum heights and areas.

Types of building codes

The codes can be either prescriptive rules and descriptions of exactly how something is to be done, or these can be based on performance requirements, leaving it to the designer to decide how this is to be achieved. The former may be effective at higher levels by enforcement through controls; however, compliance is more effective at the local level through indirect control tools such as housing loans and insurance (UNCRD, 2008).

The users

Such codes are generally developed by the government agencies and then reinforced across the country. However, in countries where the power of regulating construction and disaster preparedness is vested in local authorities, a system of model building codes is often used. These codes are a set of guidelines that have no legal status unless adopted or adapted by an authority having such jurisdiction.

The code contains regulations, which can be immediately adopted or enacted for use by various departments, municipal administrations, and public bodies. However, architects and engineers are the primary preconstruction users of the building codes. The initial designs and drawings must satisfy all the requirements embedded in these codes.

Implementing building codes

Many disaster-prone countries now have building codes that address the specific nature of disaster risks. However, their implementation is poor in many cases and many unsafe buildings are still being raised. One of the reasons is that a minimum acceptable house structure according to a building code is often beyond the means of the poor and sometimes even middle-income families. For example, this was the case in Kenya prior to the updating of the building codes in 1990, when the Intermediate Technology Development Group (ITDG) initiated participatory effort to modify the codes to reflect the local building practices while still encouraging safe building practices (Jha et al., 2010).

Although updating a building code is the first step toward improving building safety and increasing disaster resilience, other challenges such as simplifying approval processes and establishing incentives to fully apply the building codes also need to be tackled.

Professionals often understand the code's intent but do not know how to perform the necessary calculations to build hazard resistant constructions. Therefore, engineers must be given training to design structures that are resistant to specific hazards. Also it is important to bridge the gap between professional engineers, who design the structures, and the craftsmen, who put them into practice at the construction site. This is particularly a challenge in developing countries

Moreover, despite the existence of building codes and serious intention for its effective implementation, realization of safe buildings in the field requires quality material, good workmanship, and awareness in masons, builders, and house owners (UNCRD, 2008).

Also there is a constant need for the engagement of skilled and responsible experts for developing or updating the codes by adopting holistic approaches that take into account the local hazards, as well as climate, resources, cultural needs that new buildings must meet.

Summary

Building codes are a necessary instrument for reducing vulnerability to natural hazards. They should be followed by architects and engineers while designing new buildings or retrofitting existing buildings in hazard-prone areas. They must take into account the nature of building and local social, economic, and geographical context.

Effective implementation of building codes requires good institutional structure and training of engineers as well as contractors and masons, who are responsible for construction.

Bibliography

Building Code, 1996. http://media.wiley.com/product_data/excerpt/92/04717418/0471741892.pdf. Accessed April 14, 2010.

Indianetzone Construction, National Building Code, 1996. Available at http://construction.indianetzone.com/1/national_building_code.htm. Accessed April 14, 2010.

Jha, A. K., Barenstein, J. E. D. Phelps, P. M., and Pittet, D., 2010. *Safer Homes, Stronger Communities: A Handbook for Reconstructing After Natural Disasters*. Washington, DC: World Bank. Available at http://www.housingreconstruction.org/housing/sites/housingreconstruction.org/files/SaferHomesStronger-Communitites.pdf. Accessed April 14, 2010.

Subedi, J., and Mishima, N. (ed.), 2008. Handbook on Building Code Implementation. Learning from the Experience of Lalitpur Sub-Metropolitan City, Nepal. United Nations Centre for Regional Development Disaster Management Planning (UNCRD) Hyogo Office and Lalitpur Sub-Metropolitan City.

Trombly, B. 2006. The International Building Code, CMGT 564. Available at http://www.strategicstandards.com/files/InternationalBuildingCode.pdf. Accessed April 14, 2010.

UNCRD, 2008. From Code to Practice. Challenges for Building Code Implementation and the Further Direction of Housing Earthquake Safety. In *Records and outcomes of International Symposium 2008 on Earthquake Safe Housing*. Tokyo November 28–29, 2008. Available at http://www.preventionweb.net/files/10591_HESITokyoPapers.pdf. Accessed May 29, 2012.

Cross-references

Building Failure
Building, Structures and Public Safety
Damage and the Built Environment
Earthquake Resistant Design
Hazard
Mitigation
Vulnerability

BUILDING FAILURE

Tiziana Rossetto
Earthquake and People Interaction Centre (EPICentre) University College London, London, UK

Synonyms

Building collapse; Structure failure

Definition

Building failure occurs when the building loses its ability to perform its intended (design) function. Hence, building failures can be categorized into the two broad groups of physical (structural) failures (which result in the loss of certain characteristics, e.g., strength) and performance failures (which means a reduction in function below an established acceptable limit) (Douglas and Ransom, 2007).

Structural failure corresponds to the exceedance of ultimate limit state in many of the load-carrying elements, which compromise the structural stability of the building. In practice, this corresponds to extensive damage, partial or total collapse of the building, resulting in repair costs that are high relative to the replacement value of the building.

Performance failure can be induced by the failure of structural elements (as per above), nonstructural component (non-load-bearing elements or equipment) failures or combinations thereof. For example, the use of a hospital might be affected by damage to infill walls and cladding (nonstructural components) that may compromise hygiene or damage essential equipment. In the context of natural hazards, however, the established acceptable limit to function, defining the threshold of performance failure may vary according to the type of natural hazard and its frequency of occurrence. For instance, in the case of frequent events, it may not be acceptable to lose any functionality but for rare events some degree of performance failure may be deemed acceptable, (e.g., the concept of performance-based assessment in earthquake engineering, see Calvi, 1998 amongst others).

Bibliography

Calvi, G. M., 1998. Performance-based approaches for seismic assessment of existing structures. In Bisch, P., Labbe, P., and Pecker, A. (eds.), *Proceedings of the 11th European Conference on Earthquake Engineering: Invited Lectures*, pp. 3–20.

Douglas, J., and Ransom, W. H., 2007. *Understanding Building Failures*, 3rd edn. New York: Routledge. 326 pp.

Cross-references

Building Codes
Building, Structures and Public Safety
Concrete Structures
Damage and the Built Environment
Earthquake-Resistant Design
Risk Assessment
Structural Damage Caused by Earthquakes
Structural Mitigation
Unreinforced Masonry Building

BUILDINGS, STRUCTURES, AND PUBLIC SAFETY

John M. Logan
University of Oregon, Bandon, OR, USA

Definitions

Fault. A surface or zone of rock fracture along which there has been displacement from a few centimeters to kilometers.
Tsunami. A gravitational sea wave produced by any large-scale, short-duration disturbance of the ocean floor.
Marble. A metamorphic rock consisting primarily of recrystallized calcite.
Granite. An igneous rock composed primarily of quartz and feldspar with possibly lesser amounts of mica.

Introduction

The questions and issues of public safety around buildings and structures have occupied scientific and engineering research for many years and are still of major concern. Relevant questions include: (a) to what extent can buildings withstand the ground shaking associated with earthquakes or associated tsunamis, and (b) how long will building stones such as granite and marble or concrete withstand environmental weathering processes? Large buildings and structures such as major office buildings, extensive museums, libraries, and entertainment centers located in heavily populated areas are often the focus of attention because of the potential loss of life if failure occurs.

Earthquakes and tsunamis

Earthquakes occurring on land

Public safety associated with earthquakes and tsunamis in many cases is dependent upon warnings of such events (Figure 1). The anticipated ground motions depend upon not just earthquake magnitudes, but as importantly distance from the source, travel path, and near surface materials, which may produce various amplification effects. Both are caused by sudden ruptures of the Earth's crust, generally along preexisting failure surfaces called faults. The slip on the fault may be primarily horizontal (strike slip), vertical (dip slip), or a combination of both (oblique slip).

The severity of an earthquake is described by both magnitude and intensity. Magnitude, usually expressed as an "Arabic Numeral" and is not dependent upon distance from the epicenter. It is calculated by measuring the amplitude of the released seismic waves. One scale that is used to measure this magnitude is called the Richter scale. It is logarithmic assigning values from 1 to 10. As it is logarithmic, an earthquake of value 5.0 is ten times greater than one measuring 4.0 (Richter, 1958). The more recent moment magnitude scale is also used today, which gives another estimate of earthquake size (Kanamori, 1977).

The intensity of an earthquake is a measure of the ground shaking occurring at any given point on the Earth's surface. Unlike magnitude scales, they do not have a mathematical basis; instead they are a ranking based on observed effects and is contingent upon local conditions. A number of intensity scales are used throughout the world, but the Modified Mercalli Scale is commonly used in the USA (Table 1). It includes the behavior of people, responses of houses, and larger buildings, and natural phenomena. It is dependent upon distance from the earthquake epicenter, underlying rock and soil structure, and other geologic features. Earthquakes range from intensity

Buildings, Structures, and Public Safety, Figure 1 Building damage from Loma Prieta Earthquake in California, USA, of October 17, 1989. The earthquake caused 62 deaths, 3,757 injuries, 12,053 persons displaced, 18,306 homes damaged, 2,575 businesses damaged, and repairs to damaged structures over US$6 billion.

levels of I to XII on the Modified Mercalli Scale, with events not felt by most people until about level III.

It is the amplitude and duration of ground motion that pose threats to buildings, structures, and public safety. The correlation between earthquake magnitude and intensity clearly depends on local factors. The US Geological Survey shows a typical correlation in Table 2 for the USA. A local issue such as fault slip may be critical. For instance, for an event of magnitude 7t, slip along a fault segment may occur for a few tens of miles, for an event of magnitude 8, the slip may be for a few hundred miles (Bolt, 1988).

Efforts to predict earthquakes today are primarily based upon documentation of events in a specific area and the possible historic cyclic nature of such events. Research looks at not just seismic records, but rapid changes in surface elevations, inundations of coastal tidal sediments, and other geologic evidence of cyclic changes. One primary observation is that earthquakes historically have occurred along the same fault or fault array such as the San Andreas in California and the Cascadia in the Oregon-Washington region in the USA, so considerations of public safety are critical in such areas. The exact periodicity of events even in a specific region is not known except that there are cycles of seismic motion followed by periods of quiet (Witter et al., 2003). Reviewing historical records of prior earthquakes and damage, particularly with respect to specific locations, can help to mitigate public safety associated with earthquakes.

The ground motion may directly damage buildings. Damage may as well be produced by liquefaction of the wet sand at the surface to produce quicksand. Seismically induced quicksand forms when wet sand fill is sheared by the seismic waves and becomes more tightly packed. The reduced pore volume squeezes out part of the water in the pores, which migrates upward, exerting pressure on the overlying sand grains and keeping them from making contact, thus reducing the strength to near zero. In addition to potentially altering the strength of underlying material, earthquakes may also induce landslides which may pose additional threats.

Earthquakes often occur in sequences with "foreshocks" or small magnitude events preceding a large or "main shock" and frequently followed by events of lower magnitudes, called "after shocks." Presently, there are no dependable criteria to separate foreshocks from a main shock. There may only be a number of small events, or a major shock, which may not be preceded by recognized smaller events. The awareness of new small events should alert the public to the possibility of a larger one. Weakness associated with homes include foundations not anchored, unreinforced walls of any material, buildings on tall walls or posts. These issues have caused review of building codes in large cities and in smaller rural areas. It is also clear that structures such as bridges, water, and electric lines may be disrupted and poses threats to emergency routes, communications systems, and essential utilities. This may impact not just private homes but critical

Buildings, Structures, and Public Safety, Table 1 Abbreviated modified Mercalli intensity scale

I. Not felt except by a very few under especially favorable conditions
II. Felt only by a few persons at rest, especially on upper floors of buildings
III. Felt quite noticeably by persons indoors, especially on upper floors of building. Many people do not recognize it as an earthquake. Standing motor cars may rock slightly. Vibrations similar to the passing of a truck
IV. Felt indoors by many, outdoors by few during the day. At night, some are awakened. Dishes, windows, doors disturbed; walls make cracking sound. Sensation like heavy truck striking building. Standing motor cars rock noticeably
V. Felt by nearly everyone, many awakened. Some dishes, windows broken. Unstable objects overturned
VI. Felt by all, many frightened. Some heavy furniture moved; a few instances of fallen plaster. Damage slight
VII. Damage negligible in building of good design and construction, slight to moderate in well-built ordinary structures; considerable damage in poorly built or badly designed structures; some chimneys broken
VIII. Damage slight in specially designed structures; considerable damage in ordinary substantial buildings with partial collapse. Damage great in poorly built structures. Fall of chimneys, factory stacks, columns, monuments, and walls. Heavy furniture overturned
IX. Damage considerable in specially designed structures; well-designed frame structures thrown out of plumb. Damage great in substantial buildings, with partial collapse. Buildings shifted off foundations
X. Some well-built wooden structures destroyed; most masonry and frame structures destroyed with foundations. Rails bent
XI. Few, if any (masonry) structures remain standing. Bridges destroyed. Rails bent greatly
XII. Damage total. Lines of sight and level are distorted. Objects thrown into the air

Reference: US Geological Survey, Earthquake Hazards Program, (2012)

Tsunamis and fault rupture at sea

If faults rupture underwater so that the fault slip is primarily vertical and the sea floor is rapidly displaced, it produces a wave. Tsunami waves may also be produced by landslides associated with fault motion. Volcanic activity may also produce tsunami waves. Most tsunamis occur in the Pacific Ocean as very active features such as deep ocean trenches, which are fault bounded, and explosive volcanic islands, surround it. Tsunami waves differ from ordinary ocean waves as they may often measure 1,600 km from crest to crest and move at more than 100 km per hour. As the waves near shore they decrease in velocity but they increase in height dramatically. An initial apparent withdrawal after the first wave maybe followed by much larger one, minutes later (Dudley and Lee, 1988).

Buildings, Structures, and Public Safety, Table 2 Earthquake magnitude and intensity comparison. Magnitude and intensity relate different characteristics of earthquakes. Magnitude quantifies the energy released at the source of the earthquake. This is determined from measurements on seismographs. Intensity measures the strength of shaking produced by the earthquake at a particular location. Intensity is determined from effects on people, human structures, and the natural environment. The following gives intensities that are typically observed at locations near the epicenter of earthquakes of different magnitudes

Earthquake magnitude	Typical maximum modified Mercalli intensity
1.0–3.0	I
3.0–3.9	II–III
4.0–4.9	IV–V
5.0–5.9	VI–VII
6.0–6.9	VII–IX
7.0 and higher	VIII or higher

Reference: US Geological Survey, Earthquake Hazards Program, (2012)

Clearly, tsunamis cause building and structure damage, which may pose significant threats to the public. Major devastating events resulting in deaths have been recorded since about 1600 BC, with recent catastrophic ones occurring in 2004 in the Indian Ocean and the Tohoku earthquake and ensuing tsunami in 2011. Often warnings are issued for those waves traveling across long distances, but in may cases are disregarded by the public who rush to the sea shores to see the waves and are often caught by the water or are trapped by bridge and highway failures as in the 2011 Japanese earthquake and subsequent tsunami.

Buildings and structures with exteriors subject to environmental fatigue

The question of buildings and public safety raises complex issues. The most frequent issues for public safety come not from the foundation failures but from exposed portions of the structures. These areas are subject to possible deterioration from environmental factors such as heavy winds, atmospheric chemistry, and seasonal changes in temperatures. The problem of public safety arises as many buildings that potentially pose problems are often close to the center of urban areas. The price of land often leads to multi-storey, high-rise buildings. The buildings generally have steel frameworks that are covered with veneers of glass, metal, natural stones such as marble, granite or sandstone, or recently of composite materials. It is the veneers that pose the greatest issue for public safety as they often deteriorate over time with the potential for failure and pieces falling from high places on the building and endangering the public below (Toronto Star, May 17, 2007, front page, p. A1). The basic issue is the ability of the materials to withstand the wind loads generated by storms, hurricanes, and tornados.

facilities such as hospitals, ambulance services, firefighting equipment, and others.

Although there is often warning of approaching hurricanes and tornados, the high winds sometimes associated with seasonal storms are not often given as much public notice. The intrinsic strength of the material is augmented by increasing thickness, reducing the panel size, and supplementing the anchoring system.

Glass windows and panels have fractured endangering vehicles and pedestrians when they fall. In some cases, the winds will carry pieces for kilometers. Stone veneers have received the most attention and particularly those formed by marble panels. Marble panels on architectural icons such as the Amoco tower in Chicago, USA, the Finlandia Concert Hall in Helsinki, Finland, and the Grande Arche de la Defense in Paris, France all began bowing to the extent that they posed risks of fracturing and falling off the building. The potential for public injuries or deaths and the resulting wide publicity has lead to extensive research regarding the physical properties and the loss of strength of marble (Raleigh, 1934). Chemical interaction during weathering leading to mineral dissolution and fracture corrosion and subsequent loss of strength has been postulated, especially in environments of acid rain (Winkler, 1967). Cycles of freeze-thawing which could result in fractures on the microscopic or even macroscopic scale, thus reducing the marble strength, and possibly leading to panel failure have also been proposed (Bortz et al., 1993). Thermal expansion due to diurnal cycles in temperature has been extensively researched especially with respect to buildings where the vertical panels do not retain significant rain or snow fall (Logan, 2006; Koch and Siegesmund, 2004). The principal factor here is that the mineral calcite, which is the primary component of marble, is thermally anisotropic. It expands parallel to the c-crystallographic axis but contracts perpendicular to it when subjected to temperatures above the average ambient state, which makes it unique. This differential expansion during elevated temperatures followed by relaxation as the temperatures return to ambient, produces fatigue within the grain-to-grain contacts eventually weakening the marble.

With thermal cycling leading to a loss of strength, a second effect may become significant, specifically the release of residual elastic strain stored in the rock as a result of incomplete annealing upon recrystallization of the limestone to form marble (Logan, 2004; Siegsmund et al., 2000). This strain release produces increases in dimensions of the marble panel. If the panel is constrained laterally, it will bow from its original flat geometry, with the weight of the panel producing a horizontal bow. To accommodate this change in shape, microfractures will occur which further weakens the panel. The bow amplitude may become sufficient to induce bending stresses in the panel, exacerbating the condition, potentially leading to macroscopic fracture and failure of the panel at about the middle of the panel height. Additionally, as the panels are anchored to retain them on the building, the bowing may load the anchors and produce failure there. To remove the possibility of moisture accumulating behind a panel and possibly producing corrosion of the metal framework and anchors, calk is installed around the edges of the panel. Even if a horizontal fracture forms leading to complete loss of panel strength, the calk may have sufficient strength to keep the pieces from falling for a limited time.

Granite has not exhibited the extreme deterioration from environmental conditions that marble has, although laboratory studies have looked at the bowing potential (Siegsmund et al., 2007). As it is multimineralic, the major components quartz, feldspar, and micas all have different responses to the atmosphere, mitigating the effects seen in calcite marble. It does often have, however, a pronounced strength anisotropy due to the crystallization process where the crystals often have a preferred orientation parallel to each other. Although this is not an evident layering, it often results in a "rift" with the rock weaker parallel to this feature, and stronger perpendicular to it. Specifications for granite exterior panels stipulate that the rift be parallel to panel face to minimize failure due to wind loads. It is frequently presumed that the rift is consistent within a quarry so that granite may be removed with respect to the assumed rift orientation, but geological mapping of quarries has shown that the rift orientation may change considerably (Logan et al., 1993).

This variation of physical and resulting mechanical properties is common in rocks. Although buildings with marble exterior cladding showing extreme signs of weathering have received considerable attention, there are other buildings with little evidence of deterioration. Marble from Carrara, Italy, is widely used in buildings because of its white color and the fame that it achieved from sculptures, such as those produced by Michaelangelo. But there are over 50 varieties of marble all commonly known as Carrara Marble and all with different physical and mechanical properties (ERTAG, 1980). The marble types conducive to sculpture have porosities of 6% or greater which makes them very susceptible to weathering. Marble from all over the world has differing properties of composition, grain size and orientation, porosity, residual strain, and others. Variations in properties are also found in other materials such as glass due to potential differences in their processing. To ensure public safety, extensive testing of all material installed is highly desirable.

The issues of public safety become paramount when evidence of deterioration of the exterior materials becomes evident. In some cases, surveys of the exterior may determine that the problems exist locally and can be remedied by replacement in those areas. Such a step was taken on the Amoco Building when it was decided that the problems were only extreme at the corners of the building. This building was the third tallest building in the world in 1989 and the tallest marble-clad one. It was decided to install straps across each marble panel so that if failure did occur the pieces would be retained until they could be removed. In this case, unfortunately, the installation of the straps severely damaged the panels. Even thought no panels fell from the building, the result was a decision to remove all of the 43,000 marble panels from the building to avoid any panels or partial ones falling from the building and endangering the public (Williams, 2009). The building was about 15-years old at that time with cost estimated at that time to range from $60 to $100 million, to ensure public safety (Chicago Sun-Times, March 7, 1989, p. 1). It was decided to replace the marble with granite, which due to extensive quality control during replacement has remained without problems.

In contrast, 45,000 marble panels at First Canadian Place, Toronto, Canada's tallest skyscraper, were also removed as a result of two incidents in which whole or portions of panels fell from the building. The first event did not result in damage. During the second event, one panel weighing 115 kg fell from the 51st storey, and landed on a third floor roof; fortunately no one was injured. As a result of the second accident, and because the building was in the center of the downtown financial district, surrounding streets were closed to vehicle traffic for a number of days leaving some 48,000 streetcar riders and thousands of car drivers dealing with major delays (Toronto Star, May 17, 2007, front page, p. A1). The marble suffered from the same deterioration processes that affected the Amoco Building. It has been decided to remove all the marble cladding and replace it with other material.

Summary

Public safety is in most cases contingent upon proximity to buildings or structures and an awareness of conditions that may infringe on the safety. In cases of high-rise buildings and structures, the integrity of the exteriors is critical as many materials used are subject to environmental deterioration over extended periods of time. This degeneration of strength may lead to failure, especially under high wind loads with pieces falling from the buildings. In contrast, building integrity in earthquake and tsunami prone regions is dependent upon the ability to withstand sustained ground shaking or high waves. Potentially damaging earthquakes may occur without any warning, but tsunami warning systems may help to mitigate damage and public safety.

Bibliography

Bolt, B., 1988. *Earthquakes*. New York: W. H. Freeman.
Bortz, S., Stecich, J., Wonneberger, B., and Chin, I., 1993. Accelerated weathering in building stone. *International Journal of Rock Mechanics and Mining Sciences and Geomechanical Abstracts*, **30**, 1559–1562.
Dudley, W. C., and Lee, M., 1988. *Tsunami*. Honolulu: University of Hawaii Press.
Earthquake Hazards Program, 2012. USGS Informations Services. Denver: Denver Federal Center.
ERTAG, I Marmi Apuani, 1980, Nuova Grafica Fiorentina, Firenze. 55p.
Kanamori, H., 1977. The energy release in great earthquakes. *Journal of Geophysical Research*, **82**, 2981–2987.
Koch, A., and Siegesmund, S., 2004. The combined effect of moisture and temperature on the anomalous expansion behavior of marble. *Environmental Geology*, **46**, 350–363.
Logan, J. M., 2004. Laboratory and case studies of thermal cycling and stored strain of the stability of selected marbles. *Environmental Geology*, **46**, 456–467.
Logan, J. M., 2006. On-site and laboratory studies of strength loss in marble on building exteriors. In Kourkoulis, S. K. (ed.), *Fracture and Failure of Natural Building Stones: Applications in the Restoration of Ancient Monuments*. Dordrecht: Springer, pp. 345–362.
Logan, J. M., Hastedt, M., Lehnert, D., and Denton, M., 1993. Variations of rock properties within a quarry. *International Journal of Rock Mechanics and Mining Sciences and Geomechanical Abstracts*, **30**, 1527–1530.
Raleigh, L., 1934. The bending of marble. *Proceedings of the Royal Society A*, **144**, 266–279.
Richter, C. V., 1958. *Elementary Seismology*. San Francisco: W. H. Freeman.
Siegsmund, S., Ullemeyer, K., Weiss, T., and Tschegg, E. K., 2000. Physical weathering of marbles caused by thermal anisotropic expansion. *International Journal of Earth Science*, **89**, 170–182.
Siegsmund, S., Mosch, S., Scheffzuk, Ch, and Nikolayev, D. I., 2007. The bowing potential of granitic rocks: rock fabrics, thermal properties and residual strain. *Environmental Geology*, **50**, 254–265.
Williams, D. B., 2009. *Stories in Stone*. New York: Walker & Company.
Winkler, E. M., 1967. Weathering and weathering rates of natural stone. *Environmental Geology Water Science*, **9**, 85–92.
Witter, R. C., Kelsey, H. M., and Hemphill-Haley, E., 2003. Great Cascadia earthquakes and tsunamis of the past 6,700 years, Coquille River estuary, southern coastal Oregon. *Geological Society of America Bulletin*, **115**(10), 1289–1306.

C

CALDERAS

James W. Cole
University of Canterbury, Christchurch, New Zealand

Synonyms
Cauldrons

Definition
Calderas are volcanic depressions, roughly circular in surface plan, with a diameter greater than depth, and representing roof collapse into shallow underlying magma reservoirs.

Discussion
The term "caldera" comes from the Latin word "caldaria" meaning "boiling pot," and was originally used in the Canary Islands for any large "bowl-shaped" depression. Only in the last 50 years has their origin and potential hazards been fully appreciated. Calderas may occur in volcanoes of all compositions, in all tectonic environments, and show a wide range of forms. Consequently, it is difficult to classify calderas, although common collapse processes, provide "end-member" possibilities (Figure 1). The simplest form is "piston" or "plate" collapse within a cylindrical (ring) fault. This occurs within many smaller (typically basaltic) calderas, but is rare in larger (typically rhyolitic) structures. The latter are more likely to show either "piecemeal" collapse, around a number of centers in the caldera, or "downsag," where parts of the structure dip towards the center of the caldera. Regional faults can be an important boundary influence, and collapse may preferentially occur along one of these faults, with the opposite side showing "downsag," to produce a "trapdoor" caldera.

Many larger calderas have experienced multiple collapse events, often separated by tens of thousands of years. Such calderas should more correctly be called "caldera complexes". Each collapse is likely to be accompanied by explosive eruptions, usually producing pyroclastic flows, which deposit widespread ignimbrites (ash flow tuffs). Some of the ignimbrite will pond in the caldera (intra-caldera ignimbrite), whereas the remainder will be distributed radially around the caldera (outflow sheets).

While collapse is the key to caldera formation and is rapid (hours to days), it is only one phase in a process that may take tens to hundreds of thousand years. Pre-collapse volcanism is common, sometimes accompanied by uplift ("tumescence"), and most rhyolitic calderas are followed by post-collapse volcanism (usually forming lava domes and airfall tephra), often accompanied by uplift ("resurgence"). A cross section of a generalized "piston" caldera is shown in Figure 2. Hydrothermal activity and mineralization is likely to occur throughout the life of a caldera volcano, but is particularly important in the post-collapse stages. Once volcanism ceases, erosion will progressively remove much of the surface volcanism (over millions of years). This structure is called a "cauldron," when caldera-floor rocks become exposed. Once a substantial amount of the underlying magma reservoir is exposed, the term "ring-structure" is commonly used.

Calderas are a major natural hazard. The accompanying pyroclastic flows can cause total devastation for hundreds to thousands of square kilometers around the volcano. The largest of these, the "Supervolcano" eruptions, can produce >1,000 km^3 of ignimbrite and can influence climate with fine ash remaining in the atmosphere to cause a "global winter" for many years! During pre-collapse tumescence and post-collapse resurgence, ground movement is likely, which will affect structures built in the area. While there are likely to be precursor events to caldera formation (e.g., earthquake swarms, gas discharge, etc.), such

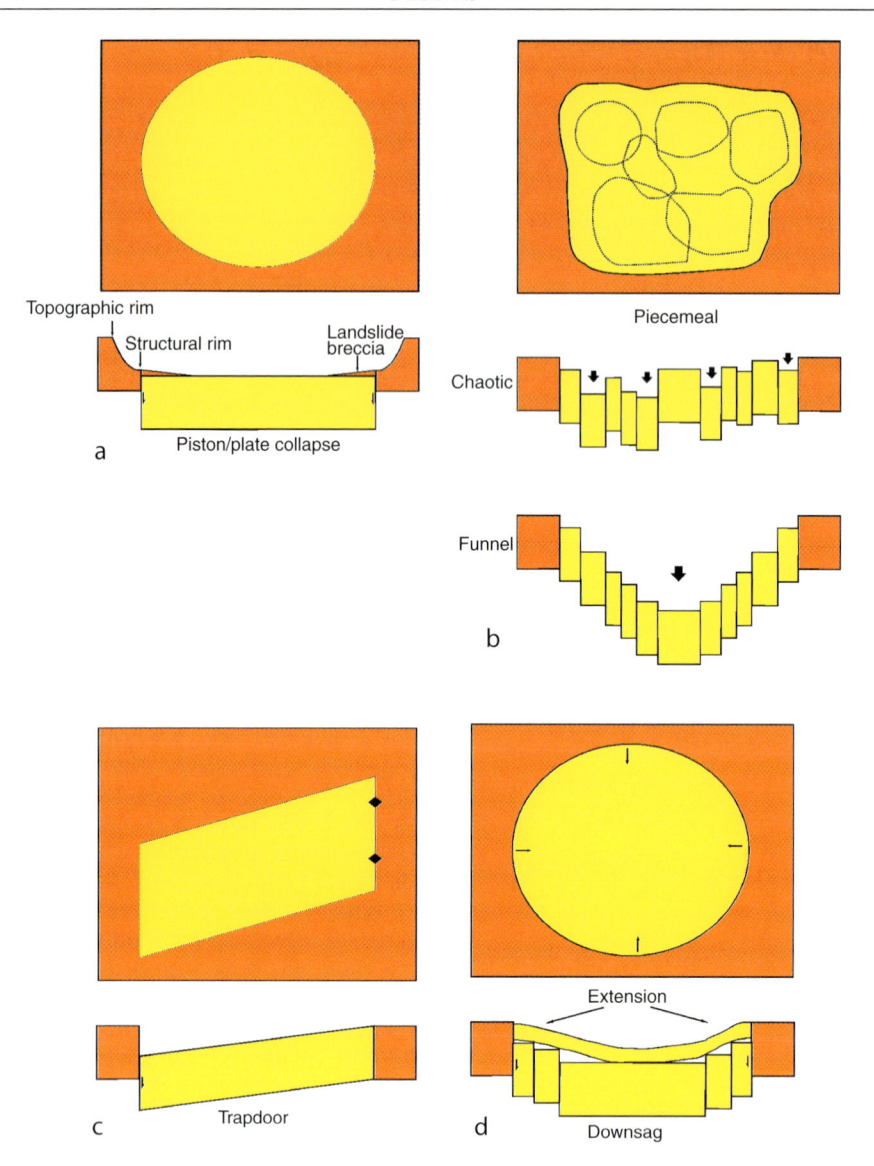

Calderas, Figure 1 Four end-member mechanisms of caldera collapse: (**a**) piston-plate, (**b**) piecemeal, (**c**) trapdoor, and (**d**) downsag (From Cole et al., 2005).

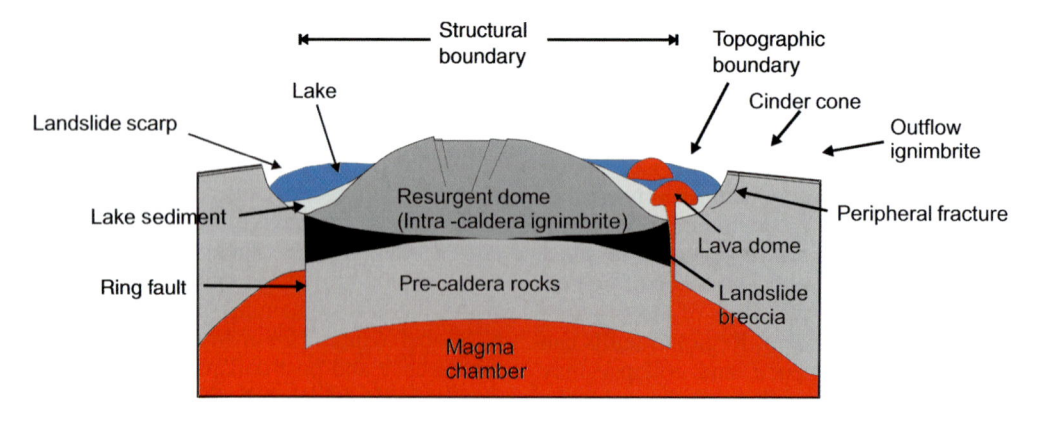

Calderas, Figure 2 Schematic block diagram of a typical resurgent piston-type caldera, showing features that may be present in the structure (From Cole et al., 2005).

events do not always culminate in an eruption. Such "false alarms" are a major problem for effective prediction.

Bibliography

Cole, J. W., Milner, D. M., and Spinks, K. D., 2005. Calderas and caldera structures. *Earth Science Reviews*, **69**, 1–26.

Lipman, P., 2000. Calderas. In Sigurdsson, H. (ed.), *Encyclopedia of Volcanoes*. San Diego: Academic, pp. 643–662.

Cross-references

Eruption Types
Krakatoa (Krakatau)
Magma
Nuee Ardente
Pyroclastic Flow
Volcanoes and Volcanic Eruptions

CASUALTIES FOLLOWING NATURAL HAZARDS

Kerrianne Watt[1], Philip Weinstein[2]
[1]Tropical Medicine & Rehabilitation Sciences; James Cook University, Townsville, QLD, Australia
[2]University of South Australia, Adelaide, SA, Australia

Synonyms

Casualties; Fatal and nonfatal injuries; Mass casualty events; Natural disasters, forces of nature

Definition

A casualty of a natural disaster can be defined as any person suffering a physical or psychological injury therefrom. Injury, in turn, is "unintentional or intentional damage to the anatomical structures or physiological processes of the body incurred from acute exposure to an exchange of energy (thermal, mechanical, electrical, or chemical), or the absence of such essentials as heat or oxygen" (p. 4, National Committee for Injury Prevention and Control, 1989; Driscoll et al., 2004). For the purposes of this contribution, a natural disaster is considered an event with one of the following: 10 or more human fatalities; 100 or more people affected; a state of emergency declared; and international assistance sought (Scheuren et al., 2008).

Introduction/background

Since the beginning of recorded history, natural disasters have been measured by the severity of their impact on human populations – beginning with the "Biblical flood" that apparently drowned all people on Earth, with the exception of one family group. More recently, it has been estimated that 6,367 natural disasters have occurred between 1974 and 2003, which involved approximately 3,135 fatalities per 100,000 population (Guha-Sapir et al., 2004). The number of people affected by natural disasters is much greater than the fatalities that occur due to such events. From 1989 to 2003, it was estimated that worldwide, approximately 13,706 persons were affected by any type of natural disaster, for every person killed (Guha-Sapir et al., 2004). Not all of the people affected by a natural disaster are physically injured – however this contribution focuses on those who experience injury as a consequence of natural disasters.

The incidence of reported natural disasters has increased substantially over the last 100 years (Figure 1). Advances in technology and communication account for some of this increase through better ascertainment, but there is also an element of increasing impact because of the rapid growth of the human population. Over the last three decades, there has been a decrease in the number of deaths caused by natural disasters, but an increase in the number of people affected by these events (Scheuren et al., 2008). These changes can be attributed to more people in bigger cities with more structures, but also to improvements in disaster preparedness, public awareness, infrastructure (e.g., anti-seismic housing and medical facilities), and our ability to manage and respond to disasters (Noji, 1992). Disaster epidemiologists, who measure and describe the health effects of disasters, and identify the factors that contribute to such adverse effects, have largely been credited with advocating for the introduction of such improvements (Noji, 1992). Whereas more people survive natural disasters, many of the survivors experience injuries and/or illness as a consequence, some of which require long-term rehabilitation and may impact on quality of life. The resultant medical burden in turn has implications for economies. The current approach to measuring the impact of disasters (the number of people affected) may therefore significantly underestimate the true public health impact of such events. An alternative approach, such as estimating Disability Adjusted Life Years (DALYS), may provide a more accurate description of the impact of natural disasters.

The severity of the impact of a natural disaster is determined by four factors: (1) population at risk (size, location, susceptibility, age distribution); (2) exposure to the effects of the disaster; (3) short-term and long-term adverse health effects resulting from such exposures; and (4) effect modifiers (building infrastructure; living conditions, communication systems including media and internet) (Dominic et al., 2005). For these reasons, the impact of a natural disaster is usually much greater in less developed regions, with higher population densities. Less infrastructure (including access to healthcare, water, electricity; financial assets and access to loans/insurance), greater communication difficulties, and less emergency response capacity exacerbate this situation (Guha-Sapir et al., 2004). Often, such regions/countries are more vulnerable to natural disasters given their geographical location (located on a flood plain or region of high seismic activity) or other environmental characteristics (e.g., soil degradation/erosion, pollution, deforestation) (Guha-Sapir et al., 2004).

Natural disasters reported 1900–2008

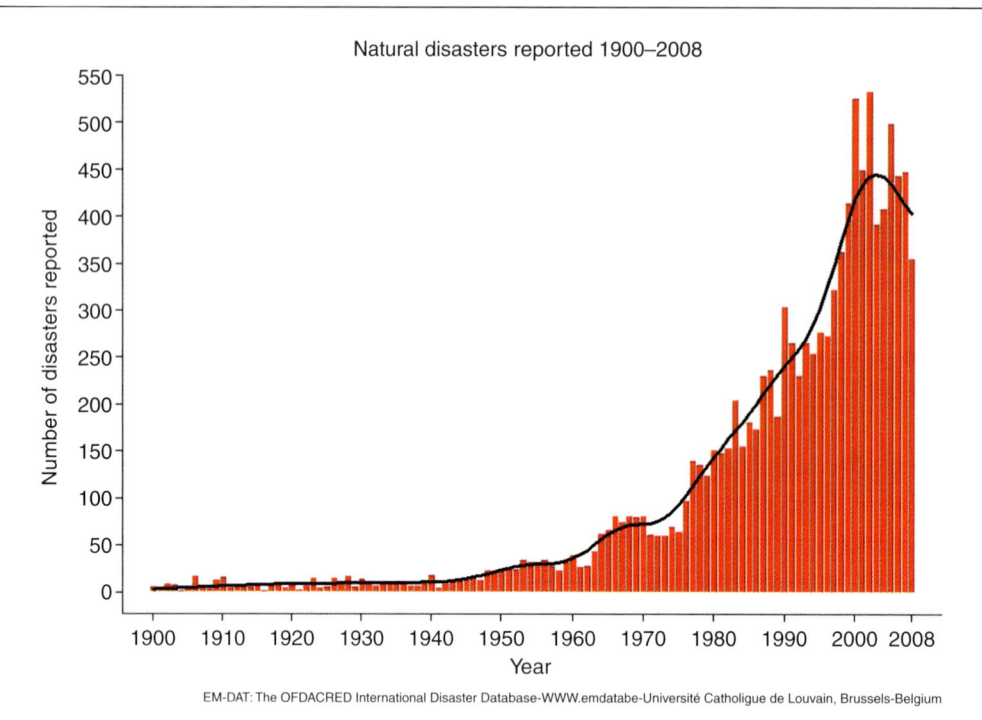

Casualties Following Natural Hazards, Figure 1 Natural disasters reported in the period 1900–2008.

The proportion of casualties (number of fatalities and severity of injuries) that occur as a consequence of natural disasters is associated with delays in reaching victims, which is in itself dependent on factors such as communication systems and the density and integrity of buildings (Chang et al., 2003). Physical location during the event has been identified as a factor that increases risk of injury and death as the result of a natural disaster. Being in a multiple unit residential or commercial structure is associated with increased risk of injury/death following an earthquake (Peek-Asa et al., 2003). Therefore, areas where there is high proportion of multilevel buildings are likely to experience higher numbers of casualties from these types of events. The very young, the elderly, and people of low socioeconomic status are also more likely to experience worse outcomes after a natural disaster (Peek-Asa et al., 2003; Milsten, 2000).

Disaster types

Natural disasters can be categorized into a range of subgroups. The classification system recently developed by EMDAT (the International Disaster database) is used for the purposes of this contribution: Climatological (droughts, extreme temperatures, wildfires); Geophysical (earthquakes, volcanoes, dry mass movements); hydrological (floods, wet mass movements); and meteorological (storms such as hurricanes, cyclones, tornadoes, etc.). This focuses on the latter three hazard types.

Geophysical

From 1974 to 2003, an estimated 767 geological disasters occurred worldwide (Guha-Sapir et al., 2004). The main subtypes of geophysical natural disasters include earthquakes and volcanoes.

Earthquakes

There were approximately 660 earthquake disasters from 1974 to 2003, which resulted in 559,608 fatalities, and affected more than 82 million people (Guha-Sapir et al., 2004). It has been estimated that over 500,000 earthquakes occur every year, and 7–11 of these cause substantial fatalities (Ramirez and Peek-Asa, 2005). Over the past 200 years, approximately 1.9 million deaths have been reported due to earthquakes (Shulz et al., 2005). Of all natural disasters, the highest rate of mortality is associated with earthquakes – 36% of deaths that have occurred due to natural disasters from 1970 to 2009 were due to earthquakes (Centre for Research on the Epidemiology of Disaster (CRED), 2010).

In addition to the generic factors that impact on injuries from natural disasters described thus far, earthquake injuries depend on several factors: number of occupants in an affected dwelling, floor surface, and time of day (Milsten, 2000). The highest rate of post-disaster suicides is associated with earthquakes (Milsten, 2000; Friedman, 1994).

Injuries associated with earthquakes commonly occur due to being trapped inside buildings, falling, or being hit by falling objects. Injuries include: asphyxiation, hemorrhage, crush syndrome, internal injuries (abdominal and pelvic), severe chest trauma, upper and lower extremity injuries, fractures, soft tissue injuries, and multiple traumatic injuries (Chang et al., 2003; Milsten, 2000). Major head injuries are usually fatal and peripheral limb injuries are also characteristic of major earthquakes, as evident in the 2003 Bam (Schnitzer and Briggs, 2004) and 2005 Kashmir (Dhar et al., 2007; Redmond, 2005) events. A significant proportion of earthquake-related injuries are sustained in the post-disaster period (i.e., "clean up") – where 22–47% of earthquake injuries have been cited as aftermath injuries (Milsten, 2000).

For instance, two catastrophic earthquakes had struck Haiti on January 12, 2010, devastating much of the country's capital city, Port-au-Prince, and surrounding regions. Most of the fatalities were due to building collapses, which were in turn influenced by structures built on unstable land (International Strategy for Disaster Reduction, 2010). Estimates by the UN in late January suggest that in excess of 80,000 were killed as a consequence of this earthquake and 200,000 injured (United Nations, 2010).

Earthquakes can also trigger other types of natural hazards, which have different patterns of injury (e.g., tsunamis, landslides, floods) (Jones, 2006). For example, the Indian Ocean Tsunami that hit several countries across southeast Asia on December 26, 2004, was triggered by an earthquake (9.3 on the Richter scale). This event is considered to be the worst natural disaster in the last decade. It resulted in 226,408 deaths in 12 countries, with injuries estimated to be in hundreds of thousands, affecting more than two million people (ISDR, 2010; Guhar-Sapir et al., 2006).

Tsunami-related injuries most frequently occur due to exposure to the extreme water forces and pressures of a tsunami, as well as oxygen deprivation, chemical reactions due to contaminants, impact from debris, and flood-related fire consequences (Guhar-Sapir et al., 2006). Suction of debris back out to sea through receding waters can also cause injury (Guhar-Sapir et al., 2006). Drowning is the most common type of injury sustained during a tsunami, but many fatalities are also caused by respiratory complications from episodes of near drowning. Traumatic injuries, contusions, open wounds, fractures, head injuries, and compression barotraumas of the tympanic membrane are common (Guhar-Sapir and van Panhuis, 2005; Fan, 2006). In Thailand, the pattern of injuries that was described following the Indian Ocean Tsunami was of small-medium multiple injuries along the head, face, and extremities, as well as the back of head, back, buttocks, and legs (Guhar-Sapir and van Panhuis, 2005).

Volcanoes
Between 1974 and 2003, 123 volcanic disasters were reported, resulting in deaths of 25,703 people, and affecting in excess of three million people (Guha-Sapir et al., 2004). The worst volcano in modern history occurred in Colombia in 1985 with the eruption of Volcano del Ruiz, which killed some 21,800 people (Guha-Sapir et al., 2004). Injuries resulting from volcanoes are influenced by four factors: (1) eruptive variables (explosive: large quantities of gas, hot ash, and dust; effusive: large lava flows; combination), which influence the duration and chemical composition of emissions, and dispersal range; (2) toxin-specific properties; (3) patterns of toxic dispersal and persistence; and (4) biological variables (Weinstein and Cook, 2005). Injuries that occur close to the site of volcanic eruption can occur due to the explosion: burns (internal and external), trauma, lacerations, asphyxiation; or due to the emission of toxic gases (asphyxiation, airway constriction, burns; ocular injuries, upper airway, and skin irritations) (Weinstein and Cook, 2005; Weinstein and Patel, 1997).

Toxic elements (e.g., Sulfur, fluoride, chlorine, carbon, silica, mercury) and compounds can be ejected to significant distances, and consequently volcanic injuries often occur well after the initial explosion (Weinstein and Cook, 2005). Injuries can occur due to resulting electrical storms, reduced visibility, water supplies contaminated with toxic substances, and air/road crashes (Jones, 2006). Other common injuries in this category are: eye/skin/airway irritations, ocular injuries (foreign bodies in eyes, corneal abrasions), and sometimes suffocation (Jones, 2006; Weinstein and Cook, 2005).

Meteorological
From 1974 to 2003, 1955 windstorm disasters (cyclones, hurricanes, tornadoes) were recorded, resulting in 293,758 fatalities and affecting an additional 557 million people (Guha-Sapir et al., 2004). Cyclone Nargis, which struck Myanmar in Burma on May 2, 2008, has been identified as the second worst natural disaster in the last decade (ISDR, 2010). The cyclone resulted in the deaths of 138,366 people, but affected many more – 19,359 were injured, and approximately 2.4 million people were severely affected (Kim et al., 2010; World Health Organization, 2008).

Injuries from windstorm events are often classified according to the disaster phase during which they occurred (Shulz et al., 2005): pre-event; event; post-event. Falls, blunt trauma, lacerations, and muscle strains are common during the pre-event phase, as people prepare their properties and communities for the destructive winds (Shulz et al., 2005). Injuries associated with evacuation also incurred during this phase, including road traffic crashes (Shulz et al., 2005; Jones, 2006).

During windstorm events, individuals are at risk of injury from the direct exposure of the forces of the event (building collapse, flying debris, falling trees, power lines, etc.) (Shulz et al., 2005; Cook et al., 2008). The three most common injuries arising from such events are lacerations, blunt trauma, and puncture wounds, the majority of which are sustained to the lower extremities (Noji, 1993). For example, over half of the disaster-related injuries that occurred after Hurricane Iniki in 1997 were open wounds

(Hendrickson and Vogt, 1996; Hendrickson et al., 1997). Other common injuries experienced during windstorm events include: asphyxiation, abdominal injuries, spinal injuries, abrasions, contusions, sprains, fractures, ocular injuries, crush syndrome, carbon monoxide poisoning, ear/nose/throat injury, burns, and electrocution (Milsten, 2000; Shulz et al., 2005; Jones, 2006; Noji, 1993).

Injuries also occur due to storm surges, which can raise coastal waters many meters above normal tide level, and heavy rainfall, which can result in flooding (Cook et al., 2008). The principal injuries reported after such events include lacerations, blunt trauma, puncture wounds (often in the feet and lower extremities), and drowning (Cook et al., 2008).

Typically, the pattern of injury presentations changes in the aftermath (and associated cleanup) of wind storm events, and these injuries are often greater in number than injuries sustained during the event (Milsten, 2000). Electrocutions due to powerlines present a problem during this phase, but injuries such as lacerations, puncture wounds, abrasions, contusions, fractures, strains/sprains, insect stings, dog bites, and dermatitis are commonly reported as a consequence of cleaning up activities involving chainsaws, falls from heights, disturbing nests, etc. (Milsten, 2000; Shulz et al., 2005; Jones, 2006). Also common are burns (from using alternative light/heat sources such as candles, open fires, portable stoves, etc.) (Shulz et al., 2005). Injuries related to suicide attempts also occur during this phase (Jones, 2006).

Floods

An estimated 206,303 fatalities occurred as a consequence of 2,553 flood disasters between 1974 and 2003, affecting more than 2.6 billion people (Guha-Sapir et al., 2004). Whereas earthquakes have been responsible for the most natural disaster mortality, floods have affected the most number of people (ISDR, 2010). Of the estimated two billion people affected by natural disasters of any kind in the last decade, 44% were affected by floods (ISDR, 2010). Floods account for approximately 40% of all natural disasters, and importantly, can occur as a consequence of several other natural disasters (volcanoes, earthquakes, wind events, and tsunamis) (Jones, 2006). Much of the projected impact of future natural disasters is likely to occur in coastal areas, due to rising sea levels that will place these regions at increased risk of storm surges and flooding (Ahern et al. 2005; Dasgupta et al., 2007; 2009; Rodriguez et al., 2009).

The most common flood injuries are drowning, near drowning, and being hit by objects in fast flowing water (Jones, 2006; Ahern et al. 2005). A significant proportion of drowning/near drowning episodes are caused by vehicles being swept away (Milsten, 2000; Ahern et al. 2005). Hypothermia as a consequence of near drowning episodes is common (Jones, 2006). In a review of flood-related injuries, the three most common injury types were identified as: sprains/strains (34%), lacerations (24%), other injuries (11%), and abrasions/contusions (11%) (Ahern et al. 2005). Other types of injuries include multiple traumas, contusions, and minor cuts (Jones, 2006).

Suicides
This entry focuses on acute injuries experienced as a consequence of natural hazards. However, it is acknowledged that depression and suicides are commonly experienced after natural disasters – most typically after hurricanes, floods, and earthquakes (Ahern et al. 2005; Krug et al., 1998), with increased rates of both for up to 4 years post-disaster (Krug et al., 1999; Galea et al., 2005; Procter, 2005).

Long-term sequelae for casualties

Although most injuries that arise from natural disasters are specific and non-disabling, recovery for many individuals is challenging. Brain injury, amputation, or paralysis may require prolonged rehabilitation and institutional care (Pan American Health Organization and Pan American Sanitary Bureau, 2000). Orthopedic services are often limited in less developed countries, as are options for postsurgical management, such as fitting of prostheses, physical and occupational therapies, and other pathways for remobilization and return to daily activities (Dhar et al., 2007; Calder and Mannion, 2005). One year after the Gujarat earthquake of 2001, which killed 13,805 people and left 166,000 injured, many thousands still required assistance for paraplegia, poorly healed fractures, amputations, and other mobility problems (Chatterjee, 2002). Organ damage may also require long-term management, such as dialysis after renal crush injuries. Following the Armenian earthquake in 1988, the medical needs of 600 cases of acute renal failure – of which at least 225 victims required dialysis – created a second catastrophe described as the "renal disaster" (Sever et al., 2006).

Management to minimize casualties

Accurately estimating the impact of natural hazards in terms of fatalities, injuries sustained, and the long-term physical, psychological, social, and economic impacts can be difficult. There is no one agency that is responsible for collecting reliable, valid disaster data (current data sources include newspapers, insurance reports, government agencies, and humanitarian agencies) (Guha-Sapir et al., 2004). There is no standardized method for assessing damage (definitions, data collection methods), verifying information, and storing data (Guha-Sapir and Below, 2002). This is compounded by difficulties associated with obtaining data on populations affected by disasters (e.g., population size; geographical boundaries). Accurate data are essential to estimate the impact of the event, and for effective disaster management, and disaster preparedness.

Despite such limitations in disaster data, a sound public health approach can still be adopted to minimize casualties from natural disasters. Such a public health approach to injury control is based on a four-stage process that includes: defining the nature and extent of the problem; identifying associated risk and protective factors; developing effective interventions; and implementing these interventions in effective programs (Sleet et al., 1998). In the traditional injury epidemiology framework, the risk factors for injury relate to host, agent, and environment (Kraus and Roberston, 1992). Natural disaster–related injuries can be examined within this context (Ramirez and Peek-Asa, 2005). Host characteristics include demographics (age, gender, etc.), individual behaviors (running out of building, heeding evacuation warnings), and resiliency. In the natural disaster model, the agent is the energy (e.g., force of wind, magnitude of earthquake). The environment in this context is the physical location, including buildings, roads, and infrastructure where the natural disaster occurs. From the public health perspective, points of intervention most likely to reduce harm arising from natural disasters should focus on host and environment characteristics. Consequences of natural disasters can be direct or indirect (Combs et al., 1998). Direct casualties are those that occur due to the physical forces associated with the event, and indirect casualties are those that occur due to unsafe or unhealthy conditions that exist in the post-disaster phase. In public health terms, indirect consequences should be the target for intervention, as it is these factors that can be altered in the preparedness phase, and through effective disaster management.

Summary/conclusion

Over five billion natural disaster events have occurred in the last three decades, yielding more than two million deaths, and affecting more than 5.1 billion people. The incidence of natural disasters has increased significantly over the last 100 years. The World Climate Change Conference recognized that during the last five decades, nine out of ten natural disasters were the result of extreme weather and climate events (World Meteorological Organization (WMO), 2009). Further, climate change models demonstrate that there will be an increase in the frequency and intensity of extreme natural hazards such as heat waves, storms, floods, wildfires, and droughts (World Meteorological Organization (WMO), 2009; IPCC, 2001; Haines and Patz, 2005). Death and injury are direct consequences of natural disasters. Although there has been a reduction in fatalities from these events, many of the survivors experience injuries and/or illness as a consequence, some of which require long-term rehabilitation, and impact on quality of life. This has important implications for public health. Many factors influence the severity of the impact of a natural disaster, including the density, geographical location, and infrastructure and disaster response capacity. The impact of a natural disaster is usually greatest in less developed regions. The disaster type itself affects the severity of the impact, and different natural hazards yield different patterns of injury. We anticipate that the increased incidence and intensity of extreme natural hazards will be reflected in changing epidemiology of disaster-related injuries (e.g., more floods will result in more drowning/near drowning). Information about any natural disaster and the devastation it brings is limited by the quality of the data relating to the event. A unified system of definitions, data collection methods, verification, and storage will significantly improve disaster preparedness, management, and recovery.

Bibliography

Ahern, M., Kovat, S., Wilkinson, P., Few, R., and Matthies, F., 2005. Global health impacts of floods: epidemiologic evidence. *Epidemiologic Reviews*, **27**, 36–46.

Calder, J., and Mannion, S., 2005. Orthopaedics in Sri Lanka posttsunami. *Journal of Bone and Joint Surgery*, **87**(6), 759–761.

Centre for Research on the Epidemiology of Disaster (CRED), 2010. Haiti Earthquake Brief. http://cred.be/sites/default/files/Haiti_Earthquake_Brief. Accessed on January 28, 2010.

Chang, C. C., Lin, Y.-P., Chen, H.-H., Chang, T.-Y., Cheng, T.-J., and Chen, L.-S., 2003. A population-based study on the immediate and prolonged effects of the 1999 Taiwan earthquake on mortality. *Annals of Epidemiogloy*, **13**, 502–508.

Chatterjee, P., 2002. One year after the Gujarat earthquake. *Lancet*, **359**(9303), 327.

Combs, D. L., Quenemoen, L. E., Parrish, R. G., et al., 1998. Assessing disaster-attributed mortality: development and application of a definition and classification matrix. *International Journal of Epidemiology*, **28**, 1124–1129.

Cook, A., Watson, J., van Buynder, P., Robertson, A., and Weinstein, P., 2008. 10th anniversary review: natural disasters and their long-term impacts on the health of communities. *Journal of Environmental Monitoring*, **10**, 167–175.

Dasgupta, S., Laplante, B., Meisner, C., Wheeler, D., and Yan, J., 2007. The impact of sea level rise on developing countries: a comparative analysis. *World Bank Policy Research Working Paper 4136*, February 2007. http://www-wds.worldbank.org/external/default/WDSContentServer/IW3P/IB/2007/02/09/000016406_20070209161430/Rendered/PDF/wps4136.pdf. Accessed on January 28, 2010.

Dasgupta, S., Laplante, B., Murray, S., Wheeler, D., 2009. Sea-level rise and storm surges: a comparative analysis of impacts in developing countries. *World Bank Policy Research Working Paper, no. WPS 4901*.

Dhar, S. A., Halwai, M. A., Mir, M. R., Wani, Z. A., Butt, M. F., Bhat, M. I., and Hamid, A., 2007. The Kashmir earthquake experience. *European Journal of Trauma Emergency Surgery*, **33**(1), 74–80.

Dominic, F., Levy, J., and Louis, T., 2005. Methodological challenges and contributions in disaster epidemiology. *Epidemiologic Reviews*, **27**, 9–12.

Driscoll, T., Harrison, J., and Langley, J., 2004. The burden of injury. In McClure, R., Stevenson, M., and McEvoy, S. (eds.), *The Scientific Basis of Injury Prevention and Control*. Melbourne: IP Communications.

Fan, S. W., 2006. Clinical cases seen in tsunami-hit Banda Aceh – from a primary healthcare perspective. *Annals of the Academy of Medicine, Singapore*, **35**(1), 54–56.

Friedman, E., 1994. Coping with calamity: how well does health care disaster planning work. *Journal of the American Medical Association*, **272**(23), 1875–1879.

Galea, S., Nandi, A., and Vlahov, D., 2005. The epidemiology of post-traumatic stress disorder after disasters. *Epidemiologic Reviews*, **27**, 78–91.

Guhar-Sapir, D., and van Panhuis, W. G., 2005. *The Andaman Nicobar Earthquake and Tsunami 2004*. Belgium: Centre for Research on the Epidemiology of Disasters (CRED).

Guhar-Sapir, D., Parry, L. V., Degomme, O., Joshi, P. C., and Saulina-Arnold, J. P., 2006. *Risk Factors for Mortality and Injury: Post-Tsunami Epidemiological Findings from Tamil Nadu*. Belgium: Centre for Research on the Epidemiology of Disasters (CRED).

Guha-Sapir, D., and Below, R., 2002. The quality and accuracy of disaster data a comparative analyses of three global data sets. Who center for research on the epidemiology of disasters, for the prevention consortium. The Disaster Management Facility, World Bank.

Guha-Sapir, D., Hargitt, D., and Hoyois, P., 2004. *Thirty Years of Natural Disasters 1974–2003: The Numbers*. Louvain: Centre for Research of the Epidemiology of Disasters. UCL Presses.

Haines, A., and Patz, J., 2005. Health effects of climate change. *Journal of the American Medical Association*, **291**(1), 99–103.

Hendrickson, L. A., and Vogt, R. L., 1996. Mortality of Kauai residents in the 12-month period following Hurricane Iniki. *American Journal of Epidemiology*, **144**(2), 188–191.

Hendrickson, L. A., Vogt, R. L., Goebert, D., and Pon, E., 1997. Morbidity on Kauai before and after Hurricane Iniki. *Preventive Medicine*, **26**(5 Pt 1), 711–716.

IPCC (Intergovernmental Panel on Climate Change), 2001. Synthesis report (ed. Watson, T. R., and Core Writing Team) *Contribution of Working Groups I, II, and III to the Third Assessment Report of the Intergovernmental Panel on Climate Change*. Cambridge University Press, Cambridge.

International Strategy for Disaster Reduction, 2010. UNISDR calls for long-term measures to rebuild a safer Haiti, January 22, 2010. http://www.unisdr.org/news/v.php?id=12398. Accessed on January 28, 2010.

ISDR, 2010. Earthquakes caused the deadliest disasters in the past decade. January 28, 2010. http://www.unisdr.org/news/v.php?id=12470. Accessed January 30, 2010.

Jones, J., 2006. Mother nature's disasters and their health effects: a literature review. *Nursing Forum*, **41**(2), 78–87.

Kim, H., Han, S. B., Ji Hye, K., Kim, J. S., and Hong, E. S., 2010. Post-Nargis medical care: experience of a Korean Disaster Relief Team in Myanmar after the cyclone. *European Journal of Emergency Medicine*, **2010**(17), 37–41.

Kraus, J. F., and Roberston, L. S., 1992. Injuries and Public Health. In Last, J. M., and Wallace, R. B. (eds.), *Public Health and Preventive Medicine*. Connecticut: Appleton and Lance, pp. 1021–1034.

Krug, E. G., Kresnow, M. J., Peddicord, J. P., et al., 1998. Suicide after natural disasters. *The New England Journal of Medicine*, **338**, 373–378.

Krug, E. G., Kresnow, M. J., Peddicord, J. P., et al., 1999. Retraction: suicide after natural disasters. *The New England Journal of Medicine*, **340**, 148–149.

Milsten, A., 2000. Hospital responses to acute-onset disasters: a review. *Prehospital and Disaster Medicine*, **15**(1), 32–40.

National Committee for Injury Prevention and Control, 1989. *Injury Prevention: Meeting the challenge*. New York: Oxford University Press.

Noji, E., 1992. Disaster epidemiology: challenges for public health action. *Journal of Public Health Policy*, **13**(3), 332–340.

Noji, E. K., 1993. Analysis of medical needs during disasters caused by tropical cyclones: anticipated injury patterns. *The Journal of Tropical Medicine and Hygiene*, **96**, 370–376.

Pan American Health Organization and Pan American Sanitary Bureau, Natural disasters: protecting the public's health, 2000, Washington, DC: Pan American Health Organization, Pan American Sanitary Bureau, Regional Office of the World Health Organization.

Peek-Asa, C., Ramirez, M., Selingson, H., and Shoaf, K., 2003. Seismic, structural and individual risk factors associated with earthquake-related injury. *Injury Prevention*, **9**, 62–66.

Procter, N., 2005. Tsunami waves of mental trauma. *Contemporary Nurse*, **18**, 215–218.

Ramirez, A., and Peek-Asa, C., 2005. Epidemiology of traumatic injuries from earthquakes. *Epidemiologic Reviews*, **27**, 47–55.

Redmond, A. D., 2005. Natural disasters. *British Medical Journal*, **330**(7502), 1259–1261.

Rodriguez, J., Vos, F., Below, R., and Guha-Sapir, D., 2009. *Annual Disaster Statistical Review 2008 the Numbers and Trends*. Belgium: Centre for Research on the Epidemiology of Disasters (CRED).

Scheuren, J.-M., Waroux, O. L., Below, R., Guha-Sapir, D., and Ponserre, S., 2008. *Annual Disaster Statistical Review: The Numbers and Trends 2007*. Belgium: Center for Research on the Epidemiology of Disasters (CRED): Jacoffset Printers.

Schnitzer, J. J., and Briggs, S. M., 2004. Earthquake relief—the US medical response in Bam, Iran. *The New England Journal of Medicine*, **350**(12), 1174–1176.

Sever, M. S., Vanholder, R., and Lameire, N., 2006. Management of crush related injuries after disasters. *The New England Journal of Medicine*, **354**(10), 1052–1063.

Shulz, J., Russell, H., and Espinel, Z., 2005. Epidemiology of tropical cyclones: the dynamics of disaster, disease, and development. *Epidemiologic Reviews*, **27**, 21–35.

Sleet, D., Bonzo, S., and Branche, C., 1998. An overview of the national center for injury prevention and control at the centers for disease control and prevention. *Injury Prevention*, **4**, 308–312.

United Nations, (2010). General Assembly GA/10913. Department of Public Information • News and Media Division • New York, January 22, 2010. http://www.un.org/News/Press/docs/2010/ga10913.doc.htm. Accessed January 29, 2010.

Weinstein, P., and Cook, A., 2005. Volcanic emissions and health. In Selinus, O., Alloway, B., Centeno, J., et al. (eds.), *Essentials of Medical Geology: Impacts of the Natural Environment on Public Health*. China: Elsevier.

Weinstein, P., and Patel, A., 1997. The Mount Ruapehu eruption, 1996: a review of potential. *Australian and New Zealand Journal of Public Health*, **21**(7), 773–777.

World Health Organization, 2008. *Post-Nargis Joint Assessment Report*. Geneva: WHO.

World Meteorological Organization (WMO), United Nations International Strategy for Disaster Reduction (UN/ISDR) and other international partners, 2009. Fact sheet #1. Climate information for reducing disaster risk www.wmo.int/wcc3 September 2009. Accessed on January 30, 2010.

Cross-references

Asteroid Impact
Avalanche
Building Failure
Coping Capacity
Damage and the Built Environment
Disaster
Earthquake
Flood Hazard
Geological/Geophysical Disasters
Hurricane
Landslide
Natural Hazard
Volcanoes

CHALLENGES TO AGRICULTURE

Julie A. March
United States Agency for International Development (USAID), Washington, DC, USA

Definitions

Food Security: When all people at all times have both physical and economic access to sufficient food to meet their dietary needs for a productive and healthy life (USAID, 1992).

Introduction

Worldwide, farmers rely on a complex combination of seed, water, sunlight, and soil nutrients to assure a season of good production. Natural hazards can disrupt the unique balance necessary for a successful harvest. In cases where farmers are able to control for some of these factors (for example, irrigation to compensate for lack of rain or fertilizer to enhance poor soil fertility), negative effects leading to a poor harvest can be mitigated. Subsistence farmers are often more susceptible to the risks associated with natural hazards as they have limited access to costly inputs to mitigate natural disasters. In some cases though, mitigation measures are not an option for any farmers, especially if the hazard is not predicted. Hazards include events such as storm surges, volcanic eruptions, droughts, and floods. The impact of natural hazards on agricultural production can be evident immediately following a disaster and recovery can take many years. Both subsistence agriculture and commercial agriculture can be damaged by extreme events. While some small-scale farmers may lose their seed stocks and their food stores for the coming season, commercial farmers may face disruption or destruction of local market and market chains for their crops and for access to agricultural inputs. Widespread damage to a region or a particular crop can affect the price of agricultural products in international markets. The extent of the damage caused to agricultural systems will depend on a variety of factors, including topography, weather, crop selection, and stage of crop growth when the hazard strikes. The speed of recovery for affected farmers will be influenced by all of those factors as well as the general resilience of the farming population.

Rainfall irregularity and drought

Many parts of the world have experienced great climatic variability over the past few decades. In parts of Africa, which depend primarily on rain-fed agriculture, these changes have had a negative impact on food security of subsistence farmers. Droughts can both decrease food security in the near term, whereas in the longer term, successive droughts can erode the ability of farmers and pastoralists to recover as recurrent shocks lead to loss of assets and erosion of coping capacity for vulnerable populations. Throughout Africa, drought has contributed greatly to large magnitude food security crises. Some examples include the famine in Ethiopia (1984), and the food insecurity in Niger (2005). These countries and many others affected by drought are still struggling with food insecurity, highlighting the need to address emergency needs related to current hazards while at the same time examining the agricultural system, farming methods, and underdevelopment related to the agriculture sector as a whole with sustainability of the system as a major objective (Trench et al., 2007).

Subsistence farmers who depend on rain-fed agriculture are often challenged by the inability to determine when rains will begin. For example, in Southern Sudan in 2010, rains began later than anticipated (WFP CFSAM, 2010). Farmers quickly planted once rains began, yet subsequently they lost the seed when the rain stopped soon after, leading to crop failure. With a shortened planting season, a short cycle crop variety may have produced a decent harvest, yet it is difficult to anticipate this prior to planting. Deciding what varieties to plant is difficult for farmers who lack reliable information on what the rainy season may bring, when it is likely to arrive, and how long it will last. Even with knowledge of weather predictions, access to or availability of preferred seed varieties to respond to the altered weather patterns can be limited.

Choosing alternative crops, irrigation, soil management, and improved information and early warning are common approaches to mitigating drought effects, yet these strategies are not simple to implement. Crop and varietal preferences develop over many years and are reflective of cultural preferences – complicating efforts to simply trade out crops for drought-resistant alternatives. For example, although orange fleshed sweet potato is vitamin rich and can withstand periods of drought, The International Potato Center (CIP) is working diligently to breed varieties that meet consumer preferences and that can compete in local markets with less resilient but preferred varieties (http://cipotato.org/research/sweetpotato-in-africa). Irrigation can be costly for automatic pump models, and labor intensive for human powered models. Additionally, if there is a drought, there may already be competing needs for water resources. Soil and watershed management for improved moisture retention can enhance water availability but they are longer-term programs that are not suited for the short time frame of many emergency programs. Early warning systems such as the Famine Early Warning System Network (FEWS NET) (www.fews.net) can be used to understand food security and weather trends by sharing information on rains, planting, market data, and general climate trends such as the presence of an El Niño or La Niña phenomenon. Ideally, by using early warning information of low rainfall or soil moisture, farmers can be proactive in selecting and implementing their mitigation strategies.

The views expressed in this chapter are those of the author and do not necessarily reflect the views of the United States Agency for International Development or the US Government.

While there are many easily predicted effects of irregular rainfall, such as crop failure and food insecurity, there can also be social effects of reduced water availability in the form of conflict over scarce resources. Climate variability can increase friction between different livelihood groups. For example, in parts of West Africa where farmers and pastoralists have established a system of resource sharing over time which is mutually beneficial, scarcity of water can stress this relationship. When the relationship is optimally beneficial, the pastoralists arrive with their animals in their seasonal migration just after the farmers harvest their crops. In exchange for the benefit the animal manure contributes to their fields, the farmers allow animals to graze on the stover from the harvested crops. In years where water and fodder are scarce, pastoralists begin moving early, and may even arrive prior to harvest. This can lead to land and resource conflict between the farmers and the pastoralists. Competition for scarce water resources and the potential for the livestock to consume the not-yet harvested crops heightens tensions, as do larger issues related to land tenure and resources (Shettima and Tar, 2008).

Water events: storm surges, floods, tsunamis

Agricultural areas bordered by rivers or oceans are at risk when weather and hazards bring too much water too quickly to be utilized by crop production or inundate areas with water that is not suitable for crop production, such as saline water.

Farmers in southern Africa are regularly challenged with growing conditions that include long periods of limited rainfall or drought, followed by inundation with rain leading to *floods*, an overflow that comes from a river or other body of water and causes damage, or any relatively high stream flow overtopping the natural or artificial banks in any reach of a stream (http://ks.water.usgs.gov/waterwatch/flood/definition.html). Floods within the Zambezi River basin have become so common as to be almost an annual event. These floods regularly claim lives and submerge crops and assets, reducing food security and resiliency to future droughts. In response, humanitarian agencies are promoting a combination of early warning and early action. Early warning against floods has proven very effective and potentially reduces the loss of human lives (http://www.usaid.gov/our_work/humanitarian_assistance/disaster_assistance/publications/prep_mit/files/fy2012/mozambique_pounds_of_prevention.pdf). Early action involves developing response and mitigation plans with local communities.

Not all floods adversely affect food security. In many riverine areas, seasonal flooding can bring much needed soil moisture, nutrients, and organic material to the banks. As the water subsides, farmers then plant on the banks of the river, taking advantage of the extra soil moisture. This recessional planting can provide an additional short season for crop production.

Storm surges involve a rise in sea level due to a hurricane or similarly intense storm. The increase in water level over the normal tide then combines with wind and waves and finally, water is forced ashore, and may proceed to infiltrate agricultural areas. December 2008 brought one such storm surge to the Federated States of Micronesia (FSM) and resulted in sea water washing over taro fields on several of the islands. Taro is a starchy tuber and a major food security staple for FSM islanders. Salt water can inflict varying degrees of damage on the taro patches depending on the length of time the water stands on the patch and the timing and duration of rainfall afterward to flush out the salt. Damage can result in total loss where the taro rots in the ground (Figure 1). Storm surges can sometimes be anticipated but island nations often have limited area to use for agricultural production. Some mitigation strategies include moving agricultural production to higher ground where possible, planting reserve plots of taro seedlings to ensure healthy planting material, planting in concrete beds or forming other barriers to prevent salt water intrusion, and diversifying crop production. Many months can pass on the island without rains which in turn does not allow recharging of the groundwater supplies, making the thin freshwater lens vulnerable to contamination by salt water. Storm surges can also hamper agricultural production by displacement of the population, destruction of crops or agricultural land through erosion or salinization, or destruction of infrastructure (docking areas, bridges, boats).

Tsunamis can wreak havoc on agricultural systems. Unlike storm surges, tsunamis are generally not caused by surface weather but rather, by earthquakes, submarine landslides, volcanic eruptions, explosions, or meteorites. The Indian Ocean Tsunami which sent a wall of water to Aceh Indonesia is estimated by FAO to have caused damage to more than 61,000 ha of agricultural land. The most common damages to agricultural land reported for this event were: "(a) Crop destruction by waves, salt poisoning, and uprooting; (b) de-surfacing of landscape as a result of erosion and sedimentation; (c) deposition of salt sediment; (d) trash and debris accumulation; (e) salt infiltration; and (f) fertility depletion." (FAO, 2005a). Ample rainfall washed away much of the surface salt in the weeks and months following. What remains is a high concentration of salts in layers of clay and silt that were deposited during the event. These layers are fairly impermeable to water, making the removal of salt through leaching when rainwater passes through, very slow.

Volcanic eruptions

In many countries agricultural production takes place in the shadow of quiescent and active volcanoes. Volcanic eruptions can lead to the displacement of populations due to the threat of various volcanic hazards, including ash fall. Displacement can last until the immediate threat of an eruption has passed, or it can persist until infrastructure and services that were damaged by the eruption are

Challenges to Agriculture, Figure 1 Rotten taro root following a storm water surge.

repaired. Damage to roads and infrastructure can affect future market access for agricultural products and inputs.

Early warning and monitoring of potential volcanic activity can help farmers take some mitigative actions, such as moving animals or choosing alternative locations for planting. Once ash falls, irrigation to settle the ash, as well as mixing the ash into the soil is a key rehabilitation strategy to aid in topsoil development. Other methods such as selecting appropriate varieties and adding lime to modify soil acidity can help reduce the negative impacts on production (http://www.maf.govt.nz/environment-natural-resources/funding-programmes/natural-disaster-recovery/volcanic-eruptions).

In addition to lava flows, volcanic activity can yield lahars, a moving fluid mass composed of volcanic debris and water, (e.g., 1993 Mt. Pinatubo eruption) and pyroclastic flows – a surface-hugging cloud of very hot gas and volcanic particles that moves rapidly across the ground surface, (http://www.geonet.org.nz/volcano/glossary.html), as well as volcanic ash falling for many months. All of these can cause extensive agricultural destruction. The effect of ash fall on agriculture and livestock can be significant and depends primarily on thickness of the ash cover, composition of the ash (the presence of soluble fluoride), weather following the eruption, and availability of feed and water for livestock.

The thickness of the ash fall can largely determine whether soil will be completely deprived of oxygen and "sterilized" or not. Ash fall thicker than 10–15 cm typically results in a complete burial of soils (Folsom, 1986). Chances of plant survival after ash fall can be improved if rain follows within 2–3 days of an eruption as the rain will wash ash from plants, compact the thickness of the ash fall, and facilitate recovery. Complete burial for several days often results in the death of the plants. Because ash composition and ash pH varies between volcanoes, the effects of ash mixing into the soil cannot be predicted. In some cases, soils will have a pH post eruption that no longer supports the crops which were previously grown. In addition to causing crop loss, livestock loss can be high when available water and fodder resources are contaminated with ash. This is especially true if fluorine is present in the ash in high concentrations, causing fluorine poisoning and death. Interestingly, the majority of livestock deaths following a major eruption are due to starvation (Wilson et al., 2011). Provision of emergency fodder and feed from unaffected areas might be a strategy for maintaining herds after an eruption.

Ash can also affect agricultural production by changing the amount of sunlight hours, altering soil properties, or damaging leaves or other crop parts. Finally, when rain mixes with volcanic gas, there is the potential to produce acid rain, which is also detrimental to crop production.

Plant pests and diseases

Some hazards directly target the crops being produced. Two examples of this type of hazard include crop pests (e.g., insects) and plant diseases. Crop pests are responsible for tremendous amounts of crop loss both in the field and during storage (post-harvest.) Monitoring of the occurrence and movement of both plant pests and diseases is an important step in controlling damage to agricultural production. In most cases, damage due to these two categories of hazard is not restricted to one farmer's field as pests do not respect property boundaries when multiple fields are planted in their preferred food source. Plant pests and diseases can wreak havoc at the local production level and potentially on an international scale unless effective control mechanisms are identified and utilized.

Some plant pests are confined at a household garden, farm or local level, restricted by available food sources, climate, and mobility. Then there are those such as the Desert Locusts, which have sufficient mobility to follow crop development and the weather pattern. Locusts travel in swarms which can vary in size from the small (hundreds of square meters) to enormous, covering 1,000 km^2. Desert locust swarms can damage 100% of the crop in a field where they land and they can fly hundreds to thousands of kilometers between their breeding sites (FAO EMPRES). The desert locust has made its way across Africa, Asia, the Middle East, and Europe and has been decimating crops and vegetation since biblical times. Control mechanisms include ground and aerial spraying with insecticides. This method must be done by trained personnel at a significant cost, and local population may not consume the locusts once sprayed. Other methods include digging trenches around fields to catch marching bands of nymphs or hoppers as they head in the direction of crops. Techniques have improved over the last decades. Emphasis on tracking the swarms increases efficiency of spraying programs, and supports a shift toward a combined approach of barrier spraying and use of less persistent and more environmentally friendly pesticides, including biological pesticides. The goal is to reach the gregarious locust populations before they reach their reproductive stage. Ideally, preventing gregarization would be the best control intervention, but this phenomenon often occurs in hard-to-reach areas. There is much support in the early warning sector for monitoring and identifying potential areas of outbreaks and subsequent invasions. Where possible, satellite imagery, field surveys, and monitoring are coordinated to predict and report the path of the swarms, providing advance notification to launch control interventions and where relief might be needed to meet food needs should the enormity of the pest invasion override control attempts.

Plant pathogens are organisms that cause a disease on a plant. As they spread and infect plants, they can significantly reduce yields. Major pathogenic outbreaks in the past include the fungus *Phytophthora infestans*, responsible for "potato blight" which culminated in the potato famine in Ireland (1845–1849). Potato blight was eventually controlled with a chemical mixture to kill the mold. More recent examples include cassava mosaic virus or cassava mosaic disease (CMD), affecting cassava crops in many African countries including Burundi, Uganda, and Democratic Republic of Congo. Cassava decimation is especially dangerous as it is a major food security staple for vulnerable populations; it is both drought resistant and able to remain in the ground for the duration of conflicts. CMD is currently managed by planting resistant varieties identified in the 1990s and, in many cases, distributed to vulnerable farmers through both emergency relief and development programs.

International agricultural research centers worldwide are challenged with new or modified pathogens. For instance, wheat stem rust can cause losses of 50% of a wheat harvest when conditions for its development are optimal. Losses of 100% are possible with susceptible cultivars. Although the Green revolution brought with it the identification of a gene with resistance to wheat stem rust, this gene was subsequently bred into most commonly grown wheat varieties over the past several decades, providing a single line of resistance against wheat rust. Then, in 1999, a strain of wheat rust arrived in Uganda (named Ug99). This strain was able to overcome the inbred resistance, attacking and decimating plants as it spread, windborne, through fields. This dispersal method facilitates spread to the many fields of wheat across the world, the vast majority of which carry no resistance to this new strain. The fungus has already made its way across Africa, Asia, and the Middle East and is particularly virulent, resulting in 100% crop loss. Scientists from more than 17 agricultural research centers and offices worldwide, including the Borlaug Global Rust Initiative, are currently committed to find new resistant varieties.

The ability of a pathogen to change and thereby render a single line of defense ineffective highlights the value of preservation of landrace and traditional crop varieties worldwide rather than conversion to one or two high performing varieties. Preservation of agro-biodiversity is a strategy for reducing the impact of pests and pathogens. When one variety is planted, a pathogen that is particularly virulent may be able to move quickly through the susceptible cultivar. When there is a wide diversity of cultivars, levels of resistance will differ, potentially slowing the spread and subsequent crop destruction. Additionally, planting multiple crops and varieties is a risk mitigating strategy for farmers against all of the hazards mentioned.

Looking forward and conclusions

Throughout history agricultural systems have faced challenges from natural hazards. Being able to anticipate hazards through monitoring or early warning systems may allow farmers to better prepare for and withstand disasters which affect agriculture. Farmers regularly employ a variety of mitigative strategies to avert disaster or enhance the speed with which they recover from disasters. These include diversification of their farms or plots and modification of planting methods to enhance sustainability and increase resiliency to hazards.

Diversification spreads risk in several ways. It provides more chances of crop survival against a particular threat. For example, if climate is not favorable for one type of crop, perhaps a different type of crop with different light, nutrient, and water requirements will survive, ensuring some measure of food security. On a global scale, crop diversity and a large genetic pool for any given crop can slow the spread of pathogens and pests. Being able to choose from a variety of characteristics provides the greatest chance to meet demands of climatic trends and pest and pathogen threats. The same is true for attack by pests or pathogens. Within crops, maintaining genetic diversity is tremendously important to long-term agricultural sustainability. Maintaining many local varieties of a given crop such as maize provides a ready supply of genetic material to better respond to evolving hazards. When only one variety is planted, the resistance to pathogens/climate stress/insect damage is limited to what is contained in that one variety. If that particular variety is susceptible to the hazard presented, a complete loss is possible.

In addition as world population growth continues and pressure on land, water, and soil resources intensifies, farmers and agricultural scientists are increasingly interested in methods such as conservation agriculture which maintain soil and water resources, and a more holistic approach to farming which considers crops as one component of the larger agricultural system. This may mitigate against many of the challenges presented by increased climate variability. The goal of this approach (ideally) is a more integrated and efficient system with less cost and waste and, hopefully, more production.

Bibliography

Bewley, M. 2009. The race against Ug99, AGWEEK (USA).
Blong, R. J., 1984. *Volcanic Hazards: A Sourcebook on the Effects of Eruptions*. Sydney: Academic, p. 424p.
Brouwer, C., and Heibloem, M. 1983, *Irrigation Water Management*. Training Manual (FAO), no. 6 Rome, Prov. ed, 60 p.
FAO, 2005a. Field Guide: 20 Things to Know About the Impact of Salt Water on Agricultural Land in Aceh Province.
FAO, 2005b. Report of the Regional Workshop on Salt-Affected Soils from Sea Water Intrusion: Strategies for Rehabilitation and Management. Bangkok: Food And Agriculture Organization of the United Nations, Regional Office for Asia and the Pacific, RAP publication 2005/11.
FAO/WFP Crop and Food Supply Assessment Mission to Southern Sudan, 2010. http://documents.wfp.org/stellent/groups/public/documents/ena/wfp217413.pdf.
Folsom, M. M. 1986. Tephra on range and forest lands of eastern Washington: local erosion and redeposition. In: *Mount St Helens: Five years later*. Cheney: Eastern Washington University Press, p. 116–119.
Francis X. 2009. *High Water in the Low Atolls*. Micronesian Counselor #76.
Koerner, B. 2010. Red menace: stop the Ug99 fungus before its spores bring starvation. *Wired Magazine*. http://www.wired.com/magazine/2010/02/ff_ug99_fungus/all/1.
Neild, J., O Flaherty, P., Hedley, P., and Underwood, R., 1998. Impact of a Volcanic Eruption on Agriculture and Forestry in New Zealand MAF Policy Technical Paper 99/2.
Shettima, A. G., and Tar, U., 2008. Farmer-pastoralist conflict in West Africa: exploring the causes and consequences. *Information, Society and Justice*, **1**(2), 163–184.
Showler, A., 2007. The desert locust in Africa and Western Asia: complexities of war, politics, perilous terrain, and development. In Radcliffe, E. B, Hutchison, W. D. and Cancelado, R. E. (eds.), *Radcliffe's IPM World Textbook*. St. Paul, MN: University of Minnesota. http://ipmworld.umn.edu.
Sivakumar, M. V. K., 1992. Climate change and implications for agriculture in Niger. *Climatic Change*, **20**, 297–312.
Trench, P., Rowley, J., Diarra, M., Sano, F., and Keita, B., 2007. *Beyond Any Drought*. London: International Institute for Environment and Development.
UN/ISDR, 2007. *Drought Risk Reduction Framework and Practices: Contributing to the Implementation of the Hyogo Framework for Action*. Geneva: United Nations secretariat of the International Strategy for Disaster Reduction (UN/ISDR), 98 + vi pp.
United States Geological Survey, 2009. Volcanic Ash – Effects on Agriculture and Mitigation Strategies. http://volcanoes.usgs.gov/ash/agric/.
USAID Policy Determination, 1992, Definition of Food Security. http://transition.usaid.gov/policy/ads/200/pd19.pdf.
Wilson, T. M., Cronin, S. J., Stewart, C., Cole, J. W., and Johnston, D. M., (2011). Impacts on agriculture following the 1991 eruption of Vulcan Hudson, Patagonia: lessons for recovery. *Natural Hazards*, **57**(2), 185–212. http://dx.doi.org/10.1007/s11069-010-9604-8.

Cross-references

Climate Change
Coping Capacity
Drought
Early Warning Systems
Flood Hazard and Disaster
Insects Hazards
Mitigation
Storm Surge
Tsunami
Volcanic Ash

CIVIL PROTECTION AND CRISIS MANAGEMENT

Scira Menoni[1], Antonio Pugliano[2]
[1]DIAP-Politecnico di Milano, Milan, Italy
[2]Lombardia Firemen Regional Headquarters, Milan, Italy

Definitions

Civil protection is a term used in several countries to indicate the institution that coordinates emergency and crisis management. This apparently simple definition hides in fact organizational complexities which in most cases stay beyond the comprehensive term "civil protection." The latter refers in several countries to a single agency which holds the responsibility of coordinating the many others which interact and intervene on the scene of a mass calamity, ranging from firemen to emergency medical doctors, to health-care departments, and several others. The coordination agency is generally lacking own resources and means while being in a strategic governmental position, close enough to the prime minister or to similar key political levels, so to have enough authority to take the lead of otherwise independent bodies and organizations.

At the European level, for example, the Community Mechanism for Civil Protection is in charge of activating aid and assistance whenever requested both inside and outside the Community's borders on a voluntary basis.

This means that member states activate their own resources to make part of a European international team or to assist another country in need of help, according to the subsidiarity principle. The latter refers to the fact that external communitarian assistance has to be asked in case national forces are overwhelmed by the crisis and cannot cope satisfactorily with their own means.

In Australia, Emergency Management (EMA) is a division of the Government Attorney General's Department, which pursues an "all agencies, all hazards" approach, with the aim of encouraging disaster preparedness, supporting states in developing their own emergency management policies and providing help in case of crises that overwhelm individual states' coping capability. Nevertheless, no national law clearly defines what the legally binding mandates of EMA are.

In Canada, in 2003, responsibilities for emergency management were assigned to Public Safety Canada, a department which coordinates other departments, through the Government Operations Center, constituting a "hub of a network of operation centers run by a variety of federal departments and agencies, including Health Canada, Foreign Affairs," the police, and others (see the website of Public Safety Canada in the references). Despite this apparently operationally centered goal, the Center holds also responsibilities regarding planning, mitigation, response, and recovery (see *Mitigation*; *Recovery and Reconstruction After Disaster*).

In the USA, FEMA (the Federal Emergency Management Agency) is in charge of coordinating the various activities necessary to face a federal emergency as well as to set guidelines, plans, and preparedness programs. Since its inclusion into the Department of Homeland Security, FEMA lost its direct contact with the White House, that is, its crucial key position that had permitted in the past the fast deployment of forces and resources. Perrow (2007) describes rather clearly the severe shortcomings and difficulties of the newly created department, as mentioned also by several observers on the occasion of the Katrina disaster. An interesting comparison among different emergency management models around the world can be found in the Fema website (see references below).

The term "civil protection" is sometimes used also to indicate in a general, comprehensive way the entire set of organizations, agencies, and forces intervening in a disaster. Following this philosophy, the public itself is part of civil protection for a number of reasons. First because peoples' coping capacity (see *Coping Capacity*) is deemed to be important in enacting self-protection. The active role the public may play in a crisis is then fully recognized and encouraged instead of condemning it to the passive role of a spectator, defying the willing to react. Second because many times laypeople are the first respondents: it is well known, for example, that in the immediate aftermath of earthquakes, those who try to rescue relatives and friends under the debris are the same escaped victims. Last but not least, the public intervenes in the form of associations of volunteers that range from rather professionalized bodies (like volunteer firemen or members of the Red Cross and International Red Crescent Movement) and NGOs to individuals who participate in various forms to emergency response (including the more recent movement of volunteers of the technical community providing free web services as described in Harvard Humanitarian Initiative, 2011).

However the term is intended, what clearly emerges is the complexity and articulation of any institutional and organizational form of emergency crisis management. The latter being an activity that significantly challenges several of the mentioned agencies, particularly those that do not tackle emergencies on an everyday basis, as will be discussed in the following part of the text.

A last consideration regarding civil protection reflects upon the boundaries of its activity. Among the functions that are generally attributed to the civil protection, when the latter identifies a specific agency or organization, besides crisis management, prevention and mitigation are contemplated as well. Problems arise when the latter must be clearly defined. In fact, the civil protection does not have an ordinary budget specifically allocated for the structural and nonstructural measures necessary for achieving risk reduction in the short and long term. Therefore, the idea of mitigation is rather broadly used to encompass the need for risk assessment and mapping as well as training and risk communication. Still, the boundaries remain somehow vague, leaving room for controversies and institutional overlapping.

Crisis management is the set of activities aimed at facing a complex, unexpected situation originated by an accident, a war, a terrorist attack, or a natural calamity. The two words actually represent almost an oxymoron, as by definition crises are characterized by high levels of uncertainty, disruption of normal life, chaotic environment, that make them hard to "manage." Nevertheless, there are better strategies than others to cope with crises, to respond to the challenges they pose, eventually to exit them in ways that not only permit a return to normalcy but exhibit high levels of resilience (see *Resilience*). In this respect, the term management refers to a set of general rules, deriving from past experience and understanding of how complex organizations behave under severe stress, that deserve to be analyzed by those whose responsibility is to provide help, rescue, and aid during emergencies.

The word "crisis" derives from the Greek verb "krino" which means "to judge." In fact, the most crucial thing during a "crisis" is the ability to judge the situation, estimate available resources, and make decisions on how to act and respond to problems. Actually, this is the most difficult task to fulfill, making decisions under the pressure of the stress provoked by crises. Lagadec (1993), for example, suggests that whenever decisions are not taken, the chaotic situation originated by the event takes over and simply annihilates the reaction potential of exposed systems and organizations. Weick (1988) observes that action itself helps in finding interpretations to the crisis condition while reshaping its feature in the meantime. Nevertheless "there is a delicate trade off between dangerous action which produces understanding and safe inaction which produces confusion."

There are many types of crises, which are classifiable also with respect to the initial event that triggers them. Actually some crises occur without any identifiable triggering event, or in circumstances where there are many events to which the crisis can be linked to, while no one stands out in a clear cut way. In this respect, some blackouts can be cited as an example, or some political breakdowns. In this contribution, only crises originated by natural hazards are discussed.

It should be noted that the term "crisis" may be considered close to others, for example, "disaster" (see *Disasters*). In fact, in the UNISDR Glossary, the term crisis is missing, whereas the term disaster has many connotations that are attributed here to "crisis." It may be held though that in the common use, the term "disaster" is broader in its covering the entire event, from impact to longer term consequences, whereas "crisis" refers more to the initial phases, and in this respect, it gets closer to "emergency" and "contingency." The term crisis expresses the type of disruption that one is faced with, the situation in which it is necessary to decide under significant stress and disruption of normal life. Following this reasoning, while the term "disaster" depicts the overall condition for the entire affected community, the term "crisis" is the disaster seen from the eyes of the interveners, of those who have responsibilities and are attributed the means and the resources to intervene and respond.

This brief discussion points out that even though terms like "crisis," "disaster," "calamity," and the like seem obvious, they cannot be accepted in an uncritical way. As an example, the book by Quarantelli (1998) titled *What Is a Disaster?* convincingly shows how difficult it may be to provide satisfactory and universally agreed upon definitions. Actually, different organizations, including EM-DAT or Munich-Re, or various national legislations set rather different thresholds to distinguish between what can be considered a disaster and what cannot. Not any landslide nor any ground shaking produces a level of damage and devastation so as to call for a disaster declaration. Furthermore, what makes a disaster in one region of the world may not in another, a death toll is considered high in one country and negligible in another.

Similarly, it is not that easy to attach the definition of a crisis due to some natural hazards to the level of disruption and losses that a specific event may provoke. In the following paragraphs, some crucial elements and factors generating a crisis and requiring specific actions for its control and management in the aftermath of a natural extreme will be discussed.

Here, it will suffice to list some specific conditions that can be considered as specifically characterizing crises linked to natural events. The verb "linked" and not due to or provoked by is used because the assumption here is that not only large magnitude events provoke crises, the latter can arise also as a consequence of somewhat medium or even minor environmental stress, depending on the weaknesses of exposed systems. In fact, a crisis may be originated either by a severe natural event, for example, a high magnitude earthquake, a fast landslide mobilizing large volumes, a strong volcanic eruption, or be the consequence of highly vulnerable exposed systems (see *Vulnerability*).

Therefore, in investigating the types of crises that may occur as consequence of some natural event be it very severe or not, both the characteristics of the threat and of exposed systems must be identified and described.

Types of crises

Following what has been stated above, types of crises will be classified according to hazard and vulnerability aspects.

With respect to the first, spatial and time factors should be considered. From a spatial point of view, hazards may generate local, regional, or multisite events. Local events, like landslides, avalanches, or tornadoes, are such that they hit a given place, provoking concentrated damages and losses. In this case, even though the event can be very severe and provoke significant local disruption, it is possible to delimitate an area, an event core, around which a corona and a periphery from which help may come and to which victims can be temporarily or permanently evacuated can be clearly drawn. In terms of crises management, a local unit to tackle the event from a close post is generally sent so as to check needs and demands arising from the field and then control the situation from a safe place at the shortest possible distance from the core area. Concentration of rescuers, teams, and support goods must be managed and organized so as to avoid congestion that may end up getting the opposite result to the intended.

Regional events, on the contrary, involve large areas, comprising different types of settlements and infrastructures, from rural/natural areas to highly urbanized to metropolitan. Large regional events may be transboundary, across several administrative borders, including regional and national. In this case, several teams will be sent to the area; a number of advanced units must be forecasted and positioned in strategic zones. Challenges are clearly larger than in the case of local events, because of the extent of territories and the expectedly larger numbers of affected people. Whenever regional events affect different jurisdictions or even nations, a complex issue of coordination among levels of government, different governments, and authorities arises, making the crisis easily escalate beyond the

Civil Protection and Crisis Management, Table 1 Issues arising in crises that are differently characterized in terms of spatial and temporal scales. Self-elaboration. Concepts in this table can be found in Chaps. 2 and 4 of Menoni and Margottini (2011)

Space / Time	Local	Regional	Multisite
Slow onset	Crisis may not be recognized by early signs	Crisis may escalate and involve large areas and more than one country	Similarities among events in different localities may not be recognized
Fast onset	Potential indirect consequences at larger scales may not be adequately foreseen	Crisis requires significant coordination in the area without any/enough prealerting time to make first common decisions	Challenge of recognizing the crisis' actual spatial extent
Long duration	Attention by the media and even by governmental offices may fade away as time passes	Turnation of a large number of officers and workers is required. Large amount of resources. Temporary solutions (like shelters) for partial return to normalcy	Problems in assuring resources to all affected sites
Short duration	When the most critical phase is over, the local community may be left alone even in cases when it does not have the resources to fully recover	Difficulties may arise to guarantee coordination particularly in cross-border and interregional crises. Long-term effects may not be adequately considered or treated differently across borders	Challenge to assess the needs in multiple locations. Challenge in dispatching forces to a variety of places simultaneously

control and management capacity of any of the involved authorities or agencies.

The adjective multisite can be attributed to otherwise local events that occur simultaneously in different places, for example, forest fires in the dry period or a storm affecting several places in the same days. Even though events like a fire or a landslide or a storm hit individual places, their contemporary occurrence puts a much stronger pressure on intervention agencies and teams. In fact, while local events, even though very severe, permit to concentrate response forces, multisite events challenge response teams, in that resources and means must be dispatched at the same time to a variety of places. The fires which occurred in Southern Europe in the summer of 2007 are a clear example of such events that distressed significantly the Community Mechanism of the European Union and required the rapid displacement of fire fighters from France to Portugal to Italy to Greece.

As far as time factors are concerned, as shown in Table 1, two criteria must be borne in mind. The first refers to the time of onset of an extreme event and the consequent crisis. Some natural events can be sudden and rather unexpected, not so much in general as for the actual circumstance, the hour and the day in which they occur. In other words, as commented by Hewitt (1983) in his *interpretation of calamities*, there is the possibility to forecast most natural hazards, such as earthquakes, floods, and landslides which occur in areas that are prone to them and historical evidence exists of their occurrence in the past. Nevertheless, the exact moment when they will occur may not be predictable, as premonitory signals are either weak, or inexistent, or highly uncertain.

Events characterized by fast onset may generate sudden, unexpected crises, particularly when mitigation has not or has been poorly carried out. Early warning and prealert is either impossible or possible with only few seconds to minutes in advance, so that crises start when most damages and losses have already occurred. Events like earthquakes, debris, and mud flows are not only sudden but also rather rapid in their development; in a short or very short time, they deploy all their destructive potential, leaving to rescuers only the possibility to respond to losses and death toll.

Events characterized by slow or relatively slow onset, like plain floods or droughts, may be predicted in advance and actions can be taken to protect people and goods as well as to secure the most critical and strategic facilities and places (see *Early Warning Systems*).

Even though examples have been provided for slow and rapid onset events, it is noteworthy that they are only indicative and cannot be considered as fully exhaustive or satisfactory. In fact, there are large earthquakes that are announced by a series of minor tremors months ahead; there are ash crises that can or cannot be followed by a big explosion; some types of landslides do show clear signals of movement, others do not. What can be said therefore is that monitoring devices, complete warning systems, comprising besides the technical component also the social and logistic aspects, may significantly change the type of crisis, from largely unexpected to highly anticipated. The ability to generate event scenarios, previous training, exercises and simulations permit to be ready for a given event in large advance, so as to downscale the magnitude of the consequent crisis.

What is crucial in most if not all instances described above is the capacity to deal with uncertainties and make decisions despite scientific and other types of uncertainties (including legal, institutional, societal, see De Marchi 1995). In fact, the classification of crises, as mentioned above, cannot be strictly associated to the characteristics of the threats. The interface between the latter and the exposed systems, considered not only as physical but also organizational and social, is equally important to determine how a crisis will look. As suggested by Sarewitz et al. (2000), not only by reducing the scientific uncertainty associated with some natural hazards one may improve the coping capacity (see *Coping Capacity*), but also by lowering all other types of uncertainties, particularly legal, institutional, and societal. By making timing and good decisions, the catastrophic potential of some events may be lowered by reinforcing the response capacity of likely to be affected systems.

In this respect, clearly, the vulnerability of the latter plays a key role.

It would be too long and perhaps beyond the scope of the present contribution to list the variety of conditions that may shape crises, according to the characteristics of the physical built environment, land use patterns, and the mode of use of buildings in exposed areas. Clearly, all those factors influence some logistics of the crises, in terms for example of accessibility to damaged zones and to resources and facilities (Ceudech and Galderisi 2010); they also strictly influence the extent of physical damage, which in turn translates into number of affected people and extent of resources to be deployed (for partial or total evacuation, etc., see also *Evacuation*).

Another vulnerability facet determining the level of crisis can be labeled as systemic or functional. How well and how long strategic facilities like lifelines can provide service is crucial to sustain help, search, and rescue activities and therefore directly influence the level of control that can be sustained by crisis managers.

Among the variables identifying communities' vulnerability and resilience, the response capacity of established organizations, like the firemen, the army, and the medical doctors, is essential. The response system constitutes a sort of standardized and predetermined body, whose preparation and training is independent from the specific features of the crisis at stake. According to the practical experience gained in the field within an operational organization like the firemen, a fundamental lesson that can be suggested refers to the importance of being able to rely on established rules and standardized procedures at least for the most repetitive tasks, for those operations and to use those devices that are most common. Formalized crisis management models may significantly improve the performance of teams, as they permit to achieve a good level of response at least for the most repetitive and trivial operations while devoting due energy to what really stands out (Wybo et al. 2001). Without standardization and preparation, coupled with strategic management, it would be extremely difficult to even recognize exceptions and surprises.

Last but not least, as for time factors, duration of crises must be accounted for. Most plain floods, for example, may affect very large portions of a given territory but are not likely to last for long. After days, people will be able to return to their houses unless severely affected or contaminated and start reconstruction (see *Recovery and Reconstruction After Disaster*). Earthquakes would require long-term stay in temporary shelters, whereas the crisis itself may last for a number of weeks. In this case, turnover among rescuers must be carefully planned and mechanisms for exchange of solutions and information must be set up.

Models of crisis management

According to common sense, crisis management requires the presence of a strong subject able to lead the team working on the disaster scene so as to achieve the best solutions in the shortest time. While this idea holds certainly some truth, particularly when the necessity to make decisions and to lead the event instead of just being at its mercy are considered, in general, some authors (Lagadec 1995; Reason 1997) contradict the idea that centralizing decisions and actions as well as making coping organization hierarchical actually improve response. In fact, the opposite has been demonstrated. Highly hierarchical organizations are not able to respond fast to changes and be flexible enough to react to surprises and unexpected situations. They require a long chain of orders and decisions to be followed and do not allow for much initiative to those in the field, who nevertheless have the direct grasp and perception of events, even though they lack a supervision of the entire scene and of the many interconnections among areas, resources, and systems.

A good balance must be sought between one person or restricted groups' ability to control and be in charge of the situation on the one side and the personnel who are at site and have a direct vision of the event on the other, so as to guarantee decisions and leadership and, in the meantime, allow for sufficient flexibility.

Often recalled in the crisis management field is also the opposition between improvisation and preparation. To a certain extent, this opposition is linked to the one discussed above between hierarchical and "democratic" organizations. In fact, hierarchical organizations tend to rely heavily on established plans, whereas local cells guaranteed enough autonomy may take fast decisions more tailored to the upcoming situation.

One way of combining the two needs, that is, take control of the entire crisis scene, particularly when the latter is complex and extended over large areas, and the need to be "close" to the site, where the incident or the natural event occurred, is constituted by a model of operation called "Incident Command System." The latter makes part of the recently reorganized "National Incident Management System" promoted within the US Homeland Security (2008) as a model for managing large emergencies. Such a model, already well established since the 1970s, has spread beyond the USA and is currently adopted, though under different names, in many countries worldwide.

"The NIMS is based on the premise that utilization of a common incident management framework will give emergency management/response personnel a flexible but standardized system for emergency management and incident response activities. NIMS is flexible because the system components can be utilized to develop plans, processes, procedures, agreements, and roles for all types of incidents; it is applicable to any incident regardless of cause, size, location, or complexity. Additionally, NIMS provides an organized set of standardized operational structures, which is critical in allowing disparate organizations and agencies to work together in a predictable, coordinated manner" (Homeland Security 2008, p. 6). According to the model, local cells are sent to the scene with the capability to guarantee information exchange among those in the disaster scene and among the various organizations present on site and their respective operation centers. The incident command system is therefore constituted by an advanced group of technically skilled personnel who are also granted the capability to decide some immediate actions on site, coordinate the various organizations, and guarantee information and exact request of resources to each operational center and to the main emergency control room.

In the case of natural disaster, this organizational mode has to be adapted to the environmental and social contexts and to the characteristics of the hazard, particularly as far as the spatial features described above are concerned. Several challenges have to be met in adapting the NIMS to individual countries' characteristics and previous mode of operating. For example, the system requires the extensive use of technical terms and standardized documents. Both have to be "translated" linguistically and also semantically in newly produced documents and then a period of extended training must be foreseen. The transition from previous models to the new, even though more efficient structure, requires planning and the provision of additional resources.

In the case of a local hazard, an advanced command post may be enough; the same cannot be held for regional or multisite events, where clearly a net of advanced command posts must be coordinated so as to guarantee the correct treatment of each site where an event has occurred.

Civil protection and crisis management in a nutshell

In the previous section, the model of intervention, whether highly centralized or distributed, whether hierarchical or flexible, has been shortly discussed. In this section, the much more complex issue of how civil protection, as initially defined, manage crisis will be addressed.

Drawing upon Cherns and Bryant's (1984) work on the construction industry, it can be held that also crisis management requires inevitably the coordination of a complex temporary multiorganization that must achieve a unique goal (facing and exiting the crisis condition) in the shortest time and reducing as much as possible losses and errors. First, because crisis management requires the presence and the action of various agencies and organizations, ranging from the firemen, to the police, to the army, to the agencies in charge of environmental assessments, health indicators, etc. Those agencies and organizations share (or should share) a common objective, solving the crisis, but are characterized by their own culture, by their political and social mission, by their means and resources. Some of the organizations involved in the crisis are dealing with "minor" or "normal" emergencies everyday on a routine basis, for example, firemen or emergency doctors. Others are involved in crisis management only occasionally, depending on the event to be tackled, for example, lifelines managing companies, public health agencies, etc.

Those organizations do not generally meet on a routine basis, hardly know each other, both as organizations and as individuals, which makes coordination and management particularly challenging, among other reasons because the one who will be in charge of coordinating must get the approval and respect of all involved parties. What Cherns and Bryant (1984) say about the construction industry perfectly fits also the crisis management arena: "Relationships [among the various bodies] are formally governed by the contract [in the case of crisis management the contingency plan or other formal governmental arrangements and protocols can be considered], but are supplemented and moderated by informal understandings and practices which have evolved to cope with the unforeseen, sometimes unforeseeable difficulties that characterize [disasters]."

Unfortunately, few studies have been devoted to analyzing the difficulties and the solutions found by the temporary multiple organizations in charge of crisis management in given circumstances and under different crisis duration. Some work that can be quoted refers to organizations under stress, how they cope and how they can be brought to react better and even with success. A recent relevant work in this direction is provided by Comfort (2007) who suggests that beyond control, coordination, and communication capabilities, the real challenge "is to build the capacity for cognition at multiple levels of organisation and action in the assessment of risk to vulnerable communities." In her contribution, Comfort stresses the importance of cognition, as the capacity of multiple organizations to build a common understanding and interpretation of the evolving emergencies and act accordingly.

In general terms, it can be said that much more research has been carried out with respect to what happens within the same organization under stress, whereas little has been done with respect to the intercorporate dimension, that is, among distinct organizations. In general it can be said that difficulties encountered within the same organizations, for example, relatively to information exchange, decision making, identification of available resources, are exacerbated when a number of organizations must work together, particularly when such circumstance is temporary and not too frequent.

A specific point should be raised with respect to international crisis management, when aid is given to poor and developing countries. In fact, such intervention often sees the convergence of massive forces from a variety of countries in the affected place. Recent examples are the intervention on the occasion of the devastating tsunami hitting Southern Eastern Asia in 2004 (see Christoplos 2006) and of the earthquake in Haiti, January 2010. As it is already very complex to achieve coordination and cooperation among different organizations of the same country, one may easily imagine the almost insurmountable difficulties when the latter must be achieved among organizations pertaining to different countries. In this case, lack of coordination may be dramatic, producing overredundancy of some goods, complete lack of others, mismanagement in the form of goods supplied where and when they are least needed, tragic delays where they are urgent, etc. In addition to the multiple temporary organizations of different countries and speaking different languages, problems of logistics, understanding of the social, political, historic, and cultural context are also crucial, leading to a variety of mistakes, sometimes severe. There are not simple solutions to those problems; nevertheless, some elements may be considered to improve current practices. On the side of aiding countries, what can be asked is a higher understanding of the social and cultural context before providing help, building on experience to avoid errors already committed in the past, avoiding putting too much emphasis on fast results to be shown to donors, in favor of deeper analysis of actual needs, and identifying where resources can be invested so as to obtain the best results for the victims. On the side of recipients, what would be clearly ideal is the training of local responsible personnel able to direct materials and goods, to dispatch help to the most affected areas, and to provide guidance to international and external agencies. At the very least, local authorities should be able to interface with international agencies so as to avoid to be completely overridden, with the uncomfortable but almost inevitable outcome of money and resources spent haphazardly.

Main characteristics of crises today and potential challenges of tomorrow

Challenges can be grouped according to whether they are intra- or interorganizational, that is, whether they refer to problems arising within the same organizations involved in crisis management or among different agencies and organizations.

Within the same organization, the following can be mentioned:

– Ability to transfer information timely and effectively among the various members and subparts. Studies have shown that organizations relying on formal systems of communication are more likely to manage effectively crises particularly when technical disturbances in communication devices may occur (see McLennan et al. 2006).

– Level of preparation and preplanning. Regarding this particular point, a rather interesting literature exists (Lagadec 1993, 1995; Roux-Dufort 2000), depicting what works well, poorly, and not at all prepared organizations. Among other criteria, the most important is the behavior and attitude of responsible managers in facing crises, as in prepared organizations the latter tend to take the lead, whereas in the least prepared they tend to retire in their own shell and protect themselves from criticism. Equally relevant is the ability of organizations to learn from experience and to successfully interface with the public and the media.

– A specifically mentioned aspect refers to decision making, that is, the ability to make decisions (possibly sound ones) in the urgency of a disaster, under the tremendous pressure of the evolving event and the concerned public(s). Lagadec, Roux Dufort, and Weick all share the conviction that crisis management is a strategic not a reactive activity.

An example of decision which is particularly hard to make in the face of natural disasters is early warning in case of large uncertainties (see *Early Warning Systems*). Specific examples are in the field of seismic risk, where early signals may be particularly difficult to interpret correctly and, in any case, leave large room for false alarm (see *Earthquake Prediction and Forecasting*). Even though other hazards, like volcanic eruptions or floods, are in general more predictable than earthquakes, they all share some basic common aspects, like the sources of uncertainty, deriving from the quality of available data, the quality of scientific explanations and models. Other types of uncertainty intertwined with the latter refer to the societal and institutional backgrounds where the decision must be taken. As Sarewitz et al. (2000) convincingly showed, sometimes improvement in scientific understanding of a given natural phenomenon may even lead to larger and deeper uncertainties. Instead, the latter may be reduced by means of strong and sound decision making rather than better science or better data.

Larger difficulties arise when multiple organizations intervene in the same crisis scene:

– Communication among different organizations which does not only imply issues of language, jargon, secrecy, willing to keep information inside each organization, but also technical aspects, for example, different radio frequencies assigned to every agency and organization, a simple fact holding heavy consequences.

– Communication with the media, when multiple actors are in theory eligible to provide information. How to agree among the police, firemen, medical doctors, etc., about the opportunity to dispatch a unique information bulletin, particularly when stakes are high and uncertainty large?

– Need to share not only material resources but also the information about the actual availability of those resources and the way to obtain them from legitimate owners. Even though the civil protection is entitled to ask for resources, conflicts among ministries and governmental agencies must be avoided; furthermore knowledge about existing resources must preexist if they are to be practically managed during the crisis.

A final word in this section must be devoted to the so-called lay or general public. Some of the latter may actually be part of the population who may be potential victim of the disaster. Social scientists have been producing thousands pages of studies describing and reasoning about the response of "people" to disasters under different circumstances and in different contexts (just as a reference, Barton 1970; Drabek 1986; Fischer 1996). Time has come to make those studies part of active crisis management, avoiding treating the public as pure recipients of somebody else's thoughts and decisions, recognizing the essential active role that the affected population actually plays in the majority of cases. Attempts to elude this reality have often turn crisis management into failure even in the presence of substantial means and well-prepared organizations. In this respect, the issue of informing the population before and during crises is clearly crucial. As Parker (1999) stated, there is often a contradiction between the requirement to keep it secret, in the fear of "panic," and the need to have the public act in an informed way (see *Risk Perception and Communication*). Considering again the example of early warning, Parker and Handmer (1998) have shown how any information, advice, or input from official sources undergoes a process of verification and analysis of costs and benefits implied in the suggested or required actions. The decision to comply with the latter depends, among other factors, on the familiarity with the hazard, on the familiarity with the authority issuing the alert, on the correspondence between the given message and the perceived threat, and, last but not least, on the tone and wording of the message itself.

Inter- and intraorganizational challenges mentioned above are limited to what is already known about past crises, emerging from experience and thinking about what happened in past events.

This is just one part of the problem at stake for today's crisis managers, the other one being future challenges, tomorrow's problems, and constraints that are not always that easy to identify and detect in advance. Scenarios of future and emerging hazards and risks must be first depicted in order to be able to answer the just asked question of how future crises will look like (see *Global Change and its Implications for Natural Disasters*).

In their book, *La fin du risqué zero*, Guilhou and Lagadec (2002) addressed those issues, pointing at two main concerns, referring to the emergence of surprises on the one hand and to the need to develop specific scientific expertise on the other. As for the first, the Authors hold that future crises will imply larger surprises and unexpected outcomes, the only way to be prepared for is training on scenarios and simulations, not because the future will be as drawn in the scenario, but because the latter helps those dealing with crises to prepare for the unexpected. As for the second, there is an increasing demand for scientific experts able to provide guidance on the basis of poor quality (and sometimes also quantity) data, making a guess informed by their knowledge and past experience in the field of concern (for example earthquakes or floods). This may be considered as a particular case of scientists advising policy makers (Jasanoff 1990), in a condition which is particularly stressful and delicate for both. As an example of tragic problems that may arise in the aftermath of a catastrophy one may recall the ongoing trial in Italy after the l'Aquila earthquake, in which scientists who worked as consultants for the civil protection are under trial for their failure in correctly communicating the risk and/or uncertainties implied in risk estimation and assessments capabilities (see Hall 2011).

In this contribution, little room has been devoted to technology, despite its omnipresence in all arenas of modern life, certainly in the field of emergency management. Computers, satellites, and cellular phones have changed substantially the conditions under which officers and civil protection servants are working (Harvard Humanitarian Initiative 2011). The increasingly extensive use of Internet has changed also victims' ability to get informed, to exchange feelings, problems, and sometime crucial information. Many technologies, starting from the GIS to several communication devices, have slowly shifted from military to civilian applications. In spite of such major influx of modern technologies, which certainly must be used to manage crises at best, a number of warnings must be raised, not with the aim to contradict the obvious potential of technologies, but rather to promote their most effective usage.

Quarantelli (1998) suggested for example that overreliance upon modern technologies should not divert attention from the need to provide backups, including manual backups, in case of technologies' failure; in any case, technologies should be looked for and developed to address actual needs rather than reshape crisis management to fit the technical features of existing devices offered in the market. Finally, the rather trivial but nonetheless important reminder of the fact that problems of interpretation and meaning cannot be solved by "more technology" (nor by more science as suggested by Sarewitz et al. 2000).

Conditions for successful crisis management: lessons learnt from successful and unsuccessful cases

There are a number of conditions that are commonly considered as keys to positive outcome of crisis management, including ability to govern complex and highly dynamic contexts and situations, ability to select the crucial information in the midst of flouring data and uncontrolled rumors, and ability to anticipate on the basis of understanding of the situation and thanks to a prior effort in designing scenarios and simulations helping to identify weak points and fragilities of both the exposed environment and communities.

One of the most crucial aspects refers to the capacity to learn upon experience, to capitalize past mistakes and successful results, and to rethink strategy and form of organizations. Well-organized agencies are able to face even failures so as to learn and question basic assumptions in an effort to be much more prepared for the next occasion; unprepared organizations do not even have the tools to analyze what went wrong and lack human, technical, and financial resources to recover in a resilient way. A resilient organization in this sense is not only able to recover after a failure but also to restructure itself so as to become stronger and take advantage of the lessons learnt. This is clearly a very demanding achievement, while most organizations tend to restore pre-event patterns, aiming at surviving, keeping the attention all focused on the specific aspects of the just occurred crisis, without questioning the entire set of organizational assumptions and fundamental beliefs.

Still the problem that remains open is how temporary multiple organizations can accomplish such learning which should not be only individual but also collective, that is, related to the entire set of different agencies, private companies, groups, and organizations making part or coordinated by civil protection and contributing to the solution of a crisis for the best or for the worst.

And even more challenging is the question of how to keep the memory of such learning, of the conditions that led to positive as well as to negative outcomes. Who should be responsible for keeping such memory and what form such memory can take. It can be suggested as a partial solution that emergency plans (see *Emergency Planning*) may be one of the material places where such memory can be kept, in the sense that the plan should constitute both a reminder of activities and procedures that proved to work well under given scenarios and may as well provide room for learning lessons from real events and simulations so as to revise the plan whenever the latter is being felt obsolete or requiring any kind of updating.

One unfortunate observation made by some authors (De Marchi 1996; Murphy 2009) is regarding the large amount of information, expertise, and know-how that is lost after some time has passed since the last crisis and several lessons must be learnt again and solutions found again, whereas they had already been achieved but not successfully transmitted in the past. In this regard, reports of emergencies that have been tackled in the recent past, in various developed and developing countries, are worthwhile reading and analyzing with the aim to build a reference archive at least for those problems and obstacles that arise over and over, which would deserve to become a common patrimony of all those in charge of crisis management at different stages and with varying levels of responsibility.

Summary

Civil protection is a term used in several countries to indicate the institution(s) that coordinates (or tackle) emergency and crisis management.

Crisis management is the set of activities aimed at facing a complex, unexpected situation originated by an accident, a war, a terrorist attack, or a natural calamity.

Crisis management is a particularly complex activity, which requires a number of qualities from those who are in charge of its solution. It is stated that crisis management as a definition holds an intrinsic contradiction in that crises are unmanageable by their very nature. They are characterized by a number of aspects, like difficulties in getting the right picture and extent of damage, disruption and resource needs, problems in communication at all levels, among stakeholders and with the public, rapid development, strong pressure on decision makers, and significant uncertainties about potential outcome of alternative decisions and consequent actions. Those and other features make the solution of crises particularly troublesome and questioning fundamental beliefs and procedures of the established organizations which are expected to deal with them effectively. Those organizations are generally grouped under the label of civil protection. The latter term may either refer to an individual organization which is in charge of coordinating the activity of the many others who intervene on the scene of a disaster or to the entire set of organizations entering in a disaster field. In both cases, crisis management often implies the establishment of a complex temporary multiorganization, comprising a variety of different agencies and organizations that meet on the occasion of a disaster and have to cooperate despite cultural, language, and mission differences. Findings of recent literature and deriving from practical cases are proposed to discuss what are the most agreed upon conditions that may lead to satisfactory crisis management solutions.

Bibliography

Barton, A. H., 1970. *Communities in disaster: A sociological analysis of collective stress situations*. Garden City: Anchor Books.

Boin, A., and Lagadec, P., 2000. Preparing for the future: critical challenges in crisis management. *Journal of Contingencies and Crisis Management*, **8**, 185–191.

Ceudech, A., and Galderisi, A., 2010. The "seismic behaviour" of urban complex systems. In Menoni, S. (ed.), *Risks challenging publics, scientists and governments*. Balkema: Leiden CRC Press.

Cherns, A. B., and Bryant, D. T., 1984. Studying the client's role in construction management. *Construction Management and Economics*, **2**, 177–184.

Christoplos, I., 2006. *Links between relief, rehabilitation and development in the tsunami response*. London: Tsunami Evaluation Coalition.

Comfort, L., 2007. Crisis management in hindsight: cognition, communication, coordination, and control. *Public Administration Review*, **67**(Suppl 1), 189–197.

De Marchi, B., 1995. Uncertainty in environmental emergencies: a diagnostic tool. *Journal of Contingencies and Crisis Management*, **3**(2), 103–112.

De Marchi, B., 1996. *Review of Chemical Emergencies Management in The EU Member States*. Luxembourg: Joint Research Centre, European Commission, Institute for Systems Engineering and Informatics. 5529-93-10 ED ISP I.

Drabek, T. E., 1986. *Human System Responses to Disaster: An Inventory of Sociological Findings*. New York: Springer.

FEMA, Comparative Emergency Management book project, at http://training.fema.gov/EMIWeb/edu/CompEmMgmtBook Project.asp.
Fischer, H. W., 1996. What emergency management officials should know to enhance mitigation and effective disaster response. *Journal of Contingencies and Crisis Management*, 4(4), 208–217.
Guilhou, X., Lagadec, P., 2002. *La fin du risque zero*. Paris: Les Èchos, Eyrolles Editeur.
Hall, S., 2011. Scientists on trial: at fault? *Nature*, 477, 264–269.
Harvard Humanitarian Initiative, 2011. *Disaster Relief 2.0: The Future of Information Sharing in Humanitarian Emergencie*. Washington, D.C. and Berkshire, UK: UN Foundation & Vodafone Foundation Technology Partnership.
Hewitt, K., 1983. *Interpretations of calamity*. Boston: Allen and Unwin Inc.
Jasanoff, S., 1990. *The Fifth Branch: Science Advisers as Policymaker*. Cambridge, MA: Harvard University Press.
Lagadec, P., 1993. *Preventing Chaos in a Crisis. Strategies for Prevention, Control and Damage Limitation*. Berkshire: McGraw Hill.
Lagadec, P., 1995. *Cellules de crise. Les conditions d'une conduite efficace.*. Paris: Les Èditions d'Organisations.
McLennan, J., Holgate, A., Omodei, M., and Wearing, A., 2006. Decision making effectiveness in wildfire incident management teams. *Journal of Contingencies and Crisis Management*, 14, 27–37.
Menoni, S., and Margottini, C. (eds.), 2011. *Inside Risk. Strategies for sustainable risk mitigation*. Milano: Springer.
Murphy, R., 2009. *Leadership in Disaster. Learning for a Future with Global Climate Change*. Canada: McGill-Queens University Press.
Parker, D., 1999. Disaster response in London. In Mitchell, J. (ed.), *Crucibles of Hazard: Megacities and Disasters in Transition*. New York: United Nations University Press.
Parker, D., and Handmer, J., 1998. The role of unofficial flood warning systems. *Journal of Contingencies and Crisis Management*, 6, 45–60.
Perrow, C., 2007. *The Next Catastrophe: Reducing Our Vulnerabilities to Natural, Industrial, and Terrorist Disasters*. New Jersey: Princeton University Press.
Quarantelli, E., 1998a. *The Computer Based Information/Communication Revolution: A Dozen Problematical Issues and Questions they Raise for Disaster Planning and Managing*. Newark: DRC. Preliminary Paper.
Quarantelli, E. (ed.), 1998b. *What is a Disaster?* UK: Routledge.
Raley, A., 1990. Artefacts, memory and a sense of the past. In Middleton, D., and Edwards, D. (eds.), *Collective Remembering*. London: Sage.
Reason, J., 1997. *Managing the Risk of Organizational Accident*. UK: Ashgate.
Roux-Dufort, C., 2000. *La gestione de crise. Un enjeu stratégique pour les organisations*. Paris-Bruxelles: DeBoek Université.
Sarewitz, D., Pielke, R. A., and Byerly, R., 2000. *Prediction: Science, Decision Making, and The Future of Nature*. Washington DC: Island Press.
United States Department of Homeland Security, 2008. *National Incident Management System*. Washington, DC: United States Department of Homeland Security. http://www.fema.gov/pdf/emergency/nims/NIMS_core.pdf.
Weick, K., 1988. Enacted sensemaking in crisis situations. *Journal of Management Studies*, 25, 305–317.
Wybo, J. L., Delaitre, S., and Therrien Euquem, M. C., 2001. Formalisation and use of experience in forest fires management. *Journal of Contingencies and Crisis Management*, 9, 131–137.
http://ec.europa.eu/echo/policies/disaster_response/mechanism_en.htm
http://www.fema.gov/
http://www.em.gov.au
http://www.publicsafety.gc.ca/prg/em/index-eng.aspx

CLASSIFICATION OF NATURAL DISASTERS

Thomas Glade[1], David E. Alexander[2]
[1]University of Vienna, Vienna, Austria
[2]Global Risk Forum, Davos Platz, Switzerland

Introduction

The question of how to define a disaster and which criteria should be applied to classify it has been the subject of vigorous debate among practitioners of the field (MunichRE, 2006; Perry and Quarantelli, 2005; Quarantelli, 1998). For example, Berren et al. (1980) offer an independent and comprehensive classification that is not limited to natural disasters and is based on type and duration of disaster, magnitude of impact, potential for occurrence, and ability to control the impact. Other classification schemes consider the differentiation by magnitude of event or consequences, by the different scales (such as individual, family, community, and region), or by speed of onset and predictability. Hence, numerous classification schemes have been proposed, and little would be gained from reviewing them all here.

Despite these reservations, there is broad consensus that a disaster is an event or situation that severely disrupts normal socioeconomic activities and causes damage and possibly casualties. Attempts to quantify the definition, for example, in terms of monetary losses and numbers of people killed (Foster, 1976; Keller et al., 1992; Munich Re, 2006), have not met with universal acceptance. Nonetheless, it is clear those disasters there is a qualitative difference between disasters and lesser events, in that they require extraordinary responses in terms of resources and organization (Kreps, 1983). A common definition of a disaster is that the coping capacities of the affected individual, group or unit (local, regional or national governments, public institutions, social groups, etc.) are exceeded and external support is likely to be required. Hence, it may be appropriate to base the classification of the magnitude of emergencies and contingencies upon ability to cope with and respond to events of a given size (Table 1).

Three global data sources for natural disasters are available. Two are data catalogs compiled by insurance companies: the Sigma database of SwissRe and NatCatService of MunichRE. However, the most widely used data bank on disasters is the OFDA/CRED International Disasters Database (EM-DAT, refer also to www.em-dat.net), maintained at the Centre for Research on the Epidemiology of Disasters (CRED) of the Catholic University of Louvain in Belgium. Besides temporal information, all entries are arranged by continent, country, and theme, as requested by the UNISDR Secretariat.

It has long been noted that the term "natural disaster" is not particularly apt (O'Keefe et al., 1976). For example, although most earthquakes are entirely natural phenomena, the root cause of seismic disasters could be regarded

Classification of Natural Disasters, Table 1 A size classification of emergencies and contingencies (Partly after Tierney, 2008)

	Incidents	Major incidents	Disasters	Catastrophes
Impact	Very localized	Generally localized	Widespread and severe	Extremely large
Response	Local efforts	Some mutual assistance	Intergovernmental response	Major international response
Plans and procedures	Standard operating procedures	Emergency plans activated	Emergency plans fully activated	Plans potentially overwhelmed
Resources	Local resources	Some outside assistance	Interregional transfer of resources	Local resources overwhelmed
Public involvement	Very little involvement	Mainly not involved	Public very involved	Extensively involved
Recovery	Very few challenges	Few challenges	Major challenges	Massive challenges

as poor construction of buildings rather than the occurrence of ground shaking. Hence, there is a motive for regarding earthquakes as human-made disasters. In fact, because so much of the impact of disasters depends upon vulnerability, a predicament that mainly depends on human decision making, "natural" disaster can be regarded as a convenience term which distinguishes one class of phenomena from others. In this case the generating mechanisms stem directly from events in the geosphere, biosphere, atmosphere, and hydrosphere.

"Natural" disasters are caused by extreme events, in the sense of large departures from long-term mean values. For instance, sudden excesses of precipitation can cause floods, whereas long drawn-out shortages can result in drought. In this respect, *speed of onset* and *duration* are important criteria in classifying events. Earthquakes and rapid debris avalanches are examples of sudden-impact disasters, whereas drought and desertification or soil erosion are examples of slow-onset events. Most earthquakes have a main shock that will last from a few tens of seconds to a couple of minutes, but the sequence of aftershocks can stretch the emergency period to hours or days. This contrasts with a drought that may be prolonged for months or years and desertification that is essentially a permanent condition, i.e., one that is technically challenging and expensive to reverse. The typology suggested by the U.S. National Research Council's Committee on Disaster Research in the Social Sciences. (US NRC, 2006) is similar to the discussion presented here, but also includes the scope of impact.

Among extreme natural phenomena there is a wide variety of speeds of onset and duration, and in turn a large variation in predictability and potential for warning. For example, tsunamis are generated abruptly by the sudden displacement of a column of ocean water. Triggers may be earthquake activity, submarine landslides, or meteorite impact. However, the very long distances that tsunamis travel allow monitoring to take place and warnings to be issued to distant coastal areas in their path. This has been successfully applied, for instance, in the tsunami generated by the Chile earthquake on February 27, 2010, and the respective precautionary response along the Western Pacific coasts including Japan and New Zealand. For the Pacific basin warning lead times may exceed, at most, 17 h, but the main problems occur with "near-field" tsunamis (those that are generated locally) in which even instant detection does not allow more than a few minutes' warning to be issued to local communities. However, one should never forget that the population at risk from a local tsunami can often recognize earthquake shaking as an environmental cue indicating a need to evacuate to higher ground (McAdoo et al., 2009).

The prospect of short-term prior warning of earthquake main shocks has long been a goal for seismologists, but has proved consistently elusive, mainly because each earthquake involves some degree of complex uniqueness. Hence, most seismic disasters occur without prior warning, other than the long-term identification of areas at risk and recurrence intervals of earthquakes of a particular maximum size.

The predictability and warning potential of volcanic eruptions is highly variable. Heat fluxes, harmonic tremor, and gas emissions all indicate the rise of molten magma close to the Earth's surface, but the exact timing of eruptions tends to defy prediction. In the mid-1980s volcanic emergencies occurred in the Caribbean and southern Italy that lasted months without any actual eruptions. In contrast, many extreme events of an atmospheric or hydrological origin have a higher degree of predictability. The preparatory phenomena can be observed by direct measurement (e.g., rain gauges and streamflow monitoring) or remote sensing (synoptic views of storms), and numerical modeling can give forecasts of pending events.

Drought is the archetype of a slow-onset, or "creeping" disaster, a category that includes desertification (the degradation of land productivity) and accelerated soil erosion. This sort of phenomenon tends to be insidious. It may go undetected until the impact is chronic, and thus the state of disaster is defined by the cumulative sum of effects.

Recurrence interval and regularity are two further elements in the classification of natural disasters. Although many anthropogenic phenomena are nonrecurrent (for example, transportation crashes and catastrophic pollution tectonically-active slopes and high rates of chemical weathering, also leads to slope weakening and high incidence of landslides, as in Hong Kong (Peart et al., 2009). These climatic conditions are also important in the transformation of mass movements and volcanic debris into highly dangerous landslides and mudflows (see *Mass Movement*).

More widely, significant future changes in low-latitude climates will be closely associated with changes in the position and dynamics of the ITCZ, which marks the equatorial convergence of trade winds. Changes in land-sea surface temperatures and moisture availability, particularly over large forest areas such as Amazonia (Huntington, 2006; Cook and Vizy, 2008), will have significant impacts on meridional energy fluxes, resulting in widening of the ITCZ and tropics with accompanying shifts in precipitation patterns and ecosystems (Seidel et al., 2008).

Mid-latitude areas

Mid-latitude areas tend to have high levels of urbanisation and population density, developed over long time periods, and as such are characterised by significant modification of landuse and water resources. Climate change impacts on the activity of the mid-latitude Ferrel cell are closely related to changes in the strength of surface ocean currents (including the North Atlantic Current (Gulf Stream), Kuroshio Current, Humboldt Current, Benguela Current etc.) that are important in meridional heat transport (Herwiejer et al., 2005). Mid-latitude oceans are also sensitive to changes in thermohaline circulation (deep water) and freshwater input from rivers and melting sea ice (surface water). As such, hazards that impact on mid-latitude coasts are linked very closely to variations in the dynamics of adjacent oceans. Strengthening zonal circulation in the North Atlantic region is associated with changes in frequency of strong westerly winds and wind gusts (Kaas et al., 1996) and with increase in mean wave height (Bacon and Carter, 1991). The net result of these is increased frequency and/or magnitude of coastal storm, storm surge, and related flood events, which is a likely outcome of climate change along these coasts (IPCC, 2007).

Other coastal impacts are a consequence of the Pleistocene glacial inheritance of many mid-latitude areas (see *Paraglacial*) which means that many of these coastlines are sediment-rich with well-developed sandy beaches, sand dunes and estuaries that are vulnerable to coastal erosion (Hansom, 2001). Enhanced coastal erosion can be related directly to climate change effects of sea-level rise and increased storminess (see *Coastal Erosion*). Coastal and river management has, in many areas, decreased sediment supply to nourish these coasts, thereby making them more vulnerable (Stive, 2004).

Inland, river engineering and management and the presence of historic settlements on floodplains make river systems vulnerable to variations in precipitation input and storage that may lead in turn to increased lowland flood frequency. This is seen clearly in some managed rivers worldwide (e.g., Wilby et al., 2008).

High-latitude areas

Climate change in high-latitude areas is causing increased glacier melt (Greenland, Antarctica) and permafrost warming (Eurasia, Canada), both of which are driven mainly by increased MAAT (Camill, 2005; Osterkamp, 2005). As a result, meltwater availability in these environments is increasing, with increasing volumes of fresh water being stored within proglacial or subglacial lakes (see *Jökulhlaups*). Increased glacier melt and ice retreat can also lead to slope instability, mass movement and land surface rebound that can contribute to increased seismicity (Hampel et al., 2010). To date, hazards associated with glacier retreat and permafrost melt in high-latitude areas have been relatively isolated in location, and subdued in magnitude (Gude and Barsch, 2005). For example, roads, buildings and oil pipelines in many high-latitude and high-altitude areas have been adversely affected by land surface subsidence as a result of permafrost melt, although some of this can be attributed directly to poor engineering practice (Harris et al., 2001; Khrustalev, 2001) (see *Cryological Engineering*). On shallow slopes, however, increased depth of the seasonally-thawed permafrost leads to meltwater pooling and mass flow hazard (including bog bursts) that decrease the stability of the land surface. More widely, permafrost melt is associated with secondary effects of increased CH_4 (methane) and CO_2 degassing where peatlands are undergoing aerobic decomposition (Schuur et al., 2009). Increased subsurface water mobility can also lead to increased contaminant transport through thawed substrates (see *Permafrost*). Hazards associated with ice retreat in mountain areas are more commonly observed (see *Paraglacial*).

Studies show that, as a result of changes in freshwater input and sea-surface warming, sea ice cover in the Arctic Ocean has decreased dramatically in recent years, continuing a trend that began in the 1970s (Serreze et al., 2007). Increased exposure of arctic coastlines due to sea ice retreat, and warming and destabilisation of coastal outcrops of permafrost, has led to a dramatic increase in coastal erosion rates across the entire Arctic Ocean, including northern Russia and Canada (Jones et al., 2009). In turn, sediment release into the nearshore zone increases turbidity and heat capacity of the water, leading to additional warming. This positive feedback effect is the basis for the arctic amplification of climate (Serreze et al., 2009). Long-term implications of climate change in the high-latitudes are likely to be significant but as yet are poorly understood, including CH_4 and CO_2 release (Schuur et al., 2009). More widespread permafrost warming will also increase the geographical area over which mass movement hazards operate, which will in turn become far more common than at present. This will be a major outcome of climate change in high-latitude areas over the next decades.

Driving factors linking hazards in different latitudes

Interrelationships between hazards that are most common in different latitudes are driven, largely, by the strength of the hydrological cycle (Huntington, 2006). This is manifested by both the vigour of macroscale atmospheric circulation patterns, including the strength and positioning of the low-latitude Hadley cell and high-latitude polar cell, and the detailed dynamical processes that take place within the atmospheric part of the hydrological cycle. This includes, critically, the role of latent heat effects which are related directly to the amount of energy (heat) moved through the system, and thus to climate change.

Water within the hydrological cycle can exist in solid, liquid and gaseous phase states. A change in phase between any of these stable states by processes of melting/freezing and evaporation/condensation is associated with the exchange of latent heat, as given by the Clausius-Clapeyron relation. Based on this relation, an airmass with a higher water vapour content and therefore associated with very vigorous evaporation/condensation processes generates a large amount of latent heat that, in turn, fuels atmospheric convection and instability (Willett et al., 2007). Increased SSTs particularly in low-latitude regions are actively contributing to increased rates of evaporation, leading to higher relative humidity and higher meridional heat flux as a result of macroscale atmospheric circulation (Lorenz and DeWeaver, 2007; Seidel et al., 2008). This process, driven largely by low-latitude warming, provides an atmospheric bridge to higher latitudes, and contributes to increased convective rainfall (thunderstorms) within the Ferrel cell. Furthermore, increased moisture content is also associated with changes in cloudiness, aerosols and atmospheric optical depth (Santer et al., 2007).

Surface and subsurface circulation patterns within the oceans are important in macroscale heat transport, and their paths depend on the three-dimensional evolution of density gradients within the oceans. As such, increased low-latitude SSTs have impacts on global temperatures: the globally-warm year of 1998 was due in large part to the strong El Nino event of that year associated with positive SST anomalies in the equatorial Pacific. Warmer surface waters have less capacity to draw down atmospheric CO_2, and are associated with lower levels of dissolved oxygen, reduced productivity where upwelling is suppressed by water stratification, and reduction in water quality, with impact on fisheries and other ecosystems. Studies of thermohaline circulation dynamics show that subsurface ocean current strength in the North Atlantic is variable spatially and temporally, which has implications for meridional heat transport (Cunningham et al., 2007). It also suggests that multiple factors can affect thermohaline circulation strength, with multiple outcomes for higher-latitude regions. Several studies discuss possible future changes in thermohaline circulation strength, including the likelihood of reversal or collapse (e.g., Vellinga and Wood, 2008), but most modelled predictions suggest that thermohaline circulation is likely to weaken rather than collapse fully (IPCC, 2007).

Impacts of climate change on the vulnerability of human systems to hazards

Climate change and climate variability give rise to changes in the frequency and magnitude of extreme events that can lead to hazards, as previously discussed. This relationship alone, however, cannot explain the impact of those hazards upon human systems (including agricultural production, socioeconomic impacts, cultural impacts, human health and wellbeing, etc.) (Hulme, 2009). The future climate changes that will frame the environmental conditions under which future hazardous events and processes will take place are considered explicitly by the IPCC (2007) (Table 1). These hazards operate over spatial and temporal scales that mean their impacts will be diverse, unpredictable, and multidisciplinary in nature. Table 1 shows that future climate change will impact in many different ways on almost all areas of human activity. Of particular concern is the role of climate change in the sustainability of food production. For example, Peng et al. (2004) show that, despite increased CO_2 which generally promotes plant growth, increased night-time temperatures will cause a decrease in rice yields.

Many natural hazards, where they impact upon human activity, have the capacity to lead to environmental or humanitarian "disasters" (see *Disasters*). The World Health Organisation-linked Centre for Research on the Epidemiology of Disasters (CRED) defines a disaster as "a situation or event which overwhelms local capacity, necessitating a request to a national or international level for external assistance; an unforeseen and often sudden event that causes great damage, destruction and human suffering" (Below et al., 2009, Annex II) (see *Casualties Following Natural Hazards*). The interlinkage of this definition to the capacity of local systems to deal with environmental variability is very similar in basis to that presented in Figure 3. CRED collates information on disasters of all types, but one such disaster category is hydrometeorological disasters that are related to floods, droughts, storms and storm surges, and landslides and avalanches (see *Hydrometeorological Hazards*).

Analysis of spatial and temporal trends in disasters and their impacts shows a very clear relationship to hydrometeorological variability (ISDR, 2009). In the period 1950–2007, two thirds of all disaster-related deaths and economic losses were caused by hydrometeorological disasters. Due to the dependence of floods, droughts and other events on atmospheric circulation patterns and local-scale factors, the disaster risk from hydrometeorological hazards is highly concentrated spatially, and disaster impacts are very variable depending on an individual country's infrastructure and wealth. For example, 75.5% of all expected global mortality from tropical cyclones (1975–2007) was in Bangladesh, but the entire south Asia region represents only 2.7% of total economic losses from these events over this time period. Patt et al. (2010) use a numerical model to predict how vulnerable developing countries will be to future climatic hazards. They show

Climate Change, Table 1 Summary of some of the major outcomes of future climate change for different aspects of earth systems (Summarised from IPCC, 2007), and their implications for hazards

IPCC subheading	Summary of likely changes (from IPCC, 2007)	Implications for hazards
Ecosystems	• Ecosystems will become more sensitive to environmental disturbance under future climate change • Ecosystems will contribute to changes in net carbon storage (to about 2050) and then outgassing (thereafter) • Increased risk of species' extinction under future climate change • Changes in ecosystem functioning, structure and biodiversity as a result of future CO_2 increase	• Risk of invasive species and pathogens that contribute to biodiversity loss and/or crop failure • On mountains, increased forest cover that will stabilise slopes • Increased CH_4/CO_2 outgassing from warming permafrost and desiccation of wetlands and peatlands
Food	• Spatial changes in agricultural productivity, with likely increase in mid-latitude areas and decrease in low-latitude areas, with impacts on food security in many regions	• Increased uncertainty in food and fibre production, particularly in agriculturally marginal areas in semi-arid regions and mountain slopes • Increased soil erosion and salinisation, loss of soil fertility • Reliance on irrigation in many areas, impacts on potable water quality; increase in water contamination/pollution and eutrophication
Coasts	• Increased risk of coastal erosion and flooding	• Increased frequency and height of flooding along coastal fringes, river estuaries and floodplains • Increased likelihood of storm surges and increased height of storm waves • Increased coastal erosion along all coastline types (including rock and sandy coasts) • Increased sediment mobility in some areas, implications for port/harbour access and navigation channel infilling
Industry, settlements and society	• Increasing vulnerability of industry, settlements and society to hydrometeorological events, particularly in low-lying and coastal regions	• Increased risk to all aspects of economic activity in coastal and low-lying areas from sea and river flooding • Increased risk of built and natural heritage loss in low-lying and coastal regions due to flood inundation and warmer sea temperatures, e.g. Venice, Great Barrier Reef
Health	• Increased risk of malnutrition and hunger as a result of variations in food production and hydrometeorological events • Increased risk of infectious and respiratory diseases and heat-related deaths, but fewer cold-related deaths	• Mass movement and flood events can cause groundwater and river water contamination
Water	• Increase in water-stressed regions, particularly in areas with high urbanisation • Reduced water availability from retreating mountain glaciers • Increased precipitation in mid-latitude areas, decreased precipitation in some low-latitude and semi-arid areas • Increased precipitation variability and seasonality, decreased water quality	• Increased mass loss from retreating glaciers, increased risk of downstream floods, jökulhlaups, landslides, mass and debris flows, rockfalls • Increased variability of precipitation at all latitudes with impacts on domestic, agricultural and industrial water supply and water quality, eutrophication

that increasing the country's resilience and adaptive capacity to such events, through socioeconomic development, is the most effective means by which hazardous impacts can be minimised (Figure 4).

Figure 5 shows that hydrometeorological disasters have increased disproportionate with respect to other types in recent decades, and are dominated by floods (30.7% of all disasters in the period 1970–2005) and wind storms (26.8%). This overall pattern is broadly similar irrespective of geographical region. The IPCC (2007) also confirms that hydrometeorological extremes are more likely under future climate change, therefore hazards of these types are likely to increase in frequency. Responses of the physical environment and human activity to these challenges are uncertain but are likely to be multidimensional and wide-ranging (McBean, 2004).

Hazards and the normal distribution

Present and future climate, in its totality, is headed on an upward trajectory, in which anthropogenic global warming is causing a directional shift in standard statistical measures of climate, for example in increased MAAT (IPCC, 2007). Other meteorological variables, however, are unlikely to show a comparable simple statistical response. For example, precipitation impacts are more complex spatially and temporally, because they involve important feedback processes via ecosystems, humidity, and latent heat (Allen and Ingram, 2002). As a result, it is unlikely that under projected climate change precipitation will follow a normal distribution but rather will show a polymodal distribution in which the likelihood of both flood and drought hazards will increase (e.g., Figure 4d). Recent patterns of precipitation changes by latitude, season, and intensity suggest that these events are now beginning to take place (Zhang et al., 2007; Müller et al., 2009). Further evidence for the inapplicability of the normal distribution to some climatic variables comes from studies of changes in strength of thermohaline circulation over time that suggest this circulation can switch between one of two modes ("on" and "off") and over very short time scales and in response largely to changes in high-latitude freshwater input (Rahmsdorf, 2002). If this dynamic behaviour characterises thermohaline circulation under conditions of rapid climate change, then it confirms the importance of non-normality, nonlinear responses, and hysteresis (Rahmsdorf, 2002). These properties are not associated with simple shifts in the normal distribution, and highlight the role of spatial and temporal climate variability.

Changes in internal properties of the normal distribution, and shifts to other distributional types, show the importance of climate variability, for two reasons. First, it reveals the internal dynamical processes that are associated with changes in individual meteorological variables. Second, it impacts on the likelihood of occurrence of the extreme events that are most often associated with hazards.

Summary and outlook for the future

Anthropogenic climate change is causing unprecedented changes in physical and human systems that, together, are converging on increased frequency and magnitude of hazardous events related explicitly to changes in meteorological processes on different spatial and temporal scales. This suggests that earth systems are experiencing unprecedented rates of change and systems' organisation is responding dynamically to meteorological drivers.

One of the major areas of uncertainty in the prediction of future climate is related to climate sensitivity. This is defined as the equilibrium temperature response to a doubling of pre-industrial values of atmospheric CO_2 (Gregory et al., 2002) and is calculated, based on climate models that consider radiative forcing only, to be in the range $+1.5–4.5°C$ (IPCC, 2007). Climate sensitivity is relevant to climate hazards because, if temperature reaches and maintains an equilibrium value, it can be inferred that other climatic variables such as precipitation must also be at equilibrium. However, much of the uncertainty associated with the calculation of climate sensitivity is due to the role of nonlinear feedbacks in the climate system, in particular atmospheric moisture content and cloudiness (Zickfeld et al., 2010). There will also be time lags and

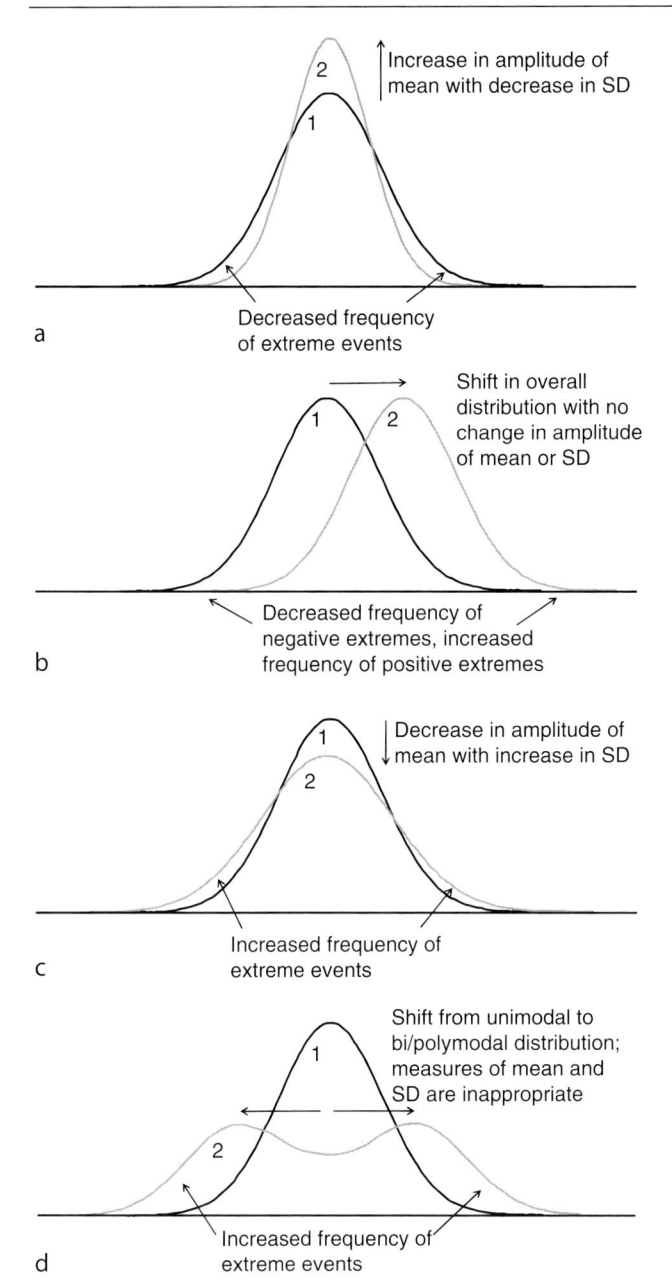

Climate Change, Figure 4 Schematic distribution plots of a meteorological variable showing four possible changes in the nature of the distribution from time 1 to time 2. (**a**) Where the mean value stays the same but where standard deviation (SD) decreases. This leads to decreased frequency of extreme events. (**b**) Where the distribution stays the same but shifts to a higher (or a lower) value with no accompanying change in SD. In the example of precipitation this results in a decreased probability of droughts (floods) and increased probability of floods (droughts). (**c**) Where the mean value stays the same but where the SD increases, giving an increased probability of both positive and negative extreme events. (**d**) Where the distribution plot changes from unimodal (normal distribution) to a bimodal or polymodal distribution. In this case there may be an increased probability of extreme events, and that standard measures of mean and SD are inappropriate descriptors of the distribution.

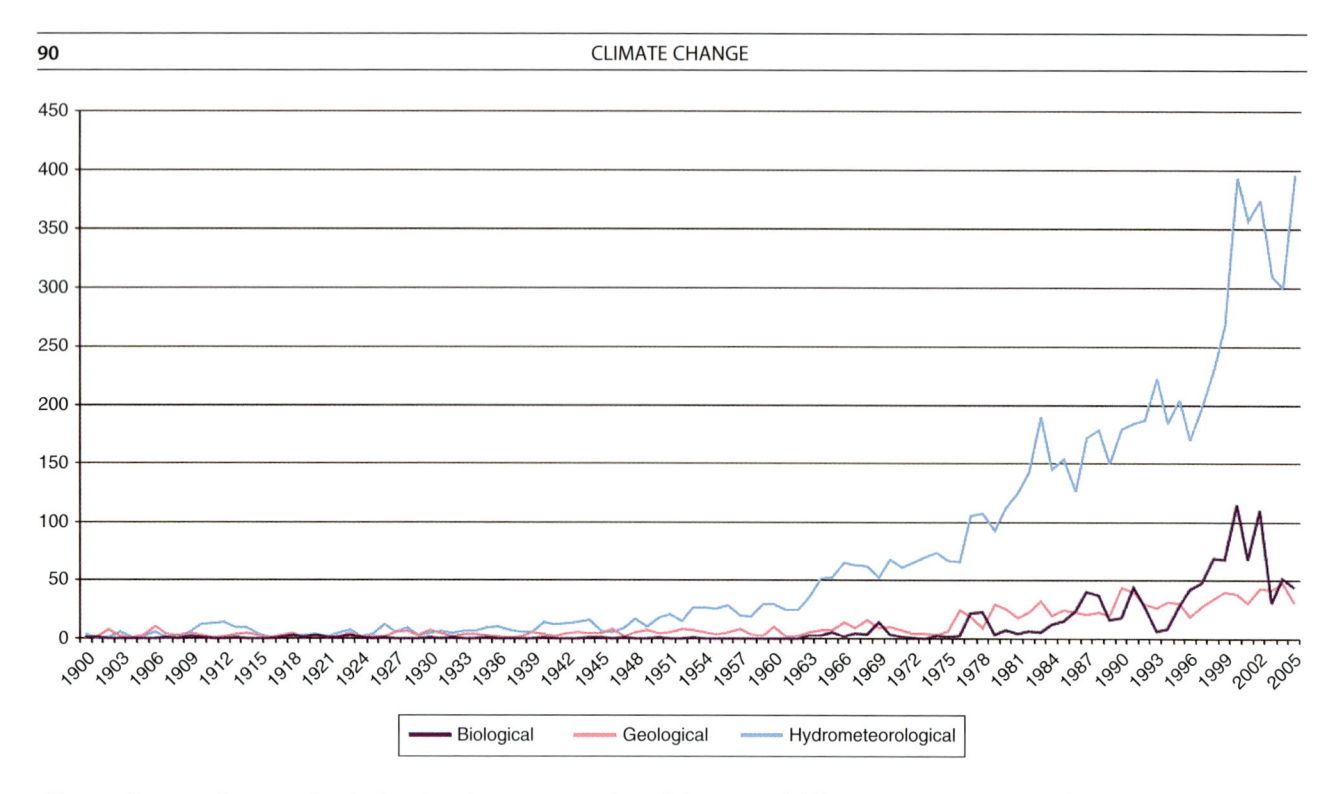

Climate Change, Figure 5 Graph showing changes in number of disasters of different types, 1900–2005 (from http://www.unisdr.org/disaster-statistics/occurrence-trends-century.htm). Note the recent increase in hydrometeorological hazards relative to other hazards types, and individual peaks that correspond to strong El Nino events in 1982 and 1998.

feedbacks associated with the long-term response of earth surface processes (Lunt et al., 2010). In combination, these factors mean that long-term climate sensitivity could well be higher than predicted by radiative forcing alone (e.g., Pagani et al., 2010). There are a number of important implications of climate sensitivity for climatic hazards. The role of feedback processes means that future patterns of precipitation are unlikely to show the same speed or magnitude of change as that of temperature. It may also mean that temperature reaches equilibrium at a different time to other meteorological variables.

Many meteorological hazards yield secondary impacts which are predominantly negative, can be long-lasting, and which affect many earth systems and the human environment. Secondary impacts include, but are not limited to, changes in fluvial sediment budgets; coastal erosion; biogeochemical cycling; chemical weathering; biodiversity; food production; and renewable energy production. The precise nature and timescale of these impacts are often poorly-defined, and although they often arise from individual hazardous events such as a single flood, such events in themselves cannot be attributed with certainty to anthropogenic climate change. Rather, the events represent one aspect of climate variability that arises out of a set of ongoing directional changes to earth and human systems that have cumulative effect (i.e., climate change). Climate change therefore provides the hydrometeorological setting from which individual hazardous events can arise. In this way, climate change (sensu lato) has potential to impact in unexpected, complex and unpredictable ways on both human activity and the physical environment (Hulme, 2009).

Individual hazardous events cannot be predicted. General circulation models (GCMs) can, however, predict global to regional scale changes in some meteorological variables (temperature, precipitation etc.) that can in turn provide input to regional ecological, geophysical and geochemical models. None of these model types downscale very well to the local level where issues of land surface stability, earth surface processes and human activity are most significant, and where real hazards take place. As such there is a mismatch between our knowledge of future climate and our knowledge of future climate responses. Impacts of future climate change on hazard characteristics (type, location, magnitude) are therefore still unknown and, as such, cannot be predicted with any certainty. This is an important area for future research.

Bibliography

Allen, M. R., and Ingram, W. J., 2002. Constraints on future change sin climate and the hydrological cycle. *Nature*, **419**, 224–232.

Allan, R. P., and Soden, B. J., 2008. Atmospheric warming and the amplification of precipitation extremes. *Science*, **321**, 1481–1484.

Bacon, S., and Carter, D. J. T., 1991. Wave climate changes in the North Atlantic and North Sea. *International Journal of Climatology*, **11**, 545–558.

Below, R., Wirtz, A., and Guha-Sapir, D. 2009. Disaster category classification and Peril terminology for operational purposes. CRED Working Paper 264. Available from http://cred.be/sites/default/files/DisCatClass_264.pdf.

Bhutiyani, M. R., Kale, V. S., and Pawar, N. J., 2008. Changing streamflow patterns in the rivers of northwestern Himalaya: implications of global warming in the 20th century. *Current Science*, **95**, 618–626.

Braganza, K., Karoly, D. J., Hirst, A. C., Stott, P., Stouffer, R. J., and Tett, S. F. B., 2004. Simple indices of global climate variability and change – Part II: attribution of climate change during the twentieth century. *Climate Dynamics*, **22**, 823–838.

Camill, P., 2005. Permafrost thaw accelerates in boreal peatlands during late-20th century climate warming. *Climatic Change*, **68**, 135–152.

Cook, K. H., and Vizy, E. K., 2008. Effects of twenty-first-century climate change on the Amazon rain forest. *Journal of Climate*, **21**, 542–560.

Cronin, T. M., 1999. *Principles of Paleoclimatology*. New York: Columbia University Press. 560.

Cunningham, S. A., Kanzow, T., Rayner, D., Baringer, M. O., Johns, W. E., Marotzke, J., Longworth, H. R., Grant, E. M., Hirschi, J. J.-M., Beal, L. M., Meinen, C. S., and Bryden, H. L., 2007. Temporal variability of the Atlantic meridional overturning circulation at 26.5°N. *Science*, **317**, 935–938.

Duan, A. M., and Wu, G. X., 2005. Role of the Tibetan Plateau thermal forcing in the summer climate patterns over subtropical Asia. *Climate Dynamics*, **24**, 793–807.

Fuchs, S., 2009. Susceptibility versus resilience to mountain hazards in Austria – paradigms of vulnerability revisited. *Natural Hazards and Earth System Sciences*, **9**, 337–352.

Gregory, J. M., Stoufer, R. J., Raper, S. C. B., Stott, P. A., and Rayner, N. E., 2002. An observationally based estimate of the climate sensitivity. *Journal of Climate*, **15**, 3117–3121.

Gude, M., and Barsch, D., 2005. Assessment of geomorphic hazards in connection with permafrost occurrence in the Zugspitze area (Bavarian Alps, Germany). *Geomorphology*, **66**, 85–93.

Hampel, A., Hetzel, R., and Maniatis, G., 2010. Response of faults to climate-driven changes in ice and water volumes on Earth's surface. *Philosophical Transactions of the Royal Society of London. Series A*, **368**, 2501–2517.

Hansom, J. D., 2001. Coastal sensitivity to environmental change: a view from the beach. *Catena*, **42**, 291–305.

Harris, C., Davies, M. C. R., and Etzelmüller, B., 2001. The assessment of potential geotechnical hazards associated with mountain permafrost in a warming global climate. *Permafrost and Periglacial Processes*, **12**, 145–156.

Hastenrath, S., 2007. Equatorial zonal circulations: Historical perspectives. *Dynamics of Atmospheres and Oceans*, **43**, 16–24.

Herwiejer, C., Seager, R., Winton, M., and Clement, A., 2005. Why ocean heat transport warms the global mean climate. *Tellus*, **Series A**(57A), 662–675.

Hilker, N., Badoux, A., and Hegg, C., 2009. The Swiss flood and landslide damage database 1972–2007. *Natural Hazards and Earth System Sciences*, **9**, 913–925.

Hulme, M., 2009. *Why We Disagree About Climate Change*. Cambridge: Cambridge University Press. 392.

Huntington, T. G., 2006. Evidence for intensification of the global water cycle: review and synthesis. *Journal of Hydrology*, **319**, 83–95.

IPCC, 2007. *Climate change 2007: synthesis report.* Contribution of Working Groups I, II and III to the Fourth Assessment Report of the Intergovernmental Panel on Climate Change. Geneva, Switzerland: IPCC, 104 pp.

ISDR, 2009. *Global assessment report and disaster risk reduction. risk and poverty in a changing climate*. United Nations, Geneva, 207 pp. Available from http://www.preventionweb.net/english/hyogo/gar/report/index.php?id=9413.

Jones, B. M., Arp, C. D., Jorgenson, M. T., Hinkel, K. M., Schmutz, J. A., and Flint, P. L., 2009. Increase in the rate and uniformity of coastline erosion in Arctic Alaska. *Geophysical Research Letters*, **36**, L03503, doi:10.1029/2008GL036205.

Kaas, E., Li, T.-S., and Schmith, T., 1996. Statistical hindcast of wind climatology in the North Atlantic and northwestern European region. *Climate Research*, **7**, 97–110.

Kehrwald, N. M., Thompson, L. G., Tandong, Y., Mosley-Thompson, E., Schotterer, U., Alfimov, V., Beer, J., Eikenberg, J., and Davis, M. E., 2008. Mass loss on Himalayan glacier endangers water resources. *Geophysical Research Letters*, **35**, L22503, doi:10.1029/2008GL035556.

Khrustalev, L. N., 2001. Problems of permafrost engineering as related to global climate warming. In Paepe, R., and Melnikov, V. (eds.), *Permafrost Response on Economic Development*. Amsterdam: Kluwer. Environmental Security and Natural Resources, pp. 407–423.

Lawford, R. G., Prowse, T. D., Hogg, W. D., Warkentin, A. A., and Pilon, P. J., 1995. Hydrometeorological aspects of flood hazards in Canada. *Atmosphere-Ocean*, **33**, 303–320.

Lorenz, D. J., and DeWeaver, E. T., 2007. The response of the extratropical hydrological cycle to global warming. *Journal of Climate*, **20**, 3470–3484.

Lunt, D. J., Haywood, A. M., Schmidt, G. A., Salzmann, U., Valdes, P. J., and Dowsett, H. J., 2010. Earth system sensitivity inferred from Pliocene modelling and data. *Nature Geoscience*, **3**, 60–64.

Mall, R., Singh, R., Gupta, A., Srinivasan, G., and Rathore, L., 2006. Impact of climate change on Indian agriculture: a review. *Climatic Change*, **78**, 445–478.

Mann, M. E., Woodruff, J. D., Donnelly, J. P., and Zhang, Z., 2009. Atlantic hurricanes and climate over the past 1,500 years. *Nature*, **460**, 880–883.

McBean, G., 2004. Climate change and extreme weather: a basis for action. *Natural Hazards*, **31**, 177–190.

Müller, M., Kaspar, M., and Matschullat, J., 2009. Heavy rains and extreme rainfall-runoff events in Central Europe from 1951 to 2002. *Natural Hazards and Earth System Sciences*, **9**, 441–450.

O'Hare, G., Sweeney, J., and Wilby, R., 2005. *Weather, Climate and Climate Change*. Harlow: Pearson.

Osterkamp, T. E., 2005. The recent warming of permafrost in Alaska. *Global and Planetary Change*, **49**, 187–202.

Pagani, M., Liu, Z., LaRiviere, J., and Ravelo, A. C., 2010. High Earth-system climate sensitivity determined from Pliocene carbon dioxide concentrations. *Nature Geoscience*, **3**, 27–30.

Patt, A. G., Tadross, M., Nussbaumer, P., Asante, M., Metzger, M., Rafael, J., Goujon, A., and Brundrit, G., 2010. Estimating least-developed countries' vulnerability to climate-related extreme events over the next 50 years. *PNAS*, **107**, 1333–1337.

Peart, M. R., Hill, R. D., and Fok, L., 2009. Environmental change, hillslope erosion and suspended sediment: Some observations from Hong Kong. *Catena*, **79**, 198–204.

Peng, S. B., Huang, J. L., Sheehy, J. E., Laza, R. C., Visperas, R. M., Zhong, X. H., Centeno, G. S., Khush, G. S., and Cassman, K. G., 2004. Rice yields decline with higher night temperature from global warming. *PNAS*, **101**, 9971–9975.

Rahmsdorf, S., 2002. Ocean circulation and climate during the past 120,000 years. *Nature*, **419**, 207–214.

Reading, A. J., 1990. Caribbean tropical storm activity over the past four centuries. *International Journal of Climatology*, **10**, 365–376.

Santer, B. D., Mears, C., Wentz, F. J., Taylor, K. E., Gleckler, P. J., Wigley, T. M. L., Barnett, T. P., Boyle, J. S., Brüggemann, W., Gillett, N. P., Klein, S. A., Meehl, G. A., Nozawa, T., Pierce, D. W., Stott, P. A., Washington, W. M., and Wehner, M. F., 2007. Identification of human-induced changes in atmospheric moisture content. *Proceedings of the National Academy of Sciences*, **104**, 15248–15253.

Schneider, S. H., Semenov, S., Patwardhan, A., Burton, I., Magadza, C. H. D., Oppenheimer, M., Pittock, A. B., Rahman, A., Smith, J. B., Suarez, A., and Yamin, F. 2007. Assessing key vulnerabilities and the risk from climate change. In: Parry, M. L.,

Canziani, O. F., Palutikof, J. P., van der Linden, P. J., and Hanson, C. E. (eds) *Climate Change 2007: Impacts, Adaptation and Vulnerability.* Contribution of Working Group II to the Fourth Assessment Report of the Intergovernmental Panel on Climate Change. Cambridge: Cambridge University Press, pp. 779–810.

Schuur, E. A. G., Vogel, J. G., Crummer, K. G., Lee, H., Sickman, J. O., and Osterkamp, T. E., 2009. The effect of permafrost thaw on old carbon release and net carbon exchange from tundra. *Nature,* **459,** 556–559.

Seidel, D. J., Fu, Q., Randel, W. J., and Reichler, T. J., 2008. Widening of the tropical belt in a changing climate. *Nature Geoscience,* **1,** 21–24.

Serreze, M. C., Barrett, A. P., Stroeve, J. C., Kindig, D. N., and Holland, M. M., 2009. The emergence of surface-based Arctic amplification. *The Cryosphere,* **3,** 11–19.

Serreze, M. C., Holland, M. M., and Stroeve, J., 2007. Perspectives on the Arctic's shrinking sea-ice cover. *Science,* **315,** 1533–1536.

Stive, M. J. F., 2004. How important is global warming for coastal erosion? An editorial comment. *Climatic Change,* **64,** 27–39.

Thorndycraft, V. R., Benito, G., and Gregory, K. J., 2008. Fluvial geomorphology: a perspective on current status and methods. *Geomorphology,* **98,** 2–12.

Vellinga, M. A., and Wood, R. A., 2008. Impacts of thermohaline circulation shutdown in the twenty-first century. *Climatic Change,* **91,** 43–63.

Wilby, R. L., Beven, K. J., and Reynard, N. S., 2008. Climate change and fluvial flood risk in the UK: more of the same? *Hydrological Processes,* **22,** 2511–2523.

Willett, K. M., Gillett, N. P., Jones, P. D., and Thorne, P. W., 2007. Attribution of observed surface humidity changes to human influence. *Nature,* **449,** 710–712.

Zhang, X., Zwiers, F. W., Hegerl, G. C., Lambert, F. H., Gillett, N. P., Solomon, S., Stott, P. A., and Nozawa, T., 2007. Detection of human influence on twentieth-century precipitation trends. *Nature,* **448,** 461–465.

Zickfeld, K., Morgan, M. G., Frame, D. J., and Keith, D. W., 2010. Expert judgments about transient climate response to alternative future trajectories of radiative forcing. *PNAS,* **107,** 12451–12456.

Cross-references

Casualties Following Natural Hazards
Coastal Erosion
Cryological Engineering
Disaster
Drought
Flood Hazard and Disaster
Hurricane
Hydrometeorological Disasters
Jökulhlaups
Mass Movement
Monsoon
Paraglacial
Permafrost

CLOUD SEEDING

Steven T. Siems
Monash University, Monash, VIC, Australia

Synonyms

Precipitation enhancement; Rain making

Definition

Cloud seeding is the practice of intentionally adding aerosols (e.g., silver iodide, common salt) or even ice itself (or dry ice) with the intent of changing the development of a cloud.

Discussion

Specifically the aim is to change either the phase of the cloud droplets/crystal (e.g., convert supercooled liquid water to ice for glaciogenic cloud seeding) or the size distribution (e.g., produce largely droplets that will coalesce more rapidly for hygroscopic seeding). Primarily cloud seeding has been used with the intent of enhancing the precipitation of a cloud, although efforts have been made to prevent precipitation, suppress hail, and burn off fog. The principles of cloud seeding have even been attempted for the suppression of hurricanes. Cloud seeding is perhaps the most common/vivid example of the practice of weather modification or weather engineering.

While cloud seeding is practiced in dozens of countries around the globe, it still remains a highly controversial practice primarily because of the lack of evidence supporting its effectiveness outside of fog dispersal. As stated in the World Meteorological Organisation Weather Modification Statement and Guidelines (2007): "Evidence for significant and beneficial changes in precipitation on the ground as a result of seeding is controversial and in many cases cannot be established with confidence." This lack of statistically significant evidence is even more apparent for the intent of hail suppression. These are the same conclusions reached the by committee established by the US Academy of Sciences in 2003.

Bibliography

Committee on the Status and Future Directions in U.S Weather Modification Research and Operations, 2003. *Critical Issues in Weather Modification Research.* Washington, DC: National Research Council.

WMO, 2007. WMO statement on the weather modification.

Cross-references

Doppler Weather Radar
Fog Hazard Mitigation
Hurricane
Lightning

COAL FIRE (UNDERGROUND)

Glenn B. Stracher
East Georgia College, University System of Georgia, Swainsboro, GA, USA

Synonyms

Burning coal; Coal combustion; Coal fire; Underground coal fire

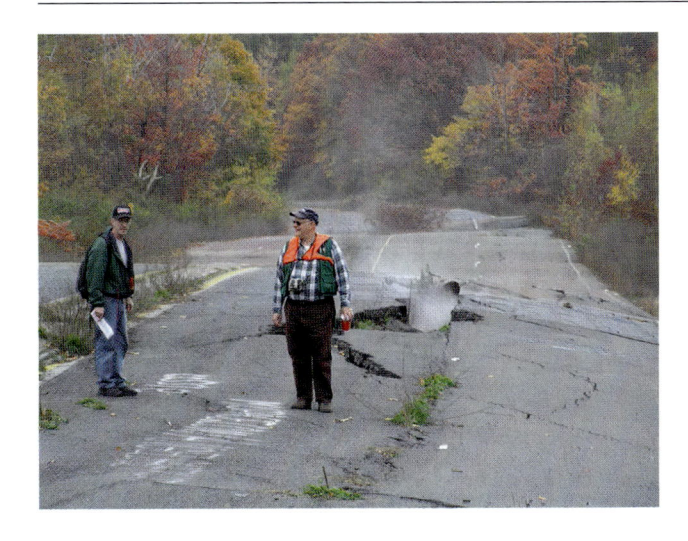

Coal Fire (Underground), Figure 1 Surface manifestation of an underground coal-mine fire includes ground subsidence and smoke, as shown here along a former section of Pennsylvania Route 61 at the Centralia mine fire (Stracher et al., 2006). The asphalt road through the town of Centralia is subsiding into abandoned coal-mine workings, where anthracite is burning underground. A detour was constructed around this section of the highway. Coal fires emit dozens of toxic gases into the atmosphere including carbon monoxide, benzene, and toluene, in addition to green-house gases. Minerals and creosote (coal tar) that nucleate at the surface from coal-fire gas may serve as vectors for the transmission of bio-accumulated pollutants, including Hg and As, from water, soil, and wind-blown dust (Stracher, 2004; Stracher et al., 2005, 2009) (Photo by Janet L. Stracher, 2006).

Definition

An underground coal fire is defined as the combustion of coal below the Earth's surface accompanied by heat-energy transfer and the emission of gas, but not necessarily flames and consequently, the emission of light. Although "fire" implies flames, coal burning underground is seldom observed, and peer-reviewed publications about underground burning or even coal burning at the surface (Stracher, 2004, 2007; Stracher et al., 2010, 2012) do not consider flames and light as a necessary criterion when describing coal fires. When a coal fire is not accompanied by flames, the terminology "smoldering" is sometimes used in reference to such fires (Hadden and Rein, 2010).

Discussion

Underground coal fires may occur just beneath, many meters below, or as is commonly the case – at an unknown depth below the Earth's surface. They may occur in association with active or abandoned coal mines (Figure 1) and also in coal seams that were never mined. Combustion is supported by the circulation of atmospheric oxygen underground to the burning coal via joints, faults, intake vents, and coal-mine portals.

Origin: Underground coal fires are started by either anthropogenic or non-anthropogenic processes. The causes include (Stracher and Taylor, 2004; Stracher et al., 2009) (1) lightning strikes; (2) forest, grass, or brush fires; (3) landfill, coal-gob pile, or coal-storage-facility fires transmitted to underground coal seams; (4) mining activities such as the use of explosives, welding or electrical work, and short circuits; (5) arson; (6) spontaneous combustion promoted by "self heating" of the coal during exothermic oxidation of sulfide minerals in the coal (e.g., pyrite); (7) the illegal distillation of liquor in underground coal-mine tunnels; and (8) smoking or other activities that may accidentally ignite hydrogen or methane gas in an underground mine.

Extinguishing: Any attempt to extinguish or contain an underground coal fire is based on available funding, and some fires are cost-prohibitive to extinguish. Fire-fighting technology includes the use of compressed-air foam, inert-gas injection, fly-ash grout, water and mud or water and fly-ash slurries, fire breaks, burial of gas vents and fissures at the surface beneath soil, remote sensing technology, and excavating the fire either by hand or mechanical digging.

Bibliography

Hadden, R., and Rein, G., 2010. Burning and water suppression of smoldering coal fires in small-scale laboratory experiments. In Stracher, G. B., Prakash, A., Sokol E. V. (eds.), *Coal and Peat Fires: A Global Perspective*, Chapter 18, Volume 1: Coal – Geology and Combustion. Amsterdam: Elsevier, pp. 317–326.

Stracher, G. B., 2004. *Coal Fires Burning around the World: a Global Catastrophe. International Journal of Coal Geology, 59*. Amsterdam: Elsevier, p. 151.

Stracher, G. B., 2007. *Geology of Coal Fires: Case Studies from Around the World. Reviews in Engineering Geology, XVIII*. Boulder: Geological Society of America, p. 283.

Stracher, G. B., and Taylor, T. P., 2004. Coal fires burning out of control around the world: thermodynamic recipe for environmental catastrophe. In Stracher, G. B. (ed.), *Coal fires Burning around the World: a Global Catastrophe, International Journal of Coal Geology, 59*. Amsterdam: Elsevier, pp. 7–17.

Stracher, G. B., Prakash, A., Schroeder, P., McCormack, J., Zhang, X., Van Dijk, P., and Blake, D., 2005. New mineral occurrences and mineralization processes: Wuda coal-fire gas vents of Inner Mongolia. *American Mineralogist, 90*. Mineralogical Society of America, pp. 1729–1739.

Stracher, G. B., Prakash, A., and Sorol, E. V., 2010. *Coal and Peat Fines: A Global perspective*, Volume 1: Coal-Geology and Combustion. pp. 357.

Stracher, G. B., Prakash, A., and Sorol, E. V., 2012. *Coal and Peat Fines: A Global perspective*, Volume 2: Photographs and Multimedia Tours. pp. 564.

Stracher, G. B., Nolter, M. A., Schroeder, P., McCormack, J., Blake, D. R., and Vice, D. H., 2006. The great Centralia mine fire: a natural laboratory for the study of coal fires. In Pazzaglia, F. J. (ed.), *Geological Society of America Field Guide 8, Excursions in Geology and History: Field Trips in the Middle Atlantic States*. Boulder: Geological Society of America, pp. 33–45.

Stracher, G. B., Finkelman, R. B., Hower, J. C., Pone, J. D. N., Prakash, A., Blake, D. R., Schroeder, P. A., Emsbo-Mattingly, S. D., and O'Keefe, J. M. K., 2009. Natural and Anthropogenic Coal Fires. In Cleveland, C. J. (ed.), *The Encyclopedia of Earth*. Washington, D.C.: Environmental Information Coalition, National Council for Science and the Environment, http://www.eoearth.org/article/Natural_and_anthropogenic_coal_fires.

Cross-references

Forest and Range Fires
Land Degradation
Land Subsidence
Lightning

COASTAL EROSION

Wayne Stephenson
University of Otago, Dunedin, New Zealand

Definition

Coastal erosion. The net landward retreat of the shoreline, as measured relative to a given datum, over a given temporal scale that is longer than cyclic patterns of coastal variability.

Introduction

Coastal erosion is the net landward retreat of the shoreline, where the shoreline is represented by a datum such as mean high water spring or a mapped feature such as a cliff, beach or road (Figure 1). The landward retreat is persistent through time as distinct from shorter term (e.g., weeks, months) fluctuations in the position of the shoreline associated with storm erosion and recovery, typical of beaches. On cliffed coasts there is no recovery from erosion events because such landforms are wholly erosional in origin. Coastal erosion is a common condition of coastlines globally, with 70% of the world's beaches thought to be eroding (Bird, 1987, 1996) and 80% of the world's total shoreline composed of eroding cliffed and rocky morphologies (Emery and Kuhn, 1982). Coastal erosion is also associated with coastal hazards when infrastructure is built on an eroding shore, coastal erosion overtakes human use systems, or when erosion increases the exposure of infrastructure to processes such as inundation or direct wave attack.

Whilst coastal erosion is considered to be a hazard and is a significant issue for coastal managers, it is also a natural geomorphic process representing the adjustment of shorelines toward a new equilibrium, often in response to sea level rise, changes in sediment supply, wave climate, or a combination of these factors. Much of the world's shoreline is configured the way it is because of erosional processes. Coastal landscapes valued for their intrinsic beauty are often erosional in origin (e.g., cliffs, arches, stacks) and contribute to economic activity, particularly tourism. Paradoxically, coastal erosion management may involve allowing erosive processes to take place in order to maintain landscape values.

Causes

There are three first order controls of coastal erosion applicable to all coastal landforms, a rise in relative sea level, a deficit in the local sediment budget, and a change in wave climate. In addition to these fundamental causes, human activities introduce a myriad of interactions that can generate coastal erosion where there was none before or make existing coastal erosion worse. Human activity usually involves interference with local sediment budgets but can also change processes such as wave energy or wave driven longshore currents that transport sediment.

Sea level rise since the last glacial maximum is one of the most important long-term causes of coastal erosion, and many shorelines are still responding to this rise. The melting of glaciers and continental ice sheets since the last glacial maximum added vast volumes of water to the world's oceans causing very rapid sea level rise. The broad configuration of the world's coastlines and coastal erosion observed over historical time frames is in response to sea level changes that have occurred over the last 6,000–7,000 years. In addition to eustatic (changes in ocean water volume) causes of sea level rise, isostatic processes can also cause a relative rise in sea level through subsidence of land masses. Isostatic processes include the addition or removal of mass from the Earth's crust, such as ice sheets, sediment, or water, and cause the crust to sink under loading and rebound after the load is removed, such as when ice sheets melt during warm phases of Earth's climate. Tectonic processes may also cause subsidence of land masses giving the appearance of sea level rising. When there is no eustatic change in sea level in this scenario, the shoreline will respond as if sea level is rising. However, a rise in sea level does not always result in coastal erosion, and in some instances, shorelines can respond by advancing when there is a supply of sediment eroded from elsewhere or transported onshore from the continental shelf, generating a positive sediment budget.

Sediment budgeting offers a useful management approach to coastal erosion by accounting for inputs and losses of sediment from a defined length of coast known as a littoral cell (e.g., Komar, 1996; Bray et al., 1995; Patsch and Griggs, 2008; Mazzer et al., 2009). Sediment budgets are usually expressed as a volume of sediment gained or lost annually from a littoral cell. A negative sediment budget (where a littoral cell loses sediment over time) equates to coastal erosion whereas a positive budget results in shoreline progradation. Littoral cells are often delimited by natural features such as headlands, river mouths, and cliffs, or artificial structures such as harbor breakwaters. Some boundaries are fixed (e.g., headlands) while others can be transient such as river mouths that migrate alongshore or the seaward limit of sediment transport that changes as wave conditions change. Cell boundaries can also be absolute, in that no sediment crosses them, or partial in that they allow some leakage of sediment.

Importantly, a properly accounted sediment budget with well-defined boundaries allows the cause of sediment loss to be identified and the amount of sediment loss to be quantified. Determining why a cell is losing sediment and the amount of the deficient in the budget allows for a better assessment of appropriate mitigation techniques (e.g., Cooper and Pethick, 2005; Bezzi et al., 2009).

Coastal Erosion, Figure 1 Coastal erosion occurring on the North Otago coast of South Island, New Zealand. The management strategy has been to abandon the road after attempts to build protective rip-rap structures failed (source of debris on the beach) (Photo: Wayne Stephenson).

Increases in wave heights have been observed in the North Atlantic (Wolf and Woolf, 2006) and the northwest Pacific over recent decades and have been linked to evidence for enhanced beach erosion (Allan and Komar, 2002, 2006). Increases in wave height and frequencies of storms are also expected to increase the incidence of coastal erosion and shoreline retreat under climate change scenarios (Zhang et al., 2004).

In addition to natural changes in sea level, sediment budgets, and wave patterns, human activities can also cause or exacerbate coastal erosion. Sea level rise from global warming will in many instances contribute to coastal erosion. Coastline development can remove, slow, or stop sediment delivery to coastal systems and prevent transport within or between littoral cells. The effect is often to shift neutral or positive sediment budgets into negative ones. Shorelines respond by eroding in order to restore the supply of sediment that has been lost. Dams on or extraction of sediment from rivers can cause coastal erosion by reducing sediment supply to the coast (Patsch and Griggs, 2008). Sediment extraction from dunes, beaches, reefs, or the nearshore for the purposes of construction may also cause coastal erosion. Seawalls built to stop cliff erosion may prevent sediment delivery to the coast and, where cliffs are important local sources of sediment for beaches, coastal erosion can result (Bird, 1987). Beach loss following the construction of a seawall is a widely recognized consequence. Wave energy is reflected by the wall, causing scour in front of the wall rather than dissipating wave energy as a beach would. Structures perpendicular to the shoreline such as groynes, built to intercept longshore sediment transport or retain sediment, may well accumulate sediment on one side (the up-drift side) but also cause erosion on the down-drift side, or farther along the shoreline by preventing sediment transport alongshore. Thus, many actions taken to prevent coastal erosion can make the situation worse or simply transfer the problem elsewhere along the coast. Breakwaters built for harbors can also inadvertently act as groyne and produce the same effects.

Measuring coastal erosion

Erosion rates are useful for planning and hazard mitigation purposes since they determine the amount of time remaining before infrastructure is overtaken or the rates at which land is being lost. Coastal erosion rates can be measured using a wide range of techniques, from regularly surveying of cross shore transects to aircraft using lasers known as LiDAR (Light Detection And Ranging). Maps, charts, and aerial photographs can be used to analyze shoreline change over longer historical time scales. Regardless of the method used, knowing the rate of shoreline retreat and/or the volume of sediment being lost from a coastal cell is critical for, first, determining that there is a real erosion problem and, second, deciding on the appropriate management response. Furthermore, the analysis needs to be over a sufficiently long timescale to remove short-term fluctuations associated with storm erosion and recovery in the case of sedimentary coasts, or noise associated with episodic failures common on cliffed coasts. Published rates of coastal erosion vary greatly, depending on shoreline type and geology and exposure. Erosion is also an episodic process both spatially and temporally. While erosion is commonly reported as a rate per year, this belies the highly episodic nature of the process.

Impacts

An obvious impact of coastal erosion is the permanent loss of land and infrastructure (Figure 1). Erosion of land can lead to loss of economic activity such as where beaches provide tourist amenity. Ecological function can also be lost as erosion progressively removes mangroves or salt marshes, and this is a significant issue when the coastal erosion is caused by human activity. Community vulnerability is increased as coastal erosion reduces the ability of shorelines to dissipate wave energy or to withstand barrier breaching. Vulnerability may also increase when artificial defenses (e.g., seawalls) are lost or damaged.

Response and mitigation

Approaches to erosion mitigation generally take one of two forms: retreat or defend. In the case of retreat, assets or activities can be abandoned or relocated to a safer position farther inland from the shore. However sociopolitical considerations work often to make this option difficult to employ. In many situations, this option is not viable because accommodation space behind the coast is unavailable to allow relocation. Defensive mitigation involves building a wide range of structures such as seawalls, breakwaters, groynes, and beach nourishment, where sediment is placed to recreate the natural buffer provided by beaches and dunes. Beach nourishment has become a popular and widely used method of erosion mitigation and is a useful approach since it addresses one of the fundamental causes of erosion – a deficit in the sediment budget. Psuty and Pace (2009) illustrate the use of sediment budgeting to better inform a beach nourishment approach to coastal erosion at Sandy Hook, New Jersey, USA. However, as with any response, it is not without problems, such as economic high costs, technical difficulties such as a lack of suitable sediment and negative ecological impacts (Grain et al., 1995; Speybroeck et al., 2006). For example, sand is often dredged from the sea floor disrupting or destroying benthic habitat and organisms. Sea turtle nesting success is known to be impacted when renourished sand is different to the original sand on the beach or when beach geometry is significantly altered (Brock et al., 2009). Alternatively well-designed beach nourishment could potentially improve nesting success by replacing lost beaches and nesting sites and restoring ecological opportunities. In situations where there is economic dependence on beach tourism, the cost (financial and ecological) of nourishment may be justifiable. Defensive structures are also expensive to build and maintain and often have detrimental impacts on shorelines, such as beach loss following seawall construction or down-drift erosion after a groyne is built. However, for high value assets, defensive structures are often necessary and beach loss is accepted. Importantly, there is no single response that is suitable for all coastal erosion problems and each case must be assessed on its own merits. The variety of responses to coastal erosion was reviewed by Pope (1997).

In situations where development has not occurred or is proposed, careful assessment of future erosion trends (based on measured or projected erosion rates) is necessary to avoid creating a new hazard. In such cases, planning schemes and zoning can be used to avoid creating erosion hazards. Predicting the position of future shorelines as sea levels rise is another approach and a wide variety of models are available for different shore types. A commonly used model to predict shoreline retreat in the face of sea level rise is the Bruun Rule, but the title is misleading since it is not a rule in a strict scientific sense but a rather simple model. The model requires a strict set of conditions to be met to successfully predict shoreline erosion, and more common than not these conditions are seldom meet (Cooper and Pilkey, 2004).

A common problem for coastal management is the view that shorelines should be rendered stable by erosion mitigation techniques, but this view is at odds with an environment that is naturally dynamic. Mitigation techniques that are also dynamic provide against current erosion problems and future climate change and sea level rise. Shoreline structures can be modified to return to dynamic behavior or built in such a way so as to allow dynamic behavior. For example, groynes can be constructed so as to be permeable to allow some sediment to continue to move alongshore (Nordstrom et al., 2007). Restoration of ecological function of mangroves, salt marshes, and coral reefs also provides protection from sea level rise, cyclones and tsunami, and hence coastal erosion. Such methods may be a more sustainable approach to erosion management than engineered structures.

Conclusions

Coastal erosion will continue to be a major hazard on coastlines into the future and in the face of projected climate change and increasing population pressure on coastal resources, it is likely to become an even more important hazard for coastal managers. Hazard management will need to come from adaptive strategies utilizing a wide range of mitigation techniques focused on managing human use systems.

Bibliography

Allan, J. C., and Komar, P. D., 2002. Extreme storms on the Pacific Northwest coast during the 1997–98 El Nino and 1998–99 La Nina. *Journal of Coastal Research*, **18**, 175–193.

Allan, J. C., and Komar, P. D., 2006. Climate controls on US West Coast erosion processes. *Journal of Coastal Research*, **22**, 511–529.

Bezzi, A., Fontolan, G., Nordstrom, K. F., Carrer, D., and Jackson, N. L., 2009. Beach nourishment and foredune restoration: practices and constraints along the Venetian Shoreline, Italy. *Journal of Coastal Research*, **SI56**, 287–291.

Bird, E. C. F., 1987. The modern prevalence of beach erosion. *Marine Pollution Bulletin*, **18**, 151–157.

Bird, E. C. F., 1996. *Beach Management*. Chichester: Wiley.

Bray, M. J., Carter, D. J., and Hooke, J. M., 1995. Littoral cell definition and budgets for central southern England. *Journal of Coastal Research*, **11**, 381–400.

Brock, K. A., Reece, J. S., and Ehrhart, L. M., 2009. The effects of artificial beach nourishment on marine turtles: differences between Loggerhead and Green Turtles. *Restoration Ecology*, **17**, 297–307.

Cooper, J. A. G., and Pilkey, O. H., 2004. Sea-level rise and shoreline retreat: time to abandon the Bruun Rule. *Global and Planetary Change*, **43**, 157–171.

Cooper, N. J., and Pethick, J. S., 2005. Sediment budget approach to addressing coastal erosion problems in St. Ouen's Bay, Jersey, Channel Islands. *Journal of Coastal Research*, **21**, 112–122.

Emery, K. O., and Kuhn, G. G., 1982. Sea cliffs: their processes, profiles, and classification. *Geological Society of America Bulletin*, **93**, 644–654.

Grain, D. A., Bolten, A. B., and Bjorndal, K. A., 1995. Effects of beach nourishment on sea turtles: review and research initiatives. *Restoration Ecology*, **3**, 95–104.

Komar, P. D., 1996. The budget of littoral sediments concepts and applications. *Shore and Beach*, **64**, 18–26.

Mazzer, A. M., Souza, C. R. G., and Dillenburg, S. R., 2009. A method to determinate coastal cells in sandy beaches at southeast coast of Santa Catarina Island, Brazil. *Journal of Coastal Research*, **SI56**, 98–102.

Nordstrom, K. F., Lampe, R., and Jackson, N. L., 2007. Increasing the dynamism of coastal landforms by modifying shore protection methods: examples from eastern German Baltic Sea coast. *Environmental Conservation*, **34**, 205–214.

Patsch, K., and Griggs, G., 2008. A sand budget for the Santa Barbara littoral cell, California. *Marine Geology*, **252**, 50–61.

Pope, J., 1997. Responding to coastal erosion and flooding damages. *Journal of Coastal Research*, **13**, 704–710.

Psuty, N. P., and Pace, J. P., 2009. Sediment management at Sandy Hook, NJ: an interaction of science and public policy. *Geomorphology*, **104**, 12–21.

Speybroeck, J., Bonte, D., Courtens, W., Gheskiere, T., Grootaert, P., Maelfait, J. P., Mathys, M., Provoost, S., Sabbe, K., Stienen, E. W. M., Van Lancker, V., Vincx, M., and Degraer, S., 2006. Beach nourishment: an ecologically sound coastal defence alternative? A review. *Aquatic Conservation: Marine and Freshwater Ecosystems*, **16**, 419–435.

Wolf, J., and Woolf, D. K., 2006. Waves and climate change in the north-east Atlantic. *Geophysical Research Letters*, **33**, L06604.

Zhang, Z., Douglas, B., and Leatherman, S., 2004. Global warming and coastal erosion. *Climatic Change*, **64**, 41–58.

Cross-references

Beach Nourishment (Replenishment)
Breakwater
Climate Change
Coastal Zone, Risk Management
Erosion
Sea Level Change
Tsunami
Zoning

COASTAL ZONE RISK MANAGEMENT

Norm Catto
Memorial University of Newfoundland, St. John's, NL, Canada

Definition

A coordinated strategy to cope with risks in the context of environmental, sociocultural, and sustainable multiple uses of the coastal zone.

Discussion

Risk management in the coastal zone requires an understanding of the evolving trends, particularly the desire for Integrated Coastal Zone Management (ICZM). A thorough understanding of the biophysical and ecoclimatic systems, using scientific principles and methods of investigation, is required. Socioeconomic concerns and maritime traditions must be incorporated into any management initiative. Management strategies must be practical, reflecting and respecting socioeconomic systems, demographic differences, existing political and regulatory systems, and biophysical conditions. They must also simultaneously respect the principles of conservation, stewardship, and sustainable development, and must acknowledge the human contribution to management.

In theory, ICZM requires:

- *Comprehensiveness* – all components of all systems must be included; all affected (or potentially affected) groups and individuals must be consulted and their views incorporated.
- *Coherence of elements* – the management plan as designed must "fit together," with individual components intermeshed; no overlapping jurisdictions or responsibilities, but perfect coordination between units with adjacent responsibilities; resource partitioning completely accomplished.
- *Consistency over time* – the plan remains in effect, not subject to political or regulatory change.
- *Cost-effectiveness of results* – the plan is economically (or socioeconomically) feasible, within the ability of the community or political jurisdiction to pay for its implementation.

In practice, this is unattainable. Not only is it impractical, but it also ignores the real nature of coastal systems: most significantly, by demanding consistency over time, it tacitly assumes that systems can reach a condition of stability. This is demographically and physically impossible. Although some degree of political consistency is desirable, any good plan must be suitably flexible to adapt to changing circumstances.

Applied specifically to risk management, this means that the hazard must be considered in the context of the coastal community or region. Coastal communities develop because they are adjacent to resources, such as fish species, convenient for transportation, attractive locations for humans to live and play, or for combinations of these reasons. Risks associated with waves and storms are inherent in any coastal setting. Some localities are also vulnerable to tsunami, slope failures, or terrestrial flooding. Communities are rarely settled in the complete absence of any perception of the hazards: although tsunami events may be rare, wave activity is a daily occurrence and is well known to maritime residents. The communities develop despite the known hazards, because the tangible benefits outweigh the hazards, both in visibility and frequency of occurrence.

Rather than thinking strictly of "cost-effectiveness" in terms of money, it is better and more productive to consider "effort-effectiveness." Does the risk management initiative make sense not only from a financial viewpoint but from the viewpoint of the amount of human effort required? Is the effort required to enforce a new regulation proportionate to the benefit? A risk management strategy in the coastal zone may affect demographic, sociocultural, or biophysical returns, in addition to fiscal ones.

These practical and conceptual issues and difficulties suggest that flexibility is the key to successful risk management in the coastal zone. The principles or theoretical ideals of Integrated Coastal Zone Management can be applied in a realistic way within the particular biophysical, political, and socioeconomic systems and subsystems.

Bibliography

Beatley, T., Brower, D. J., and Schwab, A. K., 2002. *An Introduction to Coastal Zone Management*. Washington: Island Press, 352 pp.

Dahl, E., Moksness, E., and Stottrup, J., 2009. *Integrated Coastal Zone Management*. Chichester: Wiley-Blackwell, 360 pp.

Cross-references

Beach Nourishment (Replenishment)
Breakwater
Coastal Erosion
Critical Infrastructure
Disaster Risk Management
Erosion
Erosivity
Flood Protection
Hazard and Risk Mapping
Hazardousness of Place
Ice and Icebergs
Indian Ocean Tsunami 2004
Marine Hazards
Pacific Tsunami Warning and Mitigation System (PTWS)
Red Tide
Rip Current
Sea-Level Change
Storm Surges
Storms
Tidal Bores
Tsunami
Waterspout

COGNITIVE DISSONANCE

Jaroslaw Dzialek
Jagiellonian University, Krakow, Poland

Synonyms

Contradictory beliefs; Misperception

Definition

Cognitive dissonance is an unpleasant sensation that appears when someone is confronted with two contradictory facts or ideas at the same time. People usually tend to keep consistency internalized and thus reduce the dissonance.

Overview

The cognitive dissonance theory is one of the most influential theories in social psychology. It was proposed and developed by Leon Festinger (1957). It states that people have some persistent beliefs about their physical and social environment, and try to behave in a self-consistent manner. When they encounter two cognitions (attitudes, beliefs, behaviors), which are relevant to each other, but dissonant at the same time, it generates an uncomfortable psychological tension. People are then motivated to reduce the dissonance by altering one of the causative cognitive elements. Consequently, it results in changing their attitudes, beliefs, and behaviors, or in attempts to justify or rationalize some of these. Dissonance reduction explains attitude and behavior changes of people, but fails in predicting what changes will happen in a particular situation (Glassman and Hadad, 2009). This human reaction can be interpreted as a kind of an irrational behavior as it prevents one from discovering real facts or in taking more appropriate decisions. From the psychological point of view, this mechanism serves to defend people's ego and keep a positive image of themselves (Aronson, 1979).

Cognitive dissonance processes help to understand the perception of natural hazards and disasters, and they especially explain why people living in hazardous areas tend to underestimate the risk and resign from undertaking appropriate mitigation actions. It results from misperception of natural hazards, particularly their frequency and intensity. People generally have difficulties in assessing the probability of rare phenomena and they usually think of disasters as something exceptional. They also feel they have limited control over such phenomena (National Research Council of the National Academies, 2006).

People living in hazardous areas may be confronted with dissonant ideas having both positive and negative consequences. On the one hand, they fear the eventuality that a disaster will happen that can cause losses and casualties. On the other hand, they may appreciate their living conditions or feel strongly attached to their neighborhood. Moreover, they might not have sufficient resources to move to a secure area. For example, inhabitants of volcanic slopes perceive higher benefits of living in a clean environment with opportunities to develop agriculture and tourism over the risk of eruptive and seismic activity (Chester et al., 2002). Understanding people's perception of natural hazards, particularly cognitive dissonance reduction, helps to improve hazard mitigation policies in cooperation with local population.

Bibliography

Aronson, E., 1979. *The Social Animal*. San Francisco: W. H. Freeman.
Chester, D. K., Dibben, C. J. L., and Duncan, A. M., 2002. Volcanic hazard assessment in Western Europe. *Journal of Volcanology and Geothermal Research*, **115**, 411–435.
Festinger, L., 1957. *A Theory of Cognitive Dissonance*. Stanford: Stanford University Press.
Glassman, W. E., and Hadad, M., 2009. *Approaches to Psychology*. London: McGraw-Hill Education.
National Research Council of the National Academies, 2006. *Facing Hazards and Disasters. Understanding Human Dimensions*. Washington: Committee on Disaster Research in the Social Sciences: Future Challenges and Opportunities, Division on Earth and Life Studies, National Research Council of the National Academies, The National Academies Press.

Cross-references

Insurance
Perception of Natural Hazards and Disasters
Risk Perception and Communication
Social–Ecological Systems
Sociology of Disasters

COLLAPSING SOIL HAZARDS

Andrew J. Stumpf
University of Illinois at Urbana-Champaign, Champaign, IL, USA

Synonyms

Hazards of collapsible or metastable soils

Definition

Collapsing soil hazard. A major hazard to natural land, disturbed ground, or engineered structures worldwide resulting from the structural collapse of constituents in soil. In most cases, collapse occurs following the wetting and loading of unsaturated materials (unconsolidated sediments), but soils with higher moisture content such as quick clays may undergo collapse as well. Collapsible soils also include those sediments that contain perennial ice or permafrost that has subsequently melted.

Introduction

Collapsing soils are not a local problem, but rather a worldwide phenomenon occurring on a variety of landscapes under different subsurface conditions. Soils may collapse catastrophically, but often signs of impending failure remain undetected especially in remote areas or on land modified by humans. The rate of collapsibility in soils depends on a number of factors such as their internal structure, moisture content and wetting history, degree of weathering or alteration, age, landscape position and mode of deposition, climatic conditions, mineralogy and shape of soil particles, presence of cementation, and compaction history due to loading (Dudley, 1970; Barden et al., 1973; Darwell and Denness, 1976; Rogers, 1995).

Alone, soil collapse annually results in hundreds of millions of dollars in damage to private and public property in the USA (Prior and Holzer, 1991). These collapse events may cause significant instability that engineers and

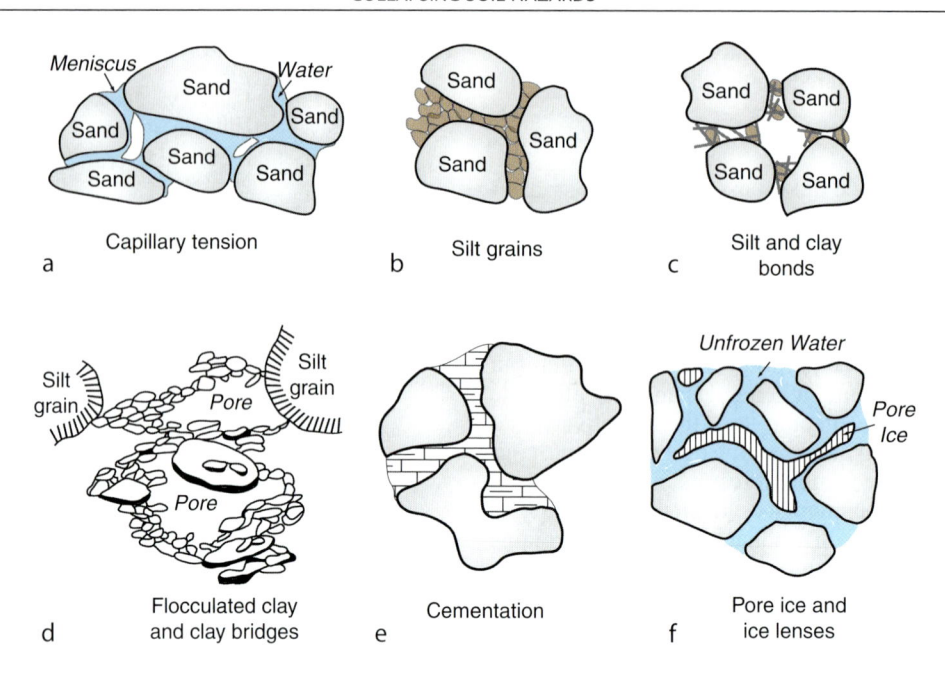

Collapsing Soil Hazards, Figure 1 Typical internal structure of collapsible soils. Soil particles and minerals are held together by: (**a**) capillary tension, (**b**) silt grains, (**c**) bonds containing silt and/or clay, (**d**) flocculated clay or clay aggregations, (**e**) mineral cements, and (**f**) pore ice or ice lenses (Modified from Collins and McGowan, 1974; Pavlik, 1980; Clemence and Finbarr, 1981; Holtz and Kovacs, 1981).

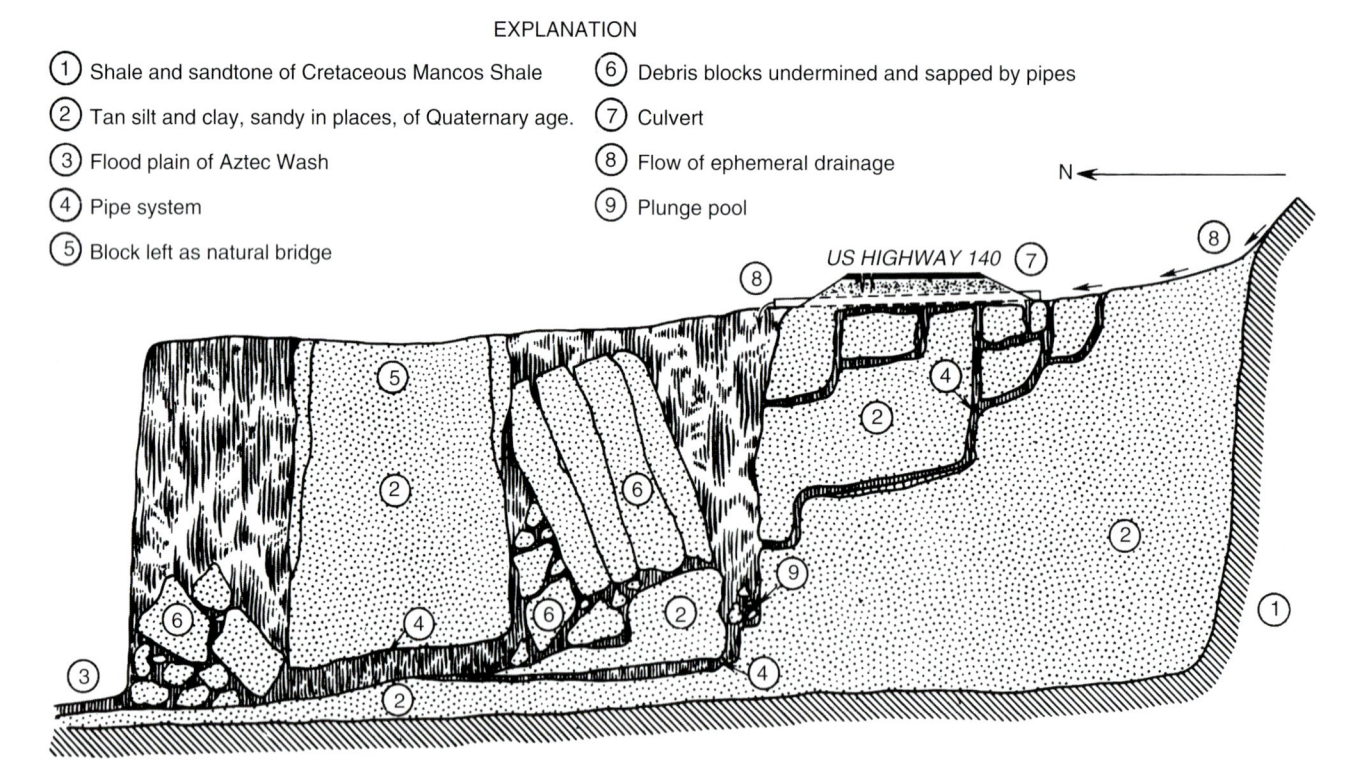

EXPLANATION

(1) Shale and sandtone of Cretaceous Mancos Shale
(2) Tan silt and clay, sandy in places, of Quaternary age.
(3) Flood plain of Aztec Wash
(4) Pipe system
(5) Block left as natural bridge
(6) Debris blocks undermined and sapped by pipes
(7) Culvert
(8) Flow of ephemeral drainage
(9) Plunge pool

Collapsing Soil Hazards, Figure 2 Idealized north to south cross-section of Aztec Wash alluvium in southwestern Colorado, USA showing incipient piping system beneath US highway 140 (Figure 18 in Parker and Jenne, 1967). As a result of culvert-concentrated drainage, a system of gullies developed by piping and subsidence that has undermined the roadbed (Reproduced with permission from the United States Geological Survey).

scientists must address in the design, construction, and maintenance of water distribution systems, pipelines, low gradient canals, power transmission lines, highways, railroads, buildings, and various aspects of land development and use (Curtin, 1973).

Properties of collapsible soils

Soils develop over a long period of time through a combination of physical, chemical, and biological processes that install distinctive internal structures and arrangement of soil particles. In situ compaction and wetting can further alter these soils. Collapsible or metastable soils are generally described as hard, dry, or partially saturated materials that have a low dry density and high porosity and will undergo an appreciable amount of volume change upon wetting, loading, or a combination of both (Sultan, 1969; Dudley, 1973; Handy, 1973; Jennings and Knight, 1975; Booth, 1977). These soils tend to be relatively young or recently altered, and have an open structure, a high void ratio, high sensitivity, and low interparticle bond strength (Derbyshire et al., 1995). Additional studies by Rosenqvist (1966), Czudek and Demek (1970), Mackay (1970), Torrance (1983), and MacKechnie (1992) provide additional information that show collapse may also occur in saturated soils, in transported or residual soils, dispersive soil, dispersive soil, and soils containing perennial ice. Soils overlying karst qv "*Karst hazard*" and soft bedrock may also collapse (Waltham et al., 2005), but these failures require the underlying rock be dissolved or fractured, processes not directly controlled by the overlying materials.

At the microscopic scale, collapse occurs when bonds between soil particles are broken and their internal structure weakened. Through this process, soil particles become further compacted producing subsidence that can be substantial and nonuniform (Barden et al., 1973; Clemence and Finbarr, 1981; Hunt, 2007). A soil is relatively stable when made up of uniform spherical particles that readily pack together to a near stable configuration reaching an optimum density (Rogers, 1995), whereas soils that contain much larger interparticle spaces between irregularly shaped grains, often of silt or fine sand size, are more prone to collapse (Clemence and Finbarr, 1981). Certain deposits containing flocculating clays may also collapse if the mineralogy and ion bonding are altered qv "*Dispersive soil hazard*" (Quigley, 1980).

Temporary strength in partially saturated, fine-grained cohesionless soils is provided through bonds between particles maintained by capillary tension, interlocking silt grains, silt and clay films and bridges, flocculated clay and clay agglomerations, chemical precipitates of iron oxide, calcium carbonate and gypsum, or pore ice and ice lenses as illustrated in Figure 1. Collapse in soil occurs once the bonds are broken particularly when: (a) water is applied increasing the degree of saturation; (b) loading exceeds the maximum strength; (c) cementing agents and salts are dissolved; and (d) permafrost is melted (Hunt, 2007; Muller et al., 2008).

Collapsing Soil Hazards, Figure 3 Complex retrogressive block collapse in thick loess on the fourth terrace of the Yellow River (Huang He) at Heifantai, ca. 60 km southwest of Lanzhou city, Gansu Province, China. Soil collapse has been accelerated by rapid growth of human settlement and resulting widespread irrigation of the terrace for agriculture. The river in the distance is the Huang Shui, here just above its junction with the Huang He. Photograph taken in 2006 (Reproduced with permission from E. Derbyshire).

Collapsing Soil Hazards, Figure 4 Photograph of the quick-clay slide at Rissa near Trondheim occurring in 1978. The largest quick-clay slide in Norway in the twentieth century, it covered a 330,000 m² area. The event began as a small failure in fill by a lake, but in a few hours approximately 6,000,000 m³ of soil had collapsed by retrogressive sliding (Gregersen, 1981). The deposits of liquid clay that spilled over the edge of the scar during failure are encircled. Note the houses for scale (Reproduced with permission from Elsevier Limited).

When the support is removed, particles are able to move and slide past one another to fill the vacant pore space (Clemence and Finbarr, 1981).

Wetting that penetrates deeply into unsaturated, low-density soils often causes interparticle bonds to be broken and washing out of silt and clay particles. This creates open pore space that promotes the denser packing of grains that can result in subsidence. This process, hydrocompaction qv "*Hydrocompaction subsidence*", is commonly observed in fine-grained soils of arid and semiarid environments (Rogers et al., 1994). Some of these soils, once wetted, are prone to significant subsurface erosion by piping qv "*Piping hazards*", which may lead to further subsidence (Paige-Green, 2008). Water entering the soil through fractures formed by desiccation or subsidence carries suspended clay particles to discharge points on slopes or permeable layers at depth. Over time, the fractures enlarge, undercutting the overlying materials (Figure 2). Piping is most prevalent in dispersive

Collapsing Soil Hazards, Figure 5 Slumping at Ester Drain on Gold Hill near Fairbanks, Alaska, USA. Following the melting of permafrost in loess, a large piping system developed causing rapid subsidence. Photograph taken by T. L. Pewe on September 14, 1949 (Reproduced with permission from the United States Geological Survey).

soils qv "*Dispersive soil hazards*"; soils that have a higher concentration of soluble salts that are developed in alluvial clays, windblown fine sand and silt (loess qv "*Loess*"), flood plain deposits qv "*Flood deposits*", and residuum on marine claystone and shale (Sherard et al., 1977). Knight (1963) and Holtz and Hilf (1961) have shown that collapse occurs in saturated and unsaturated soils following loading, particularly when the moisture content exceeds the liquid limit of the soil.

Soils containing weakly cemented particles or dispersive clays are highly susceptible to collapse after wetting (Figures 3 and 4), especially when exposed at the surface or interstratified with higher permeability sediments (Handy, 1973; Torrance, 1987). Reports by National Research Council (1985) and Wang et al. (2006) suggest that ground motions from earthquakes and underground explosions may produce enough energy to trigger collapse in these soils.

At high latitudes and altitudes, melting of perennial ice in both fine-grained and permeable soils may cause significant subsidence of the ground surface (Figure 5). In finer textured soils where drainage is poor, the melting of ice creates supersaturated conditions causing the sediment to liquefy (Andersland and Ladanyi, 2004). In more permeable soils, however, subsidence of the ground surface occurs by consolidation after the loss of excess water by drainage or evaporation (Murton, 2009).

Distribution of collapsible soils

Collapse in partly saturated or saturated soil is a phenomenon recognized throughout the world posing significant problems to engineered structures and land management. As noted previously, collapse soils are geologically young and have not undergone significant compaction or weathering by natural processes. Collapsible soils are most commonly found in Upper Pleistocene loess of the North America, central Europe, China, Africa, Russia, India, Argentina, and elsewhere (Rogers et al., 1994 and therein; Trofimov, 2001) however other soils considered prone to collapse are developed in:

1. Weathered bedrock (Brink and Kantey, 1961; Rao and Revanasiddappa, 2002; Pereira et al., 2005)
2. Aeolian sand (Knight, 1963; Amin and Bankher, 1997)
3. Glacial lake silt and clay (Clague and Evans, 2003; Kohv et al., 2009)
4. Glaciomarine and marine clay (Egashira and Ohtsubo, 1981; Geertsema et al., 2006: Hansen et al., 2007)
5. Alluvial and flood deposits (Parker and Higgins, 1990; Psimoulis et al., 2007; White and Greenman, 2008)
6. Organic deposits (Wösten et al., 1997; Haeberli and Burn, 2002)
7. Perennially frozen sediment (Demek, 1996; Jorgenson and Osterkamp, 2005)
8. Volcanic ash and dust (Wright, 2001; Iriondo and Kröhling, 2007)
9. Cemented soil (Ola, 1978; Petrukhin and Presnov, 1991)
10. Saline soil (Loveland et al., 1986; Azam, 2000)
11. Man-made materials (Booth, 1977; Herbstová et al., 2007)

Summary

The large number of studies undertaken worldwide to identify and predict the distribution of collapsing soil suggests that their presence in geologic and man-made materials is more common than originally presumed (this point is set forth in Derbyshire et al., 1995). There has been much debate between the soil scientists, geologists, geomorphologists, and geotechnical engineers on definition of collapsible soils. It is expected that with continued discussion between these academic disciplines, along with additional characterization studies of soils prone to collapse, a comprehensive criteria for identifying collapsible soils will assist local- and regional-based practitioners in determination and mitigating their hazards.

Bibliography

Amin, A., and Bankher, K., 1997. Causes of land subsidence in the Kingdom of Saudi Arabia. *Natural Hazards*, **16**, 57–63.

Andersland, O. B., and Ladanyi, B., 2004. *An Introduction to Frozen Ground Engineering*. Hoboken: Chapman and Hall.

Azam, S., 2000. Collapse and compressibility behavior of arid calcareous soil formations. *Bulletin of Engineering Geology and the Environment*, **59**, 211–217.

Barden, L., McGown, A., and Collins, K., 1973. The collapse mechanism in partly saturated soil. *Engineering Geology*, **7**, 49–60.

Booth, A. R., 1977. *Collapse Settlement in Compacted Soils*. Pretoria: National Institute for Transport and Road Research.

Brink, A. B, A., and Kantey, B. A., 1961. Collapsible grain structure in residual granite soils in South Africa. In *Proceedings Fifth International Conference on Soil Mechanics and Foundation Engineering*, Paris, pp. 611–614.

Clague, J. J., and Evans, S. G., 2003. Geologic framework of large historic landslides in Thompson River valley, British Columbia. *Environmental and Engineering Geoscience*, **9**, 201–212.

Clemence, S. P., and Finbarr, A. O., 1981. Design considerations for collapsible soils. *Journal of the Geotechnical Engineering Division*, **107**, 305–317.

Collins, K., and McGown, A., 1974. Form and function of microfabric features in a variety of natural soils. *Geotechnique*, **24**, 223–254.

Curtin, G., 1973. Collapsing soil and subsidence. In Moran, E. E., Slosson, J. E., Stone, R. O., and Yelverton, C. A. (eds.), *Geology, Seismicity, and Environmental Impact*. Los Angeles: University Publishers. Association of Engineering Geologists Special Publication, pp. 89–101.

Czudek, T., and Demek, J., 1970. Thermokarst in Siberia and its influence on the development of lowland relief. *Quaternary Research*, **1**, 103–120.

Darwell, J. L., and Denness, B., 1976. Prediction of metastable soil collapse. In *Proceedings of Anaheim Symposium*. International Association of Hydrological Sciences, Vol. 121, pp. 544–552.

Demek, J., 1996. Catastrophic implications of global climatic change in the cold regions of Eurasia. *GeoJournal*, **38**, 241–250.

Derbyshire, E., Dijkstra, T., and Smalley I. J., 1995. *Genesis and Properties of Collapsible Soils*. Proceedings of a Workshop, Loughborough, April 1994. NATO ASI Series C: mathematical and physical sciences, Vol. 468, Dordrecht: Kluwer/NATO Scientific Affairs Division.

Dudley, J. H., 1970. Review of collapsing soils. *Journal of the Soil Mechanics and Foundation Division*, **96**, 925–947.

Egashira, K., and Ohtsubo, M., 1981. Low-swelling smectite in a recent marine mud of Ariake Bay. *Soil Science and Plant Nutrition*, **27**, 205–211.

Geertsema, M., Cruden, D. M., and Schwab, J. W., 2006. A large rapid landslide in sensitive glaciomarine sediments at Mink Creek, northwestern British Columbia, Canada. *Engineering Geology*, **83**, 36–63.

Gregersen, O., 1981. *The Quick Clay Landslide in Rissa, Norway; The Sliding Process and Discussion of Failure Modes*. Stockholm: Norwegian Geotechnical Institute.

Haeberli, W., and Burn, C. R., 2002. Natural hazards in forests: glacier and permafrost effects as related to climate change. In Sidle, R. C. (ed.), *Environmental Change and Geomorphic Hazards in Forests*. Wallingford: CABI Publishing, pp. 167–203.

Handy, R. L., 1973. Collapsible Loess in Iowa. *Proceedings of the Soil Science Society of America Journal*, **37**, 281–284.

Hansen, L., Eilertsen, R. S., Solberg, I. L., Sveian, H., and Rokoengen, K., 2007. Facies characteristics, morphology and depositional models of clay-slide deposits in terraced fjord valleys, Norway. *Sedimentary Geology*, **202**, 710–729.

Herbstová, V., Boháč, J., and Herle, I., 2007. Suction and collapse of lumpy spoilheaps in northwestern Bohemia. *Experimental Unsaturated Soil Mechanics*, **112**, 293–300.

Holtz, W. G., and Hilf, J. W., 1961. Settlement of soil foundations due to saturation. In *Proceedings of the 5th International Conference on Soil Mechanics and Foundation Engineering*, Vol. 1, Paris, pp. 673–679.

Holtz, R. D., and Kovacs, W. D., 1981. *An Introduction to Geotechnical Engineering*. Englewood Cliffs: Prentice-Hall.

Hunt, R. E., 2007. *Geologic Hazards – A Field Guide for Geotechnical Engineers*. Boca Raton: Taylor and Francis Group.

Iriondo, M. H., and Kröhling, D. M., 2007. Non-classical types of loess. *Sedimentary Geology*, **202**, 352–368.

Jennings J. E., and Knight, K., 1975. Guide to construction on or with materials exhibiting additional settlement due to collapse of grain structure. In *Proceedings of Sixth Regional Conference for Africa on Soil Mechanics and Foundation Engineering*, Johannesburg, pp. 99–105.

Jorgenson, M. T., and Osterkamp, T. E., 2005. Response of boreal ecosystems to varying modes of permafrost degradation. *Canadian Journal of Forest Research*, **35**, 2100–2111.

Knight, K., 1963. The origin and occurrence of collapsing soils. In *Proceedings of the 3rd African Conference on Soil Mechanics and Foundation Engineering*, Vol. 1, Salisbury, pp. 127–130.

Kohv, M., Talviste, P., Hang, T., Kalm, V., and Rosentau, A., 2009. Slope stability and landslides in proglacial varved clays of western Estonia. *Geomorphology*, **106**, 315–323.

Loveland, P. J., Hazelden, J., Sturdy, R. G., and Hodgson, J. M., 1986. Salt-affected soils in England and Wales. *Soil Use and Management*, **2**, 150–156.

Mackay, J. R., 1970. Disturbances to the tundra and forest tundra environment of the western Arctic. *Canadian Geotechnical Journal*, **7**, 420–432.

MacKechnie, W. R., 1992. Collapsible and swelling soils. In *Proceedings of the Twelfth International Conference on Soil Mechanics and Foundation Engineering*, Vol. 12, Rio De Janeiro, pp. 2485–2490.

Muller, S. W., French, H. M., and Nelson, F. E., 2008. *Frozen in Time: Permafrost and Engineering Problems*. Reston: American Society of Civil Engineers.

Murton, J. B., 2009. Global warming and thermokarst. In Margesin, R. (ed.), *Permafrost Soils*. Berlin: Springer, pp. 185–203.

National Research Council, 1985. Liquefaction of soils during earthquakes. In *Report from the National Science Foundation Workshop on Liquefaction*. Washington: National Academy Press.

Ola, S. A., 1978. The geology and geotechnical properties of the black cotton soils of northeastern Nigeria. *Engineering Geology*, **12**, 375–391.

Paige-Green. P., 2008. Dispersive and erodible soils – fundamental differences. In *Proceedings of Problem soils in South Africa*

conference. South African Institute for Engineering and Environmental Geologists, pp. 59–65.
Parker, G. G. Jr., and Higgins, C. G., 1990. Piping and pseudokarst in drylands, with case studies by G. G. Parker, Sr. and W. W. Wood. In Higgins, C. G., and Coates, D. R. (eds.), *Groundwater Geomorphology; The Role of Subsurface Water in Earth-Surface Processes and Landforms*. Geological Society of America Special Paper, Vol. 252, pp. 77–110.
Parker, G. G. and Jenne, E. A., 1967. Structural failure of western U.S. highways caused by piping. In *Symposium on Subsurface Drainage*. National Academy of Sciences, Highway Research Board, Highway Research Record no. 203, pp. 57–76.
Pavlik, H. F., 1980. A physical framework for describing the genesis of ground ice. *Progress in Physical Geography*, **4**, 531–548.
Pereira, J. H. F., Fredlund, D. G., Cardão Neto, M. P., and De Gitirana, G. F. N., Jr., 2005. Hydraulic behavior of collapsible compacted gneiss soil. *Journal of Geotechnical and Geoenvironmental Engineering*, **131**, 1264–1273.
Petrukhin, V., and Presnov, O., 1991. Collapse deformations of gypsum sands. *Soil Mechanics and Foundation Engineering*, **28**, 127–130.
Prior, D. B., and Holzer, T. L., 1991. *Mitigating Losses from Land Subsidence in the United States*. Washington, DC: National Academy Press.
Psimoulis, P., Ghilardi, M., Fouache, E., and Stiros, S., 2007. Subsidence and evolution of the Thessaloniki plain, Greece, based on historical leveling and GPS data. *Engineering Geology*, **90**, 55–70.
Quigley, R. M., 1980. Geology and mineralogy, and geochemistry of Canadian soft soils: a geotechnical perspective. *Canadian Geotechnical Journal*, **17**, 261–285.
Rao, S. M., and Revanasiddappa, K., 2002. Collapse behavior of a residual soil. *Geotechnique*, **52**, 259–268.
Rogers, C. D. F., 1995. Types and distribution of collapsible soils. In Derbyshire, E., Dijkstra, T., and Smalley, I. J. (eds.), *Genesis and Properties of Collapsible Soils*. Dordrecht: Kluwer/NATO Scientific Affairs Division, pp. 1–17.
Rogers, C. D. F., Dijkstra, T. A., and Smalley, I. J., 1994. Hydroconsolidation and subsidence of loess: studies from China, Russia, North America and Europe. *Engineering Geology*, **37**, 83–113.
Rosenqvist, I. T., 1966. Norwegian research into the properties of quick clay – a review. *Engineering Geology*, **1**, 445–450.
Sherard, J. L., Dunnigan, L. P., and Decker, R. S., 1977. Identification and nature of dispersive soils. *Journal of the Geotechnical Engineering Division, Proceedings of the American Society of Civil Engineers*, **102**, 287–301.
Sultan, H. A., 1969. Collapsing soils. In *Proceedings Seventh International Conference on Soil Mechanics and Foundation Engineering*, Speciality Session no. 5, Mexico City: Sociedad Mexicana de Mecanica.
Torrance, J. K., 1983. Towards a general model of quick clay development. *Sedimentology*, **30**, 547–555.
Torrance, J. K., 1987. Quick clays. In Anderson, M. G., and Richards, K. S. (eds.), *Slope Stability: Geotechnical Engineering and Geomorphology*. New York: Wiley, pp. 447–474.
Trofimov, V. T., 2001. *Loess Mantle of the Earth and its Properties*. Moscow: Moscow University Press (in Russian).
Waltham, T., Bell, F. G., and Culshaw, M. G., 2005. *Sinkholes and Subsidence Karst and Cavernous Rocks in Engineering and Construction*. Berlin: Springer.
Wang, C., Wong, A., Dreger, D. S., and Manga, M., 2006. Liquefaction limit during earthquakes and underground explosions: implications on ground-motion attenuation. *Bulletin of the Seismological Society of America*, **96**, 355–363.
White J. L., and Greenman C., 2008. *Collapsible Soils in Colorado*. Denver: Colorado Geological Survey, Report 14.
Wösten, J. H. M., Ismail, A. B., and van Wijk, A. L. M., 1997. Peat subsidence and its practical implications: a case study in Malaysia. *Geoderma*, **78**, 25–36.
Wright, J. S., 2001. "Desert" loess versus "glacial" loess: quartz silt formation, source areas and sediment pathways in the formation of loess deposits. *Geomorphology*, **36**, 231–256.

Cross-references

Dispersive Soil Hazards
Flood Deposits
Hydrocompaction Subsidence
Karst Hazards
Land Subsidence
Loess
Piping Hazards

COMET

Paul R. Weissman
Jet Propulsion Laboratory/California Institute of Technology, Pasadena, CA, USA

Definition

Cometary nucleus – The solid, icy-conglomerate body that is the heart of the comet and the source of its activity.

Cometary coma – The freely outflowing atmosphere of dust and gas around the nucleus.

Comet tails – The coma materials separate and form two tails trailing behind the nucleus, one composed of dust and one composed of ionized gas molecules (see Figure 1).

Introduction

Comets are primitive bodies left over from the formation of the solar system. They were among the first solid bodies to form in the solar nebula, the collapsing interstellar cloud of dust and gas out of which the Sun and planets formed. Comets formed in the outer reaches of the planetary system where it was cold enough for volatile ices to condense. This is generally taken to be beyond 5 AU (astronomical units), or beyond the orbit of Jupiter. Because comets have been stored in distant orbits beyond the planets, they have undergone little of the modifying processes that have melted or changed most other bodies in the solar system. Thus, they retain a physical and chemical record of the primordial solar nebula and of the processes involved in the formation of planetary systems.

Cometary nuclei

Cometary nuclei are small bodies, typically only a few kilometers in diameter, and composed of roughly equal parts of volatile ices, silicate dust, and organic materials. The ices are dominated by water ice (~80% of the total ices) but also include carbon monoxide, carbon dioxide, formaldehyde, and methanol. The silicate and organic mix is similar to that found in the most primitive meteorites, carbonaceous chondrites. These materials are intimately mixed at micron scales. Images of the five

Comet, Figure 1 Comet Hale-Bopp in 1997, showing the major components that make up a typical comet. The nucleus is embedded deep within the bright coma and is too small and dark to see in this image.

cometary nuclei visited so far by interplanetary spacecraft are shown in Figure 2.

The nuclei formed in the solar nebula as dust and ice particles settled to the central plane of the nebula. When these particles collided they tended to stick. Micron-sized particles grew through this process of agglomeration and accretion to meter-sized and then kilometer-sized bodies.

When cometary nuclei come close to the Sun, the ices in them sublimate, transforming directly from the solid to the vapor phase. The evolving gas molecules flow off the nucleus surface, carrying with them silicate and organic dust particles embedded in the ices. This outflowing mix of materials then forms the cometary coma, the comet's atmosphere. Because cometary nuclei are small, their gravity is too weak to retain this atmosphere and it flows freely out into space.

Because the different ices sublimate at different temperatures, gases are liberated from different depths below the surface as the solar heat wave penetrates into the surface. So the layers closest to the surface become progressively depleted in the most volatile ices. Also, a lag deposit of nonvolatile dust develops on the surface. These are typically particles too large to be lifted by the escaping gases. This nonvolatile layer can become so thick that it effectively insulates the icy component below it, preventing further sublimation.

Another feature of cometary activity is driven by the fact that the water ice in comets condensed at very low temperatures, <100 K. At these low temperatures, ice forms in the amorphous state, a random ordering of molecules. As the amorphous water ice is warmed above $\sim$115 K, it begins to transform to crystalline ice, first in the cubic form and then the normal hexagonal ice that we are most familiar with. This transition is complete at $\sim$153 K. It is an exothermic reaction, that is, it releases energy. This energy further sustains the reaction as it warms the ice around it, but dies out because it must also heat the nonvolatile dust components of the nucleus. The amorphous-crystalline ice transition may be one source of cometary outbursts, sharp increases in cometary activity that appear to occur randomly.

The internal structure of cometary nuclei is still an area of speculation. It is generally believed that as icy planetesimals came together at low velocity in the solar nebula, there was not enough energy to melt or compress them into a single solid body. The two leading explanations suggest that cometary nuclei are "fluffy aggregates" or "primordial rubble piles" with low binding strength and high porosity. Key data supporting these models are estimates of nucleus bulk density, ranging from 0.2 to 1.5 g/cm^3, with preferred values of $\sim$0.3–0.6 g/cm^3. This suggests a combined microscopic and macroscopic porosity of $\sim$64% or more, a very high value.

Additional evidence for the rubble pile model for cometary nuclei comes from observations of split (disrupted) cometary nuclei. Observations show that nuclei can randomly break apart, shedding a few or many pieces. These pieces have typically been estimated to be between 8 and 60 m in diameter. In some cases, the entire nucleus disrupts. Disruption can also occur if the nucleus passes too close to the Sun or to a large planet like Jupiter, where gravitational tidal forces tear the weakly bound nuclei apart. This has been observed for Sun-grazing comets, comets with perihelia within one solar radius of the Sun's photosphere.

A particularly interesting case of a tidally disrupted nucleus is that of comet Shoemaker-Levy 9. This comet was discovered in 1993 as a string of 21 separate but co-moving, active nuclei. Observations showed that the

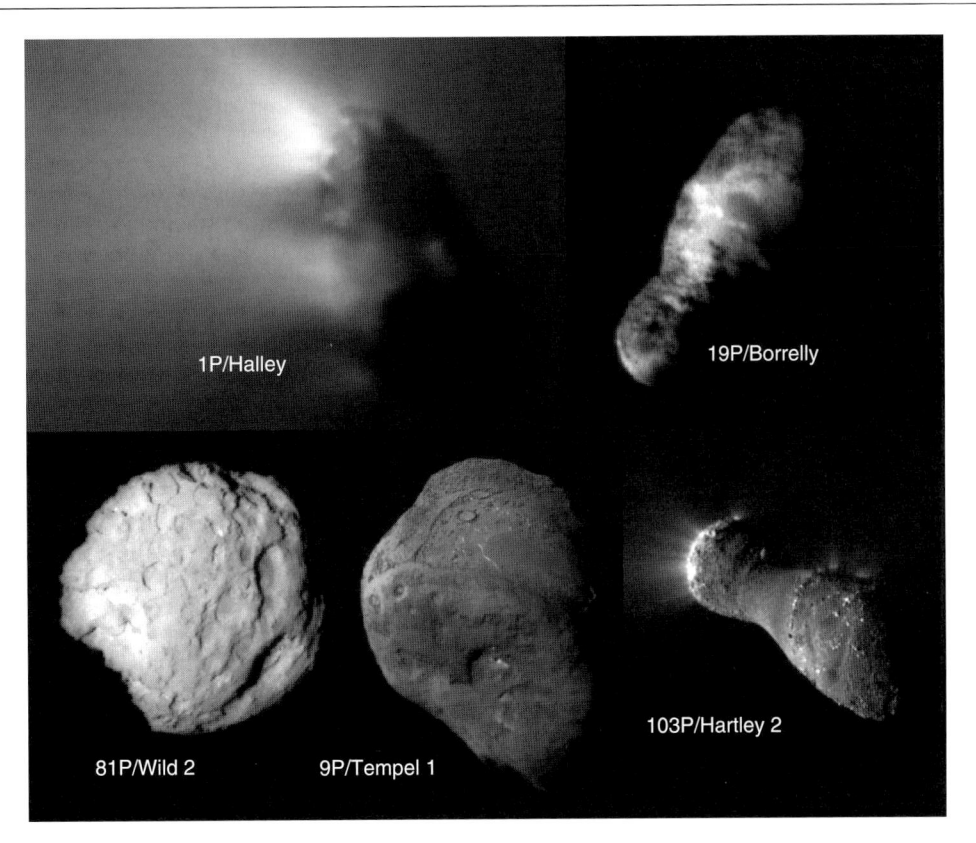

Comet, Figure 2 The five cometary nuclei imaged to date by flyby spacecraft: comet Halley (1986 by Giotto), comet Borrelly (2001 by Deep Space 1), comet Wild 2 (2004 by Stardust), comet Tempel 1 (2005 by Deep Impact), and comet Hartley 2 (2010 by EPOXI). The nuclei range in size from 16 × 8 km for Halley, down to 2 × 1 km for Hartley 2. Note the considerable differences in topographic features on each of the nuclei. These may be due to the surface features evolving as each comet makes more returns close to the Sun.

comet had been captured into orbit around Jupiter, and had passed so close to Jupiter on its last perijove passage, 1.3 Jupiter radii, that it was tidally disrupted. The nucleus appeared to have broken into thousands of separate "cometesimals." As this swarm of bodies moved away from Jupiter, their own self-gravity caused them to clump together. Interestingly, the number of final clumps was shown to be a function of the bulk density of the original nucleus, and the best fit was obtained for densities of ~ 0.6 g/cm^3. Thus, comet Shoemaker-Levy 9 is another of the proofs of a low-density, rubble pile or aggregate structure for cometary nuclei.

Cometary atmospheres

Because of the small size and low gravity of the cometary nuclei, the evolving gases from sublimating ices expand freely into the vacuum of space. Entrained in the outflowing gas are fine dust particles, typically a micron in size, composed of both silicates and organics. Because the molecules are exposed to sunlight and the solar wind they begin to disassociate, breaking up into radicals and individual atoms. The most common case of this is the water molecule, which disassociates into H and OH. Organic dust grains appear to also release radicals into the outflowing coma, most common of which are CN, C_2, and C_3. These are known as "daughter" molecules and cometary spectroscopy is used to study the chemistry that goes on in the coma as the parent and daughter molecules, radicals, and individual atoms react with each other.

The observed composition of volatiles in cometary comae is very similar to that seen in dense, cold interstellar clouds where stars and solar systems are being formed. This reinforces strongly the belief that comets are frozen remnants of the primordial solar nebula, preserving unmodified volatiles from the formation of the planetary system, 4.56 billion years ago.

Cometary comae often show geyser-like structures, or "jets," which are taken as evidence of individual active areas on the surfaces of the nuclei. As noted above, lag deposits of large dust grains can shut down sublimation on the surface. Because the nature of the source vents for the cometary activity is as yet unknown, there is no good explanation as to why some areas remain active and others do not. It is known that this is likely an ageing effect, as the "active fraction" on the nucleus is large for long-period and Halley-type comets (see below), which have made relatively few approaches close to the Sun, and very low, typically only a few percent, for Jupiter-family comets, which have made hundreds of returns, on average.

Comets also can display "outbursts," which are large, sudden releases of dust and gas. The most famous of these is comet Holmes in 2007, which brightened by 15 magnitudes (one million times brighter) in less than a day. The explanation for outbursts, and their larger cousins, disruption events, is as yet unknown, though rotational spin-up due to torques from coma outgassing, has been suggested, and the amorphous-crystalline ice transition (see above) is also likely a factor.

Comet tails

The outflowing dust and gas in the coma also interacts with the solar wind and sunlight. The fine dust is blown away from the Sun by radiation pressure on the tiny grains. This forms a broad, curved sometimes yellow-colored tail following the comet in its orbit and pointed generally away from the Sun. This is known as a Type I tail. The molecules suffer a different fate as they are ionized by charge exchange with the solar wind. Once ionized, they are caught up in the Sun's magnetic field and flow away at high velocity in the solar wind. This process forms long, narrow, straight trails that glow blue in color due to the presence of CO^+ molecules. These tails point sharply away from the Sun and are known as Type II tails. Well before the first spacecraft observations of the solar wind in 1959, the existence of the solar wind was inferred from the appearance of cometary ion tails.

Dynamics

Comets are typically in more eccentric and more inclined orbits than other bodies in the solar system. In general, comets are classified into three dynamical groups: the Jupiter-family comets with orbital periods less than 20 years, the Halley-type comets with orbital periods between 20 and 200 years, and the long-period comets with orbital periods greater than 200 years. A more formal definition involves a quantity called the Tisserand parameter:

$$T = a_J/a + 2\sqrt{(a/a_J)(1-e^2)}\cos i \qquad (1)$$

where a, e, and i are the semimajor axis, eccentricity, and inclination of the comet's orbit and a_J is the semimajor axis of Jupiter's orbit. Jupiter-family comets have Tisserand parameters between 2.0 and 3.0 and Halley-type and long-period comets have T values less than 2.0. Asteroids have T values greater than 3.0. However, there are both some comets whose orbits have evolved to T values slightly greater than 3, and some asteroids with T values slightly less than 3.

Another important difference in the dynamical groups is their orbital inclination distributions. Jupiter-family comets typically have orbits that are modestly inclined to the ecliptic (the plane of the Earth's orbit), with inclinations up to about 35°. The Halley-type comets have much higher inclinations, including retrograde orbits that go around the Sun in the opposite direction, though not totally randomized. The long-period comets have totally random inclinations and can approach the planetary system from all directions. As a result, the Jupiter-family comets are also known as "ecliptic comets," whereas the long-period comets are also known as "isotropic comets."

The inclinations of the cometary orbits provide important clues to their origin. Dynamical simulations show that the great concentration of Jupiter-family orbits close to the ecliptic can only originate from a flattened source of comets. This source has been identified as the Kuiper belt, a flattened disk of icy bodies beyond the orbit of Neptune and extending to at least 50 AU from the Sun. The Kuiper belt is analogous to the asteroid belt and is composed of ice-rich bodies that never had enough time to form into a larger planet.

The exact source of the Jupiter-family comets is called the Scattered disk, Kuiper belt comets that are in more inclined and eccentric orbits with perihelia close to Neptune. Neptune can gravitationally scatter comets from the Scattered disk inward to become Jupiter-family comets, or outward, to the Oort cloud (see below).

The origin of the long-period comets with their random inclinations was a mystery until 1950 when Dutch astronomer Jan Oort proposed that these comets came from a vast cloud of comets surrounding the solar system and stretching to interstellar distances. The key to recognizing this was the distribution of orbital energies, which showed that a large fraction of the long-period comets were in very distant orbits with semimajor axes of ∼25,000 AU or more. The orbits of comets in the "Oort cloud," as it is now known, are so distant that they are perturbed by random passing stars and tidal forces from the galactic disk. Again, dynamical simulations show that the Oort cloud is the only possible explanation for the number of comets with very distant orbits, but still gravitationally bound to the solar system.

But where did the Oort cloud comets come from? The solar nebula was too thin at those large distances for comet-sized bodies to form. Current thinking is that the Oort cloud comets are icy-planetesimals that formed in the region of the giant planets, between 5 and 30 AU, and were gravitationally ejected to distant orbits by the growing giant planets. This process is fairly inefficient and most icy planetesimals were ejected to interstellar space by the giant planets. If other forming solar systems are doing the same thing, then there is a vast swarm of comets in interstellar space. However, no comet has ever been observed entering the planetary system that was on an obviously interstellar orbit.

It is also possible that if the Sun formed in a cluster of stars, as most stars do, then it might have exchanged comets with the growing Oort clouds of those nearby stars. This could be a significant contributor to the Oort cloud population.

The source of the Halley-type comets, with their intermediate inclinations and eccentricities, is still a matter of debate. Both the Scattered disk and the Oort cloud have been suggested as sources. It may be that the explanation lies with a combination of the two cometary reservoirs.

The comet impact hazard

Comets pose a natural hazard to the Earth. This is because many of them are in orbits that cross the Earth's orbit and may collide with the Earth. Approximately 10 long-period comets, on the order of 1 km in diameter (or larger), cross the Earth's orbit each year. Because the Earth is a relatively small target and space is vast, the impact probability per comet is, on average, very low. A random long-period in an Earth-crossing orbit has an average impact probability of 2.2×10^{-9} per perihelion passage. This means that, on average, one long-period comet will strike the Earth for every 454 million comets that cross its orbit. Given the estimated rate of 10 long-period comets crossing the Earth's orbit per year, this results in a mean time between long-period comet impacts of 45 Myr.

However, because the long-period comets move on highly eccentric and highly inclined orbits, their mean impact velocities are much higher than for other celestial bodies, that is, asteroids. The average long-period comet will strike the Earth with a velocity of 51.8 km/s. If we weight the impact velocity by the probability of impact for a particular orbit, then the weighted mean impact velocity increases to 56.4 km/s. These values are much higher than that for Earth-crossing asteroids, which are typically only ~15 km/s.

An interesting case is that of Earth-crossing long-period comet Hale-Bopp (Figure 1), which passed closest to the Sun in 1997. Hale-Bopp was an unusually large and active comet, easily seen with the naked eye in evening skies. With a perihelion distance of 0.914 AU (the point in the orbit closest to the Sun), Hale-Bopp's orbit crossed inside the orbit of the Earth. Hale-Bopp was believed to have an unusually large nucleus, estimated to be 27–42 km in diameter. Taking a median value of 35 km and assuming a mean bulk density of 0.6 g/cm^3 results in an estimated mass of 1.3×10^{19} g.

The impact probability for Hale-Bopp on the Earth is 2.54×10^{-9} per perihelion passage, fairly typical for a long-period comet. Because of the comet's high orbital eccentricity, 0.9951, and inclination, 89.43°, the impact velocity would be 52.5 km/s. The resulting impact energy is equivalent to 4.4×10^9 megatons of TNT. This is ~44 times the estimated energy of the asteroid impact 65 Myr ago that killed the dinosaurs. Such an energetic impact may have the capability to completely sterilize the Earth, resulting in the extinction of all life on the planet! Fortunately, Hale-Bopp passed through the plane of the Earth's orbit on the far side of the Sun from the Earth, so there was never any possibility of an impact. Also, the average time between impacts of cometary nuclei as large as Hale-Bopp far exceeds the age of the solar system.

This illustrates an important point about the cometary impact hazard. Although asteroid impacts are far more frequent than comet impacts, some comets crossing the Earth's orbit are considerably larger than any of the known near-Earth asteroids. Thus, the largest and most devastating impacts on the Earth are likely to be of comets. Other known Earth-crossing comets with large nuclei include comet Halley, ~16 × 8 km in diameter, and comet Swift-Tuttle, ~23–30 km in diameter.

Also, the flux of long-period comets can vary over time. If a star comes close enough to the Sun to pass through the Oort cloud, in particular at distances less than 10,000 AU, then the star can cause a "shower" of comets to enter the planetary system. The rate of long-period comets crossing the Earth's orbit could increase by a factor of ~200 and the complete shower would last for about 2.5 Myr. Fortunately, such close stellar passages are rare, about one every 300 Myr.

For Jupiter-family comets, whose returns are predictable (once discovered), only 22 Earth-crossers are known (excluding the many small fragments of disrupted comet 73P/Schwassmann-Wachmann 3). Of these, four are lost, eight have only been observed on only one return, and one is no longer Earth-crossing. Their mean impact probability is 7.3×10^{-9} per orbit or 1.3×10^{-9} per year, and their mean encounter velocity with the Earth is 22.9 km/s, with a most probable encounter velocity of 19.9 km/s. The mean time between Jupiter-family comet impacts is 35 Myr

For Halley-type comets, whose returns are also predictable, another 16 Earth-crossers are known, of which 1 is lost and 6 have not yet made a second observed appearance. Their mean impact probability is 7.0×10^{-9} per orbit but only 0.16×10^{-9} per year because of their longer orbital periods. Their mean encounter velocity is 45.4 km/s, with a most probable encounter velocity of 52.3 km/s. The mean time between Halley-type comet impacts is 390 Myr. Note that the impact frequency for both Jupiter-family and Halley-type comets may be higher if there are yet undiscovered members of each group.

Summary

Comets are among the most interesting bodies in the solar system because they retain a cosmochemical record of the physical and chemical conditions at the time the planets formed. They have been kept in "cold storage" far from the Sun, during most of the solar system's history, and thus are essentially unmodified from their primitive state 4.56 billion years ago. Comets pose a small but significant part of the impact hazard to the Earth, and probably can account for the largest impacts on our planet over the last 3 billion years.

Bibliography

Festou, M. C., Keller, H. U., and Weaver, H. A. (eds.), 2004. *Comets II*. Tucson: University of Arizona Press. 733 pp.

Weissman, P. R., 2007. The cometary impactor flux at the Earth. In Valsecchi, G. B., and Vokrouhlický, D. (eds.), *IAU Symposium 236: Near-Earth Objects, Our Celestial Neighbors: Opportunities and Risk*. Cambridge: Cambridge University Press, pp. 441–450.

Cross-references

Asteroid

COMMUNICATING EMERGENCY INFORMATION*

John H. Sorensen
Oak Ridge National Laboratory, Oak Ridge, TN, USA

Definition and introduction

The empirical study of public communications in emergencies has been ongoing for almost 50 years (Perry and Mushkatel, 1986, 1984; Leik et al., 1981; Quarantelli, 1980; Baker, 1979; Mileti and Beck, 1975; Drabek and Stephenson, 1971; Lachman et al., 1961). These studies, when viewed collectively, have compiled an impressive record about how and why public behavior occurs in the presence of impending disaster or threat. For example, it is well documented that emergency warnings are most effective at eliciting public protective actions like evacuation when those warnings are frequently repeated (Mileti and Beck, 1975), are confirmatory in character (Drabek and Stephenson, 1971), make specific recommendations, and are perceived by the public as credible (Perry et al., 1981). Informal warning mechanisms (friends or relatives) are also at times very effective. In many evacuations, people leave the area at risk before an official warning is announced. Evacuation behavior is also influenced by other factors such as personal or family resources, age, and social relationships including social networks, level of education completed, experience with previous emergencies, social and environmental cues of immediate hazard, physical or psychological constraints to evacuating, as well as other more specific circumstances (such as time of day, weather conditions, etc.). Table 1 provides a list of those factors and how they have covaried with decisions to respond (Mailman School of PH @ Columbia, annual preparedness survey, focuses on why parents may not heed evacuation orders).

Studies that have used surveys of random samples of people living in or near-disaster areas have been conducted for a variety of hazard events. For hurricanes, these include Elena and Kate (Baker, 1987; Nelson et al., 1988), Eloise (Windham et al., 1977; Baker, 1979), Camille (Wilkinson and Ross, 1970), David and Frederick (Leik et al., 1981), Carla (Moore et al., 1964), Floyd (Dow and Cutter, 2002; HMG, no date), Andrew (Gladwin and Peacock, 1997), Bertha and Fran (Dow and Cutter, 1998), Georges (Dash and Morrow, 2001; Howell et al., 1998), Brett (Prater et al., 2000), Bonnie (Whitehead et al., 2000) Ivan (Howell and Bonner, 2005), and Lily (Lindell et al., 2005).

Studies of flood include Denver, CO (Drabek and Stephenson, 1971); Rapid City, SD; (Mileti and Beck, 1975); Big Thompson, CO (Gruntfest, 1977); Sumner Valley, Fillmore, and Snoqualmie, WA (Perry et al., 1981); Abilene, TX (Perry and Mushkatel, 1984);

*©United States Government

Communicating Emergency Information, Table 1 Factors associated with warning response

As factor increases	Response	Level of support
Characteristics of the warning		
Channel: Electronic	Is mixed	Low
Channel: Media	Is mixed	Moderate
Channel: Siren	Decreases	Low
Personal warning vs. impersonal	Increases	High
Proximity to threat	Increases	Low
Message specificity	Increases	High
Number of channels	Increases	Low
Frequency	Increases	High
Message consistency	Increases	High
Message certainty	Increases	High
Source credibility	Increases	High
Fear of looting	Decreases	Moderate
Time to impact	Decreases	Moderate
Source familiarity	Increases	High
Characteristics of People		
Physical cues	Increases	High
Social cues	Increases	High
Perceived risk	Increases	Moderate
Knowledge of hazard	Increases	High
Experience with hazard	Is mixed	High
Education	Increases	High
Family planning	Increases	Low
Fatalistic beliefs	Decreases	Low
Resource level	Increases	Moderate
Family united	Increases	High
Family size	Increases	Moderate
Kin relations (number)	Increases	High
Community involvement	Increases	High
Ethnic group member	Decreases	Moderate
Age	Is mixed	High
Socioeconomic status	Increases	High
Being female vs. male	Increases	Moderate
Having children	Increases	Moderate
Pet ownership	Decreases	Low

Clarksburg and Rochester, NY (Leik et al., 1981); and Denver, CO, and Austin, TX, (Hayden et al., 2007).

Studies of chemical accidents include Mississauga, Ontario, Canada (Burton, 1981); Mt. Vernon, WA; and Denver, CO (Perry and Mushkatel, 1986); Confluence and Pittsburg, PA (Rogers and Sorensen, 1989); Nanticote, PA (Duclos et al., 1989); and West Helena, AR (Vogt and Sorensen, 1999). Graniteville, SC (Mitchell et al., 2005).

Other protective action studies include the Hilo, HI, tsunami (Lachman et al., 1961); the Mt. St. Helens, WA, volcanic eruption (Perry and Greene, 1983; Dillman et al., 1984); the Three Mile Island nuclear accident, PA (Cutter and Barnes, 1985; Flynn, 1979); the World Trade Center bombing, NY, in 1993 (Aguire et al., 1998); the World Trade Center collapse, NY, in 2001 (Averill et al., 2005); SoCal wildfires in 2003 and Australian bushfires in 2005 (Proudley, 2008, AJEM) and in particular 2009 (Haynes et al., 2008, *J. Volc. Geotherm. Res.*, on volcanic risk perception; Wray et al., 2008, *Am. J. Pub. Health*, on communicating with public about health threats). The National Env. Health Assn published an excellent review of risk comm., risk perception, and loss of trust in "authorities" re post-collapse risk (Lyman, 2003, *Messages in the Dust*).

Excellent summaries of this research currently exist (Lindell and Perry, 2004; Drabek, 1986; Mileti and Sorensen, 1990; Tierney et al., 2003; National Research Council, 2006) and will not be repeated here.

Summary

Empirical studies and summaries have done much to further social scientific understanding of how people process and respond to risk communications in emergencies; it has also served to inform practical emergency preparedness efforts in this nation and abroad. Relevant research on human response to risk communications derived from the empirical research record can be summarized as follows.

Research indicates that people's decisions to respond to emergency communications are influenced by:

- The frequency and channel of communication of the warning. The most important dimensions of the warning frequency/channel are the number of different channels people hear the warning from, hearing from personal channels, and the frequency that people hear the warning.
- The content of the warning message. The most important dimensions of content are a description of the hazard and impacts, the predicted location of impacts, what actions to take, and when to take those actions.
- Observing cues. These include social cues (i.e., seeing neighbors evacuating) and physical cues (i.e., seeing flames or a smoke cloud).
- Aspects of individual status. These include socioeconomic status (i.e., income level and education completed), age, gender, and ethnicity.
- The role(s) an individual holds in society. These include having children at home, family size (i.e., larger vs. smaller), extent of kin relations, being a united family at time of the event, and greater community involvement.
- Previous experience with the hazard. People are inclined to do what they did in a previous situation.
- People's belief in the warning. Belief is not determined by the credibility of the source issuing the warning but by the frequency the message is heard.
- People's knowledge about the hazard. This includes previous information and data gained in the event or by cues.
- People's perceptions of risk. This includes perception of the threat before the event and perception of risk from the specific event.
- The extent of social interactions during the event. This includes efforts to contact others about the event, being contacted by others, and being able to confirm the message as accurate and credible.

Bibliography

Aguirre, B. E., Wenger, D., and Vigo, G., 1998. A test of emergency norm theory of collective behavior. *Sociological Forum*, **13**, 301–320.

Averill, J. D., et al., 2005. *Federal Building and Fire Safety Inspection of the World Trade Center Disaster, Project #7: Occupant Behavior, Egress and Emergency Communications.* Washington, DC: National Institute of Standards and Technology.

Baker, E. J., 1979. Predicting response to hurricane warnings: A reanalysis of data from four studies. *Mass Emergencies*, **4**, 9–24.

Baker, E. J. 1987. Evacuation in response to hurricanes Elena and Kate. Unpublished draft report. Tallahassee, FL: Florida State University.

Cutter, S., and Barnes, K., 1982. Evacuation behavior at Three Mile Island. *Disasters*, **6**, 116–124.

Dash, N., and Morrow, B. H., 2001. Return delays and evacuation order compliance: The case of Hurricane Georges and the Florida Keys. *Environmental Hazards*, **2**, 119–128.

Dillman, D., Schwalbe, M., and Short, J. 1983. Communication behavior and social impacts following the May, 18, 1980, eruption of Mt. St. Helens. In Keller, S. A. C. (ed.), *Mt. St. Helens One Year Later*. Cheny, WA: Eastern Washington University Press, pp. 191–198.

Dow, K., and Cutter, S., 1998. Crying wolf: Repeat responses to hurricane evacuation orders. *Coastal Management*, **26**, 237–252.

Dow, K., and Cutter, S., 2002. Emerging hurricane evacuation issues: Hurricane Floyd and South Carolina. *Natural Hazard Review*, **3**, 12–18.

Drabek, T., 1986. *Human system response to disaster: An inventory of sociological findings*. New York: Springer.

Drabek, T. E., and Stephenson, J. S., 1971. When disaster strikes. *Journal of Applied Social Psychology*, **1**, 187–203.

Duclos, P., Binder, S., and Riester, R., 1989. Community evacuation following the Spencer metal processing plant fire, Nanticoke, Pennsylvania. *Journal of Hazardous Materials*, **22**, 1–11.

Flynn, C., 1979. *Three Mile Island telephone survey -NUREG/CR-1093*. Washington, DC: U.S. Nuclear Regulatory Commission.

Gladwin, H., and Peacock, W. G., 1997. Warning and evacuation: A night for hard houses. In Peacock, W. G., Morrow, B. H., and Gladwin, H. (eds.), *Hurricane Andrew – Ethnicity, Gender and the Sociology of Disasters*. London, NY: Routledge, pp. 52–74.

Gruntfest, E., 1977. *What People Did During the Big Thompson Flood, Working Paper #32*. Boulder, CO: Institute of Behavioral Science, University of Colorado.

Hazards Management Group (HMG). No date. Southeast states hurricane evacuation traffic study: Floyd behavioral reports. Found at www.fhwaetis.com/etis

Howell, S. E., 1998. *Evacuation behavior in Orleans and Jefferson parishes, Hurricane Georges*. New Orleans: Survey Research Center, University of New Orleans.

Howell, S., and Bonner, D. E., 2005. *Citizen hurricane and evacuation behavior in southeastern Louisiana: A twelve parish study*. New Orleans: Survey Research Center, University of New Orleans.

Lachman, R., Tatsuoka, M., and Bonk, W., 1961. Human behavior during the tsunami of May, 1960. *Science*, **133**, 1405–1409.

Leik, R. K., Carter, T. M., Clark J. P., et al. 1981. *Community Response to Natural Hazard Warnings: Final Report*. Minneapolis, MN: University of Minnesota.

Lindell, M., and Perry, R., 1992. *Behavioral foundations of community emergency planning*. Washington, DC: Hemisphere Publishing Company.

Lindell, M., and Perry, R., 2004. *Risk communication in multiethnic communities*. Thousand Oaks, CA: Sage.

Lindell, M., Lu, J., and Prater, C., 2005. Household decision making and evacuation response to Hurricane Lily. *Natural Hazard Review*, **6**, 171–179.

Mileti, D., and Beck, E. M., 1975. Communication in crisis: Explaining evacuation symbolically. *Communication Research*, **2**, 24–49.

Mileti, D., and Sorensen, J., 1990. *Communication of emergency public warnings, ORNL-6609*. Oak Ridge, TN: Oak Ridge National Laboratory.

Moore, H. E., Bates, F. L., Layman, M. V., and Parenton, V. J., 1964. *Before the Wind: A Study of Response to Hurricane Carla, National Academy of Sciences/National Research Council Disaster Study #19*. Washington, DC: National Academy of Sciences.

National Research Council, 2006. *Facing Hazards and Disasters: Understanding Human Dimensions*. Washington DC: National Academy Press.

Nelson, C. E., Crumley, C., Fritzsche, B., and Adcock, B., 1989. *Lower Southwest Florida Hurricane Study*. Tampa, FL: University of South Florida.

Perry, R. W., and Greene, M. R., 1983. *Citizen Response to Volcanic Eruptions: The Case of Mount St. Helens*. New York: Irvington Publishers.

Perry, R. W., and Mushkatel, A., 1984. *Disaster Management: Warning Response and Community Relocation*. Westport, CT: Quorum Books.

Perry, R. W., and Mushkatel, A., 1986. *Minority Citizens in Disaster*. Athens, GA: University of Georgia Press.

Perry, R. W., Lindell, M. K., and Greene, M. R., 1981. *Evacuation Planning in Emergency Management*. Lexington, MA: Lexington Books.

Prater, C., Wenger, D., and Grady, K., 2000. *Hurricane Bret Post Storm Assessment: A Review of the Utilization of Hurricane Evacuation Studies and Information Dissemination*. College Station, TX: Texas A&M University Hazard Reduction & Recovery Center.

Quarantelli, E. L., 1980. *Evacuation Behavior and Problems: Findings and Implications from the Research Literature*. Columbus, OH: Disaster Research Center, Ohio State University.

Rogers, G., and Sorensen, J., 1989. Public warning and response to hazardous materials accidents. *Journal of Hazardous Materials*, **22**, 57–74.

Tierney, K., Lindell, M., and Perry, R., 2003. *Facing the Unexpected = Disaster Preparedness and Response in the United States*. Washington DC: Joseph Henry Press.

US Department of Interior, 1995. *Federal Wildland Fire Management Policy and Program Review: Final Report*. Washington DC: US Department of Interior.

Vogt, B., and Sorensen, J., 1999. *Description of Survey Data Regarding the Chemical Repackaging Plant Accident West, Helena, Arkansas, ORNL/TM-13722*. Oak Ridge, TN: Oak Ridge National Laboratory.

Whitehead, J. C., Edwards, B., Van Willigan, M., Maiolo, J. R., Wilson, K., and Smith, K. T., 2000. Heading for higher ground: Factors affecting real and hypothetical hurricane evacuation behavior. *Environmental Hazards*, **2**, 133–142.

Wilkinson, K., and Ross, P., 1970. *Citizens Response to Warnings of Hurricane Camille, Report #35*. Starkville, MS: Mississippi State University, Social Science Research Center.

Windham, G., Posey, E., Ross, P., and Spencer, B. (1977). Reaction to Storm Threat During Hurricane Eloise, Report #35. Starkville, MS: Mississippi State University, Social Science Research Center.

COMMUNITY MANAGEMENT OF NATURAL HAZARDS

William T. Hartwell

Desert Research Institute, Nevada System of Higher Education, Las Vegas, NV, USA

Definition

Community management of natural hazards. Community-based participation in identification, mitigation, preparedness, response, and recovery and reconstruction activities related to potential and/or experienced natural hazards.

Introduction

Natural hazard management strategies typically include several broad categories of management. They include hazard identification and mitigation, preparedness (or planning), response, and recovery and reconstruction. These strategies can be viewed as a continuum, with recovery and reconstruction activities ideally resulting in increasingly effective mitigation strategies in advance of the next hazard event.

Historically, the management of natural hazards has been viewed primarily as one of response and recovery, with the responsibility resting largely on state or national government, and with direct planning and participation at the community level largely neglected (e.g., Laughy, 1991). However, as the Organization of American States has noted in its policy series on managing natural hazard risk, natural hazard risk management efforts are most effective when they are explicitly addressed at every level, including at the community level (OAS, 2004). The strengthening of stakeholder and community involvement is viewed by some as the greatest need in the evolving area of hazard management and mitigation (King, 2008). It has become increasingly clear that there is an ongoing shift from a response and recovery approach toward a mitigative approach in the management of natural hazards, which requires the integration of management practices at the community level in order to be successful and sustainable (Pearce, 2003).

There are many challenges to implementing successful community-based participation in the management of natural hazards. Among them is the influence of previous experience with a specific hazard (or lack thereof) on how local government and community members perceive risk from that hazard, which can determine the level of public participation in preparedness and mitigation strategies (Tierney et al., 2001). The effect of previous experience on how an individual responds to future participation in mitigative programs can often be counterintuitive, and has implications for the management of hazards (McGee et al., 2009). Another challenge is the difficulty of conveying concepts of hazard risk to the public, given the often very imprecise nature of the business of hazard prediction (Alexander, 2007). Additionally, individuals may not participate in mitigative strategies due to a lack of accurate information, or the perception that a hazard is a political or ideological creation rather than a reality. They also may be moved to nonparticipation as a means to avoid unpleasant emotions about the issue or in the belief, especially in the case of global threats such as climate change, that there is really nothing they can really do about it (Norgaard, 2006). Studies showing relationships between individuals' risk-taking propensity and attitudes toward preparation for natural hazards (e.g., McClure et al., 1999), as well as the role that media portrayal of natural disasters has in influencing future social behavior and attitudes (e.g., McClure et al., 2001) have significant implications for how community management strategies may help alleviate fatalism and improve hazard preparedness. Drawing on community empowerment and engagement strategies can significantly enhance the ability of communities to promote and sustain participation in hazard preparedness (e.g., Frandsen et al., 2011).

Those living in the developing world as well as those living in poverty are particularly vulnerable to the effects of natural disasters (World Bank, 2001), largely as a result of the combination of underdevelopment, poor building construction and siting, and economic inability to adequately respond to and recover from a major disaster. The earthquake that struck the impoverished nation of Haiti on January 12, 2010 provided a stark example of the confluence of these attributes in the face of a major natural disaster. Well over 200,000 people had lost their lives as a result of this event, and at least 1.5 million were homeless. In the face of warnings from scientists as recently as 2008 that Haiti was at significant risk for a major earthquake (Manaker et al., 2008), mitigative measures to prepare were lacking, with economics likely a major factor in the lack of community preparedness for such a disaster.

Hazard identification and mitigation

Mitigation includes activities that eliminate or reduce the chance of occurrence or the effects of a disaster. Identification of potential hazards and potential vulnerabilities to hazards is the first step in this process. Communities' response to recommendations for mitigative measures may be predicated on previous personal experience with specific hazards (McGee et al., 2009). Community mitigation and preparedness for hazards that are perceived to be of low risk may not be implemented, in spite of the fact that occurrence of such hazards can result in very high consequence events (e.g., 2004 Indian Ocean Tsunami). Pre-hazard mitigation programs such as those offered by the Federal Emergency Management Agency (FEMA) in the United States have shown that, while communities may not necessarily be able to prevent disasters, they can take many proactive steps that can reduce the effects of hazards upon communities and their residents (e.g., Volunteer Florida, 2004). For example, requiring structural reinforcements to homes in areas prone to seismic activity will reduce property damage and loss of life from earthquakes. Similarly, the implementation of Early Warning Systems (e.g., Zschau and Küppers, 2003; Momani and Alzaghal, 2009) at the community level has the potential to save hundreds of thousands of lives in extreme cases, such as that which occurred as a result of the 2004 Indian Ocean Tsunami. Haque (2005) provides a range of experiences in the mitigation and management of natural hazards from an international perspective. It is important to distinguish between mitigation strategies themselves and a community's capacity to respond to them in a timely and effective manner. Assessing a community's ability to adapt and respond favorably to these strategies is key to effective implementation, whether in advance of or in response to a natural hazard (e.g., Paton and Tang, 2009).

Preparedness

Preparedness, the next aspect of hazard management, involves planning how to respond in the advent of a natural hazard, and how to activate community resources to respond effectively. Careful advance planning can help save lives and minimize property damage by preparing community members to respond in a prescribed manner when a hazard occurs. In a community-based approach, this phase involves significant public informational and educational components.

Ensuring public participation in the process of hazard management planning can be problematic at times for a variety of reasons (e.g., Chen et al., 2006). However, the importance of continued public involvement throughout the entire management cycle has direct bearing on whether or not mitigative strategies can be sustained until they are needed (e.g., Tanaka, 2009). Promoting community involvement in all aspects of preparedness can result in greater post-disaster resilience, particularly in segments of the population who are likely to be most affected by the occurrence of a natural disaster, such as children and their families (Ronan and Johnston, 2005). However, it is important to note the need to distinguish between providing information on hazards preparation and people's general ability to interpret and use such information. For example, Lindell et al. (2009) note that hazard experience, risk perception, and population demographics, among others, all can have effects on attributes related to hazard preparation adjustments. Additionally, community risk management strategies are to some extent socially constructed, and how people may act to manage their risk often encompasses both social and cultural issues (e.g., Paton et al., 2010). Finally, trust in the purveyor of the information related to hazard mitigation can influence how effective resultant strategies are (e.g., Paton, 2008).

Response

Response covers the period immediately prior to (if the hazard can be predicted in advance), during, and immediately following a disaster. Responders typically include entities such as the fire and police departments, and medical services. Depending on the magnitude of the event, however, the usual responders and local government may be ill-equipped to manage the response phase without significant assistance from the state or national government, or the international community. Involvement of local community members in the aspects of disaster awareness training can influence hazard-related cognitions and preparedness behaviors (Karanci et al., 2005), resulting in the ability of the general community to participate in the response actions completely and in more beneficial effects (e.g., Paton et al., 2001).

Recovery and reconstruction

Recovery and reconstruction represent the final part of the management cycle, though it can also be viewed as the precursor to the improvement of mitigation procedures. Recovery and reconstruction continue until community functions have returned to normal. In the early part of this phase, critical community services are restored to minimum operating conditions. Depending on the severity of the hazard's impact, recovery may go on for months or even years, as in the case of disasters with major loss of life or property. Ironically, the impact of a major disaster on a community presents the opportunity for significant improvement of infrastructure and construction practices, resulting in the incorporation of features that are less likely to be affected by future events. While the process of recovery can provide opportunities to mitigate future disasters, successful implementation of such strategies requires an understanding of changes in community contributions to recovery and rebuilding efforts over time (Paton, 2006).

The recovery and reconstruction phase following a significant natural disaster often requires significant economical resources in order to succeed, and the resilience of a community may depend largely on the strength of pre-hazard mitigative strategies that are already in place at the time of the event. Just as important as the resources that contribute to the physical recovery of a community are the services that are in place to address emotional health needs, which can often be quite severe following a natural disaster. Psychiatric disorders such as post-traumatic stress disorders can be common, and while most people are resilient and will recover, some populations may be at higher risk for more serious mental health problems (Watanabe et al., 2004; Wickrama and Wickrama, 2008). A critical aspect of community management of hazards involves planning for both physical and emotional injury that could potentially result from a natural hazard, and communities can use methods such as the formation of innovative self-help groups to ensure recovery in both areas (e.g., Kuppuswamy and Rajarathnam, 2009). While the enabling of participatory planning after the occurrence of a natural disaster bears some beneficial aspects, earlier involvement of stakeholders in the mitigation process is indicated, since many may be ill-equipped emotionally immediately following a disaster (e.g., Ganapati and Ganapati, 2009).

The role of internet technology in community management of hazards

Just as radio and television in the earlier days, the Internet has become an increasingly important resource for communities marching toward active engagement in hazard management. The ability to provide near-real-time hazard-related data (Dimitruk, 2007) and interface with Geographical Information Systems (GIS) (Raheja et al., n.d.), Global Positioning Systems (GPS), and other communications technologies aid in all areas of hazard management. As early as 1998 following the advent of Hurricane Mitch's landfall in Central America, the Internet was used intensively to post regular updates on information such as epidemiological reports and public health guidelines on topics ranging from household water quality to the prevention of measles (Bittner, 2000).

For the community in the early stages of hazard management planning, the Internet is a tremendous source of ready information on all aspects of hazard management, with some sites functioning as warehouses for relevant links, such as a site hosted by Keele University in Britain (http://www.keele.ac.uk/depts/por/disaster.htm). The added benefit of being able to store relevant information and databases on computer servers far removed from the community that they will serve in the advent of a natural disaster means that the information will still be potentially accessible to communities, responders, and other critical parties even if communications infrastructure is initially disabled or destroyed at the site of the event. In the aftermath of Hurricane Katrina's landfall near New Orleans on the south coast of the United States in 2005, and also following the earthquake that struck Haiti in early 2010, the Internet was a critical resource in aiding community members at home and abroad to track down information on the status of missing loved ones (http://guides.library.msstate.edu/content.php?pid=16013&sid=107538; http://www.google.com/relief/haitiearthquake/). Finally, the Internet has become a critical component for the affected communities in the conduct of outreach to the global community to raise funds during the response and recovery phases following the advent of a natural hazard.

Selected examples of community management of hazards

The following are examples of different types of strategies of community management of various natural hazards:

Earthquakes

Tokai Earthquake Preparedness in Shizuoka Prefecture, Japan http://www.e-quakes.pref.shizuoka.jp/english/earthquakepreparedness_in_shizuoka.pdf

This entry discusses in detail the history behind, formation of, and plans for implementing highly integrated management of a potential earthquake hazard in Shizuoka Prefecture, Japan, including community education and participation.

A Novel Strong-Motion Seismic Network for Community Participation in Earthquake Monitoring

(Cochran et al., 2009) http://qcn.stanford.edu/ (Quake-Catcher Network).

This is an innovative approach to passive community-based volunteer participation in seismic data gathering and analysis through use of distributed computing techniques, with a goal of increasing the awareness of various aspects of seismic activity to aid with the aspects of earthquake preparedness planning.

Hurricanes, coastal erosion, and coastal flooding

Sustainable Coastal Communities and Ecosystems (SUCCESS) http://seagrant.gso.uri.edu/ecosystems/hazards.html

Based out of the University of Rhode Island in the United States, this program works with governments, the private sector, and community organizations to ensure that coastal communities face and recover from hurricanes, floods, and coastal storms. The goal is to help communities achieve economic growth while reducing the potential impacts of natural hazards and maximizing public safety and public access to the shore. The SUCCESS program works to help develop strategies to prepare for natural disasters, educate disaster preparedness and response professionals, enhance evacuation preparations, and plan for expediting recovery efforts.

Landslides

Landslide Management by Community-Based Approach in the Republic of Armenia (Mori et al., 2007) http://www.n-koei.co.jp/library/pdf/forum16_017.pdf

Report discussing community-based approach toward landslide hazard identification and management.

Natural radioactivity, radon hazard

The Community Environmental Monitoring Program http://cemp.dri.edu/

While this program concentrates on the potential of releases of man-made radioactivity as a result of the past testing of nuclear weapons at the Nevada Test Site, an understanding of the potential hazards of natural radioactivity, including radon, as well as concepts of dose, is an important component of the program. The program provides a hands-on role for community members in the monitoring process and equips them with the knowledge to communicate information on the subject to their neighbors (Hartwell and Shafer, 2011). This program also provides an example of how the Internet can be an effective tool for communication data dissemination.

Tornadoes

Tornado Tabletop Exercise: Engaging Youth in Community Emergency Management http://www.unce.unr.edu/publications/files/cy/2009/cm0908.pdf

An example of a classroom curriculum designed to educate students about community emergency management, including training them to use geospatial technology to create maps with shelter locations and evacuation routes, and simulating a tornado event.

Volcanoes

Maximizing Multi-stakeholder Participation in Government and Community Volcanic Hazard Management Programs: A Case Study from Savo, Solomon Islands (Cronin et al., 2004). Report on attempt at multi-level integration of volcanic risk management strategies and challenges of involvement of certain sectors of the community populations.

Bibliography

Alexander, D., 2007. Making research on geological hazards relevant to stakeholders' needs. *Quaternary International*, **171–72**, 186–192.

Bittner, P., 2000. Disaster management in the digital age: the case of Latin America. *Humanitarian Exchange Magazine* (16). Britain: Overseas Development Institute. http://www.odihpn.org/report.asp?id=1033

Chen, L. C., Liu, Y. C., and Chan, K. C., 2006. Integrated community-based disaster management program in Taiwan: a case study of Shang-An village. *Natural Hazards*, **37**(1–2), 209–223.

Cochran, E., Lawrence, J., Christensen, C., and Chung, A., 2009. A novel strong-motion seismic network for community participation in earthquake monitoring. *IEEE Instrumentation and Measurement Magazine*, **12**(6), 8–15.

Cronin, S. J., Petterson, M. G., Taylor, P. W., and Biliki, R., 2004. Maximising multi-stakeholder participation in government and community volcanic hazard management programs; a case study from Savo, Solomon Islands. *Natural Hazards*, **33**(1), 105–136.

Dimitruk, P., 2007. Disaster management using Internet-based technology: Internet-based disaster management technology can help lighten operational challenges and financial burdens. Healthcare Financial Management. http://www.ncbi.nlm.nih.gov/pubmed/17366722. Accessed 6 June, 2012.

Frandsen, M., Paton, D., and Sakariassen, K., 2011. Fostering community bushfire preparedness through engagement and empowerment. *Australian Journal of Emergency Management*, **26**, 23–30.

Ganapati, N. E., and Ganapati, S., 2009. Enabling participatory planning after disasters: a case study of the World Bank's housing reconstruction in Turkey. *Journal of the American Planning Association*, **75**(1), 41–59.

Haque, C. E. (ed.), 2005. *Mitigation of Natural Hazards and Disasters: International Perspectives*. Dordrecht: Springer.

Hartwell, W. T., and Shafer, D. S., 2011. Community Environmental Monitoring Program: a case study of public education and involvement in radiological monitoring. *Health Physics*, **101**(5), 606–617.

Karanci, A. N., Aksit, B., and Dirik, G., 2005. Impact of a community disaster awareness training program in Turkey: does it influence hazard-related cognitions and preparedness behaviors. *Social Behavior and Personality*, **33**(3), 243–258.

King, D., 2008. Reducing hazard vulnerability through local government engagement and action. *Natural Hazards*, **47**(3), 497–508.

Kuppuswamy, S., and Rajarathnam, S., 2009. Women, information technology and disaster management: tsunami affected districts of Tamil Nadu. *International Journal of Innovation and Sustainable Development*, **4**, 206–215.

Laughy, L., 1991. *A Planner's Handbook for Emergency Preparedness*. Vancouver: Centre for Human Settlements, University of British Columbia.

Lindell, M. K., Arlikatti, S., and Prater, C. S., 2009. Why people do what they do to protect against earthquake risk: perceptions of hazard adjustment attributes. *Risk Analysis*, **29**, 1072–1088.

Manaker, D. M., Calais, E., Freed, A. M., Ali, S. T., Przybylski, P., Mattioli, G., Jansma, P., Prepetit, C., and de Chabalier, J. B., 2008. Interseismic Plate coupling and strain partitioning in the Northeastern Caribbean. *Geophysical Journal International*, **174**, 889–903.

McClure, J., Walkey, F., and Allen, M., 1999. When earthquake damage is seen as preventable: attributions, locus of control and attitudes to risk. *Applied Psychology*, **48**(2), 239–256.

McClure, J., Allen, M. W., and Walkey, F., 2001. Countering fatalism: causal information in news reports affects judgements about earthquake damage. *Basic and Applied Social Psychology*, **23**(2), 109–121.

Mcgee, T. K., McFarlane, B. L., and Varghese, J., 2009. An examination of the influence of hazard experience on wildfire risk perceptions and adoption of mitigation measures. *Society and Natural Resources*, **22**(4), 308–323.

Mercer, J., Kelman, I., Taranis, L., and Suchet-Pearson, S., 2010. Framework for integrating indigenous and scientific knowledge for disaster risk reduction. *Disasters*, **34**(1), 214–239.

Momani, N., and Alzaghal, M. H., 2009. Early warning systems for disasters in Jordan: current and future trends. *Journal of Homeland Security and Emergency Management*, **6**(1), Art. No. 75.

Mori, M., Hosoda, T., Ishikawa, Y., Tuda, M., Fujimoto, R., and Iwama, T., 2007. Landslide management by community based approach in the Republic of Armenia. Japan International Cooperation Agency, Government of Armenia.

Norgaard, K. M., 2006. "People want to protect themselves a little bit": emotions, denial, and social movement nonparticipation. *Sociological Inquiry*, **76**(3), 372–396.

Organization of American States, 2004. Managing natural hazard risk: issues and challenges. Policy Series 4.

Paton, D., 2006. Disaster resilience: building capacity to co-exist with natural hazards and their consequences. In Paton, D., and Johnston, D. (eds.), *Disaster Resilience: An integrated approach*. Springfield: Charles C. Thomas, pp. 3–10.

Paton, D., 2008. Risk communication and natural hazard mitigation: how trust influences its effectiveness. *International Journal of Global Environmental Issues*, **8**, 2–16.

Paton, D., and Tang, C. S., 2009. Adaptive and growth outcomes following tsunami: the experience of thai communities following the 2004 Indian Ocean tsunami. In Askew, E. S., and Bromley, J. P. (eds.), *Atlantic and Indian Oceans: New Oceanographic Research*. New York: Nova, pp. 125–140.

Paton, D., Millar, M., and Johnston, D., 2001. Community resilience to volcanic hazard consequences. *Natural Hazards*, **24**, 157–169.

Paton, D., Sagala, S., Okado, N., Jang, L., Bürgelt, P. T., and Gregg, C. E., 2010. Making sense of natural hazard mitigation: personal, social and cultural influences. *Environmental Hazards*, **9**, 183–196.

Pearce, L., 2003. Disaster management and community planning, and public participation: how to achieve sustainable hazard mitigation. *Natural Hazards*, **28**, 211–228.

Raheja, N., Ojha, R., and Mallik, S. R., n.d. Role of internet-based GIS in effective natural disaster management. GIS Development web site: http://www.gisdevelopment.net/technology/gis/techgi0030.htm. Accessed 6 June, 2012.

Ronan, K. R., and Johnston, D. M., 2005. *Promoting Community Resilience in Disasters: The Role for Schools, Youth, and Families*. New York: Springer.

Tanaka, N., 2009. Vegetation bioshields for tsunami mitigation: review of effectiveness, limitations, construction, and sustainable management. *Landscape and Ecological Engineering*, **5**(1), 71–79.

Tierney, K., Lindell, M. K., and Perry, R. W., 2001. *Facing the Unexpected: Disaster Preparedness and Response in the United States*. Washington, DC: Joseph Henry Press.

Volunteer Florida, 2004. *Disaster Mitigation: A Guide for Community-Based Organizations*. Florida: Florida Department of Community Affairs, Division of Emergency Management, Tallahassee. http://www.tallyredcross.org/library/DisasterMitigation-AGuideForCommunityBasedOrganizations.pdf

Watanabe, C., Okumura, J., Chiu, T. Y., and Wakai, S., 2004. Social support and depressive symptoms among displaced older adults following the 1999 Taiwan earthquake. *Journal of Traumatic Stress*, **17**(1), 63–67.

Wickrama, K. A. S., and Wickrama, K. A. T., 2008. Family context of mental health risk in Tsunami affected mothers: findings from a pilot study in Sri Lanka. *Social Science & Medicine*, **66**(4), 994–1007.

World Bank, 2001. World Development Report 2000/2001, Chapter 9. Oxford: Oxford University Press.

Zschau, J., and Küppers, A. N. (eds.), 2003. *Early Warning Systems for Natural Disaster Reduction*. Berlin/New York: Springer.

Cross-references

Casualties Following Natural Hazards
Climate Change
Coastal Erosion
Coastal Zone, Risk Management
Communicating Emergency Information
Community Management of Hazards
Coping Capacity
Costs (Economic) of Natural Hazards and Disasters
Damage and the Built Environment
Disaster
Disaster Relief
Disaster Risk Management
Dose Rate
Early Warning Systems
Earthquake
Earthquake Damage
Earthquake Prediction and Forecasting
Economics of Disasters
Education and Training for Emergency Preparedness
Emergency Management
Emergency Mapping
Emergency Planning
Epidemiology of Disease in Natural Disasters
Exposure to Natural Hazards
Federal Emergency Management Agency (FEMA)
Flood Hazard and Disaster
Forest and Range Fires
Frequency and Magnitude of Events
Geographic Information Systems (GIS) and Natural Hazards
Global Change and Its Implications for Natural Disasters
Global Positioning System (GPS) and Natural Hazards
Hazard
Hazardousness of Place
Hurricane
Information and Communications Technology
Integrated Emergency Management System
International Strategies for Disaster Reduction: The IDNDR and ISDR
Internet, World Wide Web and Natural Hazards
Landslide
Misconceptions About Natural Disaster
Mitigation
Monitoring and Prediction of Natural Hazards
Mortality and Injury in Natural Disasters
Natural Hazard
Natural Hazards in Developing Countries
Natural Radioactivity

Perception of Natural Hazards and Disasters
Planning Measures and Political Aspects
Posttraumatic Stress Disorder (PTSD)
Prediction of Hazards
Psychological Impacts of Natural Disasters
Radon Hazards
Recovery and Reconstruction After Disaster
Remote Sensing of Natural Hazards and Disasters
Resilience
Risk
Risk Assessment
Risk Perception and Communication
Seismology
Social–Ecological Systems
Structural Mitigation
Tornado
Tsunami
Volcanoes and Volcanic Eruptions
Warning Systems
Wildfire

COMPLEXITY THEORY

William H. K. Lee
U.S. Geological Survey, Menlo Park, CA, USA

Synonyms
Systems theory

Definition
A complex system consists of many interacting parts, generates new collective behavior through self organization, and adaptively evolves through time. Many theories have been developed to study complex systems, including chaos, fractals, cellular automata, self organization, stochastic processes, turbulence, and genetic algorithms.

Introduction
The classical approach to study natural phenomena is to model them as *dynamical systems* governed by differential equations, which allow the temporal evolution of many *linear* phenomena (e.g., motions of a planet around the Sun, laminar fluid flow, etc.) to be predicted with considerable accuracy. A linear system is deterministic because its output is proportional to the input. However, most natural phenomena involve *nonlinear* processes. Jules Henri Poincaré analyzed the stability of the solar system and in 1890 discovered chaotic behavior in a three-body dynamical system. Since then, many new concepts and tools have been developed for solving nonlinear problems – some are successful, but many raise more questions than provide answers.

Chaos in dynamical systems
In 1963, Edward Lorenz discovered that simple computer models of weather were very sensitive to *initial conditions*, such that a slight change at the start would give very different results. Lorenz used a simplified version of the Navier–Stokes equations (formulated in the mid-nineteenth century) for his computer models. Lorenz's discovery led to the realization that our ability to predict weather is limited at best to several days, because small measurement errors in the initial conditions grow exponentially with time, leading to predictions that deviate significantly from the actual weather conditions in just a few days. This requires repeated updating of initial conditions to extend a useful prediction.

Fractals in geology and geophysics
About 1/3 of major natural catastrophes are caused by earthquakes (the other 2/3 are mostly due to hurricanes and floods); hence their occurrences have naturally drawn attention for millennia. A prominent feature of seismicity is the Gutenberg–Richter relation derived empirically from observations (Gutenberg and Richter, 1954). It is given as $\log N(M) = a - bM$, where M is the earthquake magnitude, $N(M)$ is the number of earthquakes with magnitude greater than or equal to M, and a and b are constants. It can be rewritten as $N = \alpha A^{-\beta}$, a power law that is characteristic of *fractals*, which possess scale invariance (Turcotte, 1997). As is the case with earthquakes, faults, volcanic eruptions, landslides, floods, and many other natural phenomena also exhibit scale invariance. A fractal is commonly defined as a collection of objects that have a power-law dependence of number on size. Fractals are observed in many physical, biological, and social systems, regardless of their underlying governing processes.

Discussion
The classical, deterministic approach enjoys great success in studying some natural phenomena that can be approximated as *linear* systems. However, many natural phenomena are *nonlinear*. Complexity theory has been developing to meet this challenge, but a unified theory is not yet available for universal applications. Existing theory indicates that deterministic prediction for many phenomena (e.g., weather or earthquakes) is inherently impossible but that probabilistic forecasts are feasible. Mitchell (2009) provides a "tour" of complexity theory, and an introduction to complexity in earthquakes, tsunamis, and volcanoes is given by Lee (2009).

Bibliography
Gutenberg, B., and Richter, C. F., 1954. *Seismicity of the Earth*. Princeton: Princeton University Press.
Lee, W. H. K., 2009. Complexity in earthquakes, tsunamis, and volcanoes, and forecast, Introduction to. In Meyers, R. A. (ed.), *Encyclopedia of Complexity and Systems Science*. New York: Springer, Vol. 2, pp. 1213–1224.
Mitchell, M., 2009. *Complexity: A Guide Tour*. New York: Oxford University Press.
Turcotte, D. L., 1997. *Fractals and Chaos in Geology and Geophysics*, 2nd edn. Cambridge: Cambridge University Press.

Cross-references
Early Warning Systems
Earthquake

Earthquake Prediction and Forecasting
Earthquake Resistant Design
Extreme Value Theory
Fault
Tsunami
Volcanoes and Volcanic Eruptions

CONCRETE STRUCTURES

Murat Saatcioglu
University of Ottawa, Ottawa, ON, Canada

Synonyms

Reinforced concrete structure

Definition

A structure constructed primarily of concrete reinforced with steel.

Concrete is a commonly used construction material that is locally available throughout the world. It consists of fine aggregate (sand), coarse aggregate (crushed stone), cement (usually Portland Cement), water, and air. Cement has the appropriate chemical composition as a binding material that hydrates in the presence of water, chemically binding fine and coarse aggregate particles together to form a rock-like material called concrete (Kosmatka et al., 2008; Neville and Brooks, 2008). The aggregates account for approximately 75% of total mix by volume. Air in concrete, purposely introduced through chemical admixtures, improves resistance to freeze-thaw cycles. Water-cement (W/C) ratio by weight is the single most important parameter that affects the quality of concrete. W/C ratio of 0.5 produces sufficient workability, good performance, and an average compressive strength of approximately 30 MPa. As the W/C ratio decreases, the strength and quality of concrete (durability, abrasion resistance, freeze-thaw resistance, permeability) improves drastically. Concrete mixtures may also contain chemical admixtures for improved quality and workability.

Concrete is strong in compression for use as a structural material. However, it is generally very weak in tension. Concrete cracks at approximately 10% of its compressive strength in tension, and further breaks into pieces unless properly reinforced. Therefore, concrete is often reinforced with a material that permits resistance to tension when used for structural applications. The resulting composite material is referred to as "reinforced concrete." The most commonly used type of reinforcement is a steel bar. Typical reinforced concrete structural elements consist of beams, columns, walls, slabs, and footings. The longitudinal reinforcing bars are often placed on the tension side to control cracks and resist tension, although sometimes they may be placed in the compression zone for additional compressive capacity. Transverse reinforcement is used to control diagonal tension cracks associated with shear, to laterally restrain compression bars against buckling or to confine concrete for improved inelastic deformation capacity. The resulting structural elements form a structural framing system, consisting of moment resisting frames, structural walls (shear walls), or the combination of the two. The primary objective in structural design is to provide resistance to gravity and lateral loads, including those caused by natural hazards. Concrete structures are built either as "cast-in-place" monolithic (continuous) structures, or "precast" structures that consist of prefabricated elements that can be assembled and connected on site. A special form of reinforced concrete is "prestressed concrete." This type of construction takes advantage of eliminating or reducing tension in concrete by imposing compressive stresses prior to the application of external loads (Nawy, 2006). The prestressing operation is often done by means of high-strength steel strands, cables, or bars that are pretensioned or posttensioned to compress concrete in regions of expected tension. Concrete structures are generally favored for providing resistance to natural hazards because of their inherent mass and rigidity, which provide stability and deformation control against extreme wind effects, storm surges, and tsunamis, while also providing fire resistance. They have to be designed for continuity and inelastic deformability for improved seismic resistance (Park and Pauley, 1975). Concrete structures are often designed to experience inelastic deformations under strong earthquakes to dissipate seismic-induced energy.

Bibliography

Kosmatka, B., Kerkhoff, W., and Panarese, W., 2008. *Design and Control of Concrete Mixtures*. Skokie: Portland Cement Association. 372 p.

Nawy, E. G., 2006. *Prestressed Concrete – A Fundamental Approach*, 5th edn. Upper Saddle River: Prentice-Hall.

Neville, A. M., and Brooks, J. J., 2008. *Concrete Technology*. Harlow: Pearson Education.

Park, R., and Pauley, T., 1975. *Reinforced Concrete Structures*. New York: Wiley. 765 p.

Cross-references

Building Codes
Buildings, Structures and Public Safety
Damage and the Built Environment
Earthquake Resistant Design
High-Rise Buildings in Natural Disaster
Structural Damage Caused by Earthquakes
Unreinforced Masonry Buildings

CONVERGENCE

Ilan Kelman
Center for International Climate and Environmental Research – Oslo (CICERO), Blindern, Oslo, Norway

Synonyms

Disaster tourism; Post disaster return

Definition

Convergence refers to the spontaneous movement of people, messages, and goods – organized and unorganized – towards a disaster area.

Overview

Following a disaster, a spontaneous movement towards the disaster-affected area of people for various reasons, messages bearing different forms of information, and goods including relief supplies are frequently observed. That movement combines organized and unorganized efforts. Such activity is termed "convergence" and is a topic in disaster research.

Fritz and Mathewson (1957) articulated reasons for what they termed "informal or unofficial convergers" to disaster sites within their theory of convergence behavior in disasters. They describe five categories – still relevant and used today, as they form the basis for ongoing convergence research – that are not mutually exclusive: the returnees, the anxious, the helpers, the curious, and the exploiters.

Returnees are disaster survivors, evacuees, or those who were away from home before the disaster and who come back to their homes, with or without official sanction. Reasons for returning include property recovery, property protection, grieving, and no other place to live. The anxious refers to those individuals with a close connection to the disaster-hit community but who do not live there and who converge on the disaster site out of anxiety for friends, relatives, or their previous home.

The helpers refer to volunteers or professional assisters who wish to provide post-disaster services. Examples are rescuing trapped people, body recovery, and meeting physical and psychological needs of on-site disaster survivors or other convergers. Some helpers self-deploy which is usually not recommended because that can interfere with post-disaster resources and coordination. Donations – of cash, time, goods, and services – is a form of helper convergence. Often, problems result from poorly considered donations, such as sending food or clothes that are culturally inappropriate for the affected area. The most effective post-disaster donations tend to be cash given to credible organizations that are familiar with the location.

The curious refers to people converging on the disaster site as sightseers or spectators.

The exploiters are subdivided into looters, pilferers or souvenir hunters, relief stealers, profiteers, each of which is self-defining. Although representatives of these categories are witnessed after many disasters, widespread and systematic looting, profiteering, and mob-related crime are not common. Instead, these tend to be isolated incidents that simply receive exaggerated publicity.

The Internet has permitted online convergence behavior. Examples are online memorials, scam artists trying to defraud disaster-affected people, and Web sites dedicated to specific disasters for memorials and/or voyeurism.

Research on convergence is principally, although not entirely, derived from sociological and American perspectives. Limited work covers convergence from other geographic, cultural, and disciplinary perspectives.

Bibliography

Fritz, C. E., and Mathewson, J. H., 1957. *Convergence Behavior in Disasters: A Problem in Social Control. Disaster Study 9*, Committee on Disaster Studies. Washington, DC: National Academy of Sciences – National Research Council.

Cross-references

Myths and Misconceptions

COPING CAPACITY

Virginia R. Burkett
United States Geological Survey, Climate and Land use Change Mission Area, Many, LA, USA

Synonyms

Adaptive capacity

Definition

Coping capacity is the ability of a system (natural or human) to respond to and recover from the effects of stress or perturbations that have the potential to alter the structure or function of the system.

Discussion

The capacity of a system to cope with a natural hazard is determined by the ability of the system to adjust to a disturbance, moderate potential damage, take advantage of opportunities, and adapt to the consequences (Gallopin, 2006). The concept of coping capacity is often associated with extreme events whereas the concept of adaptive capacity generally alludes to a longer time frame and implies that some learning either before or after an extreme event or change in conditions has occurred (Smit and Wandel, 2006; Peltonen, 2010). The IPCC (2007, p. 869) defines "adaptive capacity" in relation to climate change as "the ability of a system to adjust to climate change (including climate variability and extremes) to moderate potential damages, to take advantage of opportunities, or to cope with the consequences." Turner et al. (2003) describe "adaptation" as a system's restructuring after exposure to a stress or perturbation.

Some natural hazards are considered "extreme events" because they are associated with the rapid restructuring of physical, biological, and/or societal systems. Storms, fires, volcanic eruptions, floods, landslides, avalanches, tsunamis, and other extreme events are all capable of stressing systems to a point that leads to a rapid shift from one state to another. Other natural hazards, such as subsidence of the land surface and erosion of the coastline, occur over a longer time frame. The cumulative effects of such small-scale events can perturb natural and human systems, in some cases more severely than an "extreme event" such as a storm or earthquake.

The capacity of society to cope with a natural hazard is dependent upon many variables. The following factors are considered major determinants of coping capacity, based on a review of Yohe and Tol (2002), Gallopin (2003, 2006), Armas and Avram (2009), and Gaillard and others (2008):

1. The exposure and sensitivity of the system to direct or indirect impacts of the natural hazard and the related vulnerability of social systems and the environments on which they depend.
2. The ability of decision makers to manage information, the accuracy of information, the processes by which decision makers determine which information is credible, and the credibility of the decision makers themselves.
3. The range and availability of technological options.
4. The availability of resources and their distribution across the affected population.
5. The structure and efficiency of critical institutions and decision-making authority.
6. The stock of human and societal capital, including education, personal security, strength of livelihoods, and social networks.
7. The potential for risks to be shared or spread (e.g., insurance systems).
8. The public's perception of the natural hazard and the relative significance of exposure compared to other societal challenges.

Coping capacity is an attribute of a system that exists prior to the perturbation (Gallopin, 2006). In the context of human societies, changes in coping behavior can emerge spontaneously (unplanned) or proactively (planned). Proactive coping behavior is the outcome of deliberate policy decisions that are based on an awareness of the nature of the hazard and its potential impact, coupled with actions that are required to return to, maintain, or achieve a desired state. The enhancement of coping behavior is a necessary condition for reducing vulnerability, particularly for the most vulnerable regions and socioeconomic groups (Peltonen, 2010). Human coping behaviors and factors that determine the degree to which they increase societal capacity to cope with natural hazards are illustrated in the table below.

Natural hazard	Example of coping behavior	Factors that influence how the behavior enhances coping capacity (examples)
Tsunami	Early warning system	Availability of technology; effectiveness of evacuation; availability and distribution of resources to victims
Flood	Building codes that require elevation of structures above potential flood level	Public perception of risk; efficacy of enforcement; accuracy of flood level calculation; availability of flood insurance
Earthquake	Building codes	Availability of resources needed to for compliant building construction or retrofitting; confidence in vulnerability assessments; efficiency of institutions that regulate construction
Hurricane storm surge	Business continuity planning	Public perception of risk; speed with which utilities, transportation, and other infrastructure is restored; prior experience or simulations that reveal errors or omission in planning
Wildfire	Reduction of fuel load through prescribed fire	Public acceptance of fire as a management tool; training, skill, and availability of personnel experienced in the use of prescribed fire; presence of houses and other structures that prevent the use of fire as an option for hazard reduction
Subsidence	Control of human activity that contributes to subsidence – example: reduce rate of groundwater withdrawal	Geologic setting and other antecedent conditions; availability of alternative water resources

Summary

Coping is a behavioral capacity that can reduce the adverse impacts in a system that is exposed to an extreme event or a chronic natural hazard. The capacity for coping with a natural hazard is generally inversely related to vulnerability – the higher the coping capacity, the lower the vulnerability of a system, region, community, or individual. In some cases, however, even strong coping capacities do not necessarily reduce vulnerability. For example, transportation and sewage treatment facilities constructed in a geologic floodplain by a community with high institutional and financial resources may be as physically vulnerable to the impacts of flooding as facilities constructed by a community with low coping capacity. Coping behaviors that are based on a good understanding of both the hazard and its impacts can substantially increase the resilience of human settlements, infrastructure, and economies.

Bibliography

Armas, I., and Avram, E., 2009. Perception of flood risk in Danube Delta, Romania. *Natural Hazards*, **50**, 269–287.

Gaillard, J. C., Pangilinan, M. R. M., Cadag, J. R., and Le Masson, V., 2008. Living with increasing floods; insights from a rural Philippine community. *Disaster Prevention and Management*, **17**, 383–395.

Gallopin, G. C., 2003. A systemic synthesis of the relations between vulnerability, hazard, exposure and impact, aimed at policy identification. In *Economic Commission for Latin American and the Caribbean (ECLAC). Handbook for Estimating the Socio-Economic and Environmental Effects of Disasters*. Mexico, DF: ECLAC, LC/MEX/G.S, pp. 2–5.

Gallopin, G. C., 2006. Linkages between vulnerability, resilience, and adaptive capacity. *Global Environmental Change*, **16**, 293–303.

Intergovernmental Panel on Climate Change, 2007. Climate change 2007: the physical science basis. In Solomon, S., et al. (eds.), *Contribution of Working Group I to the Fourth Assessment Report of the Intergovernmental Panel on Climate Change*. Geneva, Switzerland: Intergovernmental Panel on Climate Change.

Peltonen, L., 2010. *Coping Capacity and Adaptive Capacity*. http://www.gtk.fi/slr/article.php?id=18. Accessed January 20, 2010.

Smit, B., and Wandel, J., 2006. Adaptation, adaptive capacity and vulnerability. *Global Environmental Change*, **16**, 282–292.

Turner II, B. L., Kasperson, R. E., Matson, P. A., McCarthy, J. J., Corell, R. W., Christensen, L., Eckley, N., Kasperson, J. X., Luers, A., Martello, M. L., Polsky, C., Pulsipher, A., Schiller, A., 2003. A framework for vulnerability analysis in sustainability science. *Proceedings of the National Academy of Sciences of the United States of America*, **100**(14), 8074–8079.

Yohe, G. W., and Tol, R. S., 2002. Indicators for social and economic coping capacity – moving towards a working definition of adaptive capacity. *Global Environmental Change*, **12**, 25–40.

Cross-references

Adaptation
Antecedent Conditions
Coastal Erosion
Disaster Risk Reduction (DRR)
Earthquake Resistant Design
Emergency Planning
Flood Hazard and Disaster
Hazard and Risk Mapping
Hurricane (Cyclone, Typhoon)
Indian Ocean Tsunami, 2004
Land Subsidence
Planning Measures and Political Aspects
Risk Perception and Communication
Wildfire

COST-BENEFIT ANALYSIS OF NATURAL HAZARD MITIGATION

Sven Fuchs
University of Natural Resources and Life Sciences, Vienna, Austria

Definition

Defined in its broadest sense, cost-benefit analysis (CBA) is a tool to estimate and sum up the equivalent monetary value of the costs and benefits of alternatives in order to establish a decision context for politicians (e.g., Mishan, 2006). CBA is used for a systematic comparison of all costs and benefits that arise over (a certain) time period; and it uses discounting to make costs and benefits that arise in future comparable. Regarding natural hazard mitigation, CBA is for the most part applied with respect to permanent technical mitigation measures such as snow-supporting structures in avalanche-prone areas or dams along rivers. Costs are usually defined as expenses needed for the respective mitigation measure, such as concrete and steel necessary to build a check dam, and the labor needed for construction works. Benefits are typically defined as prevented damage which will arise in the future due to the implementation of the planned mitigation measure. CBA allows comparing different given mitigation alternatives with each other. CBA is targeted at the socially optimum level of safety, which will be when risks have been reduced by mitigation measures up to the point where the extra cost of any risk reduction equals its benefits (Marin, 1992).

Background

Costs and benefits can be determined for any goods traded on perfect and therefore efficient markets by using existing market prices; these so-called private goods include almost everything that is available in everyday life, such as food, vehicles, realties, and flight tickets. Such private goods are characterized by rivalness and excludability (e.g., Mankiw, 2008); multiple consumers compete for the use of such goods, and if someone is not willing to pay for the good, he or she can be excluded from consumption. However, it has been repeatedly argued that some goods do not have these characteristics (e.g., Samuelson, 1954). When a person cannot be excluded from consumption even if he or she is not willing to pay (non-excludability), and the individual consumption does not detract from the ability of others to consume such goods (non-rivalry) the good is considered as a public good. Typical examples of public goods include national defense, uncongested nontoll roads, and permanent constructive natural hazard protection.

Taking the latter as an example, for an inhabitant of a settlement, the quality of hazard protection does not change by the utilization of the same good by another inhabitant (Fuchs and McAlpin, 2005; Fuchs et al., 2007). The marginal costs of the utilization of the hazard protection measure by an additional user are zero and, as a consequence, there is no market price for this good. As a result, consumption of the utility from this public good is not necessarily fully valued by the users. In turn, no user can exclude, independently of the individual willingness to pay, another user from utilization. Non-excludability creates incentives for free riding because people can attain the utility of a good without paying for it. Free riding is another source of market failure because, since people pay for less than the efficient quantity of a good, the market produces less than the efficient quantity of the good and, as a result, the private sector fails to provide this good at a sufficient level for economic efficiency (Fuchs and McAlpin, 2005). Therefore, the supply must take place via the public sector in order to meet the societal demand.

However, in some cases, due to the scarcity of protected areas for development within hazardous areas, potential users could be excluded from the utilization. This scarcity would make mitigation measures common (pool) resources, for which use by some decreases the potential utility to others (Fuchs and McAlpin, 2005; Mankiw, 2008).

To facilitate the optimal supply of mitigation measures, the public sector will need, among other information, evaluations of the costs and benefits of mitigation approaches (Musgrave, 1969). Due to the characteristics of public goods stated above, such an evaluation can be made by comparing the costs of the supply of the good with an indirect measurement of the benefit for the consumer. Whereas such an attempt is relatively robust with respect to tangibles, questions related to an evaluation of intangibles have been subject to continuous discussions for decades (e.g., Adams, 1974; Green and Penning-Rowsell, 1986; Bateman, 1992; Eade and Moran, 1996).

Methodology

It is necessary to consider all relevant costs and benefits when applying CBA, including indirect costs and those costs arising later in time. Sensitivity analysis allows coping with uncertainty by analyzing the sensitivity of the results obtained under the CBA to variations in the individual factors used. The net present value to be obtained during CBA is the discounted net benefit gained or the net cost imposed on the stream of costs and benefits over time. As a consequence, the planning horizon that is considered (e.g., with respect to a planned flood retention basin) matters for the outcome of a CBA.

From a theoretical point of view, the methodology is schematically illustrated by total cost and corresponding total benefit due to the implementation of mitigation measures in Figure 1. At the level of mitigation q^0, the marginal benefit of additional mitigation is higher than the cost. Thus, further investments produce net benefits. At q^{po}, the slope of total benefit (A) and the slope of total cost (B) are equal, the marginal benefit and marginal cost per unit of mitigation are equal, and the level of mitigation is optimal. As the level of mitigation increases beyond q^{po} up to q^{pi}, where the total costs are the same as the total benefits, the total supply of mitigation still provides positive net benefits but is greater than optimal because the marginal cost of each additional unit of mitigation exceeds the corresponding marginal benefit. Beyond q^{pi}, the total supply of mitigation produces negative net benefits (adapted from Russell, 1970, 386).

Determination of costs

Economic theory suggests evaluating the costs of mitigation measures in terms of opportunity costs, which is the alternative investment of resources in the next-best alternative available to someone who has to select between several mutually exclusive choices. These costs mirror the benefit that would have resulted from an alternative appropriation of the resources. From a practical point of view, the present value of investments in mitigation measures is taken instead since it is almost unfeasible to take into account all possible other alternatives. Apart from any material and labor force needed, the investments necessary for maintenance over the life time of the structures have to be taken into account. The present value of capital expenditures for permanent mitigation measures is calculated using Equation 1, based on the real interest rate, which takes into account inflation and therefore allows comparison of expenditures in different years. Therefore, discounting may change considerably the results of a CBA depending on the choice of the discount rate. From the perspective of society, the use of low discount rates is justified with considerations on intergenerational equity and sustainability. K_n is the present value of the total capital at the expiration of the validity in monetary units, p is the real interest rate in percent, s is the interest period, n the term, and K_0 the opening capital in monetary units. The real interest rate i_{real} is typically calculated on the basis of the nominal interest rate i_{nom} and the inflation J, using Equation 2. The corresponding nominal interest rate is derived from, e.g., the average rate of interest of government bonds in the countries were the study is located.

$$K_n = \left(1 + \frac{p \cdot s}{100}\right)^n \cdot K_0 \qquad (1)$$

$$i_{real} = \left(\frac{1 + i_{nom}}{1 + J}\right) - 1 \qquad (2)$$

Determination of benefit

The accuracy of CBA depends on how accurately benefits (and costs) are estimated and that all costs and benefits are accounted for. Principally, benefits of the impacts of an intervention are evaluated in terms of the public's willingness to pay for these impacts (benefits). The benefit related to mitigation measures can be determined in different ways. However, from a methodological point of view and focusing on the application of CBA in natural hazard risk management, the evaluation is either based on an evaluation of buildings and infrastructure lifelines exposed or with respect to protected human life. Both concepts are therefore separately described below.

- The utility can be defined in the sense of prevented damage to buildings and infrastructure, the so-called method of loss expenses. Because market processes (here for real estate within hazard-prone areas) are able to reflect the real costs, market values, from an economic point of view, are particularly suitable for the determination of possible damage. If, at the time of investigation, the market demand for the buildings is high, their current value may be above the replacement value. If, for example, due to a flood event, there is no demand on the market for those buildings, their value could be zero. The societal preferences of buildings in hazard-prone areas can therefore precisely be

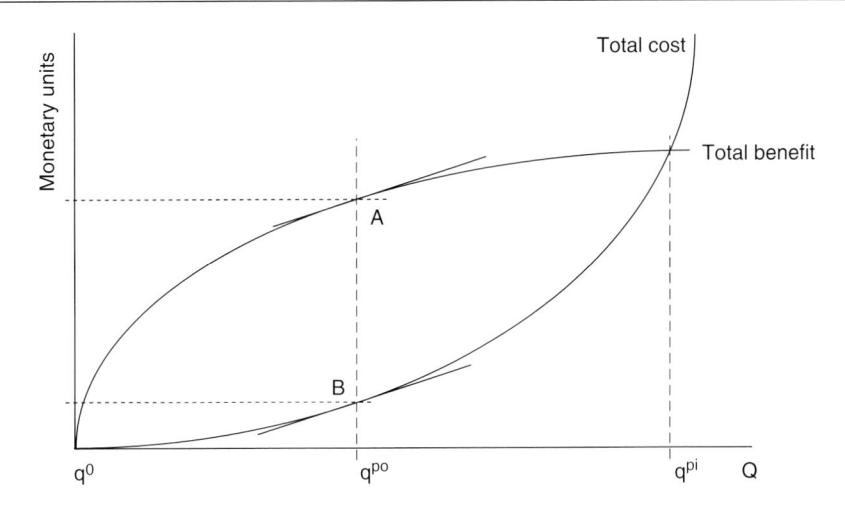

Cost-Benefit Analysis of Natural Hazard Mitigation, Figure 1 Cost-benefit analysis of natural hazard mitigation.

measured, which is the overall aim of such economic methods. However, since the investigation is exactly focusing on buildings in endangered areas, there might be a bias with respect to the socially optimum level of safety. Thus, the replacement value can be used instead as an approximation, neglecting any risk-dependent change in the demand of buildings on the market. Following this method, data concerning the number of potentially affected buildings and their respective replacement value has to be collected. With respect to infrastructure lifelines, the evaluation usually takes place by multiplying their affected length by the value per unit length. These values have to be adjusted to take into account for inflation, and the obtained sums can be directly compared to the respective year of construction.

- In a second set of calculations, the benefit can be evaluated in terms of the number of lives protected (see *Economic Valuation of Life*). The number of persons in the endangered areas is thereby determined on the basis of census data, or the number of domiciles located within areas to be protected by the mitigation measure. Subsequently, a valuation of the number of persons is undertaken in order to place monetary units on human life to be able to calculate the cost-benefit ratios. This step is solely a technical necessity and does not imply that there is a however-defined "value" of human life (which would be an ethical issue that cannot be solved by CBA, compare Adams, 1974). One possibility to achieve such a value does make use of a human capital approach. This procedure can be traced back to approaches in the insurance business, where financial compensation is paid to the immediate family upon the premature demise of the policyholder. The value of human life is calculated as follows: In the study area, the annual gross earned income per working person is identified, for example, by using available statistical information.

Subsequently, the average age of the population is achieved and compared to the mean average retirement age. By subtracting these two figures, a remaining average expectancy of working life and a corresponding expected gross income results. Equation 3 is applied to calculate the annuity value R_0 from the payment r, the factor q, and the term n. The factor q is derived summing up the rate of interest i with 1. The rate of interest is calculated by using information on the average rate of interest of government bonds in the countries where the study is located. Applying Equation 3, an annuity value with the interest paid at the end of the period results for the annuity value corresponding to the income of an average person during the remaining working life. This value is subsequently applied in the CBA in order to calculate the benefits resulting from a mitigation measure.

$$R_0 = r \cdot q^{-n} \cdot \frac{q^n - 1}{q - 1} \qquad (3)$$

Discussion

Societal and political decisions about mitigation measures concerning natural hazards are generally based on a multiplicity of interests due to the variety of parties involved. Hence, there is a particular need for methodologies ensuring the consideration of all these interests and providing simultaneously a reliable basis for the final decision maker.

However, evaluations of the net benefits of natural hazard protection measures will vary as the local context changes. The relatively high property values in the densely populated regions of central Europe and the USA and relatively high incomes of persons produce net benefits that are higher than they would be in other areas or countries with lower property values and incomes.

Although there may be potential gains in economic efficiency from changing the supply of hazard protection in some areas, the decision to supply more or less avalanche protection is a political one (Gamper et al., 2006). CBA can only inform, rather than answer, the question of how much risk protection the public sector should provide. The choice of how much to invest in mitigation measures depends on the political determination of a standard of protection. The standard may be set in terms of societal preferences such as risk reduction, the level of expenditures, as a target for the maximum number of lives lost, or in some other way. In addition to the need to incorporate CBA into a broader context of political decision making, there are still unresolved issues involved in using CBA as information for decisions about the level of protection against natural hazards. Firstly, most cost-benefit analyses assume that effects should be evaluated with respect to the preferences of individuals (Nash et al., 1975; Adams, 1993). However, since some people benefit more directly from mitigation measures than others, the preferences for the measures may be different among the group of people who live in endangered areas and among those who live outside those areas. Therefore, CBA is affected by whose preferences are used to determine the benefits of mitigation measures. In its traditional form, CBA does not consider the distribution of benefits and costs over individuals, and any increase in net benefits is desirable, regardless of to whom they occur. Secondly, while the utility from protecting property from natural hazards can be determined with relative ease and minimal debate, as the property values are already expressed in monetary terms, the valuation of protecting people from natural hazards requires placing a monetary value on each human life in the absence of objective rules for doing so (e.g., Adams, 1993; Pearce, 1998). The human capital approach presented above raises ethical issues, as it values old people less than young or middle-aged people. Thirdly, problems may arise in the aggregation of material assets and nonmaterial assets, such as an individuals' cognition of safety. Therefore, CBA seems to be an appropriate tool for a relative evaluation of different mitigation alternatives rather than for an absolute evaluation of one individual mitigation measure. CBA is simplified considerably when different alternatives attaining the same utility are evaluated against each other. This approach would apply in the situation where a level of risk acceptance has been set by the relevant community and the question is how to most effectively meet this standard. In order to determine the most competitive alternative, only relative comparisons of cost-effectiveness are necessary, which avoids the problems associated with valuing human lives.

Conclusion

Despite its limitations, economic analyses can contribute by providing information for the political choice of a standard of protection against natural hazards and on how to achieve the socially determined standard. CBA offers a tool for policy decisions because it allows for the comparison of monetary and nonmonetary factors. The comparison of economic costs and benefits is one consideration that may facilitate decision making about protection against natural hazards.

The optimal approach to natural hazards risk reduction depends on the particular hazard, the aims of the affected community, and relevant decision processes. Minimizing human fatalities may be the main priority, with an economic efficiency – thought of as a shift in welfare – as a secondary goal. Although there may be gains in economic efficiency from changing the supply of natural hazard protection, the decision to supply more or less protection is a political one. This decision is related to the society's level of risk acceptance, and should only be discussed on a participative basis.

The potential of CBA depends on its proper integration in the decision-making process as an equitable, transparent, and flexible instrument. Transparency as to assumptions used to calculate costs and benefits and the uncertainty contained in the results will increase the ability of decision makers to use findings of CBA. Decision makers have a responsibility to understand that CBA provides only part of the necessary information for natural hazards planning. Aims other than economic efficiency, such as alternative contextual factors or constraints, provide additional, necessary information for decision making.

Bibliography

Adams, J., 1974. And how much for your grandmother? *Environment and Planning*, **A6**, 619–626.

Adams, J., 1993. The emperor's old clothes: the curious comeback of cost-benefit analysis. *Environmental Values*, **2**, 247–260.

Bateman, I., 1992. Placing money values on the unpriced benefits of forestry. *Quarterly Journal of Forestry*, **86**, 152–165.

Eade, J., and Moran, D., 1996. Spatial economic valuation: benefits transfer using geographical information systems. *Journal of Environmental Management*, **48**(2), 97–110.

Fuchs, S., and McAlpin, M., 2005. The net benefit of public expenditures on avalanche defence structures in the municipality of Davos, Switzerland. *Natural Hazards and Earth System Sciences*, **5**(3), 319–330.

Fuchs, S., Thöni, M., McAlpin, M. C., Gruber, U., and Bründl, M., 2007. Avalanche hazard mitigation strategies assessed by cost effectiveness analyses and cost benefit analyses – evidence from Davos, Switzerland. *Natural Hazards*, **41**(1), 113–129.

Gamper, C., Thöni, M., and Weck-Hannemann, H., 2006. A conceptual approach to the use of cost benefit and multi criteria analysis in natural hazard management. *Natural Hazards and Earth System Sciences*, **6**(2), 293–302.

Green, C., and Penning-Rowsell, E., 1986. Evaluating the intangible benefits and costs of a flood alleviation proposal. *Journal of the Institution of Water Engineers and Scientists*, **40**(30), 229–248.

Mankiw, N., 2008. *Principles of Economics*. Mason: South-Western Cengage.

Marin, A., 1992. Cost and benefits of risk reduction. In Royal Society Study Group (ed.), *Risk: Analysis, Perception and Management*. London: The Royal Society, pp. 192–201.

Mishan, E., 2006. *Cost-Benefit Analysis*. London: Routledge.

Musgrave, R., 1969. Cost benefit analysis and the theory of public finance. *Journal of Economic Literature*, **7**, 797–806.

Nash, C., Pearce, D., and Stanley, J., 1975. An evaluation of cost-benefit analysis criteria. *Scottish Journal of Political Economy*, **22**(2), 121–134.

Pearce, D., 1998. Valuing risks. In Calow, P. (ed.), *Handbook of Environmental Risk Assessment and Management*. Malden: Blackwell, pp. 345–375.

Russell, C., 1970. Losses from natural hazards. *Land Economics*, **43**, 383–393.

Samuelson, P., 1954. The pure theory of public expenditure. *The Review of Economics and Statistics*, **36**(4), 387–389.

Cross-references

Costs (Economic) of Natural Hazards and Disasters
Damage and the Built Environment
Economic Valuation of Life
Economics of Disasters

COSTS (ECONOMIC) OF NATURAL HAZARDS AND DISASTERS

Howard Kunreuther, Erwann Michel-Kerjan
Risk Management and Decision Processes Center,
The Wharton School, University of Pennsylvania,
Philadelphia, PA, USA

Definition and introduction

Given the hundreds of billions of dollars in economic losses that catastrophes have caused in the USA since 2001, it is difficult to remember that when Hurricane Hugo hit the USA in 1989, it was the first catastrophe to inflict more than $1 billion of insured losses. But times have changed and there have been numerous large-scale natural disasters in the USA and other parts of the world in the past two decades that have been far costlier than Hugo both in terms of economic losses as well as fatalities and injuries due to the increasing concentration of population and activities in hazard-prone areas.

In Southeast Asia, the tsunami in December 2004 killed approximately ¼ million people residing in coastal areas. Cyclone Nargis, which made landfall in Myanmar in May 2008, killed an estimated 140,000 people, making it the deadliest natural disaster in the recorded history of the country. During the same month, the Great Sichuan Earthquake is estimated to have killed over 85,000, injured 374,000, and left almost five million homeless (Munich Re, 2008). Deaths from the Haitian earthquake in January 2010 are estimated at 230,000 (Insurance Journal, 2010).

Data reveals that the year 2011 is the costliest year the insurance industry has ever faced with respect to catastrophic losses. The Japan earthquake, tsunami, and nuclear power plant accident in March 2011 caused over US$210 billion in economic losses (not including nuclear-related damage), and insured losses in the range of US$35–40 billion (Munich Re, 2012). This disaster highlights the interdependencies between natural and technological accidents: the 9.0 magnitude earthquake that struck the Tohoku region of northeastern Japan caused a tsunami that hit the country's coastline within half an hour, taking the lives of nearly 20,000 people and destroying over 100,000 buildings, including the cooling system and the backup power generator of the Fukushima nuclear plant. The resulting meltdown of three nuclear reactors led to high radiation levels, which required the evacuation of more than 60,000 people (World Economic Forum Global Risk Report, 2012).

Although the USA has extensive experience with natural catastrophes and the resources to adequately prepare for them, loss-reduction measures and emergency-preparedness capacity are often inadequate to deal with large-scale natural disasters. Hurricane Katrina, which hit Louisiana and Mississippi at the end of August 2005, killed 1,300 people and forced 1.5 million people to evacuate the affected area – a historic record for the country. Economic losses from Hurricane Katrina are estimated in the range of $125–$150 billion (Munich Re, 2010).

Hurricanes in 2008 caused billions of dollars in direct economic losses along the Caribbean basin and in the USA. Hurricane Ike was the most expensive individual event in 2008, with privately insured losses estimated at $17.6 billion in addition to $2.4 billion in claims paid by the US National Flood Insurance Program for flood surge resulting from Ike (Swiss Re, 2009). Based on these figures, Hurricane Ike ranks as the third worst weather-related disaster in US history, after Hurricane Katrina and Hurricane Andrew, which hit southeast Florida in August 1992.

A new era of catastrophes

The economic and insured losses from great natural catastrophes such as hurricanes, earthquakes, and floods worldwide have increased significantly in recent years. According to Munich Re (2012), economic losses from natural catastrophes alone increased from $528 billion (1981–1990), $1.2 trillion (1991–2000) to $1.6 trillion over the period 2001–2011. During the past 10 years the losses were principally due to hurricanes and resulting storm surge occurring in 2004, 2005, and 2008. Figure 1 depicts the evolution of the direct economic losses and the insured portion from great natural disasters over the period 1970–2011.

Catastrophes since 1990 have had a more devastating impact on insurers than in the history of insurance before that time. Between 1970 and the mid-1980s, annual insured losses from natural disasters (including forest fires) were in the $3–$4 billion range. There was a radical increase in insured losses in the early 1990s, with

NatCatSERVICE
Great natural catastrophes worldwide 1950 – 2011
Overall and insured losses with trend

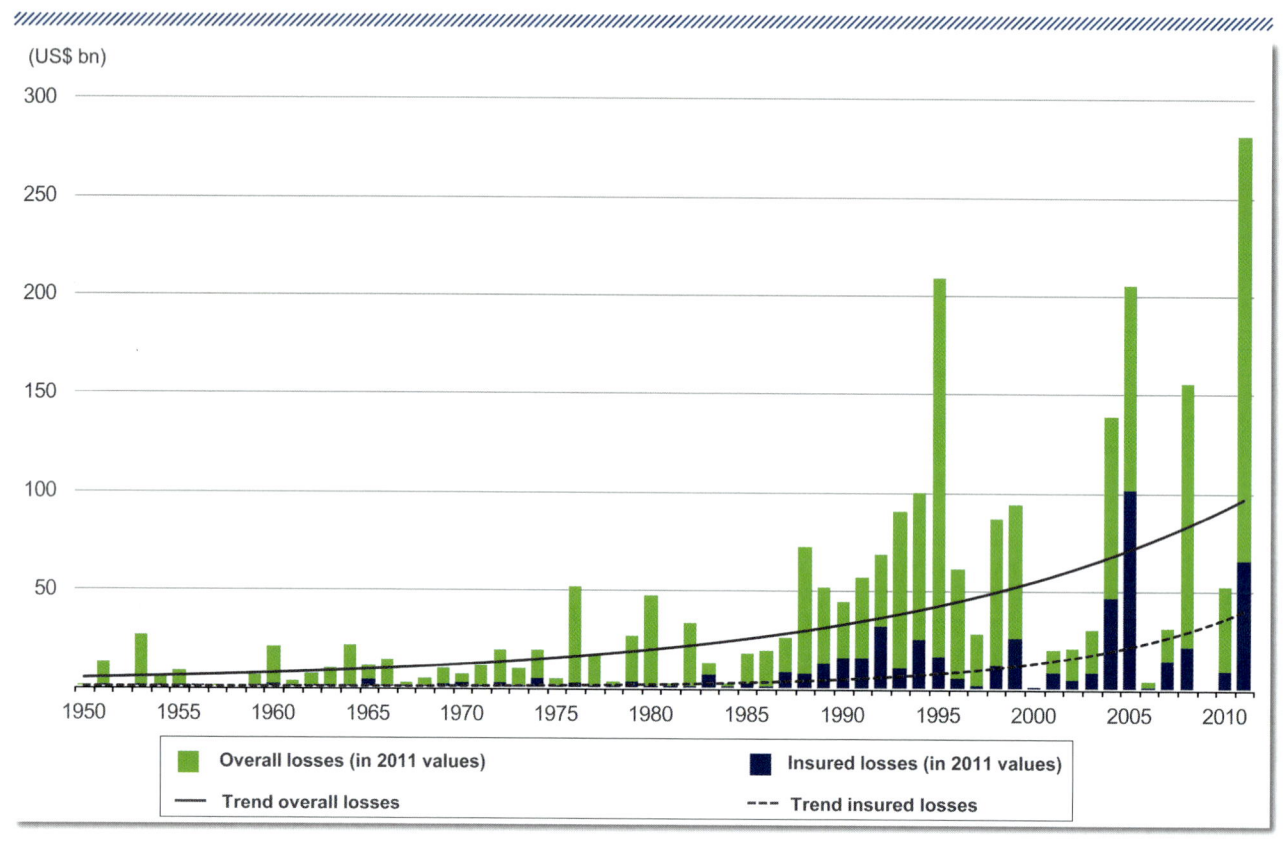

Munich RE

Costs (Economic) of Natural Hazards and Disasters, Figure 1 Natural catastrophes worldwide 1980–2011 – Overall and insured losses ($ billion) (Sources: Munich Re geo risks research).

Hurricane Andrew in Florida ($24.6 billion in 2008 dollars) and the Northridge earthquake in California ($20.3 billion in 2008 dollars). The four hurricanes in Florida in 2004 (Charley, Frances, Ivan, and Jeanne) collectively totaled almost $35 billion in insured losses. Hurricane Katrina alone cost insurers and reinsurers an estimated $48 billion, with total losses of $87 billion paid by private insurers for major natural catastrophes in 2005.

Table 1 reveals the 25 costliest insured catastrophes from 1970 to 2011 (in 2011 dollars). Of these 25 major events, 15 have occurred since 2001, 14 in the USA. With the exception of the terrorist attacks on September 11, 2001, all 25 of the costliest catastrophes were natural disasters. More than 85% of these were weather-related events – hurricanes, typhoons, storms, and floods –with nearly three quarters of the claims in the USA. Hurricane Andrew and the Northridge earthquake were the first two catastrophes that the industry experienced where losses were greater than $10 billion (designated "super-cats") and caused insurers to reflect on whether risks from natural disasters were still insurable. To assist them in making this determination, many firms began using catastrophe models to estimate the likelihood of, and consequences to, their insured portfolios from specific disasters in hazard-prone areas (Grossi and Kunreuther, 2005).

There is a very clear message from these data. Twenty or thirty years ago, large-scale natural disasters were considered to be low-probability events. Today, they not only are causing considerably greater economic losses than in the past but also appear to be occurring at an accelerating pace. In this context, it is important to understand more fully the factors influencing these changes in order to design more effective programs for reducing losses from future disasters.

The question of attribution
Several elements explain the increased costs of disasters in recent years. These include a higher degree of

Costs (Economic) of Natural Hazards and Disasters, Table 1 Twenty-five costliest insured catastrophes worldwide, 1970–2011

$ Billion	Event	Victims (dead or missing)	Year	Area of primary damage
50.1	Hurricane Katrina	1,836	2005	USA, Gulf of Mexico
38.2	9/11 Attacks	3,025	2001	USA
35–40	Earthquake and Tsunami	15,840	2011	Japan
25.6	Hurricane Andrew	43	1992	USA, Bahamas
21.2	Northridge Earthquake	61	1994	USA
18.5	Hurricane Ike	348	2008	USA, Caribbean
15.3	Hurricane Ivan	124	2004	USA, Caribbean
15.3	Hurricane Wilma	35	2005	USA, Gulf of Mexico
13.0	Earthquake	181	2011	New Zealand
11.7	Hurricane Rita	34	2005	USA, Gulf of Mexico, et al.
10.0	Floods, landslides	813	2011	Thailand
9.6	Hurricane Charley	24	2004	USA, Caribbean, et al.
9.3	Typhoon Mireille	51	1991	Japan
8.2	Maule earthquake (M_w: 8.8)	562	2010	Chile
8.2	Hurricane Hugo	71	1989	Puerto Rico, USA, et al.
8.0	Winter Storm Daria	95	1990	France, UK, et al.
7.8	Winter Storm Lothar	110	1999	France, Switzerland, et al.
7.3	Storms and tornadoes	350	2011	USA
7.0	Hurricane Irene	55	2011	USA, Caribbean
6.6	Winter Storm Kyrill	54	2007	Germany, UK, NL, France
6.1	Storms and floods	22	1987	France, UK, et al.
6.1	Hurricane Frances	38	2004	USA, Bahamas
5.5	Winter Storm Vivian	64	1990	Western/Central Europe
5.5	Typhoon Bart	26	1999	Japan
4.8	Hurricane Georges	600	1998	USA, Caribbean

Sources: Kunreuther and Michel-Kerjan (2011) with data from Swiss Re (2012).

urbanization, and an increase in the value at risk and insurance density. In 1950, approximately 30% of the world's population lived in cities. In 2000, about 50% of the world's population (six billion) resided in urban areas. Projections by the United Nations (2008) show that by 2025, this figure will have increased to 60% based on a world population estimate of 8.3 billion people.

In the USA in 2003, 53% of the nation's population, or 153 million people, lived in the 673 US coastal counties, an increase of 33 million people since 1980, according to the National Oceanic Atmospheric Administration. And the nation's coastal population is expected to increase by more than 12 million by 2015 (Crossett et al., 2004). Yet coastal counties, excluding Alaska, account for only 17% of land area in the USA. In hazard-prone areas, this urbanization and increase in population translate into greater concentration of exposure and hence a higher likelihood of catastrophic losses from future disasters. This new vulnerability is best understood in historical context – that is, compared to the cost of hurricanes in the past. It is possible to calculate the total direct economic cost of the major hurricanes affecting the USA in the past century, adjusted for inflation, population, and wealth normalization. Several studies have estimated how much previous hurricanes would have cost had they hit today. The most recent study by Pielke et al. (2008) normalizes mainland US hurricane damage for the period 1900–2005. Drawing on these data, Table 2 lists the 20 hurricanes that would have been costliest had they occurred in 2005. The estimate for each is a range based on the two approaches to normalizing losses used by the Pielke et al. study. The table provides the year when the hurricane originally occurred, the states that were the most seriously affected, and the hurricane category on the Saffir-Simpson scale. The data reveal that the hurricane that hit Miami in 1926 would have been almost twice as costly as Hurricane Katrina had it occurred in 2005, and the Galveston hurricane of 1900 would have had total direct economic costs as high as those from Katrina. We are very likely to see even more devastating disasters in the coming years because of the ongoing growth in values located in risk-prone areas.

There is another element to consider in determining how to adequately manage and finance catastrophe risks: the possible impact of a change in climate on future weather-related catastrophes. Between 1970 and 2004, storms and floods were responsible for over 90% of the total economic costs of extreme weather-related events worldwide. Storms (hurricanes in the US region, typhoons in Asia, and windstorms in Europe) contributed to over 75% of insured losses. In constant prices (2004), insured losses from weather-related events averaged $3 billion annually between 1970 and 1990 and then increased significantly to $16 billion annually between 1990 and 2004 (Association of British Insurers, 2005). In 2005, 99.7% of all catastrophic losses

Costs (Economic) of Natural Hazards and Disasters, Table 2 Twenty costliest Hurricanes, 1900–2005 (ranked using 2005 inflation, population, and wealth normalization)

Rank	Hurricane	Year	Category	Cost range in 2005 ($ billions)
1	Miami (Southeast FL/MS/AL)	1926	4	140–157
2	Katrina (LA/MS)	2005	3	81
3	North Texas (Galveston)	1900	4	72–78
4	North Texas (Galveston)	1915	4	57–62
5	Andrew (Southeast FL and LA)	1992	5–3	54–60
6	New England (CT/MA/NY/RI)	1938	3	37–39
7	Southwest Florida	1944	3	35–39
8	Lake Okeechobee (Southeast Florida)	1928	4	32–34
9	Donna (FL-NC/NY)	1960	4–3	29–32
10	Camille (MS/Southeast LA/VA)	1969	5	21–24
11	Betsy (Southeast FL and LA)	1965	3	21–23
12	Wilma	2005	3	21
13	Agnes (FL/CT/NY)	1972	1	17–18
14	Diane (NC)	1955	1	17
15	(Southeast FL/LA/AL/MS)	1947	4–3	15–17
16	Hazel (SC/NC)	1954	4	16–23
17	Charley (Southwest FL)	2004	4	16
18	Carol (CT/NY/RI)	1954	3	15–16
19	Hugo (SC)	1989	4	15–16
20	Ivan (Northwest FL/AL)	2004	3	15

Source: Pielke et al. (2008).

worldwide were due to weather-related events (Mills and Lecomte, 2006).

There have been numerous discussions and scientific debates as to whether the series of major hurricanes that occurred in 2004 and 2005 might be partially attributable to the impact of a change in climate. One of the expected effects of global warming will be an increase in hurricane intensity. This increase has been predicted by theory and modeling, and substantiated by empirical data on climate change. Higher ocean temperatures lead to an exponentially higher evaporation rate in the atmosphere, which increases the intensity of cyclones and precipitation. An increase in the number of major hurricanes over a shorter period of time is likely to translate into a greater number hitting the coasts, with a greater likelihood of damage to a residences and commercial buildings today than in the 1940s – a trend that raises issues about the insurability of weather-related catastrophes.

Conclusions

Since the 1990s, a series of large-scale catastrophes have inflicted historic economic and insured losses. Fifteen of the 25 costliest insured catastrophes worldwide between 1970 and 2011 occurred after 2001, and all were natural disasters except for the 9/11 terrorist attacks. The USA has been particularly challenged because 14 of these disasters occurred in this country. The growing concentration of population and structures in high-risk areas, combined with the potential consequences of global climate change, are likely to lead to even more devastating catastrophes in the coming years unless cost-effective risk-reduction measures are put in place.

The task facing the USA and many other countries is ascertaining how to prevent the natural disaster syndrome. Even when risk-reduction measures are available and are cost-effective, many people still do not invest in them because they are myopic and misvalue the upfront costs of these measures as much greater than the expected benefits in reduced damage from disasters in future years. Many victims of Hurricane Katrina suffered severe losses from flooding because they had not undertaken loss mitigation and did not have flood insurance. As a result, an unprecedented level of federal disaster assistance – $81.6 billion (2005 prices) – was provided to these victims and the affected communities (Kunreuther and Michel-Kerjan, 2011).

Bibliography

Association of British Insurers, 2005. *Financial risks of climate change*. London: Association of British Insurers.

Crossett, K. M., Culliton, T. J., Wiley, P. C., and Goodspeed, T. R., 2004. *Population trends along the coastal United States: 1980–2008*. Silver Spring: National Oceanic and Atmospheric Administration.

Grossi, P., and Kunreuther, H. (eds.), 2005. *Catastrophe modeling: a new approach to managing risk*. New York: Springer.

Insurance Journal, 2010. "Haiti death toll rises to 230,000; CCRIF, CIMH study future disaster mitigation efforts," February 10, 2010 http://www.insurancejournal.com/news/international/2010/02/10/107249.htm. Accessed 26 June 2012.

Kunreuther, H., and Michel-Kerjan, E., 2011. *At war with the weather: managing large-scale risks in a new era of catastrophes*. Cambridge, MA: MIT Press. Paperback edition.

Mills, E., Lecomte, E. 2006. *From risk to opportunity: how insurers can proactively and profitably manage climate change*. Ceres. August. http://eetd.lbl.gov/emills/pubs/pdf/ceres_report_090106.pdf. Accessed 26 June 2012.

Munich Re. 2008. *Catastrophe figures for 2008 confirm that climate agreement is urgently needed*. December. http://www.munichre.com/en/press/press_releases/2008/2008_12_29_press_release.aspx. Accessed 26 June 2012.

Munich Re. 2010. *Topics geo. Natural catastrophe 2009*. Munich: Munich Re. http://www.munichre.com/publications/302-06295_en.pdf. Accessed 26 June 2012.

Munich Re. 2012. *Topics geo natural catastrophes 2011, Analyses, assessments, positions*.

Pielke, R., Jr., Gratz, J., Landsea, C., Collins, D., Saunders, M., and Musulin, R., 2008. Normalized hurricane damage in the United States: 1900–2005. *Natural Hazards Review*, **9**(1), 29–42.

Swiss Re. 2009. *Natural catastrophes and man-made disasters in 2008*. Sigma, 2/2009.

Swiss Re. 2012. *Natural catastrophes and man-made disasters in 2011*. Sigma, 2/2012.

United Nations. 2008. *World population trends*, UN Population Division, Department of Economic and Social Affairs, New York.

World Economic Forum. 2012. *Global Risks 2012 (Seventh Edition)*.

CREEP

Piotr Migoń
University of Wrocław, Wrocław, Poland

Synonyms

Permafrost creep; Rock creep; Soil creep

Definition

Creep is defined as a semi-continuous, time-dependent deformation of solids which occurs at a low rate, under stress imposed by gravity. In Earth Sciences, creep of rock, soil, and frozen ground are distinguished. Not only are they different from the mechanical point of view, but they are associated with different types of natural hazards.

Overview

Soil creep is a very slow downslope movement of the near-surface part of the soil profile, at a rate usually decreasing exponentially with depth. At a depth > 50 cm, the effects of soil creep are hardly visible. It is a combination of different mechanisms, including pure shear, viscous laminar flow, expansion, and contraction. Frequent freeze/thaw and wetting/drying cycles contribute to the efficacy of creep. Creep rate, typically a few centimeters per year, is dependent on slope angle, cohesion of soil material, climatic conditions and biotic factors (vegetation cover, animal trampling). In the most favorable circumstances, e.g., on steep tropical slopes, rates approaching

Creep, Figure 1 Bent trees are commonly viewed as an evidence of soil creep.

0.5 m year^{-1} have been observed. Terracettes and bent trees (Figure 1) are noted as typical surface indicators of soil creep.

Rock creep is a unique behavior of solid rock and occurs under two circumstances. It may affect heavily fractured rock masses near the surface, which deform by joint opening and shearing along joint surfaces. Primary structural discontinuities bent downslope are the evidence of near-surface creep. Rock creep is also known to occur at great depths under considerable lithostatic stress, mainly in weak sedimentary rocks, particularly evaporates. Tunnel closures and excavation-wall buckling are typical manifestations of rock creep.

Permafrost creep is a term used to describe deformation of ice-saturated debris bodies, typically rock glaciers and protalus lobes, primarily under their own weight. However, doubts are expressed if this expression is correct, as permafrost is commonly understood as a thermal state of lithosphere.

Creep, being a deformation occurring at rather low rate, is rarely hazardous, although it may result in weakening of building foundations and tilting of trees and other vertical man-made structures (e.g., poles, masonry walls, gravestones) in the longer term, the latter leading to their collapse. However, creep may be a precursor to much more rapid deformations, in the form of either a mudslide (for soil creep) or rock slope failure (for rock creep). Rock creep may also cause severe problems in mine operations and transportation tunnels. Acceleration of creep usually occurs prior to a catastrophic yield. Therefore, in areas identified as potentially hazardous, the rate of creep should be monitored.

Bibliography

Selby, M. J., 1993. *Hillslope Materials and Processes*. Oxford: Oxford University Press.

CRITICAL INCIDENT STRESS SYNDROME

Cross-references

Landslide Types
Mass Movement

CRITICAL INCIDENT STRESS SYNDROME

Ann M. Mitchell[1], Kirstyn Kameg[2]
[1]University of Pittsburgh, Pittsburgh, PA, USA
[2]Robert Morris University, Moon Township, PA, USA

Synonyms

Acute stress disorder; Acute stress reaction; Post-traumatic stress disorder; Traumatic stress

Definition

Experiencing trauma is an essential part of being human, yet most people who experience critical incidents survive without developing psychiatric disorders. However, traumatic experiences can alter people's psychological, biological, and social equilibrium (van der Kolk et al., 1996).

Normal reactions

Most people exposed to critical incidents or traumas do not go on to develop psychiatric disorders; in fact, there is literature to support that there is the potential for post-traumatic growth following trauma (Linley and Joseph, 2004; Joseph and Linley, 2005; Paton, 2005). Natural disaster victims and their significant others who have been exposed to a sudden event that precipitates fear of injury or loss of life can respond to the traumatic event with a wide range of physical and emotional responses. Simply witnessing such an event can also produce psychological, social, and physiological dysfunction. Natural, technological, and other types of disasters (i.e., man-made, terroristic) expose innumerable people to scenes of destruction and human loss, and they can react with a classic set of symptoms similar to an acute stress reaction. Their emotional responses to disasters may be conceptualized as progressing through a number of phases. During the impact phase within the first few days, individuals often feel stunned and in shock. In these early days, individuals may also experience disbelief, numbness, fear, and confusion to the point of disorganization (Lubit, 2008). In the crisis phase, individuals may alternate between denial and intrusive symptoms with hyperarousal and may experience any number of somatic symptoms as well as irritability, apathy, and social withdrawal. Here, persons may become angry with caregivers who fail to solve problems and/or may be unable to be organized in the chaos of dealing with the crisis (Lubit, 2008). During the resolution phase, grief, guilt, and depression may be prominent and last through the first year as people continue to cope with their numerous losses, and finally, in the reconstruction phase, reappraisal, "meaning-making," and the integration of the event into a new self-concept occurs (Lubit, 2008).

Epidemiology

Epidemiological surveys of large groups of the general public have been done to determine exposure to various traumatic events. Researchers concluded that lifetime exposure to traumatic events may be as high as 73.6% for men and 64.8% for women (Solomon and Davidson, 1997). The lifetime prevalence of those individuals who will experience post-traumatic stress disorder (PTSD) at some point in their life varies from 7.8% (Kessler et al., 1995) for all to 5% of men and 10–12% of women (Solomon and Davidson, 1997). This figure jumps from 3% to 58% for "at risk individuals" (APA, 1994). Individuals may be at an increased risk for the development of PTSD if they witness an event that involves death, interpersonal violence, grotesque sensory images, or some natural disasters. It is also important to remember that critical incident stress may affect professionals (e.g., police, fire, healthcare professionals, and others) working in the field with victims of disasters (Paton and Violanti, 2011).

Complications

Two possible complications following exposure to a disaster include the development of acute stress disorder (ASD) and PTSD. Guidelines established by the Diagnostic Statistical Manual of Mental Disorders, Fourth Edition, Text Revision (DSM-IV-TR) remain the gold standard for diagnosing ASD and PTSD. For both disorders, the individual must have been exposed to or witnessed a traumatic event that involved actual or threatened death, serious injury, or a threat to physical integrity in addition to responding with fear, helplessness, or horror. Additional symptoms seen in both disorders include reexperiencing the traumatic event, avoidance of stimuli associated with the traumatic event, and increased arousal. Reexperiencing the trauma may occur through intrusive recollections, nightmares, flashbacks, hallucinations, and psychological distress/physiological reactivity upon exposure to cues that symbolize the trauma. Symptoms of avoidance include efforts to avoid thoughts, feelings, or conversations associated with the trauma, inability to remember certain aspects of the trauma, reduced interest in activities, feeling detached from others, and a sense of a foreshortened future. Symptoms of increased arousal include difficulty with sleep, irritability/angry outbursts, poor focus, hypervigilance, and exaggerated startle response. Lastly, in both disorders, the symptoms cause significant distress or impair the individual's ability to function (APA, 1994).

Differences between the two diagnoses include the time frame and the occurrence of dissociative symptoms. The onset of ASD must occur and resolve within 4 weeks of the traumatic event. Additionally, in ASD, the individual experiences dissociative symptoms during exposure to the trauma or immediately following the trauma. In PTSD, the duration of the symptoms exceeds 1 month. PTSD can also be classified as acute (duration of symptoms is less than 3 months), chronic (duration of symptoms is greater than 3 months), or delayed (duration of symptoms is at least 6 months after the stressor).

There is emerging evidence that there is a potential for positive outcomes following exposure to trauma. Paton (2005) and others (Tedeschi and Calhoun, 2003) have identified post-traumatic growth, enhanced professional capability, greater appreciation of family, and increased sense of control over significant adverse events as adaptive outcomes that may occur following a crisis. Factors that influence positive growth include personal resilience and vulnerability factors.

Treatment

Following exposure to a trauma, it is necessary to ensure a sense of safety. Provision of basic needs including food, clothing, and medical care must be met as well as ensuring that survivors are protected from reminders of the event, if possible; the onlookers; and the media. Mobilization of family members is critical, as social support networks may provide an important resource for coping with the aftermath of a traumatic event. It is also important to assist the survivor in reestablishing a sense of efficacy through education about stress responses and normal versus abnormal symptoms, as well as strategies to reduce anxiety.

Psychotropic medications should be used sparingly in the first 48 h following a natural disaster or trauma unless the individual is experiencing psychotic symptoms or their behavior is presenting a danger to themselves or others. If this is the case, a fast-acting benzodiazepine and/or an antipsychotic may be warranted as described in guidelines for managing agitation (Yildiz et al., 2003). Individuals who are experiencing acute panic symptoms and severe insomnia may benefit from a short-term (<1 week) prescription for benzodiazepines; however, early administration of benzodiazepines may be associated with a higher incidence of PTSD (Gelpin et al., 1996). According to the American Psychiatric Association (APA) practice guidelines, selective serotonin reuptake inhibitors (SSRIs) and serotonin-norepinephrine reuptake inhibitors (SNRIs) have shown superiority over placebo for noncombat-related PTSD (Benedek et al., 2009).

Conclusions

Exposure to a natural disaster or another traumatic event can disrupt an individual's physical, emotional, and psychosocial functioning. Besides providing emotional support, psychoeducation, improvement in coping skills, and reestablishment of a sense of resilience, a thorough assessment of the individual's symptoms and impairment in functioning is essential. There are numerous assessment scales available (Keane and Wilson, 2004; Norris, 1990) that can be utilized to assist in this process. In addition, screening the person for ASD and/or PTSD utilizing DSM-IV-TR diagnostic criteria is also necessary. Although medications should be used judiciously in the first days following a trauma, SSRIs have been found to be beneficial in the treatment of PTSD.

Bibliography

American Psychiatric Association, 1994. *Diagnostic and Statistical Manual of Mental Disorders*, 4th edn. Washington, DC: APA.

Benedek, D. M., Friedman, M. J., Zatzick, D., and Ursano, R. J. (2009). *Guideline Watch: Practice Guideline for the Treatment of Patients with Acute Stress Disorder and Post-Traumatic Stress Disorder*. Retrieved on February 1, 2010 from: http://www.psychiatryonline.com/content.aspx?aID=156514

Gelpin, E., Bonne, O., Peri, T., Brandes, D., and Shalev, A. Y., 1996. Treatment of recent trauma survivors with benzodiazepines: a prospective study. *The Journal of Clinical Psychiatry*, 57, 390–394.

Joseph, S., and Linley, A. P., 2005. Positive adjustment to threatening events: an organismic valuing theory of growth through adversity. *Review of General Psychology*, 9, 262–280.

Keane, J. P., and Wilson, T. M. (eds.), 2004. *Assessing Psychological Trauma and PTSD*. New York: Guilford Press.

Kessler, R., Sonnega, A., Bromet, E., Hughes, M., and Nelson, C., 1995. Post-traumatic stress disorder in the national comorbidity survey. *Archives of General Psychiatry*, 52, 1048–1060.

Linley, P. A., and Joseph, S., 2004. Positive change following trauma and adversity: a review. *Journal of Traumatic Stress*, 17, 11–21.

Lubit, R. H., 2008. *Acute Treatment of Disaster Survivors*. Retrieved on January 4, 2010 from: http://emedicine.medscape.com/article/295003-overview

Norris, F. H., 1990. Screening for traumatic stress: a scale for use in the general population. *Journal of Applied Social Psychology*, 20, 1704–1718.

Paton, D., 2005. Posttraumatic growth in protective services professional: individual, cognitive, and organizational influences. *Traumatology*, 11, 335–346.

Paton, D., and Violanti, J. M., 2011. *Working in High Risk Environments: Developing Sustained Resilience*. Springfield, IL: Charles C. Thomas.

Solomon, S. D., and Davidson, J. R. T., 1997. Trauma: Prevalence, impairment, service use, and cost. *The Journal of Clinical Psychiatry*, 58, 5–11.

Tedeschi, R. G., and Calhoun, L. G., 2003. Routes to posttraumatic growth through cognitive processing. In Paton, D., Violanti, J. M., and Smith, L. M. (eds.), *Promoting Capabilities to Manage Posttraumatic Stress: Perspectives on Resilience*. Springfield, IL: Charles C. Thomas, pp. 12–26.

van der Kolk, B. A., McFarlane, A. C., and Weisaeth, L. (eds.), 1996. *Traumatic Stress: The effects of Overwhelming Experience on Mind, Body, and Society*. New York: Guilford Press.

Yildiz, A., Sachs, G. S., and Turgay, A., 2003. Pharmacological management of agitation in emergency settings. *Emergency Medical Journal*, 20, 339–346.

Cross-references

Adaptation
Coping Capacity
Disaster Relief
Education and Training for Emergency Preparedness
Emergency Management
Federal Emergency Management Agency (FEMA)
Human Impact of Hazards
Natural Hazard
Resilience

CRITICAL INFRASTRUCTURE

Susanne Krings
United Nations University, Bonn, Germany

Synonyms
Lifelines; Lifeline utilities

Definition
The term critical infrastructure is used to cover physical and organizational structures that provide services that are estimated to be essential to the functioning of society. Hence it is feared that the unavailability of critical infrastructure services may have severe consequences for basic societal needs.

The functioning of critical infrastructures depends to varying degrees on personnel and resources (material as well as information resources). The provision of critical infrastructure services in many cases involves the private sector. Public interest in their availability has frequently been articulated, e.g., in critical infrastructure protection policies (for an overview, see Brunner and Suter, 2008). Conventionally, concrete sectors, such as communication infrastructure or energy infrastructure, are listed in these policies. As most critical infrastructures are characterized by a high degree of (inter) dependencies, a failure in one sector is likely to affect others.

Among others, natural hazards are held to be threats for critical infrastructures and the services they provide. Destruction of critical infrastructure and service outages may initially cause severe problems and/or aggravate the situation in the course of events, most notably when services are needed to carry out relief measures to mitigate the immediate impact, and during recovery and reconstruction. Vulnerability and risk assessments as well as safeguards and risk-management measures may either focus on the level of single infrastructure components and/or opt for a system perspective.

Bibliography
Brunner, E. M., and Suter, M., 2008. *The International CIIP Handbook 2008/2009. An Inventory of 25 National and 7 International Critical Information Infrastructure Protection Policies*. Zurich: Center for Security Studies, ETH. http://www.crn.ethz.ch/publications/crn_team/detail.cfm?id=90663. Accessed 2 Nov 2010.

Cross-references
Disaster
Disaster Relief
Disaster Risk Management
Exposure to Natural Hazards
Hazard
Hospitals in Disaster
Information and Communication Technology
Mitigation
Natural Hazard
Post Disaster Mass Care Needs
Recovery and Reconstruction After Disasters
Risk Assessment
Vulnerability

CRYOLOGICAL ENGINEERING

Lukas U. Arenson[1], Sarah M. Springman[2]
[1]BGC Engineering Inc., Vancouver, BC, Canada
[2]Institut für Geotechnik/Institute for Geotechnical Engineering, Zurich, Switzerland

Synonyms
Cold regions engineering; High mountain engineering; Northern engineering; Permafrost engineering

Definitions
Cryosphere. That part of the earth's crust, hydrosphere, and atmosphere subject to temperatures below 0°C for at least part of each year.
Cryology. The study of materials having a temperature below 0°C.
Geocryology. The study of earth materials having a temperature below 0°C.
Engineering. The creative application of scientific principles to design or develop structures, machines, apparatus, or manufacturing processes, or works utilizing them singly or in combination; or to construct or operate the same with full cognizance of their design; or to forecast their behavior under specific operating conditions; all to meet an intended function, economics of operation, and safety to life and property (cf. American Engineers' Council for Professional Development). Additionally, the planning for, and maintenance of, a sustainable lifetime performance of the design object and its environment is essential.

Introduction
Generally, *Cryological Engineering* can be considered as the application of scientific principles to any design assignment that is subjected to temperatures below 0°C. The engineering disciplines that are most likely to be related to cryological engineering are civil, geotechnical, and mining, in particular when linked with projects in cold regions, such as northern and southern latitudes or high elevations. In recent years, engineering projects and developments in these cryospherical environments gained in importance due to access required to enable mining of natural resources (e.g., in the high Andes and the Arctic), resource transportation (e.g., pipelines), or in improving

infrastructure and accessibility (e.g., Qinghai–Tibetan railway and railroad). But also in the aviation and space industry and science, subfreezing conditions are of importance either for the design of aircrafts, space, Moon or Mars stations, or just for the study of extraterrestrial processes.

The cryosphere

The cryosphere may be divided into the cryoatmosphere, the cryohydrosphere (snow cover, glaciers, ice caps, ice sheets and river, lake and sea ice), and the cryolithosphere (perennially and seasonally cryotic ground). Some authorities exclude the earth's atmosphere from the cryosphere (e.g., UNEP, 2007); others restrict the term "cryosphere" to the regions of the earth's crust where permafrost, that is, perennially frozen ground, exists (Baranov, 1978). However, for engineering purposes, it is important to understand the physical differences in the materials that may be encountered and used as foundations or construction materials from the cryosphere, which include snow, firn, and ice in special forms, such as sea ice, glacier ice, pore ice, segregated ice, ground ice, ice shelves or ice bergs, just to name a few.

However, good knowledge about the special conditions that prevail in the cryosphere is required for the design of conditions that are artificially induced, such as artificial ground freezing used to increase the strength of the ground temporarily to build tunnels, caverns, or shafts (e.g., Harris, 1995).

Divisions of cryological engineering

Cryological engineering is complex and can, therefore, be seen as sub-categories in a series of engineering disciplines. Figure 1 illustrates some aspects of cryological engineering that demonstrates the variety of engineering disciplines involved. *Geotechnical* engineers design foundations and dams, and assess slope stability or general geohazards due to ground ice degradation. *Road* engineers design road surfaces that resist the harsh climatic conditions, which also affect the design of towers, buildings, and bridges that have to be designed by *structural* and *civil* engineers. *Mechanical* engineers have to consider the effect of subfreezing temperatures in their designs; in particular for moving elements where freezing water or the temperature-dependent material behavior may affect a machine's performance over its lifetime, which may also be reduced by repeated cycles of freezing and thawing. These examples are not exclusive and aspects of cryological engineering can probably be found in any engineering field.

Although the engineering problems are diverse, the main challenge is similar for most disciplines, that is, the change in the mechanical behavior of unfrozen material versus frozen materials – in other words the differences in the physical response of a material containing fluid water opposed to ice.

Engineering challenges

The challenges associated with cryological engineering are as diverse as the projects. A good understanding of the *material properties* is essential. Phase changes (i.e., latent heat effects), thermal expansion, viscosity of ice, fatigue, and self-healing mechanisms as well as the temperature- and loading-dependent material stiffness are only some aspects that need to be considered. Andersland and Ladanyi (2004) or Paterson (1994) provide/provides valuable overviews on frozen ground and glacier physics. Figure 2 shows a diagram that illustrates schematically how the mechanical response of a frozen soil varies as a function of the loading conditions, temperature, and ice content. The differences in the mechanical response may result in variations of several magnitudes in the strength response and are, therefore, crucial for a sustainable foundation design. In addition, spatial variations and heterogeneities in the ground conditions pose problems in creating a standard design for linear infrastructure foundations, for example, and continuous in situ adaptations are often required in the field. Therefore, flexible engineering solutions are required.

The challenges from the material properties are, however, only one element to be considered in cryological engineering designs. Often more expensive are challenges related to the *logistics*, such as the remoteness of the construction site, available resources (e.g., gravel for concrete), or access in steep and high mountain environments. But also the effect of the harsh climatic conditions with cold temperatures and dark days at high latitude, or major diurnal air temperature variations and low oxygen levels in high mountain areas, are wearing on equipment and working crews. The logistical challenges and remoteness of some construction sites often result in limited information for the design, such as site investigation or historical climate data. The latter are important in predicting and designing for future climate change effects. Generally, the cryosphere is often found in environments that are ecologically very sensitive and it is, therefore, important to understand how a planned structure affects it and what adaptation strategies are required.

Not only local aspects are to be considered in the design process, but larger-scale effects may become important. For example, the cryogenic conditions of the surrounding landscape may change in the future, so that formerly stable, frozen slopes become unstable and transform into a potentially dangerous debris flow source zone. Hence it is important to familiarize oneself with the proximate surroundings, and with the general environment and landscape. A summary of these problems is presented in Bommer et al. (2010) for mountain permafrost environments.

Solutions

As with most engineering projects, in particular with civil projects, designs are typically prototypes, and no real testing is possible. Because of this unique situation, and the

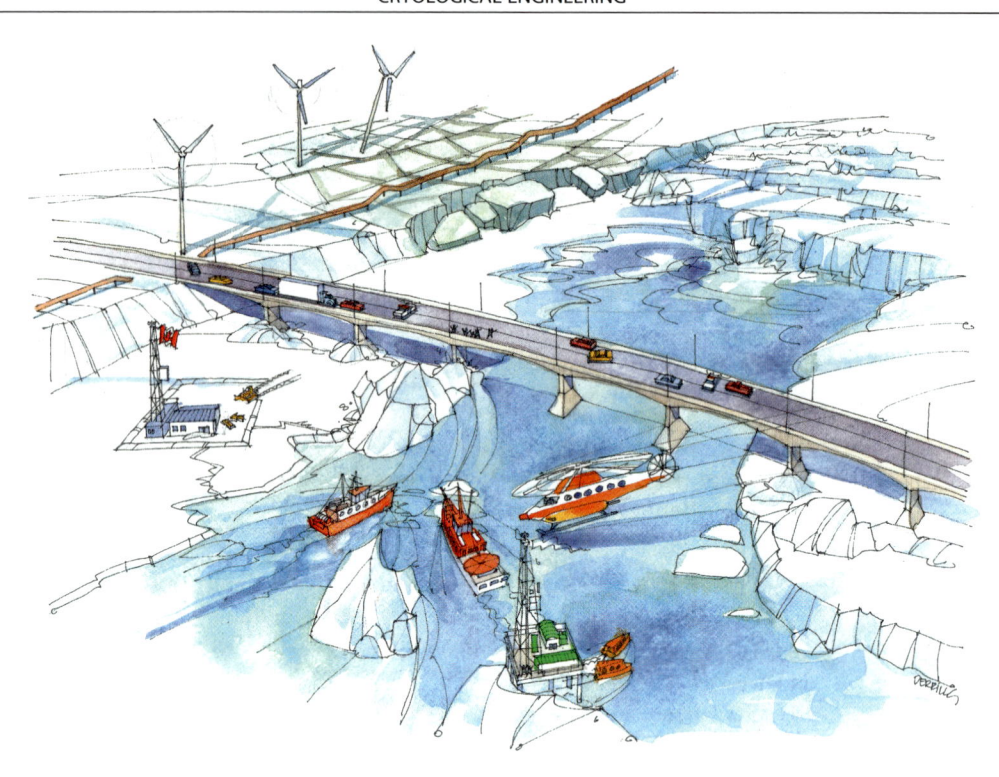

Cryological Engineering, Figure 1 An overview of the diversity of cryological engineering (Illustration by Derrill Shuttleworth).

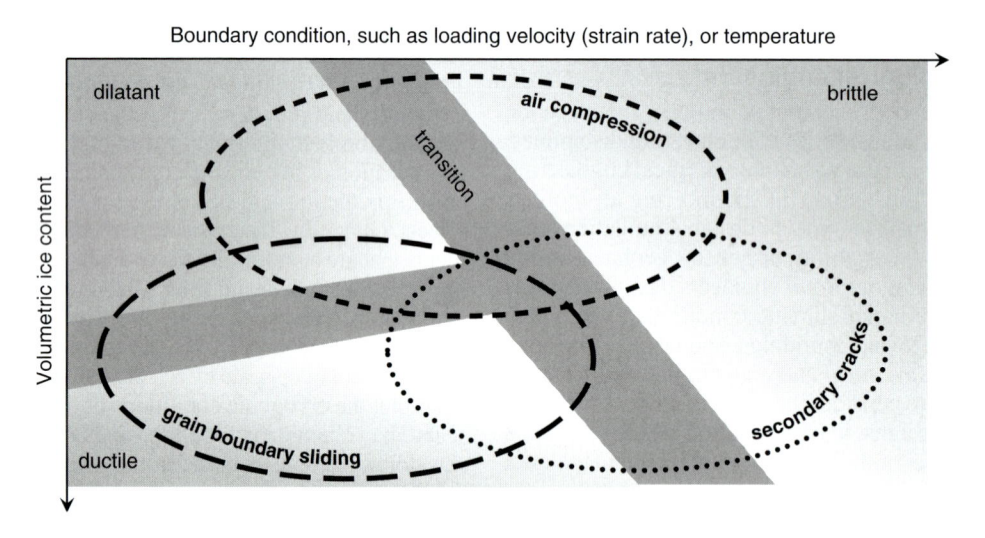

Cryological Engineering, Figure 2 Schematics of the dependency of the response mechanism of frozen soil on the boundary conditions (After Arenson and Springman, 2005).

challenges indicated above, special care is required during the design, construction, and operation of an engineered structure. Successful cryological engineering requires enough resources for planning in terms of time and financial means. It is important to have a good spatial representation of any data and long time series that allow for statistical trend analyses. Designs are to be favored that minimize the impact on the environment, notwithstanding the uncertainty about ongoing climate change over several decades. However, the environmental sensitivity and complex interactions between climate, foundation, and structure require designs to be adaptive, and the incorporation of an observational approach is essential. Ongoing data evaluation and updating predictions should be a crucial part in the structure's operation and maintenance plan to monitor a structure's performance effectively. Redundancies, designed and implemented in time, help in minimizing any operational interruption in the future due to unforeseen changes in the boundary conditions. An estimation of project vulnerability, hazard, and associated risks is critical for the decision-making process and should also be carefully planned ahead of time.

Cryological engineering projects are, therefore, often expensive and require more resources for project management than similar projects in unfrozen environments. The project lifetime is often to be chosen shorter, or design re-evaluations are required at regular intervals (e.g., 20–30 years) and are to be accounted for in the original design process. In particular, the effects of climate change, including second- and third-order effects that are almost impossible to predict, have to be analyzed regularly, especially for sensitive structures and locations.

While the challenges of cryological engineering are substantive, various innovative solutions have been presented in recent years or are currently being evaluated. These include adjustable foundation designs (e.g., Phillips et al., 2007) or standardized protocols are in development to account for the potential impacts of climate change (e.g., CSA, 2007). Additional resources are listed in the bibliography to assist in cryological engineering designs.

Summary

Cryological Engineering implies the adaptation of multidisciplinary engineering processes to account for cryological conditions in subfreezing environments. When liquid water turns into ice, several physical processes change and this has to be considered in the design. However, the effect of the structure on its environment or climate change may result in current cryogenic conditions changing into non-cryogenic ones with time. The structural integrity or the serviceability of the engineered structures may be affected by such changes in the boundary conditions. It is, therefore, critical to consider such potential changes adequately in an adaptable design that has been developed on the basis of thorough investigations and historical data analyses.

Bibliography

Andersland, O. B., and Ladanyi, B., 2004. *Frozen Ground Engineering*. Hoboken, NJ: Wiley.

Arenson, L. U., and Springman, S. M., 2005. Triaxial constant stress and constant strain rate tests on ice-rich permafrost samples. *Canadian Geotechnical Journal*, 42(2), 412–430.

Baranov, I. Y., 1978. Problems of cryology. In *Proceedings of the Second International Conference on Permafrost, Yakutsk, U.S.S.R*, July 1973, Washington, DC, U.S.S.R. Contribution, U.S. National Academy of Sciences, pp. 3–7.

Bommer, C., Phillips, M., and Arenson, L. U., 2010. Practical recommendations for planning, constructing and maintaining infrastructure in mountain permafrost. *Permafrost and Periglacial Processes*, 21(1), 97–104.

Canadian Standards Association, 2007. Climate Change and Infrastructure Engineering: Moving Towards a New Curriculum. http://www.csa.ca/koa/Climate_Change_and_Infrastructure_Engineering.pdf. Accessed 6 June 2012.

Clarkem, E. S. 2007. *Permafrost Foundations: State of the Practice*. Reston, VA: American Society of Civil Engineers.

Doré, G., and Zubeck, H. K., 2009. *Cold Regions Pavement Engineering*. Reston, VA: American Society of Civil Engineers/McGraw-Hill.

Esch, D. C. 2004. *Thermal Analysis, Construction, and Monitoring Methods for Frozen Ground*. Technical Council on Cold Regions Engineering. Reston, VA: American Society of Civil Engineers.

Harris, J. S., 1995. *Ground Freezing in Practice*. London, UK: Thomas Telford.

Paterson, W. S. B., 1994. *The Physics of Glaciers*. Burlington, MA: Elsevier Science.

Phillips, M., Ladner, F., Muller, M., Sambeth, U., Sorg, J., and Teysseire, P., 2007. Monitoring and reconstruction of a chairlift midway station in creeping permafrost terrain, Grächen, Swiss Alps. *Cold Regions Science and Technology*, 47(1–2), 32–42.

Smith, D. W., 1996. *Cold Regions Utilities Monograph*. Reston, VA: American Society of Civil Engineers and the Canadian Society for Civil Engineering.

United Nations Environment Programme (UNEP), 2007. *Global Outlook for Ice and Snow*. Nairobi, Kenya: United Nations Environment Programme, Division of Communications and Public Information (DCPI). http://www.unep.org/geo/geo_ice. Accessed 6 June 2012.

Online resources

International Permafrost Association: IPA: http://ipa.arcticportal.org. Accessed 6 June 2012.

The Cryosphere: http://nsidc.org/cgi-bin/words/glossary.pl. Accessed 6 June 2012.

Cross-references

Climate-Change
Frost Hazard
Geohazards
Glacier Hazards
Ice and Icebergs
Impact Winter
Paraglacial
Permafrost

CULTURAL HERITAGE AND NATURAL HAZARDS

Piotr Migoń
University of Wrocław, Wrocław, Poland

Definition

Cultural heritage is understood as the legacy of past generations which is maintained by the present one and intended to be passed on to future generations. It can be intangible (customs, beliefs) and tangible, the latter including various physical objects, from human-transformed landscapes (e.g., paddy rice fields on hillslopes) through places, buildings, monuments, to movable artifacts. The significance of cultural heritage may be local, regional, or global. The most valued places are those with the status of World Heritage granted by UNESCO.

Natural hazards may adversely impact tangible cultural heritage, but in specific instances, the remains of an inhabited place, or a building, may become a valuable component of cultural heritage because of their destruction by natural forces at some time in the past. Likewise, stories of ancient catastrophes have become a part of the intangible heritage of many societies.

Introduction

Relationships between cultural heritage of humankind and natural hazards are many and complex. Hence, many interrelating themes appear within the subject, including:

(a) Damage or destruction of cultural heritage due to natural catastrophic processes of various sorts
(b) Problems of adequate protection of cultural heritage sites against natural hazards
(c) The occurrence of globally or regionally significant representatives of ancient cultural heritage which have undergone catastrophic destruction by natural forces and have now become highly valued cultural properties because of their history of destruction
(d) The presence of natural hazards and catastrophes in oral folk traditions, hence a part of intangible cultural heritage

Natural hazards affecting cultural heritage properties do not form a specific category of hazards in terms of process or effect. Sites of cultural significance may become affected by catastrophic events of either endogenous (earthquakes, volcanic eruptions, tsunami) or exogenous origin (landslides, floods, ground collapses, wildfires, cyclones) (Smith, 1996), for which little or no warning has been received (particularly prior to the twentieth century). However, these sites may also suffer from processes which are not catastrophic in the conventional sense (i.e., have not appeared suddenly) but their cumulative effects in the long term may have a highly adverse impact. These include ground subsidence, especially in coastal settings, accelerated weathering of building stone, sandstorms, and recession of coastal cliffs.

Natural processes and loss of cultural heritage

Natural catastrophes have been known to affect and occasionally destroy material evidence of human activities since prehistory. Those from the most distant past are often shrouded in uncertainty and subject to scientific debate, such as the probable destruction of Sodom and Gomorrah located in the Dead Sea Graben due to an interaction of earthquakes, natural gas explosions, and fire. Volcanic eruptions were among the most devastating, able to wipe out island populations, as on the Aegean island of Santorini in the fifteenth century BC. At a more local scale, pyroclastic flows from Vesuvius were responsible for the total destruction of Pompeii and Herculaneum in 79 AD, whereas lava flows destroyed and buried the native American ceremonial center at Cuicuilco (present-day Mexico City) in the first century AD. Likewise, deteriorating environmental conditions in the longer term, particularly droughts, are often suspected as reasons for apparent declines of once mighty societies and political entities. It is widely believed that decreasing rainfall and diminishing river flows resulted in temporary or ultimate collapses of early "hydraulic" civilizations such as the Old Kingdom of Egypt around the twenty-second century BC or the great Harappan civilization of the Indus Valley, where channel changes may also have been important. More controversially perhaps, fragmentation of the Mayan states and an apparent decline of many Mayan cities in the eighth to eleventh century has been attributed to climate changes, mainly increasingly unreliable rainfall. More recent societies may also have been vulnerable, especially those living in marginal conditions. The demise of Nordic settlements in Greenland in the thirteenth/fourteenth century was influenced by climate cooling and the advent of the Little Ice Age.

The concept of cultural heritage was apparently present among ancient societies as early as the third century BC. In the Hellenistic world, its reflection was the list of Seven Wonders of the World, which was also the list of "must-see" places for ancient travelers. It included objects and sites, from Greece to Babylon, then considered absolute masterpieces of human genius. Only one of them – the Great Pyramid of Giza – has survived until today. Among the other six, three have been damaged by earthquakes. The Colossus of Rhodes tumbled down around 227 BC, the Pharos Lighthouse in Alexandria finally in the fourteenth century AD, whereas the Mausoleum in Halikarnassos (today Bodrum), destroyed by floods and earthquakes and rebuilt several times, eventually disappeared in the fifteenth century.

In recent decades, many significant sites of cultural heritage have suffered damage, occasionally irreversible, from natural processes. Mud-brick walls of an ancient fortress in Bam, Iran, largely crumbled during an earthquake on December 26, 2003, whereas numerous components of the famous Dujiangyan Irrigation System in the Sichuan province, China, dated to 256 BC, collapsed during the Wenchuan earthquake on May 12, 2008. Earthquake-induced ceiling collapse in the basilica of Assisi, Italy, in 1997 led to the destruction of unique frescoes from the thirteenth/fourteenth century. The revered pre-Columbian site of Chan Chan in northern Peru, built of dried mud brick, greatly suffered from several floods related to El Niño years. Widespread forest fires in western Peloponnese, Greece, in August 2007, put at serious risk the site of ancient games at Olympia, destroying parts of the surrounding landscape. Subsequent to fire, soil erosion from burnt slopes became an issue and a widespread erosion control project was undertaken. Floods tend to threaten historical cities located in the valley floors, late twentieth century examples being inundation of downtown Florence (Italy) in 1966 and parts of Cologne (Germany) in 1993 and 1995.

For cultural heritage sites, slow-acting natural processes may be devastating too. However, it is useful to make a distinction between those processes which are an intrinsic part of the natural environment and those which have been induced, or accelerated, by human activities. This differentiation has considerable implications on the choice of remedial solutions toward preservation of sites under threat. Many ancient temples were located along shores, on exposed cliffs and promontories. Long-term transport of salts derived from sea-spray and its subsequent crystallization has affected building stones and caused disappearance of various fine architectonic details, particularly if these were built of easily weathered limestone (e.g., the megalithic temples of Malta, and the Poseidon temple at Cap Sounion near Athens). Salt weathering is also of considerable concern at the UNESCO site of Petra, Jordan. Both scarp-foot and mid-wall weathering, influenced by capillary rise and seepage respectively, have caused widespread damage to the finely carved facades of rock-hewn tombs (Figure 1). Rock/cave art is another highly valued legacy of past cultures that is highly susceptible to weathering and in many places suffers from rapid deterioration. In Calatayud, Spain, slow gypsum dissolution underneath the medieval city results in extensive ground subsidence and building destruction. Sea level changes affected the historic Serapis temple in Pozzuoli, Italy, as well as numerous other Mediterranean examples. In urban and industrialized areas, however, damage experienced by buildings of cultural significance can be often ascribed to anthropogenic sources of salt and air pollution (Goudie and Viles, 1997). In the context of aeolian processes, human impact has been suggested as the reason behind the damage of western sections of the Great Wall in Gansu province, China. In the last 20 years more than 40 km of this unique construction disappeared or was severely reduced by sand blasting during windstorms, and destructive farming methods with consequent enhanced dust production are considered responsible (http://www.msnbc.msn.com/id/20492488/ns/world_news-world_environment/t/sandstorms-eating-away-chinas-great-wall/(retrieved 2012-03-18)).

Coastal erosion is another process to impact building constructions located on cliffs and along beaches. Usually, the coast sections under threat are those undergoing slow long-term recession, but major damage is experienced during storm episodes, when wave energy is sufficient to induce cliff undercutting, leading to rock fall or retrogressive landslides. The southern coast of England hosts many examples of cultural heritage objects affected by cliff recession, from ancient Roman forts to remains from World War II (Bromhead and Ibsen, 2006), as do Mediterranean coasts. In Tanzania, beach erosion and wave inundation threat the integrity of ancient harbors of Kilwa Kisiwani and Songo Mnara. Coastal subsidence is of major concern too, the best known example being Venice, Italy. High floods, the famous *acqua alta*, have increased

Cultural Heritage and Natural Hazards, Figure 1 The unique cultural legacy of Petra in Jordan is under threat from various geomorphic processes, including salt weathering. The picture shows two zones of accelerated rock breakdown and deterioration, caused by capillary rise (near the bottom) and seepage (in the middle of the facade) (Photo P. Migoń).

in frequency, causing weakening of building foundations and setting the stage for accelerated weathering.

Significant sites of cultural heritage: Evidence of ancient natural catastrophes

Our cultural heritage consists of objects and sites of various origin, context, and age. Many such objects are treasured because they have survived largely intact since the very distant past, occasionally even from prehistory. Their maintenance in a condition as close to original as possible is now the priority of conservation efforts and a significant constraint in access policy. Hence, any damage arising from any cause is considered highly detrimental for the integrity of a site. However, a considerable number of much valued cultural heritage sites, including many listed as UNESCO World Heritage properties, bear evidence of either natural catastrophes or slow deterioration. These natural processes once led to the abandonment of the sites, occasionally destruction, and their subsequent

Cultural Heritage and Natural Hazards, Figure 2 The ruined church tower rising from lava field north of the Paricutín volcano, Mexico, overwhelmed by lava in 1944, is already visited as a cultural heritage site (Photo P. Migoń).

disappearance from human memory. Much later archaeological work unearthed these sites as they appeared in ancient times, offering thereby unique insights into the past, not obstructed by subsequent societal and architectural developments. Examples come from different parts of the world, from the Mediterranean realm through the Middle East to the Far East, as well as from Central America.

Perhaps the best known example is Pompeii near Naples in Italy, the remains of a wealthy town buried by pyroclastic flow deposits from the eruption of Vesuvius in 79 AD. Excavations carried out since the eighteenth century, and more comprehensively since 1863, have revealed an astonishingly complete picture of daily life in the Roman Empire, with no parallels from elsewhere in the Mediterranean region. Interestingly, archaeological work has also shown evidence of earlier damage by a strong earthquake in 62 AD. Another important archaeological site is Akhrotiri on the Island of Thera (Santorini), likely abandoned just prior to the catastrophic explosion of Santorini volcano in the fifteenth century BC and then buried by many meters of pyroclastic deposits. Excavations, initiated in 1967, brought to light many details of the Minoan culture, including unique frescoes. Many ancient sites or buildings suffered from high-magnitude earthquakes, such as Kourion (Curium) in Cyprus in 365 AD, abandoned soon after. Archaeological work has not only revealed remnants of important buildings of the ancient city, but opened a window on the everyday life of this important, predominantly Christian settlement. The evidence of ancient earthquakes is common at archaeological sites in Asia Minor (e.g., Hierapolis) and along the Dead Sea Rift (e.g., Jericho). Patterns of building destruction are now used as important palaeoseismological tools (Hancock and Altunel, 1997).

River mouth siltation and channel changes have contributed to the decline and later abandonment of many important settlements of antiquity. Today many of these sites, excavated and partially reconstructed, are cherished sites of cultural heritage and important tourist attractions. They also tell instructive stories of how people interact with nature and how things can go wrong. Excellent examples are provided by ancient Greek-Roman cities in Asia Minor, along the Aegean coast, such as Miletus or Ephesus. Once important harbors and trading posts located at river/sea junctions, they declined concurrently with delta buildup, often considered a response to accelerated soil erosion in the hinterland.

Today, damage brought by natural events is usually repaired as quickly as possible. With current technological advances and international aid, the evidence of destruction may be obliterated in a few years and rebuilding sites of cultural heritage is often a priority. Very few places are left as standing memories of violent natural processes and these, over time, may join the family of cultural heritage sites. One such object is the ruined church at a site of the former town of San Juan Parangaricutiro in Mexico,

Cultural Heritage and Natural Hazards, Figure 3 The remains of a medieval church in Trzęsacz, northern Poland, atop a Baltic Sea cliff. The church was built in the fifteenth century about 1 km from the cliff edge, but long-term cliff recession resulted in a few successive collapses since 1900. Today the site is considered to be of special cultural importance and protected against further cliff erosion. However, erosion continues unabated next to the site (Photo P. Migoń).

the only survivor of a lava flow issued by the Paricutín volcano in 1944 (Figure 2). Others include sites near Pinatubo volcano in the Philippines.

Protection of cultural heritage against natural hazards

Natural hazards affecting cultural heritage properties do not form a specific category of hazards in terms of process or effect. It is the vulnerability and universal value of these properties which is decisive for the increasing risk experienced by cultural heritage sites. Mitigations and risk reduction strategies should consider characteristics of natural processes potentially affecting a site, particularly the likelihood of its occurrence in a specified period, the magnitude of expected damage, and the feasibility of preventive actions. Further, any potential human contribution to the hazard needs to be identified.

In many instances, hazards are simply unavoidable as the properties cannot be relocated to safer places. This applies to all cultural heritage sites in seismic zones and in the vicinity of active volcanoes. Many great heritage cities have been built along active fault zones and their cultural legacy is at particular risk from high-magnitude earthquakes (e.g., Istanbul, Athens, Mexico City, Kyoto). Construction strengthening is practically the only measure which can be undertaken. Others cities are located in zones prone to pyroclastic flows from volcanic eruptions (e.g., Naples, Mexico City). Large tsunamis can affect cultural heritage sites along seismogenic coasts of the Mediterranean Sea, southeast Asia, and the Pacific Ocean, as they did in Lisbon in 1755. Some sites, e.g., the Inca site of Machu Picchu, are located on hillslopes conducive to slope failures and these, if occur, may irreversibly damage the entire property.

Surface processes, such as shallow landslides, mudflows, or floods, can be predicted with more confidence and there is a choice of mitigation strategies. Landslide hazard and risk mapping is now routinely carried out and helps to identify the most vulnerable places. After these are identified, various methods of slope stabilization, depending on the type of mass movement likely to occur, may be used to protect a cultural heritage site. These include rock slope strengthening, reduction of slope angle, drainage diversion or improvement, bioengineering, and others (see Sidle and Ochiai, 2006). Flood hazard may be reduced by bank strengthening and dyke heightening adjacent to a site of concern, but these measures may not be sufficient during low-frequency, high-magnitude floods. It is also important to remember that for flood mitigation schemes to be effective, they should be designed for entire catchments and those for cultural heritage sites specifically need to be integrated within catchment-wide strategies. Valuable cliff-top sites may be protected by various coastal defenses, including sea walls, artificial boulder beaches, and concrete tetrapods (Figure 3). However, coasts are complex systems of mass transfer from one place to another and emphasis on preventing erosion and cliff recession in one locality may result in accelerated erosion in an adjacent locality. There are also different methods available to avert salt weathering and ground salinization, such as the UNESCO attempt to rescue Mohenjo Daro, Pakistan, by reducing the water table and hence the capillary rise of salt. Any preventive action is bound to be very costly (e.g., plans to build protective barriers at the entrances to the Lagoon of Venice) and often there is no guarantee that the effects will be satisfactory.

An important part of disaster prevention and risk management programs at sites of cultural significance should be adequate preparedness (Spennemann, 1999; Taboroff, 2000). Ideally, it includes components such as hazard assessments for each natural process likely to occur, individual emergency plans integrated with disaster plans for wider areas, priority lists, detailed inventories of objects, and records of past dealings with natural events, and staff specialized training. There is no doubt that cultural heritage sites will continue to suffer from natural processes, which are largely beyond our ability to control them, but accumulated knowledge from past disasters can be of considerable help to reduce negative impact of any future event.

Summary

Cultural heritage is exposed to different types of hazards and potentially devastating natural events which may result in different degrees of damage or, less commonly, total destruction. Earthquakes appear to have most serious effects and many great heritage sites are located along major fault zones in Europe, Asia, and America. Other hazards include volcanic eruptions, gravitational mass movements, flash floods, and coastal erosion. Weathering and ground subsidence are slow-acting processes whose cumulative effects may nevertheless seriously affect the stability of structures and their visual quality. Perhaps the most important aspect of relationships between cultural heritage and natural hazards is that damage or loss of properties cannot be measured in monetary units only. Their value to the humankind can be hardly expressed in this way, and some are considered of outstanding universal value, protected by international conventions. In disaster-affected areas, if objects of cultural heritage are prioritized for rebuilding, they may be used as catalysts of renewed tourism interest and, thereby, as means to improve a shaken local economy.

Bibliography

Bromhead, E. N., and Ibsen, M.-L., 2006. A review of landsliding and coastal erosion damage to historic fortifications in South East England. *Landslides*, **3**, 341–347.

Goudie, A., and Viles, H., 1997. *Salt Weathering Hazards*. Chichester: Wiley.

Hancock, P. L., and Altunel, E., 1997. Faulted archaeological relics at Hierapolis (Pamukkale), Turkey. *Journal of Geodynamics*, **24**, 21–36.

Sidle, R. C., and Ochiai, H., 2006. *Landslides: Processes, Prediction, and Land Use*. Washington: American Geophysical Union, Water Resources Monographs, p. 18.

Smith, K., 1996. *Environmental Hazards. Assessing Risk and Reducing Disaster*. London: Routledge.

Spennemann, D. H. R., 1999. Cultural heritage conservation during emergency management: luxury or necessity? *International Journal of Public Administration*, **22**, 745–804.

Taboroff, J., 2000. Cultural heritage and natural disasters: incentives for risk management and mitigation. In Kreimer, A., and Arnold, M. (eds.), *Managing Disaster Risk in Emerging Economies*. New York: World Bank. Disaster Management Risk, Vol. 2, pp. 71–79.

Cross-references

Biblical Events
Coastal Erosion
Damage and the Built Environment
Earthquake
Flood Hazard and Disaster
Geological/Geophysical Disasters
Historical Events
Santorini, Eruption
Subsidence Induced by Underground Extraction
Vesuvius
Volcanoes and Volcanic Eruptions
Wenchuan, China

D

DAMAGE AND THE BUILT ENVIRONMENT

Adriana Galderisi, Andrea Ceudech
University of Naples "Federico II", Naples, Italy

Synonyms
Losses; Urban areas; Urban environment

Definition
Damage: Losses, injuries, failures, or troubles which may occur in a given area as a consequence of the impact of a hazard on exposed and vulnerable elements and systems. Types and amount of damage depend on the features of both hazards and exposed elements and systems.

Built environment: The areas most modified by human beings, where people and all man-made structures are concentrated. It is generally opposite to the natural environment, even though close relationships between the built and natural environments are largely recognized. Built environments can greatly differ, in terms of spatial and functional patterns. The various types of built environments respond differently to hazards: that is, a densely built up historical city and a recent low-density urban area have a different susceptibility to damage from a given hazard.

Introduction

The damage due to natural hazards in built environments generally refers to the physical destruction to man-made structures and to the impacts on people. These types of damages are generally defined as direct damages and occur at the onset of a disaster or shortly thereafter. Direct damages can be due both to a primary hazard or to its impact on exposed vulnerable elements which represent, in turn, further hazard sources: for example, damages to buildings or people due to fires following an earthquake which causes, in turn, breaks in gas pipelines. Such damages are very common in the built environment, as clearly highlighted by several disasters that have occurred in the recent past. Nevertheless, in many cases, official reports do not distinguish the damages in relation to the different hazard sources.

Besides direct damages, indirect ones may occur and can result from physical damages affecting the functioning of some elements or systems (hospitals, lifelines, etc.) or inducing economic consequences (temporary unemployment due to physical damages to industries). Indirect damages may even occur over a long period of time after the event, affecting broader areas than that directly hit by the hazard. Permanent or temporary losses in relevant economic activities at local scale can reverberate on macroeconomic variables.

By their nature, indirect damages are harder to quantify than physical ones. The amount of indirect damages may constitute the majority of the total losses in large disasters affecting urban areas. Nevertheless, long-term indirect damages are frequently neglected, mainly in official reports often designed to provide governments with estimates of the amount of funds required to address emergency and reconstruction needs.

The overall damage is quantified in terms of economic losses, generally including direct damages and the most relevant economic effects, as well as distinguishing the insured losses within the overall amount. It is worth noting that in many cases, the evaluation of indirect damages is made even more difficult by the "nonmarket effects." Such effects are related to the loss in functioning of public equipment or infrastructure. Due to their public nature, such equipment provides services free of charge (Rose 2004); hence, damage cannot be easily assessed in economic terms.

Most of the available databases provide quantitative information related to damages to people (killed, injuries,

P.T. Bobrowsky (ed.), *Encyclopedia of Natural Hazards*, DOI 10.1007/978-1-4020-4399-4,
© Springer Science+Business Media Dordrecht 2013

Debris Avalanche, Figure 3 Effects of the Huascaran debris avalanches (Perù), event of May 31, 1970. (**a**) hummocky topography; (**b**) devastated buildings and (**c**) bus at Yungay; (**d**) damaged tree stump, and debris avalanche deposits (photo by W. Alberto, 2008).

somehow preserved, as suggested by Schneider et al. (1999) for the cases of Flims, Switzerland, and Blackhawk, California. In other cases, grain-size segregation during granular flow causes pseudo-stratification of debris avalanches deposits (Fineberg, 1997). Flow transformations are particularly evident in distal and lateral parts of the deposits, where the internal structures of debris accumulation disappear.

The role of basal lubricant for debris avalanches also can be achieved by other fluidized materials; sand or finer materials can be "trapped" beneath the avalanche debris after having been eroded by the moving landslide mass.

Hazards and risks

Spatial and temporal occurrences, and magnitude of instability events, are key factors controlling hazards and risks derived from debris avalanches. Hazards assessment studies should also address the recognition of possible emplacement modes and triggering mechanisms related to specific environmental contexts. As shown elsewhere large-scale volcanic debris avalanches are of higher magnitude than nonvolcanic ones. Larger mobility and volume of mobilized materials lead to wider affected areas that can be assessed by using H/L diagrams, showing the relationships between the maximum collapse height and the runout distance. By analyzing volcanic instabilities, attention has to be paid also to the evidence of transformed flows, that is, from debris avalanches to debris/mud flows or to lahars. The collapse of the north side of the volcanic edifice of Mount St. Helens during the cataclysmic eruption of the May 18, 1980, caused a massive 2.8 km^3 rock-slide/debris avalanche that traveled 28 km from its source (Voight et al., 1985). Associated debris flows (*lahars*) and muddy stream flows traveled farther, extending the damaged area northwestward, when the water-saturated parts of the massive debris avalanche deposits began to slump and flow.

Other intensity parameters, such as velocity and flow depths, should be also estimated in order to assess derived parameters such as run-up of mass from instable slopes and area of possible impacts on structures. In the case of the Huascaran debris avalanches (Perù), a first event (January 10, 1961) had limited damaging impacts due to lower magnitude and thanks to the protective effect of a 240 m-high ridge near Yungay; the subsequent event (May 31, 1970) of higher magnitude (3,800-m total height, 16-km runout; Figure 1b), overtopped the ridge, and Yungay and Ranrayrca were devastated (Figure 3a–d). Casualties were up to 18,000 people, associated debris flows filled Rio Santa and flowed 160 km downstream to the Pacific Ocean.

Bibliography

Cruden, D. M., and Varnes, D. J., 1996. Landslide types and processes. In Turner, A. K., and Schuster, R. L. (eds.), *Landslides: Investigation and Mitigation*. Washington, DC: National Research Council. Transportation Research Board Special Report 247, pp. 36–75.

Davies, T. H. R., 1982. Spreading of rock avalanches debris by mechanical fluidization. *Rock Mechanics*, **15**, 9–24.

Eisbacher, G. H., and Clague, J. J., 1984. *Destructive Mass Movements in High Mountains: Hazard and Management*. Ottawa: Energy, Mines and Resources Canada. Geological Survey of Canada Paper 84-16, 230 pp.

Fineberg, J., 1997. From Cinderella's dilemma to rock slides. *Nature*, **386**, 323–324.

Heim, A., 1932. *Bergsturz und Menschenleben*. Zurich: Wasmuth Verlag (English translation by Skermer, N., 1989. *Landslide and Human Lives*. Vancouver: Bitech).

Hsü, K. J., 1975. Catastrophic debris streams (sturzstroms) generated by rockfalls. *Bulletin Geological Society of America*, **86**, 129–140.

Hungr, O., 2005. Classification and terminology. In Jakob, M., and Hungr, O. (eds.), *Debris-Flow Hazards and Related Phenomena*. Berlin/Heidelberg: Praxis Springer, pp. 9–23.

Kent, P. E., 1966. The transport mechanism in catastrophic rock falls. *Journal of Geology*, **74**, 79–83.

Plafker, G., and Ericksen, G. E., 1978. Nevados Huascaran Avalanches, Peru. In Voight, B. (ed.), *Rockslides and Avalanches*. Amsterdam: I. Elsevier, pp. 277–314.

Schneider, J. L., Wassmerb, P., and Ledesert, B., 1999. The fabric of the sturzstrom of flims (Swiss Alps): characteristics and implications on the transport mechanisms. *C. R. Academy of Science Paris – Earth and Planetary Sciences*, **328**, 607–613.

Shaller, P. J., and Komatsu, G., 1994. Landslide on mars. *Landslide News*, **8**, 12–22.

Sharpe, C. F. S., 1938. *Landslide and related phenomena*. New York: Columbia University Press. 137 pp.

Shreve, R. L., 1968. Leakage and fluidization in air-lubricated avalanches. *Geological Society of America Bulletin*, **79**, 653–658.

Siebert, L., 1984. Large volcanic debris avalanche: characteristic of the source areas, deposits, and associated eruptions. *Journal of Volcanology and Geothermal Research*, **22**, 163–197.

Ui, T., Takarada, S., and Yoshimoto, M., 2000. Debris avalanches. In Sigurdsson, H. (ed.), *Encyclopedia of Volcanoes*. San Diego: Academic, pp. 617–626.

Varnes, D. J., 1958. Landslide types and processes. In Eckel, E. B. (ed.), *Landslide and Engineering*. Washington, DC. NAS-NRC Publication 544, Highway Research Board Special Report 29, pp. 20–47.

Varnes, D. J., 1978. Slope movement types and processes. In Schuster, R. L., and Krizek R. K. (eds.), *Landslides: Analysis and Control*. Washigton, DC: National Academy of Sciences. Transportation Research Board Special Report 176, pp. 11–33.

Voight, B., Janda, R. J., Glicken, H., and Douglass, P. M., 1985. Nature and mechanics of the Mount St Helens rockslide-avalanche of 18 May 1980. *Geotechnique*, **33**(3), 243–273.

Cross-references

Debris Flow
Landslide
Landslide Types
Mass Movement
Nuee Ardente
Rockfall
Slide and Slump
Slope Stability

DEBRIS FLOW

Oldrich Hungr
University of British Columbia, Vancouver, BC, Canada

Definition

Debris flow is an extremely rapid, flow-like mass movement, traveling in a steep, established channel and involving a saturated, unsorted mixture of granular soils, organics, and other debris (Hungr et al., 2001).

Most steep natural slopes are mantled by varying thickness of colluvial soils, disturbed by surficial phenomena known collectively as soil creep, as well as small-scale landslides. Colluvium naturally accumulates in depressions and near the thalweg of steep low-order drainage channels, where it is likely to become saturated during high infiltration periods. During the same period of high near-surface groundwater pressures, a local slope instability such as a debris slide from a steep gully headwall or an erosion-undercut side slope above the channel can occur. The moving debris impacts the steep floor of the channel, instantly increasing total stresses on the saturated colluvium, or poorly sorted channel deposits accumulated there. Given the rapid occurrence of such loading, even pervious granular materials cannot drain quickly enough, and excess pore pressures are generated. As a result of this "rapid undrained loading" mechanism, the colluvium on the gully floor is destabilized. It joins material already in motion, and a translating wave of liquefied debris continues downslope (e.g., Johnson, 1970; Sassa, 1985).

In addition to debris slides, initiation can occur as a result of rock fall impact, or by the spontaneous shear failure of the steep bed of the channel, carrying heavy flood flow. Because the initiation typically occurs during a period of unusually intensive precipitation or snowmelt, multiple, near-simultaneous initiation in several branches of a steep drainage system often takes place (Figure 1).

As the resulting flow-like mass movement progresses downslope, the rapid loading cycle is continuously repeated, as long as the slope remains sufficiently steep

Debris Flow, Figure 1 Multiple debris flow initiation points and converging channels in a Rocky Mountains, Alberta, Canada, mountain drainage.

Debris Flow, Figure 2 Bouldery levee created by a debris flow surge in Kyrghistan. The boulders in the center of the photo are approximately one meter in diameter.

and loose saturated material is available. Often, the entire length of the steep segments of zero and first- or second-order channel is scoured of loose material, and the debris flow "surge" grows to a volume far exceeding that of the initial slide (Hungr et al., 2005). While entraining saturated solid material, the debris flow surge also overtakes and incorporates surface flow from the channel, becoming progressively more diluted.

Flow of saturated liquid debris in a steep channel is not a stable process. Surges form either by longitudinal sorting and accumulation of coarse clasts (Iverson, 1997) or by increasing turbulence where the flow depth is large (Davies, 1986). Either process leads to the development of "surges," characterized by a steep front consisting either of a concentration of boulders ("a moving dam") or a zone of highly turbulent flow. The passage of a bouldery front often creates a ridge-like "levee" deposit along the margins of the channel (Figure 2). Behind the front, the surge attains a shorter or longer region of maximum depth, containing fairly dense liquefied material and often flowing in a macroscopically laminar regime and a tapering "tail" or "intersurge flow," where the flow becomes more highly diluted and turbulent (Pierson, 1980).

The surge building process is the most important aspect of debris flows, as it magnifies the flow depth and thereby peak discharge. While a debris flow surge may not travel much faster than an extreme water flood flow in the same steep channel, its peak discharge may be several tens of times greater than the peak discharge of a flood. Typical flow parameters of debris surges in the coastal region of British Columbia are peak discharge of 300–500 m^3/s, velocity of 10–15 m/s, and flow depth of 2–5 m. Thus, debris surges are capable of vastly greater impact and potential damage than even the most extreme water floods.

Due to the presence of multiple initiation points and contributing channels, as well as the erratic time development process of surge building, many debris flow "events" consist of multiple and sometimes periodic surges, separated by periods of flood flow.

Heavy water floods in steep channels can also transport large quantities of sediment, often in the form of bulk channel instability driven by the tractive forces of overflowing water. Some writers suggested using peak discharge as a criterion to distinguish between a "debris flood" and "debris flow" (Aulitzky, 1980; Hungr, 2005). A debris flood has a peak discharge of the same order as a hydrologic flood wave. It may transport large quantities of sediment but is limited in its ability to transport very large boulders and inflicting damaging impacts on trees and structures. A debris flow at peak discharge is much deeper, can transport boulders several meters in diameter, and destroys objects in its path (Figure 3). Many events contain individual surges that can be classified as debris floods, in addition to those that are true debris flow surges. Usually, the latter dominate in terms of resulting damage.

Debris surges traverse and erode material from steep channels and then gradually deposit on "debris fans." Debris fans can be recognized and distinguished from alluvial fans by high steepness, extremely poor channel stability, and the presence of very large clasts in the deposits (Figure 4).

Debris flows are ubiquitous in mountain drainages and on steep slopes everywhere, particularly where concentrated periods of high precipitation occur and good supply

Debris Flow, Figure 3 A large boulder deposited by a debris flow in Taiwan.

Debris Flow, Figure 4 A debris fan in the Coast Ranges, British Columbia, Canada.

of unconsolidated colluvium exists. Because it is such a common process, debris flow is responsible for considerable losses in terms of fatalities, destruction of structures, and, most often, interruption and damage to infrastructure such as roads and railways. This type of damage exacts a steady toll in all regions with steep slopes. Sometimes, juncture of unfavorable circumstances, especially in terms of extreme rainfall, creates clusters of debris flows that can devastate an entire region. For example, the 1999 debris flows on the northern slopes of the Aquila Range in the Venezuelan state Vargas claimed approximately 30,000 fatalities among the inhabitants of several debris fans in the area (Larsen and Wieczorek, 2006).

Bibliography

Aulitzky, H., 1980. Preliminary two-fold classification of debris torrents. In *Interpraevent*, Bad Ischl, Austria, Vol. 4, pp. 285–309 (translated into English by G. Eisbacher).
Davies, T. R. H., 1986. Large debris flows: a macroviscous phenomena. *Acta Mechanica*, **63**, 161–178.
Hungr, O., 2005 Terminology and classification. In Jakob, M., and Hungr, O. (eds.), *Debris Flow Hazards and Related Phenomena*. Heidelberg, Germany: Springer, in association with Praxis Publishing Ltd., Chapter 2, pp. 9–23.
Hungr, O., Evans, S. G., Bovis, M., and Hutchinson, J. N., 2001. Review of the classification of landslides of the flow type. *Environmental and Engineering Geoscience*, **VII**, 221–238.
Hungr, O., McDougall, S., and Bovis, M., 2005. Entrainment of material by debris flows. In Jakob, M., and Hungr, O. (eds.), *Debris Flow Hazards and Related Phenomena*. Heidelberg, Germany: Springer, in association with Praxis Publishing Ltd., Chap. 7, pp. 135–158.
Iverson, R. M., 1997. The physics of debris flows. *Reviews of Geophysics*, **35**, 245–296.
Johnson, A. M., 1970. *Physical Processes in Geology*. New York: W. H. Freeman, p. 577.
Larsen, M. C., and Wieczorek, G. F., 2006. Geomorphic effects of large debris flows and flash floods, northern Venezuela, 1999. *Zeitschrift fur Geomorphologie N.F.*, **145**, 147–175.
Pierson, T. C., 1980. Erosion and deposition by debris flows at Mt Thomas, North Canterbury, New Zealand. *Earth Surface Processes*, **5**, 227–247.
Sassa, K., 1985. The mechanism of debris flows. In *Proceedings, 11th International Conference on Soil Mechanics and Foundation Engineering*, San Francisco, Vol. 1, pp. 1173–1176.

Cross-references

Creep
Debris Avalanche
Landslide (Mass Movement)
Landslide Types
Mass Movement
Mudflow
Pore-Water Pressure
Quick Clay
Rockfall

DEEP-SEATED GRAVITATIONAL SLOPE DEFORMATION

Mauro Soldati
Università degli Studi di Modena e Reggio Emilia, Modena, Italy

Definition

A deep-seated gravitational slope deformation (DGSD) is a gravity-induced process affecting large portions of slopes evolving over very long periods of time. A DGSD may displace rock volumes of up to hundreds of millions of cubic meters, with thicknesses of up to a few hundred meters.

Introduction

Deep-seated gravitational slope deformations (DGSDs) are not considered hazardous phenomena because they evolve very slowly. However, they must not be neglected when defining slope instability in a territory and the related hazard implications. Despite their slow deformation rates, DGSDs may cause damage to surface and underground (e.g., tunnels) structures. In addition, they may evolve into faster mass movements or favor collateral landslide processes.

Causes

Deep-seated gravitational slope deformations are often recognized in formerly glaciated valleys and in seismically active mountain regions. The occurrence of DGSD is generally related to tensional stresses induced by gravity in steep-sided valleys which can be linked to (1) unloading of oversteepened slopes due to glacier retreat or fluvial downcutting, (2) presence of discontinuities of tectonic origin (joints, fractures, etc.), and (3) seismic activity (Kostak and Avramova-Taceva, 1981; Dramis and Sorriso-Valvo, 1983, 1994; Bisci et al., 1996; Pasuto and Soldati, 1996; Agliardi et al., 2001; Gutiérrez et al., 2008). Displacements related to DGSDs occur due to the presence of deformation belts at depth, where the rock mass is affected by micro-fractures (Radbruch-Hall, 1978). The main feature that distinguishes DGSDs from landslides s.s. is the absence of a continuous or well-defined sliding surface.

The term "deep-seated gravitational slope deformation" was firstly used by Malgot (1977). Alternative definitions for DGSD include *sackung* (Zischinsky, 1966), *gravity faulting* (Beck, 1968), *depth creep of slopes* (Ter-Stepanian, 1966, 1977), *gravitational slope deformations* (Nemčok, 1972), *deep-seated creep deformations* (Mahr and Nemčok, 1977), *gravitational block-type movements* (Pasek, 1974), or *gravitational spreading* and *gravitational creep* (Radbruch-Hall, 1978).

It is worth mentioning an earlier contribution to the definition of the processes involved in DGSDs by Terzaghi (1950), who showed the difference between "landslide" and "creep." According to Terzaghi, a "landslide" occurs along a well-defined sliding plane when the stress conditions for failure are satisfied, whereas "creep" is a continuous process characterized by undefined boundaries between stationary and moving material. This implies that the deformation/creep phase may be naturally followed by a sliding and much quicker phase. Though the temporal evolution from a deformation phase to a sliding phase is generally hard to predict, and possibly extremely long, the hazard implications become clear. Furthermore, as far as hazard is concerned, it should be noted that the presence of deep-seated gravitational deformations within

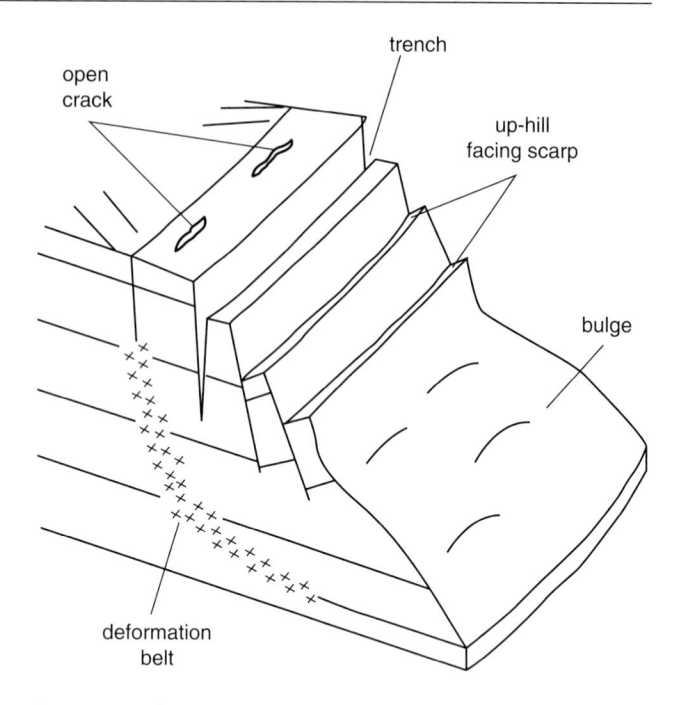

Deep-seated Gravitational Slope Deformation, Figure 1 Sackung (After Bisci et al., 1996; modified).

a rock mass also favors the onset of collateral slope movements (e.g., rock falls, rock topples, block slides, etc.), which are likely to occur suddenly. If this takes place in inhabited areas, risk issues may have to be faced.

Types

Two main types of DGSDs can be outlined: *sackung* and *lateral spreading*. Sackung (or rock flow, according to the landslide classification by Cruden and Varnes, 1996) refers to the sagging of high and steep slopes due to visco-plastic deformation occurring at depth (Zischinsky, 1969; Bisci et al., 1996). Homogeneous and jointed or stratified rock masses, characterized by a brittle mechanical behavior, are generally affected by this type of DGSD. As a result of the deformation, typical sackung morphological features are twin ridges, trenches, gulls, and uphill-facing scarps in the upper part of the slopes as well as bulges in the medium and lower parts of the slopes (Figure 1). In addition, subhorizontal joints can be found at the lower edge of the slopes. In spite of a general agreement on the morphological features which characterize a sackung, there is still some uncertainty and dispute on the kinematics of the process. However, the development of deep-seated deformations is generally referred to the rock mass behavior at depth which is different from that at surface, owing to the higher confining pressure. Two main displacement models have been proposed. One the one hand, it is assumed that a high confining pressure does not allow the formation of well-defined shear surfaces at depth (Mahr, 1977), enabling only viscous deformations to take place (nonshearing model). This generally occurs

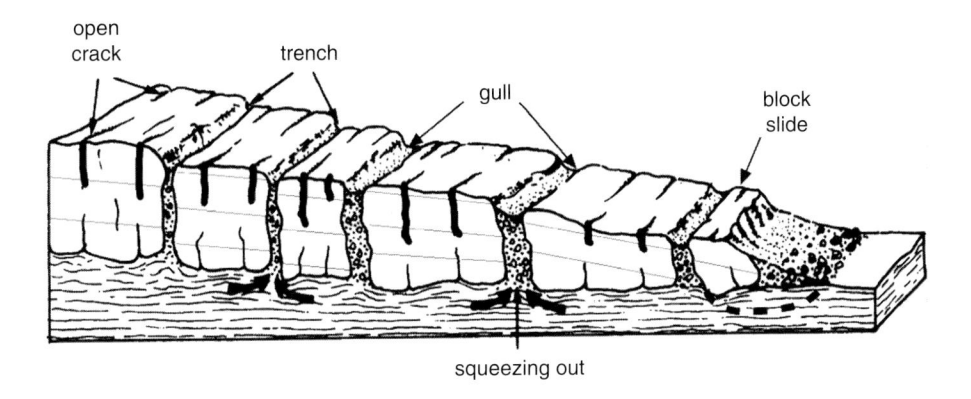

Deep-seated Gravitational Slope Deformation, Figure 2 Lateral spreading affecting brittle rock masses overlying ductile terrains (After Pasuto and Soldati, 1996; modified).

at the central parts of the slope, whereas at the top and toe of the slope, shear surfaces may still develop due to lower confining pressures. On the other hand, it is stated that the deformation zone is indeed interrupted along a shear surface located at the base of the unstable rock mass (plastic failure model) (Savage and Varnes, 1987).

Lateral spreading (or rock spreading) consists of lateral expansions of rock masses occurring along shear or tensile fractures. Two types of rock spreading can be distinguished which occur in different geological conditions (Pasuto and Soldati, 1996).

The first type refers to lateral spreading affecting brittle rock masses overlying ductile terrains. It may extensively affect rock plateaus, the effects being most evident at their edges. The spreading process is generally caused by the deformation of the underlying material and favored by slope or cliff downcutting. The deformation takes place through horizontal displacements along tensile fractures or subvertical tectonic discontinuities (Conti and Tosatti, 1996). Trenches, gulls, grabens, karst-like depressions in the brittle rocks and bulges in the weaker material are typical morphological features of lateral spreading (Figure 2). The overburden of the rock slabs is generally assumed as the cause of long-term deformation affecting the underlying terrains. As a result, squeezing out of the weaker rock types and rock block spreading due to tensile stresses take place. The process may be accelerated by water percolation through the fissures and consequent softening of the weaker materials. Lateral spreading phenomena are often accompanied by collateral slope movements occurring at the edges of the plateaus, such as rock falls and topples, block slides, and earth flows. Downcutting of valleys may favor the onset of these movements. The process may continue for long periods and cause progressive spreading and dismembering of rock plateaus.

The second type of lateral spreading occurs in homogeneous, and usually brittle, rock masses and has mostly been recognized in mountain areas with high relief energy. The spreading process generally occurs without a well-defined basal shear surface or zone of visco-plastic flow. As a result of the deformation process occurring at depth,

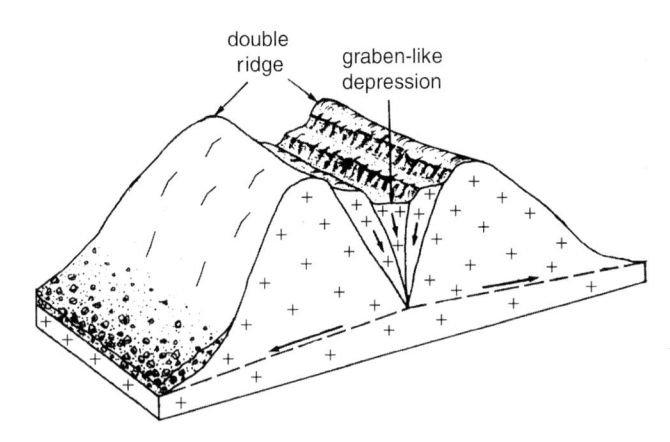

Deep-seated Gravitational Slope Deformation, Figure 3 Lateral spreading in homogeneous rock masses (After Pasuto and Soldati, 1996; modified).

double ridges, uphill-facing scarps, ridge-top depressions, and infilled troughs can be found at surface (Figure 3). The presence of joints in the rock mass is considered as a predisposing factor, but the mechanics of the deformation has not yet been well defined.

Hazard implications

Sackung and lateral spreads may evolve into faster mass movements due to various causes (e.g., earthquakes), and therefore, their presence should be adequately taken into account in landslide hazard assessment (Coltorti et al., 1985; Hewitt et al., 2008). As a matter of fact, sackung may evolve into rotational/translational slides or into rock/debris avalanches, inducing possible risk situations due to the potential high velocity of these mass movements. On the other hand, lateral spreads tend to evolve in slower movements, such as block slides occurring at the edges of the areas affected by the spreading. Nevertheless, the possible occurrence of collateral landslides favored by the lateral spreading (e.g., rock falls, rock topples, earth flows, and slides) should not be overlooked when landslide hazard has to be assessed;

these phenomena are likely to be connected with the modifications of the groundwater flow net and the stress conditions related to the active deep-seated deformations.

The awareness of the role that DGSDs can play on slope instability is fundamental when stabilization and monitoring of single landslides are to be planned. If the influence of deep-seated deformations is not considered, mitigation measures may result incomplete and noneffective in long terms. It is clear that resolving interventions aiming at the mitigation of the effects of deep-seated gravitational deformations are hard to envisage due to the large size of these phenomena. However, the application of new methods for the detection, survey, and monitoring of DGSDs is very important to understand the kinematics of the movements and foresee possible evolution scenarios. In this respect, a geomorphological survey, accompanied by high-precision topographic measurements (e.g., by means of the GPS technique), LIDAR surveys, and interferometric analyses of radar images, may provide significant outputs with respect to the areas affected and volumes involved in DGSDs, as well as on the displacement rates and trends. Structural and geomechanical analyses of the rock masses are also very important to understand the lithological and tectonic control on DGSD onset and development.

Recently, the importance of the recognition of DGSDs morphological features, especially those related to sackung, has been acknowledged in paleoseismic investigations. Actually, sackung features may represent natural archives of the recurrence of high-intensity paleoearthquakes, providing useful information for seismic hazard assessment (Gutiérrez-Santolalla et al., 2005; Gutiérrez et al., 2008). The causal and chronological relationship between sackung features and earthquakes has been shown in terms of generation and/or reactivations of sackung scarps during historic earthquakes (McCalpin and Hart, 2003; Moro et al., 2007). Therefore, timing of sackung features caused by earthquakes, as determined by trenching and radiocarbon dating of sediments, can improve existing paleoseismic catalogues. The use of this technique can be particularly important where seismogenetic faults are buried, blind, or have scarce geomorphological evidence which makes conventional paleoseismic investigations difficult. It should be added that geomorphological features generated by DGSDs, such as double ridges and scarps, are frequently interpreted as of tectonic origin. This may lead to seismic hazard overestimations with significant economic implications (Bovis and Evans, 1995) and proves the importance of detecting the role of gravity in shaping these features for a correct and reliable hazard assessment.

Bibliography

Agliardi, F., Crosta, G., and Zanchi, A., 2001. Structural constraints on deep-seated slope deformation kinematics. *Engineering Geology*, **59**, 83–102.

Beck, A. C., 1968. Gravity faulting as a mechanism of topographic adjustment. *New Zealand Journal of Geology and Geophysics*, **11**(1), 191–199.

Bisci, C., Dramis, F., and Sorriso-Valvo, M., 1996. Rock flow (sackung). In Dikau, R., Brunsden, D., Schrott, L., and Ibsen, M.-L. (eds.), *Landslide Recognition: Identification, Movement and Causes*. Chichester: John Wiley & Sons, pp. 150–160.

Bovis, M. J., and Evans, S. G., 1995. Rock slope movements along the Mount Currie "fault scarp", Southern Coast Mountains, British Columbia. *Canadian Journal of Earth Science*, **32**, 2015–2020.

Coltorti, M., Dramis, F., Gentili, B., Pambianchi, G., Crescenti, U., and Sorriso-Valvo, M., 1985. The December 1982 Ancona landslide: a case of deep-seated gravitational slope deformation evolving at unsteady rate. *Zeitschrift fuer Geomorphologie N. F.*, **29**(3), 335–345.

Conti, S., and Tosatti, G., 1996. Tectonic *vs* gravitational processes affecting Ligurian and Epiligurian units in the Marecchia valley (Northern Apennines). *Memorie di Scienze Geologiche*, **48**, 107–142.

Cruden, D. M., and Varnes, D. J., 1996. Landslides types and processes. In Turner, A. K., and Schuster, R. L. (eds.), *Landslides: Investigation and Mitigation*. Washington, DC: Transportation Research Board, National Academy of Sciences, Special Report 247, pp. 36–75.

Dramis, F., and Sorriso-Valvo, M., 1983. Two cases of earthquake-triggered gravitational spreading in Algeria and in Italy. *Rendiconti della Società Geologica Italiana*, **6**, 7–10.

Dramis, F., and Sorriso-Valvo, M., 1994. Deep-seated gravitational slope deformations, related landslides and tectonics. *Engineering Geology*, **38**, 231–243.

Gutiérrez, F., Ortuño, M., Lucha, P., Guerrero, J., Acosta, E., Coratza, P., Piacentini, D., and Soldati, M., 2008. Late Quaternary episodic displacement on a sackung scarp in the Central Spanish Pyrenees. Palaeoseismic evidence? *Geodinamica Acta*, **21**(4), 187–202.

Gutiérrez-Santolalla, F., Acosta, E., Ríos, S., Guerrero, J., and Lucha, P., 2005. Geomorphology and geochronology of sackung features (uphill-facing scarps) in the Central Spanish Pyrenees. *Geomorphology*, **69**(1–4), 298–314.

Hewitt, K., Clague, J. J., and Orwin, J. F., 2008. Legacies of catastrophic rock slope failures in mountain landscapes. *Earth-Science Reviews*, **87**, 1–38.

Kostak, B., and Avramova-Taceva, E., 1981. Propagation of coastal slope deformations at Taukliman, Bulgaria. *Bulletin of the IAEG*, **23**, 67–73.

Mahr, T., 1977. Deep-reaching gravitational deformations of high mountain slopes. *Bulletin of the IAEG*, **16**, 121–127.

Mahr, T., and Nemčok, A., 1977. Deep-seated creep deformations in the crystalline cores of the Tatry Mts. *Bulletin of the IAEG*, **16**, 104–106.

Malgot, J., 1977. Deep-seated gravitational slope deformations in neovolcanic mountain ranges of Slovakia. *Bulletin of the IAEG*, **16**, 106–109.

McCalpin, J. P., and Hart, E. W., 2003. Ridge-top spreading features and relationship to earthquakes, San Gabriel Mountain region, southern California. In Hart E. W. (ed.), *Ridge-Top Spreading in California: Contributions toward Understanding a Significant Seismic Hazard*. California Geological Survey, CD 2003–05.

Moro, M., Saroli, M., Salvi, S., Stramondo, S., and Doumaz, F., 2007. The relationship between seismic deformation and deep-seated gravitational movements during the 1997 Umbria-Marche (Central Italy) earthquakes. *Geomorphology*, **89**, 297–307.

Nemčok, A., 1972. Gravitational slope deformation in high mountains. In *Proceedings 24th International Geological Congress*, Montreal, Sect. 13, pp. 132–141.

Pasek, J., 1974. Gravitational block-type movements. In *Proceedings 2nd International Congress*. São Paulo, Brasil: IAEG, pp. V-PC-1.1–V-PC-1.9.

Pasuto, A., and Soldati, M., 1996. Rock spreading. In Dikau, R., Brunsden, D., Schrott, L., and Ibsen, M.-L. (eds.), *Landslide*

Recognition: Identification, Movement and Causes. Chichester: John Wiley & Sons, pp. 122–136.

Radbruch-Hall, D. H., 1978. Gravitational creep on rock masses on slopes. In Voight, B. (ed.), *Rockslides and avalanches.* Amsterdam: Elsevier, pp. 607–675.

Savage, W. Z., and Varnes, D. J., 1987. Mechanics of gravitational spreading of steep-sided ridges ("Sackung"). *Bulletin of the IAEG,* **35**, 31–36.

Ter-Stepanian, G., 1966. Types of depth creep of slopes in rock masses. In *Proceedings 1st Conference of International Society for Rock Mechanics,* Lisbon, Sect. 2, pp. 157–160.

Ter-Stepanian, G., 1977. Deep-reaching gravitational deformation of mountain slopes. *Bulletin of the IAEG,* **16**, 87–94.

Terzaghi, K., 1950. Mechanism of landslides. In Paige, S. (ed.), *Application of Geology to Engineering Practice (Berkey Volume).* Washington, DC: Geological Society of America, pp. 83–123.

Zischinsky, Ü., 1969. Über Sackungen. *Rock Mechanics,* **1**(1), 30–52.

Zischinsky, Ü., 1966. On the deformation of high slopes. In *Proceedings 1st Conference of International Society for Rock Mechanics,* Lisbon, Sect. 2, pp. 179–185.

Cross-references

Landslide
Lateral Spreading
Mass Movement
Sackung
Slope Movement

DESERTIFICATION

Nicholas Lancaster
Desert Research Institute, Reno, NV, USA

Synonyms

Land degradation; Loss of productivity

Definition

There are many definitions of desertification, but the most widely used is that adopted by the United Nations Convention to Combat Desertification (UNCCD) http://www.unccd.int/, which defined desertification as land degradation in drylands (arid, semiarid, and dry subhumid areas) resulting mainly from adverse human impact and climate variability (Safriel and Adeel, 2005).

Introduction

Drylands, defined as areas in which the ratio between precipitation and evapotranspiration is less than 0.75, cover as much as 47% of the world's land surface (Figure 1) and are home to over one billion people (Ezcurra, 2006). Following the UNEP definitions, drylands include areas classified climatically as hyperarid to dry subhumid. Drylands are fragile environments that experience great natural variability in climate and are easily affected by natural disturbance and anthropogenic stressors. In many areas, they are being impacted by a rapidly growing and increasingly urban population. One manifestation of natural and environmental stress in drylands is the phenomenon of desertification, a term originally used by French colonial scientists to describe environmental degradation in dry areas of West Africa. Although land degradation in drylands has been occurring locally for millennia, the problem gained prominence and recognition by the international community in the 1970s following severe drought in the Sahel region of Africa and with the United Nations Conference on Desertification held in Nairobi, Kenya, in 1977. Desertification is, however, a concept that encompasses a variety of physical and biological processes, none of which are restricted to drylands, but which take on a particular form in such areas (Thomas and Middleton, 1994).

Causes of desertification

Desertification involves the degradation of ecosystems by a combination of natural disturbance (e.g., droughts) and human stress (e.g., overgrazing, inappropriate land use). The result is a persistent reduction in the level of productivity and ecosystem services over an extended period. Desertification occurs when the natural resilience of dryland ecosystems to disturbance is impaired by anthropogenic stressors, so that it is unable to revert in the short term to its prior state when the stresses are relaxed (Whitford, 2002).

In their natural state, dryland ecosystems are resilient to natural disturbances such as drought and fire. Likewise, traditional land-use practices such as nomadic pastoralism, hunting of wildlife, and localized rain-fed agriculture have evolved to cope with a highly variable environment. Socioeconomic changes, including population growth, settling of previously mobile populations, imposition of political borders, commercial (often export-oriented) agriculture and livestock raising, provision of permanent water sources (wells and boreholes) together with poor land-use practices, such as overgrazing, fuel wood extraction, excessive irrigation, and cultivation of marginal land, impose stresses on ecosystems such that their resilience to disturbance is impaired. The result is loss of biologic and economic productivity (Adeel et al., 2005).

Occurrence of desertification

According to the UNEP, 70% of the world's drylands (excluding hyperarid deserts), or some 3,600 million hectares, are estimated to experience varying degrees of desertification. The most vulnerable areas occur where human pressures combine with fragile ecosystems and soils, and tend to be concentrated in semiarid or desert margin areas (Figure 2). Over 250 million people are directly affected by desertification. In addition, at least one billion people in over 100 countries are at risk. These people include many of the world's poorest, most marginalized, and politically weak citizens.

Ecological effects of desertification

Land degradation in drylands takes many forms, manifested by physical changes such as reduction in vegetation cover, changes in the types of plants, loss of soil

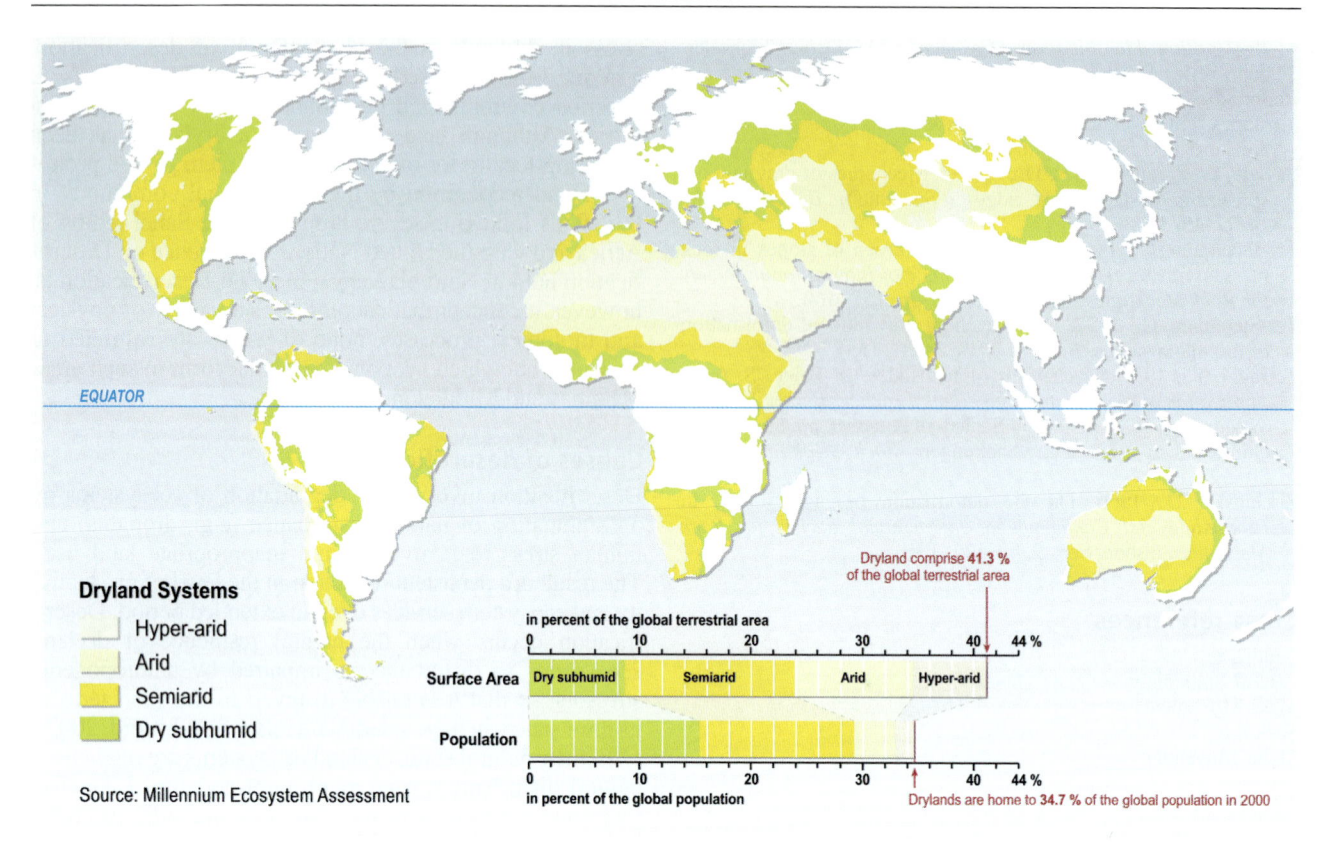

Desertification, Figure 1 Global extent of drylands From http://www.maweb.org/documents/document.355.aspx.pdf.

by wind or water erosion, and changes in plant nutrient distribution. Some of these changes can result in loss of primary productivity. In many areas, lower productivity is measured by reduced carrying capacity for livestock as a result of changes in the amount and composition of vegetation. For example, perennial grasses may be replaced by unpalatable or inedible shrubs, as documented in the Southwestern USA, southern Africa, and parts of Australia. Replacement of grassland by shrubs, partly as a result of overgrazing, may lead to a progressive and irreversible reorganization of soil and water resources at a landscape scale (Figure 3). Formerly widely distributed nutrients and water are concentrated into a series of "islands" of productivity nucleated by shrubs (Schlesinger et al., 1990). The intervening bare areas are subject to increasing erosion by wind and/or water, so creating a series of positive feedback loops that lead to a highly degraded and functionally different ecosystem.

This process is well documented from the southwestern USA, where in the Chihuahuan Desert, grasses have been gradually replaced by mesquite shrubs over the past 50–100 years, with important consequences for ecosystems and geomorphic processes (Okin et al., 2001). In the Jornada area of New Mexico, the mesquite shrubs anchor coppice dunes separated by "streets" of high sand transport rates (Okin, 2008).

Grazing also damages or destroys the biological soil crusts that are important in stabilizing soils, reducing wind and water erosion, and fixing nitrogen in many desert regions. In southeastern Utah, even after 30 years without grazing, soils in areas that have been historically grazed have much lower contents of organic matter and primary nutrients, compared to areas that have never been grazed, suggesting that livestock grazing promotes increased wind erosion which reduces soil productivity in native grasslands (Neff et al., 2008).

Land degradation as a result of fire and grazing pressure may also lead to replacement of native species by invasive or exotic species. In the Great Basin of the USA, the perennial sagebrush steppe is being invaded by exotic annual grasses such as *Bromus tectorum* (cheat grass), which leads to an increase in fire frequency and inhibits recolonization by native grasses and shrubs after fires (Smith et al., 2000).

Physical effects of desertification

Desertification rarely involves the expansion of existing deserts, represented most dramatically by encroachment of active (mobile) sand dunes. Despite reports of "spreading deserts" in the Sahel region of Africa and western China, there is little scientific evidence to support such assertions (Thomas and Middleton, 1994).

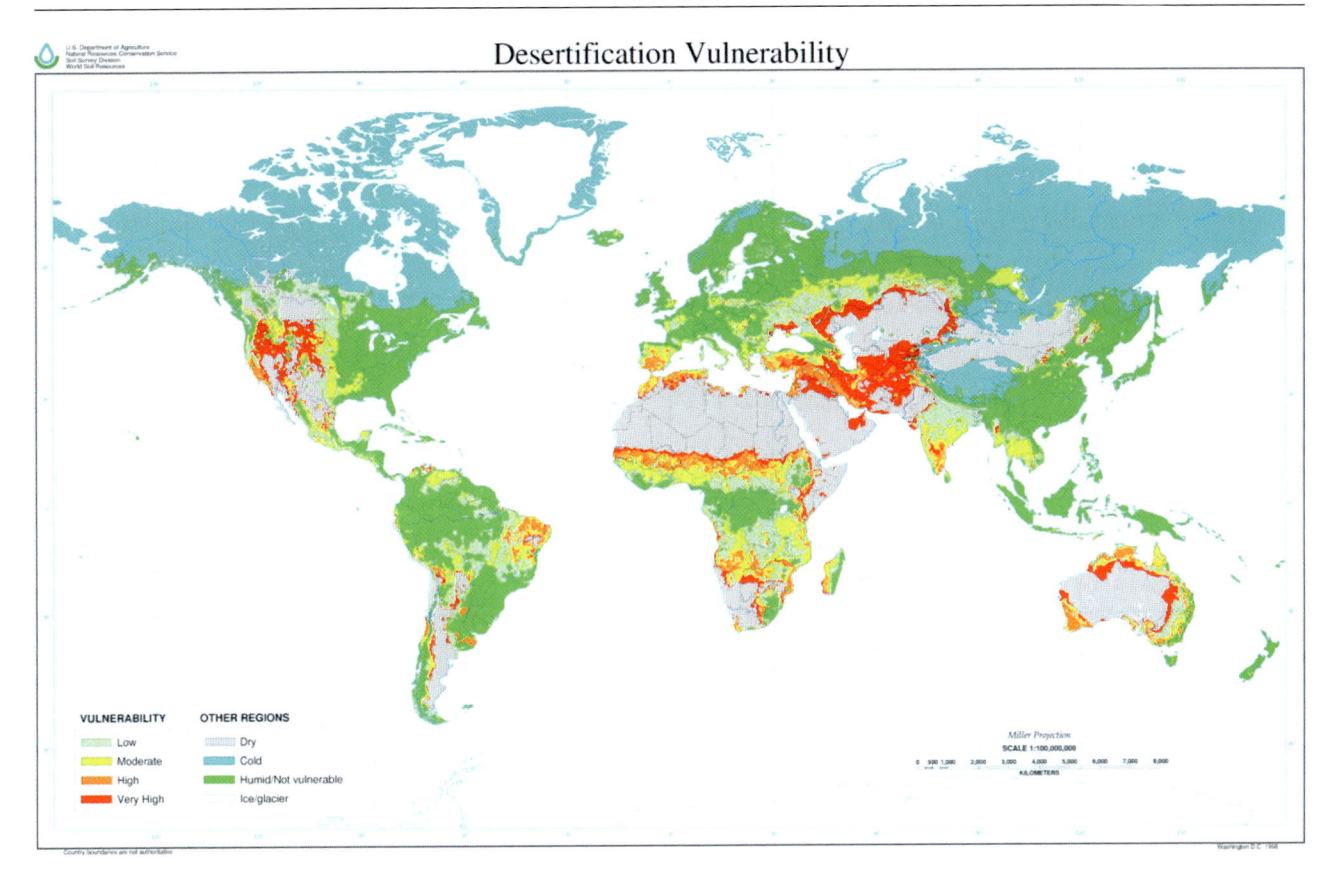

Desertification, Figure 2 Global extent of desertification vulnerability – after NRCS 2003: http://soils.usda.gov/use/worldsoils/mapindex/desert.html.

Overgrazing and fire may combine with drought to reduce vegetation cover on sand dunes so that sand transport becomes more active and more widespread. Localized reactivation of vegetation-stabilized dunes has been documented from western China (Wang et al., 2008) and the Sahel region of Africa (Niang et al., 2008).

Increased sand transport, mainly as a result of a sparse vegetation cover, may also increase the occurrence of dust storms, as saltating sand grains impact fine-grained soils. An increase in the frequency of dust storms was observed during the Sahel droughts of the 1970s (Prospero and Lamb, 2003). Areas of fine-grained soils that are subject to rain-fed agriculture are particularly susceptible to wind erosion, as in the Dust Bowl of the USA in the 1930s (Goudie and Middleton, 2006) and further in the 1970s (Lee and Tchakerian, 1995).

In many areas, reduction of vegetation cover, or replacement of grasslands by shrubs on hillslopes, reduces infiltration of rainwater and leads to increased surface runoff and loss of soil (Abrahams et al., 1995). In extreme cases, extensive networks of gullies may result.

Salinization of soils is a further physical expression of desertification, resulting mainly from poor irrigation practices. Excessive application of water leads to accumulation of salts in the soil to levels that are toxic to plants. Leakage from unlined irrigation canals contributes to a rise in the water table unless countered by drainage systems.

Social dimensions of desertification

It is widely assumed that desertification is a consequence of human activities. Reduction in the productivity of drylands has many adverse effects on the populations of these areas, leading to widespread and far reaching social, economic, and political repercussions (Adeel et al., 2005; Reynolds et al., 2007). Reduced productivity of drylands leads to lower food production, competition for scarce resources, and an increase in poverty; subsistence agriculture and cattle raising are no longer viable, so people are forced to migrate to urban areas. Lack of opportunities may lead to political instability and emigration to more developed countries. A cycle of increasing poverty and loss of livelihood ensues. Even in developed nations such as the USA and Australia, land degradation and loss of agricultural productivity in drylands have serious economic consequences.

Climate change and desertification

Hydroclimatological observations show that changes in seasonal and annual temperature, precipitation, runoff,

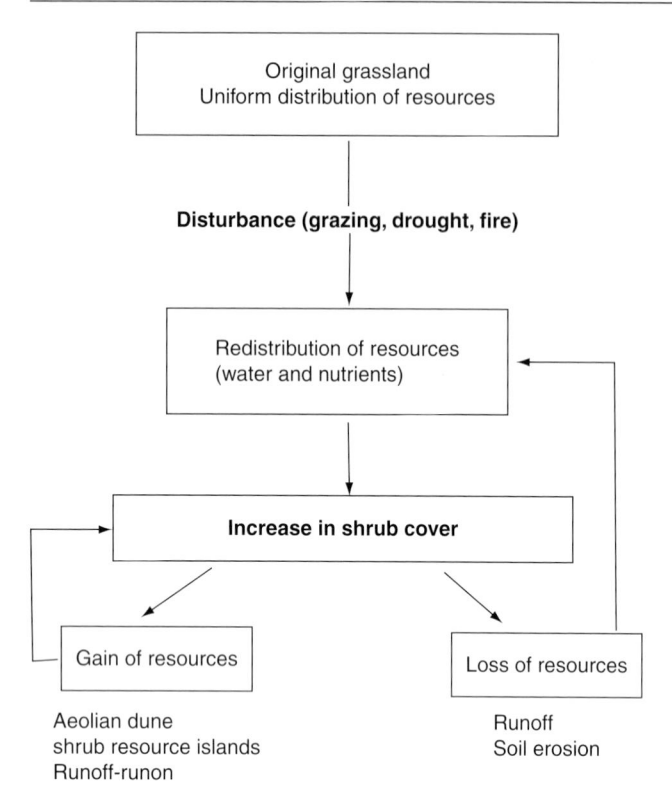

```
┌─────────────────────────────────────┐
│       Original grassland            │
│  Uniform distribution of resources  │
└─────────────────────────────────────┘
                  │
                  ▼
       Disturbance (grazing, drought, fire)
                  │
                  ▼
┌─────────────────────────────────────┐
│    Redistribution of resources      │◄──┐
│       (water and nutrients)         │   │
└─────────────────────────────────────┘   │
                  │                       │
                  ▼                       │
       ┌───────────────────────┐          │
       │ Increase in shrub cover│─────────┘
       └───────────────────────┘
          │                 │
          ▼                 ▼
   ┌──────────────┐   ┌──────────────────┐
   │Gain of       │   │ Loss of resources│
   │resources     │   └──────────────────┘
   └──────────────┘
   Aeolian dune         Runoff
   shrub resource       Soil erosion
   islands
   Runoff-runon
```

Desertification, Figure 3 Conceptual model for progressive degradation of arid ecosystems (After Whitford, 2002).

groundwater recharge, and evapotranspiration are occurring today in most deserts and models predict that they are likely to continue in the future. Many drylands are already experiencing significant increases in temperature and a reduction in rainfall over the past two decades, manifested in extended droughts. Climate models differ in their predictions for drylands: Some may enjoy increased rainfall as monsoon circulations are enhanced, but this effect is offset by higher temperatures. Increased variability of climate will likely result in more frequent and severe drought conditions. Because human pressures in many areas are increasing, dryland ecosystems will be less resilient to natural disturbance in the future, resulting in an enhanced vulnerability to desertification and its physical, biological, and human consequences (Safriel and Adeel, 2005).

Bibliography

Abrahams, A. D., Parsons, A. J., and Wainright, J., 1995. Effects of vegetation change on interrill runoff and erosion, Walnut Gulch, Arizona. *Geomorphology*, **13**(1–4), 37–48.
Adeel, Z. et al., 2005. Ecosystems and Human Well-Being: Desertification Synthesis. Millennium Ecosystem Assessment: 1–25.
Ezcurra, E. (ed.), 2006. *Global Deserts Outlook*. Nairobi, Kenya: United Nations Environment Program, 148 pp.
Goudie, A. S., and Middleton, N. J., 2006. *Desert Dust in the Global System*. Berlin/Heidelburg/New York: Springer, 287 pp.
Lee, J. A., and Tchakerian, V. P., 1995. Magnitude and frequency of blowing dust on the southern High Plains of the United States 1947–1989. *Annals of the Association of American Geographers*, **85**(4), 684–693.
Neff, J. C., et al., 2008. Increasing eolian dust deposition in the western United States linked to human activity. *Nature Geoscience*, **1**, 189–195.
Niang, A. J., Ozer, A., and Ozer, P., 2008. Fifty years of landscape evolution in southwestern Mauritania by means of aerial photographs. *Journal of Arid Environments*, **72**, 97–107.
Okin, G. S., 2008. A new model of wind erosion in the presence of vegetation. *Journal of Geophysical Research, Earth Surface*, **113**, F02S10.
Okin, G. S., Murray, B., and Schlesinger, W. H., 2001. Degradation of sandy arid shrubland environments: observations, process modeling, and management implications. *Journal of Arid Environments*, **47**, 123–144.
Prospero, J. M., and Lamb, P. J., 2003. African droughts and dust transport to the Caribbean: climate change implications. *Science*, **302**, 1024–1027.
Reynolds, J. F., et al., 2007. Global desertification: building a science for dryland development. *Science*, **316**(5826), 847–851.
Safriel, U., and Adeel, Z., 2005. Dryland systems. In Hassan, R., Scholes, R., and Ash, N. (eds.), *Ecosystems and Human Wellbeing: Current Status and Trends*. Washington, DC: Island Press, pp. 625–662.
Schlesinger, W. H., et al., 1990. Biological feedbacks in global desertification. *Science*, **247**, 1043–1048.
Smith, S. D., et al., 2000. Elevated CO_2 increases productivity and invasive species success in an arid ecosystem. *Nature*, **408**, 79–82.
Thomas, D. S. G., and Middleton, N. J., 1994. *Desertification: Exploding the Myth*. Chichester: Wiley.
Wang, X., Chen, F.-H., Hasi, E., and Li, J., 2008. Desertification in China: an assessment. *Earth Science Reviews*, **88**, 188–206.
Whitford, W. G., 2002. *Ecology of Desert Systems*. San Diego: Academic Press, 343 pp.

Cross-references

Climate Change
Drought
Dust Bowl
Dust Storm
Erosion
Land Degradation

DISASTER DIPLOMACY

Ilan Kelman
Center for International Climate and Environmental Research – Oslo (CICERO), Blindern, Oslo, Norway

Synonyms

Disaster-conflict nexus; Disaster politics; Disasters and political change

Definition

Disaster diplomacy examines how and why disaster-related activities (e.g., mitigation, prevention, response, and recovery) do and do not reduce conflict and support peace.

Overview

Disaster diplomacy research and application (see http://www.disasterdiplomacy.org) examines how and why disaster-related activities do and do not reduce conflict and support peace. Disaster-related activities include pre-disaster actions, such as prevention, mitigation, preparedness, and planning. Disaster-related activities also include post-disaster actions, such as response, recovery, and reconstruction. The range and scale of investigation and use is wide, including countries at war with each other to communities suffering internal conflict.

This field is investigated mainly through scientific research seeking policy recommendations, but practitioners have also advanced and used disaster-diplomacy knowledge (e.g., Renner and Chafe, 2007). Approaches take mainly two forms: specific case studies and overall analyses seeking patterns and predictions.

Case study examples include India-Pakistan after the 2001 Gujarat and 2005 Kashmir earthquakes, collaboration across the Middle East on seismic building codes, the internal conflicts in Sri Lanka and Aceh (Indonesia) after the 2004 tsunamis, and Cuban and American scientists collaborating on hurricane prediction and monitoring.

The case studies examined indicate that disaster-related activities do not create new approaches to conflict resolution. If, however, ongoing connections amongst antagonists exist on a solid basis, disaster-related activities in the short term can be the catalyst to turn those connections into hopes for peace. Examples of connections are peace negotiations, the desire for trade links, or a topic of common concern such as pollution or another enemy.

Over the long-term, non-disaster factors tend to overtake any disaster-related influences on peace. Examples are leadership changes, prominence of an historical conflict or grievance, or continuing distrust. As a result, disaster-related activities can exacerbate conflict, especially when a disaster has raised expectations of peace immediately following the calamity.

Work on disaster diplomacy also covers theories and patterns underlying the case study observations. Such work helps to understand why case studies do not yield disaster diplomacy. It further contributes to attempts at predicting what could occur in instances or circumstances under which disaster diplomacy might be more successful.

Explanations include why governments, organizations, or people involved in disaster diplomacy select specific pathways that try to promote or inhibit disaster-related reconciliation. Other approaches to developing typologies and explanations for disaster-diplomacy case studies cover the proximity of those involved in disaster diplomacy (e.g., if countries are involved, do they border each other?) along with their aid relationships (e.g., a donor-recipient situation or mutual aid).

The reason why disaster-related activities usually fail to support peace are complex, often involving the long histories of the parties involved and the personalities and personal interests of the current leaders. Overall, the evidence suggests that dealing with disasters appropriately and seeking peace are not always the first priorities of those involved. Consequently, disaster diplomacy has good intentions but is not likely to succeed often.

Bibliography

Renner, M., and Chafe, Z., 2007. *Beyond Disasters: Creating Opportunities for Peace.* Washington, DC: Worldwatch Institute.

Cross-references

Civil Protection and Crisis Management
Disaster Relief
Disaster Research and Policy History
Disaster Risk Reduction
Emergency Management
Federal Emergency Management Agency (FEMA)
Global Network of Civil Society Organisations for Disaster Reduction
Hyogo Framework for Action 2005–2015
Mitigation
Planning Measures and Political Aspects
Recovery and Reconstruction After Disaster
United Nations Organisations and Natural Disasters

DISASTER RELIEF

Jane Carter Ingram
Wildlife Conservation Society, Bronx, NY, USA

Synonyms

Disaster assistance; Disaster response

Definition

Disaster relief refers to interventions aimed at meeting the immediate needs of the victims of a disastrous event. Disaster relief commonly refers to aid that can be used to alleviate the suffering of national or foreign disaster victims. It often includes aid or assistance in the form of humanitarian services and transportation; the provision of food, clothing, medicine, medical services, beds, and bedding; temporary shelter and housing; and making repairs to essential services such as electricity, water supplies, and phone lines.

Discussion

Disaster relief, often commonly referred to as disaster response, is an intermediate phase of several phases in disaster management, which include (1) mitigation, (2) preparedness, (3) relief, or response, and, (4) recovery (the longest phase). Relief in this context comprises the range of actions that are necessary during the critical period immediately following a disaster when infrastructure has been damaged, communication lines have been destroyed, access to services such as electricity and water has been compromised, people have been injured and/or separated from their families, and victims' homes, assets, and/or livelihoods have been lost or damaged. Disaster relief entails ensuring that people's immediate needs such as food, shelter, and medical assistance are met, first and foremost. Relief efforts also include reconnecting severed social networks through helping people locate their family, friends, and/or loved ones if they have been separated from one another. In order to meet these basic physical and social needs, disaster relief operations may involve clearing rubble from damaged infrastructure, extinguishing fires, repairing damaged electrical lines, or reducing other potential hazards that have resulted from the disaster so that vital resources and services can be delivered efficiently to affected populations.

A major challenge during the disaster relief phase is ensuring that vulnerable people are kept safe, which often requires external support because local police or security forces may have been fragmented during a disaster. Security during the relief phase is of special concern in temporary shelters. For example, following the Indian Ocean tsunami of 2004, many people in Sri Lanka were housed in temples that were converted into temporary shelter for multiple families or were given tents in which to live until more permanent structures could be built in safe locations. Following Hurricane Katrina, thousands of people were temporarily housed in the Super Dome. While these locations may have provided relief from natural hazards, in both cases, the temporary housing options where men and women from different families were living in close quarters with each other presented security risks for vulnerable populations in society, particularly the elderly, women, and children. Furthermore, in the absence of police and security forces, looting, and other crimes may increase during the relief phase as people deal with anguish over their losses and are desperate to meet their basic needs. Thus, it is critical that relief efforts, especially those that place women, children, and the elderly in new housing situations are developed with a consideration of preexistent and new safety concerns that have arisen as a result of the disaster so that people's needs are met and their safety is not worsened by temporary conditions.

While the disaster "recovery" phase would ideally start as soon as immediate needs are met, the relief phase often lasts for quite a while due to funding delays or other constraints that hamper the recovery phase.

Bibliography

http://www.colorado.edu/hazards/resources/web/relief.html.
www.reliefweb.int/rw/dbc.nsf/doc100?OpenForm.
www.redcross.org.
http://www.ri.org/.

Cross-references

Casualties Following Natural Hazards
Civil protection and Crisis Management
Cognitive Dissonance
Coping Capacity
Critical Infrastructure
Disaster
Disaster Relief
Disaster Risk Management
Disaster Risk Reduction (DRR)
Education and Training for Emergency Preparedness
Emergency Management
Emergency Shelter
Evacuation
Federal Emergency Management Agency (FEMA)
Hospitals in Disaster
Human Impact of Disasters
Integrated Emergency Management System
International Strategies for Disaster Reduction
Livelihoods and Disasters
Natural Hazard
Posttraumatic Stress Disorder (PTSD)
Red Cross and Red Crescent
Risk
Risk Assessment

DISASTER RESEARCH AND POLICY, HISTORY

J. C. Gaillard[1], Ilan Kelman[2]
[1]The University of Auckland, Auckland, New Zealand
[2]Center for International Climate and Environmental Research – Oslo (CICERO), Blindern, Oslo, Norway

Definition

Disasters are not new phenomena. Disasters have affected humans throughout history and, consequently different forms of research (that evolved into the modern scientific method) as well as policies have been applied to deal with them. Overall, these approaches have moved from assuming a lack of control over disasters, to blaming external forces such as the "hazard" along with people affected by "hazards," to more recent approaches accepting that humans are responsible for creating the conditions that lead to disasters. To fully tackle disasters, research shows that policies should focus on reducing vulnerability which is created by humans, usually affecting others.

Introduction

Disasters have long been viewed principally as natural hazards through physical science disciplines such as seismology, volcanology, climatology, geomorphology, hydrology, and meteorology. Up until the mid-twentieth century, the responsibility for the occurrence of disasters that include physical phenomena (e.g., earthquakes or tornadoes) was therefore attributed to external natural forces or the whims of angry, vindictive, or inebriated deities.

In a 1756 letter to Voltaire regarding the 1755 Lisbon earthquake and tsunami, Rousseau was the first person known to attribute humanity's responsibility in disasters (e.g., Dynes, 1997). However, it was not until the 1940s that the human dimension of disasters began to be widely accepted. Two major research and policy paradigms developed, and were often opposed in their understanding of disasters and in the way that disaster risk should be reduced. The contemporary emphasis on climate change is a step backward in the approaches taken.

The hazard paradigm

The paradigm which has long dominated scientific studies of disasters emphasizes the importance of nature's threats, called natural hazards. This approach has been spearheaded by White's (1945) pioneering dissertation on people's adjustments to flooding in the USA. This paradigm is known as the hazard paradigm. Burton and Kates (1964, p. 413), two of White's students, define natural hazards as "*those elements in the physical environment, harmful to man and caused by forces extraneous to him.*" Frampton et al. (2000) further stress their "*uncontrollable dimension.*" Many more definitions of natural hazards similarly emphasize extreme (function of magnitude) and rare (function of time) natural phenomena that exceed humanity's ability to cope with them.

The "extraneous" and extreme dimension of natural hazards leads disasters identified with these natural phenomena to be considered out of the regular social fabric (e.g., Kates and Clark, 1996). Scientists, institutions, governments, and media thus often mention "extra-ordinary," "un-controllable," "in-credible," "un-predictable," and "un-certain" phenomena along with "un-expected" disasters and "un-scheduled" and "un-anticipated" damage (see Hewitt, 1983 for a critique). Regions affected are claimed to be unable to face such forces of nature and are often considered to be "under-developed," "over-populated," "un-informed," "un-prepared," and "un-planned" (again see Hewitt, 1983 for a critique). Therefore, a clear border is delineated between regions of the world which are often struck by disastrous events and those that are supposedly safe (see Bankoff, 2001 for a critique).

In this context, earth and climate scientists and engineers tend to focus on monitoring, predicting, and calculating probabilities and parameters for extreme natural events. In contrast, social scientists are interested in how people and societies perceive the potential danger and how they adjust to possible threats. Individuals and societies said to have a low perception of risk allegedly adjust poorly to possible threats. People and societies considered to have a high risk perception are assumed to adjust well to natural hazards (Kates, 1971; Burton et al., 1978; Fischhoff et al., 1978; Slovic, 1987). Factors that affect people's perception of risk are hazard related too (i.e., hazard magnitude, duration, frequency and temporal spacing, plus the recentness, frequency, and intensity of past personal experiences with hazards). Kates (1971, p. 441) underlines this held belief that those factors are independent from the socioeconomic environment.

In many countries, disaster policies and practices reflect the influence of the hazard paradigm. These policies are primarily geared toward the extreme dimension of natural phenomena and often reflect war strategies (Gilbert, 1995). In many countries, disaster policies are handled by the army or civil protection institutions, relying on military chains of command, and treating natural hazards as enemies to fight against (Alexander, 2002). Risk reduction strategies consequently tend to focus on technocratic, command-and-control measures such as engineering structures, technology-based warning systems activated only after a natural event, hazard-based land-use planning, and hazard-based risk awareness campaigns.

Internationally, disaster risk reduction policies have long relied on contrasting safe, affluent countries with dangerous, poor countries. That fits into wider development policies which foster top-down transfers of knowledge, technology, and experience from the rich to the poor, because the poor countries are allegedly unable to cope without external assistance (Bankoff, 2001; de Waal, 1997).

The vulnerability approach

In the 1970s, there was a critical evolution in the way disasters have been considered and faced (Waddell, 1977; Torry, 1979). Drawing on cases from around the world, scholars such as O'Keefe et al. (1976), Wisner et al. (1977) and Hewitt (1983) increasingly emphasized people's vulnerability in the face of natural hazards, an approach referred to as the "vulnerability paradigm."

Vulnerability in facing natural hazards reflects people's marginalization within society. Disaster-affected people disproportionately include those who are chronically marginalized in daily life (Wisner, 1993; 2004). These people are marginalized geographically because they live in hazardous places (e.g., informal settlers); socially because they are members of minority groups (e.g., ethnic or caste minorities, people with disabilities, prisoners, and refugees); economically because they are poor (e.g., homeless, underemployed and jobless); and politically because their voice is disregarded (e.g., women, non-heterosexuals, children, and elderly). People's vulnerability varies in time and space and is determined by mainly hazard-independent, structural constraints that are social, cultural, economic, and political (Watts and Bohle, 1993; Wisner et al., 2004; Gaillard, 2007).

Disasters thus most affect individuals with limited and fragile incomes (low wages, informal jobs, lack of savings), reducing their capability to deal with natural hazards (e.g., location of home, type of housing, knowledge of mitigation measures). Vulnerability and marginality also result from inadequate social protection (e.g., health insurance, health services, construction rules, prevention measures) and limited solidarity networks.

That does not necessarily indicate that means of protection are unavailable locally. In many instances, such as for famines (Hartmann and Boyce, 1983; Sen, 1983), the lack of access does not reflect the availability of food, knowledge, technologies, or financial capital, but rather an unequal distribution of available resources and the nature, strength, and diversity of people's livelihoods. Assets and resources essential in the sustainable or unsustainable livelihoods are conversely crucial in defining vulnerability.

Such an intimate relationship between livelihood and vulnerability supports the justification that many people have no other choice but to face natural hazards to sustain their daily needs. The difficulty of accessing sustainable livelihoods may further lead to environmental degradation, which often manifests in increasing natural hazards. For instance, the need for firewood may accelerate deforestation that in turn exacerbates the effects of landslides and floods.

People's incapacity to safely endure "natural hazards" therefore results from their inability to control their daily life and to choose the location of their home and their livelihoods (Blaikie, 1985). In that context, disasters highlight or amplify people's daily hardship and everyday emergencies (Baird et al., 1975; Maskrey, 1989). Thus, disastrous events cannot be considered as "accidents" beyond the usual functioning of the society (Hewitt, 1983; Wisner, 1993). Instead, disasters generally reflect development failures where the root causes of vulnerability have origins in other, usually contextual, development-related crises.

Some of the central ideas of the vulnerability paradigm have been progressively integrated into some, but not all, international policy documents such as United Nations strategies (United Nations, 1995; United Nations International Strategy for Disaster Reduction, 2005). Numerous NGOs focusing on development and disaster risk reduction have also adopted aspects from the vulnerability paradigm, translating them into sound policies (e.g., Global Network of Civil Society Organisations for Disaster Reduction, 2011). Conversely, some more powerful institutions like the International Monetary Fund (IMF) have not yet integrated the basic concepts of the vulnerability paradigm (Freeman et al., 2003), while other organizations such as the World Trade Organization (WTO) downplay the negative effects of disasters (World Trade Organization, 2006).

Community-based disaster risk reduction

Despite the advantages of the vulnerability approach, no single solution can address every situation. However, a general concept does seem to apply universally; reducing disaster risk requires increasing participation of local communities, a concept that has for a long time been encouraged in development research, policy, and practice (e.g., Chambers, 1983; Wisner, 1995). The actions of local communities are nearly always the first line of defense in reducing disaster risk. A wide consensus acknowledges the capacities of local communities in dealing with natural hazards on their own, as long as they are empowered with adequate resources (e.g., Quarantelli and Dynes, 1972; Delica-Willison and Willison, 2004). Community-Based Disaster Risk Reduction (CBDRR) fosters the participation of threatened communities in the evaluation and reduction of risk (including hazards, vulnerabilities, capacities, and resiliencies).

CBDRR empowers communities with self-developed and culturally, socially, and economically acceptable ways of coping with and avoiding crises related to natural hazards (e.g., Anderson and Woodrow, 1989; Maskrey, 1989). CBDRR enhances endogenous resources which prevent people from resorting to exogenous means that are often hard to access and which often create a cycle of dependency. CBDRR further aims at strengthening people's livelihoods to enable local communities to live with risk on an everyday basis (Benson et al., 2001; Cannon et al., 2003; Twigg, 2004), thus favoring the integration of disaster risk reduction into development policy and planning. It is often impossible to prevent people from settling in hazardous areas because these same locations provide resources to communities on a daily basis, such as fertile agricultural land that lies on floodplains and low-lying coastal zones with fisheries. Focusing on livelihoods simultaneously when considering disaster risk reduction will address people's ability to sustain their daily needs and their capacities to face natural hazards and other development threats.

CBDRR is also increasingly promoted among local governments and scientific communities in order to strengthen the links between top-down and bottom-up disaster-related measures (Kafle and Murshed, 2006) and to facilitate their integration into wider development policy frameworks. Top-down responsibility is not absolved by CBDRR, but instead should support, rather than substitute for, the capacities of communities. Local communities should indeed be externally assisted when large-scale measures are required, such as massive evacuations over long distances, regional settlement and infrastructure planning, and considerable debris cleaning. CBDRR accepts external assistance when appropriate on the *community's terms*, thereby also assisting in avoiding one community's measures creating or exacerbating problems for other communities.

CBDRR and other locally driven measures are crucial for development policy; however, these should be backed by a strong commitment from national and international institutions, especially those which have not yet incorporated the central arguments of the vulnerability paradigm (e.g., IMF, WTO). Too many disasters are rooted in governance issues, for instance, disobedience of laws such as building codes, corruption, the misuse of available resources, and the looting of natural and economic resources to benefit the most powerful (e.g., Lewis, 2008). In most cases, simple, affordable, and locally available measures as part of development policy would remediate such concerns and avoid disasters (Lewis, 1999; Hewitt, 2007).

Climate change and conceptual regressing?

The dominant hazard paradigm now seems to be regaining ground, fuelled by the media, political, and scientific discourse on climate change (Kelman and Gaillard, 2008). Uncertainties around the evolution of climate conditions constitute a powerful argument for the proponents of the hazard paradigm for considering Nature as the major threat (e.g., White, 2004). Uncertainties are frequently associated with the probability of occurrence of rare and extreme natural hazards which should be addressed through scientific models and statistical probabilities. The contemporary focus on climate change thus reinforces a paradigm where Nature is the danger source (even if exacerbated by human activity, as with climate change and many other hazards).

The recent interjection of climate change into disaster research is evident in the latest report of the Intergovernmental Panel on Climate Change (IPCC) (International Panel on Climate Change, 2007). In this IPCC document, vulnerability is defined as *"the degree to which a system is susceptible to, and unable to cope with, adverse effects of climate change, including climate variability and extremes. Vulnerability is a function of the character, magnitude, and rate of climate change and variation to which a system is exposed, its sensitivity, and its adaptive capacity."* This definition emphasizes extreme events and dependence on the climate systems, especially the magnitude of change to be experienced, which are both characteristics of the hazard paradigm.

In the face of such changing threats, people must change to deal with them. The analogy with the hazard paradigm is obvious. The vocabulary has shifted from adjustment to adaptation, but the meaning and scientific justification are the same as emphasized in the definition of adaptation provided in IPCC (2007): *"adjustment in natural or human systems in response to actual or expected climatic stimuli or their effects, which moderates harm or exploits beneficial opportunities."* As for the hazard paradigm, people's ability to cope with changes in climate patterns have largely been overlooked, even though their response is constrained by the same development-related factors underpinning vulnerability (Bohle et al., 1994).

Such a scientific discourse on climate change and other disasters, highly emphasized by some media, distracts national governments and international institutions from the root causes of vulnerability, making climate change a perfect scapegoat for disasters and lack of development (Kelman and Gaillard, 2008). Pinpointing a phenomenon with global scale and diffused responsibility enables governments to evade their own responsibility in addressing the root causes of vulnerability.

Therefore, there is no surprise that climate change policies and practices have reintroduced most of the actions that have failed to mitigate the occurrence of disasters over the last 60 years.

Outlook

Beyond the different, often opposing paradigms on the origins of disasters and actions to be undertaken to reduce disaster risks, there is a crucial need for increasing dialogue between the proponents of all approaches and those most affected by disasters. The top-down approach of the hazard paradigm has obviously failed to mitigate the occurrence of disasters. Consequently, it is unlikely that the similar paradigm emerging from the discourse on climate change will achieve better results.

Conversely, bottom-up actions rarely suffice to tackle the root causes of vulnerability on a large scale. Therefore, a need exists for better integration of top-down and bottom-up approaches into a single paradigm that involves working collaboratively toward the same goal, to prevent disasters (Hewitt, 2007).

Bibliography

Alexander, D., 2002. From civil defense to civil protection and back again. *Disaster Prevention and Management*, **11**(3), 209–213.

Anderson, M. B., and Woodrow, P., 1989. *Rising from the Ashes: Development Strategies in Times of Disasters*. Boulder: Westview Press.

Baird, A., O'Keefe, P., Westgate, K., and Wisner, B., 1975. *Towards An Explanation and Reduction of Disaster Proneness*. Bradford: Disaster Research Unit-University of Bradford. Occasional Paper No. 11.

Bankoff, G., 2001. Rendering the world unsafe: 'vulnerability' as western discourse. *Disasters*, **25**(1), 19–35.

Benson, C., Twigg, J., and Myers, M., 2001. NGO initiatives in risk reduction: an overview. *Disasters*, **25**(3), 199–215.

Blaikie, P., 1985. *The Political Economy of Soil Erosion in Developing Countries*. New York: Longman.

Bohle, H. G., Downing, T. E., and Watts, M. J., 1994. Climate change and social vulnerability: toward a sociology and geography of food insecurity. *Global Environmental Change*, **4**(1), 37–48.

Burton, I., and Kates, R. W., 1964. The perception of natural hazards in resource management. *Natural Resources Journal*, **3**, 412–441.

Burton, I., Kates, R. W., and White, G. F., 1978. *The Environment as Hazard*. New York: Oxford University Press.

Cannon, T., Twigg, J., and Rowell, J., 2003. *Social Vulnerability, Sustainable Livelihoods and Disasters*. London: Conflict and Humanitarian Assistance Department and Sustainable Livelihoods Support Office, Department for International Development.

Chambers, R., 1983. *Rural Development: Putting the Last First*. London: Longmans.

De Waal, A., 1997. *Famine Crimes: Politics and the Disaster Relief Industry in Africa*. Oxford, Bloomington: James Currey, Indiana University Press.

Delica-Willison, Z., and Willison, R., 2004. Vulnerability reduction: a task for the vulnerable people themselves. In Bankoff, G., Frerks, G., and Hilhorst, D. (eds.), *Mapping Vulnerability: Disasters, Development and People*. London: Earthscan, pp. 145–158.

Dynes, R. R., 1997. The Lisbon earthquake in 1755: contested meanings in the first modern disaster. Delaware: Disaster Research Center. Preliminary paper No. 255

Fischhoff, B., Slovic, P., Lichtenstein, S., Read, S., and Combs, B., 1978. How safe is safe enough? A psychometric study of attitudes towards technological risks and benefits. *Policy Sciences*, **9**(2), 127–152.

Frampton, S., McNaught, A., Chaffey, J., and Hardwick, J., 2000. *Natural Hazards*, 2nd edn. London: Hodder &Stoughton.

Freeman, P. K., Keen, M., and Mani, M., 2003. *Dealing with Increased Risk of Natural Disasters: Challenges and Options*. Washington: International Monetary Fund. IMF Working Paper 03/197

Gaillard, J.-C., 2007. De l'origine des catastrophes: phénomènes extrêmes ou âpreté du quotidien? *Natures Sciences Sociétés*, **15**(1), 44–47.

Gilbert, C., 1995. Studying disaster: a review of the main conceptual tools. *International Journal of Mass Emergencies and Disasters*, **13**(3), 231–240.

Global Network of Civil Society Organisations for Disaster Reduction, 2011. *If We Do Not Join Hands: Views from the Frontline 2011*. Teddington: Global Network of Civil Society Organisations for Disaster Reduction.

Hartmann, B., and Boyce, J. K., 1983. *A Quiet Violence: View from a Bangladesh Village*. London: Zed Books.

Hewitt, K., 1983. The idea of calamity in a technocratic age. In Hewitt, K. (ed.), *Interpretation of Calamities*. Boston: Allen & Unwin. The Risks and Hazards, Vol. 1, pp. 3–32.

Hewitt, K., 2007. Preventable disasters: addressing social vulnerability, institutional risk, and civil ethics. *Geographische Rundschau International Edition*, **3**(1), 43–52.

International Panel on Climate Change, 2007. *IPCC Fourth Assessment Report*. Geneva: Intergovernmental Panel on Climate Change.

Kafle, S. K., and Murshed, Z., 2006. *Community-Based Disaster Risk Management for Local Authorities*. Bangkok: Asian Disaster Preparedness Center.

Kates, R. W., 1971. Natural hazard in human ecological perspective: hypotheses and models. *Economic Geography*, **47**(3), 438–451.

Kates, R. W., and Clark, W. C., 1996. Environmental surprise: expecting the unexpected. *Environment*, **38**(2), 6–11. 28–33.

Kelman, I., and Gaillard, J.-C., 2008. Placing climate change within disaster risk reduction. *Disaster Advances*, **1**(3), 3–5.

Lewis, J., 1999. *Development in Disaster-Prone Places: Studies in Vulnerability*. London: IT Publications.

Lewis, J., 2008. The worm in the bud: corruption, construction and catastrophe. In Bosher, L. (ed.), *Hazards and the Built Environment: Attaining Built-in Resilience*. London: Routledge, pp. 238–263.

Maskrey, A., 1989. *Disaster Mitigation: A Community Based Approach*. Oxford: Oxfam. Development Guidelines, Vol. 3.

O'Keefe, P., Westgate, K., and Wisner, B., 1976. Taking the naturalness out of natural disasters. *Nature*, **260**(5552), 566–567.

Quarantelli, E. L., and Dynes, R. R., 1972. When disaster strikes: it isn't much like what you've heard & read about. *Psychology Today*, **5**(9), 66–70.

Sen, A. K., 1983. *Poverty and Famines: An Essay on Entitlement and Deprivation*. Oxford: Oxford University Press.

Slovic, P., 1987. Perception of risk. *Science*, **236**, 280–285.

Torry, W. I., 1979. Hazards, hazes and holes: a critique of the environment as hazard and general reflections on disaster research. *The Canadian Geographer*, **23**(4), 368–383.

Twigg, J., 2004. *Disaster Risk Reduction: Mitigation and Preparedness in Development and Emergency Programming*. London: Humanitarian Practice Network. Good Practice Review, Vol. 9.

United Nations, 1995. *Stratégie et plan d'action de Yokohama pour un monde plus sûr: directives pour la prévention des catastrophes naturelles, la préparation aux catastrophes et l'atténuation de leurs effets – Conférence mondiale sur la prévention des catastrophes naturelles – Yokohama, Japon, 23-27 mai 1994*. New York/Geneva: United Nations.

United Nations Development Programme, 2008. *Human Development Report 2007–2008 – Fighting Climate Change: Human Solidarity in a Divided World*. New York: United Nations Development Programme.

United Nations International Strategy for Disaster Reduction, 2005. *Building the Resilience of Nations and Communities to Disaster: An Introduction to the Hyogo Framework for Action*. Geneva: United Nations International Strategy for Disaster Reduction.

Waddell, E., 1977. The hazards of scientism: a review article. *Human Ecology*, **5**(1), 69–76.

Watts, M. J., and Bohle, H. G., 1993. The space of vulnerability: the causal structure of hunger and famine. *Progress in Human Geography*, **17**(1), 43–67.

White, G. F., 1945. *Human Adjustment to Floods: A Geographical Approach to the Flood Problem in the United-States*. Chicago: Department of Geography-University of Chicago. Research Paper No. 29.

White, R. R., 2004. Managing and interpreting uncertainty for climate change risk. *Building Research & Information*, **32**(5), 438–448.

Wisner, B., 1993. Disaster vulnerability: scale, power, and daily life. *GeoJournal*, **30**(2), 127–140.

Wisner, B., 1995. Bridging "expert" and "local" knowledge for counter disaster planning in urban South Africa. *GeoJournal*, **37**(3), 335–348.

Wisner, B., O'Keefe, P., and Westgate, K., 1977. Global systems and local disasters: the untapped power of peoples' science. *Disasters*, **1**(1), 47–57.

Wisner, B., Blaikie, P., Cannon, T., and Davis, I., 2004. *At Risk: Natural Hazards, People's Vulnerability, and Disasters*, 2nd edn. London: Routledge.

World Trade Organization, 2006. *World Trade Report: Exploring the Links Between Subsidies, Trade and the WTO*. Geneva: World Trade Organization.

Cross-references

Climate Change
Disaster
Disaster Risk Reduction
Exposure to Natural Hazards
Global Change and Its Implications for Natural Disasters
Hazard
Hazardousness of Place
Vulnerability

DISASTER RISK MANAGEMENT

N. Nirupama
York University, Toronto, ON, Canada

Definition

Disaster risk management is a comprehensive approach involving the identification of threats due to hazards; processing and analyzing these threats; understanding people's vulnerability; assessing the resilience and coping capacity of the communities; developing strategies for future risk reduction; and building up capacities and operational skills to implement the proposed measures.

Introduction

Disaster risk cannot be eliminated completely, but it can be assessed and managed in order to reduce the impact of disasters (Smith and Petley, 2009). The management of disaster risks has attracted much attention since the 2005 initiative of the International Strategy for Disaster Reduction (ISDR, 2004) that defined the Ten Essentials in order to empower local governments and other agencies to implement the Hyogo Framework for Action until 2015. In the twenty-first century, our understanding of disasters that are caused by natural, technological, and/or human sources has improved significantly. Both the developing and the developed worlds have made considerable progress, within their capacity and limitations, toward the development of policies and mitigation measures to reduce future disasters. However, disasters continue to harm millions of people each year worldwide. A disaster can affect, or be affected by our natural environment, social processes, psychological elements, cultural issues, historical information, and political and economic ideologies. Certain risks are often inherent within a social system or physical location, but they can also be created due to certain natural or technological hazards (Alexander, 1999). The consequences, however, can be similar in that they wreak havoc in communities and destroy social and economic systems. In order to effectively and efficiently manage disaster risks, our focus should be on addressing vulnerability (The conditions determined by physical, social, economic and environmental factors or processes, which increase the susceptibility of a community to the impact of hazards (ISDR, 2004)) and improving the resilience and coping capacity of populations (Nirupama, 2009, 2012; Twigg, 2007; Canton, 2007; Cutter, 2001).

Risk identification is a measure of individual perception – how those perceptions are understood by society as a whole, as well as an objective assessment (Cardona, 2006). The holistic approach of disaster risk management would involve risk identification and risk reduction components, a disaster management component, which is about response and recovery; and a financial protection piece that will account for institutional support, financial resources, and risk transfer tools. The shaping of risk identification, risk reduction, and risk management strategies, policies, resource allocation, and operational plans should, ideally, engage all the stakeholders in the process. A risk management team must have adequate information and understanding of high-probability/low-consequences versus low-probability/high-consequences events. A number of risk management strategies, such as education, awareness, economic incentives for individual mitigation measures, as well as legal, and legislative requirement can be considered. The process can be challenging as transfer of knowledge from science to politics is not easy (Schneider et al., 2006).

Understanding risk

Risk is defined as a function of probability of occurrence of hazardous event, and potential loss to people, property, and/or the environment (Smith, 2004; Wisner et al., 2004; ISDR, 2004; HRVA, 2004) as shown in Eq. 1. Historical records of past disasters provide reasonable estimates of the probability of occurrence of hazards, hence risk is considered to be quantifiable using probabilities and consequences (Helm, 1996; Green, 2004; Smith and Petley, 2009). Information on vulnerable populations and elements that are particularly exposed to risk can be assessed using a variety of indicators and criteria (Birkmann, 2006; Armenakis and Nirupama, 2009, 2012). Risk perception also plays a significant role in how disaster risk management is carried out in various societies and cultures (Slovic, 2000). Therefore, perception becomes a noteworthy factor to be accounted for in risk management, and risks can vary with geographic location and local conditions. Figure 1 demonstrates various perceptions of risk.

The standard risk formula is expressed as:

$$R = H \times V \qquad (1)$$

Here, R = risk; H = hazard, determined as a probability (or likelihood) of the occurrence of hazard; V = vulnerability (also loss, impact or consequences).

Several variations of standard risk formula have been proposed by experts (Table 1) and are as much practiced as the standard risk formula given in Eq. 1.

Key elements of disaster risk management

Key elements of a comprehensive disaster risk management system are shown in Figure 2. Each element is briefly explained below.

Threat recognition – risk and vulnerability identification

Identifying potential risks from natural, technological, or human-induced hazards; and recognizing vulnerable populations, such as very old, very young, single parents with young children, low income earners, unemployed, those facing language barriers, and physically and emotionally challenged people.

Risk analysis and assessment

Understanding the magnitude, frequency of occurrence, and severity of consequences and prioritization of risks. The standard risk formula is given in Eq. 1. A few risk evaluation methods are discussed here.

Qualitative and quantitative frameworks and methods have been developed to understand and evaluate disaster risk. Qualitatively speaking, all individual/institutional perceptions of risk carry equal weight as they choose to respond in a certain manner to a certain threat in certain circumstances (Nirupama and Etkin, 2009). Among the qualitative models, *Pressure and Release* (*PR*) *and Access to Resources* (*AR*) models (Wisner et al., 2004) are widely used. The *PR* is a static model, founded on the concept of progression of vulnerability by looking at how underlying causes create an environment that allows for some dynamic pressures (e.g., lack of education, land degradation, population growth) to translate into unsafe conditions (e.g., exposure to risk, lack of social network) in a given timeframe. Unlike the *PR*, the *AR* model is dynamic and community based. It focuses on access to

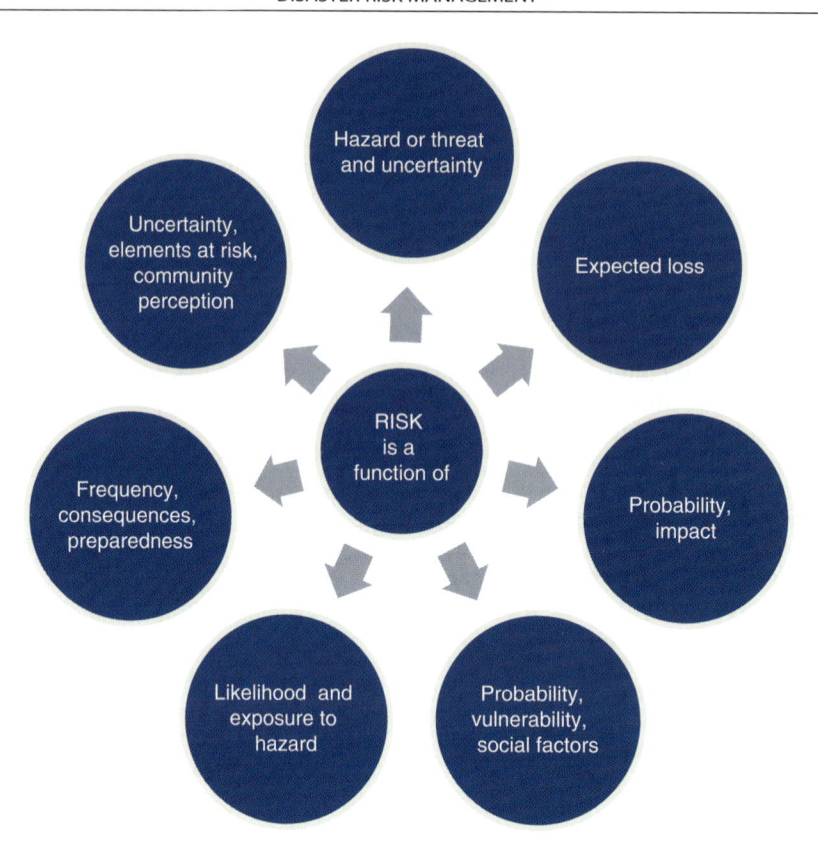

Disaster Risk Management, Figure 1 Various perceptions of hazard/disaster risk.

Disaster Risk Management, Table 1 List of disaster risk assessment approaches that are similar to the conventional approach as given in Eq. 1. Here, commonly known variables are: R = risk; p (or H) = probability; L = loss; V = vulnerability; I = impact.

Proposed risk evaluation equation	Variable other than probability and impact	Expert(s)
$R = p \cdot L^x$	$x\ (>1)$ = people's perception	Whyte and Burton (1982)
$R = P \cdot S$	S = severity	Government of Michigan (2001)
$R = p \cdot V \cdot n$	n = social consequences	Ferrier and Haque (2003)
$Risk = \frac{H \cdot L}{preparedness(mitigation)}$	Preparedness or mitigation are measurable measures	Smith (2004)
$R = p \cdot L \cdot f(x)$	$f(x)$ = risk aversion factor	Schneider et al. (2006)
$R = H \cdot V \cdot M$	M = manageability or ability of humans	Noson (2009)
$R = H \cdot Elements\ at\ Risk \cdot V$	Elements at risk = physically exposed assets	Smith and Petley (2009)
$R = H \cdot (V \cdot cp)$	cp = community perception	Nirupama (2012)

income opportunities, and the development of coping strategies during and after a disaster.

In most quantitative risk assessment methods, two variables – probability of the occurrence of hazards and their potential impact – are commonly used. A few methods are discussed here. The HRVA (Hazard, Risk, and Vulnerability Analysis) method (HRVA, 2004) of BC, Canada evaluates disaster risk based on event likelihood; assessment of vulnerability (social, physical, economic, and environmental) and severity of consequences (fatality, injury, damage, and disruption of essential services – water, electricity, communication networks, physical, and economic impact). Although the HIRA (Hazard Identification and Risk Assessment) (HIRA, 2011) of Ontario, Canada follows similar steps: Hazard identification, risk assessment, risk analysis, and monitoring/review for future revisions, it accounts for psychosocial factors, such as panic and hoarding behavior, in assessing disaster impacts.

The FEMA (Federal Emergency Management Agency) model (FEMA) was developed in the USA to provide guidance to the nation for planning and decision making

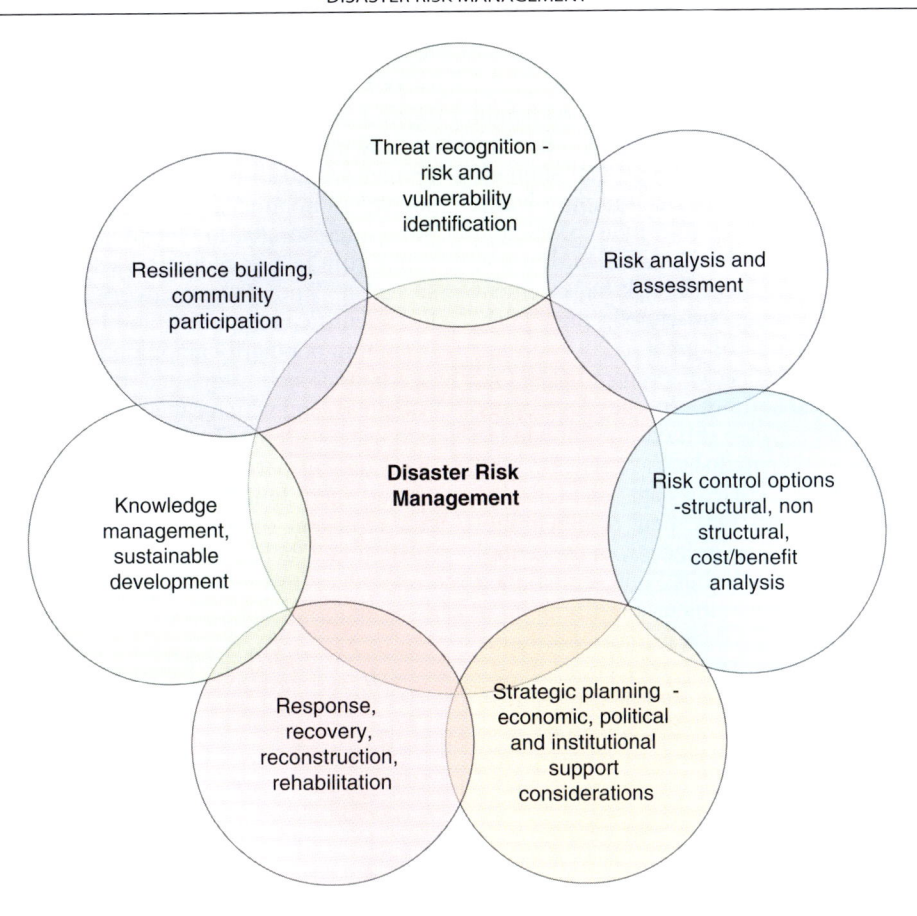

Disaster Risk Management, Figure 2 Elements of comprehensive risk management.

during disaster management through the use of mitigation. The model accounts for threat identification and rating, assessment of assets, vulnerability, risk, and mitigation options. NOAA (National Oceanic and Atmospheric Administration)'s Geographic Information System (GIS)-based vulnerability assessment tool identifies opportunities beyond the existing built environment for reducing future hazard vulnerability and identifies the large tracts of undeveloped land in communities that can be used for future land-use planning for sustainable growth.

The SMUG (Seriousness, Manageability, Urgency, and Growth) (CDEMG, 2005) model was developed by the Civil Defence Unit of Chatham Islands Council of New Zealand. The model describes the prioritization of potential hazard risks based on four criteria: Seriousness (number of lives lost, potential for injury; physical, social, and economic consequences), Manageability (ability to mitigate, both hazard and vulnerability), Urgency (measure of capability to address the hazard), and Growth (rate at which hazard risk will increase through either an increase in the probability of occurrence, in the exposure of the community, or combination of the two); and four R's (Reduction, Readiness, Response, and Recovery).

In less developed regions such as Latin America and the Caribbean and Asian countries, national governments and NGOs usually play a pivotal role in managing disasters. The concept of risk evaluation, however, is similar to that of shown in Eq. 1 and risk assessment methodologies are similar to the ones used in developed world. In an ideal disaster risk management, a hazard and vulnerability analysis would be carried out and then appropriate action would be taken based upon the analysis (NDM, 2012).

Risk control options – structural, nonstructural, cost/benefit analysis

These considerations are based on feasibility, effectiveness, and cost/benefit analysis. Structural measures may include the building of dykes, dams, and other protective structures. Nonstructural measures may include land-use planning, hazard risk zoning, early warning systems, education and awareness campaigns, affordable disaster insurance, and legal and regulatory policy. Market-like tools, such as reinsured catastrophe funds (Mexico) and mitigation-focused insurance schemes (Barbados), have been implemented in a few countries (Freeman et al., 2002).

Strategic planning – economic, political, and institutional support considerations

Financial commitment and political will are fundamental to any successful disaster management program. The allocation of resources, the building of institutional support, the creation of social programs, and community-based initiatives toward individual and collective protection measures are most important. In North America, Europe, and other developed countries, disaster risk management programs are well established, structured, and fairly funded. These regions also have great early warning systems in place, remarkable disaster preparedness, and response and recovery capabilities. In the developing world, however, the focus has shifted to knowledge dissemination, disaster preparedness awareness, and community-based programs. For example, in India, the authorities at the state level take the main responsibility for disaster relief with financial assistance from the central government. A small Calamity Relief Fund, constituted with both state and central government contributions is managed by the Disaster Management Authority of India, under the Ministry of Home Affairs (Freeman et al., 2002). In case of a major disaster, the central government provides predetermined reimbursement sums for loss of life, limb, and partial and total loss of housing and productive assets.

Response, recovery, reconstruction, and rehabilitation

Response capability and mutual agreement with neighboring regions (depending on the size and type of the event), assistance with recovery, and reconstruction would be extremely important for the impacted communities to deal with their loss and remain optimistic about their future. The rehabilitation phase provides a rare opportunity to reassess the situation, consider various options to relocate or build a better, stronger, and more resilient community. Disaster aid – internal and/or international, bilateral (government to government or through NGOs) or multilateral (through the UN agencies) must be in place to reduce the impact of a disaster. The Government of India, in partnership with the United Nations Development Program (GOI-UNDP, 2008, 2010), has developed a Disaster Risk Management Programme through disaster preparedness and vulnerability reduction. Their goal is to strengthen institutional capacity with specific emphasis on women and other marginalized groups. They have adopted a multi-hazard approach with an objective of achieving a sustainable disaster risk reduction in some of the most hazard-prone districts in selected states in India. Another example is from Fiji, where exposure to cyclones, floods, droughts, earthquakes, and tsunamis is widespread. Fiji has been able to develop a good disaster preparedness, response, and recovery plan in which NGOs are encouraged to actively participate in all the functions of disaster risk management (Freeman et al., 2002).

Knowledge management and sustainable development

Institutional knowledge must be preserved for better learning and understanding. An approach of sustainable development would allow for the use of local resources (human, social, environmental) and thus contributes to local economy. Interestingly, in developing nations, NGOs play an active role in risk reduction activities in the region. The so-called "knowledge network" involving civil society, the scientific community, and to some extent, the market is gaining popularity among people in India.

An approach suggested by Cardona (2006) for the Americas, and which can also be applied to other regions, is to use a system of indicators to measure a country's risk management performance. As shown in Eq. 2, the Risk Management Index (RMI) is based on a set of indicators that represent organization, development, capacity, and institutional actions taken to reduce vulnerability and losses, to prepare for crisis, and to recover efficiently from disasters.

$$RMI = \frac{(RMI_{RI} + RMI_{RR} + RMI_{DM} + RMI_{FP})}{4} \quad (2)$$

where

RMI_{RI} = risk identification, includes objective and perceived risks
RMI_{RR} = risk identification, includes objective and perceived risks
RMI_{DM} = measures of response and recovery
RMI_{FP} = governance and financial protection measures

Resilience building and community participation

The final element in the cycle of disaster risk management is to work toward building resilient communities with community participation and community owned programs. For an effective and helpful risk management program, it is critical that communities make risk-based choices to address vulnerabilities and mitigate disaster impact. Resilience building must become the foundation of future risk management programs. A well-designed communication strategy can be instrumental in the successful implementation of policy and other measures. In Asian countries, a communities based holistic approach is gaining popularity as people feel responsible for their safe future (Padmanabhan, 2008).

Summary

Disaster risk management involves overall understanding and realization of potential hazards, identification of vulnerable people and property, risk evaluation, institutional support, and the adoption of a culture invested in preserving institutional knowledge. Various qualitative and quantitative methods can be used for risk assessment for the purpose of the development of a disaster risk management framework. The use of indicators to capture a sense of the central components in a holistic risk management process is worth examining. It is, however, safe to say that in recent years, most nations have shown an increasing trend toward developing comprehensive disaster management programs. They have broadened their national disaster management programs to encompass risk assessment, risk control, mitigation, preparedness, political will, economic feasibility, response, recovery, resilience building, and strategic and sustainable development activities. The success of such a framework or program may depend on the commitment of stakeholders such as communities, professionals, academics, and policy and decision makers.

Bibliography

Alexander, D., 1999. *Natural Disasters*. Dordrecht: Kluwer. 632p.

Armenakis, C., and Nirupama, N., 2012. Prioritization of disaster risk in a community using GIS, special issue on sociological aspects of natural disasters Springer. *Natural Hazards*, doi:10.1007/s11069-012-0167-8.

Armenakis, C., and Nirupama, N., 2009. Vulnerability assessment using GIS: Toronto propane explosion. In *Proceedings CRHNet Symposium*, November 23–26, Edmonton.

Birkmann, J., 2006. Indicators and criteria for measuring vulnerability: theoretical bases and requirements. In Birkmann, J. (ed.), *Measuring Vulnerability to Natural Hazards: Towards Disaster Resilient Societies*. Tokyo: United Nations University Press. 524p.

Canton, L. G., 2007. *Emergency Management: Concepts & Strategies*. Hoboken: Wiley.

Cardona, O. D., 2006. A System of Indicators for Disaster Risk Management in the Americas. In Birkmann, J. (ed.), *Measuring Vulnerability to Natural Hazards: Towards Disaster Resilient Societies*. United Nations University Press, pp. 189–209.

CDEMG, 2005. Chatham Islands Civil Defense Emergence Management Plan, http://www.cic.govt.nz/CDEMPlansReports.html. Accessed 30 March 2012.

Cutter, S., 2001. The changing nature of risks and hazards. In *American Hazardscapes: The Regionalization of Hazards and Disasters*. Washington, DC: Joseph Henry Press. 211p.

FEMA, 2012. http://www.fema.gov/library/viewRecord.do?id=1938. Accessed March 2012.

Ferrier, N., and Haque, C. E., 2003. Hazards Risk Assessment Methodology for Emergency Managers: A Standardized Framework for Application. *Natural Hazards*, **28**, 271–290.

Freeman, P. K., Martin, L. A., Linnerooth-Bayer, J., Mechler, R., Pflug, G., and Warner, K. 2002. *Disaster Risk Management: National Systems for the Comprehensive Management of Disaster Financial Strategies for Natural Disaster Reconstruction*, SDD/IRPD, Regional Policy Dialogue, Washington, DC: Inter-American Development Bank.

GOI-UNDP, 2008. Through participation of communities and local self governments. National Disaster Management Division, Ministry of Home Affairs, Government of India. http://www.ndmindia.nic.in/EQProjects/goiundp2.0.pdf. Accessed March 2012.

GOI-UNDP, 2010. Disaster risk management programme: evaluation and review of lessons learnt. Ministry of Home Affairs, Government of India.

Government of Michigan, 2001. The Michigan hazard analysis. *Michigan Emergency Management*. http://www.michigan.gov/msp/0,4643,7-123-1593_3507_8948-15248--,00.html. Accessed March 2012.

Green, C., 2004. The evaluation of vulnerability to flooding. *Disaster Prevention and Management*, **13**(4), 323–329.

Helm, P., 1996. Integrated risk management for natural and technological disasters. *Tephra*, **15**(1), 4–13.

HIRA, 2011. *Emergency Management Ontario*, Ministry of Community Safety and Correctional Services, Toronto.

HRVA, 2004. http://www.pep.bc.ca/hrva/toolkit.html. Accessed February 2010.

Hyndman, D., Hyndman, D., and Catto, N., 2008. *Natural Hazards and Disasters*. Toronto: Thomson Brooks/Cole. 526p.

ISDR, 2004. *Living with Risk: A Global Review of Disaster Reduction Initiatives*. United Nations, 429p, www.unisdr.org.

NDI National Disaster Management. 2012. Ministry of Home Affairs. India. http://www.ndmindia.nic.in/. Accessed March 2012.

Nirupama, N., and Maula, A., 2012. Engaging public for building resilient communities to reduce disaster impact, special issue on sociological aspects of natural disasters Springer. *Natural Hazards*, doi:10.1007/s11069-011-0045-9.

Nirupama, N., and Etkin, D., 2009. Emergency Managers in Ontario: An Exploratory Study of Their Perspectives. *Journal of Homeland Security and Emergency Management*, **6**(1). https://mymail.yorku.ca/horde/util/go.php?url=http%3A%2F%2Fwww.bepress.com%2Fjhsem%2Fvol6%2Fiss1%2F38&Horde=25004e320ce13ac50787d7192ca13e48_blank, http://www.bepress.com/jhsem/vol6/iss1/38

Nirupama, N., 2008. Disaster risk management: theory, practice and policy. *Workshop Report: Public Safety Canada*, Government of Canada, Ottawa, 149p.

Nirupama, N., 2009. Socio-economic implications based on interviews with fishermen following the Indian Ocean tsunami. *Natural Hazards*, **48**(1), 1–9.

Nirupama, N., 2012. Risk and vulnerability assessment – a comprehensive approach. *International Journal of Disaster Resilience in the Built Environment*, Emerald, **3**(2).

NOAA, 2012. http://unfccc.int/adaptation/nairobi_work_programme/knowledge_resources_and_publications/items/5340txt.php. Accessed March 2012.

Noson, L., 2009. Hazard mapping and risk assessment. Asian Disaster Preparedness Center, http://www.adpc.net.

Padmanabhan, G., 2008. Conference on Community-based Disaster Risk Reduction, UNICEF, Kolkata, November 26–28, 2008.

Schneider, Th., Basler, E., and Partner, A. G., 2006. A delicate issue in risk assessment. In Ammann, Dannenmann, and Vulliet (eds.), *RISK21 – Coping with Risks due to Natural Hazards in the 21st Century*.

Slovic, P., 2000. *The Perception of Risk*, 1st edn. London: Earthscan Publications.

Smith, K., and Petley, D. N., 2009. *Environmental Hazards: Assessing Risk and Reducing Disaster*, 5th edn. London: Routledge.

Smith, K., 2004. *Environmental Hazards: Assessing Risk and Reducing Disaster*. London: Routledge. 306p.

Twigg, J., 2007. Characteristics of a disaster-resilient community. *DFID Disaster Risk Reduction Interagency Coordination Group*, 39p.

Whyte, A. V., and Burton, I., 1982. Perception of risk in Canada. In Burton, I., Fowle, C. D., and McCullough, R. S. (eds.), *Living with Risk*. Toronto: University of Toronto, pp. 39–69.

Wisner, B., Blaikie, P., Cannon, T., and Davis, I., 2004. *At Risk: Natural Hazards, People's Vulnerability and Disasters*. London: Routledge. 471p.

Cross-references

Disaster
Disaster Research and Policy, History
Disaster Risk Reduction
Education and Training for Emergency Preparedness

Emergency Management
Emergency Planning
Expert (Knowledge-Based) Systems for Disaster Management
Hazard and Risk Mapping
Mitigation
Resilience
Risk
Risk Assessment
Risk Governance
Risk Perception and Communication
Vulnerability

DISASTER RISK REDUCTION

Walter J. Ammann
Global Risk Forum GRF Davos, Davos Platz, Switzerland

Synonyms

Disaster reduction and recovery; Integrative risk management; Risk reduction and disaster management

Definition

Disaster risk reduction (DRR) refers to a wide range of opportunities for risk abatement and disaster management. Risk reduction includes prevention, preparedness, and part of the recovery process, and it gives particular emphasis to the reduction of vulnerability, which is defined as "the conditions determined by physical, social, economic and environmental factors or processes, which increase the susceptibility of a community to the impact of hazards" (UNISDR, 2005) Disaster management includes warning, alert, emergency response, and part of recovery. It includes a focus on methods of increasing resilience. DRR aims to limit risks – assuming that they cannot be completely avoided – and concentrates on minimizing the adverse impacts on disasters. This needs to be accomplished within the broad context of sustainable development.

Introduction

The increasing world population, coupled with globalization and urbanization, has greatly increased the risks and impacts of disasters. Climate change and land degradation aggravate the situation in terms of intensity, occurrence, and complexity. Recent disasters, such as the Asian tsunami, Hurricane Katrina, the earthquake in Haiti, and influenza pandemics confirm the global reach of disasters and the tendency for their impacts to increase over time. Trends in risk management confirm that the world we live in today is more complex, more vulnerable, and more interdependent than at any time before in history (UNISDR, 2012). When settlements or infrastructure overlap with major hazard zones, natural events can cause significant damage. Natural hazards limit the availability of living space and thus incur social costs. Studies by the World Bank (World Bank, 2005; Global Facility, 2007) show that more than 3.5 billion people are located, and about 80% of the world's gross domestic product is produced, in areas exposed to at least one natural hazard with a significant probability of occurrence.

Over the last few years numerous catastrophes have drawn attention to the fact that the extent to which life, limb, and property can be protected is limited. The protection of life is certainly the primary concern, but also economic damage has to be reduced in order to protect vital economic growth, especially in developing countries, in which large disasters can absorb more than 10% of GDP. Sustainable development and poverty reduction go hand in hand with disaster risk reduction strategies to achieve the UN Millennium Development Goals (UN MDGs). Disasters and risks should no longer be seen as a purely humanitarian affair but as an integrative part of sustainable development and adaptation to climate change (UNISDR, 2009). In fact, adaptive capacity is considered a core characteristic of a resilient socioeconomic system (CRN, 2011, p. 39).

The World Conference on Disaster Risk Reduction held in Kobe, Japan, in January 2005 (WCDR, 2005) brought consensus that to achieve risk-resilient, sustainable societies, extreme events – such as natural hazards, climate change, diseases (including pandemics), man-made hazards and terrorism – have to be managed in an integrative way. As a result, the "Hyogo Framework for Action 2005–2015: Building the Resilience of Nations and Communities to Disasters" HFA (UNISDR, 2005) was approved by the 168 government representatives gathered in Kobe. Under the Hyogo Framework, governments committed themselves to the following five priorities:

- Make disaster risk reduction a priority: ensure that it is a national and local priority with a strong institutional basis for implementation. The creation of national platforms and national focal points is strongly encouraged.
- Know the risks and take early action: identify, assess, and monitor disaster risks, as these tasks are essential components of risk reduction that will enhance early warning.
- Build an understanding of awareness: use knowledge, innovation, and education to build a culture of safety and resilience at all levels.
- Reduce risk: identify those variables latent in society and the environment that contribute to risk and ways to mitigate them.
- Be prepared and ready to act: strengthen disaster preparedness for an effective response at all levels.

Subsequently, the UNISDR produced assessment reports on disaster risk reduction (e.g., UNISDR, 2011) and a policy discussion document on the way forward after the Hyogo initiative ends in 2015 (UNISDR, 2012). DRR is clearly going to benefit from integration with the Millennium Development Goals and the Rio + 20 resolutions. At the time of writing it is unclear what instruments will replace the five Hyogo goals and whether the succeeding initiative will be merely an extension of the current one or will involve new legal and administrative instruments to induce governments to reduce the risk of disasters.

Disaster risk reduction (DRR)

Disaster risk reduction has two components:

- *Risk reduction* refers to efforts to limit risks due to hazardous situations. This can be achieved by good prevention.
- *Disaster management* signifies the need to reduce or limit the resulting damages caused by a disaster. This can be achieved by good preparedness, an efficient disaster or crisis management system and an effective recovery process.

DRR is thus a process of both, risk reduction and disaster management and is sometimes called integrative risk management (IRM–Ammann, 2006). Besides risks due to natural hazard, which is aggravated by climate change, IRM includes numerous other risks to be considered simultaneously such as those of a technical, biological, and chemical nature; pandemics, terrorism, and financial risks.

DRR requires an approach that not only tackles multiple risks, but also involves multiple stakeholders. Although the HFA recognizes that governments have the primary responsibility to guide and implement measures for achieving DRR, to create the necessary political will at the national level, a wide group of risk management experts, practitioners, scientists, and key players from civil society and other sectors with a strong emphasis on implementation at "the last mile" has to be involved and has to interact with key players from line ministries and disaster management authorities. Practice, science, policy, and decision making have to be closely linked in the search for sustainable solutions to the complex risks society is facing today. Only an interdisciplinary approach can bridge the gap between problems and their main causes on the one hand, and governance and technology perspectives for problem solving on the other. Demand-driven, practical application has to supplant purely supply-driven scientific knowledge. The task of protecting people and private and public goods has to be the central focus of this knowledge development process, and it has to be achieved in a sustainable manner.

As climate change is aggravating the meteorological hazards in terms of frequency, intensity and interdependency, measures for *climate change adaptation (CCA)* have to be closely linked to programs for DRR (UNISDR, 2009). The *harmonization of DRR and CCA* measures is already a crucial issue. This must take place through a common process of adaptation to both the effects of climate change and the increasing impacts of disasters. Common strategies of vulnerability reduction are needed. For instance, in tropical coastal areas, settlement and livelihoods need to be made resistant not only to hurricane storm surges and tsunamis, but also to potential sea-level rise and the intensification of storms that climate change may bring. Coasts are very attractive areas for settlement and are in many cases the most economically buoyant parts of countries, rich and poor alike. However, it may be necessary to manage a retreat from the coast if the worst hazards are to be avoided or reduced. This will involve both costs and economic sacrifices.

Dealing with natural hazards is not just complex, but also contradictory when technical, social, economic, and ecological aspects have to be balanced. It is no longer adequate for risk management professionals to focus solely on risk within a particular realm. Rather, in a world with interdependent systems of rapidly growing complexity (such as critical infrastructures and interdependent processes and services), risk management must have a new vision that overcomes boundaries between subject areas, one that reaches across specialisms and departments. Safety and security have to be seen as a holistic means of enabling better planning, response, and reduction of the most pressing risks.

Integrative risk management, risk culture, and governance

The key questions are: How do we create a safer world and how can our developing knowledge support this process of change? The approach must be that of integrative risk management across subject areas, professions, and sectors, encompassing natural sciences, social sciences, and engineering. Scientific understanding must be placed at the service of business, policy responses, and citizen participation. Among the risk management communities, stronger ties have to be built with private-public partnership models, and approaches need to be devised to move toward a more truly integrative way of thinking about risk: a holistic approach to risk reduction with safety, security, and sustainability at the center. This is an approach that will help policy makers and business people, risk managers and civil society to address the complex risks around them more effectively.

To be able to take effective and efficient decisions for disaster risk reduction and climate change adaptation measures, which lead to transparent and comparable results in different risk situations, a consistent and systematic risk management approach has to be followed. Hereafter, this approach will be called "integrative risk management," a process that embodies a systematic framework for risk analysis and assessment procedures, that leads eventually to consistent decisions and to the optimized, integrative planning of risk reduction measures. A consistent risk concept provides a substantial base and allows the comparison of various risk scenarios at different locations and originating from different natural disasters. Hence, the key to the future is risk-based management, rather than an approach based solely on hazard management. A significant driving force for this paradigm shift is the demand for accountability and improved effectiveness of the risk reduction measures.

The public perception of natural hazards differs from the perception of ecological, technical, and social risks leading to conflicting security philosophies, which hinders consensus on integrative measures. Different ways in which people perceive risks have an important effect on how they may or may not accept any measures that are imposed. A strategy for protection from natural disasters has to find a way to put the various risks onto a common scale to allow for comparability and that serves as a platform from which measures can be agreed upon. Any risk to humans and the environment has to be considered within the context of social, financial, and economic consequences and increased interdependencies between the various risks. The way a society handles questions of safety and security may be summarized with the term "*risk culture*." This means that security can only be gained by risk-oriented thinking.

Risk governance looks at how risk-related decision making unfolds when a multitude of stakeholders and actors is involved, requiring coordination and possibly reconciliation between a profusion of roles, perspectives, goals, and activities. Good risk governance stands for transparency in decision making, effectiveness and efficiency of the measures, accountability, a strategic focus, sustainability, equity, fairness, respect for the law, and the need for the solution to be politically attractive and legally permissible, as well as ethically and publicly acceptable.

Integrative risk management and good risk governance are complicated by the fact that in today's society many risks are not isolated, single events with limited extent, but are trans-boundary risks that affect countries with different political systems and coping strategies.

Framework for DRR and CCA

The concept of integrative risk management (i.e., DRR) is shown in Figure 1. Integrative risk management starts with the process of identifying and analyzing risks in order to answer the question "What can happen?," followed by risk assessment, which should answer the question "Is what happens acceptable?," which leads in turn to the planning of risk reduction measures. The ultimate objective is to create protective measures. The main criterion for choosing the correct protective measures is cost-effectiveness. However, DRR (and CCA) have to overcome a number of problems and obstacles:

- The risk-oriented approach and the methodology of dealing with uncertainties may determine the solution rather than the risk itself. This applies both to the analysis and the assessment of risk.
- Measures designed to promote safety may have limitations that are greater than the expectations of safety held by civil society.
- The various points of view, attitudes, and values of all stakeholders involved and affected by the risk may differ and possibly conflict.
- Disaster risk prevention and mitigation measures have to take the whole set of pre- and post-disaster measures into consideration, as well as measures during a crisis itself, and measures to transfer risk using insurance (Figure 4).

Disaster Risk Reduction, Figure 1 Framework for integrative risk management (Source: author).

- All solutions have to fulfill the criteria of sustainability, that is, a sustainable approach to disaster risk management has to be a socially, economically, and environmentally balanced and acceptable approach.
- Integrative risk management also needs a strategic and systematic process of control, including the periodic evaluation of the risk situation and a comprehensive dialogue on risk between all stakeholders.
- When setting limits for the protection and defining the processes of decision making, there is a need for dialogue and communication in order to ensure the participation of all stakeholders. Risk communication can have a major impact on how well society is prepared to cope with risks and how people react to crises and disasters.
- A balance is needed between acceptable residual risk and the economic costs of risk reduction measures.

Risk concept

To be able to compare different types of natural hazards and their related risks and to design adequate risk reduction measures, a consistent and systematic approach has to be used (Figure 1). The risk concept represents the methodological basis of several elements: integrative risk management, the decision-making process in risk reduction and mitigation, and disaster management. It serves to aid transparency in risk dialogue between all stakeholders (Ammann, 2006). The basic principles of the risk concept are represented in Figure 2 and can be summarized by the following key questions:

- How safe is safe enough?
- What can happen?
- What is acceptable (to happen)?
- What needs to be done?

The question "What can happen?" has to be answered by risk analysis, whereas the question "What is acceptable?" needs the assessment of risks. The necessary steps

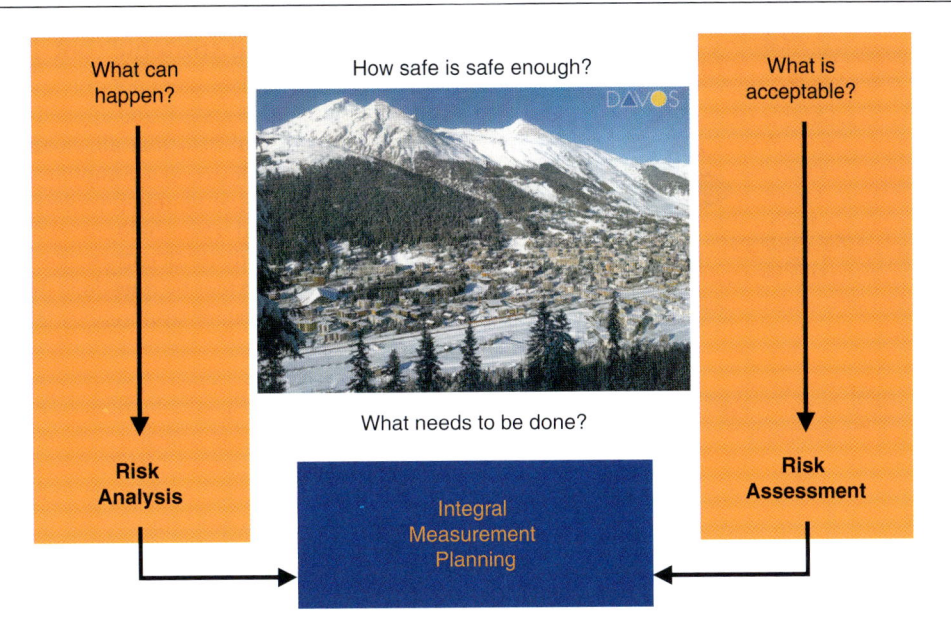

Disaster Risk Reduction, Figure 2 Basic questions and elements of the risk concept (Source: author).

Risk Anaysis	Risk Assessment	Planning of Measures

▪ Hazard Analysis
  – Event Analysis
  – Effect Analysis
▪ Exposure Analysis
▪ Impact Analysis (Vulnerability Analysis/ Robustness)
▪ Risk estimation and Risk de-scription/Visualisation
	▪ Protection Goals
	  – Life Risk
	    - Individual Risk
	    - Collective Risk
	  – Assets/Material Damage
	▪ Risk Categories
	▪ Risk Aversion
		▪ Risk- Cost- relationship
		▪ Marginal Costs
		▪ Integration of all possible measures
		▪ Comprehensive Assessment of all measures

Disaster Risk Reduction, Figure 3 Necessary steps in risk analysis, risk assessment, and the integrative planning of risk reduction measures (Source: author).

are summarized in Figure 3. The goal of a risk analysis is to achieve the most objective possible identification of the risk factors for a specific, damaging event, object, or area. The question "What can happen?" has to be answered by considering a variety of factors that influence it.

Risk assessment aims to give an explicitly subjective answer to the question "What is acceptable?". Thus, it asks how big a residual risk is acceptable. Risk assessment is by nature very complex and has to deal with the fact that risk is a mental construct but not a fully rational one. An important aspect is risk aversion, as practiced in relation to catastrophic events: people's wish to prevent large, spectacular, or particularly frightening events may be disproportionate to the event's real consequences. The acceptance of a risk also depends on whether it is given by active choice or not. Risk categories are defined to the extent that self-reliance and autonomy are possible.

Risk assessment is closely linked with the protection goals that people want to achieve. A protection goal is a set of criteria for the implementation of the primary goals of all efforts to improve safety. It represents the acceptable risk level and thus defines how far the measures should go. A protection goal has different meanings as it has to cover individual and collective perspectives. An individual's protection goal is often defined in terms of the probability of dying. The marginal costs of safety measures (Ammann, 2006) have proven to be the most useful means of defining protection goals in terms of the collective perspective of society. The marginal costs represent certain expenses per fatality avoided or per human life saved. The safety measures can be increased until the desired level of risk reduction is achieved.

Determining the marginal costs of avoiding a fatality can lead to the misunderstanding that a price can be allocated to a human life. The criterion of marginal costs

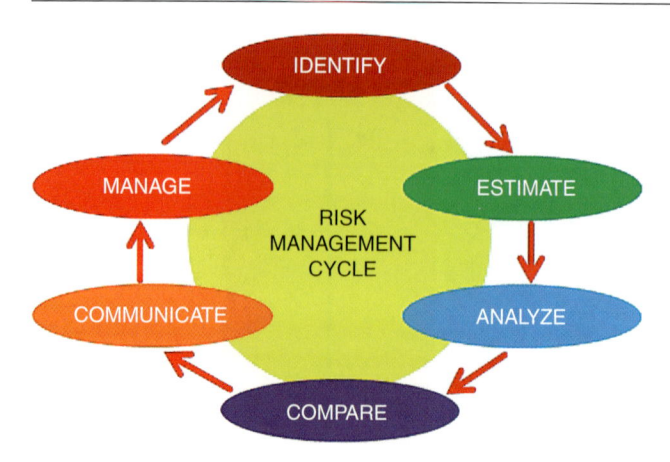

Disaster Risk Reduction, Figure 4 The risk management cycle (Source: author).

should be seen as the optimization of safety measures in terms of lives saved within the limitations of available means and resources.

Planning helps identify measures that are necessary and appropriate in order to reach the protection goals. The main function of the planning of integrative measures is to achieve the intended level of safety in the most cost-effective way. Organizational, technical, and biological protective measures must be planned, checked for effectiveness, and undertaken in concert, while keeping in mind that prevention, intervention, and reconstruction are all equally valid risk management measures (Figure 4). Whereas preventive measures serve primarily to reduce vulnerability, preparedness and intervention measures primarily serve to strengthen resilience. Further criteria such as sustainability, acceptability, feasibility, and the reliability of solutions have also to be kept in mind.

Safety measures are always accompanied by side effects. The most obvious of these is financial; however, aspects of ecology, landscape protection, and land-use planning can be of equal importance. The optimal coordination of all measures has to bear in mind that all relevant aspects and activities in the field of disaster risk reduction have to be sustainable. Measures need to be environmentally sound, to consider societal preferences and to be cost-effective. Disaster risk reduction has also to integrate with the sustainable use of natural resources and with sustainable development. This is why it is considered a cross-cutting issue.

The sociopolitical aspects of sustainability are a question of development and welfare priorities and have to be seen in context of other targets such as education and health care. Especially in developing countries, a reallocation of resources is often needed after major catastrophes for recovery purposes – resources which have been allocated originally to be used for investments in, for example, education, health care, or welfare. What is needed is a political balance between long-term investments for prevention and short-term measures for disaster response and recovery.

Risk dialogue and strategic controlling

Integrative risk management not only dictates that the measures are planned, assessed, and applied in accordance with the risk concept, but also that all those who are involved and affected are included in a comprehensive risk dialogue, and in the processes of planning protection measures. Risk communication and risk dialogue with all stakeholders have to start promptly at an early stage. They will be dominated to a greater extent by questions than by answers, and by processes rather than solutions. A continuous, comprehensive risk dialogue is therefore of vital importance, as it will help ensure that risk management becomes a transparent, understandable affair of public trust.

Active information supply and communication play a dominant role in crisis situations. A well-informed public will weather a catastrophic situation much better than an ill-informed one, and the risk of panic and long-term damage can thus be reduced.

Strategic controls should be used periodically to check the risk situation and monitor the costs and benefits of measures. It is also necessary to monitor residual risks. Integrative risk management enables the overarching aims to be reached using protection measures that can be justified in technical, economic, societal, and environment terms.

Numerous factors can increase future risks and thus create additional uncertainty. Among the most important factors to be taken into consideration, monitored, and periodically checked are globalization, mass mobility, vulnerability, the spread of urban areas, the increase in fixed capital investments, sensitivity (through increasing economic interdependencies), international leisure activities, sociopolitical changes, and changing climate and weather patterns. Developments in hazard and risk management must be followed carefully and the potential for optimization exploited. In the future, the challenge will be to understand and cope with constant change; new risk scenarios, new hazards, climate change, new sociopolitical conditions, and so on. This means that strategies for dealing with risks due to natural and anthropogenic hazards will have to undergo constant adaptation.

Conclusions

Disaster risk reduction is embodied in the combination of risk reduction and disaster management. It addresses the whole risk cycle of prevention, intervention, and recovery. In coping with natural hazards, most countries still focus on reactive disaster management, whereas proactive risk reduction using preventive measures is politically more difficult to justify and implement. To cite Kofi Annan, former Secretary General of the United Nations, "The benefits of prevention are not tangible; they are the disasters that did not happen." To strengthen prevention is only possible with a risk-related approach that needs a paradigm shift from hazard-oriented reaction to risk-related preemptive action. The benefits of prevention can only be made clear with a strict risk controlling process, and political support for prevention and climate change adaptation can only be gained with continuous activities designed to raise public awareness.

Bibliography

Ammann, W. J., 2006. Risk Concept, Integrative Risk Management and Risk Governance. In Ammann, W. J., Dannenmann, S., and Vulliet, L. (eds.), *RISK 21 Coping with Risks Due to Natural Hazards in the 21st Century*. London: Balkema, Taylor and Francis Group, pp. 3–23.

CRN, 2011. *Risk Analysis: Resilience – Trends in Policy and Research*. Focal Report no. 6, Crisis and Risk Network, ETH, Zurich, 56 pp.

GFDRR Global Facility for Disaster Reduction and Recovery, World Bank, 2007. *Committed to Reducing Vulnerabilities to Hazards by Mainstreaming Disaster Reduction and Recovery in Development*. Washington: World Bank. http://www.unisdr.org/we/inform/publications/2237. 29 March 2012.

UN Millennium Development Goals UN MGDs. http://www.un.org/millenniumgoals/. 17 March 2010.

UNISDR, 2005. *Hyogo Framework for Action 2005–2015: Building the Resilience of Nations and Communities to Disasters (HFA)*. For download at: http://www.unisdr.org/eng/hfa/docs/Hyogo-framework-for-action-english.pdf. 17 March 2010.

UNISDR, 2009. *Reducing Disaster Risks through Science: Issues and Actions – The Full Report of the ISDR Scientific and Technical Committee*. Geneva: United Nations International Strategy for Disaster Reduction. http://www.unisdr.org/we/inform/publications/11543. 29 March 2012.

UNISDR, 2011. *Revealing Risk, Redefining Development: Global Assessment Report on Disaster Risk Reduction*. Geneva: United Nations International Strategy for Disaster Reduction. http://www.preventionweb.net/english/hyogo/gar/2011/en/home/index.html. 29 March 2012.

UNISDR, 2012. *Towards a Post-2015 Framework for Disaster Risk Reduction*. Geneva: United Nations International Strategy for Disaster Reduction. http://www.unisdr.org/files/25129_towardsapost2015frameworkfordisaste.pdf. 29 March 2012.

World Bank, Columbia University, 2005. *Natural Disaster Hotspots: A Global Risk Analysis*. Washington: World Bank. http://sedac.ciesin.columbia.edu/hazards/hotspots/synthesisreport.pdf. 29 March 2012.

Cross-references

Climate Change
Cost-Benefit Analysis of Natural Hazards Mitigation
Disaster
Disaster Risk Management
Economic Valuation of Life
Emergency Planning
Frequency and Magnitude of Events
International Strategies for Disaster Reduction
Perception of Natural Hazards and Disasters
Risk
Risk Assessment
Risk Governance
Risk Perception and Communication

DISASTERS

Ian Stewart
University of Plymouth, Plymouth, UK

Introduction

In the novel Candide by Voltaire (1759), two characters debate the roots of a great earthquake that had just leveled one of Europe's greatest and most religious cities, Lisbon (Figure 1). To one, Candide, the calamity – which struck on All Saints' Day 1755 when all the pious were in church – was clearly the day of final judgment; to the other, his tutor Pangloss, it was simply Lisbon's location above a subterranean seam and an inconsequential event in what was otherwise the best of all possible worlds. The authorities of the city, a powerhouse of the inquisition, tolerated no such debate: for his sins in not recognizing divine retribution as the cause, Pangloss was rapidly hanged, whereas Candide was administered a hundred lashes for listening and watching.

Voltaire's international best seller may have been parodying the new religious optimism of the Enlightenment which saw the best in everything, but the Lisbon disaster would turn out to be a turning point in the recognition that events like earthquakes were not the result of

Disasters, Figure 1 The cathedral in Lisbon was one of many that collapsed during the 1755 earthquake in which many of the city's inhabitants perished. The result was a cathartic reinterpretation of natural hazards as acts of God or physical nature and also the emergence of a social science paradigm that viewed disasters as the consequence of human culpability.

divine wrath or an unmerciful God but instead were natural phenomena. However, amid the theological fires that the quake ignited among major Enlightenment thinkers, one of them, the philosopher Rousseau, drew attention to human culpability. After all "...nature did not construct 20,000 houses of six to seven stories there, and that if the inhabitants of this great city had been more equally spread out and more lightly lodged, the damage would have been much less and perhaps of no account" (Rousseau, translation in Dynes, 2000, p. 106). What is more, Rousseau pointed out that if the population had evacuated promptly at the first tremors, they would have been safe, but instead, "How many unfortunate people have perished in this disaster because of wanting to take his clothes, another his papers, another his money?" For Rousseau, human beings were responsible for risk because their actions, not the actions of an unmerciful god, brought consequences.

Today, the 1755 Lisbon earthquake is regarded as the world's first modern natural disaster (Dynes, 2000). But two and a half centuries on, those same questions concerning the "naturalness" of disasters, and of the competing significance of their human and physical roots, remain. At the heart of contemporary hazards, research is a clash between two broad schools of thought (Alexander, 1993). The first is fixed in the pioneering US flood hazard research of Gilbert White in the 1940s, which spawned a generation of scientists – the so-called Chicago School – convinced that scientific and technological solutions could protect society against natural disasters (Burton et al., 1978). Their belief that we can adapt to destructive natural forces and reduce their adverse impacts through engineering and planning was a mainstay of the International Decade of Natural Disaster Reduction. IDNDR strategy, developed in the 1980s, sought to transfer this knowledge on disaster reduction as practiced in developed nations to hazard-prone developing countries (Press and Hamilton, 1999). Around this time, however, an opposing school of thought was emerging, gaining ground particularly among social scientists working in the field of development studies, who saw Western technocratic methods as being inadequate for tackling the root causes of most disasters, namely, underdevelopment and the marginalization of people in poor communities. The so-called radical critics increasingly argued that if an individual or a community was already economically or ecologically marginalized, a transfer of technology would not alleviate disaster (O'Keefe et al., 1976; Hewitt, 1983).

Today, the science of natural disasters is underpinned by a broad acceptance of the paradox that while the causative events (hazards) emerge from nature, the consequent disasters are made in society (Alexander, 1993; Varley, 1994; Hewitt, 1997; Pelling, 2003; Wisner et al., 2004). In other words, although an understanding of the dynamics of the physical environment is crucial for anticipating the incidence of hazardous phenomena, equally important is an understanding of the social, economic, political, and cultural dynamics within a community or society that transform a particular hazard event into a specific disaster. In the contest between the physical and human framing of natural hazards, the latter has arguably now gained the upper hand as the dominating disaster paradigm. As the prominent hazard geographer Kenneth Hewitt (1997, p. 141) notes, "...society, rather than nature, decides who is more likely to be exposed to dangerous geophysical agents."

Modern disaster science, consequently, attempts to fuse an interdisciplinary perspective in which geoscientists and engineers seek to improve their understanding of the frequency and intensity of potentially damaging physical events (the "hazard"), and social scientists reveal the characteristics of a community or society to anticipate, cope with, resist, and recover from such events (the "vulnerability") (Smith, 2001; ISDR, 2004; Wisner et al., 2004). Taken together, analysis of both the physical hazard and the social vulnerability constitutes an assessment of "risk" – the probability of loss resulting from a specified hazard event affecting a particular societal target. It is the realization of this threat that turns a "natural hazard" into a "human disaster." An event that seriously disrupts the functioning of a community or a society (causing widespread human, material, economic, or environmental losses which would exceed the ability of the affected group to cope using it own resources) is designated a "disaster" (Smith, 2001; ISDR, 2004).

The statistics of disaster

The emergence of vulnerability-oriented disaster perspectives has reflected the perceived failure of technocratic approaches to stem the swelling tide of disaster. In 1978, when Gilbert White and colleagues ushered in modern disaster science with the classic text "Environment as Hazard," they did so with opening remarks that lamented how "... the global deathtoll from extreme events of nature is increasing. Loss in property from natural hazards is rising in most regions of the earth, and loss of life is continuing or increasing among many of the poor nations of this world" (Burton et al., 1978, p. 1). Despite three decades of scientific efforts, including an international decade – the 1990s – devoted to natural disaster reduction (Press and Hamilton, 1999), the world still confronts a soaring toll of natural crises.

The raw statistics show that the past four decades have witnessed a fourfold increase in the number of reported natural disasters, from fewer than 100 per year in the mid-1970s to around 400 in the period 2000–2007 (Guha-Sapir et al., 2004; Rodriguez et al., 2009). Since the 1990s, something of the order of 1.5 million people have been killed in natural crises, with the annual death toll averaging around 55,000–65,000 fatalities. In years blighted by major catastrophes, the toll is far greater, such as in 2008 when Cyclone Nargis killed 138,366 people in Myanmar and the Sichuan earthquake in China caused the deaths of 87,476 people, producing mortality estimates more than three times the recent average. In terms of

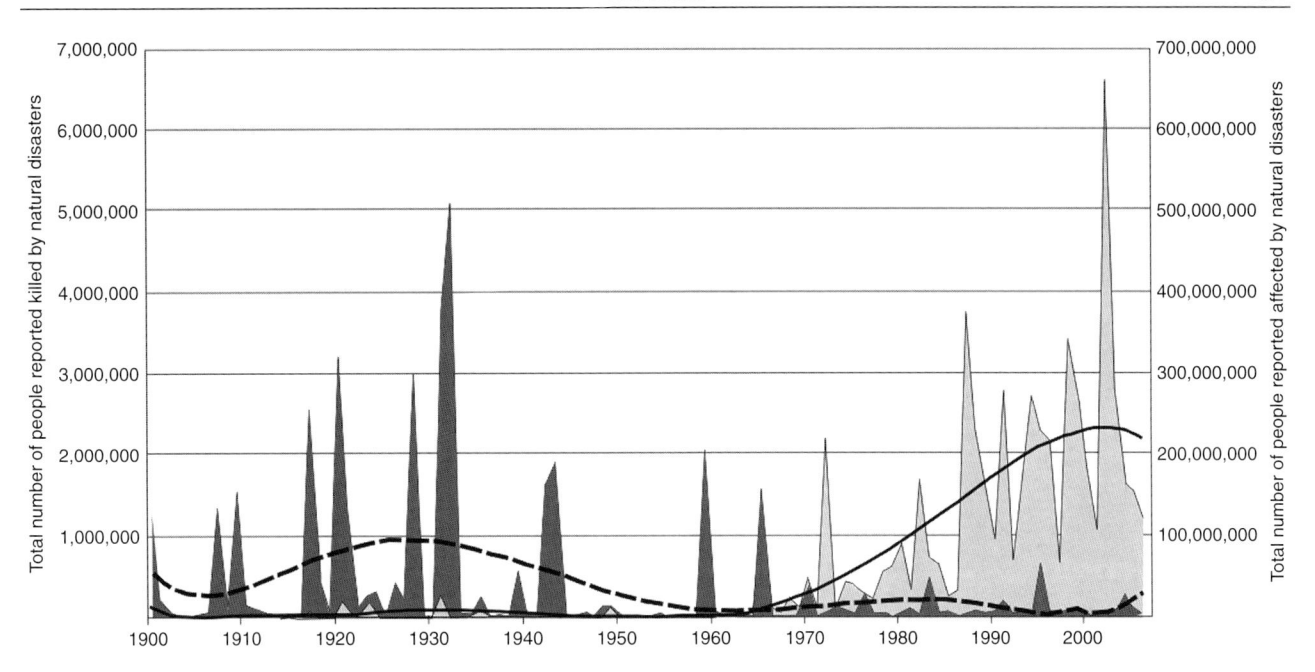

Disasters, Figure 2 Human losses due to natural disasters, 1900–2006. *Dark shading* indicates fatalities reported due to natural disasters (scale on *left*) and *light shading* indicates total number of people affected (scale on *right*). *Dashed line* shows the smoothed trend for fatality numbers and *solid line* shows smoothed trend for number affected (Source: EM-DAT – the OFDA/CRED International Disaster database, http://www.em-dat.net, Université Catholique de Louvain, Brussels, Belgium).

those affected by disasters, the situation is even more perilous – in recent times (1994–2003), more than 255 million people have been annually affected by natural calamities (EM-DAT, 2006) (Figure 2).

Such crude numbers obscure an underlying geography to disaster fatalities. For the period 1980–2004, the number of disasters and the at-risk populations of high-income and low-income countries are broadly similar (Stromberg, 2007). However, the numbers killed in disasters are over an order of magnitude lower in the wealthier nations – around 75,000 fatalities compared with over 900,000 for poorer nations. This reflects the fact that high-income countries have invariably have invested substantially in a wide range of preparedness and mitigation measures (Figure 3): buildings can be constructed of stronger and more durable materials or elevated above flood levels, farmland can be irrigated to reduce losses during droughts, warning systems for certain natural disasters, such as hurricanes, can save lives, and after a disaster strikes, mass evacuation and emergency medical care and food can limit the human toll of the disaster. Lacking the wealth, infrastructure, and institutional capacity to afford adequate protection, it is no surprise to find that over 90% of the hazard-related deaths are in less-developed nations.

The headline message of the economics of disaster is simple: the costliest collateral losses are incurred by wealthy industrialized nations, but the greatest fiscal burden of disasters (as a proportion of a country's gross domestic product) is inequitably borne by the least economically favored nations (Dilley et al., 2005).

Moreover, the financial costs are rising (Figure 4). Current economic losses are up 14-fold compared to 1950s levels, and in the mid-1990s, natural hazards in the USA alone were estimated at US$54 billion per year – or a staggering US$1 billion per week (van der Vink et al., 1998). Currently (2000–2007), the annual global average loss is thought to be around US$82 billion. Again, a contagion of disasters make some years more expensive than others. 2008, for example, was especially costly with the Sichuan earthquake in China (US$85 billion) and hurricane Ike in the USA (US$30 billion), contributing to economic losses more than double the recent average (Rodriguez et al., 2009). As the global economy grows and the number of at-risk assets swells worldwide, the cost of natural disasters in both monetary and human terms is expected to spiral higher still.

The changing face of disasters

Despite the rising incidence of disasters over recent decades, there is little sign that the physical environment that we occupy is becoming intrinsically more dangerous. There is no appreciable increase detected in the frequency or magnitude of major geophysical phenomena such as earthquakes, volcanic eruptions, and tsunamis. It is possible that anthropogenic climate change ("global warming") is invigorating the incidence and severity of tropical storms and other hydrometeorological hazards (Mitchell et al., 2006; Knutson et al., 2010; Lubchenco and Karl, 2012), but even if a heightened level of some hazardous

Disasters, Figure 3 Two contrasting views of how urban settlements face up to the earthquake threat. (*Left*) In wealthy industrial nations considerable effort has gone into engineering buildings to withstand earthquakes, evident in this Tokyo skyscraper. However, in major cities in lesser developed nations, such as Istanbul, Turkey (*right*), it is the weakly assembled building stock that is the main threat to live and livelihood in future seismic disasters. Although engineers have the technical knowledge to design buildings to withstand moderate earthquake strikes, in many countries the implementation of good construction practices and effective planning measures is hampered by weak regulatory controls and political corruption.

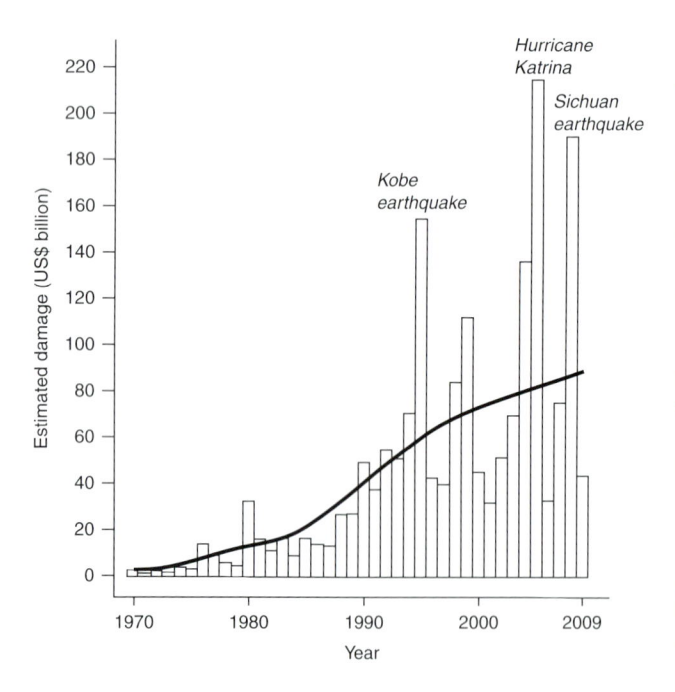

Disasters, Figure 4 Estimated damages (US$ billion) caused by reported natural disasters, 1970–2009 (Source: EM-DAT – the OFDA/CRED International Disaster database, http://www.em-dat.net, Université catholique de Louvain, Brussels, Belgium).

processes is real, it is insufficient to account for the dramatic increases in natural disasters over recent times.

Instead, for many disaster scientists, the root of our more perilous predicament lies not in the physical domain but in the human one. Specifically, it lies in the increase in the world's population, its concentration in large conurbations, the high vulnerability of modern societies and technologies, and the social and economic consequences of development in highly exposed regions, such as coastlines (Smolka, 2006). Coastalization is a trend recognized worldwide whereby more and more population, property, and infrastructure squeezed along shorelines facing rising sea levels and threatened with saltwater intrusion into groundwater aquifers and inundation from storm surges and tsunamis (Figure 5). Drawing attention to the preferential migration of the most affluent sectors in US society to the popular retirement destinations of the earthquake-prone shores of California and Washington and the hurricane-prone Gulf Coast, for example, van der Vink et al. (1998, p.537) asserted "We are becoming more ulnerable to natural disasters because of the trends in our society rather than those of nature."

Along with a move to the coast, the global shift to urban living has made many cities as dangerous as the natural environments they replace. With the rise of supercities (>2 million people) and megacities (>10 million people), human settlement has been forced into marginal,

Disasters, Figure 5 The concentration of people, infrastructure and economic development along tsunami-prone shores has changed the nature and extent of vulnerability in many coastal zones. The 2004 Indian Ocean earthquake and tsunami, for example, affected 14 Asian and African countries and killed people from 48 nationalities, 34 representing foreign tourists from around the world. Expensive beachfront tourist complexes, such as one destroyed here in Khao Lak (Thailand), greatly contributed to the human and economic losses of this calamitous event.

dangerous places, sometimes within the sprawling metropolitan areas. Alongside the physical marginality of such areas is the acute social and economic marginality of the people who must inhabit them (Wisner et al., 2004). Such a situation is tragically exemplified by the slide of solid waste from the Payatas rubbish dump in central Manila (Philippines) in July 2000 which killed 300 people in the contiguous squatter settlement (Gaillard and Cadag, 2009). Such an event illustrates how, although hazard typologies have in the past made a clear separation between "natural disasters" produced by geophysical agents and "human disasters" that arise from technological failures or human conflict, such a distinction is ever more difficult to sustain in the real world. In the real world, disasters are increasingly messy amalgams between natural processes acting on human environments.

For all their unnaturalness, disasters retain a clear natural geography to their incidence. That is because the hazard processes that underpin them tend to strike repeatedly in the same places. The hazard "hot spots" are familiar: droughts have been occurring in the Sahelian region of Africa for millennia, monsoonal storm surges annually inundate the deltaic plains of Bangladesh, hurricanes seasonally batter the Atlantic and Gulf coasts of the USA, and earthquakes and volcanic eruptions routinely plague the tectonic plate boundaries like the Pacific Ring of Fire (Dilley et al., 2005). Because geophysical phenomena are, by and large, persistent offenders, knowledge of their past incidence can provide a reasonable expectation of the physical exposure to hazard in any particular geographic area. Of course, monitoring the nascent signs of impending hazards and forecasting the likely location, size, or style of their impact remain fraught with technical difficulties and scientific uncertainties. The 2011 Tohoku (Japan) earthquake provided a telling reminder of those difficulties; the giant ($M > 9$) earthquake and its accompanying tsunami were not unexpected based on geological evidence, but that data was overlooked in subsequent hazard assessments, leading to inadequate mitigation measures (insufficiently high seawalls). Nonetheless, gauging the physical exposure to floods, hurricanes, earthquake, volcanic eruptions, and the like is often more readily constrained than assessing a hazard-prone community's capacity to resist such events (e.g., Wisner et al., 2004).

The nature of vulnerability

A fundamental challenge of disaster reduction is to anticipate the intrinsic vulnerability (or lack thereof) of communities at risk? The notion of vulnerability has been confronted by disaster researchers for decades, but it is only in the past few years that it has become an issue that is explicitly addressed, and it remains a concept that is difficult to define and quantify (Bankoff, 2004; Wisner et al.,

The Progression of Vulnerability

1 ROOT CAUSES

Unlimited access to
- Power
- Structures
- Resources

Ideologies
- Political systems
- Economic systems

2 DYNAMIC PRESSURES

Lack of
- Training
- Local markets
- Local institutions
- Press Freedom
- Local investments
- Local institutions
- Ethical standards in public life

Macro-forces
- Rapid population growth
- Rapid urbanisation
- Arms expenditure
- Debt repayment schedules
- Deforestation
- Decline in soil productivity

3 UNSAFE CONDITIONS

Fragile physical environment
- Dangerous locations
- Unprotected buildings and infrastructure

Fragile local economy
- Livelihoods at risk
- Low income levels

Vulnerable society
- special groups at risk
- Lack of local institutions

Public actions
- Lack of disaster preparedness
- Prevalence of endemic disease

DISASTER

Risk = Hazard + Vulnerability

HAZARDS

Earthquake

High winds (cyclone / hurricane / typhoon)

Flooding

Volcanic eruption

Landslide

Drought

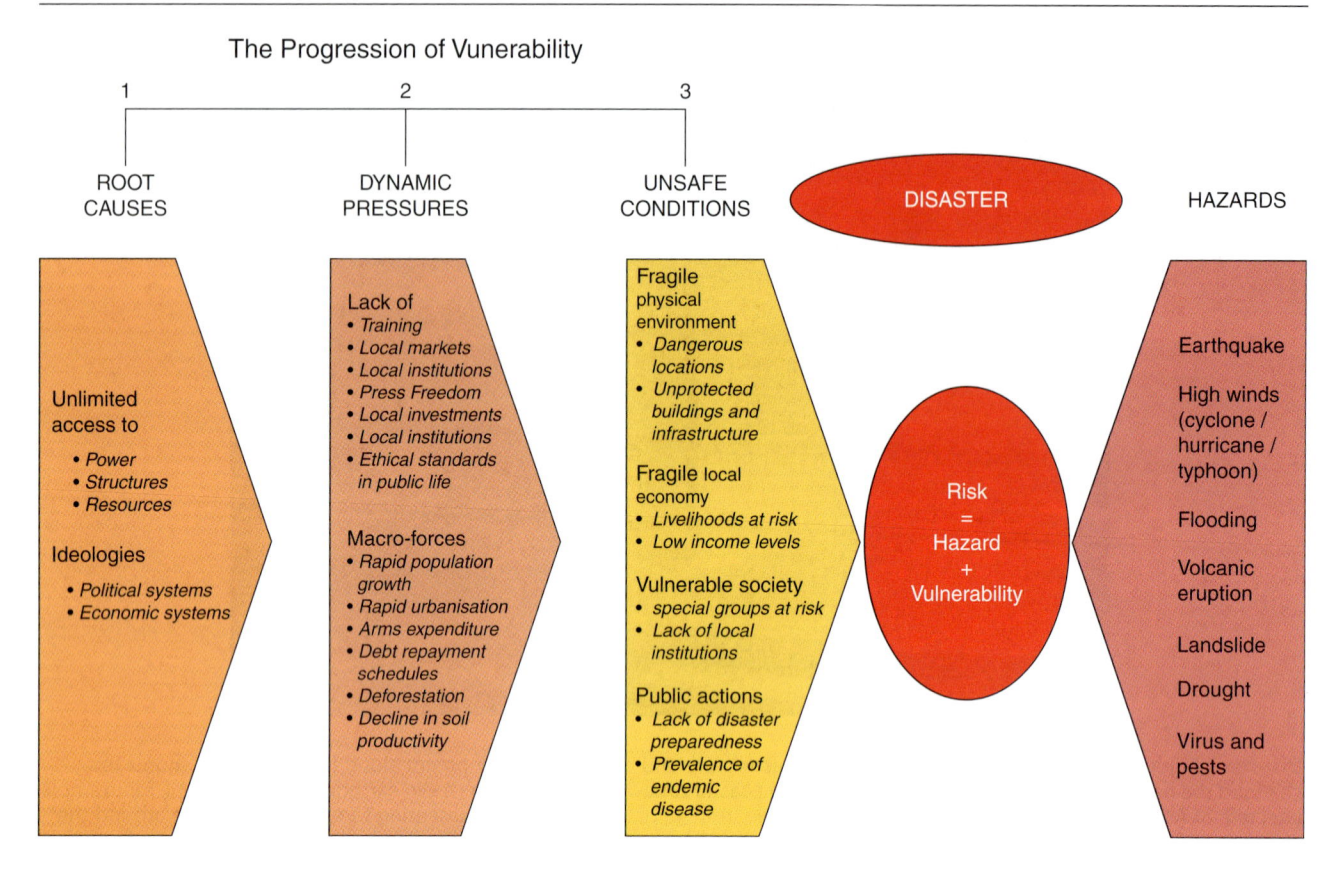

Virus and pests

Disasters, Figure 6 Pressure and Release (PAR) model of Blaikie et al. (1994) showing the progression of vulnerability. The diagram shows a disaster as the intersection between socio-economic pressures on the *left* and physical exposures (natural hazards) on the *right*.

2004; Cutter, 2006). As disasters lie at the intersection of socioeconomic pressures and physical exposures, different kinds of vulnerability prevail (Figure 6). Physical vulnerability (exposure to hazard threats) is the easiest to determine, identifying those that live in perilous places as being potential victims of disaster. Less easy to determine are those whose situation is made perilous because they are socially excluded, economically disadvantaged, and/or politically marginalized. Social, economic, and political vulnerability ensures that access to hazard mitigation measures and disaster reduction strategies are often unevenly distributed across and among at-risk communities. These different facets of vulnerability operate dynamically during hazard events, as is evident in this account of Bangladesh floods:

> On the eve of Bangladesh's massive floods in August 1988, this relatively powerless group (landless squatters] was living in an economically marginal situation but close to the city, on low-lying land prone to flooding. Their economic and political marginality meant they had few assets in reserve. It also meant that their children were unusually malnourished and chronically ill. This channelled the dynamic pressure arising out of landlessness and economic marginalization into a particular form of vulnerability: lack of resistance to diarrheal disease and hunger following the flooding in 1988. Factors involving power, access, location, livelihood, and biology mutually determined a situation of particular unsafe conditions and enhanced vulnerability. (Blaikie et al., 1994, p. 27)

For most practitioners, the first "line of defense" against vulnerability to disaster is livelihood – ensuring that individuals have appropriate assets to grow food themselves or make earnings (Cannon, 2008). Higher incomes and stable employment enable households to have livelihoods that are buffers against hazards. Livelihood presets a person's basic nutritional state, their baseline status, and their general health and welfare. Individuals with poor nutrition are generally less resistant to disease and less capable of making a good recovery when stressed by a hazard impact; morale and personal resilience, stress and general mental health are all factors that are likely to affect the ability to overcome the impact of a hazard. An individual's income determines their capacity to build a home that is safe from endemic hazards and their ability to site that home somewhere out of harm's way. Many people remain vulnerable precisely because they do not have the financial resources to live sufficiently above the regular flood levels or away from steep marginal slopes prone to landslides, excluded from safer areas by high land prices.

"Self protection" from hazards requires knowledge or skills that may be available from the local community or from outside agencies. Equally, for some vulnerable groups like children, the elderly, sick, or disabled, adequate protection from hazards can only be provided at a community or society level. Much of this "social protection" is conferred by local, regional, and national authorities, in the form of hazard-awareness programs, warning systems, emergency plans, and regulations to do with land-use management or engineering and building controls.

Yet ill-judged or inadequate social protection measures can also instigate human disasters, such as through inappropriate policies, weak infrastructure, poor governance and corruption, ineffective monitoring and communication, bad development decisions, injustice, and discrimination. Inaction, in the form of authorities not carrying out their expected regulatory functions, also aggravates disaster. For example, a failure to deal effectively with land squatting and irregular construction fuels vulnerability in the environs of many hazard-prone cities. In Istanbul (Turkey), for example, many people ignore mandatory requirements to live in homes approved as conforming to stringent earthquake design standards, preferring instead to illegally self-build. Yet they do so in part because of a suspicion that the formal approved building stock is "unsafe," having been built by people and using materials unknown to them and signed off as safe by potentially corrupt engineers or officials, all chronic failings of endemic construction practices exposed by the 1999 Izmit earthquake in which tens of thousands of people lost their lives through the collapse of improperly constructed buildings (Green, 2008). Given this deep distrust of the Istanbul's authorized, commercially built housing stock, Green (2008) suggests that bolstering the city's unauthorized self-built housing might actually be an effective means of providing protection against the future earthquake threat.

Regardless of levels of self- and social protection, in landscapes of chronic vulnerability like urban squatter settlements, disasters in some form are probably inevitable. As noted by Hewitt (1983), *"In most places and segments of society where calamities are occurring, the natural events are about as certain as anything within a person's lifetime."* The point here is that although Western disaster discourse typically depicts "disasters" as abnormal occurrences, in communities in many acutely marginalized parts of the world, vulnerability emerges from the "normal" order of things – hazards simply compound the struggles that are part of people's daily lives. Hazards are, in effect, the ordinary, not the extraordinary. Communities living on the margins will have a very low capacity to withstand even small damaging events. For that reason, basic capacity-building measures are a fundamental part of disaster reduction.

In less acute situations, vulnerability (and its alter ego, resilience) is difficult to track, being a dynamic that changes through time as individuals, groups, and institutions adapt to internal and external pressures (Oliver-Smith, 1999a; Turner et al., 2003; Berkes, 2007). Communities can become less vulnerable to hazards if they have a range of options for coping with external shocks and stresses. The key to reducing vulnerability, therefore, is to increase "resilience," a concept defined by the United Nations Strategy for Disaster Reduction as "the ability of a system, community, or society exposed to hazards to resist, absorb, or recover from the effects of a hazard in a timely and efficient manner." Most strategies for growing resilience involve reducing risks by spreading them out, thereby increasing opportunities in the face of hazards (Paton and Johnson, 2006). Ultimately, however, a measure of the success of a community's adaptations to anticipated threats is only apparent after the event (Cannon, 2008). Prior to acute environmental crises, the manifestations of vulnerability – social, economic, institutional, and infrastructure – may be hidden from view. Only when a hazard strikes do the societal and technical bonds of an at-risk community become truly tested and often found wanting (Oliver-Smith and Hoffman, 2002).

A potent example of the revelatory power of disasters was provided by Hurricane Katrina in 2005. The likely impact of major hurricane making landfall in the low-lying Mississippi delta was well known (e.g., Fischetti, 2001), and the landfall of the destructive Katrina storm in New Orleans was accurately forecasted and emergency evacuation plans were put in motion (McCallum and Heming, 2006). What surprised few was that the aging infrastructure of the Mississippi coast's flood protection levees – designed for a category three storm surge – failed under the onslaught of the storm, allowing widespread inundation of the city. What stunned many was the resulting institutional meltdown, which for several days left evacuees with no power, no drinking water, dwindling food supplies, understaffed law enforcement, and delayed search and rescue activities (Cutter et al., 2006).

Events like Hurricane Katrina throw into question how resilience is fostered in social systems. It has long been assumed that governments, from the federal to the municipal, comprise the backbone of emergency management, but increasing community organizations are shown to have a major role to play in the face of disaster (King, 2007). Community resilience takes the form of networks of strong and weak ties – families, churches, local volunteer and relief groups, hobby clubs, even neighborhood and crime watch organizations – that is referred collectively as "social capital" (Dynes, 2002). Through social capital, citizens assume roles as active agents rather than passive victims since they are able to draw upon collective strengths, assistance, and resources to deal with disasters, thereby being more proactive in decision making and effecting a more speedy recovery.

Cultures of catastrophe

While considerable attention has been devoted by hazard practitioners to elucidate and quantify the factors underpinning social vulnerability (Cutter, 2006), some argue

Disasters, Figure 7 The narrow streets, multi-story houses and tiled roofs of many Andean towns are a cultural import from Spanish Andalucia, transforming earlier Inca settlements into places of heightened seismic vulnerability.

that the whole concept of vulnerability is itself a Western ideological construct that fails to acknowledge how natural hazards are themselves a cultural driver, shaping community adaptations in ways that allow disasters to be incorporated into daily life. Bankoff (2003) recognizes this effect of the "normalization of threat" in Philippine culture, seeing it in "...the design and construction of buildings, in the agricultural system, in the constant relocation of settlements and in the frequency of migration. Filipino society has evolved certain 'coping mechanisms' to come to terms with the constancy of hazard and to mitigate the worst effects of disasters. Often, too, in the way in which people deal with emotional and psychological requirements of living with uncertainty may influence what are seen as 'Filipino' beliefs and character traits."

The loss or removal of such cultural coping mechanisms can expose communities to heightened hazard threats, even when that transformation happened decades or centuries before. According to Oliver-Smith (1999b), the calamitous May 31st 1970 Peru earthquake (M 7.7) in Peru had its roots five centuries before when local Andean resilience was replaced by imported Spanish practices; the dispersed design of Inca towns was replaced by the Andalucian-inspired new towns favoring narrow streets with multistory houses pressed close together. Sturdy monumental stonemasonry and anti-seismic wall ties were abandoned, and thatched roofs were replaced with heavy ceramic roof tiles, all of which made houses into earthquake death traps (Figure 7). For these reasons, Oliver-Smith (1999b) argues that the 1970 earthquake – an event which saw 70,000 people killed, 140,000 injured, and half a million homeless – was a calamity 500 years in the making.

Where indigenous cultural practices have persisted alongside recurrent hazard experiences, they are often in stark tension with Western scientific and social dialogues of hazards as interactions between extreme natural events and vulnerable human populations (Chester, 2005). Local knowledge, customs, and traditional beliefs can motivate a community's actions during a crisis, including their propensity to evacuate; in some cases, the cultural ties between community and hazard can have lethal consequences. In 1963, Bali's Mt. Agung erupted during the once-in-a generation Hindu rite killing 1,200 people, many of them waiting patiently and clothed in ritual dress within their temples and resisting attempts by officials and even priests to evacuate them. On the neighboring island of Java, spiritual ties with Mt. Merapi are part of the reason why communities on the perilous upper slopes have resisted efforts to evacuate during repeated volcanic crises, although socioeconomic factors also exert a strong control (Donovan, 2010) (Figure 8). Across many hazard-prone developing regions "...the battle against natural forces is often fought in the cultural arena – with religion as a backdrop" (Svensen, 2009).

Although indigenous cultural traits can at times undermine hazard science approaches to disaster reduction, traditional cultures can also reduce vulnerability by strengthening resilience and providing effective mitigation techniques (Cashman and Cronin, 2008). Perhaps

Disasters, Figure 8 (a) The fertile environs of Mount Merapi on the Indonesian island of Java is a hazardous high-population environment where lethal volcanic crises recur every few years. (b) Scientific studies of the volcano through monitoring of ground deformation, seismicity and summit gas activity give rise to early warning alerts and lead to mandatory evacuation orders by local authorities. However, routinely at-risk communities on the volcano's upper flanks refuse to evacuate, sometimes with lethal consequences. (c) During the 2006 eruption crisis, a pyroclastic flow killed 60 people attending a wedding ceremony in Turgo, a settlement located inside the high-risk exclusion zone. Reluctance to follow volcano emergency management plans reflects a complex combination of socio-economic and cultural factors.

the most dramatic example of this was the self-evacuation of coastal communities on Simeulue and Nias Islands during the 2004 Indian Ocean tsunami. Despite being close to the epicenter of the earthquake and experiencing considerable wave heights, oral traditions of calamitous tsunamis from more than a century before motivated them, at the onset of the initial tremors, to evacuate to higher ground and hardly any lives were lost (Sieh, 2006).

Along the adjacent tsunami-stricken shores of Sumatra, such cultural memories have been largely lost, eroded by economic and tourist development, and with them, have gone traditional practices that long protected communities (McAdoo et al., 2006; Gaillard et al., 2008). According to Sieh (2006, p.1947), disaster reduction in such areas "…does not necessarily involve hugely expensive or high-tech solutions such as the construction of coastal defences or sensor-based tsunami warning systems. More valuable and practical steps include extending the scientific research, educating the at-risk populations as to what to do in the event of a long-lasting earthquake (i.e., one that might be followed by a tsunami), taking simple measures to strengthen buildings against shaking, providing adequate escape routes and helping the residents of the vulnerable low-lying coastal strips to relocate their homes and businesses to land that is higher or farther from the coast."

The politics of disaster

Local disaster cultures exist because communities, and in some cases, whole societies, have coevolved with perilous nature. Strengthening or reestablishing indigenous practices may provide the means by which such communities can confront their hazard threats, but equally, the solutions may come from outside. The technical ability to construct buildings and defenses that can withstand modest hazard shocks exists, as does the scientific knowledge to identify and delineate hazard threats (e.g., Bilham, 2009). But embedding those good building practices and good land-use planning into local environments does not just require an appreciation of cultural sensitivities. Safe construction and effective planning protocols are also underpinned by robust regulatory control. Here, a very different culture can arise – a culture of ignorance, incompetence, and corruption within the authorities charged with emergency planning. With disaster reduction obligations invested in the hands of political authorities, it is the role and efficacy of the state itself that becomes the ultimate element in where and when disasters happen.

According to Berkes (2007), for example, the same hurricane striking Samoa and neighboring American Samoa in the Pacific produced markedly different results: the former was prepared and capable, whereas the latter, much less affluent and used to outside aid for disasters, had weaker institutions for response. Political environments in which there is strong linkage from local to national levels tend to withstand disasters better; Wisner (2001) has argued that so few people died when Hurricane Michelle hit Cuba severely because of the existence of strong organic links between government and people. It has also been argued that emergency crises may be less severe in countries with democratic governments (Sen, 1981), where disaster reduction measures can be more effectively monitored and made accountable through firm civil liberties and a free press (Besley and Burgess, 2002). Of course, as recent disasters in Japan and the USA testify, active democratic systems do not provide immunity from natural emergencies.

Disasters themselves are political instruments. In some cases, they can be a pretext for international political and economic "engineering," and disaster recovery is the impetus for institutional reform (Klein, 2005). In this way, governance – the manner in which power is exercised in the management of a country's economic and social resources for development – exerts a powerful influence on national and international disaster policies. After Hurricane Mitch in 1998, for example, afflicted countries in Central America agreed to a set of principles with international aid donors that included promotion of democracy and good governance, political decentralization and economic debt reduction (Wisner et al., 2004).

The realization that disasters can be significant agents of societal change leads to the paradoxical question as to whether they might, in any sense, bring positive benefits. One benign facet of natural disasters might be in aiding international diplomacy. The earthquakes that struck Istanbul and Athens, in August and September 1999, respectively, opened communication channels between feuding Greece and Turkey, whereas the Bam (southern Iran) earthquake of December 2003 prompted offers of aid from 40 countries, including the USA – the "Great Satan" – which had broken off diplomatic relations with Iran 20 years before. What patchy evidence there is on this notion of disaster diplomacy suggests that while disaster-related initiatives can be catalysts for diplomatic interchanges that have already started, they rarely cement political rapprochement, with a possible exception being the peace deal reached in Aceh after the December 26, 2004, earthquake and tsunami (Kelman, 2006).

Although natural hazards may offer up opportunities for "disaster diplomacy," they can also stir cross-border tensions. During the 2000 and 2004 floods along the India-Bangladesh border, Indian border security forces breached river embankments to allow the water to spill out, thereby ameliorating its downstream impacts in West Bengal (India) but exacerbating destruction of life, crop, and property in Bangladesh (Ali, 2007). In this instance, there was no cross-border conflict, but disasters can trigger political action. It has been argued, for example, that the cyclone and storm surge in East Pakistan in 1970 contributed to the development of the Bangladesh independence movement, while the revolutionary movement in Nicaragua from 1974 to 1979 derived some of its impetus from the effect of the Managua earthquake of 1972 (Wisner et al., 2004). Disasters striking politically peripheral regions can catalyze regional tensions, especially where existing regional deprivations are worsened by post-disaster governmental responses (Pelling and Dill, 2006). Disasters can enhance or even regain the popular legitimacy of political leaders, and many political regimes might interpret spontaneous collective actions by afflicted communities in the aftermath of a disaster as a threat and thereby respond with repression.

Overall, fractured or contested political landscapes often promote a heightened risk to disaster specifically because they sustain an inequitable distribution of resources. But economic resources too can be redefined by disasters; emergency crises can bring direct monetary gain in the form of disaster relief funds that are injected into the local economy. Following the earthquake that killed 80,000 people in China's Sichuan Province in May 2008, for example, funds allocated to rebuilding outweighed the economic loss caused by the quake, enough to raise national economic growth by 0.3% (Hewitt, 2009). Disasters may be economic catalysts at the regional level too. It has been argued that the reconstruction activities following the 1994 Northridge earthquake boosted Los Angeles economy in a similar way to Miami that benefitted after the 1992 Hurricane Andrew (Romero and Adams, 1995; Cochrane, 1997). Following the 1991 volcanic eruption of Mt. Pinatubo in the Philippines, financial resources, investment, and infrastructure poured into the area, turning Luzon into an economic hub. However, although reconstruction efforts may contribute positively to an economy (as measured by gross domestic product or GDP), the loss of productive capital may reduce it. As a result, the financial balance sheet of natural crises shows that the growth in real incomes is not significantly different in years when disasters strike than in an average year (Stromberg, 2007).

The recognition that the financial costs of natural disasters typically have comparatively little effect on most national economies is arguably less applicable to the fiscal fate of less-developed nations. Many developing nations will be hard pressed to develop economically due to recurrent hazard losses, and for many countries, probable economic losses over the next century exceeds their current financial resources (Cardona, 2005). Average losses from disasters in low-income countries (e.g., Sri Lanka, Bangladesh, Nicaragua) can be 10–20 times greater than in disasters in high-income nations (Haas et al., 1977). Whereas floods and droughts typically claim about one tenth of 1% of the GDP of industrialized countries, they cost up to 20 times more (up to 2% of GDP) in less-developed nations (Alexander, 1993).

In such a context, it is difficult to appreciate a silver lining of disasters. Indeed, most disaster scientists would contest the notion that natural calamities can be "good value" at all. A recent e-discussion on the question of whether disasters can help a country's economy drew these remarks (Hewitt, 2009): "*To say that disasters help the economy is a materialistic view…as well as loss of life, disasters entail a loss of investment in those who are killed, and have a long-term psychological impact on those who survive, affecting their capacities and capabilities, and resulting in a loss of productivity, opportunity costs, and more. Therefore, the indirect cost of a disaster is much larger than the direct cost. A loss is a loss and cannot be turned into an investment and produce income or benefits. In addition, losses are not limited to lives, materials, and animals, but also include traditional wisdom and knowledge, making future settlements more prone to natural disasters.*"

Bibliography

Alexander, D., 1993. *Natural Disasters*. London: Springer. 650p.

Ali, A. M. S., 2007. September 2004 flood event in southwestern Bangladesh: a study of the nature, causes and human perception and adjustments to a new hazard. *Natural Hazards*, **40**, 89–111.

Bankoff, G., 2003. *Cultures of Disaster: Society and Natural Hazard in the Philippines*. London: Routledge. 256 p.

Bankoff, G., 2004. The historical geography of disaster: 'vulnerability' and 'Local Knowledge' in western discourse. In Bankoff, G.,

Frerks, G., and Hilhorst, D. (eds.), *Mapping Vulnerability: Disasters, Development and People*. London: Earthscan, pp. 25–36.

Berkes, F., 2007. Understanding uncertainty and reducing vulnerability: lessons from resilience thinking. *Natural Hazards*, **41**, 283–295.

Besley, T. J., and Burgess, R., 2002. The political economy of government responsiveness: theory and evidence from India. *Quarterly Journal of Economics*, **117**(4), 1415–1452.

Bilham, R., 2009. The seismic future of cities. *Bulletin of Earthquake Engineering*, **7**(4), 839–887.

Blaikie, P., Cannon, T., Davis, I., and Wisner, B., 1994. *At Risk: Natural Hazards, People's Vulnerability and Disasters*, 1st edn. London: Routledge.

Burton, I., Kates, R. W., and White, G. F., 1978. *The Environment as Hazard*. Oxford: Oxford University Press. 258 p.

Cannon, T. 2008. *Reducing People's Vulnerability to Natural Hazards: Communities and Resilience*. UNU-WIDER Research Paper No. 2008/34.

Cardona, O. D. 2005. *Indicators of disaster risk and risk management: program for Latin America and the Caribbean*. Summary Report. Washington, DC, Inter-American Development Bank.

Cashman, K. V., and Cronin, S. J., 2008. Welcoming a monster to the world: myths, oral tradition and modern societal response to volcanic disasters. *Journal of Volcanology and Geothermal Research*, **176**, 407–418.

Chester, D., 2005. Theology and disaster studies: the need for dialogue. *Journal of Volcanology and Geothermal Research*, **146**, 319–328.

Cochrane, H. C., 1997. Forecasting the economic impact of a Mid-West earthquake. In Jones, B. G. (ed.), *Economic Consequences of Earthquakes: Preparing for the unexpected*. Buffalo: New York Center for Earthquake Engineering Research, pp. 223–248.

Cutter, S. L., 2006. *Hazards, Vulnerability and Environmental Justice*. Sterling: Earthscan. 448 p.

Cutter, S. L., Emrich, C. T., Mitchell, J. T., Boruff, B. J., Schmidtlein, M. T., Burton, C. G., and Melton, G., 2006. The long road home: race, class and recovery from Hurricane Katrina. *Environment*, **48**, 9–20.

Dilley, M., 2006. Setting priorities: global patterns of disaster risk. *Philosophical Transactions of the Royal Society A*, **364**, 2217–2229.

Dilley, M., Chen, R. S., Deichmann, U., Lerner-Lam, A. L., and Arnold, M., 2005. *Natural Disaster Hotspots: A Global Risk Analysis*. International Bank for Reconstruction and Development/The World Bank and Columbia University, Washington, DC, 145 p.

Donovan, K., 2010. Doing social volcanology: exploring volcanic culture in Indonesia. *Area*, **42**, 117–126.

Dynes, R. R., 2000. The dialogue between Voltaire and Rousseau on the Lisbon earthquake: the emergence of a social science view. *International Journal of Mass Emergencies and Disasters*, **28**, 97–115.

Dynes, R. R., 2002. *The Importance of Social Capital in Disaster Response*. University of Delaware, Disaster Research Centre, Preliminary Paper No. 327, 59 p.

EM-DAT 2006. *The OFDA/CRED International Disaster Database*. Université Catholique de Louvain, Brussels (www.cred.be/emdat). Accessed 11 May 2010.

Fischetti, M., 2001. Drowning in New Orleans. *Scientific American*, **285**, 77–85.

Gaillard, J.-C., and Cadag, J. R. D., 2009. From marginality to further marginalization: experiences from the victims of the July 2000 Payatas trashslide in the Philippines. *Journal of Disaster Risk Studies*, **3**(2), 197–215.

Gaillard, J.-C., Clare, E., Ocean, V., Azhari, D., Denain, J.-C., Efend, Y., Grancher, D., Liamzon, C. C., Sari, D. R., and Setiwan, R., 2008. Ethnic groups' response to the 26 December 2004 earthquake and tsunami in Aceh, Indonesia. *Natural Hazards*, **47**, 17–38.

Green, R. A., 2008. Unauthorised development and seismic hazard vulnerability: a study of squatters and engineers in Istanbul, Turkey. *Disasters*, **32**(3), 358–376.

Guha-Sapir, D., Hargitt, D., and Hoyois, Ph, 2004. *Thirty years of natural disasters 1974–2003: the numbers*. Louvain-la Neuve: Presses Universitaires de Louvain, p. 188.

Haas, E., Kates, R., and Bowden, M., 1977. *Reconstruction Following Disaster*. Cambridge, MA: MIT Press. 366 p.

Hewitt, K. (ed.), 1983. *Interpretations of Calamity*. Winchester: Allen and Unwin. 304 p.

Hewitt, K., 1997. *Regions of Risk: A Geographical Introduction to Disasters*. Harlow: Longman. 389 p.

Hewitt, K., 2009. Culture and Risk: Understanding the Sociocultural Settings that Influence Risk from Natural Hazards: Synthesis Report from a Global E-Conference organised by ICIMOD and the Mountain Forum, Kathmandu, 2008. http://www.mtnforum.org/sites/default/files/forum/files/participants-contributions-cartthreads1-2.pdf.

ISDR 2004. *The International Strategy for Disaster Reduction Terminology: Basic terms of disaster risk reduction*, www.unisdr.org.

Kelman, I., 2006. Acting on disaster diplomacy. *Journal of International Affairs*, **59**, 215–240.

King, D., 2007. Organizations in disasters. *Natural Hazards*, **40**, 657–665.

Klein, N., 2005. *The Rise of Disaster Capitalism*. The Nation, May 2, (www.thenation.com/article/rise-disaster-capitalism).

Knutson, T. R., McBride, J. L., Chan, J., Emanuel, K., Hollands, G., Landsea, C., Held, I., Kossin, J. P., Srivastava, A. K., and Sugi, M., 2010. Tropical cyclones and climate change. *Nature Geoscience*, **3**, 157–163.

Lubchenco, J., and Karl, T. R., 2012. Predicting and managing extreme weather events. *Physics Today*, **65**, 31–33.

McAdoo, B. G., Dengler, L., Eeri, M., Prasetya, G., and Titov, V., 2006. Smong: how an oral history saved thousands on Indonesia's Simeulue Island. *Earthquake Spectra*, **22**, 661–669.

McCallum, E., and Heming, J., 2006. Hurricane Katrina – an environmental perspective. *Philosophical Transactions of the Royal Society A*, **364**, 2099–2115.

Mitchell, J. F. B., Lowe, L. A., Wood, R. A., and Vellinga, M., 2006. Extreme events due to human-induced climate change. *Philosophical Transactions of the Royal Society A*, **364**, 2117–2133.

O'Keefe, P., Westgate, K., and Wisner, B., 1976. Taking the naturalness out of natural disasters. *Nature*, **260**, 566–567.

Oliver-Smith, A., 1999a. What is a disaster? Anthropological perspectives on a persistent question. In Oliver-Smith, A., and Hoffman, S. M. (eds.), *The Angry Earth: disaster in anthropological perspective*. New York: Routledge, pp. 18–33.

Oliver-Smith, A., 1999b. Peru's five-hundred year earthquake: vulnerability in historical context. In Oliver-Smith, A., and Hoffman, S. M. (eds.), *The Angry Earth: disaster in anthropological perspective*. New York: Routledge, pp. 74–88.

Oliver-Smith, A., and Hoffman, S. M., 2002. Why anthropologists should study disasters. In Hoffman, S. M., and Oliver-Smith, A. (eds.), *Catastrophe and Culture: The Anthropology of Disaster*. Santa Fe: School of American Research Press, pp. 3–21.

Paton, D., and Johnson, D., 2006. *Disaster Resilience: an integrated approach*. Springfield: Charles C Thomas Pub Ltd.

Pelling, M. (ed.), 2003. *Natural Disasters and Development in a Globalizing World*. London: Routledge. 272 p.

Pelling, M., and Dill, K., 2006. *'Natural' Disasters as Catalysts of Political Action*. Chatham House ISP/NSC Briefing Paper 06/01, pp. 4–6.

Press, F., and Hamilton, R. M., 1999. Mitigating natural disasters. *Science*, **284**, 1927.

Rodriguez, J., Vos, F., Below, R., and Guha-Sapir, D., 2009. *Annual Disaster Statistical Review 2008: the numbers and trends.* Brussels: Centre for Research on the Epidemiology of Disaster. 25p.

Romero, T. J., and Adams, J. L., 1995. Economic impact of the Northridge earthquake. In Woods, M. C., and Seiple, W. R. (eds.), *The Northridge, California, Earthquake of 17 January 1994.* California Department of Conservation, Division of Mines and Geology Special Publication, 116, pp. 263–271.

Sen, A. K., 1981. *Poverty and Famines – An Essay on Entitlement and Deprivation.* Oxford: Oxford University Press. 270 p.

Sieh, K., 2006. Sumatran megathrust earthquakes: from science to saving lives'. *Philosophical Transactions of the Royal Society A*, **364**, 1947–1963.

Smith, K., 2001. *Environmental Hazards: Assessing Risk and Reducing Disaster*, 3rd edn. London: Routledge. 392 p.

Smolka, A., 2006. Natural disasters and the challenge of extreme events: risk management from an insurance perspective. *Philosophical Transactions of the Royal Society A*, **364**, 2147–2165.

Stromberg, D., 2007. Natural disasters, economic development, and humanitarian aid. *Journal of Economic Perspectives*, **21**, 199–222.

Svensen, H., 2009. *The End is Nigh: A History of Natural Disasters.* London: Reaktion Books. 224p.

Turner, B. L., II, Kasperson, R. E., Matson, P. A., McCarthy, J. J., Corell, R. W., Christensen, L., Eckley, N., Kasperson, J. X., Luers, A., Martello, M. L., Polsky, C., Pulsipher, A., and Schiller, A., 2003. A framework for vulnerability analysis in sustainability science. *Proceedings of the National Academy of Sciences USA*, **100**, 8074–8079.

van der Vink, G., Allen, R. M., Chapin, J., Crooks, M., Fraley, W., Krantz, J., Lavigne, A. M., LeCuyer, A., MacColl, E. K., Morgan, W. J., Ries, B., Robinson, E., Rodriquez, K., Smith, M., and Sponberg, K., 1998. Why the United States is becoming more vulnerable to natural disasters. *EOS Transactions*, **79**, 533–537.

Varley, A. (ed.), 1994. *Disasters, Development and Environment.* Chichester: Wiley. 182 p.

Voltaire, F.-M. A. de, 1759. *Candide, ou L'Optimisme*: Cramer, Geneva, 299 p.

Wisner, B. 2001. *Lessons from Cuba? Hurricane Michelle, November.* Radix, Radical Interpretations of Disasters. http://online:northumbrian.ac.uk/geography_research/radix/cuba:html. Accessed 11 May 2010.

Wisner, B., Blaikie, P., Cannon, T., and Davis, I., 2004. *At Risk: Natural Hazards, People's Vulnerability and Disasters*, 2nd edn. Abingdon: Routledge. 496 p.

Cross-references

Adaptation
Casualties Following Natural Hazards
Classification of Natural Disasters
Coping Capacity
Cost-Benefit Analysis of Natural Hazard Mitigation
Disaster Relief
Economics of Disasters
Geohazards
Hazard
International Strategies for Disaster Reduction (IDNDR and ISDR)
Models of Hazard and Disaster
Natural Hazard
Natural Hazard in Developing Countries
World Wide Trends in Natural Disasters

DISPERSIVE SOIL HAZARDS

Andrew J. Stumpf
Institute of Natural Resource Sustainability, University of Illinois at Urbana-Champaign, Champaign, IL, USA

Synonyms
"Dispersive" soils (Volk, 1937; Fletcher and Carroll, 1948)

Definition

Some natural clay-rich soils are highly erodible by flowing water both at and below land surface. These soils contain an abundance of clay particles that disperse (slake) and deflocculate when relatively pure water is added. Such "dispersive soils" have clays with a higher exchangeable sodium percentage – the proportion of sodium cations to the total of other soluble cations (e.g., calcium and magnesium). Because of the mineralogy of their clay particles, these soils are distinct highly susceptible to erosion by gulleying, tunneling, and piping when cultivated or when disturbed to some depth below land surface.

Soil characteristics

Dispersive soils were first recognized over 120 years ago, but were not studied in depth until over 50 years later by Volk (1937) and Richards (1954), and later by Australian engineers (e.g., Aitchison and Wood, 1965). These soils contain a high proportion of clay particles that have weak electrochemical attraction to adjacent particles. These bonds are affected primarily by the type of clay minerals present, however pH, amount of organic matter, temperature, water content, thixotropy (viscosity change), and chemistry of pore water (Bell and Maud, 1994) also can affect dispersion. These soils have higher percentage of exchangeable sodium (expressed by the exchangeable sodium percentage – ESP) than most soils. Dispersive sodic soils in Australia have ESP > 6 in the top meter of the soil horizon (Northcote and Skene, 1972; Raine and Loch, 2003). Commonly, dispersive soils contain little organic matter content and have alkaline pore waters with a pH > 8.5. More recent studies have found these soils in humid tropical climates where the pore water may be acidic (Sherard et al., 1977). Often, little or no evidence of their associated instability are exposed at land surface, because the soil is covered with silty or sandy material (containing no dispersive clays) or a continuous layer of topsoil and vegetation. Since many traditional laboratory index tests, including specific gravity and Atterberg limits, fail to differentiate dispersive soils collapsible soil hazards (see entry *Collapsing Soil Hazards*) from non-dispersive soils (Sherard et al., 1972), a number of specialized experiments have been developed to measure the erodibility of dispersive soil, (Reilly, 1964; Sherard et al., 1976b, and Emerson, 2002).

Process

Erosion occurs by a process in which individual clay particles are electrochemically suspended in water ponded at the land surface or in the subsurface in soil pores. The particles are transported when the water flows. This process is significantly different than erosion taking place in other soil types, where a considerable velocity is needed for water to erode clay particles (Sherard et al., 1976a). More specifically, as water wets the soil clay particles interact with the water to weaken interparticle sodium bonds. Eventually, these bonds are broken, the sodium cations disperse, and the individual clay particles begin to deflocculate (Knodel, 1991).

Distribution

Many early studies indicated that dispersive soils form exclusively in arid and semiarid climates in alkaline soils. In alluvial floodplain deposits, slope wash, lake bed deposits, and Loess (see entry *Loess*). Some dispersed soils have been identified in residuum on marine claystone and shale, granites, and sandstone (Sherard et al., 1977; Clark, 1986). More recent field work has extended their distribution to humid climates and are now identified in United States, Venezuela, Australia, South Africa, Iran, Tasmania, Mexico, Trinidad, Vietnam, South Africa, Thailand, Israel, Ghana, and Brazil (Knodel, 1991).

Hazards

The breakdown in the internal structure of dispersive soils has led to problems such as surface crusting, reduced water infiltration, and retarded plant establishment and growth (Rengasamy et al., 1984). Dispersive soils are prone to gully erosion and piping hazards (see entry *Piping Hazard*). Damage due to subsidence has occurred in existing earth embankment dam built with dispersive soil, and in new reservoirs and buildings constructed on these soils. In the United States, the most notable failure in dispersive soils occurred at Teton Dam (Sherard, 1987). Some studies have shown that the failure of structures built on dispersive clay soils occurs after the first wetting (Knodel, 1991).

Summary

The study of dispersive soil has a long history, and a worldwide effort has been undertaken to identify and characterize them through rigorous testing, in order to develop a comprehensive classification scheme. Ancillary research has utilized this data to develop products (i.e., filters and chemical additives) used to mitigate their affect on agriculture and water-retaining structures. In many countries, major outreach programs have been developed to educate landholders, planners, and engineers to draw awareness toward the problems associated with dispersive soils.

Bibliography

Aitchison, G. D., and Wood, C. C., 1965. Some interactions of compaction, permeability, and post-construction deflocculation affecting the probability of piping failures in small dams. In *Proceedings, 6th International Conference on Soil Mechanics and Foundation Engineering*, International Society of Soil Mechanics and Foundation Engineering, 2, p. 442.

Bell, F. G., and Maud, R. R., 1994. Dispersive soils: a review from a South African perspective. *Quarterly Journal of Engineering Geology and Hydrogeology*, **27**, 195–210.

Clark, M. R. E., 1986. *Mechanics, Identification, Testing, and Use of Dispersive Soil in Zimbabwe*.

Emerson, W. W., 2002. Emerson dispersion test. In McKenzie, N., Coughlan, K., and Cresswell, H. (eds.), *Soil Physical Measurement and Interpretation for Land Evaluation*. Collingwood: CSIRO, pp. 190–199.

Fletcher, J. E., and Carroll, P. H., 1948. Some properties of soils that are subject to piping in southern Arizona. *Proceedings - Soil Science Society of America*, **13**, 545–547.

Knodel, P. C., 1991. Characteristics and problems of dispersive clay soils. Denver: United States Department of the Interior, Bureau of Reclamation, Report 91-09.

Northcote, K. H., and Skene, J. K. M., 1972. *Australian soils with saline and sodic properties*. Melbourne: CSIRO Australia, Division of Soils.

Raine, S. R., and Loch, R. J., 2003. What is a sodic soil? Identification and management options for construction sites and disturbed lands. In *Roads, Structures and Soils in Rural Queensland*. Brisbane: Queensland Department of Main Roads.

Reilly, L. A., 1964. The nature and identification of chemically unstable clays. In Contributions to Colloquium on Failure of Small Earth Dams. Melbourne: Water Research Foundation of Australia Limited/Commonwealth Scientific and Industrial Research Organization (CSIRO), Division of Soil Mechanics, Paper No. 22.

Rengasamy, P., Greene, R., Ford, G. W., and Mehanni, A. H., 1984. Identification of dispersive behaviour and the management of red-brown earths. *Australian Journal of Soil Research*, **22**, 413–431.

Richards, L. A., 1954. *Diagnosis and Improvement of Saline and Alkali Soils*. Washington: United States Department of Agriculture, Handbook No. 60.

Shainberg, I., Rhoades, J. D., and Prather, R. J., 1981. Effect of low electrolyte concentration on clay dispersion and hydraulic conductivity of a sodic soil. *Journal of the Soil Science Society of America*, **45**, 273–277.

Sherard, J. L., Decker, R. S., and Ryker, N. L., 1972. Piping in earth dams of dispersive clays. In *Proceedings, Specialty Conference on Performance of Earth and Earth-Supported Structures*. American Society of Civil Engineers, 1, pp. 584–626.

Sherard, J. L., Dunnigan, L. P., and Decker, R. S., 1976a. Identification and nature of dispersive soils. *Journal of the Geotechnical Engineering Division, Proceedings of the American Society of Civil Engineers*, **102**, 287–301.

Sherard, J. L., Dunnigan, L. P., Decker, R. S., and Steele, E. F., 1976b. Pinhole test for identifying dispersive soils. *Journal of the Geotechnical Engineering Division, Proceedings of the American Society of Civil Engineers*, **102**, 69–85.

Sherard, J. L., Dunnigan, L. P., and Decker, R. S., 1977. Some engineering problems with dispersive clays. In Sherard, J. L., and Decker, R. S. (eds.), *Dispersive Clays, Related Piping, and Erosion in Geotechnical Projects*. American Society for Testing and Materials, Special Technical Publication, 623, pp. 3–12.

Sherard, J. L., 1987. Lessons from the Teton Dam Failure. *Engineering Geology*, **24**, 239–256.

Volk, G. M., 1937. Method of determination of degree of dispersion of the clay fraction of soils. *Proceedings - Soil Science Society of America*, **2**, 561–567.

Cross-references

Collapsing Soil Hazards
Hydrocompaction Subsidence
Land Subsidence
Loess
Piping Hazard

DOPPLER WEATHER RADAR

Rodger A. Brown
NOAA/National Severe Storms Laboratory, Norman, OK, USA

Synonyms

Doppler frequency shift

Definition

Conventional weather radar. A conventional weather radar transmits a narrow pulse of electromagnetic radiation (centimeter wavelength), and then listens to see if any energy is scattered back from distant targets before the next pulse is transmitted (e.g., Rinehart, 2010). The targets of meteorological interest are hydrometeors (raindrops, hailstones, ice crystals). The time delay between the transmitted and returned pulse determines the distance to the hydrometeors and the amount of energy received (called radar reflectivity) is proportional to the size and scattering characteristics of the hydrometeors within the pulse volume. Owing to random fluctuations of hydrometeors within the pulse volume, tens of consecutive pulses are averaged together to obtain a representative measurement.

Doppler weather radar. A Doppler weather radar is a conventional weather radar that has the additional capability of detecting a slight frequency shift (Doppler shift) in the returned pulse (e.g., Rinehart, 2010). The frequency shift is caused by *the component of hydrometeor motion toward or away from the radar*. The three basic quantities measured by a Doppler radar are radar reflectivity, Doppler velocity (the mean Doppler velocity component of hydrometeor motion within the series of returned pulses), and spectrum width (the standard deviation of the velocity components within the series of pulses).

Overview

During the mid-1950s, a few research organizations around the world started to apply Doppler radar techniques to study weather phenomena (e.g., Rogers, 1990). The first radars were pointed vertically because updrafts and downdrafts in storms could be uniquely identified. However, by the late 1960s and early 1970s, researchers began to scan radars horizontally through thunderstorms. They discovered that – even though a Doppler radar measures only the single component of flow toward or away from the radar – there are unique Doppler velocity signatures of rotating and divergent/convergent flows that have warning implications. With coordinated measurements from two or more nearby Doppler radars and a few assumptions, researchers can estimate the full three-dimensional components of airflow within storms.

During the 1980s and 1990s, based on the existence of single Doppler velocity signatures, it was becoming apparent to national weather services in various countries that they could improve the timeliness and accuracy of hazardous weather warnings by replacing their conventional radar networks with Doppler radar networks (e.g., Whiton et al., 1998). With many such networks now in existence, Doppler radar data are beginning to be assimilated into numerical weather prediction models with the goal of producing more accurate short-term (1–6 h) forecasts of evolving hazardous weather conditions.

Bibliography

Rinehart, R. E., 2010. *Radar for Meteorologists*, 5th edn. Nevada: Rinehart.
Rogers, R. R., 1990. The early years of Doppler radar in meteorology. In Atlas, D. (ed.), *Radar in Meteorology*. Boston: American Meteorological Society, pp. 122–129.
Whiton, R. C., Smith, P. L., Bigler, S. G., Wilk, K. E., and Harbuck, A. C., 1998. History of operational use of weather radar by U.S. weather services. Part II: Development of operational Doppler weather radars. *Weather and Forecasting*, **13**, 244–252.

Cross-references

Cloud Seeding
Dust Storm
Hurricane
Hurricane Katrina
Ice Storm
Storms
Thunderstorms
Tornado
Waterspout

DOSE RATE*

Cathy Scheib
British Geological Survey, Nottingham, UK

Synonyms

Radiation absorbed

Definition

Dose rate is the quantity of radiation absorbed per unit time (Gy s^{-1}).

Overview

The *absorbed dose* is the amount of energy deposited by ionizing radiation in a unit mass of medium, such as tissue.

*©British Geological Survey

This dose is expressed in units of joule per kilogram (J kg^{-1}), which is called the "Gray" (Gy). The unit Gray can be used for any type of radiation, but it does not describe the biological effects resulting from different radiation types. Absorbed dose rate in air (nGy h^{-1}) is commonly used to express gamma ray intensity in the air from radioactive materials in the earth and atmosphere. *Equivalent dose* relates the absorbed dose in human tissue to the effective biological damage and is expressed in the unit Sievert (Sv) (The International System of Units, 2008). To determine equivalent dose, the absorbed dose is multiplied by a quality factor that is unique to the type of incident radiation in question (e.g., alpha particles, 20; beta particles, 1; gamma and x-rays, 1). To take account of the susceptibility of organs and tissues to radiation doses, weighted equivalent doses in all the tissues and organs of the body are summed to determine the *effective dose* (Sv) (Wrixon, 2008; US Department of Health and Human Studies).

Sources and effects of human exposure to ionizing radiation

Natural radiation contributes over 80% of the average radiation dose received; approximately half the overall dose occurring due to exposure to radon gas and its decay products (Eisenbud and Gesell, 1997). Terrestrial gamma radiation – which is largely controlled by geological variation of naturally occurring radioactive materials in rocks, soils, and building materials – and cosmic radiation, which varies with altitude and latitude, contribute on average 13% and 12%, respectively of the average annual dose to the UK population (Hughes et al., 2005). The average global annual effective dose from natural radiation is approximately 2.4 mSv. This level of exposure varies around the world, usually by a factor of 3, although at some locations it can be exceeded by more than a factor of 10 (UNSCEAR, 2000).

Anthropogenic sources of exposure to ionizing radiation include medical screening and therapeutic procedures, nuclear weapons testing, electricity generation, and accidents such as the one at Chernobyl in 1986, although the contribution to dose from these sources is small in comparison to that from natural radiation (UNSCEAR, 2000).

Damage to DNA in the nucleus is the main initiating event by which radiation causes long-term damage to organs and tissues of the body (UNSCEAR, 2006). There is no convincing scientific evidence that cancer risk from radiation exposure disappears at very low doses, and this is currently the focus of major research (e.g., US Department Of Energy – Low Dose Radiation Research Programme).

Bibliography

Eisenbud, M. and Gesell, T., 1997. *Environmental Radioactivity from Natural, Industrial, and Military Sources*, 4th Edn. San Digeo: Academic. 656 pp. ISBN 0-12-235154-1.

Hughes, J. S., Watson, S. J., Jones, A. L., and Oatway, W. B., 2005. Review of the radiation exposure of the UK population. *Journal of Radiological Protection*, **25**, 493–496.

The International System of Units (SI), Ed. by B. N. Taylor and Ambler Thompson, Natl. Inst. Stand. Technol. Spec. Publ. 330, 2008 Edition (U.S. Government Printing Office, Washington DC http://physics.nist.gov/cuu/Units/bibliography.html (last accessed at 11:24 on June 22nd 2010).

UNSCEAR, 2000 Report. Vol. 1. Annex B. New York: United Nations. http://www.unscear.org/unscear/en/publications/2000_1.html (last accessed at 11:27 on June 22nd 2010).

UNSCEAR, 2006 Report. Vol. 1. Annexes A and B. New York: United Nations. http://www.unscear.org/unscear/en/publications/2006_1.html (last accessed on 11.28 on June 22nd 2010).

US Department Of Energy – Low Dose Radiation Research Programme http://lowdose.energy.gov/default.aspx (last accessed at 11:28 on June 22nd 2010).

US Department of Health and Human Studies- Dictionary of radiological terms. http://www.remm.nlm.gov/dictionary.htm (last accessed at 11:29 on June 22nd 2010).

Wrixon, A. D., 2008. New ICRP recommendations. J. Radiol. Prot. **28**, 161–168. http://www.iop.org/EJ/abstract/0952-4746/28/2/R02 (last accessed 11:30 on June 22nd 2010).

Cross-references

Natural Radioactivity
Radon Hazards

DROUGHT

Suzanne Hollins, John Dodson
Australian Nuclear Science and Technology Organisation, Lucas Heights, NSW, Australia

Definition

Drought is a severe natural hazard that affects more people than any other natural disaster. It is usually only recognized as a natural hazard when social, economic, or environmental impacts become apparent. Drought is different from many other natural hazards in that it lacks easily identified onsets and terminations (Maybank et al., 1995). It is also unusual in that it is a hazard of scarcity rather than one of excess. Drought is a natural, recurring pattern of climate that occurs within nearly all climatic regions. However, it is not just a physical phenomenon or natural event caused by changes in climatic conditions. Rather, drought results from a connection between the natural event of lower than expected precipitation, and the demand of human usage on water supplies (Wilhite, 2000). Anthropogenic activities can exacerbate the severity and impacts of drought, but within a natural variability range.

Drought does not have a universal definition; rather it has hundreds, as Wilhite and Glantz (1985) discussed in their classification study. Despite the number of definitions, many are not useful to policy makers or scientists. This causes some uncertainty in declaring whether a region is suffering from drought, and its degree of

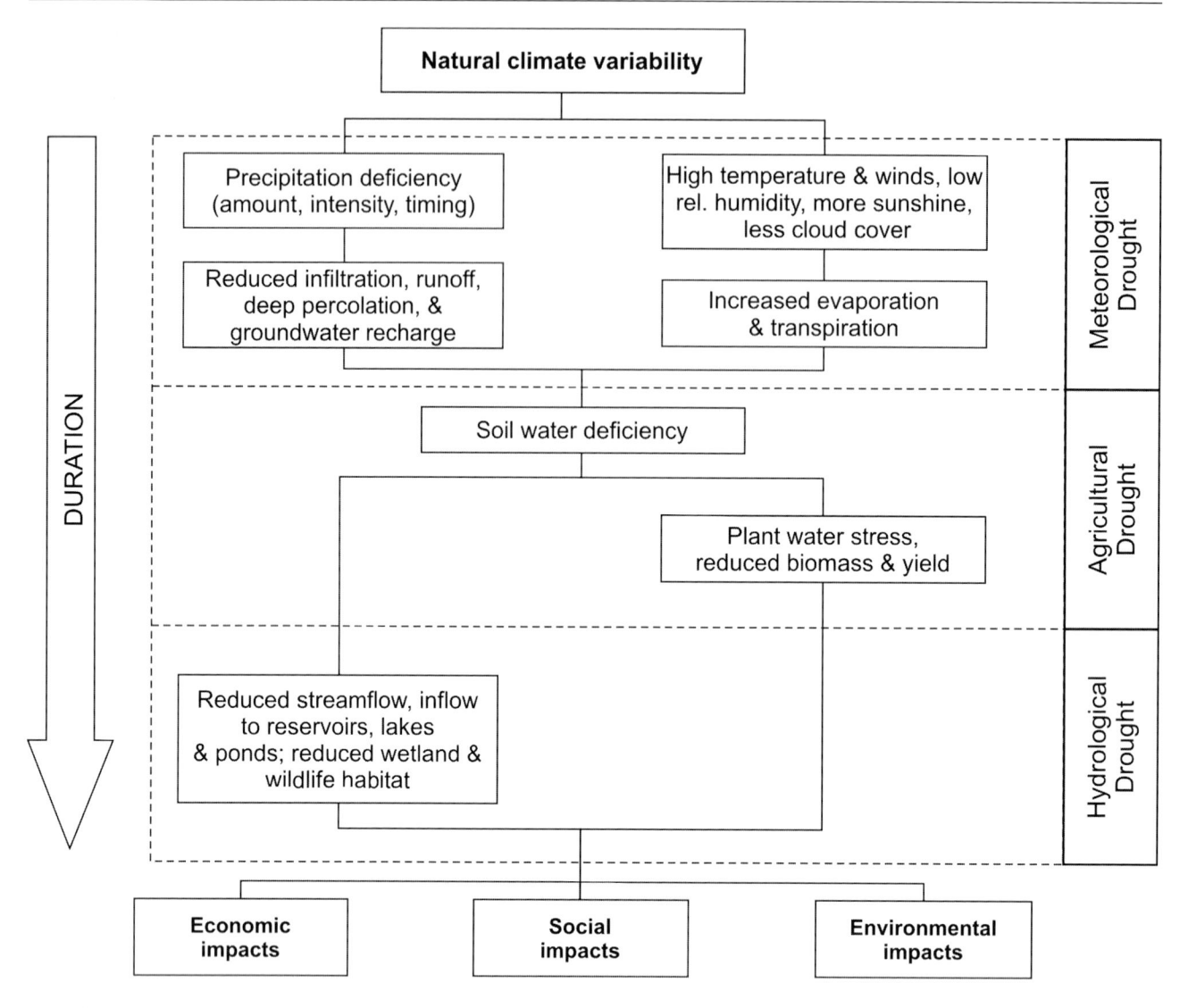

Drought, Figure 1 The relationship between types of drought and duration of drought events (Figure modified from Wilhite, 2000, p. 10.).

severity. There are three major characteristics of drought – intensity, duration, and spatial extent. Drought is never small scale or short term (by definition). The effects of drought can build up over lengthy periods of time, and these effects may be felt for years after the drought has "broken," making the onset and conclusion of drought difficult to define. "Seasonal" droughts are frequent and predictable, as distinguished from "supra-seasonal" droughts, which are aberrant and unpredictable (Bond et al., 2008). It is generally accepted that drought can be divided into four categories based on disciplinary viewpoints: meteorological, hydrological, agricultural, and socioeconomic (Wilhite and Glantz, 1985). Figure 1 (modified from Wilhite, 2000) shows the relationship between the various categories of drought and their durations. Each discipline incorporates different physical, biological, and/or socioeconomic factors in its definition (Wilhite, 2000), but common to all these is inadequate precipitation.

A working definition of meteorological drought is "an extended period (season, year, or several years) of deficient rainfall relative to the statistical multiyear mean for a region" (Druyan, 1996a). A lack of rainfall does not necessarily constitute drought. It must be distinguished from aridity, which occurs in areas where there is a high probability of low rainfall for indeterminate periods of time (Druyan, 1996b). Meteorological drought must be defined on a regional basis, as deficiencies in precipitation are specific to local atmospheric conditions. Once meteorological drought establishes itself, both agricultural and hydrological drought usually follows.

Hydrological drought is associated with the effects of a persistent scarcity in rainfall on the capacity and availability of surface water (e.g., rivers, lakes, reservoirs) and groundwater supplies. The frequency and severity of hydrological drought is often defined on a catchment or basin scale. The commencement and finishing of groundwater drought both usually lag well behind that of surface water drought (Bond et al., 2008), and both are usually out of phase with meteorological and agricultural drought. As a lack of expected rainfall continues, water levels in temporary water bodies decrease, and they eventually dry up. The drought also decreases water levels in "perennial" surface water bodies, and if it continues long enough, these may also disappear. As surface waters are depleted by ongoing drought, groundwater levels may also decrease over time in those aquifers influenced by modern recharge. This can exacerbate the effects of drought in surface water systems in which groundwater forms the base flow. After a return to normal rainfall conditions, surface water drought usually breaks well before groundwater drought.

Agricultural drought is associated with a shortage of available water for plant growth. It is assessed as insufficient soil moisture to replace evapotranspiration losses, and links meteorological and hydrological drought to impacts on agriculture. Most regions can be affected by agricultural drought, but its duration and intensity varies greatly between climatic zones (Wilhite, 2000). There are many definitions of agricultural drought, but in general they account for the varying susceptibility of crops during development to deficient topsoil moisture.

Scientists tend to frame the broad social dimensions of droughts into a general category called "socioeconomic drought" (Kallis, 2008). Socioeconomic drought associates supply and demand of economic goods with at least some elements of meteorological, hydrological, and agricultural drought (Wilhite, 2000). It can result when the demand for economic goods exceeds supply because of a shortfall in water related to variations in climate. It can also occur when the demand for goods increases due to population increase and/or per capita consumption.

Drought indices

Measurements of the frequency and severity of droughts are important in the development of mitigation strategies and preparedness plans. As in drought definitions, it is generally agreed that there can be no universal drought index or operational definition (Kallis, 2008), and so numerous indices have been developed to monitor and measure drought. Droughts differ in three major ways: intensity, duration, and spatial coverage (Wilhite, 2000). Intensity is related to the precipitation deficit, and several indices measure how precipitation has deviated from historical norms. The duration of drought is a discerning characteristic, which along with intensity and timing, is closely related to the level of impact. The spatial characteristics of drought also differ as the degree of severity evolves across areas and through seasons.

Drought and its severity can be numerically defined using indices that integrate temperature, precipitation, and other variables that effect evapotranspiration and soil moisture (IPCC, 2007a). The simplest of these assess meteorological drought using a measure of the precipitation deficit over a particular time period, whereas those more complicated use models that incorporate soil moisture conditions and land-use parameters (Oladipo, 1985).

One of the main difficulties with using indices that measure precipitation deficiency is setting the threshold below which the onset of drought is defined (Wilhite, 2000). The Palmer Drought Severity Index (PDSI; Palmer, 1965) is one of the most extensively used meteorological drought indices across the world, and particularly in the USA. PDSI is a reflection of how much soil moisture is currently available compared to that for normal or average conditions (Cook et al., 2007). The PDSI was one of the first methods to successfully quantify the severity of droughts across different climates. The index is based upon a primitive water balance model which accounts for the difference between precipitation required to maintain a water balance and the actual precipitation. The PDSI also incorporates calculations that attempt to account for climatic differences between locations and seasons of the year (Wells et al., 2004). Despite its popularity, the PDSI has been widely criticized for its empiricism (Keyantash and Dracup, 2002). It does not incorporate variables such as wind speed, water vapor, or solar radiation into its calculation of potential evapotranspiration. Commonly, it is said that calculated PDSI values are not comparable between diverse climatological regions. This led to the development of a self-calibrating version of the PDSI by Wells et al. (2004) to ensure consistency with the climate at any location. A relatively new precipitation deficit index used in the USA is the Standardized Precipitation Index (SPI) developed by McKee et al. (1993, 1995) in recognition of the impacts that precipitation deficit has on groundwater, soil moisture, streamflow, and other water resources. It was designed to quantify precipitation deficit for multiple timescales, allowing for the determination of the rarity of drought as well as the probability of precipitation necessary to break a drought. In Australia, the drought definition is based on the Rainfall Deciles method (Gibbs and Maher, 1967), chosen because it is relatively simple to calculate, and requires fewer assumptions than the PDSI (Smith et al., 1993).

Similar to the PDSI is the Palmer Hydrological Drought Severity Index (PHDI), with the primary difference being stricter criterion for the ending of a drought (or wet spell). This is considered more appropriate for hydrological drought assessment, as it is much slower to build up than meteorological drought (Keyantash and Dracup, 2002). Shafer and Dezman (1982) developed the Surface Water Supply Index (SWSI) to account for snowpack and delayed runoff, and it is useful in providing a measure of hydrological drought in areas where snow makes up a significant component of the hydrological budget.

Agricultural drought is specifically related to cultivated crops rather than natural vegetation (Keyantash and Dracup, 2002), and it is characterized by short-term changes to volumetric soil moisture in the root zone. The Crop Moisture Index (CMI) was developed by Palmer (1968) and uses a meteorological approach to monitor agricultural drought. The CMI was developed from procedures within the PDSI, but it was designed to measure short-term moisture conditions across crop-producing regions, rather than monitoring long-term meteorological drought like the PDSI (Hayes, 2009).

Many parts of the world have not adopted clear indices for agricultural drought and hydrological drought, making attempts to compare the impacts of drought between places and between times difficult (Bond et al., 2008). In the context of climate change and increasing land degradation, it is becoming increasingly important to be able to calculate drought impacts if the consequences of climate change are to be understood (Vicente-Serrano, 2007).

Impacts of drought

Drought, as one of the most complicated yet least understood natural hazards, is associated with many other kinds of hazard, and these play out in impacts on economic, social, and environmental systems (Kallis, 2008). The onset of drought is difficult to identify or even recognize, although predictive capabilities are increasing. The study of past droughts can indicate what onset might look like, how drought develops and the kinds of impacts that follow. The palaeo record shows that severe droughts of the last century were greatly eclipsed by megadroughts in the past (Maybank et al., 1995; Woodhouse and Overpeck, 1998). These will occur again, and are likely to be exacerbated by greenhouse warming.

There are methodological problems in assessing the impact of droughts due to the difficulty of defining it. However, the most obvious first-order impacts are through drought impact on agricultural production, water supply, and forestry. Forests are usually less sensitive to drought as they tend to occur in wetter regions.

Reduction in crop and animal production has secondary affects on food prices and may feed through to global markets and consumer demand (Kallis, 2008). Reduction in river flows may have consequences for water supply, hydroelectricity generation, and the amount of potable water. Poor quality water can have significant negative health outcomes for affected populations.

Drought takes a heavy toll on life in Africa, causes social disruption in Asia, and has economic impacts in Western countries. Exposure and vulnerability have strongly regionalized patterns; where drought coincides with war, poverty, or recession, the impact is magnified and exposed population are made more vulnerable (Kallis, 2008).

There is a significant difference between aridity and drought. Deserts occur in areas where there is extreme heating of the surface, and/or lack of moisture. These are created when subsiding air, which becomes compressed and thus heated, forms subtropical high-pressure zones. The deserts of Australia, Peru-Chile (Atacama), southwestern USA, Namib, Sahara, and Kalahari are of this type. Additionally, deserts occur in the lee of major mountains: Patagonia, Middle East, central Asia, Ethiopia, and the Thar (India) are examples of these.

Deserts are naturally dry most of the time, and thus drought is not a hazard in them in the strictest sense. However, droughts can be a normal weather pattern in all regions.

Projected precipitation anomalies estimated from regional climate models depend heavily on the scenario applied for the simulations. The IPCC (2007b) has applied a relatively large number of simulations (21) and these show a high degree of consistency. Drought conditions will be exacerbated whenever simulations suggest a decline in precipitation, especially outside the natural regions of aridity. Figure 2 (modified from IPCC, 2007b) shows that the main areas of predicted precipitation decrease are:

1. Annual decrease in the Mediterranean region, northern Africa, Central America, and SW USA.
2. Winter decrease in SW Australia, eastern French Polynesia, southern Africa.
3. Winter and spring decreases in southern Australia.
4. Decrease in snow season length and likely snow depth in Europe and North America.
5. Summer decrease in the southern Andes, southern SE Asia, SE South America, central Asia, central Europe, and southern Canada.

The effect of decreased precipitation will be enhanced with higher temperatures. The main regions of impact are outside the tropics and high latitude zones, and areas with winter and spring dominated rainfall patterns will be particularly disadvantaged, as will midlatitude areas dependent on snow melt for water supply. The burden of enhanced drought will fall quite unevenly across the nations of the world. These same regions will need to develop robust adaption and mitigation strategies to reduce future vulnerability to drought.

Historical impact of drought

Severe drought can have serious consequences for exposed societies. The degree of exposure depends on the kind of drought and the resilience of the society. All but the least resilient of societies can weather single seasonal droughts when they occur at some kind of recurring interval. They often do this by building reserves that can be drawn upon in times of need. Societies meet the most challenging of situations when long sequences of unexpected drought conditions occur. These may be due to rare and essentially unanticipated sequences or due to a shift in climate pattern. It is expected that climate change will, as it always has, alter the geographical patterns and severity of droughts.

In the past, many societies have encountered unexpected enduring drought conditions and they have had to adjust and adapt, migrate, or they collapse. One can wonder at

June–July–August (JJA)

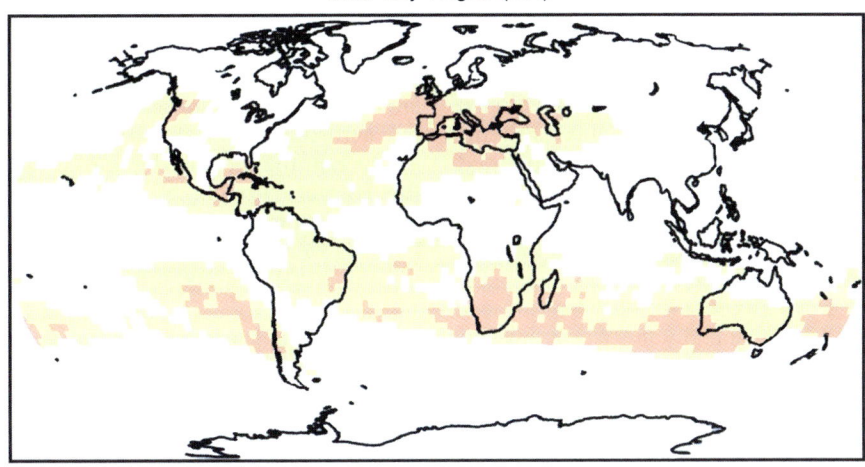

December–January–February (DJF)

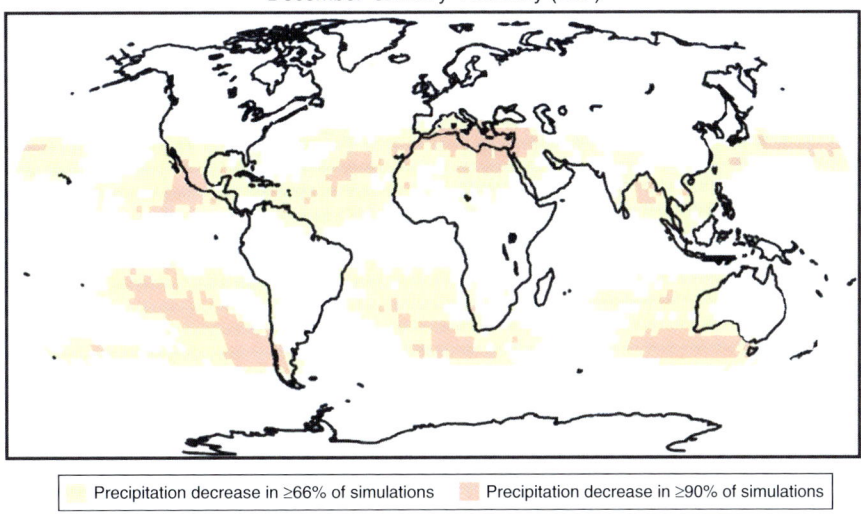

Precipitation decrease in ≥66% of simulations   Precipitation decrease in ≥90% of simulations

Drought, Figure 2 The fraction of 21 atmosphere–ocean global climate model simulations that predict a decrease in mean precipitation in a model grid cell (comparing the period 2080–2099 with control period 1980–1999) (Figure modified from IPCC, 2007a, p. 859.).

the thought processes that accompanied these circumstances. Initially, a poor season would have placed strains that would have been endured with the expectation that "normality" would return the next season. After all, this was what experience shows to be the case. A string of poor seasons challenges this experience, and the longer the sequence the more challenging this becomes. When do societies accept that conditions have indeed changed and adjustments must be made? This dilemma has been met before and will be visited upon many societies in the future.

An example of how this occurred in the past concerns the Classic Maya Civilization. The Maya occupied the Yucatan Peninsula region of Mesoamerica from 250 to 850 AD. The Late Classic culture (550–850 AD) was known for being a highly stratified society; there were vast trade networks, and widespread construction of urban centers and monuments. Complex language, belief systems, sports, and mathematics were embedded as elements of society. This all came to a sudden end when society seemed to be at its peak. Many potential factors have been cited for the collapse and include deforestation, overpopulation, warfare, and social upheaval for political reasons.

Recent research suggests prolonged drought was at least a contributing factor. Lake sediments reveal substantially lowered water levels and changes from freshwater to saline conditions (Hodell et al., 2001), reduction in forest cover according to pollen diagrams (Mueller et al., 2009), and increased soil erosion is recorded in marine sediments (e.g., Gischler et al., 2008). These indicate substantial environmental change, which coincides with the main phase of collapse of the Maya in terms of buildings and the desertion of urban centers. In the latter phase of the Maya, Sun God worship was evident, and this may have been an attempt to appease the Sun as the cause of ongoing drought.

Similar fates are thought to have transpired to the Harappan people of NW India as an arid phase developed over the region and made extensive irrigation systems become dysfunctional, and the base which supported a huge urban population was swept away (Staubwasser et al., 2003).

In some cases, human activities have appeared to exacerbate the impact of drought. In northern China, there is an environmental boundary between the loess and desert. Loess is windblown dust deposited by the Westerlies and is a highly productive soil where there is sufficient rain or irrigation that can be applied to it. In the northern region the loess gives way to desert sand-dominated soils, these are mobile and rarely watered by the monsoon rains which sweep in from the Pacific Ocean. The border region of the Northern Loess Plateau and Chinese deserts west of Beijing supported many Neolithic villages in the mid-Holocene. It appears that monsoon rain reached the region and provided sufficient water for millet-based agriculture and animal husbandry. By about 3,000–4,000 years ago villages were abandoned as desertification set in (Zhou et al., 2002). This may have been due to drought resulting from failure of the monsoon rains reaching the region, perhaps in concert with anthropogenic driven land degeneration. In any case, the desert sands shifted some hundreds of kilometers south, and so did the villages.

The observational record of drought

Observations of drought based on meteorological records indicate that they have become more intense, of longer duration, and occurred over wider areas of the tropics and subtropics since the 1970s (IPCC, 2007a). Reliable meteorological records for much of the world only exist for the last 100 years or so, but they provide a basis to investigate possible causes for drought.

Since the 1950s, the number of heat waves and warm nights has increased. These have contributed to the area under drought, although the drivers of changes in precipitation are also very important. While increases in continental temperatures are important for some regions, changes in snowpack and sea surface temperatures related to phenomena, such as El Niño–Southern Oscillation, are also strong drivers for climate in other regions. Extreme events such as the drought for western North America (Canada to Mexico) in 1999–2004 seem to be strongly related to a diminished snowpack and hence runoff (McCabe et al., 2004; Stewart et al., 2004). These in turn may be driven by sea surface temperatures in the tropical Pacific (Herweijer et al., 2007). Recent Australian droughts correlate well with higher continental temperatures, and the 15% decline in precipitation for southwestern Australia since the 1970s (mostly a failure of early winter precipitation) is related to sea surface temperature variation in the tropical Indian Ocean (Samuel et al., 2006).

Drought in the Sahel is due to failure of rainfall. Simulations have been good at reproducing the decadal variations in Sahel rainfall (Held et al., 2005) and these suggest sea surface temperatures of the Indian Oceans and Mediterranean are significant drivers of this, as is sulfate aerosol concentration (Rosenfeld et al., 2008).

Overall, the increased risk of drought, as measured by the Palmer Drought Severity Index suggests that the anthropogenic fingerprint is there, but simulations have this as a weaker component than the observed occurrence of drought (IPCC, 2007b).

Future vulnerability to drought

Observations show drought has already increased. Models can be used to simulate possible future drought intensity, frequency, and extent. In general, these suggest that the trends already seen can be expected to intensify, and the increase will be between 1% and 30% of land area in the next few decades – with greatest increase in midlatitude areas. The Mediterranean, western USA, Southern Africa, and northeastern Brazil are all expected to see intensification of drought. Russia, Mongolia, China, southern SE Asia will see drought intensification due to higher temperatures in the summer and drier months and due to changes in ENSO (IPCC, 2007b), and poleward migration of annular weather modes (Yin, 2005; Menéndez and Carril, 2010).

The impact of drought will be intensified due to human population increases. About one sixth of the world's population relies on meltwater, and reduced snowfall and snowpack will result in a reduction in delayed runoff. People in Bolivia, Ecuador, Peru, and the Hindu Kush – Himalaya are particularly vulnerable to this (Barnett et al., 2005). Soil moisture deficits will reduce pasture growth in the eastern South Island and Bay of Plenty regions of New Zealand (Mullan et al., 2005). There will be increased fire danger in seasonal environments of the midlatitudes (Gonzalez et al., 2010).

The high cost of drought, for example, in Australia in 1982–1983 ($2.3b), 1991–1995 ($3.8b), and 2002–2003 ($7.6b) (IPCC, 2007b), has already driven measures for adaption. A range of options have been used or are being considered for vulnerable areas. These include increased rainwater harvesting, adjustment of silvicultural techniques, channel and pipe leakage reduction and modifying crop planting dates, and choosing varieties which are more drought resistant.

Models also suggest precipitation extremes will be more prevalent, but the gaps between high magnitude events will increase, and so will the likelihood of drought. Of course the scale of these changes will depend on the willingness of nations to reduce the size of the anthropogenic fingerprint on global warming.

Mitigation

As a natural element of climate, the recurrence of hydrometeorological drought is inevitable. However, drought can also be exacerbated by anthropogenic influences such as rapid population growth, excessive water demand, and land degradation, and vulnerability to these impacts can be mitigated by appropriate drought plans (Rossi et al., 2005).

The uncertainty about drought definition leads to uncertainty about its characteristics and impacts, which contributes to poor drought management and mitigation across many parts of the world (Wilhite et al., 2007). A key element in any drought plan is a set of indicators that characterize drought conditions, and location-specific triggers (indicator values) which prompt some kind of response. Unfortunately, drought plans often contain ad hoc indicators and triggers that lack scientific validation or operational relevancy, and this can weaken the effectiveness of the mitigation plan (Steinemann and Cavalcanti, 2006). Other factors contributing to the difficulty of developing an effective drought plan include the changing spatial and temporal scales of drought impacts, the unique characteristics of each region or watershed, and changing societal structures and demands, to name just a few. Even though an existing drought may be of similar intensity and duration to one that has occurred in the past, changes in socioeconomic structures and environmental conditions can result in strikingly different impacts, and therefore changing vulnerability (Wilhite et al., 2007).

According to Wilhite (2000), drought mitigation is "short and long-term actions, programs, or policies implemented during and in advance of drought that reduce the degree of risk to human life, property, and productive capacity." These measures can be classified as either proactive or reactive. The proactive measures are prepared according to a planning strategy rather than in an emergency situation (Rossi et al., 2005). The most effective actions are long-term measures taken in advance of drought, such as building infrastructure to increase the reliability of water supply under increasing demand and drought conditions (Dziegielewski, 2003). Short-term measures are taken after the onset of drought, and these are aimed at mitigating impacts within existing infrastructure and management policies. An effective mitigation strategy will contain an appropriate mix of long- and short-term actions to reduce the vulnerability of human life, property, and production to future droughts.

Mitigating agricultural drought

Vulnerability, and therefore appropriate mitigation actions, differs significantly between the developing world, where drought can lead to livelihood loss, famine, and even death, and the developed world, where impacts are usually economic or asset losses. Numerous mitigation measures have been formulated to reduce the impacts of drought, and especially that of agricultural drought because of its huge environmental, economic, and social costs (Maybank et al., 1995).

Approximately, 80% of the world's agricultural land is rainfed (Rockström, 2003), so developing mitigation strategies to build ecological resilience in drought-prone and semiarid agricultural land is very important for food security. To a certain extent, water harvesting through small-scale systems such as farm ponds and subsurface tanks, can help mitigate the impacts of drought or dry spells in these areas. The building of ecological resilience to drought also requires strategies such as conservation farming (minimal or no tillage), improved crop varieties, and soil fertility management.

Decreases in agricultural production can have a roll-on effect leading to financial disaster for farmers and higher food prices for all consumers, unemployment, and even migration. Water is frequently wasted in agriculture practices through over-irrigation, poorly designed canals, and inefficient irrigation systems, and this waste can be reduced through adoption of improved channeling and irrigation practices (Le Houerou, 1996). A multidisciplinary approach of genetic improvement and physiological regulation to increase crop water productivity is another way to help achieve efficient and effective use of water (Cattivelli et al., 2008). Combining these biological water-saving measures with engineered solutions (e.g., water-saving irrigation methods) and agronomic and soil manipulation will contribute to an effective drought mitigation strategy for agriculture (Ali and Talukder, 2008).

Mitigating hydrological drought

Mitigation of hydrological drought primarily involves optimal water supply management under drought conditions, that is, making water more productive. This requires a contingency plan that includes a systematic evaluation of drought conditions with associated responses.

Traditionally, mitigation has focused on increasing water supplies through the construction of dams and reservoirs to capture and store increasing fractions of surface runoff. High levels of surface storage can effectively buffer against low runoff periods, especially in regions that experience high interannual variability in river flows (Bond et al., 2008). This practice was carried out with little analysis of how water was actually being used or of the impacts of this practice on the aquatic ecosystems. As new fresh surface water supplies for exploitation have dwindled, governments have turned to groundwater to augment supplies, especially during drought. However, the increased dependence on groundwater resources is leading to dwindling reserves and/or quality degradation. More and more countries are turning to nonconventional water sources to boost supplies. Desalination and waste water treatment and recycling are usually more expensive options than traditional water sources, but the associated environmental benefits can compensate for some of the costs.

Summary/Conclusions

Drought is a severe natural hazard that affects more people than any other natural disaster. It is difficult to define and recognizing its onset and termination is also difficult. It can be expressed in meteorological, hydrological, agricultural, and socioeconomic terms. The severity and extent of drought has increased in recent decades, and regional climate models suggest these will increase further in the future. The burden of dealing with drought will be unevenly distributed. Midlatitude regions and those heavily dependent on snow melt will have the greatest challenges. Multidisciplinary approaches will need to be developed to mitigate the extreme impacts of drought.

Bibliography

Ali, M. H., and Talukder, M. S. U., 2008. Increasing water productivity in crop production – a synthesis. *Agricultural Water Management*, **95**(11), 1201–1213.

Alley, W. M., 1984. The Palmer Drought Severity Index: limitation and assumptions. *Journal of Climate and Applied Meteorology*, **23**, 1100–1109.

Barnett, T. P., Adam, J. C., and Lettenmaier, D. P., 2005. Potential impacts of a warming climate on water availability in snow-dominated regions. *Nature*, **438**, 303–309.

Bond, N. R., Lake, P. S., and Arthington, A. H., 2008. The impacts of drought on freshwater ecosystems: an Australian perspective. *Hydrobiologia*, **600**, 3–16.

Cattivelli, L., et al., 2008. Drought tolerance improvement in crop plants: an integrated view from breeding to genomics. *Field Crops Research*, **105**(1–2), 1–14.

Cook, E. R., Seager, R., Cane, M. A., and Stahle, D. W., 2007. North American drought: reconstructions, causes, and consequences. *Earth-Science Reviews*, **81**(1–2), 93–134.

Druyan, L. M., 1996a. Drought. In Schneider, S. H. (ed.), *Encyclopedia of Climate and Weather*. New York: Oxford University Press, Vol. 1, pp. 256–259.

Druyan, L. M., 1996b. Arid climates. In Schneider, S. H. (ed.), *Encyclopedia of Climate and Weather*. New York: Oxford University Press, Vol. 1, pp. 48–50.

Dziegielewski, B., 2003. Long-term and short-term measures for coping with drought. In Rossi, G., Cancelliere, A., Pereira, L. S., Oweis, T., Shatanawi, M., and Zairi, A. (eds.), *Tools for Drought Mitigation in Mediterranean Regions*. The Netherlands: Kluwer, pp. 319–339.

Gibbs, W. J., Maher, J. V., 1967. Rainfall deciles as drought indicators. *Bureau of Meteorology Bulletin No. 48*, Melbourne, Australia, 84 pp.

Gischler, E., Shinn, E. A., Oschmann, W., Fiebig, J., and Buster, N. A., 2008. A 1500-year Holocene Caribbean climate archive from the Blue Hole, Lighthouse Reef, Belize. *Journal of Coastal Research*, **24**, 1495–1505.

Gonzalez, P., Neilson, R. P., Lenihan, J. M., and Drapek, R. J., 2010. Global patterns in the vulnerability of ecosystems to vegetation shifts due to climate change. *Global Ecology and Biogeography*, **19**, 755–768.

Hayes, M., 2009. *Drought indices*. Available from World Wide Web: www.drought.unl.edu/index.htm. Accessed December 18, 2009.

Held, I. M., Delworth, T. L., Lu, J., Findell, K. L., and Knutson, T. R., 2005. Simulation of Sahel drought in the 20th and 21st centuries. *Proceedings of the National Academy of Sciences*, **102**, 17891–17896.

Herweijer, C., Seager, R., Cook, E. R., and Emile-Geay, J., 2007. North American droughts of the last millennium from a gridded network of tree-ring data. *Journal of Climate*, **20**, 1353–1376.

Hodell, D. A., Brenner, M., Curtis, J. H., and Guilderson, T., 2001. Solar forcing of drought frequency in the Maya lowlands. *Science*, **292**, 1367–1370.

IPCC, 2007a. *Climate Change 2007: The Physical Science Basis. Contribution of Working Group I to the Fourth Assessment Report of the Intergovernmental Panel on Climate Change*. Cambridge, UK/New York: Cambridge University Press, 996 pp.

IPCC, 2007b. *Change 2007: Impacts, adaptation and vulnerability. Contribution of Working Group II to the Fourth Assessment Report of the Intergovernmental Panel on Climate Change*. Cambridge, UK/New York: Cambridge University Press.

Kallis, G., 2008. Droughts. *Annual Review of Environment and Resources*, **33**, 85–118.

Keyantash, J., and Dracup, J. A., 2002. The quantification of drought: an evaluation of drought indices. *Bulletin of the American Meteorological Society*, **83**, 1167–1180.

Le Houerou, H. N., 1996. Climate change, drought and desertification. *Journal of Arid Environments*, **34**(2), 133–185.

Maybank, J., Bonsal, B., Jones, K., Lawford, R., O'Brien, E. G., Ripley, E. A., and Wheaton, E., 1995. Drought as a natural disaster. *Atmosphere-Ocean*, **33**(2), 195–222.

McCabe, G. J., Palecki, M. A., and Betancourt, J. L., 2004. Pacific and Atlantic Ocean influences on multidecadal drought frequency in the United States. *Proceedings of the National Academy of Sciences*, **101**, 4136–4141.

McKee, T. B., Doesken, N. J., and Kleist, J., 1993. Drought monitoring with multiple time scales. In *Proceedings Ninth Conference on Applied Climatology*. Boston, MA: American Meteorological Society, pp. 179–186.

McKee, T. B., Doesken, N. J., and Kleist, J., 1995. The relationship of drought frequency and duration to timescales. In *Proceedings Eighth Conference on Applied Climatology*. Boston, MA: American Meteorological Society, pp. 233–236.

Menéndez, C. G., and Carril, A. F., 2010. Potential changes in extremes and links with the Southern Annular Mode as simulated by a multi-model ensemble. *Climatic Change*, **98**, 359–377. doi:10.1007/s10584-009-9735-7.

Mueller, A. D., Islebe, G. A., Hillesheim, M. B., Grzesik, D. A., Anselmetti, F. S., Ariztegui, D., Brenner, M., Curtis, J. H., Hodell, D. A., and Venz, K. A., 2009. Climate drying and associated forest decline in the lowlands of northern Guatemala during the late Holocene. *Quaternary Research*, **71**, 133–141.

Mullan, A. B., Porteous, A., Wratt, D., and Hollis, M. 2005. Changes in drought risk with climate change. *NIWA Report WLG2005-23*. Wellington, New Zealand, 56 pp.

Oladipo, E. O., 1985. A comparative performance analysis of three meteorological drought indices. *International Journal of Climatology*, **5**, 655–664.

Palmer, W. C., 1965. Meteorological drought. *Research Paper No. 45*. Washington, DC: U.S. Weather Bureau, 58 pp.

Palmer, W. C., 1968. Keeping track of crop moisture conditions, nationwide: the new Crop Moisture Index. *Weatherwise*, **21**, 156–161.

Rockström, J., 2003. Resilience building and water demand management for drought mitigation. *Physics and Chemistry of the Earth*, **28**, 869–877.

Rosenfeld, D., Lohmann, U., Raga, G. B., O'Dowd, C. D., Kulmala, M., Fuzzi, S., Reissell, A., and Andreal, M. O., 2008. Flood or drought: how do aerosols affect precipitation? *Science*, **321**, 1309–1313.

Rossi, G., Cancelliere, A., and Giuliano, G., 2005. Case study: multicriteria assessment of drought mitigation methods. *Journal of Water Resources Planning and Management*, **131**(6), 449–457.

Samuel, J. M., Verdon, D. C., Sivapalan, M., and Franks, S. W., 2006. Influence of Indian Ocean sea surface temperature variability on southwest Western Australian winter rainfall. *Water Resources Research*, **42**, W08402.

Shafer, B. A., and Dezman, L. E., 1982. Development of a Surface Water Supply Index (SWSI) to assess the severity of drought conditions in snowpack runoff areas. In *Proceedings of the Western Snow Conference*, Reno, NV, pp. 164–175.

Smith, D. I., Hutchinson, M. F., and McArthur, R. J., 1993. Australian climatic and agricultural drought: payments and policy. *Drought Network News*, **5**(3), 11–12.

Staubwasser, M., Sirocko, F., Grootes, P. M., and Segl, M., 2003. Climate change the 4.2 ka BP termination of the Indus valley civilization and Holocene south Asian monsoon variability. *Geophysical Research Letters*, **30**, 1425, doi:10.1029/2002GL016822.

Steinemann, A. C., and Cavalcanti, L. F. N., 2006. Developing multiple indicators and triggers for drought plans. *Journal of Water Resources Planning and Management*, **132**(3), 164–174.

Stewart, I. T., Cayan, D. R., and Dettinger, M. D., 2004. Changes in snow melt runoff timing in western North America under a "business as usual" climate change scenario. *Climate Change*, **62**, 217–232.

Vicente-Serrano, S. M., 2007. Evaluating the impact of drought using remote sensing in a Mediterranean semi-arid region. *Natural Hazards*, **40**, 173–208.

Wells, N., Goddard, S., and Hayes, M. J., 2004. A self-calibrating Palmer Drought Severity Index. *Journal of Climate*, **17**, 2335–2351.

Wilhite, D. A., 2000. Drought as a natural hazard: concepts and definitions. In Wilhite, D. A. (ed.), *Drought: A Global Assessment*. London: Routledge, Vol. 1, pp. 3–18.

Wilhite, D. A., and Glantz, M. H., 1985. Understanding the drought phenomenon: the role of definitions. *Water International*, **10**, 111–120.

Wilhite, D. A., Svoboda, M. D., and Hayes, M. J., 2007. Understanding the complex impacts of drought: a key to enhancing drought mitigation and preparedness. *Water Resources Management*, **21**, 763–774.

Woodhouse, C., and Overpeck, J., 1998. 2000 years of drought vulnerability in the central United States. *Bulletin of the American Meteorological Society*, **79**, 2693–2714.

Yin, J. H., 2005. A consistent poleward shift of the storm tracks in simulations of 21st century climate. *Geophysical Research Letters*, **32**, L18701.

Zhou, W. J., Dodson, J. R., Head, M. J., Li, Y. J., Hou, X. F., Donahue, D. J., and Jull, A. J. T., 2002. Environmental variability within the Chinese desert-loess transition zone over the last 20,000 years. *The Holocene*, **12**, 107–112.

Cross-references

Adaptation
Climate Change
Costs (Economic) of Natural Hazards and Disasters
Desertification
Disaster
Dust Bowl
Hazard
Historical Events
Land Degradation
Loess
Mitigation
Models of Hazard and Disaster
Natural Hazard
Risk
Vulnerability

CASE STUDY

DUST BOWL

Richard Seager[1], Benjamin I. Cook[2]
[1]Lamont Doherty Earth Observatory of Columbia University, Palisades Geophysical Institute, Palisades, NY, USA
[2]NASA Goddard Institute for Space Studies, New York, NY, USA

Definition

Dust Bowl. A period of drought, soil erosion, and intense dust storms that impacted the Great Plains of the United States during the 1930s.

Introduction

The Dust Bowl refers to the years of drought and dust storms that affected the Great Plains of the United States during the 1930s. The term "Dust Bowl" was proposed by a reporter writing an article 1 day after "Black Sunday" – April 14, 1935 – which was one of the worst days of dust storms. The term originally referred to some of the worst affected regions in Texas, Oklahoma, Colorado, and Kansas. "Dust Bowl" is now used to refer more generally to the entire catastrophe in the 1930s comprising drought, crop failure, soil erosion, dust storms, economic collapse, and human migration.

Meteorological origins of the dust bowl

The Dust Bowl began with drought. Rain gauge data show that average precipitation over the Great Plains was less than normal for two thirds of the seasons between 1932 and 1939. Averaged over the core years of the Dust Bowl, 1932–39, the precipitation was less than 80% of normal in most of the Great Plains (Figure 1). Droughts of this length and severity are normal features of the climate in the Plains and several had occurred since European settlement with the most recent occurring in the early to mid-1890s. What made the Dust Bowl different from these earlier droughts was the widespread soil erosion and dust storms. In the period after World War I, the Plains were transformed by the expansion of agriculture (primarily wheat, much of it for export to Europe) and the removal of drought-resistant prairie grasses (Worster, 1979). During the 1920s, adequate rains allowed for bountiful crops, thereby encouraging more new planting. When the drought struck in the early 1930s, the non-drought-resistant strains of wheat that had been planted died, exposing bare soil that was easily eroded from the surface by the wind, creating the dust storms that were characteristic of the period. The scale and magnitude of wind erosion and dust storm activity during the Dust Bowl was fairly unique and did not occur during the earlier droughts.

In the mid-2000s, computer simulations with atmosphere models forced by ship-observed historical sea

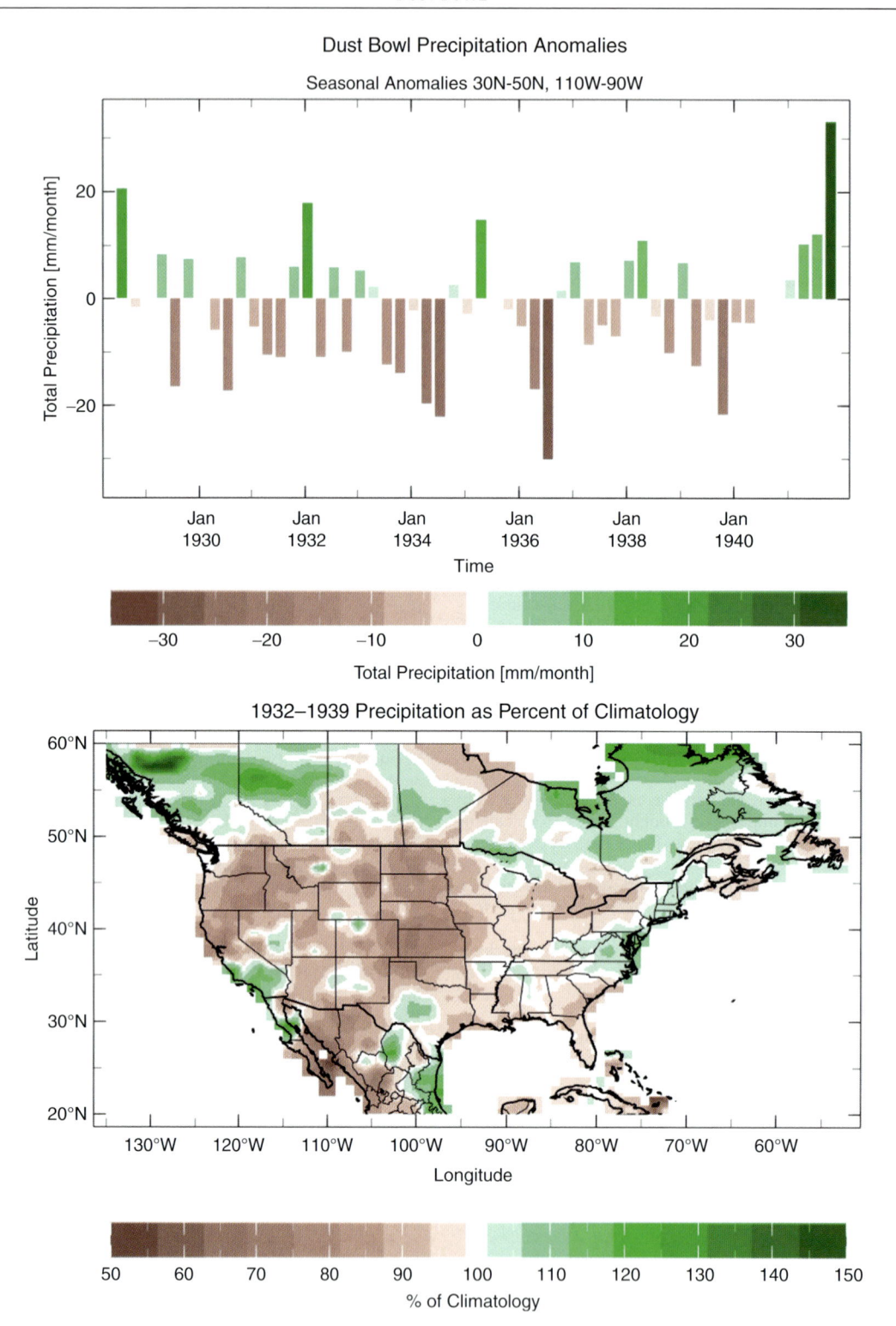

Dust Bowl, Figure 1 The precipitation anomaly (mm/day), relative to a 1900–2007 climatology, averaged over 30–50°N and 110 to 90°W, by season for the decade of the 1930s and adjacent years (*top*). The 1932–39 averaged precipitation over North America as a percent of climatology (*bottom*). Data are from the Global Precipitation Climatology Centre.

surface temperatures (SSTs) demonstrated that small variations of tropical Pacific and Atlantic SSTs forced the sequence of multiyear, persistent droughts over the Plains and Southwest North America, including the Dust Bowl drought (Schubert et al., 2004a, b; Seager et al., 2005). North American drought is particularly common when the tropical Pacific Ocean is colder than normal (referred to as a La Niña-like state) and the tropical North Atlantic is warmer than normal. This was the case during the 1930s, and again during the early to mid-1950s when a separate drought struck southwest North America (Seager et al., 2008). These SST anomalies arise naturally from ocean-atmosphere interaction and cause drought over North America through changes in the circulation and thermal structure of the atmosphere. The end result is subsidence (sinking air) over the Plains, suppressing precipitation.

The role of dust storms in modifying and intensifying the drought

Typical SST-forced droughts, however, tend to be centered in the southern Plains, southwest USA, and Mexico, whereas the Dust Bowl drought extended up into the northern Plains and the Canadian prairies. Because of this, some researchers have argued that the Dust Bowl drought was largely forced by internal atmosphere variability and not related to anomalous SSTs (Hoerling et al. 2009). An alternative theory is that the dust storms were so frequent, widespread, and intensive that they actually altered the regional climate. Climate model simulations have been performed in which an atmosphere model was forced by observed 1930s SSTs but also with bare soil placed at the surface where contemporary maps indicated wind erosion occurred. The model created dust storms that interacted with the solar radiation. By reflecting radiation to space, the dust storms induced subsiding air and suppressed conversion of water vapor into precipitation and intensified the drought (Figure 2). Since the dust transport was north and east from the Plains, the modeled drought center also shifted north with the dust, bringing the spatial pattern of the Dust Bowl drought into better agreement with observations (Cook et al., 2009).

Impacts of the Dust Bowl drought and efforts to control the soil erosion

At the peak of dust storm activity, the Plains were emitting dust at a rate equivalent to current dust emissions in the most productive areas of the Sahara (Cook et al., 2008; Laurent et al., 2008). Dust was transported eastward to coastal cities and the Atlantic Ocean, creating widespread health problems related to dust inhalation and choking of the lungs. This became known as "dust pneumonia" and led to an uncertain number of deaths (Egan, 2005). In 1935, after the wind erosion and dust storms had been ongoing for several years, the Soil Conservation Service was created specifically to address the problem of soil erosion. In 1936, the Service completed a map showing the full extent of wind and water erosion during the Dust Bowl, an area extending through the Great Plains from the Gulf of Mexico to Canada (see Hansen and Libecap, 2004).

Soil Conservation Service scientists diagnosed the cause of wind erosion to lie in a combination of drought and poor cultivation practices (i.e., lack of fallowing of land and strip cropping and the absence of shelterbelts and vegetative residue to protect soils) (e.g., Chepil, 1957). Consistently, soil erosion from cultivated land in the 1930s greatly exceeded the erosion from pasturelands (Chepil, 1957). Hansen and Libecap (2004) also noted that the small size of Dust Bowl farms encouraged farmers facing drought to plant as much area as possible to compensate for reduced yield, instead of instituting erosion control measures that would reduce erosion risk but also take land out of cultivation. In many cases, eroded soil from one farm would be transported to neighboring farms, causing a chain reaction of crop failures and wind erosion within an area. To counter such destructive practices, the Soil Conservation Service created Soil Conservation Districts that, through a mix of incentives and coercion, encouraged farmers to cooperatively practice soil conservation techniques. In addition, some marginal lands were purchased by the Federal government and allowed to return to natural grasslands. Soil conservation techniques achieved some gains against erosion, but by 1941, rains were above normal and the drought and Dust Bowl had ended.

Crop failure put many farmers into debt and forced farm sales and abandonment. According to Worster (1979), by the end of the Dust Bowl, about three million people had left their farms and about 0.5 million migrated entirely out of the affected areas, with about half of those moving to California.

Legacy of the Dust Bowl

The Dust Bowl permanently altered the agricultural economy and farming of the Plains and directly led to the widespread adoption of soil conservation techniques in the United States. Out-migration led to farm consolidation. When drought returned to the Great Plains in the 1950s, the soil erosion and dust storms were more limited than in the 1930s and the social disruption was minor by comparison. Federal farm support policies and the beginning of irrigation also helped alleviate the impact of the 1950s drought on farmers (Hansen and Libecap, 2004; Worster, 1979).

Summary

The 1930s drought was, by meteorological standards, a multiyear drought of the kind the Great Plains had experienced previously and thereafter. It was forced by a combination of cold tropical Pacific and warm tropical Atlantic sea surface temperature anomalies that in turn generated changes in atmospheric circulation that suppressed precipitation in the drought region. Poor agricultural practices, such as expansive cropping of non-drought-resistant plants with little regard for soil erosion

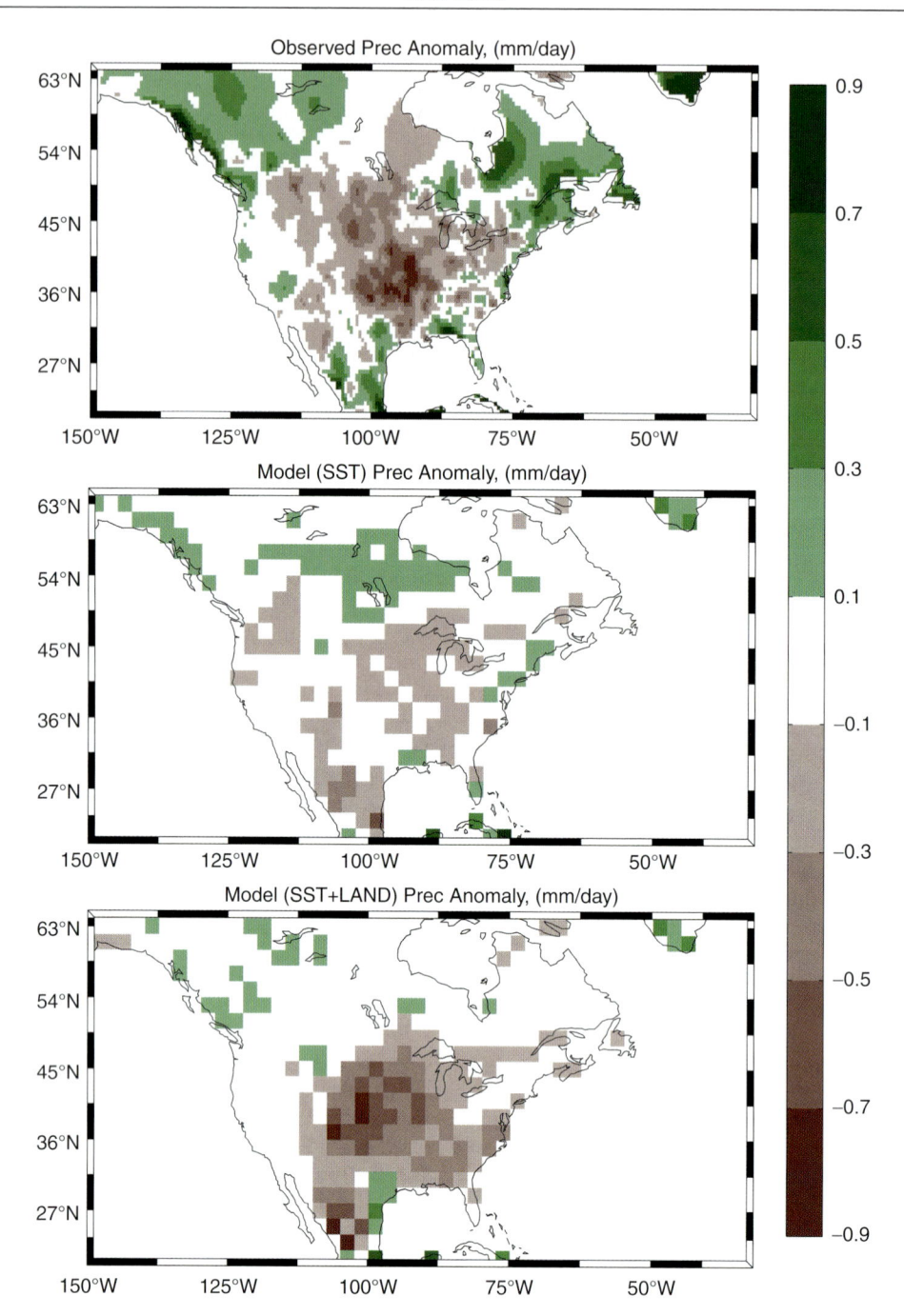

Dust Bowl, Figure 2 Observed (*top panel*) and climate model simulated warm season (April–September) precipitation anomalies for the Dust Bowl period (1932–1939). When the model is forced with observed SSTs only (*central panel*), a weak drought is simulated that is centered too far south. If the effect of the dust storms is integrated into the model in addition to the SSTs (*bottom panel*), the drought intensifies and moves northward into the central Great Plains.

potential, turned the drought into the Dust Bowl as crops failed, bare soil was exposed, and wind erosion led to dust storms. The dust storms intensified the drought and moved its center northward. In response, the Soil Conservation Service was created and put in place conservation measures to limit the erosion. The drought ended when normal levels of rainfall returned in the early 1940s. The Dust Bowl led to a permanent transformation of Plains agriculture in terms of farm size, farming practices, and farm support policies. Subsequent droughts have not led to the same scale of soil erosion because of these changes and the introduction of irrigation.

Bibliography

Chepil, W. S., 1957. Dust bowl: causes and effects. *Journal of Soil and Water Conservation*, **12**, 108.
Cook, B. I., Miller, R., and Seager, R., 2008. Dust and sea surface temperature forcing of the 1930s "dust bowl" drought. *Geophysical Research Letters*, **35**, doi:10.1029/2008GL033486.
Cook, B. I., Miller, R., and Seager, R., 2009. Amplification of the North American dust bowl drought through human-induced land degradation. *Proceedings of the National Academy of Sciences*, **106**, 4997.
Egan, T., 2005. *The Worst Hard Time*. New York: Houghton Mifflin, 352 pp.
Hansen, Z. K., and Libecap, G. D., 2004. Small farms, externalities, and the dust bowl of the 1930s. *Journal of Political Economy*, **112**, 665.
Hoerling, M. P., Quan, X.-W., and Eisched, J., 2009. Distinct causes of two principal US droughts of the 20th century. *Geophysical Research Letters*, **36**, L19708, doi:10.1029/2009GL039860.
Laurent, B., Marticorena, B., Bergametti, G., Leon, J. F., and Mahowald, N. M., 2008. Modeling mineral dust emissions from the Sahara desert using new surface properties and soil database. *Journal of Geophysical Research*, **113**, doi:10.1029/2007JD009484.
Schubert, S. D., Suarez, M. J., Pegion, P. J., Koster, R. D., and Bacemeister, J., 2004a. On the causes of the 1930s dust bowl. *Science*, **303**, 1855.
Schubert, S. D., Suarez, M. J., Pegion, P. J., Koster, R. D., and Bacemeister, J., 2004b. Causes of long term drought in the United States Great Plains. *Journal of Climate*, **17**, 485.
Seager, R., Kushnir, Y., Herweijer, C., Naik, N., and Velez, J., 2005. Modeling of tropical forcing of persistent droughts and pluvials over western North America: 1856–2000. *Journal of Climate*, **18**, 4068.
Seager, R., Kushnir, Y., Ting, M., Cane, M. A., Naik, N., and Velez, J., 2008. Would advance knowledge of 1930s SSTs have allowed prediction of the dust bowl drought? *Journal of Climate*, **21**, 3261.
Worster, D., 1979. *Dust Bowl: The Southern Plains in the 1930s*. New York: Oxford University Press.

Cross-references

Drought
Dust Storm
El Niño/Southern Oscillation
Global Dust/Aerosol Effects
Hydrometeorological Hazards
Land Degradation

DUST DEVIL

Nilton O. Rennó
University of Michigan, Ann Arbor, MI, USA

Synonyms

Convective vortex; Whirlwind

Definition

Warm-core vortices, normally less than 100 m high formed at the base of convective plumes often appearing as a well-defined dust funnel.

Overview

Dust devils are low-pressure, warm-core vortices with typical diameters ranging from 1 to 10 m, and heights of less than 100 m (Figure 1). However, occasionally, they can be larger or taller by more than an order of magnitude. Dust devils form at the bottom of convective plumes. Since their sources of angular momentum are local wind shears, caused either by the convective circulation itself or by larger scale phenomena, they can rotate clockwise or anticlockwise with equal probability. A distinctive feature of intense dust devils is their well-defined dust funnel. Theory indicates that dust is focused around the funnel by a dynamic pressure drop caused by increases in the speeds of the air spiraling toward the vortex.

Like waterspouts, tornadoes, and hurricanes, dust devils can be idealized as convective heat engines. They are the smallest and weakest members of this class of weather phenomena. The intensity of a dust devil depends on the depth of the convective plume and the transfer of heat from the ground into the air. When the surface is composed of loose particles, they might become airborne and make a dust devil visible. Dust particles are indirectly lifted from the surface by a process known as saltation. In this process, the larger particles are forced to move by the wind and bounce along the surface, lifting the smaller, harder to lift (because of strong cohesion forces) dust particles into the air. When loose particles are not present, intense vortices may exist and may not be visible to the observer. When a dust devil crosses a cold terrain, the dust column is cut off, and the vortex dissipates.

The abrupt increase in wind speed around dust devils is what creates a hazard; more than 10% of the accidents with light airplanes and helicopters are caused by visible or invisible dust-free dust devils. However, the abrupt reduction in visibility caused by them can also be a hazard.

Dust devils are more frequent in hot desert regions, but they also have been observed in colder regions such as the subarctic. Dust devils move with the ambient wind and slope with height in the wind shear direction. When the wind is strong, their diameters are biased toward large values.

The occurrence of dust devils increases abruptly from nearly zero at around 10 am to a maximum value at around 1 p.m. Then, dust devil activity slowly decreases toward nearly zero at the end of the afternoon. The abrupt increase at around 10 a.m. is due to increases in the solar radiation and abrupt increase in the depth of the boundary layer. The decrease to nearly zero at the end of the afternoon is due to the decrease in solar radiation and therefore convective activity.

Charge transfer occurs when sand and dust particles collide with each other and the surface. In this process, the smaller particles charge negatively and the surface and large particles positively. Then, the convective updrafts cause charge separation by carrying the smaller particles upward and electric fields of the order of 10,000 V/m can be generated. There are suggestions that

Dust Devil, Figure 1 Large dust devil photographed in the Nevada Desert in USA, July 2009 (Credit: University of Michigan).

these electric fields affect dust transport and, that on Mars they can even affect atmospheric chemistry.

Dust devils have been observed on Mars by almost all orbiters and landers that visited the planet. On Mars, they are ubiquitous, can form almost anywhere on the planet, and can have diameters of more than 1 km and heights of more than 10 km. There is evidence that dust devils play an important role on the global aerosol budget of both Earth and Mars.

Bibliography

Renno, N. O., Burkett, M. L., and Larkin, M. P., 1998. A simple theory for dust devils. *Journal of the Atmospheric Sciences*, **55**, 3244–3252.

Renno, N. O., Abreu, V., Koch, J., Smith, P. H., Hartogenisis, O., de Bruin, H. A. R., Burose, D., Delory, G. T., Farrell, W. M., Parker, M., Watts, C. J., and Carswell, A., 2004. MATADOR 2002: a field experiment on convective plumes and dust devils. *Journal of Geophysical Research*, **109**, E07001, doi:10.1029/2003JE002219.

Renno, N. O., 2008. A general theory for convective plumes and vortices. *Tellus*, **60A**, 688–699.

Cross-references

Dust Storm
Tornado
Waterspout

DUST STORM

Nilton O. Rennó
University of Michigan, Ann Arbor, MI, USA

Synonyms

Haboob; Sand storm

Definition

Dust storms are weather systems that lift large quantities of dust particles into the air causing extremely low visibility.

Overview

Dust storms form when the wind speed exceeds the threshold value necessary to move sand particles over plowed fields and arid terrains. Particles of diameters between about 50 and 500 μm (sand-sized particles) are the first to move as the wind speed increases. When these sand particles move, they bounce along the surface in a process known as saltation. The impact of saltating particles on the surface consequently ejects smaller, hard to lift (because of large ratio of cohesive with wind drag forces) dust particles into the air creating a dust storm when enough dust is lifted over extensive areas. Saltation is also the process by which dust particles are lifted from the surface by smaller weather systems such as dust devils.

Dust storms can be created by large-scale low-pressure weather systems or gust fronts, areas of cool dense air propagating ahead of thunderstorms. Intense gust fronts, capable of lifting significant amounts of dust, form when rain evaporates while falling through relatively dry air. Such intense gust fronts generate impressive dust storms that can lift large amounts of dust a few kilometers above the surface, and form well-defined dust fronts. The abrupt boundary between the dust fronts and the clear air ahead of them can reduce the visibility to nearly zero in a few seconds. This abrupt reduction in visibility is one of the major causes of accidents by dust storms.

The dense cool air that may form gust fronts is known as a density current. As a first approximation, the propagation speed of a density current is proportional to its depth and the difference in density between it and the surrounding air. Thus, the denser and deeper gust fronts produced by the evaporation of precipitation falling through extremely dry air are capable of producing the strongest and fastest propagating dust storms. These are the most hazardous dust storms that frequently cause accidents in arid regions.

Large-scale frontal weather systems capable of initiating saltation usually cause moderately strong dust storms that can last a day or more. Such large-scale dust storms are more common between autumn and spring. Gust fronts usually produce stronger and more dangerous dust storms lasting no more than a few hours. The smaller, but potentially more dangerous dust storms are more common in the summer. The abrupt reduction in visibility to nearly

Dust Storm, Figure 1 A dust storm in Niger. Courtesy of the University of Michigan

zero and the strong wind make dust storms extremely dangerous. They frequently cause serious aviation and automobile accidents.

Charge transfer occurs when sand and dust particles collide with each other and the surface during saltation. During this process, the smaller particles charge negatively whereas the surface and large particles charge positively. Near-surface electric fields in excess of 100,000 V/m have been measured during dust storms. There is evidence that such large electric fields affect saltation and can even directly lift dust particles from the surface.

Dust storms are ubiquitous on arid Mars. Every few years, global dust storms form and cover the entire planet, reducing the visibility and the flux of solar energy at the surface by more than an order of magnitude. The dust storms last a few weeks and can be hazardous to landers and rovers.

Bibliography

Bagnold, R. A., 1941. *The physics of blown sand and desert dunes.* New York: Methuen.
Kok, J. F., and Renno, N. O., 2008. Electrostatics in wind-blown sand. *Physical Review Letters*, **100**, 014501.
Kok, J. F., and Renno, N. O., 2009. A comprehensive numerical model of steady state saltation (COMSALT). *Journal of Geophysical Research*, **114**, D17204, doi:10.1029/2009JD011702.

Cross-references

Aviation, Hazards to
Climate Change
Dust Devils
Erosion
Fog Hazards
Global Dust
Space Weather
Storms
Tornado
Volcanic Ash

DVORAK CLASSIFICATION OF HURRICANES

Raymond Zehr
CIRA/CSU, Fort Collins, CO, USA

Definition

A method for estimating tropical cyclone intensity ranging in values from T1 to T8.

Overview

The Dvorak technique estimates tropical cyclone intensity using satellite images. Vernon Dvorak and his colleagues developed the Dvorak technique in the early 1970s. It was one of the first innovative applications of meteorological satellite imagery, and it is still widely used today at tropical cyclone forecast centers throughout the world (Velden et al., 2006).

The intensity of a tropical cyclone is generally quantified as the associated maximum surface wind speed. Near the United States, routine aircraft reconnaissance gives intensity estimates using dropsondes and flight level data. However, most of the world's tropical cyclone intensity analysis relies on satellite images and the Dvorak technique. Another indicator of tropical cyclone intensity is the minimum sea-level pressure near the tropical cyclone center. The Dvorak technique uses an intensity unit called

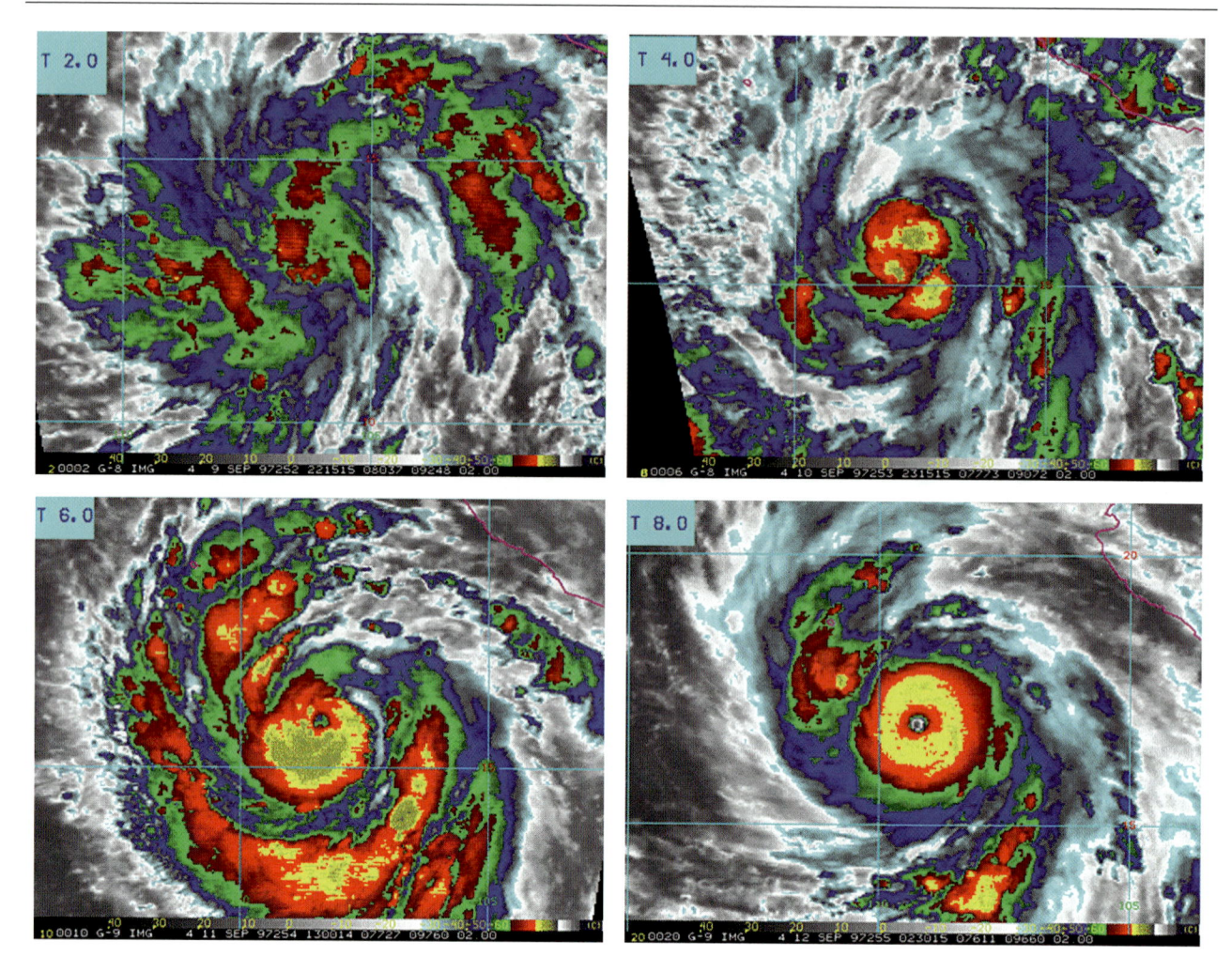

Dvorak Classification of Hurricanes, Figure 1 Enhanced IR images of Hurricane Linda at Dvorak intensities T2, T4, T6, and T8. Hurricane Linda was located in the eastern North Pacific, southwest of Mexico, during 9–17 September 1997.

a T-number in increments of ½ ranging from T1 to T8. The Dvorak T-number intensity scale is normalized according to typical observed daily changes in intensity. T2.5 is the minimal tropical storm intensity (35 knots = 18.0 m/s), whereas T4.0 is the minimal hurricane intensity (33.4 m/s). T6.0 has a wind maximum of 59.1 m/s, and T8.0 approximates record maximum intensity (87.4 m/s).

The Dvorak technique primarily uses satellite observed cloud patterns and infrared (IR) cloud top temperatures to estimate intensity, with independent methods for visible satellite images and IR images.

With weaker intensities, the Dvorak analysis is based on either the curved band pattern or the shear pattern. Using the curved band analysis, the extent to which a spiral-shaped band of deep convective clouds surrounds the tropical cyclone center determines the intensity. The shear pattern refers to the cloud pattern observed when broadscale vertical wind shear induces a distinctly asymmetric cloud pattern with respect to the tropical cyclone low-level circulation center. The degree of deep convective cloud displacement due to the vertical shear decreases with intensification.

With hurricane intensities, the cloud pattern typically evolves into what is called a central dense overcast, which describes the deep convective clouds that surround the center. As intensification proceeds, an eye is observed within this central dense overcast. The eye is the familiar cloud-free or cloud minimum area associated with the lowest pressure at the tropical cyclone center. The eye is surrounded by a circular area which has the strongest winds within very deep clouds and heavy rain, known as the eyewall. The Dvorak technique analyzes visible features and IR temperatures of the eye and the surrounding deep clouds to assign the intensity. In general, the Dvorak tropical cyclone intensity increases as they eye gets warmer and better defined, and the surrounding clouds get colder and more symmetric. A continuous, very cold circular ring of cloud tops generated by the eyewall along with a warm eye temperature indicate an intense tropical cyclone. Enhanced IR images of Dvorak intensities T2, T4, T6, and T8 are shown in Figure 1 with Hurricane Linda that was located in the eastern North Pacific in September 1997.

Dvorak (1984) gives a detailed description of the methodology and procedures of the Dvorak technique. Following Dvorak's original work, research and development efforts have been focused on replicating and refining the Dvorak approach with objective and automated routines using the IR temperatures (Velden et al., 2006). Automated Dvorak techniques give reliable results that are quickly updated as the latest IR satellite image becomes available, and the tropical cyclone intensity data supplement the general use of satellite data for analysis and forecasting.

Bibliography

Dvorak, V. F., 1984. Tropical cyclone intensity analysis using satellite data. NOAA Tech. Report NESDIS 11, U.S. Department of Commerce, Washington DC, 45 pp.

Velden, C., Harper, B., Wells, F., Beven, J., Zehr, R., Olander, T., Mayfield, M., Guard, C., Lander, M., Edson, R., Avila, L., Burton, A., Turk, M., Kikuchi, A., Christian, A., Caroff, P., and McCrone, P., 2006. The Dvorak tropical cyclone intensity estimation technique: A satellite-based method that has endured for over 30 years. *Bulletin of American Meteorological Society*, **87**, 1195–1210.

Cross-references

Airphoto and Satellite Imagery
Beaufort Wind Scale
Hurricanes
Hurricane Katrina
Storms

E

EARLY WARNING SYSTEMS

Graham S. Leonard[1], Chris E. Gregg[2], David M. Johnston[1]
[1]GNS Science, Massey University, Lower Hutt, New Zealand
[2]East Tennessee State University, Johnson City, TN, USA

Synonyms

Alerting system; Immediate warning system; Short fuse warning; Warning systems

Definition

The term early warning system often refers to the technological monitoring, telemetry, and notification aspects of warning systems. The term is also used to distinguish cases where a warning is able to be delivered in a time frame that permits protective action, such as may occur for tsunamis, volcanic eruptions, or severe weather, but not necessarily an earthquake.

Overview

Synonyms

While all aspects of a *warning system* are sometimes referred to as an early warning system, the term early warning system overlaps most in meaning with *immediate* and *short-fuse warning systems* which are warning systems designed in regard to hazardous events with very short durations of time between the onset of the event and the occurrence of a hazardous process.

Monitoring

Methods used for monitoring natural hazards vary widely, but usually revolve around one or more environmental sensors that telemeter data to communications hardware, which is capable of disseminating a warning notification. Sensors detect a variety of geological and geophysical phenomena, including, but not limited to pressure, temperature, distance, chemistry, electromagnetic radiation, and ground-shaking. Monitoring equipment is often located in remote areas, requiring special design or housing to withstand extreme environmental conditions of cold, heat, and precipitation, in addition to its own power supply, which would include batteries and solar or fuel generators. Monitoring equipment may be positioned in permanent to semipermanent fixed locations or mounted on vehicles, balloons, aircraft, or satellites and positioned in multiple areas over short time periods. Some monitoring networks are permanent, whereas others are deployed in response to initial signs of unrest.

Telemetry

Monitoring equipment is sometimes located far from population centers. This requires development of a telemetering network that can provide communications links between the monitoring equipment and scientists or emergency managers. Data are often telemetered over a fixed communication network (e.g., a phone line) or wireless (e.g., mobile phone, VHF-UHF radio, satellite uplink) technology. Regular testing is needed for networks that are not routinely used.

Automated decision-making

Warning systems need to allow for a decision to notify (or not) based upon predetermined thresholds of monitoring data provided by the monitoring equipment. For early warning systems built around hazards with short lead times, this decision is often made automatically, or at least the system automatically processes monitoring information and provides a few simple options to a decision-maker. Warning thresholds for monitored data need to be preplanned for such a decision to be made automatically or made quickly when reviewed manually. Computer-based decision support models, taking into account probabilities, uncertainty, and time frames may be used.

Dissemination of notification

Warning systems need to deliver notification of an impending hazard event and there are a wide range of options available. While the term "early warning system" may be used to refer to monitoring through notification, it is vital for an effective system that people receiving the message know what it means and act appropriately in response. Early warning systems for hazards with short lead times often rely heavily on technology for notification, including warning-dedicated hardware (e.g., sirens) and hardware used for other purpose (e.g., landline telephones, mobile phones, radio, television). NOAA weather radio in the United States is an example of hardware normally used by regular radio stations that also includes a warning-specific component that will enable the device and broadcast a warning automatically.

Bibliography

Sorensen, J. H., 2000. Hazard warning systems: review of 20 years of progress. *Natural Hazards Review*, **1**, 119–125.

UN/ISDR Platform for the Promotion of Early Waring (UN/ISDR-PPEW), 2006. Developing early warning systems: a checklist. United Nations International Strategy for Disaster Reduction (UN/ISDR), Bonn.

Cross-references

Accelerometer
Earthquake Prediction and Forecasting
Extensometers
Global Positioning System (GPS) and Natural Hazards
Global Seismograph Network (GSN)
Monitoring and Prediction of Natural Hazards
Pacific Tsunami Warning and Mitigation System (PTWS)
Piezometer
Pore-Water Pressure
Seismograph/Seismometer
Warning Systems

EARTHQUAKE

John F. Cassidy
Natural Resources Canada, Sidney, BC, Canada

Definition

Earthquake. A tectonic or volcanic phenomenon that represents the movement of rock and generates shaking or trembling of the Earth.

Introduction

Earthquakes are one of the most frightening natural phenomena that occur. They shake our very foundation – the ground beneath our feet. They almost always strike without warning. The shaking, in the form of aftershocks, can continue for days, weeks, months, or even years. The effects of earthquakes can be widespread (a tsunami caused by a large earthquake can damage regions tens of thousands of kilometers away) and wide-ranging (strong shaking, fires, landslides, liquefaction). Each year, several million earthquakes occur around the world, including about 20,000 that are large enough to be located, and, on average, more than 1,300 that are large enough (magnitude (M) greater than 5) to cause some damage. As the world's population continues to expand, losses and deaths from earthquakes are climbing rapidly, particularly in developing countries. This contribution briefly summarizes the causes of earthquakes, the history of earthquakes, the effects of earthquakes, how earthquakes are monitored and studied, and what can be done to protect ourselves from, and reduce the impact of, future earthquakes.

The causes of earthquakes

The ultimate cause of earthquakes is the interior heat of the Earth. Near the Earth's core, the temperature is estimated to be in the range of 3,500–5,700 C (De Bremaecker, 1985). This heat is trying to escape to the surface, and as it flows upward, large "circulation cells" are generated in the mantle of the Earth (Figure 1) much like convection cells in boiling water. These flow patterns help to drive the motion of the tectonic plates that make up the surface of the Earth (Figure 2). These plates, some large (like the Pacific plate),

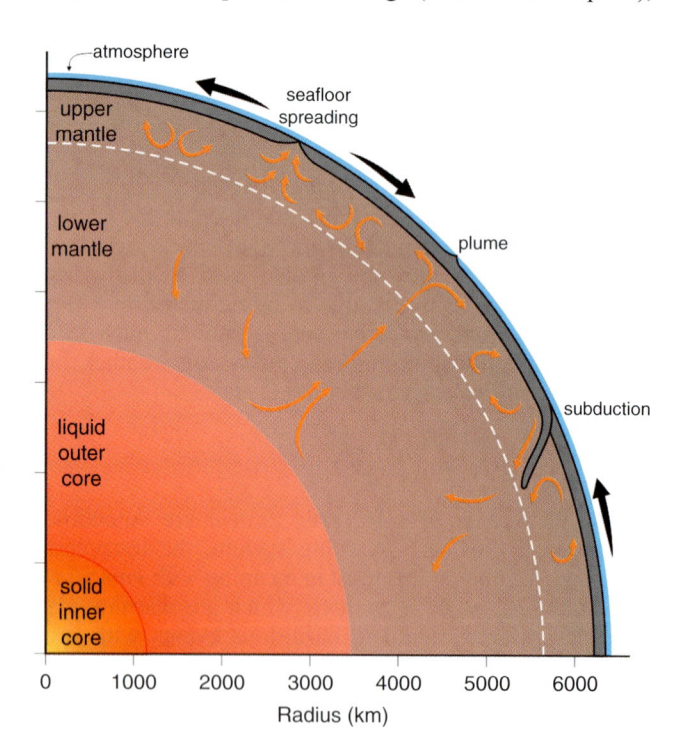

Earthquake, Figure 1 Cartoon cross section of the Earth showing the major layers (crust mantle and core), as well as the convection cells (*orange arrows*) in the mantle that help to drive motions of the tectonic plates (*dark grey*) on the surface of the Earth (illustration by Richard Franklin).

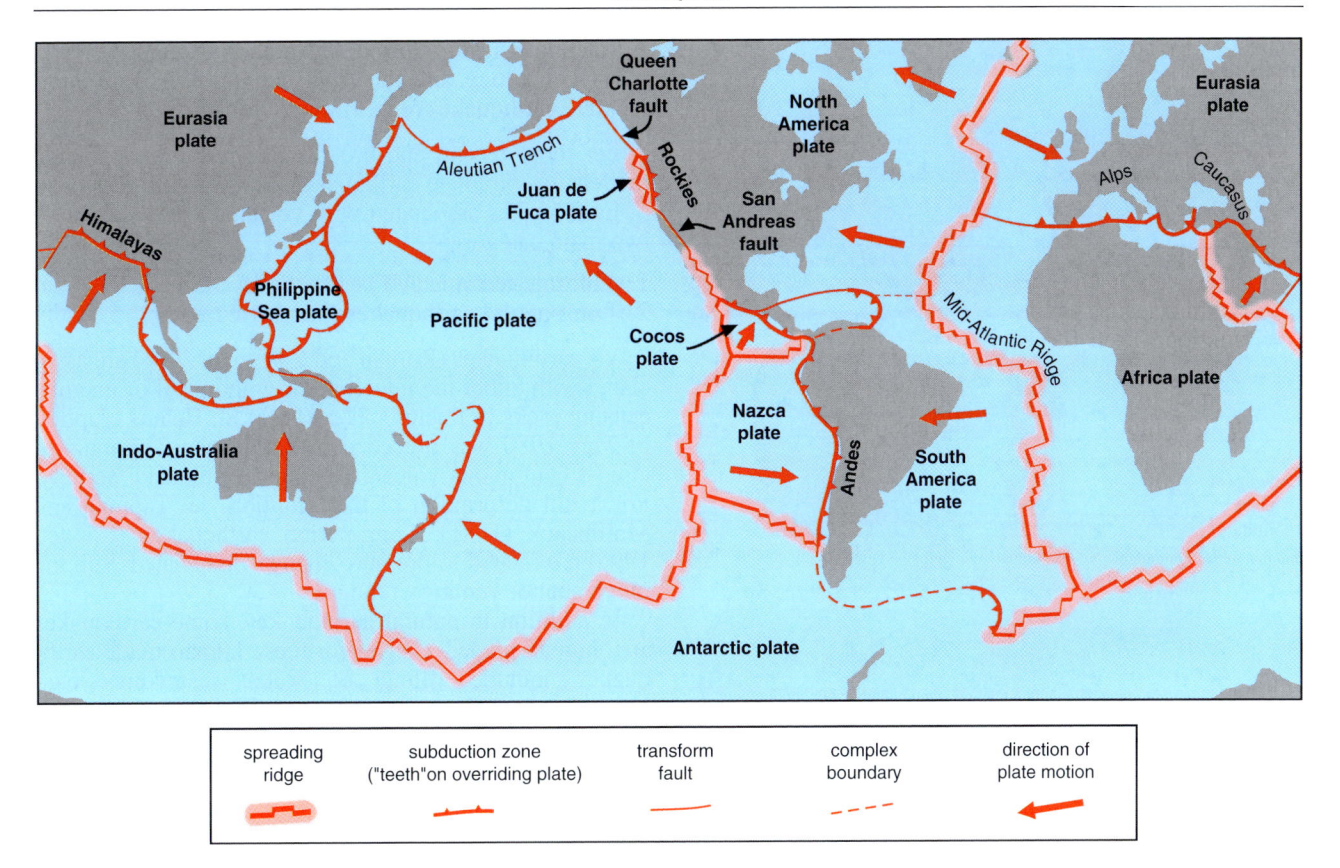

Earthquake, Figure 2 Map of the major tectonic plates of the Earth and select faults (e.g., the San Andreas Fault). The direction of plate motions is shown by the *red arrows*, and different line types are used for different types of faults (see Figure 3). Illustration reprinted from "At Risk: Earthquakes and Tsunamis on the West Coast" (With permission from Tricouni Press).

others small (Juan de Fuca plate), are moving relative to one other at speeds of about 1–10 cm/year (this is about how fast your fingernails grow). Where plates meet (active plate boundaries), one of three things can happen (Figure 3):

1. The plates can move apart (divergent boundary).
2. The plates can slide past one another (transform boundary).
3. The plates can collide (convergent boundary).

Most of the worlds earthquakes (about 90%), and the world's largest earthquakes (M > 8.5) occur at these active plate boundaries. These basic plate movements (or combination of those basic motions) result in three types of earthquakes:

1. Normal faulting (usually at divergent boundaries)
2. Strike-slip earthquakes (dominated by horizontal sliding – usually at transform boundaries)
3. Subduction earthquakes (the world's largest type of earthquake that occurs at convergent boundaries)

Tectonic plates are comprised of oceanic lithosphere and continental lithosphere. Continental areas represent the oldest rocks (up to five billion years) on Earth. These are stable regions and are typically composed of a 30–50-km-thick crust of mainly granitic rock (quartz-silicone rich) underlain by a mafic mantle. Oceanic crustal rocks are much younger (the oldest oceanic plate is about 180 Ma old) and thinner and are composed of mafic material. Oceanic plates are composed of heavier and denser material, and so when oceanic lithosphere collides with a continent at a subduction zone, the oceanic plate sinks beneath the continent and is "recycled" into the Earth. This explains why no oceanic crust is older than about 180 Ma. When two continental plates collide (like India pushing north into Asia) neither plate "sinks" and the result is very high mountains (the Himalayas) and large earthquakes. When two oceanic plates collide, subduction of one plate occurs (e.g., the Marianas region of the South Pacific).

Most (about 75%) of the world's earthquakes are "shallow" (that is less than about 60 km). The remainder are "intermediate" (60–300 km) or "deep" (>300 km). At some subduction zones earthquakes extend down to about 700-km depth (Frolich, 2006).

Although the vast majority of the world's earthquakes occur at active plate boundaries, up to about 10% of the world's earthquakes (including some very large earthquakes) occur within continental plates. These are called "intraplate" earthquakes (Talwani and Rajendran, 1991). Some notable examples of intraplate events include:

3. High heat flow – high temperatures weaken the rock, alter rheology, and can focus stresses and cause intraplate earthquakes (e.g., Liu and Zoback, 1997).

For more recent, and broadly applicable geodynamic models for intraplate earthquakes, see Mazzotti (2007).

Earthquakes associated with volcanic activity fall into two main categories:

1. Volcanic-tectonic (VT) earthquakes
2. Long-period earthquakes (harmonic tremor)

VT earthquakes (Roman and Cashman, 2006) are associated with either the injection, or withdrawal of magma, causing rocks to fracture. The Mount St. Helens eruption of May 1980 was preceded by more than 2 months of almost continuous volcanic-tectonic earthquake activity. For a full description of this eruption, see Lipman and Mullineaux (1981). Long-period earthquakes (Chouet, 1996) are related to pressure changes associated with the movement or venting of magma.

In addition to natural earthquakes, some earthquakes are "human made." Causes for these human-made earthquakes include filling of water reservoirs (e.g., Lamontagne et al., 2008), pumping fluids into the ground to aid with resource extraction (e.g., Horner et al., 1994), and mining (e.g., Gendzwill et al., 1982). For an overview of induced seismicity, see Guha (2000).

The history of earthquakes

Earthquakes have been a part of myth, legend, and history for as long as humans have been on the Earth. In Greek mythology, Poseidon (Neptune in the Roman pantheon), the "God of the Sea," was thought to be responsible for earthquakes. Some of the earliest references to earthquakes come from ancient China, including the following two descriptions taken from Loewe and Shaughnessy (1999):

1. In approximately the twenty-third century B.C. – "the three Miao tribes were in great disorder and for three nights it rained blood. A dragon appeared in the ancestral temple and dogs howled in the market place. Ice formed in summertime, *the earth split open until springs gushed forth*, the five grains (i.e. kinds of staple crop) grew all deformed, and the people were filled with a great terror."
2. In 1767 B.C. "the five planets deviated in their courses, and there were meteor showers in the night. *An earthquake occured*, and the Yi and Luo Rivers dried up."

The first earthquake described in some detail is the China earthquake of 1177 B.C. Historical earthquake databases for China are described in Wang (2004).

In Japan, one of the world's most seismically active countries, earthquakes have been described going back in time as far as 416 A.D. The earliest Japanese earthquakes are documented in the Journal of Disaster Research (2006).

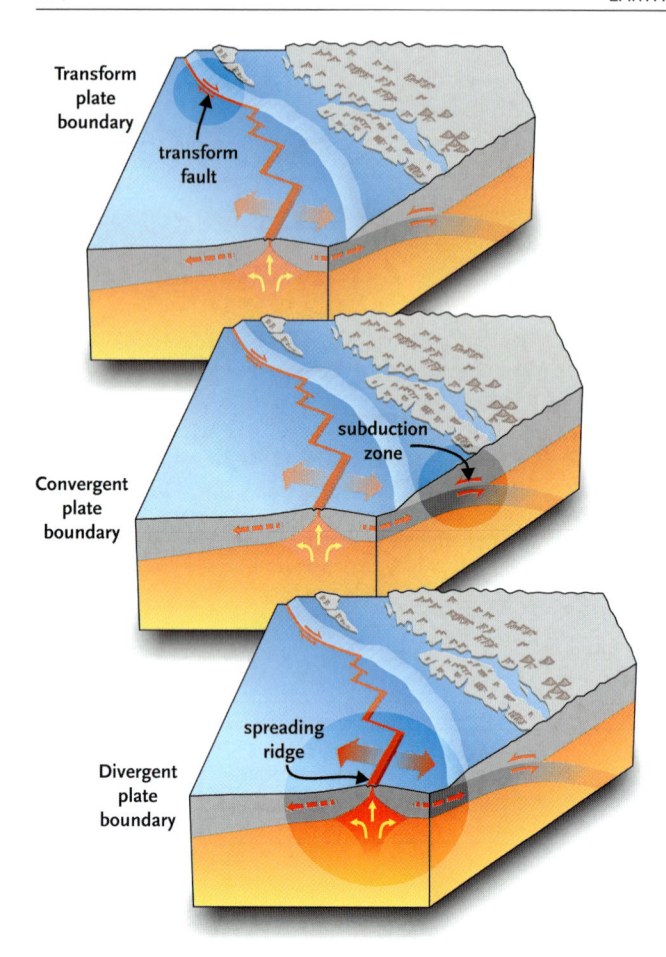

Earthquake, Figure 3 Sketches of the three primary types of plate boundaries (using offshore British Columbia, Canada, as an example): transform (*top*); convergent (*middle*); and divergent (*bottom*). Illustration reprinted from "At Risk: Earthquakes and Tsunamis on the West Coast" (With permission from Tricouni Press).

1. The M 7.2-8.1 New Madrid earthquake sequence of 1811–1812 (Johnston and Schweig, 1996; Hough, 2001).
2. The M 7.8 Kutch, India earthquake of 1819 (Rajendran and Rajendran, 2001).
3. The M 5.6 Newcastle, Australia earthquake of 1989 (Rynn et al., 1992).

Several theories have been suggested to explain the occurrence of large intraplate earthquakes, including:

1. Zones of weakness – crustal stresses are transmitted from the active plate boundaries, thousands of kilometers through continental plates, to reactivate old fault zones (e.g., Johnston and Kanter, 1990).
2. Stress concentration – inhomogeneities in the crust, such as large mafic intrusions, that concentrate crustal stresses and can cause earthquakes (e.g., Chandrasekhar et al., 2007).

Some of the world's largest and most devastating earthquakes have occurred in India. For documentation on nearly four centuries of significant earthquakes in this region, see Martin and Szeliga (2010) and Szeliga et al. (2010).

In Europe, historical earthquakes have been described as far back in time as about 2500 years (the first earthquake was mentioned in 580 B.C.). For details of historical European-Mediterranean earthquakes, the reader is referred to the European-Mediterranean Intensity Database at http://emidius.mi.ingv.it/EMID/. While much of the European earthquake hazard is concentrated in Italy and the Mediterranean region, large earthquakes have occurred throughout Europe, including the 1356 Basel, Switzerland earthquake that destroyed that city, and the 1755 Lisbon earthquake, which triggered a massive fire and an Atlantic-wide tsunami. The combination of strong shaking, tsunami, and fires destroyed most of the city of Lisbon and nearby areas. The geological causes and effects of this massive earthquake (likely between magnitude 8 and 9) are still being researched and debated (e.g., Mendes-Victor et al., 2009). For a description of some historical earthquakes in Europe, see Kozak and Thompson (1991). For details on some historical earthquakes of the Middle East, and an interpretation of how these earthquakes may have influenced history, see Nur and Burgess (2008) and Ambraseys (2008).

The main regions in Africa that experience earthquakes are the East African Rift zone (where the African plate is splitting apart) and North Africa (where the continent is colliding with Europe). A summary of earthquakes in North Africa beginning in the year 1045 is provided in Pelaez et al. (2007).

The earliest known earthquakes in the Americas occurred in Mexico in the late fourteenth century and in Peru in 1471, but descriptions of the effects were not well documented. For a description of historical earthquakes of Mexico, see Acosta (2004). Some historical earthquakes of South America (Chile and Peru) are documented by Lomnitz (2004). Charles Darwin was exploring the coast of Chile in February 1835 when a large earthquake (now known to be about magnitude 8.5) occurred off the coast near Concepcion. In a letter to his sister, Darwin describes this experience as follows:

"We are now on our road from Concepcion. The papers will have told you about the great Earthquake of the 20th of February. I suppose it certainly is the worst ever experienced in Chili (sic). It is no use attempting to describe the ruins – it is the most awful spectacle I ever beheld. The town of Concepcion is now nothing more than piles and lines of bricks, tiles and timbers – it is absolutely true there is not one house left habitable; some little hovels builts of sticks and reeds in the outskirts of the town have not been shaken down and these now are hired by the richest people. The force of the shock must have been immense, the ground is traversed by rents, the solid rocks are shivered, solid buttresses 6-10 feet thick are broken into fragments like so much biscuit. How fortunate it happened at the time of day when many are out of their houses and all active: if the town had been over thrown in the night, very few would have escaped to tell the tale. We were at Valdivia at the time. The shock there was considered very violent, but did no damage owing to the houses being built of wood. I am very glad we happened to call at Concepcion so shortly afterwards: it is one of the three most interesting spectacles I have beheld since leaving England – A Fuegian Savage – Tropical Vegetation – and the ruins of Concepcion. It is indeed most wonderful to witness such desolation produced in three minutes of time." (Darwin, 1845).

In the USA, the first documented earthquake occurred in 1638 (Coffman et al., 1982, and http://earthquake.usgs.gov/earthquakes/states/events/1638_06_11_hs.php). The first documented earthquake in eastern Canada occurred in 1663 in the vicinity of Charlevoix, Quebec (Smith, 1962). This earthquake triggered landslides, caused building damage, and was felt throughout New France and New England. Note that an earthquake reported in 1534–1535 in some catalogs (Smith, 1962) has been demonstrated to be a nonevent – see Gouin (1994). In 1811–1812 a series of major earthquakes (magnitude 7.2–8.1) struck the New Madrid region of the central USA. These earthquakes were felt across the eastern USA and southeastern Canada. They caused significant damage, including destroying the town of New Madrid and altering the course of the Mississippi river (Hough et al., 2000).

As Spanish expeditions were made through California, the first written records of earthquakes were obtained in 1769 and 1775 in the present-day Los Angeles area (Ellsworth, 1990). For a detailed description of historic earthquakes of the America's, see Kovach (2004).

The first reported earthquake in western Canada is described in Captain George Vancouver's journal. He wrote that on February 17, 1793, "a very severe shock of an earthquake had been felt" at the Spanish settlement at Nootka on the west coast of Vancouver Island (Rogers 1992). However, long before the Europeans explored the west coast of Canada, First Nations Peoples lived here. They have numerous oral traditions, masks, dances, and ceremonies that are earthquake related. Many of these traditions that are associated with giant subduction earthquakes (tsunamis, strong ground shaking, and aftershocks) are documented by Ludwin et al. (2005).

Around the world, more than one million earthquakes (almost all very small) occur each year. Of those, about 1,300 are large enough to cause some damage if they occur in a populated region, and about 17 are "major" (greater than magnitude 7). As the population of the Earth increases, and as development increases, particularly in developing countries, the impact of large earthquakes is becoming more significant. Table 1 is a list of the most destructive (deadliest) earthquakes in the world, and it is noteworthy that of these 22 events that have occurred during the past 1,154 years, 4 of these (or about 18% of the total) have occurred during the very short time period of 2004–2010. This trend is likely to continue in the future

Earthquake, Table 1 List of the world's most destructive earthquakes. Modified from the USGS webpage: (http://earthquake.usgs.gov/earthquakes/world/most_destructive.php)

Date (UTC) Year month day	Location	Magnitude	Number of deaths
1156 Jan. 23	Shaanxi, China	8	830,000
1976 Jul. 27	Tangshan, China	7.5	255,000 +
1138 Aug. 09	Syria, Alleppo	?	230,000
2004 Dec. 26	Sumatra	9.1	228,000
2010 Jan. 12	Haiti	7.0	223,000
856 Dec. 22	Iran, Damghan	?	200,000
1920 Dec. 16	Haiyuan Ningxia, China	7.8	200,000
893 Mar. 23	Ardabil, Iran	?	150,000
1923 Sep. 1	Kanto, Japan	7.9	142, 800
1948 Oct. 5	Turkmenistan, USSR	7.3	110,000
1290 Sep. 27	Chihli, China	?	100,000
2008 May 12	Sichuan, China	7.9	87,587
2005 Oct. 8	Pakistan	7.6	86,000
1667 Nov.	Caucasia	?	80,000
1727 Nov. 18	Tabriz, Iran	?	77,000
1908 Dec. 28	Messina, Italy	7.2	72,000
1970 May 31	Chimbote, Peru	7.9	70,000
1755 Nov. 1	Lisbon, Portugal	8.7	70,000
1693 Jan. 11	Sicily, Italy	7.5	60,000
1268	Asia Minor, Silicia	?	60,000
1990 Jun. 20	Western Iran	7.4	40–50,000
1783 Feb. 4	Calabria, Italy	?	50,000

A "?" symbol indicates that the magnitude is unknown

with increasing development and populations in earthquake-prone areas of the world.

Recording earthquakes

Originally, earthquakes were only "monitored" by humans. Intensity scales (see the following section) were developed to describe what people felt during an earthquake and the effects of that earthquake. Intensity scales are still in use today. The earliest known instrument to detect earthquakes was developed by Zhang Heng in China about 1800 years ago. That instrument, a heavy vase attached to a chassis to keep it from falling over, contained eight dragons located around the outside of the vase, and each with a brass ball. When earthquake waves approached, they would cause a ball to drop from one of the dragons into the mouth of a toad located below. This would show the direction of the earthquake waves. However, it was not until the mid-1800s that pendulum seismographs to record actual ground motions were developed. In the 1850s, Robert Mallet determined a method to measure the velocity of seismic waves. In Italy, Luigi Palmieri invented an electromagnetic seismograph. These seismographs were the first seismic instruments capable of routinely detecting earthquakes that were too small to be felt by humans. In 1881–1882, continuously recording pendulum seismographs were designed by John Milne and Thomas Gray. By 1900, the first global seismograph network was operational on every continent. For details on the early history of seismology, see Agnew (2002).

Real advancement in earthquake science required accurate recording of earthquakes, and this has only been the case for the past half-century. The modern era of seismology began in the early 1960s with the deployment of the World Wide Standard Seismograph Network (WWSSN). Since the early 1990s the digital Global Seismic Network (GSN) and Incorporated Research Institutions for Seismology (IRIS) – see http://www.iris.edu/hq/programs/gsn/maps – make high-quality digital seismic data freely available to anyone. This aspect of seismology, freely sharing data, is critical for studies of both earthquakes, and studies of the Earth. For details on international seismology, see Adams (2002). The advent of digital data since about the 1980s has changed the world of earthquake science. These high-quality data allow for details of the earthquake that could not be obtained using older analogue data (paper recordings).

The size of earthquakes

Earthquakes are described in one of two ways; by "intensity" or by "magnitude." These are very briefly described here.

Intensity scales

Intensity describes how an earthquake was felt, and its effects at a specific location (e.g., were people frightened? were items knocked from shelves? was there damage to buildings). For any given earthquake there is a range of intensity values. Intensity is controlled by a number of factors including the magnitude and depth of the earthquake, and it varies with distance from the earthquake source, and local soil conditions.

A number of intensity scales have been developed over time, including the Rossi-Forel scale (the first intensity scale developed in the late 1800s), the Modified Mercalli Intensity (MMI) scale (Wood and Neumann, 1931), the Omori scale (used in Japan), the USGS DYFI (Wald et al., 1999b) scale currently used in the USA, and the European Macroseismic scale that is the current standard in Europe (Grünthal, 1998).

Magnitude scales

The magnitude of an earthquake is a single number that describes the "size" of an earthquake and is directly related to the energy release. Most magnitude scales are based on the recorded amplitude of seismic waves. There are many different magnitude scales, including the most famous – the Richter Scale. Other commonly used scales today (for a summary of many magnitude scales, including primary references, see Utsu, 2002) include the surface wave magnitude (Ms) which is based on the amplitude of the seismic surface waves and moment magnitude (Mw) which is directly related to the area of the rupture and the amount of movement along a fault (Hanks and Kanamori, 1979). The magnitude scale is unbounded, but in practice earthquake magnitudes range from less than 0 to the largest earthquake ever recorded at 9.5.

The effects of earthquakes

The effects of earthquakes can range from "none" (not felt or no obvious effects) to "total destruction" (or Modified Mercalli Intensity XII). The effects of an earthquake depend on a number of factors and can vary drastically across a region. Key factors include: the size of the earthquake; the depth of the earthquake; the proximity to population centers or critical infrastructure; the rupture pattern of the earthquake; local soil conditions; and local topology.

The most common effects of earthquakes are:

1. *Strong shaking*. It is strong shaking that can damage or destroy buildings and other structures. Older, or unreinforced, masonry buildings are usually at highest risk of damage. An example of adobe structures that collapsed in Constitucion, Chile, during the 2010 M 8.8 Chile earthquake is shown in Figure 4. The strength and the duration of shaking depends upon the amount of energy released (and is directly related to the earthquake magnitude). The strength and duration of shaking also depends upon a number of other factors, including: the depth of the earthquake; distance from the earthquake; rupture style of the earthquake; and

Earthquake, Figure 4 Collapse of old adobe structures in Constitucion, Chile, caused by the M 8.8 Maule earthquake of February 27, 2010 (Photo by John Cassidy).

Earthquake, Figure 5 Tsunami damage was widespread in Constitucion, Chile, resulting from the M 8.8 Maule earthquake of February 27, 2010 (Photo by John Cassidy).

local soil conditions. Ground motion equations, which predict how ground shaking varies with distance and magnitude, are one of the most important aspects of earthquake science and have direct applications to earthquake engineering. Additional details are provided in the following section.

2. *Tsunamis*. Tsunamis can be one of the most deadly effects of earthquakes. The impact of tsunamis can reach across an ocean. For example, the tsunami generated by the 2004 Mw 9.3 Sumatra earthquake killed more than 220,000 people from Indonesia to Thailand to Africa and India. This wave was recorded around the globe (Titov et al., 2005). Large-scale (ocean-wide) tsunamis can typically occur for earthquakes larger than about Mw 8.5, that are beneath the ocean, and have a faulting mechanism that involves vertical movement. It is this vertical motion that generates the wave. Tsunami waves travel across the ocean at speeds of about 500–1,000 km/s and their amplitude is controlled by local bathymetry and topology. For the Sumatra tsunami run-up heights varied greatly over small distances, e.g., from 5 to 25 m over distances of less than 50 km (Borrero, 2005). It is important to note that localized (smaller scale) tsunamis can be generated by smaller earthquakes (e.g., M 7) or by earthquakes beneath land that may trigger landslides or underwater slumps. These tend to be localized, but can still have devastating effects (e.g., Synolakis et al., 2002). In some cases "slow earthquakes" (or silent quakes, that cannot be "felt") can generate large, localized tsunamis. An example is the Nicaragua earthquake of 1992 (Kanamori and Kikuchi, 1993). A photo of the devastating tsunami damage in Constitucion, Chile from the 2010 M 8.8 earthquake is shown in Figure 5.

3. *Landslides*. In mountainous areas large earthquakes can trigger hundreds or thousands of landslides. Earthquake-induced landslides have been documented from as early as 373 or 372 B.C. (Seed, 1968). In a study of 40 historical earthquakes, Keefer (1984) finds a relationship between the maximum area affected by landslides as a function of magnitude (0 km for M ~ 4–500,000 km^2 for M = 9.2). The number and type of landslides depends on numerous factors including the local geological conditions, type of material, and the steepness of slopes. For details, see Keefer (1984) and Rodriguez et al. (1999).

4. *Fault displacement*. During earthquakes, displacement along a fault occurs. If that rupture reaches the surface, regions on opposite sides of a fault may move relative to one another. Displacements may be as large as 20–30 m for magnitude 8+ earthquakes. Knowing the amount of displacement that might be expected for an earthquake of a given magnitude (e.g., Wells and Coppersmith, 1994) is critically important in the design of structures crossing an active fault. For an example of the success story of the design of the Alaska Pipeline and the M 7.8 Denali earthquake, see Cluff et al. (2003).

Earthquake, Figure 6 Soil liquefaction and resulting road failure along Lake Vichuquén, Chile, resulting from the M 8.8 Maule earthquake of February 27, 2010 (Photo by John Cassidy).

5. *Liquefaction.* Strong shaking during earthquake can cause some soils to liquefy and lose their strength (Seed and Idriss, 1982; Idriss and Boulanger, 2008). Widespread liquefaction has been observed in many earthquakes including 1964 Alaska (M 9), 1964 Niigata Japan (M 7.5), 1989 Loma Prieta, CA (M 7.1), and 1995 Kobe Japan (M 6.9). For details on soil failure during the Kobe earthquake, see Aguirre and Irikura (1997). Soil failure affected numerous highways in Chile after the M 8.8 earthquake – an example is given in Figure 6.
6. *Fires.* Fires triggered by earthquakes are very common. For many years the 1906 San Francisco earthquake was referred to as the "San Francisco Fire" as the fire caused more damage than the ground shaking.
7. *Psychological effects.* People can be traumatized by earthquakes, the effects of earthquakes, and the aftershocks that may continue for weeks or months. Numerous studies have documented the psychological effects of earthquakes, and that consequences of earthquake exposure are long lasting and linked to damage/loss suffered (Bland et al., 1996; Kiliç and Ulusoy, 2003).

The science of earthquakes

People began studying earthquakes long before instruments were developed to record ground shaking. For example, Aristotle created a theory on the origin of earthquakes in 330 B.C., attributing earthquakes to heavy winds within the Earth (Missiakoulis, 2008). Aristotle also classified earthquakes into different types (Oeser, 1992), he linked some earthquakes to volcanic activity, and he made a connection between soil type and strength of shaking. Our understanding of earthquakes began to significantly change after the great Lisbon earthquake of 1755. Shortly after that event, John Mitchell in England and Elie Bertrand in Switzerland began a comprehensive study of the timing and severity of earthquakes. The study of earthquakes was also hastened by Charles Darwin and his crew of the Beagle. They experienced a large earthquake (likely close to M 9) in South America, noting the very serious effects of earthquakes, including changes in the elevation of coastlines. Recordings of earthquakes, beginning in the late 1800s and early 1900s, were basic and only the largest events (M > ~7) could be recorded. As more instruments were deployed, and as they improved in quality, a much more accurate picture of earthquake distribution began to emerge.

Seismology – the study of earthquakes and seismic waves – is a relatively young science that has developed over the past approximately 100 years. During the past two to three decades, the science of earthquakes has improved dramatically, due to a number of key factors, including:

1. The advent of widespread digital seismic recordings beginning in the 1980s (digital data were available earlier than this, but was very limited). This allowed for detailed studies of the earthquake rupture process, wave propagation, ground shaking, earthquake triggering, and more.
2. New and diverse data sets, including Global Positioning Satellite (GPS) data, Light Detection and Ranging (LIDAR), high-resolution marine imaging methods, high-resolution gravity and magnetic data, paleoseismological data, and better images of Earth structure.
3. Faster and better computers that could be used to develop more accurate models of where energy is accumulating for future earthquakes, to develop detailed models of earthquake shaking, stress transfer, and much more.

These factors are contributing to greatly improved earthquake models. Specifically, identification of where future earthquakes are more likely to occur, how large they can be (and how often they may occur), time-varying earthquake hazard, earthquake triggering, and ground shaking. The discussion below is intended to briefly highlight ten activities that have significantly contributed to our understanding of earthquakes and assessment (and mitigation) of earthquake hazard.

1. Continuous GPS monitoring. With the ability to monitor crustal movements at the fraction of an mm per year level using continuous GPS data, scientists could begin to identify regions where energy is being stored for future earthquakes, and to help identify which faults are active. A few examples of how continuous GPS data are being used to contribute to improved assessments of earthquake hazards include: (a) using crustal deformation data to map the locked portion of a subduction fault (e.g., Wang et al., 2003; Sagiya, 2004); (b) constraining fault slip rates and source zone strain rates/earthquake statistics (e.g., Field et al., 2008), and (c) examining crustal deformation rates to help obtain constraints on earthquake

recurrence rates (e.g., Hyndman et al., 2003). The next generation of GPS data is real-time, high-rate (one sample per second) data that provides information on strong ground shaking and displacements.

2. Subduction zone studies. The largest earthquakes on Earth (M 9+) occur at subduction zones. This is where an oceanic plate is pushed beneath a continental plate. Some examples of this type of earthquake include the M 9.5 1960 Chile earthquake, the M 9.1 Sumatra earthquake of 2004, the M 9 1964 Alaska earthquake, and the M 8.8 Chile earthquake of February 2010. Over the past few decades our understanding of the hazards associated with subduction earthquakes has increased dramatically. This is largely due to researchers incorporating a wide variety of data sets (e.g., heat flow, gravity, seismic reflection, magnetic, earthquakes, magnetotelluric, crustal deformation measurements, and paleoseismology – see below) to identify and model subduction zone hazards. These new datasets and improved modeling techniques allow for the identification of where past subduction earthquakes have occurred, how often, how large, when the last one occurred, where energy is being stored for future subduction earthquakes, and how the ground will shake during those future earthquakes. For additional details, see (for example) Ruff and Tichelaar (1996), Hyndman and Wang (1995), Mazzotti et al. (2003), and Hyndman (1995). Faster computers, better datasets, and more sophisticated modeling algorithms are providing for a better understanding of subduction zone hazards – both earthquakes and tsunamis (e.g., see Wada and Wang, 2009 and Wang et al., 2009).

3. Episodic Tremor and Slip. This phenomenon, involving repeated episodes of slow slip along a fault zone combined with seismic tremors, was first discovered in the Cascadia subduction zone in 2003 (Rogers and Dragert, 2003). This builds upon the discovery of "slow slip" using continuous GPS data in Cascadia as described by Dragert et al. (2001). These slip events occur in the deeper portion of the subduction fault (landward of the locked portion of the subduction fault) and release a portion (perhaps up to 50%) of the strain accumulation along this part of the fault. It also likely plays a key role in adding stress to the shallower locked portion of the fault (Rogers and Dragert, 2003). ETS has also been discovered in other subduction zones of the world (e.g., Japan, Mexico, Alaska). Recent studies of ETS using dense seismic and GPS arrays are providing critical new information on this phenomenon, including propagation and source location (e.g., Ghosh et al., 2009). ETS has also been discovered along transform faults, including the San Andreas in California (e.g., Nadeau and Dolenc, 2005).

4. Paleoseismology. Seismographs provide a relatively short recorded history of earthquakes (just over 100 years). Estimates of long-term hazard require a much longer term history of earthquake activity. In many cases, paleoseismology (geological evidence of past earthquakes) can provide information on large earthquakes going back thousands of years. Paleoseismology has been extensively applied to better understand the earthquake history of the San Andreas fault in California (e.g., see Grant and Lettis (2002) and Sieh (1978)), magnitude 9 Cascadia subduction earthquakes (e.g., see Adams, 1990; Atwater et al., 1995; Goldfinger et al., 2003) and, relatively rare cratonic earthquakes in eastern North America (e.g., Atkinson and Martens, 2007; Mazzotti and Adams 2005; Fenton et al., 2006). For additional details on paleoseismology, see McCalpin (2009).

5. Wave propagation and ground motion studies. One of the most important parameters in the development of seismic hazard maps (described in detail in the next section) is the relationship between ground shaking and distance from an earthquake. The propagation of seismic waves differs significantly around the world depending on the geological and tectonic environment. For example, the hard rock of old "continental cratons" (such as eastern North America) transmits high frequency waves very efficiently, whereas those same waves are absorbed and scattered by the complicated crustal rocks of mountainous areas (e.g., western North America). Attenuation relationships have been developed in various parts of the world (e.g., Eastern North America, California, etc.) and for different types of earthquakes (transform, subduction), and different sizes of earthquakes. Some examples of recent seismic attenuation relationships include: subduction zone earthquakes (Atkinson and Macias, 2009); shallow crustal earthquakes in active tectonic regions (Abrahamson et al., 2008); and eastern North American earthquakes (Atkinson and Boore, 2006). For a summary of numerous global relationships, see Douglas (2001).

6. LIDAR. This relatively new technique (mid-1990s) provided, for the first time, a means of "seeing through the trees" in densely vegetated regions to identify active fault scarps (Harding and Berghoff, 2000). One of the most famous examples is the identification of various strands of the Seattle Fault that extends through downtown Seattle, WA, using LIDAR data (Harding and Berghoff, 2000). Geological work on portions of the Seattle fault zone had indicated a 7-m uplift (Bucknam et al., 1992) and that earthquakes as large as magnitude 6.7 have struck this fault in the past, the most recent event being in the year 900 A.D. (Bucknam et al., 1992). LIDAR is now a standard tool for identifying active faults in regions of dense vegetation (Haugerud et al., 2003) and has been used to map numerous active faults, including strands of the Seattle Fault Zone, the Tacoma Fault, the Darrington-Devil's Mountain fault zone, the Whidbey Island Fault, and the Boulder Creek Fault. These faults have been incorporated into

earthquake hazard maps to help estimate future ground shaking during earthquakes.

7. Real-time Seismology. With readily available digital data and inexpensive communications systems, "real-time seismology" is becoming increasingly important (Kanamori et al., 1997). Applications include rapid earthquake locations and magnitudes (required for tsunami warnings and stopping trains (for example)), the development of earthquake early warning systems (see the following section), "shakemaps" (Wald et al., 1999a), "shakecast" (Wald et al., 2008a), and other real-time products such as "Pager" (Wald et al., 2008b). These near-real-time information products are providing critical new information for earthquake first responders, planners, and managers.

8. Earthquake triggering and stress changes. It has long been known that earthquakes cause aftershocks in the vicinity of the earthquake rupture. Aftershocks are typically confined to the "source region" of about 1–2 rupture lengths of the earthquake. However, until recently, the remote triggering of earthquakes was not considered possible. The M 7.3 Landers earthquake of 1992 changed all of that. This earthquake triggered a sudden and well-documented increase in earthquake activity across much of the western United States to distances of 1,350 km (Hill et al., 1993). These triggered earthquakes were concentrated in regions that experience persistent seismic activity – specifically volcanic and geothermal areas. Hill et al. (1993) argue that several processes must be taking place to explain this seismicity that occurs over a period of days and weeks. Almost all of these events were small (M < 2) and were likely triggered by dynamic stresses associated with the passage of the surface waves. Some events are interpreted as triggered by other processes in the days to even weeks after the Landers earthquake. This same earthquake, its aftershocks, and smaller events prior to the mainshock, all demonstrated (Stein et al., 1992) that stress changes in crustal rocks caused by earthquakes can bring neighboring fault zones closer to failure. Since that study, stress triggering of earthquakes has been shown as an important factor in time-varying earthquake hazard. An example is the migration of large earthquakes along the North Antolian fault (Stein et al., 1997). For additional details and examples, see Stein (1999) and Steacy et al. (2005).

9. Site response studies. Detailed studies of the variation in earthquake ground shaking are rapidly increasing as the number of seismographs deployed (particularly in urban settings) increases. Some factors that make significant contributions to the strength, duration, and frequency of earthquake ground shaking include: (1) surface geology (e.g., soft soils, firm soil, rock); (2) basin edge effects (as demonstrated by the Kobe Japan earthquake of 1995 – see Kawase, 1996); (3) topographic effects (shaking can be amplified at the top of hills and ridges, see Geli and Jullien, 1988; Spudich et al., 1996); and (4) the thickness and velocity contrast of surface sediments. Detailed studies of local geology, geotechnical information, and recordings of shaking are increasingly being used to produce seismic microzonation maps around the world, but great caution is still warranted in using these maps (e.g., Mucciarelli, 2008).

10. Drilling into active fault zones. Currently, a number of studies (California, Japan, Taiwan, New Zealand) are underway to drill into active fault zones to better understand the processes that control earthquakes. Two studies briefly summarized here are: (1) the San Andreas Fault Observatory at Depth (the SAFOD project); and (2) the Nankai Subduction fault drilling project in Japan. The SAFOD project involves drilling to depths of 2–3 km into the San Andreas fault near the location of the M 6 Parkfield earthquake (Zoback, 2006). Drilling began in June 2004, and reached a depth of 3.2 km in October 2007. These new data from SAFOD are providing information on the composition and mechanical properties of rocks in the fault zone, the nature of stresses that cause earthquakes, and the role of fluids in controlling faulting. For a list of some SAFOD-related publications, see the "International Continental Scientific Drilling Project" website at: http://www.icdp-online.org/front_content.php?idcat=894

The Nankai drilling project in Japan involves drilling (beneath the ocean) through the megathrust fault – the boundary region between the oceanic plate that is being pushed beneath Japan and deploying long-term instrumentation at this interface (e.g., Tobin and Kinoshita, 2006). This region has experienced repeated large (M 8+) subduction earthquakes over the past 1,300 years, with the most recent being in 1944 (M 8.1) and 1946 (M 8.3). For details on this project, including updates, see the webpage: http://www.icdp-online.org/front_content.php?idcat=1056

This study will provide an unprecedented glimpse into an active subduction fault and may also provide additional "early warning time" for future earthquakes (e.g., an additional 30 s of "warning" before the waves from an offshore earthquake hit the population centers of Japan).

The results of earthquake research are used to reduce losses from earthquakes in numerous ways, particularly through modern design codes and standards. There is also growing interest in the applications of earthquake forecasting, and earthquake early warning. These are described in the following section.

Reducing losses from future earthquakes

While there sometimes may be a public perception that the best way to reduce the impact of future earthquakes is through earthquake prediction, at this point in time the best defense against earthquakes is through modern, robust codes and standards that protect lives, buildings, and critical infrastructure. Also, it is critical to ensure that populations living in earthquake-prone regions are aware of earthquake hazards, and that they are engaged in personal preparedness efforts. Another area of active research is in the development of "real-time early warning systems." Here, a very brief overview of these topics is presented.

Earthquake prediction

Earthquake prediction is generally defined as specifying *where*, *when*, *how big*, and *how probable* a predicted event is (and *why* the prediction was made). Predictions can be classified (Sykes et al., 1999) as short term (hours to weeks), intermediate term (1 month to 10 years), or long term (10–30 years). Attempts at earthquake prediction have existed for almost as long as earthquakes. Often these predictions are based on animal behavior, foreshock activity, "earthquake lights," electrical effects, changes in water levels, and much, much more. The most famous earthquake prediction (and the only major earthquake considered to have been successfully predicted) is that the M = 7.3 1975 Haicheng, China event. A recent reevaluation of this event (Wang et al., 2006) concluded that there was an official midterm (1–2 years) prediction. They also concluded that "the most important precursor was a foreshock sequence, but other anomalies such as geodetic deformation, changes in groundwater level, color, and chemistry, and peculiar animal behaviour also played a role." This prediction resulted in saving thousands of lives. However, a year later, an even larger earthquake occurred in China – the M 7.5 Tangshan earthquake that was not predicted and resulted in the loss of more than 250,000 lives (Table 1).

One of the most famous earthquake prediction experiments in the world was the Parkfield, California experiment – summarized in a 2006 special volume of the Bulletin of the Seismological Society of America (e.g., see Harris and Arrowsmith, 2006). The Parkfield segment of the San Andreas fault had experienced five magnitude 6 earthquakes between 1857 and 1966. Based on the regular occurrence and similarities between these earthquakes, one was expected to occur before 1993, and this segment of the fault was instrumented with numerous types of monitors in an attempt to identify precursors to the expected earthquake (Bakun and Lindh, 1985). The earthquake arrived late (very late) in September of 2004. There were no obvious precursors to this earthquake (Bakun et al., 2005) indicating that forecasting the timing of earthquakes remains a significant challenge. For a general discussion of the predictability of earthquakes, the reader is referred to Jordan (2006).

In summary, at this point in time, there is still considerable debate about the potential for earthquake prediction. Some (e.g., Geller et al., 1997) state that "reliable issuing of alarms of imminent large earthquake appears to be effectively impossible" (very-short-term prediction), while others (Sykes et al., 1999) are optimistic about the possibility of intermediate-term and long-term predictions on at least some active fault segments. Overall, the consensus appears to be that at this time earthquake prediction is still in its infancy as a scientific discipline and remains "a challenge." However, great advances are being made in the identification of regions of "higher hazard" (or long-term forecasting) based on earthquake science and the incorporation of new datasets, as described in the earthquake science section above.

Robust building codes and standards

The use of earthquake provisions in Building Codes has led to a significant reduction in earthquake-related damage. This is obvious when comparing the effects of earthquakes such as M 6.6 Bam (2003) and M 7 Haiti (2010) where codes and standards are lacking, to the M 6.7 Northridge, CA (1994) or M 6.8 Kobe (1995) earthquakes in areas with modern building codes. It is noteworthy, however, that the modern codes and standards currently in place in Japan and California are largely the result of devastating earthquakes in those countries. Earthquake provisions in building codes have generally been the result of significant earthquakes. For example, the devastating 1923 Kanto Japan earthquake (see Table 1) led to the inclusion of seismic provisions in the Japan code in 1924. Updates to these codes were made in 1950, and significant changes were made in 1981. The characteristics of building damage during the Kobe earthquake of 1995 clearly showed the improvements provided by the modern (post-1981) building code. Of the post-1981 buildings, only 3% collapsed, compared to 17% of all pre-1971 buildings. Of the post-1981 buildings, nearly 75% had no damage or little damage, compared to 26% for pre-1971 buildings.

In California, the San Francisco earthquake of 1906 led to the first inclusion of earthquake provisions in building design codes. Each of the major earthquakes in California (1925, 1933, 1971, 1989, and 1994) led to improved understanding of, and enhancements to, building codes and standards (see Cutcliffe, 2000). The 1971 San Fernando earthquake in particular led to significant improvements in the building code, and the passage of the California Hospital Seismic Safety Act.

In Canada, the first seismic provisions in the National Building Code of Canada (NBCC) were put in place in 1953. These early seismic hazard maps included four "zones," based on qualitative assessment of historic earthquake activity. Significant updates to these seismic hazard maps were made in 1970, 1985, 2005 (see Adams and Atkinson 2003), and some changes will be in the 2010 code.

The Global Seismic Hazard Assessment Program (GSHAP) was started in 1992. The purpose was to mitigate earthquake risk by providing uniform estimates of hazard on a global scale. This was a regionally coordinated effort that utilized a homogeneous approach to seismic hazard evaluation (Giardini and Basham, 1993;

Giardini, 1999). One of the key products was the GSHAP Global Seismic Hazard Map that depicts peak ground acceleration (pga) with a 10% chance of exceedance in 50 years, corresponding to a return period of 475 years. This map and all associated documentation, including regional reports, maps of seismicity, source characterization information, and GSHAP yearly reports, are available via the Internet through the GSHAP homepage, http://seismo.ethz.ch/GSHAP/.

Real-time warning systems

With advancements in instrumentation, data transmission, data processing and dissemination, there is an increased interest, and rapid advancements are being made in the area of Earthquake Early Warning (EEW). For a summary of the current state of the art, see the special volume on Earthquake Early Warning published in Seismological Research Letters (Volume 5, 2009). For earthquakes, warning times range from seconds to just over one minute, depending upon the location and depth of the earthquake and the seismic monitoring network in place in the region. A successful early warning system requires a dense seismic network. Some examples of warning systems described in this volume include the SAS system for Mexico City. Seismic stations along the west coast of Mexico (where subduction earthquakes occur) are used to provide warning to Mexico City (about 320 km inland) before the strong shaking arrives. This system, with a demonstrated ability to identify earthquakes of M > 6 and provide warning, is described in detail by Suarez et al. (2009). In Japan, data from thousands of seismic stations deployed across the country are used to feed into a public warning system for earthquakes (Allen et al., 2009). Warnings for strong shaking are issued when the Modified Mercalli Intensity is predicted to be greater than VII-VIII). These alerts are distributed to the public via the Japan Broadcasting Corporation, some cell-phone providers, radio stations, and some loud-speaker systems (Allen et al., 2009). These warnings are used to control traffic, stop trains, control elevators and critical factory systems, and more (Kamigaichi et al., 2009). The Japanese system was successfully employed during the M 8.9 2011 Tohoku earthquake. For example, 24 high-speed bullet trains were stopped without a single injury (Kanamori, 2012). Other examples of early warning systems using real-time seismic data are provided by Wu and Kanamori (2008) and Zechar and Jordon (2008). Allen et al. (2009) concluded that although there are still some misperceptions and challenges with EEW, significant progress has been made in developing and implementing warning systems.

Summary

Earthquakes pose a major threat to human life and economic development. We are reminded of this year after year. A recent example being the tragedy of the January 12, 2010, magnitude 7 earthquake in Port au Prince, Haiti, that resulted in the loss of more than 250,000 lives. That earthquake, like many earlier ones, clearly showed that the populations at most risk are those in developing countries with few or no guidelines for earthquake-resistant construction. This is the fastest growing population on Earth. The M 8.9 2012 Tohoku, Japan earthquake struck a country having modern building code, yet this earthquake still resulted in 15,854 deaths (most from the tsunami) and had an economic cost of more than $235B – making it the world's most expensive natural disaster.

However, we are making great progress in our ability to monitor earthquakes, study the details of the earthquake process, and estimate the ground shaking expected from future earthquakes.

Reducing losses requires better building codes, better construction, and a population that is aware of the risks and actively involved in mitigation efforts. The first step is in knowing the risks. I hope that this entry contributes, at least in a small way, to knowing the risks. I hope that this entry helps inspire you to mitigate the risks of earthquakes, whether that is by drawing up a personal preparedness plan for your home and family, volunteering for a community organization, conducting research into hazard assessments, utilizing hazard information as an engineer or community planner, or as a politician. Earthquakes cannot be stopped, but the effects of earthquakes can be minimized.

Acknowledgments

We greatly acknowledge the thoughtful and thorough review of this entry by Jane Wynne, Maurice Lamontagne and anonymous reviewers. This is ESS Contribution number 20100066.

Bibiliography

Abrahamson, A., Atkinson, G., Boore, D., Bozorgnia, Y., Campbell, K., Chiou, B., Idriss, I. M., Silva, W., and Youngs, R., 2008. Comparisons of the NGA ground-motion relations. *Earthquake Spectra*, **24**, 45, doi:10.1193/1.2924363.

Acosta, V. G., 2004. Historical earthquakes in Mexico. Past efforts and new multidisciplinary achievements. *Annals of Geophysics*, **47**, 487–496.

Adams, J., 1990. Paleoseismicity of the Cascadia subduction zone: evidence from turbidites off the Oregon-Washington margin. *Tectonics*, **9**(4), 569–583, doi:10.1029/TC009i004p00569.

Adams, R. D., 2002. International seismology. In Lee, W., Jennings, P., Kisslinger, C., and Kanamori, H. (eds.), *International Handbook of Earthquake and Engineering Seismology, Part A*. Amsterdam: Elsevier Press, Vol. 81A. pp. 1200.

Adams, J., and Atkinson, G. M., 2003. Development of seismic hazard maps for the proposed 2005 edition of the national building code of Canada. *Canadian Journal of Civil Engineering*, **30**, 255–271.

Agnew, D. C., 2002. History of seismology. In Lee, W., Jennings, P., Kisslinger, C., and Kanamori, H. (eds.), *International Handbook of Earthquake and Engineering Seismology, Part A*. Amsterdam: Elsevier Press, Vol. 81A. pp. 1200.

Aguirre, J., and Irikura, K., 1997. Nonlinearity, liquefaction, and velocity variation of soft soil layers in Port Island, Kobe, during the Hyogo-ken Nanbu earthquake. *Bulletin of the Seismological Society of America*, **87**, 1244–1258.

Allen, R. M., Gasparini, P., Kamigaichi, O., and Bose, M., 2009. The status of earthquake early warning around the world: an introductory overview. *Seismological Research Letters*, **80**, 682–693.

Ambraseys, N. N., 2008. Descriptive catalogues of historical earthquakes in the eastern Mediterranean and the middle east; revisited. In Fréchet, J., Meghraoui, M., and Stucchi, M. (eds.), *Historical Seismology Interdisciplinary Studies of Past and Recent Earthquakes*. Dordrecht: Springer, doi:10.1007/978-1-4020-8222-1_3.

Atkinson, G. M., and Boore, D. M., 2006. Earthquake ground-motion prediction equations for eastern North America. *Bulletin of the Seismological Society of America*, **96**, 2181–2205.

Atkinson, G. M., and Macias, M., 2009. Predicted ground motions for great interface earthquakes in the Cascadia subduction zone. *Bulletin of the Seismological Society of America*, **99**, 1552–1578.

Atkinson, G. M., and Martens, S. N., 2007. Seismic hazard estimates for sites in the stable Canadian craton. *Canadian Journal of Civil Engineering*, **34**(13), 1299–1311.

Atwater BF and 15 others, 1995. Summary of coastal geologic evidence for past great earthquakes at the Cascadia subduction zone. *Earthquake Spectra* **11**, 1–18.

Bakun, W. H., and Lindh, A. G., 1985. The Parkfield, California, earthquake prediction experiment. *Science*, **229**, 619–624, doi:10.1126/science.229.4714.619.

Bakun, W., Aagaard, B., Dost, B., Ellsworth, W., Hardbeck, J., Harris, R., Ji, C., Johnston, M., Langbein, J., Lienkaemper, J., Michael, A., Nadeau, R., Reasenburg, P., Reichle, M., Roeloffs, E., Shakai, A., Simpson, R., and Waldhauser, F., 2005. Implications for prediction and hazard assessment from the 2004 Parkfield earthquake. *Nature*, **437**, 969–974.

Bland, S. H., O'Leary, E. S., Farinaro, E., Jossa, F., and Trevisan, M., 1996. Long-term psychological effects of natural disasters. *Psychosomatic Medicine*, **58**, 18–24.

Borrero, J. C., 2005. Field survey of northern Sumatra and Banda Aceh, Indonesia after the tsunami and earthquake of 26 December 2004. *Seismological Research Letters*, **76**, 312–320.

Bucknam, R. C., Hemphill-Haley, E., and Leopold, E. B., 1992. Abrupt uplift within the past 1700 years at southern Puget Sound, Washington. *Science*, **258**, 1611–1614, doi:10.1126/science.258.5088.1611.

Chandrasekhar, D. V., Ramalingeswara Rao, B., and Singh, B., 2007. Subsurface stress analyses for the Mw 7.6 Bhuj earthquake, India: an insight from the integrated geophysical approach. *Current Science*, **92**, 75–80.

Chouet, B., 1996. Long-period volcano seismicity: its sources and use in eruption forecasting. *Nature*, **380**, 309–316.

Cluff, L. S., Page, R. A., Slemmons, D. B., and Crouse, C. B., 2003. Seismic hazard exposure for the Trans-Alaska pipeline. In *Proceedings of the Sixth U.S. Conference and Workshop on Lifeline Earthquake Engineering, ASCE Technical Council on Lifeline Earthquake Engineering*, Long Beach, CA, August 2003.

Coffman, J. L., Von Hake, C. A., and Stover, C. W., 1982. *Earthquake History of the United States*. Publication 41-1, Revised Edition (with Supplement Through 1980). Boulder, Colorado: National Oceanic and Atmospheric Administration and U.S. Geological Survey, p. 258.

Cutcliffe, S. H., 2000. Earthquake resistant building design codes and safety standards: the California experience. *GeoJournal*, **51**, 259–262, doi:10.1023/A:1017566714380.

Darwin, C., 1845. *Journal of the Researches into the Natural History and Geology of the Countries Visited During the Voyage of H.M.S. Beagle Round the World*. London: John Murray.

De Bremaecker, J., 1985. Temperature in the core. In *Geophysics of the Earth's Interior*. Wiley, New York, pp. 296–297.

Douglas, J., 2001. *A Comprehensive Worldwide Summary of Strong-Motion Attenuation Relationships for Peak Ground Acceleration and Spectral Ordinates (1969 to 2000)*. Earthquake Seismology and Earthquake Engineering Report No. 01-1, Imperial College of Science, Technology and Medicine, Civil Engineering Department, London, pp. 126.

Dragert, H., Wang, K., and James, T. S., 2001. A silent slip event on the deeper Cascadia subduction interface, science express. *Science*, **292**(5521), 1525–1528, doi:10.1126/science.1060152.

Ellsworth, W. L., 1990. Earthquake History, 1769–1989. In R. E. Wallace (ed.), *The San Andreas Fault System, California*. Washington: United States Government Printing Office. USGS Professional Paper 1515, pp. 153–187.

Fenton, C. H., Adams, J., and Halchuk, S., 2006. Seismic hazards assessment for radioactive waste disposal sites in regions of low seismic activity. *Geotechnical and Geological Engineering*, **24**, 579–592, doi:10.1007/s10706-005-1148-4.

Field, E. H., Dawson, T. E., Felzer, K. R., Frankel, A. D., Gupta, V., Jordan, T. H., Parsons, T., Petersen, M. D., Stein, R. S., Weldon, R. J., and Wills, C. J., 2008. *The Uniform California Earthquake Rupture Forecast, Version 2 (UCERF 2), 2007 Working Group on California Earthquake Probabilities*, USGS Open File Report 2007-1437 and California Geological Survey Special Report 203.

Frolich, C., 2006. *Deep Earthquakes*. Cambridge: Cambridge University Press. pp. 574.

Geli, L., Bard, P.-Y., and Jullien, B., 1988. The effect of topography on earthquake ground motion: a review and new results. *Bulletin of the Seismological Society of America*, **78**, 42–63.

Geller, R. J., Jackson, D. D., Kagan, Y. Y., and Mulargia, F., 1997. Earthquakes cannot be predicted. *Science*, **275**(5306), 1616, doi:10.1126/science.275.5306.1616.

Gendzwill, D. J., Horner, R. B., and Hasegawa, H. S., 1982. Induced earthquakes at a potash mine near Saskatoon, Canada. *Canadian Journal of Earth Sciences*, **19**(3), 466–475, doi:10.1139/e82-038.

Ghosh, A., Vidale, J. E., Sweet, J. R., Creager, K. C., and Wech, A. G., 2009. Tremor patches in Cascadia revealed by seismic array analysis. *Geophysical Research Letters*, **36**, L17316, doi:10.1029/2009GL039080, 2009.

Giardini, D., 1999. The global seismic hazard assessment program (GSHAP) – 1992/1999. *Annali di Geofisica*, **42**, 957–974.

Giardini, D., and Basham, P., 1993. The global seismic hazard assessment program (GSHAP). *Annali di Geofisica*, **XXXVI**(3–4), 3–13.

Goldfinger, C., Nelson, H., Johnson, J. E., and The Shipboard Scientific Party, 2003. Deep-water turbidites as Holocene earthquake proxies: the Cascadia subduction zone and Northern San Andreas fault systems. *Annals of Geophysics*, **46**(5), 1169–1194.

Gouin, P., 1994. About the first earthquake reported in Canadian history. *Bulletin of the Seismological Society of America*, **84**, 478–483.

Grant, L. B., and Lettis, W. R., 2002. Introduction to the special issue on Paleoseismology of the San Andreas fault system. *Bulletin of the Seismological Society of America*, **92**, 2551–2554, doi:10.1785/0120000600.

Grünthal, G. (ed.), 1998. *European Macroseismic Scale 1998*. Luxembourg: Cahiers du Centre Européen de Géodynamique et de Séismologie, Vol. 15.

Guha, S. K., 2000. *Induced Earthquakes*. Dordrecht, The Netherlands: Kluwer Academic Publishers. pp. 312. ISBN 0-412-49870-7.

Hanks, T. C., and Kanamori, H., 1979. Moment magnitude scale. *Journal of Geophysical Research*, **84**(B5), 2348–2350, doi:10.1029/JB084iB05p02348

Harding, D. J., and Berghoff, G. S., 2000, Fault scarp detection beneath dense vegetation cover: Airborne LIDAR mapping of the Seattle Fault Zone, Bainbridge Island, Washington State.

In *Proceedings of the American Society of Photogrammetry and Remote Sensing Annual Conference*, Washington, DC.

Harris, R. A., and Arrowsmith, J. R., 2006. Introduction to the special issue on the 2004 parkfield earthquake and the parkfield earthquake prediction experiment. *Bulletin of the Seismological Society of America*, **96**, S1–S10, doi:10.1785/0120050831.

Haugerud, R. A., Harding, D. J., Johnson, S. Y., Harless, J. L., Weaver, C. S., and Sherrod, B. L., 2003. High-resolution Lidar topography of the Puget Lowland, Washington – A bonanza for earth science. *GSA Today*, **13**, 4–10.

Hill, D. P., and Prejean, S. G., 2007. Dynamic triggering. In Kanamori, H. (ed.), *Earthquake Seismology Treatise on Geophysics*. Amsterdam: Elsevier.

Hill, D. P., Reasenberg, P. A., Michael, A., Arabaz, W. J., Beroza, G., Brumbaugh, D., Brune, J. N., Castro, R., Davis, S., Depolo, D., Ellsworth, W. L., Gomberg, J., Harmsen, S., House, L., Jackson, S. M., Johnston, M. J., Jones, L., Keller, R., Malone, S., Munguia, L., Nava, S., Pechmann, J. C., Sanford, A., Simpson, R. W., Smith, R. B., Stark, M., Stickney, M., Vidal, A., Walter, S., Wong, V., and Zollweg, J., 1993. Seismicity remotely triggered by the magnitude 7.3 Landers, California, earthquake. *Science*, **260**(5114), 1617–1623.

Horner, R. B., Barclay, J. E., and MacRae, J. M., 1994. Earthquakes and hydrocarbon production in the Fort St. John area, northeastern British Columbia. *Canadian Journal of Exploration Geophysics*, **30**, 39–50.

Hough, S. E., 2001. Triggered earthquakes and the 1811–1812 New Madrid, central United States, earthquake sequence. *Bulletin of the Seismological Society of America*, **91**(6), 1574–1581.

Hough, S. E., Armbruster, J. G., Seeber, L., and Hough, J. F., 2000. On the modified mercalli intensities and magnitudes of the 1811–1812 new Madrid earthquakes. *Journal of Geophysical Research*, **105**(B10), 23,839–23,864.

Hyndman, R. D., 1995. Giant earthquakes of the Pacific Northwest. *Scientific American*, **273**, 50–57.

Hyndman, R. D., and Wang, K., 1995. The rupture zone of Cascadia great earthquakes from current deformation and the thermal regime. *Journal of Geophysical Research*, **100**, 22,133–22,154.

Hyndman, R. D., Mazzotti, S., Weichert, D., and Rogers, G. C., 2003. Frequency of large crustal earthquakes in Puget Sound–Southern Georgia Strait predicted from geodetic and geological deformation rates. *Journal of Geophysical Research* **108**, 2033, pp. 12. doi: 10.1029/2001JB001710.

Idriss, I. M., and Boulanger, R. W., 2008. *Soil Liquefaction During Earthquakes*. Oakland, CA: Earthquake Engineering Research Institute. pp. 262.

Johnston, A. C., and Kanter, L. R., 1990. Earthquakes in stable continental crust. *Scientific American*, **262**(3), 68–75.

Johnston, A. C., and Schweig, E. S., 1996. The enigma of the New Madrid earthquakes of 1811–1812. *Annual Review of Earth and Planetary Sciences*, **24**, 339–384.

Jordan, T. H., 2006. Earthquake predictability, brick by brick. *Seismological Research Letters*, **77**, 3–6.

Journal of Disaster Research, 2006. Chronology of Earthquakes. *Journal of Disaster Research* **1**, 452–487.

Kamigaichi, O., Saito, M., Doi, K., Matsumori, T., Tsukada, S., Takeda, K., Shimoyama, T., Nakamura, K., Kiyomoto, M., and Watanabe, Y., 2009. Earthquake early warning in Japan: warning the general public and future prospects. *Seismological Research Letters*, **80**, 717–726.

Kanamori, H., 2012. Earthquake hazards: putting seismic research to most effective use. *Nature*, **483**, 147–148.

Kanamori, H., and Kikuchi, M., 1993. The 1992 Nicaragua earthquake: a slow tsunami earthquake associated with subducted sediments. *Nature*, **361**, 714–716.

Kanamori, H., Hauksson, E., and Heaton, T., 1997. Real-time seismology and earthquake hazard mitigation. *Nature*, **390**, 461–464.

Kawase, H., 1996. The cause of the damage belt in Kobe: "the basin-edge effect", constructive interference of the direct S wave with the basin-induced diffracted Rayleigh waves. *Seismological Research Letters*, **67**, 25–34.

Keefer, D. K., 1984. Landslides caused by earthquakes. *Geological Society of America Bulletin*, **95**, 406–421.

Kiliç, C., and Ulusoy, M., 2003. Psychological effects of the November 1999 earthquake in Turkey: an epidemiological study. *Acta Psychiatrica Scandinavica*, **108**(3), 232–238.

Kovach, R. L., 2004. *Early Earthquakes of the Americas*. Cambridge: Cambridge University Press. pp. 280. ISBN 0 521 82489 3.

Kozak, J., and Thompson, M. C., 1991. *Historical Earthquakes in Europe*. Zurich: Swiss Reinsurance.

Lamontagne, M., Hammamji, Y., and Peci, V., 2008. Reservoir-triggered seismicity at the Toulnustouc hydroelectric project, Québec North Shore, Canada. *Bulletin of the Seismological Society of America*, **98**, 2543–2552.

Lipman, P. W., and Mullineaux, D. R. (eds.), 1981. *The 1980 Eruptions of Mount St. Helens*. Washington: U.S. Geological Survey Professional Paper 1250. pp. 844.

Liu, L., and Zoback, M. D., 1997. Lithospheric strength and intraplate seismicity in the New Madrid seismic zone. *Tectonics*, **16**, 585–595.

Loewe, M., and Shaughnessy, E. L., 1999. *The Cambridge History of Ancient China from the Origins of Civilization to 221 B.C.* Cambridge: Cambridge University Press. pp. 1148. ISBN 0521470307.

Lomintz, C., 2004. Major earthquakes of Chile: a historical survey, 1535–1960. *Seismological Research Letters*, **75**, 368–378, doi:10.1785/gssrl.75.3.368.

Ludwin, R. S., Dennis, R., Carver, D., McMillan, A. D., Losey, R., Clague, J., Jonientz-Trisler, C., Bowechop, J., Wray, J., and James, K., 2005. Dating the 1700 Cascadia earthquake: great coastal earthquakes in native stories. *Seismological Research Letters*, **76**, 140–148, doi:10.1785/gssrl.76.2.140.

Martin, S., and Szeliga, W., 2010. A catalog of felt intensity data for 570 earthquakes in India, from 1636 to 2009. *Bulletin of the Seismological Society of America* **100**, 562–569, doi:10.1785/0120080328.

Mazzotti, S., 2007. Geodynamic models for earthquake studies in intraplate North America. In Stein, S., and Mazzotti, S. (eds.), *Continental Intraplate Earthquakes: Science, Hazard, and Policy Issues*. Boulder: Geological Society of America. Geological Society of America Special Paper 425, pp. 17–33. doi:10.1130/2007.2425(02).

Mazzotti, S., and Adams, J., 2005. Rates and uncertainties on seismic moment and deformation in eastern Canada. *Journal of Geophysical Research*, **110**(B09301), 16, doi:10.1029/2004JB003510.

Mazzotti, S., Dragert, H., Henton, J., Schmidt, M., Hyndman, R., James, T., Lu, Y., and Craymers, M., 2003. Current tectonics of northern Cascadia from a decade of GPS measurements. *Journal of Geophysical Research*, **108**(B12), 2554, doi:10.1029/2003JB002653, 2003.

McCalpin, J. P. (ed.), 2009. *Paleoseismology*, Second edn. San Diego: Academic. International Geophysics, Vol. 95. pp. 708. ISBN 13: 978-0-12-373576-8.

Mendes-Victor, L. A., Sousa Oliveira, C., Azevedo, J., and Ribeiro, A. (eds.), 2009. *The 1755 Lisbon Earthquake: Revisited*. Dordrecht: Springer. pp. 597. ISBN 978-1-4020-8608-3.

Missiakoulis, S., 2008. Aristotle and earthquake data: a historical note. *International Statistical Review, International Statistical Institute*, **76**(1), 130–133.

Mucciarelli, M., 2008. Codes, models and reality: reductionism vs. holism in a review of microzonation studies in the Umbria-Marche region. *Annals of Geophysics*, **2–3**(51), 491–498.

Nadeau, R. M., and Dolenc, D., 2005. Nonvolcanic tremors deep beneath the San Andreas fault. *Science*, **307**(5708), 389, doi:10.1126/science.1107142.

Nur, A., and Burgess, D., 2008. *Apocalypse: Earthquakes, Archaeology, and the Wrath of God*. Princeton: Princeton University Press. 324 pp.

Oeser, E., 1992. Historical earthquake theories from Aristotle to Kan. In Gutdeutsch, R., Grünthal, G., and Musson, R. (eds.), *Historical Earthquakes in Central Europe*. Vienna: Abhandlungen der Geologischen Bundesanstalt, pp. 11–31.

Peláez, J. A., Chourak, M., Tadili, B. A., Aït Brahim, L., Hamdache, M., López Casado, C., and Martínez Solares, J. M., 2007. A catalog of main Moroccan earthquakes from 1045 to 2005. *Seismological Research Letters*, **78**, 614–621, doi:10.1785/gssrl.78.6.614.

Rajendran, C. P., and Rajendran, K., 2001. Characteristics of deformation and past seismicity associated with the 1819 Kutch earthquake, northwestern India. *Bulletin of the Seismological Society of America*, **91**(3), 407–426.

Rodríguez, C. E., Bommer, J. J., and Chandler, R. J., 1999. Earthquake-induced landslides: 1980–1997. *Soil Dynamics and Earthquake Engineering*, **18**, 325–346.

Rogers, G. C., 1992. The history of earthquake studies in British Columbia: from indian legend to satellite technology. In *Pioneering Geology in the Canadian Cordillera, B.C. Geological Survey Branch, Open File 1992-19*, pp. 61–66.

Rogers, G., and Dragert, H., 2003. Episodic tremor and slip on the cascadia subduction zone: the chatter of silent slip. *Science*, **300**(5627), 1942–1943, doi:10.1126/science.1084783.

Roman, D. C., and Cashman, K. V., 2006. The origin of volcano-tectonic earthquake swarms. *Geology*, **34**, 457–460.

Ruff, L. J., and Tichelaar, B. W., 1996. What controls the seismogenic plate interface in subduction zones? In Bebout, G. E., Scholl, D. W., Kirby, S. H., and Platt, J. P. (eds.), *Subduction: Top to Bottom, Geophysical Monogram Series*. Washington, DC: AGU, Vol. 96, pp. 105–111.

Rynn, J. M. W., Brennan, E., Hughes, P. R., Pedersen, I. S., and Stuart, H. J., 1992. The 1989 Newcastle, Australia, earthquake: the facts and the misconceptions. *Bulletin of the New Zealand National Society for Earthquake Engineering*, **25**, 77–144.

Sagiya, T., 2004. A decade of GEONET: 1994–2003: The continuous GPS observation in Japan and its impact on earthquake studies. *Earth Planets Space*, **56**, xxix–xli.

Seed, H. B., 1968. Landslides during earthquakes due to soil liquefaction, American Society of Civil Engineers. *Journal of the Soil Mechanics and Foundation Division*, **94**, 1053–1122.

Seed, H. B., and Idriss, I. M., 1982. *Ground motions and soil liquefaction during earthquakes*. Berkeley, CA: Earthquake Engineering Research Institute, p. 134 (475 S41 1982).

Sieh, K., 1978. Pre-historic large earthquakes produced by slip on the San Andreas Fault at Pallett Creek, California. Journal of Geophysical Research **83**, 3907–3939 (Reprinted in: A.G. Sylvester (ed.), Wrench Fault Tectonics, American Assoc. Petroleum Geologists, Reprint Series No. 28, 1984 (pp. 223-275)).

Smith, W. E. T., 1962. Earthquakes of eastern Canada and adjacent areas 1534–1927. *Publications of the Dominion Observatory*, **26**(5), 269–301.

Spudich, P., Hellweg, M., and Lee, W. H. K., 1996. Directional topographic site response at Tarzana observed in aftershocks of the 1994 Northridge, California, earthquake: implications for mainshock motions. *Bulletin of the Seismological Society of America*, **86**(1B), S193–S208.

Steacy, S., Gomberg, J., and Cocco, M., 2005. Introduction to special section: stress transfer, earthquake triggering, and time-dependent seismic hazard. *Journal of Geophysical Research*, **110**, B05S01, doi:10.1029/2005JB003692.

Stein, R. S., 1999. The role of stress transfer in earthquake occurrence. *Nature*, **402**, 605–609, doi:10.1038/45144.

Stein, R. S., King, G. C. P., and Lin, J., 1992. Change in failure stress on the southern San Andreas fault system caused by the 1992 magnitude = 7.4 Landers earthquake. *Science*, **258**, 1328–1332.

Stein, R. S., Barka, A. A., and Dieterich, J., 1997. Progressive failure on the North Anatolian fault since 1939 by earthquake stress triggering. *Geophysical Journal International*, **128**, 594–604. ISSN: 0956-540X.

Suarez, G., Novelo, D., and Mansilla, E., 2009. Performance evaluation of the seismic alert system (SAS) in Mexico City: a seismological and social perspective. *Seismological Research Letters*, **80**, 707–714.

Sykes, L. R., Shaw, B. E., and Scholz, C. H., 1999. Rethinking earthquake prediction. *Pure and Applied Geophysics*, **155**, 207–232.

Synolakis, C. E., Bardet, J.-P., Borrero, J. C., Davies, H. L., Okal, E. A., Silver, E. A., Sweet, S., and Tappin, D. R., 2002. The slump origin of the 1998 Papua New Guinea tsunami. *Proceedings of the Royal Society London A*, **458**, 763–789, doi:10.1098/rspa.2001.0915.

Szeliga, W., Hough, S. E., Martin, S., and Bilham, R., 2010. Intensity, magnitude, location and attenuation in India for felt earthquakes since 1762. *Bulletin of the Seismological Society of America*, **100**(2), 570–584.

Talwani, P., and Rajendran, K., 1991. Some seismological and geometric features of intraplate earthquakes. *Tectonophysics*, **186**, 19–41.

Titov, V., Rabinovich, A. B., Mofjeld, H. O., Thomson, R. E., and Gonzalez, F. I., 2005. The global reach of the 26 December 2004 Sumatra tsunami. *Science*, **309**, 2045–2048, doi:10.1126/science.1114576.

Tobin, H., and Kinoshita, M., 2006. NanTroSEIZE: the IODP nankai trough seismogenic zone experiment. *Scientific Drilling*, **2**, 23–27, doi:10.2204/iodp.sd.2.06.2006.

Utsu,T., 2002. Relationships between magnitude scales. In Lee, W. H. K, Kanamori, H., Jennings, P. C., and Kisslinger, C., (eds.) *International Handbook of Earthquake and Engineering Seismology: Academic Press, a division of Elsevier, two volumes, International Geophysics*, London, UK. Vol. 81-A, pp. 733–746.

Wada, I., and Wang, K., 2009. Common depth of slab-mantle decoupling: reconciling diversity and uniformity of subduction zones. *Geochemistry Geophysics Geosystems*, **10**, Q10009, doi:10.1029/2009GC002570.

Wald, D. J., Quitoriano, V., Heaton, T. H., Kanamori, H., Scrivner, C. W., and Worden, C. B., 1999a. TriNet "shakemaps": rapid generation of peak ground motion and intensity maps for earthquakes in Southern California. *Earthquake Spectra*, **15**, 537–556.

Wald, D. J., Quitoriano, V., Dengler, L. A., and Dewey, J. W., 1999b. Utilization of the internet for rapid community intensity maps. *Seismological Research Letters*, **70**(6), 680–697.

Wald, D., Lin, K.-W., Porter, K., and Turner, L., 2008a. ShakeCast: automating and improving the use of shakemap for post-earthquake decision-making and response. *Earthquake Spectra*, **24**, 533, doi:10.1193/1.2923924.

Wald, D. J., Earle, P. S., Allen, T. I., Jaiswal, K., Porter, K., and Hearne M., (2008b). Development of the U.S. Geological Survey's PAGER system (prompt assessment of global earthquakes for response). In *Proceedings of the 14th World Conf. Earthquake Engineering*, Beijing, China, pp. 8.

Wang, J., 2004. Historical earthquake investigation and research in China. *Annals of Geophysics*, **47**, 831–838.

Wang, K., Wells, R., Mazzotti, S., Hyndman, R. D., and Sagaya, T., 2003. A revised dislocation model of interseismic deformation of the Cascadia subduction zone. *Journal of Geophysical Research*, **108**, 2026, doi:10.1029/2001JB001227.

Wang, K., Chen, Q.-F., Sun, S., and Wang, A., 2006. Predicting the 1975 Haicheng earthquake. *Bulletin of the Seismological Society of America*, **96**, 757–795, doi:10.1785/0120050191.

Wang, K., Hu, Y., and He, J., 2009. Wedge mechanics: relation with subduction zone earthquakes and tsunamis. In Meyer, R. (ed.), *Encyclopedia of Complexity and System Science*. New York/London: Springer, pp. 10047–10058.

Wells, D. L., and Coppersmith, K. J., 1994. New empirical relationships among magnitude, rupture length, rupture width, rupture area, and surface displacement. *Bulletin of the Seismological Society of America*, **84**, 974–1002.

Wood, H. O., and Neumann, F., 1931. Modified mercalli intensity scale of 1931. *Bulletin of the Seismological Society of America*, **21**, 277–283.

Wu, Y.-M., and Kanamori, H., 2008. Development of an earthquake early warning system using real-time strong motion data. *Sensors*, **8**, 1–9.

Zechar, J. D., and Jordon, T. H., 2008. Testing alarm-based earthquake predictions. *Geophysical Journal International*, **172**, 715–724, doi:10.1111/j.1365-246X.2007.03676.x.

Zoback, M. D., 2006. SAFOD penetrates the San Andreas fault. *Scientific Drilling*, **2**, 32–33, doi:10.2204/iodp.sd.2.07.2006.

Cross-references

Building Codes
Building Failure
Buildings, Structures and Public Safety
Costs (Economic) of Natural Hazards and Disasters
Cultural Heritage and Natural Hazards
Damage and the Built Environment
Early Warning Systems
Earthquake Damage
Earthquake Prediction and Forecasting
Earthquake Resistant Design
Economics of Disasters
Elastic Rebound, Theory of
Epicentre
Fault
Global Positioning System (GPS) and Natural Hazards
Hypocentre
Induced Seismicity
Intensity Scales
Isoseismal
Liquefaction
Macroseismic Survey
Magnitude Measures
Mitigation
Modified Mercalli (MM) Scale
North Anatolian Fault
Pacific Tsunami Warning and Mitigation System (PTWS)
Paleoseismology
Plate Tectonics
Primary Wave (P Wave)
Psychological Impacts of Natural Disasters
Remote Sensing of Natural Hazards and Disasters
Richter, Charles F.
San Andreas Fault
Secondary Wave (S wave)
Seiche
Seismic Gap
Seismograph/Seismometer
Seismology
Shear
Subduction
Tangshan, China (1976 Earthquake)
Tectonic and Tectono-Seismic Hazards
Tiltmeters
Triggered Earthquakes
Tsunami
Unreinforced Masonry Building
Warning Systems
Wenchuan, China (2008 Earthquake)
World Economy, Impact of Disasters on
World-Wide Trends in Disasters Caused by Natural Hazards

EARTHQUAKE DAMAGE

Nicolas Desramaut, Hormoz Modaressi,
Gonéri Le Cozannet
BRGM - French Geological Survey, Orléans, France

Synonyms
Earthquake destruction; Seismic damage

Definition
Damage: Physical harm reducing the value, the operation, or the usefulness of something. (Oxford English dictionary, see "*Damage and the Built Environment*").
Earthquake: A sudden violent shaking of the ground as a result of movements within the earth's crust. (Oxford English dictionary, see "*Earthquake*").
Earthquake damage: Damage or destruction of properties and good caused by an earthquake, either as a direct result of the ground shaking, or as an induced consequence.

Introduction
In the natural hazards community, different expressions are used to differentiate the domains affected by the earthquakes. "Damage" refers to the degradation or destruction of the physical assets, such as buildings, facilities, infrastructure, or nonstructural elements (see "*Damage and the Build Environment*" and "*Structural Damage Caused by Earthquakes*"). "Loss" concerns either the casualties (e.g., human losses, see "*Casualties Following Natural Hazards*") or the economical consequences (such as lost jobs, lost properties, business interruptions, repair and reconstruction costs; see "*Costs (Economic) of Natural Hazards and Disasters*"). "Impact" focuses on the repercussions on society (socioeconomical dimensions).

The term "Earthquake damage" refers to the different kinds and levels of degradations to the physical environment resulting from seismic events. Earthquake damage is the cause of the induced consequences of earthquakes that societies have to face. In particular, there is an old say stating that: "earthquakes do not kill people, buildings do."

Causes of damage
Direct causes
Ground shaking is considered to be a major cause of damage to buildings and components due to earthquakes (Noson et al., 1988). Structures are particularly vulnerable to these horizontal, oscillatory, and dynamic loads. The ground surface, moving quickly back-and-forth and

side-to-side, creates inertial forces inside the structures. These stresses might produce irreversible damage, from cracking, permanent displacement, or entire collapses of the structures (see "*Building Failure*," "*Structural Damage Caused by Earthquakes*").

Indirect causes

However, earthquakes can trigger other secondary hazards (domino effects), which could locally inflict damage more severe than the direct effects of the ground shaking. Earthquake-induced failures of the underlying sediment or rock (see "*Fault*," "*Liquefaction*," "*Lateral Spreading*," or "*Landslide*") are likely to damage underground infrastructure, resulting in potentially widespread impacts, even in areas some distance away from the epicenter (Bird and Bommer, 2004). These ground failures could also impact on the structures' foundations, which might cause a building to lean sufficiently to cantilever and fail. Another potential cause of considerable damage is due to the disturbance of water bodies, either in open seas (see "*Tsunami*") or in enclosed bodies of water (see "*Seiche*"). The consequence of such secondary events can be catastrophic (e.g., Indian Ocean Tsunami of December 2004, generated by a Magnitude 9 earthquake).

Direct earthquake damage can also result in new causes of damage (e.g., induced technological hazards). Damaged pipelines are, for example, new threats to the integrity of the remaining infrastructures. Gas leakage might ignite fires or explosions. Water-pipe breakage might cause flooding but also prevent fire-fighters from effectively intervening. The failure of the Fukushima Nuclear Plant in 2011, triggered by a tsunami which in turn was initiated by an earthquake, is an example of such an induced technological hazard.

Factors of damage

A building will endure damages that depend on the intensity of the event (e.g., for ground shaking, these loads are function of the magnitude of the earthquake, the focal depth, the distance on the surface from the epicenter, and the side effects conditioned by the surface geology and past human intervention), but also from its own physical susceptibility to the different threats triggered by the earthquake. The susceptibility to damage is a function of several intrinsic specificities of the elements, such as existence and compliance to existing building codes (see "*Building Code*," "*Earthquake Resistant Design*"), building types, material properties, construction quality, number of storeys, building positions in the site (Şengezer et al., 2008), but also age and previous stresses applied to the structures (see "*Antecedent Conditions*"). The nonstructural elements (e.g., parapets, architectural decorations, partition walls, chimneys, doors, windows, air conditioners, and heaters) are subject to other modes of damage.

They might be displaced (due to poor attachment) or distorted (due to building flexure) during ground shaking, triggering new threats for the structures and people (Fierro et al., 1994).

Measures of damage

There are several approaches to measure damage. The traditional damage assessment methods are based on surveys, resulting in an enumeration or evaluation by extrapolation of the number of damaged elements, and eventually an assessment of their value. The result will be a quantification of the property losses. However, this measure is dependent on the initial value of the stock (such as the pre-earthquake real estate value). The damage ratio (damage cost or repair cost compared to the construction cost or the replacement value) addresses this issue by evaluating the amount of degradation buildings have suffered. This method gives similar indications to the methodology which consists in estimating the proportion of damage for each component, producing classes of damage or indicators. Damage can also be assessed based on physical measures, such as Inter-Storey Drift, which provide indices correlated to the physical integrity of the buildings.

Damage scales

In order to evaluate the intensity of earthquakes based on their effects on animals, humans, and buildings, several scales have been created (Hill and Rossetto, 2008; Blong, 2003). Damage corresponding to each level is described in a qualitative way (e.g., level VII of the European Macroseismic Scale corresponds to "damaging: Most people are frightened and run outdoors. Furniture is shifted and objects fall from shelves in large numbers. Many ordinary buildings suffer moderate damage: small cracks in walls; partial collapse of chimneys" (Grünthal, 1998)). Historically, one of the first was the Rossi-Forel scale, named after the nineteenth century developers, which was divided into ten intensity levels. Damage to nonstructural objects generally relate to level IV damages, and damage to buildings, to level VIII. This scale was revised by Mercalli, and then improved successively. It is now known as the Modified Mercalli Scale (see "*Modified Mercalli (MM) Scale*"). The scale is composed of 12 levels, with 7 levels related to building damage (MM VI-XII). Other scales have been developed in order to reflect local contexts and buildings typologies. For example, the European Macroseismic Scale (EMS-98) and the Medvedev-Sponheuer-Karnik Scale (MSK-64) are used respectively in Europe and Russia, whereas HAZUS 99 or ATC-13 scales are more adapted to American buildings.

Conclusion

Earthquakes are considered to be one of the most damaging natural hazards. This is not only due to the tremendous energy released, but also due to the intrinsic uncertainties associated with this hazard. It is currently impossible to predict when earthquakes will occur. However, significant progress has been made in the comprehension of the hazards, vulnerability (see "*Vulnerability*"), and in construction practice. Hence, where resources are available, the consequences of this threat can be, if not avoided, at least minimized, and physical susceptibility reduced. This would make human infrastructure and communities less prone to damage, and thus more resilient to earthquakes.

Bibliography

Bird, J. F., and Bommer, J. J., 2004. Earthquake losses due to ground failure. *Engineering Geology*, **75**, 147–179.

Blong, R., 2003. A review of damage intensity scales. *Natural Hazards*, **29**, 57–76.

Fierro, E. A., Perry, C. A., and Freeman, S. A., 1994. *Reducing the Risks of Nonstructural Earthquake Damage*, 3rd edn. Washington, DC: Wiss, Janney, Elstner Associates. FEMA.

Grünthal, G., 1998. *European Macroseismic Scale 1998*. Luxembourg: European Seismological Commission, Subcommission on Engineering Seismology, Working Group Macroseismic Scales.

Hill, M., and Rossetto, T., 2008. Comparison of building damage scales and damage descriptions for use in earthquake loss modelling in Europe. *Bulletin of Earthquake Engineering*, **6**, 335–365.

Noson, L. J., Qamar, A., and Thorsen, G. W., 1988. *Washington State Earthquake Hazards*. Washington, DC: Washington State Department of Natural Resources, Division of Geology and Earth Resources.

Osteraas, J. (ed.), 2007. *General Guidelines for the Assessment and Repair of Earthquake Damage in Residential Woodframe Buildings*. Richmonds: Consortium of Universities for Research in Earthquake Engineering.

Şengezer, B., Ansal, A., and Bilen, Ö., 2008. Evaluation of parameters affecting earthquake damage by decision tree techniques. *Natural Hazards*, **47**, 547–568.

Cross-references

Antecedent Conditions
Building Codes
Building Failures
Building, Structures and Public Safety
Casualties Following Natural Hazards
Costs (Economic) of Natural Hazards and Disasters
Damage and the Built Environment
Earthquake
Earthquake Resistant Design
Fault
Fire and Firestorms
Human Impact of Hazards
Landslide (Mass Movement)
Lateral Spreading
Liquefaction
Mass Movement
Mitigation
Modified Mercalli (MM) Scale
Recovery and Reconstruction After Disaster
Resilience
Seiche
Structural Damage Caused by an Earthquake
Tangshan, China (1976 Earthquake)
Tsunami
Vulnerability

EARTHQUAKE PREDICTION AND FORECASTING

Alik T. Ismail-Zadeh
Karlsruher Institut für Technologie, Karlsruhe, Germany
Institut de Physique du Globe de Paris, Paris, France
Russian Academy of Sciences, Moscow, Russia

Synonyms

According to the Oxford Dictionary for Synonyms and Antonyms (ODSA, 2007), the words *prediction* and *forecasting* are synonyms. Meanwhile a part of the seismological community makes a difference between the terms *earthquake prediction* and *earthquake forecasting*. Namely, an *earthquake prediction* is referred to as a statement about a target earthquake of a certain range of magnitudes in a specified geographical region, which involves an *alarm*, a time window of the increased probability of the target earthquake. If a target earthquake occurs during the alarm, the prediction is true; if not, it is a false alarm; if a target earthquake occurs without an alarm it is called a failure to predict of the earthquake. An *earthquake forecast* is described as a statement about a future earthquake in a specified spatial-temporal window, which is based on the specified probabilities of target earthquakes. "*A time-independent forecast is one in which the subdomain probabilities depend only on the long-term rates of target events; the events are assumed to be randomly distributed in time, and the probabilities of future events are thus independent of earthquake history. In a time-dependent forecast, the probabilities depend on the information available at time when the forecast is made*" (Jordan et al., 2011).

The difference between *earthquake prediction* and *earthquake forecasting* presented above is related basically to methodologies rather than to definitions, and they are associated with two different (alarm-based and probability-based) approaches to earthquake prediction (or earthquake forecasting). In this entry, I do not distinguish between the terms *earthquake prediction* and *earthquake forecasting*, considering them as synonyms.

Definition

The Earth's lithosphere can be considered as a hierarchy of volumes of different size (from tectonic plates to the grains of rocks). Driven by thermal convection, the movement of the lithosphere is controlled by a wide variety of processes on and around the fractal mesh of boundary zones (fault zones) and does produce earthquakes. An earthquake is a sudden movement within the Earth's crust or the upper mantle usually triggered by the release of tectonic stress (and energy) along a fault segment. The hierarchy of movable lithospheric volumes composes a large nonlinear dynamical system (Keilis-Borok, 1990; Keilis-Borok et al., 2001). Prediction of such a complex system in the sense of extrapolating discrete future trajectories is futile. However, upon coarse-graining the integral empirical regularities emerge, premonitory seismicity patterns (e.g., rise of seismic activity; rise of spatial-temporal earthquake clustering; changes in earthquake frequency-magnitude distribution; rise of the earthquake correlation range) signal the approach of a large earthquake, and this opens possibilities of earthquake prediction.

Earthquake prediction is a statement about future earthquake occurrence based on the information, data, and scientific methods that are available now. To predict an earthquake, one must "*specify the expected magnitude range, the geographical area within which it will occur, and the time interval within which it will happen with sufficient precision so that the ultimate success or failure of the prediction can readily be judged. Only by careful recording and analysis of failures as well as successes can the eventual success of the total effort be evaluated and future directions charted. Moreover, scientists should also assign a confidence level to each prediction*" (Allen et al., 1976).

Introduction

"An earthquake is an evil that spreads everywhere; it is inevitable and damaging for everybody. Besides the destruction of people, houses and entire towns, it can make disappear nations and even entire nations without leaving a trace of what has been once," wrote Lucius Annaeus Seneca (1 BC–65 AD), a Roman philosopher, about AD 62 Pompeii earthquake in his Quaestiones Naturales, an encyclopedia of the natural world. The extremely destructive nature of earthquakes has been known for millennia. Although the origins of observational seismology date back to the East Han Dynasty in China, when Zhang Heng (78–139 AD) invented an earthquake detection instrument, it would be misleading to pretend that state-of-the-art physics of earthquakes is a well-developed branch of science. To the contrary, most seismologists clearly understand the grand challenges for seismology (Forsyth et al., 2009). Several important questions still remain pressing: How do faults slip? What happens during an earthquake? How do we measure the size of earthquakes? Why, where, and when do earthquakes occur? The fundamental difficulty in answering these questions comes from the fact that no earthquake has been ever observed directly and only a few were subject to an in situ verification of their physical parameters.

The mature wisdom of any science is determined by its ability to forecast phenomena under study, and in the case of seismology, to predict an earthquake. The scientific research aimed at predicting earthquakes began in the second half of nineteenth century, when seismology reached the level of a recognized scientific discipline. The desire to find tools that would permit forecasting the phenomenon under study is so natural that as early as in 1880 John Milne, a British engineer-seismologist, defined earthquake prediction as one of the pivotal problems of seismology and discussed possible precursors of large earthquakes. Over the following century earthquake prediction research experienced alternating periods of high enthusiasm and critical attitude. The abruptness, along with apparent irregularity and infrequency of large earthquake occurrences, facilitates formation of a common perception that earthquakes are random unpredictable phenomena. Earthquake prediction research has been widely debated, and opinions on the possibilities of prediction vary from the statement that earthquake prediction is intrinsically impossible (Geller et al., 1997) to the statement that prediction is possible, but difficult (Knopoff, 1999).

Earthquake prediction: Success, failure, and perspectives

Two types of earthquake prediction methods can be distinguished. The first type of the prediction methods is based on finding and monitoring of an earthquake precursor (a physical, chemical, or biological signal, which indicates that a large earthquake is approaching) and issuing an alarm at the time of the abnormal behavior of the precursor. The earthquake precursors fall into several broad categories: biological (e.g., anomalous animal behavior from dogs, snakes, horses), electromagnetic (e.g., changes in an electromagnetic signal associated with a dilatancy-induced fluid flow); geochemical (e.g., changes in radon concentration in groundwater); geodetic (e.g., strain-rate changes); geoelectrical (e.g., changes in electrical conductivities of rocks); hydrological (e.g., changes in a borehole water level); physical (e.g., changes in seismic wave velocities in the crust, Vp/Vs changes); thermal (e.g., changes in thermal infrared radiation); and others. Although many observations reveal unusual changes of natural fields at the approach of a large earthquake, most of them report a unique case history and lack a systematic description (Wyss, 1991). Unfortunately, so far the search for a precursor of a large earthquake has been unsuccessful. Moreover, perhaps such a search for a single precursor is useless, and monitoring of a set of precursors could give more reliable answer on an anticipating earthquake.

Chinese seismologists were the first to successfully predict a devastating earthquake, the M7.0 Haicheng earthquake of 1975 (Zhang-li et al., 1984). Their prediction was based on monitoring physical fields (anomalies in land elevation, in ground water level, and in seismicity prior to the large event) and on the observations of peculiar behavior of animals. The success of this prediction stimulated further design of methods for diagnosis of an approaching large quake. Most of the prediction methods suggested at that time were not confirmed in the following years. The catastrophic M7.5 Tangshan, China earthquake of 1976, which caused hundreds of thousands of fatalities, was not predicted; this failure was like a cold shower of disillusionment, not only for Chinese seismologists. The necessity of strict formulations and stringent methods of testing the complex prediction hypotheses nominated by seismology, tests that would distinguish a lucky guess from a reliable prediction, became evident as never before.

The second type of the two methods for earthquake prediction is based on statistical analysis of seismicity. A pattern of earthquakes is an important key in understanding the dynamics of the lithosphere in an earthquake-prone region. In the 1970s the progress in formalizing the description and pattern recognition of earthquake-prone areas (e.g., Gelfand et al., 1976), which indeed deliver a term-less zero-approximation prediction of large earthquakes, resulted in better understanding of certain universal aspects of seismic processes in different tectonic environments. A distinctive similarity of criteria in zero-approximation provided an encouraging foundation for a systematic search of universal patterns, for example, a unique low-magnitude seismic sequence at the approach of a large quake. Along with 20 years of accumulated global seismic data of high quality and completeness and developments in earthquake physics, a novel understanding of seismic processes basically emerged. Several earthquake prediction methods of the second type have been developed including the CN (Keilis-Borok et al., 1988), M8 (Keilis-Borok and Kossobokov, 1990), MSc (Kossobokov et al., 1990), RTL (Sobolev, 2001), the pattern informatics (Rundle et al., 2002), STEP (Gerstenberger et al., 2005), and the RTP (Shebalin et al., 2006) methods.

According to the scientific definition of earthquake prediction (see section Definition), one can identify an earthquake prediction of certain magnitude range by duration of its time interval and/or by territorial specificity. Four major stages can be distinguished in earthquake prediction: (1) long term (a decadal time scale), (2) intermediate term (one to several years), (3) short term (weeks and months), and (4) immediate (seconds to hours) (see Table 1).

Long-term prediction of earthquakes (it is also called a long-term earthquake forecasting, Kanamori, 2003) is based on the elastic rebound theory, a physical theory about increasing the shear stress on the fault that forms the boundary between crustal blocks until the fault reaches its yield stress and ruptures (Reid, 1911). The Reid's model is rather simple to explain all features of earthquake dynamics, and the use of *characteristic earthquake* (Schwartz and Coppersmith, 1984) and *seismic gap* (Fedotov, 1965; Kelleher et al., 1973) hypotheses allows for developing a long-term prediction of large earthquakes on specified faults or fault segments. Based on fault slip rates, geodetic measurements, and paleoseismological studies it is possible to determine the average strain accumulation and to estimate long-term earthquake rates in terms of probabilities. However, the long-term earthquake forecast is complicated by the irregular occurrence of large earthquakes. The long-term earthquake forecasting models were developed to determine the probabilities of active fault segments to rupture for the next few decades (e.g., Working Group on California Earthquake Probabilities, 1988, 2007). The models are based on a renewal process, in which the expected time of the next event depends only on the date of the last event. The times between successive events are considered to be independent, identically distributed, random variables. When a rupture occurs on the fault segment, it resets the renewal process to its initial state (e.g., Ellsworth et al., 1999). The advanced model (UCERF2 time-dependent forecast for California, Field et al., 2009) incorporates renewal models for the major strike-slip faults of the San Andreas fault system. The long-term earthquake forecasts can guide engineering and emergency planning measures to mitigate the impact of pending large earthquakes.

Earthquake Prediction and Forecasting, Table 1 Classification of earthquake predictions

Temporal, in years		Spatial, in source zone size L	
Long term	10–100	Long range	Up to 100
Intermediate term	1	Middle range	5–10
Short term	0.01–0.1	Narrow	2–3
Immediate	0.001	Exact	1

An intermediate-term prediction is an update of the long-term prediction brought about by some indicators (e.g., an increase in background seismicity, clustering of events in space and time, transformation of magnitude distribution, and some others). Keilis-Borok and Kossobokov (1990) proposed an alarm-based method for intermediate-term predictions of great earthquakes (magnitude 8 and greater), so-called M8 algorithm, which is based on the analysis of seismic patterns and dynamics preceding the large event. An alarm (or time of increased probability of a large earthquake) is declared, when several functions describing seismic activity (Healy et al., 1992) become "anomalous" within a narrow time window. The earthquake prediction M8 algorithm receives a fair amount of attention, because the algorithm's testing is unprecedented in rigor and coverage. The algorithm M8 is subject to ongoing real-time experimental testing by V. Kossobokov (http://www.mitp.ru/en/default.html, section Prediction), and the relevant predictions are communicated, with due discretion, to several dozens of leading scientists and administrators in many countries including China, France, Italy, Japan, Russia, and USA. The accumulated statistics of this experiment confirms intermediate-term predictability of large earthquakes with middle- to exact-range of location (Kossobokov, 2006): For the last 25 years, 13 out of 18 great earthquakes were predicted by the M8 algorithm. The ratio of the alarms to the total space-time volume is about 33%, and a confidence level is higher than 99.9% (Ismail-Zadeh and Kossobokov, 2011). To bring the confidence level down to generally accepted value of 95%, the ongoing experimental testing has to encounter eight failures to predict in a row, which appears unlikely to happen. Independent assessments of the M8 algorithm performance show that the method is nontrivial to predict large earthquakes (Zechar and Jordan, 2008; Molchan and Romashkova, 2011), although a drawback of the method is the limited probability gain (Jordan et al., 2011).

Stress-induced variations in the electric conductivity of rocks have been studied in lab experiments (e.g., Morat and Le Mouël, 1987). In the 1980s P. Varotsos, K. Alexopoulos, and K. Nomicos proposed a method of short-term earthquake prediction, the VAN method (named after the researchers' initials), which is based on detection of characteristic changes in the geoelectric potential (so-called seismic electric signals, SES) via a telemetric network of conductive metal rods inserted in the ground (e.g., Varotsos et al., 1986). The anomaly pattern is continually refined as to the manner of identifying SES from within the abundant electric noise the VAN sensors are picking up. The researchers have claimed to be able to predict earthquakes of magnitude larger than 5, within 100 km of epicentral location and within 0.7 units of magnitude (Lighthill, 1996; Uyeda et al., 2002). The feasibility of short-term earthquake prediction (days to weeks and months) is still controversial, and the major difficulty here is to identify short-term precursors in the background of intermediate-term alarms.

Another type of short-term prediction is based on calculating the probabilities of target events within future space-time domains. An example of this type of predictions is the Short-Term Earthquake Probability (STEP) method developed by Gerstenberger et al. (2005) and employed by the U.S. Geological Survey for operational forecasting in California. STEP uses aftershock statistics to make hourly revisions of the probabilities of strong ground motion. The probability-based forecasts are the mean for transmitting information about probabilities of earthquakes in the particular region under monitoring. While the probability gains of short-term forecasts can be high, the probabilities of potential destructive earthquakes remain much smaller than 0.1 as the forecasting intervals are much shorter than the recurrence intervals of large earthquakes. Moreover short-term forecasts to be of use for practitioners should cast an alarm, and in this case there is no principal difference between earthquake prediction and earthquake forecast in the terms as it was defined in the section Synonyms.

Immediate earthquake prediction for the next few hours is sometimes mixed with "early warning," which is usually based on the first arrival of seismic waves and transmission of an electronic alert within a lead-time of seconds. It is used (e.g., in Japan, http://www.jma.go.jp/jma/en/Activities/eew1.html, retrieved on April 10, 2012) to shut down nuclear reactors, gas and electricity grids, and to stop high-speed trains in the event of a strong earthquake.

Rethinking earthquake prediction, Sykes et al. (1999) wrote: "*The public perception in many countries and, in fact, that of many earth scientists is that earthquake prediction means short-term prediction, a warning of hours to days. They typically equate a successful prediction with one that is 100% reliable. This is in the classical tradition of the oracle. Expectations and preparations to make a short-term prediction of a great earthquake in the Tokai region of Japan have this flavor. We ask instead are there any time, spatial and physical characteristics inherent in the earthquake process that might lead to other modes of prediction and what steps might be taken in response to such predictions to reduce losses?*"

Following common perception, some investigators overlook spatial modes of predictions and concentrate their efforts on predicting the "exact" fault segment to rupture, which is by far more difficult and might be an unsolvable problem. Being related to the rupture size L of the incipient earthquake, such modes could be summarized in a classification that distinguishes the "exact" location of a source zone from wider prediction ranges (Table 1).

From a viewpoint of such a classification, the earthquake prediction problem might be approached by a hierarchical, step-by-step prediction technique, which accounts for multi-scale escalation of seismic activity to the main rupture. It starts with spatial recognition of the earthquake-prone zones (e.g., Gorshkov et al., 2003; Soloviev and Ismail-Zadeh, 2003) for earthquakes from a number of magnitude ranges, then follows with determination of temporal long- and intermediate-term areas and times of increased probability (Keilis-Borok and Soloviev, 2003), and, finally, may come out with a short-term or immediate alarm. "Predicting earthquakes is as easy as one-two-three. Step 1: Deploy your precursor detection instruments at the site of the coming earthquake. Step 2: Detect and recognize the precursors. Step 3: Get all your colleagues to agree and then publicly predict the earthquake through approved channels" (Scholz, 1997).

Accuracy and testing of earthquake predictions

Though the current accuracy of earthquake prediction is limited, any scientifically validated prediction can be useful for earthquake preparedness and disaster management, if the accuracy of the prediction is known, even though it is not high. In such a case an inexpensive low-key response to the prediction (e.g., to lower water level in reservoirs located in the area of a predicted earthquake, in order to prevent large flooding due to a possible damage of the reservoirs) would be well justified, if even a little part of the total damage due to a strong event is prevented.

K. Aki, one of the most distinguished seismologists of the twentieth century, stated at the workshop on Nonlinear Dynamics and Earthquake Prediction (Trieste, Italy, October 2003) that earthquake prediction would be greatly advanced, when we understand how the seismogenic part of the crust (featuring nonlinear dynamics) interacts with the underlying crust (featuring ductile behavior). A monitoring of the dynamics of the lower crust and physical parameters (e.g., attenuation of seismic waves) could significantly assist in accurate prediction of the large event.

Another approach to accurate prediction should be based on a multidisciplinary nature of earthquake prediction and requires knowledge from various disciplines in geobiosciences. For example, if the time of increased probability of a large earthquake occurrence in a specific region for a specific time interval is announced based on a statistical analysis of seismicity and pattern recognition methods, then monitoring of electromagnetic field, radon emissions, animal behavior, etc. is required to determine more precisely the time and place of the large event. Despite the large scientific and financial resources needed to proceed with the tasks, the efforts would be worthwhile and less expensive compared to a disaster, which can occur due to unpredicted event.

An earthquake prediction method can be judged based on simple statistics using the number of predicted events of a certain magnitude range, versus the total number of earthquakes of the same magnitude range that have occurred in the study area; the time of alarms compared to the total time of observations; and the confidence level of predictions. In other words, each method can be assessed based on the number of successful predictions, failures to predict, and false alarms. (It is important to mention that the probability of occurrence of a large event by a random guess should be evaluated as well, if possible.)

The performance of an earthquake prediction method actually can be evaluated using only two quantities: the rate of failures to predict and the relative alarm time (Molchan, 1990). An efficient method for testing earthquake predictions is described by Zechar and Jordan (2008). This method is based on the Molchan (1990) diagram – a plot of the rates of failures to predict versus the fraction of space time occupied by alarm – and is applicable to a wide class of predictions, including probabilistic earthquake forecasts varying in space, time, and magnitude.

Conclusion

The earthquake predictions made by seismologists have a large economic impact, and as the methods of predicting improve along with seismological data, these economic effects will increase. Meanwhile, compared to the accuracy of weather forecasting, the current accuracy of earthquake forecasting might appear low. Our knowledge of earthquake physics and earthquake dynamics is limited to predicting strong earthquakes with a high accuracy. We do not know well (1) how earthquakes, especially large events, originate; (2) when an earthquake starts, when it stops, and what magnitude could it be; (3) how and why earthquakes cluster; (4) what were the initial conditions of stress state before a large event in terms of stress transfer. Moreover, there is no quantitative description of earthquake physics, namely, no mathematical equations to describe nonlinear dynamics of fault systems and earthquake "flow." The Navier-Stokes equations in meteorology describe atmospheric flow and hence allow weather forecast with a high accuracy for timescales ranging from a few hours to a few days.

The scientific community should use the full potential of mathematics, statistics, statistical physics, and computational modeling and the data derived from seismological (monitoring of physical parameters of earthquakes and tectonic stress, fluid migration, etc.), geodetic (GPS, InSAR, and other measurements of the crustal deformation), and geological (e.g., determination of the time intervals between large earthquakes using paleoseismological tools) studies to improve intermediate- and short-term earthquake predictions (Ismail-Zadeh, 2010).

Acknowledgments

The author thanks Jim McCalpin and Volodya Kossobokov for the review of the initial manuscript and constructive comments.

Bibliography

Allen, C. R., Edwards, W., Hall, W. J., Knopoff, L., Raleigh, C. B., Savit, C. H., Toksoz, M. N., and Turner, R. H., 1976. *Predicting Earthquakes: A Scientific and Technical Evaluation – With Implications for Society*. Panel on Earthquake Prediction of the Committee on Seismology, Assembly of Mathematical and Physical Sciences, National Research Council. Washington, DC.: U.S. National Academy of Sciences.

Ellsworth, W. L., Matthews, M. V., Nadeau, R. M., Nishenko, S. P., Reasenberg, P. A., and Simpson, R. W., 1999. *A Physically-based Earthquake Recurrence Model for Estimation of Long-term Earthquake Probabilities*, U. S. Geological Survey Open-File Report, pp. 99–522.

Fedotov, S. A., 1965. Zakonomernosti raspredeleniya sil'nykh zemletryaseniy Kamchatki, Kuril'skikh ostrovov i severo-vostochnoy Yaponii (Regularities of the distribution of strong earthquakes in Kamchatka, the Kurile islands, and northeastern Japan. In *Trudy Inst. Fiziki Zemli Akad. Nauk SSSR (Proceedings of the Institute Physics of the Earth of the USSR Academy of Sciences)*, Vol. 36(203), pp. 66–93.

Field, E. H., Dawson, T. E., Felzer, K. R., Frankel, A. D., Gupta, V., Jordan, T. H., Parsons, T., Petersen, M. D., Stein, R. S., Weldon, R. J., and Wills, C. J., 2009. Uniform California earthquake rupture forecast, version 2 (UCERF 2). *Bulletin of the Seismological Society of America*, **99**, 2053–2107.

Forsyth, D. W., Lay, T., Aster, R. C., and Romanowicz, B., 2009. Grand challenges for seismology. *EOS, Transactions American Geophysical Union*, **90**(41), doi:10.1029/2009EO410001.

Gelfand, I. M., Guberman, Sh. A., Keilis-Borok, V. I., Knopoff, L., Press, F., Ranzman, E. Ya., Rotwain, I. M., and Sadovsky, A. M., 1976. Pattern recognition applied to earthquake epicenters in California. *Physics of the Earth and Planetary Interiors*, **11**, 227–283.

Geller, R. J., Jackson, D. D., Kagan, Y. Y., and Mulargia, F., 1997. Earthquakes cannot be predicted. *Science*, **275**, 1616–1617.

Gerstenberger, M. C., Wiemer, S., Jones, L. M., and Reasenberg, P. A., 2005. Real-time forecasts of tomorrow's earthquakes in California. *Nature*, **435**, 328–331.

Gorshkov, A. I., Kossobokov, V., and Soloviev, A., 2003. Recognition of earthquake-prone areas. In Keilis-Borok, V. I., and Soloviev, A. A. (eds.), *Nonlinear Dynamics of the Lithosphere and Earthquake Prediction*. Heidelberg: Springer, pp. 239–310.

Healy, J. H., Kossobokov, V. G., and Dewey, J. W., 1992. A test to evaluate the earthquake prediction algorithm, M8. U.S. Geological Survey Open-File Report, 92–401, 23 p. with 6 Appendices.

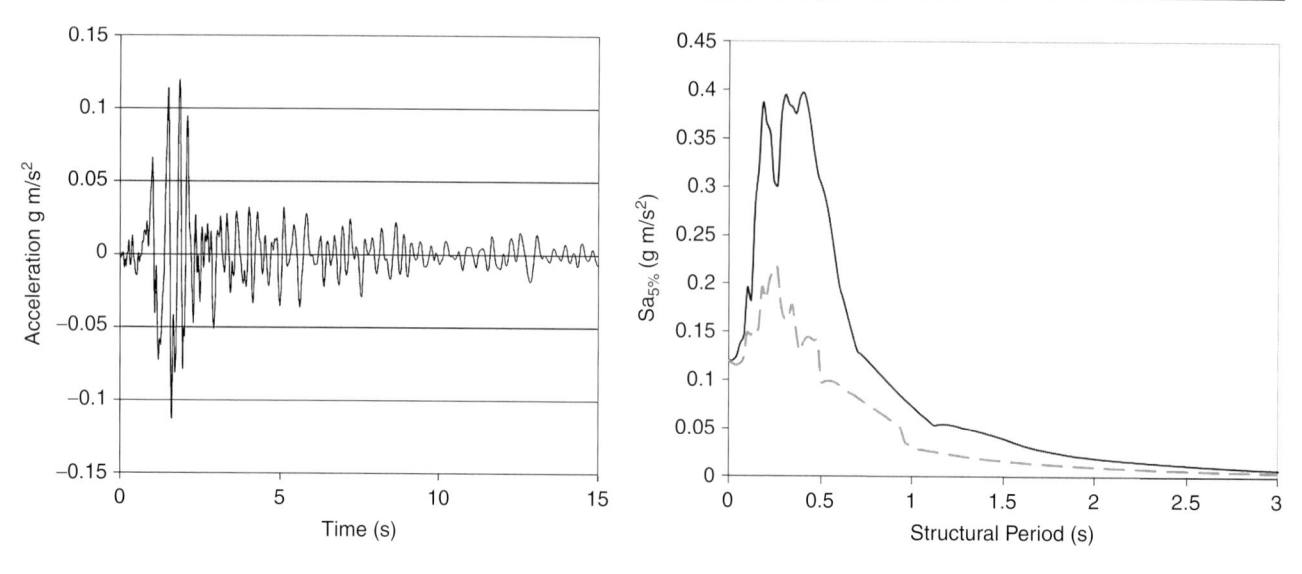

Earthquake Resistant Design, Figure 2 (a) The accelerogram from the November 28, 1974, Hollister (USA) earthquake, recorded at the City Hall recording station. The *pga* of this record is 0.12 g m/s^2 (b) the elastic ($\mu = 1$, *solid line*) and inelastic (for $\mu = 2$, *dashed line*) acceleration spectrum derived for 5% damping for the accelerogram in (a). Note: the maximum spectral acceleration is much greater than the *pga* of the record.

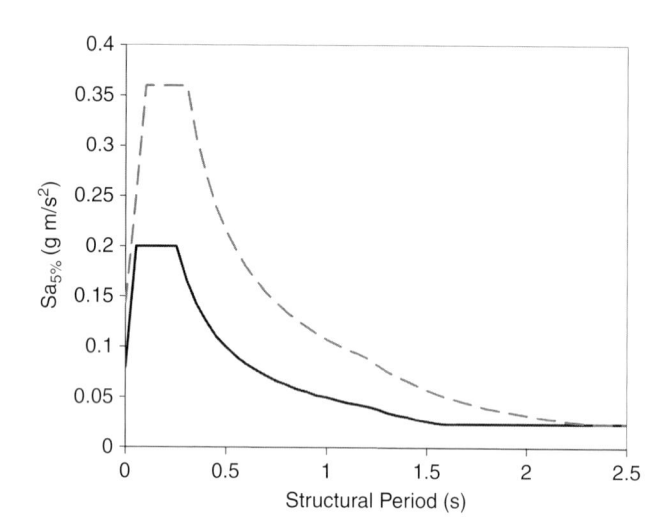

Earthquake Resistant Design, Figure 3 Comparison of elastic spectra for rock (*solid line*) and soft soil (*dashed line*) site conditions as given by Eurocode 8 (CEN, 2003). Assumptions made in drawing the graph are: "Type 2" spectral shape (for high seismicity areas), a regular importance structure, a behavior factor of 1.5 and pga of 0.12 g m/s^2, soil class A for representing rock and soil class D for representing soft soil.

Static methods are acceptable for the analysis of structures which have a dominant first mode response. If higher modes of vibration contribute in a significant manner to the overall structure response or where the structure is highly irregular, nonlinear time-history analyses are required to determine the design forces. This method of analysis usually requires the finite element modeling of the structure and uses accelerograms directly as a forcing function. The analysis involves applying the acceleration time history to the structure in time increments and resolving the equations of motion at each time step to assess the structure response. Due to the uncertainty in determining the ground motion at any particular site, the analysis is often repeated multiple times using different accelerograms (see also "*Uncertainty*"). This method of analysis is therefore only used for high importance or highly irregular structures as it is computationally and time expensive.

Effects of founding soil conditions on structure response to earthquake

Different idealized response spectra shapes are usually presented in codes for different site soil conditions (see Figure 3). This is because the accelerogram, and consequently the response spectrum, are heavily influenced by local site geology.

When seismic waves pass from the bedrock to the surface, they pass through soil layers where their velocity of propagation is less and hence they slow down. In a similar way to what happens with sea waves entering shallow waters, in order to conserve energy, the amplitudes of the seismic waves increase as they slow down. The amount of increase in amplitude depends on the relative velocities in the bedrock and overlying soil. This phenomenon is known as "impedance" and is the main way in which soil layers amplify the ground motion. Waves can also be reflected by the surface of the Earth and then as they propagate downward they are again reflected back upward at the rock face and in this way they become "trapped" in the soil layer. This has the effect of increasing the duration of the strong-ground motion.

If the dominant period of the ground motion coincides with the natural period of vibration of the underlying soil column, then resonant response can result in very high amplitudes on the spectrum in that range of vibration periods. In general, structures founded on rock will be subjected to short period ground motion, whereas soft sites result in longer period excitation. Hence all standard spectral shapes given in codes show greater spectral amplitudes at long periods on soft soils than on rock sites (see Figure 3).

A further phenomenon associated with soil at a site, which might directly cause damage to buildings is Liquefaction (see "*Liquefaction*"). In earthquake resistant design liquefiable sites can be improved by removing and replacing unsatisfactory material, densifying the loose deposits, improving material by mixing additives, through grouting or chemical stabilization or by implementing drainage solutions (e.g., Brennan and Madabhushi, 2006). Generally however, the easiest thing is not to build on a site that might potentially liquefy.

Seismic hazard maps

Seismologists have developed methods to provide probabilistic estimates of the seismic hazard at a site based on studies of regional tectonics, active faults in the area and their associated seismic activity, and local geology. The seismic hazard presents estimates of strong ground motion parameter values at a site and their associated recurrence interval (see "*Recurrence Interval*"). If seismic hazard analyses are carried out at several points on a grid covering a region, then a map can be drawn of the strong-ground motion parameter values occurring at each grid point for a given recurrence interval. By joining these points we can draw contours of the parameter values. Such hazard maps present the engineer with a graphical representation of the distribution of seismic hazard in a country or region. Many seismic hazard maps exist for countries around the world that present values of peak ground acceleration (*pga*) for set recurrence intervals. This is due to the fact that *pga* is used together with idealized spectral shapes in building codes to define the earthquake load on a structure in force-based methods of design (see above).

The recurrence interval on hazard maps present in building codes is often represented in terms of the probability that the *pga* value will be equaled or exceeded during the lifetime of the assessed structure (assumed to be 50 years). In most seismic codes of practice the design seismic ground motions (design earthquake) for residential buildings is determined from hazard maps showing *pga* values associated with a 10% probability of exceedance in 50 years. This implies a recurrence interval of 475 years (if a Poisson process is assumed for earthquake occurrence). For more important structures hazard maps for larger recurrence intervals are used, which correspond to the occurrence of larger *pga* values.

Design strategies
Force-based earthquake resistant design

Force-based design of structures for earthquake resistance is at the basis of building codes worldwide (see "*Building Code*"), and is regarded as the conventional mode of such design. The method consists in determining the applied loads to the structure from a combination of the structure's self-weight, the design earthquake (see section "*Seismic Loading*" above), and other live loads (nonpermanent actions) applied with appropriate safety factors to account for uncertainties. The resulting internal forces (flexural) moments, shear forces and tension/compression felt by each element of the structure are calculated. Then the elements are dimensioned and detailed (provided with reinforcement) to resist these internal forces. In this way building failure (see "*Building Failure*") is avoided in the case of occurrence of the design earthquake event.

Model earthquake codes such as the Uniform Building Code 1997 and later International Building Code 2009 have two main purposes:

- Provide minimum provisions for the design and construction of structures to resist earthquake effects
- "…to safeguard against major structural failures and loss of life, not to limit damage or maintain function" (UBC, 1997)

The aim of the design is therefore to provide life safety (or no collapse) under the design earthquake. Life safety practically means that heavy irreparable damage may occur in the structure but collapse and loss of life are avoided. This desired performance is assumed to coincide with achievement of the ultimate limit state in critical structural members. The ultimate limit state in a structural member is reached when it is unable to stably carry any further load or deformation, and is often characterized by a loss of 20% of its maximum strength. This state is characterized by a state of stress and strain in the element critical sections, which values are used to design the elements.

If life safety is to be achieved with no damage occurring in the structure, then the structure should be designed to remain within its elastic range of deformation during the earthquake and dissipate the input earthquake energy solely through the notional viscous damping. This is the case for structures where the consequences of damage are unacceptable (e.g., nuclear power plants). Elastic design implies high design forces with correspondingly large structural elements to resist the forces elastically, and consequently high construction costs.

However, an earthquake usually constitutes the most severe loading to which most civil engineering structures might possibly be subjected. Yet in most parts of the world, even those that are highly seismic, it is possible that an earthquake may not occur during the life of the structure. Therefore, it is generally uneconomic, and often unnecessary to design structures to respond to design-level earthquakes in the elastic range, unless it is

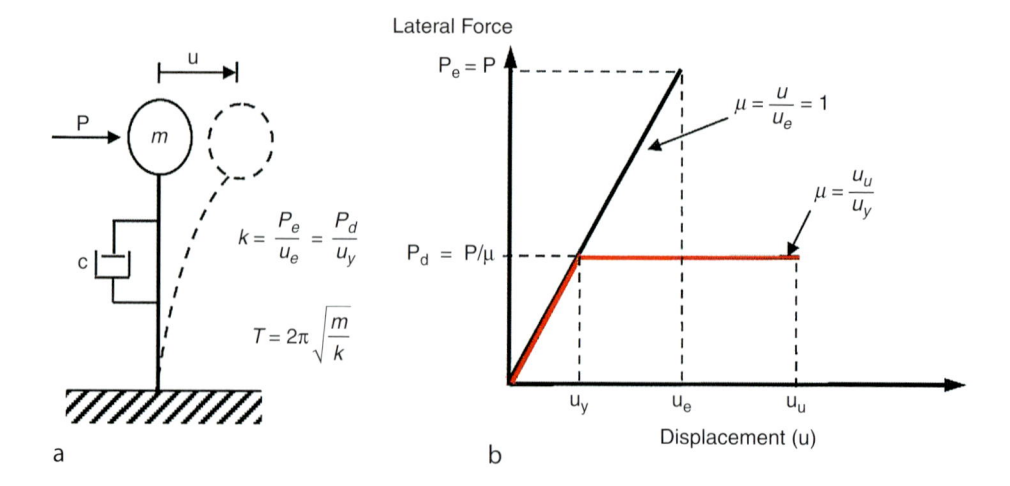

Earthquake Resistant Design, Figure 4 (a) Diagram of a single degree of freedom (SDOF) system with mass m, stiffness k, damping c, and fundamental period of vibration T. (b) The force versus top displacement diagram for the SDOF showing elastic response (*black line*) and inelastic response (*red line*). In the elastic system the SDOF resists the applied load by being designed to have a strength $P_e = P$ and shows no residual deformation after loading. In the inelastic system, the structure is designed to have a resistance $P_d = P/\mu$ but yields at a displacement of u_y and undergoes large inelastic deformations until it reaches its ultimate deflection, u_u.

absolutely critical that the structure remains functional during and immediately after the event.

In the majority of structures a degree of damage under an earthquake event is allowed, so that part of the input earthquake energy is dissipated through plastic deformation. Figure 4 shows the difference in the force – displacement response of a single degree of freedom (SDOF) system, (with short natural period of vibration) subjected to a lateral load (P). In both the elastic and inelastic cases the same amount of energy is dissipated by the structure (i.e., area under the black and red response curves is the same) but in the case of the inelastic structure, its ductility is used to reduce the design loads. The ductility of a structure is defined as its ability to withstand large deformations beyond its yield point without fracture (Williams, 2009). Ductility (μ) can be expressed as the ratio of maximum displacement (or deformation) response to the yield displacement, and is illustrated for the single degree of freedom (SDOF) system in Figure 4. In civil engineering design, high ductility is mainly achieved through the provision of good materials, connections, and detailing rather than through large member sizes. This can result in more economic designs but physically, allowing a structure to deform inelastically means that it will sustain a degree of damage, e.g., through yielding of steel and possibly concrete or masonry crushing. Provided that the strength does not degrade as a result of the inelastic deformation, acceptable resistance can be achieved.

As previously stated, where damage is acceptable, the inelasticity of the structure can be taken into account to reduce the design-level forces. Figure 2b shows the difference in the acceleration response spectrum obtained analyzing an SDOF with elastic and inelastic behavior for the same ground motion. In codes of practice a reduction factor is applied to the elastic acceleration spectrum to account for this effect and reduce the calculated applied loads (see earlier). Most buildings are generally designed for only about 10–30% of the elastic earthquake lateral loads, but the lateral resisting system, structure connections and member detailing are chosen and designed to allow the structure to undergo inelastic deformations without losing strength. In this manner the energy imparted by the earthquake is dissipated through the inelastic deformation and controlled damage (see "Capacity Design" below).

Most codes of practice lead to designs that satisfy a single limit state (ultimate) to ensure safety under the maximum expected loads but they also carry out checks under smaller loads to satisfy other limit states (e.g., serviceability, where the function of the structure is not compromised). In earthquake resistant codes, the typical design procedure is similar to that for gravity load design, where the main structural design is carried out to satisfy the ultimate limit under the design earthquake. The design earthquake is characterized by a peak ground acceleration (see later) that is associated with a given frequency of occurrence, which in most codes of practice is 10% exceedance probability in 50 years, (the typical life of a building). Following the primary design, a check is carried out to ensure the structure performance satisfies the serviceability limit state under a scaled down value of the seismic action (representative of a more frequent earthquake event occurrence). These limit states may have different names in different codes of practice.

New concepts in earthquake resistant design

Conventional earthquake resistant design allows the designer to control the structural performance (i.e., life-safety) at one limit state (ultimate) under excitation by a design earthquake event. Until recently, the general belief was that designing structures explicitly for life safety provides adequate damage protection. However, several recent earthquakes have shown that this perception is false. Although the vast majority of the structures designed to new codes met the life safety requirement, the financial cost, due to direct damage of structural or nonstructural members and business interruption, was much higher than expected. A noticeable example is the January 17, 1994, earthquake in Northridge, California. Although the death toll was not high (57 deaths) the direct financial cost amounted at US $20–40 billion. The economic consequences of the January 17, 1995, Hyogo-ken-Nanbu earthquake, Japan, were even more dramatic, affecting the economy not only in Japan but worldwide. The estimates of the repair costs only have been reported in the range of US $95–147 billion. After these two earthquakes the importance of performance goals that go beyond simple life safety, such as damage control, have been widely acknowledged. A consensus among most researchers and professional institutions has been reached that new design methodologies need to focus more on performance criteria.

Over the last 20 years there is a movement toward developing frameworks for the design of structures under multiple limit states (performance limits) for earthquake events associated with different frequencies of occurrence. The performance-based design approach was first introduced by Vision 2000 (SEOAC, 1995) which outlined a relationship between performance objectives, the type and importance of the designed facility and the probability of earthquakes of different sizes (see also "*Frequency and Magnitude of Events*"). For a normal importance structure it suggests the following limit states for design:

- No damage under a frequent earthquake event (with an exceedance probability of 68% in 50 years)
- Serviceability should be ensured under an occasional event (with an exceedance probability of 50% in 50 years)
- Damage control should be ensured under a rare earthquake event (with an exceedance probability of 10% in 50 years)
- Life safety should be ensured under a very rare earthquake event (with an exceedance probability of 5% in 50 years)

It states that the performance objective must increase (i.e., less damage is acceptable) for a high probability earthquake (one that can occur several times during the life of the structure) or for an important structure or one whose damage will have severe consequences (e.g., a hospital or nuclear power station). Conversely more damage is acceptable for a rare, severe earthquake event and for less critical or temporary facilities. The framework has been further refined by several researchers and code documents (e.g., see Ghobarah, 2001).

Performance-based design assumes engineers are able to check or carry out designs for specific damage levels other than ultimate limit state, that is collapse (see also "*Structural Damage Caused by Earthquakes*"). This is not possible using conventional force-based techniques of damage, as it has long been recognized by the earthquake engineering community that forces are poor indicators of structural damage. Once a structure passes its yield limit, the strength of a structure can remain largely unchanged throughout the inelastic range (see Figure 1b, red curve), thus changes in strength are not indicative of the achievement of different damage states between the yield limit and collapse. Moehle (1992) first proposed the concept of designing a structure directly for displacements. This differs from conventional force-based design where the primary input to the process is a set of forces, with a check on deformation only being carried out in a second stage. In displacement-based design this process is reversed and the main design quantity is a target displacement, which corresponds to a deformation value with associated damage state. Several improvements to the displacement-based design approach proposed by Moehle (1992) have been made by other authors (e.g., Kowalsky et al., 1994; Priestley et al., 2008). Displacement-based design frameworks are currently a theme of active research and have spurred the development of displacement-based parameters for earthquake load characterization (e.g., Displacement spectra, see below) and of relationships between a structural element's mechanical properties and its deformation capacity (e.g., Panagiotakos and Fardis, 2001; Rossetto, 2002). No consensus has yet been reached regarding performance-based (and displacement-based) design methods and tools.

Methods for providing earthquake resistance in structures

Seismic codes of practice

A seismic code is a building code (see "*Building Code*") that regulates the design of earthquake resistant structures in a country. They provide guidance and tools for the evaluation of the seismic loads and to provide safe and economic structural design to resist these loads. They incorporate seismic hazard maps specific to the area covered by the code. Examples of existing national codes include "IS1893 (Part1): Earthquake Resistant Design of Structures: General Provisions for Buildings" for India and the "NZS 4203:1992 Code of Practice for General Structural Design Loadings for Buildings" for New Zealand. In the USA several seismic codes exist that have been developed for each State based on the "NEHRP Recommended Provisions for Seismic Regulations for New Buildings and Other Structures." The latter is a multi-edition guidance (latest edition in 2009) that aims to aid seismic code development. Recently, the European seismic code Eurocode 8 (CEN, 2003) has been developed that regulates seismic design of structures in the European member states. Each member state is also equipped with a National Annex to the code that accounts for variations in building practice in different countries and contains a country specific hazard map.

At the basis of all these seismic codes are similar concepts of earthquake load calculation (see above), a minimum standard for building materials and basic design principles to provide life-safe structures. The latter can be summarized as:

1. Regularity of form, mass, and stiffness distribution
2. Pounding with adjacent buildings should be avoided
3. Appropriate lateral load resisting system
4. Detailing for ductility and structural integrity
5. Capacity design for a controlled failure mechanism

Structures should have a simple and regular form in plan and elevation to avoid nonuniform stress and ductility demand distributions. Where complex geometries exist in plan, the structure can be subdivided into more regular layouts using seismic joints. Height restrictions or restrictions on other structure characteristics may be imposed by codes depending on local seismicity and structure importance. Considerations for the location in plan of structural components are included in codes to ensure that the center of stiffness and mass are coincident. The center of mass is essentially where the resultant earthquake force is applied. The center of stiffness of the lateral load resisting elements instead defines the center of structural resistance. If the two are not coincident, then a couple is applied and torsion occurs in the structure, which may not have been considered in the design. Code procedures for seismic design assume a regular distribution of seismic loads over the building height. Hence, if there are severe mass or stiffness irregularities in elevation (e.g., a concentrated load or large change in lateral stiffness between stories), high demand concentrations will ensue that are not taken into account in the design.

Buildings must be protected from earthquake induced collisions with adjacent structures (pounding), by allowing adequate spacing between structures.

A clear system for resisting the vertical and lateral loads imposed by the structure's self-weight and by earthquakes should exist, where there is a good transfer of lateral loads between resisting structural elements and to the foundations. Continuity and regular transitions are essential requirements to achieving an adequate load path (Elnashai and Di Sarno, 2008). Equally, it is important that horizontal elements in the structure (e.g., beams and floor diaphragms) are sufficiently rigid to be able to transfer the seismic loads to the vertical resisting elements (that provide the main seismic resistance). Excessive deformation of the structure under the applied loads should be avoided. This can be achieved through the addition of shear walls or braces in tall buildings.

The structure should be designed so as to allow its full ductility to be reached. This is done in two ways. Firstly, since the seismic performance of the structure depends on the behavior of its critical regions, the detailing of these regions (or structural elements) must be such as to maintain under cyclic loading the ability to transmit the necessary forces and to dissipate energy. The detailing of members should ensure that they do not fail in a brittle manner (such as shear dominated failure). The structure should act as a single unit and appropriate connections between structural elements should be ensured. Failure of a building is often seen when poor connections exist that do not allow the transfer of loads between elements and for the structure to work together as a system for resisting the lateral loads. Secondly, to avoid brittle failures or unstable mechanisms of failure and to make sure that damage occurs only at the desired locations, the concept of "capacity design" is used. Once an engineer has chosen a desirable failure mechanism for the structure being designed (i.e., one that ensures life safety) they can identify those structural elements that should enter their inelastic range and act as dissipators (noncritical members) and those that instead should remain elastic in order to avoid complete collapse (critical elements). The latter are often the vertical elements in a structure as they "hold up" the building. Capacity design is the principle that critical members should be protected and designed to withstand, not the applied forces, but those commensurate with the achievement of plasticity in adjacent noncritical elements. In practical design, this means that the noncritical elements are designed first so that they are able to resist the applied earthquake loads. The critical elements are then designed to be stronger than the designed noncritical elements, taking into account any possible source of overstrength. This allows the designer to control the sequence of occurrence of plasticity and damage in the structure and avoid unstable failure mechanisms. However, it also means that a tight control must be maintained on sources of overstrength during construction, such as the use of a higher grade steel, as this can compromise the occurrence of the desired building failure mechanism. Further discussion of all the above issues can be found in Booth and Key (2006), Elnashai and Di Sarno (2008) and Elghazouli (2009).

Resisting earthquakes using devices

The design techniques discussed thus far essentially consist of carefully sizing and locating members and structural elements in buildings. A range of technological devices can be used in conjunction with code-based designs in order to relax requirements and constraints imposed on the structural elements. A comprehensive review of these types of solutions can be found in Soong and Dargush (1997). These devices essentially fall into two main categories: solely dissipative devices or base-isolation devices.

Base-isolation devices

If one could imagine a building floating in thin air like a hovercraft, then obviously it would be unaffected and remain intact through any earthquake ground motion. The inertia of the building would keep it steady while the ground underneath could slide back and forth. Such isolation could actually be achieved by basing the building on some kind of rollers. When the building is well isolated, most of the deformation induced by the ground motion occurs at the interface between the ground and the building, i.e., the isolating device suffers most of the deformation whereas the structure above remains relatively undeformed. However, this presents a number of difficulties. Although a building can be effectively isolated from the ground structurally (at least in any horizontal direction), service supplies (gas, water supply, and waste) still need to be connected to it in such a way that they survive earthquake events. To achieve this, special flexible connections must fit to pipes or cables connecting the building to utilities networks. Access to the building must also be carefully considered in order to allow large relative displacements between the building and surrounding pavements and access routes. In practice, this means that isolated buildings must be surrounded by a clear gap all around. If the building is completely free to move sideways then any moderate lateral load could push it and offset its position. Hence, mechanisms need to be introduced that restore the building to its initial position after an earthquake.

Two main types of devices can be used to isolate a structure from the ground: (1) laminated rubber bearings and (2) friction-pendulum bearings. Laminated rubber bearings are weak in shear (laterally) but strong enough vertically to resist the weight of the building above. The natural elasticity of the rubber solves any issue of re-centring and will also resist wind loading. However they constitute a lateral connection to the ground – albeit a weak one – so they will transmit some ground motion to the structure. The rubber is usually treated so as to provide large energy dissipation through deformation thereby releasing some the input energy as heat rather than transmitting it to the building. Friction-pendulum bearings, in their simplest form, rely on a top part and a bottom part sliding along a curved interface as shown diagrammatically in Figure 5.

The bottom part is attached to the ground (foundation) whereas columns supporting the building rest above the top part. The middle crescent shaped element allows the top and bottom part to move sideways without actually rotating. The friction between the two parts dissipates some of the input energy. Re-centering and resistance to winds is ensured by the shape of the surface, which allows gravity to bring the system back to its lower point. These devices can be effective for earthquakes of moderate amplitude but can lead to large sudden accelerations being transmitted to the structure if the ground motion is so large that the relative displacement exceeds the range of the device. Double friction pendulums are improvements on the simplest version and are designed especially to increase this range. Both single and double friction pendulums present intrinsic nonlinear behavior as they rely on friction. This makes the behavior of structures fitted with them difficult to model and analyze (Zayas et al., 1990).

Dissipative devices
The next class of devices are dissipative devices. Most of these are composed of some type of ram or piston that dissipates energy using fluid viscosity. They often consist of long members that are bolted diagonally between floors so as to be worked to their maximum when the building is oscillating in an earthquake and its stories undergo relative displacements. Unlike base isolation, they rely on the building deforming to work effectively but the energy they dissipate also reduces the extent of this deformation to acceptable limits. They effectively play the role of ductility in conventional design by focusing the dissipation of energy to specific areas of the structure. These devices are also used to dampen wind-induced oscillation in tall buildings.

Tune-mass dampers (TMD) are another category of dissipative device. They consist of a mass-spring system whose natural frequency is tuned to that of a particular problematic resonant frequency of the structure. As both systems are tuned, the device is set in resonance at the same time as the building. This causes the mass of the TMD to undergo large displacements when the building resonates. By immersing the damper's mass in oil, its displacement is converted into heat due to the fluid viscosity. This effectively drains a significant amount of energy out of the system. A single TMD must be fine-tuned to work well and it can only be effective at a specific frequency.

Summary

Earthquake resistant design is carried out to ensure a desired structural behavior in the case of a seismic event with given recurrence interval. The seismic excitation of a structure depends on the characteristics of the earthquake ground

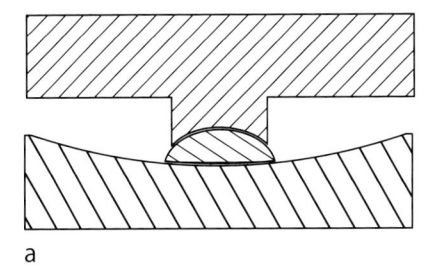

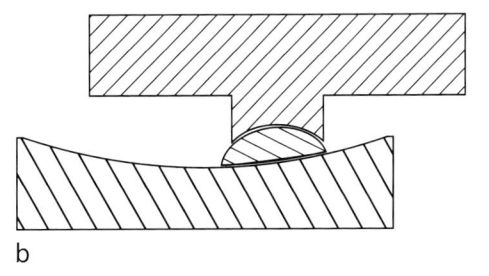

Earthquake Resistant Design, Figure 5 Schematic representation of a friction pendulum in neutral (**a**) and shifted position (**b**). Supporting columns of the building rest on top while the lower part of the device rests on the foundation. The curved shape allows the structure to slide back in neutral position under gravity.

motion and on the inertia of the structure. The stiffness, strength, damping, and ductility of the structure interact with the ground motion to amplify or reduce the structural response to the earthquake, and hence the design forces and displacements. Earthquake resistant design can be achieved through the direct design of structural elements to resist the applied earthquake loads coupled with requirements for regularity of structural form, good interconnection of elements, and capacity design principles for the control of the building failure mechanism. Alternatively, devices can be used to reduce seismic actions on the structure by isolating it from the ground or to enhance its energy dissipation capacity.

Bibliography

Bisch, P., 2009. Introduction: seismic design and eurocode 8. In Elghazouli, A. Y. (ed.), *Seismic Design of Buildings to Eurocode 8*. London/New York: Spon Press, p. 318. ISBN 978-0-415-44762-1.

Booth, E., and Key, D., 2006. *Earthquake Design Practice for Buildings*. London: Thomas Telford.

Brennan, A. J., and Madabhushi, S. P. G., 2006. Liquefaction remediation by vertical drains with varying penetration depths. *Soil Dynamics and Earthquake Engineering*, **26**(5), 469–475.

CEN, 2003. *Eurocode 8: Design Provisions for Earthquake Resistance of Structures*. European Committee for Standardisation, Brussels.

Clough, R. W., 1960. The finite element method in plane stress analysis. In *Proceedings of 2nd ASCE Conference on Electronic Computation*, Pittsburgh, September 8–9, 1960.

Elghazouli, A. Y. (ed.), 2009. *Seismic Design of Buildings to Eurocode 8*. London/New York: Spon Press, p. 318. ISBN 978-0-415-44762-1.

Elnashai, A. S., and Di Sarno, L., 2008. *Fundamentals of Earthquake Engineering*. Chichester: Wiley, p. 347. ISBN 978-0-470-01483-6.

EQE international, 1995. *The January 17, 1995 Kobe Earthquake*. An EQE Summary Report, April 1995. http://www.eqe.com/publications/kobe/kobe.htm.

Fralleone A., and Pizza A. G. 2000. *Sintesi dei cambiamenti più significativi della normativa italiana per le costruzioni in zone sismiche*. Italian National Seismic Survey Report, Italy.

Ghobarah, A., 2001. Performance-based design in earthquake engineering: state of development. *Engineering Structures*, **23**(8), 878–884.

International Building Code, 2009. Whittier: International Code Council.

Kowalsky, M. J., Priestley, M. J. N., MacRae, G. A. 1994. *Displacement-Based Design, a Methodology for Seismic Design Applied to Single Degree of Freedom Reinforced Concrete Structures*. Report No. SSRP-94/16. Structural Systems Research, University of California, San Diego.

Lazan, B. J., 1986. *Damping of Materials and Members in Structural Mechanics*. Oxford: Pergamon.

Moehle, J. P., 1992. Displacement based design of reinforced concrete structures. In *Proceedings of the 10th World Conference on Earthquake Engineering*. Rotterdam: A.A.Balkema, pp. 4297–4302.

Panagiotakos, T. B., and Fardis, M. N., 2001. Deformations of reinforced concrete at yielding and ultimate. *ACI Structural Journal*, **98**(2), 135–147.

Priestley, M. J. N., Calvi, G. M., and Kowalsky, M. J., 2008. Displacement-based seismic design of structures. *Earthquake Spectra*, **24**(2), 555–557.

Rossetto, T., 2002. Prediction of deformation capacity of non-seismically designed reinforced concrete members. In *Proceedings of the 7th U.S. National Conference on Earthquake Engineering*, Boston.

SEOAC, 1995. *Performance Based Seismic Engineering of Buildings*. Sacramento: Vision 2000 Committee, Structural Engineers Association of California.

Soong, T. T., and Dargush, G. F., 1997. *Passive Energy Dissipation Systems in Structural Engineering*. Chichester: Wiley.

Uniform Building Code, 1997. Sacramento: California Building Standards Commission.

Williams, M. S., 2009. Structural analysis. In Elghazouli, A. Y. (ed.), *Seismic Design of Buildings to Eurocode 8*. London/New York: Spon Press, p. 318. ISBN 978-0-415-44762-1.

Zayas, V. A., Low, S. S., and Mahin, S. A., 1990. A simple pendulum technique for achieving seismic isolation. *Earthquake Spectra*, **6**, 317.

Cross-references

Building Codes
Building Failure
Frequency and Magnitude of Events
Liquefaction
Recurrence Interval
Seismology
Structural Damage Caused by Earthquakes
Uncertainty

ECONOMIC VALUATION OF LIFE

Mohammed H. I. Dore, Rajiv G. Singh
Brock University, St Catharines, ON, Canada

Synonyms

Indirect Methods of Placing a Monetary Value on the Whole Life of an Individual

Definition

Methods employed to assign a monetary value on an individual's life which may be useful for insurance purposes, or for assessing compensation or damages due to loss of life, as a result of accidents or as a result of natural disasters.

Economic valuation

There are at least four methods that can be used to determine the economic value of human life. These are the: (1) human capital approach (HK), (2) willingness-to-pay approach (or WTP), (3) actuarial basis, and (4) the value of a statistical life (VSL). The human capital approach considers the value of a human life as the present value of an individual's net earnings discounted over his/her lifetime; net earnings are defined as an individual's income over his/her useful lifetime less the portion which he/she spends on consumption. This approach requires an individual to have a lifetime income stream or requires some measure of potential earnings. Problems may arise in evaluating the earnings of the elderly, children, the unemployed, and the handicapped since they may have no income stream. This method of valuing a human life omits the value placed by individuals on leisure time activities as well as periods of ill health which can affect lifetime earnings. Due to HK's heavy reliance on income in its evaluation, low-income earners are technically of "lower value" than high-income earners as they have less potential future income, which is ethically debatable. Also, the discount rate chosen in the HK calculation is arbitrary, and difficult to justify, which raises another ethical problem. The monetary valuation of human life depends heavily on the rate of discount chosen.

The willingness-to-pay method places a value on life according to how much an individual is willing to pay to reduce the probability of death or is willing to pay for an improvement in health. The WTP approach can be calculated in two ways: (1) the contingent valuation method and (2) the revealed preference method. In the contingent valuation method, individuals are asked to supply information, usually via a survey or questionnaire, on how much they are willing to pay in monetary terms to receive a decrease in the probability of death or ill health. This would involve the construction of hypothetical or contingent scenarios where respondents will in effect estimate their willingness to pay. The disadvantages of this method are all of those which are associated with surveys: respondents may over- or understate their willingness to pay; an individual's willingness to pay in the hypothetical scenario may differ from his/her real reaction to when a particular contingency is in fact a reality; and individuals may interpret and answer abstract hypothetical questions in a manner very different from the way it was intended by the designers of the survey. However, once properly designed, the contingent approach can provide estimates on the valuation of life or at the least a decrease in a unit probability of death from respondents. On the other hand, the revealed preference method is based on individuals revealing their preferences implicitly or explicitly during actual market behavior. For instance, in wage-risk studies, a wage premium acts as an indicator for the trade-off between willingness to pay for an increase in a unit of risk which causes harm (e.g., workplace injury or even death). This can explain why high-risk occupations correlate well with high wage premia. In consumer-market studies, a similar trade-off is found between willingness to pay and consumption goods which reduce the probability of injury or death. For instance, the value of a human life can be calculated from the reduction in the probability of death arising from the purchase of vehicle safety features or fire alarms (Dionne & Lanoie, 2002). Both the contingent valuation method as well as the revealed preference method are *ex ante* (before the fact) and reflect an individual's valuation given current conditions. Individuals may change their preferences *ex post* (or after the fact), for example, an injury sustained from a vehicular accident leaves an individual out of work for the remainder of his/her life; as a result income levels, level of risk aversion, and consumption preferences would also change.

The third method to the value of life is based on actuarial estimation. Actuaries use elements of the HK and WTP approach in their valuation of human life. Economic and demographic assumptions are the two key elements in the valuation of human life under the actuarial basis. For the economic aspect, the present value of future liabilities and potential earnings are calculated. For the demographic assumptions, factors such as employment status, gender, occupation, disability, and health status (whether the person is a smoker, is a diabetic, or has cancer), are used by the actuary to determine the net effect on mortality rate for an individual. Many insurance companies use an actuarial (also called mortality or life) table to compute the probability of mortality given the economic status and demographic factors associated with a particular individual. This information is then used to compute the value of a life insurance policy which is in effect the economic valuation of the individual's life.

Finally, the value of a statistical life (VSL) is often associated with the economic value of human life. If individuals were asked to attach a numerical monetary value on their lives, many would respond by saying that life is "priceless" or "of infinite value." However, under this method of valuation, the cost–benefit analysis of projects that reduce the risk of death by a very small amount would be deemed worthwhile regardless of its cost (Erickson and Moskalev, 2009). To avoid this, a quantifiable and justifiable method of evaluating a human life within practical monetary boundaries is required; hence a statistical approach to valuing life. The term statistical emphasizes the nature in which human life is evaluated: via a methodology including calculations. The value arrived at represents a monetary trade-off between accepting an extra unit of risk associated with death and an increase in wealth. VSL represents the summation of such trade-offs for an entire population; hence the VSL for an individual or for a population would on average be the amount the entire population would pay to reduce the risk of death by a certain increment. In order to estimate VSL, Viscusi and Aldy (2003) analyzed data on risk premia from ten countries and 60 studies as well as current estimates of injury risk premiums from 40 studies. Their study indicates a contribution of around $0.50–$0.60 to the value of a statistical life for every dollar increase in income. They found that the value of statistical life for an individual in the UK ranges from US $4.2 to $21.7 millions in constant 2,000 dollars whereas for India this value lies between US $1.2 and $4.1 m. In Canada, some studies have reported the VSL to be as low as US $2.2 m and as high as US $21.7 m whereas for Japan, Korea, Austria, and Australia, the estimated VSL can be as high as US $9.7 m, $0.89 m, $6.5 m, and $19.1 m, respectively. The VSL for the US varies quite considerably, mainly due to the number of studies conducted in the country. However, VSL can range from a low of $0.5 m to a high of $20.8 m constant 2,000 dollars. Other studies such as Murphy and Topel (2006) found that due to improvements in health care in the US, a 1% decline in cancer mortality can contribute up to a $500bn gain to society. In fact increased life expectancy contributed $3.2 m per year to national wealth since 1970 (Murphy and Topel, 2006).

Bibliography

Dionne, G., and Lanoie, P., 2002. How to make a public choice about the value of a statistical life: the case of road safety. *HEC Montreal, Risk Management Chair Working Paper Series*, No. 02-02.

Erickson, C., and Moskalev, R., 2009. Economic value of a human life. Retrieved Feb 8, 2010, from: http://www.ecosys.com/spec/ecosys/download/UniFr/Travaux/Life.pdf

Murphy, K. M., and Topel, R. H., 2006. The value of health and longevity. *Journal of Political Economy*, **114**(5), 871–904.

Viscusi, W. K., and Aldy, J. E., 2003. The value of a statistical life: a critical review of market estimates throughout the world. *NBER Working Paper*, No. W9487.

Cross-references

Casualties Following Natural Hazards
Cost-Benefit Analysis of Natural Hazard Mitigation
Costs (Economic) of Natural Hazards and Disasters
Disaster Relief
Economics of Disasters
Federal Emergency Management Agency (FEMA)
Hyogo Framework for Action 2005–2015
Insurance
Livelihoods and Disasters
Red Cross and Red Crescent
World economy, Impact of Disasters on

ECONOMICS OF DISASTERS

Pierre-Alain Schieb
OECD SGE/AU IFP, Paris, France

Definition

The economics of disasters. An emerging branch of economics which studies the economic dimensions of disasters. Its primary domain encompasses the economic consequences of disasters, be they direct or indirect, short-term or long-term, and negative or positive, and on how to deal with them. Its whole domain entails all the economic dimensions associated with the different stages of the risk management cycle: identification, assessment, prevention, mitigation, emergency, rescue and recovery activities, post-impact issues, and reconstruction.

Disasters: definition and scope

A disaster can be defined as an event that adversely affects a significant number of people, devastates a significant geographical area, and taxes the assets and resources of local communities and governments (adapted from Gad-el-Hak, 2008).

- A disaster is not an event/accident impacting an isolated victim: a single family house fire, a car accident, a boat or someone drowning in the sea, a tree falling on a house, a lightning strike impacting a farm, are all isolated events and do not fit with the concept of a disaster. Of course, in the eyes of the victims, it can be a disaster.
- According to the definition, human, but also social or environmental consequences are discussed with monetary or financial indicators (losses, liabilities, costs, budgets, ...) even if the primary data are in physical terms: number of casualties, job losses, and area of polluted sea.
- The definition starts by identifying the trigger, an "event." In the case of natural disasters, examples include earthquakes, floods, hurricanes, tsunamis, forest fires, etc. Disasters can also be provoked by man-made events such as wars, financial crises due to economic cycles or bubble bursts, industrial accidents (Bhopal, Seveso, Chernobyl, oil spills from tankers, etc.), and even terrorist acts.
- The definition could be extended to also cover "silent phenomena." A silent risk can go unnoticed for a long period before the negative consequences take a significant toll: for example, asbestos with its health, social, and economic impacts in the twentieth century is often described as leading to one of the biggest economic disasters of the twenty-first century. The potential impacts of the rising obesity rates among humans could be another example of a silent upcoming risk in the twenty-first century.
- Although the common understanding is that "disasters" have mostly negative economic consequences, it should be noted that some economic consequences can be "positive" even though it does not look morally correct to highlight the "benefits" from disasters. For example, a destroyed manufacturing plant can be rebuilt with the latest technological equipment and then an unprecedented level of productivity achieved: new and more appropriate legislation can be passed; regenerative medicine could progress; and disaster resilience could be improved.
- A disaster is the result of the occurrence of a damaging event. In insurance terms, a risk is quantified as the probability of occurrence of a damaging event multiplied by the estimated potential damages. A risk becomes a disaster only when the event has occurred with a significant economic impact after which the actual damages can be recorded.
- A disaster also has to be distinguished from a catastrophe. "Catastrophe" is professionally used in two different ways: (a) to refer to casualties, human losses; and often to the case of extreme events characterized by low probability, high loss, and less empirical data and (b) as a subset of large-scale disasters, relative to the resources of the impacted community, whereby both the public authorities, the first responders and the victims, are overwhelmed by the damages and are not in a position to cope with the situation in an orderly fashion.

- Intrinsic to the definition of disasters lays the question of scale and proportions. An earthquake of magnitude 7.2 for a country such as Turkey, or Chinese Taipei in 1999 for example, may seem to be a disaster; however, a similar event can be a devastating catastrophe for a more fragile country, such as Haiti in 2010, without the resources to cope.

What can be learned from the economics of disasters?

To what extent do disasters affect the national GDP?

The economic context of disasters in industrialized OECD countries and developing countries is not the same. OECD countries are often equipped with a set of prevention and insurance tools that make the situation much easier to cope with; for instance, the economic losses are much larger in OECD countries than in developing countries in absolute numbers but quite less in relative percentage of the GDP.

To what extent do disasters affect longer-term growth?

The negative impact on GDP is usually a very short one for OECD countries, a matter of months quite often, as opposed to other countries for which the impacts can be very long lasting. The Chernobyl nuclear accident in Ukraine is supposed to have had a negative impact on their GDP for 20 years (OECD, 2004).

However, Chicago recovered from the Great Fire of 1871 in approximately one and one-half years (Macaulay, 2005). San Francisco was almost completely rebuilt within 3 years after the 1906 earthquake. Economic activity in Kobe, Japan, returned to normal 19 months after the massive earthquake in 1995 (Macaulay, Philips, OECD 2008).

Unexpectedly, in OECD countries, after a disaster, the trend is for the GDP to rise, due to the massive injection of recovery funds by government and insurance. For example, it is estimated that 40% of the losses were covered by insurance in the September 11, 2001 terrorist attack in New York. Of course, this process can create a transfer of public debt to the next generation since the budget of the government could then suffer a deficit. It also affects the balance sheet of insurance companies, which could then be forced to sell assets to be able to pay the indemnities. Overall, because of potential public deficit and increase of insurance premiums, this process can lead to rising interest rates in the country (Table 1).

To what extent does the number of damaging events grow over time and why?

The number of damaging events is globally increasing over time. This trend can be attributed to two sources: (a) with modern communication tools, natural and man-made disasters are much more recorded than in the past, creating a possible bias in the trend; and (b) man-made footprints can create conditions which are more favorable to the repetition of disasters (e.g., anthropic emissions and climate change) (Figure 1).

To what extent are damages from disasters growing over time?

It is now understood that as urbanization increases (51% of global population, at a fast rate in non-OECD countries, and at a slower rate in OECD countries) the assets of society (human, real estate, industrial, and financial, ...) are increasingly concentrated in cities or even megacities. Therefore, even at a constant probability of an earthquake over time, any occurrence will create more damages in economic terms. Increased human development along the coastal margin has lead to an increase in hurricane and typhoon losses (Figure 2).

What are the questions/issues that economists of disasters have to answer?

Typical questions/issues: In what ways do disasters impact the economy (impacts on GDP, production, consumption, scarcity of resources, savings, investment, productivity, interest rates, public or private debt, ...)? To what extent are these impacts a function of the stock of capital (more or less recent, in compliance with building codes and other safety regulations), of the level of technology involved, or the level and quality of maintenance? What kinds of incentives for individual or community protection,

Economics of Disasters, Table 1 Impact of large-scale disasters on GDP (Source: Adapted from OECD, 2004)

Year	Disaster	Economic cost (estimates)	% of GDP (estimates)
1995	Kobe earthquake, Japan	USD 130 billion	Over 2%
2001	September 11	USD 120 billion	1.2%
Since 1986	Bovine spongiform encephalopathy (BSE) in Europe	EUR 92 billion	1% (EU15)
2003	Severe acute respiratory syndrome (SARS) in E/SE Asia	USD 60 billion	2%
1992	Hurricane Andrew	USD 25 billion	About 0.5%
2002	Central European flooding (Germany, Austria, Czech Republic)	EUR 15 billion	About 0.75%
1999	Marmara earthquake, Turkey	USD 9–13 billion (SPO)	About 6%
		USD 6–10 billion (World Bank)	About 4%
2000	"I love you" computer virus	USD 8 billion	–
2000	Foot and mouth disease, UK	£6 billion	0.6%

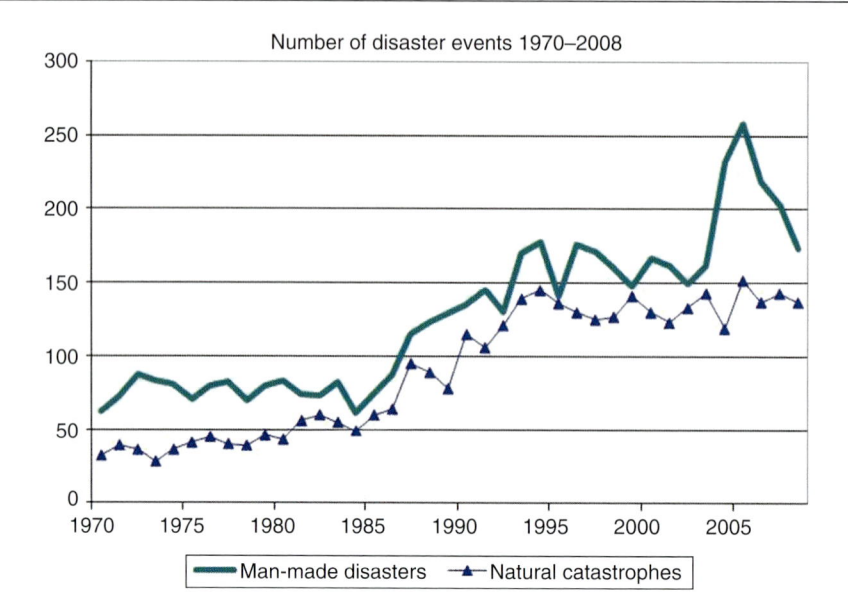

Economics of Disasters, Figure 1 Number of disaster events 1970–2008 (Source: Swiss, 2009).

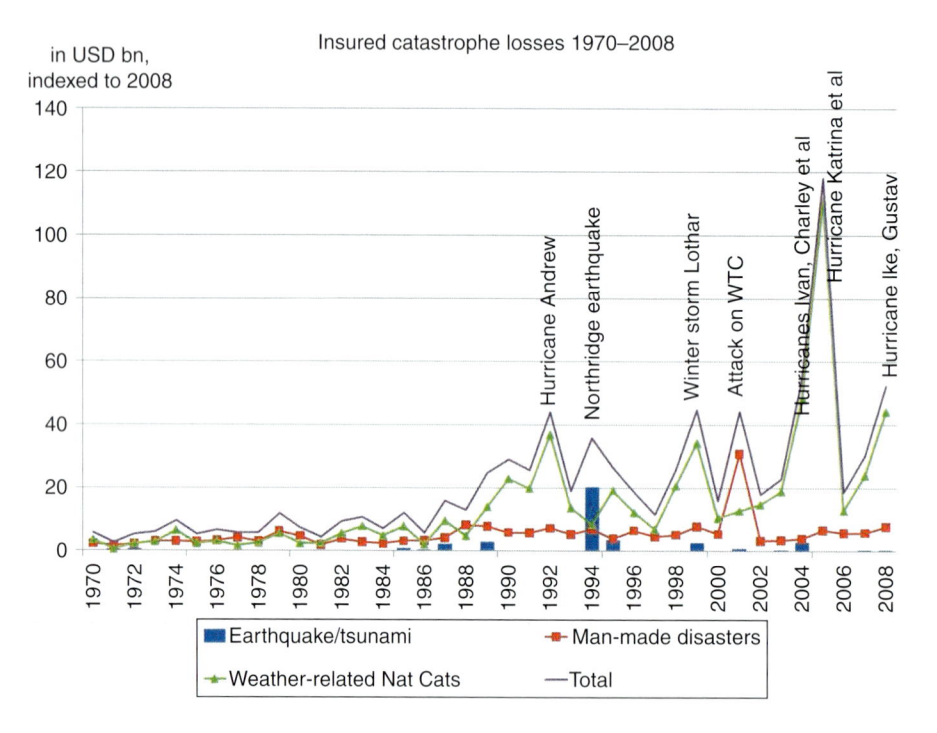

Economics of Disasters, Figure 2 Insured catastrophe losses 1970–2008 (Source: Swiss, 2009).

mitigation, and insurance are provided within existing institutions? In short, how can or will the markets react? What can governments do in terms of economic and other policies (public expenditures, incentives, regulations, monetary policies, ...)?

Other issues that the economists have to contribute to are as follows: Trade-offs between investments in prevention versus recovery; Actors' level of risk tolerance or preference; Channels for transfer of risks; Role of State as last resort guarantor of asymmetry of information; and Opportunities for predatory behavior and/or free-riding impacts on labor markets, and price mechanisms such as impacts on interest rates or insurance premiums, and intergenerational debts.

Status as a branch of economics

Economics of disasters can be defined as an emerging branch since:

- In the past, it was not a case for scientific investigation. In the western world at least, disasters were often perceived by monotheist religions as revenge by god or "acts of god," more than a scientific phenomena that could be investigated.
- The economic dimension of disasters has only recently been studied. Of course, numerous records have been written on such large disasters, such as the Black Plague (1347–1350) or more recently the Spanish Flu (1918–1919), but they were not trying to identify "economic laws." Early work on economics can be traced to the 1950s with work by Jack Hirshleifer of the RAND Corporation (Hirshleifer, 1987) and in the 1960s when Douglas C. Dacy and Howard Kunreuther published *The Economics of Natural Disasters* (1969). The relevant facts and databases are slow to emerge in a consistent way and are still not at the desired level of accuracy, comprehensiveness, or comparability. There is still no reliable database aimed at collecting the data and the common definitions of the parameters accounting for economic disasters, besides the few and confidential databases of insurers. Few inroads have since been made by other players, such as International Governmental Organizations, the research community, or NGOs.

Example of "laws" or working hypothesis in the field of economics of disasters

- The more resources allocated for the recovery and construction after a disaster, the faster is the speed of recovery (Okuyama and Chang, 2004).
- Due to the replacement of older equipment with new equipment based on newer technologies after a disaster, it is assumed that the rate of technological progress in the stressed area will increase but on temporary basis only (Okuyama and Chang, 2004).

However, still in its infancy, it can be claimed that the economics of disasters already has the attributes of a branch of economics: a separate/specific scope and domain; produced laws and principles, if not yet a general body of knowledge; and the adaption to its own domain of some specific concepts and methodological tools. Not unrelated is the fact that interest in disaster management during the beginning of twenty-first century is increasing among all stakeholders: governments; the private sector; and taxpayers and citizens as well, for the following reasons: (a) Disasters are not acceptable by populations in "welfare states." (b) They should not "happen again." (c) The repetition of shocks within the process of globalization is a threat (the last frontier) to the human well-being.

Related branches

Economics of disasters as a branch ought to be distinguished from the economics of insurance, although insurance economics is older and there are overlaps between the two. For instance, mandatory insurance can lead to a disincentive for the risk-bearer to act prudently (a case of moral hazards): this issue could be discussed under the two branches. However, the question of economic impact of a disaster on GDP, on interest rates, or even on the insurance industry, following a disbursement of funds, is more a question of economics of disaster than insurance.

Economics of disasters has a strong link to economic and other policy areas which is specific to its scope. A disaster quite often has a direct effect on policy shaping and decision making processes, such as special programs for restoring the capacity of public services, or changing the policies related to building codes in case of floods, or changing the regulations of safety features of industrial plants.

Related sciences

As is the case for most branches of this discipline, economics of disasters can use a number of tools or principles that are generic to the other branches of economics; this is the case of probabilistic or mathematical approaches to risks, the concept of extreme events and all tools linked to cost/benefit approaches in decision making, for instance.

Social sciences such as philosophy, history, psychology or sociology, organizational theory, and political science can be useful to economists when dealing with disasters (Slovic, 1993; 2001) (e.g., the issues of risk perception, willingness to be insured, moral hazards, and long-term social psychological consequences of disasters all have economic consequences).

Hard sciences are extremely useful for the economics of disasters as they provide the basis for the assessment of potential risks: estimates of probabilities of occurrence and of magnitude of possible damaging events in the case of natural disasters or man-made events are indispensable for economists as a foundation for their own estimate of economic consequences as well as their analysis of options for prevention, mitigation, emergency or recovery processes, and as a prerequisite for cost/benefit analysis.

Toolbox and methodological tools specific to the economics of disasters

Approaches from the perspective of conventional economic methodologies are used in economics of disasters. This is particularly true when one tries to measure and understand the impacts of disasters. Input/output tables, social accounting, and general equilibrium models are generally found in the literature (Cochrane, 2004; Okuyama and Chang, 2004).

More adaptations of the methodologies to the specific nature of disasters can be found. A recent example is CATSIM, which is a model originally started by the Inter-American Development Bank for Latin American countries and enhanced by the IIASA on estimates of impacts of disasters on public finance. The concept of financial resilience of public finance is a major factor which can encompass the degree of preparation of the public sector as well as the potential access to funds both ex ante and ex post disaster.

A number of methodologies at the interface of macro- and microeconomics as well as probabilistic and stochastic decision processes can also be found under loss estimation models, risk mitigation decisions under uncertainty, and modeling mitigation strategies with cost/benefit analyses.

At the forefront of the improvement of economic tools lies the convergence of interest from multiple stakeholders in implementing GIS databases, fed by satellites and other remote sensing measurement, census information, and databases from the private sector. These databases will increasingly provide data that disaster economists need to help decision and policy makers, as well as investors, to make informed decisions.

Conclusions

Although, economics of disasters is an emerging branch of economics, international governmental organizations (such as the OECD, the World Bank, specialized UN agencies), the European Union, regional development banks, and national governments, as well as universities, research centers, consultants have started in the early twenty-first century to increase activities in the domain of risk management policies, risk indicators, and databases, as well as in economics of disasters.

Another branch of economics is trying to link knowledge of economics and physics, under the name of "econophysics." It mainly tries to use inductive or experimental methods from physics, as well as concepts from thermodynamics, to learn more on the economics particularly of financial processes including bubbles. Some of the potential benefits are to avoid being constrained by the previous assumptions of economists, such as general equilibrium conditions, and also to improve the "detrending" capacity in database analysis to overcome the weaknesses of older multiple linear regression tools and therefore, better identify the risks of disruption, cycles and crises, etc. These developments could be also helpful in the economics of disasters.

A parallel and convergent effort is underway with the emerging body of knowledge on complex systems, which also uses interdisciplinary concepts and tools from ecology, biology, mathematics, and physics in trying to identify universal features of complex systems and particularly their endogenous dynamics. An example is the concept criticality in economics of disasters which has been defined as the impact of a disaster which drives the economy at a critical state at macroeconomic level, together with strong localized interactions between individual elements at the microeconomic level. An illustration would be the fall of the Iron Curtain in 1989 (a disaster for some, a shock in more neutral terms) with macroeconomic impact on the East German economy and local impacts on the labor markets of Western Germany (Reggiani and Nijkamp, 2004).

Summary

A few centuries ago, disasters were "acts of gods." During the twentieth century, disasters progressively became an important theme for hard sciences (e.g., seismology for earthquakes, hydrology for floods) as well as, slowly, a theme for economics. It is now recognized that potential negative damages from natural and man-made hazards are large enough to justify efforts to better understand the laws and principles that govern the economic dimensions of disasters: primarily assessing their costs and impacts. Even more importantly, economic questions are linked to the decisions that governments, firms, and citizens have to make about potential disasters: invest in research or not, and if yes, to what extent? In prevention or not? In early warning or not? In insurance or not? And, if yes, up to what extent? As efforts to understand these laws are growing, economics of disasters is becoming a branch of economics.

Bibliography

Bostrom, A., French, S., and Gottlieb, S. (eds.), 2008. *Risk Assessment, Modeling and Decision Support: Strategic Directions*. Berlin, Heidelberg: Springer.

Cochrane, H. C., 2004. Chapter 3, Indirect losses from natural disasters: measurement and myth. In Okuyama, Y., and Chang, S. E. (eds.), *Modeling the Spatial and Economic Impacts of Disasters*. Berlin: Springer.

Dacy, D., and Kunreuther, H., 1969. *The Economics of Natural Disasters: Implications for Federal Policy*. New York, NY: The Free Press.

Gad-el-Hak, M., 2008. *Large Scale Disasters*. Cambridge/New York: Cambridge University Press.

Hirshshleifer, J., 1987. *Economic Behavior in Adversity*. Chicago: University of Chicago Press.

Macaulay, D., 2005. *The Chicago Fire of 1871: An Empirical Analysis*. Unpublished paper. University of Chicago.

OECD, 2003. *Emerging Risks in the 21st Century*. Paris: OECD.

OECD, 2004. *Large Scale Disasters: Lessons learned*. Paris: OECD.

OECD, 2005. *Catastrophic Risks and Insurance*. Paris: OECD.

OECD, 2009. *Japan: Large-Scale Floods and Earthquakes*. Paris: OECD.

Okuyama, Y., and Chang, S. (eds.), 2004. *Modeling the Spatial Economic Impacts of Disasters*. Berlin: Springer.

Philillips, M., and Crossen, C., 2005. Will New Orleans Rebound?. *The Wall Street Journal Online*. September 1, 2005.

Reggiani, A., and Nijkamp, P., 2004. Fall of iron curtain and German regional labour markets. In Okuyama, Y., and Chang, S. E. (eds.), *Modeling the Spatial and Economic Impacts of Disasters*. Berlin: Springer.

Slovic, P., 1993. Perceived risk, trust, and democracy: a systems perspective. *Risk Analysis*, **13**, 675–682.

Slovic, P., 2001. *Emerging Systemic Threats: Risk Assessment, Public Perception, and Decision Making, Contribution to the OECD International Futures Project on Emerging Systemic Risks*.

Swiss Re, 2009. *Natural Catastrophes and Man-Made Disasters 2008*. Sigma No 2/2009, Swiss Re.

Cross-references

Classification of Natural Disasters
Complexity Theory
Cost-Benefit Analysis
Costs (Economic) of Natural Hazards and Disasters
Economic Valuation of Life
Exposure to Natural Hazards
Risk Assessment
World Economy (Impact of Disasters on)

EDUCATION AND TRAINING FOR EMERGENCY PREPAREDNESS

Kevin R. Ronan
CQUniversity Australia, North Rockhampton, QLD, Australia

Synonyms

Emergency readiness

Definition

Emergency Preparedness: Actions taken in anticipation of an emergency to facilitate rapid, effective, and appropriate response to the situation (Inter-Agency Contingency Planning Guidelines for Humanitarian Assistance, 2001).

This entry is based on Ronan and Johnston (2005). Portions are reprinted from that book, also published by Springer.

Education and training for emergency preparedness

The effects of natural hazards are many. One classification system considers effects at three levels: primary effects, secondary effects, and higher order effects (Petak and Atkisson, 1982; see also Ronan and Johnston, 2005). Primary effects include direct effects (e.g., death and injury; physical damage; changes to river flows, flood plains, and landscapes; and trauma); secondary effects include the need for evacuation, lifeline disruptions, industry, business and school closure; financial expenditure for response and recovery; insurance; and secondary stressors). Higher order effects include flow on effects such as alteration of population trends, unemployment, loss of income, capital, savings, land values, changes to socioeconomic trends, and increased cost and tax burden.

Given the many and potentially large-scale effects of natural hazards, societies must undertake certain activities if they are to coexist with hazardous events. Emergency preparedness is one of those activities. The other tasks include risk reduction (or mitigation), response, and recovery. These tasks are linked together across the phases of a hazardous event and include those activities that can be done prior to an event (preparedness, risk reduction), during the event itself (response), and following the event (recovery). Emergency managers emphasize prevention in the form of risk reduction and preparedness activities as the key to effective response and recovery (Tierney et al., 2001).

Research has supported the idea that the effects of a hazardous event can be reduced through sound decision making and planning (e.g., Peek and Mileti, 2002). On the other hand, some forms of planning can be intended to assist emergency preparedness but inadvertently create other problems. For instance, costly solutions such as construction of levees or flood banks and raising sea walls may divert attention from more effective social solutions. Extending this example, putting in flood banks aimed at 100-year floods may then also lead to population growth that is then at risk in the event of a flood that exceeds 100-year return periods.

In light of this backdrop, education and training for emergency preparedness needs to consider both physical and social solutions to assist a community mitigate risk as well as be prepared to respond more effectively. Strategies available for emergency preparedness include (Burby et al., 2000): (1) building standards and codes, (2) development regulations, (3) policies for public and critical facilities, (4) good fiscal policy including taxation, (5) property management and *Land-Use Planning*, and (6) dissemination of information through various channels. The dissemination of information includes through training and educating government, organizations, and the public about the value of emergency preparedness, including specific risk reduction and readiness activities that can be undertaken.

Thus, the goal in education and training for emergency preparedness is to assist all sectors of the community to understand risk in relation to natural (and other) hazards, how to reduce risk and how to prepare to respond more effectively (Johnston and Ronan, 2000). Activities related include creating response plans, undertaking training, including simulation exercises and drills, and engaging in necessary physical and social activities that will reduce risk and increase preparedness to respond. Of course, all of these activities can, and do, occur at various levels within a community, including at community level, organization level, and household level. Communities in many parts of the world have a statutory obligation to have an *Emergency Management* preparedness and response plan. However, many organizations also engage in some forms of education and training in emergency preparedness, including having written emergency plans. At the household level, research has demonstrated that in general, most communities have low rates of household planning and preparedness, even in high hazard zones (Mileti and Darlington, 1997; Ronan and Johnston, 2005).

In addition, even where emergency plans are available, they will not be effective if they are not disseminated or applied by stakeholders. Thus, a written plan is a necessary but by no means sufficient component of an adequate emergency preparedness education and training program. In fact, prior to a written plan, the more input and buy-in from several sectors of a community that there is, the more likely that it will then be adopted and applied when necessary. This implicates a need for developing relationships across several sectors of a community, initiated by those with the most vested interests in emergency preparedness. Those with the most vested interests of course will include local emergency planners and other government agencies but may also come from several other community sectors, including schools, non-governmental organizations, scientists, business and industry, neighborhood, community, service and volunteer groups, and the media. Research has demonstrated that it does not merely have to be officials that initiate and raise the level of community concern about education and training, but it can include "emergent groups" of concerned citizens (Quarantelli, 1985).

The written plan itself should serve two major functions (Tierney et al., 2001): (1) an agreement that is signed off by various organizations and documents specific roles and related functions for promoting training and education, preparedness activities and activation thresholds for specific response, and recovery related activities and (2) a template for continuing training and education. The importance of this second function is to buffer against the chances that the plan itself is not "filed, shelved, and forgotten" (see also Ronan and Johnston, 2005).

With a template for ongoing training and education, practice, simulations, and important community linkages, an integrated approach to community-based emergency preparedness can then be a continuing process of keeping community members up-to-date with risk-related information. Ensuring community members have up-to-date knowledge will have members informed on what to do to be prepared for, and respond to hazardous events. This includes not only raising awareness but providing specific guidance across multiple media, all endorsed by multiple, linked, and trusted organizations (Mileti, 1999). This up-to-date knowledge should come from different trusted information sources (e.g., from school education programs, emergency management, businesses, scientists, and government officials); emphasize important content (e.g., specific guidance on protective and response actions by households and organizations, physical effects and features of an expected hazard, and anticipating and responding effectively to warnings); and use a range of education delivery modes (education through schools, public media, brochures, demonstrations, community, organizational, and household-response exercises) aimed at different target audiences (youth, adults, and those in higher risk groups including those in low socioeconomic groups; see also Lindell and Perry, 2007; Ronan and Johnston, 2005).

A final point on the motivation to engage in emergency preparedness and training: owing to community emergency preparedness levels typically being low, motivation to prepare is quite obviously an issue that is worth keeping in mind when planning emergency preparedness training and education programs. Developing multiple relationships and linkages through community partnerships is one key to raising community motivation to engage in a variety of activities aimed at strengthening communities, including emergency preparedness (Ronan and Johnston, 2005). Another is to locate "motivational reservoirs" within communities, including schools and schoolchildren and concerned citizens' groups, and include them in training and education efforts.

Summary and outlook for emergency preparedness

Research indicates that continuing research and efforts on the parts of localities to encourage emergency preparedness has merit. For example, documented preparedness for earthquakes and other natural hazards in California has seen quite a significant improvement in the last 3 decades of research done in that US state (Lindell and Perry, 2000). However, there also appears to be a certain "half-life" to community-focused education and training efforts, where effectiveness can diminish over time. Thus, a "spaced" effort across time rather than a one-off "massed" effort appears to be the more fruitful pathway for helping a community prepare more effectively for a range of hazardous events (Ronan and Johnston, 2005).

Bibliography

Burby, R. J., Deyle, R. E., Godschalk, D. R., and Olshansky, R. B., 2000. Creating hazard resilient communities through land-use planning. *Natural Hazards Review*, **1**, 99–106.

Inter-Agency Standing Committee Reference Group on Contingency Planning and Preparedness (2001). Inter-agency contingency planning guidelines for humanitarian assistance: Recommendations to the IASC. United Nations.

Johnston, D., and Ronan, K., 2000. Risk education and intervention. In Sigurdsson, H., Houghton, B., McNutt, S. R., Rymer, H., and Stix, J. (eds.), *Encyclopedia of Volcanoes*. San Diego, CA: Academic.

Lindell, M. K., and Perry, R. W., 2000. Household adjustment to earthquake hazard: A review of research. *Environment & Behavior*, **32**, 590–630.

Lindell, M. K., and Perry, R. W., 2007. Planning and preparedness. In Tierney, K. J., and Waugh, W. F., Jr. (eds.), *Emergency Management: Principles and Practice for Local Government*, 2nd edn. Washington DC: International City/County Management Association, pp. 113–141.

Mileti, D., 1999. *Disasters by Design: A Reassessment of Natural Hazards in the United States*. Washington, D.C.: Joseph Henry Press.

Mileti, D. S., and Darlington, J. D., 1997. The role of searching in shaping reactions to earthquake risk information. *Social Problems*, **44**, 89–103.

Peek, L. A., and Mileti, D. S., 2002. The history and future of disaster research. In Bechtel, R. B., and Churchman, A. (eds.), *Handbook of Environmental Psychology*. New York: John Wiley & Sons, Inc.

Petak, W. J., and Atkisson, A. A., 1982. *Natural Hazard Risk Assessment and Public Policy: Anticipating the Unexpected*. New York: Springer-Verlag.

Quarantelli, E. L., 1985. What is disaster? The need for clarification in definition and conceptualization in research. In Sowder, B. J. (ed.), *Disasters and Mental Health: Selected Contemporary Perspectives*. Rockville, MD: National Institute of Mental Health.

Ronan, K. R., and Johnston, D. M., 2005. *Community Resilience to Disasters: The Role for Schools, Youth, and Families*. New York: Springer.

Ronan, K. R., Crellin, K., Johnston, D. M., Finnis, K., Paton, D., and Becker, J., 2008. Promoting child and family resilience to disasters: Effects, interventions, and prevention effectiveness. *Children, Youth, and Environments*, **18**(1), 332–353.

Tierney, K. J., Lindell, M. K., and Perry, R. W., 2001. *Facing the unexpected: Disaster response in the United States*. Washington, D.C.: Joseph Henry Press.

Cross-references

Building Codes
Cognitive Dissonance
Communicating Emergency Information
Community Management of Hazards
Cost-Benefit Analysis of Natural Hazard Mitigation
Costs (Economic) of Natural Hazards and Disasters
Damage and the Built Environment
Disaster Risk Reduction (DRR)
Early Warning Systems
Emergency Management
Emergency Planning
Expert (Knowledge-Based) Systems for Disaster Management
Federal Emergency Management Agency (FEMA)
Integrated Emergency Management System
International Strategies for Disaster Reduction (IDNDR and ISDR)
Land-Use Planning
Livelihoods and Disasters
Mass Media and Natural Disasters
Nevado del Ruiz
Perception of Natural Hazards and Disasters
Psychological Impacts of Natural Disasters
Risk Perception and Communication
Social-Ecological Systems
Structural Mitigation
Vulnerability
Warning Systems

ELASTIC REBOUND THEORY

John Ristau
GNS Science, Avalon, Lower Hutt, New Zealand

Synonyms

Earthquake cycle; Strain accumulation

Definition

The gradual accumulation of elastic strain on either side of a locked fault.

Overview

Elastic rebound theory is one of the central ideas to the mechanics of earthquakes and was originally proposed by Reid (1910) following the 1906 San Francisco earthquake on the San Andreas Fault. The theory involves the slow build up of elastic strain due to large-scale stresses in the crust on either side of a locked fault (i.e., a fault on which steady movement is not occurring). Once the strain accumulation exceeds the strength of the locked fault, the stored energy is abruptly released by rapid displacement along the fault in the form of an earthquake, whereby the rocks return to their original state, and the cycle repeats. Elastic rebound theory fits in well with the theory of plate tectonics and helps explains the cyclical nature of many earthquakes including why earthquakes repeatedly occur in the same regions. It is now known that the actual accumulation of strain does not follow a simple, steady-strain accumulation, and that the concept of a single breaking point is overly simplistic; however, as a general concept, elastic rebound theory works well (Figure 1).

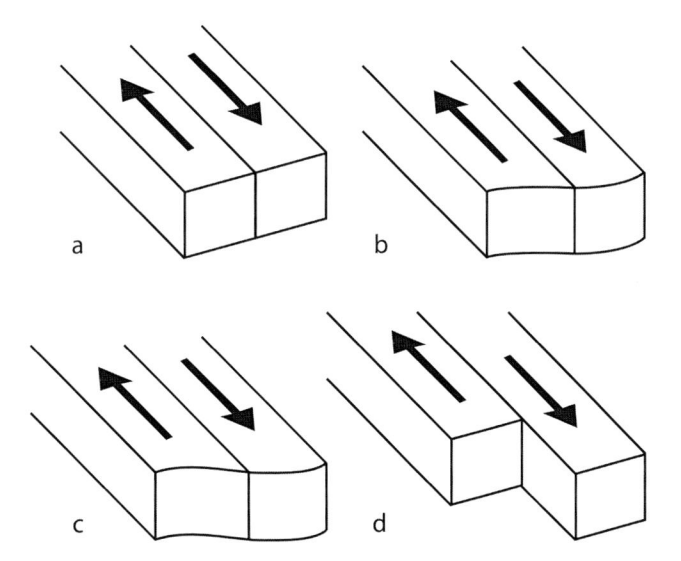

Elastic Rebound Theory, Figure 1 Due to large-scale stress in the crust strain is built up along a fault from the unstrained state (**a**), through (**b**) to the fully strained state (**c**) at which point sudden displacement occurs along the fault to release strain and return to the original state (**d**).

Bibliography

Lay, T., and Wallace, T. C., 1995. *Modern global seismology*. San Diego: Academic, p. 521.

Reid, H. F., 1910. The mechanism of the earthquake. In *The California Earthquake of April 19, 1906, Report of the State Earthquake Investigation Commission, 2*. Washington, DC: Carnegie Institution, p. 192.

Scholz, C. H., 1990. *The Mechanics of Earthquakes and Faulting*. Cambridge: Cambridge University Press, p. 439.

Cross-references

Earthquake
Earthquake Prediction and Forecasting
Epicenter
Fault
Hypocenter
San Andreas Fault
Seismology

ELECTROMAGNETIC RADIATION (EMR)

Norman Kerle
Faculty of Geo-Information Science and Earth Observation (ITC), University of Twente, Enschede, The Netherlands

Synonyms

Electromagnetic waves; Radiant energy

Definition

Energy propagating through space at the speed of light in the form of sine-shaped electromagnetic waves, composed of perpendicularly arranged electric and magnetic fields. EMR ranges from gamma rays with very short wavelength to long radio waves. The shortest wavelengths can also be modeled as particles (photons). The interaction of EMR with matter forms the basis for remote sensing.

Overview

Electromagnetic radiation (EMR) is composed of sine-shaped waves that propagate through space at the speed of light (approximately 300,000 km s^{-2}), characterized by electrical and magnetic fields that are arranged perpendicular to each other (Lillesand et al., 2004). The central property of EMR is wavelength, inversely proportional to frequency. It ranges from high-frequency gamma rays (with picometer [10^{-16} m] wavelength and that are better thought of as particles or photons) to radio waves many kilometers long and with low frequencies, collectively known as the electromagnetic spectrum (EMS). Wave energy is also proportional to frequency.

EMR forms the basis for remote sensing (RS), which has gained great relevance in studying and monitoring of hazards (Tralli et al., 2005). RS is divided into passive and active methods: reflected or emitted radiation is recorded (passive), or the response of an artificial signal is received (active, for example radar). To detect or monitor phenomena related to hazards, a careful selection of the appropriate part of the EMS is critical. Most Earth observation instruments, such as regular cameras, passively record EMR in the visible part of the spectrum (approximately 0.4–0.7 μm [10^{-6} m]), and in the adjacent near-infrared (NIR, 0.7–1.4 μm). This is ideal to detect the state of vegetation, as the cell structure of healthy green leaves strongly reflects NIR energy, which declines in stressed leaves. Vegetation stress possibly leading to crop failure can thus be detected early.

Less common are detectors that record thermal infrared (TIR) radiation (8–14 μm), for example, to measure surface temperatures. The main forms of active RS are lidar (laser scanning), radar, and sonar (*l*ight/*ra*dio/*so*und *d*etection *a*nd *r*anging, respectively). Lidar uses very short waves between about 400 nm and 1 μm, whereas radar waves range between approximately 0.1–1 m. Sonar uses acoustic waves several meters long. An advantage of all active sensors is that they are largely weather-independent and may also be applied at night.

EMR is also the basis for other tools important in hazard work, for example, GPS, which uses radio waves of about 20 cm, marginally more than other important communication systems, such as wireless networks.

EMR itself can constitute a hazard to living organisms. Well-known examples of radiation to which exposure should be minimized or avoided are X-rays (wavelength of a few nm), ultraviolet rays than cause sunburn (about 0.3–0.4 μm), but also microwaves (wavelength of about 12 cm).

Bibliography

Lillesand, T. M., Kiefer, R. W., and Chipman, J. W., 2004. *Remote Sensing and Image Interpretation*. New York: Wiley.

Tralli, D. M., Blom, R. G., Zlotnicki, V., Donnellan, A., and Evans, D. L., 2005. Satellite remote sensing of earthquake, volcano, flood, landslide and coastal inundation hazards. *ISPRS Journal of Photogrammetry and Remote Sensing*, **59**, 185–198.

Cross-references

Global Positioning System and Natural Hazards
Remote Sensing of Natural Hazards and Disasters

EL NIÑO/SOUTHERN OSCILLATION

Michael Ghil[1,2], Ilya Zaliapin[3]
[1]Ecole Normale Supérieure, Paris, France
[2]University of California, Los Angeles, CA, USA
[3]University of Nevada Reno, Reno, NV, USA

Synonyms

El Niño; Southern oscillation; Tropical pacific warming

Definitions

Easterlies. Low-latitude trade winds that blow from east to west and extend from the Galapagos Islands to Indonesia.

Kelvin wave. A nondispersive subsurface oceanic wave, several centimeters high and hundreds of kilometers wide, that balances the Earth's Coriolis force against the equator.

El Niño. A recurrent increase of the ocean surface temperature across much of the tropical eastern and central Pacific; the term was introduced by Peruvian fishermen.

La Niña. The opposite to El Niño – cooling of the ocean surface temperature across much of the tropical eastern and central Pacific.

Rossby wave. A planetary-scale wave caused by the variation in the Coriolis force with latitude, discovered by Carl-Gustaf Rossby in 1939.

Southern oscillation. Coupled changes in surface air pressure between the eastern and western Pacific associated with El Niño and La Niña events; the term was introduced by Sir Gilbert Walker in 1924.

Southern Oscillation Index (SOI). A scalar measure of the strength of Southern Oscillation, calculated as the difference in surface air pressure between Tahiti and Darwin, Australia. El Niño episodes correspond to negative SOI values, La Niñas to positive ones.

Thermocline. A thin layer of ocean water that divides a layer of relatively warm water just below the surface from colder, deeper, nutrient-rich waters. The thermocline is usually a few tens of meters deep in the eastern Tropical Pacific, and can reach depths of hundreds of meters in the western Tropical Pacific. The thermocline depth is heavily affected by El Niño dynamics.

Thermohaline circulation. Large-scale circulation throughout the world's oceans that transforms low-density upper-ocean waters to higher-density intermediate and deep waters, and returns the latter back to the upper ocean.

Walker circulation. Zonal flow in a roughly longitude-altitude plane near the Equator, caused by differences in heat distribution between ocean and land, and described by Sir Gilbert Walker during the 1920s.

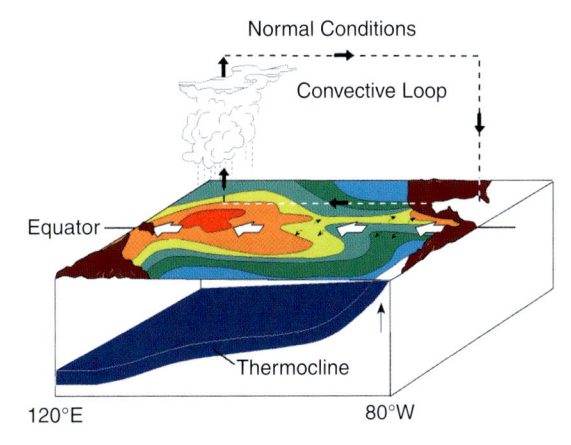

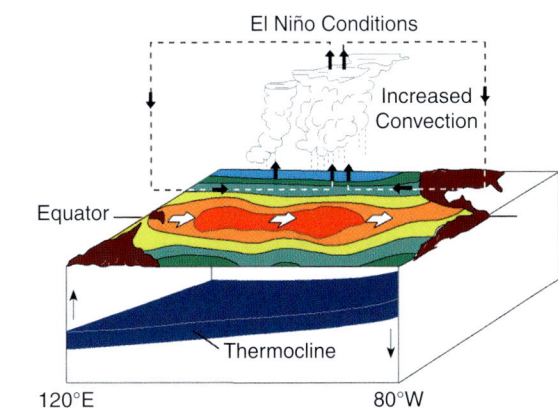

El Niño/Southern Oscillation, Figure 1 Schematic diagram of the atmospheric and oceanic circulation in the Tropical Pacific. *Upper panel*: climatological mean ("normal"), *lower panel*: El Niño (warm) phase. The three-dimensional diagrams show the deepening of the thermocline near the coast of Peru during the warm phase, accompanied by anomalous surface winds (*heavy white arrows*), modified Walker circulation (*lighter black arrows*), and a displacement and broadening of the warmest SSTs from the "warm pool" in the western Tropical Pacific, near Indonesia, toward the east (After McPhaden et al., 1998, with permission of the American Geophysical Union AGU.).

Introduction

The El-Niño/Southern Oscillation(ENSO) phenomenon is the most prominent signal of global seasonal-to-interannual climate variability. It was known for centuries to fishermen and sailors along the west coast of South America, who witnessed a seemingly sporadic and abrupt warming of the cold, nutrient-rich waters that support the food chains in those regions; these warmings caused havoc to the fish harvests (Diaz and Markgraf, 1993; Philander, 1990). The common occurrence of such warmings shortly after Christmas inspired Peruvians to name it El Niño, after the "Christ child" (*el niño* is Spanish for *little boy*). The phenomenon was discussed at the Geographical Society of Lima meeting in 1892, and El Niño became the official name of this phenomenon.

The ENSO phenomenon manifests itself in both atmospheric and oceanic processes (Figure 1). The terms El Niño and La Niña refer to the temperature state of the oceanic surface. An *El Niño* (or warm ENSO phase) represents the warming of waters across much of the tropical eastern and central Pacific. A *La Niña* is the opposite, cooling phase (*la niña* is Spanish for *little girl*). The warm phase occurs typically, to some extent at least, every boreal winter. More significant warm episodes, as well as not so warm ones, occur roughly every 2 years: this is the so-called quasi-biennial oscillation (QBO). In the climate literature, differences between the instantaneous, or short-term average map, and the climatological or *normal* values associated with the mean seasonal cycle are called *anomalies*; in the context of interannual variability, El Niño and La Niña represent, respectively, warm and cold anomalies of sea surface temperature (SST).

These changes in the ocean temperature are directly related to the atmospheric pressure; the changes in pressure between the eastern and western Pacific associated with El Niño and La Niña events are referred to as the *Southern Oscillation*; the term was coined by Sir Gilbert Walker in 1924. El Niño favors higher pressure in the western Pacific; La Niña favors higher pressure in the eastern Pacific.

ENSO dynamics is commonly monitored by averaging SST anomalies over some portion of the Tropical Pacific. The *normal* SST field is defined here in terms of a long-term average (often taken to be 30 years) of mean-monthly maps. There are four regions commonly used for ENSO monitoring:

- Niño 1 + 2 (0°–10°S, 80°–90°W) – the region that typically warms first when an El Niño event develops
- Niño 3 (5°S–5°N; 150°–90°W) – the region with the largest variability in SST on El Niño time scales
- Niño 3.4 (5°S–5°N; 170°–120°W) – the region that is most important for monitoring global climate variability, because the SST variability in this region has the strongest effect on shifting rainfall in the western Pacific and elsewhere
- Niño 4 (5°S–5°N; 160°E–150°W) – the region with average SST close to 27.5°C, which is thought to be an important threshold in producing rainfall

As is typical for threshold-crossing episodes in a continuous field, there is no objective, quantitative definition of ENSO events. Trenberth (1997) addressed the problem of formally defining El Niño and La Niña by suggesting that "an El Niño can be said to occur if 5-month running means of sea surface temperature (SST) anomalies in the Niño 3.4 region exceed 0.4°C for 6 months or more." Table 1 lists all the ENSO events since 1950 according to this definition.

Global climatic and socioeconomic impacts

Starting in the 1970s, El Niño's climatic, and hence socioeconomic, effects were found to be far broader than its manifestations off the shores of Peru (Glantz et al., 1991; Diaz and Markgraf, 1993). This realization was triggered in particular by the strong El Niño episode of 1976–1977 that coincided with a so-called global "climate shift" (Miller et al., 1994) in the Pacific Ocean, and it led to a global awareness of ENSO's significance and to an increased interest in modeling (Cane and Zebiak, 1985; Philander, 1990; Neelin et al., 1992, 1994, 1998), as well as in monitoring and forecasting (Barnston et al., 1994; Latif et al., 1994; Ghil and Jiang, 1998) exceptionally strong El Niño and La Niña events.

Affected by numerous competing mechanisms, the remote effects of ENSO, also called *teleconnections*, do vary from one El Niño or La Niña event to another. Nevertheless, it has been observed that significant ENSO events are consistently related to particular weather anomalies around the globe. In many regions, these anomalies

El Niño/Southern Oscillation, Table 1 El Niño and La Niña years, after IRI (2010)

El Niño	Year	La Niña	Year
1	1951	1	1950–1951
2	1953	2	1954–1956
3	1957–1958	3	1964–1965
4	1963–1964	4	1967–1968
5	1965–1966	5	1970–1972
6	1968–1970	6	1973–1976
7	1972–1973	7	1984–1985
8	1976–1977	8	1988–1989
9	1977–1978	9	1995–1996
10	1982–1983	10	1998–2000
11	1986–1988	11	2000–2001
12	1990–1992		
13	1993		
14	1994–1995		
15	1997–1998		

present the second largest contribution to climate variability after the seasonal cycle (IRI, 2010). Seasonal climate changes, in turn, affect the flood and landslide frequencies, air quality, forest fire likelihood, agricultural production, disease outbreaks, fisheries catches, energy demand and supply, as well as food and water availability. Accordingly, both damages and benefits of ENSO's impact present an important component of the socioeconomical and political life of the affected regions. This impact is generally more severe in developing countries, where people's lives are heavily dependent on agricultural production and natural water sources. ENSO effects, as well as any other climate impacts, can be amplified or reduced by such factors as government policies; infrastructure resilience; crop and human diseases; current military, economic, or political conflicts; disruption of storage and shipping facilities; mitigation policies for natural hazards; and many others (IRI, 2010).

In the physical climate system, ENSO affects SSTs, wind, pressure, and rainfall patterns: it does so directly in the Tropical Pacific and via teleconnections in many other parts of the globe. The key observation in understanding ENSO climatic patterns is that the SST field plays a key role in determining rainfall intensity: the higher the SSTs, the higher are the rainfalls. Under *normal conditions* (absence of El Niño), the largest SSTs and greatest rainfall intensity is found over the "warm pool" in the western Tropical Pacific, whereas the eastern Tropical Pacific and the west coast of South America enjoy cold, nutrient-rich waters; the prevailing winds are the *easterly* trade winds. *El Niño conditions* shift the highest SSTs and rainfall eastward and weaken the easterlies; *La Niña conditions* shift high SSTs and rainfall farther west, extend the cold waters to the central Pacific, and strengthen the easterlies. ENSO teleconnections manifest the close relationship between the tropical rainfall and prevailing winds on the one hand and the global atmospheric wind patterns on the other.

A large El Niño event affects, within the next year, the temperature and precipitation patterns along both coasts of South America, the Caribbean, the Equatorial Pacific, Southeast Asia, India, Southeast and West Africa, Australia, and both coasts of North America (Ropelewski and Halpert, 1987; New and Jones, 2000). Hurricanes, typhoons, and tropical cyclones are also affected by ENSO, either directly (in the Pacific) or via teleconnections (in the Indian and Atlantic Oceans): the changes may be seen in the frequency of events or in the initial location of these storms (Landsea, 2000).

The most important impact of El Niño events on local hydroclimatic conditions and thus on regional ecology and economics can be seen in floods and landslides caused by high rainfalls in Southern California and Peru, forest fires and air pollution in Indonesia, crop failures and famine due to droughts in southern Africa, as well as in the collapse of Peruvian anchovy fisheries due to the warming of coastal waters. At the same time, ENSO affects in one way or another all continents and diverse sectors of socio-economic systems around the globe. Table 2 illustrates the global ENSO impacts according to the data from the International Research Institute for Climate and Society (IRI, http://portal.iri.columbia.edu).

What causes ENSO?

The following conceptual elements play a determining role in ENSO dynamics.

The Bjerknes hypothesis. Jacob Bjerknes (1897–1975) laid the foundation of modern ENSO research. Bjerknes (1969) suggested a *positive feedback* as a mechanism for the growth of an internal instability that could produce large positive anomalies of SSTs in the eastern Tropical Pacific. Using observations from the International Geophysical Year (1957–1958), he realized that this mechanism must involve air-sea interaction in the tropics. The "chain reaction" starts with an initial warming of SSTs in the "cold tongue" that occupies the eastern part of the equatorial Pacific. This warming causes a weakening of the thermally direct Walker cell circulation; this circulation involves air rising over the warmer SSTs near Indonesia and sinking over the colder SSTs near Peru. As the trade winds blowing from the east weaken and give way to westerly wind anomalies, the ensuing local changes in the ocean circulation encourage further SST increase. Thus the feedback loop is closed and further amplification of the instability is triggered.

Delayed oceanic wave adjustments. Compensating for Bjerknes positive feedback is a *negative feedback* in the system that allows a return to colder conditions in the basin's eastern part. During the peak of the cold-tongue warming, called the *warm* or El Niño phase of ENSO, westerly wind anomalies prevail in the central part of the basin. As part of the ocean's adjustment to this atmospheric forcing, a Kelvin wave is set up in the tropical wave guide and carries a warming signal eastward. This signal deepens the eastern-basin thermocline, which separates the warmer, well-mixed surface waters from the colder waters below, and thus contributes to the positive feedback described above. Concurrently, slower Rossby waves propagate westward, and are reflected at the basin's western boundary, giving rise therewith to an eastward-propagating Kelvin wave that has a cooling, thermocline-shoaling effect. Over time, the arrival of this signal erodes the warm event, ultimately causing a switch to a *cold* or La Niña phase.

Seasonal forcing. A growing body of work (Chang et al., 1994; Jin et al., 1994; Tziperman et al., 1994; Ghil and Robertson, 2000) points to resonances between the Pacific basin's intrinsic air-sea oscillator and the annual cycle as a possible cause for the tendency of warm events to peak in boreal winter, as well as for ENSO's intriguing mix of temporal regularities and irregularities. The mechanisms by which this interaction takes place are numerous and intricate and their relative importance is not yet fully understood (Battisti, 1988; Tziperman et al., 1994; Dijkstra, 2005).

Time series that depict ENSO dynamics

An international 10-year (1985–1994) Tropical-Ocean–Global-Atmosphere (TOGA) Program greatly improved the observation (McPhaden et al., 1998), theoretical modeling (Neelin et al., 1994, 1998), and prediction (Latif et al., 1994) of exceptionally strong El Niño events. It has confirmed, in particular, that ENSO's significance extends far beyond the Tropical Pacific, where its causes lie.

An important conclusion of this program was that – in spite of the great complexity of the phenomenon and the

El Niño/Southern Oscillation, Table 2 Global impacts of ENSO (according to the IRI data base http://iri.columbia.edu/climate/ENSO/)

Region	Impact
Africa	Changes in land use
Asia	Changes in available water resources; droughts and floods; changes in river discharge; outbreaks of wheat stripe rust disease, cholera, hemorrhagic fever, dengue; interannual variability of ozone; influence on rice production
Australia and the Pacific	Changes in alpine-lake inflow; changes in river discharge; amount of rainfall; outbreaks of encephalitis; availability of banana prawns
Central America and the Caribbean	Floods; changes of a coastal fish assemblage; variations of annual maize yields; coral bleaching; mortality rates
Europe	Agricultural yields; wine production and quality
North America	Variations in the occurrence of wildfires; annual runoff; insect population
South America	Invertebrate behavior; maize, grain production; precipitation and streamflow; river discharge

differences between the spatiotemporal characteristics of any particular ENSO cycle and other cycles – the state of the Tropical Pacific's ocean-atmosphere system could be characterized, mainly, by either one of two highly anticorrelated scalar indices. These two indices are an SST index and the Southern Oscillation Index (SOI): they capture the East-West seesaw in SSTs and sea level pressures, respectively.

A typical version of the SST index is the so-called Niño-3.4 index, which summarizes the mean anomalies of the spatially averaged SSTs over the Niño-3.4 region (Trenberth, 1997). The evolution of this index since 1900 is shown in Figure 2: it clearly exhibits some degree of regularity on the one hand as well as numerous features characteristic of a deterministically chaotic system on the other. The regularity manifests itself as the rough superposition of two dominant oscillations, the quasi-biennial (QBO) and quasi-quadrennial (Jiang et al., 1995c; Ghil et al., 2002) mode, by the phase locking of warm events to boreal winter that gives El Niño its name and by a near-symmetry of the local maxima and minima (i.e., of the positive and negative peaks). The lack of regularity has been associated with the presence of a "Devil's staircase," which is discussed in further detail below (Chang et al., 1994; Jin et al., 1994; Tziperman et al., 1994), and may be due to stochastic effects as well (Ghil et al., 2008a).

A hierarchy of climate models

Climate dynamics has emerged as a modern scientific discipline about a half-century ago (Pfeffer, 1960), and it is within this broader framework that ENSO variability should be considered. The climate system is highly complex, its main subsystems have very different characteristic times, and the specific phenomena involved in various climate problems are quite diverse. It is inconceivable, therefore, that a single model could successfully be used to incorporate all the subsystems, capture all the phenomena, and solve all the problems.

Hence the concept of a *hierarchy of models*, from the simple to the complex, had been developed almost four decades ago (Schneider and Dickinson, 1974). Climate models can be divided into *atmospheric*, *oceanic*, and *coupled*. Each group is characterized, in addition, by the model dimension, where the number of dimensions, from zero to three, refers to the number of independent space variables used to describe the model domain, i.e., to physical-space dimensions. Coupled atmosphere-ocean models – from the simplest zero-dimensional (0-D) ones to three-dimensional (3-D), general circulation models (GCMs) – might be better able to model ENSO dynamics than other climatic phenomena, because ENSO is generally thought to operate through atmosphere-ocean coupling.

A fairly well-developed hierarchy of coupled ocean-atmosphere models has been applied to the problem of seasonal-to-interannual variability in the Tropical Pacific ocean, directly related to ENSO dynamics (Neelin et al., 1994). Its most important rungs are, in ascending order: essentially 0-D simple models, like the delay-oscillator model of Suarez and Schopf (1988); essentially 1-D intermediate coupled models (ICMs: Cane and Zebiak, 1985; Jin et al., 1994); essentially 3-D hybrid coupled models, in which an ocean GCM is coupled to a much simpler, diagnostic atmospheric model (Neelin, 1990; Barnett et al., 1993); and fully coupled GCMs (Neelin et al., 1992; Robertson et al., 1995). Hybrid models of this type have also been applied to climate variability for the mid-latitude (Weng and Neelin, 1998) and global (Chen and Ghil, 1996; Wang et al., 1999) coupled system.

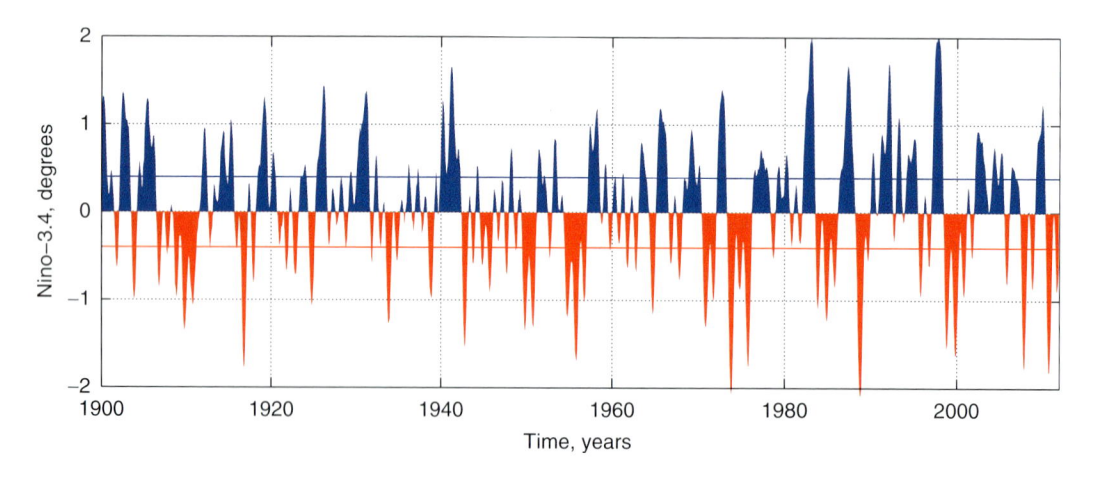

El Niño/Southern Oscillation, Figure 2 Time evolution of the Niño-3.4 index during 1900–2012. The index depicts the sea surface temperature (SST) anomalies (deviations from the climatological mean) in the Niño-3.4 region, between 170°W–120°W and 5°S–5°N. Horizontal lines are drawn at ±0.4°; according to Trenberth (1997), El Niño (warm) and La Niña (cold) events can be defined as a 6-month exceedance of these thresholds.

Modeling ENSO: Goals and approaches

ENSO modeling is focused on a broad twofold goal: (1) to depict the essential mechanisms behind the observed ENSO variability and (2) to forecast future ENSO dynamics, in particular large El Niño and La Niña events that have potential to impact on human activity. The modeling efforts are fostered by understanding the basic physical principles behind the examined phenomenon (see the section above) as well as confronting the existing models with observations.

There are two main paradigms in modeling ENSO variability (Neelin et al., 1994, 1998; Ghil and Robertson, 2000). A *deterministically chaotic, nonlinear* paradigm explains the complexities of ENSO dynamics by the nonlinear interplay of various internal driving mechanisms. For instance, the complex evolution of the SST and thermocline depth can be simulated by the interplay of the two basic ENSO oscillators: an internal, highly nonlinear one, produced by a delayed feedback of the oceanic wave propagation, and a forced, seasonal one (Tziperman et al., 1994; Zaliapin and Ghil, 2010). A *stochastic, linear paradigm*, on the other hand, attempts to explain characteristic features of ENSO dynamics by the action of fast *weather noise* on a linear or weakly nonlinear slow system, composed mainly on the upper ocean near the equator. Boulanger et al. (2004) and Lengaigne et al. (2004), among others, provide a comprehensive discussion of how weather noise could be responsible for the complex dynamics of ENSO, and, in particular, whether westerly wind bursts trigger El Niño events. It seems that any successful modeling effort should combine these two paradigms to obtain richer and more complete insights into climate dynamics in general (e.g., Ghil et al., 2008a).

Much of our theoretical understanding of ENSO comes from relatively simple, essentially 0-D and 1-D coupled models, consisting of a shallow-water or two-layer ocean model coupled to steady-state, shallow-water-like atmospheric models with heavily parameterized physics; the more complete ones among these models are the previously mentioned ICMs (Neelin et al., 1994). In these models, ENSO-like variability results from an oscillatory instability of the coupled ocean-atmosphere's annual-mean climatological state. Its nature has been investigated in terms of the dependence on fundamental parameters, such as the coupling strength, oceanic adjustment time scale, and the strength of surface currents (Jin and Neelin, 1993).

The growth mechanism of ENSO is fairly well established, arising from positive atmospheric feedbacks on equatorial SST anomalies via the surface wind stress, as first hypothesized by Bjerknes (1969). The cyclic nature of the unstable mode is subtler and depends on the time scales of response within the ocean. The next section reviews the deterministic, nonlinear paradigm in understanding ENSO's quasi-periodic behavior; the section emphasizes a toy-modeling approach, which is sufficient to capture the main ENSO-driving mechanisms and, unlike GCMs, can be reviewed here in sufficient detail.

ENSO as a coupled oscillator

The 1980s and 1990s saw the development of a dynamical theory that explains ENSO variability via the interaction of two oscillators: *internal*, driven by the negative feedback associated with oceanic wave propagation, and *external*, due to the seasonal cycle.

Schopf and Suarez (1988), Battisti (1988), Suarez and Schopf (1988), and Battisti and Hirst (1989) demonstrated that ENSO's complex dynamics can be studied using the formalism of *delayed differential equations* (DDE). The first attempts dealt with autonomous DDEs, without seasonal forcing, and with a linear delayed part:

$$\frac{dT}{dt} = -\alpha T(t-\tau) + T \quad (1)$$

Here, T represents the SSTs averaged over the eastern equatorial Pacific. The first term on the right-hand side of Equation 1 mimics the negative feedback due to the Kelvin and Rossby waves, while the second term reflects Bjerknes's positive feedback.

The delay equation idea happens to be very successful in explaining the recurrent nature of ENSO events in easily intelligible mathematical settings. Indeed, the delayed negative feedback does not let a solution of Eq. 1 converge to zero or go to infinity as it would go in the ordinary differential equation case with $\tau = 0$: the delay effect thus creates an internal oscillator whose period depends on the delay and the particular form of the equation's right-hand side. Thus, a simple DDE like Eq. 1 has reproduced some of the main features of a fully nonlinear, coupled atmosphere-ocean model of ENSO dynamics in the tropics (Zebiak and Cane, 1987; Battisti, 1988; Battisti and Hirst, 1989). DDE modeling has also emphasized the importance of nonlinear interactions in shaping the complex dynamics of the ENSO cycle.

At the same time, many important details of ENSO variability still had to be explained. First, a delayed oscillator similar to Eq. 1 typically has periodic solutions with a well-defined period of about 4τ. However, the occurrence of ENSO events is irregular. Second, the period suggested by delay equations deviates significantly from the actual recurrence time of warm events, which is about 2–7 years. The delay τ, which is the sum of the basin-transit times of the westward Rossby and eastward Kelvin waves, can be roughly estimated to lie in the range of 6–8 months. Accordingly, model (1) suggests a period of 24–32 months, at most, for the repeating warm events; this period lies at the lower end of the range of recurrence times. Finally, El Niño and La Niña events always peak during the Northern Hemisphere (boreal) winter, hence their name; such phase locking does not exist in a purely internal delayed oscillator.

The next important step in developing ENSO modeling in the DDE framework was made by Tziperman et al. (1994), who demonstrated that the above discrepancies can be removed by considering nonlinear interactions between the internal oscillator and the external periodic forcing by the seasonal cycle. These authors also introduced a more realistic nonlinear coupling between atmosphere and ocean to reflect the fact that the delayed negative feedback saturates as the absolute value of the key dependent variable T increases.

Munnich et al. (1991) made a detailed comparison between cubic and sigmoid nonlinearities in an iterated map model of ENSO. As a result, the sigmoid type of nonlinearity was chosen in Tziperman et al. (1994), resulting in the periodically forced, nonlinear DDE:

$$\frac{dT}{dt} = -\alpha \tanh[\kappa T(t-\tau_1)] + \beta \tanh[\kappa T(t-\tau_2)] + \gamma \cos(\omega t) \quad (2)$$

Here, the first term on the right represents the westward-traveling Rossby wave, the second term represents the eastward Kelvin wave, and the last one is a seasonal forcing. The parameters α, β, and γ represent the relative strengths of these three driving forces; τ_1 and τ_2 are Rossby and Kelvin wave delays, respectively; ω determines the period of the seasonal forcing; and κ represents the strength of the atmosphere-ocean coupling.

Depending on the parameter values, this model has solutions that possess an integer period, are quasi-periodic, or exhibit chaotic behavior. The increase of solution complexity – from period one, to integer but higher period, and on to quasi-periodicity and chaos – is caused by the increase of the atmosphere-ocean coupling parameter κ. Tziperman et al. (1994) also demonstrated that this forced DDE system exhibits period locking, when the external "explicit" oscillator wins the competition with the internal delayed one, causing the system to stick to an integer period; dependence of the system's period on model parameters is realized in the form of a *Devil's staircase* (see below).

These and other ENSO studies with DDE models have been limited to (1) the linear stability analysis of steady-state solutions, which are not typical in forced systems; (2) case studies of particular trajectories; or (3) one-dimensional scenarios of transition to chaos, where one varies a single parameter, while the others are kept fixed. A major obstacle for the complete bifurcation and sensitivity analysis of such DDE models lies in the complex nature of DDEs, whose numerical and analytical treatment is harder than that of models with no delays.

Zaliapin and Ghil (2010) took several steps toward a comprehensive analysis, numerical as well as theoretical, of DDE models relevant for ENSO phenomenology. These authors considered a simplified version of Eq. 2:

$$\frac{dT}{dt} = -\tanh[\kappa T(t-\tau)] + b\cos(2\pi t) \quad (3)$$

and, for the first time, performed its analysis in the complete 3-D space of the physically relevant parameters: strength of seasonal forcing b, ocean-atmosphere coupling κ, and transit time τ of oceanic waves across the Tropical Pacific. This model reproduces many scenarios relevant to ENSO phenomenology, including prototypes of El Niño and La Niña events; intraseasonal activity reminiscent of Madden-Julian oscillations (Madden and Julian, 1994) or westerly wind bursts; and spontaneous interdecadal oscillations.

The model also provided a good justification for the observed QBO in Tropical Pacific SSTs and trade winds (Philander, 1990; Diaz and Markgraf, 1993; Jiang et al., 1995b; Ghil et al., 2002), with the 2–3-year period arising naturally as the correct multiple (four times) of the sum of the basin transit times of Kelvin and Rossby waves. Zaliapin and Ghil (2010) found regions of stable and unstable solution behavior in the model's parameter space; these regions have a complex and possibly fractal distribution of solution properties. The local continuous dependence theorem (Zaliapin and Ghil, 2010) suggests that the complex discontinuity patterns indicate the presence of a rich family of unstable solutions that point, in turn, to a complicated attractor.

The simple DDE model (3), with a single delay, does reproduce the Devil's staircase scenario documented in other ENSO models, including ICMs and GCMs, as well as in observations (Jin et al., 1994; Tziperman et al., 1994; Ghil and Robertson, 2000). The latter result suggests that interdecadal variability in the extratropical, thermohaline circulation (Dijkstra, 2005; Dijkstra and Ghil, 2005) might interfere constructively with ENSO's intrinsic variability on this time scale. Zaliapin and Ghil (2010) found that model (3) is characterized by *phase locking* of the solutions' local extrema to the seasonal cycle; in particular, solution maxima – i.e., model El Niños – tend to occur in boreal winter. These authors also found multiple solutions coexisting for physically relevant values of the model parameters.

Figure 3 illustrates the model's sensitive dependence on parameters in a region that corresponds roughly to actual ENSO dynamics. The figure shows the behavior of the period P of model solutions as a function of two parameters: the propagation period τ of oceanic waves across the Tropical Pacific and the amplitude b of the seasonal forcing; for aperiodic solutions one takes $P = 0$. Although the model is sensitive to each of its three parameters (b,κ,τ), sharp variations in P are mainly associated with changing the delay τ, which is plotted on the ordinate. This sensitivity is an important qualitative conclusion since in reality the propagation times of Rossby and Kelvin waves are affected by numerous phenomena that are not related directly to ENSO dynamics. The sensitive dependence of the period on the model's parameters is consistent with the irregularity of occurrence of strong El Niños, and can help explain the difficulty in predicting them (Latif et al., 1994; Ghil and Jiang, 1998).

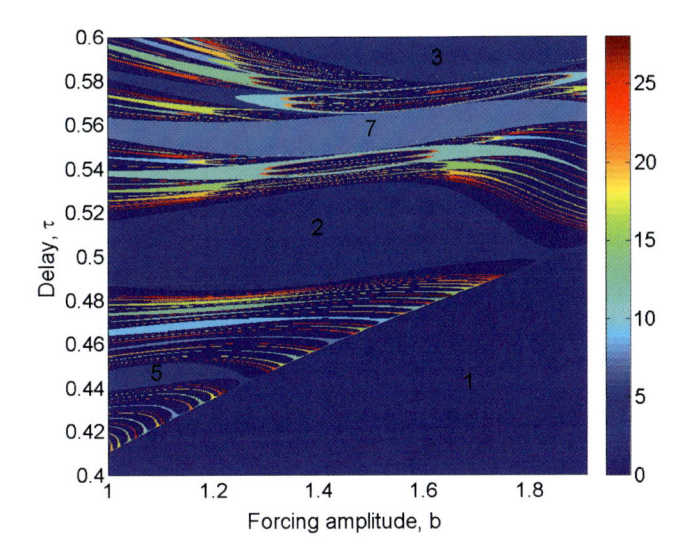

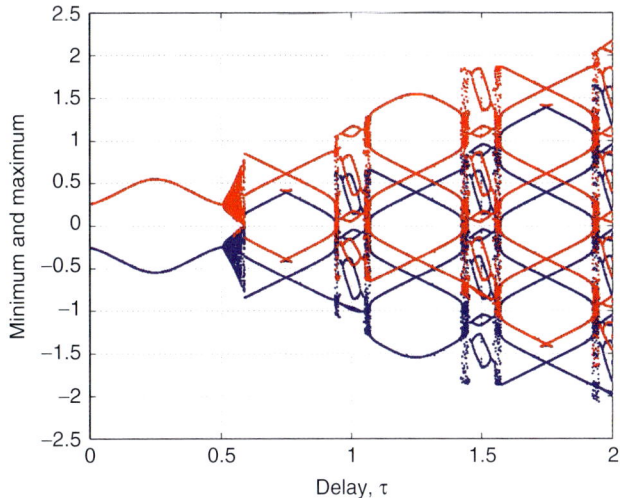

El Niño/Southern Oscillation, Figure 3 Period map for the delayed coupled oscillator of Eq. 3. The figure shows the period P as a function of two model parameters: amplitude b of seasonal forcing and delay τ of the oceanic waves; the ocean-atmosphere coupling strength is fixed at $\kappa = 10$. Aperiodic solutions correspond to $P = 0$. Numbers indicate the period values within the largest constant-period regions.

El Niño/Southern Oscillation, Figure 4 Local maxima (red) and minima (blue) of solutions of Eq. 3 as a function of delay τ; the other parameter values are fixed at $\kappa = 11$ and $b = 2$. Note the aperiodic regimes between periodic windows of gradually increasing period. This figure corresponds to the rightmost vertical section of the region shown in Figure 3 (From Zaliapin and Ghil, 2010).

The model's instabilities disappear and the dynamics of the system becomes purely periodic, with a period of 1 year (not shown), as soon as the atmosphere-ocean coupling $\kappa \tau$ vanishes or the delay τ decreases below a critical value. Figure 4 illustrates this effect in greater detail: the period P of model solutions increases with τ in discrete jumps, $P = 2k + 1$, $k = 0, 1, 2,\ldots$, separated by narrow, apparently chaotic "windows" in τ. This increase in P is associated with the increase of the number of distinct local extrema, all of which tend to occur at the same position within the seasonal cycle. This resembles in fact the behavior of chaotic dynamical systems in discrete time (Kadanoff, 1983) and suggests that the model's aperiodic dynamics is in fact chaotic. This chaotic behavior implies, in particular, that small perturbations in the model parameters or in initial states may lead to significant changes of the model dynamics. Due to this sensitive dependence, forecasting the model's behavior, as well as that of the related natural phenomenon, is a hard problem.

The boundary between the domains of stable and unstable model behavior is clearly visible at the lower right of Figure 3. The period-1 region below and to the right of this boundary contains simple solutions that change smoothly with the values of model parameters. The region above and to the left is characterized by sensitive dependence on parameters. The range of parameters that corresponds to present-day ENSO dynamics lies on the border between the model's stable and unstable regions. Hence, if the dynamical phenomena found in the model have any relation to reality, Tropical Pacific SSTs and other fields that are highly correlated with them, inside and outside the Tropics, can be expected to behave in an intrinsically unstable manner; they could, in particular, change quite drastically with global warming.

Quasi-periodic behavior

The ENSO phenomenon dominates interannual climate variability over the Tropical Pacific. Figure 5, top panel, shows the power spectrum of the monthly SSTs averaged over the eastern equatorial Pacific's Niño-3 area, for the time interval 1960–1997 (Ghil and Robertson, 2000). The observed SST time series contains a sharp annual cycle, together with two broader interannual peaks centered at periods of 44 months (the so-called quasi-quadrennial or low-frequency ENSO cycle) and 28 months (the QBO cycle).

This power spectrum provides a fine example of the distinction between the sharp lines produced by purely periodic forcing and the broader peaks resulting from internal climate variability or from the interaction of the latter with the former. The sharp annual peak reflects the seasonal cycle of heat influx into the tropical Pacific and the phase locking of warm events to boreal winter that gives El Niño its name. The two interannual peaks correspond to the quasi-quadrennial and QBO components of ENSO, as identified by a number of authors (Rasmusson et al., 1990; Allen and Robertson, 1996). These components play a determining role in the ENSO dynamics: Jiang et al. (1995b) have demonstrated that these two

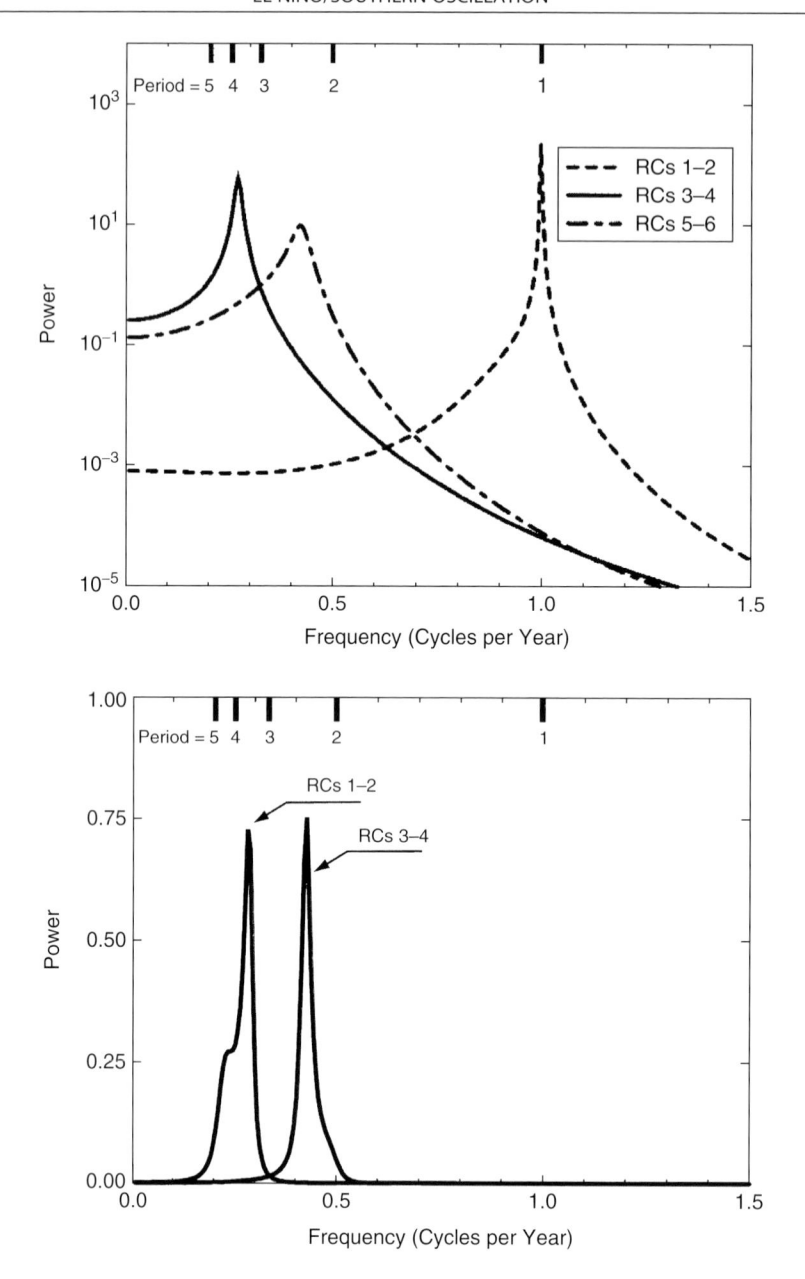

El Niño/Southern Oscillation, Figure 5 *Top*: Power spectrum of the leading reconstructed components (RCs) of the Niño-3 SSTs for the time interval 1960–1997, using monthly data from the Climate Prediction Center of the National Centers for Environmental Prediction (NCEP). An SSA analysis with a window width of 72 months was used to derive the RCs, whose power spectra were then computed using the maximum entropy method, with 20 poles. RCs (1,2) capture the annual cycle, RCs (3,4) the quasi-quadrennial oscillation, and RCs (5,6) the QBO. *Bottom*: Power spectrum of Niño-3 SST anomalies from a 60-year integration of NCEP's coupled GCM, with the seasonal cycle removed (Ji et al., 1998).

components account for about 30% of the variance in the time series analyzed in Figure 5; accordingly, the major El Niño (warm) and La Niña (cold) events during the time interval 1950–1990 can be well reconstructed from ENSO's quasi-quadrennial and QBO components.

The existence of both these oscillatory components has been established in coupled GCMs. The University of California Los Angeles (UCLA) atmospheric GCM, coupled to a tropical-Pacific basin version of the Geophysical Fluid Dynamics Laboratory (GFDL) ocean GCM (Mechoso et al., 2000), is characterized by ENSO-like quasi-quadrennial and QBO modes, but with weaker variability than that of the observed modes (Robertson et al., 1995). A 100-year-long simulation with NASA Goddard's Aries-Poseidon coupled GCM exhibits both quasi-quadrennial and QBO spectral peaks of a strength very similar to that in observations (not shown). These results are further confirmed by a 60-year run of National Center for Environmental Prediction's (NCEP) coupled GCM (Ji et al., 1998).

Devil's staircase

Various toy and intermediate models (Chang et al., 1994; Jin et al., 1994; Tziperman et al., 1994; Ghil et al., 2008b) have demonstrated that annual forcing causes the internal ENSO cycle to lock to rational multiples of the annual frequency in a Devil's staircase. The period map for the DDE ENSO model of Eq. 3, as shown in Figure 3, is a representative example of this behavior. The Devil's staircase is a function that exhibits peculiar properties, which challenge our intuition about the concept of continuity: it is continuous on the interval [0, 1]; its values span the entire range between 0 and 1; and at the same time it is constant almost everywhere. In other words, the total length of the point sets over which the function increases equals zero!

A classical example of a Devil's staircase, related to the celebrated Cantor set, is shown in Figure 6. Despite its seemingly unnatural features, this type of behavior is commonly seen in coupled, oscillatory mechanical systems, as well as in phase-locked electronic loops (Rasband, 1990). In these systems, the Devil's staircase depicts the relations between the phase of a driven nonlinear oscillatory system and the phase of the external driving force. This staircase represents a generic scenario of transition to deterministic chaos via a subharmonic resonance (e.g., period doubling). Such a scenario involves two parameters: the phase of the external force and the degree of coupling to the driving force.

The period staircase of ENSO also involves two parameters: one governs the period of the intrinsic ENSO instability (i.e., the propagation time of oceanic Kelvin and Rossby waves), whereas the other is the coupling strength between the model's ocean and atmosphere. As the intrinsic period increases, a subharmonic resonance causes frequency locking to successively longer rational multiples of the annual cycle. As the coupling strength increases, the steps on the staircase broaden and begin to overlap, and the model's ENSO cycle becomes irregular, due to jumps between the steps.

The complete Devil's staircase scenario, in fact, calls for successively smaller peaks associated with the harmonics of the 4-year (quasi-quadrennial) mode, at $4/1 = 4$, $4/2 = 2$, and $4/3$ years. Both the QBO and the $4/3$-year $= 16$-month peak are present in observed SST data (Jiang et al., 1995b; Allen and Robertson, 1996). There is a smaller and broader 18–20 month peak present in the UCLA coupled GCM, which can be interpreted as a merged version of these two peaks (Robertson et al., 1995).

Thus, the results of GCM simulations, along with existing observational data, provide reasonable support to the following conjecture: the interaction of the seasonal cycle and the fundamental ENSO mode nonlinearly

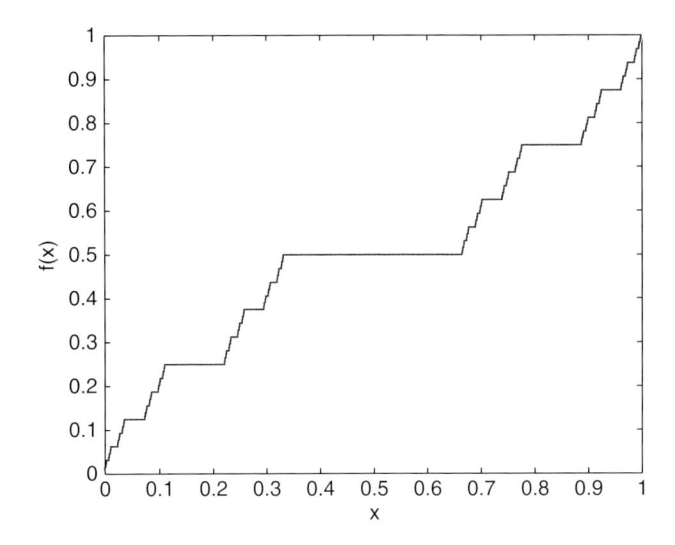

El Niño/Southern Oscillation, Figure 6 Devil's staircase $f(x)$: a continuous function that takes all the values between 0 and 1 and at the same time is constant almost everywhere. There is solid evidence that period locking phenomena in the ENSO system are organized according to a Devil's staircase.

entrain this mode to a rational multiple of the annual frequency, and produce additional peaks, according to a Devil's staircase. Still, it is possible that different frequency peaks, in particular the quasi-quadrennial and the QBO peaks, could represent separate oscillations, generated by different mechanisms, each with an independent frequency (see Neelin et al., 1998, and references therein).

Forecasts

Forecasts of ENSO dynamics are commonly based on modeling and monitoring the evolution of the equatorial Pacific SST. There are two main types of such models: *dynamical* and *statistical*. Roughly speaking, a dynamical model extrapolates the current oceanic and atmospheric conditions into the future by using deterministic principles of atmosphere-ocean interaction. A statistical model uses the past observations to identify conditions that statistically favor occurrence of an El Niño or a La Niña event. The dynamical and statistical models can span the entire modeling hierarchy: from toy to intermediate to fully coupled.

ENSO forecasts are facilitated by (1) the dominant ENSO regularities, mainly the quasi-quadrennial and QBO modes described above, and (2) the persistent nature of ENSO warm and cold episodes, each of which lasts for a few months up to a year. Hence, episode initiation in April–May does facilitate forecasting an event's peak in December–January; likewise, the absence of an episode by June is a very reliable signal for normal conditions in the next year.

In spite of these marked regularities, forecasting El Niño and La Niña events has met with mixed success, even for short, subannual lead times (Latif et al., 1994). The authors of over a dozen models – all of which routinely produce ENSO forecasts on a monthly or quarterly basis – have noticed year-to-year variations in forecast skill at a lead time of 6–12 months. In addition, forecasts from individual models may significantly vary from one to another.

We illustrate the latter statement with an example of SST forecasts by 15 coupled GCMs (dynamical) and 7 statistical models for the time interval from December 2009 to November 2010, summarized by the International Research Institute for Climate and Society (IRI, 2010); they are shown in Figure 7. All the models start from the observed Niño-3.4 SST anomaly of about 1.8° in December 2009 and try to extrapolate it into the future. Strikingly, this "forecast plume" is characterized by overall regression to the mean of most forecasts: many of the models, dynamical as well as statistical, regress to the no-anomaly mean in about a year. This regression is due to the well-known statistical principle that the best long-term prediction of an ergodic process is its mean. Even so, one observes a persistent 2-degree spread in individual forecasts at all lead times, a spread that characterizes the existing "state-of-the-art" uncertainty in the ENSO prediction problem.

Barnston et al. (1994) and Ghil and Jiang (1998) formally assessed the 6-month lead forecast skills of six ENSO models, three dynamical and three statistical. They used the following two measures of the forecast skill: (1) the Pearson correlation r between the monthly forecast and actual SST anomaly values and (2) the root-mean-squared error (RMSE) of the forecast versus actual SST normalized by the variation of the actual values. This assessment, illustrated in Table 3, suggests that different models have comparable forecast skills with $r \approx 0.65$; these skills are of intermediate quality (RMSE ≈ 0.9), and most probably different models will show different local performance under different circumstances.

Summary and outlook

El Niño/Southern Oscillation (ENSO) is a prominent climatic phenomenon that affects Earth's atmosphere and oceans, as well as their interactions, on time scales up to several years and influences the global climate. ENSO creates anomalies in the sea-surface temperature (SST), thermocline depth, and atmospheric pressure across the Tropical Pacific and affects most immediately the waters off Peru, where this phenomenon was noticed centuries ago. ENSO's oceanic manifestations are called El Niño (warmer waters in the eastern Pacific) and La Niña (colder

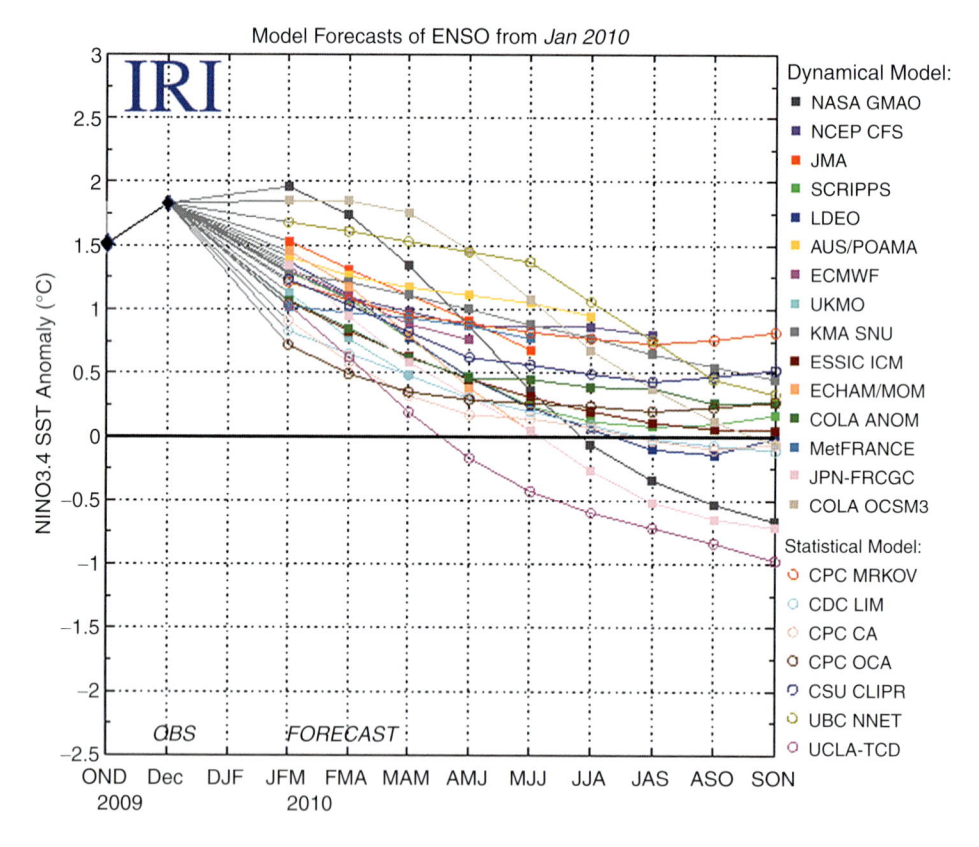

El Niño/Southern Oscillation, Figure 7 "Forecast plume": juxtaposition of SST forecasts for the year 2010, made in December 2009 by 15 dynamical and 7 statistical models (IRI, 2010). Most forecasts regress to the no-anomaly mean within a year, while still giving a persistent 2-degree spread of the individual values at all lead times larger than 3 months.

El Niño/Southern Oscillation, Table 3 Forecast skill (6-month-lead) of six ENSO models (After Ghil and Jiang, 1998, Table 1)

Authors	Zebiak and Cane (1987)	Barnett et al. (1993)	Ji et al. (1994)	Barnston and Ropelewski (1992)	Van den Dool (1994)	Jiang et al. (1995)
Model type	Dynamical			Statistical		
Forecast region 5°S–5°N	Niño-3 90°–150°W	Central Pacific 140°–180°W	Niño-3.4 120°–170°W			Niño-3 90°–150°W
Period	1970–1993	1966–1993	1984–1993	1956–1993	1956–1993	1984–1993
Skill: Correlation	0.62	0.65	0.69	0.66	0.66	0.74
Skill: RMSE	0.95	0.97	0.83	0.89	0.89	0.50

waters there), whereas its atmospheric part is referred to as the Southern Oscillation. The most prominent natural hazards caused by ENSO are felt in all parts of the World (see Table 2); they include local hydroclimatic extremes and affect the regional ecology and economy.

The physical growth mechanism of ENSO is due to the positive atmospheric feedbacks on equatorial SST anomalies via the surface wind stress, cf. Bjerknes (1969). Still, its unstable quasi-periodic behavior prevents robust ENSO predictions, even at subannual lead times. Numerical modeling plays a prominent role in understanding ENSO variability and developing forecasts. There are two main paradigms in ENSO modeling. A *deterministically chaotic, nonlinear* paradigm tries to explain the complexities of ENSO dynamics by the nonlinear interplay of various internal mechanisms. A *stochastic, linear paradigm* approaches this problem via the action of fast *weather noise* on an essentially linear, slow system, composed mainly of the upper ocean near the Equator. Despite the existence and importance of comprehensive numerical models, much of our theoretical understanding of ENSO comes from relatively simple models. Initiated in the 1980s, the study of such conceptual models has significantly contributed to shedding new light on many aspects of ENSO, including its quasi-periodic behavior, onset of instabilities, phase locking, power spectrum, and interdecadal variability; some of the most interesting simple models involve delay effects and are summarized herein.

The easiest to forecast are the large-scale SST and sea level patterns in the Tropical Pacific; even here, forecast skill rarely extends beyond 6–12 months. Beyond this ocean basin, atmospheric teleconnections provide some skill in certain parts of the world where ENSO effects are statistically significant, especially during major warm or cold events. Enhanced probabilities for local and regional hazards can be inferred from the larger-scale atmospheric anomalies via downscaling, but such probabilistic forecasts are clearly less reliable than the large-scale patterns on which they are based. Due to the importance of the associated natural hazards, considerable effort is invested in improving these forecasts.

Bibliography

Allen, M. R., and Robertson, A. W., 1996. Distinguishing modulated oscillations from coloured noise in multivariate datasets. *Climate Dynamics*, **12**, 775.

Barnett, T. P., Latif, M., Graham, N., Flügel, M., Pazan, S., and White, W., 1993. ENSO and ENSO-related predictability. Part I: Prediction of equatorial Pacific sea surface temperature with a hybrid coupled ocean-atmosphere model. *Journal of Climate*, **6**, 1545–1566.

Barnston, A., van den Dool, H., Zebiak, S., et al., 1994. Long-lead seasonal forecasts – Where do we stand? *Bulletin of the American Meteorological Society*, **75**, 2097.

Barnston, A. G., and Ropelewski, C. F., 1992. Prediction of NESO episodes using canonical correlation analysis. *Journal of Climate*, **5**, 1316.

Battisti, D. S., 1988. The dynamics and thermodynamics of a warming event in a coupled tropical atmosphere/ocean model. *Journal of the Atmospheric Sciences*, **45**, 2889.

Battisti, D. S., and Hirst, A. C., 1989. Interannual variability in a tropical atmosphere-ocean model – influence of the basic state, ocean geometry and nonlinearity. *Journal of the Atmospheric Sciences*, **46**, 12, 1687.

Bjerknes, J., 1969. Atmospheric teleconnections from the equatorial Pacific. *Monthly Weather Review*, **97**, 163.

Boulanger, J. P., Menkes, C., and Lengaigne, M., 2004. Role of high- and low-frequency winds and wave reflection in the onset, growth and termination of the 1997-1998 El Nino. *Climate Dynamics*, **22**, 267.

Cane, M., and Zebiak, S. E., 1985. A theory for El Niño and the Southern Oscillation. *Science*, **228**, 1084.

Chang, P., Wang, B., Li, T., and Ji, L., 1994. Interactions between the seasonal cycle and the Southern Oscillation: frequency entrainment and chaos in intermediate coupled ocean-atmosphere model. *Geophysical Research Letters*, **21**, 2817.

Chen, F., and Ghil, M., 1996. Interdecadal variability in a hybrid coupled ocean-atmosphere model. *Journal of Physical Oceanography*, **26**, 1561.

Diaz, H. F., and Markgraf, V. (eds.), 1993. *El Niño: Historical and Paleoclimatic Aspects of the Southern Oscillation*. New York: Cambridge University Press.

Dijkstra, H. A., 2005. *Nonlinear Physical Oceanography: A Dynamical Systems Approach to the Large Scale Ocean Circulation and El Niño*, 2nd edn. New York: Springer.

Dijkstra, H. A., and Ghil, M., 2005. Low-frequency variability of the ocean circulation: a dynamical systems approach. *Reviews of Geophysics*, **43**, RG3002.

Ghil, M., 2002. Natural climate variability. In Munn, T. (ed.), *Encyclopedia of Global Environmental Change*. Chichester/New York: Wiley, Vol. 1, pp. 544–549.

Ghil, M., and Jiang, N., 1998. Recent forecast skill for the El Niño/Southern Oscillation. *Geophysical Research Letters*, **25**, 171.

Ghil, M., Allen, M. R., Dettinger, M. D., Ide, K., Kondrashov, D., Mann, M. E., Robertson, A. W., Saunders, A., Tian, Y., Varadi, F., and Yiou, P., 2002. Advanced spectral methods for climatic time series. *Reviews of Geophysics*, **40**, 1003.

Ghil, M., Chekroun, M. D., and Simonnet, E., 2008a. Climate dynamics and fluid mechanics: natural variability and related uncertainties. *Physica D*, **237**, 2111.

Ghil, M., and Robertson, A. W., 2000. Solving problems with GCMs: general circulation models and their role in the climate modeling hierarchy. In Randall, D. (ed.), *General Circulation Model Development: Past Present and Future*. San Diego: Academic, pp. 285–325.

Ghil, M., Zaliapin, I., and Coluzzi, B., 2008b. Boolean delay equations: a simple way of looking at complex systems. *Physica D*, **237**, 2967.

Glantz, M. H., Katz, R. W., and Nicholls, N. (eds.), 1991. *Teleconnections Linking Worldwide Climate Anomalies*. New York: Cambridge University Press.

IRI: The International Research Institute for Climate and Society, 2010. Resources on El Niño and La Niña, http://iri.columbia.edu/climate/ENSO/.

Ji, M., Behringer, D. W., and Leetmaa, A., 1998. An improved coupled model for ENSO prediction and implications for ocean initialization, Part II: the coupled model. *Monthly Weather Review*, **126**, 1022.

Jiang, N., Ghil, M., and Neelin, J. D., 1995a. Forecasts of equatorial Pacific SST anomalies by an autoregressive process using similar spectrum analysis. *Experimental Long-Lead Forecast Bulletin (ELLFB)*, **4**, 24. National Meteorological Center, NOAA, U.S. Department of Commerce.

Jiang, S., Jin, F.-F., and Ghil, M., 1995b. Multiple equilibria, periodic, and aperiodic solutions in a wind-driven, doublee-gyre, shallow-water model. *Journal of Physical Oceanography*, **25**, 764.

Jiang, N., Neelin, J. D., and Ghil, M., 1995c. Quasi-quadrennial and quasi-biennial variability in the equatorial Pacific. *Climate Dynamics*, **12**, 101.

Jin, F.-F., and Neelin, J. D., 1993. Modes of interannual tropical ocean-atmosphere interaction – a unified view. Part III: analytical results in fully-coupled cases. *Journal of the Atmospheric Sciences*, **50**, 3523.

Jin, F-f, Neelin, J. D., and Ghil, M., 1994. El Niño on the Devil's Staircase: Annual subharmonic steps to chaos. *Science*, **264**, 70.

Kadanoff, L. P., 1983. Roads to chaos. *Physics Today*, **12**, 46.

Landsea, C. W., 2000. El Niño/Southern Oscillation and the seasonal predictability of tropical cyclones. In Diaz, H. F., and Markgraf, V. (eds.), *El Niño and the Southern Oscillation: Multiscale Variability and Global and Regional Impacts*. Cambridge: Cambridge University Press, pp. 149–181.

Latif, M., Barnett, T. P., Flügel, M., Graham, N. E., Xu, J.-S., and Zebiak, S. E., 1994. A review of ENSO prediction studies. *Climate Dynamics*, **9**, 167.

Lengaigne, M., Guilyardi, E., Boulanger, J. P., et al., 2004. Triggering of El Nino by westerly wind events in a coupled general circulation model. *Climate Dynamics*, **23**, 601.

Madden, R. A., and Julian, P. R., 1994. Observations of the 40–50-day tropical oscillation – a review. *Monthly Weather Review*, **122**, 814.

Mechoso, C. R., Yu, J.-Y., and Arakawa, A., 2000. A coupled GCM pilgrimage: from climate catastrophe to ENSO simulations. In Randall, D. A. (ed.), *General Circulation Model Development: Past, Present and Future: Proceedings of a Symposium in Honor of Professor Akio Arakawa*. New York: Academic Press, p. 539.

McPhaden, M. J., Busalacchi, A. J., Cheney, R., Donguy, J. R., Gage, K. S., Halpern, D., Ji, M., Julian, P., Meyers, G., Mitchum, G. T., Niiler, P. P., Picaut, J., Reynolds, R. W., Smith, N., and Takeuchi, K., 1998. The tropical ocean-global atmosphere observing system: a decade of progress. *Journal of Geophysical Research*, **103**, 14169.

McWilliams, J. C., 1996. Modeling the oceanic general circulation. *Annual Review of Fluid Mechanics*, **28**, 215.

Miller, A. J., et al., 1994. The 1976-77 climate shift of the Pacific Ocean. *Oceanography*, **7**, 21.

Mitchell, J. M., Jr., 1976. An overview of climatic variability and its causal mechanisms. *Quaternary Research*, **6**, 481.

Munnich, M., Cane, M., and Zebiak, S. E., 1991. A study of self-excited oscillations of the tropical ocean-atmosphere system Part II: nonlinear cases. *Journal of the Atmospheric Sciences*, **48**, 1238.

Neelin, J. D., 1990. A hybrid coupled general circulation model for El Niño studies. *Journal of the Atmospheric Sciences*, **47**, 674.

Neelin, J. D., Latif, M., Allaart, M. A. F., Cane, M. A., Cubasch, U., Gates, W. L., Gent, P. R., Ghil, M., Gordon, C., Lau, N. C., Mechoso, C. R., Meehl, G. A., Oberhuber, J. M., Philander, S. G. H., Schopf, P. S., Sperber, K. R., Sterl, A., Tokioka, T., Tribbia, J., and Zebiak, S. E., 1992. Tropical air-sea interaction in general circulation models. *Climate Dynamics*, **7**, 73.

Neelin, J. D., Latif, M., and Jin, F.-F., 1994. Dynamics of coupled ocean-atmosphere models: the tropical problem. *Annual Review of Fluid Mechanics*, **26**, 617.

Neelin, J. D., Battisti, D. S., Hirst, A. C., Jin, F.-F., Wakata, Y., Yamagata, T., and Zebiak, S., 1998. ENSO theory. *Journal of Geophysical Research*, **103**, 14261.

New, M. G., and Jones, P. D., 2000. Representing twentieth-century space-time climate variability Part II: development of a 1901-96 mean monthly grid of terrestrial surface climate. *Journal of Climate*, **13**, 2217.

Pfeffer, R. L. (ed.), 1960. *Dynamics of Climate*. New York: Pergamon Press.

Philander, S. G. H., 1990. *El Niño, La Niña, and the Southern Oscillation*. San Diego: Academic.

Rasband, S. N., 1990. *Chaotic Dynamics of Nonlinear Systems*. New York: Wiley.

Rasmusson, E. M., Wang, X., and Ropelewski, C. F., 1990. The biennial component of ENSO variability. *Journal of Marine Systems*, **1**, 71.

Robertson, A. W., Ma, C.-C., Ghil, M., and Mechoso, R. C., 1995. Simulation of the Tropical-Pacific climate with a coupled ocean-atmosphere general circulation model. Part II: interannual variability. *Journal of Climate*, **8**, 1199.

Ropelewski, C. F., and Halpert, M. S., 1987. Global and regional scale precipitation patterns associated with the El Niño/Southern Oscillation. *Monthly Weather Review*, **115**, 1606.

Schneider, S. H., and Dickinson, R. E., 1974. Climate modeling. *Reviews of Geophysics and Space Physics*, **25**, 447.

Schopf, P. S., and Suarez, M. J., 1988. Vacillations in a coupled ocean-atmosphere model. *Journal of the Atmospheric Sciences*, **45**, 549.

Suarez, M. J., and Schopf, P. S., 1988. A delayed action oscillator for ENSO. *Journal of the Atmospheric Sciences*, **45**, 3283.

Trenberth, K. E., 1997. The definition of El Niño. *Bulletin of the American Meteorological Society*, **78**, 277.

Tziperman, E., Stone, L., Cane, M., and Jarosh, H., 1994. El Niño chaos: overlapping of resonances between the seasonal cycle and the Pacific ocean-atmosphere oscillator. *Science*, **264**, 272.

Van den Dool, H. M., 1994. Searching for analogues, how long must we wait? *Tellus*, **46A**, 314.

Wang, X., Stone, P. H., and Marotzke, J., 1999. Global thermohaline circulation, Part II: sensitivity with interactive atmospheric transports. *Journal of Climate*, **12**, 83.

Weng, W., and Neelin, J. D., 1998. On the role of ocean-atmosphere interaction in midlatitude interdecadal variability. *Geophysical Research Letters*, **25**, 170.

Zaliapin, I., and Ghil, M., 2010. A delay differential model of ENSO variability, Part 2: phase locking, multiple solutions, and dynamics of extrema. *Nonlinear Processes in Geophysics*, **17**, 123–135.

Zebiak, S. E., and Cane, M. A., 1987. A model for El Niño oscillation. *Monthly Weather Review*, **115**, 2262.

Cross-references

Climate Change
Complexity Theory
Cultural Heritage and Natural Hazards
Disaster
Drought
Global Change and Its Implications for Natural Disaster
Monitoring and Prediction of Natural Hazards
Natural Hazard in Developing Countries

EMERGENCY MANAGEMENT

Michael K. Lindell
Texas A&M University, College Station, TX, USA

Definition

Emergency management. The process by which communities identify the hazards to which they are exposed and the physical (casualties and damage) and social (psychological, demographic, economic, and political) impacts these hazards might inflict, as well as assess and develop their capabilities to mitigate, prepare for, respond to, and recover from these impacts.
Hazard mitigation. Preimpact activities that provide passive protection during disaster impact by eliminating the causes of a disaster, reducing the likelihood of its occurrence, or limiting the magnitude of its impacts if it does occur.
Disaster preparedness. Preimpact activities that provide the trained personnel plans and procedures, facilities, and equipment needed to support active response at the time of disaster impact.
Emergency response. Activities conducted between the detection of hazardous conditions and the stabilization of the situation following impact whose goal is to minimize physical and social impacts.
Disaster recovery. Activities that restore a community's buildings (residential, commercial, and industrial), physical infrastructure (water, waste disposal, electric power, fuel, telecommunication, and transportation), and social, economic, and political activities to at least the same level as they were before disaster impact.

Introduction

Losses from disasters, in the United States and the rest of the world, have been growing over the years and are likely to continue to grow (Berke, 1995; Bourque et al, 2006; Mileti, 1999; Noji, 1997). Losses can be measured in a variety of ways – with physical impacts (casualties and property damage) being the most common indexes. The 2004 Indian Ocean earthquake and tsunami is estimated to have killed more than 258,000 people, and the 2010 Haiti earthquake is estimated to have killed more than 230,000. Recent economic losses have also been staggering; Hurricane Katrina cost over US$100 billion, and disaster costs are rising exponentially (Mileti, 1999).

Communities can cope with the potential for disasters through emergency management, which is the process by which communities identify the hazards to which they are exposed and the physical (casualties and damage) and social (psychological, demographic, economic, and political) impacts these hazards might inflict, as well as assess and develop their capabilities to mitigate, prepare for, respond to, and recover from these impacts. Although emergency management is often conceived as the responsibility of government, households, neighborhoods, and private business organizations are taking an increasing interest in reducing their hazard vulnerability. As a consequence, community emergency management should comprise a network of organizations at all levels of government and throughout the broader social and economic sectors – including for-profit and nonprofit organizations.

Emergency management is necessary because people occupy physical environments that consist of natural (geophysical, meteorological, and hydrological) and technological (energy producing and materials handling) systems that pose a variety of risks to the people, property, and the natural environment. The term *hazard* refers to the potential for variations in natural and technological processes to produce extreme events having very negative consequences (Burton et al., 1993). Communities can adjust to hazards by modifying human behavior (including land use and building construction practices) or modifying environmental systems to enable people to live in greater safety (Lindell and Perry, 2004). Thus, an event that is extremely hazardous to one community might not be so hazardous to another. For example, the 2010 Chile earthquake (M_W 8.8) released substantially more energy than the 2010 Haiti earthquake (M_W 7.0), but the death toll in Chile (approximately 500 persons) was less than one percent of that in Haiti.

The term *emergency* is commonly used in two slightly different but closely related ways. One usage of the term refers to an event involving a minor consequences for part of a community – perhaps a few casualties and a limited amount of property damage. Thus, emergencies are events that are "routine" in the sense that they are frequently experienced, relatively well understood, and can be managed successfully with local resources – sometimes with a single local government agency using standardized response protocols and specialized equipment (Quarantelli, 1987). Nonetheless, it is important to understand that each emergency can present unusual elements. For example, there is no such thing as a "routine" house fire; the belief that each new fire will be like all the previous ones has a high probability of producing firefighter deaths and injuries (Brunacini, 2002).

Another usage of the term emergency refers to the imminence of an event rather than the severity of its consequences. In this context, an emergency is a situation in which there is a higher than normal probability of an extreme event occurring, a meaning that is more appropriately designated by the term *crisis*. For example, a hurricane approaching a coastal community creates an emergency because the probability of casualties and damage is much greater than it was when the hurricane was far offshore. The urgency of the situation requires increased attention and, at some point, action to minimize the impacts if the hurricane should strike. Unlike the previous usage of the term emergency, the event has not occurred, the consequences are not likely to be minor, and routine methods of response by a single agency are unlikely to be effective if the event does occur.

The term *disaster* is reserved for the actual occurrence of events that produce substantial casualties, damage, and social disruption. Unlike the uncertain time of impact associated with a hazard (whether or not the impact would exceed community resources), a disaster reflects the actuality of an event whose consequences exceed a community's resources. Unlike crises, the consequences have occurred; unlike routine emergencies, disasters involve severe consequences for the community. By extension, a *catastrophe* is an event that exceeds the resources of many local jurisdictions – in some cases crippling those jurisdictions' emergency response capacity and disrupting the continuity of other local government operations.

Characterizing emergency management activities

Emergency management is a local responsibility because few counties are small enough for national authorities to initiate an immediate disaster response. Indeed, local volunteers and emergent groups are often the first responders in a disaster (Fischer, 2008). Moreover, in some countries such as the United States, local jurisdictions establish land use planning and building construction policies that determine the extent of their hazard vulnerability. As will be discussed in more detail below, communities can manage their hazard vulnerability through one of four major types of emergency management strategies – hazard mitigation, disaster preparedness, emergency response, and disaster recovery. Hazard mitigation comprises preimpact actions that protect passively against casualties and damage at the time of hazard impact (as opposed to an active emergency response to reduce those casualties and damage). Disaster preparedness consists of preimpact actions that provide the human and material resources needed to support active responses at the time of hazard impact, whereas emergency response comprises the planned and improvised actions implemented at the time of disaster impact to limit physical (casualties and damage) and social (psychological, demographic, economic, and political) impacts. Disaster recovery is the emergency management function that seeks to reestablish normal social, economic, and political routines once an incident has been stabilized – that is, after the immediate threats to human safety and property resulting from the physical impacts of the primary and secondary (e.g., fires following earthquakes) hazard agents have been resolved.

Community hazard management strategies can be individually implemented by households and businesses or collectively implemented by government agencies acting on behalf of the entire community. The individual strategies, which only reduce the vulnerability of a single household or business, generally involve simple measures such as elevating structures above expected flood heights, developing emergency response plans, and purchasing emergency supplies and hazard insurance. The collective strategies are generally complex – and expensive – technological systems that protect entire communities. Thus, they mitigate hazards through community protection works such as dams and levees and prepare for hazard impacts through measures such as installing warning systems and expanding highways to facilitate rapid evacuation.

Collective hazard adjustments are relatively popular because they permit continued development of hazard-prone areas and yet do not impose any constraints on individual households or businesses. In addition, their cost is spread over the entire community and often buried in the overall budget. Indeed, the cost is often unknowingly subsidized by taxpayers in other communities. For this reason, these collective hazard adjustments are often called "technological fixes." By contrast, individual hazard adjustment strategies require changes in households' and businesses' land use practices and building construction practices. Such changes require one of three types of motivational tactics – incentives, sanctions, or risk communication. Incentives provide *extrinsic rewards* for compliance with community policies. That is, they offer positive inducements that add to the inherent positive consequences of a hazard adjustment or offset the inherent negative consequences of that hazard adjustment. Incentives are used to provide immediate extrinsic rewards when the inherent rewards are delayed or when people must incur a short-term cost to obtain a long-term benefit. For example, incentives are used to encourage people to buy flood insurance by subsidizing the premiums.

Sanctions provide *extrinsic punishments* for noncompliance with community policies. That is, they offer negative inducements that add to the inherent negative consequences of a hazard adjustment or offset the inherent positive consequences of that hazard adjustment. Sanctions are used to provide immediate extrinsic punishments when the inherent punishments are delayed or when people incur a short-term benefit that results in a long-term cost. For example, sanctions are used to prevent developers from building in hazard-prone areas or using unsafe construction materials and methods. The establishment of incentives and sanctions involves using the political process to adopt a policy, and the enforcement of incentives and sanctions requires an effective implementation program (Lindell et al., 2006). By contrast, risk communication seeks to change households' and businesses' practices for land use, building construction, and contents protection by pointing out the *intrinsic* consequences of their behavior. That is, risk communication explains specifically what are the personal risks associated with risk area occupancy and also the hazard adjustments that can be taken to reduce hazard vulnerability.

Principles of community emergency planning

Over the years, researchers have identified eight fundamental principles of community emergency planning that can be used to increase the level of community preparedness regardless of the amount of available funding (Lindell and Perry, 2007):

1. Anticipate both active and passive resistance to the planning process and develop strategies to manage these obstacles.
2. Identify and address all hazards to which the community is exposed.
3. Include all response organizations, seeking their participation, commitment, and clearly defined agreement.
4. Base preimpact planning on accurate assumptions about the threat, about typical human behavior in disasters, and about likely support from external sources such as state and federal agencies.
5. Identify the types of emergency response actions that are most likely to be appropriate, but encourage improvisation based on continuing emergency assessment.
6. Address the linkage of emergency response to disaster recovery and hazard mitigation.
7. Provide for training and evaluating the emergency response organization at all levels – individual, team, department, and community.
8. Recognize that disaster planning is a continuing process.

Emergency management functions

Emergency management involves six functions – community preparedness analysis, hazard mitigation, disaster preparedness, emergency response, disaster recovery, and evaluation of the emergency management system.

Community preparedness analysis

Community preparedness analysis involves hazard/vulnerability analysis, hazard operations analysis, population protection analysis, and incident management analysis. Hazard/vulnerability analysis identifies the natural and technological hazards to which the community is exposed, the locations that would be affected, and the amount of damage that could be expected from events of various intensities. It also assesses the community's structures (residential, commercial, and industrial buildings) and infrastructure systems (water, waste disposal, electric power, fuel, telecommunication, and transportation) in terms of their ability to withstand the events identified in the hazard analyses. Finally, it assesses the community's susceptibility to psychological, demographic, economic, and political impacts. Both the physical and social impacts are examined to determine the degree to which demographic segments and economic sectors differ in their susceptibility to physical and social impacts. This social vulnerability analysis (Wisner et al., 2004) represents an important extension of previous theories of hazard vulnerability (Burton et al., 1993). Whereas people's physical vulnerability refers to their susceptibility to biological changes (i.e., impacts on anatomical structures and physiological functioning), their social vulnerability refers to limitations in their physical assets (buildings, furnishings, vehicles) and psychological (knowledge, skills, and abilities), social (community integration), economic (financial savings), and political (public policy influence) resources. The central point of the social vulnerability perspective is that just as people's occupancy of hazard-prone areas and the physical vulnerability of the structures in which they live and work are not randomly distributed, neither is social vulnerability randomly distributed – either geographically or demographically. Thus, just as variations in structural vulnerability can increase or decrease the effect of hazard exposure on physical impacts (property damage and casualties), so too can variations in social vulnerability (Bolin, 2006; Enarson et al., 2006).

Hazard operations analysis and population protection analysis identify alternative responses to disaster demands and evaluate them in terms of their effectiveness in protecting persons and property and their resource requirements – the amount of time, effort, money, and organizational cooperation needed to adopt and implement them (Lindell et al., 2006). For some hazard agents, especially technological hazards, it is possible to intervene into the hazard generating process to prevent a disaster from occurring. Thus, hazard operations analysis seeks to identify the hazard source control, community protection works augmentation, building construction augmentation, and building contents protection actions that can protect property from destruction. In the case of hazardous materials releases, source control involves actions such as patching holes or replacing leaking valves in tank cars. Community protection works augmentation can be accomplished by adding sandbags to increase the height of levees or cutting fire breaks to isolate wildfires. Building construction augmentation can be achieved by strengthening building soft spots, as when storm shutters are installed to protect against high wind. Finally, building contents protection can be accomplished by moving furniture and appliances to higher floors when flooding is forecast.

Population protection analyses identify the actions that can reduce casualties, as well as the facilities, equipment, and training that is needed to implement these actions. Thus, population protection analyses address systems for hazard detection and notification, warning, evacuation traffic management and transportation support, search and rescue, and emergency medical transportation and treatment. Finally, incident management assessment determines whether households, businesses, government agencies, and nongovernmental organizations have the capacity (i.e., resources) and commitment (i.e., motivation) needed to implement the hazard operations and population protection actions.

Hazard mitigation

FEMA defines mitigation as "any action of a long-term, permanent nature that reduces the actual or potential risk of loss of life or property from a hazardous event" (Federal Emergency Management Agency, 1998, p. 9). This definition is rather ambiguous because it encompasses the development of forecast and warning systems, evacuation route systems, and other preimpact actions that are designed to develop a capability for active response to an imminent threat. Thus, Lindell and Perry (2000) contended that the defining characteristic of hazard mitigation was that it provides passive protection at the time of disaster impact, whereas emergency preparedness measures develop the capability to conduct an active response at the time of disaster impact. Thus, hazard mitigation should be defined as activities that provide passive protection during disaster impact by eliminating the causes of a disaster, reducing the likelihood of its occurrence, or limiting the magnitude of its impacts if it does occur.

Since 1995, FEMA has emphasized mitigation as the most effective and cost-efficient strategy for dealing with hazards. Indeed, a recent study by the Multihazard Mitigation Council (2005) concluded that investments in hazard mitigation return four dollars in losses averted for every dollar invested. The ways in which mitigation activities can reduce hazard losses can best be understood by recognizing that natural hazards arise from the interaction of natural event systems and human use systems (Burton et al., 1993). Thus, the potential human impact of an extreme natural event such as a flood, hurricane, or earthquake can be altered by modifying either the natural event system or the human use system or both. In the case of floods, for example, the natural event system can be modified by dams or levees that confine floodwater. The human use system can be modified by land use practices that limit development of the flood plain or building construction practices that flood-proof structures.

Attempts to mitigate natural hazards, or events over which there is little human control, involve controlling human activities in ways that minimize hazard exposure. Thus, land use practices restricting residential construction in floodplains are important mitigation measures against riverine floods. The Hazard Mitigation and Relocation and Assistance Act of 1993, for example, allows FEMA to purchase homes and businesses in floodplains and remove these structures from harm's way. Although moving entire communities involves considerable stress for all concerned, an intense and systematic management process – characterized especially by close coordination among federal, state, and local agencies – can produce successful protection of large numbers of citizens and break the repetitive cycle of "flood-rebuild-flood-rebuild" that is so costly to the nation's taxpayers (Perry and Lindell, 1997). Likewise, building code requirements are used to restrict construction to those designs that can better withstand the stresses of hurricane force winds or earthquake shocks. However, the adoption of land use and building construction policies is an intensely political process (Prater and Lindell, 2000; Stallings, 1995).

Although the amount of control that human societies can exercise over natural event systems is often limited, technological hazards are inherently susceptible to such controls. Chemical, biological, radiological/nuclear, and explosive/flammable materials can all be produced, stored, and transported in ways that minimize adverse effects to plant workers, local residents, and the public at large. However, this control can be lost, resulting in releases to the air or to surface or groundwater. It is possible to control the hazard agent by locating the system away from populated areas, designing it with diverse and redundant components, or by operating it with smaller quantities of hazardous materials, lower temperatures and pressures, safer operations and maintenance procedures, and more effective worker selection, training, and supervision. Alternatively, one can control the human use system by preventing residential and commercial development – especially schools and hospitals – near hazardous facilities and major hazardous materials transportation routes. As is the case with natural hazards, the choice of whether to mitigate technological hazards by controlling the hazard agent or the human use system depends upon political and economic decisions about the relative costs and benefits of these two types of control.

Disaster preparedness

Disaster preparedness activities are undertaken to protect human lives and property in response to threats that cannot be controlled by means of mitigation measures or from which only partial protection is achieved. Thus, preparedness activities are based upon the premise that disaster impact will occur and that plans, procedures, and response resources must be established in advance. These are designed not only to support a timely and effective emergency response to the threat of imminent impact but also to guide the process of disaster recovery. A jurisdiction's disaster preparedness program needs to be defined in terms of:

- Which agencies will participate in disaster preparedness and the process by which they will plan
- What emergency response and disaster recovery actions are feasible for that community
- How the emergency response and disaster recovery organizations will function and what resources they require
- How disaster preparedness will be established and maintained

Emergency managers can address the first of these issues – which agencies will participate and what will be the process for developing disaster preparedness – by promoting the development of a local emergency management committee. As will be discussed later, this requires identifying the emergency management stakeholders in the community and developing a collaborative structure within which they can work effectively. It also requires ensuring an adequate statutory basis for disaster preparedness and administrative support from senior elected and appointed officials. Moreover, they can address the second issue – what are the feasible response and recovery actions – by means of analyses conducted to guide the development of major plan functions. These include, for example, evacuation analyses to assess the population of the risk areas, the number of vehicles that will be taken in evacuation, when people will leave, and what is the capacity of the evacuation route system.

Emergency managers can address the third issue – how the response and recovery organizations will function – in the emergency operations plan (EOP) and the recovery operations plan (ROP), which might be combined in an emergency management plan (EMP). These documents define which agencies are responsible for each of the functions that must be performed in the emergency response and disaster recovery phases. Some of the generic emergency response functions include emergency assessment, hazard operations, population protection, and incident management (Lindell and Perry, 1992; Perry and Lindell, 2007). While developing the plans and procedures, emergency managers also need to identify the resources required to implement them. Such resources include facilities (e.g., mobile command posts and emergency operations centers – EOCs), trained personnel (e.g., police, fire, and EMS), equipment (e.g., detection systems such as river gages and chemical sensors, siren systems, pagers, emergency vehicles, and radios), materials and supplies (e.g., traffic barricades, chemical detection kits, and self-contained breathing apparatus), and information (e.g., chemical inventories in hazmat facilities, congregate care facility locations and capacities, and local equipment inventories).

Emergency managers can also address the fourth issue – how disaster preparedness will be established and maintained – in the EOP and ROP. Sections of these plans should define the methods and schedule for plan maintenance, training, drills, and exercises. Training should always be conducted for emergency responders in fire, police, and EMS. In addition, training is needed for personnel in special facilities such as hospitals, nursing homes, and schools.

Emergency response

Emergency response activities are conducted during the time period that begins with the detection of hazardous conditions and ends with the stabilization of the situation following impact. For some hazards, monitoring systems ensure authorities are promptly alerted to disaster onset either by means of systematic forecasts (e.g., hurricanes) or prompt detection (e.g., flash floods detected by stream gages), so there is adequate forewarning and sufficient time to activate the emergency response organization. For other hazards such as earthquakes, preimpact prediction is not available, but prompt assessment of the impact area is feasible within a matter of minutes to hours and can quickly direct emergency response resources to the most severely affected areas.

The actual performance of individuals and organizations in disasters can be characterized by four basic emergency response functions – emergency assessment, hazard operations, population protection, and incident management (Lindell and Perry, 1992). *Emergency assessment* comprises diagnoses of past and present conditions and prognoses of future conditions that guide the emergency response. *Hazard operations* refers to expedient hazard mitigation actions that emergency personnel take to limit the magnitude or duration of disaster impact (e.g., sandbagging a flooding river or patching a leaking railroad tank car). *Population protection* refers to actions – such as sheltering in-place, evacuation, and risk area access control – that protect people from hazard agents. *Incident management* consists of the activities by which the human and physical resources used to respond to an emergency are prioritized, mobilized, and directed to accomplish the goals of the emergency response organization. These emergency response functions provide a useful framework for summarizing and evaluating existing research on disaster preparedness and response. These functions are similar, but not identical, to the Incident Command System structure of *command, operations, planning, logistics* and *finance and administration*. Emergency assessment is performed within the planning section, hazard operations and population protection are performed within the operations section, and incident support is performed within the command, planning, logistics, and finance and administration sections.

Emergency response activities are usually accomplished through the efforts of diverse groups – some formally constituted, others volunteer – coordinated through an EOC. Usually, local emergency responders dominate the response period. These almost always include police, firefighters, and EMS personnel, and often include public works and transportation employees. Uncertainty and urgency – less prevalent in mitigation, preparedness, and recovery – are important features of the response period. In the world of disaster response, minutes of delay can cost lives and property, so speed is typically essential. However, speed of response must be balanced with good planning and intelligent assessment to avoid actions that are impulsive and possibly counterproductive. Finally, emergency response actions need to be coordinated with disaster recovery. That is, life and property are priorities, but response actions foreshadow recovery actions. For example, damage assessments are later used to support requests for presidential disaster declarations and debris removal might be concentrated on roadways that are essential for restoring infrastructure.

Disaster recovery

Disaster recovery activities begin after disaster impact has been stabilized and extends until the community's citizens have returned to their normal activities. The immediate objective of recovery activities is to reconstruct damages or destroyed buildings and restore the physical infrastructure – water, waste disposal, electric power, fuel (e.g., natural gas), telecommunication, and transportation – but the ultimate objective is to return the community's quality of life to at least the same level as it was before the disaster. Recovery has been defined in terms of short-range (relief and rehabilitation) measures versus long-range (reconstruction) measures. Relief and rehabilitation activities usually include clearance of debris and restoration of access to the impact area, reestablishment of economic (commercial and industrial) activities, restoration of essential government or community services, and provision of an interim system for caring for victims – especially housing, clothing, and food. Reconstruction activities tend to be dominated by the rebuilding of major structures – buildings, roads, bridges, dams, and such – and by efforts to revitalize the area's economic system. In some communities, leaders view the reconstruction phase as an opportunity to institute plans for change that existed before the disaster or to introduce mitigation measures into reconstruction that would reduce the community's preimpact hazard vulnerability.

One important finding from disaster research is that the speed, efficiency, and equity of community recovery depend significantly upon local government's ability to improvise effective recovery strategies (Rubin et al., 1985). That is, communities recover more quickly and effectively if they can identify and respond to the specific problems that arise from its unique circumstances. More recently, practitioners and researchers have begun to agree that community disaster recovery is even faster and more effective if it is based on a recovery plan that has been developed prior to disaster impact (Schwab et al., 1998; Wu and Lindell, 2004). The recovery plan needs to establish clear goals and an implementation strategy (Smith and Wenger, 2006), preferably one that does not reproduce the community's preimpact hazard vulnerability. Of course, disaster recovery requires a substantial amount of resources, most of which (particularly in reconstruction) are derived from extracommunity sources. In the United States, these sources include insurance proceeds as well as grants and loans from private organizations and state governments. However, most of the resources for recovery from major disasters come from the federal government.

Evaluation of the emergency management system

In the past, jurisdictions focused most of their preimpact disaster planning on their preparedness programs and evaluated their emergency management systems by having the staff of the local emergency management agency compare the emergency operations plan and emergency response resources (equipment, facilities, and personnel) to the requirements identified in the community preparedness analyses, although the audit was sometimes conducted by outside consultants from private industry or from state or federal agencies. In addition, they conducted performance evaluations by conducting drills, exercises, and critiques. More recently, the National Fire Protection Association developed Standard 1600 which, in turn, led to the development of the Emergency Management Accreditation Program (EMAP). In addition to providing more systematic methods for auditing the local emergency preparedness program, NFPA 1600 and EMAP serve as a basis for evaluating the entire emergency management program – including hazard mitigation and disaster recovery preparedness.

Performance evaluations
Drills are techniques by which an individual emergency responder is given an emergency response task and his or her performance is evaluated by a qualified observer. Exercises are similar in concept but are broader in scope – generally involving entire teams or multiple organizations. In either case, the observers assess responders' ability to perform the necessary tasks effectively. Oral and written performance critiques are integral components of drills and exercises because they help the participants to identify deficiencies in plans, procedures, training, equipment, and facilities. In turn, the deficiencies can serve as the basis for specific, measurable, and achievable objectives for revising these emergency response resources.

National fire protection association standard 1600
The National Fire Protection Association (NFPA) Standards Council established a disaster management committee in 1991 that developed standards for preparedness, response, and recovery from the entire range of disasters. The current version of NFPA 1600 (National Fire Protection Association, 2007) defines a set of criteria for all emergency management programs, including business continuity programs. The standard requires a public or private sector organization to have a documented emergency management program with an adequate administrative structure, an identified coordinator, an advisory committee, and procedures for evaluation. NFPA 1600 can be used in self-assessment and also by external evaluators. The program must address the elements identified in Table 1.

The emergency management accreditation program
The Emergency Management Accreditation Program (EMAP) is based on NFPA 1600 (Emergency Management Accreditation Program, 2004), but has language that is specifically appropriate for state and local emergency management agencies (EMAs). An EMA that submits an application for EMAP accreditation must conduct a self-assessment that includes a proof of compliance record for each EMAP standard. Following a review of the self-assessment, the EMAP Commission dispatches an assessor team to conduct an onsite assessment that examines the jurisdiction's written documents, interviews local personnel, and inspects facilities, equipment, materials, and supplies to verify their adequacy. If accredited, the

Emergency Management, Table 1 Emergency management program elements

Element	Title
1	General
2	Laws and authorities
3	Risk assessment
4	Incident prevention
5	Mitigation
6	Resource management and logistics
7	Mutual aid/assistance
8	Planning
9	Incident management
10	Communications and warning
11	Operational procedures
12	Facilities
13	Training
14	Exercises, evaluations, and corrective actions
15	Crisis communications and public information
16	Finance and administration

applicant is issued a certificate that is valid for five years, subject to continuing compliance with the EMAP Standard, continuing documentation of compliance, and filing an annual report with the EMAP Commission.

Emergency management policy development

All public policy is determined by stakeholders, which are people who have, *or think they have*, a personal interest in the outcome of a policy. This interest motivates them to attempt to influence the development of that policy. Community stakeholder groups can be divided into three different categories – social groups, economic groups, and political groups. In turn, each of these types of groups can be characterized by its horizontal and vertical linkages (Berke et al., 1993). Horizontal linkages are defined by the frequency and importance of contacts with other groups of the same type; vertical linkages consist of ties with larger groups at regional, national, or international levels.

Social groups comprise households and other groups such as neighborhoods, religious organizations, service organizations, environmental organizations, and other nongovernmental organizations (NGOs), nonprofit organizations (NPOs), and community-based organizations (CBOs). All of these groups vary widely in size, level of organizational complexity, and amount of resources available to them. They also vary based on the functions they perform in society and, thus, varying levels of interest in community emergency management activities. Nonetheless, all are potential partners in formulating emergency management policies as well as participating in hazard mitigation, disaster preparedness, emergency response, and disaster recovery activities.

The principal type of economic group, the business, is an important stakeholder because these organizations are responsible for most of the flow of goods and services in society. One especially important type of business that is a stakeholder in emergency management is the public utility provider, whether privately or publicly owned. This includes the providers of electricity, fuel, water, sewer services, solid waste management, and communications such as telephone, television, and Internet access. Another set of businesses – the news media – is especially important to the success of emergency management programs because their coverage of all phases of emergency management can be an important way to educate the public about hazards that might strike the community, not just to inform them of an imminent disaster.

Finally, there are various types of governmental stakeholders. The lowest levels of organization, the municipality (i.e., town or city) and the county, have varying levels of power because states differ in the powers that they grant to their political subdivisions. Much emergency management policy is set at the state level, and the federal government has historically been seen as a supporter to local and state efforts.

Social, economic, and political stakeholders are all involved in developing emergency management policy, which can be explained by the *policy process model*. This model, which is adapted from Anderson (1994), lists five stages through which policies move – agenda setting, policy formulation, policy adoption, policy implementation, and policy evaluation (see Table 2). Those who are concerned about reducing their communities' hazard vulnerability need to understand the policy process thoroughly so they can be effective advocates. In particular, they need to use the occurrence of a natural or technological disaster in their own or another jurisdiction as a *focusing event* to draw public attention to the need for local disaster planning and hazard mitigation (Birkland, 1997). The focus of public and official attention on a particular hazard for some period of time provides a *window of opportunity* for policy change (Kingdon, 1984).

Institutionalizing emergency management networks

The achievement of valued organizational outcomes such as high-quality plans, procedures, training, equipment, and facilities is substantially determined by two factors – the quality of the emergency planning process and the individual outcomes experienced by each LEMA or LEMC member (Lindell and Perry, 2007). Important outcomes for individual LEMA and LEMC members include job satisfaction, organizational commitment, organizational attachment behaviors, and organizational citizenship behaviors. Positive individual outcomes are important because the time people must contribute to the activities of LEMCs is often unpaid. Consequently, people are more likely to commit their time and energy when they perceive social and environmental problems within their community, they are committed to the success of these communities, and they expect the LEMC to be successful in solving these problems.

Emergency Management, Table 2 The policy process model

Policy terminology	Stage 1: Agenda setting	Stage 2: Policy formulation	Stage 3: Policy adoption	Stage 4: Policy implementation	Stage 5: Policy evaluation
Definition of policy stage	Establishing which problems will be considered by public officials	Developing pertinent and acceptable proposed courses of action for dealing with a public problem	Developing support for a specific proposal so that a policy can be legitimized or authorized	Applying the policy by using government's administrative machinery	Determining whether the policy was effective and what adjustments are needed to achieve desired outcomes
Typical objective	Getting the government to consider action on a problem	Generating alternative solutions to the problem	Getting the government to accept a particular solution to the problem	Applying the government's policy to the problem	Evaluating effectiveness and identifying improvements

Source: Lindell et al. (2006)

An effective planning process is characterized by the involvement of key personnel from diverse agencies in a participative and consensus-oriented process that acquires critical resources such as emergency personnel, facilities, and equipment. In addition, these organizations create positive organizational climates that have five major facets (Lindell et al., 2006, Chap. 3). The *leadership* facet involves being clear about what tasks are to be performed, as well as recognizing individual members' strengths and weaknesses and being supportive of their needs. The *team* facet is characterized by task (rather than purely social) orientation, coordination among members, a high level of inclusiveness that produces team cohesion, and a belief in excellence of their organization (team pride). An organization with a high quality *role* facet avoids member role ambiguity (uncertainty about what to do), role conflict (disagreement about what to do), and role overload (too much work to do). When the *job* facet has high quality, members have independence (personal autonomy) in the way they exercise a variety of significant skills (skill variety) to perform a "whole" piece of work that provides a meaningful contribution to the group product (task identity). Finally, a positive *reward* facet is characterized by members having opportunities to perform new and challenging tasks (member challenge), opportunities to work with other people (social contacts), and being told that other people appreciate their work (social recognition).

LEMAs and LEMCs also must make strategic choices. These strategies include a *resource building strategy*, which emphasizes acquisition of human, technical, and capital resources needed for effective agency performance, and an *emergency resource strategy*, defined by securing the participation of emergency-relevant organizations in emergency planning and response. An *elite representation strategy* involves placing members of a focal organization (in this case, the LEMA) in positions where they can interact with influential members of other emergency-relevant organizations; the *constituency strategy* consists of establishing a symbiotic relationship between two organizations, whereby both benefit from cooperation. The *co-optation strategy* consists of absorbing key personnel, especially those from other organizations, into the focal organization's formal structure as directors or advisors, whereas the *audience strategy* focuses on educating community organizations and the public at large about the importance of community emergency preparedness. Finally, there is the *organizational development strategy*, when local emergency managers actively try to increase the resource base of all local agencies, not just their own, in order to foster predisaster relationships among organizations that must respond to a disaster. Emergency managers can pursue this strategy by relying on committees and joint ventures to involve other community organizations and by having frequent contacts and formalized interagency agreements (e.g., memoranda of understanding), especially with other emergency-relevant agencies.

Finally, LEMAs and LEMCs can obtain the resources they need to support an effective planning process by increasing community support through an effective risk communication program and acquiring extracommunity resources by developing horizontal ties with neighboring jurisdictions and vertical ties with state and federal agencies. LEMAs and LEMCs can also enhance their staffing and organization by establishing defined roles for elected officials, clear internal hierarchy, good interpersonal relationships, commitment to planning as a continuing activity, member and citizen motivation for involvement, coordination among participating agencies, and public/private cooperation. Ultimately, though LEMAs and LEMCs do need money to accomplish objectives, there are many things they can do to increase their effectiveness that can be achieved at little or no cost (Lindell and Perry, 2007).

Summary

Emergency management is the process by which communities identify the hazards to which they are exposed and the potential impacts these hazards might inflict, as well as assess and develop their capabilities to respond to these impacts. In particular, communities implement hazard mitigation, disaster preparedness, emergency response, and disaster recovery actions to manage their hazard vulnerability. Communities can evaluate their emergency management systems by conducting performance evaluations as well as standardized audits using the National Fire Protection Association's Standard 1600. Finally, communities can maintain their emergency management agencies through an effective planning process that involves the development of an effective organizational climate, the implementation of strategic choices, and acquisition of resources from inside and outside the jurisdiction.

Bibliography

Anderson, J. E., 1994. *Public Policymaking: An Introduction.* Boston, MA: Houghton Mifflin Company.

Berke, P. R., 1995. Natural-hazard reduction and sustainable development: a global assessment. *Journal of Planning Literature*, **9**, 370–382.

Berke, P. R., Kartez, J., and Wenger, D. E., 1993. Recovery after disaster: achieving sustainable development, mitigation and equity. *Disasters*, **17**, 93–109.

Birkland, T. A., 1997. *After Disaster: Agenda Setting, Public Policy and Focusing Events.* Washington, DC: Georgetown University Press.

Bolin, B., 2006. Race, Class, Ethnicity, and Disaster Vulnerability. In Rodríguez, H., Quarantelli, E. L., and Dynes, R. R. (eds.), *Handbook of Disaster Research.* New York: Springer, pp. 113–129.

Bourque, L. B., Siegel, J. M., Kano, M., and Wood, M. M., 2006. Morbidity and Mortality Associated With Disasters. In Rodríguez, H., Quarantelli, E. L., and Dynes, R. R. (eds.), *Handbook of Disaster Research.* New York: Springer, pp. 97–112.

Brunacini, A. V., 2002. *Fire Command: The Essentials of IMS.* Quincy, MA: National Fire Protection Association.

Burton, I., Kates, R., and White, G. F., 1993. *The Environment as Hazard*, 2nd edn. New York: Guildford Press.

Emergency Management Accreditation Program, 2004. *EMAP Standard.* Lexington, KY: Emergency Management Accreditation Program.

Enarson, E., Fothergill, A., and Peek, L., 2006. Gender and disaster: Foundations and directions. In Rodríguez, H., Quarantelli, E. L., and Dynes, R. R. (eds.), *Handbook of Disaster Research.* New York: Springer, pp. 130–146.

Federal Emergency Management Agency, 1998. *Introduction to Mitigation, IS-393.* Emmitsburg, MD: FEMA Emergency Management Institute.

Fischer, H. W., III, 2008. *Response to Disaster: Fact Versus Fiction and Its Perpetuation*, 3rd edn. Lanham, MD: University Press of America.

Kingdon, J. W., 1984. *Agendas, Alternatives and Public Policy.* Boston, MA: Little, Brown.

Lindell, M. K., and Perry, R. W., 1992. *Behavioral Foundations of Community Emergency Planning.* Washington, DC: Hemisphere.

Lindell, M. K., and Perry, R. W., 2000. Household adjustment to earthquake hazard: a review of research. *Environment and Behavior*, **32**, 590–630.

Lindell, M. K., and Perry, R. W., 2004. *Communicating Environmental Risk in Multiethnic Communities.* Thousand Oaks, CA: Sage.

Lindell, M. K., and Perry, R. W., 2007. Planning and Preparedness. In Tierney, K. J., and Waugh, W. F., Jr. (eds.), *Emergency Management: Principles and Practice for Local Government*, 2nd edn. Washington, DC: International City/County Management Association, pp. 113–141.

Lindell, M. K., Prater, C. S., and Perry, R. W., 2006. *Fundamentals of Emergency Management.* Emmitsburg, MD: Federal Emergency Management Agency Emergency Management Institute. Available from World Wide Web: www.training.fema.gov/EMIWeb/edu/fem.asp or www.archone.tamu.edu/hrrc/Publications/books/index.html.

Mileti, D. S., 1999. *Disasters by Design: A Reassessment of Natural Hazards in the United States.* Washington, DC: Joseph Henry Press.

Multihazard Mitigation Council, 2005. *Natural Hazard Mitigation Saves: An Independent Study to Assess the Future Savings from Mitigation Activities.* Washington, DC: Multihazard Mitigation Council.

National Fire Protection Association, 2007. *Standard on Disaster/Emergency Management and Business Continuity Programs.* Boston, MA: National Fire Protection Association.

Noji, E. K., 1997. The Nature of Disaster: General Characteristics and Public Health Effects. In Noji, E. K. (ed.), *The Public Health Consequences of Disasters.* New York: Oxford University Press, pp. 3–20.

Perry, R. W., and Lindell, M. K., 1997. Principles for Managing Community Relocation As a Hazard Mitigation Measure. *Journal of contingencies and crisis management*, **5**, 49–60.

Perry, R. W., and Lindell, M. K., 2007. *Emergency Planning.* Hoboken, NJ: John Wiley.

Prater, C. S., and Lindell, M. K., 2000. Politics of hazard mitigation. *Natural Hazards Review*, **1**, 73–82.

Quarantelli, E. L., 1987. What should we study? *International Journal of Mass Emergencies and Disasters*, **5**, 7–32.

Rubin, C.B., Saperstein, M.D., and Barbee, D.G., 1985. *Community Recovery from a Major Natural Disaster.* Monograph # 41. Boulder, CO: University of Colorado, Institute of Behavioral Science.

Schwab, J., Topping, K.C., Eadie, C.C., Deyle, R.E., and Smith, R.A., 1998. *Planning for Post-disaster Recovery and Reconstruction*, PAS Report 483/484. Chicago IL: American Planning Association.

Smith, G. P., and Wenger, D. E., 2006. Sustainable Disaster Recovery: Operationalizing an Existing Agenda. In Rodríguez, H., Quarantelli, E. L., and Dynes, R. R. (eds.), *Handbook of Disaster Research.* New York: Springer, pp. 234–257.

Stallings, R. A., 1995. *Promoting Risk: Constructing the Earthquake Threat.* New York: Aldine de Gruyter.

Wisner, B., Blaikie, P., Cannon, T., and Davis, I., 2004. *At Risk: Natural Hazards, People's Vulnerability and Disasters*, 2nd edn. London: Routledge.

Wu, J. Y., and Lindell, M. K., 2004. Housing reconstruction after two major earthquakes: the 1994 Northridge earthquake in the United States and the 1999 Chi-Chi earthquake in Taiwan. *Disasters*, **28**, 63–81.

Cross-references

Building Codes
Civil Protection and Crisis Management
Coastal Zone, Risk Management
Community Management of Hazards
Disaster Relief
Disaster Risk Reduction (DRR)
Emergency Planning
Integrated Emergency Management System
Land-Use Planning
Recovery and Reconstruction After Disaster
Risk Assessment
Worldwide Trends in Disasters Caused by Natural Hazards

EMERGENCY MAPPING

Frank Fiedrich[1], Sisi Zlatanova[2]
[1]Wuppertal University, Wuppertal, Germany
[2]Delft University of Technology, Delft, BX, The Netherlands

Synonyms
Crisis mapping

Definition
The term "Emergency mapping" refers to the creation and use of maps – paper or digital – before, during or after emergencies and disasters. While "hazard and risk mapping" is primarily used to visualize the hazards and risks during the pre-event phase, "emergency mapping" focuses on supporting response and relief efforts. Nevertheless, both types of maps are closely related to one another since hazard and risk maps can be included into emergency maps as important components. Currently "Geographic (al) Information Systems" (GIS) play a critical role in the development and use of these maps. GIS-based emergency maps are often an integral part of web-enabled crisis information management systems.

Introduction
Successful emergency management would not be possible without maps. Emergency maps visualize vital spatial information for planning and response through an easily understandable mean. One of the well-known early examples of emergency maps is the mapping of the cholera outbreak in London around 1850. During the cholera outbreak, Dr. John Snow plotted the observed cholera deaths on a hand-drawn map. He realized that many deaths occurred in the immediate vicinity of a specific water pump. By examining this pump it became obvious that it drew polluted water from the sewage system. Dr. Snow simply recommended to remove the handle. The Cholera outbreak stopped soon afterward (Snow, 1855). In the course of time, emergency maps became more sophisticated and with the advent of Geographical Information Systems in the 1980s the creation of these maps and the analysis of emergency management-related information became easier and more efficient. Today emergency managers rely heavily on map products created before and after a disaster. Recent experiences during the response to the World Trade Center attacks (2001), the Indian Ocean tsunami (2004), Hurricane Katrina (2005), and the recent Haiti earthquake (2010) show that emergency maps were used very successfully (Committee on Planning for Catastrophe, 2007).

Categories of emergency maps
Up to now, no overall accepted classification of emergency maps exists. One commonly used classification is based on the usage of the maps related to an incident. Whereas some emergency maps are used prior to an incident (pre-emergency maps) others are used during the response to and recovery from an event (post-emergency maps). Both categories are discussed below in more detail.

Pre-event emergency maps
Pre-event emergency maps are typically used for emergency planning. They are integral part of emergency plans and are either publicly available or attached to planning documents of response agencies or high-risk industrial facilities. The goal of pre-event maps is to improve speed and efficiency in case of an actual event as they provide guidelines on intended behavior or desired response activities. Pre-event emergency maps may exist for any type of possible incident, including evacuation of buildings and areas, large events, or possible emergencies due to natural and technological hazards. They should be updated frequently and should reflect the most current information about the potential hazards and risks.

Evacuation maps are often included in public information materials related to possible threats, like building fire or industrial accidents. In many countries, evacuation maps have to be highly visible in public use buildings, including hotels, shopping malls, or stadiums. The goal of these maps is to provide guidance to the best possible evacuation routes under the assumption that people are unfamiliar with the area and may be under stress.

Pre-emergency maps for large areas, like maps for accidents of nuclear power plants or snow emergency maps are mainly addressed to local residents. These maps have slightly different design requirements because people usually have a longer time available to familiarize themselves with the map. Different zones for different alert levels are frequently used in these large-scale maps (e.g., evacuation zones for storm surges).

Pre-event maps addressed to responders serve a different purpose and are dependent on the domain. These maps may include information about locations of possible response resources, detailed hazard and risk maps, and detailed estimates for possible event scenarios.

Post-event emergency maps
Although post-event emergency maps are sometimes created to guide and support the affected population, emergency managers are the key users of this type of maps. Post-event maps are used to support any kind of emergency management function. Among others, some of the most critical functions using emergency maps include:

- Damage and needs assessment
- Emergency logistics and resource tracking
- Mass care and shelter
- Search and rescue
- Fire fighting
- Health and medical care
- Evacuation
- Hazardous material response

- Forecasts (e.g., storm and plume modeling)
- Public safety and security support, including crowd control
- Critical infrastructure repair and recovery

While each function has its own relevant map data and design requirements, the common denominator is that the maps are used in time-critical decision environments. Therefore, these maps are created to answer one or more operational questions through visual representation of key information.

Symbology

As with any other map, symbols are essential components of emergency maps. Symbols are abstract graphical objects used for the representation of natural or artificial features. Although the use of self-developed sets of emergency symbols is still very common among response organizations, several standardization approaches try to create a set of standard cartographic symbols. The main goal of these initiatives is to facilitate information sharing between involved response agencies. One of these standards is the US Homeland Security Mapping Standard (ANSI/INCITS, 2006). It includes a set of common symbols for different types of incidents and natural events, infrastructures, and operational data. In Germany, the responding organizations use a standardized set of tactical symbols (SKK, 2003). In Netherlands, a set of symbols was designed for emergency response sector (Heide and Hullenaar, 2007). Some of these symbols, especially for resources, can be rather complex since they include information about organizational affiliation, resource type, unit size, direction, and time.

Geospatial data needs

The information needed for emergency response can be grouped into two large clusters, dynamic information (situational and operational) and static (existing) information. Data collected during a disaster are denoted as dynamic data, whereas the information existing prior the disaster is named static information. Static information provides also the basis for pre-event emergency maps. For both categories, the collected information can either apply to all hazards or it is specific to a single hazard or type of event. Typical information needs are, for example, published in (Board on Natural Disasters, 1999) or (US DHS et al., 2008).

Some examples of dynamic information are:

- Incident: location, nature, scale
- Effects/consequences: affected and threatened area, predictive modeling results
- Damages: damaged objects, damaged infrastructure
- Casualties: dead, injured, missing, and trapped people and animals
- Accessibility: building entrances, in- and out-routes, traffic direction, blocked roads

- Temporary centers: places for accommodating people (and animals), relief centers, morgues
- Meteorological information: wind direction, humidity, temperature
- Remote sensing imagery of the affected area
- Up-to-date data about involved response personnel and resources
- Hazard-specific information: for example, in case of flood – velocity and water depth, flood pattern

The most commonly used static (existing) information used for emergency maps includes

- Reference data: topographic maps, aerial photographs (orthophoto images), satellite images, cadastral maps and data
- Managerial and administrative data: census data, administrative borders, risk objects (gas stations, storage places of dangerous goods, etc.), vulnerable objects (schools, nursing homes, etc.)
- Infrastructure: road network, utility networks (gas, water, electricity), parking lots, dykes, etc.
- Building catalogs: high/low-rise, material, number of floors, usage (residential, industrial), presence of hazardous materials, owners, cables and pipes, etc.
- Accessibility maps: for buildings, industrial terrains, etc.
- Locations of preplanned resources
- Planned evacuation routes and shelters
- Water sources: fire hydrants, uncovered water, drilled water well, capacity, etc.
- Hazard-specific information: Hazard and risk maps, calculated event scenarios

Existing data is usually available from a variety of sources, including local authorities, national mapping agencies, cadastre, and private companies.

Emergency mapping and remote sensing

In emergency mapping, the use of sensors is very important for collecting dynamic data before and shortly after a disaster. The information derived from sensor products is critical for monitoring the natural hazard and ensuring a better situational and operational awareness.

Remote sensing refers to the entire suite of sensors that allow the collection of data from various platforms. Relevant sensors (optical, thermal, range, radar, acoustic, temperature, water level, humidity, etc.) can be deployed on the ground, in the air, or in space. Some sensors (water level gauges, seismic, air quality sensors, etc.) can be mounted on stationary platforms (near rivers, volcanoes, chemical and nuclear plants) while others (optical, thermal, acoustic, range, etc.) are often mounted on moving platforms (satellites, aircrafts, helicopters, unmanned aerial vehicles, cars, etc.). To be able to estimate which technology is appropriate for a specific disaster a number of technical, cost, and usability aspects have to be considered. Some of the most important technical aspects related to emergency maps are spatial resolution, spatial coverage, and deployment time (Kerle et al., 2008). Examples of usability factors include:

- *Availability of software for data processing*: For example, software packages for raster image processing are widely available for all major GIS systems whereas software for laser scanner data processing is still subject to extensive research.
- *Required expertise*: Some products such as images (satellite or airborne) and videos do not require a specific expertise and can easily be used. However, many products are either not human-readable or require processing to derive the needed information. Expert knowledge may be required for the interpretation of this data.
- *Required post-processing time*: Some products like image classifications, feature extraction, creating digital terrain models, and creation of damage assessment maps can require days or even weeks.
- *Sensor suitability for different emergencies*: Different sensors have strengths for different emergencies: radar sensors are appropriate for mapping flooded areas, damage detection, land subsidence; laser scanners are useful for mapping emergencies with height differences before and after an event; thermal and infrared images are appropriate for fire monitoring.

Sensors and their products can also be used simultaneously. The integration requires geo-referencing to one predefined coordinate reference system. For example, using frequent snapshots from several sensors contributed greatly to the emergency response following the World Trade Center collapse (Rodarmel et al., 2003). For comparative analyses before and after a disaster, sensor products are regularly overlaid with existing maps and imagery. Examples for large-scale disasters like the Indian Ocean tsunami and Hurricane Katrina are published, for example, by Brecht (2008) and Kevany (2008).

Interaction with emergency maps and visual analytics

Many classifications for interactions with digital maps exist, but most commonly they can be grouped in:

- Animation/video
- Interaction (navigate, zoom, manipulate)
- Query (explore)
- Feedback
- Change (edit)

Amongst all the interactions, emergency mapping benefits greatly from animation, query, and change/edit.

Animation is a dynamic visualization of a series of images. The images can be snapshots of an area (or a specific object of the area) with different time stamps or a walk-through/flyover of a given area. Whereas the first technique is used mostly for the simulation of hazardous events, the second is used for orientation, path finding, and navigation. Although visually "dynamic," the user can only observe, but not change the sequence of visualized materials. Video recording falls in the same group due to the same characteristics. Animations are widely used to represent expected flooded areas, plume spread, forest fires, or tsunamis. For example, Jern et al. (2010) describe how animations can be applied as final visualization technique in different stages of flood management.

Querying objects allows the user to obtain additional information about an object on the map or information about new maps. The additional information can be simple text, explaining a characteristic, or the query may execute an animation or voice recording. All major Geographic Information Systems allow for rather complex querying including selection by location or by attribute (e.g., buffer analysis) or spatial joins and relates.

Change (edit) is the highest level of interaction. It allows users to invoke changes in the shape of an object or its attributes. This is usually the most critical functionality for a successful collaboration during emergencies. The editing could either be temporal (to explore different options) or permanent (persistently recording the changes in the map). Many of the emergency response systems developed in the last years largely rely on such change/edit functionalities (e.g., Eagle, IBridge in the Netherlands).

Visual Analytics is yet another emerging technology, which is defined as the science of analytical reasoning facilitated by interactive visual interfaces (Andrienko and Andrienko, 2005; Thomas and Cook, 2005). Visual Analytics introduces a new level of indigence in the visualization, by finding specific patterns in a data set or after integration with other portions of information. Visual Analytics can be seen as fusion of visualization techniques with other areas such as data mining, databases, and spatial analysis. Advanced emergency response systems apply Visual Analytics to support the decision making process. Examples are published by Todin et al. (2004) and Jern et al. (2010).

Innovative systems for emergency mapping

In the past, many systems for emergency mapping of different disasters have been developed as specialized centralized systems (e.g., desktop systems) in which the data are available in a single repository (Amdabl, 2001; Greene, 2002). In centralized systems, data are constantly accessible; however, they might be easily outdated and due to proprietary data formats the integration of new data sets might be problematic. Related to centralized systems is the notion of scenario-based systems, in which complex models can be used to create realistic predictions and simulations.

In contrast, distributed systems rely on access and integration of data from different repositories. Since emergency mapping is highly dependent on the dynamics of the disaster, it is difficult to predict which information is actually needed in a specific situation. Generally, emergency mapping has to fulfill two premises: (1) ensure supply of sufficient data from the field and (2) discover, access, and fetch the most appropriate data from existing information sources. Consequently, emergency mapping can be also seen as an on-demand system. One of the main challenges of on-demand systems is the design and implementation of well-defined standardized services for discovery and exchange of existing information. Such services are closely related to the development of a Spatial Information Infrastructure (SII) for local, regional, national, and international levels. As of 2010 a number of SII-initiatives are in progress worldwide, including INSPIRE in Europe (www.ec-gis.org/inspire). Those initiatives are further enriched with specific services for the emergency management sector. Large international projects like ORCHESTRA (www.eu-orchestra.org), OASIS (www.oasis-fp6.org), and WIN (www.win-eu.org) have reported valuable results.

Most of the technology that is required for access and exchange of 2D spatial information is available as implementation standards (e.g., WFS, WMS, WCS, WPS, WCPS, OpenLS, SFS, SOS, and GML), or as concepts (e.g., OGC Abstract specifications for open distributed management of geographical imagery, GeoRSS). Many extensions of existing standards are proposed for further discussions and new ones have also been developed now for 3D (CityGML, Web3D Service). The third dimension is also considered with respect to indoor modeling and integration with Building Information Models (BIM) for evacuation and navigation (Lapierre and Cote, 2008; Lee, 2007).

Systems for emergency mapping have been traditionally developed by and for specialists involved in relief operations. However, experiences from recent disasters have clearly shown that information provided by local citizens and volunteers could be of great help especially in the first critical hours. For example, Ushahidi, Google Map, Open Street map, MS Bing Maps have been successfully used in Afghanistan, Pakistan, and Haiti to share logistical and recue information. Although there are many open issues (reliability, security, accuracy, etc.), such technologies should be further investigated and developed.

Conclusions

Emergency maps are an essential component of effective emergency management during planning, response, and recovery. Obtaining the best possible information about potential or ongoing emergencies is vital for emergency managers and the involved public. Since access to a disaster area is frequently limited the use of remote sensing technologies is often the primary way to receive initial information about the affected area, but the applicability of different sensor technologies and platforms often depends on the actual type event. Collected data can be integrated, manipulated, and analyzed via GIS and other related systems. Because of the complexity resulting from data diversity, sophisticated analysis tools, and due to the involvement of multiple stakeholders, new approaches for collaboration using emergency maps are currently being developed. Although the research in this field is still very young, recent disasters proof the value of the integration of emergency maps with distributed, collaborative systems.

Bibliography

Andrienko, G., and Andrienko, N., 2005. Visual exploration of the spatial distribution of temporal behaviors. In *Proceedings of the International Conference on Information Visualisation*. Los Alamitos: IEEE Computer Society, pp. 799–806.

Amdahl, G., 2001. *Disaster Response: GIS for Public Safety*. Redlands, CA: ESRI press. 108p.

American National Standards Institute, International Committee for Information Technology Standards (ANSI/INCITS), 2006. *Homeland Security Mapping Standard – Point Symbology for Emergency Management* ANSI/INCITS 415-2006.

Board on Natural Disasters, National Research Council, 1999. *Reducing Disaster Losses Through Better Information*. Washington, DC: National Academy Press.

Brecht, H., 2008. The application of geo-technologies after the hurricane Katrina. In Nayak, S., and Zlatanova, S. (eds.), *Remote Sensing and GIS Technologies for Monitoring and Prediction of Disasters*. Berlin: Springer, pp. 25–36.

Committee on Planning for Catastrophe, National Research Council, 2007. *Successful Response Starts with a Map: Improving Geospatial Support for Disaster Management*. Washington, DC: Committee on Planning for Catastrophe: A Blueprint for Improving Geospatial Data, Tools, and Infrastructure, The National Academy Press.

Dymon, J. D., and Winter, N. L., 1993. Evacuation mapping: the utility of guidelines. *Disasters*, 17(1), 12–24.

Greene, R. W., 2002. *Confronting Catastrophe: A GIS Handbook*. Redlands, CA: ESRI Press. 140 p.

Heide, J. van der, and van 't Hullenaar, B., 2007. *Simbolenset voor rampenbestrijding and grootschalig optreden*, Eindrapport RGI-210, 33 p. Avaible from World Wide Web: http://kennis.rgi.nl (in dutch)

Jern, M., Brezzi, M., and Lundblad, P., 2010. Geovisual analytics tools for communicating emergency and early warning. In Konecny, M., Zlatanova, S., and Bandrova, T. L. (eds.), *Geographic Information and Cartography for Risk and Crisis Management: Towards better solutions*. Berlin: Springer, pp. 379–394.

Kevany, M., 2008. Improving geospatial information in disaster management through action on lessons learned from major events. In Zlatanova, S., and Li, J. (eds.), *Geospatial Information Technology for Emergency Response*. London/Leiden: Taylor & Francis. ISPRS book series, pp. 3–19.

Kerle, N., Heuel, S., and Pfeifer, N., 2008. Real-time data collection and information integration using airborne sensors. In Zlatanova, S., and Li, J. (eds.), *Geospatial Information Technology for Emergency Response*. London, UK: Taylor & Francis, pp. 43–74.

Konecny, M., Zlatanova, S., and Bandrova, T. (eds.), 2010. *Geographic Information and Cartography for Risk and Crisis Management: Towards Better Solutions*. Heidelberg/Dordrecht/London/New York: Springer, p. 446.

Lapierre, A., and Cote, P., 2008. Using Open Web Services for urban data management: a testbed, resulting from an OGC initiative for offering standard CAD/GIS/BIM services. In Coors, V., Rumors, M., Fendel, E. M., and Zlatanova, S. (eds.), *Urban and Regional Data Management, UDMS Annual 2007*. London: Taylor & Francis, pp. 381–393.

Lee, J., 2007. A three-dimensional navigable data model to support emergency response in microspatial built-environments. *Annals of the Association of American Geographers*, **97**(3), 512–529.

Li, J., Zlatanova, S., and Fabbri, A. (eds.), 2007. *Geomatics Solutions for Disaster Management*. Berlin/Heidelberg: Springer. 444 p.

Nayak, S., and Zlatanova, S. (eds.), 2008. *Remote Sensing and GIS Technologies for Monitoring and Prediction of Disasters*. Berlin/Heidelberg: Springer. 271 p.

Oosterom, P., Zlatanova, S., and Fendel, E. M. (eds.), 2005. *Geo-information for Disaster Management*. Berlin/Heidelberg/New York: Springer. 1434 p.

Rodarmel, C., Scott, L., Simerlink, D., and Walker, J., 2003. Multisensor fusion over the World Trade Center disaster site. *Optical Engineering*, **41**(9), 2120–2128.

Snow, J., 1855. *On the Mode of Communication of Cholera*. London: John Churchill.

SKK (Ständige Konferenz für Katastrophenvorsorge und Katastrophenschutz), 2003. *Taktische Zeichen: Vorschlag einer Dienstvorschrift DV 102, SKK*, Cologne.

Thomas, J., and Cook, K., 2005. *Illuminating the Path: The Research and Development Agenda for Visual Analytics*. Available from World Wide Web: http://nvac.pnl.gov/.

Todin, E., Catelli, C., and Pani, G., 2004. FLOODSS, flood operational DSS. In Balabanis, P., Bronstert, A., Casale, R., and Samuels, P. (eds.), *Ribamod: River Basin Modelling, Management and Flood Mitigation*. Luxembourg: European Commission, Directorate-General Science, Research and Development, Environment and Climate Programme.

U.S. Department of Homeland Security (U.S. DHS), Federal Emergency Management Agency Region IX and Governor's Office of Emergency Services, 2008. *California Catastrophic Incident Base Plan: Concept of Operations*.

Zhang, X., Zhang, J., Kuenzer, C., Voigt, S., and Wagner, W., 2004. Capability evaluation of 3-5 micrometer and 8-12.5 micrometer airborne thermal data for underground coal fire detection. *International Journal of Remote Sensing*, **25**(12), 2245–2258.

Zhang, Y., and Kerle, N., 2008. Satellite remote sensing for near-real time data collection. In Zlatanova, S., and Li, J. (eds.), *Geospatial Information Technology for Emergency Response*. London: Taylor & Francis, pp. 75–102.

Zlatanova, S., and Li, J. (eds.), 2008. *Geospatial Information Technology for Emergency Response*. London: Taylor & Francis. 381 p.

Cross-references

Airphoto and Satellite Imagery
Disaster Relief
Emergency Management
Geographic Information Systems (GIS) and Natural Hazards
Geographical Information System
Hazard and Risk Mapping
Landsat Satellite
Remote Sensing of Natural Hazards and Disasters

EMERGENCY PLANNING

Scira Menoni
DIAP-Politecnico di Milano, Milan, Italy

Definition

Emergency or contingency planning is the activity aimed at preparing all concerned organizations to face a given crisis when an accident or a natural extreme occurs in a given area, provoking victims, damage, and various degrees of disruption of everyday life. According to Perry and Lindell (2003), emergency planning, training, and exercising are key aspects of emergency preparedness. The latter is to be intended as "the readiness" of social systems and governmental organizations to respond to environmental stress, minimizing negative consequence in terms of health and safety for people and avoiding the breakdown of fundamental community functions. The most important point raised by the two authors refers to the fact that emergency planning must be looked at more like a process rather than a product. Emergency plans, in fact, have to be updated often, according to changes in the environment to be protected, to the kind and severity of threats, to the amount and quality of available personnel, resources, and means. Contingency plans are closely related to those who developed it, they constitute a sort of predefined agreement on how certain problems and situations will be tackled by concerned actors rather than a consultation document. In this respect, updating does not only mean that data and information may/have to change but also that the same stakeholders who are in charge of emergency management should continuously check the efficacy of the plan or get acquainted to it if they have not participated to its development. In other words, whenever a new officer or a new disaster manager enters the teams in charge of contingency planning and management, he/she must learn not only the details of the plan but also the reasoning behind it.

In general terms, it can be said that contingency plans provide reference guidance to those who will be caught in an emergency and should work as a platform allowing for adjustments while making decisions in a turbulent environment. In this respect, emergency planning must start with a careful and deep analysis of crises, of weaknesses that past crises have manifested, so as to develop a tool that will be actually tailored to the specific context rather than being an act of formal compliance with some law. Therefore, understanding what crises exist, what their main features are and what specific character they may assume in a given area is crucial before any emergency planning is initiated (see *Civil Protection and Crisis Management*). The latter in fact should respond to problems and challenges posed by crises rather than constitute a predetermined list of functions or actions to be delivered but detached from the real potential development of a disaster.

While this is true in general, in the following some aspects of emergency planning specific to natural hazards will be treated.

Types of plans

As emergency planning is a process aimed at tackling crises in the best way possible, plans can be also grouped into categories depending on which types of crises they should be prepared for.

As for spatial and geographical aspects, the first aspect to be considered is whether in the area of concern one or more natural hazards exist. When the second option is confirmed, a further analysis must be carried out to understand whether existing hazards may or may not be linked in one single complex event (this would be the case of landslides triggered by earthquakes or lahars triggered by volcanic explosions, etc.). While independent events can be treated separately, even though the unfortunate co-occurrence of the latter can be a concern, it is certainly much more appropriate to be prepared for linked events, one triggered by the other. In this respect, in highly urbanized areas, na-techs (technological disasters initiated by a natural hazard, see Showalter and Fran Myers, 1992) must be always taken into consideration and prepared for in order to disentangle the dramatic sequence of a fire induced by damages to plants or lifelines or contamination of water as a consequence of flooding in areas where hazardous material are stored.

In general terms, multi-hazard plans provide a complete reference for rescuers and concerned authorities as well as permit economies of scale in identifying stakeholders, personnel, and means and resources to be deployed in case of need.

As far as the spatial dimension is concerned, the scale factor is crucial in contingency planning. In most countries in fact, there is some subsidiarity principle according to which emergencies must be dealt with by the governmental level closest to the extreme event. The intervention of higher levels, going as far as regional, national, or even international levels must be sought for when the extent of damage, disruption, and magnitude of the disaster goes far beyond local, provincial, regional, or national coping capacities. This implies that all levels, at least up to the national, must prepare contingency plans responding to the requirements and responsibilities that are applicable to the scale of concern. The concept of scale then becomes extremely important, as at each scale different problems, means, and opportunities must be considered, as well as different responsibilities. Nevertheless, because temporary multiple organizations will work in the crisis management, plans must be conceived in a way and shaped in a format best fit to provide guidance to all involved organizations and to constitute a common, shared, and agreed platform for cooperation.

It must also be remembered that many of the involved organizations, for example, the firemen or medical doctors organisations, are also split among various levels of government and are often organized in territorial districts and units. This means that there may even be a situation in which a good level of cooperation among agencies exists at the regional but not at the provincial level, at the local but not at the national scale, with all the consequent problems that such a situation may entail.

As for the time dimension, clearly the characteristic of the threat and the vulnerability of places imply varieties of options for detailed planning. An alert phase must be planned for those hazards that allow for prediction and forecasting, whereas it is useless when forecasting is not possible and for fast onset crises. In the meantime, it should be reminded that an emergency plan should cover the entire extent of a crisis, be it short or long, including the monitoring of crucial variables in the aftermath. Such monitoring may regard health issues (e.g., after a volcanic eruption with toxic gases), the quality of water in cases of contamination, and even psychological distress due to the event itself and to the traumatic consequences it may imply.

The likely duration of crises determines the type of decisions and resource deployment that must be set, which clearly varies in case of short or long crises, in case a large or small number of people are potentially affected.

Mode of development and structure of emergency plans

Following what has been said about emergency planning being a process rather than a product and the variety of organizations that are inevitably involved in its development and updating, the mode of development is a fundamental aspect to be carefully designed. As Lagadec (1993) correctly noted, coordination cannot be obtained by decree, it has to be constructed with the people who are in charge of crisis management. A minimum number of organizations, particularly those with a key role during emergencies must be involved in all stages of the plan development. Although this seems rather obvious and trivial, often it is not the case. Plans are prepared by a limited number of stakeholders and then disseminated, with the inevitable result that those who did not participate to its development will not use it and sometimes will even forget about its existence. In the meantime, it is also true that there are time constraints in the number of plans which each stakeholder can contribute to, particularly when the same people are mandated to represent an authority in different arenas. Countries have identified different ways to cope with such difficulty, some more efficient than others. What can be certainly said is that the problem of how to harmonize and integrate the various levels of emergency plans has still to be solved in a satisfactory way. In many instances and in many organizations, this is not even recognized as a relevant problem. The criticalities clearly increase when larger scales are concerned, reaching the maximal complexity in disasters in poor or developing countries where international assistance is required (see *Natural Hazards in Developing Countries*). Abstracting from individual occasions and controversies, there are growing concerns about the failure and partial failure of international aid and assistance during crises and in supporting reconstruction in poor countries. The lack of common plans and common understanding, protocols about what is really needed and how operations should be carried out results in conflicts between the military and nonmilitary bodies, between governmental agencies and NGOs, and between those who are willing to assist the population in the immediate aftermath of a disaster leaving reconstruction to local forces and those who are mandated of development and rehabilitation missions. The main problem of such international intervention has to be indicated in the complete lack of useful information about the vulnerability of places and the specific factors that make a community in a given context vulnerable (see *Vulnerability*). Needs are often guessed but not really identified, whereas tools for rapid assessment are also lacking and in any case not agreed upon before missions start.

The knowledge of the vulnerability of places, as well as their strengths and opportunities for coping and rebuilding are seldom accounted for in emergency plans, even in developed countries (see *Resilience*; *Coping Capacity*). Instead, it is essential, as plans should be tailored not only to respond to specific threats and their spatial and time features but also to the specific weaknesses and resilience factors of both the built environment and the population.

Structure and main ingredients of emergency plans

As for the structure of an emergency plan, there are several forms that it can take, some more efficient than others. Plans that rely too much on scientific reports about risk areas are in general poorly operational; plans that do not contain logistics information and maps (see *Emergency Mapping*) provide little guidance to those who will come from outside to provide help. In general terms, it can be suggested that emergency plans must be synthetic documents, organized in working sheets and forms, containing the crucial information mostly needed to handle a crisis:

- Who are the people in charge of what and what are the basic actions they are expected to take, particularly at the beginning, when confusion is sovereign?
- What are the means and resources that are available to face the problems created by the crisis, where can these be found and how can they be obtained?
- What are the main features of the involved areas, in terms of maps, logistics, any information that can be relevant for managing the crisis in a specific context?

It is a challenging task to identify the key components of an emergency plan in abstraction from a given hazard and a specific area. There are nevertheless some crucial parts that should be never neglected.

The first relates to the event scenarios to which the plan is aimed to respond. Event scenarios are images of what may happen when a given natural extreme hits a specific context. The latter is made of a spatial setting, a given built environment including the strategic facilities and infrastructures that will be used to tackle the crisis, of a community, comprised by individuals, families, and social groups of different age, culture, origin, etc. Scenarios should depict how a given stress (provided by the occurrence of a natural hazard) is likely to affect constructions, roads, networks, people, etc. Further, the scenarios should consider how damage and losses may trigger secondary and induced consequences, given a certain level of systemic vulnerability, to be understood as level of interdependence and structure making a facility and/or a service able to keep functioning even in case some physical damage has occurred. Scenarios are essential in that they show how a natural phenomenon may transform itself into a severe challenge to a given environment, as a consequence of the specific and particular features of the latter. This means that in contrast to what is often done, emergency plans are unique to the area of interest, they respond to that particular community only, they cannot be transferred from one place to another without carefully evaluating the local conditions from a variety of points of view (geographical, natural, social, economic).

The development of scenarios is a critical stage in emergency plans development as they are an expression of the knowledge existing in a place regarding the combination of hazards, vulnerabilities, resilience of the natural, the built environment and of the settled population.

A second fundamental component of any emergency plan relates to the resources, both human and material, that can be deployed in case of need. The "needs" that may arise are partially indicated by scenarios, which are there also to provide hints about the necessities that would probably arise under given circumstances in a given place. A structured survey of available resources, including indication of proprietary usage conditions, is the part of the plan to which concerned agencies will contribute more enthusiastically. In fact, the development of an emergency plan may prove to be a unique occasion for identifying crucial requirements in terms, for example, of protection means for search and rescue personnel, tools designed to look for people under debris, and for monitoring given environmental conditions. This is also a part of the plan that must be constantly updated, not only because new means can be obtained by various organizations but also because new devices may be introduced in the market. The prior development of scenarios, with the complete description of the features of the area of interest is relevant also for selecting means that can be effectively used under specific environmental conditions. For example, heavy material like cranes are difficult to transport on an island; large trucks are not likely to make it in narrow mountain lanes, etc., in other words, resources must fit the environmental conditions of places for which the emergency plan is developed.

A third part, often the most problematic, refers to the stakeholders who must agree on actions and responsibilities in the emergency plan. The complete and exact description of tasks that will be accomplished, even though always cited as one of the crucial thing to do, proves to be unrealistic and unfeasible, particularly as multiple organizations, each with a definite hierarchy, structure, and mode of operation will have to meet and cooperate in the emergency situation. In part, the exact forces, agencies, and organizations that will be asked to intervene depend on the scenario itself as they cannot be completely imagined before the event strikes. Furthermore, political external concerns may override the plan and call for stronger interference of forces that were not necessarily involved at the beginning. Given as a general rule that the largest number possible of organizations likely to be involved should be involved in the plan development, what can be suggested is that the latter constitute a reference, a guidance for making coordinated decisions responding to the problems raised by the crisis and for renegotiating the use of resources in case of need. Actually this is not too far from what happens in the field when collaborative organizations work together during a contingency: they often share resources and expertise improvising new forms of mutual cooperation. As those are the situations generating the best outcomes, it can be suggested that the plan contains elements, information, and tools to facilitate such process of mutual exchange and aid that often takes place in real emergencies. Predefining on paper what each organization will do will probably work only for the most obvious operations (first aid, search and rescue, etc.) but will not help in ambiguous or unexpected situations.

Finally, the emergency plan, being a process, should contain indications for its own upgrading, like sections where lessons learned from real events, simulations, and exercises can be archived.

Summarizing, it can be held that emergency plans are a tool to keep collective memory alive (Middleton and Edwards, 1990), to store it and make it available to the largest number of stakeholders in charge of a crisis. Such a tool maintains its validity as long as concerned actors see it as a reference and a guidance of their activity, recognizing it as a repository of information and criteria to make decisions.

The importancce of communication in emergency planning

The Canadian Roundtable on Crisis Management (2003), in a rather interesting booklet of guidance for emergency managers indicates that contingency plans are made of two main parts: the operational and the communicational. The latter must be granted the same attention as the previous, as information, knowledge, and perceptions exchange is crucial for a successful exit from crises.

Within the emergency planning process, communication not only holds a central role but must be carefully designed according to the involved actors and stakeholders (see *Communicating Emergency Information*).

As for the latter, there are at least three main areas for which communication processes must be carefully designed and maintained: among the temporary multiorganizations that will take part in the contingency management operations, between the decision makers, the various organizations, and the public. In the case of a disaster, further distinctions should be drawn between the population directly affected and the more general public, as well as between scientists and decision makers, when the intervention of the latter is deemed essential.

As for the time dimension, communication during the event is clearly different from the information that must be guaranteed in "normal" time, so as to prepare people to react correctly and to apply the parts of the plan that are relevant to them in case of an extreme event.

The communication plan is therefore a significant chapter of the emergency plan itself, be it separated or nested into it. In any case, if people in different governmental positions, pertaining to various organizations and the same public are not fully aware about the content and the implications of the contingency plan, the latter is likely to fail, no matter how well conceived and developed it is.

Lagadec (1993) stresses the importance to have the media as part of the contingency management framework, because media are better equipped to transfer information to people, because in any case they will be there whenever and wherever a disaster occurs. As a suggestion, instead of trying to avoid contacts with the media, that will be inevitably there, one should allocate time, resources, and tools to accommodate the media within the entire process of emergency management, so as to be as transparent as possible in conveying news in a turbulent situation and perhaps even be able to exploit media's capability to dispatch information for the victims (like facilitating contacts among relatives and friends, disseminating information about health care, shelters, etc.).

Major challenges of emergency planning

Emergency planning is successful if involved parties and stakeholders take actually part in the process and if documents that are produced are known and used by the largest number possible of agencies and organizations. Clearly, in all those cases where plans are forgotten in the drawers or are not used as a reference guidance to prepare for and during emergencies, it means the process has failed in one or more steps and components. It can be held that the major challenge for emergency planning is therefore to generate a process of mutual respect, trust, and cooperation among the multiorganizations in charge of civil protection and produce documents that constitute a reference for operations (see *Civil Protection and Crisis Management*).

Future challenges imply the starting of processes of preparation and production of plans for the crises of the future. In this respect, increased probability of na-tech in urban and metropolitan environments are to be considered as more likely than in the past and therefore requiring specific intervention. The effects of climate change as a potential trigger of changes in other hydrometeorological hazards might be also accounted for (see *Global Change and Its Implications for Natural Disasters*). Multi-hazard emergency plans and stronger cooperation among different levels of government and among countries to deal with likely transboundary threats are part of the actions that can be suggested for the future. Certainly one major aspect is rethinking the linkages among different scales, including the global one, when global threats are at stake and whenever international intervention as part of humanitarian aid is sought in the aftermath of a calamity. Arrangements found until now proved often of limited efficacy: while intervention on the concrete action is always local, the implications, when regional, multisite events are implied, must be carefully envisaged, as well as when local events become international, because of political, economic, or even humanitarian reasons.

Summary

Emergency or contingency planning is the activity aimed at preparing all concerned organizations to face a given crisis when an accident or a natural extreme occurs in a given area, provoking victims, damage, and various degrees of disruption of everyday life.

Emergency planning is fundamentally a multidisciplinary activity, requiring the active participation and contribution of several experts and particularly of those stakeholders who will have the responsibility to implement the plan.

Emergency plans should constitute a valid reference for action and intervention in case of a natural disaster for all agencies and forces implied in emergency management and in search and rescue activities. It must respond to apparently two contradictory requirements: be the basis of standardization on the one hand (for all those tasks and procedures that have to be carried out more frequently and repeated in almost all contingencies) and create the condition for good improvisation and collaboration on the other. Such a balance is needed as emergency plans have to be used not only to tackle "usual" contingencies but also "exceptional" threats and conditions, typically associated with crisis management.

Therefore emergency planning should not focus only on the expected product (the contingency plan) but rather be viewed as a process aimed at finding solutions and creating the conditions for negotiating and renegotiating tasks and use of available resources in case of need, whenever surprises challenge the most standardized procedures.

Bibliography

Canadian Centre for Management Development Roundtable on Crisis Management (Canada), 2003. *Crisis and Emergency Management: A Guide for Managers of the Public Service of Canada*, Canadian Centre for Management Development.

Lagadec, P., 1993. *Preventing Chaos in a Crisis. Strategies for Prevention, Control and Damage Limitation*. Berkshire: McGraw Hill.

Middleton, D., and Edwards, D. (eds.), 1990. *Collective Remembering*. Newbury Park, CA: Sage.

Perry, W., and Lindell, M., 2003. Preparedness for emergency response: guidelines for the emergency planning process. *Disasters*, **27**, 336–350.

Showalter, P., and Fran Myers, M., 1992. *Natural Disasters as the Cause of Technological Emergencies: A Review of the Decade 1980–1989*. Boulder, CO: Natural Hazard Research and Applications Center, University of Colorado.

Cross-references

Civil Protection and Crisis Management
Communicating Emergency Information
Coping Capacity
Emergency Mapping
Global Change and Its Implications for Natural Disasters
Natural Hazards in Developing Countries
Resilience
Vulnerability

EMERGENCY SHELTER

Camillo Boano, William Hunter
University College London, London, UK

Synonyms

Emergency relief; Immediate relief phase

Definition

Emergency shelter: The initial, immediate phase of crisis recovery development; a type of physical structure used to accommodate people in the immediate aftermath of a crisis or disaster situation; accommodation for actual or potential disaster victims seeking quarters outside of their own permanent homes for short periods: hours in many cases, overnight at most (Quarantelli, 1995:45).

Transitional shelter/settlements: Accommodation resulting from conflict and natural disasters, ranging from emergency response to durable solutions (Corsellis and Vitale, 2005).

Introduction

In the wake of a disaster or crisis, whether man-made or the result of natural occurrence, or as often the case, a combination of these and other factors, shelter remains a sometimes elusive and consistently controversial and challenging element of sustainable recovery efforts. Compounding the challenge is the delineation of terminology (Quarantelli, 1995; Zetter, 1995; Zetter and Boano, 2008) and physical attributes assigned to autonomous and overlapping phase shifts within disaster relief planning and implementation.

If temporary accommodation fills the gap between immediate relief and later reconstructive stage, the initial emergency shelter phase of the recovery process is fundamental and requires strategic interdisciplinary collaboration and decision-making between diverse actors and expertise. Emergency shelter involves different scales regarding time, space, and resources.

Each disaster situation is unique, and thus, it begs for a unique set of appropriate actions rather than prescriptive standardized solutions. Determining a "best-fit" solution for eventual shelter means that emergency relief, rehabilitation, and development mechanisms, including preparedness planning and immediate disaster assessment, must be produced in a coordinated manner. Thus, emergency sheltering should be conceived as an act of providing suitable habitat that fosters security and protection rather than mere logistics and object distribution or at the scale, encampment (Zetter, 1995; Zetter and Boano, 2008; Babister and Kelman, 2002; Boano, 2009). Hence, the ever-changing nature of emergencies and the complex multi-faced shelter process render emergency shelter an intricate, debatable topic of discussion among professionals and institutions.

Emergency shelter and rights to shelter: toward adequate housing

The provision of adequate shelter designed to satisfactory and appropriate physical and cultural standards during a disaster triggered by natural hazards is obligatory for governments, agencies, and institutions in order to offer protection and relief and could be conceived as a basic human right concept. Although it is not directly grounded in international human rights law, humanitarian law, or in codes of conduct (see Zetter and Boano, 2008), it can be recalled indirectly in a rich body of laws specifically referring to an adequate standard of living and right to housing (UN, 1948).

The UN ECOSOC (1992) provides the most authoritative legal interpretation of the right to adequate housing, suggesting that shelter is not seen exclusively as a commodity. Rather it views that it is the right "to live somewhere in security, peace, and dignity" stressing different integral components of the right (para 8.) as "legal security of tenure; availability of services, materials, facilities, and infrastructure; affordability; habitability; accessibility; location; and cultural adequacy." Although emergency shelter by definition normally does not meet the criteria of "adequate housing," a number of minimum human requirements are still applicable in such context (UN, 1948, 1966).

Protection related to shelter, housing, and adequate living conditions is acknowledged through *The Guiding Principles on Internal Displacement* as the right to be protected if displaced from "home or place of habitual residence" (UN, 1998). Principle 7(1) expresses the need to explore alternative solutions to displacement in order to minimize its adverse effect, whereas 7(2) affirms the need for proper accommodation in satisfactory conditions of safety.

In order to protect an individual from the elements, shelter must be constructed in a fashion appropriate to the effects of its environment. Described as protection from human elements, preservation of dignity is a less-tangible issue. Dignity demands an understanding of how shelter can temper relationships between displaced individuals or between the migrant and host communities. The restoration of an individual's dignity in an emergency situation involves the construction and provision of a place where the person can enjoy privacy and safety. This often has as much to do with the layout and location of shelter units as the type of shelter unit itself.

For this reason a house can be a keystone for recovery, providing a physical support mechanism, where other resources cannot. Though different shelter types breed different cultural associations (Rapoport, 1969), if provided, *temporary shelter* is often considered best to reduce the risk of future disasters. However, this is difficult to implement. Unsuitable sites can lead to lost livelihoods, lost sense of community and social capital, cultural alienation, poverty, and people abandoning new sites and returning to their original community (Cernea, 1997).

Emergency shelter in discourse and practice

In the aftermath of disasters, international agencies, governments, and NGOs often emphasize the quick delivery of emergency shelters to show progress results and to alleviate the suffering and psychological stress caused by devastation and homelessness and faster transition to a "normal" situation (Davis, 1978; UN, 2008). Even though this objective seems correct initially, vast literature confirms continuous failures in addressing such transition (Zetter, 1995; Lizarralde et al., 2010). Building affordable housing is a complex process that, even in regular circumstances, consumes great amounts of time and resources, requiring complex logistics, administrative innovation, and careful management (an argument studied by Keivani and Werna, 2001). Thus, providing emergency shelter raises a variety of corollary issues such as availability, affordability, location and use of land, scale, technology, and participation – all themes that converge on the issue of displacement.

The processes that people go through after a disaster to stabilize their housing situation can be quite lengthy and convoluted. People affected by the same disaster are affected differently and respond differently. Some will begin repairs of their damaged houses in the first days after the disaster, whereas others will be displaced for a period of time, even finding their situation changing weekly, monthly, or yearly. Because of such diversity, it is important to understand the range of options people face without imposing artificial "phases."

Shelter literature is based on the assumption that three levels of solutions are normally employed (Quarantelli, 1995) – emergency sheltering, temporary housing, and permanent housing. In this logic, emergency sheltering corresponds to the immediate protection against natural elements for the first few days after the disaster. This type of sheltering often includes handing out tents, plastics, corrugated iron sheets, among other materials to affected families. Emergency sheltering is essentially provisional and the agencies that provide it do not usually build houses. Therefore, their intervention is limited in time. Once the emergency assistance phase has been completed, the problem of housing is still present but permanent solutions often seem far away.

Adding to confusion among the sector is considerable variation in the terms used to describe shelter in humanitarian crises and these vary from location to location:

transitional shelter "provides a habitable, covered living space...over the period between a disaster and achieving a durable shelter solution" as Corsellis and Vitale (2005) suggests; emergency shelter typically involves the supply of temporary shelter materials such as tents and plastic sheeting, as UNHCR (2006:7) argues; temporary shelter/ mass shelter/collective building may include the use of public buildings such as mosques, churches, and schools, empty binding (Holzman and Nezam, 2004), and specially built temporary living centers (Lambert and Pougin de la Maisonneuve 2007:12).

Following disasters, it is imperative to minimize the distance and duration of displacement, while keeping safety in mind. This allows people to better maintain their livelihoods and allows households to protect their land, property, and possessions. Displacement can continue long after risks have receded, due to (1) the inability of households to document their property rights, which may be a prerequisite to reconstruction; (2) inappropriate reconstruction strategies, such as one that ignores the variety of needs within the affected population; or (3) the lack of resources and capacities of government and agencies to assist the displaced population.

People displaced from their original location have different sheltering options as recognized internationally by UN (2008):

- *Host families*: The displaced are sheltered within the households of local families, or on land or in properties owned by them.
- *Urban self-settlement*: The displaced settle in an urban area, occupying available public or private property or land.
- *Rural self-settlement*: The displaced settle in a rural area, occupying available public or private property or land.
- *Collective centers*: The displaced shelter in collective centers, or mass shelters, often-transitory facilities housed in pre-existing structures.
- *Self-settled camps*: The displaced settle independently in camps, often without services and infrastructure.
- *Planned camps*: The displaced settle in purposely built sites, where services and infrastructure are offered by government or the humanitarian community.

During the 1970s and 1980s, temporary housing was often delivered by special contractors with high-tech industrialized methods (UNDRO, 1982). These solutions usually implied industrialization and standardization and resulted in repetition of a universal solution that rarely responded to the specifics of climate, topography, local customs, and local forms of living. Most recent practices include the construction of shack-type temporary units made of timber and/or corrugated iron sheets. Usually located in public or vacant land and built with perishable materials, this form of housing usually has primitive infrastructure and is made by organizations that are rather transitional solution providers and not permanent (regular) housing builders (Boano, 2009) but (Corsellis and Vitale, 2005).

Camps, just as they are manifested spatially between the open and the closed, exist somewhere between the complex and difficult categorization of what is temporary and what is permanent. Essentially a time-based concept, "camps are understood as having a limited, although sometimes indeterminate, duration" (Hailey, 2009: 4). Perhaps the most obvious camp in terms of emergency is the *refugee camp* – designated zones for displaced persons and more specifically as a mandated space of protection. The UNHCR objective is to administer camps that work between safety and assistance, though it is not always able, or required to assist refugees (Hailey, 2009). Stemming from debates in the 1990s regarding refugee camp management, *Transitional Settlement* declared that "camps are not intended to be sustainable settlements, but every effort should be made to create and support livelihood opportunities for displaced populations, to empower them by increasing their self-sufficiency, and to reduce demands upon the aid community" (Hailey, 2009: 325).

This seemingly contradictory assumption of definitive limitation versus expectant-sustained independency led to the concept of "neighborhood planning" (Hester, 1984; Jones, 1990) that should be adopted in the design and layout of camps and settlements to promote a sense of community and reinforcement of community-based protection (IASC, 2005:54; Goethert and Hamdi, 1989) while also preserving the privacy of the family unit (Corsellis and Vitale, 2005) in order to avoid the camp becoming a "zone of indistinction" (Hailey, 2009: 325). Specifically, the *Camp Management Toolkit* suggested to "start planning from the family unit" (NRC, 2008: 201) and then expand to whom they normally relate to and live near. While this may seem like the description of a more long-term permanent situation, the fact remains that in many emergency situations, the established camps existed and developed for much longer than expected (Zetter, 1995; Boano and Floris, 2005).

Fred Cuny, a pioneer in the development of holistic approaches to planning and organization of refugee camps, in *Refugee camps and Camp Planning: The State of The Art* (1977) proposed a system of organization and management based on the full participation of the refugees themselves (UN IASC, 2006:18). The first edition of the UNHCR emergency manual came out in 1982 and is mainly based on Cuny's work. Further steps to define standards, founded on a rights-based perspective were further developed and introduced by the Sphere Charter for Humanitarian Intervention (Sphere Project, 2004) where basic dimensional standards were suggested. More recently, The IASC Emergency Shelter Cluster has outlined a set of general principles that it claimed had assisted in widening the community of practitioners, strengthened coordination systems in the global and national level, and more importantly encouraged concerned agencies to target alternative and more appropriate methods for ensuring "integrated and robust humanitarian programming" (ISAC, 2008).

Currently Humanitarian community and practitioners are more inclined to work around the notion of transitional shelters as used to house affected households with habitable, covered living space and a secure, healthy living environment with privacy and dignity during the period between a natural disaster and the availability of a permanent shelter solution (Corsellis and Vitale, 2005; UN, 2008). This approach to sheltering provides incremental support from the moment recovery begins, and gives households mobility and autonomy.

Conclusion

Emergency shelter and settlement should provide protection and privacy, for "human dignity and to sustain family and community life as far as possible in difficult circumstances" (Sphere Project, 2004:208). Far from being merely a design exercise, or logistic distribution, the practice of such spatialities and the notion of *camp* render evident that the design of shelters and settlements, which are responsive to the wide range of needs and values which housing serves, is a complex task. Complexity exacerbated by the political economy contours in which "makeshift architecture and emergency urbanism" (Lewis, 2008) take place.

The central concern is that the conceptualization of shelter, notably around the principles of space and place, is a critical factor in addressing the widely documented shortcomings of shelter and settlement responses. As advocated elsewhere (Zetter and Boano, 2008; Boano, 2009) post-disaster shelter inventions should be constructed around (1) a coherent understanding of the space and place nexus which embodies a pluralistic perception – as physical commodity, as the expression of social processes, cultural relations, and as an economic resource; (2) a clearer articulation of the processes linking relief, rehabilitation, and development in the production of space and place; (3) addressing national- and international-level institutional constraints to achieving better integrated responses; and (4) recognition of rights-based approaches.

Bibliography

Babister, E., and Kelman, I., 2002. The emergency shelter process with application to case studies in Macedonia and Afghanistan. *Journal of Humanitarian Assistance*. Available at: http://sites.tufts.edu/jha/files/2011/04/a092.pdf.

Boano, C., 2006. From Terra Nullius to Aidland: different geographies in post-tsunami recovery. Paper presented at International Aid Ideologies and Policies in the Urban Sector. In *Seventh N-Aerus Annual Conference*, September 8–9, 2006, Darmstadt, Germany.

Boano, C., 2009. Housing anxiety, paradoxical spaces and multiple geographies of post tsunami housing intervention in Sri Lanka. *Disasters*, **34**(3), 762–785.

Boano, C., and Floris, F., 2005. *Citta Nude. Iconografia dei Campi Profughi*. Milano: Franco Angeli.

Cernea, M., 1997. The Risks and Reconstruction Model for Resettling Displaced Populations. *World Development*, **25**(10), 1569–87.

Comerio, M., 1998. *Disaster Hits Home: New Policy for Urban Housing Recovery*. Berkeley, CA: University of California Press.

Corsellis, T., and Vitale, A., 2005. *Transitional Settlement Displaced Populations*. Oxford: University of Cambridge Shelterproject/Oxfam.

Cuny, F. C., 1977. Refugee camps and camp planning: the state of the art. *Disasters*, **1**(2), 125–143.

Davis, I., 1978. *Shelter after Disaster*. Oxford: Oxford Polytechnic Press.

ECOSOC, 1992. UN Doc. E/1992/23-E/C.12/1991/4, Annex III. All General Comments and Recommendations. UN Doc. HRI/GEN/1.

European Commission, 1996. Linking relief, rehabilitation and development – communication from the commission of April 30, 1996. European Commission. Available at: http://europa.eu.int/comm/development/body/legislation/recueil/en/en16/en161.htm.

Geipel, R., 1991. *Long-Term Consequences of Disasters: The Reconstruction of Friuli, Italy in Its International Context, 1976–1988*. New York: Springer.

Goethert, R., and Hamdi, N., 1989. *Refugee Settlements, A Primer for Development*. Unpublished preliminary outline prepared for the United Nations High Commissioner for Refugees, Geneva.

Hailey, C., 2009. *Camps: A Guide to 21st-Century Space*. Cambridge, MA: The MIT Press.

Harrell-Bond, B., 1998. *Camps: Literature Review, Forced Migration Review*. Oxford: Refugee Studies Centre, Vol. 2, pp. 22–23.

Hester, R. T., 1984. *Planning Neighborhood Space with People*. New York: Van Nostrand Reinhold.

Holzman, S. B., and Nezam, T., 2004. *Living in Limbo. Conflict-Induced Displacement in Easter Europe and Central Asia*. Washington, DC: World Bank.

IASC, 2005. Cluster Working Group on Protection Progress Report, December 12, 2005. Geneva: Palais des Nations. Available at: http://ocha.unog.ch/ProCapOnline/docs/library/Report%20of%20IASC%20PWG_Dec%202005.doc.

Johnson, C., 2007. Strategic planning for post-disaster temporary housing. *Disasters*, **31**, 435–458.

Johnson, C., 2008. Strategies for the reuse of temporary housing. In Ruby, I., and Ruby, A. (eds.), *Urban Transformation*. Berlin: Ruby Press.

Jones, B., 1990. *Neighborhood Planning*. Chicago, IL: American Planning Association.

Keivani, R., and Werna, E., 2001. Refocussing the housing debate in developing countries from a pluralist perspective. *Habitat International*, **25**(2), 191–208.

Kennedy, J., et al., 2007. Post-tsunami transitional settlement and shelter: field experience from Aceh and Sri Lanka. *Humanitarian Exchange Magazine*, **37**, 28–31.

Lambert, B., and Pougin de la Maisonneuve, C., 2007. *UNHCR's response to the tsunami emergency in Indonesia and Sri Lanka, December 2004–November 2006*. An independent evaluation PDES/2007/01. Geneva: UNHCR.

Lewis, J., 2008. The exigent city. *The New York Times*, June 8, 2008.

Lizarralde, G., Johnson, C., and Davidson, C. (eds.), 2010. *Rebuilding after Disasters. From Emergency to Sustainability*. London: Taylor and Francis.

Norwegian Refugee Council (NRC), 2008. *The camp management project*. Edition May, 2008. Available at: http://www.humanitarianreform.org/humanitarianreform/Portals/1/cluster%20approach%20page/clusters%20pages/CCm/CampMgmtToolKit.pdf.

Quarantelli, E. L., 1995. Patterns of shelter and housing in US disasters. *Disaster Prevention and Management*, **4**(3), 43–53.

Rapoport, A., 1969. *House, Form and Culture*. Englewood Cliffs, NJ: Prentice-Hall.

Sphere Project, 2004. *Humanitarian Charter and Minimum Standards in Disaster Response.* Geneva: Sphere Project. Available at: http://www.sphereproject.org/.

UN, 1948. Universal Declaration of Human Rights, G.A. res. 217A (III), U.N. doc. A/810 at 71.

UN, 1966. International covenant on civil and political rights. Adopted and opened for signature, ratification and accession by General Assembly resolution 2200A (XXI) of December 16, 1966.

UN, 1998. *Guiding Principles on Internal Displacement.* Presented by the UN Secretary-General Francis M. Deng to the United Nations Commission on Human Rights, UN doc. E/CN.4/1998/53/Add.2.

UNDRO, 1982. *Shelter after Disaster: Guidelines for Assistance.* New York: United Nations.

UNHCR, 2006. *Operational Protection in Camps and Settlements. A Reference Guide of Good Practices in the Protection of Refugees and Other Persons of Concerns.* Geneva: UNHCR.

United Nations, Inter-Agency Standing Committee (UN IASC), 2006. *Operational Guidelines on Human Rights Protection in Situations of Natural Disasters, with Particular Reference to the Persons who are Internally Displaced (Guidelines on Human Rights and Natural Disasters).* Geneva: UN.

United Nations, Inter-Agency Standing Committee (UN IASC), 2008. *Shelter Project 2008.* Nairobi: UN-Habitat. Available at: http://www.unhabitat.org/pmss/getPage.asp?page=bookView&book=2683.

Zetter, R. W., 1995. *Shelter Provision and Settlement Policies for Refugees. A State of the Art Review.* Uppsala: Nordiska Afrikainstitutet. Studies on Emergencies and Disaster Relief, Vol. 2.

Zetter, R. W., and Boano, C., 2008. *Protection during and after Displacement: Basic Shelter and Adequate housing in Brookings-Bern Institute, Protecting Internally Displaced Persons: Manual for Law and Policymakers.* Washington, DC: Brookings-Bern Institute, p. 129. Available at: http://www.brookings.edu/papers/2008/1016_internal_displacement.aspx.

Cross-references

Disaster Relief
Emergency Planning
Land-Use Planning
Recovery and Reconstruction

EPICENTER

Valerio Comerci
ISPRA - Institute for Environmental Protection and Research, Roma, Italy

Definition

The epicenter is the point on the earth's surface which lies vertically above the *hypocenter*.

Overview

The epicenter was originally defined before the development of *seismographs*: it was commonly taken to be near the center of the meizoseismal area (the area within the *isoseismals* of higher intensity). Subsequently, instrumental records proved that in most cases the epicenter was at one side of the meizoseismal area (Richter, 1958, 17–18, 144). Therefore, macroseismic and instrumental epicenters seldom coincide.

Today, epicenters (Figure 1) are generally determined based on the travel time of seismic waves, such as *primary waves* (and sometimes *secondary waves*), from the *hypocenter* to the *seismographs*.

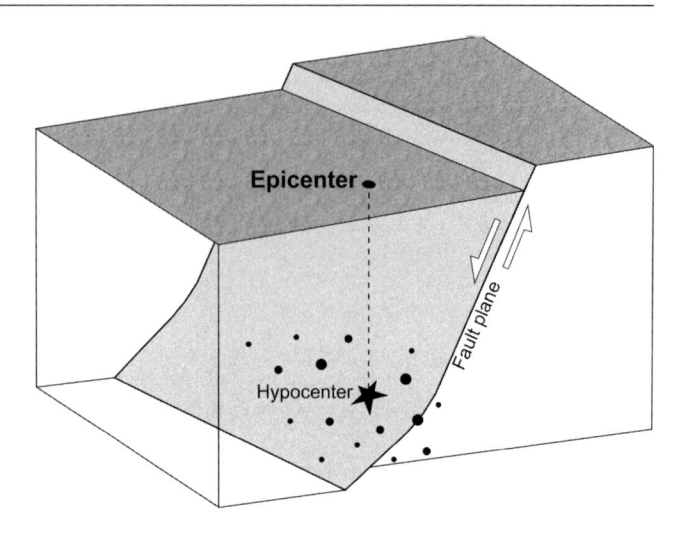

Epicenter, Figure 1 The epicenter is the point on the surface vertically above the *hypocenter*. The star represents the point where the rupture along the *fault* starts, while the dots represent smaller earthquakes generally occurring before and after the main shock (foreshocks and aftershocks).

Bibliography

Bolt, B. A., 2006. *Earthquakes: 2006 Centennial Update.* New York: W. H. Freeman and Company.

Richter, C. F., 1958. *Elementary Seismology.* San Francisco: W. H. Freeman and Company.

URLs

http://earthquake.usgs.gov/learn/glossary/?term=epicenter

Cross-references

Earthquake
Fault
Hypocenter
Intensity Scales
Isoseismal
Macroseismic Survey
Primary Wave
Secondary Wave
Seismograph/Seismometer
Seismology

EPIDEMIOLOGY OF DISEASE IN NATURAL DISASTERS

Gilbert M. Burnham
The Johns Hopkins Bloomberg School of Public Health, Center for Refugee and Disaster Response, Baltimore, MD, USA

Definition

The epidemiology of disease in natural disasters encompasses both *epidemic diseases*, defined as a group of illnesses of similar nature, clearly in excess of normal expectancy and derived from a common or propagated sources, and *endemic diseases* which are those usually or commonly present in a population (Gordis, 2004). Whereas epidemic diseases are commonly thought of as communicable diseases, epidemics of non-communicable diseases can occur with a disaster. Examples include increases in injuries or mental health conditions arising from a disaster, the common source in the Gordis definition. Endemic conditions can become epidemic following a disaster, directly related to the event, or indirectly, arising from the collapse of health services and other support services. Examples include communicable diseases such as respiratory infections, scabies, hepatitis, or sexually transmitted diseases.

Introduction

Epidemics of communicable diseases are less a feature of natural disasters than they are of Complex Humanitarian Emergencies (CHE), where populations are displaced, food insecure, and with limited access to health services (Watson et al., 2007). Outbreaks of communicable disease and natural disasters are often thought to go hand in hand. The news media continues to believe that epidemics of cholera and natural disasters are inextricably linked, even where cholera is not normally present. Although rare, conditions such as cholera, meningitis, or dysentery when they do occur after natural disasters can overwhelm health services and create widespread fear. More common infections are outbreaks of conjunctivitis among people temporarily housed in schools and public buildings or head lice among children temporarily displaced. There are many factors which contribute to the increased transmission of disease after natural disasters. Each of these needs to be considered in estimating potential risk to a disaster-affected population so that limited resources can focus on potential response to those events which are most probable. Although the nature of the disaster does affect the type and nature of disease transmission, the extent and duration of population displacement from the disaster is often a more powerful determinant of risk. Communicable disease outbreaks are more a feature of countries in epidemiological transitions where the major causes of illness and death are infectious agents commonly present before disaster.

In more developed countries, where populations live longer, complications of chronic and degenerative diseases may increase with disasters, as patients are cut off from diabetes and cardiovascular medications or cannot access dialysis for kidney failure. The importance of understanding the epidemiology of these conditions and how persons with these conditions are affected by lack of access to services is only recently being appreciated (Chan, 2009; Howe et al., 2008).

Factors contributing to disease after disasters
Nature and extent of the disaster

Not all disasters generate the same type or level of risk. Whereas earthquakes may create a great demand for injury care, and sometimes for complex orthopedic procedures, in themselves they cause little direct increase in disease transmission. The preoccupation with dead bodies as a potential source of epidemic disease continues in spite of multiple publications to the contrary (PAHO, 2005). Hasty mass burials carried out because of the fear of disease outbreaks can have negative psychological effect on those surviving (Morgan et al., 2006).

Flooding or other disasters affecting water and sanitation are disasters most likely to be associated with increased transmission of disease (Ahern et al., 2005). Wells or water sources can be directly contaminated since contamination can occur through disruption to water reticulation systems, as Gomez et al. (1987) noted in the 1985 Mexico City earthquake. Contamination with human fecal material is generally, but not always of more risk than animal fecal contamination, *Cryptosporidium* being an exception (MacKenzie et al., 1994).

More indirectly, flooding may increase the breeding sites for vectors of disease or bring populations and vectors into closer proximity. Outbreaks of malaria and dengue fever have occurred after flooding. Rodents dislocated by flooding may introduce leptospirosis into a human population. Those in suboptimal conditions post disaster may have increased exposure to disease vectors.

At the other extreme, drastic reductions in water available for personal hygiene can increase prevalence of skin conditions as well as transmission of fecal-oral infections.

Existing endemic disease patterns

The disease pattern post disaster is likely to represent alterations or variations from the patterns observed before disaster. In locations where cholera or meningococcal meningitis are endemic, such conditions might appear with the population displacement and collapse of health service. When new epidemic conditions are introduced into a community with a disaster, it is usually associated with population displacement (Jawaid, 2008). The pattern of care-seeking for chronic diseases in a community can be expected to change following a natural disaster (Guha-Sapir et al., 2008) and this may be a function of the priorities of both the community and those selected by emergency health services.

Health of the population

Where widespread acute malnutrition exists, illness and death from common conditions are likely to increase if the equilibrium between the status of the population and their environment is worsened. This is a common pattern seen where there is prolonged food insecurity, particularly among children. Populations debilitated by physical exhaustion may be more susceptible to disease. In countries with low immunization coverage, events associated with a disaster may increase the risk of a measles outbreak among young children. More than 18,000 cases of measles were diagnosed among the population displaced by the Mt Pinatubo eruption in 1991 (Surmieda et al., 1992). Such events can have a devastating impact, especially if there is preexisting malnutrition and vitamin A deficiency. Following a disaster, injuries may be extensive in the cleanup process. Poor immunization coverage led to an outbreak of tetanus following the tsunami in Indonesia (Jeremijenko et al., 2007).

Some fragile states have experienced a series of interspersed conflicts and natural disasters. This series of events can leave a population physically and psychologically debilitated and with increased susceptibility to disease even before the next event occurs.

Population characteristics

Populations with a high dependency ratio have increased disease risks. In some high income countries, this would be mostly elderly, and in high fertility countries, this would be mostly young children. Vulnerability is increased if the household head is physically or psychologically impaired or the household is headed by a single parent or a child. Populations with a high density may be at an increased risk of injury and disability both from the direct effects from the natural disaster as well as from communicable diseases. A population which is forced from their location of usual residence is likely to be at increased risk of illness and death (Wilder-Smith, 2005). For example, during the 1992 Somalia famine, households that were not displaced had a lower mortality than those who were forced to move (Moore et al., 1993). Households which sought shelter in camps had the highest mortality. The extent of crowding in temporary housing also contributes to the risk of communicable diseases such as measles, meningitis, and tuberculosis. The physical location where a population lives will also influence the epidemiology of injury and the illness patterns resulting from the disaster.

Response capacity

The epidemiology of diseases and conditions which follow a natural disaster is heavily influenced by the capacity of health services. At the same time that disasters increase health needs, they often dramatically decrease in the capacity of the health services to provide services. Both structural and nonstructural components of health services are at risk. In the 2003 Algeria earthquake, half of health facilities could not function (Lancet, 2008). Similarly, Indonesia lost 122 hospitals and health centers during the 2004 tsunami. Even when facilities are adequately protected or left intact by a disaster, the health workforce may be decimated or incapacitated because of their internal personal losses. Conditions which could be otherwise managed or contained by a functioning health system may worsen and potentially spread. Although most countries have national disaster plans, the presence of district, provincial, or facility disaster response plans is variable. Following the 2007 Ica earthquake in Perú, Chapin found those facilities which had developed emergency response plans were able to more effectively provide services than those without plans.

Where present, voluntary and auxiliary groups can provide community assistance which will help to contain outbreaks of disease and help to meet the needs of those with disabilities or chronic conditions (Kilby, 2008).

Prevention and control

Managing disease in disaster is based on the basic prevention concepts of public health which start with efforts to prevent disease and limit the spread or deterioration of disease once it has occurred.

Primary prevention is the prevention of disease through preventing exposure of a population to potential pathogens. Examples are ensuring water is safe, and sanitation maintained, and that immunization coverage is adequate to prevent infection. Prompt information to those at risk about preventing exposure or infection is another example. Primary prevention is also an approach to prevent exposure to hazardous substance released during disasters, and to prevent complications in persons requiring regular treatment of chronic diseases such as insulin-dependent diabetes. Many primary prevention actions are part of preparedness planning, but others can be done to avoid specific threats as a disaster occurs.

Secondary prevention is the prompt treatment or prompt action to prevent a disease having serious consequences. In the first instance, this is the control of outbreaks so that the spread of infection is limited. Once a disease or condition is established in a population, it also means good case management. This can include rescue and transport for persons with surgical injuries, as well as proper treatment of diseases such as malaria, meningitis, or tetanus. Establishing standard procedures or protocols is an early step in disease outbreak control as well as clinical management of disease. During disaster preparedness, creation of short just-in-time training courses materials can make implementation more rapid.

Although outbreaks of communicable diseases may occupy health services immediately post disaster, the demands for care of chronic and routine conditions soon become predominant. Understanding the epidemiology of chronic disease in communities, and prevention of deterioration in existing conditions is important to ensure health needs of all are being met. This is especially true for populations with large numbers of elderly populations, or where many people are receiving treatment for TB or HIV.

Tertiary prevention is the long-term treatment of the complications which arise from disasters. This is especially applicable after earthquakes where physical rehabilitation of those injured may extend for years.

Surveillance systems

An effective surveillance system is central to understanding the evolving epidemiology of disease in a community affected by a disaster. Surveillance provides the ongoing systematic collection of data which are then analyzed and interpreted to provide the basis for public health interventions and also used for evaluating the success of efforts (Connolly, 2005; WHO, 2005). Establishing a surveillance system is one of the immediate public health tasks in emergencies. Surveillance systems may develop as an outgrowth of initial rapid assessment surveys or may start with organizing reporting from functioning health facilities. This initial information helps identify the common conditions or potential epidemic diseases for a population. Establishing case definitions for potentially epidemic conditions is an early activity, if they do not exist already. Although information is important from all health units for a comprehensive understanding of disease epidemiology, using sentinel sites may be an interim step before full coverage of health services can be achieved. Initially the conditions tracked by a surveillance system may be limited to those of major public health importance, because of limited resources. Surveillance systems for cases of cholera, measles, or meningitis are examples of such priority conditions where these are likely threats. It is critical that information gathered through surveillance be quickly transferred into public health action. This part may break down if there is weak implementation capacity or a lack of coordination among agencies or groups assisting a population.

As services are restored, surveillance systems must mature to better monitor the health status of the affected population. This means including data from community sources, not just facility data, as much morbidity and mortality occurs outside the health system in many developing countries. In addition to tracking incidence of routine conditions, a well-functioning surveillance system can identify new or unexpected conditions. Identification of "tsunami lung" in Aceh is one example (Allworth, 2005). Surveillance systems can include verbal autopsy methods to establish cause-specific mortality, which is particularly important where many deaths occur outside the health system. As assistance programming has become increasingly rights based, the utilization of health services by age and sex is now carefully tracked almost everywhere, as measures of equal access and utilization.

Establishing accurate population denominators, stratified by age and sex, is critical to the estimation of the rates and ratios needed to measure effectiveness of programs for populations affected. This is often very difficult in the rapidly changing environment post disaster where substantial population shifts are common. Without these population estimates, programs may use resources inefficiently, leaving some vulnerable groups unassisted.

Summary

Natural disasters are not commonly associated with large outbreaks of epidemic diseases, with the possible exception of flooding. Once a population is displaced by a disaster, the risks of epidemic disease rise. Natural disasters may affect the epidemiology of endemic diseases or chronic diseases present in a community through reducing or changing the access a population has to health services. Preventing or limiting the public health impact of a disaster requires an effective disease surveillance system that is a critical early measure to establish in any emergency. Acting on surveillance data can not only prevent or limit epidemics from developing, but is also a means of evaluating the public health response to disasters.

Bibliography

Ahern, M., Kovats, R. S., Wilkinson, P., Few, R., and Matthies, F., 2005. Global health impact of floods: epidemiological evidence. *Epidemiological Reviews*, 27, 36–46.

Allworth, A. M., 2005. Tsunami lung: a necrotizing pneumonia in survivors of the Asian tsunami. *Medical Journal of Australia*, 182, 364.

Chan, E. Y., and Griffiths, S., 2009. Comparison of health needs of older people between affected rural and urban areas after the 2005 Kashmir, Pakistan earthquake. *Prehospital and Disaster Medicine*, 5, 365–371.

Chapin, E., Daniels, A., Elias, R., Aspilcueta, D., and Doocy, S., 2009. Impact of the 2007 Ica earthquake on health service provision in southern Peru. *Pre-hospital and Disaster Medicine*, 24, 4. http://pdm.medicine.wisc.edu. February 2, 2010.

Connolly, M. A. (ed.), 2005. *Communicable Disease Control in Emergencies, a Field Manual*. Geneva: WHO, pp. 90–106.

Gomez, V., Cerillo, P., Amor, S., Ortega, D., Amor, C., and Jimenez, A., 1987. Quality of drinking water in Mexico City in relation to the earthquakes of September 1985. *Salud Pública de México*, 29, 412–420.

Gordis, L., 2004. *Epidemiology*, 3rd edn. Philadelphia: Elsevier Saunders, p. 18.

Guha-Sapir, D., van Panjuis, W. G., and Lagoutte, J., 2008. Patterns of chronic and acute diseases after natural disasters-a study from the ICRC field hospital in Banda Aceh after the 2004 Indian Ocean tsunami. *Tropical Medicine & International Health*, 12, 1338–1341.

Howe, E., Victor, D., and Price, E. G., 2008. Chief complaints, diagnoses, and medications prescribed seven weeks post-katrina in New Orleans. *Prehospital and Disaster Medicine*, 23, 41–47.

Jawaid, A., and Zafar, A. M., 2001. Disease and dislocation, the impact of refugee movements on the geography of malaria in NWFP, Pakistan. *Social Science & Medicine*, 52, 1042–1055.

Jeremijenko, A., McLaws, M. L., and Kosasih, H., 2007. A tsunami related tetanus epidemic in Aceh, Indonesia. *Asia Pacific Journal of Public Health*, 19, Spec no. 40–44.

Kilby, P., 2008. The strength of networks: the local NGO response to the tsunami in India. *Disasters*, **32**, 120–130.

Lancet, 2008. Keeping hospitals safe from all types of disasters. *Lancet*, **371**, 448.

MacKenzie, W. R., Hoxie, N. J., Proctor, M. E., Gradus, S., Blair, K. A., Peterson, D. E., Kazmierczak, J. J., Addiss, D. G., Kim, R., Fox, K. R., Rose, J. B., and Davis, J. P., 1994. A massive outbreak in Milwaukee of cryptosporidium infection transmitted through the public water supply. *The New England Journal of Medicine*, **331**, 161–167.

Moore, P. S., Marfin, A. A., Quenemoen, L. E., Gessner, B. D., Ayub, Y. S., Miller, D. S., Sullivan, K. M., and Toole, M. J., 1993. Mortality rates in displaced and resident populations of central Somalia during 1992 famine. *Lancet*, **341**, 395–398.

Morgan, O. W., Sribanditmongkol, P., Perera, C., Sulasmi, Y., van Alphen, D., and Sondorp, E., 2006. Mass fatality management following the South Asian tsunami disaster: case studies in Thailand, Indonesia, and Sri Lanka. *PLoS Medicine*, **3**(6), e195.

PAHO, 2005. *Management of Dead Bodies in Disaster Situations*. Washington: Pan American Health Organization. PAHO Disaster Manuals and Guidelines on Disasters Series No. 5.

Surmieda, M. R., Lopez, J. M., Abad-Viola, G., Miranda, M. E., Abdllanosa, I. P., Sadang, R. A., et al., 1992. Surveillance in evacuation camps after the eruption of Mt Pinatubo, Philippines. *MMWR*, **41**, 963.

Watson, J. T., Gayer, M., and Connolly, M. A., 2007. Epidemics after natural disasters. *Emerging Infectious Diseases*, **13**, 1–5.

WHO, 2005. Epidemic-prone disease surveillance and response after the tsunami in Aceh province, Indonesia. *Weekly Epidemiological Record*, **80**(18), 160–164.

Wilder-Smith, A., 2005. Tsunami in South Asia: what is the risk of post-disaster infectious disease outbreaks? *Annals of the Academy of Medicine, Singapore*, **34**, 625–631.

Cross-references

Earthquakes
Floods
Hospitals
Water

EROSION

Matija Zorn, Blaž Komac
Anton Melik Geographical Institute, Ljubljana, Slovenia

Definition

Erosion is a geomorphic process that detaches and removes material (soil, rock debris, and associated organic matter) from its primary location by some natural erosive agents or through human or animal activity.

Overview

The term is derived from the Latin *erodere*, to gnaw away, and is often used for overall exogenic processes in contrast to endogenic processes that build up. In this wide meaning, it includes also transportation and deposition processes. In the most broad and common meaning, erosion includes all exogenic processes, in the absence of weathering (which causes the breakdown of rock material) and mass movements.

Natural agents of erosion include water, wind, glaciers, snow, sea/lake waves, and gravity (as a constant force on unstable slopes). Besides these, human activity can also be the cause for erosion processes. Erosion can be either mechanical or chemical (e.g., corrosion). Erosive agents transport material either by traction (rolling, sliding, pushing, jumping), suspension (moving material in turbulent flows either in running water or wind) or solution ("chemical" transportation in running water).

Erosion is a function of the erosivity of erosion agents and erodibility of the ground. Connected to these are various erosion factors (e.g., relief, bedrock geology, climate, vegetation, humans). For example, bedrock geology controls erosion as erodibility depends on it. Erodibility also depends on tectonic deformation of bedrock; the higher the deformation, the greater the erodibility.

Climate influences weathering and influences several erosive agents (e.g., rainfall, wind). It also influences vegetation that controls the erosivity of some erosive agents.

The principal types of erosion related to single erosion agents are water erosion, wind erosion, glacial erosion, snow erosion, sea/lake erosion, and anthropogenic/animal erosion. The most important erosion connected to several erosive agents is soil erosion.

Water erosion is connected to running water, ranging from raindrops that cause splash erosion, to sheet, rill, and gully erosion that occur because of surface runoff. The former prior to and the latter after surface runoff merges into trickles and begins to erode vertically to form erosion rills and eventually erosion gullies. In river channels, water erosion is referred to as river or stream erosion. When a river/stream erodes a riverbank, this is referred to as lateral or bank erosion, which is often accelerated by flooding.

Sea and lake erosion can also be considered as water erosion. It is a consequence of wave action affecting coasts.

Wind erosion is the consequence of wind action usually connected to climatic aridity and the absence of vegetation cover.

Snow erosion is usually connected to erosive effects of avalanches, whereas glacial erosion is connected to the erosion effects of glaciers, and it is considered the most powerful type of mechanical erosion.

Soil erosion is any removal of soil particles and regolith by natural agents that is often accelerated by the activity of humans (clear-cutting, overgrazing, road construction) and animals, and which is more intensive than soil formation.

Today, the most visible and economically serious is anthropogenic erosion. It is commonly expressed as accelerated soil erosion and connected to the destruction of natural vegetation thus exposing the bare soil or bedrock.

In common conditions, erosion is usually a low-magnitude, high-frequency process, e.g., water erosion may be reflected in a gradual change of hydrological response and hence a change in flood frequency. On the other hand, erosion may occur over the short term through catastrophic loss in specific situations.

Erosion hazard in a landscape is usually significant only in the long-term perspective and can be determined by social and economic losses.

Bibliography

Alcántra-Ayala, I., and Goudie, A. (eds.), 2010. *Geomorphological Hazards and Disaster Prevention*. Cambridge: Cambridge University Press.

Boardman, J., and Poesen, J. (eds.), 2006. *Soil Erosion in Europe*. Chichester: Wiley.

Fairbridge, R. W., 2008. Erosion. In Chesworth, W. (ed.), *Encyclopedia of Soil Science*. Dordrecht: Springer, pp. 216–221.

Flanagan, D. D., 2006. Erosion. In Lal, R. (ed.), *Encyclopedia of Soil Science*. New York: Taylor & Francis, pp. 523–526.

Hole, F. D., 1968. Erosion. In Fairbridge, R. W. (ed.), *Encyclopedia of Geomorphology*. New York: Reinhold, pp. 317–320.

Lupia-Palmieri, E., 2004. Erosion. In Goudie, A. S. (ed.), *Encyclopedia of Geomorphology*. London: Routledge, pp. 331–336.

Skinner, B. J., Porter, S. C., and Park, J., 2004. *Dynamic Earth – An Introduction to Physical Geology*. New York: Wiley.

Stallard, R. F., 2000. Erosion. In Hancock, P. L., and Skinner, B. J. (eds.), *The Oxford Companion to the Earth*. Oxford: Oxford University Press, pp. 314–318.

Summerfield, M. A., 1996. *Global geomorphology: An Introduction to the Study of Landforms*. Burnt Mill: Longman.

Zorn, M., 2008. *Erosion Processes in Slovene Istria*. Ljubljana: ZRC Publishing.

Cross-references

Avalanches
Coastal Erosion
Desertification
Erosivity
Flash Flood
Land Degradation
Mass Movement
Universal Soil Loss Equation (USLE)

EROSIVITY

Matija Zorn, Blaž Komac
Anton Melik Geographical Institute, Ljubljana, Slovenia

Definition

Erosivity is a measure of the potential ability of soil, regolith, or other weathered material to be eroded by rain, wind, or surface runoff.

Overview

Historically, the term erosivity was first associated with an *R*-factor (rainfall-runoff erosivity factor) in the Universal Soil Loss Equation (USLE). *R*-factor as used in the USLE and the revised USLE (RUSLE) relates to the mean annual sum of EI_{30} values. EI_{30} is the most commonly used rainfall erosivity index, where *E* is the total kinetic energy per unit area during a precipitation event (MJ·ha^{-1}) and I_{30} is its peak 30-min intensity (mm·ha^{-1}). Thus, erosivity of precipitation events is a function of their intensity and duration, and of the mass, diameter, and velocity of the raindrops. In principle, each detachment-transport system can be represented by an equation that has an erosivity term.

Wind erosivity has often been determined using indices based on wind velocities and durations above certain threshold velocities, precipitation and potential evaporation.

Erosivity is one of the influential factors of erosion hazard which is influenced by both, physical and social factors. Physical factors are represented by changing climate-influenced erosivity factors, such as intensive precipitation or strong winds in storm events. Highly erosive precipitation is a key climate variable that determines the type and magnitude of water erosion and some mass movements.

The effects of vegetation on erosivity depend on type and density of vegetation. It reduces the intensity of precipitation (or kinetic energy of raindrops) and its erosivity. Vegetation also influences runoff erosivity by restraining the action of running water and wind erosivity with obstruction of winds. Drought, dry soil surface, and sparse vegetation cover are necessary for strong winds to become highly erosive.

Erosivity is also influenced by climate patterns and land use changes that are controlled also by human activity, leading to increased hazard especially in agricultural areas.

A term connected to erosivity is erodibility as both are functions of erosion.

Bibliography

Bofu, Y., 2008. Erosion and Precipitation. In Trimble, S. W. (ed.), *Encyclopedia of Water Science*. Boca Raton: CRC Press, pp. 214–217.

Goudie, A. S., 2004. Erosivity. In Goudie, A. S. (ed.), *Encyclopedia of Geomorphology*. London: Routledge, p. 336.

Kinnell, P. I. A., 2006. Erosivity and erodibility. In Lal, R. (ed.), *Encyclopedia of Soil Science*. New York: Taylor & Francis, pp. 653–656.

Morgan, R. P. C., 1995. *Soil Erosion and Conservation*. Harlow: Longman.

Renard, K. G., Foster, G. R., Weesies, G. A., McCool, D. K., and Yoder, D. C., 1997. Predicting soil erosion by water: A guide to conservation planning with the revised universal soil loss equation (RUSLE). Agricultural Handbook No. 703. Washington DC: U.S. Department of Agriculture.

Wischmeier, W. H., and Smith, D. D., 1978. Predicting rainfall erosion losses–A guide to conservation planning. Agricultural Handbook No. 537. Washington DC: U.S. Department of Agriculture.

Cross-references

Avalanches
Coastal Erosion
Desertification
Erosion

Flash Flood
Land Degradation
Mass Movement
Universal Soil Loss Equation (USLE)

ERUPTION TYPES (VOLCANIC ERUPTIONS)

Catherine J. Hickson[1,2], T. C. Spurgeon[2], R. I. Tilling[2,3]
[1]Magma Energy Corp., Vancouver, BC, Canada
[2]Alterra Power Corp., Vancouver, BC, Canada
[3]Volcano Science Center, U.S. Geological Survey, Menlo Park, CA, USA

Synonyms

Magmatic eruptions; Volcanic explosions

Definition

Volcanic Eruptions. The expulsion of liquid rock (*magma*) – explosively or effusively – onto the earth's surface, either above or below water, through a vent. During a volcanic eruption, lava, tephra (ash, lapilli, rocks, pumice), and various gases are expelled. The following are the main eruption types:
Hawaiian low viscosity lava, high effusion rates, passive venting to fire-fountaining; low level to no plume
Strombolian moderate viscosity lava, high effusion rates, vigorous fire-fountaining; low level plume
Vulcanian moderate viscosity lava, moderate effusion rates, fire-fountaining to explosive ejection; low-mod level plume; early phases sometime phreatomagmatic
Peléan moderate viscosity lava, low-moderate effusion rates, explosive ejection; low-mod level plume
Plinian high viscosity lava, moderate effusion rates, explosive ejection; moderate-high level plume
Ultra-Plinian high viscosity lava, high effusion rates, explosive ejection; very high level plume
Surtseyan low viscosity lava, moderate-high effusion rates, explosive ejection with water interaction; moderate plume
Phreatoplinean low-high viscosity lava, moderate-high effusion rates, explosive ejection with water interaction; very high level plume

In addition, another eruption type, called *phreatic* (or *steam-blast*), involves explosive activity that only ejects nonmagmatic materials (preexisting volcanic or country rock) but no new magma. A variant involves magma-water interaction, *phreatomagmatic*, or *hydrovolcanic*, with its most violent variant being Phreatoplinean.

Volcanic eruptions range from passive to explosive, benign to catastrophic, and can have local to global impacts. The "style" of eruption is a fundamental characteristic that influences the eruption and its impact. The morphological form as well as the severity and impact of the hazards posed by a volcano are all related to the characteristics of the eruption and have been classified into eight main categories: *Hawaiian, Strombolian, Vulcanian, Peléan, Plinian, Ultra-Plinian, Surtseyan*, and *Phreatoplinian*. Some rarely used variants of categories include *Bandaian, Icelandic*, and *Katmaian*.

These names were for the most part derived from the respective volcanoes where characteristic eruptions had been observed. Early workers realized that there was consistency between certain types of eruptions, so they began to describe eruptions at other places in terms of those seen at the "type" locality. For example, Mount Pelée has lent its name to "Peléan" eruptions because eruptions there early in the twentieth century were well documented (c.f. Fisher and Heiken, 1982). Similar eruptions at other volcanoes could be described within the context of what was understood at Mount Pelée.

The Volcanic Explosivity index (VEI; Newhall and Self, 1982) provides a numerical and graphical, semiquantitative, logarithmic measure of *volcanic eruption's* magnitude or size, determined by the force of the eruption and volume of erupted (ejected) material (Figure 1). It is a methodology to better quantify eruptions rather than referring back to the "type locality." It seeks to understand the commonality of eruptions of similar size, style, etc. Accordingly, a rough correlation exists between eruptive style, type of eruptive vent, erupted volume, and the estimated VEI values (Figure 1). VEI 1–2 eruptions generally form only small cinder cones (*Hawaiian* to *Strombolian*), whereas with increasing eruptive energy the amount of erupted material increases. VEI 3 eruptions are typically *Strombolian* to *Vulcanian* through *Peléan*. VEI 4–5 are typical of major eruptions at stratovolcanoes (*Peléan* to *Plinian*), and VEIs > 5 (*Plinian* to *Ultra-Plinian*) often results in caldera forming eruptions (c.f. Miller and Wark, 2008; Sparks et al., 2005). Nonexplosive eruptions, regardless of size, are assigned VEI = 0. If sufficient data exist, VEI rankings can be estimated for eruptions during the Holocene (Simkin and Siebert, 1994).

Passive *volcanic eruptions* (*Hawaiian;* VEI 0–1) are dominated by the effusion of lava of a viscosity sufficiently low such that the fluid behaves as a liquid and flows under the influence of gravity. Flowage will stop when the lava crosses the liquid-solid boundary (crystallizes), or when gravity is insufficient to overcome obstacles along the flow surface or channels; effusive flows pond in topographic lows. Depending on viscosity, lavas may form thin flows less than a meter in thickness, or thick tabular bodies, sometimes referred to as flow domes. For more detail about the wide variety of effusive volcanic activity, see, for example, Tilling (2009) and papers in Sigurdsson et al., 2000: Part III.

In explosive *volcanic eruptions* (*Peléan, Plinian, Ultra-Plinian;* VEI > 3) magma has sufficient viscosity such that it moves only with great difficulty. Nonetheless, density contrasts with surrounding wall rock allow viscous magma to continue to ascend. At the same time, as pressure decreases progressively during ascent, fluids begin to exsolve from the rising magma, forming bubbles. With continued pressure decline the bubbles expand until they rupture, breaking their surrounding lava carapace.

VEI	Ejecta Volume	Eruption Type	Description	Plume	Frequency	Dispersal area (km²)	Fragmentation
0	< 10,000 m³	Hawaiian	non-explosive	< 100 m	constant	< 0.05 km; nil to very small	None to low
1	> 10,000 m³	Hawaiian/Strombolian	gentle	100-1000 m	daily	0.05 – 5; small	Low
2	> 1,000,000 m³	Strombolian/Vulcanian	explosive	1-5 km	weekly	0.05 – 5; small to moderate	Moderately low
3	> 10,000,000 m³	Vulcanian/Peléan	severe	3-15 km	yearly	0.05 – 5; moderate	Moderate
4	> 0.1 km³	Peléan/Plinian/Surtseyan	cataclysmic	10-25 km	≥ 10 yrs	5 – 500; moderate to large	High
5	> 1 km³	Plinian/Surtseyan	paroxysmal	> 25 km	≥ 50 yrs	500 – 5000; large	Very high
6	> 10 km³	Plinian/Ultra-Plinian/Phreatoplinian	colossal	> 25 km	≥ 100 yrs	5000; very large	Extreme
7	> 100 km³	Plinian/Ultra-Plinian/Phreatoplinian	super-colossal	> 25 km	≥ 1000 yrs	>5000; extensive (continental scale)	Extreme
8	> 1,000 km³	Ultra-Plinian/Phreatoplinian	mega-colossal	> 25 km	≥ 10,000 yrs	>5000; continental to global scale	Extreme

Eruption Types (Volcanic Eruptions), Figure 1 Adapted from Cas and Wright (1987); Newhall and Self (1982) and Walker (1973).

The expanding fluid, whose volume increases by orders of magnitude during the phase change from water to steam, is initially confined by the volcano's conduit and overburden pressure but then ultimately is released suddenly to produce an explosive discharge of fragmented lava (Cashman et al., 2000). Such discharges can be extremely vigorous and will propel eruption plumes of particles and gases many kilometers into the air. The variation in explosiveness and fragmentation is shown in Figure 1 (Cas and Wright, 1987; Newhall and Self, 1982; Walker, 1973).

The most explosive *eruption types* are those involving mixing of magma and water. If there is mixing of magma with subsurface water, the resulting phreatomagmatic (hydromagmatic) explosion can create an extremely destructive to catastrophic eruptions (*Surtseyan, Phreatoplinian; VEI 4>*; Figure 1). These types of eruptions create erosive pyroclastic surges (Moore et al., 1966a, b; Valentine and Fisher, 2000). The surges can sculpt the landscape by being highly erosive near source, stripping and scouring underlying soils and vegetation and infrastructure, yet leave relatively thin (or no) stratigraphic marker horizons in the geologic record. This lack of preserved "marker" horizons makes their frequency and intensity sometimes difficult to determine, impacting the validity of hazard assessments at volcanoes prone to these types of eruptions. More distally, pyroclastic surges are less destructive, depositing material, but they still have environmental consequences that takes years to decades before recovering.

Eruptions involving intermediate to higher silica composition magmas (e.g., andesitic dacitic, and rhyolitic) typically are moderately to highly explosive. However, basaltic eruptions, normally benign (*Hawaiian; VEI 0 – 1*), can on occasion become explosive and highly destructive (Fiske et al., 2009). Well-documented phreatomagmatic basaltic eruptions in Hawaii devastated an area of 420 km² with a highly explosive eruption between 2,800 and 2,100 ^{14}C years ago (Dzurizin et al., 1995). The violent basaltic subsea eruptions of Surtsey, Iceland, during 1963–1967, which formed a new volcanic island, lent their name to the

Surtseyan eruption type. Even stratovolcanoes subject to *plinian* eruptions can erupt more violently when water mixes with magma as in the May 18th, 1980, eruption of Mount St. Helens. This eruption proved to be far more explosive than anticipated. A growing lava dome inside the volcano ("cryptodome") failed with spectacular and devastating results. The hydrothermal system developed within the volcano over past eons, combined with water melted from the overlying glaciers by the cryptodome, created a phreatomagmatic eruption upon the virtually instantaneous depressurization due to gravitational failure of the volcanoes north flank (flank collapse). The blast from the mixing of the depressurizing cryptodome with ice and expanding steam from the hydrothermal system (Lipman and Mullineaux, 1981) destroyed more than 650 km^2 leaving a thin, characteristic pyroclastic surge deposit (Hickson et al., 1982; Moore and Albee, 1981). The thickness of the deposit is not indicative of the force of the eruption and is something that must be carefully analyzed when carrying out hazard assessments (MAP:GAC, 2007).

The rate and volume of expelled magma dictate the size of the *volcanic eruption*. If explosive, high discharge rates will create soaring eruption plumes (*Plinian*), the effects of which can impact areas of 100 s of square kilometers (Figure 1). High plumes (>30 km; *Ultra-Plinian*) from large explosive eruptions can affect global weather by injecting acidic aerosols into the atmosphere along with particles of ash (Mills, 2000). The airborne particles typically cause global cooling for several years following a very large eruption. Drifting eruption clouds containing abrasive ash particles also present a significant hazard to aviation (Casadevall, 1994; Miller and Casadevall, 2000). The biggest eruptions create subsurface void space that often collapses, forming a large crater or depression referred to as a caldera (Lipman, 2000).

If the magma is sufficiently fluid, high volumes of erupted lava – effusively or explosively erupted – can overwhelm rivers and snow fields, substantially altering the existing landforms by infilling topography and also by creating lahars or floods. The interaction of lava and rivers has been the focus of recent investigations (e.g., Grant and Cashman, 2007). Unstable dams of lava or other volcanic debris can fail catastrophically, leading to significant downstream hazards and morphological changes in the river channel (Hickson et al., 1999).

In addition to eruption rate and duration, conduit diameters and the rheology of the erupting material from a volcano also results in *volcanic eruptions* of widely differing scales.

Summary

Volcanic eruptions range from benign to highly destructive events with devastating consequence for people, flora, fauna, and infrastructure. The style of eruption is dependent on the viscosity, chemical composition, morphology, magma discharge rate, and a number of other factors.

Bibliography

Cas, R. A. F., and Wright, J. V., 1987. *Volcanic successions: modern and ancient: a geological approach to processes*. London: Allen and Union. 520 pp.

Casadevall, T. J. (Ed.), 1994. *Volcanic Ash and Aviation Safety: Proceedings of the First International Symposium on Volcanic Ash and Aviation Safety*. U.S. Geological Survey Bulletin 2047, Government of the United States of America, Washington, DC, 450 pp.

Cashman, K. V., Sturtevant, R., Papale, P., Navon, O., et al., 2000. Magmatic fragmentation. In Sigurdsson, H. (ed.), *Encyclopedia of Volcanoes*. San Diego: Academic, pp. 421–430.

Dzurizin, D., Lockwood, J. P., Casadevall, T. J., and Rubin, M., 1995. The Uwekahuna ash member of the Puna Basalt: product of violent phreatomagmatic eruptions at Kilauea volcano, Hawaii, between 2800 and 2100 ^{14}C years ago. *Journal of Volcanology and Geothermal Research*, **66**, 163–184.

Fisher, R. V., and Heiken, G., 1982. Mt. Pelée, Martinique: May 8 and 20, 1902, pyroclastic flows and surges. *Journal of Volcanology and Geothermal Resources*, **13**, 339–371.

Fiske, R. S., Rose, T. R., Swanson, D. A., Champion, D. E., and McGeehin, J. P., 2009. Kulanaokuaiki Tephra (ca. A.D. 400–1000): newly recognized evidence for highly explosive eruptions at Kīlauea Volcano, Hawai'i. *GSA Bulletin*, **121**-5(6), 712–728.

Francis, P., and Oppenheimer, C., 2004. *Volcanoes*, 2nd edn. Oxford: Oxford University Press. 521 pp.

Grant, G. E., and Cashman, K. V., 2007. Fire and water; interactions between lava flows and rivers during eruptive events. *Geological Society of America*, Abstracts with Programs – **39** (6), 180 pp.

Hickson, C. J., Hickson, P., and Barnes, W. C., 1982. Weighted vector analysis applied to surge deposits from the May 18, 1980 eruption of Mount St. Helens, Washington. *Canadian Journal of Earth Sciences*, **19**, 829–836.

Hickson, C. J., Russell, J. K., and Stasiuk, M. V., 1999. Volcanology of the 2350 B.P. eruption of Mount Meager Volcanic Complex, British Columbia, Canada: implications for hazards from eruptions in topographically complex terrain. *Bulletin of Volcanology*, **60**, 489–507.

Lipman, P. W., and Mullineaux, D. R., (Eds), 1981. *The 1980 Eruptions of Mount St. Helens, Washington*. USGS Professional Paper 1250, Government of the United States of America, Washington, DC, 844 pp.

Lipman, P. W., et al., 2000. Calderas. In Sigurdsson, H. (ed.), *Encyclopedia of Volcanoes*. San Diego: Academic, pp. 643–662.

MAP: GAC: 2007. Lineamientos para la preparación, representación y socialización de mapas de amenazas/peligros geológicos. In: *The Andean Experience in the Mitigation of Geological Risk/Experiencias Andinas en Mitigación De Riesgos Geológicos*. Proyecto Multinacional Andino: Geociencias para las Comunidades Andinas. Publicación Geológica Multinacional, No. 6, 2007, ISSN 0717-3733.

Miller, C. F., and Wark, D. A., 2008. Supervolcanoes and their explosive supereruptions. *Elements*, **4**, 11–15.

Miller, T. P., Casadevall, T. J., et al., 2000. Volcanic ash hazards to aviation. In Sigurdsson, H. (ed.), *Encyclopedia of Volcanoes*. San Diego: Academic, pp. 915–930.

Mills, M. J., et al., 2000. Volcanic aerosol and global atmospheric effects. In Sigurdsson, H. (ed.), *Encyclopedia of Volcanoes*. San Diego: Academic, pp. 931–943.

Moore, J. G., and Albee, W. C., 1981. Topographic and structural changes, March–July 1980; photogrammetric data. In Lipman, P.W., Mullineaux, D.R. (eds.), *The 1980 Eruptions Mount St. Helens, Washington*. USGS Professional Paper 1250, Government of the United States of America, Washington, DC, pp. 123–134.

Moore, J. G., Nakamura, K., and Alcaraz, A., 1966a. The September 28–30, 1965 eruption of Taal volcano, Philippines. *Bulletin of Volcanology*, **29–1**, 75–76.

Moore, J. G., Nakamura, K., and Alcaraz, A., 1966b. The 1965 eruption of Taal volcano. *Science*, **151**(3713), 955–960. New Series.

Newhall, C. G., and Self, S., 1982. The volcanic explosivity index (VEI): An estimate of explosive magnitude for historical volcanism. *Journal of Geophysical Research*, **87**, 1231–1238.

Sigurdsson, H., Houghton, B. F., McNutt, S. R., Rymer, H., and Stix, J. (eds.), 2000. *Encylopedia of Volcanoes*. San Diego: Academic. 1417 pp.

Simkin, T., and Siebert, L., 1994. *Volcanoes of the World: A Regional Directory, Gazetteer, and Chronology of Volcanism During the Last 10,000 Years*, 2nd edn. Washington, DC/Tucson, Arizona: Smithsonian Institution/Geoscience Press. 349 pp.

Sparks, S., Self, S., Grattan, J., Oppenheimer, C., Pyle, D., and Rymer, H., 2005. Supereruptions: global effects and future threats. In *Report of a Geological Society of London Working Group*. London: Geological Society of London, 28 pp.

Tilling, R. I., 2009. *Volcanoes*, Online Edition. United States Geological Survey, http://pubs.usgs.gov/gip/volc/,accessed22/11/2009.

Valentine, G. A., and Fisher, R. V., 2000. Pyroclastic surges and blasts. In Sigurdsson, H., et al. (eds.), *Encyclopedia of Volcanoes*. San Diego: Academic, pp. 571–580.

Walker, G. P. L., 1973. Explosive volcanic eruptions – a new classification scheme. *Geologische Rundschau*, **62**, 431–466.

Cross-references

Base Surge
Magma
Pyroclastic Flow
Pyroclastic Surge

EVACUATION

Graham A. Tobin[1], Burrell E. Montz[2], Linda M. Whiteford[1]
[1]University of South Florida, Tampa, FL, USA
[2]East Carolina University, Greenville, NC, USA

Synonyms

Emergency migration; Flight; Mass departures; Relocation

Definition

Evacuation: The temporary or permanent relocation of people from hazardous environments to minimize injuries and deaths from disasters.

Evacuation context

In 1999, approximately 25,000 people were evacuated from communities around the Tungurahua volcano in Ecuador. Fears of a major eruption had spurred authorities to declare an emergency and enforce a mandatory evacuation. People relocated to homes of families and friends and to apartments in nearby towns, or they were sent to temporary shelters in old schools. While some residents returned to their homes after 2 months, others remained in the shelters for up to a year waiting for the volcano to stop erupting (Whiteford and Tobin, 2004). In 2005, as Hurricane Katrina approached land along the Gulf coast of the USA, warnings were issued for residents to move inland away from storm surge areas (see *Hurricane Katrina*). Thousands of people left, with some moving to homes of family and friends, and others to hotels. Still others evacuated to *Emergency Shelter* such as the New Orleans sports' arena. However, many individuals stayed in their homes, whereas others were unable to evacuate. Some of those who evacuated are scattered in cities across the USA and will never return (Brinkley, 2006). Another example, the 2009 bushfires in Victoria, Australia tested the national policy of "prepare, stay and defend, or leave early". More than 170 people died in the fires illustrating the importance of the timing of evacuations, which in turn, is dependent on sufficient warnings.

These experiences are emblematic of the complexities of evacuation, no matter what the scale or location of the disaster, and illustrate some of the successes and failures of evacuation strategies. Certainly, the evacuation of people from hazardous areas in the face of impending disaster appears to be a sensible policy to follow if lives are to be saved. Indeed, getting out of harm's way and relocating to a safe place is logical. However, evacuation is not simple; it is embedded with many conflicting pressures that affect both efficiency and effectiveness of evacuation strategies. Not all individuals will evacuate, for instance, and even those who do, may suffer untoward difficulties. The concept of evacuation, therefore, requires an understanding of those factors that influence calls for evacuation and response, including characteristics of the impending event and the socio-economic context of the location at risk.

Evacuation can occur without formal notification, but more typically, comes as a result of warnings about an imminent event from official sources such as emergency managers (see *Early Warning Systems*). Thus, evacuation is part of an emergency management (see entry *Emergency Management*) plan that addresses critical issues including the timing of evacuations, modes of transport and routes, alternative safe sheltering sites, and return. Evacuation, of course, is just one of many possible responses to warnings, but it is one that requires active participation on the part of those at risk. However, getting out of harm's way may not be perceived as necessary, desirable, or even viable. Understanding this decision-making process, then, is important and requires consideration of the human dimensions associated with forecasting, warning, and response (Sorenson, 2000).

Evacuation is the result of a complex of decisions by many parties (from public officials to those at various levels of risk) regarding the location and nature of an event's potential impacts as well as the balance of costs and benefits, in both real dollars and unquantifiable factors, including loss of social networks. The effectiveness for example, it was found that characteristics of social networks affect the levels of support (Murphy et al., in press). The ramifications for evacuation, especially if long-term, could be devastating.

External networks can be affected, too. Evacuees may move en masse to homes of families and friends in other areas, thus putting stress on recipient communities especially when this entails long stays (Whiteford and Tobin, 2004). In addition, a large influx of evacuees can change the dynamics of receiving communities creating new challenges. Long-term evacuation, relocation projects, and permanent resettlements can all have such impacts. For example, a resettlement project in Ecuador has more than doubled the size of a village near Tungurahua, changing social interactions within the community. Thus, it is important to look at the effects of evacuation on the receiving area and on the people and networks of evacuees.

Summary and conclusion

Evacuation is a wise choice to avoid loss of life and injuries when faced with severe geophysical events. However, this wisdom is constrained by numerous factors that relate to the nature of the impending event, the political context, and various social and individual variables that influence response. Perception and social and family networks affect decision-making whereas temporal and spatial elements define evacuation opportunities and constraints. An over-riding theme is one of complexity that embeds evacuation planning and management with individual decision-making in different social and economic situations, leading to differential *Vulnerability* and, thus, unequal access to the benefits of evacuation.

Bibliography

Atwood, L. E., and Major, A. M., 1998. Exploring the "Cry Wolf" hypothesis. *International Journal of Mass Emergency Disasters*, **16**, 279–302.

Bea, K., 2005. *Disaster Evacuation and Displacement Policy: Issues for Congress*. Congressional Research Service, Report for Congress, USA. Government and Finance Division, September 2nd. Washington DC: Library of Congress.

Bell, H. M., 2007. *Situating the Perception and Communication of Flood Risk: Components and Strategies*. Doctoral Dissertation, Tampa, FL, Department of Geography, University of South Florida.

Brinkley, D., 2006. *The Great Deluge: Hurricane Katrina, New Orleans, and The Mississippi Gulf Coast*. New York: Harper Collins.

Dash, N., and Gladwin, H., 2007. Evacuation decision making and behavioral responses: individual and household. *Natural Hazards Review*, **8**(3), 69–77.

Dash, N., and Morrow, B. H., 2000. Return delays and evacuation order compliance: the case of hurricane Georges and the Florida Keys. *Environmental Hazards*, **2**(3), 119–128.

Dow, K., and Cutter, S. L., 2002. Emerging hurricane evacuation issues: Hurricane Floyd and South Carolina. *Natural Hazards Review*, **3**(1), 12–18.

Dixit, V. V., Pande, A., Radwan, E., and Abdel-Aty, M., 2008. Understanding the Impact of a Recent Hurricane on Mobilization Time During a Subsequent Hurricane. *Transportation Research Record*, **2041**, 49–57.

Fu, H., Wilmot, C. G., Zhang, H., and Baker, E. J., 2007. Sequential logit dynamic travel demand model and its transferability. *Transportation Research Record*, **1882**, 19–26.

Gladwin, H., and Peacock, W. G., 1997. Warning and evacuation: a night for hard houses. In Morrow, B. H., and Gladwin, H. (eds.), *Hurricane Andrew: Gender Ethnicity and the Sociology of Disasters*. New York: Routledge, pp. 52–74.

Hughey, E. P., 2008. *A Longitudinal Study: The Impact of a Comprehensive Emergency Management System on Disaster Response in The Commonwealth of the Bahamas*. Doctoral dissertation, Tampa, FL, Department of Geography, University of South Florida.

Hurlbert, J. S., Beggs, J. J., and Haines, V. A., 2001. Social networks and social capital in extreme environments. In Lin, N., Cook, A., and Burt, S. K. (eds.), *Social Capital: Theory and Research*. New Jersey: Transaction Publishers, pp. 209–232.

Lindell, M. K., Lu, J., and Prater, C. S., 2005. Household decision making and evacuation in response to hurricane Lili. *Natural Hazards Review*, **6**(4), 171–179.

Lindell, M. K., Prater, C. S., and Peacock, W. G., 2007. Organizational communication and decision making in hurricane emergencies. *Natural Hazards Review*, **8**, 50–60.

Murphy, A. D., Jones, E. C., Whiteford, L. M., Tobin, G. A., Faas, A. J., Vargas, I. P., and Guevara, F. J., (in press). Factores en el Bienestar de Personas Bajo Situaciones de Riesgos Crónicos. In Lopez Garcia, A., Fuentes, A. F., Sanchez, S. C., and Ramon, J. C., (eds.), *Encuentros Sobre el Volcán Popocatépetl: A 15 Años de su Erupción*. Puebla, Mexico: Universidad de Puebla, CUPREDER-Benemérita.

Perry, R. W., 1979. Incentives for evacuation in natural disaster research based community emergency planning. *Journal of the American Planning Association*, **45**(4), 440–447.

Rashid, H., Haider, W., and McNeil, D., 2007. Urban riverbank residents' evaluation of flood evacuation policies in Winnipeg, Manitoba, Canada. *Environmental Hazards*, **7**, 372–382.

Ruch, C., Miller, C., Haflich, M., Farber, N., Berke, P., and Stubbs, N., 1991. *The Feasibility of Vertical Evacuation*. Boulder: University of Colorado Institute of Behavioral Science.

Simmons, K. M., and Sutter, D., 2008. Tornado warnings, lead times, and tornado casualties: an empirical investigation. *Weather and Forecasting*, **23**, 246–258.

Slovic, P. (ed.), 2000. *The Perception of Risk*. London: Earthscan.

Sorenson, J. H., 2000. Hazard warning systems: Review of 20 years of progress. *Natural Hazards Review*, **1**(2), 119–125.

Tobin, G. A., and Montz, B. E., 1997. *Natural Hazards: Explanation and Integration*. New York: Guilford Press.

Urbina, E., and Wolshon, B., 2003. National review of hurricane evacuation plans and policies: a comparison and contrast of state practices. *Transportation Research Part A*, **37**, 257–275.

Whiteford, L. M., and Tobin, G. A., 2004. Saving lives, destroying livelihoods: emergency evacuation and resettlement policies. In Castro, A., and Springer, M. (eds.), *Unhealthy Health Policies: A Critical Anthropological Examination*. Walnut Creek, California: AltaMira Press, pp. 189–202.

Zeigler, D. J., Brunn, S. D., and Johnson, J. R. Jr., 1981. Evacuation from a nuclear technological disaster. *Geographical Review*, **17**, 1–16.

Cross-references

Community Management of Hazards
Early Warning Systems
Emergency Management
Emergency Shelters
Hazard and Risk Mapping
Hurricane Katrina
Monitoring and Prediction of Natural Hazards

Myths and Misconceptions
Perception of Natural Hazards and Disasters
Risk Perception and Communication
Vulnerability
Warning Systems

EXPANSIVE SOILS AND CLAYS

Ghulappa S. Dasog[1], Ahmet R. Mermut[2]
[1]University of Agricultural Sciences, Dharwad, India
[2]Harran University, Şanlıurfa, Turkey

Synonyms
Cracking clay soils; Smectitic soils; Vertisols

Definition
Expansive soils and clays: Soils (or clay minerals) that exhibit considerable volume change with change in moisture content. Expansive soils contain appreciable amounts of swelling clay minerals.

Introduction
Expansive soils, known by various nomenclatures worldwide, are a unique group of soils that expand when wetted and shrink when dried. These include predominantly Vertisols and vertic intergrades in US soil taxonomy (Soil Survey Staff, 2006). The extent of Vertisols alone is estimated at 350 million ha worldwide. The pressure these soils exert can be strong enough to crack driveways, floors, walls and foundations. The American Society of Civil engineers estimates that 25% of all homes in the United States have some damage caused by expansive soils. Expansive soils cause substantial damage and yet few people are aware of this hazard. It is estimated that shrinking and swelling soils cause about $2.3 billion damage annually in the United States alone (Holtz and Kovacs, 1981) which is more than twice the annual cost of damage from floods, hurricanes, tornadoes, and earthquakes combined!

Swell-shrink process
A soil may be described as a three phase system: solid, liquid, and gaseous. Under ideal conditions, about half of the volume is occupied by the solid phase and the other half is shared in equal proportion by liquid and gaseous phases. In non-expansive soils, the bulk volume would not change whether hydrated or dehydrated. The liquid and gaseous phases would interchange without significantly affecting the volume and other soil properties. However, in expansive soils considerable volume changes occur upon hydration and dehydration due to changes in porosity and water content (Figure 1). When the soil is rewetted, part of the volume occupied by air is replaced by water, but the pore volume of the soil also increases and, consequently, its water content (Coulombe et al., 1996a).

When the soil dehydration process starts, three stages are usually observed in the shrinking phenomenon. The first one termed "structural shrinkage" corresponds to a slight volume change when water is removed from larger pores (Figure 2). The second stage represents a volume reduction which is proportional to water loss in the system. This stage is termed "linear shrinkage" and occurs around 0.03–1 MPa, and even down to 1.5 MPa of suction in the case of smectitic soils (Coulombe et al., 1996a). With further drying, the decrease in volume is much less than the water lost and is termed "residual shrinkage." Further collapse is prevented by electrostatic repulsion, strongly bound water, and particle rigidity.

Interparticle and intraparticle porosity of the microstructures is largely responsible for the shrink-swell phenomena in soils. The other theories of expansion/collapse of the interlayer space of clay minerals and diffuse double layer, reportedly, have a slight influence on swelling under very specific conditions (Coulombe et al., 1996b).

The three groups of factors that influence the shrink-swell potential of the soil are soil properties, environmental factors, and the state of stress. The soil properties include clay mineralogy, fabric, and dry density. The initial moisture content, moisture fluctuations induced by climate, ground water, drainage, and vegetation are factors related to environment (Nelson and Miller, 1992). Everything else being equal, smectites swell more than illites, which swell more than kaolinites. Soils with random fabrics tend to swell more than soils with oriented fabrics and monovalent cation clays swell more than divalent clays. Cementation and organic substances tend to reduce swelling (Holtz and Kovacs, 1981).

Characterization of expansive soils
Expansive soils are characterized by a number of test parameters and the popular ones are liquid limit (LL), plastic limit (PL), and plasticity index (LL-PL). Skempton (1953) defined activity(A) as the ratio of plasticity index to clay fraction (percent of soil particles finer than 2 μm) and is given as:

$$\text{Activity, A} = \frac{plasticity\ index}{(\% < 2\mu m) - 5}$$

Soils are considered *inactive* if A is less than 0.75, normal for activities between 0.75 and 1.25 and active with values greater than 1.25.

Free swell test is one of the simple swelling tests performed by slowly pouring 10 cm^3 of dry soil, which had passed the No. 40 sieve, into a 100 cm^3 graduated cylinder filled with water, and observing the equilibrium swelled volume. Free swell is defined as:

$$\text{Free swell} = \frac{(final\ volume) - (initial\ volume)}{initial\ volume} \times 100(\%)$$

Highly swelling bentonites (Na-montmorillonite) are known to have free swell values of greater than 1,200%.

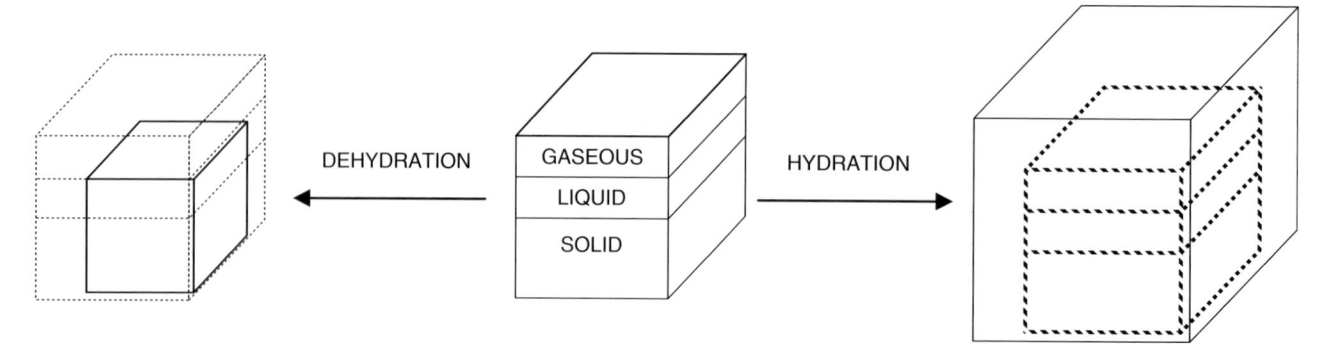

Expansive Soils and Clays, Figure 1 Schematic representation of soil volume change in an expansive soil (Modified from Coulombe et al., 1996a).

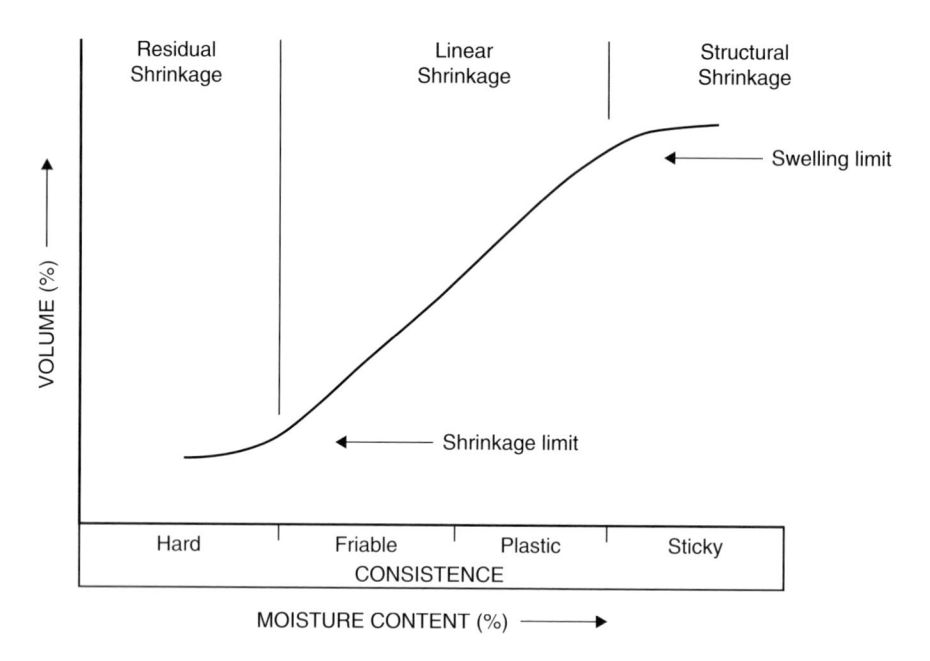

Expansive Soils and Clays, Figure 2 Schematic representation of shrinkage curve and consistence as a function of moisture content (From Coulombe et al., 1996a).

Even soils with free swells of 100% may cause damage to light structures when they become wet and soils with free swells less than 50% have been found to exhibit only small volume changes. The degree of expansion in relation to various soil test parameters is suggested in Table 1.

COLE (Coefficient of Linear Extensibility) is a measure of the change in sample dimension from the moist to dry state and is estimated from the bulk densities of the soil clod at a suction of 33 kPa (BDm) and oven dry moisture conditions (BDd).

The value of $COLE = (BDd/BDm)^{-1/3} - 1$

COLE, widely used in soil classification, is influenced by clay content and clay mineralogy.

Clay content, nature of clay minerals, CEC, and surface area which are all determined in the laboratory also help to identify expansion in soils. Thomas et al. (2000) developed Expansive Soil Index (ESI) by summing the absolute values of swelling 2:1 minerals, swell index, liquid limit, and CEC.

Potential volume change (PVC), Expansion index (EI), and California bearing ratio (CBR) are other tests performed in soil mechanics laboratories to measure one-dimensional swell, under a surcharge pressure, of compacted remolded soils under varying moisture conditions.

Active zone

Expansion in soils occurs as a result of an increase in water content in the upper few meters from ground surface influenced by climatic conditions and environmental factors. The zone affected by seasonal fluctuation of moisture is termed active zone as illustrated in Figure 3.

If the moisture content and/or soil type differs at various locations under the foundation of a structure, localized or nonuniform settlement may occur in the structure. This differential settlement of sections of the structure can cause damage to the foundation and framing, evidenced by cracking of the slab or foundation, cracking in the exterior and/or interior wall covering (indicating movement of the framing) uneven floors and/or misaligned doors and windows (Nelson and Miller, 1992).

A second effect of expansive soils is additional horizontal pressure applied to foundation walls found in basements and crawlspaces. Increased moisture in the soils adjacent to the foundation wall will cause the soils to expand and increase the lateral pressure applied to the foundation wall. If the foundation wall does not have sufficient strength, minor cracking, bowing or movement up to serious structural damage to, or even failure of, the wall may occur. If the soil is located on a slope, the top layer of soil can creep (slow movement) downhill or even cause a landslide.

The depth of surface cracks in expansive soils determine the depth of the active zone in many cases (Picornell and Lytton, 1989). Rainfall and surface runoff can fill these cracks, and the water in the cracks can travel, wherever the crack goes. If the water travels beneath a pavement, it will remain there, soaking into the soil on each side of the crack, and cause swelling. Thus, the depth of the surface cracks determines the depth to which a vertical moisture barrier should be placed in order to control moisture beneath a pavement.

Solutions

Prior to building a structure, a soil test will assist in determining if the soils are capable of properly supporting the structure. In addition, information on the soil properties can ensure that the foundation is designed to withstand the effects of the existing soil conditions.

For structures already in existence, common preventative solutions include proper soil maintenance such as maintaining a uniform and constant moisture level in the soil. This may involve introducing moisture into the soils continually and uniformly to prevent shrinking; and/or

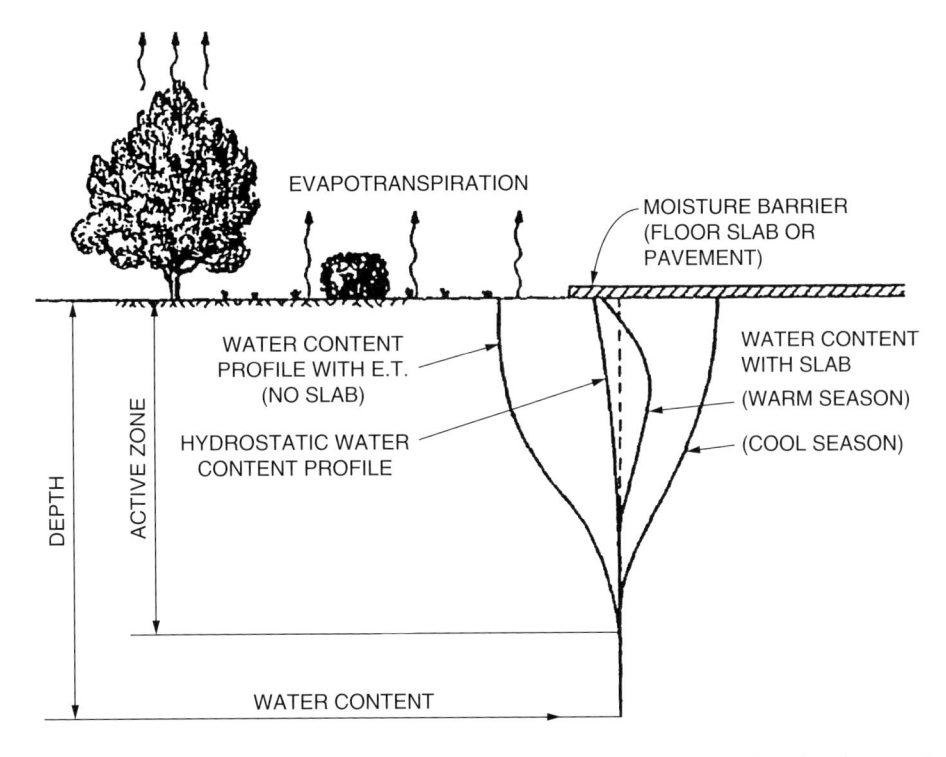

Expansive Soils and Clays, Table 1 Probable expansion as estimated from classification test data

Degree of expansion	Probable expansion as a % of the total volume change	Colloidal content (% < 1 μm)	Plasticity index PI	Shrinkage limit SL
Very high	>30	>28	>35	<11
High	20–30	20–31	25–41	7–12
Medium	10–20	13–23	15–28	10–16
Low	<10	<15	<18	>15

From Holtz and Kovacs (1981)

Expansive Soils and Clays, Figure 3 Figure showing the moisture content in active zone with and without moisture barrier (From Nelson and Miller, 1992).

preventing excessive or isolated saturation of the soil through proper drainage and grading techniques that prevent swelling. For structures affected by expansive soils, further movement can be prevented by providing additional strength and support to the foundation. This may include preventing vertical movement and/or sliding and/or reinforcing of the foundation walls to withstand lateral pressure. Addition of lime, gypsum, sand, and cohesive nonswelling (CNS) material are known to decrease the swelling pressure and are prescribed to mitigate swelling.

Significance of shrinkage and swelling

The effects of shrinkage of expansive soils can be of considerable significance from an engineering point of view. Shrinkage cracks can occur locally when the capillary pressures exceed the cohesion or the tensile strength of the soil. These cracks reduce the overall strength of a soil mass and affect the stability of clay slopes and the bearing capacity of foundations. The desiccated and cracked dry crust usually found over deposits of soft clay affects the stability of, for example, highway embankments constructed on these deposits. The volume changes resulting from both shrinkage and swelling of fine-grained soils are often large enough to seriously damage small buildings, highway pavements, and canal linings. An ordinary building is believed to weigh about 10 kPa per story. An embankment of 5 or 6 m would be required to prevent all swelling of subgrade with swelling pressures of 100 or 200 kPa (Holtz and Kovacs, 1981).

Summary

Expansive soils are known to cause damage to driveways, floors, and foundations and in monetary terms this damage annually amounts to billions of dollars. Structural, linear, and residual shrinkage are the three distinct stages in soil shrinkage of which linear shrinkage is predominant over a wide moisture range and is directly related to moisture loss. Interparticle and intraparticle porosity is largely considered responsible for the shrink-swell phenomenon in these soils.

Soil characteristics like clay mineralogy and dry density and environmental factors such as moisture content, moisture fluctuation range induced by climate, ground water, and vegetation, and the state of stress are known to influence the shrink-swell potential of the soil. These soils are characterized mainly by liquid limit and plastic limit, clay content, or COLE (Coefficient of Linear Extensibility), a term used in soil classification by USDA. The swell-shrink is limited to what is termed as an active zone in the upper few meters affected by seasonal moisture fluctuations and is related to the depth of surface cracks in these soils. To overcome the hazardous effects due to expansion, a soil test is recommended prior to construction of a structure. Treatment of soil with lime, gypsum, sand, and other non-swelling materials is known to mitigate swelling.

Bibliography

Coulombe, C. E., Wilding, L. P., and Dixon, J. B., 1996a. Overview of vertisols: characteristics and impacts on society. *Advances in Agronomy*, **57**, 289–375.

Coulombe, C. E., Dixon, J. B., and Wilding, L. P., 1996b. Mineralogy and chemistry of vertisols. In Ahmed, N., and Mermut, A. (eds.), *Vertisols and technologies for their management*. Amsterdam: Elsevier.

Holtz, R. D., and Kovacs, W. D., 1981. *An introduction to geotechnical engineering*. Englewood Cliffs: Prentice-Hall, p. 733.

Nelson, J. D., and Miller, D. J., 1992. *Expansive soils*. New York: Wiley, p. 259.

Picornell, M., and Lytton, R. L., 1989. Field measurement of shrinkage crack depth in expansive soils. *Transportation Research Record*, **1219**, 121–230.

Skempton, A. W., 1953. The colloidal activity of clays. In *Proceedings of the third International conference on soil mechanics and foundation engineering*, Zurich, Switzerland, Vol. 1, pp. 57–61.

Soil Survey Staff, 2006. *Keys to soil taxonomy*, USDA, Natural Resources Conservation Service, Washington, DC.

Thomas, P. J., Baker, J. C., and Zelazny, L. W., 2000. An expansive soil index for predicting shrink- swell potential. *Soil Science Society of America Journal*, **64**, 268–274.

Cross-references

Collapsing Soil Hazards
Damage and the Built Environment
Dispersive soil hazards
Erosion
Hydrocompaction Subsidence
Karst Hazards
Piping Hazards
Quick Clay
Quick Sand
Solifluction
Universal soil loss equation (USLE)

EXPERT (KNOWLEDGE-BASED) SYSTEMS FOR DISASTER MANAGEMENT

Jean-Marc Tacnet[1], Corinne Curt[2]
[1]Irstea, UR ETGR, Unité Erosion Torrentielle Neige et Avalanches (Snow Avalanche Engineering and Torrent Control Research Unit), Saint Martin d'Hères, France
[2]Irstea, UR OHAX, Hydraulic Engineering and Hydrology Research Unit, Aix-en-Provence, France

Synonyms

Decision support systems (DSS); Information systems (IS)

Definition

Natural disaster: Serious disruption triggered by natural hazards (earthquakes, floods, droughts, landslides, volcanoes, tsunamis, wildfires, avalanches, storms, cyclones, and tornadoes) causing human, material, economic, or environmental losses which exceed the ability of those affected to cope (ISDR, 2004).

Disaster management: Systematic management of administrative decisions, organization, operational skills, and abilities to implement policies, strategies, and coping capacities of the society or individuals to lessen the impacts of natural and related environmental and technological hazards (ISDR, 2004).

Expert system: An intelligent computer program that uses knowledge and inferences procedures to solve problems that are difficult enough to require significant human expertise for their solution. The knowledge of an expert system consists of facts and heuristics. The "facts" constitute a body of information that is widely shared, publicly available, and generally agreed upon by experts in the field (Harmon and King, 1985).

Introduction

Throughout the year, the world is subjected to natural hazards of a variety of types, sizes, and impacts. They may be geological hazards such as avalanches, earthquakes or volcanic eruptions, hydrological hazards such as floods or tsunamis, or climatic and atmospheric hazards such as cyclonic storms or tornadoes. Occasionally, they are particularly devastating: the tsunami in Indonesia in December 2004 claimed some 230 000 victims and the earthquake in Haiti in January 2010 may have had a comparable death total with virtual complete destruction of local infrastructures. Disaster management consists in coping with such major crises by planning, arranging, and controlling activities and resources in order to minimize the impact of these inherently uncertain and catastrophic phenomena.

As with any decision situation, disaster management requires one to obtain and use information for decision making. Not only expert systems but also Information systems are therefore essential tools to provide real-time information and help decision at the different crisis and disaster management phases.

Crisis and disaster management phases

Several spatial and temporal steps can be identified in a disaster management (cf. Figure 1):

- Crisis management and response consist in taking emergency measures to save and rescue people and infrastructure at risk.
- Recovery is the postcrisis step.
- Prevention and mitigation aim at avoiding damage through structural (protection works) or nonstructural measures (land-use planning).
- Preparation aims at anticipating the crisis through forecasting, rescue training, and resources positioning and inventory.

The management of these different steps involves many different activities and tools (cf. Figure 1). Decision Support Systems (DSS) are developed to help governments and public bodies to manage the impact of natural catastrophes and mitigate (ideally prevent) their disastrous consequences (Wallace and De Balogh, 1985). Some of them are expertise or knowledge based.

Expertise and expert systems for disaster management

Expert knowledge encompasses what qualified individuals know with respect to their technical practices, training, and experience. It is related to an individual and is acquired by studies or practical experience. Expert knowledge can be categorized into "expertise" and "expert judgment." Expertise refers to the tacit thinking processes used in everyday decision making and problem solving. The definition and structure of the problem, its representation and scope, determining relevant information, information organization, and flow through the problem are aspects of the expertise. In contrast, expert judgment, expert estimates, or expert opinion refer to the contents of the problem (Booker and McNamara, 2004).

Expertise is not always clearly described and explained. Knowledge or formal description is an important step for traceability and quality improvement (cf. Figure 2). Expert (or knowledge-based) systems constitute a method to formalize the available knowledge retained by experts. This supposes that knowledge is available.

An expert system or knowledge-based system is able to propose a conclusion through an automated reasoning process. Rule-based systems, which are the oldest systems, consist in a knowledge base, a rule-based inference engine, and a user interface. The knowledge base is set of rules and facts: rules are chained together and executed according to logic statements to provide a solution. In case-based systems, experiences are used instead of rules. New problems are solved through adaptation of old solutions (Bossé et al., 2007). Expert systems are not the only techniques of artificial intelligence techniques, which also include pattern matching algorithms and self-learning systems such as neural networks (Woo, 1999).

Expert systems are very interesting tools but present some limits. In addition to those involved in any software systems, expert system failures result (Bell, 1985) both from:

- Lack of knowledge about the problem to solve.
- Lack of awareness from the designers, users about this missing, incomplete, or imperfect knowledge.
- Difficulties or impossibilities to test the systems.
- Lack of confidence from the users due to partial understanding of used rules and reduced traceability of reasoning processes involved in those systems.

From the point of view of disaster management, expert (knowledge-based) systems can concern:

- Prevention:
  - Phenomenon analysis, hazard and risk assessment such as the forecasting of avalanches (Buisson and Charlier (1989); Buisson and Giraud, 1995; McClung, 1995; Naresh and Pant, 1999; Zischg et al., 2005; Singh and Ganju, 2008) or the

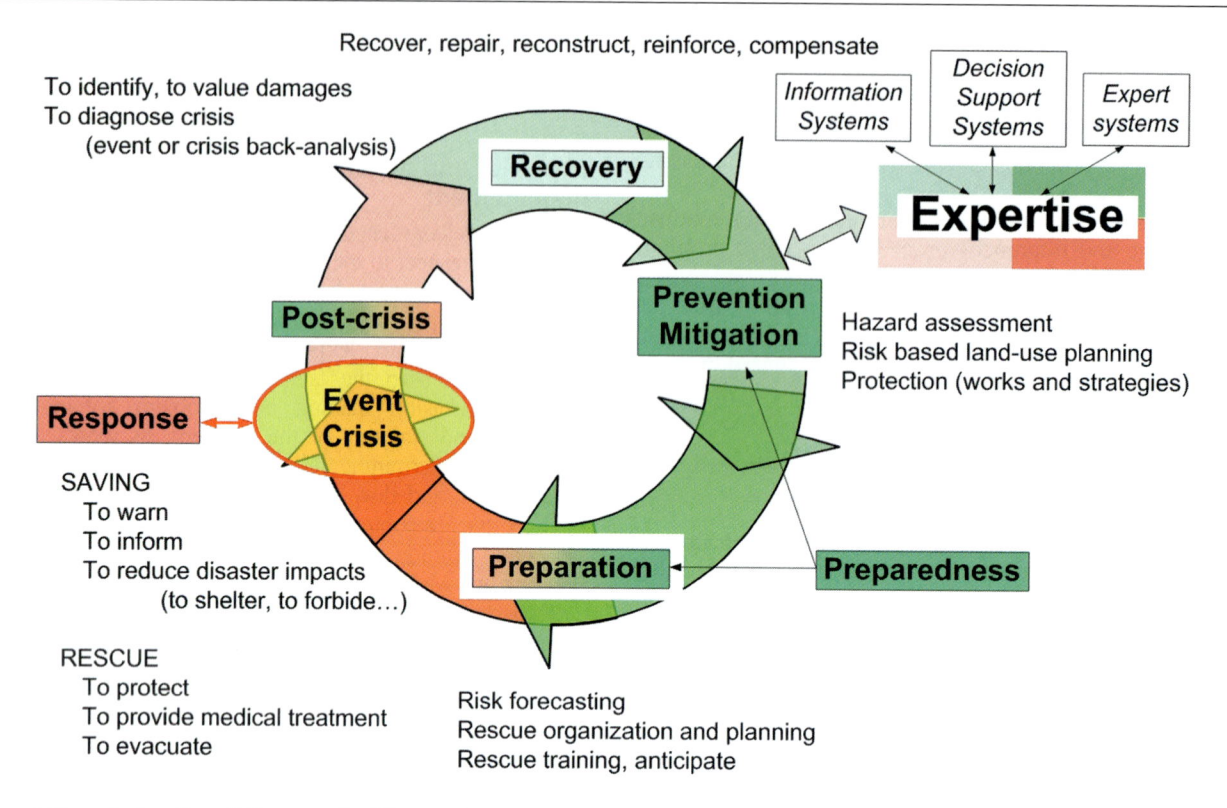

Expert (Knowledge-Based) Systems for Disaster Management, Figure 1 Description of main steps of natural hazards and risks management.

- landslide-hazard mapping (Muthu and Petrou, 2007; Chmelina and Eichhorn, 2008);
- Land-use planning;
- Protection works (description, design, security, and reliability assessment) such as in the case of floods (Ahmad and Simonovic, 2001) or the design of hazard resistant infrastructures such in the case of earthquake (Syrmakezis and Mikroudis, 1997; Berrais, 2005).
• Crisis: development of an expert system combined with an information system supporting the onsite search and rescue (Schweier and Markus, 2009), early warning and decision (Woo, 1999).
• Recovery: development of an expert system combined with an information system for the evaluation of buildings' states after earthquake (Schweier and Markus, 2009).

Particularities of knowledge-based systems for disaster management

In the context of natural hazards, knowledge and information are often imperfect meaning either incomplete, imprecise, uncertain, inconsistent, or conflicting. Expertise is therefore required to produce phenomenon maps, to interpret numerical modeling results and/or to produce hazards or risk maps (cf. Figure 2). At each step, expertise using and interpreting available information is required to take decisions. Finally, to apply their expertise, experts use different systems including (cf. Figure 2):

• Databases and GIS (Geographical Information Systems)-based information systems. Databases are essential to trace the geography of risk and factors producing hazard and vulnerability (UNDP, 2004). Some expert (or knowledge-based) systems can be coupled with a GIS when spatial representation or decision are needed to produce estimates of hazard-related damage before, or after, a disaster occurs: HAZUS-MH is an example of a methodology for analyzing potential losses from floods, hurricane winds, and earthquakes (FEMA, 2010). Others are linked to information systems.
• Numerical modeling.
• Decision support systems based either on Artificial Intelligence or Multicriteria decision frameworks.
• Expert systems (automatic reasoning applications).

Conversely, Information Systems collect information resulting both from measures, including sensors but also from expert analysis. Therefore, information is not only a simple collection of data but also includes some expert interpretation. For that reason, information systems can often be considered in risk management and for a better disaster management, the classical vision of expert systems should be extended to the information systems used to make decisions (Tacnet et al., 2009).

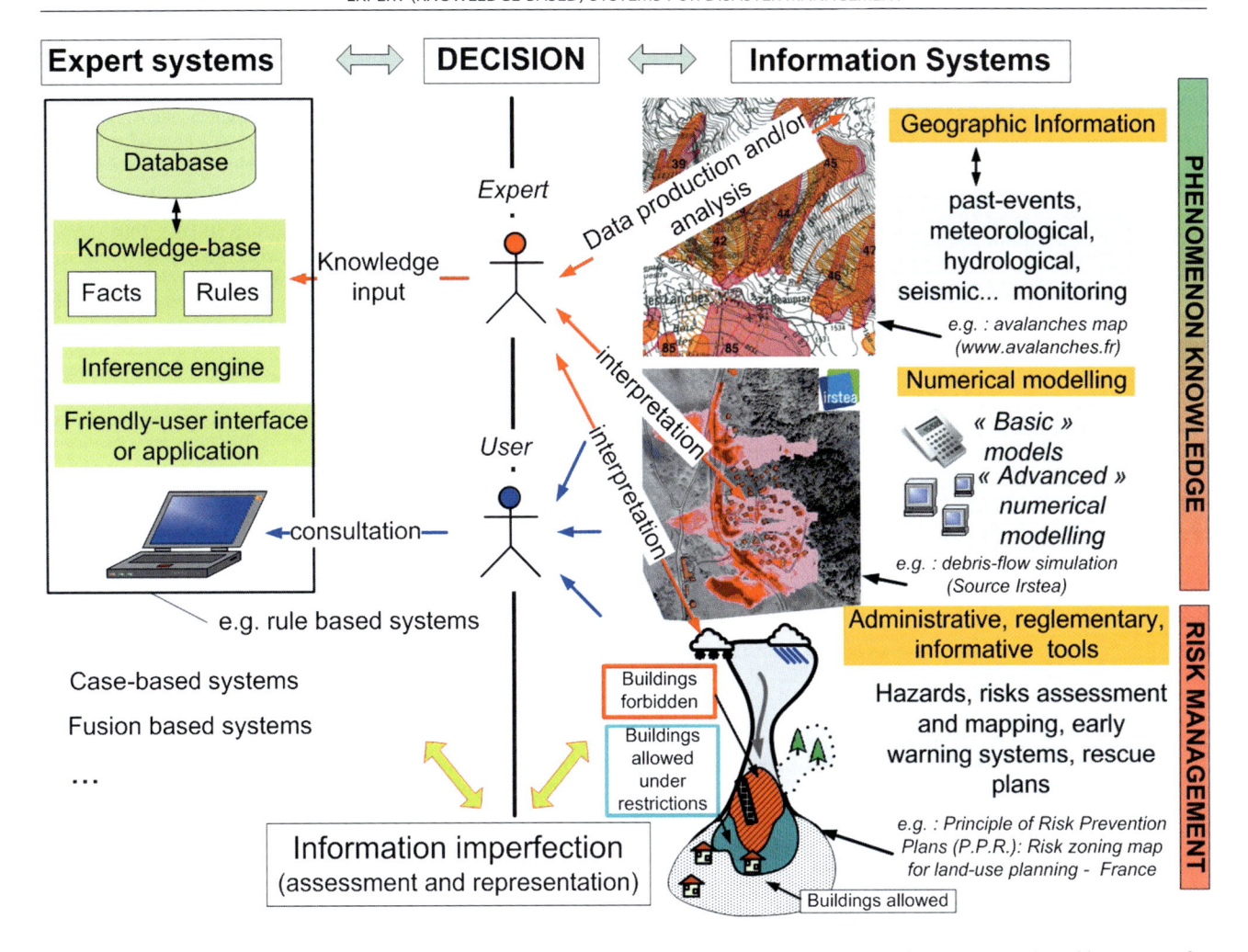

Expert (Knowledge-Based) Systems for Disaster Management, Figure 2 Expert systems and information produced by experts for risk management.

A typology can be proposed according to the added value:

- Systems corresponding to data collection and provision: information is stored and available when needed. This can be basic information (phenomenon maps, exposed buildings, or infrastructures), preprocessed information (modeling results);
- Systems designed to help and propose solutions through an automated and preprocessed way.

Data are frequently incomplete, outdated, or unusable for a variety of reasons (Guha-Sapir and Below, 2002). Thus, experts are needed to cope with all of these information imperfections and to help with decisions. As a consequence, decision, expertise and imperfect information are closely linked together in the natural hazards context whereas decisions have to be provided quickly. Methods that prevent information overload but also take into account imprecise, uncertain, contradicting expert analysis have to be used in combination with expert systems. Information fusion techniques based on quantitative theories such as fuzzy sets, possibility or evidence theory are particularly interesting to represent imperfect knowledge and take decisions in the case of natural hazards. The older theory developed to deal with uncertainty is the probability theory. Indeed, it cannot always be used because the probability laws are not always known. Moreover, information is often vague, imprecise, or incomplete (Bouchon-Meunier, 1999; Ben Armor and Martel, 2004). This lack of knowledge may stem from a partial lack of data, either because collecting this data is too difficult or costly, or because only experts can provide certain items of imprecise information. In this case, the evidence theory (Dempster, 1967; Shafer, 1976; Smarandache and Dezert, 2004–2009) or the possibility theory (Zadeh, 1965; Zadeh, 1978; Mauris et al., 2000; Baudrit et al., 2004; Dubois, 2006) provide tools that are better adapted to represent imperfections. In the natural hazards context, expertise can be considered as a combination of decisions based on multiple criteria imprecisely evaluated by more or less conflicting and reliable sources: multicriteria decision analysis and information fusion are combined to help decisions (Tacnet et al., 2009).

Summary and conclusions

In the context of natural hazards, expert systems must be considered in a broader sense than in classical industrial applications due to the role of expert in information production. Information systems and expert systems have an essential contribution to improve the organizations and society resilience (ability to withstand and minimize human and financial losses to natural disasters).

Due to imperfections of knowledge and information, disaster management cannot be operated through the exclusive use of experts systems. Operational Research (OR) and Management Science (MS) can also provide valuable tools to make decisions for disaster management (Altay and Green, 2006).

Bibliography

Ahmad, S., and Simonovic, S. P., 2001. Integration of heuristic knowledge with analytical tools for the selection of flood damage reduction measures. *Canadian Journal of Civil Engineering*, **28**, 208–221.

Altay, N., and Green, W. G., 2006. OR/MS research in disaster operations management. *European Journal of Operational Research*, **175**, 475–493.

Baudrit, C., Dubois, D., and Fargier, H., 2004. Practical representation of incomplete probabilistic information. In *Proceedings of 2nd International Conference on Soft Methodology and Random Information Systems*, Oviedo, Spain.

Bell, M. Z., 1985. Why experts systems fail. *Journal of the Operational Research Society*, **36**(7), 613–619.

Ben Armor, S., and Martel, J. M., 2004. Le choix d'un langage de modélisation des imperfections de l'information en aide à la décision. In *Proceedings of Congrès de l'ASAC*, Québec, Canada.

Berrais, A., 2005. A knowledge-based expert system for earthquake resistant design of reinforced concrete buildings. *Expert Systems with Applications*, **28**, 519–530.

Booker, J. M., and McNamara, L. A., 2004. Solving black box computation problems using expert knowledge theory and methods. *Reliability Engineering and System Safety*, **85**, 331–340.

Bossé, E., Roy, J., and Wark, S., 2007. *Concepts, Models and Tools for Information Fusion*. London, Boston: Artech House.

Bouchon-Meunier, B., 1999. *La Logique Floue*. Paris: Presses Universitaires de France.

Buisson, L., and Charlier, C., 1989. Avalanche starting-zone analysis by use of a knowledge-based system. *Annals of Glaciology*, **13**, 27–30.

Buisson, L., and Giraud, G., 1995. Two examples of expert knowledge based system for avalanche forecasting and protection. *Surveys in Geophysics*, **16**, 603–619.

Chmelina, K., and Eichhorn, A., 2008. Applications of knowledge-based systems in technical surveying. *Journal of Applied Geodesy*, **2**, 31–38.

Dempster, A. P., 1967. Upper and lower probabilities induced by a multivalued mapping. *Annals of Mathematical Statistics*, **38**, 325–339.

Dubois, D., 2006. Possibility theory and statistical reasoning. *Computational Statistics & Data Analysis*, **51**, 47–69.

FEMA (Federal Emergency Management Agency), 2010. *Hazards U.S. Multi-Hazard (HAZUS-MH)* (see http://www.fema.gov/plan/prevent/hazus/hz_overview.shtm).

Guha-Sapir, D., and Below, R., 2002. *The quality and Accuracy of Disaster Data: A Comparative Analyses of Three Global Data Sets*. Washington, DC: Provention Consortium, The Disaster Management Facility, The World Bank.

Harmon, P., and King, D., 1985. *Expert Systems*. New-York: Wiley.

International Strategy for Disaster Reduction (ISDR), 2004. *Living with Risk: A Global Review of Disaster Reduction Initiatives*. New York: ISDR Secretariat publication, United Nations.

Mauris, G., Lasserre, V., and Foulloy, L., 2000. Fuzzy modeling of measurement data acquired from physical sensors. *IEEE Transactions on Measurement and Instrumentation*, **49**, 1201–1205.

McClung, D. M., 1995. Use of expert knowledge in avalanche forecasting. *Defence Science Journal*, **45**, 117–123.

Muthu, K., and Petrou, M., 2007. Landslide-hazard mapping using an expert system and a GIS. *IEEE Transactions on Geoscience and Remote Sensing*, **45**, 522–531.

Naresh, P., and Pant, L. M., 1999. Knowledge-based system for forecasting snow avalanches of Chokibal-Tangdhar axis (J&K). *Defence Science Journal*, **49**, 381–391.

Schweier, C., and Markus, M., 2009. Expert and information systems for techncal SAR measures and buildings' state evaluation. *Natural Hazards*, **51**, 525–542.

Shafer, G., 1976. *A Mathematical Theory of Evidence*. Chichester: Princeton University Press.

Singh, D., and Ganju, A., 2008. Expert system for prediction of avalanches. *Current Science*, **94**, 1076–1081.

Smarandache, F., and Dezert, J., 2004–2009. *Advances and Applications of DSmT for Information Fusion (Collected Works)*, Vol. 1–3. Rehoboth: American Research Press (see http://www.gallup.unm.edu/smarandache/DSmT.htm).

Syrmakezis, C. A., and Mikroudis, G. K., 1997. ERDES – An expert system for the aseismic design of building. *Computers and Structures*, **63**, 669–684.

Tacnet, J. M., Batton-Hubert, M., and Dezert, J., 2009. Information fusion for natural hazards in mountains. In Dezert, J., and Smarandache, F. (eds.), *Advances and Applications of DSmT for Information Fusion – Collected works*. Rehoboth: American Research Press, Vol. 3, pp. 565–659.

United Nations Development Programme (UNDP), 2004. *A Global Report – Reducing Risk Disaster: A Challenge for Development*. New-York: Bureau for Crisis Prevention and Recovery (see www.undp.or/bpcr).

Wallace, W. A., and De Balogh, F., 1985. Decision support systems for disaster management. *Public Administration Review – Special Issue: Emergency management: A challenge for Public Administration*, **45**, 136–146.

Woo, G., 1999. *The Mathematics of Natural Catastrophes*. London: Imperial College Press.

Zadeh, L., 1965. Fuzzy sets. *Information and Control*, **8**, 338–353.

Zadeh, L., 1978. Fuzzy sets as a basis for a theory of possibility. *Fuzzy Sets and Systems*, **1**, 3–28.

Zischg, A., Fuchs, S., Keiler, M., and Meisi, G., 2005. Modelling the system behaviour of wet snow avalanches using an expert system approach for risk management on high alpine traffic roads. *Natural Hazards and Earth System Sciences*, **5**, 821–832.

Cross-references

Avalanches
Disaster Risk Management
Earthquake Resistant Design
Flood Hazard and Disaster
Geographic Information Systems (GIS) and Natural Hazards
Hazard

Hazard and Risk Mapping
Land-Use Planning
Mitigation
Natural Hazard
Prediction of Hazards
Risk
Risk Assessment
Slope Stability
Uncertainty

EXPOSURE TO NATURAL HAZARDS

Jörn Birkmann
United Nations University, Bonn, Germany

Synonyms

Elements at risk; Risk inventory

Definition

Exposure encompasses all elements, processes, and subjects that might be affected by a hazardous event. Consequently, exposure is the presence of social, economic, environmental or cultural assets in areas that may be impacted by a hazard.

Overview

Exposure is most commonly understood, in risk and natural hazard research, as the social and material context represented by persons, resources, infrastructure, production, goods, services, and ecosystems that may be affected by a hazardous event. Hence, the term is often used to characterize, in a more general sense, the components of *society* and *environment* and respective processes that are exposed to a specific hazard or a variety of potential hazardous phenomena. That means exposure can further be qualified by focusing on the spatial and temporal patterns of exposure of the element at risk.

Temporal aspects of exposure deal with the questions of when and how long an element at risk might be exposed to a hazard or stressor.

For example, in Indonesia, in coastal communities potentially exposed to tsunami hazards, it is noteworthy that the male population in contrast to female population, children, and elderly (more than 65 years) have a different spatial exposure at different times of the day, related to their respective activities and mobility patterns. Setiadi et al. (2010) analyzed temporal exposure patterns in the coastal city of Padang and found that the more vulnerable population groups (female, children, and elderly) to tsunamis have a different spatiotemporal exposure to potential tsunami risks in the high-risk zone as compared to the male population (see Setiadi et al., 2010; see also Figure 1).

The *spatial dimension of exposure* deals with the question of how much territory is exposed to a certain hazard phenomena as compared to the total territory of the entity exposed such as a community, city, or country. This spatial exposure provides a rough overview of how strongly an exposed entity would be affected. If a coastal storm hits a larger country, recovery processes and losses often remain relatively small, since the level of exposure is low compared to the overall population and values of the country. In contrast, small islands, where nearly 100% of

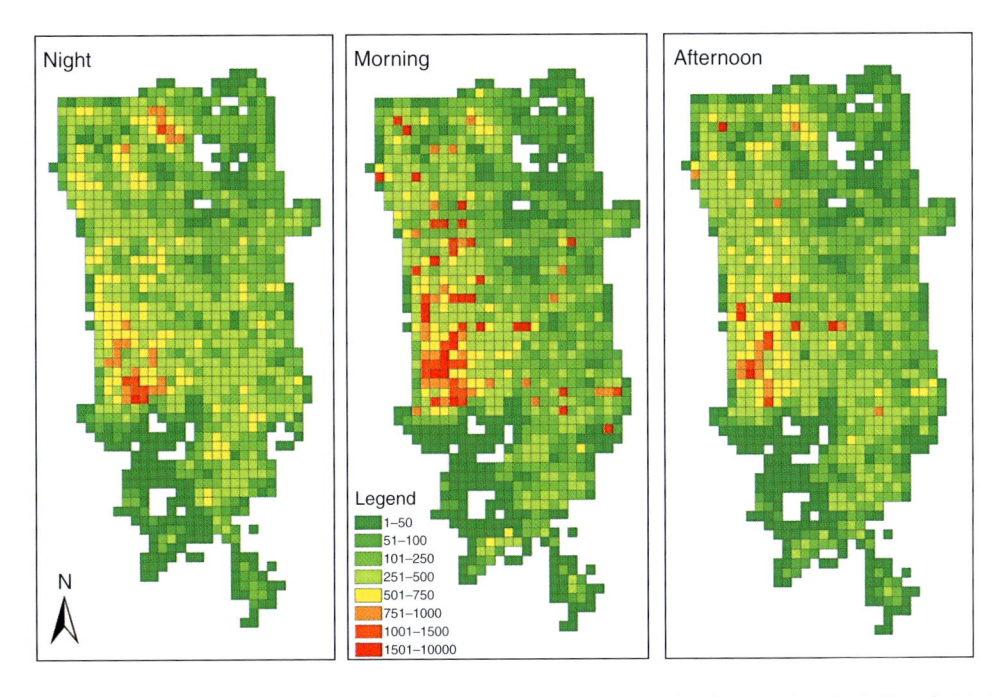

Exposure to Natural Hazards, Figure 1 Female, elderly, and children concentration by daytime (see Setiadi et al., 2010; Setiadi, 2011).

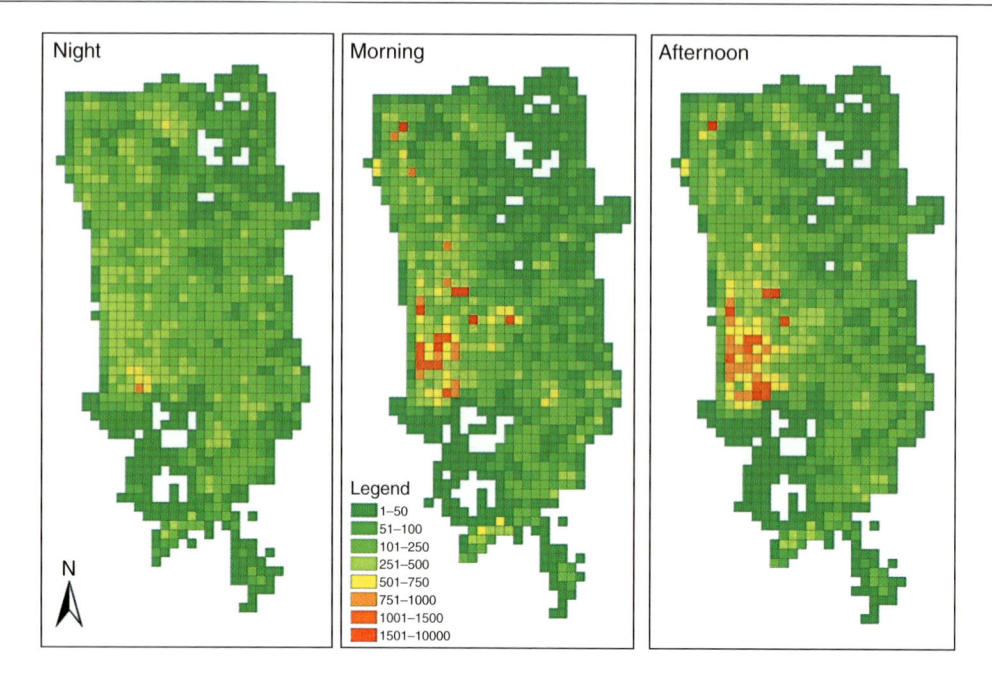

Exposure to Natural Hazards, Figure 2 Male population concentration by daytime (see Setiadi et al. 2010; Setiadi 2011).

the territory and population is exposed and could be impacted by a storm surge, would, consequently, already be more at risk. Exposure is in some concepts seen as a component of vulnerability and respectively risk, since it contributes in the degree of impacts brought by a hazard event (see Birkmann, 2006).

Risk reduction, in terms of exposure, may encompass different strategies and measures. Spatial and urban planning, for example, can potentially restrict growth of exposure of people and settlements in high-risk zones. In addition, the consideration of temporal and spatial exposure calls for measures, such as the construction of safe places in areas where many people are or will be exposed. Furthermore, awareness and understanding is needed that people might be exposed differently to certain hazards and stressors during different phases of the day. Based on the improved understanding of dynamic exposure patterns, differences between women and men, in terms of spatial exposure could also be used to develop specific awareness, early warning, and preparedness strategies (Figures 1 and 2).

Bibliography

Birkmann, J., 2006. Measuring vulnerability to promote disaster-resilient societies: conceptual frameworks and definitions. In Birkmann, J. (ed.), *Measuring Vulnerability to Natural Hazards: Towards Disaster Resilient Societies*. Tokyo: United Nations University Press, pp. 9–54.

Setiadi, N., Taubenboeck, H., Raupp, S., and Birkmann, J., 2010. Integrating socio-economic data in spatial analysis: an exposure analysis method for planning urban risk mitigation, In Schrenk, M., Popovich, V., Engelke, D., and Elisei, P. (eds.), *Proceedings of the 15th International Conference on Urban Planning and Regional Development in the Information Society GeoMultimedia 2010*, Vienna.

Setiadi, N., 2011. Daily Mobility, Excursus – Padang, Indonesia, In Chang Seng, D., and Birkmann, J. (eds.), *Early Warning in the Context of Environmental Shocks, Demographic Change, Dynamic Exposure to Hazards, and the Role of EWS in Migration Flows and Human Displacement*. Background Paper for the Foresight Project-State of Science, Bonn, pp. 41–45

Cross-references

Hazardousness of Place
Human Impact of Hazards
Land use, Urbanization and Natural Hazards
Natural Hazards in Developing Countries
Risk
Risk Assessment
Risk Perception and Communication
Susceptibility
Vulnerability

EXTENSOMETERS

Erik Eberhardt
University of British Columbia, Vancouver, BC, Canada

Synonyms

Convergence gauges; Crack gauges; Crackmeters; Distometers; Jointmeters; Sliding micrometers; Strainmeters

Definition

Extensometers are devices used to measure the changing distance between two points. They are commonly used in the monitoring of landslides.

Overview

Measurement points may be located on the surface of a landslide to measure ground movements, for example, spanning a tension crack to monitor its rate of opening, or in a borehole to measure differential displacements at depth, for instance to identify active landslide shear surfaces. Extensometers vary in type between those that involve manual measurements and those that are automated using vibrating wire electronics, differential transducers, or more recently fiber optics. Another variant involves the use of mechanical or electrical probes (i.e., probe extensometers) to monitor changing distances between fixed points with depth along a borehole as determined by the probe position. Measurement accuracy and repeatability depend on the type of sensing device and the distance between the monitoring points. Typical accuracies range from submillimeter to millimeter over distances of less than a meter when using probe extensometers or stiff sliding rods fixed between the monitoring points, to millimeter to centimeter over distances of meters or several tens of meters when extending a flexible tape or wire between the monitoring points.

Extensometers are frequently employed in landslide investigations to determine the active boundaries of the landslide (on the surface and at depth), correlate landslide movements to external environmental factors (e.g., precipitation events), and for early warning of accelerating landslide behavior. Recent examples of their use for landslide hazard investigations include those reported by Corominas et al. (2000), Coe et al. (2003), Willenberg et al. (2008), and Wang et al. (2008).

Bibliography

Coe, J. A., Ellis, W. L., Godt, J. W., Savage, W. Z., Savage, J. E., Michael, J. A., Kibler, J. D., Powers, P. S., Lidke, D. J., and Debray, S., 2003. Seasonal movement of the slumgullion landslide determined from global positioning system surveys and field instrumentation, July 1998–March 2002. *Engineering Geology*, **68**, 67–101.

Corominas, J., Moya, J., Lloret, A., Gili, J. A., Angeli, M. G., Pasuto, A., and Silvano, S., 2000. Measurement of landslide displacements using a wire extensometer. *Engineering Geology*, **55**, 149–166.

Wang, F., Zhang, Y., Huo, Z., Peng, X., Araiba, K., and Wang, G., 2008. Movement of the Shuping landslide in the first four years after the initial impoundment of the three Gorges Dam reservoir, China. *Landslides*, **5**, 321–329.

Willenberg, H., Evans, K. F., Eberhardt, E., Spillmann, T., and Loew, S., 2008. Internal structure and deformation of an unstable crystalline rock mass above Randa (Switzerland): Part II – three-dimensional deformation patterns. *Engineering Geology*, **101**, 15–32.

Cross-references

Creep
Deep-Seated Gravitational Slope Deformations
Dispersive Soil Hazards
Early Warning Systems
Expansive Soils and Clays
Hydrocompaction Subsidence
Inclinometer
Land Subsidence
Lateral Spreading
Mass Movement
Monitoring and Prediction of Natural Hazards
Piezometer
Slope Stability

EXTINCTION

Ross D. E. MacPhee
American Museum of Natural History, New York, NY, USA

Synonyms

Species loss

Definition

Complete loss of a species, which occurs when all individuals comprising the species' populations disappear without issue. Loss of one or more (but not all) constituent populations (subspecies) of a species constitutes *extirpation*, since in fact the species still exists and may possibly recover.

Introduction

The *IUCN Red List* (published under various extended titles, now available online at www.iucnredlist.org) has provided a biennial snapshot of the state of the world's biota for more than five decades, with particular attention given to identifying endangered species. The list's purpose is to provide (optimally) the conservation status of every species within a given taxonomic group, rated according to a series of criteria. From its inception, the *Red List* has published compendia of extinct species (and often subspecies), initially relying on expert opinion alone but in recent years moving toward a more transparent and quantitative system of analysis through its network of Species Survival Commission (SSC) Specialist Groups. The current *Red List* (2008) provides status assessments for nearly 45,000 species, some 800 of which are classified as extinct.

Extinction: an overview

More than 99% of all species that have ever lived since organized life first appeared more than 2 billion years ago are now extinct (Eldredge, 1999). One implication of this figure is that extinction is and always has been a ubiquitous feature of life on Earth, which means that extinction is an entirely natural process. Indeed, it is inconceivable that life as we know it could have evolved to its present tremendous complexity were it not for the fact that species have always come and gone, with the disappeared being succeeded by others having different adaptations, different lifestyles, and different capacities to handle the effects of natural change over time. However, the concern at present is whether humans, with their unprecedented capacity to alter habitats and degrade ecosystems, are now influencing this pattern in a way no other species ever has.

From the standpoint of extinction biology, this pattern has two primary features. Over geologically long periods of time, the vast majority of species losses have tended to occur in clumps (*mass extinctions*), prompted by unusual forcing functions such as the bolide impact 65 million years ago that wiped out non-avian dinosaurs and perhaps as much as 70% of other life. It is also clear from the fossil record that a continuous but relatively low turnover of species, not related to massive episodic losses, is another long-standing feature. This turnover, measured in a variety of ways, can be used to formulate a *background extinction rate* (see below), against which, for example, the much larger rates of loss characteristic of mass extinctions can be compared.

The relevant question from a natural hazards standpoint, of course, is whether the current rate of species loss qualifies as a mass extinction. Many scientists believe that, within the next few centuries, the number of losses will place modern extinctions among the six greatest episodes of extinction of all time (Eldredge, 1999; Vié et al., 2009). This makes an understanding of the biology of extinction one of the critical issues of our time. This entry is devoted to three fundamental questions: what is biological extinction, how is it measured, and how is it caused?

Biological extinction and related concepts

For the sake of simplicity, discussion in this text is framed around the formal categories "extinct" and "possibly extinct" as developed or endorsed by member units of the International Union for the Conservation of Nature (IUCN) (www.iucnredlist.org).

At first glance, determining whether the complete loss of a given species has occurred would appear to be a strictly empirical issue – a species either is, or is not, in existence. However, making realistic, verifiable determinations can be complicated (Diamond, 1989; Butchart et al., 2006). Actual census taking to establish whether any viable populations or individuals persist is often not practicable, for a host of reasons. In the case of well-characterized taxa that are highly visible or "charismatic" (at least to humans), making a plausible inference of complete extinction may be straightforward, as in the case of species formerly numerous in a particular area that are now no longer seen (passenger pigeon scenario). However, a finding of "not seen," even over an appreciable interval such as 50 years may nevertheless be insufficient to conclude that final extinction has actually occurred (MacPhee and Flemming, 1999; MacPhee, 2008). For example, surviving populations may have withdrawn to a new range outside of the census area. Or overlooked individuals may survive in unexpected pockets, so that extreme rarity is misinterpreted as final disappearance (e.g., *Campephilus principali*, ivory-bill woodpecker, possibly surviving in Cuba and Arkansas; Fitzpatrick et al., 2005).

Erroneous claims of extinction are, however, comparatively rare, and little is gained scientifically or by requiring potentially unattainable levels of certainty before declaring a species to be extinct. A typical problem is the difficulty of detecting *cryptic survivorship*, as in the situation in which a few individuals from a severely endangered species, perhaps rarely or never seen, manage to survive in a small part of their original range or at greatly reduced densities. If this possibility seems real in a given case, some other designation such as Birdlife International's (www.birdlife.org) category of *possibly extinct* – species likely to be extinct, but for which there is a small chance that of survival – is much to be preferred over any other.

Some other extinction-related concepts and definitions merit brief mention. A species that is possibly extinct is probably already *functionally extinct*: The taxon is probably no longer a viable entity and may well completely disappear in short order. The concept of *ecologically extinct* is essentially identical: A species whose populations have collapsed down to a few individuals is unlikely to be an effective participant in normal ecological processes in any significant way, and from that standpoint might as well be regarded as nonexistent (Purvis et al., 2000). More problematic is *Extinct in the Wild*, an IUCN status that is assigned to species (or subspecies) which have been so reduced that they survive only in captivity or outside their historic range in very small numbers. This category, however, is semantically confusing because it applies to taxa that are acknowledged not to be completely extinct in any formal sense. While it is doubtless true that very few species currently considered extinct in the wild will ever return to non-managed, natural conditions, there are examples of recovery to normal levels in original habitats even after stupendous losses – once, of course, the agent of loss is removed (e.g., northern elephant seal, *Mirounga angustirostris*; (Hedrick, 1995), reduced to a few tens of individuals by overhunting in the nineteenth century, now returning to pre-seal-hunting levels).

Despite difficulties in application, the concepts just described are useful in conservation education, as they emphasize the close connection between species welfare and population size, structure, and geographic range.

Rates of extinction

Measurement of extinction is conventionally presented in the form of an *extinction rate*. Several rate-estimation methods are currently in use (for an extensive treatment, see papers collected by Lawton and May, 1995).

Paleontological studies reveal that rates have varied widely across space, time, and taxa, even when losses due to mass extinctions are excluded. According to prevailing interpretations of the fossil record, on a per-taxon basis, extinction outside of periods of mass extinction is comparatively rare. However, paleontological evidence is of very uneven quality for many groups, and doubtless many instances of extinction have been missed even for comparatively well-researched groups with extensive fossil records, such as many marine groups and land vertebrates.

Nevertheless, estimates of "normal" or background rates developed using paleontological data have assumed particular prominence in contemporary conservation biology because of the need to provide a baseline for evaluating the severity of the current biodiversity crisis (e.g., Gaston and Spicer, 2004; Pimm et al., 2006). In light of growing human impacts on biodiversity, it is crucial that we find ways to accurately forecast species extinctions in order to minimize biodiversity loss and maintain critical ecosystem functions. From the basic science standpoint, it is important to note that, because forecasted losses have not actually occurred (yet), rate estimates cannot be based on direct counts of the disappeared but depend instead on reasonable proxies for species viability (e.g., likelihood that very small populations will disappear in a specific envelope of years; Caughley and Gunn, 1996). Thus although IUCN currently records only ~80 instances of recent extinction among mammals (1% of total number of assessed species, about 5,500), the model projection is that 1 in 4 species (25%) are threatened to some degree. How many of these threatened species will actually go extinct within the next century or so cannot be known in advance, but the real point is that having some idea of which species are more likely to disappear provides a needed focus for directing mitigation efforts. Much work is now being directed toward establishing better methodologies for using IUCN categories of information (Butchart et al., 2005).

Losses on islands in the last several centuries have been investigated with particular intensity, in part because such contexts often feature (at least for land vertebrates) relatively low population numbers and restricted habitats, two constraints that are widely regarded as factors predisposing species to extinction (MacPhee, 2008). Harrison and Stiassny (1999) have pointed out that freshwater habitats, which for a given species might consist of a single watershed, stream, or lake, function in an island-like manner from a probability-of-extinction standpoint. And indeed, "aquatic islands" account for a large proportion of recent extinctions among all freshwater fishes. In the case of birds and mammals, the correlation between the first arrival of humans, as determined archeologically, and evidence of catastrophic species losses tends to follow a particularly strong pattern of "dreadful syncopation" (MacPhee and Flemming, 1999; MacPhee, 2008): When the humans come, the animals disappear shortly thereafter. On some Pacific islands extinctions of large endemic birds appear to have occurred within a few decades of human colonization (Worthy and Holdaway, 2002), leading in the end to losses that have been conservatively estimated at several hundred species. The endemic mammals of the West Indies were also hit hard, and in similar fashion. Because more than 80% of the land mammals (excluding bats) that lived on these islands during the Holocene (last 10,000 years) are now extinct, the West Indies is not normally thought of as having once had a significant fauna. On most continents with the exception of Australia, mammal and bird losses have not been nearly as severe as island losses during the past 500 years. However, because of forest clear-cutting, rampant urbanization, and ever-higher levels of resource exploitation, many parts of the world – including those with the highest concentrations of biodiversity at the species level – exhibit extensive areas of forest that are beginning to be cut to pieces by exploitation, becoming in the process "virtual islands." There is an immense danger in this; there is probably no single factor more important than contiguous, untransformed habitat in providing for species welfare (Gaston and Spicer, 2004).

Causation of extinctions

Although extinctions may theoretically have many (including multiple) causes, most attention at present is directed toward defining so-called anthropogenic causes. These are too numerous to document here and are individually the subject of much current research: Climate change, invasive species, forest loss, emerging diseases, pollution, and by-catch as causative agents of species declines and extinction are among the topics being most actively investigated at present (www.redlist.org).

A particularly interesting and useful approach is to compare extinctions in particular groups (when possible) across a much longer timeline than is usually the case, for example, the entire Holocene (i.e., the past 10,000 years). The perspective that is gained by these means is to place the last several hundred years in proper perspective (see Turvey, 2009), by revealing periods and places in which extinction rates notably climbed and attempting to account for them with the kinds of sophisticated methods now available. However, this can only be meaningfully done for taxa that leave behind hard tissues (bones, shells) as documents of their former existence. For the vast majority of living things, especially invertebrates, the unseen departed will likely remain among the least documented.

Summary

Biological extinction is a natural process; indeed, the natural loss of species over time needs to be thought of as the complement to, not the antithesis of, the production of new species by natural selection. The objective pattern of life on Earth is that species are continually going extinct, but not in large numbers (at a so-called background rate). However, mass extinctions, or short periods of greatly enhanced loss, are also part of this pattern. A question for our time is whether species loss rates have surged to such a great degree in the last several hundred years that it is meaningful to identify the modern era as another of these mass extinction events. For more than 50 years, the International Union for the Conservation of Nature (IUCN) has tracked endangerment and loss with increasing sophistication. Although no one can predict the exact form the future will take, there is no doubt that public education efforts are key to conservation efforts, and that the IUCN is in the forefront of organizations performing this essential task.

Bibliography

Butchart, S. H. M., Stattersfield, A. J., Baillie, J., Bennun, L. A., Stuart, S. N., Akçakaya, H. R., Hilton-Taylor, C., and Mace, G. M., 2005. Using red list indices to measure progress towards the 2010 target and beyond. *Philosophical Transactions of the Royal Society of London*, **B360**, 255–268.

Butchart, S. H. M., Stattersfield, A. J., and Brooks, T. M., 2006. Going or gone: defining "possibly extinct" species to give a truer picture of recent extinctions. *Bulletin of the British Ornithological Club*, **12A**, 7–24.

Caughley, G., and Gunn, A., 1996. *Conservation Biology in Theory and Practice*. Cambridge, MA: Blackwell Science.

Diamond, J., 1989. The past, present, and future of human-caused extinctions. *Philosophical Transactions of the Royal Society of London*, **B325**, 469–477.

Eldredge, N., 1999. Cretaceous meteor showers, the human ecological "niche", and the sixth extinction. In MacPhee, R. D. E. (ed.), *Extinctions in Near Time: Causes, Contexts, and Consequences*. New York: Kluwer Academic/Plenum Press, pp. 1–16.

Fitzpatrick, J. W., Lammertink, M., Luneau, D. L., et al., 2005. Ivory-billed woodpecker (*Campephilus principalis*) persists in continental North America. *Science*, **308**, 1460–1462.

Gaston, K., and Spicer, J., 2004. *Biodiversity: An Introduction*, 2nd edn. Cambridge, MA: Blackwell Science.

Harrison, I. J., and Stiassny, M. L. J., 1999. The quiet crisis: a preliminary listing of the freshwater fishes of the world that are extinct or "missing in action". In MacPhee, R. D. E. (ed.), *Extinctions in Near Time: Causes, Contexts, and Consequences*. New York: Plenum Press/Kluwer Academic, pp. 271–331.

Hedrick, P. W., 1995. Elephant seals and the estimation of a population bottleneck. *Journal of Heredity*, **86**, 232–235.

Lawton, J. H., and May, R. M. (eds.), 1995. *Extinction Rates*. Oxford: Oxford University Press.

MacPhee, R. D. E., 2008. *Insulae infortunatae*: establishing the chronology of late quaternary mammal extinctions in the West Indies. In Haynes, G. (ed.), *American Megafaunal Extinctions at the End of the Pleistocene*. Dordrecht: Springer, pp. 169–193.

MacPhee, R. D. E., and Flemming, C., 1999. *Requiem aeternum*: the last five hundred years of mammalian species extinctions. In Cracraft, J. L., and Grifo, F. T. (eds.), *The Biodiversity Crisis: Science, Policy, and Society*. New York: Columbia University Press, pp. 73–124.

Pimm, S., Raven, P., Peterson, A., Sekercioglu, C. H., and Erlich, P. R., 2006. Human impacts on the rates of recent, present, and future bird extinctions. *Proceeding of the National Academy of Sciences, USA*, **103**, 10941–10946.

Purvis, A., Gittleman, J. L., Cowlishaw, G., and Mace, G. M., 2000. Predicting extinction risk in declining species. *Proceedings of the Royal Society of London*, **B325**, 469–477.

Turvey, S. (ed.), 2009. *Holocene Extinctions*. Oxford: Oxford University Press.

Vié, J. C., Hilton-Taylor, C., and Stuart, S. N. (eds.) 2009. *Wildlife in a Changing World – An Analysis of the 2008 IUCN Red List of Threatened Species*. Gland, Switzerland: IUCN. (http://data.iucn.org/dbtw-wpd/edocs/RL-2009-001.pdf).

Worthy, T., and Holdaway, R. N., 2002. *The Lost World of the Moa: Prehistoric Life of New Zealand*. Bloomington, IN: Indiana University Press.

www.birdlife.org (for catalog of world's threatened and endangered birds; Accessed May 1, 2010).

www.iucnredlist.org (for catalog of world's threatened and endangered species; Accessed May 1, 2010).

Cross-references

Asteroid Impact Predictions
Climate Change
Exposure to Natural Hazards
Humanity as an Agent of Natural Disaster
Impact Fireball
Land Use, Urbanization and Natural Hazards
Overgrazing
Sea Level Change

EXTREME VALUE THEORY

Gianfausto Salvadori
Università del Salento, Lecce, Italy

Synonyms
Theory of outliers

Definition
The Extreme Value Theory deals with the study of the limiting behavior of nondegenerate probability distributions of vectors of componentwise maxima (or minima) of random variables.

Overview

Extreme Value Theory can be further examined as follows. Since $\min\{X_1, \ldots, X_n\} = -\max\{-X_1, \ldots, -X_n\}$, usually only the laws of maxima are investigated.

Two complementary methods are used to study extremes: the "Blocks" and the "Peaks-Over-Threshold" approach. The former involves a partition of the sample into arbitrary subsets (or *blocks*), and maxima are extracted from the population in each block. The latter requires one to fix an arbitrary high threshold, and only the *excesses* above it are therefore analyzed.

In the univariate case (Coles, 2001), the *Generalized Extreme Value* distribution provides the probability law of the "blocks" maxima, whereas the *Generalized Pareto* distribution models the statistics of the excesses. In the multivariate case, universal laws do not exist. A general representation of a Multivariate Extreme Value distribution F can be given (Salvadori et al., 2007) in terms of its univariate marginals F_i's (which must be Generalized

Extreme Value laws), and a suitable multivariate *copula* (or *dependence function*) C satisfying the Max-Stable property, i.e., $C(u_1^t, \ldots, u_d^t) = C^t(u_1, \ldots, u_d)$ for all $t > 0$ and $u_i \in [0, 1]$. In fact, for all $x_i \in \Re$, it holds $F(x_1, \ldots, x_d) = C(F_1(x_1), \ldots, F_d(x_d))$.

The Extreme Value Theory is a fundamental tool in applications, since it provides the theoretical framework for performing risk assessment and rational decision-making in all areas of geophysics.

Bibliography

Coles, S., 2001. *An Introduction to Statistical Modeling of Extreme Values*. Berlin: Springer.
Salvadori, G., De Michele, C., Kottegoda, N. T., and Rosso, R., 2007. *Extremes in Nature. An Approach Using Copulas*. Dordrecht: Springer. Water Science and Technology Library series, Vol. 56.

Cross-references

Biblical Events
Disaster
Frequency and Magnitude of Events
Hazard
Models of Hazard and Disaster
Prediction of Hazards
Risk

CASE STUDY

EYJAFJALLAJÖKULL ERUPTIONS 2010

Freysteinn Sigmundsson
Institute of Earth Sciences, University of Iceland,
Reykjavík, Iceland

Definition

Eyjafjallajökull. An ice-capped volcano in South Iceland that erupted in 2010 and disrupted air traffic. The ice cap covering the volcano has the same name as the volcano itself.

Eyjafjallajökull eruptions 2010. Two eruptions at Eyjafjallajökull in 2010. An explosive summit eruption beginning 14 April, with sustained activity until 22 May. Ash transported toward mainland Europe led to closure of large part of European airspace for many days, with global disruption of air traffic and economic influence at an unprecedented scale. A preceding effusive eruption occurred 20 March–12 April at the eastern flank of the volcano, in the Fimmvörðuháls area just east of the Eyjafjallajökull ice cap. The effusive eruption is also referred to as the Fimmvörðuháls eruption and the explosive summit eruption as the Eyjafjallajökull eruption of 2010. The two eruptions are also sometimes referred to as a single eruption with different phases of activity.

Introduction

Iceland is a subaerial part of the divergent plate boundary between the North American and Eurasian plates, with over 30 active volcanic systems and eruptions occurring an average every 3–4 years (Thordarson and Larsen, 2007; Sigmundsson, 2006). Basaltic eruptions are common, but explosive eruptions of more evolved magma do also occur. Some of the most active volcanoes are subglacial, leading to generation of eruption plumes due to explosive phreatomagmatic fragmentation of magma even in basaltic eruptions that would, without ice cover in the source area, have generated lava flows. Although the effects of many eruptions in Iceland remain local, there is the potential for widespread influence from long-range transport of ash and gas from Icelandic eruptions. This was the case for the explosive eruption of Eyjafjallajökull volcano that began on 14 April 2010 and directly influenced more people on Earth than any preceding eruption. Advice of the Volcanic Ash Advisory Centre in London (London VAAC) formed the basis for closure of large part of European airspace 15–21 April, leading to cancellation of over 100,000 European flights (Oxford Economics, 2010). Additional airspace closure occurred in early May in a limited area.

Tectonic setting and volcanic unrest

Mt. Eyjafjallajökull (~1,660 m.a.s.l.) is a moderately active ice-capped volcano with eruptive periods producing both tephra (airborne eruptive material) and lava, separated by hundreds of years of dormancy. The volcano is located south of the spreading part of the Eastern Volcanic Zone of Iceland and has a tectonic setting comparable to an intraplate volcano within the Eurasian Plate (Figure 1). The volcano is aligned in an east–west direction, with a shape intermediate between that of a stratovolcano and an elongated shield, with a small summit caldera. Its edifice links to the edifice of the neighboring Katla volcano, one of the most active volcanoes in Iceland that has had ten explosive eruptions breaking through the ice cover of the overlying Mýrdalsjökull since 1500 A.D. (Larsen, 2000; Eliasson et al., 2006). Volcanic hazards due to eruptions at Eyjafjallajökull include tephra fallout and short lava flows, but the main local threats of eruptions of Eyjafjallajökull are jökulhlaups, sudden glacial outburst floods that come swiftly down the slopes of the volcano when eruptions melt their way through overlying ice (Gudmundsson et al., 2008).

Eyjafjallajökull had a prolonged summit eruption 1821–1823, when a short phreatomagmatic phase in December 1821 was followed by a yearlong period of intermittent magmatic/phreatomagmatic activity and flooding (Larsen, 1999). After that, no activity is known at Eyjafjallajökull until unrest began in 1992 with increased seismicity. Intrusive episodes occurred in 1994 and 1999–2000, witnessed as ground deformation and seismicity interpreted in terms of $\sim(10-30) \times 10^6$ m^3 sill intrusions at 4.5–6.5 km depth (Sigmundsson et al., 2010;

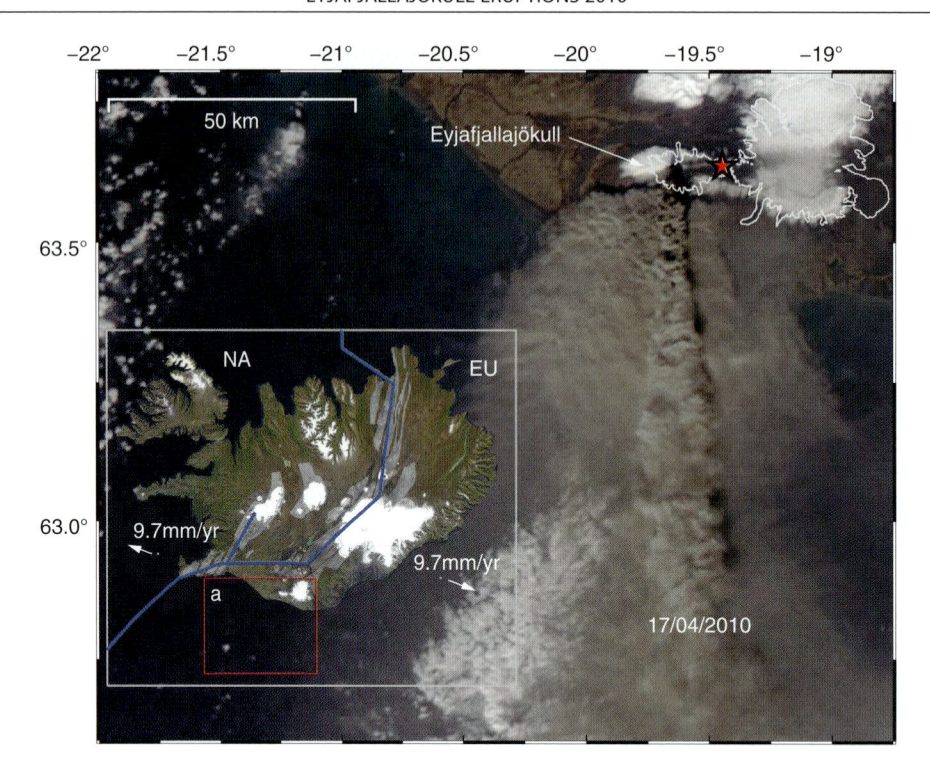

Eyjafjallajökull Eruptions 2010, Figure 1 Satellite image (MODIS) of the eruption plume emanating from Eyjafjallajökull main crater on 17 April 2010. Outline of ice caps shown with *gray line*. *Red star*, location of the preceding flank eruption. *Inset*, satellite image of Iceland overlain by fissure swarms in rift zones (*shaded grey*) and showing schematically in *blue* the main axes of the plate boundary between the North American (*NA*) and Eurasian (*EU*) plates. Half-spreading rate is 9.7 mm/year. (Modified from Sigmundsson et al., 2010). The MODIS image is courtesy of NASA/GSFC, MODIS Rapid Response (http://rapidfire.sci.gsfc.nasa.gov/).

Hjaltadóttir et al., 2009; Pedersen and Sigmundsson, 2004, 2006; Sturkell et al., 2003, 2010; Hooper et al., 2009). A peak in seismicity occurred also in 1996, without observed deformation. After 2000, no deformation was detected until mid-2009, when a period of elevated seismic activity was associated with 10–12-mm displacement at a continuously recording GPS geodetic station at the base of the volcano, related to minor intrusive activity.

Activity in 2010

In January 2010, deformation and elevated seismicity indicated the beginning of a new intrusive episode in the roots of Eyjafjallajökull that escalated until 20 March when an eruptive fissure opened at the eastern slopes of the volcano. The unrest signals were stronger than in the earlier intrusion periods, in particular in the last 3 weeks preceding the flank eruption, with more than 100 earthquakes recorded on some days and observed deformation rates of more than 5 mm/day. The deformation was well mapped with both GPS geodesy and InSAR (interferometric analysis of satellite synthetic aperture radar images). Interpretation of these data suggests evolution of a complicated intrusion in the volcano roots, with magma flow rates into the intrusion of 30–40 m^3/s after 4 March (Sigmundsson et al., 2010). Detailed seismic studies also reveal the evolution of the intrusive complex (Hjaltadottir and Vogfjord, 2010; Tarasewicz et al., 2012). The eruptive activity that began on 20 March ended 18 years of intermittent precursory activity of the volcano, associated with main intrusions formed in 1994 and 1999 in addition to the 2010 intrusion that evolved in the volcano roots for 3 months prior to delivering magma toward the surface.

From 20 March to 12 April, an effusive basaltic eruption at a short fissure complex at the eastern flank of the Eyjafjallajökull, just east of the ice cap, produced ~ 0.02-km^3 lava field (Edwards et al., 2012; Eyjólfur Magnússon, personal communication, 2012) (Figure 2). Activity on a short eruptive fissure focused quickly onto several craters; in early May, another short fissure opened up directly northwest of the initial one. During this eruption, lava flowed toward north into steep canyons producing lava falls and steam plumes originating at the lava fronts, as the advancing lava melted snow. Plume originating at the eruptive craters, and tephra produced, was miniscule. The volcano practically stopped deforming during this eruption, signifying magma drainage from a large depth through the intrusive complex formed in the preceding months, rather than draining and collapse of the intrusive complex (Sigmundsson et al., 2010). When the eruptive fissure opened in late evening of 20 March, the earthquake activity had been declining and the onset of eruption tremor was very gradual. Seismic signals of the immediate precursors in the preceding few hours and the

Eyjafjallajökull Eruptions 2010, Figure 2 The Eyjafjallajökull flank eruption during its later stages in early April 2010 (Photo: Freysteinn Sigmundsson.).

eruption beginning were so minor that the exact timing of the eruption was not forecast; the eruption was visually confirmed before it was recognized on instrument recordings. The eruption attracted a large crowd of tourists who enjoyed spectacular lava fountaining at close distance, as well as the lava falls into canyons. Lava expelled during the eruption is olivine- and plagioclase-bearing mildly alkali basalt with SiO_2 content about 46 wt% (Sigmarsson et al., 2011; Moune et al., 2012).

In late evening of 13 April, seismic activity renewed, this time under the ice-capped summit, signified new propagation of magma toward the surface (Tarasewicz et al., 2012). At about 1:30 h on 14 April, a new eruption broke out at the ice-capped summit of Eyjafjallajökull, with initial phase of the eruption (few hours) being fully subglacial. Ice cauldrons melted by heat of eruptive products, creating floods of meltwater rushing down the slopes of the volcano in jökulhlaups. The catastrophic explosive phase of the eruption began when the eruption had melted its way through the overlying ice cap. A complete cloud cover hindered initially visual observations in the summit region, but airborne SAR instrument on board an airplane of the Icelandic Coast Guard allowed mapping of the development of the ice cauldrons.

The height of the eruption plume was monitored throughout the eruption with weather radar at the Keflavik airport, at a distance of 155 km (Arason et al., 2011), as well as web cameras, and visual evaluations from ground and overview flights. The eruption was highly explosive on 14–18 April and again after 4 May, with

Eyjafjallajökull Eruptions 2010, Figure 3 The Eyjafjallajökull explosive eruption on 17 April 2010 (Photo: Eyjólfur Magnússon.).

a 5–10-km-high eruption plume (Figure 3). In the intervening interval, the plume was mostly below 5 km height. During that interval, a narrow lava stream flowed northward from the crater area, at the Gígjökull outlet glacier. It produced a massive white steam plume due to ice melt, in addition to the ash-loaded eruption plume. Westerly and northerly winds prevailed during most of the eruption (Petersen et al., 2012), carrying ash quickly toward mainland Europe (Prata and Prata, 2012). The latter explosive phase followed renewed flow of basaltic magma from depth into the volcano as witnessed by seismicity and deformation. Intense ash fallout occurred in some inhabited areas in south Iceland both in the initial as well as the latter explosive phase of the eruption, associated with total darkness during the periods of most intense tephra fall.

This eruption was the first in Iceland to be monitored with web cameras. They provided valuable timely information on pulsating activity of the plume and allowed the general population easy access to monitor the progress of the eruption. A number of other techniques were also used for the first time to monitor Icelandic eruptive activity, including acoustic observations of infrasound and lidar measurements of the arrival of the ash cloud in Europe.

Magma erupted during the summit eruption was benmoreitic to trachytic in composition with SiO_2 content in the range of 55–61 wt% (Sigmarsson et al., 2011). This silicic magma was mingled with basalt, revealed by a variety of basaltic, intermediate, and silicic glass in the eruption fallout. The geochemical studies thus show that the summit eruption was triggered by injection of basalt into silicic magma. This was also suggested by Sigmundsson et al. (2010) based on the pattern of deformation associated with the eruptions and intrusions. The deflation source associated with the explosive eruption under the summit of the volcano had not inflated during the preceding unrest. Rather, the recharging of the volcano was associated with inflow of basaltic magma under its eastern flank into an intrusive complex that later started to "leak" magma toward the surface in the flank eruption. The meeting of the two different magmas, silicic magma residing under the summit area and newly arrived basaltic magma, was a key feature triggering the summit eruption.

Ash generated in the explosive eruption was unusually fine grained, fragmented by two processes. Purely brittle magmatic fragmentation due to gas exsolution in the eruption conduit occurred throughout the eruption. This fragmentation process was augmented by fragmentation due to magma-water interaction at least in the initial phase of the eruption; the eruption plume was rich in water vapor during the initial explosive phase but dry during the later one (Dellino et al., 2012). A study using nanotechniques demonstrated that ash particles from the initial explosive phase were especially sharp and abrasive over their entire size range from submillimeter to tens of nanometers

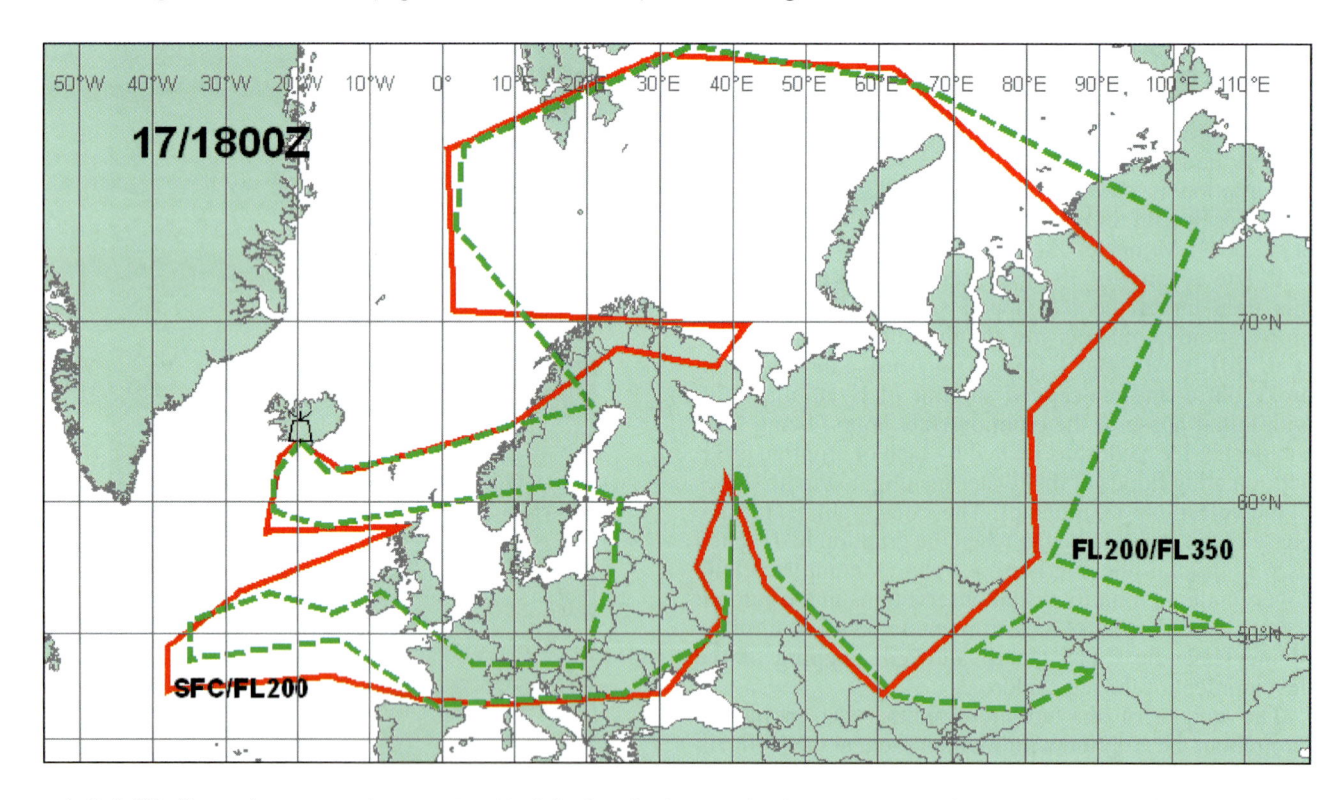

Eyjafjallajökull Eruptions 2010, Figure 4 Graphical display of volcanic ash advisory on 17 April 2010 at 18:00 GMT from the London VAAC. *Red and green outlines* delimit airspace with predicted contamination of ash. *Red outline* from the Earth's surface to flight levels (FL) 200, corresponding to about 20,000 ft. *Dashed line* for FL200-350, or from about 20,000 to 35,000 ft.

(Gislason et al., 2011). Long-range ash transport was facilitated by the fine-grained ash, but aggregation of ash particles was an important process contributing to the fallout (Taddeucci et al., 2011). The amount of erupted material during the summit eruption is estimated to corresponding to dense rock equivalent of 0.18 ± 0.05 km^3 (Gudmundsson et al., submitted manuscript).

Eruption response and impact

Monitoring of eruptive activity was led by the Icelandic Meteorological Office and the Institute of Earth Sciences at University of Iceland (Gudmundsson et al., 2010) that provided advice to civil protection authorities. A National Crisis Coordination Centre operated by Icelandic civil protection authorities, hosted by the National Commissioner of the Icelandic Police, coordinated communication with the public, local, national, and international authorities; scientists; and the media and was the hub for emergency operations. Good response plans were in place at a local level because of prior evaluation of the volcanic hazards and training of population in the area. During the 2010 eruptions, the area surrounding the volcano was temporarily evacuated for short periods during times of uncertain development of eruptive activity and jökulhlaups.

The catastrophic effect of the explosive eruption was the closure of large part of European airspace for many days, based on the guidance of the International Civil Aviation Organization (ICAO) that ash in the atmosphere should be avoided by aircraft. Information on eruption plume height, translated into estimates of mass eruption rate using an empirical relation, was used by the London VAAC to run models of transport and dispersion of the eruption cloud that formed the basis for their advisories of regions where ash was forecasted to be present (Webster et al., 2012). With this approach, a large part of European airspace became closed (Figure 4). During the eruption, a revised procedure was, however, agreed on in Europe, allowing aircraft to fly within the predicted volcanic ash cloud in regions with low levels of ash. The new limits are organized into three levels with a "no fly" high contamination zone for concentrations greater than 4 mg/m^3 (see, e.g., London VAAC, 2011; Hooper et al., in press). These zones apply in European airspace and have not been accepted for global use. These new rules imply that ash concentrations can both be measured, e.g., by using multispectral satellite measurements (Prata and Prata, 2012) and forecast, a major challenge for the response to future volcanic eruptions.

Bibliography

Arason, P., Petersen, G. N., and Björnsson, H., 2011. Observations of the altitude of the volcanic plume during the eruption of Eyjafjallajökull, April–May 2010. *Earth System Science Data*, 3, 9–17.

Dellino, P., Gudmundsson, M. T., Larsen, G., Mele, D., Stevenson, J. A., Thordarson, T., and Zimanowski, B., 2012. Ash from the Eyjafjallajökull eruption (Iceland): Fragmentation processes and aerodynamic behavior. *Journal of Geophysical Research*, 117, B00C04, doi:10.1029/2011JB008726.

Economics, O., 2010. *The Economic Impacts of Air Travel Restrictions Due to Volcanic Ash, Report*. Oxford, UK: Abbey House, 12 pp.

Edwards, B., Magnússon, E., Thordarson, T., Gudmundsson, M. T., Höskuldsson, A., Oddsson, B., and Haklar, J., 2012. Interactions between lava and snow/ice during the 2010 Fimmvörðuháls eruption, south-central Iceland. *Journal of Geophysical Research*, 117, B04302, doi:10.1029/2011JB008985.

Eliasson, J., Larsen, G., Gudmundsson, M. T., and Sigmundsson, F., 2006. Probabilistic model for eruptions and associated flood events in the Katla caldera, Iceland. *Computational Geosciences*, 10, 179–200.

Gislason, S. R., Hassenkam, T., Nedel, S., Bovet, N., Eiriksdottir, E. S., Alfredsson, H. A., Hem, C. P., Balogh, Z. I., Dideriksen, K., Oskarsson, N., Sigfusson, B., Larsen, G., and Stipp, S. L. S., 2011. Characterization of Eyjafjallajökull volcanic ash particles and a protocol for rapid risk assessment. *Proceedings of the National Academy of Sciences*, 108, 7307–7312.

Gudmundsson, M. T., Larsen, G., Höskuldsson, Á., and Gylfason, Á. G., 2008. Volcanic hazards in Iceland. *Jökull*, 58, 251–268.

Gudmundsson, M. T., Thordarson, T., Höskuldsson, Á., Larsen, G., Björnsson, H., Prata, F. J., Oddsson, B., Magnusson, E., Högnadóttir, Th., Petersen, G. N., Hayward, C. L., Stevenson, J. A., and Jónsdóttir, I., (submitted manuscript). Ash generation and distribution from the April–May 2010 eruption of Eyjafjallajökull, Iceland, *Scientific Reports*.

Gudmundsson, M. T., Pedersen, R., Vogfjörd, K., Thorbjarnardóttir, B., Jakobsdóttir, S., and Roberts, M. J., 2010b. Eruptions of Eyjafjallajökull volcano. *EOS Transactions, American Geophysical Union*, 91, 190–191.

Hjaltadottir, S., and Vogfjord, K., 2010. Seismic evidence of magma transport in Eyjafjallajökull during 2009–2010, Abstract V21F-02 presented at 2010 Fall Meeting, AGU, December 13–17, San Francisco, CA.

Hjaltadóttir, S., Vogfjörð, K. S., and Slunga, R., 2009. *Seismic Signs of Magma Pathways Through the Crust in the Eyjafjallajökull Volcano, South Iceland*, Rep. VI 2009-13. Reykjavik: Icelandic Meteorological Office.

Hooper, A., Pedersen, R., and Sigmundsson, F., 2009. Constraints on magma intrusion at Eyjafjallajökull and Katla volcanoes in Iceland, from time series SAR interferometry. In Bean, C. J., et al. (eds.), *The VOLUME Project – Volcanoes: Understanding Subsurface Mass Movement*. Dublin: University College, pp. 13–24.

Hooper, A., J., Prata, F., and Sigmundsson, F., (in press). Remote sensing of volcanic hazards and their precursors. In *IEEE Proceedings, Special issue on Remote Sensing of Natural Disasters*.

Larsen, G., 1999. Gosið í Eyjafjallajökli 1821–23 (The Eyjafjallajökull eruption 1821–23, In Icelandic), Rep. RH-28-99. Reykjavík: Sci. Inst., Univ. of Iceland, 13 pp.

Larsen, G., 2000. Holocene eruptions within the Katla volcanic system, south Iceland: characteristics and environmental impact. *Jökull*, 49, 1–28.

London VAAC, 2011. Volcanic Ash Concentration Forecasts – Specifications of Data Formats for Data and Graphic Files – Implementation Date 31st March 2011 (memorandum). Downloaded in 2011 from http://www.metoffice.gov.uk/aviation/vaac/Changes_To_Ash_Concentration_Forecasts_V13web.pdf

Moune, S., Sigmarsson, O., Schiano, P., Thordarson, T., and Keiding, J. K., 2012. Melt inclusion constraints on the magma source of Eyjafjallajökull 2010 flank eruption. *Journal of Geophysical Research*, 117, B00C07, doi:10.1029/2011JB008718.

Pedersen, R., and Sigmundsson, F., 2004. InSAR based sill model links spatially offset areas of deformation and seismicity for the 1994 unrest episode at Eyjafjallajökull volcano, Iceland. *Geophysical Research Letters*, 31, L14610, doi:10.1029/2004GL020368.

Pedersen, R., and Sigmundsson, F., 2006. Temporal development of the 1999 intrusive episode in the Eyjafjallajökull volcano, Iceland, derived from InSAR images. *Bulletin of Volcanology*, **68**, 377–393.

Petersen, G. N., Bjornsson, H., and Arason, P., 2012. The impact of the atmosphere on the Eyjafjallajökull 2010 eruption plume. *Journal of Geophysical Research*, **117**, D00U07, doi:10.1029/2011JD016762.

Prata, A. J., and Prata, A. T., 2012. Eyjafjallajökull volcanic ash concentrations determined using Spin Enhanced Visible and Infrared Imager measurements. *Journal of Geophysical Research*, **117**, D00U23, doi:10.1029/2011JD016800.

Sigmarsson, O., Vlastelic, I., Andreasen, R., Bindeman, I., Devidal, J.-L., Moune, S., Keiding, J. K., Larsen, G., Höskuldsson, A., and Thordarson, Th, 2011. Remobilization of silicic intrusion by mafic magmas during the 2010 Eyjafjallajökull eruption. *Solid Earth*, **2**, 271–281.

Sigmundsson, F., 2006. *Iceland Geodynamics, Crustal Deformation and Divergent Plate Tectonics*. Chichester: Praxis/Springer, 209 pp.

Sigmundsson, F., Hreinsdóttir, S., Hooper, A., Árnadóttir, Th, Pedersen, R., Roberts, M. J., Óskarsson, N., Auriac, A., Decriem, J., Einarsson, P., Geirsson, H., Hensch, M., Ófeigsson, B. G., Sturkell, E., Sveinbjörnsson, H., and Feigl, K. L., 2010. Intrusion triggering of the 2010 Eyjafjallajökull explosive eruption. *Nature*, **468**, 426–430, doi:10.1038/nature09558, 2010.

Sturkell, E., Sigmundsson, F., and Einarsson, P., 2003. Recent unrest and magma movements at Eyjafjallajokull and Katla volcanoes, Iceland. *Journal of Geophysical Research*, **108**, B8–B13, doi:10.1029/2001jb000917.

Sturkell, E., Einarsson, P., Sigmundsson, F., Hooper, A., Ófeigsson, B. G., Geirsson, H., and Ólafsson, H., 2010. Katla and Eyjafjallajökull volcanoes. *Developments in Quaternary Sciences*, **13**, 5–21.

Taddeucci, J. P., Scarlato, P., Montanaro, C., Cimarelli, C., Del Bello, E., Freda, C., Andronico, D., Gudmundsson, M. T., and Dingwell, D. B., 2011. Aggregation-dominated ash settling from the Eyjafjallajökull volcanic cloud illuminated by field and laboratory high-speed imaging. *Geology*, **39**, 891–894.

Tarasewicz, J., Brandsdóttir, B., White, R. S., Hensch, M., and Thorbjarnardóttir, B., 2012. Using microearthquakes to track repeated magma intrusions beneath the Eyjafjallajökull stratovolcano, Iceland. *Journal of Geophysical Research*, **117**, B00C06, doi:10.1029/2011JB008751.

Thordarson, T., and Larsen, G., 2007. Volcanism in Iceland in historic time: volcano types, eruption styles and eruptive history. *Journal of Geodynamics*, **43**, 118–152.

Webster, H. N., Thomson, D. J., Johnson, B. T., Heard, I. P. C., Turnbull, K., Marenco, F., Kristiansen, N. I., Dorsey, J., Minikin, A., Weinzierl, B., Schumann, U., Sparks, R. S. J., Loughlin, S. C., Hort, M. C., Leadbetter, S. J., Devenish, B. J., Manning, A. J., Witham, C. S., Haywood, J. M., and Golding, B. W., 2012. Operational prediction of ash concentrations in the distal volcanic cloud from the 2010 Eyjafjallajökull eruption. *Journal of Geophysical Research*, **117**, D00U08, doi:10.1029/2011JD016790.

Cross-references

Aviation, Hazards to
Caldera
Emergency Management
Eruption Types (volcanic)
Galeras Volcano, Colombia
Jökulhlaup
Karakatoa (Krakatau)
Lava
Magma
Montserrat Eruptions
Nevado del Ruiz, Colombia
Plate Tectonics
Remote Sensing of Natural Hazards and Disasters
Santorini
Stratovolcano
Vesuvius
Volcanic Ash
Volcanic Gas
Volcanoes and Volcanic Eruptions

FAULT

William A. Bryant
California Geological Survey, Sacramento, CA, USA

Synonyms
Seismic source; Shear

Definitions
Fault – a tectonic fracture in the earth's crust along which displacement (horizontal, vertical, or diagonal) of one side relative to the other has taken place. The fracture may be either a discrete plane or a zone containing multiple fracture planes. Cumulative displacement may be measurable from millimeters to kilometers.

Slip – distance, measured on the fault plane, between two originally adjacent points situated on opposite sides of the fault. It would be represented by a straight line on the fault plane connecting these two points after displacement.

Introduction
The word fault was derived from a late eighteenth-century mining term used to describe a surface, or plane, across which coal layers were displaced (Twiss and Moores, 1992). Realization that sudden displacement along a fault produces earthquakes evolved from observations of four large events that ruptured the ground surface in the late nineteenth and early twentieth centuries: the 1888 Marlborough (New Zealand), 1891 Nobi (Japan), 1892 Tapanuli (Indonesia); and 1906 San Francisco (California) earthquakes (Yeats et al., 1997). Earthquakes are the dominant natural hazard associated with faults. Although strong ground shaking accounts for the vast majority of potential damage from earthquakes, significant damage to structures and infrastructure can result when the rupture of a fault at depth extends to and offsets the ground surface. Surface displacement can also disrupt drainage patterns, such as by damming streams, with resultant flooding and/or catastrophic breaching. Rapid uplift of the seafloor resulting from surface fault rupture is one of the ways seismic sea waves, or tsunamis, are formed.

Types of faults
Faults can be recognized by the juxtaposition of dissimilar rock types across a generally planar surface exhibiting textures and structures characteristically produced by shearing. A fault plane can be vertical, or inclined at an angle (dip) less than 90°. Dip is the acute angle between the fault plane and a horizontal surface. High-angle faults dip greater than 45°; low-angle faults dip less than 45°. Faults are classified according to the direction the bounding crustal blocks have been displaced parallel to the fault plane (Figure 1a–e). Dip-slip displacement is vertical movement oriented parallel to the dip of the fault plane. Horizontal offset parallel to the fault plane is known as strike-slip displacement. Oblique-slip faults will have components of both dip-slip and strike-slip displacement.

Dip-slip faults
Relative movement of crustal blocks bounding a fault plane determines whether a dip-slip fault is described as normal or reverse. The hanging wall is the structurally upper block bounding a dipping fault; the structurally lower block is the footwall. If the hanging wall has been displaced down relative to the footwall, the fault is considered a normal fault (Figure 1a). A reverse fault has the opposite sense of displacement: The hanging wall has moved up relative to the footwall (Figure 1b).

Normal faults
Normal faults generally form in response to extension of the earth's crust, and are found in regions such as oceanic

DIP-SLIP FAULTS

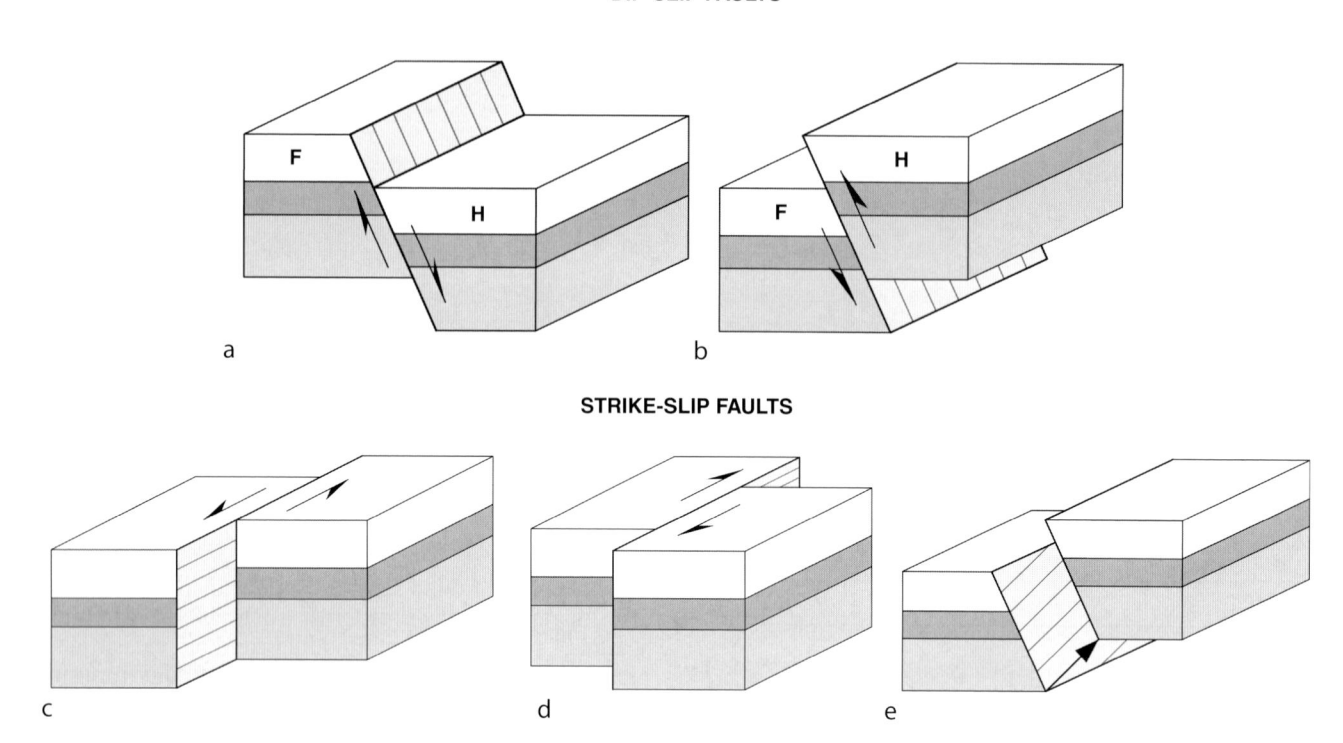

STRIKE-SLIP FAULTS

Fault, Figure 1 *Block* diagrams showing types of fault displacement. *Arrows* indicate direction of movement; *H* hanging wall, *F* footwall. Striations, or slickensides, on the fault plane generally denote direction of last displacement. (**a**) normal fault; (**b**) reverse fault; (**c**) left-lateral (sinistral) fault; (**d**) right-lateral (dextral) fault; (**e**) oblique-slip (sinistral-reverse) fault.

spreading centers and continental rift zones. The maximum principal stress axis (σ_1) is vertical and the minimum stress axis (σ_3) is horizontal. Although normal faults typically are steeply dipping (50°–70°), the dip angle of some normal faults decreases with increasing depth below the earth's surface (listric faults or low-angle detachment faults). Some earthquakes have been associated with low-angle normal faults, but there is still debate within the geologic community whether listric faults commonly produce earthquakes (Yeats et al., 1997; McCalpin, 1996). Normal faults vertically offset the ground surface and commonly are delineated by cliff-like features referred to as scarps. Examples of earthquakes produced by normal faults are listed in Table 1.

Reverse faults

Reverse faults are the result of compressive forces where the maximum compressive stress axis (σ_1, typically perpendicular to the fault's strike) is horizontal and the minimum compressive stress axis (σ_3) is vertical. Thrust faults are a subset of reverse faults where the fault plane dips less than 45°. Convergent plate margins (subduction zones and continent-continent collision zones) are the most significant and widespread compressional tectonic environments. Most major earthquakes have occurred along these convergent plate margins, contributing about 90% of the total energy released by earthquakes (seismic moment) in the twentieth century (Pacheco and Sykes, 1992). The devastating 2004 Mw 9.2 Sumatra-Andaman earthquake, and associated Indian Ocean tsunami, were the result of displacement along a 1,600-km section of the Sunda Megathrust Fault, an east-dipping subduction zone. The Himalayan Frontal Fault System is the locus of uplift of the massive Himalayan block resulting from the collision of the Indian and Eurasian plates (Nakata, 1989) (see Table 1 for selected reverse fault earthquakes). Surface traces of reverse faults typically are highly complex, irregular, and often are difficult to detect. This is especially true of thrust faults, which may form complex fault scarps, surface folding, and warping. Gordon and Lewis (1980) documented examples of variations in thrust fault scarp morphology.

Blind thrust faults

The blind thrust fault is a low-angle structure recognized as a significant seismic source in areas of active folding, including parts of Taiwan, Iran, Argentina, India, Algeria, and California (Stein and Yeats, 1989). The upper extent of the fault plane may terminate several kilometers below the ground surface. Surface expression of a blind thrust fault often is delineated by young folded terrain. Displacement along a blind thrust fault can deform overlying strata

Fault, Table 1 Representative historical earthquakes with surface faulting. Slip types: *RL* right-lateral, *LL* left-lateral, *R* reverse, *N* normal. Number in parentheses in Maximum displacement column is average displacement

Date and earthquake name	Fault	Slip type	Magnitude	Maximum displacement (m)	Surface rupture length (km)	Reference
1959 Hebgen Lake, Montana	Red Canyon, Hebgen	N	Ms 7.6	5.4 (2.5)	25	Myers and Hamilton (1964)
1987 Edgecumbe, New Zealand	Edgecumbe	N	Ms 6.6	2.9 (1.7)	18	Beanland et al. (1989)
1980 El Asnam, Algeria	El Asnam	R	Ms 7.3	6.5 (1.5)	31	Philip and Meghraoui (1983)
1999 Chi Chi, Taiwan	Chelungpu	R	Mw 7.6	12.7 (3.5)	72	Lin et al. (2001)
1857 Fort Tejon, California	San Andreas	RL	Ms ~8.3	9.1 (4.7)	360	Sieh (1978)
1920 Haiyuan, China	Haiyuan	LL	Ms 8.5	10 (7.3)	220	Zhang et al. (1987)

in a number of ways, including fault-bend folds and fault propagation folds (Suppe, 1985). Recent blind thrust fault earthquakes include the 1983 Mw 6.4 Coalinga, California earthquake, and the 1994 Mw 6.7 Northridge, California earthquake.

Strike-slip faults

Strike-slip displacement results from a horizontal maximum compressive axis (σ_1) oriented at an acute angle to the fault's strike, vertical intermediate compressive stress axis (σ_2), and minimum compressive stress axis (σ_3) oblique to the fault's strike. By convention, a strike-slip fault exhibits left-lateral, or sinistral, displacement if the crustal block opposite the block on which an observer stands is offset to the observer's left (Figure 1c). A right-lateral, or dextral, offset occurs when the opposite crustal block is displaced to the observer's right (Figure 1d). Strike-slip faults often are the longest faults on continental landmasses, extending for hundreds to thousands of kilometers, and usually are delineated by prominent geomorphic expression (Weldon et al., 1996). Because strike-slip faults have steep to vertical dips, their surface trace is relatively straight, even through high-relief topography. Some of the most hazardous strike-slip faults form transform plate boundaries, such as the San Andreas Fault in California and the Alpine Fault in New Zealand. Table 1 lists representative strike-slip fault earthquakes.

Fault creep

Displacement that occurs in the uppermost part of the earth's crust without the occurrence of large earthquakes is known as fault creep. Fault creep was first recognized on the central part of the San Andreas Fault at a winery about 17 km south of San Juan Bautista, California (Tocher, 1960). Additional strike-slip faults of the San Andreas Fault System (Hayward, Calaveras, Concord-Green Valley, Maacama faults), the North Anatolian Fault in Turkey, and some reverse faults, such as the Buena Vista Thrust in central California and the Chihshang Fault in the Longitudinal Valley of eastern Taiwan, also exhibit fault creep. Displacement due to fault creep can be steady state, episodic, or episodic with steady-state background. Of critical importance is whether fault creep extends down along the entire fault plane, or whether the fault is locked below a certain depth. A fault that is locked through most of the seismogenic zone (typically 10–20 km below earth's surface) will be able to store seismic moment at a much faster rate, and can be more hazardous than a fault that slips freely from the surface to the base of the seismogenic zone. Some faults, such as the Hayward Fault in California, produce both fault creep and large earthquakes.

Active fault

The definition of "active fault" usually implies likelihood of future displacement that would constitute a geologic hazard. Wallace (1986) introduced the term "active tectonics" to denote "tectonic movements that are expected to occur within a future time span of concern to society." Most definitions of active fault have legal significance and vary according to the type of structure and the degree of acceptable risk. Jennings (1985) discusses the various definitions of active faults.

Seismic hazard assessment

The behavior of faults must be characterized in order to understand and mitigate earthquake hazard. Specific fault characteristics to consider include fault geometry (length, slip type, and dip), slip rate, recurrence interval of significant earthquakes, age of most recent event, and maximum earthquake magnitude (typically inferred from amount of displacement per event and expected rupture area of the fault). Slip rate is the displacement on a fault averaged over a time period usually involving several large earthquakes, and is typically expressed in millimeters per year. Recurrence interval is the elapsed time between earthquakes within a given magnitude range and can be measured by identifying geologic evidence of past seismic events (paleoseismology), or estimated using a fault's slip rate and expected slip per event. Ideally, paleoseismic investigations of faults can reveal the fault's geometry, slip rate, recurrence interval (and dates of past earthquakes, including the most recent event), and amount of displacement during past earthquakes. Realistically, these data often are difficult to obtain and it is extremely unlikely, if not impossible, to obtain all of this information at an individual study site. The difficulty of characterizing blind thrust faults is much greater than surface faults due to the concealed nature of this type of fault.

Designing buildings to withstand surface fault rupture is particularly challenging and generally not cost effective. Mitigation of surface fault rupture over the past several decades has been by avoiding the placement of buildings across faults that have potential for surface rupture. California enacted the Alquist-Priolo Earthquake Fault Zoning Act in 1972 following the destructive 1971 Mw 6.6 San Fernando earthquake (Bryant and Hart, 2007). This law prohibits constructing most buildings for human occupancy across active fault traces (faults with evidence of surface offset in the last 11,000 years). Nevada (Nevada Earthquake Safety Council, 1998), Utah (Christenson et al., 2003), and New Zealand (Kerr et al., 2003) have developed similar guidelines for mitigating the hazard of surface faulting.

Summary

Understanding faults was one of the earlier concerns of applied geology that originated from the practical problem of tracing coal beds. During the nineteenth century, geologists recognized a correlation between large earthquakes and ground displacement caused by faults. The question of whether earthquakes produced faults, or faults caused earthquakes, was answered following the 1906 San Francisco earthquake. Investigation of faults remains an important task for geologists in order to better understand and quantify seismic hazards caused by sudden, catastrophic fault displacement. In the past 20 years several countries, including the United States (Machette et al., 2004), New Zealand (GNS Science, 2010), Italy (Basili et al., 2008), and Japan (AFRC, 2009), have compiled databases of essential fault parameters necessary for seismic hazard assessment. These databases are useful not only for cataloging data that can be used in earthquake forecast models, but also to quantify what geologists and seismologists know and do not know about potential seismic sources. This information can be used to better direct future paleoseismology research to address the uncertainties of critical seismic source parameters such as slip rate, timing of past earthquakes, and expected earthquake magnitude. Geologists also need to continue to develop geomorphic and stratigraphic models for active folds in order to better describe paleoearthquakes on blind thrust faults (McCalpin, 2009).

Bibliography

AFRC, 2009. Active fault database of Japan. *National Institute of Advanced Industrial Science and Technology Active Fault Research Center.* Internet database, (http://riodb02.ibase.aist.go.jp/activefault/index_e.html). Accessed April 19, 2012.

Basili, R., Valensise, G., Vannoli, P., Burrato, P., Fracassi, U., Mariano, S., Tiberti, M. M., and Boschi, E., 2008. The database of individual seismogenic sources (DISS), version 3: summarizing 20 years of research on Italy's earthquake geology. *Tectonophysics*, **453**, 20–43.

Beanland, S., Berryman, K. R., and Blick, G. H., 1989. Geological investigations of the 1987 Edgecumbe earthquake, New Zealand. *New Zealand Journal of Geology and Geophysics*, **32**, 73–91.

Bryant, W.A., and Hart, E.W., 2007. *Fault-Rupture Hazard Zones in California*. Sacramento, CA: California Geological Survey Special Publication 42. Digital version only, electronic document available at ftp://ftp.consrv.ca.gov/pub/dmg/pubs/sp/Sp42.pdf

Christenson, G. E., Batatian, L. D., and Nelson, C. V., 2003. *Guidelines for evaluating surface-fault-rupture hazards in Utah*. Salt Lake City, Utah: Utah Geological Survey Miscellaneous Publication 03–06.

GNS Science, 2010. *New Zealand Active Fault Database*. Lower Hutt, New Zealand: Institute of Geological and Nuclear Science. Internet database, (http://data.gns.cri.nz/af/index.jsp)

Gordon, F. R., and Lewis, J. D., 1980. The Meckering and Calingiri earthquakes, October 1968 and March 1970. *Geological Survey of Western Australia, Bulletin*, **126**, 1–229.

Jennings, C. W., 1985. *An Explanatory Text to Accompany the 1:750,000 Scale Fault and Geologic Maps of California*. Sacramento, CA: California Department of Conservation, Division of Mines and Geology Bulletin 201.

Kerr, J., Nathan, S., Van Dissen, R., Webb, P., Brunsdon, D., and King, A., 2003. *Planning for development of land on or close to active faults*. Wellington, New Zealand: Ministry for the Environment, Publication Number ME 483.

Lin, A., Ouchi, T., Chen, A., and Maruyama, T., 2001. Co-seismic displacements, folding and shortening structures along the Chelungpu surface rupture zone that occurred during the 1999 Chi Chi (Taiwan) earthquake. *Tectonophysics*, **350**, 225–244.

Machette, M., Haller, K., and Wald, L., 2004. *Quaternary Fault and Fold Database for the Nation. U.S. Geological Survey Fact Sheet 2004–3033*. Reston, Va: U.S. Department of the Interior, U.S. Geological Survey. Internet database, (http://earthquake.usgs.gov/hazards/qfaults/)

McCalpin, J. P., 1996. Paleoseismology in extensional tectonic environments. In McCalpin, J. P. (ed.), *Paleoseismology*. San Diego: Academic, pp. 85–146.

McCalpin, J. P., 2009. Application of Paleoseismic data to seismic hazard assessment and neotectonic research. In McCalpin, J. P. (ed.), *Paleoseismology*, 2nd edn. San Diego: Academic.

Myers, W.B., and Hamilton, W., 1964. *Deformation Accompanying the Hebgen Lake Earthquake of August 17, 1959. U.S. Geological Survey Professional Paper 435-I*, pp. 55–98.

Nakata, T., 1989. Active faults of the Himalaya of India and Nepal. *Geological Society of America Special Paper 232*, pp. 243–264.

Nevada Earthquake Safety Council, 1998. Guidelines for evaluating potential surface fault rupture/land subsidence hazards in Nevada. Digital version only, electronic document available at http://www.nbmg.unr.edu/nesc/guidelines.htm.

Pacheco, J. F., and Sykes, L. R., 1992. Seismic moment catalog of large shallow earthquakes, 1900 to 1989. *Seismological Society of America Bulletin*, **82**, 1306–1349.

Philip, H., and Meghraoui, M., 1983. Structural analysis and interpretation of the surface deformations of the El Asnam earthquake of October 10, 1980. *Tectonics*, **2**, 17–49.

Sieh, K. E., 1978. Slip along the San Andreas fault associated with the great 1857 earthquake. *Seismological Society of America Bulletin*, **68**, 1421–1448.

Stein, R. S., and Yeats, R. S., 1989. Hidden earthquakes. *Scientific American*, **260**, 48–57.
Suppe, J., 1985. *Principles of Structural Geology*. Englewood Cliffs: Prentice-Hall.
Tocher, D., 1960. Creep on the San Andreas fault- creep rate and related measurements at Vineyard, California. *Seismological Society of America Bulletin*, **50**, 396–404.
Twiss, R. J., and Moores, E. M., 1992. *Structural Geology*. New York: W.H. Freeman and Company.
Wallace, R. E. (ed.), 1986. *Studies in Geophysics – Active Tectonics*. Washington, DC: National Academy Press.
Weldon, R. J., II, McCalpin, J. P., and Rockwell, T. K., 1996. Paleoseismology of strike-slip tectonic environments. In McCalpin, J. P. (ed.), *Paleoseismology*. San Diego: Academic, pp. 271–329.
Yeats, R. S., Sieh, K., and Allen, C. R., 1997. *The Geology of Earthquakes*. New York: Oxford University Press.
Zhang, W., Jiao, D., Zhang, P., Molnar, P., Burchfiel, B. C., and Deng, Q., 1987. Displacement along the Haiyuan fault associated with the great 1920 Haiyuan, China, earthquake. *Seismological Society of America Bulletin*, **77**, 117–131.

Cross-references

Creep (Fault)
Earthquake
Mitigation
Neotectonics
Paleoseismology
Plate Tectonics
Tectonic and Tectono Seismic Hazards
Tsunami
Zoning

FEDERAL EMERGENCY MANAGEMENT AGENCY (FEMA)

Vincent R. Parisi
Park Ridge, IL, USA

Definition

The Federal Emergency Management Agency (FEMA) is part of the U.S. Department of Homeland Security (DHS) and coordinates response and recovery to disasters that occur in the USA and its territories. FEMA's mission is to support citizens and first responders to build, sustain, and improve capabilities to prepare for, protect against, respond to, recover from, and mitigate all hazards.

In 1979, FEMA was established by an executive order that merged many separate disaster-related responsibilities into a single agency, and on March 1, 2003, FEMA became part of the DHS. As of December 2009, FEMA has responded to more than 2,900 presidentially declared disasters. FEMA has more than 3,700 full time employees and over 4,000 on-call employees available to support disaster field operations. FEMA headquarters are located in Washington, D.C., with ten regional offices across the country (see Figure 1).

FEMA coordinates federal response and recovery efforts to disasters under the Robert T. Stafford Disaster Relief and Emergency Assistance Act (Robert T. Stafford Disaster Relief and Emergency Assistance Act, as amended, and

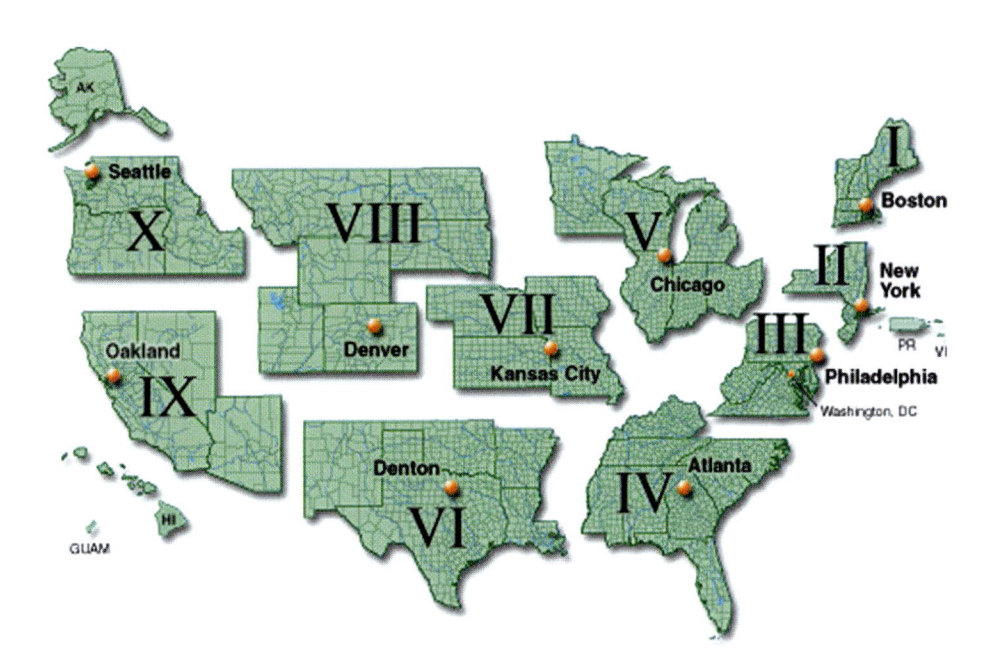

Federal Emergency Management Agency (FEMA), Figure 1 Regional office map.

Related Authorities as of June 2007). Once the President declares a major disaster, FEMA is authorized to provide disaster assistance and works in partnership with state and local emergency management agencies, numerous federal agencies, and the American Red Cross. FEMA provides assistance to individuals through grants for temporary housing and repair assistance when losses are not covered by insurance or other aid programs (Disaster Assistance Available from FEMA). FEMA also provides grant assistance to state, local, and tribal governments that is used to repair or rebuild public facilities damaged or destroyed by a disaster and to pay for debris removal and other emergency services (FEMA: Public Assistance Grant Program).

FEMA's other responsibilities include:

- Incident Management – FEMA works to respond swiftly and decisively to all hazards with around-the-clock support. The agency continues to professionalize its workforce by training and certifying staff in emergency management skills and techniques. FEMA also works closely with external partners to improve and update standards, and support the efforts of first responders (http://www.fema.gov/emergency/nims/index.shtm).
- Operational Planning – FEMA's Operational Planners assist state, local, and tribal jurisdictions with developing planning capabilities as well as write area- and incident-specific operational plans that will guide local response activities (FEMA: Plan).
- Integrated Preparedness – FEMA works closely with federal, tribal, state, and local governments; voluntary agencies; private sector partners; and the public to ensure the nation is prepared to respond to and recover from terror attacks, major disasters, and other emergencies. Preparedness information and tips can be found online at www.ready.gov.
- Hazard Mitigation – FEMA works proactively to reduce the physical and financial impact of future disasters through improved risk analysis and hazard mitigation planning, risk reduction, and flood insurance. FEMA helps implement effective hazard mitigation practices in order to create safer communities, promote rapid recovery from disasters, and reduce the financial impact at the federal, tribal, state, and local levels. Information about FEMA's mitigation efforts can be found online at www.fema.gov/government/mitigation.shtm.

Cross-references

Civil Protection and Crisis Management
Community Management of Hazards
Disaster Diplomacy
Disaster Relief
Disaster Risk Reduction (DRR)
Education and Training for Emergency Preparedness
Emergency Management
Emergency Shelter
Integrated Emergency Management System
Pacific Tsunami Warning and Mitigation System (PTWS)
Red Cross and Red Crescent
Risk Assessment

FETCH

Norm Catto
Memorial University, St. John's, NL, Canada

Synonyms
Fetch length

Definition
Fetch is the linear distance of open, unobstructed water lying beneath a wind traveling with a constant direction.

Overview

Fetch is calculated by measuring the distance from a point across open water of suitable depth for allowing wave travel, unobstructed by land masses including islands and shallow banks. Fetch is directional, and hence should include both a distance and orientation. Commonly, the fetch specified for a port is the maximum open water distance associated with a prevailing wind direction.

The fetch length is directly related to the height of waves, which can be generated by wind. Formation of waves requires the wind to frictionally interact with the water, transferring energy and overcoming both the inertia of the water and any existing waves traveling in different directions. Thus, the height of the waves generated increases with the fetch distance, as the duration of the wind–water interaction increases.

The maximum height of waves in open water potentially generated by a wind of given velocity traveling across a fetch of given length can be calculated, giving an estimate of anticipated wave heights. However, the actual wave heights measured may differ considerably from this theoretical value, depending upon local weather conditions and nearshore bathymetry. Generation of maximum wave heights also depends upon maintenance of constant wind velocity, for periods ranging from hours for meter-high waves to several days for the largest waves.

For a specific wind velocity, greater fetch will result in higher, more powerful waves. A coast exposed to a long fetch is thus more susceptible to high energy wave conditions, posing a greater hazard to infrastructure and residents.

Bibliography

Woodroffe, C. D., 2002. *Coasts: Form Process, Evolution*. Cambridge: Cambridge University Press.

Cross-references

Beaufort Wind Scale
Breakwater
Coastal Erosion
Coastal Zone, Risk Management
Displacement Wave
Hurricane

Ice and Icebergs
Levee
Rip Current
Rogue Wave
Seiche
Tsunami

FIRE AND FIRESTORMS

John Radke
University of California, Center for Catastrophic Risk Management, Berkeley, CA, USA

Synonyms

Catastrophic wildfire; Conflagration

Definition

Fire is the result of combustion (a series of chemical reactions) between a fuel (an organic compound such as wood) and an oxidant (oxygen source) producing heat, light, and often sound. A firestorm occurs when uncontrollable combustion, a conflagration, impacts the wind system, exacerbates the flow of oxygen to the fire, and sustains the risk to humans and their habitat.

Overview

Fire is a natural phenomenon that controls vegetation growth in healthy landscapes. It becomes a natural hazard when it negatively impacts humans and their habitat. As human habitat continues to expand globally, the expansion of human communities into the surrounding landscapes serves to exacerbate conditions for this type of hazard.

The nature of humans to control and manage their environment has led to the evolution of a new class of landscape on the urban edge, the wildland–urban interface WUI (Radke, 1995; Cova, 2005; Radeloff et al., 2005; Stewart et al., 2006). Here, houses and wildland vegetation coincide to form a heterogeneous environment where fire suppression policies and misguided fuel management strategies have often resulted in extremely hazardous fire conditions (Russell and McBride, 2003). Such conditions have recently given rise to a new class of conflagrations that appear to be on the rise in number and intensity (Pagni, 1993; Hartzell, 2001; Rey, 2003; Radke, 2007).

Typical wildland fires consist of rapidly heated vegetation that undergo pyrolysis, emit flammable gases, mix with oxygen, combust, and spread. The speed and intensity of these fires vary as a function of vegetation type (fuel), topography, and weather conditions (Rothermel, 1972; Burgan and Rothermel, 1984). Such conditions are typically modeled and attempts are made to predict and mitigate fires (Radke, 1995; Finney, 1998; Finney et al., 2005). However, in the WUI, the mixing of houses and vegetation creates a new fuel condition where ignition and rate of spread is enhanced by vegetation, and fire temperature and intensity is exacerbated by the addition of extreme fuel loads in the form of houses and other structures. Add to this the decoration of individual properties with exotic plants to suit individual tastes and one gets a very unique, complex, and dangerous landscape where new predictive fire models are needed for mitigation (Radke, 2007).

Firefighting involves modeling, classifying, preventing, and ultimately extinguishing fire. Proficient firefighting in both urban and wildland environments requires very different strategies and skills. Although success at fighting fires has occurred in both urban and wildland regions, the WUI presents a hybrid environment where strategies and techniques are still evolving. WUI fires can quickly become catastrophic, escalating out of control and seemingly driven by, and sustaining their own wind systems. Such extreme fire behaviors, or uncontrollable firestorms (Radke, 2007; Cohen, 2008; Sommers, 2008), require a new breed of firefighters and fire models (Mell et al., 2007; Rehm and Mell, 2009), if we are to successfully control them in the rapidly expanding WUI (Stewart et al., 2006).

Fire knows no political or administrative boundary yet humans attempt to control and manage fire and their landscapes within politically defined or administrative boundaries. The WUI is no exception. As this human urge spreads globally, we continue to enhance this natural hazard.

Bibliography

Burgan, R. E., and Rothermel, R. C., 1984. *BEHAVE: Fire behavior prediction and fuel modeling system-FUEL subsystem. Gen. Tech. Rep. INT-167*. Ogden: U.S. Department of Agriculture, Forest Service, Intermountain Research Station, p. 126.

Close, K. R., 2005. Fire behavior vs. human behavior: why the lessons from Cramer matter. In *Eighth International Wildland Fire Safety Summit*, April 26–28, Missoula.

Cohen, J., 2008. The wildland-urban interface fire problem: a consequence of the fire exclusion paradigm. Forest History Today, **Fall**, 20–26.

Cova, T., 2005. Public safety in the urban–wildland interface: should fire-prone communities have a maximum occupancy? *Natural Hazards Review*, 6(3), 99–108.

Finney, M. A., 1998. *FARSITE: fire area simulator-model development and evaluation. Res. Pap. RMRS-RP-4*. Ogden: U.S. Department of Agriculture, Forest Service, Rocky Mountain Research Station, p. 47.

Finney, M. A., McHugh, C. W., and Grenfell, I. C., 2005. Stand and landscape effects of prescribed burning on two Arizona wildfires. *Canadian Journal of Forest Resources*, **35**, 1714–1722.

Hartzell, T., 2001. Implementation of the National Fire Plan. A statement before the Sub- committee on Forests and Forest Health, Committee on Resources, United States House of Representatives, March 8, 2001. Hills Emergency Forum (HEF). Retrieved from http://www.hillsemergencyforum.org.

Mell, W. E., Jenkins, M. A., Gould, J. S., and Cheney, N. P., 2007. A physics-based approach to modeling grassland fires. *International Journal of Wildland Fire*, **16**, 1–22.

Pagni, P., 1993. Causes of the 20th October 1991 Oakland Hills conflagration. *Fire Safety Journal*, **21**, 331–340.

Radeloff, V. C., Hammer, R. B., and Stewart, S. I., 2005. The wildland urban interface in the United States. *Ecological Applications*, **15**(3), 799–805.

Radke, J., 1995. Modeling urban/wildland interface fire hazards within a geographic information system. *Geographic Information Science*, **1**, 7–20.

Radke, J., 2007. Modeling fire in the wildland–urban interface: directions for planning. In Troy, A., and Kennedy, R. (eds.), *Living on the Edge: Economic, Institutional and Management Perspectives on Wildfire Hazard in the Urban Interface*. Amsterdam: Elsevier Science.

Rehm, R. G., and Mell, W., 2009. A simple model for wind effects of burning structures and topography on wildland–urban interface surface-fire propagation. *International Journal of Wildland Fire*, **18**, 290–301.

Rey, M., 2003. 2002 Wildfire season and wildfire threats for the 2003 season. A statement before the Committee on agriculture. U.S. House of Representatives, July 23, 2003.

Rothermel, R. C., 1972. *A mathematical model for predicting fire spread in wildland fuels*. Ogden: U.S. Department of Agriculture, Forest Service, Research Paper INT-115. Intermountain Forest and Range Experiment Station.

Russell, W. H., and McBride, J. R., 2003. Landscape scale vegetation-type conversion and fire hazard in the San Francisco bay area open spaces. *Landscape and Urban Planning*, **64**, 201–208.

Sommers, W. T., 2008. The emergence of the wildland-urban interface concept. Forest History Today, **Fall**, 12–19.

Stewart, S. I., Radeloff, V. C., and Hammer, R. B., 2006. The wildland–urban interface in the United States. *Ecological Applications*, **15**(3), 799–805.

Cross-references

Coal Fire (Underground)
Forest and Range Fires (Wildfire)
Impact Fireball
Lightening
Mega-Fires in Greece (2007)
Nuee Ardente
Pyroclastic Flow
Wildfire

FLASH FLOOD

Yang Hong[1], Pradeep Adhikari[2], Jonathan J. Gourley[3]
[1]Center for Natural Hazards and Disaster Research, National Weather Center, University of Oklahoma, Norman, OK, USA
[2]Atmospheric Radar Research Center, University of Oklahoma, Norman, OK, USA
[3]NOAA National Severe Storms Laboratory, Norman, OK, USA

Synonyms

Freshet; Huayco

Definition

A flash flood is a rapid flooding of water over land caused by heavy rain or a sudden release of impounded water (e.g., dam or levee break) in a short period of time, generally within minutes up to several hours, a timescale that distinguishes it from fluvial floods. It also includes freshet, which is a great rise or overflowing of a stream caused by heavy rains or melted snow and huayco or huaico (Quechua term waygu meaning "depth, valley") which are Peruvian terms referring to flash floods caused by torrential rains in high mountains.

Overview

Characteristics: Flash floods are typically characterized by raging torrents in response to heavy rainfall that rip through riverbeds, urban streets, or mountain canyons sweeping away large debris and sediment with them. The Federal Emergency Management Agency (FEMA) characterizes it as a flood with a quick rise in water surface elevation with abnormally high water velocity often creating a "wall" of water moving down the channel and floodplain (FEMA, 1981).

Causes: Heavy rainfall from convective thunderstorms and tropical cyclones are the main causes of flash floods in most parts of the world. The rainfall-producing storms are often intense and impact a given region repeatedly due to stationarity of the system, backward regeneration, or enhancement by steep relief. Impermeable surfaces in urban regions and hydrophobic soils from wildfire as well as high antecedent soil moisture conditions exacerbate flash flood impacts.

Occurrence: Flash floods typically occur in tropical, arid and semiarid zones. They can occur in developed urban zones, steep mountainous regions, canyons, dry washes in deserts, transitional areas, and coastal zones. The climatology of their occurrence is related to the seasonality of the rainfall-producing storms.

Impact on life and property: Flash floods are a major natural hazard throughout the world. In the United States, flash floods are the number one killer among all weather-related hazards with approximately 140 lives lost each year. A vast majority of deaths from flash floods, as high as 90% in tropical countries, are due to drowning from victims being caught by rapidly rising waters (Smith and Petley, 2008).

Prediction: Predicting flash floods requires accurate detection and estimation of rainfall events, which are typically intense and very localized. Operational prediction methods include flash flood monitoring and prediction algorithms (FFMPA) used in Europe and the United States. FFMPA alerts forecasters when flash flooding is imminent based on radar-estimated rainfall amounts compared to hydrologic model–based rainfall thresholds. Advancements in forecasting convective rainstorms help to improve the performance of FFMPA by providing a longer lead time of impending flash floods.

Mitigation: Reducing societal exposure to flash floods is a key to any mitigation measure. Preventive activities including open space preservation, planning and zoning, property protection such as relocation and insurance, public education, emergency services in the form of warning and evacuation, and structural engineering projects such as levees, diversions, and channel improvements are some of the actions that mitigate the societal impacts of flash floods (Colombo et al., 2002).

Bibliography

Colombo, A. G., Hervás, J., and Vetere-Arellano, A. L., 2002. *Guidelines on Flash Flood Prevention and Mitigation.* Italy: European Commission Joint Research Center.

FEMA, 1981. *Design Guidelines for Flood Damage Reduction.* Washington, DC: Federal Emergency Management Agency.

Smith, K., and Petley, D. N., 2008. *Environmental Hazards: Assessing Risk and Reduction Disaster,* 5th edn. London and New York: Routledge.

Cross-references

Flood Deposits
Flood Hazard and Disaster
Flood Protection
Flood Stage
Floodplain
Floodway

FLOOD DEPOSITS

János Kovács
University of Pécs, Pécs, Hungary

Synonyms

Vertical-accretion deposits

Definition

Flood deposits are sediments derived from a running water source (river) and can consist of clay, gravel, sand, and silt. The sediments deposited are in an unconsolidated form until lithification.

Overview

Flood deposits are represented by channel deposits and overbank deposits associated with seasonal increases in river discharge caused by prolonged periods of snow melt, excessive rain, or monsoon conditions (Aslan, 2003). Lasting upward of several months, such "seasonal floods" are most typical of medium to large size rivers (e.g., Mississippi, Amazon, and Danube).

Subglacial outburst floods (jökulhlaup) that are episodic flood events accumulate *fluvial outburst flood deposits* (Duller et al., 2008). The structureless and diffusely stratified characteristics of depositional units are interpreted as the result of rapid deposition following flow deceleration at the stoss side of major antidunes, as well as at hydraulic jumps. This sedimentary evidence indicates that an outburst flood is a highly turbulent, sediment-charged flow (Duller et al., 2008).

According to Benito et al. (2003) and McKee (1938), *high-stage flood deposits* are characterized by a sheet-like layer of poorly sorted and relatively fine-grained material (pebble to silt size), directly overlain along a sharp contact by a well-sorted layer of cobbles and boulders. These two layers are interpreted to have been deposited in quick succession. The poorly sorted basal unit is deposited by a forward advancing stream flow that is losing turbulent energy and therefore also losing its suspended sediment load. This layer is in turn overlain by a slower advancing traction load of coarse boulders and cobbles. Late stage erosion, reworking of sediments, plus removal of fines typically occurs as progressively clearer flood waters wash over and dissect earlier deposited sediments.

In general, many floodplain deposits (channel and overbank deposits, or alluvium) are planar-stratified and small-scale cross-stratified, with fine to very fine sand interbedded with silt and clay (Aslan, 2003; Bridge and Demicco, 2008; Miall, 1996). The planar laminae are formed on upper-stage plane beds, and the small-scale cross strata are formed by current ripples. Small-scale cross strata are commonly of the climbing-ripple type and associated with convolute lamination. Ripples formed at the top of the sandy part of a strataset may be draped with mud that can show desiccation cracks. Wave-ripple marks can be superimposed on the ripples, representing the ponding stage of the waning flood. The upper parts of such stratasets are commonly bioturbated with root casts and animal burrows (Bridge and Demicco, 2008).

Bibliography

Aslan, A., 2003. Floodplain sediments. In Middleton, G. V. (ed.), *Encyclopedia of Sediments and Sedimentary Rocks.* Dordrecht: Springer, pp. 286–287.

Benito, G., Sánchez-Moya, Y., and Sopeña, A., 2003. Sedimentology of high-stage flood deposits of the Tagus River, Central Spain. *Sedimentary Geology,* **157,** 107–132.

Bridge, J., and Demicco, R., 2008. *Earth Surface Processes, Landforms and Sediment Deposits.* New York: Cambridge University Press.

Duller, R. A., Mountney, N. P., Russell, A. J., and Cassidy, N. C., 2008. Architectural analysis of a volcaniclastic jokulhlaup deposit, southern Iceland: sedimentary evidence for supercritical flow. *Sedimentology,* **55,** 939–964.

McKee, E. D., 1938. Original structures in Colorado River flood deposits of Grand Canyon. *Journal of Sedimentary Petrology,* **8,** 77–83.

Miall, A. D., 1996. *The Geology of Fluvial Deposits.* Berlin: Springer.

Cross-references

Avulsion
Flash Flood
Flood Hazard and Disaster
Floodplain
Flood Protection
Flood Stage
Floodway
Jökulhlaup (Débâcle)
Paleofloods
Sedimentation of Reservoirs

FLOOD HAZARD AND DISASTER

Yang Hong[1], Pradeep Adhikari[2], Jonathan J. Gourley[3]
[1]Center for Natural Hazards and Disaster Research, National Weather Center, University of Oklahoma, Norman, OK, USA
[2]Atmospheric Radar Research Center, University of Oklahoma, Norman, OK, USA
[3]NOAA National Severe Storms Laboratory, Norman, OK, USA

Synonyms

Deluge; Freshet; Spate; Inundation. The word flood comes from the Old English *flod*, a word common to Germanic languages (compare German *Flut*, Dutch *Vloed* from the same root as seen in *flow, float*). *Deluge* myths are stories of a great flood sent by a deity to destroy civilizations as an act of divine retribution and are featured in the folklore of many cultures.

Introduction

Definition

Floods are an overflow or inundation that comes from a river or other body of water and often threatens lives and properties. Therefore, any relatively high streamflow overtopping natural or artificial banks (e.g., levees) in any reach of a stream can be termed a flood. Floods can happen when the flow capacity of river channels, streams, or coastal areas is exceeded due to heavy, intense, or continuous rainfall or when the absorptive capacity of the soils is exceeded. This causes water in a river channel to overflow its banks onto adjoining land area, known as a floodplain. Floodplains are, therefore, highly prone to flood. In addition, coasts and deltas, areas directly below dams, inland shorelines, and alluvial fans are vulnerable to floods (Smith and Petley, 2008).

Flooding creates a significant threat to life and property. Communities situated near riverbanks or coastal zones are most vulnerable to flood, but many historic cities and towns have been built in such areas due to the conveniences of transportation, commerce, recreation, and pleasant scenery. Occasionally, flood hazard and flood risk are used interchangeably. Flood risk has the additional implication of the statistical chance of a particular flood actually occurring, whereas hazard is a naturally occurring or human-induced process which has the potential to create loss (Smith and Petley, 2008).

Generally, flooding invokes images of destruction and catastrophe. In contrast, floods also play an important role in the functioning of ecosystems, analogous to wildfire. Floodwaters bring many benefits such as increasing soil fertility by depositing nutrients from upstream and recharging ground water. Many aquatic species depend on normal flooding to wash debris into the water, which they subsequently use for shelter and food. Periodic floods also transport eroded soil and other materials that are essential for delta areas and coastal marshes to persist over time and sustain the wetland ecosystems. Freshwater floods in particular play an important role in maintaining ecosystems in river corridors and are a key factor in maintaining floodplain biodiversity (APFM, 2006). Many ecosystems develop with regular flooding as one of the key components to their existence. For example, the Sundarbans is the largest single block tidal forest of the world, located in the Ganges river delta. Similarly, it has been found that flooding was the key to the well-being and prosperity of ancient communities along the Nile, the Tigris-Euphrates, the Yellow, and the Ganges Rivers.

Flood generation

There are various factors that contribute to flood generation, with the primary ones being precipitation and snowmelt. In regions without extended periods of below-freezing temperatures, floods usually occur in the season of highest rainfall. The flood season is spring and early summer in regions where floods result from snowmelt, often accompanied by rainfall. Some other causes of floods include impediments to flood paths due to natural or man-made structures (e.g., ice jams, log jams, bridges, weirs), sudden dam failure and rapid snowmelt in the catchment basin (UNESCO, 2001). Because floods can transport large volumes of sediment and debris, they are often associated with landslides, mudflows, and debris flows. Figure 1 summarizes the causes of flood generation.

Flood measurement

Hydrologists generally quantify floods in two different ways. One is in absolute quantitative terms including discharge (m^3/s or ft^3/s), specific discharge (i.e., discharge normalized by catchment area or unit width of channel) or the height (stage) that the floodwaters reach (in m or ft). In contrast, a relative approach of quantifying flood is to express the magnitude of flood discharge or water depth based on the probability that the discharge or stage height will equal or exceed a certain level on a recurring basis, in terms of their statistical frequency. A "100-year flood" describes an event with a magnitude corresponding to a 1% probability of that flood occurring in any given year. This concept does not mean such a flood will occur only once in 100 years. Whether or not it occurs in a given year has no bearing on the fact that there is still a 1% chance of a similar occurrence in the following year. Any other statistical frequency of a flood event may be chosen depending on the degree of risk that is selected for evaluation; for example, 5-year, 20-year, 50-year, 500-year floodplain (DSD, http://www.oas.org/USDE/).

Frequency of flooding depends on the climate, the conditions of the banks of the stream, and the channel slope. Where substantial rainfall occurs in a particular season each year, or where the annual flood is derived principally from precipitation and snowmelt, the floodplain may be inundated nearly every year, even along large streams with

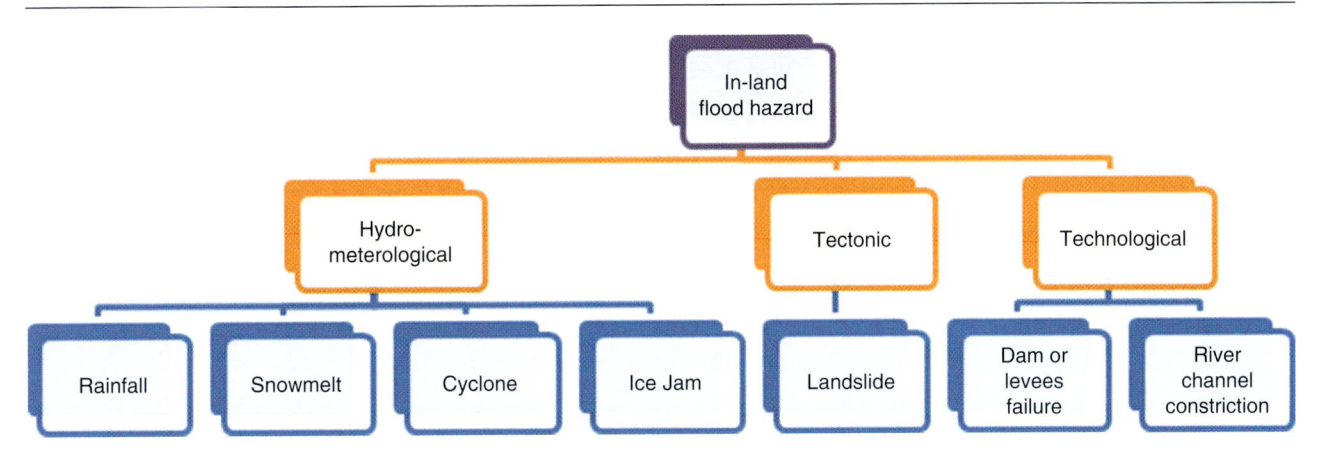

Flood Hazard and Disaster, Figure 1 Physical causes of in-land floods (Modified from Smith and Petley, 2008).

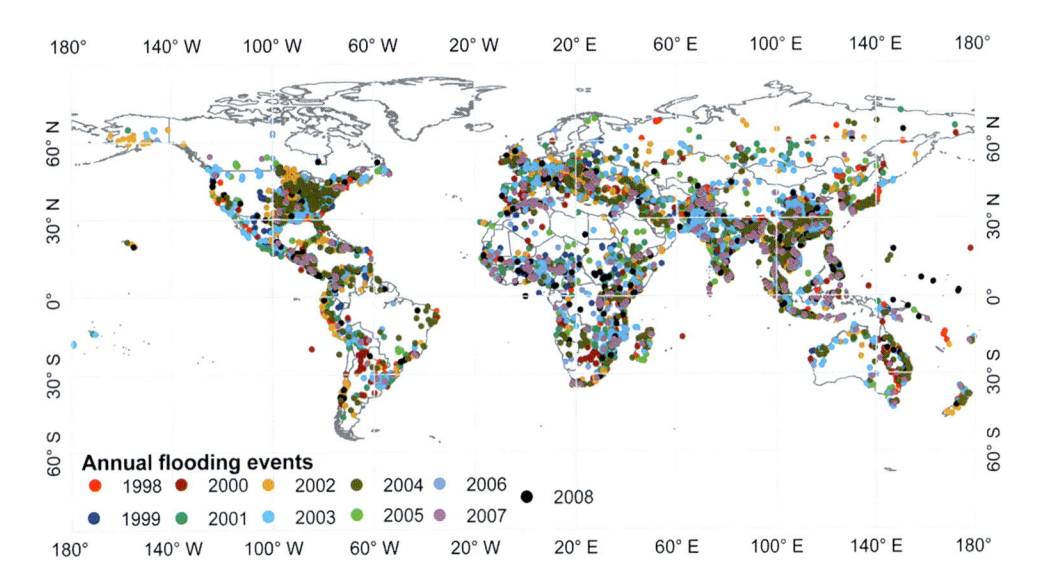

Flood Hazard and Disaster, Figure 2 Global flooding events 1998–2008 (Source: Adhikari et al., 2010).

very small channel slopes. The Yangtze River in China, the Ganges River in India and Bangladesh, and the Mekong River in Vietnam are some examples where flooding occurs annually, causing significant losses of life and property.

Spatiotemporal distribution of flood

Floods occur frequently in most parts of the populated world, inundating land and killing thousands of people annually. Figure 2 presents the distribution of annual floods for 1998–2008 based on the Global Flood Inventory (Adhikari et al., 2010).

A number of flood databases exist. The Emergency Disasters Data Base (EM-DAT) managed by the Centre for Research on the Epidemiology of Disasters (CRED) at the Catholic University of Louvain, Belgium, is a publicly accessible international database with information on natural and technological disasters including floods. The United Nations Office for the Coordination of Humanitarian Affairs (OCHA) through its portal ReliefWeb provides current disaster information on humanitarian emergencies and disasters relief works (ReliefWeb, http://www.reliefweb.int/). It does not intend to provide a comprehensive database of disaster events, but can serve as a valuable resource to verify current events and obtain additional disaster details for rapid response and humanitarian support. The Dartmouth Flood Observatory (DFO) compiled the Global Archive of Large Flood Events, which covers events from 1985 to the present in a simple Microsoft Excel spreadsheet format (DFO, http://www.dartmouth.edu/~floods/). This database also includes links to high-quality maps for selected events since 2006, showing the entire affected region. This database is exclusively dedicated for flood hazards. It has

Flood Hazard and Disaster, Table 1 Flood databases showing number of events and direct fatalities. Note that the variations among the database for the selected base period, 1998–2008

DATA Base	Events		Fatality	
	No. of events	Variation with respect to GFI (±)%	No. of fatality	Variation with respect to GFI (±)%
1. EM-DAT	1,796	−33	95,823 (96,000)	−61
2. DFO	2,132	−21	391,814 (392,000)	+60
3. GFI	2,700	N/A	246,087 (246,000)	N/A

Source: (Adhikari et al., 2010; DFO; and EM-DAT)

about 3,400 events recorded for 1985–2008. The Global Flood Inventory (GFI) has compiled global flood data for 1998–2008 from publicly available online resources, irrespective of their scale of impacts so as to have a comprehensive flood database. It has geo-referenced locations of all the flooding events for a period of 11 years. The Global Flood Inventory has record of approximately 2,700 events in total for the period, which is about 250 per annum (Adhikari et al., 2010). The discrepancies in the database discussed above are mainly due to the inherited biases as a result of the scope of the database. The entry criteria, sources of data, and definition of specific hazard terms differ amongst the databases. Table 1 shows an example of such variation in reporting the number of events and fatalities for the same time period of 1998–2008. It is obvious that analyses based on different database entries might result in contradicting conclusions.

It is important to note that some databases are impact-based making them potentially biased toward reporting more events in populated areas, whereas others are compiled from publicly available resources. The latter are prone to undermine the number of actual flooding events because only those floods that have significant effects on the community, local government, or national government are recorded. Therefore, a very careful interpretation of flood data is necessary before reaching conclusions regarding the number of events or their impacts to the people and the community. A more elaborate analysis on the different types of disaster databases and their scope and limitations can be found at Tschoegl (2006). Based on the GFI, it is found that the USA has recorded the highest number of flood events followed by China and India for the period 1998–2008. Table 2 gives the top ten countries which have witnessed the maximum number of flood events for the same period. In addition, the GFI shows a seasonal pattern in flooding, with the number of events increasing in May and peaking in the months of July to August (Figure 3).

In terms of the spatial distribution of flooding, Table 3 shows the percentage of reported flooding events for each region, sorted by year. Asia and Africa continuously recorded the highest percentage of events throughout the globe, followed by Southeast Asia, Central America, and the Caribbean.

Flood Hazard and Disaster, Table 2 Top ten countries with the most flooding events reported for 1998–2008

Rank	Name	No. of events
1	USA	216
2	China	193
3	India	126
4	Indonesia	120
5	Philippines	92
6	Vietnam	80
7	Australia	79
8	Russia	76
9	Afghanistan	63
10	Thailand	53

Source: Adhikari et al. (2010)

Socioeconomic impact of floods

Flooding is one of the most destructive types of natural disasters. In the twentieth century, out of the top 100 most fatal natural disasters, 12 were associated to flooding. Floods resulted in the death of seven million people (EM-DAT; The Disaster Center, http://www.disastercenter.com). Out of the 12 most fatal flood events, 10 occurred in China alone (Table 4).

Natural disasters such as flooding have the potential to eradicate decades of investments in infrastructure and the personal wealth of a country, in addition to the insurmountable loss of lives. The situation is much worse for developing regions. The poor are the most severely affected by natural disasters including floods. According to UN/ISDR (2004), people from the low-income category accounted for approximately half of those killed in floods from 1975 to 2000.

For 1998–2008, North America, Europe, and Oceania experienced much fewer fatalities compared to the rest of the world (Figure 4), with the notable exception of North America in 2005 due to Hurricane Katrina. It is observed that almost 75% of the fatalities worldwide occurred in Asia and Southeast Asia combined. Specifically, Southeast Asia had the highest number of flood fatalities, mainly as a result of cyclone Nargis that hit Myanmar in May 2008. Totally, 247,000 people were killed worldwide from 1998 to 2008 (Figure 5). The seasonal trend of flood fatalities generally follows that of

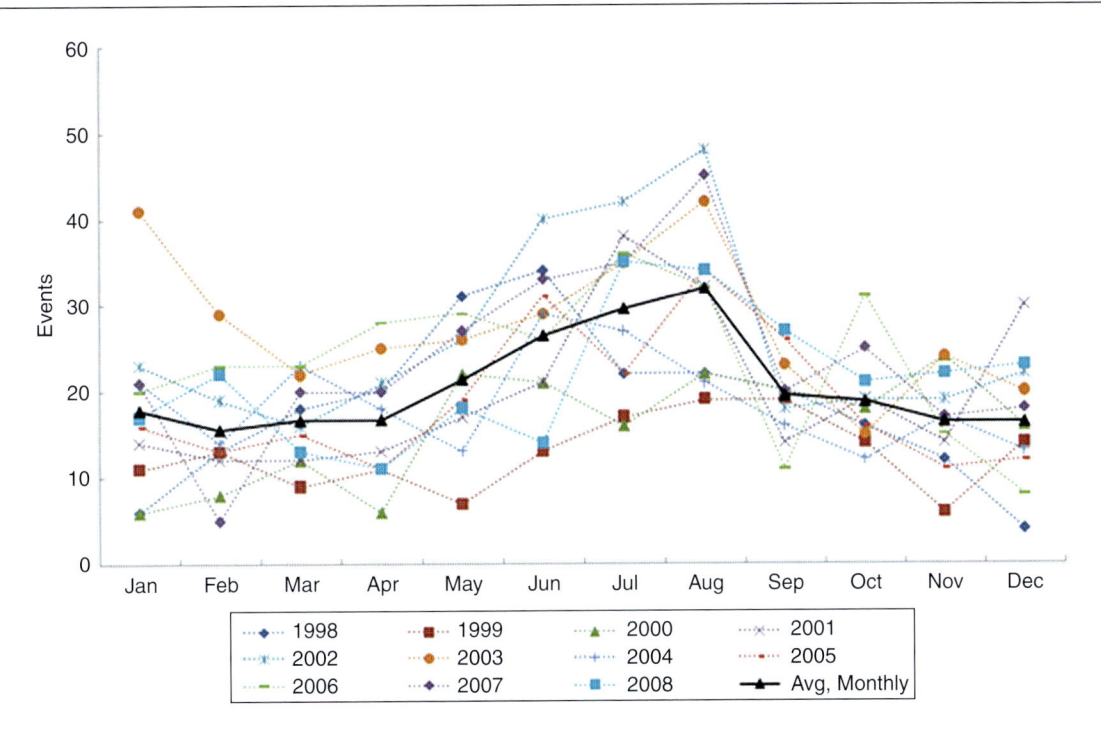

Flood Hazard and Disaster, Figure 3 Seasonal variation flood events for each year of the GFI record.

Flood Hazard and Disaster, Table 3 Flooding events recorded for 1998–2008 (in percentage) across the regions

Region	1998	1999	2000	2001	2002	2003	2004	2005	2006	2007	2008	Yearly average
Africa	17	17	18	23	24	22	14	33	29	25	23	23
Asia	32	22	25	20	23	20	27	20	21	23	20	17
Central America and Caribbean	7	14	8	18	13	14	7	15	15	10	10	12
Europe	14	17	18	16	13	12	12	8	10	9	7	11
North America	13	6	5	10	9	11	11	8	10	5	10	7
Oceania	5	3	3	5	7	9	7	6	5	5	5	6
South America	4	8	11	5	7	6	8	5	5	5	7	8
Southeast Asia	7	13	13	2	3	6	13	4	5	18	18	9

Source: Adhikari et al. (2010)

flooding events, but deaths extend into August through November. The month of May exhibited the highest fatality values; however, this may be skewed by the previously described tropical cyclone in Myanmar in 2008 (Figure 6).

Historical records show that the world has witnessed about 3,500 floods since 1900 till 2008 killing about seven million people globally. China is the country that witnessed the highest number of flood-related fatalities, a staggering 6.6 million during that time period (EM-DAT). Analyzing impacts for a longer duration from 1970 to 2008 based on the data from EM-DAT, it is observed that the world experienced total damages around 1.8 trillion USD from natural hazards, out of which 23% is attributed to floods alone. For the same time period, about three million people lost their lives; 8% of those were killed by floods (Figure 7).

It is clear from Figure 7 that, in terms of total fatalities for the 1970–2008 period, floods killed much less people in comparison to other natural hazards like storms, earthquakes, and droughts. But for the period 1985–1999 floods were the deadliest natural disaster followed by the earthquakes (Abramovitz, 2001). Irrespective of the discrepancies in the number of people killed or damages to local economies, the propensity of floods to harm human beings is very high and therefore requires attention from researchers, planners, and governments alike.

It is difficult to calculate an estimate of global economic impacts from floods alone. There is some information about the impact from all catastrophic natural hazards in which flooding is one of the major components. For 1985–1999, a total of US$920 billion were estimated to

Flood Hazard and Disaster, Table 4 Historical top 12 most fatal flood events

S. No.	Country	Year	Region	Continent	Fatality
1	China	1931	E. Asia	Asia	3,700,000
2	China	1959	E. Asia	Asia	2,000,000
3	China	1938	E. Asia	Asia	500,000
4	China	1939	E. Asia	Asia	500,000
5	China	1935	E. Asia	Asia	142,000
6	China	1908	E. Asia	Asia	100,000
7	China	1911	E. Asia	Asia	100,000
8	China	1949	E. Asia	Asia	57,000
9	Guatemala	1949	C. America	Americas	40,000
10	China	1954	E. Asia	Asia	30,000
11	Bangladesh	1974	S. Asia	Asia	28,700
12	China	1933	E. Asia	Asia	18,000

Source: EM-DAT; http://www.disastercenter.com/disaster/TOP100K.html

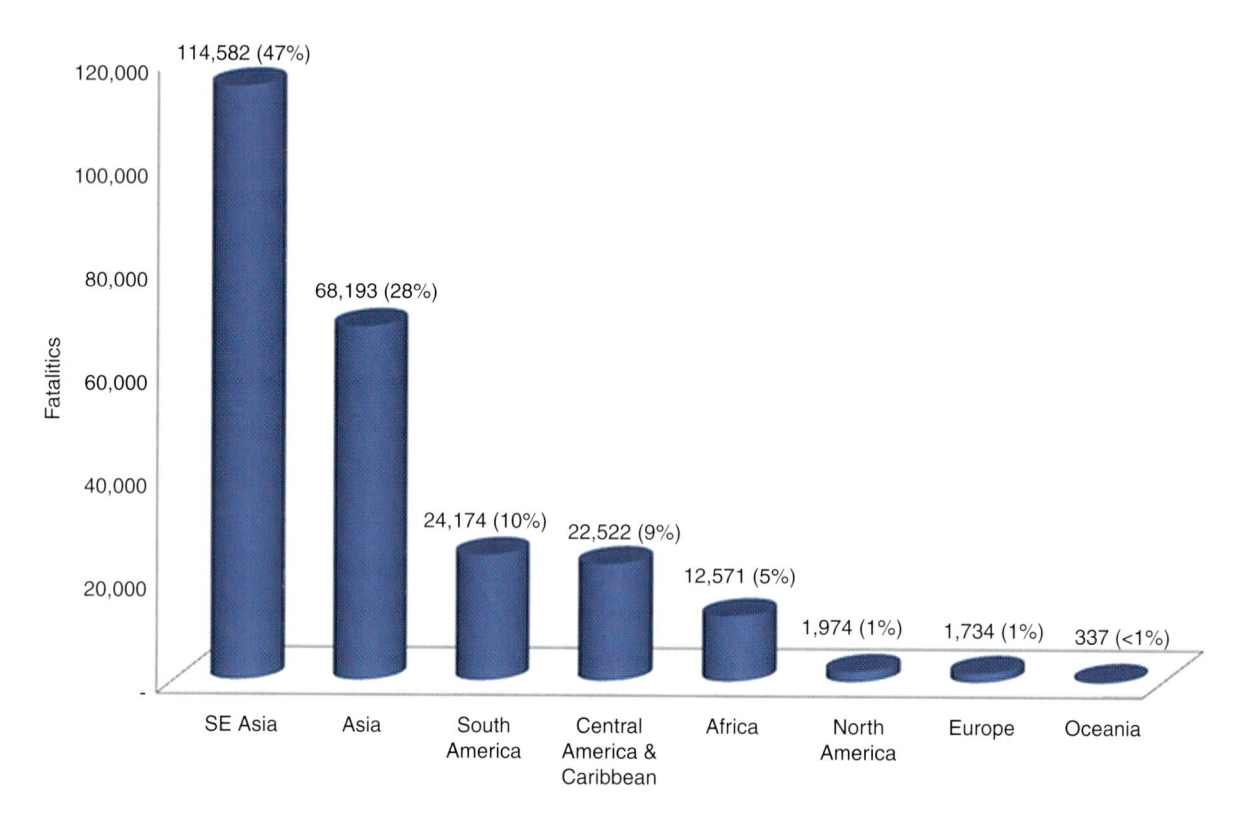

Flood Hazard and Disaster, Figure 4 Total fatalities for 1998–2008 segregated into regions. Southeast Asia witnessed the maximum fatality followed by Asia and South America.

have been lost from the natural hazards. The loss in richest and poorest countries is 57.4% and 24.4%, respectively. However, the loss in terms of Gross Domestic Product is 2.5% and 13%, respectively, in richest and poorest countries. This shows that disasters make the poorest countries even more vulnerable to future disasters (Abramovitz, 2001). The difference in loss from the disasters can be due to the phenomenal difference between rich and poor countries in infrastructure setup and better insurance coverage, which is the only reliable source of information to estimate the damages due to hazards.

Flood management

The ultimate objective of flood management is to minimize the destruction and damages that flood hazards can

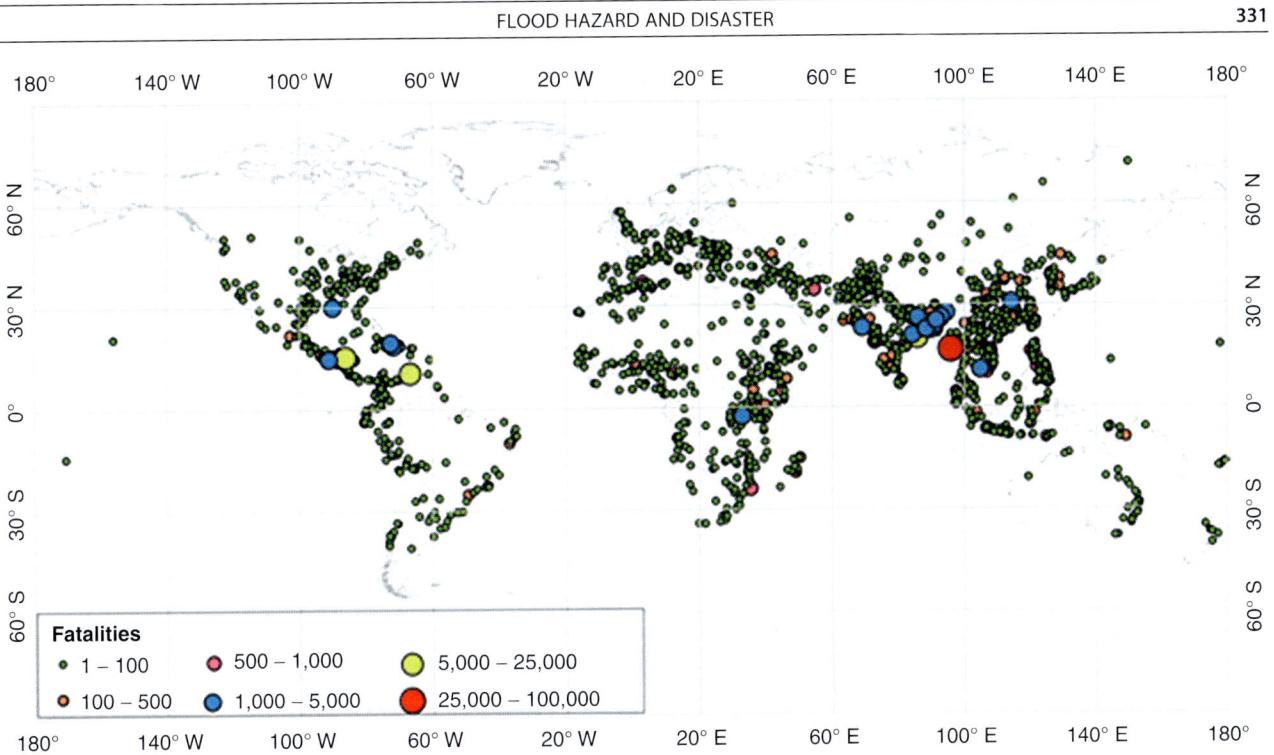

Flood Hazard and Disaster, Figure 5 Total number of fatalities for 1998–2008. Results show that Asia, Southeast Asia, Eastern Africa, Central America, and the Caribbean and South America regions experienced the most fatal floods.

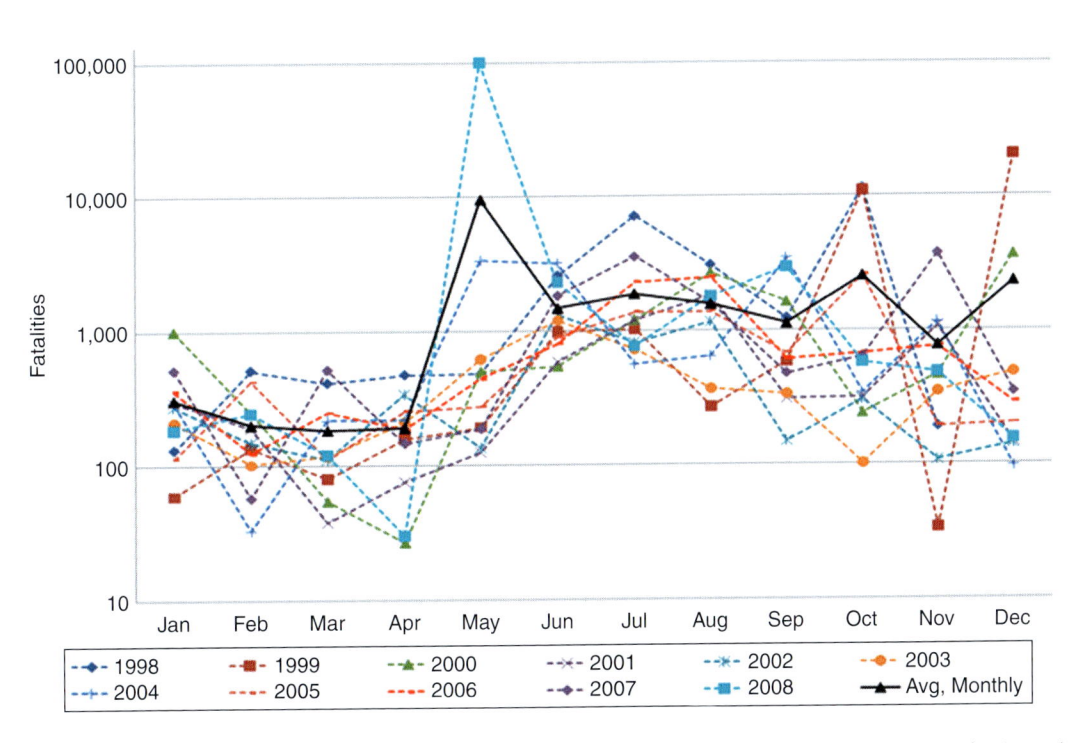

Flood Hazard and Disaster, Figure 6 Seasonal variation in flood fatalities for 1998–2008. Fatalities on vertical axis are in log scale.

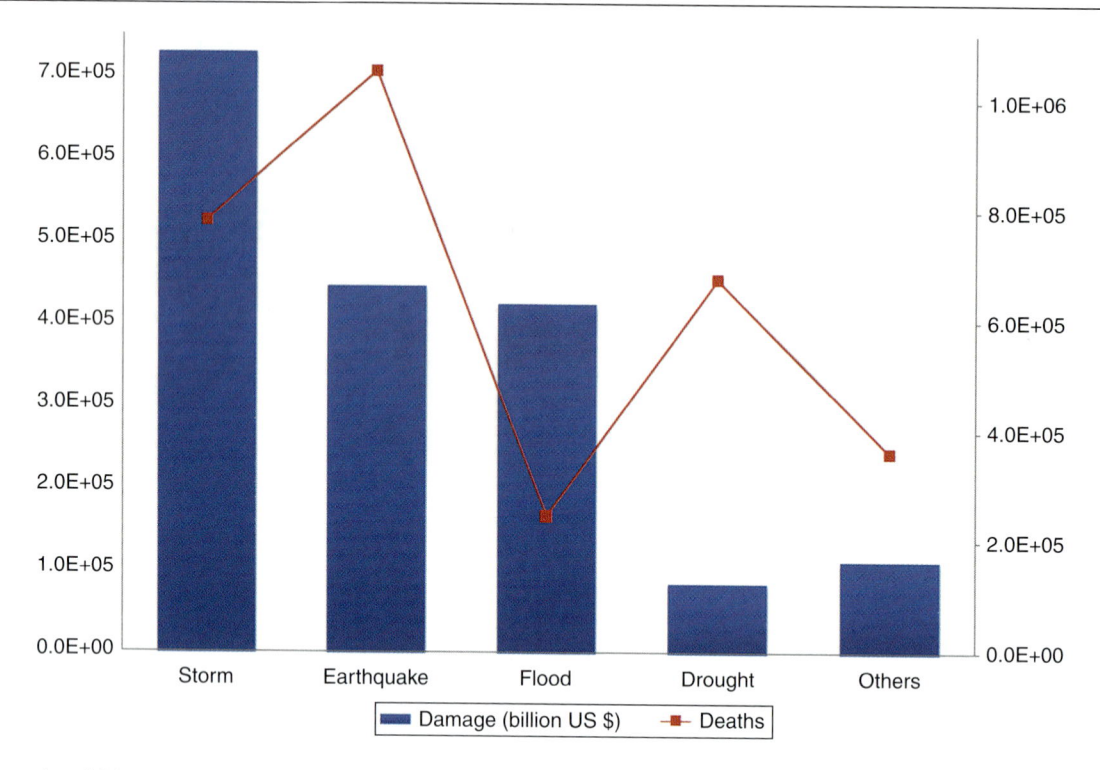

Flood Hazard and Disaster, Figure 7 Flood impacts for 1970–2008. The primary y-axis on *left* shows the damage in billion US dollar while the secondary y-axis on *right* shows the deaths as a result of natural hazards (Source: EM-DAT).

inflict on human lives and society. Flood management comprises three major components: adaptation, protection, and mitigation. These can be categorized into structural and nonstructural measures. These measures have their own advantages and disadvantages.

Structural measures

Some of the major structural measures to control floods and protect communities are dams, levees, and river channel improvements. They have played a key role in protecting the people and property around the globe from floods for centuries.

Dams

Dams are constructed on many rivers around the world. They are constructed for irrigation, to store runoff and reduce flooding downstream. Dams attenuate flood peaks by storing water in reservoirs and releasing it when river levels have fallen (Figure 8).

There is no agreement on the number of existing dams in the world. The World Commission on Dams reports 45,000 large dams worldwide but the Global Reservoir and Dam database mentions only about 33,000 large dams (WCD, 2000; Lehner et al., 2011). Three Gorges in China, Syncrude Tailings in Canada, and Chapetón in Argentina are some of the largest dams in the world. Dams can serve multiple purposes in addition to flood control, such as water supply, irrigation, hydroelectric power generation, recreation, and fishing. Most dams are multipurpose

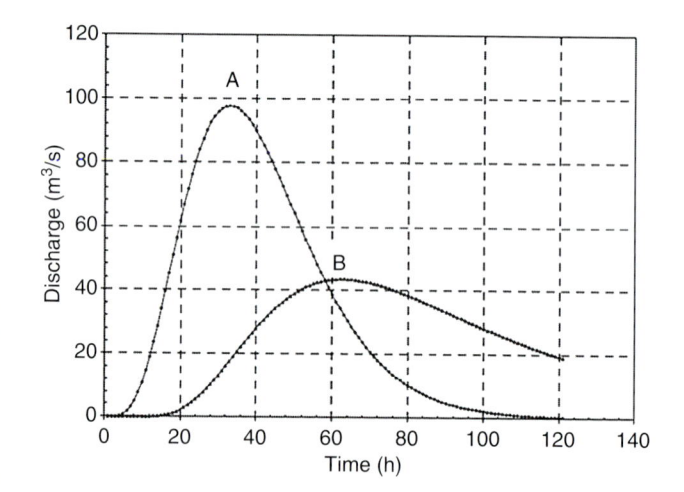

Flood Hazard and Disaster, Figure 8 Flood hydrograph. The point A is the peak inflow to the reservoir dam. Peak B is the regulated outflow from the reservoir. The difference in peak A and B is called attenuation which essentially reduces the flood peak downstream of the dam.

making them an economically attractive investment, although meeting the demand of competing uses makes their operation a very challenging task. At the same time, one should not minimize the negative impacts dams and levees have on riparian areas, forest land, and above all the displacement and resettlement of millions of people (WCD, 2000; Smith and Petley, 2008; Dixit, 2003;

Mishra, 2003). The World Commission on Dams (WCD) notes that large dams worldwide have led to the loss of aquatic biodiversity; upstream and downstream fisheries; and impacted the health of downstream floodplains, wetlands, estuaries, and adjacent marine ecosystems. Another negative consequence of dams is the displacement of people at the dam site and areas upstream that become inundated upon the completion of the dam. Globally, the total displacement from large dams could range from 40 to 80 million (WCD, 2000). In addition, dams have been connected to systematic changes in large-scale land use and land cover type, as they serve their multiple purposes possibly altering extreme precipitation patterns (Hossain et al., 2009).

Levees

Levees or dykes are artificial embankments built to control the spread of flood waters and to guide the river. Whereas dams are constructed across the river to impound the river water upstream and to regulate the downstream flow, levees are constructed laterally along the river banks. In low-lying areas or near deltas, levees are the main "protection" structure from floods. Historically, levees were constructed in the Indus Valley Civilization, as well as ancient Egypt, Mesopotamia, and China.

In modern times, levees play a major role to control floods. In the Upper Mississippi River basin of the USA alone there are more than 8,000 miles (~13,000 km) of levees (Interagency Floodplain Management Task Force, 1994). Levees are often part of a much larger project, such as the system of floodways along the lower Mississippi. However, environmentalists often challenge the notion of "safety" from the levees. Levees failed to protect vast areas in the Mississippi and Missouri River valleys during the record-setting floods of 1993 in which 1,600 levees were damaged (Wright, 2000). In 2005, levee and floodwall failures were blamed for the catastrophic inundation of New Orleans by Hurricane Katrina. Recently, in 2008, levees were breached by large floods in the Koshi River on the Nepalese-Indian boarder inundating vast settlements and killing hundreds of people.

River channel improvement

River channel improvement essentially enlarges and stabilizes the channel increasing the carrying capacity of the river and containing the floodwaters within its banks. During the channel improvement process, the channel bed is smoothened, which helps to clear the floodwaters faster downstream preventing floods in the upper reaches. But this type of intervention in the river channel potentially results in increased bank erosion raising the risk of embankment collapse, thus further exacerbating the problem (Smith and Petley, 2008). Sometimes channelization or dredging will aggravate flood problems downstream.

Therefore, a careful analysis of the situation is required before adopting suitable structural measures for flood control. Striking a balance between possible benefits and costs is a challenge to the water resources planning and engineering communities.

Nonstructural measures

In general, nonstructural measures of flood protection are of two types. The first one reduces the susceptibility to flood impacts including actions such as flood monitoring, forecasting and warning, and floodplain management. The second one reduces the impacts on individuals and the communities they live in. This includes flood insurance and relief and recovery aspects after the flood occurs.

Flood monitoring

Flood monitoring plays an important role in flood mitigation measures. Although flood monitoring cannot reduce flood incidence, it greatly reduces the loss to life and property. Monitoring river water levels and discharge are the major components of flood monitoring. Traditionally, river water levels were recorded only at limited in situ measurement stations, which failed to capture the spatiotemporal variability of a river's flow. Flood extents were usually derived from historical analysis of maps of past events or from model simulations based on in situ measurements. But with the latest communication technology, both recording and communicating a real-time water level or discharge measurements have become conventional. In addition, operational networks of stations are capable of capturing the spatiotemporal variability of river water levels along its reaches, significantly improving the efficiency of flood monitoring. Today, most of the developed countries have sophisticated systems of river water level and discharge measurements in place. For example, the United States Geological Survey (USGS) records and maintains a real-time water level and discharge network across the country. Similar examples can be found in Europe and Australia. Advanced remote-sensing technology has created a new opportunity for efficient flood monitoring even in ungauged basins of the least developed countries. Space-based rainfall monitoring has paved the way toward a global flood monitoring system (Hossain et al., 2007; http://trmm.gsfc.nasa.gov).

Flood forecasting

Flood forecasting is the capability to estimate, in real time, the future state of discharge and water levels at various sites in a river network as a result of observed or forecast rainfall, or from upstream streamflow measurements. An important aspect in flood forecasting is the lead time, which is the time between issuance of the forecast and the occurrence of the event. More lead time means ample opportunity to prepare and mobilize resources so as to evacuate people to a safer place and minimize the loss of life. Flood forecasting essentially includes three steps: getting real-time rainfall (and/or snowmelt) estimates or forecasts and streamflow observations; using these data in hydrological (rainfall-runoff and streamflow routing) simulation models; and forecasting discharge and water levels for periods ranging between a few hours and days, depending on many factors such as the size and shape of the watershed. The rainfall and streamflow observations are collected using appropriate networks of instruments, through microwave, radio, or satellite communication systems. Numerical weather prediction models provide rainfall forecasts with global availability.

Over the last one and a half decade, the significant advancement of remote-sensing capabilities onboard space-borne platforms has yielded precipitation estimates and subsequent flood forecasts on a quasi-global scale. This major development has been made possible largely from the launch of the Tropical Rainfall Measuring Mission (TRMM) by NASA in 1997. These remotely sensed data have made it possible to predict floods even in the ungauged areas of the world.

Following estimation of rainfall, hydrological simulation models are needed to estimate the flood timing and severity. There are numerous hydrologic simulation models available that can be used for flood forecasting. Sometimes such models can be as simple as a statistical rainfall-runoff relation model or they can be much more sophisticated. Hydrological models have been classified according to their complexity in representing physical processes and by their handling of the spatial distribution of physical processes and parameters as either basin-integrated (or lumped), spatially distributed, or semi-distributed. The selection of a model often depends on the availability of input, output, basin characteristics data, the modeling objective, and the experience of the modeler. A lumped model generalizes a given watershed as a single unit for calculating runoff in response to basin-averaged forcing (i.e., rainfall) and rainfall-runoff processes. The calculations are statistically based and relate to the underlying hydrological processes as a spatially averaged process. Depending on forecast needs and the characteristics of the watershed, a lumped model may be sufficient. A distributed model simulates the key hydrological processes that occur in a watershed using spatially variable data inputs, parameters, and process representations. Distributed hydrologic models commonly use rainfall and evapotranspiration estimates as inputs and parameterize processes (e.g., rainfall-runoff generation, infiltration, channel routing, interflow, baseflow) using remotely sensed observations and in situ measurements of soil types and depths, land use and land cover types, and digital elevation models (and their derivatives). Distributed models require more data and knowledge of watershed processes than lumped models. Despite the increased complexity of physical process representation of a distributed model over a lumped one, data have often been the limiting factor to accurately calibrate and verify simulation results. Also, both types of models often require observations of system behavior (i.e., rainfall and runoff) for accurate estimation of model parameters and improved model skill.

Flood-forecasting systems are more common to developed countries like in Europe and the USA and have been found to be effective in large river basins like the Danube and Mississippi (Smith and Petley, 2008). Reasons for the lack of flood forecasting systems in lesser-developed countries can be attributed to the affordability of the technology and availability of precipitation and streamflow data in real time. Space-based rainfall, however, has paved the way for a global system of flood monitoring; results have shown improved flood forecasts and flood vulnerability reduction can now become a reality (Hong et al., 2007; Hong and Adler, 2008; Li et al., 2008). At present, the latest capabilities in remotely sensed forecasting and monitoring of floods have been in use in some of the most vulnerable areas of the world like Africa, Asia, and South America (http://www.servir.net; http://eos.ou.edu/; Begkhuntod, 2007; Islam and Sado, 2000).

Floodplain and wetland management
Floodplain management is the implementation of corrective and preventative measures for reducing flood damage to structures and properties in close proximity to a river. These measures generally include requirements for zoning, subdivision, or building, particularly applicable to new construction projects. According to the Federal Emergency Management Agency (FEMA) in the USA, over 20,100 communities voluntarily adopt and enforce local floodplain management ordinances that provide building standards for new and existing development to minimize flood losses. Although attempts are made to protect floodplains from encroachment (e.g., urbanization or agriculture) through regulations, it has a mixed success (Pinter, 2005). In less-developed countries, high population growth and poverty force people to encroach on the floodplains for shelter and cultivation, making flood mitigation a challenging prospect.

Wetlands, such as swamps, marshes, and lowlands along the rivers and coastal areas have saturated soils and support a wide diversity of vegetation and wildlife. Wetlands also serve as natural buffers to floods, especially in coastal regions where hurricanes make landfall. They store water from intense rainfall and release it more gradually into the groundwater or surface and thus attenuate the peak flows that would otherwise occur. Active management and conservation practices of wetlands benefits the water quality, local environment, and prevents impacts from floods.

Providing flood insurance, community involvement and increasing the resilience are all equally important aspects of flood preparedness and mitigation. Examples from Bangladesh, Philippines, and Indonesia where community involvement has been the core of the nonstructural measures at local levels have significantly improved the flood mitigation efforts to protect the lives of the people (Adikari and Yoshitani, 2009; Perez et al., 2007; BUDMP, undated).

Summary and conclusions

Flooding arises as a result of extreme hydrological events like heavy rainfall or snowmelt. It is one of the major natural hazards affecting the lives of millions of people around the world. Events such as the 1931 and 1938 flooding of the Yellow River in China, the great floods of the Mississippi in 1993, and flooding in Myanmar from Cyclone

Nargis in May 2008 serve as sobering testimonies to the significant effect floods can inflict on populations worldwide. Historically, it has been observed that floods have wiped out not only decades of investments for a country but they also have the capability to destabilize the economy of a nation and make them more vulnerable to future disasters. Flood impacts are most felt by the poor and marginalized communities in developing nations.

A spatial analysis of flood events indicates that almost all countries have experienced flood impacts unless they were situated at latitudes greater than approximately 60°. When flooding reports were grouped by region, Asia and Africa regularly had the highest percentage of flood events per year, followed by Southeast Asia, Central America, and the Caribbean. While considering flood fatalities, almost 75% of the total worldwide deaths from floods occurred in Asia and Southeast Asia combined.

Due to the impact on human lives and property, investment in flood management strategies is a worthy enterprise. Flood forecasting and mitigation measures have contributed to the reduction in fatalities and economical damage as evidenced in Bangladesh (Adikari and Yoshitani, 2009). Globally, it is observed that each dollar spent on disaster preparedness has saved $7 in disaster-related economic losses, an impressive return on the investment (Abramovitz, 2001). Flood prediction and forecasting of impending floods at national and global levels look even more promising due to innovation in information technology and remote-sensing techniques over the past one and half decades. However, continued investments to reduce the flood vulnerability of the world's poorest and marginalized people are needed given anticipated hydrologic responses to climate change.

Bibliography

Abramovitz, J., 2001. Averting unnatural disasters. In: Brown L, Flavin C, French Hillary (eds.), State of the World 2001, World Watch Institute 2001. W.W. Norton, New York, pp 123–142.

Adhikari, P., Hong, Y., Douglas, K. R., Kirschbaum, D. B., Gourley, J., Adler. R., Brakenridge, G. R., 2010. A digitized global flood inventory (1998–2008): compilations and preliminary results. *Natural Hazards*, **55**, 405–422.

Adikari, Y., and Yoshitani, J., 2009. Global Trends in Water-Related Disasters: An Insight for Policymakers, International Centre for Water Hazard and Risk Management (ICHARM). The United Nations World Water Development Report 3, Water in a Changing World.

Associated Program on Flood Management (APFM), 2006. Environmental Aspects of Integrated Flood Management, APFM Technical Document No. 3, Flood Management Policy Series, World Meteorological Organization.

Bangladesh Urban Disaster Mitigation Project (BUDMP), undated. Hazard Mapping and Vulnerability Assessment for Flood Mitigation. Bangkok: Asian Disaster Preparedness Center/AUDMP.

Begkhuntod, P., 2007. Application of satellite information for flood risk reduction in Mekong River Basin. *Paper Presented in Regional Workshop on Innovative Approaches to Flood Risk Reduction in the Mekong Basin*, October 17–19, 2007, Khon Kaen, Thailand. Available online from: http://www.aprsaf.org/.

Dixit, A., Adhikari, P., Bishangkhe, S. (eds.), 2004. Constructive Dialogue on Dams and Development in Nepal. IUCN–The World Conservation Union Nepal and Nepal Water Conservation Foundation, Kathmandu.

Dixit, A., 2003. Floods and Vulnerability: Need to rethink flood. *Natural Hazards*, **28**, 155–179.

EM-DAT: The OFDA/CRED International Disaster Database – Université catholique de Louvain, Brussels, Belgium. www.emdat.net

Hong, Y., Adler, R., 2008. Predicting global landslide spatiotemporal distribution: Integrating landslide susceptibility zoning techniques and real-time satellite rainfall estimates. *International Journal of Sediment Research*, **23**(3), 249–257.

Hong, Y., Adler, R. F., Hossain, F., Curtis, S., Huffman, G. J., 2007. A first approach to global runoff simulation using satellite rainfall estimation. *Water Resources Research* **43**(8), W08502. doi:10.1029/2006WR005739.

Hossain, F., Katiyar, N., Hong, Y., Wolf, A., 2007. The emerging role of satellite rainfall data in improving the hydro-political situation of flood monitoring in the under-developed regions of the world. *Natural Hazards*, **43**, 199–210.

Hossain, F., Jeyachandran, I., Pielke, R., 2009. Have large dams altered extreme precipitation pattern? *Eos*, **90**(48), 453–454, American Geophysical Union.

Interagency Floodplain Management Task Force, 1994. Report to the Administration Floodplain Management Task Force-Sharing the Challenge: Floodplain Management into the 21st Century, p.191.

Islam, M. M., Sado, K. 2000. Flood hazard assessment in Bangladesh using NOAA AVHRR data with geographical information system. *Hydrological Processes*, **14**, 605–620.

Lehner, B., Liermann, C. R., Revenga, C., Vorosmarty, C., Fekete, B., Crouzet, P., Doll, P., Endejan, M., Frenken, K., Magome, J., Nilsson, C., Robertson, J. C., Rodel, R.,Sindorf, N., and Wisser, D., 2011. Global Reservoir and Dam (GRanD) Database, Technical Documentation, Version 1.1.

Li, L., Hong, Y., Wang, J., Adler, R. F., Policelli, F.S., Habib, S., Irwn, D., Korme, T., and Okello, L., 2008. Evaluation of the real-time TRMM-based multi-satellite precipitation analysis for an operational flood prediction system in Nzoia Basin, Lake Victoria, Africa. *Journal of Natural Hazards*, doi: 10.1007/s11069-008-9324-5. (2007).

Mishra, D., 2003. Life Within the Kosi Embankments. *Water Nepal* (Eds. Ajaya Dixit), **10** (1), 277.

Perez, R. T., Espinueva, S. R., and Hernando, H., 2007. Community-based flood early warning systems. *Briefing Paper: Workshop on The Science and Practice of Flood Disaster Management in Urbanizing Monsoon Asia*/April 4–6, 2007, Chiang Mai, Thailand.

Pinter, N., 2005. One step forward, two steps back on US floodplains. *Science*, **308**, 207–208.

Smith, K., Petley, D. N., 2008. Environmental Hazards, Assessing Risk and Reducing Disaster, 5th edn. Routledge, London.

Tschoegl, L., Below R., and Guha-Sapir D., 2006. An analytical review of selected data sets on natural disasters and impacts. *Paper Prepared for UNDP/CRED Workshop on Improving Compilation of Reliable Data on Disaster Occurrence and Impact*, April 2–4, 2006, Bangkok, Thailand.

UN/ISDR, 2004. Living with Risk: A Global Review of Disaster Reduction Initiatives. UN, Geneva.

UNESCO, 2001. *Guidelines on Non-Structural Measures in Urban Flood Management, Technical Documents in Hydrology, No. 50.* Paris: UNESCO.

Word Commission on Dams (WCD), 2000. Dams and Development: A New Framework for Decision Making. Earthscan, London/Sterling, VA, 2000.

Wright, J. M., 2000 The Nation's Responses to Flood Disasters: A Historical Account. Association of State Floodplain Managers, Madison.

Web sites

http://www.oas.org/USDE/publications/Unit/oea66e/ch08.htm. Accessed December 2009–April 2011
http://www.disastercenter.com. Accessed December 2009–April 2011
http://www.servir.net. Accessed December 2009–April 2011
http://eos.ou.edu/. Accessed December 2009–April 2011
http://www.reliefweb.int/. Accessed December 2009–April 2011
http://www.emdat.be/. Accessed December 2009–April 2011
http://www.dartmouth.edu/~floods/. Accessed December 2009–April 2011

Cross-references

Avulsion
Coastal Zone Risk Management
Disaster
Disaster Relief
Disaster Risk Management
Flash Flood
Floodplain
Hazard
Hazard and Risk Mapping
Landslide Dam
Levees
Risk
Risk Assessment
Susceptibility
Tsunami
Vulnerability
Warning Systems
Worldwide Trends in Natural Disasters

FLOOD PROTECTION

Fernando Nardi
University of Tuscia, Viterbo (VT), Italy
Hydraulics Applied Research & Engineering Consulting S.r.l., Rome, Italy

Synonyms

Flood control; Flood defense; Flood management

Definition

Flood protection refers to the human-made and natural structures, strategies, and actions undertaken to prevent, manage, control, and mitigate the social, economical, and physical impact of floods.

Overview

The protection of a territory from floods is achieved by means of structural and nonstructural measures. Structural measures are physical features, devices, and operations that directly impact flood waters dynamics at different scales – from a river basin to river transects and a single building. Nonstructural measures are legislative and regulatory policies formulated to plan, manage, and identify the flood risk – *floodplain* and *floodways* – at the river basin scale.

Structural flood protection is generally achieved by means of artificial hydraulic structures, land surface modifications, and maintenance activities specifically designed to control, store, convey, and attenuate the inundation process as well as the adoption of flood proofing techniques to prevent storm waters from entering predefined urban areas and properties. Nonstructural flood protection is achieved by strategic activities and, in particular, by river basin zoning and planning, actions aimed at increasing humans preparedness and/or to minimize the presence of humans and valuables in flood-prone urban and rural areas. In other words, structural methods interact with the inundation phenomenon by interventions that directly modify the flood whereas nonstructural methods act indirectly by means of spatial organization of urban and rural areas, steering toward safer land use planning and managing real-time flood forecasting-warning systems with the objective of avoiding people and properties to be exposed to flood-induced hazard.

Flood protection structural systems include flood and *debris flow* barriers, *levees*, floodwalls, dams, embankment and dykes, detention basins and reservoirs, artificial channels, diversions, and bypasses. Structural measures also include the human-made landscape modifications such as channel geometry reshaping and floodplain maintenance operations (e.g., vegetation cleaning) that modify river hydraulics and surface roughness conditions, and also the flood proofing activities that are part of interventions aimed to prevent (dry flood proofing) or control (wet flood proofing) the damages due to the passage of flood flows through private properties.

Flood protection nonstructural systems include flood-induced hazard and risk assessment, hydrologic and hydraulic modeling for floodplain mapping, law and regulations by policy makers, land use planning, soil and vegetation management, real-time online flood forecasting and warning systems, as well as human preparedness to floods.

Cross-references

Debris Flow
Floodplain
Floodways
Levees

FLOOD STAGE

Fernando Nardi
University of Tuscia, Viterbo (VT), Italy
Hydraulics Applied Research & Engineering Consulting S.r.l., Rome, Italy

Synonyms

Flood (plain) flow height; Flood (water) level

Definition

Flood stage is the depth of the inundated areas in fluvial and coastal environments that is generally originated by an unusual and abrupt increase of river flow and tidal level during severe storm events.

Discussion

In river basins, the flood stage is that specific *floodplain* flow depth that corresponds to the exceeding of the maximum conveyance and storage capacity of natural or artificial water systems – channels and water bodies – which results in the flooding of usually dry lands.

At the watershed scale, the concept of flood stage is often associated to a predefined *floodplain* cross section of the river reach, where a flow gauge is usually located, in order to monitor, identify, and map the corresponding downstream flood hazard. The analysis of downstream inundated areas corresponding to the increasing flood stage of the upstream floodplain river station is the core information of nonstructural *flood protection* methods (e.g., flood hazard mapping, flood forecasting and warning systems). In fact, the real-time monitoring and forecasting of the flood stage of gauged river stations, located upstream of flood-prone areas, is regularly used for preparing and organizing the evacuation plan for a population situated in downstream valleys/regions.

Cross-references

Floodplain
Flood Protection

FLOODPLAIN

Klement Tockner
Leibniz Institute of Freshwater Ecology and Inland Fisheries, IGB, Berlin, Germany

Synonyms

Riparian zone

Definition

Floodplains are low-relief Earth surfaces positioned adjacent to freshwaters and subject to flooding (fringing floodplains of lakes and rivers, internal and river deltas). Similarly, riparian zones are transitional semi-terrestrial areas that extend from the edge of permanent water bodies to the edge of uplands and are influenced by freshwater (Junk et al., 1989; Naiman et al., 2005; Tockner et al., 2002, 2008).

Overview

Floodplains are distributed worldwide and cover about 2% of the land surface. Floodplains are particularly abundant adjacent to large rivers including their deltaic areas. The Amazon, the Brahmaputra, the Nile, the Niger, the Zaire, and the Siberian Rivers, among others, are still fringed by extensive floodplains that are regularly inundated. Floodplains are disturbance-dominated ecosystems tightly linked to fluvial and geomorphic dynamics. As expanding and contracting ecosystems, they are among the most complex, dynamic, and diverse ecosystems globally. Because of their unique position at the lowest location in the landscape, floodplains integrate and accumulate upstream processes. They offer a remarkable diversity of ecosystem services including flood retention, recharge of ground water, biomass production, nutrient removal, as well as aesthetic and cultural values. Worldwide, floodplains provide approximately 25% of all continental ecosystem services, more than any other ecosystem type. Such multiple services favored the development of modern civilizations along the Nile, Euphrates, Indus, Amazon, Yangtze, or Mississippi rivers; and many present indigenous human societies are well adapted to the unique conditions of floodplains. Today, about half of the human population of Europe and Japan live on (former) floodplains. About 80% of Bangladesh, the most densely populated country worldwide, is covered by floodplains. At the same time, floods are among the most costly natural disasters worldwide. Between 1900 and 2004, almost 3 billion people were affected by floods, causing 2.9 million human deaths, and making more than 130 million other people homeless. At the same time, floodplains are among the most threatened ecosystems. In Europe and North America, more than 90% of the former floodplains are functionally extinct or have been converted into cropland and urban areas. Habitat degradation, species invasion, pollution, and climate change are among the most important pressures threatening floodplain ecosystems and their rich biodiversity. Therefore, conserving the remaining intact floodplains as strategic global resources and restoring degraded floodplains poses highest priority for future ecosystem management. However, floodplains designated for restoration and conservation must be large enough to support its native plant and animal communities and to perform key ecosystem functions.

Bibliography

Junk, W. J., Bayley, P. B., and Sparks, R. E., 1989. The flood pulse concept in river-floodplain systems. *Canadian Special Publication of Fisheries and Aquatic Sciences*, **106**, 110–127.

Naiman, R. J., Decamps, H., and McClain, M. E., 2005. *Riparia*. San Diego: Elsevier/Academic.

Tockner, K., and Stanford, J. A., 2002. Riverine floodplains: present state and future trends. *Environmental Conservation*, **29**, 308–330.

Tockner, K., Bunn, S. E., Quinn, G., Naiman, R., Stanford, J. A., and Gordon, C., 2008. Floodplains: critically threatened ecosystems. In Polunin, N. C. (ed.), *Aquatic Ecosystems*. Cambridge: Cambridge University Press, pp. 45–61.

Cross-references

Flash Flood
Flood Deposits

Flood Hazard and Disaster
Flood Protection
Flood Stage
Floodway

FLOODWAY

Armand LaRocque
University of New Brunswick, Fredericton, NB, Canada

Synonyms
Regulatory floodway

Definition
A floodway is defined as a man-made diversion channel that is created for the discharge of river flow during flood events, without increasing the water surface elevation more than a particular reference level, usually the 100-year flood level. For the US Federal Emergency Management Agency (FEMA), the floodplain includes also the channel of a river or another watercourse, with the adjacent portion of the floodplain being used for the discharge of flood water (FEMA, 2002).

Floodways are among the protection systems that are built to avoid the catastrophic floods that occur annually in many densely populated floodplains. A floodway includes not only the diversion channel, but also any adjacent areas of the floodplain that can carry overflow. The floodway can be delineated by dikes that are built on both sides of the adjacent overbanks and by other control structures that regulate the water flow, at both ends of the diversion channel. At the inlet of the floodway, the control structures may include control gates which are built to redirect the overflow from the river to the diversion channel, at the beginning of flood occurrences. The implementation of a floodway can be a very complex project, including excavation of earth, construction of dikes, building or redesign of highway and railway bridges, and modifications to utilities and services (e.g., power and sewer lines) which are already in place.

One example of a floodway is the Red River Floodway that was built to protect the city of Winnipeg (Manitoba, Canada) against catastrophic floods that occur regularly in the Red River valley. The floodway consists of an artificial 47 km (29 mi) long channel that diverts a portion of the Red River's flow during flood periods along the eastern edge of the city of Winnipeg. This system also includes a network of dikes and hydraulic structures controlling the water flow. Constructed between 1962 and 1968 at a total cost of $63 millions (CAD), this floodway was used over 20 times until 2005 and has resulted in savings estimated to be $10 billion (CAD). The original floodway was built following the 1950 flood, but the "Flood of the Century" (Spring, 1997) showed that there was a need to improve the flood protection of the valley.

Once completed in 2010, the new expansion of this floodway will prevent floods having a 700-year probability of recurrence, by increasing the capacity of the floodway from 1,700 m^3 (60,000 ft^3) of water per second to 4,000 m^3 (140,000 ft^3) of water per second.

Bibliography
Federal Emergency Management Agency (FEMA), 2002. Emergency management and assistance. Federal Emergency Management Agency, U.S. Department of Homeland Security, Washington (DC), Title 44, 499p, http://www.access.gpo.gov/cgi-bin/cfrassemble.cgi?title=200244.
Federal Emergency Management Agency (FEMA), 1979. The floodway: A guide for community permit officials. Federal Emergency Management Agency (FEMA), Federal Insurance Agency, Washington (DC), Community Assistance Series No. 4, 17 p. http://www.arkansasfloods.org/afma/docs/cfm/Floodway-Guide-for-Permit-Officials.pdf
Jones, D. E. Jr., and Jones, J. E., 1987. Floodway delineation and management. *Journal of Water Resources Planning and Management*, 113(2), 228–242.

Cross-references
Flash Flood
Flood Deposits
Flood Hazard and Disaster
Flood Protection
Flood Stage
Floodplain

FOG HAZARD MITIGATION

Steve LaDochy[1], Michael R. Witiw[2]
[1]California State University, Los Angeles, Los Angeles, CA, USA
[2]Embry-Riddle Aeronautical University Worldwide, Everett Campus, Everett, WA, USA

Synonyms
Haze; Mist; Smog

Definition
Fog is a cloud consisting of minute droplets or ice crystals that is in contact with the surface of the Earth. Horizontal visibility is reduced to less than 1 km, according to international standards (George, 1951).

Dense or heavy fog. When visibility is less than 5/16ths of a mile (less than 400 m) fog is dense or heavy.

Mitigation is efforts to lessen the negative effects of a natural hazard such as fog on society and the natural environment.

Introduction
Despite its beneficial impacts including moisture for forests and plants and the ability to collect fog to be used as a water source in otherwise arid regions, fog has been

a hazard probably for as long as humans have inhabited Earth. It is the reduced visibility which makes fog a hazard. The hazards of fog can be epitomized by the sinking of the RMS Titanic in 1912, when on its maiden voyage it became enshrouded in fog, hit an iceberg, and sank into the cold North Atlantic (see *Marine Hazards*). In some years fog causes more deaths in the USA than tornadoes and hurricanes combined. Most of these deaths are caused by auto accidents in poor visibility. For these reasons, in many areas, humans have made the attempt to mitigate the effects of fog, by either modifying the fog itself or devising ways to cope with it.

Road transport
Fog lights
A good fog lamp or lights produces a wide, bar-shaped beam of light with a sharp horizontal cutoff (dark above, bright below) at the top of the beam, and minimal upward light above the cutoff. Studies show that in North America, more people inappropriately use their fog lamps in dry weather than for poor weather (Sivak et al., 1997). The effectiveness of fog lights decreases when drivers do not slow down.

Fog visibility sensors
A nephelometer is an instrument that measures light scattering in the forward direction by fog particles. Normally found with weather stations at an airport, these instruments can also detect patchy or dense fog along a highway (FHA, 1999). In 2006, the USA employed over 7,000 closed circuit televisions (CCTV) to monitor traffic nationwide. The Federal Highway Administration uses CCTV along with nephelometers in developing road weather management strategies aimed at mitigating weather impacts by advising motorists of prevailing and predicted conditions. Part of this program employs the 511 national travelers information telephone number (DOT, 2009).

Fog road warning systems
In 1995, the California Department of Transportation (Caltrans) installed an automated fog warning system that uses nine roadside weather and visibility monitoring stations and 36 detectors lodged in the pavement. Motorists are advised of prevailing conditions via flashing beacons atop static signs, information broadcast over advisory radio frequencies, and messages posted on dynamic message signs. Success was seen in a 70% reduction in incidents where the system was installed. Because of this success, the system was expanded in 2005 (Traffic Technology Today, 2010). A similar system was installed in Tennessee after a chain-reaction collision of 99 vehicles occurred in 1990 (Robinson et al., 2002).

Driver's behavior
Even with advanced warning systems, it is often driver's behavior that determines accidents. In a recent study in California, it was noted that although the advised speed during fog was lowered to 30 mph, the mean speed of drivers averaged 61 mph (MacCarley et al., 2007).

Widespread dense fog can have large socioeconomic impacts through disruption of commerce as well as in injury and death. Data obtained from the Illinois Department of Transportation indicate that between 1975 and 1995 some 4,000 collisions occurred annually under foggy conditions in Illinois, excluding the city of Chicago, resulting in an average of 30 deaths per year and millions of dollars in damage (Westcott, 2007). Accident reports from 1992 to 1995 indicate that, on average, at least two fog-related traffic accidents occur somewhere in Illinois, excluding Chicago, on 288 days per year.

Dense fog with reduced visibility slows traffic and causes costly accidents. Long traffic delays may be harmful to perishable food, especially fresh fruit, vegetables, and sushi. A 1960s report from the British Transport Commission, London, said that one fog day caused them additional costs of about $10,000 due to loss of income or pay for extra personnel.

The National Transportation Safety Board found that between 1981 and 1989, accidents where fog was present on highways resulted in more than 6,000 deaths in the USA (NCHRP, 1998). On average, dense fog is present during 680 fatalities per year in the USA (NTSB Highway Accident Report, 1992). In 2001, fog was associated with 43,792 crashes; over 670 people were killed and more than 19,000 people were injured (Goodwin, 2003). In Canada, approximately 50 people per year die because of motor vehicle accidents in which fog was a contributing factor (Wiffin et al., 2004). In Europe, a fog chamber, with the capability of simulating different visibilities, was built in Clermont-Ferrand, France. This chamber is used to simulate fog with a goal of improving transport safety (Andre et al., 2004; Colomb et al., 2004).

Marine transport
Fog horns
Lighthouses use high-powered lenses to warn ships of coastlines or islands. But as in the case of Alcatraz Island, the San Francisco fog made the light ineffective and a fog bell was needed (Veronico, 2008).

A foghorn or "fog signal" or "fog bell" is a device that uses sound to warn vehicles of hazards (or of the presence of other vehicles) in foggy conditions. The term is most often used in relation to marine transport. When visual navigation aids such as lighthouses are obscured, foghorns provide an audible warning of rocks, shoals, headlands, or other dangers to shipping.

Since automation of lighthouses became common in the 1960s and 1970s, most older, manual foghorn installations have been removed and have been replaced with electrically powered diaphragms or compressed air horns. Activation is completely automated using a laser or photo beam that reflects back due to the fog. The sensor tells a computer to activate the foghorn. In many cases, modern navigational aids have rendered large, long-range foghorns completely unnecessary (Noble, 2004).

Foghorns have very low pitches because sounds with low pitches have a long wavelength. This is important because a long wavelength means that the sound wave can pass around barriers, like rocks, easily (Young, 2010).

Radar
Modern shipping relies less on lighthouses and fog horns and more on technology, such as sonar and radar to avoid objects in dense fog (Amichai, 1978).

Aviation
Fog also greatly impacts the aviation community, that is, closing some airports as well as keeping VFR (Visual Flight Rules) pilots on the ground (see *Aviation (Hazards to)*). Aviation is directly impacted by fog as visibility is critical for landing and takeoffs. Elevated risk of a fog event is defined as ceiling/visibility less than 200 feet/½ mile (60m/800m). In New Delhi, where pollution has led to increasing fog, flight delays and cancellations due to fog during the winter cost airlines millions of dollars while stranding thousands (Air Transport Association, 2002). Efforts are being made on improving forecasting techniques (Whiffen, 2001).

Fog seeding
At Fairchild Air Force Base, near Spokane, Washington, cold fog (that which forms at subfreezing temperatures) has been routinely dispersed through the use of liquid propane dispensers for years. Liquid propane is dispensed into the foggy air. The vaporized propane results in very low temperatures (<-40 C). The super-cooled water droplets become glaciated. The resultant ice crystals initiate the ice crystal process, causing the fog to dissipate. Visibility improves as the ice crystals grow and fall out as snow (Vardin et al., 1971).

Cold fog is easier to disperse, using seeding techniques than warm fog (which forms at temperatures above freezing). But most mid-latitude airports have to deal with warm fog. Since fog droplets are found in a narrow size range of about 4–10 μm, dissipation can be achieved through coalescence processes. Houghton and Radford (1938) were the first to use hygroscopic particle (NaCl, $CaCl_2$) seeding to dissipate fog droplets. Giant hygroscopic nuclei seeding, NaCl, was used by Jiusto et al. (1968) to improve visibility.

Successful programs of fog dissipation have been carried out in Korea (Oh et al., 2008) and Germany (Moller et al., 2003) and Iran using liquid carbon dioxide (Jamali et al., 2005).

Heat
FIDO (Fog Intensive Dispersal Of) was a successful program using burners to heat the air in fog during WWII in England. A similar program was installed at Paris. (Fabre, 1971) showed that the use of heat and turbulence from jet engine exhaust could improve visibility on runways in Paris.

Turbulent mixing
Another method for dispersing ground fog is by turbulent mixing of saturated surface air with drier air aloft. Airports have used helicopters to mix dry air into the fog layer by the helicopter's downwash (Plank et al., 1971). Heat from jet engines also produces convection and turbulent mixing.

For aviation, fog is the only weather hazard that can be mitigated by technical means. Enhanced vision systems and space and ground-based navigation and augmentation systems will allow automated landings and takeoffs also under very reduced visibility conditions.

Forecasting
Better forecasting/nowcasting skills can help reduce financial losses arising from road, air, and marine transport systems that can be significantly impacted by unpredicted low-visibility conditions (Gultepe et al., 2009). In Europe, a major fog project called Cooperation in Field of Scientific and Technical Research (COST-722), with objectives of reducing economic loss and fatalities, was also created to develop advanced methods for very short-range forecasts of fog and low clouds (Jacobs et al., 2007).

Remote sensing
Satellites are used operationally to monitor developing fog and to verify forecasts. The use of satellite observations for fog studies has been discussed in many earlier studies (Ellrod, 1995; Bendix, 2002). In these studies, differencing between channel 2 (ch2; 3.9 μm) and channel 4 (ch4; 11.2 μm) from Geostationary Operational Environmental Satellite (GOES) observations, and between the channel-3 IR component (3.7 μm) and ch4 (11.0 μm) data from National Oceanic and Atmospheric Administration Advanced Very High Resolution Radiometer observations, has been used to detect fog regions. The discrimination between fog and other surfaces is due to the low emissivity of fog in channel 3 at night and the high reflectivity in channel 3 during daylight (Bendix, 2002). In Switzerland, ground-based cloud radars and ceilometers were used to detect low clouds and fog (Nowak et al., 2008).

Laser telecommunications
Since the 1990s, laser telecommunications, also called free space optical (FSO) communications have been used to transmit large amounts of data in a line of site fashion over distances that range up to 2,000 m. Wavelengths that are less susceptible to atmospheric attenuation are chosen. However, attenuation due to absorption by fog droplets remains an issue. To mitigate this problem, laser communications companies have done studies on attenuation caused by fog. Basically, to maintain an average downtime

of the communications equipment due to low visibility to less than 0.1%, a climatological study of the fog regime at the location is made. Receiver and transmitter are then spaced apart accordingly. In a relatively foggy area, like Seattle, depending on the power and wavelength chosen, transmitter and receiver may need to be separated by no more than 500 m. On the other hand, a place like Phoenix, Arizona, may sustain downtime less than 0.1% with transmitter and receiver separated by a distance of 2,000 m (Fischer et al., 2008, 2004).

Health

Air pollutants, especially particulate matter, are also known to enhance the reduction in visibility resulting from fog. The famous Killer Smog (smoke and fog) of London in December of 1952 not only reduced visibility but resulted in an estimated 4,000 deaths (Bach, 1972). The Donora Pennsylvania fog of 1948 also led to many deaths but occurred in a much less populated area (Pennsylvania Department of Environmental Protection, 1998). These two events helped lead to the clean air acts that were subsequently passed in the UK and USA. Largely as a result of the clean air acts, the health hazards posed by fogs of the past have greatly diminished. Visibility has also improved. For example, since 1980, Los Angeles International Airport (LAX) has only experienced 1 year where dense fog occurrence exceeded 50 h and, in fact, in 1997, no dense fog was reported at LAX. This contrasts to the years prior to 1980, when more than 50 h of dense fog per year were routinely reported, peaking at over 300 h in 1950 (Witiw and LaDochy, 2008) (see *Acid Rain*).

Summary

Fog is an underrated natural hazard that takes a large toll in lives every year from accidents mainly in transportation (road, air, and marine). Fog also disrupts transport, communications, and construction on a global scale impacting economies with billions of dollars (US) in losses, although total impacts may not be calculable. Mitigation efforts at dissipating fog, providing accurate and early forecasts, and devising successful warning systems have a long and successful history. The continued list of international accidents shows that further efforts are needed to make these efforts universal.

Bibliography

Air Transport Association, (2002). *2002 State of the U.S. Airline Industry: A Report on Recent Trends for U.S. Carriers*. Available at http://www.airlines.org/public/industry/display1.asp?nid=1026. Accessed 2010.
Amichai, Y., 1978. *The Use of Radar at Sea*. London: Naval Institute Press.
Andre, P., Silva, C. A., Balocco, E., Boreux, J. J., Cavallo, V., Colomb, M., Dore, J., Dufour, J., Hannay, J., Hirech, K., Kelly, N., Lacote, P., and Mealares, L., 2004. The main results of a European research project: Improvement of transport safety by control of fog production in a chamber ("FOG"). In *Proceedings of the Third International Conference of Fog, Fog Collection and Dew*, Capetown.
Bach, W., 1972. *Atmospheric Pollution*. New York: McGraw-Hill.
Bendix, J., 2002. A satellite-based climatology of fog and low-level stratus in Germany and adjacent areas. *Atmospheric Research*, **64**, 3–18.
Clark, R. S., Wright, T. L., Evert, R. W., Loveland, R. B. and Northrup, R. E., 1973. Project Foggy Cloud V: Panama Canal Warm Fog Dispersal Program, NWC TP 5542, 92 pp.
Colomb, M., Hirech, K., Morange, P., Boreux, J. J., Lacote, P., and Dufour, J., 2004. Innovative artificial fog production device, a technical facility for research activities. In *Proceedings of the Third International Conference of Fog, Fog Collection and Dew*, Capetown.
Department of Transport, 2009. Department of transport weather programs. Available at http://www.ofcm.gov/fp-fy09/pdf/3Sec3i-DOT.pdf. Accessed 2010.
Ellrod, G. P., 1995. Advances in the detection and analysis of fog at night using GOES multispectral infrared imagery. *Weather Forecasting*, **10**, 606–619.
Ellrod, G. P. and Lindstrom, S., 2006. Performance of satellite fog detection techniques with major, fog-related highway accidents. National Weather Association. Available at http://www.nwas.org/ej/pdf/2006-EJ3.pdf. Accessed 2010.
Fabre, R., 1971. Airport D'Orly-installation de Denebulation Turboclair. Report by the Aeroport de Paris and Societe Bertin et Cie.
Federal Highway Administration, 1999. Highway Fog Warning System. FHA, DOT Pub. No. FHWA-RD-99-110. Available at http://www.fhwa.dot.gov/tfhrc/safety/pubs/its/ruralitsandrd/tb-hwyfog.pdf. Accessed 2010.
Fischer, K. E., Witiw, M. R., and Eisenberg, E., 2008. Attenuation in fog at a wavelength of 1.55 millimeters. *Atmospheric Research*, **87**, 252–258.
Fischer, K. E., Witiw, M. R., Baars, J. A., and Oke, T. R., 2004. Atmospheric laser communications: new challenges for applied meteorology. *Bulletin of the American Meteorological Society*, **8**(5), 725–732.
George, J. J., 1951. Fog. In T.F. Malone, ed., Compendium of Meteorology. Boston, MA: American Meteorological Society, 1179–1189.
Goodwin, L., 2003. *Weather-related crashes on U.S. highways in 2001*. Available at http://www.ops.fhwa.dot.gov/weather/docs/2001CrashAnalysisPaperV2.doc. Accessed 2010.
Gultepe, I., et al., 2009. The fog remote sensing and modeling field project. *Bulletin of the American Meteorological Society*, **90**, 341–359.
Houghton, H. G., and Radford, W. H., 1938. On the local dissipation of warm fog. *Papers in Physical Oceanography and Meteorology*, **6**(3), 63.
Jacobs, W., Nietosvaara, V., Bott, A., Bendix, J., Cermak, J., Silas, M. and Gultepe, I., 2007. Short-range forecasting methods of fog visibility and low clouds. *Final Report on COS-722 Action*. Available from COST Office, Avenue Louise 149, B-1050, Brussels.
Jamali, J. B., Javanmard, S., and Heydari, M., 2005. Role of low level cloud seeding, aimed at fog dispersion for the purpose of safety upgrading in mountainous roads. *Journal of Transportation Research*, **1**(2), 49–63.
Jiusto, J. E., Pilie, R. J., and Kocmond, W. C., 1968. Fog modification with giant hygroscopic nuclei. *Journal of Applied Meteorology*, **7**, 860–869.
MacCarley, C. A., Ackles, C., and Watts, A., 2007. Highway traffic response to dynamic fog warning and speed advisory messages. *Transportation Research Record: Journal of the Transportation Research Board*, **1980**, 95–104.
Moller, P., Wieprecht, W., Hofmeister, J., Kalass, D., Elbing, F. and Ulbricht, M., 2003. Fog dissipation by nucleation scavenging

using particle blasting. In *8th WMO Scientific Conference on Weather Modification*, Casablanca, WMO Report No. 39, pp. 389–392.

National Cooperative Highway Research Program (NCHRP), 1998. *Reduced Visibility Due to Fog on the Highway, A Synthesis of Highway Practice. National Cooperative Highway Research Program (NCHRP) Synthesis of Highway Practice 228*.

Noble, D. L., 2004. *Lighthouses and Keepers*. Annapolis, MD: Naval Institute Press.

Nowak, D., Ruffieux, D., and Agnew, J. L., 2008. Detection of fog and low cloud boundaries with ground-based remote sensing systems. *Journal of Atmospheric and Oceanic Technology*, **25**, 1357–1368.

Oh, S., Chang, K. and Lee, M., 2008. The blanket effect: clearing fog in South Korea. Available at: http://theblanketeffect.blogspot.com/2008/02/clearing-fog-in-south-korea.html. Accessed 2010.

Oliver, J. E., Jr., 1973. *Climate and Man's Environment: An Introduction to Applied Climatology*. New York: Wiley.

Pennsylvania Department of Environmental Protection, 1998. *Donora Smog*. Available at http://www.portal.state.pa.us/portal/server.pt/community/events_that_shaped_our_environment/13894/donora_smog_/588401. Accessed 2010.

Plank, V. G., Spatola, A. A., and Hicks, J. R., 1971. Summary results of the Lewisburg fog clearing program. *Journal of Applied Meteorology*, **10**, 763–779.

Robinson, M., et al., 2002. Safety applications of ITS in rural areas: section 3.2.8 fog detection warning system – Tennessee, EDL No. 13609, Prepared by SAIC for the U.S. DOT. Washington, DC.

Sivak, M., Flannagan, M. J., Traube, E. C. and Kojima, S., 1997. Fog lamps: frequency of installation and nature of use. SAE International. Available at http://www.sae.org/technical/papers/970657. Accessed 2010.

Traffic Technology Today, 2010. Caltrans unveils life-saving Fog Pilot project. Available at http://www.traffictechnologytoday.com/news.php?NewsID=10958#loaded. Accessed 2010.

Vardin, L., Figgins, E. D., and Appleman, H. S., 1971. Operational dissipation of supercooled fog using liquid propane. *Journal of Applied Meteorology*, **10**, 515–525.

Veronico, B. S., 2008. *Lighthouses of the Bay Area*. Charleston, SC: Arcadia Publishing.

Westcott, N. E., 2007. Some aspects of dense fog in the Midwestern United States. *Weather and Forecasting*, **22**, 457–465.

Whiffen, B., 2001. Fog: impact on aviation and goals for meteorological prediction (2001). In *Proceedings of the Third International Conference of Fog, Fog Collection and Dew*, St. John's, Newfoundland.

Wiffin, B. P., Delannoy, P. and Siok, S., 2004. Fog: impact on road transportation and mitigation options. In *National Highway Visibility Conference*, Madison, WI, May 18–19, 2004.

Witiw, M. R., and LaDochy, S., 2008. Trends in fog frequencies in the Los Angeles Basin. *Atmospheric Research*, **87**, 293–300.

Young, T., 2010. Why do foghorns always have very low pitches? Do they have high or low amplitude? Phys-Link.com. Physics and astronomy online. Available at http://www.physlink.com/Education/AskExperts/ae264.cfm?CFID=28338689%26CFTOKEN=4e53d03bedd5ec40-5C48084B-15C5-EE01-B901EBCBA290AFDC. Accessed 2010.

Cross-references

Aviation (Hazards to)
Coastal Hazards
Fog Hazards
Marine Hazards
Mitigation

FOG HAZARDS

Paul J. Croft
Kean University, Union, NJ, USA

Synonyms

Cloud; Ice particles (suspended); Mist; Nebulosus; Rime; Water droplets

Definition

Fog. Cloud at or near the surface. Word origin may be primarily from northern latitude countries to describe small/fine rain sometimes with wind obscuring visibility. After the Greek (nephele) or Latin (nebula) for cloud.

Nebulosus. Veil of clouds without definition as a stratus layer or sublayer.

Water droplets. Suspended above or near the surface.

Ice particles. Suspended above or near the surface.

Mist. Suspended and falling liquid hydrometeors.

Rime. Suspended and falling solid hydrometeors.

Internationally, fog is known by a variety of names (see Table 1) and is reported operationally (and for climatic purposes) according to WMO criteria (e.g., see WMO-No. 485, Geneva; World Meteorological Organization, 1992) and coding symbols (i.e., 10–12, 28, 40–49).

Fog impact

Fog is observed worldwide at all locations and elevations under substantially variant synoptic, mesoscale, and microscale conditions of the atmosphere. As such, its formation and intensity – and thus hazards posed – may occur any time or day of the year, and its coverage may be widespread, regional, or highly localized. Although fog is often very shallow in the vertical layer, it is very often in close proximity to the surface (e.g., see the COMET MetEd online learning module "Fog: Its Processes and Impacts to Aviation and Aviation Forecasting" revised 2010; https://www.meted.ucar.edu/ – user registration is required). The production, maintenance, and dissipation of fog pose significant risks and hazards for multiple modes of transportation, and thus these show the greatest impacts in terms of total populations affected (i.e., by populations and costs; e.g., see Encyclopedia of Atmospheric Science, 2003). For example, the Bosphorus and Dardanelles Straits experience closures in Turkey due to dense fog, and the closures impact the flow of oil and grain from ports in the Russian Black Sea.

Multiple vehicle collisions and chain-reaction crashes are the most common and may be exacerbated when ice fog (or riming) also takes place. Marine, rail, and aviation impacts tend to produce more of an economic impact due to delays but on occasion, have led to serious accidents (or significant delays) with loss of life and property (e.g., Croft et al., 1997; Chagnon, 2006). This is also true in some instances of search and rescue operations. Notable transportation-related disasters have included the highway chain-reaction crash on the Mobile Bayway (1993) in

Fog Hazards, Table 1 Fog terms by country or language of origin with some locally known names

Fog Name/Definition	Origins/Language	Local names and meanings
la niebla	Coastal Chile/ Spanish	Gorua
le brouillard (or brouiller)	French	To blur; cloud over
nebel	German	
megla	Slovene	
雾	Chinese	"Fog City" (Chongqing)
안개	Korean	
la niebla (a neblina)	Spanish	Chilean coastal town
Tage	Danish	
الضباب	Arabic	
Dimma	Swedish	
mgla	Polish	
mist	Dutch	
霧	Japanese	
nevoeiro	Portugese	
sumutusjärjestelmiä	Finnish	
nebbia	Italian	
tåke	Norwegian	

Mobile, Alabama; the shipping collision of the Andrea Doria with the MS Stockholm (1956); and the Tenerife airport disaster (1977). Other collisions with loss of property and life, or with significant damages, have also occurred between transportation vehicles and fixed structures (e.g., buildings, bridges) and even mountains. The majority of transportation impacts have resulted from obstructed visibility and the lack of adequate instrumentation or warning systems. However, while such systems have been used, fog-related accidents still occur often due to human error or poor judgment of the conditions or their seriousness.

In combination with other atmospheric constituents (natural or anthropogenic), fog also has the potential to damage the health and welfare of human, plant, and animal communities as well as insect populations. These may occur in the air, within the ground surface layers, or the water cycle components. When these constituents include noxious (and/or irritant) aerosols and chemicals, the impact may be deadly (e.g., the Donora Episode of 1948 and the London Fog Event of 1952) or lead to chronic illnesses and thus represents a significant but less common and more isolated hazard. The occurrence of smog, a term first used circa 1905 in London, dates back to the burning of coal reserves to at least the thirteenth century. Similar smog episodes occur around the world including Mexico City, Tehran, Los Angeles, and in many portions of Southeast Asia.

In cases where fog is harvested, risks exist in the use of fog-water in terms of its composition, contamination, and storage. It has been suggested that the generation of mists by fog machines for entertainment purposes may also pose health risks. There is also some concern for the use of various chemicals used for fog dispersal (seeding) that may have short-term impacts on those communities exposed to the application and dispersion of the chemical treatments. Fog has also been popularized with regard to film (often as mystical and menacing), music, and artwork, as well as for industrial, commercial purposes, and marketing (e.g., clothing or manufacturing, anti-fog devices). In some instance, fog has had a negative, but not critical impact, with regard to sporting and other outdoor events (e.g., the "Fog Bowl" in the United States, or impacts on ski slope operations).

Fog investigation

The advent of a more scientific and structured observational (or empirical) study of fog followed cloud classification schema and the development of more sophisticated instrumentation for study of the atmosphere, particularly on more localized scales (e.g., Sutton, 1953; Geiger, 1965). This phase of exploration led to improved understanding of the observed properties and behaviors of fog based on weather observation data and fog's associated characteristics as a function of dynamic processes in the atmosphere. This linkage afforded the first efforts at fog prediction in order to provide opportunity for the avoidance, mitigation, or prevention of its impacts. Based upon numerous studies in this regard, fog has since typically been classified by its formative method (e.g., radiative), location (e.g., sea fog), or atmospheric processes (e.g., frontal fog) and has led to many types being recognized internationally as well as locally named fogs (Table 2). In the latter case, these processes provide explanation of the movement or evolution of fog and thus allow for an "entity-following" method to study an evolving phenomenon.

Further development of these methods have included statistical and time series analyses to produce simple conditional climatologies of fog occurrence, intensity, and tendencies, as well as Markov Chain and other analytic techniques to more accurately assess fog occurrence and probability distributions for specific locations or regions (and intensity). In any of these cases, the specificity of the information is limited to the period of record and how representative each location is with regard to its surrounding area. Further specification of fog occurrence is possible when these techniques are linked to the synoptic setting or when statistical or climatic values are mapped across a region in order to infer the frequency of occurrences, intensities, and fog coverage or behavior. However, while such findings may be applied in a predictive (or diagnostic) manner, they are limited according to an observational framework and time scale (e.g., hours or days) without full consideration of the process-oriented nature of fog formation, maintenance, evolution, and dissipation.

As the observational study of fog has matured, the study of cloud physics and the microphysical processes

Fog Hazards, Table 2 Fog types, names, or classifications and key formation mechanisms and common locations

Classification/type of fog	Formation	Locations
Radiative fog	Thermal loss	Particularly low elevations
Advective fog	Transport and/or mixing	Ocean/coastal;
Advective-radiative fog	Delayed cooling modified air	Inland and isolated areas
Precipitation fog	Associated with synoptic pattern	Anywhere
Steam fog	Air-sea differential	Ocean, lakes, rivers
High-altitude fog	Cloud formation processes	Mountains
Upslope fog	Mechanical lift	Mountains
Frontal fog	Mechanical lift and instability	
Mixing fog	Convective/advective mixing	
Ocean fog	Evaporation, cooling	Fog bank formation
Valley fog	Cold air drainage	Depressions
Freezing fog	Thermal loss	
Hail fog	Rapid cooling (thunderstorm)	
Ground fog	Moist ground and rapid cooling	

involved in cloud growth provided distinct thermodynamic information as to the internal mechanisms and factors associated with fog formation (and dissipation) as well as its radiative impacts (as recognized many years before). This much more comprehensive insight of fog as an "entity" phenomenon to observe and follow was also assisted by the modern development of more sophisticated instrumentation (e.g., transmissometer) to measure cloud (or fog) properties and to monitor subsequent development (or dissipation) for shorter time scales and across shorter distances. These help to establish the sensing of fog's presence and the impact of fog intensity according to visibility criteria as linked to arbitrary interrogations versus one based on more defined impacts or meaningful thresholds. This provided more distinctive identification of fog and its variations in terms of visibility, or depth/thickness, beyond simple measurements of visual range according to predetermined target ranges. Microphysical studies have also provided sufficient evidence to explain variations in fog behavior based upon its phase composition (i.e., liquid, solid, or mixed) and its subsequent evolution.

Fog drop-size distributions have been determined to range from 1 to 10 μm according to the aerosol (or particulate) matter and the availability of moisture (e.g., mixing ratio of the atmosphere) and condensational processes. Larger sizes are preferred as fog increases during condensation processes, while smaller drops dominate in dissipation. While these also have obvious synoptic signatures as found through observational studies, they are also driven by interactions between fog and the local landscape, including local sources and sinks acting on the micro- and mesoscale. These act to contribute to locally preferred drop-size distributions with specific precipitation or deposition (or terminal velocity) tendencies. An understanding of the dependence and interaction of fog on local features has been aided by air quality and atmospheric chemistry studies (for near-surface locations) as well as the emergent air-surface and air-sea studies (over the last 20 years) that focus on boundary layer exchanges and flux behaviors. These allow a more precise explanation of fog's physical and chemical characteristics (or the statistical family of these) which modulates evolution of fog in time and space – and that create regional and local differences. Due to the physical-chemistry associated with fog events, their electrical, radiative, optical, and acoustical properties also necessarily show variation.

Fog mitigation and applications

More recent investigations have focused on the physical-chemistry involved in cloud (and thus fog) processes as well as techniques for fog dissipation or dispersal based on the foregoing knowledge of fog's physical, chemical, radiative, and internal properties. Other work has been focusing on the chemical mixture of species present in fog in terms of its impacts on human (or animal/plant) respiration (e.g., acute or chronic, mortality), external contact (e.g., surface of the skin or leaf contact), and ingestion (e.g., mucous membranes or eyes, fog-water consumption, or irrigation). In tandem with prior empirical and statistical study, these are assisting in the development of a more comprehensive conceptual model of fog in terms of its characteristics, physical and chemical attributes, and behaviors. Together these provide for integrated numerical modeling of fog with regard to its formation mechanisms, maintenance, intensity, evolution, and dissipation – particularly as related to fog's physical-chemistry and its attendant optical and radiative effects. The ability to quantify these will provide an enhanced ability to model the interactions between atmospheric systems producing fog and the systems impacted by fog (i.e., human and other populations). To best represent and visualize such systems and their interactions, current work has focused on artificial intelligence, impact-response models, and the use of GIS systems.

Presently, fog is known to be formed through three basic mechanisms, each of which are found to a lesser or greater extent in the creation of each fog type or classification. The basic processes are radiative (cooling), advective (cooling and/or lift), and mixing (thermodynamic). Making use of the historical empirical evidence, statistical evaluations, and the principles of physical-chemistry, researchers and application-specialists have used this information to enact two basic types of mitigation or prevention strategies with regard to fog hazards and a third option for use in other cases: (1) fog dispersion, (2) air quality management, and (3) special cases. In the first case fog dispersion has focused on relatively costly mechanical means (i.e., bulk mixing) as opposed to thermodynamic and physical methods (i.e., heating and/or seeding) that have been applied predominantly at aerodrome locations. For air quality issues, emphasis has been placed on reduced exposure (e.g., avoidance by remaining indoors or masking through the use of personal filters), reduction of contributing sources, or ventilation (similar to bulk mixing) and is often a function of the population affected (i.e., receptors) and the resources required for cost-effective implementation and the likely effectiveness.

The third aspect (special cases) typically offers a mix of the foregoing methods, requires other techniques, or is not well known. These would include situations of ice or acid fog, fog-water collection, smog, or fogs occurring with smoke and haze. Related to these anthropogenic issues are the creations of artificial fogs by cooling towers and other industrial processes or those fogs made for use "on stage" or for entertainment purposes. While most of these make use of simple ventilation techniques or equipment, these environments are often complicated due to the presence of other airborne aerosols which may be more readily deliquesced and thus affective for the population that is present. Impacts may range from ear, eye, nose, or throat ailments or chronic exposure that may need to be treated on an individual basis. Secondary impacts may be wet or dry deposition from fog on plants, buildings, statues, and other surfaces causing damage due to chemical interactions (or scalding).

Operational support

In support of the avoidance, mitigation, or prevention of fog hazards, a variety of operational support and decision-making tools and information is available to forecasters as well as the affected and responding communities (e.g., aerodromes, emergency managers, or similar personnel). In some cases, these are also linked to, or provide data ingest for, numerical or statistical guidance packages. These are particularly effective when tied to GIS databases and decision-support software (or artificial intelligence) and are more commonly used in a military (e.g., transmittal or blockage of select frequencies of communication, visible light and infrared/thermal sensors) or disaster-related type of response (e.g., applications in catastrophe modeling). The additional deployment of meso-and microscale surface-based observing networks will increase the spatial and temporal acuity of data as related to fog occurrence and its evolution so that it may be more accurately detected, assessed, and compared with forecasts (e.g., the National Digitized Forecast Database as per the United States: NOAA/NWS). In some instances, highway systems have been deployed to provide "instant" warning signs to alert drivers to rapidly changing visibility in fog-prone regions.

Satellite and similar remote sensing platforms offer a variety of products (e.g., the website of "Nighttime Fog and Low Cloud Images" as produced by NOAA-NESDIS, USA: http://www.orbit.nesdis.noaa.gov/smcd/opdb/aviation/fog.html) and provide a gross estimate of fog occurrence, intensity, and coverage by channel differencing as well as through examination of sounder data to construct vertical and near-surface profiles of temperature and moisture in the atmosphere. While useful, these products only provide information after fog has formed and thus allow tracking its movements and evolution. While other parameter fields are available through satellite imagery, none presently provide estimates of fog precursors for predictive purposes (or for the assessment of the pre-fog environment). Although microwave sensors and ground-based radar, lidar, and profiler platforms may offer additional information and operational support, none are presently suited to fog detection or prediction (with the exception of automated highway alert systems). Additional information on atmospheric chemistry and structure is now available through several remote sensing platforms, and these will provide a real-time observation profile of the physical-chemistry of the atmosphere as related to fog and other phenomena or conditions in the atmosphere (including the diagnosis of the pre-fog environment).

Summary

Continued progress in understanding the physical and chemical attributes and behaviors of natural and artificial fogs will provide for localized modeling. When these are integrated with community or population-impact models the prediction of fog occurrence, intensity, evolution, coverage, and impacts will be further enhanced. These will require the use of emerging decision-making tools, visualization environments, additional observational networks, and the development of advanced warning systems that monitor the pre-fog environment in order to provide real-time updates and automated advisories. Further work on the mitigation and prevention of fog will proceed from these efforts in order to enhance operations while reducing impacts and costs to the safety and health of select populations as well as economic sectors.

Bibliography

Chagnon, S. A., 2006. *Railroads and Weather*. Boston: American Meteorological Society.

Croft, P. J., 2003. Fog. In Holton, J. R., Curry, J. A., and Pyle, J. A. (eds.), *Encyclopedia of Atmospheric Sciences*. New York: Academic.

Croft, P. J., Pfost, R. L., Medlin, J. M., and Johnson, A. J., 1997. Fog forecasting for the southern region: a conceptual model approach. *Weather Forecast*, **12**(3), 545–556.

Geiger, R., 1965. *The Climate Near the Ground*. Cambridge: Harvard University Press. 611 pp.

Sutton, O. G., 1953. *Micrometeorology*. Maidenhead: McGraw-Hill. 333 pp.

World Meteorological Organization, 1992. *Manual on the Global Data-Processing and Forecasting System*. WMO-No. 485; 153 pp.

Cross-references

Climate Change
Cloud Seeding

Doppler Weather Radar
Fog Hazards Mitigation
Fohn

FÖHN

Anita Bokwa
Jagiellonian University, Kraków, Poland

Synonyms

Aspre; Austru; Berg; Chinook; Favogn; Halny; Nor-Wester; Samun; Santa Ana; Zonda

Definition

Föhn is a warm, dry, strong, and gusty wind descending down on the leeward side of a mountain range, as a result of a synoptic-scale, cross-barrier flow over the mountain range.

Föhn winds occur along the leeward slopes of mountains in various regions of the world and they have regional names, such as *zonda* in Argentina, *halny* in Poland, *austru* in Romania, *favogn* in Switzerland, *aspre* in France, *chinook* and *Santa Ana* in USA, *berg* in South Africa, *Nor-wester* in New Zealand, and *samun* in Iran. The particular term *föhn* is associated with the Alps. When moist air is forced to rise over the mountains, it cools at the saturated adiabatic lapse-rate ($0.5°C \cdot 100$ m^{-1}). The humidity increases until the condensation begins. Precipitation may occur and föhn clouds can be formed. Föhn clouds are either orographic clouds or mountain wave clouds. Orographic clouds may include crest clouds and the föhn wall (föhn bank), that is, clouds that form along a mountain range, either on the ridge or slightly above and to the lee of it, and which present, to an observer downwind from the ridge, the appearance of a vertical wall. The most distinctive wave clouds are altocumulus lenticularis clouds. Subsequently the air descends along the leeward side of the ridge and warms at the dry adiabatic lapse-rate ($1.0°C \cdot 100$ m^{-1}), which generates a rapid rise of air temperature, as high as 20°C (36°F) in 1 h and a decrease in relative air humidity, possibly to less than 5%. Föhn can cause dangerous avalanches and sudden snowmelt in the mountains, which may result in floods. The extreme dryness of air can lead to dangerous fire weather conditions. Föhn has a negative impact on bioclimatic conditions for humans; it increases cardiovascular problems and affects the symptoms of some mental illnesses.

Sometimes before the occurrence of a föhn wind, a high föhn (free föhn) can be observed. The latter are föhn-like conditions at higher elevations, while the lower elevations are under a cold air mass. The mountains are then often warmer than the lowlands. The warming is attributed to subsiding air in a synoptic anticyclone above the cold surface air. The high föhn does not necessarily have strong, gusty winds.

Bibliography

Ahrens, C. D., 1999. *Meteorology Today: An Introduction to Weather, Climate, and the Environment*. Pacific Grove: Brooks/Cole.
Glossary of Meteorology, 2009. American Meteorological Society, http://amsglossary.allenpress.com/glossary
International Meteorological Vocabulary, 1992. Geneva: WMO.
Whittow, J. B., 1984. *The Penguin Dictionary of Physical Geography*. London: Penguin Books.

Cross-references

Climate Change
Cloud Seeding
Doppler Weather Radar
Dust Storm
Fire and Firestorms
Heat Wave
Ice Storm
Storms

FOREST AND RANGE FIRES

George Eftychidis
Algosystems S.A, Kallithea, Greece
Pangaiasys Ltd., Pikermi, Greece

Synonyms

Forest fires; Range fires; Wildfires

The forest fire problem

Fires have been a major ecological factor that shaped and continue shaping the earth landscape and vegetation along centuries and across the entire globe (Trabaud and Prodon 1992). There have been wildfires long before *Homo sapiens* evolved. One main component of carboniferous and north hemisphere coal is charcoal left over by forest fires. Being an intrinsic factor of the ecosystem, fire occurs where and when the conditions favor ignition. Wildfires are common in many places around the world, where the climates are sufficiently moist to allow the growth of trees, but feature extended dry, hot periods when fallen branches, leaves, and other material dry out and become highly flammable. Wildfires are also common in grasslands and scrublands.

In the Mediterranean region, as well as in areas with a Mediterranean climate type (hot dry summer combined with cold and rainy winter) worldwide, vegetation is considered the result of competition and adaptation to fire of the forest species. Plants use a variety of strategies from possessing reserve shoots that sprout after a fire, to fire-resistant seeds for surviving fire, and they even encourage fire (e.g., eucalypts contain flammable oils in the leaves) as a way to eliminate competition from less fire-tolerant species (http://www.sciencedaily.com/articles/b/bushfire.htm).

In 2004, researchers discovered that exposure to smoke from burning plants actually promotes germination in other types of plants by inducing the production of the chemical butenolide (Dawsa et al., 2004). Most native animals in fire-prone regions are also adapted to surviving wildfires.

The ecological role of the fire in the natural ecosystems can be compared to the role of fever as a protective response to human diseases. According to the circumstances both can be a healthy systemic reaction or a lethal upshot. Thus, small-scale fires or fires in old, mature forest stands can be part of the natural dynamics of the ecosystem and, therefore, beneficial to the health of the vegetation from the ecological point of view. On the contrary, frequent, repetitive fires which prevent the stands from maturity and suppress the regeneration capability of the forest or large-scale fires that consume the tree seeds due to high fire intensity can be catastrophic or leading to a permanent impair of the ecosystem.

Although the vast majority of fire ignition is due to human activity or is intentionally set, there are fires that start from natural causes such as lightning or self-ignition. Rapid land-use changes, socioeconomic conflicts, and competing interests, which are quite typical in the regions threatened by forest fires, contribute to increasing their occurrence and tend to aggravate their effects. With extensive urbanization of wild lands, fires often involve destruction of suburban homes located in the surroundings of large cities and in the zone of transition between developed areas and undeveloped wild land, the so-called wildland urban interface. The problem of fire in the wildland urban interface increased greatly over the last decades due to lack of management of nonproductive forests and the extensive use of forests as recreation and residence areas. The combination of flammable vegetation and human activity in the same area fanned most of the large disastrous fires in Greece in 2007 (Peloponnesus) and 2009 (Attica), Portugal in 2003, Australia in 2009 (Victoria), USA in 2003 and 2008 (California), etc.

Every year, 9,200 million tons of biomass are burnt globally for energy generation. Wildfires consume over half of this – some 5,130 million tons of biomass. At the same time, they release 3,431 million tons of CO_2 into the atmosphere, contributing to global greenhouse gas emissions and thus to climate change (FAO, 2007).

Although the main reason for increased fire occurrence over the last decades is land-use change, climate change plays a very important role in modifying forest fire regimes worldwide. The fire-prone Mediterranean landscapes in Europe are likely to be affected by regional climate change that may exacerbate the current fire conditions. Recent assessments on climatic changes in the Mediterranean Basin show that air temperature increases while summer rainfall is reduced, which leads to an increase in the soil water deficit and the vegetation water stress (Dimitrakopoulos and Mitsopoulos, 2006). As vegetation burns, it releases stored-up carbon into the atmosphere, speeding global warming and thereby exacerbating conditions that may lead to extreme behavior and generate a greater incidence of wildfires in the coming years (Flannigan et al., 2005). This phenomenon can be further engraved by the use of fire on a massive scale as a cheap and efficient way to clear forests for agriculture or development. In fact, deforestation fires alone have contributed 20% of the total greenhouse gases that humans have contributed to the atmosphere since industrialization (Scientific American, 2009).

The fire season

Fire season is defined as the period or periods of the year during which wild land fires are likely to occur, spread, and do sufficient damage to warrant organized fire-suppression activities.

Naturally caused fires are more frequent as the dry and warm season advances, due to the cumulative decrease of moisture in the soil and the limited rainfall. Geographically the number of fires increases as we move from smaller to greater latitudes and from higher altitudes to the sea level. The fire season typically starts after the end of the winter rainfall and finishes following the first rains in autumn.

Types of fire

It is quite usual to describe the type of a fire burning in natural surroundings in terms of the main form of supporting vegetation. A forest fire, often termed as "wildfire" in the US literature, and "vegetation fire" or "bushfire" in Australasia, is an uncontrolled fire in tall, closed forests, ignited by natural or human causes. The term is commonly used referring to any fire event burning in wildland areas. Grass fire is a relative term referring to fire burning herbaceous vegetation, which may be substituted by the terms prairie, rangeland, or veldt (similar to savanna areas in African regions) fire (Luke and McArthur, 1986). There can be also tundra fires, moor or muir (poorly drained areas with patches of heath and peat bogs) fires, and peat fires which usually burn underground. In North America, the term "brushfire" is used if the main fuel is a low, scrubby vegetation, whereas in Australia such type of fire is called scrub fires. Rural fire is also used to describe fires burning in the countryside in a mosaic of agricultural cultivations.

All grass species have a lifestyle where the grass grows, flowers, and dies. This process is called curing. Due to curing, grasslands and rangeland fires can occur earlier than fires in shrub lands and forest stands, according to the degree of curing of grass along the fire season. Brush and chaparral (maquis) fires usually start somewhat later in the season than the first grass fires, and they normally have more intense, longer lasting impacts.

Furthermore fires can be characterized based on the vegetation layer in which they burn or propagate. We may then distinguish ground, surface, and crown fires, whereas the term "spot fire" is used to denote a secondary fire which has been ignited ahead of the main fire by an airborne firebrand or ember.

Overall, the types characterized by the vegetation layer carrying the fire include the following:

- *Ground fires* which burn below the surface or the ground in deep layers of organic material such as peat bogs, fed by subterranean roots, duff, and other buried organic matter. This fuel type is associated with smoldering combustion and is particularly susceptible to ignition due to spotting. Ground fires can be caused due to inadequate vegetation management, such as peat fires in Kalimantan and Eastern Sumatra, Indonesia, which resulted from a rice land creation project that unintentionally drained and dried the peat.
- *Crawling or surface fires* are fueled by low-lying vegetation such as leaf and timber litter, debris, grass, and low-lying shrubbery without extending into the tree crowns. In this category we include grass, brush, and shrub fires.
- *Crown, canopy, or aerial fires* burn tree crowns and suspended material – called ladder fuels – at the canopy level, such as tall trees, vines, and mosses. The ignition of a crown fire, termed "crowning," is dependent on the density of the suspended material, distance of the canopy base from ground, canopy continuity, and sufficient surface and ladder fuels for the flames to reach the tree crown.

Crown fires evolve often to extremely violent fires that are described as conflagrations or blow-up fires, meaning fires having "assumed three-dimensional characteristics" (Luke and McArthur, 1986).

Fires are also characterized according to the combustion phase as flaming, creeping, and smoldering. Flaming is normally associated with the front of the fire propagation. A creeping fire is a low-intensity fire with a negligible rate of spread whereas a fire burning without flame and barely spreading is characterized as smoldering. Smoldering combustion of biomass can linger for days or weeks after flaming has ceased without having any visual impact; however, it has important consequences for the forest ecosystem resulting in large quantities of fuels consumed and becoming a global source of emissions to the atmosphere. The slow propagation during the smoldering phase of combustion leads to prolonged heating and might cause sterilizations of the soil or the killing of roots, seeds, and plant stems at the ground level.

Detection

Fast and effective detection is a key factor in wildfire fighting. Early detection and relevant warning of the protection authorities aim to early response for limiting the fire size and thus require accurate location of the fire spot in relation to topography, day and nighttime operation, and ability to prioritize fire danger (Hirsch, 1964).

Fires were detected systematically from observers in lookout towers since the beginning of the twentieth century. Most of these towers were more than 30 m tall with views reaching up to 50 km. However, forest authorities limited such high towers later being concerned that people might climb up and injure themselves (CBC Digital archives). Before radio contact was possible, observers reported fires using telephones, carrier pigeons, and heliographs (a heliograph operator uses Morse code to communicate by tilting a mirror to reflect sunlight). Aerial and land photography using instant cameras were used in the 1950s until infrared scanning was developed for fire detection in the 1960s. However, information analysis and delivery was often delayed by limitations in communication technology. Early satellite-derived fire analyses were hand-drawn on maps at a remote site and sent via overnight mail to the Fire Manager.

In the last decades, public hotlines, fire lookouts in towers, and ground and aerial surveillance platforms have been used as means of early detection of forest fires. However, since accurate human observation may be influenced by operator fatigue, time of day, time of year, and geographic location, electronic systems have gained popularity in recent years as a possible alternative. The layout of a network of fire detection stations is defined based on the results of visibility analysis and fire risk assessment of the envisaged area performed using GIS technology. Alerts from fire detection systems are normally merged by the Fire Manager together with satellite data, aerial imagery, and personnel position, available via Global Positioning System (GPS), into a response module of an Integrated Forest Fire Management system. There are currently systems under development which consider broadcasting of fire detection alerts to local population using SMS messaging (ESS project, 2009).

Small, high-risk areas that feature thick vegetation, a strong human presence, or that are close to a critical urban area can be monitored using a local wireless sensor network in order to overcome privacy problems associated with the use of visual cameras (Dubbeld, 2005). These networks act as specific automated weather systems: detecting temperature, humidity, and smoke. This is a modern approach for detecting fire ignition and monitoring its propagation using dispersed wireless temperature sensors, fixed on trees (Figure 1), which communicate to each other their status and measurement, broadcasting alerts to long distances if a fire occurs.

Wireless sensors are normally powered by batteries, since their size is rather small and forest conditions are not favorable for using solar energy. A major environmental issue regarding this technology is the dispersal of the batteries in the forest. This problem requires the monitoring of the sensors at the physical level, which in fact downgrade the performance of the solution and prevent its safe use in natural environments. Recent developments consider the use of bioenergy-harvesting technology (tree-supplied energy) for recharging batteries using converted metabolic plant energy into usable electricity, which can be a relevant solution in the future (Figure 2).

Larger, medium-risk areas are usually monitored by scanning towers that incorporate fixed cameras and sensors to detect smoke or additional factors such as the

Forest and Range Fires, Figure 1 Wireless fire sensors (Photo courtesy Enveve SA).

infrared signature of carbon dioxide produced by fires. Brightness and color change detection and night vision capabilities may be incorporated also into sensor arrays.

A particular solution is considered to be the use of mobile fire detection and monitoring stations which can be deployed in an area only during the fire season and moved easily in other areas during the rainfall period in order to be used, for example, for flood monitoring (Figure 3).

Satellite and aerial monitoring can provide a wider view and may be efficient to monitor very large, low-risk areas. Properly equipped unmanned aerial vehicles (UAV) are used for such purpose. These more sophisticated systems employ GPS and aircraft-mounted infrared or high-resolution visible cameras to identify and target wildfires (Figure 4).

Satellite-mounted sensors, such as Envisat's Advanced Along Track Scanning Radiometer and European Remote-Sensing Satellite's Along-Track Scanning Radiometer, can measure infrared radiation emitted by fires, identifying hot spots greater than 39°C. The National Oceanic and Atmospheric Administration's Hazard Mapping System in the USA combines remote-sensing data from satellite sources, such as Geostationary Operational Environmental Satellite (GOES), Moderate-Resolution Imaging Spectroradiometer (MODIS), and Advanced Very High Resolution Radiometer (AVHRR), for detection of fire and smoke plume locations. However, satellite detection is limited so far for operational use in the conditions of the European forests. This is due to the fact that satellite detection of fires is prone to offset errors, anywhere from 2 to 3 km for MODIS and AVHRR data and up to 12 km for GOES data. Satellites in geostationary orbits may become disabled, and satellites in polar orbits are often limited by their short window of observation time. Cloud cover and image resolution may also limit the effectiveness of satellite imagery. On the contrary,

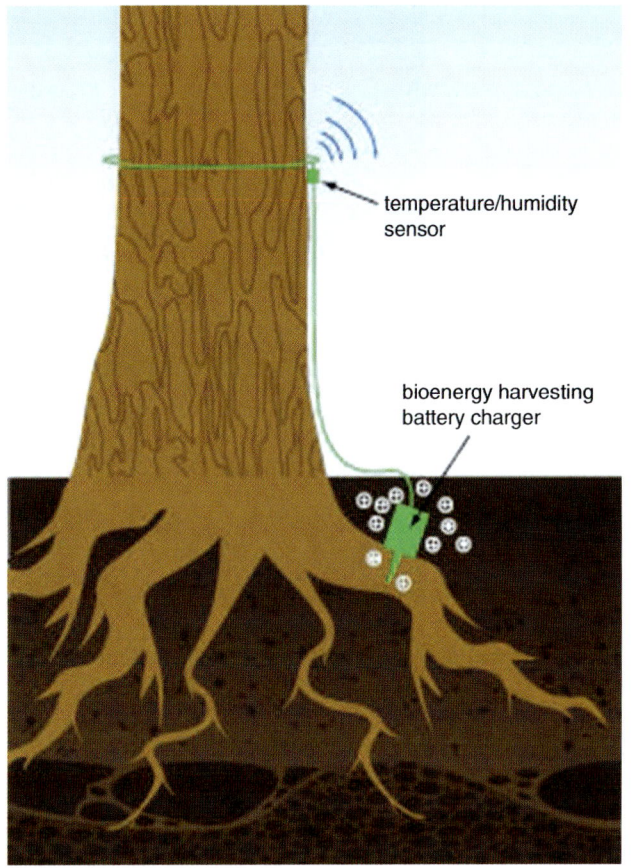

Forest and Range Fires, Figure 2 Voltreebioenergy harvesting system (Photo courtesy Voltree power).

satellite technology can be very helpful for monitoring very large fires, detecting fires in isolated vast forest areas as well as for supporting post-fire management decisions (Figure 5).

Prevention

Since most of the fire events recorded in the various regions of the planet are caused by human activity it is expected that fire ignition can be prevented. Similarly measures can be taken for mitigating the impact of a fire in order to keep its size within the boundaries of the fire control mechanism. This is the main objective of the fire prevention planning worldwide.

While considered global the wildfire problem appears different due to the diversity of the ecological factors and the existing variety of land-use patterns, the environmental awareness, the fire culture, and the forest management practices across the globe. Thus, the percentage of human-caused fires can differ from as low as 60% in the Northern American continent (except Mexico) up to 95% in the Mediterranean region (FAO, 2007). However in Mexico and Central America, this percentage raises to 97% since several fires are intentionally set for

Forest and Range Fires, Figure 3 Mobile fire detection station installed in the island of Karpathos in Greece (Photo courtesy Faenzi Srl).

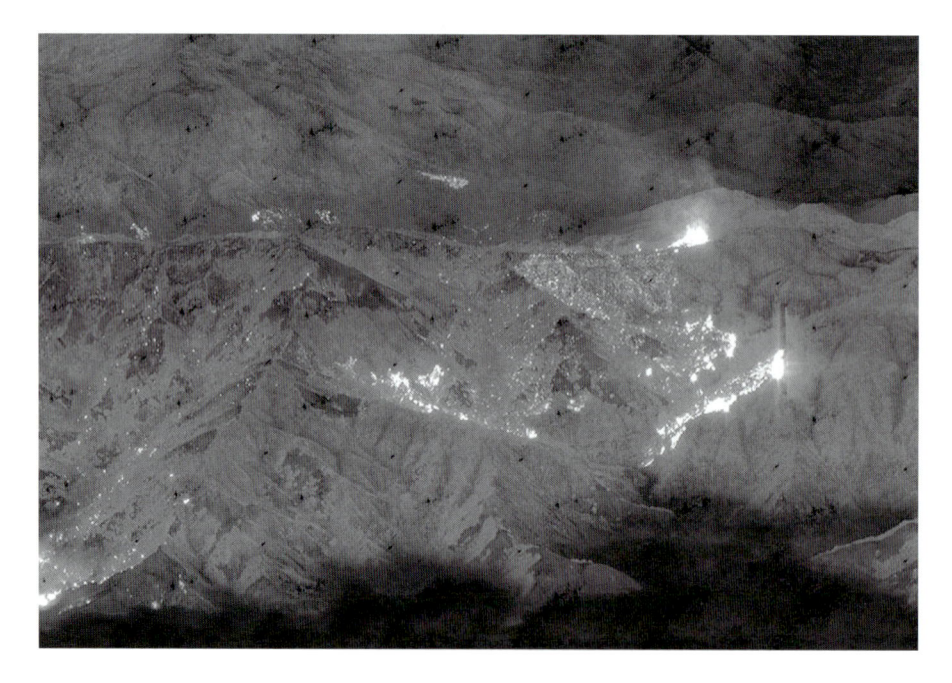

Forest and Range Fires, Figure 4 A photo of the wildfires in Northern California taken by the RQ-4 Global Hawk (UAV).

agricultural purposes. In countries such as Canada and the Russian Federation, lightning is a major cause of starting a fire in the forest. In Canada, for instance, 35% of the total number of fires is caused by lightning compared with the 60% of fires ignited by human activity. On the contrary, in southern EU countries (Portugal can be an exception here), although being the main natural factor, lightning causes less than 5% of the fire starts (Kettlunen et al., 2008). Fire is also used as a forest management tool for eliminating competition between species, reducing fire hazard, improving the recycling of nutrients of the soil, etc. Although prescribed fire is normally practiced since

Forest and Range Fires, Figure 5 Wildfires across the Balkans in late July 2007 (MODIS image).

several decades in the USA and Australia, it is slowly progressing in Europe as a Forest and Fire Prevention Management practice.

Wildfire prevention refers to the preemptive methods of reducing the risk of fires as well as lessening its severity and spread. Effective prevention techniques allow supervising agencies to manage air quality, maintain ecological balances, protect resources, and limit the effects of future uncontrolled fires. However, prevention policies must consider the role that humans play in wildfires, since, for example, only 5% of forest fires in Europe are not related to human involvement. Sources of human-caused fire may include arson, accidental ignition, or the uncontrolled use of fire in land-clearing and agriculture such as the slash-and-burn farming in Southeast Asia.

Fire danger rating systems are used by national authorities in most countries for producing qualitative and/or numeric indices of fire potential, based mainly on fuels, topography, and weather. These rating systems allow Fire Managers to estimate present and future fire danger for a given area.

From the management point of view, the risk of major wildfires can be reduced by reducing the amount of fuel present. In wild land, this can be accomplished by either conducting "controlled burns" – deliberately setting areas ablaze under less-dangerous weather conditions in spring or autumn – or physical fuel removal by removing some trees or shrubs beneath the forest stands. Multiple fuel treatments are often needed to influence future fire risks, and wildfire models may be used to predict and compare the benefits of different fuel treatments on future wildfire spread. Both approaches are controversial with some environmentalists, who regard them as tampering with the forest ecosystem.

Beyond the operational management of fires, major issues for preventing catastrophic events are education of the population and reasonable planning of land development in particular in the wildland urban interface.

People living in fire-prone areas typically take a variety of precautions, including building their homes out of flame-resistant materials following particular national codes, reducing the amount of fuel within a prescribed distance from their home or property (including *firebreaks* – their own miniature control lines, in effect), and investing in their own firefighting equipment. Communities in the Philippines maintain fire lines 5–10 m wide between forests and villages, and patrols along these lines are set during summer months or seasons of dry weather. In Switzerland, agricultural zones are planned between settlements and forests for reducing the risk and providing defensible space to address a fire approaching the village.

Rural farming communities are rarely threatened directly by wildfire. These types of communities are usually located in large areas of cleared, usually grazed, land, and during the drought conditions that prevail in the wildfire years, there is often very little grass left on such grazed areas. Hence the risk is minimized. However, urban fringes spread in and communities have literally built themselves in the middle of highly flammable forests. These communities are at high risk of destruction in the case of forest fires.

Wildfire mitigation is accomplished through the assessment of fire risks, the reduction of fire risks, and the prevention of unwanted fires. Loss of life, property, and resources can be reduced if planners, developers, fire agencies, and homeowners work together to define, enforce, and maintain reasonable fire safety standards.

The fire paradox

Since a great percentage of fires are due to the human activity and carelessness, for many decades the fire management policy was to suppress all fires. In 1937, US President Franklin D. Roosevelt initiated a nationwide fire prevention campaign, highlighting the role of human carelessness in forest fires. Sixty years after implementing aggressive fire-suppression programs they realized that no new sequoias had been grown in the redwood forests of California, because fire is an essential part of their life cycle. This situation led to including controlled burns in the fire management policy for reducing much of the forest undergrowth which is the main stratum supporting the propagation and growth of the fire. Therefore in North America, current firefighting policies may permit naturally caused fires to burn to maintain their ecological role, so long as the risks of escape onto high-value areas are mitigated.

It was in the mid-1800s when explorers from the HMS Beagle (the ship of Charles Darwin mission) observed Australian Aborigines using fire for ground clearing, hunting, and regeneration of plant food in a method called fire-stick farming. Such careful use of fire has been employed for centuries in the lands protected by Kakadu National Park in Australia to encourage biodiversity.

In Europe, fire management practice continues to be closer to the fire suppression, although there are efforts to consider the use of fire as a tool for managing wild land fires. However, the human activity continues to be the center of any fire management policy, trying to combine easier access to the forest while reducing the risk of dangerous high-intensity fires caused by many years of fuel buildup. Although not a European peculiarity, there are fire management policies that are still missing which can address fire prevention with the objective to ensure sustainability, biodiversity, and community-based protection of the forests (Figure 6).

The fire-exclusion policy results also in the buildup of forest fuel resulting in large and severe fires in all countries where the policy proved to be successful for several years in the past. This is known as the fire paradox, which put on evidence that efficient fire suppression leads after years to bigger fires if not accompanied with other forest management measures. Large fires, the so-called Megafires, were the result of this policy in the Yellowstone National Park in 1988, the fires in Greece during the summer of 2007, and many others. Urbanization of forested areas is another factor resulting in fuel buildup and devastating fires, such as those within the California cities of Oakland and Berkeley (1991, 2003), all over Colorado (2002), in Attica region (Greece 2009), in Guadalajara (Spain 2005), etc.

Forest fuels and fuel management

In wild land fires, fuel corresponds to all combustible plant-derived material including grass, litter, duff, down dead woody debris, exposed roots, plants, shrubs, and trees. This plant-derived material can be dead or alive. Thus, a significant difference between fuel and vegetation is that fuel refers mainly to small- and medium-sized combustible material including dead vegetative material, while vegetation refers to the living plant components of any size. Plant parts that are not consumed, such as the trunks of live trees, are not considered fuel.

The forest fuel concept, as used in fire behavior modeling, includes further the horizontal and vertical distribution of the combustible material in the forest. Thus, the spread of wildfires varies based on the quantity and distribution of the flammable material present in the forest fuel bed and its vertical arrangement.

Fuel status is influenced not only by the weather but also due to the topographic conditions in the fire scene. For example, fuels uphill from a fire front are more readily dried and warmed by the fire than those downhill, yet burning logs can roll downhill from the fire to ignite other fuels. Fuel arrangement and density is governed in part by topography, as land shape determines factors such as available sunlight and water for plant growth.

Fuels can be produced or modified by the human activity in the forest. Fuels resulting from, or altered by,

Forest and Range Fires, Figure 6 Controlled burning in Portugal.

forestry practices such as timber harvest or thinning, as opposed to naturally created fuels are called Activity fuels. Branches and other noncommercial wood left on site after timber harvesting (slash) is the quintessential activity fuel. Other land management and activity, such as pruning, herbicide application, noncommercial thinning of forest stands, and even controlled burning (killing tree branches or whole trees) for silvicultural purpose, can create activity fuels.

A series of fuel classification systems have been historically developed taking into account the needs of fire management and fire modeling. In the traditional sense, fuel models are a set of parameters that describe vegetation in terms that are required for the solution of a mathematical rate-of-spread model. Fuel models were thus defined initially as input to fire growth simulation models, such as Rothermel's surface fire behavior and spread model (Rothermel, 1972), BEHAVE (Burgan and Rothermel, 1984; Andrews et al., 2005), FARSITE (Finney, 2006), FlamMap fire potential simulator (Finney, 2006), and Canadian Wild land Fire Growth Model (CWFGM) Prometheus (Prometheus, 2000). Nowadays, it is recognized that solid fuel classification systems are also essential tools for fuel mapping, hazard assessment, evaluation of fuel treatment options, and monitoring fire effects (Sandberg et al., 2001).

Fuels are the main component of the fire triangle which fire management can influence directly. Fuel management projects are prepared on a regular base, including thinning and pruning of vegetation, partial removal of understory shrubs, construction of fuelbreaks and control lines, etc. The objective of the fuel management is to keep fire behavior corresponding to predefined environmental scenarios within the limits of the fire-suppression mechanism. Fuel typology is useful for supporting planning of forest fuel management projects.

Fire behavior

Wildfires occur when fuel, heat, oxygen, and the chain reaction occur in a wooded area. The fire occurrence is modeled as a tetrahedron having in each of its sides one of the aforementioned elements as shown in the "Figure 7".

The time of burning is considered an additional dimension to the traditional fire triangle since longer fire duration can release additional energy, increase the potential of the fire, and cause extreme or erratic behavior.

Wildfire behavior and severity result from the combination of factors such as available quantity of burnable material (fuel load); physical setting, that is, topographic conditions; and weather. Based on these conditions the size of the fire can vary greatly. Wildfires can be kept small if wind strength, fuel quantity, and distribution (sparse vegetation) prevent its propagation. Fires are also kept small if they are suppressed at a very early stage before gaining significant size and releasing energy that supports their propagation and outweighs the suppression efforts.

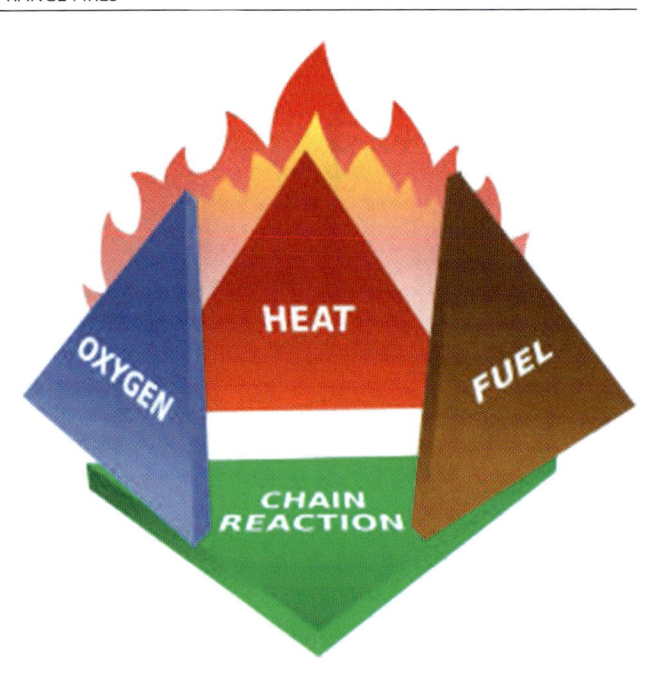

Forest and Range Fires, Figure 7 The fire tetrahedron.

An ignition source is brought into contact with a combustible material such as vegetation, which is subjected to sufficient heat and has an adequate supply of oxygen from the ambient air. High quantity of water in the burnable material (fuel moisture content) usually prevents ignition and slows propagation, because higher temperatures are required to evaporate any water within the material and heat further the material to its ignition point. Dense forests usually provide more shade, resulting in lower ambient temperatures and greater humidity, and are therefore less susceptible to wildfires. Less-dense materials, such as grasses and leaves, are easier to ignite because they contain less water than denser materials, such as branches and trunks. Plants continuously lose water by evapotranspiration, but water loss is usually balanced by water absorbed from the soil, humidity, or rain. When this balance is not maintained, plants dry out and are therefore more flammable, often a consequence of droughts.

Occasionally fires spread outward at an equal rate in all directions and assume a circular form. This is more probable in grass fires burning a homogeneous fuel bed, in calm conditions and in rather plain areas. Forest fires burn and grow according to the availability of the material that they consume, the prevailing wind, and the landscape topography. Therefore, they normally become irregular or elongated in shape when spreading. For supporting fire suppression, specific names are assigned to the different parts and sectors of a spreading fire as shown in the Figure 8.

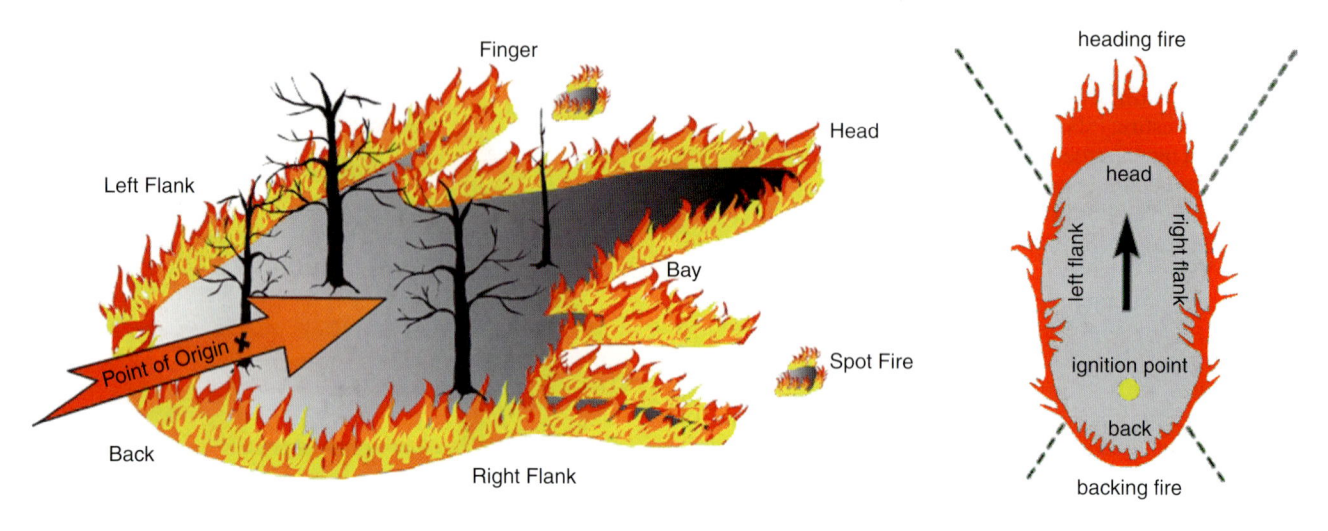

Forest and Range Fires, Figure 8 The parts of a forest fire.

A wildfire front is the portion sustaining continuous flaming combustion, where unburned material meets active flames, or the smoldering transition between unburned and burned material. As the front approaches, the fire heats both the surrounding air and woody material through convection and thermal radiation. First, wood is dried as water is vaporized at a temperature of 100°C. Next, the pyrolysis of wood at 230°C releases flammable gases. Finally, wood can smolder at 380°C or, when heated sufficiently, ignite at 590°C. Even before the flames of a wildfire arrive at a particular location, heat transfer from the wildfire front warms the air to 800°C, which preheats and dries flammable materials, causing materials to ignite faster and allowing the fire to spread faster. High-temperature and long-duration surface wildfires may encourage flashover or torching: the drying of tree canopies and their subsequent ignition from below.

Wildfires have a rapid forward rate of spread when burning through dense, uninterrupted fuels. They can move as fast as 10.8 km/h in forests and 22 km/h in grasslands. Wildfires may spread by jumping or spotting as winds and vertical convection columns carry firebrands (hot wood embers) and other burning materials through the air over roads, rivers, and other barriers that may otherwise act as firebreaks. Torching (an intermittent ignition of tree crowns as fire advances) and fires in tree canopies encourage spotting, and dry ground fuels that surround a wildfire are especially vulnerable to ignition from firebrands. Spotting can create spot fires as hot embers, and firebrands ignite fuels downwind from the fire. In Australian bushfires, spot fires are known to occur as far as 10 km from the fire front.

Large wildfires may particularly affect air currents in their immediate vicinities by the stack effect: Air rises as it is heated, and large wildfires create powerful updrafts that will draw in new, cooler air from surrounding areas in thermal columns. Great vertical differences in temperature and humidity encourage the formation of pyrocumulus clouds, strong winds, and fire whirls with the force of such tornadoes to reach speeds of more than 80 km/h. Rapid rates of spread, prolific crowning or spotting, the presence of fire whirls, and strong convection columns define extreme conditions of fire behavior.

Fire intensity increases during daytime hours. Burn rates of smoldering logs are up to five times greater during the day due to lower humidity, increased temperatures, and increased wind speeds. Sunlight warms the ground during the day which creates air currents that travel uphill. At night the land cools, creating air currents that travel downhill. Wildfires are fanned by these winds and often follow the air currents over hills and through valleys.

From the chemical point of view, fire is an oxidation process that rapidly transforms the potential energy stored in chemical bonds of organic compounds into the kinetic energy forms of heat and light. Like the much slower oxidation process of decomposition, fire destroys organic matter, creating a myriad of gases and ions, and liberating much of the carbon and hydrogen as carbon dioxide and water. A large portion of the remaining organic matter is converted to ash which may go up in the smoke, blow or wash away after the fire, or, like the humus created by decomposition, be incorporated into the soil.

Fires in shrub lands compared to grasslands burn with much greater thermal output because their fuel loading can be 5–50 times greater than that in grasslands. Similarly, the fuel in a mature forest, which may be 100 times greater than that of a medium density shrub land, requires more time to dry and become available for combustion.

The state of grassland curing is critical to the initiation and spread of grass fires. The proportion of dead grass in a pasture has a major effect on grassfire behavior. Thus, the fire is spread fast in grasslands even in cases where the grass is cured barely more than 50% and despite the presence of several green shoots.

Fire is traditionally used worldwide as grassland management tool and several wildfires start from grasslands burning for soil improvement (fertilization).

Fire modeling

Fire modeling is concerned with numerical simulation of wildfires in order to comprehend and predict fire behavior. Models are helpful tools for scientists and engineers. Fire models are used for organizing prevention, training firefighters, and supporting the coordination and decisions during the fire-suppression operations.

Existing models used for fire modeling are usually classified into the following: empirical models which are based primarily on statistics collected by observation of experimental or historical fires; physical models based on physical principles of fluid dynamics and laws of conservation of energy and mass; and semiempirical models based on physical laws, but enhanced with some empirical factors.

Various wildfire propagation models have been proposed in the past, including simple ellipses and egg- and fan-shaped models. Early attempts to determine wildfire behavior assumed terrain and fuel uniformity in space. However, the exact behavior of a wildfire's front is dependent on a variety of factors, including wind speed and slope steepness. Modern growth models utilize a combination of past ellipsoidal descriptions and Huygens' Principle to simulate fire growth as a continuously expanding polygon (Finney, 1998).

Computer models help predicting the propagation of a fire under varying fuel, weather, and topographic conditions. The most known modeling system for assessing wildfire is BEHAVE, which is a wild land fire behavior and fuel modeling system developed by the US Department of Agriculture Forest Service (Andrews et al., 2003). Numerous information systems have been developed based on BEHAVE and are currently in use for supporting operational fire management decisions worldwide (Farsite, Flamap, CARDIN, GFMIS, Fire tactics, Fire station, etc.) (Figure 9).

Fire suppression

Although fire is vital to the long-term health and sustainability of many ecosystems, wildfires take numerous human lives and cause damages in the range of several millions of euro per year.

Suppression includes all the procedures which start on, or after the fire alarm. Fire-suppression efforts are based on the fact that any fire requires three things: heat, fuel, and oxygen. Together, these make up the three legs of the so-called fire triangle known to all firefighters. The strategy in all firefighting is to extinguish the blaze by breaking one of the legs of this triangle. An entire science has developed around fire behavior and the effects of changing weather, topography, and fuels on that behavior.

There are several techniques for controlling the fire propagation in the forest. Fuelbreaks are 20–300-m-wide strips of land on which the natural vegetation has been permanently modified so that fires burning into it can be more readily controlled (Heikkila et al., 1993). Fuelbreaks are created as a prevention and mitigation measure before the start of any fire event in the area. They can include roads, streams, or rivers for extending the width of the fire barrier. During the fire event, firebreaks, which are strips from which all fuel is removed, are created and they are used as a control line for suppressing an approaching fire front. Existing natural or constructed fire barriers in the direction of the fire propagation are used as firebreaks as well.

Most fire-prone areas are served by fire authorities aiming to protect them from forest fires. Similarly to the water-spraying fire trucks used in urban fires, forest fire services use a variety of alternative firefighting techniques. These may refer to a variety of tools and technologies including chain saws and hand tools, all terrain fire-engines using water and foam, water-bombing air planes using sea- or lake-water and fire-retardants, helicopters with water bucket, etc. Sophisticated firefighting technology can consider edge techniques, such as the use of cloud seeding with silver iodide, full-scale aerial assaults by UAVs, establishment of fire control lines using water mist, use of explosives, etc. However, large fires are often of such a size that no conceivable firefighting service could attempt to douse them efficiently, and so alternative techniques are used.

Normally forest firefighters attempt to control the fire by controlling the area where it is spread, by creating firebreaks as control lines. These control lines can be produced by physically removing fuel (e.g., using bulldozers), or by "backburning," which refers to small, low-intensity fires that are set by the fire authorities in order to burn the flammable material between the fire front and the control line in a (hopefully) controlled way. These fires may then be extinguished by firefighters, or, ideally, directed in such a way that they meet the main fire front, at which point both fires run out of flammable material and are thus extinguished (Figure 10).

Unfortunately, such methods can fail in the face of wind shifts causing fires to miss control lines or to jump over them (e.g., because a burning tree falls across a line, burning embers are carried by the wind over the line, or burning tumbleweeds cross the line).

The actual goals of firemen during wildfire fighting vary. Protection of life (both of the firefighters and the "civilians") is given top priority, then follows private property according to economic and social value and also to its "savability" (structure triage), that is, to give an example, in a wildland urban interface fire, more effort will be dedicated towards saving a house with a tile roof than one with a wooden shake roof.

The usual firefighting tactic is the creation of control lines, that is, to create a defense line by removing fuels very far from the moving fire front. The control line is decided based on the estimation of the fire propagation according to the wind, topography, and forest fuels ahead

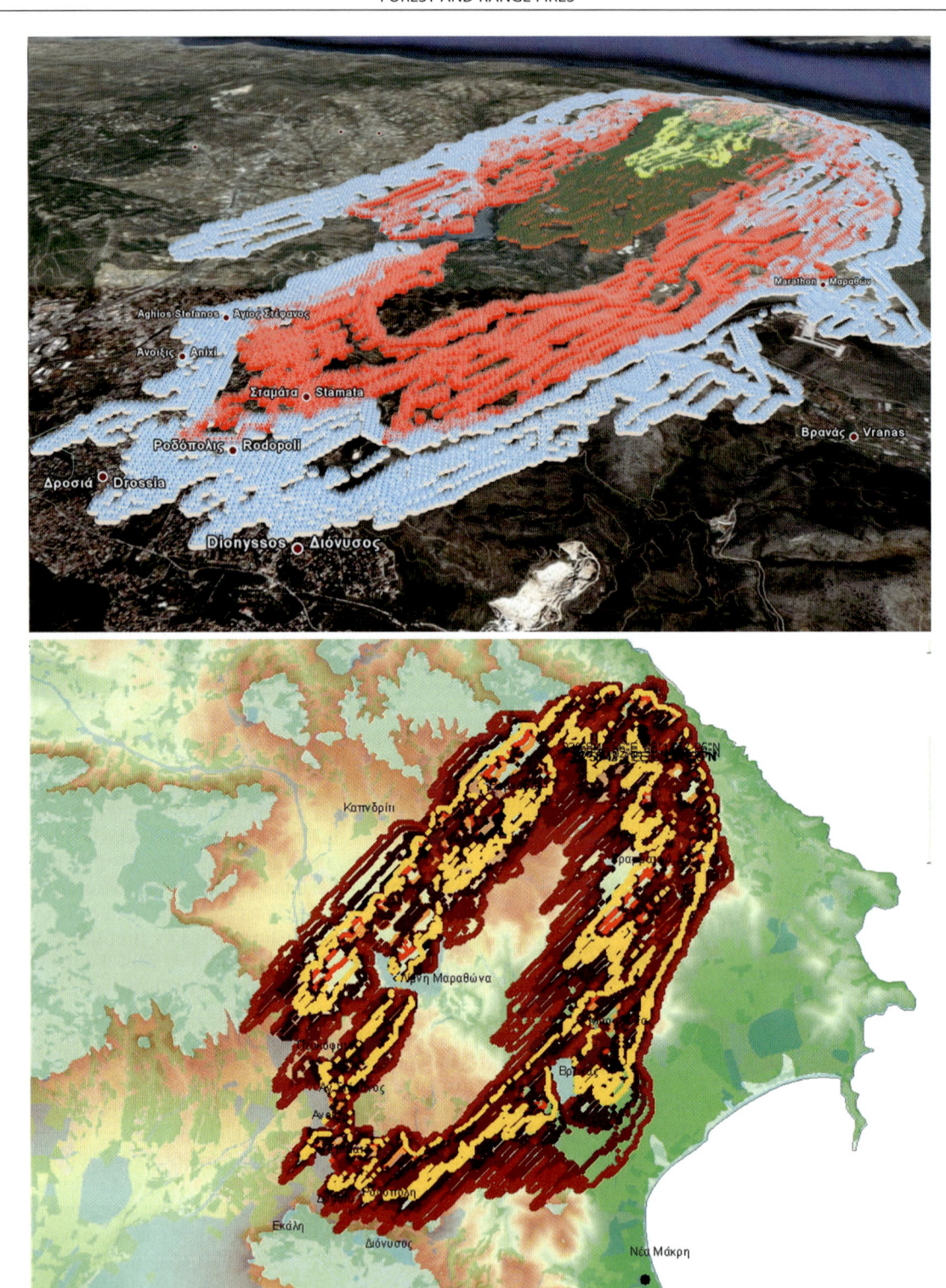

Forest and Range Fires, Figure 9 Simulation of a fire propagation using the GFMIS forest fire simulator.

Forest and Range Fires, Figure 10 Backburning.

of the front. Preventing the burning of publicly owned forested areas is in some countries (e.g., in Australia) of least priority, and it is quite common for firefighters to simply observe a fire burning toward control lines through forest rather than attempt to put it out more quickly – it is, after all, a natural process. This is not, however, the case in wildland urban interface or rural fires where the firemen have to protect the residents, their property, and the threatened infrastructures. In such cases, life protection comes first, whereas protecting houses is regarded as more important than, say, farming machinery sheds, although firefighters, if possible, try to keep fires off farmland to protect stock and fences (steel fences are destroyed by the passage of fire, as the wire is irreversibly stretched and weakened by it).

During the first stage of the fire evolution suppression can be performed using back-pack pumps and hand tools. As the fire grows mechanical means like chain saws and equipment (dozers, etc.) are required. The last stage of a controllable wildfire is addressed using aerial means. Despite the fact that aerial firefighting is considered by the public the most efficient practice for extinguishing wildfires, the fire propagation can be efficiently stopped by ground crews creating appropriate control lines (removing fuels). It is true that large fires and conflagrations cannot be distinguished from the ground. However, very large fires (so-called Megafires) burning with extreme weather conditions cannot be extinguished by aerial means either.

Fire effects

The aftermath of a wildfire can be as disastrous if not more so than the actual fire itself. A particularly destructive fire burns away plants and trees that prevent the forest soil from erosion. If heavy rains occur after such a fire, landslides, debris flows, and flash floods can occur. This can result in property damage in the immediate fire area and can affect the quality of water in streams, rivers, and lakes. Thus, soil erosion and flooding are the main effects expected following a wildfire.

Fires of both natural and anthropogenic origin have generated biotic and abiotic effects on ecosystem properties and the environment. The biotic effects of fire include changes in vegetation composition and subsequent impact on wildlife. Abiotic effects include changes in soil properties, nutrient cycling, water quality, and air quality. The quality of life experienced by human populations near the burned area is also impacted.

When the forest burns, the ground may or may not be intensely heated, depending on the arrangement of fuels from the ground to the forest canopy. Under a hot burn with the heavy ground fuel found in some forests, heat can penetrate mineral soil to a depth of tens of centimeters, significantly altering the physical, chemical, and biological properties of the soil. When the soil is heated, water is driven out; soil structure, which is the small aggregations of sand, silt, clay, and organic matter, may be destroyed, leaving a massive soil condition to a depth of several inches.

Forest and shrub lands soils often become hydrophobic or water repellent after the fire passage. A hydrophobic layer of a few centimeters thick commonly develops just below the burned surface. This condition is created because the fire's heat turns organic matter into gas and drives it deeper into the soil, where it then condenses on cooler particle surfaces. Under severe conditions, water simply beads and runs off this layer, like water applied to a freshly waxed car. Soil above the hydrophobic layer is highly susceptible to erosion during the first rains after a fire. Fortunately, the hydrophobic layer is soon broken by insects and burrowing animals, which have survived the fire by going underground; they penetrate the layer, allowing water to soak through.

Forest fires decrease soil acidity, often causing pH to increase by three units before and after the fire (e.g., from 5 to 8). Normally, conditions return to pre-fire levels in less than a decade. Fires also transform soil nutrients, most notably converting nutritive nitrogen into gaseous forms that go up with the smoke. Some of the first plants (legumes) to recolonize a hotly burned area are those whose roots support specialized bacteria that replenish the lost nutritive nitrogen through a process called nitrogen fixation. Large amounts of other nutrients, including phosphorus, potassium, and calcium, remain on the site, contributing to the so-called ashbed effect. Fires convert through burning organic vegetative material that could not before be used by the plants to a fertile ashbed. Plants that colonize these fertile ashbeds tend to be more vigorous than those growing outside of them. When heating has been prolonged and intense in areas such as those under burning logs, stumps, or debris piles, soil color can change from brown to reddish. Fires hot enough to cause these color changes are hot enough to sterilize the soil, prolonging the time to recovery. In the absence of heavy ground fuels, so much of the energy of an intense forest fire may be released directly to the atmosphere that soils will be only moderately affected.

Competing conceptions of the costs and benefits of wildfires have led to conflicting fire management and suppression objectives, most notably in the national parks and national forests. The debate on which fires should be allowed to burn and which should be suppressed undoubtedly will continue for some time.

Summary and conclusions

Forest fires are a major environmental issue worldwide. Despite the progress of the fire-suppression technology and the advances in fire science, the results in terms of burned area and damages still remain poor. People have now exposed a wealth of assets in fire-prone forest areas which competes with forest downgrading its environmental services. The policy of aggressive fire suppression is to limit the number of fires and to control the small- to medium-sized fires. However, the size of large and very large fires, fanned by global warming, exceeded all the previous records in the last decade and threatens the forest sustainability in the future if economic growth will continue feeding human activity in the forest without revising citizen's fire culture and creating rational plans of land development.

Bibliography

Andrews, P. L., Bevins, C. D., and Seli, R. C., 2003. *BehavePlus Fire Modeling System, Version 2.0:User's Guide*. USDA Forest Service Gen. Technical Report RMRS-GTR-106WWW. 132 pp.

Andrews, P. L., Bevins, C. D., and Seli, R. C., 2005. *BehavePlus Fire Modelling System, Version 4.0: User's Guide*. Ogden, UT: Department of Agriculture, Forest Service. Rocky Mountain Research Station. General Technical Report RMRS-GTR-106WWW Revised.

Burgan, R. E., and Rothermel, R. C., 1984. *BEHAVE: Fire Behavior Prediction and Fuel Modeling System–FUEL Subsystem*. USDA Forest Service General Technical Report INT-167. 126 pp.

Dawsa, M. I., Pritcharda, H. W., and Van Stadenb, J., 2004. Butenolide from plant-derived smoke functions as a strigolactone analogue: evidence from parasitic weed seed germination. *South African Journal of Botany*, **74**(1), 116–120.

Dimitrakopoulos, A. P., and Mitsopoulos, I. D., 2006. *Global Forest Resources Assessment 2005 – Thematic report on forest fires in the Mediterranean region*. FAO Fire Management Working Paper 8.

Dubbeld, L., 2005. Protecting personal data in camera surveillance practices. *Surveillance & Society*, **2**(4), 546–563. 'People Watching People' (ed. Wood).

ESS project, 2009. Public website: http://ess-project.eu

FAO, 2007. Global forest resources assessment 2010: options and recommendations for a global remote sensing survey of forests, Rome.

Finney, M. A., 1998. *FARSITE: Fire Area Simulation Model Development and Evaluation*. Fort Collins, CO: USDA Forest Service, Rocky Mountain Research Station. Research Paper RMRS-4.

Finney, M. A., 2006. An overview of FlamMap fire modeling capabilities. In Andrews, P. L., and Butler, B. W. (eds.), *Fuels Management-How to Measure Success: Conference Proceedings*. Fort Collins, CO: U.S. Department of Agriculture, Forest Service, Rocky Mountain Research Station, Portland, OR, pp. 213–220.

Flannigan, M. D., Amiro, B. D., Logan, K. A., Stocks, B. J., and Wotton, B. M., 2005. Forest fire and climate change in the 21st century. *Mitigation and Adaption Strategies for Global Change*, **11**(4), doi:10.1007/s11027-005-9020-7.

Heikkila, T. V., Gronovist, R., and Jurvelius, M., 1993. *Handbook of Forest Fire Control*. Forestry Training Programme, Publication 21. Helsinki: National Board of Education of the Government of Finland. ISBN 951 47 7459 0. Reprinted 1998.

Hirsch, S. N., 1964. Forest fire detection systems. In *Western Forest Fire Research Council Proceedings*, pp. 3–5.

http://archives.cbc.ca/environment/natural_disasters/topics/849/
http://en.wikipedia.org/wiki/Wildfire
http://wildfiremag.com/tactics/grassy_knowns/
http://www.bookrags.com/research/wildfire-enve-02/
http://www.forestencyclopedia.net
http://www.knowledgerush.com/kr/encyclopedia/Forest_fire/
http://www.sciencedaily.com/articles/b/bushfire.htm

Kettlunen, M., Bassi, S., Kampa, E., and Cavalieri, S., 2008. Forest fires: causes and contributing factors in Europe. Institute for European Environmental Policy, London (UK). Report published by the European Parliament's Committee on the Environment.

Luke, R. H., and McArthur, A. G., 1986 [1978]. Bushfires in Australia. Dept Primary Industry. Forestry and Timber Bureau. CSIRO Division of Forest Research.ISBN 0 642 02341 7. Reprinted with corrections 1986. xii+359pp. Canberra: AGPS.

Nyland, R. D., 1996. *Silviculture Concepts and Applications*. New York: McGraw-Hill. 633 pp.

Prometheus, S. V., 2000. Management techniques for optimization of suppression and minimization of wildfire effects. System Validation. European Commission – Contract number ENV4-CT98-0716.

Rothermel, R. C. 1972. A mathematical model for predicting fire spread in wildland fuels. Ogden, UT: U.S. Department of Agriculture, Forest Service, Intermountain Forest and Range Experiment Station. 40 p. Res. Pap. INT-115.

Sandberg, D. V., Ottmar, R. D., and Cushon, G. H., 2001. Characterizing fuels in the 21st century. *International Journal of Wildland Fire*, **10**, 381–387.

Scientific American, 2009. In *Depth Report: Wildfires and Climate Change*. http://www.scientificamerican.com/.

Trabaud, L., and Prodon, R. (eds.) 1992. *Fire in Mediterranean Ecosystems*, Banyuls-sur-Mer, France, September 21–25, 1992. Commission of the European Communication, Ecosys. Res. Rep., 5.

Cross-references

Forest Fire Effects
Forest Fires
Wildfires

FREQUENCY AND MAGNITUDE OF EVENTS

Lionel E. Jackson, Jr.
Geological Survey of Canada, Vancouver, BC, Canada

Definitions

Frequency. The frequency of a natural hazard event is the number of times it occurs within a specified time interval.

Magnitude. The magnitude of a natural hazard event is related to the energy released by the event. It is distinguished from *intensity* which is related to the effects at a specific location or area.

Introduction

The magnitude of a natural hazard event varies in its frequency of occurrence over time in an inverse power relationship. The relationship is often depicted as log-normal (Figure 1) where the magnitude increases linearly (e.g., 1, 2, 3, ...) whereas the frequency decreases as an inverse power function (e.g., 1/3, 1/9, 1/81) with increasing magnitude (Keller et al., 2008, p. 23). In other words, the larger and the more energetic the event, the rarer it is in time.

Linear magnitude scales or proxy indicators are logarithmically related to exponential increases in energy as events increase in size. Log-normal plots are useful to engineers and scientists because, given a linear magnitude scale, magnitude-frequency data approximate a straight line on this type of plot and are relatively easy to visualize and compare. Magnitude values are explicitly or implicitly related to the energy released during an event. The Moment Magnitude (M_W) scale is used by seismologists to rate earthquake magnitude (Hanks and Kanamori, 1979). It increases linearly (i.e., 1, 2, 3, ...) in relation to the exponential increases in the energy released by earthquakes. For example, an earthquake of M_W 6 releases an energy equivalent to approximately 32 times more than the energy released by an earthquake of M_W 5 and 1,024 (32×32) times more than that of an earthquake of M_W 4 (Figure 2). Proxy indicators of magnitude may have various dimensions or be dimensionless. Peak stream discharge in cubic metres/second (m^3/s) is used as a measure of magnitude in flood frequency analysis (Figure 1). The total volume discharged (m^3) is used as a measure of magnitude in the case of the debris flow magnitude scale of Jakob (2005) or for other types of landslides. The moment magnitudes of earthquakes (M_w) are dimensionless as are the units (0-7) used in volcanic explosivity index (0-7) (Newhall and Self, 1982) are dimensionless but are related to energy released in Joules and volumes of tephra erupted in cubic kilometres, respectively. The relationship between magnitude and frequency is not open-ended: magnitude of events is limited by physical limits to the energy that can be liberated during an event within the context of such variables as plate interactions in the solid earth (earthquakes), the relative relief between mountains and valleys (landslides), and ocean-atmospheric circulation (floods).

Global earthquake magnitude and frequency records 2000–2009

Earthquakes are among the best phenomena to illustrate the magnitude-frequency concept as they occur frequently and are globally monitored. Also the largest earthquakes can be expected to occur several times along earth's seismically active belts during an average human lifespan. Figure 2 is a plot of the frequency of earthquakes above M_w 4.0 for the decade 2000–2009 from data posted on the United States Geological Survey website http://earthquake.usgs.gov/ (February 2010). The log-log plot groups earthquakes in M_w classes 4.0–4.9, 5.0–5.9, etc. up to 9.0+ against their frequency over the decade. The data points are plotted at the lower limit of their class range. The period includes the great Andaman-Nicobar earthquake (26 December, 2004) that was rated as M_w 9.1 (the only one in the 9.0+ class that occurred during the observation period). It is interesting to note that this value approaches the upper limit of earthquake magnitude: the energy released during an earthquake is determined by area of the rupture along the source fault. Subduction zones, where one tectonic plate overrides another, such as those along the coasts of Chile, Sumatra, and northwestern North America, are regarded as producing the largest earthquakes because they have the largest potential rupture areas with lengths over 1,000 km and down-dip widths up to 100 km (Wells and Coppersmith, 1994). The great 1960 earthquake that occurred along the subduction zone along the coast of Chile ($M_w \sim 9.5$) is regarded as being close to the ultimate limit of earthquake magnitude. This is because the rupture surface area approached the largest potential rupture surface existing in any subduction zone on the planet.

The frequencies of earthquakes increase exponentially with decreasing magnitude in Figure 2. The asymptotic trends of the curve at extreme values of magnitude and frequency give it a sinusoidal shape. This is typical of the relationship between magnitude in terms of energy released and frequency of natural hazards. No data were

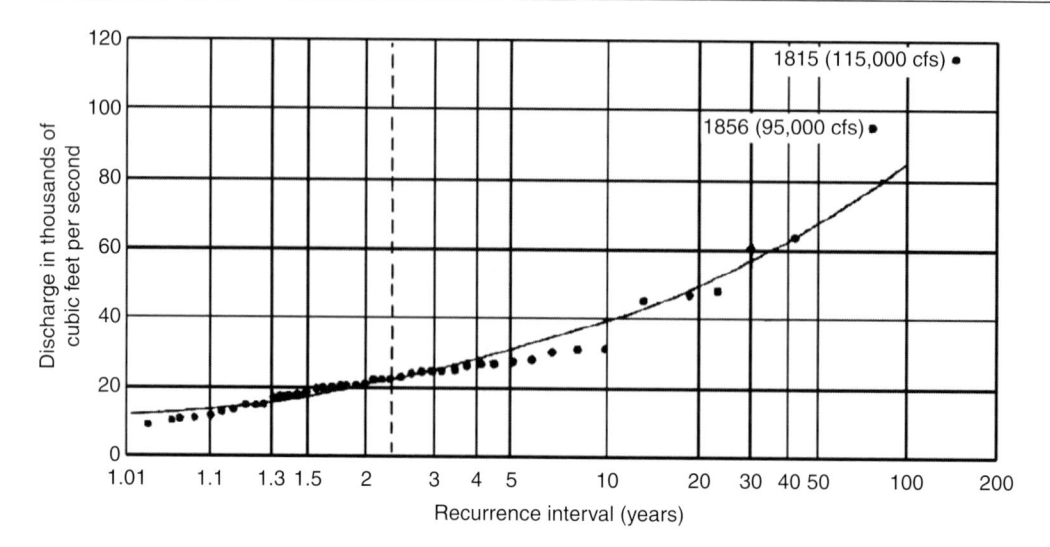

Frequency and Magnitude of Events, Figure 1 A log-normal frequency plot showing variation between maximum instantaneous stream discharge and flow recurrence in years for the Skagit River at Newhalem, Washington (From Stewart and Bondhaine, 1961). The *vertical dashed line* is mean annual discharge. The *trend line* is hand-fitted to values from the instrumental gauge record. This plot was used by Baker (2008) to illustrate the nonconformity of extreme flood values (1815 and 1856) with the extrapolation of more frequent peak flows. The extreme flood values were defined by paleohydrologic flood-stage indicators identified in the field.

plotted for earthquakes with M_w values below 4.0 in Figure 2. Had they been plotted, they would have taken several more log cycles and accentuated the asymptotic shape of the curve for high frequency events.

Applications and limitations of magnitude-frequency in flood frequency analysis

The magnitude-frequency relationship of natural events is most widely applied in the field of flood frequency analysis which spans fields of hydrology, civil engineering, and geomorphology. Water and suspended sediment discharge data are routinely collected through stream gauging and water sampling worldwide. These data are used to construct stream flow frequency plots (Figure 1) and sediment transport plots (Figure 3). Magnitude-frequency plots are valuable in understanding the relationship between magnitude and frequency of relatively common peak flows and the amount of sediment transported by them. This led Wolman and Miller (1960) to conclude that intermediate ranges of stream discharge (intermediate magnitudes and frequency) can be more significant for transporting sediment and shaping river channels over long periods of time than those of low frequency/high magnitude discharge rates (see example of the Eel River, Figure 3). Over the 10-year record period (1957–1967), the stream transported an average suspended load of 2.8×10^7 tons per year past the Scotia gauging station with 50% of the sediment transported by high flows during 6 or fewer days per year. The period of record included a very high magnitude low frequency flood in 1964 (probability of occurring in a given year has been estimated as low as 0.003 (Helly and LaMarche, 1968)). This transported 51% of the entire suspended sediment for the 10-year period. However, because of the rarity of such a flood event, more frequent yearly peak stream flows will transport much more sediment in the course of a few decades.

Extremely low frequency/extreme magnitude floods bring effects that are not associated with more frequent peak flows: they can cause major and permanent changes to stream channels and landscapes through permanent channel modification, establishment of new stream channels, and abandonment of old ones (stream avulsion). In mountainous landscapes, prolonged and intense rainfall associated with extreme floods can trigger landslides over large parts of drainage basins which in turn create debris flows. Such extreme floods are capable of transporting large boulders that are stable during all but the most extreme flood events. The disastrous flood in the Vargas area along the northern coast of Venezuela in December 1999 is a good example of such an extreme flood (García-Matinez and López, 2005). It killed tens of thousands of people and boulders driven by torrents severely damaged multistory buildings.

Attributes and understanding of extreme flood events have not been reliably captured or their frequencies reliably estimated by extrapolating flood frequency from intermediate and low frequency events (Baker, 1994). Baker (2008) originally used the plot in Figure 1 (after Stewart and Bodhaine, 1961) to illustrate this point: the outlying values in the extreme upper right of the plot represent peak discharges reconstructed for ungauged historical floods of 1815 and 1856 on Skagit River, Washington State, USA. They lie outside of the trend line based on stream gauge records. Understanding of extreme

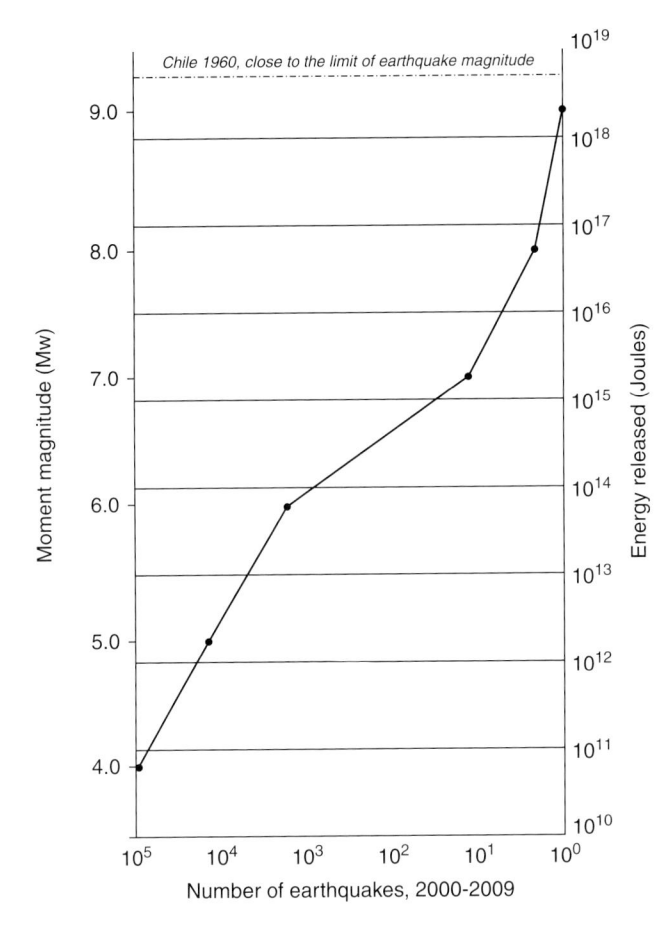

Frequency and Magnitude of Events, Figure 2 Log-log plot of earthquake energy and frequency for global earthquakes >M_w 4.0 for the decade 2000–2001. The moment magnitude (M_w) scale is shown along the left side of the plot to illustrate the relationship between the moment magnitude scale and energy released by earthquakes (scale on the right side). The earthquakes are grouped in M_w classes 4.0–4.9, 5.0–5.9, etc. up to 9.0+. Points are plotted at the lower limit of their class for the sake of explanation. The great 1960 earthquake along coastal Chile approached the upper limit of the energy that can be released in an earthquake. The sinusoidal shape of the curve is typical of the magnitude-frequency relationship for natural hazard events.

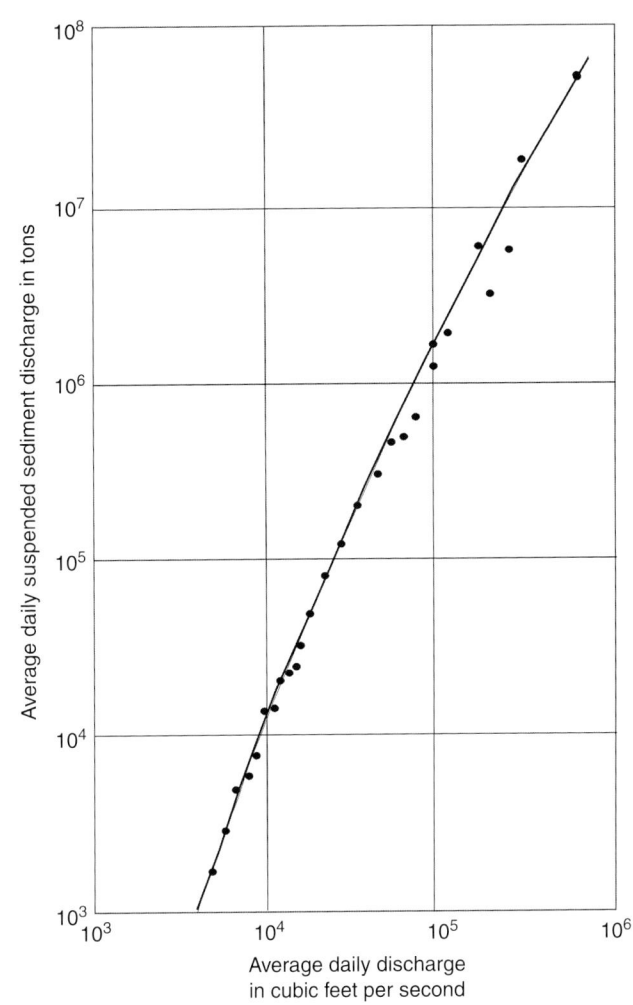

Frequency and Magnitude of Events, Figure 3 Log-log plot of average daily water discharge versus average daily discharge of sediment for the Eel River at Scotia, California, for the period October 1957–September 1967 (Redrawn from Brown and Ritter, 1971). Such plots are used when two variables change exponentially. Data spanning many orders of magnitude can be depicted in a single plot. The period includes the great flood of December 1964 that was estimated to have a 400-year return period based upon paleoflood indicators analysis (Helly and LaMarche, 1968).

past flood events requires their reconstruction and modeling based upon geological evidence. This has led to the development of paleohydrology as a discipline (Baker, 2008).

Applications of magnitude and frequency to landslide-type hazards

The magnitude-frequency concept can be applied regionally to landslides (Guthrie and Evans, 2005) and is the basis of landslide risk management (Fell et al., 2005). Landslides encompass a spectrum of mass wasting phenomena including translational or rotational failures and mass flow (Cruden and Varnes, 1996; Hungr et al., 2001). Separate magnitude-frequency relationships will exist for each type of landslide within a given region with similar climate, physiography, and geological substrate (Picarelli et al., 2005). For example, the magnitude and frequency of debris flows (lahars) in volcanic terrain are typically much higher than in non-volcanic mountainous areas (Vallance, 2005).

The frequency of an event of a given magnitude may vary with time where climate-driven cycles operate on the landscape. For example, based upon indicators of relative rates of sedimentation on flood plains and alluvial fans, Jackson et al. (1982) concluded that the frequency of debris flows in the southern Canadian Rocky Mountains was greater following deglaciation at the end of the last glaciation about 10,000 years ago than they have been during the last 5,000 years. An analogue is occurring globally in contemporary mountain-glacier regions in response to climatic warming: rapid retreat of glaciers is leaving unstable moraines and moraine-dammed lakes resulting in landslide-failure of slopes and outburst floods (Jackson et al., 1989; Evans and Clague, 1994). The magnitude-frequency relationship for debris flow and outburst flood hazards in a given mountain-glacier region would have been different before the current period of climatic warming that began in the latter part of the nineteenth century (Mann et al., 1999). Concerning landslide frequency over a similar time period, Jackson (2002) postulated that the frequency of large rockslides and rock avalanches in glaciated mountains is highest following deglaciation and decreases and becomes asymptotic toward a basal rate determined by structural and lithologic architecture and regional rates of uplift and stream incision. Similar relationships may prevail in the magnitude and frequency of submarine landslides: the Storrega landslide west of Norway on the floor of the North Atlantic Ocean is one of the largest known on earth (volume >3,000 km^3; Nadim and Locat, 2005). Its failure 8,500 years ago (which produced a tsunami) was conditioned partly by rapid deposition of sediment along the edge of the Norwegian continental shelf during and immediately after the last glaciation. Such high rates of sedimentation associated with deglaciation are no longer in operation today in that region. Consequently, the frequency of events of that size in shelf areas is likely less at present than it was the end of the last glaciation.

Summary

The frequencies and magnitudes of potentially hazardous geophysical and hydro-meteorological events have an inverse power relationship: the larger and more energetic the event, the rarer it is in time. The relationship is not open-ended but is limited by physical limits dictated by such factors as plate tectonics and climate. The magnitude-frequency relationship can imply that events of intermediate frequency and magnitude may be the most significant in shaping aspects of the earth's surface. However, rare extreme events can cause permanent changes to fluvial systems and other aspects of the landscape that would not have occurred under the regime of more frequent and less energetic events. The attributes and frequencies of extreme floods are unlikely to be predicted by analysis of more frequent and lower magnitude stream flow data. For natural hazard processes that are influenced by climatic change, the frequency of an event of a given magnitude can vary depending on the length of time and the specific time interval considered with respect to climatic variation.

Bibliography

Baker, V. R., 1994. Geomorphological understanding of floods. *Geomoprphology*, **10**, 139–156.
Baker, V. R., 2008. Paleoflood hydrology: origin, progress, prospects. *Geomorphology*, **101**, 1–13.
Brown, W. M. III, and Ritter, J. R., 1971. Sediment transport and turbidity in the Eel River basin, California. United States Geological Survey, Water Supply Paper 1986, 71p.
Cruden, D. M., and Varnes, D. J., 1996. Landslide types and processes. In Turner, K., and Schuster, R. L. (eds.), Landslides investigation and mitigation. *Transportation Research Board Special Report*, 247, Washington, DC: National Academy Press, pp. 36–75.
Evans, S. G., and Clague, J. J., 1994. Recent climatic change and catastrophic geomorphic processes in mountain environments. *Geomorphology*, **10**, 107–108.
Fell, R., Ho, K. K. S., Lacasse, S., and Leroi, E., 2005. A framework for landslide risk management. In Hungr, O., Fell, R., Couture, R., and Eberhardt, E. (eds.), *Landslide Risk Management*. Leiden: A.A. Balkema, pp. 3–26.
García-Matinez, R., and López, J. L., 2005. Debris flows of December 1999 in Venezuela. In Jakob, M., and Hungr, O. (eds.), *Debris Flow Hazards and Related Phenomena*. Chichester: Springer-Praxis, pp. 411–443.
Guthrie, R. H., and Evans, S. G., 2005. The role of magnitude-frequency relations in regional landslide risk analysis. In Hungr, O., Fell, R., Couture, R., and Eberhardt, E. (eds.), *Landslide Risk Management*. Leiden: A.A. Balkema, pp. 375–380.
Hanks, T. C., and Kanamori, H., 1979. Moment magnitude scale. *Journal of Geophysical Research*, **84**(B5), 2348–2350.
Helly, E. C., and LaMarche, V. C., Jr., 1968. December 1964, a 400 year flood in northern California. United States Geological Survey, Professional Paper 600-D, pp. D34–D37.
Hungr, O., Evans, S. G., Bovis, M. J., and Hutchinson, J. N., 2001. A review of the classification of landslides of the flow type. *Environmental and Engineering Geoscience*, **7**, 221–238.
Jackson, L. E., Jr., 2002. Regional landslide activity and Quaternary landscape evolution, Rocky Mountain Foothills, Alberta, Canada. In Evans, S. G., and DeGraff, J. V. (eds.), *Catastrophic Landslides; Effects, Occurrence, and Mechanisms*. Boulder, CO: Geological Society of America. Reviews in Engineering Geology, Vol. XV, pp. 325–344.
Jackson, L. E., Jr., MacDonald, G. M., and Wilson, M. C., 1982. Paraglacial origin for terrace river sediments in Bow Valley, Alberta. *Canadian Journal of Earth Sciences*, **19**, 2219–2231.
Jackson, L. E., Jr., Hungr, O., Gardner, J. S., and Mackay, C., 1989. Cathedral mountain debris flows, Canada. *Bulletin of the International Association for Engineering Geology and the Environment*, **40**, 35–54.
Jakob, M., 2005. Debris-flow hazard and analysis. In Jakob, M., and Hungr, O. (eds.), *Debris Flow Hazards and Related Phenomena*. Chichester: Springer-Praxis, pp. 411–443.
Keller, E. A., Blodgett, R. H., and Clague, J. J., 2008. *Natural Hazards: Earth's Processes as Hazards, Disasters, and Catastrophes*. Toronto: Pearson-Prentice Hall. 421p.
Mann, M. E., Bradley, R. S., and Hughes, M. K., 1999. Northern hemisphere temperatures during the past millennium. *Geophysical Research Letters*, **26**, 759–762.
Nadim, F., and Locat, J., 2005. Risk assessment for submarine slides. In Hungr, O., Fell, R., Couture, R., and Eberhardt, E. (eds.), *Landslide Risk Management*. Leiden: A.A. Balkema, pp. 321–333.
Newhall, C. G., and Self, S., 1982. The volcanic explosivity index (VEI): an estimate of explosive magnitude for historical volcanism. *Journal of Geophysical Research*, **87**(C2), 1231–1238.

Picarelli, L., Oboni, F., Evans, S. G., Mostyn, G., and Fell, R., 2005. Hazard classification and quantification. In Hungr, O., Fell, R., Couture, R., and Eberhardt, E. (eds.), *Landslide Risk Management*. Leiden: A.A. Balkema, pp. 27–61.

Stewart, J. E., and Bodhaine, G. L., 1961. Floods in the Skagit River basin. United States Geological Survey, Water Supply Paper 1527, 57p.

Vallance, J. W., 2005. Volcanic debris flows. In Jakob, M., and Hungr, O. (eds.), *Debris Flow Hazards and Related Phenomena*. Heidelberg: Springer, pp. 247–254.

Wells, D. L., and Coppersmith, K. J., 1994. New empirical relationships among magnitude, rupture length, rupture width, rupture area and surface displacement. *Bulletin of the Seismological Society of America*, **84**, 974–1002.

Wolman, M. G., and Miller, J. P., 1960. Magnitude and frequency of forces in geomorphic processes. *Journal of Geology*, **68**, 54–74.

Cross-references

Avulsion
Climate Change
Disaster Risk Management
Debris Flow
Earthquake
Flood Hazard and Disaster
Hazard
Lahar
Landslide (Mass Movement)
Landslide Types
Natural Hazard
Paraglacial
Probable Maximum Flood (PMF)
Storms

FROST HAZARD

Leanne Webb[1], Richard L. Snyder[2]
[1]Institute of Land and Food Resources, University of Melbourne, CSIRO Division of Marine and Atmospheric Research, Aspendale, Vic, Australia
[2]University of California, Davis, CA, USA

Synonyms

Freeze hazard; Freezing injury; Frost damage

Definition

A frost event occurs when air temperature falls to 0°C or lower, measured at a height of between 1.25 and 2.0 m above soil level, inside an appropriate weather shelter. In meteorology, *frost* refers to the formation of ice crystals on surfaces due to deposition, that is, a phase change from vapor to ice. In agriculture or biology, however, *frost* refers to an event where temperature falls to the point where ice forms inside plant tissues and causes damage to the cells. *Frost hazard* is the potential for damage or negative impact caused by freezing temperature or frost.

Introduction

Frost hazard, as it relates to the potential for plant damage (or injury) caused by freezing temperature is discussed in the following description. Although other impacts through freezing conditions may occur, for example, road or infrastructure damage or soil heaving (e.g., Knollhoff et al., 2003; Little et al., 2003; Greenfield and Takle, 2006), these have not been addressed in detail here.

Frost events can be described as advective or radiative (Kalma et al., 1992). Advective frosts are associated with large-scale incursions of cold air with a well-mixed, windy atmosphere and a temperature that is often below zero, even during the day. Radiative frosts are associated with clear night-time skies and calm winds, where more heat is radiated away from the surface than is received, so that the temperature drops. In some cases, a combination of both advective and radiative conditions occurs. For example, it is not uncommon to have advective conditions bring a cold air mass into a region, resulting in an advection frost. This may be followed by several days of clear, calm conditions that are conducive to radiation frosts.

There are two subcategories of radiation frosts. A "hoar frost" occurs when water vapor deposits onto the surface and forms a white coating of ice that is commonly called "frost." A "black" frost occurs when temperature falls below 0°C and no ice forms on the surface. If the humidity is sufficiently low, then the surface temperature does not reach the ice point temperature and no frost forms. When the humidity is high, ice is more likely to deposit and a "hoar frost" can occur. Because heat is released during the ice deposition process, hoar frosts usually cause less damage to vegetation than black frosts.

The average length of the frost-free period, measured in days, varies across the world through latitudes, continentally and topographically (Kalma et al., 1992). The length of the frost-free period can be used as a general guide to the agro-climatic suitability of a region, though as explained subsequently, this also depends on the crop sensitivity.

Frost conditions can result in losses to crops through damage to plant tissue. The extent of damage can be significant and is influenced by many factors. These include the location of the crop, the time of the frost event in relation to the developmental stage of the plant, the severity of the frost event, plant nutrition, the type of plant, and the conditions leading up to the event (e.g., cold hardening or acclimation). Ice-nucleating bacteria promote the initiation of freezing so their presence also influences damage levels.

Frost causes significant damage to agricultural products in both commercial and noncommercial enterprises. Freezing accounts for greater economic losses of fruits and vegetables than any other environmental or biological hazard (Rieger, 1989). Frosts severely affected coffee in Brazil and Africa in the 1960s and 1970s (Hewitt, 1983), and, with losses amounting to US$3.5 billion in 1989/90, citrus in Florida (Kalma et al., 1992). Annual crops (e.g., wheat, barley, sorghum) are also vulnerable; though sowing dates can be selected to reduce risk of frost exposure (Gomez-Macpherson and Richards, 1995).

Depending on the type of frost event (advective or radiative) either passive or active protection methods are employed to reduce the potential hazard (Kalma et al., 1992). The risk and the relative benefit and cost need consideration to ensure effective strategies are implemented for hazard reduction.

Frost hazard: biological impact

Low temperature freeze injury can occur in all plants, but the mechanisms and types of damage vary considerably. Crop plants that develop in tropical climates often experience serious frost damage when exposed to temperature slightly below zero, whereas most crops that develop in colder climates often survive with little damage if the freeze event is not too severe. The temperature at which plant tissue damage occurs is correlated with air temperatures called "critical temperatures" and these are specific to particular plant species. In addition to species differences in susceptibility to low temperatures, some species or varieties can exhibit different frost damage at the same temperature and phenological stage, depending on antecedent weather conditions. During cold periods prior to the frost event, plants tend to harden against freeze injury, and they lose the hardening during warm spells.

Direct frost damage can occur when ice crystals form inside the protoplasm of cells (intracellular freezing), whereas indirect damage can occur when ice forms inside the plants but outside of the cells (i.e., extracellular freezing). Levitt (1980) reports that, in nature, freeze injury results from extracellular ice crystal formation and there is usually no evidence of intracellular freezing. It is not cold temperature but ice formation that actually injures the plants. The extent of damage due to extracellular freezing depends mainly on how fast the temperature drops and to what level it supercools before freezing. There is little or no evidence that the duration of the freezing affects damage levels (Levitt, 1980). In fact, Levitt (1980) states that extent of freeze injury is independent of time, at least for short periods (e.g., 2–24 h).

Levitt (1980) proposed that cells were gradually damaged as a result of growth of the extracellular ice mass. As a result of extracellular ice formation, water evaporates from the liquid water inside the cells passing through the semipermeable cell membranes, potentially causing cell desiccation, to deposit on the ice crystals outside of the cells. As water is removed from the cells, the solute concentration increases and reduces the chances of cell freezing. Therefore, the main cause of frost damage to plants in nature is extracellular ice crystal formation that causes secondary water stress to the surrounding cells. In fact, there is a close relationship between drought-tolerant plants and freeze-tolerant plants.

Protection from frost hazard

Farmers can elect to passively protect their crop from frost or employ more active protection measures to reduce frost hazard. Passive methods are usually less costly than active methods and often the benefits are sufficient to eliminate the need for active protection. The overall aim of both types of strategy is to maintain plant temperatures high enough to minimize extracellular ice formation. Critical air temperatures, which provide a guideline when frost damage is likely to occur, vary with crops described as either tender, slightly hardy, moderately hardy, and very hardy (Levitt, 1980). These temperatures have been quantified in growth chamber studies for many crops (Snyder and Paulo de Melo-Abreu, 2005).

Energy transfer rates determine how cold it will get and the effectiveness of frost protection methods. The four main forms of energy transfer that are important in frost protection are radiation; conduction (or soil heat flux); convection (i.e., fluid transfer of sensible heat), and phase changes associated with water (latent heat). For all nonbiological methods of frost protection, the goal is to minimize losses or maximize gains of energy from one or more of the energy balance components (Figure 1).

Humidity is an important factor in freeze protection because of phase changes which convert sensible to latent (evaporation) or latent to sensible (condensation) heat and because moist air absorbs more long-wave radiant energy. When the surface temperature drops to near the dew point temperature, condensation can occur releasing latent heat and reducing the rate of temperature drop. Also air with higher humidity cools more slowly than drier air (as it absorbs more long-wave radiation).

Both passive and active protection methods (Table 1) are more effective during radiative than advective frost events. Active methods, for example, heaters and wind machines, often use the presence of a temperature inversion near the ground to enhance heat transfer to the plants. Inversions, however, tend to be weak or absent during advective freezes. Protection methods that use water are

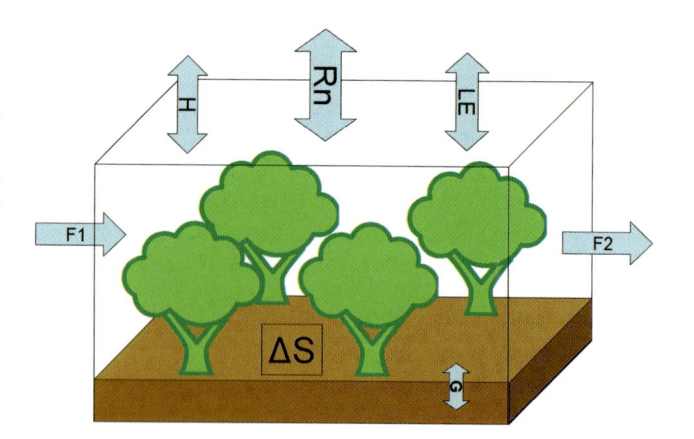

Frost Hazard, Figure 1 A *box* energy diagram showing possible sources and losses of energy from a crop represented by the *box*. Net radiation (Rn), sensible heat flux (H), latent heat flux (LE), soil heat flux or conduction (G), sensible heat advection in (F1) and out (F2), and energy storage in the crop (ΔS) (Adapted from Barfield and Gerber, 1979).

Frost Hazard, Table 1 Frost protection options indicating which part of the energy balance (Figure 1) is influenced to reduce frost hazard (*italicized options are generally less effective for advective frost hazard avoidance*) (Adapted from Snyder and Paulo de Melo-Abreu, 2005)

	Frost event avoidance	Reduce radiation loss (Rn)	Maximize sensible heat gain (H, G,)	Maximize latent heat gain (LE, Fl)	Ice formation within plant
Passive					
Site selection	X	X			
Managing cold air drainage	X	X			
Plant selection	X				X
Canopy trees		X			
Plant nutritional management					X
Plant pruning	X	X	X		
Plant covers		X			
Avoiding soil cultivation			X		
Irrigation			X	X	
Removing cover crops			X		
Soil covers			X		
Trunk painting and wraps		X			
Bacteria control					X
Planting date for annual crops	X				
Active					
Heaters			X		
Wind machines			X		
Helicopters			X		
Sprinklers				X	
Surface irrigation			X	X	
Foam insulation		X			

more effective when evaporation rates are low, that is, when the humidity is higher and the wind speed is lower. Advective frosts are associated with low humidity and high wind speeds, so using water for protection is more problematic during advective frost events. Passive methods, for example, cover cropping and soil moisture management, are beneficial for both types of frosts, but they are less effective for advective frosts because air turbulence tends to dominate energy transfers. Avoiding low spots, where cold air accumulates, is advisable for locations with radiation frost, but avoiding hilltops is more prudent in regions characterized by advective frost. If the risk is high for advective frosts, it might be best to select a different region for growing sensitive crops. A good example is the citrus industry in the southeastern USA that, 200 years ago, extended north to South Carolina but is now restricted to southern Florida (Attaway, 1997) because of advective frost. While most protection methods attempt to optimize energy transfer and maintain higher plant temperatures, passive methods also include biological factors such as plant selection, hardening, and reduction of ice nucleating bacteria concentration (Lindow, 1983), which reduce the chances for ice formation and/or cell damage.

A wide range of simple to sophisticated frost protection technologies are used around the world. The main determining factors depend on local availability and costs. For example, liquid-fuel heaters are widely used in Mexico because low-cost fuel is available, but not used where costs are higher. Of course, protection methods vary depending on the size and wealth of the farming operation as well as government support. Each protection method must be considered on its own merits and an economic evaluation should be performed to determine whether or not the method is cost-effective (Snyder et al., 2005).

Probability and risk

Remote sensing, for example, aircraft or satellite thermal imagery, can be employed to produce both frost risk maps and to improve frost prediction services (Kalma et al., 1992). Maps indicating frost risk incorporate both climate and topography data (Laughlin and Kalma, 1990) and, more recently, have been produced through GIS-based studies, for example (Lindkvist et al., 2000; Geerts et al., 2006). Due to the importance of frost hazard to agricultural economies, frost risk maps are produced by many governmental meteorological departments around the world, for example, http://www.ncdc.noaa.gov/oa/climate/freezefrost/frostfreemaps.html.

As frost damage can occur in almost any location in temperate and arid climates, and even in tropical zones at elevated sites, minimum temperature forecasting techniques were developed and are used extensively by farm managers (Kalma et al., 1992). Empirical formulae employing meteorological variables such as air temperature, various measures of humidity, wind speed, and cloud cover are used in combination with local conditions and topography to develop these forecasts.

Hazards: The hazards related to fumaroles are similar to those presented by volcanic gases in that they can be harmful to humans, animals, plants, agricultural crops, and property. An example is the death of four skiers (via asphyxiation) exposed to fumarole emissions on Mammoth Mountain, California, USA, and the death of two children and animal stock in Italy (Hill, 2000; Beaubien et al., 2003; Cantrell and Young, 2009). Central Italy (Alban Hills area), in particular, is an example of one region where a large human population is threatened by fumarole emissions. Elevated SO_2 levels near fumaroles in the Azores have also been identified as a potential hazard for those with asthma (Baxter et al., 1999).

Bibliography

Baxter, P. J., Baubron, J. C., and Coutinho, R., 1999. *Heath Hazards and Disaster Potential of Ground Gas Emissions at Furnas Volcano*. Azores: São Miguel.

Beaubien, S. E., Ciotoli, G., and Lombardi, S., 2003. Carbon dioxide and radon gas hazard in the Alban hills area (Central Italy). *Journal of Volcanology and Geothermal Research*, **123**, 63–80.

Cantrell, L., and Young, M., 2009. Fatal fall into a volcanic fumarole. *Wilderness & Environmental Medicine*, **20**, 77–79.

Heggie, T. W., 2009. Geotourism and volcanoes: health hazards facing tourists at volcanic and geothermal destinations. *Travel Medicine and Infectious Disease*, **7**, 257–261.

Hill, P. M., 2000. Possible aspyxiation from carbon dioxide of a cross-country skier in eastern California: a deadly volcanic hazard. *Wilderness & Environmental Medicine*, **11**, 192–195.

Cross-references

Aa Lava
Caldera
Eruption Types (Volcanic)
Galeras Volcano, Colombia
Krakatoa
Methane Release from Hydrate
Montserrat Eruptions
Mt. Pinatubo
Nevado del Ruis Volcano, Colombia (1985)
Nuee Ardente
Pahoehoe Lava
Pyroclastic Flow
Stratovolcano
Surge
Vesuvius
Volcanic Gas
Volcanoes and Volcanic Eruptions

G

CASE STUDY

GALERAS VOLCANO, COLOMBIA

Barry Voight[1,2], Marta L. Calvache[3]
[1]Penn State University, University Park, PA, USA
[2]Cascades Volcano Observatory, Vancouver, WA, USA
[3]INGEOMINAS, Bogota, Colombia

Summary points

Galeras Volcano, Colombia: Volcano Crises, 1988–2010+: Adverse Impacts without a Major Eruption

- Lava dome growth with strong seismic, deformation and gas precursors
- Vulcanian explosions with very subtle precursors and strong shocks
- Fears of population early in crisis magnified by 1985 Nevado del Ruiz catastrophe
- Lack of a pre-crisis hazard map led to a *chain-reaction* of events that culminated in an economic crisis, hostility of authorities and loss of credibility of the scientists
- Important lessons for safety of volcanologists from 1993 tragedy
- Continuing difficulty of compliance to evacuation recommendations

Introduction

Located in southwest Colombia, 4,270-m high Galeras Volcano rises 1,600 m above Pasto, the capital city and economic nucleus of Narino Province (Figure 1). It is the most active volcano in Colombia. Pasto is located on the eastern slope of the volcano, about 7–11 km from the crater. Its population is about 350,000, and over 50,000 more people reside in neighboring villages.

Activity at Galeras Volcano extends back about 4,500 years and included six major eruptions before the most recent episode of activity (Calvache, 2000; Stix et al., 1997). It consists of an active cone some 100–150 m high and 500 m diameter, nested inside a horseshoe-shaped amphitheater created by a prehistoric cone collapse.

About 50 periods of unrest have been described in the 500 years of recorded history at Galeras, of which 27 have proceeded into eruptions. Twelve of these episodes were brief and included small explosions, and the remainder were more notable but still modest. Pyroclastic currents of mixtures of blocks, ash, and gas generated by a collapsing eruption column descended beyond the crater rim in 1580, 1616, 1641–43, and 1936, and represent the most serious current potential hazard. The effects on Pasto in historic events include ash deposits of 4 cm or less and shock waves breaking windows. Incandescent ballistic blocks have reached about 3 km from the crater rim. Some building damage and a few fatalities have been caused by strong volcanic earthquakes. The towns of Consaca and Bambona, 12 km west of the active cone, have reported heavier ash falls and stronger earthquakes. Historic loss of life has been limited to tourists and scientists in and near the active cone (Munoz et al., 1993), and several deaths in Pasto due to earthquake-induced building collapse.

This case history reviews activity since 1989, and underscores the severe and complex problems that can emerge when no hazards assessments, emergency preparedness plans, or mitigation plans are available at the start of a crisis – such as at Galeras in 1989 and early 1990s. The contribution also examines background information on nine deaths within the crater from an explosion in January 1993.

Post-1988 activity and deadly explosions

In general, volcano monitoring is carried out to improve understanding of volcanic processes and to identify times of enhanced risk of eruption. Most techniques utilize

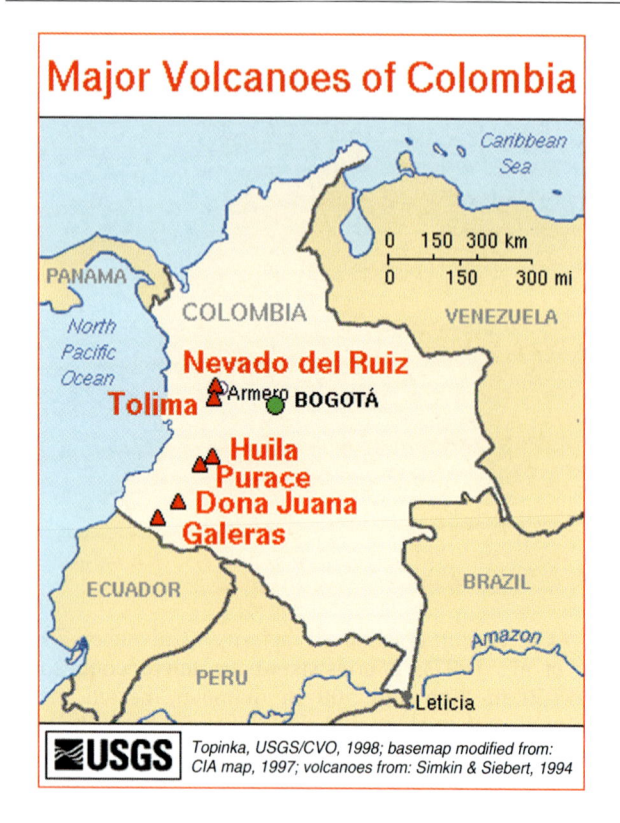

Galeras Volcano, Colombia, Figure 1 Location map of Galeras and other major volcanoes in Colombia (After USGS).

sophisticated equipment and require trained specialists to acquire, process, and interpret the data. Typical indicators include seismicity, changes in gas flux or temperature, and deformation of the edifice linked to observations of activity. Seismic indicators are particularly important, with one family of events consisting of volumetric sources involving flowing gas and/or magma, and another family involving rock breakage (Chouet, 1996). *Long-period (LP)* events and *tremor* are examples of the former, and *volcano-tectonic (VT)* events are an example of the latter. These types of events were involved in the debates which occurred after the fatal incidents in 1993.

Following recognition of anomalous fumarole activity in 1987–1988, continuous monitoring began in February 1989 (Cortes and Raigosa, 1997). Seismic activity accelerated in the next months, and between 4 and 9 May 1989 explosions occurred and eruption plumes rose as much as 3.5 km high (Figure 2). Ash deposits around the crater were as much as 6 m thick and a thin ash blanket covered a region 35 km^2.

After the May eruption, seismicity continued at a moderate level and SO_2 emissions increased, several times exceeding 5,000 t/day (Zapata et al., 1997). After June 1990, an increase in CO_2/SO_2 ratio was observed, suggestive of rising magma. After July 1991, an extended period of long-period (fluid-flow) seismicity occurred, with very large tilt inflation, and in October fresh andesite lava extruded. The lava was stiff and developed the shape of a dome, and grew for about a month, with 400,000 cubic meters emplaced in the first 2 weeks. Throughout this active period, VT (rock breakage) seismicity was low. From December 1991 through June 1992, LP seismicity reduced to moderate levels, tilt stabilized, gas emissions remained moderate, and VT seismicity remained very low (Figure 3).

Daily occurrences of LP events halved by June 1992, and continued to decrease in early July. However, a relatively new type of long-period signature (a few had been noted in March 1989) was observed on 11 July (Narvaez M. et al., 1997; Gomez and Torres, 1997), characterized by low-frequency, decaying signal with long duration (Figure 4). This type was described as a *tornillo*, Spanish for *screw*, and nine such events occurred between 11 and 16 July. On 15 July a swarm of small high-frequency events occurred and were interpreted as VTs (Cortes and Raigosa, 1997); much later they were reinterpreted as events of intermediate (hybrid) character between LP and VT events (Gil Cruz and Chouet, 1997). The last tornillo event occurred 1 h before an explosion on 16 July 1992 which destroyed about 90% of the lava dome.

Other possible eruption precursors for 16 July included SO_2 gas flux, variable for a month prior to the eruption, from ~300 to 3,500 t/day (Zapata et al., 1997), and minor tilt changes possibly indicating localized pressure buildup. The explosion generated a shock wave felt in local villages (Cortes and Raigosa, 1997), ejected blocks over 3 m diameter, and produced a 4-km high plume; its signal lasted about 9 min.

Monitored activity remained at low levels into 1993, when five small (Volcano Explosivity Index, VEI 1) explosions occurred on 14 January, 23 March, 4 and 13 April, and 7 June. The explosions were similar in style, volume of incandescent lava blocks and ash ejected, and similar activity preceding and following each explosion (Cortes and Raigosa, 1997; Zapata et al., 1997).

The 14 January explosion, which lasted 15 min, caused the deaths of six scientists participating in a scientific workshop, and three others, all near the active cone inside the amphitheater (Munoz et al., 1993; Baxter and Gresham, 1997). The imminence of the explosion was not recognized by the scientists. Seismicity was low before the explosion (Cortes and Raigosa, 1997). A total of 12 tornillo signals were recorded from 23 December to 14 January (about one event per day), but no swarm of high-frequency events occurred as in July 1992. Gas flux had been low in December although no data were recorded for 3 weeks prior to the explosion (Zapata et al., 1997). No precursory deformation was detected. An intense LP swarm occurred for 18 h after the eruption, but thereafter the rate declined to <10 events per day, a background level comparable to that before the eruption.

The other explosions were similar to the one of 14 January (Cortes and Raigosa, 1997), and by then the tornillo signals had acquired importance as an eruption precursor.

Galeras Volcano, Colombia, Figure 2 Galeras volcano, showing large amphitheater-like crater, and the active vent inside the crater. The amphitheater gives a measure of protection to Pasto and neighboring villages, because pyroclastic flows must surmount the rim to descend into the nearest populated areas.

Seismicity was low before the explosions (<100 long-period events/month), with a sparse number of tornillos generally recorded, ranging from zero before 4 April, to 103 in the 50 days preceding 7 June. Intense LP swarm activity and tremor occurred for a few days immediately following the explosion but seismicity then returned to low levels. Deformation showed no significant changes before, during or after the explosions. Gas flux (SO_2) was generally low before and after the explosions, apart from a short-lived increase associated with the explosions themselves. An exception is for the period preceding 7 June, when variable SO_2 flux was recorded, reminiscent of 16 July 1992. Authorities were alerted on 17 May for a possible sudden explosion (Narvaez et al., 1997), and the explosion occurred on 7 June, expelling 1.3 million cubic meters of material. This was the largest of the 1992–1993 explosions (Cortes and Raigosa, 1997). The various explosion signals had lasted 2–17 min, and eruption plumes reached 2–8 km above the vent.

Between July and November 1993, 94 tornillo-type events were recorded, without being followed by an explosion (Cortes and Raigosa, 1997). Likewise sporadic tornillos occurred in January, March, and May 1994, without explosions. Between 9 August and 23 September 1994, 31 tornillos preceded a long-period swarm with a minor gas emission; after 20 October 1994, 89 tornillos occurred, prior to a minor gas release on 9 January associated with LP events. The event type continues to be recognized in sparse numbers and is postulated to reflect pressurized fluids at shallow depth.

Activity and monitoring continues to present (e.g., Smithsonian BGVN, 2009).

Crisis response and management
Institutional structure and political reactions

The catastrophe caused by the Nevado del Ruiz eruption in 1985 (*see entry Nevado del Ruiz Volcano, Colombia 1985*) led the Government of Colombia in 1988 to create a national system for risk mitigation and disaster preparedness. Its responsibilities were defined at national, regional, and local levels. It incorporated public and private organizations that already existed and was designed such that decisions would be decentralized through Regional and Local Risk Committees, headed by the Governor at regional level and the mayors of municipalities at local level (Cardona, 1997). Coordination was provided by a National Office for Emergency Prevention and Assistance of the Office of the President (ONAD).

The study of Galeras by INGEOMINAS (the geological survey of Colombia) was initiated in 1988, in response to a request by ONAD to evaluate all active volcanoes of Colombia. In February 1989, enhanced fumarolic activity at Galeras was reported to the newly created Regional Risk Committee for Narino Province, and ONAD then provided funding for continuous monitoring by INGEOMINAS. With the first eruptive events in May 1989, intensive work began at all Risk Committee levels to provide emergency and contingency plan, prompted by fear based particularly on the recent horrible outcome at Nevado del Ruiz and Armero. Although significant progress was achieved in this planning, the "community was in large part isolated from the preparedness process" (Cardona, 1997). For example, videos from the 1980

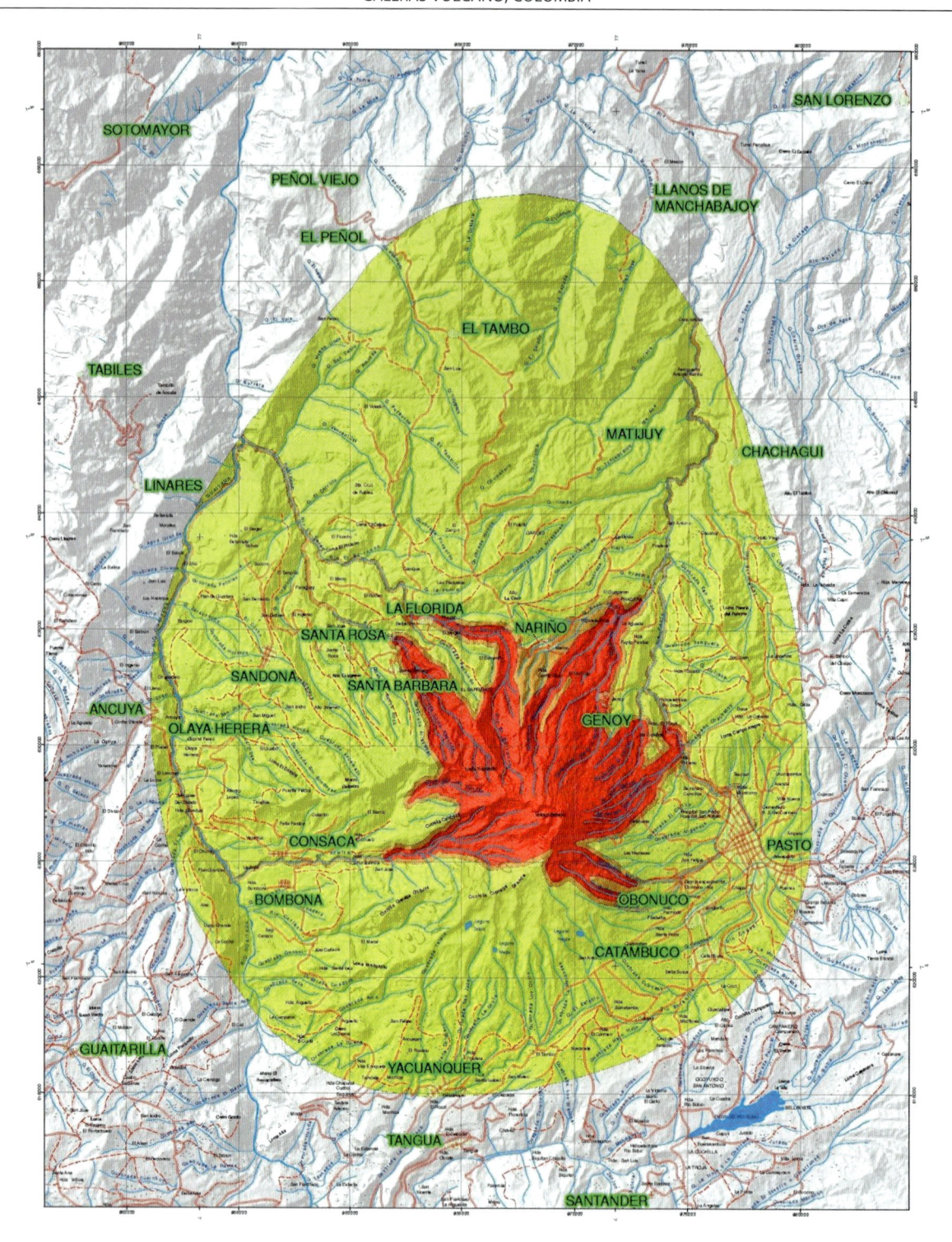

Galeras Volcano, Colombia, Figure 3 Hazard map of Galeras in 2009, relation to surrounding region. Pasto is located to the east. *Red* indicates flowage hazards; *yellow* is area of lesser hazard including ash fall (INGEOMINAS).

Mount St Helens eruption were used for public education, but these were criticized for generating undue fear and alarm, because some of the processes illustrated were different from what was possible at Galeras.

In 1989, some unfortunate events generated strong dissatisfaction within economic organizations and the population. For example, the Mayor of Pasto promoted "voluntary preventive evacuations" that alarmed

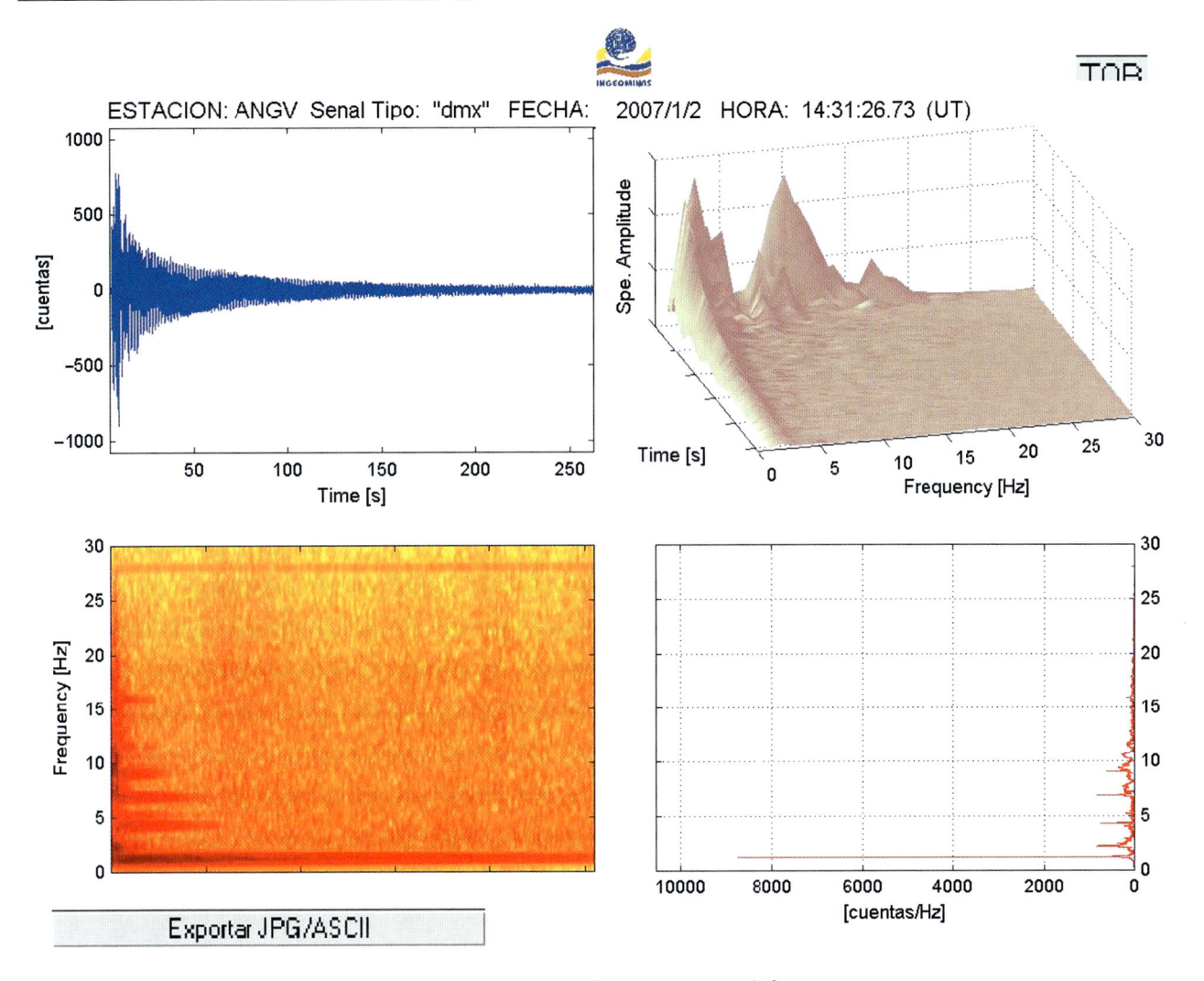

Galeras Volcano, Colombia, Figure 4 Seismic record with tornillo-type event, with frequency spectra.

inhabitants (Cardona, 1997). On 8 May, during the eruption, an evening "emergency meeting" was held in Pasto to advise the Governor and Mayor; the meeting was attended by INGEOMINAS staff, the Director of ONAD, and several volcanologists acting as advisors to INGEOMINAS. The possibility of cyclic eruptive activity, the gradational nature of hazard zone boundaries, the uncertainties regarding precursors, and the uncertain probabilities of pyroclastic flowage hazards were discussed, and the government officials displayed impatience and frustration concerning the uncertain information received by them. A few days later, on 10 May 1989, taking into account the concern of volcanologists regarding the status of activity, ONAD in Bogota unilaterally declared an "orange warning" (second-highest warning level). The volcanologists, influenced by the uncertain and insufficient knowledge of Galeras, were concerned about potential large eruptions and the potential for pyroclastic currents in some drainage channels. However the "orange warning" promoted sensationalism by the media, particularly at the national level, in exaggerated reports on the evolution of activity at Galeras. These media pressures aggravated the relationships between scientists and the new Mayor of Pasto and the Governor of Narino (Cardona, 1997).

Also, on several occasions when changes in activity were noted, the national media used archival eruption images of other volcanoes that were spectacular but unrepresentative of Galeras activity. Such instances enhanced concerns and stress levels for local inhabitants and relatives living in other areas. The national government held seminars for the media with the aim to minimize distortional reporting and to emphasize the social responsibility of media toward inhabitants in affected areas, but these goals were not achieved and sensationalized reporting continued. The problem was exacerbated when regional and local officials impeded the flow of information in an attempt to exert control on media.

In late 1991, about the time the new lava dome was emplaced, a new Governor and Mayor were elected for the term of office 1992–1994. The election campaigns featured criticism of the previous administration's management of the May 1989 volcanic crisis and the subsequent directly related economic crisis. The governments had been asked several times by ONAD to restrict tourism near the crater and to establish better security against theft or damage for volcano monitoring equipment. These requests to the old governments had been ignored, and the new governments proved even more unwilling to support hazards work. After the explosion of July 1992, the national government was asked by the Governor to restrict information releases to the public from national offices, including ONAD, about the activity of Galeras – despite the explosive destruction of the dome in July 1992 that hurled ballistic fragments several kilometers, damaged telecommunication facilities high on Galeras, and generated a strong shock wave. The minister reiterated the constitutional obligation of authorities to inform citizens in risk zones. The same Governor ridiculed the activities and ignored the monitoring of the Pasto observatory scientists, and appointed an independent geologist as his scientific advisor. During this period, local media began to claim that information was being concealed. The Governor's attitude finally led to an investigation by the Attorney General of Colombia.

The deaths to scientists and tourists in January 1993 generated a confrontation between national and regional authorities, each blaming the other (Cardona, 1997). The national level of government had to address the emergency, as the local and regional governments through lack of preparation exhibited inadequate procedures and coordination. Although the disaster generated new meetings at local and regional level, the debate instead led to requests to the President of the Republic to declare an economic emergency, and to invest resources from the national budget into the local economy. Although not unsuitable, the requests bore no direct relation to improving risk management.

During 1994 and 1995, seismic swarms and damage-causing earthquakes occurred. The emergency caused by the March 1995 earthquake, which resulted in six deaths, was dealt with by yet again a new Mayor and Governor. Despite their positive attitudes toward preparedness, resources available for mitigation were limited. Yet between 1989 and 1994, almost two million dollars was made available through ONAD and related entities, and used for volcano monitoring, hospitals, communications, and public education.

Hazards assessments and the role of scientists

Most Colombian volcanologists assigned to Pasto in 1989 had some experience from the Nevado del Ruiz emergency (there were *no* trained "volcanologists" in Colombia before Ruiz). Despite this limited experience, they were disconcerted by the rapid buildup of volcanic activity, and were under pressure from national and regional authorities to accurately forecast future developments, and simultaneously to determine zones of volcano hazard to aid emergency preparations.

The first official hazard map, released 3 May 1989 by the Regional Risk Committee, was essentially a "bullseye" map with five concentric zones. Its necessarily rapid preparation, the lack of basic knowledge about Galeras, and the Ruiz experience, collectively contributed to an overestimation of the extent of hazardous zones. Western Pasto was mapped as high hazard, and the main part of the city was in a medium-high hazard zone.

In addition to the map, alert levels concerning the volcano's activity were being discussed. In late March 1989, INGEOMINAS volcanologists recommended consideration of issuing a "white warning" of increasing activity, with the intention of using this unconventional color code to avert undue alarm. Days later, a "yellow warning" was declared by the Regional Risk Committee. However, concern at national level prompted ONAD in Bogota to declare an "orange warning" to the media, without the input of the regional and local committees who should have been the source of information (Cardona, 1997). The orange alert was noted on the front-page headline in the Pasto newspaper *Diario del Sur* on 10 May, along with the colored hazard map. This action led to some exaggerated media reports and a deterioration in scientist-government relations at regional and local levels, as mentioned earlier.

Meanwhile several experts from the USA arrived in early May to assist the Colombians, and a second, unofficial hazard map was prepared and with a summary distributed to authorities and DN-PAD by 30 May 1989; a report was issued the next month (Janda and Voight, 1989). The map considered the influence of valley topography on potential flowage hazards and showed that Pasto was under much less risk than indicated in the first map.

An international workshop on volcano emergencies was held in Pasto between 8 and 28 May 1989, sponsored by UNESCO, USGS, and USAID, and attended by foreign scientists and trainees. The hazards were being discussed at this time, and more monitoring instruments were installed by USGS. At about the same time, a decision was made by the National Emergency Committee (NEC) to completely and quickly evacuate Pasto and neighboring towns in order to conservatively prevent loss of life. The decision was reversed when the eruptive scenarios were presented to NEC in probabilistic terms, along with concepts of acceptable risk, "that should be considered to avoid a social and economic disaster in the absence of a volcanic event" (Cardona, 1997). This information was received by local and regional governments as more suitable than the existing official hazards map for planning purposes.

Other problems were recognized. Maintaining an "orange warning" for a long time had led to loss of its effectiveness, and the population drifted back to the same mind-set as had existed before the warning. Thus some thought was given to restructuring the criteria for warnings and the actions implied.

The second "official" hazard map was issued December 1989. This version considered the eruptive potential during the last 10,000 years, the runouts of hazardous flows, and their probabilities. The result was three zones with Pasto in a low hazard zone. But in a contrary development in 1990, the Pasto Chamber of Commerce and Mayor encouraged the involvement in the crisis of the director of the geophysical institute at a local university. The institute had no credentials in volcanology, but their optimistic and reassuring opinions on the volcano were desired by the Chamber of Commerce. For example, the institute denied that pyroclastic currents had occurred in the past, despite proof to the contrary recognized by INGEOMINAS.

When the lava dome appeared in 1991 the observatory in Pasto was strengthened further, and a two-man USGS seismology team provided technical assistance on 8–15 November. Because of the political situation, the team was instructed to maintain a low profile and not to offer public opinions on volcano status or other actions. A meeting that included regional and local authorities and INGEOMINAS was held on 17 November, and appeared to be positive in the limited sense that the Governor and Mayor agreed to the necessity of having an emergency plan (USGS file information).

Clearly the nadir for those who had insisted on strengthening preparedness was the eruption of 14 January 1993. Not only were valued colleagues lost, but some inhabitants were shocked by the accident and later ridiculed it, deteriorating the credibility of the scientists still further (Cardona, 1997). Some recognized that the eruption was proof that the volcano was dangerous and a threat to the community, but some poorly educated persons believed that the volcano was venting anger only at foreigners that disturbed it (Calvache, 2000). Years later this accident was sensationalized in nonscientific reporting, with personality issues emphasized and the field trip leader vilified. But before 14 January there was a paucity of experience with which to assess a conjectured link between tornillos and dome-destroying eruptions at Galeras, and other potential precursors were then at a low level. In the few days preceding the eruption, "It is important to mention that at that time no special or clear indications of these signals as precursors was seen…in the sense to have a real premonition of the incoming eruption" (R. Torres and D. Gomez, Observatory Seismologists, written communication to BV). Several experienced volcanologists have also judged the January 1993 situation as manifestly unclear: from S. McNutt, "…based on my experience elsewhere, it was not obvious"; and from B. Voight, "…before hindsight entered the picture, there was plenty of room for alternative interpretations regarding the hazard implications of the signals" (Monastersky, 2001). In a detailed assessment of the precursor evidence using probability theory, Aspinall et al. (2003) concluded: "With the benefit of hindsight, it would be invidious to criticize an expedition leader for taking with him or her other scientists who were keen to enter the crater and whose posterior odds of imminent explosive activity were probably 'professionally acceptable.'"

The incident also prompted consideration of general safety recommendations for volcanologists and the public (Aramaki et al., 1994). Precautions had been followed by the Pasto observatory during (and even before) January 1993, limiting the personnel entering the crater area, minimizing exposure time, and keeping radio contact with Pasto. Certainly, their standard of safety practice was no worse than that observed at most other observatories in the 1990s, but a raised worldwide standard was deemed desirable.

Only after four eruptions and two seismic crises (clearly felt by the population) had occurred in 1993 was it finally accepted by government officials and the community that Galeras was a genuine threat. The seismic crises particularly affected the population, and even the Governor "decided to believe" the volcano was active (Calvache, 2000). Gradually, after political resistance waned, the observatory scientists began to develop more advanced procedures for communicating information to authorities and the community. Several strategies were used to make the scientist's work and tools more accessible to the public, including student tours to the observatory, scientist interviews on local television, preparation of educational materials using easy-to-understand language, and production of videos on volcano hazards and monitoring. Nevertheless, as recently as 2009, recommended evacuations have not been generally met with compliance.

Social and economic impacts and feedback to the crisis

Many people left Pasto c.1990 and some property was sold at very low prices. A recession occurred in construction, because the Mayor prohibited construction in western Pasto (where the hazard was originally perceived as high), and in turn this fuelled unemployment. Financial and business concerns suspended or tightened credit and debtors failed their obligations. Transportation companies were unwilling to send freight and passengers to the area, leading to a severe problem of supplies and food. Tourism ceased and businesses were significantly affected.

When the volcano failed after many months to generate a significant eruption, doubts of a large eruption arose among both authorities and scientists. The events that had occurred were within the range encompassed by oral tradition and history, over a time span that seemed long to nonscientists but brief to volcanologists. The political campaigns of 1990–91 focused on meeting the economic crisis, and on criticism of the incumbent authorities' management of it.

Acceding to pressure from the economic guilds, the Mayor of Pasto promoted a "Meeting of the Volcanic Risk Situation and its Socioeconomic Consequences" in July 1990, whose purpose was the "*de-galerasization*" of

Pasto. At this meeting, attended by representatives of INGEOMINAS, ONAD, the Regional Risk Committee, and the economic guilds, it was suggested to de-emphasize the importance of volcanic activity to the city. The main speaker was the director of the Jesuit geophysical institute mentioned earlier, and, with misleading evidence and logic, he minimized the risk posed by Galeras. As a result, information concerning volcanic activity was almost totally restricted. There were no interinstitutional, educational, or public information actions, and information issued by scientists were undermined. Statements made by ONAD at national level were ignored.

The Governor and several parliament members later (in 1992) asked the President of the Republic to declare an economic emergency, but this was denied. The attitudes of most inhabitants evolved from one of extreme stress to complete apathy or denial of risk (Cardona, 1997). The lack of an alternative to coexistence with the risk prompted a collective psychological attitude of simply ignoring the volcano. Elements of that attitude exist to the present time and are a factor in poor compliance to evacuation recommendations.

Summary of crisis management issues

Because of a lack of available detailed knowledge about the volcano and no previous hazards mapping or emergency preparedness plans, the initial, necessarily rushed hazards assessments by Colombian scientists overestimated the extent of the hazard zones, and the first official hazard map in 1989 showed the densely populated city of Pasto within a zone of high to medium-high hazard.

This initial, appropriately cautious attitude of scientists on the risk posed by Galeras was translated, in the post-Armero environment, by political authorities at local, regional, and national levels into extreme alarm – because if a serious eruption were to occur, they would be held responsible due to a lack of recommendation for evacuation. This alarm overrode considerations of social impacts which might be produced by a false alarm or an unnecessary evacuation. Although a formal evacuation was not instituted, the way information was managed also contributed to problems in the fragile economic environment of southern Colombia – even though it would have been difficult to eliminate such problems.

Given the negative implications from the 1989 crisis, the local and regional political authorities focused on the economic problems and were thereafter reluctant to promote risk reduction activities. Further, since the Governor and Mayor were automatically the designated chairpersons for the regional and local risk committees, respectively, progress on these committees was negligible. Exemplified by the "*de-galerasization*" of Pasto, and fearing that the media would spread alarm and thereby affect the economy, information concerning volcanic activity was almost totally restricted. There were no inter-institutional, educational, or public information actions, information issued by scientists were consistently undermined, and statements from ONAD at national level were ignored. More than four Governors and four Mayors had dealt with the crisis in the period to c.2000, and such lack of continuity (and related politicization of the crisis) affected hazards management. The problem of continuity is typical for long-lasting crises, e.g., at Montserrat.

Conclusions – the lessons learned

1. The lack of precrisis scientific knowledge of the volcano contributed not unreasonably to an assessment that overemphasized the hazards, but this in turn led to a *chain-reaction* of serious problems – undue extreme alarm by civil authorities, the spreading of alarm nationally by media, deterioration of a fragile regional economy, loss of credibility of the scientific team with authorities and the public, and adverse politicization of the crisis which lasted many years.
2. The Galeras example shows how a potentially active volcano can lead to serious economic and social problems without accompanying serious volcanic activity, and indeed even without an evacuation. Much depends upon the way the emergency is managed, although some aspects of the problem can develop irrespective of management skills. Such situations are dangerous because they may lead to a state of denial of risk, with potentially disastrous consequences.
3. Scientists, risk committees, and civil authorities must learn how to work together, in order to maintain credibility of the public and to protect the public interest. Lacking such credibility, effective communications with the public will be impossible, and the implementation of effective mitigation measures will be severely hindered.
4. Despite management problems, the scientific achievements from Galeras have been notable in multiparameter monitoring, detection of strong precursors to lava dome emplacement, LP seismicity as indicators of gas pressurization, and interpretation of gas flux fluctuations. Interpreting explosion precursors was for a time problematic, given the subtle nature and limited understanding (pre-January 1993) of the precursors.

Acknowledgments

We thank our highly dedicated colleagues at OV Pasto for much support, notably F. Munoz, H. Cepeda, E. Parra., D. Gomes, R. Torres. The 1989 mission to Galeras by BV, R. Janda and D. Harlow was supported by USGS and United Nations (UNDRO) funds. C. Cardenas and O.D. Cardona of ONAD provided assistance and insight. We recall the memory of colleagues and friends killed in the 1993 eruption, Geoff Brown, Fernando Cuenca, Nestor Garcia, Igor Menyailov, Carlos Trujillo, and Jose Arles Zapata, and of Bruno Martinelli, dear friend and wise advisor to the Pasto observatory scientists.

Bibliography

Aramaki, S., Barberi, F., Casadevall, T., and McNutt, S., 1994. Safety recommendations for volcanologists and the public. *Bulletin of Volcanology*, **56**, 151–154.

Aspinall, W. P., Woo, G., Voight, B., and Baxter, P. J., 2003. Evidence-based volcanology: application to eruption crises. *Journal of Volcanology and Geothermal Research*, **128**, 273–285.

Baxter, P. J., and Gresham, A., 1997. Deaths and injuries in the eruption of Galeras volcano, Colombia, 14 January 1993. *Journal of Volcanology and Geothermal Research*, **77**, 325–338 (causes of casualties and advice on safety).

Calvache, M. L. V., 2000. Mt. Galeras: activities and lessons to be learned. *Philosophical Transactions of the Royal Society of London*, **358**, 1607–1617 (summary of activity and management problems).

Cardona, O. D., 1997. Management of the volcano crises of Galeras volcano: social, economic and institutional aspects. *Journal of Volcanology and Geothermal Research*, **77**, 313–324 (insights on management).

Chouet, B. A., 1996. New methods and future trends in seismological volcano monitoring. In Scarpa, R., and Tilling, R. (eds.), *Monitoring and Mitigation of Volcano Hazards*. Berlin: Springer, pp. 23–97 (seismicity types).

Cortes, G. P., and Raigosa, J. J. A., 1997. A synthesis of the recent activity of Galeras volcano: seven years of continuous surveillance, 1989–1995. *Journal of Volcanology and Geothermal Research*, **77**, 101–114 (summary of monitoring over full crisis period).

Gil Cruz, F., and Chouet, B., 1997. Long-period events, the most characteristic seismicity accompanying the emplacement and extrusion of a lava dome in Galeras volcano, Colombia. *Journal of Volcanology and Geothermal Research*, **77**, 121–158 (detailed study of long-period seismicity).

Gomez, D. M., and Torres, R. A., 1997. Unusual low-frequency volcanic seismic events with slowly decaying coda waves observed at Galeras and other volcanoes. *Journal of Volcanology and Geothermal Research*, **77**, 173–194 (compares 'tornillo'-type events at Galeras and other volcanoes).

Hurtado Artunduaga, A. D., and Cortes Jimenez, G. P., 1997. Third version of the hazard map of Galeras volcano. *Journal of Volcanology and Geothermal Research*, **77**, 89–100.

Janda, R. J., and Voight, B., 1989. *Mission to Galeras volcano, Colombia, April–May 1989: Results and recommendations*. Report to USGS, United Nations and Colombian officials, 49 pp.

Monastersky, R., 2001. Under the volcano. *The Chronicle of Higher Ed.*, A18-22 (March 30) (review of contrasting opinions of 1993 disaster).

Munoz, F. A., et al., 1993. Galeras volcano: international workshop and eruption. *Eos Transactions, American Geophysical Union*, **74**(6), 286–287.

Narvaez, L. M., Torres, R. A., Gomez, D. M. M., Cortes, G. P. J., Cepeda, H. V., and Stix, J., 1997. 'Tornillo'-type seismic signals at Galeras volcano, Colombia. *Journal of Volcanology and Geothermal Research*, **77**, 159–172.

Smithsonian Bull. Global Volcanism Network (BGVN), 2009. **34**(12), 9–10.

Stix, J., Calvache, M. L. V., and Williams, S. N., 1997. Galeras Volcano, Colombia: interdisciplinary study of a Decade volcano. *Journal of Volcanology and Geothermal Research*, **77**(1–4), 338. (comprehensive scientific treatise on Galeras).

Zapata, J. A., et al., 1997. SO$_2$ fluxes from Galeras volcano, Colombia, 1989–1995: progressive degassing and conduit obstruction of a Decade Volcano. *Journal of Volcanology and Geothermal Research*, **77**, 195–208.

Cross-references

Casualties Following Natural Hazards
Disaster Diplomacy
Disaster Risk Management
Emergency Management
Emergency Planning
Evacuation
Hazard and Risk Mapping
Magma
Mass Media and Natural Disasters
Montserrat Eruption
Mt Pinatubo
Nevado del Ruiz Volcano
Pyroclastic Flow
Stratovolcano
Uncertainty
Volcanoes and Volcanic Eruptions

GAS-HYDRATES

Harsh K. Gupta[1,2], Kalachand Sain[2]
[1]Government of India, NDMA Bhawan, New Delhi, India
[2]National Geophysical Research Institute, Hyderabad, India

Synonyms

Clathrates; Methane-hydrates

Definition

Gas-hydrates, also called clathrates, are ice-like crystals of hydrocarbons (mainly methane) and water, which are found in shallow sediments of deepwater and permafrost regions, controlled by high-pressure and low-temperature regimes.

Discussion

The origin of methane can be either biogenic or thermogenic or a mixture of both. The estimated global reserve of methane (2×10^{16} m^3) within gas-hydrates is believed to exceed the equivalent amount of gas in total fossil fuels, and is roughly 3,000 times the amount present in the atmosphere. The carbon in gas-hydrates is speculated to be 10,000 gigatons, which is roughly two times the carbon content in global fossil fuels (Milkov, 2004).

Methane is the cleanest fuel of all hydrocarbons and is 20 times more potent as a greenhouse gas than carbon dioxide. Therefore, a large release of methane from this source may have a significant impact on climate change. Destabilization of gas-hydrates weakens the host sediments influencing the occurrence of submarine *Landslide*. The study of gas-hydrates has attracted the attention of researchers because of their (1) potential submarine geohazards, (2) likely role in global climate change, and (3) future fuel resource. The first two aspects are dealt with here.

Periodic tidal loading and unloading, and an increase in bottom-water temperature can change the pressure–temperature regime of a gas-hydrate stability field thereby leading to dissociation of gas-hydrates, and consequently causing sediment instability. The tectonic disturbances may lead to explosive dissociation of massive gas-hydrates, which can trigger large submarine landslides that can damage oil platforms, coastal installations, and fishing networks. Due to a decrease in permeability, the hydrate-bearing sediments can trap free-gas underneath. The association of gas-hydrates with active mud volcanoes enhances the chances for flaming eruptions.

Deepwater drilling operations may melt the gas-hydrates that can cause sediment mass movement, borehole collapse, gas blowouts, etc. The clogging of pipelines by the formation of gas-hydrates also poses a threat to transmission. Since a depressurizing hydrate plug can travel at ballistic speeds, it has the potential to injure workers and rupture the pipelines, and thus this type of dislodging can be dangerous.

The dissociation of gas-hydrates may add a large volume of methane to the atmosphere, which is known to contribute to climate change. This also causes an uncontrolled release of over-pressured gas trapped underneath, which also represents a significant hazard. Global warming could melt gas-hydrates, causing methane release into the atmosphere and enhanced warming. This raises the risk of more extreme events such as floods, droughts, and wildfires.

There is a need to develop an environment-safe technology for commercial production of gas-hydrates, avoiding possible hazards and meeting the enormous global energy deficit. It is essential to recognize the prospective places of gas-hydrates' occurrences for evaluating the resource potential and impact of environmental hazard for further advancement. The identification and quantification (Sain and Gupta, 2008) of gas-hydrates is a prerequisite for exploration of conventional oil and gas in the deepwater regions.

Bibliography

Milkov, A. V., 2004. Global estimates of hydrate-bound gas in marine sediments: how much is really out there? *Earth Science Reviews*, **66**, 183–197.
Sain, K., and Gupta, H. K., 2008. Gas hydrates: Indian scenario. *Journal of the Geological Society of India*, **72**, 299–311.

Cross-references

Coastal Zone Risk Management
Critical Infrastructure
Damage and the Built Environment
Economics of Disasters
Expert (Knowledge-Based) Systems for Disaster Management
Hazardousness of Place
Induced Seismicity
Landslide
Marine Hazards
Volcanic Gas

GEOGRAPHIC INFORMATION SYSTEMS (GIS) AND NATURAL HAZARDS

Paolo Tarolli[1], Marco Cavalli[2]
[1]Department of Land, Environment, Agriculture and Forestry, University of Padova, Legnaro, Padova, Italy
[2]National Research Council – Research Institute for Geo-Hydrological Protection, Padova, Italy

Definitions

Geographic Information Systems (GIS). GIS is a computer-based information system designed for capturing, storing, analyzing, managing, and displaying spatial data representing human and natural phenomena from the real world. It may include application to remote sensing, land surveying, mathematics, and geography.
Natural Hazard. Any natural phenomenon that poses a threat to human life or properties.

Introduction

Natural hazards include geological (e.g., earthquakes and landslides) and meteorological events such as cyclones, tornadoes, hailstorms, floods, droughts, and wildfires (Chen et al., 2003). Natural hazards are deeply linked to the concepts of *magnitude*, *geographical location*, and *time recurrence* which denote intensity, place of potential occurrence, and frequency of the natural phenomenon, respectively. Since many factors can play an important role in the occurrence of a natural disaster and their spatial information is crucial to risk assessment and management, the analysis of natural hazards greatly benefits from GIS. The use of GIS to manage and understand complex natural hazards in spatial and temporal contexts is nowadays essential (e.g., Chen et al., 2001; Zerger, 2002; Gamper et al., 2006; Fernández and Lutz, 2012), since many issues related to hazard assessment and mitigation have a strong spatial component and their combination can provide valuable information for decision making. For example, in the case of hazard assessment and risk mitigation of diseases such as malaria and cholera, mapping the spatial distribution of incidence or space-time evolution, and merging these maps with land-use and land-cover layers and population density is really useful and represents a precious tool for decision making, method preparedness, and early warning system for malaria or cholera control (Bertuzzo et al., 2008; Hanafi-Bojd et al., 2012). Even in the case of destructive phenomena, such as hurricanes and wildfire, GIS tools are essential to store and manage geographical data in order to study effective mitigation measures (Castrillón et al., 2011; Krishnamurthy et al., 2011). In the frame of a joint analysis of several natural hazards via GIS, the combination of multiple hazard maps related to the individual phenomena can lead to the creation of a composite "multi-hazards" map (Schneider and Schauer, 2006; Reese et al., 2007; Kappes et al., 2012).

In recent years, GIS technology has allowed individuals to store and manipulate earth surface data in more efficient and innovative ways. Considering the topography, morphological factors, such as slope, landform curvature, flow direction, and drainage area, play a key role in flood and landslide hazard assessment. During the last decade, a range of new remote sensing techniques have led to a dramatic increase in terrain information, both in terms of quality and amount of topographic data (Tarolli et al., 2009). Both Terrestrial Laser Scanner (TLS) and Airborne Laser Swath Mapping technology (ALSM), using LiDAR (Light Detection And Ranging) technology, now provide high-resolution topographic data with notable advantages over traditional survey techniques (Slatton et al., 2007; Tarolli et al., 2009). A valuable characteristic of these technologies is their capability to produce sub-meter resolution Digital Terrain Models (DTMs), and high-quality land cover information (Digital Surface Models, DSMs) over large areas (Tarolli et al., 2009; Cavalli and Tarolli, 2011). With the use of GIS, it is possible to combine topographic information with other land surface data (e.g., geology, land use, location of buildings and roads) to improve the natural hazards assessment (Ayalew et al., 2004; Ho and Umitsu, 2011; Guzzetti et al., 2012), and to increase the effectiveness of decision-makers for natural hazards mitigation and preparedness (Rashed and Weeks, 2003; Malczewski, 2006; Ballesteros-Cánovas et al., 2012; Yang et al., 2012). In order to assure the effectiveness of a GIS model, all the data layers must be periodically updated and the analysis repeated with new data (Cubellis et al., 2004; Tropeano and Turconi, 2004).

At present, there is a great variety of GIS software, such as packages developed by academic institutions, commercial software, and, especially in the last few years, many widely spread open-source software systems (Steiniger and Bocher, 2009), that can be used in different operating systems (OS). In addition, they are developed in various languages: Java, C++, C#, VB.Net, etc. (Lei et al., 2011). In the last few years, the Web-based GIS have become largely widespread, with an increasing use of the related applications in natural hazards assessment (Nappi et al., 2008; Gitis et al., 2012). A Web-based GIS combines Internet and Geographic Information System environments; GIS provides the capability for storing and managing large amounts of spatial data, while Internet technology allows easy access to the geospatial information (Nappi et al., 2008).

GIS and landslide hazard

The distribution of landslides and the resulting hazards has been estimated following both statistical and physically-based or process-based approaches. Landslide mapping through GIS is an important part of landslide research because maps showing observed landslides are necessary to calibrate and validate the landslide susceptibility models (Guzzetti et al., 2006; Chang et al., 2007; Rossi et al., 2010). Different techniques are used to compile landslide event-inventory maps where GIS plays an important role (Guzzetti et al., 2012): interpretation of stereoscopic aerial photographs taken shortly after an event (Guzzetti et al., 2004), using single or multi-temporal aerial photographs (Brardinoni et al., 2003; Prokesova et al., 2010); satellite images (e.g., Nichol et al., 2006; Hong et al., 2007; Mondini et al., 2011); synthetic aperture radar (SAR) (Czuchlewski et al., 2003; Hilley et al., 2004; Bulmer et al., 2006; Roering et al., 2009); and analysis of high-resolution DTMs obtained from airborne Lidar (McKean and Roering, 2004; Van Den Eeckhaut et al., 2007; Kasai et al., 2009; Tarolli et al., 2012). All these methodologies are based on the detection of landslides, through digital quantification of the DTM, or statistical interpretation of images. In addition to these, the physically-based or process-based approaches (Hammond et al., 1992; Montgomery and Dietrich, 1994; Crosta and Frattini, 2003) were largely used for landsliding hazard mapping in the last few decades. They consist in integrating hydrologic and geotechnical models for evaluating landsliding susceptibility maps. These maps usually comprised a stability index spatially distributed and expressed as Factor of Safety FS and Stability Index SI in the case of Pack et al. (1998), and Tarolli and Tarboton (2006) respectively, or as critical rainfall (Montgomery and Dietrich, 1994; Borga et al., 2002; Tarolli et al., 2011) needed to cause instability for each topographic element. Susceptibility maps need to be evaluated and validated (Dietrich et al., 2001; Tarolli et al., 2011). The evaluation of the effectiveness of susceptibility maps can be carried out by mapping all observed landslides, and comparing them with the map of predicted stability index. Figure 1 shows an example of such approach where Most Likely Landslide Initiation Points (MLIPs) (Tarolli and Tarboton, 2006), landslide locations, and high-resolution aerial photographs are superimposed in a GIS environment. Similar landslide susceptibility maps can be of immense support to decision-makers and environmental planners.

Several studies have demonstrated the potential of GIS for hazard assessments in the particular case of debris flow (Huggel et al., 2003; Medina et al., 2007; Cavalli and Marchi, 2008; Carrara et al., 2008; Stolz and Huggel, 2008; Scheidl and Rickenmann, 2010). Many different runout prediction methods can be applied to estimate the mobility of future debris flows during hazard assessment (Hürlimann et al., 2008). Using a statistical approach, He et al. (2003) developed a quantitative model of hazards assessment and zonation through synthesis analysis of basin areas, gradients, and the relative reliefs of these debris flow sites. The location of each debris flow site and its physical and activity characteristics were entered into the GIS-based database. Cavalli and Marchi (2008) used LiDAR data in GIS environment to examine the morphology and to characterize surface complexity of the alluvial fan of an alpine debris-flow stream.

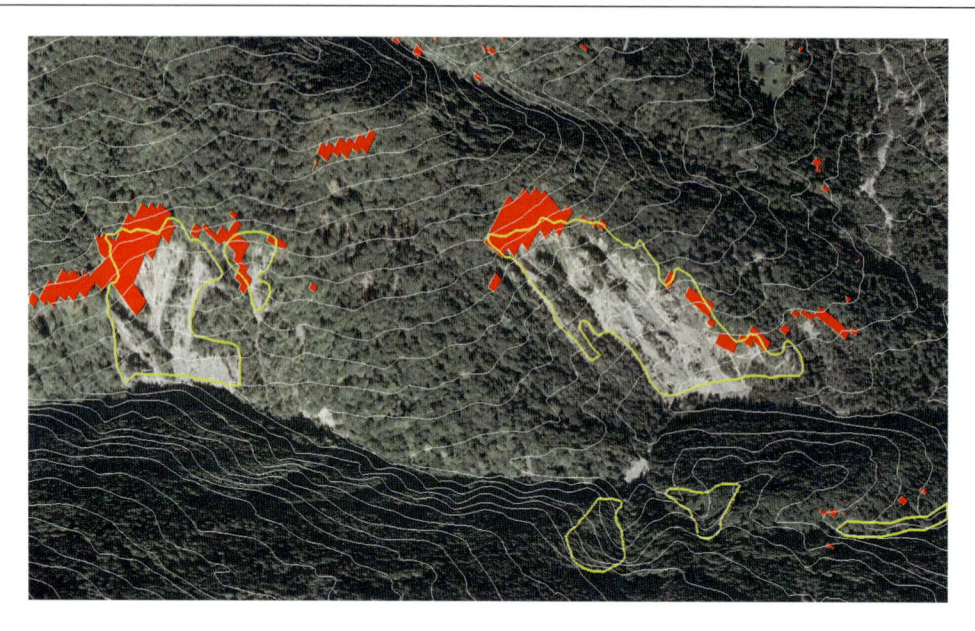

Geographic Information Systems (GIS) and Natural Hazards, Figure 1 Most likely landslide intiation points and landslide locations (Tarolli and Tarboton, 2006).

Carrara et al. (2008) in order to predict location of the debris-flow source areas proposed two multivariate procedures (discriminant and logistic regression) and a physically-based approach. Berti and Simoni (2007) and Scheidl and Rickenmann (2010) proposed two empirical approaches for prediction of debris flow inundation areas and prediction of debris-flow mobility and deposition on fans respectively. Medina et al. (2007) and Stolz and Huggel (2008) proposed 2D model analysis of debris-flow hazard zones which can be easily implemented in GIS software.

GIS and flood hazard

Floods are one of the major natural hazards. Predicting an inundation in an urban area is a crucial tool for the mitigation of flood risk, in order to be prepared for the flood, and to minimize its impact and for providing the basis for management decision, flood protection system, and a suitable urban planning. Flood hazard mapping can be performed using different methods that can be grouped into two main categories: geological-geomorphological and hydrological-hydraulic methods (Lastra et al., 2008). Geomorphological analysis has been carried out for flood-risk prevention plans in the Mediterranean regions (Guzzetti and Tonelli, 2004; Guzzetti et al., 2005). For regional planning purposes, the prediction of flood processes can be carried out by means of GIS analysis of topographic and remote sensing data. Simple morphometric properties (i.e., Melton ratio) can help in identifying watersheds prone to flooding (Wilford et al., 2004), and the analysis of remote sensing images taken during a flood and in a dry season allows to estimate flooded area and therefore, to create accurate flood hazard map (Islam and Sado, 2000). Conventional image processing techniques for extracting flooded areas from SAR imagery have been used with varying accuracy (depending on spatial resolution) (Kiage et al., 2005; Schumann et al., 2009), since SAR imagery presents advantages over optical instruments, especially in flood management applications (Matgen et al., 2007). Combining supervised land cover classifications from LANDSAT and Shuttle Radar Topographic Mission Digital Elevation Model (SRTM DEM) seemed an efficient method for mapping coastal flood risk (Wang et al., 2002; Demirkesen et al., 2006). Umitsu et al. (2006) demonstrated the utility of SRTM incorporated with GIS data for studies on flooding and micro-landforms. Ho and Umitsu (2011) developed an integrated method for classifying micro-landforms and flood hazard zones based on a geomorphological approach utilizing in GIS environment SRTM and LANDSAT ETM (Enhanced Thematic Mapper Plus) data combined with field investigation. Lastra et al. (2008) combined geomorphological and hydrological methods for flood hazard delineation. Several other approaches included also the hydrological-hydraulic approach for flood hazard mapping (Horritt et al., 2001; Horritt and Bates, 2001; Xia et al., 2011; Stephens et al., 2012). The results are flood inundation maps visualized in GIS environments, wherein it is possible to recognize and locate, in detail, the places with different water depth and extension. A good risk analysis of such phenomena is given by combining flood risk maps with other data such as streets, buildings, land use, etc. Such maps play an important role in flood management (Plate, 2002). GIS flood applications depend on resolution of topographic data used in flood simulations. New remote sensing technologies such as LiDAR measure high-resolution terrain (elevation) data that help in measuring surface morphology and roughness (Cavalli et al., 2008). This is advantageous when small-scale processes have a significant effect on model, for example, where the extent of inundation is controlled by predictions of small topographic features (Horritt and Bates, 2001). Fernandéz and Lutz (2012) proposed a GIS-aided urban flood hazard zoning, applying multicriteria decision analysis and to evaluate it by means of uncertainty and sensitivity analysis.

The use of GIS for decision makers to limit or analyze the impact of natural coastal hazards, such as sea level rise, storm surges, and coastal erosion given by tsunamis, and thus, protect the exposed assets (population, property, settlements, communications networks, etc.) (Desprats et al., 2010) is worth to be mentioned. By the analysis of the coastal risk scenarios within GIS environment, it is possible to identify "weak points" in human settlements and test the exposure of their future development (Desprats et al., 2010). This is indeed a critical issue if we consider the recent devastating 2004 Indian Ocean and 2011 Japan tsunamis. The analysis of tsunami hazards and mapping, and thus, the evaluation of population exposure can greatly benefit from the use of GIS which helps to improve the tsunami hazards assessment and mitigation (Løvholt et al., 2012).

GIS and seismic hazard

Managing seismic hazard requires different and independent observations and techniques (Salvi et al., 1999): epicenter distribution of recent and historical seismicity, geological and geomorphologic data, geophysical analysis, analysis and monitoring of deformation by geodetic data, photo interpretation of aerial and satellite images, and fluid geochemistry monitoring (e.g., Quattrocchi et al., 2000). The approach to seismic hazard evaluation is a multidisciplinary approach, wherein a GIS is strongly required for managing multi-scaled datasets in the same geographically referenced view (Salvi et al., 1999). The geologic and seismologic maps can be easily combined, and integrated in a GIS environment (Nath, 2005). This allows to overlay and analyze different thematic layers, and merge raster and vectorial data (Cubellis et al., 2004; Pal et al., 2008). With the overlay procedure of different thematic fields it is possible to derive new integrated maps of potential damage that may then be used in hazard assessment for a suitable planning in an urbanized area (Cubellis et al., 2004). Nath (2005) proposed a seismic microzonation model through thematic mapping and GIS integration of geological and motion features through GIS. The microzonation maps may be useful for land use planning or for making hazard mitigation decisions. Selcuk and Yucemen (2000) and Selcuk-Kestel et al. (2012) proposed a GIS-based software for lifeline reliability analysis under seismic hazard. The developed GIS-based software imports seismic hazard and lifeline network layers. It adopts a network reliability algorithm to calculate the upper and lower bounds for system reliability of the lifeline under seismic hazard. Zolfaghari (2009) proposed an approach to provide practical facility for better capture of spatial variations of seismological and tectonic characteristics, which allows better treatment of their uncertainties. In the proposed approach, GIS raster-based data models are used in order to model geographical features in a cell-based system. The cell-based source model provides a framework for implementing many geographically referenced seismotectonic factors into seismic hazard modeling.

As with the greater part of geographical data, even seismic information is currently shared with a large target of end-users (Pessina and Meroni, 2009). Due to this reason Web-GIS should be a strategic tool for sharing information on seismic hazard and to develop real-time warning systems. Martinelli and Meletti (2008) proposed a WebGIS application for rendering seismic hazard data in Italy, whereas Pessina and Meroni (2009) proposed a WebGIS tool for seismic hazard scenarios which consists in ground shaking scenarios associated with the repetition of the earthquake that struck an area.

GIS and volcanic hazard

Volcanic hazards have important scientific, economic, and political implications, especially for densely populated areas close to volcanoes (a clear example is Naples in Italy). Volcanic hazards require accurate assessment and mitigation plans, the development of suitable tools for prediction and management of critical situations, and the promotion of sustainable development within volcanic hazard areas (Felpeto et al., 2007). The use of GIS is crucial in this aspect. Mapping and studying of lava flows have been the subject in the last few years of recent advances where the combination of GIS and remote sensing techniques played a key role (Tarquini and Favalli, 2011). Digital elevation models, satellite images, hazards maps, vector data on natural and artificial features, and population density are the main input information that can be used in GIS for volcanic risk management and mitigation (Pareschi et al., 2000). Volcanic hazard maps and eruption scenarios can be developed from past eruptions and the geologic records (Felpeto et al., 2007). Several authors proposed GIS modeling and spatial analysis in order to produce maps for prevention and emergency management (Iverson et al., 1998; Gomez-Fernandez, 2000; Gaspar et al., 2004). Iverson et al. (1998) express the importance of defining lahar (i.e., a landslide of volcanic fragments mixed with water flowing on the flanks of a volcano) inundation zones from past events. They developed a model for lahar inundation based on the assumption that past inundation zones provide a basis for predicting future inundation areas. Renschler (2005) emphasizes the importance of considering spatial and temporal variability in volcanic processes and representing it in GIS-based volcano hazard simulations. Toyos et al. (2007) presented a methodology implemented within a GIS for hazard mapping of small volume pyroclastic density currents.

The results achieved highlight the applicability of this approach for hazard mitigation and real-time emergency management. Tarquini and Favalli (2011) introduced a GIS to obtain high-resolution boundaries of lava flow fields. This technique is mainly based on the processing of LIDAR-derived maps and DEMs.

GIS and drought hazard

Droughts are among the most important natural disasters, particularly in the arid and semiarid regions of the world (Núñez et al., 2011), especially in recent years due to climate change (Hao et al., 2012). Drought hazards maps combined with other layers in the GIS environment, especially those related to irrigated and food production areas, represent a useful tool for drought risk management and plan. Detailed analyses of spatial and temporal drought dynamics during monsoon and non-monsoon seasons have been carried out through drought index maps generated in GIS environment (Bhuiyan et al., 2006). Vicente-Serrano (2007) analyzed, in a semi-arid region, monthly and spatial differences in the effects of drought on the natural vegetation and agricultural crops by means of the joint use of vegetation indexes, a drought index (standardized precipitation index), and GIS. Shahid and Behrawan (2008) developed a standardized precipitation index method in a GIS environment for mapping the spatial extents of drought hazards in different time steps.

Summary

Understanding the natural hazards is one of the most important issues in the earth science field. This is useful not only for analyzing a natural process from the scientific point of view, but also for understanding the link between the human system and natural and potentially dangerous phenomena. In the last two decades, the GIS technology (commercial, open source, and also web-based) offered to the scientific and technical community a powerful tool to analyze natural processes. Natural hazards, such as landslides, debris flows, floods, volcanic and seismic activity, hurricanes, coastal erosion, and wildfires, but also droughts and human diseases (e.g., malaria and cholera), are now more easier to manage and understand, thanks to the introduction of GIS. In the field of natural hazards, the increasing availability of GIS coupled models and decision-support tools offers even more efficient support for governmental agencies and decision makers to help in risk management and protection activities.

Bibliography

Armanini, A., Fraccarollo, L., and Rosatti, G., 2009. Two-dimensional simulation of debris flows in erodible channels. *Computer & Geosciences*, **35**, 993–1006, doi:10.1016/j.cageo.2007.11.008.

Ayalew, L., Yamagishi, H., and Ugawa, N., 2004. Landslide susceptibility mapping using GIS-based weighted linear combination, the case in Tsugawa area of Agano River, Niigata Prefecture, Japan. *Landslides*, **1**, 73–81.

Ballesteros-Cánovas, J. A., Díez-Herrero, A., and Bodoque, J. M., 2012. Searching for useful non-systematic tree-ring data sources for flood hazard analysis using GIS tools. *Catena*, **92**, 130–138.

Berti, M., and Simoni, A., 2007. Prediction of debris flow inundation area using empirical mobility relationship. *Geomorphology*, **90**, 144–161, doi:10.1016/j.geomorph.2007.01.014.

Bertuzzo, E., Azaele, S., Maritan, A., Gatto, M., Rodriguez-Iturbe, I., and Rinaldo, A., 2008. On the space-time evolution of a cholera epidemic. *Water Resources Research*, **44**, W01424, doi:10.1029/2007WR006211.

Bhuiyan, C., Singh, R. P., and Kogan, F. N., 2006. Monitoring drought dynamics in the Aravalli region (India) using different indices based on ground and remote sensing data. *International Journal of Applied Earth Observation and Geoinformation*, **8**(4), 289–302.

Borga, M., Dalla Fontana, G., and Cazorzi, F., 2002. Analysis of topographic and climatic control on rainfall-triggered shallow landsliding using a quasi-dynamic wetness index. *Journal of Hydrology*, **268**, 56–71.

Brardinoni, F., Slaymaker, O., and Hassan, M. A., 2003. Landslide inventory in a rugged forested watershed: a comparison between air-photo and field survey data. *Geomorphology*, **54**, 179–196.

Bulmer, M. H., Petley, D. N., Murphy, W., and Mantovani, F., 2006. Detecting slope deformation using two-pass differential interferometry: implications for landslide studies on earth and other planetary bodies. *Journal of Geophysical Research*, **111**, E06S16, doi:10.1029/2005JE002593.

Carrara, A., Crosta, G., and Frattini, P., 2008. Comparing models of debris-flow susceptibility in the alpine environment. *Geomorphology*, **94**, 353–378.

Castrillón, M., Jorge, P. A., López, I. J., Macías, A., Martín, D., Nebot, R. J., Sabbagh, I., Quintana, F. M., Sánchez, J., Sánchez, A. J., Suárez, J. P., and Trujillo, A., 2011. Forecasting and visualization of wildfires in a 3D geographical information system. *Computers and Geosciences*, **37**(3), 390–396.

Cavalli, M., and Marchi, L., 2008. Characterisation of the surface morphology of an alpine alluvial fan using airborne LiDAR. *Natural Hazards and Earth System Sciences*, **8**(2), 323–333.

Cavalli, M., Tarolli, P., Marchi, L., and Dalla Fontana, G., 2008. The effectiveness of airborne LiDAR data in the recognition of channel bed morphology. *Catena*, **73**, 249–260.

Cavalli, M., and Tarolli, P., 2011. Application of lidar technology for river analysis. *Italian Journal of Engineering Geology and Environment, Special Issue*, **1**, 33–44, doi:10.4408/IJEGE.2011-01.S-03.

Chang, K.-T., Chiang, S.-H., and Hsu, M.-L., 2007. Modeling typhoon- and earthquake-induced landslides in a mountainous watershed using logistic regression original research article. *Geomorphology*, **89**(3–4), 335–347.

Chen, K., Blong, R., and Jacobson, C., 2001. MCE-RISK: integrating multicriteria evaluation and GIS for risk decision-making in natural hazards. *Environmental Modelling and Software*, **16**(4), 387–397.

Chen, K., Blong, R., and Jacobson, C., 2003. Towards an integrated approach to natural hazards risk assessment using GIS: with reference to bushfires. *Environmental Management*, **31**(4), 546–560.

Crosta, G. B., and Frattini, P., 2003. Distributed modelling of shallow landslide triggered by intense rainfall. *Natural Hazards and Earth System Sciences*, **3**, 81–93.

Cubellis, E., Carlino, S., Iannuzzi, R., Luongom, G., and Obrizzo, F., 2004. Management of historical seismic data using GIS: the Island of Ischia (Southern Italy). *Natural Hazards*, **33**, 379–393.

Czuchlewski, K. R., Weissel, J. K., and Kim, Y., 2003. Polarimetric synthetic aperture radar study of the Tsaoling landslide generated by the 1999 Chi-Chi earthquake, Taiwan. *Journal of Geophysical Research*, **108**(F1), 7.1–7.11.

Demirkesen, A. C., Evrendilek, F., Berberoglu, S., and Kilic, S., 2006. Coastal flood risk analysis using Landsat-7 ETM imagery and SRTM DEM: a case study of Izmir, Turkey. *Environmental Monitoring and Assessment*, **131**, 293–300.

Desprats, J. F., Garcin, M., Attanayake, N., Pedreros, R., Siriwardana, C., Fontaine, M., Fernando, S., and De Silva, U., 2010. A "coastal-hazard gis" for sri lanka. *Journal of Coastal Conservation*, **14**(1), 21–31, doi:10.1007/s11852-009-0084-5.

Dietrich, W.E., Bellugi, D. and Real de Asua, R. 2001. Validation of the shallow landslide model SHALSTAB for forest management. In M.S. Wigmosta and S.J. Burges (eds.), Land Use and Watersheds: Human influence on hydrology and geomorphology in urban and forest areas. American Geophysical Union, Water Science and Application, vol. 2, pp. 195–227.

Felpeto, A., Martí, J., and Ortiz, R., 2007. Automatic GIS-based system for volcanic hazard assessment. *Journal of Volcanology and Geothermal Research*, **166**, 106–116.

Fernández, D. S., and Lutz, M. A., 2012. Urban flood hazard zoning in Tucumán Province, Argentina, using GIS and multicriteria decision analysis. *Engineering Geology*, **111**, 90–98.

Gamper, C., Thöni, M., and Weck-Hannemann, H., 2006. A conceptual approach to the use of cost benefit and multi criteria analysis in natural hazard management. *Natural Hazards and Earth System Sciences*, **6**(2), 293–302.

Gaspar, J. L., Goulart, C., Queiroz, G., Silveira, D., and Gomes, A., 2004. Dynamic structure and data sets of a GIS database for geological risk analysis in the Azores volcanic islands. *Natural Hazards and Earth System Sciences*, **4**(2), 233–242.

Gitis, V., Derendyaev, A., Metrikov, P., and Shogin, A., 2012. Network geoinformation technology for seismic hazard research. *Natural Hazards*, **62**(3), 1021–1036, doi:10.1007/s11069-012-0132-6.

Gomez-Fernandez, F., 2000. Application of a GIS algorithm to delimit the areas protected against basic lava flow invasion on Tenerife Island. *Journal of Volcanology and Geothermal Research*, **103**(1–4), 409–423.

Guzzetti, F., and Tonelli, G., 2004. Information system on hydrological and geomorphological catastrophes in Italy (SICI): a tool for managing landslide and flood hazards. *Natural Hazards and Earth System Sciences*, **4**, 213–232.

Guzzetti, F., Cardinali, M., Reichenbach, P., Cipolla, F., Sebastiani, C., Galli, M., and Salvati, P., 2004. Landslides triggered by the 23 November 2000 rainfall event in the Imperia Province, Western Liguria, Italy. *Engineering Geology*, **73**, 229–245.

Guzzetti, F., Stark, C. P., and Salvati, P., 2005. Evaluation of flood and landslide risk to the population in Italy. *Environmental Mangement*, **36**(1), 15–36.

Guzzetti, F., Reichenbach, P., Ardizzone, F., Cardinali, M., and Galli, M., 2006. Estimating the quality of landslide susceptibility models. *Geomorphology*, **81**, 166–184.

Guzzetti, F., Mondini, A., Cardinali, M., Fiorucci, F., Santangelo, M., Chang, K.-T, 2012. Landslide inventory maps: New tools for an old problem. *Earth-Science Reviews*, **112**(1–2), 42–66, doi:10.1016/j.earscirev.2012.02.001.

Hammond, C. J., Prellwitz, R. W., and Miller, S. M., 1992. Landslides hazard assessment using Monte Carlo simulation. In Bell, D.H. (ed.) *Proceedings of 6th International Symposium on Landslides*, Christchurch, New Zealand, Balkema, 2, pp. 251–294.

Hanafi-Bojd, A. A., Vatandoost, H., Oshaghi, M. A., Charrahy, Z., Haghdoost, A. A., Zamani, G., Abedi, F., Sedaghat, M. M., Soltani, M., Shahi, M., and Raeisi, A., 2012. Spatial analysis and mapping of malaria risk in an endemic area, south of Iran: a GIS based decision making for planning of control. *Acta Tropica*, **121**, 85–92.

Hao, L., Zhang, X., and Liu, S., 2012. Risk assessment to China's agricultural drought disaster in county unit. *Natural Hazards*, **61**, 785–801, doi:10.1007/s11069-011-0066-4.

He, Y. P., Xie, H., Cui, P., Wei, F. Q., Zhong, D. L., and Gardner, J. S., 2003. GIS-based hazard mapping and zonation of debris flows in Xiaojiang Basin, southwestern China. *Environmental Geology*, **45**, 286–293.

Hilley, G. E., Bürgmann, R., Ferretti, A., Novali, F., and Rocca, F., 2004. Dynamics of slow-moving landslides from permanent scatterer analysis. *Science*, **304**, 1952–1955, doi:10.1126/science.1098821.

Ho, L. T. K., and Umitsu, M., 2011. Micro-landform classification and flood hazard assessment of the Thu Bon alluvial plain, central Vietnam via an integrated method utilizing remotely sensed data. *Applied Geography*, **31**(3), 1082–1093.

Hong, Y., Adler, R., and Huffman, G., 2007. Use of satellite remote sensing data in the mapping of global landslide susceptibility. *Natural Hazards*, **43**, 245–256.

Horritt, M. S., and Bates, P. D., 2001. Effects of spatial resolution on a raster based model of flood flow. *Journal of Hydrology*, **253**, 239–249.

Horritt, M. S., Mason, D. C., and Luckman, A. J., 2001. Flood boundary delineation from synthetic aperture radar imagery using a statistical active contour model. *International Journal of Remote Sensing*, **22**(13), 2489–2507.

Huggel, C., Kääb, A., Haeberli, W., and Krummenacher, B., 2003. Regional-scale GIS-models for assessment of hazards from glacier lake outbursts: evaluation and application in the Swiss Alps. *Natural Hazards and Earth System Sciences*, **3**, 647–662.

Hürlimann, M., Rickenmann, D., Medina, V., and Bateman, A., 2008. Evaluation of approaches to calculate debris-flow parameters for hazard assessment. *Engineering Geology*, **102**, 152–163.

Islam, M. M., and Sado, K., 2000. Flood hazard assessment in Bangladesh using NOAA AVHRR data with geographical information system. *Hydrological Processes*, **14**, 605–620.

Iverson, R. M., Schilling, S. P., and Vallance, J. W., 1998. Objective delineation of lahar-inundation hazard zones. *Geological Society of America Bulletin*, **110**(8), 972–984.

Kappes, M. S., Gruber, K., Frigerio, S., Bell, R., Keiler, M., and Glade, T., 2012. The MultiRISK platform: the technical concept and application of a regional-scale multihazard exposure analysis tool. *Geomorphology*, **151–152**, 139–155.

Kasai, M., Ikeda, M., Asahina, T., and Fujisawa, K., 2009. LiDAR-derived DEM evaluation of deep-seated landslides in a steep and rocky region of Japan. *Geomorphology*, **113**, 57–69.

Kiage, L. M., Walker, N. D., Balasubramanian, S., Babin, A., and Barras, J., 2005. Applications of radarsat-1 synthetic aperture radar imagery to assess hurricane-related flooding of coastal Louisiana. *International Journal of Remote Sensing*, **26**, 5359–5380.

Krishnamurthy, P. K., Fisher, J. B., and Johnson, C., 2011. Mainstreaming local perceptions of hurricane risk into policymaking: a case study of community GIS in Mexico. *Global Environmental Change*, **21**(1), 143–153.

Lastra, J., Fernández, E., Díez-Herrero, A., and Marquínez, J., 2008. Flood hazard delineation combining geomorphological and hydrological methods: an example in the Northern Iberian Peninsula. *Natural Hazards*, **45**, 277–293, doi:10.1007/s11069-007-9164-8.

Lei, X., Wang, Y., Liao, W., Jiang, Y., Tian, Y., and Wang, H., 2011. Development of efficient and cost-effective distributed hydrological modeling tool MWEasyDHM based on open-source MapWindow GIS. *Computers and Geosciences*, **37**(9), 1476–1489.

Løvholt, F., Glimsdal, S., Harbitz, C. B., Zamora, N., Nadim, F., Peduzzi, P., Dao, H., and Smebye, H., 2012. Tsunami hazard and exposure on the global scale. *Earth-Science Reviews*, **110**, 58–73.

Malczewski, J., 2006. GIS-based multicriteria decision analysis: a survey of the literature. *International Journal of Geographical Information Science*, **20**(7), 703–726.

Martinelli, F., and Meletti, C., 2008. A WebGIS application for rendering seismic hazard data in Italy. *Seismological Res Lett*, **79**(1), 68–78.

Matgen, P., Schumann, G., Henry, J.-B., Hoffmann, L., and Pfister, L., 2007. Integration of SAR-derived river inundation areas, high-precision topographic data and a river flow model toward near real-time flood management. *International Journal of Applied Earth Observation and Geoinformation*, **9**, 247–263.

McKean, J., and Roering, J., 2004. Objective landslide detection and surface morphology mapping using high-resolution airborne laser altimetry. *Geomorphology*, **57**, 331–351.

Medina, V., Hurlimann, M., and Bateman, A., 2007. Application of FLATModel, a 2D finite volume code to debris flows in the northeastern part of the Iberian Peninsula. *Landslides*, **5**, 127–142, doi:10.1007/s10346-007-0102-3.

Mondini, A. C., Guzzetti, F., Reichenbach, P., Rossi, M., Cardinali, M., and Ardizzone, F., 2011. Semi-automatic recognition and mapping of rainfall induced shallow landslides using satellite optical images. *Remote Sensing of Environment*, **115**(7), 1743–1757, doi:10.1016/j.rse. 2011.03.006.

Montgomery, D. R., and Dietrich, W. E., 1994. A physically based model for the topographic control on shallow landsliding. *Water Resources Research*, **30**, 1153–1171.

Nappi, R., Alessio, G., Bronzino, G., Terranova, C., and Vilardo, G., 2008. Contribution of the SISCam Web-based GIS to the seismotectonic study of Campania (Southern Apennines): an example of application to the Sannio-area. *Natural Hazards*, **45**(1), 73–85.

Nath, S. K., 2005. An initial model of seismic microzonation of Sikkim Himalaya through thematic mapping and GIS integration of geological and strong motion features. *Journal of Asian Earth Sciences*, **25**(2), 329–343.

Nichol, J. E., Shaker, A., and Wong, M. S., 2006. Application of high-resolution stereo satellite images to detailed landslide hazard assessment. *Geomorphology*, **76**, 68–75.

Núñez, J. H., Verbist, K., Wallis, J. R., Schaefer, M. G., Morales, L., and Cornelis, W. M., 2011. Regional frequency analysis for mapping drought events in north-central Chile. *Journal of Hydrology*, **405**(3–4), 352–366.

Pack, R. T., Tarboton, D. G., and Goodwin, C. N., 1998. The SINMAP Approach to Terrain Stability Mapping, *8th Congress of the International Association of Engineering Geology*, Vancouver, British Columbia, Canada.

Pal, I., Nath, S. K., Shukla, K., Pal, D. K., Raj, A., Thingbaijam, K. K. S., and Bansal, B. K., 2008. Earthquake hazard zonation of Sikkim Himalaya using a GIS platform. *Natural Hazards*, **45**, 333–377.

Pareschi, M. T., Cavarra, L., Favalli, M., Giannini, F., and Meriggi, A., 2000. GIS and volcanic risk management. *Natural Hazards*, **21**(2–3), 361–379.

Pessina, V., and Meroni, F., 2009. A WebGis tool for seismic hazard scenarios and risk analysis. *Soil Dynamics and Earthquake Engineering*, **29**, 1274–1281.

Plate, E. J., 2002. Flood risk and flood management. *Journal of Hydrology*, **267**, 2–11.

Prokesova, R., Kardos, M., and Medvedova, A., 2010. Landslide dynamics from high resolution aerial photographs: a case study from the Western Carpathians, Slovakia. *Geomorphology*, **115**, 90–101.

Quattrocchi, F., Pik, R., Pizzino, L., Guerra, M., Scarlato, P., Angelone, M., Barbieri, M., Conti, A., Marty, B., Sacchi, E., Zuppi, G. M., and Lombardi, S., 2000. Geochemical changes at the Bagni di Triponzo thermal spring during the Umbria-Marche 1997–1998 seismic sequence. *Journal of Seismology*, **4**(4), 567–587.

Rashed, T., and Weeks, J., 2003. Assessing vulnerability to earthquake hazards through spatial multicriteria analysis of urban areas. *International Journal of Geographical Information Science*, **17**, 547–576.

Reese, S., Bell, R., and King, A., 2007. RiskScape: a new tool for comparing risk from natural hazards. *Water & Atmosphere*, **15**, 24–25.

Renschler, C. S., 2005. Scales and uncertainties in using models and GIS for volcano hazard prediction. *Journal of Volcanology and Geothermal Research*, **139**, 73–87.

Roering, J. J., Stimely, L. L., Mackey, B. H., and Schmidt, D. A., 2009. Using DInSAR, airborne LiDAR, and archival air photos to quantify landsliding and sediment transport. *Geophysical Research Letters*, **36**, L19402, doi:10.1029/2009GL040374.

Rossi, M., Guzzetti, F., Reichenbach, P., Mondini, A., and Peruccacci, S., 2010. Optimal landslide susceptibility zonation based on multiple forecasts. *Geomorphology*, **114**(3), 129–142.

Rott, H., 2009. Advances in interferometric synthetic aperture radar (InSAR) in earth system science. *Progress in Physical Geography*, **33**, 769–791.

Salvi, S., Quattrocchi, F., Brunori, C. A., Doumaz, F., Angelone, M., Billi, A., Buongiorno, F., Funiciello, R., Guerra, M., Mele, G., Pizzino, L., and Salvini, F., 1999. A multidisciplinary approach to earthquake research: implementation of a geochemical geographic information system for the Gargano site, Southern Italy. *Natural Hazards*, **20**(1), 255–278.

Scheidl, C., and Rickenmann, D., 2010. Empirical prediction of debris flow mobility and deposition on fans. *Earth Surface Processes and Landforms*, **35**, 157–173, doi:10.1002/esp. 1897.

Schneider, P., and Schauer, B., 2006. HAZUS — its development and its future. *Natural Hazards Review*, **7**, 40–44.

Schumann, G., Di Baldassarre, G., and Bates, P. D., 2009. The utility of spaceborne radar to render flood inundation maps based on multi-algorithm ensembles. *IEEE Transactions on Geoscience and Remote Sensing*, **47**(8), 2801–2807.

Selcuk, A. S., and Yucemen, M. S., 2000. Reliability of lifeline networks with multiple sources under seismic hazard. *Natural Hazards*, **21**, 1–18.

Selcuk-Kestel, A. S., Duzgun, H. S., and Oduncuoglu, L., 2012. A GIS-based software for lifeline reliability analysis under seismic hazard. *Computers and Geosciences*, **42**, 37–46.

Shahid, S., and Behrawan, H., 2008. Drought risk assessment in the western part of Bangladesh. *Natural Hazards*, **46**, 391–413, doi:10.1007/s11069-007-9191-5.

Slatton, K. C., Carter, W. E., Shrestha, R. L., and Dietrich, W., 2007. Airborne Laser Swath Mapping: achieving the resolution and accuracy required for geosurficial research. *Geophysical Research Letters*, **34**, L23S10, doi:10.1029/2007GL031939.

Steiniger, S., and Bocher, E., 2009. An overview on current free and open source desktop GIS developments. *International Journal of Geographical Information Science*, **23**(10), 1345–1370, doi:10.1080/13658810802634956.

Stephens, E. M., Bates, P. D., Freer, J. E., and Mason, D. C., 2012. The impact of uncertainty in satellite data on the assessment of flood inundation models. *Journal of Hydrology*, **414–415**, 162–173.

Stolz, A., and Huggel, C., 2008. Debris flows in the Swiss National Park: the influence of different flow models and varying DEM grid size in modelling results. *Landslides*, **5**, 311–319, doi:10.1007/s10346-008-0125-4.

Tarolli, P., and Tarboton, D. G., 2006. A new method for determination of most likely landslide initiation points and the evaluation of digital terrain model scale in terrain stability mapping. *Hydrology and Earth System Science*, **10**, 663–677.

Tarolli, P., Arrowsmith, J. R., and Vivoni, E. R., 2009. Understanding earth surface processes from remotely sensed digital terrain models. *Geomorphology*, **113**, 1–3.

Tarolli, P., Borga, M., Chang, K.T., and Chiang, S.H., 2011. Modeling shallow landsliding susceptibility by incorporating heavy

rainfall statistical properties. *Geomorphology*, **133**, 199–211, doi:10.1016/j.geomorph.2011.02.033.
Tarolli, P., Sofia, G., and Dalla Fontana, G., 2012. Geomorphic features extraction from high-resolution topography: landslide crowns and bank erosion. *Natural Hazards*, **61**, 65–83, doi:10.1007/s11069-010-9695-2.
Tarquini, S., and Favalli, M., 2011. Mapping and DOWNFLOW simulation of recent lava flow fields at Mount Etna. *Journal of Volcanology and Geothermal Research*, **204**(1–4), 27–39.
Toyos, G. P., Cole, P. D., Felpeto, A., and Martí, J., 2007. A GIS-based methodology for hazard mapping of small pyroclastic density currents. *Natural Hazards*, **41**(1), 99–112.
Tropeano, D., and Turconi, L., 2004. Using historical documents for landslide. Debris flow and stream flood prevention. Applications in Northern Italy. *Natural Hazards*, **31**(3), 663–679.
Umitsu, M., Hiramatsu, T., and Tanavud, C., 2006. Research on the flood and micro landforms of the Hat Yai plain, southern Thailand with SRTM data and GIS. *Transaction, Japanese Geomorphological Union*, **27**(2), 205–219.
Van Den Eeckhaut, M., Poesen, J., Verstraeten, G., Vanacker, V., Nyssen, J., Moeyersons, J., Van Beek, L. P. H., and Vandekerckhove, L., 2007. Use of LIDAR-derived images for mapping old landslides under forest. *Earth Surface Processes and Landforms*, **32**, 754–769.
Vicente-Serrano, S. M., 2007. Evaluating the impact of drought using remote sensing in a mediterranean, semi-arid region. *Natural Hazards*, **40**, 173–208, doi:10.1007/s11069-006-0009-7.
Wang, Y., Colby, J. D., and Mulcahy, K. A., 2002. An efficient method for mapping flood extent in a coastal flood using Landsat TM and DEM data. *International Journal of Remote Sensing*, **23**(18), 3681–3696.
Wilford, D. J., Sakals, M. E., Innes, J. L., Sidle, R. C., and Bergerud, W. A., 2004. Recognition of debris flow, debris flood and flood hazard through watershed morphometrics. *Landslide*, **1**, 61–66.
Xia, J., Falconer, R. A., Lin, B., and Tan, G., 2011. Numerical assessment of flood hazard risk to people and vehicles in flash floods. *Environmental Modelling and Software*, **26**(8), 987–998.
Yang, B., Madden, M., Kim, J., and Jordan, T. R., 2012. Geospatial analysis of barrier island beach availability to tourists. *Tourism Management*, **33**(4), 840–854.
Zerger, A., 2002. Examining GIS decision utility for natural hazard risk modelling. *Environmental Modelling and Software*, **17**, 287–294.
Zolfaghari, M. R., 2009. Use of raster-based data layers to model spatial variation of seismotectonic data in probabilistic seismic hazard assessment. *Computers and Geosciences*, **35**, 1460–1469.

Cross-references

Coastal Erosion
Debris Flow
Disaster Risk Management
Disease
Earthquake
Emergency Management
Flood Hazard and Disaster
Geographical Information System
Hazard and Risk Mapping
Hydrometeorological Hazards
Lahar
Landslide (Mass Movement)
Landslide Inventory
Models of Hazard and Disaster
Natural Hazard
Remote Sensing of Natural Hazards and Disasters
Risk Assessment
Risk Assessment Remote Sensing of Natural Hazards and Disasters
Tsunami
Volcanoes and Volcanic Eruptions
Zoning

GEOGRAPHIC INFORMATION TECHNOLOGY

Brigitte Leblon
University of New Brunswick, Fredericton, NB, Canada

Synonyms

Environmental Information System; Geographic Information System (GIS); Geographically Referenced Information System (GIRS); Geo-Information System; Land Information System (LIS); Land Management System; Land Resources Information System; Spatial Information System (SIS)

Definition

Geographic information technology uses computer-based tools to analyze spatial information into a geographic information system (GIS). In a GIS, data of the real world are stored into a georeferenced database, which can be displayed via maps. There is a dynamic link between the displayed maps and the stored georeferenced data, that is, a change on the maps leads to a change in the database and vice versa.

Use of geographic information technologies

There are a number of applications that use geographic information technologies in numerous economical sectors, such as transportation, medicine, agriculture, mineral exploration, forestry, and governance. Geographic information technologies are used in natural resource management, environmental assessment and planning, demographical studies, land use management and planning, emergency management, transportation management, logistics, and so on.

Geographic information system components

A GIS is mainly used for storing, accessing, manipulating, transforming, analyzing, and displaying spatial information (georeferenced data). It can create new spatial information from those stored in the database. GIS outputs are maps, tables, or charts. A GIS allows study of spatial relationships, patterns, and trends among data, for example, searching for the best soil type for a given crop. Real world objects, such as streets, streams, cities, forest stands, and others, are defined in a GIS using three main characteristics:

1. Geographic features, which define geometric shapes and geographic positions of world objects. This characteristic is used in the map display.
2. Attributes, which are variables or parameters describing each geographic feature. They are stored into an attribute table.

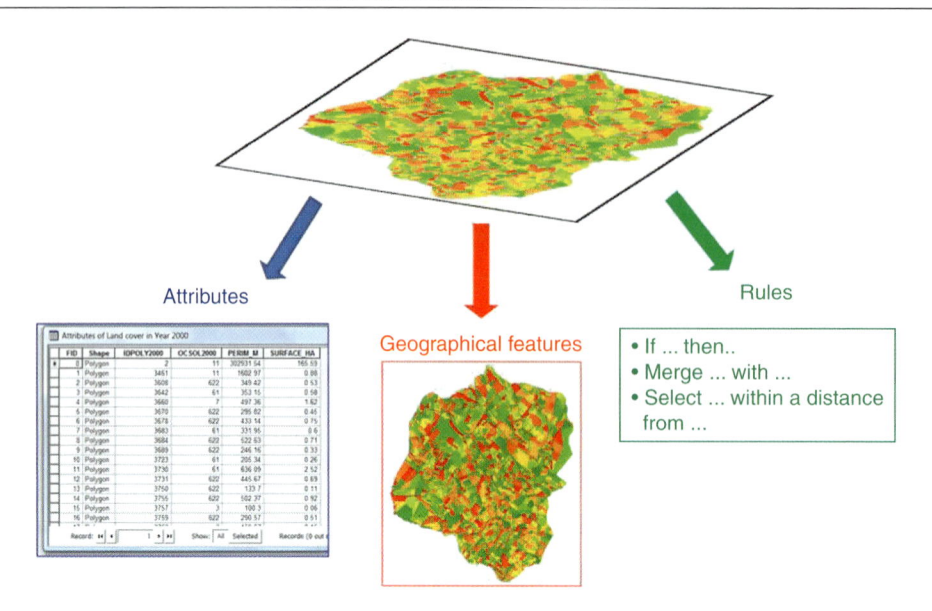

Geographic Information Technology, Figure 1 Three characteristics used in GIS to represent real world objects.

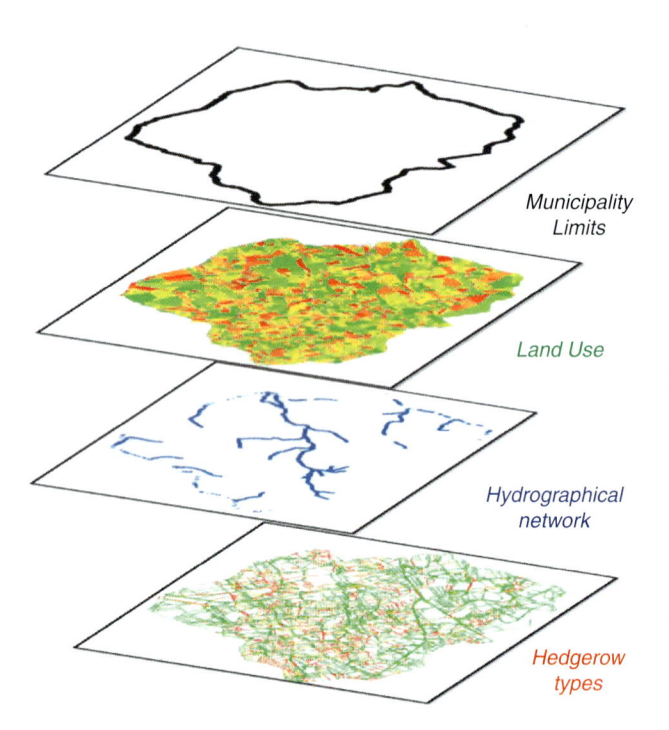

How spatial information is stored in a GIS database?

In a GIS database, spatial information from the real world is stored as thematic layers that are geographically linked. Each layer has features with similar attributes. The following GIS database has as layers the municipality limits, the land use, the hydrographical network, and the hedgerow type.

Data in a GIS database can be of two main formats: vector and raster. Vector data are either point layers (made of punctual data), line layers (made of linear data like the above hydrographical network layer), and polygon layers (made of data having an area like the above land use layer). Raster (or matrix-type) data are either digital images or grid data (Figure 2).

Bibliography

Leblon, B., and LaRocque, A. ENR 4286 geographic information system, online course 8 chapters, http://learning.unb.ca.

Geographic Information Technology, Figure 2 Thematic layers used to store specific spatial information in a GIS database.

3. Rules, which allow analyzing or displaying spatial data.

In the illustration example, agricultural plots are represented as polygons. Each plot is described by attributes, that is, municipality where it is located, crop type in the year 2000, and plot area. Rules such as "*locate the corn plots that are close to wheat plots*" can be defined (Figure 1).

Cross-references

Communicating Emergency Information
Early Warning Systems
Education and Training for Emergency Preparedness
Emergency Mapping
Geographic Information Systems (GIS)
Global Positioning Systems (GPS)
Hazard and Risk Mapping
Information and Communications Technology
Integrated Emergency Management System
Land-Use Planning
Planning Measures and Political Aspects
Zoning

GEOHAZARDS

Blaž Komac, Matija Zorn
Anton Melik Geographical Institute, Ljubljana, Slovenia

Synonyms

Geological Hazards; Geomorphological Hazards

Definition

Geohazard is a relatively new scientific term related to *Natural Hazard* studies. It indicates geomorphological, geological, or environmental processes, phenomena, and conditions that are potentially dangerous or pose a level of threat to human life, health, and property, or to the environment.

Discussion

Geohazards include subaerial and submarine processes, such as *Earthquake*, volcanic eruptions, floods, erosion, debris flows, rockfalls, and other types of *Landslide* and *Tsunami*. Human-induced processes may also be considered as geohazards.

Since the term *Hazard* represents a particular state of the geomorphic system that may develop further into a situation leading to damage or uncontrolled risk, studies in this field treat the changes in nature that may affect society. Geohazards research may be considered a fully geographical topic, since it includes both natural and social dimensions and is related to the complex of activities related to assessment, prevention, and mitigation, including the adjustments of society to natural hazards.

From the geomorphological perspective, innumerable studies on natural systems, their processes, and phenomena have been elaborated until now. They have focused on the geomorphic processes, dynamics, space, and timescales with special regard to frequency and magnitude. Statistical analysis of the geomorphic processes and evolution of the geographic information systems make it possible to predict the occurrence of certain natural events. Geomorphologists have increasingly become aware of the societal dimensions of the problem by implementing vulnerability analyzes and hazard and risk assessment.

The majority of countries have evolved several complex physical, social, and normative responses to the most frequent natural processes, raising their resilience and thus enabling them to absorb the effects of potential geohazards. In many cases, the effects of geohazards are multiplied by the results of deforestation and unsuitable land-use practices, often related to human activities. This is one of the reasons why the European Commission has lately put more emphasis on social aspects of geohazards. In response to the increasing threat of geohazards to global society, several international organizations and initiatives have been struggling to cope with geohazards; among them are the United Nations International Strategy for Disaster Reduction (UN ISDR), the EUR-OPA Major Hazards Agreement, GeoHazards International, and several scientific organizations, such as the Japan Landslide Society, International Centre for Geohazards, and the Platform on Natural Hazards of the Alpine Convention – PLANALP.

Bibliography

Alcántra-Ayala, I., and Goudie, A. (eds.), 2010. *Geomorphological Hazards and Disaster Prevention*. Cambridge: Cambridge University Press.
Bell, F. G., 2003. *Geological Hazards: Their Assessment, Avoidance and Mitigation*. Oxford: Taylor & Francis.
Coch, N. K., 1995. *Geohazards: Natural and Human*. Englewood Cliffs: Prentice Hall.
Embleton, C., Federici, P. R., and Rodolfi, G., 2010. Geomorphological Hazards. *Supplementi di Geografia Fisica e Dinamica Quaternaria, 2*. Torino: Comitato Glaciologico Italiano.
MacCall, G. J. H., Laming, D., and Scott, S. (eds.), 1992. *Geohazards: Natural and Man-Made*. London: Chapman & Hall.

Cross-references

Disaster
Earthquake
Geological/Geophysical Disasters
Hazard
Landslide
Natural Hazard
Tsunami
Volcanoes and Volcanic Eruptions

GEOLOGICAL/GEOPHYSICAL DISASTERS

Richard Guthrie
MDH Engineered Solutions, SNC Lavalin Group, Calgary, AB, Canada

Definition

Disaster

A disaster is an event that causes hardship, loss of life, or damage to infrastructure, the environment, the economic well-being, or other things that humans value. It is a cultural construct that refers to the relative impact of a hazard. It is important to recognize that a disaster would not exist without some direct impact on humans. A volcanic event, for instance, is an occurrence of a volcano. A volcanic hazard is the threat to humans from impacts of this event. A volcanic disaster is the negative impact of a volcanic hazard on things humans value. Disasters are therefore heavily influenced by a community's vulnerability to a particular hazard or group of hazards.

As a cultural construct, a disaster may be perceived differently by individual persons or groups of people, and there is no absolute disaster scale. The magnitude of a disaster may not be related to the magnitude of a hazard, highlighting, once again, the important role of vulnerability by humans and human communities.

A large magnitude earthquake in a remote region, for instance, may have no impact to humans and be considered an event, but a more moderate earthquake in a vulnerable and heavily populated city that results in widespread loss of life is very likely to be considered a disaster. A disaster typically implies that external assistance is required to facilitate recovery or that the effects of a hazard will be particularly persistent.

Disaster research is a relatively new but rapidly growing science with professionals publishing in several key scientific journals and research that spans almost all types of human endeavor. On December 11, 1987, the 42nd session of the General Assembly of the United Nations declared the International Decade for Natural Disaster Reduction (IDNDR) from 1990 to 1999 inclusive. In 1994, the Yokohama strategy was declared, and in the resolution, it was stated, "Natural disasters continue to strike and increase in magnitude, complexity, frequency and economic impact. While the natural phenomena causing disasters are in most cases beyond human control, vulnerability is generally a result of human activity. Therefore, society must recognize and strengthen traditional methods and explore new ways to live with such risk, and take urgent actions to prevent as well as to reduce the effects of such disasters. The capacities to do so are available" (ISDR, 2004, p. 9).

The IDNDR ended at the turn of the twenty-first century, however, buoyed by its success in engaging both scientists and policy makers toward reducing disasters and recognizing that the need for disaster reduction had arguably grown, the United Nations created the International Strategy for Disaster Reduction (ISDR), an institution that is still active today.

The IDSR definition of disaster states, "A serious disruption of the functioning of a community or a society causing widespread human, material, economic or environmental losses which exceed the ability of the affected community or society to cope using its own resources" (ISDR, 2004, p. 17).

In 1988, the World Health Organization (WHO) collaborated with the Belgian government to form the Centre for Research on the Epidemiology of Disasters (CRED). CRED collects and maintains EM-DAT (a database of worldwide disasters), which, in addition to the above criteria, includes any of the following situations:

- Ten or more people are killed.
- One hundred or more people are affected.
- A state of emergency is declared.
- A call for international assistance is made.

Geological/geophysical disasters

Geological disasters are those destructive events that originate within or are caused by the processes of the earth. Geological disasters include:

- Earthquakes
- Tsunamis
- Volcanoes
- Landslides and other mass movements
- Floods (including glacial lake outburst floods and landslide dam failures)
- Subsidence

Geophysical disasters include a slightly broader range of events that involve the physics of the earth that function principally in five areas:

- Electricity
- Gravity
- Magnetic
- Seismic
- Thermal

Therefore, geophysical disasters include all the geological disasters as well as:

- Cyclones
- Lightning storms
- Drought
- Ice storms
- Fire
- Floods (due to high runoff)

The latter categories of geophysical disasters are typically considered climatological or hydrometeorological disasters. Climatological and hydrometeorological events are both subcategories of geophysical events; however, they are not always recognized as such and may instead be considered categories in their own right.

There is, in addition, a class of geological/geophysical hazards for which the potential for geological/geophysical disasters is sufficient that it bears mentioning. These hazards have not yet occurred in modern human history, but have occurred repeatedly in geological time and include the terrestrial collision of near-Earth objects (NEOs) such as asteroids, volcanoes of sufficient explosive capacity to be popularly called "supervolcanoes," and magnetic pole reversals.

The global occurrence of geological/geophysical disasters

Geological and geophysical disasters affect every continent substantially populated by humans. History contains a long record of geophysical disasters that both capture our imagination and give us pause: the 1556 Shaanxi earthquake in China that killed 830,000, the 1755 tsunami in Portugal that killed 60,000, the explosion of Vesuvius in 79 AD that buried Pompeii and Herculaneum, and the landslide in Tajikistan in 1949 that killed 12,000. Disaster reporting has increased dramatically in recent decades (Figure 1) as a result of increased awareness and technology, as well as increased populations, for some, increasing their vulnerability. In contrast, however, the trend for numbers of deaths reported in disasters has remained steady over the same time period (Figure 2), suggesting that major events were well recorded historically as they are now. Variability in the reporting of geophysical disasters appears to have diminished since the latter half of the

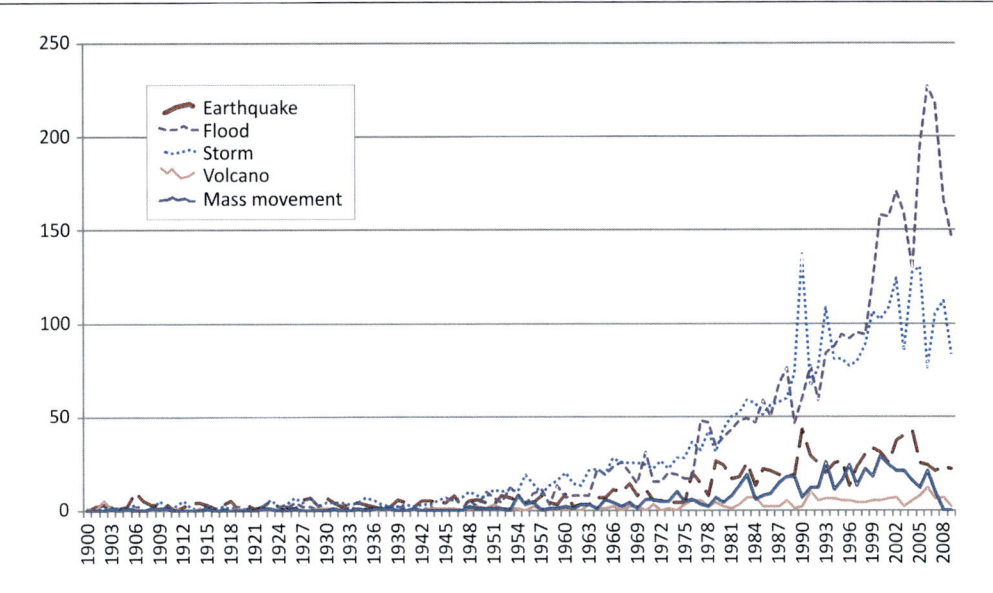

Geological/Geophysical Disasters, Figure 1 The number of geophysical disasters reported per year over the twentieth century (Data from EM-DAT, 2010).

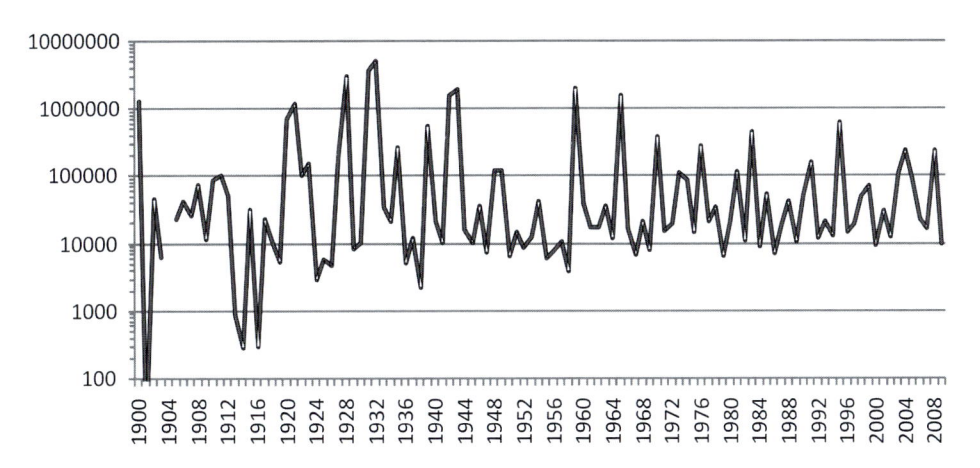

Geological/Geophysical Disasters, Figure 2 The number of reported deaths due to geological and geophysical disasters worldwide 1900–2009 (Data from EM-DAT, 2010).

twentieth century, reflecting perhaps more accurate characterization of geophysical disasters and more consistent reporting of "smaller" events. Problems of reporting remain. If the cost of damage in terms of dollars is considered a benchmark, then developed industrialized countries tend to be the most heavily impacted by geological disasters (Table 1 and Figure 3). However, if one reports the cost in terms of displaced persons or number of people affected, geological disasters are most severe in densely populated developing countries. An argument could be made that the fundamental measure of quality of life begins with life itself. While it may overlook millions of affected individuals, loss of life is a relatively straightforward measure of disaster magnitude. Once again, the impacts are most severe in developing countries that have close proximity to abundant geophysical hazards (Table 1 and Figure 4). The countries located in an area, known as the Pacific or Asian "Ring of Fire," are particularly vulnerable to the full suite of geophysical hazards. These areas are affected by the tectonic and volcanic events occurring along active plate margins, landslides in steep terrain among the highest mountains in the world, as well as cyclone and high temperatures and humidity characteristic of the tropical location.

Earthquakes

Earthquakes are caused by the quick release of stored potential energy into kinetic energy of motion in the form of seismic waves and the propagation of fractures in the

Geological/Geophysical Disasters, Table 1 Top 20 countries affected geophysical/geological disasters for the cumulative period 1900–2009 by cost (in thousands of US$) and by number of individuals killed or presumed dead (EM-DAT, 2010). Highlighted countries are on both lists

Country	Cost (millions of US$)	Country	Number killed (thousands of deaths)
United States	558,359	China	11,149
China	323,928	Soviet Union	6,368
Japan	214,259	India	4,570
Italy	67,036	Bangladesh	2,589
India	48,781	Korea	612
Korea	38,847	Ethiopia	404
Germany	35,302	Indonesia	236
France	33,252	Japan	227
Australia	29,294	Pakistan	170
United Kingdom	29,286	Iran	155
Spain	26,590	Sudan	151
Mexico	26,551	Myanmar	146
Turkey	25,166	Italy	140
Indonesia	23,553	Mozambique	103
Iran	21,500	Turkey	91
Soviet Union	20,713	Niger	85
Taiwan	19,695	Cape Verde Is	85
Bangladesh	18,171	Peru	84
Canada	17,218	Guatemala	83
Brazil	12,677	Chile	61

Earth's crust. Tectonic earthquakes (the usual case) are the release of elastic strain along plate boundaries, in turn caused by frictional resistance between plates as they slide against one another. The three main fault types associated with earthquakes are strike-slip (two plates slide horizontally against one another along the fault plane), thrust (crustal shortening that causes one plate to be pushed over another along the fault plane), and normal (crustal lengthening that causes one plate to slide against another as it drops down along the fault plane). These faults occur along transform, convergent, and divergent plate boundaries, respectively, and often, a combination of vertical (normal or thrust) and horizontal (strike-slip) movement results in oblique faults. Crustal deformations are ubiquitous around the globe, and small earthquakes are almost constant, though often go unnoticed.

Earthquakes may also be caused by man-made explosions, volcanic activity, landslides and outburst floods, and mining practices. Many of the natural hazards (and subsequent disasters) are themselves related to earthquake activity including landslides and tsunamis.

The magnitude of earthquakes is measured by the moment magnitude (M_W) scale (Hanks and Kanamori, 1979), a logarithmic scale that describes the amount of slip along a fault and the area of fault that slipped, which is a successor to the earlier Gutenberg-Richter scale (1954, 1956). In the moment magnitude scale, each increasing whole number represents an earthquake amplitude ten times greater than the previous number (M_W of 7 is 100 times greater than M_W of 5) and occurs ten times less frequently. The energy released during an earthquake, however, is about 32 times the earthquake amplitude, so an earthquake of M_W 7 releases about 1,000 times more energy than an earthquake of M_W 5. It is the energy release that ultimately causes damage and results in disasters.

Globally, the most frequent earthquakes occur along convergent boundaries of the Pacific Plate in what is known as the Pacific "Ring of Fire" or the Asian "Ring of Fire" along convergent and transform boundaries of the Eurasian Plate and along the Nazca Plate as it collides with South America (Figure 5). Not surprisingly, the greatest losses of life from earthquake disasters have been in countries of Eastern Asia, followed closely by countries of South America (Table 2, Figure 6).

The deadliest earthquake on record occurred on January 23, 1556, in the Shaanxi province of China. An estimated 830,000 people died in one of the largest geophysical disasters ever recorded. The earthquake, an estimated M_W 8, was particularly devastating because a large percentage of the population lived in "yaodongs" or caves carved into cliffs composed of loess. Loess is wind-transported silt that is very unstable and may collapse when wetted or seismically shaken (e.g., Derbyshire, 2001). In this province, the loess cliffs collapsed or were buried in landslides, killing up to 60% of the population (cultural China website, 2010). In this same province of China, there have been at least four great earthquakes in the last 10,000 years, including one that occurred in 1,303 killing an estimated 270,000–470,000 people (Guogan et al., 1984).

Earthquakes remain a significant problem in China, and the country has experienced the greatest loss of life as compared to other countries in the twentieth century (Table 2). However, earthquake disasters occur elsewhere in the world, including the recent Haitian, New Zealand, and Japanese earthquakes.

On January 12, 2010, an earthquake of M_W 7 occurred 25 km west of the capital of Haiti along a strike-slip fault and caused widespread building collapse (Figure 7), the deaths of approximately 230,000 people, and the displacement of over 1,000,000 residents (USGS, 2010a). The financial cost of the disaster may reach as much as 14 billion dollars (IDB, 2010).

The largest magnitude earthquake recorded globally was a Chilean earthquake of magnitude M_W 9.5 that occurred in Chile in 1960. This earthquake caused a transpacific tidal wave that eventually reached the island of Hilo, Hawaii (USGS, 2010b).

Tsunamis

A tsunami is a wave or a series of waves that form from the displacement of a large body of water. The term, literally meaning "harbor wave," originated in Japan where over 340 tsunamis have been recorded since 684 AD

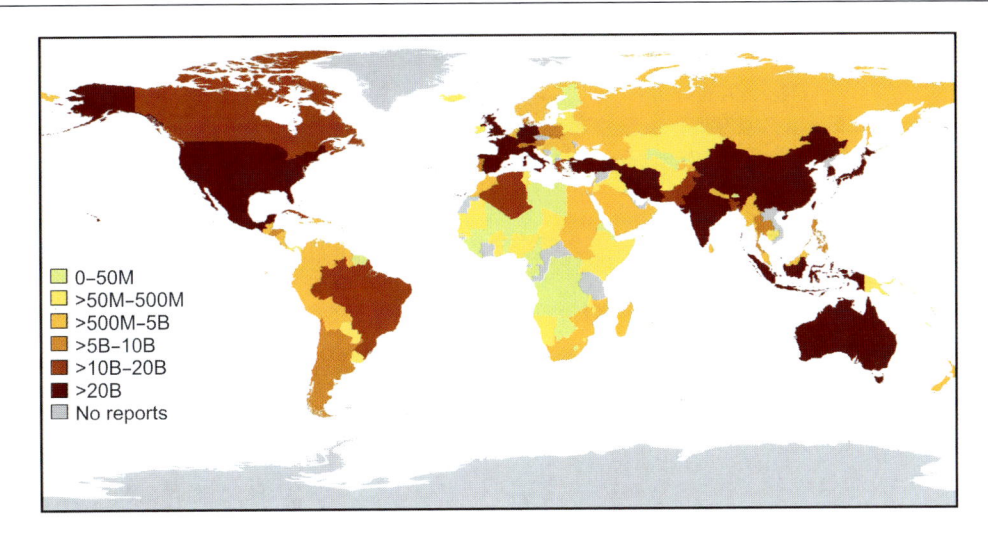

Geological/Geophysical Disasters, Figure 3 The total cost of geophysical disasters reported over the period 1900–2009 by country (Data from EM-DAT, 2010).

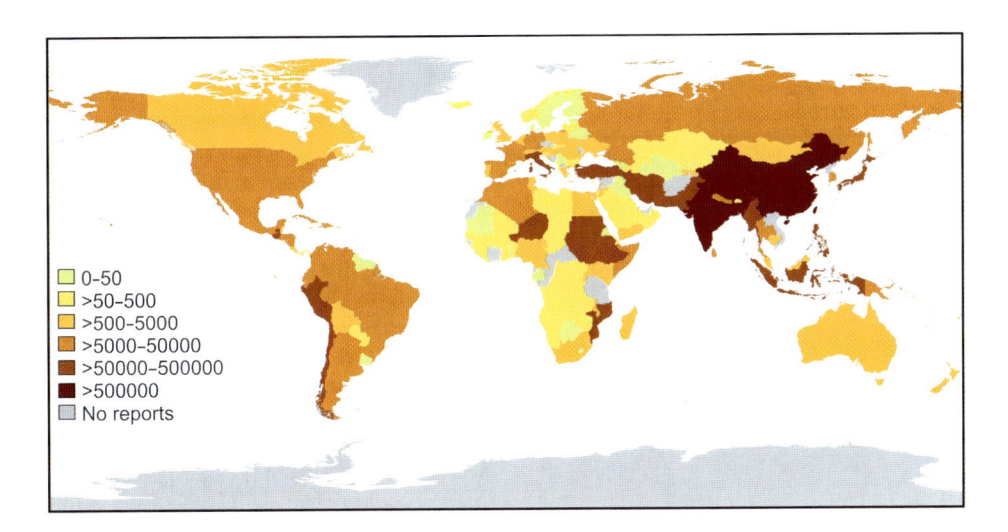

Geological/Geophysical Disasters, Figure 4 The global distribution of geophysical disasters reported for the years 1900–2010. *Colors* represent the number of deaths by country for the entire period. The greatest numbers of deaths (over 11 million) last century were in China, due in a large part to flooding, earthquakes, and drought (Data from EM-DAT, 2010).

(NGDC, 2010). Other words to describe tsunamis have included tidal waves. However, this terminology is now discouraged as tides are predictable and driven by the gravitational attraction of the sun and the moon. While both tides and tsunamis can produce big waves, tsunamis have much greater potential for sudden and unexpected damage.

Tsunamis are caused by the displacement of a large body of water, which then travels outward from the center of the generating event like ripples in a pond. The most common source of a tsunami is sudden vertical movement in the Earth's crust along a convergent plate boundary, which, in turn, displaces a large part of the overlying water column in the ocean. Subduction zones, where oceanic plates are being pushed below an overriding continental plate, are the primary tectonic locations that generate tsunamis, and consequently, tsunamis are most common along plate boundaries rimming the Pacific Ocean. The displaced column of water may travel away from the source of seismic activity at speeds in excess of 1,000 km/h (NGDC, 2010). In the deep ocean, tsunami wavelengths may be hundreds of kilometers long, and their height may only be a few meters, making them hard to detect. However, as they approach land, shallower water slows the waves down, compressing them dramatically and causing a marked increase in amplitude

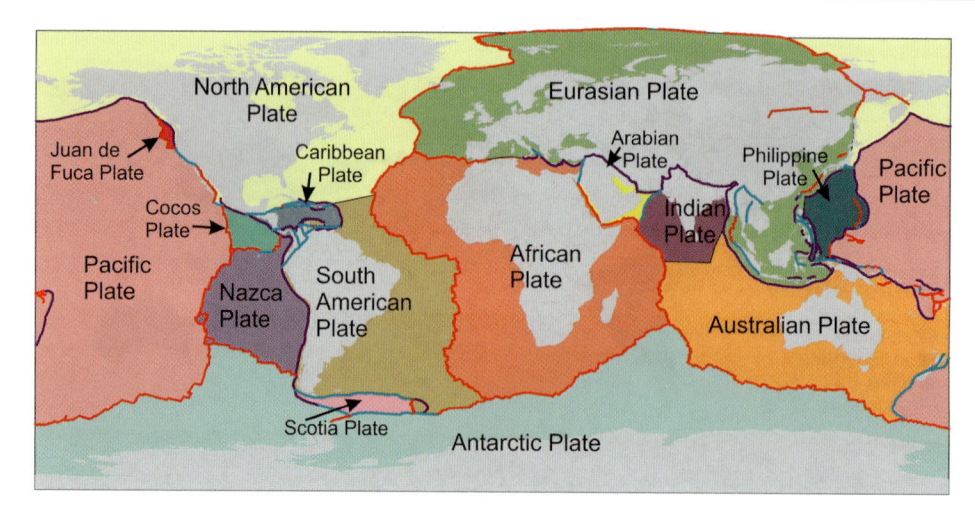

Geological/Geophysical Disasters, Figure 5 The global distribution of major plates. Convergent boundaries are indicated by *purple lines*, transform (strike-slip) boundaries by *blue*, and divergent boundaries by *red*. The mid-Atlantic ridge is a divergent boundary, and few earthquakes occur there as a result. In contrast, the convergent margins of the Pacific Plate as it collide with Asia produce significant earthquakes on a regular basis.

Geological/Geophysical Disasters, Table 2 The top 20 cumulative deaths from earthquakes reported by country for the period 1901–2009 (EM-DAT, 2010)

Country	Number killed
China	872,366
Indonesia	197,847
Japan	172,637
Former Soviet Union	155,400
Iran	147,106
Pakistan	142,978
Italy	115,621
Turkey	88,538
India	78,094
Peru	70,709
Chile	58,955
Sri Lanka	35,399
Guatemala	27,697
Taiwan	15,801
Morocco	12,728
Nicaragua	12,686
Ecuador	11,336
Mexico	10,677
Argentina	10,076
Nepal	9,929

(wave height). Tsunamis may also be generated by earthquakes, volcanoes, underwater explosions, and landslides.

Though approximately ten tsunamis occur each year, only one, on average, takes human lives (WHOI, 2010). The first recorded tsunami disaster was in Syria in 2000 BC and caused an estimated 100–1,000 deaths (the median number for this category of damage is 300 based on other events worldwide). Most of the early tsunami records have come from the Mediterranean countries. However, with better recording systems developed over the last several decades, more accurate information has been collected. The countries hardest hit by tsunami disasters are in the Asian Pacific region and countries along the west coast of South America, specifically Chile and Peru (Table 3 and Figure 8).

One of the worst geophysical disasters globally occurred on December 26, 2004, a result of the Sumatra-Andaman megathrust (Mw 9.2) earthquake between the subducting oceanic Indian Plate and the thicker Burma microplate. The earthquake released the energy equivalent of 100 gigatons of dynamite and created a sudden vertical displacement of 4–5 m, moving an estimated 30 km^3 of water and creating devastating waves up to almost 50 m high (Kawata et al., 2007). Nearly 230,000 lives were lost with the greatest devastation occurring along the shores of Indonesia, Sri Lanka, India, Thailand, and the Maldives (Figure 9).

The tallest tsunami wave ever recorded was in 1958 in Lituya Bay, Alaska, by a landslide-generated tsunami. The tsunami removed forest and soil from the bay to a height of 524 m above sea level (for comparison, the Eiffel Tower in Paris is 300.5 m high and the Petronas Twin Towers in Malaysia are 452 m high).

Other notable tsunamis include the wave that devastated Lisbon, Portugal, on All Saints Day in 1755, the 1960 Chilean tsunami that was generated by the greatest earthquake ever recorded (M_W 9.5) causing a transpacific wave of water that killed people as far away as the island of Hilo, Hawaii (USGS, 2010b), and the most recent 2011 Tohoku-Oki earthquake and tsunami events in Japan that displaced over 400,000 residents and reconstruction is estimated to be several hundred billions of dollars.

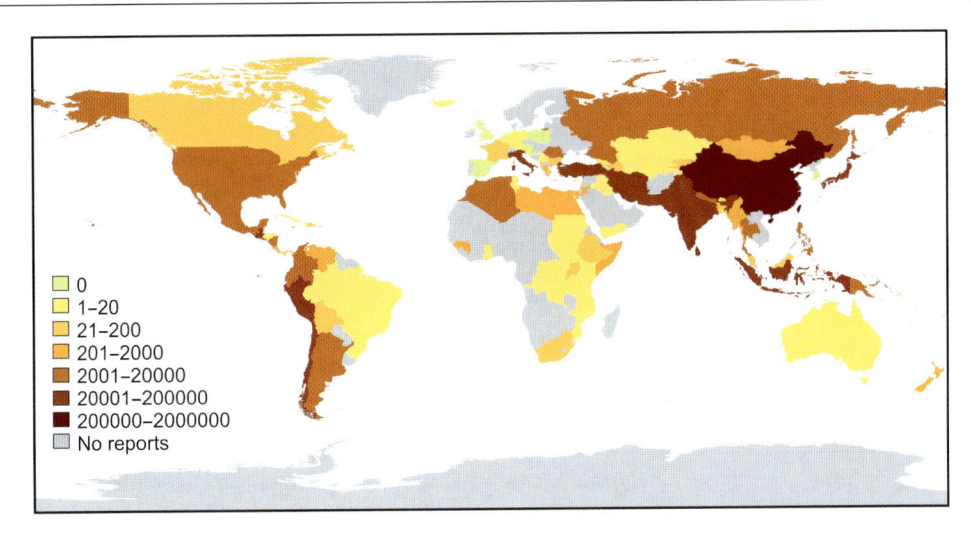

Geological/Geophysical Disasters, Figure 6 The global distribution of earthquake-related disasters due to earthquakes reported for the period 1901–2009. Colors represent the number of deaths by country for the entire period (Data from EM-DAT, 2010).

Geological/Geophysical Disasters, Figure 7 Collapsed buildings in Port-au-Prince, the capital of Haiti, following the January 2010 earthquake (Public domain photograph by Marco Dormino).

Volcanoes

Volcanoes are the extrusion of molten rock, gas, and ash from within the Earth's crust onto the land surface. Volcanoes typically occur along plate margins at convergent and divergent boundaries (Figure 10). Consequently, volcanoes are particularly active along subducting portions of the plates and especially along the Pacific "Ring of Fire." Volcanoes may erupt explosively, effusively, as flows, or as clouds of superheated or poisonous gases; all the eruption styles can pose a threat to humans. Explosive volcanoes are particularly troublesome, and the volcanic explosivity index (VEI), a logarithmic index that describes the amount of material ejected into the atmosphere, the height of the explosion, and the qualitative size of the event, was developed to compare volcanoes to one another (Newhall and Self, 1982).

Volcanic disasters are associated with humans and human infrastructure that are in the path of extruded lava and gases, particularly during explosive events from stratovolcanoes or with lahars, volcanic mudflows, which are generated by the volcanic activity. The top ten volcanic disasters that have occurred globally in the last century (1901–2009) are listed in Table 4. Only the Columbia disaster of 1985 is listed near the top of both lists (most expensive and most deadly). The reporting of cost from the last half of the twentieth century compared to lives lost over the first half of the century reveals that volcanoes are somewhat predictable and that loss of life is largely avoidable.

The deadliest and most expensive disaster in the United States (VEI 5) was the 1980 eruption of Mount St. Helens (Figure 11), a stratovolcano that lies along the eastern margin of the Pacific Ring of Fire along the Cascadia subduction zone. Mount St. Helens ejected about 1 km^3 of volcanic material into the atmosphere. While 57 people were killed, many more were evacuated safely from the area. Although the volcanic eruption was predicted, both the blast impact and subsequent debris flows from Mount St. Helens travelled farther than predicted.

History is replete with famous volcanic disasters, including the 79 AD eruption of Mount Vesuvius (VEI 5) in Italy; ash and pumice from the volcano buried the cities of Pompeii and Herculaneum killing more than 10,000. Today, about 3,000,000 people live in the shadow of Vesuvius, and it is considered to be one of the most dangerous volcanoes on the planet.

Krakatau was a stratovolcano that erupted in 1883 expelling 21 km^3 of volcanic material (VEI 6), killing approximately 36,417 people. The explosion was the most violent in recorded history and was heard about 3,500 km away.

Mount Pinatubo erupted in 1991 in the Philippines, ejecting 10 km^3 of magma and 20 million tons of sulfur dioxides into the atmosphere thus making it the second biggest eruption in the twentieth century (VEI 6). The eruption caused the deaths of more than 350 people (landslides and disease increased the death toll to 722) and was responsible for a maximum temporary cooling of global temperatures of about 1.5 °C (NOAA, 2010).

Another class of volcanoes exists that have the potential to cause a geological disaster of enormous magnitude. Supervolcanoes are those that eject over 1,000 km^3 of material (VEI 8) into the atmosphere with such enormous

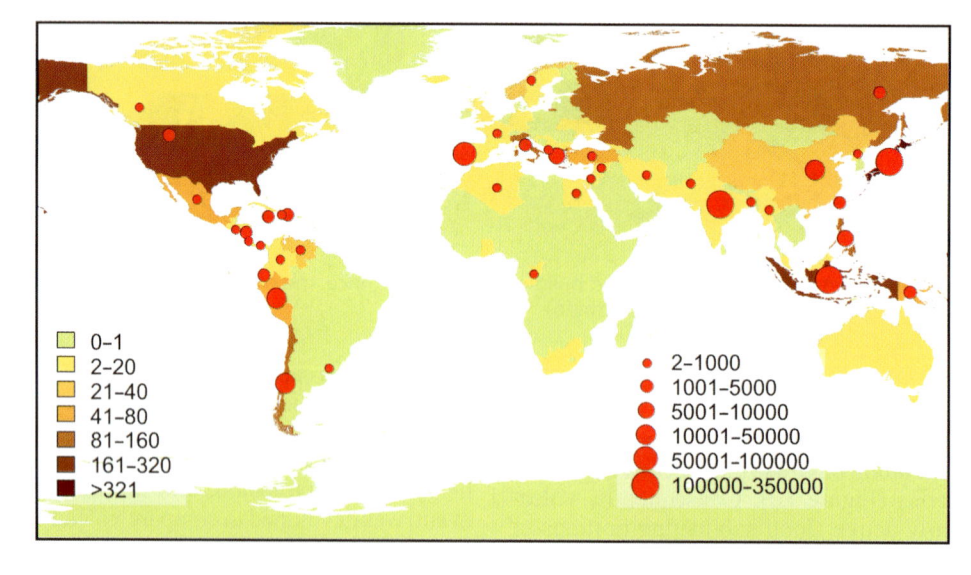

Geological/Geophysical Disasters, Table 3 The 20 most severe tsunami disasters by country and year recorded for the period between 2000 BC and 2010 (NGDC, 2010)

Country	Year	Number killed
India	1737	300,000
Indonesia	2004	227,898
Portugal	1755	60,000
Indonesia	1883	36,000
Japan	1498	31,000
Japan	1707	30,000
Japan	1896	27,122
Chile	1868	25,000
Japan	1771	13,486
China	1765	10,000
Japan	1586	8,000
Greece	365	5,700
Japan	1703	5,233
Japan	1605	5,000
Japan	1611	5,000
Peru	1687	5,000
Peru	1746	4,800
Philippines	1976	4,456
Japan	1792	4,300
Japan	1512	3,700

Geological/Geophysical Disasters, Figure 8 The global distribution of tsunamis reported for the period 2000 BC to 2010. *Country colors* represent the number of reported tsunamis, while the *circles* represent lives lost by country (Data from NGDC, 2010).

Geological/Geophysical Disasters, Figure 9 Tsunami wave hitting Ao Nang, Thailand caused by the December 26 Sumatra-Andaman megathrust (M 9.2) earthquake (Public domain photograph by David Rydevik).

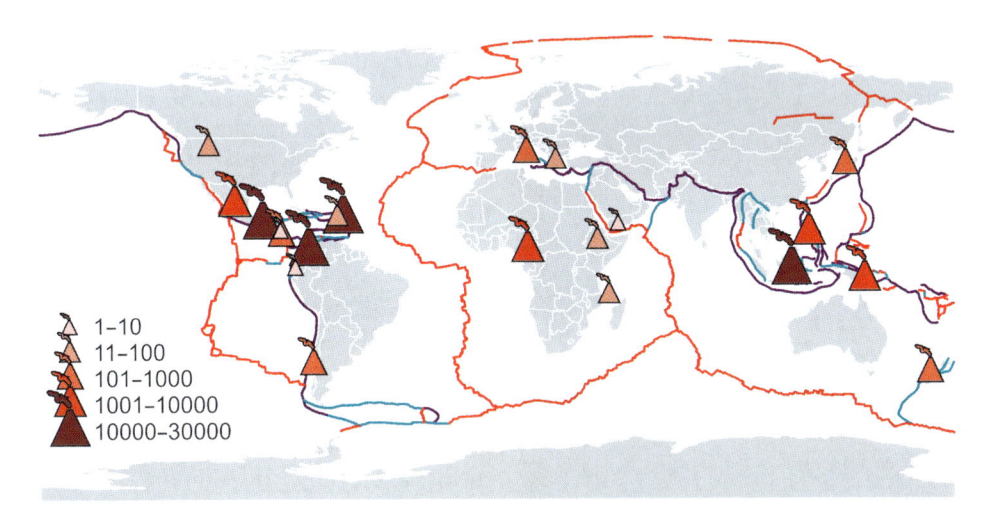

Geological/Geophysical Disasters, Figure 10 The global distribution of volcanic disasters from 1901 to 2009. Note that the volcanoes are largely concentrated along the convergent margins of the Pacific Plate (Data from EM-DAT, 2010).

impact that they change long-term weather patterns and cause extinctions of species. Examples of supervolcanoes include Lake Toba in Indonesia, Yellowstone in the United States, and Aira Caldera in Japan. Supervolcanoes often form as the result of rising magma beneath the crust that form hot spots, but initially are unable to reach the land surface. The underground pool of magma, therefore, continues to grow and increase pressure on the crust. Eruptions are infrequent; Yellowstone, for example, erupted 2.1 million, 1.3 million, and 640,000 years ago (Newhall and Dzurisin, 1988). However, the colossal size of such an event warrants recognition. Previous eruptions ejected between 1,000 and 2,500 km^3 of material into the air.

Landslides

Landslides are the downslope movement of rock, sediment, earth, and debris under the influence of gravity.

Geological/Geophysical Disasters, Table 4 The ten deadliest and ten most expensive volcanoes for the period between 1901 and 2009. Note that the most expensive volcanoes have all occurred since 1980 except for Japan (1945), whereas the deadliest volcanoes are mostly in the first half of the century. Only the Columbia disaster (1985) is on both lists. This suggests improvements in hazard zoning, awareness, and prediction, but at a substantial economic cost. Serious hazards remain, however, for dense settlements that are currently threatened by volcanic hazards

Country	Year	Deaths	Country	Year	Cost (000's of USD)
Martinique	1902	30,000	Colombia	1985	1,000,000
Colombia	1985	21,800	United States	1980	860,000
Guatemala	1902	6,000	Philippines	1991	211,000
Indonesia	1909	5,500	Indonesia	1982	160,000
Indonesia	1919	5,000	Ecuador	2006	150,000
Guatemala	1929	5,000	Indonesia	1983	149,690
Papua New Guinea	1951	3,000	Mexico	1982	117,000
Cameroon	1986	1,746	Papua New Guinea	1994	110,000
Indonesia	1963	1,584	Japan	1945	80,000
St. Vincent	1902	1,565	Indonesia	1983	25,500

Geological/Geophysical Disasters, Figure 11 The 1980 eruption of Mount St. Helens, the deadliest and most expensive volcano in the US history (Public domain photograph by Austin Post and the USGS).

They include a range of events including rock falls, debris flows, slumps, earth flows, and rock avalanches. Landslides may travel great distances at incredible speeds or more slowly along predictable paths. Landslide disasters have taken lives, caused untold injuries, and resulted in massive economic losses. Some records of landslide disasters date as far back as 373 BC (Seed, 1968) with the total loss of the Greek town of Helice. However, landslide disasters are notoriously underreported because such events frequently occur in conjunction with other natural hazards such as earthquakes, volcanoes, and floods. The deadly 1556 earthquake in Shaanxi province of China, for instance, had a massive death toll (~830,000 people), many a direct result from landslides. Similarly, volcanic debris flows may extend many kilometers beyond the eruption center of a volcano and threaten lives and property. The deaths in the Colombia volcanic disaster of 1985 (totally 21,800) were largely attributable to the volcanic debris flow associated with the event (Evans, 2006). Floods are especially linked to and often confused with landslides. High rainfall produces both floods and landslides, and they may grade into one another as part of a continuum depending on the concentration of sediment in the flow. In addition, many floods have been caused by the failure of a landslide-dammed lake, often to devastating effect. For example, in 1786, a landslide dam was breached on the Dadu River in the Sichuan province of China, causing flooding for 1,400 km downstream and the deaths of about 100,000 people (Schuster and Wieczorek, 2002).

Despite the problem of landslide reporting, global reports of landslide disasters do provide a sense of the distribution of landslide disasters in countries around the world (Figure 12). The distribution of landslide disasters relates to both the proximity of hazards, but also to the concentrations of people. The Canadian Cordillera, for example, has relatively high landslide hazards, but a low population density (~4 persons/km^2) and consequently

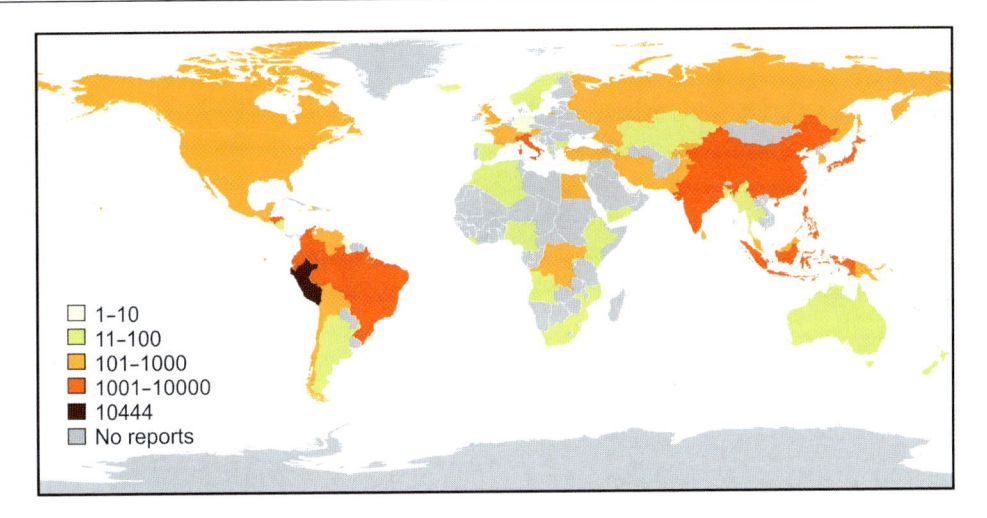

Geological/Geophysical Disasters, Figure 12 The cumulative global distribution of deaths by country for landslide disasters not otherwise recorded as another disaster type (flood, volcano, earthquake).

few disasters. In contrast, Peru has a much higher population density of 23 persons/km^2. Many people live at the base of the Andes Mountains, part of the South American Cordillera, and the populations are situated between the mountains and the ocean. Disasters are consequently more frequent and of considerably higher magnitude.

Landslide hazards may be substantially reduced or exacerbated by the activities of humans. Many landslides are readily monitored, hazard zonation is particularly effective, and there are numerous mitigation techniques that reduce the likelihood of a hazard becoming a disaster. Nevertheless, landslide disasters continue to occur.

One of the most famous landslide disasters of all time began high on Huascarán Mountain in the Yungay province of Peru in 1962 and again in 1970. Both events began as rock/ice falls and transformed into high-velocity (up to 85 m/s in 1970) debris flows by incorporating sediment and ice from the glacier and moraines below. About 7,000 people died (Evans et al., 2009a); earlier estimates were substantially higher. The two events and associated debris floods triggered by the 1970 landslide continued downslope for an additional 180 km.

On February 17, 2006, 1,221 people were killed and 19,000 displaced by a rockslide-debris avalanche (landslide) that buried the village of Guinsaugon in the Philippines (Figure 13). A rock mass slid down an 800-m-high escarpment (the surface expression of the slip-strike Philippine fault zone), entraining millions of cubic meters of colluvial debris that spread rapidly across rice paddies in the fields below. The total volume of debris was 15 Mm^3, and it was the deadliest single landslide in a decade (Guthrie et al., 2009). Landslides remain an ongoing hazard in the Philippines.

On July 10, 1949, a massive (245 Mm^3) loess flow (landslide) travelled through the Yasman valley in Tajikistan on slopes of only 2° following the M_W 7.4 Khait earthquake. The landslide travelled over 20 km and buried 20 villages in its path. The death toll from that single event was about 4,000 people, but landslides in adjacent valleys raised the regional magnitude of the disaster, increasing the death total to over 7,000 people, making this event one of the most devastating landslides of the twentieth century (Evans et al., 2009b).

Floods

Floods occur when land that is normally dry becomes saturated, submerged, and underwater. Floods are caused by rainfall (convective as well as cyclonic), snowmelt, tidal and storm surges, dam creation (natural and man-made), dike and dam breaches (both natural and man-made), and glacial lake outburst floods (GLOFs). Floods are one of the most widespread hazards, and flood disasters have affected nearly every country at one time or another (Figure 14). The Pacific Asian region is particularly vulnerable to flood disasters, both because of constant exposure to tropical cyclonic weather patterns and because of high population densities in flood-prone regions.

Flood disasters in China have been especially destructive; in the twentieth century alone, almost seven million residents have been reported killed by floods. In fact, the deadliest geophysical disaster on record is the 1931 flooding of the Yellow River, which occurred between July and November of that year, killing 3.7 million people (EM-DAT, 2010). China had repeated flood disasters in the twentieth century, including 1911 (100,000 killed), 1931 (3.7 million killed), 1933 (18,000 killed), 1935 (142,000 killed), 1939 (500,000 killed), 1949 (57,000 killed), 1954 (31,000 killed), and 1959 (two million killed). The 1959 flood, also along the Yellow River, coincided with experimental (and ultimately unsuccessful) changes in agricultural practices; in addition to the two million persons lost directly to flooding, the agricultural losses contributed to an ongoing famine that killed between 15 and 43 million

Geological/Geophysical Disasters, Figure 13 A cross marking the dead following the Guinsaugon landslide disaster on Leyte island in the Philippines (Photo by R.H. Guthrie).

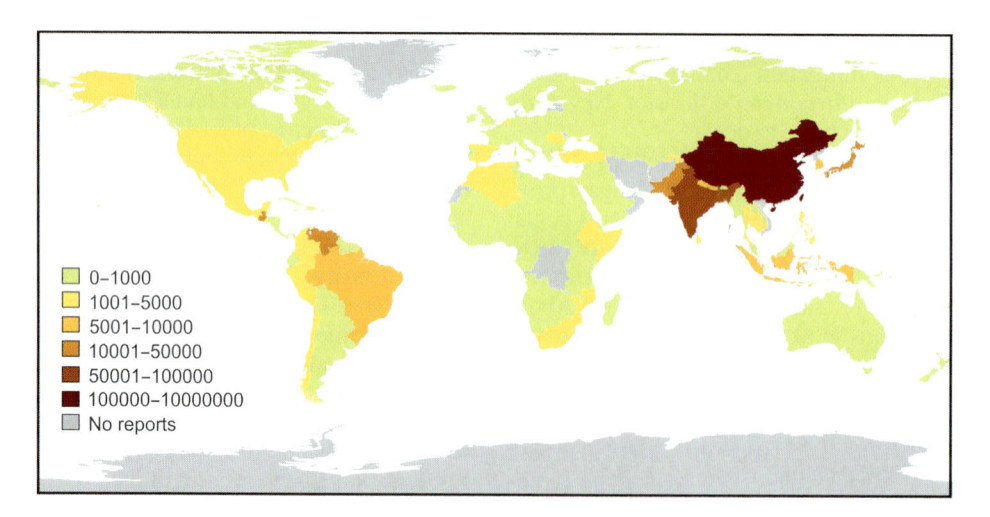

Geological/Geophysical Disasters, Figure 14 The total number of reported deaths due to flooding disasters worldwide (1900–2010) (Data from EM-DAT, 2010). Most countries are affected by some level of flood disasters. Canada, for example, reports 168 flood disasters between 1900 and 1997, killing at least 198 and costing several billion dollars (Brooks et al., 2001). However, the devastation due to floods in China is particularly severe. The total number of deaths reported for China for that period is about 6.7 million.

people. Such unintended consequences are not limited to developing countries during certain periods of history.

The 2005 flooding in New Orleans, United States was made worse by the catastrophic failure of a man-made dike, normally used to protect houses from just such an event. The total damage from Hurricane Katrina and subsequent floods was over 80 billion dollars, making it the most expensive disaster in the country to date.

Glacial lake outburst floods (GLOFs), also known as jokulhlaups, occur when there is a sudden release of water from beneath or behind a glacier. The outflow suddenly and often catastrophically can incorporate sediment and often become hyper-concentrated flows. In 1954, for example, a GLOF occurred in the Yarlung Zangbo basin of Tibet, releasing up to 10,000 m^3/s of water and damaging cities up to 200 km away (SAARC, 2010). As glaciers retreat worldwide, thousands of glacial lakes are forming and growing in size. The threat of GLOF disasters is increasing as a result.

Landslide dams create flooding in watersheds where the impound lakes and streams form, and when breached, they may fail catastrophically devastating everything in their path, similar to GLOFs. For example, in the United States in 1925, the 38,000,000 m^3 Gros Ventre landslide created a 60-m-high dam that backed up the Gros Ventre River and created a temporary lake. Two years later, a partial failure of the dam generated a flood that travelled for 40 km downstream, destroying the town of Kelly, 10 km along its path.

In 1786, in Sichuan, China, a landslide dam was breached on the Dadu River, flooding lands 1,400 km downstream and killing approximately 100,000 people.

At the time of writing, workers are trying to construct a spillway and mitigate the potential threats caused by a massive landslide that dammed the Hunza River in Pakistan. The current water level is 86 m above the valley floor and rising. Overtopping of dam is expected to occur on or about June 8, 2011. Downstream resources include a man-made dam and populated communities.

Subsidence

Subsidence is the lowering of land relative to some datum such as sea level, due to the removal of underlying support or by compression of the soil. Subsidence may produce an overall lowering (the sinking of Venice for example), or it may produce sudden catastrophic sinkholes. Sinkholes and subsidence may result from mining, the dissolution of limestone or underground salt, the removal of sediment by water, melting of permafrost, or any other mechanism that removes underlying support. Fortunately, there are relatively few instances of disasters related to ground subsidence.

In February 2007, a large (100 m deep) sinkhole appeared in Guatemala City, resulting in the evacuation of 1,000 people and the death of three people.

In the state of Florida, United States, public work officials struggle with sinkholes on a regular basis including a large (50 m deep) sinkhole that appeared beside a gypsum stack in 1994. Hazardous waste from the failure contaminated 90% of the drinking water in Florida.

In 1986, a sinkhole appeared in Berezniki, Russia at the location of 10% of the world's potash production. The sinkhole has been growing wider every year. The hole is currently 200 m deep, 80 m long by 40 m wide and threatens the nearby rail system.

Asteroids and near-earth objects

The understanding that humans are potentially threatened by comets and asteroids (near-Earth objects) careening through space is a fairly recent occurrence; however, it is not without precedent. In the geologic record, impacts of near-Earth objects (NEOs) have changed the global climate, food supply, and ultimately preceded global mass extinctions. The consequences are so potentially devastating that NASA maps all NEOs that approach the earth and ranks them accordingly to the Torino scale (a 1–10 logarithmic scale of probability and magnitude; Binzel, 2000). Should a NEO of a category 8 or higher be on a collision course with the Earth, it is likely that humans would use technology to intercept the object. In this manner, NEOs are the one potential devastation that is expected to be fully preventable.

Other disasters

There are other geophysical hazards that may lead to a disaster. In 1998, the North American ice storm killed 35 and cost several billion dollars in damages. Snow avalanches in 218 BC killed 20,000 soldiers and numerous elephants of Hannibal's army as they tried to cross the Alps. Lightning during electric storms kills people and starts fires that do millions of dollars of damage on a yearly basis. In 1281, 100,000 people died in a typhoon (cyclone) that inundated Japan. In March 2010, over 6,000,000 people were impacted by drought in Thailand.

Summary

Geophysical disasters will occur whenever humans and things that human's value are exposed to geophysical hazards. In the end, it is a matter of scale: Humans and human structures are small relative to the workings of the planet, and even small perturbations in the geophysical world can have devastating effects to humankind. However, humans have a responsibility for awareness, prevention, and mitigation. Much of the devastation from the world's most calamitous disasters is ultimately avoidable through good planning, engineering, avoidance, and improved geophysical science. As we move forward into the twenty-first century, humans will face new geophysical challenges and disasters will become more expensive. However, enacting procedures and policies to mitigate these disasters will ultimately save lives.

Bibliography

Binzel, R. P., 2000. The Torino impact hazard scale. *Planetary and Space Science*, **48**, 297–303.

Brooks, G. R., Evans, S. G., and Clague, J. J., 2001. Flooding. In Brooks, G. R. (ed.), *A Synthesis of Natural Geological Hazards in Canada*, Geological Survey of Canada Bulletin 548, Ottawa, pp. 101–143.

Cultural China, 2010. *1556 Shaanxi Earthquake (Jiajing earthquake) – The Deadliest Earthquake on Record*. http://history.cultural-china.com/en/34History6357.html. Accessed November 25, 2010.

Derbyshire, E., 2001. Geological hazards in loess terrain, with particular reference to the loess regions of China. *Earth-Science Reviews*, **54**, 231–260.

EM-DAT, 2010. *The OFDA/CRED International Disaster Database*. www.emdat.be. Université Catholique de Louvain, Brussels.

Evans, S. G., 2006. Single-event landslides resulting from massive rock slope failure: characterizing their frequency and impact on society. In Evans, S. G., Mugnozza, G., Strom, A., and Hermanns, R. L. (eds.), *Landslides from Massive Rock Slope Failure*. NATO Science Series, Earth and Environmental Sciences, Vol. 49, pp. 53–73.

Evans, S. G., Bishop, N. F., Smoll, L. F., Murillo, P. V., Delaney, K. B., and Oliver-Smith, A., 2009a. A re-examination of the mechanism and human impact of catastrophic mass flows originating on Nevado Huascarán, Cordillera Blanca, Peru in 1962 and 1970. *Engineering Geology*, **108**, 96–118.

Evans, S. G., Roberts, N. J., Ischuk, A., Delaney, K. B., Morozova, G. S., and Tutubalina, O., 2009b. Landslides triggered by the 1949 Khait earthquake, Tajikistan, and associated loss of life. *Engineering Geology*, **109**, 195–212.

Guogan, Y., Yun, J., and Xueming, Y., 1984. Investigation on the 1303 Zhaocheng, Shanxi, earthquake (M = 8) and its parameters concerned. *Journal of Seismological Research*, 1984–2003.

Gutenberg, B., and Richter, C. F., 1954. *Seismicity of the Earth and Associated Phenomena*. Princeton: Princeton University Press.

Gutenberg, B., and Richter, C. F., 1956. Magnitude and energy of earthquakes. *Annali di Geofisica*, **9**, 1–15.

Guthrie, R. H., Evans, S. G., Catane, S. G., Zarco, M. A. H., and Saturay, R. M., Jr., 2009. The catastrophic February 17, 2006 rockslide-debris avalanche at Guinsaugon, Philippines: a synthesis. *Bulletin of Engineering Geology and the Environment*, **68**, 201–213.

Hanks, T. C., and Kanamori, H., 1979. A moment magnitude scale. *Journal of Geophysical Research*, **84**, 2348–2350.

IDB, 2010. *Haiti Reconstruction Cost May Near $14 Billion, IDB Study Shows*. http://www.iadb.org/features-and-web-stories/2010-02/english/haiti-reconstruction-cost-may-near-14-billion-idb-study-shows-6528.html. Accessed November 25, 2010. Inter-American Development Bank News.

ISDR, 2004. *Living with Risk: A Global Review of Disaster Reduction Initiatives*. Geneva: United Nations.

Kawata, Y., Hayashi, H., Imamura, F., Koshimura, S., Statake, K., Tsuji, Y., Fujima, K., Hayashi, I., Matsutomi, H., Takahshi, T., Tanioka, Y., Nishimura, Y., Matsuyama, M., Maki, N., Horie, K., Koike, N., Harada, K., and Suzuki, S., 2007. *Comprehensive Analysis of the Damage and Its Impact on Coastal Zones by the 2004 Indian Ocean Tsunami Disaster*. Research Group on the December 26, 2004 Earthquake Tsunami Disaster of Indian Ocean, Japan. http://www.drs.dpri.kyoto-u.ac.jp/sumatra/index-e.html. Accessed November 25, 2010.

Newhall, C. G., and Dzurisin, D., 1988. *Historical Unrest at Large Calderas of the World*. United States Geological Survey Bulletin 1855, 1108 p.

Newhall, C. G., and Self, S., 1982. The volcanic explosivity index (VEI): an estimate of explosive magnitude for historical volcanism. *Journal of Geophysical Research*, **87**, 1231–1238.

NGDC, 2010. *National Geophysical Data Center, Tsunami Event Database*. http://www.ngdc.noaa.gov/nndc/struts/results?t=102564%26s=207%26d=207. Accessed November 25, 2010. US National Oceanic and Atmospheric Administration.

NOAA, 2010. *Teachers Guide to Stratovolcanoes of the World*. http://www.ngdc.noaa.gov/hazard/stratoguide/strato_home.html. Accessed November 25, 2010. US National Oceanic and Atmospheric Administration.

SAARC, 2010. *SAARC Disaster Management Centre*. http://saarc-sdmc.nic.in/home.asp. Accessed November 25, 2010. South Asia Association of Regional Cooperation.

Schuster, R. L., and Wieczorek, G. F., 2002. Landslide triggers and types. In Rybář, J., Stemberk, J., and Wagner, P. (eds.), *Landslides: Proceedings of the First European Conference on Landslides*. A.A. Balkema Publishers, Lisse, pp. 59–78.

Seed, H. B., 1968. Landslides during earthquakes due to soil liquefaction. *Journal of the Soil Mechanics and Foundations Division, ASCE*, **94**, 1055–1122.

USGS, 2010a. *Magnitude 7.0 – Haiti region, 2010 January 12 21:53:10 UTC*. http://earthquake.usgs.gov/earthquakes/recenteqsww/Quakes/us2010rja6.php#summary. Accessed November 25, 2010. United States Geological Survey Earthquake Hazards Program.

USGS, 2010b. *Historic Earthquakes, Chile 1960, May 22, 19:11:14 UTC, Magnitude 9.5*. http://earthquake.usgs.gov/earthquakes/world/events/1960_05_22.php. Accessed November 25, 2010. United States Geological Survey Earthquake Hazards Program.

WHOI, 2010. *Tsunami: An Interactive Guide that Could Save Your Life*. http://www.whoi.edu/home/interactive/tsunami/. Accessed November 25, 2010. Woods Hole Oceanographic Institution, Massachusetts.

Cross-references

Cyclones
Disaster
Drought
Earthquakes
Fire
Floods
Hazard
Ice storms
Landslide Dams
Landslides
Land Subsidence
Lightning Storms
Mass Movements
Natural Disaster
Tsunamis
Volcanoes
Vulnerability

GLACIER HAZARDS

John J. Clague
Simon Fraser University, Burnaby, BC, Canada

Definition

Glacier hazards are natural Earth processes associated with alpine glaciers, ice caps, or ice sheets that threaten people or property.

Introduction

About 10% of Earth's surface is covered by glacier ice, and about 99% of this glacier cover is in Greenland and Antarctica. The other 1% comprises ice fields, ice caps, and cirque, valley, and piedmont glaciers, mainly in mountains of northwest North America, Arctic Canada, and Asia. Glaciers provide many benefits or natural service functions. For example, meltwater from alpine glaciers augments stream runoff during summer, which is important for agriculture, municipal water supply, and hydroelectric power generation. However, processes associated with glaciers can also be hazardous. In addition, some glacier hazards may be amplified by climate change.

Glaciers, climate change, and sea-level rise

Glaciers around the world have grown and receded in response to climate change throughout the Holocene. Most alpine glaciers achieved their maximum Holocene size during the late stage of the "Little Ice Age" in the eighteenth and nineteenth centuries (Grove, 1988). In marked contrast, alpine glaciers have thinned and receded during the twentieth century and the first part of the current century, a response to the warming that has occurred over the past 150 years. Today, ice cover in most mountain ranges is one-half to two-thirds of what it was in the mid-nineteenth century. The response of the large ice sheets in Greenland and Antarctica to climate warming is less clear, although large amounts of ice are now being lost in Greenland due to accelerated flow of outlet glaciers, and ice shelves in Antarctica are showing signs of instability (Howat et al., 2008; Rignot et al., 2008).

Sea-level rise, which is a hazard in its own right, is being caused by glacier melt and thermal expansion of upper ocean waters, both driven by climate warming. Some of the current global sea-level rise of about 3 mm/year is due to retreat of alpine glaciers, mainly in western North America and Eurasia (Larsen et al., 2007; Leclercq et al., 2011), and ice loss in Greenland and Antarctica is increasingly contributing to the rise in sea level (Rignot et al., 2011). If all alpine glaciers were to disappear sea level would only rise several tens of centimeters, but even that rise would displace people living along some shorelines and would threaten or damage infrastructure. A large reduction in ice cover in Greenland or Antarctica would be catastrophic. Enough freshwater is currently stored in glacier ice in Greenland and Antarctica to raise sea level, respectively, about 7 m and 70 m (National Snow and Ice Data Center).

Landslides and other mass movements

Thinning and retreat of alpine glaciers appear to be responsible for some catastrophic rock-slope failures in high mountains (Evans and Clague, 1994, 1999; Holm et al., 2004). Many marginally stable slopes that were buttressed by glacier ice during the Little Ice Age failed after they became ice-free in the twentieth century (Figure 1). An ancillary factor that may have contributed to the failures is steepening of rock slopes by cirque and valley glaciers during the Pleistocene Epoch.

Landslides related to glacier debuttressing are most common in mountain ranges with the greatest ice cover (Himalaya, St. Elias Mountains, and Coast Mountains), because those ranges have experienced the largest ice losses in the twentieth century. An extreme example is Glacier Bay, Alaska, which until the end of the eighteenth century was completely occupied by glacier ice. Since then, the bay has become deglaciated, with the loss of over 1,000 km^3 of ice. Ice loss in Glacier Bay is so great that the land is rising due to glacio-isostatic rebound (Larsen et al., 2005).

Other climate-related mechanisms besides debuttressing cause or trigger landslides. Temperatures in high-glacierized mountains are normally below freezing. They may rise rapidly, however, during warm spells, accompanied by melting snow and ice. The water generated by melting may infiltrate fractures in marginally stable rock masses (Davies et al., 2001). When temperatures again fall below freezing, water within the fractures freezes or freezing at the surface may seal the fractures. In either case, pore water pressures rise, producing extensional forces in an already weak rock mass. A related process is thawing of permafrost beneath high-elevation rock slopes (Gruber et al., 2004; Fischer et al., 2006; Gruber and Haeberli, 2007). The lower limit of alpine permafrost has risen in most or all mountains in the twentieth century due to climate warming (Zhang et al., 2006). Thawing of previously frozen slopes has lowered the strength of stratified or fractured rock, because water along bedding planes or in the fractures melts (Deline et al., 2011).

The importance of these processes is illustrated by the unusually large number of rock falls in the European Alps during the summer of 2003, which was one of the warmest summers in European history. The rockfalls posed a significant hazard to climbers, and they attracted the attention of scientists and the public. Larger landslides have also occurred more recently, for example in 2004 at Thurwieser Peak in the central Italian Alps and in 2007 at Monte Rosa in the northwestern Italian Alps, both areas heavily trafficked by tourists.

Another glacier hazard with possible links to climate is ice avalanches. The heavily crevassed snouts of glaciers that terminate on steep slopes are prone to avalanching, in part because of the large gravitational forces exerted on the ice and in part because rates of ice flow on steep slopes are high and accompanied by enhanced basal sliding. In such environments, ice avalanches can occur any time. Many ice avalanches occur during warm or wet weather, when there is abundant water flowing at the base of the glacier. This water can accelerate glacier flow and locally elevate pore pressures. It is no coincidence, then, that many notable ice avalanches have occurred during summer. Examples include a snow and ice avalanche from Glaciar 511 on the summit of Nevados Huascaran, Peru, in January 1962, which destroyed the town of Yungay, killing an estimated 4,000 people; an avalanche from the toe of Allalin Glacier in August 1965, which killed 12 men at a dam construction site in Switzerland; an avalanche of 2–3 million cubic meters of ice from the snout of Diadem Glacier in British Columbia in July 2007; and a 1.1 million cubic meters ice avalanche on the east face of Monte Rosa in August 2005.

Glacial outburst floods

Many glaciers impound large water bodies that drain suddenly, causing downvalley floods, termed jökulhlaups, that are far larger than normal nival and rainfall-triggered floods (Costa and Schuster, 1988; Clague and Evans, 1994). Water may be trapped on top of, within, beneath, or at the margins of glaciers. Some lakes at the margins of alpine glaciers may hold 100,000,000 m^3 of water or more.

Glacier Hazards, Figure 1 Jubilee rock avalanche, New Zealand. This landslide, which occurred in the austral summer of 2008, is an example of a landslide that may have been caused, in part, by glacial debuttressing.

The largest of these lakes, and commonly those most prone to sudden emptying, are situated in low-gradient trunk valleys at the toes of glaciers that flow out of tributary valleys, and in a few fiords at the margins of calving tidewater glaciers. Subglacial lakes form beneath Vatnajökull, an ice cap in Iceland, during eruptions of Grimsvötn and other active volcanoes, and generally drain suddenly to produce large jökulhlaups. When climate is stable and glaciers neither advance nor retreat, few glacier-dammed lakes drain. The situation is different, however, when climate changes over periods of decades or longer. During the Little Ice Age, for example, new lakes formed when glaciers advanced across streams and blocked drainage (Figure 2). When these lakes first formed, their glacier dams may have been weak and many of the lakes probably drained repeatedly. As these glaciers continued to advance, both the dams and the impounded lakes stabilized. Most Little Ice Age lakes drained one or more times in the twentieth century when their dams weakened due to glacier thinning and retreat (Figure 3; Costa and Schuster, 1988; Clague and Evans, 1994; and references therein). In each case, a critical threshold of instability was reached, at which time the dam failed. The most common drainage mechanism is rapid development of one or more subglacial channels that serve as conduits for outflowing water. Glacier dams may also fail by mechanical collapse after glacier surges that block streams or fiords. Many former glacier-dammed lakes no longer exist, because the glaciers that dammed them have wasted so much that they no longer impound water. However, lakes have formed in new locations at the margins of some receding glaciers, typically at higher elevations than former lakes, and pose new risks to downvalley areas (Geertsema and Clague, 2005).

Lakes also developed behind Little Ice Age terminal moraines as glaciers retreated in the late nineteenth and early twentieth centuries (Costa and Schuster, 1988; O'Connor and Costa, 1993; Clague and Evans, 1994, 2000). Some of these moraine dams are unstable and vulnerable to failure because they are steep-sided and consist of loose sediment. Irreversible rapid incision of moraine dams may be caused by a large overflow triggered by an ice avalanche or a rockfall (Figure 3). Other failure mechanisms include earthquakes, slow melt of buried ice, and removal ("piping") of fine sediment from the dam.

Many outburst floods from glacier and moraine-dammed lakes display an exponential rise in discharge, followed by an abrupt decrease to background levels when the water supply is exhausted. Peak discharges are controlled by lake volume, dam height and width, the material properties of the dam, failure mechanism, and downstream topography and sediment availability. Floods from glacier-dammed lakes tend to have lower peak discharges than those from moraine-dammed lakes of similar size because enlargement of tunnels in ice is a slower process than overtopping and incision of sediment dams.

Floods from glacier and moraine-dammed lakes may transform into debris flows as they travel down steep valleys. Such flows can only form and be sustained on slopes greater than 10–15° and only where there is an abundant supply of sediment in the valley below the dam (Clague and Evans, 2000). Entrainment of sediment and woody plant debris by floodwaters may elevate peak discharges, which has important implications for hazard appraisal, because debris flows are more destructive than floods of the same size.

Outburst floods from lakes dammed by glaciers and moraines erode, transport, and deposit huge amounts of sediment over distances of tens of kilometers (Figure 4). They alter river floodplains far from the flood source. They broaden floodplains, destroy preflood channels, and create a new multichannel, braided planform. The changes can persist for decades after the flood, although rivers quickly reestablish their preflood grades by incising the flood deposits.

A relation exists between climate and the stability of moraine and glacier dams. Most existing moraine-dammed lakes formed in the twentieth century when glaciers retreated from bulky terminal moraines constructed during the Little Ice Age. The lakes soon began to fail as climate warmed. If warming and glacier retreat continue, the supply of moraine-dammed lakes susceptible to failure in most mountain ranges will decrease and the threat they pose will diminish (Clague and Evans, 2000). The relation is different for glacier-dammed lakes. Typically, a glacier-dammed lake goes through a period of cyclic or sporadic outburst activity, lasting up to several decades. The cycle of outburst of floods from a lake ends when either the glacier advances and forms a stronger dam or retreats so much that it can no longer trap water behind it.

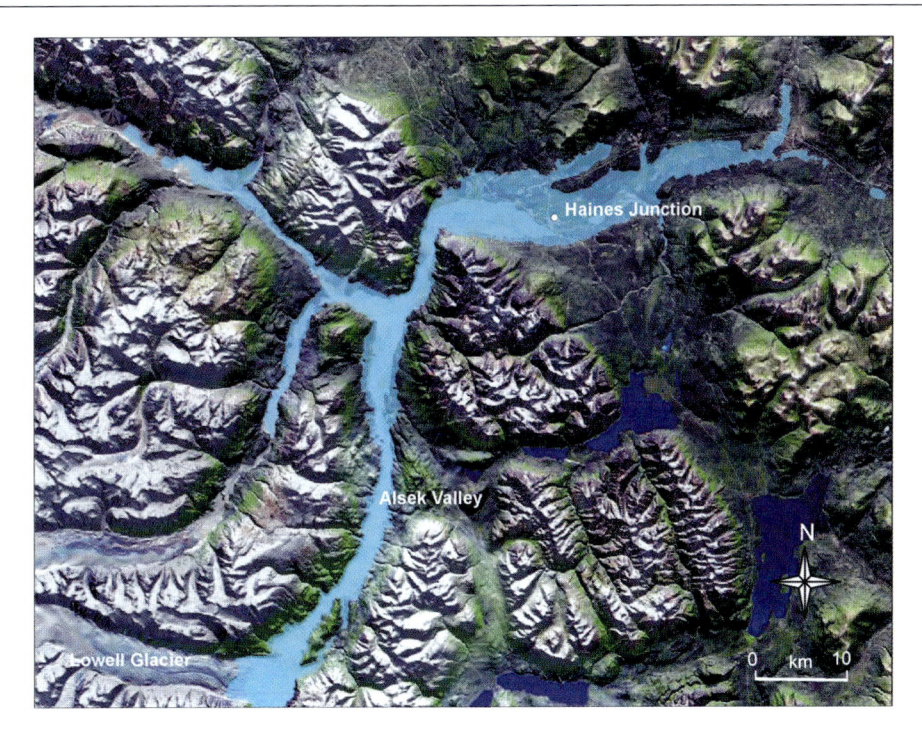

Glacier Hazards, Figure 2 Glacial Lake Alsek formed repeatedly during the Little Ice Age when Lowell Glacier, a large valley glacier in the St. Elias Mountains of southwest Yukon Territory, advanced across Alsek River alley and blocked the flow of the river. This figure depicts the lake in the mid-nineteenth century, when it reached to the present site of Haines Junction on the Alaska Highway, and earlier when the site of Haines Junction was inundated by water about 70 m deep (Courtesy of Jeff Bond, Yukon Geological Survey).

Glacier Hazards, Figure 3 Breached Little Ice Age terminal moraine of Diadem Glacier, British Columbia. About 6 million m^3 of water flowed out of moraine-dammed Queen Bess Lake on August 12, 1997, after an ice avalanche from the toe of Diadem Glacier entered the lake and produced a displacement wave that overtopped and incised the moraine (Photo courtesy of Interfor).

Glacier Hazards, Figure 4 Floor of the west fork of Nostetuko River valley, British Columbia several years after the 1997 outburst flood from moraine-dammed Queen Bess Lake. The flood transported and deposited large amounts of coarse sediment and destroyed the preexisting river channel and floodplain on the valley floor (Photo: J. J. Clague).

Changes to streams

A sustained advance or retreat of a glacier will change the amount of water and sediment delivered to its snout. These changes impact the stream in the valley below the glacier snout. During glacier advance, for example, meltwater streams aggrade their channels as sediment supply increases (Maizels, 1979). Sediment stored within and beneath the glacier, and that previously stored in the valley, is delivered at an increasing rate to the fluvial system as the glacier advances (Karlén, 1976; Maizels, 1979; Leonard, 1986, 1997; Karlén and Matthews, 1992; Lamoureux, 2000). At the same time, the areas of subglacial erosion and rockfall onto the glacier increase and meltwater may carry more sediment into the river valley than at times when the glacier is more restricted. A sediment pulse may propagate rapidly downstream in a narrow mountain valley as the glacier approaches and reaches its maximum extent.

Such climatically induced changes in stream discharge and sediment yield have been documented in the Coast Mountains of British Columbia. Streams responded to increased sediment delivery during the Little Ice Age by aggrading their channels and braiding over distances up to tens of kilometers down valley from glaciers in their headwaters (Church, 1983; Gottesfeld and Johnson-Gottesfeld, 1990; Wilkie and Clague, 2009). Subsequently, during the twentieth century, the streams incised their Little Ice Age deposits and reestablished single-thread channels characteristic of periods of lower sediment flux.

Summary

Warming of climate during the twentieth century has caused massive deglacierization of high mountains around the world. The loss of glacier ice, in tandem with permafrost thaw, has destabilized many steep slopes and triggered rockfalls, rock slides, and ice and rock avalanches. Glacier downwasting and retreat also have triggered outburst floods from glacier- and moraine-dammed lakes. These floods, which have peak discharges much larger than normal nival and rain-on-snow floods, pose a serious threat to downvalley communities and infrastructure in many mountain ranges. Melting glaciers are also contributing to sea-level rise, which will accelerate in the future if warming continues.

Bibliography

Church, M., 1983. Pattern of instability in a wandering gravel bed channel. In Collinson, J. D., and Lewin, J. (eds.), *Modern and Ancient Fluvial Systems*. Oxford: Blackwell Scientific Publications. International Association of Sedimentologists Special Publication, Vol. 6, pp. 169–180.

Clague, J. J., and Evans, S. G., 1994. *Formation and Failure of Natural Dams in the Canadian Cordillera*. Geological Survey of Canada Bulletin 464.

Clague, J. J., and Evans, S. G., 2000. A review of catastrophic drainage of moraine-dammed lakes in British Columbia. *Quaternary Science Reviews*, **19**, 1763–1783.

Costa, J. E., and Schuster, R. L., 1988. The formation and failure of natural dams. *Geological Society of America Bulletin*, **100**, 1054–1068.

Davies, M. C. R., Hamza, O., and Harris, C., 2001. The effect of rise in mean annual temperature on the stability of rock slopes containing ice-filled discontinuities. *Permafrost and Periglacial Processes*, **12**, 137–144.

Deline, P., Alberto, W., Broccolato, M., Hungr, O., Noetzli, J., Ravanel, L., and Tamburini, A., 2011. The December 2008 Crammont rock avalanche, Mont Blanc massif area, Italy. *Natural Hazards and Earth System Sciences*, **11**, 3307–3318.

Evans, S. G., and Clague, J. J., 1994. Recent climatic change and catastrophic geomorphic processes in mountain environments. *Geomorphology*, **10**, 107–128.

Evans, S. G., Clague, J. J., 1999. Rock avalanches on glaciers in the Coast and St. Elias Mountains, British Columbia. In *Proceedings, 13th Annual Vancouver Geotechnical Conference*. Richmond, BC: Vancouver/BiTech Publishers, pp. 115–123.

Fischer, L., Kääb, A., Huggel, C., and Noetzli, J., 2006. Geology, glacier retreat and permafrost degradation as controlling factor of slope instabilities in a high-mountain rock wall: the Monte Rosa east face. *Natural Hazards and Earth System Sciences*, **6**, 761–772.

Geertsema, M., and Clague, J. J., 2005. Jökulhlaups at Tulsequah Glacier, northwestern British Columbia, Canada. *The Holocene*, **15**, 310–316.

Gottesfeld, A. S., and Johnson-Gottesfeld, L. M., 1990. Floodplain dynamics of a wandering river, dendrochronology of the Morice River, British Columbia, Canada. *Geomorphology*, **3**, 159–179.

Grove, J. M., 1988. *The Little Ice Age*. London: Methuen Press.

Gruber, S., and Haeberli, W., 2007. Permafrost in steep bedrock slopes and its temperature-related destabilization following climate change. *Journal of Geophysical Research*, **112**, F02S18, doi:10.1029/2006JF000547.

Gruber, S., Hoelzle, M., and Haeberli, W., 2004. Permafrost thaw and destabilization of alpine rock walls in the hot summer of 2003. *Geophysical Research Letters*, **31**(13), doi:10.1029/2004GL020051.

Holm, K., Bovis, M., and Jakob, M., 2004. The landslide response of alpine basins to post- Little Ice Age glacial thinning and retreat in southwestern British Columbia. *Geomorphology*, **57**, 201–216.

Howat, I. M., Smith, B. E., Joughin, I., and Scambos, T. A., 2008. Rates of southeast Greenland ice volume loss from combined ICESat and ASTER observations. *Geophysical Research Letters*, **35**, L17505, doi:10.1029/2008GL034496.

Karlén, W., 1976. Lacustrine sediments and tree limit variations as evidence of Holocene climatic variations in Lappland, northern Sweden. *Geografiska Annaler*, **58**, 1–34.

Karlén, W., and Matthews, J. A., 1992. Reconstructing Holocene glacier variations from glacial lake sediments: studies from Nordvestlandet and Jostedalsbreen-Jotunheimen, Southern Norway. *Geografiska Annaler*, **74**, 327–348.

Lamoureux, S., 2000. Five centuries of interannual sediment yield and rainfall-induced erosion in the Canadian High Arctic recorded in lacustrine varves. *Water Resources Research*, **36**, 309–318.

Larsen, C. F., Motyka, R. J., Freymueller, J. T., Echelmeyer, K. A., and Ivins, E. R., 2005. Rapid viscoelastic uplift in Southeast Alaska caused by post-Little Ice Age glacial retreat. *Earth and Planetary Science Letters*, **237**, 548–560.

Larsen, C. F., Motyka, R. J., Arendt, A. A., Echelmeyer, K. A., and Geissler, P. E., 2007. Glacier changes in southeast Alaska and Northwest British Columbia and contribution to sea level rise. *Journal of Geophysical Research*, **112**, F01007, doi:10.1029/2006JF00586.

Leclercq, P. W., Oerlemans, J., and Cogley, J. G., 2011. Estimating the glacier contribution to sea-level rise for the period 1800–2005. *Surveys in Geophysics*, **32**, 519–535.

Leonard, E. M., 1986. Varve studies at Hector Lake, Alberta, Canada, and the relationship between glacial activity and sedimentation. *Quaternary Research*, **25**, 199–214.

Leonard, E. M., 1997. The relationship between glacial activity and sediment production: Evidence from a 4450-year varve record of neoglacial sedimentation in Hector Lake, Alberta, Canada. *Journal of Palaeolimnology*, **17**, 319–330.

Maizels, J. K., 1979. Proglacial aggradation and changes in braided channel patterns during a period of glacier advance: an Alpine example. *Geografiska Annaler*, **61**, 87–101.

O'Connor, J. E., and Costa, J. E., 1993. Geologic and hydrologic hazards in glacierized basins in North America resulting from 19th and 20th century global warming. *Natural Hazards*, **8**, 121–140.

Rignot, E., Bamber, J. L., van den Broeke, M. R., Davis, C., Li, Y., van de Berg, W. J., and van Meijgaard, E., 2008. Recent Antarctic ice mass loss from radar interferometry and regional climate modelling. *Nature Geoscience*, **1**, 106–110.

Rignot, E., Velicogna, I., van den Broeke, M. R., Monaghan, A., and Lenaerts, J., 2011. Acceleration of the contribution of the Greenland and Antarctic ice sheets to sea level rise. *Geophysical Research Letters*, **38**, L05503, doi:10.1029/2011GL046583.

Wilkie, K., and Clague, J. J., 2009. Fluvial response to Holocene glacier fluctuations in the Nostetuko River valley, southern Coast Mountains, British Columbia. In Knight, J., and Harrison, S. (eds.), *Periglacial and Paraglacial Processes and Environments*. London: Geological Society. Geological Society Special Publications, Vol. 320, pp. 199–218.

Zhang, Y., Chen, W., and Riseborough, D. W., 2006. Temporal and spatial changes of permafrost in Canada since the end of the Little Ice Age. *Journal of Geophysical Research*, **111**(D22), doi:10.1029/2011GL046583.

Cross-references

Avalanche
Climate Change
Jökulhlaup
Landslide
Permafrost Thaw
Rock Avalanche
Sea-Level Rise

GLOBAL CHANGE AND ITS IMPLICATIONS FOR NATURAL DISASTERS

Gonéri Le Cozannet, Hormoz Modaressi, Nicolas Desramaut
BRGM – French Geological Survey, Orléans, France

Definition

Global change refers to environmental and societal transformations, including those affecting the Earth system as a whole such as climate change (Field et al., 2012) or ozone hole depletion (Farman et al., 1985), but also more direct impacts of human activities that affect environmental systems worldwide, such as urbanization (see entry "*Urban Environments and Natural Hazards*") and environmental degradation (Verstappen, 2011). Hence, global change includes the dynamic of society and socioeconomic factors (Clark, 1988; Gallopin, 1991) and their interactions with environmental changes (Slaymaker et al., 2009). A *disaster* is defined by the International Strategy for Disaster Reduction (ISDR) as "a serious disruption of the functioning of a community or society, causing widespread human, material, economic or environmental losses which exceed the ability of the affected community or society to cope using its own resources" (ISDR, 2004). The term "natural" can be disputed because human vulnerability and exposures are essential components of disasters induced by natural phenomenon (O'Keefe et al., 1976; Turner, 1976; Pelling, 2001). *Global change implications of natural disasters* (Pelling, 2003) (see "*Natural Hazard*") thus include all components of risk: hazard, exposure, vulnerability as well as the communities' capacity to respond. Responses to global change as a factor of aggravation of natural disasters (Alexander, 1993) address at least three major policies with implications from local to global decision levels: disaster risk management (see "*Disaster Risk Management*"), adaptation to climate change (see "*Adaptation*"), and sustainable development as a response to environmental degradations (UN (United Nations), 2009).

Discussion

Global change affects natural disasters through three major components: (1) climate change, which is influenced by human activities (Hegerl et al., 2007); (2) more local human-induced pressures on the environment that constitute a global issue either because they occur worldwide, or because the magnitude of change is increasing (Turner et al., 1990); and (3) global societal transformations (see "*Humanity as an Agent of Natural Disaster*").

Global change affects or is expected to affect the following components of natural disasters:

- Hazard, which may be affected by climate change (Field et al., 2012) and inappropriate land use planning, such as deforestation affecting landslides (Glade, 2003; Huggel et al., 2012).
- Exposure, which has considerably grown recently because of population growth and urbanization (see "*Urban Environments and Natural Hazards*").
- Vulnerability (see "*Vulnerability*"): as an example, disorganized urbanization that does not take into account building regulation contributes to increasing physical vulnerability (as defined in Douglas, 2007). Another example is economic development causing environmental degradations that reduce the capacity of natural systems to protect against disasters (World Bank, 1992). This is the case, for example, for the mangrove that plays a role in protecting coastal areas against marine submersions and coastal erosion (Othman, 1994).
- The capacity to respond to disasters may be worsened by environmental degradation and population growth pressure (Adger et al., 2005), as there may already be a stress on basic resources such as food, water, or even land (see "*Land Use, Urbanization, and Natural Hazards*").

1. *Widespread anthropogenic environmental changes and their influence on natural disasters:* Anthropogenic changes such as land use (Dale, 1997), urbanization (Parnell et al., 2007), and some agriculture practices may modify all components of risk, including the hazard itself in certain cases. However, they primarily affect exposure, vulnerability, as well as, the capacity to respond of communities. EM-DAT (www.emdat.net) reports a significant increase in the number of natural disasters since 1900 (See "*Worldwide Trends in Natural Disasters*"). These data should be considered cautiously because of observational bias and the improvements in disaster reporting. However, such a trend would be the expected consequence of increased vulnerability and exposure in the context of social and economic development barely taking into account disaster risk prevention and mitigation.

2. *Climate change* (see "*Climate Change*") *influence on natural disasters:* Climate change is expected to primarily affect the heat wave hazard and secondarily hydrometeorological hazards (see "*Hydrometeorological Hazards*"), such as heavy precipitation, flooding, and drought. Finally, it is assumed that climate change will modify hazards that are related to hydrometeorological processes (Meehl et al., 2007) such as landslides. Coastal risks, in particular coastal erosion (see "*Coastal Erosion*") and storm surge hazard (see "*Storm Surges*"), should increase as a consequence of sea-level rise (Nicholls and Cazenave, 2010). Whereas the consequences of climate change on hazard are expected to be important in the future, observations of current extreme events are rarely attributed unambiguously to climate change. Recently, a notable advance in this field has been to investigate how heat waves can be attributed to climate change (Yiou et al., 2007). However, many of the significant trends that have been observed in recent decades are the manifestation of decadal or multi-decadal climate variability (Schneider et al., 2007). Moreover, the assessment of climate change influence on natural disasters is limited by a scaling issue: Global climate models produce projections at spatial scales and timescales that are barely appropriate for many extreme events. Thus, projections from global models are downscaled to identify or track indices of extreme events trends for climate change scenarios. This regionalization adds an additional layer of uncertainties to those of global climate models

(Van Aalst, 2006). Climate change influences on cyclonic hazard in the Caribbean are an example of these difficulties. The scientific community not only debates the possibility of attributing recent cyclonic patterns changes to climate change (Holland and Webster, 2007), but they are also debating what could finally be the impact of climate change on cyclonic (see "*Hurricane (Typhoon, Cyclone)*") patterns and genesis in the future. For many hazards such as landslides (see "*Landslide (Mass Movement)*"), the relative importance of temperature and hydrometeorological forcing with respect to other processes and features, such as geomechanics, is still a research topic. Because of this limited knowledge of the influence of climate change on natural disasters, methods are being developed within climate change adaptation strategies to support decision making in an uncertain future (Hallegatte, 2009).

3. *Links between climate change adaptation, disaster reduction, and sustainable development:* The question of climate change influence on natural disasters includes several major issues: disaster risk management, adaptation to climate change, and sustainable development as a response to environmental degradations. Connecting these policies also increases the complexity of finding suitable strategies because of different understanding of concepts such as vulnerability (see "*Vulnerability*"): For climate change communities, vulnerability is defined as the losses to be expected because of global changes minus the benefits of adaptation, whereas in communities concerned with disaster risk management, it is a component of risk (Romieu et al. 2010). Beyond the semantic difficulties, the connection of these three major policies raises major questions for decision-making authorities. For example, it is presently often unclear whether short-term reduction of present risks is a sufficient measure of adaptation to climate change or if new approaches need to be developed.

4. *Extreme events and long-term trends:* Over a certain threshold, the poorest societies are unable to manage the consequences of extreme events, recovery is limited, and so these societies are kept in the loop of underdevelopment (Schipper and Pelling, 2006). Since the influence of climate change on natural disasters remains uncertain, there is no evidence whether the worldwide impact of global change will be limited to part of the gross domestic product or if a threshold may be passed leading to a new pathway of underdevelopment (Hallegatte et al., 2007). A relatively open research issue is the question of which kind of interactions between short-term effects, such as extreme events, and long-term trends, such as climate and global changes, are implied in terms of human development and livelihood.

5. *Global change as a hazard in itself:* The above highlighted examples also lead to consideration of global change as a hazard in itself (ISDR, 2008). This is supported by possible future abrupt changes of climate, such as collapse of large Greenland or Antarctic ice sheets, major changes in the oceanic thermohaline circulation, or major changes in the carbon cycle (Schneider et al., 2007). This raises again the question of the robustness of human societies to global changes. The collapse of the Akkadian empire in the context of Mesopotamia becoming more arid (4000 BC) (De Menocal, 2001) and the rise of the Egyptian civilization in the context of Sahara's desertification and population growth around the Nile (5300 BC to 3500 BC) (Kuper and Kröpelin, 2006) are two of many historical examples of major environmental changes affecting human societies. In these two examples, environmental stress creates a crisis in a society, which may adapt to the new conditions or not, for example, by creating a more complex and structured culture (Brooks, 2006). While such research provides a perspective on the ongoing global change consequences, there are major differences between the present situation and historical analogues. Despite the uncertainties, there is presently some knowledge of the expected changes. In addition, while the expected climax of the environmental crisis has probably not been already reached, human societies are now investigating how they can anticipate adverse effects of contemporary global change and reduce their vulnerability to the forecasted future disasters.

Summary or conclusion

The environmental impact of human development and climate change are two major components of global change. It is widely recognized that fast population growth and disorganized urbanization has led to increasing vulnerability and exposure to natural disasters during the twentieth century. With respect to climate change, our ability to understand and predict its actual impact to natural disasters is limited, for example, by the difficulties in capturing natural disasters triggering factors in global climate models. In addition, Roe and Baker (2007) showed that our ability to predict future changes due to climate change is limited by intrinsic uncertainties of climate models. This supports the idea that the suspected implications of global change for natural disasters should be anticipated through mitigation and adaptation measures to climate change (see "*Climate Change*"), which should be consistent with disaster risk reduction policies (see "*Disaster Risk Reduction*") and sustainable development policies (ISDR, 2005; Adger et al., 2007; Field et al., 2012).

Bibliography

Adger, W. N., Hughes, T. P., Folke, C., Carpenter, S. R., and Rockstrom, J., 2005. Social-ecological resilience to coastal disasters. *Science*, **309**, 1036–1039.

Adger, W. N., Agrawala, S., Mirza, M. M. Q., Conde, C., O'Brien, K., Pulhin, J., Pulwarty, R., Smit, B., and Takahashi, K., 2007. Assessment of adaptation practices, options, constraints and capacity.

In Parry, M. L., Canziani, O. F., Palutikof, J. P., van der Linden, P. J., and Hanson, C. E. (eds.), *Climate Change 2007: Impacts, Adaptation and Vulnerability*. Cambridge: Cambridge University Press. Contribution of Working Group II to the Fourth Assessment Report of the Intergovernmental Panel on Climate Change, pp. 717–743.

Alexander, D., 1993. *Natural Disasters*. London: UCL Press.

Brooks, N., 2006. Cultural responses to aridity in the Middle Holocene and increased social complexity. *Quaternary International*, **151**, 29–49.

Clark, W. C., 1988. *The Human Dimension of Global Environmental Change, Toward an Understanding of Global Change*. Washington, DC: National Academy Press.

Dale, V. H., 1997. The relationship between land-use change and climate change. *Ecological Applications*, **7**, 753–769.

De Menocal, P., 2001. Cultural responses to climate change during the late Holocene. *Science*, **292**(5517), 667.

Douglas, J., 2007. Physical vulnerability modelling in natural hazard risk assessment. *Natural Hazards and Earth System Sciences*, **7**, 283–288.

Farman, J. C., Gardiner, B. G., and Shanklin, J. D., 1985. Large ozone losses of total onze in Antarctica reveal seasonal ClOx/NOx interaction. *Nature*, **315**, 207–210.

Field, C. B., Barros, V., Stocker, T. F., Qin, D., Dokken, D. J., Ebi, K. L., Mastrandrea, M. D., Mach, K. J., Plattner, G.-K., Allen, S. K., Tignor, M., and Midgley, P. M., (eds.), 2012. Summary for policymakers. In: *Managing the Risks of Extreme Events and Disasters to Advance Climate Change Adaptation*. Cambridge/New-York: Cambridge University Press.

Gallopin, G. C., 1991. Human dimensions of global change: linking the global and the local processes. *International Social Science Journal*, **130**, 707–718.

Glade, T., 2003. Landslide occurrence as a response to land use change: a review of evidence from New Zealand. *Catena*, **51**(3–4), 297–314.

Hallegatte, S., 2009. Strategies to adapt to an uncertain climate change. *Global Environmental Change*, **19**, 240–247.

Hallegatte, S., Hourcade, J.-C., and Dumas, P., 2007. Using climate analogues for assessing climate change economic impacts. *Climatic Change*, **82**, 47–60.

Hegerl, G. C., Zwiers, F. W., Braconnot, P., Gillett, N. P., Luo, Y., Marengo Orsini, J. A., Nicholls, N., Penner, J. E., and Stott, P. A., 2007. Understanding and attributing climate change. In Solomon, S., Qin, D., Manning, M., Chen, Z., Marquis, M., Averyt, K. B., Tignor, M., and Miller, H. L. (eds.), *Climate Change 2007: The Physical Science Basis*. Cambridge/New York: Cambridge University Press. Contribution of Working Group I to the Fourth Assessment Report of the Intergovernmental Panel on Climate Change.

Holland, G. J., and Webster, P. J., 2007. Heightened tropical cyclone activity in the North Atlantic: natural variability or climate trend? *Philosophical Transactions of the Royal Society A*, **365**, 2695–2716.

Huggel, C., Clague, J. J., and Korup, O., 2012. Is climate change responsible for changing landslide activity in high mountains? *Earth Surface Processes and Landforms*, **37**, 77–91.

ISDR (International Strategy for Disaster Reduction), 2004 (rev. 2009). *Terminology of Disaster Risk Reduction*. United Nations, Geneva, Switzerland, May 2009. Available at http://www.unisdr.org/files/7817_UNISDRTerminologyEnglish.pdf (Accessed 18 April 2012).

ISDR (International Strategy for Disaster Reduction), 2005. *Hyogo Framework for Action: Building the Resilience of Nations and Communities to Disasters*. Extract from the final report of the World Conference on Disaster Reduction (A/CONF.206/6); United Nations, Geneva, Switzerland, 2005. Available at: http://www.preventionweb.net/files/1037_hyogoframeworkforactionenglish.pdf (Accessed 18 April 2012).

ISDR (International Strategy for Disaster Reduction), 2008. *Environment and Disaster Risk: Emerging Perspectives*. ISDR Working Group on Environment and Disaster Reduction; United Nations Environment Programme (UNEP); Geneva, Switzerland and Nairobi, Kenya, July 2008, ISBN: 978-92-807-2887-3. Available at http://www.preventionweb.net/files/624_EnvironmentanddisasterriskNov08.pdf (Accessed 18 April 2012).

Kuper, R., and Kröpelin, S., 2006. Climate-controlled holocene occupation in the Sahara: motor of Africa's evolution. *Science*, **313**(5788), 803.

Meehl, G. A., Stocker, T. F., Collins, W. D., Friedlingstein, P., Gaye, A. T., Gregory, J. M., Kitoh, A., Knutti, R., Murphy, J. M., Noda, A., Raper, S. C. B., Watterson, I. G., Weaver, A. J., and Zhao, Z.-C., 2007. Global climate projections. In Solomon, S., Qin, D., Manning, M., Chen, Z., Marquis, M., Averyt, K. B., Tignor, M., and Miller, H. L. (eds.), *Climate Change 2007: The Physical Science Basis*. Cambridge/New York: Cambridge University Press. Contribution of Working Group I to the Fourth Assessment Report of the Intergovernmental Panel on Climate Change.

Nicholls, R. J., and Cazenave, A., 2010. Sea-level rise and its impact on coastal zones. *Science*, **328**(5985), 1517–1520.

O'Keefe, P., Westgate, K., and Wisner, B., 1976. Taking the naturalness out of natural disasters. *Nature*, **260**(5552), 566–567.

Othman, M. A., 1994. Value of Mangrove in coastal protection. *Hydrobiologia*, **285**, 277–282.

Parnell, S., Simon, D., and Vogel, C., 2007. Global environmental change: conceptualising the growing challenge for cities in poor countries. *Area*, **39**, 357–369.

Pelling, M., 2001. Natural disasters? In Castree, N., and Braun, B. (eds.), *Social Nature*. London: Blackwells, pp. 170–188.

Pelling, M. (ed.), 2003. *Natural Disasters and Development in a Globalizing World*. London: Routledge.

Roe, G., and Baker, M., 2007. Why is climate sensitivity so unpredictable? *Science*, **318**(5850), 629.

Romieu, E., Welle, T., Schneiderbauer, S., Pelling, M., and Vinchon, C., 2010. Vulnerability assessment within climate change and natural hazard contexts: revealing gaps and synergies through coastal applications. *Sustainability Science*, **5**, 159–170.

Schipper, L., and Pelling, M., 2006. Disaster risk, climate change and international development: scope for, and challenges to, integration. *Disasters*, **30**, 19–38.

Schneider, S. H., Semenov, S., Patwardhan, A., Burton, I., Magadza, C. H. D., Oppenheimer, M., Pittock, A. B., Rahman, A., Smith, J. B., Suarez, A., and Yamin, F., 2007. Assessing key vulnerabilities and the risk from climate change. Climate change 2007: impacts, adaptation and vulnerability. In Parry, M. L., Canziani, O. F., Palutikof, J. P., van der Linden, P. J., and Hanson, C. E. (eds.), *Contribution of Working Group II to the Fourth Assessment Report of the Intergovernmental Panel on Climate Change*. Cambridge: Cambridge University Press, pp. 779–810.

Slaymaker, O., Spencer, T., and Embleton-Hamann, C., 2009. *Geomorphology and Global Environmental Change*. Cambridge: Cambridge University Press.

Turner, B. A., 1976. The organizational and interorganizational development of disasters. *Administrative Science Quarterly*, **21**, 378–397.

Turner, B. L., Kasperson, R. E., Meyer, W. B., Dow, K. M., Golding, D., Kasperson, J. X., Mitchell, R. C., and Ratick, S. J., 1990. 2 types of global environmental change – definition and spatial scale issues in their human dimensions.

Global Environmental Change -Human and Policy Dimensions, **1**, 14–22.

UN (United Nations), 2009. *Global Assessment Report on Disaster Risk Reduction: Risk and Poverty in a Changing Climate – Invest Today for a Safer Tomorrow*. Geneva: UNISDR.

Van Aalst, M. K., 2006. The impacts of climate change on the risk of natural disasters. *Disasters*, **30**(1), 5–18.

Verstappen, H. T., 2011. Natural disaster reduction and environmental management: a geomorphologist's view. *Geografia Fisica E Dinamica Quaternaria*, **34**, 55–64.

World Bank, 1992. *World Development Report 1992: Development and the Environment*. New York: Oxford University Press.

Yiou, P., Vautard, R., Naveau, P., and Cassou, C., 2007. Inconsistency between atmospheric dynamics and temperatures during the exceptional 2006/2007 fall/winter and recent warming in Europe. *Geophysical Research Letters*, **34**, L21808.

Cross-references

Adaptation
Climate Change
Coastal Erosion
Disaster Risk Management
Disaster Risk Reduction (DRR)
Humanity as an Agent of Geological Disaster
Hurricane (Cyclone, Typhoon)
Hydrometeorological Hazards
Land Use, Urbanization, and Natural Hazards
Landslide (Mass Movement)
Natural Hazard
Storm Surge
Vulnerability
Worldwide Trends in Natural Disasters

GLOBAL DUST

Edward Derbyshire
Cheltenham, Gloucestershire, UK
Royal Holloway, University of London, Egham, Surrey, UK

Definition

Global dust refers to palls or plumes of solid particles suspended in the troposphere, to be transported for thousands of kilometers by the action of strong, turbulent winds in source regions characterized by dry land surfaces with little or no vegetation cover. Other notable sources include the volcanic ash (tephra) emitted during eruptions and organic matter and elemental carbon derived from forest fires and industrial sources (Gieré, 2010). A large proportion of global dust particles have aerodynamic diameters finer than 20 micrometers (μm).

Introduction

Erosion, airborne transport and deposition of solid dust particles have occurred throughout geological time. Evidence of such events includes wind transport and deposition of mineral dust over the past 2.6 Myr in the form of the thick clayey-silt formation, known as *loess*, found on all inhabited continents. Long range (hemispherical and global) transport of dust involves atmospheric turbulence in source areas such as occurs along air-mass frontal systems, the finer dust particles being lifted into the upper troposphere (up to about 10 km). The stronger volcanic dust plumes, ejected at speeds greater than those of the local winds, reach the upper atmosphere (the stratosphere) (Durant et al., 2010). Information on the relative proportions of global dust emitted by three of the primary processes (dust storms, volcanic eruptions, and forest fires) is sparse but fine volcanic ash and dust from biomass burning may make up only 12% and 3% by weight, respectively, of the dust storm total (Gieré and Querol, 2010). The written record of visible atmospheric dust dates back to some 3,000 years in China (Zhang, 1984), since classical and medieval times in the Mediterranean region (e.g., Homer's "blood rain" referred to in the *Iliad* (eighth century BCE); Georgius Agricola (1556), cited in Guthrie, 1997) and since the eighteenth century in North Atlantic ships' logs (e.g., Darwin, 1846).

Airborne dust and its entrainment

Mineral particles, a major component of global dust, are products of physical and chemical weathering of rocks, biochemical processes involved in soil formation, and volcanic eruptions. Detachment of loose mineral dust from such dry surfaces (deflation) varies with particle size, degree of particle exposure, soil surface moisture content and roughness, lower atmosphere dynamics, and the wind speed (threshold velocity) required to detach the dust. Some dust particles, including sulfates, form within the atmosphere.

Global dust consists mainly of silt-size particles (>2 μm); its mineral composition is dominated by oxides of silicon, aluminum, iron, and calcium, as well as carbonates, the relative proportions reflecting the average composition of the upper continental crust. However, composition varies with dust source. Fibrous materials (including the fibrous amphibole minerals and organic fibers) can be transported long distances. The finer mineral fractions (clay-size grains: <2 μm) are not easily detached from soil surfaces as single particles because of their high interparticle cohesive charges; most detachment occurs as a result of impact on soil surfaces by larger, sand-size particles (the process of saltation). Clay-size material commonly occurs as attachments to silt-sized grains and also as silt-sized aggregates.

The entrainment of dust from barren or sparsely vegetated terrain varies with numerous factors, but the particle size, density, and shape exert a strong influence on the process of entrainment of particles by a critical wind speed known as the threshold velocity at the surface of the dust source. Dust particles held in suspension in the atmosphere over global distances are generally finer than 20 μm.

Volcanoes with silica-rich magma tend to have more violent eruptions, in which the thrust of volcanic gases may reach a velocity of 600 m s^{-1}, slowing with increase of altitude and with progressively finer particles in the upper part of the plume. About one quarter of such emissions reach the stratosphere and strong, high magnitude plumes can sustain global atmospheric impacts for months to years (Sparks, 1986; Durant et al., 2010).

Dust sources

Deflation of mineral dust occurs on all continents, but the drylands of the Earth (semiarid and arid) provide the main sources, notably in rainstorm channels (*wadis*), seasonally dry alluvial fans, and ephemeral and former lake depressions (*playas*). The most prolific sources are found in a 10,000 km tract extending across the Northern Hemisphere from the North Africa/Middle East region through to central and eastern Asia, which makes up one third of Earth's land area. Dust sourced in North Africa and the Middle East constitutes up to 70 wt.% of total global dust emissions, with Central and East Asia contributing 15% (Tanaka and Chiba, 2006). Other important sources include Australia (6%), southern Africa (3%), South America (2%), and the western U.S.A. (ca. 0.1%).

Eruptions of the world's volcanoes also contribute to the global dust flux, as do natural and human-induced wildfires (bush fires). Smoke plumes from burning vegetation provide a regional contribution to global dust emissions in the form of black carbon and silica particles (Le Blond et al., 2008), a product resulting from incomplete combustion of biomass able to absorb solar radiation. While volcanic eruptions remain largely unpredictable, natural biomass burning and some human-induced wildfires, including those arising from some agricultural practices, tend to be seasonal.

Dust pathways and deposition

The atmospheric dust pathways from Earth's two dominant dust-source regions (North Africa/Middle East and central and eastern Asia) are global in distribution and seasonally variable under the influence of three global wind systems – the subtropical trade winds, the Westerlies, and the monsoons. Other transport pathways include the Middle East (including north-westerly *shamal* winds), Australia (seasonally variable owing to southern monsoon influence but including the Westerlies), southern South America (Westerlies), southern Africa (southern Trade Winds), and the Great Basin of the USA (Westerlies) (Figure 1). Modeling studies have suggested that composition of far traveled dust in the northern hemisphere is a mixture of particles derived from all three major sources (North Africa, the Middle East, and central and eastern Asia: Uno et al., 2009).

In North Africa during the northern winter (February and March), mobile anticyclonic (high) pressure cells strengthen the NE trade winds (*Harmattan*) that raise long-lasting, highly concentrated mineral dust palls into the lower troposphere, carrying dust into the Atlantic basin (Figure 1). In summer (June–August), Saharan dust outbreaks are lower in frequency and density but may persist for several weeks. Summer heating generates a thermal depression over the Sahara, which raises the regional anticyclone to altitudes above 1,500 m, causing summer dust outbreaks at higher tropospheric levels. This semipermanent Saharan Air Layer (Prospero and Carlson, 1981) sustains a westward flow of dust across the Atlantic that reaches South and Central America and southeastern North America. In the autumn-winter transition (October and November), low temperatures and increased rainfall in North Africa's Sahel zone inhibit dust release.

Global dust pathways from sources in central and eastern Asia are also largely driven by the Westerlies. Although more seasonal than the Saharan system, with dust storm frequency peaking between winter and late spring, dust release may occur in all seasons in north and west China. In winter, the presence of a large and stable high pressure cell (the Siberian High) maintains very cold winters, but it becomes smaller and less stable as spring approaches. Vigorous Westerly pre-frontal low pressure troughs track along the Siberian High's retreating flanks, raising dust from two main dryland sources (Mongolia/northeast China and Taklimakan Desert/northern Tibetan plateau) into the upper troposphere. These Asian pathways impact upon western, northern, and eastern China, Korea, Japan, Pacific islands, and North America. Dust from both sources traveling within both lower and upper troposphere reaches the Greenland ice sheet and the French Alps (Bory et al., 2003: Grousset et al., 2003).

Environmental effects of airborne dust

The presence of dust particles in the atmosphere affects Earth environments in several ways. These include impact on climate by changes in the radiation budget of the atmosphere, lowering of air quality, changes in the biogeochemical cycles including enrichment of soils, delivery of vital nutrition for, but also inflicting damage upon marine systems, and impacts upon human health by inhalation.

Atmospheric effects

The radiation budget is an expression of the state of balance between scattering and absorption of shortwave radiation received from the sun and long-wave (infrared) radiation emitted from Earth's surface (Figure 2). This balance is affected by the presence and characteristics of airborne dust including the load (concentration) and vertical thickness of the dust pall, and the properties of the dust particles including size, shape (and, hence, surface area), density, and mineralogical/chemical composition. Optical properties of dust, notably the ability to absorb and scatter light, are influenced by particle size, shape, and chemical composition. By reducing the solar radiation that reaches Earth's surface, dust concentration

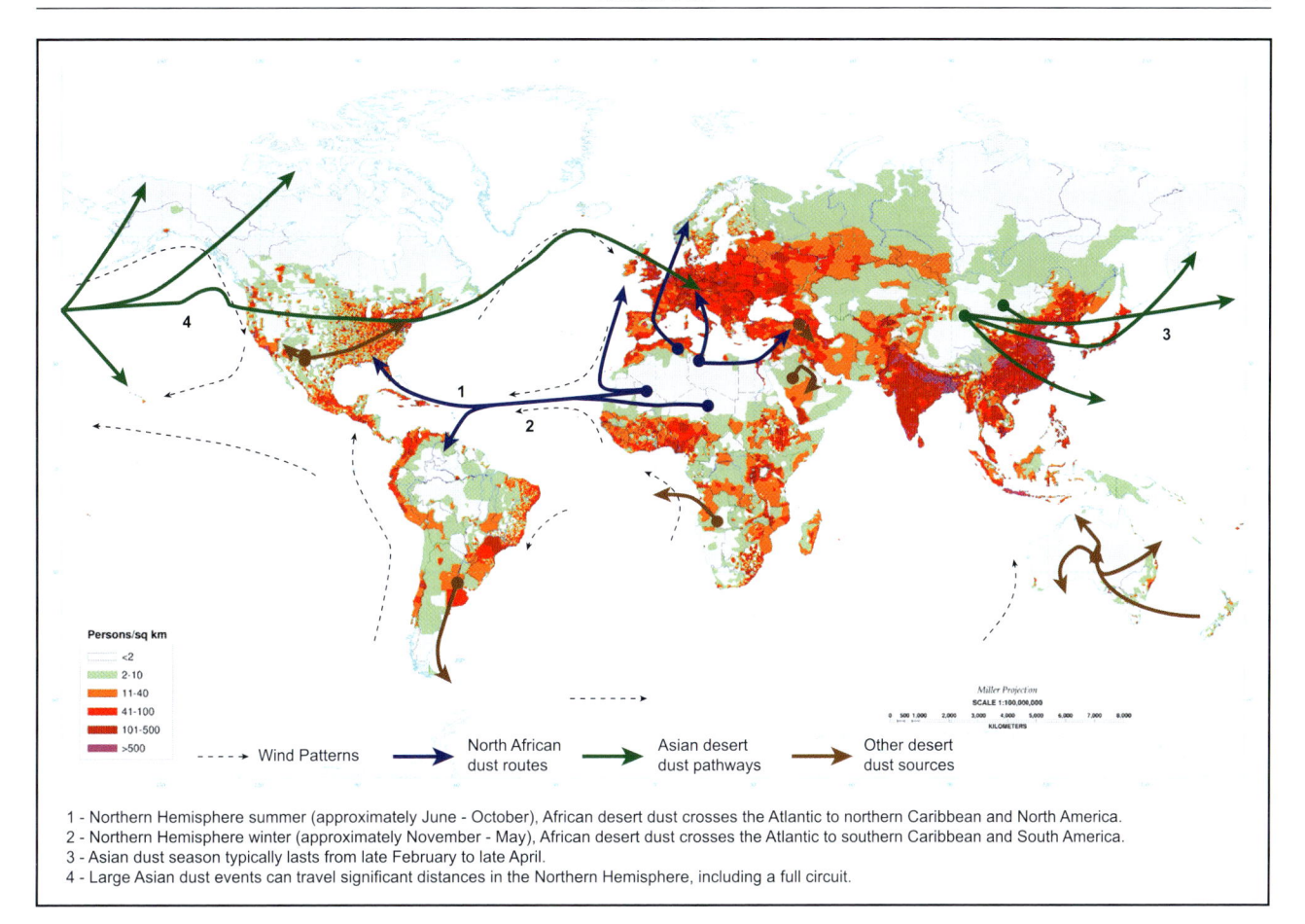

Global Dust, Figure 1 Primary sources of mineral dust and their generalized atmospheric pathways (Modified after Griffin, 2007), and world population density 1994 (United States Department of Agriculture). Global pathways of Asian dust derived from two major sources traverse areas with population densities of over 100 to more than 500 people per km². Smaller regions with significantly dense populations traversed by Saharan global dust palls include the greater Caribbean (from southeast USA to northern South America), while both Asian and Saharan global dust palls traverse moderately dense populations (>41 per km² and locally >101 per km²) in Europe and North America.

causes deterioration of air quality, warming of the aerosol layer, and cooling of Earth's surface.

Large atmospheric dust loads increase energy absorption levels. This is also the case with palls of coarse-grained dust commonly found close to dust sources, but the fining of mean particle size with distance along transport pathways and the increase in the proportion of small, disk- or blade-shaped mineral particles (Lawrence and Neff, 2009) increases energy scattering. The albedo of the Earth's surface (the ratio of light reflected from a surface compared to light received from the sun) beneath a dust pall and certain atmospheric dynamics can also influence the radiation balance. Surfaces with high albedo (i.e., high reflectivity), including snow and deserts, are warmed beneath a dust pall, while darker surfaces (low albedo), such as forested terrain and oceans, are cooled.

Mineral dust also affects the atmosphere's chemical composition by adsorption onto dust particles of atmospheric compounds, including those of nitrogen and sulfur. Dust particles are commonly coated with nitrates, involving heterogeneous processes in which reaction rates rise with increasing relative humidity when initially dry dust particles enter more humid air masses. Nitric acid reacts with atmospheric ammonia, forming ammonium nitrate particles, which have the potential to scatter solar radiation. Dust particles are also frequently coated with sulfur compounds which, attached as sulfur dioxide, rapidly convert to sulfuric acid. Airborne sulfate is important in scattering incoming solar radiation. At least 10% of atmospheric sulfate is associated with dust particles, the figure reaching 50% over extensive parts of the globe (Dentener et al., 1996). Reactions between acids and soluble dust particles diminish atmospheric acidity and remove aerosols with the potential to affect the radiation balance (Arimoto, 2001). The smallest dust fractions also serve as condensation nuclei in the process of cloud formation; cloud cover scatters solar radiation and thus cools the surface of land and ocean.

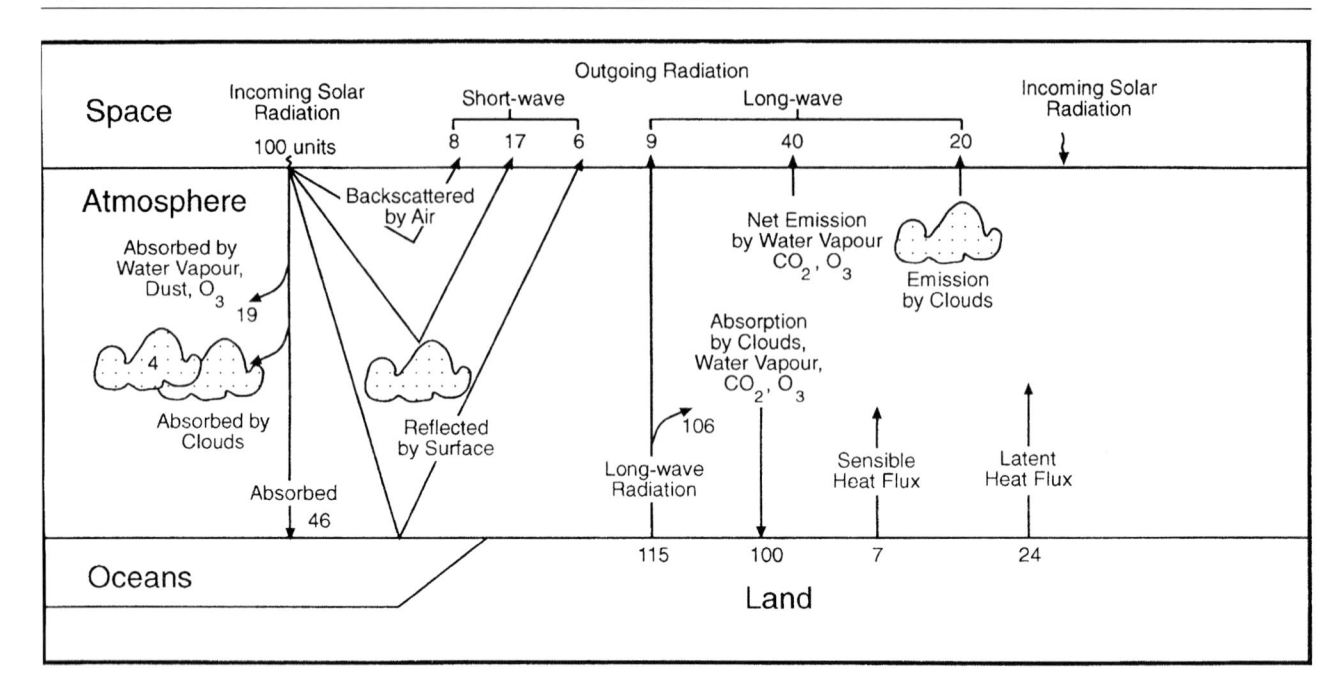

Global Dust, Figure 2 The radiation balance of the Earth. Numbers refer to arbitrary units of radiation (After Pickering and Owen, 1997, page 105. Re-drawn after Our future world: global environmental change (1989)).

Volcanic eruptions release mineral particles and compounds rich in sulfur and fluorine. Plume dust confined to the troposphere warms the higher altitudes and cools the Earth's surface, but the effects of many volcanic eruptions are relatively short lived. In contrast, major explosive eruptions from which ash, mineral compounds, and gaseous aerosols reach the stratosphere can prolong the "solar dimming" process and associated surface cooling effects for years. On June 15, 1991, a cataclysmic eruption of Mount Pinatubo volcano (Philippine Islands) ejected debris over 5 km^3 in volume that reached 35 km into the atmosphere (Newhall et al., 1997). The fine-grained ash cloud from this second largest eruption of the twentieth century encircled the globe.

Some wild (or bush) fires can give rise to dust and smoke plumes that are generally local to regional in scale rather than global. The environmental impact of dense pollutants arising from such fires can be intense although they are frequently short lived (Morawska and Zhang, 2002). However, the fires in the Indonesian forests in 1997–98 burned about eight million hectares of land, inflicting huge damage on the tropical ecosystem and a range of health problems across the region, as well as suffering a severe economic downturn across the whole region (Newhall et al., 1997). In the dry summer of 1998, extensive fires broke out in the boreal forest zone of both western and eastern Siberia, reaching the far east of Russia by October. The estimate of the area of forest burning is 1.1×10^{7} with carbon (as CO_2, CO, CH_4, non-methane hydrocarbons, and C particles (as smoke)) entering the atmosphere (Kajii et al., 2002).

Soils and ecosystems

The transportation of dust by wind action, as well as being a degradation process involving soil erosion and nutrient loess, also plays a positive role in adding vital nutrients to ecosystems following deposition on both land and the open ocean. Actions that can have effects on the biogeochemical cycles include enrichment of soils on land and phytoplankton growth in the ocean.

It is well known that African dust reaches the SE of North America and NE South America as well as the Caribbean and other islands (Prospero and Lamb, 2003), but soils on the limestone bedrock of this region have been regarded as the result of long-term residual dissolution of the carbonate-rich bedrock. In a study of the clay-rich soils of Barbados, the Florida Keys, and the Bahamas, Muhs et al. (2007) found that certain rare earth and other trace elements are present in only very low concentrations in the local coral. Also, while tephra from the Lesser Antilles volcanic arc includes a broad range of mixed oceanic and continental crustal components, African dust and the fine-grained (<20 μm) fraction of the lower Mississippi valley loess have a narrow range of such components. Thus, it appears that North African dust is the source of the dominant parent materials in the soils of the Florida Keys and the Bahamas and an important source material in the soils of Barbados. It was also conjectured that the lower Mississippi loess may have been an important source of the regional soils for the past 700,000 years.

Wind-transported mineral dust, as a vital source of nutrition in marine ecosystems, can also influence the nutrient dynamics and biogeochemical cycles in the world's oceans. For example, deposition of iron in oceanic waters stimulates nitrogen fixation by plankton, thus enhancing productivity; the dominant input of iron is provided by wind-blown dust from the drylands of the world (Jickells et al., 2005). In contrast, influxes of airborne dust may have deleterious effects on ocean ecosystems. For example, in addition to transporting nutrients to the Caribbean and Gulf of Mexico region, Saharan dust also deposits harmful microorganisms, pathogens, organic pollutants, and metals (Shinn et al., 2000; Garrison et al., 2003, 2006; Griffin, 2007). Studies in the Caribbean and elsewhere have suggested that changes in the level of nutrients affect coral reef health because dust can enhance the effect of rapid growth of algae, which causes changes in marine microbial diversity (Griffin and Kellogg, 2004).

Impacts of mineral particles on human health

Many inhaled mineral particles are intrinsically toxic and/or carcinogenic and, together with the dose (dust concentration plus length of exposure), are prime causal factors in respiratory diseases. The potential of inhaled mineral particles to raise morbidity and mortality rates varies with particle properties including mineralogical and geochemical composition, degree of crystallinity, size, shape, density, solubility, and reactivity with human lung tissue and fluids; other factors affecting variability include pathogens attached to the minerals, dust concentration (the load), and the natural vulnerability of individual human subjects.

Human lung airways are susceptible to the reactivity of fine mineral dust, which is conventionally divided into the coarse (<10 μm, or PM10), fine (<2.5 μm, or PM 2.5) and ultrafine fractions (<100 nm). Dust particles with diameters < 4 μm can reach the lung's distal bronchioles and alveoli (Figure 3). Macrophage cells (mononuclear phagocytes) patrol the fluids that line the respiratory tract and engulf (i.e., phagocytise) foreign particles. Alveolar macrophages are a first line of defense in the clearing of particles from the alveoli (Plumlee et al., 2006). Acidic lysosomes and digestive enzymes within alveolar macrophages aid in the dissolution of soluble particles while, at the same time, releasing chemicals (cytokines) that attract other macrophages to assist in clearing additional foreign particles (Lehnert, 1992, cited in Plumlee et al., 2006). Macrophages, with diameters in the range 5–10 μm, are able to engulf only small particles (<2–5 μm in size (Fubini and Areán, 1999)). Particles that avoid being phagocytised and escape other clearance processes injure lung tissue. PM10 particles are usually cleared by coughing, but upper airway morbidity may result if mineral particles are sufficiently toxic and inhalation exposure is prolonged. Nanoparticles are released into the environment by natural processes and by industrial activity including manufacturing and engine exhausts, but little is known about their concentrations in global dust and their likely health impact therein.

Mineral particle toxicity

Important naturally occurring and potentially toxic particles that may be carried in global dust include crystalline silica, fibrous amphiboles, trace elements, and alkaline salts. Prolonged inhalation of silica (SiO_2: commonest form – quartz) can cause chronic (long-term) inflammation and fibrosis of the lung leading to silicosis, with potential progression to lung cancer, as known from occupational health studies. Quartz pathogenicity is influenced by the condition of particle surfaces (e.g., freshly fractured or presence of mineral coatings) and the presence on or within them of toxins. For example, aluminum salts and metallic iron can reduce the toxicity of quartz and so inhibit lung inflammation but, on the other hand, trace amounts of ferric or ferrous iron can stimulate surface oxidative stress leading to epithelium cell damage (Donaldson and Borm, 1998).

Emissions of silica, silicates, and trace metals during volcanic eruptions can have acute (short-term) effects on exposed subjects, especially those with preexisting respiratory and cardiac conditions. However, opinion remains divided as to whether exposure to volcanic ash leads to chronic diseases such as silicosis (Horwell and Baxter, 2006). The impact on lung tissue of toxic trace elements adsorbed onto mineral particles is also not fully understood, but epidemiological evidence points to an association with human airways inflammation, both acute (asthma, bronchitis, rhinitis) and chronic (fibrosis).

Dust pathways: particle size, mineral dust composition, and potential health impact

Decline in mean particle size and bulk mineral compositional changes, owing to both dry and wet deposition *in transit*, results in progressive increase in the fine silt and clay grades with distance from source (Aluko and Noll, 2006). Compositional changes can include a decline in primary mineral content, such as quartz, feldspars, and carbonates, and a proportionate increase in the phyllosilicates (e.g., clay minerals). Trace element proportions may also be raised by size fractionation during deflation and deposition. Thus, source-proximal dust (local and regional, <10 km and <1,000 km from source, respectively) may differ in composition from its distal equivalent (global dust: >1,000 km from source: Lawrence and Neff, 2009); however, this progression can be complicated by mixing of dusts derived from additional sources along the dust pathway.

The health risk to communities living close to major dust sources is likely to be higher than for more distal populations. Both acute and chronic morbidity can arise from dust inhalation close to global dust sources, as recorded in the Sahara and in western China, but information on health impacts of global-scale mineral dust is sparse and opinion is more contentious

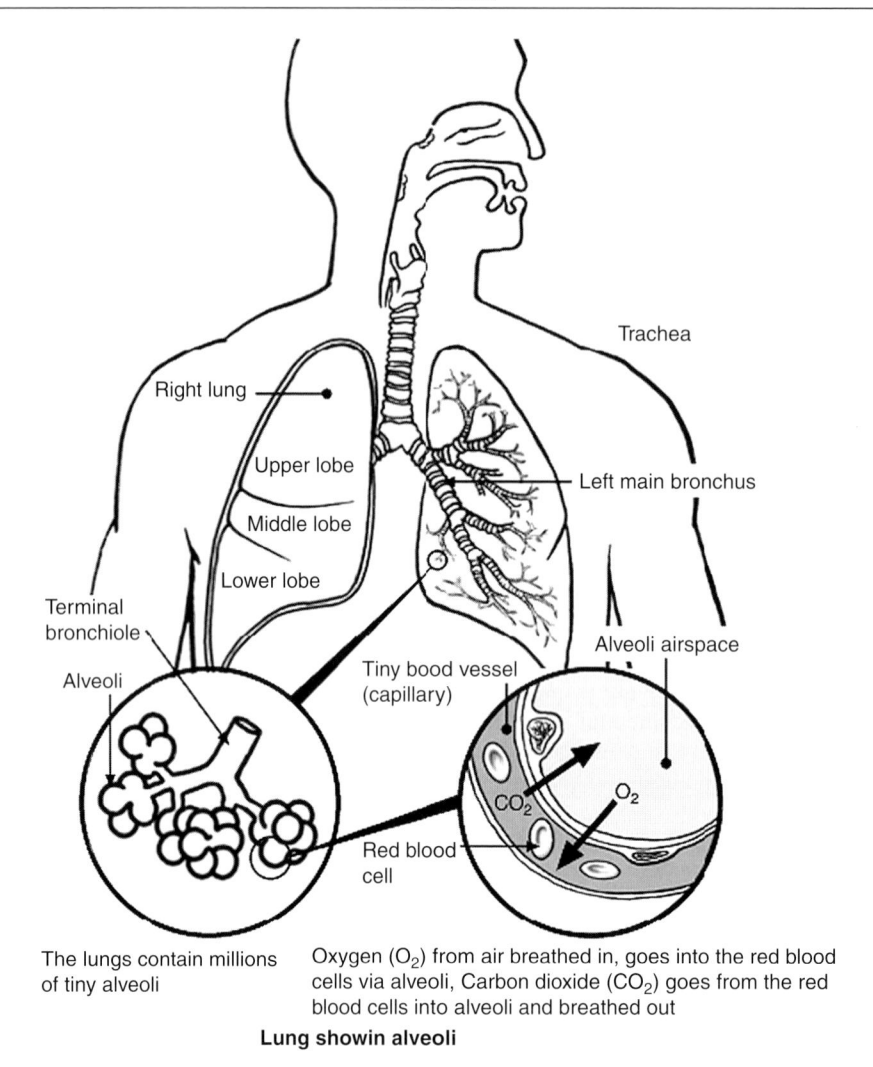

Global Dust, Figure 3 Diagram showing lungs, alveoli, and oxygen-carbon dioxide gas exchange (From http://www.patient.co.uk/diagram/Lungs-and-alveoli.htm).

(e.g., Monteil and Antoine, 2009). Cartographic comparison of the major global dust pathways and global population density suggests that the regions of greatest potential risk from exposure to global dust may be western, northern, and eastern China, Korea, and Japan (Figure 1).

Broader agreement on causal links between distal dust loads and human morbidity rests on the presence within dust particles of microorganisms (bacteria, fungi, pollen, spores, and viruses: e.g., Griffin, 2007). Health threats include the meningitis pathogen (*Neisseria meningitidis*) (Molesworth et al., 2003), but the role of airborne pathogens in epidemics is not clearly defined. Concentrations of potentially harmful elements and compounds released by human action sometimes exceed those in natural dust palls (Holmes and Miller, 2004), but estimates of the magnitude of anthropogenic pollutant input to the global dust flux vary widely, mainly because of the sparseness of measured observational data.

Summary

Natural mineral particles, a major component of global dust, are products of rock weathering and erosion, soil-forming processes, and volcanic eruptions. Principal global dust sources are located in arid and semiarid terrains, mainly in the Northern hemisphere. Lofted into the atmosphere by strong winds, dust palls frequently travel for thousands of kilometers, sometimes circling the Earth. Dust palls impact upon the atmosphere's radiation budget and so affect the climate and biogeochemical cycles. Airborne dust can also affect human health, inhalation of toxic minerals and pathogens contributing to human morbidity and mortality. People living close to global dust sources may be subject to health risks greater than in communities located further along a dust pathway, but opinion is divided. The magnitude and impact of anthropogenically released toxins and pathogens in global dust palls are poorly known.

Bibliography

After Pickering, K. T., and Owen, L. A., 1997. *An introduction to global environmental issues*, 2nd edn. London/New York: Routledge. 512 pp.

Aluko, O., and Noll, K. E., 2006. Deposition and suspension of large, airborne particles. *Aerosol Science and Technology*, **40**(7), 503–513.

Arimoto, R., 2001. Eolian dust and climate: relationships to sources, tropospheric chemistry, transport and deposition. *Earth-Science Reviews*, **54**, 29–42.

Bory, A. J. M., Biscaye, P. E., and Grousset, F. E., 2003. Two distinct seasonal Asian source regions for mineral dust deposited in Greenland (NorthGRIP). *Geophysical Research Letters*, **30**, 1167, doi:10.1029/2002GL016446, 2003.

Darwin, C., 1846. An account of the fine dust which often falls on vessels in the Atlantic Ocean. *Quarterly Journal of the Geological Society of London*, **2**, 26–30.

Dentener, F. J., Carmichael, G. R., Zhang, Y., Lelieveld, J., and Crutzen, P. J., 1996. The role of mineral aerosols as a reactive surface in the global troposphere. *Journal of Geophysical Research*, **101**, 22869–22889.

Donaldson, K., and Borm, P. J. A., 1998. The quartz hazard: a variable entity. *Annals of Occupational Hygiene*, **42**, 287–294.

Durant, A. J., Bonadonna, C., and Horwell, C. J., 2010. Atmospheric and environmental impacts of volcanic particulates. *Episodes*, **6**, 235–240.

Fubini, B., and Areán, C. O., 1999. Chemical aspects of the toxicity of inhaled mineral dusts. *Chemical Society Reviews*, **28**, 373–381.

Garrison, V. H., Shinn E. A., Foreman, W. T., Griffin, D. W., and six others. 2003. African and Asian dust: from desert soils to coral reefs. *Bioscience* **53**, 469–480.

Garrison, V. H., Foreman, W. T., Genualdi, S., Griffin, D. W., Kellogg, C. A., Majewski, M. S., Mohammed, A., Ramsubhag, A., Shinn, E. A., Simonich, S. L., and Smith, G. W., 2006. Saharan dust – a carrier of persistent organic pollutants, metals and microbes to the Caribbean? *Revista de Biologia Tropical*, **54**(Suppl. 3), 9–21.

Gieré, R., and Querol, X., 2010. Solid particulate matter in the atmosphere. *Elements*, **6**, 215–222.

Gieré, R., 2010. Atmospheric particles. *Elements*, **6**, 215–252.

Griffin, D. W., 2007. Atmospheric movement of microorganisms in clouds of desert dust and implications for human health. *Clinical Microbiology Reviews*, **20**, 459–477.

Griffin, D. W., and Kellogg, C. A., 2004. Dust storms and their impact on ocean and human health: dust in earth's atmosphere. *EcoHealth*, **1**, 284–295.

Grousset, F. E., Ginoux, P., Bory, A., and Biscaye, P. E., 2003. Case study of a Chinese dust plume reaching the French Alps. *Geophysical Research Letters*, **30**, 6, doi:10.1029/2002GL016833.

Guthrie, G. D., 1997. Mineral properties and their contributions to particle toxicity. *Environmental Health Perspectives*, **105**, 1–16.

Holmes, C. W., and Miller, R., 2004. Atmospherically transported metals and deposition in the southeastern United States: local or transoceanic? *Applied Geochemistry*, **19**, 1189–1200.

Horwell, C. J., and Baxter, P. J., 2006. The respiratory health hazards of volcanic ash: a review for volcanic risk mitigation. *Bulletin of Volcanology*, **69**, 1–24.

Jickells, T. D., An, Z. S., Andersen, K. K., Baker, A. R., Bergametti G and 14 others, 2005. Global iron connections between desert dust, ocean biogeochemistry, and climate. *Science* **308**, doi: 10.1126/science.1105959.

Kajii, Y., Cato, S., Streets, D. G., Tsai, N. Y., Shvidenko, A., Nilsson, S., McCallum, I., Minko, N. P., Abushenko, N., Altyntsev, D., and Khodzer, T. V., 2002. Boreal forest fires in Siberia in 1998: estimation of area burned and emissions of pollutants by advanced very high resolution radiometer satellite data. *Journal of Geophysical Research*, **107**(D24), 4745, doi:10.1029/2001JD001078.

Lawrence, C. R., and Neff, J. C., 2009. The contemporary physical and chemical flux of aeolian dust: a synthesis of direct measurements of dust deposition. *Chemical Geology*, **267**, 46–63.

Le Blond, J. S., Williamson, B. J., Horwell, C. J., Monro, A. K., Kirk, C. A., and Oppenheimer, C., 2008. Production of potentially hazardous respirable silica airborne particulate from the burning of sugar cane. *Atmospheric Environment*, **42**(22), 5558–5568.

Lehnert, B. E., 1992. Pulmonary and thoracic macrophage subpopulations and clearance of particles from the lung. *Environmental Health Perspectives*, **97**, 17–46.

Molesworth, A. M., Cuevas, L. E., Connor, S. J., Morse, A. P., and Thomson, M. C., 2003. Environmental risk and meningitis epidemics in Africa. *Emerging Infectious Diseases*, **9**, 1287–1293.

Monteil, M. A., and Antoine, R., 2009. African dust and asthma in the Caribbean: medical and statistical perspectives. *International Journal of Biometeorology*, **53**, 379–381.

Morawska, L., and Zhang, J., 2002. Combustion sources of particles. 1. Health relevance and source signatures. *Chemosphere*, **49**, 1045–1058.

Muhs, D. R., Budahn, J. R., Prospero, J. M., and Carey, S. N., 2007. Geochemical evidence for African dust inputs to soils of western Atlantic islands: Barbados, the Bahamas, and Florida. *Journal of Geophysical Research*, **112**, F02009, doi:10.1029/2005JF000445, 2007.

Newhall, C., Hendley, J. W., II and Stauffer, P. H., 1997. The Cataclysmic 1991 Eruption of Mount Pinatubo, Philippines, U.S. Geological Survey Fact Sheet, pp. 113–197.

Plumlee, G. S., Morman, S. A., and Ziegler, T. L., 2006. The toxicological geochemistry of earth materials: an overview of processes and the interdisciplinary methods used to understand them. *Reviews in Mineralogy and Geochemistry*, **64**, 5–57, doi:10.2138/rmg.2006.64.2.

Prospero, J. M., and Carlson, T. N., 1981. Saharan air outbreaks over the tropical North Atlantic. *Pure and Applied Geophysics*, **119**, 677–691.

Prospero, J. M., and Lamb, P. J., 2003. African droughts and dust transport to the Caribbean: climate change implications. *Science*, **302**, 1024–1027.

Shinn, E. A., Smith, G. W., Prospero, J. M., Betzer, P., Hayes, M. L., Garrison, V. H., and Barber, R. T., 2000. African dust and the demise of Caribbean coral reefs. *Geophysical Research Letters*, **27**, 3029–3032.

Sparks, R. S. J., 1986. The dimensions and dynamics of volcanic eruption clouds. *Bulletin of Volcanology*, **48**, 3–15.

Tanaka, T. Y., and Chiba, M., 2006. Transport of Saharan dust from the Bodele depression to the Amazon Basin: a case study. *Atmospheric Chemistry and Physics Discussions*, **10**, 4345–4372.

Uno, I., Eguchi, K., Yumimoto, K., Takemura, T., Shimizu, A., and 5 others, 2009. Asian dust transported one full circuit around the globe. *Nature Geoscience* **2**, 557–560.

Zhang, D., 1984. A preliminary analysis of synoptic climatology of falling dust since the historical period in China. *Scientia Sinica*, **Serie B 3**, 278–288. 825–836.

Cross-references

Albedo
Dust Storm
Epidemiology of Disease in Natural Disasters
Exposure to Natural Hazards

Loess
Mortality and Injury in Natural Disasters
Mt Pinatubo
Risk
Volcanic Ash
Wildfire

GLOBAL NETWORK OF CIVIL SOCIETY ORGANIZATIONS FOR DISASTER REDUCTION

Terry Gibson
Global Network of Civil Society Organisations for Disaster Reduction (GNDR), Teddington, Middlesex, UK

Definition

International network of nongovernmental organizations collaborating in the field of *Disaster Risk Reduction*.

Discussion

The Global Network of Civil Society Organizations for Disaster Reduction (GNDR) is an international network of civil society organizations working to influence and implement disaster risk reduction policies and practice around the world. It relies on the commitment, diversity of skills, knowledge, and extensive reach of its membership at all levels (particularly local) across virtually every region of the world. Members work for a broad range of organizations including large national and international NGOs, local or community-based organizations, and academic and/or research institutions. Many within the Network serve on their own national and regional networks, alliances, and associations.

The stated Goal of GNDR is to harness the potential of global civil society to engage strategically in disaster risk reduction policy and practice – placing the interests and concerns of vulnerable people at the heart of policy formation and implementation.

The need for the Global Network was identified by a number of civil society organizations at the World Conference for Disaster Reduction in 2005. It was initiated with the close support of the United Nations International Strategy for Disaster Reduction (UN-ISDR) Secretariat, in collaboration with the Special Unit for South–South Cooperation of the United Nations Development Program (UNDP). It was officially launched in Geneva during the first session of the Global Platform for Disaster Risk Reduction (GP-DRR) in June 2007.

GNDR works closely with civil society organizations, government bodies, UN agencies, and international institutions, with the intention of amplifying the voice and influence of disaster-prone communities at the national, regional, and international levels through sharing of learning and experiences, building consensus, and supporting collaborative approaches and joint actions.

A key project based on these principles is the "Views from the Frontline" (VFL) participative monitoring project (http://www.globalnetwork-dr.org/VFLreports.htm), which was undertaken by GNDR in 2009. The VFL project involved network members in a face-to-face survey of stakeholders involved in Disaster Risk Reduction at the local level, in order to establish what progress had been made in terms of the United Nations "Hyogo Framework for Action" on Disaster Risk Reduction, which had been established in 2005. The survey generated data from over 7,000 respondents at a local level, including local government officials, civil society members, and community members. There was a special focus on women and young people.

The VFL report was presented at the second session of the GP-DRR in June 2009 with the aim of promoting stronger linkages between policy established at the national level based on the HFA, and practical implementation and resourcing at the local level. The VFL program will be continued to compare results with the 2009 "baseline" survey.

Bibliography

www.globalnetwork-dr.org

Cross-references

Civil Protection and Crisis Management
Communicating Emergency Information
Community Management of Hazards
Disaster Diplomacy
Disaster Research and Policy History
Disaster Risk Management
Disaster Risk Reduction
Education and Training for Emergency Preparedness
Emergency Management
Federal Emergency Management Agency (FEMA)
Hyogo Framework for Action 2005–2015
Information and Communications Technology
Integrated Emergency Management System
International Strategies for Disaster Reduction (IDNDR and ISDR)
Red Cross/Red Crescent, International Federation of Vulnerability

GLOBAL POSITIONING SYSTEMS (GPS) AND NATURAL HAZARDS

Norman Kerle
University of Twente, Faculty of Geo-Information Science and Earth Observation (ITC), Enschede, The Netherlands

Synonyms

Global Satellite Positioning System; GPS; Sat Nav

Definition

Global Positioning Systems (GPS) are all-weather systems based on constellations of satellites that allow worldwide terrestrial, maritime, and aerial navigation and position determination, using passive receivers that triangulate a three-dimensional position based on radio signals received from at least four satellites.

Discussion

Although Global Positioning System (GPS) is a generic term, it has become synonymous with the US-operated NAVSTAR GPS. With the first satellites launched in 1978, it became fully operational in 1995. Developed primarily for military purposes, it remains the only complete global satellite-based positioning system. It employs 24 active satellites orbiting at approximately 20,200 km altitude, ensuring that between 4 and 12 satellites are within direct line of sight of anywhere on Earth. Three-dimensional positions are determined using passive devices that receive encoded signals from at least four satellites, triangulating the position based on distances to the satellites, calculated from the message travel time (see *Electromagnetic Radiation (EMR)*). Different signals are encoded for precision military positioning and coarser civilian uses. Early handheld GPS receivers achieved positional accuracies of about 100 m. After discontinuation in 2000 of "Selective Availability," which introduced an artificial error in the civilian signal, this increased to approximately 20 m. In 2005, a signal with a different frequency was added, allowing errors due to frequency-dependent ionospheric signal delay to be calculated, improving horizontal accuracies to about 5 m. By using data from a base station with well-known coordinates, the measurements from a mobile (rover) GPS can be corrected (differential GPS), leading to mm-accuracies (see Kennedy, 2002).

GPS has become vital in disaster risk applications, since all aspects of disaster risk (see *Remote Sensing of Natural Hazards and Disasters*) have a certain location and extent that GPS allows to locate or track. Differential GPS is frequently used to collect precise ground control points needed for photogrammetric applications, but also to monitor mm-scale movements of tectonic plates or deforming volcanoes. It is also essential for airborne laser scanning as it provides precise positioning of the aircraft. GPS is the backbone of car navigation used in rapid emergency response, and reduces the hazards of air travel and maritime navigation. More recently GPS-based applications have been developed that allow multi-agent emergency response, based on knowledge of real-time positions of all response personnel. Modern GPS receivers are tiny and cheap, and increasingly common in cameras and mobile phones.

Although weather-independent, GPS requires line of sight; thus, under heavy tree cover or indoors use is limited. Development of pseudolites that broadcast GPS-like signals from fixed positions in urban areas, or linking a GPS framework with other radio networks for indoors navigation, look promising. In addition to NAVSTAR, Europe is developing the competing and more accurate Galileo system, whereas Russia has been reviving the GLONASS system (originally also commissioned in 1995), expecting to operate a global system with 24 satellites (Cojocaru et al., 2009). The Chinese Beidou system, planned as a regional service, will morph into the global Compass system with 35 satellites by 2015.

Bibliography

Cojocaru, S., Birsan, E., Batrinca, G., and Arsenie, P., 2009. GPS-GLONASS-Galileo: a dynamical comparison. *Journal of Navigation*, **62**, 135–150.

Kennedy, M., 2002. *The Global Positioning System and GIS*. London and New York: Taylor & Francis.

Cross-references

Electromagnetic Radiation
Remote Sensing of Natural Hazards and Disasters

GLOBAL SEISMOGRAPH NETWORK (GSN)

Allison Bent
Geological Survey of Canada, Natural Resources Canada, Ottawa, ON, Canada

Synonyms

Global Digital Seismograph Network (GSN); Global Telemetered Seismograph Network (GTSN)

Definition

The Global Seismograph Network (GSN) is a high quality, standardized, digital, real-time seismograph network consisting of more than 150 stations distributed globally.

Discussion

The GSN provides free, open access data via the Incorporated Research Institutions for Seismology Data Management Center (IRIS DMC), which is also responsible for archiving the data from the GSN and many other seismograph stations. The data may be used for research, monitoring, or outreach purposes. The network is operated by IRIS and the United States Geological Survey (USGS) with additional funding from the National Science Foundation (NSF) and through coordination with the international seismological community. The GSN is an American contribution toward the International Federation of Digital Seismograph Networks' goal of uniform Earth coverage by broadband, three component seismographs. The GSN represents the most recent version of several global seismograph networks operated principally by the United States starting with the Worldwide Standardized Seismograph Network (WWSSN) in the 1960s. As seismograph, digital archiving, and telecommunications technology improved, the network was upgraded to the Digital Worldwide Standardized Seismograph Network (DWWSSN), then the Global Digital Seismograph Network (GDSN), and finally the GSN. The Global Telemetered Seismograph Network (GTSN) is a smaller, also US initiated, network aimed at filling in gaps in global coverage and consists of 13 seismograph stations primarily in the southern hemisphere. A more thorough description of some of the earlier networks may be found in Engdahl et al. (1982) and the references therein.

Bibliography

Butler, R., and Anderson, K., 2009. GSN. *IRIS Annual Report 2008*, pp. 6–7.

Engdahl, E. R., Peterson, J., and Orsini, N. A., 1982. Global digital networks-current status and future directions. *Bulletin Seismological Society of America*, **72**, S243–S259.

Incorporated Research Institutions for Seismology, 2010. Global Seismograph Network, http://www.iris.edu/hq/programs/gsn.

Cross-references

Seismograph/Seismometer

H

CASE STUDY

HAITI EARTHQUAKE 2010: PSYCHOSOCIAL IMPACTS

James M. Shultz[1], Louis Herns Marcelin[2], Zelde Espinel[1], Sharon B. Madanes[3], Andrea Allen[4], Yuval Neria[5]
[1]Center for Disaster & Extreme Event Preparedness (DEEP Center), University of Miami Miller School of Medicine, Miami, FL, USA
[2]Interuniversity Institute for Research and Development (INURED), Port-au-Prince, Haiti, and University of Miami, Coral Gables, FL, USA
[3]Columbia University, New York, NY, USA
[4]Barry University, Miami Shores, FL, USA
[5]Columbia University, The New York State Psychiatric Institute, New York, NY, USA

Synonyms

Disaster behavioral health; Disaster health; Disaster mental and behavioral health; Disaster mental health

Overview

At 4:53 PM local time (21:53 GMT) on January 12, 2010, a magnitude 7.0 earthquake devastated the Haitian capital city of Port-au-Prince and surrounding area, killing an estimated 222,570–316,000 persons (Inter Agency Standing Committee, 2010; CBC News, 2011). The Haiti 2010 earthquake is noteworthy among natural disasters that subject the affected human population to intense psychological stressors (Table 1).

Norris and colleagues (2002) posit that disasters that possess two or more of the following four characteristics are likely to create significant mental health consequences for the affected community: (1) large numbers of deaths and/or injuries, (2) widespread destruction and property damage, (3) disruption of social support and ongoing economic problems, and (4) "human" contribution to the disaster's causation. By the numbers, the Haiti 2010 earthquake possesses all of these attributes: (1) mass mortality, (2) near-total physical destruction over large areas, (3) social dislocation to a point bordering on societal collapse, and (4) ample evidence of human amplification of the earthquake's harm (preventable deaths and severe injuries attributable to dangerous and derelict standards for housing construction; uncommonly widespread and brutal interpersonal violence directed against earthquake survivors in the aftermath) (Shultz et al., 2011).

Profiling the Haiti 2010 earthquake will provide insights into the psychologically traumatizing capacity of this exceptional event (Shultz et al., 2011). The psychological consequences of the Haiti 2010 earthquake can be portrayed across five salient dimensions: (1) disaster type, (2) severity, (3) duration, (4) mortality, and (5) scope and social context of psychological impact (Table 2).

Disaster type: event description and disaster consequences

Haiti occupies the western third of the island of Hispaniola, one of the Greater Antilles islands, located between Puerto Rico and Cuba. The Haiti 2010 earthquake occurred south of the east-west trending "strike-slip" fault zone separating the massive North American tectonic plate (to the north) from the Caribbean plate (to the south). Relative to the North American plate, the Caribbean plate "slips" about 20 mm to the east each year. In Haiti, the plate boundary between these two plates presents a more complex scenario; historically, shearing pressures have created an additional fracture along the ragged juncture. The resulting tectonic shard, "the Gonave microplate," on which Haiti rides, is bracketed by two fault zones running east-west in northern and southern Haiti. The Haiti 2010 earthquake occurred along the Enriquillo-Plantain

Haiti Earthquake 2010: Psychosocial Impacts, Table 1 Haiti 2010 earthquake: disaster stressors

Impact phase stressors	Post-impact phase stressors
Exposure to severe shaking	Exposure to ongoing aftershocks
Exposure to multiple strong aftershocks	Physical injury and personal harm
	Pain, debility, loss of function
	Loss of limb
Physical injury and personal harm	Rampant infectious diseases
Entrapment in collapsed structures	Searching for missing loved ones
	Caring for injured loved ones
	Grief for lost loved ones
Perception of risk of death/extreme harm	Lack of access to emergency care
	Lack of access to primary medical care
Panicked flight	Lack of access to mental health care
Loss of loved one	Lack of survival needs (food, water)
Separation from loved ones during event	Lack of utilities and essential services
	Lack of sanitation, public health
Witnessing harm to others	Lack of communications
Witnessing death of others	Lack of transportation
Witnessing grotesque, troubling scenes	Lack of personal security
	Rumors of additional shocks to come
Damage/destruction of home	Damage/destruction of home
	Loss of personal possessions
Displacement from home	Displacement
Community-wide physical destruction	Living in temporary camps
	Witnessing dead bodies
	Witnessing severely harmed persons
	Witnessing traumatized/bereaved survivors
	Looting and gang activities
	Interpersonal and gender-based violence
	Damage to worksite
	Loss of employment
	Financial hardship
	Damage to place of worship
	Absence of government infrastructure
	Damage to schools
	Earthquake trauma reminders from aftershocks and scenes of destruction
	Family distress

Garden Fault Zone (EPGFZ) to the south. Along the EPGFZ, the plates had been "locked" in position for approximately 250 years, with escalating stress prior to the moment of release. At that moment, 4:53 PM, January 12, 2010, along 65 km of the EPGFZ, the Caribbean Plate jolted violently eastward an average distance of 1.8 m (with amplitudes as large as 4 m in some locales). As precisely described by Calis et al. (2010), the earthquake occurred on a previously unmapped steeply dipping fault that makes a high angle with the Enriquillo-Plantain fault.

With a rupture that was as close as 17 km from downtown Port-au-Prince, Haiti's capital and major population center, this near-surface earthquake shattered the city and terrified its population. At Moment Magnitude 7.0, the Haiti 2010 earthquake was very intense, but for perspective, the February 2010 earthquake in Maule, Chile, registered Moment Magnitude 8.8, generating 500 times the energy of its Haitian counterpart. Although the 2010 Chilean earthquake ranks high among the most powerful seismic events, its memory was short lived, whereas the weaker 2010 Haitian earthquake, by decimating a densely packed and structurally vulnerable population, has made its mark on human history and on the psyche of the Haitian people.

The entire nation of Haiti experienced the sensations of ground movement during the initial shock, with millions in the throes of significant shaking. The death toll, in the range of 222,570–316,000 deaths (Inter Agency Standing Committee, 2010; CBC News, 2011), places this event among the deadliest sudden-impact natural disasters on record. More than 300,000 survivors were seriously injured. Among these, an estimated 4,000 survivors sustained amputating injuries. Millions experienced excruciating pain from personal injury, or directly observed gruesome harm or brutal death as buildings toppled, or witnessed piles of decomposing bodies strewn throughout the streets. In the aftermath, with hundreds of thousands of collapsed or uninhabitable dwellings, 1.3 million persons were internally displaced, eventually finding temporary lodging in the myriad impromptu camps that were hastily erected. In the weeks following the earthquake, displaced persons, deprived of secure shelter and personal privacy, were easy prey to looting, gang violence, and gender-based violence including rape. Environmental hazards posed risks for falls, punctures, abrasions, and lacerations. Destruction of infrastructure, leading to absence of clean water, sanitation, and hygienic healthcare services, produced a cascade of untoward outcomes: infectious disease outbreaks, life-threatening wound infections, and malnutrition contributing to preventable deaths among infants and young children. Population-wide vulnerability following the earthquake exacerbated the spread and severity of a raging outbreak of cholera that caused illness in 470,000 persons and killed almost 7,000 in 1 year (CDC, 2011) (Table 3).

Disaster severity and psychological impact

The confluence of multiple physical and psychological consequences distinguishes the Haiti 2010 earthquake as singularly destructive. In a disaster, one of the strongest predictors of the severity of psychological effects is the degree of exposure to the forces of harm (Neria et al., 2008; Shultz et al., 2007, 2012a). The absolute magnitude of the forces of harm largely determines the extent of damage, destruction, displacement, death, and injury, as well as the extent of stress and trauma experienced by the disaster-affected population (Shultz et al., 2011).

At the moment of impact, exposure to the Haiti 2010 earthquake was experienced throughout the nation as ground shaking. Both the objective intensity and the subjective experience of ground shaking relate directly to the distance from the earthquake's epicenter. Persons closest to the strike zone were subjected to the kinesthetics of extreme shaking, compounded by the multisensory

Haiti Earthquake 2010: Psychosocial Impacts, Table 2 Haiti 2010 earthquake: hazard profile

Disaster type	*Definition*	Catastrophe/complex emergency
	Initial event classification	Natural disaster/geophysical/earthquake: left-lateral strike-slip faulting at angle with the Enriquillo-Plantain Garden fault system
	Ongoing event classification	Complex emergency: natural disaster followed by infrastructure collapse, massive loss of life, large-scale displacement, and interpersonal violence
	Forces of harm	Strike-slip faulting on a fault associated with a tectonic plate boundary
		Severe ground shaking
		Structural collapse of buildings
Magnitude and severity	*Moment magnitude*	Initial earthquake: M 7.0
		First 4 weeks: 16 aftershocks $\geq$ M5.0
		First 4 weeks: 59 aftershocks $\geq$ M4.5
	Modified Mercalli Intensity	Epicenter: X (on a 12-point scale: I–XII)
		MMI intensity generally diminishes as a function of distance from epicenter
Place dimension	*Epicenter*	A 40-km long rupture; shortest distance is 17 km from Haitian capital of Port-au-Prince
	Hypocenter	Rupture from near surface to 13-km depth
	Geography/geology	The region is near a tectonic plate boundary region separating the Caribbean plate and the North America plate. The main shock mostly represents faulting on an unmapped steeply dipping fault at angle with the Enriquillo-Plantain Garden fault zone (EPGFZ).
	Scale/scope	Intense shaking/destruction throughout southeastern Haiti, particularly in the capital of Port-au-Prince and surrounding areas.
		Shaking of varying intensity felt throughout entire nation of Haiti (27,750 km^2) and adjoining portions of Dominican Republic on island of Hispanola.
		Psychological impact extends to Haitian "diaspora" in United States, Canada, Caribbean, responders from many nations
	Built environment	Poorest nation in Western Hemisphere
		Rank 147/182 on Human Development Index
		55% below extreme poverty index
		Widespread structural deficiencies
		High-density urban crowding
Time dimension	*Initial earthquake strike*	January 12, 2010, 16:53 local time (21:53 GMT)
	Duration	Initial earthquake: 15–35 s of strong shaking (estimate)
	Frequency	Strong aftershocks of gradually decreasing frequency over several months (59 aftershocks $\geq$ M4.5 in first 4 weeks)
	Duration of life/health risk	Months: cholera outbreak caused illness in 470,000 with 6,631 deaths through October 2011
	Duration of disruption	Years

experience of structural disintegration occurring in all directions around them. In some areas of particularly intense tumult, "liquefaction" occurred as the sediment lost its weight-bearing properties, causing structures to sink and crumble (Shultz et al., 2011). Punctuating the jarring, disorienting, fear-triggering sensations of tremulousness and vibration, many persons were physically injured, pinned by falling rubble, or entrapped in collapsed structures. Bodily pain, injury, immobilization, trapped confinement, and perceived life threat are psychological stressors (Shultz et al., 2011). Haitian survivors engaged in frenetic, bare-handed scramble to find and extricate family members and neighbors who had disappeared amid the heaving earth and caving debris.

Assuming an earthquake source of given orientation and rupture characteristics, it is possible to estimate the levels of shaking over an area and convert them to ratings on the Modified Mercalli Intensity Scale. The Modified Mercalli Intensity Scale (MMI) is a 12-point rating scale (I–XII). Each point on the scale is composed of an estimate of shaking intensity and a corresponding description of structural damage. The MMI estimates how strongly an earthquake is felt by people in a geographical area, and the corresponding amount of damage to buildings and dwellings.

For the Haiti 2010 earthquake, the United States Geological Survey (USGS, 2010a) has created a "shakemap" displaying the geographic areas, and estimated population numbers, for each of the MMI categories. For example, more than 1.0 million Haitians experienced violent or extreme shaking (MMI Scale IX or higher), with heavy to catastrophic damage all around. An additional 3.5 million residents experienced strong, very strong, or severe ground shaking (MMI Levels VI–VIII). Together, these data provide clear evidence that almost 4.6 million Haitians experienced ground movement at the level of

Haiti Earthquake 2010: Psychosocial Impacts, Table 3 Haiti 2010 earthquake: disaster consequences

Disaster consequence	Haiti 2010 earthquake description
Exposure to ground shaking	MMI[a] ≥ IX: violent/extreme intensity: 1.0 million persons MMI[a] VI–VII: strong/very strong/severe intensity: 3.5 million persons
Mortality	222,570–316,000 deaths
Bereavement	Millions have lost a primary family member
Morbidity: injury	300,000 injuries requiring medical care 4,000 injuries requiring amputation
Morbidity: infectious disease	Hundreds of thousands of disease cases
Internally displaced persons (IDPs) due to earthquake	1,300,000 earthquake IDPs
Damage to homes	Destroyed homes: 97,000 Severely damaged homes: 188,000
Gender-based violence	High rate of reported rapes and assaults on women
Children without caregiver	Separation from caregiver Earthquake orphans
Lack of access to clean water	Millions of persons
Lack of access to food	Malnutrition risk: 495,000 children and 198,000 pregnant/lactating women
Lack of access to sanitation	Millions of persons

[a]MMI: Modified Mercalli Index

Haiti Earthquake 2010: Psychosocial Impacts, Table 4 Haiti 2010 earthquake: persons exposed to ground shaking by MMI level

MMI	Perceived shaking	Estimated population AT MMI level	Estimated population AT or ABOVE MMI level
X–XII	Extreme	118,000	118,000
IX	Violent	908,000	1,026,000
VIII	Severe	2,030,000	3,056,000
VII	Very strong	598,000	3,654,000
VI	Strong	926,000	4,580,000
V	Moderate	6,361,000	10,941,000
IV	Light	7,468,000	18,409,000
I–III	Weak to imperceptible	50,000	18,459,000

Source
United States Geological Survey, 2010a.

MMI VI or higher. The entire Haitian population of approximately 10 million persons experienced some degree of perceptible ground shaking, as did millions more in neighboring Dominican Republic and nearby Caribbean islands (Table 4, Figure 1).

Duration and frequency of psychological stressors

During exposure to threat or overt forces of harm, survival stress responses are activated. Moreover, the frequency of discrete disaster impacts relates to psychological trauma; a sequence of multiple strikes, and exposure to both stress and loss, tends to be more alarming than a single, discrete event (Neria and Litz, 2004). In the case of the Haiti 2010 earthquake, both duration and frequency operated in tandem to exacerbate fear and distress.

While the duration of the initial earthquake was estimated in the range of 15–35 s, repeated strong aftershocks triggered fear reactions and acted as powerful psychological reminders of the initial "mainshock" experience. Aftershocks began immediately. The first powerful aftershock (M 6.0) occurred 7 min after the "mainshock." The second aftershock (M 5.5) was experienced 12 min later. According to the United States Geological Survey (2010b), in precisely 4 weeks following the earthquake's origin, 16 aftershocks of magnitude 5.0–6.0, and 43 aftershocks of magnitude 4.5–4.9, occurred close to the original epicenter.

Disaster stress persisted long after the aftershocks tapered. Hardships in the aftermath were severe and prolonged with near-complete destruction of infrastructure, lack of survival supplies, absence of basic services, and sporadic episodes of violence on the streets and in the camps.

Mass mortality in relation to psychological impact

Sudden, concentrated, mass mortality distinguishes the Haiti 2010 earthquake from past earthquakes of similar magnitude. Most of the deaths occurred instantly or within minutes or hours of the M 7.0 mainshock. Some injured victims suffered a more protracted death over a period of days due to the circumstances of injury, entrapment, lack of access to medical care, or absence of basic survival needs.

Only one earthquake in recorded history significantly exceeded the numbers of deaths experienced in the Haiti 2010 earthquake; an estimated 830,000 persons died in the catastrophic earthquake that leveled Shaanxi, China, on January 23, 1556. Among ground-shaking seismic incidents in modern times, the Tangshan, China earthquake on July 28, 1976, with 255,000 deaths, rivaled the Haiti death toll. On December 26, 2004, a tsunami-generating seismic event killed an estimated 280,000 coastal dwellers distributed across more than one dozen nations encircling the Indian Ocean. Apart from these two earthquakes and the Indian Ocean tsunami, no other earthquake in recorded history has equaled the magnitude of mortality experienced in Haiti on January 12, 2010.

Beyond the exceptional numbers of deaths, a related feature of Haiti 2010 is the geographic concentration of mortality in a densely populated urban capital city and surrounding towns and communities. Large numbers of Haitians lost multiple family members.

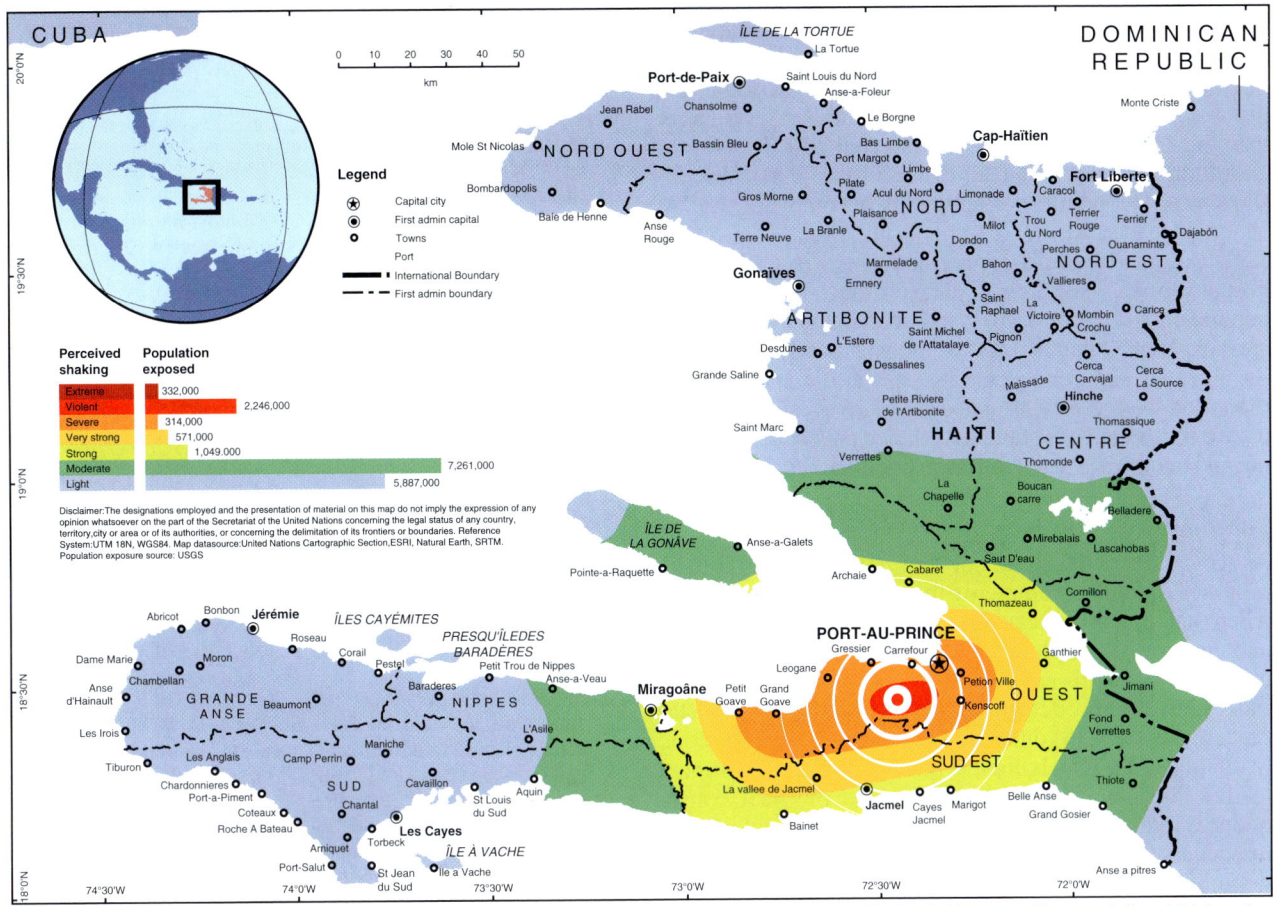

Haiti Earthquake 2010: Psychosocial Impacts, Figure 1 Map of Haiti showing the epicenter of the January 12, 2010, earthquake and population affected.

The Centre for Research on the Epidemiology of Disasters (CRED) in Brussels, Belgium, is a WHO Coordinating Center and serves as the international repository for information on all reported disasters worldwide. CRED maintains an international disaster database dating from 1900 to the present. During the 110 years of tracking international disasters, a total of 1,095 damaging earthquakes have been reported including the early 2010 earthquakes in Haiti and Chile. Collectively, across these earthquakes, 2,323,000 persons were killed. The death toll in Haiti alone accounted for 10% of all earthquake deaths in 110 years of surveillance.

Loss of a loved one in a natural disaster is one of the most psychologically devastating experiences. Traumatic bereavement, leading to complicated grief, is associated with a host of psychiatric disorders including posttraumatic stress disorder (PTSD) and depression (Neria et al., 2007). Complicated (or prolonged) grief disorder (PGD) is a relatively new diagnosis, and different from normal grief in its extended duration and symptom profile (Horowitz et al., 1997). Correlates of PGD include severe functional impairment, decreased productivity, suicidality, and physical health problems (Neria et al., 2007). Yet to be explored is the interaction between disaster trauma and loss of a loved one (Neria and Litz, 2004); many survivors of the 2010 Haiti earthquake experienced both. More research is needed to fully understand the relations between PTSD and complicated grief, and whether they differ in their risk and protective factors.

Scope and social context of psychological impact

Psychosocial consequences of disasters are wide ranging and pervasive (Shultz et al., 2012a). More persons are affected psychologically than are harmed physically (Shultz et al., 2007). The Haiti 2010 earthquake has psychologically affected not only the persons within the immediate strike zone but the entire Haitian population (10 million persons), the expansive diaspora of Haitian emigrants to the United States, Canada, and throughout the Caribbean (Shultz et al., 2012b), and large numbers of persons who have engaged in the earthquake response efforts. According to CRED data, only two earthquakes in the past 110 years affected more persons: The Haiti 2010 earthquake provides strong evidence that, in natural disasters, the "psychological footprint" of disaster is larger than the "medical footprint" (Shultz et al., 2007).

Adding to the widespread distress from physical hardships was the public perception of a complete void of national leadership to communicate with the public (INURED, 2010). In a survey conducted in Cité Soleil, Port-au-Prince, the key phrase that captured the population's sense of despair and abandonment, was, "We are on our own." This sentiment was repeatedly documented and echoed through the press and research reports. A lead researcher stated, "Failure to rally the country at a time when it is gravely wounded added to the suffering and trauma of the overall population" (INURED, 2010).

Summary

As we have previously stated, "the 2010 Haiti earthquake provides a potent example of the rare catastrophic event where all major risk factors for psychological distress and impairment are prominent and compounding," (Shultz et al., 2011). In this case example, we have illustrated the links between the descriptors of the event: (1) disaster type, (2) severity, (3) duration, (4) mortality, and (5) scope in relation to the psychological impact. We have attempted to describe how the psychological and physical consequences interplay and synergize in a seismic event of such impact severity marked by quantum loss of life, obliteration of infrastructure, and barbarous hardship in the aftermath. In fact, our detailed assessment of this event became the basis for introducing "trauma signature analysis" (TSIG) to the field (Shultz et al., 2011).

Bibliography

Calis, E., Freed, A., Mattioli, G., Amelung, F., et al., 2010. Transpressional rupture of an unmapped fault during the 2010 Haiti earthquake. *Nature Geoscience*, **3**, 794–799, doi:10.1038/ngeo992. Available at: http://www.nature.com/ngeo/journal/v3/n11/full/ngeo992.html. Accessed April 22, 2012.

CBC (Canada) News (2011). Haiti raises quake death toll on anniversary. Available at: http://www.cbc.ca/news/world/story/2011/01/12/haiti-anniversary-memorials.html. Accessed April 22, 2012.

Centers for Disease Control and Prevention (CDC) (2011). *Cholera in Haiti: One Year Later*. Available at: http://www.cdc.gov/haiticholera/haiti_cholera.htm. Accessed April 22, 2012.

Horowitz, M. J., Siegel, B., Holen, A., et al., 1997. Diagnostic criteria for complicated grief disorder. *The American Journal of Psychiatry*, **154**, 904–910.

Inter Agency Standing Committee (IASC)/United Nations Office for the/Coordination of Humanitarian Assistance (2010). *Response to the Humanitarian Crisis in Haiti Following the 12 January 2010 Earthquake*. Available at: www.humanitarianinfo.org/iasc/downloaddoc.aspx?docID=5366&type. Accessed April 22, 2012.

INURED, 2010. *Voices from the Shanties: A Post-Earthquake Rapid Assessment of Cite Soleil, Port-au-Prince*. Port-au Prince, Haiti: Interuniversity Institute for Research and Development (INURED), pp. 1–24.

Neria, Y., and Litz, B., 2004. Bereavement by traumatic means: the complex synergy of trauma and grief. *Journal of Loss and Trauma*, **9**, 73–87.

Neria, Y., Gross, R., Litz, B., et al., 2007. Prevalence and psychological correlates of complicated grief among bereaved adults 2.5–3.5 years after 9/11 attacks. *Journal of Traumatic Stress*, **20**, 251–262.

Neria, Y., Nandi, A., and Galea, S., 2008. Post-traumatic stress disorder following disasters: a systematic review. *Psychological Medicine*, **38**(4), 467–480.

Norris, F. H., Friedman, M., and Watson, P., 2002. 60,000 disaster victims speak. Summary and implications of the disaster mental health research. *Psychiatry*, **65**, 240–260.

Shultz, J. M., Espinel, Z., Galea, S., and Reissman, D. B., 2007. Disaster ecology: implications for disaster psychiatry. In Ursano, R.J., et al., (eds.) *Textbook of Disaster Psychiatry*. Cambridge, UK: Cambridge University Press, pp. 69–96.

Shultz, J. M., Marcelin, L. H., Madanes, S., Espinel, Z., and Neria, Y., 2011. The trauma signature: understanding the psychological consequences of the Haiti 2010 earthquake. *Prehospital and Disaster Medicine*, **26**(5), 353–366.

Shultz, J. M., Neria, Y., Allen, A., and Espinel, Z., 2012a. Psychological impacts of natural disasters. In Bobrowsky, P. (ed.), *Encyclopedia of Natural Hazards*. Heidelberg: Springer, pp. xx–xx.

Shultz, J.M., Besser, A., Kelly, F., Allen, A., Schmitz, S., Hausmann, V., Marcelin, L.H., and Neria, Y., 2012b. Psychological consequences of indirect exposure to disaster due to the Haiti earthquake: Six hundred nautical miles from danger. *Prehospital and Disaster Medicine* (in press).

United States Geological Survey, 2010a. Pager – M 7.0 – Haiti region. Available at: http://earthquake.usgs.gov/earthquakes/pager/events/us/2010rja6/index.html. Accessed April 22, 2012.

United States Geological Survey, 2010b. Magnitude 7.0 – Haiti region. Available at: http://earthquake.usgs.gov/earthquakes/recenteqsww/Quakes/us2010rja6.php. Accessed April 22, 2012.

Cross-references

Casualties Following Natural Hazards
Community Management of Hazards
Critical Incidence Stress Syndrome
Disaster Relief
Disaster Risk Reduction (DRR)
Earthquake
Education and Training for Emergency Preparedness
Emergency Shelter
Epicentre
Epidemiology of Disease in Natural Disasters
Federal Emergency Management Agency (FEMA)
Hazardousness of Place
Hypocentre
Indian Ocean Tsunami 2004
Modified Mercallit (MM) Scale
Mortality and Injury in Natural Disasters
Post Disaster Mass Care Needs
Post-Traumatic Stress Disorder (PTSD)
Psychological Impacts of Natural Disasters
Recovery and Reconstruction After Disaster
Risk
Seismology
Tangshan, China (1976 earthquake)

HARMONIC TREMOR*

Melanie Kelman
Natural Resources Canada, Geological Survey of Canada, Vancouver, BC, Canada

Definition

Harmonic tremor is defined as seismicity consisting of a continuous low frequency, often single frequency, sine wave with smoothly varying amplitude, and is caused by complex interactions between magma (see *Magma*), exsolved gases, and bedrock at volcanoes (McNutt, 2000).

Harmonic tremor comprises a sequence of seismic waves whose dominant frequencies typically range from 1 to 5 Hz. Harmonic tremor and spasmodic bursts (also called spasmodic tremor) are specific variants of more general volcanic tremor. However, harmonic bursts have a relatively uniform signal, whereas spasmodic bursts have a higher frequency, pulsating, irregular signal.

Volcanic tremor, including harmonic tremor, may persist for minutes to days or longer, may lack a distinct beginning or end, frequently precedes and almost always accompanies eruptions, and was the most consistent short-term indicator of impending eruptions in a study of 200 calderas worldwide (McNutt, 2000) (see also *Volcanoes and Volcanic Eruptions*). Detection of tremor is a major part of most volcano-monitoring programs (see *Monitoring Natural Hazards*). However, intrusion events can also produce similar seismicity, even though no eruption occurs.

Models for the origin of harmonic tremor involve oscillations of fluid-filled cracks (e.g., Julian, 1994; Chouet 1988), boiling of water (Leet, 1988), water or gas motions in a volcanic conduit (Hellweg, 2000; Maryanto et al., 2008), or thermoacoustic instabilities (Busse et al., 2005). Spasmodic tremor may be associated with modification of the volcanic plug or vent (Lesage et al., 2006; Jolly et al., 2010) or other brittle failure related to the movement of magma, rather than being caused by the movement of magma itself (Nishi et al., 1996). Variations in depth, duration, and amplitude of volcanic tremor suggest that multiple mechanisms may be responsible for its generation, even at a single volcano.

Bibliography

Chouet, B., 1988. Resonance of a fluid-driven crack: radiation properties and implications for the source of long-period events and harmonic tremor. *Journal of Geophysical Research*, 93, 4375–4400.
Hellweg, M., 2000. Physical models for the source of Lascar's harmonic tremor. *Journal of Volcanology and Geothermal Research*, 101, 183–198.
Jolly, A. D., Sherburn, S., Jousset, P., and Kilgour, G., 2010. Eruption source processes derived from seismic and acoustic observations of the 25 September 2007 Ruapehu eruption – North Island, New Zealand. *Journal of Volcanology and Geothermal Research*, 191, 33–45.
Julian, B. R., 1994. Volcanic tremor: nonlinear excitation by fluid flow. *Journal of Geophysical Research*, 99, 11859–11877.
Leet, R. C., 1988. Saturated and subcooled hydrothermal boiling in groundwater flow channels as a source of harmonic tremor. *Journal of Geophysical Research*, 93, 4835–4849.
Lesage, P., Mora, M. M., Alvarado, G. E., Pacheco, J., and Métaxian, J.-P., 2006. Complex behavior and source model of the tremor at Arenal volcano, Costa Rica. *Journal of Volcanology and Geothermal Research*, 157, 49–59.
Maryanto, S., et al., 2008. Constraints on the source mechanism of harmonic tremors based on seismological, ground deformation, and visual observations at Sakurajima volcano, Japan. *Journal of Volcanology and Geothermal Research*, 170, 198–217.
McNutt, S. R., 2000. Seismic monitoring. In Sigurdsson, H. (ed.), *Encyclopedia of Volcanoes*. San Diego: Academic Press, pp. 1095–1119.
Nishi, Y., Sherburn, S., Scott, B. J., and Sugihara, M., 1996. High-frequency earthquakes at White Island volcano, New Zealand: insights into the shallow structure of a volcano-hydrothermal system. *Journal of Volcanology and Geothermal Research*, 72, 183–197.

Cross-references

Magma
Monitoring and Prediction of Natural Hazards
Volcanoes and Volcanic Eruptions

HAZARD

Farrokh Nadim
Norwegian Geotechnical Institute, Oslo, Norway

Synonyms

Danger; Frequency of danger; Frequency of threat; Threat

Definition

Hazard is an event, phenomenon, process, situation, or activity that may potentially be harmful to the affected population and damaging to the society and the environment. A hazard is characterized by its location, magnitude, geometry, frequency or probability of occurrence, and other characteristics.

Introduction

Hazard involves something which could potentially be harmful to a person's life, health, and property; to the society as a whole; or to the environment. There is a lack of a common ground among different disciplines and schools of thought regarding what hazard means and different disciplines apply the term in different ways (e.g., Alexander, 2002; EEA (European Environment Agency), 2006; Emergency Management, 1998; ISSMGE (International Society for Soil Mechanics and Geotechnical Engineering), 2004; TRADE (The Training Resources and Data Exchange), 1999; UNISDR (United Nations

---

*©Her Majesty the Queen in right of Canada

International Strategy for Disaster Reduction), 2009). However, there is general consensus that hazard is not synonymous with risk (*risk*), but rather an important determinant of the latter.

The disaster risk community, geoscientists, and economic risk analysts often view hazard as a potentially damaging event, phenomenon, situation, or human activity that may have a negative impact on cultural, economic, environmental, institutional, physical, or social assets. Hazard may be a dormant or latent condition that represents a potential future threat such as an earthquake (*Earthquake*), or it may be an existing threat such as a creeping slope (*Slope Stability, Creep*). A hazard is characterized by its location, magnitude, geometry, mechanical, and other characteristics (*Frequency and Magnitude of Events*).

Dual meaning of hazard

One of the reasons for the ambiguity in the definition of hazard is that the term is also used to describe the temporal probability of occurrence of the event or situation in question. Whether hazard refers to the event or to its probability of occurrence within a given period of time depends on the context of its usage. As a scientific concept, this ambiguity in the meaning of hazard is not desirable. To avoid this problem, Technical Committee 32 of the International Society for Soil Mechanics and Geotechnical Engineering (ISSMGE), for instance, suggested using the term "danger" or "threat" to refer to the phenomenon that could lead to damage, and the term "hazard" to the probability that a particular danger (threat) occurs within a given period of time. In this definition, the characterization of danger or threat does not include any forecasting, whereas the characterization of hazard includes the assessment of frequency or temporal probability of occurrence. The ISSMGE definition of hazard is fine for scientific discussions. It is, however, not entirely consistent with the common usage of the term. For example, to be consistent with the ISSMGE terminology, one should refer to earthquakes (*earthquake*) and floods (*flood*) as natural threats, and not natural hazards.

Summary

Hazard is an event, phenomenon, process, situation, or activity that may potentially be harmful to the affected population and damaging to the society and the environment. The term is also used to describe the temporal probability of occurrence of the event or situation in question. The dual meaning of hazard creates ambiguity in scientific discussions and different disciplines apply the term in different ways. Considering the pros and cons of different definitions, the general definition of hazard given above is recommended.

Bibliography

Alexander, D. E., 2002. *Principles of Emergency Planning and Management*. New York: Oxford University Press.

Emergency Management Australia, 1998. *Australian Emergency Manuals Series Manual 3: Australian Emergency Management Glossary*. Canberra: Emergency Management Australia. Compiled and typeset by Peter Koob, Tasmania State Emergency Service. ISBN 0 7246 4675 2.

European Environment Agency (EEA) 2006. *Multilingual Environmental Glossary*. http://glossary.eea.eu.int/EEAGlossary/.

International Society for Soil Mechanics and Geotechnical Engineering (ISSMGE) – Technical Committee 32 (Risk Assessment and Management), 2004. *Glossary of Risk Assessment Terms – Version 1*. http://www.engmath.dal.ca/tc32/2004Glossary_Draft1.pdf

The Training Resources and Data Exchange (TRADE) 1999. *Glossary and Acronyms of Emergency Management Terms*, 3rd edn. Prepared for the Office of Emergency Management U.S. Department of Energy. http://orise.orau.gov/emi/prod-trng/files/glossary-emt.pdf

United Nations International Strategy for Disaster Reduction (UNISDR) 2009. *UNISDR Terminology on Disaster Risk*. http://www.unisdr.org/eng/terminology/terminology-2009-eng.html

Cross-references

Creep
Earthquake
Flood
Frequency and Magnitude of Events
Risk
Slope Stability

HAZARD AND RISK MAPPING

Brian R. Marker
London, UK

Synonyms

Hazard and risk maps are sometimes included in sets of maps under names such as "Engineering geology maps," "Environmental geology maps," "Applied geological maps" or "Maps for planning, development, or conservation."

Definitions

Hazard – the probability of occurrence within a specific period of time and within a given area from a natural phenomenon of given magnitude.
Susceptibility – the extent to which a causal mechanism might affect and destabilize a potentially hazardous system (but sometimes used as a synonym for vulnerability).
Vulnerability – the sensitivity of exposure of people and assets (buildings, infrastructure, and other property) to a hazard of a given magnitude (usually expressed as a percentage loss or as a value between 0 and 1).
Risk – the likely impacts or costs of hazardous events in terms of likely deaths, injuries, damage to properties, direct financial costs, and economic disruption due to hazardous events.

Hazard mapping – the assembly of information from field and archival sources, remote sensing and, in some cases, ground investigation and sample testing on the nature, distribution, causes, magnitude, and frequency of hazardous events.

Risk mapping – the process of estimating and depicting the likely impacts or costs of hazardous events in terms of likely deaths, injuries, damage to properties, and economic disruption due to hazardous events.

Introduction

Hazard maps showing the nature and extent of actual and potential hazards have been prepared for many years, initially manually but more recently by using Geographic Information Systems (GIS). These are used for several purposes including land-use planning, insurance, environmental protection, and disaster management. Recently increasing emphasis has been placed on risk mapping which describes potential losses from hazardous events. The flexibility and ease of amending maps using GIS is valuable because many hazard and risk maps should be revised fairly often as social, economic, and environmental circumstances change and new data and methods become available. A wide range of topics are addressed including: seismic and volcanic phenomena, tsunami, landslides, rockfall, avalanche, subsidence, flooding, storm damage, dust, wildfire, and potentially harmful substances. The variety makes it difficult to generalize about the approaches to hazard and risk mapping and types of maps. Each investigation must reflect the nature of the specific hazard so the reader should refer to other entries of this encyclopedia for details of those characteristics.

Some studies consider a single hazard but that is not always appropriate because:

(a) In addition to primary damage and injuries some types of event, especially earthquakes, trigger secondary hazards such as landslides, rockfall, subsidence, and tsunami so integrated assessment can be more useful in informing decision makers of necessary actions.
(b) Many areas are subject to several potential hazards and therefore require maps for each or a multi-hazard map.

Mapping of some hazards requires consideration of ecological, meteorological, and hydrological information and risk assessment requires social and economic information. Therefore it is sometimes necessary to organize multidisciplinary teams of researchers.

The level of detail of assessments depends on the location concerned and the accessibility and amount of available data. Thus, seismic hazard maps are often more generalized than, for example, landslide hazard maps because of the uncertainty as to where a seismic event of given magnitude and depth might have impacts compared, for example, the relatively more localized effects of slope instability.

Although many studies of hazard, vulnerability and risk have been undertaken there are still many difficulties in comparing these in terms of potential damage other than in a simplified way (Carpignano et al., 2009). Also, many countries have not yet undertaken strategic hazard mapping initiatives for large areas but there are notable exceptions such as the maps for land-use planning prepared in Norway over the past 30 years (Hestnes and Lied, 1980).

Preparation of hazard maps

The initial stage in hazard mapping is the collection, collation, and interpretation of data on the nature, frequency, and magnitude of past events. Depending on the type of hazard to be considered this may include:

(a) Field mapping and/or use of air photograph and remote sensing data to establish the geology, geomorphology, topographical characteristics, vegetation cover, etc.: use of remote sensing is important where studies are of large or remote areas.
(b) Collecting samples of soils or water to determine engineering or geochemical characteristics.
(c) Examination of historical documentary data (e.g., topographical maps and air photographs of various ages, site investigation reports, media reports, historical books, and manuscripts) on hazardous events to determine the magnitudes and frequencies.
(d) For some hazards, particularly anthropogenic contamination, mapping of former or current uses of land.
(e) Assembly of information on factors that trigger or exacerbate hazards (e.g., meteorological records in relation to ground instability, or rates of erosion that might affect landslide potential).

Databases are assembled that can be drawn upon to plot a variety of maps on cadastral and topographical base maps. Topography is usually drawn from a digital elevation model (Doornkamp, 1989).

Some illustrative, but not exhaustive, examples of types of information are given in Table 1.

Caution is needed in assembling and interpreting data because:

(a) Previously mapped data may have been plotted at a number of scales and, although GIS is scale independent, these may need to be sampled and normalized to a common level of detail (see Briggs, 2000).
(b) A long historical data record is available for some hazards in certain areas (e.g., parts of Europe and China) but is sparse in other places, in which case dating techniques (e.g., radiometric, lichenometry, tree ring analysis, thermoluminescence, and cosmogenic radionuclide) must be used to establish the chronology of events (see, for instance, Smith and Petley, 1991).

Hazard and Risk Mapping, Table 1 Some factors that are taken into account in mapping of various types of hazard and risk and selected references to examples of studies

Hazard	Comments	Example references
Seismic	Identification of the maximum credible seismic event that might affect an area, likely peak particle acceleration, and likely depth ranges of events. In order to assess impacts it is also necessary to characterize the distribution of poorly consolidated overburden and bed rock, since impacts on structures are generally greater on the former, and of areas subject to liquefaction hazards. It is also prudent to assess other hazards that may be triggered by earthquakes (secondary seismic hazards) such as landslides, rockfall, and subsidence.	Adams and Atkinson (2003); Jimenez et al. (1999); Kayabali and Akin (2003); Musson and Sargeant (2007); Sanabria and Dhu (2005).
Volcanic	Identification of volcanic centers and the types of eruption that tend to occur so that the potential effects of pyroclastic gas falls and flows, lahars and lava flows, gas emissions, and the potential for large-scale cone collapse can be assessed. Many of these require assessment of directions that materials may take in leaving the volcanic center and the topography that they may flow across. Dispersal of higher level dust clouds however requires consideration of prevailing wind directions and potential for rainfall at various times of year.	Favalli et al. (2009); Hickson (2002); Hoblitt et al. (1998); Pomonis et al. (1999); Simpson and Shepherd (2001); Wadge and Isaacs (1998).
Tsunami[a]	Prediction of the likely maximum heights reached by tsunami taking account of likely directions of origin, orientations of shorelines that may be affected, sea- or lake-floor morphology, especially gradient and the occurrence of features such as reefs, locations, and orientations of constrained straits and bays, and the distribution of adjacent low lying land.	Burbidge et al. (2008); Clague et al. (2003); Study Committee on Storm Surge and Hazard Maps (2004).
Landslide	Identifying the types and extents of various types of landslides in relation to the nature of bedrock and superficial deposits, geological structures, slope angles, morphologies and aspects, slope, vegetation and groundwater behavior, as well as the potential for reactivation of earlier landslides.	Ardizzone et al. (2002); Keaton and Rinne (2002); Chacon et al. (2006); Fell et al. (2008a, b); Glade et al. (2005); Lee and Jones (2004); Malet and Macquaire (2008).
Rockfall	Angles and roughness of slopes, block sizes, strength and morphology, heights of falls and run-out distance, dip of strata, nature, spacing and infillings of joints and other discontinuities, weathering and groundwater behavior.	Abbruzzese et al. (2009); Agliardi et al. (not dated); Ayala-Carcedo et al. (2003); Irigaray et al. (2003).
Avalanche	Tracks left by past avalanches, likely extent of snow and ice cover, slope steepness, morphology and roughness, isolation, meteorological conditions, and vegetation cover.	Arnalds et al. (2004); Barbolini et al. (2011); Gruber and Bartelt (2007); Keylock et al. (1999).
Subsidence (natural cavities)	Recorded extent and nature of caves and sinkholes, bedrock characteristics including patterns of faults, joints and other discontinuities, depth to bedrock and characteristics of superficial deposits, and groundwater behavior.	Forth et al. (1999); Edmonds (2001a, b); Marker (2010); Ragozin and Yolkin (2003); Simón-Gómez and Soriano (2002).
Subsidence (mined ground)	Subsidence associated with mined ground – recorded extent, sizes, and depths of mined ground and shafts, and of associated collapses, the type of mining undertaken, the extent of mineral deposits that could contain unrecorded mined ground, the thickness and depth of overburden, patterns of faults, joints and other discontinuities, and groundwater conditions	Choi et al. (2010); Lee and Sakalas (2001); Marker (2010); Oh and Lee (2011); Statham and Treharne (1991).
Fluvial flooding	Levels, volumes, and frequencies ("return periods") of floods of specific sizes, topography of drainage basins, precipitation patterns, zones of erosion and deposition, channel morphology, surface runoff, soil characteristics, infiltration, vegetation, urbanization, defenses, and drainage.	Buchele et al. (2006); Rango and Anderson (2007); Schwab et al. (2002)
Coastal flooding	Nature, scale and frequency of recorded events, extent of coastal lowlands, coastal aspect relative to potential storm surge directions and onshore winds, coastal impediments to surges including coastal defenses	Cendrero (1989); Monirul Islam and Sado (2000); Study Committee on Storm Surge and Hazard Maps (2004).

Hazard and Risk Mapping, Table 1 (Continued)

Hazard	Comments	Example references
Storm	Consideration of the nature, scale, frequency, and impacts of significant events in relation to meteorological records of seasonal prevailing wind directions and strengths, precipitation, and potential exposure to these slopes and structures.	Heneker et al. (2006); Hofherr (2007); Watson (1995).
Dust[b]	Particle size and moisture content of soils, and vegetation cover, in areas liable to wind erosion. Prevailing wind directions and strengths. Expected patterns of rainfall (see also volcanic ash above).	Afifi and Gad (2011); McLaurin et al. (2011).
Wildfire	Vegetation cover, type, and combustibility; topography especially slope aspects; precipitation patterns; atmospheric and vegetation humidity; and prevailing wind strengths and directions.	Chuvieco and Congleton (1989); Mermoz et al. (2005); Romme et al. (2006); Sampson et al. (2008).
Potentially harmful substances	Concentrations of potentially harmful inorganic and organic elements and compounds (natural or anthropogenic) in soils, surface water or groundwater compared with concentrations that are believed to be harmful or of concern. Concentrations of substances in vegetation, bones, teeth etc. Comparison of concentrations with legal and regulatory limits.	Birke et al. (2011); Demetriades (2011); Ducci (2000); Lahr and Kooistra (2010).

[a]Tsunami are caused by earthquakes, but also by some volcanic eruptions and landslides.
[b]Dust hazards arise from soil erosion, volcanic emissions, wildfire, and many human activities.

(c) Documents may be uneven over time, or in scope, and older records are often less reliable. Also these are not necessarily comprehensive. For instance, plans of old mines and associated shafts may not show all workings, and some may have been inaccurately plotted (Freeman Fox Ltd, 1988; Marker, 2002).

(d) Some hazard phenomena cover appreciable areas (e.g., large landslides are extensive) and can be plotted accurately but others are very localized (e.g., small sinkholes and mine subsidence crown holes) and can only be shown by point symbols. In the latter case it may be necessary to represent the occurrences as, for instance, numbers occurring per unit area (Marker, 2010).

If parts of an area have more comprehensive and accurate information than others, it is necessary to extrapolate information to the latter (e.g., if, in the case of landslides, a particular geological horizon associated with slopes exceeding a particular angle is known to have a high frequency of instability, that information can be used to judge landslide potential in other parts of the outcrop). Results can be validated to some extent by developing an assessment for one part of an area and then comparing it with another part, or with a different area (see, for instance, Chung and Fabbri, 2003). A systematic and structured approach is required when considering each type of hazard. An example is given in Figure 1.

Information is usually depicted as shaded or colored zones, contours, or pixels that are classified according to presumed levels of hazard. Weightings can be allocated from quantitative scores, be expressed by relative values on a numerical scale, or indicated by general terms such as high, medium, or low. If levels of hazard are uncertain due to uneven base data gradational boundaries between zones may needed. GIS can provide auto-contouring but the defined rules are applied strictly and results can be misleading unless these are moderated using professional judgment. A further step is to attempt to predict the nature and extent of future occurrences. This requires examination of likely frequency or probability of damaging events through statistical estimation and/or modelling (Doornkamp 1989).

Modelling simulates a hazard, in many cases by expressing the physics of a process, through either of the following:

(a) Deterministic mathematical models which provide results for a chosen scenario, often the worst case, but do not provide information on uncertainty of results.
(b) Probabilistic mathematical models which use random variables to allow for uncertainties. These are now generally preferred to deterministic approaches.

There are two basic types of simulation: event based and hazard based. The event based approach calculates, for example, the damage that would be caused at each individual building by a particular type of event and is time consuming. The hazard based method is used to estimate effects from a hazard map rather than by considering specific events, but tends to overestimate consequences. Uncertainties either from the natural variability and randomness of complex phenomena, which can be estimated but not reduced, or inadequate data or

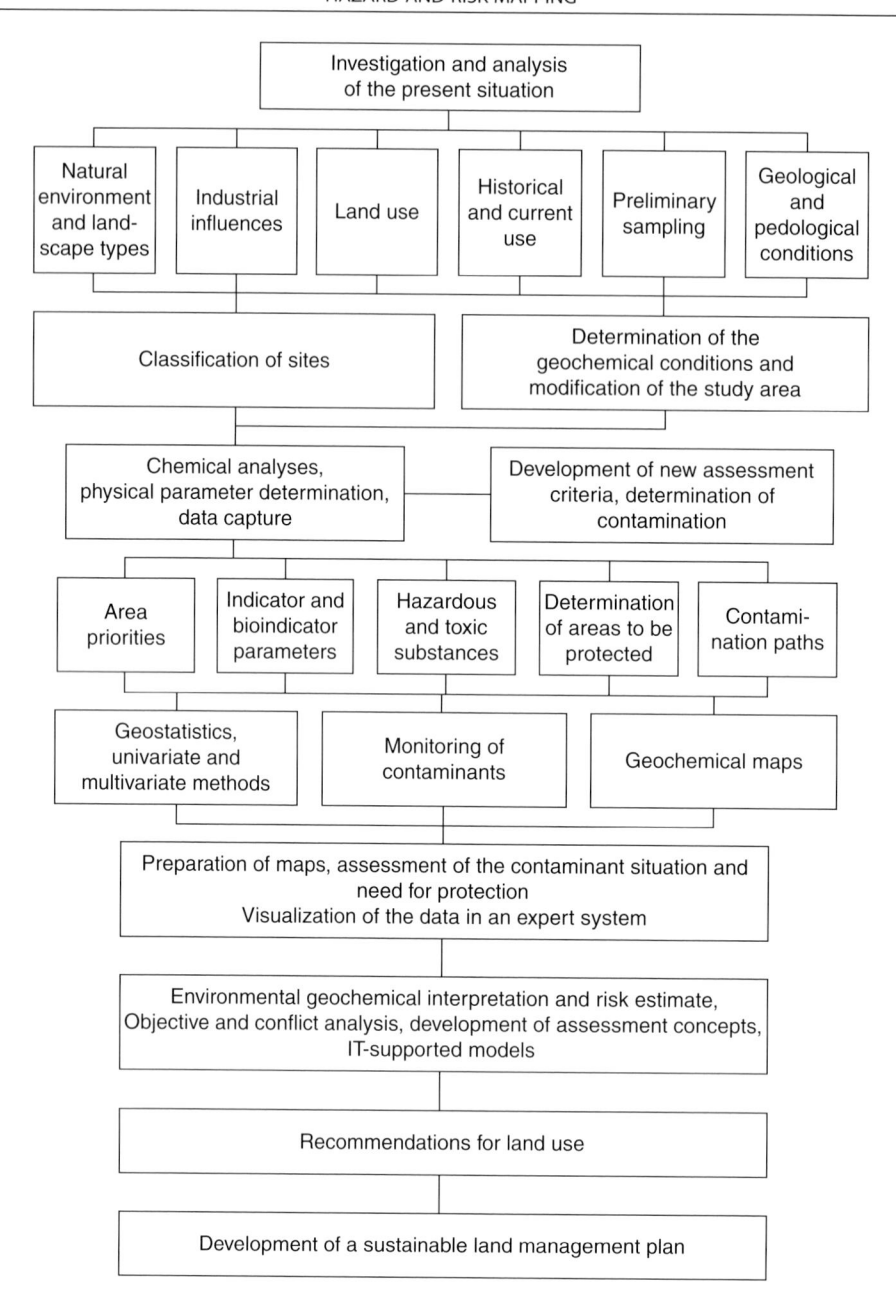

Hazard and Risk Mapping, Figure 1 Geochemical flow chart for investigating industrial influences and mining damage in urban regions (the flow of investigation is from the top to bottom (Birke et al., 2011)).

incomplete models due to limitations on understanding which can be reduced as data and understanding improve. Several models may be developed to allow for the hazardous events of various magnitudes. Alternatively a model may be used to consider several sets of originating and triggering conditions (Sanabria and Dhu, 2005). The design of models varies between types of hazard. Some illustrative examples are:

(a) Seismic hazard – Source zone characteristics can be defined and assumptions made about the origin and magnitude of a range of possible events; energy transmission and attenuation for events of specific magnitudes and distances estimated and evaluated in terms of subareas based on regolith characteristics; and building damage fragility curves used to assess impacts (Sanabria and Dhu 2005).

(b) Landslide hazard – A model can be developed by examining the proportion of rainfall that percolates down into soils and underlying bedrock, groundwater levels and flows allowing groundwater levels to be

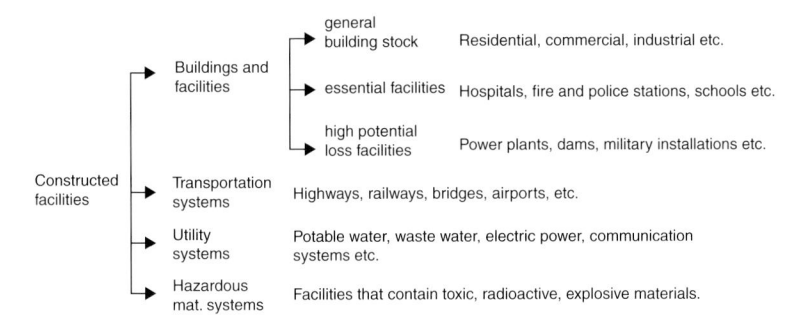

Hazard and Risk Mapping, Figure 2 A structural inventory classification system (FEMA, 1999).

estimated for different time intervals and related to slope characteristics that determine stability thus enabling the estimation of safety factors for various slope profiles (Terlien et al., 1995).
(c) Flood hazard – Hydrological frequency analysis can be used to consider the magnitude of extreme events based on the historical record, and combined with results of hydraulic modelling taking account of channel characteristics at various points down-stream and resistance to flow to establish river channel discharges followed by sensitivity analysis to identify the key influencing factors (Mosquera-Machado and Ahmad, 2007).
(d) Wildfire – After comparing vegetation, slope characteristics, and meteorological information fuel models can be used to evaluate fire heat release, flame length, and rates of spreading for extreme but realistic circumstances (Romme et al., 2006).

In addition to mapping potential levels of hazards within an area, it is important to understand the extent to which causal mechanisms might exacerbate or destabilize a potentially hazardous system (susceptibility or sensitivity), for instance, if excavation or loading of slopes might lead to landsliding, to show where particular precautions are likely to be needed.

The reliability of hazard maps depends on the amount and accuracy of original data and the validity of prediction and modelling. While maps can provide a reasonably good understanding of what has happened in the past and an indication of where future problems might be expected, they are not necessarily an accurate guide because: some past events may have obscured effects of hazards; future events may take place where hazards have not previously occurred; and environmental change and land use may cause different patterns of hazard potential and risk in future (e.g., Ardizzone et al., 2002).

The presence of a hazard in a particular locality does not necessarily mean that it will cause significant adverse effects. A major hazardous event in a fairly sparsely populated area might have few consequences a moderately sized hazardous event in a densely populated area can lead to severe impacts on people and property. Therefore it is necessary to assess the people and assets that might be vulnerable to, and at risk from, likely levels of hazards.

Preparation of vulnerability and risk maps

Risk maps are essentially vulnerability maps enhanced to estimate the consequences in terms of lives, injuries, and/or financial impacts of events of given magnitudes. This requires the examination of the susceptibility of assets and populations and the losses. Layers of information are needed on matters such as:

- Buildings, constructions, and infrastructure, including the likelihood of damage to particular categories of structure and structural elements and the financial costs of replacing these (see Federal Emergency Management Agency (FEMA), 1999).
- Locations of key facilities such as evacuation and aid routes, hospitals, fire and police stations, and water supply facilities (see FEMA op cit).
- The distribution of people, sometimes including factors such as social circumstances, ages of potential victims or population distributions at different times of day since this influences exposure to hazards (e.g., see Ragozin and Yolkin, 2003).
- Management of responses to disasters including emergency and aid services and consequent costs (Commission of the European Communities (CEC), 2010; Morrow, 2002; Tran et al., 2008).

An illustrative example of a structural inventory classification system for constructed facilities that can be used to consider data on, and estimate financial costs of, seismic events is shown in Figure 2.

The mapping of risk has recently extended into attempts to provide multi-risk maps for substantial areas, for instance, the Risk Map Germany initiative that aims to draw on earthquake, storm, flood, and other disaster expertise (Müller et al., 2006).

Communication of results

It is important that hazards are taken properly into account when developing policies for land use and development. The extent to which that is done varies in different parts of the World, thus in some areas:

(a) There are legal and regulatory requirements to make and use maps for relevant types of hazard, for example: Regulations are in place in Germany, Austria, Switzerland (see, for example, Aulitzky, 1994; Petrascheck and Keinholz, 2003; Bründl et al., 2009), and Iceland (Iceland Meteorological Office, 2000); in the European Union steps are being taken to develop an integrated risk management strategy (European Environment Agency (EEA), 2010); and, in the USA State, Indian Tribal and local governments are required to develop a hazard mitigation plan (FEMA, 2011).

(b) Maps are prepared as necessary and are used as guidance rather than being mandatory, for instance, most types of hazard maps in the United Kingdom (see, for instance, DTLR, 2002).

(c) Maps are prepared in response to events but are not yet formally required either legally or for guidance.

Where legal and regulatory provisions exist, planners are aware of hazard and risk maps and have professional support whereas elsewhere, some planners do not have experience in considering natural hazards. Therefore a class of simplified interpretative maps is sometimes developed for public communication purposes. These focus on the ways in which land should be considered for development using categories such as "unsuitable for built development," "areas for which full site investigation by a competent person is required with any planning application," "areas within which a walk over survey by a competent person is required to determine whether detailed site investigation is needed," and "areas where the hazard is not thought to be present" (see, for instance, Thompson et al., 1998).

Sometimes a "traffic light" approach is used (i.e., red for hazardous, amber for caution, and green for safe). This has the attraction of being simple but caution is needed because, for many types of hazard, development may still be possible in red areas if suitable precautions are taken. Also areas defined as of high relative hazard in one study are not necessarily equivalent to those defined as of high relative hazard in another. It is desirable to use similar criteria in all areas so that hazards are not exaggerated with consequent loss of confidence in investment in new development, effects of value and sales potential of existing development, or increases in costs or loss of cover of insurance. Care is needed, therefore, in developing the wording of the map legend and in disseminating the findings to politicians, administrators, financial interests, international aid organizations, and communities. While planning maps provide a basis for decision making, even where the significance of the underlying technical information is not appreciated, it is best for experts to advise alongside those maps (Marker, 1998).

Conclusion

Hazard and risk mapping are essential, steps in safe and cost-effective planning for development and for protection of people but is currently undertaken unevenly, if at all. Commonly hazard mapping is first undertaken after a damaging event takes place and causes strong public concern, but does not extend to all potential hazards and is not updated as time passes and memory of the event fades. In many cases hazard maps are prepared but are not carried through to risk and planning maps even though those are of greater assistance to practitioners in development and emergency planning. However an increasing number of countries are now incorporating hazard and risk maps into legal and regulatory processes. It is hoped that others will follow suit in future.

Bibliography

Abbruzzese, J. M., Sauthier, C., and Labiouse, V., 2009. Considerations on Swiss methodologies for rock fall hazard mapping based on trajectory modelling. *Natural Hazards and Earth Systems Science*, **9**(4), 1095–1109.

Adams, J., and Atkinson, G., 2003. Development of seismic hazard maps for the proposed 2005 edition of the National Building Code of Canada. *Canadian Journal of Civil Engineering*, **30**(2), 255–271.

Afifi, A. A., and Gad, A. A., 2011. Assessment and mapping areas affected by soil erosion and desertification in the north coastal part of Egypt. *International Journal of Water Resources and Arid Environments*, **1**(2), 83–91.

Ardizzone, F., Cardinali, M., Carrara, A., Guzetti, F., and Reichenbach, P., 2002. Impact of mapping errors on the reliability of landslide hazard maps. *Natural Hazards and Earth System Sciences*, **2**(1–2), 3–14.

Arnalds, B., Jonasson, K., and Sigurdsson, S., 2004. Avalanche hazard zoning in Iceland based on individual risk. *Annals of Glaciology*, **38**(1), 285–290.

Aulitzky, H., 1994. Hazard mapping and zoning in Austria: methods and legal implications. *Mountain Research and Development*, **14**(4), 307–313.

Ayala-Carcedo, F. J., Cubillo-Nielsen, S., Alvarez, A., Dominguez, M.J., Lain, L., and Ortiz, G., 2003. Large scale rock-fall reach susceptibility maps in La Cabrera Sierra (Madrid) performed with GIS and dynamic analysis at 1:5,000. In Chacon, J., Corominas, J. (Eds) Special Issue on Landslides and GIS. *Natural Hazards*, **33**(3), 325–340.

Barbolini, M., Pagliardi, M., Ferro, F., and Corradeghini, P., 2011. Avalanche hazard mapping over large undocumented areas. *Natural Hazards*, **56**(2), 451–464.

Birke, M., Rauch, U., and Chmieleski, J., 2011. Environmental geochemical survey of the city of Stassfurt: an old mining and industrial urban area in Sachsen-Anhalt, Germany. In Johnson, C. C., Demetriades, A., Locutura, J., and Ottesen, R. T. (eds.), *Mapping the chemical environment of urban areas*. Chichester: Wiley, pp. 269–306.

Briggs, D., 2000. *Environmental health hazard mapping for Africa*. Harare, Zimbabwe: WHO-AFRO. 140 pp.

Bründl, M., Romang, H. E., Bischof, N., and Rheinberger, C. M., 2009. The risk concept and its application in natural hazard risk management in Switzerland. *Natural Hazards and Earth Systems Science*, **9**(3), 801–813.

Buchele, B., Kreibich, H., Kron, A., Thieken, A., Ihringer, J., Oberle, P., Merz, B., and Nestmann, F., 2006. Flood-risk mapping: contributions towards an enhanced assessment of extreme events and associated risks. *Natural Hazards and Earth System Sciences*, **6**(4), 485–503.

Burbidge, D., Cummins, P. R., Mleczko, R., and Hong Kie, T., 2008. A probabilistic tsunami hazard assessment for Western Australia. *Pure and Applied Geophysics*, **165**(11–12), 2059–2088.

Carpignano, A., Golia, E., di Mauro, C., Bouchon, S., and Nordvik, J. P., 2009. A methodological approach for the definition of multi-risk maps at regional level: first application. *Journal of Risk Research*, **12**(3–4), 513–534.

Cendrero, A., 1989. Mapping and evaluation of coastal areas for planning. *Coastal and Shoreline Management*, **12**(5–6), 427–462.

Chacon, J., Irigaray, C., Fernandez, T., and el Hamdouni, R., 2006. Engineering geology maps: landslides and geographic information. *Bulletin of Engineering Geology and the Environment*, **65**(4), 341–411.

Choi, J-K., Kim, K-D., Lee, S., and Won, J-S., 2010. Application of a fuzzy operator to susceptibility estimations of coal mine subsidence in Taebaek City, Korea. *Environmental Earth Sciences*, **59**(5), 1009–1022.

Chung, C.-J. F., and Fabbri, A., 2003. Validation of spatial prediction for landslide hazard mapping. *Natural Hazards*, **30**(3), 451–472.

Chuvieco, E., and Congleton, R. G., 1989. Application of remote sensing and geographic information systems to forest fire hazard mapping. *Remote Sensing of Environment*, **29**, 147–159.

Clague, J. J., Munro, A., and Murty, T., 2003. Tsunami hazard and risk in Canada. *Natural Hazards*, **28**(2–3), 435–463.

Commission of the European Communities (CEC), 2010. *Risk Assessment and Mapping Guidelines for Disaster Management*. Commission Staff Working Paper SEC (2010)1626 final, 42 pp.

Demetriades, A., 2011. Understanding the quality of chemical data from the urban environment – Part 2: measurement uncertainty in the decision making process. In Johnson, C. C., Demetriades, A., Locutura, J., and Ottesen, R. T. (eds.), *Mapping the chemical environment of urban areas*. Chichester: Wiley, pp. 77–98.

Department for Transport, Local Government and the Regions (DTLR), 2002. *Planning Policy Guidance Note 14 Development on Unstable Land Annex 2 Subsidence and planning*. London: DTLR. 56 pp.

Doornkamp, J. C., 1989. Techniques of map presentation. In McCall, G. J. H., and Marker, B. R. (eds.), *Earth Science Mapping for Planning, Development and Conservation*. London: Graham and Trotman, pp. 15–28.

Ducci, D., 2000. GIS techniques for mapping groundwater contamination risk. *Natural Hazards*, **20**, 279–294.

Edmonds, C. N., 2001. Subsidence hazard in Berkshire in areas underlain by chalk karst. In Griffiths, J. S. (ed.), *Land surface evaluation for engineering practice*. London: Geological Society. Engineering Geology Special Publication, Vol. 18, pp. 97–106.

European Environment Agency (EEA), 2010. *Disasters in Europe: More Frequent and Causing More Damage*. http://www.eea.europa.eu/highlights/natural-hazards-and-technological-accidents. Accessed 17 Mar 2012

Favalli, M., Tarquini, S., Fornaciai, A., and Boschi, E., 2009. A new approach to risk assessment of lava flow at Mount Etna. *Geology*, **37**(12), 1111–1114.

Federal Emergency Management Agency (FEMA), 1999. *HAZUS Technical Manual*. http://www.fema.gov/hazus/dl_st2.shtm. Accessed 23 Mar 2012

Federal Emergency Management Agency (FEMA), 2011. *Mitigation Planning Laws, Regulations and Guidance*. http://www.fema.gov/plan/mitplanning/guidance.shtm. Accessed 23 Mar 2012

Fell, R. J., Corominas, J., Bonnard, C., Cascini, L., Leroi, E., and Savage, W. Z., 2008a. Guidelines for landslide susceptibility, hazard and risk zoning for land-use planning. *Engineering Geology*, **102**(3–4), 85–98.

Fell, R. J., Corominas, J., Bonnard, C., Cascini, L., Leroi, E., and Savage, W. Z., 2008b. Commentary on guidelines for landslide susceptibility, hazard and risk zoning for land-use planning. *Engineering Geology*, **102**(3–4), 99–111.

Forth, R. A., Butcher, D., and Senior, R., 1999. Hazard mapping of karst along the coast of the Algarve, Portugal. *Engineering Geology*, **52**, 67–74.

Freeman Fox Ltd, 1988. *Methods of Compilation, Storage and Retrieval of Data on Disused Mine Openings and Workings*. London: HMSO. 18 + 58 + 52 pp.

Glade, T., Anderson, M., and Crozier, M. (eds.), 2005. *Landslide Hazard and Risk*. Chichester: Wiley-Blackwell. xix + 824 pp.

Gruber, U., and Bartelt, P., 2007. Snow avalanche hazard modelling of large areas using shallow water numerical methods and GIS. *Environmental Modelling and Software*, **22**(10), 1472–1481.

Heneker, P., Hoffherr, T., Ruck, B., and Kottmeier, C., 2006. Winter storm risk and residential structures – model development and application to the German state of Baden-Wüttemberg. *Natural Hazards Earth Systems Science*, **6**, 721–733.

Hestnes, E., and Lied, K., 1980. Natural-hazard maps for land-use planning in Norway. *Journal of Glaciology*, **6**(94), 331–343.

Hickson, C. J., 2002. An overview of volcanic hazard maps: past, present and future. In Bobrowsky, P. T. (ed.), *Geoenvironmental Mapping: Methods, Theory and Practice*. AA Balkema: Lisse, pp. 557–576.

Hoblitt, R.P., Walder, J.S., Dreidger, C.L., Scott, K.M., Pringle, P. T., and Vallance, J.W. 1998. *Volcano Hazards from Mount Rainier, Washington*. Revised 1998. Open file report 98-428. Denver:USGS, 11 pp.

Hofherr, T. 2007. Countryside storm hazard for Germany In *Proceedings of Forum DKKV/CEDIM – Disaster Reduction in Climate Change* 2007, October 16, Karlsruhe, 4 pp.

Icelandic Meteorological Office, 2000. *Regulations on Hazard Zoning Due to Snow- and Landslides, Classification and Utilisation of Hazard Zones, and Preparation of Provisional Hazard Mapping*. Regulation Nr. 505/2000. Reykjavik: The Ministry of the Environment.

Irigaray, C., Fernandez, T., and Chacon, J., 2003. Preliminary rock-slope-susceptibility assessment using GIS and the SMR classification. In Chacon, J., and Corominas, J. (eds.) *Special Issue on Landslides and GIS. Natural Hazards*, **33**(3), 309–324.

Jimenez, M.J., Garcia-Fernandez, M., and the GSHAP Ibero-Maghreb Working Group 1999. Seismic hazard assessment in the Ibero- Maghreb Region. *Annali di Geofisica* **42**(6), 1057–1065.

Kayabali, K., and Akin, M., 2003. Seismic hazard map of Turkey using the deterministic approach. *Engineering Geology*, **69**(1–2), 127–137.

Keaton, J. R., and Rinne, R., 2002. Engineering-geology mapping of slopes and landslides. In Bobrowsky, P. (ed.), *Geoenvironmental mapping: methods, theory and practice*. Lisse: AA Balkema, pp. 9–28.

Keylock, C., McClung, D. M., and Magnusson, M. M., 1999. Avalanche risk mapping by simulation. *Journal of Glaciology*, **45**(150), 303–314.

Lahr, J., and Kooistra, L., 2010. Environmental risk mapping of pollutants: state of the art and communication aspects. *Science of the Total Environment*, **408**(18), 3899–3907.

Lee, E. M., and Jones, D. K. C., 2004. *Landslide Risk Assessment*. London: Thomas Telford. 454 pp.

Lee, E. M., and Sakalas, C. F., 2001. Subsidence map development in an area of abandoned salt mines. In Griffiths, J. S. (ed.) *Land Surface Evaluation for Engineering Practice*. London: Geological Society. Geological Society Engineering Group Special Publication, 18, pp. 29–38.

Malet, J.P., and Macquaire, O., 2008. *Risk Assessment Methods of Landslides. Risk Assessment Methodologies for Soil Threats*. European Union Sixth Framework Programme, Scientific Support to Policies. Project report 2.2. Deliverable 2.3.2.4, 29 pp.

Marker, B. R., 1998. The incorporation of information on geohazards into the planning process. In Maund, J. G., and Eddleston, M., (eds.) *Geohazards in Engineering Geology*. London: Geological Society. Geological Society Engineering Group Special Publication, 15, pp. 385–389.

Marker, B. R., 2002. The development and current state of land instability mapping. In McInnes, R. and Jakeways, J. (eds) *Instability – Planning and Management*. London: Thomas Telford, pp. 97–108.

Marker, B. R., 2010. Review of approaches to mapping of hazards arising from subsidence into cavities. *Bulletin of Engineering Geology and the Environment*, **69**(2), 159–183.

McLaurin, B. T., Goossens, D., and Buck, B. J., 2011. Combining surface mapping and process data to assess, predict, and manage dust emissions from natural and disturbed land surfaces. *Geosphere*, **7**(1), 260–275.

Mermoz, M., Kitzberger, T., and Veblen, T., 2005. Landscape influences on occurrence and spread of wildfires in Patagonian forests and shrublands. *Ecology*, **86**(10), 2705–2715.

Monirul Islam, M. D., and Sado, K., 2000. Development of flood hazard maps of Bangladesh using NOAA-AVHRR images with GIS. *Hydrological Sciences*, **45**(3), 337–355.

Morrow, B. H., 2002. Identifying and mapping community vulnerability. *Disasters*, **23**(1), 1–18.

Mosquera-Machado, S., and Ahmad, S., 2007. Flood hazard assessment of the Atrato River in Colombia. *Water Resource Management*, **21**, 591–609.

Müller, M., Vorogushyn, S., Maier, P., Thieken, A. H., Petrow, T., Kron, A., Buchele, B., and Wachter, J., 2006. CEDIM risk explorer – a map server solution in the project "risk map Germany". *Natural Hazards and Earth System Sciences*, **6**(5), 711–720.

Musson, R.M.W., and Sargeant, S.L. 2007. *Eurocode 8 Seismic Hazard Zoning Maps for the UK*.Technical report CR/07/125. Keyworth,UK:British Geological Survey,spp. 70.

Oh, H.-Y., and Lee, S., 2011. Integration of ground subsidence hazard maps of abandoned coal mines in Samcheok. *Korea International Journal of Coal Geology*, **86**(1), 58–72.

Petrascheck, A., and Kienholz, H., 2003. Hazard assessment and mapping of mountain risks in Switzerland. In Rickenmann, D., and Chen, C. M. (eds.), *Debris-Flow Hazards Mitigation: Mechanics, Prediction, and Assessment 1*. Rotterdam, The Netherlands: Millpress, pp. 25–38.

Pomonis, A., Spence, R., and Baxter, P., 1999. Risk assessment of residential buildings for an eruption of Furnas Volcano, São Miguel, the Azores. *Journal of Volcanology and Geothermal Research*, **92**(1–2), 107–131.

Ragozin, A. I., and Yolkin, V. A., 2003. *Individual karst risk qualitative assessment at a regional level (an example of Tartar Republic)*. Portsmouth: Proceedings Annual Conference International Association Mathematical Geology, pp. 1–6.

Rango, A., and Anderson, A. T., 2007. Flood hazard studies in the Mississippi River Basin using remote sensing. *Journal of the American Water Resources Association*, **10**(5), 1060–1081.

Romme, W. H., Barry, P. J., Hanna, D. D., Floyd, M. L., and White, S., 2006. A wildfire hazard assessment and map for La Plata County, Colorado. *USA Fire Ecology*, **2**(1), 7–30.

Sampson, R. N., Atkinson, R. D., and Lewis, J. W., 2008. *Mapping Wildfire Hazards and Risks*. Boca Raton, USA: CRC Press. 343 pp.

Sanabria, L.A., and Dhu, T. 2005. A methodology for consistent modelling of natural hazards. In *Proceedings of Internat Congress on Modelling and Simulation*, University of Melbourne, pp. 765–771.

Schwab, J. W., Hogan, D. L., and Weiland, I., 2002. Flood plain hazard assessment: application to forest land management in British Columbia, Canada. In Bobrowsky, P. T. (ed.), *Geoenvironmental Mapping: Methods, Theory and Practice*. Lisse: A A Balkema, pp. 343–368.

Simón-Gómez, J. L., and Soriano, M. A., 2002. Actual and potential doline subsidence hazard mapping: case study in the Ebro Basin, Spain. In Bobrowsky, P. T. (ed.), *Geoenvironmental mapping: methods, theory and practice*. AA Balkema: Lisse, pp. 649–666.

Simpson, K., and Shepherd, J. B., 2001. *Volcanic-Hazard Assessment for St Kitts, Lesser Antilles*. St Augustine, FL: Seismic Research Unit, The University of the West Indies. 15 pp.

Smith, K., and Petley, D., 1991. *Environmental Hazards: Assessing Risk and Reducing Disaster*, 5th edn. Abingdon: Routledge. xxv + 383 pp.

Statham, I., and Treharne, G., 1991. Subsidence due to abandoned mining in the South Wales Coalfield, UK: causes, mechanisms and environmental risk assessment. In: Johnson, A. I. (ed.) *Land subsidence: Proceedings of the 4th International Conference on Land Subsidence*, Houston, Texas 12–17 May 1991. IAHS Publication No 200, London, Thomas Telford 143–152.

Study Committee on Tsunami and Storm Surge Hazard Maps, 2004. *Tsunami and Storm Surge Hazard Map Manual*. Tokyo: Cabinet Office-Disaster Management. 113 pp.

Terlien, M.T.J., van Westen, C.J., and van Asch, T.W.J., 1995. Deterministic modelling in GIS-based landslide hazard assessment. In Carrara, A. and Guzzetti, F. (Eds) *Geographic Information Systems in Assessing Natural Hazards*. Dordrecht: Kluwer, pp. 57–78.

Thompson, A., Hine, P., Peach, D., Frost, L., and Brook, D. 1998. Subsidence hazard assessment as a basis for planning guidance in Ripon. In Maund, J.G., and Eddleston, M. (eds.) *Geohazards in Engineering Geology*. London: Geological Society. Geological Society London Engineering Geology Special Publications 15, pp. 415–426.

Tran, P., Shaw, R., Chantry, G., and Norton, J., 2008. GIS and local knowledge in disaster management: a case study of flood risk mapping in Viet Nam. *Disasters*, **33**(1), 152–169.

Wadge, G., and Isaacs, M.C., 1988. Mapping the volcanic hazards from Soufriere Hills volcano, Montserrat, West Indies using an image processor. *J Geol Soc*, **145**(4), 541–555.

Watson, C. C., 1995. Arbiter of storms: a high resolution GIS-based system for integrated storm hazard modelling. *National Weather Digest*, **20**, 1–9.

Cross-references

Avalanches
Building Codes
Casualties Following Natural Hazards
Coastal Zone Risk Management
Cost (Economic) of Natural Disasters
Damage and the Built Environment
Earthquake Prediction and Forecasting
Flood Hazard and Disaster
Frequency and Magnitude of Events
Geographic Information Systems (GIS) and Natural Hazards
Karst Hazards
Lahar
Land Subsidence
Landslide
Liquefaction
Monitoring and Prediction of Natural Hazards
Probable Maximum Flood
Pyroclastic Flow
Recurrence Interval
Remote Sensing of Natural Hazards and Disasters
Risk
Risk Assessment
Risk Perception and Communication
Rockfall
Sinkhole
Slope Stability
Storm Surge
Storms
Subsidence Induced by Underground Extraction

Surge
Tsunami
Uncertainty
Vulnerability
Zoning

HAZARDOUSNESS OF A PLACE

Netra Raj Regmi[1], John Rick Giardino[2], John D. Vitek[2]
[1]Desert Research Institute, Reno, NV, USA
[2]Texas A&M University, College Station, Texas, USA

Definition

Hazard: Probability of occurrence of a certain natural or human-induced phenomenon that poses threat to life, property, and the environment within a certain period of time and within a given area (Varnes, 1984).
Hazardous place: A location with a high probability of experiencing a natural event that may be hazardous.

Introduction

Hazards broadly can be classified into three types: (a) hazards related to geophysical events, such as mass movement, land subsidence, earthquake, volcanic eruption, flood, tsunami, and fire; (b) atmospheric events, such as climate change, geomagnetic storms, winds (including hurricanes, tornadoes, and straight winds), drought, hail, freeze, and lightning; and (c) hazards created by anthropogenic activities, such as pollution, deforestation, use of herbicides and pesticides, chemical spillage, dam and reservoir failure, and fire. Every location is unique with regard to the events that can occur and those events only become hazards if people are impacted by the event; otherwise the event is simply a natural event that has occurred on Earth for eons.

In this contribution, we briefly describe geophysical and anthropogenic events and associated hazards, fundamental techniques of assessment of the hazards and associated risks, and common approaches to the mitigation of hazards.

Events and related hazards
Geophysical events
Hazards related to the geophysical events can be classified into: (1) movement of slope materials under the effect of gravity; (2) subsidence of land associated with karst terrain and the overdraft of groundwater and oil; (3) volcanic eruptions and associated movement of lahars and slope materials; (4) earthquake shaking and tsunamis; (5) floods associated with large rainstorms, snowmelt, and storm surges; (6) coastal erosion; and (7) sea level rise.

Landslides
The term landslide includes many downslope movements of soil, rock, or other Earth materials. Landslides can be activated by earthquakes, rapid snowmelt, intense rainstorms, groundwater rise, slope toe cutting by rivers, or volcanic eruptions, in conjunction with gravity and occur when driving forces, such as gravity, exceed the frictional strength of the slope materials. Landslides include: (1) rockfalls and topples, characterized by free-falling of rocks from overlying cliffs; (2) slides, slumps, and avalanches, a displacement of overburden material because of shear failure along a structural feature; (3) flows and lateral spreads of unconsolidated material associated with a shallow water table; (4) complex landslides, combination of two or more types of landslides (Varnes, 1978). In a rockfall, rocks may fall, bounce, or roll down the slope. In a topple, rocks on a steep slope break loose and rotate forward. Flows can be categorized as: (1) debris flow, (2) debris avalanche (3) earthflow, (4) mudflow, (5) lahar, and (6) creep, based on the rate of movement and the types of unconsolidated materials involved. A debris flow, characterized by a combination of water-saturated loose soil, rock, organic matter, and air, with materials varying in size from clay to large boulders, forms when loose masses of unconsolidated wet debris become unstable. A fast moving debris flow is known as a debris avalanche. Earthflows consist of saturated soil or fine-grained rock deposits and flow downhill usually on moderate slopes. A mudflow is an earthflow consisting of material that is wet enough to flow rapidly. A lahar is a special type of debris flow that originates from the slopes of a volcano. Creep is an imperceptibly slow, steady, downward movement of slope-forming soil or rock. Creep can occur seasonally, where the movement is within the depth of soil affected by seasonal changes in soil moisture and soil temperature.

The hazards and risks from landslides depend on the type of landslides. Rockfalls pose a localized threat to life and property because of limited runout distances. But slides, avalanches, flows, and lateral spreads can have great runout distance and can result in significant loss of life and property. Similarly, mudflows, associated with volcanic eruptions, can travel great distances with high speeds and may be a quite destructive phenomena. Worldwide, landslides constitute a major geophysical hazard. A most probable location for landslides is a moderate to steep slope where the soil is saturated, an area susceptible to earthquakes, volcanic activities, fluctuation in groundwater, and human-induced disturbances, or any combination of these factors. A variety of other natural causes may also result in landslides.

Some examples of recent, large and most hazardous landslides worldwide are: a volcanic eruption-induced landslide that occurred at Mount St. Helens, USA on May 18, 1980; a landslide occurred in Usoy, Tajikistan, in 1911; a landslide that caused Vaiont dam disaster in Italy on October 9, 1963; an earthquake-induced landslide occurred in Yungay, Peru in 1970; and a rain-induced landslide occurred on January 10, 2005 in La Conchita, California (Figure 1). These landslides killed many people, countless animals and destroyed great amounts of property. The landslide at Mt. St. Helens killed 50

Hazardousness of a Place, Figure 1 A landslide reactivated on January 10, 2005 in La Conchita, Callifornia. The landslide killed 10 people and damaged 36 houses (Photo by Mark Reid, U.S. Geological Survey).

people and countless animals (Ylvisaker, 2003). The landslide in Usoy killed almost all people of the Usoy village. The Vaiont dam disaster killed ∼1,900 people (Coch, 1995). The landslide in Yungay killed ∼20,000 people. But the earthquake and landslides together in this event killed ∼70,000 people. The landslide in La Conchita damaged 36 houses and killed 10 people (Jibson, 2005).

Land subsidence
Land subsidence is a gradual sinking or sudden downward settling of land with little or no horizontal motion. Subsidence can occur at or near the surface. This type of subsidence occurs mostly in areas underlain by compressible deposits, expandable clays, and organic soils. Natural and anthropogenic activities can cause land subsidence when the loss of subsurface support occurs from mining, pumping of oil or groundwater, and dissolving of carbonate rocks. Subsidence and collapse are serious hazards because they can pose significant risks to human health and safety, cracking or complete damage of structures, and interruption to transportation and other services.

Some examples of subsidence are the overdraft of groundwater and petroleum in the USA in the Houston-Galveston region of Texas; Las Vegas, Nevada; and areas in Santa Clara Basin, California (Coch, 1995). Many parts of the Japan are experiencing land subsidence because of the withdrawal of groundwater (Yamamoto, 1995).

Earthquakes
Earthquakes, most common along tectonic plate boundaries, are caused by the sudden releases of strain energy stored along a fault in the bedrock. Earthquakes are perhaps the most terrifying hazards; they can destroy life and property in a very short period of time and usually occur without warning. If other hazardous events occur in conjunction with an earthquake, the severity of the hazard increases. Hazards related to an earthquake can be: (1) ground shaking and structural dislocation, (2) faulting or breaking of surface materials, (3) mass movement, (4) liquefaction of gently sloping unconsolidated materials, (5) subsidence or surface depressions resulted from the settling of loose or unconsolidated sediments, (6) tsunamis generated by seismic activities under the ocean floor, (7) flooding, and (8) fire (Figure 2).

Many metropolitan areas, such as Tokyo, San Francisco, Los Angles, and Mexico City, are in seismically active zones. These cities have been struck by earthquakes in the past, and they still have high probability for future earthquakes. Table 1 lists some of the deadliest earthquakes.

Volcanoes
Volcanoes are vents in the crust through which molten rock and associated gases escape. Volcanic eruptions, dramatic and violent events, influence large areas, can continue for a long period of time, and pose significant

Hazardousness of a Place, Figure 2 Photographs showing destruction caused by April 18, 1906 San Francisco earthquake shaking and fire; (a) Financial district, (b) Mission District (Source: Museum of the City of San Francisco).

Hazardousness of a Place, Table 1 Some of the deadliest earthquakes in history

Location	Time	M	Loss of life and property	Other hazards	Reference
Shaanxi (Shensi), China	January 23, 1556	8.0	Killed ~830,000		USGS, 2010
Lisbon, Portugal	November 1, 1755	8.7	Killed more than 60,000 people and countless animals and damaged large property	Tsunami, fire	Gutscher, 2004
San Francisco, USA	April 18, 1906	7.9	Damaged structures throughout the city, death may have exceeded 3,000 people.		Hansen and Condon, 1989
Valdivia-Puerto Montt, Chile	May 22, 1960	9.5	Killed 1,655–5,700 people	Tsunami	USGS, 2010
Prince William Sound, Alaska	March 27, 1964	9.2	Damaged many towns and cities and killed 128 people	Tsunami	USGS, 2010
Bam, Iran	December 26, 2003	6.6	Killed 26,000–40,000 people		Jackson et al., 2006
Coast of northern Sumatra	December 26, 2004	9.1–9.3	More than 283,000 people were killed and countless people were displaced in 14 countries in South Asia, and East Africa.	Tsunami	Lay et al., 2005
Port-au-Prince, Haiti	January 12, 2010	7.0	111,481 confirmed death, possibly more		CNN, January 23, 2010

hazards to people. Volcanic eruption can be very explosive or quiet. Explosive eruptions pose a risk by scattering rock blocks, fragments, and lava at varying distances from the source. Explosive eruptions can spread lava, gas, and other materials over a wide area, and may significantly modify the landscape. In quiet eruptions, materials flow rather than explode from the vent. The composition of flow varies; because it may originate from multiple sources.

Major hazards associated with volcanic eruptions are: lava flows; pyroclastic flows; tephra and ash fall; lahar, landslides, or mudflows; and emanations of toxic gases. Volcanic activity may also trigger other hazardous events including tsunamis, deformation of the landscape, floods, and tremor-provoked landslides. Lava may flow quickly or slowly, but destroys everything in its path. Lava entering the sea poses risks to life in the sea by increasing the local temperature of seawater. Lahar, a mixture of volcanic ash, rock, debris, and water, occurs when a high volume of hot or cold water mixes with ash and rock. The water may come from melting snow or ice, rainfall, or lakes. As a lahar moves downslope its size grows significantly because of the erosion and addition of water. The lahar will damage everything until the sediment load is deposited. Pyroclastic flow, a rapidly moving mixture of hot and dry rock fragments, ash, and hot gases, also destroys everything in its path by burying and burning. Other events associated with a volcanic eruption, such as earthquake and rainfall, can trigger landslides on the slope of the volcanic cone. Tephra, fragments of volcanic rock and lava blast into the air, creating hazards in a local extent as well as in a large region. Large tephra typically falls back to the ground near the volcano whereas smaller fragments are carried away by the wind. Volcanic ash can travel hundreds to thousands of miles downwind from a volcano and disrupt life and property over a wide region.

Hazardousness of a Place, Figure 3 Landslides and lahar caused by Mt. St. Helens Volcano on May 18, 1980. (a) The cone of the volcano showing landslides and a lahar. (b) Destruction caused by a lahar (Photographs were taken on June 1, 1980).

Gases and acid aerosol particles released from a volcanic eruption, fumaroles, and hydrothermal systems also create hazards. The volcanic gases, such as sulfur dioxide, carbon dioxide, and hydrogen fluoride, pose the greatest risks to human health, animals, and plants. Sulfur dioxide and hydrogen fluoride gases can lead to acid rain and air pollution downwind from a volcano. Concentrations of carbon dioxide gas can be lethal to human, animals, and vegetations.

Most volcanoes have erupted at the convergent plate boundaries that surround the Pacific Ocean. Far more unobserved eruptions occur on the ocean floor. Indonesia, Japan, and the USA experience volcanic events. Some of the deadliest volcanoes occurred at Mt. Tambora, Indonesia; Mt. Krakatau, Indonesia; Mt. Ruiz, Colombia; Mt. Pelee, Martinique; Mt. Vesuvius, Italy; Mt. St. Helens, USA; Mount Pinatubo, Philippines; and Mt Etna, Italy. Loss of life because of these volcanoes is very high, for example, the Mt. Tambora volcano killed ~88,000 people in 1815 (Stothers, 1984); Krakatau volcano along with a volcano-induced tsunami killed ~36,000 people in 1883 (Dörries, 2003); Mount Pelee volcano along with an ash flow killed at least 29,000 people in 1902 (Witham, 2005); Mt. Ruiz volcano along with volcano-induced mudflows killed more than 23,000 people in 1985 (Pierson et al., 1990); Mt. St. Helens volcano (Figure 3) along with heat and hot volcanic ash released from the volcano killed 58 people in 1980 (Witham, 2005); Mt. Pinatubo volcano along with volcano-induced avalanches of hot ash and gas, giant mudflows, and a cloud of volcanic ash killed ~ 800 people and displaced more than 100,000 people in 1991.

Flooding

Two types of flooding can be distinguished: (1) land-borne floods, or river flooding, caused runoff from large amounts of rain, and (2) sea-borne floods, or coastal flooding, caused by storm surges.

Land-borne floods occur when the capacity of stream channels to conduct water is exceeded and water overflows banks. The severity of the river flooding increases by the conditions within the drainage basin, such as excessive rainfall, low rate of infiltration, steep slopes, and lack of vegetation in conjunction with human activities, such as urbanization, improper practice in agriculture and timber production. Destruction by river flooding is attributable to: (1) residential and agricultural areas in close proximity to river,; (2) stream habitat, and (3) failure of hillslope triggered by undercutting of slope toe. The major cause of the coastal flooding is storm surge. Tsunamis are a special type of coastal flood. Most often, destruction resulted by the coastal flooding is associated with: (1) wave impact; (2) hydrostatic/dynamic forces; (3) the impact of the objects carried by the wave; (4) flooding of deltas and low-lying coastal areas by the influence of tidal action, storm waves, and frequent channel shifts; and (5) weakening of major infrastructures, such as highways and railroads. The most significant damage often results from the direct impact of waves on fixed structures.

Worldwide, the deadliest floods reported include: Yellow River floods in 1887 and 1931 (China), Mississippi River flood in 1927 (USA), and hurricane and tide-induced flood of the Netherland in 1953. The Yellow River flood of 1887 killed between 900,000 and 2,000,000 people (Kynge, 2007), countless animals and destroyed a large property. It was one of the deadliest natural disasters ever recorded. The Mississippi River flood broke out of its levee system and flooded 13 million acres of land in 1927. The flood caused over US$236 million in damage and killed more than 246 people, and left more than 700,000 people homeless (The American Institutes for Research et. al., 2005). The Yellow River flood of 1931 killed as many as four million people (Famighetti, 1995), countless animals and destroyed a large property. The Netherland flood of 1953 killed ~1,836 people and many animals and damaged 136,500 ha of land (Gerritsen, 2005).

Tsunamis

Tsunami, a series of large destructive waves that are caused by the sudden movement of a large area of the sea floor, are mostly caused by earthquakes, volcanic activity, and landslides. As the water is displaced, it surges outward in all directions in a large wave. The wave can travel rapidly across oceans and cause destruction far from the location where they were generated. Most of the tsunamis are caused by faulting associated with earthquakes. Some tsunamis are caused by volcanic eruptions, and submarine landslides, and very rare tsunamis are caused by meteorite impact in the ocean.

Tsunami waves have exceptionally long wavelengths that measure tens of kilometers, even up to 100 km. The height of a tsunami wave is as low as 0.5 m in the open ocean with a deep water speed up to 700 km/h. As it enters shallow coastal waters, speed decreases and height dramatically increases. This water rises up offshore to form the first tsunami wave to strike the coast. Successive waves may attain several tens of meters in height, depending on local conditions. Such huge waves can dramatically alter shorelines by flooding shore and far inland areas, eroding and depositing large amounts of debris and destroying buildings and other structures.

All oceanic regions of the world experience tsunamis. Tsunamis in the Atlantic, Mediterranean, and Caribbean, however tend to be smaller and less destructive and less frequent than those in the Pacific and Indian Oceans. The Pacific Ocean is surrounded by a geologically active series of mountain chains, deep ocean trenches, and island arcs, sometimes called "the ring of fire." The earthquakes and volcanic eruptions that occur in the "ring of fire" are the source of many tsunamis. An example of the violent and largest tsunami over 40 years is the tsunami that occurred in Pacific and Indian Ocean on December 26, 2004 because of an earthquake of magnitude 9.1–9.3 near Sumatra. The tsunami killed at least 283,000 people (Lay et al., 2005). An example of volcanic eruption-induced tsunami is the tsunami induced by Krakatau volcanic eruption on August 26 and 27, 1883. The tsunami was very violent, killed thousands of people, and wiped out numerous coastal villages.

Fire
Well known causes of fire are: (1) volcanic lava; (2) earthquake tremor; (3) hot, dry, and windy days or nights; (4) electric failure, broken gas pipes, as well as lack of proper management of the fire; (5) accidental spreading of fire by people; and (6) lightning. Fire can be a severe hazard depending on the location of occurrence and the weather condition.

An example of a wind-induced fire is the one that occurred in Matheson, Canada on July 29, 1916. The fire burned an area of approximately 2,000 km^2 (490,000 acres) and killed 223 people. An example of the earthquake-induced fire is the Great San Francisco fire of April 18, 1906. The event destroyed most parts of the city and left 250,000 people homeless. An example of fire generated during an extremely hot day is the 1918 Cloquet fire. On October 12, 1918, during dry condition, sparks on the local railroads in northern Minnesota generated the fire which took 453 lives. The fire also injured 52,371 people and damaged US$73 million property (Carrol and Raiter, 1983). An example of fire induced by lightning is the Yellowstone National Park fire of 1988. Several fires burned for several months and destroyed 793,880 acres (3,213 km^2) area of the park. The fires destroyed structures, killed many animals and cost US$120 million to control the fire (National Park Service, 2008).

Coastal erosion
Areas along coastal regions are subject to erosion from actions of adjacent water bodies. Seaward sides of beaches, sand spits, and uplifted marine terraces of dunes and bluffs are the most susceptible areas. Major causes of this erosion are continuous impact of waves, currents, tides, storms, high winds, rain, runoff, increased water level, and ocean conditions caused by periodic El Niños. The impact may be gradual over a season or many years or drastic during the course of a single storm. The impact might also create landslides and floods. Coastal erosion creates special challenges for people living near the ocean and requires thoughtful planning to minimize the potential dangers to life and property. Attempts to stabilize the shoreline or beach are often futile because the forces that shape the coast are persistent and powerful.

Examples of the extensive coastal erosion occur in the coastal regions of Louisiana, California, and Shishmaref, Alaska.

Atmospheric events
Climate change
Variations in the global and regional climate have changed the size, shape, and volume of glaciers, snow, and ice sheets. Hazards associated with climate change are: (1) permafrost melting, (2) GLOF (Glacial Lake Outburst Flood) and debris flows, (3) ice avalanches from steep glaciers, (4) destabilization of frozen or unfrozen debris slopes, (5) destabilization of rock walls, and (6) freeze and drought.

Glacier and permafrost-related hazards represent a continuous threat to human life and property in mountainous regions. Related disasters can cause large damage to property and kill hundreds or even thousands of people and animals at once. Permafrost occurs on one fourth of the land in the Northern Hemisphere and is also common within continental shelves of the Arctic Ocean. Permafrost in these regions is melting because of climatic warming. This process might be a serious danger because of very large quantities of methane, a very potent greenhouse gas for global warming, stored in the permafrost. Furthermore, permafrost melting reduces the ability of the ground to support large structures.

A GLOF occurs when a lake, dammed by a glacier or moraine, fails. Main reasons for the break out of the glacial lake are: a buildup of water pressure in the lake, an avalanche of rock or heavy snow, an earthquake, intense erosion by the lake water, and breaking off a large portion of glacier. The failure of the lake causes floods and debris flows which are very hazardous downstream. Major GLOFs have been reported from Iceland, USA, Canada, and Himalayan countries. GLOFs were frequently released from Vatnajökull Ice Cap, Iceland (Scharrer et al., 2007). Cathedral Glacier and Farrow Creek areas are known as major GLOF susceptible areas in Canada. In 1978, debris flows triggered by a GLOF from Cathedral Glacier destroyed part of the Canadian Pacific railway track, derailed a freight train, and buried parts of the Trans Canada Highway (Jackson et al., 1989). Knik River and Abyss Lake areas in Alaska and Wind River Mountains in Wyoming are major GLOF susceptible areas in the USA. Knik River had large annual outbreaks from 1918 to 1966. Almost every year, GLOF occurs in Abyss Lake in southeastern Alaska. Likewise, in early September 2003, a GLOF occurred from Grasshopper Glacier in the Wind River Mountains (Oswald and Wohl, 2007). The 12-ha lake drained an estimated 3.2 million cubic meters of water downslope into the Wind River valley.

Many GLOFs are reported from Himalayas. GLOFs regularly occur in the valleys and rivers of Bhutan. For example, in October 1994, a GLOF occurred in Pho Chhu River which caused a massive flood that destroyed many lives and property. Of the 2,674 glacial lakes in Bhutan, 24 have been identified dangerous (Srivastava et al., 2008). The higher Himalaya in Nepal is also a susceptible region for GLOF. Dig Tsho, Imja, Lower Barun, Tsho Rolpa, and Thulagi regions are potentially dangerous for GLOF. For example, Dig Tsho glacier initiated a GLOF in 1985. A recent inventory carried out by ICIMOD and UNEP/EAP-AP shows that 2,315 glacial lakes exist in Nepal and 26 are potentially dangerous (Srivastava et al., 2008). Similarly, many parts of the Chinese Himalaya are susceptible for GLOF. Longbasaba and Pida lakes of Chinese Himalaya are very dangerous. GLOF from these lakes can affect people living in 23 towns and villages downstream (Xin et al., 2008).

Geomagnetic storms
Geomagnetic storm is a rapid variation in the geomagnetic field, caused either by a gust in the solar wind or by a temporary linking of the magnetic field of the Sun with the geomagnetic field. Variation in the magnetic field does not directly impact human health, but it can disturb long-range radio communication, degrade global positioning systems, damage satellites, affect long-distance pipelines, and produce surges in electric power grid. For example, when in September 1859 the largest recorded geomagnetic storm occurred, telegraph wires in the USA and Europe experienced induced emf (Boteler, 2006). In some cases the event shocked telegraph operators and caused fires. Similarly, an electromagnetic storm disrupted power throughout most of Quebec, Canada, in 1989 (Boteler et al., 1998). The event collapsed the Hydro-Québec power grid. Geomagnetic storms can expose astronauts and high-altitude pilots to increased levels of radiation.

Winds, hurricanes, storm surge, and large rainfalls
High-speed winds can damage structures like power and communication lines, residential homes, and sometimes people. As winds exceed 45 mph, damage begins on human constructions, such as traffic signals and trees may be toppled.

A hurricane is a massive tropical storm system with rotary winds that exceed 119 km/h (74 mi/h) blowing counterclockwise around a central area of very low pressure (Coch, 1995). The formations of hurricanes begin with the disturbance in the westward-flowing air directly north of the equator. The disturbance sets up a vertical air movement that draws heat laden water vapor upward from the warm ocean. As the vapor rises, it condenses and releases great amounts of heat energy which drives the wind system into a cyclonic spiral system. When the storm intensifies, it develops hurricanes, with its spiraling arms of thunderstorms. A mature hurricane has a series of rainbands, spiral bands of high wind, and torrential rain that surround a relatively calm area of low pressure known as the eye of the hurricane. The greatest speed of winds of a hurricane is in the eyewall. Hurricane winds can extend more than 120 km (75 mi) out from the eye. Destructive tornadoes can also develop away from the storm's center during landfall. Gulf and the Atlantic coasts are the most susceptible areas to hurricanes. Similar storms in the northwestern Pacific are called typhoons, whereas cyclones occur in the Indian Ocean. Failure of coastal engineering structures increases the severity of the hurricane hazard along these coastal areas.

Hurricanes are particularly dangerous because of the large zone of influence, spontaneous generation, and erratic movement. Hazardous phenomena associated with hurricanes are: (1) winds exceeding 119 km/h (74 mi/h), (2) intense rainfall which commonly precedes and follows hurricanes for up to several days, (3) storm surge, and (4) impact on structures by wind-borne objects. The quantity of rainfall is dependent on the amount of moisture in the air, and the size and speed of hurricane. In extreme storms (stronger than category 3), the force of wind alone can cause tremendous devastation, such as toppling of trees and power lines, destruction of weak elements of homes and buildings. Likewise, intense rainfall can saturate soils and cause flooding and landslides.

Storm surge is a dome of water that moves ashore near the hurricane eyewall. As a storm makes landfall, tides can inundate most populated areas along the shore. Storm surge causes salt water flooding, which can contaminate supplies of surface and subsurface drinking water. Furthermore, storm surge floods wash out roads and leave streets filled with sand and debris.

Each year, an average of ten tropical storms, six of which become hurricanes, develop over the Atlantic Ocean, Caribbean Sea, or Gulf of Mexico. Whereas many of these remain over the ocean, about five hurricanes strike the USA coastline every 3 years (Schmidt, 2001). The loss of life and property associated with hurricanes

Hazardousness of a Place, Figure 4 Oblique aerial photographs of Bolivar Peninsula, Texas, (a) before (September 9, 2008) and (b) after (September 15, 2008) the Hurricane Ike. The Hurricane hit Bolivar hit Bolivar Peninsula on September 23, 2008 (Photo source: USGS).

and storm surges are very large. The Texas Hurricane of 1900 (Category 4) killed at least 8,000 individuals of Galveston City (Blake et al., 2007). The Okeechobee Hurricane of 1928 (Category 4) killed at least 2,500 individual (Blake et al., 2007). The Bangladesh Cyclone of 1970 killed at least 300,000 individual (Berz, 1988). Some examples of recent deadliest hurricanes are: Hurricane Andrew, Hurricane Katrina, Hurricane Rita, and Hurricane Ike. Hurricane Andrew (Category 5), Atlantic Hurricane Season of 1992 struck the Bahamas, south Florida and southwest Louisiana. The hurricane killed 40 individuals and caused US$26.5 billion in damage (Blake et al., 2007). Hurricane Katrina (Category 3) hit southern coast of the USA (south Florida, parts of Miami and Gulf of Mexico) in Atlantic season of 2005. The hurricane killed at least 1,500 individuals and damaged US$81 billion (Blake et al., 2007). Hurricane Rita (Category 3) made landfall on the US Gulf of Coast in Atlantic hurricane season of 2005, killed 120 individual (direct and indirect impact) and caused damages of US $11.3 billion (Blake et al., 2007). Hurricane Ike caused extensive destruction in Haiti, Dominican Republic, Turks and Caicos Islands, southeastern Bahamas, Cuba, and parts of southeastern Texas (Figure 4), western Louisiana, and Arkansas (USA) in 2008. The event killed 195 individuals and damages were US$29.5 billion (Blake et al., 2011).

Intense rainfall could also be very hazardous to life and property. Torrential rains associated with slow moving or stationary tropical weather systems can produce enough water to drown animals and destroy agriculture. Freshwater flooding can cause as many drowning deaths as storm surge flooding. Besides flooding, intense rain can have a disastrous effect on agriculture by drowning crops and increasing the probability of disease and pest infestations in surviving crops.

Tornadoes

Mixing warmer, humid air with cold, dry air can generate thunderstorm, tornadoes, lightning, winds, hailstorms, intense rain, and flash floods. Tornadoes are violently rotating (counterclockwise) columns of air that form a funnel shape with very low interior pressure and very high velocities in the wall. A tornado often is encircled by cloud of debris and dust. The speed of a tornado ranges from stationary to 110 km/h (68 mi/h), with average of 50 km/h (30mi/h) (Coch, 1995). Tornadoes form in the rainbands of a hurricane and cause significant damage to life and property.

Tornadoes are serious local hazards because of an extraordinarily low air pressure and very high winds. Tornadoes cause damage in various ways. The rapid, rotating winds knock down weaker structures. The wind force picks up smaller objects and can transport them to a very long distance. Tremendous amounts of debris are mobilized in a tornado and can destroy larger structures.

Many of the deadliest tornadoes were recorded with the USA, and Australia. Areas in Texas and surrounding states in the USA, known as "Tornado Alley," frequently experience tornadoes. Tornadoes hit Montana, Illinois, Indiana, and Kentucky on March 18, 1925 and killed 734 people and damaged US$16.2 million in property (Brooks and Doswell, 2000). On April 5 and 6, 1936, tornadoes hit Mississippi and Georgia, which killed about 419 people and damaged US$16 million (Brooks and Doswell, 2000). A tornado on May 11, 1953 hit Waco, Texas, and killed 114 people, and damaged US$41 million (Brooks and Doswell, 2000). On June 8 of the same year another tornado hit Flint and Beecher communities of Michigan and killed 116 people, injured hundreds and destroyed US$19 million property (Hanson et al., 1979). On April 11, 1965, one of the worst tornados outbreak in the USA history hit many areas in Michigan, Ohio, and Indiana, leaving more than 128 people dead, many people injured, and more than US$105 million property damaged (Brooks and Doswell, 2000). A super tornado on April 3, 1974 hit many areas in Ohio, Alabama, Indiana, and Kentucky and killed 183 people, injured thousands of people, and damaged US$262 million property (Brooks and Doswell, 2000).

Many parts of Australia frequently get tornadoes, also known as Willi Willi or Dust Devil.

Lightning

Lightning is a large charge of electricity that can descend from clouds to the ground or to other clouds. Lightning initiates fires and sometimes can hurt or kill people. Worldwide, eight million lightning flashes occur every day and kill typically 300–600 people each year (Glunčič et al., 2001). The chances of being struck by lightning depend on the location. Lightning can strike a long distance from a storm, which means a wide region around a storm area can be a hazardous region of lightning. Lightning tends to travel the path of least resistance and often seeks out tall or metal objects. In the USA, lightning strikes are most prevalent in the south, southwest, and mid-western regions of the country. An average of 62 people are killed each year by lightning in the USA (Adekoya and Nolte, 2005). The deadliest lightning strike ever hit a Boeing 707 near Elkton MD on December 8, 1963. The plane crashed and killed 81 people (Uman and Rakov, 2003). Another deadly lightning strike in 1993 hit an Egyptian army fuel depot and killed 430 people.

Hail

Hail, a type of precipitation, occurs when updrafts in thunderstorms carry raindrops upward into areas of the atmosphere with temperatures below freezing. Hail is only produced by thunderclouds, usually at the front of the storm system. Hail forms on condensation nuclei, such as a dust particle or ice crystal, when supercooled water freezes in contact. The size of the hailstone ranges from very tiny (mm in diameter) to several centimeters. It can damage aircrafts, structures, and vehicles. It also can be deadly to livestock and people. For example, on April 30, 1888, hailstorms killed more than 250 people and many cattle in India (Wylie, 1936). On July 19, 1932 hailstorms killed 200 individuals in China. Similarly, hailstorms killed 56 individuals in Nebraska, USA, in 1991 and 39 individuals in Oklahoma, USA, in 1995 (Rocke et al., 2005).

Freeze

Freeze, a winter hazard, can cause substantial damage and affect facilities in the entire temperate zone (middle latitudes), and sometimes in subtropical regions where freeze is thought to be rare. In areas where freeze is common, the typical freeze incident results from lack of proper management of the structures designed to prevent freeze, for example, leaving a door or window open. In areas where freeze is infrequent and rare, the problems result from the lack of adequate insulation. Freeze directly impacts a facility by breaking or impairing the fire protection system and water system. In addition, freeze can impact instruments in industries and can interrupt production processes. Other causes of the freeze incidents are the lack of proper monitoring of forecasting system of the temperature and lack of waning system.

Drought/desertification

Drought is a prolonged, abnormally dry period, from moisture deficiency. Drought creates adverse impacts on water supply, water quality, hydropower, vegetation, soils, animals, and people. It results from large-scale disruptions of atmospheric circulation patterns that may persist for months or years. All droughts originate from too little precipitation (i.e., meteorological drought), but vulnerability increases as patterns of land and water use change. Drought occurs with the fluctuations in global weather. The most widely known recurring climatic irregularity every few years is known as the El Niño. El Niño is an extreme swing in a recurring air pressure shift across the Pacific Ocean called the Southern Oscillation. Many droughts affecting eastern and northern Australia are considered as direct result of a strong swing in the Southern Oscillation.

In last 60 years droughts affected the USA, Canada, Australia, China, and Magnolia. Although the location varies, drought occurs each year in the USA and results in serious economic, social, and environmental impacts. Based on the study of FEMA in 1995, annual losses associated with drought in the USA have been estimated to be between US$6 billion and US$8 billion (FEMA, 1995). During the 1950s, many areas of the USA, extending from the mid-west to the Great Plains and southward into New Mexico, experienced severe drought conditions. The drought subsided in most areas with the spring rains of 1957. The drought was characterized by low rainfall and excessively high temperatures. The drought devastated agricultural production and degraded the land in these regions. Again, in the 1980s (1987–1989), drought impacted many areas of the USA. The drought intensified over the northern Great Plains, spread across much of the eastern half of the USA, and affected the primary corn and soybean growing areas. The summer of 1988 is well known for the extensive forest fires that burned across western North America, including the catastrophic Yellowstone fire. In addition to dry conditions, heat waves during the summer of 1988 broke long-standing temperature records in many mid-western and northeastern metropolitan areas. Combining the losses in energy, water, ecosystems, and agriculture, the total cost of the drought was estimated at US$39 billion (Coch, 1995). Western Canada also experienced drought in 1988 and the loss in this event was estimated more than US$1.8 billion.

In the 1980s and 1990s, many parts of Australia experienced a drought. During 1979–1983, very large areas of central and eastern Australia had record of low rainfall. Although no known human deaths from starvation or disease occurred, the total impact of the event was estimated at US$7 billion at that time. The drought affected production of cereal grain, cotton, and sugar and caused loss of livestock as well as tons of topsoil. Again, from 1991 to late 1995, a prolonged and severe drought occurred in most parts of northeastern New South Wales and Queensland. Water levels lowered in major reservoirs of the area because of the low rainfall. The deficiency in water significantly affected agriculture to a loss of US$5 billion.

Desertification, the gradual transformation of arable and habitable land into desert, is one of the most important environmental challenges in the future. Climate change and improper use of land are contributing factors. Each year, desertification and drought account for a loss of US $42 billion in food productivity worldwide (UNEP, 2004; Sivakumar, 2006). In China, nearly 20% of land area is desert. As a result of a combination of poor farming practices, drought, and increased demand for groundwater, desertification has become arguably the most important environmental challenge in China. Because of the effects of increasing desertification, farmers are abandoning their land, and the intensity of sandstorms is increasing.

Anthropogenic events

The interaction of human activities with natural geologic systems sometimes results in a hazard. Hazards created by the human activities are attributable to pollution and contamination of food, air, and water because of increase in industries, mining, and inappropriate dumping of toxic wastes; results of conflicts and wars; dam failure; deforestation; and climate change because of the excessive production of greenhouse gases.

Air and water pollution
Air pollution is becoming a serious environmental problem worldwide. The rapid rate of urbanization, industrialization, and utilization of fossil fuels are affecting air quality worldwide. Furthermore, the rise in the use of old and outdated vehicles, mostly in developing countries, is also increasing air pollution. The results of these activities are increasing the concentrations of toxic metal and dusts in air and water. Lead pollution from automobiles and cottage industries represents a major health hazard.

The sites of the disposal of the toxic waste material, such as raw sewage, sludge, incinerated ashes, contaminated soils, nuclear materials, acids, and poisonous solvents ejected by chemical, pharmaceutical, and fertilizer-producing plants, contaminate and pollute surface and subsurface water storage. The localities become a hazardous environment to the health of the local people.

Seepage of the chemicals and oils from industries and mining areas as a result of accidents, natural disaster, or human recklessness contaminate food, air, and water, sometimes in local area but occasionally over a wide region. Spills of chemicals into coastal waters can harm people, environment, and aquatic life, and can cause substantial disruption of marine transportation and economy.

Mining and gas flaring
Mining contributes to the pollution of water and air. Basic environmental impacts of intensive mining operations include: accumulation of sediments in rivers and lakes, land degradation by mining-induced deforestation, erosion and mass movement, contamination of air and water by the accumulations of dust and heavy metals (lead, arsenic, cadmium, mercury). One of the major causes of health hazards for the people is working in mines and living around mining areas. The main causes of health hazards among miners include: inhaling large amounts of siliceous dust, careless handling of mercury during gold panning, and sharing of poor quality air in the mines by large numbers of individuals.

Burning of natural gas during crude oil production is the most common gas flaring. Gas flaring also emits carbon dioxide, hydrogen sulfide, sulfur dioxide, and small quantities of toxic compounds, such as carbonyl sulfide, carbon disulphide, nitrogen oxides, and volatile organic compounds. These gases are hazardous for the health of the people living around the locations of oil productions.

Conflict-related hazards
Famine, mental and physical illness, and rapid spreading of diseases are major hazards to people created by conflict, war, and terrorism. Other hazards related to conflict, war, and terrorism include an increase in rate of mortality, morbidity and disability, degradation of land and environment, degradation of social life, and increase in crime rate. Recently, many Asian countries, such as Afghanistan, Nepal, Iraq, Sri Lanka, and African countries, such as Cote d'Ivoire, Guinea, Liberia, Nigeria, Sierra Leone, Togo, Eritrea, Ethiopia, Somalia, Sudan, Uganda, Burundi, Democratic Republic of the Congo, Rwanda, Algeria, Angola, and Zimbabwe, are experiencing these hazards because of the political conflicts, civil war, and terrorism.

Deforestation and construction
Deforestation is a process of degradation of forests resulting from various factors, such as climatic variations, and human activities. Other reasons of the deforestation are erosion, landsliding induced by road cutting, forest fire, and poor agricultural practices.

Forests, the most biologically diverse regions of the world, yield millions of species of plants and animals. Forests are also a source of medicine, money, food, shelter, recreation, and livelihood for many indigenous and nonindigenous people. Deforestation destroys this unique environment and also causes change in climate. Deforestation results in desiccation of the forest soil and leads to an increase in temperature, decrease in water storing capacity, and increase in soil erosion, which ultimately converts a moist humid region into a desert. Furthermore, deforestation results in less of an exchange of carbon dioxide and nitrogen.

Dam failure
Dam, an artificial barrier to impound water, wastewater, or any liquid-borne material for the purpose of storage or control, can fail and damage life and property downstream. Reasons for the dam failure can be: (1) overtopping caused by floods, (2) deliberate acts of sabotage, (3) structural failure of materials used in dam construction, (4) movement and/or failure of the foundation supporting the dam, (5) settlement and cracking of concrete or embankment dams, (6) piping and internal erosion of soil in embankment dams, (7) inadequate maintenance and improper operation, and (8) vandalism or terrorism. A well-known example of dam failure is the landslide-induced Vaiont dam disaster on October 9, 1963, which killed ~1,900 people living downstream of the dam.

Assessment of hazards and risks

Identification and analysis of hazards is the first step necessary to reduce the impact of potential hazards and to remain prepared for an emergency situation. Although it is difficult to address all potential hazards of a place, assessment of hazards and risks can follow priorities so that the most likely and dangerous situations are addressed first and those least likely to occur and less likely to cause major problems can be considered later. Hazards should be prioritized based on the location and the frequency and magnitude of the events. Large magnitude events cause greater damage but fortunately they generally occur with low frequency. Events with small magnitudes typically cause smaller damage but are very frequent. Some hazards, such as landslides, usually have a local effect, others including earthquakes and hurricanes have a regional effect and some, such as volcanic dust clouds can have a global effect.

The major hazards in a place can be identified by the geological records of the historical events, survey of the recent events, current scientific knowledge, and the sensing of the environmental changes, such as change in transportation, industrial and urban development, and population densities. Then hazards can be evaluated in terms of the likelihood that a problem may occur and the damage it would cause. This information can be used to identify and prioritize major hazards. After data for all kinds of hazards are collected, a rating or weighting approach can be used to prioritize different hazards (Odeh, 2002; Li et al., 2009). Then a qualitative or quantitative approach can be used to determine the spatial distribution of a selected hazard. The result helps decision makers design and institute approaches of hazard mitigation or emergency preparedness. Hazards that are very likely to happen and would do considerable damage to people and property should be addressed first. Hazards that are less likely or that would do less damage need to be identified for attention after the more serious hazards have been addressed. Assessment requires identifying populations at risk, structures, critical facilities, and natural resources to estimate deaths and injuries, property damage, and economic losses. With such information, geographical information systems and computer models can be used to integrate hazard and vulnerability assessments to carry out the assessment of risks. Assessment of risk is necessary to make rational decisions about investments in mitigation, emergency preparedness, and warning systems.

Mitigation of hazards

Over the past few decades, science has greatly advanced the understanding of hazardous events through new theoretical approaches, experiments, computer simulations, remote sensing, and monitoring. Some events, such as volcanic eruptions, hurricanes and storms, tornadoes, and floods, are now predictable. Although prediction of some events, such as earthquakes and tsunamis, remains elusive, an early warning system based on the monitoring of the related activity can provide some time prior to the emergency situation.

Mitigation of hazards is an activity taken to reduce or eliminate the risk to life and property from hazards. An assumption of mitigation strategy is that the economy invested in mitigation will significantly reduce the economy needed for emergency recovery, repair, and reconstruction after a hazardous event. The first focus of any mitigation plan should be human settlement areas, engineering structures, and ecologically sensitive areas, such as wetlands, floodplains, lakes, and natural features which significantly reduce the impact of the hazardous events. Furthermore, natural resources and precious cultural treasure should be protected.

A mitigation approach should plan for long and short-term perspectives. Some examples of the long-term mitigation approach are adopting codes for buildings, schools, hospitals, bridges, power plants and dams; and preventing construction in areas susceptible to hazardous events, such as vulnerable seismic zones, fire-prone areas, flood plains, and low coastal areas. Another broad way of mitigating the impact of hazards is preventing and controlling the hazards by engineering methods. Events like floods, landslides, or erosion can be prevented by this method. Finally, the third approach of reducing the impact of hazards is public awareness. Expert studies are of limited value unless public officials and the public are educated about hazards, risks, and warnings of hazardous events. These approaches can produce remarkable savings of life and property.

Emergency preparedness, including evacuation before a hazardous event, is the final stage to be safe from a hazardous event. Ignoring warnings of a hazardous event could be fatal. Two examples will contrast the results of preparedness versus ignoring the warnings. Hurricane Gilbert in the Caribbean (Category 5) in 1988 was one of the most powerful storms, but only 318 people died (Lawrence and Gross, 1989). The low death toll occurred because people and governments were well prepped for Hurricane Gilbert. Weather reports were timely and reasonably accurate. The public was educated to heed these warnings and take appropriate actions. Land use restrictions and building codes minimized loss of life and property. Governments had emergency teams waiting to provide assistance. In contrast, the 1985 eruption of Nevado del Ruiz, a volcano in Colombia, buried more than 23,000 individuals in a mudflow (Pierson et al., 1990). Scientists warned the government of Colombia about the possibility of eruption. Unfortunately, people near the volcano were not alerted and remained in their homes at the time of eruption.

Summary

The first step in the assessment of hazards is to understand major hazards associated with an area, to develop an appropriate approach of hazard mitigation, and to be prepared for emergency situations. Hazards can be evaluated in terms of the likelihood that a problem may occur and the damage it would cause. Magnitude and the frequency of the events are always necessary to understand in the analysis of severity of a potentially damaging event. Some hazards, such as landslides, have a local effect; some such as earthquakes and hurricanes, have a regional effect; and other hazards, such as volcanic dust clouds, have a hemispheric or global effect. Prioritizing hazards based on the records of the historical events, survey of the recent events, current scientific knowledge, and the sensing of the environmental changes is an appropriate way to identify the major hazards in an area.

Mitigation of hazards is an activity taken to reduce or eliminate the risk to life and property. The first choice of any mitigation plan should focus on human settlement areas, engineering structures, and ecologically sensitive areas. Any mitigation approach should be a plan with a long- and short-term perspective. Adopting codes for buildings, schools, hospitals, dams, bridges, and power plants are the long-term approach. Preventing and controlling the hazards by engineering methods can be adopted as a short-term approach.

Emergency preparedness, including evacuation before a hazardous event, is the final stage to be safe from a hazardous event. Many hazardous events are preceded by warning precursors. Educating the public about these warnings and appropriate actions needed during a hazardous event would be one of the best approaches of emergency preparedness.

Bibliography

Adekoya, N., and Nolte, K. B., 2005. Struck-by-lightning deaths in the United States. *Journal of Environmental Health*, **67**(9), 45–50.

Berz, G., 1988. List of major natural disasters, 1960–1987. *Natural Hazards*, **1**, 97–99.

Blake, E. S., Rappaport, E. N., Jarrell, J. D., and Landsea, C. W., 2007. The deadliest, costliest, and most intense United States tropical cyclones from 1851 to 2006 (and other frequently requested hurricane facts). *NOAA Technical Memorandom NWS TPC-5*, 45 pp.

Blake, S. E., Landsea, C. W., and Gibney, E. J., 2011. The deadliest, costliest, and most intense United States tropical cyclones from 1851 to 2010 (and other frequently requested hurricane facts). *NOAA Technical Memorandum NWS NHC-6*, 47 pp.

Boteler, D. H., 2006. The super storms of August/September 1859 and their effects on the telegraph system. *Advances in Space Research*, **38**(2), 159–172.

Boteler, D. H., Pirjola, R. J., and Nevanlinna, H., 1998. The effects of geomagnetic disturbances on electrical systems at the earth's surface. *Advances in Space Research*, **22**(1), 17–27.

Brooks, H., and Doswell, C. A., 2000. Normalized damage from major tornadoes in the United States: 1890–1999. *Weather and Forecasting*, **16**, 168–176.

Carrol, F. M., and Raiter, R. F., 1983. At the time of our misfortune: relief efforts following the 1918 Cloquet Fire. *Minnesota History*, **48**(7), 270–282.

CNN, January 23, 2010. *Survivor Found Amid Rubble in Haiti After 11 days*, http://www.cnn.com/2010/WORLD/americas/01/23/haiti.earthquake/index.html.

Coch, N. E., 1995. *Geohazards: Natural and Human*. Englewood Cliffs: Prentice-Hall. 481 pp.

Dörries, M., 2003. Global science: the eruption of Krakatau. *Endeavour*, **27**(3), 113–116.

Famighetti, R., 1995. *The World Almanac and Book of Facts 1996*. Mahwah, NJ: The World Almanac.

Federal Emergency Management Agency (FEMA), 1995. *National Mitigation Strategy: Partnerships for Building Safer Communities*. Washington, DC: Federal Emergency Management Agency.

Gerritsen, H., 2005. What happened in 1953? The big flood in the Netherlands in retrospect. *Philosophical Transactions of the Royal Society*, **363**, 1271–1291.

Glunčić, I., Roje, Ž., Glunčić, V., and Poljak, K., 2001. Ear injuries caused by lightning: report of 18 cases. *The Journal of Laryngology and Otology*, **115**, 4–8.

Gutscher, M. A., 2004. What caused the Great Lisbon earthquake? *Science*, **305**(5688), 1247–1248.

Hansen, G., and Condon, E., 1989. *Denial of Disaster*. San Francisco: Cameron. 150 pp.

Hanson, P. O., Vitek, J. D., and Hanson, S., 1979. Awareness of tornadoes: the importance of an historic event. *Journal of Geography*, **78**(1), 22–25.

Holle, R. L., and Lopez, R. E., 2003. A comparison of current lightning death rates in the U.S. with other locations and times. In *International Conference on Lightning and Static Electricity*, September 16–18, 2003, Paper 103-34 KMS, 7 pp.

Jackson, L. E., Hungr, O., Gardner, J. S., and Mackay, C., 1989. Cathedral Mountain debris flows, Canada. *Bulletin of Engineering Geology and the Environment*, **40**(1), 35–54.

Jackson, J., Bouchon, M., Fielding, E., Funning, G., Ghorashi, M., Hatzfeld, D., Nazari, H., Parsons, B., Priestley, K., Talebian, M., Tatar, M., Walker, R. and Wright, T., 2006. Seismotectonic, rupture process, and earthquake-hazard aspects of the 2003 December 26 Bam, Iran, earthquake. *Geophysical Journal International*, **166**, 1270–1292.

Jibson, R.W., 2005. *Landslide hazards at La Conchita, California*. United States Geological Survey, Open-File Report 2005-1067, 12 pp., http://pubs.usgs.gov/of/2005/1067/508of05-1067.html.

Kynge, J., 2007. *China Shakes the World*. New York: Houghton Miffin. 257 pp.

Lawrence, M. B., and Gross, J. M., 1989. Atlantic Hurricane season of 1988. *Monthly Weather Review*, **117**, 2248–2259.

Lay, T., Kanamori, H., Ammon, C. J., Nettles, M., Ward, S. N., Aster, R. C., Beck, S. L., Bilek, S. L., Brudzinski, M. R., Butler, R., DeShon, H. R., Ekstrom, G., Satake, K., and Sipkin, S., 2005. The great Sumatra-Andaman earthquake of 26 December 2004. *Science*, **308**(5725), 1127–1133.

Li, H., Apostolaskis, E. G., Gifun, J., VanSchalkwyk, W., Leite, S., and Barber, D., 2009. Ranking the risks from multiple hazards in a small community. *Risk Analysis*, **29**, 438–456.

National Park Service, 2008. *The Yellowstone Fires 1988*. Yellowstone National Park, 7 pp.

Odeh, J. D., 2002. Natural hazards vulnerability assessment of statewide mitigation planning in Rhode Island. *Natural Hazards Review*, **3**, 177–187.

Oswald, L., and Wohl, E., 2007. Jökulhlaup in the wind river mountains, Shoshone National Forest, Wyoming. In Furniss, M., Clifton, C., and Ronnenberg, K. (eds), *Advancing the Fundamental Sciences: Proceedings of the Forest Service National Earth Sciences Conference*, San Diego, CA, October 18–22, 2004, pp. 363–367.

Pierson, T. C., Janda, R. J., Thouret, J. C., and Borrero, C. A., 1990. Perturbation and melting of snow and ice by the 13 November 1985 eruption of Nevado del Ruiz, Colombia, and consequent mobilization, flow and deposition of lahars. *Journal of Volcanology and Geothermal Research*, **41**, 17–66.

Rocke, T., Converse, K., Meteyer, C., and McLean, B., 2005. The impact of disease in the American White Pelican in North America. *Waterbirds*, **28** (sp1), 87–94.

Scharrer, K., Mayer, C., Martinis, S., Münzer1, U., and Gudmundsson, Á., 2007. Ice dam fluctuations at the marginal lake Grænalón (Iceland) before and during a GLOF. In *Proceeding in Envisat Symposium 2007*, Montreux, Switzerland, April 23–27, 2007 (ESA SP-636, July 2007).

Schmidt, L. J., 2001. *Forecasting Fury*. NASA Earth Observatory, http://earthobservatory.nasa.gov/Features/Seawinds/.

Sivakumar, M. V. K., 2006. Interactions between climate and desertification. *Agricultural and Forest Meteorology*, **142**(2–4), 143–155.

Srivastava, S., Ray, P. K. C., Shakya, B., Joshi, D. D., and Kumar, R., 2008. *South Asian Disaster Report 2007*. New Delhi: SAARC Disaster Management Centre. 169 pp.

Stothers, R. B., 1984. The great Tambora Eruption in 1815 and its aftermath. *Science*, **224**(4654), 1191–1198.

The American Institutes for Research, The Pacific Institute for Research and Evaluation and Deloitte & Touche LLP, 2005. *A Chronology of Major Events Affecting the National Flood Insurance Program*. The American Institutes for Research, Washington, DC, 86 pp.

Uman, M. A., and Rakov, V. A., 2003. The interaction of lightning with airborne vehicles. *Progress in Aerospace Sciences*, **39**, 61–81.

United Nations Environment Programme, 2004. Global Environment Outlook. Progress Press LTD, Valletta, Malta. http://www.unep.org/geo/GEO4/report/GEO-4_Report_Full_en.pdf.

US Geological Survey (USGS), 2010. *Historic Worldwide Earthquakes*. http://earthquake.usgs.gov/earthquakes/world/events.

Varnes, D. J., 1978. Slope movement types and processes. In Schuster, R. L., and Krizek, R. J. (eds.), *Landslides: Analysis and Control*. National Research Council, Washington, DC, Transportation Research Board, Special Report 176, pp. 11–33.

Varnes, D. J., 1984. *Landslide Hazard Zonation: A Review of Principles and Practice*. IAEG Commission on Landslides and other Mass movements on Slopes, UNESCO, Paris, 63 p.

Witham, C. S., 2005. Volcanic disasters and incidents: a new database. *Journal of Volcanology and Geothermal Research*, **148**(3–4), 191–233.

Wylie, C. C., 1936. The Lord cast down great stones from heaven. *Popular Astronomy*, **44**, 378.

Xin, W., Shiyin, L., Wanqin, G., and Junli, X., 2008. Assessment and simulation of Glacier Lake outburst floods for Longbasaba and Pida Lakes, China. *Mountain Research and Development*, **28**(3/4), 310–317.

Yamamoto, S., 1995. Recent trend of land subsidence in Japan. In *Land Subsidence. Proceedings of the Fifth International Symposium on Land Subsidence*, The Hague, October 1995. IAHS Publication No. 234, pp. 487–492.

Ylvisaker, A., 2003. Landslides. In Doeden, M., and Risch, K. (eds.), *Natural Hazard Disaster*. Minnesota: Capstone Press. 48 pp.

Cross-references

Aa Lava
Avalanches
Caldera
Casualties Following Natural Hazards
Classification of Natural Disasters
Climate Change
Coastal Erosion
Coastal Zone, Risk Management
Creep
Cryological Engineering
Debris Avalanche (Sturzstrom)
Debris Flow
Deep-Seated Gravitational Slope Deformations
Disaster
Earthquake
Earthquake Damage
Earthquake Prediction and Forecasting
Erosion
Eruption Types (Volcanic)
Fault
Flash Flood
Flood Deposits
Flood Hazard and Disaster
Flood Protection
Flood Stage
Floodplain
Frequency and Magnitude of Events
Geohazards
Geological/Geophysical Disasters
Glacier Hazards
Hazard
Hazard and Risk Mapping
History of Natural Disasters
Human Impact of Hazards
Humanity as an Agent of Geological Disaster
Hurricane (Cyclone, Typhoon)
Jökulhlaup (Débâcle)
Krakatoa (Krakatau)
Lahar
Land Subsidence
Landslide (Mass Movement)
Landslide Dam
Landslide Inventory
Landslide Types
Lava
Mass Movement
Mt Pinatubo
Mudflow
Natural Hazard
Pahoehoe Lava
Paraglacial

Permafrost
Pyroclastic Flow
Risk Assessment
Rock Avalanche (Sturzstrom)
Sea Level Change
Slide and Slump
Slope Stability
Storm Surges
Stratovolcano
Structural Damage Caused by Earthquakes
Tornado
Tsunami
Vajont Dam, Italy
Volcanic Ash
Volcanoes and Volcanic Eruptions
Warning Systems
Wildfire

HEAT WAVES

Gerd Tetzlaff

Institut für Meteorologie, Universität Leipzig, Leipzig, Germany

Definition

Heat is a feeling of discomfort. Continuous heat increases the discomfort and may cause adverse effects on health particularly combined with high levels of humidity or when exacerbated by a person's physical condition. If heat lasts for more than a couple of days it is called a heat wave. In the mid and high-latitudes, heat waves are embedded in the usual course of summer weather; in the tropics they are endemic.

Introduction

Most individuals feel comfortable with respect to thermal conditions, if little or no thermal-regulatory activities are necessary to keep the body core temperature at the level of 36.8°C. This generally is the case in conditions of slight air movement, no solar radiation, low humidity, and the air temperature in the lower to mid-20s given in °C, for indoor and outdoor conditions (Fanger, 1972). There are many additional factors involved when considering the felt heat comfort of a concrete individual, such as the general health and living conditions, the current disposition of an individual, etc. Higher air temperatures often do come together with additional effects of one or several of the mentioned parameters and then act together in causing discomfort called heat, if lasting for a period of several days, a heat wave. The feeling of discomfort increases not only with the increase of the deviation from comfortable conditions but also with time spent in such conditions.

Weather and environmental conditions are far from being constant and therefore, *almost permanently* activate the thermoregulatory system of the human body. In a wide range of environmental conditions this happens rather smoothly and unnoticed. The feeling of discomfort from exposure to heat conditions is a warning to the body to discontinue the actual exposure and/or bodily activity, otherwise adverse effects in health have to be expected. Particularly severe effects occurred in the European heat wave in August 2003 causing many thousand fatalities (e.g., Luterbacher et al., 2004; EM-DAT, 2012).

The activation of the temperature control mechanisms depends on the energy balance of the body (Jendritzky et al., 1979; US Army 2003). The temperature of the body is so warm, that in most natural environments energy is transported from the core parts of the body to the surface. The source of the energy is food transformed by the body's metabolism. There are quite different environmental conditions, climate zones, distributed over the globe. In the warmer parts of the world people do live permanently in a wide range of heat conditions and have acclimated to their endemic environmental heat. The concept of acclimation also applies for non-indigenous individuals, with several phases of the actual degree of acclimation taking time periods from days to years. The quantity and rate of the energy flux from the core to the surface depends on the structure of the thermoregulation system and available fluids, the size and the shape of the body, the quantity of the internal energy production, clothing, and the above mentioned conditions of the environment. The energy balance of the body's surface comprises the body's internal energy production as a source; and the sinks are the pulmonary energy flux, the radiation budget, the convective fluxes of sensible and latent heat, the energy conduction, and the storage. The storage is determined by the increase or decrease of the body temperature.

In heat conditions, there is a deviation from heat comfort conditions, bringing imbalances to the body's energy budget and thermoregulatory activities, going with enhanced energy transport from the core of the body to the surface enforcing increased blood circulation trying to keep thermal comfort. Additionally, often external mechanical work is performed, exercising the muscles. These then need an increased energy supply, enhancing the overall energy production in the core of the body. To keep the core temperature of the body constant when doing such external work requires the transport of the work-related extra energy to the surface. In this whole process, the muscles need increased amounts of oxygen supplied by the circulating blood, whereas on the other side the blood is responsible for the increased transport of energy from the core body to its surface.

The energy transport from the core of the body to the surface acts based upon a gradient between the core temperature and the surface temperature. The surface temperature and the surface energy fluxes, however, are largely influenced by the complex set of environmental parameters. The key parameters allowing to measure environmental conditions are air temperature, humidity, solar radiation, long wave radiation, wind speed, and air quality. Often temperature is taken as key indicator, although the energy fluxes from the body's surface depend on the combined effect of all these parameters. Certain parameter combinations result in noticeable energy storage in the body's core, with overheating bringing adverse effects to any affected individual.

Any deviation from the felt thermal comfort conditions leads to imbalances of the body's energy budget and thermoregulatory activities. In the past, some attention was given to the development of objectively measurable indicators to characterize comfort in warm environments. The number of indices is quite large, because it proved to be difficult to quantify the whole complex cooling process of the body. This is particularly true in the absence of sufficient, or representative, empirical investigation and sampling (of select populations) from both a statistical and significance point of view. It proved to be especially difficult to quantify the individually varying properties of the body, but also the effects of bodily exercise. One of the most widely used indicators is the WBGT (Yaglou and Minard, 1957; Gaspar and Quintela, 2009), the wet-bulb global temperature. It consists of three quantities (see equation).

$$WBGT = 0.7\, T_{nw} + 0.2\, T_g + 0.1\, T_a$$

T_{nw} is the wet-bulb temperature, T_g the globe temperature, and T_a the air temperature. The wet-bulb temperature is taken from a thermometer that is cooled by a water-soaked wick, the globe temperature from a thermometer surrounded by a black globe exposed to solar and thermal or long wave radiation, and the air temperature from a shaded thermometer. T_{nw} represents the effects of humidity, air temperature, and ventilation; T_g the effects of solar and thermal radiation, air temperature, and ventilation; and T_a the air temperature (representing a source or sink of heat with regard to the person). In addition to the definition of WBGT, some empirical thresholds were found and used to form a five-step scale for comfort and discomfort, respectively (see table below).

The WBTG concept is valid for stationary, long-term conditions. The effects of short, intermittent variations might result in differences between the measured and the individually felt comfort values, resulting in a rather general recommendation to take any heat-related discomfort feelings as a warning to prepare for action. Furthermore, comfort is to be differentiated for age, body shape and health state, and clothing. To allow for the effects of these parameters, empirically based correction factors are in use.

The reference values of WBGT heat stress index related to a body core temperature of 38°C are presented in the table. The range is from no heat stress (1) to extreme heat stress (5), valid for indoor and outdoor conditions.

Metabolic rate	According to WBGT value in W/m^2 for acclimatized persons in °C
1 (rest; ~65)	33
2 (light exercise; 65–130)	30

Metabolic rate	According to WBGT value in W/m^2 for acclimatized persons in °C
3 (medium; 130–200)	28
4 (heavy; 200–260)	25
5 (extreme; >260)	23 (still air), 25 (sensible air movement)

The reaction of the body to defend itself from overheating happens through three mechanisms: increase of blood flow, breathing, and perspiration (i.e., more generally transpiration). The first reaction is to increase the blood flow to bring more energy to the body's surface. Increased breathing follows as an effect. In continuing heat stress conditions sweating begins. All of these require use of the body's water and salt reserves and its ability to pump enough energy from the body's core to the surface. Sufficient water supply, thus, is a major precondition to cope with heat conditions.

If cooling mechanisms are not efficient enough to keep the core temperature at constant optimum level, or respond too slowly, a condition of heat stress arises. This can lead to problems, like heat exhaustion, heat cramps, fainting, or even heat stroke. There is abundant information available on the symptoms of these health problems and on how to respond, if they occur. In recent years, public authorities have taken efforts to also mitigate, prevent, or avoid the heat effects available to the public (Bureau of Meteorology of Australia, 2010).

Heat exhaustion is the most common heat-related ailment occurring when doing strenuous work or exercise for the environmental conditions, indicated by WBGT. The elderly or the very young may already suffer exhaustion even being at rest. This usually means that the individual sweats a lot and does not drink enough water, or does not take enough salt, or both. The symptoms are sweating, feeling weak or tired, and sometimes giddy or nauseous. The face looks pale, the skin feels clammy. In uncomplicated cases, it is sufficient to bring the individual to rest in shady and cooler place and administer some appropriate drink.

Heat cramps do occur when an individual works or exercises in a warm environment, and sweats a lot without replacing the salts lost from sweating. In cases with no further health problems, relief is achieved by drinking electrolyte solutions and resting in cool shady places. Fainting usually happens to individuals not acclimatized to a warm environment while standing still. In the majority of cases, relief is achieved if the individual lies down in a shady, cooler place.

A heat stroke is the most serious health problem caused by heat stress. It is caused by a rise of the body core temperature to about 41°C or higher. The symptoms are mental confusion, delirium, or seizures. The skin is hot and dry, often red. Such a condition requires immediate medical intervention. Rapid cooling by soaking with cold water or fanning with cool air can give relief.

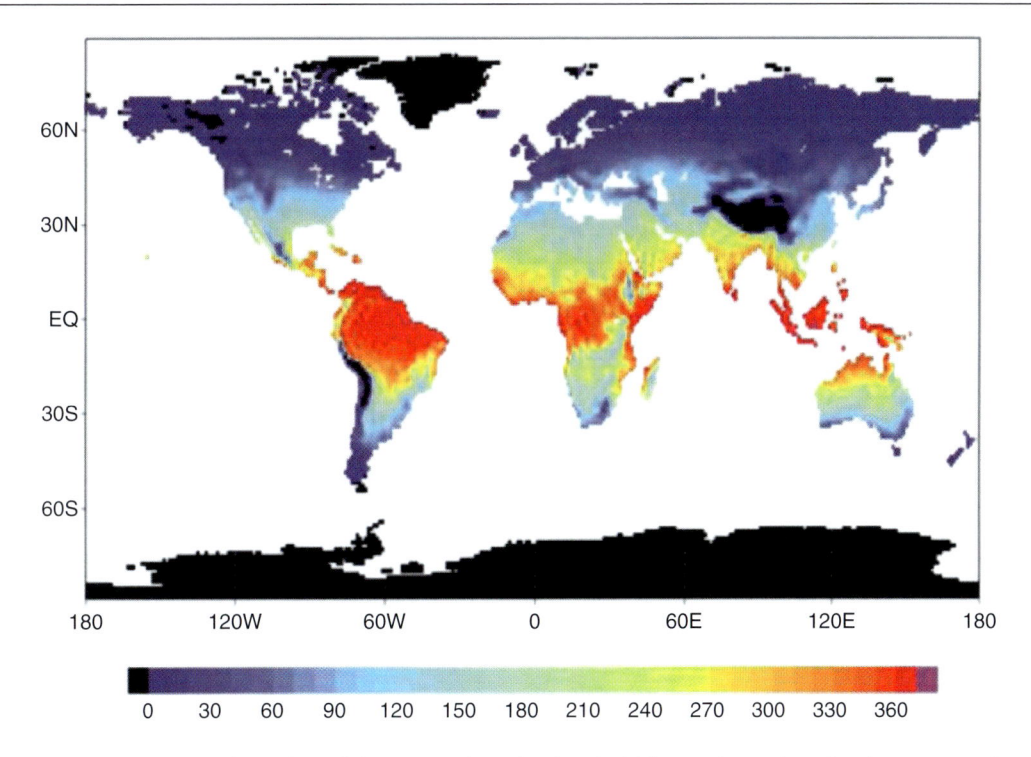

Heat Waves, Figure 1 Mean annual number of days exceeding the threshold for moderate heat load at 12 mean local time taking acclimatization into account with no or light exercise (Jendritzky and Tinz, Global Health Action 2009).

As already shown in the formulation of WBGT, heat is acquired by the body by several physical properties, such as temperature, wet-bulb temperature, ventilation, solar radiation, long wave radiation, etc. For the use in weather forecasting and climatology the statistics of these parameters are mostly evaluated separately. Statistics of all parameters contributing to the formation of heat stress were merged by Jendritzky and Binz (2009). The map (Figure 1) shows the global distribution of the number of days with moderate heat load (see the aforementioned table on WBGT heat stress index). As to be expected, the humid tropical regions show an occurrence of more than 300 days per year, permanent residents mostly being acclimated and prepared to cope with the endemic heat conditions. In the mid-latitudes, the occurrence of such days ranges between about 20 and 100.

In the mid-latitudes, the days with heat stress mainly occur in the summer season and are then grouped in sequences of days, in heat waves, mostly exceeding long-term mean temperatures. Following the mid-latitude weather patterns often heat waves last 2 to about 5 days. This duration is closely connected to the basic properties of mid-latitude weather, dominated by transient, high- and low-pressure systems. The intensity of any particular heat wave depends on the temporal and local peculiarities of the pressure systems, in rare cases allowing a heat wave to last for weeks and to bring particularly hot conditions. The duration is of some relevance, because it takes a few days for the heat to penetrate the walls of most types of standard non-air-conditioned buildings and to raise the indoor temperature to heat stress conditions.

The occurrence of heat waves and their adverse effects is globally wide spread. In the last century, about 150 heat waves were documented to reach disaster level (EM-DAT, 2012). The deadliest heat waves on record globally since 1900 occurred in Europe in the years 2003 and 2010. Europe was hit almost unaware by the heat and the large number of fatalities reached more than 66,000 in the year 2003 and 55,000 in 2010 (EM-DAT, 2012). Both heat waves were accompanied by an enormous amount of material damage, mostly in forestry and agriculture. Both heat waves were caused by a long-lasting mid-latitude air flow pattern, a so-called blocking action. Such weather situations are in general well-known; however, two factors, high temperature and low precipitation, combine in a particularly unfavorable way. In both cases, the monthly average maximum temperature (July 2010 in Russia, August 2003 in France) was about 10 K higher than the long-term average. In both cases the number of days with extremely high daily maximum temeprature was unprecedented (the thresholds were 36°C in western Europe in 2003, and 30°C in Russia in 2010; Schönwiese et al., 2003; Trigo et al., 2005). In the 2003 heat wave, a large part of the victims were from the elderly (Robine et al., 2007). Several factors added to the heat stress for this population group, comprising reduced mobility, health

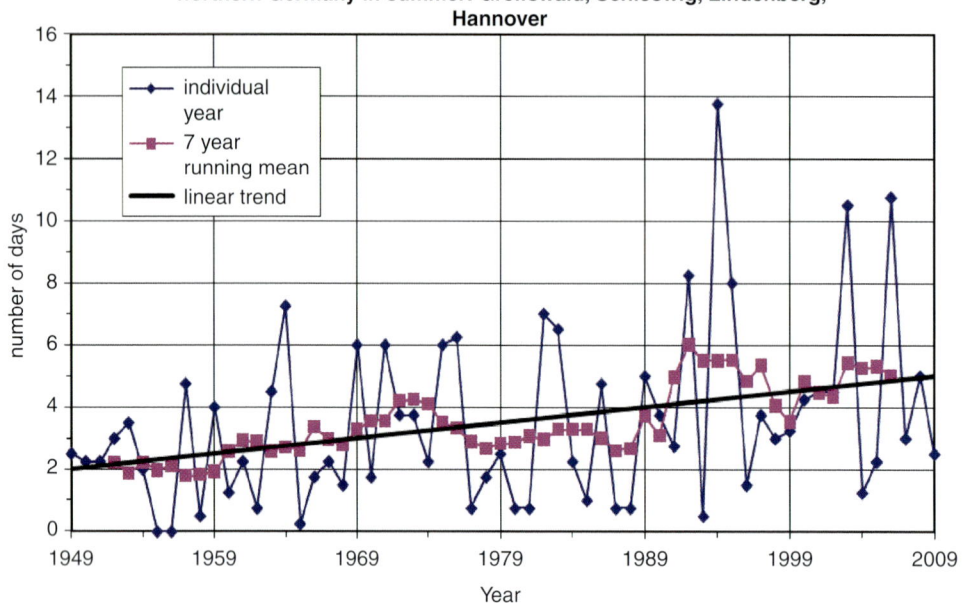

Heat Waves, Figure 2 Number of days with maximum temperature exceeding 30°C in northern Germany (data: Deutscher Wetterdienst).

problems, reduced ability and/or damage of the cardiovascular system, living alone, etc.

The frequency of heat waves is expected to change in the future (IPCC, 2012). Observations for Central Europe already show, that the temperature changes went with a drastic increase of the high (>30°C) and very high temperatures (Schär et al., 2004). In the last decades, the number of days with high temperature more than doubled while the increase of the average temperature was about 1 K (Figure 2).

Summary

Thermal comfort is felt individually. Heat generally means thermal discomfort activating the thermoregulatory system of the body. Heat conditions are weather-determined and if lasting for several days or longer are called heat wave. Quantification of heat needs the total energy budget of the human body. This energy budget is influenced by air temperature, humidity, ventilation, radiation, bodily activity, clothing, mass and shape of the body, health state, age, and also by individual predisposition. In the case of overheating the body's thermoregulatory system failed to keep the temperature in the tolerable range of temperature in the core parts of the body. Indicators are in use allowing a simplified description of the energy budget of the human body, so capturing effects of heat as well. These indicators classify the environmental conditions by either recommending the type of feasible bodily activity, or by requiring the cooling of the human body, both in order to avoid health damage due to overheating. The climatic conditions show a global distribution of heat stress. In some parts of the tropics, heat stress occurs on more than 300 days per year, and in the mid-latitudes, heat stress occurs on about 20–100 days. Indigenous and non-indigenous populations do acclimate to endemic or periodical heat conditions, helping to better cope with these. Heat waves can develop more frequently, when on average there are more such heat days. In recent years, the number of days with heat stress, and of days occurring in heat waves, increased in the mid-latitudes. The forecast climate change points to further increase of heat stress in many parts of the world.

Bibliography

Bureau of Meteorology of Australia, 2010. *About the WBGT and Apparent Temperature Indices*. Commonwealth of Australia 2010, Bureau of Meteorology, ABN 92 637 533 532.

EM-DAT, 2012. *The OFDA/CRED International Data Base*, Université catholique de Louvain, Brussels, Belgium. http://www.emdat.be.

Fanger, P. O., 1972. *Thermal Comfort. Analysis and Applications in Environmental Analysis*. New York: Mc Graw-Hill.

Gaspar, A., and Quintela, D., 2009. Physical modeling of globe and natural wet bulb temperatures to predict WBGT heat stress index in indoor environments. *International Journal of Biometeorology*, **53**, 221–230.

IPCC, 2012. Summary for policymakers. In *Managing the Risks of Extreme Events and Disastersto Advance Climate Change Adaptation*. Cambridge, UK/ NewYork: Cambridge University Press.

Jendritzky, G., and Birger, T., 2009. The thermal environment of the human being on the global scale. *Global Health Action* **2**. Online : globalhealthaction.net. doi:10.3402/gha.v2/0.2005.

Jendritzky, G., Sönning, W., Swantes, H.-J., and Jendritzky, G., 1979. *Ein objectives Bewertungsverfahren zur Beschreibung des thermischen Milieus in der Stadt- und Landschaftsplanung (Klima-Michel-Modell)*. Hannover: H. Schroedel. Beitr. Akad. F. Raumforschung und Landesplanung, Vol. 28. 85.

Luterbacher, J., Dietrich, D., Xoplaki, E., Grisjean, M., and Wanner, H., 2004. European seasonal and annual temperature variability, trends and extremes since 1500. *Science*, **303**, 1499–1503.

Robine, J.-M., Cheung, S., Le Roy, S., van Oyen, H., and Herrmann, F., 2007. *Report on Excess Mortality in Europe During Summer 2003* EU community Action Programme for Public Health, p.15, Health, grant agreement 2005114; ec.europa.eu/health/ph-projects/2005/action1/docs/action1_2005_ a2_15_en.pdf.

Schär, C., Vidale, P., Lüthi, D., Frei, C., Häberli, C., Liniger, M., and Appenzeller, C., 2004. The role of increasing temperature variability in European summer heat waves. *Nature*, **427**, 332–336.

Schönwiese, C., Staeger, T., Trömel, S., Jonas, M., 2003. *Statistisch-klimatologische Analyse des Hitzesommers 2003 in Deutschland*. Deutscher Wetterdienst Klimastatusbericht 2003, pp. 123–132.

Trigo, R. M., Garcia-Herrera, R., Diaz, J., Trigo, F., and Valente, M., 2005. How exceptional was the early August heat wave in France? *Geophysical Research Letters*, **32**, L10701.

US Army and Air Force Department, 2003. Heat stress control and heat casulaty management. Technical Bulletin TB MED 507/AFPAM 48–152 (1), Washington, DC: USA, 72 p.

Yaglou, C., and Minard, D., 1957. Control of heat casualties at military training centres. *American Medical Association Archives of Industrial health*, **16**, 302–316.

Cross-references

Adaptation
Building Codes
Civil Protection and Crisis Management
Climate Change
Community Management of Hazards
Drought
Urban Environment
Warning System

HIGH-RISE BUILDINGS IN NATURAL DISASTER

Murat Saatcioglu
University of Ottawa, Ottawa, ON, Canada

Definition

There is no precise definition of what constitutes a high-rise building. A building code definition includes buildings of over 23 m in height or approximately 6 stories high (IBC, 2009). Emporis Standards (2010) define high-rise buildings as multistory structures that are 35–100 m tall, or 12–39 stories high. Sometimes special service requirements dictate the definition. A building that is tall enough to necessitate an elevator, or as in the case of a fire code, "any structure where the height can have a serious impact on evacuation," may be classified as high-rise. In the context of natural hazards, two aspects distinguish high-rise buildings from others: (1) consequence of damage and (2) structural properties and their interactions with effects of natural hazards. High-rise buildings provide increased density of occupancy per structure, providing concentration of people in one building. Therefore, the consequence of damage tends to be more severe than that of a low-rise building. Furthermore, evacuation of upper floors of a tall building after a natural disaster may pose challenges during relief and response operations.

High-rise buildings are generally built using reinforced concrete or structural steel, and possess the required strength and stiffness against potential natural hazards. Geotechnical risk factors, including soft and compressible soils, as well as potential for landslide and liquefaction cause serious threats to high-rise buildings, which consequently require high soil-bearing capacities. Structural response of high-rise buildings gains a new dimension if the natural hazard induces dynamic forces, as in the case of wind pressures and earthquakes. The natural frequency of tall buildings tends to become lower with increased height, increasing their vulnerability to low-frequency wind and earthquake effects. High-rise buildings whose height-to-width ratios exceed approximately 4, or whose height is approximately in excess of 120 m, tend to become susceptible to vibration by extreme wind effects. These buildings may experience human perception problems, causing discomfort to the occupants, while increasing risk for potent damage to windows, external cladding elements, and other nonstructural components.

Most high-rise buildings have a fundamental period of longer than 1 s. Seismic hazard generally decreases with fundamental period. Figure 1 illustrates the Uniform Hazard Spectrum on firm ground for the city of Vancouver in Canada (NBCC, 2010). In this example, a building with a fundamental period of 1 s has approximately ½ the earthquake spectral acceleration for a low-rise building with a fundamental period of 0.5 s. Taller buildings usually attract much less seismic forces than low-rise buildings. When soil conditions are unfavorable, however, the amplification of earthquake accelerations can be multiple times higher for long-period structures. The 1985 Mexico city earthquake is a good example of the vulnerability of

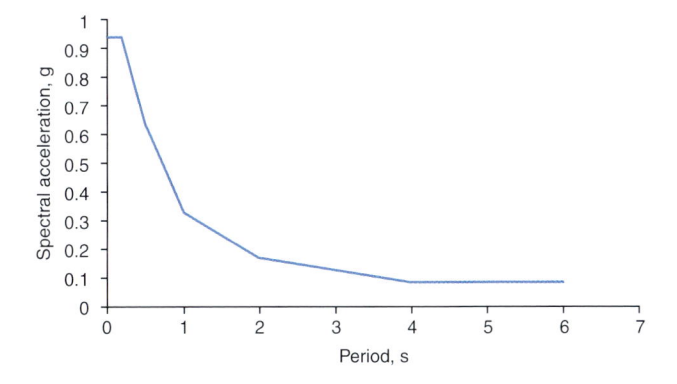

High-Rise Buildings in Natural Disaster, Figure 1 Uniform hazard spectrum for Vancouver, BC, Canada.

high-rise buildings to seismic excitations when located on soft soils. In this earthquake, buildings that were 7–13 stories suffered much more damage and collapses when compared with short-period and very long-period buildings, underlining the importance of interaction between the natural frequency of buildings and the frequency of earthquake excitation.

Bibliography

Emporis Standards, 2010. Datas Standards (ESN 18727), http://standards.emporis.com/?nav=realestate&lng=3&esn=18727

IBC, 2009. International Building Code. International Code Council.

NBCC, 2010. National building code of Canada. Associate Committee on the National Building Code, National Research Council of Canada, Ottawa, ON.

Cross-references

Building Code
Building Failure
Buildings, Structures and Public Safety
Concrete Structures
Damage and the Built Environment
Earthquake Resistant Design
Structural Damage Caused by Earthquakes
Structural Mitigation
Unreinforced Masonry Buildings

HISTORICAL EVENTS

Suzanne A. G. Leroy[1], Raisa Gracheva[2]
[1]Brunel University, Uxbridge (London), UK
[2]Institute of Geography of RAS, Moscow, Russia

Synonyms
Historical natural hazards

Definition
Natural hazards that occurred in the historical past.

Introduction

This entry deals with historical natural hazards, with some reference to the ensuing disasters (see *Hazard, Geohazards, Disasters, Natural Hazard*). Therefore, pollution, famines, and epidemics and natural hazards caused by humans (e.g., desertification) are not considered here. Typical fast events, such as earthquakes, and slower changes, such as climatic change that affected vast areas, and slower pervasive processes (e.g., karst dissolution) with an abrupt end (e.g., collapse of infrastructure), are included. Some of these fast and slower phenomena may have a significant geomorphological-environmental impact with long-lasting repercussions on civilizations (Trifonov and Karakhanyan, 2004; Leroy, 2012, in press).

This entry focuses on historical natural disasters which have: (1) affected the environment, resources, or material culture so much that they caused long-lasting changes in human society at a regional or global scale, (2) led to new research and involved the scientific community in scientific, and sometimes less scientific, discussions, or (3) left lasting images (Figure 1) of catastrophes that have become "imprinted" in human history (Table 1). For example, when the eruption of Mt Vesuvius is mentioned, one remembers the annihilation of the city of Pompeii.

The start of history is by definition the beginning of the written record. The appearance of the first written evidences of natural hazards and disasters varies considerably between different parts of the world. The oldest are found in Mesopotamia on clay tablets in the fourth millennium before Christ, with perhaps much earlier findings in China on turtle bones. Very early, people felt the need to commit to writing certain recurring or exceptional natural observations. The chosen period for this contribution mostly extends from 2,200 years BC to the twentieth century, with most examples from the nineteenth and twentieth centuries (Table 1). Whenever possible, the age annotation has been given in years before Christ (BC) and after Christ (Anno Domini or AD). Radiocarbon dates presented in Years BP, that is, before 1950, are given when necessary.

Techniques

Techniques to reconstruct historical natural hazard events are abundant and drawn from a very wide range of scientific fields. For the time period in this text, historical and phenological documents are supplemented by high resolution, or otherwise well-constrained, earth sciences data and archaeological records.

Historical documents and phenological indicators

In ancient history, various forms of written texts on natural catastrophes can be found. Biblical stories, such as Noah's flood or the destruction of Jericho, have a high level of uncertainty (see *Biblical Events, Religion and Hazards*). Chronicles usually describe events post-factum such as those in the Viking sagas, which were recorded by individuals other than direct witnesses. Contemporary detailed accounts, such as by Josephus Flavius or Pliny the Younger, are often rich in descriptions, but may be rather subjective and incomplete.

Historical observations and documents may provide valuable information about past temperatures and other features of the environment, but commonly are seasonally specific or cover a short time span. In Europe, prior to about AD 1700, the evidence weakened and often became discontinuous. China, having had written documents for much longer, has a wealth of diaries kept by individuals, many of which still remain to be deciphered.

Historical Events, Figure 1 Eye-catching view of collapsed temple in a pancake pile, following the JiJi (or Chi-Chi) earthquake in Taiwan in 1999 (Photo: K. Arpe 2000).

Ship logs have been used for detecting extreme conditions at sea (Figure 2). García et al., (2000) were able, for example, to obtain a reconstruction of the North Atlantic atmospheric circulation during the sixteenth, seventeenth, and eighteenth centuries using documentary sources of the Spanish voyages to the Americas kept in the "Archivo General de Indias." Twelve possible hurricanes were identified in this way.

April to August temperature anomalies in Burgundy, France, were reconstructed from grape-harvest dates from AD 1370 to 2003, carefully registered in parish and municipal archives. Temperatures as high as those reached in the 1990s have occurred several times since AD 1370. However, the heat wave of summer AD 2003 appears to have been extraordinary, with temperatures that were probably higher than in any other year since AD 1370 (Chuine et al., 2004). The WHO has estimated that the extreme heat caused more than 15,000 excess deaths in France, Portugal, and Italy alone (Fink et al., 2004) (see *Heat Waves*).

Zerefos et al., (2007) analyzed paintings created by artists representing sunsets throughout the period AD 1500–1900 to detect the effect of volcanic eruptions on the atmosphere color.

Fluctuations of Nile flood discharge during the period from the ninth to the fifteenth century AD are documented in historical measurements of the Nile flood levels at the Roda Nilometer, opposite Old Cairo. Special attention was focused on two yearly events – one when the Nile was at its lowest point and the other when it reached its peak (Hassan, 2007). Excessively low or high Nile floods led to famines. The results indicate that there was a sharp increase in extreme low floods in the tenth century AD and a rise in the frequency of both extreme high and low Nile floods in the fourteenth century AD.

Archaeological evidence

Geohazards are recorded in many ways by archaeological evidence. New fields are emerging such as archaeoseismology and archaeo-volcanology. Investigations focus on the geohazard effects on ancient structures, uncovered by means of archaeological excavations or pertaining to the monumental heritage (Galadini et al., 2006). Archaeological evidence records periods of abandonment and collapse of infrastructure. Building damage has been used to provide information on the location and magnitude of past earthquakes (Figure 3) (Marco 2008) (see *Earthquake Damage, Paleoseismology, Building Failure*). Houses, personal belongings, temples, citadels, and even whole towns and cities can be preserved intact beneath the volcanic ash or mud that engulfed them (Figure 4). Studies of eruptive deposits, particularly those associated with human habitation, document the nature of the event itself, including precursor activity, the eruptive sequence and the interaction of volcanic flows with human structures (Cashman and Giordano, 2008).

Paleoanthropology based on the interpretation of oral traditions and myths may also provide information on palaeohazards, such as volcanic eruption (Cashman and Giordano, 2008).

Historical Events, Table 1 List of the natural hazards discussed in the text, in detail (bold), briefly mentioned (normal) and as other entries in the encyclopedia (italics)* indicates unconfirmed events

	Eruptions	Earthquakes	Landslides	Floods	Droughts	Other Meteorological	Rare
Early twenty-first		2010 Maule & Haiti *2004 Indian Ocean*		2001 Lensk		2010 Russia fire 2005 Katrina hurricane 2003 hot summer, France	
Late twentieth	*1991 Pinatubo* **1980 Mt St. Helens**	*1976 Tangshan* 1966 Tashkent **1960 Valdivia** **1952 Severo-Kurilsk**	1991 Switzerland **1970 Huascarán**	**1996 Vatnajökull**	1970–1980s Sahel **1959–1961 China**	1972 & 2002 Russian fires **1998 Bola cyclone** **1970 Bhola cyclone**	**1988 Nimes karst** **mid-1980s Cameroon burst** **early 1980s Dead Sea sinkholes** **1978 Lithuania karst** **1972 Calera karst** **1947 Sikhote-Alin meteorite**
Early twenty-first	1906 Chile	1948 Ashgabat **1927 Palestine & Yalta** **1908 Messina** **1906 San Francisco**	1920 Gansu	**1953 The Netherlands** **1931 China**	*1930 Dust Bowl*		1923 Tokyo fire **1908 Tungushka object**
1800–1900	*1883 Krakatau* **1815 Tambora**		1881 Elm	**1887 Huang He**	**1876–1879 N. China and elsewhere**	**1839 Coringa cyclone**	**1871 Peshtigo & Chicago fires**
1500–1800	1783 Laki	**1755 Lisbon**	**1618 Piuro** **1556 Shaanxi**	1556 Shaanxi		Sixteenth–Eighteenth C. Tropical cyclones	1700? Wabar craters
1–1500	**969 Baitoushan** **79 Vesuvius**	**749 Jerash** **363 & 365 Crete & E. Mediterranean**		Fourteenth C. irregular Nile Tenth C. low Nile	Twelfth–Fifteenth C. Puebloans 900 Maya 600 Moche		**1237 Kitezh karst*** 536 dust veil
						850 Global climate	c. 300 Rio Cuarto craters
BC	*1,600–1,500? Santorini*	31 Jericho	Saidmarreh		2,200 Harappans and Akkadians	2,200 Global climate	**1,690–1,510 Kaali craters** c. 2000 Campo del Cielo craters

Historical Events, Figure 2 Ship log from a ship sailing near the Philippines on a day with a thunderstorm from August 5–6, 1796 (Photo: David Gallego).

Historical Events, Figure 3 Convento do Carmo, a building which survived the earthquake and the fire of Lisbon in AD 1755. The damage above the red door is probably caused by earthquake shaking (Photo: P. Costa 2010).

Earth sciences

The tectonic and geological structure of a region, including the occurrence of faults, rock movement, soft sediment formation-deformation, and seismic and geodynamic observations, may all provide information on past physical events and their damaging power (Marshak, 2004). Past landslides, lahars and floods, and any rapid and massive sediment or water movement can be reconstructed by studying landforms and sediment (Huggett, 2007).

Buried soils are specific archives of past environmental changes, both of slow and fast events (Yaalon, 1971; Alexandrovskiy, 1996). Preservation of microfossils within buried soils has proved to be an effective long-term recorder of a range of environmental parameters, such as moisture and temperature and have thus been widely used as proxy records of climatic change (Meunier and Colin, 2001; Golyeva and Terhorst, 2003). The stratigraphic sequence of buried soils and overlying or interstratified deposits provide evidence of interruptions of soil formation by rapid sedimentation. The radiocarbon analysis of organic matter buried in soils allows the dating of the accumulative or destructive phenomena (Alexandrovskiy et al., 2004).

Continuous proxy records

Lake sediments generally accumulate on a continuous basis, and, where obtained in outcrops or by coring/drilling, are archives for environmental changes including rapid events (Figure 5). Natural hazards are marked by anomalies, such as changes in grain size, disturbances to fine structures, and changes in sediment origin seen in, for example, geochemistry, pollen content, and charcoal concentration. Lake sediments have provided past records of floods (Arnaud et al., 2005), earthquakes (Schwab et al., 2009), hurricanes (Besonen et al., 2008), and volcanic eruptions (Bertrand et al., 2008a; Costa et al., 2012).

Deep marine sediment with a high sedimentation rate may be considered to provide comparable information to that in lake sediment; shallow marine sediment may, in addition, record phenomena such as tsunamis (see *North Anatolian Fault*) (Bertrand et al., 2011).

Corals contain annual growth bands, and therefore provide the possibility of high resolution using isotopic and geochemical analyses that may contribute to a history of flood events (Shen et al., 2005), extreme sediment flux events, changes in upwelling, sea surface temperature, and El Niño Southern Oscillation (ENSO) (Grottoli and Eakin, 2007). Ice cores may provide a fairly complete record of volcanic eruptions and comet/asteroid impacts through their chemical record, especially acid layers (Baillie, 2008) (see *Comet*). Tree-rings, in addition to the traditional dendrochronology and dendroclimatology, provide evidence for local and regional fires (Grissino-Mayer and Swetnam, 2000), hurricanes (Miller et al., 2006), and volcanic eruptions (Pearson et al., 2005).

It is worth noting the benefit of working on annually laminated sediment, tree-rings, coral annual bands, and ice core annual layers leading to strong age-depth models that allow constraining the event age, the duration of the disturbance, and their recurrence times. For example, in the annually laminated sediment of the Dead Sea, Holocene earthquakes and disturbance of lake sedimentation have been recorded by seismites. The associated palynology has been interpreted to show a decline in agriculture for 4–5 years following the two earthquakes analyzed at 31 years BC and AD 363 (Leroy et al., 2010).

Palaeoecology is a technique that is applied to the study of natural hazard reconstruction in continuous records: shells for high-energy events (Morales et al., 2008); foraminifera for seawater invasions (Mamo et al., 2009); ostracods for changes in salinity (Ruiz et al., 2010); diatoms for emergence-submergence (Sawai et al. 2004); pollen, spores, and non-pollen palynomorphs for soil wash in lakes (Leroy et al., 2009), amongst others, as these techniques record the impact of natural hazards on ecology.

Historical Events, Figure 4 House buried by an Etna eruption (Photo: V. Targulian 2007).

Historical Events, Figure 5 Coring in Lake Sapanca on the North Anatolian Fault, using the Reasoner technique (Photo: I. Stewart 2003).

Case studies

This section provides an overview of some notable natural disasters, beginning with classical geohazards, then hazards related to water and weather, followed by large-scale climatic events, and finally rarer events.

Instantaneous and destructive geohazards

This section is concerned with classical geohazards such as volcanoes, earthquakes, and landslides.

Volcanic eruptions and lahars

The AD 79 eruption of Mount Vesuvius (Italy) is one of the most famous catastrophic events in human history because of the vivid description recorded by Pliny the Younger, eyewitness of the event (see *Vesuvius*). The eruption buried the towns of Pompeii and Herculaneum (Figure 6). It was only at the end of the sixteenth century that Pompeii was rediscovered, buried below ash and rocks. Some 1,150 full casts and skeletons of people have been discovered during subsequent archaeological excavations of Pompeii (Luongo et al., 2003; Italian National Institute of Geophysics and Volcanology, 2006).

The destructive power of volcanic eruptions, strong in itself, is often exacerbated by subsequent lahars, landslides, and atmospheric emissions causing long-term consequences for humans. The AD 1815 eruption of Mount Tambora, Indonesia, is the largest and the most deadly volcanic eruption in recorded history (Oppenheimer, 2003). The mountain erupted, and ash fell 1,300 km away (Stothers, 1984). The death toll was at least 71,000 people; 11,000–12,000 were killed directly

Historical Events, Figure 6 Pompeii ruins with the Vesuvius volcano in the background (Photo: V. Targulian 2009).

Historical Events, Figure 7 Mount St Helens with trees felled in the same direction by the blast of the eruption in 1980 (Photo: K. Arpe 2002).

by the eruption, and other deaths were from starvation and disease (Oppenheimer, 2003). The eruption released sulfur into the stratosphere, causing global climate anomalies. AD 1816 became known as the "Year Without a Summer" because of the effect on North American and the European weather. Agricultural crops failed and livestock died in much of the Northern Hemisphere, resulting in the worst famine of the nineteenth century (Oppenheimer, 2003).

The relatively large scale of the May 18, 1980, eruption of Mount Saint Helens, USA, was totally unexpected despite regular monitoring (Figure 7). It claimed 57 lives and affected several hundred thousand people (Major and Scott, 1988). A large lateral blast sent a large mass failure down the mountain, sending ash up to 18 km into the atmosphere and causing a large landslide, which buried the surroundings under 45–180 m of debris. The lahars of rocks, ice, and mud rushed into river valleys and destroyed

27 bridges, 200 homes, 300 km of roadway, and 25 km of railway (Major and Scott, 1988; Major, 2003) (see *Lahar*). Spirit Lake was dammed with debris over 100 m deep, and the navigation channel of the Columbia River was reduced overnight in depth from 12 to just 4.2 m.

One of the world's largest historical explosive eruptions took place at about AD 969 from Baitoushan (Changbaishan or Paektusan) volcano, on the Chinese-North Korean border. The eruption column is estimated to have reached 25 km and thus entered the stratosphere; the eruption is thought to have had a substantial, but possibly short-lived effect on climate (Horn and Schmincke, 2000). The volcano now has a large lake in its caldera considered sacred by the Korean population living in the region: Lake Tianchi or Sky Lake.

Other extreme events not detailed here are the Santorini eruption (Greece) in the mid-second millennium BC with its tempting to link to the cultural downfall of the Minoans (see *Santorini, Eruption*), the Laki eruption (Iceland) in AD 1783 which caused a widespread dry fog and weather extremes (Leroy, 2012, in press), the Krakatau eruption (Indonesia) in AD 1883 in which more than 50,000 people perished (see *(Krakatoa) Krakatau*), many in a way similar to the Indian Ocean tsunami of 2004, and finally the Pinatubo eruption (Philippines) in AD 1991 (see Mt Pinatubo). The Smithsonian Institution has produced a catalog of world-wide volcanic eruptions, including historical ones, which can be searched by eruption date, region, and name: http://www.volcano.si.edu/world/. The Institute of Volcanology and Seismology of the Far Eastern Branch of the Russian Academy of Sciences has prepared a catalog of volcanoes and volcanic eruptions for Kamchatka and the Kuril Islands, which can be searched by name: http://www.kscnet.ru/ivs/volcanoes/holocene/ and http://www.kscnet.ru/ivs/kvert/volcanoes/.

Earthquakes and tsunamis

The strongest earthquake (see *Earthquake*) ever recorded occurred in Chile on May 22, 1960, off Valdivia. This 9.5 magnitude (M_w) earthquake killed more than 2,000 people, left two million homeless, caused heavy damage in several coastal cities, and triggered a tsunami in the Pacific Ocean that caused destruction as far as Hawaii and Japan (USGS, 2010b) (see *Tsunami*). Inland, this earthquake triggered numerous landslides in the Southern Cordillera of the Andes. It was also assumed to be responsible for the eruption of the Puyehue–Cordón de Caulle volcano (Bertrand et al., 2008b). On February 27, 2010, another very large earthquake, M_w 8.8, occurred offshore from Maule, only 273 km north of Valdivia.

The deadliest and most destructive earthquake was in AD 1556 when a M_w 8 earthquake hit the province of Shaanxi, and surrounding provinces in China, and caused 830,000 deaths (Hou et al., 1998). The death toll was especially high as many people lived in caves in loess cliffs. These easily collapsed in the shaking and created many deadly landslides.

The tectonic uplift of western Crete (Greece) of up to 9 m, a cluster of coastal uplifts in the East Mediterranean, as well as historical and archaeological data are evidence for a major seismic destruction on a nearly Mediterranean scale in AD 365 (Stiros, 2010). The earthquake affecting western Crete was associated with a reverse fault offshore of southwestern Crete, whose minimum magnitude was 8.5, hence putting it among the largest ever earthquakes. Stiros, (2010) suggested that the AD 365 destruction included at least two other major events, with epicenters close to Cyprus and Sicily and probably other areas.

Lisbon (Portugal) was destroyed in AD 1755 by earthquake shocks, a resulting tsunami, and fire (Fonseca, 2004). The M_w 8.5 earthquake and perhaps a second event occurred offshore near the Gorringe Bank and in Lisbon, respectively (Vilanova et al., 2003). Their impact extended from the Algarve to North Africa and Southeast Spain. Detailed and accurate historical accounts of the disaster have been kept by individuals who were in the city. Also, the first minister of Portugal sent a questionnaire to every parish priest requesting an exact description of what happened, leading to an exceptional scientific documentation. From these accounts, it appears that three shocks caused fires, which commonly started from stoves in the houses or from candelabra in churches, and those fires were then propagated by a northeast wind. Many people sought refuge on the banks of the Tagus River. However, many were drowned when a tsunami inundated the area 75–90 min after the earthquake (Fonseca, 2004). The Lisbon earthquake was the first disaster in which the state took on the responsibility for organizing the response to the emergency (Fonseca, 2004). This action greatly reduced hunger, spread of disease, and looting. When the time for reconstruction came, several proposals were made to reconstruct the city in a safer place; but the Portuguese king and the people insisted on occupying the lower area of Lisbon with buildings despite the engineers' advice. The King decided on razing the whole central area using rubble to make it flat. The only mitigation measures that were successfully implemented were the wider streets and the reinforcement of the house walls with a mesh to make the structures more resistant to shaking (Fonseca, 2004).

The area along the Dead Sea Fault Zone, which has been inhabited since antiquity, has a long and well-documented history of seismic activity. For example, in 31 BC, a major upheaval on the Jericho fault destroyed the town of Qumran (Karcz, 2004). A detailed description of this event appears in the writing of the historian, Josephus Flavius, who records the death from falling houses of 10,000 men. In AD 749, a major earthquake destroyed the ancient city of Jerash (Jordan) and its surroundings (Ambraseys, 2005) (Figure 8). On July 11, 1927, a M_w 6.5 earthquake hit the central, populated area of Palestine. Three hundred people were killed and about 1,000 homes were destroyed. Rock slides triggered by the earthquake stopped the flow of the Jordan River

Historical Events, Figure 8 Fallen columns at the classical Greco-Roman city of Susita, overlooking the Sea of Kinneret, collapsed during the AD 749 earthquake on the Dead Sea Fault (Photo: S. Leroy 2004).

for a day. The earthquake was large enough to be recorded at seismological stations in Europe, South Africa, North America, and the USSR (Avraham et al., 2005).

In the same year, on September 11, 1927, a M_w 8 earthquake with an epicenter under the Black Sea affected the Crimean Peninsula, USSR (now Ukraine), near the city of Yalta. Seventy percent of the town buildings were destroyed (Kondorskaya and Shebalin, 1977). During this event, flares erupted from the Black Sea ranging from 20 to 500 m high and visible to a distance of 60–70 km. These flames were most likely caused by the combustion of methane gas eruptions triggered by the earthquake (Nikonov, 2009).

The AD 1908 Messina earthquake (Italy) was one of the most destructive to hit Europe. This M_w 7.5 earthquake occurred along the Strait of Messina between the island of Sicily and mainland Italy. It triggered a local tsunami, which struck within minutes of the earthquake. The cities of Messina and Reggio di Calabria were completely destroyed; the death toll was estimated between 60,000 and 120,000 (Barbano et al., 2005). The Italian government relocated many of the Messina survivors to other cities within Italy, others choose to emigrate to America.

The AD 1906 San Francisco (USA) earthquake (see *San Andreas Fault*) and the resulting fire is, as arguably, the worst natural disaster in the history of the USA. It ranks as one of the most significant earthquakes in history with M_w estimated between 7.8 and 8.5 (USGS, 2009). The death toll from the earthquake and fires was over 3,000, and at least 225,000 people were left homeless, out of a population of about 410,000 (USGS, 2009). About 90% of the city was destroyed as a result of the subsequent fires, many of them caused by ruptured gas mains.

In the Russian Far-East, tsunamis have a recurrence interval of 50–100 years (Vorobyev et al., 2006). The most destructive tsunami on record in this area took place on November 1952. This tsunami was generated by a M_w 8.5 earthquake in the Pacific Ocean. The city of Severo-Kurilsk was almost totally destroyed by three waves; the second was up to 20 m high; the cold water killed people who were swept into the sea. The death toll was estimated between 5,000 and 14,000, although the exact information was kept secret by the authorities at that time. The remaining survivors were evacuated to the interior; later Severo-Kurilsk was rebuilt in a more elevated location. After this event a tsunami warning system was established in the USSR (Piyp, 2005; Vorobyev et al., 2006).

Two other highly destructive earthquakes, the Tangshan earthquake of AD 1976 (China) and the Haiti earthquake of AD 2010, are among those having caused the highest numbers of victims. Both occurred in highly populated and poor areas (see *Tangshan, China (1976 Earthquake)*). Several catalogs of historical earthquakes have been prepared by historians, especially for the Mediterranean Region (Guidoboni and Ebel, 2009).

Landslides

Mass movements (see *Landslide*) affect people and their environment, sometimes leading to secondary disasters, such as the bursting of temporary lakes dammed by the landslide (Schuster and Highland, 2007). Heavily populated mountainous regions in particular suffer from landslides including extremely destructive events, such as the earthquake-generated landslide in AD 1556 in China, described above (Hou et al., 1998).

In the Alps, a number of landslide events were recorded in various chronicles. A historically well-documented landslide occurred on September 4, 1618 in the Chiavenna Valley, Northern Italy. About 2,420 people from villages, especially Piuro, were killed in the landslide and subsequent flooding (Guzzetti, 2000) (Figure 9). The business activity of the area, which was thriving at an international level, was totally lost and many of the surviving rich local families had to emigrate elsewhere in Europe.

The Elm landslide in Switzerland occurred on September 11, 1881, likely as a consequence of open

Historical Events, Figure 9 The Piuro landslide (N. Italy) that buried the ancient village below several debris flows (Photo: S. Leroy 2005). The *red line* shows the extent of the landslide.

slate mining undercutting a large bedding plane on the side of Plettenbergkopf Mountain. The landslide lasted 40 s and traveled about 2 km. The total volume is estimated to have been about 0.01 km^3. Some 115 fatalities were recorded and 800,000 m^2 of the village of Elm were covered with rubble reaching 3–6 m (Heim, 1882; Warburton, 2007).

Probably the greatest number of deaths from a series of landslides in recorded history occurred in Gansu (Kansu), China, on December 16, 1920. The landslides, triggered by a M_w 7.8 earthquake, occurred over an area of about 48,000 km^2, moving into valleys and covering villages, fields, and roads. An estimated 180,000 people were killed from the earthquake, landslide, and cold weather. Thousands of people living in caves carved in the high loess cliffs were buried by landslides.

A large mountain landslide disaster occurred in the Peruvian Andes following an offshore M_w 7.7 earthquake on May 31, 1970. A debris and snow avalanche slid from the summit of the Huascarán Mountain (6,654 m). By the time it reached the river, it had changed into a mudflow of over 1 km width (see *Mudflow*). Eyewitnesses describe the flow as a huge wave 80 m high (Whittow, 1980). The flow buried settlements in its path up to 10 m deep, leaving 18,000 people dead or missing (Warburton, 2007).

One of the largest landslides, or rock avalanches (see *Rock Avalanche (Sturzstrom)*), was the Saidmarreh (or Seymareh) landslide in southwest Iran, which occurred at 10,370 ± 120 ^{14}C years BP (Shoaei and Ghayoumian, 1998; Schuster and Highland, 2004). It is highly likely that this landslide was earthquake triggered. A mass of debris collapsed from a height of about 1,600 m. The landslide had an exceptionally high volume, of about 20 km^3, a depth of 300 m, a travel distance of 14 km, and a width of 5 km. It blocked two rivers, forming lakes, now drained away, and at present the lake sediment provides fertile farmland (Shoaei and Ghayoumian, 1998).

The USGS has created a list of the twentieth-century landslides that occurred in the western hemisphere on http://pubs.usgs.gov/of/2001/ofr-01-0276/.

Hazards linked to water and weather

This section includes extreme water-related and weather events which had an influence on a large area, including floods, droughts, and cyclones (see *Hydrometeorological Hazards*).

Floods

Rivers may become a serious hazard by threatening human life or property with inundation and sediment erosion or deposition (Davies, 1991) (see *Flood Hazard and Disaster*; *Paleoflood Hydrology*). China has suffered numerous floods in its history. In AD 1887, the Huang He River (Yellow River) flood damaged thousands of settlements with a combined population of seven million, of which at least one million perished (Gunn, 2007). The Yangtze (Chang Jiang), the Huang He, and the Huai rivers in AD 1931 affected more than 40 million people, with at least one million fatalities (Mingteh, 2003). This flood was preceded by a prolonged drought in China from 1928 to 1930 (Mingteh, 2003). The Yellow River is commonly called "River of Sorrow," or "China's sorrow." These Chinese floods are arguably thought to be the deadliest natural disasters ever recorded.

The Netherlands also suffered serious floods, including in AD 1953 when the North Sea flood caused one of the worst natural disaster in northern Europe over the past two centuries. A storm surge (see *Surge*), in combination with a high tide, struck the southwest coast of the Netherlands overwhelming flood defenses. The water level locally exceeded 5.6 m above mean sea level. The damage was significant and more than 1,800 people lost their lives (Glaser and Stangl, 2003). In order to prevent future catastrophes, the coastline was protected by the construction of larger dams along the North Sea coast, one of the most massive water defense systems in the world. The event also had serious effects in eastern England (over 300 deaths) and ultimately led to building the Thames barrier in 1984.

A catastrophic flood on the Lena River at Lensk City, eastern Siberia, in May 2001 was caused by an ice jam. Lensk and adjacent villages were submerged by cold water containing large blocks of floating ice. Some 30,800 people were affected, with six fatalities; several villages and more than 3,000 houses were completely destroyed (Buzin et al., 2007). Lensk is currently being reconstructed on a more elevated site.

Glacial Lake Outburst Floods (GLOFs) occur when lakes, which have developed behind unstable ice-cored moraines, burst catastrophically because the terminal moraine dam fails. Flood triggers are ice avalanche in the lake, or breach of the dam. Within the last century, GLOFs have been recorded in Oregon, Austria, Nepal, Peru, and Argentina (Clague and Evans, 2000). Fear of glacial lake outbursts is mounting in the Himalayas for glaciers showing signs of retreat. Iturrizaga (2005) analyzed the risk in a valley of the eastern Karakoram based on recent changes and past outbursts. In the last 150 years, on average every 5 years, glacier lake floods occurred in the upper Shimshal Valley (Pakistan) with impacts sometimes as far as 150 km downstream. Traces of older and larger floods can be observed.

A jökulhlaup is a specific type of GLOF, associated with subglacial lakes (see *Jökulhlaups*). Its trigger may be, for example, subglacial volcanic activity, such as in Iceland. A large jökulhlaup occurred from the Vatnajökull glacier in Iceland, due to an eruption of the Grímsvötn volcano (Björnsson, 2002). As most of the volcano lies underneath Vatnajökull, most of its eruptions have been subglacial. Eruptions may melt enough ice to fill the Grímsvötn caldera with water, and the pressure may become sufficient to suddenly lift the glacier, allowing water to escape as an outburst flood. As a result, the Grímsvötn caldera is carefully monitored. When a large eruption occurred in AD 1996, geologists knew well in advance that a GLOF was imminent. It did not occur until several weeks after the eruption finished, but the monitoring allowed the Icelandic ring road to be closed when the burst occurred. The total volume of water released from the glacier was estimated at 3.5 km^3 (Magilligan et al., 2002). A section of road across the Skeidara plain was washed away in the ensuing flood, but no one was hurt. The Grímsvötn volcano erupted in AD 1783, that is, at the same time as the Laki volcano, and continued to erupt until AD 1785; it is likely that a jökulhlaup occurred during this time.

Drought

The archaeological record contains numerous instances of civilization collapse due to periods of drought (see *Drought*) exceeding societal resilience (Fagan, 2000; Diamond, 2005; Leroy, 2012, in press): for example, the Harappan and the Akkadian civilizations 2,200 years BC, the Moche in AD 600, the classical Maya

Historical Events, Figure 10 Medieval Jeffrey pine stumps (AD 1112 and 1350) in the West Walker River Canyon (Sierra Nevada). They grew in the valley during the droughts that contributed to the collapse of the Puebloan Indian civilization. Now the valley is a treeless place because of sudden floods (Photo: S. Leroy 2003).

period in AD 900, and the various groups of Pueblo Indians between the twelfth and the fifteenth century AD (Figure 10).

The deadliest drought in recorded history was in Northern China, called the Garden of China, between AD 1876 and 1879. Rivers were completely dry; so most crops and livestock died. Food production could not take place in a 1 million km^2 area of nine provinces, roughly equal to the size of France. The drought caused the death of an estimated nine million people in China. During these years, India also suffered a loss of eight million people, with simultaneous droughts in north Africa and Brazil, indicating a quasi-global impact. Such a large-scale phenomenon is explained by a strong El-Niño event, upwelling along the coast of Peru, with worldwide teleconnections (Aceituno et al., 2009) (see *El Niño/Southern Oscillation*).

Several other important droughts hit China with several tens of millions of victims until as recently as AD 1959–1961. This last drought, which occurred in a non-El-Niño year, was kept secret for several decades by the communist government.

The Dust Bowl (see *Dust Bowl*) in the American Midwest in the 1930s and the Sahel drought in the 1970s and 1980s are the results of a lack of rain combined with, respectively, farming malpractice and overgrazing (Fagan, 2000).

Other meteorological events

In November 1970, the most severe cyclone in the century, with respect to storm surge height, inundation, and loss of life occurred in Bangladesh, then East Pakistan. The Bhola cyclone, a category 3 hurricane, produced a surge height of up to 10 m, which hit the coast at high spring tide, causing inundation of nearly every low-lying coastal area. The loss of life was estimated to be 500,000 (Madsen and Jakobsen, 2004). The northern Bay of Bengal has been hit by many cyclones of similar magnitudes over the years. The main factors leading to disaster are the low-lying continental shelf of the Ganges delta, the large tidal ranges, a funneling coastal configuration, frequent storms, and the densely populated coastline (Frank and Husain, 1971). The cyclone that hit the City of Coringa in the Indian section of the Bay of Bengal, in AD 1839, caused the deaths of 300,000 people. The 12 m-high storm surge caused such destruction that the harbor city, which was well known for ship building and repairing at an international level, was never completely rebuilt.

One of the well-investigated damaging cyclones was Cyclone Bola, which hit the northwest of the North Island of New Zealand in 1998. It was the largest storm event that had occurred since European settlement in the 1830s. Between March 6 and 9, 917 mm of rain fell; three people were killed due to flooding, and hundreds were evacuated when a swollen river threatened a town. Heavy rain caused massive landslides and erosion along the valley systems causing severe damage to the economy of the Gisborne Region (Page et al., 1999).

Global climatic events

The focus is here on large-scale climatic changes (see *Climate Change*), as well as relatively rapid events. Two such events concurrent with widespread cultural declines have been recorded in both palaeoecological and archaeological records at 2,200 years BC (Weiss and Bradley, 2001; Leroy, 2012, in press) and at 850 years BC (Coombes and Barber, 2005).

The latter case, which is part of the time period we examine here, corresponds to the 850 years BC shifts in settlement patterns in Western Europe (van Geel et al. 1996). It has been assigned to a change to cooler and wetter conditions at the Sub-boreal/Sub-Atlantic transition or Late Bronze Age-early Iron Age transition in Europe. A synchronous change has been found in South America (van Geel et al., 2000). This global climatic change has been attributed to altered production of ^{14}C in the atmosphere. Chambers et al., (2007) have underlined that global climatic changes at that time could be detrimental to ancient societies, for example, in the abandonment of some low-lying areas in Netherlands due to a considerable rise in ground water level (van Geel et al., 1996). At the same time, climatic changes could be beneficial to others, for example, the acceleration of the cultural development and the increase in the nomadic population of the Scythians in south-central Siberia (van Geel et al., 2004).

Rare events

This section deals with rarer events such as karst dissolution impact, bolides, limnic eruptions, lake bursts, and wildfires.

Sinkholes

Sinkholes of various origins (see *Sinkhole, Karst Hazards*) are a phenomenon occurring commonly at a local scale only. Sinkholes cause significant damages in karst areas where limestone bedrock dissolves. Sinkholes are often unexpected natural events because of their sudden and rapid occurrence in the absence of any visible external forces. Features such as sinkholes can be attributed to various myths and legends, for example, the Legend of the Invisible City of Kitezh in Russia. According to this legend, the city of Kitezh when attacked by the Mongols in AD 1237 was submerged in the waters of a bottomless lake, which suddenly appeared (Likhachev et al., 1997). It is supposed that it is Svetloyar Lake, in the Upper Volga region not far from the city of Nizhny Novgorod, where the event occurred. The origin of the lake is debatable, but carbonate and sulfate karsts are widely distributed in the area and deep lakes of karst origin are common (Bayanov, 2007).

Large sinkholes in the area of the Dead Sea, probably caused by the dissolution of underlying salt layers, pose a permanent hazard to people and the environment. Sinkholes started appearing in the Dead Sea region in the early 1980s, and the danger may only become worse as the water level of the Dead Sea continues to fall at its current rate of 1 m per year (Abelson et al., 2006). Karst in limestone was responsible for a flash flood in AD 1988 in Nîmes (France) when the homes of 45,000 people were damaged, and 9 deaths were reported (Maréchal et al., 2008). Thousands of sinkholes are known to have formed recently in the USA. Perhaps one of the most spectacular impacts of dissolving limestone was the "Golly Hole" collapse whose surface expression appeared in a matter of hours on December 2, 1972 in Calera, Alabama, USA. It remains among the largest sinkholes in the USA with the size of a football field: 130 × 100 m in area and 50 m deep (Lamoreaux and Newton, 1986). Much of the county where Calera is located is affected by sinkhole formation, which has increased in relation to droughts and human activities, including mining and groundwater withdrawal.

Karst sinkholes associated to gypsum layers have developed extensively in Lithuania since AD 1978, damaging crop areas and communications. It is supposed that this phenomenon is correlated with climate change, which impacted local hydrological and hydrogeological regimes (Satkunas et al., 2007).

Bolides and near-earth objects

Relatively few impact craters have been discovered dating to the most recent millennia, reflecting perhaps an underestimation of their true frequency (see *Asteroid Impact, Comet*). Among them are the Wabar craters in Saudi Arabia dated at c. 290 years ago (Prescott et al., 2004), the Kaali craters in Estonia at 3,305 ± 65 ^{14}C years ago (or 1690–1510 BC) (Veski et al., 2004) (Figure 11), and the Rio Cuarto impacts at 2,300 ± 1,600 ^{14}C years ago (Schultz et al., 2004) and the Campo del Cielo craters at 3,945 ± 84 ^{14}C years ago, both in Argentina (Cassidy and Renard, 1996).

None are known to have caused a major impact on civilization. The pollen diagram from Piila peat bog, 6 km away from the Kaali impact crater, however shows a reduction of cereal pollen grains, reflecting the cessation of human activities in the region for perhaps 100 years and a dramatic drop in the concentration of arboreal taxa, reflecting the destruction of the woodland in the region (Veski et al., 2004). The mysterious dry fog of AD 536 reconstructed from various proxies, including tree-rings from both the northern and southern hemispheres, at times attributed to a possible cometary impact, at other times to a possible volcanic eruption, has been linked to several civilization downturns quasi worldwide (Baillie, 1999, 2008).

One of the largest meteorites in recorded history is the Sikhote-Alin iron meteorite, which fell in AD 1947 in the forests of the Sikhote-Alin Mountains, Far East, Russia (see *Meteorite*). The bright flash and the sound of the fall were observed by eyewitnesses for 300 km around the point of impact; a long smoke trail remained in the sky for several hours. This meteorite broke up in the atmosphere and fell as a meteoritic shower spreading fragments over an area of 48 km^2 and forming a number of craters. The initial size of the meteorite has been estimated at 900,000 kg, and the post-atmospheric mass is estimated between 70,000 and 100,000 kg (Krinov, 1966).

An object struck the sparsely populated taiga of the Podkamennaya Tunguska River basin, not far from the Siberian settlement of Vanavara, Russia on June 30, 1908. It is generally accepted that the event resulted from the catastrophic explosion of a large meteorite high above the ground. A brilliant, sun-like fireball was seen hundreds of kilometers away. A powerful explosion devastated more than 2,150 km^2 of taiga, flattening by blast several tens of millions of trees (Figure 12). At distances around 60 km, people were thrown to the ground or even knocked unconscious. The first study of the Tunguska event began with the scientist, Leonid Kulik, who carefully documented eye witnesses' descriptions in AD 1921. In AD 1927, investigating damaged taiga, he discovered that fallen trees within a huge area lay with their tops radially pointing outward from the center; in the center of this forest there were standing trunks charred and stripped of their branches; in the very center, he described a peat marsh blasted and tortured into a wavy landscape. At present

Historical Events, Figure 11 The main crater (58° 22′ 22″ N, 22° 40′ 08″ E) of the Kaali meteorite crater field that has nine identified craters located on the Island of Saaremaa, Estonia (Photo: S. Veski 2008).

Historical Events, Figure 12 Roots of trees that were pulled from the ground by the blast of the Tunguska event in AD 1908, 5 km northeast from the explosion epicenter. These trees are the only remaining witnesses of the blast. These roots can still be found 100 years after Tunguska event (Photo: V. Bidyukova, on 2 July 2008).

the "Kulik's path" has become a popular route for scientists from around the world. The Tunguska event has been reconstructed in detail (Rubtsov, 2009). This Siberian phenomenon nevertheless continues to generate scientific discussions and a number of hypotheses, including methane explosion (Kundt, 2001) or a geophysical, non-extraterrestrial origin (Ol'khovatov 2003).

Limnic eruptions and lake bursts

In Africa, a series of lakes are known to accumulate CO_2 in their lower water strata. Some of them, such as Lakes Nyos and Monoun (Cameroon), erupted in the mid-1980s and caused thousands of deaths due to carbonic acid burn and poisoning. Other volcanic lakes worldwide have dissolved CO_2 although at lower levels; but historically their water level has been known to fluctuate considerably, and at times even overflow. Lake Albano, which is the deepest (167 m) crater lake in Italy, shows frequent seismic activity and gas emissions (Anzidei et al., 2008). Historical and archaeological data were recently reassessed due to the extension of urbanization in this area close to Rome; they indicate rapid lake level changes and overflows (Funiciello et al., 2003). This reached such a dangerous situation that in 396 year BC Romans decided to dig a tunnel through the wall of the maar in order to maintain the lake at a low level.

Wildfires

Global fires and even continent-wide fires are not documented to have occurred in the recent millennia (see *Forest and Range Fires, Wildfire*). One of the largest wildfires in the US history is that of AD 1871 in the logging community of Peshtigo, Wisconsin, when a massive forest fire destroyed much of the state: nine towns and more than 1,500 people perished (Brown, 2004). This was caused by drought, high temperatures, unusual weather conditions, and by the lumbering practices of the time that left a quarter of the felled trees in the forest and large piles of sawdust near the mills.

Fires may also occur as secondary effects of other natural disasters such as in Lisbon in AD 1755, San Francisco in AD 1906, and Tokyo in AD 1923 following earthquakes.

Some periods are more favorable than others for fires. Biomass burning has broadly increased in the Northern and Southern hemispheres throughout the second half of the Holocene associated with changes in human practices and climate (Carcaillet et al., 2002). Climate, indeed, is a powerful driver of wildfire frequency. This has been especially well demonstrated in western North America (Fauria and Johnson, 2008). Moreover, the fire frequency changes in the boreal forests during the Holocene have been triggered by climate rather than by vegetation types, as demonstrated in eastern Canada (Carcaillet et al., 2001). Climatic factors, such as ENSO and insolation, have played a key role in Amazonia (Bush et al., 2008). Changes in the ENSO and the Pacific Decadal Oscillation (PDO) system affected fire frequencies all across the boreal forest of North America (Fauria and Johnson, 2008).

In summer 2010, with a major drought and prolonged high temperatures, forest and peatland fires occurred in vast areas of Russia. About 15 million hectares of forests were damaged and more than 7,000 people lost their homes. The net effect of extreme temperature, smoke, and fires resulted in 55,800 deaths. For almost 2 months, smoke affected Moscow and its densely populated surrounding districts resulting in perhaps the most significant exposure to smoke of a major population center in recorded history. These fires as well as fires in 1972 and 2002 in Central Russia can be considered as mega-fire events with long-term consequences for atmosphere and vegetation (Chubarova et al. 2008). Data on the world fires can be found on http://www.fire.uni-freiburg.de and http://www.emdat.be.

Discussion

Why are historical events relevant to our present and future?

Many natural hazards are comprised of a series of events that are of short duration, with a long recurrence interval. Therefore, in order to have a relatively full view of the potential range of hazards in a region, a first step toward mitigation, it is essential to look back into the past using geological and historical records.

In AD 1989, as a result of the study of historical earthquakes, geodynamic and geological evidence, including radiocarbon dating of buried soils, the maximal expected earthquake intensity in the Crimean Peninsula was reestimated and raised from M_w 6.0 to 9.0. This study resulted in the cancellation of the construction plans of a Nuclear Power Plant in East Crimea (Nikonov, 2000).

The study of past wildfires is becoming increasingly relevant for our near future, as fires generate greenhouse gases (van der Werf et al., 2008), and global warming is suspected to cause droughts in some regions, which would in turn be favorable to fire increase (Torn and Fried, 1992).

Limnic eruptions are rare natural hazards, but they are interesting analogs to what could happen to geological sinks used for carbon storage, which are being put in place in order to counteract the release of CO_2 from anthropogenic activities.

Factors leading to significant influences on human environment and human society

Four factors need to combine in order to transform a natural hazard into a disaster, or even a catastrophe: time, space, type of society, and accumulation of events (Leroy, 2006; Leroy, 2012, in press) (see *Disasters*).

1. Time: An absence of warning and a spread of the hazards over several weeks or months enhance the disaster: for example, earthquakes commonly have a series of aftershocks.
2. Space: The larger the area affected, the worse the impact. Global or hemispheric events are however relatively rare. Some of them stand out, that is, global climatic changes at 2,200 and 850 years BC, tropical volcanic eruptions with an impact in the stratosphere (e.g., Tambora in AD 1815), and the ENSI event of AD 1876–79.
3. Type of society: The nature of society plays an essential role in the preparedness and response to disasters. Typically, planning is left in the hands of politicians who have short-term goals and therefore hesitate in engaging public money for long-term mitigation. Two radically different sociopolitical approaches have been recognized as efficient: either a centralized country such as the Soviet Union or the Kingdom of Portugal of AD 1755 with top-down actions to impose their authority, or bottom-up action when citizens are able to put pressure on local governments, such as in The Netherlands after the flood of AD 1953.
4. Accumulation of events: The combined effects and successive hazards can be found in the examples we have just used, such as earthquakes combined with landslides in AD 1920 in China, or subglacial eruptions with a jökulhlaup in Iceland. An analysis of historical volcanic eruptions after great earthquakes in Chile showed more volcanic eruptions than normally expected in the 12 months following the earthquakes. This was especially true for the two large earthquakes in AD 1906 and AD 1960 (Watt et al., 2009).

Are there any catastrophes in the twentieth century?

At historical and geological timescales, the twentieth century has not been subjected to many extreme events. The largest GLOFs were during the melting of the Late Pleistocene icecap and one of the largest landslides was 10,000 years ago in Iran. Few natural hazards have reached more than one million deaths in the twentieth century. The largest mega-hazards were mostly floods, for example, the China floods of AD 1931, where several million victims were counted, as these floods took place in a very densely populated region. On the scale of the number of deaths, the next largest twentieth-century hazards were cyclones and earthquakes (Wikipedia, 2010).

How different is the twentieth century from the past?

More understanding of the mechanisms behind natural hazards and more monitoring should bring us to a situation where we are better protected. However the impact of natural hazards increased dramatically during the twentieth century as shown by the trends in the frequency of natural disasters and in insured catastrophe losses (ICSU, 2005; SwissRe, 2009). Two main reasons explain this. First, more people are put at risk as the world population has increased exponentially. More people live in cities where damaged buildings may become a source of hazards. More people have been forced to settle in more marginal and risky areas. The second reason is that our modern societies increasingly depend on technology in order to function normally (Leroy, 2012, in press). Natural hazards can easily damage the infrastructure we so heavily depend on, such as telecommunication, electricity, and transport networks. This dependency on technology may have the opposite effect than expected: it may actually make us more vulnerable.

An illustration of this in the twenty-first century is that the most recent hurricanes/cyclones have become the most expensive ones: for example, the costliest hurricane so far is Hurricane Katrina in AD 2005 (see *Hurricane Katrina*).

Historical changes and their impact: Not only negative

Despite the damage to the environment and society, many natural disasters have had long-term positive sociopolitical effects. It is clear that rapid landscape changes initiate chaos, from which, in the best cases, innovation may emerge. This surprising beneficial outcome has been recognized for many civilizations in the Holocene (Brooks, 2006). The AD 1970 Bhola cyclone in East Pakistan had a positive sociopolitical outcome, in that it led in the following year to the creation of a new country, Bangladesh. The Scythians were able to derive benefits from the 850 years BC global climatic change: they expanded the area dominated by their culture and their nomadic population density increased as more semidesert was turned into steppe (van Geel et al., 2004).

Positive scientific consequences of some twentieth- and twenty-first-century events

Extreme natural events may become a trigger for deepening of the scientific understanding of nature and may be an impetus for new ideas and technology.

A warning system for tsunami has been put in place in the Indian Ocean following the large-scale disaster caused by the earthquake and tsunami of December 2004 (see *Indian Ocean Tsunami, 2004, Warning Systems*). The system linking about 20 countries became active in late June 2006.

The AD 1948 Ashgabat earthquake comfortably sits in the list of the most unforgiving earthquakes in history. On the night of the October 5–6, in a matter of seconds, Ashgabat city, USSR (now Turkmenistan), with a population of 130,000, was completely devastated by a M_w 7.3 earthquake. An estimated 110,000 people perished under the ruins, and others became homeless (Nikonov, 1998). The Tashkent earthquake on April 26, 1966, USSR (now Uzbekistan) with a M_w 5.2 did not cause many human losses, but 36,000 buildings disappeared, and about 300,000 people lost their homes (Ulomov, 1971). Both towns were quickly rebuilt, taking into account the regional seismicity. These events were significant in the history of the USSR because thousands of specialists and workers from all over the USSR participated in providing aid to earthquake victims and cities' recovery.

The AD 1991 Randa rockslide, which occurred on April 18, in the tourist valley in southern Valais, Switzerland, caused disruption and devastation to the valley infrastructure and loss of livestock; 31 chalets were buried in the debris (Quanterra, 2003). On May 9, a series of slides over a period of 7 h caused a dam across the Vispa River, just below the village of Randa, resulting in 30 houses being flooded. Dust was deposited to a depth of 10–40 cm in a 1 km radius and some local housing was buried in debris up to 60 m deep (Gotz and Zimmermann 1993). The first event in April had prompted the authorities to install a seismographic and geodetic warning system. As a result, the May event was forecasted and the area was successfully evacuated (Warburton, 2007).

Finally, our society cannot stop natural hazards from happening but learning from the past may reduce the resulting disaster.

Summary

Historical natural hazards contain a wealth of information worth studying when dealing with the improvement of present mitigation plans. The instrumental record of many natural hazards is very short and it is only by looking at historical and geological events that the full range of natural forces can be estimated. Moreover, lessons may be derived from the way our ancestors have reacted to past events; we could learn from the best cases and avoid repeating the worst.

Bibliography

Abelson, M., Yechieli, Y., Crouvi, O., Baer, G., Wachs, D., Bein, A., and Shtivelman, V., 2006. Evolution of the Dead Sea sinkholes. In Enzel, Y., Agnon, A., and Stein, M. (eds.), *New Frontiers in Dead Sea Paleoenvironmental Research*. Boulder, CO: Geological Society of America, Special papers, 401, pp. 241–253.

Aceituno, P., Prieto, Md. R., Solari, M. E., Martínez, A., Poveda, G., and Falvey, M., 2009. The 1877–1878 El Niño episode: associated impacts in South America. *Climatic Change*, **92**, 389–416.

Alexandrovskiy, A. L., 1996. Natural environment as seen in soil. *Eurasian Soil Science*, **29**(3), 245–254.

Alexandrovskiy, A. L., Glasko, M. P., Krenke, N. A., and Chichagova, O. A., 2004. Buried soils of floodplains and paleoenvironmental changes in the Holocene. *Revista Mexicana de Ciencias Geológica*, **21**, 9–17.

Ambraseys, N. N., 2005. The seismic activity in Syria and Palestine during the middle of the 8th century; an amalgamation of historical earthquakes. *Journal of Seismology*, **9**, 115–125.

Anzidei, M., Carapezza, M. L., Esposito, A., Giordano, G., Lelli, M., and Tarchini, L., 2008. The Albano Maar lake high-resolution bathymetry and dissolved CO_2 budget (Colli Albani volcano, Italy): constrains to hazard evaluation. *Journal of Volcanology and Geothermal Research*, **171**, 258–268.

Arnaud, F., Revel, M., Chapron, E., Desmet, M., and Tribovillard, N., 2005. 7200 Years of Rhône river flooding activity in lake Le Bourget: a high-resolution sediment record of NW Alps hydrology. *The Holocene*, **15**(3), 420–428.

Avraham, Z.-V., Lazar, M., Schattner, U., and Marko, Sh, 2005. The Dead sea fault and its effect in civilization. In Wenzel, F. (ed.), *Perspectives in modern seismology*. Berlin Heidelberg: Springer, pp. 145–168.

Baillie, M. G. L., 1999. *Exodus to Arthur: Catastrophic Encounters with Comets*. London: B.T. Batsford.

Baillie, M. G. L., 2008. Proposed re-dating of the European ice core chronology by seven years prior to the 7th century AD. *Geophysical Research Letters*, **35**, L15813.

Barbano, M. S., Azzaro, R., and Grasso, D. E., 2005. Earthquake damage scenarios and seismic hazard of Messina, north-eastern Sicily (Italy) as inferred from historical data. *Journal of Earthquake Engineering*, **9**(6), 805–830.

Bayanov, N. G., 2007. Preliminary results of palaeolimnological study of Svetloyar Lake and stages of its limnogenesis. *Izvestia RGO*, **139**(5), 73–80 (in Russian).

Bertrand, S., Castiaux, J., and Juvigné, E., 2008b. Tephrostratigraphy of the late glacial and Holocene sediments of Puyehue lake (southern volcanic zone, Chile, 40°S). *Quaternary Research*, **70**(3), 343–357.

Bertrand, S., Charlet, F., Chapron, E., Fagel, N., and De Batist, M., 2008a. Reconstruction of the Holocene seismotectonic activity of the Southern Andes from seismites recorded in Lago Icalma, Chile, 39°S. *Palaeogeography, Palaeoclimatology, Palaeoecology*, **259**, 301–322.

Bertrand, S., Doner, L., Akçer On, S., Sancar, U., Schudack, U., Mischke, S., Cagatay, N., and Leroy, S. A. G., 2011. Sedimentary record of coseismic subsidence in Hersek coastal lagoon (Izmit Bay, Turkey) and the Late Holocene activity of the North Anatolian Fault. *Geochemistry Geophysics Geosystems*, **12**(6), Q06002, doi:10.1029/2011GC003511. 17 pages.

Besonen, M. R., Bradley, R. S., Mudelsee, M., Abbott, M. B., and Francus, P., 2008. A 1,000-year, annually-resolved record of hurricane activity from Boston, Massachusetts. *Geophysical Research Letters*, **35**, L14705.

Björnsson, H., 2002. Subglacial lakes and jökulhlaups in Iceland. *Global and Planetary Change*, **53**, 255–271.

Brooks, N., 2006. Cultural responses to aridity in the middle Holocene and increased social complexity. *Quaternary International*, **151**, 29–49.

Brown, H., 2004. "The air was fire": fire behavior at Peshtigo in 1871. *Fire Management Today*, **64**(4), 20–30.

Bush, M. B., Silman, M. R., McMichael, C., and Saatchi, S., 2008. Fire, climate change and biodiversity in Amazonia: a Late-Holocene perspective. *Philosophical Transactions of the Royal Society B*, **363**, 1795–1802.

Buzin, V. A., Klaven, A. B., and Kopaliani, Z. D., 2007. Laboratory modelling of ice jam floods on the Lena River. In Vasiliev, O. F., van Gelder, P. H. A. J. M., Plate, E. J., and Bolgov, M. V. (eds.), *Extreme Hydrological Events: New Concepts for Security*. NATO Science Series: IV: Earth and Environmental Sciences, 78, pp. 269–277.

Carcaillet, C., Almquist, H., Asnong, H., Bradshaw, R. H. W., Carrión, J. S., Gaillard, M.-J., Gajewski, K., Haas, J. N., Haberle, S. G., Hadorn, P., Müller, S. D., Richard, P. J. H., Richoz, I., Rösch, M., Sánchez Goñi, M. F., von Stedingk, H., Stevenson, A. C., Talon, B., Tardy, C., Tinner, W., Tryterud, E., Wick, L., and Willis, K. J., 2002. Holocene biomass burning and global dynamics of the carbon cycle. *Chemosphere*, **49**, 845–863.

Carcaillet, C., Bergeron, Y., Richard, P. J. H., Fréchette, B., Gauthier, S., and Prairie, Y. T., 2001. Change of fire frequency in the eastern Canadian boreal forests during the Holocene: does vegetation composition or climate trigger the fire regime? *Journal of Ecology*, **89**, 930–946.

Cashman, K. V., and Giordano, G., 2008. Volcanoes and human history. *Journal of Volcanology and Geothermal Research*, **176**, 325–329.

Cassidy, W. A., and Renard, M. L., 1996. Discovering research value in the campo del cielo, Argentina, meteorite craters. *Meteoritics &Planetary Science*, **31**, 433–448.

Chambers, F. M., Mauquoy, D., Brain, S. A., Blaauw, M., and Daniell, J. R. G., 2007. Globally synchronous climate change 2,800 years ago: proxy data from peat in South America. *Earth and Planetary Science Letters*, **253**, 439–444.

Chubarova, N. Yu., Prilepsky, N. G., Rublev, A. N., and Riebau, A. R., 2008. Chapter 11 a mega-fire event in central Russia: fire weather, radiative, and optical properties of the atmosphere, and consequences for subboreal forest plants. *Developments in Environmental Sciences*, **8**, 247–264.

Chuine, I., Yiou, P., Viovy, N., Seguin, B., Daux, V., and Le Roy Ladurie, E., 2004. Grape ripening as a past climate indicator. *Nature*, **432**, 289–290.

Clague, J. J., and Evans, S. G., 2000. A review of catastrophic drainage of moraine-dammed lakes in British Columbia. *Quaternary Science Reviews*, **19**, 1763–1783.

Coombes, P., and Barber, K., 2005. Environmental determinism in Holocene research: causality or coincidence? *Area*, **37**(3), 303–311.

Costa, P., Leroy, S. A. G., Dinis, J. L., Dawson, A., and Kortekaas, S., 2012. Recent high-energy marine events in the sediments of the Lagoa de Óbidos and Martinhal (Portugal): recognition, age and likely causes. *Natural Hazards & Earth System Science*, **12**, 1367–1380.

CRED, 2009. EM-Dat, the international disaster database. http://www.emdat.be/. Last accessed 16 March 2010.

Davies, T. R. H., 1991. Research of fluvial processes in mountains – a change of emphasis. In Armanini, A., and Di Silvio, G. (eds.), *Fluvial Hydraulics of Mountain Regions*. Berlin: Springer, pp. 251–260.

Diamond, J., 2005. *Collapse: How Societies Choose to Fail or Succeed*. New York: Viking.

Fagan, B., 2000. *Floods, Famines and Emperors*. London: Pimlico.

Fauria, M. M., and Johnson, E. A., 2008. Climate and wildfires in the North American boreal forest. *Philosophical Transactions of the Royal Society B*, **363**, 2317–2329.

Fink, A. H., Brücher, T., Krüger, A., Leckebusch, G. C., Pinto, J. G., and Ulbrich, U., 2004. The 2003 European summer heatwaves and drought – synoptic diagnosis and impacts. *Weather*, **59**(8), 209–216.

Fonseca, J. D., 2004. *1755 The Lisbon earthquake*. Lisbon: Argumentum.

Frank, N. L., and Husain, S. A., 1971. The deadliest tropical cyclone in history? *Bulletin of the American Meteorological Society*, **52**(6), 438–444.

Funiciello, R., Giordano, G., and De Rita, D., 2003. The Albano maar lake (Colli Albani volcano, Italy): recent volcanic activity and evidence of pre-roman Age catastrophic lahar events. *Journal of Volcanology and Geothermal Research*, **123**, 43–61.

Galadini, F., Hinzen, K.-G., and Stiros, S., 2006. Archaeoseismology: methodological issues and procedure. *Journal of Seismology*, **10**, 395–414.

García, R., Gimeno, L., Hernández, E., Prieto, R., and Ribera, R., 2000. Reconstructing the North Atlantic atmospheric circulation in the 16th, 17th and 18th centuries from historical sources. *Climate Research*, **14**, 147–151.

Glaser, R., and Stangl, H., 2003. Historical floods in the Dutch Rhine delta. *Natural Hazards and Earth System Sciences*, **3**, 605–613.

Golyeva, A., and Terhorst, B., 2003. Biomorphic analysis of paleosols in the upper Pleistocene loess-paleosol sequence of gunderding (upper Austria). *Tübinger Geowissenschaftliche Arbeiten, D*, **9**, 106–115.

Gotz, A., and Zimmermann, M., 1991. The 1991 rock slides in randa: causes and consequences. *Landslide News*, **7**(3), 22–25.

Grissino-Mayer, H. D., and Swetnam, T. W., 2000. Century-scale climate forcing of fire regimes in the American southwest. *The Holocene*, **10**(2), 213–220.

Grottoli, A. G., and Eakin, M. C., 2007. A review of modern coral $\delta^{18}O$ and $\Delta^{14}C$ proxy records. *Earth-Science Reviews*, **81**, 67–91.

Guidoboni, M., and Ebel, J. E., 2009. *Earthquakes and Tsunamis in the Past. A Guide to Techniques in Historical Seismology*. Cambridge: Cambridge University Press.

Gunn, A. M., 2007. Yellow River China flood 1887. In *Encyclopedia of Disasters: Environmental Catastrophes and Human Tragedies*. Westport, Conn: Greenwood Press, pp. 141–144.

Guzzetti, F., 2000. Landslide fatalities and the evaluation of landslide risk in Italy. *Engineering Geology*, **58**(2), 89–107.

Hassan, F. A., 2007. Extreme Nile floods and famines in medieval Egypt (AD 930–1500) and their climatic implications. *Quaternary International*, **173–174**, 101–112.

Heim, A., 1882. Der Bergsturz von Elm. *Deutsche Geologische Gesellschaft für Zeitschrift*, **34**, 74–115.

Horn, S., and Schmincke, H.-U., 2000. Volatile emission during the eruption of Baitoushan Volcano (China/North Korea) ca. 969 AD. *Bulletin of Volcanology*, **61**(8), 537–555.

Hou, J.-J., Han, M.-K., Chai, B.-L., and Han, H.-Y., 1998. Geomorphological observations of active faults in the epicentral region of the Huaxian large earthquake in 1556 in Shaanxi Province, China. *Journal of Structural Geology*, **20**(5), 549–557.

Huggett, R., 2007. *Fundamentals of Geomorphology*. London: Routledge.

ICSU, 2005. Science and natural hazards. www.icsu.org/Gestion/img/ICSU_DOC_DOWNLOAD/865_DD_FILE_Hazards_Report_Final.pdf. Last accessed 16 March 2010.

Italian National Institute of Geophysics and Volcanology, 2006. Summary of the eruptive history of Mt. Vesuvius Osservatorio Vesuviano. http://www.ov.ingv.it/inglese/vesuvio/storia/storia.htm. Last accessed 10 March 2010.

Iturrizaga, L., 2005. New observations on present and prehistorical glacier-dammed lakes in the Shimshal valley (Karakoram mountains). *Journal of Asian Earth Sciences*, **25**, 545–555.

Karcz, I., 2004. Implications of some early Jewish sources for estimates of earthquake hazard in the holy land. *Annals of Geophysics*, **47**, 759–792.

Kondorskaya, N. V., and Shebalin, N. V., 1977. *New Catalog of Strong Earthquakes on the Territory of the USSR*. Moscow: Nauka (in Russian).

Krinov, E. L., 1966. The sikhote-aline iron meteorite shower. In Beynon, M. M. (ed.), *Giant Meteorites*. New York: Pergamon Press, pp. 266–376.

Kundt, W., 2001. The 1908 Tunguska catastrophe: an alternative explanation. *Current Science*, **81**(4), 399–407.

Lamoreaux, Ph. E., and Newton, J. G., 1986. Catastrophic subsidence: an environmental hazard, Shelby county, Alabama. *Environmental Geology*, **8**(1–2), 25–40.

Leroy, S. A. G., 2006. From natural hazard to environmental catastrophe, past and present. *Quaternary International*, **158-1**, 4–12.

Leroy, S. A. G., 2012. Natural hazards, landscapes, and civilizations. In Shroder, J., Jr., James, L. A., Hardon, C., and Clague, J. (eds.), *Treatise on Geomorphology*. San Diego, CA: Academic Press, Vol. 13. 14 pages (in press).

Leroy, S. A. G., Boyraz, S., and Gürbüz, A., 2009. High-resolution palynological analysis in Lake Sapanca as a tool to detect earthquakes on the North Anatolian fault. *Quaternary Science Reviews*, **28**, 2616–2632.

Leroy, S. A. G., Marco, S., Bookman, R., and Miller, Ch. S., 2010. Impact of earthquakes on agriculture during the roman-byzantine period in the Dead Sea laminated sediment. *Quaternary Research*, **73**, 191–200.

Likhachev, D. A., Dmitriev, L. A., Alekseev, A. A., and Pnyrko, N. V. (eds.), 1997. *Biblioteka Literatury Drevnei Rusi*, T. 5: XIII vek. *Bibliotheca of Literature of Ancient Russia*, Vol. 5: XIII c. Nauka: St-Petersburg (in Russian).

Luongo, G., Perrotta, A., Scarpati, C., De Carolis, E., Patricelli, G., and Ciarallo, A., 2003. Impact of the AD 79 explosive eruption on Pompeii II. Causes of death of the inhabitants inferred by stratigraphic analysis and areal distribution of the human casualties. *Journal of Volcanology and Geothermal Research*, **126**, 169–200.

Madsen, H., and Jakobsen, F., 2004. Cyclone induced storm surge and flood forecasting in the northern Bay of Bengal. *Coastal Engineering*, **51**, 277–296.

Magilligan, F. J., Gomez, B., Mertes, L. A. K., Smith, L. C., Smith, N. D., Finnegan, D., and Garvin, J. B., 2002. Geomorphic effectiveness, sandur development, and the pattern of landscape response during jökulhlaups: skeiðoarsandur, southeastern Iceland. *Geomorphology*, **44**(1–2), 95–113.

Major, J. J., 2003. Post-eruption hydrology and sediment transport in volcanic river system. *Water Resources Impact*, **5**, 10–15.

Major, J. J., and Scott, K. M., 1988. Volcaniclastic sedimentation in the Lewis River Valley, Mount St. Helens, Washington; processes, extent, and hazards. *U.S. Geological Survey Bulletin*, 1383-D.

Mamo, B., Strotza, L., and Dominey-Howes, D., 2009. Tsunami sediments and their foraminiferal assemblages. *Earth-Science Reviews*, **96**(4), 263–278.

Marco, S., 2008. Recognition of earthquake-related damage in archaeological sites: examples from the Dead Sea fault zone. *Tectonophysics*, **453**, 148–156.

Maréchal, J. C., Ladouche, C., and Dörfliger, N., 2008. Karst flash flooding in a Mediterranean karst, the example of Fontaine de nîmes. *Engineering Geology*, **99**(3–4), 138–146.

Marshak, S., 2004. *Essentials of Geology*. New York: W.W. Norton & Company.

Meunier, J. D., and Colin, F. (eds.), 2001. *Phytoliths: Applications in Earth Sciences and Human History. Proceedings of the 2nd International Meeting on Phytolith Research*. Lisse: A.A. Balkema Publishers.

Miller, D. L., Mora, C. I., Grissino-Mayer, H. D., Mock, C. J., Uhle, M. E., and Sharp, Z., 2006. Tree-ring isotope records of tropical cyclone activity. *Proceedings of the National Academy of Sciences*, **103**(39), 14294–14297.

Mingteh, Ch, 2003. *Forest Hydrology: An Introduction to Water and Forests*. Boca Raton, FL: CRC Press.

Morales, J. A., Borrego, J., San Miguel, E. G., López-González, N., and Carro, B., 2008. Sedimentary record of recent tsunamis in the Huelva estuary (southwestern Spain). *Quaternary Science Reviews*, **27**(7–8), 734–746.

NERC, 2010. Natural hazards. http://www.nerc.ac.uk/research/issues/naturalhazards/. Last accessed 16 March 2010.

Nikonov, A. A., 1998. Chronicle of Askhabat catastrophe. *Herald of the DGGGMS RAS*, 2(4). http://www.scgis.ru/russian/cp1251/h_dgggms/ca_2-1998.htm#1. Last accessed 2 February 2010.

Nikonov, A. A., 2000. Seismic potential of Crimean region: comparison of regional maps and parameters of identified events. *Fizika Zemli*, **7**, 53–62 (in Russian).

Nikonov, A. A., 2009. *Earthquakes...: The Past, The Present, and Prediction*. URSS (in Russian): Moscow.

NOAA, 2010. Images of geologic hazards. http://www.ngdc.noaa.gov/mgg/image/hazardsimages.html. Last accessed on 16 March 2010.

Ol'Khovatov, A. Yu., 2003. Geophysical circumstances of the 1908 Tunguska event in Siberia, Russia. *Earth, Moon, and Planets*, **93**(3), 163–173.

Oppenheimer, C., 2003. Climactic, environmental, and human consequences of the largest known historic eruption: Tambora volcano (Indonesia) 1815. *Progress in Physical Geography*, **27**, 230–259.

Page, M. J., Reid, L. M., and Linn, I. H., 1999. Sediment production from cyclone Bola landslides, Waipaoa catchment. *Journal of Hydrology (New Zealand)*, **38**(2), 289–308.

Pearson, C., Manning, S. W., Coleman, M., and Jarvis, K., 2005. Can tree-ring chemistry reveal absolute dates for past volcanic eruptions? *Journal of Archaeological Science*, **32**, 1265–1274.

Piyp, B. I., 2005. Secret tsunami. *Priroda*, **5**, 36–43.

Prescott, J. R., Robertson, G. B., Shoemaker, C., Shoemaker, E. M., and Wynn, J., 2004. Luminescence dating of the Wabar meteorite craters, Saudi Arabia. *Journal of Geophysical Research*, **109**, E01008.

Quanterra, 2003. Short guide about slope instabilities between Lausanne and Zermatt (Switzerland). (18) Randa. http://www.quanterra.org/guide/guide1_18.htm. Last accessed 26 February 2010.

Rubtsov, V., 2009. *The Tunguska Mystery*. New York: Springer.

Ruiz, F., Abad, M., Cáceres, L. M., Vidal, J. R., Carretero, M. I., Pozo, M., and Gonzáles-Regalado, M. L., 2010. Ostracods as tsunami tracers in Holocene sequences. *Quaternary Research*, **73**, 130–135.

Satkunas, J., Marcinkevicius, V., Mikulenas, V., and Taminskas, J., 2007. Rapid development of karst landscape in North Lithuania – monitoring of denudation rate, site investigations and implications for management. *GFF*, **129**(4), 345–350.

Sawai, Y., Horton, B. P., and Nagumo, T., 2004. The development of a diatom-based transfer function along the pacific coast of eastern Hokkaido, northern Japan — an aid in paleoseismic studies of the Kuril subduction zone. *Quaternary Science Reviews*, **23**(23–24), 2467–2483.

Schultz, P. H., Zárate, M., Hames, B., Koeberl, C., Bunch, T., Storzer, D., Renne, P., and Wittke, J., 2004. The quaternary impact record from the Pampas, Argentina. *Earth and Planetary Science Letters*, **219**, 221–238.

Schuster, R. L., and Highland, L. M., 2004. Impact of landslides and innovative landslide-mitigation measures on the natural environment. In *International Conference on Slope Engineering*, Hong Kong, China, December 8–10, 2003, keynote address, Proceedings 29.

Schuster, R. L., and Highland, L. M., 2007. Overview of the effects of mass wasting on the natural environment. *Environmental and Engineering Geoscience*, **13**(1), 25–44.

Schwab, M. J., Werner, P., Dulski, P., McGee, E., Nowaczyk, N., Bertrand, S., and Leroy, S. A. G., 2009. Palaeolimnology of lake sapanca and identification of historic earthquake signals, northern Anatolian fault zone (Turkey). *Quaternary Science Reviews*, **28**, 991–1005.

Shen, C.-C., Lee, T., Liu, K.-K., Hsu, H.-H., Edwards, R. L., Wang, C.-H., Lee, M.-Y., Chen, Y.-G., Lee, H.-J., and Sun, H.-T., 2005. An evaluation of quantitative reconstruction of past precipitation records using coral skeletal Sr/Ca and $\delta^{18}O$ data. *Earth and Planetary Science Letters*, **237**(3–4), 370–386.

Shoaei, Z., and Ghayoumian, J., 1998. Seimareh landslide, the largest complex slide in the world. In *Proceedings Eight International Congress of the International Association for Engineering Geology and the Environment*, 1–5, pp. 1337–1342.

Stiros, S. C., 2010. The 8.5+ magnitude, AD365 earthquake in Crete: coastal uplift, topography changes, archaeological and historical signature. *Quaternary International*, **216**(1–2), 54–63.

Stothers, R. B., 1984. The great tambora eruption in 1815 and its aftermath. *Science*, **224**(4654), 1191–1198.

SwissRe, 2009. Natural catastrophes and man-made disasters in 2008: North America and Asia suffer heavy losses. Sigma, 2. www.swissre.com/sigma. Last accessed 16 March 2010.

Torn, M. S., and Fried, J. S., 1992. Predicting the impacts of global warming on wildland fire. *Climatic Change*, **21**(3), 257–274.

Trifonov, V. G., and Karakhanyan, A. S., 2004. *Geodynamics and the History of Civilization*. Nauka (in Russian): Moscow.

Ulomov, V. I., 1971. Deformation of rocks in the source area of the Tashkent, April 26, 1966 earthquake. Izvestia Akademii Nauk SSSR, Ser. *Fizika Zemli*, 9: 22–30. (in Russian).

USGS, 2009. Science for a changing world. The Great 1906 San Francisco Earthquake. http://earthquake.usgs.gov/regional/. Last accessed 4 March 2010.

USGS, 2010a. Natural hazards gateway. http://www.usgs.gov/hazards/. Last accessed 16 March 2010.

USGS, 2010b. Historic earthquakes http://earthquake.usgs.gov/earthquakes/world/events/1960_05_22_articles.php

van der Werf, G. R., Dempewolf, J., Trigg, S. N., Randerson, J. T., Kasibhatla, P. S., Giglio, L., Murdiyarso, D., Peters, W., Morton, D. C., Collatz, G. J., Dolman, A. J., and DeFries, R. S., 2008. Climate regulation of fire emissions and deforestation in equatorial Asia. *Proceedings of the National Academy of Sciences*, **105**(51), 20350–20355.

van Geel, B., Buurman, J., and Waterbolk, H. T., 1996. Archaeological and palaeoecological indications of an abrupt climate change in the Netherlands, and evidence for climatological teleconnections around 2650 BP. *Journal of Quaternary Science*, **11**, 451–460.

van Geel, B., Heusser, C. J., Renssen, H., and Schuurmans, C. J. E., 2000. Climatic change in Chile at around 2700 BP and global evidence for solar forcing: a hypothesis. *The Holocene*, **10**, 659–664.

van Geel, B., Bokovenko, N. A., Burova, N. D., Chugunov, K. V., Dergachev, V. A., Dirksen, V. G., Kulkova, M., Nagler, A., Parzinger, H., van der Plicht, J., Vasiliev, S. S., and Zaitseva, G. I., 2004. Climate change and the expansion of the Scythian culture after 850 BC, a hypothesis. *Journal of Archaeological Sciences*, **31**, 1735–1742.

Veski, S., Heinsalu, A., Lang, V., Kestlane, ü, and Possnert, G., 2004. The age of the kaali meteorite craters and the effect of the impact on the environment and man: evidence from inside the kaali craters, island of Saaremaa, Estonia. *Vegetation History and Archaeobotany*, **13**, 197–206.

Vilanova, S. P., Nunes, C. F., and Fonseca, J. F. B. D., 2003. Lisbon 1755: a case of triggered onshore rupture? *Bulletin of the Seismological Society of America*, **93**(5), 2056–2068.

Vorobyev, Yu. L., Akimov, B. A., and Sokolov, Yu. I., 2006. *Tsunami: Warning and Protecting*. Vector TiS (in Russian): MChS Russia. Moscow.

Wang, G., and Xu, B., 1984. Brief introduction of landslides in loess in China. In *Proceedings 4th International Symposium on Landslides*, 2. Toronto: Canadian Geotechnical Society, pp. 197–207.

Warburton, J., 2007. Mountain environments. In Perry, C., and Taylor, K. (eds.), *Environmental Sedimentology*. Oxford: Blackwell Publishing, pp. 32–74.

Watt, S. F. L., Pyle, D. M., and Mather, T. A., 2009. The influence of great earthquakes on volcanic eruption rate along the Chilean subduction zone. *Earth and Planetary Science Letters*, **277**, 399–407.

Weiss, H., and Bradley, R. S., 2001. What drives societal collapse? *Science*, **291**, 609–610.

Whittow, J., 1980. Landslides and avalanches – avalanches. In *Disasters: The Anatomy of Environmental Hazards*. Harmondsworth: Penguin, pp. 163–170.

Wikipedia, 2010. List of natural disasters. http://en.wikipedia.org/wiki/List_of_natural_disasters. Last accessed 16 March 2010.

Yaalon, D. H., 1971. *Paleopedology – Origin, Nature and Dating of Paleosols*. Jerusalem: ISSS and Israel University Press.

Zerefos, C. S., Gerogiannis, V. T., Balis, D., Zerefos, S. C., and Kazantzidis, A., 2007. Atmospheric effects of volcanic eruptions as seen by famous artists and depicted in their paintings. *Atmospheric Chemistry and Physics*, **7**, 4027–4042.

Cross-references

Asteroid
Asteroid Impact
Biblical Events
Building Failure
Climate Change
Comet
Disaster
Drought
Dust bowl
Earthquake
Earthquake Damage
El Niño/Southern Oscillation
Flood Hazard and Disaster
Forest and Range Fires (wildfire)
Geohazards
Hazard
Heat Wave
Hurricane Katrina
Hydrometeorological Hazards
Indian Ocean Tsunami
Jökulhlaup (Débâcle)
Karst Hazards
Krakatau
Lahar
Landslide
Meteorite
Mt Pinatubo
Mudflow
Natural Hazard
North Anatolian Fault
Paleofloods
Paleoseismology
Religion and Hazards
Rock avalanche
San Andreas Fault
Santorini Eruption
Sinkhole
Surge
Tangshan 1976 Earthquake (China)
Tsunami
Vesuvius
Warning System
Wildfire

HOSPITALS IN DISASTER

Jeffrey N. Rubin
Tualatin Valley Fire & Rescue, Tigard, OR, USA

Synonyms

Hospital crisis management; Hospital disaster preparedness; Hospital emergency management; Hospital emergency preparedness

Definition

Hospital emergency management. A combination of actions, programs, processes, and capabilities that allows a hospital organization and its constituent departments to prepare for, respond to, and recover from major emergencies and disasters while maintaining critical functions.

Introduction

Hospitals are an essential community resource on a daily basis and even more so during and after a disaster (e.g., Ardagh et al., 2012). A hospital that cannot withstand the effects of a disaster not only cannot provide its critical services at the time of greatest need, it adds to the burden of the disaster (e.g., Kirsch et al., 2010): its skilled staff and vulnerable patients and visitors are likely to become victims as well. As succinctly stated by the Hospitals Safe from Disasters program (WHO and ISDR, 2008), "A safe hospital: will not collapse in disasters, killing patients and staff; can continue to function and provide its services as a critical community facility when it is most needed; and is organized, with contingency plans in place and health workforce trained to keep the network operational." This is the most basic expression of expectations of hospitals in disaster.

Standards and critical elements

There are numerous hospital preparedness standards, ranging from consensus to regulatory (e.g., Sauer et al., 2009); the latter tend to be national, the former of broader scope. Most pertain to hospitals in industrialized societies, particularly the United States. ASTM International (2009) has developed intentionally broad standards for hospital preparedness and response. The National Fire Protection Association's Standard 101 (2012), the Life Safety Code, addresses a variety of occupancies; Standard 99 (2012) covers health care facilities. The Joint Commission (2012) has long enumerated emergency preparedness standards for hospitals seeking accreditation; these standards have evolved from "disaster plans" to full-fledged emergency management programs. In terms of depth and breadth, these standards are broadly applicable, even if accreditation is not an issue. Current standards drive toward establishing preparedness as a sustainable organizational component: developing and maintaining a hazard vulnerability analysis (HVA) and an emergency operations plan (EOP). Once in place, sustainability is achieved by ongoing evaluation and update, including equipment procurement, development of procedures and training, testing through exercise and actual event, and appropriate revision (e.g., Cosgrove et al., 2008; Rubin, 2006). These criteria form the essence of effective emergency management programs in multiple disciplines and their achievement makes any organization, including a hospital, more likely to maintain function in disaster.

Hazard vulnerability analysis (HVA)

An HVA allows a hospital to prioritize preparedness and mitigation activities, including training and exercises (e.g., McLaughlin, 2001; Joint Commission, 2001; Briggs and Brinsfield, 2003; Jenkins et al., 2009). It comprises identifying hazards relevant to the area in question and their potential impact on a hospital's ability to provide services: Ideally it combines event history, probability, and severity, so that a hospital can differentiate between likelihood and severity. As most hospitals are not stand-alone facilities, but dependent on community services, utilities, and the ability of their employees to arrive for work, a hospital HVA should consider impacts on external (community) as well as internal stakeholders. Kaiser Permanente has developed the most widely used HVA tool for hospitals (CHA, 2010); the World Health Organization's guide (2006) is applicable across a broader range of development and infrastructure status.

Emergency operations plan (EOP)

Whether called a disaster plan, emergency management plan, or emergency operations plan (EOP), the concept applies to all users. An effective EOP is not a "cookbook" or a set of instructions, but it is an important tool in all phases of disaster impact. In its essence, an EOP does the following: identifies roles, responsibilities, and authorities among the hospital staff and community partners before, during, and after an emergency; indicates incident management structures for emergency operations (e.g., Hospital Incident Command System (California EMSA, 2006) or something similar) and identifies how those structures integrate with community and larger-scale structures (e.g., Barbera and Macintyre, 2002); establishes protocols and capabilities for notifying and communicating with staff, patients, families, external agencies, media, and other stakeholders; establishes non-routine response procedures and their activation thresholds for addressing specific concerns (e.g., evacuation, safety and security, staffing, utilities, curtailing services); identifies capabilities and procedures for maintaining critical services for at least 96 h in the absence of external support (e.g., Wagner et al., 2008; CDC and AWWA, 2011), including resource management procedures for requesting, assigning, tracking, and replenishing medical and nonmedical supplies; establishes protocols for expanding staff with volunteer health care providers; and addresses post-disaster recovery and restoration. Emergency operation plans can take multiple forms, with numerous models, content guides, and checklists available (e.g., Belmont et al., 2004; GNYHA, 2006; WHO, 2006; Joint Commission, 2008; WHO and ISDR, 2010), particularly for areas of special concern or capability, but the form is less important than the quality of the underlying assumptions (e.g., Auf der Heide, 2006; Barbera et al., 2009), HVA, internal adoption, and integration with local and higher-level planning.

Surge

A major component of hospital disaster preparedness is identifying and addressing medical surge capacity and capability. Medical surge "describes the ability to provide adequate medical evaluation and care during events that exceed the limits of the normal medical infrastructure of an affected community" (CNA, 2004, 1–5). Medical surge capacity describes the ability to handle increased patient volume; surge capability describes the ability to provide unusual or specialty care (e.g., pediatric patients (Boyer et al., 2009; Kissoon et al., 2011)) or other specialized services (CNA, 2004). The concept of medical surge recognizes that disasters may compromise a hospital's ability to provide critical services not only by impact to facility and staff, but by extraordinary patient needs. Whether due to catastrophic physical impact (e.g., hurricane or earthquake) or mass illness (e.g., pandemic), medical surge needs must be assessed and incorporated into the emergency management process, identifying baseline capacity and capabilities (e.g., Rubinson et al., 2010), available (and nontraditional) resources, and an ethical process of scarce resource allocation (e.g., Christian et al., 2008; Devereaux et al., 2008a, b; Rubinson et al., 2008a, b; Hick et al., 2009; IOM, 2010).

Challenges

The most common barrier to hospital preparedness is one of economics (e.g., Barbera et al., 2009), as preparing for a future event competes with daily needs, but in many industrialized countries, such barriers may relate as much to executive prioritization as they do to actual lack of resources. The fact that most communities have not endured a catastrophe means that motivation must be driven by something other than direct experience. Inaccurate perception of risk, whether related to hazards, their physical impact, or their economic and legal impact, can lead to unrealistic planning assumptions, and in turn to insufficient preparedness (e.g., Barbera et al., 2009). Lack of sufficient resource allocation is commonly manifested in inadequate emergency equipment (particularly backup communications systems), insufficient training, and drills and exercises that are too short, too small in scope, or too optimistic (e.g., Rubin, 2006). Since the beginning of the millennium, substantial progress has been made in awareness and tangible improvements, particularly in the industrialized world, but the very scope of major disasters means that few if any regions are truly prepared for a catastrophe.

Conclusion

Hospitals are generally acknowledged as among the most critical resources for sustaining a population on a daily basis, let alone in a disaster. Although the past several years have seen greater awareness, more encompassing standards, and greater access to preparedness guidance and resources (e.g., AHRQ, 2011; CHA, 2010; ASPR, 2012a, b), preparedness is rarely considered a daily priority, resulting in chaotic disaster response and resultant increased morbidity and mortality. Even where awareness is sufficient and preparedness considered a high priority, progress is variable in industrial nations and many facilities in developing countries (and the societies in which they function) have scant resources for daily operations.

The WHO/ISDR (2008) recognition of the importance of safe hospitals is a necessary step but far from a sufficient one. Preparedness is far more a journey than a destination, and "true" preparedness more of a goal than a realistic target, but there is still ample room to expand on the progress made to date.

Bibliography

Agency for Healthcare Research and Quality, 2011. Public Health Emergency Preparedness Archive website: http://archive.ahrq.gov/prep. Accessed May 27, 2012.

Ardagh, M. W., Richardson, S. K., Robinson, V., Than, M., Gee, P., Henderson, S., Khodaverdi, L., McKie, J., Robertson, G., Schroeder, P. P., and Deely, J., 2012. The initial health-system response to the earthquake in Christchurch, New Zealand, in February, 2011. *Lancet*, **379**, 2109–2115.

Assistant Secretary for Preparedness and Response, 2012a. *Healthcare Preparedness Capabilities: National Guidance for Healthcare System Preparedness.* Washington: US Department of Health and Human Services.

Assistant Secretary for Preparedness and Response, 2012b. Public Health Emergency Preparedness website: http://www.phe.gov. Accessed May 27, 2012.

ASTM International, 2009. *E2413-04: Standard Guide for Hospital Preparedness and Response.* West Conshohocken: ASTM International.

Auf der Heide, E., 2006. The importance of evidence-based disaster planning. *Annals of Emergency Medicine*, **47**, 34–49.

Barbera J. A., and Macintyre A. G., 2002. *Medical and Health Incident Management (MaHIM) System: A Comprehensive Functional System Description for Mass Casualty Medical and Health Incident Management.* Washington: Institute for Crisis, Disaster, and Risk Management, The George Washington University.

Barbera, J. A., Yeatts, D. J., and Macintyre, A. G., 2009. Challenge of hospital emergency preparedness: analysis and recommendations. *Disaster Medicine and Public Health Preparedness*, **3**, S74–S82.

Belmont, E., Fried, B. M., Gonen, J. S., Murphy, A. M., Sconyers, J. M., and Zinder, S. F., 2004. *Emergency Preparedness, Response, and Recovery Checklist: Beyond the Emergency Management Plan.* Washington: American Health Lawyers Association.

Boyer, E. W., Fitch, J., and Shannon, M., 2009. *Pediatric Hospital Surge Capacity in Public Health Emergencies.* AHRQ Publication No. 09-0014. Rockville: Agency for Healthcare Research and Quality.

Briggs, S. M., and Brinsfield, K. H., 2003. *Advanced Disaster Medical Response. Manual for Providers.* Boston, MA: Harvard Medical International Trauma and Disaster Institute.

California Emergency Medical Services Authority, 2006. *Hospital Incident Command System.* State of California: Sacramento.

California Hospital Association, 2010. Emergency Preparedness website: http://www.calhospitalprepare.org. Accessed May 27, 2012.

Centers for Disease Control and Prevention and American Water Works Association, 2011. *Emergency Water Supply Planning Guide for Hospitals and Health Care Facilities.* Atlanta: US Department of Health and Human Services.

Christian, M. D., Devereaux, A., Dichter, J. R., Geiling, J. A., and Rubinson, L., 2008. Definitive care for the critically ill during a disaster: current capabilities and limitations. *Chest*, **133**, 8S–17S.

CNA Corporation, 2004. *Medical Surge Capacity and Capability: A Management System for Integrating Medical and Health Resources During Large-Scale Emergencies.* Alexandria: CNA Corporation.

Cosgrove, S. E., Jenckes, M. W., Wilson, L. M., Bass, E. B., and Hsu, E. B., 2008. *Tool for Evaluating Core Elements of Hospital Disaster Drills.* AHRQ Publication No. 08-0019. Rockville: Agency for Healthcare Research and Quality.

Devereaux, A., Christian, M. D., Dichter, J. R., Geiling, J. A., and Rubinson, L., 2008a. Summary of suggestions from the Task Force for Mass Critical Care summit, January 26–27, 2007. *Chest*, **133**, 1S–7S.

Devereaux, A., Dichter, J. R., Christian, M. D., Dubler, N. N., Sandrock, C. E., Hick, J. L., Powell, T., Geiling, J. A., Amundson, D. E., Baudendistel, T. E., Braner, D. A., Klein, M. A., Berkowitz, K. A., Curtis, J. R., and Rubinson, L., 2008b. Definitive care for the critically ill during a disaster: a framework for allocation of scarce resources in mass critical care. *Chest*, **133**, 51S–66S.

Greater New York Hospital Association, 2006. *Recovery Checklist for Hospitals After a Disaster.* http://www.gnyha.org. Accessed May 27, 2012.

Hick, J. L., Barbera, J. A., and Kelen, G. D., 2009. Refining surge capacity: conventional, contingency, and crisis capacity. *Disaster Medicine and Public Health Preparedness*, **3**, S1–S9.

Institute of Medicine, 2010. *Medical Surge Capacity: Workshop Summary.* Washington: National Academies Press.

Jenkins, J. L., Kelen, G. D., Sauer, L. M., Fredericksen, K. A., and McCarthy, M. L., 2009. Review of hospital preparedness instruments for National Incident Management System compliance. *Disaster Medicine and Public Health Preparedness*, **3**, S83–S89.

Joint Commission, 2001. Analyzing your vulnerability to hazards. *Joint Commission Perspectives*, **21**(12), 8–9.

Joint Commission, 2012. *Comprehensive Accreditation Manual for Hospitals.* Oakbrook Terrace: The Joint Commission.

Kirsch, T. D., Mitrani-Reiser, J., Bissell, R., Sauer, L. M., Mahoney, M., Holmes, W. T., Santa Cruz, M., and de la Maza, F., 2010. Impact on hospital functions following the 2010 Chilean earthquake. *Disaster Medicine and Public Health Preparedness*, **4**, 122–128.

Kissoon, N.; for the Task Force for Pediatric Mass Critical Care, 2011. Deliberations and recommendations of the Pediatric Emergency Mass Critical Care Task Force: executive summary. *Pediatric Critical Care Medicine*, **12**, S103–S108.

McLaughlin, S. B., 2001. *Hazard Vulnerability Analysis.* Chicago: AHA Publishing.

National Fire Protection Association, 2012a. *Standard 99: Standard for Health Care Facilities.* Quincy: NFPA.

National Fire Protection Association, 2012b. *Standard 101: Life Safety Code.* Quincy: NFPA.

Rubin, J. N., 2006. Recurring pitfalls in hospital preparedness and response. In McIsaac, J. H., III (ed.), *Preparing Hospitals for Bioterror: A Medical and Biomedical Systems Approach.* Burlington: Academic.

Rubinson, L., Hick, J. L., Curtis, J. R., Branson, R. D., Burns, S., Christian, M. D., Devereaux, A., Dichter, J. R., Talmor, D., Erstad, B., Medina, J., and Geiling, J. A., 2008a. Definitive care for the critically ill during a disaster: medical resources for surge capacity. *Chest*, **133**, 32S–50S.

Rubinson, L., Hick, J. L., Hanfling, D. G., Devereaux, A., Dichter, J. R., Christian, M. D., Talmor, D., Medina, J., Curtis, J. R., and Geiling, J. A., 2008b. Definitive care for the critically ill during a disaster: a framework for optimizing critical care surge capacity. *Chest*, **133**, 18S–31S.

Rubinson, L., Vaughn, F., Nelson, S., Giordano, S., Kallstrom, T., Buckley, T., Burney, T., Hupert, M., Mutter, R., Handrigan, M., Yeskey, K., Lurie, N., and Branson, R., 2010. Mechanical ventilators in US acute care hospitals. *Disaster Medicine and Public Health Preparedness*, **4**, 199–206.

Sauer, L. M., McCarthy, M. L., Knebel, A., and Brewster, P., 2009. Major influences on hospital emergency management and disaster preparedness. *Disaster Medicine and Public Health Preparedness*, **3**, S68–S73.

Wagner, W. M., Jansen-Adams, L., and Bartels, R. H., 2008. A process for determining sustainability during emergencies: dealing with the 96-hour rule. *Inside ASHE*, **16**(5), 26–31.

World Health Organization and International Strategy for Disaster Reduction, 2008. *Hospitals Safe from Disasters Field Kit*. Geneva: World Health Organization.

World Health Organization and International Strategy for Disaster Reduction, 2010. *Hospitals Safe from Disasters bibliography*: http://safehospitals.info/index.php?option=com_content&task=blogcategory&id=21&Itemid=198. Accessed May 27, 2012.

World Health Organization Regional Office for the Western Pacific, 2006. *Field Manual for Capacity Assessment of Health Facilities in Responding to Emergencies*. Manila: World Health Organization.

Cross-references

Building, Structures and Public Safety
Casualties Following Natural Hazards
Classification of Natural Disasters
Community Management of Hazards
Disaster
Diseases and Disasters
Earthquake
Education and Training for Emergency Preparedness
Emergency Management
Emergency Planning
Epidemiology of Disease in Natural Disasters
Hazard
Integrated Emergency Management System
ISDR
Mitigation
Natural Hazards in Developing Countries
Planning Measures and Political Aspects
Resilience

HUMAN IMPACTS OF HAZARDS

Douglas Paton[1], David M. Johnston[2], Sarb Johal[3]
[1]University of Tasmania, Launceston, TAS, Australia
[2]GNS Science/Massey University, Lower Hutt, New Zealand
[3]GNS Science/Massey University, Wellington, New Zealand

Synonyms

Disaster preparedness; Disaster resilience; Disaster vulnerability; Response and recovery; Risk management

Definition

The human impact of hazards is the product of the interaction between hazard characteristics and the personal, community and societal factors implemented to influence people's capacity to cope, adapt, and recover from hazard effects.

Introduction

Throughout human history, people have established and developed communities and societies in locations that allowed them to take advantage of the resources and amenities (e.g., fertile soils, natural harbors, navigable rivers that serve as commercial highways, and coastal scenery) afforded by the action of natural and geological processes. However, periodically the processes that create such benefits (e.g., tectonic activity) become hazards (e.g., tectonic activity can create earthquakes, volcanic eruptions, and tsunami). For instance, the volcanic processes that contribute to creating fertile soils and scenic amenities can, during volcanic eruptions, expose those living within the surrounding area (which can have a radius of several hundred miles) to hazards such as ashfall. Similarly, the rivers that provide a resource for commerce, recreation, and tourism can channel heavy rainfall in ways that floods those living on their banks.

While natural hazards have the *potential* to create substantial loss and destruction, the nature, magnitude, and duration of hazard impacts is influenced by the choices people make regarding where and how societal development occurs (e.g., not building on flood plains), and what people do to enhance their capacity to cope with hazard impacts, adapt to their consequences, and recover from the loss and disruption experienced (Paton and Johnston, 2006; Tobin, 1999). To identify how hazards impact on human populations it is thus necessary to define the hazards that exist in a given area and the actions people and societies can take to influence the impact of hazards on people and communities.

Hazards

Most places are susceptible to experiencing multiple hazards. The term used to describe this is "hazard-scape." The hazard-scape describes not only the hazard (e.g., earthquake, wildfire) but also the characteristics and behavior of the hazards that can occur. For example, for volcanic events, hazard characteristics include ashfall (tephra), lava, ballistic material (e.g., rocks ejected by the explosive forces of the eruption), lahars, and gases. Seismic hazard characteristics include, for example, ground acceleration and liquefaction. Hazard behavior includes, for example, the return (how frequently an event occurs over time) period, speed of onset, intensity, and duration. The hazard-scape (the hazard, its characteristics and behavior) contributes to defining the risk hazards pose to the physical, social, and administrative infrastructure of human settlements (Lindell and Perry, 1992; Paton and Johnston, 2006).

For instance, hazard characteristics such as the ground shaking that accompanies an earthquake, volcanic ashfall, and tsunami inundation can cause damage to buildings, homes, roads, and other infrastructure. Hazard activity can thus injure or kill people, leave them homeless, and limit their access to the resources normally required to sustain everyday life. Hazard impacts on infrastructure can have secondary consequences for human populations as a result of, for example, loss of utilities (e.g., water, power, and roads) and societal functions (e.g., welfare, local government, and financial services). The particular mix of hazard characteristics and behavior will differ from place to place, as will the communities they impact. These examples introduce the fact that the impact of a hazard is defined not by its hazard-scape per se, but by the consequences hazard activity has for people, the buildings that house them, the infrastructure that supports activities, and the societal functions that impose structure on everyday life. Drawing a distinction between hazard characteristics and the consequences of hazard activity is important.

How people respond to hazards

Although the causes of hazard events (e.g., earthquake) are beyond human control, people can influence the consequences (e.g., level of destruction, time to recover) that hazard events create. Thus, the impact of hazards on human populations is defined by the actions people undertake to influence the nature, severity, and duration of the consequences they experience (Lindell and Perry, 1992; Paton and Johnston, 2006). People can make choices about the consequences they may experience. If people decide not to make these choices, they increase their vulnerability to experiencing adverse impacts from hazard events. Vulnerability can also be affected by factors such as demographics (e.g., proportion of elderly people and ethnic minorities in affected populations) and socio-economic factors such as levels of poverty and quality of housing stock (Wisner et al., 2004). If people choose to, they can increase their ability to cope, adapt, and recover should a hazard event occur (Tobin, 1999). The choices that influence hazard impacts occur at several levels (person, community, societal, and environmental (natural and built)) and fall into two general categories.

The first concerns the actions that societies and their members undertake to mitigate the effects of hazards and increase their ability to cope with and adapt to hazard impacts. Since these activities are undertaken prior to hazard events occurring, they are called *preparation* or *readiness strategies*. The second relates to the procedures and resources put in place to facilitate community and societal recovery. These response and recovery strategies are undertaken during and after a period of hazard activity respectively.

The impact of hazard events is a function of the actions taken to safeguard the physical integrity of the built environment (e.g., *Land-Use Planning*, building codes, lifeline engineering, and retrofitting buildings). However, because of a public desire to use hazardous areas (e.g., adjacent to rivers or at the coast) and the relatively long return period of some damaging events, building codes, regulations and land use planning cannot provide effective mitigation for the entire risk. For hazard events that can be foreseen, an effective integrated warning system can address the residual risk (Sorensen, 2000).

Hazard impact is also influenced by the degree to which people and households prepare. Strategies must also encompass the social (including ensuring economic, business, and administrative continuity), cultural, and environmental contexts within which societal activities occur (Paton and Johnston, 2006). However, the complex nature of hazard activity makes it impossible to prepare for every eventuality. Since some level of physical and social impact is inevitable, strategies designed to facilitate physical and social recovery must be developed and be capable of implementation in a timely manner. Examples of preparation and recovery strategies are described in the next section.

Reducing the impact of hazards on infrastructure

Natural hazard activity affects specific geographical areas. As a consequence, one strategy for reducing the human impact of hazards utilizes land use planning to prevent or restrict development and population growth in areas susceptible to hazard impacts (Burby et al., 2000). While this approach can help mitigate future impacts (e.g., not developing in areas susceptible to hazard activity, ensuring post-disaster rebuilding occurs in less risky locations), it does not apply to all circumstances.

Hazard events are generally infrequent, with return periods often measured in decades or centuries. As a consequence, much economic, infrastructure, and social development has occurred in areas susceptible to experiencing hazard activity. In this context, strategies such as building design, retrofitting buildings, and strengthening essential lifelines (e.g., water, power, and roads) in ways that reduce the likelihood of their being damaged or destroyed are essential. If buildings and infrastructure remain intact and functioning, the impact of the hazard event on human populations is reduced.

While planning, regulatory, and legislative processes can facilitate the implementation of this element of a *Disaster Risk Management* strategy, the action taken to safeguard domestic housing stock and their inhabitants, requires that people take steps to reduce the impact that hazard events have on them. This level of understanding is also important from the perspective of understanding the psychological and social impacts that hazard events have on people and how they can be managed (Dass-Brailsford, 2010).

Reducing the impact of hazards on people

Hazard events expose people and communities to demands and consequences that fall outside the usual realm of human experience and increase the risk of significant loss, death, and destruction. As a result, experience of hazard activity can be accompanied by Acute Stress and, on occasion, Posttraumatic Stress Disorder (PTSD).

A comprehensive account of the psychological impact on people and its implications can be found in Norris et al. (2002a, b). People also face increased levels of stress from the temporary loss of societal functions and services. However, the impact is not the same for all people. Diversity in psychological impacts reflects demographic, psychological, and social factors.

From a demographic perspective, the psychological impacts of hazards are greater amongst children, elderly individuals, people who live alone, and those with a history of prior trauma, mental illness, and disability. Psychological impact is also a function of the distribution of event-related factors such as being injured, losing family members, loss of property and/or livelihood, and being displaced from home and losing normal support, networks, and resources. On the other side of the psychological impact coin, however, are the strengths (e.g., spiritual and cultural resources, active coping, collective efficacy, and sense of community) that enhance people's ability to cope and recover (Paton and Johnston, 2006). A critical influence on the impact of the hazard is the psychological support and recovery resources available to survivors (Dass-Brailsford, 2010).

Hazard events impact people by presenting them with demands that are atypical, challenging, and unavoidable, which creates significant change to their environment, and which overwhelms their usual coping and support resources. The demands people experience are strongly influenced by the physical impacts (e.g., ground shaking, liquefaction, wind damage, and flooding levels) that directly affect levels of damage to infrastructure and the number of casualties (Lindell et al., 2006). Not only do these physical impacts have implications for the nature of people's response to hazards, the associated loss of normal infrastructure increases the need for improvised disaster response and recovery intervention (Lindell et al., 2006). One way of doing this involves mobilizing the community response.

Effective intervention involves working *with* community members to help them regain equilibrium by recognizing and mobilizing existing strengths and capabilities in a way that facilitates their coping and adaptation (Dass-Brailsford, 2010). In the short term, the impact can be ameliorated by providing concrete help (e.g., safety, food/water, and shelter). In the medium term, intervention aims to, for example, increase peoples' understanding of the stress response, mobilizing or developing coping strategies to help them reduce their stress, and facilitating the development of social support relationships with other survivors and responders. While strategies such as these facilitate recovery in most people, some may experience more significant mental health problems. Mitigating the psychological impact of hazard experience for the latter requires the availability of mental health professionals to provide counseling and therapeutic services.

Another factor influencing the psychological impact of hazards is the degree to which people possess the resources required to respond (Hobfoll, 2001). Thus, encouraging people to develop their resources (e.g., knowledge, beliefs, coping strategies, capacity for self-reliance, household emergency plans and resources, and ability to work with others to confront local problems) will influence the nature and extent of the impact of hazards on human populations. The issue here is the generally low level of people's preparedness (Tierney et al., 2001).

Contrary to what might be expected, neither living in a high-risk area nor being informed, via public education or risk communication programs, of this risk influences people's levels of preparedness. However, public education programs that assist people to make sense of hazards and that demonstrates how and why preparing is effective in mitigating hazard consequences facilitates preparedness and, should a hazard event occur, reduce its impact on the individuals. Several models, such as the Person-Relative-to-Event-Model (Mulilis and Duval, 1995), the Protective Action Decision model (Lindell and Perry, 1992), and the Social Context model (Paton, 2008), were developed to guide the process of tailoring public education in ways that facilitate preparing. The Protective Action Decision model and the Social Context model highlight the role community processes play in facilitating both sustained hazard preparedness and the availability of support resources for recovery.

Reducing the impact of hazards on communities

Faced with uncertainty, people turn to others who share their interests and values to help them reduce uncertainty and decide how to manage their risk (Paton, 2008). Family and members of the communities (e.g., workplaces, social clubs, churches, etc.) with whom people interact regularly play a central role in helping people understand hazards and formulating and implementing their hazard management plans. The levels of collective preparedness that results from this collaboration can mitigate hazard impacts in the immediate aftermath of a disaster.

When hazard activity creates a disaster, the loss, destruction, and disruption overwhelms emergency response resources. This means that help and recovery efforts in the immediate aftermath of the hazard event come from neighbors and others within the proximal community. The level of, for example, informational and emotional social support that comes from membership of a cohesive social network of people helps reduce the psychological impact on community members. However, because it is impossible to prepare for all eventualities, communities benefit from intervention to help rebuild its support resources and networks.

The impact of a hazard event on a community is reduced and recovery facilitated by working with community members to help them regain access to and use their own resources (Dass-Brailsford, 2010). This reduces dependency on external resources, sustains community and cultural identity, and maintains social support. In so doing, it is important to accommodate community diversity (e.g., the coexistence of groups with different needs, etc.) within an affected area. One important aspect of diversity is culture. Cultural beliefs and practices can influence the human impact of hazards in several ways.

For example, in the context of the possibility of using engineering solutions to divert lava flows from future volcanic eruptions in Hawaii, Gregg et al. (2008) found that native Hawaiians' belief that interfering with lava flows disrespects the wishes of Pele (the Goddess of volcanoes) reduced their willingness to support mitigation actions (building walls to divert lava flows, bombing lava flows) that were inconsistent with their cultural beliefs. Thus, certain cultural beliefs can increase people's vulnerability to experiencing loss should a hazard event occur. Cultural factors can also increase people's ability to cope with hazard impacts. For example, the Hakka Spirit, a cultural characteristic of the Hakka people in Taiwan, characterized by reciprocal support practices and collaborative problem solving evolved to limit the impact of typhoons and other natural hazards on people's livelihood (Paton and Jang, 2010).

Societal strategies for reducing hazard impacts

In most environments hazard events occur infrequently. This limits the opportunities people have to gain experience of hazard consequences or the effectiveness of mitigation measures for themselves. To identify what they may have to contend with and what they might do to manage their risk, people rely on information from expert scientific and emergency management sources through public education programs. Societal-level factors influence the impact of hazards in other ways.

Government plays a role by influencing the policies and resourcing that contributes to readiness and recovery efforts. They provide direction for planning, land use strategies and structural and engineering mitigation measures. Factors such as disaster business continuity planning play an important role in minimizing economic impacts of hazard impacts and ensure that continued availability of employment that makes an important contribution to recovery (Paton and Johnston, 2006).

Summary

The impact of hazards on people is a function of the interaction between the hazard and the characteristics of the people, communities, and society that lives in harm's way. From a social perspective, hazard impacts are influenced by several factors. Community members, businesses, and societal institutions that are prepared (e.g., household emergency plans and business continuity plans) in ways that enhance their safety and facilitate their continued functioning even when experiencing disruptive hazards consequences (e.g., ground shaking, volcanic ashfall, and flood inundation) will experience less severe impacts and will recover more quickly. Since people cannot prepare for all eventualities, the severity and duration of hazard impacts will be influenced by the availability of strategies that facilitate social and physical recovery and that accommodate the community diversity (e.g., demographics, ethnicity, and culture) that affects the distribution of hazard impacts within affected communities. Finally, strategies must accommodate the fact that the hazards that communities may face will change over time. For example, changes in land use patterns (e.g., land clearance, industrial development, and pressure for affordable housing) and factors such as climate change are changing the hazard-scape people will have to contend with. Understanding and managing the human impacts of hazards is thus a dynamic and iterative process.

Bibliography

Burby, R. J., Deyle, R. E., Godschalk, D. R., and Olshansky, R. B., 2000. Creating hazard resilient communities through land-use planning. *Natural Hazards Review*, **1**, 99–106.

Dass-Brailsford, P., 2010. *Crisis and Disaster Counselling*. Los Angeles, CA: Sage.

Gregg, C. E., Houghton, B. F., Paton, D., Swanson, D. A., Lachman, R., and Bonk, W. J., 2008. Hawaiian cultural influences on support for lava flow hazard mitigation measures during the January 1960 eruption of Kilauea volcano, Kapoho, Hawai'i. *Journal of Volcanology and Geothermal Research*, **172**, 300–307.

Hobfoll, S. E., 2001. The influence of culture, community, and the nested-self in the stress process: advancing conservation of resources theory. *Applied Psychology: An International Review*, **50**, 337–421.

Lindell, M. K., and Perry, R. W., 1992. *Behavioral Foundations of Community Emergency Management*. Washington, DC: Hemisphere Publishing Corporation.

Lindell, M. K., Prater, C. S., and Perry, R. W., 2006. *Fundamentals of Emergency Management*. Emmitsburg MD: Federal Emergency Management Agency Emergency Management Institute. Available at www.training.fema.gov/EMIWeb/edu/fem.asp or archone.tamu.edu/hrrc/Publications/books/index.html.

Mulilis, J.-P., and Duval, S. T., 1995. Negative threat appeals and earthquake preparedness: A Person-Relative-to-Event (PrE) model of coping with threat. *Journal of Applied Social Psychology*, **25**, 1319–1339.

Norris, F. H., Friedman, M. J., Watson, P. J., Byrne, C. M., Diaz, E., and Kaniasty, K., 2002a. 60,000 Disaster victims speak: Part I. An empirical review of the empirical literature, 1981–2001. *Psychiatry*, **65**(3), 207–239.

Norris, F. H., Friedman, M. J., and Watson, P. J., 2002b. 60,000 Disaster victims speak: Part II. Summary and implications of the disaster mental health research. *Psychiatry*, **65**(3), 240–260.

Paton, D., 2008. Risk communication and natural hazard mitigation: How trust influences its effectiveness. *International Journal of Global Environmental Issues*, **8**, 2–16.

Paton, D., and Jang, L., 2010. Disaster resilience: Exploring all-hazards and cross cultural perspectives. In Miller, D., and Rivera, J. (eds.), *Community Disaster Recovery and Resiliency: Exploring Global Opportunities and Challenges*. London: Taylor & Francis.

Paton, D., and Johnston, D., 2006. *Disaster Resilience: An integrated approach*. Springfield, IL: Charles C. Thomas.

Sorensen, J. H., 2000. Hazard warning systems: review of 20 years of progress. *Natural Hazards Review*, **1**, 119–125.

Tierney, K. J., Lindell, M. K., and Perry, R. W., 2001. *Facing the Unexpected: Disaster Preparedness and Response in the United States*. Washington, DC: Joseph Henry Press.

Tobin, G. A., 1999. Sustainability and community resilience: The holy grail of hazards planning. *Environmental Hazards*, **1**, 13–26.

Wisner, B., Blaikie, P., Cannon, T., and Davis, I., 2004. *At risk: Natural hazards, people's vulnerability and disasters*, 2nd edn. London: Routledge.

Cross-references

Disaster Risk Management
Disaster Risk Reduction (DRR)
Evacuation
Land-Use Planning
Resilience
Structural Mitigation
Warning Systems

HUMANITY AS AN AGENT OF NATURAL DISASTERS

Thomas Glade[1], Andreas Dix[2]
[1]University of Vienna, Vienna, Austria
[2]Otto-Friedrich-University Bamberg, Bamberg, Germany

Definition and introduction: humans and natural disasters

Humans have relied on the Earth's landscapes for thousands of years. Only in the last few centuries have people become more independent from the natural environment. With the advent of specialized tools and technology, they have become the primary contributors to change in the environment. For example, people applied fire for widespread forest clearance, they have developed unique methods to modify the soil for agricultural use, they routinely construct water reservoirs and divert rivers, they actively modify landscapes directly through excavation, and other land modifications or indirectly through groundwater extraction. Such advanced change and development in human capacity has led to an interesting phenomenon. On the one hand, people have become better and more efficient in their use of the environment, but on the other hand they have become much more reliant on the Earth as a whole. Within the vulnerability debate, this is referred to as the "vulnerability paradox" (e.g., Burton and Cutter, 2008), which suggests that the more efficient humans become, the more vulnerable they are toward disturbances such as natural hazards.

In terms of natural hazards and disasters, societies' exposure to the vulnerability paradox has important consequences. Numerous studies are available which try to prove that natural processes have changed dramatically over the several last decades. Indeed, it is correct that some process have actually changed in their magnitude and frequency. However, what is the driver of these changes? We have to ask: are these still considered natural processes, ones which act independent of human influence or society? What do we actually have to measure when we want to better understand and manage natural hazards? In short, this leads to the final question: what actually causes the disaster?

There are a number of arguments that support the premise of independency that humanity can influence natural processes. This is in fact true for geophysical hazards including earthquakes, volcanic eruptions, or tsunamis. However, the consequent disaster is more dependent on many more factors than just the natural elements (Kennedy, 2006; Wisner et al., 2004).

It is clear that humans are heavily dependent on nature. Although some authors claim that humanity is independent from nature and actually influenced by the human actions, this strong position is clearly questioned within sociology (e.g., Lidskog, 1998). It is possible that society considers itself independent and self-determined, but it is evident from many recent natural disasters, such as the December 2004 tsunami or the earthquakes in Pakistan 2005, in China 2007, or in Chile and Haiti 2010, that humans do rely heavily on nature and the natural processes affecting the Earth.

There is indeed a large need to cope with a suite of natural processes, in terms of management or, as often discussed, as risk governance. Besides the assessment of the social system, ranging from personal matters to national interests and global involvements over a wide variety of institutions, it is most important to effectively assess the natural processes. One of the main problems currently facing the natural science-based analyses and assessments are the limited availability of data. Continuous and comparable measurements of environmental parameters are available for some decades (e.g., water levels, volcanic activities, earthquakes, etc.) or centuries (e.g., single climate parameters, water stages, ocean tides, etc.). Unfortunately to allow the application of sustainable land use and secure developments, there is a need to extract information to longer time periods in order to recognize extreme events with returning periods of hundreds or thousands of years. Since we are now reaching the maximum limit of recording periods we should consider and assess other sources of information on such data – besides historical archives. Such information is relevant as it includes the knowledge gained on former natural physical processes as well as of the coping response of prior societies. Both types of information (society and nature) is only available through historical archives.

Historical perspectives

After an event there is a wide range of sources and methods for collecting available information. In the case of historical events the information is limited to written records. These records were commonly and intentionally produced for purposes other than reporting on the disasters. Consequently, one of the major methodological hurdles is to locate relevant and informative files. The respective archives and files have to be analyzed in the appropriate context (e.g., socioeconomic, political). This is even more important for earlier periods such as the fourteenth or fifteenth century where only a few written records survived or the social and historical context is unclear.

Nevertheless, historical events involving disasters can be reconstructed, if the historian is able to carefully reconstruct the available and often sparse information. For example, Scanlon (2002) successfully reconstructed one catastrophe, the explosion in the harbor of Halifax, Canada (December 6, 1917), where two vessels collided, one of them transporting ammunition. This was one of the largest nonnuclear explosions in human history where more than 1,600 people died, and much of the city was destroyed. Scanlon (2002) shows how poor and inaccurate the contemporary reports are and then shows how it is possible to successfully reconstruct the explosion and aftermath differently and much more precisely (Scanlon, 2002). In this example, additional historical information included pictures and maps. The use of many different information sources is really the only way of collecting data to ensue a successful study (Dix, 2008). A clear advantage is that we are able to learn more about distinct events as well as the societal reactions, specifically how societies deal with a given. With this knowledge base, it is possible to learn from previous experiences of human societies that cope with natural risks.

Interestingly, it is also possible to study how such societies calculated natural risks and how they perceived vulnerability in the past. Bankoff (2003, 2004), for instance, successfully shows how the historical perspective can be used within the analysis of vulnerability issues in a non-Western society, in this case the Philippines (Bankoff, 2003, 2004).

A major problem is the extensive time required to adequately collect sufficient and accurate information to reconstruct a good (quality, reliability, quantity) data and time series of an event or events (Dix, 2012). In the case of climate history research based on historical archives, it is apparent that there are good quality data (Glaser, 2001). Unfortunately, this is not the case in regard to the quality of data for other topics. To emphasize this point, it is clear that prior to the Indian Ocean tsunami of 2004, the global level of knowledge concerning tsunami threats was generally poor in this region but the level of awareness increased dramatically thereafter. Unfortunately, the situation regarding detailed knowledge on the frequency and magnitude of paleo-tsunamis remains poor over many parts of the world (e.g., Mediterranean).

Humanity as a major force for disasters

Results of analysis of some historical archives demonstrates that, indeed in the case of some past societies which have experienced major disasters, human intervention of the natural processes was not a factor (all geophysical hazards). Major historic examples include volcanic eruptions, such as Vesuvius on April 12, AD 79, which buried Pompeii (Luongo et al., 2003) or the three major volcanic eruptions of the nineteenth and twentieth century; Tambora 1815 (Oppenheimer, 2003), Krakatau 1883 (Simkin and Fiske, 1984), and Mount Pelée on Martinique 1902 (Scarth, 2002). Similarly, many earthquakes have caused severe damage without the influence of human societies including the 1348 earthquake in Southern Germany and Switzerland (Borst 1981) or the Lisbon earthquake on November 1, 1755, which caused more than 10,000 fatalities (Chester, 2001).

Non-geophysical hazards such as storm surges are different. As with most hazards, the spatial exposition to the natural environment is the main factor affecting damages. But in the case of events such as storm surges, issues surrounding the quality of adjustments toward natural hazards are relevant, for example, dykes in case of storm surges, as realized as early as the late medieval period (Allemeyer, 2012). The recent case of the flooding of New Orleans and wide parts of the coastal areas of Louisiana in the aftermath of Hurricane Katrina demonstrates that existing institutions and measures of security like storm warning, hurricane shelters, and dykes may fail (Brinkley, 2006). In comparison, the indirect human influence resulting from extensive peat mining and the leveling of marshland surfaces contributed to the tremendous loss of life and cultivated areas especially in the case of the so-called "Grote Mandränken" (directly translated: "large drowning of population") storm surges on the coast of the North Sea in 1362 and 1634 which totaled several thousand victims (Glaser, 2001).

Another case of indirect influence is the wide range of excessive erosion events. Bork (1988) discusses the major influence of severe heavy rainfalls in 1342 in parts of Central Europe and the consequent gully erosion and appearance of deserted villages in the 1340s (Bork, 1998). He also collected a number of examples of the interference of land use and excessive erosion from all the continents (Bork, 2006).

There are also historic disasters which resulted from a direct interference by humans to the landscape, such as the Elm rockfall in the Canton Glarus, Switzerland on September 11, 1881 which was triggered by mining excavations of the slope toe (Hauer, 2001). Another example is the Vaiont landslide on October 9, 1963, a tragic case of reactivation of unstable slopes and old landslide areas following the construction of a dam (Kiersch, 1980; Genevois and Ghirotti, 2005). This latter disaster shows clearly that a historical survey in advance of human construction would have had indicated old landslide events in this area.

In all cases, intensified land use of any type (e.g., agriculture, built-up area) contributes to the major adverse forces related to disasters. These are, in most cases, unintentional responses. In the "industrial period," most people did not consider the potential consequences of large magnitude events. There is often a general belief that people can adequately manage their environment. Society commonly believes that to be aware of potential nature hazards is sufficient and that their limited past experience with potential natural hazards will suffice. Most societies are not aware that the Earth surface was also formed by very large events, generally considered as catastrophic. When focusing on long-term landform development, numerous landforms only exist because of these large-scale events. Hence, societies tend to manage natural processes and make decisions based on their limited knowledge of temporal and spatial patterns for potential hazardous processes. In this context, historical information can significantly contribute to this knowledge because it can demonstrate the possible effects of previous larger events on humanity.

Summary

It is evident, that natural processes will always occur and that disasters are simply a result of the inappropriate adaptation of society to live with such processes. However, it is also evident that occasionally the anthropogenic actions or inactions enhance the disaster potential. Extensive use of natural resources changes the environment as we know it, for example, extensive groundwater extraction caused land subsidence of up to 3.9 m in some Chinese cities (Xu et al., 2008) and geothermal fluid extraction in New Zealand led to a subsidence of about 15 m since 1970 (Allis et al., 2009). Inappropriate agricultural practices can result in extensive soil erosion (Uri, 2000), and artificial water reservoirs may destabilize slopes and trigger possible failures that could produce disasters (Kiersch, 1980; Genevois and Ghirotti, 2005). As early as 1999, Hooke showed that for the USA, more sediment had been moved by humans than under natural conditions.

The key issue for the future is not to assign responsibility or blame to specific groups or individuals in the discussion on the impact of humanity on the environment. It is most important that we better understand the environment, but is also important to understand the social systems and in particular the interaction between society and environment. In order to make sustainable decisions, we have to be fully aware of potential consequences, to weigh the negative and positive responses of development based not only on short-term considerations, but also reflecting upon long-term environmental changes. It is essential to find a consensus within all affected parties, that is, by applying risk governance principles. Herein, the monitoring of our current environment, the modeling of acting processes, and the definition of scenarios helps society to live with the risk. Our current knowledge on previous disasters is very limited, but largely important to determine the actual role of humanity as an agent of natural disaster. Historical information on former natural disasters will help us to understand the processes involved and the related consequences.

Bibliography

Allemeyer, M. L., 2012. The struggle against the sea: an early modern coastal society between metaphysical and physical attempts to control nature. In Janku, A., Schenk, G. J., and Mauelshagen, F. (eds.), *Historical Disasters in Context. Science, Religion, and Politics*. London: Routledge, pp. 75–93.

Allis, R., Bromley, C., and Currie, S., 2009. Update on subsidence at the Wairakei–Tauhara geothermal system, New Zealand. *Geothermics*, **38**(1), 169–180.

Bankoff, G., 2003. *Cultures of Disaster. Society and Natural Hazard in the Philippines*. London: Routledge.

Bankoff, G., 2004. The historical geography of disaster. Vulnerability and local knowledge. In Bankoff, G., Frerks, G., and Hilhorst, D. (eds.), *Mapping Vulnerability. Disasters, Development and People*. Sterling: Earthscan Publications, pp. 25–36.

Bork, H.-R., 1998. *Landschaftsentwicklung in Mitteleuropa*. Gotha: Klett-Perthes.

Bork, H.-R., 2006. *Landschaften der Erde unter dem Einfluss des Menschen*. Darmstadt: Wissenschaftliche Buchgesellschaft.

Borst, A., 1981. Das Erdbeben von 1348. Ein historischer Beitrag zur Katastrophenforschung. *HistorischeZeitschrift*, **233**, 529–569.

Brinkley, D., 2006. *The Great Deluge. Hurricane Katrina, New Orleans and the Mississippi Gulf Coast*. New York: Morrow.

Burton, C., and Cutter, S. L., 2008. Levee failures and social vulnerability in the Sacramento-San Joaquin delta area, California. *Natural Hazards Rev.*, **9**(3), 136–149.

Chester, D. K., 2001. The 1755 Lisbon Earthquake. *Progress in Physical Geography*, **25**(3), 363–383.

Dix, A., 2008. Historische Ansätze in der Hazard- und Risikoanalyse. In Felgentreff, A., and Glade, T. (eds.), *Naturrisiken und Sozialkatastrophen*. Berlin/Heidelberg: Spektrum, pp. 201–211.

Dix, A., 2012. Forgotten risks. Mass movements in the mountains. In Janku, A., Schenk, G. J., and Mauelshagen, F. (eds.), *Historical Disasters in Context. Science, Religion, and Politics*. London: Routledge, pp. 140–152.

Genevois, R., and Ghirotti, M., 2005. The 1963 Vaiontlandslide. *Giornale di Geologia Applicata*, **1**, 41–52.

Glaser, R., 2001. *Klimageschichte Mitteleuropas – 1000 Jahre Wetter, Klima, Katastrophen*. Darmstadt: WBG.

Hauer, K., 2009. *Der plötzliche Tod: Bergstürze in Salzburg und Plurs kulturhistorisch betrachtet*. Wien: Lit Verlag.

Hooke, R. L., 1999. Spatial distribution of human geomorphic activity in the United States: comparison with rivers. *Earth Surface Processes and Landforms*, **24**, 687–692.

IPCC, 2007. *Fourth Assessment Report: Climate Change 2007*. Cambridge: Cambridge University Press.

Janku, A., and Schenk, G. J., Mauelshagen, F. (eds.), Historical Disasters in Context. Science, Religion, and Politics. London: RoutledgeCurzon.

Kennedy, B. A., 2006. *Inventing the earth*. Oxford: Blackwell Publishing.

Kiersch, G. A., 1980. Vaiontreservoirdisaster. *Civil Engineering*, **34**, 32–39.

Lidskog, R., 1998. Society, space and environment. Towards a sociological re-conceptualisation of nature. *Journal Housing, Theory and Society*, **15**(1), 19–35.

Luongo, G., Perrotta, A., and Scarpati, C., 2003. Impact of the AD 79 explosive eruption on Pompeii, I. Relations amongst the depositional mechanisms of the pyroclastic products, the framework of the buildings and the associated destructive events. *Journal of Volcanology and Geothermal Research*, **126**(3), 201–223.

Oppenheimer, C., 2003. Climatic, environmental and human consequences of the largest known historic eruption. Tamboravolcano (Indonesia) 1815. *Progress in Physical Geography*, **27**(2), 230–259.

Scanlon, T. J., 2002. Rewriting a living legend. Researching the 1917 Halifax explosion. In Stallings, R. A. (ed.), *Methods of Disaster Research*. Philadelphia: Xlibris, pp. 266–301.

Scarth, A., 2002. *La Catastrophe – the Eruption of Mount Pelee, the Worst Volcanic Eruption in the Twentieth Century*. Harpenden: Oxford University Press.

Simkin, T., and Fiske, R. S., 1984. *Krakatau 1883 – The Volcanic Eruption and its Effects*. Washington, DC: Smithsonian Institution Press.

Uri, N. D., 2000. Agriculture and the environment – The problem of soil erosion. *Journal of Sustainable Agriculture*, **16**, 71–94.

Wisner, B., Blaikie, P., Cannon, T., and Davis, I., 2004. *At Risk – Natural Hazards, People's Vulnerability and Disasters*. London: Routledge.

Xu, Y.-S., Shen, S.-L., Cai, Z.-Y., and Zhou, G.-Y., 2008. The state of land subsidence and prediction approaches due to groundwater withdrawal in China. *Natural Hazards*, **45**(1), 123–135.

Cross-references

Risk Governance

HURRICANE (TYPHOON, CYCLONE)

Robert Korty
Texas A&M University, College Station, TX, USA

Definition

Tropical cyclone. An organized, cyclonically rotating system of convection driven by fluxes of heat derived from the ocean. Tropical cyclones are classified by their intensities using nomenclature that varies regionally. Tropical storms are tropical cyclones that have maximum sustained winds between 17 and 32 m/s (34–63 kts); intense tropical cyclones – those with winds of at least 33 m/s (64 kts) – are called *hurricanes* in the Atlantic and eastern North Pacific basins and *typhoons* in the western North Pacific. Typhoons whose maximum sustained 1-minute surface winds exceed 65 m/s (130 kts) are called *super typhoons*, and the strongest hurricanes (Category 3 or higher on the Saffir-Simpson Hurricane Wind Scale; see Tables 1 and 2) are classified as major hurricanes in the Western Hemisphere. In the Indian Ocean and South Pacific, tropical cyclones are referred to as *cyclones*.

Hurricanes

Introduction

Tropical cyclones are low-pressure systems driven by fluxes of enthalpy, a thermodynamic variable that consists of sensible and latent heat components, from the ocean to the atmosphere. In a tropical cyclone, air-sea fluxes of enthalpy principally arise in the form of evaporation from the ocean. These cyclones are characterized by deep convection near their center and 1-min sustained surface speeds of at least 17 m/s. Figure 1 shows a satellite image of Hurricane Mitch (1998) in the Caribbean Sea near the time of its peak intensity. The storm exhibits the classic hallmarks of intense hurricanes: a ring of deep convection surrounding a comparatively calm eye in which clouds are mostly absent. Outer bands of convection spiral into the core of the storm, but the most violent winds are confined to a tight radius around the center. Upper level outflow carries moist air away from the center at the top of the storm, which fans high-level clouds out to radii several hundred kilometers away from the eye.

To aid warnings and communication, intense tropical cyclones – those with wind speeds of at least 33 m/s – are further classified in the Western Hemisphere using the Saffir-Simpson Hurricane Wind Scale, which is based on the damage they are capable of causing (Table 1 reports the original scale from Simpson and Riehl (1981) whereas Table 2 shows the current definitions). This scale bins hurricanes by strength into categories numbered from 1 (weakest) to 5 (most severe). Cyclones in the Southern Hemisphere affecting Australia are rated using a similar but slightly different scale (Table 3). Whereas the Saffir-Simpson scale uses the highest sustained wind to establish categorization, the Australian scale rates storms by their highest wind gust. Note that in any event the damage potential does not increase linearly with the wind speed – the kinetic energy of the wind increases as the square of the wind speed, and the damage that can result is also a nonlinear function of wind speed.

Tropical cyclones form principally in low latitudes around the globe. Figure 2 shows the points of origin of

Hurricane (Typhoon, Cyclone), Table 1 Original Saffir-Simpson scale of hurricane intensity (Simpson and Riehl, 1981). Classification is determined by wind speed alone, and slight modifications to the scales have been made in 2012; see Table 2 for current definitions. The minimum central pressure and storm surge in the original scale are estimates of common values associated with storms of that strength

Category	Wind (m/s)	Pressure (hPa)	Storm surge (m)	Damage
1	33–42	>980	1.0–1.7	Minimal
2	43–49	965–979	1.8–2.6	Moderate
3	50–58	945–964	2.7–3.8	Extensive
4	59–69	920–944	3.9–5.6	Extreme
5	≥70	<920	≥5.7	Catastrophic

Hurricane (Typhoon, Cyclone), Table 2 Saffir-Simpson Hurricane Wind Scale. This scale underwent a minor alteration in 2012 to alleviate awkwardness associated with conversions between various wind units. Hurricane intensities are estimated in 5 kt intervals, and the more commonly used units of mph or km/h had been previously rounded in the tables

Category	Wind (mph)	Wind (kts)	Wind (km/h)
1	74–95	64–82	119–153
2	96–110	83–95	154–177
3	111–129	96–112	178–208
4	130–156	113–136	209–251
5	≥157	≥137	≥252

Hurricane (Typhoon, Cyclone), Figure 1 Satellite image of Hurricane Mitch taken at 12:45 UTC on October 26, 1998. At the time this image was taken, the storm was bordering on Category 5 intensity with maximum sustained winds estimated to be 155 mph and a minimum central pressure of 923 mb. (Image courtesy of NOAA Operational Significant Event Imagery: http://www.osei.noaa.gov/Events/Tropical/Atlantic/1998/Mitch_10/TRCmitch299B_G8.jpg).

Hurricane (Typhoon, Cyclone), Table 3 Australian classification of tropical cyclone severity

Category	Strongest wind gust
1	<125 km/h (< 34.7 m/s)
2	125–170 km/h (34.7–47.2 m/s)
3	170–225 km/h (47.2–62.5 m/s)
4	225–280 km/h (62.5–77.8 m/s)
5	>280 km/h (>77.8 m/s)

all tropical cyclones between 1979 and 2008. They are common in the tropical North Atlantic, Caribbean Sea, Gulf of Mexico, tropical North Pacific, tropical Indian Ocean, and the central and western tropical South Pacific. They are rare in the South Atlantic, where only one hurricane was observed during this time period, and none has been observed in the eastern South Pacific. Storms do not form on the equator, where the vertical component of Earth's planetary vorticity (measured by the Coriolis parameter) vanishes. Storms most commonly form during the summer and autumn months in each hemisphere (June to November in the Northern Hemisphere; December to May in the Southern), but can form outside of this window, particularly in the western North Pacific, where some tropical storm activity is common throughout the year. The distribution as a function of month is plotted in Figure 3, which shows the late summer/early autumn peaks in each hemisphere.

Approximately 90 storms form around the world each year. This entry reviews the features common to these storms around the world: their structure and factors that influence formation and govern their intensity. More recent studies have probed a relationship between tropical cyclone climatology and climate, and a discussion of these ongoing efforts is included in the "Summary" section.

Structure and size

In mature storms, the axisymmetric part of the circulation is so much stronger than the asymmetric motions that it is often helpful to view these separately. The primary circulation is very nearly in gradient balance, a balance between pressure gradient, Coriolis, and centripetal accelerations, although winds in the surface boundary layer often exceed the balanced value (see Bui et al., 2009 for further discussion). The primary circulation evolves when heat and momentum sources drive a secondary circulation, which in turn change the distribution of heat and momentum.

Figure 4 shows a modeled cross section in the radial-height plane of a mature, axisymmetric storm. Air flowing

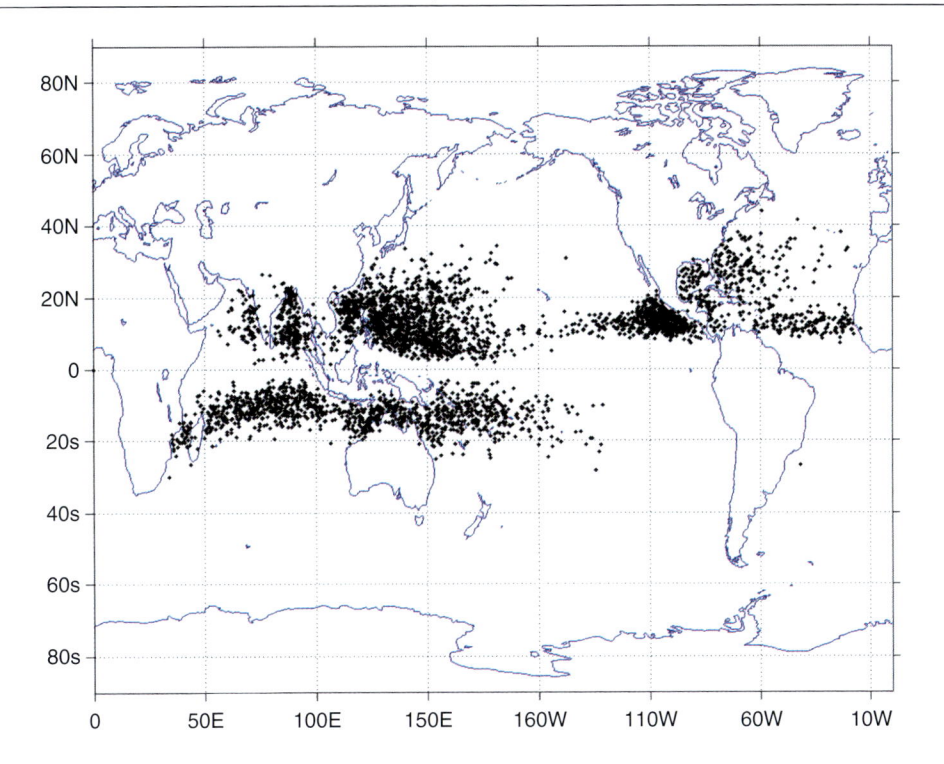

Hurricane (Typhoon, Cyclone), Figure 2 Locations of tropical cyclone genesis for all storms from 1979 to 2008.

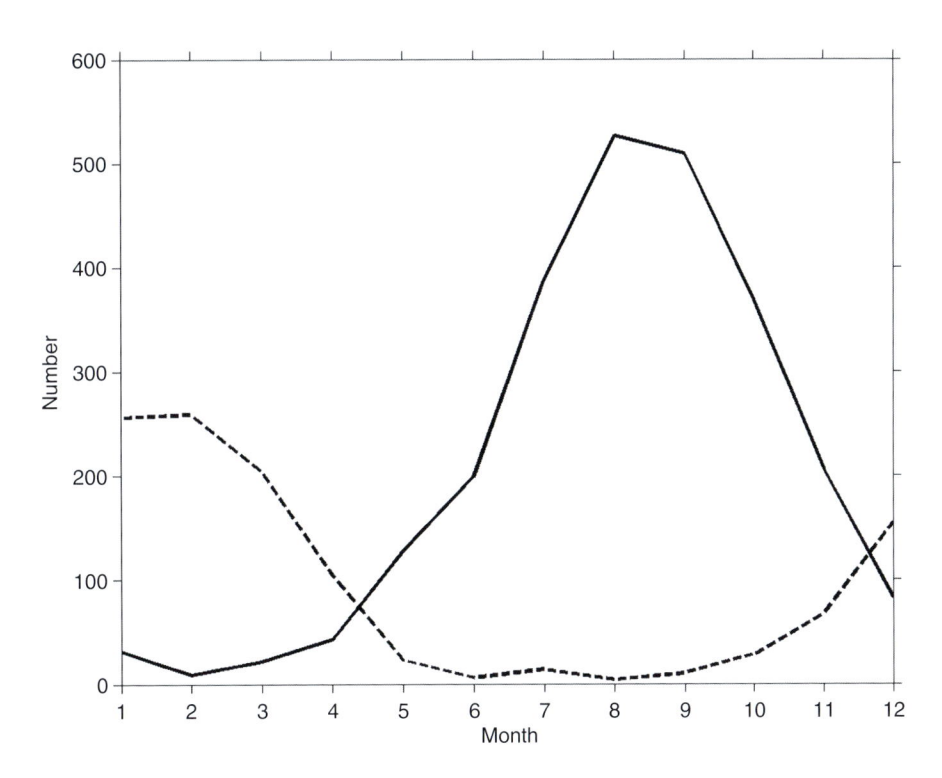

Hurricane (Typhoon, Cyclone), Figure 3 Monthly distribution of storm formation. Total number of storms that formed between 1979 and 2008 during each month of the year in the Northern Hemisphere (*solid curve*) and Southern Hemisphere (*dashed curve*).

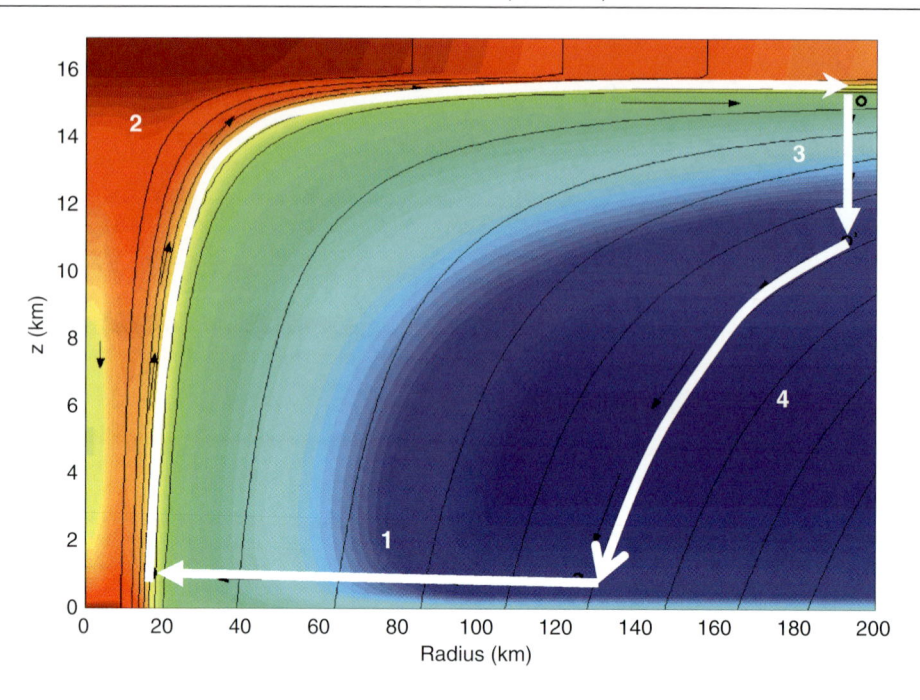

Hurricane (Typhoon, Cyclone), Figure 4 Carnot heat engine in a radial-height cross section of an axisymmetric, mature hurricane. Colors show θ_e, which is proportional to moist entropy (*blue* is low, *red* high), and *solid black curves* are surfaces of constant angular momentum.

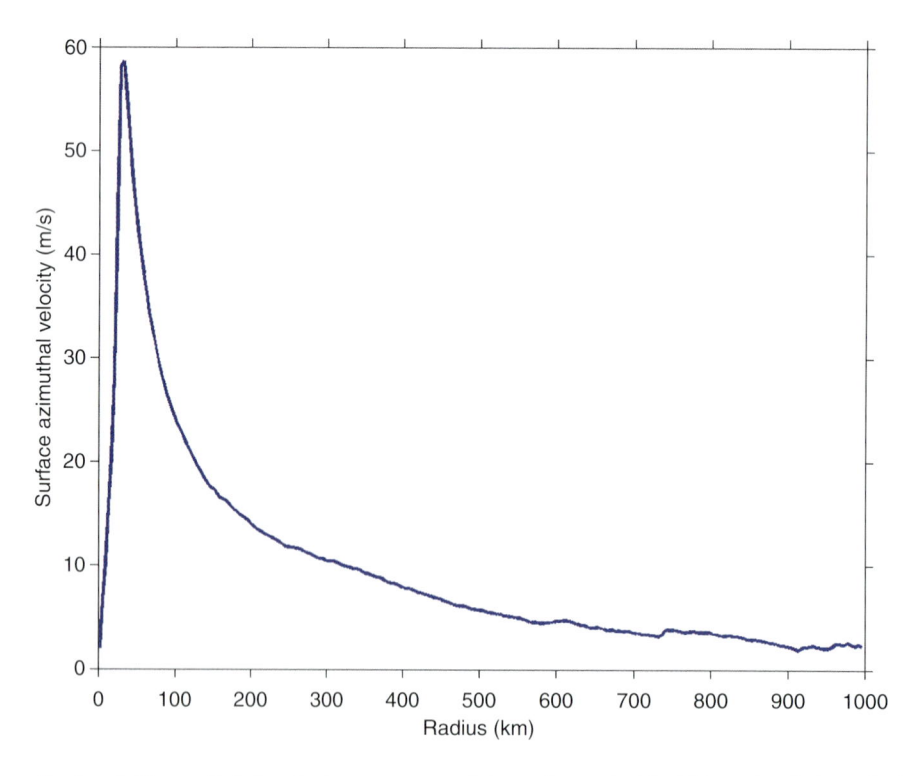

Hurricane (Typhoon, Cyclone), Figure 5 Azimuthal component of the surface wind (m/s) in a simulation of a hurricane using a model similar to that described by Rotunno and Emanuel (1987).

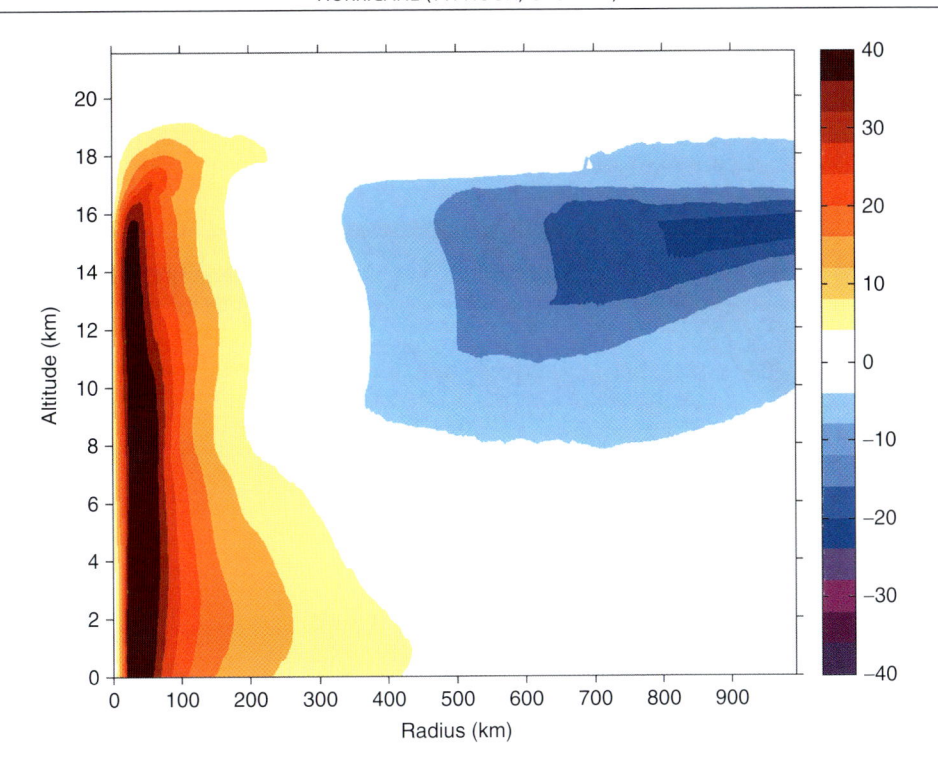

Hurricane (Typhoon, Cyclone), Figure 6 Radial-height cross sections of the azimuthal component of velocity (m/s) from an updated version of the model first used by Rotunno and Emanuel (1987).

into the circulation in the boundary layer spirals in, rotating around the vortex center while migrating toward the center via a radially inward component of the surface wind field. In the eyewall, saturated air rises along angular momentum surfaces that are tilted away from the center with height. Thus, the eye of a hurricane is shaped like a football stadium: deep convection that usually towers through the full depth of the troposphere (~15 km high) that tilts at an angle of up to 45°.

Surface winds are highest in this narrow region, and radial pressure gradients are correspondingly tight. As one moves outward from the eyewall, surface winds decrease at a rate roughly inversely proportional to the square root of the distance from the eyewall (see, e.g., Kossin et al., 2000; Mallen et al., 2005). Figure 5 shows this profile is captured in models of idealized hurricanes, such as those described by Rotunno and Emanuel (1987); note the rapid drop in the azimuthal component of the surface wind with radius moving away from the eyewall.

Figures 6–9 show the modeled radial-height cross sections of several meteorological variables from an updated version of the Rotunno and Emanuel (1987) model. Figure 6 shows the concentration of strongest azimuthal winds close to the center, with values tapering off at greater radii. Figure 7 shows the radial component of the wind, which is inward in the boundary layer and outward aloft. Figure 8 completes the mass circulation with strong upward velocities in the eyewall and gradual descent occupying a much area outside of the eyewall (including in the eye itself). Liquid water content (Figure 9) is highest in the boundary layer, but it is elevated at great heights in the eyewall and generally high in the upper-level outflow between 12 and 15 km. These model cross sections are consistent with measurements taken directly in the field. Figure 10 shows radar images taken during an Air Force Reconnaissance mission into a mature Hurricane Floyd in 1999. The convection is deep and tilted in the eyewall, and generally shallower at greater radii, as can be seen in Figure 11, which shows a radar image on a vertical cross section through the center of Floyd. The actual three-dimensional structure can have important azimuthal asymmetries, even in the most mature storms, which feature organized convective rainbands; see Houze (2010) for a good review of their structure.

The size of a tropical cyclone varies considerably from one storm to another, and considerably less research has been devoted to studying the factors that characterize the size distribution of storms. Dean et al. (2009) found that the size distribution in Atlantic hurricanes follows a roughly log-normal distribution, which would suggest that the size of a given storm may be primarily a function of the geometry of the initial disturbance, but Hill and Lackmann (2009) hypothesize that environmental relative humidity may play a role. Typical eye diameters are O(10 km), though some reach O(100 km). Typical radii for tropical storm force winds are a few

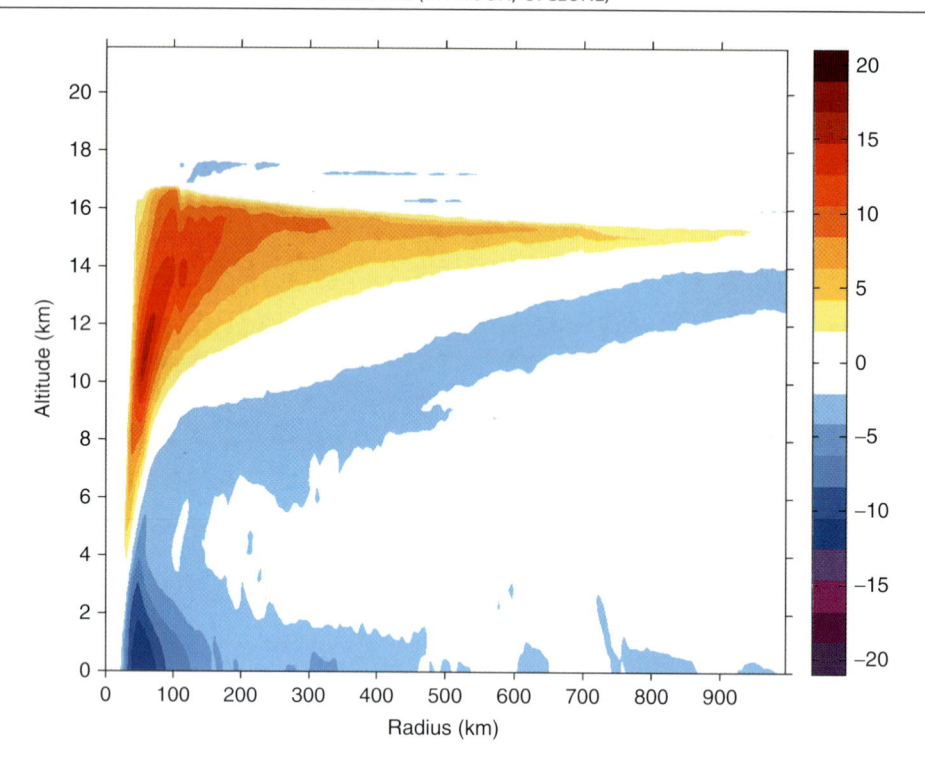

Hurricane (Typhoon, Cyclone), Figure 7 As in Figure 6, but for the radial component of velocity (m/s).

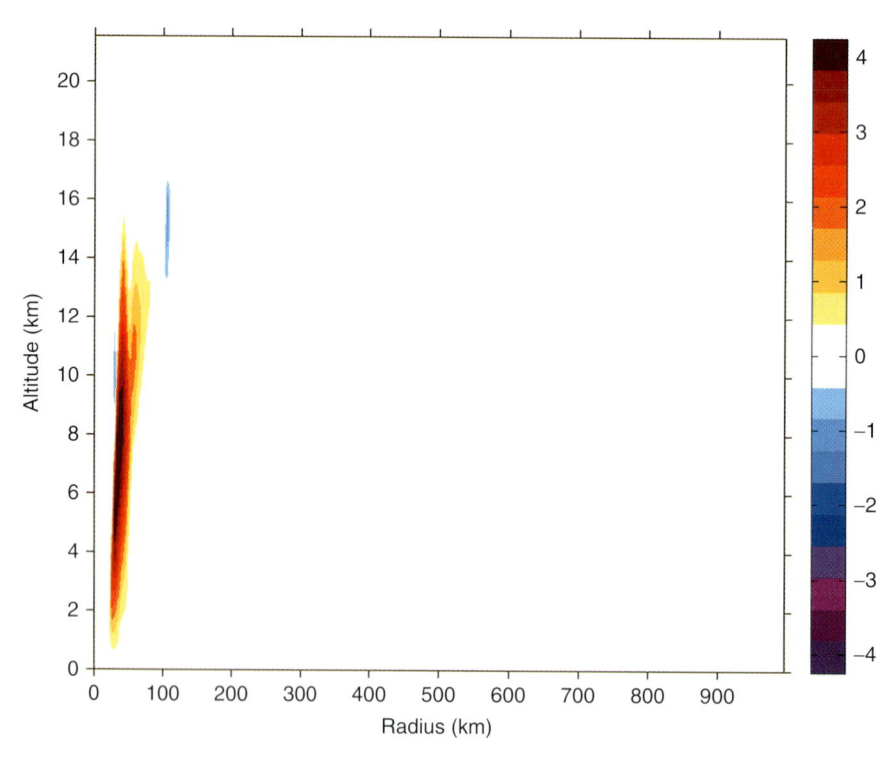

Hurricane (Typhoon, Cyclone), Figure 8 As in Figure 6, but for the vertical component of velocity (m/s).

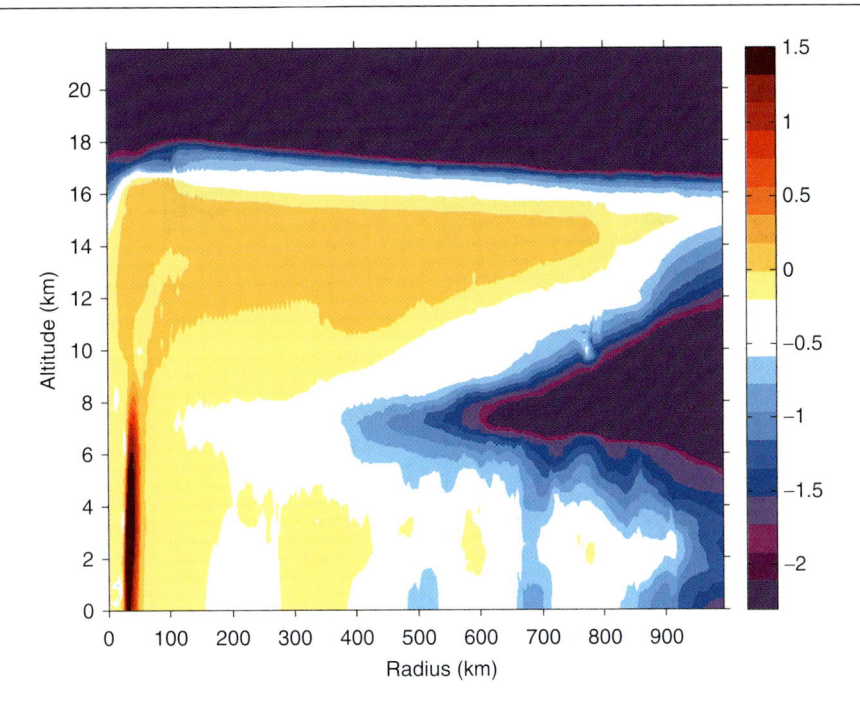

Hurricane (Typhoon, Cyclone), Figure 9 As in Figure 6, but for the natural logarithm of the liquid water content (g/kg).

hundred kilometers, though some storms – particularly those transitioning into extratropical cyclones – are considerably larger in size.

Genesis and climatology

As seen in Figure 2, tropical cyclones form over a wide swath of the world's tropical and subtropical oceans, and are principally born in one of three belts. One lies in low latitudes of the Northern Hemisphere and stretches from the coast of Africa across the Atlantic Ocean, Caribbean Sea, and Gulf of Mexico into the eastern Pacific Ocean. A second Northern Hemisphere cluster begins in the central and western Pacific and extends westward to the northern Indian Ocean, the Bay of Bengal, and the Arabian Sea. In the Southern Hemisphere, genesis is concentrated in the Indian and Pacific Oceans from Madagascar eastward to the central South Pacific. Tropical cyclones in the South Atlantic are rare. Storms form most often during late summer and early autumn in their respective hemispheres (the peak is August–October in the Northern Hemisphere and January–March in the Southern Hemisphere).

William Gray indentified a number of environmental conditions common to genesis in regions around the world in pioneering work (Gray, 1968, 1979). These include a propensity to form over oceans with surface layers that are warm and deep. The structure of the upper tropical ocean includes a well-mixed surface layer that occupies the topmost 25–200 m in which temperatures vary little; below this is the thermocline, a region in which temperatures begin a sharp decline toward a cold abyss. Surface water temperatures at the point of tropical cyclone genesis are usually >26°C, although each year a small number of storms form over water colder than this. This isotherm is not a required threshold for genesis, but rather a coincident indicator of regions in which genesis is common.

Gray (1968) also identified the importance of temperature and moisture profiles in the lower atmosphere to developing tropical cyclones. The role has been clarified by the subsequent theoretical development of potential intensity theory by Emanuel (1986, 1988) and Holland (1997). Tropical cyclones are driven by positive fluxes of enthalpy from the ocean to the atmosphere, but this heat must be transported vertically by deep convection. If the atmosphere's thermodynamic profiles contain stable layers that inhibit convection, then a tropical cyclone will be unable to generate the deep layer structure characteristic of intense systems.

Other factors that are common at points of tropical cyclone genesis include low vertical wind shear (usually defined as the vector difference between the 850 and 200 mb winds), and some incipient cyclonic vorticity from which an initial disturbance can grow. Tropical cyclones do not form on the equator, where the vertical component of the planetary vorticity vanishes.

These criteria, first catalogued by Gray (1968, 1979), are the basis for a genesis potential index used in more recent studies by Camargo et al. (2007). The presence of a warm, deep layer of surface water, convectively neutral (or unstable) profiles, vorticity, and the absence of wind shear are clearly conducive to genesis, but this collection of conditions is common in the tropics, while genesis remains sporadic and rare. Thus, while these factors are

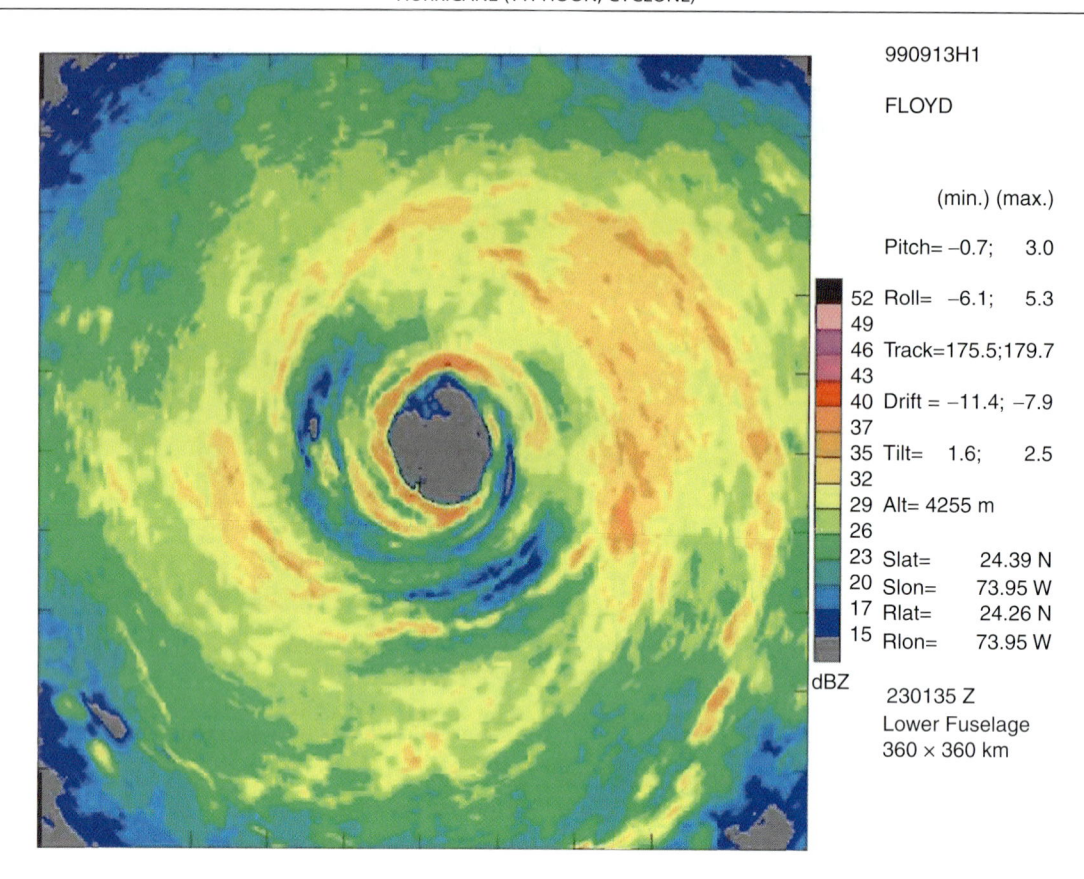

Hurricane (Typhoon, Cyclone), Figure 10 Radar image in Hurricane Floyd taken at 23:01 UTC on September 13, 1999.

important and appear to be necessary, their presence never guarantees formation of a tropical cyclone.

Part of the difficulty in forming tropical cyclones is that they are unable to spontaneously generate when the large-scale conditions are conducive; they require a trigger to develop. The trade wind regions of the tropics are most often characterized by a moist marine boundary layer topped by an inversion in temperature that arises from a gradual subsidence in the free troposphere. Thus the moisture content of a column of atmosphere in this region is confined to very near the surface, and it falls off sharply with height across the inversion. Deep convection that has enough buoyancy to punctuate through the inversion layer and tap unstable air aloft occurs in regular but isolated bursts, but these convective cells are unable in our present climate to spawn tropical cyclones on their own. The convective towers quickly consume the supply of moist enthalpy stored in the boundary layer, which leaves in their wake cool, dry air that is brought down to the surface by downdrafts generated by the convection, choking off a continued supply of fuel. Evaporation from the ocean resumes, but under normal wind speeds this process occurs slowly: the timescale to restore the moisture levels to their normal boundary layer values is a day or so. If winds are stronger (owing to an incipient vortex, for example), the evaporative process is accelerated.

Thus tropical cyclogenesis can occur only where evaporation is augmented by stronger surface winds; they require a trigger to start. Nascent clusters of convection are inhibited from growing into tropical cyclones because downdrafts bring down drier air from the middle troposphere, which is usually quite dry in the tropical atmosphere. As air parcels ascend in convective updrafts, excess water content is removed by precipitation. When air sinks back to the surface, it is dry, although mixing from shallow convection and evaporation of falling precipitation partially rehydrates parcels. Thus cyclogenesis will be favored when the middle troposphere is more humid than average, as it appears that nascent disturbances must first saturate the troposphere before genesis can occur (Bister and Emanuel, 1997).

Tropical cyclones also require vorticity. As shown in Figure 2, while cyclogenesis is common throughout the tropics, these storms do not form on the equator itself, where the vertical component of planetary vorticity goes to zero. Genesis is also preferred in regions where the atmosphere develops cyclonic spin with respect to the earth (cyclonic relative vorticity). These factors that favor genesis have been known for several decades (Gray, 1979;

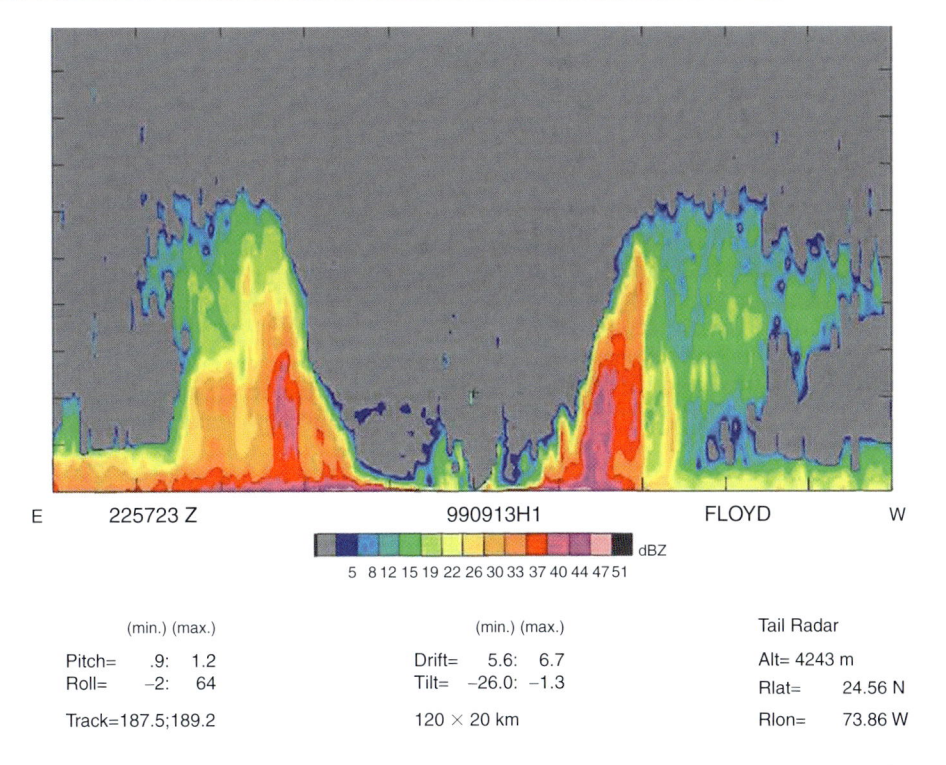

Hurricane (Typhoon, Cyclone), Figure 11 Radar image taken along a vertical cross section inside the center of Hurricane Floyd taken at 22:57 UTC on September 13, 1999.

McBride and Zehr, 1981), but their presence is insufficient to guarantee development. Undoubtedly, there are several routes by which genesis can occur, and these details remain the subject of active research studies.

Tracks and motion

Once formed, tropical cyclones follow a path guided by both external, environmental factors and internal dynamics. Steering by the background wind patterns are often of primary importance in advecting a storm downstream, and the vertically averaged lower-tropospheric flow is most often a principal driver. Absent any strong steering, storms tend to move poleward and westward owing to some internal dynamical constraints.

Earth's planetary vorticity f increases with the sine of the latitude heading northward. Consider a storm in the Northern Hemisphere that is being advected to the west by a background flow of easterly winds. Because the storm has cyclonic vorticity (relative vorticity ζ and f have the same sign), air circulates around its center in a counterclockwise fashion in the Northern Hemisphere. As air rotates northward up the eastern side of the storm, it moves into an area where the planetary vorticity is higher than it was where it originated. Thus it decreases in relative vorticity to compensate, and it acquires an anomalous clockwise spin. Conversely, air rotating southward along the western side of the storm moves to an area where planetary vorticity is lower than it was where it originated, and thus the air acquires additional counterclockwise spin to compensate. The net effect of both of these anomalous spins is to induce a poleward push to the trajectory of the storm, which is embedded in an otherwise east-to-west flow. This is known as the β-effect, after the rate at which Earth's planetary vorticity changes with latitude (Smith, 1993).

Tracks generally follow a slightly poleward off due west trajectory in the deep tropics, but as a storm ventures toward the subtropics, interactions with middle latitude weather systems being advected from west-to-east can turn the trajectories in a more pronounced poleward direction before recurving them in the westerly middle latitude flow.

Intensity

Tropical cyclones have their strongest winds near the surface, a consequence of being warm-cored systems. As noted previously, wind speeds are highest in the eyewall of a hurricane and decrease inversely with distance from the eyewall (see Figure 5). Thus the most violent part of a storm is confined to a smaller fraction of the storm near its center, although as noted earlier, the size of both the eye and the storm itself can vary substantially from one hurricane to another. In the strongest hurricanes, the maximum winds can reach 85 m/s, making these among the fiercest weather phenomena.

As high as the surface wind speeds are, they are not unbounded by nature. In fact, there is a thermodynamic limit on how strong storms can be. The factors generally needed for genesis outlined in the last section show the importance of both warm surface waters and a favorable atmospheric temperature and moisture profile. The relationship between these imposes a constraint on a cyclone's maximum wind speed, analogous to a Carnot heat engine's constraint on the thermodynamic efficiency of an engine.

Early researchers recognized that hurricanes derive their energy from fluxes of heat from the ocean (Riehl, 1950; Kleinschmidt, 1951). Most of this comes in the form of evaporation of ocean water into the boundary layer of the hurricane, and these surface fluxes become stronger as the wind speed increases. Emanuel (1986) showed that a mature tropical cyclone can be viewed as a steady, axisymmetric flow with an energy cycle that resembles that of an ideal Carnot heat engine, which is diagrammed in Figure 4. On the first leg, as a parcel of air spirals toward the center of a hurricane in the atmospheric boundary layer, fluxes of enthalpy from the ocean and dissipative heating (Bister and Emanuel, 1998) add heat to the hurricane while surface temperatures remain nearly constant. This causes pressures to fall as entropy rises and angular momentum decreases as a result of the loss of kinetic energy through frictional dissipation with the sea surface. As temperatures are nearly constant along this journey, the pressure drop is analogous to the first leg in a Carnot cycle: one of isothermal expansion. Once the parcel reaches the eyewall, it turns upward as it rises along a surface of constant angular momentum under slantwise convective ascent (leg 2 of Figure 4); this ascent is one in which entropy remains nearly constant, and so this is analogous to adiabatic expansion to the lower pressures in the upper troposphere.

In nature, where actual storms are not bounded to recirculate every parcel of air, the outflow from a hurricane is mixed with its environment far removed from a hurricane's core, which results in an open energy cycle. Numerical models, however, can be enclosed with artificial walls, which allow the cycle to be closed conceptually (Emanuel, 2003). In the third leg of Figure 4, where temperatures in the lower stratosphere or upper troposphere are around 200 K, a parcel of air loses heat to space by radiation as it sinks to higher pressures; this is isothermal expansion. In the final leg, although air continues to lose heat to space just as in leg 3, this amount is very nearly equal to the amount that would have been gained if all of the rainwater had remained suspended in the air and evaporated as the parcel descends rather than fall out of the system. So while entropy is lost and gained in this leg, little or none is available for kinetic energy production (see Emanuel, 2003 for a full discussion). Thus, this final leg is nearly one of adiabatic compression.

Heat input and dissipation of kinetic energy occur through heat and momentum fluxes in the boundary layer. These are usually quantified using bulk aerodynamic formulae:

$$F_p = -C_D \rho |\vec{V}| \vec{V} \tag{1}$$

and

$$F_k = C_k \rho |\vec{V}| (k_0^* - k). \tag{2}$$

Here F_p and F_k are the fluxes of momentum and heat, respectively, V is the near-surface vector wind (usually taken to be at 10 m), ρ is air density, k is the specific enthalpy of air near the surface, k_0^* is the saturation enthalpy of air that is saturated with water at the ocean temperature (i.e., a parcel of air in contact with the ocean), and C_D and C_k are nondimensional drag and enthalpy exchange coefficients. Both C_D and C_k depend on sea state, but their precise values are uncertain in the complex sea states and high wind environments characteristic of a hurricane.

The ocean to atmosphere enthalpy flux that feeds a hurricane is augmented through dissipative heating in the atmospheric boundary layer by an amount

$$D = C_D \rho |\vec{V}|^3 \tag{3}$$

(see Bister and Emanuel, 1998). Thus the total heat put into a hurricane, integrated over all radii and azimuthal angle θ along leg 1 of Figure 4, is

$$Q = \int_0^{2\pi} \int_0^{r_o} \left[C_k \rho |\vec{V}| (k_0^* - k) + C_D \rho |\vec{V}|^3 \right] r dr d\theta. \tag{4}$$

The work done by the hurricane's Carnot cycle is simply the total heat input given by Eq. 4 multiplied by the thermodynamic efficiency, which is the ratio of the difference between the hot input (SST, T_s) and cold output (T_o, the temperature at the top of the convection) divided by T_s. This work must balance the dissipation of momentum given by Eq. 3 integrated over all radii and azimuthal angles. Combining these, and assuming that the integrals are dominated by the values near the radius of maximum wind where these quantities spike, the maximum wind speed possible in a hurricane is given by

$$|V_{\max}|^2 = \frac{C_k}{C_D} \frac{T_s - T_o}{T_o} (k_0^* - k). \tag{5}$$

This derivation parallels Emanuel (2003); an alternate, more technical derivation is presented in Bister and Emanuel (1998), but yields the same result. This equation shows that the maximum possible wind speed in a hurricane is a function not merely of the SST (T_s), but rather a more complex function that couples it to the temperature at which convection outflows in the atmosphere (T_o) – that value will be established by the temperature and moisture profiles in the atmosphere that dictate how high a parcel of air will rise before it is no longer buoyant. Thus assessing how intensities could change in other climates requires not just knowing how SST changes, but rather how it changes in relation to the atmosphere that is in equilibrium with it.

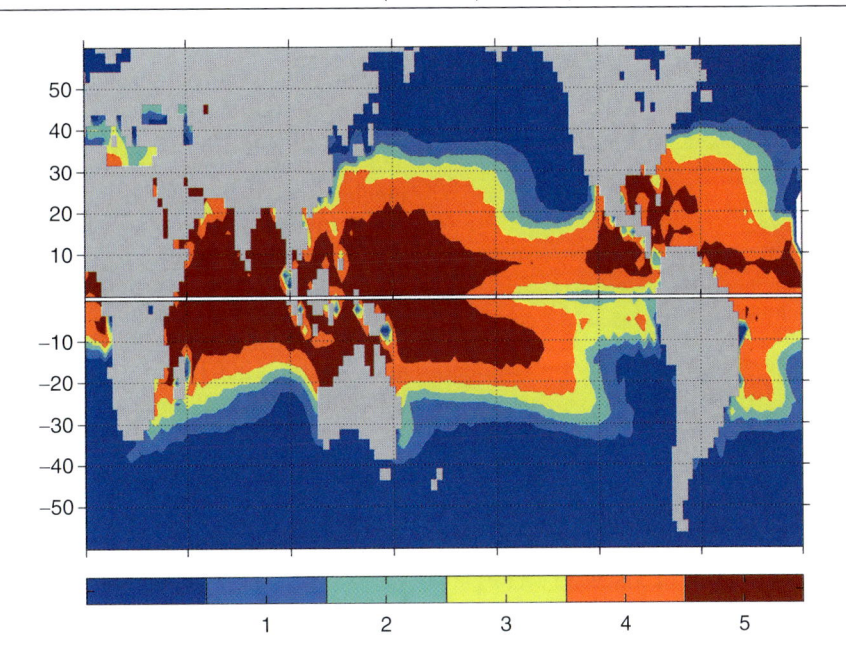

Hurricane (Typhoon, Cyclone), Figure 12 Peak value of the maximum potential intensity at any point during the annual cycle (Values were computed using the Bister and Emanuel, 1998) algorithm from data in the NCEP/NCAR reanalysis fields between 1979 and 2008. These were then averaged over each of the 30 years to form a single mean-annual cycle. The highest values from this cycle are shown here.).

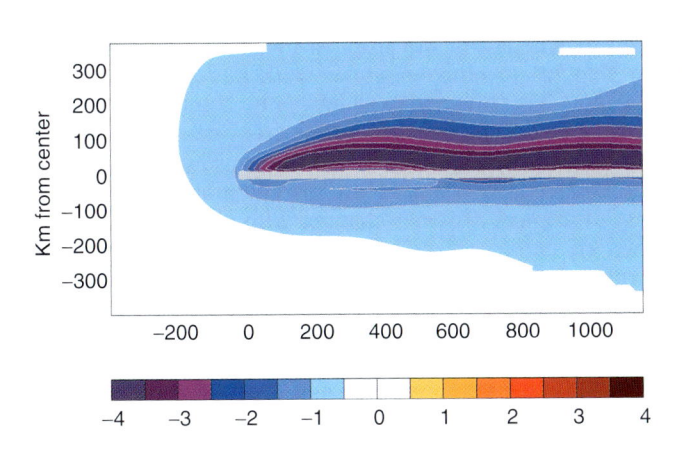

Hurricane (Typhoon, Cyclone), Figure 13 SST change in the wake of a westward-moving storm in a coupled hurricane-ocean model. The track of the storm is shown by the solid gray line, and its current location is centered at (0, 0). Cooling of several degrees Celsius is pronounced along and to the right of the track in its wake.

As seen in Figure 12, potential intensities equal to the strength of a Category 4 or 5 hurricane are common over a wide part of the globe, yet very few of the storms that form reach their upper bound (Emanuel, 2000). An individual storm will be subjected to specific conditions that often limit its ability to reach its full potential. These include wind shears, dynamic interactions with neighboring meteorological features, and interactions with the underlying ocean itself.

Price (1981) described the mechanism by which tropical cyclones cool the surface of the ocean. Figure 13 shows the deviation from initial sea surface temperature (SST) in the wake of a westward moving hurricane output from a simple coupled model. Water temperatures are cooler by several degrees along and north of the track (the right hand side of this hurricane). While some heat is lost to the atmosphere during the passage of a storm through air-sea fluxes, the amount is quite negligible compared to heat content of the upper ocean; if these were the only cause of cooling, SSTs would be only a fraction of a degree lower behind a hurricane. The principal reason why water temperatures are lower by several degrees Celsius after the passage of a hurricane is that cooler water from the thermocline has been mixed into the surface layer by turbulent processes instigated by the hurricane itself.

The strong surface winds in a hurricane act as a stress on the upper ocean, and this stress imparts momentum to the ocean currents in the mixed layer. As currents here build in strength under a hurricane, shear increases at the interface between the surface layer and thermocline. When this shear becomes sufficiently strong, turbulent mixing ensues, and colder, denser water from the upper thermocline is entrained into the surface layer, resulting in a drop in temperature.

As can be seen in Figure 13, there exists a pronounced bias in the location of maximum cooling: it is much larger to the north of the track, while SSTs are nearly unchanged to the south. This is a result of the action of the Coriolis force on a temporally varying wind stress. When the winds blow on the ocean, water is initially pushed in the direction of the wind, but the Coriolis force turns currents to the

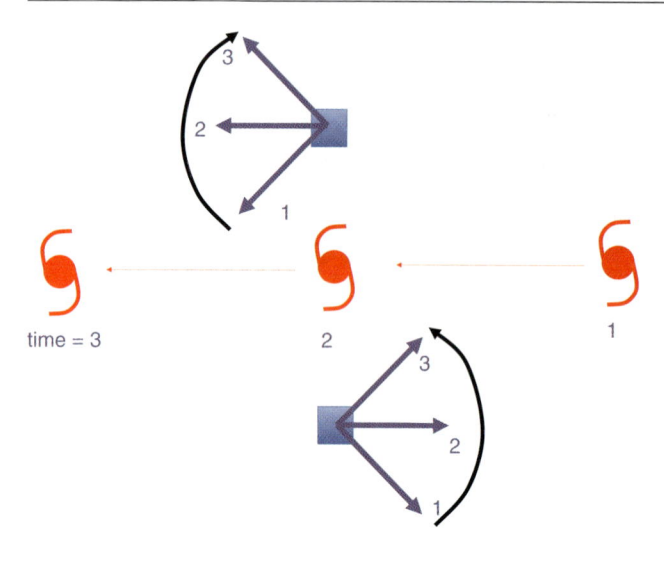

Hurricane (Typhoon, Cyclone), Figure 14 The wind vectors at two points north and south of a westward-moving Northern Hemisphere hurricane are shown for three times in gray. As a hurricane moves west, the winds at a fixed point turn as the cyclonic wind field moves by. The direction the wind vector turns is different on the right and left sides of the track: to the right, the wind vector turns clockwise or to its right with time, while it turns counterclockwise or to its left on the left side of the track. The temporal turning of the wind vector augments the torque exerted by the Coriolis force on parcels of water to the right of the track, while it opposes it on the left. As a result, currents in the upper ocean will be stronger in magnitude on the right than on the left of a track.

right in the Northern Hemisphere, so that transport is ultimately at a right angle to the wind. In the case of a moving hurricane, the wind stress itself is turning with time, and as diagrammed in Figure 14, the wind stress vector rotates clockwise on the right side of a track, and counterclockwise on the left. This means that on the right side of the track, the wind stress vector will rotate in the same direction that the Coriolis force is turning the currents, augmenting the strength of the currents here. To the left of the track, the reverse is true, and currents are weaker. As noted above, the strength of the currents determines the strength of the shear that builds on the base on the mixed layer; thermocline water is much more readily entrained on the right side of the track where currents have the largest velocities (Price, 1981, 1983).

While much of the cooling occurs in the wake of a storm, this process develops quickly enough to influence the air-sea fluxes that feed the hurricane itself. Thus a hurricane induces cooling of the surface water, which acts as a negative feedback on hurricane intensity by reducing the strength of the very fluxes on which a storm relies. Storms that move rapidly, travel over deep layers of warm water, and those that pass over parts of the ocean with a weaker thermocline stratification will be less affected by this interaction than those that do not (Schade and Emanuel, 1999).

One factor that is often responsible for inhibiting intensification well short of the thermodynamic limit is vertical wind shear (DeMaria and Kaplan, 1994; Emanuel, 2000). While a direct, negative link between vertical wind shear and tropical cyclone intensification exists, the physical mechanisms that cause a reduction in intensity remain under investigation. Several mechanisms have been proposed; shear may reduce intensity by ventilating the vortex core, increasing tropospheric stability, and eddy momentum fluxes, which decelerate tangential and radial winds (Simpson and Riehl, 1958; DeMaria, 1996; Frank and Ritchie, 2001; Wu and Braun, 2004; Tang and Emanuel, 2010).

Summary and open questions

Tropical cyclones rank among the most intense weather events in the world, although most fail to reach the full limits set by thermodynamics. They are sporadic events by nature, with only about 90 storms forming worldwide each year: they require a trigger to form, as even when conditions are supportive they do not spontaneously generate. While they form mostly in low latitudes of the world, they can affect subtropical and higher latitudes as well; tracks are biased to turn poleward owing to variations in Earth's planetary vorticity.

Recently, there has been renewed interest in the question of whether or not tropical cyclones are responding to changes in climate. Some work has focused on the Atlantic basin, which contains only about 15% of the storms that form worldwide, but has one of the longest and most reliable records of observations owing to regular aircraft reconnaissance. How the climatology and intensity of tropical cyclones changes with climate remains an area of active research. Historical records of tropical cyclones provide the most obvious avenue for investigation, but they are in many respects ill suited for analyzing how characteristics respond to climate because they are relatively short and have varying quality over their length (Landsea et al., 2004). Global satellite coverage was unavailable before the 1970s, meaning that records in remote parts of the world are likely incomplete before then. Another avenue is to use global climate models (e.g., Bengtsson et al., 2007), but even as resolution improves, these remain unable to resolve the inner core of tropical cyclones given the fine scales of the physics involved (Rotunno et al., 2009). Downscaling – where output from global climate models is then used to run a full physics hurricane model – offers a way to study these questions with models without succumbing to the computational costs of running global-scale models at very high resolutions (Emanuel, 2010).

One other avenue of research comes from the nascent field of paleotempestology. Liu and Fern (1993) showed that overwash from intense tropical cyclones is recorded in the sediment record on the inland side of barrier islands, and some of these records cover thousands of years. Each of these cores might be viewed as a single paleoweather station, but as the number of cores grows around the world, it may be possible to get some important information about how the climatology of landfalling, intense tropical cyclones varied over the Holocene.

Bibliography

Bengtsson, L., Hodges, K. I., Esch, M., Keenlyside, N., Kornbleuh, L., Luo, J.-J., and Yamagata, T., 2007. How may tropical cyclones change in a warmer climate? *Tellus*, **59**, 539–561.

Bister, M., and Emanuel, K., 1997. The genesis of hurricane Guillermo: TEXMEX analyses and a modeling study. *Monthly Weather Review*, **125**, 2662–2682.

Bister, M., and Emanuel, K., 1998. Dissipative heating and hurricane intensity. *Meteorology and Atmospheric Physics*, **65**, 233–240.

Bui, H. H., Smith, R. K., Montgomery, M. T., and Peng, J., 2009. Balanced and unbalanced aspects of tropical cyclone intensification. *Quarterly Journal of the Royal Meteorological Society*, **135**, 1715–1731.

Camargo, S. J., Emanuel, K. A., and Sobel, A. H., 2007. Use of a genesis potential index to diagnose ENSO effects on tropical cyclone genesis. *Journal of Climate*, **20**, 4819–4834.

Dean, L., Emanuel, K. A., and Chavas, D. R., 2009. On the size distribution of Atlantic tropical cyclones. *Geophysical Research Letters*, **36**, L14803, doi:10.1029/2009GL039051.

DeMaria, M., 1996. The effect of vertical shear on tropical cyclone intensity change. *Journal of the Atmospheric Sciences*, **53**, 2076–2087.

DeMaria, M., and Kaplan, J., 1994. Sea surface temperature and the maximum intensity of Atlantic tropical cyclones. *Journal of Climate*, **7**, 1324–1334.

Emanuel, K. A., 1986. An air-sea interaction theory for tropical cyclones. Part I: steady-state maintenance. *Journal of the Atmospheric Sciences*, **43**, 585–604.

Emanuel, K. A., 1988. The maximum intensity of hurricanes. *Journal of the Atmospheric Sciences*, **45**, 1143–1155.

Emanuel, K. A., 2000. A statistical analysis of tropical cyclone intensity. *Monthly Weather Review*, **128**, 1139–1152.

Emanuel, K. A., 2003. Tropical cyclones. *Annual Review of Earth and Planetary Sciences*, **31**, 75–104.

Emanuel, K., 2010. Tropical cyclone activity downscaled from NOAA-CIRES reanalysis, 1908-1958. *Journal of Advances in Modeling Earth Systems*, **2**, doi:10.3894/JAMES.2010.2.1.

Frank, W., and Ritchie, E., 2001. Effects of vertical wind shear on the intensity and structure of numerically simulated hurricanes. *Monthly Weather Review*, **129**, 2249–2269.

Gray, W. M., 1968. A global view of the origin of tropical disturbances and storms. *Monthly Weather Review*, **96**, 669–700.

Gray, W. M., 1979. Hurricanes: their formation, structure and likely role in the tropical circulation. In Shaw, D. B. (ed.), *Meteorology Over Tropical Oceans*. Bracknell, Berkshire: James Glaisher House, Royal Meteorological Society, pp. 155–218.

Hill, K. A., and Lackmann, G. M., 2009. Influence of environmental humidity on tropical cyclone size. *Monthly Weather Review*, **137**, 3294–3315.

Holland, G., 1997. The maximum potential intensity of tropical cyclones. *Journal of the Atmospheric Sciences*, **54**, 2519–2541.

Houze, R. A., Jr., 2010. Clouds in tropical cyclones. *Monthly Weather Review*, **138**, 293–344.

Kleinschmidt, E., Jr., 1951. Gundlagen einer theorie des tropischen zyklonen. *Archives for Meteorology Geophysics Bioklimatology Series A*, **4**, 53–72.

Kossin, J. P., Schubert, W. H., and Montgomery, M. T., 2000. Unstable interactions between a hurricane's primary eyewall and a secondary ring of enhanced vorticity. *Journal of Atmospheric Sciences*, **57**, 3893–3917.

Landsea, C. W., co-authors, 2004. The Atlantic hurricane database re-analysis project: Documentation for the 1851-1910 alterations and additions to the HURDAT database. In Murnane, J., and Liu, K.-B. (eds.), *Hurricanes and Typhoons: Past, Present, and Future*. New York: Columbia University Press, pp. 177–221.

Liu, K.-B., and Fearn, M. L., 1993. Lake-sediment record of late Holocene hurricane activities from coastal Alabama. *Geology*, **21**, 793–796.

Mallen, K. J., Montgomery, M. T., and Wang, B., 2005. Reexamining the near-core radial structure of the tropical cyclone primary circulation: implications for vortex resiliency. *Journal of Atmospheric Sciences*, **62**, 408–425.

McBride, J. L., and Zehr, R., 1981. Observational analysis of tropical cyclone formation. Part II: comparison of non-developing versus developing systems. *Journal of Atmospheric Sciences*, **38**, 1132–1151.

Price, J. F., 1981. Upper ocean response to a hurricane. *Journal of Physical Oceanography*, **11**, 153–175.

Price, J. F., 1983. Internal wave wake of a moving storm. Part I: scales, energy budget, and observations. *Journal of Physical Oceanography*, **13**, 949–965.

Riehl, H., 1950. A model for hurricane formation. *Journal of Applied Physics*, **21**, 917–925.

Rotunno, R., and Emanuel, K. A., 1987. An air-sea interaction theory for tropical cyclones. Part II: evolutionary study using axisymmetric nonhydrostatic numerical model. *Journal of Atmospheric Sciences*, **44**, 542–561.

Rotunno, R., Chen, Y., Wang, W., Davis, C., Dudhia, J., and Holland, C. L., 2009. Large-eddy simulation of an idealized tropical cyclone. *Bulletin of the American Meteorological Society*, **90**, 1783–1788.

Schade, L. R., and Emanuel, K. A., 1999. The ocean's effect on the intensity of tropical cyclones: results from a simple coupled atmosphere-ocean model. *Journal of Atmospheric Sciences*, **56**, 642–651.

Simpson, R. H., and Riehl, R., 1958. Mid-tropospheric ventilation as a constraint on hurricane development and maintenance. In *Technology Conference on Hurricanes*. American Meteorological Society, Miami Beach, FL, pp. D4-1–D4-10.

Simpson, R. H., and Riehl, H., 1981. *The hurricane and its impact*. Baton Rouge: Louisiana State University Press. 398 pp.

Smith, R. B., 1993. A hurricane beta-drift law. *Journal of Atmospheric Sciences*, **50**, 3213–3215.

Tang, B., and Emanuel, K., 2010. Mid-level ventilation's constraint on tropical cyclone intensity. *Journal of Atmospheric Sciences*, **67**, 1817–1830.

Wu, L., and Braun, S., 2004. Effects of environmentally induced asymmetries on hurricane intensity: a numerical study. *Journal of Atmospheric Sciences*, **61**, 3065–3081.

Cross-references

Damage and the Built Environment
Disaster
Doppler Weather Radar
Dvorak Classification of Hurricanes
Hurricane Katrina
Intensity Scales
Marine Hazards
Natural Hazards
Saffir-Simpson Hurricane Intensity Scale
Storm Surges
Storms
Surge
Warning Systems

CASE STUDY

HURRICANE KATRINA

Joann Mossa
University of Florida, Gainesville, FL, USA

Background

Hurricane Katrina of the Atlantic Ocean in August 2005 is an important natural hazard case study because it was the costliest hurricane as well as one of the five deadliest in the history of the United States (http://www.katrina.noaa.gov/; Beven et al., 2008). Although it affected several states, the most severe damages occurred in southeastern Louisiana and along the Mississippi coast, the former receiving the most media attention. The Gulf Coast of Mississippi experienced the most severe quadrant of the storm and suffered near total devastation on August 28–29 with a maximum 9 m (27.8-ft) storm surge along the shore but an overall surge that reached up to 19 km (12 miles) inland and wiped out towns and forests, leaving 238 people dead (Knabb et al., 2006). There were an estimated 1,577 deaths in Louisiana directly or indirectly related to Katrina because the heavily populated Greater New Orleans area was affected (Knabb et al., 2006). Many citizens, politicians, and scientists considered the preparation, flooding, and response to the storm in New Orleans a debacle, and as a result, it spurred many investigations, reports, documentaries, and books (e.g., Horne, 2006; van Heerden and Bryan, 2006; U.S. House of Representatives, 2006).

Environmental vulnerability and engineering problems

The vulnerability and risk of living in southeastern Louisiana and the other Gulf areas impacted by Hurricane Katrina has been long recognized by many (e.g., Kelley et al., 1984; Colten, 2006). Located in a deltaic setting, on a natural levee and backswamp between the Mississippi River and several large lakes, elevations in New Orleans are low and large portions of the city are below sea level (McCullough et al., 2006). Engineering structures and modifications are widespread; the metropolitan area is surrounded by artificial levees and floodwalls, plus numerous canals and other waterways creating multiple inroads of vulnerability. The area surrounded by levees requires a system of pumps to remove rainfall from lowlands following rainstorms because there is nowhere else for water to drain. During the past century, wetlands throughout southeastern Louisiana have been disappearing at an accelerated rate. One reason is the maintenance of the Mississippi River in its present course by multiple control structures near the juncture with the Atchafalaya, a site of potential avulsion, which results in sediment resources going off the continental shelf. Land loss is also due to the lack of wetland deposition from Mississippi River floodwaters due to presence of artificial levees and creation of numerous canals through wetlands to support extraction of petroleum, other resources, and navigation. The region is also experiencing subsidence related to compaction and settling of recently deposited sediments and movement of growth faults (Dixon et al., 2006). Subsidence has been accelerated by oxidation and shrinkage of organic-rich drained soils, associated mostly with development of areas surrounded by artificial levees; the loading of sediments with structures, buildings, and levees; and pumping of groundwater.

Flooding preparation, impacts, and response

The challenges associated with protecting this type of terrain are numerous, as the region has become more vulnerable with time; yet, ultimately, engineering failures caused the flooding associated with Hurricane Katrina. Portions of levees and floodwalls were overtopped or bent by the incoming storm surge. Other sections developed leaks and breaches or collapsed due to problems with the materials or foundation. Ultimately, about 80% of New Orleans was flooded, with low areas inundated with waters 6 m (20 ft) and higher for several days to weeks (http://www.katrina.noaa.gov/maps/maps.html). Levee problems prompted external reviews of levee design and construction, which is the responsibility of the U.S. Army Corps of Engineers, and maintenance, which is the purview of local Levee boards. It was concluded that both system design flaws and lack of adequate maintenance were problematic (Andersen et al., 2007).

Coupled with the environmental and engineering problems were historical, behavioral, and socioeconomic factors. There had not been a storm of this severity for decades; thus, officials lacked experience in planning for and recovering from major storms, and some residents did not take this potential threat seriously. The social vulnerability of poor and older residents with limited access to information and transportation was also a major problem. Although the National Hurricane Center and National Weather Service provided accurate forecasts and abundant lead time, many inhabitants could not or did not evacuate. After the flooding, many were without basic provisions, some were rescued by helicopter off rooftops, and others found shelter in places like the Superdome or Morial Convention Center, which were not staged to handle a problem of this scope and magnitude. These places were not adequately stocked with water and food and had insufficient personnel for emergency assistance. Dissemination of information and emergency response coordination of local, state, and federal agencies were made difficult by confusion regarding the chain of command, disruption of communications infrastructure, and other information flow impediments (Day et al., 2009). The U.S. Coast Guard seemed best prepared of the agencies involved in the emergency response and rescued most of the roughly 60,000 people stranded in New Orleans (U.S. Government

Accountability Office, 2006). Until transported to shelters in other states for assistance, people suffered for days in difficult conditions. About 50–70% of the housing experienced severe flooding; financial aspects of this debacle were magnified by the fact that many homeowners lacked federal flood insurance (U.S. Department of Housing and Urban Development, 2006). There were many problems with efforts by individuals and city, state, and federal authorities to evacuate, plan, communicate, provide relief, and assist in recovery. Ultimately, investigations resulted in the resignations of the Federal Emergency Management Agency director Michael D. Brown and the New Orleans Police Department Superintendent Eddie Compass and widespread criticisms of President George W. Bush, Governor Kathleen Blanco, and Mayor Ray Nagin.

Aftermath

Southeastern Louisiana and the Gulf Coast of Mississippi will likely be forever changed from this event. Years after the storm, the population of New Orleans is much lower than prior to Hurricane Katrina. People of different backgrounds have different levels of place attachment and reasons for returning (Li et al., 2010). The levees were repaired prior to the subsequent storm season, but the city is still vulnerable to storms of this magnitude or larger. Redevelopment is being planned more carefully than in the past; some areas will not be rebuilt, and in other areas, buildings will need to be elevated (U.S. Army Corps of Engineers, New Orleans District, 2009). Conflicts are continuing over levee improvements and who should be responsible for funding them. The U.S. Government, (2006) identified 125 specific recommendations from studying what went wrong with Hurricane Katrina. Legal battles of citizen plaintiffs against the United States Army Corps of Engineers have been fought and won (U.S. District Court, Eastern District of Louisiana, 2009). It remains unknown how much learning has occurred but it is now better understood that the city and its people and others in low-lying areas of the Gulf Coast are vulnerable and that the numerous engineering structures built to protect them are not infallible.

Bibliography

Andersen, C. F., Battjes, J. A., Daniel, D. E., Edge, B., Espey, W., Jr., Gilbert, R. B., Jackson, T. L., Kennedy, D., Mileti, D. S., Mitchell, J. K., Nicholson, P., Pugh, C. A., Tamaro, G., Jr., and Traver, R., 2007. *The New Orleans Hurricane Protection System: What Went Wrong and Why.* American Society of Civil Engineers Hurricane Katrina External Review Panel. http://www.asce.org/files/pdf/ERPreport.pdf.

Beven, J. L., II, Avila, L. A., Blake, E. S., Brown, D. P., Franklin, J. L., Knabb, R. D., Pasch, R. J, Rhome, J. R., and Stewart, S. R., 2008. *Annual Summary Atlantic Hurricane Season of 2005 Tropical Prediction Center,* NOAA/NWS/National Hurricane Center, Miami, FL, pp. 1109–1173. www.aoml.noaa.gov/general/lib/lib1/nhclib/mwreviews/2005.pdf

Colten, C. E., 2006. *An Unnatural Metropolis: Wresting New Orleans from Nature.* Baton Rouge: Louisiana State University Press.

Day, J. M., Junglas, I., and Silva, L., 2009. Information flow impediments in disaster relief supply chains. *Journal of the Association for Information Systems,* **10**, 637–660.

Dixon, T. H., Amelung, F., Ferretti, A., Novali, F., Rocca, F., Dokka, R., Sellall, G., Kim, S.-W., Wdowinski, S., and Whitman, D., 2006. Subsidence and flooding in New Orleans. *Nature,* **441**, 587–588.

Horne, J., 2006. *Breach of Faith: Hurricane Katrina and the Near Death of a Great American City.* New York: Random House.

Kelley, J. T., Kelley, A. R., and Pilkey, O. H., 1984. *Living with the Louisiana Shore.* Durham, NC: Duke University Press.

Knabb, R., Rhome, J., and Brown, D., 2006. *Tropical Cyclone Report.* Retrieved November 2009, from http://www.nhc.noaa.gov/pdf/TCR-AL110005_Katrina.pdf.

Li, W., Airriess, C. A., Chen, A. C.-C., Leong, K. J., and Keith, V., 2010. Katrina and migration: evacuation and return by African Americans and Vietnamese Americans in an Eastern New Orleans Suburb. *Professional Geographer,* **62**, 103–118.

McCullough, R. P., Heinrich, P. V., and Good, B., 2006. Geology and hurricane-protection strategies in the Greater New Orleans Area, *Louisiana Geological Survey Public Information Series 11.* http://www.lgs.lsu.edu/deploy/content/PUBLI/contentpage14.php.

U.S. Army Corps of Engineers, New Orleans District, 2009. *LACPR (Louisiana Coastal Protection and Restoration) Final Technical Report,* 3280 pp. http://lacpr.usace.army.mil/default.aspx?p=LACPR_Final_Technical_Report.

U.S. Department of Housing and Urban Development, 2006. *Current Housing Unit Damage Estimates.* http://www.dhs.gov/xlibrary/assets/GulfCoast_HousingDamageEstimates_021206.pdf.

U.S. District Court, Eastern District of Louisiana, 2009. *Katrina Canal Breaches Consolidated Litigation,* Colleen Berthelot et al v. Boh Brothers Construction Co. LLC et al 05–4182. http://www.laed.uscourts.gov/CanalCases/CanalCases.htm.

U.S. Government, 2006. *The Federal Response to Hurricane Katrina: Lessons Learned.* http://georgewbush-whitehouse.archives.gov/reports/katrina-lessons-learned.pdf.

U.S. Government Accountability Office, 2006. *Coast Guard: Observations on the Preparation, Response, and Recovery Missions Related to Hurricane Katrina.* http://www.gao.gov/new.items/d06903.pdf.

U.S. House of Representatives, 2006. *A Failure of Initiative: Final Report of the Select Bipartisan Committee to Investigate the Preparation for and Response to Hurricane Katrina.* Washington, DC: Government Printing Office. http://www.gpoaccess.gov/katrinareport/fullreport.pdf.

van Heerden, I., and Bryan, M., 2006. *What Went Wrong and Why During Hurricane Katrina.* New York: Viking.

Cross-references

Avulsion
Coastal Erosion
Disaster Relief
Evacuation
Flood Deposits
Flood Hazard and Disaster
Flood Protection
Floodplain
Hurricane (Cyclone, Typhoon)
Levee
Storms and Storm Surge
Urban Environments and Natural Hazards

HYDROCOMPACTION SUBSIDENCE

Andrew J. Stumpf
Prairie Research Institute, University of Illinois at Urbana-Champaign, Champaign, IL, USA

Synonyms

Hydroconsolidation; Near-surface subsidence; Sagging; Shallow subsidence

Definition

A process of collapse and compaction that occurs in silty to sandy sediment (soil) having a low bulk density, is saturated for sustained periods, but the water is subsequently removed (Krynine and Judd, 1957; Dudley, 1970). Infiltrating water enters the porous structure of the sediments weakening interparticle bonds of clay if present and reducing capillary tension between coarser soil particles that provided strength when unsaturated. Removal of these structural bonds causes overall consolidation and is manifested by settlement under loads and subsidence at the ground surface (Inter-Agency Committee on Land Subsidence in the San Joaquin Valley, 1958; Lofgren, 1960; Sajgalik, 1990).

Soil characteristics

Soils susceptible to hydrocompaction are geologically immature deposits or reclaimed soils that are uncompacted or unconsolidated predominantly composed of silt and sand grains with a small amount of clay, usually less than 12% (Bull, 1964). Likely soils typically have a water content below 10% (Waltham, 1989), loose, bulk density <1.3 g/cm^3, and a high void ratio or porosity ranging from 38% to 60% (Feda, 1966; Dudley, 1970). The soils also have sufficient dry strength because of interparticle bonds that resist compacting and readily absorb large load stresses (Waltham, 1989). Hydrocompaction is common in soils containing montmorillonite clay (Rogers et al., 1994) and other swelling clay minerals.

Process

Hydrocompaction occurs after saturating and drying, silty to sandy soils that contain weak structural bonds formed between particles by capillary tension, clay, or soluble precipitants (Prokopovich, 1963, 1986; Lofgren, 1969). As the soil becomes saturated, the structural support provided by both capillary action and clay bonds are weakened. The soil then undergoes deformation or a change in volume to a denser structure under its own weight or the weight of an overlying structure. The areal subsidence or settlement underlying a structure has been triggered by large-scale irrigation projects, canal leakage, waste water disposal, and other modifications to drainage and evaporation (Lofgren, 1969).

Distribution

Collapsible soils qv "*Collapsing Soil Hazards*" prone to hydrocompaction are present in wind-blown silt (loess qv "*Loess*") found in the United States, Europe, Russia, Ukraine, Afghanistan, Thailand, and China, and alluvial silt on fans and mudflows in the southwestern United States (Lofgren, 1969; Waltham, 1989). Hydrocompaction may also occur in volcanic ashes, colluvial deposits, and organic-rich sediment (Stephens et al., 1984; Waltham, 1989).

Hazards

Hydrocompaction may cause rapid subsidence of the ground surface causing significant damage to property. Subsidence as large as 5 m can occur over a wide area (Waltham, 1989) and can promote further collapse by sapping and piping (see entry *Piping Hazard*) in the subsurface.

Summary

Hydrocompaction is a serious land subsidence issue where unsaturated, fine-grained soils are modified by human interactions. The extent over which this collapse process has been identified indicates its importance as a natural hazard. Rogers et al., (1994) provides a thorough review of the problem in a global perspective and encourage discussion between the scientists and engineers to overcome their misunderstandings about the process that have resulted from differences in terminology used by the varied disciplines involved with the process and consequence.

Bibliography

Bull, W. B., 1964. *Alluvial Fans and Near-Surface Subsidence in Western Fresno County, California*. Washington: United States Geological Survey, Professional Paper 437-A.

Dudley, J. H., 1970. Review of collapsing soils. *Journal of the Soil Mechanics and Foundation Division*, **96**, 925–947.

Feda, J., 1966. Structural stability of subsident loess soil from Praha-Dejvice. *Engineering Geology*, **1**, 201–219.

Inter-Agency Committee on Land Subsidence in the San Joaquin Valley, 1958. *Progress Report on Land-Subsidence Investigations in the San Joaquin Valley, California, Through 1957*. Sacramento, CA: United States Geological Survey/Inter-Agency Committee on Land Subsidence in the San Joaquin Valley.

Krynine, D. P., and Judd, W. R., 1957. *Principles of Engineering Geology and Geotechnics*. New York: McGraw-Hill.

Lofgren, B. E., 1960. Near-surface subsidence in western San Joaquin Valley California. *Journal of Geophysical Research*, **65**, 1053–1062.

Lofgren, B. E., 1969. Land subsidence due to the application of water. In Varnes, D. J., and Kiersch, G. (eds.), *Reviews in Engineering Geology*. Boulder, CO: The Geological Society of America, Vol. 2, pp. 271–303.

Prokopovich, N. P., 1963. Hydrocompaction of soils along San Luis Canal alignment, western Fresno County, California. In *Program 58th Cordilleran Section Annual Meeting of the Geological Society of America*, Abstract.

Prokopovich, N. P., 1986. Origin and treatment of hydrocompaction in the San Joaquin valley, CA, USA. In Johnson, I., Carbognin, L., and Ubertini, L. (eds.), *Land Subsidence*. Proceedings of the Third

International Symposium on Land Subsidence. Wallingford, CT: International Association of Hydrological Sciences, 151, pp. 537–546.
Rogers, C. D. F., Dijkstra, T. A., and Smalley, I. J., 1994. Hydroconsolidation and subsidence of loess: studies from China, Russia, North America and Europe. *Engineering Geology*, **37**, 83–113.
Sajgalik, J., 1990. Sagging of loess and its problems. *Quaternary International*, **14**, 314–316.
Stephens, J. C., Allen, L. H., Jr., and Chen, E., 1984. Organic soil subsidence. In Holzer, T. L. (ed.), *Man-Induced Land Subsidence*. Boulder, CO: The Geological Society of America. Reviews in Engineering Geology, Vol. 6, pp. 107–122.
Waltham, A. C., 1989. *Ground Subsidence*. New York: Chapman and Hall, pp. 156–164.

Cross-references

Collapsing Soil Hazards
Land Subsidence
Loess
Piping Hazards

HYDROGRAPH, FLOOD

Fernando Nardi
University of Tuscia, Viterbo (VT), Italy
Hydraulics Applied Research & Engineering Consulting S.r.l, Rome, Italy

Synonyms

Flood time–discharge relationship; Flood wave; Storm hydrograph

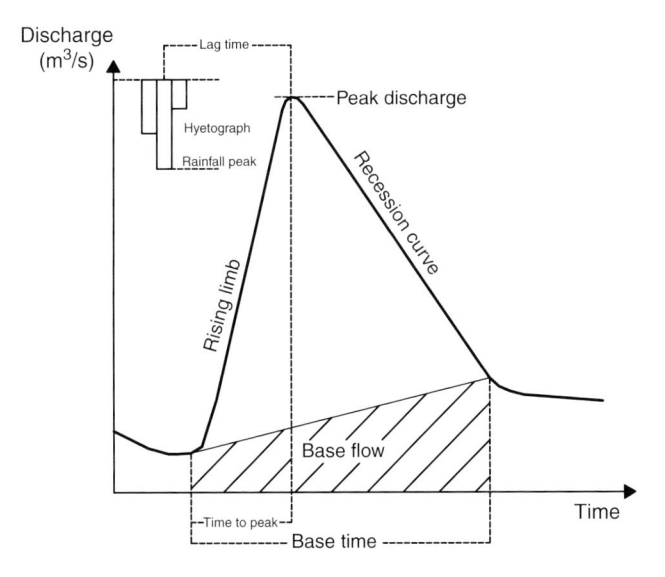

Hydrograph, Flood, Figure 1 Schematic representation of the flood hydrograph.

Definition

The hydrograph is the plot of water-flow discharge (runoff) versus time at a given river station (where a flow gauge might be located). The storm hydrograph represents the discharge record of river flow during extreme events, as the flood volume exceeds the maximum conveyance capacity of the river channels (see also *Flood Stage*) and the floodwaters – overtopping the banks – propagate through the floodplain (see entry *Floodplain*).

Discussion

The flood hydrograph is indirectly linked to a wide range of heterogeneous, spatially distributed factors pertaining to the upstream areas of the watershed. Those factors include the climatic, hydrologic, geomorphologic, geologic, hydraulic, and ecological conditions that, together with artificial features (e.g., levees, reservoirs, bridges, etc.), characterize the flood generation, propagation, and inundation phenomena, and range from the hydrologic forcing (precipitation regime, represented by the rainfall hyetograph) to the hillslope and river network morphology, the land use, soil, vegetation, and surface roughness properties up to the human-made interventions by means of civil infrastructures and structural *flood protection* systems.

Figure 1 shows a schematic representation of the flood hydrograph and its main features: the rising and falling limb (recession curve), the base flow and time, the lag time (time interval between the rainfall and discharge peaks), the peak discharge, and the total volume (estimated as the area under the curve that is the integral of the hydrograph function for the given time period). The shape of the flood hydrograph curve provides direct information for understanding the dynamics of the flood hydraulics and its evolution in time with specific regard to the volume and the maximum (peak) discharge.

Cross-references

Flood Protection
Flood Stage
Floodplain

HYDROMETEOROLOGICAL HAZARDS

Gordon McBean
The University of Western Ontario, London, ON, Canada

Synonyms

Climate and related hazards; Flood; Water; Weather

Definition

Process or phenomenon of atmospheric, hydrological or oceanographic nature that may cause loss of life, injury or other health impacts, property damage, loss of livelihoods and services, social and economic disruption, or environmental damage. (United Nations International Strategy for Disaster Reduction, UNISDR, 2009a).

Introduction

Throughout the history of this planet, hydrometeorological hazards have had great impact. From the prehistoric to biblical floods to the tragic events of recent times, humanity has been afflicted by floods, storms and droughts. In 2008, Cyclone Nargis swept through the Bay of Bengal and caused over 130,000 deaths in Myanmar (Center for the Research on the Epidemiology of Disasters, CRED, 2010); 17 years earlier Cyclone Gorky had resulted in a similar death toll in Bangladesh. Drought in sub-Saharan Africa has been a frequent occurrence and in 1983 the death tolls in Ethiopia and Sudan totaled close to half a million people. About 30,000 people died in the floods in Venezuela in 1999. The European death toll due to the summer of 2003 heat wave is now estimated at 70,000 people (Robine et al., 2008).

The exact direct and indirect economic costs (lives, infrastructure, etc.) associated with hydrometeorological hazards is not readily quantifiable. An estimated total loss of $125 Billion (CRED, 2010) (all values in $US) was attributed to Hurricane Katrina in 2005 whereas Hurricanes Charley and Wilma, before and after Katrina, added another $30B in losses. The losses due to the Yangtze flood of 1998 in China and Hurricane Andrew of 1992 in the USA were also close to $30B. The wild fires in Indonesia in 1997 had estimated losses of $17B.

This entry will address different types of hydrometeorological hazards with respect to their past and projected changes with a changing climate. This will be followed with a more in-depth discussion of the impacts of hydrometeorological hazards. Reducing impacts due to these events is very important; one of the ways of reducing impacts is to advise people in advance, the role of scientific predictions, so they can prepare and take appropriate actions. A description of the current research investigations related to gaps in current knowledge will follow and then a summary.

Types of hydrometeorological hazards

Hydrometeorological hazards include a wide range of phenomena and can be generally thought to include storms, floods, droughts, and temperature extremes and their associate phenomena. Hydrometeorological hazards are a major part of what is called the climate. Climate is a broad term but in the minds of most people it is the major hydrometeorological events that play a critical part in defining the climate. The category of storms includes tropical cyclones, which are known in some regions as typhoons and in others as hurricanes; and higher-latitude, extra-tropical cyclones, including blizzards and heavy snowfalls which sometimes cause avalanches. In coastal regions, storms can result in storm surges resulting in high waves and temporarily high sea level leading to coastal flooding. Other storm examples are smaller-scale events such as tornadoes, thunderstorms, and heavy precipitation, sometimes causing flash floods. Warming events leading to rapidly melting snow or the breakup of ice jams on rivers can also result in flooding. Hazards can also result from significant temperature variations. Heat waves, together with reduced precipitation, can lead to droughts and wildfires. Cold spells create havoc for human health and some ecosystems, for example icing, black ice, frost damaged plants, wet snow and downed trees. Hydrometeorological conditions also can be a factor in other hazards such as air pollution, landslides, locust plagues, epidemics, and in the transport and dispersal of food, water, vector and rodent-borne diseases, toxic substances, and material from volcanic eruptions. Hydrometeorological hazards directly and indirectly impact human health.

The Center for the Research on the Epidemiology of Disasters (CRED) (Rodriguez et al., 2009) defines: meteorological disasters as "events caused by short-lived/small to mesoscale atmospheric processes (in the spectrum from minutes to days) (such as a storm)"; climatological disasters as "events caused by long-lived/meso to macroscale processes (in the spectrum from intra-seasonal to multi-decadal climate variability) (such as extreme temperature, drought and wildfire)"; and hydrological disasters as "events caused by deviations in the normal water cycle and/or overflow of bodies of water caused by wind set-up (such as flood and mass movement (wet)." Hydrometeorological hazards, as defined above (from UNISDR), include meteorological, climatological and hydrological hazards and disasters as defined by CRED.

Observed and projected changes in hydrometeorological hazards

The Intergovernmental Panel on Climate Change (IPCC) 2007 Fourth Assessment Report has concluded that the "warming of the climate system is unequivocal" (IPCC, 2007, p. 5). With a warming climate, unprecedented hydrometeorological extreme events can occur as the trend in the mean variable, such as temperature, leads to a situation beyond the previous range. For some variables there may be critical thresholds that cause hazardous conditions and these thresholds may be breached more frequently with the shift in the mean. Changes can also be caused by changes in the variability. With the observed warming there have been increases in the frequency and/or intensity of hydrometeorological hazards as tabulated in Table 1. The IPCC used the following terms to indicate the assessed likelihood, using expert judgment, of an outcome or a result: *Virtually certain* > 99% probability of occurrence, *Extremely likely* > 95%, *Very likely* > 90%, *Likely* > 66%, *More likely than not* > 50%, *Unlikely* < 33%, *Very unlikely* < 10%, *Extremely unlikely* < 5% (IPCC, 2007, Box TS.1, p. 3). With respect to the risks of extreme weather events, the IPCC, in 2007, found that responses to some recent extreme events reveal higher

Hydrometeorological Hazards, Table 1 Recent trends, assessment of human influence on the trend and projections for extreme weather events for which there is an observed late-twentieth century trend. (IPCC, 2007 – Table SPM.2. p. 8)

Phenomenon and direction of trend	Likelihood that trend occurred in late twentieth century (typically post 1960)	Likelihood of a human contribution to observed trend	Likelihood of future trends based on projections for twenty-first century using SRES scenarios
Warmer and fewer cold days and nights over most land areas	Very likely	Likely	Virtually certain
Warmer and more frequent hot days and nights over most land areas	Very likely	Likely (nights)	Virtually certain
Warm spells/heat waves. Frequency increases over most land areas	Likely	More likely than not	Very likely
Heavy precipitation events. Frequency (or proportion of total rainfall from heavy falls) increases over most areas	Likely	More likely than not	Very likely
Area affected by droughts increases	Likely in many regions since 1970s	More likely than not	Likely
Intense tropical cyclone activity increases	Likely in many regions since 1970s	More likely than not	Likely
Increased incidence of extreme high sea level (excludes tsunamis)	Likely	More likely than not	Likely

levels of vulnerability than found earlier. It is important to examine decadal statistics as it is difficult, if not impossible, to attribute a specific, single extreme event due to a changing climate. The impacts of an extreme event may also depend on antecedent conditions. For example, a heavy rainfall event that occurs after a long period of drought may create more of a hazard than situations without these antecedent conditions.

The IPCC projections, also in Table 1, are that there will be very likely more heat waves and heavy precipitation events and likely increases in intense tropical cyclone activity. The IPCC (2007) report notes higher confidence than in earlier assessments with respect to projected increases in droughts, heat waves, and floods, as well as their adverse impacts. They comment that the "altered frequencies and intensities of extreme weather, together with sea level rise, are expected to have mostly adverse effects on natural and human systems" (IPCC, 2007, Sect. (3.3.5).

The IPCC (2007) assessment was largely based on 2006 and earlier scientific publications. A new IPCC report "Managing the Risks of Extreme Events and Disasters to Advance Climate Change Adaptation" has been published IPCC (2012). This report is primarily focused on hydrometeorological hazards, their changes with climate and disaster risk reduction – climate change adaptation strategies, and will provide an authoritative update. Newer research has confirmed the role of human activities in the global hydrological and atmospheric moisture cycles (Stott et al., 2009; Willett et al., 2007) and short-duration extreme precipitation (Zhang et al., 2010). Extreme precipitation would be expected to increase at a rate higher than changes in mean precipitation due to the nonlinear relationship between moisture content and temperature (Pall et al., 2007; Kharin et al., 2007) but the relationships are complicated for both mid to higher-latitudes (Meehl et al., 2005) and the tropics (Emori and Brown, 2005).

As the ocean surface temperature increases, it is likely that there will be an increase in intense tropical cyclone activity (CCSP, 2008; Gillett et al., 2008; Wing et al., 2007; Elsner et al., 2008) with the strongest tropical cyclones becoming stronger. Associated rainfall is also expected to increase (Nolan et al., 2007; Knutson et al., 2008). Due to the large amount of the interannual variability, detectable increases in tropical cyclone intensities may not be clearly manifest for decades to come (Bender et al., 2010). Storms at mid to higher latitudes will also be influenced by changes in climate but average cyclone activity may not be expected to change much (Bengtsson et al., 2009).

The general definition of drought is "a period of abnormally dry weather sufficiently prolonged for the lack of precipitation to cause a serious hydrological imbalance" (Heim, 2002). Drought also has several alternative definitions depending on the perspective. A meteorological drought is defined in terms of the magnitude of precipitation deficit. An agricultural drought depends on root zone soil water balance and a hydrological drought is related to stream flow, lake and groundwater levels (Heim, 2002). The duration of the precipitation deficit is also important from the point of view of impacts (Nicholls and Alexander, 2007). Drought indices such as the Palmer Drought Severity Index (PDSI) (Palmer, 1965) or the Standard Precipitation Index (Lloyd-Hughes and Saunders, 2002) are often used due to the lack of direct measurements (Trenberth et al., 2007; Seneviratne et al., 2010). Extreme droughts have had extensive impacts (Kallis, 2008; Beniston, 2009; Burke et al., 2006; Alexander and Arblaster, 2009; Easterling et al., 2008; Wang et al., 2009). Heat waves, such as that in Europe in 2003, are often related to drought (Fischer et al., 2007a, 2007b). In some regions, the changes in drought frequency have been difficult to document since one climate-induced change, the increase in precipitation, may be countering the tendency for more droughts (Easterling et al., 2007). There has also been more interannual variability in the Sahel drought in recent years and more spatial variations (Greene et al., 2009; Ali and Lebel, 2009). Projections on a regional basis are summarized in Christensen et al. (2007).

Global average sea level is now rising faster than earlier predictions (Copenhagen Diagnosis, 2009; Rahmstorf et al., 2007) such that by 2100, global sea level rise in a world of unmitigated greenhouse emissions may well exceed 1 m with the upper limit ~2 m (Note these quoted estimates exceed the 0.6 m given in the IPCC report). Sea level will also continue to rise for centuries after global temperatures have been stabilized, and several meters of sea level rise must be expected over the next few centuries. With increases in storm intensity, there will likely be higher precipitation and winds, leading to higher storm surges flooding coastal regions and risks of river flooding as well. Many low-lying coastal, river-delta megacities, already stressed by rapid population growth and economic, social, health, and cultural difficulties, are now increasingly vulnerable due to hydrometeorological hazards associated with climate change leading to increased risk of disasters that will affect not only the cities but the regions. An OECD report has ranked cities (Nichols et al., 2008) in terms of population and other exposures. The projected increases in hydrometeorological hazards with sea level rise will impact coastal megacities (Nichols et al., 2007).

Impacts of hydrometeorological hazards

A hazard, be it hydrometeorological or otherwise, is defined as a "potentially damaging physical event, phenomenon or human activity that may cause the loss of life or injury, property damage, social and economic disruption or environmental degradation." (UNISDR, 2009a) Hazards by themselves do not create disasters which are defined as a "serious disruption of the functioning of a community or a society involving widespread human, material, economic or environmental losses and impacts, which exceeds the ability of the affected community or society to cope using its own resources" or alternatively the characteristic of a disaster is the sense of overwhelming the capacity of communities to respond, and cause "extensive loss or disruption to the physical, social and administrative infrastructure" of a nation (Paton and Johnston, 2006). Disasters actually are the result of the situation when a hazard impacts a vulnerable community, ecosystem, or other region, where a vulnerability is defined as the "conditions determined by physical, social, economic, and environmental factors or processes, which increase the susceptibility of a community to the impact of hazards." This leads to the formal definition of disasters as social phenomena which stem from interaction between two key elements: hazards–triggering agents stemming from nature, as well as from human activity–and vulnerabilities–susceptibility to injury or loss influenced by physical, social, economic and cultural factors (Alexander, 1997; Mileti, 1999; McEntire, 2001; Paton et al., 2001).

The impacts of hydrometeorological hazards are immense (McBean and Ajibade, 2009). The number of disasters impacting global society has been increasing rapidly from about 150 per year in the 1980s, to over 200 per year in the 1990s to almost 1 per day for the period 2000–2008 (Table 2) (Rodriguez et al., 2009). In 2008 the two most disastrous events were Cyclone Nargis with major impacts in Myanmar, and the Sichuan earthquake in China. During the period 2000–2008, 356 were hydrometeorological disasters compared to 36 geophysical disasters (Table 3) (Rodriguez et al., 2009). Annually, these hydrometeorological disasters affected 220 million people as victims (deaths plus people affected) and caused about $82B in damages. The hydrometeorological disasters were predominantly hydrological (54% of which 44% were floods) and meteorological (30% of which 27% were storms). Although these figures are staggering, both insurers and scientists expect that climate change will bring more frequent and intense extreme hydrometeorological hazards, potentially resulting in more costly disasters in years to come. "In view of continued global warming, we anticipate a long-term increase in severe, weather-related natural catastrophes." (Topics Geo, 2006).

One of the challenges for analyzing trends in disaster statistics is that organizations use different methodologies for tabulating the information (Sapir and Vos, 2009). For 2008, the Centre for Research on the Epidemiology of Disasters (CRED) recorded 354 natural disasters of which 322 were hydrometeorological with about 166 million

Hydrometeorological Hazards, Table 2 Number of hydrometeorologicaly related disasters by decade and per year (with number of geophysical for comparison) (Rodriguez et al., 2009)

	1900–1909	1910–1919	1920–1929	1930–1939	1940–1949	1950–1959	1960–1969	1970–1979	1980–1989	1990–1999	2000–2008
Hydrometeorological	28	72	56	72	120	232	463	776	1,498	2,034	3,202
Per year (unit)	3	7	6	7	12	23	46	78	150	203	356
Geophysical	40	28	33	37	52	60	88	124	232	325	328

Hydrometeorological Hazards, Table 3 Average number of occurrences, victims and damages per year for period 2000–2008 for climatological, hydrological and meteorological and their total – of hydrometeorological origin, with statistics for geophysical events for comparison (Rodriguez et al., 2009)

	Occurrences (per year)	Number of Victims (Million per year)	Damages ($Billion per year)
Climatological	54	82	9
Hydrological	194	99	20
Meteorological	108	39	53
Total hydrometeorological	356	220	82
Geophysical	36	9	20

victims (people killed or affected) and economic costs exceeding US$ 104B. In contrast, Munich Re Data Service (Topics Geo, 2008) recorded a total of 750 disasters with the total cost of hydrometeorologicaly related events being US$ 134B. Although CRED and the MunichRe Data Service (NatCatService, 2010) recorded different disaster occurrences and economic losses, their ratios of hydrometeorological events to the total were about the same. Also, analyses of their data show similar rising trends (Gall et al., 2009; Rodriguez et al., 2009). With the upward trend with regard to the scope and magnitude of disaster impacts, it is important to note that the losses are highly influenced by the occurrence of "mega-disasters" affecting tens of millions of people and/or causing billion of dollars worth of economic damage. Topics Geo (2006) categorizes disasters on a scale of 1–6, with Category 5 events, called devastating catastrophes, causing more than 500 deaths and/or overall losses of more than $US 500 million, The number of these Category 5 events has increased from 5 to 15 events per year in the 1980s to 15–25 events per year in the period 1990–2005 to 28–41 events per year in the 2006–2008 period (2008 had 41 devastating catastrophes, the largest number ever). Between 1950 and 2009, 41% of disasters were due to meteorological hazards; 28% due to geophysical hazards; 25% due to hydrological hazards; and 6% due to climatological hazards such as droughts and heavy rains affected by El Nino. Overall losses (US$ 2,000B) are distributed similarly to the number of events but insured losses (about US$ 415B) are predominantly (80%) due to meteorological hazards (storms). This reflects the impacts of storms more on developed countries where there is extensive insurance compared to earthquakes that often happen with greater impact on people and uninsured property in the developing world. Although geophysical events are less frequent, they cause more of fatalities (53% of the two million fatalities) compared to 36% for the meteorological events.

Although all countries have been impacted by natural disasters, the relative impacts usually are larger in human lives in developing countries and larger in economic costs in developed countries (Mileti, 1999; Mizra, 2003). In highly developed countries, the average number of deaths per disaster is 23, whereas the number increases dramatically to about 150 deaths per disasters in medium and to over 1,000 deaths per disaster in less developed countries (Mutter, 2005). While the absolute dollar costs of disasters in highly-developed countries are large, the damage as a percentage of Gross Domestic Production (GDP) is much larger in the developing country (Handmer, 2003). Among the key findings and recommendations of the UN ISDR Global Assessment Report on Disaster Risk Reduction (2009b) were the following:

- Global disaster risk is highly concentrated in poorer countries with weaker governance.
- Weather-related disaster risk is expanding rapidly both in terms of the territories affected, the losses reported and the frequency of events.
- Climate change is already changing the geographic distribution, frequency, and intensity of weather-related hazards and threatens to undermine the resilience of poorer countries and their citizens to absorb and recover from disaster impacts.
- The governance arrangements for disaster risk reduction in many countries do not facilitate the integration of risk considerations in development.

The highest number of hydrometeorological disastrous events is most often recorded in Asia. Of the four category 6 catastrophes of 2008, for example, two were in Asia, one in the USA and one in the Caribbean. This trend is especially important due to the population density of the most vulnerable nations. An event in India or China is likely to affect more people than in one of the smaller, less densely populated nations since there is likely to be fewer inhabitants in any given site. However, the number of victims per 100,000 inhabitants list was led by Djibouti, Tajikistan, Somalia and Eritrea. This demonstrates that, in addition to population density and area of vulnerability, the economic ability of a nation to respond is an important factor in assessing the potential impact of any natural hazard. Developing nations often have minimal preventative measures and are unable to respond adequately in the immediate aftermath. Additionally, the attempt to recover from such events may be economically debilitating as well. For instance, the events in Myanmar and Tajikistan resulted in damages exceeding 20% of their Gross Domestic Production (GDP).

Disaster risk reduction

Historically, public policy in disaster risk reduction has been heavily concentrated on responding to the disasters after they have happened, reflecting a belief that disasters are unfortunate but random calamities beyond our control (Henstra and McBean, 2005). The present approaches used to address hazards and disasters are built around the key themes of traditional disaster management: mitigation (lessening or limitation of adverse impacts), preparedness (capacities to effectively anticipate, respond to, and recover from the impacts), response (services and public assistance during or immediately after a disaster), and recovery (restoration, and improvement where appropriate, of communities) (Paton and Johnston, 2006; Godschalk, 1991; Godschalk and Brower, 1985). While all four themes are necessary, most national disaster management plans focus on response and recovery. Primarily, governments are often more willing to put funding into response instead of prevention and preparedness because, while they must be seen to be responding to tragic events that have occurred, investments to prevent future events that may not occur are more politically risky. Further, developing countries are often limited to these strategies since while aid for response efforts is often available, funding for prevention measures is more difficult to obtain (UNISDR, 2005; Mileti, 1999). There will, unfortunately, always be need for response and recovery but the focus on disaster management is and should be shifting toward the inclusion of mitigation and preparedness efforts. Mitigation and preparedness both rely in substantial part on knowledge of what is or may happen – seeing the future.

Seeing the future: the role of scientific predictions

Prediction is used across the natural, environmental, social and economic sciences. "Prediction is a statement or claim that a particular event will occur in the future. Narrowing the sense of prediction it may be added that the place and time of event are known as well (Mesjasz, 2005)." The Oxford Dictionary defines the verb to predict as to *"foretell, prophesy."* The noun *forecast* is defined as: "conjectural estimate of something future, especially, of coming weather." *Conjecture* is the "formation of opinion on incomplete grounds." The sense of estimate, future and incomplete information is certainly consistent with the sense of prediction of natural and human systems. Prediction is the process of looking ahead on the basis of incomplete knowledge of the present and with incomplete understanding of how the system works.

Prediction can play a role in better planning for the future to reduce the impacts of hydrometeorological hazards (McBean, 2007). For natural physical systems including hydrometeorological systems, there are sets of physical "laws" such as those of Newton that are well tested and provide a basis for prediction. For weather forecasting, "progress in understanding and predicting weather is one of the great success stories of twentieth century science" (National Research Council, 1998). The skill of weather forecasts on all scales has been improving (Jolliffe and Stephenson, 2003; Nichols, 2001). Simonovic (2009) has recently reviewed flood modeling approaches. Predictions of the occurrence of other geophysical hazards, such as earthquakes, show increasing skill but due to the complexities of the relationships and the difficulties in observing the details of the present state, it is not yet possible to make predictions of the timing and magnitude of an event with high skill (Nigg, 2000, pp. 135–156).

Since disasters result from the intersection of hydrometeorological events and vulnerable communities, there is also need for predictions of human actions which is often more difficult. The World Summit on Sustainable Development (WSSD, 2002) recommended that an integrated, multi-hazard, inclusive approach be taken to address vulnerability, risk assessment and disaster management, and that among the necessary actions was to strengthen early warning systems. The World Conference on Disaster Reduction (UNISDR, 2005) called for "people-centred early warning systems" to be a key part of the response to the tragedies of natural hazards. The international community has clearly identified early-warning systems, which must be based on predictions, as a key response to global concerns.

Hydrometeorological events can occur on time scales from minutes to days to seasons to decades to centuries. There is a general relationship between physical and temporal scales of the events. Small physical-scale events generally have short lifetimes and larger events last longer. Thus a tornado, very small-scale event, is formed, travels over a short distance, and then disappears in less than a few hours. Storms that generate high winds and precipitation amounts typically cover a 100 km area and track over half a continent or more and last for days. The accuracy of prediction of hydrometeorological hazards will be dependent on the cumulative uncertainties in each component of the prediction system; this has been referred to as the "cascade of uncertainty" (IPCC, 2004). For floods, the skill depends on first the skill of the weather forecast and then of the hydrological prediction system (McBean, 2002). The skill of predictions (Murphy, 1997) depends on the characteristics of the phenomenon. There are several types of predictions of hazardous natural events.

Deterministic prediction uses information on the observed state of the system at an initial time to predict future successive states to some future time. The sequence events are "determined," to the extent the predictive model has skill. The coupled atmosphere-water-ocean system is what mathematicians call a dynamic, non-linear, chaotic system in which small differences in the initial state, time t_o, amplify with time (Lorenz, 1993, pp. 102–110) and this sets the limit of system predictability since the initial state cannot be known exactly. For small-scale, short-lived phenomena like tornadoes and thunderstorms, skilful deterministic predictions are only for minutes to hours. However, for major weather systems, there is skill for several days although it decreases as the length of the forecast increases. Lorenz (1993) has demonstrated that the theoretical limit for deterministic weather predictions is about 2 weeks.

Beyond the deterministic limit, predictions of statistical quantities or probabilities are possible using statistical approaches based on the longer deterministic prediction timescales of the components of the coupled atmosphere-ocean system that naturally change more slowly than others. Ensemble prediction technique is now being widely used in weather predictions, for days through seasons, to improve skill and also to provide the user with information on the probable skill of the prediction. Since tornadoes are embedded in and largely determined by the large-scale weather system, a probabilistic forecast can be made that gives an increased risk of a tornado for the next day, then clarifies and refines the forecast as the risk becomes clearer and eventually a deterministic forecast can be made. Prediction of these small-scale phenomena and their possible changing characteristics with climate must be approached by risk management techniques (McBean, 2005). The large-scale features of the atmosphere adjust more slowly than the smaller-scale cloud-weather systems, and the oceans, due to their large thermal capacity, adjust much more slowly than the atmosphere. Thus, one can use this information to extend beyond the atmospheric deterministic limit to provide predictions of the statistical occurrence of events. The ensemble approach is now widely used in seasonal predictions. These could be that a region will be warmer or colder than "normal" or wetter with risk of floods, or drier with risk of drought. This provides useful information but does not predict the sequence of events over the prediction period just that more warm days will occur, for example. Some skill in predictions for several seasons is possible for some events. This approach of cascading forecasts and increasing clarity of the risk of a hazardous event needs to be part of the prediction system for reducing the impacts of natural hazards. The concepts of a seamless prediction system are outlined in Shapiro et al. (2010).

One of the important challenges for national meteorological-hydrological services (NMHS) is to develop better their prediction skills and their approaches to risk management, recognizing there will always be some uncertainty in their predictions. Understanding risk communications and the relationship between warnings of risk, risk communication and society's response to risk is essential when dealing with hydrometeorological hazards (Leiss, 2001). Scientific assessment of risk and the public perception of risk are probably different, unless a good job is done of risk communication, connecting the two. Predictions need to better translated, including the statements of probabilities and risk assessment, into understandable terms. There is need to understand better how the public sees a risk, especially its qualitative dimensions (Descurieux, 2010).

Building capacity for knowledge-based disaster risk reduction

Of particular concern is the fact that capacity building for disasters has not been implemented in all nations. There are many barriers that account for this trend. First, as the Global Assessment Report on Disaster Reduction- 2009 asserts, disaster risk is "highly concentrated in poorer countries with weaker governance." (UNISDR, 2009b). Other barriers in the design and implementation of strategies for capacity development include how priorities should be set, what role each level of government should play, how strategies should be coordinated and how outcomes should be evaluated. Finally, the substantial amount of uncertainty surrounding hazards, vulnerability and the prediction of future events, proves to be a significant barrier to global capacity building (McBean, 2009b; Burton et al., 1993). The impacts of a changing climate or disaster events are largely determined by a society's or community's vulnerability; a function of its exposure to climate and other hazards, its sensitivity to the stresses they impose, and its capacity to adapt to these stresses. Vulnerability can be reduced through actions to minimize exposure, reduce the sensitivity of people and systems, and strengthen the community's adaptive capacity (McBean and Rodgers, 2010). Each of these actions requires an integrated approach and their implementation will necessitate surmounting barriers and constraints.

Research investigations to support disaster risk reduction

On December 22, 1989, the United Nations General Assembly (1989) declared the decade of the 1990s as the International Decade for Natural Disaster Reduction. The objective of the International Decade for Natural Disaster Reduction was to "reduce through concerted international action, especially in developing countries, the loss of life, property damage and social and economic disruption caused by natural disasters such as earthquakes, windstorms, tsunamis, floods, landslides, volcanic eruptions, wildfires, grasshopper and locust infestations, drought and desertification and other calamities of natural origin." Among the goals were: "(c) To foster scientific and engineering endeavours aimed at closing critical gaps in knowledge in order to reduce loss of life and property;" and "(e) To develop measures for the assessment, prediction, prevention and mitigation of natural disasters through programmes of technical assistance and technology transfer, demonstration projects, and education and training, tailored to specific disasters and locations, and to evaluate the effectiveness of those programmes." This was the first major international research program focussing on disaster risk reduction, including that due to hydrometeorological hazards. At the World Conference on Natural Disaster Reduction, Yokohama, 1994 (UNISDR, 1994), hydrometeorological hazards and the need for research were a focus of the resulting Yokohama Strategy and Plan of Action for a Safer World.

The World Climate Research Programme was originated in 1980 and is now co-sponsored by the International Council for Science (ICSU), the World Meteorological Organization (WMO) and the Intergovernmental Oceanographic Commission (IOC) of the United Nations Educational, Scientific and Cultural Organization (UNESCO). The two overarching objectives of the WCRP are: *to determine the predictability of climate*; and *to determine the effect of human activities on climate*. To achieve its objectives, the WCRP adopts a multidisciplinary approach, organizes large-scale observational and modeling projects, and facilitates focus on aspects of climate too large and complex to be addressed by any one nation or single scientific discipline. One of the challenges in climate prediction is in improving forecasts of climate extreme events with time-scales from several weeks, to seasons, years and even decades. Statistics of extreme events such as their probability of occurrence, duration, and intensity are very important for assessing the risks of hydrometeorological hazards. Linking the ongoing research on global energy and water cycles, the cryosphere and climate variability, the WRCP has identified and is now focusing on these extreme events.

Whereas the World Climate Research Program has focussed on hydrometeorological hazards within the climate time–scale context, the new World Weather Research Programme of the World Meteorological Organization is addressing these events on the weather time–scale. The Program is working to advance society's ability to cope with high impact weather through research focused on improving the accuracy, lead time and utilization of weather prediction. A major component of WWRP is the THORPEX (THe Observing Research and Predictability Experiment) program which is a global international research program focusing on high-impact weather with the research priorities of addressing: Global-to-regional influences on the evolution and predictability of weather systems; Global observing system design and demonstration; Targeting and assimilation of observations; and societal, economic, and environmental benefits of improved forecasts. The scope of the research includes mesoscale weather forecasting; nowcasting; weather modification assessment; tropical meteorology; and verification research. The primary research strategy is to promote, initiate, coordinate or manage: field campaigns, long-term research projects, and programs that are well suited for international collaboration and are designed to advance the underlying science of weather forecasting, to use research to advance forecasting techniques and to enhance the utilization of weather information; the establishment of archive centers that bring together international data sets that would not be easily accessible through other means and to set up reference datasets that form the basis for testing, comparing, and improving modeling and data assimilation strategies; end-to-end Research and Development Projects (RDPs) to advance understanding of weather processes, improve forecasting techniques and increase the utility of forecast information with an emphasis on high-impact weather; and Forecast Demonstration Projects (FDPs) to evaluate research techniques, tools, and concepts in an operational setting to facilitate the transfer of research results into operational practice.

When the International Council for Science (2003) undertook a Priority Area Assessment on Environment and its Relation to Sustainable Development and reviewed strategic options for future ICSU activities related to environmental research, it was concluded that research on "Natural and human-induced hazards" was one of four possible new fields of work. The resulting ICSU Planning Group concluded that, "despite all the existing or already planned activities on natural hazards, an integrated research programme on disaster risk reduction, sustained for a decade or more and integrated across the hazards, disciplines and geographical regions, is an imperative. The value-added nature of such a programme would rest with the close coupling of the natural, socio-economic, health and engineering sciences."

As a result a new research program Integrated Research on Disaster Risk – addressing the challenge of natural and human-induced environmental hazards (IRDR) was created (McBean, 2009a; ICSU, 2008) with the co-sponsorship of ICSU, the International Social Sciences Council and the United Nations International Strategy for Disaster Reduction (ISDR). The Science Plan of the proposed IRDR Programme would focus on hazards related to geophysical, oceanographic, and hydrometeorological trigger events; earthquakes; volcanoes; flooding; storms (hurricanes, typhoons, etc.); heat waves; droughts and fires; tsunamis; coastal erosion; landslides; aspects of climate change; space weather; and impact by near-Earth objects. The effects of human activities on creating or enhancing hazards, including land-use practices, would be included. The IRDR Programme would deal with epidemics and other health-related situations only where they were consequences of one or more of the aforementioned events. Technical and industrial hazards and warfare and associated activities would not be included per se. The focus on risk reduction and the understanding of risk patterns and risk-management decisions and their promotion would require consideration of scales from the local through to the international level.

Focusing on disaster risk reduction, the research will be aimed at integrated risk analysis, including consideration of relevant human behavior and decision-making processes in the face of risk. The IRDR is guided by three broad research objectives:

1. Characterization of hazards, vulnerability, and risk; with sub-objectives: identifying hazards and vulnerabilities leading to risks; forecasting hazards and assessing risks; and dynamic modeling of risk.
2. Understanding decision making in complex and changing risk contexts with sub-objectives: identifying relevant decision-making systems and their interactions; understanding decision-making in the context of environmental hazards; improving the quality of decision making practice.

Reducing risk and curbing losses through knowledge-based actions. The IRDR research program fulfills the need for an international, multidisciplinary, and an all hazard research program emphasized in the Hyogo Framework for Action. The added value of such a research program lies in its coupling of natural sciences' examination of hazards with socio-economic analysis of vulnerability and mechanisms for engaging the policy decision-making process. The IRDR will draw upon the expertise and scientific outputs of many partners in research. Specifically, it is hoped that IRDR will be able to catalog and analyze successful capacity building systems and strategies for resilient communities in order to benefit those threatened by climate-related hazards.

Summary

Hydrometeorological hazards, mainly floods, storms, and droughts, are the trigger mechanism for most of the "natural" disasters around the world. These events affect hundreds of millions of people and create major economic and social hardships. Scientific research has led to much improved knowledge of these events so they can, with some confidence, be predicted in the short term. With a changing climate, the characteristics of these hazards are changing, mostly in ways to cause more impacts, and these changes raise major scientific challenges that, while being addressed, still need more emphasis. Further, since the disasters result from the impact of a hazard on vulnerable communities and systems, there is a strong need for an integrated approach to understanding and addressing these hazards, vulnerabilities and resultant disasters to reduce the impacts in the future.

Bibliography

Alexander, D., 1997. the study of natural disasters, 1977–1997: some reflections on a changing field of knowledge. *Disasters*, **21**(4), 284–304.

Alexander, L. V., and Arblaster, J. M., 2009. Assessing trends in observed and modelled climate extremes over Australia in relation to future projections. *International Journal of Climatology*, **29**(3), 417–435.

Ali, A., and Lebel, T., 2009. The Sahelian standardized rainfall index revisited. *International Journal of Climatology*, **29**(12), 1705–1714.

Bender, M. A., Knutson, T. R., Tuleya, R. E., Sirutis, J. J., Vecchi, G. A., Garner, S. T., and Held, I. M., 2010. Modeled impact of anthropogenic warming on the frequency of intense Atlantic hurricanes. *Science*, **327**(5964), 454–458.

Bengtsson, L., Hodges, K. I., and Keenlyside, N., 2009. Will extratropical storms intensify in a warmer climate? *Journal of Climate*, **22**(9), 2276–2301.

Beniston, M., 2009. Trends in joint quantiles of temperature and precipitation in Europe since 1901 and projected for 2100. *Geophysical Research Letters*, **36**, L07707, doi:10.1029/2008GL037119.

Breda, N., and Badeau, V., 2008. Forest tree responses to extreme drought and some biotic events: towards a selection according to hazard tolerance? *Comptes Rendus Geoscience*, **340**(9–10), 651–662.

Burke, E. J., Brown, S. J., and Christidis, N., 2006. Modeling the recent evolution of global drought and projections for the twenty-first century with the Hadley centre climate model. *Journal of Hydrometeorology*, **7**(5), 1113–1125.

Burton, I., Kates, R. W., and White, G. F., 1993. *The Environment as Hazard*. New York: The Guilford Press.

CCSP, 2008. *Weather and climate extremes in a changing climate. Regions of focus: North America, Hawaii, Caribbean, and U.S. Pacific Islands*. In Karl, T. R., et al. (eds.), A Report by the U.S. Climate Change Science Program and the Subcommittee on Global Change Research. Department of Commerce, NOAA's National Climatic Data Center: Washington, DC, 164 pp.

Christensen, J. H., Hewitson, B., Busuioc, A., Chen, A., Gao, X., Held, I., Jones, R., Kolli, R. K., Kwon, W.-T., Laprise, R., Magaña Rueda, V., Mearns, L., Menéndez, C. G., Räisänen, J., Rinke, A., Sarr, A., and Whetton, P., 2007. Regional climate projections. In Solomon, S., Qin, D., Manning, M., Chen, Z., Marquis, M., Averyt, K. B., Tignor, M., and Miller, H. L. (eds.), *Climate Change 2007: The Physical Science Basis. Contribution of Working Group I to the Fourth Assessment Report of the Intergovernmental Panel on Climate Change*. Cambridge/New York: Cambridge University Press.

Copenhagen Diagnosis, 2009. Updating the World on the latest climate science. I. Allison, et al. The University of New South Wales Climate Change Research Centre (CCRC), Australia, 60 pp

CRED, 2010. Center for the research on the epidemiology of disasters. http://www.emdat.be.

Descurieux, J., 2010. Post Hoc evaluation of hazardous weather: snowstorms in the Montréal, Québec, area in March 2008. *Weather, Climate, and Society*, **2**, 36–43.

Easterling, D. R., Wallis, T. W. R., Lawrimore, J. H., and Heim, R. R., 2007. Effects of temperature and precipitation trends on U.S. drought. *Geophysical Research Letters*, **34**, L20709, doi:10.1029/2007GL031541.

Easterling, D. R., et al., 2008. Measures to improve our understanding of weather and climate extremes. In Karl, T. R., et al. (eds.), *Weather and Climate Extremes in a Changing Climate. Regions of Focus: North America, Hawaii, Caribbean, and U.S. Pacific Islands*. A Report by the U.S. Climate Change Science Program and the Subcommittee on Global Change Research, Washington, DC.

Elsner, J. B., Kossin, J. P., and Jagger, T. H., 2008. The increasing intensity of the strongest tropical cyclones. *Nature*, **455**(7209), 92–95.

Emori, S., and Brown, S. J., 2005. Dynamic and thermodynamic changes in mean and extreme precipitation under changed climate. *Geophysical Research Letters*, **32**, L17706.

Fischer, E. M., Seneviratne, S. I., Lüthi, D., and Schär, C., 2007a. The contribution of land-atmosphere coupling to recent European summer heatwaves. *Geophysical Research Letters*, **34**, L06707.

Fischer, E. M., Seneviratne, S. I., Vidale, P. L., Lüthi, D., and Schär, C., 2007b. Soil moisture – atmosphere interactions during the 2003 European summer heatwave. *Journal of Climate*, **20**, 5081–5099.

Gall, M., Borden, K. A., and Cutter, S. L., 2009. When do losses count? Six fallacies of natural hazards loss data. *Bulletin of the American Meteorological Society*, **90**, 799–809.

Gillett, N. P., Stott, P. A., and Santer, B. D., 2008. Attribution of cyclogenesis region sea surface temperature change to anthropogenic influence. *Geophysical Research Letters*, **35**, L09707, doi:10.1029/2008GL033670.

Godschalk, D. R., 1991. Disaster mitigation and hazard management. In Drabek, T. E., and Hoetmer, G. J. (eds.), *Emergency Management Principles and Practice for Local Government*. Washington: International City Management Association Washington.

Godschalk, D. R., and Brower, D. J., 1985. Mitigation strategies and integrated emergency management. *Public Administration Review*, **45**, 64–71.

Greene, A. M., Giannini, A., and Zebiak, S. E., 2009. Drought return times in the Sahel: a question of attribution. *Geophysical Research Letters*, **36**, L12701, doi:10.1029/2009GL038868.

Handmer, J., 2003. Adaptive capacity: what does it mean in the context of natural hazards. In Smith, J. B., Klein, R. J. T., and Huq, S. (eds.), *Climate Change, Adaptive Capacity and Development*. London: Imperial College Press, pp. 51–70.

Heim, R. R., Jr., 2002. A review of twentieth-century drought indices used in the United States. *Bulletin of the American Meteorological Society*, **83**, 1149–1165.

Henstra, D., and McBean, G., 2005. Canadian disaster management policy: moving toward a paradigm shift? *Canadian Public Policy*, **31**(3), 303–318.

International Council for Science, 2003. Priority area assessment on environment and its relation to sustainable development. http://www.icsu.org/Gestion/img/ICSU_DOC_DOWNLOAD/58_DD_FILE_ICSU_PAA_REPORT.pdf.

International Council for Science, 2008. *A Science Plan for Integrated Research on Disaster Risk: Addressing the Challenge of Natural and Human-Induced Environmental Hazards.* ISBN 978-0-930357-66-5.

IPCC, 2004. *IPCC Workshop on describing scientific uncertainties in climate change to support analysis of risk and of options.* 11–13 May 2004, Ireland, 2004. Workshop Report. See http://ipcc-wg1.ucar.edu/meeting/URW/product/URW_Report_v2.pdf.

IPCC, 2007. Summary for policymakers. In Solomon, S., Qin, D., Manning, M., Chen, Z., Marquis, M., Avery, K. B., Tignor, M., and Miller, H. L. (eds.), *Climate Change 2007: The Physical Science Basis. Contribution of Working Group 1 to the Fourth Assessment Report of the Intergovernmental Panel on Climate Change.* Cambridge/New York: Cambridge University Press.

IPCC, 2012. Managing the Risks of Extreme Events and Disasters to Advance Climate Change Adaptation. In Field, C.B., Barros, V., Stocker, T. F., Qin, D., Dokken, D. J., Ebi, K. L., Mastrandrea, M. D., Mach, K. J., Plattner, G. -K., Allen, S. K., Tignor, M., and Midgley, P. M. (eds.). *A Special Report of Working Groups I and II of the Intergovernmental Panel on Climate Change.* Cambridge/New York: Cambridge University Press, pp 582.

Jolliffe, I. T., and Stephenson, D. B., 2003. *Forecast Verification: A Practitioner's Guide in Atmospheric Science.* Chichester: Wiley.

Kallis, G., 2008. Droughts. *Annual Review of Environment and Resources*, 33, 85–118.

Kharin, V., Zwiers, F. W., Zhang, X., and Hegerl, G. C., 2007. Changes in temperature and precipitation extremes in the IPCC ensemble of global coupled model simulations. *Journal of Climate*, 20, 1419–1444.

Knutson, T. R., Sirutis, J. J., Garner, S. T., Vecchi, G. A., and Held, I. M., 2008. Simulated reduction in Atlantic hurricane frequency under twenty-first- century warming conditions. *Nature Geoscience*, 1(6), 359–364.

Leiss, W., 2001. *Understanding Risk Controversies.* Montreal: McGill-Queen's University Press.

Lloyd-Hughes, B., and Saunders, M. D., 2002. A drought climatology for Europe. *International Journal of Climatology*, 22, 1571–1592.

Lorenz, E., 1993. *The Essence of Chaos.* Seattle: University of Washington Press.

McBean, G. A., 2002. Prediction as a basis for planning and response. *Water International*, 7(1), 70–76.

McBean, G. A., 2005. Risk mitigation strategies for tornadoes in the context of climate change and development. *Mitigation and Adaptation Strategies for Global Change*, 10(3), 357–366.

McBean, G. A., 2007. Role of prediction in sustainable development and disaster management. In Brauch, H. G., Grin, J., Mesjasz, C., Dunay, P., Chadha Behera, N., Chourou, B., Oswald Spring, U., Liotta, P. H., and Kameri-Mbote, P. (eds.), *Globalisation and Environmental Challenges: Reconceptualising Security in the 21st Century.* Berlin/Heidelberg/New York/Hong Kong/London/Milan/Paris/Tokyo: Springer. Hexagon Series on Human and Environmental Security and Peace, Vol. 3.

McBean, G., 2009a. Introduction of a new international research program: integrated research on disaster risk – the challenge of natural and human-induced environmental hazards. In Beer, T. (ed.), *Geophysical Hazards: Minimizing Risk, Maximizing Awareness.* Berlin: Springer. International Year of Planet Earth series.

McBean, G., 2009b. Coping with global environmental change: need for an interdisciplinary and integrated approach. In Brauch, H. G., Oswald Spring, U., Mesjasz, C., Grin, J., Kameri-Mbote, P., Chourou, B., Dunay, P., and Birkmann, J. (eds.), *Coping with Global Environmental Change, Disasters and Security Threats, Challenges, Vulnerabilities and Risks.* Berlin/Heidelberg/New York: Springer. Hexagon Series on Human and Environmental Security and Peace.

McBean, G. A., and Ajibade, I., 2009. Climate change, related hazards and human settlements. *Current Opinion in Environmental Sustainability*, 1(2), 179–186.

McBean, G. A., and Rodgers, C. 2010. Climate hazards and disasters: the needs for capacity building. Wiley Interdisciplinary Reviews: Climate Change. 1–6, 871–884.

McEntire, D. A., 2001. Triggering agents, vulnerabilities and disaster reduction: towards a holistic paradigm. *Disaster Prevention and Management*, 10(3), 189–196.

Meehl, G. A., Arblaster, J. M., and Tebaldi, C., 2005. Understanding future patterns of increased precipitation intensity in climate model simulations. *Geophysical Research Letters* 32, L18719.

Mesjasz, C., 2005. Prediction in security, theory and policy. Paper presented at the First World International Studies Conference at Bilgi University, Istanbul, Turkey, 24–27 August.

Mileti, D. S., 1999. *Disasters by Design: A Reassessment of Natural Hazards in the United States.* Washington, DC: Joseph Henry Press.

Mirza, M. M. Q., 2003. Climate change and extreme weather events: can developing countries adapt? *Climate Policy*, 3, 233–248.

MunichRe Group, 2007. Press release. http: www.munichre.com.

Murphy, A. H., 1997. Forecast verification. In Katz, R. W., and Murphy, A. H. (eds.), *Economic Value of Weather and Climate Forecasts.* Cambridge, UK: Cambridge University Press, pp. 19–74.

Mutter, J. C., 2005. The Earth sciences, human well-being, and the reduction of global poverty. In *EOS*, 86,16, 19 April, vol 157, pp. 164–165.

NatCatService, 2010. Geo Risks Research – Long-term statistics since 1950. http://www.munichre.com/en/ts/geo_risks/natcatservice/long-term_statistics_since_1950/default.aspx.

National Research Council, Board on Atmospheric Sciences and Climate, 1998. The atmospheric sciences entering the twenty-first century. Washington: National Academy Press, Vol. 364, p. 169.

Nicholls, N., and Alexander, L., 2007. Has the climate become more variable or extreme? Progress 1992–2006. *Progress in Physical Geography*, 31, 77–87.

Nicholls, R. J., Wong, P. P., Burkett, V. R., Codignotto, J. O., Hay, J. E., McLean, R. F., Ragoonaden, S., and Woodroffe, C. D., 2007. Coastal systems and low-lying areas. In Parry, M. L., Canziani, O. F., Palutikof, J. P., van der Linden, P. J., and Hanson, C. E. (eds.), *Climate Change 2007: Impacts, Adaptation and Vulnerability. Contribution of Working Group II to the Fourth Assessment Report of the Intergovernmental Panel on Climate Change.* Cambridge, UK: Cambridge University Press, pp. 315–356.

Nicholls, R. J., et al., 2008. Ranking port cities with high exposure and vulnerability to climate extremes: exposure estimates. Organisation for Economic Co-operation and Development, ENV/WKP(2007)1, 62 pp.

Nichols, N., 2001. Atmospheric and climatic hazards: improved monitoring and prediction for disaster mitigation. *Natural Hazards*, 23(2–3), 137–155.

Nigg, J. M., 2000. Predicting earthquakes: science, pseudoscience, and public policy paradox. In Sarewitz, D., Pielke, R. A., Jr., and Byerly, R. (eds.), *Prediction: Science Decision Making and the Future of Nature.* Washington, DC: Island Press, pp. 135–158.

Nolan, D. S., Rappin, E. D., and Emanuel, K. A., 2007. Tropical cyclogenesis sensitivity to environmental parameters in radiative-convective equilibrium. *Quarterly Journal of the Royal Meteorological Society*, **133**(629), 2085–2107.

O'Gorman, P. A., and Schneider, T., 2008. Energy of midlatitude transient eddies in idealized simulations of changed climates. *Journal of Climate*, **21**(22), 5797–5806.

Osborn, T. J., and Hulme, M., 1997. Development of a relationship between station and grid-box rainday frequencies for climate model evaluation. *Journal of Climate*, **10**(8), 1885–1908.

Pall, P., Allen, M. R., and Stone, D. A., 2007. Testing the Clausius-Clapeyron constraint on changes in extreme precipitation under CO_2 warming. *Climate Dynamics*, **28**, 351–363.

Palmer, W. C., 1965. *Meteorological drought*. Report 45, US Weather Bureau, Washington, DC.

Paton, D., and Johnston, D., 2006. *Disaster Resilience: An Integrated Approach*. Springfield: Charles C Thomas/Springfield. 6.

Paton, D., Johnston, D., Smith, L., and Millar, M., 2001. Responding to Hazard Effects: Promoting Resilience and Adjustment Adoption. *Emergency Management Australia*, p. 47

Rahmstorf, S., Cazenave, A., Church, J. A., Hansen, J. E., Keeling, R. F., Parker, D. E., and Somerville, R. C. J., 2007. Recent climate observations compared to projections. *Science*, **316**(5825), 709–709.

Robine, J. M., Cheung, S. L. K., Le Roy, S., Van Oyen, H., Griffiths, C., Michel, J. P., et al., 2008. Death toll exceeded 70,000 in Europe during the summer of 2003. *Comptes Rendus Biologies.*, **331**(2), 171–U5.

Rodriguez, J, Vos, F, Below, R., and Guha-Sapir, D., 2009. *Annual Disaster Statistical Review 2008 – The numbers and trends*. Centre for Research on the Epidemiology of Disasters. http://www.emdat.be

Sapir, D. G., and Vos, F., 2009. Quantifying global environmental change impacts: methods, criteria and definitions for compiling data on hydrometeorological disasters. In Brauch, H. G., et al. (eds.), *Coping with Global Environmental Change, Disasters and Security: Threats, Challenges, Vulnerabilities and Risks*. Berlin/Heidelberg/New York: Springer. Hexagon Series on Human and Environmental Security and Peace, Vol. 5.

Schipper, L., and Burton, I., 2009. *Adaptation to Climate Change: the Earthscan Reader*. London: Earthscan.

Seneviratne, S. I., Corti, T., Davin, E. L., Hirschi, M., Jaeger, E., Lehner, I., Orlowsky, B., and Teuling, A. J., 2010. Investigating soil moisture-climate interactions in a changing climate: a review. *Earth Science Reviews*, **99**(3–4), 125–161.

Shapiro, M. A., Shukla, J., Brunet, G., Nobre, C., Béland, M., Dole, R., Trenberth, K., Anthes, R., Asrar, G., Barrie, L., Bougeault, P., Brasseur, G., Burridge, D., Busalacchi, A., Caughey, J., Chen, D., Church, J., Enomoto, T., Hoskins, B., Hov, Ø., Laing, A., Le Treut, H., Marotzke, J., McBean, G., Meehl, G., Miller, M., Mills, B., Mitchell, J., Moncrieff, M., Nakazawa, T., Olafsson, H., Palmer, T., Parsons, D., Rogers, D., Simmons, A., Troccoli, A., Toth, Z., Uccellini, L., Velden C., and Wallace, J., 2010. An Earth-system prediction initiative for the 21st century. *Bulletin American Meteorological Society*, **91**, 1377–1388.

Simonovic, S. P., 2009. *Managing Water Resources: Methods and Tools for a Systems Approach*. Paris/London: UNESCO/Earthscan/James & James, p. 576.

Stott, P. A., Gillett, N. P., Hegerl, G. C., Karoly, D., Stone, D., Zhang, X., and Zwiers, F. W., 2009. Detection and attribution of climate change: a regional perspective. *Wiley Interdisciplinary Reviews: Climate Change*, **1**, 192–211.

Topics Geo, 2006 – Natural catastrophes 2006, analyses, assessments, positions. Copyright 2007 Münchener Rückversicherungs-Gesellschaft, Königinstrasse 107, 80802 München, Germany http://www.Munichre.com.

Topics Geo, 2008. Natural catastrophes 2008, analyses, assessments, positions. Copyright 2009 Münchener Rückversicherungs-Gesellschaft, Königinstrasse 107, 80802 München, Germany. http://www.munichre.com.

Trenberth, K. E., Jones, P. D., Ambenje, P., Bojariu, R., Easterling, D., Klein Tank, A., Parker, D., Rahimzadeh, F., Renwick, J. A., Rusticucci, M., Soden B., and Zhai, P., 2007. Observations: surface and Atmospheric climate change. In Solomon, S., Qin, D., Manning, M., Chen, Z., Marquis, M., Averyt, K. B., Tignor M., Miller H. L. (eds.), *Climate Change 2007: The Physical Science Basis*. Contribution of Working Group I to the Fourth Assessment Report of the Intergovernmental Panel on Climate. Cambridge, UK/New York, Cambridge University Press.

UN General Assembly, 1989. Resolution A/RES/44/236 http://www.un.org/documents/ga/res/44/a44r236.htm.

UNISDR, 1994. Yokohama strategy. http://www.unisdr.org/eng/about_isdr/bd-yokohama-strat-eng.htm.

UNISDR, 2005. Hyogo Framework for Action 2005-2015: Building the Resilience of Nations and Communities to Disasters. World Conference on Disaster Reduction. 2005, p. 6.

UNISDR, 2009a. United Nations international strategy for disaster reduction, terminology on disaster risk reduction http://www.unisdr.org/terminology.

UNISDR, 2009b. Global assessment report on disaster risk reduction. United Nations, Geneva, Switzerland, 2009. ISBN/ISSN: 9789211320282, 207 pp http://www.preventionweb.net/english/hyogo/gar/report/index.php?id=1130&pid:34&pih:2.

Wang, X. M., Yang, Y., Dong, Z. B., and Zhang, C. X., 2009. Responses of dune activity and desertification in China to global warming in the twenty-first century. *Global and Planetary Change*, **67**(3–4), 167–185.

Willett, K. M., Gillett, N. P. Jones, P. D., and Thorne, P. W., 2007. Attribution of observed surface humidity changes to human influence. *Nature*, **449**(7163), 710–712.

Wing, A. A., Sobel, A. H., and Camargo, S. J., 2007. Relationship between the potential and actual intensities of tropical cyclones on interannual time scales. *Geophysical Research Letters*, **34**, L08810.

World climate research programme: http://wcrp.wmo.int/wcrp-index.html.

World weather research programme: http://www.wmo.int/pages/prog/arep/wwrp/new/wwrp_new_en.html.

WSSD, 2002. *World summit on sustainable development*. Report of the World Summit for Sustainable Development, Johannesburg, South Africa, 26 Aug– 4 Sept 2002, A/CONF.199/20, See at: www.un.org.

Zhang, X., Wang, J., Zwiers, F. W., and Groisman, P. Y., 2010. The influence of large scale climate variability on winter maximum daily precipitation over North America. *Journal of Climate*, **23**, 2902–2915.

Cross-references

Avalanches
Beaufort Wind Scale
Climate Change
Cloud Seeding
Coastal Erosion
Coping Capacity
Cryological Engineering
Desertification
Doppler Weather Radar
Drought
Dust Bowl
Dust Storm
Dvorak Classification of Hurricanes
Fire and Firestorms
Flash Flood

Forest and Range Fires
Fujita Tornado Scale
Global Dust
Heat Waves
Hurricane
Hurricane Katrina
Ice Storms
International strategies for Disaster Reduction (IDNDR and ISDR)
Mega-Fires in Greece (2007)
Ozone
Permafrost
Queensland floods (2010–2011) and "Tweeting"
Sea Level Change
Snowstorm and Blizzard
Space Weather
Tornadoes
Wildfire

HYOGO FRAMEWORK FOR ACTION 2005–2015

Pedro Basabe
United Nations Complex, Nairobi, Kenya

Definition

Resilience: "The ability of a system, community, or society exposed to hazards to resist, absorb, accommodate to and recover from the effects of a hazard in a timely and efficient manner, including through the preservation and restoration of its essential basic structures and functions." UNISDR. Geneva 2009.

Introduction

Every year, millions of people, their livelihood, and assets are affected by droughts, floods, cyclones, earthquakes, volcanic eruptions, wildland fires, and other hazards. In early 2010, the earthquakes in Haiti, Chile, China, caused more than 300,000 casualties, and the volcanic eruption of Iceland's Eyjafjallajökull volcano paralyzed the air traffic and trade in Europe for almost 1 week. This reminds all how vulnerable people are to natural hazards and related disasters, and the respect that society needs to give to nature and its related hazardous phenomena.

Increased population densities, environmental degradation, and global warming adding to poverty will make the impacts of natural hazards even worse.

Whereas many know the human misery and crippling economic losses resulting from disasters, what few realize is that this devastation can be prevented through comprehensive disaster risk reduction (DRR) policies, mechanisms, programs, and measures on the ground.

This contribution summarizes the agreed framework to build the resilience of nations and communities to disasters as well as existing strategy and program for DRR in Africa which includes strategic areas of intervention, major areas of activities, expected results, and indicators to monitor progress.

Framework for disaster risk reduction

Governments around the world have committed to take action to reduce disaster risk, and adopted, in January 2005 at the World Conference on Disaster Reduction held in Kobe Japan, a guideline to reduce vulnerabilities associated with natural hazards, called the Hyogo Framework for Action (HFA). This was accomplished under the auspices of the United Nations International Strategy for Disaster Reduction (UNISDR).

The Hyogo Framework is the key instrument for implementing disaster risk reduction, it assists the efforts of nations and communities to become more resilient to, and cope better with, the hazards that threaten their development gains.

Its overarching goal is to build resilience of nations and communities to disasters, by achieving substantial reduction of disaster losses by 2015 – in lives, and in the social, economic, and environmental assets of communities and countries.

To attain this expected outcome, the HFA has three strategic objectives (Figure 1):

1. The more effective integration of disaster risk considerations into sustainable development policies, planning, and programming at all levels, with a special emphasis on disaster prevention, mitigation, preparedness and vulnerability reduction.
2. The development and strengthening of institutions, mechanisms, and capacities at all levels, in particular at the community level that can systematically contribute to building resilience to hazards.
3. The systematic incorporation of risk reduction approaches into the design and implementation of emergency preparedness, response, and recovery programs in the reconstruction of affected communities.

The HFA offers the following five areas of priorities for action, guiding principles, and practical means for achieving disaster resilience for vulnerable communities in the context of sustainable development.

1. Ensure that disaster risk reduction is a national and a local priority with a strong institutional basis for implementation.
2. Identify, assess, and monitor disaster risks and enhance early warning.
3. Use knowledge, innovation, and education to build a culture of safety and resilience at all levels.
4. Reduce the underlying risk factors.
5. Strengthen disaster preparedness for effective response at all levels.

In their approach to disaster risk reduction, States, regional and international organizations, and other actors concerned should take into consideration the key activities listed under each of these five priorities and should implement them, as appropriate, to their own circumstances and capacities.

Hyogo Framework for Action 2005–2015, Figure 1 Hyogo framework for action 2005–2015: building the resilience of nations and communities.

Since the adoption of the HFA, many global, regional, national, and local efforts have addressed disaster risk reduction more systematically. The UNISDR has facilitated guidelines, indicators, and a number of good practices, and promote implementation at all levels (Figure 2). Nevertheless much remains to be done.

One of the key achievements has been the establishment by Member States of multisectoral national platforms to coordinate disaster risk reduction programs and activities in countries. Many regional and subregional intergovernmental organizations have adopted strategies and programs for disaster risk management (Figure 3).

Hyogo Framework for Action 2005–2015, Figure 2 Words into action: a guide for implementing the hyogo framework.

The Africa strategy and program of action for disaster risk reduction (DRR)

Africa was the first continent that showed political commitment to disaster risk reduction by formulating in 2004 the "Africa Regional Strategy for Disaster Risk Reduction," "Programme of Action 2005–2010" for its implementation and "Guidelines for Mainstreaming Disaster Risk Assessment in Development," under the aegis of the African Union Commission, the NEPAD Secretariat, African Development Bank, and with support of the UNISDR and UN partners (Figures 4, 5, and 6). These documents were officially adopted at the First African Ministerial Conference on DRR in Addis Ababa in 2005.

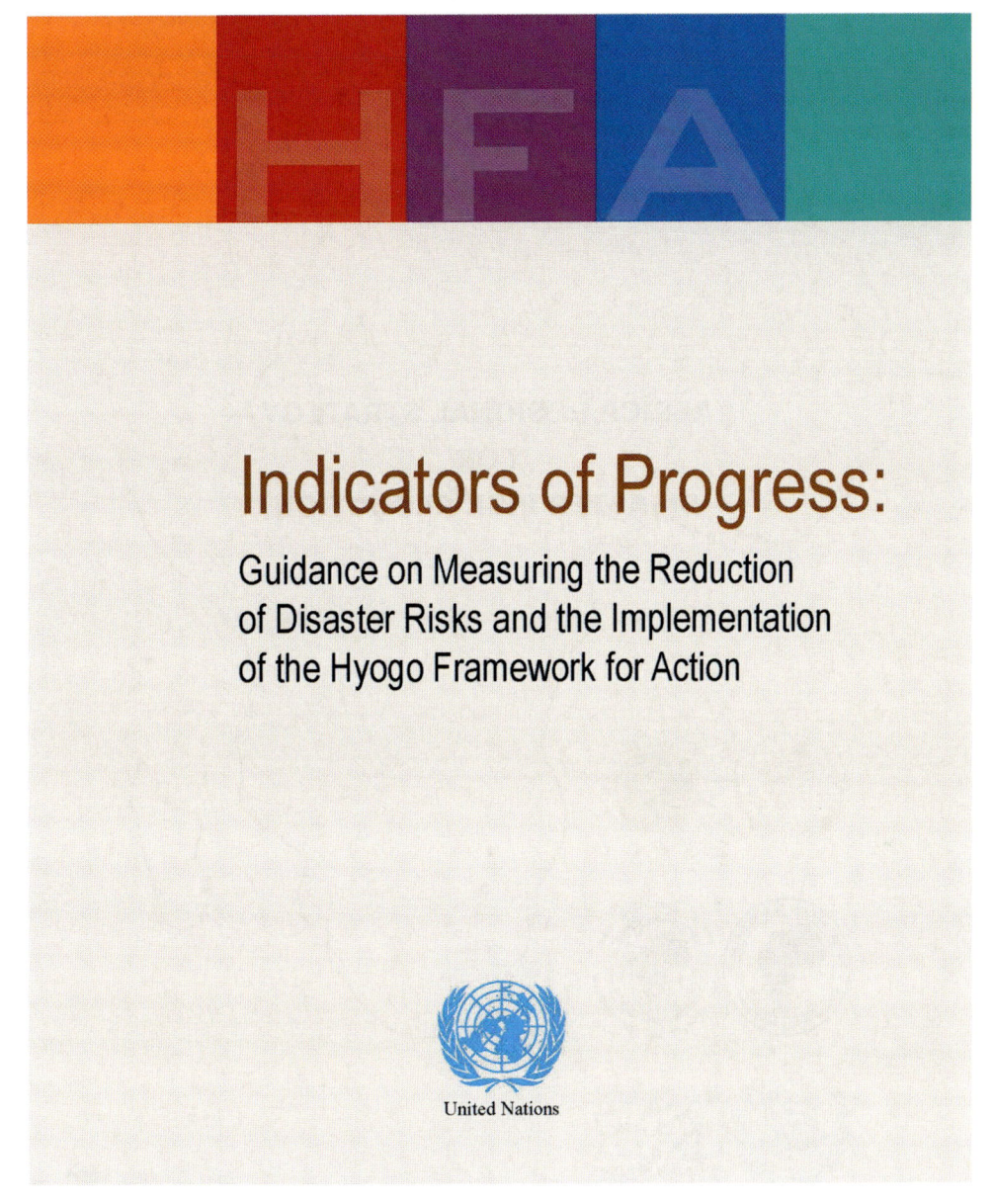

Hyogo Framework for Action 2005–2015, Figure 3 Indicators of progress: guidance on measuring the reduction of disaster risks and the implementation of the hyogo framework for action.

Africa has also advanced the implementation of the HFA and the Africa Strategy and Programme of Action at the subregional level. Several Regional Economic Communities (RECs) have engaged with DRR issues.

The Intergovernmental Authority on Development (IGAD), the Southern African Development Community (SADC), The Economic Community of Central African States (ECCAS), the Economic Community of West African States (ECOWAS), and the East African Community (EAC) and the Indian Ocean Commission (IOC) have developed strategies, policies, or programs for disaster risk reduction based on the priorities for action of the HFA and the Africa Regional Strategy for DRR. There have also been initiatives for South-South cooperation to build on successful experiences from across the region.

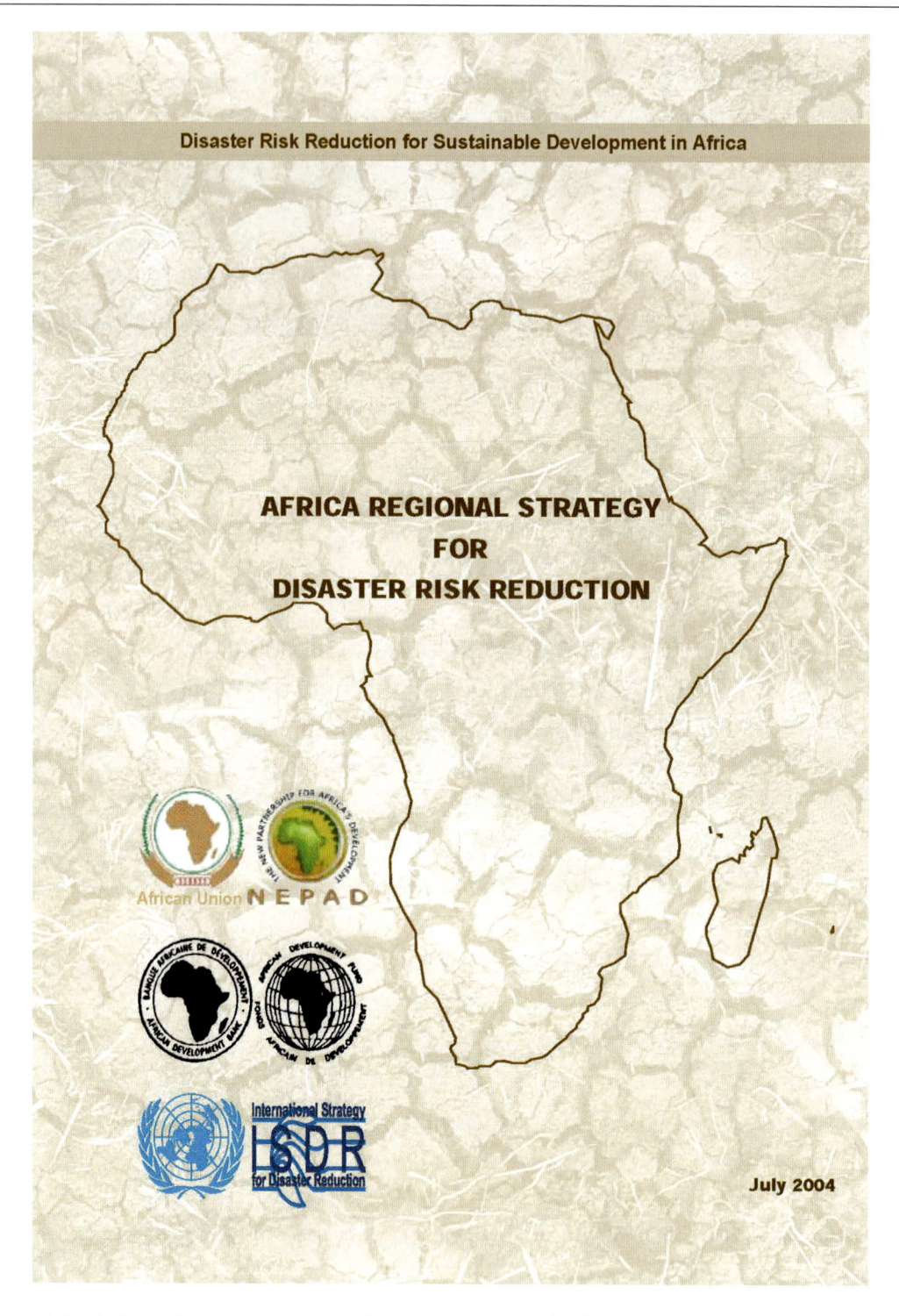

Hyogo Framework for Action 2005–2015, Figure 4 Africa regional strategy for disaster risk reduction.

In addition, specialized subregional institutions such as the IGAD Climate Prediction and Applications Centre (ICPAC), the Southern African Development Community's Climate Services Centre (SADC CSC), the AGRHYMET Regional Centre (ARC), and the African Centre of Meteorological Application for Development (ACMAD) are responding to a major regional and global challenge through enhanced services for DRR and climate change adaptation.

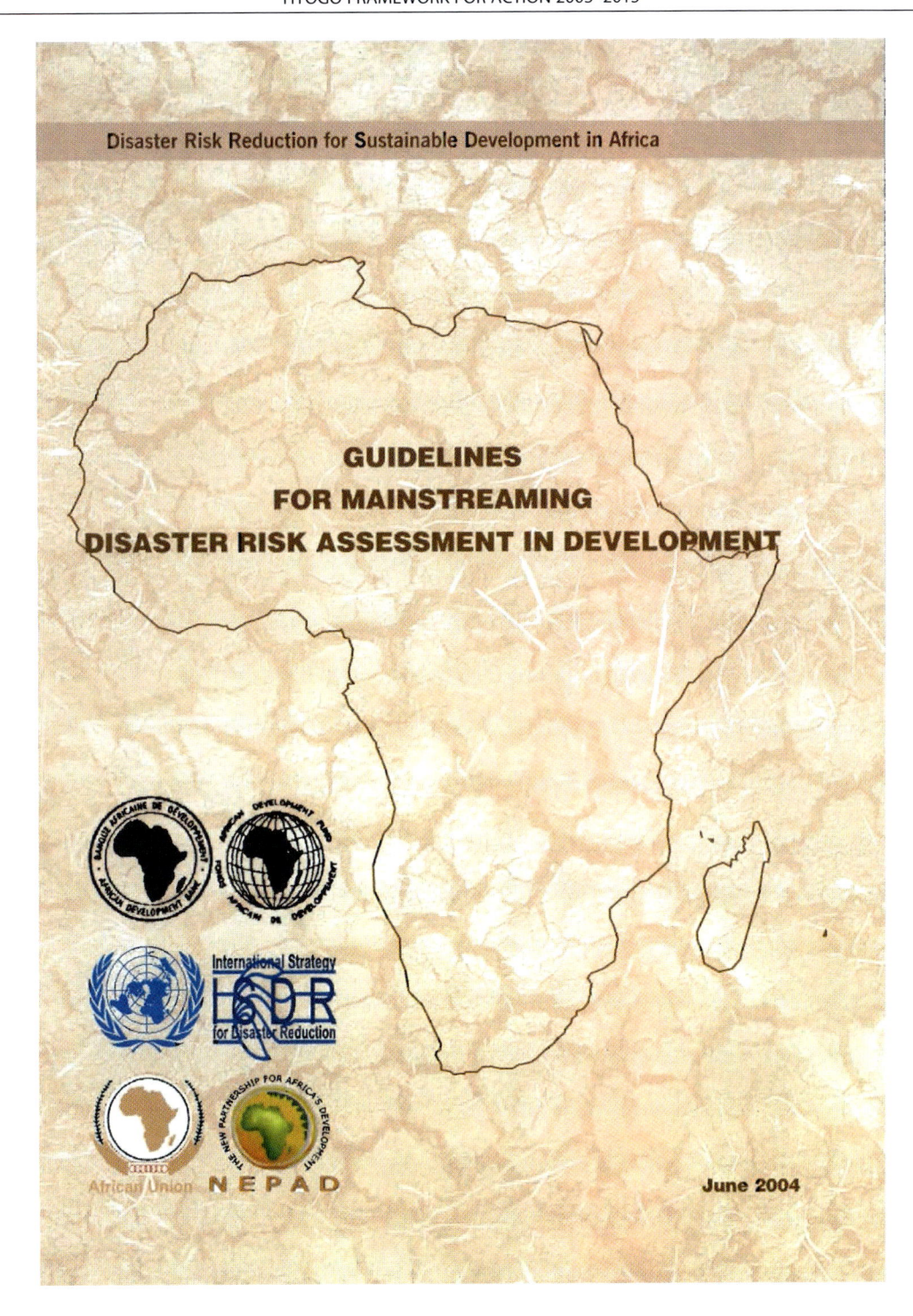

Hyogo Framework for Action 2005–2015, Figure 5 Guidelines for mainstreaming disaster risk assessment in development.

At the national level, governments in Africa have moved forward with the implementation of the HFA priorities for action and related regional objectives. Until the end of 2011, 35 countries have established or strengthened National Platforms for DRR or similar coordinating mechanisms. Several of them are working to develop legal frameworks and national plans or have included DRR topics in National Poverty Reduction Strategic Papers (PRSPs) and considered linkages between Climate change and DRR in the National Adaptation Programmes of Action

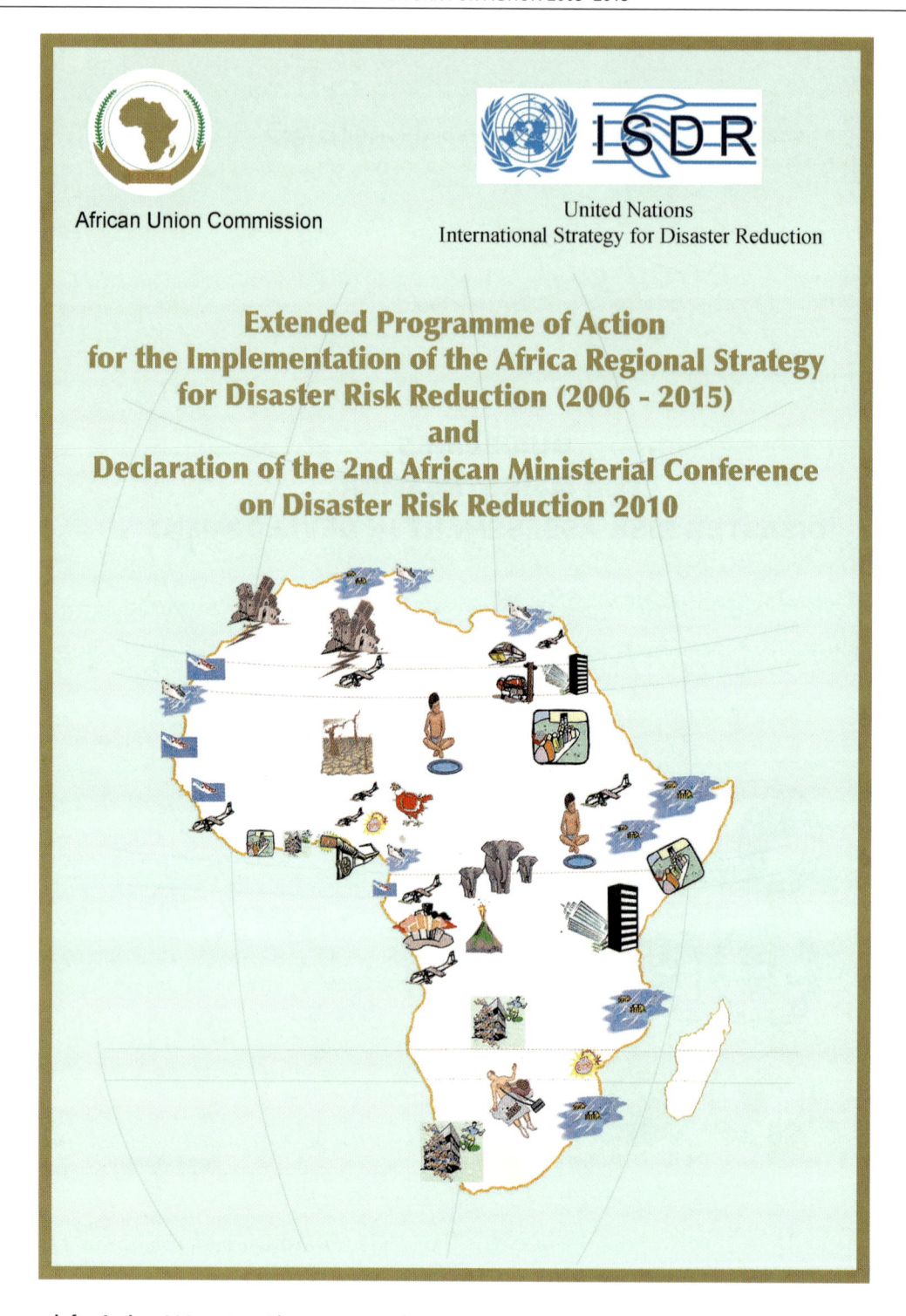

Hyogo Framework for Action 2005–2015, Figure 6 Extended program of action for the implementation of the Africa regional strategy for disaster risk reduction (2006–2015).

(NAPAs). More specifically, six African countries are working in multiyear strategic planning with UN assistance, including disaster risk management (Figures 7 and 8).

At the Second Africa Ministerial Conference held in April 2010 in Nairobi, Kenya, it was agreed to revise and extend the "Africa Programme of Action for the Implementation of the Africa Strategy for Disaster Risk Reduction till 2015 and better align with the HFA.

The extended Africa Programme of Action considers emerging challenges and the most frequent natural

Hyogo Framework for Action 2005–2015, Figure 7 Floodwaters in Madagascar after Cyclone Bingiza struck the Indian Ocean island of Madagascar on February 14, 2011 © Hannah McNeish/IRIN.

Hyogo Framework for Action 2005–2015, Figure 8 Dry earth in the desert plains of the Danakil depression in northern Ethiopia. © Siegfried Modola/IRIN.

hazards in Africa, such as drought and floods currently more frequent due to climate change factors and growing vulnerability. The program defined concrete activities, expected results and indicators to monitor progress.

Ministers also agreed to accelerate the implementation of the African Strategy and Programme of Action, including the officialization of mechanisms at national, subregional, and regional level to support implementation and concrete recommendations to invest in DRR and carry out systematic activities (Figure 9).

Extracts of the Africa ministerial declaration on DRR, 2010

Recommendation 7:
 To strongly urge Member States to increase their investments in disaster risk reduction through the allocation of a certain percentage of their national budgets and other revenue dedicated to disaster risk reduction and report to the next Ministerial Conference, considering other related African Ministerial resolutions;

Recommendation 8:
 To call upon development and humanitarian partners to ensure that disbursement of one percent (1%) of development assistance and ten percent (10%) of humanitarian assistance, in line with the Chair's Summary of the Second Session of the Global Platform, supports disaster risk reduction, preparedness and recovery, including from violent conflicts and/or severe economic difficulties;

Recommendation 12:
 To call upon Member States to undertake vulnerability assessments of schools, health facilities and urban centres, and develop and implement plans to ensure their safety and resilience;

Hyogo Framework for Action 2005–2015, Figure 9 A man carts a precious drum of water through a sandstorm in Tillaberi region in southwestern Niger © Jaspreet Kindra/IRIN.

Under priority one:
  4. *Compared with 2005 baseline, an increased number of countries have DRR in their PRSPs, NAPAs, and other relevant development plans.*

Under priority two:
  3. *At least 2 RECs have sub-regional hazard risk early warning systems and protocols for sharing such early warning information with countries.*
  4. *Number of cities with policies and/or strategies on Safer Cities and multi-dimensional vulnerability reduction, etc., etc.*

Under priority four:
  8. *Compared with the 2005 baseline, an increased number of countries with plans to ensure the safety of schools and hospitals.*

Bibliography

African Development Bank, African Development Fund, UNISDR, African Union, NEPAD, 2004a. *Africa Regional Strategy for Disaster Risk Reduction*.

African Development Bank, African Development Fund, UNISDR, African Union, NEPAD, 2004b. *Guidelines for Mainstreaming Disaster Risk Assessment in Development*.

African Union Commission, UNISDR, 2005. *Extended Programme of Action for the Implementation of the African Regional Strategy for Disaster Risk Reduction (2006–2015) and Declaration of the 2nd Ministerial Conference on Disaster Risk Reduction 2010*.

http://www.unisdr.org/africa/

http://www.unisdr.org/news/v.php?id=13655

UNISDR, 2005. *Hyogo Framework for Action 2005–2015: Building the Resilience of Nations and Communities to Disasters.* Extract from the final report of the World Conference on Disaster Reduction (A/CONF.206/6).

UNISDR, 2007. *Words into Action: A guide for Implementing the Hyogo Framework - Hyogo Framework for Action 2005–2015: Building the Resilience of Nations and Communities to Disasters*.

UNISDR, 2008. *Indicators of Progress: Guidance on Measuring Reduction of Disaster Risks and the Implementation of the Hyogo Framework for Action*.

HYPOCENTER

Maurice Lamontagne
Natural Resources Canada, Ottawa, ON, Canada

Synonyms

Earthquake focus; Focal point

Definition

A point beneath the Earth's surface where the vibrations of an earthquake originates and which corresponds to the location where the motion on a fault surface starts. The focal depth of earthquake hypocenters may vary from the surface down to nearly 700 km in subduction zones.

Discussion

Earthquake magnitude, and more precisely the Moment Magnitude, is function of the amount of slip and area of the fault surface that ruptures during an earthquake. For small earthquakes, such as magnitude 4 or less, or for earthquakes displayed on maps and cross-sections at a regional scale, the hypocenter is essentially a point with three coordinates (latitude, longitude, and focal depth). As the magnitude increases, the earthquake rupture is no longer a point but is rather a surface. In this latter case, the hypocenter is defined as the point where the rupture started.

The traditional method used to calculate the earthquake hypocenter is similar to that used for determining the epicenter, that is, by triangulation of epi- or hypo-central distances from recording stations. These distances are derived from arrival times of seismic phases assuming a velocity model for the seismic phases. The simplest method to calculate these distances relies on the difference between the arrival times of the S-wave and the P-wave at each station. The philosophy behind calculating an epicenter versus hypocenter is similar, the main exception being that the third dimension (i.e., the focal depth) is also considered in the triangulation.

To calculate the location of an earthquake, one has to determine the optimum values of four earthquake parameters: the latitude, the longitude, the depth of the hypocenter, and the time of occurrence of the event. The best estimates of these parameters are those that minimize the differences between the observed and the calculated arrival times (i.e., the residuals) for each phase recorded at seismograph stations (Kissling, 1988). The focal depth, essential to the calculation of a hypocenter, can also be calculated or estimated by other methods. A traditional one is to recognize reflected phases in teleseismic records of an earthquake, such as the pP phase (P phase reflected off the surface in the region of the epicenter) (Kulhánek, 2002). This can also be done for local earthquakes recorded at regional distances if intracrustal phases can be recognized or their arrival times measured. For local earthquakes, it is advisable to use near-field recordings of P and S phases and find the solution that minimizes the residuals (time difference between the recorded phase and the estimated phase assuming a given hypocenter and origin time). Crustal phases and the modeling of surface waves can also provide estimates of focal depths (Kulhánek, 2002). The presence of Rayleigh waves can indicate a near surface (less than 15 km depth) for a small earthquake (Kulhánek, 2002). Finally, macroseismic data can be used to infer focal depths (Musson and Cecić, 2002).

Bibliography

Kissling, E., 1988. Geotomography with local earthquake data. *Review of Geophysics*, **26**, 659–698.

Kulhánek, O., 2002. The structure and interpretation of seismograms. In Lee, W. H. K., Kanamori, H., Jennings, P. C., and Kisslinger, C. (eds.), *International Handbook of Earthquake and Engineering Seismology*. San Diego: Academic Press, Vol. 81A, pp. 333–348.

Musson, R. M. W., and Cecić, I., 2002. Macroseismology. In: Lee W. H. K., Kanamori, H., Jennings, P. C., and Kisslinger C. (eds.), *International Handbook of Earthquake and Engineering Seismology*. San Diego: Academic Press, Vol. 81A, pp. 807–825.

Cross-references

Body Wave
Earthquake
Epicenter
Global Seismograph Network (GSN)
Intensity Scales
Isoseismal
Magnitude Measures
Modified Mercalli (MM) Scale
Primary Wave (P Wave)
Secondary Wave (S Wave)
Tectonic Tremor

ICE AND ICEBERGS

Norm Catto
Memorial University of Newfoundland, St. John's, NL, Canada

Definition
An iceberg is a floating mass of ice.

Discussion

Ice poses hazards to mariners in three ways: as floating sea ice, icebergs, and riming. Floating ice and icebergs continue to pose hazards, despite advances in meteorological and ice forecasting, radar, and communication. Since the sinking of *Titanic* in 1912, more than 500 ice–ship collisions have occurred in the North Atlantic. Worldwide, 5–10 collisions occur annually.

Sea ice

Floating sea ice poses a hazard to ship traffic in several shipping lanes, including the North Atlantic, Arctic, and Southern Oceans. Sea ice is composed of freshwater, formed from surface freezing in coastal areas and embayments, ice lifted by waves and tides from beaches, and pieces which break off (calve) from glaciers and ice shelves that reach tidewater. Sea ice gradually accumulates throughout the winter. Although some areas of the Arctic and Antarctic support multiyear ice, which have survived the previous summer, ice which poses a hazard to mariners is dominantly single-year. The hazard is thus at its greatest from midwinter towards the end of winter and early spring, where the maximum amount of ice has accumulated and has moved toward the shipping lanes. Sea ice transported by currents flowing toward the Equator, notably the Labrador Current off eastern Canada, may extend more than 1,000 km seaward. In this area, the maximum southern extent typically is reached in early March, with ice extending to 44°N. Sea ice in this area may require the services of icebreakers to keep shipping lanes clear. Sea ice typically recedes from the northern tip of Labrador, Cape Chidley, in early July, opening Hudson Strait for navigation by vessels plying the Churchill (Manitoba, Canada)–Europe grain shipping route. The autumn sea ice expands southward to close off Hudson Strait to navigation in September.

For northern and Arctic coastal communities, the development of sea ice is critical for winter transportation and subsistence hunting. Unexpected variations in ice thickness, or breakup due to strong wave action, can pose hazards for travelers. Climate warming has resulted in inconsistencies in the timing, persistence, and thickness of sea ice in many Arctic areas. Open leads (*polyna*; *ashkui*) present in coastal areas can also pose hazards for travelers, although they are also suitable areas for harvesting fish and marine mammals.

Icebergs

Icebergs are composed of freshwater generated from calving of glaciers, which reach tidewater in the Arctic and Antarctic, in addition to isolated glaciers in mountainous coastal areas.

Icebergs are concentrated by the major ocean surface currents. In southernmost Davis Strait, one of the major areas of iceberg concentration, iceberg numbers have varied between 1,800 and 2,000 annually since 1990. Traveling with velocities of 1–4 km/h, icebergs can persist for several years from the point of calving, and reach latitudes of 40°N or S. However, icebergs melt rapidly in water warmer than 2°C. An iceberg 40 m high above the water line and 100 m long (approximate volume 2,000,000 m^3 and mass approximately 10 million ton) will completely ablate in 24 days in 2°C water, and requires only 15 days in 4°C water. Such large icebergs make up approximately

Ice and Icebergs, Figure 1 This iceberg drifted south to St. John's, NL, Canada.

Ice and Icebergs, Figure 2 Floating Sea Ice, Cape Kellett, Banks Island, Arctic Canada.

20% of the iceberg population in the southern Davis Strait, and about 5–6% at the mouth of the Strait of Belle Isle. Smaller fragments ("growlers") increase in proportion southwards as the larger icebergs disintegrate, but an average "growler" will endure for less than 5 days in 4°C water. Less than 10% of the icebergs that enter Labrador waters from the southern Davis Strait survive to 52°N; less than 1 in 10,000 potentially would reach the latitudes of the northernmost trans-Atlantic shipping routes.

Riming

Riming occurs where super-cooled spray from waves freezes on contact with objects. The process is analogous to the occurrence of freezing rain over land, but the proximity of the waves can result in much more rapid riming, causing vessels to become top-heavy and capsize. Ships and oil platforms operating in northern waters are vulnerable to sea-spray riming. Crews on vessels encountering freezing sea spray need to work rapidly to remove the ice in any way possible to prevent riming. Riming can also occur along the coastline, causing damage to property.

Bibliography

C-CORE, 2005. *Calculation of iceberg collision risk during ice-free season*. Canadian Centre for Cold Ocean Research, St. John's, NL, Canada, Report R-04-093-341.

Drinkwater, K. F., 2004. Atmospheric and sea-ice conditions in the Northwest Atlantic during the decade, 1991–2000. *Journal of Northwest Atlantic Fishery Science*, **34**, 1–11.

Hyndman, D., Hyndman, D., and Catto, N. R., 2008. Winter hazards. In *Natural Hazards and Disasters*. Toronto: Nelson, pp. 349–352. Chap. 13.

Cross-references

Beach Nourishment (Replenishment)
Breakwater
Challenges to Agriculture
Climate Change
Coastal Erosion
Critical Infrastructure
Early Warning Systems
Fog Hazards
Ice Storm
Marine Hazards
Permafrost
Sea Level Change
Snowstorm and blizzard
Storm Surges
Storms
Tidal Bores
Warning Systems

ICE STORMS

Ronald E. Stewart
University of Manitoba, Winnipeg, MB, Canada

Synonyms

Freezing precipitation storms; Freezing rain storms; Icing events

Background

Winter storms in which the temperature falls below 0°C commonly occur over many regions of the world. The form of precipitation within such storms is often variable and may include snow (dry or wet), freezing rain, ice pellets, and rain (Stewart, 1992). These various forms of precipitation can occur simultaneously in a storm.

Winter precipitation that contains some liquid may freeze when it strikes a cold surface; such an event is referred to as an ice storm. The result of this icing is hazardous and sometimes catastrophic (Henson et al., 2007).

Definitions
There are several types of winter precipitation (<0°C) that contain or are composed entirely of liquid. Terms to describe some of these are defined formally by the World Meteorological Organization, and are used in all countries, including Canada; however, some are not defined formally. For example, terms such as slush and liquid core pellets were only coined by researchers in the last few years. The precipitation types containing liquid that can occur below 0°C are summarized below. Freezing Rain – Rain that falls in liquid form but freezes upon impact to form a coating of glaze upon the ground and on exposed objects (Environment Canada, 1992). Ice Pellets – A type of precipitation consisting of transparent or translucent pellets of ice, 5 mm or less in diameter (Environment Canada, 1992). Wet Snow – Snow that contains a great deal of liquid water (Environment Canada, 1992). Slush – Precipitation composed of a mixture of liquid and ice in which the original snowflake's shape is not discernible. (Theriault et al., 2006). Liquid Core Pellets: Partially refrozen drop with an ice shell and liquid water within it (Theriault and Stewart, 2007).

Formation
Winter precipitation is formed through several different processes. A particular type of temperature profile, consisting of a warm layer of air (>0°C) aloft and a cold layer below (<0°C), is required to form some of the hazardous types of freezing precipitation (e.g., Zerr 1997). These two atmospheric layers are called the melting and the refreezing layers, respectively. Snowflakes falling through a melting layer either melt or partially melt, depending on their size and atmospheric conditions of the layer. Depending on the degree of melting, these particles may or may not refreeze completely when falling through the refreezing layer before reaching the surface. In addition, freezing drizzle (small supercooled drops) can be formed by all-liquid processes within an atmosphere completely below 0°C.

Several types of precipitation containing liquid can occur at temperatures below 0°C and/or coexist because of the varying precipitation size distribution aloft. For example, when falling through a given melting layer, smaller particles will be more likely to melt completely, whereas larger ones will only partially melt. The completely melted particles will reach the surface as liquid (freezing rain) if the refreezing region is not cold enough (approximately >−5°C) to initiate freezing through ice nucleation. Larger particles may melt only partially and these will at least begin to freeze in the refreezing layer; if freezing is complete they would normally fall as ice pellets and if incomplete, as liquid core pellets. It may also be that the largest particles only melt a small amount in the melting layer, resulting in wet snow falling into the refreezing layer. These particles may not refreeze entirely prior to reaching the ground, where they do finally freeze. Such explanations highlight how icing at the surface can be due to the freezing of several types of particles. Freezing precipitation is a catch-all phrase to describe such particles.

Collisions between the different types of precipitation can alter their relative amounts. For instance, Hogan (1985) states that the freezing of a falling supercooled drop can be initiated by a collision with an ice crystal. In this case, it will decrease the amount of supercooled rain and ice crystals and ice pellets tend to be formed. As well, Stewart et al. (1990) showed that collisions between supercooled liquid drops and ice pellets could significantly decrease the amount of freezing rain at the surface.

Freezing precipitation is normally associated with a surface warm front, although under some conditions it may also be associated with a cold front. The warm frontal circulation with warm air rising above cold air is conducive to generating the necessary temperature profile for freezing precipitation. The ensuing band of freezing precipitation generally varies from about 10 to 100 km in width and can extend hundreds of kilometers along a front. More intense storms with stronger warm and cold air circulations generally lead to more favorable conditions for more intense and/or wider bands of freezing precipitation.

There is often a well-defined organization to the evolution of the forms of precipitation in the transition region. Stewart and King (1987) showed that a five-step evolution from snow to ice pellets and then to freezing rain is expected under simple conditions and one of these steps actually includes all three precipitation types occurring simultaneously (freezing rain, ice pellets or liquid core pellets, and wet snow). This progression arises because of the systematic warming aloft and cooling below as a warm front passes.

The actual form of precipitation also depends upon local features. Ascending and descending air induced by local features leads to adiabatic cooling and heating, respectively, and this can alter the form of precipitation through changes in the vertical temperature profile. The presence of valleys and surface topographic effects can, for example, channel low-level air into particular locations and subsequently alter the forms of precipitation. This process contributed to the 1998 ice storm over Montreal and surrounding areas, one of Canada's largest disasters (Henson et al., 2007).

Summary
Freezing precipitation is a hazardous form of precipitation that causes major impacts whenever and wherever it occurs. It is a natural feature of winter weather and there is also growing concern as to how it may change in occurrence within a future climate. If winter storms become more intense, this may well lead to more ice storms.

Bibliography

Canada, E., 1992. *Manual of Surface Weather Observations (MANOBS), User's Manual*. Ontario: Meteorological Service of Canada, Downsview.

Henson, W. L., Stewart, R. E., and Kochtubajda, B., 2007. On the precipitation and related features of the 1998 Ice Storm in the Montreal area. *Atmospheric Research*, **83**, 36–54.

Hogan, A. W., 1985. Is sleet a contact nucleation phenomenon? In *Proceeding of the Eastern Snow Conference*, Montreal, pp. 292–294

Stewart, R. E., 1992. Precipitation types in the transition region of winter storms. *Bulletin of the American Meteorological Society*, **73**, 287–296.

Stewart, R. E., and King, P., 1987. Freezing precipitation in winter storms. *Monthly Weather Review*, **115**, 1270–1279.

Stewart, R. E., Crawford, R. W., Donaldson, N. R., Low, T. B., and Sheppard, B. E., 1990. Precipitation and environmental conditions during accretion in Canadian East Coast winter storms. *Journal of Applied Meteorology*, **29**, 525–538.

Theriault, J., and Stewart, R. E., 2007. On the effect of vertical air motion on winter precipitation types. *Natural Hazards and Earth System Sciences*, **7**, 231–242.

Theriault, J., Stewart, R. E., and Mildebrandt, J. A., 2006. On the simulation of winter precipitation types. *Journal of Geophysical Research*, **111**, D18202, doi:10.1029/2005JD006665.

Zerr, R. J., 1997. Freezing rain: an observational and theoretical study. *Journal of Applied Meteorology*, **36**, 1647–1661.

Cross-references

Avalanches
Climate change
Cloud Seeding
Cryological Engineering
Ice and Icebergs
Snowstorm and Blizzards
Storms
Thunderstorms

IMPACT AIRBLAST

Natalia Artemieva
Planetary Science Institute, Tucson, AZ, USA

Definition

Impact airblast is a shock wave in the atmosphere caused either by a cosmic body entering the Earth with hypersonic velocity or by an expanding impact plume.

Discussion

Damage effects of an airblast depend on the peak overpressure (the maximum pressure in excess of the ambient atmospheric pressure) and the gas velocity behind the shock (commonly known as the wind speed). The effects of impact-generated airblasts are based on data from US nuclear explosion tests (Glasstone and Dolan, 1977). Some objects are sensitive to overpressure: 0.03–0.07 bar is enough to shatter a glass window, 0.2–0.7 bar – to destroy a brick wall panel; other objects are sensitive to the wind speed: ~30% of trees are blown down at 40 m/s and 90% at 60 m/s (a *tornado* wind speed is usually lower than 30 m/s). The user-friendly Web site: www.lpl.arizona.edu/impacteffects provides estimates of impact airblast effects (Collins et al., 2005).

Although any hypervelocity cosmic body hitting Earth produces shock waves, the amount of damage on the surface depends on body size. Small bodies (diameters <50 m) are efficiently decelerated in the upper atmosphere; shock waves decay quickly and reach the surface as a package of acoustic waves. These waves may be used to estimate the meteoroid's parameters. Larger bodies (50–300 m) are decelerated at lower altitudes and strong shock waves may reach the surface. A classic example of such an airblast is the Tunguska explosion in 1908 (Vasilyev, 1998). The ground effect of this event is a butterfly-shaped region of devastated forest extending over 2,000 km^2. Numerical models (Artemieva and Shuvalov, 2007; Boslough and Crawford, 2008) show that the Tunguska airblast was caused by a 50–100 m diameter *asteroid* or *comet* releasing an energy of 5–20 Mton TNT (1 ton TNT is equivalent to the energy of $4.18 \cdot 10^9$ J) at altitudes of 5–15 km. No crater or projectile material has been confirmed at the impact site so far. The largest objects (>300 m) reach the surface and excavate an impact crater (*asteroid impact*). The hypervelocity expansion of an impact plume and *impact ejecta* also creates shock waves in the atmosphere. The damage effects are combined with other environmental consequences of impacts – *impact wildfires, earthquakes, tsunami, and impact ejecta*.

Bibliography

Artemieva, N., and Shuvalov V., 2007. 3D effects of Tunguska event on the ground and in atmosphere. In *Proceeding of Lunar and Planetary Science Conference*, abstract 1537.

Boslough, M. B. E., and Crawford, D. A., 2008. Low-altitude airbursts and the impact threat. *International Journal of Impact Engineering*, **35**, 1441–1448.

Collins, G. S., Melosh, H. J., and Marcus, R. A., 2005. Earth Impact Effects Program: a Web-based computer program for calculating the regional environmental consequences of a meteoroid impact on Earth. *Meteoritics & Planetary Science*, **40**, 817–840.

Glasstone, S., and Dolan, P. J., 1977. *The effects of nuclear weapons*, 3rd edn. Washington: United States Department of Defense and Department of Energy.

Vasilyev, N. V., 1998. The Tunguska meteorite problem today. *Planetary and Space Science*, **46**, 129–150.

Cross-references

Asteroid
Asteroid Impact
Comet
Earthquakes
Impact Ejecta
Impact Wildfires
Tornado
Tsunami

IMPACT EJECTA

Christian Koeberl
University of Vienna, Vienna, Austria
Natural History Museum, Vienna, Austria

Definition

Impact ejecta are sediments comprising material that was thrown out from an impact crater during its formation and deposited around it. Impact ejecta are in general rather heterogeneous due to variations in target composition, and often include glassy materials (impact glasses) and shocked minerals, and, in some cases, a minor meteoritic component.

Discussion

In contrast to many other planets (and moons) in the solar system, the recognition of impact craters on the Earth is difficult, because active geological and atmospheric processes on our planet may obscure or erase the impact record in geologically short times. Impact craters are recognized from the study of actual rocks – remote sensing can only provide supporting information. Petrographic studies of rocks at impact craters can lead to the discovery of impact-characteristic, shock metamorphic effects, whereas geochemical studies may yield information on the presence of meteoritic components in these rocks.

Apart from studying meteorite impact craters, significant information can also be gained from the study of impact ejecta. Such ejecta are found within the normal stratigraphic record, where they can provide excellent time markers, and allow one to relate an impact event directly to possible biological effects. Impact ejecta are commonly divided into two groups – proximal ejecta (those that are deposited closer than 5 crater radii from the crater rim) and distal ejecta. In some cases, impact events have been identified solely from the discovery and study of regionally extensive or globally distributed impact ejecta.

A well-known case in point is the Cretaceous–Tertiary (K-T, now usually called the K-Pg = Cretaceous-Paleogene boundary), where the discovery of an extraterrestrial signature, together with the presence of shocked minerals, led not only to the identification of an impact event as the cause of the end-Cretaceous mass extinction, but also to the discovery of a large buried impact structure about 200 km in diameter, the Chicxulub structure (Yucatan peninsula, Mexico).

The first physical evidence pointing to a contribution of extraterrestrial material that was discovered was the presence of anomalously high PGE abundances in K-T boundary clay in Italy and other locations around the world. The contents of Ir and other PGEs were found to be enriched in these K-T boundary clay layers by up

Impact Ejecta, Figure 1 The figure shows the famous K-T boundary location at Gubbio, Italy, where the initial evidence was discovered that a large-scale impact event occurred at the end of the Cretaceous. Within a succession of layered limestone a thin (1–2 cm) clay layer was found, which contains shocked minerals (such as shocked quartz, similar to the upper inset, which shows a secondary electron microphotograph of an etched quartz grain with shock lamellae or planar deformation features), as well as distinct enrichments in the element iridium and other platinum group elements of extraterrestrial origin (*lower inset*).

to four orders of magnitude compared to average terrestrial crustal abundances; inter-element ratios of the PGEs in K-T boundary clay samples are very similar to the values observed in chondritic meteorites, and osmium and chromium isotopic studies of the K-T boundary provided further evidence of an extraterrestrial component. Further evidence for impact are: shocked minerals (including shocked quartz and shocked zircon), impact glass (some fresh, some devitrified), impact-derived diamonds, and spinel. The source crater of the K-T boundary ejecta, the ca. 200 km-diameter Chicxulub structure in Mexico (also subject of ICDP drilling projects) was only discovered in the early 1990s and is now firmly linked to the K-T ejecta by geochemical evidence.

Another well-known ejecta layer occurs in late Eocene marine sediments around the world that contain evidence for at least two closely spaced impactoclastic layers, which are linked to the Popigai and Chesapeake Bay impact craters (100 and 85 km diameter, respectively).

Tektites are another form of distal impact ejecta, the source craters of which have long remained elusive. To date only three of the four known Cenozoic tektite-strewn fields have been connected to source craters. Distal ejecta ("impactoclastic layers") can be used as markers for impact events in the stratigraphic record. "Impact markers" are a variety of chemical, isotopic, and mineralogical species derived from the encounter of cosmic bodies (such as cometary nuclei or asteroids) with the Earth. Distal ejecta layers can be used to study a possible relationship between biotic changes and impact events, because it is possible to study such a relationship in the same outcrops, whereas correlation with radiometric ages of a distant impact structure is always associated with larger errors. The discovery and detailed study of distal ejecta layers have led to the discovery of previously unknown large impact structures (e.g., Chicxulub and Acraman).

Bibliography

Montanari, A., and Koeberl, C., 2000. *Impact Stratigraphy: The Italian Record. Lecture Notes in Earth Sciences.* Heidelberg: Springer Verlag, Vol. 93, 364 pp., ISBN 3-540-66368-1.

Simonson, B. M., and Glass, B. P., 2004. Spherule layers – records of ancient impacts. *Annual Review of Earth and Planetary Sciences*, **32**, 329–361.

Cross-references

Asteroid
Asteroid Impact
Asteroid Impact Mitigation
Asteroid Impact Predictions
Comet
Impact Airblast
Impact Fireball
Impact Firestorms
Impact Tsunami
Impact Winter

IMPACT FIREBALL

Peter Brown
Centre for Planetary Science and Exploration (CPSX), University of Western Ontario, London, ON, Canada

Definition

Impact fireball is the collective light, heat, and shock phenomena created when a relatively large solid object ("meteoroid") encounters the atmosphere at high speed (>11 km/s).

Discussion

Surface damage from fireballs surviving atmospheric passage to the ground with a substantial portion of their original kinetic energy is usually confined to larger stony objects (the precise cutoff is poorly known, but believed to be in the range of >100 m) or small (down to a few meters in size) iron meteoroids (which comprise about 3% of the impacting population (Ceplecha et al., 1998)). Damage at ground level may also result from smaller fireballs that are stopped higher in the atmosphere due to airblast effects of the shock propagating to the ground.

The direct danger from fireballs increases with size – a 100 m diameter object (producing a very large fireball) is believed to impact the Earth roughly once per 7,000 years whereas a 2 m iron object collides with the Earth about once every 5 years (Brown et al., 2002).

In addition to the danger at ground level caused by direct impact effects, it has been proposed that impact fireballs of sufficiently large energy may produce ionospheric and magnetospheric disturbances, which could affect radio propagation. The debris trail left by very large fireballs consists of significant amounts of micron-sized meteoric dust that may cause short-lived localized cooling at the Earth's surface while production of NO at the impact site for very large fireballs may affect ozone on regional scales (Adushkin and Nemchinov, 1994). Finally, impact fireballs may be misinterpreted as nuclear airbursts and could accidentally trigger hostile responses from nation states unable to distinguish between impact fireballs of natural origin and nuclear detonations (Tagliaferri et al., 1994).

Bibliography

Adushkin, V. V., and Nemchinov, I. V., 1994. Consequences of impacts of cosmic bodies on the surface of the Earth. In Gehrels, T. (ed.), *Hazards due to Comets and Asteroids.* Tucson: University of Arizona Press, pp. 721–778.

Brown, P., Spalding, R. E., ReVelle, D. O., Tagliaferri, E., and Worden, S. P., 2002. The flux of small near-earth objects colliding with the Earth. *Nature*, **420**, 294–296.

Ceplecha, Z., Borovička, J., Elford, W. G., Revelle, D. O., Hawkes, R. L., Porubčan, V., and Šimek, M., 1998. Meteor phenomena and bodies. *Space Science Reviews*, **84**, 327–471.

Tagliaferri, E., Spalding, R. E., Jacobs, C., Worden, S. P., and Erlich, A., 1994. Detection of Meteoroid impacts by optical sensors in Earth orbit. In Gehrels, T. (ed.), *Hazards due to Comets and Asteroids*. Tucson: University of Arizona Press, pp. 199–220.

Cross-references

Asteroid
Asteroid Impact
Comet
Earthquakes
Impact Airblast
Impact Ejecta
Impact Wildfires
Space Weather
Tornado
Tsunami

IMPACT FIRESTORMS

Tamara Goldin
Center for Earth Sciences, University of Vienna, Vienna, Austria

Synonyms

Impact wildfires

Definition

Impact firestorms are a possible environmental effect of large meteorite impacts and a consequence of the large amount of heat released by an impact event, which may ignite wildfires. Locally, impact firestorms can result due to heat radiated near the impact site by the ablating impactor and the rising fireball (Toon et al., 1997). On a larger scale, impact firestorms can result from heat radiated by high-speed distal ejecta as they reenter, decelerate, and heat up in the Earth's atmosphere.

Discussion

The discovery of soot in distal ejecta deposits (Wolbach et al., 1985) from the Chicxulub impact at the end of the Cretaceous (65 Ma) first suggested that global wildfires raged on the continents immediately following the impact event. Calculations suggested that the ejecta, transported around the globe at hypersonic speeds (5–10 km/s in distal localities), radiated enough thermal energy to ignite the world's forests (Melosh et al., 1990) and thus would have been a major contributor to the environmental catastrophe following Chicxulub or any impact of this magnitude. However, the impact firestorm hypothesis for Chicxulub has faced recent criticism. New soot analyses support an origin from fossil hydrocarbons in the target rocks at Chicxulub (Harvey et al., 2008), not woody biomass. Furthermore, numerical modeling of the atmospheric reentry of ejecta suggests that ejecta particles settling through the atmosphere shield the Earth's surface from much of the downward thermal radiation emitted from later-entering ejecta and reduce the thermal pulse at the Earth's surface below the limits for wood ignition (Goldin and Melosh, 2009). Although the heat pulse would still cause significant environmental damage, global impact firestorms may not result from a Chicxulub-sized impact. However, wildfires in the vicinity of the impact will still occur, as was observed following the 1908 Tunguska airburst in Siberia.

Bibliography

Goldin, T. J., and Melosh, H. J., 2009. Self-shielding of thermal radiation by Chicxulub impact ejecta: firestorm or fizzle? *Geology*, **37**, 1135–1138.
Harvey, M. C., Brassell, S. C., Belcher, C. M., and Montanari, A., 2008. Combustion of fossil organic matter at the Cretaceous-Paleogene (K-P) boundary. *Geology*, **36**, 355–358.
Melosh, H. J., Schneider, N. M., Zahnle, K. J., and Latham, D., 1990. Ignition of global wildfires at the Cretaceous/Tertiary boundary. *Nature*, **343**, 251–254.
Toon, O. B., Zahnle, K., Morrison, D., Turco, R. P., and Covey, C., 1997. Environmental perturbations caused by the impacts of asteroids and comets. *Reviews of Geophysics*, **35**, 41–78.
Wolbach, W. S., Lewis, R. S., and Anders, E., 1985. Cretaceous extinctions; evidence for wildfires and search for meteoritic material. *Science*, **230**, 167–170.

Cross-references

Asteroid
Asteroid Impact
Asteroid Impact Mitigation
Comet
Impact Airblast
Impact Ejecta
Impact Fireball
Impact Tsunami
Meteorite
Wildfires

IMPACT TSUNAMIS

Galen Gisler
University of Oslo, Oslo, Norway

Synonyms

Meteor impact wave; Water crater

Definitions

Impact tsunami. A water wave produced by the impact of a large meteor (asteroid or comet) into a large body of water, the ocean, for example. Impact tsunamis may prove destructive to persons or property on coastlines near the impact site.

Introduction

Collisions between bodies in the Solar System occur very frequently. When an asteroid or comet collides with a rocky planet (e.g., Mercury, Venus, the Earth, the Moon, and Mars), it can produce an impact crater. On the Moon, craters have been recognized since the beginnings of the telescopic age. Mars and Mercury were found to have craters early in the space age, and craters on Venus have been seen using high-resolution radar observations through its dense cloud cover. There are now roughly 200 impact craters recognized on Earth, all on dry land or on continental shelves. For reviews, see Holsapple (1994) and Melosh (1989). A list of terrestrial impact craters is maintained on the website http://www.lpi.usra.edu/science/kring/epo_web/impact_cratering/World_Craters_web/intromap.html.

Because the Earth's surface is covered 70% by ocean, most asteroid or comet impacts have occurred in the ocean, and will do so in the future. Unfortunately, up to now, no definite crater sites on the abyssal plains of deep oceans have been identified and confirmed. Tentative identifications in the literature have been accompanied by intense debate (Masse et al., 2006; Pinter and Ishman, 2008). It is much more difficult to survey the seafloor to the same extent as dry land, of course. But because water absorbs so much of a projectile's energy by moving out of the way and vaporizing, very little energy is left to penetrate the seafloor unless the projectile's size is a substantial fraction of the water depth. In contrast, a few underwater craters are known to exist on the continental shelf. One is the Mjølnir crater, off the northern coast of Norway (Tsikalas et al., 2002) and another is the Montagnais crater, off Nova Scotia (Jansa et al., 1989).

When an impact occurs in the deep ocean, a transient crater is produced as the water is pushed out of the way. Some water is vaporized, and some water is lifted into a "crown splash" high up into the atmosphere immediately around the transient crater, much like the splash made when a rock is thrown into a pond. Because a water crater is unstable, it soon closes in on itself and produces a vertical jet from its center, which can also be seen in the pond's rock splash.

The collapse of the crown splash and the subsequent collapse of the central vertical jet both produce waves that emanate from the impact site. These are often referred to as *impact tsunamis*, but they are very different in many important respects from classical tsunamis caused by earthquakes or underwater landslides.

Probability of impact

In the 4.6 billion year history of the Earth and Moon, these bodies have been struck by asteroids or comets many thousands of times. Most impacts are small ones, and the big ones are very rare. The rate of impact has decreased dramatically over time. Early on, the solar system was cluttered with millions of multi-kilometer-sized bodies. Collisions between the major planets and smaller bodies gradually cleaned out the vast majority of the latter. The probability of an Earth impact by a body of a particular size in the future can be estimated from the present number of bodies of that size in the asteroid population. According to Morrison et al. (2002), Earth impacts of kilometer-sized bodies occur once in a million years on average, whereas bodies of a hundred meter diameter strike the Earth every few centuries. Bodies smaller than a hundred meters in size enter the Earth's atmosphere more frequently, but most of these burn up as meteors or explode high in the atmosphere, occasionally leaving fragments on the ground as meteorites.

Physics of water impacts

The physics of impact cratering in general is covered extensively in the book *Impact Cratering* by Melosh (1989). Cratering in rock differs from cratering in water only in the late stages, when the reduced mobility of rock results in a permanent change to the landscape. A high-speed impact into water initiates a shock wave that compresses and heats the water in an expanding hemispherical region around the projectile. In the rarefaction wave behind the shock, a significant amount of water is vaporized, as is much of the projectile itself. Water expands explosively when suddenly vaporized, to a volume a thousand times greater than it occupied when still liquid, and the pressure of this expansion drives the development of the water crater. The transient crater expands to a diameter twenty or more times the diameter of the asteroid within a few seconds. This crater is very symmetrical even for fairly high-angle impacts, though the crown splash may not be (see Figure 1).

The pressure of the water surrounding the transient crater forces a refilling of the crater. Water rushes into the center from all directions and collides with itself at high speed, producing a jet that can rise upward many kilometers, even into the stratosphere. Meanwhile, the collapse of the crown splash develops into a rim wave that moves outward in all directions. The central jet falls back into the center and produces a second, smaller, transient crater, which in turn rebounds into a second jet. This may go on for several cycles for a very large impact. Each rebound produces a new wave.

Wave production in impacts

The waves that are produced in impacts result from the collapse of the crown splash and the collapse, possibly multiple times, of the central jet. These waves can be of very high amplitude near the impact point, but they decay rapidly away from the impact site. These waves feel the ocean bottom because of their amplitude, much as near-shore waves do. They therefore *break* in deep water, and lose a great deal of their energy to surf, local turbulence in air and water, and interference. The remaining weaker waves propagate out in all directions, spreading out energy as they go.

Differences from classical tsunamis

Classical tsunamis are produced when large earthquakes or landslides happen under water. In both of these cases, the *source region* is much larger than an impact crater. The source region for the 2004-12-26 Sumatran tsunami

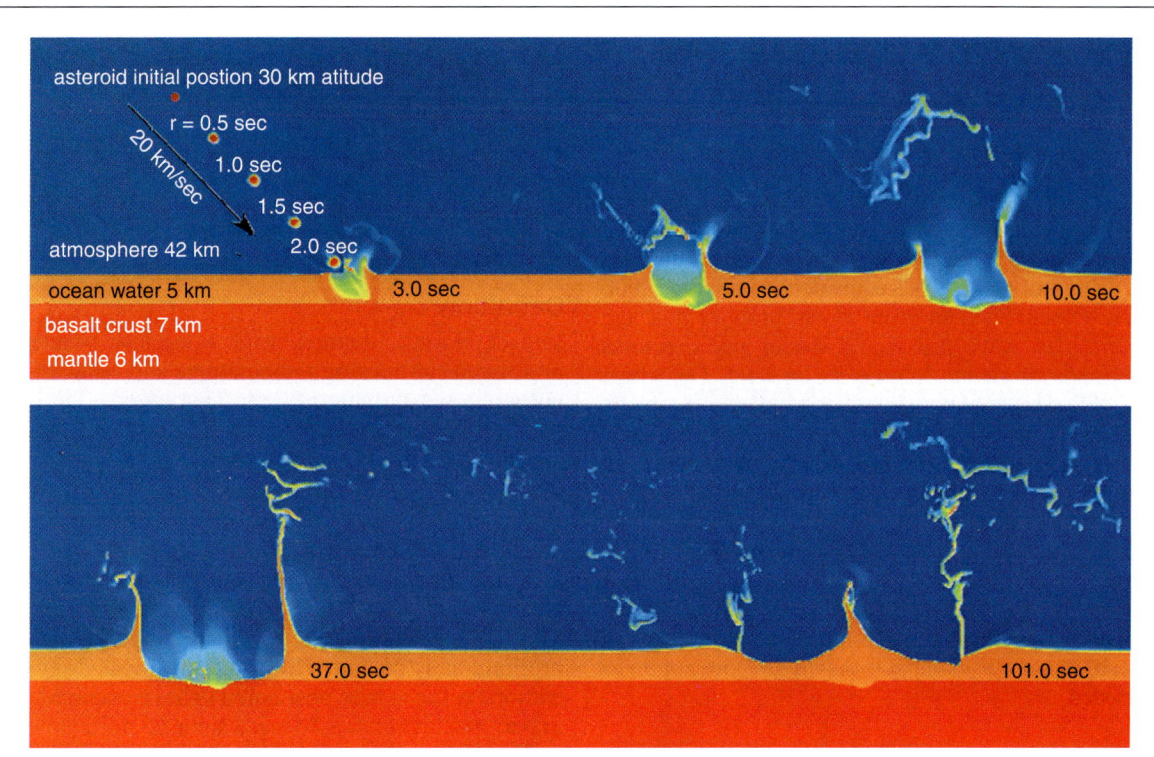

Impact Tsunamis, Figure 1 Montage of frames from a three-dimensional simulation of an impact of a 1 km diameter asteroid into an ocean of 5 km depth. The asteroid is assumed to come in at a 45-degree angle at a speed of 20 km/s. As it penetrates the water, it is vaporized, along with a comparable quantity of the ocean water, and an asymmetric transient crater is produced. The explosive expansion of the vaporized water leads to the cavity's becoming more symmetric, and large amounts of water are lofted tens of kilometers up into the atmosphere in the crown splash and central jet. The subsequent collapse of these two features leads to the setup of impact waves.

(a 300-year event) was roughly 1200-km long by 50-km wide, whereas a 1 km asteroid impact (a million-year event) will produce a transient crater of only 20-km diameter. The much larger source region for a classical tsunami means that the wavelength is considerably longer, roughly 100 km for the Sumatran tsunami. Calculations indicate a wavelength of about 40 km for a 1 km asteroid impact (Gisler et al., 2010).

Because the wavelength of the Sumatran tsunami, like other classical tsunamis, is so much greater than the water depth in mid-ocean (about 5 km), it propagates as a *shallow-water wave*, with high speed and low amplitude in mid-ocean but piling up as it moves into shallower water. The wave speed for a shallow-water wave is proportional to the square root of the water depth, so the front of the wave slows down while the rest catches up as the depth decreases. The asteroid impact tsunami is not a shallow-water wave, but an intermediate wave, whose speed is less strongly dependent on the water depth. It therefore experiences less pileup on entering shallower water. Since the wavelength is considerably shorter, a smaller volume of water is delivered to the shoreline.

The Sumatran tsunami was a *line source*, so it propagated perpendicular to that line with little angular dispersion, whereas an asteroid impact is a *point source*, so it will propagate in all directions with considerable dispersion. Other geological sources of tsunamis like underwater landslides have different directionality, some being quite highly focused.

All of these considerations indicate that the amount of water delivered per meter of affected shoreline is much less for an asteroid impact than for an earthquake tsunami of comparable energy. The total earthquake energy released by the Sumatran event at the Earth's surface has been estimated at 26 Mt TNT equivalent (USGS, http://neic.usgs.gov/neis/eq_depot/2004/eq_041226/neic_slav_e.html). The same energy would be released by a stony asteroid of 75-m diameter entering Earth's atmosphere at 20 km/s. Such an event would occur on average once in 300 years, coincidentally similar to the estimated return time of the Sumatran earthquake.

Recent calculations (Gisler et al., 2010) show that ocean impacts of less than about 300-m diameter would have dangerous effects near the impact site, but negligible ocean-wide effects.

For these most probable impacts, the effects on distant shores would be similar to the effects of severe storm surges rather than tsunamis. Asteroids larger than about

700 m in diameter could indeed produce dangerous waves at great distances, but even for these cases, the near-field and atmospheric effects would be of greater severity.

Summary: dangers from oceanic impacts

The real dangers from likely asteroid impacts in the ocean are, just as for land impacts, the atmospheric effects. A 100-m asteroid striking the earth at 20 km/s has a kinetic energy of 66 Mt TNT equivalent, similar to 3,000 Hiroshima bombs. Such an event would produce a severe atmospheric blast wave that would shatter any structures over thousands of square kilometers on land. On water, but within a few tens of kilometers of a populated shoreline, this could result in tens of thousands of deaths. Hurricane force winds, high atmospheric temperatures, and the fallout of large quantities of water from the crown splash and the central jet will also produce dangers for coasts within range. Larger impacts will inject significant amounts of water vapor into the stratosphere, and could thereby produce a significant change in Earth's climate.

But long distance tsunamis – impact tsunamis – are not a significant danger from impacts that might come upon us unexpectedly.

Bibliography

Gisler, G. R., Weaver, R. P., and Gittings, M., 2010. Calculations of asteroid impacts into deep and shallow water, Novosibirsk Impact Symposium. *Pure and Applied Geophysics*, **168**, 1187–1198, doi:10.1007/s00024-010-0225-7.

Holsapple, K. A., 1994. Catastrophic disruptions and cratering of solar system bodies: a review and new results. *Planetary and Space Science*, **42**, 1067–1078.

Jansa, L. F., Pe-Piper, G., Robertson, P. B., and Freidenreich, O., 1989. Montagnais, a submarine impact structure on the Scotian Shelf, eastern Canada. *Geological Society of America Bulletin*, **101**, 450–463.

Kenkmann, T., Hörz, F., and Deutsch, A., 2005. *Large Meteorite Impacts III*. Boulder, CO: Geological Society of America Special Paper 384, 476p.

Masse, W., Bryant, E., Gusiakov, V., Abbott, D., Rambolamana, G., Raza, H., Courty, M., Breger, D., Gerard-Little, P., Burckle, L., 2006. Holocene Indian Ocean cosmic impacts: the megatsunami chevron evidence from Madagascar*EOS Transactions AGU* **87** (52), Fall meeting supplement, Abstract PP43B-1244.

Melosh, H. J., 1989. *Impact Cratering*. New York: Oxford University Press.

Morrison, D., Harris, A. W., Sommer, G., Chapman, C. R., and Carusi, A., 2002. Dealing with the impact hazard. In Bottke, W. F., Cellino, A., Paolicchi, P., and Binzel, R. P. (eds.), *Asteroids III*. Tucson: University of Arizona Press, pp. 739–754.

Pinter, N., and Ishman, S. E., 2008. Impacts, megatsunami, and other extraordinary claims. *GSA Today*, **18**, 37–38.

Tsikalas, F., Gudlaugsson, S. T., Faleide, J. I., and Eldholm, O., 2002. The Mjølnir marine impact crater porosity anomaly. *Deep-Sea Research II*, **49**, 1103–1120.

Cross-references

Asteroid Impacts
Earthquakes
Storm Surges
Tsunamis

IMPACT WINTER

Owen Brian Toon
University of Colorado, Boulder, CO, USA

Synonyms

Nuclear winter

Definition

Impact winter, named after "nuclear winter," has prolonged subfreezing temperatures and reduced precipitation caused by stratospheric dust, smoke, and sulfates resulting from an impact.

Discussion

Following a large impact the vaporized impactor, containing iridium, as well as melted target material rises in a hot fireball and reenters globally like an immense swarm of shooting stars, heating the upper atmosphere. The hot upper atmosphere radiates light to the surface, broiling many creatures living at the time and setting vegetation on fire as was postulated to have occurred after the impact at Chicxulub in Mexico some 65 million years ago.

The fires generate sooty smoke, which joins dust recondensed from the vapor, and sulfate from the impactor in the atmosphere. Each of these materials has a separate origin, and evolution, and each is capable of causing an impact winter (Toon et al., 1997). The particles absorb sunlight, and reflect sunlight back to space, so less light reaches the surface. Hence, surface temperatures begin to drop and, with a cooled ocean, precipitation also begins to decrease. Low temperatures, precipitation, and sunlight eventually kill many of the animals that escaped that original broiling to death and can cause mass extinctions of many species in the oceans. Even modest smoke injections, such as would follow a nuclear conflict, may cause a "nuclear winter" under which mid-latitudes would remain below freezing for several years, and global precipitation could decline by 90% (Robock et al., 2007). Similar "anti-greenhouse effects" have been observed on Saturn's moon Titan and on Mars after global dust storms.

Bibliography

Robock, A., Oman, L., and Stenchikov, G. L., 2007. Nuclear winter revisited with a modern climate model and current nuclear arsenals: still catastrophic consequences. *Journal of Geophysics Research*, **112**, doi:10.1029/2006JD008235.

Toon, O. B., Zahnle, K., Morrison, D., Turco, R. P., and Covey, C., 1997. Environmental perturbations caused by the impacts of asteroids and comets. *Reviews of Geophysics*, **35**, 41–78.

Cross-references

Asteroid
Asteroid Impact
Asteroid Impact Mitigation
Comet

Impact Airblast
Impact Ejecta
Impact Fireball
Impact Firestorm
Impact Tsunami
Meteorite
Wildfires

INCLINOMETERS

Erik Eberhardt
University of British Columbia, Vancouver, BC, Canada

Synonyms
Slope indicators; Transverse deformation gauges

Definition
Inclinometers are devices used to monitor subsurface landslide movements through a probe or fixed transducer designed to measure inclination with respect to vertical (see Dunnicliff, 1993 for a detailed description).

Discussion
Operation using a probe inclinometer involves lowering the probe down a borehole and measuring its inclination at a number of fixed points as it is pulled back to surface (Figure 1). Comparison of repeated periodic surveys provides an indication of active shear surfaces at depth as a function of time. Operation can also be carried out by fixing a probe, or series of in-place inclinometer probes, at desired depths to monitor continuous movements across known active shear surfaces.

Bibliography
Dunnicliff, L., 1993. *Geotechnical Instrumentation for Monitoring Field Performance.* New York: Wiley.

Cross-references
Creep
Extensometers
Landslide (Mass Movement)
Piezometer

CASE STUDY

INDIAN OCEAN TSUNAMI, 2004

Franck Lavigne[1], Raphaël Paris[2], Frédéric Leone[3], J. C. Gaillard[4], Julie Morin[5]
[1]UMR 8591 CNRS, Paris 1 Panthéon-Sorbonne University, Meudon, France
[2]Clermont University, Clermont-Ferrand, France
[3]University of Montpellier and GESTER Laboratory, Montferrier-sur-Lez, France
[4]The University of Auckland, Auckland, New Zealand
[5]La Reunion University, Saint Denis, France

Definition
Tsunami: a series of traveling ocean waves of extremely long length generated primarily by earthquakes occurring

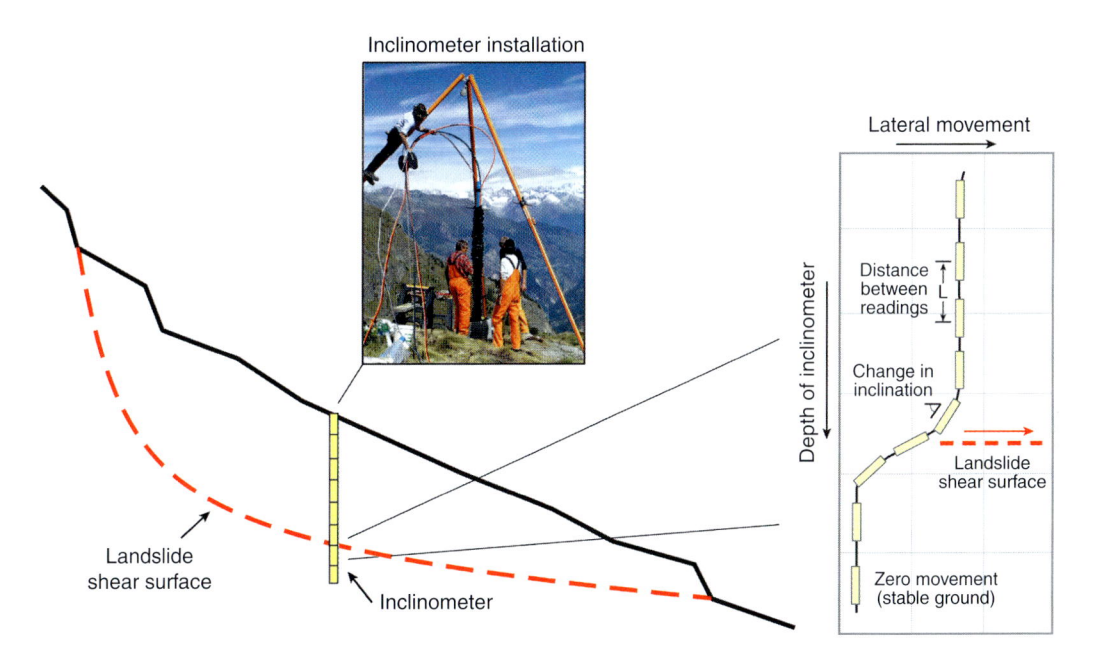

Inclinometers, Figure 1 Use of inclinometer to detect depth of landslide basal shear surface, with insets showing photo of inclinometer installation and schematic of operation measuring cumulative lateral movements.

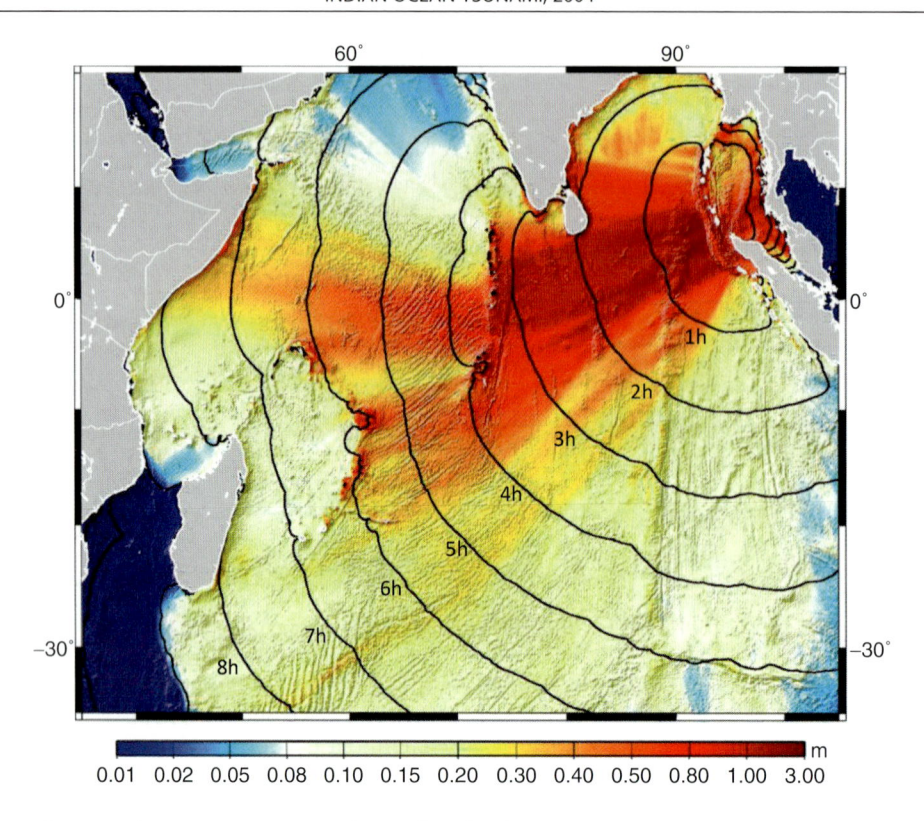

Indian Ocean Tsunami, 2004, Figure 1 Tsunami travel time chart for the 2004 Sumatra tsunami across the Indian Ocean (H. Hébert, Commissariat Energie Atomique, France).

below or near the ocean floor. Volcanic eruptions, landslides, and asteroids can also generate tsunamis.

Introduction

The December 26, 2004, tsunami was the most deadly tsunami and which led to one of the greatest disasters in historical times. Some 280,000 people were killed in South Asia, Southeast Asia, and East Africa. The Aceh province in the Sumatra Indonesian Island was the most affected area with about 167,000 killed. The December 26, 2004, tsunami was unusually violent, but deadly tsunamis are frequent in Indonesia. This country may have faced more than 250 tsunamis during the last four centuries, with over one third being deadly. This contribution focuses on the December 26, 2004, tsunami impacts in the Aceh province, in Indonesia, which was the most damaged area.

Tsunami sources and offshore tsunami propagation

The 2004 Indian Ocean tsunami was triggered by a 9.15 magnitude earthquake (Meltzner et al., 2006; Chlieh et al., 2007) that occurred at 0:58:53 GMT, 7:58:53 LT (t_{EQ}). The epicenter was located at 3.3 N, 95.8 E with a focal depth of approximately 30 km. The earthquake was responsible for a sudden fault slip estimated on average from 12 to 15 m (Synolakis et al., 2005; Lay et al., 2005) to 20 m (Fu and Sun, 2006). In the model suggested by Chlieh et al. (2007), the latitudinal distribution of released moment has three distinct peaks at about 4°N, 7°N, and 9°N, which compares well to the latitudinal variations evident in the seismic inversion and of the analysis of radiated T waves.

The tsunami waves propagated across the Indian Ocean (Figure 1) at an average velocity of 800 km/h (Lay et al., 2005), impacting the Sumatran coast between 20 and 40 min after the main shock of the earthquake (Lavigne et al., 2009).

Precursory signs of the tsunami

In the Banda Aceh area, the great earthquake generated ground subsidence (Meltzner et al., 2006), with amplitudes ranging from a few centimeters to about 2 m. Such large subsidence along the coast could have suggested to an informed observer the occurrence of a vertical displacement of the seafloor as a result of the earthquake, which was one natural warning signal of the incoming tsunami.

Local people reported hearing three detonations similar to bomb explosions that sounded between the main tectonic shock and the tsunami arrival. These bangs are probably peculiar to earthquakes caused by the breaking of a sinking plate (Kato and Tsuji, 1995). However, Shuto (1997) suggested that "thunder-like" sounds are generated and heard at distant places when tsunamis higher than 5 m hit coastal cliffs.

Another preliminary sign of the impending tsunami was a withdrawal of the ocean waters near the shore after the first shock of the earthquake (Yalciner et al., 2005). The extent of the withdrawal exceeded 1 km off Banda Aceh and Lhok Nga. The corresponding lowering of the sea level has been estimated at 5 m +/− 1 m by local fishermen (Lavigne et al., 2009).

The last warning sign of the tsunami arrival was the massive migration of bird colonies flying landward from the open sea. Numerous eyewitnesses reported – after the disaster – to hear birdcalls, which were interpreted by some villagers as a warning for people threatened by the tsunami (Lavigne et al., 2009).

Overland tsunami propagation and its geomorphological effects

Based on eyewitness accounts, about ten separate waves affected the region; this indicates a high-frequency component of the tsunami wave energy in the extreme near field. The largest tsunami wave heights were on the order of 35 m (Figure 2) with a maximum run-up height of 51 m (Lavigne et al., 2009). This run-up related to earthquake-induced tsunami has been exceeded at least four times previously: Indonesia in 1674 (80 m), Kamchatka in 1737 (63 m), Japan in 1771 (85 m), and Krakatoa in 1883. Evidence of a significant discontinuity in the tsunami wave heights and flow depths was noticed along a line approximately 3 km inland (Figure 2), which is interpreted to be the location of the collapse of the main tsunami bore caused by sudden energy dissipation (Lavigne et al., 2009).

The geomorphologic impacts of the tsunami waves include (Paris et al., 2009) beach erosion; destruction of the sand barriers; numerous erosion escarpments up to 2 m in the dunes; bank erosion in the river beds, with local retreats exceeding 30 m; and plurimetric scars typically 20–50 cm deep on the slopes. The upper limit of erosion appears as a continuous trimline at 20–30 m a.s.l. The erosional imprints of the tsunami extend to 500 m from the shoreline and exceed 2 km along riverbeds. The fringing reefs were not efficient in reducing the erosional impact of the tsunami.

The recognition and analysis of tsunami deposits provide clues to better understand the sedimentary signature, sediment transport, and deposition of past tsunamis and to better assess tsunami hazard. Normally, graded couplets or triplets of layers were used to identify the run-up of the three main waves (Paris et al., 2007). The local effects of the topography could be identified: thickest deposits in the topographic lows (50–80 cm), great spatial variations in thickness, landward coarsening, and very poor sorting at the wave breaking point, bimodal

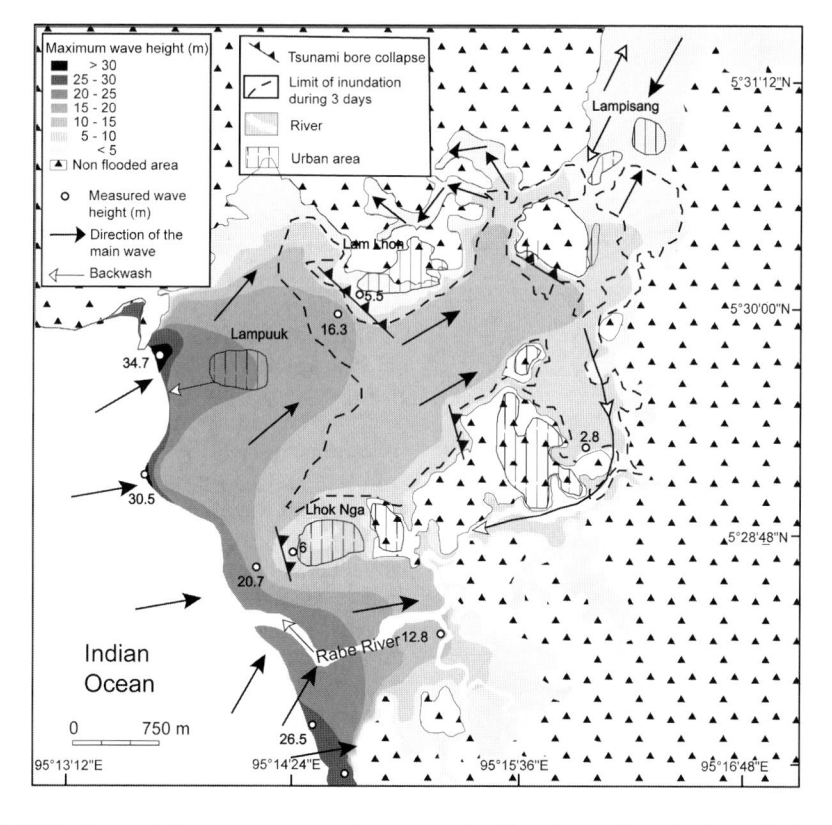

Indian Ocean Tsunami, 2004, Figure 2 Tsunami propagation across the West Coast of Banda Aceh (Lhok Nga subdistrict) (Source: Lavigne et al., 2009).

grain-size distributions reflecting different sources of sediments.

Although the deposition by the tsunami is mainly represented by extensive sheets of sand up to 5 km inland (Paris et al., 2007), megaclasts of soil, road, boulders of coral, and beach rock were also deposited onshore. The tsunami was able to detach and transport coral boulders in excess of 10 t over 500–700 m and megaclasts of the platform in excess of 85 t over a few meters. The fraction of boulders transported from offshore and deposited inland represent only 7% of the total number of boulders moved during the tsunami. Almost 1,800 boulders were identified offshore (Paris et al., 2010). However, the boulder deposits do not appear, at present, as powerful indicators to reconstitute palaeo-tsunami magnitudes.

Tsunami-induced damages

In Indonesia, 654 villages and 63,977 families were affected by the tsunami (Republic of Indonesia, 2005). As a whole, the December 26, 2004, disaster in Aceh and North Sumatra is estimated to have resulted in a loss of about US$4.5 billion (Leitmann, 2007), representing 2.7% of the national GDP or more than 97% of the Aceh province's GRDP. In addition to casualties, the tsunami also caused damage to various sectors. For example, it is estimated that 1,168 schools experienced the impact, or 16% of the schools existing before the disaster. Six hospitals and 6 polyclinic units were damaged (Figure 3). Damaged places of worship in Aceh comprised 1,069 mosques and musholla praying rooms. In the industrial sector, the extent of damage to small and medium industries was estimated at an average of 65%, and large industry at 60%. The impact of the disaster on the infrastructure sector includes the damage suffered by the sectors of housing (about 252,000 houses totally or partially damaged), transportation, energy and electricity, postal and telecommunications service, drinking water and sanitation, water resources, as well as other facilities.

A spatial analysis of tsunami-induced damages in the northwestern suburbs of Banda Aceh city (Leone et al., 2011) pointed out that nearly all of the buildings suffered grade 5 damage (i.e., destruction/collapse), and only a few reinforced concrete buildings (e.g., big mosque, hospital, and school) suffered from very heavy damage to structures (grade 4). No substantial to heavy damage (grade 3) has been observed. The tsunami also damaged port breakwaters, destroyed or washed away small vessels, and violently moved large vessels ashore. Figure 4 underlines a steep drop in the damage gradient around 2.7 km from the coast. Patterns of building damage are related to the location of the propagating bore with overall less damage to buildings beyond the line where the bore collapsed (Leone et al., 2011; Lavigne et al., 2009). The final shape of the line outlines digitations that can be associated with different wave heights or roughness variations of the topography.

The root causes of the disaster: People's vulnerability in the face of the hazard

During the last decade, several deadly tsunamis occurred in Indonesia. One thousand nine hundred sixty people perished in Flores in 1992 (Imamura et al., 1995), 238 in East Java in 1994 (Maramai and Tinti, 1997), 110 in Irian Jaya in 1996 (Matsutomi et al., 2001), and 733 in South Java in 2006 (Lavigne et al., 2007). The recurrence of tsunami-related disasters increased the awareness of scientists and authorities and emphasized the need for improved tsunami disaster risk reduction policies. In response, the TREMORS seismic network was created in 1996. The Meteorological and Geophysical Agency of Indonesia (BMG) has been managing this network which is operating 24/7. On December 26, 2004, an earthquake warning was issued and transmitted to Indonesian authorities. It was further broadcast on Metro TV national channel but only five minutes before the tsunami struck the city of Banda Aceh in Northern Sumatra. The time lead was, however, insufficient to prevent a major disaster.

The tsunami further hit mostly unprepared local communities who were not covered by tsunami disaster risk reduction programs unlike other communities throughout the country (Morin et al., 2008). Indeed, only a few people spontaneously ran to higher ground following the withdrawal of the sea. In some instances, many locals actually rushed to the beaches to gather fish grounded on the beaches before being swept away by the tsunami. Significant differences were, however, observed among the different ethnic groups who live in the area (Gaillard et al., 2008a). About 170,000 Acehnese and Minangkabau people died in the Northern tip of Sumatra whereas only 44 people passed away in the neighboring Simeulue Island located near the earthquake epicenter. Such a difference in the death toll does not lie in the nature of the hazard but in different human

Indian Ocean Tsunami, 2004, Figure 3 Destruction of the Uleelheue hospital, northwest suburb of Banda Aceh (Photo: F. Lavigne).

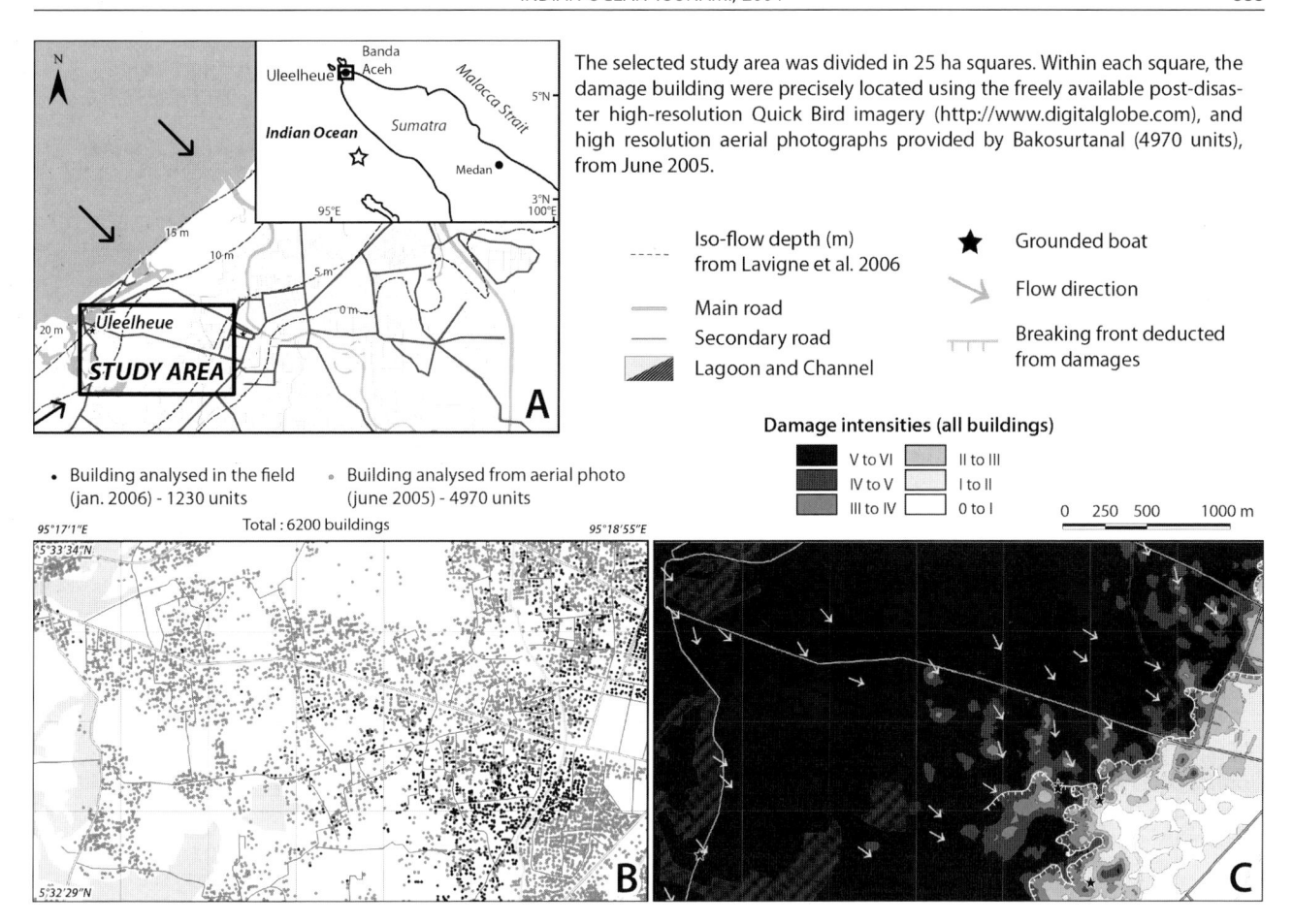

Indian Ocean Tsunami, 2004, Figure 4 Investigation of tsunami-induced damage on buildings after the December 26, 2004 tsunami. (a) Studied area between Uleelheue and downtown Banda Aceh. (b) Map of studied buildings. (c) Map of interpolated damage intensity for all building types. (Source: Leone et al., 2011).

behaviors, which were deeply influenced by the cultural, social, economic, and political context. Seeing the sea's withdrawal, Simeulue Island inhabitants immediately escaped toward surrounding mountains. The accounts passed from generation to generation of the deadly 1907 tsunami enabled them to understand what was happening. Simeulue inhabitants even have their own word to name the phenomenon: *smong*. Conversely to Simeulue communities, Acehnese and Minangkabau people, respectively, in the cities of Banda Aceh and Meulaboh, did not anticipate the phenomenon and were thus caught by the waves. Eventually, the Indonesian government fostered the use of the word *smong* through its integration into the official Indonesian language. This helped in increasing national awareness of tsunami hazards, all the more so as *smong* creates a very helpful acronym: SeMua Orang Naik Gunung ("Everybody moves up on the hills") (Morin et al., 2008). Another similar account of indigenous knowledge in the face of the 2004 tsunami was reported by Arunotai (2008) in Surin Island in Thailand where Moken fishing communities, attentive to nature forewarnings, avoided the tsunami.

The Simeulue and Surin islands stories suggest that loss of life is easily avoidable in the face of tsunami hazard. However, in the event of a warning signal, whether based on local or scientific knowledge, people's behaviors and the capacity to protect oneself may be further hindered by a deep tangle of structural factors. Among these factors, political and economic constraints played a great role in explaining why Acehnese people were unable to face the tsunami. At the time of the disaster, an armed conflict had been affecting the province for more than 30 years (Gaillard et al., 2008b). This led to the impoverishment of the region and the fear of the people to evacuate inland where violence was raging. The armed conflict also contributed to the progressive deconstruction of the social and political organization of the villages, including the lingering erosion of local knowledge, by the Java-based Indonesian government (Gaillard et al., 2008a).

What can we henceforth expect if a tsunami hit again regions which are now covered by the Indian Ocean Warning System? This system was actually already working when a tsunami hit Pangandaran, Java, in July 2006, but hundreds of people were still killed. Although warning was issued on time, it was not transmitted down to local authorities for political reasons, allegedly connected to avoid panic movements among locals. Moreover, people did not systematically recognize natural forewarning signs. Both the December 26, 2004, catastrophe and the July 2006 disaster revealed the pressing need for implementing an efficient warning system and empowering local communities and authorities with adequate resources to face the threat of tsunami hazard. Tsunami preparedness is a key component of disaster risk reduction policies. It should be included in multi-hazard risk reduction programs which emphasize local knowledge and socially and economically acceptable actions to mitigate people's vulnerability. It should further involve a large range of stakeholders which span from local communities, local and national authorities, NGOs, international organizations, scientists, and the media.

Summary

The December 26, 2004, tsunami was the most deadly tsunami (280,000 people killed) in recent history which led to one of the greatest disasters in historical times. This entry explores a broad range of scientific investigations in the Aceh province, Indonesia, from geophysical aspects (tsunami sources, offshore and overland tsunami propagation, and its geomorphological effects) to social aspects (damages, people's vulnerability, interpretations of precursory signs, etc.).

Bibliography

Arunotai, N., 2008. Saved by an old legend and a keen observation: the case of Moken Sea Nomads in Thailand. In Shaw, R., Uy, N., and Baumwoll, J. (eds.), *Indigenous Knowledge for Disaster Risk Reduction: Good Practices and Lessons Learnt from the Asia-Pacific Region*. UNISDR Asia and Pacific, United Nations for Disaster Risk Reduction, Bangkok, pp. 73–78.

Chlieh, M., Avouac, J.-P., Hjorleifsdottir, V., Song, T.-R. A., Sieh, K., Sladen, A., Hébert, H., Prawirodirdjo, L., Bock, Y., and Galetzka, J., 2007. Coseismic slip and afterslip of the great (Mw 9.15) Sumatra–Andaman Earthquake of 2004. *Bulletin of the Seismological Society of America*, **97**, 152–173.

Fu, G., and Sun, W., 2006. Global co-seismic displacements caused by the 2004 Sumatra–Andaman earthquake (Mw 9.1). *Earth Planets Space*, **58**, 149–152.

Gaillard, J.-C., Clavé, E., and Kelman, I., 2008a. Wave of peace? Tsunami disaster diplomacy in Aceh, Indonesia. *Geoforum*, **39**(1), 511–526.

Gaillard, J.-C., Clavé, E., Vibert, O., Azhari, D., Denain, J.-C., Efendi, Y., Grancher, D., Liamzon, C. C., Sari, D. S. R., and Setiawan, R., 2008b. Ethnic groups' response to the 26 December 2004 earthquake and tsunami in Aceh, Indonesia. *Natural Hazards*, **47**(1), 17–38.

Imamura, F., Gica, E., Takahashi, T., and Shuto, N., 1995. Numerical simulation of the 1992 Flores tsunami: interpretation of tsunami phenomena in northeastern Flores Island and damage at Babi Island. *Pure and Applied Geophysics*, **144**(3–4), 555–568.

Kato, K., and Tsuji, Y., 1995. Tsunami of the Sumba earthquake of August 19, 1977. *Journal of Natural Disaster Science*, **17**(2), 87–100.

Lavigne, F., Gomez, C., Giffo, M., Wassmer, P., Hoebreck, C., Mardiatno, D., Prioyono, J., and Paris, R., 2007. Field observations of the 17th July 2006 tsunami in Java. *Natural Hazard and Earth Sciences Systems*, **7**, 177–183.

Lavigne, F., Paris, R., Grancher, D., Wassmer, P., Brunstein, D., Vautier, F., Leone, F., Flohic, F., De Coster, B., Gunawan, T., Gomez, C., Setiawan, A., Cahyadi, R., and Fachrizal, 2009. Reconstruction of tsunami inland propagation on December 26, 2004 in Banda Aceh, Indonesia, through field investigations. *Pure and Applied Geophysics*, **13**(166), 259–281.

Lay, T., Kanamori, H., Ammon, C. J., Nettles, M., Ward, S. N., Aster, R. C., Beck, S. L., Bilek, S. L., Brudzinski, M. R., Butler, R., Deshon, H. R., Ekström, G., Sakate, K., and Sipkin, S., 2005. The great Sumatra–Andaman earthquake of 26 December 2004. *Science*, **308**, 1127–1133.

Leitmann, J., 2007. Cities and calamities: learning from post-disaster response in Indonesia. *Journal of Urban Health*, **84**(Suppl. 1), 144–153.

Leone, F., Vinet, F., Denain, J.-C., and Bachri, S., 2011. L'analyse spatiale des dommages sur le bâti: contribution méthodologique et enseignements pour les futurs scénarios de risque tsunami. In Lavigne, F., and Paris, R. (eds.), *Tsunarisque: le tsunami du 26 décembre 2004 à Aceh, Indonésie*. Paris: Publications de la Sorbonne, pp. 77–96.

Maramai, A., and Tinti, S., 1997. The 3 June 1994 Java tsunami: a post-event survey of the coastal effects. *Natural Hazards*, **15**(1), 31–49.

Matsutomi, H., Shuto, N., Imamura, F., and Takahashi, T., 2001. Field Survey of the 1996 Irian Jaya earthquake tsunami in Biak Island. *Natural Hazards*, **24**(3), 119–212.

Meltzner, A. J., Sieh, K., Abrams, M., Agnew, D.-C., Hudnut, K.-W., Avouac, J.-P., and Natawidjaja, D. H., 2006. Uplift and subsidence associated with the great Aceh-Andaman earthquake of 2004. *Journal of Geophysical Research*, **111**, B02407.

Morin, J., De Coster, B., Flohic, F., Lavigne, F., Le Floch, D., and Paris, R., 2008. Assessment and prevention of tsunami risk in Indonesia. *Disaster Prevention and Management*, **17**(3), 430–446.

Paris, R., Fournier, J., Poizot, E., Etienne, S., Morin, J., Lavigne, F., and Wassmer, P., 2010. Boulder and fine sediment transport and deposition by the 2004 tsunami in Lhok Nga (western Banda Aceh, Sumatra, Indonesia): a coupled offshore – onshore model. *Marine Geology*, **268**, 43–54.

Paris, R., Lavigne, F., Wassmer, P., and Sartohadi, J., 2007. Coastal sedimentation associated with the December 26, 2004 tsunami in Lhok Nga, West Banda Aceh (Sumatra, Indonesia). *Marine Geology*, **238**, 93–106.

Paris, R., Wassmer, P., Sartohadi, J., Lavigne, F., Barthomeuf, B., Desgages, E., Grancher, D., Baumer, P., Vautier, F., Brunstein, D., and Gomez, C., 2009. Tsunamis as geomorphic crisis: lessons from the December 26, 2004 tsunami in Lhok Nga, West Banda Aceh (Sumatra, Indonesia). *Geomorphology*, **104**, 59–72.

Republic of Indonesia, 2005. Main book of rehabilitation and reconstruction. In *Master Plan for Rehabilitation and reconstruction for the Regions and People of the Province of Nanggroe Aceh Darussalam and Nias Islands of the Province of North Sumatra*, Badan Rehabilitasi dan Rekonstruksi, Banda Aceh, Vol. 1 (13 Vol).

Shuto, N., 1997. A natural warning of tsunami arrival. In Hebenstreit, G. T. (ed.), *Perspectives on Tsunami Hazard Reduction, Observation Theory and Planning*. Dordrecht: Springer. Advances in Natural and Technological Hazards Research, pp. 157–173.

Synolakis, C. E., Okal, E. A., and Bernard, E. N., 2005. The Megatsunami of December 26 2004. *The Bridge*, **35**(2), 26–35.

Yalciner, A. C., Perincek, D., Ersoy, S., Presateya, G. S., Hidayat, R., and McAdoo, B., 2005. December 26, 2004 Indian Ocean tsunami field survey (January 21–31, 2005) at north of Sumatra Island. Available from World Wide Web: http://yalciner.ce.metu.edu.tr/sumatra/survey/.

Cross-references

Disaster Relief
Disaster Risk Reduction (DRR)
Early Warning System
Earthquake
Earthquake Damage
Earthquake Prediction and Forecasting
Exposure to Natural Hazards
Geohazards
Geological/Geophysical Disasters
Natural Hazard in Developing Countries
Pacific Tsunami Warning and Mitigation System (PTWS)
Tsunami
Vulnerability
Warning Systems

INDUCED SEISMICITY

Maurice Lamontagne
Natural Resources Canada, Ottawa, ON, Canada

Synonyms

Anthropogenic seismicity; Stimulated seismicity; Triggered seismicity

Definition

Induced seismicity is a general expression that includes all earthquake activity brought about by man-made perturbations of the effective shear stress existing on faults and fractures. This can arise by changing the acting stresses or the pore-fluid pressure. These perturbations can be caused by the impounding of water reservoirs, mining activity either underground or at the surface, high pressure injection of fluids for hydrothermal power generation or oil production, removal of underground fluids such as oil, gas, and water, and underground explosions (Simpson, 1986).

Induced seismicity

Strictly speaking, the expression "induced seismicity" is correctly used if the seismic activity causes a stress change comparable in magnitude to the ambient shear stress acting on a fault to cause slip (McGarr et al., 2002). These authors define "triggered seismicity" if the stress change is only a small fraction of the ambient level. In other words "induced seismicity" applies if the stress changes brought about by the change in conditions are causing the seismicity whereas the other definition implies that it is the perturbation that "advanced the clock" of a seismic event that would have eventually occurred. It should be noted that this distinction in terms is not followed by all.

Water reservoir impounding

Notwithstanding the distinction above, the most famous types of "induced seismicity" are related to the impounding of large reservoirs (more correctly referred to as "reservoir-triggered seismicity"; RTS). At least 70 cases of RTS are recognized worldwide (Gupta, 1992) but more cases may have escaped detection depending on the magnitude detection threshold of the local seismograph networks. Damaging RTS has occurred at Hsinfengkiang (China), Kariba (Zaire), Kremasta (Greece), Koyna (India), Oroville (USA), and Aswan (Egypt) (Gupta, 1992). This phenomenon is well documented and it is recommended that reservoir impounding be monitored by an array of local seismographs a few years prior to impounding (ICOLD, 2008).

Mining and quarrying activity

Mining-induced seismicity, sometimes called rockbursts, because of the mining safety aspects is probably the best studied type of induced activity. Because the activity is taking place very close to areas subject to direct observation and measurement, mining-induced activity offers the best opportunity to examine the relative importance of depth, scale, state of stress, pore-fluid pressure, and mining technique (McGarr et al., 2002). The safety considerations of mining-induced activity have also led to advanced monitoring of the activity itself as well as to the potential causative factors. Surface mining is also reported to have caused induced activity (McGarr et al., 2002).

Injection of fluids

There are a number of cases of seismicity caused by the injection of fluids. In 1960s, the injection of fluids in the Rocky Mountain Arsenal well near Denver Colorado was the first time when a clear relationship was established between the fluid injection and the triggered seismicity. One of the best studied case is the KTB well of Germany where fluids under pressure were injected at a depth of 9.1 km (Zoback and Harjes, 1997). Through relatively modest increase in pore pressure, hundreds of microearthquakes were triggered. In 2006, injection of fluids near Basel, Switzerland, has triggered seismic activity with a magnitude 3.4 earthquake that has caused some three million dollars of damage to houses. This is a reminder that the physical processes and parameters that control injection-induced seismicity are still poorly understood (Kraft et al., 2009).

Summary

Due to the increase in exploration and exploitation of deep wells, the phenomenon of induced seismicity may well become subject to increased interest by scientists and, if this seismicity is felt at the surface, by the public. The possibility of induced seismicity should be considered in the design phase of large-scale projects such as large reservoirs (ICOLD, 2008).

Bibliography

ICOLD (International Commission on Large Dams), 2008. *Reservoir and seismicity: state of knowledge" bulletin* (International Commission on Large Dams); Rough version of Bulletin 137, 50 p.

Kraft, T., Mai, P. M., Wiemer, S., Deichmann, N., Ripperger, J., Kästli, P., Bachmann, C., Fäh, D., Wössner, J., and Giardini, D., 2009. Enhanced geothermal systems: mitigating risk in urban areas. *Eos, Transactions, American Geophysical Union*, **90**(32), 273.

McGarr, A., Simpson, D., and Seeber, L., 2002. Case histories of induced and triggered seismicity. In Lee, W., Kanamori, H., Jennings, P., and Kisslinger, C. (eds.), *International Handbook of Earthquake and Engineering Seismology*. New York: Academic, pp. 1253–1259.

Simpson, D. W., 1986. Triggered earthquakes. *Annual Review of Earth and Planetary Sciences*, **14**, 21–42.

Cross-references

Earthquake
Pore-Water Pressure
Reservoir Dams and Natural Hazards
Seismology
Triggered Earthquakes

INFORMATION AND COMMUNICATION TECHNOLOGY

Peter S. Anderson
Simon Fraser University, Burnaby, BC, Canada

Definition

Information and communication technology (ICT) refers to the advancement and use of contemporary electronic information systems, brought about by the convergence of computer and telecommunications technology. ICT is also concerned with improving human and organizational problem solving through use of technologically based processes and systems that enhance the efficacy of information in a variety of tactical and strategic operational situations. ICT provides a common intellectual framework to interconnect all of the stages of hazard phenomena and resultant human and environmental interaction.

Introduction

The healthy functioning of modern societies is effectively based on the high-quality and expeditious circulation of information. This is even more germane in the case of societies vulnerable to natural hazards. There, accurate information and its reliable and timely communication are of importance, as people's lives, assets, and environments are at stake. However, with the increasing complexity of modern society, no single organization involved in disaster risk reduction practices has the requisite knowledge to understand and address all of the consequences of hazards.

A fundamental problem in dealing with natural hazards, especially atmospheric and oceanographic hazards, is that they respect no boundaries. Their effects can cut across social, cultural, economic, geopolitical, and ecological domains and the adequacy of countermeasures is greatly influenced by the degree to which actions can be coordinated and integrated among the relevant disciplines, professions, and jurisdictions within all phases of hazard phenomena (Anderson, 1991). The systematic organization of these activities forms the basis for an emerging discipline, often referred to as "emergency management," "disaster management" or, more recently, "disaster risk reduction." In general, it aims to foster a continuous process by which individuals, groups, and communities of interest share responsibility for helping to avoid or ameliorate the impacts of disasters resulting from hazards. This more comprehensive and collaborative approach marks a paradigm shift from a traditional fixation on relief and recovery to incorporate a broader focus on risk and vulnerability management to help make communities more resilient to disasters.

Information creation and exchange underpin these activities and form crucial interdependent links in the chain of sound natural disaster risk reduction measures to bring together those who have knowledge about hazards and their characteristics with those who must deal with their consequences from local to global scales (NRC, 1998). This chain of measures includes hazard, vulnerability, and risk analysis (HVRA), prevention and mitigation, preparedness, response, recovery, and rehabilitation.

"HVRA" is a three-stage process that enables communities to: (1) collect and analyze information on the probable location and severity of dangerous natural phenomena and likelihood of their occurrence within a specific period in a given area (OAS, 1991); (2) as determined by physical, social, economic, and environmental factors or processes, estimate the vulnerability – the degree of loss or harm that would result from the occurrence of a natural phenomena of given severity, and (3) for a given hazard event, determine the nature and extent of risk or exposure by analyzing potential hazards and evaluating existing conditions of vulnerability (UNISDR, 2004). The results of the HVRA process establish the social and operational contexts for all subsequent measures. "Prevention and mitigation" involve longer term and sustainable structural (technological) and nonstructural (social) measures to prevent hazards from developing into disasters altogether, or to reduce the effects of disasters when they occur. They include activities such as hazard and risk mapping, risk communication and education, land-use planning and zoning, regulation, structural fortification, and insurance. "Preparedness" is aimed at minimizing the loss of life and property during an actual hazard event, and includes developing and adopting strategic and tactical plans to identify and effectively mobilize human and material resources in case of a disaster. Early warning is a key component that facilitates delivery of timely information through authorized institutions and available communication channels, to enable individuals and organizations, about to be exposed to a hazard, to take steps to avoid or reduce their risk and to initiate follow-on response actions. "Response" includes actions employed during or immediately after a hazard event to meet the urgent life preservation and basic subsistence needs of those affected. "Recovery and Rehabilitation" is concerned with immediate and long-term decisions and actions taken after a disaster to restore or improve the pre-disaster conditions of the stricken society and environment, while encouraging and facilitating necessary adjustments within the other risk reduction measures to reduce future disaster risk (UNISDR, 2004).

Each measure requires information drawn from multiple sources to support decision making at many different levels and scales (NRC, 1998). Local authorities need information that is sufficiently detailed to be useful in planning all aspects of disaster risk reduction – information such as hazard and risk maps; land-use planning; locations of vulnerable populations; community hazard awareness; critical infrastructure; transportation routes and emergency resources. Senior level authorities need aggregated regional and national level information to support policymaking, program development, multijurisdictional disaster response coordination, and recovery financial and other assistance. International bodies require national and regional level information to formulate and stimulate global disaster reduction strategies and partnership programs. Utilities and essential service providers need information to develop and implement business continuity and restoration plans. Individuals and community organizations need information to support personal and neighborhood preparedness and risk reduction initiatives. Sometimes, the same information, such as real-time atmospheric and flood hazard information, is required at all levels.

Role of information systems

A major challenge to adopting an effective and comprehensive disaster risk reduction approach is that crucial information is gathered and held by a range of individuals, organizations, and disciplines through complex and varied processes (NRC, 1998). However, information systems can assist collective planning and decision making by providing a standardized means to systemically gather, organize, and present data from a wide assortment of sources. The advancement and use of contemporary electronic information systems, brought about by the convergence of computer and telecommunications technology, commonly referred to as Information and Communication Technology (ICT), is unparalleled for gathering, storing, retrieving, processing, analyzing, and transmitting information. ICT is also concerned with improving human and organizational problem solving through use of technologically based processes and systems that enhance the efficacy of information in a variety of tactical and strategic operational situations. ICT provides a common framework to interconnect all of the stages of hazard phenomena and human and environmental interaction. However,

to be effective, countries need to develop the necessary human resources and strengthen the capacities of institutions and communities to use and integrate ICT applications (UNESCAP, 2009).

Today, hazard data and derived information is readily available across all types of ICT networks and user devices. This trend is evident in the expanded use of: space and terrestrially based remote sensing to detect and monitor hazards; global positioning systems (GPS) and geographical information systems (GIS) to locate and map hazards, associated risk, vulnerable populations, and critical infrastructures; complex computer modeling to predict and forecast extreme hazard events; and an explosive growth in interconnected communication channels that enable millions of people to share data and information concurrently and to collaborate and problem solve in ways not previously possible (NRC, 1998). Using Internet and data warehousing techniques, ICT tools are also being widely used to build knowledge warehouses to support planning and policy decisions for mitigation, preparedness, response, and recovery at all levels.

ICT assists scientists to systematically collect, manage, analyze, and interpret baseline data. For example, field data can be remotely collected through broad sensor arrays and measurement devices and instantaneously relayed over networks. Using a multitude of interoperable space and terrestrial communication methods, the data can be simultaneously gathered from numerous locations and quickly processed by high capacity computers to construct predictive models to improve forecasting and help users interpret and understand the hazard's physical processes. Increasingly, through specialized national and international networks linked to the Internet, derived hazard occurrence information can be viewed almost immediately after its data collection, enabling notification of seismic events around the world, the tracking of hurricanes and tsunamis waves, monitoring of stream flows and water levels, forecasting of extreme weather, etc. The derived information can be communicated to planners, managers, responders, and other stakeholders to enhance and improve their understanding of extreme hazard events and to trigger appropriate warnings and response.

Some information systems integrate current with archived information to enable dynamic prediction modeling of hazard behavior. For example, regional tsunami warning centers use forecast modeling to generate estimates of wave arrival time, wave height, and likely coastal inundation areas immediately following large coastal or submarine earthquakes. Given time constraints of running very complex tsunami forecast models in real time, wave generation, propagation, and inundation computation can now be expedited by accessing a database of precomputed scenarios containing information about tsunami propagation from a multitude of sources. As the actual waves propagate across the ocean and successively reach preinstalled sea level gauges and sophisticated pressure gauges set on the ocean floor, the detected and recorded sea changes are transmitted back to the warning center which, in turn, processes the information and produces new and refined estimates of the tsunami characteristics. The result is a progressively more accurate forecast of the tsunami that can be used to help identify threatened coastal areas, calculate estimated times of wave arrival, and formulate and issue targeted watches, warnings, and/ or recommended evacuation instructions (IOC, 2008).

Remote and in situ sensing and data acquisition

One of the most important tools available to the risk reduction community is remote sensing of the environment. Remote sensing is commonly referred to as the science of obtaining and interpreting information about area, object, or phenomena from a distance, using sensors that are not in physical contact with the object or phenomena being observed. In its broadest sense, remote sensing includes aerial, satellite, and spacecraft observations of the Earth's surfaces. Remote sensors measure different wavelengths, from optical to radar, of electromagnetic radiation that have interacted with the Earth's surface. Different sensors aboard airplane or satellites are used extensively in providing imaging, ortho-photographic, and mapping information to support both detailed micro- and regional level studies of land surface, marine, and atmospheric hazard interactions. It is particularly valuable in detecting and mapping many types of natural hazards that were previously undetected or recorded such as landslides, volcanoes, tornados, and earthquake-induced fault lines.

Earth observation is also useful in viewing the same area over long periods of time and, as a result, make it possible to monitor natural processes, environmental change, and human interaction and impacts to model and simulate past, present, and future trends and projections. If susceptibility to natural hazards can be identified in the early stages, measures can be introduced to reduce the social and economic impacts of potential disasters through better mitigation, preparedness, and adaptation strategies (Wattegama, 2007). Examples include using remote sensing to observe and monitor: coastal shores, snow cover regimes, glaciers, and sea ice surface features for climate change prediction; precipitation, soil water, and crop lands for food production and famine prediction, and land transformation before and after earthquakes for infrastructure design and engineering.

Remote sensing has become a valuable resource during and after hazard events to support early warning, response, and rescue and post-event recovery. Meteorological satellites enable officials to observe atmospheric patterns and visualize the formation and intensity of cyclones and other extreme weather events, track their progress across regions and predict impact of locations at-risk in time to warn residents and deploy vital resources. Other systems enable similar observation of flooding and tracking of major wildfires. Post-event observation has permitted rapid damage assessment mapping of tsunami (Bonda Ache 2004, Japan 2011), flood (Nepal 2008) and earthquake (Haiti 2010, Japan 2011) hazard impacts.

Remote sensing is often supplemented by the integration and networking of surface-based in situ sensing in which measuring devices are either immersed in, or at least touch, the object(s) of observation. These devices when connected to satellite or terrestrial telecommunications networks permit real-time observation. For example, sensors can record and transmit remote watershed snow pack data to enable hydrologists to predict annual flood and drought conditions and, when supplemented by real-time stream flow gauges, support flood warning and response. Wireless networks incorporating seismometers, temperature probes and cameras are now routinely used to remotely monitor high-risk volcanoes. Both remote and in situ sensing systems have become critically important in the monitoring of locations where the hazards pose extreme physical risk to observers, such as volcanoes and landslides.

Geographic information systems

With the availability of large volumes of remotely sensed data, computing systems and specialized applications, such as Geographic Information Systems (GIS), permit geo-referenced data to be stored, retrieved, manipulated, mapped, and visualized in combination with data collected from other sources such as human settlement, critical infrastructure, evacuation routes, and public assembly locations. By utilizing online GIS, planners, responders, and the public can pool and share information through computer-generated databases and maps. GIS provides the mechanism to standardize, integrate, centralize, and visually display both static and dynamic critical information before, during, and after hazard occurrences. In recent years, the explosion of Internet mapping applications, such as Google Earth, Google Maps, and Bing Maps, has given the public access to ever-expanding amounts of geographic data and allowed users to annotate and share maps with others. GPS enabled mobile smartphones and personal data assistants (PDAs) incorporating GIS can display their locations in relation to fixed facilities (nearest hotel, gas station, fire hall), mobile assets (vehicle, people), or communicate their coordinates back to central servers and Web sites for display or other processing.

Decision support and incident management systems

ICT provides a framework for integrating database management systems with analytical models, GIS, statistical databases, and tabular reporting capabilities into decision support and incident management systems (DSIMS) that can be used to reduce the time needed to make crucial decisions regarding task assignment and resource mobilization during response phases. Among other features, these systems can also incorporate event reporting, resource and critical assets management, interagency situation reporting, duty logging, call center tracking, contact databases, infrastructure status reporting, action planning, impact sighting and damage assessment, and public and interagency notification. Through secure wide-area network connections, system information can be shared on an interagency basis to enhance situational awareness and provide a common operational perspective. DSIMS also provides a stored memory of events to assist in recovery and longer-term mitigation efforts.

Early warning systems

For many natural hazards, the knowledge about their current and forecasted states often resides beyond the geographical locations of those who could be affected. ICT provides a unique means to ensure the timely and accurate delivery of this information in the form of routine bulletins, special notifications, or warnings. A warning can be defined as the timely communication of information about a hazard or threat to people at risk in order for them to take necessary actions to mitigate any potentially negative impacts on themselves, those in their care and their property (Samarajiva et al., 2005). Warning, therefore, must be viewed not simply as a technology, but rather as a unified system made up of five critical and interrelated elements: (1) HVRA; (2) detection and monitoring; (3) emergency management structure; (4) local notification and dissemination, and (5) public education and community capacity (Anderson, 2006). No single technology is sufficient to warn everyone, and ICT must encompass both traditional media (radio, television, loud speakers, sirens, amateur radio) as well as new media (text messaging, cell broadcasting, Internet, satellite radio, automated telephone notification) in order to reach all segments of the public, day, or night.

Internet and new social media

Since the early 1990s, the Internet has become one of the world's most important social networking tools for stimulating and enhancing virtually all aspects of risk reduction across disciplines, jurisdictions, genders, and cultures. More recently, an even newer generation of interworking has begun to take shape that allows increasingly larger portions of the population to play a more self-determinist role in these activities. The emergence of Web 2.0 – SMS, blogs, social networks (YouTube, Facebook, Twitter), Flickr, RSS feeds, Voice-over-IP, and Open Source Software represents a culture shift from the Internet being largely shaped by storage and retrieval applications toward a real-time interactive environment that cultivates users becoming active participants rather than passive viewers. The Internet now offers rich user experiences including Web video and audio streaming, interactive maps, timely content, and virtual worlds that can be used not only for online entertainment, but also for practical risk reduction purposes both for the public-at-large and risk managers (NRC, 2009).

Even in the absence of official warning and communication, people affected by natural hazards now communicate directly with each other and the world-at-large. The first reported impacts of the 2004 Indian Ocean Tsunami were sent via Short Message Service (SMS) and posted on blogs in Sri Lanka. During the 2005 Hurricane Katrina disaster, many New Orleans residents of affected coastal areas were able to communicate with friends and relatives via SMS when traditional landlines were unavailable. Similarly, Haitians buried under debris from the 2010 earthquake were able to text for help and be located through their mobile phones. Crowdsourcing, crisismapping and resource management tools such as Ushahidi and Sahana enable citizens to report incidents, self-organize and collectively respond to local relief and recovery needs.

These same applications allow risk managers to shift focus from what needs to be done for the community, to what can be done with the community (NRC, 2009). Electronic social networks enable emergency responders and support agencies to assess continuously social impacts of extreme hazard events and observe self-organized community responses. They can also supplement scientific observation and analysis of the same events. For example, while seismic detection systems help to generate numeric models to identify the source and estimated magnitude of the earthquakes, social feedback systems such as email, Twitter, and Facebook, enable rapid subjective assessment of how and where they were felt. In the near future, smart phones and other PDAs will serve as sensor platforms to instantly record and relay hazard-related data from millions of user locations globally, enabling more precise modeling of natural hazards and effects. This data tied into GIS systems and land type and land-use databases will enable rapid impact and damage predictions to enable emergency teams to plan initial response.

Summary

In summary, ICT can assist collective planning and decision making by providing a standardized means to systemically gather, organize, and present data from a variety of sources. Today, hazard data is readily available across all types of ICT networks and user devices to enable millions of people to share data and information concurrently and to collaborate and problem solve in ways not previously possible. The derived information can be communicated to planners, managers, responders, and other stakeholders to enhance and improve their understanding of extreme hazard events and to trigger appropriate mitigation, warning, response, and recovery actions. Newer social media versions enable emergency responders and support agencies to assess continuously social impacts of extreme hazard events and observe self-organized community responses. However, to be effective, countries need to continuously foster and support the development of human, institution, and community capacities to use and integrate ICT applications.

Bibliography

Anderson, P. S., 1991. *Toward an Integrated Australian Disaster-Management Information System; Challenges and Prospects for the 1990s*. Melbourne: Centre for International Research on Communication and Information Technology.

Anderson, P. S., 2006. *British Columbia Tsunami Warning Methods: A Toolkit for Community Planning*. Burnaby: Telematics Research Lab, Simon Fraser University.

International Oceanic Commission (IOC), 2008. *Indian Ocean Tsunami Warning and Mitigation System (IOTWS): Implementation Plan for Regional Tsunami Watch Providers*. IOC Information Series. No. 81. Paris: UNESCO.

National Research Council (NRC), 1998. *Reducing Disaster Losses Through Better Information*. Washington, DC: National Academy Press.

National Research Council (NRC), 2009. *Applications of Social Network Analysis for Building Community Disaster Resilience: Workshop Summary*. Washington, DC: National Academy Press.

Organization of American States (OAS), 1991. *Primer on Natural Hazard Management in Integrated Regional Development Planning*. Washington, DC: Organization of American States. Available on-line at: http://www.oas.org/usde/publications/unit/oea66e/begin.htm. Accessed 16 January 2010.

Samarajiva, R., Knight-John, M., Anderson, P. S., Zainudeen, A., 2005. *National Early Warning System: Sri Lanka – A Participatory Concept Paper for the Design of an Effective All-Hazard Public Warning System*, Colombo: LIRNEasia. Available on-line at: http://www.lirneasia.net/2005/03/national-early-warning-system. Accessed 12 January 2010.

United Nations Economic and Social Commission for Asia and the Pacific (UNESCAP), 2009. *Information and Communication Technology For Disaster Risk Reduction*. Policy Brief on ICT Applications in the Knowledge Economy. No. 4. Bangkok: UNESCAP.

United Nations International Strategy for Disaster Reduction (UNISDR), 2004. *Terminology: Basic Terms of Disaster Risk Reduction*. Available on-line at: http://www.unisdr.org/eng/library/lib-terminology-eng%20home.htm. Accessed 21 February 2010.

Wattegama, C., 2007. *ICT for Disaster Management*. Bangkok: Asian and Pacific Training Centre for Information and Communication Technology for Development and United Nations Development Programme-Asia-Pacific Development Information Programme.

Cross-references

Early Warning Systems
Emergency Mapping
Geographic Information Systems (GIS) and Natural Hazards
Geographical Information System
Global Positioning System (GPS) and Natural Hazards
Hazard and Risk Mapping
Internet, World Wide Web and Natural Hazards
Landsat Satellite
Mass Media and Natural Disasters
Remote Sensing of Natural Hazards and Disasters
Seismograph/Seismometer
Warning Systems

INSECT HAZARDS

Philip Weinstein
University of South Australia, Adelaide, SA, Australia

Definition

Insects have been an integral part of human culture since the beginning of recorded history, and most likely much earlier. The world's few remaining hunter-gatherer societies eat insects as well as their products, suggesting that insects have formed part of human ecology throughout our evolutionary history. Then as now, insects posed a hazard to people impinging on their domain, for example, in harvesting honey from bees – the latter being the biggest insect killer of humans in the developed world. The relationship between humans and insects is thus a complex one, and they have been revered (sacred dung beetles of Egypt) and feared (insect phobias) with equal passion. This contribution focuses on insect hazards that can be defined in the same way as insect pests – that is, insects that are "judged by man to cause harm to himself, his crops, animals, or his property" (Dent, 1993, p. 1). Such harm can be readily discussed in the following categories: stings and allergies, bites and disease, and physical and imagined hazards.

Stings and allergies

The Hymenoptera (bees, ants, and wasps) arguably present the greatest direct hazard to humans of any insects, because their stings are both venomous and allergenic – meaning that people are at risk not only of the acute pain of the venom injected with a bee sting, but also of anaphylactic shock (and potentially death) if they have developed an allergy to the same venom. The Hymenoptera possess a highly modified ovipositor, which, instead of serving the original function of egg laying, has developed an associated venom gland and thus, can be used defensively. It is important to appreciate that with the exception of hematophagous (blood-feeding) insects discussed elsewhere, no insect actively pursues humans. Thus, bites and stings are largely administered in self-defense, sometimes to victims who accidentally disturb insects in the course of other duties (stepping on a foraging bee), and other times in response to deliberate provocation (removing honey from a hive), but never proactively in the way that a predator seeks prey. Having said that, swarms of bees protecting their hive can rarely cause death in humans, by the sheer number of stings (even in the absence of allergy); particularly worrisome are aggressive varieties commonly called "killer bees." In Australia, where underground nests of the imported European wasp can grow to over a meter, an unusual hazard is caused for people who step on and fall into such nests. Some ant venoms can be as potent as bee or wasp venom, but again the major threat is also from allergic reactions (Goddard, 1996). Many insects cause or exacerbate respiratory allergies in humans, hence fly and cockroach antigens (often in feces) are likely to be as potent in asthma etiology as are the better recognized house dust mite allergens. Occupational exposure to particular species of insects can lead to any number of unexpected allergic reactions, from skin irritation to anaphylaxis. By far the majority of morbidity (disease) and mortality (death) cases from insect stings is a result of allergic reactions – only insect relatives (spiders, scorpions) possess venom potent enough for a single animal to cause death by envenomation.

Bites and disease

By analogy with stings and allergies, it is generally not the bite of an insect that poses a hazard to humans, but more so the diseases that can be transmitted by biting insects. It is true that larger individuals of many species (beetles, soldier ants) are capable of breaking human skin to cause severe pain to a probing finger – but again this behavior is purely defensive and does not constitute a significant hazard from a public health perspective. By contrast, disease transmission by biting insects does constitute a public health problem, with more people dying annually of mosquito-borne disease than of any other disease. For the continued existence of their species, blood-feeding insects are dependent on at least the female finding a blood meal: egg maturation, and therefore the next generation, is dependent on the availability of protein from such a meal. By the early Cretaceous, mosquitoes and other biting flies were "on the lookout for the fresh blood of every [vertebrate] from small frogs to gigantic dinosaurs" (Poinar and Poinar, 2008, p. 33). As mammals evolved, the feeding preferences of many hematophagous insects changed to capitalize on the increasing availability of mammalian blood – culminating in the ready and regular availability of human blood once agricultural communities appeared. The latter were often situated near water bodies where mosquitoes breed, and hungry swarms can potentially make life outdoors a veritable misery in such situations.

Every year, close to 300 million people catch mosquito-transmitted malaria, and about a million people – mostly children – die (WHO, 2009). Malaria parasites (protozoa of the genus *Plasmodium*) are transmitted to a new host from the mosquito's saliva when an infected mosquito takes a blood meal. The parasites go on to multiply in the host, causing the symptoms of infection and sometimes death. Add to that problem a dozen other diseases transmitted by mosquitoes (including dengue fever and West Nile virus), and it is clear that mosquitoes are, globally, by far the most hazardous insects to humans. Some other insects (tsetse flies, sand flies, reduviid bugs, fleas) and insect relatives (ticks) have a similar ecology, actively pursuing hosts to obtain blood meals, and incidentally infecting the latter with pathogens that include the causative organisms of sleeping sickness, viral encephalitis, plague, and other potentially lethal diseases. Hematophagous insects can, therefore, be legitimately feared in areas that harbor such diseases, and precautions should be taken against being bitten. Generally the risk can be dramatically reduced by simple measures: avoiding peak biting times (dawn and dusk); wearing light colored, loose, long-sleeved clothing; using personal insect repellents outdoors and knockdown sprays indoors; and sleeping under insecticide-impregnated bed nets.

Finally, insect parasites rate a mention in this category – insects that spend their lives on humans, feeding on blood. Lice are virtually the only insects that have adapted to the relatively hairless state of humans, and not surprisingly, the ones that do infest us have species specifically adapted to heads, pubic hair, and clothing. Only the latter, body lice, pose a significant threat, and again it is because of their potential to transmit disease. During the First World War, more soldiers died from louse-borne typhus than died in battle.

Physical impacts

Insects are at their most spectacular when in large numbers in outbreaks and swarms, and can pose direct hazards that result from the sheer mass of their bodies, as well as indirect hazards that result from losses in human productivity. An outbreak can generally be conceived of as any situation where insect numbers reach unacceptable levels (as judged by economic, medical, or aesthetic criteria), and outbreaks are driven by both abiotic (nonliving environment) and biotic (living environment) triggers.

The interaction between climate and locusts provides a classic example of an abiotically driven insect hazard. Locusts, like all insects, are cold blooded, and the rate of development is, therefore, temperature-dependent. In many cases, it is also humidity or rainfall dependent and female locusts require soft wet soil in order to successfully lay their eggs underground. When these conditions are perfect (from the insect's perspective!), nymphs (young locusts) hatch in such large numbers that a physiological change is triggered: The high frequency of contact with their fellow nymphs causes these locusts to grow into swarming, migratory adults. Swarms can contain millions of individuals, and have plagued man since our earliest recorded history: "[The locusts] covered the face of the whole earth, so that the land was darkened; and they did eat every herb of the land, and all of the fruit of the trees..." (Exodus 10: 14). Swarms have been recorded that are tens of kilometers long and traveling at over 100 km per day, stripping the ground of all vegetation when landing to feed. Food security is obviously threatened by such large masses of herbivorous insects, and historical famines have led to the entrenchment of such events in our culture and language (as reflected in everyday expressions of overconsumption such as "like a plague of locusts"). In the modern era, plagues are arguably less frequent as a result of monitoring and control efforts, and famines can sometimes be avoided by the remedial redistribution of international food reserves. Locusts, nevertheless, also continue to affect human well-being indirectly, for example by interfering with modern means of transport such as cars, trains, and aeroplanes by overwhelming the mechanical operation with masses of (crushed) bodies.

Turning to biotic triggers of outbreaks, it is fair to say that many of these are anthropogenic. Natural predators of insects may be inadvertently removed by changing land use, as when bat habitats are destroyed. The elimination of such useful creatures removes their nightly mosquito-control service, thereby producing an increased risk of disease transmission as discussed elsewhere. Other common examples include the spraying of agricultural insecticides, when predatory spiders are often also killed, and their prey can then "rebound" to create worse pest problems: The indirect hazard created is then one of threatening food security or other industries such as timber production. Importation of potential pest species into new environments can also pose a hazard, particularly when these environments are devoid of the insects' usual predators (as with the American cockroach, now a globally distributed threat to hygiene in urban kitchens). In anthropogenically simplified ecosystems, there is generally a lack of biodiversity, and insect pests are, therefore, more likely to escape from the biological control normally provided not only by predation, but by interspecific competition (other species competing effectively for the same resources, such as food or nesting areas). Abiotic (climatic) variables often enhance the effect of such biological mechanisms because pest species are generally the best adapted to exploiting environments with rapidly changing conditions – and it is, therefore, likely that the number and extent of insect hazards will increase with global climate change.

Imagined hazards

Insects have profoundly influenced Western culture through time, permeating our language, arts, history, philosophy, and religion. However, as human society has become progressively more urbanized, insects have become progressively more estranged: There is now generally an aversion to insects, with much misplaced fear resulting from inadequate information and lack of more regular contact. The range of such aversions spans the full spectrum of appropriate apprehension when faced with the possibility of a bee sting, through subclinical and clinical insect phobias, to full-blown psychotic delusions of insect attacks and infestations. Phobias involve an irrational fear of insects without the insect bites or infestation actually being experienced, and phobic objects are often totally harmless insects (e.g., dragonflies, moths, and crickets); delusions involve the patient believing that the bites or infestations are actually occurring (Weinstein and Slaney, 2004).

Outlook

It is worth concluding here with consideration of the paradoxical nature of attributing a hazardous nature to insects, both real and imagined. On the one hand, we have seen that mosquito-borne disease kills over a million people every year: In countries with these diseases, people generally do not fear insects, but accept them as part of everyday life (and in some places even depend on them as food). On the other hand, in most developed countries, there are few or no diseases transmitted by insects, and the risk of dying from a sting is less than one in a million (in the same order as being struck by lightning and much less likely than being murdered). Yet in these developed countries, insects are feared to the point of featuring in psychiatric disorders alongside devils and alien abductions. As increasingly intangible elements of urbanized Western culture, it seems likely that insects will continue to be over-rated as a hazard to our health and wealth.

Bibliography

Dent, D., 1993. *Insect Pest Management*. Wallingford, UK: CAB International.
Exodus 10:14. The Holy Bible, King James Version. Meridian, USA, 1974
Goddard, J., 1996. *Physician's Guide to Arthropods of Medical Importance*. Boca Raton: CRC Press.
Poinar, G., and Poinar, R., 2008. *What Bugged the Dinosaurs?* Princeton: Princeton University Press.
Weinstein, P., and Slaney, D., 2004. Psychiatry and insects: phobias and delusions of insect infestations in humans. In Capinera, J. L. (ed.), *Encyclopeadia of Entomology*. Dordrecht: Kluwer, pp. 1845–1849.
WHO, (2009). *Malaria*. Fact Sheet No. 94, January. http://www.who.int/mediacentre/factsheets/fs094/en/index.html. Accessed Jan 2009.

Cross-references

Aviation (Hazards to)
Biblical Events
Global Change
Myths and Misconceptions
Natural Hazards
Risk Perception

INSURANCE

Jaroslaw Dzialek
Jagiellonian University, Krakow, Poland

Synonyms

Assurance; Guarantee; Policy; Protection; Security

Definition

Insurance is a risk management tool that redistributes a disaster risk among a high number of insured individuals or businesses, and enables one to cover large and accidental losses incurred by them.

Overview

Insurance is a key element of a disaster management cycle, especially as part of the last stage of recovery and reconstruction. It allows each individual to transfer the risks they face to an intermediary that mutualizes these risks among a large number of the insured. When a disaster occurs in one area, they obtain a premium to cover their losses. However, insurance systems in different countries face a number of difficulties related to low insurability rates and the growth of insured damages.

According to the German reinsurance company Munich Re, between 2000 and 2008 natural catastrophe losses reached USD 898 billion, of which USD

287.6 billion (32%) were insured. In 2008, an earthquake in Sichuan, China, the largest natural disaster that year, caused USD 85 billion in economic losses, but only USD 300 million were actually insured. The same year, hurricane Ike destroyed parts of the Caribbean and southern United States, with overall estimated losses of USD 38 billion, but in this case almost half of the losses (USD 15 billion) were insured. Munich Re notices a continuous increase of extreme weather-related events likely linked to climate change (Natural catastrophes, 2009). This growth in claims puts a serious strain on insurance systems and, in consequence, some of the largest disasters in the United States have led to the bankruptcy of a number of small insurers and cessation of disaster insurance offers in hazard-prone areas (Godschalk et al., 1999).

Policyholders, even in the developed countries, are concerned about a low level of insurability (Paklina, 2003). This reluctance is in many cases the result of the human perception of the risk that neglects events with very low probabilities of occurrence, like natural hazards. People do not perceive insurance in the manner that economists assume, but treat it rather as a form of investment. They rather prefer to insure against high-probability, low-consequence events to those with low probability, but with high consequences (Slovic et al., 2005). Cognitive dissonance reduction, on the other hand, may lead to denying the risk and renouncing of mitigation actions (Facing Hazards and Disaster, 2006). Another explanation is that individuals expect to obtain governmental and private aid when a disaster happens. In that case, they perceive insurance as an unnecessary expense (Raschky and Weck-Hannemann, 2007). In Japan, where the national government does not generally authorize financial aid to individuals, the insurability rate is considerably higher (Japan Large-Scale Floods and Earthquakes, 2009).

Imperfections of the insurance market lead some governments to the introduction of new schemes, increasing market penetration, for example, compulsory insurance or government-backed reinsurance mechanisms (France Policies for Preventing and Compensating Flood Related Damage, 2006). In hazard-prone areas, two types of insurance exist: optional and a package system (Paklina, 2003). The first leads to very low penetration rates (in case of flood insurance between 5% and 10% in countries like Austria, Belgium, Germany, Italy, Mexico, or the Netherlands) as the demand for insurance coverage is high in flood-prone areas. Insurance rates are then expensive due to repeated damages. In the package system, the insurance covers the combination of such risks as floods, fires, hurricanes, storm surges, and earthquakes. The risk is spread over a higher number of the insured, across a number of areas with different levels of various disaster risks.

In 1968, the United States adopted the National Flood Insurance Program (NFIP), a unique system providing federally backed basic flood insurance for residential and commercial areas. Its main aim was to reduce the risk of flood losses, reduce demand for federal assistance, and to preserve and restore natural functions of floodplains (Wetmore et al., 2006). Its characteristic feature is linking of the insurance system with other flood mitigation measures. The Community Rating System (CRS) provides discounts on flood insurance for individuals in the communities that establish floodplain management programs, including public information, mapping and regulatory activities, flood damage reduction, and flood preparedness (Burby, 2001). Still, NFIP effectiveness is described as limited with a penetration rate growth from 13% in 1990 to an estimated 22–25% in 2002 (Paklina, 2003).

Bibliography

Burby, J. R., 2001. Flood insurance and floodplain management: the US experience. *Environmental Hazards*, **3**, 111–122.
Committee on Disaster Research in the Social Sciences: Future Challenges and Opportunities, Division on Earth and Life Studies, National Research Council of the National Academies, 2006. *Facing Hazards and Disasters. Understanding Human Dimensions*. Washington: The National Academies Press.
Godschalk, D., Beatley, T., Berke, P., Brower, D., and Kaiser, E., 1999. *Natural Hazard Mitigation. Recasting Disaster Policy and Planning*. Washington: Island Press.
OECD, 2009. *Japan Large-Scale Floods and Earthquakes*. Paris: OECD Publications. OECD Reviews of Risk Management Policies.
OECD, 2006. *France Policies for Preventing and Compensating Flood Related Damage*. Paris: OECD Publications. OECD Studies in Risk Management.
Munich Re Group, and Munich Re Group, 2009. *Natural catastrophes 2008. Analyses, assessments, positions*. München: Münchener Rück, Munich Re Group. Topics Geo Knowledge Series.
Raschky, P. A., and Weck-Hannemann, H., 2007. Charity hazard. A real hazard to natural disaster insurance? *Environmental Hazards*, **7**, 321–329.
Slovic, P., Peters, E., Finucane, M. L., and Macgregor, D. G., 2005. Affect, risk, and decision making. *Health Psychology*, **24**, 35–40.
Paklina, N., 2003. *Flood Insurance*. Paris: OECD.
Wetmore, F., Bernstein, G., Conrad, D., DiVincenti, C., Larson, L., Plasencia, D., Riggs, R., Monday, J., Robinson, M. F., and Shapiro, M., 2006. *An Evaluation of the National Flood Insurance Program. Final Report*. Washington: American Institutes for Research.

Cross-references

Civil Protection and Crisis Management
Cognitive Dissonance
Communicating Emergency Information
Community Management of Hazards
Cost (Economic) of Natural Hazards and Disasters
Cost-Benefit Analysis of Natural Hazard Mitigation
Disaster Relief
Disaster Risk Management
Economics of Disasters
Emergency Management
Emergency Planning
Federal Emergency Management Agency (FEMA)
Flood Protection
Hazardousness of Place
Livelihoods and Disasters
Perception of Natural Hazards and Disasters

Post Disaster Mass Care Needs
Recovery and Reconstruction After Disaster
Rights and Obligations in International Assistance
World Economy, Impact of Disasters

INTEGRATED EMERGENCY MANAGEMENT SYSTEM

Frank Fiedrich
Wuppertal University, Wuppertal, Germany

Synonyms
IEMS

Definition
The Integrated Emergency Management System (IEMS) refers to an all-hazard approach to the coordination, direction and control of disasters independent of their type, origin, size, and complexity. In the early 1980s, this term was coined by the Federal Emergency Management Agency, FEMA, of the United States.

Today the term Integrated Emergency Management System is sometimes also used for computer-based emergency management systems that allow managing all scales of incidents by fostering collaboration and information sharing. These systems are discussed elsewhere (see section Cross-references).

Elements of the integrated emergency management system
FEMA's Integrated Emergency Management System is a framework for effective emergency management that integrates partnerships on local, state, and federal level, including the collaboration of government agencies, private sector, and media. IEMS covers all phases of emergency management, namely mitigation, preparedness, response, and recovery. The IEMS framework describes different elements, processes, and principles for these emergency management phases. Important key elements of the IEMS are:

- **Hazard analysis**: An all-hazard analysis provides the basis for the other IEMS processes since it leads to a better understanding of the consequences and needs of different types of disasters.
- **Capability assessment**: All currently available response capabilities for different emergency response functions, like resource management, mass care, or communication need to be documented and assessed in order to identify shortfalls.
- **Emergency planning**: Emergency plans are based on the hazard analysis and capability assessment. They include a capability inventory, discuss possible consequences for different events, and list the required actions. Plans are important for training and response.
- **Capability maintenance and development**: Even if the current response capabilities are considered to be adequate the available capabilities have to be maintained by regular plan updates, resource maintenance, and training. In case of insufficient capabilities, short-, medium-, and long-term capability development plans have to be created.
- **Emergency response**: Emergency response activities should be based on the emergency plans and modified as necessary. Evaluations of after-action reviews need to be integrated into the existing plans.
- **Recovery and mitigation efforts**: The goal of the recovery efforts is to restore the community function to normal as soon as possible. Recovery itself can be viewed as an opportunity to create a more resilient community. Therefore mitigation efforts, like for example, strengthening infrastructures, can either be an integral part of the recovery phase or they can be initiated prior to a disaster.

Bibliography
McLoughlin, D., 1986. A framework for integrated emergency management. *Public Administration Review*, **45**, 165–172.

Cross-references
Civil Protection and Crisis Management
Community Management of Hazards
Critical Infrastructure
Disaster Relief
Disaster Risk Management
Emergency Management
Emergency Mapping
Emergency Planning
Evacuation
Federal Emergency Management System (FEMA)
Geographic Information Systems (GIS) and Natural Hazards
Hazard and Risk Mapping
Recovery and Reconstruction After Disaster
Risk Governance

INTENSITY SCALES

David Giles
University of Portsmouth, Portsmouth, UK

Synonyms
Earthquake measure; Earthquake severity; Earthquake size

Definition
Intensity Scales. A scale to measure the effects and degree of damage caused by an earthquake to the local environment and buildings affected by the seismic event using descriptive evidence to categorize the severity of the damage caused.

Introduction

The size and damaging effects or severity of an earthquake are described by measurements of both magnitude and intensity. In seismology (the study of earthquakes), scales of seismic intensity are used to measure or categorize the effects of the earthquake at different sites around its epicenter. Various seismic scales can be used to measure and compare the severity of the seismic event. The amount of elastic energy released by an earthquake is measured on a magnitude scale (see *Magnitude Measures*) whereas the effects of intensity of ground motion or "shaking" occurring at a given surface point are measured on an *intensity scale*.

Seismologists use earthquake intensity as their most widely applicable measure of the size of an earthquake. Intensity is measured by means of the degree of damage to structures of human origin, the amount of disturbance to the surface of the ground and the extent of human and animal reaction to the shaking (Bolt, 2006).

Intensity can therefore be defined as a classification of the strength of shaking at any place during an earthquake, in terms of its observed effects (Musson, 2002).

A series of *intensity scales* have been developed to semi-quantify this degree of damage and disturbance. These scales use descriptive evidence in order to establish the probable size of the earthquake causing the damage. This assessment of earthquake intensity depends on the macroseismic observations (observations of the actual effects of the earthquake), in what is termed the "meizoseismal zone" (the area of maximum disturbance during an earthquake where there is observable damage to buildings). It is not based on measuring the ground motion with instruments (microseismic observations). The descriptive scales have important uses, firstly in determining earthquake size in areas where there are limited seismographs to measure strong ground motion and secondly to retrospectively determine the size of historical earthquakes from contemporary accounts taken at the time of the seismic event. This method can be subjective and is very much dependent on the underlying geological ground conditions which may dampen or exaggerate the effects of the earthquake, but it does provide some limited data on the distribution of ground shaking during a particular event and gives an approximate estimate of the epicenter of the earthquake. It is therefore important that correlations can be made between the various scales used over historical time (Musson et al., 2009). The intensity at a point not only depends on the strength of the earthquake (magnitude) but also on the distance from the earthquake to that point and on the local geology at that point.

Intensity studies enable the macroseismic field of historical and contemporary earthquakes to be reconstructed and through this reconstruction it is often possible to identify the seismogenic source (Panza et al., 1991; Gasperini et al., 1999). The intensity parameter allows for a comparison to be made between more recent earthquakes and historical ones. This is based on the destructive effects described and detailed in the intensity scale used.

Macroseismic scales

A considerable number of macroseismic scales have been developed over the last 200 years in order to try and semi-quantify and describe the effects of an earthquake via a measure of intensity (Musson, 2002; Musson et al., 2009). The first such scale was considered to have been developed by the Italian Jacopo Gastaldi in 1564 (Hao et al., 2005; Xie, 1958). About 8 or so such scales have been more widely adopted, evolving over time and usage. The first widely adopted scales were developed by de Rossi (1874) and Forel (1881) leading to the Rossi-Forel Scale (de Rossi, 1883). The inadequacies of these early scales led to the development of more modern versions together with entirely new scales, an evolving

Intensity Scales, Table 1 Nonprescriptive guidelines to conversion from five major scales to EMS-98 (Musson et al., 2009)

RF	EMS 98	MCS 30	EMS 98	MMI 56	EMS 98	MSK 64	EMS 98	JMA 96	EMS 98
								0	1
1	1	1	1	1	1	1	1	1	2 or 3
2	2	2	2	2	2	2	2	2	4
3	3	3	3	3	3	3	3	3	4 or 5
4	4	4	4	4	4	4	4	4	5
5	5	5	5	5	5	5	5	5L	6
6	5	6	6	6	6	6	6	5U	7
7	6	7	7	7	7	7	7	6L	8
8	7 or 8	8	8	8	8	8	8	6U	9 or 10
9	9	9	9	9	9	9	9	7	11
10	Note 1	10	10	10	10	10	10		
		11	11	11	Note 1	11	11		
		12	Note 1	12		12	Note 1		

Note 1: This intensity is defined in such a way that it relates to phenomena that do not represent strength of shaking, e.g., those due to surface faulting, or reaches a saturation point in the scale where total damage refers to total damage to buildings without antiseismic design
Note 2: *RF* Rossi-Forel Scale, *MCS* Mercalli-Cancani-Sieberg Scale 1930, *EMS* European Macroseismic Scale 1998, *MMI* Modified Mercalli Intensity Scale 1956, *MSK* Medvedev-Sponheuer-Karnik Scale 1964, *JMA* Japanese Meteorological Agency 1996

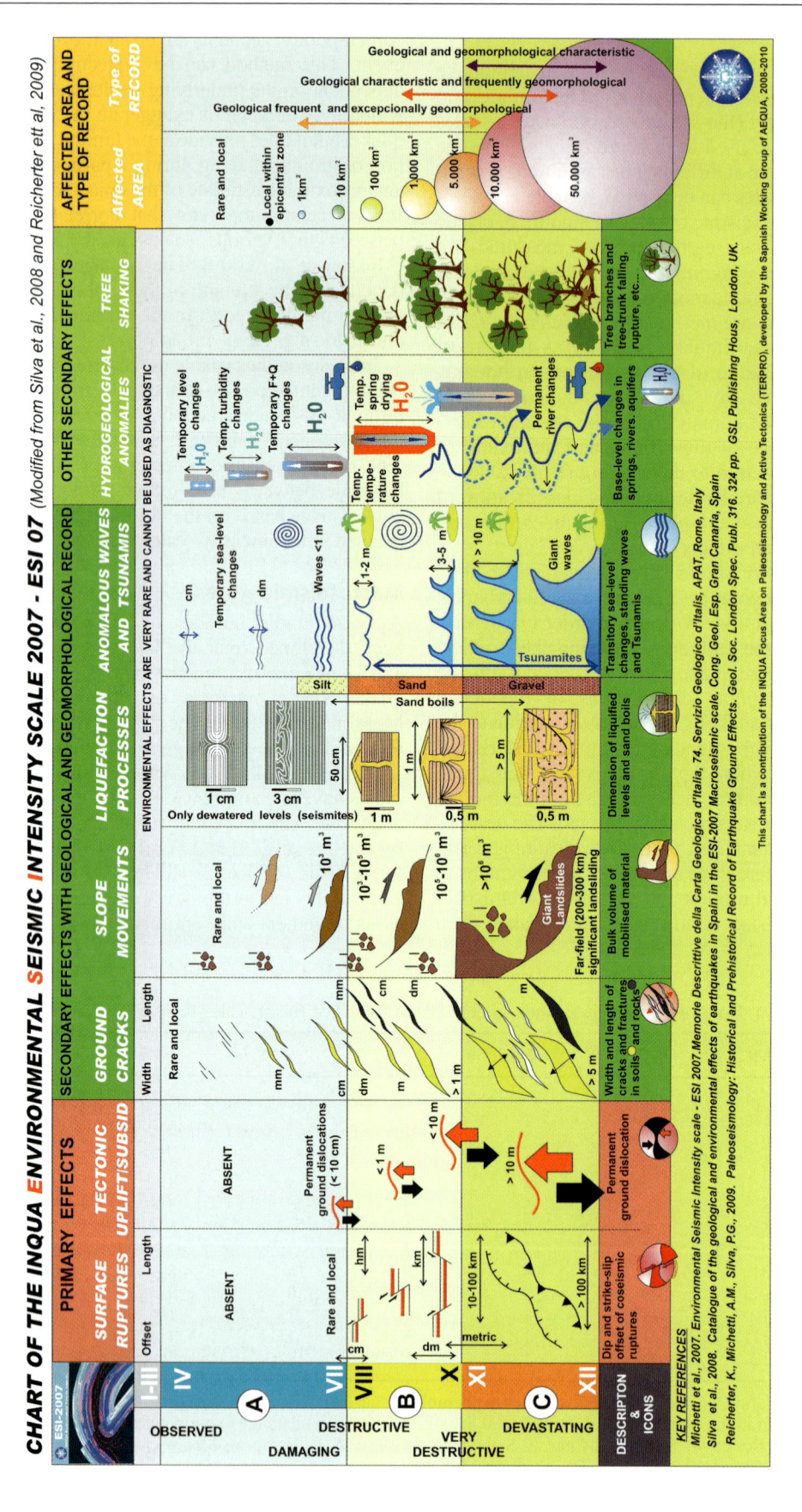

Intensity Scales, Figure 1 The environmental Seismic Intensity scale, ESI-07 (Reicherter et al., 2009; Silva et al., 2008).

Intensity Scales, Table 2 Modified Mercalli Scale after Richter (Richter, 1958)

Intensity level	Description
I	Not felt. Marginal and long period effects of large earthquakes
II	Felt by persons at rest, on upper floors, or favorably placed
III	Felt indoors. Hanging objects swing. Vibration like passing light trucks. Duration estimated. May not be recognized as an earthquake
IV	Hanging objects swing. Vibration like passing of heavy trucks; or sensation of a jolt like a heavy ball striking the walls. Standing motor cars rock. Windows, dishes, doors rattle. Glasses clink. Crockery clashes. In the upper range of IV, wooden walls and frame creak
V	Felt outdoors; direction estimated. Sleepers wakened. Liquids disturbed, some spilled. Small unstable objects displaced or upset. Doors swing, close, open. Shutters, pictures move. Pendulum clocks stop, start, change rate
VI	Felt by all. Many frightened and run outdoors. Persons walk unsteadily. Windows, dishes, glassware broken. Knickknacks, books, etc., off shelves. Pictures off walls. Furniture moved or overturned. Weak plaster and masonry D cracked. Small bells ring (church, school). Trees, bushes shaken (visibly, or heard to rustle)
VII	Difficult to stand. Noticed by drivers of motor cars. Hanging objects quiver. Furniture broken. Damage to masonry D, including cracks. Weak chimneys broken at roof line. Fall of plaster, loose bricks, stones, tiles, cornices (also unbraced parapets and architectural ornaments). Some cracks in masonry C. Waves on ponds; water turbid with mud. Small slides and caving in along sand or gravel banks. Large bells ring. Concrete irrigation ditches damaged
VIII	Steering of motor cars affected. Damage to masonry C; partial collapse. Some damage to masonry B; none to masonry A. Fall of stucco and some masonry walls. Twisting, fall of chimneys, factory stacks, monuments, towers, elevated tanks. Frame houses moved on foundations if not bolted down; loose panel walls thrown out. Decayed piling broken off. Branches broken from trees. Changes in flow or temperature of springs and wells. Cracks in wet ground and on steep slopes
IX	General panic. Masonry D destroyed; masonry C heavily damaged, sometimes with complete collapse; masonry B seriously damaged. (General damage to foundations.) Frame structures, if not bolted, shifted off foundations. Frames racked. Serious damage to reservoirs. Underground pipes broken. Conspicuous cracks in ground. In alluvial areas sand and mud ejected, earthquake fountains, sand craters
X	Most masonry and frame structures destroyed with their foundations. Some well-built wooden structures and bridges destroyed. Serious damage to dams, dikes, embankments. Large landslides. Water thrown on banks of canals, rivers, lakes, etc. Sand and mud shifted horizontally on beaches and flat land. Rails bent slightly
XI	Rails bent greatly. Underground pipelines completely out of service
XII	Damage nearly total. Large rock masses displaced. Lines of sight and level distorted. Objects thrown into the air
Masonry A	Good workmanship, mortar, and design; reinforced, especially laterally, and bound together by using steel, concrete, etc.; designed to resist lateral forces
Masonry B	Good workmanship and mortar; reinforced, but not designed in detail to resist lateral forces
Masonry C	Ordinary workmanship and mortar; no extreme weaknesses like failing to tie in at corners, but neither reinforced nor designed against horizontal forces
Masonry D	Weak materials, such as adobe; poor mortar; low standards of workmanship; weak horizontally

process that continues to this day (Musson et al., 2009). The comparison of different intensity scales is thus an important issue when trying to understand and cross-correlate historical seismic events (Table 1).

One of the first intensity scales describing earthquake effects on the environment is given in Annex C to the European Macroseismic Scale (EMS-98) by Grünthal (Grünthal, 1998). One of the first proposals of an intensity scale based on effects on rocks and considering terrain vulnerability, thus complementing the EMS-98 environmental scale, is presented by Vidrih (Vidrih et al., 2001). In 2007, another scale was constructed resulting from a series of international meetings (Michetti et al., 2004, 2007) (Figure 1). INQUA (International Union for Quaternary Research) ratified the Environmental Seismic Intensity Scale (ESI-2007) which is now being widely adopted (Reicherter et al., 2009).

All of these *intensity scales* use 10, or more commonly 12, degrees or classes of earthquake effects to describe and define the intensity of the earthquake and the consequent severity of its effects.

Historical development of macroseismic scales

The first widely accepted *intensity scale* was developed by P N C Egen in 1828 (Egen, 1828) followed by the work of de Rossi and Forel (Musson et al., 2009). In 1921 Charles Davison (Davison, 1921) identified 27 different *intensity scales* which increased to 39 in 1933 (Davison, 1933). Giuseppe Mercalli, born in Italy, was the principal developer of two key scales: the Mercalli, 1883 Scale and a second scale published in 1902 (Mercalli, 1883, 1902). These were principally modifications to the Rossi-Forel Scale. In 1904 another Italian Adolfo Cancani proposed a 12-point scale (Cancani, 1904) which added two extra degrees of intensity to deal with the reporting of very strong earthquakes. In 1912, August Sieberg, a German geophysicist, revised the scale definitions which involved a considerable expansion of the description entries (Sieberg, 1912). He continued to develop this scale and published the Mercalli-Cancani-Sieberg Scale in 1923 (Sieberg, 1923). This scale was further modified in 1931 by two American seismologists, Wood and Neumann,

Intensity Scales, Table 3 Ground effects in the MCS-1930, MM-1931, MSK-1964, and Japanese (JMA) intensity scales (After Esposito et al., 1997; Serva, 1994)

Ground effect	Scale equivalent
Cracks in saturated soil and/or loose alluvium:	
Up to 10 mm:	MSK: VI
A few cm:	MSK: VIII: MM: VIII; MCS: VIII
Up to l00mm:	MSK: IX; MM: IX
A few dm up to 1 m	MSK: X; MCS: X
Cracks on road backfills and on natural terrigenous slopes over 100 mm	MSK: VII, VIII, IX; MM: VIII; MCS: VIII
Cracks on dry ground or on asphalted roads	MSK: VII, IX, XI: MCS: X, XI; JMA: VI
Faults cutting poorly consolidated Quaternary sediments	MSK: XI; MCS: XI
Faults cutting bedrock at the surface	MSK: XII; JMA: VII
Liquefaction and/or mud volcanoes and/or subsidence	MSK: IX, X; MM: IX, X; MCS: X, XI
Landslides in sand or gravel artificial dykes	MSK: VII, VIII, X; MM: VII; MCS: VII
Landslides in natural terrigenous slopes	MSK: VI, IX, X, XI; MM: X; MCS: X, XI; JMA: VI, VII
Rockfalls	MSK: IX, XI, XII; MM: XII; MCS: X, XI
Turbulence in the closed water bodies and formation of waves	MSK: VII, VIII, IX; MM: VII; MCS: VII, VIII
Formation of new water bodies	MSK: VIII, X, XII; MCS: XII
Change in the direction of flow in watercourses	MSK: XII; MCS: XII
Flooding	MSK: X, XII; MM: X; MCS: X
Variation in the water level of wells and/or the flow rate of springs	MSK: V, VI, VII, VIII, IX, X; MM: VIII; MCS: VII, X
Springs which dry out or are starting to flow	MSK: VII, VIII, IX

MSK Medvedev-Sponheuer-Karnik 1964 Scale, *MCS* Mercalli-Cancani-Sieberg 1930, *MM*: Modified Mercalli 1931 Scale, *JMA* Japanese Meteorological Agency 1996 Scale

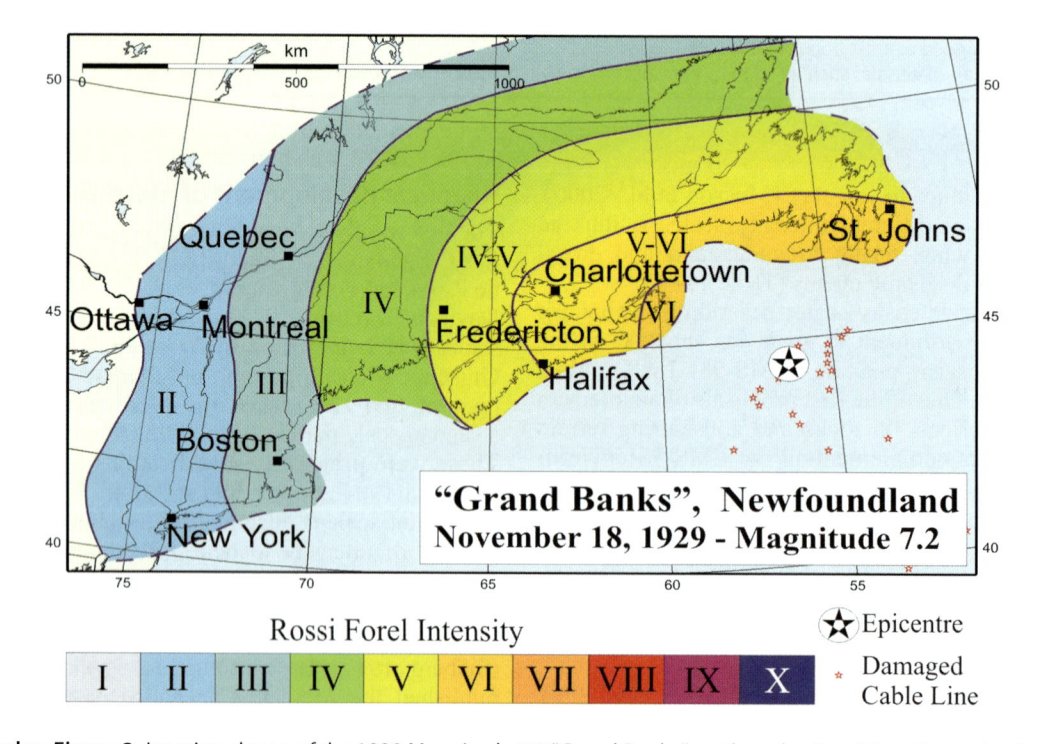

Intensity Scales, Figure 2 Isoseismal map of the 1929 Magnitude 7.2 "Grand Banks" earthquake, Rossi-Forel Intensity Scale (Halchuk, S., Geological Survey of Canada).

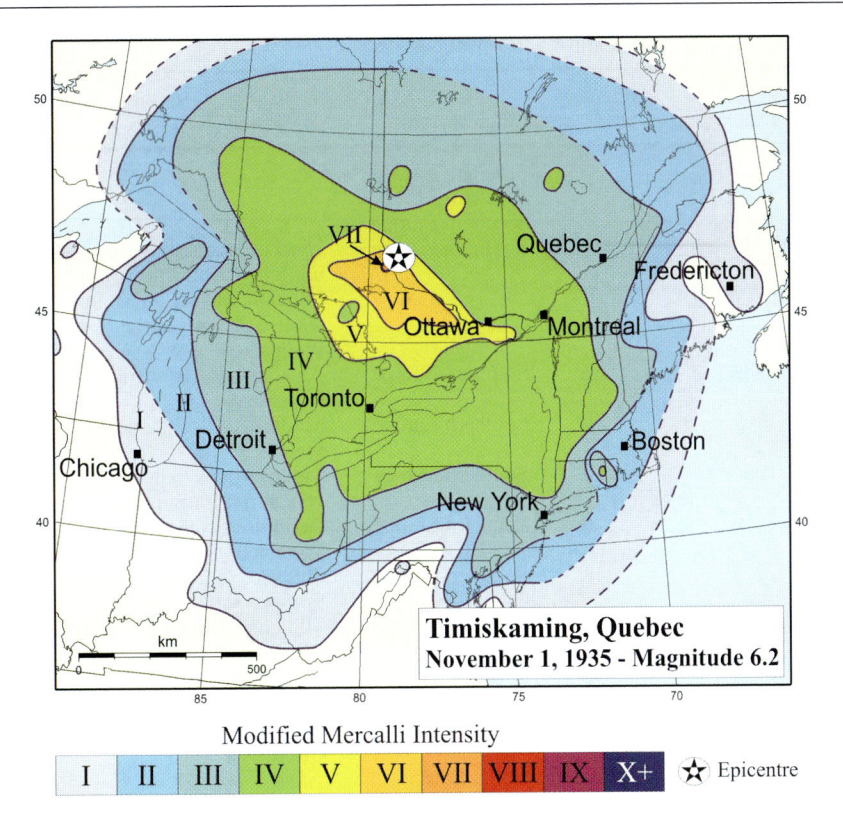

Intensity Scales, Figure 3 Isoseismal map of the Timiskaming earthquake 1935, Modified Mercalli Intensity Scale (Halchuk, S., Geological Survey of Canada).

where the original work was translated into English with some changes and published as the Modified Mercalli Scale (Wood and Neumann, 1931). This was completely reworked in 1956 by Charles Richter (Richter, 1958) to become the Modified Mercalli Scale of 1956 (Musson, 2002) (Table 2).

In Europe initial work by Sergei Medvedev, a Russian, together with a Czechoslovak Vit Kárník and an East German Wilhelm Sponheuer led to another 12-point scale being published in 1964 known as the Medvedev-Sponheuer-Kárník or MSK-64 Scale (Medvedev et al., 1964). In 1988, the European system was again further developed to include aspects of more modern building types and designs which resulted in the European Macroseismic Scale, EMS-98 (Grünthal, 1998).

The Japanese Meteorological Agency published a very detailed *intensity scale* which was idiosyncratic to aspects of Japanese buildings and environment (JMA, 1996). This *intensity scale* is measured in units of *shindo* which literally means "degree of shaking." The present use of the JMA scale converts instrumental ground motion readings into these intensity values (Musson et al., 2009). Cross-correlation with previous versions of the scale and other scales is difficult.

In China, the China Seismic Intensity Scale (CSIS) is a 12-degree system created in 1954 by Prof. Li Shan-Bang (Li, 1954) and rewritten by Prof. Xie Yu-sou in 1957 (Xie, 1957). The China Seismic Intensity Scale was supplemented and simplified by Prof. Liu Hui-Xian in 1978 (Liu, 1978), taking account of new kinds of modern buildings, which had been destroyed by strong earthquakes in recent years both in China and elsewhere in the world (Wang, 2004). The scale is sometimes referred to as the Lièdù Scale, literally "degrees of violence."

Environmental seismic intensity scale (ESI-2007)

The INQUA Environmental Seismic Intensity Scale is a 12-degree macroseismic scale which follows the same basic structure of the other historical 12-point scales (Michetti et al., 2007; Reicherter et al., 2009). The main advantage of the ESI 2007 scale is the classification, quantification, and measurement of several known geological, hydrological, botanical, and geomorphological features for different intensity degrees (Reicherter et al., 2009). The scale differentiates two main categories of earthquake effects on the environment:

Primary
Fault surface ruptures
Tectonic uplift/subsidence

Intensity Scales, Table 4 Summary of significant macroseismic scales

Scale name	Acronym	Date	Intensity levels	Original usage	Comments	Reference
Egen		1828		Southern Europe		Egen (1828)
de Rossi		1874		Southern Europe		de Rossi (1874)
Forel		1881		Southern Europe		Forel (1881)
Rossi-Forel	RF	1883	10	Southern Europe	Used for about two decades until the introduction of the Mercalli intensity scale in 1902. Still used in the Philippines	de Rossi (1883)
Mercalli		1883	6	Southern Europe	An adaptation of the de Rossi scale.	Mercalli (1883)
Mercalli		1902	10	Southern Europe	Modification of the de Rossi scale.	Mercalli (1902)
Cancani		1904	12	Southern Europe	Modifications to deal with very strong earthquakes.	Cancani (1904)
Sieberg		1912	6	Southern Europe	Definitions considerably expanded.	Sieberg (1912)
Mercalli-Cancani-Sieberg	MCS	1923	12	Southern Europe Global	Still used today in Southern Europe.	Sieberg (1923, 1930)
Modified Mercalli	MM-31	1931	12	Southern Europe USA	MCS translated into English with some additions.	Wood and Neumann (1931)
Modified Mercalli of 1956	MM-56	1956	12	Southern Europe USA	Complete overhaul of the 1931 version.	Richter (1958)
Medvedev		1953	12	Russia	Also known as the GEOFIAN scale.	Medvedev (1953)
Medvedev-Sponheuer-Karnik	MSK-64	1964	12	USSR Europe, India, Russia, CIS		Medvedev et al. (1964)
	MSK-81	1981	12		Minor modifications made. Became EMS-92	Grünthal (1993)
European Macroseismic Scale	EMS-98	1998	12	Europe, Global	Employs vulnerability classes and involve construction type.	Grünthal (1998)
China Seismic Intensity	CSIS	1954	12	China	Lièdù Scale	Li (1954)
		1957	12	Hong Kong		Xie (1957)
		1980	12			Liu (1978)
Japan Meteorological Agency Seismic Intensity	JMA	1884	4	Japan, Taiwan	Shindo scale	Japanese Meteorological (1996)
		1898	8		Changed to a numerical system	
		1908	8		Levels given descriptions	
	JMA-96	1996	10		Following the Great Hanshin earthquake in 1995, Shindo was further expanded to a total of 10 levels	
Environmental seismic intensity	ESI-2007	2007	12	Europe, Global	Ratified following the 17th INQUA Congress.	Michetti et al. (2004)

Secondary
Ground cracks
Slope movements
Liquefaction processes
Anomalous waves and tsunamis
Hydrogeological anomalies
Tree shaking

Primary effects triggered by surface faulting can be further sub categorized:

Almost absent for intensity degrees below VIII
Characteristic but moderate for intensities between VIII and X
Diagnostic for stronger intensities of XI and XII

Table 3 highlights the ground effects and the associated intensity degree from the various macroseismic scales. Figure 1 summarizes the ESI-2007 scheme and highlights the main diagnostic characteristics of the different effects resulting from a seismic event. The chart provides a qualitative framework to assess affected areas, the geological and geomorphological setting of the area and their respective degree of presentation through time. Reicherter et al. (2009) provides a detailed appendix, which outlines the definitions of the intensity degrees used in the ESI-2007 scale. Various case studies are available (Mosquera-Machado et al., 2009; Papanikolaou et al., 2009; Tatevossian et al., 2009; Ota et al., 2009) which have been used to benchmark this *intensity scale*.

Intensity and isoseismal maps

The presentation of intensity data is usually done in the form of a map (Musson, 2002). By drawing lines on a map between places of equal damage (hence equal intensity) a series of isoseismal (term coined by Robert Mallet in 1862) contours can be created. This isoseismal line can be defined as a line bounding the area within which the intensity is predominantly equal to, or greater than, a given value. Figures 2 and 3 show examples of isoseismal maps for two different *intensity scales*.

Arias intensity

In 1970, Arturo Arias, a Chilean engineer, proposed a way to determine objectively the intensity of shaking during an earthquake by measuring the acceleration of transient seismic waves (Arias, 1970). It is an important measure of the strength of ground motion. The time-integral of the square of the ground acceleration became known as the Arias Intensity (Equation 1), which represents the square root of the energy per mass thus having units of "m/s." This intensity must not be confused with the macroseismic intensity scales, which describe the subjective intensity of shaking as reported by people and building damage.

$$\textbf{Arias Intensity } I_A = \frac{\pi}{2g} \int_0^{T_d} a(t)^2 dt (\text{m/s}) \quad (1)$$

Where g is the acceleration due to gravity and T_d is the duration of signal above the threshold level.

Summary

A variety of *Intensity Scales* have been developed to measure the degree of damage caused by an earthquake to the local environment and buildings affected during a seismic event. A variety of descriptive evidence is used to categorize the severity of the damage caused. Many scales have been developed and modified for local site conditions (Table 4).

Bibliography

Arias, A., 1970. A measure of earthquake intensity. In Hansen, R. J. (ed.), *Seismic Design for Nuclear Power Plants*. Cambridge, MA: MIT Press, pp. 438–483.

Bolt, B. A., 2006. *Earthquakes*. New York: W.H. Freeman and Company.

Cancani, C., 1904. Sur l'emploi d'une double echelle seismique des intensites, empirique et absolue. *Gerlands Beiträge für Geophysik*, **2**, 281–283.

Davison, C., 1921. On scales of seismic intensity and on the construction of isoseismal lines. *Bulletin of the Seismological Society of America*, **11**, 95–129.

Davison, C., 1933. Scales of seismic intensity: supplementary paper. *Bulletin of the Seismological Society of America*, **23**, 158–166.

de Rossi, M. S. L., 1874. Bibliografia con annotazione. *Bullettino del Vulcanismo Italiano*, **1**, 46–56.

de Rossi, M. S. L., 1883. Programma dell'Osservatorio ed Archivio Centrale Geodinamico presso il R. Comitato Geologico d'Italia. *Bollettino del Vulcanismo Italiano*, **10**, 3–128.

Egen, P. N. C., 1828. Über das Erdbeben in den Rheinund Niederlanden vom. *Annales de Physique*, **13**, 153–163.

Esposito, E., Porfido, S., Mastrolorenzo, G., Nikonov, A. A., and Serva, L., 1997. Brief review and preliminary proposal for the use of ground effects in the Macroseismic Intensity Assesment. In *Proceedings. 30th International Geological Congress, Beijing, China, Vol 5. Contemporary Lithospheric Motion and Seismic Geology*, pp. 233–243.

Forel, F. A., 1881. Intensity scale. *Archives des Sciences Physiques et Naturelles*, **6**, 465–466.

Gasperini, P., Bernardini, F., Valensise, G., and Boschi, E., 1999. Defining seismogenic sources from historical earthquake felt reports. *Bulletin of the Seismological Society of America*, **89**, 94–110.

Grünthal, G. (ed.), 1993. *European Macroseismic Scale 1992. (Updated MSK Scale)*. Luxembourg: Centre Européen de Géodynamique et de Séismologie. Cahiers du Centre Européen de Géodynamique et de Séismologie, Vol. 7.

Grünthal, G. (ed.), 1998. *European Macroseismic Scale 1998. (EMS-98). Cahiers du Centre Européen de Géodynamique et de Séismologie 15*. Luxembourg: Centre Européen de Géodynamique et de Séismologie, p. 99.

Hao, M., Xie, L., and Xu, L., 2005. Some considerations on the physical measure of seismic intensity. *Acta Seismologica Sinica*, **18**(2), 245–250.

Japanese Meteorological Agency, 1996. Explanation Table of JMA Seismic Intensity Scale.

Li, S. B., 1954. On the application of seismic scales. *Acta Geophysica Sinica*, **3**(1), 35–54.

Liu, H. X., 1978. On the concept and application of earthquake intensity. *Acta Geophysica Sinica*, **21**(4), 340–351.

Mallet, R., 1862. *Great Neapolitan Earthquake of 1857: The First Principles of Observational Seismology*. London: Royal Society.

Medvedev, S. V., 1953. A new seismic scale. *Trudy Geofizicheskogo Instituta, Akademiya Nauk SSSR*, **21**, 148.

Medvedev, S., Sponheuer, W., and Karnik, J., 1964. Neue seismische skala. In Veröff, J. (ed.), *Institut für Bodendynamik und Erdbebenforschung in Jena*. Berlin: Deutsche Akademie der Wissenschaften zu, Vol. 77, pp. 69–76.

Mercalli, G., 1883. Vulcani e fenomeni vulcanici in Italia. In Negri, G., Stoppani. A., Mercalli. G. (eds.), *Geologia d'Italia*. Vallardi, pp. 217–218.

Mercalli, G., 1902. Sulle modificazioni proposte alla scala sismica De Rossi–Forel. *Bollettino della Società Sismologica Italiana*, **8**, 184–191.

Michetti, A. M., Esposito, E., et al., 2004. The INQUA Scale. An innovative approach for assessing earthquake intensities based on seismically-induced ground effects in natural environment. In Vittori, E., and Comerci, V. (eds.), *Special Paper, APAT, Memorie Descrittive della Carta Geologica d'Italia*. Roma: SystemCart Srl, Vol. LXVII.

Michetti, A. M., Esposito, E., et al., 2007. Environmental Seismic Intensity scale 2007 – ESI 2007. In Vittori, E., and Guerrieri, L. (eds.), *Memorie Descrittive della Carta Geologica d'Italia. Dipartimento Difesa del Suolo, APAT, SystemCart Srl*. Roma: Servizio Geologico d'Italia, Dipartimento Difesa del Suolo, APAT, SystemCart Srl, Vol. LXXIV, pp. 7–54.

Mosquera-Machado, S., Lalinde-Pulido, C., Salcedo-Hurtado, E., and Michetti, A. M., 2009. Ground effects of the 18 October 1992, Murindo earthquake (NW Colombia), using the Environmental Seismic Intensity scale (ESI 2007) for the assessment of intensity. In Reicherter, K., Michetti, A. M., and Silva, P. G. (eds.), *Palaeoseismology: Historical and Prehistorical Records of Earthquake Ground Effects for Seismic Hazard Assessment*. London: Geological Society, Special Publications, 316, pp. 123–144.

Musson, R., 2002. Intensity and intensity scales. In Bormann, P. (ed.), *IASPEI New Manual of Seismological Observatory Practice*. Potsdam: GeoForschungs Zentrum Potsdam, Vol. 12, pp. 1–20, doi:10.2312/GFZ.NMSOP_r1_ch12.

Musson, R. M. W., Grünthal, G., and Stucchi, M., 2009. The comparison of macroseismic intensity scales. *Journal of Seismology*, doi:10.1007/s10950-009-9172-0.

Ota, Y., Azuma, T., and Lin, Y. N., 2009. Application of INQUA environmental Seismic Intensity scale to recent earthquakes in Japan and Taiwan. In Reicherter, K., Michetti, A. M., and Silva, P. G. (eds.), *Palaeoseismology: Historical and Prehistorical Records of Earthquake Ground Effects for Seismic Hazard Assessment*, 316th edn. London: Geological Society, pp. 55–71. Special Publications.

Panza, G. F., Craglietto, A., and Suhadolc, P., 1991. Source geometry of historical events retrieved by synthetic isoseismals. In Stucchi, M., and Postpischl, D., (eds.), *Multidisciplinary Evaluation of Historical Seismicity, Tectonophysics*, Vol. 192, pp. 173–184.

Papanikolaou, D., Papanikolaou, D. I., and Lekkas, E. L., 2009. Advances and limitations of the Environmental Seismic Intensity scale (ESI 2007) regarding near-field and far-field effects from recent earthquakes in Greece: implications for the seismic hazard assessment. In Reicherter, K., Michetti, A. M., and Silva, P. G. (eds.), *Palaeoseismology: Historical and Prehistorical Records of Earthquake Ground Effects for Seismic Hazard Assessment*. London: Geological Society, Vol. 316, pp. 11–30. Special Publications.

Reicherter, K., Michetti, A. M., and Silva Barroso, P. G., 2009. Palaeoseismology: historical and prehistorical records of earthquake ground effects for seismic hazard assessment. In Reicherter, K., Michetti, A. M., and Silva, P. G. (eds.), *Palaeoseismology: Historical and Prehistorical Records of Earthquake Ground Effects for Seismic Hazard Assessment*. London: The Geological Society, Vol. 316, pp. 1–10. Special Publications.

Richter, C. F., 1958. *Elementary Seismology*. San Francisco: W.H. Freeman, p. 768.

Serva, L., 1994. Ground effects in the intensity scales. *Terra Nova*, 6, 414–416.

Sieberg, A., 1912. Über die Makroseismische Bestimmung der Erdbebenstrke. *Gerlands Beiträge für Geophysik*, 11, 227–239.

Sieberg, A., 1923. *Geologische, Physikalische und Angewandte Erdbebenkunde*. G. Fischer, Jena.

Sieberg, A., 1930. Geologie der Erdbeben. *Handbuch der Geophysik*, 2(4), 552–555.

Silva, P. G., Rodriguez Pascua, M. A., et al., 2008. Catalogación de los efectos geológicos y ambientales de los terremotos en España en la Escala ESI-2007 y su aplicación a los estudios paleosismológicos. *Geotemas*, 6, 1063–1066.

Tatevossian, R. E., Rogozhin, E. A., Arefiev, S. S., and Ovsyuchenko, A. N., 2009. Earthquake intensity assessment based on environmental effects: principles and case studies. In Reicherter, K., Michetti, A. M., and Silva, P. G. (eds.), *Palaeoseismology: Historical and Prehistorical Records of Earthquake Ground Effects for Seismic Hazard Assessment*. London: Geological Society, Vol. 316, pp. 73–91. Special Publications.

Vidrih, R., Ribičič, M., and Suhadolc, P., 2001. Seismogeological effects on rocks during the April 12, 1998 Upper Soča Territory Earthquake (NW Slovenia). *Tectonophysics*, 330, 153–175.

Wang, J., 2004. Historical earthquake investigation and research in China. *Annals of Geophysics*, 47(2/3), 831–838.

Wood, H. O., and Neumann, F., 1931. Modified Mercalli Intensity scale of 1931. *Bulletin of the Seismological Society of America*, 21, 277–283.

Xie, Y. S., 1957. A new scale of seismic intensity adapted to the conditions in Chinese territories. *Acta Geophysica Sinica*, 6(1), 35–48.

Xie, Y. S., 1958. The seismic intensity scale. *Chinese Journal of Civil Engineering*, 5(2), 73–85.

WebLinks

British Geological Survey. http://www.earthquakes.bgs.ac.uk
Earthquakes Canada. http://earthquakescanada.nrcan.gc.ca
European Macroseismic Scale EMS-98. http://www.gfz-potsdam.de/portal/gfz/home
IASPEI New Manual of Seismological Observatory Practice, Chapter 12. dx.doi.org/10.2312/GFZ.NMSOP_r1_ch12
Japanese Meteorological Agency Explanation Table of JMA Seismic Intensity Scale. http://www.jma.go.jp/jma/en/Activities/inttable.html
The INQUA Scale: Environment Intensity Scale ESI-2007. http://www.isprambiente.gov.it/site/en-gb/Projects/INQUA_Scale/Environmental_Seismic_Intensity_Scale_-_ESI_2007
United States Geological Survey. http://earthquake.usgs.gov

Cross-references

Accelerometer
Damage and the Built Environment
Earthquake
Earthquake Damage
Epicentre
Federal Emergency Management Agency
Harmonic Tremor
Isoseismal
Macroseismic Survey
Magnitude Measures
Modified Mercallit (MM) Scale
Primary wave (P wave)
Richter, Charles Francis (1900–1985)
Secondary wave (S wave)
Tectonic and Tectono-Seismic Hazards
Tectonic Tremor

INTERNATIONAL STRATEGIES FOR DISASTER REDUCTION (IDNDR AND ISDR)

Karl-Otto Zentel[1], Thomas Glade[2]
[1]Deutsches Komitee Katastrophenvorsorge, Bonn, Germany
[2]University of Vienna, Vienna, Austria

Terminology

Disaster Risk Reduction: The concept and practice of reducing disaster risks through systematic efforts to analyze and manage the causal factors of disasters, including through reduced exposure to hazards, lessened vulnerability of people and property, wise management of land and the environment, and improved preparedness for adverse events.

Early Warning: The set of capacities needed to generate and disseminate timely and meaningful warning information to enable individuals, communities, and organizations threatened by a hazard to prepare and to act appropriately and in sufficient time to reduce the possibility of harm or loss.

ENSO-El Nino Southern Oscillation Phenomenon: A complex interaction of the tropical Pacific Ocean and the global atmosphere that results in irregularly occurring episodes of changed ocean and weather patterns in many parts of the world, often with significant impacts over many months, such as altered marine habitats, rainfall changes, floods, droughts, and changes in storm patterns.

National Platform for Disaster Risk Reduction: A generic term for national mechanisms for coordination and policy guidance on disaster risk reduction that are multi-sectoral and interdisciplinary in nature, with public, private, and civil society participation involving all concerned entities within a country (UNISDR, 2009, Terminology of Disaster Risk Reduction).

Background

During the 1970s and 1980s more than 800 million people were affected by natural disasters. More than 23 billion US$ damage were caused and 3 million people killed (General Assembly A/RES/43/202). The worst disasters were the droughts in sub-Saharan Africa region and floods in Southeast Asia, each of which claimed several hundred thousand victims.

These developments led to an increasing attention to the subject of disaster reduction. The President of the American Society of Science, Frank Press, started an initiative to declare the 1990s a Decade for Disaster Reduction. The aim of this initiative was to integrate already existing scientific knowledge in the field of disaster reduction into development decisions and projects.

The increasing losses caused by natural disasters led to a number of developments in the humanitarian sector. The tragic events of the 1980s led to the decisions in 1992 at the European Commission to establish the European Commission Humanitarian Office (ECHO) and at the United Nations to improve coordination of the work of the UN Agencies by a central coordinating organization with a humanitarian mandate (today), the Office for the Coordination of Humanitarian Affairs (OCHA).

In its Report "Our Common Future," the World Commission on Environment and Development (so-called Brundtland Commission) made a clear link between the need for implementation of disaster reduction measures and sustainable development (Report of the World Commission on Environment and Development "Our Common Future").

> The Commission has sought ways in which global development can be put on a sustainable path into the 21st century.

During the 1970s, twice as many people suffered each year from "natural" disasters as during the 1960s. The disasters most directly associated with environment/development mismanagement – droughts and floods – affected the most people and increased most sharply in terms of numbers affected. Some 18.5 million people were affected by drought annually in the 1960s. 24.4 million in the 1970s. There were 5.2 million flood victims yearly in the 1960s. 15.4 million in the 1970s. Numbers of victims of cyclones and earthquakes also shot up as growing numbers of poor people built unsafe houses on dangerous grounds.

The results are not in for the 1980s. But we have seen 35 million affected by drought in Africa alone and tens of millions affected by the better managed and thus less- publicized Indian drought. Floods have poured off the deforested Andes and Himalayas with increasing force. The 1980s seem destined to sweep this dire trend on into a crisis-filled 1990s (Report of the World Commission on Environment and Development "Our Common Future". p 23).

The Commission identified the link between the negative effects of natural disasters and poverty: "*Such disasters claim most of their victims among the impoverished in poor nations, ...*" (Report of the World Commission on Environment and Development "Our Common Future"., p 42).

Reasons for the increasing negative effects of natural disasters were seen in unsolved development problems: "*All major disaster problems in the Third World are essentially unsolved development problems. Disaster prevention is thus primarily an aspect of development, and this must be a development that takes place within sustainable limits.*" (Grann et al., 1985, p 43)

The start

In its 42nd General Assembly in 1987 the United Nations – based on the Report of the World Commission on Environment and Development – already made important and guiding decisions to establish an International Decade on natural disaster reduction in the 1990s. It was expected that special attention should be given to less developed countries as those are the most affected by natural disasters. Goals for the Decade were identified and member states were asked to establish national committees in order to support the goals. The main emphasis was given towards the transfer of technology and scientific knowledge in the field of disaster reduction. Resolutions (refer to Box 1).

An Ad Hoc Expert Group, chaired by Frank Press, was established to further define the role and structure of the Decade. On April 11, 1989, this expert group provided a joined declaration the "Tokyo Declaration" in which the experts clearly identified the root causes of the increasing losses, identified the important role of the international system, and underlined the need for national activities to support the decade.

Within the Tokyo Declaration it was stated:

> Vulnerability to natural disasters is rising due to population growth, urbanization, and the concentration of industry and infrastructure in disaster-prone areas....
> We believe that the Decade is a moral imperative. It is the first coordinated effort to prevent the unnecessary loss of life from natural hazards. ...

The Group calls for all countries to form national committees to plan for and coordinate national efforts (Tokyo Declaration, 1989, p.1).

In its letter to the Secretary General dated June 1, 1989, Frank Press gave a strong statement toward the need for action, for integrated approaches combining different constituencies as well as a continuum approach throughout the disaster management cycle.

> Fatalism is not longer acceptable; it is time to bring the full force of scientific and technological advancement to reduce the human tragedy and economic loss from natural disaster. We must take an integrated approach to disaster reduction, bringing new emphasis to pre disaster planning, preparedness, and prevention, while sustaining our post disaster relief capabilities. Our humanitarian efforts must be broadened to encompass disaster- resistant instrument as well as timely warnings in which people at risk receive, understand and act upon the information conveyed (ref. June 1, 1989 p iii).

At the 44th session the General Assembly of the United Nations the members proclaimed IDNDR.

para1"Proclaims the International Decade for Natural Disaster Reduction, beginning on 1 January 1990"; (GA 44/236; 85th Plenary meeting; 22 December 1989)

The goals defined in the Resolution of 1987 were kept as objectives for the Decade. The member states agreed:

> The objective of the International Decade for Natural Disaster Reduction is to reduce through concerted international action, especially in developing countries, the loss of life, property damage and social and economic disruptions caused by natural disasters such as earthquakes, windstorms, tsunamis, floods, landslides, volcanic eruptions, wildfires, grasshopper and locust infestations, drought and desertification and other calamities of natural origin (ref. GA/RES/44/236 Annex A).

Member states were asked to establish National Committees and inform the Secretary General about their initiatives. IDNDR became through this request one of the very few United Nations organizations/structures linked to corresponding national structures. UNESCO and WMO are examples of similar protocols. A special high-level council should be established to provide advice to the Secretary General. A Scientific and Technical Committee (STC) to develop overall programs, identify gaps in technical knowledge, and to assess and evaluate activities carried out was foreseen.

A secretariat was established – attached to the United Nations Disaster Relief Coordinator - to manage and coordinate the day-to-day work. The United Nations Development Programme (UNDP) became the focal point to oversee the programs. The resident coordinators were asked to integrate the goals of the Decade into their work. The World Meteorological Organisation (WMO) and the United Nations Education and Scientific Organisation (UNESCO) became major contributors to the Decade. In order to fund the secretariat and programs, member states were asked for voluntary contributions to a trust fund. Already in this declaration a review of the Decade was foreseen for 1994.

Based on the broadly defined goals, the STC identified a number of specific areas of activities which would mark progress to be achieved at the end of the Decade period (ref. Plate/Merz Naturkatastrophen p. 33). Herein it was claimed that by the year 2000, all countries should have in place:

1. Comprehensive national assessments of risks from natural hazards, with these assessments taken into account in development plans
2. Mitigation plans at national and/or local levels, involving long-term prevention and preparedness and community awareness
3. Ready access to global, regional, national, and local warning systems and broad dissemination of warnings

---

Box 1

"The General Assembly . . .

Recognizing the responsibility of the United Nations system for promoting international cooperation in the study of natural disasters of geophysical origin and in the development of techniques to mitigate risks arising there from, as well as for co-ordinating disaster relief, preparedness and prevention, including prediction and early warning.

para 3. Decides to designate the 1990s as a decade in which the international community, under the auspices of the Unites Nations, will pay special attention to fostering international co-operation in the field of natural disaster reduction.

para 4. Goals

(a) To improve the capacity of each country to mitigate the effects of natural disasters expeditiously and effectively, paying special attention to assisting developing countries in the establishment, when needed, of early warning systems;
(b) To devise appropriate guidelines and strategies for applying existing knowledge, taking into account the cultural and economic diversity among nations;
(c) To foster scientific and engineering endeavours aimed at closing critical gaps in knowledge in order to reduce loss of life and property;
(d) To disseminate existing and new information related to measures for the assessment, prediction, prevention and mitigation of natural disasters;
(e) To develop measures for the assessment, prediction, prevention and mitigation of natural disasters through programmes of technical assistance and technology transfer, demonstration projects, and education and training, tailored to specific hazards and locations, and to evaluate the effectiveness of those programmes;

para 7. Calls upon all governments to participate during the decade in concerted international action for the reduction of natural disasters and, appropriate, to establish national committees, in co-operation with the relevant scientific and technological communities with a view to surveying available mechanisms and facilities for the reduction of natural hazards, ..." (42/169, 1987 96th plenary meeting, 11. December 1987)

The midterm review "World Conference on Natural Disaster Reduction, Yokohama, Japan, 23–27 May 1994"

In the beginning years of the Decade, the efforts made were very scientifically oriented, giving special emphasis on technical possibilities in the sectors of prevention and mitigation. The Yokohama conference 2004 clearly recognized the still existing gaps between the vision – as formulated at the beginning of the Decade – and reality. This reality was that the goals and targets were far from being achieved as one would have expected after half of the Decade had passed by: Many of the delegates of the 147 nations represented at the conference commented that 95% of the Decade's work needs to be done in its second half. In the "Assessment of the status of disaster reduction midway into the Decade/Yokohama Strategy" it was concluded that *"Awareness of the potential benefits of disaster reduction is still limited to specialized circles and has not been successfully communicated to all sectors of society, in particular policy makers and the general public"* (Yokohama Strategy and Plan of Action, 1994, p 7).

The conference came to the conclusion that the intended impact of the technical solutions could only achieve the given goals if they were integrated in an appropriate socioeconomic and political framework. This can be evaluated as a major turnover for problem-oriented thinking and management. Suddenly, the involved parties realized that structural measurements indeed help to address the problems; however, they were far from solving the issues.

The conference unanimously accepted the declaration of the "Yokohama Strategy." The 18-page document gives clear outlines and a plan of action although no specific disaster or action is mentioned in detail. The targets of IDNDR were broadened in the "Yokohama Strategy" to integrate cultural, socioeconomic, and political aspects into disaster reduction programs of the Decade.

End of the IDNDR decade

The international community was increasingly aware that natural disasters are a major threat to social and economic stability and that disaster prevention is the main long-term solution to this threat. The largest challenge of the Decade laid, therefore, in the creation of a global culture of prevention as the Secretary General Kofi Annan stressed: *"We must, above all, shift from a culture of reaction to a culture of prevention. Prevention is not only more human cure; it is also much cheaper ... Above all, let us not forget that disaster prevention is a moral imperative, no less than reducing risks of war."* (International Decade for Natural Disaster Reduction, Programme Forum, 1999, Proceedings p 13) The IDNDR Secretariat in the United Nations organized the "IDNDR International Programme Forum – Towards Partnerships for Disaster Reduction in the 21st Century" as the closing event of the Decade.

The Geneva Programme Forum ended with the "Geneva Mandate", a joint statement of the participants of the Forum. Furthermore, the strategy "A Safer World in the 21st Century: Disaster and Risk Reduction" was developed. In this strategy the following goals were formulated:

(a) To increase public awareness of the risks that natural, technological, and environmental hazards pose to modern societies.
(b) To obtain commitment by public authorities to reduce risks to people, their livelihoods, social and economic infrastructure, and environmental resources.
(c) To engage public participation at all levels of implementation so as to create disaster-resistant communities through increased partnerships and expanded risk reduction networks at all levels.
(d) To reduce economic and social losses caused by disasters as measured, for example, by gross domestic product. (A Safer World in the 21st Century: Disaster and Risk Reduction, http://www.unisdr.org/eng/about_isdr/bd-safer-world-eng.htm)

These goals were the basis for the successor arrangement "International Strategy for Disaster Reduction (ISDR)".

During the Decade tragically the number of major natural disasters rose by nearly a third, compared to the 1980s, more than two and a half times as many people died, and economic damage tripled, according to statistics provided by Munich Reinsurance Company. Some suggested that this showed the failure of the IDNDR Decade. However, it has to be emphasized that first of all any action taken on the political level needs time to be implemented and therefore, visible also in decreasing consequences. Additionally one could also argue: If the IDNDR had not been in place, the losses would have been even greater. In any case, these developments made it evident how necessary further efforts in the area of disaster reduction were and how important it was that the findings from the Decade become implemented. The need for a continuation was obvious. However the discussion about if and how to organize a follow-up to the IDNDR was very much dominated by diverting interests of agencies and member states on structural and administrative aspects.

The Decade contributed very much to an increased understanding for the need of disaster reduction in order to achieve sustainable development. One of the major achievements of the Decade was the clear recognition that hazards are only one factor which may cause disasters and that human activities are at least equally important. Thus, the humanity is not the victim of environmental conditions, it is also responsible and even partly the driver of this adverse development. However, the focus of the Decade on natural disasters was based on several reasons:

(a) Larger losses and more negative effects by natural disasters in developing countries
(b) The traditional point of view by developed countries with regard to vulnerabilities, different levels of

development, coping capacities, and the need for a transfer of technology
(c) Avoiding diverting discussions about real causes of susceptibility to disasters
(d) Developing a program which could be acceptable to the majority of member states

From the political point of view, the shift of real causes from poverty and underdevelopment to natural hazards proved to be successful in order to reach an agreement for the IDNDR.

Continuation and the International Strategy for Disaster Reduction (ISDR)

The negative trends in disasters throughout the 1990s clearly underlined the need to continue with international coordinated efforts to reduce the impacts of natural disasters. The Secretary General of the United Nations, Kofi Annan, proposed to the member states to continue the successful work of IDNDR in form of an International Strategy for Disaster Reduction. The objectives and goals are based on the outcomes of the Geneva Forum of IDNDR. They show a shift from protection to the management of disasters. Emphasis is placed on the resilience of societies.

> para 6 The main objectives of the Strategy are: (a) to enable communities to become resilient to the effects of natural, technological and environmental hazards ... (b) to proceed from protection against hazards to the management of risk, by integrating risk prevention strategies into sustainable development activities. (International Decade for Natural Disaster Reduction: successor arrangements (Report of the SG) A/54/497, 1 November 1999)

The strategy is based on four main goals derived from the above objectives taken from the Geneva Mandate.

In its resolution 1999/63, the Economic and Social Council requested the Secretary-General to:

> *(a) Establish, as at January 2000, an inter-agency task force, with representation from all relevant United Nations bodies and members of the scientific and technical community, including regional representation, to serve as the main forum within the United Nations for continued and concerted emphasis on natural disaster reduction, in particular for defining strategies for international cooperation at all levels in this field, while ensuring complementarity of action with other agencies; (b) Maintain the existing inter-agency secretariat function for natural disaster reduction as a distinct focal point for the coordination of the work of the task force, to place the inter-agency task force and inter-agency secretariat under the direct authority of the Under-Secretary-General for Humanitarian Affairs and to finance it from extrabudgetary resources through a specific trust fund.* (International Decade for Natural Disaster Reduction: successor arrangement, A/54/497)

At the same time member states were asked to maintain their IDNDR Committees in the form of National Platforms for disaster reduction. This change from Committees to Platforms was based on the recognition of disaster reduction being a cross-cutting issue which needs to involve many stakeholders. The term "Platform" was seen as being the appropriate one to express the model of a place where all stakeholders – despite their background – could meet at the same level.

The General Assembly agreed to this proposal and by Resolution A/RES/54/219 the International Strategy for Disaster Reduction became established.

By accepting the by ECOSOC proposed structure the secretariat kept its status equal to OCHA. The structure was chosen in order to ensure the universal, interdisciplinary, and intersectoral nature of the secretariat instead of integrating disaster reduction into the structures of the United Nations Development Programme (UNDP) or the Office for the Coordination of Humanitarian Affairs.

However, with the structures of an Inter-Agency-Secretariat and an Inter-Agency Task Force the contact to structures outside the United Nations like the scientific community and member states was weakened.

World Summit on Sustainable Development, South Africa, 2002

The first important World Conference following the establishment of ISDR was the World Summit for Sustainable Development (WSSD) in Johannesburg, 2002, which provided the floor for the follow-up to the decisions taken at the United Nations Conference on Environment and Development (UNCED), Rio de Janeiro, 1992. In the outcomes of the WSSD member states of the United Nations recognized disaster reduction as a prerequisite to achieve sustainable development. In Chapter IV member states agreed on:

> "An integrated, multi-hazard, inclusive approach to address vulnerability, risk assessment and disaster management, including prevention, mitigation, preparedness, response and recovery, is an essential element of a safer world in the twenty-first century. Actions are required at all levels to:
>
> (a) Strengthen the role of the International Strategy for Disaster Reduction and encourage the international community to provide the necessary financial resources to its Trust Fund;
> (b) Support the establishment of effective regional, subregional and national strategies and scientific and institutional support for disaster management;
> ...
> (f) Encourage the dissemination and use of traditional and indigenous knowledge to mitigate the impact of disasters and promote community-based disaster management planning by local authorities, including through training activities and raising public awareness; (Report of the World Summit on Sustainable Development, Annex, Plan of Implementation of the World Summit on Sustainable Development. (Document A/CONF.199/20). Chapter IV: Protecting and managing the natural resource base of economic and social development, No. 37, pp. 27–28.)

These references were clear improvements with regard to the recognition of disaster reduction compared to the Outcomes of United Nations Conference on Environment and Development (UNCED), Rio de Janeiro 1992.

The Outcomes of the WSSD led to a number of resolutions on ISDR with the title "Natural Disasters and

Vulnerability." The first of this resolutions A/RES/58/215 (2003) referred to the Johannesburg Declaration on Sustainable Development and:

> ...*urges* the international community to continue to address ways and means, including through cooperation and technical assistance, to reduce the adverse effects of natural disasters, including those caused by extreme weather events, in particular in vulnerable developing countries, through the implementation of the International Strategy for Disaster Reduction, and encourages the Inter-Agency Task Force for Disaster Reduction to continue its work in this regard; (A/RES/58/215 para2)

A link to climate negotiations was made the first time and kept in the following resolutions until today:

> ...*encourages* the Conference of the Parties to the United Nations Framework Convention on Climate Change and the parties to the Kyoto Protocol to the United Nations Framework Convention on Climate Change to continue to address the adverse effects of climate change, especially in those developing countries that are particularly vulnerable, in accordance with the provisions of the Convention, and also encourages the Intergovernmental Panel on Climate Change to continue to assess the adverse effects of climate change on the socio-economic and natural disaster reduction systems of developing countries; (A/RES/58/215 para´6)

Since A/RES/60/196 (2005) a reference to the WCDR and the Hyogo Framework for Action was included into the Resolution as well as to the role of National ISDR Platforms.

> *Reaffirming also* the Hyogo Declaration and the Hyogo Framework for Action 2005–2015: Building the Resilience of Nations and Communities to Disasters, as adopted by the World Conference on Disaster Reduction, held at Kobe, Hyogo, Japan, from 18 to 22 January 2005, (A/RES/60/196) *Encourages* Governments, through their respective International Strategy for Disaster Reduction national platforms and national focal points for disaster risk reduction, in cooperation with the United Nations system and other stakeholders, to strengthen capacity-building in the most vulnerable regions, to enable them to address the socio-economic factors that increase vulnerability, and to develop measures that will enable them to prepare for and cope with natural disasters, including those associated with earthquakes and extreme weather events, and encourages the international community to provide effective assistance to developing countries in this regard; (A/RES/60/196 para5)

These references were kept (WSSD, WCDR, National Platforms and the Climate negotiations) in A/RES/61/200 (2006) and in the Resolution of 2008 A/RES/63/217.

World Conference on Disaster Reduction (WCDR), Kobe 2005

However, despite these resolutions, the review of the Yokohama Strategy for Action (2004) undertaken in the preparation to the World Conference for Disaster Reduction (WCDR) in January 2005 provided quite disillusioning findings:

In its report to the General Assembly the Secretary-General pointed out:

Para 14. Since the Yokohama Strategy was adopted, there have been about 7,100 disasters resulting from natural hazards around the world. They have killed more than 300,000 people, and caused more than US$ 800 billion in losses. The UN Under-Secretary General for Humanitarian Affairs has indicated "that on average, with well over 200 million people affected every year by 'natural' disasters since 1991, this is seven times more than the average of 30 million people affected annually by conflict."

...

Para 22. While only 11% of people exposed to natural hazards live in low human development countries, they account for more than 53% of total recorded deaths. - The Hyogo Framework for Action 2005

...

Para 79. Particular significance has been given to the sociology of disasters and other human dimensions that highlight the relevance of vulnerability in conditioning people's exposure to risk. (Draft Review of Yokohama Strategy and Plan of Action for a Safer World A/CONF.206/PC(II)/3 8 September 2004)

In the General Assembly Resolution A/RES/58/214, the United Nations decided to convene a World Conference on Disaster Reduction in 2005 (para7).

The WCDR was the second World Conference on Disaster Reduction following Yokohama. Taking place in January 2005, still under the full impressions of the Indian Ocean Tsunami from December 26, 2004, the conference got the full attention of politicians and international media; 168 member states and more than 4,000 participants were present in Kobe. The member states voluntarily agreed on the Hyogo Framework for Action which provided the workplan for the period 2005–2015. The Document is structured around three strategic goals and five Priorities for Action which are summarized in Box 2.

The Hyogo Framework for Action provided the workplan for the ISDR systems for the period 2005–2015. The clear structure of the documents and the assignments of tasks in the implementation of the Framework made it necessary to analyze the existing structures and identify arrears where they could be improved. The need for a better interlinkage with the scientific community and a closer involvement of member states into the ISDR structures were identified. In order to address these topics it was decided to establish a Scientific Committee. Additionally the General Assembly agreed that every 2 years a Global Platform would be organized as a Forum for member states, UN agencies, scientific community, and civil society to meet.

In order to move from a focus on hazards and losses and to address the dynamic nature of risk UNISDR started a biannual publication "Global Assessment Report" (GAR) which was presented for the first time at the Global Platform 2009. This publication addresses the changing nature of risk. It builds on studies undertaken among others by UNDP and the World Bank.

While a number of advancements were reported in the 2009 GAR, major obstacles to improve disaster risk reduction were identified in the underlying root causes.

Poverty, lack of integration of disaster risk reduction into development planning, and weak governance structures were recognized as the main barriers. At the same time it became evident that it needs additional efforts to integrate disaster risk reduction firmly into the international debate on the negative effects of Climate Change. The decision of the Intergovernmental Panel on Climate Change (IPCC) to prepare a special report on "Managing the Risks of Extreme Events and Disasters to Advance Climate Change Adaptation" and its findings presented in November 2011 is a milestone in this direction.

The Global Platform 2009 meeting and the launch of the GAR were the major ISDR events in 2009. While they stimulated the discussions and contributed to a further integration of disaster risk reduction in the development agenda, a number of gaps were identified. The issue of disaster statistics is not sufficiently addressed. Until today the disaster reduction community relies on three main sources, the sigma database from SwissRE, the NATCAT Service of Munich Reinsurance, and EMDAT, the database of the Center for the Research and Epidemiology of Disasters. The same is true for the important topic of cost-benefits of disaster reduction. A study of the World Bank addressing this issue "Natural Hazards, UnNatural Disasters – The Economics of Effective Prevention" was published in 2010.

> **Box 2**
>
> **Strategic goals**
>
> 12. To attain this expected outcome, the Conference resolves to adopt the following strategic goals
>
> (a) The more effective integration of disaster risk considerations into sustainable development policies, planning, and programming at all levels, with a special emphasis on disaster prevention, mitigation, preparedness, and vulnerability reduction
>
> (b) The development and strengthening of institutions, mechanisms, and capacities at all levels, in particular at the community level, that can systematically contribute to building resilience to hazards
>
> (c) The systematic incorporation of risk reduction approaches into the design and implementation of emergency preparedness, response, and recovery programs in the reconstruction of affected communities
>
> **Priorities for action**
>
> 14. Drawing on the conclusions of the review of the Yokohama Strategy, and on the basis of deliberations at the World Conference on Disaster Reduction and especially the agreed expected outcome and strategic goals, the Conference has adopted the following five priorities for action:

> 1. Ensure that disaster risk reduction is a national and a local priority with a strong institutional basis for implementation.
> 2. Identify, assess, and monitor disaster risks and enhance early warning.
> 3. Use knowledge, innovation, and education to build a culture of safety and resilience at all levels.
> 4. Reduce the underlying risk factors.
> 5. Strengthen disaster preparedness for effective response at all levels (Hyogo Framework for Action, 2005).

The midterm review of the Hyogo Framework for Action

In order to investigate the effectiveness of the Hyogo Framework for Action, a midterm review was carried out. It was stated that

> ..."We are still far from having empowered individuals to adopt a disaster risk reduction approach in their daily lives and demand that development, environmental and humanitarian policies and practices be based on sound risk reduction measures" (HFA MTR, Foreword the Special Representative of the Secretary-General for the implementation of the Hyogo Framework for Action, p 9)

In the document of the Hyogo Framework for Action it was declared that its implementation "will be appropriately reviewed" (HFA, IV Implementation and follow-up, A. General Considerations, para 29, p 14). The terms of reference of a Midterm Review were discussed in three Plenary sessions of the Global Platform 2009. It was decided to carry out the Midterm Review based on the following five analytical tools:

1. A Literature Review
2. Outcomes of structured workshops held at regional and national level
3. Selected in-depth studies
4. One-on-one interviews with key policy makers
5. Online debates

To provide advice on the MTR an Advisory Group of senior experts in disaster risk reduction, donor representatives, evaluation experts, and civil society were established. One of the challenges identified in the review process was: "Measuring progress against the expected outcome of reducing the loss of lives and assets due o disasters is difficult in the absence of a commonly agreed baseline at the time of the HFA adoption and of regular, standardized data collection by governments on disaster losses." (HFA Midterm Review, 2.1 Challenges, p 18)

While the Midterm Review noted that: "...significant progress has been made over the past five years in disaster risk reduction and that the adoption of the HFA 2005 has played a determinant role in pushing this process across international, regional and national agendas ..." it also was concluded "... that these connections, strongly driven by the disaster risk reduction community, have not been fully internalized in the ways in which international development assistance agencies, some government institutions, and the United Nations are institutionally and financially organized to manage disaster risk reduction." (HFA MTR, Suggestions for accelerating implementation of the Hyogo Framework for Action, p 55)

Based on these findings an institutional reassessment is suggested: "...where disaster risk reduction is placed within the international national and regional agencies to ensure that critical functions such as mainstreaming for sustainable development, strategic advice, monitoring of implementation, and reporting on impacts can effectively influence development policies and plans." (HFA MTR, Suggestions for accelerating implementation of the Hyogo Framework for Action, p 56)

The conclusion and recommendations address the key point of a further implementation of disaster risk reduction. In order to be able to measure progress they ask for targets to be set, "...standards to ensure quality in the delivery..." and "International, national and local level accountability mechanisms should be encouraged and developed to help measure action taken and progress achieved..." (HFA MTR, Conclusions and recommendations for the way forward, p 70)

In order to strengthen the ISDR system the improvement of government for disaster risk reduction at international and national level and an assessment of the effectiveness of National Platforms is proposed. The MTR of the HFA started the debate about the expected outcomes of the HFA by 2015 and actions to be taken. At the same time it already launched the debate about a Post 2015 Disaster Risk Reduction Framework.

Special thematic foci throughout IDNDR and ISDR

Starting with the IDNDR two selected core topics were put prominently on the agenda of disaster reduction. The following points will introduce how Early Warning and the El Nino phenomenon have been addressed.

Early Warning was seen as one of the most effective instruments to save lives and reduce losses caused by disasters. Thus, the development and implementation of local, regional, and global early warning centers was formulated as one of the goals of IDNDR by the Scientific and Technical Committee.

At the World Conference 2004 in Yokohama, a specific technical committee session was devoted to this subject.

> Subsequently, at its forty-ninth session, the General Assembly called for improvements and better coordination within the United Nations system with regard to natural disasters and similar disasters with an adverse effect on the environment. The General Assembly, in its resolution 49/22 B, placed this initiative distinctly within the concerted efforts of implementing the Yokohama Strategy and Plan of Action, and thus within the framework of IDNDR (A/54/132 – E/1999/80, para 22).

In order to address this complex issue scientifically and to get a State of the Art on Early Warning, the First International Conference on Early Warning EWCI was organized in 1998 in Potsdam, Germany. In its final declaration participants emphasized:

> The Potsdam Early Warning Conference has identified major strengths and weaknesses in early warning capacities around the world. Participants repeatedly emphasized the multidisciplinary and multi-sectoral character of the early warning process. Although based on scientific and technology, early warning must be tailored to serve people's needs, their environments, and their resources. Successful early warning requires unrestricted access to data that is freely available for exchange. Ultimately, all resulting information must be credible, and emanate from a single officially designated authority (Final Declaration EWC I, http://www.geomuseum.com/ewc98/).

The conclusions of the conference highlighted the importance of Early Warning in the framework of disaster reduction, put special emphasis on the multi-sectoral and interdisciplinary character of Early Warning, and the need to support the implementation of early warning systems on the local level.

1. Early warning represents a cornerstone of disaster reduction. It should, therefore, become a key element of future disaster reduction strategies for the twenty-first Century that are to be formalized in the conclusion of the IDNDR.
2. Effective early warning depends upon a multi-sectoral and interdisciplinary collaboration among all concerned actors, as demonstrated during the Potsdam Early Warning Conference.
3. While early warning capabilities must continue to be strengthened at the global level, it is important that greater emphasis be given to developing capacities that are relevant, and responsive to, the needs of local communities (Conclusion EWC I).

The Final report of the Scientific and Technical Committee of the International Decade for Natural Disaster Reduction (A/54/132/Add.1-E/1999/80/Add.1) shared these opinions (para 37–38).

Following the end of IDNDR and the establishment of ISDR and 5 years after EWCI, the Second International Conference on Early Warning (EWCII) took place in 2003 in Bonn, Germany. The subtitle of EWCII "Integrating Early Warning into Public Policy" expressed the concept of EWCII. It was based on the recognition of increasing scientific knowledge on early warning but a gap on the side of integration into public policy. Thus, the conference aimed to establish links between science and policy. The conference statement emphasized this goal and asked for the establishment of an early warning platform to sustain the dialogue.

> Calls for the integration of early warning systems into government policies and requests the organizers to disseminate widely to authorities at all levels the relevant guidelines recommended by the Conference, governments and relevant

organizations including the private sector to support the implementation of the early warning programme as recommended by the Conference and to integrate the programme into disaster reduction strategies at all levels, the early warning programme to focus on: (i) integration of early warning into relevant development policies and programmes; (ii) improvement of data collection, facilitating access to relevant data and forecasting; (iii) enhancement of capacities; (iv) people centered warning systems in particular ensuring gender balance and a; (v) platform to sustain the early warning dialogue. "Conference Statement EWCII"

The international community followed this suggestion and in early 2004 an Early Warning Platform (later on renamed to: Platform for the Promotion of Early Warning "PPEW") was established. As one of the first tasks, PPEW developed the International Early Warning Programme (IEWP), which was launched at the WCDR 2005 in Kobe.

The program is a vehicle by which partner organizations cooperate and develop shared and systematic approaches to advancing early warning systems worldwide. IEWP aims to:

- Develop international dialogue and a common framework for action, and promote early warning in policy debates and as a development priority
- Collate and disseminate good practices and other information on early warning systems
- Define and support capacity building projects in priority areas of need, involving humanitarian and development communities
- Develop improved tools and techniques, including guidelines and performance standards for early warning systems, and formulate priorities for further research and development (International Early Warning Programme)

Almost at the same time, following the Indian Ocean Tsunami, December 2004, the Secretary General asked the ISDR secretariat to carry out a Global Survey of Early Warning Systems.

In his March 2005 report to the Summit on the Implementation of the Millennium Declaration *In larger freedom: towards development, security and human rights for all*, he requested that a global survey of capacities and gaps for early warning systems be undertaken:

> The countries of the Indian Ocean region, with the help of the United Nations and others, are now taking steps to establish a regional tsunami early warning system. Let us not forget, however, the other hazards that people in all regions of the world are exposed to, including storms, floods, droughts, landslides, heat waves and volcanic eruptions. To complement broader disaster preparedness and mitigation initiatives, I recommend the establishment of a worldwide early warning system for all natural hazards, building on existing national and regional capacity. To assist in its establishment, I shall be requesting the International Strategy for Disaster Reduction secretariat to coordinate a survey of existing capacities and gaps, in cooperation with all United Nations system entities concerned, and I look forward to receiving its findings and recommendations (In larger freedom: towards development, security and human rights for all Report of the Secretary-General (A/59/2005)).

The Survey coincided with the preparation to a Third International Conference on Early Warning (EWCIII) 2006 in Bonn, Germany, an initiative which was also fueled by the tragic event of the 2004 Tsunami. The subtitle "From knowledge to action" underlined the intention to move to the implementation of early warning systems.

The basic aims of the conference were:

- To identify unused potentials in all areas of early warning
- To identify and launch specific early warning projects of high priority and illustrate the bridging of gaps
- To stimulate discussion and action toward concrete follow-up projects
- To discuss proposals for global integration of early warning systems whenever feasible and useful

The outcomes, a project portfolio of 100 peer-reviewed early warning projects and guidelines for the implementation of early warning systems, were very practically oriented in order to support these aims.

The Global Early Warning Systems Survey was presented at EWCIII. The survey report recommends the development of a globally comprehensive early warning system, rooted in existing early warning systems and capacities. It also recommends a set of specific actions toward building national people-centered early warning systems, filling in the main gaps in global early warning capacities, strengthening the scientific and data foundations for early warning, and developing the institutional foundations for a global early warning system. (Global Early Warning Systems Survey, Summary)

The Survey reached the following overall conclusion:

> Nevertheless, there are significant inadequacies in existing early warning systems, as illustrated by the experience of the Indian Ocean tsunami in late 2004, Hurricane Katrina in the Gulf of Mexico in 2005 and other recent events such as heat waves, droughts, famine, wildfires, tsunami, floods and landslides. Early warning systems especially in developing countries lack basic equipment, skills and financial resources and are for certain hazards even nonexistent. A major challenge is to integrate the knowledge and insight of relevant social and economic communities into the predominantly technically based existing systems.
> 13. One of the survey's key findings is that the weakest elements in early warning systems are the dissemination of warnings and the preparedness to respond. This is true for developing and developed nations alike. Warnings may fail to reach all those who need to take action, including local authorities, community-based organizations and the public at large, and often the warnings are not properly understood or may not be taken seriously. A good understanding by the public and by community organizations of their real vulnerabilities and the risk posed by an event is often lacking. Root causes of such failures appear to be inadequate political commitment, weak coordination among an often-diverse group of actors, and insufficient public awareness and participation in the development and operation of early warning systems. (Global Early Warning Systems Survey, Summary)

Following these major initiatives the work on early warning obtained momentum in a sense that major players of the UN family like the World Meteorological Organization (WMO) became more active. The PPEW became the focal point for the Indian Ocean Tsunami Consortium in order to coordinate the implementation of a Tsunami Early Warning system in the Indian Ocean together with international key players. Even national funding bodies incorporated these developments into their funded thematic themes. For example, the BMBF (German Federal Ministry of Education and Research) launched a key research program on "Early Warning Systems in Earth Management" within their research program "Geotechnologies."

At the fifty-second session in 1998 with Resolution A/RES/52/200 (2 March, 1998) the first time a special resolution on: "International cooperation to reduce the impact of the *El Niño* phenomenon" was adopted by the General Assembly of the United Nations. The resolution made reference to a number of Resolutions on IDNDR, established a clear link to the Yokohama Strategy for a Safer World, and requested the Secretary General to develop a strategy within the IDNDR to integrate the El Niño into the International Framework of Action for the Decade.

Recalling its resolutions 44/236 of 22 December 1989, 48/188 of 21 December 1993, 49/22 A of 2 December 1994, 49/22 B of 20 December 1994, 50/117 A and B of 20 December 1995 (A/RES/52/200)

> 2. Calls upon States, relevant intergovernmental bodies and all others involved in the International Decade for Natural Disaster Reduction to participate actively in the financial and technical support for Decade activities, including those related to international cooperation to reduce the impact of the El Niño phenomenon, in order to ensure the implementation of the International Framework of Action for the Decade, in particular with a view to translating the Yokohama Strategy for a Safer World: Guidelines for Natural Disaster Prevention, Preparedness and Mitigation and its Plan of Action into concrete disaster reduction programmes and activities; (A/RES/52/200 para2)
> 4. Requests the Secretary-General to facilitate, within the framework of the Decade, an internationally concerted and comprehensive strategy towards the integration of the prevention, mitigation and rehabilitation of the damage caused by the El Niño phenomenon, including the development of long-term strategies which take into due consideration the need for technical cooperation, financial assistance, the transfer of appropriate technology and the dissemination of existing scientific knowledge, as part of the Decade's activities, the International Framework of Action for the Decade and the Yokohama Strategy for a Safer World: Guidelines for Natural Disaster Prevention, Preparedness and Mitigation and its Plan of Action, and taking into account the relevant parts of the Programme of Action for the Sustainable Development of Small Island Developing States (A/RES/52/200 para 4)

In the follow-up to this resolution the government of Ecuador established together with WMO and ISDR the International Center for the study of the El Niño phenomenon at Guyaquil. A development which was welcomed by the Resolution on El Niño in 2005.

> Welcomes the efforts of the Government of Ecuador, the World Meteorological Organization and the inter-agency secretariat for the International Strategy for Disaster Reduction which led to the establishment of the International Centre for the Study of the El Niño Phenomenon at Guayaquil, Ecuador, and to its opening in February 2003, and encourages those parties to continue their efforts for the advancement of the Centre; (A/RES/59/232)

This positive comment was accompanied by a request to governments and international organizations to support the work of the center.

> Calls upon States, relevant intergovernmental bodies and all others involved in the International
> Decade for Natural Disaster Reduction to participate actively in the financial and technical support for Decade activities, including those related to international cooperation to reduce the impact of the El Niño phenomenon, in order to ensure the implementation of the International Framework of Action for the Decade, in particular with a view to translating the Yokohama Strategy for a Safer World: Guidelines for Natural Disaster Prevention, Preparedness and Mitigation and its Plan of Action into concrete disaster reduction programmes and activities; (A/RES/59/232)

In the latest resolution on the "International cooperation to reduce the impact of the El Niño phenomenon" of 2007 (A/RES/61/199) the support provided to the center was positively welcomed together with a strong plea to maintain the center.

> Welcomes the activities undertaken so far to strengthen the International Centre for the Study of the El Niño Phenomenon, through collaboration with international monitoring centres, including the national oceanographic institutions, and efforts to enhance regional and international recognition and support for the Centre and to develop tools for decision–makers and Government authorities to reduce the impact of the El Niño phenomenon; (A/RES/61/199)
> Underscores the importance of maintaining the El Niño/Southern Oscillation observation system, continuing research into extreme weather events, improving forecasting skills and developing appropriate policies for reducing the impact of the El Niño phenomenon and other extreme weather events, and emphasizes the need to further develop and strengthen these institutional capacities in all countries, in particular in developing countries; (A/RES/61/199 para 7)

The International Center for the Study of the El Niño Phenomenon is still operational and continues to work today.

Conclusion

The international community undertook an important move when it agreed to address the issue of increasing losses caused by hazards through development and implementation of disaster reduction strategies. The initiative was borne out of the sustainable development discussion. The focus on technical solutions was enlarged in the mid-1990s and sociocultural aspects were integrated and considered equally important. Consequently the conceptual discussion moved from preparedness and response to long-term aspects. Until today, however, disaster risk reduction is very much rooted in the humanitarian community. It is only there where solid commitments with regard to a percentage of funding earmarked for disaster risk reduction activities are made. The integration of disaster risk reduction into development policies still needs to be improved. A reason for this situation might be the decision of the United Nations to place IDNDR and later on ISDR under the umbrella of the Under Secretary General for Humanitarian Affairs.

The structure which links UN level to the national level through National Platforms provides a number of opportunities for the integration of disaster risk reduction in a concerted way through different levels. However, it has to be stated that the available synergies and added values of this structure have not been utilized to the extent possible.

The current debate about negative effects caused by climate change and the recognition of adaptation as equally important to mitigation became a new driving force. Although links have been made in a number of Resolutions on "Natural Disasters and Vulnerability" by the General Assembly following the WSSD in 2002, disaster risk reduction as being one important part of adaptation to climate change needs to be further promoted in order to ensure the necessary integration. The IPCC Special Report is an important step in this direction.

Since the beginning of the debate major improvements have been made. However a number of important issues still need to be addressed.

- The conceptual evolution to capture the development of risk and its dynamics illustrates an important improvement made. However, following disasters still need to be assessed in detail and comprehensive vulnerability assessments are required in order to be able to build back better.
- The existing data basis needs to be improved in order to base the analysis of risk and its development on a sound foundation. Statistical agencies should be encouraged to collect systematically risk relevant data.
- Decision makers need to be convinced to invest money into risk reduction scientific information on cost-benefit of risk reduction. This would allow to implement an extremely useful tool. The study undertaken by the World Bank may become a reference point for this discussion.
- The GAR identified the underlying root causes as major problems for the reduction of risk. In order to overcome this burden it is needed to establish firm links between the different stakeholders (politicians, decision makers, scientists, civil society actors, disaster managers, and private sector) and establish cross-border and international cooperation.
- The debate about a Post 2015 agreement on disaster risk reduction needs to take place in the context of the Millennium Development goals and the sustainable development agenda in order to ensure that DRR remains visible in the international agenda.

Abbreviations

ECOSOC–Economic and Social Council
EWC–International Conference on Early Warning
GAR–Global Assessment Report
IDNDR–International Decade for Natural Disaster Reduction (Decade of the United Nations 1990–1999)
IEWP–International Early Warning Programme
IPCC–Intergovernmental Panel on Climate Change
ISDR–International Strategy for Disaster Reduction (successor arrangement for IDNDR starting from 2000)
PPEW–Platform for the Promotion of Early Warning
STC–Scientific and Technical Committee within the IDNDR
UNCED–United Nations Conference on Environment and Development
WCDR–World Conference on Disaster Reduction
WSSD–World Summit for Sustainable Development

Bibliography

Addressing the Challenge (German Committee for Disaster Reduction), 2009.

Closing the Gaps (Commission on Climate Change and Development), 2009.

Federal Foreign Office (publisher), Are Disasters inevitable? The disaster reduction strategy of the Federal Foreign Office, 2004.

General Assembly Resolution A/42/169 International Decade for natural disaster reduction.

General Assembly Resolution A/44/236 International Decade for natural disaster reduction.

General Assembly Resolution A/RES/52/200; International cooperation to reduce the impact of the El Niño phenomenon.

General Assembly Resolution A/RES/54/219; International Decade for Natural Disaster Reduction: successor arrangements.

General Assembly Resolution A/RES/56/195; International Strategy for Disaster Reduction.

General Assembly Resolution A/RES/58/215; Natural Disasters and Vulnerability.

General Assembly Resolution A/RES/59/232; International cooperation to reduce the impact of the El Niño phenomenon.

General Assembly Resolution A/RES/60/196; Natural Disasters and Vulnerability.

General Assembly Resolution A/RES/61/199; International cooperation to reduce the impact of the El Niño phenomenon.

General Assembly Resolution A/RES/61/200; Natural Disasters and Vulnerability.

General Assembly Resolution A/RES/63/217; Natural Disasters and Vulnerability.

Global Survey of Early Warning Systems, A/C.2/61/CRP.1.

Grann, O., 1985. Secretary General, Norwegian Red Cross; WCED Public Hearing Oslo, 24–25 June 1985 published in: GA/A/42/427; Report of the World Commission on Environment and Development.

International Decade for Natural Disaster Reduction, Report of the Secretary-General Addendum, Final report of the Scientific and Technical Committee of the International Decade for Natural Disaster Reduction (A/54/132/Add.1 – E/1999/80/Add.1).

ISDR, 2002. *Living with Risk, A Global Review of Disaster Reduction Initiatives*. Geneva: United Nations.

ISDR, 2004. *Living with Risk, A Global Review of Disaster Reduction Initiatives*. Geneva: United Nations, Vol. I.

ISDR, 2009. *Global Assessment Report on Disaster Risk Reduction*. Geneva, Switzerland: United Nations.

Plate, E. J., and Merz, B. (eds.), 2001. *Naturkatastrophen: Ursachen, Auswirkungen, Vorsorge*. Stuttgart: Schweizerbart'sche Verlagsbuchhandlung. 475 pp.

Report of the Secretary-General; International Decade for Natural Disaster Reduction: Successor Arrangements (A/54/497).

Report of the Secretary-General, Implementation of the International Strategy for Disaster Reduction (A/56/68-E/2001/63).
Report of the World Commission on Environment and Development, Note by the Secretary-General A/42/427.
Schmitt, A., Bloemertz L., and Macamo E. (eds.), 2005. *Linking Poverty Reduction and Disaster Management*. Eschborn: GTZ.
Tokyo Declaration on the International Decade for natural Disaster Reduction, April 11 1989.
UNISDR, 2005. *Hyogo Framework for Action*. Geneva: UNISDR.
Yokohama Strategy and Plan for a Safer World, World Conference on Natural Disaster Reduction, Yokohama, Japan, May 23–27 1984.
Yokohama Strategy and Plan of Action for a Safer World - Guidelines for Nattural Disaster Prevention, Preparedness and Mitigation World Conference on Natural Disaster Reduction, Yokohama, Japan, May 23–27, 1994. http://www.preventionweb.net/english/professional/publications/v.php?id=8241

Cross-references

Casualties Following Natural Hazards
Climate Change
Communicating Emergency Information
Disaster
Disaster Diplomacy
Disaster Relief
Disaster Research and Policy, History
Disaster Risk Reduction (DRR)
Early Warning Systems
History of Natural Disasters
Hyogo Framework for Action
Indian Ocean Tsunami, 2004
Natural Hazard
Perception
World-Wide Trends in Natural Disasters

INTERNET, WORLD WIDE WEB AND NATURAL HAZARDS

Lucy Stanbrough
University College London, London, UK

Synonyms

The net; Web; World Wide Web

Definition

Internet. An electronic communications network that connects computer networks and organizational computer facilities around the world

Introduction

Increasing the usage and accessibility of information is widely recognized as a key requirement: before, during, and post natural hazard events, and with an increasing global population and encroachment into hazardous areas there is a greater emphasis on reducing the risks generated, of which the Internet is well placed to be involved in as a tool in this process.

The Indian Ocean Tsunami (2004) and Hurricane Katrina (2005) revealed the coming of age of the internet as an effective tool for natural hazard information, facilitating the exchange of information and increasing the speed of communication, sparking a wave of realization to the capabilities of internet application and *applications* within the field. Free and easy access to a wealth of satellite imagery, made available by various groups during these events, raised expectations and awareness to the internet's applicability, both to the public and the scientific community, and the topic continues to grow with these realizations.

Natural hazards and the internet

In the last 20 years, the internet has grown from a simple group work tool for scientists at CERN (European Organization for Nuclear Research) into a complex global information space with an estimated two billion users worldwide. With multiple connection points ranging from desktop computers to laptops, mobile phones to data warehouses, the internet has rapidly become one of the most effective methods for the distribution and coordination of information, and has become firmly integrated into everyday life; whether it is sending emails to your research partners or reading the latest journal article online, internet use has become a seamless part of facilitating information across networks.

It is this ability to facilitate information that is a useful component in the field of natural hazards, and was formally identified as such in a report in 1995 by the IDNDR (International Decade for Natural Disaster Reduction), who raised a call for better application of current information technologies to enhance the accessibility and to increase the understanding of warnings by a greater number of people. This point has since been widely enforced and repeated by various international and national reports and policy documents, and has spread to all areas of activity within the field of natural hazards, whether it is pure scientific research or ground-based field work dealing with communities and policy makers.

Growth of the internet and global access has allowed the reachability for natural hazard information and the introduction of high-speed data transfer systems and processors, overall reductions in costs, and the rapid development of computer technologies has resulted the capabilities and applications we see today – collaboration between individuals or research groups can be instant, secure, and timely, regardless of distance constraints. Users from anywhere around the world can link to up-to-date feeds of earthquake activity at the United States Geological Survey, browse and download vast directories of remotely sensed global data at NASA's Jet Propulsion Laboratory, or view Italy's Istituto Nazionale di Geofisica e Vulcanologia volcano database, all at the click of a button. Hundreds of thousands of websites currently exist, with more and more created every day, allowing users to track hurricanes, pinpoint earthquake locations, and read reports from the field or laboratory in real time.

While the use of the internet has spread, it is important to note however that less affluent countries, which are often those in greatest need of assistance, are technologically vulnerable and have problems with technical security, often lagging behind in technical expertise, computer literacy, and often, basic literacy. Rural communities are also experience issues, as the internet can be an unattainable commodity in remote regions and areas with poor technological infrastructure; although the combination of satellite and wireless mesh networks guaranteeing broadband communications is reducing this, it is not without cost, and therefore out of reach to those unable to meet the associated financial burdens. There are also issues with usability for those with disabilities, especially blindness, as the internet is primarily a visual medium, although there are software suites, browser add-ons, and website design guidelines that can be employed to make the internet suitable to such needs; this can especially be an issue in event reactionary postings, where following guidelines and establishing accessible sites or applications might be sacrificed to allow accessibility to the wider audiences.

Development and growth

Like many tools available for assisting with natural hazard processes, the internet is relatively young and is in a constant state of evolution in order to meet user's needs as the technology and tools available develop – by the time you read this there will already be numerous new examples of its application. Browser technology development has allowed the growth of these kinds of applications with the introduction of what are known as Rich Internet Applications (RIA). Currently, the main technology for delivering RIAs is AJAX (Asynchronous JavaScript and XML), although there are some alternatives which are mainly based on Flash technology. Two new technologies, AJAX and image tiling have also improved the performance and response times of internet applications significantly, resetting the "gold standard" and expectations of users. These have been responsible for the explosion of mapping products illustrating natural hazard information as developers are able to combine vast catalogs of data with available applications, but only serve up the immediate area a user is interested in viewing.

The success of the web has turned the browser into a central application, whose standardized capabilities can be exploited by applications such as mapping, video and audio streaming, funding generation, and blogs; creating a useable platform to seamlessly integrate and distribute information. Compared to print-based encyclopedias and traditional libraries, the internet has enabled a sudden and extreme decentralization of information and data; making tools and databases increasingly accessible to stakeholders at all levels, and removing the cost and time frame issues of traditional print.

Examples of this can be seen in services like ReliefWeb, which collates information from websites of international and nongovernmental organizations, governments, research institutions, and the media for news, reports, press releases, appeals, policy documents, analysis, and maps related to humanitarian emergencies worldwide, and then delivers it in a variety of formats for use. Data and news can be published instantly and made globally accessed at the push of a button, providing a common virtual "space" and "place" for discussion and collaboration, eliminating physical limitations and boundaries – meetings no longer need to be physical constructs, and can take place through instant voice or video communication. There are increasing numbers of online journal versions where content can be read in advance of print versions, often featuring interactive content that could not exist in another format.

Where traditionally internet applications only provided one-way information sharing from servers to clients, the introduction of natural hazard-specific standards like CAP and the wiki-like applications, allow multiple users to upload their own information and share it with other users. Common Alerting Protocol (CAP) is an XML-based data format for exchanging public warnings and emergencies between alerting technologies, allowing a warning message to be consistently disseminated simultaneously over many warning systems to many applications. Organizations are now able to provide feeds of standardized data that can be utilized easily by external users, the *USGS Recent Earthquake* RSS feed is a prime example, which can be seen embedded as news sites, within maps using background coordinate fields, or pulled into other applications to run models.

Virtual communities

Online communities have grown from the internet, easily connecting people worldwide, they exist in many different sizes, with some communities supporting the communication of a small circle of close friends, whereas others support tens of thousands of people in a single group. These range from mailing list discussion groups to fund-raising drives, to those involved broadly in natural hazards down to the smallest of niche areas of research, and allow connections to be easily made in a virtual environment between those who may never meet face-to-face to connect with one another. Twitter and Facebook are becoming major players in this area, with Facebook trialling services such as an "I'm Safe" button for their users in event-specific locations to let others in their social connections know they are alive after an event, and scientists able to track #hashtags or words on Twitter through data mining to get damage assessments based off users posts with location information. Social networks, while much more ad hoc and unstructured, contain a wealth of data; Facebook has some 800 million users and an estimated 250 million tweets are sent every day.

Web 2.0

Increasing technical mobility and familiarity of users is driving changes in the content and features offered, forcing content providers to reevaluate the services they are offering. The revolution of Web 2.0 is defined by Tim O'Reilly as the network as platform, spanning all connected devices; Web 2.0 applications are those that make the most of the intrinsic advantages of that platform: delivering software as a continually updated service that gets better the more people use it, consuming and remixing data from multiple sources, including individual users, while providing their own data and services in a form that allows remixing by others, creating network effects through an "architecture of participation," and going beyond the page metaphor of Web 1.0 to deliver rich user experiences.

Chief among these rules are the principals to build applications that harness network effects to get better the more people use them, and the importance of democracy. Crowdsourcing, the process of outsourcing tasks to a distributed group of people that can take place offline and online, has emerged out of this principal with efforts like those seen post-Christchurch in the comparison of pre- and post-earthquake satellite imagery to create damage assessments through the online Tomnod Disaster Mapper. Users were given two images to compare with different building damage estimates to categorize structures based on satellite image interpretation. While the validity of such results could always be called into question due to poor data or malicious efforts, the important point is that it can work with pools of verified experts and that there are vast networks of potential workers that can be connected through the internet.

The most dramatic example of Web 2.0 democracy is not in the selection of ideas, but their production and the ability to engage with users. With little to no knowledge of internet technologies, users are increasingly able to generate their own content, whether this be a simple Google Map with pinpoint locations denoting field sites, or a personal blog tracking a research area. People appropriated the technologies such as forums, bulletin boards, blogs, and online donation options, and personal websites are also being used to coordinate activities ranging from academic conferences to grassroots responses to disasters. Adoption of these resources has been widespread and has seen various National scientific groups and global organizations such as the United Nations disseminating information via services such as Twitter, and even the generation of podcasts by universities on iTunes that users can subscribe to.

Summary

Through continuous development and application, space and place are becoming increasingly conceptual notions, with the ease of developing communication connections and links worldwide. The combination of the internet and applicable applications offers a treasure trove of approaches, paradigms, and methods, with which hazards and disasters can be explored, modeled, and analyzed to assist with the mammoth task of bringing together global natural hazard efforts. It has the potential to be a focal linchpin for connecting the various arms of research organizations and interests that this subject encompasses.

Bibliography

Butler, D., 2006. Virtual globes: the web-wide world. *Nature*, **439**(7078), 776–778.

Hamilton, R. M., 2000. Science and technology for natural disaster reduction. *Natural Hazards Review*, **1**(1), 56–60.

Laituri, M., and Kodrich, K., 2008. On line disaster response community: people as sensors of high magnitude disasters using internet GIS. *Sensors*, **8**, 3037–3055.

Maguire, D. J., and Longley, P. A., 2005. The emergence of geoportals and their role in spatial data infrastructures. *Computers, Environment and Urban Systems*, **29**(1), 3–14.

Nourbakhsh, I., Sargent, R., Wright, A., Cramer, K., McClendon, B., and Jones, M., 2006. Mapping disaster zones. *Nature*, **439**(7078), 787–788.

O'Reilly, T., 2005. *Web 2.0: Compact Definition?* O'Reilly. http://radar.oreilly.com.

OASIS, 2007. *Common Alerting Protocol*. Oasis Emergency. http://www.oasis-emergency.org/cap.

Stephenson, R., and Anderson, P. S., 1997. Disasters and the information technology revolution. *Disasters*, **21**(4), 305–334.

UN, 1995. *A/50/526: SG report on early-warning on natural disasters (9 Oct)*. New York: United Nations.

Cross-references

Communicating Emergency Information
Community Management of Hazards
Early Warning Systems
Risk Perception and Communication

ISOSEISMAL

Valerio Comerci
ISPRA – Institute for Environmental Protection and Research, Roma, Italy

Synonyms

Isoseismal line

Definition

An isoseismal (line) is a contour or line on a map connecting points of equal intensity relating to a specific earthquake and confining the area within which the intensity is the same.

Discussion

When an *earthquake* occurs, its intensity at a given site is determined through the classification of observed effects on humans, buildings, and on the natural environment. Today, these kinds of data are acquired from *macroseismic surveys* and, in some countries, also from standard questionnaires distributed to the public. The results are then compared to the effects classified in an *intensity scale*, in order to assign intensity values. All of the assigned values are subsequently placed onto a map; the zones with the same intensities are then bounded by the isoseismal lines and marked with a roman number (corresponding to the degree of intensity). The significance of isoseismals depends on the number of observations in each place and on the close distribution of the places. If the number of recorded data at one's disposal is considerable, the isoseismals depict the variation of intensity throughout the whole area affected by the *earthquake* (Davison, 1921, 124, 127).

When an *earthquake* occurs inland, generally the isoseismals are closed curves, irregularly shaped and concentric, with the intensity degrees increasing as one moves toward the inner curve; when an *earthquake* occurs in a coastal region or offshore, the isoseismals open out toward the area, in the sea, where no data exist.

The first occasion on which the variation in the intensity was represented by isoseismal lines was the investigation, conducted by Robert Mallet (1862), of the great 1857 Neapolitan *earthquake*, in southern Italy (Davison, 1921, 99). The isoseismals are currently employed for evaluating the severity of *earthquakes* occurred before the availability of instrumental records, and for comparing these *historical events* with more recent seismic events. Moreover, the isoseismals allow one to locate an approximate *epicenter* of pre-instrumental *earthquakes*, besides indicating the direction of the seismogenic fault (Davison, 1921, 127): the elongation axis of the inner lines can be oriented according to the strike of the fault.

Finally, the isoseismals are useful for *earthquake hazard* assessment and for communicating *earthquake risk* to the population.

Bibliography

Bolt, B. A., 2006. *Earthquakes: 2006 Centennial Update*. New York: W. H. Freeman and Company.
Davison, C., 1921. On scales of seismic intensity and on the construction of isoseismal lines. *Bulletin Seismological Society of America*, **11**, 95–129.
Mallet, R., 1862. *Great Neapolitan Earthquake of 1857*. London: Chapman and Hall.
Reiter, L., 1990. *Earthquake hazard analysis: issues and insights*. New York: Columbia University Press.
Richter, C. F., 1958. *Elementary seismology*. San Francisco: W. H. Freeman and Company.

URLs

http://earthquake.usgs.gov/learn/glossary/?term=isoseismal (line)

Cross-references

Earthquake
Earthquake Damage
Epicenter
Fault
Hazard
Hazard and Risk Mapping
Historical Events
Intensity Scales
Macroseismic Survey
Modified Mercalli (MM) Scale
Seismology

J

JÖKULHLAUPS

Marten Geertsema
British Columbia Forest Service, Prince George, BC, Canada

Synonyms
Glacial lake outburst floods; Glacier burst

Definition
Catastrophic floods resulting from the breaching of glacier-dammed lakes occur in many regions of the world with glaciers. Some lakes drain and fill frequently, whereas others remain empty for years after draining, or never fill again. The resulting floods are referred to as jökulhlaups, an Icelandic term meaning *glacier burst*. Jökulhlaups are orders of magnitude larger than normal nival (snow melt) floods in the same basins – they can be very destructive. They are also called GLOFs, short for Glacial Lake Outburst Floods.

Discussion
Most glacier-dammed lakes occur at the margins of glaciers, but some are located beneath, within, or on top of glaciers (Figures 1 and 2). Glacier-dammed lakes usually drain through subglacial tunnels. The tunnels rapidly grow in size as the flowing water melts the surrounding ice. When the outflow ceases, the tunnels close by plastic flow of ice, allowing the lake to refill. Many lakes drain during or after the melt season in the summer or early fall, but drainage can occur at any time of the year. The hydrograph of jökulhlaups show an exponential increase in discharge followed by an abrupt decline (Figure 3). Water temperature decreases markedly during the jökulhlaup because the water discharging from the lake is colder than that of the background flow of the river.

Glacier-dammed lakes may go through cycles of growth and decay. Lakes will drain once a threshold of glacier thinning and retreat is reached. Lakes may also form when an advancing glacier dams a tributary or trunk valley. As glaciers continue to thin and retreat, both lake volumes and flood magnitudes decrease. Jökulhlaups can be expected to increase in frequency initially with global warming, as new lakes form, and eventually taper off as ice thins and retreats. Because glacier-dammed lakes can be in various stages of development at the same time, one can expect more jökulhlaups in different locations as most of the Earth's glaciers continue to thin and retreat. Mountain villages and infrastructure in the Himalayan

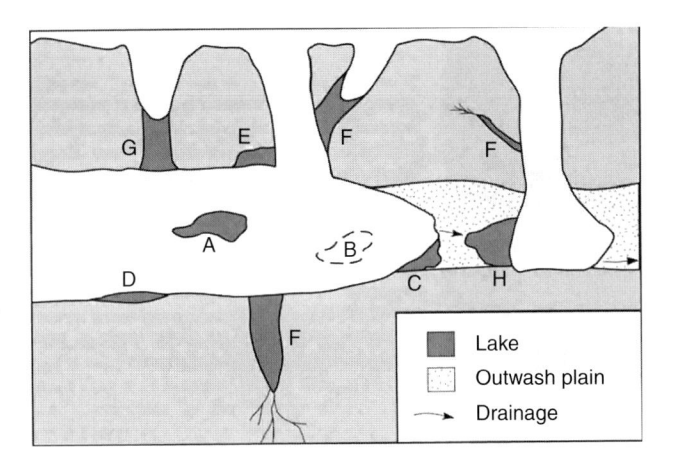

Jökulhlaups, Figure 1 Schematic diagram showing locations of different types of glacier-dammed lakes. (**a**) Supraglacial; (**b**) subglacial; (**c**) proglacial; (**d**) embayment in slope at glacier margin; (**e**) area of coalescence between two glaciers; (**f**) tributary valley adjacent to a trunk or tributary glacier; (**g**) same as F except glaciers dam both ends of lake; (**h**) main valley adjacent to a tributary glacier. Light toned area is land, white area is ice (After Clague and Evans, 1994).

P.T. Bobrowsky (ed.), *Encyclopedia of Natural Hazards*, DOI 10.1007/978-1-4020-4399-4,
© Springer Science+Business Media Dordrecht 2013

Jökulhlaups, Figure 2 Lake no lake (foreground) drains subglacially under Tulsequah Glacier up to two times per year. Note the stranded ice bergs on the drained lake bottom (Photo by Marten Geertsema, BC Forest Service).

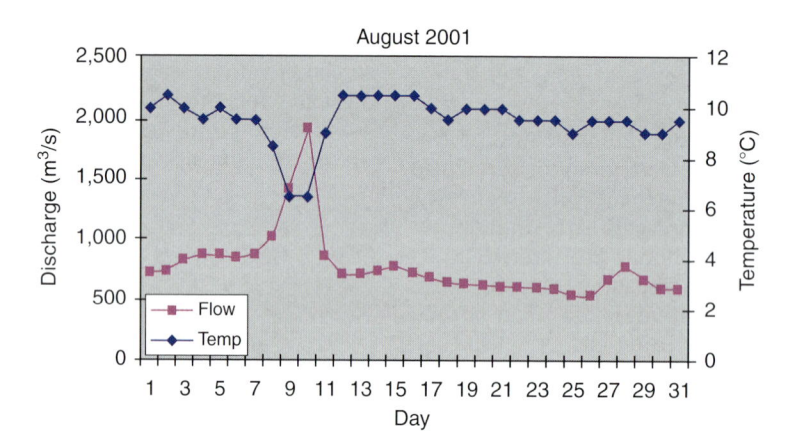

Jökulhlaups, Figure 3 Graph showing relationship between water temperature and discharge during a jökulhlaup at Tulsequah Glacier, British Columbia.

regions are especially at risk from destructive outburst floods.

Jökulhlaups may also be caused by subglacial volcanic eruptions. This happened in Iceland in 1996 when Gjálp volcano erupted beneath Vatnajökull glacier. The resultant flood was second in discharge only to that of the Amazon River.

During deglaciation several thousand years ago, several enormous jökulhlaups are known to have occurred. Ice-dammed Glacial Lake Agassiz (North America) drained through various outlets at different times. Its final outburst flood occurred about 8,200 years ago when it released enough freshwater into the North Atlantic to raise global sea level by 40 cm. This event changed the thermohaline circulation of the Atlantic Ocean, and caused global cooling for many decades. Glacial Lake Missoula also drained catastrophically on a number of occasions. The Missoula floods carved deep channels into basalt and impacted Idaho, Washington, and Oregon. There is also evidence for large-scale sheet flooding beneath Pleistocene ice sheets in various regions of the world.

Bibliography

Bretz, J. H., 1969. The Lake Missoula floods and the channelled scabland. *Journal of Geology*, **77**, 505–543.

Clague, J. J., and Evans, S. G., 1994. Formation and failure of natural dams in the Canadian Cordillera. Geological survey of Canada. *Bulletin*, **464**, 35–35.

Geertsema, M., and Clague, J. J., 2005. Jökulhlaups at Tulsequah Glacier, northwestern British Columbia. *Holocene*, **15**, 310–316.

Gudmundsson, M. T., Sigmundsson, F., and Björnsson, H., 1997. Ice-volcano interaction of the 1996 Gjálp subglacial eruption, Vatnajökull, Iceland. *Nature*, **389**, 954–957.

Shaw, J., 2002. The meltwater hypothesis for subglacial bedforms. *Quaternary International*, **90**, 5–22.

Teller, J. T., Leverington, D. W., and Mann, J. D., 2002. Freshwater outbursts to the oceans from glacial Lake Agassiz and their role in climate change during the last deglaciation. *Quaternary Science Reviews*, **21**, 879–887.

Cross-references

Flash Flood
Flood Hazard and Disaster
Flood Protection
Flood Stage
Floodplain
Floodway
Glacier Hazards
Hydrograph, Flood
Ice and Icebergs

K

KARST HAZARDS

Viacheslav Andreychouk, Andrzej Tyc
University of Silesia, Sosnowiec, Poland

Synonyms
Hazards in karst areas

Definition
Karst. All processes, forms, and landscape related to dissolution and efficient underground drainage in soluble rocks (e.g., limestone, dolomite, marbles, chalk, gypsum, or salt).
Karst hazards. Natural and human-induced hazards in karst areas, connected with the nature of karst.
Collapse. The gradual or rapid failure of roof rock or caprock into an underground cavity manifested on the surface by collapse (natural) or subsidence (human-induced) sinkholes.
Subsidence. The process of gentle and continuous surface deformation, manifested on the surface by shallow depressions.

Introduction
Karst hazards are an important example of natural hazards. They occur in areas with soluble rocks (carbonates, mostly limestone, dolomite, and chalk; sulfates, mostly gypsum and anhydrite; chlorides, mostly rock salt and potassium salt; and some silicates, quartzite and amorphous siliceous sediments) and efficient underground drainage. Karst is one of the environments in the world most vulnerable to natural and human-induced hazards. *Karst hazards* involve fast-acting processes, both on the surface and underground (e.g., collapse, subsidence, slope movements, and floods) and their effects (e.g., sinkholes, degraded aquifers, and land surface). They frequently cause serious damage in karst areas around the world, particularly in areas of intense human activity. *Karst threat* is the potential hazard to the life, health, or welfare of people and infrastructure, arising from the particular geological structure and function of karst terrains. The presence of underground cavities in the karst massif masks the threat from the hazards of collapse. This means that in some instances, the potential threats from karst, which are inherent features of the karst environment, become hazards. They range in category from potential to real.

The term (*karst hazards*) is related to two other terms, used mostly in applied geosciences, particularly engineering geology – *risk assessment* and *mitigation*. Risk is the probability of an occurrence, and the consequential damages are defined as hazards. Risk assessment is the determination of quantitative or qualitative value of risk related to a concrete situation and a recognized hazard. Quantitative *risk assessment* requires calculations of two components: the magnitude of the potential loss and the probability that the loss will occur. Risk assessment is a step in a *risk management*. *Mitigation* may be defined as the reduction of risk to life and the environment by reducing the severity of collapse or subsidence, building subsidence-resistant constructions, restricting land use, etc.

Classification of karst hazards
Karst hazards can be divided into two main groups: *gravidynamic* (a wide group of gravitational processes in karst) and *hydrodynamic* (relating to water circulation and violent changes of the water regime in karst) (Figure 1) (Andreychouk and Tyc, 2005). Both kinds of hazards can occur on the surface and/or underground. Natural or human-induced collapses and subsidence as well as mass movements often accompanying these processes are examples of *superficial gravidynamic hazards*. *Underground gravidynamic hazards* include cave or cavity breakdown, collapse of consolidated internal cave

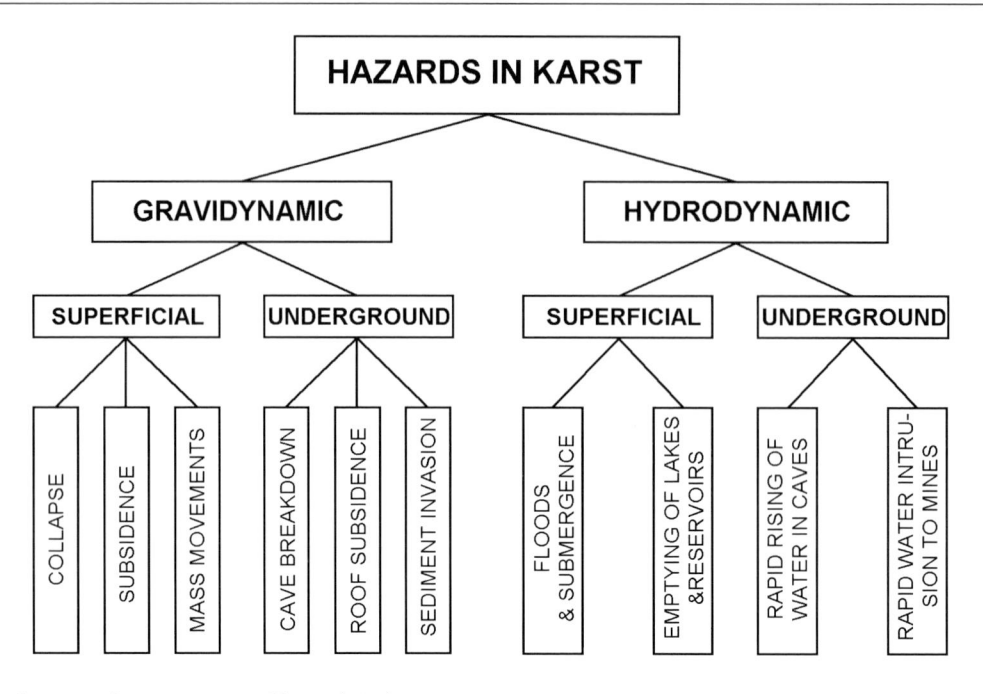

Karst Hazards, Figure 1 The main types of hazards in karst terrains.

sediments, and invasion cavities by unconsolidated sediments. Hydrodynamic hazards connected with the surface of karst terrains (*superficial hydrodynamic hazards*) include floods (e.g., in poljes), submergence of karst springs, and emptying of lakes or reservoirs. Rapid karst water intrusion into mine galleries and sudden rising of waters in caves (a hazard for cave exploration and show caves) are examples of *underground hydrodynamic hazards*. Hazards in karst terrains are usually rapid and frequently catastrophic, but there is also a group of evolutionary, slow-acting hazards such as saltwater intrusions into karst aquifers and pollution of karst waters (Andreychouk and Tyc, 2005; Parise and Gunn, 2007).

Gravidynamic karst hazards

The collapse of bedrock into underlying cavities is one of the most serious and common hazards in karst areas. Although large collapse sinkholes are known from many karst terrains in the world, the largest and most spectacular, called tiankeng (sky holes), occur in tropical karst (Ford and Williams, 2007). Rock collapse as a result of the failure of the cave roof is a rare event, with few recent cases reported. Nearly all recent cases of collapse are subsidence sinkholes (e.g., dropout, suffosion, caprock, or cover-collapse sinkholes) induced by human activity in karst (Waltham et al., 2005).

Human-induced dewatering of unconsolidated sediments covering karstified rock is the most important cause of subsidence sinkholes. Lowering of the groundwater table due to intense water pumping for water supplies and draining the rock mass for mining or quarrying produces a cone of hydraulic depression below the rockhead of the karstified rock and consequently loss of buoyant support from the water. Human-induced changes to the hydrodynamic conditions of groundwater circulation in karst aquifers produce conditions favorable for collapse and subsidence. As a result, numerous subsidence sinkholes, mainly caprock sinkholes, occur in karst areas affected by water exploitation or mining. Their size depends on dimensions of the underground cavity, the type and thickness of the caprock, and they vary widely in diameter and depth, from a few meters to hundreds of meters (Tyc, 1999; Parise and Gunn, 2007; Waltham et al., 2005). The large subsidence sinkhole in Winter Park, Florida, is one of the best known and most studied in the world (e.g., Beck, 2005; Waltham et al., 2005; Ford and Williams, 2007). Nearly 150,000 m^3 of unconsolidated cover sediments have disappeared into underground cavities after collapse (Waltham et al., 2005).

Due to the high rate of dissolution of gypsum, collapse and subsidence sinkholes are a far greater hazard in gypsum karst than in carbonate karst (e.g., Pinega or Perm regions in Russia, Podolia and Bukovina in Ukraine). In the gypsum karst of the Urals, there are regions with density of sinkholes up to 200 per km^2 and very high rate of sinkhole appearance up to 3,0 per km^2 per year (Klimchouk and Andreychouk, 1996; Andreychouk, 1996, 1999). Rapid and extensive dissolution occurs when groundwater circulation reaches salt deposits. This can result in extremely dangerous hazards. Removal of salt by solution mining commonly induces development of regional subsidence (continuous deformations on a regional scale). Salt mining and leakage from the rock mass overlying the salt deposits can cause the development of enormous underground cavities, consequently replaced by breccia pipes and finally, manifested on the surface by spectacular large subsidence sinkholes. One of the best examples is the sinkhole that developed in 1986 over the world's largest potash mine at Bereźniki (Urals, Russia) (Figure 2) (Andreychouk, 2002).

Numerous subsidence sinkholes in areas of mining and water extraction cause hazards to engineering, construction, and sometimes to human life. The high density of sinkholes transforms some regions into *karst badlands* (e.g., some industrial areas in pre-Urals – Russia and western Ukraine, Figure 3). Collapse of bedrock in karst areas is of a rapid and unforeseen character and is usually accompanied by a range of physical phenomena including acoustic, luminous, pneumatic, seismic, and hydrodynamic events. These phenomena can constitute an additional hazard – slight earthquakes (see *Induced Seismicity*) or explosions of rock fragments.

Another *karst hazard* of the *gravidynamic* type is the process of ground subsidence, the gradual and gentle lowering of the surface without any distinct breaking of the caprock. Shallow depressions resulting from subsidence develop in gypsum karst, mostly, where there is solution of rocks at shallow depth (Ford and Williams, 2007). The intensity of the subsidence varies from several millimeters to several meters per year. Depressions are sometimes of large dimensions, up to several square kilometers across. Ground subsidence in karst is less dangerous than sinkhole failure and is a more predictable hazard. The gradual and gentle development of subsidence potentially makes it possible to develop a protection strategy. In karst areas with a shallow groundwater table, collapse and subsidence sinkholes as well as larger subsidence depressions can fill with water and form lakes (Figure 4).

Limestone and gypsum are very prone to different types of landslides or rockslide avalanches – mass movements that cause serious hazards in karst areas. Mass movements can occur on slopes and in the sides of large collapse or subsidence sinkholes. The most hazardous occur where thick limestone strata overlay slick, impermeable strata such as shales that are dipping downslope. Such settings are favorable for downslope detachment, the movement of limestone, and can cause landslides. Heavy rain or earthquakes are important triggers for such mass movements.

Cave roof failure (breakdown) and sediment invasion of cave passages are the most dangerous *underground gravidynamic hazards*. Susceptibility to ceiling breakdown is related to the structural integrity of the cave roof (thickness of bedding, density of fractures, etc.) and the width of its unsupported span. Breakdown stabilizes when the span of the bed is less than the critical width for its thickness and a tension dome is formed. In caves located below perched aquifers, separated from the cave space by poorly permeable sediments, invasion of liquefied unconsolidated sediments can occur. Cave roof failure and sediment invasion can cause damage to the cave space. Both hazards are a danger to cavers during exploration. Permanent monitoring of roof stability and other geotechnical parameters are used to minimize both hazards in show caves. Liquefaction and consequent sediment invasion can cause damage to the infrastructure of show caves.

Hydrodynamic karst hazards

Floods and *submergence* are periodic processes in karst that pose superficial hydrodynamic hazards. *Floods* and *submergence* mostly occur in flat karst depressions such as poljes (e.g., Dinaric Karst in Mediterranean Europe) or depressions in tropical or subtropical karst (e.g., south China, Vietnam, or Cuba). Karst is susceptible to rapid changes and the rising of water levels, resulting from overflowing of groundwater onto the karst surface through karst forms in the depression bottom flooding farmlands and roads. In the past, boats were typical equipment on farms located in frequently flooded poljes such as those in the Dinaric Karst. Flooding and submergence of flat karst depressions is a natural occurrence in karst, so flood hazard management systems are in place in areas where this occurs. Much more hazardous and less predictable than floods are rapid *emptying of lakes* and *reservoirs*. This is quite common in areas where karst lakes or artificial reservoirs are perched above the regional karst water table. Development of subsidence sinkholes in the lake bottom can cause rapid emptying of the lake over periods ranging from minutes or hours to several days (Figure 5).

Karst Hazards, Figure 2 Catastrophic subsidence sinkhole over the world's largest potash mine in Bereźniki area, Urals, Russia (Photo V. Andreychouk).

Karst Hazards, Figure 3 High density of subsidence sinkholes in the area of Olkusz lead and zinc mines, South Poland – example of karst badlands (Photo M. Dobrzański).

Karst Hazards, Figure 4 Large-scale subsidence in settled region of potash mine, vicinity of Kalush, western Ukraine (Photo V. Andreychouk).

Karst Hazards, Figure 5 Rapid emptying of artificial lake as a result of collapse of bottom sediments and its drainage by subsidence sinkhole, Bukovina, Ukraine (Photo V. Andreychouk).

Sudden emptying of lakes or reservoirs is dangerous to fish and other aquatic animals. Such phenomena are known primarily from flat and lowland karst terrains on Russian Plain and Wolyn and Bukovina in Ukraine.

Rapid rising of water in caves caused by heavy rains is a serious underground hydrodynamic hazard in karst areas. Depending on the rainfall and the morphology of the cave system, the water level can rise rapidly from several to several tens of meters and, on rare occasions, even several hundreds of meters. In flat areas, whole horizontal cave systems and, in alpine karst, some horizontal passages can be completely flooded. Rapid rising of water is very dangerous for cavers exploring deep cave systems in high mountains and cave systems with active water flows in flat areas. Quite a large number of accidents involving cavers being trapped by underground flooding in caves in the Alps, Pyrenees, and Caucasus have been fatal. Flooding of show caves can cause damage to their infrastructure.

Rapid water intrusions into mines exploiting karst deposits are very dangerous hazards. These can occur in bauxite, nickel, lead, zinc, iron, gypsum, and salt mines in the karst and in coal or other mines adjacent to the karst (e.g., Kizelovski coal basin; Severouralsk bauxite deposits and Berezniki potassium salt deposits in Urals, Russia; Olkusz lead-zinc ore deposits in Poland; Tyc, 1999; Andreychouk, 2002). Water-filled cavities or watering zones in the karst rock commonly drain into galleries during the mining operation (Figure 6). Rapid, sometimes violent, inflow of large amounts of water into the mine gallery causes great risk for miners and machinery.

Gasodynamic karst hazards

Concentrations of carbon dioxide (CO_2) are often considerably higher in caves than in the open air with caves having CO_2 levels of 2–6% and higher. Such high concentrations can result from CO_2 migration from a deeper part of the Earth's crust, the activity of microorganisms in the caves, or from the diffusion of soil CO_2 into the caves. High concentrations (over 5–7%) in cave air can pose a serious danger to human life (*gasodynamic hazards*). Fatal CO_2 intoxication of cavers in caves with high concentrations of CO_2 has been reported. Some caves with poor ventilation are hazardous due to relatively high concentrations of radon, sometimes also found in basements and crawl spaces of houses built on karst.

Summary

Karst hazards are an important example of natural hazards, which occur in areas with soluble rocks (carbonates, sulfates, or chlorides). Karst hazards can be divided into two main groups: *gravidynamic* and *hydrodynamic*. Both kinds of hazards can occur on the surface and/or underground. Additionally, high concentrations (over 5–7%) of CO_2 in cave air can pose a serious danger to human life and can be called as *gasodynamic hazards*.

Karst Hazards, Figure 6 Large amount of water inflow from karst conduit into mine gallery, Olkusz lead and zinc mine, south Poland (Photo J. Jackowski).

Karst hazards affecting karst areas all over the world can be fully appreciated only through a multidisciplinary approach, combining expertise from different fields, including speleology, karstology, engineering geology, hydrology, hydrogeology, geophysics, geochemistry, and biology.

Bibliography

Andreychouk, V., 1996. *Bereznikovski collapse*. Perm: UrO RAN [In Russian], 133 pp.
Andreychouk, V., 1999. *Collapses above Gypsum Labyrinth Caves and Stability Assessment of Karstified Terrains*. Chenovtsy: Prut [In Russian], 52 pp.
Andreychouk, V., 2002. Collapse above the world's largest potash mine (Ural, Russia). *International Journal of Speleology*, **31**(1–4), 137–158.
Andreychouk, V., and Tyc, A., 2005. Hazards and risks in karst terrains – definitions and classification. *Geophysical Research Abstracts*, **7**, 10080.
Beck, B. F., (ed.) 2005. *Sinkholes and the Engineering and Environmental Impacts of Karst*. Reston: Geotechnical Special Publication, Vol. 144, American Society of Civil Engineers, 677 pp.
Ford, D., and Williams, P., 2007. *Karst Hydrogeology and Geomorphology*. Chichester: Wiley, p. 562.
Klimchouk, A., and Andreychouk, V., 1996. Breakdown development in cover beds and landscape features induced by interstratal gypsum karst. *International Journal of Speleology*, **23**(3–4), 127–144.
Parise, M., and Gunn, J., (eds.) 2007. *Natural and Anthropogenic Hazards in Karst Areas: Recognition, Analysis and Mitigation*. The Geological Society of London, Special Publication, Vol. 279, 202 pp.
Tyc, A., 1999. Collapse and piping induced by human activity in the Olkusz lead-zinc exploitative district of the Silesian Upland, Poland. In Drew, D., and Hötzl, H. (eds.), *Karst Hydrogeology and Human Activities: Impacts, Consequences and Implications*. Rotterdam: A.A. Balkema, pp. 215–217.
Waltham, T., Bell, F., and Culshaw, M. (eds.), 2005. *Sinkholes and Subsidence: Karst and Cavernous Rocks in Engineering and Construction*. Chichester: Springer &Praxis Publishing, p. 382.

Cross-references

Disaster Risk Management
Flood Hazard and Disaster
Induced Seismicity
Land Subsidence
Mitigation
Monitoring and Prediction of Natural Hazards
Prediction of Hazards
Radon Hazards
Risk
Risk Assessment
Sinkhole

CASE STUDY

KRAKATOA (KRAKATAU)

Bill McGuire
Aon Benfield UCL Hazard Research Centre, University College London, London, UK

Synonyms
Krakatau

Introduction

The Indonesian volcano of Krakatoa is one of the best known on Earth, primarily as a consequence of a catastrophic explosive eruption in 1883. This event took an estimated 36,417 lives, mainly due to associated tsunamis, and was detected on tide gauges and barometers around the world.

Discussion

Krakatoa is an island volcano located in the Sunda Strait between the south coast of Sumatra and east coast of Java (lat: 6.102°S long: 105.423°E). The volcano is one of more than 130 active volcanoes in Indonesia, which form an arc stretching from Sumatra in the west to New Guinea in the east. The volcanic arc is fed by magma formed as a consequence of the northeastward subduction of the Indo-Australian Plate beneath the Sunda Plate. Indonesian volcanoes, including Krakatoa, are typically characterized by the violent eruption of viscous magmas, generating extensive ash-fall, hot ash and debris flows, volcanic mud-flows (lahars), and, where adjacent to water, tsunamis.

The current height of Krakatoa is 813 m. Prior to the 1883 eruption, Krakatoa island comprised three volcanic cones, Rakata, Danan, and Perboewatan, with Rakata the highest at 820 m. The 1883 event destroyed the latter two cones, leaving only a remnant of Rakata. Currently, Krakatoa comprises four islands, Sertung, Panjang, a remnant of the original Krakatoa, and Anak Krakatoa (Child of Krakatoa). Located centrally between the other three islands, Anak Krakatoa is the new active center, building itself up to a height of 300 m since breaching the sea-surface in 1927. Recently, relatively minor eruptions have occurred from 2000 to 2001 and from 2007 to 2009, adding – on average – around 5 m to the height of the cone each year.

Prior to the 1883 eruption, the activity of Krakatoa is not well documented. Eight mild to moderate eruptions are recorded in the 1,000 years before 1883, the latest almost 200 years earlier in 1684. There is serious speculation, based on historical records, of a cataclysmic eruption in 416 AD, leading to the collapse of an "ancestral" Krakatoa and the formation of a 7 km-wide caldera, leaving as remnants the Verlaten, Lang, and Krakatoa islands that existed prior to the 1883 eruption. The Javanese *Book of Kings* notes that in the year 338 Saka (416 AD), "a thundering sound" was heard from a volcano in the Sunda Strait that eventually "burst into pieces" causing the land to be inundated by the sea, which swept away people and property. An alternative version of events, which also remains unverified, argues for this major eruption occurring in 535 AD and holds it responsible for a significant deterioration of the global climate in 535–536 AD.

Reports of seismic activity in the vicinity of Krakatoa in the years leading up to 1883 argue for magma accumulation and ascent over a long period. The first evidence at the surface took the form of steam venting in May 1883, three months before the climactic eruption. By mid-June, eruption columns were carrying ash and pumice several kilometers into the atmosphere, and activity continued to escalate through July. The paroxysmal phase commenced on August 26, with a 27 km high eruption column leading to extensive hot ash and pumice fall across the region and onto ships 20 km distant in the Sunda Strait. The climax came the following day, with four cataclysmic explosions loud enough to be heard 4,653 km away at Rodriguez Island in the Indian Ocean, and 4,600 km distant at Alice Springs in Australia. While the series of explosions was clearly associated with the destruction of the volcano, the precise course of events and the cause of the extreme explosions remain a matter for conjecture. Possibilities include partial submarine collapse of the edifice, exposing the pressurized magma reservoir to seawater, or violent magma mixing due to the emplacement of hot, basaltic magma into the main body of cooler, more silica-rich, and more viscous dacite magma.

Determining the sources of the catastrophic tsunamis that inundated the shores of Sumatra and Java has also proven problematical, although these are now generally regarded to have been formed by the entry into the sea of massive hot pumice flows (known as ignimbrites) generated by the gravitational collapse of the eruption column. The tsunamis had run-up heights of up to 46 m on the Java coast, and carried sufficient strength to loft 600 ton coral boulders far onto the shore and to destroy the 40 m high lighthouse at Anjer on the west coast of Java. Although most deaths (in excess of 34,000) resulted from the tsunamis destroying 165 communities, more than a thousand lives were lost in the vicinity of Ketimbang (southern Sumatra) by surges of hot ash and gas that traveled the 40 km across the sea surface before climbing the southern slopes of the island.

The legacy of the eruption was the destruction of much of the original island of Krakatoa, the formation of a 250 m deep caldera, and a dramatically altered sea floor topography, which included an area of more than 1.1 million km^2 covered by ignimbrite deposits. Further afield, loading of the atmosphere by dust and sulfuric acid aerosols led to meteorological effects worldwide, including spectacular sunsets, rings around the Sun, and "blue" moons. Global average temperature fell by 1.2°C for a number of years, while the cooling effect could be detected in the oceans well into the twentieth century.

Summary

After the 1815 eruption of Tambora, also in Indonesia, Krakatoa hosted the second most lethal eruption of the last 250 years. The total volume of material erupted is estimated at between 22 and 25 km^3, corresponding to a score of six on the Volcano Explosivity Index. The eruption was also important in another context, with the advent of undersea telegraph cables making it possible, for the first time, for news of a major catastrophe to spread across the world in just a few days. The eruption, therefore, was a key media event as well as a major expression of volcanic power.

Bibliography

Self, S., and Rampino, M. R., 1981. The 1883 eruption of Krakatau. *Nature*, **294**, 699–704.

Simkin, T., and Fiske, R. S., 1980. *Krakatau, 1883: The Volcanic Eruption and Its Effects*. Washington: Smithsonian Institution Press, p. 464.

Winchester, S., 2003. *Krakatoa: The Day the World Exploded: August 27, 1883*. New York: HarperCollins, p. 448.

Cross-references

Base Surge
Caldera

Eruption Types (Volcanic Eruptions)
Galeras Volcano, Colombia
Lahar
Magma
Montserrat Eruptions
Mt. Pinatubo
Nevado del Ruiz Volcano, Colombia (1985)
Nuee Ardente
Pyroclastic Flow

Santorini
Shield Volcano
Stratovolcano
Tsunami
Vesuvius
Volcanic Ash
Volcanic Gas
Volcanoes and Volcanic Eruptions

L

LAHAR

Richard B. Waitt
U.S. Geological Survey, Vancouver, WA, USA

Synonyms

Volcanic mudflow; Debris flow

Definition

A lahar is a flowing slurry of rock debris and water originating on the slopes of a volcano. The term may also mean the deposit of such a flow.

Discussion

Characteristics: Lahars contain grains from clay to large boulders. The flowing material is water-lubricated sand or mud, but the whole mixture maybe more than half of incorporated cobbles and boulders. Volumes can reach 10^9 m^3 and peak discharge may exceed 10^7 m^3/s. Depending on proportions of freshly erupted volcanic debris and snow or water, temperatures range from nearly 100°C to 0°C but are typically below 50°C. Flowing lahars peak swiftly and wane more slowly, passing any one place within minutes to a couple hours but sometimes lasting several hours.

Origin: Lahars can originate by eruption of hot fragmental debris onto snow or ice. A melting mass then flows swiftly downslope, incorporating more and more ash and rock from the volcano's slopes. Some large lahars initiate from volcanic debris avalanches. If the avalanche consists of rock highly altered to clay, almost all the flowing mass becomes lahar, for instance, the Osceola Mudflow off Mount Rainier, USA. But if the avalanche is rocky, its groundwater can escape, incorporate loose sand, and flow downvalley – as in May 1980 at Mount St. Helens, USA. Lahars may initiate by eruption through a crater lake, as at Kelut in Indonesia or by breakout of a crater lake, as at Ruapehu in New Zealand. Lahars may originate from torrential rainfall during or after eruption of voluminous ash, as at Merapi in Indonesia, Pinatubo and Mayon in the Philippines, and Unzen in Japan. Large lahars occasionally originate by breakout of a debris-dammed lake on a volcano's lower flank.

Downstream evolution: An initially watery lahar may incorporate more debris from the channel and increase the initial flow volume by many times. Lahars moving down a volcano's flanks flow at 10–40 m/s but slow downvalley. A dense lahar may at first overrun river water, incorporating enough to dilute the mass into an intermediate (hyperconcentrated) flow or even muddy-water flow.

Distribution: Lahars radiate from a volcano. They scour and drape steep valley reaches and accumulate in gentler downstream reaches far beyond the volcano's flanks. Here they may spread widely and inundate areas far beyond the delivering valley. Empirical runout models can estimate areas inundated by future lahars.

Hazard: A lahar can sluice down a valley and burst fast and deep upon a town, smashing or removing almost all structures. Tens of thousands of people have perished from lahars, most infamously at Nevado del Ruíz in Colombia in 1985. Close monitoring of an impending eruption, assessing probable runout in valleys at risk, and effective and timely communication to people in the way can mitigate the hazard to human life if not to infrastructure.

Bibliography

Major, J. J., and Newhall, C. G., 1989. Snow and ice perturbation during historical volcanic eruptions and the formation of lahars and floods. *Bulletin of Volcanology*, **52**, 1–27.

Pierson, T. C., and Scott, K. M., 1985. Downstream dilution of a lahar – transition from debris flow to hyperconcentrated streamflow. *Water Resources Research*, **21**, 1511–1524.

Vallance, J. W., 2005. Volcanic debris flows. In Jakob, M., and Hungr, O. (eds.), *Debris-Flow Hazards and Related Phenomena*. Berlin: Springer, pp. 247–274.

Cross-references

Base Surge
Caldera
Eruption Types (Volcanic Eruptions)
Galeras Volcano, Colombia
Krakatoa (Krakatau)
Lava
Montserrat Eruptions
Mt Pinatubo
Nevado del Ruiz Volcano, Colombia (1985)
Nuee Ardente
Shield Volcano
Stratovolcano
Surge
Vesuvius
Volcanic Ash
Volcanic Gas
Volcanoes and Volcanic Eruptions

LAND DEGRADATION

Matija Zorn, Blaž Komac
Anton Melik Geographical Institute, Ljubljana, Slovenia, Slovenia

Definition

Land degradation is a natural or human-induced process (Figure 1) that negatively affects the land to function effectively within an environmental system and can be defined as a process of degrading land from a former state. Land degradation is closely related to sensitivity, resilience, and carrying capacity of land, as well as to vulnerability of people living on and from these lands. It may be defined as the loss of utility or potential utility, or reduction, loss, or change of features or organisms which cannot be replaced (Barrow, 1991). A pure anthropogenic definition of land degradation is the loss of a sustained economic, cultural, or ecological function due to human activity in combination with natural processes (Bush, 2006).

Mechanism

Mechanisms that initiate land degradation include physical, chemical, and biological processes. Important among physical processes are a decline in soil structure leading to crusting, compaction (including oxidation of organic soils leading to subsidence), hard-setting (crusting, induration, formation of plinthite and duripans), *erosion, natural disasters*, desertification, anaerobism, environmental pollution, and unsustainable use of natural resources. Significant chemical processes include acidification, leaching, salinization, decrease in cation retention capacity, fertility depletion, and accumulation of toxic metals. Biological processes include reduction in total and biomass carbon, and decline in land biodiversity. The latter comprises important concerns related to eutrophication of surface water, contamination of groundwater, and emissions of trace gases (CO_2, CH_4, N_2O, NO_x) from terrestrial/aquatic ecosystems to the atmosphere. Thus, land degradation may be defined as a biophysical process driven by socio-economic and political causes. Some lands or landscape units are affected by more than one process.

Depending on their inherent characteristics and the climate, lands vary from highly resistant, or stable, to those that are vulnerable and extremely sensitive to degradation. Fragility, which is sensitivity to degradation processes, may refer to the whole land, a degradation process (e.g., *erosion*) or a property (e.g., soil structure). Stable or resistant lands do not necessarily resist change since fragile lands may degrade and become less capable of performing environmental regulatory functions.

Terms commonly associated with land degradation are soil degradation and desertification. While there is a clear distinction between "soil" and "land" (the term "land" refers to an ecosystem comprising land, landscape, terrain, vegetation, water, climate), there is often no clear distinction between the terms "land degradation" and "desertification." Usually desertification refers to land degradation in arid, semiarid, and sub-humid areas due to anthropogenic activities. Some 33% of the Earth's land surface (42 million km^2) is subject to desertification.

Because of different definitions and terminology, a large variation in the available statistics on the extent and rate of land degradation exists. Two principal sources of data include the global estimates of land degradation in drylands (desertification) by Dregne and Chou (1994), and of land degradation by the "Global Assessment of Human Induced Soil Degradation" (GLASOD assessment) (Oldeman, 1994). According to the first source degraded lands in dry areas of the world amount to 3.6 billion hectares or 70% of all global drylands. According to the second source the global extent of degraded land (by all processes and all ecoregions) is about 2 billion hectares, i.e., 562 million hectares of arable lands, 685 million hectares of permanent pastures, and 719 million hectares of forests or woodlands. Both assessments, however, are qualitative and do not refer to a quantitative and reliable database (Eswaran et al., 2001; Bush, 2006).

Some causes of land degradation

The anthropogenic causes of land degradation are mainly agriculture-related (agricultural intensification, cropland expansion, livestock extension, shifting cultivation without adequate fallow periods, absence of soil conservation measures, cultivation of fragile or marginal lands, unbalanced use of fertilizers and pesticides, possible problems arising from faulty planning or management of irrigation, use of high-yield hybrid crops, etc.). It is estimated that up to 40% of the world's agricultural land can be considered as degraded (Kertész, 2009). The anthropogenic causes of

Land Degradation, Figure 1 Land degradation can be either human induced, e.g., as a result of mining activity (on the *left*; Cave del Predil, NE Italy), or induced by natural processes, e.g., Stovžje landslide (around 2 million cubic meters) (on the *right*; Stovžje, NW Slovenia) (Photographer: Matija Zorn).

land degradation are as follows: land clearing (deforestation, clear-cutting, and overlogging), agricultural depletion of soil nutrients, urban sprawl, irrigation, land pollution, quarrying and mining, other industrial causes and vehicle off-roading. Other socioeconomic driving forces include population growth and density, migration patterns, land-use policies, lack of education, conflicts, wars, poverty, rapid technological change, etc. The results are accelerated *erosion*, removal of nutrients, soil acidification or alkalinization, salinization, destruction of soil structure, loss of organic matter, etc.

Severe land degradation affects a significant portion of the Earth's arable lands, decreasing the wealth and economic development of nations. The link between a degraded environment and poverty is direct and intimate. As the land resource base becomes less productive, food security is compromised and competition for dwindling resources increases.

Land degradation affects also waters (rivers, wetlands, and lakes) because soil, along with nutrients and contaminants associated with soil, are delivered in large quantities to environments that respond detrimentally to their input. Land degradation, therefore, has potentially disastrous effects on lakes and reservoirs that are designed to alleviate flooding, provide potable water, irrigation, and generate hydroelectricity.

Overcutting of vegetation occurs when people cut forests, woodlands, and shrublands to obtain timber, fuelwood, and other products at a pace exceeding the rate of natural regrowth. This is frequent in semiarid environments, where fuelwood shortages are often severe.

Overgrazing is the grazing of natural pastures at stocking intensities above the livestock-carrying capacity; the resulting decrease in the vegetation cover is a leading cause of *erosion*.

Soil compaction is a worldwide problem, especially with the adoption of mechanized agriculture. It has caused yield reductions of 25–50% in some regions of Europe and North America, and between 40% and 90% in West African countries. On-farm losses through land compaction in the USA have been estimated at US$1.2 billion per year (Eswaran et al., 2001).

Nutrient depletion as a form of land degradation has a severe economic impact at the global scale, especially in sub-Saharan Africa. In Zimbabwe, soil erosion results in an annual loss of N and P alone, totaling to US$1.5 billion. In South Asia, the annual economic loss is estimated at US $600 million for nutrient loss by erosion, and US$1,200 million due to soil fertility depletion (Eswaran et al., 2001).

An estimated 950 million ha of salt-affected lands occur in arid and semiarid regions, nearly 33% of the potentially arable land area of the world. Productivity of irrigated lands is severely threatened by buildup of salt in the root zone. In South Asia, annual economic loss is estimated at US $500 million from waterlogging, and US $1,500 million due to salinization. Potential and actual economic impact globally is not known (Eswaran et al., 2001).

Land degradation due to population growth occurs as a consequence of land shortage. Population pressure also operates through other mechanisms. Improper agricultural practices, for instance, occur only under constraints such as the saturation of good lands under population pressure which investment that is essential for reclamation and regeneration. Unfortunately, while the potential for many types of subsidence can be identified, the precise locations and times of many types of damaging events cannot usually be predicted confidently.

Types of land subsidence

Land subsidence is most commonly and geographically widespread associated with:

- Underground cavities, both natural (Figure 1) and man-made (Figures 2 and 3)
- Vertical displacement of the ground due to fault activation by earthquakes
- Withdrawal of fluids or gas from the ground
- Compression of weak and/or water-logged soils under superimposed loads, because of vibrations, or withdrawal of support (Figure 4)

Less frequent or extensive causes are linked to lateral movements of the ground, permafrost, and subsurface erosion within unconsolidated soils (Table 1).

Effects vary greatly. Some types of subsidence are localized, affecting only a few square meters. But some events affect square kilometers (e.g., general mining subsidence) and even tens of square kilometers (e.g., some earthquake subsidence). Displacements vary between many meters and a few millimeters but even modest displacements can be sufficient to disrupt foundations of buildings. Some events are sudden (e.g., from earthquakes, mine and cave collapses) leading to injuries and loss of life, while others involve slow down-warp over periods of years (e.g., slow dissolution of strata, pumping of fluids and gas). In these, the ground surface may be gradually depressed below local ground water level or beneath peak flood levels. This is a particular problem where extraction of fluids takes place in low-lying river deltas and coast plains. Sudden subsidence is often the final result of processes that have progressed over a long time. Many subsidence events are triggered by increased inputs of water (e.g., rainfall, melt water, or leaking pipes) into potentially unstable strata.

Voids often form and subsidence processes take place beneath natural or man-made superficial deposits. Differential subsidence may occur where a structure or infrastructure is built straddling boundaries between contrasting materials, for instance, at the margin of an infilled pit or quarry, or across an active geological fault. Therefore, caution is needed when evaluating sites in subsidence-prone areas.

Investigation

Most subsidence problems affecting recent and new development result from failure to recognize, properly investigate, or to fully evaluate subsidence potential often because:

- This type of problem is not expected in the area concerned
- The site investigation is not appropriate for the circumstances or is inadequately funded
- Precautionary or remedial measures that are adopted are not appropriate for dealing fully with the hazard

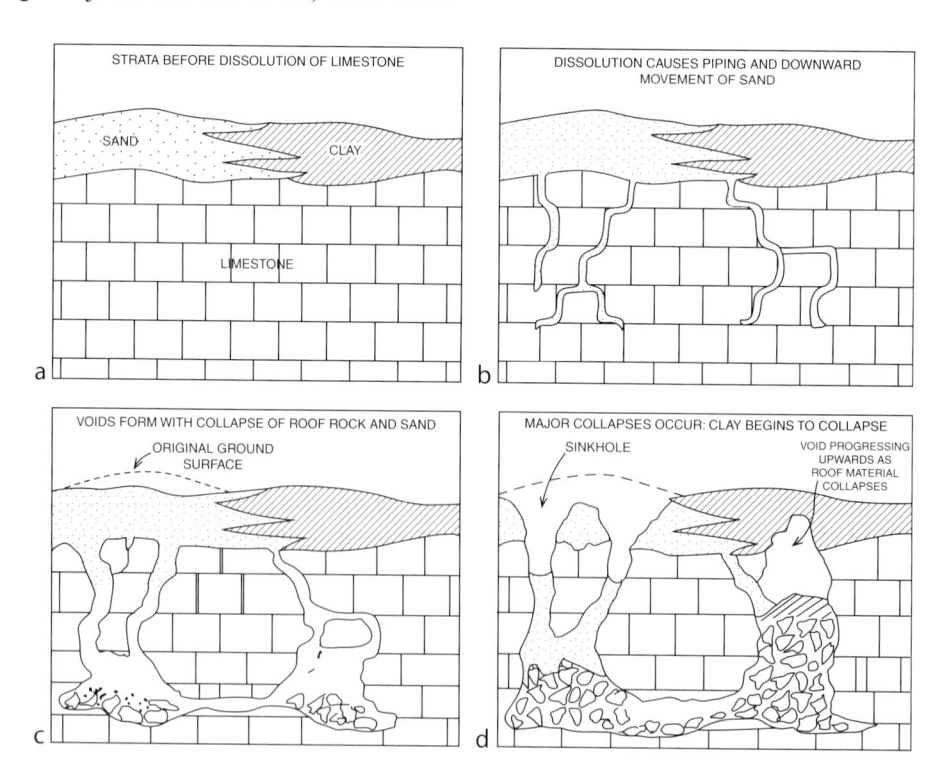

Land Subsidence, Figure 1 Development of land subsidence due to dissolution of bedrock.

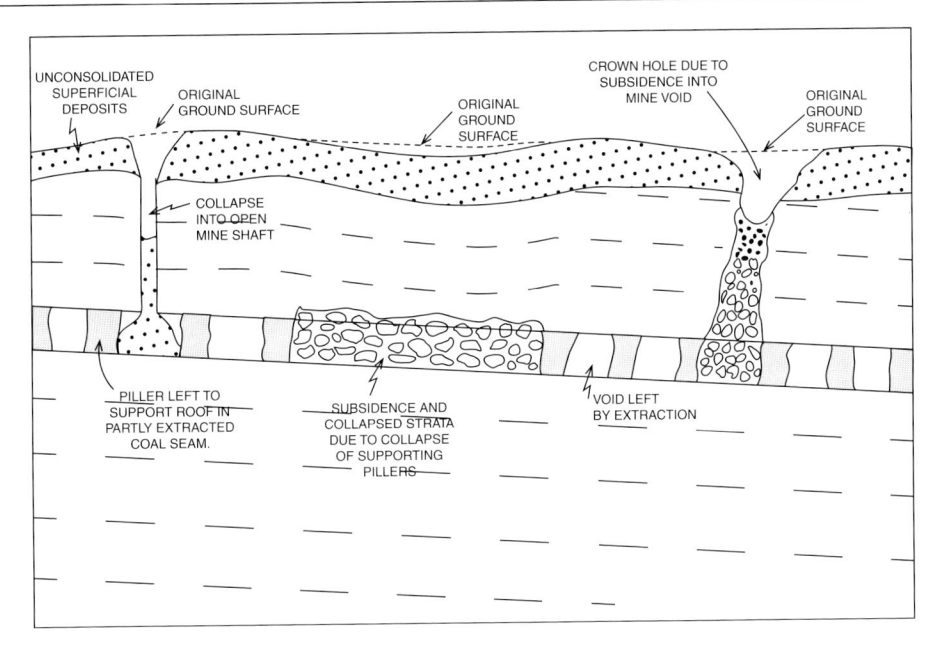

Land Subsidence, Figure 2 Subsidence over mined ground in an area of partial extraction.

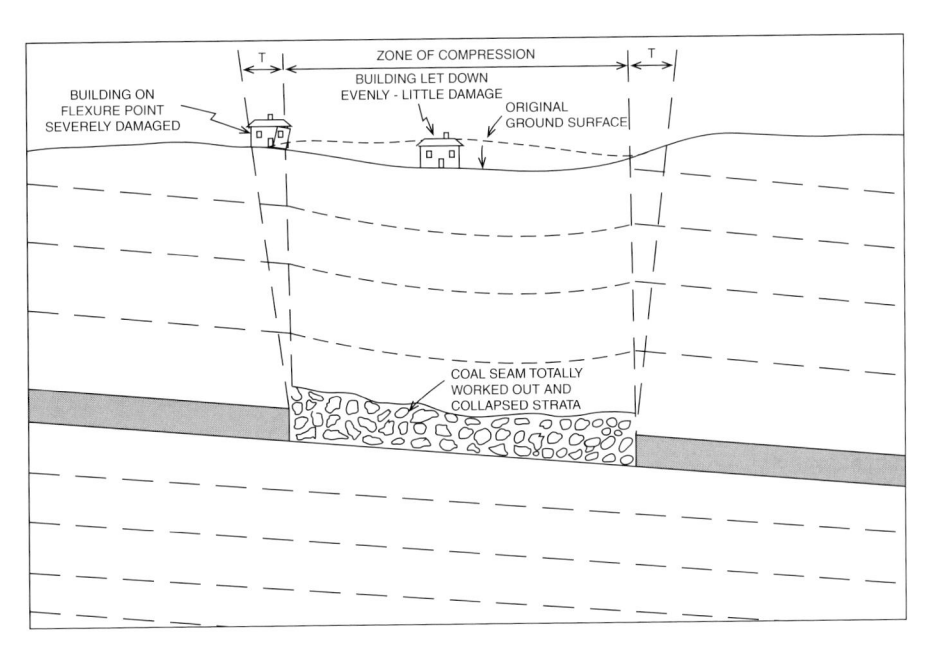

Land Subsidence, Figure 3 Subsidence over mined ground in an area of total extraction. *T* zone of tension and surface flexuring.

Problems can be reduced by ensuring that: geological conditions are properly investigated; ground subsidence is taken into account when developing planning policies and determining planning applications; site investigations are properly designed and provide all of the information needed for planning decisions; and adequate planning conditions and building control measures are imposed.

Careful site investigation including walk over surveys, direct ground investigations including drilling, trial pits, and trenches (Bell, 1975) and indirect methods such as ground geophysics (Carpenter et al., 1995) and remote sensing (Donnelly and McCann, 2000; Strozzi et al., 2003; Ge et al., 2007; Wright and Stow, 1999), supported, if available, by archive information (Freeman Fox, 1988; Howard Humphreys and Partners, 1993), usually provide

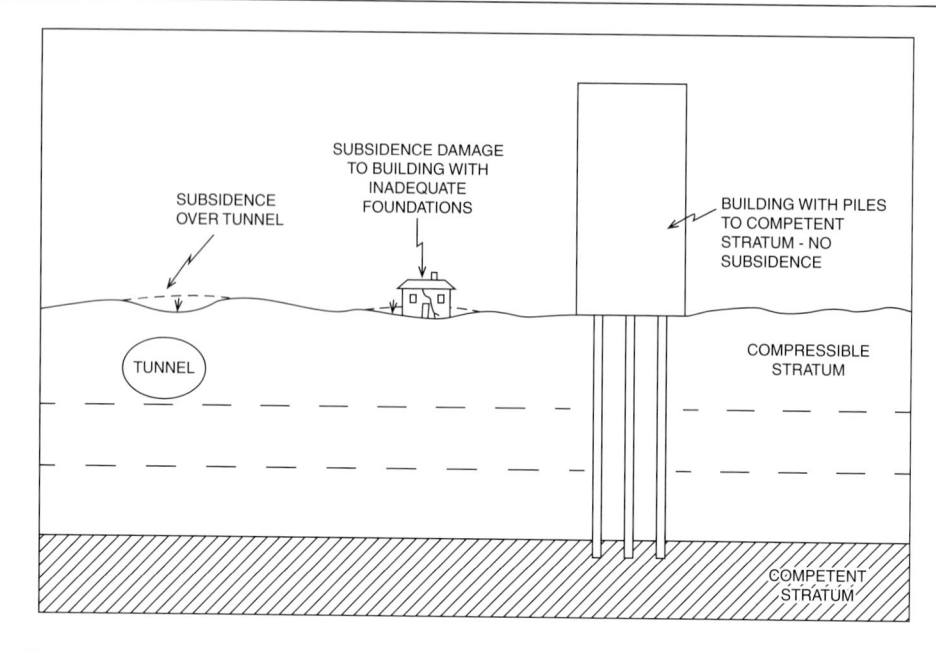

Land Subsidence, Figure 4 Subsidence due to building of surface structures on, and tunneling in, compressible strata.

a reasonable definition of potential problems and basis for designing solutions against all but the most extreme events.

There is an important role for generalized information contained in databases and hazard and risk maps (e.g., Kim et al., 2006) if these are made easily available to those who are considering development of land. Information pitched at this broader level can also assist land use planners in making decisions on allocations of land (zoning) for specific purposes and deciding what information is needed to make sound decisions on individual planning applications (DTLR, 2002).

Remediation and treatment

Some subsidence takes place on such a large scale and so suddenly that little can be done, except to provide help. Large sinkholes may measure many tens of meters across, or seismic subsidence may be appreciable. However, for many types of subsidence, sound investigation and precautions normally allow development to proceed safely although the cost of necessary work on a few sites may make small developments economically unviable. Ground subsidence can be dealt with in a number of ways depending on the cause and scale (see, for instance, National Coal Board, 1982; Kratsch, 1983; Healy and Head, 1984; Driscoll and Skinner, 2007):

- Localized areas of potential instability can be fenced off and the land above can be used for agriculture or nature conservation purposes or can be left vacant.
- Shallow voids can be excavated and backfilled.
- Voids of modest size can be infilled by injection of cement or rock paste, or by underground emplacement of granular fill.
- Shafts and wells can be capped using a concrete plug or a wire cage filled with aggregate.
- Foundations can be designed and constructed to bridge across fairly small cavities.
- Compressible soils can be strengthened.
- Minor subsidence can be dealt with by underpinning foundations.

It is sensible to record treatment that has been undertaken in a readily accessible place for use by redevelopers of sites.

Responsibility and liability

If subsidence results from human activity, such as mining, the responsibility for dealing with problems and, if appropriate, paying compensation usually rests with the mining company or its successors. In many countries, the responsibility for dealing with other types of subsidence rests with the land or property owner. However, they may have a claim against consultants if ground investigations were inadequate and hazards have not been properly defined. Ground investigation firms usually have professional indemnity insurance to cover any possible claims against them. In most developed countries, if no responsibility or liability can be proved, costs usually fall on the relevant local government authority either directly or through, for example, re-housing displaced persons. Insurance against costs of subsidence is widely available, but insurance companies may charge unaffordable premiums or

Land Subsidence, Table 1 Factors associated with land subsidence

Subsidence-prone situations		Comments	Example references
Underground cavities	Dissolution of soluble strata	The most geographically and geologically widespread soluble strata consist of limestone and dolomite. These frequently contain caves, tunnels, and fissures, formed during percolation and flow of groundwater, that may remain as empty voids or infilled partly or wholly with sediments, collapse debris, or water. Similar features develop, less extensively, in gypsum deposits. Sudden subsidence into voids forms sink holes. Slow dissolution may let the surface down by gentle down-warping. Salt deposits are also soluble but are generally absent near the ground surface, except in arid areas, due to past dissolution and are not usually a direct cause of subsidence.	Beck, 2003; Cooper, 1998; Gutiérrez, 1996; Milan Vic, 2003; Waltham et al., 2004
	Lava tubes and tunnels	Some flows develop hard surface crusts beneath which molten lava flows and, if drained away, leaves tubes, tunnels, and gas cavities.	Waltham et al., 2004, 129–137
	Mines and shafts	Voids associated with mining, particularly shallow mining, are major causes of subsidence. The type of mining influences the nature of subsidence. Partial extraction mines, leaving pillars supporting the mine roof, can remain stable for long periods, but mechanical failure of pillars, withdrawal of hydrostatic support when water is withdrawn from mine voids, or erosion of fissures and joints in the mine roof can lead to sudden subsidence. In long-wall mines (usually for coal), the whole seam is extracted and controlled subsidence is allowed behind the progressing working face leading to relatively gentle surface down-warping. Particular problems may arise where differential subsidence is caused by mining through geological faults. Ground around mine shafts can collapse into the void, especially if affected by downward percolation of water, if these have not been properly sealed.	Arup Geotechnics, 1992; Bell et al., 2000; Kratsch, 1983; National Coal Board, 1982; Ren et al., 1987
	Cellars, tunnels and wells	The excavation of tunnels, particularly in unconsolidated or poorly consolidated soils, can cause surface displacement. Other man-made cavities in previously developed land can become buried and forgotten following demolition, disasters or war damage can also present a localized hazard.	Rodriguez-Roa, 2002
Withdrawal of fluids	Extraction of water, oil, gas or brine	Extraction of fluids and gases under natural or induced pressure allows compaction of unconsolidated or poorly consolidated strata with consequent general subsidence of the ground surface and associated fissuring.	Addis, 1988; Davis and Boling, 1983; Donaldson et al., 1995; Holzer and Johnson, 1985; Milliman and Haq, 1996; Phien-wej et al., 2006
Vertical displacement	Faulting or down-warping associated with seismic events	Earthquakes can cause vertical downward displacement of sections of the Earth's crust. These may be localized or, sometimes, more extensive. Instances are known that have affected tens of square kilometers.	Shennan and Hamilton, 2008
	Salt, clay, and shale	Salt deposits deform and flow in response to pressure and can cause subsidence. Clay and shale can also behave plastically.	Jackson et al., 1994
Lateral movements	Soil creep and landslides	Gradual downward movement under the force of gravity of soils on slopes and mass movements in landslides can open fissures in the ground that may persist and become concealed.	Humpage, 1996

Land Subsidence, Table 1 (Continued)

Subsidence-prone situations		Comments	Example references
Compressible, collapsing and shrinking soils	Loading of soils	Unconsolidated or poorly consolidated superficial deposits (e.g., loess, some sands and silts) become compacted if placed under a load such as a building, tip, or reservoir.	CUR Centre for Civil Engineering, 1996; Dudley, 1970
	Vibration	Vibrations from earthquakes, machinery, or activities such as pile driving can cause liquefaction of water-logged unconsolidated sediments that can be associated with subsidence if the resulting water-sediment mixture can either flow away laterally (e.g., "running sands") or be ejected from the ground (e.g., "mud or sand volcanoes").	Berrill and Yasuda, 2002; Jeffries and Been, 2006; Seed and Idriss, 1982
	Shrinking clays	Some clays swell in the presence of water and shrink when these dry out (e.g., in drought conditions or when water is withdrawn by trees). Shrinking causes subsidence of the ground and can cause significant damage to overlying constructions.	Doornka, 1993; Hammer and Thompson, 1966
	Peat	Peat consists of water-logged fibrous plant material that is compressible under load. If peat is drained, the plant debris rapidly dry out and collapse from a fibrous to a granular state with substantial loss of volume causing substantial surface subsidence.	Gambolati et al., 2006; Wösten et al., 1997
	Infilled sites	Depressions in the ground including valleys, quarries, and ponds are commonly infilled prior to construction. Fills may sometime be poorly compacted and variable in physical properties. Landfill sites containing substantial quantities of putrescible material are constructed to allow for reductions of volume, and consequent subsidence, as organic materials decompose.	Charles, 2001; Emberton and Parker, 1987; Suter et al., 1992
Permafrost		Frozen ground in polar or mountain regions contains ice in layers and masses as well as in pore spaces between grains in soils. Summer melting affects the surface layer leaving marshy conditions above ground that remains frozen but the extent of permafrost has been decreasing in recent years. This melting and drainage can give rise to subsidence as can foundations beneath heated buildings if precautions are not taken.	McFadden and Bennett, 1991; Nelson et al., 2001
Underground fires		Coal and peat deposits can burn underground either as a result of spontaneous combustion or human activities with consequent surface subsidence and fissuring.	Ide et al., 2010; Kuenzer et al., 2007
Subsurface erosion	Piping	Flowing groundwater or leakage from water pipes can form erosion cavities in unconsolidated sediments beneath resistant overlying material if the water and entrained sediments flow away.	Bonelli et al., 2006; Parker et al., 1990
	Burrows	Animal burrows can occasionally be extensive or large enough to cause very localized ground subsidence.	Dolbeer et al., 1994

withdraw cover in areas where subsidence is frequent and expensive (Edwards, 1995).

Conclusion

A wide variety of mechanisms cause land subsidence events sometimes large scale, causing major disasters, whereas others are gradual or localized, but even these can lead to significant economic losses as time passes. The extent to which events can be predicted and taken into account in planning for development varies. Sound ground investigation normally establishes whether a subsidence hazard is present allowing suitable precautionary or remedial measures to be identified although sometimes it can be imprudent or uneconomic for development to proceed. For existing development, treatment often is only undertaken after the event. While effects of minor subsidence on buildings can be treated, major events are likely to lead to demolition, major works on the site or, in some cases, cannot be remedied. Where human actions lead to subsidence, some compensation may be obtainable, but for other circumstances, redress can be obtained but only if there is adequate insurance cover. Subsidence hazard mapping is an important step in planning for development or safeguarding existing development in any subsidence-prone area.

Bibliography

Addis, M. A., 1988. *Mechanisms of Sediment Compaction Responsible for Oilfield Subsidence*. London: University of London Press, p. 1122.

Arup Geotechnics, 1992. *Mining Instability in Great Britain – Summary Report*. London: Department of the Environment, p. 22.

Beck, B. F., 2003. *Sinkholes and the Engineering and Environmental Impacts of Karst*. Reston: American Society of Civil Engineers, p. 737.

Bell, F. G., 1975. *Site Investigation in Areas of Mining Subsidence*. Oxford: Butterworth-Heinemann. 168 pp.

Bell, F. G., and Donnelly, L. J., 2006. *Mining and Its Impact on the Environment*. Abingdon: Taylor and Francis. 547 pp.

Bell, F. G., Stacey, T. R., and Genske, D. D., 2000. Mining subsidence and its effect on the environment: some differing examples. *Environmental Geology*, **40**(1–2), 135–152.

Berrill, J., and Yasuda, S., 2002. Liquefaction and piled foundations: some issues. *Journal Earthquake Engineering*, **6**(1), 1–41.

Bonnelli, S., Brivois, O., Borghi, R., and Benahmed, N., 2006. On the modelling of piping erosion. *Comptes Rendu Mécanique*, **334**(8–9), 555–559.

Bruhn, R. W., Magnusson, M. O., Gary, R. E., 1978. Subsidence over the mined out Pittsburgh coal. Proceedings American Society of Civil Engineers Convention, Pittsburgh, pp. 26–55.

Carpenter, P. J., Booth, C. J., and Johnston, M. A., 1995. *Application of Surface Geophysics to Detection and Mapping of Mine Subsidence Fractures in Drift and Bedrock*. Champaign: Illinois State Geological Survey. 21 pp.

Charles, J. A., 2001. *Building on Fill: Geotechnical Aspects BR424*, 2nd edn. Abingdon: Taylor and Francis (HIS-BRE Press). 208 pp.

Cooper, A. H., 1998. Subsidence hazards caused by the dissolution of Permian gypsum in England: geology, investigation and remediation. In: Maund, J. G., Eddleston, M. (eds.), *Geohazards in Engineering Geology*. Special publication in engineering geology 15 London. Geological Society., pp 265–275.

CUR Centre for Civil Engineering, 1996. *Building on Soft Soils: Design and Construction of Earth Structures Both on and into Highly Compressible Subsoils of Low Bearing Capacity*. Abingdon: Taylor and Francis, 500 pp.

Davis, S. N., and Boling, J., 1983. *Measurement, Prediction and Hazard Evaluation of Earth Fissuring and Subsidence due to Groundwater Overdraft*. Tucson: University of Arizona Press. 88 pp.

Department for Transport, Local Government and the Regions (DTLR) 2002. *Planning Policy Guidance Note 14 Development on Unstable Land, Annex 2 Subsidence and Planning*. London: The Stationery Office, 57 pp.

Dolbeer, R. A., Holler, R. N., Hawthorne, D. W., 1994. Identification and assessment of wildlife damage: an overview. In: Hygnstrom, S. E., Timm, R. M., Larson, G. E. (eds.), *The Handbook: Prevention and Control of Wildlife Damage*. Lincoln: University of Nebraska, 18 pp.

Donaldson, E. C., Chilingarian, G. V., and Yen, T. F., 1995. *Subsidence due to Fluid Withdrawal*. Amsterdam: Elsevier Science. Developments in Petroleum Science, Vol. 41. ix+498 pp.

Donnelly, L. J., and McCann, D. M., 2000. The location of abandoned mine workings using thermal techniques. *Engineering Geology*, **57**(1–2), 39–52.

Doornkamp, J. C., 1993. Clay shrinkage induced subsidence. *Geographical Journal*, **159**(2), 196–202.

Driscoll, R., and Skinner, H., 2007. *Subsidence Damage to Domestic Buildings: A Guide to Good Technical Practice*. Abingdon: Taylor and Francis (HIS-BRE Press), p. 66.

Dudley, J. H., 1970. Review of collapsing soils. *Journal of the Soil Mechanics and Foundations Division, American Society of Civil Engineers*, **96**(3), 925–947.

Edwards, G. H., 1995. *Subsidence, Landslip and Groundheave, with Special Reference to Insurance 2nd Edition*. London: Chartered Institute of Loss Adjusters. Wetherby. 363 pp.

Emberton, J. R., and Parker, A., 1987. The problems associated with building on landfill sites. *Waste Management and Research*, **5**(4), 473–482.

Freeman Fox Ltd, 1988. *Methods of Compilation, Storage and Retrieval of Data on Disused Mine Openings and Workings*. London: HMSO. 18 + 58 + 32 pp.

Gambolati, G., Putti, M., Teatini, P., and Stori, G. G., 2006. Subsidence due to peat oxidation and impact on drainage infrastructures in a farmland catchment south of the Venice Lagoon. *Environmental Geology*, **49**(6), 814–820.

Ge, L., Chung, H., and Rios, C., 2007. Mine subsidence monitoring using multi-source satellite SAR images. *Photogrammetric Engineering and Remote Sensing*, **73**(3), 259–266.

Gutiérrez, F., 1996. Gypsum karstification: effects on alluvial systems and derived geohazards. *Geomorphology*, **16**(4), 277–293.

Hammer, M. J., and Thompson, O. B., 1966. Foundation clay shrinkage caused by large trees. *Journal of Soil Mechanics and Foundations Division, American Society of Civil Engineers*, **92**, 1–17.

Healy, P. R., Head, J. M., 1984. *Construction over abandoned mine workings*. CIRIA Special Publication 32; PSA Civil Engineering Technical Guide 34. London: CIRIA, 94 pp.

Howard Humphreys and Partners, 1993. *Subsidence in Norwich*. London: HMSO. 112 pp.

Holzer, T. L., and Johnson, A. I., 1985. Land subsidence caused by ground water withdrawal in urban areas. *Geojournal*, **11**(3), 245–255.

Humpage, A. J., 1996. *Cambering and valley bulging in Great Britain – a review of distribution, mechanisms of formation, and the implications for ground movements*. British Geological Survey Onshore Geology Series Technical Report WA/96/104 Keyworth: British Geological Survey. 7 pp.

Land Use, Urbanization, and Natural Hazards, Table 1 Selected urban population statistics

	Total population 2010 (millions)	Urban populations 2010 (millions)	Urban population 2010 (%)	Average annual growth of urban population 2005–2010 (%)	Urban population living in slums 2005–2007 (%)
More developed countries[a]	1,234	929	75.3	3.8	35.0
Less developed countries[b]	5,660	2,551	45.1	2.0	–[c]
World	6,896	3,480	50.5	2.7	35.0

Source: UN DESA Population Division (2011a)
Notes:
[a]Europe, Northern America, Australia, New Zealand, Japan
[b]Other countries
[c]Very minor

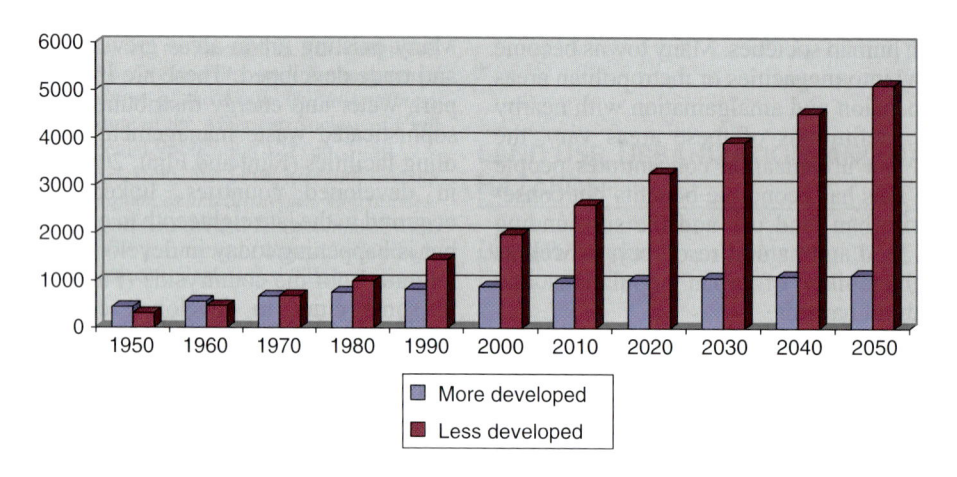

Land Use, Urbanization, and Natural Hazards, Figure 1 Rates of urbanization since 1950 projected forward to 2050 (Source: UN DESA (2011b). Note: scale in millions of people).

settlements located on coasts, deltas, and rivers are often susceptible to major floods (Berz, 2000). Hazards also affect the nearby countryside but the potential severity of impacts is greater in urban areas because:

(a) Urbanization concentrates population, development, and infrastructure leading to greater potential losses and, therefore, higher risks
(b) Rapid growth may outstrip the capacity of infrastructure to cope, for instance, drainage may be inadequate increasing flood potential, and may be accompanied by poor standards of construction making buildings more liable to damage with consequently higher casualties (Mitchell, 1999).

In developed countries, financial losses due to hazard events in urbanized areas are often large but losses of life and injuries are fewer but, in developing countries, there is often greater loss of life and more injuries compared with financial losses for a given magnitude of event. That is because, in developed countries, land and real estate prices are higher, building codes are generally stricter and better enforced, and transport and other infrastructure is often better able to cope with post-event rescue and relief operations and reconstruction (Mitchell, 1995). Although impacts on urban areas tend to be greater in human and financial terms, impacts outside urban areas are also important because of loss of crops that are needed to feed towns and cities, and disruption to transport routes. Sensible land use planning and management is important in both.

Urban management should include assessment of all potential hazards and appropriate steps at planning policy and site development stages. Risks can be reduced if buildings are constructed to survive expected events or if potentially hazardous ground is reserved for open space uses. But, even if that is done, the intention can be frustrated. Especially in developing countries, inward migrants to cities often set up home in poorly constructed slums on land that has been deliberately kept vacant, placing the most vulnerable members of society at the greatest risk (Marker 2009). It is also important to plan and practice emergency responses; site essential facilities such as hospitals and fire stations in least vulnerable areas (Boullé et al., 2002); and ensure that key evacuation and access routes are, as far as possible, protected to ensure resilience to adverse events (Godschalk, 2003; Institution of Civil Engineers, 1995; Kreimer et al., 2003; United Nations, 2002; Valentine, 2003).

Planning on the basis of past experience is insufficient because of environmental, social, and economic changes. It is important to consider changes to vulnerability, risk, and resilience for a possible range of future conditions (Kraas, 2007; Romero Lankao and Qin, 2011) so that key threats, adaptation measures, interactions between strategies and measures, and limits to adaptation can be identified (Birkmann et al., 2010) and governance arrangements can be improved (Tanner et al., 2009).

Conclusions

It is important to take account of natural hazards when planning and managing development, whether in urban areas or the surrounding countryside, and to minimize risk and to maximize the effectiveness of post-event responses. This can be done through hazard assessment and appropriate zoning of land uses, ensuring that development is designed and constructed to survive likely events, and allowing for likely changes in the coming years. It is also important that attention should continue to be paid to the issue even if there have been no recent damaging events because complacency is likely to lead to future damage or disaster.

Bibliography

Arias, E. H., Asai, Y., Chen, J. C., Cheng, H. K., Ishii, N., Kinugasa, T., Ko, P. C., Murayama, Y., Kwong, P. W., and Ukai, T., 2001. Sharing Pacific-Rim experiences in disasters: summary and action plan. *Prehospital Disaster Medicine*, **16**(1), 29–32.

Berz, G., 2000. Flood disasters: lessons from the past – worries for the future. *ICE Proc. Water and Maritime Engineering*, **142**, 3–8.

Birkmann, J., Garschagen, M., Kraas, F., and Quang, N., 2010. Adaptive urban governance: new challenges for the second generation of urban adaptation strategies to climate change. *Sustainability Science*, **5**(2), 185–206.

Boullé, P., Vrolijks, L., and Palm, E., 2002. Vulnerability reduction for sustainable urban action. *Journal of Contingencies and Crisis Management*, **5**(3), 179–188.

Chester, D. K., Degg, M., Duncan, A. M., and Guest, J. E., 2000. The increasing exposure of cities to the effects of volcanic eruptions: a global survey. *Global Environmental Change Part B Environmental Hazards*, **2**(3), 89–103.

Culshaw, M. G., Reeves, H. J., Jefferson, I., and Spink, T. W. (eds.), 2009. *Engineering Geology for Tomorrow's Cities*. London: Geological Society. Geological Society Engineering Group Special Publication 22, 303pp.

Flint, C., and Flint, D., 2001. *Urbanisation: Changing Environments. Landmark Geography*, 2nd edn. London: Collins Educational, p. 192.

Fuchs, R. K., 2002. *Mega-City Growth and the Future*. New Delhi: Bookwell Publications, p. 436.

Fujita, M., 1991. *Urban Economic Theory: Land Use and City Size*. Cambridge: Cambridge University Press, p. 380.

Genske, D. D., and Ruff, A., 2009. Expanding cities, shrinking cities, sustainable cities: challenges, opportunities and examples. CD insert. In Culshaw, M. G., Reeves, H. J., Jefferson, I., and Spink, T. (eds.), *Engineering Geology for Tomorrow's Cities*. London: Geological Society, Engineering Geology Special Publication 22.

Godschalk, D. R., 2003. Urban hazard mitigation: creating resilient cities. *Natural Hazards Review*, **4**(3), 136–143.

Institution of Civil Engineers, 1995. *Megacities: Reducing Vulnerability to Natural Disasters*. London: Thomas Telford, p. 170.

Kraas, F., 2007. Megacities and global change: key priorities. *Geographical Journal*, **173**(1), 79–82.

Kreimer, A., Arnold, M., and Carlin, A., 2003. *Building Safer Cities: The Future of Disaster Risk. Disaster Risk Management*. Washington, DC: World Bank Publications, p. 320.

Marker, B. R. 2009. Geology of megacities and urban areas. In Culshaw, M. G., Reeves, H. J., Jefferson, I., and Spink, T. (eds.), *Engineering geology for tomorrow's cities*. London: Geological Society. Engineering Geology Special Publication 22, pp. 33–48.

Marker, B. R., Pereira, J. J., and de Mulder, E. F. J., 2003. Integrating geological information into urban planning and management: approaches for the 21st century. In Heiken, G., Fukundiny, R., and Sutter, J. (eds.), *Earth Sciences in the Cities: A Reader*. Washington, DC: American Geophysical Union, pp. 379–411.

McCall, G. J. H., de Mulder, E. F. J., and Marker, B. R. (eds.), 1996. *Urban geoscience*. London: Taylor and Francis. AGID Special Publication Series No 20 in association with Cogeoenvironment. 280 pp.

Mitchell, J. K., 1995. Coping with natural disasters in megacities: perspectives of the twenty first century. *GeoJournal*, **37**(3), 303–311.

Mitchell, J. K., 1999. *Crucibles of Hazard: Megacities and Disasters in Transition*. Washington, DC: Brookings Institution, p. 450.

Pelling, M., 2003. *The Vulnerability of Cities: Natural Disasters and Social Resilience*. London: Earthscan Publications. xi+212pp.

Romero Lankao, P., and Qin, H., 2011. Conceptualising urban vulnerability to global climate and environmental change. *Current Opinions on Environmental Sustainability*, **3**, 142–149.

Tanner, T., Mitchell, T., Polack, E., and Guenther, B., 2009. *Urban Governance for Adaptation: Assessing Climate Change Resilience in Ten Asian Cities*. Brighton: Institute of Development Studies, p. 47.

United Nations 2002. *Living with risk: a global review of disaster reduction initiatives*. Inter-Agency Secretariat International Strategy for Natural Disaster Reduction (ISDR). Preliminary report. Geneva: United Nations. www.unisdr.org.

United Nations Department Economic and Social Affairs (DESA) Population Division, 2011b. *World urbanisation prospects, the 2011 revision*. http://www.un.org/unpd/wup/CD-ROM/Urban-rural-Population.htm.

United Nations Department of Economic and Social Affairs (DESA) Population Division, 2011a. *Urban population, development and the environment* (wall chart and data table). http://www.un.org/esa/population/publications/2011UrbanPopDevEnvchart/urbanpopdevenv2011wallchart.html.

Valentine, G. A., 2003. Towards integrated natural hazard reduction in urban areas. In Heiken, G., Fakundiny, R., and Sutter, J. (eds.), *Earth Science in the City: A reader*. Washington, DC: American Geophysical Union, pp. 63–73.

Cross-references

Building Codes
Buildings, Structures and Public Safety
Damage and the Built Environment

Hazard and Risk Mapping
Land-Use Planning
Megacities and Natural Hazards
Planning Measures and Political Aspects
Risk Assessment
Risk Perception and Communication
Urban Environment and Natural Hazards

LANDSAT SATELLITE

María Asunción Soriano
Universidad de Zaragoza, Zaragoza, Spain

Synonyms
ERTS

Definition
Landsat is the name applied since 1974 to a program of unmanned satellites whose main objective is the study of the surface of the Earth.

Discussion
The Landsat program includes seven satellites. Before 1974 the program was known as the ERTS (Earth Resources Technology Satellite). The satellites were launched in 1972, 1975, 1978, 1982, 1984, 1993 (failed), and 1999 (Landsat 7). Two of them, Landsat 5 and 7, are still active. The missions and commercialization of results of these civilian satellites have generally been managed by agencies of the US Government (NASA, NOAA and the US Geological Survey) although between 1984 and 2001 a private company was contracted for the purpose (NASA, 2010).

The orbit of Landsat satellites is Sun-synchronous near polar at an altitude of 918 km for Landsat 1, 2, 3 and 705 km for the remaining satellites. The spacecraft completes just over 14 orbits per day, covering the entire Earth between 81°N and S latitude every 18 days for the three first Landsats and 16 days for the rest. Some side lap of orbits occurs especially in north latitudes (around 80%). There is a Worldwide Reference System for the images where each scene is designated by path and raw numbers.

The systems on board Landsat satellites have been modified over time, increasing their resolution. The best known are the Multispectral Scanner System (MSS), the Thematic Mapper (TM) and the Enhanced Thematic Mapper Plus (ETM+). All are cross-track scanning systems with an oscillating mirror that scans a width of 185 km on the ground. In the MSS (Landsat 1–5), four different wavelength intervals (four bands) of the electromagnetic spectrum are detected, two included in the visible and two in the reflected infrared interval. Their spatial resolution is 79 by 79 m. The TM (Landsat 4 and 5) detects seven bands, three in the visible, three in the reflected infrared and one in the thermal infrared interval. The spatial resolution is 30 by 30 m, except in the thermal infrared (120 by 120 m). The main modifications in the ETM+ (Landsat 7) are the addition of a panchromatic band with a 15 m spatial resolution and the improvement of the resolution of the thermal infrared band (60 m) (NASA, 2010). The system detectors sense the energy from the ground and transform it into electrical signals that are codified in a binary system and later transformed into gray levels with an intensity proportional to the energy received (i.e., no energy is black and the maximum energy is white), generating an image similar in appearance to black and white photographs (Sabins, 1987). The combination of three bands enables color images to be obtained (false color only in the case of the MSS because of the absence of the blue wavelength interval, and both true and false in the TM and ETM+).

The Landsat Program has been providing data about the Earth's surface that have proved very useful in the fields of Geology, Forestry, Agriculture, Cartography, Urban change, dynamic environments, etc. The main customers of these images are researchers, governments, commercial users, and educators.

Bibliography
Sabins, F. F., 1987. *Remote Sensing. Principles and Interpretation*. New York: Freeman. 449 p.
NASA, 2010. http://landsat.gsfc.nasa.gov/

Cross-references
Airphoto and Satellite Imagery
Remote Sensing of Natural Hazards and Disasters

LANDSLIDE

John J. Clague
Simon Fraser University, Burnaby, BC, Canada

Definition
A *landslide* is the failure and movement of a mass of rock, sediment, soil, or artificial fill under the influence of gravity. (*Soil*, as used here, is the thin layer that directly underlies the land surface formed by pedogenic processes.)

Introduction
Globally, landslides kill thousands of people each year and cause tens of billions of dollars in damage (Centre for Research on the Epidemiology of Natural Disasters, 2012). One of the most deadly landslides on record occurred during an earthquake in Peru on May 31, 1970. A streaming mass of blocky debris produced by a failure on Nevados Huascaran killed 8,000 inhabitants of the town of Yungay (Figure 1; Plafker and Ericksen, 1978;

Landslide, Figure 1 Oblique view of Nevados Huascaran and the path of the debris flow that overran the town of Yungay on May 31, 1970, killing about 8,000 people. (George Plafker, U.S. Geological Survey).

Evans et al., 2009). Landslides were also responsible for thousands of the more than 70,000 fatalities of the Sichuan (Wenchuan) earthquake in southwest China in 2008. Large landslides can also block rivers, creating impoundments and upstream flooding. When overtopped after the reservoirs reach full-pool level, the dams may fail, producing devastating downstream floods. In some cases, however, seepage through the dams is greater than inflow into the reservoir and the dams and the impoundments may persist indefinitely.

Landslides range in volume from less than one cubic meter to tens of cubic kilometers. They are primarily associated with mountainous terrain but also occur in areas of low relief, for example, in roadway and building excavations, mine-waste piles, and river bluffs. Landslides may travel only short distances and leave their deposits at the base of the source slope. In some cases, however, they can run-out far from their sources; in the case of some lahars, the travel distance may be tens of kilometers.

Types of landslides

Landslides are classified on the basis of their velocity, the type of movement, and the source material (Figure 2; Cruden and Varnes, 1996). They move at rates ranging from millimeters per year to more than 100 m per second. Source materials range from unconsolidated sediments to lithified rocks. Perhaps the most landslide-prone unconsolidated sediments are silts and clays, but all other types of sediments may fail under certain circumstances. Crystalline and metamorphic rocks that are massive and have few joints or fractures are most resistant to slope failure; these rocks include some granitoid rocks and most non-foliated metamorphic rocks. In contrast, strongly foliated metamorphic rocks, including phyllites, schists, gneisses, and bedded sedimentary rocks, and any rock that is strongly jointed, fractured, or faulted, may fail where structural elements or stratification dip in the same direction as the slope. Some of the most failure-prone rocks are pyroclastic deposits and lavas found on the slopes of stratovolcanoes. Topography is, of course, also important. All other things being equal, steeper slopes are more likely to fail than gentler ones, because gravitational driving forces increase as a slope becomes steeper. The presence or absence of water can be critically important because it exerts a strong influence on the type of movement. The classification scheme shown in Figure 2, although simplified, provides an understanding of the range and complexity of landslides.

Falls involve rolling and bouncing of rock and, less commonly, sediment from cliffs or down steep slopes. Initial failure occurs along steeply inclined fractures or other discontinuities in rock or sediment. This process is responsible for cones and aprons of talus, which are common landforms in mountains. Large blocks and boulders sometimes roll or bound beyond the foot of the talus slope, causing loss of life and property damage (Holm and Jakob, 2009). Numerical models have been created to estimate, for a given slope, the "rockfall shadow" – the zone vulnerable to boulder run-out. Rockfall is also common along some roads and railways (Figure 3).

Although relatively small, rockfall is among the most costly of all types of landslides. In addition to economic losses due to traffic delays, the costs of scaling, blasting, and grouting threatening rock faces and removing debris from roads and rail lines are considerable.

Topples involve the forward rotation of rock or sediment about a pivot point under the influence of gravity. Movement occurs along steeply inclined fractures. Topples range from shallow movements to deep-seated displacements of large volumes of rock. The process operates almost imperceptibly, with movement rates commonly in the range of millimeters to centimeters per year, but a threshold of stability may be reached, at which time the material suddenly fails, producing a fall or a slide.

Slides involve the downslope translational movement of rock or sediment along a discrete surface (Voight, 1978).

They are subdivided into translational and rotational types, although many slides are complex phenomena, involving both types of movements.

Translational sliding takes place on planar or undulating surfaces dipping in the direction of the slope. The basal failure plane may be a bedding plane, foliation, a fault, or a tight set of joints. The basal failure plane is bordered by the head and marginal scarps, which may be joints or faults. The slide mass commonly disintegrates as it moves downslope, but the fragments tend to retain their positions with respect to one another.

Rotational sliding involves translation of rock or sediment along a curved, concave-upward failure surface, producing what are termed slumps (Figure 4). This style of movement involves extension at the head of the slump and compression at its toe. The result is one or more steep down-stepping scarps at the headwall and bulging of the toe. Movement rates range from millimeters per hour to meters per second.

Flows are a large and varied group of landslides that share one similarity – the failed material moves in the manner of a fluid. In wet flows, rock fragments are partly supported by interstitial water. Debris flows are the most common type of wet flow; they consist of mixtures of water, rock fragments, and plant detritus that move down steep stream courses or ravines as slurries and, less commonly, down open slopes without lateral confinement (Figure 5; Jakob and Hungr, 2005). Most debris flows are triggered by heavy rainfall or by rain on snow. Those that move down the flanks of volcanoes are called lahars. This group includes very large flows caused by melt of snow and ice during eruptions glacier-clad volcanoes. The Osceola lahar, which traveled down valleys on the north and east flanks of Mount Rainier in Washington state about 5,600 years ago, covered an area of about 550 km^2 and had a volume in excess of 3 km^3 (Crandell, 1971; Scott and Vallance, 1995). The smaller Electron lahar, which is about 500 years old, traveled more than 50 km from its source on the west flank of Mount Rainier. Mudflows are similar to debris flows, but the solid fraction consists of sand, silt, and clay, with little or no gravel or coarser material. Debris flows and mudflows can travel at speeds of up to a few tens of meters per second.

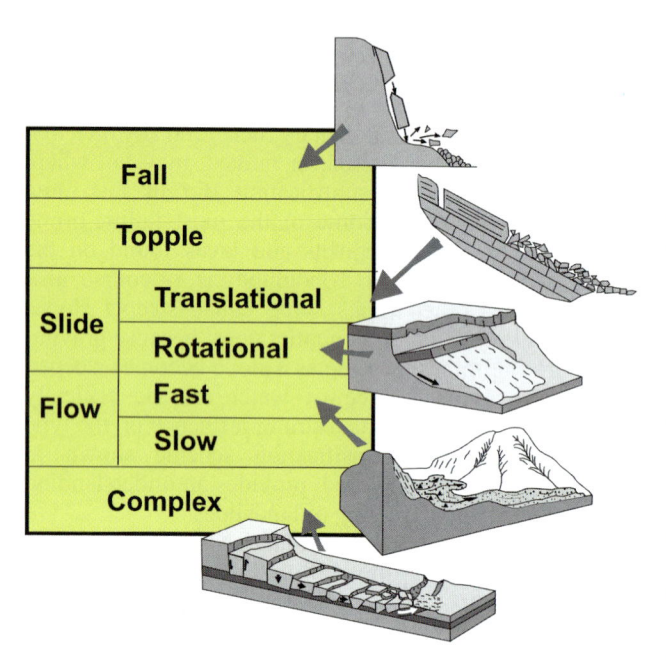

Landslide, Figure 2 Simplified classification of landslides.

Landslide, Figure 3 Rockfall on the Trans-Canada Highway near Yale, British Columbia (Duncan Wyllie).

Landslide, Figure 4 Slump in unconsolidated sediments along Thompson River at Ashcroft, British Columbia. The slump happened in 1897 (J. J. Clague).

Landslide, Figure 5 Debris flow tracks on steep slopes above Wahleach Lake in the Cascade Range, British Columbia (J. J. Clague).

Flows of water-saturated sediment are also common on slopes underlain by permafrost in the Arctic (McRoberts and Morgenstern, 1974). The failure plane is shallow, at the contact between the permanently frozen ground and the overlying active layer. These flows can occur on slopes as low as a few degrees and are especially common along river and coastal bluffs underlain by ice-rich sediments. As climate warms, as is expected through the remainder of this century, the active layer in permafrost terrain may thicken, leading to an increase in the numbers of these flows.

Sediment flows also happen in oceans and lakes, especially off deltas, at the heads of submarine canyons, and at the seaward edges of continental shelves. Those that travel down submarine canyons into deep ocean waters are termed turbidity currents. Submarine landslides are especially common on the foreslopes of rapidly prograding deltas, such as those at the mouths of fiords in Norway, British Columbia, and Greenland (Prior and Bornhold, 1988), and off the mouths of some of the world's largest rivers, for example, the Mississippi, Congo, Ganges, and Indus rivers.

Rock avalanches (also known as sturzstroms) are large flows of fragmented rock that contain little or no water. They flow due to the release of energy by particle interactions and particle comminution. Rock avalanches are the fastest of all landslides, in some cases achieving speeds of 100 m per second or more. They also travel long distances where unimpeded by topography. Rock avalanches are far less common than other types of

Landslide, Figure 6 The Frank Slide, which occurred in April 1903, is a classic rock avalanche. It involved the failure of about 30 million cubic meters of limestone. The rock rapidly fragmented and streamed onto the valley floor at high velocities. The debris was deposited as a sheet over an area of over 2 km^2 (Geological Survey of Canada).

landslides, but are important and of scientific interest because of their long travel distances and the destruction and death they cause. One of the most famous examples is the Frank Slide, which killed about 70 people in the mining town of Frank, Alberta, in April 1903 (Figure 6; Cruden and Krahn, 1978). Many mechanisms have been proposed to explain the high mobility and long run-out of rock avalanches (see recent review by Hewitt et al., 2008). Examples include travel of the debris on a thin layer of air, generation of high pore pressures at the base of the flow by entrainment of water, and forces generated by rapid and extreme comminution of particles. The discovery of rock avalanche deposits on other planets suggests that explanations involving trapped air and water are likely invalid.

Many landslides do not fit comfortably into existing classification schemes. Prominent examples include *sackung* and *lateral spreads*. Sackung, a German verb meaning "to sag," is deep-seated downslope movement of large, internally broken rock masses, with no single, well-defined basal failure plane (Zischinsky, 1969). Movement is manifested at the land surface by cracks, trenches, and scarps at mid and upper slope positions, and by bulging of the lower slope. Lateral spreading involves extension of a slab of earth material above a nearly flat shear plane. The moving slab may subside, rotate, disintegrate, or flow. Lateral spreading in silts and clays is commonly progressive – failure starts suddenly in a small area, but spreads rapidly, ultimately affecting a much larger area. Lateral spreading commonly results from liquefaction of a subsurface sand layer. An unusual form of lateral spreading is the failure of what are termed "quick clays." Thick glaciomarine clays were deposited on isostatically depressed coastal lowlands in some areas of Canada, Scandinavia, and Alaska near the end of the Pleistocene and subsequently elevated above sea level due to rebound of the crust. Quick clays are sensitive, meaning that they may liquefy and flow when disturbed (Rankka et al., 2004). A famous example is the Leda clay, deposited in lowlands bordering the St. Lawrence River and its tributaries in southern Quebec and Ontario (Eden and Mitchell, 1970; Smalley, 1980). Landslides in the Leda clay have caused much damage and loss of life during the past few centuries (Evans et al., 1997).

Finally, many landslides, including most large ones, are *complex*, that is, they involve more than one type of movement. A rockslide, for example, may evolve into a debris flow by entraining water or saturated sediments along its path. The 1970 Nevados Husacaran landslide, mentioned above, is an example of a complex landslide. It started as a large slide or fall of rock and ice, but soon transformed into a debris flow.

Causes and triggers

When considering landslides, it is important to understand the difference between their causes and triggers, although in practice, the two concepts define end points on a continuum. The *cause* of a landslide is the combination of external and internal factors that, over time, leads to failure. The main causes of landslide are geological (lithologic and structural), steep topography, weathering, erosion, subsurface solution, depositional loading at the top of an unstable slope, a change in climate, and human disturbance. In contrast, a landslide is *triggered* by a single event such as an earthquake, rainstorm, volcanic eruption, or a series of freeze-thaw cycles (Wieczorek, 1996).

Moderate and large earthquakes commonly trigger rockfalls, slides, and debris avalanches. Very large (magnitude >7) earthquakes in mountainous terrain may trigger rockslides, large slumps, and rock avalanches. Debris flows and small rockfalls are commonly triggered by intense rain. Rainwater infiltrates sediment and fractures in rock, raising water pressures in these materials and inducing failure. Many rockfalls are triggered by frequent freeze-thaw cycles (Figure 7). When water in fractures freezes, its volume increases by about 9%, sufficient to induce large tensile forces on the rock bounding the fractures. Volcanic eruptions trigger lahars (Figure 8) and, in some cases, huge flank collapses, such as happened at the onset of the cataclysmic eruption of Mt. St. Helens in May 1980.

Many landslides, including some large ones, occur without known triggers. Slopes may slowly deteriorate over centuries or millennia to the point that they fail of their own accord. An example of a large landslide without a known trigger is the Hope slide, which occurred in

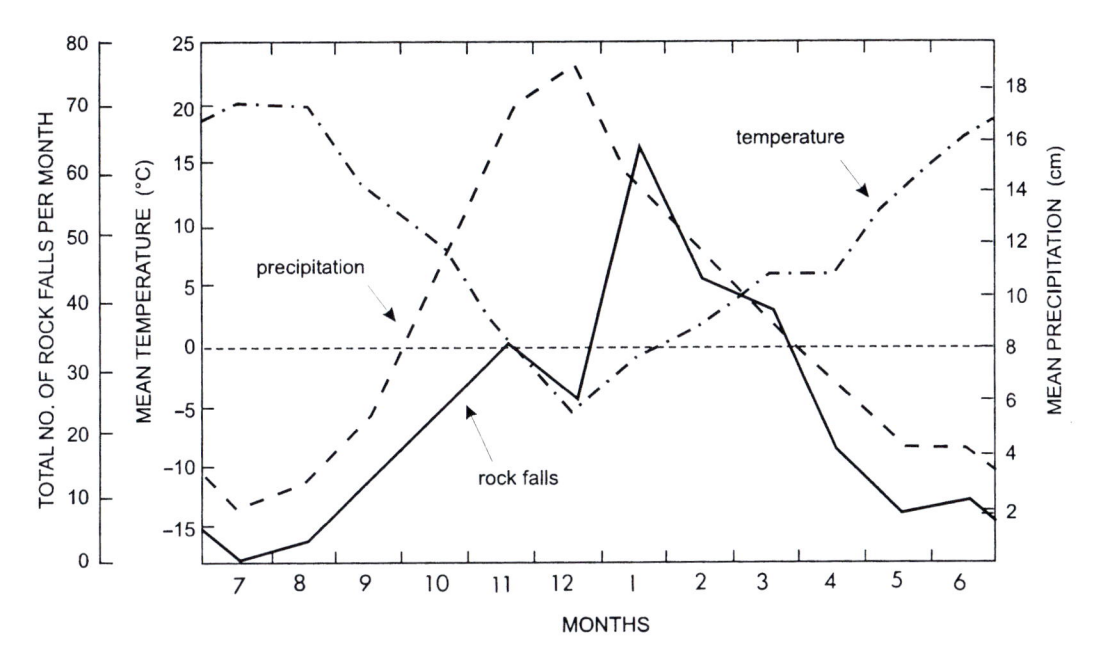

Landslide, Figure 7 Average monthly rockfall frequency and weather over the period 1933–1970 in the Fraser Canyon (From Evans and Savigny, 1994; Modified from Peckover and Kerr, 1977).

Landslide, Figure 8 Aftermath of a huge lahar that swept through the town of Armeio in Colombia on November 13, 1985, killing 20,000 of its 29,000 inhabitants. The lahar was triggered by an eruption of Nevado del Ruiz, which melted snow and glacier ice on the summit.

southwest British Columbia in January 1965 (Mathews and McTaggart, 1978). Weather conditions in the lead-up to the landslide were not unusual and no earthquakes occurred at the time of failure. Interestingly, however, a large (magnitude 7) earthquake, with an epicenter approximately 50 km south of the site of the landslide, occurred in 1872 (Bakun et al., 2002). Although the seismic acceleration at the landslide site during this earthquake may have been greater than 0.2 g, the slope did not fail. Rather, the unstable slope deteriorated over the next 93 years sufficiently to reach the threshold of failure without the requirement of a significant trigger.

The water content of slope materials is an important factor in their stability. Slopes that are stable in arid environments may fail in humid ones, and some rock slopes in areas where temperatures frequently fluctuate above and below freezing are vulnerable to rockfall. Even in arid environments, however, heavy rainfall can trigger widespread slope instability. In any case, water enters slopes along fractures, faults, and permeable strata. Under conditions of poor drainage, pore pressures may rise within the slope, increasing the likelihood of failure. Vegetation reduces this effect by intercepting rainfall and, through evapotranspiration, reducing the amount of water in slope materials. Tree roots also act to stabilize near-surface earth materials on slopes.

Climate warming, which is expected later in this century, may increase annual or seasonal precipitation in some areas (Solomon et al., 2007), thereby increasing the frequency of rockfalls, debris flows, debris avalanches, and perhaps other types of landslides there. Of critical importance will be changes in the frequency of extreme rainfall events. A rise in extreme storms in an area would certainly increase the frequency of debris flows and probably rockfalls. Similarly, areas that become more arid are likely to experience more frequent or more intense wildfires, with an attendant loss in vegetation leading to increased soil failure. The incidence of rockfalls from high mountain slopes may also increase in the future due to thaw of alpine permafrost (Noetzli and Gruber, 2009). Ice within fractures in rock that is now below 0°C will melt if air temperatures rise above the freezing point over much of the year, possibly destabilizing already marginally stable rock slopes (Fischer et al., 2006; Huggel, 2009).

Recognition and mitigation of landslide hazards

The first step in landslide hazard mitigation is to identify and assess the hazard (Schuster and Kockelman, 1996). This assessment involves: (1) determining the age and frequency of past landslides from geologic evidence and historical records and (2) assessing ground conditions where failure might occur. The second step is mitigation to minimize risk to people and property. Mitigation measures are of three types: (1) restrictions on land use, (2) monitoring and early warning, and (3) corrective and defensive works.

Land-use restrictions

If a site is deemed too hazardous and the risk cannot be reduced to an acceptable level, development may be disallowed or restricted. Such determinations must be made carefully and with a sense of what is acceptable because limitations on land use commonly have economic consequences – the full value of the land may not be realized. Whether a risk is perceived as acceptable or unacceptable depends on social and economic factors. In general, less risk is tolerated in wealthy countries than in poor ones. In wealthy countries, knowledge of past landslides is generally taken into account during development. For this reason, land-use restrictions are more commonly accepted in these countries than in poor ones. The high level of risk that people in some poor countries accept, largely because they have few or no options, contributes to the greater loss of life from landslides in these countries than elsewhere.

Monitoring and early warning

Rockfalls, small slumps, opening and widening of ground cracks, and tilting of trees precede many large landslides. Observation or instrumentation of unstable slopes may provide early warning of their imminent catastrophic failure.

Advanced technologies are providing new opportunities to monitor potentially hazardous slopes. Very small movements of the ground can be detected using satellite remote sensing tools, notably InSAR (Interferometric Synthetic Aperture Radar). A powerful tool for producing highly precise digital elevation models of unstable slopes is LIDAR (Light Detection and Ranging), which uses laser pulses to measure the time delay between transmission of a pulse and detection of the reflected signal. LIDAR uses millions of pulses to produce a model of the Earth surface in an area of interest. Comparison of images acquired over time may reveal slow slope movements that could precede catastrophic failure (Oppikofer et al., 2008).

Corrective and defensive works

Corrective and defensive measures include reforestation, control works, and protective structures. Careful management of forests reduces the likelihood of some types of landslides. Reforestation of logged or burned slopes, in combination with other corrective measures, may stabilize debris source areas. Control works, including dykes along stream channels, deflection dams and dykes, and debris retention basins, provide protection against debris flows. Steep rock slopes along highways and railways can be stabilized with retaining walls, anchored beams, rock bolts, bulkheads, toe buttresses, metal nets and fences, and ditches (Wyllie, 1991). Unstable slopes can be dewatered with tunnels or permeable pipes. Stabilization of large bedrock slides and sagging slopes requires extensive surface and subsurface drainage.

Summary

Landslides are a near-global hazard, but are most common in areas of high relief where steep and moderate slopes are common. Landslides are classified according to type of movement (falls, topples, slides, slumps, flows, sags), type of material (rock, unconsolidated sediments), and velocity (slow, fast). The most important cause of landslides is geology – failure generally occurs on weak planes in rock, including stratification, foliation, joints, or faults; and poorly lithified rocks and non-lithified sediments are generally more prone to failure than lithified materials. Almost as important is water, which increases the mass of the earth material and therefore the gravitational driving force; water also penetrates joints, fractures, and permeable strata, possibly elevating pore-water pressures to the threshold of failure. Landslide triggers are related to, but distinct from, causes. They are the phenomena that raise the slope above the threshold of failure. The most common triggers of landslides are severe storms, earthquakes, volcanic eruptions, and human activity. In the case of human activity, loading the top of an unstable slope or removing material from its base may trigger a landslide, as both increase the gravitational driving force on the slope. Human activity, however, can also increase the stability of a slope, for example, by loading the toe of the slope or by dewatering unstable materials. Landslide hazards can be mitigated by land-use restrictions, monitoring, warning systems, and corrective and defensive structures.

Bibliography

Bakun, W. H., Haugerud, R. A., Hopper, M. G., and Ludwin, R. S., 2002. The December 1872 Washington State earthquake. *Bulletin of the Seismological Society of America*, **92**, 3239–3258.

Centre for Research on the Epidemiology of Disasters, 2012. EM-DAT, The International Disaster Database. Universite catholique de Louvain, Brussels, Belgium. http:www.emdat.be. Accessed 7 April 2012.

Crandell, D. R., 1971. *Postglacial Lahars from Mount Rainier Volcano, Washington*. Washington, DC: U.S. Government Printing Office. U.S. Geological Survey Professional Paper, Vol. 677.

Cruden, D. M., and Krahn, J., 1978. Frank rockslide, Alberta, Canada. In Voight, B. (ed.), *Rockslides and Avalanches*. Amsterdam: Elsevier. Natural Phenomena, Vol. 1, pp. 97–112.

Cruden, D. M., and Varnes, D. J., 1996. Landslide types and processes. In Turner, A. K., and Schuster, R. L. (eds.), *Landslides: Investigation and Mitigation*. Washington, DC: National Academy Press. National Research Council, Transportation Research Safety Board Special Report, Vol. 247, pp. 36–75.

Eden, W. J., and Mitchell, R. J., 1970. The mechanics of landslides in Leda clay. *Canadian Geotechnical Journal*, **7**, 285–296.

Evans, S. G., and Savigny, K. W., 1994. Landslides in the Vancouver–Fraser Valley–Whistler region. In Monger, J. W. H. (ed.), *Geology and Geological Hazards of the Vancouver Region, Southwestern British Columbia*. Ottawa: Natural Resources Canada. Geological Survey of Canada Bulletin, Vol. 481, pp. 251–286.

Evans, S. G., Couture, R., and Chagnon, J. Y., 1997. Notes on major Leda Clay landslides of the St. Lawrence Lowlands of eastern Canada, 1615–1996. In *50th Canadian Geotechnical Conference of the Canadian Geotechnical Society*. Alliston, ON: Canadian Geotechnical Society. Geological Survey of Canada Contribution Series, pp. 839–846.

Evans, S. G., Bishop, N. F., Fidel Smoll, L., Valderrama, M. P., Delaney, K. B., and Oliver-Smith, A., 2009. A re-examination of the mechanism and human impact of catastrophic mass flows originating on Nevado Huascaran. *Engineering Geology*, **108**, 96–118.

Fischer, L., Kääb, A., Huggel, C., and Noetzli, J., 2006. Geology, glacier retreat and permafrost degradation as controlling factor of slope instabilities in a high-mountain rock wall: the Monte Rosa east face. *Natural Hazards and Earth System Sciences*, **6**, 761–772.

Hewitt, K., Clague, J. J., and Orwin, J. F., 2008. Legacies of catastrophic rock slope failures in mountain landscapes. *Earth Science Reviews*, **87**, 1–38.

Holm, K., and Jakob, M., 2009. Long rockfall runout, Pascua Lama, Chile. *Canadian Geotechnical Journal*, **46**, 225–230.

Huggel, C., 2009. Recent extreme slope failures in glacial environments; effects of thermal perturbation. *Quaternary Science Reviews*, **28**, 1119–11130.

Jakob, M., and Hungr, O. (eds.), 2005. *Debris-Flow Hazards and Related Phenomena*. Berlin: Springer.

Mathews, W. H., and McTaggart, K. C., 1978. Hope rockslides, British Columbia, Canada. In Voigt, B. (ed.), *Rockslides and Avalanches*. Amsterdam: Elsevier. Natural Phenomena, Vol. 1, pp. 259–275.

McRoberts, E. C., and Morgenstern, N. R., 1974. Stability of slopes in frozen soil. *Canadian Geotechnical Journal*, **11**, 554–573.

Noetzli, J., and Gruber, S., 2009. Transient thermal effects in Alpine permafrost. *The Cryosphere*, **3**, 85–99.

Oppikofer, T., Jaboyedoff, M., and Keusen, H. R., 2008. Collapse of the eastern Eiger flank in the Swiss Alps. *Nature Geosciences*, **1**, 531–535.

Peckover, F. L., and Kerr, J. W. G., 1977. Treatment and maintenance of rock slopes on transportation routes. *Canadian Geotechnical Journal*, **14**, 487–507.

Plafker, G., and Ericksen, G. E., 1978. Nevados Huascaran avalanches, Peru. In Voight, B. (ed.), *Rockslides and Avalanches*. Amsterdam: Elsevier. Natural Phenomena, Vol. 1, pp. 277–314.

Prior, D. B., and Bornhold, B. D., 1988. Submarine morphology and processes of fjord fan deltas and related high-gradient systems: modern examples from British Columbia. In Nemec, W., and Steel, R. J. (eds.), *Fan Deltas: Sedimentology and Tectonic Settings*. London: Blackie and Son, pp. 125–143.

Rankka, K., Andersson-Sköld, Hultén, C., Larsson, R., Leroux, V., and Dahlin, T., 2004. *Quick clay in Sweden*. Swedish Geotechnical Institute Report, 65.

Schuster, R. L., and Kockelman, W. J., 1996. Principles of landslide hazard reduction. In Turner, A. K., and Schuster, R. L. (eds.), *Landslides: Investigation and Mitigation*. Washington, DC: National Academy Press. National Research Council, Transportation Research Safety Board Special Report, Vol. 247, pp. 91–105.

Scott, K. M., and Vallance, J. W., 1995. *Debris Flow, Debris Avalanche, and Flood Hazards at and Downstream from Mount Rainier, Washington*. Reston, VA: U.S. Geological Survey. U.S. Geological Survey Hydrologic Investigations Atlas, Vol. HA-729.

Smalley, I., 1980. Factors relating to the landslide process in Canadian quickclays. *Earth Surface Processes and Landforms*, **1**, 163–172.

Solomon, S., Qin, D., Manning, M., Chen, Z., Marquis, M., Averyt, K. B., Tignor, M., and Miller, H. L. (eds.), 2007. *Climate Change 2007: The Physical Science Basis. Contribution of Working Group I to the Fourth Assessment Report of the Intergovernmental Panel on Climate Change*. Cambridge: Cambridge University Press.

Voight, B. (ed.), 1978. *Rockslides and Avalanches (2 Vols.)*. Amsterdam: Elsevier.

Wieczorek, G. F., 1996. Landslide triggering mechanisms. In Turner, A. K., and Schuster, R. L. (eds.), *Landslides: Investigation and Mitigation*. Washington, DC: National Academy Press. National Research Council, Transportation Research Safety Board Special Report, Vol. 247, pp. 76–90.

Wyllie, D., 1991. Rock slope stabilization and protection measures. In *Proceedings of a National Symposium on Highway and Railroad Slope Maintenance*, Chicago, IL, pp. 41–63.

Zischinsky, U., 1969. Über sackungen [Subsidence]. *Rock Mechanics*, **1**, 30–52.

Cross-references

Climate Change
Debris Flow
Earthquake
Hazard
Lahar
Landslide Dam
Landslide Types
Mass Movement
Mitigation
Mudflow
Permafrost
Rock Avalanche
Rockfall
Rockslide
Sackung
Slide and Slump
Slope Stability
Solifluction
Warning Systems

LANDSLIDE DAM

Reginald L. Hermanns
International Centre for Geohazards, Geological Survey of Norway, Trondheim, Norway

Synonyms

Debris dam; Quake lake

Definition

Landslide dams are formed by landslide (see entry *Landslide*) deposits or moving landslides which block a permanent or ephemeral water course leading to the formation of a natural reservoir which fills with water and/or sediments. The term quake lake is used in the case that the landslide was triggered seismically, this term became established as a name for a landslide-dammed lake in southwestern Montana triggered by an earthquake on August 17, 1959; it then became more widely used after the 2008 Wenchuan China earthquake that triggered hundreds of large valley-damming landslides.

Introduction

Landslide dams range in size from a few cubic meters in volume and a few decimeters high (e.g., that can block a drainage ditch beside a road), to a dam several cubic kilometers in volume and several hundreds of meters high that can block an entire mountain valley. In all cases, the damming adds to the hazard (see entry *Hazard*) of the landslide due to combinations of flooding of the valley upstream of the dam, diversion of the water course, and catastrophic flooding of downstream areas if the dam fails. However, in the *natural hazard* literature, the term "landslide dam" is most often used to describe larger landslide blockages that form lakes more than several meters deep (Schuster, 2006).

Large landslides in all mountain environments have formed natural dams which have resulted in upstream lakes (Costa and Schuster, 1988; Evans et al., 2011a). Often these are related to earthquake triggering of landslides, and single earthquakes have formed more than hundreds of dams (quake lakes) (Evans et al., 2011a). Impounded lakes can have capacities of more than 10 km^3, as in the case of lake Sarez which is impounded by the Usoi dam (see entry *Usoi Landslide and Lake Sarez*) in Tajikistan; with a volume of 16 km^3 this is the largest rockslide dam reservoir on Earth (Alford and Schuster, 2000; Ischuk, 2011; Stone, 2009). The height of this natural dam is 670 m and its volume is 2.2 km^3. Despite the large volume of such dams, they are not necessarily stable. Half of all dams formed in the twentieth century with a volume in excess of 20×10^6 m^3 have breached within the following 7–3,435 days (Evans, 2006), often releasing much of the dam water and causing catastrophic floods downriver that causes significant destruction and loss of life (e.g., Groeber, 1916; Zevallos et al., 1996; Evans et al., 2011b). In history the *landslide dam* with highest reported death toll (100,000) was the result of the breach of a dam on the Dadu river in Sichuan, China, on June 11, 1786; this had formed only 10 days earlier after being triggered by an earthquake (see entry *Earthquake*) of ~M 7.8 (Dai et al., 2005). In general, there is an agreement that those dams which fail catastrophically, are usually new and ~80% of catastrophic dam failures occur within the first year of dam existence (Costa and Schuster, 1988; Ermini and Casagli, 2003; Evans et al., 2011a). However, landslide dams which survive their first years of existence can become stable landforms that control the geomorphic developments of valleys for thousands of years (e.g., Hermanns et al., 2004; Schuster, 2006; Hewitt, 2006). Such dams and their remains have to be recognized as such in the geological record in order to not misinterpret these landforms with landforms related to tectonic or climate-driven landscape evolution (Hewitt et al., 2011). Due to the large volume of water stored and the steep gradient caused in the river profile these dams have often been used for hydropower generation (e.g., Duman, 2009; Hermanns et al., 2009). Based on this experience, rockslides have been triggered artificially by explosives in the former Soviet Union to construct dams in remote areas rapidly and at low costs (e.g., Aduschkin, 2011; Korchevskiy et al., 2011) for hydropower generation, but also for debris flow protection. Nevertheless, much older dams have also breached and have caused catastrophic floods (e.g., González Díaz et al., 2001; Hermanns et al., 2004).

Classification of landslide dams

The first widely used classification of landslide dams emphasized the relation of the *landslide* deposit to valley and impoundment morphology. It proposed six different types of dams, based mainly on the two-dimensional distribution of the *landslide* deposit in a single valley and the dam plan form (Costa and Schuster, 1988). The recent increase in identification of landslide dam deposits (Evans et al., 2011b) has revealed some limitations of this classification. For example, landslide dams are not two-dimensional; they have a three-dimensional distribution in relation to valley morphology (Dunning et al., 2005; Strom and Pernik, 2006) which affects the formation and long-term stability of impoundments. A new classification system (Hermanns et al., 2011a, Figure 1) not only takes into account the three-dimensional relation of *landslide* bodies to valley morphology, but also the common phenomena of (a) lakes formed on the landslide deposit itself, (b) formation of multiple landslide dams in a valley, (c) landslide dams at confluences of two or more river valleys that may give rise to more than one lake, and (d) the rarer but special case of landslide dams affecting drainage divides.

Landslide dam failure and outburst floods

Most landslide dams fail by overtopping (Costa and Schuster, 1988; Evans et al., 2011a) when the landslide-dammed lake fills with water and *erosion* of the dam crest starts downcutting into the deposit. However, overtopping by displacement waves (see entry *Landslide Triggered Tsunami, Displacement Wave*) caused by a mass movement (see entry *Mass Movement*) into the landslide-dammed lake has also been reported and is considered a serious threat when large unstable slope areas or glaciers exist above the lake (Hermanns et al., 2004; Stone, 2009). Other failure modes are piping (e.g., Meyer et al., 1994; Quenta et al., 2007), progressive upstream erosion (see entry *Erosion*) (Hancox et al., 2005), and sliding collapse of the downstream face of the dam (Dunning et al., 2006). All of these processes are self-accelerating because the more the water escapes growth, the higher the outflow velocity becomes and the faster the material is eroded. Hence, landslide dam *erosion* often results in breaching. Peak discharges during such failures can be several times the seasonal peak discharge of a river and have volumes of several tens or more than 100,000 m^3/s (e.g., Abbott, 1848; Zevallos et al., 1996; González Díaz et al., 2001; Evans et al., 2011a). Such discharges can flood large portion of the river valleys downstream and lead to the damage and destruction of bridges, villages, agricultural land, and hydropower facilities in the days that the wave moves downriver (Evans et al., 2011b). More than 1,000 km long valleys can strongly be affected by these masses of water and solids (González Díaz et al., 2001; Schuster, 2006; Evans et al., 2011a) flushing downstream. This downstream flush of sediments is not restricted to the breach event itself but lasts over years and leads to aggradations through the river system (Hancox et al., 2005; Davies and Korup, 2007).

Hazard assessment of landslide dams

Hazard (see entry *Hazard*) assessment of landslide dams includes assessment of flooding of the upriver area, stability assessment of the dam, assessment of a potential catastrophic outburst flood, and assessment of the longer-term river aggradation and flood hazard downstream. While the first step is straightforward and restricted to outline the surface area below the minimum dam crest level, stability assessment of the dam itself and of potential outburst floods is a severe challenge. Geotechnical stability calculations similar to stability calculations for artificial earth- and rock-filled dams are especially challenging because they require knowledge of parameters such as grain size distribution, permeability, and groundwater flow which are difficult and expensive to obtain and have been used only in limited occasions (e.g., Meyer et al., 1985, 1994; Bianchi Fasani et al., 2011). Limited data exists for grain size distribution of eroded rockslide dams in the Italian Apennine (Casagli et al., 2003); however, landslide dam material varies significantly depending on the properties of the initial slide material and the landslide type (see entry *Landslide Types*) (Weidinger, 2011). This difference in material can significantly influence the life span of a dam. Therefore, geomorphic parameters have been used to group dams in stability domains (Ermini and Casagli, 2003; Korup, 2004), for example, height of the dam and size of drainage system and the upstream basin. However, although these domains are a good approximation, they cannot be used as a rule (Hermanns et al., 2011b) and the three-dimensional relation between dam form and valley must also be assessed (Hermanns et al., 2009, 2011a).

The breach of a dam can also be analyzed by hydraulic models that base upon principles of hydraulics, sediment transport, and soil mechanics used in software simulation programs (e.g., Fread, 1993) and also give the breach-outflow hydrograph (see entry *Hydrograph, Flood*). Physical hydraulic modeling has also been used to assess peak outflows and test numerical models (Davies et al., 2007).

The downstream flooding of a potential dam breach can be assessed using empirical peak discharge estimation (e.g., Evans, 1986; Walder and O'Connor, 1997); hydraulic models (see above); and simulation of downstream propagation of outburst floods, debris floods, and debris flows (see entry *Debris Flow*) using GIS-based programs (e.g., O'Brien, 2003; Iverson et al., 1998).

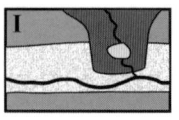

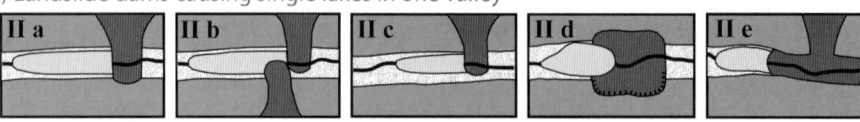

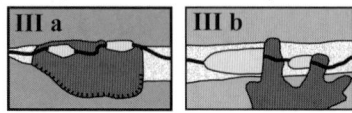

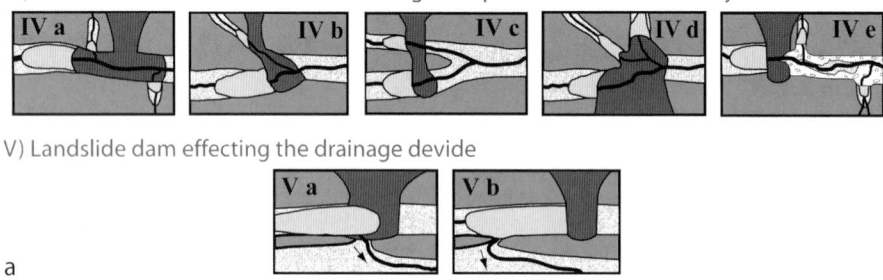

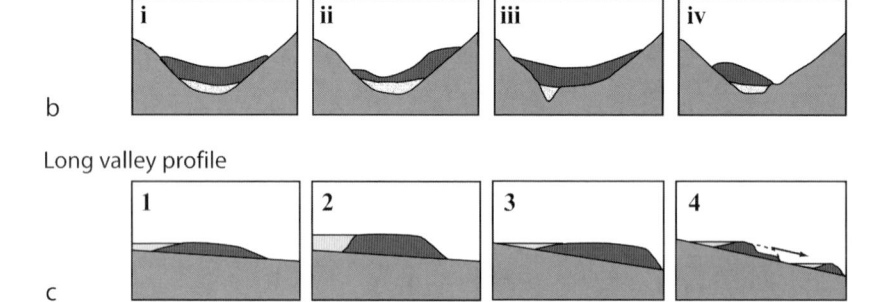

Landslide Dam, Figure 1 Classification system for rockslide dams based upon the three-dimensional distribution of rockslide debris (*dark gray*) within the valley (*light gray* = valley slopes, *dotted fill* = valley fill) with (**a**) showing the plan view distribution, (**b**) the cross valley profile, and (**c**) the long valley profile (Summarized after Hermanns et al., 2011a).

Risk management of landslide dams

Often, due to the size of outburst floods and the length of valleys which can be affected, a landslide dam failure cannot be mitigated with flood protection (see entry *Flood Protection*) structures. In order to allow for more time-consuming prevention measures, high-capacity pumping is often a first step to reduce an imminent hazard of dam breaching (Evans et al., 2011a; Schuster and Evans, 2011). Long-term stability of dams is best sought by constructing by-pass tunnels and spillways (Evans et al., 2011a; Schuster and Evans, 2011). Spillways may not always prevent the dam from failure (Zevallos et al., 1996; Schuster and Evans, 2011) because of retrogressive erosion (see entry *Erosion*) due to the fast outflow; however, they have been shown to reduce the upstream and downstream impacts by reducing the crest level at failure and thus, the water volume. In any case, monitoring of the inundation of the valley upriver the dam and of the dam itself has to start immediately after dam formation and early warning systems (see entry *Early Warning Systems*) might be the only cost-effective nonstructural mitigation (see entry *Mitigation*) measure against significant loss of life during outburst floods (Droz and Spasic-Gril, 2006). In the very recent past, a combination of such measures have been applied at both the Hattian Bala landslide dam formed by a rock avalanche triggered by the M 7.6 October 2005 earthquake and the Hunza dam formed on January 4, 2010, both in Pakistan. Spill ways were dug at both sites to lower the lake level, early warning systems got installed, and people were evacuated from potential flood areas (Evans et al., 2011a; Delaney and Evans, 2011). At both sites no breaching occurred after dam overtopping along the spill ways, and outflow stabilized only shortly after overtopping and erosional modification of the spill way. The Hattian landslide dam breached unfortunately on February 9 in 2010 after heavy rainfall and a slide on the downstream side of the dam (Konagai and Sattar, 2011). At the Hunza site stabilization work was ongoing in 2011 focusing on widening the spill way and lowering of the lake level (Delaney and Evans, 2011).

Summary

Landslide dams occur in all mountain terrains and form lakes with water volumes which can be in excess of 1 km^3. They often form in mountain regions when earthquakes trigger large landslides. Due to the relative short life of most of these dams and often catastrophic failure, urgent has to be given including assessment of upriver inundation and catastrophic downstream flooding. As dam stabilization methods are time-consuming, expensive, and often impractical, the only feasible method to save lives may be the immediate evacuation (see entry *Evacuation*) of the downstream valley. Slower-filling landslide-dammed lakes may be drained by diversion tunnels, and natural landslide dams can even be used for hydropower generation assuming they are confidently assessed as stable (which can only be the case after they have been stable for many years). Landslide dams have also been constructed artificially by triggering large landslides in narrow valley sections with explosives for hydropower generation but also for debris flow protection.

Bibliography

Abbott, J., 1848. Inundation of the Indus, taken from the lips of an eye-witness A. D., 1842. *Journal of the Asiatic Society of Bengal*, **17**, 230–232.

Aduschkin, V. V., 2011. Russian experience with blast-fill dam construction. In Evans, S. G., Hermanns, R. L., Strom, A. L., and Scarascia Mugnozza, G. (eds.), *Natural and Artificial Rockslide Dams*. Berlin: Springer. Lecture Series in Earth Sciences, pp. 595–616.

Alford, A., and Schuster, R. L., 2000. *Usoi Landslide Dam and Lake Sarez*. New York: United Nations Publication. ISDR Prevention Series.

Bianchi Fasani, G., Esposito, C., Petitta, M., ScaraciaMugnozza, G., Barbierei, M., Cardarelli, E., Cercato, M., and Di Fillipo, G., 2011. The importance of the geological models in understanding and predicting the life span of rockslide dams. In Evans, S. G., Hermanns, R. L., Strom, A. L., and ScarasciaMugnozza, G. (eds.), *Natural and Artificial Rockslide Dams*. Berlin: Springer. Lecture Series in Earth Sciences, pp. 323–346.

Casagli, N., Ermini, L., and Rosati, G., 2003. Determining grain size distribution of material composing landslide dams in the Northern Apennine: sampling and processing methods. *Engineering Geology*, **69**, 83–97.

Costa, J. E., and Schuster, R. L., 1988. The formation and failure of natural dams. *Geological Society of America Bulletin*, **100**, 1054–1068.

Dai, F. C., Lee, C. F., Deng, J. H., and Tham, L. G., 2005. The 1786 earthquake-triggered landslide dam and subsequent dam-break flood on the Dadu river, Southwestern China. *Geomorphology*, **65**, 205–221.

Davies, T. R. H., and Korup, O., 2007. Persistent alluvial fanhead trenching resulting from large, infrequent sediment inputs. *Earth Surface Processes and Landforms*, **32**, 725–742.

Davies, T. R. H., Manville, V., Kunz, M., and Donadini, L., 2007. Modelling landslide dambreak flood magnitudes: case study. *Journal of Hydraulic Engineering*, **133**, 713–720.

Delaney, K. B., and Evans, S. G., 2011. Rockslide dams in the northwest Himalayas (Pakistan, India) and the adjacent Pamir Mountains (Afghanistan, Tajikistan), Central Asia. In Evans, S. G., Hermanns, R. L., Strom, A. L., and ScarasciaMugnozza, G. (eds.), *Natural and Artificial Rockslide Dams*. Berlin: Springer. Lecture Series in Earth Sciences, pp. 205–241.

Droz, P., and Spasic-Gril, L., 2006. Lake Sarez mitigation project: a global risk analysis. In *Proceedings 22nd Congress on Large Dams*, Barcelona, Q36-R75.

Duman, T. Y., 2009. The largest landslide dam in Turkey: Tortum landslide. *Engineering Geology*, **104**, 66–79.

Dunning, S., Petley, D., Rosser, N., and Strom, A., 2005. The morphology and sedimentology of valley confined rock-avalanche deposits and their effect on potential dam hazard. In Hungr, O., Couture, R., Eberhardt, E., and Fell, R. (eds.), *Landslide Risk Management*. Amsterdam: Balkema, pp. 691–701.

Dunning, S. A., Rosser, N. J., Petley, D. N., and Massey, C. R., 2006. Formation and failure of the Tsatichhu landslide dam, Bhutan. *Landslides*, **3**, 107–113.

Ermini, L., and Casagli, N., 2003. Prediction of the behaviour of landslide dams using a geomorphologic dimensionless index. *Earth Surface Processes and Landforms*, **28**, 31–47.

Evans, S. G., 1986. The maximum discharge of outburst floods caused by the breaching of man-made and natural dams. *Canadian Geotechnical Journal*, **23**, 385–387.

Evans, S. G., 2006. The formation and failure of landslide dams: an approach to risk assessment. *Italian Journal of Engineering Geology and Environment*, **1**(Special issue), 15–20.

Evans, S. G., Delaney, K. B., Hermanns, R. L., Strom, A. L., and Scarascia-Mugnozza, G., 2011a. The formation and behavior of natural and artificial rockslide dams: Implications for engineering performance and hazard management. In Evans, S. G., Hermanns, R. L., Strom, A. L., and ScarasciaMugnozza, G. (eds.), *Natural and Artificial Rockslide Dams*. Berlin: Springer. Lecture Series in Earth Sciences, pp. 1–75.

Evans, S. G., Hermanns, R. L., Strom, A. L., and Scarascia-Mugnozza, G., 2011b. *Natural and Artificial Rock Slide Dams*. Berlin: Springer. Lecture Notes in Earth Sciences.

Fread, D. L., 1993. NWS FLDWAV Model: the replacement of DAMBRK for dam-break flood prediction. In *Proceedings: The 10th Annual Conference of the Association of State Dam Safety Officials*. Kansas City, MO, pp. 177–184.

González Díaz, E. F., Giaccardi, A., and Costa, C., 2001. La avalancha de rocas del río Barrancas (Cerro Pelán), norte del Neuquén: su relación con la catástrofe del río Colorado (29/12/1914). *Revista de la Asociación Geológica Argentina*, **56**, 466–480.

Groeber, P., 1916. Informe sobre las causas que han producido las crecientes del río Colorado (Provincia de Neuquén, Argentina). *Dirección General de Minas, Geología e hidrogeología*, **11**, 1–29.

Hancox, G. T., McSaveney, M. J., Manville, V. R., and Davies, T. R., 2005. The October 1999 Mt, Adams rock avalanche and subsequent landslide dam-break flood and effects in Poera River, Westland, New Zealand. *New Zealand Journal of Geology and Geophysics*, **48**, 683–705.

Hermanns, R. L., Blikra, L. H., and Longva, O., 2009. Relation between rockslide dam and valley morphology and its impact on rockslide dam longevity and control on potential breach development based on examples from Norway and the Andes. In Bauer, E., Semprich, S., and Zenz, G. (eds.), *Long Term Behavior of Dams: Proceedings of the 2nd International Conference*. Graz: Verlag der Technischen Universität Graz, pp. 789–794.

Hermanns, R. L., Folguera, A., Penna, I., Fauqué, L., and Niedermann, S., 2011b. Landslide dams in the central Andes of Argentina (Northern Patagonia and the Argentine Northwest). In Evans, S. G., Hermanns, R. L., Strom, A. L., and ScarasciaMugnozza, G. (eds.), *Natural and Artificial Rockslide Dams*. Berlin: Springer. Lecture Series in Earth Sciences, pp. 147–176.

Hermanns, R. L., Hewitt, K., Strom, A. L., Evans, E. G., Dunning, S. A., and ScarasciaMugnozza, G., 2011a. The classification of rock slide dams. In Evans, S. G., Hermanns, R. L., Strom, A. L., and ScarasciaMugnozza, G. (eds.), *Natural and Artificial Rockslide Dams*. Berlin: Springer. Lecture Series in Earth Sciences, pp. 581–593.

Hermanns, R. L., Niedermann, S., Ivy-Ochs, S., and Kubik, P. W., 2004. Rock avalanching into a landslide-dammed lake causing multiple dam failures in Las Conchas valley (NW Argentina) – evidence from surface exposure dating and stratigraphic analyses. *Landslides*, **1**, 113–122.

Hewitt, K., 2006. Disturbance regime landscapes: mountain drainage systems interrupted by large rockslides. *Progress in Physical Geography*, **30**, 365–393.

Hewitt, K., Gosse, J., and Clague, J. J., 2011. Rock avalanches and the pace of late Quaternary development of river valleys in the Karakoram Himalaya. *Geological Society of America Bulletin*, **123**, 1836–1850.

Ischuk, A. R., 2011. Usoi rockslide dam and lake Sarez, Pamir mountains, Tajikistan. In Evans, S. G., Hermanns, R. L., Strom, A. L., and ScarasciaMugnozza, G. (eds.), *Natural and Artificial Rockslide Dams*. Berlin: Springer. Lecture Series in Earth Sciences, pp. 423–440.

Iverson, R. L., Scilling, S. P., and Vallance, J. W., 1998. Objective delineation of lahar-inundation hazard zones. *Geological Society of America Bulletin*, **110**, 972–984.

Konagai, K., and Sattar, A., 2011. Partial breaching of Hattian Bala landslide dam formed in the 8th October 2005 Kashmir Earthquake, Pakistan. *Landslides*, doi:10.1007/s10346-011-0280-x.

Korchevskiy, V. F., Kolichko, A. V., Strom, A. L., Pernik, L. M., and Abdrakhmatov, K., 2011. Utilisation of data derived from large-scale experiments and study of natural blockages for blast fill dam design. In Evans, S. G., Hermanns, R. L., Strom, A. L., and ScarasciaMugnozza, G. (eds.), *Natural and Artificial Rockslide Dams*. Berlin: Springer. Lecture Series in Earth Sciences, pp. 617–637.

Korup, O., 2004. Geomorphometric characteristics of New Zealand landslide dams. *Engineering Geology*, **73**, 13–35.

Meyer, W., Sabol, M. A., Glicken, H. X., and Voight, B., 1985. *The Effects of Groundwater, Slope Stability, and Seismic Hazard on the Stability of the South Fork Castle Creek Blockage in the Mt. St. Helens area*, Washington. USGS Professional Paper, 1345, pp. 1–42.

Meyer, W., Schuster, R. L., and Sabol, M. A., 1994. Potential for seepage erosion of landslide dam. *Journal of Geotechnical Engineering*, **120**, 1211–1229.

O'Brien, J. S., 2003. Reasonable assumptions in routing a dam break mudflow. In *Proceedings of the 3rd International Conference on Debris Flow hazard Mitigation: Mechanics, Prediction, and Assessment*. Davos, 1, pp. 683–693.

Quenta, G., Galaza, I., Teran, N., Hermanns, R. L., Cazas, A., García, H., 2007. Deslizamiento traslacional y represamiento en el valle de Allpacoma, ciudad de La Paz, Bolivia. In: *Proyecto Multinacional Andino: Geosciencias para las Communidades Andinas*. Servicio Nacional de Geología y Minería, Publicación Multinacional, 4, pp. 230–234.

Schuster, R. L., 2006. Impacts of landslide dams on mountain valley morphology. In Evans, S. G., ScarasciaMugnozza, G., Strom, A. L., and Hermanns, R. L. (eds.), *Landslides from Massive Rock Slope Failures Earth and Environmental Sciences*. Dodrecht: Springer. NATO Science Series IV, Vol. 49, pp. 591–618.

Schuster, R. L., and Evans, S. G., 2011. Risk reduction measure for landslide dams. In Evans, S. G., Hermanns, R. L., Strom, A. L., and ScarasciaMugnozza, G. (eds.), *Natural and Artificial Rockslide Dams*. Berlin: Springer. Lecture Series in Earth Sciences, pp. 77–100.

Stone, R., 2009. Peril in the Pamirs. *Science*, **326**(5960), 1614–1617.

Strom, A. L., and Pernik, L., 2006. Utilization of the data on rockslide dams formation and structure for blast-fill dams design. *Italian Journal of Engineering Geology and Environment*, 1(Special Issue), 133–136.

Walder, J. S., and O'Connor, J. E., 1997. Methods for predicting peak discharge of floods caused by the failure of natural and constructed earthen dams. *Water Resources Research*, **33**, 2337–2348.

Weidinger, J. T., 2011. Stability and life span of landslide dams in the Himalayas (India, Nepal) and the Qin Ling mountains (China). In Evans, S. G., Hermanns, R. L., Strom, A. L., and Scarascia-Mugnozza, G. (eds.), *Natural and Artificial Rockslide Dams*. Berlin: Springer. Lecture Series in Earth Sciences, pp. 243–278.

Zevallos, O., Fernandez, M. A., Plaza Nieto, G., Klinkicht Sojos, S., 1996. Sin plazo para la esperanza, reporte sobre el desastre de La Josephina – Quito: Escuela Politécnica Nacional.

Cross-references

Debris Flow
Disaster Risk Management
Displacement Wave
Early Warning System
Earthquake
Erosion
Evacuation
Flood Protection
Hazard
Hydrograph
Landslide
Landslide Type
Mitigation
Natural Hazard
Rock Avalanche
Usoi Landslide and Lake Sarez

LANDSLIDE IMPACTS

Michael James Crozier[1], Nick Preston[1], Thomas Glade[2]
[1]Victoria University of Wellington, Wellington, New Zealand
[2]University of Vienna, Vienna, Austria

Definition

Consequences experienced by natural or human systems as a result of landslide activity are referred to as *landslide impacts*. These can result from direct physical damage or indirect disruption of economic and social activities. Impacts may be experienced immediately at the time of the initial event or be manifest some time afterward. They may be confined solely to the landslide site or be experienced off-site, at some distance from the landslide. The degree of impact can range in magnitude and in the extreme cases can result in human disasters and catastrophes.

Physical and human context

Landslides are defined here according to Cruden and Varnes (1996) and Dikau et al. (1996) as "the movement of a mass of rock, earth, or debris down a slope." However, *landslide impact* is a much more difficult concept to define because it is subjective and depends to a large extent on the system under consideration (whether physical or human) and the values placed on components of those systems. For example, impacts in the social context may range through loss of life, community disruption, damage to buildings and infrastructure, and disruption to means of production and wider economic activity. On the other hand, impacts to elements of the physical system can be viewed as part of a natural geomorphic process and from this perspective are less likely to be construed as *negative* impacts. In terms of geomorphic landscape development, landslides contribute significantly to landform evolution in a number of ways. Landslide processes transport sediments and rocks downslope and are, thus, a major contributor to the overall sediment cycle: erosion – transport – accumulation – compaction – uplift – and again erosion (Hancox et al., 2005).

The degree of impact produced by landslides depends partly on factors such as their volume, velocity of movement, depth of displaced mass, and the extent of disruption of the displaced mass, which together constitute landslide *intensity* (Glade and Crozier, 2005). The other major factor that determines the degree of impact is the sensitivity of the receiving system. In social systems, this is reflected by the concepts of vulnerability and resilience (Hufschmidt and Glade, 2010) whereas in natural systems, sensitivity is influenced by such factors as connectivity between hillslope and fluvial systems (Korup et al., 2004), vegetation cover, and other geological and geomorphological characteristics. The spatial and temporal occurrence of landslides also influences the magnitude of impact in an area over time. Landslides can occur in widely different configurations. They can be single events ranging from small magnitude (e.g., minor volumes with short displacements) to large magnitudes (e.g., collapses of complete slope segments with extensive travel distances). Alternatively, landslides that are triggered by large storms or earthquakes (Keefer, 1984) may involve the simultaneous occurrence of numerous landslides, ranging from localized multiple events involving tens to hundreds of landslides occurring over a few square kilometers, to large regional landslide events producing 10,000 failures or more over hundreds of square kilometers (Crozier, 2005). Some research on spatial landslide data using geospatial analytical techniques suggests that magnitude-frequency distributions have a common form irrespective of the type of landslide and the triggering agent, for example, landslides triggered by rainstorms or earthquakes (Guzzetti et al., 2002; Malamud et al., 2004).

Assessing and managing impacts

The probability of consequences (expected impacts) arising from landslide activity is referred to as landslide-related *risk* (see entry *Risk*). Hazard and risk management is now a well established practice and depends primarily on the assessment and evaluation of risk. Conventionally landslide risk is treated as a function of hazard (the frequency of a given magnitude of landslide), the elements at risk (e.g., people or structures), and their vulnerability. Risk determined in this way is evaluated in terms of its level of tolerability and potential treatment options available to reduce risk.

As with all hazards, the ability to predict occurrence in time and space has a bearing on the ability to determine risk and take mitigating or evasive measures and ultimately to prevent or reduce impact. If the physical system is well understood, susceptibility maps or even hazard maps can be made that will identify the location and probability of occurrence respectively (Guzzetti et al., 2006). For certain location-specific landslides such as debris flows and lahars which tend to reoccur along the same pathways, warning systems can be developed to allow evasive action and reduce impact (Bell et al., 2009). However, the record of landslide activity used to determine frequency and magnitude relationships has often been established on the basis of historic natural conditions and does not reflect more recent destabilizing effects of human activity, such as creation of reservoirs, deforestation, and surcharging of slopes with water from leaking utilities or disposal of waste in unstable localities.

All landslides involve some degree of impact. Significant impacts change landscapes and can affect humanity. Effects vary from surface modifications resulting from slow creep processes up to fatalities caused by mobilization of large masses of sediment or rocks. The nature of impact and consequences for society are discussed below.

Physical and human controls of impact severity

The impacts from landslides and associated consequences are closely related to landslide intensity, magnitude, and frequency of occurrence as well as the pattern of occurrence. Glade and Crozier (2005) discussed the impacts with respect to displacement and movement mechanisms and were able to draw some broad correlations with landslide type (such as fall, topple, rotational slide, translational slide, and flow). However, individual characteristics of landslide behavior, such as velocity (Table 1) and contextual factors, such as degree of exposure of individuals (Table 2) are generally much more

Landslide Impacts, Table 1 Classification of speed of movement and related impacts, according to Cruden and Varnes (1996) and Australian Geomechanics Society (2002)

Speed class	Description	Velocity (mm/s)	Typ. Velocity	Probable destructive significance
7	Extremely fast	5×10^3	5 m/s	Disaster of major violence, buildings destroyed by impact of displaced material, many deaths, escape unlikely
6	Very fast	5×10^1	3 m/min	Some lives lost; velocity too great to permit all persons to escape
5	Fast	5×10^{-1}	1.8 m/h	Escape evacuation possible; structures; possessions and equipment destroyed
4	Moderate	5×10^{-3}	13 m/month	Some temporary and insensitive structures can be temporarily maintained
3	Slow	5×10^{-5}	1.6 m/year	Remedial construction can be undertaken during movement; insensitive structures can be maintained with frequent maintenance work if total movement is not large during a particular acceleration phase
2	Very slow	5×10^{-7}	16 mm/year	Some permanent structures undamaged by movement
1	Extremely slow			Imperceptible without instruments, construction possible with ***precautions***

Landslide Impacts, Table 2 Vulnerability of a person being affected by a landslide in open space, in a vehicle and in a building, based on Wong et al. (1997)

Location	Description	Data range	Recommended value	Comments
Open space	Struck by rockfall	0.1–0.7	0.5	May be injured but unlikely to cause death
	Buried by debris	0.8–1	1	Death by asphyxia
	Not buried, but hit by debris	0.1–0.5	0.1	High chances of survival
Vehicle	Vehicle is buried/crushed	0.9–1	1	Death almost certain
	Vehicle is damaged only	0–0.3	0.3	High chances of survival
Building	Building collapse	0.9–1	1	Death almost certain
	Inundated building with debris and person is buried	0.8–1		Death is highly likely
	Inundated building with debris, but person is not buried	0–0.5	0.2	High chances of survival
	Debris strikes the building only	0–0.1	0.05	Virtually no danger

important determinants of the degree of impact than landslide type alone.

Direct impacts lead to immediate consequences. Houses and infrastructure might be displaced, damaged, or destroyed, farmland might be removed or covered by debris and/or people might be injured or killed. Indirect effects, on the other hand, include disruption of the normal socioeconomic activities, for example, production losses may occur if people are prevented from attending their workplace or road blockages prevent supply of raw material for manufacturing or export of products.

Short-term impact causes only temporary interruption or distortion of human activity. In contrast, some effects can last for extended periods, or indeed the consequences may be permanent. For example, a major landslide into a river may cause avulsion and the creation of a new channel or even reversal of drainage direction. Similarly, constant creep of a slope may require continuous adaptation to ongoing movement, for example, the city of Ventnor, Isle of Wight is located on a large landslide and requires regular maintenance (e.g., Ibsen and Brunsden, 1996). On the other hand societies might be affected so badly, that the affected people suffer from continuous trauma (Catapano et al., 2001) or preexisting industrial activity cannot be resumed. The duration of social impact depends on community resilience, which in turn is related to a number of factors including the available resources, the strength of community networks, the level of insurance, and the external support.

Scenarios of system change

The wide range of landslide impacts both in type and severity demand intensive research in order to offer sustainable solutions. Of particular importance is assessment of the possible increase of landsliding driven by environmental changes, including climate change and potential human interventions associated with development. Examples of such interventions are direct slope modifications (e.g., undercutting, leveling, changed drainage pattern, etc.) or indirect effects through vegetation change. Development, however, is driven by population increase, accelerating rates of urbanization and demand for increased living standards. These demographic and social trends have been closely linked to increase in risk from landslides and other hazards (Cendrero et al., 2006).

In this respect, process-based landslide studies play a major role in estimating effects in scenarios of future environmental and social changes. If these studies focus on single and distinctive failures, applicable geotechnical and engineering methods range from rockfall modeling based on lump mass approaches and debris flow routing based on modified flow laws to landslide failures applying soil mechanical principles. If the problem is widespread and distributed over a large region, then totally different, spatially focused methods have to be applied to assess the future likelihood of failure (Glade and Crozier, 2005). Recent examples of such widespread failures are landslides resulting from the 2008 Wenchuan, China or 2005 Pakistan earthquake (e.g., Sato et al., 2007) or failures following a thunderstorm such as that occurred in1999 in Venezuela (Larsen et al., 2001) or in the Collazzone area, Umbria, Central Italy (Guzzetti et al., 2006).

Irrespective of the type of investigation, landslide scenarios only have relevance for stakeholders and decision makers if they are coupled with information on possible consequences, including the implications of social trends. These consequences can be assessed based on a variety of methods ranging from classic heuristic approaches such as expert opinion or detailed landslide risk analysis involving modeling (Risk).

Summary

Landslides are part of normal landform evolution. Our landscapes would not appear as they are, and sediment flux on hillslopes and within river systems would not function as effectively without landslides. Impacts of landslides occur on various levels. Impacts have significance for both natural environments (e.g., removal of soil from the upper slopes and accumulation on the slope foot or in the adjacent fluvial system) and for human systems (e.g., human lives, economic activity, infrastructure, and the built environment). In a strict philosophical sense, "negative" consequences for the natural environment do not exist. However, it is evident, that society has to assume responsibility for many of the negative consequences associated with landslides. To judge the extent of exposure to landslides is crucial, but it is also important to employ appropriate adaptation and coping strategies.

Future work on landslide impact has to be threefold. First, more research is required on the full range of landslide types, associated triggering mechanisms, and the controlling factors such as slope geometry, material properties of lithology and regolith, hydrological patterns, and vegetation. Second, there is a need to fully understand the role of human intervention and social systems such as coping strategies in influencing the degree of impact. Third, and most importantly, the linkage between the landslide system and the societal dimensions (Bell and Glade, 2004), its change over time (e.g., Hufschmidt and Crozier, 2008), and the possible consequences have to be explored. Only then, are we able to understand our coupled natural environment and slope systems better and are able to reduce the impact of landslides.

Bibliography

Australian Geomechanics Society 2002. Landslide risk management concepts and guidelines. *Australian Geomechanics*, **37**(1), 51–70.

Bell, R., and Glade, T., 2004. Landslide risk analysis for Bíldudalur, NW-Iceland. *Natural Hazard and Earth System Science*, **4**, 1–15.

Bell, R., Glade, T., Thiebes, B., Jaeger, S., Krummel, H., Janik, M., and Holland, R., 2009. Modelling and web processing of early warning. In Malet, J. P., and Bogaard, T. (eds.), *Landslide Processes: From Geomorphologic Mapping to Dynamic Modelling*. Strasbourg: European Centre on Geomorphological Hazards, pp. 249–252.

Castellanos Abella, E.A., 2008. *Multi-scale Landslide Risk Assessment in Cuba*. ITC Dissertation 154. University of Utrecht, Utrecht, pp. 273.

Catapano, F., Malafronte, R., Lepre, F., Cozzolino, P., Arnone, R., Lorenzo, E., Tartaglia, G., Starace, F., Magliano, L., and Maji, M., 2001. Psychological consequences of the 1998 landslide in Sarno, Italy: a community study. *Acta Psychiatrica Scandinavica*, **104**(6), 438–442.

Cendrero, A., Remondo, J., Bonachea, J., Rivas, V., and Soto, J., 2006. Sensitivity of landscape evolution and geomorphic processes to direct and indirect human influence. *Geographia Fisica e Dinamica Quaternaria*, **29**, 125–137.

Crozier, M. J., 2005. Multiple-occurrence regional landslide events: hazard management perspectives. *Landslides*, **2**(4), 245–256.

Cruden, D. M., and Varnes, D. J., 1996. Landslide types and processes. In Turner, A. K., and Schuster, R. L. (eds.), *Landslides: Investigation and Mitigation*. Washington, DC: National Academy Press. Special Report 247, pp. 36–75.

Dikau, R., Brunsden, D., Schrott, L., and Ibsen, M. (eds.), 1996. *Landslide Recognition*. Chichester: Wiley, p. 251.

Glade, T., and Crozier, M. J., 2005. The nature of landslide hazard impact. In Glade, T., Anderson, M. G., and Crozier, M. J. (eds.), *Landslide Hazard and Risk*. Chichester: Wiley, pp. 43–74.

Guzzetti, F., Malamud, B. D., Turcotte, D. L., and Reichenbach, P., 2002. Power-law correlations of landslide areas in central Italy. *Earth and Planetary Science Letters*, **195**, 169–183.

Guzzetti, F., Galli, M., Reichenbach, P., Ardizzone, F., and Cardinali, M., 2006. Landslide hazard assessment in the Collazzone area, Umbria, Central Italy. *Natural Hazard and Earth System Science*, **6**, 115–131.

Hancox, G. T., McSaveney, E. R., and Manville, V., 2005. The October 1999 Mt Adams rock avalanche and subsequent landslide dam-break flood and effects in Poerua River, Westland, New Zealand. *New Zealand Journal of Geology and Geophysics*, **48**, 1–22.

Hufschmidt, G., and Crozier, M., 2008. Evolution of natural risk: analysing changing landslide hazard in Wellington, Aotearoa/New Zealand. *Natural Hazards*, **45**(2), 255–276.

Hufschmidt, G., and Glade, T., 2010. Vulnerability analysis in geomorphic risk assessment. In Alcántara-Ayala, I., and

Goudie, A. S. (eds.), *Geomorphological Hazards and Disaster Prevention*. New York: Cambridge University Press, pp. 233–243.

Ibsen, M.-L., and Brunsden, D., 1996. The nature, use and problems of historical archives for the temporal occurrence of landslides, with specific reference to the south coast of Britain, Ventnor, Isle of Wight. *Geomorphology*, **15**(3–4), 241–258.

Keefer, D. K., 1984. Landslides caused by earthquakes. *Geological Society of America Bulletin*, **95**(4), 406–421.

Korup, O., McSaveney, M., and Davies, T. R. H., 2004. Sediment generation and delivery from large historic landslides in the Southern Alps, New Zealand. *Geomorphology*, **61**, 189–207.

Larsen, M. C., Wieczorek, G. F., Eaton, S., Sierra, H. T. 2001. The Venezuela landslide and flash flood disaster of December 1999. In Mugnai, A. (ed.), *2nd Plinius Conference on Mediterranean Storms*, October 16–18, 2000. Siena, Italy: EGS.

Malamud, B. D., Turcotte, D. L., Guzzetti, F., and Reichenbach, P., 2004. Landslide inventories and their statistical properties. *Earth Surface Processes and Landforms*, **29**(6), 687–711.

Sato, H., Hasegawa, H., Fujiwara, S., Tobita, M., Koarai, M., Une, H., and Iwahashi, J., 2007. Interpretation of landslide distribution triggered by the 2005 Northern Pakistan earthquake using SPOT 5 imagery. *Landslides*, **4**(2), 113–122.

van Westen, C. J., van Asch, T. W. J., and Soeters, R., 2006. Landslide hazard and risk zonation – why is it still so difficult? *Bulletin of Engineering Geology and the Environment*, **65**(2), 167–184.

Wong, H. N., Ho, K. K. S., and Chan, Y. C., 1997. Assessment of consequences of landslides. In Cruden, D. M., and Fell, R. (eds.), *Landslide Risk Assessment – Proceedings of the Workshop on Landslide Risk Assessment*, Honolulu, Hawaii, USA, February 19–21, 1997. Rotterdam: A. A. Balkema, pp. 111–149.

Cross-references

Antecedent Conditions
Coping Capacity
Disaster
Exposure to Natural Hazards
Humanity as an Agent for Natural Disasters
Risk
Wenchuan, China (2008 Earthquake)

LANDSLIDE INVENTORY

Javier Hervás
Joint Research Centre, European Commission, Ispra (Va), Italy

Synonyms

Inventory of slope movements; Landslide archive; Landslide database; Landslide register; Mass movement inventory

Definition

A landslide inventory is a detailed register of the distribution and characteristics of past landslides.

Discussion

For each landslide recorded in an inventory, core information usually includes a unique identification code, landslide site name (for major landslides), location (geographical coordinates, municipality, province or county, region or state), type of landslide, date of occurrence if known or last reactivation, state of activity, and volume (or surface extent). Additional information may include landslide geometry (surface dimensions, depth of failure surface), geology (lithology, structure, material properties), hydrogeology, land cover or use, slope geometry, triggering cause, impact (e.g., casualties, damage expressed in economic value or in descriptive terms), remedial measures, surveying methods and date, and surveyor's name and bibliographical references. Complementary data such as illustrations (ground or aerial photographs, drawings) and monitoring data (type of instrumentation, rate of movement) can sometimes be found in an inventory. However, since much of the above-mentioned data are rarely available or expensive to collect, most existing landslide inventories only contain a subset of these data. Moreover, they do not usually include the same level of information for all landslides.

Landslide inventory data can be collected by aerial photointerpretation, field surveys and instrumentation, bibliographical research (e.g., scientific publications, technical reports, newspapers, historical chronicles, previous inventories, and geological maps), satellite and airborne remote sensing techniques, and interviews, depending on the scope and scale of the inventory and available resources.

Landslide inventories usually consist of a spatial component (i.e., the inventory map), showing landslide spatial distribution, and an associated alphanumeric component including the above-mentioned landslide-related information. On the maps, individual (sometimes clustered) landslides can be represented as dots, lines, or closed lines (polygons), depending mainly on the surface extent and shape of the landslides in relation to the map scale. Occasionally, large-scale inventory maps may also differentiate landslide source and deposit areas and depict features such as scarps, ridges, troughs, and ponds for large landslides. Today, digital landslide inventories are built on spatial databases using geographical information systems (GIS) technology for relatively simple inventories and relational database management systems (RDBMS) with geospatial data management capabilities or combined with GIS technology for comprehensive inventories.

Landslide inventories provide useful although spatially limited information on landslide distribution and occurrence for scientific, planning, decision-making, and other purposes. They are particularly valuable to generate landslide density maps and, especially, susceptibility, hazard, and risk maps, which are essential tools for devising risk reduction measures. Landslide inventories should be periodically updated, especially after a major landslide-triggering event such as a big rainstorm or earthquake.

Bibliography

Galli, M., Ardizzone, F., Cardinali, M., Guzzetti, F., and Reichenbach, P., 2008. Comparing landslide inventory maps. *Geomorphology*, **94**, 268–289.

Hervás, J., and Bobrowsky, P., 2009. Mapping: inventories, susceptibility, hazard and risk. In Sassa, K., and Canuti, P. (eds.), *Landslides – Disaster Risk Reduction*. Heidelberg Berlin: Springer, pp. 321–349.

Cross-references

Airphoto and Satellite Imagery
Community Management of Hazards
Debris Avalanche
Debris Flow
Deep-seated Gravitational Slope Deformations
Geographic Information Systems (GIS)
Global Positioning System (GPS) and Natural Hazards
Land-Use Planning
Landslide Types
Mass Movement
Mudflow
Rock Avalanche
Rockfall
Slope Stability

LANDSLIDE TRIGGERED TSUNAMI, DISPLACEMENT WAVE

Reginald L. Hermanns[1], Jean-Sébastien L'Heureux[2], Lars H. Blikra[3]
[1]Geological Survey of Norway, International Centre for Geohazards, Trondheim, Norway
[2]Norwegian Geotechnical Institute (NGI), Trondheim, Norway
[3]Åknes/Tafjord Early-Warning Centre, Stranda, Norway

Synonyms

Landslide-triggered tsunami (see entry *Tsunami*), displacement wave, non-seismic tsunami, and other terms such as surface wave, mega-tsunami, giant wave (see also titles in reference list) or even *seiche* have been used (e.g., Stone, 2009), although these waves are not standing oscillations in a (semi)closed basin.

The term "landslide-triggered tsunami" is a widespread term and has been used for waves caused by the impact of a landslide (see entry *Landslide*) into a water body or caused by a subaqueous landslide, whereas the term "displacement wave" has been used mainly for rock and/or ice avalanche/fall triggered waves in mountain lakes.

Definition

The term "*tsunami*" is Japanese and translates as "sudden wave in a harbor," referring to waves not visible on the open water which build up near the shore. It is used for earthquake-triggered waves (seismic tsunami) and waves triggered by the displacement of water by mass movements (see entry *Mass Movement*) or asteroid impact (non-seismic tsunami). In the following, we use the term "non-seismic tsunami" as a general term for waves triggered by a mass movement and specifically the term "displacement wave" for waves triggered by subaerial mass movements, whereas the term "landslide-triggered tsunami" is used for those waves triggered by subaqueous landslides. The initial wave amplitude (wave height) for displacement waves measures often several meters, tens or hundreds of meters which is in contrast to landslide-triggered tsunami that are in general in the order of a few meters or less.

Introduction

Non-seismic tsunamis occur in a multitude of environments, including the continental margins, ocean islands, fjords, natural and artificial lakes and rivers (Moore and Moore, 1984; Hendron and Patton, 1987; Grimstad and Nesdal, 1990; Evans, 2001; Tappin, 2009; L'Heureux et al., 2011, 2012; Harbitz et al., 2012). A number of records show that devastating displacement waves with a large loss of life may occur from subaerial *landslides* (e.g., Sicily, Calabria in 1783, Graziani et al., 2006; Mt. Unzan, Japan in 1792, Siebert et al., 1987; Vajont, Italy in 1963, Hendron and Patton, 1987). In some regions of the world, such as the fjords of western Norway, displacement wave sourced by subaerial rock avalanches (see entry *Rock Avalanche* (*Sturzstrom*)) have taken place several times during the last century often resulting in large loss of life (e.g., Blikra et al., 2006; Hermanns et al., 2012).

Until recently, and despite landslide-triggered tsunamis such as the Grand Banks in 1929 (e.g., Heezen et al., 1954; Piper and Asku, 1987) and those associated with the Good Friday 1964 earthquake in Alaska (e.g., Lee et al., 2003), historic submarine landslides were rarely identified as a source for devastating landslide-triggered tsunamis. It was not until 1998, when a submarine slump (see entry *Slide and Slump*) caused the devastating tsunami in Papua New Guinea in which 2,200 people died, that the threat from submarine mass movement was fully recognized (Tappin et al., 2001, 2008).

Displacement waves caused by rockslides, or collapse of a glacier into a water body

The highest displacement wave ever recorded during a *historical event* was observed at Lituya Bay, Alaska, on July 9, 1958 (Miller, 1960). Here the 524-m-high wave was sourced by a 4.3-km^3 rockslide previously triggered by an *earthquake*. In 1783, at Calabria in Sicily, a cliff collapsed into the sea and created a relatively small (8.3 m run-up) but devastating displacement wave that killed 1,500 people (Graziani et al., 2006). Run-up heights in the order of a several tens of meters are common for these types of events (e.g., Blikra et al., 2006; Hermanns et al., 2006; Naranjo et al., 2009). *Ice avalanches* from collapsing glaciers into moraine lakes are considered as one of the major *glacier hazards* in the Peruvian Andes, as displacement waves overtop and erode the moraine dams leading to catastrophic downstream *flash floods* (Reynolds, 1992). Similarly, the importance of *landslides* along reservoirs became especially obvious during the *disaster* of the *Vajont dam*, Italy, on October 9, 1963, where a *landslide* caused a 250-m-high displacement wave overtopping the dam. The *flash flood* down valley destroyed several villages and killed nearly 2,000 people (Hendron and Patton, 1987). Since then such displacement waves have been recognized as a serious *geohazard* for dams, whether they are man-made or not (e.g., Hermanns, et al., 2004).

Displacement waves in the ocean and on lakes surrounding volcanoes

The earliest record of displacement waves from volcanic flank collapse is from the *eruption* at the island of Thera, also known as *Santorini*, in the Cycladic Islands, Greece, about 3,600 years B.P. during the Late Bronze Age. This event had profound impacts on civilization in the Aegean and eastern Mediterranean region (McCoy and Heiken, 2000). Early records are also known from Japan where catastrophic volcanic collapse events at Hokkaido in 1640 and 1741 resulted in over 700 and 2,000 fatalities, respectively (Nishimura et al., 1999; Satake, 2007). In 1792, the *hazard* posed by such events became clear with the collapse of Mt. Unzen and the accompanied *debris avalanche* that swept into the Ariaka Sea. The 100-m-high wave resulted in the loss of more than 10,000 lives (Siebert et al., 1987). More recently, the flank collapse of Mount St. Helens on May 18, 1980 led to an improved understanding of the processes of flank failure and their potential to create devastating displacement waves. During this event a lobe of the *debris avalanche* entered into Spirit Lake and caused a 260-m-high displacement wave (Voight et al., 1983). During the past 20 years, flank collapses causing *debris avalanches* have been documented on multiple *volcanoes* around the world and are also mapped in the surrounding of most island *volcanoes* (e.g., Moore, 1964; Labazuy, 1996; Ward and Day, 2001; Krastel et al., 2001; Paras-Carayannis, 2004). These partial collapses can have volumes in the order of several hundred km^3 and modeled displacement wave heights can be several hundred or up to 1,000 m high (Ward and Day, 2001; McMurtry et al., 2003; Løvholt et al., 2008). Conspicuous deposits containing marine sediments were found far above sea level on multiple volcanic islands (Moore and Moore, 1984; Tanner and Calvari, 2004). On the island of Lanai, Hawaii, these date back to 101–134 ka (Moore and Moore, 1988) indicating the order of *recurrence intervals* of this particular *marine hazard*.

Landslide-triggered tsunamis caused by submarine landslides

It is well known today that a sudden displacement of the seafloor through catastrophic sliding or slumping has the potential to displace large volumes of water, generating *tsunami*s that can affect coastal areas and offshore infrastructure (Murty, 1979; Jiang and LeBlond, 1992). Landslide-triggered tsunamis have large run-up heights close to their source, but have far-field effects that are smaller relative to the initial wave height than *earthquake*-generated *tsunami*s (Okal and Synolakis, 2004). The 1998 Papua New Guinea landslide-triggered tsunami, which resulted in over 2,200 casualties, showed how catastrophic such types of events can be (Imamura and Hashi, 2003). There are numerous submarine *landslides* at all scales mapped along continental margins, fjords, and lakes. One of the most studied examples is the giant Storegga slide which occurred 8,200 years ago off the edge of the Norwegian shelf in the North Atlantic (Bugge et al., 1988; Bryn et al., 2005). The 3,500 km^3 Storegga slide caused a landslide-triggered tsunami wave up to more than 12 m high along the Norwegian coast (Harbitz, 1992; Bondevik et al., 2003), 20 m high on the Shetland Islands (Bondevik et al., 2005), 3–5 m high in Scotland (Smith et al., 2004) and even reached the coast of Greenland in at least four waves (Wagner et al., 2007). The evolution of the *landslide* is probably representative of similar slides along the Norwegian margin (Solheim et al., 2005). Failure took place after the end of the last deglaciation by translational and retrogressive sliding along specific marine clay layers. Destabilization prior to failure was associated to rapid loading from glacial deposits and generation of excess *pore-water pressure* considerably reducing the effective strength of the sediments. In this case, climatic processes led to a preconditioning of slope instability and a subsequent failure was most likely triggered by an *earthquake* due to isostatic rebound of the crust following the ice retreat (Bryn et al., 2005).

Hazard assessment of non-seismic tsunamis

Historical examples of non-seismic tsunamis in different environments demonstrate the potential catastrophic *hazard* of such events on coastal populations and infrastructures, but also on offshore installations (e.g., Miller, 1960; Siebert et al., 1987; Hendron and Patton, 1987; Grimstad and Nesdal, 1990; Blikra et al., 2006; Naranjo et al., 2009). Assessment of the *hazard* and *risk* posed by non-seismic tsunami waves is at an early stage, although our understanding of the processes has advanced significantly over the past decades. The scientific community has now a fairly good understanding of *landslide* processes affecting coastal and submarine areas, but the precise forecasting of *landslides* is still a challenging task. Also, there is still a paucity of data to fully understand how slope failures relate to the generation of displacement waves. Some attempts to evaluate the *hazard* and *risk* posed by non-seismic tsunamis have been performed (e.g., Blikra et al., 2005; Moscardelli et al., 2009; Lacasse and Nadim, 2009). A first task in such a study is to locate and identify a potential volume and geometry of unstable sediments or rocks which could affect a significant body of water (e.g., Day et al., 1999; Blikra et al., 2005). In a next step, the displacement wave propagation can be modeled to assess the area of impact (e.g., Harbitz, 1992; Tinti et al., 1999; Ward, 2001; Løvholt et al., 2008). Here, several parameters must be taken into account since the magnitude of displacement wave generated by a *landslide* depends on (1) volume of moving material, and especially the morphology of the front area, (2) water depth of the water body, and (3) velocity, while the magnitude of a landslide-triggered tsunami rather depends on (4) volume of moving material, (5) water depth where the *landslide* occurs, (6) acceleration and initial velocity of the *landslide*, (7) rheology of the failed sediments and the dynamics, and (8) distance to shore and the seafloor morphology (e.g., Pelinovsky and Poplavsky, 1996; Sælevik et al., 2009; L'Heureux et al., 2011).

Risk management for displacement waves

Disastrous non-seismic tsunamis are caused by large subaerial or submarine *landslides* where physical *mitigation* is difficult or sometimes impossible. Some large *deep-seated gravitational slope deformations* have been stabilized by large-scale drainage systems in order to avoid destructive displacement waves, as in the *mitigation* related to hydropower reservoirs in British Columbia, Canada (Watson et al., 2006). The *mitigation* of displacement waves is so far mainly handled by passive or non-physical *mitigation* measures, including the implementation of *monitoring* systems, *early warning systems*, and *evacuation* plans. However, sea walls and dikes used to protect coastal areas against *storm surges* might also protect coastal areas against non-seismic tsunamis. *Monitoring* and *early warning systems* of *landslides* related to the generation of displacement waves have been implemented in reservoir in British Columbia (Watson et al., 2006) and at the Åknes rockslide site in western Norway (Blikra, 2008). Recent advances in deep-ocean *tsunami* measurement technology (e.g., DART buoys) coupled with *tsunami* forecast model have demonstrated that *tsunami* impact can be predicted before it reaches the affected coastlines (Percival et al., 2011). Such systems were developed to detect *earthquake*-triggered *tsunamis* but could also be used for the early warning of non-seismic tsunamis. *Hazard* maps, pre-defined *evacuation* routes, and preparedness on all society levels and sectors are essential for the total *early warning system*.

Summary

Displacement waves caused by the impact of a *landslide* into a water body and landslide-triggered tsunamis caused by submarine *landslides* pose a major *hazard* along the surrounding coast lines and lake shores. Displacement waves related to volcanic collapses have, in general, a long *recurrence interval* of several thousand up to millions of years. Submarine slides and subaerial *landslides* into water bodies are more frequent, and in some regions several can take place in a century. With the exception of narrow mountain valleys below artificial and natural mountain lakes where displacement waves can affect tens of kilometer long mountain valleys, these waves affect only coastal areas and offshore installations. Depending on the magnitude of the event and the geological settings, results may be catastrophic with loss of lives and total destruction of infrastructure. Similar to *earthquake*-triggered *tsunami* waves, hazardous displacement waves are difficult to mitigate. However, our increasing knowledge of *landslide* processes coupled with the validation of our *tsunami* modeling codes from *historical events* and improved *risk assessment* strategies gives the opportunity to elaborate *hazard mapping* that can be used in *risk mitigation* strategies.

Bibliography

Blikra, L. H., 2008. The Åknes rockslide; monitoring, threshold values and early-warning. In Zuyu, C., Jian-Min, Z., Ken, H., Fa-Quan, W., and Zhong-Kui, L. (eds.), *Landslides and Engineered Slopes. From the Past to the Future. Proceedings of the 10th International Symposium on Landslides and Engineered Slopes*, June 30-July 4, Xi'an, China, Taylor and Francis, 1850 pp. ISBN 9780415411967.

Blikra, L., Longva, O., Harbitz, C., and Løvholt, F., 2005. Quantification of rock-avalanche and tsunami hazard in Storfjorden, western Norway. In Senneset, K., Flaate, K., and Larsen, J. O. (eds.), *Landslides and Avalanches*. London: Taylor & Francis, pp. 57–63.

Blikra, L. H., Longva, O., Braathen, A., Anda, E., Dehls, J. F., and Stalsberg, K., 2006. Rock slope failures in Norwegian fjord areas: examples, spatial distribution and temporal pattern. In Evans, S. G., Scaraascia Mugnozza, G., Strom, A., and Hermanns, R. L. (eds.), *Landslides from Massive Rock Slope Failures*. Dodrecht: Springer, pp. 475–496.

Bondevik, S., Mangerud, J., Dawson, S., Sawson, A., and Lohne, O., 2003. Record-breaking height for 8000-year-old tsunami in the North Atlantic. *EOS, Transactions, American Geophysical Union*, **84**, 289.

Bondevik, S., Løvholt, F., Harbitz, C., Mangerud, J., Dawson, A., and Svendsen, J. I., 2005. The Storegga slide tsunami; comparing field observations with numerical simulations. *Marine and Petroleum Geology*, **22**, 195–208.

Bryn, P., Berg, K., Forsberg, C. F., Solheim, A., and Lien, R., 2005. Explaining the Storegga slide. *Marine and Petroleum Geology*, **22**, 11–19.

Bugge, T., Belderson, R. H., and Kenyon, N. H., 1988. The Storegga slide. *Philosophical Transactions of the Royal Society of London*, **325**, 357–388.

Day, S. J., Carracedo, J. C., Guillou, H., and Gravestock, P., 1999. Recent structural evolution of the Cumbre Vieja Volcano, La Palma, Canary Islands; volcanic rift zone reconfiguration as a precursor to volcano flank instability? *Journal of Volcanology and Geothermal Research*, **94**, 135–167.

Evans, S. G., 2001. Landslides. In Brooks, G. R. (ed.), *A Synthesis of Geological Hazard in Canada*. Ottawa: Geological Survey of Canada. Geological Survey of Canada Bulletin, Vol. 548, pp. 151–177.

Graziani, L., Maramai, A., and Tinti, A., 2006. A revision of the 1783–1784 Calabrian (southern Italy) tsunamis. *Natural Hazards and Earth System Science*, **6**, 1053–1060.

Grimstad, E., and Nesdal, S., 1990. The Loen rockslides – a historical review. In Barton, M., and Stephansson, W. (eds.), *Rock Joints*. Rotterdam: Balkema, pp. 1–6.

Harbitz, C. B., 1992. Model simulations of tsunamis generated by the Storegga slides. *Marine Geology*, **105**, 1–21.

Harbitz, C. B., Glimsdal, S., Bazin, S., Zamora, N., Løvholt, F., Bungum, H., Smebye, H., Gauer, P., and Kjekstad, O., 2012. Tsunami hazard in the Caribbean: regional exposure derived from credible worst case scenarios. *Continental Shelf Research*, **38**, 1–23, doi:10.1016/j.csr.2012.02.006.

Heezen, B. C., Ericsson, D. B., and Ewing, M., 1954. Further evidence of a turbidity current following the 1929 grand banks earthquake. *Deep Sea Research*, **1**, 193–202.

Hendron, A. J., Jr., and Patton, F. D., 1987. The Vaiont Slide; a geotechnical analysis based on new geologic observations of the failure surface. *Engineering Geology*, **24**, 475–491.

Hermanns, R. L., Niedermann, S., Ivy-Ochs, S., and Kubik, P. W., 2004. Rock avalanching into a landslide-dammed lake causing multiple dam failure in Las Conchas valley (NW Argentina) – evidence from surface exposure dating and stratigraphic analyses. *Landslides*, **1**, 113–122.

Hermanns, R. L., Blikra, L. H., Naumann, M., Nilsen, B., Panthi, K. K., Stromeyer, D., and Longva, O., 2006. Examples of multiple rock-slope collapses from Köfels (Ötz valley, Austria) and western Norway. *Engineering Geology*, **83**, 94–108.

Hermanns, R. L., Hansen, L., Sletten, K., Böhme, M., Bunkholt, H., Dehls, J. F., Eilertsen, R., Fischer., L., L'Heureux, J. -S., Høgaas, F., Nordahl, B., Oppikofer, T., Rubensdotter, L., Solberg, I. -L., Stalsberg K., and Yugsi Molina, F. X., 2012. Systematic geological mapping for landslide understanding in the Norwegian context. In Eberhardt, E., Froese, C., Turner, A. K., and Leroueil, S. (eds.), *Landslides and Engineered Slopes, Protecting Society through improved understanding*. London: Taylor and francis Group, pp. 265–271.

Imamura, F., and Hashi, K., 2003. Re-examination of the source mechanism of the 1998 Papua New Guinea earthquake and tsunami. *Pure and Applied Geophysics*, **160**, 2071–2086.

Jiang, L. C., and LeBlond, P. H., 1992. The coupling of a submarine slide and the surface waves which it generates. *Journal of Geophysical Research*, **97**, 12731.

Krastel, S., Schmincke, H.-U., Jacobs, C. L., Rihm, R., Le Bas, T. P., and Alibes, B., 2001. Submarine landslides around the Canary Islands. *Journal of Geophysical Research*, **106**, 3977–3997.

L'Heureux, J. S., Glimstad, S., Longva, O., Hansen, L., and Harbitz, C. B., 2011. The 1888 shoreline landslide and tsunami in Trondheimsfjorden, central Norway. *Marine Geophysical Researches*, **32**, 313–329.

L'Heureux, J. S., Eilertsen, R. S., Glimstad, S., Issler, D., Solberg, I.-L., and Harbitz, C. B., 2012. The 1978 quick clay landslide at Rissa, mid-Norway: subaqueous morphology and tsunami simulations. In Yamada, Y., et al. (eds.), *Submarine Mass Movements and Their Consequences*. Dordrecht: Springer Science + Business Media B.V. Advances in Natural and Technological Hazards Research, Vol. 31, doi:10.1007/978-94-007-2162-3_45.

Labazuy, P., 1996. Recurrent landslides events on the submarine flank of Piton de la Fournaise Volcano (Reunion Island). *Geological Society Special Publications*, **110**, 295–306.

Lacasse, S., and Nadim, F., 2009. Landslide risk assessment and mitigation strategy. In Sassa, K., and Canuti, P. (eds.), *Landslides – Disaster Risk Reduction*. Berlin: Springer, pp. 31–61.

Lee, H. J., Kayen, R. E., Gardner, J. V., and Locat, J., 2003. Characteristics of several tsunamigenics submarine landslides. In Locat, J., and Mienert, J. (eds.), *Submarine Mass Movements and Their Consequences*. Dordrecht: Kluwer Academic. Advances in Natural and Technological Hazards Research, pp. 357–366.

Løvholt, F., Pedersen, G., and Gisler, G., 2008. Oceanic propagation of a potential tsunami from the La Palma Island. *Journal of Geophysical Research*, **113**, C09026.

McCoy, F. W., and Heiken, G., 2000. Tsunami generated by the Late Bronze Age eruption of Thera (Santorini), Greece. *Pure and Applied Geophysics*, **157**, 1227–1256.

McMurtry, G. M., Watts, P., Fryer, G. J., Smith, J. R., and Imamura, F., 2003. Giant landslides, mega-tsunamis, and paleo-sea level in the Hawaiian Islands. *Marine Geology*, **203**, 219–233.

Miller, D. J., 1960. *Giant Waves in Lituya Bay, Alaska*. Washington: GPO. U.S. Geological Survey Professional Paper, Vol. 354 C, pp. 51–86.

Moore, J. G., 1964. *Giant Submarine Landslides on the Hawaiian Ridge*. Reston, Virginia: U.S. Geological Survey, pp. D95–D98.

Moore, J. G., and Moore, G. W., 1984. Deposit from a giant wave on the island of Lanai, Hawaii. *Science*, **226**, 1312–1315.

Moore, G. W., and Moore, J. G., 1988. Large-scale bedforms in boulder gravel produced by giant waves in Hawaii. Special Paper – Geological Society of America, 229, pp. 101–110.

Moscardelli, L., Hornbach, M., and Wood, L., 2009. Tsunamigenic risks associated with mass transport complexes in offshore Trinidad and Venezuela. In Mosher, D. C., Shipp, R. C., Moscardelli, L., Chaytor, J. D., Baxter, C. D. P., Lee, H. L., and Urgeles, R. (eds.), *Submarine Mass Movements and Their Consequences*. Dordrecht: Springer. Advances in Natural and Technological Hazards Research, Vol. 28, pp. 733–744.

Murty, T. S., 1979. Submarine slide-generated water waves in Kitimat Inlet, British Columbia. *Journal of Geophysical Research*, **84**, 7777–7779.

Naranjo, J. A., Arenas, M., Clavero, J., and Muñoz, O., 2009. Mass movement-induced tsunamis: main effects during the Patagonian Fjordland seismic crisis in Aisén (45º25'S), Chile. *Andean Geology*, **36**, 137–145.

Nishimura, Y., Miyaji, N., and Suzuki, M., 1999. Behavior of historic tsunamis of volcanic origin as revealed by onshore tsunami deposits. *Physics and Chemical of the Earth, Part A: Solid Earth and Geodesy*, **24**, 985–988.

Okal, E. A., and Synolakis, C. E., 2004. Source discriminants for near-field tsunamis. *Geophysical Journal International*, **158**, 899–912.

Paras-Carayannis, G., 2004. Volcanic tsunami generating source mechanisms in the Eastern Caribbean region. *Science of Tsunami Hazards*, **22**, 74–114.

Pelinovsky, E., and Poplavsky, A., 1996. Simplified model of tsunami generation by submarine landslides. *Physics and Chemistry of the Earth*, **21**, 13–17.

Percival, D. B., Denbo, D. W., Eble, M. C., Gica, E., Mofjeld, H. O., Spillane, M. C., Tang, L., and Titov, V. V., 2011. Extraction of tsunami source coefficients via inversion of DART® buoy data. *Nat. Hazards*, **58**(1), 567–590. doi: 10.1007/s11069-010-9688-1.

Piper, D. J. W., and Asku, A. E., 1987. The source and origin of the 1929 Grand Banks turbidity current inferred from sediment budgets. *Geo-Marine Letters*, **7**, 177–182.

Reynolds, J. M., 1992. The identification and mitigation of glacier-related hazards; examples from the Cordillera Blanca, Peru. In McCall, G. J. H., Laming, D. D. C., and Scott, S. C. (eds.), *Geohazards; Natural and Man-Made*. London: Chapment and Hall, pp. 143–157.

Sælevik, G., Jensen, A., and Pedersen, G., 2009. Experimental investigation of impact generated tsunami; related to a potential rock slide, Western Norway. *Coastal Engineering*, **56**, 897–906.

Satake, K., 2007. Volcanic origin of the 1741 Oshima-Oshima tsunami in the Japan Sea. *Earth Planets Space*, **59**, 381–390.

Siebert, L., Glicken, H., and Ui, T., 1987. Volcanic hazards from Bezymianny- and Bandai-type eruptions. *Bulletin of Volcanology*, **49**, 435–459.

Smith, D. E., Shi, S., Cullingford, R., Dawson, A., Firth, C., Foster, L., Fretwell, P., Haggart, B., Holloway, L., and Long, D., 2004. The Holocene Storegga slide tsunami in the United Kingdom. *Quaternary Science Reviews*, **23**, 2291–2321.

Solheim, A., Berg, K., Forsberg, C. F., and Bryn, P., 2005. The Storegga slide complex: repetitive large scale sliding with similar cause and development. *Marine and Petroleum Geology*, **22**, 97–107.

Stone, R., 2009. Peril in the pamirs. *Science*, **326**(5960), 1614–1617.

Tanner, L. H., and Calvari, S., 2004. Unusual sedimentary deposits on the SE side of Stromboli Volcano, Italy; products of a tsunami caused by the ca. 5000 years BP Sciara del Fuoco collapse? *Journal of Volcanology and Geothermal Research*, **137**, 329–340.

Tappin, D. R., 2009. Mass transport events and their tsunami hazard. In Mosher, D. C., Shipp, R. C., Moscardelli, L., Chaytor, J. D., Baxter, C. D. P., Lee, H. L., and Urgeles, R. (eds.), *Submarine Mass Movements and Their Consequences*. Dordrecht: Springer. Advances in Natural and Technological Hazards Research, Vol. 28, pp. 667–684.

Tappin, D. R., Watts, P., McMurtry, G. M., Lafoy, Y., and Matsumoto, T., 2001. The Sissano Papua New Guinea tsunami of July 1998 – offshore evidence on the source mechanism. *Marine Geology*, **175**, 1–23.

Tappin, D. R., Watts, P., and Grilli, S. T., 2008. The Papua New Guinea tsunami of 17 July 1998: anatomy of a catastrophic event. *Natural Hazards and Earth System Science*, **8**, 1–24.

Tinti, S., Bortolucci, E., and Armigliato, A., 1999. Numerical simulation of the landslide-induced tsunami of 1988 on Vulcano Island, Italy. *Bulletin of Volcanology*, **61**, 121–137.

Voight, B., Jandra, J. R., Glicken, H., and Douglass, P. M., 1983. Nature and mechanics of the Mount St. Helens rockslide-avalanche of 18 May 1980. *Geotechnique*, **33**, 243–273.

Wagner, B., Bennike, O., Klug, M., and Cremer, H., 2007. First indication of Storegga tsunami deposits from East Greenland. *Journal of Quaternary Science*, **22**, 321–325.

Ward, S. N., 2001. Landslide tsunami. *Journal of Geophysical Research B, Solid Earth and Planets*, **106**, 11,201–11,215.

Ward, S. N., and Day, S., 2001. Cumbre Vieja Volcano; potential collapse and tsunami at La Palma, Canary Islands. *Geophysical Research Letters*, **28**, 3397–3400.

Watson, A. D., Derik Martin, C., Moore, D. P., Stewart, T. W. G., and Lorig, L. J., 2006. Integration of geology, monitoring and modeling to assess rockslide risk. *Felsbau*, **24**, 50–58.

Cross-references

Asteroid Impact
Avalanches
Debris Avalanche
Deep-Seated Gravitational Slope Deformation
Disaster
Disaster Risk Management
Early Warning System
Earthquake
Evacuation
Flash Flood
Geohazard
Glacier Hazards
Hazard
Hazard Mapping
Historical Event
Landslide
Marine Hazard
Mass Movement
Mitigation
Monitoring
Pore-Water Pressure
Recurrence Interval
Risk
Risk Assessment
Rock Avalanche
Santorini, Eruption
Seiche
Slide and Slump
Surge
Tsunami
Vajont Dam
Volcanoes

LANDSLIDE TYPES

David Cruden
University of Alberta, Edmonton, AB, Canada

Synonyms

Landslide classification; Landslide description; Landslide names

Definition

Landslides, movements of masses of rock, earth, or debris down slopes, have observable characteristics (activity, rate of movement, moisture content, material, type of movement). Landslides with similar characteristics belong to the same landslide type.

Introduction

What can you usefully observe about a landslide? How can these observations be succinctly and unambiguously described? These are questions which have found answers in classifications of landslides. The International Union of Geological Sciences Working Group on Landslides has developed an international consensus on landslide classification which has been summarized in the Multilingual Landslide Glossary (WP/WLI, 1993b). This classification, the Working Classification, has been used in the latest edition of the Transportation Research Board's Special Report (Turner and Schuster, 1996) to update Varnes' (1978) widely used classification. Highland and Bobrowsky (2008) provide an accessible version.

The criteria used in the description of landslides (Cruden and Varnes, 1996) follow Varnes (1978) in emphasizing type of movement and type of material. A landslide can be described by a word describing the material and a second word describing the type of movement. The divisions of materials are unchanged from Varnes (1978): rock, debris, and earth. Movements have been divided into five types: falls, flows, slides, spreads, and topples. The sixth type proposed by Varnes (1978), complex landslides, has been dropped from the formal classification, though the term "complex" has been retained as a description of the style of activity of a landslide. Complexity can also be indicated by combining the five types of landslide in the ways suggested below.

The name of a landslide can become more elaborate as more information about movement becomes available. Adjectives can be added in front of the two nouns to build up the description of the movement. A preferred sequence of terms in naming the movement, a progressive narrowing of the focus of the descriptors, first in time, then in space, from a view of the whole landslide to parts of the movement and to the materials involved, would follow a typical landslide reconnaissance. The recommended sequence, Activity, Rate of Movement, Moisture Content, Material, Type of Movement, is the sequence of sections in this entry.

Activity

Under activity, broad aspects of landslides are described that should focus the initial reconnaissance of movements before more detailed examination of materials displaced (WP/WLI, 1993a, b). The terms Varnes (1978) considered relating to age and state of activity, with some of the terms from sequence or repetition of movement, have been regrouped under three headings: State of Activity, which describes what is known about the timing of movements; Distribution of Activity, which describes where the landslide is moving; and Style of Activity, which indicates how different movements contribute to the landslide.

State of activity

Active landslides are those that are currently moving. Landslides which have moved within the last annual cycle of seasons but which are not moving at present are described as suspended. A landslide which is again active after being inactive may be called reactivated.

Inactive landslides are those which have last moved more than one annual cycle of seasons ago. This state can be subdivided. If the causes of movement apparently remain, then the landslides are dormant. Perhaps, however, the river which had been eroding the toe of the moving slope has itself changed course and the landslide is abandoned. If the toe of the slope had been protected against *erosion* by bank armoring or other artificial remedial measures have stopped the movement, the landslide can be described as stabilized. Landslides often remain visible in the landscape for thousands of years after they have initially moved. Landslides which have clearly developed under different geomorphological or climatic conditions perhaps thousands of years ago can be called relict.

Distribution of activity

Varnes (1978) defined a number of terms that can be used to describe the activity distribution in a landslide. Movement may be limited to the displacing material or the rupture surface may be enlarging, continually adding to the volume of displacing material. If the rupture surface is extending in the direction opposite to the movement of the displaced material, the landslide is said to be retrogressing. If the rupture surface is extending in the direction of movement, the landslide is advancing. If the rupture surface is extending at one or both lateral margins, the landslide is widening. Confined movements have a scarp but no rupture surface visible in the foot of the displaced mass; displacements in the head of the displaced mass are taken up by compression and slight bulging in the foot of the mass. If the rupture surface of the landslide is enlarging in two or more directions, Varnes (1978, p. 23) suggested the term "progressive" for the landslide while noting this term had also been used for both advancing and retrogressing landslides. This term is also current for describing the process, progressive failure, by which the rupture surface in some slides extends. The possibility of confusion seems sufficient now to abandon "progressive" in favor of describing the landslide as enlarging. To complete the possibilities, terms are needed for landslides in which the volume of displacing material can be seen to be reducing with time and for those landslides where no trend is obvious.

Movements such as rotational slides and topples may stop naturally after substantial displacements because the movements themselves reduce the gravitational forces on the displaced masses. Alternatively, rock masses may be dilated by movements that rapidly increase the volumes of cracks in the masses and cause decreases in fluid pore pressures within these cracks. However, to conclude that the displacing mass is stabilizing because its volume is decreasing may be premature; the activity of rotational slides caused by *erosion* at the toe of slopes in cohesive soils is often cyclic. The term "diminishing" for a landslide whose displacing material is decreasing in volume seems free of undesired implications. Landslides whose displaced materials continue to move but whose rupture surfaces show no visible changes can be simply described as moving.

Style of activity

The way in which different movements contribute to the landslide, the style of the landslide activity, can be described by terms from Varnes (1978, p. 23). There, complex landslides are defined as exhibiting at least two types of movements. The term is limited here to movements in which the types are in sequence: a topple in which some of the displaced mass subsequently slid is a complex rock topple, rock slide. Not all the toppled mass slid, but no significant part of the displaced mass slid without first toppling. Notice that some of the displaced mass may be still toppling whereas other parts are sliding.

We can use a former synonym of complex, composite, to describe landslides in which different types of movement occur in different areas of the displaced mass, sometimes simultaneously. These different areas of the displaced mass show different sequences of movements. WP/WLI (1993a, b) adopted the convention of treating the higher of the two movements as the first movement and the lower of the two movements as the second movement in the absence of more definite information.

A multiple landslide shows repeated movements of the same type, often following the enlargement of the rupture surface. The newly displaced masses are in contact with previously displaced masses and often share a rupture surface with them. In a retrogressive, multiple, rotational slide, two or more blocks have each moved on curved rupture surfaces tangential to a common generally deep rupture surface.

A successive movement is identical in type to an earlier movement but, in contrast to a multiple movement, does not share displaced material or a rupture surface with it.

Single landslides consist of a single movement of displaced material often as an unbroken block. They contrast with the other styles of movement which require disruption of the displaced mass or independent movements of portions of the mass.

Rate of movement

The IUGS Working Group (1995) modified the rate of movement scale given in Varnes (1978, Fig. 2:1 u). The seven divisions of the scale are now adjusted to increase in multiples of 100 by slightly increasing the uppermost limit of the scale and decreasing the lowest limit of the scale. These two limits span ten orders of magnitude (from 0.5×10^{-6} to 5×10^3 mm/s).

The important division between very rapid and extremely rapid movement approximates the speed of a person running (5 m/s.). Some extremely rapid movements have been called *avalanches*. Another important boundary is between the slow and very slow classes (1.6 m/year), below which some structures on the landslide are undamaged. Terzaghi (1950, p. 84) identified slope movements "proceeding at an imperceptible rate" as "*creep*." The many uses of "creep" have been discussed by Varnes (1978, p. 17); the term should be replaced by either very slow or extremely slow, applied to the other landslide descriptors.

Materials

Varnes (1978) suggested four terms derived from simple observations of the water content of the displaced material: (1) dry, no moisture; (2) moist, contains some water but no free water – the material may behave as a plastic solid but does not flow; (3) wet, contains enough water to behave in part as a liquid, has water flowing from it, or supports significant bodies of standing water; (4) very wet, contains enough water to flow as a liquid under low gradient.

These terms should be used to describe the masses displaced by the landslide. The water content of the displaced masses may give useful guidance for assumptions about the water content of the displacing materials while the materials were displacing. However, soil or rock masses may drain quickly after displacement, and individual rock or soil masses may have water contents which differ considerably from the average water content of the displacing material. In some fine-grained soils, the boundaries between the terms may correspond approximately with Atterberg limits – the shrinkage, plastic, and liquid limits separating dry, moist, wet, and very wet soils, respectively.

We can follow Varnes (1978) in describing materials in landslides as either rock, a hard or firm mass that was intact and in its natural place before the initiation of movement, or soil, an aggregate of solid particles generally of minerals and rocks which has either been transported or formed by the weathering of rock in place. Gases or liquids filling the pores of the soil form part of the soil.

Soil is further divided into debris and earth. Debris contains a significant proportion of coarse material: 20–80% of the particles are larger than 2 mm and the remainder are less than 2 mm. Earth describes material in which 80% or more of the particles are smaller than 2 mm; it includes a range of materials from nonplastic sand to highly plastic clay.

Types of movement

In this section, the five kinematically distinct types of landslides are described in the sequence fall, topple, slide, spread, and flow.

A soil or rock *fall* starts with detachment from a steep slope along a surface on which little or no shear displacement takes place. The material then descends largely through the air by falling, saltation, or rolling. Movement is very rapid to extremely rapid. Except when the displaced mass has been undercut, falling is preceded by small sliding or toppling movements which separate the displacing material from the undisturbed mass. Undercutting typically occurs at the toe of cliffs undergoing wave attack or in the eroding banks of rivers.

A *topple* is the forward rotation out of the slope of a mass of soil or rock about a point or axis below the center of gravity of the displaced mass. Toppling is sometimes driven by gravity exerted through material upslope of the displaced mass and sometimes through water in cracks in the mass. Topples may lead to falls or slides of the displaced mass depending on the geometry of the moving mass, the geometry of the surface of separation, and the orientation and extent of the kinematically active discontinuities. Topples range from extremely slow to extremely rapid, sometimes accelerating throughout the movement.

A *slide* is a down slope movement occurring dominantly on surfaces of rupture or relatively thin zones of intense shear strain. Movement is usually progressive; it does not initially occur simultaneously over the whole of what eventually becomes the surface of rupture. Often the first signs of ground movement are cracks in the original ground surface along which the main scarp of the slide will form. The displaced mass may slide beyond the toe of the surface of rupture covering the original ground surface of the slope which then becomes a surface of separation.

In a *(lateral) spread*, there is an extension of a cohesive mass combined with a general subsidence of the fractured mass of cohesive material into softer underlying material. The rupture surface is not a surface of intense shear. Spreads may result from *liquefaction* or flow and extrusion of the softer material. Varnes (1978) distinguished spreads, typical of rock, which extended without forming an identifiable rupture surface from movements in cohesive soils overlying liquefied materials or materials which are flowing plastically. The cohesive materials may also subside, translate, rotate, disintegrate, or liquefy and flow.

Clearly these movements are complex, but they are sufficiently common in certain materials and geological situations that a separate type of movement is worth recognizing.

A *flow* is a spatially continuous movement in which surfaces of shear are short-lived, closely spaced, and not usually preserved. The distribution of velocities in the displacing mass resembles that in a viscous liquid. The lower boundary of the displaced mass may be a surface along which appreciable differential movement has taken place or a thick zone of distributed shear. There is then a gradation from slides to flows depending on the water content, mobility, and evolution of the movement. Debris slides may become extremely rapid *debris flows*, *debris avalanches*, as the displaced material loses cohesion, gains water, or encounters steeper slopes.

Summary

An initial reconnaissance of a landslide might be expected to describe the activity and the materials displaced in this particular type of landslide. This format lends itself to the creation of simple databases suited to much of the database management software now available (WP/WLI, 1990). The information collected can be compared with summaries of other landslides (WP/WLI, 1991) and used to guide further investigations and mitigative measures. Further investigation increases the precision of estimates of the dimensions and confidence in the descriptions of activity and material and in the hypotheses about causes of the movement. The new information may finally be added to the database to influence the analysis of further landslides. These databases can be expected to form the foundations of systems for landslide *risk assessment and management* (Cruden and Fell, 1997).

Bibliography

Cruden, D. M., and Fell, R. (eds.), 1997. *Landslide Risk Assessment*. Rotterdam: Balkema, p. 370.

Cruden, D. M., and Varnes, D. J., 1996. Landslide types and processes. In Turner, A. K., and Schuster, R. L. (eds.), *Landslides: Investigation and Mitigation*. Wasington, DC: National Academy Press, pp. 36–75. Transportation Research Board Special Report 247, National Research Council.

Highland, L. M., and Bobrowsky, P., 2008. *The Landslide Handbook – A Guide to Understanding Landslides*. Reston, VA: US Geological Survey. United States Geological Survey Circular 1325. 129 p.

IAEG Commission on Landslides, 1990. Suggested nomenclature for landslides. *Bulletin International Association of Engineering Geology*, **41**, 13–16.

International Union of Geological Sciences Working Group on Landslides, 1995. A suggested method for describing the rate of movement of a landslide. *Bulletin International Association of Engineering Geology*, **52**, 75–78.

Terzaghi, K., 1950. Mechanism of landslides. In Paige, S. (ed.), *Application of Geology to Engineering Practice*. New York: Geological Society of America, pp. 83–123.

Turner, A. K., and Schuster, R. L. (eds.), 1996. *Landslides: Investigation and Mitigation*. Washington, DC: National Academy Press. Transportation Research Board Special Report 247, National Research Council.

Varnes, D. J., 1978. Slope movement types and processes. In Schuster, R. L., and Krizek, R. J. (eds.), *Landslides: Analysis and Control*. Washington, DC: Transportation Research Board, National Academy of Sciences.

WP/WLI (International Geotechnical Societies' UNESCO Working Party on World Landslide Inventory), 1990. A suggested method for reporting a landslide. *Bulletin International Association of Engineering Geology*, **41**, 5–12.

WP/WLI (International Geotechnical Societies' UNESCO Working Party on World Landslide Inventory), 1991. A suggested method for a landslide summary. *Bulletin International Association of Engineering Geology*, **43**, 101–110.

WP/WLI (International Geotechnical Societies' UNESCO Working Party on World Landslide Inventory), 1993a. A suggested method for describing the activity of a landslide. *Bulletin International Association of Engineering Geology*, **47**, 53–57.

WP/WLI (International Geotechnical Societies' UNESCO Working Party on World Landslide Inventory), 1993b. *Multilingual Landslide Glossary*. Richmond, British Columbia: Bitech Publishers. 59 p.

WP/WLI (International Geotechnical Societies' UNESCO Working Party on World Landslide Inventory), 1994. A suggested method for describing the causes of a landslide. *Bulletin International Association of Engineering Geology*, **50**, 71–74.

Cross-references

Avalanches
Creep
Debris Avalanches (Sturzstrom)
Debris Flow
Disaster Risk Management
Lateral Spreading
Liquefaction
Risk Assessment
Rockfall
Slide and Slump

LAND-USE PLANNING

Stefan Greiving[1], Philipp Schmidt-Thomé[2]
[1]TU Dortmund University, Dortmund, Germany
[2]Geological Survey of Finland (GTK), Espoo, Finland

Synonyms

City planning; Regional planning; Spatial planning; Territorial development; Town and country planning

Definition

Land-use planning is defined as the whole comprehensive, coordinating planning at all scales (from national to local), which aims at an efficient and balanced territorial development.

Introduction: natural hazards and their relevance for land-use planning

Natural hazards are usually defined as extreme natural events that have the potential to damage societies and individuals. These extreme events occur in closed time spans of seconds or weeks, after which the initial state before the extreme event is sometimes reached again. Longer lasting natural processes, such as climate change and desertification, might pose certain threats or trigger hazards, but do not belong to hazards sensu stricto. Most natural hazards arise from the normal physical processes operating in the Earth's interior, at its surface, or within its enclosing atmosphere (Schmidt-Thomé, 2006).

"Land-use planning operates on the presumption that the conscious integration of (particularly public) investment in sectors such as transport, housing, water management, etc., is likely to be more efficient and effective than uncoordinated programmes in the different sectors" (ODPM 2005). Thus, the core element of land-use planning is to prepare and make decisions about future land use. This can be specified for different scales as follows:

Regional planning/development: the task of settling the land use and development by drawing up regional plans or programs. Regional planning is required to specify aims of land-use planning which are drawn up for an upper, overall level and sets a framework for decisions on land use taken at the local level within land-use planning of the municipalities. Depending on the planning system of a country, there might even be textual and cartographic determinations and information which typically range on a scale from 1:50,000 to 1:100,000.

Urban land-use planning: creation of policies at local/municipal level to guide land and resource uses. The main instrument of land-use planning is zoning or zoning ordinances, respectively. Local land-use planning normally consists of two stages with specific planning instruments on each of these: first, a general or preparatory land-use plan (scale from 1:5,000 to 1:50,000) for a whole municipality and second, a detailed land-use plan for small part of it, mostly legally binding (scale 1:500 to 1:5,000).

Each hazard has a spatial dimension (it takes place somewhere). Space can therefore be defined as an area where human beings and their artifacts are threatened by spatially relevant hazards. However, spatially relevant does not mean land-use planning relevance. The reaction of tolerating or altering risk can be understood as an integrated part of the given socioeconomic structures with land-use planning as a certain part of a reaction. Land-use planning makes decisions for society regarding if and how certain spaces will be used. Therefore, land-use planning more or less influences vulnerability with regard to natural (and technological) hazards.

Land-use planning is responsible for the development of a particular land-use area (where the sum of hazards and vulnerabilities defines the overall land-use risk) and not for a particular object or thread (e.g., sectoral engineering sciences). Land-use planning must adopt a multi-hazard approach in order to appropriately deal with risks and hazards in a land-use context (Greiving, 2002). There is a tradition of land-use planning research for single hazards (coastal flooding, river flooding, earthquakes, nuclear power plants), an integrated research approach to land-use relevant hazards has only recently been undertaken by a few authors in Europe (Egli, 1996; Greiving, 2002), whereas in the USA the role of land-use planning has been highlighted in several publications (e.g., Godschalk et al., 1999 or Burby, 1998; both with further references).

The land-use character of a hazard can either be defined by land-use effects that might occur if a hazard turns into a disaster or by the possibility of an appropriate land-use planning response. This also opens up questions about the relevance of different levels of land-use planning as well as the relationship to sectoral planning. Relevant for land-use planning are those hazards where frequency and/or magnitude of the event itself can be influenced by land-use planning or where mitigation actions influence land use. If this is not the case, like for meteorite impacts or pandemics, nevertheless it might be of interest for a sectoral planning division or an emergency response unit.

The normative dimension of natural hazards

One of the most serious problems in the context of dealing with natural hazards in land-use planning is represented by external effects: a land use and temporal inconsistency between chances and risks which are related with every decision making about a future land use or a concrete investment at a certain location. A classic example for this planning problem is represented by the (intra-generational) conflict between actors which are located upstream and downstream: A municipality located upstream might profit from the chances of a suitable location for an industrial area located on the flood plains of a river and could protect this area by means of a dike. The direct consequence of this action would be an increased flood risk for downstream located areas because of the reduced flood plain capacities in combination with flood waves that would occur faster and with a higher peak.

In terms of sustainable development, this conflict can be described as an intra-generational conflict. Aside from this, intergenerational aspects have to be taken into consideration. Intergenerational justice can be understood as a second prerequisite for reaching a balance of chances and risks. The "Theory of Justice" based the necessity of a consensus about normative regulations on a consensus with the righteous interests of future generations instead of just a consensus of people who are actually alive now. The "Veil of Ignorance" or the view of short-term chances hinders an appropriate estimation of long-term negative affects that might threaten mainly future generations (Rawls, 1971, p. 328). The greater the persistence of possible harmful impacts on an event or decision, the greater the importance and problems related to a decision that accepts consequences from hazardous events (Berg et al., 1995, p. 30). For example, the Chief Building Inspector had justified a governmental responsibility for building safety standards after the Loma-Pieta-earthquake as follows: "I represent, in absentia, the unknown future user" (Godschalk et al., 1999, p. 494). This example indicates that planning-related decisions based on a consensus of all stakeholders could fail

Land-Use Planning, Table 1 Strength and weaknesses of land-use planning in the context of dealing with natural hazards.

Task	Milestones	Potential of land-use planning	Description
Risk assessment	Assessment and appraisal of long-term impacts on the human-environmental-system such as climate change	Fair	Based on impact studies, regional planning is essential. A strength of comprehensive planning is the traditionally integrated view on different change processes (demography, economy, environment, climate)
	Identification of interaction between land uses and hazards	Good	Assessments can easily be integrated in the strategic environmental assessment which is obligatory for any land-use plan or program
	Assessment of frequency and magnitude of extreme events (exposure)	Poor	This is clearly a task for specialized authorities, like water management and terrain mapping, where land-use planning does not have competence in assessment
Risk management	Adaptation of existing land-use structures (settlements, infrastructure)	Poor	For regulatory planning, adaptation of existing structures is difficult because of given private property rights. Requires suitable approaches based on incentives and communication to private households
	Avoidance of non-adapted developments	Good	This focus of planning is very much about future developments. The effectiveness of actions depends partly on existing regulatory frameworks (zoning instruments)
	Keeping disaster prone areas free of further development	Good	If conforming planning systems have regulatory zoning instruments at hand it is possible to keep free of areas prone to extreme events
	Differentiated decisions on land use: Acceptable land-use types according to the given risk	Fair	Possible, but usually not effective with regard to existing settlement structures
	Relocation/retreat from threatened areas	Poor	Again in conflict with property rights. Full compensation is normally needed, which fails mostly due to the lack of financial resources. Possible in areas with shrinking population where the existing building stock will be (partly) deconstructed based on planning strategies (e.g., Eastern Germany)

in relation to the temporal and, as mentioned above, land-use dimensions. The same decision is possibly based on free market transactions. Even if all participants of a transaction of land designated for construction would come to an agreement, they might fail in relation to an unacceptable use of common pool goods.

The core elements of sustainable development were laid out in the Rio Declaration in 1992. The development of societies cannot be sustainable in view of increasing risks from natural and technological hazards (Lass et al., 1998). The US National Science and Technology Council states, "Sustainable development must be resilient with respect to the natural variability of the earth and the solar system. The natural variability includes such forces as floods and hurricanes and shows that much economic development is 'unacceptable brittle and fragile'" (FEMA, 1997, p. 2). A resilient community is one that lives in harmony with nature's varying cycles and processes including earthquakes, storms, and floods as natural events, which cause harm only for a non-sustainable society (Godschalk et al., 1999, p. 526). A fourth criterion should be added to sustainability's economic, social, and ecologic aspects (Greiving, 2002): Sustainability can be understood as a mission for the development of mechanisms for adaptation of societies to future consequences of present processes.

Decisions based on normative findings, made by supranational (like the European Union) or national policies as a framework for regional and local weighting-up processes within land-use planning, can take the interests of future generations into account.

Risk management and land-use planning

Risk management is defined as adjustment policies that intensify efforts to lower the potential for loss from future extreme events, i.e., risk management is characterized by decisions of stakeholders. Decision making is a normative, politically influenced strategy about

Land-Use Planning, Table 2 Contribution of land-use planning and supporting instruments to risk management strategies.

Risk management strategy	Regional planning/development	Local land-use planning	Supporting instruments
Long-term prevention	Fostering resilience as planning strategy, i.e., by following the robustness principle		Tax system; strategies for reducing greenhouse gas emissions
Mitigation of hazard impacts (nonstructural)	Maintenance or reinforcement of protective features that absorb or reduce hazard impacts (mangroves, retention areas, etc.)	E.g., local rain water infiltration adapted to land cultivation	Economic incentives; communication strategies
Mitigation of hazard impacts (structural)	Secure the availability of space for protective infrastructure	Protective infrastructure; Obligations for the design of individual buildings; Retrofitting of existing buildings	Communication strategies
Vulnerability reduction	Spatial development concepts like decentralized concentration	Keeping hazard-prone areas free of further developments	Financial incentives for reallocation of threatened objects
Preparedness, response, recovery	–	Allocation of critical infrastructure outside hazard-prone areas; Rebuilding planning	Emergency plans; Information and training; Risk awareness

Source: Based on Greiving and Fleischhauer, 2006, p. 119

tolerating or altering risks. The authority in charge (democratically legitimized) has to decide the main planning goals to deal with hazards.

Risks due to natural hazards mean a certain challenge for many stakeholders involved in risk governance. The International Risk Governance Council defines risk governance as a "process by which risk information is collected, analysed and communicated and management decisions are taken" (IRGC, 2005). The assessment and management of risks are embedded in a communication process.

As risk assessment and management can be interpreted as an ongoing process, it is often illustrated as the disaster management cycle by which public and private stakeholders plan for and reduce the impact of disasters in the pre-emergency phase (mitigation and preparedness), react in the emergency phase (response) and in the post-emergency phase (recovery).

At all points of the cycle, appropriate actions lead to a reduction of damage potential, reduced vulnerability, or a better prevention of disasters. Land-use planning most likely does not play a decisive role in all phases of the disaster management cycle but it has some specific functions in risk management.

The action decided upon is the result of a weighting process between different management options that can be structured along the triangle "resistance-resilience-retreat" (Greiving, 2004; Greiving and Fleischhauer, 2006): Resistance is the protection against (all) hazards by means of structural measures. Resilience can be defined as minimization of the risk to life and property when a disaster occurs and retreat is the abandonment of risky areas.

Even though land-use planning is considered to be an important instrument to cope with climate change induced impacts, it is limited in its powers and can only solve parts of the problem (Schmidt-Thomé and Greiving, 2008). The following table indicates to what extent land-use planning is able to handle natural hazards. It is divided into assessment and management (Table 1).

Risk assessment is a task for sectoral planning authorities. Land-use planning plays a minor role in this context and can be understood as an important end user of hazard related information provided by sectoral planning. Hazard maps with a scale of about 1:2,000 to 1:10,000 are necessary for the enforcement of restrictions of land use at the municipal land-use planning level.

Due to its coordinative role and responsibility, land-use planning is relevant and responsible for nonstructural adaptation measures as part of risk management strategies. There are several zoning-related instruments that can improve nonstructural mitigation and some supporting instruments that promote planning initiatives (Table 2).

Climate change in general, but particularly as a triggering factor for many natural hazards, is particularly troublesome for Europe with its existing settlement structures, cultural landscapes and infrastructures which have been developed over centuries. Mitigation and prevention actions, carried out by spatial planning, are under such circumstances less effective than in countries which are still growing rapidly in terms of population and the built environment. Here, disaster prone areas can be kept free from further development whereas most of these areas are in Europe already built-up. However, this problem calls for authorities to improve public risk awareness and to look for means to mitigate this problem due to cooperative solutions (Fleischhauer et al., 2006).

The Strategic Environmental Assessment (SEA), which came into force through EU Directive 2001/42/EC in 2001, offers a suitable procedural frame for risk assessment and embedding risk management in decision making by land-use planning. The use of impact assessment methodologies encourages a more informed approach to planning and regional development. The identification of cumulative impacts highlights areas where land-use planning needs to focus on adaptation measures designed to deal with several impacts similarly. Planning is mainly able to guide future developments. Adapting existing settlement structures can be seen as the main challenge for regulatory land-use planning because of given private property rights. Risk governance is regarded as an important success factor for the development of adaptation strategies, and in this context, land-use planning has an important role to play (see EC, 2009, Ribeiro et al., 2009; Swart et al., 2009).

Summary

Land-use planning prepares and endorses decisions about future land use. It is specified in different scales from regional planning/development and in most countries sets the framework for detailed planning at a local, i.e., municipal level. One of the main instruments of local land-use planning is zoning or zoning ordinances, respectively. Local land-use planning normally consists of two stages: a general, or preparatory land-use plan; and detailed land-use plans. Land-use planning is responsible for the development of a particular land-use area (where the sum of hazards and vulnerabilities defines the overall land-use risk) and not for a particular object or thread (e.g., sectoral engineering sciences). The involvement of land-use planning is only institutionalized in some countries. Currently, single hazard concepts dominate in land-use planning. Multi-hazard approaches to appropriately deal with the potential sum and interaction of hazards and risks are seldom found in practice. Land-use planning is limited in its powers and can only solve parts of the problem. Its coordinative role and responsibility is relevant and responsible for nonstructural adaptation measures as part of risk management strategies. The adaptation of existing settlement structures is a main challenge for regulatory land-use planning, e.g., due to private property rights. Therefore risk governance is regarded as a tool to support land-use planning in the development of adaptation strategies.

Bibliography

Berg, M., et al., 1995. *Was ist ein Schaden? Zur normativen Dimension des Schadensbegriffs in der Risikowissenschaft.* Zürich: Verlag der Fachvereine.

Burby, R. J. (ed.), 1998. *Cooperating with Nature: Confronting Natural Hazards with Land-Use Planning for Sustainable Communities.* Washington, DC: Joseph Henry Press.

Egli, T., 1996. *Hochwasserschutz und Raumplanung: Schutz vor Naturgefahren mit Instrumenten der Raumplanung – dargestellt am Beispiel von Hochwasser und Murgängen.* vdf – Hochschulverlag, ORL-Bericht: Zürich, Vol. 100.

European Commission (EU) 2009. *Staff working document accompanying the white paper adapting to climate change: towards a European framework for action impact assessment.* Brussels.

Federal Emergency Management Agency (FEMA) (ed.), 1997. *Strategic Plan – Partnership for a Safer Future.* Washington, DC: FEMA.

Fleischhauer, M., Greiving, S., and Wanczura, S. (eds.), 2006. *Natural Hazards and Spatial Planning in Europe.* Dortmund: Dortmunder Vertrieb für Bau- und Planungsliteratur.

Godschalk, D. R., et al., 1999. *Natural Hazard Mitigation – Recasting Disaster Policy and Planning.* Washington, DC: Island Press.

Greiving, S., 2002. *Räumliche Planung und Risiko.* München: Gerling Akademie Verlag.

Greiving, S., 2004. Risk assessment and management as an Important Tool for the EU Strategic Environmental Assessment. *DISP*, **157**, 11–17.

Greiving, S., and Fleischhauer, M. 2006. Spatial planning response towards natural and technological hazards. In Schmidt-Thomé, P. (ed.), *Natural and technological hazards and risks affecting the spatial development of European regions.* Geological Survey of Finland, Special Paper 42, Espoo 2006.

International Risk Governance Council (IRGC), 2005. *Basic concepts of risk characterisation and risk governance*, IRGC: Geneva. Available at: (http://www.irgc.org/_cgidata/mhscms/_images/12395-3-1.pdf). 4 pp.

Lass, W., Reusswig, F., and Kühn, K. D., 1998. Disaster Vulnerability and "Sustainable Development". In *Integrating Disaster Vulnerability in the CSD's List of Indicators: Measuring Sustainable Development in Germany.* Bonn: IDNDR, p. 14e.

Office of the Deputy Prime Minister (OPDM), 2005. *Polycentricity scoping study.* Glossary. Available at: http://www.odpm.gov.uk/index.asp?id=1145459.

Rawls, J., 1971. *A Theory of Justice.* New York: Harvard University Press.

Ribeiro, M., Losenno, C., Dworak, T., Massey, E., Swart, R., Benzie, M., and Laaser, C. 2009. Design of guidelines for the elaboration of regional climate change adaptations strategies. *Study for European Commission – DG Environment – Tender DG ENV. G.1/ETU/2008/0093r.* Vienna: Ecologic Institute.

Schmidt-Thomé, P. (ed.), 2006. *Natural and technological hazards and risks affecting the spatial development of European regions.* Geological Survey of Finland, Special Paper 42.

Schmidt-Thomé, P., and Greiving, S., 2008. Response to natural hazards and climate change in Europe. In Faludi, A. (ed.), *European Spatial Planning and Research.* Cambridge: Lincoln Institute for Land Policy.

Swart, R., Biesbroek, R., Binnerup, R., Carter, T. R., Cowan, C., Henrichs, T., Loquen, S., Mela, H., Morecroft, M., Reese, M., and Rey, D., 2009. *Europe Adapts to Climate Change – Comparing National Adaptation Strategies.* Online available: http://peer-initiative.org/media/m256_PEER_Report1.pdf.

Cross-references

Adaptation
Climate Change
Disaster Risk Management
Risk Governance
Uncertainty
Zoning

LATERAL SPREADING

Steven L. Kramer
University of Washington, Seattle, WA, USA

Synonyms

Liquefaction

Definition

Lateral spreading is the finite, lateral movement of gently to steeply sloping, saturated soil deposits caused by earthquake-induced liquefaction.

Discussion

The movement of soil deposits undergoing lateral spreading can range from a few centimeters to a few meters, and can cause significant damage to buildings, bridges, pipelines, and other elements of infrastructure. Lateral spreading often occurs along riverbanks and shorelines where loose, saturated sandy soils are commonly encountered at shallow depths. Structures supported on shallow foundations, pavements, and buried pipelines are particularly susceptible to damage from lateral spreading.

Lateral spreading occurs as the generation of porewater pressure in the soil resulting from earthquake shaking reduces the stiffness and strength of the soil. Under the action of the static stresses required to maintain equilibrium under sloping ground conditions, each cycle of seismic stress causes incremental deformation of the soil. The deformations generally originate in soils at shallow depths beneath the ground surface, and can cause cracking and severe disruption, as well as horizontal and vertical displacements, of the ground surface (Figure 1).

The level of ground deformations caused by lateral spreading is influenced by the ground slope, depth of the water table, density of the soil beneath the water table, and the strength of the ground shaking caused by the earthquake. The mechanics of lateral spreading are quite complex, so ground surface deformations are currently predicted using empirical models calibrated against lateral spreading behavior observed in past earthquakes. Subsurface ground deformations can also be important, particularly for structures such as bridges supported on piles that extend through liquefiable soils into underlying stable deposits.

Mitigation of lateral spreading hazards usually involves densification, reinforcement, or cementation of the liquefaction-susceptible soil. A wide variety of construction methods are available to reduce lateral spreading hazards.

Bibliography

Idriss, I. M., and Boulanger, R. W., 2008. *Soil Liquefaction During Earthquakes*. Oakland: Earthquake Engineering Research Institute.
Kramer, S. L., 1996. *Geotechnical Earthquake Engineering*. Englewood Cliffs: Prentice-Hall.

Cross-references

Collapsing Soil Hazards
Dispersive Soil Hazards
Expansive Soils and Clays
Hydrocompaction Subsidence
Land Subsidence
Landslide (Mass Movement)
Landslide Types
Liquefaction
Mass Movement
Pore Water Pressure
Quick Clay
Sinkhole

Lateral Spreading, Figure 1 Lateral spreading damage at Capital Lake in Olympia, Washington, following the 2001 Nisqually earthquake.

LAVA

Robert Buchwaldt
Massachusetts Institute of Technology, Cambridge, MA, USA

Synonyms

Magma; Molten rock

Definition

Lava is molten rock that reaches the Earth's surface through a volcano or fissure. When the molten rock solidifies the resulting rock is called igneous rock. The igneous rock can originate from different sources. Some rocks originate deep in the Earth's mantle whereas others can originate high within the Earth's crust directly underneath a vent. Dependent on the origin, the composition of the crystallized rock is different and therefore provides some hint where the rock came from and what the interior of the Earth looks like.

The erupted lavas usually can be distinguished into three major types: basaltic, andesitic, and rhyolitic lava. Basaltic lava is an extrusive rock of "mafic" composition (high in iron, magnesium, and calcium) with relatively low silica content, andesitic lava has an intermediate silica content, and finally rhyolitic lava has a "felsic" composition (high in sodium and potassium) with silica content greater than 68 vol%.

The vast majority of lava on Earth, more than 90% of the total volume, is estimated to be basaltic in composition (that includes most of our ocean floors). Andesites and other lavas of intermediate composition account for most of the rest, whereas silica rich rhyolitic flows make up about 1% of the total. In recent times, the volume of lava flows, for example, in Hawaii ranges up to 0.5 km^3. One of the largest lava eruptions in historic times is reported from the Laki fissure in Iceland in 1783, which had a volume of 12 km^3 and traveled up to 88 km from its source. There are prehistoric basaltic lava eruptions like the material that formed the Columbia River plateau called Large Igneous Provinces, which produced a volume that exceeded 1,200 km^3.

Because of their lower silica content, the basaltic lavas are usually very fluid. In Hawaii, one measured lava flow had a speed of 30 km/h. In contrast, the movement of silica-rich lava may be too slow to perceive. In addition, because of the high viscosity of rhyolitic lava the flow is usually not more than a few kilometers from their vents.

Bibliography

Francis, P., and Oppenheimer, C., 2003. *Volcanoes*. New York: Oxford University Press.
Schmincke, H.-U., 2004. *Volcanism*. Berlin: Springer.

Cross-references

Aa Lava
Pahoehoe Lava
Volcanoes and Volcanic Eruptions

LEVEE

Joann Mossa
University of Florida, Gainesville, FL, USA

Synonyms

Dike (dyke); Embankment

Definition

Levee an embankment produced naturally by river sedimentation or constructed by humans to prevent flooding.

Natural levees

Natural levees are ridges formed by overbank flooding, which deposits sand and silt-size sediments adjacent to the river channel (Brierley et al., 1997). In low-lying areas, natural levees are the highest topographic features and thus were preferential locations for both prehistoric and more recent settlement (Hudson, 2005).

Artificial levees

Artificial levees are often built upon natural levees for purposes of flood protection and damage reduction. They are generally built from sediment, and may be reinforced with concrete, rock, and/or vegetation. Artificial levee construction dates back thousands of years in the valleys of the Indus and Nile rivers, Mesopotamia, and China. By confining the flow of the river, artificial levees produce higher water levels and velocities (see Zong and Chen, 2000). If levees are set back from the river, there is more capacity for floodwaters and greater potential for flood protection. Artificial levees provide primary or secondary protection to a lowland, and are used with structures to route waters through floodways away from populated areas. Locations surrounded by a ring of levees, such as New Orleans, Louisiana, U.S.A. are vulnerable to flooding from many directions.

Levee hazards

Flood hazards still occur despite the presence of levees. Levees may be overtopped during extreme floods, during wind-driven surges, or because environmental conditions such as sea level or land level have changed since the levee was constructed. Breaches (crevasses) can develop in weaker portions of both natural and artificial levees. Locations with seepage and sand boils, especially in weak materials, are often where these breaches occur. If the overbank flow is of sufficient magnitude and floodplain conditions are appropriate, breaches may lead to avulsions.

Levees have also been intentionally destroyed for a variety of reasons. Some examples include the downstream dynamiting of the levee near Caernarvon to protect New Orleans in the Mississippi River flood of 1927, selective blasting of levees during the Mississippi River flood of 1993, and dynamiting of upstream levees in 1998 to protect Wuhan, a city of over seven million along the Yangtze River in China, where flooding occurred nonetheless. The Chinese destroyed levees on the Yellow River in 1938 to disrupt the invading Japanese, whereas the Germans blew up levees in the Netherlands during wartime in 1945.

Thus, although levees are built to protect lowlands from flood hazards, these areas are vulnerable nonetheless to both storms and human decisions. The presence of artificial levees allows for development and more intense land use, but the price is a false sense of security and a lifetime of maintenance and repair.

Bibliography

Brierley, G. J., Ferguson, R. J., and Woolfe, K. J., 1997. What is a fluvial levee? *Sedimentary Geology*, **114**, 1–9.
Hudson, P. F., 2005. Natural levees. In Trimble, S. (ed.), *Encyclopedia of Water Science*. Boca Raton: Taylor & Francis.
Zong, Y., and Chen, X., 2000. The 1998 flood on the Yangtze. *Natural Hazards*, **22**, 165–184.

Cross-references

Avulsion
Flood Deposits
Flood Hazard and Disaster
Flood Protection
Floodplain
Floodway
Hurricane Katrina

LIGHTNING

Leopoldo C. Cancio
Colonel, Medical Corps, U.S. Army, Fort Sam Houston, TX, USA

Definition

Lightning is a sudden, massive discharge of electrical current, most commonly arising from thunderstorms (but rarely from forest fires, volcanic eruptions, or dust storms). Lightning discharges can be classified as cloud-to-ground, cloud-to-cloud, cloud-to-air, or cloud-in-cloud. Lightning is the second leading cause of weather-related death, as well as a fascinating (and incompletely understood) natural phenomenon.

History

Since ancient times, lightning has both stimulated awe and symbolized the divine. Lightning is the instrument by which Zeus, in the *Iliad* Chap. VIII, acts against the Achaeans: "Then he thundered aloud from Ida, and sent the glare of his lightning upon the Achaeans; when they saw this, pale fear fell upon them and they were sore afraid." Among the most beautiful passages in the Bible are references to lightning. Psalm 144:6 depicts it as a manifestation of God's power, and the psalmist prays, "Cast forth lightning, and scatter them: shoot out thine arrows, and destroy them." In Matthew 24:27 Jesus foretells his second coming: "For as the lightning cometh out of the east, and shineth even unto the west; so shall also the coming of the Son of man be." In Luke 10:18, He gives the 70 disciples power over evil, saying: "I beheld Satan as lightning fall from heaven." In Surah 13 of the Quran (Al Ra'ad, "The Thunder"), lightning is a manifestation sent by God by which He reveals Himself: "It is He Who doth show you the lightning, by way both of fear and of hope: It is He Who doth raise up the clouds, heavy with rain. Nay, thunder repeateth His praises, and so do the angels, with awe..."

Our understanding of lightning as the result of electrical activity within clouds, and more specifically our ability to protect ourselves by means of the lightning rod, are attributed to Benjamin Franklin. In 1751, a series of five letters written by Franklin to a fellow of the Royal Society of London were published as *Experiments and Observations on Electricity, Made at Philadelphia in America*, describing various electrical experiments. In the fourth letter, he hypothesized: if two electrified gun barrels "will strike at two Inches distance, and make a loud Snap; to what great a Distance may 10,000 Acres of Electrified Cloud strike and give its Fire, and how loud must be that Crack!" In the fifth letter, he proposed the lightning rod, "to fix on the highest Parts of those Edifices upright Rods of Iron...Would not these pointed Rods probably draw the Electrical Fire silently out of a Cloud before it came nigh enough to strike, and thereby secure us from that most sudden and terrible Mischief!" In the fifth letter he also proposes an experiment: a man would stand in a sentry box on a high tower. A lightning rod would extend skyward from the box. During a rainstorm, then, the man "might be electrified, and afford Sparks, the rod drawing fire to him from the Cloud."

In his 1767 review, *The History and Present State of Electricity*, Joseph Priestley recounts the brave enactment of that proposal by two French philosophers, assisted by a priest and an artisan, in May 1752. The priest "drew sparks from the bar of a blue colour, an inch and a half in length, and which smelled strong of sulphur. He repeated the experiment at least six times in the space of about four minutes...each experiment taking up the time, as he, in the stile of a priest expresses himself, of a *Pater* and an *Ave*. In the course of these experiments he received a stroke on his arm...such as might have been made by a blow with the wire on his naked skin..." Priestley also describes Franklin as carrying out his famous kite-flying experiment a month later: "Dr. Franklin, astonishing as it must have appeared, contrived actually to bring lightning from the heavens, by means of an electrical kite, which he raised when a storm of thunder was perceived to be coming on." In Franklin's 1753 description of and advocacy for the lightning rod in *Poor Richard's Almanac* he sought to reconcile faith and reason – the sense of lightning as a manifestation of God's power and the new concept of lightning as an understandable and controllable natural force. "It has pleased God in his Goodness to Mankind, at length to discover to them the Means of securing their Habitations and other Buildings from Mischief by Thunder and Lightning." (Krider, 2006).

Lightning physics

According to a hypothesis advanced by C.T.R. Wilson in 1920, thunderstorms play a central role in energizing the Earth and its atmosphere, which together can be viewed as a *global electric circuit*. Even during fair weather, the Earth's surface is charged negatively, and the air positively; lightning serves to deliver negative charge to the ground. The atmosphere above 60 km is conductive due to the presence of free electrons, a zone sometimes called the *electrosphere*. Lateral currents flow both in the

---

The opinions or assertions contained herein are the private views of the author, and are not to be construed as official or as representing the views of the Department of the Army or Department of Defense.

electrosphere and across the highly conductive surface of the planet (Rakov and Uman, 2003).

Lightning is generated when a voltage difference develops within a thundercloud, or between a thundercloud and the ground. A conventional view is that these conditions are created when ice particles (*hydrometeors*) in thunderclouds form and then fracture. Through a poorly understood process, smaller crystals tend to develop a positive charge, and larger crystals a negative charge. Gravity causes positively charged crystals to move upward within the cloud and negatively charged crystals to move downward. This gives rise to a potential difference between the lower and upper portions of the cloud, and between the lower portion of the cloud and the ground. When the voltage exceeds 2–3 million Volt per meter, arcing occurs. This model is problematic, however, because the strongest fields actually observed within thunderclouds are about ten times lower. An alternative hypothesis, *runaway breakdown*, was advanced by Gurevich and colleagues. Cosmic ray particles impacting the upper thundercloud generate electrons with sufficient energy (*runaway electrons*) that the drag or braking force exerted by air molecules diminishes, in essence serving as a catalyst for a lightning stroke (Dwyer, 2005).

The initial pathway through which current travels is called the *leader*, is relatively invisible, and opens an ionized channel between cloud and ground in a series of steps. Just before this *stepped leader* reaches the ground (at a height of 50–100 m), an *upward streamer* propagates from the ground to the leader. Once this channel is opened between the cloud and the ground, a much greater current passes through it called the *return stroke*, which results in the visible lightning bolt. The electric current involved in lightning strikes is direct (DC). The amount of DC delivered by a lightning strike, on the order of 30,000–50,000 amp, is far greater than that inflicted by the usual man-made electrical contact. The duration of exposure is, on the other hand, much shorter, approximately 10–100 ms. This brief current causes the release of a large amount of heat, raising temperatures to approximately 30,000 K, which in turn causes a "thermoacoustic blast wave" or thunder. The overpressure generated by thunder at the source may approach 100 atm. (Lee et al., 2000).

Lightning may follow one of several paths upon interacting with a body. Casualties may sustain a *direct strike*, a *contact injury* (lightning strikes a conductive object touching the casualty), a *side flash* (lightning splashes from a nearby object onto the casualty), *ground current* (lightning travels through the ground then into the casualty), or an *upward streamer* (rarely, lightning passes upward from the casualty). In addition, thunder may cause *blast injury*, manifested, for example, by tympanic membrane rupture or by blunt trauma when the casualty is thrown (Ritenour et al., 2008).

During a direct strike, the primary current arc travels outside rather than through the body, a phenomenon known as *flashover*. This would seem protective, but the immense current likely generates large magnetic fields perpendicular to the body surface, which in turn induce secondary electric currents within the body. These secondary currents may cause cardiac arrest and other internal injuries, even without external evidence of injury. When lightning hits the ground causing a ground current, current spreads out from the contact point such that if a casualty is standing nearby with feet apart, the potential difference between the feet may approximate 1,500 V. As a result, injuries are more severe when a person's feet are apart than when they are together. When lightning directly strikes the upper body, a very large potential difference between the upper and lower body is established. A brief, large current flow will result. The duration is generally not sufficient to cause Joule heating, but can damage muscle and nerve cells by mechanisms such as electroporation (Bier et al., 2005).

Electroporation, also known as *electropermeabilization*, features reorganization of lipids in the cell membrane into "pores" as a result of an imposed transmembrane potential. This results in a large increase in membrane permeability that significantly augments the work necessary to maintain transmembrane concentration gradients. When cellular metabolic energy stores become depleted, ATP-driven ion pumps can no longer compensate for the rapid diffusion of ions through the damaged cell membrane. If the membrane does not then seal itself, cell death will occur. Skeletal muscle and nerve cells are especially susceptible to electroporation because of their length, which is directly proportional to their transmembrane potential. Some suggest that the delayed neurological sequelae following lightning injury are due largely to the gradual effects of electroporation.

Epidemiology

Today, given the unchanged destructive power carried by lightning, and its daily frequency, it is extraordinary that deaths due to lightning are not more common. Remarkably, at any moment in time, about 2,000 thunderstorms are occurring over about 10% of the planet's surface (Rakov and Uman, 2003). Space-based optical sensors have made it possible to document lightning strike frequency with accuracy. Worldwide, strike density is highest in central Africa (over 50 flashes per km^2 per year) and rare over the open ocean. In the USA, central Florida is the leading region for lightning strikes.

Lightning is the second leading cause of weather-related death in much of the world. The epidemiology of lightning injury is well described only for certain areas of the world, and significant underreporting of both injuries and deaths has been demonstrated by comparing disparate databases. The annual number of deaths reported in the USA is approximately 60, and given underreporting, may be as high as 70. With a USA population of 300 million, this gives a death rate of 0.20–0.23 per million. A South African study noted a much higher rate of 6.3 deaths per million inhabitants for the Highveld, a region consisting predominantly of urban poor – suggesting the influence not only of strike frequency, but also of factors such as building construction and availability of safety information. In the USA, the largest number of deaths occurs in two southern states: Texas and Florida. On the other hand, during 1968–1985, the highest number of deaths per inhabitants in the USA was reported for the rural Rocky Mountain states of Wyoming, at 1.96 per million, and New Mexico at 1.70. At the same time, the national average was 0.61 per million.

The month of July consistently features the largest number of casualties in the USA. In India, the peak months are the monsoon season of June to September. The opposite pattern is noted south of the equator in Australia and South Africa. In Singapore, an equatorial country, two peaks are observed in April and November. Worldwide, most injuries take place in the afternoon (1,200–1,800 h local times) (Ritenour et al., 2008).

Despite rising population, the number of lightning deaths in the USA decreased from 377 during 1891–1894 to 239 during 1991–1994 (Holle et al., 2001). From the 1950s until the 1990s, there was a slight decrease in the per person risk of lightning injury or death (Curran et al., 2000). A likely explanation for this finding is a decrease in the number of individuals involved in farming, and an increase in the proportion of the population living in an urban setting. Similar long-term decreases in the number of lightning injuries and fatalities were observed in England and Wales over the period 1852–1990. In that region, the mean annual number of lightning deaths decreased from 20.5 for 1852–1859 to 4.2 for 1980–1989. This occurred despite a doubling of the population, implying an eightfold decrease in the risk of death (Elsom, 1993). Improved medical care is likely improving survival following lightning strikes. One paper based on the US *Storm Data* database concluded that the ratio of injuries to deaths increased from two in 1959 to about seven in 1994 (Curran et al., 1997).

The typical lightning casualty is a young man who is engaging in outdoor work (such as farming or construction) or recreation. Men are five times more likely to be struck by lightning than are women (Center for Disease Control and Prevention, 1998). Young people (ages 10–29 years) are at greatest risk in several series. Over time, some regions have noted a decrease in outdoor-work-related injuries, and an increase in outdoor-recreation-related injuries. However, a significant fraction of lightning injuries in the USA takes place indoors. Improvements in building design in developed countries likely reduced the number of indoor injuries between the 1890s and 1990s. Lightning injury may afflict individuals riding bicycles, motorcycles, or boats. Lightning has also struck aircraft in flight, resulting in fatalities; engineering improvements have made aircraft safer from this threat.

Prevention of injury

The first key to prevention of lightning injury is adherence to the building codes which have evolved since the invention of the lightning rod. The primary reference in the USA is the National Fire Protection Association's Standard for the Installation of Lightning Protection Systems (NFPA-780).

Equally important is a high level of awareness, particularly for those working or recreating outdoors during the thunderstorm season. Lightning is often associated with cumulonimbus rain clouds, but may precede the rainstorm and may even strike with blue skies overhead (a "bolt from the blue"). Lightning may rarely occur during snowstorms, presaged by *graupel* (soft hail or snow pellets). The "30-30 rule" states that a flash-to-thunder interval of less than 30 s places one at risk of lightning strike, and mandates shelter for 30 min after the last strike is seen or heard.

A third key to prevention is to seek shelter in a safe place during a thunderstorm. When thunder is heard, personnel should seek shelter in a building or enclosed vehicle. Alternatively, there is a relatively safe triangle near walls, no closer than 1 m to the wall, and no farther from the wall than the wall's height. Trees or tall objects, high ground, water, open spaces, metal objects, and ungrounded buildings, such as shacks and huts, should be avoided. Instinctively, persons may seek shelter under isolated trees during thunderstorms, but this increases strike risk. Holding a metal object during a storm is particularly dangerous, and was associated with over 60% of Florida lightning injuries. Indoors during a thunderstorm, appliances should be turned off and the telephone should not be used.

A final key to prevention is appropriate immediate action. When stranded in the open, it is best to crouch with the feet and knees together rather than lying flat. An impending hit may be signaled by a crackling sound, a visible glow (*St. Elmo's Fire*), a tingling sensation, and/or hair standing on end. The correct response is to crouch immediately with the feet together.

Several popular lightning myths are false. These include the notion that lightning never strikes in the same place twice or always hits the tallest object; and that it is dangerous to touch a lightning victim after a strike (Ritenour et al., 2008).

Injury and treatment

Lightning usually affects multiple organs simultaneously, and has several unique effects on the body. The combination of acute life-threatening multisystem injuries and long-term neuropsychiatric sequelae make the multidisciplinary team approach championed by burn centers appropriate for lightning survivors. Thus, lightning injury is one of the burn center referral criteria published by the Committee on Trauma of the American College of Surgeons (Ritenour et al., 2008).

Cardiopulmonary resuscitation

First responders must approach lightning casualties as if they had sustained high-energy blunt trauma. Airway, breathing, and circulation should be rapidly assessed, and full spine immobilization should be strongly considered. Immediate cardiopulmonary resuscitation (CPR) may be required. The most common cause of death is cardiac arrest (asystole or ventricular fibrillation) at the moment of injury, caused by a massive DC "countershock." Respiratory arrest may also occur due to both chest muscle paralysis and suppression of respiratory centers in the brainstem. Thus, lightning victims may face two highly lethal "hits": primary cardiac arrest, followed by respiratory arrest and secondary cardiac ischemia. Those patients who do not sustain cardiopulmonary arrest at the time of injury may subsequently die of various complications, but this is less likely. The cardiovascular response also features a massive catecholamine release through an unknown mechanism, which may be manifested (in patients who do not arrest) by hypertension, tachycardia, nonspecific electrocardiographic (EKG) changes, and contraction-band myocardial necrosis. In a few patients, fatal lightning injury may occur in the absence of any obvious external or internal injury. This finding has been attributed to the induction of current by strong magnetic fields, sufficient to cause cardiac arrest.

The American Heart Association (AHA) and the European Resuscitation Council have published recommendations for CPR of lightning victims. Advanced Cardiac Life Support (ACLS), and defibrillation should be performed as needed. Because respiratory arrest may persist even after the return of spontaneous circulation, advanced airway management and bag-valve ventilation may be required. Lightning patients may respond to resuscitation even when they appear dead, and even when the interval between injury and resuscitation is prolonged. Thus, the AHA recommends "vigorous resuscitative measures...even for those who appear dead on initial evaluation." (Anonymous, 2000).

Lightning may strike several people at once, creating a mass casualty situation that requires the sorting of casualties (*triage*). Under other circumstances, it might be reasonable to triage patients who appear dead to the "expectant" category during a mass casualty scenario. In the case of lightning, however, apparently dead patients should probably be treated first ("immediate" category), since they may respond well to CPR and may only require ventilation.

Lightning casualties may require ongoing intravenous fluid resuscitation, particularly if hypotension or muscle breakdown (*rhabdomyolysis*) is present. Hypotension should also prompt a search for sources of bleeding secondary to blunt trauma.

Skin

There are four types of skin lesions that can result from lightning: linear, punctate, "feathering," and thermal. *Linear burns* tend to follow areas of high sweat concentration, such as under the breasts and arms, and down the middle of the chest. They are generally small, from 1 to 4 cm in diameter. They may be present initially, or develop over several hours, and are thought to be due to vaporization of water on the skin's surface. *Punctate burns* are small, multiple, closely spaced circular burns. *"Feathering" lesions* do not represent a burn, and the epidermis and dermis are normal. Lichtenberg figures, also known as keraunographic markings, are one example. These fern-like, branching, arborizing, serpiginous, or fractal patterns are pathognomonic for lightning, but are often not present. Upon pathophysiologic examination they consist of extravasation of blood in the subcutaneous tissues. Lichtenberg figures are evanescent, and usually disappear after several hours without known residua. *Thermal injury* may occur if the patient is wearing metal objects (e.g., zippers), or if clothing ignites. Patients with lightning injury rarely suffer from extensive tissue destruction or large cutaneous burns. Thus, lightning burns – in contrast to other electric injuries – should generally be treated conservatively as they are usually superficial, and tend to heal quickly. Standard wound care procedures should be employed.

Muscle

Extremities may appear cool, blue, or pulseless due to transient vasospasm; this phenomenon may also feature keraunoparalyisis (see below). Emergent fasciotomy and debridement should be considered for patients with elevated compartment pressures or other clear evidence of intramuscular compartment syndrome. However, steady improvement of the cool extremity, with subsequent return of pulses, is the more likely outcome. Although extensive muscle damage following lightning strike is unusual, patients with myoglobinuria should be treated as for high-voltage electric injury, to include aggressive fluid resuscitation.

Central nervous system

Central nervous system (CNS) injury is common in lightning victims. According to Cherington's (2003) classification, the four groups of CNS injury include: Immediate and Transient Effects, Immediate and Prolonged Effects, Possible Delayed Neurological Syndromes, and Trauma from Falls or Blast. *Immediate and Transient Effects* are common, and include loss of consciousness, confusion, amnesia, and headaches, paresthesias, weakness, and *keraunoparalysis*. Keraunoparalysis (Charcot's paralysis) is specific to lightning injury. It features transient paralysis and loss of sensation, especially in the lower extremities; it lasts 1 to several hours and then resolves. Keraunoparalysis may be a result of intense catecholamine release because it is often accompanied by pallor, vasoconstriction, and hypertension. However, patients with such neurologic deficits should be assumed to have spinal injury until proven otherwise. *Immediate and Prolonged Effects* feature the sequelae of significant neurological injury, such as postarrest cerebral anoxia. *Possible Delayed Neurological Syndromes* are those which may be related to lightning, but which present in delayed fashion. These include motor neuron diseases and movement disorders. *Trauma from Falls or Blast* include closed head injuries such as subdural or epidural hematomas and subarachnoid hemorrhages. In addition, lightning injury may be associated with long-term neuropsychological impairment. Common complaints include fatigue, lack of energy, poor concentration, irritability, and emotional lability. Post-traumatic stress disorder is another common problem, occurring in about 30% of survivors. Cognitive testing may reveal memory, attention, and visual-reaction-time abnormalities. Some patients will meet criteria for depression. Of course, these effects may cause significant vocational and interpersonal difficulty, and early neuropsychiatric intervention is urged.

Eye and ear

Because of the blast overpressures generated by a lightning strike, patients frequently present with tympanic membrane rupture, and more rarely with more severe otologic injuries such as sensorineural deafness or vestibular injury. A wide variety of eye injuries can be caused by lightning strike. The most common of these is "lightning cataract," the presentation of which may be delayed by years. Dilated or nonreactive pupils are not considered a reliable sign of brain death in the early postinjury period.

Summary

Despite man's ancient fascination with lightning, our understanding of lightning physics remains incomplete. Improved building codes and a transition away from agricultural work for the majority of persons in developed countries has led to a decrease in the number of lightning deaths per year. Nevertheless, lightning remains one of the leading causes of weather-related death. Awareness of lightning hazards is the key to prevention.

Bibliography

Anonymous, 2000. Part 8: advanced challenges in resuscitation.- Section 3:special challenges in ECC. 3G: electric shock and lightning strikes. European Resuscitation Council. *Resuscitation*, 46(1–3), 297–299.

Bier, M., Chen, W., et al., 2005. Biophysical injury mechanisms associated with lightning injury. *Neurorehabilitation*, 20(1), 53–62.

Center for Disease Control and Prevention, 1998. Lightning-associated deaths – United States, 1980–1995. *MMWR: Morb Mort Wkly Rpt*, 4, 391–394.

Cherington, M., 2003. Neurologic manifestations of lightning strikes. *Neurology*, 60, 182–185.

Curran, E. B., Holle, R. L., and Lopez, R. E., 1997. *Lightning Fatalities, Injuries, and Damage Reports in the United States: 1959–1994. Technical Memorandum NWS SR-193*. Washington, DC: National Oceanic and Atmospheric Administration.

Curran, E. B., Holle, R. L., and Lopez, R. E., 2000. Lightning casualties and damages in the United States from 1959 to 1994. *Journal of Climate*, 13, 3448–3453.

Dwyer, J. R., 2005. Out of the blue. *Scientific American*, 292(5), 65–71.

Elsom, D. M., 1993. Deaths caused by lightning in England and Wales, 1852–1990. *Weather*, 48, 83–90.

Holle, R. L., Lopez, R. E. et al., 2001. U.S. lightning deaths, injuries, and damages in the 1890s compared to the 1990s (NOAA Technical Memorandum OAR NSSL-106).

Krider, E. P., 2006. Benjamin Franklin and lightning rods. *Physics Today*, 59, 42.

Lee, R. C., Zhang, D., et al., 2000. Biophysical injury mechanisms in electrical shock trauma. *Annual Review of Biomedical Engineering*, 2, 477–509.

Rakov, V. A., and Uman, M. A., 2003. *Lightning: Physics and Effects*. Cambridge, UK: Cambridge University Press.

Ritenour, A. E., Morton, M. J., et al., 2008. Lightning injury: a review. *Burns*, 34(5), 585–594.

Cross-references

Aviation, Hazards to
Building Codes
Casualties Following Natural Hazards
Hydrometeorological Hazards
Misconceptions About Natural Disasters (Physical Processes)
Monitoring and Prediction of Natural Hazards
Mortality and Injury in Natural Disasters
Myths and Misconceptions
Natural Hazard
Perception of Natural Hazards and Disasters
Snowstorm and Blizzard
Storms
Thunderstorms

LIQUEFACTION

Steven L. Kramer
University of Washington, Seattle, WA, USA

Definition

Liquefaction is a term used to describe the loss of soil strength and/or stiffness due to the generation of porewater pressure in saturated soil subjected to rapid loading. Liquefaction is most commonly triggered by earthquake ground shaking, but may also be caused by non-seismic loading (e.g., train traffic, rapid deposition of sediment, or construction vibrations).

Damage

Liquefaction has caused extensive damage in many historical earthquakes. Liquefaction damage is usually caused by the excessive ground deformations that result from the weakening and/or softening of liquefied soil. Liquefaction is frequently accompanied by the development of sand boils, small to large piles of ejecta brought to the ground surface by pressurized groundwater. Extensive weakening due to porewater pressure generation can cause soils that were stable prior to earthquake shaking to become unstable. When such soils support a building (Figure 1) significant weakening can cause foundation failure. When they underlie a slope, weakening can cause landsliding (Figure 2) to occur. Even under level ground, liquefied soils can densify as porewater pressure dissipates leading to significant post-earthquake settlement

Liquefaction, Figure 1 Liquefaction-induced foundation failure of Kawagishi-cho apartment buildings in 1964 Niigata earthquake.

Liquefaction, Figure 2 Flow slide along bank of Lake Merced, San Francisco.

(Figure 3). Ground deformations associated with liquefaction can be predominantly horizontal or vertical, or may include both components. The deformations can range from centimeters to hundreds of meters and can develop slowly or very rapidly. The evaluation and mitigation of liquefaction hazards is an important part of geotechnical engineering practice in seismically active areas of the world.

Liquefaction hazard evaluation

Evaluation of liquefaction hazards at a particular site generally involves three primary activities: evaluation of the susceptibility of the in situ soil to liquefaction, evaluation of the potential for initiation of liquefaction under the levels of ground shaking anticipated at the site, and evaluation of the expected effects of liquefaction.

Susceptibility

A number of factors control the susceptibility of soil to liquefaction, and not all soils are susceptible to liquefaction. Liquefaction results from the generation of high porewater pressure, which can only occur when the soil is saturated; dry and partially saturated soils, therefore, are not susceptible to liquefaction. Also, the high permeability of gravelly soils will generally not allow them to sustain high porewater pressure, so gravels are generally

Liquefaction, Figure 3 Liquefaction-induced settlement of Hotel Sapanca following 1999 Kocaeli earthquake.

not susceptible to liquefaction unless bounded by lower-permeability zones that impede the drainage of porewater. The generation of high porewater pressure is also retarded by soil plasticity, so clayey soils are generally not susceptible to liquefaction (although some clay-rich sediments can exhibit macro-behavior that shares certain characteristics of the behavior of liquefiable soils). The most susceptible sediments are non-plastic fine-grained (silty) materials with contrasted granulometry, such as alternately homogeneous fine sands or silts. Indeed, most occurrences of liquefaction in historical earthquakes have been observed to occur in such materials.

Liquefaction-susceptible soils are found in a relatively narrow range of geological environments. Processes that sort soils into uniform particle sizes and deposit them in loose states produce soils with high liquefaction susceptibility. Fluvial deposits, and colluvial and aeolian deposits, when saturated, are frequently susceptible to liquefaction. Alluvial, lacustrine, and estuarine deposits can also be susceptible to liquefaction. The combinations of soil type and groundwater conditions required for liquefaction susceptibility are commonly found in and along rivers and shorelines. Since important transportation and lifeline facilities such as bridges and ports are also located in such areas, they are frequently impacted by liquefaction in strong earthquakes.

Initiation

The initiation of liquefaction requires rapid loading of sufficient amplitude and duration to produce high porewater pressure in the soil. The level of porewater pressure required to trigger liquefaction depends on the initial (pre-earthquake) density and stress conditions in the soil. If the initial shear stress required to maintain static equilibrium is greater than the shear strength of the soil after it has liquefied (the residual strength of the soil), flow liquefaction can occur; if not, liquefaction can still occur through a mechanism known as cyclic mobility.

Flow liquefaction is triggered when rapid loading of sufficient amplitude brings the stress conditions in the soil to a critical point at which the structure of the soil skeleton becomes unstable and rapidly breaks down. As the structure collapses, compressive stresses are transferred from the soil skeleton to the porewater so the intergranular, or effective, stress decreases. As a result, the shear strength, which is proportional to the effective stress, also decreases. The extent of the strength loss depends on the density of the soil; if the soil is very loose, the residual strength may be extremely low. Large, unidirectional ground movements are then driven by the difference between the shear stress required for static equilibrium and the residual strength. If the residual strength drops to a value so low that the stress difference is large, the deformations will be both rapid and large.

The phenomenon of cyclic mobility in liquefiable soils is quite complicated. The cyclic shear stresses induced in the soil by earthquake shaking cause an incremental rise of porewater pressure. If the shaking is strong enough and of sufficient duration, the porewater pressure may instantaneously reach the level of the initial effective stress, at which point the effective stress will be zero. At that point, the stiffness of the soil is extremely low, allowing it to strain significantly in response to the initial and earthquake-induced shear stresses. Dilation eventually causes the stiffness to increase, but the initial static shear stress will cause strain to accumulate preferentially in one direction. The amplitude of the final, permanent strain will depend on the level and duration of ground shaking. Integrating these strains over the thickness of the liquefied soil layer yields the permanent displacement of the ground surface. These displacements develop in a series of increments and usually cease to accumulate after earthquake shaking has ended. In some cases, however, redistribution of porewater pressure following earthquake shaking can lead to delayed ground movements.

The resistance of a soil to liquefaction depends most strongly on its in situ density. Because in situ density is very difficult to measure, liquefaction resistance is usually correlated to in situ test indices such as standard penetration resistance (Seed et al., 1985; Idriss and Boulanger, 2008), cone penetration resistance (Robertson and Wride, 1997), or shear wave velocity (Andrus and Stokoe, 2000). These indices, which are readily and commonly measured in the field, serve as proxies for density. Liquefaction potential is usually expressed in terms of a factor of safety against liquefaction, which is computed as a ratio of capacity (liquefaction resistance) to demand (liquefaction loading). Design factors of safety, which reflect both uncertainty in the evaluation process and the potential consequences of liquefaction, are typically on the order of 1.2–1.5; higher design factors of safety provide greater conservatism than lower values. Probabilistic liquefaction potential analyses are becoming more commonly used in practice.

Effects of liquefaction

The initiation of liquefaction can have a number of damaging consequences. The extent of the damage caused by liquefaction is usually related to the density of the soil and strength of the ground motion, with all deleterious effects increasing with decreasing density and increasing ground motion level.

Ground shaking

Liquefaction can have a strong effect on earthquake ground motions, and hence on the seismic performance of structures founded on liquefiable soil deposits. Prior to the initiation of liquefaction, a soil deposit will typically be able to transmit both high and low-frequency components of ground motions to the surface. The softening of the soil that occurs upon initiation of liquefaction, however, causes high-frequency components to be reflected downward from the bottom of the liquefied layer rather than being transmitted to the surface. At the same time, the softening of the profile causes low-frequency components to be amplified. The sudden change in soil profile characteristics causes a change in the character of the surface motion – acceleration levels generally drop and displacement levels increase following the initiation of liquefaction.

Instability

The weakening and softening that occurs following initiation of liquefaction can lead to mass movement of sloping soil deposits. When the residual strength of a liquefied soil is lower than the shear stress required to maintain equilibrium, flow liquefaction can lead to the occurrence of flow slides. Such slides occur suddenly with the unstable soils moving at high velocities over distances of tens to hundreds of meters. Although they are not common, flow slides can cause disastrous damage and loss of life.

When the ground slope is flat enough and/or the soil density high enough, the residual strength will exceed the shear stress required to maintain equilibrium. Under such conditions, flow sliding is not possible but damaging deformations associated with cyclic mobility can occur. These deformations, referred to as lateral spreading, result from the softening of a liquefied soil, which allows the incremental development of horizontal and vertical soil movements. Lateral spreading displacements can range from a few centimeters to several meters. Whereas the movements caused by lateral spreading are much smaller than those associated with flow sliding, lateral spreading occurs much more commonly than flow sliding and is responsible for significantly greater total losses in most earthquakes.

Settlement

After earthquake shaking has ended, the high porewater pressure in a liquefied soil will dissipate resulting in volumetric compression, which ultimately leads to ground surface settlement. This settlement typically occurs relatively quickly – within hours to a couple days following the earthquake. It can occur in an irregular pattern, however, thereby causing damage to structures supported on shallow foundations, as well as pavements, buried pipelines, and other elements of infrastructure. The use of deep foundations, such as piles that extend through the liquefied zone and derive their support from underlying non-liquefiable soils, can prevent the structure itself from settling. Settlement-related building damage can still occur, however, if utilities entering the building from the surrounding ground are not designed with flexible connections that can accommodate the settlement. Settlement of the approaches can render a bridge impassable if hinged approach slabs do not provide a transition from the approach embankment to the bridge deck. The amount of post-liquefaction settlement depends on the density and thickness of the liquefied soil and on the strength of the earthquake shaking.

Mitigation of liquefaction hazards

Since liquefaction is caused by the buildup of high porewater pressure, techniques for mitigation of liquefaction hazards have focused on reducing the tendency of the soil to generate high porewater pressure. That tendency is most closely related to the density of the soil; as a result, most liquefaction hazard–mitigation techniques focus on soil densification. The soils that are most susceptible to liquefaction tend to be densified efficiently by vibration. Vibro techniques involve the insertion and removal of a torpedo-shaped vibrating probe, often accompanied by the addition of gravel or crushed rock, on a grid pattern across a site. The stone columns left behind in such a process provide reinforcement and drainage benefits in addition to the densification associated with their installation. Dynamic compaction involves repeatedly dropping heavy (6–30 t) weights from heights of 10–30 m on a grid pattern across a site. The combination of impact stress and vibration can densify the soil to depths of 9–12 m. Blasting with time-delayed charges placed at multiple depths in multiple boreholes has also been used successfully to densify liquefiable soils. Liquefaction hazards can also be mitigated by injecting or mixing cementitious material into the soil. Permeation grouting injects low-viscosity liquid grout (aqueous suspensions of micro-fine cement, silica and lignin gels, phenolic and acrylic resins, or other materials) into the voids of the soil without disturbing the soil skeleton. Intrusion grouting injects stronger and more viscous grout materials under pressure sufficient to fracture the soil, leaving behind a network of intersecting lenses of hardened grout. Soil mixing and jet grouting use mechanical and hydraulic means to mix the in situ soil with cement grout leaving behind columns of hard, strong "soilcrete" that can resist liquefaction and the ground deformations it can cause.

Summary

Liquefaction is an important seismic hazard that has produced significant damage to the natural and built environments in past earthquakes. Although the phenomenon is quite complex, progress has been made in the evaluation and mitigation of liquefaction hazards. Nevertheless, it remains an active area of research as geotechnical engineers seek more reliable and economical ways to evaluate and mitigate those hazards.

Bibliography

Andrus, R. D., and Stokoe, K. H., 2000. Liquefaction resistance of soils from shear wave velocity. *Journal of Geotechnical and Geoenvironmental Engineering, ASCE*, **126**, 929–936.

Idriss, I. M., and Boulanger, R. W., 2008. *Soil liquefaction During Earthquakes*. Oakland: Earthquake Engineering Research Institute.

Robertson, P. K., and Wride, C. E., 1997. Evaluating cyclic liquefaction potential using the cone penetration test. *Canadian Geotechnical Journal*, **35**, 442–459.

Seed, H. B., Tokimatsu, K., Harder, L. F., and Chung, R., 1985. Influence of SPT procedures in soil liquefaction resistance evaluations. *Journal of Geotechnical Engineering, ASCE*, **111**, 1425–1445.

Cross-references

Building Codes
Building Failure
Collapsing Soil Hazards
Dispersive Soil Hazards
Earthquake
Hazard and Risk Mapping
Land Subsidence
Lateral Spreading
Pore-Water Pressure
Primary Wave
Quick Clay
Quick Sand
Secondary Wave
Structural Mitigation
Zoning

LIVELIHOODS AND DISASTERS

J. C. Gaillard
The University of Auckland, Auckland, New Zealand

Definition

The concept of livelihood reflects the ability of people to sustain their daily needs and draws on the combination of a large array of resources which are natural, physical, human, social, financial, and political in nature. These resources strongly interplay with the ability of people to face the threat of and recover from the impact of natural hazards. Therefore strengthening livelihoods and making them sustainable is a crucial component of disaster risk reduction.

Defining livelihoods

The concept of livelihood emerged in the 1980s as an alternative to the technocratic concept of "employment" to better describe how people struggle to make a living (Chambers and Longhurst, 1986; Swift, 1989). It emphasizes people's view of their own needs. According to Chambers and Conway (1991, p. 1) sustainable livelihoods comprise "people, their capabilities and their means of living, including food, income and assets. Tangible assets are resources and stores, and intangible assets are claims and access. A livelihood is environmentally sustainable when it maintains and enhances the local and global assets on which livelihoods depend, and has net beneficial effects on other livelihoods. A livelihood is socially sustainable which can cope with and recover from stress and shocks, and provide for future generations." Livelihoods thus refer to the means and capacities required to sustain durably people's basic needs. Basic needs are vitally linked to food, but also include shelter, clothing, cultural values, and social relationships.

The capacity to meet food and other basic needs depends on assets and capitals (Scoones, 1998). The use of the term "capital" has however been criticized for its economic nature which does not reflect the entire range of resources upon which people resort to make a living (e.g., de la Peña, 2008). As part of its sustainable livelihood framework, the Department For International Development (1999) distinguishes five types of resources (hence replacing "capital" in view of the foregoing criticism): natural resources (land, water, forest, air, and other natural resources), human resources (health, skills, and knowledge), social resources (kinship, social networks, and associations), financial resources (cash, saving, credit, jewelry, and other valuables) and physical resources (housing, infrastructures, work implements, livestock, and domestic utensils). Wisner (2009) and Gaillard and Cadag (2009) further identify institutional and political resources which include the interface with formal governance and access to government-linked services, information, and overall to the larger political scene. The extent, strength, and diversity of resources condition people's capacity to produce their own food. It also commands the capacity to purchase food should it not be supplied by the household itself. In the latter case, the availability of food depends on the larger political economy framework (Start and Johnson, 2004). The availability and extent of resources is indeed deeply dependent on claims and access. Claims refer to rights and capacities/power to ask for some external support to sustain basic needs should people be unable to meet them by themselves. Claims thus depend on the extent of people's social, economic, and political networks and relationships. It is complemented by access which is the opportunity to use available stores and resources or obtain food, employment, technology, and information (Chambers and Conway, 1991). As underlined by Sen (1981, 1986), people's claims for and access to livelihoods thus go beyond the specific availability or unavailability of resources but encompass the capability or entitlement to use available resources. Watts and Bohle (1993) emphasize that entitlement reflects people's empowerment evident in class relationships and the larger distribution of economic wealth, social opportunities, and political power within the society.

Livelihoods and people's ability to face natural hazards

People's ability to face natural hazards depends on their vulnerabilities and capacities (e.g., Wisner et al., 2004; Gaillard, 2010). Vulnerability in facing natural hazards reflects their susceptibility to be harmed should the threatening phenomenon occur. It basically reflects people's ability to live in safe places and, if they are compelled to settle in a hazardous area, on their awareness of, access to, and ability to successfully apply means of protection. Capacities refer to the resources people possess to resist and cope with disasters. As for livelihoods, capacities encompass the ability to either use and access needed resources and thus goes beyond the sole availability of these resources. All these factors are closely related to people's livelihoods and everyday life (Davis et al., 2004; Wisner et al., 2012).

The nature, strength, and diversity of livelihoods are crucial in defining people's vulnerabilities and capacities in facing natural hazards (e.g., Twigg, 2001; Cannon, 2003; Wisner et al., 2004; Gaillard et al., 2009). People whose livelihoods are sustainable in the face of natural hazards prove to be less vulnerable and equipped with capacities to face environmental shocks. Resources essential in the sustainability of livelihoods are crucial in defining vulnerability too. People's ability to live in hazard-safe places depends on access to land (natural resources). Skills and knowledge (human resources) enable the diversification of activities and thus lessen households' dependence on natural resources in the event of an adverse climatic or geologic event. Incomes and savings (financial resources) are obviously important to purchase food in time of scarcity but also to build resistant houses and to access other means of protection. Furthermore, pawning or selling of valuable belongings often allows to generate additional cash should required. Social networks and kinship (social resources) are critical in providing alternative support in time of crisis. The fragility of physical resources, i.e., infrastructures (including public buildings, hospitals, schools, and housing) and working implements (such as boats and farming implements), is another crucial factor in the face of natural hazards. Ultimately people's vulnerability and capacity in facing natural hazards is tied to powerlessness, i.e., the lack of political resources, as it prevents access to other forms of resources. People's vulnerability can therefore not be dissociated from livelihoods sustainability. On the other hand, livelihood sustainability is similarly tied to people's vulnerability to natural hazards. Disasters often destroy the environment, damage physical resources, kill relatives, and drain savings, therefore endangering people's livelihoods and ability to sustain their everyday needs on the long run.

The concept of sustainability implies that basic needs are met on a quotidian basis. Considering everyday life is therefore crucial in understanding both livelihoods sustainability and vulnerability in facing natural hazards. Factors which determine both sustainability of livelihoods and vulnerability to natural hazards are similarly rooted in daily life. Many people deliberately choose to face natural hazards to sustain the daily needs of their household. Indeed, the threat related to food insecurity always weight heavier than the threat linked to natural hazards. Sustaining one's minimum food intake is the human most basic need and is rooted in daily life. Threats to everyday needs, especially to food security, are almost always more pressing than threats from rare or seasonal natural hazards. Strategies to cope with natural hazards are also often anchored in daily life. Most are adjustments in everyday activities of the affected people rather than extraordinary measures adopted to face extreme and rare natural events (Gaillard et al., 2009).

Livelihoods and people's ability to recover from disasters

The ability to recover in the aftermath of a disaster further reflects the nature, strength, diversity, and sustainability of people's livelihoods (e.g., Gaillard and Cadag, 2009). The aptitude of disaster survivors to recover is first dependent on the nature and diversity of their pre-disaster livelihoods. Those who struggle to recover are often those who extensively rely on one form of resource which is heavily impacted by the disaster, e.g., natural resources.

The strength of livelihoods is another critical factor of recovery. Important financial resources enable people to save money, which may be tapped in time of hardship to fasten recovery. Similarly people with alternative skills and knowledge have an easier time to adjust to changing

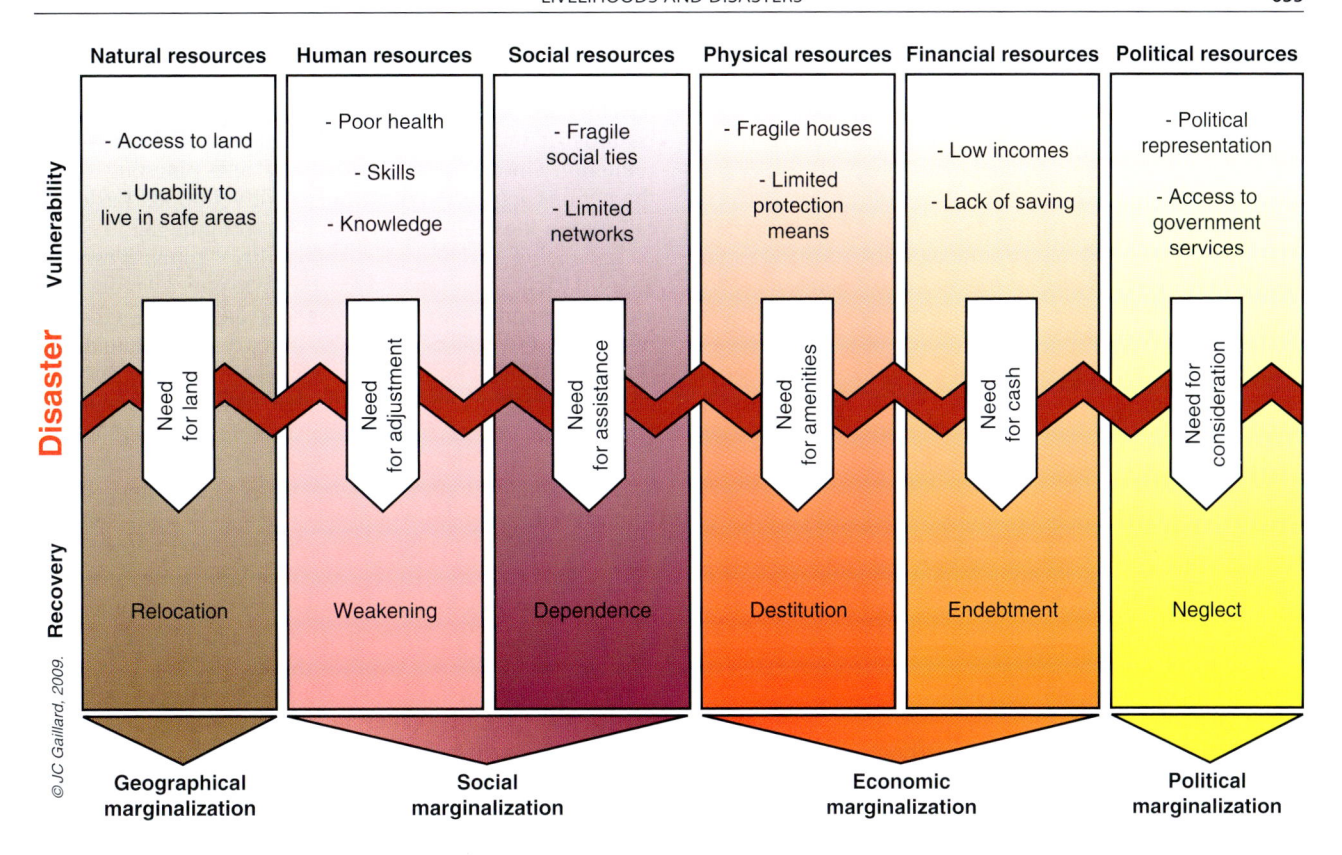

Livelihoods and Disasters, Figure 1 Interactions between livelihoods, pre-disaster vulnerability, and post-disaster recovery.

social and economic environments. Both physical and mental health also matters as disasters most frequently aggravate pre-event fragility for those most frequently affected, for example, the children, elderly, people with mental and physical disabilities. In time of disasters, alternative sources of support such as loaning money depend on the extent of social networks, that is, social resources, and the ability to pay back, financial resources, too. Entitlement to land ownership is of primary importance when disasters force people to relocate. Access to political resources and representation is also essential to benefit from post-disaster recovery programs provided by the government, nongovernment organizations (NGOs), and international institutions. The needs of those groups and survivors which are invisible on the everyday political scene, i.e., illegal settlers, ethnic and gender minorities, people with disabilities, are often neglected in disaster recovery.

Sustainability of livelihoods also turned out to be essential to the ability of the survivors to overcome disasters. Stability in livelihoods prevents a sharp decrease in households' incomes, thus preventing them to plunge into chronic endebtment. The loss of one or more relatives and lingering long-term decrease in available social resources often turned out to be a key determinant of people's ability to recover from a disaster. Those with some physical resources may sell or pawn them but on the long run, this strategy endangers their ability to sustain their daily life, especially when assets sold or pawned are cattle, farming implements, or fishing boats. Post-disaster relocation also affects the sustainability of livelihoods. It is often impossible to rely on the same resources than in the area of origin because land is lacking for farmers and skills are insufficient for urban settlers relocated in rural areas or fishermen relocated in mountainous locations.

The ability of those affected by disaster to recover is actually strongly dependant on their pre-disaster vulnerability. Disasters are amplifiers of everyday hardship (Baird et al., 1975; Maskrey, 1989). They do not level down people resources so that all survivors are equal in recovery. Following disasters, the rich are still rich, and sometimes richer, and the most able to recover quickly (Quarantelli and Dynes, 1972) while the poor are often poorer and struggle to recover. Disasters basically increase the everyday need for resources which make up people livelihood (Figure 1). For the most vulnerable coping with these increasing needs means falling in further marginalization.

Assessing the sustainability of livelihoods in the face of natural hazards

Livelihoods rarely refer to a single activity. It includes complex, contextual, diverse, and dynamic strategies developed by households to meet their needs (Chambers, 1995; Scoones, 1998, 2009). Furthermore, livelihoods and livelihood resources and strategies vary in time and space from one place to another, and from one season to another. For these reasons, understanding and assessing the sustainability of livelihoods is a challenge for researchers and practitioners engaged in development and disaster risk reduction.

Traditional research methods such as short interviews with key informants and questionnaire-based surveys are usually of limited help as they fail to encapsulate the complexity of livelihoods. Interestingly, the concept of livelihood emerged among the same group of researchers and practitioners who foster the use of participatory action research methods (e.g., Chambers, 1994). These methodologies encourage the participation of those who are most concerned, the people, in the evaluation of their own resources and strategies to assess strength and sustainability. Participatory methods encompass a large array of tools which range from listing, ranking, profiles, and Venn diagrams to transect walk, community drama, and participatory mapping, which are always conducted as part of group discussions to foster dialogue and exchange of ideas.

A number of analytical frameworks provide useful approaches to sustainable livelihoods (e.g., Hoon et al., 1997; Department For International Development, 1999) but only a few tools have been developed with the specific objective to assess the vulnerability and sustainability of those livelihoods in the face of natural hazards. Existing tools include the Community-based Risk Screening – Adaptation and Livelihoods tool or CRiSTAL developed by a consortium of NGOs and international research institutes (International Institute for Sustainable Development, 2007). CRiSTAL consists in a series of tables which integrate both hazardous phenomena, including changing climatic patterns, and people's resources. It thus provides a useful overview of the potential impact of hazards on the overall livelihood strategies.

Participatory 3-Dimensional Mapping or P3DM (Gaillard and Maceda, 2009) has also recently been used to provide a spatial analysis of livelihood resources and strategies in the face of natural hazards. All forms of resources may be plotted on the map but some which are not location based such as interpersonal and power relationships are more difficult to capture. Yet P3DM proves to be a very useful tool which helps people in visualizing intangible threats to their resources. Furthermore, it turns out to be a very powerful tool for participatory planning and strengthening of livelihoods.

Outlook: Reinforcing livelihoods to reduce disaster risks

Strengthening people's livelihoods is crucial to sustainable disaster risk reduction as it enables local communities to live with risk on an everyday basis (Benson et al., 2001; Cannon et al., 2003; Twigg, 2004). It is actually often impossible to prevent people from settling in hazardous areas, because these same locations often provide resources on a daily basis, as in the case of fertile floodplains and coastal zones with fisheries. Focusing on livelihoods simultaneously addresses people's ability to sustain their daily needs and their capacities to face natural hazards. It further favors the integration of disaster risk reduction into development policy and planning since the actions required for reinforcing livelihoods basically fall within the realm of development programs.

Enhancing livelihood sustainability emphasizes five areas of focus: creation of working days, poverty reduction, well-being and capabilities, livelihood adaptation, vulnerability and resilience, natural resource base sustainability (Scoones, 1998). Strategies to enhance livelihood sustainability should thus be people centered, multilevel and holistic, dynamic and sustainable (Department For International Development, 1999). The sustainable livelihood approach is being widely used by government agencies and NGOs to foster development both in urban and rural settings (e.g., Chambers, 1995; Hoon et al., 1997; Devereux, 2001; Scoones, 2009). It is now applied to disaster risk reduction and post-disaster recovery (e.g., Sanderson, 2000; Twigg, 2001; Cannon, 2003; Cannon et al., 2003; Wisner et al., 2004; Kelman and Mather, 2008).

Reinforcing livelihoods to reduce disaster risks required both action from the bottom up and measures from the top down. Bottom-up actions refers to community-based disaster risk reduction or CBDRR. CBDRR spurs the participation of local communities in the assessment and reduction of disaster risk in links with their daily livelihoods (e.g., Anderson and Woodrow, 1989; Maskrey, 1989). On the other hand, top-down actions from national authorities and international institutions should facilitate people's access to a large range of resources to reinforce their livelihoods, including those resources which would protect these livelihoods from the harm of natural hazards. Such actions necessitate a political will and commitment on the side of the governments to blend development policies with disaster risk reduction. Focusing on livelihoods in disaster risk reduction and development is a long-term investment as it enables to equally address poverty and vulnerability and locate both within the context of everyday life. The two approaches mutually benefit from each other as development contributes to reducing vulnerability and vulnerability reduction participates in the reinforcement of livelihoods.

Bibliography

Anderson, M. B., and Woodrow, P., 1989. *Rising from the Ashes: Development Strategies in Times of Disasters*. Boulder: Westview Press.

Baird, A., O'Keefe, P., Westgate, K., and Wisner, B. 1975. *Towards an explanation and reduction of disaster proneness*. Bradford: Disaster Research Unit, University of Bradford. Occasional Paper No. 11.

Benson, C., Twigg, J., and Myers, M., 2001. NGO initiatives in risk reduction: an overview. *Disasters*, **25**(3), 199–215.

Cannon, T., 2003. *Vulnerability Analysis, Livelihoods and Disasters Components and Variables of Vulnerability: Modelling and Analysis for Disaster Risk Management*. Manizales: Inter-American Development Bank//Instituto De Estudios

Ambientales, Program on Indicators for Disaster Risk Management, Universidad Nacional de Colombia.

Cannon, T., Twigg, J., and Rowell, J., 2003. *Social Vulnerability, Sustainable Livelihoods and Disasters*. London: Conflict and Humanitarian Assistance Department and Sustainable Livelihoods Support Office, Department for International Development.

Chambers, R., 1994. The origins and practice of participatory rural appraisal. *World Development*, **22**(7), 953–969.

Chambers, R., 1995. Poverty and livelihoods: whose reality counts? *Environment and Urbanization*, **7**(1), 173–204.

Chambers, R., and Conway, G. R., 1991. *Sustainable Rural Livelihoods: Practical Concepts for the 21st Century*. Brighton: Institute of Development Studies. IDS discussion paper 296.

Chambers, R., and Longhurst, R., 1986. Trees, seasons and the poor. *IDS Bulletin*, **17**(3), 44–50.

Davis, I., Haghebeart, B, and Peppiatt, D., 2004. *Social Vulnerability and Capacity Analysis*. Geneva: ProVention Consortium. Discussion paper and workshop report.

de la Peña, A., 2008. *Evaluating the World Bank's concept of social capital: a case study in the politics of participation and organization in a rural Ecuadorian community*. Ph.D. dissertation, Gainesville, University of Florida.

Department for International Development, 1999. *Sustainable Livelihoods Guidance Sheets*. London: Department for International Development.

Devereux, S., 2001. Livelihood insecurity and social protection: a re-emerging issue in rural development. *Development Policy Review*, **19**(4), 507–519.

Gaillard, J. C., 2010. Vulnerability, capacity, and resilience: perspectives for climate and development policy. *Journal of International Development*, **22**(2), 218–232.

Gaillard, J. C., and Cadag, J. R., 2009. From marginality to further marginalization: experiences from the victims of the July 2000 Payatas trashslide in the Philippines. *Jàmbá: Journal of Disaster Risk Studies*, **2**(3), 195–213.

Gaillard, J. C., and Maceda, E. A., 2009. Participatory 3-dimensional mapping for disaster risk reduction. *Participatory Learning and Action*, **60**, 109–118.

Gaillard, J. C., Maceda, E. A., Stasiak, E., Le Berre, I., and Espaldon, M. A. O., 2009. Sustainable livelihoods and people's vulnerability in the face of coastal hazards. *Journal of Coastal Conservation*, **13**(2–3), 119–129.

Hoon, P., Singh, N., and Wanmali, S., 1997. *Sustainable Livelihoods: Concepts, Principles and Approaches to Indicator Development*. New York: United National Development Program.

International Institute for Sustainable Development, InterCooperation, International Union for Conservation of Nature, Stockholm Environment Institute, 2007. *Community-Based Risk Screening – Adaptation and Livelihoods – CRiSTAL v.3.2*. Winnipeg: International Institute for Sustainable Development.

Kelman, I., and Mather, T., 2008. Living with volcanoes: the sustainable livelihoods approach for volcano-related opportunities. *Journal of Volcanology and Geothermal Research*, **172**(3–4), 189–198.

Maskrey, A., 1989. *Disaster Mitigation: A Community Based Approach*. Oxford: Oxfam. Development Guidelines No, 3.

Quarantelli, E. L., and Dynes, R. R., 1972. When disaster strikes: it isn't much like what you've heard and read about. *Psychology Today*, **5**(9), 66–70.

Sanderson, D., 2000. Cities, disasters and livelihoods. *Environment and Urbanization*, **12**(2), 93–102.

Scoones, I. 1998. *Sustainable Rural Livelihoods: a Framework for Analysis*. Brighton: Institute of Development Studies. IDS working paper 72.

Scoones, I., 2009. Livelihoods perspectives and rural development. *Journal of Peasant Studies*, **36**(1), 171–196.

Sen, A., 1981. *Poverty and Famines: An Essay on Entitlement and Deprivation*. Oxford: Oxford University Press.

Sen, A., 1986. *Food, Economics and Entitlements*. Helsinki: World Institute for Development Economics Research, United Nations University. WIDER working paper 1.

Start, D., and Johnson, C., 2004. *Livelihood Options? The Political Economy of Access, Opportunity and Diversification*. London: Overseas Development Institute. Overseas Development Institute working paper 233.

Swift, J., 1989. Why are rural people vulnerable to famine? *IDS Bulletin*, **20**(2), 8–15.

Twigg, J., 2001. *Sustainable Livelihoods and Vulnerability to Disasters*. London: Benfield Hazard Research Centre. Working Paper No 2.

Twigg, J. 2004. *Disaster Risk Reduction: Mitigation and Preparedness in Development and Emergency Programming*. London: Humanitarian Practice Network. Good Practice Review No 9.

Watts, M. J., and Bohle, H. G., 1993. The space of vulnerability: the causal structure of hunger and famine. *Progress in Human Geography*, **17**(1), 43–67.

Wisner, B., 2009. *SHINK & Swim: Exploring the Link Between Capital (Social, Human, Institutional, Natural), Disaster, and Disaster Risk Reduction*. Global Facility for Disaster Reduction and Recovery. Washington: World Bank.

Wisner, B., Blaikie, P., Cannon, T., and Davis, I., 2004. *At Risk: Natural Hazards, People's Vulnerability, and Disasters*. London: Routledge.

Wisner, B., Gaillard, J. C., and Kelman, I. (eds.), 2012. *Handbook of Hazards and Disaster Risk Reduction*. London: Routledge.

Cross-references

Adaptation
Coping Capacities
Disaster Research and Policy Paradigms
Hazardousness of Place
Marginality
Perception of Natural Hazards and Disasters
Vulnerability

LOESS

János Kovács, György Varga
University of Pécs, Pécs, Hungary

Synonyms

Bluff formation (Mississippi Valley region); Lehm (Alsace, France); Löss (Germany)

Definition

Loess is a homogeneous, typically nonstratified, porous, friable, slightly coherent, often calcareous, fine-grained, silty, pale yellow or buff, windblown (aeolian) sediment.

Loess consists mainly of quartz particles predominantly of silt with subordinate grain sizes ranging from clay to fine sand (Muhs and Bettis, 2003; Pye, 1995; Smalley, 1975). It generally occurs as a widespread blanket deposit that covers areas of hundreds of square kilometers and tens of meters thick. Loess covers areas extending from north-central Europe to eastern China as well as the Mississippi Valley and Pacific Northwest of the USA; and the Pampas in South America (Muhs and Bettis, 2003). Loess is generally buff to light yellow or yellowish brown, often contains shells, bones, and teeth of mammals, and is traversed by networks of small narrow vertical tubes (frequently lined with calcium-carbonate concretions) left by successive generations of grass roots, which allow the loess to stand in steep or nearly vertical faces (Smalley et al., 2001). Loess is now generally believed to be wind-blown dust of Pleistocene age carried from desert surfaces, alluvial valleys, and outwash plains, or from unconsolidated glacial or glaciofluvial deposits uncovered by successive glacial recessions but prior to invasion by a vegetation mat (Muhs and Bettis, 2003; Smalley et al., 2001). The mineral grains composed mostly of quartz and associated heavy minerals, feldspars, and clay minerals are fresh and angular, and are generally held together by calcareous cement. In some regions, for example, Moravia, Tajikistan, and China, more than ten successive loess formations are separated by red to dark brown paleosols. Etymology: German Löss, from dialectal (Switzerland) lösch, "loose," so named by peasants and brick workers along the Rhine valley where the deposit was first recognized.

Because the grains are angular, with little polishing or rounding, loess will often stand in banks for many years without slumping (Smalley and Derbyshire, 1991). The thickness of collapsible loess is as much as 20 m in the loess terrains worldwide. Dry loess can sustain nearly vertical slopes, being perennially undersaturated. However, when locally saturated, it disaggregates instantaneously. Such hydrocompaction is a key process in many slope failures, made worse by an underlying terrain of low-porosity rocks. Gully erosion of loess may yield very high sediment concentrations. Characteristic vertical jointing in loess influences the hydrology. Enlarged joints develop into natural subsurface piping systems (subsidence), which following collapse produce a "loess karst" terrain. Foundation collapse and cracked walls are common, many rapid events following periods of unusually heavy rain. Slope failure is a major engineering problem in thick loess terrain, flow-slide and spread types being common (Derbyshire, 2001). The results are often devastating in both urban and rural areas. An associated hazard is the damming of streams by landslides.

Bibliography

Derbyshire, E., 2001. Geological hazards in loess terrain, with particular reference to the loess regions of China. *Earth–Science Reviews*, **54**, 231–260.

Muhs, D. R., and Bettis, E. A. III, 2003. Quaternary loess-paleosol sequences as examples of climate-driven sedimentary extremes. In Chan, M. A., and Archer, A. W. (eds.), *Extreme Depositional Environments: Mega End Members in Geologic Time*. Boulder, CO: Geological Society of America Special Paper 370, pp. 53–74.

Pye, K., 1995. The nature, origin and accumulation of loess. *Quaternary Science Reviews*, **14**, 653–657.

Smalley, I. J. (ed.), 1975. *Loess: Lithology and Genesis*. Stroudsburg: Dowden, Hutchinson and Ross, Benchmark Papers in Geology 26.

Smalley, I. J., and Derbyshire, E., 1991. Large loess landslides in active tectonic regions. In Jones, M., and Cosgrove, J. (eds.), *Neotectonics and Resources*. London: Belhaven Press, pp. 202–219.

Smalley, I. J., Jefferson, I. F., Dijkstra, T. A., and Derbyshire, E., 2001. Some major events in the development of the scientific study of loess. *Earth–Science Reviews*, **54**, 5–18.

Cross-references

Collapsing Soil Hazards
Dust Storm
Expansive Soils and Clays
Global Dust/Aerosol Effects
Landslide (Mass Movement)
Landslide Types
Piping Hazards
Pore Water Pressure
Sinkholes
Subsidence Induced by Underground Extraction

M

MACROSEISMIC SURVEY

Roger M. W. Musson
British Geological Survey, Edinburgh, UK

Definition

The term "macroseismic survey" refers to the process of gathering information on how strongly an earthquake was felt in different places.

Discussion

It has long been standard practice in earthquake investigation to gather information on the distribution of effects of any recent earthquake. Indeed, before the introduction of reliable seismometers, this was really the only way to study an earthquake. Generally, the results of such a study are presented as a map of intensity, often contoured as isoseismals. A macroseismic survey generally comprises two parts. The most heavily damaged area needs to be examined firsthand, and the damage to individual buildings recorded. This task ideally should be conducted in collaboration with engineers qualified to assess the original strength of the damaged buildings. This is referred to as a field investigation of the earthquake. Data collection for the wider felt area of the earthquake, at non-damaging intensities, is usually done via questionnaires. Various strategies for the dissemination of questionnaires have been practiced in the past, including appeals for information published in newspapers, sending questionnaires to local officials, and maintaining a network of volunteer observers who can be relied on to fill in details after an earthquake has occurred. Today, the dominant method of collecting questionnaire data is over the internet. After even a moderate-sized event in a populated area, tens of thousands of responses can be collected very quickly via an institute's web site, and these can then be processed in real time using an automatic intensity assessment algorithm. This also has the great advantage that the results of the survey are visible immediately on the web site, rather than appearing only in a journal paper or bulletin some months later, and this is an excellent method of conveying seismological data to the general public in a timely and informative way.

Bibliography

Musson, R. M. W., 2002. Intensity and intensity scales. In Bormann, P. (ed.), *New manual of seismological observatory practice (NMSOP)*. Potsdam: GFZ.

Musson, R. M. W., and Cecić, I., 2002. Macroseismology. In Lee, W. H. K., Kanamori, H., Jennings, P. C., and Kisslinger, C. (eds.), *International Handbook of Earthquake and Engineering Seismology*. San Diego: Academic, pp. 807–822.

Wald, D. J., Quitoriano, V., Dengler, L. A., and Dewey, J. W., 1999. Utilization of the Internet for rapid community intensity maps. *Seismological Research Letters*, **70**(6), 680–697.

Cross-references

Intensity Scales
Internet, World Wide Web and Natural Hazards
Isoseismal
Magnitude Measures
Seismograph/Seismometer

MAGMA

Catherine J. Hickson[1,2], T. C. Spurgeon[2], R. I. Tilling[2,3]
[1]Magma Energy Corp., Vancouver, BC, Canada
[2]Alterra Power Corp., Vancouver, BC, Canada
[3]Volcano Science Center, U.S. Geological Survey, Menlo Park, CA, USA

Synonyms

Liquid rock; Molten rock

Definition

Magma is liquid or molten "rock."

Discussion

Magma is liquid rock which is a fluid comprised of a mixture of crystals and gas. When solidified it becomes an igneous rock. It is magma when below ground and lava when above ground. The chemical composition of magma/lava plays a major role in determining eruption characteristics and the hazard potential of a volcano. Magmas vary in composition dependent on a number of factors, in particular their plate tectonic affinity (Perfit and Davidson, 2000). Basaltic magmas are common along ocean ridges, hot spots, and continental plateaus. Magmas with higher silica contents (Andesite, Dacite, and Rhyolite) are common along subduction zones and intra-plate tectonic settings. The composition, along with crystal and gas content, controls the viscosity, temperature, and explosivity of the magma. Composition combined with pressure dictates the proportions of liquid, gases, and solids. These proportions have a strong controlling influence on the style of eruption. Basaltic (or mafic) lavas have low viscosity and are the least explosive, except in certain circumstances where there is interaction with water. As magma increases in silica content (referred to as felsic or sometimes siliceous magmas [reflecting high silica content], for example, Gillespie and Styles, 1999; Rogers and Hawkesworth, 2000; Thorpe and Brown, 1993), the explosivity tends to go up because the rise in silica creates an attendant rise in viscosity. As the magma rises to the surface (and as it crystallizes with lowering temperatures, exsolving fluids), the fluid phase (dominated by H_2O and CO_2) within the magma begins to exert pressure on the liquid phase. The exsolving bubbles, expanding as the magma rises, combined with growing crystals, increase the pressures within the magma (Scandone et al., 2007), causing a decrease in the density of the melt resulting in more rapid rise. The culmination of the ascent, high fluid pressures, and high viscosity magmas is an explosive eruption, common at stratovolcanoes. High silica, low fluid pressure magmas flow sluggishly with little or no explosive activity. Such magmas often "stall" at high crustal levels forming small stocks or sills, or larger plutons. If they egress to the surface, they flow only with great difficulty, forming domes or flow domes.

Bibliography

Gillespie, M. R., and Styles, M. T., 1999. *BGS Rock Classification Scheme*, Volume 1, Classification of igneous rocks. British Geological Survey, Research Report Number RR 99-06, 154 pp.

Perfit, M. R., and Davidson, J. P., 2000. Plate tectonics and volcanism. In Sigurdsson, H., et al. (eds.), *Encyclopedia of Volcanoes*. New York: Academic Press, pp. 89–113.

Rogers, N., and Hawkesworth, C., 2000. Composition of magma. In Sigurdsson, H., et al. (eds.), *Encyclopedia of Volcanoes*. New York: Academic Press, pp. 115–131.

Scandone, R., Cashman, K. V., and Malone, S. D., 2007. Magma supply, magma ascent and the style of volcanic eruptions. *Earth and Planetary Science Letters*, **253**, 513–529.

Thorpe, R., and Brown, G., 1993. *The Field Description of Igneous Rocks*. Chichester, England: Wiley. Geological Society of London Handbook. 154 pp.

Cross-references

Aa Lava
Lava
Pahoehoe Lava
Plate Tectonics
Shield Volcano
Stratovolcanoes
Volcanoes and Volcanic Eruptions

MAGNITUDE MEASURES

David Giles
University of Portsmouth, Portsmouth, UK

Synonyms

Earthquake measure; Earthquake severity; Earthquake size; Magnitude scale

Definition

Magnitude Measures. A variety of scales and calculations to measure, characterise and catalogue the size of an earthquake in terms of the seismic waves generated and energy released by the event.

Introduction

The size and damaging effects or severity of an earthquake are described by measurements of both magnitude and intensity. The quantification of the size of an earthquake has been considered by seismologists for many decades. A variety of different measures have been produced to estimate and report the magnitude of a seismic event. Many attempts have been made to develop a uniform scale to measure earthquake magnitude (Kanamori, 1983) but this goal has not always been achievable due to the changes in instrumentation used over time, changes in seismic data processing techniques as well as developments in the distribution of seismic monitoring stations. As a result of these influences a variety of *magnitude scales/measures* have been developed and reported which have been used at various times and locations around the world. As the science of earthquakes (seismology) has developed further advances have been made in the quantification of a seismic event. In order to provide a historical continuity of the measurements made relationships needed to be developed between the various earthquake size measuring schemes. As earthquakes are the result of complex geophysical processes it is not a simple matter to find a single measure of the size of an earthquake (Kanamori, 1978).

There are two fundamental parameters that can be used to describe the size of an earthquake. The *magnitude* of a seismic event characterises the relative size of the earthquake. It can be considered as a measure of the amount of energy released during the seismic event. For each earthquake there is only one magnitude. The *intensity* of a seismic event describes the severity of the earthquake in terms of the physicals effects on the ground, people and buildings in the area affected. For each earthquake there are many intensities depending on the location and distance from the epicentre, underlying geology, types and styles of buildings and structures present in the affected zone.

Magnitude is a logarithmic measure of the size of an earthquake based on instrumental data (Bormann et al., 2002). The measurement of magnitude is based on the amplitude of the resulting seismic waves recorded on a seismogram once the amplitudes are corrected for the decrease with distance due to geometric spreading and attenuation (Stein and Wysession, 2003).

Seismic waves

The fault rupturing process that takes place during an earthquake generates elastic waves within the earth which propagate away from the rupture front. Different types of seismic waves are generated each with different velocities and travel paths. Two fundamental types of waves are created; compressional, longitudinal waves and shear, transverse waves. The fastest P or Primary Waves travel through the body of the earth together with the slower S or Secondary Shear Waves. At the surface of the earth these two types of motion can combine to form complex surface waves. These surface waves have much higher amplitudes than the P and S waves and are therefore much more destructive as their energy is concentrated near the earth's surface. Such surface waves can be further subdivided into Rayleigh or Love Waves which both have longer periods and arrive after the P and S waves on the seismogram. Rayleigh Waves have an elliptical motion similar to that of water waves whereas Love Waves have a motion that is horizontal and perpendicular to the direction of propagation. Near the earthquake epicentre the largest recorded wave is the short period S Wave. At greater distances the longer period Surface Waves become dominant. The various magnitude scales set out to measure the fundamental properties of these different waves in order to estimate the magnitude of the seismic event.

Quantification of earthquake size

Earthquakes can be quantified with respect to various physical properties of the source site. These include the length of the fault that ruptures, the area of the fault, the fault displacement, particle velocity and acceleration of the fault motion, duration of faulting, amount of radiated energy as well as the complexity of the fault motion (Kanamori, 1983). It is not possible to represent all of these parameters by a single number such as the magnitude of the earthquake but the magnitude of a seismic event does have value in allowing an initial analysis and cataloguing of an earthquake to be undertaken.

The majority of *magnitude measure* scales that are in use are empirical in nature. A magnitude M is determined from the amplitude A and period T of the various seismic waves detected by a seismometer, recorded by a seismograph on a seismogram. The formulas used to derive an estimate of the earthquake magnitude contain constraints such that magnitude value scales can be correlated over a certain magnitude range (Kanamori, 1983).

The first widely used *magnitude measure* or scale was developed by Charles Richter in 1935 (Richter, 1935). This work was further developed with Beno Gutenberg in 1945 (Gutenberg, 1945a). Initially the magnitude scale was calculated on the maximum amplitude of the largest waveform detected from the seismic event. Subsequently the use of surface waves was included and then measurements of the body wave. Since this initial work many other magnitude scales have been developed for both local and global application utilising differing aspects of the seismic signal generated during an earthquake.

In order to overcome some of the localised issues of the early magnitude scales and their inability to differentiate larger magnitude earthquakes, a magnitude measure was developed that was based on a key seismic parameter, the Seismic Moment. The Seismic Moment is related to some of the key physical parameters of the fault which has ruptured during the seismic event. This Seismic Moment has been incorporated into a Moment Magnitude Scale (M_W) by considering the seismic energy radiated during the earthquake. The Moment Magnitude Scale is now the most frequently quoted scale in describing the size of an earthquake along with the corresponding Seismic Moment of the event.

Seismic moment

One of the major advances in the development of magnitude scales was the concept of 'seismic moment' (Kanamori, 1978). The Seismic Moment is considered to be the most accurate and comparable measure of an earthquake and can be considered as a measure of the irreversible inelastic deformation in the fault rupture area (Kanamori, 1977). The measure is completely independent of the type of seismograph used to record the seismic event. The Seismic Moment is a parameter that measures the overall deformation at the source of the seismic event (Kanamori, 1977). It has an important bearing on global phenomena such as tectonic plate motion, polar motion and on the rotation of the earth. The Seismic Moment can be interpreted in terms of the strain energy released in an earthquake. It measures the amount of energy released rather than the size of the seismic waves which are affected by the depth of the event and the geology of the rocks that the waves pass through. The Seismic Moment is related to the final static displacement after the earthquake. The Seismic Moment M_0 is defined thus:

$$M_0 = \mu \bar{D} A \tag{1}$$

Where:

M_0 = Seismic moment (*measured in* dyn. cm *or* N.m)
μ = *Rigidity or shear modulus of the rock at the source (fault) depth*
$\bar{D}$ = *Average slip or displacement on the fault after rupture*
A = *Surface area of the fault rupture zone*

It is termed Seismic Moment as Area × Stress gives a Force, and Force × Distance gives a Moment.

Seismic energy

Conventionally the energy E released by an earthquake has been estimated via the magnitude – energy relationship developed by Gutenberg and Richter (Gutenberg, 1956):

$$Log\, E_S = 1.5\, M_S + 11.8\; (E_S\; in\; Ergs) \tag{2}$$

$$Log\, E_S = 2.45\, m_B + 5.8\; (E_S\; in\; Ergs) \tag{3}$$

These equations hold well for most earthquakes but tend to underestimate for very large earthquakes which have a fault rupture length of 100 km or greater. Kanamori (1977, 1994) considered the change in strain energy during a seismic event with a fault rupturing. He stated that if the stress drop during an earthquake is complete the following equation holds:

$$E_S \approx \frac{\Delta\sigma}{2\mu} M_O \tag{4}$$

Where:

E_S = *Seismic energy radiated by the seismic source as seismic waves*
M_O = *Seismic Moment*
$\Delta\sigma$ = *Stress drop*
μ = *Rigidity or shear modulus of the rock at the source (fault) depth*

The relationship between the slip or displacement in an earthquake, its fault dimensions and its Seismic Moment is closely tied to the magnitude of the stress released by the earthquake. This is known as the stress drop, the difference between the stress before and after fault rupture. The earthquake releases the strain energy that has accumulated over time around the fault area (Stein and Wysession, 2003). The stress drop, averaged over the fault can be approximated:

$$\Delta\sigma \approx \frac{\mu \bar{D}}{L} \tag{5}$$

Where:

$\bar{D}$ = *Average slip or displacement on the fault after rupture*
L = *Fault characteristic dimension of the fault rupture*

The average slip on the fault that ruptures can also be estimated from the Seismic Moment where:

$$\bar{D} \approx \frac{c\, M_0}{\mu L^2} \tag{6}$$

Where:

c = *Fault shape factor.*

The specific relationship and values of c depend on the fault shape and fault rupture direction. This allows the stress drop to be calculated for a variety of fault morphologies.

For a Circular Fault:

$$\Delta\sigma \approx \frac{7}{16} \frac{M_0}{R^3} \tag{7}$$

For a Rectangular Fault (Strike Slip):

$$\Delta\sigma \approx \frac{2}{\pi} \frac{M_0}{w^2 L} \tag{8}$$

For a Rectangular Fault (Dip Slip):

$$\Delta\sigma \approx \frac{8}{3\pi} \frac{M_0}{w^2 L} \tag{9}$$

Where:

R = *Fault radius*
W = *Fault width*

Kanamori (1983) stated that by utilising the relationship between Seismic Moment and seismic wave energy the energy can be estimated thus:

$$E_S \approx \frac{M_O}{2 \times 10^4} \;\; as \;\; \frac{\Delta\sigma}{\mu} \sim 10^{-4} \tag{10}$$

The conventional magnitude scales discussed in detail elsewhere are said to saturate when the rupture dimensions of the earthquake exceeds the wavelength of the seismic waves used for the magnitude determination, usually between 5 and 50 km (Kanamori, 1977). This saturation leads to an inaccurate estimate of the energy released in very large earthquakes. The energy can however be estimated from the calculated Seismic Moment as it is possible to correlate the seismic energy with the Moment Magnitude, M_w:

$$E_S \approx \frac{M_O}{2 \times 10^4} \tag{11}$$

$$Log\, E_S = Log\, (M_0) - 4.3 \tag{12}$$

And:

$$M_W = \frac{2}{3} Log\, M_0 - 10.7 \tag{13}$$

So:

$$M_W = \frac{2}{3} Log\, (E_S . 20000) - 10.7 \tag{14}$$

$$Log\, (E_S) = \frac{3}{2} M_W + 11.8\; (E_S\; in\; ergs) \tag{15}$$

To illustrate that Seismic Moment and seismic energy are different, Seismic Moment is quoted in dyn.cm (CGS units) or N.m (SI units) and seismic energy in Erg (CGS) or Joules (SI), even though the units are equivalent (Stein and Wysession, 2003). 1 erg = 1 dyn.cm and 1 erg = 10^{-7} J. The radiated energy is only $1/2 \times 10^4$ or 0.00005 of the Seismic Moment released. This is because the Seismic Moment is not energy per se but is related to the stress change over the earthquake source region which gives the Seismic Moment dimensions of dyn.cm:

$$\frac{dyn}{cm} \cdot cm^3 = dyn.cm \qquad (16)$$

Note however that E_S is not the total energy released by an earthquake. It is only the estimated amount of energy radiated as seismic waves. Other energy is released as gravitational, frictional or heat energy. E_S only represents this small fraction of the total energy release during a seismic event.

Moment magnitude scale, M_W

The key concept of Seismic Moment led to the development of a Moment Magnitude Scale, M_W (Hanks and Kanamori, 1979) which more closely relates the measure of size to the tectonic effects of an earthquake. Traditional *magnitude measure* scales, discussed elsewhere, are said to saturate at large magnitudes leading to considerable underestimation of the size of very large earthquakes. These magnitude scales tend to only measure the localised failure along the crustal fault zone rather than the gross wide scale fault characteristics (Hanks and Kanamori, 1979). In order to represent the size of an earthquake as a dislocation phenomenon along a fault the Seismic Moment M_0 is considered to be the most adequate measure (Utsu, 2002). It is the most fundamental parameter that can be used to measure the strength of an earthquake caused by fault slip.

Kanamori (1977) compared the earthquake energy-moment relationship with the magnitude-energy relationship developed by Gutenberg and Richter (Gutenberg, 1956) where E_S is expressed in ergs and M_0 in dyne.cm:

$$E_S = \frac{\Delta \sigma}{2\mu} M_O \qquad (17)$$

$$Log\, E_S = 1.5\, M_S + 11.8 \qquad (18)$$

As $\Delta\sigma/\mu \approx 10^{-4}$ (Kanamori, 1983):

$$Log\, M_0 = 1.5\, M_S + 16.1 \qquad (19)$$

As has been stated previously M_S values saturate for great earthquakes with $M_0 > 10^{29}$ dyn.cm or more such that Eqs. 2 and 3 do not hold for such large earthquakes. Kanamori (1977) and Hanks and Kanamori (1979) proposed a new Moment Magnitude Scale, M_W which overcame these issues of saturation by the incorporation of the calculated Seismic Moment:

$$M_W = \frac{2}{3}\, Log\, M_0 - 10.7 \; (M_0 \,in\, \text{dyn.cm}) \qquad (20)$$

$$M_W = \frac{2}{3}\, Log\, M_0 - 6.1 \; (M_0 \,in\, \text{N.m}) \qquad (21)$$

The Seismic Moment does not saturate. For example the Great Alaskan Earthquake of 1964 was recorded as $M_S = 8.4$ whereas on the Moment Magnitude Scale as $M_W = 9.2$.

Other significant magnitude scales

Magnitude scales general form

When attempting to estimate the magnitude of a seismic event the amplitude of the seismic wave is used to determine the earthquake size once the amplitudes have been corrected for the decrease with distance from the epicentre due to geometric spread and attenuation. *Magnitudes scales* thus have the following general form:

$$M = Log\, \frac{A}{T} + F(\Delta, h) + C_S + C_R \qquad (22)$$

Where:
M = Estimated magnitude of earthquake
A = Amplitude of the signal recorded on the seismogram
T = Dominant period of the signal recorded on the seismogram
F(Δ, h) = A calibration function used for the correction of the variation of amplitude with the earthquakes depth (h) and distance in degrees or kilometres (Δ) from the epicentre to the seismometer recording station
C_S = Station correction factor
C_R = Region correction factor

Magnitude measurements scales are thus logarithmic in nature. A unit increase in magnitude will correspond to a 10-fold increase in seismic wave amplitude and a 32-fold increase in associated seismic energy. Various scales have been developed for local or teleseismic (distant) events. Distance measurements for local events are usually quoted in kilometres and in degrees for more distant events (1° = 111.19 km).

Local wave magnitude scale, M_L

The earliest *magnitude measurement* scale was introduced by Charles Richter in 1935 to assess the size of earthquakes occurring in Southern California (Richter, 1935). He developed a *local* magnitude scale (M_L) which is often referred to as the 'Richter Scale'. The magnitude of the earthquake was calculated from the amplitude of the seismic waves measured on a specific seismograph, the Wood Anderson Torsion Instrument. Equation 23 details the formula used along with calibration charts to calculate M_L. This equation is only applicable to shallow earthquakes measured in Southern California occurring within 600 km of the Wood Anderson instrument. Richter's original magnitude scale was further developed in 1945 by

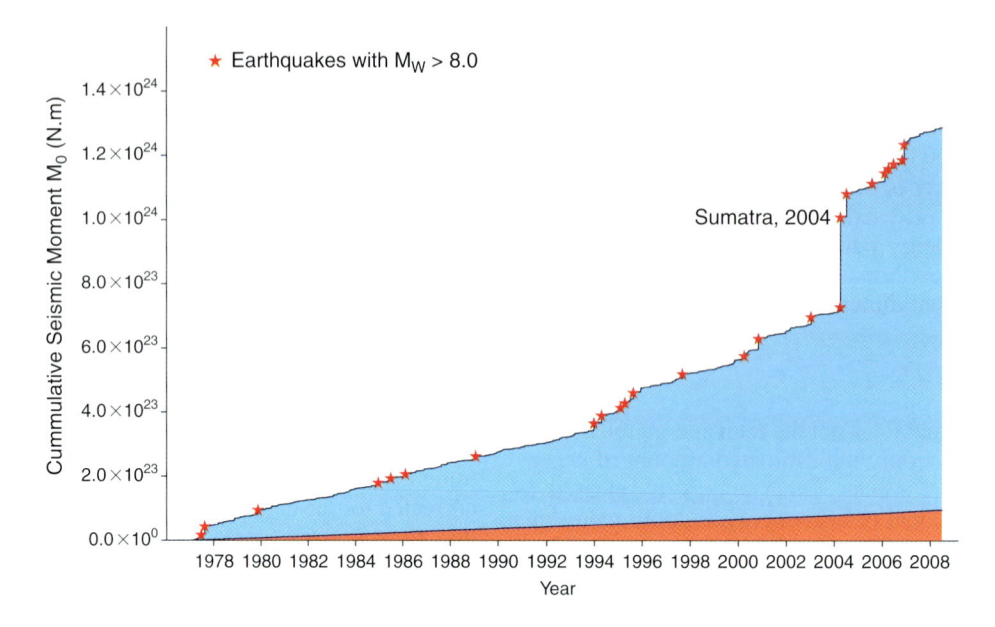

Magnitude Measures, Figure 1 Cumulative moment of all earthquakes in the Harvard University CMT catalogue from the Global Seismographic Network between 1977 and 2009. The field shaded light blue reflects the cumulative moment of earthquakes with $M_W \geq 6.5$. The field shaded *orange* reflects the cumulative moment of earthquakes with $M_W \geq 5.0$ to < 6.5. *Red stars* indicate the dates of earthquakes with $M_W \geq 8.0$. The contribution of the December 2004 Sumatra earthquake to the total cumulative moment is the largest step in the curve.

Gutenberg (Gutenberg, 1945a) to include seismic events of any epicentral distance from the recording station and for deeper focal depths as well as not being dependant on the type of seismograph used to record the event. A further two magnitude scales were developed from this early work, one dealing with *surface waves*, M_S, and another with *body waves*, M_B, (seismic waves that travel into and through the body of the Earth). Richter magnitudes in their original form are no longer quoted as most earthquakes do not occur in California and today Wood Anderson seismographs are rare (Stein and Wysession, 2003). M_L is a good indication of the structural damage that an earthquake can cause due to the recording frequency of the Wood Anderson seismograph being close to the resonant frequency (the frequency most likely to cause damage) of many buildings at around 1 Hz.

$$M_L = Log\, A_{Max} - Log\, A_0 (Richter, 1935) \tag{23}$$

To allow for possible local recording station effects (Hutton and Boore, 1987; Boore, 1989) a 'station term' is introduced:

$$M_L = Log\, A + 2.76\, Log\, \Delta - 2.48$$

Where:

A_{Max} = *Peak motion on a specific instrument (Wood Anderson seismograph)*

A_0 = *curve correction factor for the effect of distance, tabulated in Richter (1958)*

These correction factors are only truly valid for southern California. Other site specific correction factors have been developed for other 'local scales' around the world. In the UK the British Geological Survey uses the Hutton and Boore (1987) distance correction factor when estimating M_L for local UK earthquakes (Booth, 2007).

Surface wave magnitude scale, M_S

The M_S scale (Gutenberg, 1945a) use the amplitude of the surface seismic waves for earthquakes that are located between 2° and 160° epicentral distance from the recording station, with wave periods between 18 and 22 s and where the epicentre depth is less than 50 km. This scale will saturate at $M_S \geq 8$. A significant step in the development of the M_S scale was the publication of what was termed the Moscow-Prague Formula (Karnik et al., 1962). For shallow earthquakes where surface waves are generated, the magnitude of the event can be derived thus:

$$M_S = Log\, \frac{A}{T} + 1.66\, Log\, \Delta + 3.3 \tag{24}$$

Where:

A = *Maximum amplitude of the Rayleigh Wave*

Δ = *Distance in degrees between* 2° *and* 160°, $h \geq$ 50 km

Alternatively M_S can be calculated from the Rayleigh Waves with a period of 20 s, wave forms which often have the largest amplitude (Stein and Wysession, 2003):

$$M_S = Log\, A_{20} + 1.66\, Log\, \Delta + 2.0 \tag{25}$$

The Surface Wave Scale has sometimes been referred to as the Rayleigh Wave Scale (Marshall and Basham, 1973).

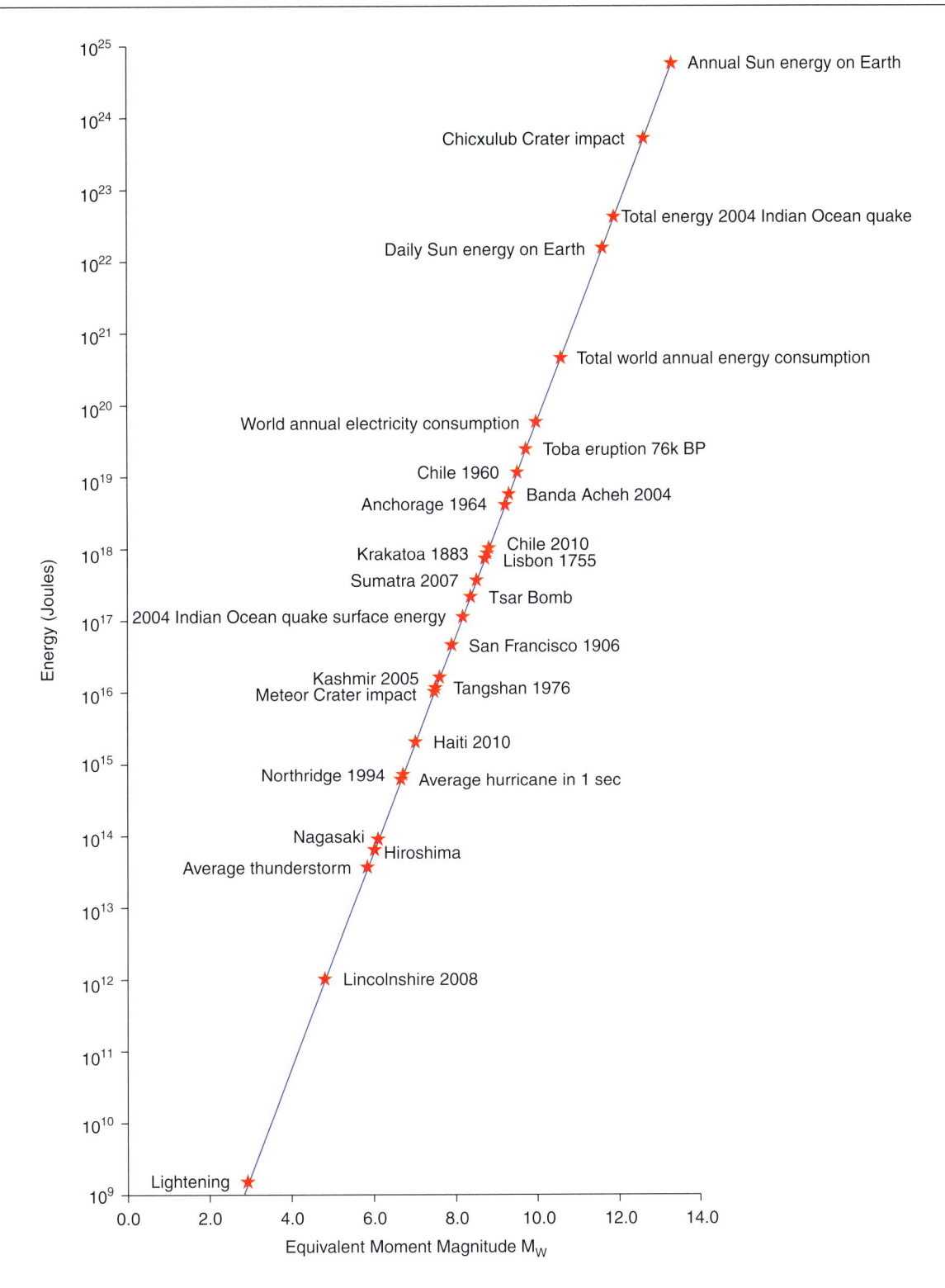

Magnitude Measures, Figure 2 A graph illustrating the equivalent Moment Magnitude M_W with respect to energy released by earthquakes and other phenomena.

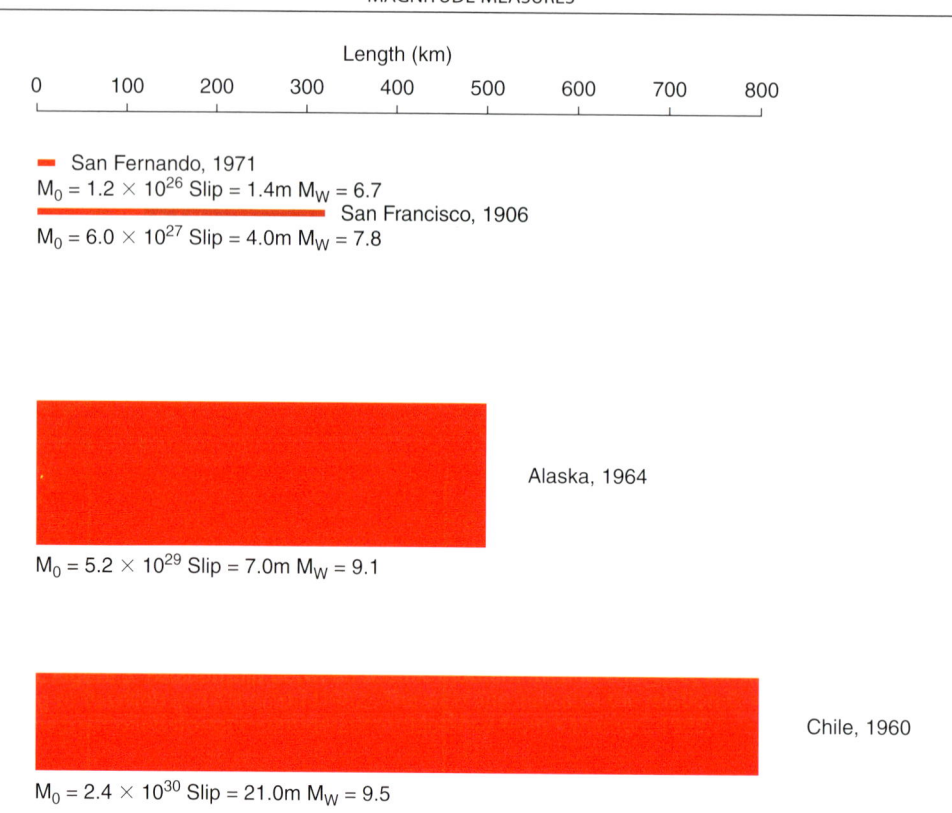

Magnitude Measures, Figure 5 Comparison of the magnitude of some significant earthquakes (After Stein and Wysession, 2003).

Where:
 d = Event duration (seconds)
 a_0, a_1, a_2 = Site specific coefficients

Aki and Chouet (1975) demonstrated that for earthquakes at epicentral distances shorter than 100 km the total duration of a seismogram is almost independent of distance and azimuth. Thus quick magnitude estimates from local events are feasible without knowing the exact distance of the stations to the source with the removal of the distance term from the equation. For example the Northern California Seismic Network calculates M_D thus (Lee et al., 1972):

$$M_D = 2.00 \, Log \, d + 0.0035 \, \Delta - 0.87 \qquad (33)$$

The scale can seriously underestimate magnitudes for events $M_L > 3.5$.

Nuttli magnitude scale, M_N

The M_N scale developed by Nuttli (1973) has been used in eastern North America and in particular Canada. The scale is based on the maximum amplitude of the Rayleigh surface waves for a frequency of 1 Hz:

$$M_N = Log \, \frac{A}{KT} + 1.66 \, Log \, R - 0.1 \qquad (34)$$

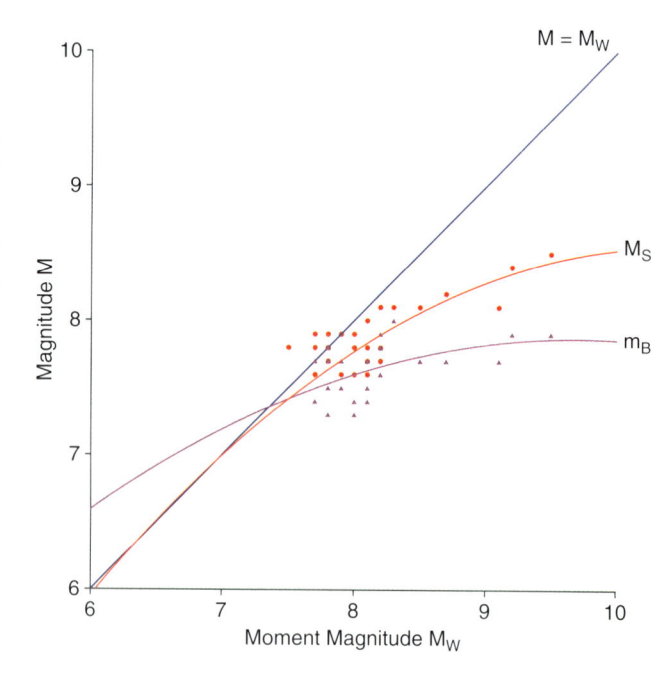

Magnitude Measures, Figure 6 Relationship between magnitude scales illustrating saturation at higher magnitudes (Data from Abe and Kanamori, 1980; Kanamori, 1983).

Magnitude Measures, Table 1 Summary of various magnitude measurement scales

Symbol	Magnitude scale	Reference/source
M_L	Local magnitude	Richter (1935)
M_S	Surface wave magnitude	Gutenberg (1945a), Moscow-Prague formula (Karnik et al., 1962)
M_B	Body wave magnitude	Gutenberg (1945b) and Gutenberg and Richter (1956, 2010)
m_B	Body wave magnitude	Gutenberg and Richter (1956, 2010)
M_D	Duration magnitude	Herrmann (1975)
M_E	Energy magnitude	Choy and Boatwright (1995) and Aki and Chouet (1975)
M_N	Nuttli magnitude	Nuttli (1973)
M_{JMA}	Japan Meteorological Agency magnitude	Magnitude used by Japan Meteorological Agency
M_W	Moment magnitude	Hanks and Kanamori (1979)
M_{GR}	Gutenberg-Richter magnitude	Magnitude used in *Seismicity of the Earth*, Gutenberg and Richter (1954)
M_R	Rothe magnitude	Magnitude used in *The Seismicity of the Earth, 1953–1965*, Rothe (1969)
$M_{S\ PDE}$	Surface wave magnitude	Magnitude used in USGS preliminary determinations of epicentres catalogue
$M_{S\ ISC}$	Surface wave magnitude	Magnitude used in International Seismological Centre catalogue
$m_{B\ PDE}$	Body wave magnitude	Magnitude used in USGS preliminary determinations of epicentres catalogue
$m_{B\ ISC}$	Body wave magnitude	Magnitude used in International Seismological Centre catalogue
M_T	Tsunami magnitude	Abe (1989)
M_K	Kawasumi's magnitude	Kawasumi (1951)
M_U	Utsu magnitude	Magnitudes for earthquakes in Japan, 1885–1925, Utsu (1982)
M_C	Large earthquake magnitude	Purcaru and Berckhemer (1978)
M_N	Mantle wave magnitude	Brune and Engen (1969)

Modified from Kanamori (1983) and Utsu (2002)

Magnitude Measures, Table 2 Source parameters for some significant earthquakes

Earthquake	Date	Body wave magnitude m_B	Surface wave magnitude M_S	Fault area Length × Width (km^2)	Average dislocation (m)	Seismic moment M_0 (dyn.cm)	Moment magnitude M_W
San Fernando	1971	6.2	6.6	20 × 14 = 280	1.4	1.2×10^{26}	6.7
Loma Prieta	1989	6.2	7.1	40 × 15 = 600	1.7	3.0×10^{26}	6.9
San Francisco	1906		8.2	320 × 15 = 4,800	4.0	6.0×10^{27}	7.8
Alaska	1964	6.2	8.4	500 × 300 = 150,000	7.0	5.2×10^{29}	9.1
Chile	1960		8.3	800 × 200 = 160,000	21.0	2.4×10^{30}	9.5

After Stein and Wysession (2003)

Where:

R = Epicentral distance
A = Wave amplitude
K = Amplitude of the seismogram
T = Natural period of the seismogram

The Nuttli Magnitude Scale is used for epicentral distances >50 km and for instruments with a natural period smaller than 1.3 s. The scale has been used in preference to M_W for small to moderate earthquakes as the Moment Magnitude Scale is more difficult to estimate these low magnitude events.

Magnitude of Japanese earthquakes, M_{JMA}

The Japanese Meteorological Agency (JMA) has estimated the magnitude of shallow Japanese earthquakes utilising the following formula (Tsuboi, 1954):

$$M_{JMA} = Log\,(A_N^2 + A_E^2) + 1.73\,Log\,\Delta - 0.83$$

Where:

AN, AE = Maximum ground amplitude measured on the N – S and E – W components of horizontal Wiechert seismographs in JMA recording stations.

For deeper focus earthquakes in and around Japan Katsumata (2001) proposed a magnitude determination utilising regional velocity-amplitude data.

Relationship between scales

The vast majority of magnitude scales in use today stem from the one introduced by Richter in 1935. This scale has been extended by many seismologists to apply to data produced by various observational environments (Utsu, 2002). As new scales were developed they were in principle to provide equal value estimates to the same earthquakes or to the same earthquakes which radiated equal amounts of energy. However, systematic bias exists in the newly created scales when compared to the original

Richter model. Studies have demonstrated that there are systematic differences between M_L, M_S and m_B. A variety of scale interrelationship curves have been produced on order to compare and correlate various described and catalogued magnitudes. Utsu (2002) and Kanamori (1983) undertook a much more detailed analysis of various intra scale relationships (Figures 1–6, Tables 1, 2).

Bibliography

Abe, K., 1981. Magnitudes of large shallow earthquakes from 1904 to 1980. *Physics of the Earth and Planetary Interiors*, **27**, 72–92.

Abe, K., 1989. Quantification of tsunamigenic earthquakes by the Mt scale. *Tectonophysics*, **166**, 27–34.

Abe, K., and Kanamori, H., 1980. Magnitudes of great shallow earthquakes from 1953 to 1977. *Tectonophysics*, **62**, 191–203.

Aki, K., and Chouet, B., 1975. Origin of coda waves: source, attenuation and scattering effects. *Journal of Geophysical Research*, **80**, 3322–3342.

Boatwright, J., and Choy, G., 1986. Teleseismic estimates of the energy radiated by shallow earthquakes. *Journal of Geophysical Research*, **91**(B2), 2095–2112.

Bolt, B. A., 2006. *Earthquakes*. New York: W.H. Freeman and Company.

Boore, D. M., 1989. The Richter scale: its development and use for determining earthquake source parameters. *Tectonophysics*, **166**, 1–14.

Booth, D. C., 2007. An improved UK local magnitude scale from analysis of shear and Lg- wave amplitudes. *Geophysical Journal International*, **169**(2), 593–601.

Bormann, P., Baumbach, M., Bock, G., Grosser, H., Choy, G. L., and Boatwright, J., 2002. Seismic sources and source parameters, chapter 3. In Bormann, P. (ed.), *IASPEI New Manual of Seismological Observatory Practice*. Potsdam: GeoForschungs Zentrum Potsdam, pp. 1–94.

Brune, J. N., and Engen, G. R., 1969. Excitation of mantle Love waves and definition of mantle wave magnitude. *Bulletin of the Seismological Society of America*, **59**, 923–933.

Choy, G. L., and Boatwright, J. L., 1995. Global patterns of radiated seismic energy and apparent stress. *Journal of Geophysical Research*, **100**(B9), 18205–18228.

Choy, G. L., Boatwright, J. L., and Kirby, S., 2001. *The radiated seismic energy and apparent stress of interplate and intraplate earthquakes at subduction zone environments: implications for seismic hazard estimation*, USGS Open-File Report, 01–005, 10 pp.

Gutenberg, B., 1945a. Amplitudes of surface waves and magnitudes of shallow earthquakes. *Bulletin of the Seismological Society of America*, **35**, 3–12.

Gutenberg, B., 1945b. Amplitudes of P, PP, and S and magnitude of shallow earthquakes. *Bulletin of the Seismological Society of America*, **35**, 57–69.

Gutenberg, B., 1956. The energy of earthquakes. *Quarterly Journal of the Geological Society of London*, **112**, 1–14.

Gutenberg, B., and Richter, C. F., 1954. *Seismicity of the Earth*, 2nd edn. Princeton: Princeton University Press, 310 pp.

Gutenberg, B., and Richter, C. F., 1956. Magnitude and energy of earthquakes. *Annali di Geofisica*, **9**, 1–15.

Gutenberg, B., and Richter, C. F., 2010. Magnitude and energy of earthquakes. *Annals of Geophysics*, **53**, 7–12.

Hanks, T., and Kanamori, H., 1979. A moment magnitude scale. *Journal of Geophysical Research*, **84**(B5), 2348–2350.

Herrmann, R. B., 1975. The use of duration as a measure of seismic moment and magnitude. *Bulletin of the Seismological Society of America*, **65**, 899–913.

Hutton, L. K., and Boore, D. M., 1987. The M_L scale in Southern California. *Bulletin of the Seismological Society of America*, **77**(6), 2074–2094.

Kanamori, H., 1977. The energy release in great earthquakes. *Journal of Geophysical Research*, **82**, 2981–2987.

Kanamori, H., 1978. Quantification of earthquakes. *Nature*, **271**(5644), 411–414.

Kanamori, H., 1983. Magnitude scale and quantification of earthquakes. *Tectonophysics*, **93**, 185–199.

Kanamori, H., 1994. Mechanics of earthquakes. *Annual Review of Earth and Planetary Sciences*, **22**, 207–237.

Karnik, V., Kondorskaya, N. V., Riznichenko, Y. V., Savarensky, Y. F., Soloviev, S. L., Shebalin, N. V., Vanek, J., and Zatopek, A., 1962. Standardisation of the earthquake magnitude scales. *Studia Geophysica et Geodaetica*, **6**, 41–48.

Katsumata, A., 2001. Magnitude determination of deep-focus earthquakes in and around Japan with regional velocity-amplitude data. *Earth Planets Space*, **53**, 333–346.

Kawasumi, H., 1951. Measures of earthquake danger and expectancy of maximum intensity throughout Japan as inferred from the seismic activity in historical times. *Bulletin of the Earthquake Research Institute, University of Tokyo*, **29**, 469–482.

Lee, W. H. K., Bennett, R., and Meagher, K., 1972. A method of estimating magnitude of local earthquakes from signal duration. *USGS Open File Report*, 28 pp.

Marshall, P. D., and Basham, P. W., 1973. Rayleigh wave magnitude scale M_S. *Pure and Applied Geophysics*, **103**, 406–414.

Nuttli, O. W., 1973. Seismic wave attenuation and magnitude relations for eastern North America. *Journal of Geophysical Research*, **78**, 876–885.

Purcaru, G., and Berckhemer, H., 1978. A magnitude scale for very large earthquakes. *Tectonophysics*, **49**, 189–198.

Richter, C., 1935. An instrumental earthquake magnitude scale. *Bulletin of the Seismological Society of America*, **25**, 1–32.

Richter, C. F., 1958. *Elementary Seismology*. San Francisco/London: W. H. Freeman and Company. 768pp.

Rothe, J. P., 1969. *The Seismicity of the Earth 1953–1965*. Paris: Unesco.

Stein, S., and Wysession, M., 2003. *An Introduction to Seismology, Earthquakes, and Earth Structure*. Malden: Blackwell Publishing.

Tsuboi, C., 1954. Determination of the Gutenberg-Richter's magnitude of earthquakes occuring in and near Japan. *Journal of the Seismological Society of Japan, II*, **7**, 185–193.

Utsu, T., 1982. Relationships between magnitude scales. *Bulletin of the Earthquake Research Institute, University of Tokyo*, **57**, 465–497.

Utsu, T., 2002. 44 Relationships between magnitude scales. In Lee, W. H. K., Kanamori, H., Jennings, P. C., and Kisslinger, C. (eds.), *International Geophysics, International Handbook of Earthquake and Engineering Seismology*. London: Academic Press, Vol 81, Part 1, pp. 733–746, DOI:10.1016/S0074-6142(02)80247-9.

Vassiliou, M. S., and Kanamori, H., 1982. The energy release in earthquakes. *Bulletin of the Seismological Society of America*, **72**, 371–387.

Web Links

British Geological Survey
http://www.earthquakes.bgs.ac.uk/
http://www.bgs.ac.uk/schoolSeismology/
Natural Resources Canada
http://earthquakescanada.nrcan.gc.ca/index-eng.php
Japan Earthquake Information
http://www.jma.go.jp/en/quake/
United States Geological Survey

http://earthquake.usgs.gov/
IASPEI New Manual of Seismological Observatory Practice, Chapter 3, Seismic Sources and Source Parameters. 10.2312/GFZ.NMSOP_r1_ch3
IASPEI New Manual of Seismological Observatory Practice, Glossary
http://ebooks.gfz-potsdam.de/pubman/item/escidoc:4141:2
Hiroo Kanamori
John E. and Hazel S. Smits Professor of Geophysics, California Institute of Technology
http://web.gps.caltech.edu/faculty/kanamori/kanamori.html
Earthquake Seismometer Equations and Formulas Calculator
http://www.ajdesigner.com/phpseismograph/earthquake_seismometer_richter_scale_magnitude.php
International Seismology Centre
http://www.isc.ac.uk/

Cross-references

Accelerometer
Building Codes
Earthquake
Epicentre
Haiti Earthquake 2010 Psychosocial Impacts
Harmonic Tremor
Hypocentre
Indian Ocean Tsunami
Intensity Scales
Isoseismal
Mercalli, Giuseppe (1850–1914)
Primary Wave
Richter, Charles (1900–1985)
Secondary Wave (S Wave)
Seismograph/Seismometer
Seismology
Tangshan China (1976 Earthquake)
Tectonic Tremor
Tohoku, Japan, Earthquake, Tsunami and Fukushima Accident (2011)
Wenchuan, China (2008 Earthquake)

MARGINALITY

Ben Wisner
Oberlin College, Oberlin, OH, USA
University College London, UK

Synonyms

Discrimination; Exclusion

Definition

Marginality is a socio-spatial process of great importance in understanding and combating vulnerability to natural hazards. It severely limits the political voice and participation, economic and livelihood options, access to resources and information, as well as locational decisions of sub-groups within society. Caste, class, religious minority, and immigration status are often underlying causes of marginality.

Discussion

Groups in society may live in places that are spatially peripheral to the majority or live in conditions that severely limit their participation in decisions that affect their lives as well as their access to resources and information. Such conditions are sometimes invisible to the majority. In a disaster, such groups often suffer greater death, injury, and economic loss (as a proportion of their already limited assets), and experience difficulty recovering. In 1978, Wisner used the term eco-demographic marginality to describe the situation of semi-pastoral people on the lower slopes of Mt. Kenya, who were politically powerless, lived in an environment undergoing degradation, and whose livelihoods depended on crops and animals of low and fluctuating value in the market. Blaikie and Brookfield (1987) adopted and subsequently expanded Wisner's notion of marginality.

Marginality is a concept with considerable utility in vulnerability assessment and planning for disaster risk reduction as well as recovery planning. Because it embraces numerous aspects of situations "on the edge," both professional planners and focus groups composed of lay people may use it to identify groups and situations that would normally not receive attention when policy, plans, and projects are focused on the needs and capabilities of the "average" person or household. Many methods such as wealth ranking exist that facilitate focus group discussion of marginality (ProVention, 2010), and this kind of situation-specificity is vital to effective project planning and programming (Wisner, 2004). It also provides understanding of what Chambers (1983) called the "deprivation trap," and thus may add a degree of reality to sometimes overly optimistic interventions that assume, for example, that everyone has time to volunteer in self-help activities or that every adult understands what it is to lobby government. Reasons for social marginality include caste, occupational, class status; religion and ethnicity; immigration status; disability; sexual orientation; and in some societies, gender and age. Political marginality may overlap with the social, but may also reflect favoritism practiced by ruling parties and historically developed center–periphery divisions of national territory. Economic marginality may be due to land and resource allocations and market dynamics that exclude or burden some, while benefiting others. However, as Perlman noted (1976), this does not imply that an economy is "dual" – a modern economy side by side with pre-modern. Indeed, in many places, marginal people are exploited for their cheap labor or commodities, and this is a reason why marginality persists and underlies much of what the United Nations (2009) has called "extensive risk" in the face of extreme natural events.

Bibliography

Blaikie, P., and Brookfield, H., 1987. *Land Degradation and Society*. London: Routledge Kegan and Paul.
Chambers, R., 1983. *Rural Development: Putting the Last First*. London: Longman.

Perlman, J., 1976. *The Myth of Marginality*. Berkeley, CA: University of California Press.

ProVention Consortium, 2010. *Community Risk Assessment Tool Kit* http://www.proventionconsortium.org/?pageid=39.

United Nations Intergovernmental Secretariat for Disaster Reduction, 2009. *Global Disaster Assessment 2009*. Geneva: UN-ISDR http://www.preventionweb.net/english/hyogo/gar/report/index.php?id=1130&pid:34&pih:2.

Wisner, B., 1978. *The Human Ecology of Drought in Eastern Kenya*. PhD dissertation, Worcester, MA, Clark University.

Wisner, B., 2004. Assessment of capability and vulnerability. In Bankoff, G., Frerks, G., and Hilhorst, T. (eds.), *Vulnerability: Disasters, Development and People*. London: Earthscan, pp. 183–193.

Cross-references

Crtical Incidence Stress Syndrome
Disaster Diplomacy
Disaster Relief
Disaster Risk Management
Disaster Risk Reduction (DRR)
Emergency Management
Emergency Planning
Exposure to Natural Hazards
Global Network of Civil Society Organizations for Disaster Reduction
Human Impact of Hazards
International Strategies for Disaster Reduction (IDNDR and ISDR)
Planning Measures and Political Aspects
Post-traumatic Stress Disorder (PTSD)
Psychological Impacts of Natural Disasters
Red Cross/Red Crescent, International Federation of
Risk
Sociology of Disasters
Susceptibility
Vulnerability

MARINE HAZARDS

Tore Jan Kvalstad
Norwegian Geotechnical Institute, Oslo, Norway

Synonyms

Offshore geohazards; Submarine hazards

Definition

Marine geohazard. Geological site and soil conditions in the ocean bottom representing a potential source of harm.

Introduction

Marine "geo*hazards*" (see entry *Geohazards*) are related to geological processes in the marine environment that have created regional or local site and soil conditions with a potential of developing into failure events that could cause loss of life or damage to health, environment, or assets. The failure events can be tectonic seabed displacements, seabed accelerations, and seabed instabilities ranging from local slumping to large-scale slope instability involving mass movement and debris flow and turbidity currents. Rapid, large-scale seabed displacements and downslope mass transport may generate tsunamis. Failure events where expulsion of gas, oil, water, and mud may flow uncontrolled from overpressured submarine reservoirs are often related to oil and gas production, but may also occur naturally through fractures and seeps to seabed and submarine mud volcanoes.

The event-triggering sources can be ongoing geological processes or human activities that change the seabed conditions or affects deeper strata mechanically or by pressure and temperature changes.

Marine hazards are of concern for the offshore petroleum industry with huge investments in wells, offshore structures, flowlines, and pipelines, but may also affect infrastructure related to telecommunications and electric energy transmission cables, the rapidly growing offshore wind power industry as well as fisheries. Also communities, industries, and infrastructure in the near-shore and shoreline area can be affected by submarine slide events reaching the shoreline, by earthquake or slide-generated tsunamis, and also by pollution from natural seeps and uncontrolled expulsion of oil.

Geological processes

Consideration of large-scale geological processes like "*plate tectonics*" and long-term climate changes are important for evaluation of marine "*hazards*." The major part of subduction zones where the oceanic crust is underthrusting continental plates is located in the oceans. This is where the most destructive earthquakes occur and the associated change in seabed level may generate tsunamis.

Long-term climate changes, especially during the last part of the Pleistocene, led to repeated "*sea level changes*" of more than 100 m. This affected the coastal zones and the continental shelves and margins on a global basis. Glacial erosion and transport of terrigenous sediments to the shelves and over the shelf edge by grounded glaciers to the continental slopes led to rapid progradation of the continental shelves along northern part of the Atlantic Ocean during glacial periods. The continental shelves and shallow water areas elsewhere were severely affected by changes in water depth and shoreline position, leading to wave, current, and river erosion and suspension of sediments. The finer fractions were transported seaward with tidal and wind-driven currents and the coarse grained sediments as hyperpycnal and turbid flows toward and locally over the shelf edge to the continental slopes.

Regional geological conditions and processes control the sedimentation rate, the thickness, and the type of marine sediments. The major river deltas of the world and the glacial fans on the margins along the North Atlantic and Arctic Seas are areas dominated by high sediment input that may lead to a combination of sloping seabed and overpressured sediments prone to slope instability and also representing a hazard for drilling operations for the petroleum industry. (In overpressured sediments, the ground water pressure is higher than hydrostatic pressures.)

Overpressures may also be generated by diagenetic changes of minerals under increased pressure and temperature transforming the mineral structure into a denser configuration under expulsion of excess water. Overpressured clayey sediments have generally lower strength, are less dense, and are more easily deformed than fully consolidated sediments.

Earthquakes

Major "*earthquakes*" originating in the oceanic subduction zones may generate enormous tsunami catastrophes like the December 26, 2004, Sumatra event (see entry *Tsunami*) and the March 11, 2011, Tohoku events in the west cost of Japan.

For marine structures and installations, the "*earthquake*" generated ground accelerations may cause damage in the same way as for buildings and structures on land. Severe earthquakes may also trigger submarine slope failures as the sediment strength can be reduced due to cyclic stress variations during the earthquake shaking. In a worst-case scenario, the slide event may transform to a tsunami generating mass flow and cause damage to marine installations and infrastructure in the slide initiation area and in the pathway of mass flow.

Earthquake-induced fault displacements may deform and damage well casings, pipelines, cables, and structures located at or crossing the fault.

An induced earthquake is a term that is assigned to human-induced seismicity. In the marine environment, this is mainly connected to microseismicity caused by extraction of oil and gas leading to reservoir compaction, changes in the stress conditions in the reservoir and overburden sediments and along faults. With increasing reservoir compaction the likelihood of larger displacements and damage to well casings increase.

Sediment strength and pore water pressure
Slope stability

The stability of the seabed depends on the strength of the sediments relative to the destabilizing forces. In a slope the shear strength of the soil will have to exceed the downslope component of gravity to prevent slope failure. If other external forces (like inertia forces under earthquake loading) are acting, even higher strength will be required. Submarine slide events can be initiated either by increased downslope loading, steepening of the slope by top accumulation or toe erosion, and reduction of the shear strength of the sediments under monotonic or cyclic shear stress variations.

Soils most susceptible to large-scale instability are marine sediments with a loose mineral grain structure. These sediments are typically hemipelagic clays and sands deposited at high sedimentation rates causing overpressure generation, lower effective stresses, and thus lower strength. These soils are susceptible to increase in pore water pressure and reduced strength when subjected to rapid changes in shear stress. The combination of excess pore pressure from rapid sedimentation and pore water pressure increase during undrained shearing is the main factor in development of submarine slide events.

Enormous submarine slide area have been mapped on the continental slopes, especially in and near in the major river deltas, Nile, Niger, Amazon, etc. and glacial fans. The slope angle is typically very low, from less than 1° to a few degrees. The understanding of the geomechanical processes involved in the triggering and development of these slide events is a key element in evaluation of marine slide hazards.

Submarine landslides are generally much larger than onshore landslides (Brunetti et al., 2009). While the larger terrestrial landslides are found to fall in the range 10^6–10^7 m^3, the larger submarine slide events are reported to have volumes of several 1,000 km^3. This is due to the long-term sedimentation under stable conditions not affected by yearly climate variations, but more dependent on the major sea level variations over 100,000 years.

The Storegga Slide is one of the largest submarine slide events worldwide. It is located at the mouth of the Norwegian trench next to and partly cutting into the North Sea Fan, a major glacial depocenter. The upper slide scar has a length of about 300 km, the downslope extension of the slide area is about 250 km, and the run-out distance of slide debris and turbidites is about 800 km. The estimated slide volume is in the range 3,000–3,500 km^3. The Storegga Slide was mapped and investigated in much detail as the Ormen Lange gas field was located in the slide scar (Solheim et al., 2005). The slide event took place about 8,200 calendar years before present and generated a major tsunami hitting the coastline of Norway, Scotland, the Faeroes, and Shetland (Bondevik et al., 2005).

The average slope angle from the toe area to the top of the upper slide scar is about 0.6°, and the slide event can be explained by existence of overpressures, a retrogressive slide process, and the sensitivity of the marine clays that formed the preferred slip planes (Kvalstad et al., 2005).

The long run-out distance of submarine landslides leads to extensive hazard zones in downslope direction and is a major source of concern for subsea installations, pipelines, and cables located below potential slide areas.

Retrogressive slide development is also observed, where the slide scar progressively moves upslope over distances of tens of kilometers.

Mud diapirs and mud volcanoes

Overpressured soils will typically have lower strength than soils that are fully consolidated under the weight of the overburden sediments, i.e., hydrostatic pore pressure conditions. This may lead to development of deep-seated failure processes (*Deep-seated Gravitational Slope Deformation*) under the delta front where there is a decrease in overburden stress in seaward direction. This gradient in overburden stress leads to compression and formation of anticlines in the toe area of the delta and growth fault over the effects of global warming have renewed interest in solifluction processes.

Surficial mass movements

Even though dry ravel and dry creep are surface processes, they are considered mass movements because they are driven by gravity. These surficial mass movements involve the rolling, sliding, and bounding of surface soil grains, aggregates, and coarse fragments down steep hillsides, often forming talus cones at slope breaks (Sidle and Ochiai, 2006). Dry ravel and creep mainly initiate during active freeze-thaw periods and wetting-drying cycles. During such natural perturbations a loss of interlocking frictional resistance among soil aggregates or grains occurs loosening the material and subjecting it to downslope gravitational transport. While dry ravel and dry creep typically transport much less sediment to streams compared to other mass movements in steep terrain, they can be significant surficial processes on steep slopes with sparse vegetation covers, thin organic horizons, and/or soils that have been disturbed (particularly by fire). Under such conditions, slope gradients that approach or exceed the internal angle of friction of surface materials ($\approx 38°-41°$) typically experience active dry ravel and creep; on gentler slopes, ravel rates diminish substantially. The impacts of these surficial mass movements are generally restricted to maintenance requirements along road cuts, but in cases of extreme and widespread fire they can contribute significant sediment pulses to streams. Additionally, dry ravel is often an important infilling process after evacuation of geomorphic hollows by shallow landslides.

Summary

Mass movements are largely episodic processes driven by gravity that can severely impact people, property, and the environment depending on their location, size, and rate of movement. These are largely triggered by rainfall, but devastating mass movements are sometimes caused by earthquakes and volcanic activity. Certain mass movement processes like soil creep can exacerbate other processes like large-scale landslides, and combination mass movements (e.g., debris slides-avalanches-flows) are common occurrences. Land use practices can exacerbate particularly shallower mass movements, with road and other excavations into hillsides being particularly problematic.

Bibliography

Chigira, M., 2002. The effects of environmental changes on weathering, gravitational rock deformation and landslides. In Sidle, R. C. (ed.), *Environmental Change and Geomorphic Hazards in Forests*. Wallingford, Oxen: CABI Publishing. IUFRO Research Series, Vol. 9, pp. 101–121.

Cruden, D. M., and Varnes, D. J., 1996. Landslide types and processes. In Turner, A. K., and Schuster, R. L. (eds.), *Landslides – Investigation and Mitigation*. Washington, DC: National Academic Press. Special Report 247, pp. 36–75.

Sidle, R. C., and Ochiai, H., 2006. *Landslides: Processes, Prediction, and Land Use*. American Geophysical Union, Washington, DC. American Geophysical Union, Water Resource Monograph 18, 312 pp.

Varnes, D. J., 1978. Slope movement types and processes. In Clark M. (ed.), *Landslide Analysis and Control*. Washington, DC: Transportation Research Board, National Academy of Science, National Research Council. Special Report 176, pp. 11–33.

Cross-references

Casualties Following Natural Hazards
Collapsing Soil Hazards
Creep
Debris Avalanche (Sturzstrom)
Debris Flow
Deep-Seated Gravitational Slope Deformations
Geohazards
Lahar
Land Use, Urbanization and Natural Hazards
Landslide Impacts
Landslide (Mass Movement)
Landslide Inventory
Landslide Types
Lateral Spreading
Liquefaction
Mudflow
Pore-Water Pressure
Quick Clay
Rock Avalanche (Sturzstrom)
Rockfall
Slide and Slump
Slope Stability

MEGACITIES AND NATURAL HAZARDS

Norman Kerle[1], Annemarie Müller[2]
[1]Faculty of Geo-Information Science and Earth Observation (ITC), University of Twente, Enschede, The Netherlands
[2]Helmholtz-Centre for Environmental Research (UFZ), Leipzig, Germany

Synonyms

Megalopolis; Megapolis

Definition

Megacities are typically defined as metropolitan areas with more than ten million inhabitants, which show high growth dynamics and a high speed of change and development. While not necessarily facing substantially different or more severe hazards than other settlement types, their high concentration of administrative and economic functions can lead to risk that extends to the national level. However, megacities also have the best resource base to mitigate risk, and prepare for and recover from disaster events.

Introduction

The global population has been growing continuously since about the fourteenth century (Raleigh, 1999), though by no means evenly. An abrupt increase in what had been a relatively low and steady growth rate only began toward the middle of the twentieth century, reaching a brief peak increase of some 2.2% in the early 1960s. While it had taken some 160 years for the population to grow from one to three billion (by 1960), this doubled over the next 40 years. Today, the number stands at 7 billion, and is projected to grow to 9.1 billion by 2050. This rapid increase coincided with a second trend – people moving into cities. While in 1800 about 3% of the global population was urbanized, a strong acceleration began by the beginning of the twentieth century (ca. 9%), reaching 50% by 2008 (United Nations Population Division (UNPD), 2006). Within this broad urbanization process, a number of individual cities grew disproportionally. In 1900, only 15 cities had more than one million inhabitants, three of them (London, Paris, New York) above three million (Wenzel et al., 2007). Today, more than 300 cities house in excess of one million people. Adjusting for the trend, we now define megacities as agglomerations of more than ten million people (Thouret, 1999), though at times a threshold of eight million is used (Wenzel et al., 2007). By the year 2000, already 19 cities with populations of more than ten million existed. Only 5 years later already 25 such cities were identified (Brinkhoff, 2010), with greater Tokyo (approximately 34 million people in 2010) being the largest.

The global population and urbanization growth rates over the last century (Grimm et al., 2008) also show similarities with an increasing number of disasters associated with natural hazards (e.g., Guha-Sapir et al., 2004; Figure 1) that are marked by an even more pronounced increase since about the 1960s. This suggests a relationship between population growth and urbanization with disaster incidence and damage. Considered in general terms, the coincidence is readily explained by disaster risk theory – more elements at risk (people, infrastructure, assets), exposed to (even unchanging) hazards, will likely lead to more frequent damaging events and higher losses. At a detailed level, the picture is more complicated, as specific hazard exposures and vulnerabilities have to be considered (e.g., Kerle and Alkema, 2012; see entry *Risk*). In this entry, the particular relationship between natural hazards and disasters and megacities is described, considering the role of actual hazard exposure, vulnerability, and resilience and capacity, also in light of global climate change. It particularly highlights that the absolute number of inhabitants is less relevant; instead, the functional value, as well as the political, administrative, and economic importance of megacities in their respective countries now define megacities and strongly influence the hazard risk.

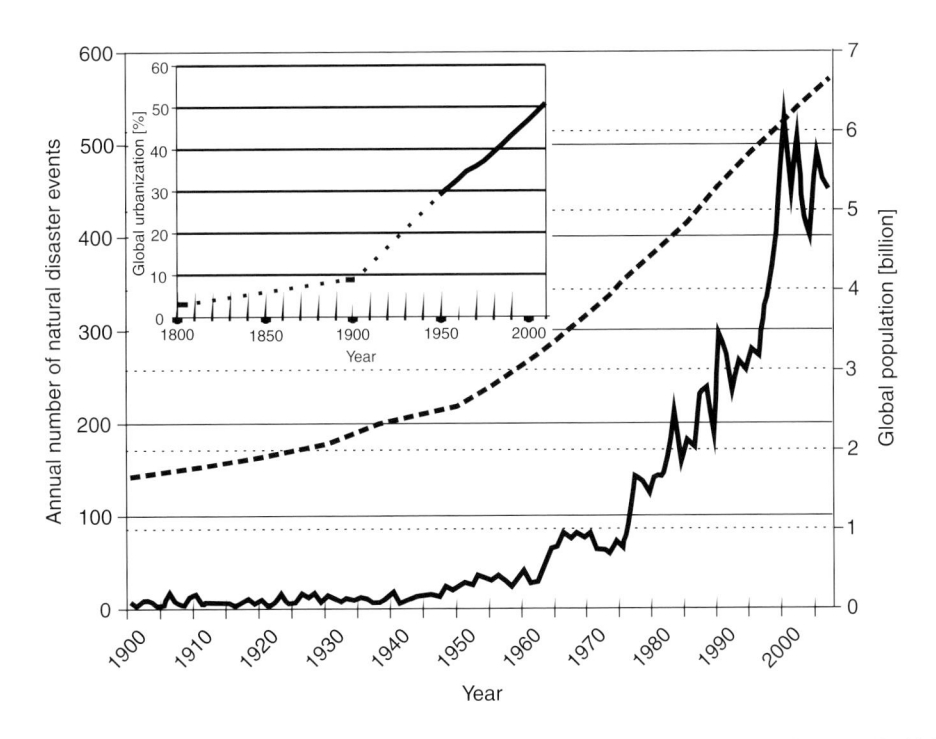

Megacities and Natural Hazards, Figure 1 Number of annual natural disasters between 1900 and 2007 in bold (primary axis, CRED 2009), and global population for the same period in hatched line (secondary axis). Inset shows estimated global urbanization rates between 1800 and today (United Nations Population Division (UNPD) 2006).

Hazard exposure of megacities

Of the 25 currently existing megacities, only six are not located in economically less developed countries (LDC). About half are exposed to substantial seismic hazard (Jackson, 2006), and all except six are situated in coastal areas (Figure 2). Those hazardous locations, however, they share with many smaller population centers. Megacities tend to occupy large areas (e.g., the Los Angeles metropolitan area covers more than 12,500 km^2). As such, given a comparable hazard setting, they are statistically more likely to get affected by an event than smaller cities or even rural communities. At the same time, a given event will likely affect a smaller fraction of a megacity area than it would in smaller cities or communities (Cross, 2001). Thus, in terms of direct exposure to environmental hazards, megacities do not show characteristics that significantly differ from smaller settlement types.

Disaster damage and the number of people killed or affected have been increasing in recent decades. Disaster statistics show that events affecting megacities have led to the highest monetary damages, such as the 1995 Kobe earthquake (part of greater Osaka; losses of > US$130 billion), reflecting the high accumulation of wealth. While some of these events have also killed many people (more than 6,000 during the Kobe earthquake), disasters outside megacities have been more devastating. The 1965–1967 drought in India caused some 1.5 million fatalities, whereas in 1970, a cyclone inundating coastal areas of Bangladesh killed an estimated 500,000 people. Disaster numbers, however, are strongly dependent on the specific location and extent of a disaster, and include an element of chance. With the exception of the 1923 Tokyo, 1976 Tangshan, and 1980 Mexico City earthquakes, major urban agglomerations have so far been spared by seismic events with a magnitude >7.5. However, an eventual direct hit of a megacity is seen as inevitable (Jackson, 2006), and capable of causing more than one million fatalities (Bilham, 2009). Whether a direct tsunami hit on a coastal megacity will lead to high fatality numbers or mostly infrastructure damage largely depends on the warning time. For tropical cyclones and impending volcanic eruptions, the other environmental hazards with destructive potential in megacities, the time to prepare is usually sufficient.

Damage is more usefully considered in relative rather than absolute terms. While perhaps causing less absolute physical damage in rural areas, the destroyed assets nevertheless often constitute a significant share of all possessions, especially in LDCs. Thus, in terms of economic consequence, less costly disasters outside megacities frequently have more severe and lasting effects than in large urban agglomerations that have broader means for rapid recovery.

The effect of megacities on hazard exposure

In addition to megacities encroaching on hazardous terrain, a range of environmental changes has been documented. Ongoing and projected climate changes strongly affect various aspects of the environmental system, with consequences for hazards levels. They relate in particular to hydrometeorological hazards, such as stronger windstorms, flooding, and general precipitation

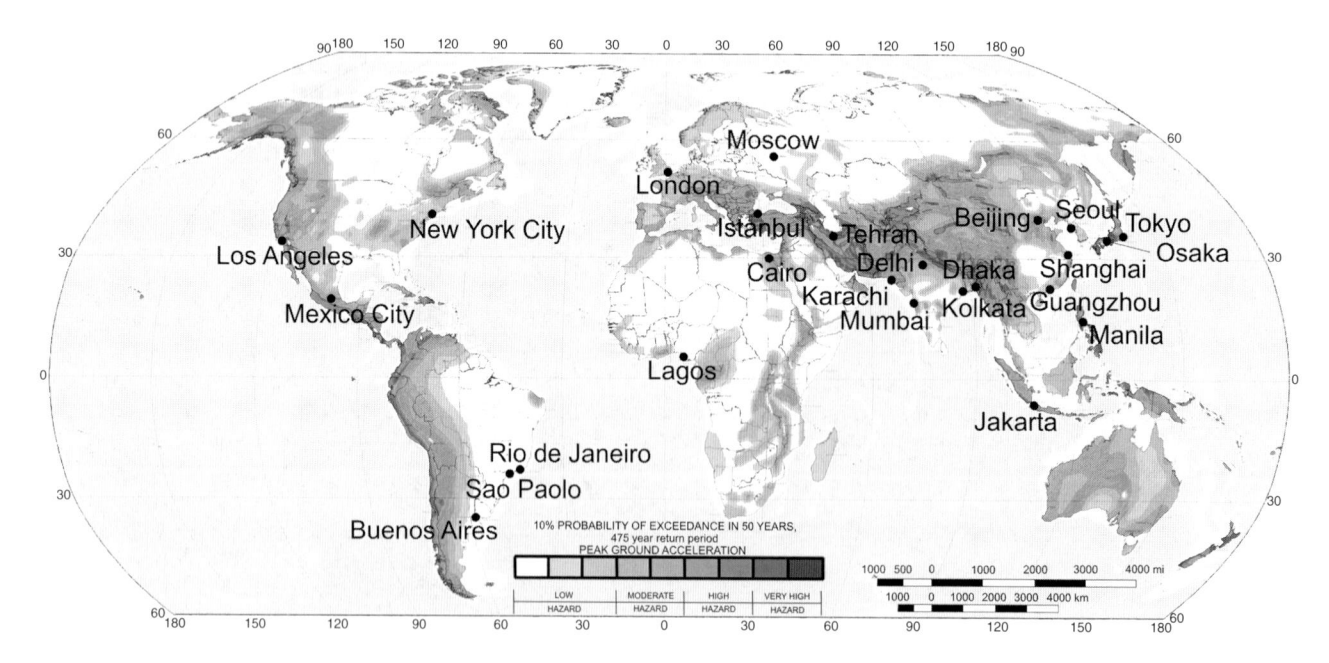

Megacities and Natural Hazards, Figure 2 Global seismic hazard map, adapted from Global Seismic Hazard Assessment Program (GSHAP), and current megacities. About half of those are exposed to substantial seismic hazard (Jackson, 2006), and all except Sao Paulo, Mexico City, Delhi, Beijing, Moscow, and Tehran are located in coastal areas.

regime changes. Megacities themselves can also have effects on the hazards they are exposed to. Those can be effectively considered in the framework of urban ecology, which displays strong similarities with disaster risk theory (Kerle and Alkema, 2012). Several observations from an urban ecological perspective offer insights in the hazard exposure of large urban areas: (1) cities are seen as both the cause and the principal victim of environmental degradation (Weiland and Richter, 2009). As a major source of pollution, and due to their extensive resource requirements and energy consumption, they contribute to global climate change. However, with their high concentration of elements at risk and frequent location in coastal areas, these cities are also poised to be most affected by sea-level rise or stronger windstorms (Klein et al., 2003). (2) Global environmental changes are outpaced by local changes (Grimm et al., 2008). For example, urban temperature increases (urban heat islands) are faster than global warming rates, leading to rapidly rising secondary hazards (e.g., new disease vectors spreading, or increased ozone concentrations). (3) Major urban areas have ecological footprints hundreds of times their size, typically also evidenced by changes in their surrounding land cover and land use (Grimm et al., 2008). Hazard sources can be potentially far away, and the characteristics of the area in between strongly affect not only the hazard, but also vulnerability and capacity (e.g., widespread deforestation or river straightening versus comprehensive floodplain management). The threat of projected sea-level rises endangering megacities in coastal areas is at times compounded by large-scale subsidence, typically resulting from excessive groundwater extraction, such as in Jakarta or Bangkok.

Do megacities face megarisks?

Whether megacities automatically face disproportionate disaster risks has been a matter of intense scientific debate. If megacities are not exposed to exceptional hazards compared to smaller settlements in comparable locations, what else determines their risk? Risk is principally a function of all present hazards and their potential interactions and amplifications, and the type, value, and vulnerability of all elements at risk (see entry *Risk*). Vulnerability, that is, the susceptibility to suffer loss (see entry *Vulnerability*), which differs for physical assets, people, and their social structures, and economic and environmental systems, is further offset by capacity. This is defined as "the combination of all the strengths and resources available within a community, society or organization that can reduce the level of risk, or the effects of a disaster" (UN/ISDR 2004, p. 430). Here, it becomes apparent that risk in megacities is much less a function of the absolute population number, but more of its complexity and development level (Hansjürgens et al., 2008). While a city such as Tokyo, with an exceptional physical asset base, faces a high seismic hazard, the actual risk is limited because of great efforts to reduce vulnerability (e.g., by imposing strict building codes), and to increase the capacity of the city (e.g., by empowering the population on how to respond in a seismic situation). In particular, megacities in LDCs face higher risks (Cross, 2001; Wenzel et al., 2007). This is not only due to these cities being located in poorer countries with fewer means for risk mitigation measures. Instead, the trajectory of urban development is of major importance. While most megacities in richer countries grew over centuries, allowing time for support infrastructure to develop, those in LDCs experienced their most rapid growth in recent decades. For example, while the population of greater London already exceeded six million by 1900 and since then remained largely unchanged, Mumbai grew from some 800,000 to over 23 million in the same period. This led to infrastructure and functional development drastically lagging behind urban expansion, and explains why some 60% of Mumbai's residents live in informal settlements (Wenzel et al., 2007), which are widely considered to be more vulnerable to hazards. Another point influencing risk is the exceptional importance of megacities in LDCs. While all western megacities are important economic, political, and administrative centers in their respective countries, they are not primate cities comparable to Manila, Lagos, or Jakarta. This, in turn, influences risk positively and negatively. While a disaster in a western megacity would lead to substantial damage, and potentially national and international repercussions, it is unlikely to compromise the ability of the respective country to function economically or administratively, as critical functions are decentralized and some level of redundancy exists. Megacities in LDCs tend to have far higher concentrations of economic, political, and administrative power, and as such are more vulnerable to disruption affecting the entire country (Hansjürgens et al., 2008). The risks such megacities face are, therefore, to some extent nationwide risks. On the other hand, their singular importance also facilitates acquisition of resources needed for disaster response and reconstruction, at the expense of the rest of the country.

The assets of megacities

Many megacities, especially those that grew rapidly in recent years, are characterized by haphazard construction, insufficient infrastructure, unhygienic environments, and inadequate administrative and medical services, all with negative effects on vulnerability and capacity. Those limitations, however, are in part counterbalanced. In addition to the comparatively high ability to obtain resources for disaster response and reconstruction, their status as primate city leads to accumulation of knowledge and expertise, and a comparatively better knowledge of the existing hazards and risk. In addition, they allow an easier early warning of the population, a more timely response following an event (both with national means and international assistance), and in principle are better equipped to empower people on vulnerability reduction and disaster preparedness. The per-person cost of any risk reduction approach, be it engineering measures or installation of early warning infrastructure, is also much lower than in smaller settlements. Creating more effective disaster risk management strategies, which have to draw on all elements of the political, administrative, and societal fabric of a city, is also facilitated by the high concentration of these elements in megacities.

Megacities and future disaster risks

The trend toward more and larger megacities is clear, with positive and negative consequences for disaster risk, posing especially high challenges for large agglomerations in LDCs. Given the generally high disaster risk, what can be effectively done to reduce it? Any form of risk mitigation and management is contingent on a solid understanding of existing risk. This is difficult as it has to include all present hazards and vulnerability types, as well as account for any present trends related to environmental degradation or climate change. This risk knowledge then forms the basis for sustainable urban development. Such planning has been performed for several megacities, such as Dhaka (Roy, 2009), Santiago de Chile (Heinrichs et al., 2012), or Istanbul (Wenzel et al., 2007), and broad recommendations for climate change adaptation in such settings have been made (Klein et al., 2003). The planning has to be integrative and consider the wider geographic setting. Given the reliance of resilient megacities on a healthy hinterland (Cross, 2001), the focus must not only be on reducing risk within the cities themselves. It is equally important to take measures that reduce the massive rural–urban migration that has been leading to a demographic imbalance that endangers the rural resource supply megacities depend on. The urban agglomerations also have to be surrounded by healthy ecosystems. Overall resilience, that is, the capacity to absorb shocks from disasters and recover, relies on proper functioning and interlinking of both human and ecological systems (Cross, 2001). As such, urban ecology considers integrative, transdisciplinary analysis of the diverse environmental, social, and political aspects as being central to urban disaster risk management, especially in megacities.

Bibliography

Bilham, R., 2009. The seismic future of cities. *Bulletin of Earthquake Engineering*, **7**, 839–887.

Brinkhoff, T., 2010. *The Principal Agglomerations of the World*. Available from World Wide Web: http://www.citypopulation.de/World.html.

CRED, 2009. *EM-DAT: The OFDA/CRED International Disaster Database*. Available from World Wide Web: www.em-dat.net.

Cross, J. A., 2001. Megacities and small towns: different perspectives on hazard vulnerability. *Environmental Hazards*, **3**, 63–80.

Grimm, N. B., Faeth, S. H., Golubiewski, N. E., Redman, C. L., Wu, J. G., Bai, X. M., and Briggs, J. M., 2008. Global change and the ecology of cities. *Science*, **319**, 756–760.

Guha-Sapir, D., Hargitt, D., and Hoyois, P., 2004. *Thirty Years of Natural Disasters 1974–2003: The Numbers*. Brussels: University of Louvain Presses. Center on Epidemiology of Disasters (CRED).

Hansjürgens, B., Heinrichs, D., and Kuhlicke, C., 2008. Mega-urbanization and social vulnerability. In Bohle, H. G., and Warner, K. (eds.), *Megacities. Resilience and Social Vulnerability*. Bonn, Germany: United Nations University – Institute for Environment and Human Security (UNU-EHS).

Heinrichs, D., Krellenberg, K., Hansjürgens, B., and Martinez, F., 2012. *Risk Habitat Megacity. The Case of Santiago de Chile*. Berlin: Springer.

Jackson, J., 2006. Fatal attraction: living with earthquakes, the growth of villages into megacities, and earthquake vulnerability in the modern world. *Philosophical Transactions of the Royal Society A – Mathematical Physical and Engineering Sciences*, **364**, 1911–1925.

Kerle, N., and Alkema, D., 2012. Multi-scale flood risk assessment in urban areas – a geoinformatics approach. In Richter, M., and Weiland, U. (eds.), *Applied Urban Ecology: A Global Framework*. Oxford, UK: Blackwell.

Klein, R. J. T., Nicholls, R. J., and Thomalla, F., 2003. Resilience to natural hazards: how useful is this concept? *Environmental Hazards*, **5**, 35–45.

Raleigh, V. S., 1999. Trends in world population: how will the millenium compare with the past? *Human Reproduction Update*, **5**, 500–505.

Roy, M., 2009. Planning for sustainable urbanisation in fast growing cities: mitigation and adaptation issues addressed in Dhaka, Bangladesh. *Habitat International*, **33**, 276–286.

Thouret, J. C., 1999. Urban hazards and risks: consequences of earthquakes and volcanic eruptions: an introduction. *GeoJournal*, **49**, 131–135.

UN/ISDR (United Nations/International Strategy for Disaster Reduction), 2004. *Living with Risk: A Global Review of Disaster Reduction Initiatives*. New York: UN/ISDR.

United Nations Population Division (UNPD), 2006. *World Urbanization Prospects: The 2005 Revision*. New York: United Nations.

Weiland, U., and Richter, M., 2009. Lines of tradition and recent approaches to urban ecology, focussing on Germany and the USA. *GAIA – Ecological Perspectives for Science and Society*, **18**, 49–57.

Wenzel, F., Bendimerad, F., and Sinha, R., 2007. Megacities – megarisks. *Natural Hazards*, **42**, 481–491.

Cross-references

Building Codes
Buildings, Structures, and Public Safety
Climate Change
Coastal Zone, Risk Management
Costs (Economic) of Natural Hazards and Disasters
Damage and the Built Environment
High-Rise Buildings in Natural Disasters
Integrated Emergency Management System
Resilience
Risk Assessment
Tangshan, China (1976 Earthquake)
Vulnerability
Worldwide Trends in Disasters Caused by Natural Hazards

CASE STUDY

MEGA-FIRES IN GREECE (2007)

George Eftychidis
Algosystems S.A., Kallithea, Greece
Pangaiasys Ltd., Pikermi, Greece

Synonyms

Greek fires; Mega-fires; Very large wildfires

The fire management policy in Greece toward the summer of 2007

Forest fire is a major natural hazard in southern Europe, which is often directly related to climate change and anomalies of meteorological conditions, in particular increased temperature and scarcity of rainfall. Long dry periods combined with other extreme weather conditions contribute to the development of forest fires that in most cases originate by anthropogenic activity and often turn into very large conflagrations. Such fires can easily burn down large forest areas, as evident in particular in the Mediterranean region.

Greece is one of the EU countries most affected by the forest fires. Areas approximately 1,850,000 ha in size have been burned between 1955 and 2007, out of which 30% was burned during the last 7 years of this period. Up to 1973, fires used to occur with a relative low frequency and the average per annum area burned was 11,500 ha. One third of this area was classified as tall forests, mainly pine stands, whereas the remaining area was shrublands, pastures, and grasslands. Starting from 1974, the annually burned area increased rapidly peaking every 3–4 years (influenced by the combination of periodic favorable climatic conditions and societal fire causes). For instance, the area burned in 1974 was 36,000 ha, in 1977 some 49,000 ha, in 1985 about 80,000 ha, and in 1985 >100,000 ha (Eftichidis, 2007).

A significant increase of the burned area was recorded following the 1998 policy shift for fire suppression to the fire brigades from the forest service. This decision marked a clear change in fire management policy in Greece. Aggressive fire suppression succeeded the preventive forest management strategy previously applied with the objective of mitigating fire behavior and impact. Unfortunately, fires continued making new national records in the years 1998 (102,000 ha) and 2000 (157,000 ha). For a period of 6 years following the record year 2000, forest fires have been controlled effectively by applying a focused and aggressive fire suppression policy, giving the impression to the citizens that the problem was being managed properly. Figure 1 summarizes the statistics for fire suppression in the Mediterranean.

During the summer of 2007, following a long dry season, a series of fires started burning the unmanaged shrublands and pine forests in southern Greece and Peloponnese.

Fanned by favorable weather conditions and a significant volume of accumulated biomass, the 2007 fires in Peloponnese evolved to catastrophic mega-fires that burned >180,000 ha in 1 week with intensities far exceeding the capabilities of the firefighting infrastructure, including the addition of an unprecedented number of resources offered to the Greek government by several other countries.

In total, more than 3,000 fires were recorded over Greece, ravaging approximately 270,000 ha of forest,

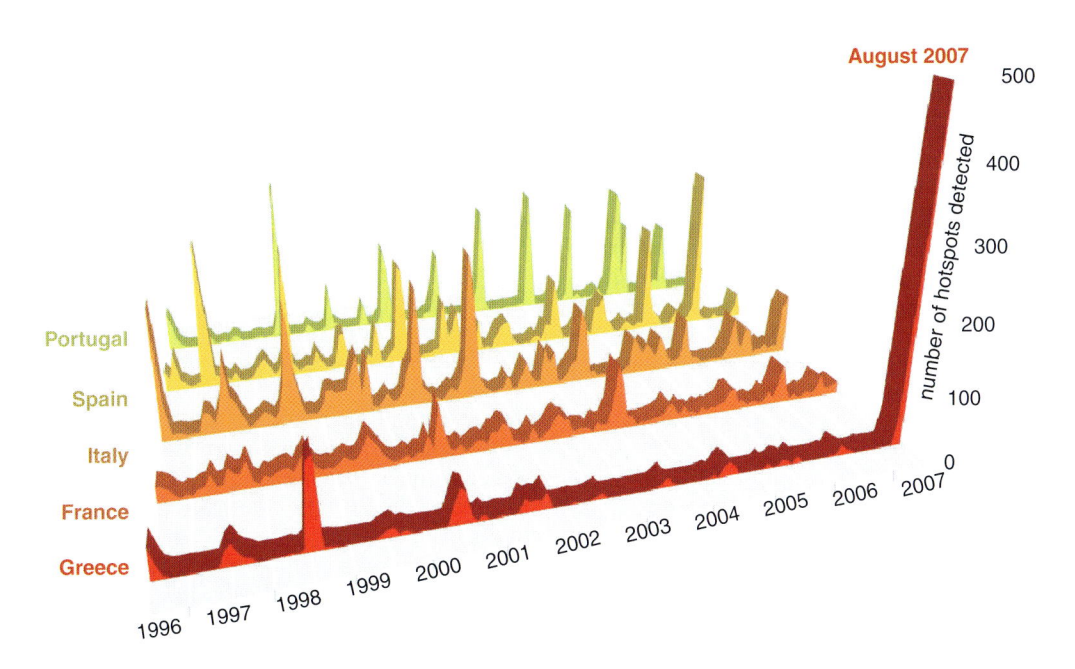

Mega-Fires in Greece (2007), Figure 1 Forest areas burnt in the Mediterranean countries of the EU between 1996 and 2007 (Source ATSR World Fire Atlas).

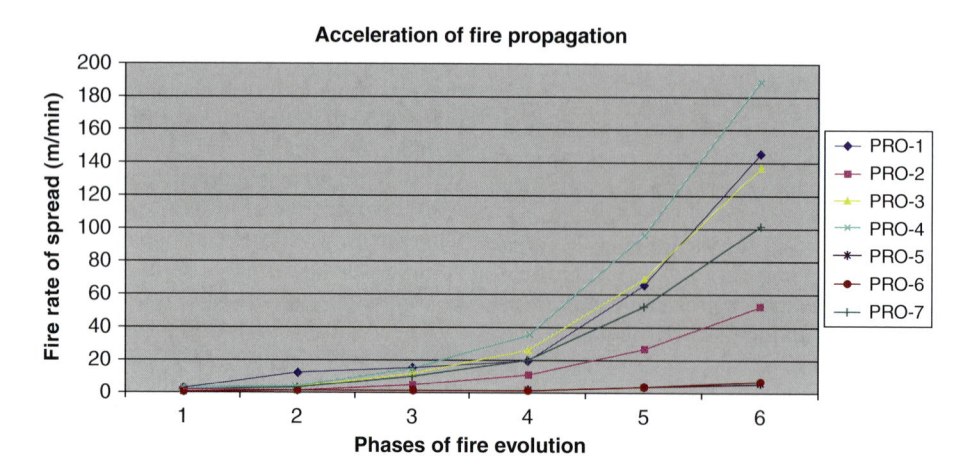

Mega-Fires in Greece (2007), Figure 2 Local acceleration of fire propagation during very large fires.

olive groves, and farmland, according to data of the European Forest Fire Information System (EFFIS) of the JRC Ispra. On the Peloponnese, 177,265.4 ha was destroyed, consisting of 55% forests and natural lands, 41.1% agricultural lands, and 0.9% built-up areas (WWF Hellas, 2007).

The special characteristics of the 2007 forest fires, which distinguish them from past forest fires in Greece, can be summarized as follows (Xanthopoulos, 2007):

- Although the number of fires recorded was not remarkable, the extent of the burnt area was very large compared to previous years.
- Many fire episodes occurred at the same time in several locations.
- There was frequent restart of already suppressed fires.

The mega-fires issue in Greece

Forest fires can be classified according to the suppression effort needed to contain them into initial attack, extended attack, large fires, and mega-fires. These four types cover the continuum of severity that runs from very small, short-duration, and noncomplex events to extraordinarily large, long-duration, and very complex fires. The difficulty of managing forest fires changes dramatically moving from a normal accident to a serious event (extended attack fire) or an ultracatastrophe arises and a mega-fire emerges. Mega-fires occur when multiple fire spots and individually propagating fronts of flames merge into a superfront (Brooking Institution, 2005).

In order to depict the relation of the classification with the number of fire events, we can consider that the majority of fires (approx. 95%) are suppressed during the "initial attack," whereas 4% usually evolve and require "extended attack" operations. Therefore, only 1% of the total number of fires evolves to large fires and only few of these become mega-fires.

The main physical reason for the occurrence of mega-fires is the buildup of dead woody material and accumulation of live biomass in fire-dependent forest ecosystems that can fuel high-intensity events. It is quite common to have such a fire regime following long periods of drought and repeated heat waves during the summer. The situation can be worst due to insect infestations and diseases. Mega-fires create their own local wind field which sustains their propagation, independently of the weather conditions prevailing in the area. Since spotting (starting of new fires by flying embers) is common, mega-fires combined with extreme weather conditions burn out of control and continue burning until relief in the weather or a break in fuel source occurs. Firefighting mechanisms can manage fronts with fire line intensity to 2,500 kW/m, whereas mega-fires often reach intensities to 100,000 kW/m. Therefore, efforts to extinguish such fire fronts are quite futile (Viegas and Eftichidis, 2007).

Fire behavior is normally defined by the topography, the meteorological conditions, and the type of vegetation burned. However, the time since fire ignition is another factor that contributes to extreme fire behavior. Since mega-fires are characterized by their long duration, the time lapse is responsible for eruptive behavior of the fire in many cases. Considering that fires evolve differently through time, we can define a series of six phases for describing this evolution. This consists of (1) the starting condition, (2) the phase of reduction of the fuel moisture, (3) the phase of vegetation dehydration, (4) the phase of wind generation, (5) the phase of wind flow, and (6) the phase of the fire eruption. Time evolution of these phases is different for various forest fuels as shown in Figure 2 (Viegas and Eftichidis, 2007), using the Prometheus (Riaño et al., 2002) classification of forest vegetation to fuels.

The above observation is particularly important in cases of fires approaching villages in case the fire accelerates and surprises the inhabitants without giving them time for evacuation.

Mega-fires are not defined by their physical attributes (e.g., by their size). Instead, they are recognized as "headline fires" in which operational limitations, public anxiety, media scrutiny, and political pressures collide. Beyond their impressive size, they are characterized by their complexity, their potential to overwhelm the capabilities and capacities of the fire suppression forces, and their extreme intensity and long duration. Due to the costs and damages associated with such events, mega-fires are often followed by policy or procedural changes. However, such changes are usually limited in improving firefighting operations and their sustainable hazard mitigation measures.

The 2007 forest fires of Greece record as the most catastrophic fire event in the country's history and the most catastrophic of the last few decades in Europe. The devastation includes the forests and agricultural lands, entire villages, infrastructure, and a large toll on human life (WWF Hellas, 2007). These fire events have been cited in the press as the fourth worst disaster due to forest fires worldwide since 1871 and by far the deadliest for humans in recent years.

Causes of the mega-fires: season of 2007 in Greece

The extended Greek forest fires of 2007 took place in a summer of three continuous heat waves. The exceptionally high summer temperatures, following a winter drought, made the resinous pine forests more flammable than usual and created very favorable conditions for extensive fires.

In the search for the underlying causes of the 2007 Greek forest fires, discussions most often lead to weaknesses in Greek physical planning and development regulations, which inadvertently encourage criminal actions such as arson.

Greek officers concluded that at least some of the fires of 2007 could be attributed to arson. In the Peloponnese, suspicions of arson were reinforced by the fact that dozens of fire episodes started at the same time. Evidence suggests that the 2007 fires broke out due to a combination of criminal intent, carelessness, and accidents. In addition to arson, the lack of maintenance of the electricity pylon network; carelessness of local farmers, villagers, and forest visitors who started fires on hot days; illegal landfills left unguarded; and the inability of elderly farmers to control fires they started to maintain grazing land are frequent cited causes of fires (Xanthopoulos, 2007).

Despite significant investments and an increase of the fire suppression budget since 1998, the Greek forests suffered record-setting forest fires in which the death toll, costs, losses, and damages involved have been staggering.

However, a "successful" 6-year period of firefighting in Greece, which was due to a number of factors, was interpreted as efficiency of the fire management system based on fire suppression. Thus, the fire problem was considered finally solved or at least under control. Vegetation management programs have been ignored, and the forests were left to accumulate billions of tonnes of biomass. In addition, the high temperatures, even during the winter months, extended the growth period of the vegetation and increased the production of biomass. Due to the alternating moist and dry periods, increased volumes of cured vegetation accumulated in the forests (Eftichidis, 2007).

Given the change in the live and dead fuel moisture conditions, the fires moved to sites that in the past were less dry and where the fire used to burn surface fuels with low intensity. Currently, fires in these sites burn intensively and develop large dimensions due to high accumulation of dead vegetative material. Furthermore, the fires tend to invade areas occupied by forest species that have become more flammable and less fire-adapted in the face of worsening climatic conditions. Fir and black pine forests are good examples of this situation in Greece.

The above-mentioned conditions eventually led to a series of mega-fires in south and southwestern Greece that burned 250,000 ha, 72% of which burned during the last week of August 2007 in five adjacent fires in the region of Peloponnese (Eftichidis, 2007).

The year of 2007 was particularly dry for Greece. Measurements from the National Observatory of Athens show that high temperatures are recorded not only in the summer but during the winter months as well. A report by the National Technical University of Athens describes the winter of 2007 as the warmest in 100 years of collected data. The summer was affected by three heat waves with continuous temperatures as high as 42–45°C for several days at a time.

During the first heat wave, in the eastern part of Greece, the weather station of the city of Pyrgos, one of the most affected areas in Peloponnese, recorded for the first time in its history maximum temperatures of 38.5°C and 41.1°C, respectively, for the 24th and 25th of June. The second heat wave was worse and lasted 10 days from July 17–26, with two peaks according to the Pyrgos meteo station data, first the 18/7 (39.7°C) and second the 25/7 (43.4°C which was also a historical record for the last 50 years). The last heat wave (22–25 August with temperatures ranging from 38°C up to 42.3°C) occurred just before the firestorm started. These persistent heat waves dehydrated the forest vegetation and prepared the environment for the mega-fire that followed (Eftichidis, 2007).

The wind speed in the area of Pyrgos during the dates of the fire (24–27 August) reached 30.6 km/h, whereas day temperature was constantly above 40°C. The humidity of the air fell below 12% during the warmer hours of the day, reaching 40% after midnight.

The majority of the fires in Peloponnese started the night of 23 August and involved several parts of south and west Peloponnese, including the regions of Messinia, Arcadia, Laconia, Ilia, and Achaia. The 24th of August was the 80th day without rain in the area of Pyrgos. On the 25th of August, a state of emergency was declared, and international assistance was requested to fight the fires. On the 29th of August, due to the change of the weather, the fires began to die, and the fire brigades succeeded to contain most of them within the next 2 days. A distribution of the burned areas in the region of Peloponnese is shown in Figure 3. According to the calculations made by the Remote Sensing Laboratory of the

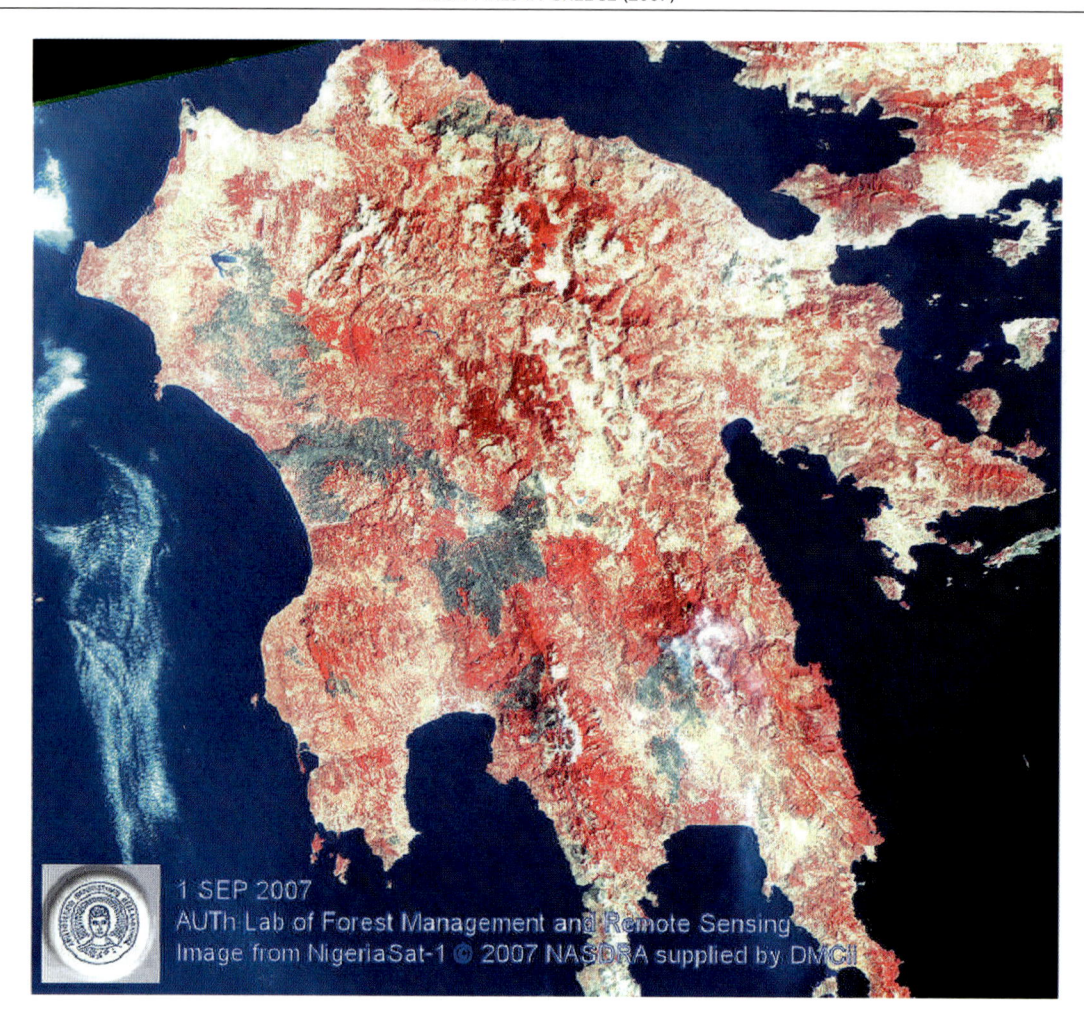

Mega-Fires in Greece (2007), Figure 3 Footprint of the areas burnt in Peloponnese during August 2007 (Source: Remote sensing Laboratory of the Aristotelian University of Thessaloniki).

Aristotelian University of Thessaloniki using satellite data of resolution 30 × 30 m, the total burned area is 177,265 ha. An area of 78,104 ha was agricultural land, whereas 1,634 ha corresponded to structures and infrastructures (villages, roads, installations, etc.).

The evolution of the fire ignitions during this period is shown in Figure 4. It is evident that most of the fire ignitions occurred in the first 2 days of the firestorm (24 and 25 August), whereas significant new fires continued to start until 28 August (Eftichidis, 2007). The situation far exceeded the capabilities of the Greek firefighting forces. Reinforcements and help provided by several other countries for the firefighting operations was not able to control the high-intensity fires in progress.

The data of the mega-fires of Peloponnese are shown in the next table (Table 1).

The fires burned hundreds of square kilometers of pine and fir forests, open forested areas, shrublands, olive groves, vineyards, as well as vast number of isolated residences, installations, and houses in the villages. Several regions faced breakdowns in telecommunications, electricity, and water supplies.

As can be seen from the data in Table 1, the fires of Megalopoli, Zacharo, and Pyrgos and the fires in dry sites were more extensive than the fires that burned in higher altitudes and more humid sites such as the mountains of Taygetos and Parnon. The extent of the fires is also related to the forest species of the regions. In Taygetos and Parnon, stands of fir trees burned more slowly than pine stands that burned elsewhere.

The extent of the 2007 situation was completely new in comparison to the historic forest fire patterns. The extreme intensity of these fires made their control impossible even when they reached areas that are normally used as fire control points.

Damages were unprecedented and of extreme severity. Many people evacuated their homes to move to safer places. Unfortunately, most of the inhabitants of the villages, in particular aged people, refused to leave their houses and belongings, and a number of individuals died as a consequence.

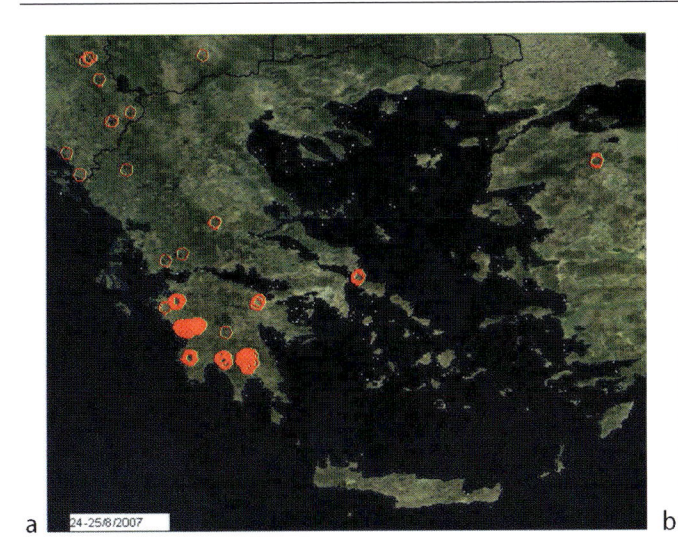

Mega-Fires in Greece (2007), Figure 4 Fire ignitions in Peloponnese during the dates 24–28/8/2007 (Source NASA).

Mega-Fires in Greece (2007), Table 1 Burned areas by mega-fires in Greece, summer 2007 (Source: MODIS burned area products)

Fire name	Burned area (ha)	Growth duration
Mnt. Taygetos	11,357	24–27/8/2007
Mnt. Parnon	20,681	23–30/8/2007
Megalopoli	42,350	24–27/8/2007
Zacharo	45,809	24–30/8/2007
Pyrgos	42,652	24–30/8/2007
Total	*162,849*	

There were also cases of people who did not evacuate in a timely manner, due to the unpredictable speed of the fire propagation as well as due to lack of coordination of the evacuation operations during the first days of the fires. Some of these people were trapped and killed in car accidents while trying to escape from the burning villages. The death toll of the mega-fires of the summer 2007 in southern Greece was more than 70 people, which is a high number of victims in worldwide wildfire history.

Flames engulfed the archeological site of Olympia, home of the first Olympic Games, and the temple of Apollo Epikourios, a 2,500-year-old monument near the town of Andritsaina in southwestern Peloponnese. Thus, the situation was made extremely complex, requiring the authorities to evacuate villages, save archeological sites, and protect human property rather than just extinguishing the flames.

Effects of the 2007 fires

The mega-fires of the summer 2007 in Greece had significant environmental impact due to the large extent and the erratic behavior of the fire. Biodiversity in several protected areas belonging to the Natura 2000 network seismologists but Mercalli's name was maintained: the *Modified Mercalli scale* is today used worldwide.

Mercalli studied some Etna eruptions and the Eolian islands, in particular Stromboli and Vulcano, but the main subject of his investigations was certainly *Vesuvius*, to which he dedicated over 20 years of his life. Moreover, he summed up his ponderous studies on active *volcanoes* of the world in the volume "I vulcani attivi della Terra", printed in 1907, which actually represents the first Italian treatise on volcanology. He not only provided a precise description of the observed phenomena but also introduced classifications, stating the specific characteristics of the different eruptive apparatus and their manifestations.

During the night of March 18–19, 1914, a fire put an end to Mercalli's life, one that had been entirely dedicated to science.

Bibliography

Baratta, M., 1915. L'opera scientifica di Giuseppe Mercalli. *Bollettino Società Geologica Italiana*, **34**, 343–419.

Galli, I., 1915. Il professore Giuseppe Mercalli. Elogio e Bibliografia *Memorie Pontificia Accademia Romana Nuovi Lincei*, s. **2**(1), 40–80.

Mariani, E., 1915. Giuseppe Mercalli. Cenni biografici. *Società Italiana di Scienze Naturali, Atti*, **54**, 1–6.

Cross-references

Earthquake
Eruption Types (Volcanic)
Intensity Scales
Modified Mercalli (MM) Scale
Seismology
Vesuvius
Volcanoes and Volcanic Eruptions

METEORITE

Jay Melosh
Purdue University, West Lafayette, IN, USA

Synonyms

Asteroid; Bolide; Meteor; Meteoroid

Definition

A meteorite is a mass of solid material (either stony or metallic) on the surface of the Earth that came from space.

Discussion

The word meteorite is used for such an object on the surface of the Earth. In space, it is called a meteoroid if small or an asteroid if large (there is no strict dividing line between a meteoroid and an asteroid: typically, a diameter of about 1 km is used, but usage varies within wide limits). A meteor is the bright streak in the sky that accompanies the entry of a meteoroid into the Earth's atmosphere. A meteor that exhibits one or more bright explosions is called a bolide.

Most meteorites originate in the asteroid belt between Mars and Jupiter, but a few come from the surfaces of larger planets, such as Mars or the Moon. Some volatile-rich types may come from comets. Meteorites are classified as stony, iron (metallic), and stony-iron. Stony meteorites, which are about 40 times more abundant in space than irons, are further classified as either chondrites (the most abundant type, with many subclasses of chondrite) or as achondrites. Chondrites contain small, mm to cm diameter, spherical inclusions that are more or less distinct in the body of the meteorite. Freshly fallen meteorites are enclosed in a glassy crust of melted material, the fusion crust, which forms by friction with the air as the meteoroid enters the Earth's atmosphere at high speed.

Meteorites are described as either finds or falls, depending upon the circumstances of their discovery. They are conventionally named after the post office nearest to the point at which they are recovered, such as Allende (fell in 1969 near the town of Allende, Chihuahua, Mexico). In the case of the recently discovered meteorites in Antarctica, names are given that refer to the location, year, and order in which they were cataloged, such as ALH84001 (found near the Allen Hills Moraine in 1984 and the first to be cataloged).

Bibliography

Dodd, R. T., 1981. *Meteorites: A Petrologic-chemical Synthesis*. Cambridge: Cambridge University Press. 368 pp.

Lauretta, D. S., and McSween, H. Y. Jr. (eds.), (2006). *Meteorites and the Early Solar System II*. Tucson: University of Arizona Press. 943 pp.

Wasson, J. T., 1985. *Meteorites: Their Record of Early Solar-system History*. New York: Freeman. 267 pp.

Cross-references

Asteroid
Asteroid Impact
Asteroid Impact Mitigation
Asteroid Impact Predictions
Comet
Impact Airblast
Impact Ejecta
Impact Fireball
Impact Firestorms
Impact Tsunami
Impact Winter

METHANE RELEASE FROM HYDRATE

Graham Westbrook
University of Birmingham, Edgbaston, Birmingham, UK

Synonyms

Climate-induced dissociation of methane hydrate; Release of methane from hydrate caused by global warming

Definition

The release of methane gas from methane hydrate, which is a clathrate (a solid in which water molecules form a cage enclosing methane molecules), occurs when an increase in pressure or a decrease in pressure create conditions that cause hydrate to break down into its separate constituents of water and gas. A natural increase in temperature can be caused by a warming climate, and reduction in pressure, for hydrate beneath the seabed, by a fall in sea level.

Discussion

Methane hydrate is stable under conditions of low temperature and high pressure such as those found on land in regions of permafrost or under the ocean in water deeper than 300–600 m, depending on the water temperature. The concentration of methane in the ocean is usually far too low for hydrate to form, but in the sediment and rocks beneath the seabed, methane concentration can be high enough to form hydrate. The thickness of the gas hydrate stability zone (GHSZ), in which hydrate can form and exist stably, is limited by the increase of temperature with depth within the Earth. Methane from deeper hydrocarbon reservoirs or generated by bacteria from the organic material in the sediment migrates upward, as free gas or dissolved in water, into the GHSZ, where it forms hydrate. The amount of carbon in hydrate beneath the seabed is probably equal to the carbon in all other sources of natural gas and petroleum in the Earth.

An increase in seabed temperature reduces the extent of the GHSZ. In deep water, the seabed remains in the GHSZ, whereas the downward propagation of the temperature increase causes the base of the GHSZ to migrate upward, releasing methane, which may re-enter the GHSZ and form hydrate again, limiting the amount that may escape into the ocean. Where the GHSZ in shallower water is removed completely by warming, the methane released is free to migrate through the sediment to the seabed. The upper continental slope is most prone to methane release by this mechanism, because temperature change is greatest in the upper water column. Although hydrate is absent from most continental shelves, because they are too shallow for the GHSZ to occur, it exists in rocks and sediment beneath the shelf in the Arctic because of the low temperature caused by the presence of permafrost created during the last glacial period when large parts of the shelf were subaerial. There, sea-level rise reinforces the effect of increasing water temperature by flooding low-lying land with water that is warmer than the average temperature of the land surface. Permafrost retards the escape of methane released from hydrate, because the extra heat required to melt the ice slows down the increase of temperature, and because ice impedes the flow of gas. This can impose time lags of hundreds of years between the onset of warming and methane escape.

Over recent years, there has been increasing evidence that methane released from hydrate as a consequence of warming enters the ocean, but little evidence that much of it enters the atmosphere to contribute to global warming. It appears that the rate of release of methane is generally too slow to overcome its solution in the ocean, where, after oxidation, it contributes to ocean acidification. Catastrophic gas venting or submarine landslides of hydrate-rich sediment might, however, be effective in releasing large amounts of methane over short periods of time. Submarine slides have been widely cited as an agent of ancient increases in atmospheric methane but their potency has still to be proven. It has been proposed that the release of gas from rapid dissociation of hydrate creates zones of over-pressured gas in sediment beneath continental slopes, reducing sediment strength and increasing the likelihood of submarine slides, which can cause tsunamis.

Bibliography

Archer, D., Buffett, B. and Brovkin, V., 2008. Ocean methane hydrates as a slow tipping point in the global carbon cycle. In *Proceedings of the National Academy of Science*, www.pnas.org_cgi_doi_10.1073_pnas.0800885105.

Kennett, J. P., Cannariato, K. G., Hendy, I. L., and Behl, R. J., 2003. *Methane Hydrates in Quaternary Climate Change: The Clathrate Gun Hypothesis*. Washington: American Geophysical Union.

Westbrook, G. K., Thatcher, K. E., Rohling, E. J., Piotrowski, A. M., Pälike, H., Osborne, A. H., Nisbet, E. G., Minshull, T. A., Lanoisellé, M., James, R. H., Hühnerbach, V., Green, D., Fisher, R. E., Crocker, A. J., Chabert, A., Bolton, C. T., Beszczynska-Möller, A., Berndt, C., and Aquilina, A., 2009. Escape of methane gas from the seabed along the West Spitsbergen continental margin. *Geophysical Research Letters*, **36**, L15608, doi:10.1029/2009GL039191.

Cross-references

Climate Change
Displacement Wave, Landslide Triggered Tsunami
Gas-Hydrates
Marine Hazards
Permafrost
Release Rate
Sea Level Change
Tsunami

MINING SUBSIDENCE INDUCED FAULT REACTIVATION

Laurance Donnelly
Wardell Armstrong LLP, Greater Manchester, UK

Synonyms

Break lines; Fault steps

Definition

Faults are naturally occurring discontinuities in rock or soil where there has been observable and measurable displacement by shearing and/or dilation. Faults located in areas prone to mining subsidence, caused by the longwall extraction of coal, are susceptible to reactivation. This may result in the generation of a fault scarp along the ground surface (also referred to by some mining and subsidence engineers as a "step" or "break line").

Summary

Mining subsidence-induced fault reactivation may generate a scarp, graben, fissure, or zone of compression along the ground surface (Figures 1 and 2). This is significant because it may cause physical damage to structures (buildings, houses, industrial premises, bridges, dams, pylons, and towers), services and utilities (sewers, water conveyances, gas mains, pipelines, and communications cables), and transport networks (tracks, roads, motorways, railways, rivers, and canals) (Figures 3–6).

The topographic expression of reactivated faults may vary considerably from subtle deflections and flexures barely recognizable across agricultural land or road side verges, to distinct, high-angled fault scarp walls, upto approximately 3–4 m high and 4 km long. In areas of high relief, reactivated faults may influence the first time failure of slopes and the reactivation of landslides (Figures 7–9). More commonly, fault scarps are less than a meter high, less than a meter wide, and vary in length from just a few meters to a few hundreds of meters long.

Reactivated faults do not always appear at their expected outcrop position as inferred on geological maps. This may be attributed to the acceptable mapping tolerances (since geological maps provide an estimate of their likely outcrop position on the ground surface). This is often complicated by the variable nature of the strata, surficial deposits, or made ground, which a fault displaces. Greater thicknesses of surficial

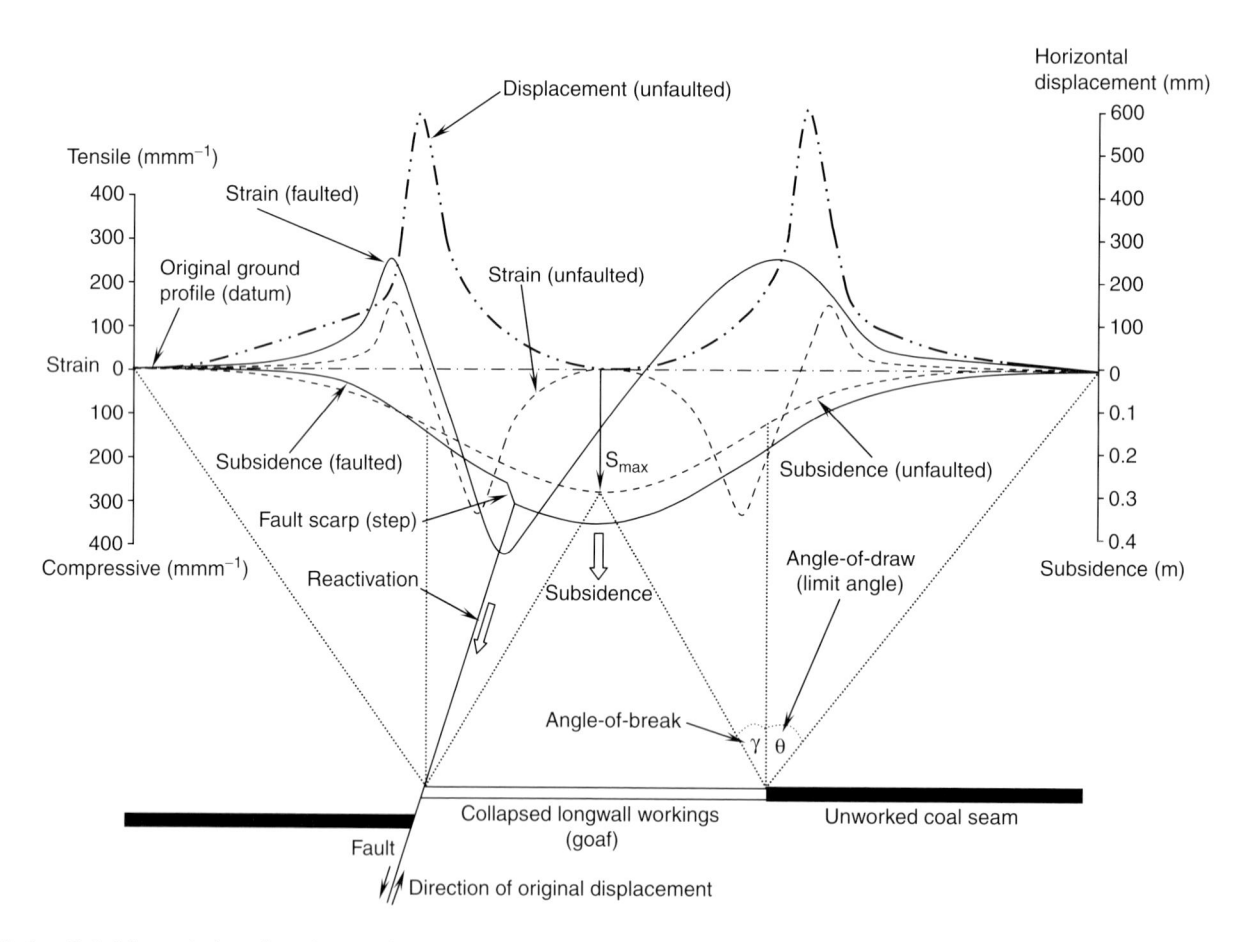

Mining Subsidence Induced Fault Reactivation, Figure 1 Schematic model to illustrate fault reactivation during the mining of a horizontal coal seam by the longwall-mining method. Fault reactivation generates a fault scarp (or step) in the subsidence profile (trough) and disrupts the distribution of the horizontal displacements and strains. High compressive ground strains tend to occur at the fault scarp (but not always, these may also be tensile, generating a fissure) (After Donnelly, 2009).

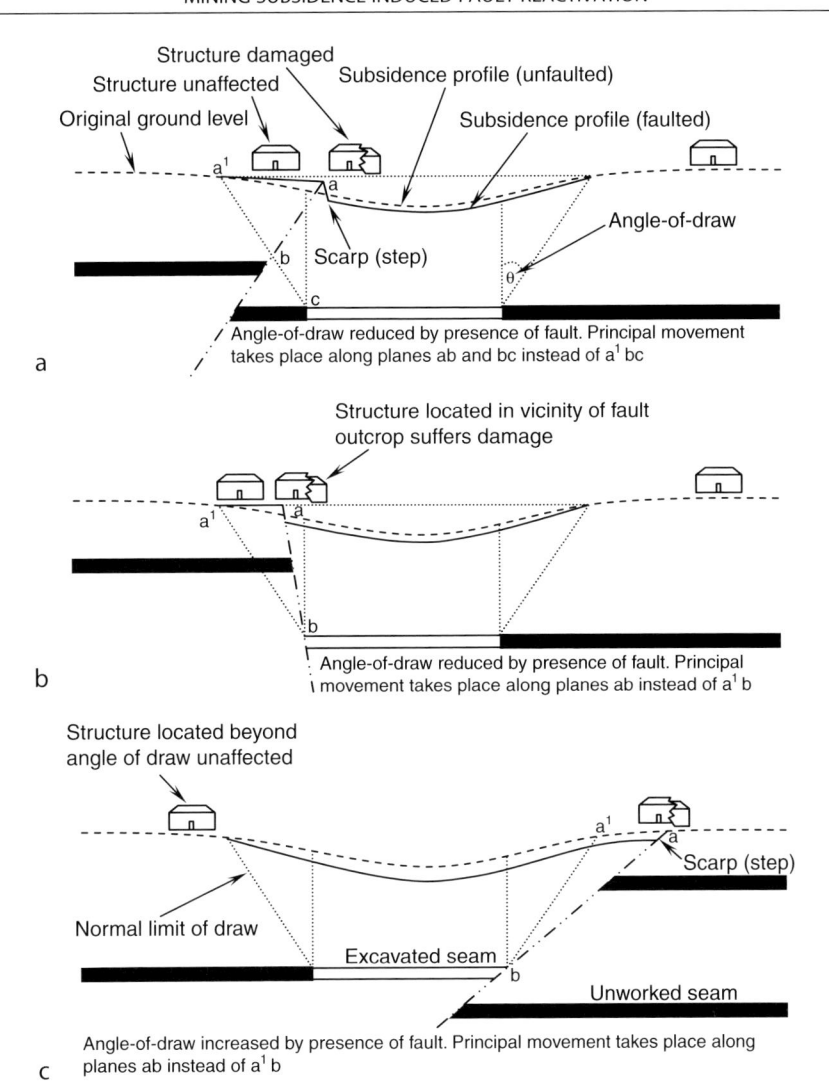

Mining Subsidence Induced Fault Reactivation, Figure 2 The influence of faults on mining subsidence and the angle-of-draw. (a and b) Any structures located in the vicinity of fault outcrops during their reactivation will almost certainly physical damage. When workings are located in the footwall of the fault, any structure located in the hanging wall may be safeguarded as the fault absorbs most of the ground strains (although this is not always the case). In examples (a and b), the presence of the fault has reduced the angle-of-draw (and therefore area-of-influence) in the hanging wall. (c) Faults may also extend the angle-of-draw, beyond that which would otherwise prevail in the absence of any faults (After Donnelly, 2009).

deposits tend to reduce the severity of a fault scarp, but influence a much broader area. Where the cover is thin or absent a distinct, high-angled fault scarp may develop, but where these are thicker a less distinct, broad, open flexure will be generated. Fault scarps are normally temporary features of the ground surface and may be destroyed soon after their generation by, for example, repairs to roads and structures, the ploughing of agricultural land, or by processes of weathering and erosion. In some instances, reactivated faults have reduced the amount of subsidence on the unworked side of a fault by absorbing ground strains and safeguarding houses, structures, and land that may have been otherwise damaged.

Faults are capable of several phases of reactivation each time they are influenced during longwall coal mining operations, separated by periods of relative stability. Fault reactivation has been documented since the middle part of the nineteenth century throughout the United Kingdom and in many other coal mining regions around the world.

Although fault reactivation, in certain circumstances, may continue for periods of time (weeks to several years) after "normal" subsidence has been completed, movements along most faults does eventually stop in the majority of cases investigated.

The mechanisms of mining subsidence-induced fault reactivation are only partially understood. Since ground movements along faults have been observed and recorded to take place over weeks, months, and years, aseismic creep appears to be the dominant mechanism. However, brittle shear failure may be possible where the fault displaces strong sandstone or limestone. There is currently no strong evidence to suggest that coal mining-induced fault reactivation induces seismicity (earthquakes), although this is difficult to prove.

Mining Subsidence Induced Fault Reactivation, Figure 3 Damage to houses caused by the mining-induced reactivation of the Hopton Fault, Oulton, Staffordshire, UK (Photograph © Laurance Donnelly).

Mining Subsidence Induced Fault Reactivation, Figure 5 Barlaston church, Staffordshire, UK, was severely damaged by mining-induced fault reactivation. (Photograph © Laurance Donnelly).

Mining Subsidence Induced Fault Reactivation, Figure 4 Compression to a 5.0 m high retaining wall, caused by fault reactivation and subsidence, Eastwood Hall, Nottinghamshire, UK (Photograph after Whittaker and Reddish, 1989).

Mining Subsidence Induced Fault Reactivation, Figure 6 Reactivation of the Inkersall Fault, Derbyshire, generating a graben, which caused widespread damage to two schools, houses, roads, and walls in the late 1980s and 1990s (Photograph © Laurance Donnelly).

Mining Subsidence Induced Fault Reactivation, Figure 7 Air Photograph to demonstrate how the reactivation of the Tableland fault and associated network of complex fissures can influence the geomorphology of entire moorland slopes, South Wales Coalfield (after Donnelly, 1994).

Mining Subsidence Induced Fault Reactivation, Figure 8 The 3–4 m high and 4 km long Tableland Fault scarp, which has influenced the Darren Goch landslide and displaced stream valleys, South Wales Coalfield (Photograph © Laurance Donnelly).

Mining Subsidence Induced Fault Reactivation, Figure 9 A typical South Wales fault scarp representing several phases of reactivation, probably initiated by valley deglaciation and exacerbated by mining subsidence. These form distinct, extensive topographic features, which may reach at least 4 m high and 3–4 km long. These influence surface drainage and groundwater flow and landsliding (including first time failures and reactivation of existing landslides) (Photograph © Laurance Donnelly).

It would be prudent on all engineering sites containing geological faults in active and former mining areas to investigate their potential effects on ground stability, mine gas emissions, or groundwater/mine water discharges, before development and construction is carried out. It is recommended that this be undertaken at the desk study and ground investigation stage of a project to reduce the risks for unforeseen ground conditions. The ground may then be suitably treated, or appropriate foundations designed, prior to any construction or developments taking place. Further information on mining-induced fault reactivation is present in Donnelly,

2006, 2009; Donnelly and Rees, 2001; Bell and Donnelly, 2006.

Bibliography

Bell, F. G., and Donnelly, L. J., 2006. *Mining and Its Impact on the Environment*. London: Taylor/Francis (Spon).
Donnelly, L. J., 2006. A review of coal mining-induced fault reactivation in Great Britain. *Quarterly Journal of Engineering Geology and Hydrogeology*, **39**, 5–50.
Donnelly, L. J., 2009. A review of international cases of fault reactivation during mining subsidence and fluid abstraction. *Quarterly Journal of Engineering Geology and Hydrogeology*, **42**, 73–94.
Donnelly, L. J., and Rees, J., 2001. Tectonic and mining-induced fault reactivation around Barlaston on the Midlands Microcraton. *Quarterly Journal of Engineering Geology and Hydrogeology*, **34**, 195–214.
Whittaker, B. N., and Reddish, D. J., 1989. Subsidence: occurrence, predicition and control. Amsterdam: Elsevier.

Cross-references

Creep
Critical Infrastructure
Fault
Land Subsidence
Landslide
Mass Movement
Risk Assessment
Subsidence Induced by Underground Extraction

MISCONCEPTIONS ABOUT NATURAL DISASTERS

Timothy R. H. Davies
University of Canterbury, Christchurch, New Zealand

Definitions

Adaptability. The ability to adapt – in this context, to unexpected or altered behavior of natural systems.
Mitigation. Measures taken by society to reduce the consequences of a disaster.
Natural disaster. An event in which the behavior of part of Earth's natural systems causes severe consequences to society, usually greater than local in scale.
Natural hazard. A natural system with the potential to damage society; alternatively, any natural process with the ability to damage society *even if society is not yet present in the area*.
Resilience. The ability to resume normal functioning after a disaster.
Risk. (noun) Probability; probability multiplied by consequence; (verb) to take a chance.
Sustainability. The ability to be sustained – requires specification of *what* is to be sustained, at what *level of intensity*, for what specified *time period*, and what are the *indicators of unsustainability*. Needless to say, these requirements are usually ignored.
Vulnerability. The degree to which society can be affected by disasters.

Introduction

As a result of many years of thinking about natural disasters, teaching students about natural disasters and trying to help communities avert natural disasters, I have come to a number of realizations about the nature and causes of hazards and disasters. These can be summarized as follows:

(a) People cause natural disasters by behaving in ways that make society vulnerable to infrequent high-magnitude natural events.
(b) More people and more development means more and bigger natural disasters.
(c) "Natural hazards" can usefully be defined as processes of nature with the potential to cause damage to society.
(d) Altering the behavior of natural systems usually results in increasing the probability of a natural disaster.
(e) Maintaining altered behavior of natural systems creates significant long-term costs to society.
(f) People behave according to their world views.
(g) Scientists are often poor communicators, especially with nonscientists.
(h) Whatever can happen, will happen one day; that could be today.
(i) Scientists should do what they are good at – science.
(j) Communities must make their own disaster-management decisions.

These realizations sometimes conflict with more conventional thinking about disaster mitigation, among both scientists ("experts") and lay people. I make no claim whatsoever that my views are "right" for anyone else – but I do think that, even if they are wrong, they are at the very least a useful set of discussion points to stimulate fundamental thinking about how to better reduce the impacts of natural disasters. The following list of "misconceptions" – perhaps better thought of as challenges – sets my realizations (which are at present "true" for me) against the background of conventional or traditional practices and thinking.

Misconceptions

That we know what we are talking about

Discussions about natural disasters are frequently plagued by the different meanings that different people attach to words such as "hazard," "disaster," "risk," and so on. The word "hazard" is particularly broadly interpreted; to some, hazard is synonymous with risk as the numerical probability of a specific event happening in a specified time interval; to others, it is synonymous with "natural process," such as a landslide occurring on an uninhabited island. Many other interpretations occur between (and even beyond) these extreme examples. Again, "risk" is sometimes defined as the product of probability and consequence, whereas to others it is simply numerical probability as noted above. Similar confusion is possible with the terms "vulnerability," "catastrophe," "disaster," "resilience," and many others.

This is not the place to propose specific meanings for words (with the exception of two examples suggested below); it is, however, appropriate to note that in order to make substantial progress in mitigating inevitable future natural disasters, the meanings of words used either in print or orally must be made completely clear by the user. If this is not done, audiences should ask for it to be done. Experience has shown that such requests are often a complete surprise to the user of the words, and indeed may be treated as an insult; this probably indicates that the user is not clear about the meaning. In any case, continued discussion in the presence of unresolved conflicting interpretations of word meanings is usually unproductive and thus a waste of time.

As examples of how it is possible to unequivocally define potentially confusing words, I offer the following:

"Sustainable" – a specified activity is sustainable at a specified level for a specified time if it does not result in unacceptable consequences (to whom?).
"Natural hazard" – a natural process that currently has the potential to be deleterious to society.

That natural disasters are caused by misbehavior of nature

Natural disaster is a term commonly used to describe severe damage and/or deaths in communities affected by events such as tornadoes, earthquakes, tsunami, volcanic eruptions, storms, etc. It is important to understand that the *events* are simply part of the normal behavior of the Earth's natural systems; they were going on for billions of years before humans evolved, and will continue for billions of years into the future. There is no element of natural misbehavior involved. Events that cause natural disasters are usually somewhat rare on the timescales commonly considered by communities, and are therefore sometimes unexpected, but the only element of misbehavior that can be identified is that the communities did not expect the event to occur and were therefore unprepared – i.e., human misbehavior.

That natural disasters can be prevented by altering the behavior of nature

This misconception arises from the idea that nature misbehaves; if it does then its behavior can be corrected. It is telling to note that the German language term for river engineering is "*Flußkorrektion*" – literally "river correction," implying that the form of the river prior to engineering was incorrect. This is undoubtedly a consequence of the definition of Civil Engineering up until the 1970s: "Harnessing the great powers of nature for the benefit of man," reflecting the idea that "man" has dominion over nature.

Certainly engineering has been vital in developing resources for (hu)mans' use, and modification to the everyday behavior of natural systems can be sustainable. To modify the infrequent events that are the usual trigger for natural disasters, though, is a much more challenging task for a number of reasons:

(a) Data describing infrequent events are usually sparse, so those events are poorly known and understood and the design of control measures to that extent is unreliable.
(b) These infrequent events are characterized by greater magnitude and power than the more frequent, lower-intensity events to which communities are accustomed, so control is correspondingly more difficult.
(c) Fiscal constraints commonly limit the magnitude of event that is able to be "controlled"; but a greater (superdesign) event can occur at any time, and when it does occur it will cause a natural disaster in spite of engineering controls.
(d) Implementation of works to alter the behavior of nature inevitably generates the public perception that there can be no more disasters in that place, so development accelerates, leading to greater costs when the inevitable superdesign event occurs.

Natural disasters cannot be prevented; given Earth's ever-increasing population and occupancy of available land, natural disasters will increase in frequency. The impacts of future natural disasters can be reduced only by better knowledge of their trigger events and careful preparation by communities to reduce their own vulnerability.

That humans are powerless against nature

This is a more recent misconception than most of the others. It is a reaction to the realization that, in many places where geological activity is intense, the forces involved are simply too large for humans to counter. It appears to follow that there is nothing we can do to prevent natural disasters.

It is certainly true that little or nothing can be done to alter the behavior of earthquakes, volcanoes, glaciers, and other large-scale physical processes, and, as outlined elsewhere in this entry, *reliable* modification of infrequent, intense natural process is not achievable. The natural processes that trigger disasters will therefore continue to occur. This does not make society powerless to reduce the impacts of disasters, however. A disaster occurs when a community is affected by an extreme natural process; but there is nothing to prevent the community from modifying *its own* behavior so as to become less vulnerable to the disaster. What is required is that society becomes aware of the likely consequences of the trigger event, and is prepared to adapt its own behavior in the light of those effects. We may in principle be unable to control nature, but we are in principle able to control ourselves.

The increase in meteorological disasters is caused by climate change

There seems to be a general awareness that natural disasters triggered by storms are increasing in frequency and magnitude (e.g., Hurricane Katrina, recent storms in Japan and the Philippines). This has been cited as evidence that anthropogenic climate change is both real and rapid, and is causing extreme meteorological events to increase in intensity. The hard factual evidence for this is pretty much nonexistent; the storm sequences of recent years lie within the natural event variability that would be expected with a stable climate, even if they may be unusual in that context. Recent storminess is not yet evidence for climate change.

What is clearly evident is the exponential increase in damage costs of weather-related disasters from the 1960s to the present day, as evidenced in many reports. This coincides with the dramatic rise in human population and investment value over the same period – the more there is to lose, the greater will be the losses.

Statistical data on natural behavior can be used to design reliable disaster countermeasures based on cost-benefit analysis

This is the classical natural hazard management concept; if we design to manage the most likely event, then over time net benefit will be maximized. There are a number of flaws in this concept. For example, the probability of the most likely event is much less than the sum of the probabilities of the other events, so the most likely event is in fact *un*likely to occur – it is much more likely that some other event will occur. For example, in 1,000 throws of a six-sided die, the most likely number of sixes is 166.7 (1,000/6). This is of course impossible, because 0.7 of a "six" cannot occur; the most likely possible number (167) of sixes is also much less likely to occur than some other result. Many of these flaws result from the fact that in dealing with disaster-triggering events, we are always dealing with a small sample. This is not only a small dataset describing the infrequent (and therefore few) recorded trigger events, but also the small number of events that will occur in the future in the timescale of relevance.

Events capable of causing disasters are by definition infrequent; if they were frequent, humans would alter their behavior so that the natural hazards were not disastrous. Thus, if we are planning to mitigate disasters at a given site over, say, the next hundred years, we can expect a small number of trigger events – certainly fewer than five, possibly none at all. Here is the point: *Statistical predictions about a small sample of events are intrinsically imprecise*. If a dice is rolled 6,000 times, we expect close to 1,000 sixes; say between 950 and 1,050, which is 1,000 ± 5%. If the same dice is rolled six times, however, we expect close to 1 six; so the best-case scenario is 1 ± 1 or 1 ± 100%. Even if we have a million years of event data from the past, the fact that we are predicting *into* a small sample space makes the prediction intrinsically imprecise.

The other fact that makes cost-benefit analysis of doubtful value is that net benefit equals unmitigated damage cost minus mitigated damage cost. Now both unmitigated and mitigated damage costs are large and imprecise numbers; this means that subtracting one from the other to get net benefit gives a much smaller and much more imprecise number – so imprecise in fact that using it as a design discriminator is often unrealistic.

Natural disasters are always big

The word "disaster" intrinsically implies something big – bigger than an "incident," say, or a "mishap" (but smaller than a "cataclysm" or a "catastrophe"). As with much terminology, however, its meaning depends on one's point of view. A minor mudslide that kills an unemployed peasant is completely unworthy of notice to the vast majority of a population, but to the close relatives of the dead man it is clearly an event that will change their lives, and could realistically be called catastrophic; for the local community in which the man had lived for many years it is a disaster.

People resist hazard mitigation because they are ignorant

It is a common experience among hazard managers that persuading people to take sensible precautions against disasters is difficult. Even persuading them to accept the fact of the existence of a hazard of which they were previously unaware can be tremendously difficult. In such cases an easy solution to the problem is to label the people stupid; but this is both untrue and unproductive.

People usually behave according to what they think is the right thing to do; their view of the right thing to do may be the result of ignorance or prejudice, but it is not the result of stupidity. Ignorance and prejudice can be altered by good communication; but by definition, stupidity cannot.

At another level, peoples' behavior aligns with their view of how the world operates. Hence, before they have experienced a natural disaster, people will resist being required to carry out hazard assessments and mitigation measures – whereas after the disaster they may blame the authorities who failed to protect them. This is not stupidity, it is the result of a change of world view.

The point of this is that informing people about potential natural disasters is always unwelcome, and the information will be resisted. In order to communicate it effectively, the "expert" needs to understand the world view of the people, and to be overtly empathetic about the psychological impact the information can have. Such empathy is not possible with people one has (even to oneself) labeled "stupid."

Worst-case scenarios are scaremongering and problematic

It is not uncommon for natural hazards scientists to be accused of scaremongering when outlining the potential impacts of extreme natural events on communities, together with the comment that this is not a constructive way to go about communicating science to society. It is indeed the case that simply stating that a community has a 1% per year chance of being devastated by a landslide is likely to create a situation where further communication is difficult; nevertheless, if that information is correct then it needs to be made available so that the community can make decisions about how to manage the situation. The reality is that

(a) Every worst-case scenario can occur, and given long enough will occur.
(b) The worst-case scenario can occur tomorrow.

Thus any disaster-management planner who does not convey such information to a community is not carrying out their duty – in fact any official whose estimate of the likely disaster magnitude is exceeded has failed.

How, then, can such information be conveyed without engendering a non-constructive reaction? This needs forethought – it is too late when standing before the microphone in the Community Hall. It is necessary to establish mutual trust between the community and the official before real communication can occur, so considerable groundwork is required. The whole topic of effectively communicating hazards science to communities and their leaders, so that it can be useful in decision making, is being seriously addressed nowadays (e.g., http://www.usgs.gov/science_impact/index.html) and is possibly one of the most important factors in advancing disaster management worldwide.

A useful aspect of considering a worst-case scenario is that any action a community takes to mitigate its impact will be much more effective against any (much more likely) lesser event. It also has the effect of making a community actively aware of the nature of the landscape they use.

Scientists know best

Reducing the impact of a potential disaster is a task that requires knowledge of the physical aspects of the disaster and knowledge of the social functioning of the community it impacts. Scientists acquire the former through research, but they do not have the latter; I would even venture to suggest that the people with the best potential knowledge of how the community functions are not sociologists or social scientists, but *the community itself*. It is not uncommon, especially where less-developed communities are receiving aid to reduce disasters, to find that scientists exceed their brief of understanding and communicating science, and carry on to state what actions the community should take to mitigate disasters.

I submit that this is not the best way to operate. Especially in dealing with communities of people whose culture is not that of science, or even that of the land of origin of the scientists, all that scientists can usefully do is make information easily available; how that information is used by the community is a decision that can only be made by the community. In doing this, the community accepts responsibility for its resilience to disaster. The community may choose to seek further advice from the scientists, but the latter group should, in my opinion, refrain at all times from trying to influence decision making (difficult though this may be).

This is not just cultural correctness: It is a pragmatic way of ensuring that the disaster-management decisions made are acceptable to the community, and therefore are carried out. There is a long list of situations where solutions have been imposed on communities, found not to be acceptable by the communities and simply not implemented; or, if implementation was part of the job, the works or procedures put in place were not maintained and lapsed through neglect. Rarely does the agency responsible for the solution return to assess its effectiveness. By contrast, when the community is the decision maker, the community will ensure implementation goes ahead and that maintenance occurs.

Conclusions

As noted at the outset, these "misconceptions" are both personal to myself and intended for discussion; however, the purpose is very serious. Of all the tertiary programs I have been involved in, disaster management is the one with far and away the greatest potential to benefit society – and, if it is done poorly, to do the opposite. *Disaster mismanagement kills people*. Natural processes do not obey the theories of scientists; if the theories are sound, they more or less represent natural processes. In disaster management the best possible information is always required; nature cannot be influenced by theory, policy, or blind faith. Hence, it is imperative that we think deeply about the behavior of nature and of communities; we take nothing on trust, however eminent the source; and we are open to admitting that our present ideas might be wrong.

Bibliography

Mileti, D., 1999. *Disasters by Design: A Reassessment of Natural Hazards in the United States*. Washington, DC: Joseph Henry Press. 371p.

Cross-references

Civil Protection and Crisis Management
Classification of Natural Disasters
Community Management of Hazards
Disaster Risk Reduction
Education and Training for Emergency Preparedness
Emergency Planning
Exposure to Natural Hazards
Frequency and Magnitude of Events
Hazardousness of Place
Humanity as an Agent of Geological Disaster
Land-Use Planning
Land Use, Urbanization and Natural Hazards
Mitigation
Myths and Misconceptions
Natural Hazard
Perception of Natural Hazards and Disasters
Recurrence Interval
Resilience
Risk Assessment
Sociology of Disasters
Uncertainty
Vulnerability

MITIGATION

Farrokh Nadim
Norwegian Geotechnical Institute, Oslo, Norway

Synonym

Risk reduction

Definition

Mitigation is the planning and execution of measures designed to reduce the risk to acceptable or tolerable levels.

Introduction

Risk mitigation is an important component of risk management. To develop effective risk mitigation measures, one should understand the key determinants of risk; that is hazard and vulnerability.

Risk mitigation strategies

Risk mitigation strategies for natural hazards aim at either reducing the hazard, or reducing the vulnerability and exposure of the population, infrastructure, and other elements at risk. They can broadly be categorized into the following groups:

- Physical measures to reduce the frequency and/or severity of the hazard
- Land-use planning
- Early warning systems (and emergency evacuation plans) (*early warning systems*)
- Risk communication (*risk perception/communication*) and public awareness campaigns
- Legislation and enforcement of building codes
- Measures to pool and transfer the risks such as natural hazard insurance

Public awareness campaigns are effective in reducing the vulnerability of the exposed population for all types of natural hazards.

Physical measures may be used to stop, delay, or reduce the impact of certain types of natural hazards such as debris flow, flash flood, river flood (*flood protection*), storm surge, and tsunami. On land, these may include "soft" measures in the form of drainage, erosion protection, vegetation, ground improvement; or "hard" structures like dikes, embankments, and vertical concrete or stone block wall. Offshore, the man-made physical barriers like jetties, moles or breakwaters, or even submerged embankments could be constructed to reduce the impact of cyclone, storm surge, and tsunami.

A well functioning and efficient early warning system, including well-designed escape routes and safe areas, is probably the best way to prevent loss of life due to tsunami, flood, storm surge, cyclone, volcanic eruption, and certain classes of landslides. To develop a reliable early warning system, the physical processes and mechanisms need to be understood and methods need to be developed for measuring, modeling, and predicting the natural hazard of concern, for example, landslide or tsunami. Design of functional networks of escape routes and safe places is strongly dependent on the local context.

The most effective method for mitigating the earthquake risk is to construct buildings and other infrastructure to withstand the earthquake-induced load effects. In seismically active regions, important structures should not be placed in areas that are exposed to earthquake-induced landslides and ground failure, unless measures to improve the ground and/or stabilize the slope(s) are implemented. Obviously relevant legislation and enforcement of building codes must be in place for this mitigation strategy to be successful.

Identification of appropriate mitigation strategy

For a given hazard and element at risk, a number of viable mitigation measures may be available. The identification of the optimal risk mitigation strategy involves:

1. Identification of possible hazard scenarios and hazard levels
2. Analysis of possible consequences (loss of life, monetary losses, damage to the environment, etc.) for the different scenarios (*risk assessment*)
3. Assessment of possible measures to reduce the hazard
4. Assessment of possible measures to reduce or eliminate the potential adverse consequences
5. Recommendation of specific measure(s) on the basis of technical evaluations and discussions with the stakeholders
6. Transfer of knowledge and communication with authorities and society

Any mitigation strategy needs to be part of a community's integrated land-use planning and subjected to analyses that assess and circumvent its potential negative environmental impacts. The optimal risk mitigation strategy is not always the most appropriate one. The exposed population and other stakeholders must be involved in the decision-making process that leads to the choice of the most appropriate risk mitigation strategy.

Summary

Mitigation is an important component of risk management and it refers to the planning and execution of measures designed to reduce the risk. Risk mitigation strategies for natural hazards may focus on reducing the hazard, or on reducing the vulnerability and exposure of the population, infrastructure, and other elements at risk. To identify the most appropriate risk mitigation strategy, the exposed population and other stakeholders must be involved in the decision-making process.

Cross-references

Breakwater
Building Codes
Debris Flow
Disaster Risk Management
Early Warning Systems
Flash Flood
Flood Protection
Hazard
Insurance
Land-Use Planning
Risk
Risk Assessment
Risk Perception and Communication
Surge
Tsunami
Volcanoes and Volcanic Eruptions
Vulnerability

MODIFIED MERCALLI (MM) SCALE

Valerio Comerci
Geological Survey of Italy, Rome, Italy

Definition

The Modified Mercalli Scale is one of the several scales used in the world to estimate the intensity of earthquakes (see entry *Intensity Scales*). It is a tool to evaluate the severity of historical earthquakes in many regions of the world, and it is currently adopted in the USA and other countries for *macroseismic surveys*. Note that there are different versions of MM Scale, all with 12 degrees. The first one was devised by Wood and Neumann in 1931 (see Table 1), modifying and condensing the Mercalli-Cancani scale, as formulated by Sieberg in 1923. This scale is a hierarchical classification of observed effects; the diagnostic effects for the lower degrees are essentially those on people, for the intermediate and higher degrees those on objects and buildings, and for the highest degrees (XI and XII) those on the environment.

Afterward, *Richter* proposed a new version, the MM Scale of 1956 (Richter, 1958), which takes into account four different classes of masonry, defined according to quality of workmanship, construction materials employed, and resistance against lateral forces. Later on, other MM scales have been produced, such as the versions by Brazee (1979) and Stover and Coffman (1993), the variant by Dengler and McPherson (1993) addressed to sparsely populated areas, or the revisions carried out by Dowrick (1996) and Hancox et al. (2002) for New Zealand, etc.

Therefore, when using MM intensity values, it is necessary to specify the scale version.

Modified Mercalli (MM) Scale, Table 1 Modified Mercalli intensity scale of 1931 (From Wood and Neumann 1931)

I *Not felt* – or, except rarely under especially favorable circumstances
Under certain conditions, at and outside the boundary of the area in which a great shock is felt:
Sometimes birds, animals, reported uneasy or disturbed;
Sometimes dizziness or nausea experienced;
Sometimes trees, structures, liquids, bodies of water, may sway –doors may swing, very slowly

II *Felt indoors by few, especially on upper floors*, or by sensitive, or nervous persons
Also, as in grade I, but often more noticeably:
Sometimes *hanging objects may swing*, especially when delicately suspended;
Sometimes trees, structures, liquids, bodies of water, may sway; doors may swing, very slowly;
Sometimes birds, animals, reported uneasy or disturbed;
Sometimes dizziness or nausea experienced

III *Felt indoors by several, motion usually rapid vibration*
Sometimes not recognized to be an earthquake at first
Duration estimated in some cases
Vibration like that due to passing of light, or lightly loaded trucks, or heavy trucks some distance away
Hanging objects may swing slightly
Movements may be appreciable on upper levels of tall structures. Rocked standing motor cars slightly

IV *Felt indoors by many, outdoors by few*
Awakened few, especially light sleepers
Frightened no one, unless apprehensive from previous experience. Vibration like that due to passing of heavy, or heavily loaded trucks. Sensation like heavy body striking building, or falling of heavy objects inside
Rattling of dishes, windows, doors; glassware and crockery clink and clash
Creaking of walls, frame, especially in the upper range of this grade
Hanging objects swung, in numerous instances
Disturbed liquids in open vessels *slightly*
Rocked standing motor cars noticeably

V *Felt indoors by practically all, outdoors by many or most: outdoors direction estimated*
Awakened many or most
Frightened few – slight excitement, a few ran outdoors
Buildings trembled throughout
Broke dishes, glassware, to some extent
Cracked windows – in some cases, but not generally
Overturned vases, small or unstable objects, in many instances, with occasional fall
Hanging objects, doors, swing generally or considerably
Knocked pictures against walls, or swung them out of place. Opened, or closed, doors, shutters, abruptly
Pendulum clocks stopped, started or ran fast, or slow
Moved small objects, furnishings, the latter to slight extent. *Spilled liquids* in small amounts from well-filled open containers. *Trees, bushes, shaken slightly*

VI *Felt by all*, indoors and outdoors
Frightened many, excitement general, some alarm, many ran outdoors. *Awakened all*
Persons made to move unsteadily
Trees, bushes, shaken slightly to moderately
Liquid set in strong motion
Small bells rang – church, chapel, school, etc
Damage slight in poorly built buildings
Fall of plaster in small amount
Cracked plaster somewhat, especially fine cracks in *chimneys* in some instances
Broke dishes, glassware, in considerable quantity, also some windows
Fall of knickknacks, books, pictures
Overturned furniture in many instances
Moved furnishings of moderately heavy kind

VII *Frightened all* – general alarm, all ran outdoors
Some, or many, found it difficult to stand
Noticed by persons driving motor cars
Trees and bushes shaken moderately to strongly
Waves on ponds, lakes, and running water
Water turbid from mud stirred up
Incaving to some extent of sand or gravel stream banks
Rang large church bells, etc
Suspended objects made to quiver
Damage negligible in buildings of good design and construction, *slight* to moderate in well-built ordinary buildings, *considerable* in poorly built or badly designed buildings, adobe houses, old walls (especially where laid up without mortar), spires, etc
Cracked chimneys to considerable extent, *walls* to some extent. *Fall of plaster* in considerable to large amount, also some stucco. Broke numerous windows, furniture to some extent

Modified Mercalli (MM) Scale, Table 1 (Continued)

	Shook down loosened brickwork and tiles
	Broke weak chimneys at the roofline (sometimes damaging roofs). *Fall of cornices* from towers and high buildings
	Dislodged bricks and stones
	Overturned heavy furniture, with damage from breaking
	Damage considerable to concrete irrigation ditches
VIII	*Fright general* – alarm approaches panic
	Disturbed persons driving motor cars
	Trees shaken strongly – branches, trunks, broken off, especially palm trees
	Ejected sand and mud in small amounts
	Changes: temporary, permanent; in flow of springs and wells; dry wells renewed flow; in temperature of spring and well waters
	Damage slight in structures (brick) built especially to withstand earthquakes
	Considerable in ordinary substantial buildings, partial collapse: racked, tumbled down, wooden houses in some cases; threw out panel walls in frame structures, broke off decayed piling
	Fall of walls
	Cracked, broke, solid stone walls seriously
	Wet ground to some extent, also ground on steep slopes
	Twisting, fall of chimneys, columns, monuments also factory stacks, towers
	Moved conspicuously, overturned, very heavy furniture
IX	Panic general
	Cracked ground conspicuously
	Damage considerable in (masonry) structures built especially to withstand earthquakes:
	Threw out of plumb some wood-frame houses built especially to withstand earthquakes;
	Great in substantial (masonry) buildings, some collapse in large part; or wholly shifted frame buildings off foundations, racked frames; serious to reservoirs; underground pipes sometimes broken
X	*Cracked ground*, especially where loose and wet, up to widths of several inches; fissures up to a yard in width ran parallel to canal and stream banks
	Landslides considerable from river banks and steep coasts
	Shifted sand and mud horizontally on beaches and flat land
	Changed level of water in wells
	Threw water on banks of canals, lakes, rivers, etc
	Damage serious to dams, dikes, embankments
	Severe to well-built wooden structures and bridges, some destroyed
	Developed dangerous cracks in excellent brick walls
	Destroyed most masonry and frame structures, also their foundations
	Bent railroad rails slightly
	Tore apart, or crushed endwise, pipe lines buried in earth
	Open cracks and broad wavy folds in cement pavements and asphalt road surfaces
XI	Disturbances in ground many and widespread, varying with ground material
	Broad fissures, earth slumps, and land slips in soft, wet ground. Ejected water in large amounts charged with sand and mud
	Caused sea waves ("tidal" waves) of significant magnitude
	Damage severe to wood-frame structures, especially near shock centers
	Great to dams, dikes, embankments, often for long distances
	Few, if any (masonry), structures remained standing
	Destroyed large well-built bridges by the wrecking of supporting piers, or pillars
	Affected yielding wooden bridges less
	Bent railroad rails greatly, and thrust them endwise
	Put pipe lines buried in earth completely out of service
XII	*Damage total* – practically all works of construction damaged greatly or destroyed
	Disturbances in ground great and varied, numerous shearing cracks. Landslides, falls of rock of significant character, slumping of river banks, etc., numerous and extensive
	Wrenched loose, tore off, large rock masses
	Fault slips in firm rock, with notable horizontal and vertical offset displacements
	Water channels, surface and underground, disturbed and modified greatly
	Dammed lakes, produced waterfalls, deflected rivers, etc
	Waves seen on ground surfaces (actually seen, probably, in some cases). Distorted lines of sight and level
	Threw objects upward into the air

Bibliography

Brazee, R. J., 1979. Reevaluation of modified Mercalli intensity scale for earthquakes using distance as determinant. *Bulletin of the Seismological Society of America*, **69**, 911–924.

Dengler, L., and McPherson, R., 1993. The 17 august 1991 Honeydew earthquake, North Coast California: a case for revising the Modified Mercalli scale in sparsely populated areas. *Bulletin of the Seismological Society of America*, **83**, 1081–1094.

Dowrick, D. J., 1996. The modified Mercalli earthquake intensity scale; revisions arising from recent studies of New Zealand earthquakes. *Bulletin of the New Zealand National Society for Earthquake Engineering*, **29**(2), 92–106.

Hancox, G. T., Perrin, N. D., and Dellow, G. D., 2002. Recent studies of historical earthquake-induced landsliding, ground damage, and MM intensity in New Zealand. *Bulletin of the New Zealand Society for Earthquake Engineering*, **35**, 59–95.

http://pubs.usgs.gov/gip/earthq4/severitygip.html

Richter, C. F., 1958. *Elementary Seismology*. San Francisco: W. H. Freeman.

Sieberg, A., 1923. *Geologische, Physikalische und Angewandte Erdbebenkunde*. Jena: G. Fisher.

Stover, C. W., and Coffman, J. L., 1993. *Seismicity of the United States, 1568–1989 (Revised)*. Washington: United States Government Printing Office.

Wood, H. O., and Neumann, F., 1931. Modified Mercalli intensity scale of 1931. *Bulletin of the Seismological Society of America*, **21**, 277–283.

Cross-references

Building Failure
Earthquake
Earthquake Damage
Intensity Scales
Isoseismal
Macroseismic Survey
Mercalli, Giuseppe
Richter, Charles F.
Seismology

MONITORING NATURAL HAZARDS

Michel Jaboyedoff, Pascal Horton, Marc-Henri Derron, Céline Longchamp, Clément Michoud
University of Lausanne, Lausanne, Switzerland

Synonyms

Observation; Surveillance; Watching

Definition

The verb "to monitor" comes from the Latin "monere" which means to warn. In geosciences, it means to watch carefully at a hazardous situation and to observe its evolution and changes over a period of time. It is also used to define the activity of a device that measures periodically or continuously sensitive states and specific parameters.

Introduction

Hazard monitoring is based on the acquisition and the interpretation of a signal indicating changes in behavior or properties of a hazardous phenomenon or the occurrence of events. This ranges from acquiring basic meteorological data to advanced ground movement measurements. Hazards monitoring began sometime ago, when the Babylonians first tried to forecast weather. When Aristotle wrote his treatise *Meteorologica*, the Chinese were also aware of weather observations (NASA, 2012a). Pliny the Elder studied in details the eruption of the Vesuvius in August 79 AD, providing one of the first scientific observations of a natural catastrophe. Presently, the evolution and the precision of monitoring are closely linked to the development of new technologies. A very interesting example highlighting the importance of technological development is provided by hurricane statistics. The number of hurricanes had often been underestimated because of the lack of information prior to the appearance of satellite imagery: many hurricanes that did not reach the coasts were simply not registered (Landsea, 2007). Today, the development of telecommunications and electronics has made easier the adoption of monitoring systems. In addition, satellite remote sensing has improved greatly the detection of changes at Earth surface. Nevertheless, monitoring remains a costly activity, implying that actually only few hazard types and locations are monitored. Moreover, as dangerous phenomena are usually complex, several parameters have to be monitored, and in most cases one single variable is not a sufficient criterion to provide reliable warnings.

Monitoring can be either linked to an early warning system, leading to act directly within the society, or used to record hazardous events to provide data for hazard assessment and a better understanding of the phenomenon. Some of the monitoring results are public and accessible at no cost, such as earthquake data, whereas meteorological data are often sold because they are profitable due to their direct impact on society (such as agriculture, air traffic, news, and tourism). In any case, with the boom of Internet, more and more free data is accessible in many countries.

In the following, we describe briefly the most common sensor types used for monitoring several hazards and further discuss monitoring aspects.

Instruments and measured variables

Originally, monitoring was mainly done by simple human observations or with limited devices, and some were performed manually, such as the first rain gauges. Now, even if some monitoring is still based on observations, as for snow avalanches, it is mainly instrumented, and many sensors are also used for remote-sensing techniques. The great advance in computer sciences and communication technologies has increased the accessibility to instruments, by improving technology and reducing costs.

Climatic variables are monitored by satellite and meteorological stations. According to the World Meteorological Organization (WMO, 2012a), the global observing system (GOS) acquires *"meteorological, climatological, hydrological and marine and oceanographic data from more than 15 satellites, 100 moored buoys, 600 drifting buoys, 3,000 aircraft, 7,300 ships and some 10,000 land-based stations."*

Hazard monitoring consists primarily of treating a signal in order to obtain information about movement, moisture, temperature, pressure, or physical properties (Table 1). A monitoring sensor is local when it records properties at its own location (thermometer, rain gauge, etc.). Remote sensors are used to collect properties of distant objects. Remote-sensing techniques can be active (a signal is sent and received) or passive (only receiving). For instance, InSAR (interferometric synthetic aperture radar) is an active remote-sensing method to detect ground movement, whether Earth surface temperatures can be measured from satellites by passive remote sensing analyzing specific bands of the electromagnetic spectrum (Jensen, 2007). Currently, satellites using microwaves or bands in the visible and infrared spectra permit one to quantify environmental variables such as rainfall, CO_2, water vapor, cloud fraction, and land temperature (NASA, 2012b).

Two important advances in the last 20 years now allow one to measure ground movements, one key factor for many natural hazards: (1) the GNSS (Global Navigation Satellite System), which allows measuring 3D displacements, and (2) the satellite and terrestrial InSAR techniques that permit one to map very accurate displacements using two successive radar images by comparing the phase signal. Of course, local direct measurements of displacements such as extensometers, tide float gauges, or inclinometers are still very much used and complement these recent techniques.

The final goal of hazard monitoring is to provide information about physical parameters directly or indirectly interpreted in order to evaluate the level of risk. The following presents some of the most current methods used to monitor the main hazards affecting human activities.

Meteorological monitoring

Monitoring meteorological variables is mainly dedicated to weather forecasting but also to the understanding of climate change. It covers phenomena from local to global scale. Spatial and temporal scales of the phenomena are linked. Local and extreme events, such as tornadoes, hail, or thunderstorms, last only a few minutes to hours, and their location and intensity cannot be forecasted in advance. These kind of events are the topics of short-range forecasting, or nowcasting, that rely on observations and measurements of the phenomena after its initiation, as, for instance, by means of satellite or ground-based radar data. Regional events, such as heavy precipitation over a mountain range, strong winds over a country, or hurricanes, can usually be foreseen a few days in advance. These are forecasted at medium range by numerical weather forecast models that rely on the actual state of the atmosphere, assessed by radiosounding balloons, meteorological stations, or satellite images. The global scale is related to climate changes and is monitored by temperature measurements (Figure 1), sea level rise tracking, and various other indices.

Weather monitoring is thus dedicated to forecasting but also to increase the knowledge about the phenomena. Most of the data acquired during an event are then used by the scientific community for various applications, such as statistical analyses, improvement of the understanding of the processes, or development of more reliable models.

Monitoring of local extreme events

The short-range forecasting, often referred to as nowcasting, focuses on the pending few hours and the local scale. It strongly relies on monitoring to anticipate the displacements of the occurring hazard.

Thunderstorms with intense precipitation or hail are usually tracked by means of ground-based precipitation radars. The returning radar pulses provide the spatial distribution of the hydrometeors and so the intensity of the precipitation. The diameter of the raindrops or the hail may be approximated based on the reflectivity factor or the signal attenuation. The main advantage of radar measurements is that it provides real-time precipitation information on a large area, but there are several issues for precipitation estimation. The first one is that the drops are detected on a wide range of altitudes and the calculated intensity may not match ground observations due to wind or evaporation (Shuttleworth, 2012). Another issue is for mountainous regions, as mountain ranges are responsible for beam shielding (Germann et al., 2006). However, various algorithms and correction methods exist to make the radar data valuable for nowcasting. The goal of such forecasting is to assess the motion and the evolution of precipitation patterns (Austin and Bellon, 1974). While it was initially just an extrapolation of the patterns, it is becoming more sophisticated by use of numerical forecasting models that are initialized with radar data (Wilson et al., 1998).

Tornado detection is possible using a Doppler radar, which uses the Doppler effect on the reflected pulse to assess the velocity of hydrometeors, according to the radial axis. By displaying the motion within a storm, it becomes possible to identify a tornado vortex signature (Donaldson, 1970; Brown et al., 1978), which is characterized by an intense and concentrated rotation. With this approach, the presence of tornado genesis can be identified before a tornado touches the ground. The US government deployed a network of 158 Doppler radars for tornadoes monitoring between 1990 and 1997 (NOAA website).

Monitoring of regional meteorological variables

Today's weather forecasts are mainly based on numerical weather prediction (NWP) models. However, these models rely on data assimilation, which is a statistical

Monitoring Natural Hazards, Table 1 Description of the most common sensors used to monitor natural hazards

Sensors	Monitored variables	Principles	Monitored phenomenon
Pressure measurement	Pressure (air, water), in situ stress measurement	Barometer: used a height of fluid in vacuum to compensate the atmospheric pressure. Pressure transducer: convert a material deformation electrical signal	Atmospheric circulation, water table, Earth crust deformations
Radar (RAdio Detecting And Ranging)	Distance to a hard object and velocity	Reflection of an emitted microwave by an object and received by an antenna. The Doppler effect permits to estimate the speed of an object	Precipitation imaging, river discharge (velocity), sea level rise, tornadoes
Laser (Light Amplification by Stimulated Emission of Radiation) and *Lidar* (Light Detection And Ranging)	Distance to a surface and orientation	The Laser consists in amplifying coherent light by using the principle of stimulated emission, creating a narrow beam that can be reflected by surfaces. The Lidar uses the principle of range finder by evaluation of the distance by the time of flight or the phase comparison. The direction of the beam is recorder in order to obtain the 3D coordinates. Information on the reflectivity can be also obtained	Landslides movements and characterization, local atmospheric circulations
Thermometer	Temperature	The measurement is realized using changes of the properties of materials under temperature variations such as volume (mercury), or the electric resistance such as thermistors or thermocouple which produce a current proportional to the temperature between two different materials	Climate, weather forecasts, volcano
Accelerometer and seismometer	Acceleration, velocity, displacement	Measurement of ground acceleration using transforming movement into electrical signal	Earthquake, surface deformation (landslides)
Wind sensor	Wind speed and direction	Anemometer is a rotating device entrained by wind such as cup. Anemometers usually use three half spheres like rotating along a vertical axe. The windvanes is a device which is orientated parallel to the wind. Measurement of ultrasonic wave by several sensors permits to obtain the wind velocity and direction	Weather, hurricanes, tornadoes
Rain gauge	Amount of precipitation throughout time	The traditional rain gauges are tipping-bucket, like a container that is emptied each time the unitary volume that can be measured is reached. Precipitations can also be measured using rain drop impact counts	Weather, bad weather
InSAR (interferometric synthetic aperture radar)	Topography, small surface displacement using radar	By using ground-based or satellite InSAR images, it is possible to extract a distance to the ground and a very accurate changes between two images down to millimeter resolution in the direction of line of site. This is based on microwave interference	Earth surface deformation: Earthquakes, volcanoes, landslides, subsidence
GNSS (Global Navigation Satellite System)	Ground position	The principle is to acquire several highly precise travel times of microwaves from at least two satellites (with highly precise positions) and to compute the distance and location to calculate the best position (can be improved include the phase information). Highest accuracy is obtained by using differential GNSS method which computes difference with a well-known GNSS position. This remove several error such atmospheric and ionosphere one. The position resolution reaches a few millimeters	Earthquakes, volcanoes, landslides, subsidence

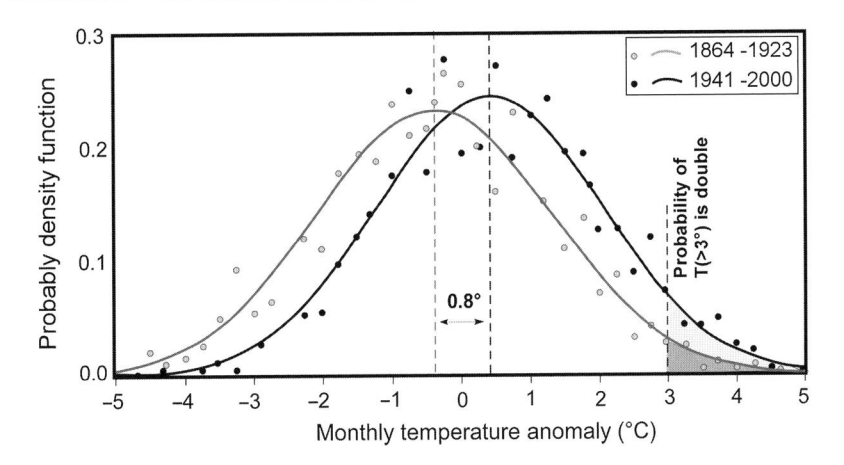

Monitoring Natural Hazards, Figure 1 Statistics of Swiss monthly temperature differences to the average over the whole period. This shows a shift of 0.8°C. The probability to get a monthly temperature 3°C greater than the average temperature is at least twice for the period 1941–2000 compare to 1864–1923 (Modified from Schär et al., 2004).

combination of observations and short-range forecasts, to adjust the initial conditions to the current state of the atmosphere (Daley, 1993; Kalnay, 2003). Data such as temperature, pressure, humidity, and wind are acquired by weather stations, or radiosounding balloons to get a profile of the troposphere (Malardel, 2005).

Air temperature, barometric pressure, wind speed, and direction are commonly measured at weather stations, but also with costal or drifting weather buoys. Some boats and aircrafts are also equipped with sensors acquiring various atmospheric variables.

Rain gauge stations provide point precipitation measurement. It is the first and most common way to measure precipitation, and so it has the advantage that long time series exist. However, these are subject to systematic errors (values lower by about 5–10%) related to the wind and to the choice of the gauge site (over exposure to the wind in open area or shade effect from obstacles around) and gauge design (Shuttleworth, 2012). The height of the gauge is a defined parameter and balances the effect of the wind that decreases closer to the ground, and of the splash-in that increases nearer to the ground. The rain gauges evolved to reduce errors linked to the wind, to evaporation, and to condensation, and changed from manual measurements toward automatic recording.

Weather station networks are organized at a national or regional scale. In 1995, the World Meteorological Organization proposed a resolution (Resolution 40) to "facilitate worldwide co-operation in the establishment of observing networks and to promote the exchange of meteorological and related information in the interest of all nations" (WMO, 2012b). This database contains time series from all over the world.

Precipitation assessment by remote sensing is not as accurate as ground-based measurements, but it provides information in area where no or few observations exist. It is likely to be the only way for precipitation measurement to be possible at a global scale (Shuttleworth, 2012). The Tropical Rainfall Measuring Mission (TRMM) satellite with precipitation radar onboard allows measuring the vertical structure of precipitation (Iguchi et al., 2000; Kawanishi et al., 2000). Precipitation can also be derived from visible and infrared satellite data (Griffith et al., 1978; Vicente et al., 1998).

In addition, the meteorological satellites such as meteosat-9 (www.eumetsat.int) deliver images in visible or infrared spectra providing important data to the meteorologist. It is also a very important source of information in case of the development of severe hazards, such as hurricanes.

Monitoring of climate and climate change

Climate studies rely on long series of high-quality climate records (Figure 1). The most analyzed parameter is the air temperature. Scientists use data recorded at weather stations over decades and employ different methods to reconstruct past data before the beginning of the measurements. Data reconstruction, rescue, and homogenization are still important topics today.

Some satellites have radiometers on board to monitor clouds and thermal emissions from the Earth and Sea Surface Temperature (SST) (NASA, 2012a). For instance, SST can be measured using the calibrated infrared Moderate Resolution Imaging Spectroradiometer (MODIS) installed on Observing System satellites Terra (Minnett et al., 2002). The sea level can be measured using a Radar altimeter of the Jason-2 satellite, which permits one to provide inputs for El Niño or hurricane monitoring. Sea level rise is mainly caused by climate change and is currently about 3.4 ± 0.4 mm/year (Nerem et al., 2010).

Floods monitoring

Floods have several origins often linked to intense precipitation, massive snowmelt, tsunamis, hurricanes, or storm surges, but several are related to other hazards like landslides and rockfalls. The main instrumental setups to forecast floods are weather stations, with a particular emphasis on the rain gauge, weather radars, and meteorological models.

The direct monitoring of floods is done by measuring rivers discharge and/or lakes and sea level. The river discharge is linked to the measurement of the stage (or level), which is the water height above a defined elevation, by a stage-discharge relation. The stages of rivers or lakes are measured by float, ultrasonic, or pressure gauges (Olson and Norris, 2007; Shaw, 1994). The stage-discharge relation has to be updated frequently because of erosion and deposition problems. This relationship is established using current-meters based on rotor or acoustic Doppler velocimeter which establishes the velocity contours of the river section (Olson and Norris, 2007; Shaw, 1994). Radars are also used and seem to be a promising way to obtain discharge (Costa et al., 2006), by using ground-penetrating radar (GPR; the echo of emitted microwave permit to get the river bed profile) coupled with a Doppler velocimeter in order to get the discharge estimation.

In several lowland areas, flood monitoring includes the embankment monitoring that means stability analysis as for landslides. The survey of affected flood area is performed by man-made mapping, aerial photography, or satellite imaging when the flood area is wide, as in Bihar (India) in August 2008 (UNOSAT, 2012).

Earthquake monitoring

Earthquakes monitoring has two objectives: one to provide data for hazard assessment and the other to develop some aspects for prediction. The main recent technologic advances are GNSS and InSAR techniques that allow one to observe the deformation of the Earth's crust before (interseismic), during (coseismic), and after (post-seismic) an earthquake (Figure 2). This permits, for instance, to expect large earthquakes like in the Cascadia Subduction Zone (Hyndman and Wang, 1995), California, and Turkey (Stein et al., 1997).

The displacements recorded by several seismometers provide the necessary information to estimate the location of an earthquake, its magnitude or the energy released. The statistics of the magnitude for defined zones lead to define the Guntherber-Richter law which may be used to obtain the probability of occurrence for earthquakes of a magnitude larger than a given value. In addition, fine analysis using inversion methods of wave signal provides information to characterize the surface of failure (Ji et al., 2002).

The use of monitoring to predict events within a few days or hours is not yet possible because of the variability of geodynamical contexts. For example, a monitored variable may display opposite signals depending on the context, such as radon which can increase before earthquakes as in Kobe in 1995 (Igarashi et al., 1995) but which can also decrease (Kuo et al., 2006). The amplitude of the signal is thus not significant. The observation of an enhanced activity close to a fault (foreshocks) can be used as signal, but this activity increase does not necessarily lead to earthquakes.

The forecast is still not accurate, but observed ground deformations coupled with history of earthquakes permit one to estimate the probability that large earthquakes occur at a location within a period of time (Stein et al., 1997). The two most promising methods are the following: (1) The first is to characterize the ground mechanical properties using ambient seismic noise. The post-seismic period leads to significant seismic velocity changes (Brenguier et al., 2008), indicating most probably stress field modification, but it seems from recent results that it can also be observed before the earthquake. (2) The second is to analyze ionospheric anomalies of the total electron content that are detected before earthquakes by GNSS systems (Heki, 2011).

Tsunamis monitoring

Tsunamis can have different origins including earthquakes, large volcanic eruptions, submarine landslides, rock falling into water, etc. The indirect monitoring is related to the triggering factors of the phenomenon, which are mainly earthquakes or landslides. The Åknes rockslide in Norway is an example of indirect monitoring applied to mountainside instability of significant volume that can fall into a fjord and generate a tsunami. The monitoring of the instability is part of a full early warning system including the evacuation of villages located on the coast within a few minutes (Blikra, 2008).

The direct monitoring of tsunamis is the record of the wave propagation and can be fundamental for different reasons: a large earthquake does not lead necessarily to a tsunami, then the alarm should be canceled if the closest gauges do not indicate any wave (Joseph, 2011); the wave can occur later than expected; the occurrence of landslides (submarines or not) are not always detected. In addition to tide gauges, several seafloor sensors (pressure) are located near the coastal areas of continents and islands, but also in the middle of the ocean (Joseph, 2011). The most advanced monitoring system is the Deep-Ocean Assessment and Reporting of Tsunamis (DART II), and it consists of a surface buoy localized by GNSS and communicating the pressure recorded at the bottom of the ocean by a pressure sensor. The communication with a satellite is bidirectional (Meinig et al., 2005). Such devices are being deployed all over the world (NOAA, 2012) showing great results, like the satellite altimeters that

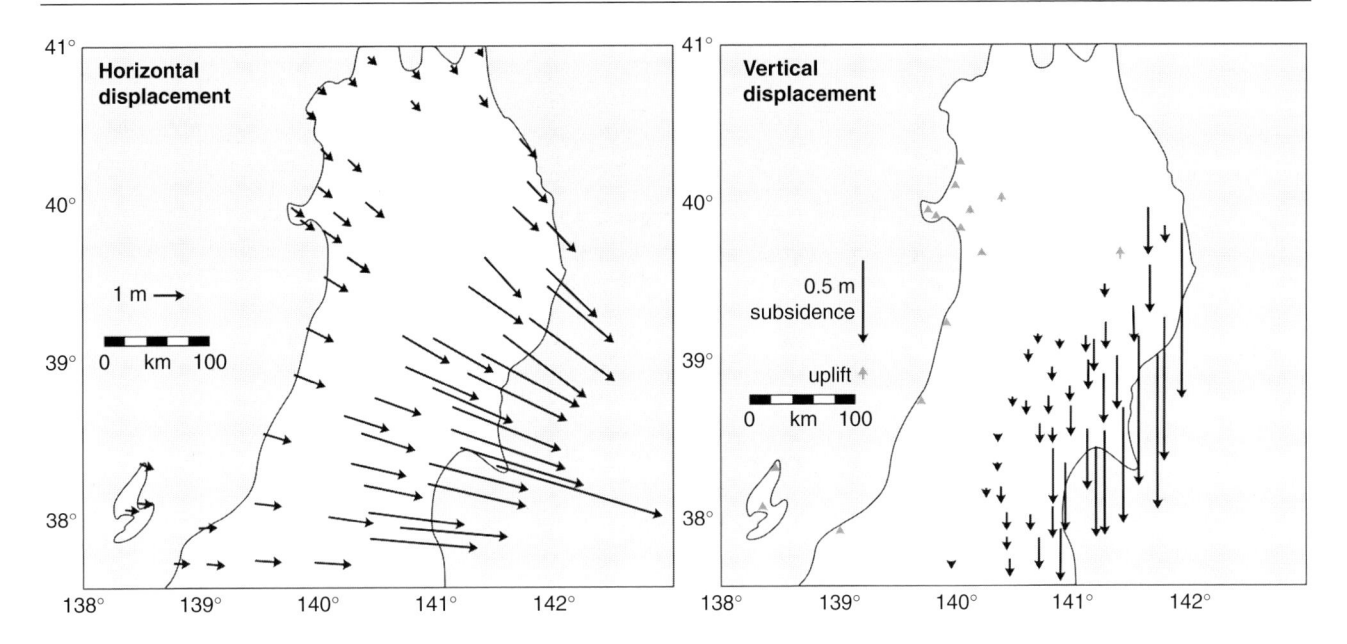

Monitoring Natural Hazards, Figure 2 Coseismic crustal deformation of the Tohoku Earthquake. Horizontal and vertical displacement. These displacements are defined by the difference between the positions on the day before the mainshock (March 10) and those after the mainshock, March 11 (Modified and simplified after RCPEVE, 2012).

recorded accurately the 2004 Sumatra tsunami wave all around the world (Smith et al., 2005).

Volcanoes monitoring

Volcanoes are one of the most spectacular natural hazards on Earth and can be the most disastrous. As an example, the eruption of the Krakatau (Indonesia) in 1883 killed some 30,000 people, releasing a significant volume of ash that briefly affected climate (Durant et al., 2010) and generated a large tsunami wave (Gleckler et al., 2006). As eruption types are so diverse, their monitoring is not easy. Several activities can provide precursory signs, linked to magma movements which change the properties of the ground. The first activity signs that are usually monitored by seismographs are tremors indicating stress adjustments. These stress changes induce ground deformations that can be observed by high precision tiltmeters, indicating changes in slope of the surface. Currently, GNSS are commonly used (Figure 3); they can provide continuous 3D displacements and have partially replaced the electronic distance meter (EDM) laser beam. In addition, since the early works of Massonnet et al. (1995), the InSAR technique allows one to observe deformation of volcanoes, providing information on their behaviors. Any change in the ground can influence measurable parameters such as gravity, temperature, and magnetic field. All those variables can be monitored. The change in gas composition in fumaroles is frequently reported, especially an increase in CO_2 content or a change in the ration F/Cl. Nevertheless, it is quite difficult to monitor gases because they follow preferential paths up to the surface that can change during a precursory period (McNutt et al., 2000). At Etna volcano, ambient seismic noise signature has been recognized as a potential precursor that can be monitored in order to forecast an eruption (Brenguier et al., 2008).

The monitoring of volcanoes does not only involve the volcano itself, but also ash that can disturb aerial traffic or have an impact on the agriculture. Sulfur dioxide, ash, and aerosols (sulfuric acid) are mostly monitored by satellite imaging (ultraviolet and infrared sensors) which is not designed directly for that purpose (Prata, 2009). As those processes are closely linked to atmosphere movements, many of the monitoring techniques of weather forecasting are also used.

Landslides monitoring

Landslides are easily observed because they are moving masses affecting and deforming the relief. As a consequence, the main variables to monitor are the movement and parameters that are modifying the stress or the properties of the material that is under deformation (SafeLand, 2010). Except in the case of earthquakes or exceptional precipitation, the displacement is the main parameter to monitor. In most of the cases, the failure is preceded by an acceleration of movements. Depending on the material geometry and the volume involved, the failure may be forecasted (Crosta and Agliardi, 2003), and this acceleration can sometime be directly correlated with groundwater level using a mechanical model (Corominas et al., 2005).

Two types of landslides must be distinguished: shallow and deep-seated landslides. The first are too small and too localized to be easily monitored, but today several

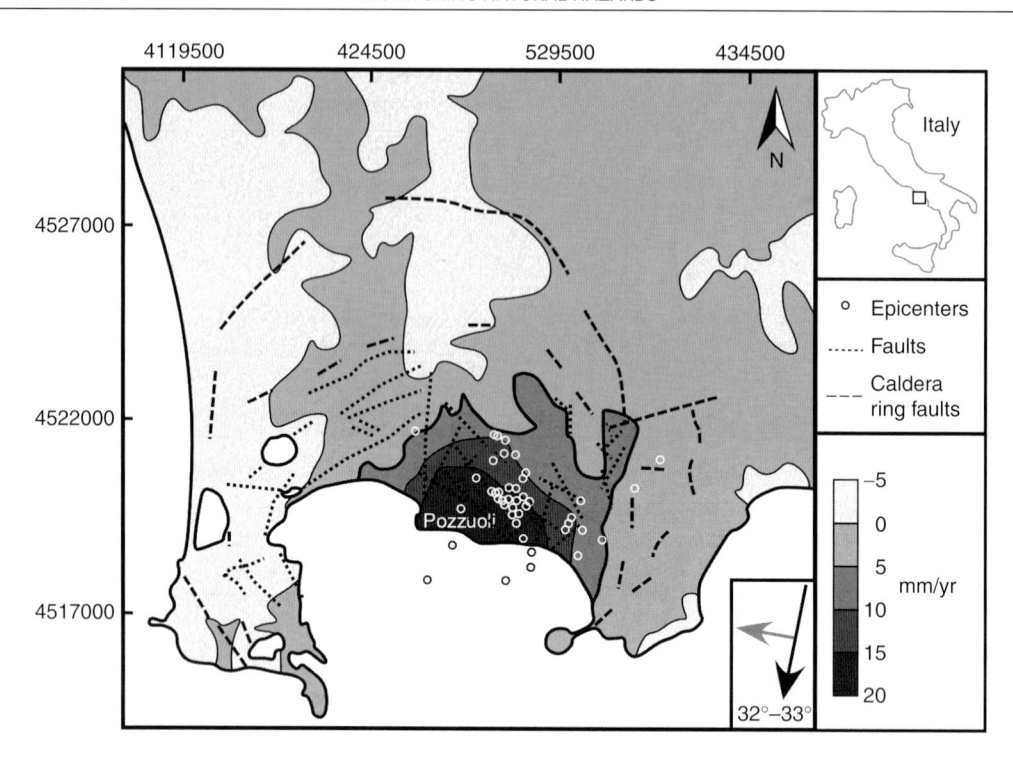

Monitoring Natural Hazards, Figure 3 PS-InSAR™ showing uplift along the line of sight with data from descending orbit on October 2005–November 2006. Observe the correlation between uplifts, structures, and seismic activity (Modified and synthetized after Vilardo et al., 2010).

attempts are made to create early warnings for shallow landslides (Sassa et al. 2009). The deep-seated failures are usually sufficiently large to display significant movements before catastrophic failure.

Large landslides monitoring

The main instruments used to monitor large landslides are dedicated to movements. Physically, extensometers can be used to measure displacements and crack meters can be used to observe the opening of cracks. When boreholes are available, manual inclinometer or permanent inclinometer columns may be used, providing the deformation profiles and often the failure surface where most of the deformation concentrates. These devices are often used for early warning system, as for the site of Åknes (Norway) (Blikra, 2008). As water plays an important role in controlling movements of a landslide, boreholes can be used to measure the level of the water table (manually or by measuring the groundwater pressure).

Surface movements can be followed using targets and total station (laser distance meter), but today, if the required conditions of visibility are appropriate, permanent GNSS can be used for a permanent monitoring of the movements (Gili et al., 2000). The disadvantage of these methods is that they are point measurements only. By using advanced satellite InSAR techniques (PS-InSAR, SBAS, etc.), a significant percentage of landslides can be imaged and monitored. In addition, time series of displacement of ground reflectors can be obtained. One of the last evolutions of the InSAR is the SqueeSAR™ method that enhances significantly the capability of tracking ground displacement (Ferretti et al., 2011). Unfortunately, satellite InSAR is not suitable for early warning because satellites take several days to pass over an area a second time. If no appropriate reflective object exists on the monitored surface (for instance due to forest cover), the InSAR method can be applied only if corner reflectors are installed on the ground, providing movements on selected points only (Singhroy et al., 2011). With ground-based InSAR (GB-InSAR), it is possible to follow the movements of the surface of a landslide or rockslide, when it is visible in the direction of the line of sight. This is very useful to observe the deformation evolution of the front of landslides (Tarchi et al., 2003).

The Lidar technique provides full 3D point clouds in the case of terrestrial Laser scanner (TLS), which allows characterizing rock slopes and landslides (Safeland, 2010; Jaboyedoff et al., 2012). It permits one to monitor and to follow the full evolution of a landslide surface that is moving, to understand mechanisms of failure (Oppikofer et al., 2008) and also to monitor rock fall by comparison of successive acquisitions (Figure 4). The airborne Laser scanner (ALS) is less accurate but

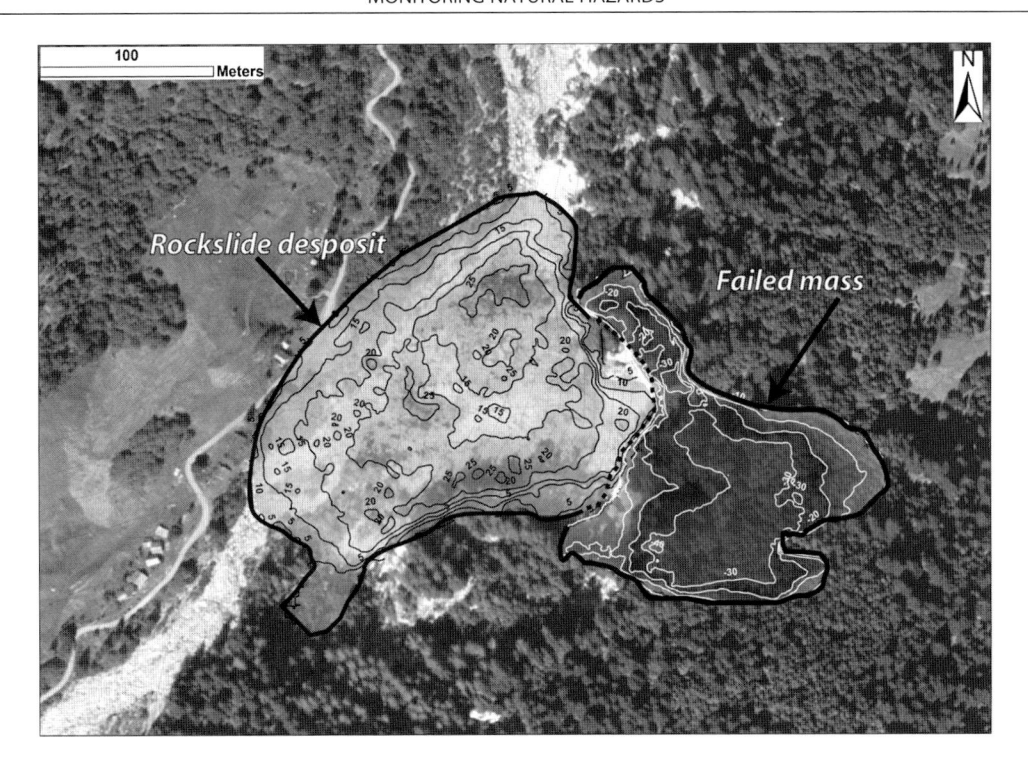

Monitoring Natural Hazards, Figure 4 Map of the deposit and failed mass thickness of the of the Val Canaria rockslide (Ticino, Southern Swiss Alps). This map based on the comparison of the airborne and terrestrial Lidar digital elevation model taken before and after the 27.10.2009 rockslide event (modified after Pedrazzini et al., 2011; the aerial picture and airborne Lidar are provided by swisstopo).

permits one to estimate differences between digital elevation models.

For most landslides, several different sensors are required to establish an early warning system (Blikra, 2008; Froese and Moreno, 2011). Since a few years ago, photogrammetry and image correlation have developed, leading to very promising results (Travelletti et al., 2012). Geophysics methods are also improving their capabilities to image the underground. One of the most interesting recent developments is ambient seismic noise analysis. For a rock mass, it indicates a decrease of the natural frequency before failures and for landslides, a decrease of the surface wave velocity (Mainsant et al., 2012).

Debris flow and shallow-landslides monitoring

Shallow landslides and debris flow landslides are mostly dependent on precipitation. As a consequence, the main monitored variables are precipitation intensity, and duration (Baum and Godt, 2010; Jakob et al., 2011). Saturation, soil moisture, and antecedent precipitation are variables that are also often monitored. In the case of shallow landslides, the exact location cannot be determined, thus the entire area is considered as hazardous if some thresholds are exceeded. It must be noted that an early warning system designed for rainfall-induced landslides is operational in Hong Kong and has been continuously improved since 1977 (Chan et al., 2003; Sassa et al., 2009).

In the case of debris flows, sensitive catchments can be equipped in order to issue warnings. The seismic sensors and ultrasonic gauges permit one to deduce velocity and peak discharge (Marchi et al., 2002).

Monitoring shallow landslides and debris flows is still a topic of research under development because the triggering and the localization of such phenomena are not yet well understood.

Snow avalanche monitoring

Snow avalanches are seasonal events and depend strongly on climate variables such as previous precipitation, snowpack depth and strength, and temperatures. As a consequence, snow avalanches monitoring concentrates essentially on hazard level quantification. This is mainly performed using human observations (SLF, 2012) and weather stations equipped by ultrasonic snow depth sensors. Observed variables are strongly dependant on local physiographic conditions. In addition to monitored data, the observers perform snow hardness tests in order to detect the potential mechanical weakness in the snowpack (Pielmeier and Schneebeli, 2002). The conditions for avalanches are so diverse (wet snow, large amount of fresh snow, etc.), that up to now, human intervention in the monitoring remains the main method to monitor and forecast this hazard.

Other monitoring

There are other hazards to monitor. Some require the integration of meteorological data in the monitoring design. For instance, a drought corresponds to a period of abnormally dry weather leading to a deficit of water in the hydrologic cycle and finally leading to problems (but the definition of drought is not unique). Forest fires are consequences of dryness, with origins that are often not natural, but anthropogenic. Hail storms are also hazardous phenomenon that can lead to serious damage; hail monitoring is mainly based on human observation and meteorological radar. Lightnings are monitored using an electromagnetic sensors network. All the sensors detecting one specific lightning provide the distance to it. The location is then deduced by searching the best agreement between all the detected distances to sensors.

Future of monitoring as a demand of the society

The monitoring of natural hazards is often a tedious task because if the physics well describes the single phenomenon, in natural environments, the occurrence of an event is controlled by several simultaneous phenomena. It implies that, for the analysis and prediction of events, a number of different variables are required to be able to describe all possible cases.

The power of computer science, communication technologies, and the improving quality of sensors, combined with decreasing prices, make the monitoring of environmental data more precise and easy. This leads to new understanding of natural hazards and also to the implementation of early warning systems that will permit one to manage territories in a safer way. In addition, nowcasting, as proposed by World Meteorological Organization, is now an objective of this organization to provide forecasts in less than 6 h. Such developments are mainly possible because of computer power available almost everywhere and a generalized ability to communicate rapidly by anybody with the "smartphone" technology.

Bibliography

Austin, G. L., and Bellon, A., 1974. The use of digital weather radar records for short-term precipitation forecasting. *Quarterly Journal of the Royal Meteorological Society*, **100**(426), 658–664.

Baum, R. L., and Godt, J. W., 2010. Early warning of rainfall-induced shallow landslides and debris flows in the USA. *Landslides*, **7**, 259–272, doi:10.1007/s10346.

Blikra, L. H., 2008. The Åknes rockslide; monitoring, threshold values and early-warning. In Chen, Z., Zhang, J., Li, Z., Wu, F., Ho, K. (eds.), *Landslides and Engineered Slopes, From Past to Future, Proceedings of the 10th International Symposium on Landslides*. Taylor and Francis Group. pp. 1089–1094.

Brenguier, F., Shapiro, N., Campillo, M., Ferrazzini, V., Duputel, Z., Coutant, O., and Nercessian, A., 2008. Towards forecasting volcanic eruptions using seismic noise. *Nature Geoscience*, **1**, 126–130.

Brown, R., Lemon, L., and Burgess, D., 1978. Tornado detection by pulsed Doppler radar. *Monthly Weather Review*, **106**, 29–38.

Chan, R. K. S., Pang, P. L. R., and Pun, W. K., 2003. Recent developments in the landslips warning system in Hong Kong. In Ho, K. K. S, Li, K. S. (eds.) *Geotechnical engineering – meeting society's needs, Proceedings of the 14th Southeast Asian Geotechnical Conference*. Hong Kong. Balkema, Rotterdam, pp. 219–224.

Corominas, J., Moya, J., Ledesma, A., Lloret, A., and Gili, J. A., 2005. Prediction of ground displacements and velocities from groundwater level changes at the Vallcebre landslide (Eastern Pyrenees, Spain). *Landslides*, **2**, 83–96.

Costa, J. E., Cheng, R. T., Haeni, F. P., Melcher, N., Spicer, K. R., Hayes, E., Plant, W., Hayes, K., Teague, C., and Barrick, D., 2006. Use of radars to monitor stream discharge by noncontact methods. *Water Resources Research*, **42**, W07422, doi:10.1029/2005WR004430.

Crosta, G., and Agliardi, F., 2003. Failure forecast for large rock slides by surface displacement measurements. *Canadian Geotechnical Journal*, **40**, 176–191.

Daley, R., 1993. *Atmospheric Data Analysis*. Cambridge: Cambridge University Press.

Donaldson, R. J., 1970. Vortex signature recognition by a Doppler radar. *Journal of Applied Meteorology*, **9**, 661–670.

Durant, A. J., Bonadonna, C., and Horwell, C. J., 2010. Atmospheric and environmental impacts of volcanic particulates. *Elements*, **6**, 235–240.

Ferretti, A., Fumagalli, A., Novali, F., Prati, C., Rocca, F., and Rucci, A., 2011. A new algorithm for processing interferometric data-stacks: SqueeSAR. *IEEE Transactions on Geoscience and Remote Sensing*, **49**, 3460–3470.

Froese, C. R., and Moreno, F., 2011. Structure and components for the emergency response and warning system on Turtle Mountain. *Natural Hazards*, doi:10.1007/s11069-011-9714-y.

Germann, U., Galli, G., Boscacci, M., and Bolliger, M., 2006. Radar precipitation measurement in a mountainous region. *Quarterly Journal of the Royal Meteorological Society*, **132**, 1669–1692.

Gili, J. A., Corominas, J., and Rius, J., 2000. Using global positioning system techniques in landslide monitoring. *Engineering Geology*, **55**, 167–192.

Gleckler, P. J., Wigley, T. M. L., Santer, B. D., Gregory, J. M., AchutaRao, K., and Taylor, K. E., 2006. Volcanoes and climate: Krakatoa's signature persists in the ocean. *Nature*, **439**, 675.

Griffith, C., Woodley, W., Grube, P., Martin, D., Stout, J., and Sikdar, D., 1978. Rain estimation from geosynchronous satellite imagery-visible and infrared studies. *Monthly Weather Review*, **106**(8), 1153–1171.

Heki, K., 2011. Ionospheric electron enhancement preceding the 2011 Tohoku- Oki earthquake. *Geophysical Research Letters*, **38**, L17312.

Hyndman, R. D., and Wang, K., 1995. The rupture zone of Cascadia great earthquakes from current deformation and the thermal regime. *Journal of Geophysical Research*, **100**(B11), 22133–22154.

Igarashi, G., Saeki, S., Takahata, N., Sumikawa, K., Tasaka, S., Sasaki, Y., Takahashi, M., and Sano, Y., 1995. Ground-water radon anomaly before the Kobe earthquake in Japan. *Science*, **269**, 60–61.

Iguchi, T., Meneghini, R., Awaka, J., Kozu, T., and Okamoto, K., 2000. Rain profiling algorithm for TRMM precipitation radar data. *Advances in Space Research*, **25**(5), 973–976.

Jaboyedoff, M., Oppikofer, T., Abellán, A., Derron, M.-H., Loye, A., Metzger, R., and Pedrazzini, A., 2012. Use of LIDAR in landslide investigations: a review. *Natural Hazards*, **61**, 5–28, doi:10.1007/s11069-010-9634-2.

Jakob, M., Owen, T., and Simpson, T., 2011. A regional real-time debris-flow warning system for the district of North Vancouver. Canada. *Landslides*, doi:10.1007/s10346-011-0282-8.

Jensen, J. R., 2007. *Remote Sensing of the Environment: An Earth Resource Perspective*, 2nd edn. Upper Saddle River, NJ: Prentice Hall.

Ji, C., Wald, D. J., and Helmberger, D. V., 2002. Source description of the 1999 Hector Mine, California, earthquake, part I: wavelet domain inversion theory and resolution analysis. *Bulletin of the Seismological Society of America*, **92**(4), 1192–1207.

Joseph, A., 2011. *Tsunamis: Detection, Monitoring, and Early-Warning Technologies*. Amsterdam: Academic.

Kalnay, E., 2003. *Atmospheric Modeling, Data Assimilation, and Predictability*. Cambridge: Cambridge University Press.

Kawanishi, T., Kuroiwa, H., Kojima, M., Oikawa, K., Kozu, T., Kumagai, H., Okamoto, K., Okumura, M., Nakatsuka, H., and Nishikawa, K., 2000. TRMM precipitation radar. *Advances in Space Research*, **25**(5), 969–972.

Kuo, T., Fan, K., Kuochen, H., Han, Y., Chu, H., and Lee, Y., 2006. Anomalous decrease in groundwater radon before the Taiwan M6.8 Chengkung earthquake. *Journal of Environmental Radioactivity*, **88**, 101–106.

Landsea, C. W., 2007. Counting Atlantic tropical cyclones back to 1900. *EOS Transactions, American Geophysical Union*, **88**(18), 197–202.

Mainsant, G., Larose, E., Brönnimann, C., Jongmans, D., Michoud, C., and Jaboyedoff, M., 2012. Ambient seismic noise monitoring of a clay landslide: toward failure prediction. *JGR-ES*, **117**, F0103, 12 pp, doi:10.1029/2011JF002159.

Malardel, S., 2005. *Fondamentaux de météorologie. À l'école du temps*, Toulouse: Cépaduès.

Marchi, L., Arattano, M., and Deganutti, A. M., 2002. Ten years of debris-flow monitoring in the Moscardo Torrent (Italian Alps). *Geomorphology*, **46**, 1–17.

Massonet, D., Briole, P., and Arnaud, A., 1995. Deflation of Mount Etna monitored by spaceborne radar interferometry. *Nature*, **375**, 567–570.

McNutt, S. R., Rymer, H., and Stix, J., 2000. Synthesis of volcano monitoring, Chapter 8 of *Encyclopedia of Volcanoes*, San Diego: Academic Press, pp. 1165–1184

Meinig, C., Stalin, S. E., Nakamura, A. I., González, F., and Milburn, H. G., 2005. Technology developments in real-time tsunami measuring, monitoring and forecasting. In *Oceans 2005 MTS/IEEE, 19–23 September 2005*, Washington, DC.

Minnett, P. J., Evans, R. H., Kearns, E. J., and Brown, O. B., 2002. Sea-surface temperature measured by the Moderate Resolution Imaging Spectroradiometer (MODIS) Geoscience and Remote Sensing Symposium, 2002. IGARSS'02. 2002 IEEE, Vol. 2, pp. 1177–1179.

NASA, 2012a. Temperature. *National Aeronautics and Space Administration*, http://science.nasa.gov/earth-science/oceanography/physical-ocean/temperature, visited in Mai 2012.

NASA, 2012b. http://earthobservatory.nasa.gov/, visited in Mai 2012.

Nerem, R. S., Chambers, D., Choe, C., and Mitchum, G. T., 2010. Estimating mean sea level change from the TOPEX and Jason Altimeter Missions. *Marine Geodesy*, **33**, 435–446.

NOAA 2012. http://www.ndbc.noaa.gov/dart.shtml, visited in Mai 2012.

Olson, S. A., and Norris, J. M., 2007. U.S. Geological Survey Streamgaging. USGS-Fact Sheet 2005–3131.

Oppikofer, T., Jaboyedoff, M., and Keusen, H.-R., 2008. Collapse of the eastern Eiger flank in the Swiss Alps. *Nature Geosciences*, **1**, 531–535.

Pedrazzini, A., Abellan, A. Jaboyedoff, M., and Oppikofer, T., 2011. Monitoring and failure mechanism interpretation of an unstable slope in Southern Switzerland based on terrestrial laser scanner. *14th Pan-American Conference on Soil Mechanics and Geotechnical Engineering*, Toronto.

Pielmeier, C., Schneebeli, M., 2002. Snow stratigraphy measured by snow hardness and compared to surface section images. In *Proceedings of the International Snow Science Workshop 2002*, Penticton, BC, Canada, pp. 345–352.

Prata, A. J., 2009. Satellite detection of hazardous volcanic clouds and the risk to global air traffic. *Natural Hazards*, **51**, 303–324.

RCPEVE, 2012. The 2011 off the pacific coast of Tohoku Earthquake (M9.0). *Research Center for Prediction of Earthquakes and Volcanic Eruptions*, http://www.aob.geophys.tohoku.ac.jp/aob-e/info/topics/20110311_news/index_html, visited in Mai 2012.

SafeLand, 2010. Deliverable 4.1 – Review of techniques for landslide detection, fast characterization, rapid mapping and long-term monitoring. Edited for the SafeLand European project by Michoud, C., Abellán, A., Derron, M.-H., and Jaboyedoff, M. Available at http://www.safeland-fp7.eu.

Sassa, K., Picarelli, L., and Yueping, Y., 2009. Monitoring, prediction and early warning. In: Chapter 20 in Sassa, K., and Canuti, P. (eds.) Landslides- disaster risk reduction. Springer, pp. 351–375.

Schär, C., Vidale, P. L., Lüthi, D., Frei, C., Häberli, C., Liniger, M. A., and Appenzeller, C., 2004. The role of increasing temperature variability in European summer heatwaves. *Nature*, **427**(6972), 332–336.

Shaw, E., 1994. *Hydrology in Practice*, 3rd edn. London: Chapman & Hall.

Shuttleworth, W. J., 2012. *Terrestrial Hydrometeorology*. Chichester: Wiley-Blackwell.

Singhroy, V., Charbonneau, F., Froese, C., and Couture, R., 2011. Guidelines for InSAR Monitoring of Landslides in Canada. *14th Pan-American Conference on Soil Mechanics and Geotechnical Engineering*, Toronto.

SLF, 2012. http://www.slf.ch/lawineninfo/zusatzinfos/howto/index_EN, visited in Mai 2012.

Smith, W. H. F., Scharroo, R., Titov, V. V., Arcas, D., and Arbic, B. K., 2005. Satellite altimeters measure tsunami. *Oceanography*, **18**, 10–12.

Stein, R. S., Barka, A. A., and Dieterich, J. H., 1997. Progressive failure on the North Anatolian fault since 1939 by earthquake stress triggering. *Geophysical Journal International*, **128**, 594–604.

Tarchi, D., Casagli, N., Fanti, R., Leva, D., Luzi, G., Pasuto, A., Pieraccini, M., and Silvano, S., 2003. Landslide monitoring by using ground-based SAR interferometry: an example of application to the Tessina landslide in Italy. *Engineering Geology*, **68**, 15–30.

Travelletti, J., Delacourt, C., Allemand, P., Malet, J.-P., Schmittbuhl, J., Toussaint, R., and Bastard, M., 2012. Correlation of multi-temporal ground-based optical images for landslide monitoring: application, potential and limitations. *ISPRS Journal of Photogrammetry and Remote Sensing*, **70**, 39–55.

UNOSAT, 2012. http://www.unitar.org/unosat/node/44/1259, visited in Mai 2012.

Vicente, G., Scofield, R., and Menzel, W., 1998. The operational goes infrared rainfall estimation technique. *Bulletin of the American Meteorological Society*, **79**(9), 1883–1898.

Vilardo, G., Isaia, R., Ventura, G., De Martino, P., and Terranova, C., 2010. InSAR permanent scatterer analysis reveals fault re-activation during inflation and deflation episodes at Campi Flegrei caldera. *Remote Sensing of Environment*, **114**, 2373–2383.

Wilson, J., Crook, N., Mueller, C., Sun, J., and Dixon, M., 1998. Nowcasting thunderstorms: a status report. *Bulletin of the American Meteorological Society*, **79**(10), 2079–2099.

WMO, 2012a http://www.wmo.int/pages/themes/weather/index_en.html, visited in Mai 2012.

WMO, 2012b. http://www.wmo.int/pages/about/Resolution40_en.html, visited in Mai 2012.

Cross-references

Accelerometer
Airphoto and Satellite Imagery
Avalanches
Climate Change
Debris flow
Deep-Seated Gravitational Slope Deformations
Doppler Weather Radar
Earthquake
Earthquake Prediction and Forecasting
El Niño/Southern Oscillation
Eruption Types (Volcanic)
Flash Flood
Flood Hazard and Disaster
Hurricane (Cyclone, Typhoon)
Hydrograph, Flood
Inclinometer
North Anatolian Fault
Piezometer
Pore-water Pressure
Remote Sensing of Natural Hazards and Disasters
Rock Avalanche
Rockfall
San Andreas Fault
Santorini
Seismic Gap
Seismograph/Seismometer
Slope Stability
Tiltmeters
Tohoku, Japan, Earthquake, Tsunami and Fukushima Accident (2011)
Tsunami

MONSOONS

Song Yang, Viviane Silva, Wayne Higgins
Climate Prediction Center, NCEP/NWS/NOAA, Camp Springs, MD, USA

Synonyms

Mausam; Rainy season; Wet season

Definition

The term "monsoon" is derived from the Arabic word "mausam," which means season. Halley (1686) defined monsoon as the seasonal reversal of steady and sustained surface winds, which blow from the northeast during winter and from the southwest during summer. In spite of this original definition rooted in atmospheric circulation, rainfall is another variable that has been widely used to define monsoon.

Discussion

Although there is no universal definition, monsoons are atmospheric systems with certain well-defined characteristics (Webster 1987). All monsoons have a life cycle characterized by distinct onset, maintenance, and demise phases. They feature abundant rainfall during summer and dry conditions during winter. The strongest monsoon, the Asian summer monsoon (Ramage 1971), affects about half of the world's population. Monsoons are also found in other tropical–subtropical land areas, including Australia, Africa, South America, and North America (Webster 1987; Nogues-Paegle et al., 2002; Sultan et al., 2003; Higgins et al., 2006).

Monsoon variability is influenced by various weather and climate phenomena, including synoptic-scale disturbances, tropical waves and cyclones, and tropical intraseasonal variations that contribute to active and break periods. Interannual and longer variations of monsoons are due to both internal dynamics of the coupled atmosphere–ocean–land system and interactions of monsoons with other climate phenomena such as El Niño-Southern Oscillation, snow cover, and the Pacific Decadal Oscillation.

Although the major cause of monsoons is the thermal contrast between land and ocean, the discernable features of monsoons vary from region to region. The monsoon climate over many Asian countries is characterized by wet and hot conditions in summer but dry and cold conditions in winter, corresponding to a pronounced seasonal reversal of surface winds. However, regions close to the equator usually experience two rainy seasons. Over eastern Africa, the monsoon rainfall is characterized by "long rain" in March–May and "short rain" in October–December. The North American monsoon is characterized by distinct rainfall maxima over western Mexico and the southwestern United States and by an accompanying upper-level anticyclone over the higher terrain of northwestern Mexico. The South American monsoon features a pronounced wet season (November–March) and a dry season (April–September) over central Brazil. An intense upper-tropospheric anticyclonic circulation, located over eastern Bolivia, appears during the wet season.

Monsoon variability is often related to floods, drought, and other hazardous extreme weather and climate events. Excessive monsoon rainfall causes floods and landslides and hence considerable social and economic impacts. Alternately, insufficient monsoon rainfall leads to drought, and therefore scarcer fresh water supplies. Monsoon depressions and tropical storms with high winds and tidal surges are often embedded within the large-scale monsoon circulation, posing threats to human lives and property. Monsoon behavior, such as the intensity and duration, influences economic planning and development, water resource management, agriculture (planting and harvesting), and emergency response. Because of the significant societal and economic impacts of monsoons, it is important to continue to improve understanding towards more realistic simulation and prediction of monsoons.

Bibliography

Halley, E., 1686. Historical account of the trade winds and monsoons. *Philosophical Transactions of the Royal Society London*, **16**, 153–168.

Higgins, W., Ahijevych, D., Amador, J., and coauthors, 2006. The NAME 2004 field campaign and modeling strategy. *Bulletin of the American Meteorological Society*, **87**, 79–94.

Nogués-Paegle, J., Mechoso C. R., and coauthors, 2002. Progress in Pan American CLIVAR research: Understanding the South American monsoon. *Meteorologica*, **27**, 3–30.

Ramage, C. S., 1971. *Monsoon Meteorology*. New York: Academic, p. 296.

Sultan, B., Janicot, S., and Diedhiou, A., 2003. The West African monsoon dynamics. Part I: documentation of intraseasonal variability. *Journal of Climate*, **16**, 3389–3406.

Webster, P. J., 1987. The elementary monsoon. In Fein, J. S., and Stephens, P. L. (eds.), *Monsoons*. New York: Wiley, pp. 3–32.

Cross-references

Challenges to Agriculture
Cloud Seeding
Doppler Weather Radar
Drought
El Niño-Southern Oscillation
Erosion
Flash Flood
Hydrometeorological Hazards
Storm Surges

CASE STUDY

MONTSERRAT ERUPTIONS

Katherine Donovan
University of Oxford, Oxford, Oxfordshire, UK

Montserrat is a small volcanically active island in the Caribbean situated on the Lesser Antilles island arc. The island's main volcano is called the Soufrière Hills and this volcano has been erupting since 1995.

1995–1998

In 1995 after 40 years of quiescence a relatively small lava dome was extruded. This dome grew at 4 m^3/s until 1997 when the dome collapsed producing multiple pyroclastic flows. These burning clouds of ash destroyed the previously evacuated capital city of Plymouth in March 1997 and killed 19 people in June 1997. The volcano continued to erupt until 1998 showing a cyclic seismic and dome growth behavior that was used by scientists at the newly established Montserrat Volcano Observatory (MVO) to provide short-term forecasts (McNutt et al., 2000). This initial period of activity changed Montserrat dramatically, destroying the prosperous south and forcing residents to relocate to the rugged and difficult north (Figure 1). In 1998 the pre-eruption population of 11,000 had reduced to just 4,000 as long-term evacuations, loss of livelihoods, and personal danger forced the people of Montserrat to transmigrate, mainly to the United Kingdom (Aspinall and Cooke, 1998).

1998–2003

As the people gradually abandoned hope the volcano continued to erupt. Between 1998 and 2003, Andesitic lava domes continued to grow and collapse, for example, in 2000 a 29 million m^3 dome collapsed generating a magmatic eruption and over 40 pyroclastic flows (Carn et al., 2004). In 2003, the volcano produced the largest dome collapse ever recorded in historical time with 210 million m^3 of material giving way, and 170 million m^3 collapsing in just 2 hours of activity (Herd et al., 2005). Figure 2 shows the smoking crater that was left behind. This major collapse followed 2 years of dome growth, caused a tsunami, a previously unrecorded pressure wave, a shock wave, and tephra fall that caused extensive damage on Montserrat and neighboring islands (Herd et al., 2005).

2003-Onwards

The Soufriere Hills is now the best monitored volcano complex in the Caribbean with an array of technologically advanced monitoring equipment and a permanent scientific team. But recent changes in seismicity, which previously aided eruption forecasts, have led to changes of procedure at the MVO and increased pressure to find more accurate precursors (Luckett et al., 2008).

As the physical monitoring of the volcano continues, so does the struggle of the Montserrat people (Figure 3). Relocation to the northern regions caused long term social issues, including a lack of cultural building considerations and inferior agricultural land causing residents to return to the dangerous regions to farm. Transmigration also caused multiple stresses and unanticipated concerns, for example, there was a lower standard of schooling in the UK compared with pre-eruption standards on the island (Kelman and Mather, 2008). As the eruption continues the future of the remaining Montserratians is unclear, they require a sustainable livelihood in order to remain on the island but with limited space and imminent danger this may be difficult to achieve. Scientists and local authorities are under extreme pressure to protect the remaining Montserratians from further suffering.

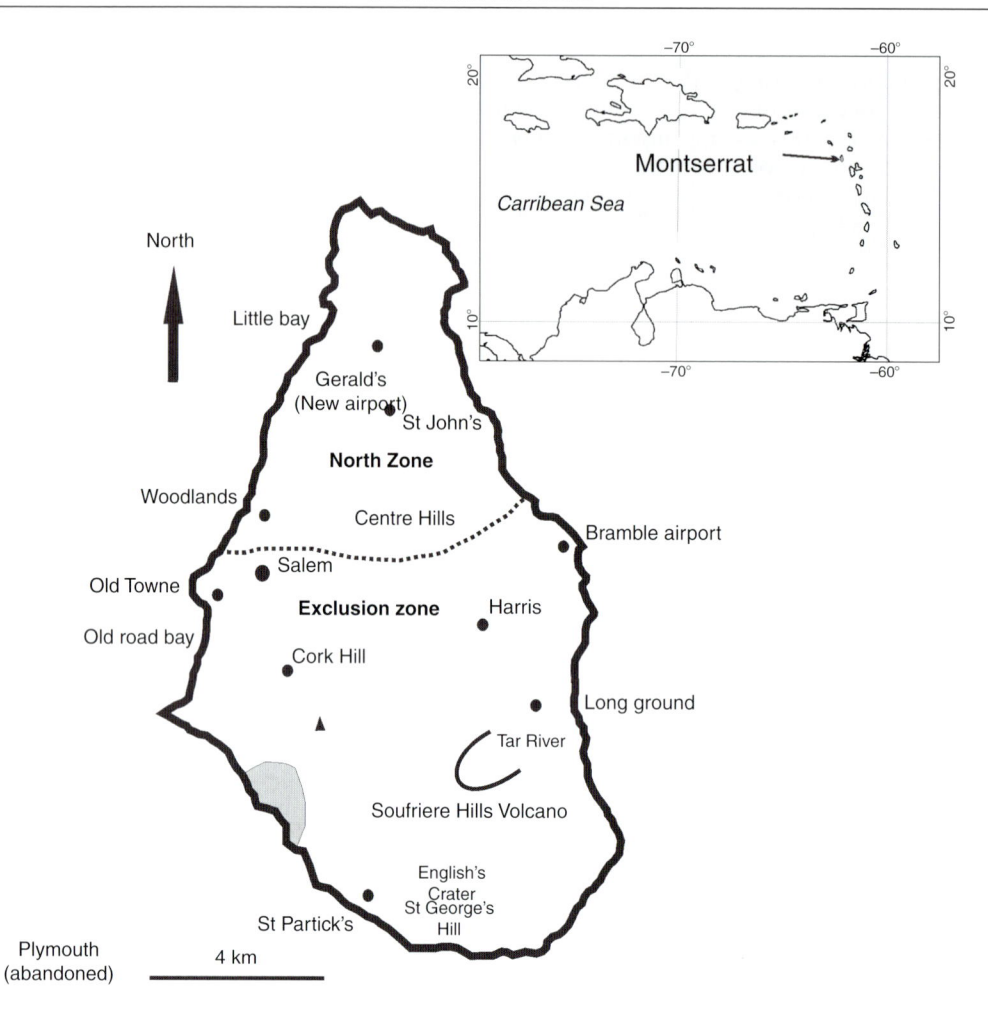

Montserrat Eruptions, Figure 1 Location of Montserrat Island and the Soufriere Hills Volcano. This map also marks the exclusion zone that covers the majority of the southern island.

Montserrat Eruptions, Figure 2 A view of the crater taken in December 2004 (*Catherine Lowe*).

Montserrat Eruptions, Figure 3 A minibus used for tourism is caught in a lahar in November 2004. This image demonstrates the difficulties in maintaining a sustainable livelihood on an active volcanic island (*Catherine Lowe*).

Bibliography

Aspinall, W., and Cooke, R. M., 1998. Expert judgement and the Montserrat Volcano eruption. In Mosleh, A., and Bari, R. A. (eds.), *Proceedings of the 4th International Conference on Probabilistic Safety Assessment and Management PSAM4*, September 13–18, 1998, New York City, USA, Vol. 3, pp. 2113–2118.

Carn, S. A., Watts, R. B., Thompson, G., and Norton, G. E., 2004. Anatomy of a lava dome collapse: the 20th March 2000 event at Soufrière Hills Volcano, Montserrat. *Journal of Volcanology and Geothermal Research*, **131**, 241–264.

Herd, R. A., Edmonds, M., and Bass, V. A., 2005. Catastrophic lava dome failure at Soufrière Hills Volcano Montserrat, 12–13 July 2003. *Journal of Volcanology and Geothermal Research*, **148**, 234–252.

Kelman, I., and Mather, T. A., 2008. Living with volcanoes: The sustainable livelihoods approach for volcano-related opportunities. *Journal of Volcanology and Geothermal Research*, **172**, 189–198.

Luckett, R., Loughlin, S., De Angelis, S., and Ryan, G., 2008. Volcanic seismicity at Montserrat, a comparison between the 2005 dome growth episode and earlier dome growth. *Journal of Volcanology and Geothermal Research*, **177**, 894–902.

McNutt, S. R., Rymer, H., and Stix, J., 2000. Synthesis of volcano monitoring. In Sigurdsson, H. (ed.), *Encyclopedia of Volcanoes*. London: Academic, pp. 1165–1184.

Cross-references

Base Surge
Civil Protection and Crisis Management
Community Management of Hazards
Disaster Risk Reduction
Early Warning Systems
Eruption Types (Volcanic)
Evacuation
Galeras Volcano, Colombia
Human impact of hazards
Krakatoa (Krakatau)
Magma
Mt. Pinatubo
Nevado del Ruiz Volcano, Colombia
Nuee Ardente
Pyroclastic Flow
Santorini
Tsunami
Volcanoes and Volcanic Eruptions

MORTALITY AND INJURY IN NATURAL DISASTERS

Shannon Doocy
Johns Hopkins Bloomberg School of Public Health, Baltimore, MD, USA

Synonyms

Casualties; Fatalities

Definition

Disaster. An event that causes significant damage, destruction, or loss of life where local response capacity is overwhelmed and outside assistance is required.

Natural disaster. Disasters resulting from the effects of naturally occurring hazards such as earthquakes, volcanoes, floods, or extreme climatic events.

Natural disaster mortality. Deaths resulting from a natural disaster, most often those that are immediate and directly attributable to the event.

Natural disaster injury. Physical damage or harm to the body caused by a natural disaster.

Natural disaster mortality and injury in the twentieth century and beyond

Since the beginning of the twentieth century, natural disasters have resulted in over 22.6 million deaths and 6.6 million injuries, and have affected the lives of more than 5.4 billion people (CRED, 2010). While the number of natural disasters reported and the size of populations affected have followed an increasing trend, fatalities have declined as a result of advances in early warning systems, disaster preparedness, and improvements in emergency management and response. However, human vulnerability to natural hazards is escalating, primarily due to the increasing population density and land use change which suggests that the human toll of future natural disasters will rise (Huppert and Sparks, 2006; United Nations, 1988). Poverty is a major risk factor for mortality and injury in natural disasters, and the size of impoverished populations in high-risk areas is likely to increase in future years (Eshghi and Larson, 2008).

A rapid-onset natural disaster is an event that is triggered by an instant shock. Most natural disasters are classified as rapid-onset events though it is important to note that in some cases there is enough warning time to allow for evacuation and other mitigation measures. In contrast, a slow-onset natural disaster unfolds over a longer time period where the hazard is felt as an ongoing stress over days, months, or even years (UNDP, 2004). Natural disaster impacts on human populations from 1900 to date are summarized in Table 1. More than half (52%) of reported deaths in natural disasters since the beginning of the twentieth century are attributable to drought. The significance of drought-related deaths is historically underappreciated where many casualties are secondary or indirect and are uncounted. Floods and earthquakes are also large contributors to natural disaster mortality, accounting for 31% and 10% of deaths, respectively. Natural disaster injuries were overwhelmingly caused by three types of events: earthquakes (33%), extreme temperature events (28%), and floods (20%).

Drought

More than half of disaster-related deaths since the beginning of the twentieth century are attributed to drought, a slow-onset natural disaster that has devastating long-term effects on communities. Drought is a frequent phenomenon that is sometimes associated with famine; however, famines are rare, complex, and often the result of multiple underlying causes including chronic poverty, economic inequalities, and conflicts (Sen, 1982).

Some of the worst famines in the recent history include the 1943 Bengal famine, the Great Leap Forward famine in China from 1958 to 1961, the 1974 famine in Bangladesh, and regional famines in the Sahel during the mid-1970s and mid-1980s (CRED, 2006). In recent decades, drought-related mortality has been concentrated in Africa where in many cases drought-related impacts are exacerbated by conflict and other preexisting cultural and political tensions. Both starvation and disease epidemics are primary causes of drought mortality; however, many secondary deaths where drought is a causal factor go unreported (CRED, 2010). While drought-related mortality is complex, multicausal, and likely to be underestimated, numerous methodologies and long-term development strategies exist that seek to reduce the impacts of drought (Dreze and Sen, 1990; FEWS, 2010). Compared to other types of natural disasters, droughts clearly resulted in the greatest number of deaths in the past century, however, drought-related mortality has substantially decreased in recent history where between 1990 and 2009, there were 37 droughts with a total of 4,472 deaths reported (CRED, 2010).

Mortality and injury in rapid-onset natural disasters, 1980–2009

Rapid population growth and changing trends in natural disasters over time suggest that earthquakes and storms will have the greatest impacts on human populations in the coming decades. Rapid-onset natural disasters, including earthquakes, volcanoes, meteorological events, floods, mass movements, and wildfires, caused over 1.4 million deaths and 5.0 million injuries within the past three decades. Deaths and injuries in rapid-onset natural disasters in the past 30 years are summarized in Figure 1 and Table 2. Earthquakes, which accounted for only 10% of events, resulted in 43% of deaths and 28% of injuries. Storms, including cyclones and hurricanes, comprised 33% of events and were the cause of 30% of deaths and 12% injuries. The most common event, floods, was associated with 16% of mortality and 23% of injuries. Extreme temperature events, which accounted for 5% of rapid-onset natural disaster events, resulted in 7% of deaths and 37% of injuries. Injury reporting is likely more complete in extreme temperature events than other types of disasters, particularly those in the middle- and low-income countries where the majority of mortality and injury occur, because the vast majority of extreme temperature events are in high-income countries where better health information systems ensure more accurate reporting. Other disaster types, including volcanic eruptions, mass movements, and wildfires, accounted for 12% of events collectively but contributed only 4% of mortality and <1% of reported injuries (CRED, 2010).

Earthquakes

Earthquakes are concentrated in Asia which is the most populous continent with approximately 60% of the world's population (UN, 2010). Over the past century,

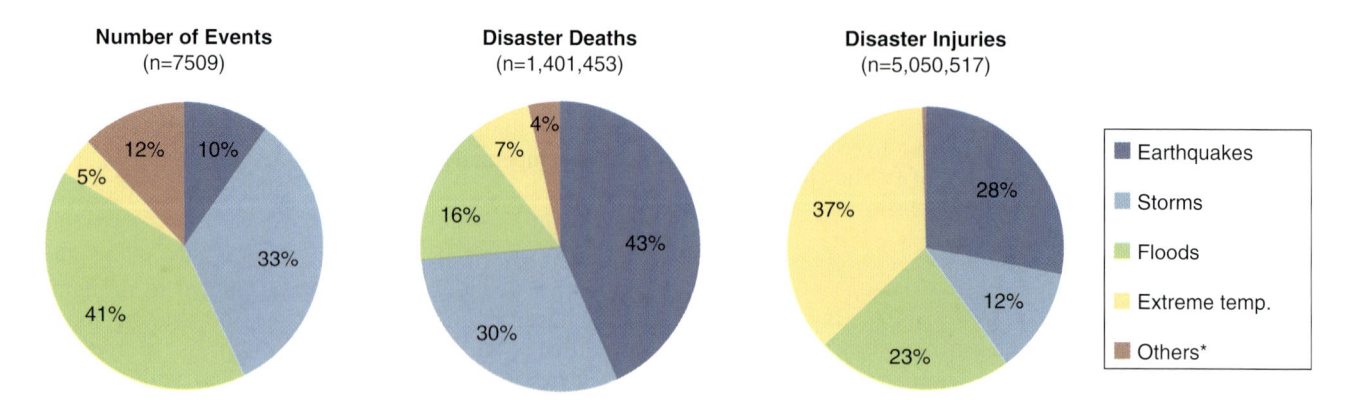

Mortality and Injury in Natural Disasters, Table 1 Mortality and injury associated with natural disasters, 1900–2009[a]

Hazard type	Mortality N	%	Injuries N	%
All geophysical events	2,414,208	10.7	2,191,887	33.0
Earthquake[a]	2,313,294	10.2	2,180,226	32.8
Volcano	95,979	0.4	11,152	0.2
Mass movement dry	4,935	0.0	509	0.0
Meteorological events (storms)	1,374,993	6.1	1,294,556	19.5
All hydrological events	6,968,301	30.9	1,303,199	19.6
Flood	6,913,134	30.6	1,293,919	19.5
Mass movement wet	55,167	0.2	9,280	0.1
All climatological events	11,821,088	52.4	1,856,696	27.9
Drought	11,708,271	51.9	–	0.0
Extreme temperature	109,344	0.5	1,852,761	27.9
Wildfire	3,473	0.0	3,935	0.1
Total	22,578,590	100	6,646,338	100

Source: CRED, 2010
[a]Includes mortality and injury from earthquake-induced tsunamis

Mortality and Injury in Natural Disasters, Figure 1 *Rapid-onset natural disasters and their impact on human populations, 1980–2009 (Source CRED, 2010. *Others include volcanoes, wet and dry mass movements, and wildfires).*

Mortality and Injury in Natural Disasters, Table 2 Casualties in rapid-onset disasters, 1980–2009

Hazard type	Total number of casualties		Average per event	
	Deaths	Injuries	Deaths	Injuries
Earthquakes	617,201	1,412,010	827	1,893
Storms	430,131	611,538	174	247
Floods	199,481	1,155,699	66	380
Extreme temperature	103,475	1,852,161	307	5,496
Others[a]	51,165	19,079	56	21
Overall	1,401,453	5,050,517	187	673

Source: CRED, 2010
[a]Include volcanoes, wet and dry mass movements, and wildfires

53% of earthquakes and 75% of earthquake mortality were in Asia, and their impact in this region has been increasing in the recent decades in parallel with rapid population growth and industrialization. In the past 30 years, 86% of earthquake deaths were in Asia. An average of 827 deaths and 1893 injuries were reported per earthquake disaster between 1980 and 2009. Earthquake-induced tsunamis, which were reported in 3% of earthquakes in the past 30 years, contributed 60% of all earthquake-related fatalities, primarily due to the catastrophic Indian Ocean tsunami in 2004 which resulted in 227,000 deaths and affected over 2.4 million people in the coastal areas of Indonesia, Sri Lanka, India, and Thailand (CRED, 2010). Other recent devastating earthquakes include the 2008 Sichuan earthquake which killed an estimated 87,476 Chinese and the 2010 Haiti earthquake where mortality estimates range from 45,000 to above 300,000 with CRED reporting 22,570 deaths (CRED, 2010; BBC, 2011). The primary cause of death in earthquakes is building collapse, and direct mortality is both rapid, occurring within hours, and delayed where deaths occur within several days of the earthquake (Kunii et al., 1995). Instantaneous deaths are caused by severe crush injuries or trauma-induced hemorrhage; other causes of rapid death include asphyxia from dust inhalation or chest compression, hypovolemic shock, or drowning in earthquake-induced tsunamis. Delayed deaths can be caused by hypothermia, hyperthermia, dehydration, crush syndrome, and sepsis (Safar et al., 1988). In the aftermath of most earthquakes, the majority of people requiring medical assistance have minor injuries including superficial lacerations, sprains, and bruises; fractures and injuries requiring surgery and hospitalization are less common (Noji, 1997). The greatest demand for emergency medical services is within the day following the earthquake, and most of the injured can be treated on an outpatient basis; within 3–5 days, the demand for medical attention at hospital emergency departments usually returns to normal (Schultz et al., 1996; Oda et al., 1997).

Storms

Meteorological events, which include hurricanes, tropical cyclones, local storms, and winter storms, occurred predominantly in Asia (39% of events) and the Americas (32% of events) over the past 30 years; however, their impact is concentrated in Asia where 90% of deaths were reported. An average of 174 deaths and 247 injuries were reported due to meteorological events between 1980 and 2009. Tropical cyclones are by far the most deadly type of meteorological event and accounted for 94% of storm fatalities, or an estimated 428,734 deaths, in the past 30 years. There were an average of 342 deaths per tropical cyclone, and mortality was concentrated in Asia where more than 91% of tropical cyclone deaths were reported. The most devastating recent tropical cyclones include the 1991 Bangladesh cyclone which killed 138,866 people and cyclone Nargis which resulted in an estimated 138,366 deaths in Myanmar in 2008 (CRED, 2010). The majority of storm-related deaths are drownings associated with storm surges; other causes of mortality and injury include burial in collapsed structures, blunt trauma, and storm-induced mudslides (French, 1989; Noji, 2000). Most care seekers after floods suffer from lacerations and can be treated on an outpatient basis; closed fracture and other penetrating injuries are also common (Noji, 1993, 2000).

Floods

Geographically, floods in the past 30 years have been concentrated in Asia (42%), the Americas (23%), Africa (21%), and Europe (14%). However, flood mortality occurred predominantly in Asia and the Americas, which accounted for 65% and 25% of flood deaths, respectively; India and Bangladesh have particularly high levels of flood mortality (NRC, 1987; CRED, 2010). The average flood between 1980 and 2009 resulted in 66 deaths and 380 injuries. General floods accounted for 59% of floods and 48% of flood mortality. Flash floods, which comprised 14% of flood events, were the most deadly type of flood, and accounted for 27% of flood mortality (50,764 deaths) with an average of 121 deaths per flash flood. The deadliest recent flash flood in Venezuela killed an estimated 30,000 people in 1999. Other recent high-mortality floods were in China (3,656 deaths in 1998 and 2,755 deaths in 1996), Haiti (2,665 deaths in 2004), Somalia (2,311 deaths in 1997), and India (1,811 deaths in 1998 and 2001 deaths in 1994) (CRED, 2010). Flood deaths and injuries are primarily the result of fast-flowing water that is laden with debris. The main cause of deaths is drowning, followed by combinations of trauma, hypothermia, and drowning (Beinin, 1985). Among flood survivors, a very low proportion of victims require emergency medical care (Noji, 2000). Injuries from floods are generally minor and include lacerations, infection of wounds, skin rashes, and ulcers (PAHO, 1981).

Extreme temperature events

Extreme temperature events which include heat waves and also extreme winter conditions and cold waves are mostly frequently reported in Europe (45%), Asia (31%), and the Americas (20%). The average extreme temperature event during the past 30 years resulted in 307 deaths and 5,496 injuries. Extreme heat events, which accounted for 36% of extreme temperature events, accounted for 89,046 deaths (87%), while more common extreme cold events (64% of events) resulted in 13,755 deaths (13%). There were an average of 707 deaths per extreme heat event and 62 deaths per extreme cold event over the past 30 years. Overall, 87% of heat wave fatalities were concentrated in Europe, while cold wave deaths were prevalent in both Asia (53%) and Europe (32%). It is important to consider that reporting of deaths and injuries may be more complete in extreme temperature events because extreme temperature events are predominantly reported in more developed countries with better information systems; furthermore as compared to other types of rapid-onset natural disasters, they do not cause infrastructure damage and widespread societal disruption. In extreme temperature events, hyperthermia (heat) and hypothermia (cold) deaths are either direct or indirect causes of mortality and injury. Hyperthermia cases are likely to be underreported because heat-related illness can exacerbate the existing medical conditions and can be difficult to identify; in addition there is variation in criteria used to identify heat-related deaths (MMWR, 2006). In heat waves, where risk of mortality is greater, numerous underlying demographic and physiological characteristics have been identified as risk factors for death, and risk of respiratory death is increased (Davido et al., 2006; Hertel et al., 2009).

Conclusion

One of the difficulties in assessing natural disaster injury and mortality is that information for many events is unreported or casualty estimates are inaccurate; this is particularly true for injuries that are undocumented for a majority of events. As a result, the true impact of natural disasters on human populations is likely to be substantially greater than the recorded impact, especially in events that occurred before substantial improvements in natural disaster reporting that were observed in the 1970s (CRED, 2010; Ehsghi and Larson, 2008). Understanding the causes of death and injury in natural disasters is important for planning disaster response. Morbidity and mortality patterns for certain types of natural disasters have been identified and can be used to plan the type of relief supplies, equipment, and personnel that will be required in the early stages of disaster response (Noji, 2000). Many factors contribute to the outcome of a natural hazard, including if the hazard evolves into a disaster and the resulting level of impact on human populations. While all natural disasters are unique and require a response that is tailored to the specific event, mortality and injury patterns can be anticipated and used to inform emergency medical relief and the planning and management of the ensuing humanitarian response.

Bibliography

Beinin, L., 1985. *Medical Consequences on Natural Disasters*. Berlin: Springer.

British Broadcasting Service (BBC), (2011). Report challenges Haiti earthquake death toll. Archived from the original on June 1, 2011. Retrieved April 2, 2012.

Center for Research on the Epidemiology of Disasters, 2006. *CRED Crunch Newsletter*, #7. Brussels: Center for Research on the Epidemiology of Disasters, Catholique Universite de Louvain.

Center for Research on the Epidemiology of Disasters, 2010. EM-DAT Emergency Events Database. http://www.emdat.be/Database/terms.html. Accessed on January 22, 2010.

Davido, A., Patzak, A., Dart, T., et al., 2006. Risk factors for heat related death during the August 2003 heat wave in Paris, France, in patients evaluated at the emergency department of Hospital European Georges Pompidou. *Emergency Medicine Journal*, **23**(7), 515–518.

Dreze, J., and Sen, A., 1990. *The Political Economy of Hunger: Famine Prevention*. Oxford: Oxford University Press, Vol. 2.

Eshghi, K., and Larson, R. C., 2008. Disasters: lessons from the past 105 years. *Disaster Prevention and Management*, **17**(1), 62–82.

Famine Early Warning System, 2010. www.fews.net. Accessed January 22, 2010.

French, J., 1989. Hurricanes. In Gregg, M. D., and Gregg, M. D. (eds.), *The Public Health Consequences of Disasters*. Atlanta: Centers for Disease Control.

Hertel, S., Le Terte, A., Karl-Heinz, J., and Hoffman, B., 2009. Quantification of the heat wave effect on cause-specific mortality in Essen, Germany. *European Journal of Epidemiology*, **24**, 407–414.

Huppert, H. E., and Sparks, R. S. J., 2006. Extreme natural hazards: population growth, globalization and environmental change. *Philosophical Transactions of the Royal Society*, **364**, 1875–1888.

Kunii, O., Akagi, M., and Kita, E., 1995. Health consequences and medical and public health response to the great Hanshin-Awaji earthquake in Japan: a case study in disaster planning. *Medicine and Global Survival*, **2**, 32–45.

Morbidity and Mortality Weekly Report (MMWR), 2006. Heat-related deaths – United State, 1999–2003. *Morbidity and Mortality Weekly Report*, **56**(29), 796–798.

National Research Council (NRC), 1987. *Confronting Natural Disasters: An International Decade for Disaster Reduction*. Washington, DC: National Academy Press.

Noji, E., 1993. Analysis of medical needs in disasters caused by tropical cyclones: the need for a uniform injury reporting scheme. *The Journal of Tropical Medicine and Hygiene*, **96**, 370–376.

Noji, E., 1997. Earthquakes. In Noji, E. (ed.), *The Public Health Consequences of Disasters*. New York: Oxford University Press.

Noji, E., 2000. The public health consequences of disasters. *Prehospital and Disaster Medicine*, **15**(4), 147–157.

Oda, Y., Shindoh, M., Yukioka, H., et al., 1997. Crush syndrome sustained in the 1995 Kobe, Japan earthquake: treatment and outcome. *Annals of Emergency Medicine*, **30**, 507–512.

PAHO, 1981. *Emergency Health Management After Natural Disaster*. Washington, DC: PAHO Office of Emergency Preparedness and Disaster Relief Coordination. Scientific publication 407.

Safar, P., Pretto, E., Bircher, N., 1988. Disaster resuscitology including the management of severe trauma. In: Baskett, P., Weller, R., (eds). *Medicine for Disasters*. London, England: Wright-Butterworth Publishers, pp. 36–86.

Schultz, C., Koenig, K., and Noji, E., 1996. A medical disaster response to reduce immediate mortality after an earthquake. *New England Journal of Medicine*, **334**, 438–444.

Sen, A., 1982. *Poverty and Famines: An Essay on Entitlement and Deprivation*. Oxford: Oxford University Press.
UNDP, 2004. *Reducing Disaster Risk: A Challenge for Development*. New York: United Nations Development Program.
United Nations, 1988. International decade for natural disaster reduction: report of the secretary general. United Nations General Assembly, 43rd Session, October 18, 1988. New York: United Nations. Agenda Item 86. A/43/723.
United Nations, (2010). World population prospects: The 2010 revision population database. www.esa.un.org. Retrieved April 2, 2012.

Cross-references

Casualties Following Natural Hazards
Drought
Economic Valuation of Life
Flood Hazards and Disaster
Galeras Volcano, Colombia
Geological/Geophysical Disasters
Haiti Earthquake 2010 Psychosocial Impacts
Heat Wave
Hurricane Katrina
Hydrometeorological Hazards
Indian Ocean Tsunami 2004
Nevado del Ruiz, Colombia (1985)
Storms
Tangshan, China (1976 Earthquake)
Tohoku, Japan, Earthquake, Tsunami and Fukushima Accident (2011)
Tornadoes
Tsunamis
Vaiont Landslide, Italy
Vesuvius
Wenchuan, China (2008 Earthquake)

CASE STUDY

MT PINATUBO

Katherine Donovan
University of Oxford, Oxford, Oxfordshire, UK

Mt Pinatubo is an active stratovolcano located in Central Luzon, Philippines, that has had significant global impacts.

The 1991 eruption

Mt Pinatubo had been quiescent for 500 years until 1990 when a nun working with traditional Aetas people living high on the volcano reported unusual activity, such as steaming cracks in the ground, to the Philippine Institute of Volcanology and Seismology (PHILVOLCS). Once these and other volcanic precursors were confirmed, PHILVOLCS in collaboration with a team from the US Geological Survey (USGS) started to monitor the activity and collate all existing geological data on previous activity.

Mt Pinatubo, Figure 1 The pre-climactic eruption column of Mt Pinatubo taken on 12 June 1991 from Clark Air Base (This photograph has been reproduced courtesy of the Philippine Institute of Volcanology and Seismology).

Previous eruptions had been recorded in local oral histories and warned of caldera forming eruptions lasting up to 3 days (Rodolfo and Umbal, 2008), yet geological surveys for the volcano were scarce and the international team of volcanologists had very little time to estimate potential eruption size and impact. Despite this, a 5-level warning system was implemented and evacuation zones were delineated (Newhall 2000). Fortunately, because of the quick actions of the scientists and government some 85,000 people were evacuated just before one of the most powerful eruptions of the twentieth century took place on 15 June 1991 (Leone and Gaillard, 1999; Gaillard, 2008) (Figure 1).

The eruption caused widespread destruction ejecting 5 km^3 of magmatic material and leaving behind a 2.5 km wide caldera (see Table 1). This eruption affected 2.1 million people and despite the added danger from a coinciding typhoon, approximately, only 300 people were directly killed and the management of this eruption was considered a success. The eruption was recorded in detail within the text *Fire and Mud: Eruptions and lahars of Mount Pinatubo* edited by C. Newhall and R.S Punongbayan (1996).

The secondary hazard and long-term social impact

The combination of widespread volcanic deposits and seasonal rains caused a secondary hazard known as lahars

Mt Pinatubo, Table 1 A statistical summary of the main products produced by the 1991 Mt Pinatubo eruption (Source: Wolfe and Hoblitt, 1996)

Hazard type	In detail	Size	Impact area
Tephra	Thickness deposited	1 cm < thick	75,000 km^2
		10 cm < thick	2,000 km^2
	Total bulk volume	3.4–4.4 km^3	
Pyroclastic flows	Distance traveled from source	12–16 km	5–6 km^2
Magma	Total volume	5 km^3	
Total ejecta	Total bulk volume	8.4–10.4 km^3	
Gas	SO$_2$	17 Mt	

Mt Pinatubo, Figure 2 The top image shows a lahar watch point in the Sacobia-Bamban River on Mt Pinatubo in 1991 and the image below was taken from the same position in 1992. In the lower image, only the roof of the watch point can be seen amongst the lahar deposits (These photographs have been reproduced courtesy of the Philippine Institute of Volcanology and Seismology).

(or volcanic mudflows) that annually threatened an area of 770 km^2 (Figure 2). This dangerous long-term hazard is responsible for killing nearly twice as many people than the actual eruption (Gaillard, 2008). Despite efforts to relocate residents out of lahar-prone regions some people still remain. Gaillard (2008) discusses the push and pull factors that motivate these at-risk communities to live in potentially dangerous regions on Mt Pinatubo. Push factors include victims having to pay for their new homes and services despite having no means of income in the long term evacuation centers. Pull factors include historical and cultural attachments, some declaring, "we are dead and drowned but we will never leave" (Gaillard, 2008, 323). Social and political issues caused increased difficulties in the management of lahars and relocation of evacuees, but the scale of recovery efforts indicate the difficulties faced by disaster managers. By 1997, 42,396 families had been re-homed in 23 centers around the volcano. This required 6,000 ha of land in addition to over 300 km of roads and electrical networks (Leone and Gaillard, 1999).

The economic costs of the eruption and lahar activities over the ensuing 2 years has been estimated at 11 billion pesos with 600,000 people losing their sources of income (Tayag and Punongbayan, 1994). This event demonstrates the potential physical and social impacts of such a large eruption and highlights the need for long-term disaster management in volcanic regions.

Bibliography

Gaillard, J.-C., 2008. Alternative paradigms of volcanic risk perception: the case of Mt Pinatubo in the Philippines. *Journal of Volcanology and Geothermal Research*, **172**, 315–328.
Leone, F., and Gaillard, J.-C., 1999. Analysis of the institutional and social responses to the eruption and the lahars of Mount Pinatubo volcano from 1991 to 1998 (Central Luzon, Philippines). *GeoJournal*, **49**, 223–238.
Newhall, C. G., 2000. Volcano warnings. In Sigurdsson, H. (ed.), *Encyclopedia of Volcanoes*. London: Academic, pp. 1185–1197.
Newhall, C. G., and Punyongbayan, R. S., 1996. *Fire and Mud: Eruptions and lahars of Mt Pinatubo*. Philippines/London: University of Washington Press.
Rodolfo, K. S., and Umbal, J. V., 2008. A prehistoric lahar-dammed lake and eruption of Mount Pinatubo described in a Philippine aborigine legend. *Journal of Volcanology and Geothermal Research*, **176**, 432–437.
Tayag, J. C., and Punongbayan, R. S., 1994. Volcanic disaster mitigation in the Philippines: experience from Mt Pinatubo. *Disasters*, **18**(1), 1–15.
Wolfe, E. W., and Hoblitt, R. P., 1996. Overview of the eruptions. In Newhall, C. G., and Punyongbayan, R. S. (eds.), *Fire and Mud: Eruptions and lahars of Mt Pinatubo*. Philippines/London: University of Washington Press.

Cross-references

Aviation, Hazards to
Base Surge
Caldera
Early Warning Systems
Evacuation
Galeras Volcano (Colombia)
Krakatoa (Krakatau)
Lahar
Magma
Nuee Ardente
Pyroclastic Flow
Stratovolcano
Volcanoes and Volcanic Eruptions

MUD VOLCANOES

Behruz M. Panahi
Azerbaijan National Academy of Sciences, Baku, Azerbaijan

Synonyms
Gas-oil volcano; Mud dome; Sedimentary volcano

Definition
Mud volcano was defined by Kopf (2002) as a surface expression of mud that originated from depth. Depending on the geometry of the conduit and the physical properties of the extrusive, the feature may be a dome or a pie with low topographic relief (Figure 1). Mud volcanoes may be the result of a piercing structure created by a pressurized mud diapir, which breaches the Earth's surface or ocean bottom.

Discussion
The connotations relate to formations created by geoexcreted liquids and gases; from extruded mud, and liquid. The parent material is rapidly deposited, overpressured, commonly thick argillaceous sequences of mostly Tertiary age.

The depositional environment includes ridges, plains, and intermountain falls and hollows occupied with temporary salty lakes, and plateaus with an abundance of mud domes and cones extruding mud and rock fragments (gryphons), and water-dominated pools with gas seeps (salses); offshore mud volcanoes form islands and banks on the sea floor that alter the topography and shape of the coastline.

In terms of origin, mud volcanoes are mainly present all over subduction zones and orogenic belts, where rapidly buried sediment overthrusts deeper stratum. With an increase of burial stress and temperature, a decrease in porosity and maturation of organic material are favored (Hedberg, 1974). In these conditions, trapped pore water and forming hydrocarbon gas may cause overpressure of the mud at depth (Judd and Hovland, 1997). The mud, depending on the magnitude of the buoyancy, either slowly ascends through the overburdened rock and forms mud diapirs or extrudes vigorously along zones of structural weakness such as faults and fractures and forms mud volcanoes (Brown, 1990). During rapid ascent, self-ignition of emanating methane may cause flaming eruptions and a societal hazard (Bagirov and Lerche, 1998; Ismail-Zadeh, 2006).

Mud volcanoes are used as source of natural gas. Clay from volcanoes can be used as raw material for production of ceramics and bricks. Mud from volcanoes contains medical qualities and is widely used in local spas and perfumery.

Bibliography
Bagirov, E., and Lerche, I., 1998. Flame hazards in the South Caspian Basin. *Energy Exploration & Exploitation*, **16**, 373–397.

Mud Volcanoes, Figure 1 One of the spectacular world mud volcanoes – Bahar (Located in Azerbaijan, photo of B. Panahi).

Brown, K. M., 1990. The nature and hydrogeologic significance of mud diapirs and diatremes for accretionary systems. *Journal of Geophysical Research*, **95**, 8969–8982.

Hedberg, H., 1974. Relation of methane generation to undercompacted shales, shale diapirs and mud volcanoes. *American Association of Petroleum Geologists Bulletin*, **58**, 661–673.

Ismail-Zadeh, A. T., 2006. Geohazard, georisk and sustainable development: multidisciplinary approach. In Ismail-Zadeh, A. T., (ed.), *Recent Geodynamics, Georisk and Sustainable Development in the Black Sea to Caspian Sea Region*, AIP Conference Proceedings 825.

Judd, A. G., Hovland, M., Dimitrov, L. I., Garcia Gil, S., and Jukes, V., 2002. The geological methane budget at continental margins and its influence on climate change. *Geofluids*, **2**, 109–126.

Kopf, A. J., 2002. Significance of mud volcanism. *Reviews of Geophysics*, **40**(2), 52.

Cross-references

Dispersive Soil Hazards
Expansive Soils and Clays
Liquefaction
Quick Clay
Quick Sand
Volcanoes and Volcanic Eruptions

MUDFLOW

Christophe Ancey
Laboratoire Hydraulique Environnementale
ENAC/ICARE/LHE, Lausanne, Switzerland

Synonyms

Debris flows; Lahars; Mudslides

Definition

There is a wide spectrum of natural processes that take the form of a rapid mass movement of saturated soil or sediment under the action of gravitational acceleration; here the adjective "rapid" means that the typical velocity is within the 1–25 m/s range (Iverson, 1997). These mass movements are often referred to as "debris flows." Mudflows constitute an end-member of this large family: when the sediment is rich in clayey materials and poor in coarse particles (note that there is no consensus in literature on classification and the definition of mudflows may vary depending on the authors), the sediment looks like a muddy fluid (Coussot and Meunier, 1996).

Discussion

In most cases, mudflows are initiated after long or heavy rainfalls over mountain slopes or result from the acceleration of a landsliding mass. Bank and bed erosion may also result in mudflows, in particular in rivers whose channel incises soils made up of loose soils (e.g., loess or volcanic-ash deposits). Once set in motion, mudflows can travel large distances (mostly in the 1–100 km range) and spread over gentle slopes; the mean slope gradient for observing mudflows is usually in excess of 10%, but on some occasions, mudflows were observed on shallow slopes (less than 1%).

The capacity of mudflows to travel large distance has been ascribed to their viscous behavior. There is still a vivid debate on the origins of mudflow fluidity (Ancey, 2007). Some authors provided evidence that the mud behaves like a viscoplastic fluid, that is, like a solid when the shear-stress level is low and like a viscous fluid for shear stresses in excess of a critical value (called the *yield stress*) (Coussot and Meunier, 1996). Other authors consider mud as a liquefied soil, that is, a soil within which pore pressure is sufficiently high to reduce shear strength resulting from particle friction (Iverson, 1997, 2005). Both theories have been implemented in generalized hydraulic models; a set of equations that describe flow evolution in terms of flow-depth and velocity (Huang and García, 1998).

Mudflows are a major threat in mountainous and volcanic areas, claiming thousands of lives and millions of dollars in lost property each year (e.g., Sarno and Quindici in southern Italy in May 1998, where approximately 200 people were killed).

Bibliography

Ancey, C., 2007. Plasticity and geophysical flows: a review. *Journal of Non-Newtonian Fluid Mechanics*, **142**, 4–35.

Coussot, P., and Meunier, M., 1996. Recognition, classification and mechanical description of debris flows. *Earth Science Review*, **3–4**, 209–227.

Huang, X., and García, M. H., 1998. A Herschel-Bulkley model for mud flow down a slope. *Journal of Fluid Mechanics*, **374**, 305–333.

Iverson, R. M., 1997. The physics of debris flows. *Reviews of Geophysics*, **35**, 245–296.

Iverson, R. M., 2005. Debris-flow mechanics. In Jakob, M., and Hungr, O. (eds.), *Debris-Flow Hazards and Related Phenomena*. Berlin: Springer, pp. 105–134.

Cross-references

Debris Flow
Flash Flood
Lahar
Landslide Types
Mass Movements

MYTHS AND MISCONCEPTIONS IN DISASTERS

Alejandro López Carresi
Centre of Studies on Disasters and Emergencies,
Madrid, Spain

Definition and introduction

Humankind has always relied on myths to provide an answer for the unknown. Fact and imagination are interwoven to account for uncertainties. Disasters are prototypical representations of uncertain situations that must be dealt with. What is real in a myth can play its role in coping with a disaster. Equally important, though, is to separate what is imaginary about myths in disaster management. The debate on disaster myths and misconceptions has been a recurrent issue among scholars, practitioners, and other actors involved. This entry does not intend to be a comprehensive account and explanation of all myths in disaster management. This is just an introduction to a few of those myths: epidemics, looting and antisocial behavior, massive population movements, and goods donations.

One of the most popular lists of myths and misconceptions was first published by the Pan-American Health Organization in the 1980s and has been widely used, modified, and adapted ever since (Table 1).

The first 18 items were used by Alexander (2007) in a survey about disaster myths among disaster management students in Italy and USA. Despite the differences in the country of origin, background, training, and field experience, the students gave similar answers. This is illustrative of how deeply rooted some of these wrong assumptions are and how persistent in time they prove to be.

Dead bodies, epidemics and disease: always predicted, hardly ever materialized

Shortly after nearly every disaster, the news headlines alert people to the risk of major epidemics of communicable diseases. For example, after the 2004 Asia tsunami, there were widespread fears that a second wave of deaths was to be expected, and that disease and epidemics may cause as many casualties as the tsunami itself. The forecasted disease mortality and epidemics, as in most previous similar occasions, failed to materialize.

Despite the widespread idea that dead bodies can generate epidemics after disasters, there is no evidence to support that myth (Morgan and de Ville de Goyet, 2005). The health risks of dead bodies resulting from natural hazards are very few because the immediate cause of death is trauma, not infectious disease. Dead bodies can transmit a number of diseases for a limited period of time (Morgan, 2004), and only if those diseases were already present in the host *before* death takes place.

The overrated risk of major epidemics after disasters does not imply that all health concerns in disaster response are irrelevant. Frequently, local health services are affected or destroyed, interrupting the provision of adequate community care. Also, some common diseases are frequent, but rarely in epidemic proportions. What can be expected is an increase in gastrointestinal diseases, respiratory diseases, and some vector-borne diseases such as malaria. There are no indications of massive mortality increases or large epidemic outbreaks in any of those cases. Besides, the lack of crucial data about the number of previous cases of the detected disease prevents any comparison and the determination of the trend as increasing or decreasing.

Myths and Misconceptions in Disasters, Table 1 List of myths and misconceptions from PAHO, 2000

Myth: Disasters are truly exceptional events
Reality: They are a normal part of daily life and in very many cases are repetitive events
Myth: Disasters kill people without respect for social class or economic status
Reality: The poor and marginalized are more at risk of death than are rich people or the middle classes
Myth: Earthquakes are commonly responsible for very high death tolls
Reality: Collapsing buildings are responsible for the majority of deaths in seismic disasters. Whereas, it is not possible to stop earthquakes, it is possible to construct anti-seismic buildings and to organize human activities in such a way as to minimize the risk of death. In addition, the majority of earthquakes do not cause high death tolls
Myth: People can survive for many days when trapped under the rubble of a collapsed building
Reality: The vast majority of people brought out alive from the rubble are saved within 24 or perhaps even 12 h of impact
Myth: When disaster strikes panic is a common reaction
Reality: Most people behave rationally in disaster. While panic is not to be ruled out entirely, it is of such limited importance that some leading disaster sociologists regard it as insignificant or unlikely
Myth: People will flee in large numbers from a disaster area
Reality: Usually, there is a "convergence reaction" and the area fills up with people. Few of the survivors will leave and even obligatory evacuations will be short-lived
Myth: After disaster has struck survivors tend to be dazed and apathetic
Reality: Survivors rapidly start reconstruction. Activism is much more common than fatalism (this is the so-called therapeutic community). Even in the worst scenarios, only 15–30% of victims show passive or dazed reactions
Myth: Looting is a common and serious problem after disasters
Reality: Looting is rare and limited in scope. It mainly occurs when there are strong preconditions, as when a community is already deeply divided
Myth: Disease epidemics are an almost inevitable result of the disruption and poor health caused by major disasters
Reality: Generally, the level of epidemiological surveillance and health care in the disaster area is sufficient to stop any possible disease epidemic from occurring. However, the rate of diagnosis of diseases may increase as a result of improved health care
Myth: Disasters cause a great deal of chaos and cannot possibly be managed systematically
Reality: There are excellent theoretical models of how disasters function and how to manage them. After >75 years of research in the field, the general elements of disaster are well-known, and they tend to repeat themselves from one disaster to the next
Myth: Any kind of aid and relief is useful after disaster provided it is supplied quickly enough
Reality: Hasty and ill-considered relief initiatives tend to create chaos. Only certain types of assistance, goods, and services will be required. Not all useful resources that existed in the area before the disaster will be destroyed. Donation of unusable materials or manpower consumes resources of organization and accommodation that could more profitably be used to reduce the toll of the disaster
Myth: In order to manage a disaster well it is necessary to accept all forms of aid that are offered
Reality: It is better to limit acceptance of donations to goods and services that are actually needed in the disaster area

Myths and Misconceptions in Disasters, Table 1 (Continued)

Myth: Unburied dead bodies constitute a health hazard
Reality: Not even advanced decomposition causes a significant health hazard. Hasty burial demoralizes survivors and upsets arrangements for death certification, funeral rites, and, where needed, autopsy
Myth: Disasters usually give rise to widespread, spontaneous manifestations of antisocial behavior
Reality: Generally, they are characterized by great social solidarity, generosity, and self-sacrifice, perhaps even heroism
Myth: One should donate used clothes to the victims of disasters
Reality: This often leads to accumulations of huge quantities of useless garments that victims cannot or will not wear
Myth: Great quantities and assortments of medicines should be sent to disaster areas
Reality: The only medicines that are needed are those used to treat specific pathologies, have not reached their sell-by date, can be properly conserved in the disaster area, and can be properly identified in terms of their pharmacological constituents. Any other medicines are not only useless, but potentially dangerous
Myth: Companies, corporations, associations, and governments are always very generous when invited to send aid and relief to disaster areas
Reality: They may be, but in the past disaster areas have been used as dumping grounds for outdated medicines, obsolete equipment, and unusable goods, all under the cloak of apparent generosity
Myth: Technology will save the world from disaster
Reality: The problem of disasters is largely a social one. Technological resources are poorly distributed and often ineffectively used. In addition, technology is a potential source of vulnerability as well as a means of reducing it
Myth: There is usually a shortage of resources when disaster occurs, and this prevents them from being managed effectively
Reality: The shortage, if it occurs, is almost always very temporary. There is more of a problem in deploying resources well and using them efficiently than in acquiring them. Often, there is also a problem of coping with a superabundance of certain types of resources

As rare as they may be, it is still worth commenting on the occasional outbreaks of disease after disasters caused by natural hazard. According to the Center for Disease Control and Prevention (CDC), an epidemic is the occurrence of more cases of disease than expected in a given area or among a specific group of people over a particular period of time. Floret and colleagues found that in only 3 out of more than 600 geophysical disasters recorded worldwide from 1984 to 2004, there were epidemic outbreaks: measles after the Pinatubo eruption in Philippines in 1991, coccidioidomycosis (a fungal infection caused by inhalation of spores) after an earthquake in California in 1994, and malaria after earthquake and heavy rains in Costa Rica in 1991 (Floret et al., 2006).

The cholera epidemic which developed in the aftermath of the devastating January 12, 2010 Haiti earthquake is instructive. Although perceptually linked with the disaster, it is clear that the epidemic itself was the product of a set of unusual circumstances more closely aligned with an external input and preexisting sanitary conditions (Piarroux et al., 2011).

Decision makers should keep in mind that infectious disease epidemics after disasters are very rare and that massive and indiscriminate actions to prevent unfounded health risks are not recommended. Health and disease after disasters are a major issue, and undoubtedly some illnesses increase and public health deteriorates (Noji, 1997). But the presence of infectious diseases does not justify unfounded fears of major epidemics (WHO, 2006).

Looting and social unrest: the augmented perception of exceptional events

According to the most widespread expectation, looting is frequent after disasters, and preventative measures must be taken immediately. This perception is based on the idea that disasters change societies and communities, triggering negative actions and antisocial collective behavior. But the reality is that looting is the exception and not the norm (Auf der Heide, 2004), and when it does happen, it follows different patterns than looting associated with riots and civil unrest crises. Pro-social adaptive behavior and the willingness to help others is generally the collective reaction to be expected.

First of all, a distinction needs to be made between looting and taking essential items for survival. While looting may be considered the illicit taking of nonessential items with the sole purpose of obtaining personal profit, many researchers use the term "appropriation" when the goods taken are used to cover basic needs, such as the need for food, water, and shelter (Quarantelli, 1994). However, most of these actions are perceived and reported by media, law enforcement, or casual observers as examples of social disorder, violent behavior, and looting. Unconfirmed rumors are also assumed as proof of looting.

When actual looting occurs in disasters, it is commonly undertaken by people from outside the community, frequently by people usually involved in criminal activities, individually or in small groups, taking advantage of the sudden opportunity (Quarantelli, 1994). By contrast, looting in riots and situations of civil unrest is enacted by normally law-abiding people from the community, in a collective manner and openly undertaken with wide social support. Most of those who loot and steal after disasters also do it before disasters. The disaster itself does not act as a social transformer that triggers deep changes or significantly increases antisocial behavior.

In summary, while detailed observation of disasters and the vast majority of the scientific literature indicate that widespread looting and social disorder is a myth and actual looting is truly exceptional, the number of disasters with actual looting and its precise extent remains unclear.

Displacement and disaster-stricken populations: the unexisting exodus

After disaster, the myth perception is that a massive displacement of those affected will follow. However, massive population displacements are not a common feature after disasters caused by natural hazards. It is in wars and armed conflicts where it is possible to find this type of exodus, with thousands or even hundreds of thousands of people painfully walking roads and paths, carrying their scarce belongings by any possible means. These displaced people will travel long distances, usually up to the first secure place they may find, and settle in quite large camps for extended periods of time.

The situation in disasters caused by natural hazards is quite otherwise. Some people may seek help from relatives outside the affected areas or in assistance camps, but most will not leave the area, or at least they will not be displaced very far away. In disasters, people will try to stay as close as possible to their homes, their neighborhoods, their villages, etc. In fact, the typical population movement more frequently observed is toward the disaster area. As early as the 1950s, this feature was identified and named "convergence behavior" (Fritz and Mathewson, 1957). People going toward the disaster-stricken zones will include concerned relatives seeking news about missing family members or aid workers. As an example, the Haiti earthquake in 2010 produced plenty of news headlines reporting massive population exodus from the capital toward the Dominican Republic by road and the USA by boats. While indeed some people attempted to reach those destinations, these actions were already commonplace in Haiti before the earthquake. And even though the difficult situation in some cities in the aftermath of the disaster may have increased attempts to leave, the reality was far from the massive exodus many predicted.

Donations: received versus needed

Donation of all kind of commodities is indeed a very typical image after disasters. All kind of goods are donated, boxed and shipped to disaster areas. But the reality is that most such donations cause significant problems. First, there are costs linked to the logistics involved in the process: reception, classification, boxing, handling, transportation, distribution, and other related logistical elements.

Second, many donated items are inappropriate or unusable: expired medicines, unpaired shoes, extremely dirty clothes, culturally unacceptable food, winter clothes to tropical areas (or the opposite), etc. All of these situations and many others have been observed regarding donations to disaster-affected countries. The consequence is that despite the intention to help, these donations compound the situation by forcing the diversion of human resources from other essential tasks into the classification and storage of the donations.

In most occasions, the mere cost of transportation will exceed by far the value of the donated goods. Although a donated blanket seems free of charge, by the time that blanket reaches the target beneficiary, particularly if it is shipped from a long distance, the final costs will be far higher than purchasing that blanket locally. The farther the donation travels from the destination country, the higher the costs will be. Besides, massive influx of external goods, if that keeps happening beyond the first days of the disaster response, may affect local markets negatively. No one will purchase in local markets goods that the aid agencies distribute for free. Even in disasters with high levels of destruction, there will be always less-affected or unaffected neighboring areas with available sellers of basic products such as clothes, blankets, and cooking items. Certainly, price inflation may affect certain local products in disaster areas in the initial stages of an emergency. But aid organizations must strive to reject unwanted donations in kind and encourage individual donors and institutions to donate cash to well-established and recognized organizations involved in the response; the cash will be used to purchase locally as many products as possible to support the recovery of the area.

Conclusions

Education about disasters for the public, the media, and above all, the professionals is critical for increasing our awareness about the consequences of distorted information.

Also, a new approach may be needed. Just denouncing the inappropriateness of mass burials will not solve the problem faced by authorities when they have many thousands of bodies to bury. There is a need to obtain basic data from the bodies (estimated age, clothes, old scars, taking digital pictures, etc.) and to keep records for future possible identification by relatives, or addressing the cultural and religious sensitivities through mass funerals or rituals. After these or other palliative measures have been taken, mass burials may still be hard to avoid. But certainly, authorities are better served by concentrating their efforts on activities that reduce fear, which could eventually bring some closure for the survivors.

Finally, better organized relief operations would contribute to reduce social problems caused by unsatisfied basic needs. Social unrest caused by poor access to essential items has been recorded in post-hurricane Katrina in New Orleans in 2006 and post-earthquake Haiti in 2010. Better disaster response and better organized relief distribution, which is based on better disaster preparedness, may contribute to solve this problem.

The struggle to debunk disaster myths was initiated long ago and it will not be won in the short-term. The final objective is not to destroy the myth itself but a reduction in human suffering. Myths persist because they give answers in uncertain situations. If disaster responders and societies learn to better provide certainties, explanations, and an organized response in a disaster situation, the myths will go back to being just imaginary stories.

Acknowledgment

The author would like to acknowledge the invaluable help of Marta Cabarcos-Traseira, Deputy Director of CEDEM, in writing this entry.

Bibliography

Alexander, D. E., 2007. Misconception as a barrier to teaching about disasters. *Prehospital and Disaster Medicine*, **22**(2), 95–103.

Auf der Heide, E., 2004. Common misconceptions about disasters: panic, the "disaster syndrome", and looting. In O'Leary, M. (ed.), *The First 72 Hours: A Community approach to Disaster Preparedness*. New York: Lincoln iUniverse Publishing.

Floret, N., Viel, J. F., Mauni, F., Hoen, B., and Piarroux, R., 2006. Negligible risk for epidemics after geophysical disasters. *Emerging Infectious Diseases*, **12**(4), 543–548.

Fritz, C. E., and Mathewson, J. H., 1957. *Convergence Behavior in Disasters: A Problem in Social Control. Committee on Disaster Studies*. Washington, DC: National Academy of Sciences, National Research Council.

Morgan, O., 2004. Infectious disease risks from dead bodies following natural disasters. *Revista Panamericana de Salud Pública/Pan American Journal of Public Health*, **15**(5), 307–312.

Morgan, O., and De Ville de Goyet, C., 2005. Dispelling disaster myths about dead bodies and disease: the role of scientific evidence and the media. *Revista Panamericana de Salud Pública/Pan American Journal of Public Health*, **18**(1), 33–36.

Noji, E. (ed.), 1997. *The Public Health Consequences of Disasters*. New York: Oxford University Press.

PAHO, 2000. *Natural Disasters: Protecting the Public's Health*. Washington, DC: Pan American Health Organisation. Scientific Publication, Vol. 575.

Piarroux, R., Barrals, R., Faucher, B., Haus, R., Piarroux, M., Gaudart, J., et al., 2011. Understanding the cholera epidemic, Haiti. *Emerging Infectious Diseases*, **17**(7), 1161–1167.

Quarantelli, E. L., 1994. *Looting and Antisocial Behavior in Disasters*. Newark: University of Delaware Disaster Research Center. Preliminary Paper, Vol. 205.

World Health Organisation, 2006. *Communicable Diseases Following Natural Disasters*. www.who.int/diseasecontrol_emergencies/en. Accessed January 2010.

Cross-references

Integrated Emergency Management System
Mass Media and Natural Disasters
Perception of Natural Hazards and Disasters
Recovery and Reconstruction After Disaster

N

NATURAL HAZARD

Anita Bokwa
Jagiellonian University, Kraków, Poland

Definition

Natural hazard is an unexpected and/or uncontrollable natural event of unusual magnitude that might threaten people.

The concept of natural hazard

A *hazard* is a source of potential harm or a situation with a potential to cause loss. It may also be referred to as a potential or existing condition that may cause harm to people or damage to property or the environment (Middelmann, 2007). A *natural hazard* is associated with geophysical processes that are an integral part of the environment and involves the potential for damage or loss that exists in the presence of a vulnerable human community (Stillwell, 1992); it is an unexpected threat to humans and/or their property (Mayhew, 1997). These definitions indicate that natural hazards have not only natural, but also social, technological, and political aspects. Natural hazards include geophysical hazards, i.e., hazards where the principal causal agent is climatic and meteorological (e.g., floods, hurricanes, and droughts) or natural hazards where the principle causal agent is geological and geomorphological (e.g., landslides, tsunamis, and earthquakes). They do not include biological hazards, both floral and faunal, such as fungal diseases, poisonous plants, viral diseases, and infestations or locusts (Geophysical hazard, 2010).

Classification of natural hazards

Natural hazards are usually grouped according to the causative element of the Earth's geosystem (Graniczny and Mizerski, 2007, modified; Karst as Geologic Hazard, 2006; Glacier Hazards, 2010):

1. Meteorological hazards: for example, thunderstorm, tornado, tropical cyclone (hurricane, typhoon), cold and heat waves, fog, hail, drought, dust storm
2. Hydrological hazards: for example, flood, snow avalanche, glacier hazards (e.g., ice avalanches or debris flows from outbursts of subglacial water reservoirs and periglacial lakes)
3. Oceanographical hazards: for example, storm surge, sea-level change
4. Geological hazards: for example earthquake, volcanism, mass movements (e.g., landslide, rock fall, debris avalanche), karst hazards (e.g., cover-collapse sinkholes and sinkhole flooding), tsunami
5. Hazards connected with vegetation: for example, wildfire, bushfire
6. Extraterrestrial hazards: for example, meteorite strike

This classification is a simplification and only one of many possible schemes used to organize the complicated issue of natural hazards. For example, within the SHIELD Project (SHIELD, 2009), natural hazards are divided into only two groups: geological hazards (earthquakes, volcanoes, floods, slope failures, tsunamis) and atmospheric hazards (fires, thunderstorms, snow and ice, fog). In fact, most natural hazards are the result of several contributing processes, for example, floods are caused by prolonged rainfalls; tsunami waves follow earthquakes; bushfires are one of the outcomes of drought. Likewise, Stillwell (1992) classified natural hazards on the basis of their origin: within the Earth, such as earthquakes and volcanoes; on the Earth's surface, such as landslides and subsidence; and above the Earth, such as violent storms and fog. The author additionally points to the fact that multiple causes also involve a human element, which is of particular importance in developing nations. The conditions of poor people living on dangerous ground, in unsafe buildings, and with fragile lifelines can magnify the impact of disasters.

Natural hazards may be also classified by:

1. Time of occurrence: Some have the potential to occur at any time of year (e.g., tsunami), whereas others are often seasonal (e.g., thunderstorm).
2. Impact: From frequent moderate impacts (e.g., bushfire) through to rare but potentially catastrophic impacts (e.g., earthquake).
3. Predictability: Some hazards may occur suddenly (e.g., rockfall), whereas in the case of others, the threat may be identified in advance and a warning provided (e.g., flood).

The spatial distribution of natural hazards is influenced by region and topography, and they also vary in the size of the geographical area affected (Middelmann, 2007). Additionally, apart from "traditional" natural hazards, new items are added to the list as the human civilization develops. For example, Beer (2001) argues that air pollution should be viewed as a natural hazard of meteorological origin, because it is caused by meteorological factors that are sporadic in nature and has the potential to cause damage to life or property, or both. Omitting important natural hazards, for example, heat waves or frosts, in some classifications is another issue.

As presented by Wilhite (1996), it is common also for droughts to be omitted from various assessment figures because they differ from other natural hazards by their slow onset, and because they seldom result in structural damage or loss of life. Drought is also one of the most underreported natural disasters, because the sources of most of the statistics are international aid or donor organizations. Unless the countries afflicted by drought request assistance from the international community or donor governments, droughts are not reported. Thus, severe droughts such as those that occurred in Australia, Uruguay, Brazil, Canada, Spain, Italy, and the United States in 1990s are not included in these statistics. Drought is considered by many to be the most complex, but the least understood of all natural hazards, affecting more people than any other hazard. For example, the droughts of the early 1980s to mid-1980s in sub-Saharan Africa are reported to have adversely affected more than 40 million persons. Drought differs from other natural hazards in several ways. Firstly, it is a "creeping phenomenon," making its onset and end difficult to determine. The effects of drought accumulate slowly over a considerable period of time, and may linger for years after the termination of the event. Secondly, the absence of a precise and universally accepted definition of drought adds to the confusion about whether a drought exists or not and, if it does, what is its severity. Thirdly, drought impacts are less obvious and spread over a larger geographical area than are damages that result from other natural hazards. Drought seldom results in structural damage.

For these reasons, the quantification of impacts and the provision of disaster relief is a far more difficult task for drought than it is for other natural hazards. Because drought affects so many economic and social sectors, scores of definitions have been developed by a variety of disciplines. In addition, because drought occurs with varying frequency in nearly all regions of the globe, in all types of economic systems, and in developing and developed countries alike, the approaches taken to define it should be impact and region specific.

Effects of natural hazards

Natural hazards have the potential to cause a number of primary and secondary phenomena. The secondary phenomena produced by a natural hazard vary with event, as does their severity. Tropical cyclones bring strong winds and heavy rains which cause secondary hazards such as floods, storm tides, landslides, and water pollution. Floods inundate areas, which in turn may trigger landslides, erosion, water quality deterioration, or turbidity, as well as sediment deposition. Severe storms range from isolated thunderstorms to intense low-pressure systems producing phenomena such as severe winds, heavy rain, lightning, floods, storm tides, hail, and coastal erosion. Secondary effects of bushfires include water pollution, erosion, and reduced water catchment yield. A landslide may block a watercourse, leading to flooding and debris flows upstream. Earthquakes may also bring fire, flooding, water pollution, landslides, tsunamis, and soil liquefaction that can be as devastating as the primary hazard. Each of these phenomena may produce physical, social, and economic effects.

Physical effects on the built infrastructure may involve structural and nonstructural damage and/or progressive infrastructure deterioration. They may also result in the release of hazardous materials such as chemicals which are usually stored in a safe environment. Social effects may include fatalities, injuries, homelessness, or loss of income, or secondary effects such as psychological impact, disease, or loss of social cohesion. Economic effects may include business disruption; disruption to the supply of power, water, and telecommunications; and the cost of response and relief operations. Secondary economic impacts, such as insurance losses and rising premiums, loss of investor confidence, and costs of providing welfare and medical assistance, may also result. However, a natural hazard is not inherently negative, as hazards produce a disaster only when they impact adversely on communities. Natural hazards can bring positive environmental and social benefits. Bushfires, for example, can stimulate growth and regenerate forest ecology, as the heat from fire is required for some seeds to germinate. Floodplains are picturesque places for recreational activity, and floods can bring welcome relief for people and ecosystems suffering from prolonged drought (Middelmann, 2007).

Natural hazard and natural disaster

A key distinction exists between what is termed a "natural hazard" and what is referred to as a "natural disaster." A natural hazard is a serious disruption to a community or region caused by the impact of a naturally occurring process, occurring as a rapid onset event that threatens or causes death, injury, or damage to property or the environment, and which requires significant and coordinated multi-agency and community response. Such serious disruption can be caused by one, or a combination, of natural hazards (Middelmann, 2007; Newton, 1997). So hazards might lead to disasters; a disaster is the impact of a hazard on a community/society: For example, a tornado is an example of a natural hazard; a tornado disaster occurs when a severe tornado destroys part of a town, causing significant loss of life and property, often beyond the ability of the local community to recover from, without assistance (White and Etkin, 1997). Scheidegger (1997) points to an important cognitive aspect, already mentioned above: What one calls a "disaster" is in any case only an anthropocentric valuation: If there are many *human* casualties, one speaks of a disaster, otherwise merely of a natural event (cf. the impact of the Tunguska meteorite in 1908, which would have been a catastrophic disaster if the object had hit Tokyo, St. Petersburg, Berlin, London, or New York).

Economical aspects of natural hazards and natural disasters

An economic framework is often used to calculate the cost of natural disasters. Each decade, property damage from natural hazards events doubles or triples. Japan and the United States are the countries where economic damages resulting from natural disasters are the highest in the world (Natural Hazards – A National Threat, 2007). In the period 1970–1990, almost 3 million lives were lost, 820 million people were affected, and up to U.S.$100 billion worth of property was damaged by various natural catastrophes worldwide (Stillwell, 1992). From 1991 to 2000, some 1.5 billion people were affected by floods alone (Natural Hazard, 2010). In 2001–2010, every year on average almost 107, 000 people were killed due to various disasters and almost 232, 000, 000 were affected (Disasters in numbers, 2011). However, the difficulty of measuring the actual impact of a natural disaster on the community continues to be a major challenge because of the complexities in assessing loss. Intangible losses, such as the destruction of personal memorabilia and the effects of post-disaster stress, are particularly difficult to measure. Though insured losses are the most easily captured, they represent only a small proportion of the total loss. These complexities need to be kept in mind when measuring and communicating the concept of "impact" (Middelmann, 2007).

The effects of urbanization and increasing population growth and density, most notable in big cities and coastal regions, have led to greater demand for and concentration of infrastructure and a higher potential exposure to natural hazards (Middelmann, 2007). White and Etkin (1997) distinguished four human-induced reasons of growing natural hazards risk:

1. Fast global population growth
2. Continued growth of material possessions by households
3. Urbanization
4. Coastalization

Nearly 100 million people per year are born, mostly in developing countries. Inevitably this leads to greater population density and hence more people exposed to perils in any particular area. Despite entrenched poverty in some countries and intermittent recession in many others, the global economy is still growing, and one result of this growth is that households and companies acquire more goods – especially expensive and fragile consumer durables. In addition to the single radio a Western middle class family might have possessed 50 years ago, we have added two or three automobiles, many radios, several televisions, a refrigerator and a freezer, cameras, mobile phones, and several computers. This Western pattern of material acquisition is now spreading rapidly to the emerging economies of Latin America, East and South Asia, and to the economies in transition in central Europe. As agriculture reduces its need for labor, and as the coal mines lose their attraction for manufacturing enterprises, there are fewer and fewer reasons for people to inhabit the interiors of continents.

The coastal movement began in the United States and Western Europe early in twentieth century, and the same trend is now gathering momentum in China and Southeast Asia. Thus, it is a phenomenon which is prevalent in both the richest and the poorest countries. Coasts are areas of a relatively high risk for natural hazards because of their vulnerability to storms, tsunamis, and flooding (as well as earthquakes). If sea-level rise continues (as is predicted by greenhouse climate models), this will further increase the risk. Canada is something of an exception to this global trend in that the Atlantic coast is not a magnet for population. In terms of economic opportunity, the draw in central and eastern Canada is the St. Lawrence Valley, focused on the Greater Toronto Area, and this is not an area which is particularly susceptible to natural hazards. However, the same dangers of concentration of insured wealth exist, given the high rate or urbanization and the concentration of Canada's population in just three urban areas. The Pacific Northwest is a classic case of concentrated coastal risk: The Greater Vancouver area is situated in an active fault zone and at risk from earthquakes, volcanic eruptions, landslides, liquefaction, tsunamis, and sea-level change. In search of recreation and retirement, Canadians conform to the global trend in seeking coastal zones with a warmer climate, especially Florida, the Caribbean Islands, and Mexico. These trends are producing a greater number of affluent people living in coastal cities and other built-up areas, and to a large extent account for the fact that billion dollar losses have become commonplace in recent years. To some extent, this trend could be reversed through land-use planning (by not allowing people to live in harm's way), or by ensuring that people assume their own risk. Although a great deal has been written on the problems of managing such huge urban agglomerations, very little has been produced specifically on the management of urban risks. The professions of "risk management and insurance" and "urban management" have yet to form any significant partnerships.

Risk, vulnerability, and mitigation of natural hazards

Natural hazards are closely linked to risk and vulnerability issues. Risk refers to the chance of something happening that will have an impact on objectives, whereas vulnerability is the degree of susceptibility and resilience of the community and environment to hazards. A risk is often specified in terms of an event or circumstance and the consequences that may stem from it. Risk is measured in terms of a combination of the consequences of an event and their likelihood (Middelmann, 2007). Overpopulation, unemployment, poverty, decreasing self-reliance through urbanization, as well as misuse of capital and natural resources, all contribute to vulnerability (Stillwell, 1992). A good understanding of hazard, exposure, and vulnerability is fundamental in any rigorous analysis of the risk posed by natural hazards, as the assessment of risk is only as good as the data used. Knowledge of the elements likely to be exposed to the impact of the hazard phenomena is vital in determining the potential impact or consequence of any hazard on a community or society. This includes information on the people, buildings, and infrastructure potentially exposed to a hazard impact. Such data are fundamental to any analysis of risk, regardless of the hazard. It is also important to consider the potential impacts of climate change on the future risk. The study of prehistoric impacts of natural hazards can also be useful in extending historical knowledge for application today (Middelmann, 2007).

Climate change scenarios indicate that the risk of natural hazards are supposed to increase. According to the IPCC Fourth Assessment Report of 2007 (IPCC, 2007), observed global climate changes have already caused increasing ground instability in permafrost regions and rock avalanches in mountain regions. In the twenty-first century, it is very likely that hot extremes, heat waves, and heavy precipitation events will become more frequent. Based on a range of models, it is likely that future tropical cyclones (typhoons and hurricanes) will become more intense, with larger peak wind speeds and heavier precipitation associated with ongoing increases of tropical sea-surface temperatures. There is less confidence in projections of a global decrease in numbers of tropical cyclones. The apparent increase in the proportion of very intense storms since 1970 in some regions is much larger than simulated by current models for that period. Extratropical storm tracks are projected to move poleward, with consequent changes in wind, precipitation, and temperature patterns.

By the 2080s, many millions more people than today who live in coastal areas are projected to experience floods every year due to sea-level rise. The numbers affected will be largest in the densely populated and low-lying megadeltas of Asia and Africa and small islands are especially vulnerable. Available research suggests a significant future increase in heavy rainfall events in many regions, including some in which the mean rainfall is projected to decrease. The resulting increased flood risk poses challenges to society, physical infrastructure, and water quality. It is likely that up to 20% of the world population will live in areas where river flood potential could increase by the 2080s. Increases in the frequency and severity of floods and droughts are projected to adversely affect sustainable development. Some regions of the world are predicted to be especially endangered by the possible future changes. Coastal areas, especially heavily populated megadelta regions in South, East, and Southeast Asia, will be at the greatest risk due to increased flooding from the sea and, in some megadeltas, flooding from the rivers.

By 2050, ongoing coastal development and population growth in some areas of Australia and New Zealand are projected to exacerbate risks from a sea-level rise and increases in the severity and frequency of storms and coastal flooding. Climate change is expected to magnify regional differences in Europe's natural resources and assets. Negative impacts will include an increased risk of inland flash floods and more frequent coastal flooding and increased erosion (due to storminess and sea-level rise). In southern Europe, climate change is projected to worsen conditions (high temperatures and drought) in a region already vulnerable to climate variability, and to reduce water availability, hydropower potential, summer tourism, and, in general, crop productivity. Climate change is also projected to increase the health risks due to heat waves and the frequency of wildfires. Cities of North America that currently experience heat waves are expected to be further challenged by an increased number, intensity, and duration of heat waves during the course of the century, with a potential for adverse health impacts.

As many natural hazards are connected with the atmospheric processes, the World Meteorological Organization (WMO) undertakes many disaster risk reduction activities. They are integrated and coordinated with other international, regional, and national organizations. WMO coordinates the efforts of national meteorological and hydrological services to mitigate human and property losses through improved forecast services and early warnings, as well as risk assessments, and to raise public awareness. Emphasis is on disaster risk reduction: One dollar invested in disaster preparedness can prevent seven dollars' worth of disaster-related economic losses – a considerable return on investment. WMO's objective is to reduce by 50%, by 2019, the associated 10-year average fatality of the period 1994–2003 for weather, climate and water-related natural disasters (Natural hazard, 2010).

Natural disasters have a significant economic, social, environmental, and political impact on societies. While some of the impact of natural disasters can be mitigated, the risk cannot be completely eliminated. Growing economic and technological advances may assist in managing disasters, but they also make communities more vulnerable to potential impacts of hazards. This occurs through the increase in numbers and concentration of people and other assets exposed to hazards, and greater reliance on infrastructure such as power and water supplies (Middelmann, 2007). As far as management strategies are concerned, for example, satellite surveillance and monitoring shows a great promise in identifying hazards and assessing damage from disasters. Brazil is using satellite imagery to follow drought developments and deforestation. An early warning system offers the chance to prepare for a disaster such as a hurricane, although it has little use against earthquakes or volcanic eruptions. In 1988, early warning minimized loss of life during Hurricane Hugo. Some regions are prepared for disaster with hazard zone maps and evacuation plans, but these measures may not always work. For example, even though hazard zone maps had been prepared for the area around Nevado del Ruiz, Colombia, residents failed to respond and were caught by the eruption of 1985. Behind much of the loss from mass movement and flooding lies overuse or deforestation of land. These are critical problems in much of Latin America which land-use regulations could help to reduce. Recognizing vulnerability to a natural hazard and avoiding settlement on hazardous sites can also reduce the effects of natural disasters. Slope of active volcanoes and floodplains of major rivers are fertile sites for crop production, but they hold latent danger (Stillwell, 1992).

According to Quarantelli (Newton, 1997), the implications of rethinking "natural" disasters as having social causation are fourfold:

1. Mitigation of disasters must stress social rather than physical approaches.
2. These approaches must place emphasis on proactive rather than reactive actions.
3. Such actions need to focus on internal flaws in society rather than external forces.
4. Reduction of vulnerability to disasters must be integrated as part of ongoing policies and programs of societal development.

If we accept that full prevention is unattainable, then our rethinking of disasters leads us toward a policy of long-term loss reduction – mitigation. Local institutional involvement is crucial for the adoption and implementation of hazard mitigation. Factors such as local leadership, locally devised rules and strategies, adaptation to dynamic conditions, recognition of local rights, monitoring and compliance, and linkage help local governments to promote recovery and mitigation efforts (Reddy, 2000).

Public perception of natural hazards

The concept of natural hazard involves assignation of values (e.g., dangerous, harmful, and bad) to natural phenomena, depending on their potential impact on humans. That anthropocentric perspective, predominantly exploited by mass media, is one of the reasons of flawed public understanding of the Earth's geosystem functioning. The term "natural hazard" refers to one of the aspects of human perception of the natural environment. The relations between human beings and environment began about 5 million years ago and from the beginning people have attempted to become as independent from the environment as possible. Nature is generous and a friendly source of food, building materials, energy, etc., but also a hostile and mysterious power bringing earthquakes, floods, tornadoes, etc.

Natural hazards have impacted humans since they first walked on the Earth, influencing, shaping, and modifying human behavior, gradually or catastrophically changing the way people live with and respond to the environment (Middelmann, 2007). Milestones in civilization development were marked by inventions of new technologies of energy production, first from biomass and later from various kinds of fuels. One of the consequences was changing public attitude to the environment. The most recent trend is based on the idea of individual and collective responsibility for the environment, followed by actions aimed to mitigate negative environmental changes (Mannion, 1997). The technological achievements of the last 200 years, followed by significant social transformations on a global scale, have created an illusion of the potential independence of humans from the environment. About 50% of the world population lives in cities, but the figures vary from over 80% in USA and Australia to 38% in Africa (United Nations, 2008). In USA, people spend over 90% of their lifetime in closed spaces, for example, homes, office buildings, cars, shopping centers (Jacobson, 2002).

Although significantly affected by humans, the environment is still ruled by the same natural processes that have operated since before the dawn of the human existence (e.g., plate tectonics, water cycle). The Earth's ecosystem remains in a dynamic equilibrium and extreme events, potentially dangerous for humans, are regular elements of this state (e.g., earthquakes or tornadoes). Scheidegger (1997) goes further and says that a landscape is, in fact, an open, nonlinear, dynamic system where tectonic uplift and seismic activity represent the input: mass wastage and relief degradation, the output. The apparent "stability" is due to the fact that open, nonlinear dynamic systems tend to develop into relatively stable, self-organized ordered states "at the edge of chaos." Short of complete breakdown, such systems reestablish order in steps of various magnitudes.

Over time, attribution of natural hazards and disasters has shifted from supernatural or mystical forces, to nature (physical forces in natural systems), and with some reluctance, to humans who have made changes to natural systems. In this transition of thinking, we have added to the list of causation, but never fully abandoned the earlier sources. Quarantelli notes the inherent danger in this approach when he observes, "the distinction often drawn between so called Acts of God (or Nature) and Acts of Men and Women is both a useless and false one. There also lurks in the distinction a supposition that one kind of disaster is more directly controllable than other ones" (Newton, 1997, p. 222). Moreover, a natural hazard or disaster, in a pure sense, does not exist; rather there is an interaction of changes in physical systems with existent social conditions. Hazards and disasters are therefore more accurately seen as social phenomena. Additionally, natural hazards are not always initiated through only natural means, and human activity can sometimes exacerbate their occurrence. For example, deforestation in mountainous areas contributes to flood occurrence by diminishing the natural water absorption. In Australia, arson is a common source of ignition for bushfires. Still, the potential impact of a hazard is the same regardless of its origin (Middelmann, 2007). Natural hazard can also depend upon the organization and values of society that control the degree to which risk may be reduced (Stillwell, 1992).

At present, two major approaches seem to dominate in the public attitude to the causes of natural hazards. One approach, called "the dominant view," emphasizes the geophysical processes underlying natural disasters. It involves monitoring and predicting, risk assessment and zoning, and emergency planning and relief. Expertise is provided by the physical sciences and engineering, such as work of Scheidegger (1997). This approach assumes that technology can solve most hazard/disaster problems, but it is restricted mainly to the developed countries. Another approach emphasizes the human/environment relation of natural disasters as seen by cultural geographers and other social scientists. It is based on the assumption that a natural hazard can only exist in the presence of a vulnerable human community and that natural disasters are characteristic rather than accidental features of places and societies. This approach also recognizes that recent disasters occur in conjunction with major social change and environmental impact, and therefore it is most applicable to the developing countries (Stillwell, 1992).

Communicating natural hazards

Information on natural hazards can often be seen as controversial or having the potential to cause panic if not adequately communicated. Natural hazards and disasters have been recently receiving much attention in books, periodicals, and the news media. This may be due to improved monitoring and communications rather than to an increase in natural phenomena. A growing concern over this problem prompted the United Nations General Assembly to designate the 1990s as the International Decade for Natural Disaster Reduction. Since 1965, the International Geographical Union's Commission on Man and Environment has been playing a central role in stimulating and coordinating natural hazard and disaster research activities of geographers worldwide (Stillwell, 1992). But as shown by Nicholls (2001), even if the forecaster-media interactions work well, there may still be problems with the understanding of forecast information. Australian experiences during the 1997/98 El Niño indicated that many users had difficulties interpreting the uncertainties inherent in a climate forecast.

There is considerable evidence in the psychological literature that people do *not* handle uncertainty and probabilities at all well. A group of psychological factors called cognitive illusions confound the attempts to communicate and understand uncertainties. Slovic summarizes the nature and effects of these illusions: "...research on basic perceptions and cognitions has shown that difficulties in understanding probabilistic processes, biased media coverage, misleading personal experiences, and the anxieties generated by life's gambles causes uncertainty to be denied, risks to be misjudged (sometimes overestimated and sometimes underestimated), and judgements of fact to be held with unwarranted confidence. Experts' judgments appear to be prone to many of the same biases as those of the general public, particularly when experts are forced to go beyond the limits of available data and rely on intuition. Strong initial views are resistant to change because they influence the way that subsequent information is interpreted. New evidence appears reliable and informative if it is consistent with one's initial beliefs; contrary evidence tends to be dismissed as unreliable, erroneous, or unrepresentative. When people lack strong prior opinions, the opposite situation exists – they are at the mercy of the problem formulation. Presenting the same information about risk in different ways (for example, mortality rates as opposed to survival rates) alters people's perspectives and actions" (Nicholls, 2001, p. 149).

The difficulties people have in dealing with probabilities and uncertainties, as summarized by Slovic, have clear implications to attempts to have climate predictions (which are inherently uncertain and probabilistic) used in an optimal fashion. Since climate forecasts must be delivered in terms of probabilities because of the chaotic nature of the climate, scientists must also learn how people interpret and misinterpret these probabilities. Cognitive biases affected the way El Niño forecasts were received and interpreted in Australia during 1997. For instance, two biases are "availability" and "anchoring." During 1997 many press articles on the El Niño described the severe impacts of the 1982/83 event. Users then had a great difficulty adjusting their expectations of the impacts of the 1997 El Niño away from what they had experienced during 1982. This occurred even when they were reminded that the 1982 impacts were very extreme, compared with historical impacts of the El Niño on Australia. The "availability" of the reports about the 1982/83 impacts led users to "anchor" to the 1982 impacts, and they subsequently could not "adjust" away from that anchor sufficiently. Nicholls (2001) discusses the ways of avoiding these cognitive biases – the "anchoring" problem, for instance, could be reduced by ensuring that a variety of El Niño events, with varying degrees of impact, are discussed in the context of a forecast based on El Niño. Scientists need to work on, not just improving our monitoring and predicting of the next El Niño and other climate and weather hazards, but on innovative ways to present these predictions to the public.

There appears to have been somewhat more attention paid to the difficulties of communication in short-range predictions; but even here, much could be gained from increased communication between atmospheric scientists and those dealing with other forms of disasters. The development of appropriate policies and communication strategies to deal with sensitive situations is therefore essential. Equally important is instilling a culture of safety and local participation in the community. The preparedness of a community for a natural hazard can reduce the impact of an event of natural disaster and allow for more rapid recovery. Therefore, a key to reducing the overall risk of natural disasters is for those who play a role in the management of natural disasters to work closely with the wider community (Middelmann, 2007). One of the obstacles that must be overcome is the communication strategy.

Creating closer links between policy, research, and practice is central to reducing the impact of natural hazards and natural disasters. Communication across these domains provides an appreciation, understanding, and involvement across interrelated areas and is of high importance in reducing risk. However, for science and research to effectively influence policy development, information must be clearly communicated to government in a timely and understandable manner. This is vital in ensuring scientific research reaches its full potential and assists policy makers to make informed and relevant decisions using the best information available. As the Centre for European Flood Research observes: "If scientists really want to influence policy more, researchers need to become more visible, and clearer about the kind of changes they are aiming for, and are able to achieve" (Middelmann, 2007, p. 61). Practitioners need to communicate effectively to those whose role is to develop a policy. Similarly, any policy which is developed needs to be coherent in "whole-of-government" terms. It is also vital that those involved in policy development seek the expertise of those working "on the ground." Researchers need to liaise with practitioners to find out what their needs are, and work toward developing relevant methodologies and techniques which can be easily applied and communicated to effectively inform policy makers. Successful linking of policy and research requires an open and continuous dialogue. Where this relationship is effective and natural hazard impacts are minimized, the benefit is felt by politicians, policy makers, researchers, practitioners, and the community (Middelmann, 2007).

Apart from uncertainty, the lack of a precise and objective definition of a phenomenon may be a serious obstacle to understanding it. A good example is the case of drought. It must be accepted that the importance of drought lies in its impacts. Institutional, political, budgetary, and human resource constraints often make drought planning difficult. One major constraint that exists worldwide is a lack of understanding of drought by politicians, policy makers, technical staff, and the general public. Lack of communication and cooperation among scientists and inadequate communication between scientists and policy makers on the significance of drought planning also complicates efforts to initiate steps toward preparedness. Because drought occurs infrequently in some regions, governments may ignore the problem or give it low priority. Inadequate financial resources to provide assistance and competing institutional jurisdictions between and within levels of government may also serve to discourage governments from undertaking planning.

Other constraints include technological limits (such as difficulties in predicting and detecting drought), insufficient databases, and inappropriate mitigation technologies. Policy makers and bureaucrats need to understand that droughts, like floods, are a normal feature of the climate. Their recurrence is inevitable. Although we cannot influence the occurrence of the natural event (i.e., meteorological drought), we can lessen vulnerability through more reliable forecasts, improved early warning systems, as well as appropriate and timely mitigation and preparedness measures. Drought manifests itself in ways that span the jurisdiction of numerous bureaucratic organizations (e.g., agricultural, water resources, health, and so forth) and levels of government (e.g., national, state, and local). Competing interests, institutional rivalry, and the desire to protect their agency missions (i.e., "turf protection") impede the development of concise drought assessment and response initiatives. To solve these problems, policy makers and bureaucrats, as well as the general public, must be educated about the consequences of drought and the advantages of preparedness. Drought is an example of an interdisciplinary problem that requires input by many disciplines and policy makers (Wilhite, 1996).

Summary

Natural hazards are phenomena resulting from the geophysical processes that are an integral part of the environment. People cannot control natural hazards, but can be significantly affected by them if a natural disaster (i.e., the concrete realization of a hazard in a populated area) occurs. There are many direct and indirect effects of natural hazards, but most of them are associated with great economic and nonmaterial losses. Therefore, those phenomena are characterized not only by natural but also social, economic, and political aspects. As the climate change scenarios indicate that the risk of natural hazard is supposed to increase, it is of utmost importance to improve the mitigation strategies. The basic problem to be solved is better communication together with better information and knowledge transfer from the scientific community to

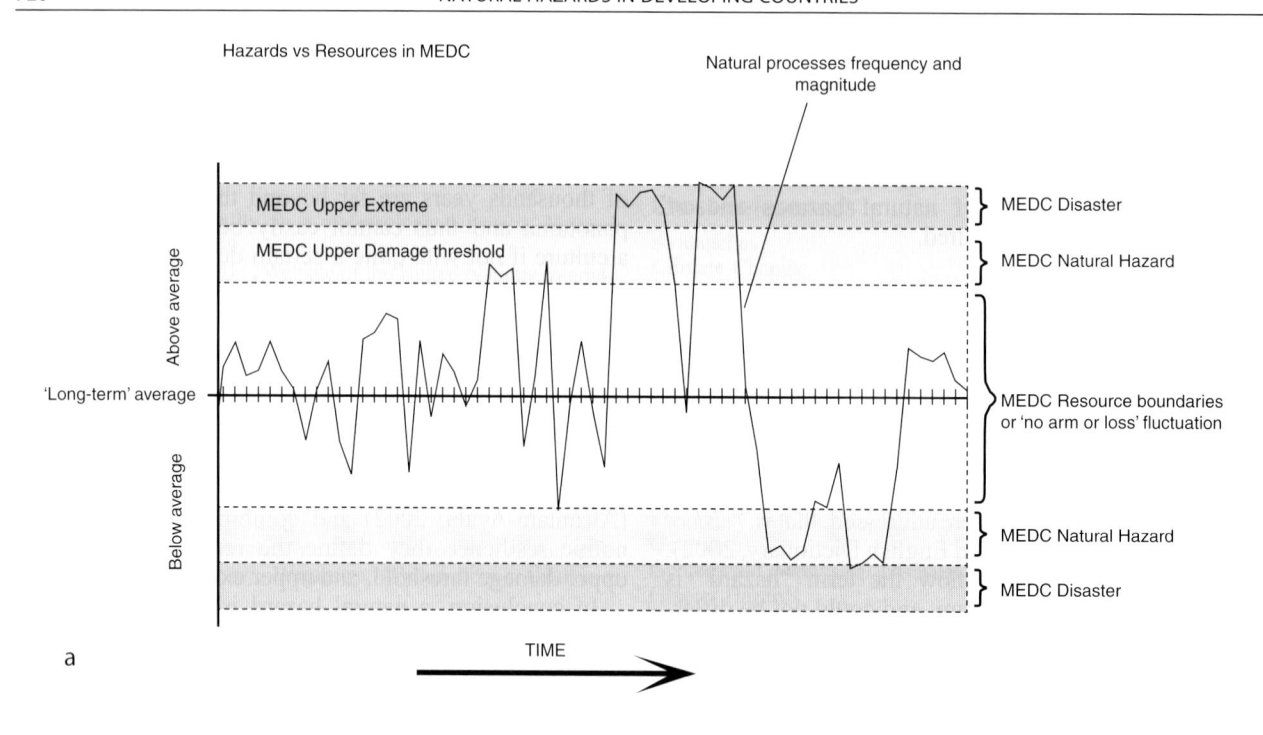

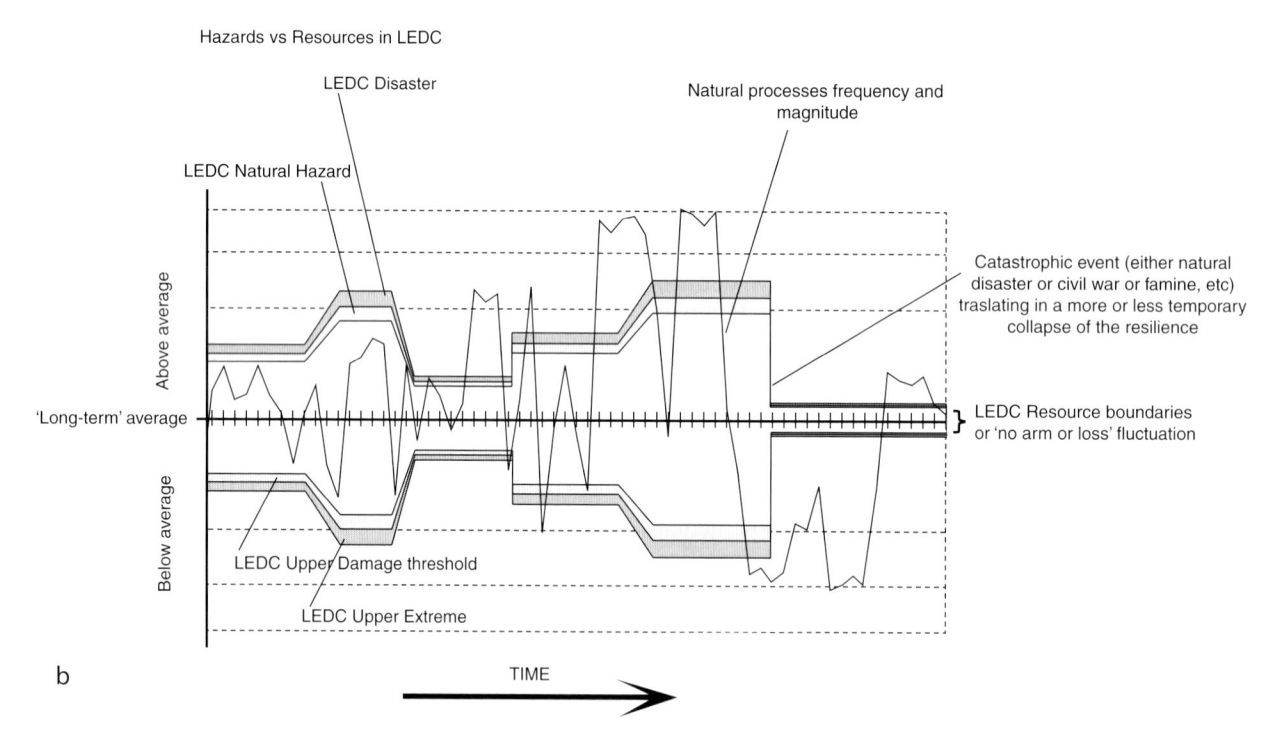

Natural Hazards in Developing Countries, Figure 1 Comparison between MEDC and LEDC thresholds for the onset of Natural Hazard and Disaster. The thin lined curve in both diagrams indicates a natural process, for example, rainfall, with its frequency and magnitude in time. The midline stands for the long-term average. (**a**) In the MEDC thresholds are quite stable through time if not becoming wider with time. Also they show a quite wide range of tolerance or resilience that defines the resource boundaries. A natural hazard impacts on society only when the MEDC Upper Damage threshold is surpassed (Modified after Smith and Petley, 2009). (**b**) In LEDC, on the other hand, all boundaries and thresholds are highly variable in time depending on socioeconomic instability of these countries. The resource boundaries are in general smaller than the one of MEDC. Furthermore in LEDC, the thresholds for Upper Damage and Upper Extreme are not very far one from the other, while in MEDC, they are more widely spaced.

NATURAL HAZARDS IN DEVELOPING COUNTRIES

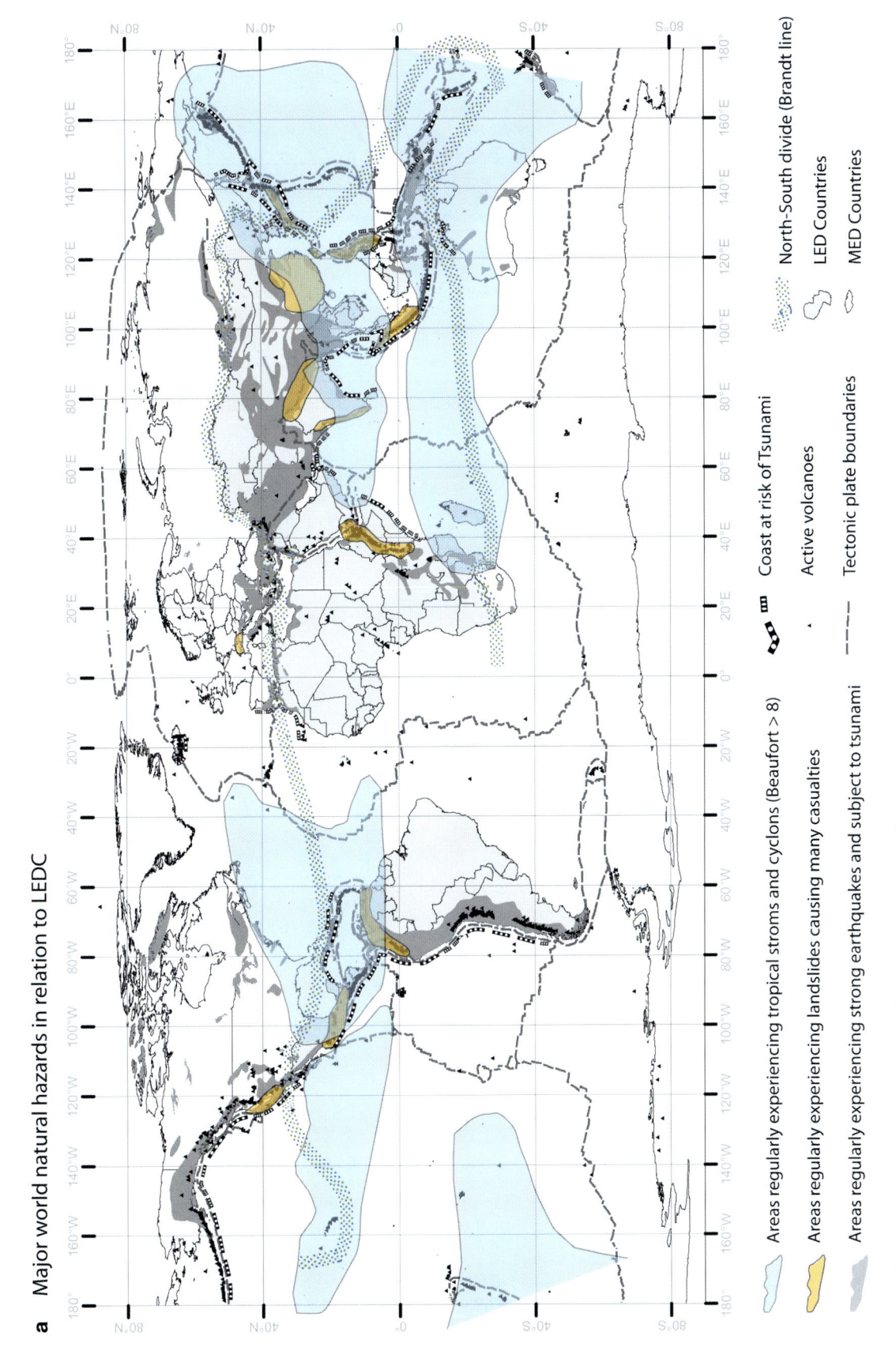

a Major world natural hazards in relation to LEDC

Natural Hazards in Developing Countries, Figure 2 (Continued)

Disaster Type Proportions by United Nations Sub-Regions: 1974–2003

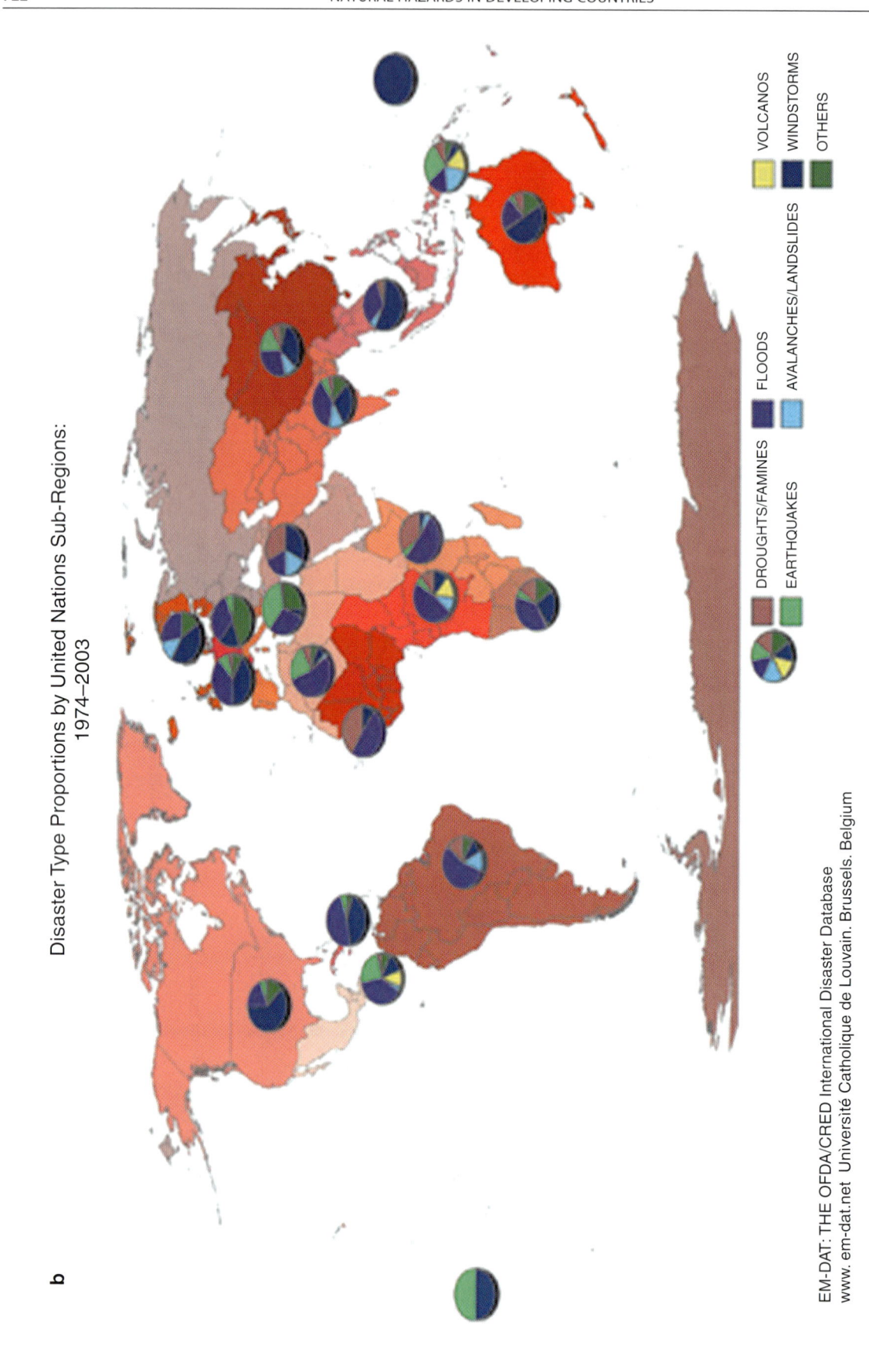

EM-DAT: THE OFDA/CRED International Disaster Database
www.em-dat.net Université Catholique de Louvain. Brussels. Belgium

Natural Hazards in Developing Countries, Figure 2 (a) Location of the LEDC and of some of the main natural hazards in the world. The LEDC regions are: Central and South America, Africa (excluding South Africa), Asia (Mid, Central, South East, Far, excluding Japan South Korea, Taiwan, and Russia). (b) Disaster type proportion by UN macro regions 1974–2003 (From CRED database).

Natural Hazards in Developing Countries, Table 1 Environments and their main locations in the LEDC regions

Environment	Occurrence
Monsoonal	Indian Ocean realm
Humid tropic	Central America, South of Mexico; Amazon basin; Central-West Africa,
Arid and semiarid (cold and hot)	Mexico; parts of South America; North, Central, East and Southern Africa; Asia and Middle East (excluding South East Asia)
Deserts	Mexico; parts of South America; North Africa (Saharan region), East and South Africa; Asia and Middle East (excluding South East Asia)
Mountain (including high mountain and associated Plateau)	Central and South America (Andes); Asia (Alpine-Himalayan mountains)
Coastal	All regions
Islands	Mainly South East Asia and Pacific realm
Glacial and periglacial	Western and Southern part of South America (Andes); Asia (Alpine-Himalayan mountains)
Technoscapes or anthropic environments (megacities)	All regions, mainly on coastal environments

Natural Hazards in Developing Countries, Table 2 Examples of natural hazards, their rapidity, and the environment they mainly occur in, with reference to Table 1

Type of natural hazard	Rapid	Slow	Environment
Avalanches	X		Mountain, glacial, periglacial
Epidemics diseases		X	All (with much less influence on the glacial and periglacial)
Coast = erosion		X	Coastal, technoscapes
Desertification		X	Arid and semiarid
Drought		X	Arid and semiarid
Dust storm	X		Arid and semiarid (with global effects)
Dzud (also spelled Zud)		X	Arid and semiarid (cold, especially Mongolia plains)
Earthquakes	X		All
Coastal Floods	X	X	Coastal, technoscapes
Floods	X	X	All (excluding glacial and periglacial)
Flash floods	X		Arid and semiarid
Fog		X	Mountain
Glacier surges		X	Glacial
Hail	X		Mountain, desert (coastal)
Hurricanes/typhoons	X		Coastal, islands (mainly between +20° and −20° of latitude)
Plant and livestock pests		X	All (with much less influence on the glacial and periglacial)
Sea level rise		X	Coastal, technoscapes
Sinkholes (karst)		X	All (if limestone or dolomite or evaporitic rocks are present)
Slope instability	X	X	Mountain, humid tropics, monsoonal
Soil erosion (including gully erosion)		X	Most, especially with human interference
Storms	X		All
Thermokarst		X	Glacial and periglacial
Tornadoes	X		Arid and semiarid, mountain
Tsunamis	X		Coastal, technoscapes
Volcanic	X		All
Wildfires	X		Arid and semiarid

Derived from Alexander (1993), Whittow (1996), Alcantara-Ayala (2002), Goudie (2002), and Smith and Petley 2009

thresholds of GNI each year on the first of July. With this classification, the bank defines the following category of countries: Low-Income, with $975 or less per capita/year; Lower-Middle Income, $976–$3,855 per capita/year; Upper-Middle Income, $3,856–$11,905 per capita/year; and High Income, $11,906 or more. The first two groups are usually referred to as developing countries, but the bank also specifies that this system is not always related to the degree of development of a nation (http://go.worldbank.org/K2CKM78CC0).

An alternative to these definitions is given by the term Less Economically Developed Countries (LEDC), as opposed to More Economically Developed Countries (MEDC). LEDC have high birth rates (>20%), death rates (>30%), and infant mortality (>30%); more than half of their workforce are involved in agriculture and have a low level of nutrition, secondary schooling, literacy, electricity consumption per head and GDP is usually less than 1,000 USD/capita/year (Mayhew, 2009).

Visually, the so-called North–south divide, separating MEDC and LEDC countries, is also shown in Figure 2 with "South" countries in gray and "North" countries in white (Brandt, 1980).

Type of natural hazards in developing countries

Natural hazards are zonal phenomena (Green, 2007) and as such they are not evenly distributed on the entire planet. Figure 2 shows the distribution of several hazards and it emerges how much more LEDC are exposed to natural hazards than MEDC.

The environments found in the LEDC countries are summarized in the following Table 1, where under the field Occurrence is listed the main geographical distribution within the LEDC.

Each of the environments of Table 1 has its own specific type of geophysical and biological processes. Table 2 presents most of the natural hazards occurring by environment, that is far from being exhaustive.

Some geophysical hazards are not following the same zonation of Table 1; instead they are located at tectonic plates boundaries (see Figure 2). Notably earthquakes, tsunamis, and volcanic activity are mainly found in the so-called circum Pacific fire belt, encompassing all the continental areas facing the

Natural Hazards in Developing Countries, Table 3 Examples of complex interaction between natural and anthropic causes of some natural hazards

Natural hazard	Anthropic component	Natural component
Desertification in semiarid area	Clearing, overgrazing, wood collection, charcoal burning, etc.	Drought and climate change
Soil erosion in a valley bottom	Runoff from new roads, removal of protective vegetation for overgrazing, increased cropping areas, plowing perpendicular to the contour lines, etc.	Change in climate or in base level
Coastal erosion	Side effect of groynes up the coast, or of a reduced sediment supply due to damming of the rivers	Higher intensity and/or frequency of storms, increase in sea level
Flooding (river)	Removal of natural vegetation, urbanization, overgrazing, mismanaged dam regulation, land encroachment	Higher intensity and may be amount of rainfall

Adapted from Goudie (2002), p. 507

Pacific Ocean, the Caribbean Sea, the East Indian Ocean and the Philippine Sea, as well as onshore Central and South Asia.

Traditionally, natural hazards are defined as a "sudden release of energy and/or matter in that specific system" (Smith and Petley, 2009), with an accent on their velocity of development. This is true for the most striking processes like earthquakes, tsunamis, volcanic eruptions, rapid landslides, and flash floods, but in fact the release of energy and/or matter can also be slowly onsetting instead of rapid. Drought, dzud (or zud), soil erosion, dust storms, desertification, coastal erosion, and salinization are some of the slow developing processes. Table 2 distinguishes between the different velocities of each natural hazard in LEDC. Although it could be noticed that most of the events are rapid in their final expression, it is important to identify the development processes of any specific natural hazard in order to be able to identify the most appropriate type of monitoring and early warning systems as well as preparedness plans that should be adopted for each specific hazard.

Most of the biological hazards are slowly on setting if not chronically endemic and in most cases it is difficult to identify them as purely natural. In countries where the human population is below the average nutritional and health status like, for example, in the bottom billion population (Collier, 2008), the impact of a virus/infection could be devastating, while in better-off countries the impacts of the same infection would be much lower. An example is given by malaria infection that is weakening most of African population but not its richer tiny percentage, because they have access to better health care, sanitation, and nutrition. Similarly some geomorphological hazards are the result of interplay between human and natural causes, as shown in Table 3.

The global death toll due to natural disasters in the developing countries can be as high as 95% of the total (Alexander, 1993). There are three indicators usually adopted for defining the magnitude of a natural hazard or disaster: number of deaths; number of people affected; and economic loss (CRED database). It has been shown (Pielke and Pielke, 2000) that there is an inverse correlation between the economic loss and the death toll: The more economic loss is suffered the less death is registered. This means that when high income countries are hit, they lose mainly economically while when a low income country is hit, the death rates are higher.

For this reason in LEDC, the most accepted indicator for magnitude of natural hazard is mortality and using this indicator it results that the top 25 countries affected by multi-hazard mortality are all in the LEDC (Mosquera-Machado and Dilley, 2009).

Vulnerability has been recognized as the most important factor in calculating risk, both to population and goods (Wisner et al., 2004). Nevertheless vulnerability is calculated in many different ways and still there is no standard practice (Alexander, 1993; Wisner et al., 2004; Mosquera-Machado and Dilley, 2009). Despite the variability in assessing vulnerability in LEDC, some factors leading to lower resilience are common: Lack of institutional organization, lack of sound early warning systems, low preventive capacity at both structural and cultural levels, low awareness and education of the population at risk, absence of effective civil protection are some of the causes of higher exposure to risk, not to mention the more general –but still relevant to the issue – poverty trap (Alexander, 1993; Smith, 2009; Wisner et al., 2004, among others).

Role of international agencies dealing with natural hazards in developing countries

The international community, led by a move of the United Nations in early 1990s, started addressing these issues especially for LEDC. The International Decade for Natural Disaster Reduction (IDNDR), which then evolved into the ISNDR (International Strategy for Natural Disaster Reduction, http://www.unisdr.org/) started with a strong engineering paradigm, top-bottom approach (Hamilton, 2000; Smith, 2009; Chester, 2002), and then, with the ISNDR, evolved into a more development/complexity type of approach.

Following the 1994 devastating famine in sub-Saharan Africa, a group of humanitarian agencies launched the

Sphere Project, with the aim of coordinating and improving the professionalism, effectiveness, and accountability of the aid actions in disaster contexts. The Sphere Humanitarian Charter and Minimum Standards in Disaster Response sets out for the first time what people affected by disasters have a right to expect from humanitarian assistance. The aim of the Project is to improve the quality of assistance provided to people affected by disasters, and to enhance the accountability of the humanitarian system in disaster response (http://www.sphereproject.org/content/view/27/84/lang, english/). Sphere standards set a new paradigm in humanitarian emergencies: Despite being a nonmandatory set of standards, most agencies/organization comply with them. Sphere standard undergoes also periodical participatory review.

Conclusions

Despite a great deal of studies concerned with the explanation and forecast of global atmospheric and volcanic hazards, addressing also the developing countries, there is still a desperate need for detailed identification of natural hazards in developing countries (especially the more complex, slow developing ones). Capacity building at academic and professional level should be supported and strengthened by the international community. There is a big opportunity here also for philanthropic donations by wealthy Africans, South Americans, and Asians who can probably better select and direct their generous efforts to deserving compatriots/research centers.

One of the biggest challenges to reduce natural hazard vulnerability especially in LEDC for policy makers, land planners, international and local community leaders is well spelled out by Smith and Pedley (2009, p. 339) "Any improvement in the connectivity between people and their environment depends on assisting all community exposed to risk to develop their own hazard-reducing capabilities and local self-reliance following disasters. This is not always an easy task because it depends, to some extent, on external inputs. For example, the construction of rural roads in landslide-prone terrain is doomed to failure if no provision is made for the use of appropriate engineering measures. Once again there is a need for integrated approaches in which sensitive external assistance is deployed to help build community skills for the anticipation of hazards and the mitigation of their impacts."

Bibliography

Alcantara-Ayala, I., 2002. Geomorphology, natural hazards, vulnerability and prevention of natural disasters in developing countries. *Geomorphology*, **47**, 107–124.
Alexander, D., 1993. *Natural Disasters*. London: UCL Press. 632 p.
Brandt, W., 1980. *North–South. A Program for Survival*. Cambridge: MIT Press. 304 p.
Chester, D. K., 2002. Overview: hazard and risk. In Allison, R. J. (ed.), *Applied Geomorphology: Theory and Practice*. Chichester: Wiley, pp. 251–263.
Collier, P., 2008. *The Bottom Billion*. Oxford: OUP. 224 p.
Concise Oxford English Dictionary, 2002. Oxford University Press, 1728 p.
CRED database, 2010. http://www.emdat.be/world-map
Goldman Sachs, 2003. *Dreaming with BRICs*. Global Economics Report, 99, p. 23. (http://www2.goldmansachs.com/ideas/brics/book/99-dreaming.pdf). Accessed 08 March 2010.
Goudie, A. S., 2002. *The Nature of the Environment*, 4th edn. Oxford: Blackwell, p. 544.
Green, C., 2007. Natural hazards. In Douglas, I., Hugget, R. J., and Perkins, C. (eds.), *Companion Encyclopedia of Geography*. London: Routledge, pp. 645–661.
Hamilton, R. M., 2000. Science and technology for natural disaster reduction. *Natural Hazards Review*, **1**(1), 56–60.
Jain, S. C., 2006. *Emerging Economies and the Transformation of International Business*. Cheltenham: Edward Elgar Publishing, p. 384.
Mayhew, S. 2009. Less economically developed country. In OUP (ed.), *Oxford Reference Online*. Oxford, UK: University of Oxford. *A dictionary of Geography*. http://www.oxfordreference.com/pages/Subjects_and_Titles__2E_PS04. Accessed 09 March 2010.
Mosquera-Machado, S., and Dilley, M., 2009. A comparison of selected global disaster risk assessment results. *Natural Hazards*, **48**, 439–456.
Pielke, R. A., Jr., and Pielke, R. A., Sr., 2000. *Storms*. Reutledge: London/New York.
Smith, K., 2009. Natural hazards. In Cuff, D., and Goudie, A. S. (eds.), *The Oxford Companion to Global Change*. New York: Oxford University Press.
Smith, K., and Petley, D. N., 2009. *Environmental Hazards. Assessing Risk and Reducing Disaster*, 5th edn. London/New York: Reutledge. 383 p.
Varnes, D. J., 1984. Landslide hazard zonation: a review of principles and practice. *International Association of Engineering Geologists, Commission on Landslides and other Mass Movements on Slopes*. Paris: Unesco. 60 p.
Whittow, J., 1996. Environmental hazards. In Douglas, I., Hugget, R. J., and Robinson, M. (eds.), *Companion Encyclopedia of Geography: The Environment and Humankind*. London: Routledge, pp. 620–650.
Wisner, B., Blaikie, P., Cannon, T., and Davis, I., 2004. *At Risk: Natural Hazards, People's Vulnerability, and Disasters*. London/New York: Routledge. 471 p.

Cross-references

Civil Protection and Crisis Management
Classification of Natural Disasters
Community Management of Hazards
Coping Capacity
Costs (Economic) of Natural Hazards and Disasters
Disaster
Disaster Risk Management
Disaster Risk Reduction (DRR)
Early Warning Systems
Education and Training for Emergency Preparedness
Exposure to Natural Hazards
Geological/Geophysical Disasters
Global Change and its Implications for Natural Disasters
Global Network of Civil Society Organisations for Disaster Reduction
Hazard
Human Impact of Hazards
Humanity as an Agent of Geological Disaster

Hyogo Framework for Action
International Strategies for Disaster Reduction (IDNDR and ISDR)
Livelihoods and Disasters
Megacities and Natural Hazards
Mitigation
Perception of Natural Hazards and Disasters
Red Cross and Red Crescent
Resilience
Risk
Time and Space in Disaster
United Nations Organisation and Natural Disasters
Vulnerability
Warning Systems
Worldwide Trends in Natural Disasters

NATURAL RADIOACTIVITY*

Cathy Scheib
British Geological Survey, Nottingham, UK

Definition
Natural radioactivity originates from two primary sources: cosmic radiation and radioactive elements in the earth's crust. All the elements from polonium (atomic number 84) to uranium (atomic number 92) are radioactive. Radioisotopes of some lighter elements are also found in nature (e.g., ^{40}K).

Introduction
Many atoms are unstable and will change quite naturally into atoms of another element accompanied by the emission of ionizing radiation. Unstable atoms that change through radioactive decay to form other nuclides are said to be radioactive and are referred to as radionuclides or radioisotopes. The rate of change or decay of an unstable radionuclide is indicated by its half-life, which is the period of time during which half the original number of atoms would have decayed. The radioactivity of the earth includes three major categories: *primordial radionuclides*, which have very long half-lives, were created in stellar processes before the earth was formed and are still present in the earth's crust; *secondary radionuclides*, which are decay products of primordial radionuclides that are themselves radioactive and will decay to other secondary radionuclides or to stable isotopes; and *cosmogenic radionuclides* which are continuously produced by bombardment of stable nuclides by cosmic rays, primarily in the atmosphere. Natural radionuclides are ubiquitous in the environment and make a major contribution to background radiation (see *Dose Rate*).

Cosmic radiation
The atmosphere is continuously exposed to primary cosmic radiation that originates in outer space. This cosmic radiation comprises predominantly protons (about 87%) and alpha-particles (about 11%), with a smaller fraction of nuclei and very high energy electrons comprising the remainder. The interactions of these primary particles with atmospheric nuclei produce electrons, gamma rays, neutrons, and mesons.

The amount of cosmic radiation increases with altitude and with polar latitudes. The annual cosmic-ray dose equivalent is about 0.3 mSv at sea level. In Leadville, Colorado (altitude 3,200 m), for example, the residents receive around 1.25 mSv year^{-1}, which is more than four times the annual dose from cosmic radiation at sea level (Eisenbud and Gesell, 1997). Because of this effect of altitude, passengers and crew of high-flying aircraft are subject to additional dose from cosmic rays. Solar activity affects the effective dose from cosmic radiation received during aviation; the amount of cosmic radiation produced by the sun varies with an approximately 11-year cycle (see *Solar Flares*). Years with increased levels of solar activity translate to a higher frequency of solar flares, some of which increase the amount of cosmic radiation in the earth's atmosphere resulting in higher annual effective doses to aircrews during those years (UNSCEAR, 2000).

The nuclear reactions initiated by cosmic particles in the atmosphere give rise to a number of cosmogenic radionuclides, such as ^{14}C, which is used to date relics containing naturally carbonaceous material.

Terrestrial natural radioactivity
The naturally occurring primordial radionuclides of the earth can be divided into those that occur singly and those that are the components of three decay chains. The uranium series originates with the most abundant uranium isotope, ^{238}U (Figure 1), and accounts for the largest proportion of human exposure to ionizing radiation due to radon gas (^{222}Rn). The actinide series begins with ^{235}U, which comprises only 0.72% of total uranium, and the thorium series originates with ^{232}Th.

In a closed system, the daughter nuclides produced by radioactive decay in each series eventually achieve a state called secular equilibrium with their parent radionuclide. This state is achieved when the half-life of the parent nuclide is much longer than those of the succeeding species, such that there is no significant change in the concentration of the parent during the time interval over which its shorter-lived descendants attain equilibrium. When this state is achieved, all nuclides within a given decay chain decay at the same rate. The ^{232}Th series comes to equilibrium in about 70 years, in contrast to the ^{238}U chain, which takes longer than 10^6 years to reach equilibrium. This state of secular equilibrium only occurs in a truly closed system, so disequilibrium can occur if the system changes and is no longer "closed." For example, if the members of the decay chain are being transported by ground water, the differing physicochemical behavior of each element in the chain may lead to differing migration rates or the precipitation or dissolution of the different decay chain members, thus leading to disequilibrium. As ^{226}Ra is chemically very different from ^{238}U, it is possible

---
*©British Geological Survey

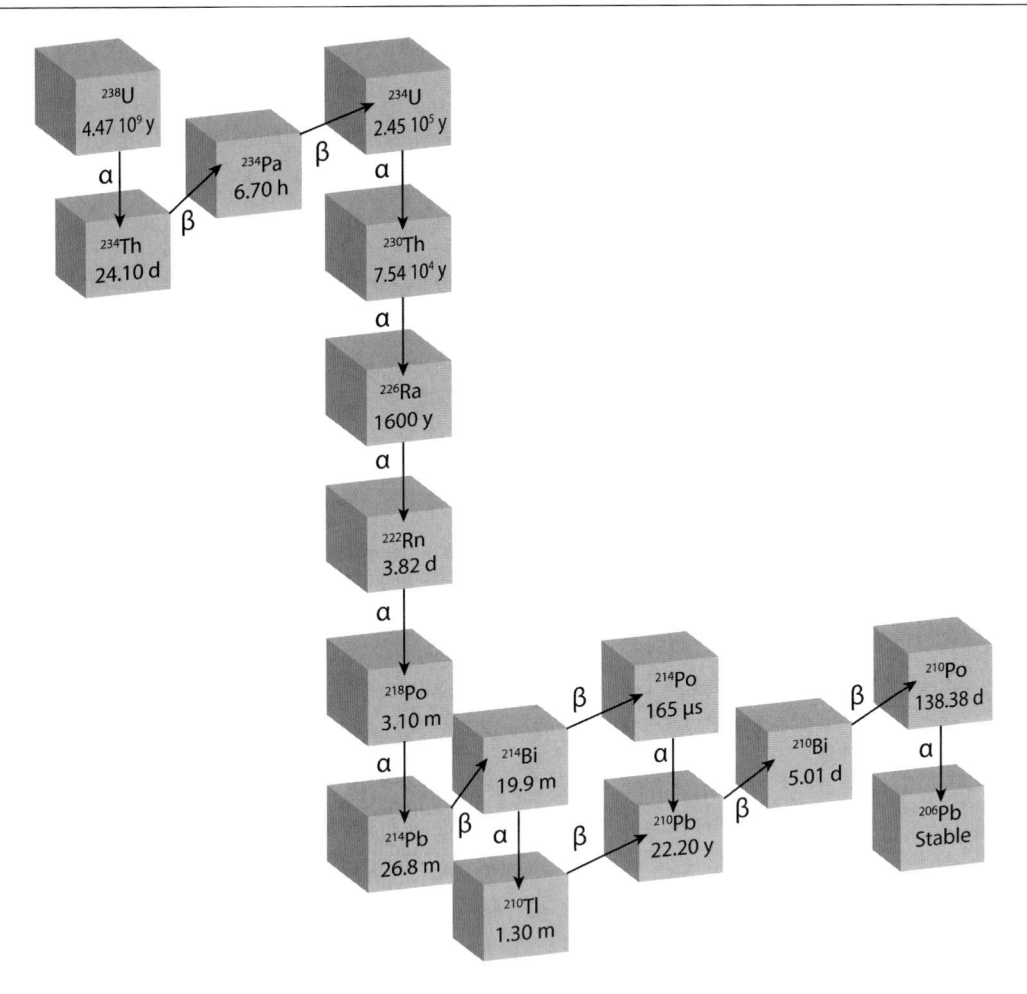

Natural Radioactivity, Figure 1 The uranium series. Half-lives and major decay radiation (α – alpha decay; β – decay (National Nuclear Data Centre, Brookhaven Laboratory). Often the transitions are accompanied by the emission of gamma radiation. y = years, d = days, h = hours, m = minutes.

in natural processes for the two to become separated so that the ^{226}Ra and its daughter products are unsupported by the parent ^{238}U. Thus, there may not be a simple relationship between measurements of secondary radionuclides beyond ^{226}Ra in the uranium series and ^{238}U. Decay series disequilibria can be used as an investigatory tool in earth and environmental sciences.

An example of a non-series primordial radionuclide is ^{40}K. Potassium-40 comprises only 0.0119% of total K and undergoes branched decay producing ^{40}Ca and ^{40}Ar, the latter producing high-energy gamma rays. The K-Ar ratio is often used in geochronology. Potassium-40 is easily the predominant radioactive component in normal foods and human tissue.

Terrestrial gamma rays originate chiefly from the radioactive decay of the natural K, U, and Th which are widely distributed in terrestrial materials including rocks, soils, and building materials extracted from the earth. In general, the gamma radiation dose at any location is proportional to the amount of K, U, and Th in the ground and in building materials. ^{214}Bi contributes most of the gamma activity of the uranium decay series (Figure 1) and ^{208}Tl is the main gamma active daughter product derived from the ^{232}Th series. In airborne gamma-ray surveying, a technique which provides spatially integrated gamma-ray data over large areas, these daughter products are used to estimate equivalent uranium (eU) and equivalent thorium (eTh), respectively. Estimates of eU (Figure 2), in addition to K, eTh, and dose rate determined by a national-scale aerogeophysical survey across Northern Ireland, showed that gamma-emitting radionuclide distribution was closely related to bedrock (Figure 2) and surficial geology and, to a lesser extent, to technological enhancement of naturally occurring materials, for example, by power production (Appleton et al., 2008; Beamish and Young, 2009).

Soils developed over radioactive rocks generally have a much lower gamma radioactivity than the rock substrate. Radioactive elements in the rock fragments and derived minerals in the weathered overburden are diluted with

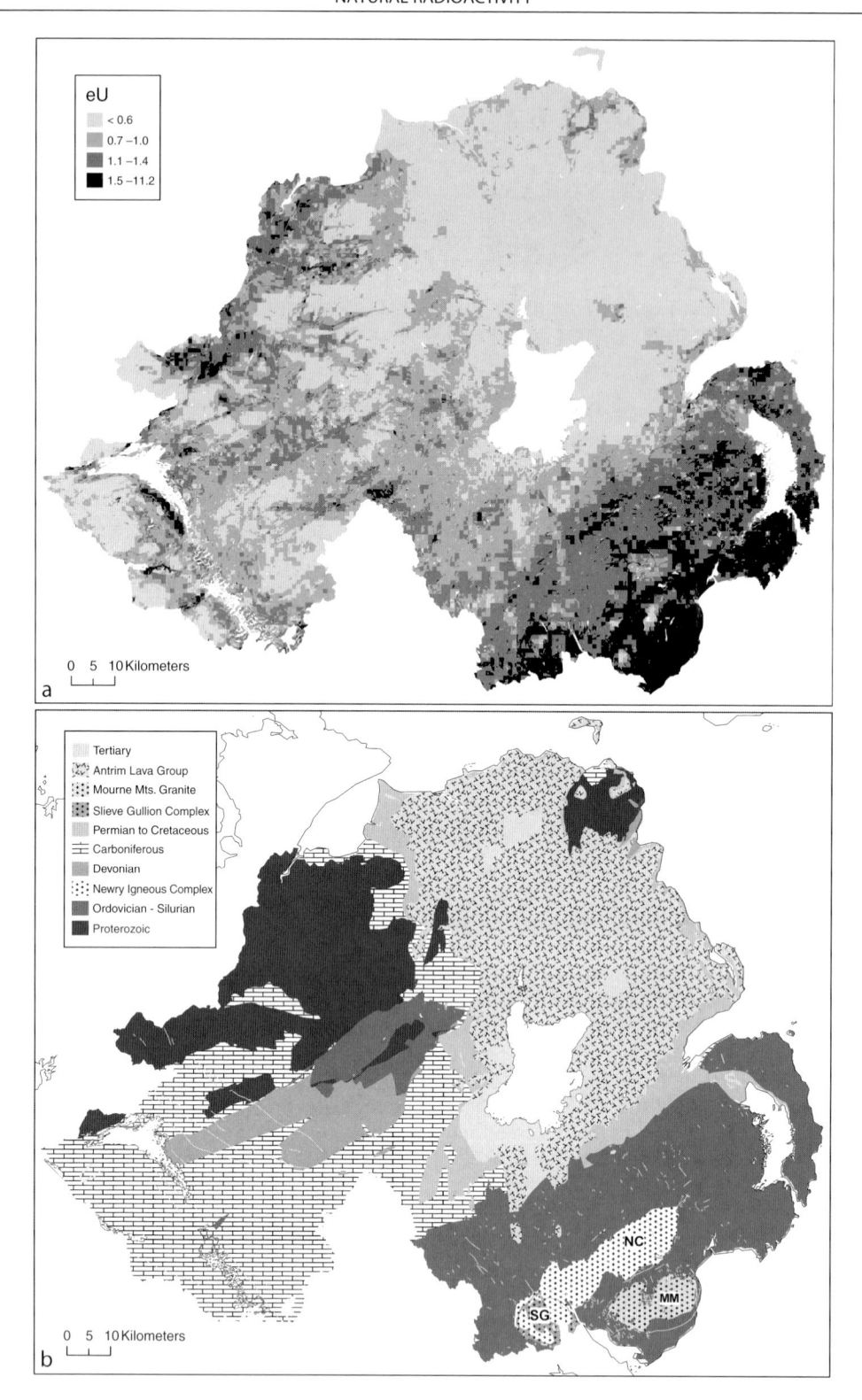

Natural Radioactivity, Figure 2 (a) Distribution of average eU (mg kg^{-1}) by airborne gamma-ray survey, grouped by 1-km grid squares and geology combination (bedrock and superficial geology type) for Northern Ireland, UK; and (b) Simplified bedrock geology of Northern Ireland. NC Newry igneous complex, SG Slieve Gullion complex, MM Mourne Mountain Complex; the igneous complexes where the highest eU in the region is found. © Crown Copyright 2010, published with the permission of the Geological Survey of Northern Ireland.

Natural Radioactivity, Table 1 Ranges and averages of the concentrations of ^{40}K, ^{232}Th, and ^{238}U in typical rocks and soil (Eisenbud and Gesell, 1997)

Material	Potassium-40 % total K	Potassium-40 Bq Kg^{-1}	Thorium-232 mg kg^{-1}	Thorium-232 Bq Kg^{-1}	Uranium-238 mg kg^{-1}	Uranium-238 Bq Kg^{-1}
Basalt (crustal average)	0.8	300	3–4	10–15	0.5–1	7–10
Granite (crustal average)	>4	>1,000	17	70	3	40
Shale sandstone	2.7	800	12	50	3.7	40
Carbonate rocks	0.3	70	2	8	2	25
Continental crust (average)	2.8	850	10.7	44	2.8	36
Soil (average)	1.5	400	9	37	1.8	22

organic matter and water. Average concentrations of ^{40}K, ^{232}Th, and ^{238}U in typical rocks and soil are given in Table 1.

Areas of high natural radioactivity

The level of natural background radiation varies depending on location, and in some areas, the level is significantly higher than average. Areas of high natural radioactivity include: Ramsar in Iran due to hot-springs containing high levels of ^{226}Ra and ^{222}Rn; areas of thorium-bearing monazite sands in Brazil, China, Egypt, and India; volcanic rocks in Brazil and Italy; uranium mineralization in France, the UK, and the United States; and radium enriched karst soils developed over limestones in Switzerland, the UK, and the United states (UNSCEAR, 2000).

Technologically enhanced naturally occurring radioactive materials (TENORM)

Human activity may lead to what is known as "technologically enhanced naturally occurring radioactive materials" (TENORM). TENORM industries may release significant amounts of radioactive material into the environment resulting in the potential for extra or enhanced exposure to ionizing radiation. These industries include mining, phosphate processing, metal ore processing, heavy mineral sand processing, titanium pigment production, fossil fuel extraction and combustion, manufacture of building materials, aviation, and scrap metal processing (Vearrier et al., 2009). Workers in TENORM-producing industries may be occupationally exposed to ionizing radiation (UNSCEAR, 2000).

Oklo natural nuclear reactor

Fossil remains of a 2-billion-year-old nuclear "reactor" were discovered in 1972 in Oklo, Gabon. The percentage of ^{235}U present in the environment was much greater 2 billion years ago than it is now, and along with oxidizing conditions which allowed the uranium to dissolve, be transported, and preferentially concentrated, a "critical mass" was attained. Criticality was sustained for on the order of 10, 000 years, during which time an estimated 15, 000 MW-years of energy was released by the consumption of ~6,000 kg of ^{235}U (Eisenbud and Gesell, 1997). Studies of the migration of the fission products from the Oklo reactors provide a natural analogue for radioactive waste management (Brookins, 1990).

Summary

Natural radioactivity originates from two primary sources: cosmic radiation and radioactive elements in the earth's crust. Natural radionuclides are ubiquitous in the environment and make a major contribution to background radiation. The level of natural background radiation varies depending on location. ^{238}U is the most abundant uranium isotope and its decay products account for the largest proportion of human exposure to ionizing radiation, primarily due to radon gas (^{222}Rn). Terrestrial gamma rays originate chiefly from the radioactive decay of natural K, U, and Th which are widely distributed in terrestrial materials including rocks, soils, and building materials extracted from the earth. In general, the gamma radiation dose at any location is proportional to the amount of K, U, and Th in the ground and in building materials. Industrial processes can concentrate naturally occurring radioactive materials, which can enhance the ionizing radiation exposure to workers or exposed populations.

Bibliography

Appleton, J. D., Miles, J. C. H., Green, B. M. R., and Larmour, R., 2008. Pilot study of the application of Tellus airborne radiometric and soil geochemical data for radon mapping. *Journal of Environmental Radioactivity*, **99**, 1687–1697.

Beamish, D., and Young, M. E., 2009. Geophysics of Northern Ireland: the Tellus effect. *First Break*, **27**, 43–49.

Brookins, D. G., 1990. Radionuclide behaviour at the Oklo nuclear reactor Gabon. *Waste Management*, **10**(4), 285–296.

Eisenbud, M., and Gesell, T., 1997. *Environmental Radioactivity from Natural, Industrial, and Military Sources*, 4th edn. San Diego, CA: Academic. 656 pp. ISBN 0-12-235154-1.

National Nuclear Data Centre. Brookhaven National Laboratory http://www.nndc.bnl.gov/ (last accessed at 14:25 on February 10, 2010).

UNSCEAR 2000 Report. Volume 1: Annex B. New York: United Nations. http://www.unscear.org/unscear/en/publications/2000_1.html (last accessed at 11.12 on February 12, 2010).

Vearrier, D., Curtis, J. A., and Greenberg, M. I., 2009. Technologically enhanced naturally occurring radioactive materials. *Clinical Toxicology*, **47**, 393–406.

Cross-references

Dose Rate
Radon Hazards
Solar Flares

NEOTECTONICS

James P. McCalpin
GEO-HAZ Consulting Inc., Crestone, CO, USA

Definition

Neotectonics. Any Earth movements or deformations of the geodetic reference level, their mechanisms, their geological origin, and their implications for various practical purposes and their future extrapolations (Mörner, 1978).

The study of young tectonic events (deformation of upper crust), which have occurred or are still occurring in a given region after its final orogeny (at least for recent orogenies) or more precisely after its last significant reorganization (Pavlides, 1989).

The study of the post-Miocene structures and structural history of the Earth's crust (AGI, 2009).

Introduction

From the three definitions given above, it is clear that neotectonics is the study of "young" tectonic movements, but there is disagreement on exactly how young they must be to qualify as "neotectonic." Hancock and Williams (1986), following Blenkinsop, suggested that neotectonics commenced "when the contemporary stress field of a region was established." The age of this establishment differs in various regions of the world, but is generally between the Oligocene and Miocene periods. Such an initiation time conforms to Vita-Finzi's (1986) succinct definition of neotectonics as "late Cainozoic tectonics."

Neotectonics has a broad spatial extent as well as broad temporal extent. At the macroscale, it describes the current and geologically recent movements of the Earth's tectonic plates (see *Plate Tectonics*). At the mesoscale, it describes vertical and lateral movements of mountain chains such as the Himalaya, and vertical isostatic movements. At the microscale, it deals with the movement on individual faults (see *Fault*) and folds, with dimensions as small as a few km.

Detecting areas of neotectonic deformation

Neotectonic movements are commonly associated with areas of active seismicity and active faulting, such as plate margins (see *Plate Tectonics*). For example, linear bands or belts of earthquakes (see *Earthquake*) typically indicate zones of active faulting and folding. However, some late Cainozoic (neotectonic) structures have not generated significant seismicity in historic time, because they have either become inactive, or because the recurrence interval (see *Recurrence Interval*) between earthquakes is longer than the period of historic record. To locate these more subtle neotectonic faults/folds, geologists look for traces of their deformation expressed as tectonic landforms, a field of study known as tectonic geomorphology. Examples of tectonic landforms are faceted spurs on mountain fronts, created by young normal and reverse faulting; deflected drainages, shutter ridges, sag ponds, and other disrupted topography along strike-slip faults (see *San Andreas Fault*; *North Anatolian Fault*); and raised marine terraces or drowned coastal forests along actively subducting coasts (see *Subduction*). At regional scales such landforms are identified by satellite imagery (Landsat, ASTER, Google Earth, synthetic aperture radar), whereas at local scales, they are normally recognized on aerial photographs or in the field. In many cases, detailed field studies may yield the number, displacement, and timing of prehistoric earthquakes from such landforms (see *Paleoseismology*).

Neotectonic movements are not limited to individual faults and folds, but may also affect broader areas of the crust via post-earthquake crustal warping, isostatic rebound, or epeirogenic uplift. In coastal zones regional uplift and subsidence create geomorphic evidence such as emergent or submergent shorelines (respectively), or stratigraphic evidence such as recent coastal, lacustrine, or colluvial deposits (e.g., Bertrand et al., 2011). In continental areas, regional uplift may be reflected by rejuvenation of drainage networks, tilting of drainage networks creating asymmetry, or tilting of lake basins. Neotectonic structures are also associated with young geologic basins (onshore or offshore), that contain Neogene and Quaternary sediments. Neotectonic faults and folds are normally located at the margins of such basins, but they may also lie hidden beneath the basin fill, in which case geophysical surveys are required to locate and characterize them.

Finally, neotectonic movements affect areas of active volcanism, including vertical deformation due to loading by volcanic eruptions, magma movements (inflation, deflation, bulging); movement on volcano-tectonic faults; rifting; and crustal-scale landsliding, and volcano flank collapse.

Measuring neotectonic motions (slip amount, slip rate)

Neotectonic studies seek to measure the rate and direction of crustal movement at all scales, either as relative or absolute motions. Prior to the 1970s, almost all measurements were of relative motion. An example is using the offsets of landforms across faults or folds to measure the fault slip amount and rate (see *Paleoseismology*). Such offsets measure the motion of the two fault blocks relative to each other, rather than the absolute motion of either block compared to some larger reference frame. In coastal areas, it is possible to measure neotectonic vertical slip amounts and rates in relation to sea level, an absolute measure, but lateral offsets across faults are only measured as relative motions. After large historic earthquakes, first-order level lines were often surveyed into the epicentral area from 50–100 km away, so at sites within 50–100 km of the coast the vertical motions could be related to sea level. Beyond 100 km, vertical motions could only be measured as relative.

With the advent of radio astronomy, satellite geodesy, and laser altimetry in the 1980s, it became possible to measure neotectonic motions relative to fixed points on the Earth's surface, to the geoid, or to any external reference frame (Vita-Finzi, 2002). The first real-time, absolute measurement of horizontal plate motions was made using VLBI (very long baseline interferometry), which measures the time difference between the arrival at two Earth-based antennas of a radio wavefront emitted by a distant quasar (see http: //cddis.nasa.gov/vlbi_summary.html). At present the VLBI method has been replaced by global positioning satellites (GPS), which measure horizontal motions in relation to an Earth centric reference frame.

Although vertical neotectonic motions can also be measured via GPS, the accuracy and precision is lower than for horizontal measurements. Fortunately, vertical crustal motions can also be measured via laser altimetry (satellite laser ranging) and by interferometry of synthetic aperture radar (InSAR). The latter technique has revealed patterns of cm-scale uplift and subsidence over hundreds to thousands of square kilometers associated with large earthquakes, with non-tectonic mechanisms such as fluid withdrawal, and also vertical movements of unknown origin. Like many new tools developed in science, InSAR is demonstrating that deformation of the Earth's surface is more complex, subtle, and widespread than previously thought.

Modeling neotectonic motions

Neotectonic motions can be modeled qualitatively or quantitatively. The qualitative models aim at establishing a region's stress regime, that is, the orientation of the horizontal maximum and least principal stresses responsible for neotectonic deformation. This type of "stress inversion analysis" can be performed on static geological data, such as the orientation and slip sense of faults. For example, a system of parallel normal faults indicates an extensional stress regime with the least principal horizontal stress perpendicular to fault strike. Neotectonic stress regimes have also been inferred from the orientation of joints. The key to inferring a neotectonic stress regime from "static" geologic fault or joint data is to measure only neotectonic faults and joints, rather than older faults and joints produced by earlier stress regimes.

Stress regimes can also be inverted from more direct data such as focal plane mechanisms of earthquakes; paleomagnetic rotations of Neogene deposits; and well-bore breakouts and other direct measurements of contemporary stress fields.

More refined, quantitative neotectonic models rely on finite-element simulations of plate or microplate motion (e.g., the NeoKinema model of Liu and Bird, 2008). Such models attempt to reproduce the observed plate motions as measured from GPS surveys, by specifying in the model an arrangement of crustal layers of a given shape/thickness/rheology, and by assigning stresses to them. These models may also include simulated erosion and the isostatic effect of that erosion, inasmuch as the isostatic effect perturbs the evolving stress field, and thus may conceivably control the locations, styles, and rates of tectonic deformation (feedback).

Distinguishing between seismogenic and non-seismogenic neotectonic deformation

It is important from a hazards viewpoint to know whether the observed, measured, or modeled neotectonic deformation is being produced by seismogenic or non-seismogenic processes. As explained elsewhere (see *Tectonic and Tectono-Seismic Hazards*), most hazards to human life and property arise from tectono-seismic (earthquake-related) processes. However, not all neotectonic deformation is associated with earthquakes. For example, some tectonic fault types produce surface deformation but do not generate earthquakes, such as bending-moment faults and flexural-slip faults. Likewise, processes such as fault creep, deep-seated gravitational spreading (sackung), gravity sliding, evaporate-related faulting and folding, karst, glacio-isostatic faulting, and dike-related faulting can deform the geoid and are thus neotectonic, but are non-seismogenic. Because these structures/processes do not produce earthquakes, they pose much less of a hazard than seismogenic structures.

In practice, it is sometimes difficult to distinguish neotectonic deformation from seismogenic sources versus non-seismogenic sources, because the surface manifestations are very similar. This dilemma has become more acute now that systems such as GPS and InSAR are able to measure hitherto-undetectable patterns and rates of surface deformation. Hanson et al. (1999) provide a treatise for specialists on how to distinguish these fault types.

Summary

Neotectonics is the study of "young" tectonic movements, subsequent to the establishment of the contemporary stress (or seismotectonic) regime in the area of study. This means that neotectonics covers the tectonics of currently active structures, as well as some Neogene structures that may no longer be active. At the macroscale, neotectonics describes the Neogene movements of the Earth's tectonic plates (see *Plate Tectonics*). At the mesoscale, it describes vertical and lateral movements of mountain chains such as the Himalaya, and vertical isostatic movements. At the microscale, it deals with the movement on individual faults (see *Fault*) and folds, with dimensions as small as a few kilometers.

Bibliography

AGI, 2009. *Glossary of geology.* Washington, D.C.: American Geological institute, online version, www.agiweb.org, Accessed November 2009.

Bertrand, S., Doner, L., Akçer On, S., Sancar, U., Schudack, U., Mischke, S., Cagatay, N., and Leroy, S. A. G., 2011. Sedimentary record of coseismic subsidence in Hersek coastal lagoon (Izmit Bay, Turkey) and the Late Holocene activity of the North Anatolian Fault. *Geochem Geophys Geosyst,* **12**(6), 17, doi:10.1029/2011GC003511.

Hancock, P. L., and Williams, G. D., 1986. Neotectonics. *J Geol Soc London,* **143**, 325–326.

Hanson, K. L., Kelson, K. I., Angell, M. A. and Lettis, W. R., 1999. *Techniques for identifying faults and determining their origins.* Contract Rept. NUREG/CR-5503, Washington, DC: U.S. Nuclear Reg. Comm., 504 p.

Liu, Z., and Bird, P., 2008. Kinematic modelling of neotectonics in the Persia-Tibet-Burma orogen. *Geophys J Int,* **172**, 779–797.

Mörner, N. A., 1978. Faulting, fracturing, and seismic activity as a function of glacial isostasy in Fennoscandia. *Geology,* **6**, 41–45.

Pavlides, S. B., 1989. Looking for a definition of neotectonics. *Terra Nova,* **1**, 233–235.

Vita-Finzi, C., 1986. *Recent Earth Movements-An Introduction to Neotectonics.* Orlando, FL: Academic Press. 226 p.

Vita-Finzi, C., 2002. *Monitoring the Earth.* New York: Oxford University Press. 189 p.

Cross-references

Earthquake
Fault
Paleoseismology
Plate Tectonics
Recurrence Interval
Subduction
Tectonic and Tectono-seismic Hazards

CASE STUDY

NEVADO DEL RUIZ VOLCANO, COLOMBIA 1985

Barry Voight[1,2], Marta L. Calvache[3], Minard L. Hall[4], Maria Luisa Monsalve[3]
[1]Penn State University, University Park, PA, USA
[2]Cascades Volcano Observatory, Vancouver, WA, USA
[3]INGEOMINAS, Bogota, Colombia
[4]Escuela Politecnica, Quito, Ecuador

Definition and facts

- Modest eruption on summit ice pack generates lethal lahar.
- 23,000 killed in second-worst volcano disaster of the twentieth century.
- Factors include delayed hazard map, inadequately prepared local authorities, unprepared populace, and refusal to accept false alarm.
- *Lessons from Armero* provide many guidelines for future emergencies.

Introduction

Nevado del Ruiz Volcano, 5,370-m high, is located near Manizales in the Central Cordillera of Colombia between the Magdalena Valley to the East and the Cauca Valley to the West (Figure 1). Although Colombia has several active volcanoes, prior to the eruption of Ruiz in 1985, the country had virtually no first-hand experience in dealing with active volcanoes, nor were Colombian scientists specialized in volcanology. For most Colombians, tales by grandparents about minor eruptions at Galeras and Puracé volcanoes in the south of the country provided the only vague impressions of active volcanism. This inexperience played a major role in the events described. The impressions of Colombians changed on November 13, 1985, when a minor magmatic eruption at Ruiz generated the worst volcanic debris-flow (lahar) disaster in recorded history worldwide, and the second-worst volcanic disaster of the twentieth century.

The small but explosive eruption at the summit crater produced pyroclastic (mixed pumice and gas) currents that scoured and melted much of the summit's snow and ice cap (Figure 2), sending torrents of meltwater and pyroclastic debris down the volcano's flanks, where they coalesced in channels and entrained additional debris and water to form large *lahars*. On the western slope, overbank flooding by lahars caused some 1,900 fatalities and destroyed 200 houses. On the eastern slope, successive lahar waves obliterated 5,000 buildings and killed over 21,000 people in the town of Armero, located 50 km east of the volcano's summit (Figures 1 and 3). The death toll ranks fourth in history, behind only Tambora in 1815 and Krakatau in 1883, both in Indonesia, and Mount Pelee, Martinique, in 1902.

Yet the eruption was not a surprise, and neither were its effects. Persistent fumarolic, seismic, and phreatic activity had served as precursors for almost a year. Colombian workers were assisted by international specialists from the USA, Switzerland, Ecuador, France, Italy, New Zealand, and Costa Rica.

Despite these circumstances, the emergency management system failed to avert catastrophe. The consensus of those who have carefully studied the case history is that many of the casualties could have been prevented by improved hazards management practices (Voight, 1988, 1990, 1996a, b; Hall, 1990, 1992; Giesecke et al., 1990; Barberi et al., 1990; Miletti et al., 1991). There are important lessons to be learned from the Ruiz disaster pertinent to the scientific and public responses and adjustments to volcanic hazards – lessons that will provide valuable guidelines for future emergencies.

The re-awakening of Nevado del Ruiz

The Nevado del Ruiz complex has a 1.8 million year eruptive history, but most relevant is its behavior over the past 11,000 years. That history had been studied sufficiently to establish that voluminous and hazardous lahars were generated during relatively small eruptive events that

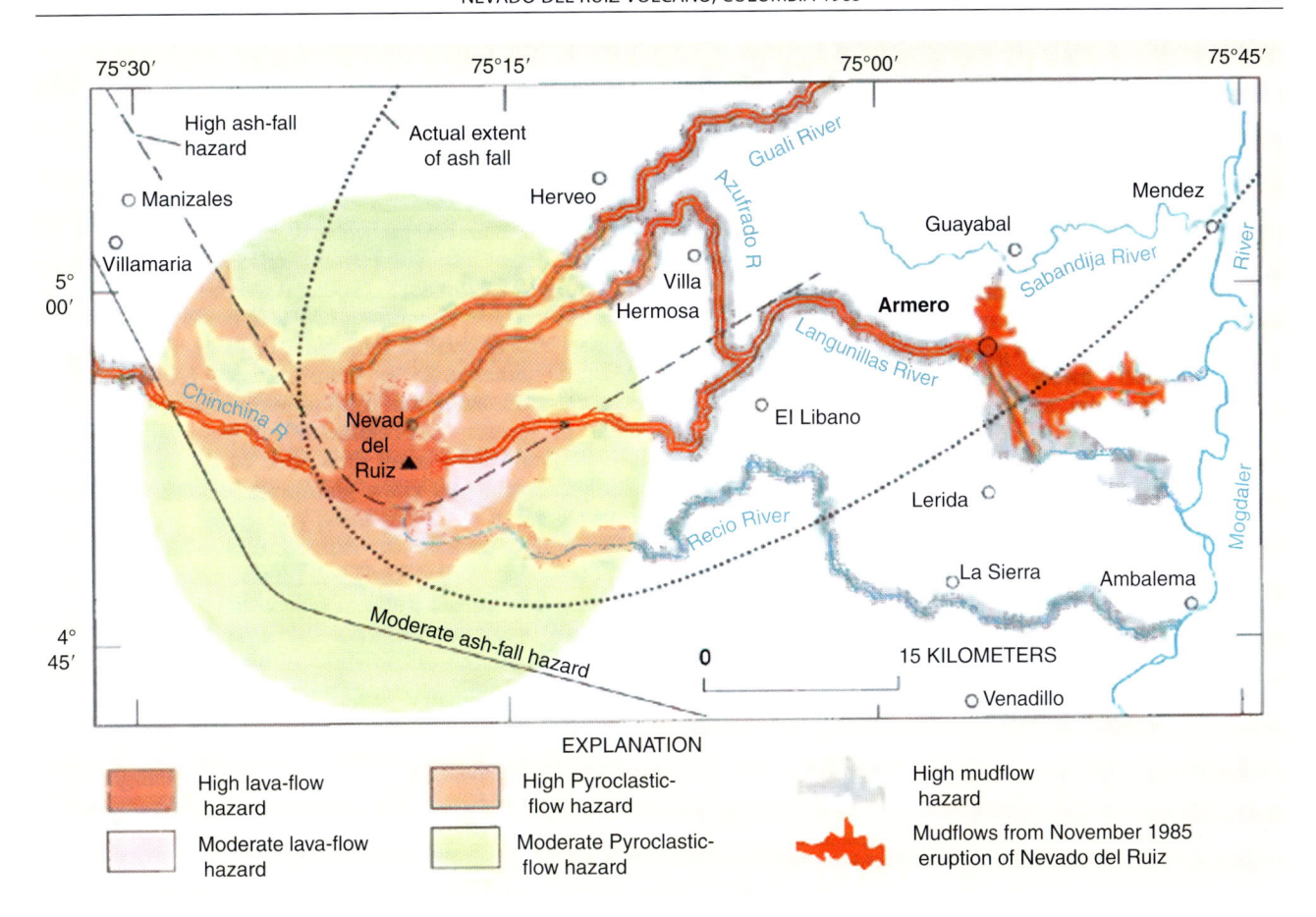

Nevado del Ruiz Volcano, Colombia 1985, Figure 1 Map of Nevado del Ruiz and surrounding region, showing volcanic hazards, and river valleys affected by lahars (mudflows) in the November 13, 1985, eruption (*red*) (Hazard map data from Parra and Cepeda (1990), redrafted by USGS).

melted only part of the summit icecap (Figure 2). The historical eruptions of 1595 A.D. and 1845 A.D., both of which produced voluminous lahars, are described by Fray Pedro Simón (in Acosta, 1850). The eruption of 1595 A. D. caused 636 fatalities (Simón 1625, as cited in Acosta, 1850), and about 1,000 were killed in the eruption of 1845 A.D. (Arboleda, 1918), during a time when the region was much less densely populated than in 1985.

The first signs of unrest at Ruiz related to the 1985 catastrophe were detected in November and December 1984. What happened next is summarized in the timeline below (Voight, 1990; Hall, 1990):

Time line

November, December 1984. Unusual fumarolic activity within the crater and earthquakes near summit.

January 6, 1985. Geologists visit summit crater, notice a new small crater within and conclude monitoring needed.

February, March 1985. Fumarolic activity at crater, felt earthquakes. Civic committee formed in Caldas Province and feature article with photographs appears in key newspaper.

April 1985. Strong fumarolic activity, felt earthquakes; UNESCO nominates an expert team in anticipation of assisting Colombians.

May 1985. Visiting UNDRO scientist recommends hazard map, civil defense planning, monitoring.

Colombian geology-mines bureau INGEOMINAS requests USGS for expertise and geophones.

June 1985. USGS supplies hardware but declines to send expert. UNESCO offers expert team and equipment to Colombian Foreign Affairs Minister, but letter is "lost" for 2 months.

July 1985. Continued fumarolic activity and earthquakes. Caldas Province, concerned by lack of progress at national level, forms risk committee and requests foreign assistance. Meanwhile INGEOMINAS installs provisional array of four portable seismographs, although initial array is unsuitable, and analysis is delayed. First long-period event recorded on July 23.

Nevado del Ruiz Volcano, Colombia 1985, Figure 2 The town of Armero at the mouth of the Lagunillas River, after impact by the lahar generated by the November 13, 1985, eruption. Lahar deposits in *brown* to *gray* color, reflectant where wet (R.J. Janda (USGS) photo).

Nevado del Ruiz Volcano, Colombia 1985, Figure 3 Nevado del Ruiz and its summit glacier ice cap, overlain by a veneer of brown pyroclastic current deposits. The pyroclastic currents rapidly eroded snow and ice at the summit, creating a deluge of water that resulted in lahars in several river systems surrounding the volcano (J. Marso (USGS) photo).

August 1985. Swiss seismologist arrives to aid Caldas Province and installs three-component short-period seismometer. Two independent groups are now operating seismometers.

September 6, 1985. Intense seismicity with tremor begins, with small ash emissions on September 8.

September 11, 1985. Strong phreatic explosion generates ash rain in population centers and 27 km-long lahar in Azufrado river. Concerns and emergency work set in motion, although provinces decide to manage emergency plans independently.

September 18, 1985. Work on a hazard map begins by local geologists. USGS and UNDRO send scientists to support efforts. Ash emissions on September 23, 24, and 29; fumarole samples taken.

October 1985. UNESCO scientist guides combining the two seismic networks.

October 7, 1985. Preliminary hazard map and report presented to government; first INGEOMINAS seismic report issued (despite operation since 20 July).

October 9, 1985. UNDRO reports ashfalls and lahars most likely and within "very near future, the necessary measures will have been taken to protect the population." Reporter's version of hazard map published with errors.

October 21, 1985. Deformation studies began at Nevado del Ruiz, too late to detect signal.

November 1985. Public presentation of revised hazard map postponed from November 12 to 15.

Meeting of mayors to review emergency plans scheduled for November 15. Nevertheless, plans to remove seismic network discussed, due to operational costs. Gas analyses suggest magmatic signature, and on November 10, 3 days of tremor began, less pronounced that in September.

November 13, 1985. Phreatic eruption begins 15:06 h. Regional emergency committee, in Tolima Province at prescheduled planning meeting, issued preliminary warnings but conditions returned to normal. Magmatic eruption begins 21:08 h; *23,000 die in Armero and Chinchiná.*

Slow progress in managing the volcanic *unrest*

Monitoring and planning activities had only gradually developed and were not merged in a coherent effort. Notably lacking was a resident scientist with significant volcanic eruption experience. A preliminary hazard map and report was released October 7. The report noted the very high probability for lahars, "with great danger for Armero, Mariquita, Honda, Ambalema, and the lower part of the Rio Chinchina." But "the government, skeptical of an impending eruption and worried more about the consequences of the map itself" (Hall, 1992), ordered that it be rechecked and resubmitted by November 12. Meanwhile, Civil Defense compiled field counts of the population-at-risk along the river systems, and scheduled meetings at national, provincial, and local levels to disseminate risk and preparedness information. The information transfer to the local level was coordinated by the emergency committees in each province with variable success.

On October 22, 1985, an advisory team representing the National Volcanological Group of Italy emphasized the lahar risk and need for a place of refuge for the populace in river towns (Barberi et al., 1990). The Caldas Emergency Committee met again in Manizales on November 12, 1985. They noted that gas analyses suggested a magmatic origin, the hazard map revision was still incomplete, and seismic analyses had not been issued since October 10. Tremors had been occurring for several days, but less pronounced than in early September. Visual observations yielded no signs of an impending eruption. But on November 13, a phreatic eruption began at 15:06 h that resulted in a light ashfall in Armero. The Tolima Provincial Emergency Committee at a prescheduled planning meeting issued preliminary warnings from its distant capitol, but conditions appeared to return to normal. During the evening, heavy rains and an international soccer game kept most Armerians inside at home. Then a magmatic eruption began 21:08 h, creating collapsing currents of hot pumice-gas mixtures that melted the snowpack and generated a flood of water and debris. The latter breached a landslide-blocked drainage and lake that had formed upriver from Armero during 1984–1985, exacerbating the subsequent lahar impact.

The day Armero died

The eruption-generated lahars raced down three of the volcano's major drainage systems. Around 22:40 h, the riverbank village of Chinchiná on the western slope was struck. Hundreds of homes had been evacuated because of the general alert by Civil Defense, but still 1,927 persons died (Gueri and Perez, 1986).

From 21:45 h to 22:00 h, officials in Tolima attempted to inform Armero of the situation, but power and communication difficulties were experienced. No evacuation orders, neither specific nor systematic, were issued, although some representatives of various agencies took individual action (Voight, 1988, 1990, 1996a, b). Disgorged from the Lagunillas canyon at 23:35 h as a wave nearly 40 m high (Figure 2), the muddy boulder-laden torrent crushed Armero (Pierson et al., 1990). A second major pulse struck at 23:50 h, followed by a number of smaller and finer-grained pulses, the last at 01:00 h. Many survivors took flight only after the first flood waves struck the town, and although many escaped, over 21,000 people perished. Thousands had managed to reach islands of high ground, but many experienced difficulty in extracting themselves from, or crossing, the soft mud of the final lahar pulses. A national TV station had broadcast news of the eruption, although many remember the message as advising no cause for alarm – a message singularly inappropriate for Chinchiná or Armero.

Analysis of a catastrophe

In early 1985, risk evaluations by UNDRO found a receptive audience in Manizales and Caldas Province near the volcano, but not in distant Tolima Province which marginally included the high-risk area of Armero. National agencies were not fully committed to the task, partly because of limited funds, experience, and equipment, but also from a lack of conviction that the threat was real. USGS managers were reluctant to send volcanologists and equipment to Ruiz beforehand, largely "because of skepticism (particularly before September 1985) that the crisis would actually culminate in significant activity." With the perception of disinterest at the national level, the Caldas group sought international assistance, creating rivalries that developed between seismic groups. The volcano hazard map requested as early as March 1985 had not been worked on until September, with the result that the existing management plans lacked the necessary understanding and focus toward areas targeted for high risk.

With the September 11, 1985, phreatic eruption and ashfall in Manizales, the gap of credibility closed especially in Caldas Province, and regional agencies accelerated their efforts and requested national assistance. By October, the management structure appeared to be gaining effectiveness, but when put to the test a month later – the system failed. Some of the factors included:

1. Although the risks were becoming known, provincial and national government made the conscious decision not to evacuate the villages until and unless the danger could be *guaranteed*. Evacuation before the event would have caused economic, political, and law-enforcement problems, and no official was willing to accept this responsibility. *Thus, the authorities on the whole acted rationally in the short term but were unwilling to bear the economic or political costs of early evacuation or a false alarm. Scientific studies accurately foresaw the hazards but were insufficiently precise to provide a reliable prediction of an uncertain but possibly devastating event. Therefore, catastrophe had to be accepted as a calculated risk, and this combination – the limitations of prediction/detection, the refusal to accept a false alarm, and the lack of will to act on the uncertain information available – provided the immediate and most obvious cause of the catastrophe"* (Voight, 1996a).
2. Under the above circumstances, mitigation would have required a supremely efficient and tested disaster-prevention system, with unerring detection (with around-the-clock well-instrumented monitoring) and data interpretation, followed by instantaneous decision making, effective alert communications, and a swift response by a thoroughly prepared population. Such elements did not exist.
3. The local authorities at Armero were inadequately prepared and equipped, and had insufficient information of what the eruption might hold for the town. "What-to-do, where-to-go" information had not been communicated to the population. The situation was different at Chinchiná in Caldas Province, where officials had made an effort to educate the population, and where an alarm was given as early as 21:30 h. Nevertheless, casualties here were also high, due to the shorter lead time and the difficulty in spreading a night-time warning to the riverbank community.
4. The response of civil defense could not be prioritized due to the lack of specific hazard maps prior to October 7.
5. A tragic aspect of this catastrophe was the fidelity of the hazard maps in relation to the events of 1985 (Parra and Cepeda, 1990; Voight, 1990). The written descriptions of the 1845 lahar (Acosta, 1850) precisely mirror those of 1985. Further, had this information been utilized earlier for city planning, catastrophe could have been wholly averted.
6. Communication breakdowns are common during volcanic emergencies. However, a review of the communications carried out on November 13, 1985, reveals that the system for information transfer had redundant elements and worked reasonably well (Voight, 1988, 1990). However, the system was not organized to make rapid decisions nor give alerts. It had been decided earlier that only the President of Colombia could authorize the needed evacuations, a situation that required more than 2 h (at the inconvenient time, 21:00–23:00 h) of ascending and subsequent descending approvals along the bureaucratic chain-of-command, in order to permit the local authorities to give the alarm. A local authority should have been charged with this responsibility. The primary problem was not information dissemination, but of lack of readiness and the will to act decisively.
7. The media played an important role (Hall, 1992), and its impact reflected the culture and socioeconomic conditions in the two affected provinces. Due to isolation from Bogota, Caldas boasted a strong local newspaper in Manizales that gave commendable coverage of the developing crisis. Tolima, on the other hand, depended on Bogota's daily newspapers which carried fewer and shorter volcano articles. As a result, Caldas' authorities and population were better prepared and more effective in instigating government action.
8. What was the role of technology? Would telemetered seismographs and real-time analysis have prevented disaster? The data suggest that there were no definitive short-term warnings of the November eruption (Martinelli, 1990, 1991). To offset the lack of telemetry, the scientific staff visually monitoring the volcano, and tending to a seismometer 9 km from the crater observed the onset of the eruption and so advised Manizales by radio with sufficient time to save most lives, *if appropriate further communication transfer and decisions had followed and the population-at-risk had been prepared*. Neither Armero nor Chinchiná directly received this early warning. And what of lahar detectors? Farmers living along the probable lahar paths were instructed to be watchful; with the eruption, they heard the approaching lahars but had no means of communicating a warning downstream to Armero. At the time lahar detectors were not off-the-shelf items, but as a result of Ruiz, acoustic debris-flow detectors with telemetry are now standard equipment and have been deployed at numerous sites worldwide.

Lessons from Armero

The *Lessons from Armero* (Voight, 1988, 1990, 1996a, 1996b) are fundamental and have aided the management of volcanic emergencies in other regions.

Recognize and document the hazards. Hazard maps play a crucial role in identifying the likely types of volcanic phenomena and the areas of high risk. The maps should also indicate areas that are safe, in order to assist emergency managers to define refuge for evacuation planning. Of course, hazard maps must not only be made, they need to be *used*. A very early priority is therefore *to produce a timely and well-distributed map and document that will allow mitigation planning to proceed effectively and without delay.*

Rank site vulnerability. Hazards and vulnerability have different degrees of severity, whereas mitigation planning and response is time consuming. Thus, the most vulnerable sites should be mitigated first. This procedure was not followed at Armero. *The lesson is that systematic attention must be given to identifying and resolving local problems of the most threatened communities at a very early stage.*

Modern volcano monitoring techniques (seismic, geodetic, infrasound, chemical), installed on the high hazard volcanoes long before any crisis, have the ability to provide timely warnings of changing activity and pre-eruption events. In parallel fashion, fixed and portable communications nets must be established in order to assure the effectiveness and diffusion of such warnings.

Accept the social responsibility to communicate the risk to the public. There are two categories. The first is public awareness and education, in which the authorities, citizens, and communities are informed about the character of volcanic risk and appropriate responses. The second is the warning, in which an endangered public is alerted to the problem and the relevant protective actions. Armero revealed failures in both categories. In regard to the first, volcanologists have produced compelling videos of volcanic hazards, and social scientists have developed risk communication guidelines. In regard to the second, scientists should try to make sure their warnings reach the affected people. S*cientists, who have the best appreciation of the true hazard, can often do more by taking on the social responsibility as well as a technical one, and can be a strong and persistent voice within the hazard management structure to ensure that the mitigation process truly reaches the affected people.*

Time is the most important variable. At Armero, an important contribution to the failure was to wait (either delay or postponement) until the last possible minute. One cannot expect that emergency management and public response operate efficiently on an extremely short time scale. Delays in hazard recognition and mitigation can reflect also the reluctance of a national government to invest time, money, and foreign expertise toward assessment of an uncertain hazard.

Plan critical decisions in advance. In moments of crisis, complex decision-making processes that involve a bureaucratic chain-of-command, or hours of committee discussion, or that assume rapid, unstressed communications linkages are not effective nor dependable. The decision-making process associated with warnings should be simplified by advanced consideration and agreement of the decision criteria.

Is the warning system tested and is it reliable? It is possible to design a risk communication and emergency warning system that maximizes the probability of sound public advice and also minimizes the potential negative impact of some personal characteristics of the public (see Voight, 1996a). The system should be tested by drills in advance of crisis.

Anticipate warning communications problems. Satellite and radio telemetry, microwave radio systems, cellular phones, standby power sources, and redundancy, all have a place in volcano emergency management.

Attention to the media is not a casual responsibility, but ranks in importance to the scientific work. Media reporting powerfully interacts with the public and officials with printed messages that are superior in providing detailed information essential for preparedness, whereas the electronic media are superior in *conveying alarms* to the public. Conflicting coverage and some distortions are inevitable, personal and commercial interests may use the media to aid mitigation efforts, or retard them, and the attention of the media to hazards may depend less on whether the underlying issues have been resolved than on whether they unfold in newsworthy fashion. In some cases, adversarial relations can develop between scientists and media producers, and such situations are clearly not helpful but can be difficult to resolve. Scientists dealing with risk and safety issues have to cope with strong uncertainty and ignorance, and yet are called upon to express judgments on ambiguous facts. Scientists have the obligation to supply balanced, reliable information to the population in a timely manner and in a form that can be widely understood, to further the mitigation process.

Do not underestimate lahar hazards. The public and decision makers need to know that lahars are not just muddy water but can include fast boulder-laden flows capable of demolishing concrete steel-reinforced buildings.

Conclusions

At Ruiz most of the elements desirable for successful hazards management had been in place. The youthful geology and geochronology of the cone had been studied, and postglacial flowage deposits had been mapped. Key historical events were studied. Advice was available from foreign experts, equipment had been provided, a hazards map was produced a month before the crucial event, and national, provincial, and local governments showed concern. The magmatic eruption was small and its effects had precedents. Thus, the catastrophe was not produced by technological defectiveness, nor by an overpowering eruption, nor by overwhelming bad luck. Instead Armero was created by cumulative human error, indecision, and shortsightedness in response to uncertainty. In the event the authorities were unwilling to bear the economic or political costs of early evacuation or a false alarm, and they delayed action to the last possible minute. Catastrophe was the calculated risk, and Nature cast the die.

And so the lessons of Armero are not new lessons; they are old lessons forged in human behavior that once again required the force of catastrophe to drive them home.

Acknowledgments

We thank our colleagues on the Comité de Estudios Volcanologicos in Manizales. Special mention is due to Bruno Martinelli, the Swiss seismologist in our paper, and the Colombians H. Cepeda, P. Medina, E. Parra, F. Munoz, A. Nieto, and N. Garcia, later killed at Galeras. We also acknowledge our USGS colleagues, particularly Dick Janda, D. Harlow, N. Banks, R. Norris, J. Lockwood, and J. Zollweg. Andinista Bis, who was killed on Ruiz, and Oscar Ospina, both made great contributions.

Bibliography

Acosta, J., 1850. Sur les montagnes de Ruiz et de Tolima (Nouvelle Grenade) et les eruptions boueuses de la Magdalena (deux lettres a Elie de Beaumont) (in French). *Bulletin de la Societe Geologique de France*, **21**, 489–496.

Arboleda, G., 1918. Historia Contemporanea de Colombia (in Spanish). In *Tomo II: Administración de Herran y Mosquera*. Bogotá: Casa Editorial Arboleda y Valencia. 474 pp.

Barberi, F., Martini, M., and Rosi, M., 1990. Nevado del Ruiz volcano (Colombia): pre-eruption observations and the November 13, 1985 catastrophic event. In Williams, S. N. (eds), Nevado del Ruiz volcano, Colombia II. *Journal of Volcanology Geothermal Research*, **42**, 1–12.

Giesecke, A., Anzola, P., Fernandez, B., Gonzalez-Ferran, O., Hall, M., Podesta, B., Rodriguez, A., and Sarria, A., 1990. *Riesgo Volcanico. Evaluacion y mitigation en America Latina*. Lima: Centro Regional de Sismologia para America del Sur-CERESIS, 288 p. (detailed analysis in Spanish).

Gueri, M., and Perez, L. J., 1986. Medical aspects of the "El Ruiz" avalanche disaster, Colombia. *Disasters*, **10**, 150–157.

Hall, M. L., 1990. Chronology of the principal events before the eruption of November 13 1985. *Journal of Volcanology and Geothermal Research*, **41**, 101–115 (concise listing of key events).

Hall, M. L., 1992. The 1985 Nevado del Ruiz eruption: scientific, social and governmental responses and interaction before the event. In McCall, G. J. H., Laming, D. J. C., and Scott, S. C. (eds.), *Geohazards: Natural and Man-Made*. London: Chapman and Hall, pp. 43–52 (best analysis of roles of culture and media at Ruiz).

Martinelli, B., 1990. Seismic patterns observed at Nevado del Ruiz volcano, Colombia, during August-September, 1985 and October, 1986. *Journal of Volcanology and Geothermal Research*, **41**, 297–314.

Martinelli, B., 1991. Understanding triggering mechanisms of volcanoes for hazard evaluation. *Episodes*, **14**(1), 19–25.

Mileti, D. S., and Fitzpatrick, C., 1991. Communication of public risk: its theory and application. *Sociological Practice Review*, **2**(1), 20–28.

Parra, E., and Cepeda, H., 1990. Volcanic hazard maps of the Nevado del Ruiz volcano, Colombia. *Journal of Volcanology and Geothermal Research*, **41**, 117–128.

Pierson, T. C., Janda, R. J., and Borrero, C. A., 1990. Origin, flow behavior, and deposition of eruption-triggered lahars on 13 November 1986, Nevado del Ruiz volcano, Colombia. *Journal of Volcanology and Geothermal Research*, **41**, 17–66.

Voight, B., 1988. Countdown to catastrophe. *Earth and Mineral Sciences*, **57**(2), 17–30.

Voight, B., 1990. The 1985 Nevado del Ruiz volcano catastrophe: anatomy and retrospection. *Journal of Volcanology and Geothermal Research*, **44**, 349–386 (Analyzed chronology more detailed than Voight 1996).

Voight, B., 1996a. The management of volcano emergencies: Nevado del Ruiz. In Scarpa, R., and Tilling, R. I. (eds.), *Monitoring and Management of Volcano Hazards*. Berlin/Heidelberg: Springer, pp. 719–769, 841 pp. (some analysis and lessons more updated than Voight 1990).

Voight, B., 1996b. Cuenta regresiva a la catástrofe. Revista Semestral de la Red de Estudios Sociales en Prevención de Desastres en América Latina (*La Red*, Lima), No. 6/Año 4, 117–136 (Spanish language version of Voight 1988).

Cross-references

Galeras Volcano

NORTH ANATOLIAN FAULT

Thomas Rockwell
San Diego State University, San Diego, California, CA, USA

Definition

The North Anatolian Fault in Turkey is a major, continental transform system that connects the compressional system of faults in eastern Turkey, the Caucasus and the Zagros to the extensional regime in the Aegean Sea region. This fault zone, which extends 1,400 km from its juncture with the East Anatolian fault near Karliova east of Erzincan westward into the Gulf of Saros in the Aegean Sea, has a long and complex history, with the modern trace superposed on a Jurassic Neothethyan suture zone. The modern North Anatolian fault was reactivated in late Miocene time as a strike-slip zone to accommodate the westward extrusion of the Anatolian Plate resulting from the northward collision of the Arabian Plate with Asia.

Geodetic studies indicate that the fault is loading at a rate of about 25 mm/year along its entire trace. Several geologic studies suggest a lower rate of 15–20 mm/year from dating of offset late Quaternary geologic markers, and this lower rate is consistent with the production rate of large earthquakes for the length of the fault for the past 1500 years, as determined from paleoseismic and historical data. The geologic rates may not account for off-fault deformation, whereas the geodetic rates may not account for a persistent transient or visco-elastic relaxation after rupture of nearly the entire plate boundary. Consequently, the actual loading rate for large earthquake production along the North Anatolian fault remains controversial.

The fault has experienced a remarkable westward progression of large earthquakes during the past century, beginning with the great 1939 M7.9 Erzincan earthquake, which resulted in 360 km of surface rupture and right-lateral displacements of up to 10 m. Subsequent earthquakes in 1942 (M7.1; 50 km of rupture), 1943 (M7.7; 280 km of rupture), 1944 (M7.4; 165 km of rupture), 1951 (6.8; 30 km of rupture), 1957 (M7; 30–50 km of rupture), 1967 (M7.2; 80 km of rupture), 1999 (M7.5; 130 km of rupture), and 1999 (M7.1; 50 km of rupture) essentially unzipped the fault westward for over 1,000 km, nearly to the Marmara Sea, which has been attributed to progressive failure due to Coulomb stress loading.

The long historical record of earthquakes in Turkey, along with extensive paleoseismic work at various sites along the fault, has led to an understanding of prior earthquake sequences, and an assessment of local and regional seismic risk. The previous earthquake sequence played out over the period of a century, and initiated with a very large earthquake in 1668 along the same faults that ruptured in 1943 and 1944. This was followed by large earthquakes in 1719, 1754, 1766, and 1766, the last two of which ruptured through the Marmara Sea and the Gallipoli peninsula. A major concern is that the current sequence has not played itself out and that another earthquake may occur in the Marmara Sea next to Istanbul, one of the great cities of the World with a population of over 13 million people.

Bibliography

Barka, A. A., 1992. The North Anatolian fault zone. *Annales Tectonicae*, **6**, 164–195.

Flerit, F., Armijo, R., King, G., and Meyer, B., 2004. The mechanical interaction between the propagating North Anatolian Fault and the back-arc extension in the Aegean. *Earth and Planetary Science Letters*, **224**(3), 347–362, doi:310.1016/j.epsl.2004.1005.1028.

Hartleb, R. D., Dolan, J. F., Akyuz, H. S., and Yerli, B., 2003. A 2000-year-long paleoseismologic record of earthquakes along the central North Anatolian fault, from trenches at Alayurt, Turkey. *Bulletin of the Seismological Society of America*, **93**(5), 1935–1954.

Hartleb, R. D., Dolan, J. F., Kozaci, O., Akyuz, H. S., and Seitz, G. G., 2006. A 2500-yr long paleoseismologic record of large, infrequent earthquakes on the North Anatolian fault at Cukurcimen, Turkey. *Bulletin of the Geological Society of America*, **118**(7–8), 823–840, doi:810.1130/B25838.25831.

Hubert-Ferrari, A., Armijo, R., King, G., Meyer, B., and Barka, A., 2002. Morphology, displacement, and slip rates along the North Anatolian Fault, Turkey. *Journal of Geophysical Research*, **107**, doi: 10.1029/2001JB000393.

Klinger, Y., Sieh, K., Altunel, E., Akoglu, A., Barka, A., Dawson, T., Gonzalez, T., Meltzner, A., and Rockwell, T., 2003. Paleoseismic evidence of characteristic slip on the western segment of the North Anatolian fault, Turkey. *Bulletin of the Seismological Society of America*, **93**(6), 2317–2332.

Kozaci, O., Dolan, J. F., Finkel, C. F., and Hartleb, R., 2007. Late Holocene slip rate for the North Anatolian fault, Turkey, from cosmogenic 36Cl geochronology: Implications for the constancy of fault loading and strain release rates. *Geology*, **35**(10), 867–870, doi:810.1130/G23187A.23181.

Kozaci, Ö., Dolan, J. F., and Finkel, R. C., 2009. Late Holocene slip rate for the central North Anatolian fault, Tahtakorpru, Turkey, from Cosmogenic 10Be Geochronology: Implications for the constancy of fault loading and strain release rates. *Journal of Geophysical Research*, **114**, doi:10.1029/2008JB005760.

Pantosti, D., Pucci, S., Palyvos, N., De Martini, P. M., D'Addezio, G., Collins, P. E. F., and Zabci, C., 2008. Paleoearthquakes of the Düzce fault (North Anatolian Fault Zone): Insights for large surface faulting earthquake recurrence. *Journal of Geophysical Research*, **113**, doi:10.1029/2006JB004679.

Reilinger, R., McClusky, S., Vernant, P., Lawrence, S., Ergintav, S., Cakmak, R., Ozener, H., Kadirov, F., Guliev, I., Stepanyan, R., Nadariya, M., Hahubia, G., Mahmoud, S., Sakr, K., ArRajehi, A., Paradissis, D., Al-Aydrus, A., Prilepin, M., Guseva, T., Evren, E., Dmitrotsa, A., Filikov, S. V., Gomez, F., Al-Ghazzi, R., and Karam, G., 2006. GPS constraints on continental deformation in the Africa-Arabia-Eurasia continental collision zone and implications for the dynamics of plate interactions. *Journal of Geophysical Research*, **111**, 1–26, doi:10.1029/2005JB004051.

Rockwell, T., Ragona, D., Seitz, G., Langridge, R., Aksoy, M. E., Ucarkus, G., Ferry, M., Meltzner, A. J., Klinger, Y., Meghraoui, M., Satir, D., Barka, A., and Akbalik, B., 2009. *Palaeoseismology of the North Anatolian Fault Near the Marmara Sea: Implications for Fault Segmentation and Seismic Hazard*. London: The Geological Society. Special Publications, 316(1), pp. 31–54, doi:10.1144/SP1316.1143.

Sengor, A. M. C., Gorur, N., and Saroglu, F., 1985. Strike-slip faulting and related basin formation in zones of tectonic escape: Turkey as a case study. In Biddle, K. T., and Christie-Blick, N. (eds.), *Strike-slip Deformation, Basin Formation, and Sedimentation*. Society of Economic Paleontologists and Mineralogists, Tulsa, Special Publication, No. 37, pp. 227–264.

Sengor, A., Tuysuz, O., Imren, C., Sakinc, M., Eyidogan, H., Gorur, N., Le Pichon, X., and Rangin, C., 2005. The North Anatolian fault: a new look. *Annual Review of Earth and Planetary Sciences*, **33**(1), 37–112, doi:110.1146/annurev.earth.1132.101802.120415.

Stein, R. S., Barka, A. A., and Dieterich, J. H., 1997. Progressive failure on the North Anatolian fault since 1939 by earthquake stress triggering. *Geophysical Journal International*, **128**(3), 594–604.

Sugai, T., Awata, Y., Tooda, S., Emre, O., Dogan, A., Ozalp, S., Haraguchi, T., Takada, K., and Yamaguchi, M., 2001. Paleoseismic investigation of the 1999 Duzce earthquake fault at Lake Efteni, North Anatolian fault system, Turkey. *Annual Report on Active Fault and Paleoearthquake Researches*, **1**, 339–351.

Cross-references

Earthquake
Fault
Intensity Scales
Magnitude Measures
San Andreas Fault

NUÉE ARDENTE

Catherine J. Hickson[1,2], T. C. Spurgeon[1], R. I. Tilling[1,3]
[1]Alterra Power Corp., Vancouver, BC, Canada
[2]Magma Energy Corp., Vancouver, BC, Canada
[3]Volcano Science Center, U.S. Geological Survey, Menlo Park, CA, USA

Synonyms
Pyroclastic flow

Definition
Nuée Ardente is a "glowing cloud" of superheated (often incandescent) particles and gases formed by the collapse of an explosive eruption column.

Discussion
Nuée Ardente is an old term that has been largely replaced by "pyroclastic flow" or more recently "pyroclastic density current" (PDC) (cf. Sulpizio et al., 2008). The name *Nuée Ardente* is derived from observations of Mount Pelée in 1902 (cf. Fisher and Heiken, 1982). Pyroclastic flows (and surges) are extremely hazardous and destructive processes that can have a number of origins (Nakada, 2000). The most well known are those that originate from the collapse of a rising eruption column. This collapse is usually instigated by a change in the eruption velocity caused by changing vapor/particle concentrations at the vent. If the expansion of the gas content of the erupting column can no longer support the lower ballistically expelled parts of the column, it will collapse. The collapse produces a density current of hot gases and particles. The heat is sufficient enough that during flowage, the cloud of particles and gas glows, hence the French name for a glowing cloud, "Nuée Ardente." The particles range from fine grained ash size to blocks tens of centimeters in diameter, depending on the disaggregation of the magma in the vent. Mixed with the clasts and being expelled from them, are hot gases that help fluidize the pyroclastic flow. Depending on the size of the eruption, height of the collapsing column, and topography, pyroclastic flows can travel tens to hundreds of kilometers from the vent. They also vary significantly in characteristics depending on the magma type and size of the eruption. The most common pyroclastic flows involve dacitic to rhyolitic ejecta erupted during Peléan, sub-plinian to plinian eruptions, but basaltic pyroclastic flows such as at Villarrica (Witter et al., 2004:305) and Kilauea (Fiske et al., 2009) have also been reported, though they are rare.

Bibliography
Fisher, R. V., and Heiken, G., 1982. Mt. Pelée, Martinique: May 8 and 20, 1902, pyroclastic flows and surges. *Journal of Volcanology and Geothermal Resources*, **13**, 339–371.

Fiske, R. S., Rose, T. R., Swanson, D. A., Champion, D. E., and McGeehin, J. P., 2009. Kulanaokuaiki Tephra (ca. A.D. 400–1000): newly recognized evidence for highly explosive eruptions at Kīlauea Volcano, Hawai'i. *Geological Society of America Bulletin*, **121–5/6**, 712–728.

Nakada, S., 2000. Hazards from pyroclastic flows and surges. In Sigurdsson, H., et al. (eds.), *Encyclopedia of Volcanoes*. San Diego: Academic Press, pp. 945–955.

Sulpizio, R., Dellino, P., Mele, D., La Volpe, L., 2008. Generation of pyroclastic density currents from pyroclastic fountaining or transient explosions: insights from large scale experiments. IOP Conference Series: Earth and Environmental Science 3 (2008) 012020, doi:10.1088/1755-1307/3/1/012020.

Witter, J. B., Kress, V. C., Delmelle, P., and Stix, J., 2004. Volatile degassing, petrology, and magma dynamics of the Villarrica Lava Lake, Southern Chile. *Journal of Volcanology and Geothermal Research*, **134**, 303–337.

Cross-references
Base Surge
Eruption Types (Volcanic Eruptions)
Lahar
Lava
Magma
Montserrat Eruptions
Mt. Pinatubo
Vesuvius
Volcanic Ash
Volcanic Gas
Volcanoes and Volcanic Eruptions

O

OVERGRAZING

Norm Catto
Memorial University of Newfoundland,
St. John's, NL, Canada

Definition
Overgrazing is the excessive consumption of vegetation by livestock.

Discussion
Overgrazing, and the consequent food shortage for livestock, represents a hazard directly for human populations dependent on the livestock. It also is a contributing cause to soil and sediment erosion, as material unprotected by vegetation is subject to removal by wind, running water, and mass movements.

Overgrazing can be a major contributing cause of desertification, where formerly arable land or pasture is converted to unproductive terrain. Overgrazing may result in desertification, even though there has been no significant change in the amount of rainfall or the temperature. However, desertification is a complex process resulting from the interplay of numerous factors, and many instances of desertification are not directly linked to overgrazing.

Human populations that have increased rapidly due to increased livestock production may be at risk if overgrazing reduces food supply. Overgrazing is a significant risk in areas subject to drought or strong winds, particularly steppes and areas adjacent to deserts, including Australia, Patagonia, northern China, India, Pakistan, the prairies of North America, and drier regions of northern and southern Africa.

As overgrazing and its effects can be very localized, detailed vegetation cover and land use assessments should be conducted, and local and regional factors considered prior to the implementation of changes in agricultural policy. Effective management of a potential overgrazing problem requires careful consideration of the human cultural factors, in addition to physical analysis.

Bibliography
Magole, L., 2009. The 'shrinking commons' in the Lake Ngami grasslands, Botswana: The impact of national rangeland policy. *Development in Southern Africa*, **26**, 611–626.

Williams, M., 2003. Desertification in Africa, Asia and Australia: Human impact or climatic variability? *Annals of the Arid Zone*, **42**, 213–230.

Cross-references
Challenges to Agriculture
Desertification
Disaster Relief
Dust Bowl
Dust Storm
Erosion
Erosivity
Insurance
Land Degradation
Universal Soil Loss Equation (USLE)

OZONE

Tom Beer
Centre for Australian Weather and Climate Research,
Energy Transformed Flagship, Aspendale, VIC, Australia

Definition
Ozone (O_3) is a tri-atomic form of oxygen that constitutes a minor and variable constituent of the atmosphere. Most of the ozone is found in the stratosphere (which is the

region from approximately 15 to 45 km altitude). It can also be found at ground level.

Discussion

In the lower part of the atmosphere, ozone is formed by chemical reactions between the hydrocarbons and oxides of nitrogen in industrial and automotive emissions that are initiated by sunlight. High ozone concentrations at ground level are an indicator of a type of air pollution called photochemical smog. Air pollution, as noted by Beer (2001), constitutes a meteorological hazard.

In the stratosphere, ozone is concentrated in a layer centered at approximately 35 km altitude. It is an efficient absorber of ultraviolet radiation and thus acts to protect living systems from the harmful effects of such radiation.

Certain anthropogenic chemicals containing chlorine and bromine, known as ozone depleting substances (ODS), react with ozone and contribute to the destruction of the ozone layer. These chemicals are able to diffuse into the stratosphere and produce a hole in the ozone layer in the polar regions at certain times (especially during spring). The Montreal Protocol was introduced to ban emissions of ozone depleting substances.

The ozone hole and global warming are separate atmospheric issues that are not necessarily related to each other.

Urban air quality

High ozone concentrations and the resulting photochemical smog are of particular concern in localities with plentiful sunshine that do not control their automobile and industrial emissions. Ozone shows average concentrations that exceed guideline values in cities on all continents, demonstrating that it is a global problem (Baldasano et al., 2003). In particular, megacities in the developing world have high ozone concentrations resulting from unregulated emissions and large numbers of emitters (industries and automobiles). For instance, Mexico City in 1999 is believed to have recorded the highest ever one-hourly ozone concentration (491 $\mu g\,m^{-3}$). For comparison, the EU has an alert threshold of 240 $\mu g\,m^{-3}$ for a 1-h ozone concentration, whereas Canada has a maximum desirable concentration of 100 $\mu g\,m^{-3}$. Australia has a maximum 1-h allowed concentration of 211 $\mu g\,m^{-3}$ (0.10 ppm), but as shown in Figure 1 the city of Sydney regularly exceeds this concentration.

Stratospheric ozone

The total amount of ozone-depleting gases in the stratosphere is measured by a variable known as Total Stratospheric Chlorine. Accumulation of Total Stratospheric Chlorine slowed during the early 1990s and levels are now declining slowly (Figure 2). Although there is no evidence of a significant reduction in the Antarctic ozone hole in recent years (e.g., the 2006 ozone hole was close to or possibly the largest ever), it stopped increasing significantly in the mid-to-late 1990s and a polynomial fit (shown in blue in Figure 3) indicates that the size of the

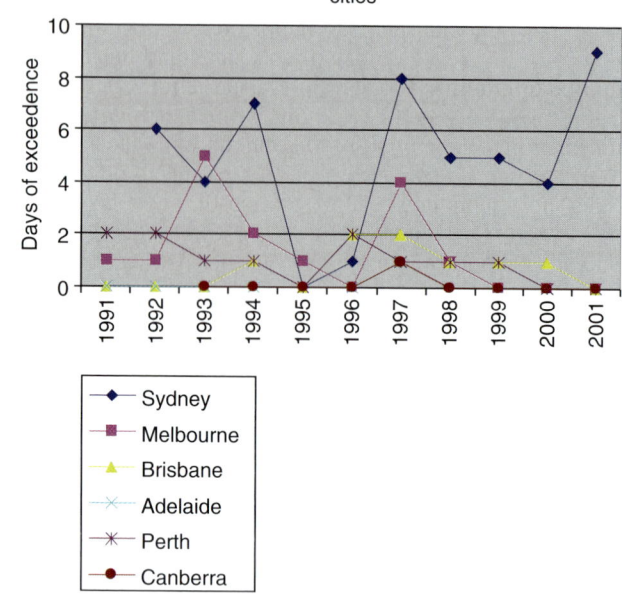

Ozone, Figure 1 Changes in lower-atmosphere ozone concentrations in Australia vary substantially from city to city and from year to year (From Beeton et al., 2006).

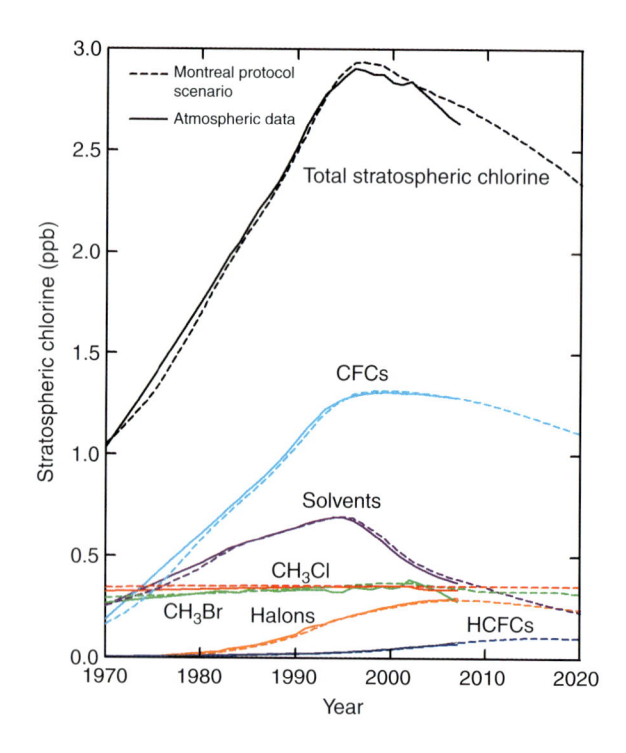

Ozone, Figure 2 Stratospheric "chlorine" from the worldwide use of major ozone depleting substances. Solid lines are data collected at Cape Grim, Tasmania; dashed lines are model calculations based on past and future emissions of ODS (Montzka and Fraser, 2003).

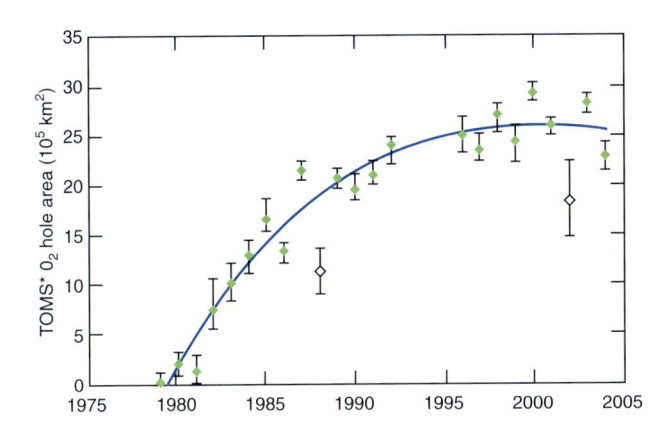

Ozone, Figure 3 The area of the Antarctic ozone hole as measured by the TOMS satellite sensor has, since 2000, started to reduce (Beeton et al., 2006).

Antarctic ozone hole started to lessen around the turn of the millennium.

The ozone hole has been linked to increased ultraviolet radiation and this in turn results in increased incidence of skin cancer. This link has become more widely known and public action in avoiding excessive ultraviolet radiation has increased significantly. Unfortunately, all other factors being equal, skin cancer incidence may be expected to continue to increase until about 2050, even though ozone levels in the stratosphere have started to recover. The reason for this is that, even though exposure to ultraviolet radiation (as a result of ozone depletion) causes skin cancer, there is about a 50-year time lag between the two.

Bibliography

Baldasano, J. M., Valera, E., and Jimenez, P., 2003. Air quality data from large cities. *The Science of the Total Environment*, **307**, 141–165, doi:10.1016/S0048-9697(02)00537-5.

Beer, T., 2001. Air quality as a meteorological hazard. *Natural Hazards*, **23**, 157–169.

Beeton, R. J. S., Buckley, K. I., Jones, G. J., Morgan, D., Reichelt, R. E., and Trewin, D., 2006. *Australia-State of the Environment 2006*. Canberra: Australian State of the Environment Committee.

Montzka, S. A., and Fraser, P. J., 2003. *Controlled Substances and Other Gas Sources in Scientific Assessment of Ozone Depletion*. Global Ozone Research and Monitoring Project Report No. 47. Geneva, Nairobi; Washington, D.C., Brussells, WMO, UNEP, NOAA, NASA, EC: 1.1–1.83.

Cross-references

Albedo
Climate Change
Cloud Seeding
Dose Rate (of risk)
Exposure to Natural Hazards
Fog Hazards
Gas-Hydrates
Global Dust
Methane Release from Hydrate
Natural Radioactivity
Ozone Loss
Radiation Hazards
Release Rate
Solar Flare
Space Weather
Sunspots
Supernova

OZONE LOSS

Mary J. Thornbush
University of Birmingham, Edgbaston, Birmingham, UK

Synonyms

(Stratospheric) Ozone depletion; (Stratospheric) Ozone hole

Definition

Involving the depletion of ozone (O_3) in the stratosphere (located 10–50 km in the atmosphere above the Earth's surface), where it protects organisms living on the Earth's surface. This is caused by chlorine molecules in ozone-depleting substances that react catalytically with ozone, destroying it (Tsai, 2002).

Background to risks

Whereas the accumulation of ground-level (tropospheric) ozone can be harmful to human health, stratospheric ozone protects human health and the environment from ultraviolet (UV) radiation. The ozone layer acts as a shield by filtering out ultraviolet B (UVB), which is linked to increased skin cancer and reduced immunity in humans, DNA damage in other animals, and impacts crop productivity (Roscoe, 2001).

Chlorofluorocarbons

Ozone depletion is caused by reactive chlorine and bromine compounds that are derived from human-made ozone-depleting substances, such as chlorofluorocarbons (CFCs) and halons (volatile bromine-containing organic substances) (Roscoe, 2001). Concern for the destruction of stratospheric ozone due to CFCs in the atmosphere developed toward the middle of the 1970s. CFCs were mainly used as refrigerants, cleaning solvents, foam-blowing agents, and aerosol propellants (Tsai, 2005). They were banned in the USA in 1978 as propellants in most aerosol uses.

Clean Air Act

More than 190 countries signed the 1987 Montreal Protocol on Substances that Deplete the Ozone Layer to protect the ozone layer. Developed countries, including the USA, committed to limiting the production and use of chemicals (such as chlorinated and brominated compounds) that are harmful to stratospheric ozone. The Montreal Protocol reduced chlorine loading, which is expected to reach 2 ppbv in 2050 – the amount above which the Antarctic ozone hole was noted (Roscoe, 2001). US Congress added provisions to the Clean Air Act of 1970 for the protection of the stratospheric ozone layer since 1990; for instance, terminating the production of harmful chemicals (such as CFCs, halons, and methyl chloroform) in 1996. Other steps are needed to protect the ozone layer, including the development of "ozone-friendly" substitutes for ozone-destroying chemicals, and the reformulation of products and processes to be more "ozone-friendly" (such as refrigerators that no longer rely on the use of CFCs). However, some substitutes are still harmful, including the use of methyl bromide (MeBr) as a pesticide in the absence of an alternative. MeBr is increasingly used in the treatment or prevention of pest infestation at international borders (Norman et al., 2008).

Scientific monitoring

Satellite data have been obtained from the Total Ozone Mapping Spectrometer (TOMS), but ground-based data derived from Dobson spectrophotometric measurements have demonstrated more long-term stability (Straehelin et al., 2002). Using a Dobson spectrophotometer, for example, it was possible to observe an elongation and splitting into two of the ozone hole over Antarctica in late September 2002 probably due to a polar stratospheric major warming (Varotsos, 2004). Ozone loss has also been observed during winter and spring months in mid-latitudes since the beginning of the 1970s, when a large fleet of civil airplanes were planned to fly in the lower stratosphere (Straehelin et al., 2002). Only later in 1985, was an "ozone hole" confirmed over Antarctica, where ozone depletion was evident that could pose a potential health and ecological threat from increased solar UV radiation onto the Earth's surface. According to the National Oceanic and Atmospheric Administration's (NOAA's) Earth System Research Laboratory (Global Monitoring Division), three gases significantly contribute to stratospheric ozone depletion, including CFC-11, CFC-12, and nitrous oxide (N_2O). Since the Montreal Protocol and its Amendments, there has been a peak in Equivalent Effective Stratospheric Chlorine (EESC) around 2000 (Straehelin et al., 2002). National Aeronautics and Space Administration's (NASA's) Laboratory for Atmospheres maintains an Ozone Hole Watch, conveying the largest ozone hole ever was observed over the South Pole on September 24, 2006. An image-based comparison of total ozone in January 2010 of images displayed by the Climate Prediction Center of NOAA's National Weather Service shows greater concentrations in the Northern Hemisphere over the Arctic than in the Southern Hemisphere over Antarctica. This is linked to the heterogeneous activation of ozone-depleting substances on surfaces (such as polar stratospheric clouds or PSCs), occurring at very low temperatures during winter in the polar stratosphere (Straehelin et al., 2002). A planetary-scale comparison of weather systems over Antarctica on September 23, 2002 with September 23, 1999 shows small losses in the 2002 event of a smaller ozone hole, with warmer temperatures, as well as a distorted shape with splitting that could be attributed to planetary waves (Varotsos, 2003). According to this author, the smaller size of the ozone hole over Antarctica in 2002 is similar to 1988, when a strong sudden stratospheric warming occurred. The Antarctic stratosphere has been warming since 1979, particularly in September and October, when most ozone-hole recovery is expected with the reduction of PSCs and a weakened Antarctic polar vortex with increased wave activity in the Southern Hemisphere (Hu and Fu, 2009).

Ozone repair

The Montreal Protocol brought about international cooperation that successfully addressed ozone depletion as a serious global environmental hazard (Norman et al., 2008). The success of the Montreal Protocol is evident in reductions in the use of ozone-destroying substances worldwide since 1986. A more recent meeting in September 2007 has led to the accelerated phaseout of hydrochlorofluorocarbons (HCFCs), which are used especially in cleaning solvents, completely by 2030 (Norman et al., 2008). Like CFCs, HCFCs share the potential for stratospheric ozone depletion and, together with hydrofluorocarbons (HFCs), they also contribute to global warming (Restrepo et al., 2008). HFCs are replacing CFCs and HCFCs, which cause considerably more stratospheric ozone depletion and global warming (Tsai, 2005). Commercial uses of HFCs are as cleaning solvents in electronic components, blowing agents in foamed plastics, refrigerants (in refrigerators and air conditioners), fire suppression agents, propellant in metered dose inhalers, and dry-etching agents in the manufacture of semiconductors (Tsai, 2005). HCFCs still contain chlorine, but release less atomic chlorine into the atmosphere compared to CFCs; also, HCFCs have a short atmospheric lifetime and relatively less potential for ozone depletion than CFCs (Tsai, 2002). It will be decades, however, before the stratospheric ozone layer heals due to the presence of ozone-destroying chemicals already implanted, which are stable compounds (such as chlorine) that can remain in the atmosphere for many years until they decay or are precipitated out of the atmosphere. For example, the Antarctic ozone hole was of record size for 3 months in 2006 (Norman et al., 2008). There are implications for significant amounts of BrO and Cl-atoms found outside the Arctic and Antarctic boundary layer, and that MeBr is ubiquitous in the troposphere at all latitudes (Platt and Hönninger, 2003).

Summary

The problem of ozone thinning has been observed since the finding in 1985 over Antarctica of an "ozone hole." This is an established connection of the disappearance of ozone in cold (polar) regions with airborne CFCs, which are concentrated in the atmosphere by the human use of refrigerants, foam products, and so on. The 1987 Montreal Protocol established strict regulations on the use of CFCs that has lessened this air pollutant in developed countries. Reductions of CFC use have resulted in ozone recovery in the stratosphere, which helps to protect humans from exposure to UV radiation that is harmful to cells in the human body (such as the skin). Other biological life is also affected by ozone loss, including other animals and plant tissues.

Bibliography

Hu, Y., and Fu, Q., 2009. Antarctic stratospheric warming since 1979. *Atmospheric Chemistry and Physics Discussions*, **9**, 1703–1726.

Norman, C. S., DeCanio, S. J., and Fan, L., 2008. The Montreal protocol at 20: ongoing opportunities for integration with climate protection. *Global Environmental Change*, **18**, 330–340.

Platt, U., and Hönninger, G., 2003. The role of halogen species in the troposphere. *Chemosphere*, **52**, 325–338.

Restrepo, G., Weckert, M., Brüggemann, R., Gerstmann, S., and Frank, H., 2008. Ranking of refrigerants. *Environmental Science and Technology*, **42**, 2925–2930.

Roscoe, H. K., 2001. The risk of large volcanic eruptions and the impact of this risk on future ozone depletion. *Natural Hazards*, **23**, 231–246.

Straehelin, J., Mäder, J., Weiss, A. K., and Appenzeller, C., 2002. Long-term ozone trends in Northern mid-latitudes with special emphasis on the contribution of changes in dynamics. *Physics and Chemistry of the Earth*, **27**, 461–469.

Tsai, W. T., 2002. A review of environmental hazards and adsorption recovery of cleaning solvent hydrochlorofluorocarbons (HCFCs). *Journal of Loss Prevention in the Process Industries*, **15**, 147–157.

Tsai, W. T., 2005. An overview of environmental hazards and exposure risk of hydrofluorocarbons (HFCs). *Chemosphere*, **61**, 1539–1547.

Varotsos, C., 2003. What is the lesson from the unprecedented event over Antarctica in 2002? *Environmental Science and Pollution Research*, **10**, 80–81.

Varotsos, C., 2004. Atmospheric pollution and remote sensing: implications for the southern hemisphere ozone hole split in 2002 and the northern mid-latitude ozone trend. *Advances in Space Research*, **33**, 249–253.

Cross-references

Ozone

PACIFIC TSUNAMI WARNING AND MITIGATION SYSTEM (PTWS)

Laura S. L. Kong
UNESCO/IOC-NOAA International Tsunami Information Center, Honolulu, HI, USA

Definition

The Intergovernmental Coordination Group for the Pacific Tsunami Warning and Mitigation System (ICG/PTWS) was started in 1965 as a subsidiary body of the Intergovernmental Oceanographic Commission of the United Nations Educational, Scientific, and Cultural Organization (UNESCO/IOC). It is comprised of Member States bordering and within the Pacific Ocean, and other interested Member States. The ICG/PTWS acts to coordinate international tsunami warning and mitigation activities. One of its most important activities is to ensure the timely issuance of tsunami warnings in the Pacific. This requires cooperation in sea level and seismic networks and data sharing, standardization and understanding of tsunami threat criteria, and effective dissemination of useful tsunami information. Comprehensive tsunami mitigation programs require complementary and sustained activities in tsunami hazard risk assessment, tsunami warning and emergency response, and preparedness. Stakeholder involvement and coordination is essential, and community-based, people-centered mitigation activities will help to build tsunami resiliency. The IOC (Paris, France) serves as the PTWS Secretariat, and coordinates the overall implementation of the global tsunami warning and mitigation system. After the devastating 2004 Indian Ocean Tsunami, the PTWS played a large role in guiding the development of regional systems in the Indian Ocean, Caribbean, and Mediterranean.

The International Tsunami Information Centre (ITIC, Hawaii, USA, started in 1965), a partnership of UNESCO/IOC and the USA National Oceanic and Atmospheric Administration (NOAA), serves as a technical and capacity building resource. The Director is provided by USA and since 1998, the Associate Director by Chile. The ITIC monitors the PTWS's effectiveness in order to recommend and facilitate improvements in the timeliness and accuracy of tsunami advisories, works closely with the international tsunami warning centers to enhance PTWS operations and directly with Member States to strengthen their national systems, facilitates technology transfer through training, workshops, and other capacity building, and acts as an information resource on historical tsunamis, education, and awareness.

NOAA's Pacific Tsunami Warning Center (PTWC, Hawaii, USA, started in 1949 after the 1946 Aleutian Islands tsunami) serves as the operational headquarters of the tsunami warning system. The Northwest Pacific Tsunami Advisory Center operated by the Japan Meteorological Agency (JMA, started in 1952 as a national center and as NWPTAC in 2005) and NOAA's West Coast/Alaska Tsunami Warning Center (WC/ATWC, started in 1967 after the 1964 Alaska tsunami) work with PTWC to provide international tsunami alerts for the Pacific and its marginal seas. The messages are advisory to designated national authorities, as each country is individually responsible for issuing warnings and public safety information to its population. National Tsunami Warning Centres are active in a number of countries. The oldest are in the Russian Federation (Sakhalin and Kamchatka Tsunami Warning Centres started after 1952 Kamchatka tsunami), in Chile (Servicio Hidrográfico y Oceanográfico de la Armada de Chile (SHOA) Sistema Nacional de Alarma de Maremotos (SNAM), started in 1964 after 1960 Chilean tsunami), and in Tahiti, France (Centre Polynésien de Prévention des Tsunamis (CPPT) started in 1965 after 1964 Alaska tsunami). For more information, visit http://ioc-tsunami.org/. For general information on tsunamis, visit http://www.tsunamiwave.info.

P.T. Bobrowsky (ed.), *Encyclopedia of Natural Hazards*, DOI 10.1007/978-1-4020-4399-4,
© Springer Science+Business Media Dordrecht 2013

Cross-references

Communicating Emergency Information
Disaster Risk Reduction (DRR)
Early Warning Systems
Earthquake
Emergency Management
Federal Emergency Management Agency (FEMA)
Global Network of Civil Society Organizations for Disaster Reduction
Hyogo Framework for Action 2005–2015
Indian Ocean Tsunami 2004
Tsunami
Tsunami Loads on Infrastructure
United Nations Organization and Natural Disasters

PAHOEHOE LAVA

Robert Buchwaldt
Massachusetts Institute of Technology, Cambridge, MA, USA

Synonyms

Ropy lava; Shelly lava; Slabby lava

Definition

Pahoehoe (the word is Hawaiian for ropy) lava is the term for cooling textures of a highly fluid, gas-charged lava flows and was introduced as a technical term by Clarance Dutton, 1883. The surface of Pahoehoe lava is usually smooth, undulant, or ropy (Figure 1). These textures are created by deformation of the flow. As the lava cools, the molten rock becomes more viscous (resistant to flow) and behaves more like a plastic. The moving flow cools from the outside toward the inside and a skin is created at the atmosphere-lava interface, whereas the center is still moving and flowing and tries to escape the surrounding skin. Due to this behavioral difference between the center and the skin, the skin can bunch upward or create wrinkles which can sometimes look like ropes and therefore is called *ropy pahoehoe*. In cases where the surface is creating more shelly textures the pahoehoe is called *Shelly pahoehoe* and where the surface is broken with closely spaced slabs, it is called *Slabby pahoehoe*.

Due to the intrinsic relationship between viscosity and temperature (the higher the temperature the lower the viscosity) the pahoehoe lavas are usually the first to erupt from a vent. The higher the volume of lava emitted the faster the current. Pahoehoe flows move forward in tongues or lobes and are characterized by a glassy, plastic skin. They may embrace obstacles at a rate of about 50 m/h.

Because of the cooling, gas loss, and crystallization of minerals Pahoehoe lava changes its viscosity over time and converts into an Aa lava flow. This conversion only occurs in this direction and it is impossible to convert lava flow occurrences from Aa to Pahoehoe (Peterson and Tilling, 1980).

Pahoehoe Lava, Figure 1 A Pahoehoe lava flow surface from a basalt flow on Isabella one of the islands of the Galapagos archipelago, Ecuador.

Bibliography

Dutton, C. E., 1883. 4th Annual Report of the U.S. Geological Survey, 95 pp.
Peterson, D. W., and Tilling, R. I., 1980. Transition of basaltic lava from Pahoehoe to Aa Kilauea volcano Hawaii: field observations and key factors. *Journal of Volcanology and Geothermal Research*, **7**, 271–293.

Cross-references

Aa Lava
Eruption Types (Volcanic)
Lahar
Shield Volcano
Stratovolcano
Volcanoes and Volcanic Eruptions

PALEOFLOOD HYDROLOGY

Gerardo Benito
CSIC-Centro de Ciencias Medioambientales, Madrid, Spain

Synonyms

Ancient floods

Definition

Paleoflood hydrology is the reconstruction of the magnitude and frequency of recent, past, or ancient floods using geological evidence (Kochel and Baker, 1982).

Discussion

Paleoflood hydrology is a multidisciplinary research field that involves expertise from geomorphology, sedimentology, hydrology, modeling, and statistics. The primary goal of paleoflood hydrology is to extend flood records over periods of time ranging from decades to millennia. Past flood data are derived from the lasting traces left by physical indicators such as sediments (e.g., slack water flood deposits, coarse clasts), erosional landforms (stripped soils, flood scarps, high-flow secondary channels), high-water marks (drift wood), tree impact scarps, and damage to vegetation (dendrogeomorphological paleoflood evidence).

The most successful paleoflood hydrology approach is based on paleostage indicators from flood-induced erosion or deposition near maximum water levels of large floods (Baker, 2008). In bedrock canyon margins and during flood stages, eddies, back-flooding and water stagnation occur, significantly reducing flow velocities and favoring the deposition of suspended clay, silt, and sand, known as slack water flood deposits (SWD). Successive layers, or sedimentary units, record a sequence of individual floods over periods of centuries to millennia. Flood timing can be derived from a variety of geochronological procedures including: (1) radiocarbon dating and (2) optically stimulated luminescence from quartz and feldspar grains (OSL).

Water stages derived from elevations of flood layers (paleostages) can be converted into flood discharge. Flood discharge is obtained by a trial-and-error procedure using a hydraulic model, and comparing and contrasting the observed water levels with the simulated ones. The estimated discharges correspond to minimum discharge values since the maximum water depth at the deposition site is unknown. Common one-dimension discharge calculation from a known water surface elevation includes: (1) slope-conveyance, (2) slope-area, (3) step-backwater, and (4) critical-depth methods.

Flood frequency analysis (FFA) may incorporate paleoflood information (non-systematic data) together with systematic (annual maximum discharge) data. It is assumed that the number of k observations exceeding an arbitrary discharge threshold (X_T) in M years is known (censored data as an analogue to partial-duration series). Maximum likelihood estimators, expected moment algorithm, and fully Bayesian approach have been shown to be efficient in the estimation of statistical parameters of traditional flood distribution functions (e.g., Gumbel, LP3, Generalized Extreme Value, Benito and Thorndycraft, 2005).

Paleoflood hydrology has been applied in many regions of the world in: (1) flood-risk estimation; (2) determination of the maximum limit of flood magnitude and non-accedences as a check of the probable maximum flood (PMF); (3) flood response to climatic variability; and (4) assessing sustainability of water resources in dryland environments where floods are an important source of water to alluvial aquifers (Benito and Thorndycraft, 2005).

Bibliography

Baker, V. R., 2008. Paleoflood hydrology: origin, progress, prospects. *Geomorphology*, **101**(1–2), 1–13.
Benito, G., and Thorndycraft, V. R., 2005. Paleoflood hydrology and its role in applied hydrological sciences. *Journal of Hydrology*, **313**, 3–15.
Kochel, R. C., and Baker, V. R., 1982. Paleoflood hydrology. *Science*, **215**, 353–361.

Cross-references

Flash Flood
Flood Deposits
Flood Hazard and Disaster
Flood Stage
Floodplain
Floodway

PALEOSEISMOLOGY

Alan R. Nelson
U.S. Geological Survey, Golden, CO, USA

Synonyms

Ancient earthquakes; Prehistoric earthquakes

Definition

Paleoseismology is the study of the location, timing, and size (magnitude) of past, usually prehistoric, earthquakes.

Discussion

Paleoseismology is a subdiscipline of the broader fields of neotectonics, active tectonics, and earthquake geology (Yeats et al., 1997; Burbank and Anderson, 2001; Keller and Pinter, 2002; Bull, 2007; McCalpin and Nelson, 2009). Paleoseismology differs from these more general studies of the deformation of Earth's crust during the past few million years in its focus on the almost instantaneous deformation of landforms and sediments during earthquakes. Paleoseismology adapts many concepts from seismology, structural geology, and tectonics, but its field methods and techniques are derived from Quaternary geology (the past 2 million years of Earth history) and related disciplines, such as geomorphology, sedimentology, archeology, paleoecology, soil science, soil mechanics, imagery analysis, and age dating. Most paleoseismic field studies require extensive training or experience in Quaternary geology, itself a highly interdisciplinary field.

Generally, paleoseismic evidence records only large (magnitude > 6.7) earthquakes because evidence of lower magnitude earthquakes is rarely created at the surface. Even where evidence of past earthquakes is preserved, much is eroded or obscured within tens to hundreds of years of an earthquake. Evidence includes rupture or warping of the ground surface along a fault, (e.g., a vertical fault scarp or a laterally offset stream valley), landforms showing sudden uplift or subsidence of large regions above a tectonic plate-boundary fault (broadly warped river terraces or uplifted or subsided shorelines), and features recording the effects of strong ground shaking or tsunamis tens to thousands of kilometers from the fault on which a large earthquake occurred (unusual landslides, liquefaction structures, or tsunami deposits).

Historical records of large earthquakes are short compared to the complete geologic history of most active faults. Even in parts of China and the Middle East where earthquake catalogs extend back 1000s of years, written records of earthquakes do not identify all faults capable of generating large earthquakes. Archeology helps document the history of large earthquakes in some regions, but much of the earthquake history of most major faults is accessible only through the techniques of paleoseismology. Paleoseismic records may extend over tens of thousands of years particularly in regions where earthquakes are infrequent or where earthquake evidence is unusually well preserved. Understanding regional patterns of large earthquakes in space and the earthquake histories of individual faults are critical in defining regional earthquake hazard, forecasting future damaging earthquakes, and mitigating their effects.

Bibliography

Bull, W. B., 2007. *Tectonic Geomorphology of Mountains*. Malden: Blackwell, 316 p.
Burbank, D. W., and Anderson, R. S., 2001. *Tectonic Geomorphology*. Malden: Blackwell, 274 p.
Keller, E. A., and Pinter, N., 2002. *Active Tectonics: Earthquakes, Uplift, and Landscape*, 2nd edn. Upper Saddle River: Prentice-Hall, 362 p.
McCalpin, J. P., and Nelson, A. R., 2009. Introduction to paleoseismology. In McCalpin, J. P. (ed.), *Paleoseismology*, 2nd edn. San Diego: Academic. International Geophysics Series, Vol. 95, pp. 1–27.
Yeats, R. S., Sieh, K. E., and Allen, C. A., 1997. *The Geology of Earthquakes*. New York: Oxford University Press, 576 p.

Cross-references

Earthquake Prediction and Forecasting
Fault
Liquefaction
Neotectonics
Seismology
Tectonic and Tectono-Seismic Hazards
Tsunami

PARAGLACIAL

Jasper Knight
University of the Witwatersrand, Johannesburg, South Africa

Definition

Paraglacial. "Nonglacial processes that are directly influenced by glaciation" (Church and Ryder 1972, p. 3059).

Paraglaciation. (1) (v.) The geomorphological regime under which landscapes undergo modification by paraglacial processes. (2) (n.) The time period during which paraglacial processes operate in a given area, starting from the time of land surface exposure upon deglaciation, to the time at which regional sediment yield reaches average interglacial background values. At this point, the effects of paraglacial processes are indistinguishable from background, and the period of paraglaciation has ceased.

Introduction and development of the paraglacial concept

The term "paraglacial" was first used by Ryder (1971) and Church and Ryder (1972) to refer to the changes in sediment yield that take place as a result of the presence, and then retreat, of glaciers in a landscape. Specifically, Church and Ryder (1972) looked at changes in fluvial sediment yield during late Pleistocene and present-day glacier retreat in Baffin Island and in British Columbia, Canada. In this paper, they argued that loose, surficial glacigenic sediment, deposited during ice retreat, was actively reworked by outflowing rivers and led to a period of enhanced sediment yield from such glaciated catchments. Ballantyne (2002a) describes this as "glacially conditioned sediment availability." Following glacier retreat, sediment yield decreases exponentially over time as the glacial sediment store is progressively reworked, reaching background interglacial values over a period of anything up to 10,000 years following initial ice retreat (Figure 1).

In the development of the paraglacial concept since the 1970s, the term has been used to include other earth surface processes including slope and periglacial weathering processes; mass movements; and lacustrine, coastal, and nearshore processes (Ballantyne, 2002a). As such, therefore, the paraglacial concept has gained wider currency as an umbrella term for all those processes that are enhanced or modified by the climatic disturbance associated with the presence, or development, of glaciers (see *Glacier Hazards*). Paraglacial processes and sediment yield should be distinguished from periglacial processes (sensu stricto), which merely imply the operation of a cold climate, not necessarily the presence of glaciers themselves (Slaymaker, 2009). Also implicit in this broader definition is the relationship between paraglacial processes and the wider glacial context within which these processes are found: "It refers both to proglacial processes, and to those occurring around and within the margins of a former glacier" (Church and Ryder, 1972, p. 3059).

From this broad definition of the term *paraglacial*, therefore, it is logical to make two key inferences. First, that the domain within which paraglacial processes operate waxes and wanes in location and size and changes in tempo with glacier extent, in particular over large spatial scales and on long, glacial-interglacial timescales (see *Climate Change*). Second, that a wide range of earth

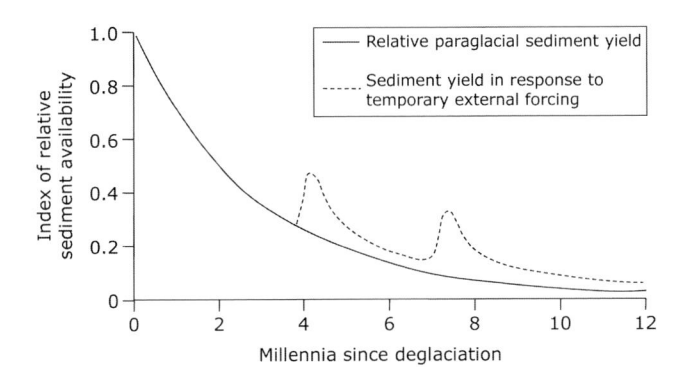

Paraglacial, Figure 1 Schematic graph showing the exponential decrease in sediment yield typical of paraglacial environments, and temporary changes in sediment yield in response to external forcing by events such as changes in base level or landsliding into valley floors (After Slaymaker, 2009).

processes that typically change when in the presence of glaciers (including subaerial and subsurface weathering, groundwater flow, geochemical pathways, and far-field effects on coastal sediment supply and sea level) should also be included under the paraglacial umbrella (Knight and Harrison, 2009a). This is a somewhat wider definition than that proposed by Ballantyne (2002a), but follows on from well-established principles of equilibrium-disturbance models in ecosystems and in systems theory (e.g., Seneviratne et al., 2006; Lenton et al., 2008). This theme of landscape disturbance highlights the important role that paraglacial processes exert, during major phases of ice retreat, on the types, locations, frequency, magnitude, and landscape impact of earth surface processes, particularly in landscapes where physical environments are changing rapidly because of glacier retreat (see *Geohazards*). It is in these environments, therefore, where a range of natural hazards occur. Typical paraglacial processes and the natural hazards that are associated with them are now examined in detail.

Evidence for paraglacial processes in past and present landscapes

Because the processes associated with paraglaciation are not exclusive to that setting, the landforms resulting from these processes are also not unique (Whalley, 2009). Some landforms that result from the paraglacial enhancement of these processes, however, can be linked morphogenetically and by radiometric dating to periods of glacier retreat and glacigenic sediment release, particularly in mountain environments which are sensitive to climate-driven changes in glacier extent (see *Climate Change*). Examples of specific physical processes and environments that respond to paraglaciation are now described. These correspond to both late Pleistocene and present-day glaciers, and in mountain and lowland glacial settings.

River responses to ice retreat

Glacier retreat is associated with the release into the proglacial environment of both loose glacigenic sediment and meltwater (Orwin and Smart, 2004). As such, glacier retreat is genetically associated with invigoration of river discharge and changes in river geomorphology, and therefore with periods of river system instability (Church and Ryder, 1972; Juen et al., 2007; Huss et al., 2008). Many rivers in alpine and mountain settings exhibit these characteristic paraglacial properties, even if these rivers are not specifically termed "paraglacial" (see *Flood Deposits*). Over short (decadal to centennial) timescales when glaciers are in retreat, increased sediment supply leads to river gravel aggradation within mountain and piedmont mid and upper-catchments, nearest to where sediment is being released (Wilkie and Clague, 2009). Sediment supply into valley bottoms can also take place from steep, unstable, glaciated slopes, particularly those that comprise glacial sediment rather than bare rock (Figure 2). Slope failure associated with slope collapse contributes a "slug" of loose sediment into these river systems, which has implications for sediment budgets and processes by which this sediment is transferred downstream (Bartsch et al., 2009). Paraglacial river responses are also spatially complex, in which barriers to downstream sediment transport (such as overdeepened, glacially scoured valleys) inhibit sediment yield to lowland floodplains. Worldwide, many paraglacial sediments today are trapped within rock-bounded upland basins and within enclosed mountain blocks, and may be unlikely to contribute to downstream sediment supply (Hewitt, 2006).

Today, meltwater production and sediment release from the margins of lowland outlet glaciers in Iceland are important in the maintenance of fronting outwash plains (sandar) (Marren, 2005). Hydrographic records of outflowing rivers from these glaciers show high temporal variability of river discharge (Old et al., 2005) (see *Flood Hazard and Disaster*). These rivers are also geomorphically unstable, and are characterized by rapid aggradation, avulsion, and bar and channel migration (see *Avulsion*; *Flood Deposits*), particularly when associated with subglacial drainage events (jökulhlaups) (Marren, 2005; Marren et al., 2009) (see *Jökulhlaups*). Several studies from SE Iceland highlight the fact that sediment delivery to the sandar of Vatnajökull has helped shape sediment supply and the dynamics of the paraglacial SE Iceland coast to which these rivers flow (Maria et al., 2000). Future changes in glacier-fed sediment supply have major implications for the maintenance of this fronting coastline (see *Coastal Erosion*).

The presence of snow/ice water sources in glaciated mountains also has implications for the temporal dynamics of outflowing rivers (Messerli et al., 2004). Where glaciers are present, glacier melt dominates mountain hydrographs through the summer period, culminating in the late summer when water from other sources is lowest. By contrast, rivers from areas dominated by winter snow cover are most vigorous during spring snowmelt (freshets)

Paraglacial, Figure 2 Photo of catastrophic rockslide/rockfall event caused by valley side debuttressing following ice retreat, Southern Alps, New Zealand (Photo: Jasper Knight).

(Vanham et al., 2008). Long-term changes in glacier and snow cover extent and duration, caused by climate change, have major implications for the sustenance of these river systems (Viviroli and Weingartner, 2004; Kehrwald et al., 2008) and the timing of consequent flood events (Vanham et al., 2008) (see *Hydrograph, Flood*).

Mass movements

Mass movement processes dominate in areas of high relief where steep slopes relax under gravity as a consequence of debuttressing as valley glaciers retreat (Fischer et al., 2006; Hewitt, 2006; Iturrizaga, 2008). Whilst mass movement processes can take place in many different physical settings, they have a greater frequency and/or higher magnitude in glaciated mountains where, first, the erosive effects of glaciers lead to the creation of high relief by valley overdeepening; second, steep-sided and unvegetated landforms comprised of glacial and periglacial debris accumulate at the margins and termini of valley glaciers; and third, bedrock slopes and glacial landforms are exposed to subaerial processes following ice retreat (Iturrizaga, 2008; Korup and Clague, 2009). Evidence for mass movement processes in glaciated lowlands is more subdued, mainly because shallower slopes mean that these processes work more slowly and the land surface is more easily stabilized by vegetation (see *Mass Movement*).

The mass movement processes and resultant landforms are morphologically diverse and not unique to paraglacial settings, and are controlled by slope, aspect, lithology, sediment supply, water content, and nature of the forcing factor that triggers the event (Korup and Clague, 2009). Specific mass movement types found in paraglacial environments include the following:

Rockslides, rock avalanches, and rock falls

Rockslides and rock avalanches are typical landscape relaxation processes in mountain areas subject to glaciation (Hewitt, 2006). They represent pressure unloading of steep rocky slopes, usually as a result of ice retreat or ice/permafrost melt, but also where overlying rock has been rapidly removed by erosion or catastrophic mass movement (e.g., Figure 2) (see *Rock Avalanche (Sturzstrom)*). There is a strong geologic control on the susceptibility of rock slopes to failure by sliding, falling, or avalanching (Jarman, 2006; Cooper, 2007), dependent on rock type, joint density, and relationship of joints to the land surface. These geologic factors also influence the geomorphological impact of failure events, including block size and runout length (Hewitt, 2006). Rocksliding (where intact slabs of rock fail and move in a coherent fashion), rock falls (where blocks are detached singly), and rock avalanches or sturzstroms (where single blocks interact with each other during movement) are all enhanced in both their frequency and magnitude in paraglacial settings (Hewitt, 2006; Cossart et al., 2008). Where available, radiocarbon and cosmogenic dating shows that high magnitude rockslide events occur most frequently in association with the climatic disturbances of the last deglaciation and Little Ice Age, and in recent times by deforestation and land use change (Curry et al., 2006; Hewitt, 2006; Cooper, 2007). Fischer et al. (2006) show that recent mass movements in the European Alps are triggered by a combination of glacier retreat and permafrost warming, which weakens the internal cohesion of rock layers and encourages failure (see *Permafrost*). These processes are typical of areas subject to increased temperature and/or decreased precipitation and where the zone in which periglacial processes operate migrates to higher-altitude locations with climate warming. Hales and Roering (2005) showed that physical weathering processes, contributing material to scree slopes in the Southern Alps (New Zealand), is strongly controlled by temperature and thus elevation. Here, climate warming results in an elevational shift in the location where rock slope failure hazards are most common, tracking the zone in which glacial and periglacial processes operate. Landslides, rockslides, and other mass movement processes are also important in overall rockwall retreat in mountain areas (Curry and Morris, 2004; Hales and Roering, 2005), which tends toward the development of equilibrium profiles that represent transient landscape stability (Roering et al., 2005). Rockslides, rockfalls, and rock avalanches are therefore common events in steep, sediment-poor areas around the time of and immediately after ice retreat.

Other slope processes
Incipient slope mass movement, that strongly contributes toward the development of these equilibrium profiles, takes place and is maintained in mountain and paraglacial landscapes by solifluction (Ballantyne and Benn, 1994; Matsuoka, 2001). Solifluction, which refers to the process of accelerated soil or sediment creep that takes place in cool climate settings under gravity, is particularly active where water is present at the ground surface (by seasonal melting of snow or permafrost) and where vegetation cover is low (see *Creep*). Solifluction causes a redistribution of unconsolidated or low-strength material from upper to lower slope positions, thereby extending slope length and decreasing slope angle (Curry, 1999). A number of landforms result from solifluction and related slope processes in paraglacial environments, including debris cones (Ballantyne, 1995; Curry et al., 2006) and alluvial fans (Ryder, 1971; Lian and Hickin, 1996; Owen and Sharma, 1998). Development of these landforms is often episodic, driven by high-magnitude events that may be seasonal (by spring snowmelt or summer rainstorms), or more infrequent and associated with particularly unusual climatic conditions.

An example of the enhancement of paraglacial slope processes under climate forcing is the period August 20–23, 2005, in which unusually heavy summer rainfall was received over the European Alps. This event resulted in extremely high rainfall intensity and volume over mountain catchments, particularly in Switzerland, Austria, and Germany, which in turn led to major river erosion and flooding events in alpine mountain valleys (Beniston, 2006). The rainfall weakened glacial sediment on steep slopes, triggering debris flows, landslides, and rockfalls (Hilker et al., 2009). In Switzerland, 25% of all damage by floods and landslides in the period 1972–2007 took place during the August 2006 event alone (ibid.).

In total, these mass movements are important in supplying material to nourish downslope debris cones and scree slopes (e.g., Ballantyne, 1995; Curry et al., 2006; Hales and Roering, 2005). In constrained valleys, this slope material can coalesce, forming a valley fill that over time stabilizes slopes and decreases available accommodation space in lower parts of mountain valleys (Ballantyne and Benn, 1994; Anderson and Harrison, 2006; Straumann and Korup, 2009) (Figure 3). The sediment yield generated by these slope processes therefore decreases over time through the paraglacial period, with accompanying decrease in frequency and magnitude of their

Paraglacial, Figure 3 Photo of paraglacial valley fill by solifluction processes that has been subsequently downcut and reworked by river rejuvenation during the Holocene, County Donegal, northwest Ireland (Photo: Jasper Knight).

Paraglacial, Table 1 Typical range of natural hazards associated with paraglacial processes that operate over short (seconds to months) and longer (years to centuries) timescales

	Drivers of the process	Associated hazardous processes and events
Short timescale paraglacial processes		
Proglacial lake outburst floods	Ice margin retreat; collapse of supporting moraine dam; landslide into the proglacial lake	Sandur erosion and aggradation; downstream flooding; debris flows; mud flows
Jökulhlaups	Ice margin retreat; subglacial volcanic eruption	Sandur erosion and aggradation; downstream floods; debris flows
Rockfalls, rockslides (sturzstrom)	Rock slope debuttressing; permafrost melt	May transform in a downslope direction into a debris flow
Landslides	High precipitation; river undercutting; slope debuttressing	Often leads to formation of landslide-dammed lakes; increased sediment input into rivers; distal debris flows and mud flows
Debris flows, mass flows, mud flows	High precipitation; permafrost melt	Formation of landslide-dammed lakes; enhanced sediment input into rivers
Longer timescale paraglacial processes		
Solifluction, gravity-driven flows	High precipitation; permafrost melt; sediment availability; steep and unstable slopes. Often results in slope stabilization and formation of valley fill	Land surface instability and movement in permafrost regions; episodic flows downslope
Sediment yield of outflowing rivers	Related to upslope sediment supply by a range of processes. Often results in sediment storage within parts of the river basin	Rapid changes in river geomorphology by erosion and deposition; increased flood frequency; mass flows and mud flows
Coastal sediment supply	Paraglacial rivers transport sediment to coasts, therefore coastal sediment budgets are strongly dependent on this sediment source	Decreased river sediment yield over time leads to sediment starvation along paraglacial coasts, and increased coastal erosion

accompanying natural hazards (Table 1). These variations in different processes and the timescales over which they operate reflect the response time of paraglacial landscapes, also called their relaxation time, to climate change and ice retreat.

Paraglacial response times

The response time of any paraglacial landscape is quite variable, depending on its location, the geomorphological process under consideration, and the baseline value of the process against which its response time to climatic disturbance is measured (Ballantyne, 2002b). The most rapid paraglacial responses will be those that take place nearest to the glaciated area, and also in areas of high relief when unstable slopes are exposed (Table 1). The processes involved here will include large-scale mass movements, in particular landslides and rockslides, which are reported from many paraglacial mountain blocks worldwide, in particular the Himalayan ranges (Hewitt, 1998) (see *Landslide*). Here, large-scale avalanches into overdeepened valleys take place shortly after (within few ka of) major ice retreat, and can often lead to development of landslide-impounded glacial lakes (Hubbard et al., 2005) (see *Landslide Dam*). Development and drainage of these lakes is a major hazard in rapidly changing glaciated landscapes. Similar processes also took place throughout the smaller-scale glaciated landscapes of Scotland (e.g., Jarman, 2006). Dating and other evidence here suggests that many of these events can take place for a long time after initial ice retreat, and sediment yield can be reinvigorated by glacial lake drainage, river flood events, or rapid climate changes (e.g., Ballantyne, 2002b). The response of paraglacial landscapes in their totality, and therefore the nature of the processes and geomorphological hazards that take place during this relaxation time, varies according to geographic setting, landscape relief, process type, and timescale (Table 1).

Outlook: the future of paraglacial environments and hazards under climate change

Today, due to global warming, most mountain glaciers worldwide are in retreat (e.g., Haeberli et al., 2007), meaning that these mountain landscapes are in transition from glacial to paraglacial in terms of the major process domains that operate in those environments. This has important implications in turn for individual geomorphological processes and natural hazards (see *Glacier Hazards*). For example, recent ice retreat in the European Alps has led to increased rockfall and debris flow hazards from surrounding slopes, increased sediment yield to outflowing rivers, and changes in water flow contribution to these rivers by glacier melt and changes in snow cover depth and duration, giving increased flood frequency in adjacent valleys (Fischer et al., 2006; Stoffel and Beniston, 2006; Fuchs, 2009). These multiple and wide-ranging impacts in the Alps are a preview of what glacier retreat will likely mean for glaciated mountains worldwide in future decades.

This present mountain glacier retreat is causing a renewed phase of enhanced paraglacial sediment delivery to valley bottoms (Knight and Harrison, 2009b). This is associated with increased magnitude and frequency of debris flows, landslides, and rockfalls, which transport sediment into mountain catchments (Stoffel and Beniston, 2006). The increased sediment supply is then available to be reworked downstream during flood events (which are themselves increasing in frequency), resulting in major changes in river geomorphology and generation of mass and mud flows. Monitoring of mountain river systems is only just now starting to identify these trends (Beylich et al., 2006; Warburton, 2007) but such studies are needed in order to identify the strength and longevity of the paraglacial signal. Paraglacial processes are extremely important in understanding contemporary mountain geomorphology, and paraglacial hazards are those that will dominate mountain settings in the next decades to centuries (Knight and Harrison, 2009b).

Bibliography

Anderson, E., and Harrison, S., 2006. Late quaternary paraglacial sedimentation in the Macgillycuddy's Reeks, southwest Ireland. *Irish Geography*, **39**, 69–77.

Ballantyne, C. K., 1995. Paraglacial debris-cone formation on recently deglaciated terrain, western Norway. *The Holocene*, **5**, 25–33.

Ballantyne, C. K., 2002a. Paraglacial geomorphology. *Quaternary Science Reviews*, **21**(18–19), 1935–2017.

Ballantyne, C. K., 2002b. A general model of paraglacial landscape response. *The Holocene*, **12**(3), 371–376.

Ballantyne, C. K., and Benn, D. I., 1994. Paraglacial slope adjustment and resedimentation following recent glacier retreat, Fåbergstølsdalen, Norway. *Arctic and Alpine Research*, **26**, 255–269.

Bartsch, A., Gude, M., and Gurney, S. D., 2009. Quantifying sediment transport processes in periglacial mountain environments at a catchment scale using geomorphic process units. *Geografiska Annaler*, **91A**(1), 1–9.

Beniston, M., 2006. August 2005 intense rainfall event over Switzerland: not necessarily an analog for strong convective events in a greenhouse climate. *Geophysical Research Letters*, **33**, L05701, doi:10.1029/2005GL025573.

Beylich, A. A., Sandberg, O., Molau, U., and Wache, S., 2006. Intensity and spatio-temporal variability of fluvial sediment transfers in an Arctic-oceanic periglacial environment in northernmost Swedish Lapland (Latnjavagge catchment). *Geomorphology*, **80**, 114–130.

Church, M., and Ryder, J. M., 1972. Paraglacial sedimentation: a consideration of fluvial processes conditioned by glaciation. *Bulletin of the Geological Society of America*, **83**, 3059–3071.

Cooper, R. G., 2007. *Mass Movements in Great Britain*. Peterborough: JNCC. Geological Conservation Review Series 33, 348pp.

Cossart, E., Braucher, R., Fort, M., Bourles, D. L., and Carcaillet, J., 2008. Slope instability in relation to glacial debuttressing in alpine areas (upper durance catchment, southeastern France): evidence from field data and Be-10 cosmic ray exposure ages. *Geomorphology*, **95**, 3–26.

Curry, A. M., 1999. Paraglacial modification of slope form. *Earth Surface Processes and Landforms*, **24**(13), 1213–1228.

Curry, A. M., and Morris, C. J., 2004. Lateglacial and Holocene talus slope development and rockwall retreat on Mynydd Du, UK. *Geomorphology*, **58**, 85–106.

Curry, A. M., Cleasby, V., and Zukowskyj, P., 2006. Paraglacial response of steep, sediment-mantled slopes to post-'Little Ice Age' glacier recession in the central Swiss Alps. *Journal of Quaternary Science*, **21**, 211–225.

Fischer, L., Kääb, A., Huggel, C., and Noetzli, J., 2006. Geology, glacier retreat and permafrost degradation as controlling factors of slope instabilities in a high-mountain rock wall: the Monte Rosa east face. *Natural Hazards and Earth System Sciences*, **6**, 761–772.

Fuchs, S., 2009. Susceptibility versus resilience to mountain hazards in Austria – paradigms of vulnerability revisited. *Natural Hazards and Earth System Sciences*, **9**, 337–352.

Haeberli, W., Hoelzle, M., Paul, F., and Zemp, M., 2007. Integrated monitoring of mountain glaciers as key indicators of global climate change: the European Alps. *Annals of Glaciology*, **46**, 150–160.

Hales, T. C., and Roering, J. J., 2005. Climate-controlled variations in scree production, Southern Alps, New Zealand. *Geology*, **33**, 701–704.

Hewitt, K., 1998. Catastrophic landslides and their effects on the Upper Indus streams, Karakoram Himalaya, northern Pakistan. *Geomorphology*, **26**, 47–80.

Hewitt, K., 2006. Disturbance regime landscapes: mountain drainage systems interrupted by large rockslides. *Progress in Physical Geography*, **30**(3), 365–393.

Hilker, N., Badoux, A., and Hegg, C., 2009. The Swiss flood and landslide damage database 1972–2007. *Natural Hazards and Earth System Sciences*, **9**, 913–925.

Hubbard, B., Heald, A., Reynolds, J. M., Quincey, D., Richardson, S. D., Luyo, M. Z., Portilla, N. S., and Hambrey, M. J., 2005. Impact of a rock avalanche on a moraine-dammed proglacial lake: Laguna Safuna Alta, Cordillera Blanca, Peru. *Earth Surface Processes and Landforms*, **30**, 1251–1264.

Huss, M., Farinotti, D., Bauder, A., and Funk, M., 2008. Modelling runoff from highly glacierized alpine drainage basins in a changing climate. *Hydrological Processes*, **22**, 3888–3902.

Iturrizaga, L., 2008. Paraglacial landform assemblages in the Hindukush and Karakoram Mountains. *Geomorphology*, **95**, 27–47.

Jarman, D., 2006. Large rock slope failures in the Scottish Highlands: characterisation, causes and spatial distribution. *Engineering Geology*, **83**, 161–182.

Juen, I., Kaser, G., and Georges, C., 2007. Modelling observed and future runoff from a glacierized tropical catchment (Cordillera Blanca, Perú). *Global and Planetary Change*, **59**, 37–48.

Kehrwald, N. M., Thompson, L. G., Tandong, Y., Mosley-Thompson, E., Schotterer, U., Alfimov, V., Beer, J., Eikenberg, J., and Davis, M. E., 2008. Mass loss on Himalayan glacier endangers water resources. *Geophysical Research Letters*, **35**, L22503, doi:10.1029/2008GL035556.

Knight, J., and Harrison, S., 2009a. Periglacial and paraglacial environments: a view from the past into the future. In Knight, J., and Harrison, S. (eds.), *Periglacial and Paraglacial Processes and Environments*. London: Geological Society Special Publication, Vol. 320, pp. 1–4.

Knight, J., and Harrison, S., 2009b. Sediments and future climate. *Nature Geoscience*, **3**(4), 230.

Korup, O., and Clague, J. J., 2009. Natural hazards, extreme events, and mountain topography. *Quaternary Science Reviews*, **28**, 977–990.

Lenton, T. M., Held, H., Kriegler, E., Hall, J. W., Lucht, W., Rahmstorf, S., and Schellnhuber, H. J., 2008. Tipping elements in the Earth's climate system. *PNAS*, **105**(6), 1786–1793.

Lian, O., and Hickin, E. J., 1996. Early postglacial sedimentation of lower Seymour valley, southwestern British Columbia. *Géographie Physique Et Quaternaire*, **50**, 95–102.

Maria, A., Carey, S., Sigurdsson, H., Kincaid, C., and Helgadottir, G., 2000. Source and dispersal of jokulhlaup sediments discharged to the sea following the 1996 Vatnajökull eruption. *Geological Society of America Bulletin*, **112**, 1507–1521.

Marren, P. M., 2005. Magnitude and frequency in proglacial rivers: a geomorphological and sedimentological perspective. *Earth-Science Reviews*, **70**, 203–251.

Marren, P. M., Russell, A. J., and Rushmer, E. L., 2009. Sedimentology of a sandur formed by multiple jökulhlaups, Kverkfjoll, Iceland. *Sedimentary Geology*, **213**, 77–88.

Matsuoka, N., 2001. Solifluction rates, processes and landforms: a global review. *Earth-Science Reviews*, **55**, 107–134.

Messerli, B., Viviroli, D., and Weingartner, R., 2004. Mountains of the world: vulnerable water towers for the 21st century. *Ambio*, **13**(Supplement), 29–34.

Old, G. H., Lawler, D. M., and Snorrason, A., 2005. Discharge and suspended sediment dynamics during two jokulhlaups in the Skafta river, Iceland. *Earth Surface Processes and Landforms*, **30**, 1441–1460.

Orwin, J. F., and Smart, C. C., 2004. The evidence for paraglacial sedimentation and its temporal scale in the deglacierizing basin of Small River Glacier, Canada. *Geomorphology*, **58**(1–4), 175–202.

Owen, L. A., and Sharma, M. C., 1998. Rates and magnitudes of paraglacial fan formation in the Garhwal Himalaya: implications for landscape evolution. *Geomorphology*, **26**, 171–184.

Roering, J. J., Kirchner, J. W., and Dietrich, W. E., 2005. Characterizing structural and lithologic controls on deep-seated landsliding: Implications for topographic relief and landscape evolution in the Oregon Coast Range, USA. *GSA Bulletin*, **117**, 654–668.

Ryder, J. M., 1971. Some aspects of the morphometry of paraglacial alluvial fans in south-central British Columbia. *Canadian Journal of Earth Sciences*, **8**, 1252–1264.

Seneviratne, S. I., Lüthi, D., Litschi, M., and Schär, C., 2006. Land–atmosphere coupling and climate change in Europe. *Nature*, **443**, 205–209.

Slaymaker, O., 2009. Proglacial, periglacial or paraglacial? In Knight, J., and Harrison, S. (eds.), *Periglacial and Paraglacial Processes and Environments*. London: Geological Society Special Publications. Geological Society, Vol. 320, pp. 71–84.

Stoffel, M., and Beniston, M., 2006. On the incidence of debris flows from the early Little Ice Age to a future greenhouse climate: A case study from the Swiss Alps. *Geophysical Research Letters*, **33**, L16404, doi:10.1029/2006GL026805.

Straumann, R. K., and Korup, O., 2009. Quantifying postglacial sediment storage at the mountain-belt scale. *Geology*, **37**, 1079–1082.

Vanham, D., Fleischhacker, E., and Rauch, W., 2008. Seasonality in alpine water resources management – a regional assessment. *Hydrology and Earth System Sciences*, **12**, 91–100.

Viviroli, D., and Weingartner, R., 2004. The hydrological significance of mountains: from regional to global scale. *Hydrology and Earth System Sciences*, **8**, 1016–1029.

Warburton, J., 2007. Sediment budgets and rates of sediment transfer across cold environments in Europe: a commentary. *Geografiska Annaler*, **89A**, 95–100.

Whalley, W. B., 2009. On the interpretation of discrete debris accumulations associated with glaciers with special reference to the British Isles. In Knight, J., and Harrison, S. (eds.), *Periglacial and Paraglacial Processes and Environments*. London: Geological Society. Geological Society Special Publication, Vol. 320, pp. 85–102.

Wilkie, K., and Clague, J. J., 2009. Fluvial response to Holocene glacier fluctuations in the Nostetuko River valley, southern Coast Mountains, British Columbia. In Knight, J., and Harrison, S. (eds.), *Periglacial and Paraglacial Processes and Environments*. London: Geological Society. Geological Society Special Publication, Vol. 320, pp. 199–218.

Cross-references

Avulsion
Climate Change
Coastal Erosion
Creep
Flood Deposits
Flood Hazard and Disaster
Geohazards
Glacier Hazards
Hydrograph, Flood
Jökulhlaup
Landslide
Landslide Dam
Mass Movement
Permafrost
Rock Avalanche

PERCEPTION OF NATURAL HAZARDS AND DISASTERS

Jaroslaw Dzialek
Jagiellonian University, Krakow, Poland

Definition

Perception of natural hazards and disasters involves intuitive judgments, beliefs, and attitudes adopted by individuals and groups of people about the likelihood of occurrence and course and mechanisms of development of such phenomena. The subjective nature of the understanding of natural hazards influences people's decisions.

Discussion

Risk perception is influenced by many factors, which are, firstly, related to the nature of the hazard itself, and, secondly, to psychological, social, and cultural components and their mutual interactions. In the case of natural hazards, risk perception analysis has to take into account that they vary immensely from sudden events such as flash floods, avalanches, or earthquakes to long lasting phenomena such as heat waves or droughts. Most comparative studies focus on human-induced dangers, whereas risk perception of natural hazards has been researched less often. It is somehow consistent with the fact, that people are usually much less concerned about natural hazards when compared to other health, safety, environmental, or social risks (Fischer et al., 1991), and that technological threats are often perceived as more dreadful than natural ones (Wachinger and Renn, 2010).

The way an average person perceives and understands natural hazards and disasters is different from that of an expert who uses scientific methods to assess the risks involved (Smith, 2001). A non-expert's knowledge about the natural environment tends to be limited, fragmentary, and uncertain (Slovic, 2000). It must be noted though that over time some indigenous populations inhabiting an area for many generations have gathered valuable knowledge to cope with disasters (Shaw et al., 2008), with an interesting example of the Solomon Islands, where immigrant populations have suffered much higher losses during 2007 tsunami than indigenous inhabitants living in the same areas (McAdoo et al., 2009).

The study of human perception of natural hazards and disasters can bring significant benefits for disaster mitigation policies because people's behavior is largely dependent on their perceptions. Individuals rely on their beliefs, which tend to differ from objective knowledge, in making decisions in the real world, including making choices where to settle, taking measures to prevent a disaster, taking out insurance and determining their behavior during the course of a disaster such as whether to evacuate. The ubiquitous subjective element is there even if an objective risk assessment is undertaken. For this reason it is important to understand factors that bias human perception of such phenomena towards underestimating or overestimating them. Based on this information, risk communication strategies can be devised between experts and non-experts that will bring the beliefs of the latter to the results of the former (Tobin and Burrel, 1997; Smith, 2001).

Experts and non-experts assess similar components of the natural environment, but their outcomes may differ. Both have to answer the same set of questions: (1) Is a certain phenomenon possible at all in a given area? (2) What is the likelihood of an occurrence within the person's lifetime? (3) What might be the scale of the phenomenon? (4) What casualties and damage can it cause? (5) How can one prepare for such an event? An important part of an individual's perception process is understanding how natural phenomena develop and, consequently, how they can be prevented or mitigated.

While assessing a threat some people may tend to overestimate it, whereas others may be prone to underestimate it. Social amplification and attenuation of risk perception are two concepts that try to explain how the initial risk is altered through social interactions in the risk communication process (Wachinger and Renn, 2010). K. Smith (2001, p. 72) and V. Covello (1991, p. 112) presented sets of factors increasing or decreasing the perception of risk associated with various hazards, including environmental hazards. People tend to be more concerned with hazards that produce an immediate and direct impact and involve numerous fatalities in a short span of time and space. According to this view, when considering only natural hazards, people may be less concerned with, for example, a drought, the effects of which are delayed and spread over large areas, than with an earthquake, which can cause catastrophic damage within a limited area. People also tend to be more anxious about phenomena whose origin they understand less (e.g., tsunami), or those that are more complicated (e.g., earthquakes), or that seem less controllable (e.g., hurricanes). Both authors point to the role of the media, which may stoke the sense of danger by increasing attention even if the objective likelihood of an event is relatively low.

Individual perception of hazards is mostly based on past experience. Familiarity with a disaster is one of the very few strong predictors for higher awareness of the possible danger (Wachinger and Renn, 2010). Many case studies show that people who have experienced a certain phenomenon are more likely to take action to mitigate future hazards (Smith, 2001). Some of this attitude may be down to what is known as the crisis effect where preventive activity is intensive immediately after a disaster, but soon dwindles and gives way to a daily routine (Bell et al., 2005). Attitudes may also differ between communities affected by a disaster depending on the nature of the event. For example in areas where minor floods are a frequent experience, people tend to prepare themselves, whereas in areas affected by a single catastrophic flood it may be perceived as something extraordinary and will not contribute to increasing a local community's resourcefulness (Tobin and Burrel, 1997; Biernacki et al., 2008). That was the case of those sites in Germany affected by a "100-year flood," where people believed that a similar event would not be possible during next 100 years, or during their lifetime (Renn, 2008). Other personal factors such as age, gender, or education level are usually tested as less important or insignificant in influencing risk perception of natural hazards (Wachinger and Renn, 2010). However, gender perspective is one of the most promising research areas as women generally rate various risks as more serious than men do (Finucane et al., 2000; Gustafson, 1998).

Some limitations of human risk perception are related to psychological mechanisms. Natural disasters may cause unpleasant feelings of anxiety and a lack of a sense of security. For this reason each individual's emotional and psychological traits have an impact on their perception. This may lead to an effect known as cognitive dissonance whereby a person is confronted by information that undermines the integrity of their beliefs about the world and causes an inner tension. An example of such cost-benefit consideration would be a strong attachment to a home that is located in a hazard zone. Rather than taking preventive action or even moving out, the predominant response to this situation observed in people is a denial of the threat and refraining from taking any active measures.

Past experience may be modified by another psychological phenomenon describing people's beliefs about what causes good and bad events in their lives and is known as the locus of control (Bell et al., 2005). People with an external locus of control perceive natural disasters as divine retribution or ill fate, whereas individuals with an internal locus of control will tend to take personal responsibility for such events. The former group will have a sense of hopelessness leading to a fatalistic attitude and a complete lack of preparation, despite their past experience of a natural disaster (Tobin and Burrel, 1997; Lin et al., 2008). The latter, on the other hand, espouse a belief that damage and harm can be prevented by active measures (McClure et al., 1999).

When a direct personal experience is not available people have to rely on indirect sources, such as media reports or their previous education. In the case of the media, their potential to contribute to a misperception of a risk mentioned above stems from a certain bias since they normally tend to focus on the news, which increases the sense of anxiety and threat. However, there is no evidence that media shape opinions. On the contrary, people tend to select elements from media coverage that are consistent with their beliefs and confirm them (Breakwell, 2007). Risk communication and risk education policies in the area of natural hazards are more and more often developed and implemented in many countries to exchange information, knowledge, and opinions between risk managers and those who may be affected by the disasters.

Different types of communities, such as rural and urban, may have different views of natural hazards. City-dwellers are further removed from the natural world and may have a lesser awareness of the hazards involved in it. They may also perceive these threats similarly to technology-related hazards, which in their view can be easily mitigated with technical means. Rural populations, on the other hand, may have a better awareness of natural threats because of their closer integration with the natural environment. Additionally, natural disasters may pose a considerable threat to their income if derived from agriculture or tourism, which would provide an additional stimulus to learn more about the topic (White, 1974; Bell et al., 2005).

The attitudes of individuals can, in general, be defined with regard to three aspects: knowledge, emotions, and actions. This can also be applied to attitudes toward natural disasters. The cognitive dimension involves knowledge (not always objective) and awareness of local hazards; the emotional aspect involves feelings (typically concern and anxiety); and finally the instrumental aspect defines actions taken in response to a potential natural disaster using the knowledge and feelings already gained. These are the elements of a decision making process in the context of a hazard in accordance to the perception-adjustment paradigm. An individual "(1) appraises the probability and magnitude of extreme events; (2) canvasses the range of possible alternative actions; (3) evaluates the consequences of selected actions; (4) chooses one or a combination of actions" (Burton et al., 1993, p. 101). A combination of the three dimensions produces various types of perception and behavior toward the perceived hazard (Raaijmakers et al., 2008):

- Ignorance when the local population is unaware of a threat and, therefore, develops no concern and takes no preventive actions
- Safety when the local population is aware of a threat, but regards its level as either low or acceptable and is, therefore, not concerned with the threat and makes no preparations for a disaster
- Risk reduction when a high level of awareness and concern produces the previously described mechanism of reducing the cognitive dissonance and denial of a disaster threat; the local population resigns from taking protective action or passes the responsibility on to the authorities
- Control when an aware population takes preventive action that helps reduce their concern

Smith (2001) proposes a different set of basic perceptions of natural hazards adopted by hazard perceivers:

- Determinate perception is demonstrated by people who, faced with a natural disaster, are trying to rationalize it and look for a pattern of occurrence of the phenomenon (e.g., how often it occurs). They will aim to expand technical measures of protection, which may lead them to be satisfied with their level of preparedness and give them a false sense of security that the hazard has been eliminated (Tobin and Burrel, 1997). This behavior is also known as the "levee effect" (Bell et al., 2005).
- Dissonant perception is typical of people who, when faced with a potential hazard, go into denial. In contrast to the previous type these people believe that previous disasters were exceptional and for this reason will rather not repeat themselves and certainly not within their lifetimes. This perception type represents a psychological adaptation strategy (Bell et al., 2005) to living in high-risk areas where, despite the presence of the hazard, people tend to care more about their daily matters than about phenomena that are largely out of their control (Tobin and Burrel, 1997).
- Probabilistic perception is typical of people who accept both the randomness and the likelihood of a natural disaster within their lifetime. Researchers point to the fact that a tendency to transfer the responsibility for dealing with the hazard on to government agencies and organizations is representative of this type of perception. Such people manifest attitudes typical for an external locus of control, that is, a fatalistic approach toward disasters as acts of God and resignation from preventive measures.

Summary

There are numerous factors influencing the perception of natural hazards and disasters, and these factors can be broken down into situational (physical and socioeconomic environment) and cognitive factors (psychological and attitudinal variables) (Tobin and Burrel, 1997). Perception studies should deal with the characteristics of the phenomena and communities as proposed by G.F. White (1974): the magnitude and frequency of an event, recency and frequency of personal experience, significance of the hazard to income interest, and personality traits. The range of variables that needs to be taken into account is large and, unfortunately, "their modeling has proved somewhat elusive in that particular variables can sometimes have different effects under different situations" (Tobin and Burrel, 1997, p. 164). Researchers are still looking for increasingly detailed explanatory models for the human perception of natural threats within existing theoretical frameworks.

Bibliography

Bell, P. A., Greene, T., Fisher, J., and Baum, A. S., 2005. *Environmental Psychology*. Orlando: Harcourt College.
Biernacki, W., Bokwa, A., Domański, B., Działek, J., Janas, K., and Padło, T., 2008. Mass media as a source of information about extreme natural phenomena in Southern Poland. In Carvalho, A. (ed.), *Communicating Climate Change: Discourses, Mediations and Perceptions*. Braga: Centro de Estudos de Comunicação e Sociedade, Universidade do Minho, pp. 190–200. Available from: http://www.lasics.uminho.pt/ojs/index.php/climate_change.
Breakwell, G. M., 2007. *The Psychology of Risk*. Cambridge: Cambridge University Press.
Burton, I., Kates, R. W., and White, G. F., 1993. *The Environment as Hazard*. New York: Guilford Press.
Covello, V. T., 1991. Risk comparison and risk communication: issues and problems in comparing health and environmental risk. In Kasperson, R. E., and Stallen, P. J. M. (eds.), *Communicating Risks to the Public*. Dordrecht: Kluwer, pp. 79–118.
Finucane, M. L., Slovic, P., Mertz, C. K., Flynn, J., and Satterfield, T. A., 2000. Gender, race, and perceived risk: the "white male" effect. *Health, Risk and Society*, 2(2), 159–172.
Fischer, G. W., Morgan, M. G., Fischhoff, B., Nair, I., and Lave, L. B., 1991. What risks are people concerned about? *Risk Analysis*, 11(2), 303–314.
Gustafson, P. E., 1998. Gender differences in risk perception: theoretical and methodological perspectives. *Risk Analysis*, 18(6), 805–811.
Lin, S., Shaw, D., and Ho, M.-C., 2008. Why are flood and landslide victims less willing to take mitigation measures than the public? *Natural Hazards*, 44(2), 305–314.
McAdoo, B. G., Moore, A., and Baumwoll, J., 2009. Indigenous knowledge and the near field population response during the 2007 Solomon Islands tsunami. *Natural Hazards*, 48(1), 73–82.
McClure, J., Walkey, F., and Allen, M., 1999. When earthquake damage is seen as preventable: attributions, locus of control and attitudes to risk. *Applied Psychology: An International Review*, 48, 239–256.
Raaijmakers, R., Krywkow, J., and van der Veen, A., 2008. Flood risk perceptions and spatial multi-criteria analysis: an exploratory research for hazard mitigation. *Natural Hazards*, 46, 307–322.
Renn, O., 2008. *Risk Governance. Coping with uncertainty in a complex world*. London: Earthscan.
Shaw, R., Uy, N., and Baumwoll, J. (eds.), 2008. *Indigenous Knowledge for Disaster Reduction: Good Practices and Lessons Learned from Experiences in the Asia-Pacific Region*. Bangkok: United Nations International Strategy for Disaster Reduction. Available from www.unisdr.org/eng/about_isdr/isdr-publications/19-Indigenous_Knowledge-DRR/Indigenous_Knowledge-DRR.pdf.
Slovic, P., 2000. *The Perception of Risk*. London: Earthscan.
Smith, K., 2001. *Environmental Hazards: Assessing Risk and Reducing Disaster*. London/New York: Routledge.
Tobin, G. A., and Burrel, E. M., 1997. *Natural Hazards: Explanation and Integration*. New York: Guilford Press.
Wachinger, G., Renn, O., 2010. *Risk perception and natural hazards*. CapHaz-Net WP3 report. Stuttgart: DIALOGIK. Available from: http://caphaz-net.org/outcomes-results
White, G. F. (ed.), 1974. *Natural Hazards: Local, National, Global*. New York: Oxford University Press.

Cross-references

Adaptation
Cognitive Dissonance
Mass Media and Natural Disasters
Risk Assessment
Risk Perception and Communication
Sociology of Disasters

PERMAFROST

Julian B. Murton
University of Sussex, Brighton, UK

Definition

Permafrost. Permafrost is ground that remains at or below 0°C for 2 years or more.
Active layer. The active layer represents the upper layer of ground subject to annual freezing and thawing in areas underlain by permafrost. The base of the active layer occurs at the depth of maximum seasonal penetration of the 0°C isotherm into the ground.
Ground ice. Ground ice denotes all types of ice contained in freezing and frozen ground.
Thermokarst. Thermokarst denotes the processes, landforms, and sediments associated with ablation – usually by thawing – of ice-rich permafrost.
Yedoma deposits. Ice-rich silts that contain on average 2–5% carbon by weight – often in the form of grass rootlets – that form a layer commonly 10–40-m thick. They underlie a region of more than 1 million km^2, mostly in central and northern Siberia. They represent the buried relict soils of the mammoth steppe-tundra ecosystem that formed during Pleistocene glacial periods.

Introduction

Permafrost on the Earth's crust forms a layer or isolated bodies that range in thickness from centimeters to ~1,500 m, and in temperature from 0°C to about −20°C. In the Northern Hemisphere, the permafrost region occupies up to ~23 × 10^6 km^2 (~24%) of the exposed land area (Zhang et al. 2008). Within this region, permafrost is classified in terms of its spatial extent, varying – in a transect from colder northern areas to warmer southern areas – from *continuous* (where it underlies 90–100% of the land surface), through discontinuous (50–90%), to sporadic (10–50%) or isolated (0–10%).

As a result, the actual area of exposed land in the Northern Hemisphere that is underlain by permafrost is significantly smaller than that of the overall permafrost region, and it is estimated to be $\sim 12 \times 10^6$ km^2 to 17×10^6 km^2 ($\sim$13–18%) (Zhang et al. 2000). Geographically, permafrost can be classified into four main types:

1. *Polar permafrost* (e.g., in Arctic and Subarctic lowlands)
2. *Mountain permafrost* (e.g., in the Rocky Mountains of Canada and the USA)
3. *Plateau or montane permafrost* (e.g., the Qinghai-Xizang Plateau of China)
4. *Subsea permafrost* (e.g., within the continental shelves of the Beaufort, Laptev, and East Siberian Seas) (French 2007)

Climatically, permafrost is either in equilibrium with current atmospheric and environmental conditions, or it is adjusting to them. Much of the polar and subsea permafrost in the Arctic developed during glacial periods of the Pleistocene, when climate tended to be colder than present and lower sea levels exposed large areas of continental shelves to atmospheric conditions. Today, such old (relict) permafrost tends to be in various stages of warming or thawing, with some decoupled from the atmosphere by a deepening (residual) thaw layer above the degrading permafrost. A second period of permafrost growth has occurred in the last $\sim$100–200 years, after the Little Ice Age. Such young permafrost is currently forming where sediments are freshly deposited in the Arctic deltas of rivers such as the Lena and Mackenzie, or where lakes drain and the thawed sediments beneath them become directly exposed to cold air temperatures.

Ground ice and carbon

Permafrost often contains ground ice and is a globally significant store of carbon. The ground ice varies from ice cements and ice lenses or veins to bodies of massive ice, meters thick and hundreds of meters or more in lateral extent (Mackay 1972; Murton and French 1994). The ice is often concentrated in the upper few meters of permafrost developed in silt-clay soils (Shur et al. 2005) and fine-grained porous bedrocks (Murton et al. 2006), forming a widespread ice-rich layer in Arctic regions that is susceptible to thaw-related disturbance. The total volume of ground ice in the Northern Hemisphere is estimated at $\sim 11 \times 10^3$ km^3 to 35×10^3 km^3, equivalent to $\sim$3–9 cm of sea-level rise (Zhang et al. 2000). These authors estimate that the area of ice-rich permafrost – ground-ice content >20% by volume – is between $\sim 1.3 \times 10^6$ km^2 and $\sim 1.6 \times 10^6$ km^2. The age of the ground ice varies from a few years to perhaps >740,000 years (Froese et al. 2008).

The carbon is stored mainly in Cryosols (permafrost-affected soils), yedoma deposits, and deltaic deposits, primarily in Arctic and Subarctic lowlands. The Cryosols comprise both organic soils (peatlands) and cryoturbated (i.e., frost-churned) mineral soils (Kimble 2004; Margesin 2009). Cryosols are estimated to contain $\sim$1,024 PgC [1 petagram $\equiv$ 1 gigatonne] at a depth of 0–3 m, whereas at depths greater than 3 m, yedoma deposits are estimated to contain 407 PgC, and deltaic deposits 241 PgC (Tarnocai et al. 2009). This gives a total below-ground carbon pool in permafrost regions of $\sim$1,672 PgC, of which 1,466 Pg occurs in perennially frozen soils and deposits. This 1,672 Pg pool of organic carbon is more than double the current amount of carbon in the atmosphere ($\sim$730 PgC), about half of the estimated global below-ground carbon pool, and more than three times greater than the total global forest biomass ($\sim$450 PgC) (Kuhry et al. 2009). Less is known about the size and distribution of the N pool in permafrost regions, although it is likely that substantial quantities are stored in permafrost peatlands (Repo et al. 2009) and other C-rich deposits.

Organic carbon accumulates in permafrost regions because cold temperatures and impeded drainage retard decomposition of plant materials, and cryoturbation moves some of this carbon into deeper and colder soil layers (Tarnocai 2009). In addition, upward growth of permafrost – as occurs in accumulating peatlands or in steppe-tundra episodically buried by windblown or alluvial sediments (Zimov et al. 2006) – freezes and preserves the carbon. In effect, the permafrost acts as a giant freezer that stores old carbon and nitrogen, sometimes for tens of thousands of years. The freezer, however, is vulnerable to permafrost thaw.

Permafrost thaw and C/N release

As permafrost is a product of climate (Burn 2007), the ice, C, and N within it are sensitive to climate or microclimate change. Thaw of ice-rich permafrost (thermokarst) results from a variety of factors, ranging from regional climate warming to site-specific disturbances (Mackay 1970; Murton 2009). In turn, a variety of permafrost-related hazards arise (Table 1). Of these, the most significant global hazard is that of climate warming and C/N release to the atmosphere. Local and regional permafrost-related hazards are reviewed by Davis (2001), French (2007), and Whiteman (2011).

Current warming and thawing of permafrost soils (ACIA 2004) results in increased rates of microbial decomposition of organic carbon and subsequent release of carbon dioxide or methane. Carbon dioxide tends to be released where the soils are affected by drying, whereas methane release is favored by wet, anaerobic conditions. Methane is also released from thawing yedoma deposits near the edges of growing thaw lakes. In North Siberia, methane bubbling in thaw lakes is estimated to emit $\sim$3.8 teragrams (0.0038 Pg) of methane per year, and lake expansion during the last quarter of the twentieth century increased methane emission by $\sim$58% (Walter et al. 2006). The Pleistocene age of this methane indicates that old carbon previously stored in permafrost is now being released to the atmosphere, where it constitutes

Permafrost, Table 1 Permafrost-related hazards

Scale	Hazard	Example
Global	Carbon release	Thaw of the permafrost carbon pool
		Thaw of permafrost associated with methane hydrates
Local to regional	Thaw subsidence	Subsidence of pipelines, roads, buildings built on ice-rich soil or bedrock
		Land-surface collapse
		Initiation and growth of thaw lakes
		Drowning of forest and conversion to wetland
	Mass movement	Enhanced rockfalls and landslide activity during thaw of ice-rich mountain permafrost
		Enhanced solifluction, active-layer detachment, and thaw slumping during degradation of ice-rich soil
		Bog bursts and mass flows
	Reduced ice strength	Reduced adfreeze strength (bonding) and bearing capacity between piles and surrounding ice as permafrost warms
		Increased velocity of rock glaciers as permafrost warms and ice weakens
	Frost heave	Enhanced frost heave as wet, silty active layers thicken during permafrost thaw
		Enhanced frost heave when chilled gas pipelines are buried in unfrozen ground in discontinuous permafrost terrain
	Erosion	Enhanced erosion as ground ice thaws within bluffs along sea coasts, lakes, and rivers
	Water supply	Contamination of water above permafrost

a powerful positive feedback to high-latitude climate warming. Methane bubbling associated with development of thaw lakes may also have contributed significantly to global warming during the Pleistocene-Holocene transition (Walter et al. 2007). N_2O emissions from unvegetated areas of subarctic peaty tundra (Repo et al. 2009) and thawing permafrost (Elberling et al. 2010) may also contribute to atmospheric greenhouse gas (GHG) concentrations, although as yet few data are available on the fluxes and sources.

The strength of the feedback between permafrost C/N and climate change depends on both the amount of organic C/N stored in permafrost and on the rate of C/N release to the atmosphere (Schuur et al. 2008). The rate of release in turn depends on the balance between (a) increased uptake of atmospheric C by enhanced plant growth and (b) increased loss of old permafrost C by increased microbial decomposition (Mack et al. 2004). Consistent with the inferred positive feedback effect, one recent study from an upland thermokarst site in Alaska found that ~15 years after permafrost thaw began, losses of old soil C began to offset increased C uptake by shrubby plants, with the result that this study area became a net source of C to the atmosphere (Schuur et al. 2009). Further studies from other permafrost regions, however, are now needed to quantify the mass balance between C/N in permafrost and the atmosphere. In particular, it is timely and important to focus such studies on C/N-thermokarst terrains that are likely to be hotspots of GHG generation in a warming world.

Thermokarst activity and global warming

Global warming is one of many factors that initiate thermokarst activity. But it is undoubtedly a key factor, based on the abundant evidence in the geological record for pan-Arctic thermokarst activity during the Last Glacial-to-Interglacial Transition (Murton 2001) and the signs of intensifying thermokarst activity in recent decades.

Last 100–150 years: Thermokarst activity during the last 100–150 years has undoubtedly spread and intensified, although not all of it can be attributed to global warming. One of the clearest examples where climate warming has exacerbated thermokarst is an abrupt increase in ice-wedge melting since 1982 in continuous permafrost of northern Alaska (Jorgenson et al. 2006). The melting probably resulted from record high summer temperatures between 1989 and 1998, and was initiated by extreme hot and wet summer weather in 1989, leading to unusually deep thaw of the active layer. This thermokarst activity coincided with a 2–5°C increase in mean annual ground temperature, partially melting ice wedges that had previously been stable for thousands of years.

Farther south, in the warm discontinuous permafrost of Subarctic regions, increased thermokarst activity has taken place since the Little Ice Age (Jorgenson and Osterkamp 2005), but the causes are complex. For example, thermokarst activity in the Tanana Flats, central Alaska, has transformed large area of birch forest into fens and bogs (Jorgenson et al. 2001). Thermokarst here probably began in the mid-1700s, associated with climate warming. But thermokarst activity during the succeeding ~250 years has been enhanced in part by (1) convective heat transfer by movement of relatively warm (2–4°C year-round) groundwater through the fens and underlying outwash gravel, (2) fires, and (3) increased snow depths. Isolating the influence of fire, snow, and climate warming is difficult because an increase in fire frequency may correlate with an increase in summer temperatures (Jorgenson and Osterkamp 2005), and because warmer winter temperatures may correlate with increased snowfall and therefore warmer mean annual ground temperatures (cf. Osterkamp 2007).

In western Siberia, climate warming since the early 1970s is thought to have driven thermokarst activity in two different ways (Smith et al. 2005). In the continuous permafrost zone, the number of lakes has increased substantially, whereas in discontinuous, isolated, and sporadic permafrost zones, it has decreased. This disparity supports a conceptual model in which initial warming of cold, continuous permafrost favors thermokarst activity and lake expansion, followed by lake drainage as permafrost degrades further. In central Siberia, increases in mean annual air temperature and summer air temperatures between 1992 and 2001 at Yakutsk have coincided with (1) thermokarst subsidence beneath stable inter-alas meadows and (2) flooding of young thermokarst basins, enhancing thermokarst activity at a nearby permafrost monitoring site (Fedorov and Konstantinov 2003).

Next 100–150 years: With climate warming predicted to continue during the next century, amplified in Arctic and Subarctic regions (ACIA 2004), thermokarst activity will generally spread and intensify still more. Thawing of permafrost is projected to be concentrated in the current discontinuous permafrost zone during the next 100 years (Delisle 2007). This is of particular concern in Subarctic Alaska, where ~40% of the area may be susceptible to thermokarst (Jorgenson et al. 2007). Thus global warming at high latitudes is putting large areas of ice-rich permafrost at risk of thermokarst subsidence and related disturbances (Nelson et al. 2001). Although projected climate warming in the Twenty-first Century will lead to deeper ground thaw in many permafrost regions, its impacts will be modulated by site-specific conditions. For example, peat and vegetation cover may buffer permafrost from severe degradation, whereas local disturbance of the ground cover or fires in the boreal forest or tundra may accelerate permafrost thaw (Yi et al. 2007). Thus caution is needed in generalizing between projected changes in atmospheric climate and permafrost responses, and modeling of projected responses must consider a buffer layer between the atmosphere and the permafrost.

Buffer layer

A buffer layer of vegetation, snow, and organic material thermally modulates the coupling between atmospheric and permafrost temperatures (Figure 1a). As a result, air temperatures are often offset from ground surface and permafrost temperatures by a few to several degrees °C (Figure 1b). As the buffer layer changes seasonally – for example, with leaf growth in summer, snowfall in winter, and year-round changes in the moisture content of the organic layer – its thermal properties (e.g., albedo, thermal conductivity, latent heat) change. In turn, this alters the surface energy exchanges and ground temperatures. Thus it is essential for permafrost models to parameterize the seasonal contrasts in the different components of the buffer layer. Additionally, models must consider ground ice within permafrost beneath the buffer layer (Figure 1a), because the high latent heat effects of the ice retard permafrost thaw.

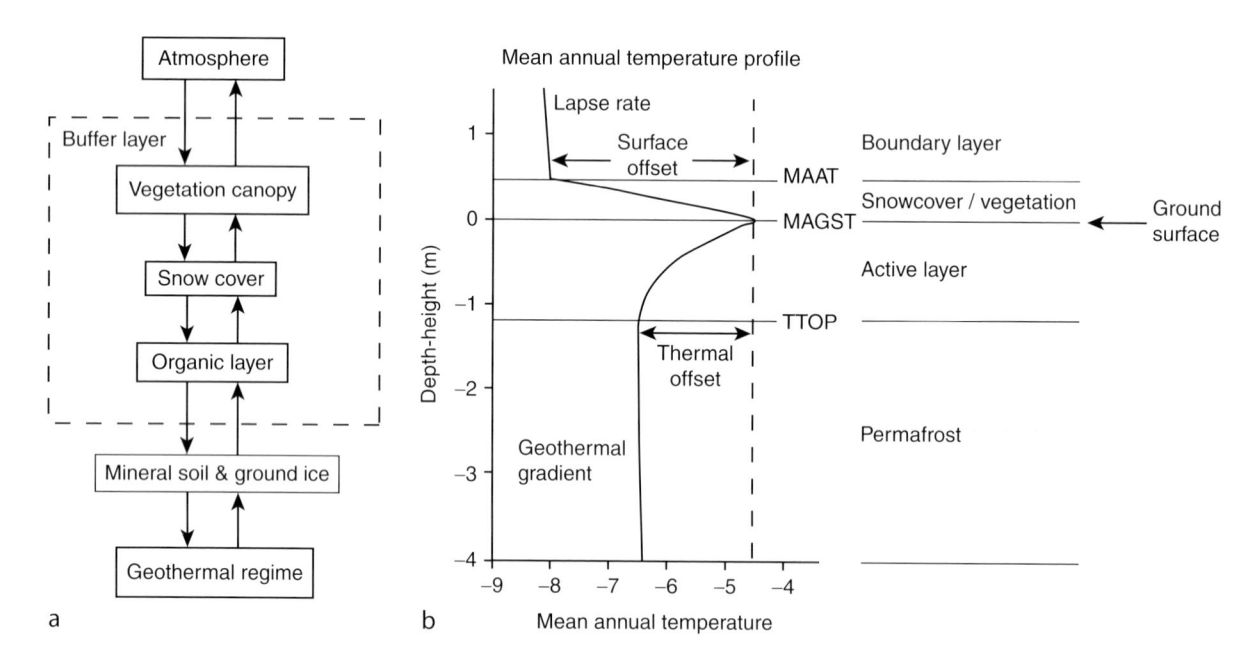

Permafrost, Figure 1 Atmosphere-ground thermal coupling in permafrost regions. (**a**) Buffer layer model, showing conceptually the thermal buffering of the mineral soil from the atmosphere by layers of vegetation, snow, and organic material (Adapted from Williams and Smith 1989). (**b**) Schematic mean annual temperature profile through surface boundary layer, showing relation between air temperature and permafrost temperature. MAAT: mean annual air temperature; MAGST: mean annual ground surface temperature; TTOP: temperature at the top of permafrost (Adapted from Smith and Riseborough 2002).

Bibliography

ACIA, 2004. *Impacts of a Warming Arctic: Arctic Climate Impact Assessment*. Cambridge: Cambridge University Press.

Burn, C. R., 2007. Permafrost. In Elias, S. A. (ed.), *Encyclopedia of Quaternary Sciences*. Amsterdam: Elsevier Science Publishers B.V. (North-Holland), pp. 2191–2199.

Davis, N. T., 2001. *Permafrost: A Guide to Frozen Ground in Transition*. Fairbanks: University of Alaska Press.

Delisle, G., 2007. Near-surface permafrost degradation: how severe during the 21st century? *Geophysical Research Letters*, **34**, L09503, doi:10.1029/2007GL029323.

Elberling, B., Christiansen, H. H., and Hansen, B. U., 2010. High nitrous oxide production from thawing permafrost. *Nature Geoscience*, doi:10.1038/NGEO803.

Fedorov, A., and Konstantinov, P., 2003. Observations of surface dynamics with thermokarst initiation, Yukechi site, central Yakutia. In Phillips, M., Springman, S. M., and Arenson, L. U. (eds.), *Proceedings of the 8th International Conference on Permafrost*. Lisse: Swets & Zeitlinger, Vol. 1, pp. 239–242.

French, H. M., 2007. *The Periglacial Environment*. Chichester: Wiley.

Froese, D. G., Westgate, J. A., Reyes, A. V., Enkin, R. J., and Preece, S. J., 2008. Ancient permafrost and a future, warmer arctic. *Science*, **321**, 1648, doi:10.1126/science.1157525.

Jorgenson, M. T., and Osterkamp, T. E., 2005. Response of boreal ecosystems to varying modes of permafrost degradation. *Canadian Journal of Forest Research*, **35**, 2100–2111, doi:10.1139/X05-153.

Jorgenson, M. T., Racine, C. H., Walters, J. C., and Osterkamp, T. E., 2001. Permafrost degradation and ecological changes associated with a warming climate in central Alaska. *Climatic Change*, **48**, 551–579.

Jorgenson, M. T., Shur, Y. L., and Pullman, E. R., 2006. Abrupt increase in permafrost degradation in Arctic Alaska. *Geophysical Research Letters*, **33**, L02503, doi:10.1029/2005GL024960.

Jorgenson, M. T., Shur, Y. L., Osterkamp, T. E., and George, T., 2007. Nature and extent of permafrost degradation in the discontinuous permafrost zone of Alaska. In: Kokelj, S., and Walters, J. (eds.), *Proceedings of the 7th International Conference on Global Change: Connection to the Arctic (GCCA-7)*. International Arctic Research Center, University of Alaska Fairbanks.

Kimble, J. M. (ed.), 2004. *Permafrost-Affected Soils*. Berlin: Springer.

Kuhry, P., Ping, C.-L., Schuur, E. A. G., Tarnocai, C., and Zimov, S., 2009. Report from the International permafrost association: carbon pools in permafrost regions. *Permafrost and Periglacial Processes*, **20**, 229–234.

Mack, M. C., Schuur, E. A. G., Bret-Harte, M. S., Shaver, G. R., and Chapin, F. S., III, 2004. Ecosystem carbon storage in arctic tundra reduced by long-term nutrient fertilization. *Nature*, **431**, 440–443.

Mackay, J. R., 1970. Disturbances to the tundra and forest tundra environment of the western Arctic. *Canadian Geotechnical Journal*, **7**, 420–432.

Mackay, J. R., 1972. The world of underground ice. *Annals of the American Association of Geographers*, **62**, 1–22.

Margesin, R. (ed.), 2009. *Permafrost Soils*. Berlin/Heidelberg: Springer. Soil Biology, Vol. 16.

Murton, J. B., 2001. Thermokarst sediments and sedimentary structures, Tuktoyaktuk Coastlands, Western Arctic Canada. *Global and Planetary Change*, **28**, 175–192.

Murton, J. B., 2009. Global warming and thermokarst. In Margesin, R. (ed.), *Permafrost Soils*. Berlin/Heidelberg: Springer. Soil Biology, Vol. 16, pp. 185–203.

Murton, J. B., and French, H. M., 1994. Cryostructures in permafrost, Tuktoyaktuk Coastlands, Western Arctic Canada. *Canadian Journal of Earth Sciences*, **31**, 737–747.

Murton, J. B., Peterson, R., and Ozouf, J.-C., 2006. Bedrock fracture by ice segregation in cold regions. *Science*, **314**, 1127–1129, doi:10.1126/science.1132127.

Nelson, F. E., Anisimov, O. A., and Shiklomanov, N. I., 2001. Subsidence risk from thawing permafrost. *Nature*, **410**, 889–890.

Osterkamp, T. E., 2007. Characteristics of the recent warming of permafrost in Alaska. *Journal of Geophysical Research*, **112**, F02S02, doi:10.1029/2006JF000578.

Repo, M. E., Susiluoto, S., Lind, S. E., Jokinen, S., Elsakov, V., Biasi, C., Virtanen, T., Pertti, J., and Martikainen, P. J., 2009. Large N_2O emissions from cryoturbated peat soil in tundra. *Nature Geoscience*, **2**, 189–192.

Schuur, E. A. G., Bockheim, J., Canadell, J. G., Euskirchen, E., Field, C. B., Goryachkin, S. V., Hagemann, S., Kuhry, P., Lafleur, P., Lee, H., Mazhitova, G., Nelson, F. E., Rinke, A., Romanovsky, V., Shiklomanov, N., Tarnocai, C., Venevsky, S., Vogel, J. G., and Zimov, S. A., 2008. Vulnerability of permafrost carbon to climate change: implications for the global carbon cycle. *Bioscience*, **58**, 701–714.

Schuur, E. A. G., Vogel, J. G., Crummer, K. G., Lee, H., Sickman, J. O., and Osterkamp, T. E., 2009. The effect of permafrost thaw on old carbon release and net carbon exchange from tundra. *Nature*, **459**, 556–559, doi:10.1038/nature08031.

Shur, Y., Hinkel, K. M., and Nelson, F. E., 2005. The transient layer: implications for geocryology and climate-change science. *Permafrost and Periglacial Processes*, **16**, 5–17.

Smith, L. C., Sheng, Y., MacDonald, G. M., and Hinzman, L. D., 2005. Disappearing Arctic lakes. *Science*, **308**, 1429.

Smith, M. W., and Riseborough, D. W., 2002. Climate and the limits of permafrost: a zonal analysis. *Permafrost and Periglacial Processes*, **13**, 1–15.

Tarnocai, C., 2009. Arctic permafrost soils. In Margesin, R. (ed.), *Permafrost Soils*. Berlin/Heidelberg: Springer. Soil Biology, Vol. 16, pp. 3–16.

Tarnocai, C., Canadell, J. G., Schuur, E. A. G., Kuhry, P., Mazhitova, G., and Zimov, S., 2009. Soil organic carbon pools in the northern circumpolar permafrost region. *Global Biogeochemical Cycles*, **23**, GB2023, doi:10.1029/2008GB003327.

Walter, K. M., Zimov, S. A., Chanton, J. P., Verbyla, D., and Chapin, F. S., III, 2006. Methane bubbling from Siberian thaw lakes as a positive feedback to climate warming. *Nature*, **443**, 71–75, doi:10.1038/nature05040. September 7, 2006.

Walter, K. M., Edwards, M. E., Grosse, G., Zimov, S. A., and Chapin, F. S., III, 2007. Thermokarst lakes as a source of atmospheric CH_4 during the last deglaciation. *Science*, **318**, 633–636, doi:10.1126/science.1142924.

Whiteman, C. A., 2011. *Cold Climate Hazards and Risks*. Chichester: Wiley-Blackwell.

Williams, P. J., and Smith, M. W., 1989. *The Frozen Earth. Fundamentals of Geocryology*. Cambridge: Cambridge University Press.

Yi, S., Woo, M., and Arain, M. A., 2007. Impacts of peat and vegetation on permafrost degradation under climate warming. *Geophysical Research Letters*, **34**, L16504, doi:10.1029/2007GL030550.

Zhang, T., Barry, R. G., Knowles, K., Heginbottom, J. A., and Brown, J., 2008. Statistics and characteristics of permafrost and ground-ice distribution in the northern hemisphere. *Polar Geography*, **31**, 47–68.

Zhang, T., Heginbottom, J. A., Barry, R. G., and Brown, J., 2000. Further statistics on the distribution of permafrost and ground ice in the northern hemisphere. *Polar Geography*, **24**, 126–131.

Zimov, S. A., Schuur, E. A. G., and Chapin, F. S., III, 2006. Permafrost and the global carbon budget. *Science*, **312**, 1612–1613, doi:10.1126/science.1128908.

Cross-references

Climate Change
Cryological Engineering
Gas-Hydrates
Glacier Hazards
Ice and Icebergs
Ice Storm
Methane Release from Hydrate
Paraglacial
Snowstorm and Blizzard
Solifluction

PIEZOMETER

Sylvi Haldorsen
Norwegian University of Life Sciences, Aas, Norway

Synonyms

Standpipe piezometer; Pneumatic piezometer; Vibrating wire piezometer

Definition

A piezometer is a piece of equipment installed to measure the hydraulic head (pore pressure) in a groundwater aquifer.

Discussion

In an unconfined aquifer the measured hydraulic head is equal to the groundwater level (elevation head) and in a confined aquifer the total hydraulic head is equal to the elevation head plus the pressure head (Fetter, 2001). The piezometers define the hydraulic head at the specific site where it is installed. A number of different piezometers are applied, and the following are the most common:

Standpipe piezometers are cased tube wells with a limited diameter. There is a short screen or only a short slotted section at the lower end of the casing. The tube above the filter acts as a riser pipe. The water level in the riser pipe defines the hydraulic head, and is commonly measured manually. This simple instrument is inexpensive and does not depend on electronic installations.

Pneumatic piezometers record the pore pressure automatically by a pneumatic filter tip. Via a cable, the record is transferred to a pneumatic reader, which may be connected to a readout unit or a data logger. The filter tip may be installed inside a cased tube well or at the end of a separate fitted tube. A pneumatic piezometer facilitates continuous readings and makes readings from remote locations possible.

Vibrating wire piezometers record the hydraulic head by a vibrating wire pressure transducer and signal cable. The wire pressure transducer is commonly installed in a cased tube well. The readings are recorded by a portable readout or a data logger.

Measuring water level underground is an important source of information in hazards studies, especially those involved with landslide monitoring and evaluation.

Bibliography

Fetter, C. W., 2001. *Applied Hydrogeology*, 4th edn. Upper Saddle River: Prentice-Hall.

Cross-references

Antecedent Conditions
Collapsing Soil Hazards
Debris Flow
Early Warning Systems
Flash Flood
Landslide Types
Liquefaction
Mass Movement
Vaiont Landslide

PIPING HAZARD

Michael James Crozier[1], Nick Preston[1], Thomas Glade[2]
[1]Victoria University of Wellington, Wellington, New Zealand
[2]University of Vienna, Vienna, Austria

Synonyms

Sapping; Seepage Erosion; Tunneling

Definition

Piping and tunneling represent distinct subsurface, linear erosion mechanisms but they are often functionally indistinguishable and generally grouped under the process term *piping* (Bryan, 2000).

Discussion

Essentially *true piping* (referred to as *seepage erosion*, by Dunne (1990)) involves the localized dislodgment and ejection of particles at an outlet in response to seepage forces produced by a positive water potential and high hydraulic gradient, leading to erosion and headward development of a conduit. By contrast, *tunneling* exploits existing macropores such as cracks and root channels, and involves hydraulic entrainment forces and hence is referred to by Dunne (1990) as *tunnel scour*. Both processes may operate in subsurface conduits and the consequent undermining effect and extension of the cavity is referred to as *sapping*.

The criteria for pipe and tunnel development include: a pathway for water entry and concentration of subsurface flow (e.g., cracks, root channels, animal burrows, and other macropores), a decrease in permeability with depth that promotes slope parallel subsurface flow, the presence of erodible material (especially silts and fine sand), space for eroded material to be evacuated (coarse porous material or an outlet from the slope), and a water supply under positive water potential and high hydraulic gradient. Susceptibility to piping in some regions has also been linked to soil and soil water chemistry.

The development of pipes in an area enhances slope drainage, increases runoff coefficients, decreases concentration time (Anderson and Burt, 1990), limits soil water storage, and consequently increases peak flow during storm events (Jones, 1990).

Pipes may develop into gully systems as a result of conduit enlargement and roof collapse (for a benchmark paper on this process see Laffan and Cutler (1977)) and hence further accelerate storm runoff, leading to localized flooding. Piping and gully development can occur within the space of a few days to years and, as a result, buildings initially thought to be in safe locations can experience flash flooding during storm rainfall. The development of pipes can also affect buildings by undermining foundations and removing structural support, through tunnel enlargement and tunnel roof collapse.

Resistance to landsliding can be affected by piping, through removal of underlying support and changes to subsurface water flow. Although pipes can stabilize slopes through rapid drainage, pipe blockage can allow localized build up of pore water pressures and significantly lower slope resistance (Hardenbicker and Crozier, 2002).

Piping represents a serious potential hazard to dams, particularly earth dams, in some cases leading to serious dam failure (UNEP, 2001). Initiation of pipes can be caused through development of stress cracks but more commonly as a result of the nature of fill material and the integrity of contact between fill and foundations or other structures. A well-known example of the destructive effect of piping is the Teton dam failure, June 5, 1976, which killed 14 people and resulted in millions of dollars of property damage downstream (Independent panel, 1976).

Bibliography

Anderson, M. G., and Burt, T. P., 1990. Subsurface runoff. In Anderson, M. G., and Burt, T. P. (eds.), *Process Studies in Hillslope Hydrology*. Chichester: Wiley, pp. 365–400.

Bryan, R. B., 2000. Soil erodibility and processes of water erosion on hillslope. *Geomorphology*, **32**(3), 385–415.

Dunne, T., 1990. Hydrology, mechanics, and geomorphic implications of erosion by subsurface flow. In Higgins, C. G., and Coates, D. R. (eds.), *Groundwater Geomorphology: The Role of Subsurface Water in Earth-surface Processes and Landforms*. Geological Society of America Special Paper 252, pp. 1–28.

Hardenbicker, U., and Crozier, M. J., 2002. Soil pipes and slope stability. In Rybár, J., Stemberk, J., and Wagner, P. (eds.), *Landslides. Proceedings of the First European Conference on Landslides*. Balkema: Prague, pp. 565–570.

Independent panel, 1976. *Independent Panel to Review Cause of Teton Dam Failure*. Report to the US Department of interior and state of Idaho on failure of the Teton dam, Idaho Falls.

Jones, J. A. A., 1990. Piping effects in humid lands. In Higgins, C. G., and Coates, D. R. (eds.), *Groundwater Geomorphology: The role of Subsurface Water in Earth-surface Processes and Landforms*. Geological Society of America Special Paper 252, pp. 111–138.

Laffan, M. D., and Cutler, E. J. B., 1977. Landscape, soils, and erosion of a catchment the Wither Hills, Marlborough. *New Zealand Journal of Science*, **20**, 279–289.

UNEP, 2001. *Tailings Dams – Risk of Dangerous Occurrences, Lessons Learnt from Practical Experiences*. United Nations Environmental Programme, Division of Technology, Industry and Economics; International Commission on Large Dams, Paris.

Cross-references

Building Failure
Collapsing Soil Hazards
Damage and the Built Environment
Dispersive Soil Hazards
Erosion
Erosivity
Expansive Soils and Clays
Hydrocompaction Subsidence
Land Subsidence
Lateral Spreading
Mining Subsidence Induced Fault Reactivation
Piezometer
Pore-Water Pressure
Sinkhole
Subsidence Induced by Underground Extraction
Universal Soil Loss Equation (USLE)

PLANNING MEASURES AND POLITICAL ASPECTS

Brian R. Marker
London, UK

Definitions

Government consists of the bodies (international, national, regional, and local) responsible for administering a state or area including developing policies for, and the regulation of, human activities and resolution of potential disputes between activities and interest groups. The term also applies to the processes of securing outcomes through legislation, regulation, guidance, the coordination of actions, and enforcement of provisions to meet policy objectives.

Land-use planning is the process of identifying the suitability of land for particular uses, setting out policies for land use and determining planning applications in accordance with those policies.

Spatial planning extends the principles of land-use planning by taking account of a full range of spatial factors whether social, economic, or environmental, in order to secure sustainable development.

Emergency planning is the process of identifying potential hazards, coordinating response systems and preparing, practicing and promulgating plans for dealing with the effects of these.

Environmental impact assessment is the process of assembling and interpreting the environmental information that is pertinent to the potential impacts of a proposed development that is considered likely to cause significant adverse environmental effects and to set out proposals for mitigation of these. The results are presented in an environmental statement.

Sustainability appraisal is a process of assembling the social, economic, and environmental information that is pertinent to assessing the acceptability of alternative policy options intended to secure sustainable development.

Strategic environmental assessment is a process of assembling the environmental information that is pertinent to assessing the acceptability of alternative proposed policy options for sustainable development and forms part of sustainability appraisal.

Introduction

This entry is necessarily generalized because of the great variety of circumstances around the World; but some general principles relevant to natural hazards can be summarized.

Political systems vary from place to place, and time to time, from democratic to centralist and, therefore, from extensive to limited public participation in political, policy, and decision making processes. They also vary in transparency and application. Cultural and other diversity issues also influence actions and outcomes. In general, the scale and complexity of problems is such that the political focus tends to be on current key issues and imperatives rather than all factors that should be taken into account when formulating policies and making decisions. The tendency is to identify short-term tactical "solutions" rather than undertaking strategic actions for the longer term especially where financial resources are limited and governance structures are weak. The result, in relation to natural hazards, is often to carry out work after damaging events rather than taking advance precautions to minimize adverse impacts. The results are higher levels of deaths, injuries, and economic losses than might be the case if societies are, as far as possible, adequately protected and prepared.

Understanding of potential hazards

Most of the World's major populated areas are subject to at least one or two natural hazards and some to several. There is often a lack of widespread appreciation of the potential for these unless damaging events occur frequently. Precautions are not taken unless potential is recognized but all too often precautions are not taken even though the potential has been recognized. Logically, all populated areas should be evaluated for locations, nature, mechanisms, and frequency of all significant potential hazards so that planning and management decisions can be made on an informed basis. In practice, that is often not done until after a damaging event has occurred. If at all, consideration is usually limited to the type of event that has taken place rather than the fuller range of potential hazards. This is partly because of a lack of awareness of potential hazards amongst administrators and decision makers and also because other calls on resources are perceived to be of greater priority. Even where hazards have been identified, problems arise if systems are not developed to provide adequate responses and there is insufficient awareness of provisions.

Government and governance

Government essentially provides the conditions and circumstances required for economic growth, social improvement, and environmental protection and enhancement. Potential conflicts between these aspects are addressed through attempts to secure sustainable development – delivering social, economic, and environmental policies with the least adverse effects for present and future generations (WCED and Commission for the Future, 1987). The extent to which this is done depends on the situation of the nation concerned. In a wealthy developed country it is possible to devote land and resources to environmental protection but in a poor developing country the emphasis is understandably on the immediate economic and social survival of the population. While there is some redistribution of wealth from rich to less rich countries through international aid, this often relates to development of infrastructure, notably water supply, and to health improvement, rather than disaster preparedness and precautionary measures. Also some aid does not reach its intended targets due to inefficiency or corruption.

Governance operates at several levels – international (e.g., the European Union), national, regional, and local. Systems exist at each level for politicians to formulate policies with advice from government officials such as civil servants, and for decision makers to determine planning applications having regard to those policies. All necessarily engage with other sections of society such as industry, commerce, nongovernmental organizations, and the general public. But structures for communication, debate, exchange of ideas and education need to be robust if sound and effective actions are to be secured.

Responsible authorities, at all levels, set out policies that decision makers and other members of society are expected to observe. Implementation depends on legislation to control the development, application of policies and procedures, and behavior of institutions and populations. It provides that certain things must be done at the risk of penalties that are imposed by government bodies or through the legal system. Legislation is usually pitched at a general level with more detailed aspects being contained in secondary regulations that are often administered by government agencies, such as environmental protection agencies, rather than government departments. In addition to the body of law, rights are due, to a greater or lesser extent, to citizens and interest groups. These range from the Universal Declaration of Human Rights, to those of owners of land and property and, in some countries, to the rights of indigenous peoples. Redress for infringements of rights can usually be sought through independent courts. The complexity of systems makes it difficult to secure optimum outcomes. For example, different government departments, levels of government and government agencies have different aims and objectives to meet as do nongovernmental interest groups. Effective debate and coordination is essential if issues are to be address in a balanced and proportionate manner.

The role of politicians depends on the nature of the political system. Where they are subject to election they must understand public aspirations and persuade people to support their policies. In less democratic systems, the public has no option but to follow government policies but may do so in a half-hearted manner if they are not content with these. In both types of system, however, politicians need to engage with and agree policies with their colleagues and supporters (Hamilton, 2000). But the extent to which those policies are implemented by regulators, decision makers, and the public is also variable.

It has to be accepted that geoscience issues, including hazards, are not key concerns of many politicians, decision makers or the public, until specific damaging events occur. They have to be persuaded to take note. When disaster strikes there may be may be promises about assistance and better future planning and construction and, sometimes, a short-term increase in resources that can detract from preparations for dealing with other relevant hazards. But often the promised aid and planning measures are not implemented (Lewis and Mioch, 2005).

Communication

If natural hazards are to be addressed adequately by politicians, administrators, the media and the public, sound communication with hazard specialists is needed. It is essential to secure adequate attention, resources, and actions before damaging events occur. This is not easy due to a widespread lack of appreciation of the technical issues and because these are often not expressed in terms that can be widely understood. Among many nonscientists there is a weak appreciation of scientific method and of concepts such as hazard, vulnerability, risk, frequency, uncertainty, and probability. Necessary scientific caution may be misinterpreted as meaning that predictions have little value. So can inability to indicate exactly when and where the next damaging event may occur. Also any warning that is not followed by an actual event may cause a loss of confidence. Poor communication can cause under or over-reactions leading to unnecessarily restrictive or excessively lax policies, unwise investment, financial losses, or failure to react. Therefore information is not necessarily received gladly particularly if it raises concerns about hazards and possible associated costs and impacts asset values. It is also difficult to convince stakeholders that hazard and risk assessment needs to be repeated at intervals because of environmental, social, and economic changes.

Overall, scientific conclusions need to be carefully and clearly framed in plain language and dissemination of information needs to be carefully planned and handled. But this is not a one way process. Hazard specialists need to understand: the complexity of administrative structures and procedures; interactions with and between different levels of administration, the wider public and the media; and when and how best to engage with these. They also need to be comfortable collaborating with other relevant disciplines, including sociologists and economists, when addressing hazards (Forster and Freeborough, 2006; Liverman et al., 2008).

Spatial and land-use planning

It is prudent, where possible, to direct development toward the least vulnerable locations. Spatial and land-use planning are systems that control the use of land in the public interest. Developed and many developing countries have systems for regulating the use of land involving the definition of what types of development and conservation might be appropriate in particular locations. Policies are expressed in some form of development planning document. In some systems development plan policies are mandatory – they have to be observed – whereas in others provisions are discretionary – they should be observed unless there are good reasons to do otherwise. Commonly guidance is issued to explain how provisions should be applied but, even so, discretionary responses can be misjudged. Development plan documents provide the opportunity for ensuring that development is undertaken prudently but a key problem is that much existing development is already vulnerable to natural hazards and it is simply impractical to move large settlements to other locations. Also, after damaging events, people usually look to governments to redevelop in the same familiar location, even though they might be safer elsewhere. The extent to which planning policies are complied with depends on the effectiveness of enforcement by the responsible authorities. Enforcement is strong in some countries but weak in others (Greiving and Fleischhauer, 2006; Stallworthy, 2002; Allmendinger, 2009). Spatial and land-use planning therefore play an important part in addressing potential natural hazards (Schmidt-Thomé, 2006) but must be accompanied by appropriate emergency planning procedures.

Evaluation of policy options and development proposals

Policy options are now widely subjected to sustainability appraisal: evaluation of policy options in terms of sustainable development – the economic, social, and environmental implications – to secure the option that will not have unacceptable adverse effects for present and future generations. Within this overall process rests strategic environmental assessment: the process of evaluating the environmental implications of policies and plans (Jones et al., 2005; Helming et al., 2008; Tang, 2008).

Developments that are likely to have adverse environmental effects are widely subjected to environmental impact assessment (EIA). Assessments address the implications of the development for the environment in detail but are often less thorough in considering how the environment might impact on the development. For most significant developments, whether an EIA is required or not, site investigations are required. The results are taken into account, alongside other factors, when determining planning applications or considering legislation for major infrastructure developments that may fall outside planning systems (Therival et al., 2005; Morris and Therival, 2009).

Site investigations should take full account of natural hazards but there can be a tendency to focus on the characteristics within the site rather than the broader regional context (Weltman and Head, 1983; ASCE, 1996). When an EIA or site investigation is undertaken planning authorities may not be fully equipped to evaluate the quality of the work or the findings, sometimes leading to poor decisions, unless they have access to independent experts.

Emergency planning

When dealing with disasters it is essential to be well prepared with resilient systems and plans. Emergency planning involves preparation of effective responses to likely emergencies, natural and man-made, through civil contingencies procedures. It requires identification of key roles and responsibilities of government, at all levels, emergency services (police, fire, medical, military), utilities and NGOs, as well as public education, assessment of the vulnerability of key services to specific hazards, and training and exercises based on credible scenarios. Key aspects are securing communications and routes that can be kept open during emergencies, safe access to rescue equipment and to supplies and medicine, and, where possible, well understood warning systems (Sinha, 2006; Fleischhauer, 2008; Moore, 2008). The Hyogo Framework of Action was adopted by 168 Member States of the United Nations at the World Disaster Conference in 2005 (UN/ISDR, 2005) to improve performance on disaster reduction and preparedness over the period 2005–2015. This set out five priority actions:

- Ensure that disaster reduction is a national and local priority with a strong institutional basis for implementation.
- Identify, assess, and monitor disaster risks and enhance early warning.
- Use knowledge, innovation, and education to build a culture of safety and resilience at all levels.
- Reduce the underlying risk factors.
- Strengthen disaster preparedness for effective response at all levels.

Conclusions

If potential problems are properly understood and there is a political will to support and fund the necessary actions it is possible to significantly reduce the impacts of hazardous events through planning, coordination, and education. It is useful to consider steps in terms of "the disaster cycle": preparedness, disaster event, relief, restoration and rebuilding, and risk reduction.

Preparedness

Risk can be reduced by:

– Planning for development in the least vulnerable locations and using potentially more hazardous sites for open-space purposes (e.g., wildlife refuges, playing fields), low-density occupation, or space to accommodate effects of hazards (e.g., flood water storage).
– Ensuring, where possible, that precautionary and preventive works are undertaken prior to specific development.
– Ensuring that foundations, constructions, and infrastructure are designed and undertaken so as to minimize the risk of damage and failure.
– Evaluating the location and vulnerability of key services and infrastructure such as power and water supply, communications, access and evacuation roads, fire and police stations, hospitals, and military bases since all have a part to play in bringing aid.
– Ensuring access to supplies through storage or appropriate delivery arrangements that can be drawn on when an event occurs.
– Where appropriate, monitoring potential hazards to give early warning of impending events.
– Developing and practicing emergency responses and warning procedures so that the relevant agencies, including nongovernmental aid organizations, know what to do and are properly coordinated.
– Informing the public of the issue and increasing awareness of early warning signs of hazardous events and of the actions that they should take to minimize risks to themselves and others.

This requires good organization and adequate funding. These are often, but not always, available in developed countries, if the will and understanding is there, but are often uneven in developing countries. Also, the scale and frequency of potential hazards varies greatly. Some cannot be avoided or addressed through preventive work – for instance, a major volcanic eruption, earthquake, or tsunami. For these, adequate design of emergency and relief responses, and, if practicable, monitoring facilities are essential.

During and immediately after the hazard event

Following a hazardous event quick action is needed to:

– Implement search and rescue plans, ensure public security, and, as far as possible, stabilize structures and clear key access routes.
– Deploy national and international aid (food, water, medical supplies, temporary shelter) where it is most urgently needed and to place adequate medical care.

Restoration and rebuilding

As soon as possible thereafter, it is necessary to:

– Undertake remedial work on constructions and infrastructure where possible and to replace it where not.
– Implement plans for clearance and replacement of damaged constructions.

Risk reduction

It is important to learn from the event to identify necessary improvements to reduce future risk and, thus, preparedness. This requires consideration of matters such as:

- The effectiveness of emergency responses.
- The possible need for changes to building codes.
- Improvements to the resilience of infrastructure.
- Changes to land-use planning policies and practices.

However, depending on the severity of the event and the availability of finance and resources recovery may take years.

The development of effective responses requires hazard specialists to understand governance systems, to communicate effectively, and to work closely with experts from other disciplines, for instance, sociologists, economists, and professional communicators. The principal constraint, particularly in developing countries, is the capacity to respond in terms of funds and structures. There is a need for international aid to focus on developing the capacity to respond to emergencies as well as responding to current problems.

Bibliography

Allmendinger, P., 2009. *Planning Theory (Planning, Environment, Cities)*. Basingstoke: Palgrave McMillan.
American Society of Civil Engineers (ASCE), 1996. *Environmental Site Investigation Guidance Manual*. Reston: American Society of Civil Engineers.
Fleischhauer, M., 2008. The use of spatial planning in strengthening urban resilience. In Pasman, H. J., and Kirillov, I. A. (eds.), *Resilience of Cities to Terrorist and Other Threats*. Amsterdam: Ios Press. NATO Science for Peace and Security Series, Series C Environmental security.
Forster, A., and Freeborough, K., 2006. A guide to the communication of geohazards information to the public. *British Geological Survey (BGS) Urban Science and Geohazards Programme Internal Report IR/06/009*. Keyworth: British Geological Survey.
Greiving, S., and Fleischhauer, M., 2006. Spatial planning response towards natural and technological hazards. In Schmidt-Thomé, P. (ed.), *Natural and Technological Hazards and Risks Affecting the Spatial Development of European Regions*. Espoo: Geological Survey of Finland. Geological Survey of Finland Special Publication, Vol. 42, pp. 109–123.
Hamilton, D. K., 2000. Organising government structure and governance functions in metropolitan areas in response to change: a critical overview. *Journal of Urban Affairs*, **22**(1), 65–84.
Helming, K., Perez-Soba, M., and Tabbush, P., 2008. *Sustainability Impact Assessment of Land Use Changes*. Heidelberg: Springer.
Jones, C., Baker, M., Carter, J., and Jay, S. (eds.), 2005. *Strategic Environmental Assessment and Land Use Planning: An International Evaluation*. London: Earthscan.
Lewis, D., and Mioch, J., 2005. Urban vulnerability and good governance. *Journal of Contingencies and Crisis Management*, **13**(2), 50–53.
Liverman, D. G. E., Pereira, C., and Marker, B. R. (eds.), 2008. *Communicating Environmental Geoscience*. Geological Society London Special Publication SP305. London: Geological Society.
Moore, A., 2008. *Disaster and Emergency Management Systems*. Teddington: British Standards Institute.
Morris, P., and Therival, R. (eds.), 2009. *Methods of Environmental Impact Assessment*. London: Routledge. Natural and Built Environment Series.
Schmidt-Thomé, P., 2006. *Integration of Natural Hazards, Risk and Climate Change into Spatial Planning Practices*. Espoo: Geological Survey of Finland.
Sinha, P. C. D., 2006. *Disaster Relief: Rehabilitation and Emergency Humanitarian Assistance*. New Dehli: SBS Publishers.
Stallworthy, M., 2002. *Sustainability, Land Use and the Environment*. London: Routledge-Cavendish.
Tang, Z., 2008. *Integrating Strategic Environmental Assessment into Local Land Use Plans: Thinking Globally, Acting Locally*. Saarbrucken: VDM Verlag Dr Müller Aktiengeschellschaft and Co KG.
Therival, R., Glasson, J., and Chadwick, A., 2005. *Introduction to Environmental Impact Assessment*. London: Routledge. Natural and Built Environment Series.
United Nations International Strategy for Disaster Reduction (UN/ISDR), 2005. *Hyogo Framework for Action: Extract from the Final Report of the World Conference on Disaster Reduction (A/CONF.206/6) UN/ISDR*, Geneva, pp. 28.
United Nations, 1948. *Universal Declaration of Human Rights United Nations*, New York, pp. 1. http://secint50.un.org/en/documents/udhr/. Accessed 15 Mar 2012.
Weltman, A. J., and Head, R., 1983. *Site Investigation Manual*. London: Construction Industries Research Information Association.
World Commission on Environment and Development (WCED) and Commission for the Future, 1987. *Our Common Future*. Oxford: Oxford University Press.

Cross-references

Building Codes
Buildings, Structures, and Public Safety
Casualties Following Natural Hazards
Civil Protection and Crisis Management
Community Management of Hazards
Damage and the Built Environment
Disaster Relief
Economics of Disasters
Emergency Management
Land-use Planning
Megacities and Natural Hazards
Recovery and Reconstruction After Disaster

PLATE TECTONICS

John Ristau
GNS Science, Avalon, Lower Hutt, New Zealand

Definition

Tectonic Plate: a large slab of rock composed of part of the Earth's crust and upper mantle.
Plate Tectonics: theory which describes the history, motions, and tectonic activity of Earth's tectonic plates.

Introduction

Plate tectonics is a theory which describes the motions of the Earth's tectonic plates over the Earth's surface. The Earth consists of a thin outer shell of solid rock (the crust), a thick viscous layer (the mantle), a liquid outer core, and a solid inner core. The mantle is divided into two layers – a colder and more solid upper mantle and a hotter and more liquid lower mantle. Together, the crust and upper mantle are called the *lithosphere* which can be thought of as floating on the lower mantle (the *asthenosphere*). The lithosphere is broken into large pieces ~100–200 km thick which make up the Earth's tectonic plates. The plates slowly drift and collide with, pull apart from, and slide past one another.

History

The history of the theory of plate tectonics provides an interesting and important example of a modern scientific revolution where a new theory replaces an existing theory due to the accumulation of new evidence which is not consistent with the existing theory. The history of plate tectonics essentially started in 1915 when Alfred Wegener argued that the fit of the coastlines of the continents, the distribution of coal and glacial deposits, and the distribution of flora and fauna suggested that the continents were once joined as a supercontinent and later drifted apart (*continental drift*). This directly opposed the accepted thinking that the positions of the continents were fixed. At the time, the idea was rejected for a number of reasons with one of the main problems being that there was no acceptable mechanism to explain how the continents moved.

Throughout the first half of the twentieth century, a vast amount of evidence in support of continental drift began to accumulate. Seismic data showed that the spatial distribution of earthquakes is not random but rather that earthquakes mainly occur in the vicinity of what are now known as plate boundaries. Earthquake sources were found to occur in dipping seismic zones around the Pacific, and studies of the transmission of seismic waves showed that the dipping zones were thick mantle slabs with high seismic velocities. These zones mark where one plate is pushing beneath another and led to the concept of *subduction zones*. Studies of the propagation of seismic waves gave important insights into the structure of the Earth's interior.

Paleomagnetism showed strip-like magnetic anomalies related to magnetic reversals along the ocean floor. These strips are offset symmetrically from mid-ocean ridges and increase in age with distance from the ridges with the youngest rocks always having the present-day magnetic polarity. Regular movement of the Earth's magnetic pole (*polar wander*) was found to be different in Europe and North America, suggesting that the two continents had moved relative to one another. The term "seafloor spreading" was introduced in the early 1960s as part of the explanation of how oceanic crust is formed at mid-ocean ridges by volcanic activity and gradually moves away from the ridge. The discoveries of spreading at mid-ocean ridges and subduction at plate boundaries complemented one another perfectly. The concept of upwelling of hot mantle material from the asthenosphere which pushes up through weak parts of the lithosphere provided the driving mechanism for plate tectonics.

By the end of the 1960s, the theory of plate tectonics became firmly established, replacing the fixist paradigm. Confirmation of plate tectonic theory continues through to the present day with modern Global Positioning System (GPS) measurements which measure the current movements of tectonic plates with centimeter scale precision. GPS measurements also show how the edges of the plates deform from collisions between plates while the plate interiors remain relatively undeformed.

How it works

There are three types of boundaries between plates:

1. Divergent boundaries where plates move away from one another
2. Convergent boundaries where plates collide with one another
3. Transform boundaries where plates slide past one another

Divergent boundaries are areas where new lithosphere is formed and provide the driving mechanism for plate tectonics. At places where the lithosphere is thin and weak, hot mantle material from the asthenosphere pushes up and through, forcing the lithosphere apart (Figure 1). It then cools, forming new lithosphere material. Divergent boundaries can occur at either thinner oceanic crust such as mid-ocean ridges, or at thicker continental crust such as the East African Rift Zone.

Convergent boundaries provide the complement to divergent boundaries and are areas where lithosphere material is recycled back into the mantle. Oceanic crust is denser than continental crust, and in regions where oceanic and continental plates collide, the oceanic plate

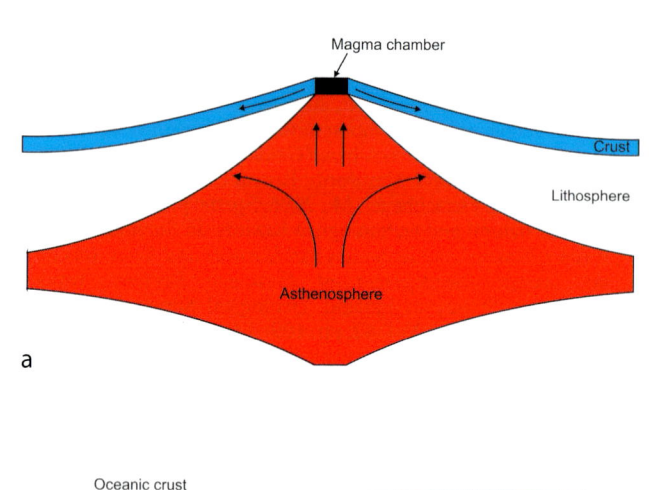

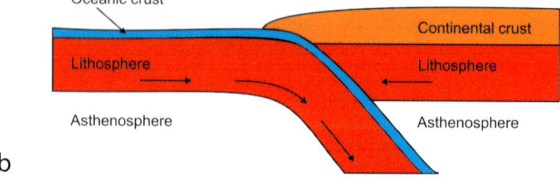

Plate Tectonics, Figure 1 (**a**) Divergent plate boundary formed by mantle material from the asthenosphere pushing through and forcing the lithosphere apart. (**b**) Convergent plate boundary where oceanic crust and lithosphere push beneath continental crust and lithosphere and down into the mantle.

pushes beneath the continental plate and down into the mantle (Figure 1). These areas are known as subduction zones. The world's largest earthquakes are those that occur on the boundary between the subducting and overriding plates such as the 1960 magnitude 9.5 Chilean earthquake or the more recent 2004 magnitude 9.3 Indian Ocean earthquake. The majority of subduction zones are along the boundaries of the Pacific and Indian Oceans and occur beneath the west coasts of North and South America, Japan, the Philippines, Taiwan, Indonesia, New Guinea, Fiji, and New Zealand. In areas where two continental plates collide, neither plate subducts, but instead the lithosphere thickens at the collision zone and has a major effect on topography. The primary example of continent-continent collision is the collision between the Indian and Eurasian plates which produced the Himalayas and the Tibetan Plateau.

Transform boundaries are areas where lithosphere is neither created nor destroyed. These are regions where one plate is sliding past another without convergence or divergence. Examples of transform boundaries include the San Andreas Fault in California and the Queen Charlotte Fault of the west coast of Canada where the Pacific Plate is sliding past the North American Plate, and the Alpine Fault in New Zealand where the Pacific Plate is sliding past the Australian Plate.

Outlook: present day

Over the history of the Earth, the tectonic plates have collided together, forming supercontinents, and moved apart several times. The last supercontinent was known as *Pangaea* which broke apart ~180 million years ago into two smaller supercontinents called *Gondwana* and *Laurasia*. These in turn broke apart into the present-day layout of tectonic plates. There are currently seven or eight major tectonic plates (depending on whether the Indian and Australian plates are considered one plate or separate plates) which cover most of the Earth's surface and a large number of minor plates (Figure 2). Many of the minor plates contribute significantly to seismic hazard in various countries, such as the Philippine plate for Japan, the Nazca plate for the west coast of South America, and the Juan de Fuca plate for the west coast of North America.

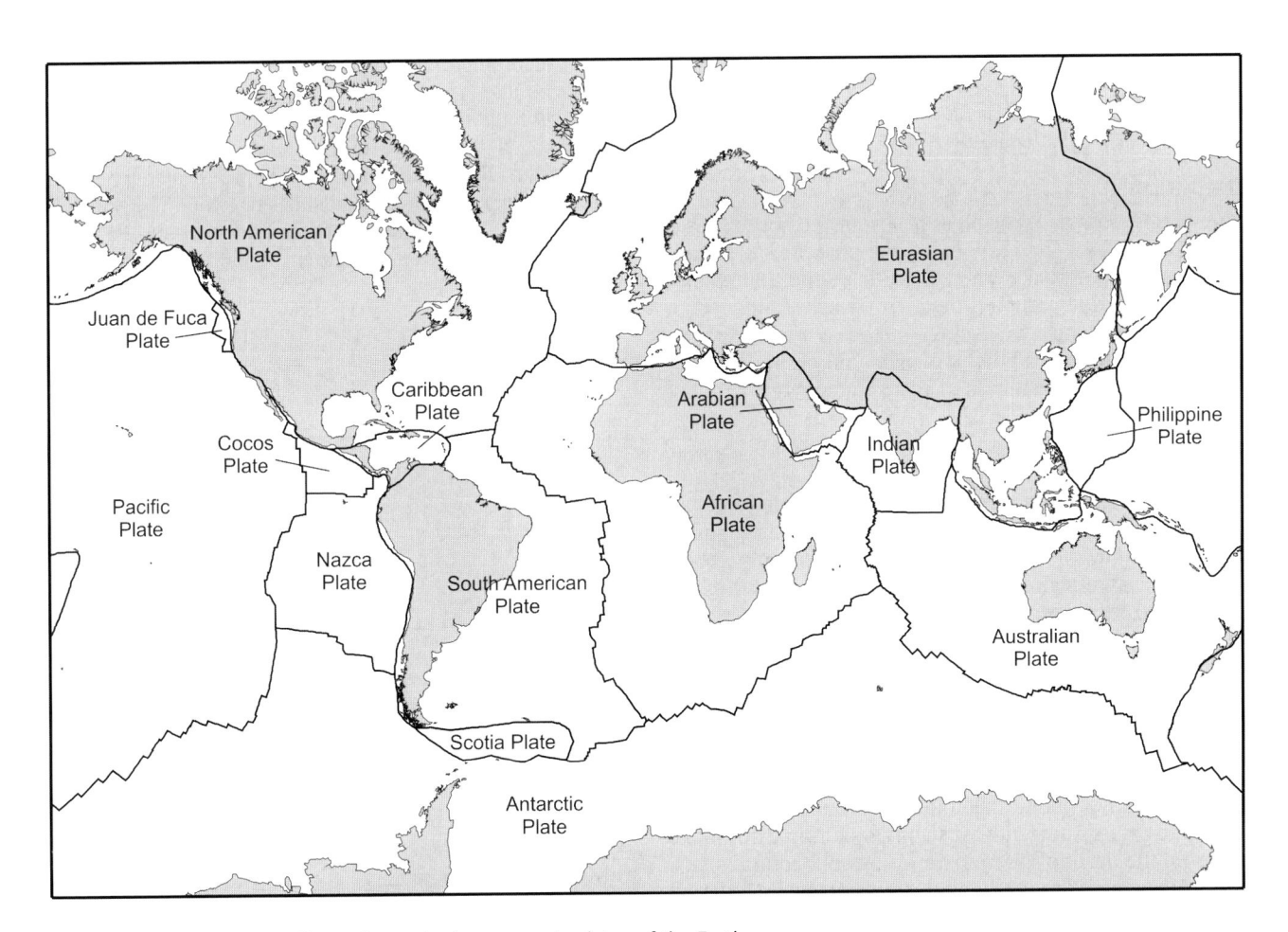

Plate Tectonics, Figure 2 The major and minor tectonic plates of the Earth.

Bibliography

DeMets, C., Gordon, R. G., Argus, D. F., and Stein, S., 1990. Current plate motions. *Geophysical Journal International*, **101**, 425–478.

Kearey, P., and Vine, F. J., 1996. *Global Tectonics*, 2nd edn. Malden: Blackwell Sciences. 348 p.

Moores, E. M., and Twiss, R. J., 1995. *Tectonics*. New York: W.H. Freeman and Company. 415 p.

Vine, F. J., and Matthews, D. H., 1963. Magnetic anomalies over oceanic ridges. *Nature*, **199**, 947–949.

Wilson, J. T., 1965. A new class of faults and their bearing on continental drift. *Nature*, **207**, 343–347.

Cross-references

Magma
Neotectonics
Subduction
Tectonic and Tectono-Seismic Hazards
Volcanoes and Volcanic Eruptions

PORE-WATER PRESSURE*

Mark E. Reid
U.S. Geological Survey, Menlo Park, CA, USA

Synonyms

Pore pressure; Pore-fluid pressure; Pore-water stress

Definition

Pore-water pressure is the pressure (isotropic normal force per unit area) exerted by the fluid phase in a porous medium (soil or rock) composed of a solid framework and pores filled or partially filled with water or other fluid. Pore-water pressure is averaged over a representative elementary volume containing many pores, rather than an individual pore. SI units of measurement are N/m^2 or Pa (Pascals).

Discussion

Pore-water pressure is commonly measured relative to an ambient atmospheric reference pressure; positive pressures (measured using piezometers) are typically found in saturated materials whereas negative pressures or suctions (measured using tensiometers) are typically found in partially saturated materials. In groundwater systems, the surface of zero pressure (relative to atmospheric) defines a water table.

The distribution and magnitude of pore-water pressures play key roles in the deformation and failure of porous earth materials in a tremendous variety of settings. These pressures can provoke the initiation of rainfall-induced landslides, the liquefaction of sediment during earthquake shaking, the onset of seismic tremor in subduction zones, the displacement of faults, and the hydrofracturing of rock around fluid injection wells. The widespread influence of pore-water pressure arises from several phenomena: (1) porous earth materials typically contain pore fluids as groundwater or soil moisture; (2) mechanical deformation is coupled between the fluid and solid phases (stress and strain in one phase affects the other phase); (3) pore-water pressure can reduce the frictional shear strength of earth materials; and (4) pore-water pressures commonly vary in response to dynamic hydrologic drivers (e.g., infiltration from precipitation, gravity drainage) and geologic events (e.g., earthquake shaking, sedimentary basin compaction, plate subduction, thermal pressurization). Spatial and temporal variations in pore-water pressure, transmitted through groundwater, can provide the trigger to induce deformation or failure.

Pore-water pressure is integral to soil and rock mechanics and geotechnical engineering. Analyses of deformation and failure (such as poro-elastic continuum models or limit-equilibrium approaches) account for the ability of pore water to support normal stress and its inability to resist shear. When earth materials deform more rapidly than induced pore-water pressures can dissipate, they behave quite differently from materials undergoing slower deformation with free drainage. Terzaghi proposed the effective stress principle to separate stress in the solid framework from stress supported by pore water. According to this principle, solid-phase normal stress is reduced as pore-water pressure increases. Thus, the frictional strength of soil or rock (which is proportional to solid-phase normal stress on potential failure surfaces) can be reduced by locally increasing pore-water pressure, thereby inducing failure in landslides or faults. The effective stress principle has been extended to account for fluid suction stress that can strengthen partially saturated earth materials.

Bibliography

Bear, J., 1972. *Dynamics of Fluids in Porous Media*. New York: Dover Publications.

Ingebritsen, S. E., Sanford, W. E., and Neuzil, C. E., 2006. *Groundwater in Geologic Processes*, 2nd edn. Cambridge: Cambridge University Press.

Terzaghi, K., Peck, R. B., and Mesri, G., 1996. *Soil Mechanics in Engineering Practice*, 3rd edn. New York: Wiley.

Cross-references

Collapsing Soil Hazards
Creep
Debris Avalanche (Sturzstrom)
Debris Flow
Deep-seated Gravitational Slope Deformations
Earthquake
Hydrocompaction Subsidence
Induced Seismicity
Lahar
Land Subsidence
Landslide (Mass Movement)
Landslide Types
Lateral Spreading
Liquefaction

---

*United States Government

Mass Movement
Mud Volcano
Mudflow
Piezometer
Piping Hazards
Quick Clay
Quick Sand
Rock Avalanche (Sturzstrom)
Slide and Slump
Slope Stability
Solifluction
Subduction
Tectonic Tremor
Triggered Earthquakes

POST DISASTER MASS CARE NEEDS

Frank Fiedrich[1], John R. Harrald[2], Theresa Jefferson[3]
[1]Wuppertal University, Wuppertal, Germany
[2]Virginia Tech, Arlington, VA, USA
[3]Loyola University Maryland, Baltimore, MD, USA

Definition

Postdisaster mass care needs include all needs of disaster victims resulting from a disaster. Traditional mass care needs include medical care, sanitation, temporary shelter, food, water, clothing, and other emergency items, as well as collecting and providing information on victims to family members. In a wider sense, further emergency assistance services, housing support, and human services can also be included.

Introduction

According to Smith (1957), traditional mass care needs are concerned with the urgent needs of a large number of displaced and homeless people independent of the cause of displacement. Therefore, typical traditional mass care needs include mainly temporary shelter, clothing, food, as well as medical, nursing, and hospital care. Over the course of time, the concept has been extended, and today, mass care needs can be defined more generally as all needs of disaster victims resulting from a disaster. The mass care annex of the current National Response Framework of the United States (FEMA, 2008b) lists also services and special topics closely related to the traditional mass care needs. Among others, this includes:

- Emergency assistance. This includes family reunification; services and aid for special needs populations, like elderly or handicapped people; medical and special shelters (e.g., pet sheltering); donation management; and volunteer management.
- Housing. Support of the affected population related to housing options, including rental assistance, replace and repair, and loan assistance.
- Human services. These services are related to recovery from nonhousing issues. Examples are programs to replace destroyed personal property and help with disaster loans, and strategies against disaster unemployment.

This chapter focuses on the traditional mass care needs and the major organizations involved in meeting these needs.

International response to mass care events

Once a disaster has affected a country, the local government is responsible for the management of the event. Local laws, plans, and guidelines define the processes and agencies involved in different aspects of response and recovery. Nevertheless, many major events exceed the capability of the local organizations, and international assistance becomes necessary. Among the most important international bodies supporting the response to large-scale disasters are the United Nations (see *United Nations Organizations and Natural Disasters*), the European Union, and the Red Cross and Red Crescent Movement (for Red Cross, see *Red Cross and Red Crescent*). The United Nations and the European Union do not provide mass care assistance themselves, but they provide help in coordinating the international response activities through various programs. In the following sections, the UN and EU mechanisms will be discussed in more detail. Further information about international disaster management can be found, for example, in Coppola (2007).

United Nations (UN)

The UN assists disasters and humanitarian crisis through various programs and funds, specialized organizations, and other bodies. Among the most important organization to meet the initial needs is the Office for the Coordination of Humanitarian Affairs (OCHA). OCHA belongs to the UN Secretariat and has a coordinating role in humanitarian assistance which is then delivered through various local and international organizations. OCHA staffs a United Nations Disaster Assessment and Coordination (UNDAC) standby team which can swiftly be deployed upon request to an affected area. The UNDAC team supports the initial damage assessment and helps to estimate the immediate needs, including mass care needs. OCHA also offers an On-Site Operations Coordination Center (OSOCC) which works closely with the local government and acts as an information hub for the international assistance.

The United Nations' response to the event is based on a cluster system with initially 11 identified sectors. Based on the magnitude of the disaster and the conducted needs assessment, the UN initiates different clusters with predefined organizations – mainly from the UN – as cluster leads. Among the most important clusters related to mass care needs are the camp coordination/management and emergency shelter clusters; health; water, sanitation, and hygiene (WASH); food aid and nutrition; as well as logistics. Important UN organizations in this field include the World Food Program (WFP), the World Health Organization (WHO), the United Nations Children's Fund

(UNICEF), and the Office of the United Nations High Commissioner for Refugees (UNHCR).

European Union (EU)

The two most relevant EU initiatives related to mass care are the Community Civil Protection Mechanism and the Commission Humanitarian Aid department (ECHO).

The Community Civil Protection Mechanism is comparable to UN-OCHA as it plays a major role in coordinating international response efforts provided through the member states of the EU (plus Iceland, Liechtenstein, and Norway). Its Monitoring and Information Centre (MIC) serves as a communication hub for information related to the current situation and the ongoing relief efforts. Once the mechanism is activated through an appeal from the affected country, the MIC facilitates and matches the disaster needs of the affected country with the offers from the participating states. It also offers field experts who can be deployed to support damage and needs assessment. Since the mechanism is typically activated during the most urgent disaster phase, it also provides assistance in the traditional mass care fields.

ECHO's mandate includes to save and preserve life after disasters and to provide assistance to the people in need. ECHO is largely a donor organization and focuses on funding short- and longer term activities and programs. ECHO monitors the progress of the funded activities and promotes the coordination among response organizations. Providing food assistance after disasters has a high priority for ECHO, but ECHO also offers support through other mass care items like toilets, water sanitation equipment, or building materials. For example, in the aftermath of cyclone Nargis in 2008, ECHO provided 10 million euro for food aid and 29 million euro for other humanitarian assistance (EU, 2009).

Estimating mass care needs

In order to provide fast and efficient help to the affected population, it is necessary to have a good understanding of the damages and the resulting mass care needs for goods and services. Until now, the needs estimates are largely based on damage and needs assessment missions after an event, but some comprehensive models to support the development of pre-event plans do also exist.

Post-disaster needs assessment

Based on experiences from past disasters and the setup/operation of camps in conflict zones, a variety of guidelines and estimates are available. The Sphere Project (2004), for example, provides guidelines, estimates, and leading principles for camp management. USAID's Field Operations Guide (OFDA, 2005) and UNDAC's Field Handbook (United Nations, 2000; NRC, 1989; IFRC, 2000; State of Florida, 2005) provide also valuable information related to mass care needs. According to these sources, some important planning figures related to mass care are:

- Total shelter space per person: 480 square feet per person (this includes space for all shelter-related infrastructures).
- Sleeping space: 60 square feet per person.
- Cots and blankets: 1 per person.
- Toilets: 1 toilet per 20–40 persons within a maximum walking distance of 1 min.
- Sinks: 1 sink/tap per 80–200 persons within 2-min walking distance.
- Garbage: 1 30-gallon refuse container for every 50 persons.
- Ice: dependent on the weather eight pounds of ice per person.
- Minimum water requirements: 5 gallons per person per day (1 gallon of drinking water, 2 gallons of water for washing and personal hygiene, 2 gallons of water for other requirements like cooking), for feeding centers approximately 9–10 gallons per inpatient per day, and for hospitals 11–20 gallons per inpatient per day.
- Food: approximately 2,100 calories per person per day. This equals two meals ready to eat (MRE) or approximately three pounds of fresh food (dependent on type of food).
- Shelter staff: it includes staff to run the shelter and to feed people. This number is dependent on the size and the purpose of the shelter.
- Medical supplies: various prepackaged sets exist, like WHO's NEHK 98, which includes medicines, disposables, and instruments, sufficient to support 10,000 people during a 3-month period.
- Clothing needs depend very much on social and cultural factors and the weather.

Scenario-based models for mass care planning

Prior to a disaster, models and scenarios can be used to estimate the impact of a disaster, including estimates for mass care needs. Once realistic scenarios have been identified and calculated, the resulting needs estimates can be included in response plans. During the immediate aftermath of an event, these initial numbers can be used as initial planning figures (Harrald et al., 2007).

While many models and methods exist for different aspects of scenario modeling, not many comprehensive frameworks exist. Among those, HAZUS is probably the most advanced system to date. HAZUS is the *Federal Emergency Management Agency (FEMA)*'s damage and loss estimation methodology for natural disasters. HAZUS is implemented as a software system based on ESRI's *Geographic Information Systems (GIS)* ArcGIS. Originally developed for earthquake risk assessment in the United States, the current release HAZUS-MH4 covers methodologies for earthquakes, hurricanes, and floods. Dependent on the hazard, it includes detailed scientific modules for:

- Hazards (e.g., earthquake hazard)
- Direct damages (e.g., building and infrastructure damages)

- Indirect damages (e.g., fire and hazardous material release)
- Direct losses (e.g., repair costs, casualties, and short-term sheltering needs)
- Indirect losses (e.g., economic losses) (FEMA, 2008a)

HAZUS allows the development of disaster scenarios, and although the focus is on planning and mitigation, it can also be used during response and recovery. While HAZUS is currently only available for the US territories, initiatives like the Global Earthquake Model (GEM) aim to create comparable global scale models and software systems (GEM, 2009).

Of specific interest for mass care planning are the casualty and short-term shelter models since they allow the estimation of scenario-specific mass care needs.

HAZUS's earthquake casualty model estimates casualty numbers due to earthquake damages in four different casualty severity classes: slightly, moderately, severely, and killed/mortally injured. The calculated number and types of casualties are largely based on the time of day of the event, the population distribution, and the estimated structural damages. Although HAZUS does not distinguish between detailed injury types (e.g., blunt trauma), the estimated numbers provide a very good basis for the resulting medical needs. Since HAZUS also calculates damages to hospitals, electricity, and water networks, the scenarios allow the performance of gap analyses to compare the available response capacities with the estimated medical needs.

HAZUS's displaced households and short-term shelter needs model is based on a multi-attribute model originally developed by Harrald et al. (1992) for the American Red Cross. The HAZUS model assumes that displacement is solely based on structural damage to residential buildings. Since a subset of the displaced population may stay with friends and family or rent apartments, the model assumes that only a portion of the displaced households will seek public shelter. While this decision depends on a variety of socioeconomic and demographic variables (Harrald et al., 2000), HAZUS uses ethnicity and income as major determining factors (FEMA, 2008a). The shelter population requires the full range of mass care, but other displaced population may only need a subset of mass care goods and services. On the other side, it must also be noted that severe and long-term damage to water and electricity networks may not allow for timely repair and force people to leave their homes and stay in shelters. Therefore, a recent FEMA-funded study on catastrophic earthquakes in the New Madrid Seismic Zone extended the HAZUS methodology and uses the following population categories for mass care planning (Elnashai et al., 2009, pp 37):

- Shelter-seeking population: this includes the people who are seeking shelter directly after the event. In the days following the event – dependent on socioeconomic and demographic factors – this number is increased by a subset of the population without access to water and electricity. Shelter-seeking population requires water, food, medical care, and shelter-related items and services (blankets, toilets, etc).
- At-risk population: the at-risk population includes the displaced population immediately after the event. During the following days, the people who still live without access to water and electricity can be added.

Once the shelter and at-risk populations are estimated, the post-event expert rules of thumb can be used as an approximation for planning (see section *Post Disaster Mass Care Needs*).

Missing persons registries

Catastrophic incidents often disrupt families. As a consequence, family members do not know the current status and location of their relatives. One of the key problems in mass care is to help people inside and outside the affected area to find information about the status of friends and family members. Shelters often use bulletin boards to publish names and additional information of the people living in the camps. Morgues provide lists of photos and – if available – names of the dead. Additional bulletin boards allow family members to inquire about missing people. Although paper lists are still frequently used, nowadays, a variety of Web-based systems allows searching as well as publishing and requesting status information of missing persons through one single system. The International Committee of the Red Cross (ICRC) provides probably the most widely accepted system. Their "Family Links" Web site helps those separated by conflict or disaster to find information about their loved ones in order to restore contact (ICRC, 2010). In case of a major event, the IRCR sets up a subsite which allows searching and publishing of information related to a specific event. A similar approach is used by Google. Their "Person Finder" Web site setup after the 2010 Haiti earthquake lists more than 55,000 entries (Google, 2010). Google initiates comparable sites for other major events.

Conclusions

Successful response to large-scale and catastrophic events depends very much on how fast professional mass care is available. While the domain of mass care is still dominated by national and international response organizations, field experts, and emergency management professionals, some scientific approaches exist or are currently under development. Reliable damage and loss estimation methodologies provide the framework for the development of mass care-related plans. In addition, the emerging research field of humanitarian logistics seeks to provide methods to improve the distribution of goods during the response to complex disasters.

Bibliography

Coppola, D. P., 2007. *Introduction to International Disaster Management*. Amsterdam: Butterworth-Heinemann.

Elnashai, A., Jefferson, T., Fiedrich, F., Cleveland, L. J., and Gress, T., 2009. *Impact of New Madrid Seismic Zone*

Earthquakes on the Central USA, Vol. I, MAE Center Report No. 09-03, Mid-America Earthquake Center, Urbana, IL.

European Union (EU), 2009. *Annual Report on Humanitarian Aid*. DG for Humanitarian Aid (ECHO), Brussels.

Federal Emergency Management Agency (FEMA), 2008a. *HAZUS-MH MR3 Technical Manual*. Washington, DC.

Federal Emergency Management Agency (FEMA), 2008b. *Emergency Support Function #6 – Mass Care, Emergency Assistance, Housing, and Human Services Annex*. Washington, DC.

Global Earthquake Model, 2009. *Global Earthquake Model: A Uniform, Independent Standard to Calculate and Communicate Earthquake Risk Worldwide*. GEM Foundation, Pavia.

Google, 2010. *Person Finder: Haiti Earthquake*. http://haiticrisis.appspot.com.

Harrald, J. R., Fouladi, B., and Al-Hajj, S. F., 1992. Estimates of demand for mass care services in future earthquakes affecting the san francisco bay region. Prepared by George Washington University for the American Red Cross Northern California Earthquake Relief and Preparedness Project (NCERPP), 41 pp. plus appendices.

Harrald, J. R., Renda-Tanali, I., Bettridge, M., and Perkins, J. B., 2000. Cost estimate model of initial mass care needs following catastrophic earthquakes affecting the San Francisco bay area. In Perkins, J. B. (ed.), *Preventing the Nightmare: Post Earthquake Housing Issue Papers*. Oakland, CA: Association of Bay Area Governments.

Harrald, J. R., Jefferson, T. I., Fiedrich, F., Sener, S., and Mixted-Freeman, C., 2007. A first step in decision support tools for humanitarian assistance during catastrophic disasters: modeling hazard generated needs. In *Proceedings of the 5th ISCRAM Conference*, Washington, DC, pp. 51–56.

International Committee of the Red Cross (ICRC), 2010. *Family Links*. http://www.familylinks.icrc.org.

International Federation of Red Cross and Red Crescent Societies (IFRC), 2000. *Disaster Emergency Needs Assessment*, Geneva.

National Research Council (NRC), 1989. *Recommended Dietary Allowances*, 10th edn. Subcommittee on the Tenth Edition of the Recommended Dietary Allowances, Food and Nutrition Board, Commission on Life Sciences, National Academy Press, Washington, DC.

Office of Foreign Disaster Assistance (OFDA), 2005. *Field Operations Guide for Disaster Assessment and Response*, version 4, Washington, DC.

Smith, D. W., 1957. Emergency mass care. *The Annals of the American Academy of Political and Social Science*, **308**, 118–131.

Sphere Project, 2004. *Humanitarian Charter and Minimum Standards for Disaster Response*. The Sphere Project, Geneva.

State of Florida, Unified Logistics Section, State Emergency Response Team, 2005. *Bulk Distribution of Resources*. Logistics Technical Bulletin, 1:1, p. 3.

United Nations, 2000. *UNDAC Field Handbook*. Office for the Coordination of Humanitarian Affairs (OCHA), United Nations Disaster Assessment and Coordination, 3rd edn.

Cross-references

Casualties Following Natural Hazards
Civil Protection and Crisis Management
Damage and the Built Environment
Disaster Relief
Emergency Management
Federal Emergency Management Agency (FEMA)
Hospitals in Disasters
Hurricane Katrina
Megacities and Natural Hazards
Natural Hazard in Developing Countries
Red Cross, Red Crescent, International Federation of
United Nations Organisation and natural disasters

POSTTRAUMATIC STRESS DISORDER (PTSD)

Fran H. Norris
Dartmouth Medical School, National Center for PTSD, White River Junction, VT, USA

Definition

Posttraumatic stress disorder (PTSD) is a psychiatric condition that occurs following a traumatic event and is characterized by a complex constellation of reexperiencing, avoidance/numbing, and arousal symptoms.

Discussion

Although the term is often used loosely to describe various manifestations of disaster-related distress, PTSD is actually a highly specific anxiety disorder that is diagnosed according to criteria established by the American Psychiatric Association (1994). Criterion A, the trauma criterion, requires that the individual: (1) has experienced, witnessed, or been confronted with an event that involves actual or threatened death or physical injury and (2) responded to the event with intense fear, helplessness, or horror. Whereas all disasters may be stressful, they may not necessarily be traumatic according to this definition. Criterion B, reexperiencing, requires that the person has experienced recurrent and intrusive recollections of the trauma, distressing dreams, subjective feelings of reliving the event, or psychological or physiological distress upon exposure to reminders. Criterion C, avoidance and numbing, requires the presence of at least three of the following: efforts to avoid thoughts or feelings associated with the trauma, efforts to avoid activities, people or places associated with the trauma, inability to recall, diminished interest in activities, estrangement from others, restricted affect, or sense of foreshortened future. Criterion D, arousal, is indicated by difficulty sleeping, irritability, difficulty concentrating, hypervigilance (a feeling of being on-guard), or exaggerated startle response; at least two of these symptoms must be present. In addition, the symptoms must last for at least one month (Criterion E) and result in clinically significant distress or impairment in social, occupational, or other important areas of functioning (Criterion F). Despite the complexity of PTSD, well-validated tools for assessing it exist (Wilson and Keane, 2004).

PTSD is the most commonly studied and observed mental health problem associated with major disasters. Reexperiencing and arousal are very common among disaster survivors, but numbing symptoms and functional impairment are much less so (Norris et al., 2002). The prevalence of PTSD varies widely across disasters depending on the extensiveness of injury, death, and destruction but, on average, it may be experienced by 30–40% of highly exposed individuals, 10–20% of rescue workers and first-responders, and 5–10% of general populations who encompass levels of exposure ranging from modest to severe (Galea et al., 2005). PTSD often co-occurs with other psychiatric conditions, such as depression and substance abuse.

Disaster-related PTSD often remits on its own. In general, PTSD takes a chronic course in about one third of persons who develop it (Kessler et al., 1995). PTSD can be treated with current evidence favoring cognitive behavioral therapies as the most effective approach (Foa et al., 2009).

Bibliography

American Psychiatric Association, 1994. *Diagnostic and Statistical Manual of Mental Disorders*, 4th edn. Washington, DC: Author.

Foa, E., Keane, T., Friedman, M., and Cohen, J., 2009. *Effective Treatments for PTSD: Practice Guidelines from the International Society for Traumatic Stress Studies*, 2nd edn. New York: Guilford Press.

Galea, S., Nandi, A., and Vlahov, D., 2005. The epidemiology of post-traumatic stress disorder after disasters. *Epidemiologic Reviews*, **27**, 78–91.

Kessler, R., Sonnega, A., Bromet, E., Hughes, M., and Nelson, C., 1995. Posttraumatic stress disorder in the National Comorbidity Survey. *Archives of General Psychiatry*, **52**(12), 1048–1060.

Norris, F., Friedman, M., Watson, P., Byrne, C., Diaz, E., and Kaniasty, K., 2002. 60,000 disaster victims speak, Part I: an empirical review of the empirical literature, 1981–2001. *Psychiatry*, **65**, 207–239.

Wilson, J., and Keane, T. (eds.), 2004. *Assessing Psychological Trauma and PTSD: A Practitioner's Handbook*, 2nd edn. New York: Guilford Press.

Cross-references

Casualties Following Natural Hazards
Cognitive Dissonance
Coping Capacity
Critical Incidence Stress Syndrome
Federal Emergency Management Agency (FEMA)
Hospitals in Disaster
Human Impact of Hazards
Integrated Emergency Management System
Livelihoods and Disasters
Marginality
Mortality and Injury in Natural Disasters
Perceptions of Natural Hazards and Disasters
Post Disaster Mass Care Needs
Psychological Impacts of Natural Disasters
Red Cross/Red Crescent
Sociology of Disasters
Uncertainty
Vulnerability

PRIMARY WAVE (P-WAVE)

Allison Bent
Natural Resources Canada, Ottawa, ON, Canada

Synonyms

Compressional wave; Longitudinal wave; P Wave

Definition

P waves are compressional elastic waves that travel through the Earth.

Discussion

P waves are seismic body waves meaning that they travel through the Earth's interior. The name primary waves stems from the fact that they are normally the first waves recorded by a seismograph. P waves typically travel at velocities of 6–7 km/s in the Earth's crust and at higher velocities in the mantle. The particle motion associated with P waves is the same as for sound waves in that it consists of a series of compressions and dilatations parallel to the direction of propagation of the wavefront. P waves are recorded by the vertical and radial components of seismographs. They are able to propagate through liquids and gases but at much slower speeds than through solids. P waves are usually small in amplitude relative to S waves and surface waves.

Bibliography

Aki, K., and Richards, P. G., 1980. *Quantitative Seismology Theory and Methods*. San Francisco: W. H. Freeman and Company.

Bolt, B. A., 1993. *Earthquakes*. New York: W. H. Freeman and Company.

Cross-references

Body Wave
Earthquake
Secondary Wave
Seismograph/Seismometer

PROBABLE MAXIMUM FLOOD (PMF)

Armand LaRocque
University of New Brunswick, Fredericton, NB, Canada

Synonyms

Extreme Flood; Maximum Flood

Definition

According to the US Federal Energy Regulatory Commission (2001), the Probable Maximum Flood (PMF) is the theoretically largest flood resulting from a combination of the most severe meteorological and hydrologic conditions that could conceivably occur in a given area. PMF is mainly used as a security assessment for existing dams and other impounding structures, and as a design criterion for proposed similar hydraulic structures, in order to avoid dam failures and catastrophic floods.

Discussion

Dams are often built to control floods in areas occupied by people and must be able to store floodwaters, in order to avoid costly disasters in life and economy. Computer models are used to assess the water storage capabilities of these hydraulic structures, without overtopping under flood conditions. The height of future floods is often predicted in reference with a time interval. For example, a 100-year flood level is considered as the level of a large flood having the chance of occurring once in a period of 100 years. However, historical records of significant floods since one century show that these unusual events may happen more often and cause dam failures and catastrophic destructions in floodplains farther downstream. For this reason, the PMF is currently more frequently used as a design criterion for more important dams, although it is related to a very rare and unlikely occurring event.

The computation method of PMF is based on the integration of the probable maximum precipitation (PMP), or the probable maximum storm (PMS), and the characteristics of the watershed upstream. The PMP is the theoretically largest height of precipitation that may fall on a particular location and for a given duration. The value of PMP can be calculated using rainfall records or estimated from meteorological models. In mountainous areas, the PMS is more frequently used than PMP. PMS is defined as the maximum precipitation that may occur if weather conditions, mainly winds and atmospheric moisture contents, are maximal, for a given location. Other meteorological factors, such as the maximum accumulation of snow and the fastest rate of snow melt, can also be considered. The highest value among PMP or PMS is then used to estimate the maximum water inflow in the drainage area. The hydrograph for PMF can then be compiled, taking into account some important characteristics of the drainage basin, including the soil type, the land use, the size and the shape of the watershed, and the average watershed slope.

Bibliography

Shalaby, A. I., 1994. Estimating probable maximum flood probabilities. *Journal of the American Water Resources Association*, **30**(2), 307–318.

United States Federal Energy Regulatory Commission, 2001. Determination of the probable maximum flood (Chap. VIII). In *Engineering Guidelines for the Evaluation of Hydropower Projects*. Washington (DC): United States Department of Energy, p. 121.

http://www.ferc.gov/industries/hydropower/safety/guidelines/eng-guide/chap8.pdf

Cross-references

Damage and the Built Environment
Evacuation
Flash Flood
Flood Deposits
Flood Hazard and Disaster
Flood Protection
Flood Stage
Floodplain
Floodway
Hydrograph, Flood
Jokulhlaup
Levee
Monsoon
Paleoflood Hydrology
Reservoir Dams and Natural Hazards
Usoi Landslide and Lake Sares
Vaiont Landslide, Italy

PROBABLE MAXIMUM PRECIPITATION (PMP)

Gerd Tetzlaff[1], Janek Zimmer[2]
[1]Universität Leipzig, Leipzig, Germany
[2]GFZ German Research Centre for Geosciences, Potsdam, Germany

Synonyms

Continuous Heavy Rain; Flash Flood; Rainstorm

Definition

The WMO defines Probable Maximum Precipitation (PMP) as "... the greatest depth of precipitation for a given duration meteorologically possible for a given size storm area at a particular location at a particular time of year, with no allowance made for long-time climatic trends." (WMO, 1986).

Discussion

Heavy precipitation and its adverse effects, mostly by floods, are reported from most parts of the world. The reduction of such effects, e.g., through engineering (dams, etc.), requires data on the magnitude and frequency of heavy precipitation events. The information is usually derived from a time series of observed past events. By their nature, such events are infrequent; often a frequency of one such event per 100 years is applied. Observational periods often comprise periods of only a few decades.

Therefore, methods are needed to estimate the probable maximum precipitation occurring in any given set of time-invariant climatic conditions. The approaches to obtain such PMP values comprise two main methods. The first method (WMO, 1986) uses the maximum observed precipitation over a river catchment area and the maximum water vapor content in the same area. The dew point temperature is often extrapolated to an event size with a frequency of one per 100 years. The vertically integrated water vapor content, as deduced from the dew point temperature, is then compared to the value measured during the precipitation event. The resulting adjustment factor is applied to the observed, hitherto maximum, precipitation value. The result is called PMP. The method can only give an estimate, because the processes contributing the second key factor in the formation of precipitation, the vertical velocity, are not considered other than in the nonquantitative method for the observed precipitation event.

To overcome this limitation, it is necessary to extend the concept of PMP to both key factors, which means addressing also the responsible lifting mechanism. By using complex, state-of-the-art numerical weather prediction models with idealized initial conditions, which should represent the maximum combination of water vapor and a suitable pressure and wind field, it is feasible to estimate the vertical velocity distribution in time and space. This is especially complicated within areas of convection, due to the vast diversity of possible size and organization. Given the 3D-distribution of water vapor and vertical velocity, a diagnostic approach assuming moist-adiabatic ascent yields the maximum rain rates for the given initial conditions (Tetzlaff, 2009).

Among the heaviest precipitation events recorded on Earth, orographic enhancement is frequently the cause. In this case, the maximum precipitation can be deduced quite reliably because of robust estimates of the vertical velocity that originates from forced upslope motion.

Bibliography

Tetzlaff, G., 2009. *Extreme Rain and Wind Storms in Mid-Latitudes.* Spring School on Fluid Mechanics and Geophysics of Environmental Hazards, Singapore (April 19–May 2, 2009).

World Meteorological Organization, 1986. *Manual for Estimation of Probable Maximum Precipitation*, 2nd edition, Operational Hydrology Report No. 1, WMO – No. 332, Geneva, ISBN 92-63-11332-2.

Cross-references

Challenges to Agriculture
Climate Change
Cloud Seeding
Debris Flow
Drought
El Nino/Southern Oscillation
Erosion
Flash Flood
Hurricane
Hydrograph, Flood
Lightning
Monsoon
Probable Maximum Flood
Storms
Thunderstorms
Waterspout

PSYCHOLOGICAL IMPACTS OF NATURAL DISASTERS

James M. Shultz[1], Yuval Neria[2], Andrea Allen[3], Zelde Espinel[1]
[1]University of Miami Miller School of Medicine, Clinical Research Building Suite 1512, Miami, FL, USA
[2]Columbia University, New York State Psychiatric Institute, New York, NY, USA
[3]Barry University, Miami Shores, FL, USA

Synonyms

Disaster behavioral health; Disaster mental and behavioral health; Disaster mental health

Definition

Disaster Mental and Behavioral Health: Professionals in the rapidly emerging field of *disaster mental and behavioral health*, focus on the interconnected psychological, emotional, cognitive, developmental, and social influences on behavior and mental health and the impact of those factors on preparedness, response, and recovery from disasters and traumatic events.

Introduction

Reverberating mental and behavioral health consequences occur when the physical forces of natural disaster collide with a vulnerable human population. In fact, the psychological impacts are more expansive in scope, more extended in time, and frequently more debilitating in severity than the injurious physical impacts of natural disaster (Shultz et al., 2007b).

The importance of "psychological impacts of natural disasters" becomes immediately apparent by examining four attributes of mental and behavioral health consequences. Aligned with the theme of this *Encyclopedia*, the interconnection between hazards, psychological stress, and mental health outcomes is explored. Next, considering persons "in harm's way," populations at elevated risk for mental health impacts are differentiated by disaster phase. Psychological reactions and psychiatric disorders that may occur following exposure to a disaster are described. Given the significant psychiatric disease burden in the wake of major natural disasters, approaches to behavioral triage, referral, and psychological treatment are presented. The range of possible psychological trajectories for survivors is examined.

Mental and behavioral health consequences in disasters

In the Disaster Ecology Model (Shultz et al., 2007b), "a disaster is characterized as an encounter between forces of harm and a human population in harm's way, influenced by the ecological context, that creates demands exceeding the coping capacity of the affected community." This model serves as the basis for integrating physical and psychological dimensions of natural disaster impact and illustrates the intriguing interplay between hazards and the human population grappling with the rampaging forces of harm (Figure 1).

Psychological consequences of natural disasters can be portrayed across four salient dimensions of scope, severity, duration, and disaster type.

Widespread Scope. Psychosocial consequences of disasters are wide-ranging and pervasive. More persons are affected psychologically than are harmed physically. The "psychological footprint" of disaster is larger than the "medical footprint" (Shultz et al., 2007a).

While many persons in the strike zone will escape physically unharmed from the ravages of a natural disaster, they will nevertheless suffer stress reactions, distress, fear, and possibly bereavement and grief. Persons who are

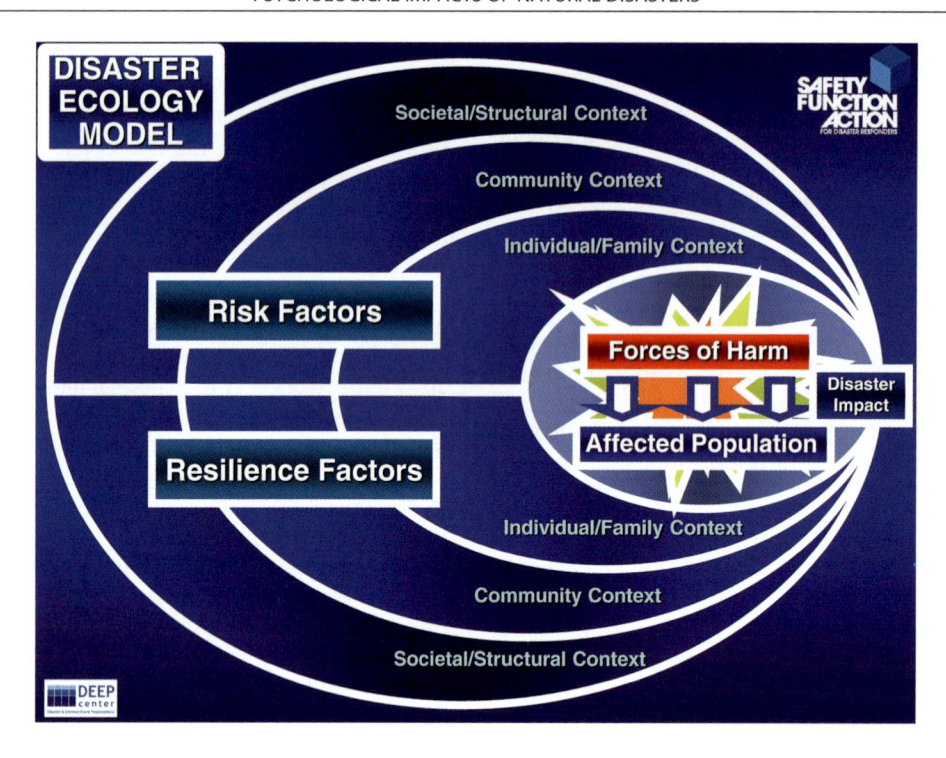

Psychological Impacts of Natural Disasters, Figure 1 Disaster ecology model: forces of harm and the human population in harm's way.

physically injured will also experience a psychological overlay of "injury-related distress" and elevated risk for subsequent development of posttraumatic stress disorder (PTSD) (Zatzick, 2007). In natural disasters characterized by high mortality, such as the 2004 Southeast Asia tsunami or the 2010 Haiti earthquake, the disaster-affected community will experience prolonged grief, with many survivors dealing with the loss of multiple close friends and family members. Under circumstances of extensive destruction, population-wide distress will arise from population displacement, loss of resources, and scarcity of basic needs.

Spectrum of Severity of Psychological Reactions. Almost all persons exposed to a natural disaster will experience increased levels of fear and distress during times of overt danger (Butler et al., 2003). In fact, the threat of disaster can trigger stress even in the absence of impact. Consider the fear responses and frantic preparatory activities in coastal communities during the warning phase for an approaching hurricane that ultimately remains over water without making landfall (Shultz et al., 2007b). In many cases, these psychological reactions are brief and relatively mild, allowing persons to quickly rebound to full functioning without need for psychological support. In contrast, a subset of disaster-exposed persons will be distressed to the point of making detrimental behavior changes, such as surging the local healthcare system (Ursano et al., 2007). Some individuals may develop psychiatric disorders following encounter with a natural disaster. Best studied is posttraumatic stress disorder (PTSD), but other psychiatric disorders may co-occur with PTSD ("comorbidities"). Among these, major depressive disorder (MDD) and generalized anxiety disorder (GAD) are commonly observed among disaster survivors.

The proportion of disaster-affected persons at each point along this spectrum of severity – from disaster stress to behavior change to psychiatric disorder – generally corresponds to the intensity of exposure to the "forces of harm" in a natural disaster.

Range of Duration. Psychological stress reactions and fear-driven behavioral responses ramp up during the disaster warning phase and escalate during disaster impact, the period of overt danger when destructive forces of harm are operating. However, in the aftermath, psychological reactions do not disappear even when physical danger ceases. The reason is that loss and change are prominent features of the post-disaster environment. The hardships of enduring physical destruction, scarcity of basic needs, displacement, community-wide disruption of services, loss of resources, and painful rehabilitation from physical injury collectively act to maintain or amplify the stress level. For large-scale disasters, the protracted period of reconstruction perpetuates chronic stress.

Type of Disaster. Natural disasters are notable for their global frequency and diversity, generating both physical and psychological effects for millions of world citizens annually. Natural disasters are relatively common, familiar, and predictable. The degree of psychological distress and the extent of mental health consequences found in survivors of these "acts of nature" tend to be less compared with victims of human-generated disasters, particularly human-perpetrated intentional acts of violence (Norris et al., 2002).

Forces of harm: Hazard characteristics related to psychological impacts

In this *Encyclopedia*, dedicated to the panoply of natural hazards, it is instructive to examine how hazard characteristics commingle and synergize to influence the extent, severity, and duration of psychological reactions (Shultz et al., 2007a, b). Predictably, the degree of psychosocial impact is directly proportional to the degree of physical harm and destruction. Five hazard descriptors predict the extent of mental and behavioral health ramifications.

First, the *absolute magnitude* (or *intensity*) of the forces of harm largely determines the extent of damage, destruction, displacement, death, and injury, as well as the amplitude of stress and trauma experienced by the disaster-affected population. Second, the *duration* of exposure to threat or overt forces of harm predicts the span of time during which survival stress responses are activated. Third, the *frequency* of discrete disaster impacts relates to psychological trauma; a sequence of multiple strikes tends to be more devastating than a single event. For example, repeated strong aftershocks following a major earthquake can provoke widespread fear reactions. As an illustration of multiple sequential strikes, in 2004, a series of four hurricanes made landfall in the state of Florida within a period of 3 months, potentiating stress levels and inducing statewide "hurricane fatigue." Fourth, *proximity* to the geographical "epicenter" of destruction forecasts the severity of exposure and the attendant psychological effects. Investigations of natural disasters commonly reveal gradations of impact, closely corresponding to the extremity and extent of physical and psychological harm sustained. Fifth, the *geographic scope and scale* defines both the expanse of territory impacted and the numbers of persons comprising the disaster-affected population; psychological consequences are experienced both individually and collectively.

Hazards as disaster stressors

Exposure to natural hazards during disaster impact, followed by the confrontation with loss and change in the wake of disaster, presents survivors with an array of disaster stressors. These "forces of harm" – exposure, loss, and change – trigger stress reactions, alter behavior patterns, and may lead to severe psychiatric outcomes for a subset of disaster-affected persons. Table 1 provides examples of disaster stressors in each of these three categories (Shultz et al., 2007a, b).

Hazard conditions that generate severe psychological impacts

According to Norris and colleagues (2002), the majority of declared disasters do not produce significant psychological repercussions because numbers of injuries and deaths are limited, the degree of destruction is not overwhelming, and community social structures remain intact. These investigators assert that pronounced disaster mental and behavioral health impacts are generally restricted to the subset of high-profile disasters that possess two or more of the following four characteristics: (1) large numbers of injuries and/or deaths, (2) widespread destruction and property damage, (3) disruption of social support and ongoing economic problems, and (4) intentional human causation. For the current focus on natural disasters, intentional causation is not applicable, so some combination of mass mortality, extreme damage, and social dislocation must occur to catapult psychological consequences to the forefront.

Psychological Impacts of Natural Disasters, Table 1 Disaster stressors associated with exposure to hazard, loss, and change

Exposure to hazards	Loss	Change
Perceived threat of harm	Bereavement	Disruption of services
Disaster warning (or) Lack of warning	Separation from loved ones	Physical displacement
Shopping/stockpiling	Physical harm, debility, pain	Separation from essential health services/medications
Evacuation	Loss of function	Lack of utilities
Sheltering	Loss of home	Lack of transportation
Perception of threat to life	Loss of worksite	Lack of communications
Exposure to physical forces of disaster impact	Property damage	Unemployment, job change
Personal physical harm	Lack of basic necessities	School closure
Witnessing	Loss of valued possessions	Disruption of community
Widespread destruction	Loss of social support	Community-wide grief
Mass casualties	Resource loss	Shortages, rationing
Death/injury to others	Financial loss	Refugee conditions
Exposure to	Loss of employment	Social violence
Grotesque scenes	Loss of independence	Poverty
Noxious agents	Loss of personal control	Disease outbreaks

Threat, harm, loss, change: Psychological impacts throughout the disaster life cycle

Disaster stress permeates all phases of the disaster life cycle. Threat alone is capable of stimulating the stress response even if physical harm does not materialize. Consider hundreds of county fair-goers frantically fleeing as a tornado funnel descends from a cloud bank overhead. Before reaching the ground, the funnel hovers, then retracts and dissipates. No touchdown, no damage, no physical injury ensues, yet the threat alone was able to set off palpable psychological stress reactions and provoke potentially injurious crowd behavior.

When disaster actually strikes a human population, physical harm and destruction is guaranteed to be accompanied by psychological distress. In the post-disaster environment, ongoing exposure to hazards, compounded by the profound realization of loss and change, combine to produce powerful psychological effects. Postimpact adversities are experienced by some survivors as more difficult than the disaster event itself. Mental health sequelae typically persist long after the physical threats abate in the postimpact phase. Table 2 contrasts physical harm and psychological impact in relation to the phases of the disaster life cycle.

In harm's way: Populations affected psychologically in disasters

During a natural disaster, the degree of psychological impact generally relates to the extent of exposure to the physical forces of harm, but a further distinction comes into play. "Direct victims" experience intense exposure to the forces of harm (Galea and Resnick, 2005; Norris and Wind, 2009). Some direct victims are physically injured. During impact, many direct victims perceive an imminent threat to life ("I thought I was going to die."). Survivors bear witness to scenes of massive physical destruction, death, and egregious harm to others. Posttrauma memories may be studded with images that are grotesque and deeply troubling. Survivors may experience traumatic bereavement due to the death, on-scene, of one or more loved ones.

Yet psychological distress also extends to "indirect victims" (Galea and Resnick, 2005), persons who are typically outside the geographical perimeter of the disaster footprint but are socially connected to the direct victims. While the mental and behavioral health impact tends to be less severe, the number of indirect victims far exceeds numbers of direct victims. During the 2010 Haiti earthquake, direct victims were concentrated around the focus of destruction, the capital city of Port au Prince. Indirect victims extended throughout the entire island nation and beyond, encompassing large Haitian immigrant communities in Miami and New York City, and Caribbean nations.

Spirals of psychological impact enfold direct impact victims and many rings of indirect victims. Indirect victims include family members, friends, and neighbors of the direct victims; colleagues and coworkers; and those who witnessed harm, death, and disturbing scenes. Moreover, both professional and volunteer disaster responders will experience psychological repercussions. The "Population Exposure Model" (USDHHS, 2004) aptly captures the concentric nature of the psychological impact of disasters. In general, the closer the person is to "ground zero" in a natural disaster, both physically and socially, the greater will be the psychological distress.

Populations at higher risk for psychiatric disorders by disaster phase

When dealing with disaster-affected populations, one important consideration is defining who is most likely to suffer serious psychological consequences (Shultz et al., 2007a; Watson and Shalev, 2005). Individuals and populations at elevated risk can be identified at each phase in the disaster life cycle.

Some risk factors that predict increased likelihood of unfavorable psychological outcomes following disaster are clearly definable in the pre-event period. Among these risk factors are demographic descriptors such as gender (higher risks for women), race/ethnicity (higher rates for ethnic and marginalized minorities), and socioeconomic status (increasing risk with decreasing SES). Pre-disaster functioning and psychiatric history are among the most forceful predictors of post-disaster symptoms (Norris et al., 2007). History of pre-disaster psychiatric diagnosis relates directly to risk for post-disaster PTSD. Past history of trauma or substance use elevates the postimpact risk for psychological consequences. Likewise, chronic physical health conditions and disability status add to risks for psychological distress.

However, many disaster survivors with no pre-event risk factors will experience significant psychological consequences. The traumatizing experiences of living through

Psychological Impacts of Natural Disasters, Table 2 Forces of harm: physical and psychological effects by disaster phase

Disaster phase	"Forces of harm"	Physical harm/Destruction	Psychological impact
Inter-disaster	Exposure to nonspecific threat (even without impact)	NO	YES
Preimpact/Warning	Exposure to specific threat	NO	YES
Impact	Exposure to hazard, harm, loss	YES	YES
Early postimpact	Exposure to hazard, loss, change	YES	YES
Late postimpact	Exposure to loss and change	NO	YES

Psychological Impacts of Natural Disasters, Table 3 High-risk groups for psychological consequences by disaster phase

Pre-disaster phase

Population Demographics
Female gender
Children
Older adults
Low socioeconomic status
   Minimal education
   Economic disadvantage
   Unemployment
 	
Culture, race, ethnicity, language
Minority status
   Minority culture, race ethnicity
   Marginalized culture
Recent immigrant
Limited proficiency in dominant language
 	
Family context characteristics
Adult with children
Single head of household
Child with dysfunctional parent
Family instability
Domestic violence
 	
Social support/Adaptive skills
Limited coping skills
Limited social support network
	Psychiatric/Psychological Health
	Serious and persistent mental illness
	History of trauma
	Psychiatric diagnosis
	Substance abuse diagnosis
	
	Physical health
	Pregnancy
	Pre-existing chronic disease
	Disease requiring life-sustaining treatment
	Disease requiring essential medications
	Immunosuppression
	
	Disability status
	Physical limitation and disability
	Learning/language disability
	Limitation of intellectual skills

Disaster impact phase

Direct victims Experiencing the physical forces Physical injury Perception of threat to life Extreme fear, horror, trauma Witnessing destruction Witnessing injury or death of others Witnessing grotesque scenes	*Indirect victims* Connection to disaster-affected community Inability to help Survivor guilt Secondary trauma *Bereaved victims* Death of loved one Death of close friends, neighbors

Post-disaster phase

Injured Pain, rehabilitation, physical disability *Disaster-impacted community at large* Massive destruction Desperate search for missing persons Lack of basic needs Loss of community function Disruption of services Diminished social support Marital stress *Unemployed* Economic hardship Lack of career identity Inability to provide for dependents	*Bereaved (loss of loved one in disaster)* Complicated grief *Disaster-displaced persons* Loss of home Lack of shelter and safety Relocation stress Temporary housing Crowding *Psychiatric "Peritraumatic" symptoms* Dissociation Panic, extreme distress Depression

the onslaught of disaster and encountering the extreme adversities in the aftermath tend to "reshuffle the deck." Planning for mental and behavioral health support for disaster-affected populations must consider the reality that some survivors with no salient preimpact risk indicators will urgently require focused help and psychological care following the disaster. These individuals emerge as members of a disaster-created special needs population (Table 3).

Psychological reactions to natural disasters

As first described under the topic of "spectrum of severity," from a public health, population-focused vantage point, no one goes through a traumatic event unchanged (Butler et al., 2003). Almost all persons exposed to a disaster are affected psychologically. Survivors will exhibit an array of distress reactions across all domains of human function (Table 4).

The intensity and variety of reactions vary based on individual differences as well as the nature of the exposure to disaster. Psychological reactions are time-phased and transient. For example, a person whose primary emotional response is extreme fear at the moment of forceful earthquake tremor may exhibit rage and anger later that same day, and possibly helplessness and despair several days into the postimpact phase.

For most persons, disaster stress responses will be relatively brief and transient, followed by return to pre-disaster levels of functioning (Butler et al., 2003). However, some persons will be distressed to the point of making behavioral changes such as avoidance of sleeping indoors after an earthquake, flocking to local healthcare centers for vague complaints, or increasing use of alcohol.

Finally, for a minority of exposed persons, symptoms will progress and persist to the point where a psychiatric diagnosis is warranted. For these individuals, the emotional sequelae of disasters may be enduring. A range of post-disaster mental health problems has been documented including posttraumatic stress disorder, major depressive disorder and complicated grief disorder, substance abuse, and physical illness.

Posttraumatic stress disorder (PTSD). PTSD is a common, frequently debilitating, psychiatric disorder among trauma-exposed populations. PTSD is classified as an anxiety disorder with three defining symptom domains: reexperiencing, avoidance, and hyperarousal (APA, 2004).

Reexperiencing entails terrifying flashbacks, disturbing thoughts, and intrusive memories of the disaster event. Avoidance involves both emotional "numbing" and behavioral choices to refrain from visiting locations or coming upon reminders of the trauma. Avoiding trauma reminders requires constant focus and can detour the survivor's lifestyle by limiting mobility, shrinking the social network, and diminishing the enjoyment of usual activities. Hyperarousal symptoms include a racing, fight-or-flight physiology; inability to calm or rest; extreme alertness; and an exaggerated startle reflex.

Psychological Impacts of Natural Disasters, Table 4 Disaster-induced stress and distress reactions across six dimensions of human function

Physical	Emotional	Cognitive
Adrenalin "rush"	Fear and terror	Disorientation
Increased heart rate	Horror and dread	Confusion
Increased respirations	Anxiety, emotional distress	Decreased concentration
Sweating	Excessive worry	Distractibility
"Butterflies" in stomach	Anger or rage	Memory problems
Muscle tension	Irritability	Reduced attention span
Headaches	Sadness, crying	Impaired problem solving
Dizziness	Depression	Difficulty making decisions
Heart palpitations	Helplessness	Difficulty setting priorities
Gastrointestinal problems	Hopelessness	Calculation impairment
Exaggerated startle reflex	Grief	Loss of objectivity
Tremors, muscle twitching	Guilt	Disbelief
Shortness of breath	Overwhelm	Distorted thinking
Visual disturbances	Apathy, denial	Inaccurate perceptions
Chronic fatigue	Numbing, shutting down	Alternatively:
	Alternatively:	Increased attention
Behavioral	Feeling heroic, euphoric	Heightened focus on tasks
Changes in sleep habits	Feeling "invulnerable"	
Inability to rest and relax		Spiritual
Changes in diet	Social	Crisis of faith
Weight loss or weight gain	Withdrawal	Questioning values
Hypervigilance	Isolation	Directing anger toward God
Impaired job performance	Interpersonal conflict	Cynicism
Academic problems	Hostility	Loss of meaning
Absenteeism	Aggression	Interpreting disaster as punishment
Inappropriate humor	Blaming	Alternatively:
Increased alcohol use	Dependency	Religious conversion
Increase medication use	Greater need for comfort	Increased reliance on faith
Decreased interest in pleasurable activities	Difficulty giving support	Increased religiosity
Avoidance of activities or places that trigger memories of disaster	Difficulty receiving support	Increased use of prayer
		Increased use of ritual
		Bargaining with God

To receive a definitive PTSD diagnosis, all three symptom clusters must be present for a period of 30 days following the traumatic event and cause impaired functioning. Up to 3 months posttrauma, the diagnosis is termed "acute PTSD" and thereafter, beyond 3 months, "chronic PTSD." For some disaster survivors, posttraumatic stress symptoms are evident across multiple domains within the first 30 days; some of these individuals meet criteria for a diagnosis of acute stress disorder (ASD), a strong predictor of future progression to PTSD beyond the 30 day threshold.

PTSD is the most investigated, and the most central, psychopathology in the aftermath of disasters (Norris et al., 2002; Breslau et al., 2004; Galea et al., 2005; Neria et al., 2008). Strong predictors of developing PTSD include serious physical injury, imminent threat to life, severe property damage, and high death toll.

The burden of PTSD among persons who were exposed to disasters is significant. Overall, studies of natural disasters report PTSD prevalence rates ranging from 3.7% to 60% in the first 2 years after the disaster, with most studies reporting prevalence estimates in the lower half of this range (Neria et al., 2008). Fortunately, PTSD symptoms and disease burden generally decrease over time (Galea et al., 2003).

The risk of PTSD has been repeatedly shown to be associated with the severity of exposure to the disaster. The prevalence of PTSD among direct victims of disasters ranges from 30% to 40%. PTSD prevalence among rescue workers is lower (10–20%), while the range of PTSD rates in the general population is the lowest (5–10%). These differentials in prevalence rates correspond to the intensity of exposure, evidence of a dose-response relationship. To validate this dose-response relationship, researchers have compared groups with quantifiably different levels of exposure. Studies that compared survivors based on proximity to the "ground zero" point of maximum impact consistently found highest PTSD rates in persons closest to the epicenter (Schlenger et al., 2002; Neria et al., 2006).

Prevalence rates of PTSD are higher for human-generated acts of mass violence compared with natural disasters. This partially reflects distinguishing features of mass violence that make such events especially troubling psychologically; these acts are intentional, perpetrated, and unpredictable (Shultz et al., 2007a). But the observed lower rates of PTSD in natural disasters may also reflect differences in study design. Compared to focalized human-made events, major natural disasters create destruction over a sweeping expanse of territory. Studies of natural disasters tend to enroll survivors with varying gradients of exposure and a mix of direct and indirect victims, thus diluting the average "dose" of exposure and thus, potentially underestimating PTSD prevalence rates (Galea et al., 2005).

Mental health consequences, most notably PTSD, are not restricted to direct victims alone. Large numbers of indirectly exposed persons may also be affected. Though not exposed onsite, indirect victims may have sustained the loss of family members or close friends, experienced the destruction of personal property or homestead, or viewed intense coverage of the event through the media. While the dose of exposure is diminished, and symptoms tend to be less severe, indirect victims greatly outnumber direct victims. Therefore, a substantial burden of mental health impact be found within the large, broadly distributed population of those who were indirectly exposed (Galea et al., 2005).

Major Depressive Disorder (MDD). MDD is frequently diagnosed in individuals exhibiting persistent feelings of deep sadness, accompanied by additional symptoms such as loss of interest in activities they once enjoyed, decreased self-worth, guilt, sleep problems, and changes in appetite (APA, 2004), together leading to impaired function. While feelings of sadness and related symptoms of MDD are expectable following a disaster, for most individuals these reactions spontaneously remit as lost resources are replenished, daily activities are resumed, and future expectations brighten.

Several disaster-related stressors increase the risk for MDD in disaster survivors: death of a loved one, displacement, relocation, lack of social support, and being alone (Ahern and Galea, 2006; Kilic et al., 2006; van Griensven et al., 2006; Tak et al., 2007). Estimates of post-disaster MDD vary and frequently do not exceed expected rates of depression in the general population. MDD prevalence is higher for persons with more intense disaster exposure. For example, 14 months following an earthquake in Turkey, MDD prevalence was 16% for persons near the epicenter compared to 8% for persons away from the center (Başoğlu et al., 2007). Displacement away from home and community support also predicts MDD. Two months following the 2004 Southeast Asia tsunami, MDD prevalence for Thai survivors was 30% for displaced survivors and 21% for non-displaced individuals (van Griensven et al., 2006).

Complicated grief. Loss of a loved one in a natural disaster is one of the most psychologically devastating experiences. Traumatic bereavement, leading to complicated grief, is associated with a host of psychiatric disorders including PTSD and depression (Neria et al., 2007). Complicated (or prolonged) grief disorder (PGD) is a relatively new diagnosis, and different from normal grief in its extended duration and symptom profile (Horowitz et al., 1997). Correlates of PGD include severe functional impairment, decreased productivity, suicidality, and physical health problems (Lichtenthal et al., 2004; Neria et al., 2007). Yet to be explored is the interaction between disaster trauma and loss of a loved one (Neria and Litz, 2004);

many survivors of the 2010 Haiti earthquake experienced both. More research is needed to fully understand the relations between PTSD and complicated grief, and whether they differ in their risk and protective factors.

Substance use. Trauma exposure is often associated with increased substance abuse, either directly or indirectly, through increased substance use associated with PTSD. Existing research does not indicate that exposure to disasters results in a substantial increase in substance use. Reported post-disaster increases in substance use (tobacco, alcohol, drugs) are typically restricted to persons who were using these substances before the disaster. Rates of substance use post-disaster generally decline over time. The field would greatly benefit from well designed, prospective examination of the relationships between substance use and disaster-related psychiatric disorders.

Triage, screening, and referral

Indicators for referral to mental health evaluation

Among the range of responses to disasters, mental health consultation and evaluation is warranted for survivors displaying any of the following symptoms (Shultz et al., 2007b; Reissman et al., 2010):

- Inability to perform necessary everyday functions
- Disorientation (confused, unable to give name/date/time/place)
- Suicidal or homicidal thoughts, plans, or actions
- Domestic violence
- Acute psychosis (hearing voices, seeing visions, delusional thinking)
- Significant disturbance of memory
- Severe anxiety, extreme fear of another disaster
- Problematic use of alcohol, prescription or illicit drugs
- Depression (hopelessness, despair, withdrawal)
- Hallucinations, paranoia
- Serious developmental regression

Triage and screening

Disaster survivors who are most likely to experience severe psychological reactions, impairment of function, and potential psychopathology are those who have experienced: (1) intense exposure, (2) loss of a loved one, (3) major disruption of basic needs and services, (4) prior trauma, and (5) major life stressors (Reissman et al., 2010). Even small increments in prevalence rates of psychiatric disorders will stimulate a daunting surge of demand for mental health services. Lacking mechanisms for triage, screening, and coordinated care systems, distress will create impairment and impairment will progress unnecessarily to psychiatric disorders and dysfunction across life roles (Reissman et al., 2010).

A psychological triage system has been developed that uses color-coded triage tags and data transmission devices, and importantly, connects persons in need of more intensive mental health evaluation to community-based "disaster systems of care." This system has been applied during the postimpact phase following the 2004

Southeast Asia tsunami and the Laguna Beach, California wildfires (Reissman et al., 2010). In practice, this mental health triage system operates separately from medical triage at a later stage in response.

Expert panels have convened to devise and standardize efficient triage systems for emergency and disaster victims. Consensus has emerged regarding the five critical features that exemplify the "ideal mass casualty triage system" for use during disaster impact and in the immediate aftermath.

The ideal triage system possesses these characteristics:

1. *Rapid* (less than 1 min per patient)
2. *Scalable* (saving time by triaging large groups collectively as numbers of incoming casualties mount)
3. *Recurring* (triage must be updated with every patient encounter)
4. *Integrates both medical and behavioral triage*
5. *Includes resource-based criteria for exclusion from care* (accounting for the possibility of insufficient medical/behavioral resources to treat all critical patients immediately)

Recently introduced as the "SALT" protocol (Sort, Assess, Lifesaving Intervention, Treatment/Transport), the Model Uniform Core Criteria for Mass Casualty Triage were developed as a consensus guideline for mass casualty triage to achieve standardization when responding to a catastrophic health event (Lerner et al., 2011). This well-publicized model matches 4-for-5 on the key triage attributes. What is lacking in the SALT structure is the ability to seamlessly integrate real-time medical and behavioral triage.

Disaster mental and behavioral health intervention

Prevention

From the vantage of public health, the greatest leverage for decreasing the burden of injured, traumatized, and bereaved disaster survivors resides in the domain of prevention. Citizen disaster preparedness carries great potential for diminishing disaster impact in all aspects. Preventive behaviors include development of family, school, worksite, and community disaster plans with regular updates and periodic drills (Schmitz et al., 2009). Stockpiling and rotation of disaster supplies and creation of Go-Kits for home, automobile, and worksite will provide a critical supply of basic needs. Citizen participation on volunteer community disaster teams will enhance the repertoire of preparedness and response skills and build social support and community cohesiveness. When disaster is approaching, heeding disaster warnings and engaging in appropriate response actions (including preparation of home site, notification of key contacts, timely evacuation, and safe sheltering) have life-saving implications (Shultz et al., 2009). Psycho-education on expectable psychological reactions and positive coping strategies for persons exposed to disasters, combined with acquisition of effective stress management skills, are important components of psychological preparedness.

Despite the wisdom and utility of disaster preparedness and the potential for significantly decreasing physical and psychological harm when disaster strikes, the vast majority of citizens do not engage in these preventive behaviors. A critical future direction for psychological research is to examine strategies to motivate citizens to participate actively in family and community disaster preparedness.

Post-disaster

Post-disaster mental and behavioral health interventions can be classified according to timing and intention. As Bryant and Litz (2009) point out, it is important to sensitively gauge receptivity to receiving services in persons exposed to the effects of trauma. Therefore, a critical factor to be considered when choosing interventions post-disaster is when to implement intervention efforts so as to avoid unnecessary taxing of already scarce resources. Further, when choosing interventions post-disaster a clear delineation according to the intervention's intention is important. One of the major aims of early post-disaster intervention is to reestablish a sense of safety and calm, whereas intermediate and long-term interventions focus on acquiring coping skills and ameliorating psychopathological presentations.

Early intervention

Early intervention for psychological impacts of disasters is presently in a quandary. The longtime standard bearer, psychological debriefing, has been scientifically discredited. The putative successor, psychological first aid, has not gained traction with disaster responders and its efficacy, to date, remains unevaluated in the field (Litz, 2008).

Psychological debriefing. Introduced in the 1980s under the name "critical incident stress debriefing" (CISD), psychological debriefing was promoted as an intervention to decrease acute stress and to eliminate or inhibit delayed stress reactions among emergency and disaster response personnel (Mitchell, 1983). Authors later championed CISD as effective for use with a broad spectrum of survivors of trauma (Everly and Mitchell, 1999). CISD was repackaged as a pivotal component among a suite of related interventions collectively titled critical incident stress management or CISM. Popularity escalated with broad adoption by first responders worldwide. Litz (2008) describes CISD as the modal intervention strategy in the immediate posttrauma phase prior to 2002.

CISD is intended for use within 48 h of the traumatic incident and features brief education about trauma reactions. This is followed by a stepwise process in which survivors are prompted to disclose cognitive and emotional aspects of the traumatic event. Participants are explicitly asked to describe "the worst part" of the experience and their accompanying psychological and physical reactions

(Bryant and Litz, 2009). Widely adopted by emergency responders and military organizations internationally, the intuitive appeal of CISD is based on its easy-to-learn, easy-to-apply, peer-provided, protocol-driven format that fits well within responder work cultures (Litz, 2008).

Despite favorable perceptions among both practitioners and persons receiving CISD, when subjected to scientific scrutiny, CISD has not been shown to prevent PTSD nor to confer any favorable benefits upon recipients when compared to those who do not receive this intervention (McNally et al., 2003; Roberts et al., 2009). The scientific critiques of CISD focus on the potential for re-traumatization of persons exposed to disaster by requiring them to recount their own vivid experiences and to listen to the oftentimes horrific stories of others (McNally et al., 2003; Roberts et al., 2009; Bryant and Litz, 2009). The process and timing involved in application of CISD may short-circuit the natural healing process, lead to ruminating thought processes, and provoke psychological hyperarousal in a manner that locks-in, or consolidates, the traumatic memories.

Under certain circumstances, such as mandatory, single-session CISD, there appears to be a low-level potential for harm and possibly elevated rates of PTSD. The absence of benefit, coupled with the potential for detrimental outcomes, violates the fundamental precept of "do no harm" (Bryant and Litz, 2009). Nevertheless, in apparent defiance of the scientific concerns, passionate advocates practice CISD undeterred, responder work-cultures continue to mandate debriefing, and teams using this approach are ever-present at disaster scenes.

Empirically informed early intervention. The demise of debriefing as the psychosocial standard created a void in the early psychological intervention armamentarium. International experts in disaster mental and behavioral health were convened by the National Institute of Mental Health in 2001 to develop consensus on early posttrauma intervention (NIMH, 2002). Their findings endorsed a movement away from debriefing and toward a more flexible, nonprescriptive, multifaceted approach. Rather than focusing narrowly on a single intervention technique, the committee recommended that a spectrum of actions be considered as components of early intervention: securing basic needs, applying the principles of psychological first aid, conducting needs assessments, monitoring the rescue and recovery environment, providing outreach and information, fostering resilience and recovery, conducting triage and referral, and providing psychiatric treatment for an identified subset of trauma survivors (NIMH, 2002).

This roster of recommendations emphasized strategies that can be applied at the community level through the concerted efforts of many responders, not just mental health professionals. In the earliest moments of response, actions directed toward reestablishing safety and security, coupled with provision of basic survival needs – mainstays of traditional disaster response – also provide a beneficial psychological effect. Trained mental health professionals are encouraged to activate an effective triage system as a safety net to identify those persons needing referral to expert psychological evaluation and possible treatment. Following initial screening, these professionals also deliver the appropriate level of intervention necessary to restore pre-disaster levels of functioning; options range from education, to psychotherapy, to prescription of therapeutic medications, to psychiatric hospitalization. Technical assistance and consultation to leadership was also included in the package of recommended early interventions. While some of these tactics are provided one-on-one, others are amenable to community-wide outreach and education.

Psychological first aid. The term "psychological first aid" was originally introduced in 1954 (Drayer et al., 1954) and models of psychological first aid had been in active use since the 1980s particularly by international relief agencies in Europe and Australia. Psychological first aid is rooted in the scientific evidence base supporting the assertion that a return to pre-disaster levels of functioning is the expected outcome for most disaster survivors. One of the major aims of psychological first aid is to draw upon survivors' strengths, and in the process help increase resiliency (Pynoos and Nader, 1988; Young, 2006). In contrast to CISD, psychological first aid allows flexible application of strategies tailored to each survivor's needs. Moreover, discussion of the trauma experience is not prompted, although such discussion is not precluded if the survivor chooses to talk about the event.

The release of the 2001 NIMH panel recommendations spurred a flurry of activity; new models of "PFA" proliferated within several years of the conference. In short order, psychological first aid was heralded by its proponents as the "acute intervention of choice" (NCTSN/NCPTSD, 2006). With multiple models in circulation and use, Bryant and Litz (2009) explored for common themes when they specified these three goals of psychological first aid: (1) reestablish safety, (2) reduce acute stress reactions, and (3) guide the survivor to access resources.

In 2004, world experts were convened to solidify the science behind early psychological support and intervention. Their primary output was a landmark publication describing the five essential elements of early intervention, defined as *safety, calming, connectedness, self-efficacy*, and *hope* (Hobfoll et al., 2007). These five principles are known to be psychologically beneficial based on scientific evidence and collectively they assist disaster survivors to cope with the stressors and challenges of disasters.

Actions taken to reestablish the physical *safety* of survivors also decrease perceptions of threat and vulnerability (Ozer et al., 2003). Efforts aimed at promoting *calming*, such as relaxation breathing, are beneficial psychologically by decreasing hyperarousal, a risk factor for PTSD. Endeavors to achieve *connectedness*, such as reuniting separated loved ones, draw upon the well-documented protective effects of social support (Norris et al, 2002). Actions that empower survivors to participate actively in recovery promote *self-efficacy* and help reestablish a degree of personal control for survivors in the post-disaster environment. Rekindling *hope* can be beneficial psychologically based on studies that relate optimism and a hopeful appraisal style to favorable outcomes among survivors of trauma (e.g., Antonovsky, 1979).

Reissman and colleagues (2010) envision these five principles as pathways to guide survivors back from the traumatizing disaster experience to favorable adaptation. Effective early intervention, then, should move survivors (1) from risk to *safety*, (2) from fear to *calming*, (3) from loss to *connectedness*, (4) from helplessness to *self-efficacy*, and (5) from despair to *hope*. These five principles align with common sense disaster response actions throughout history. While many of the response strategies that embody these principles are not new, recently acquired scientific support indicates that they can be both helpful in a practical sense as well as beneficial psychologically.

Multiple models of psychological first aid coexist, each applying a variation of the empirically informed principles of early intervention. The most literal example comes from the Australian Psychological Society (2009) which explicitly organized its psychological first aid response to the Victorian bushfires around five "core components." Not surprisingly, these are: (1) *promote safety* by providing basic needs and emergency medical services, (2) *promote calming* by listening to the stories of survivors who wish to share their experiences and providing accurate information, (3) *promote connectedness* by keeping families together and helping separated family members make contact, (4) *promote self and community efficacy* by engaging people in meeting their own needs, and (5) *promote hope* by enhancing people's natural resilience.

During the 2000s, the use of psychological first aid received endorsement from a series of expert panels (NIMH, 2002; Hobfoll et al., 2007; National Biodefense Science Board, 2008). However, it has not been scientifically tested for efficacy in real disaster field applications (Litz, 2008; Raphael and Maguire, 2009). "Psychological first aid does not purport to prevent the future onset of psychiatric disorders, such as PTSD. Instead, its goals are more modest, decrease distress, and promote adaptive coping skills in post-disaster environment," (Bryant and Litz, 2009).

In summary, early intervention models continue to evolve. In the earliest moments after disaster strikes, what is needed is neither formal psychological treatment nor therapy. In fact, the greatest psychological good comes from actions that are not primarily "psychological," but draw on time-tested disaster response tactics. Moving survivors to safety, calming them by supplying basic needs, connecting survivors to missing loved ones, actively involving survivors in helping themselves, and maintaining a positive, optimistic presence – all mainstays of traditional disaster response – are now known to also be beneficial psychologically.

Intermediate-term intervention

In the intermediate aftermath of a disaster (i.e., weeks and months post-disaster), more specific psychological intervention may be appropriate for those persons exhibiting continued high levels of disaster distress (e.g., high levels of anxiety, high levels of physiological arousal, lack of coping skills) that markedly interfere with daily functioning.

Skills for Psychological Recovery. As an intermediate intervention, Skills for Psychological Recovery (Australian Psychological Society, 2009) focuses on providing survivors of disasters with appropriate coping skills aimed at alleviating persistent disaster distress. As such, SPR builds on psychological first aid, emphasizing the importance of gathering information and providing support. More specifically than psychological first aid, though, SPR delves into managing physiological arousal, modulating emotions and cognitions, increasing problem-solving skills, and scheduling positive activities.

Long-term intervention

In the long-term post-disaster aftermath, as disaster distress continues to persist, PTSD is the most commonly seen psychiatric disorder. The intervention of choice for PTSD is a cognitive-behaviorally based psychotherapy approach. The overall aims of such an approach are the restructuring of dysfunctional cognitions, redressing of problematic behaviors, and modulating affective responses.

Different variants of PTSD-specific cognitive behavioral therapy approaches have been developed, such as exposure therapy, stress inoculation therapy, systematic desensitization, cognitive processing therapy, cognitive therapy, assertiveness training, and biofeedback/relaxation training (Rothbaum et al., 2001). The exposure-based approach has been found to be most efficacious in treating PTSD (e.g., Foa and Meadows, 1997). An exposure-based approach addresses the pervasive avoidant behavioral patterns that often result from trauma exposure and lead to significant impairment in daily functioning. The approach seeks to reexpose survivors to the disaster experience in a therapeutic and safe manner.

Psychological trajectories following disaster

The severity and duration of psychological outcomes is highly variable for individuals even when the severity of disaster exposure is approximately equal. For persons who are psychologically impacted by exposure to a natural disaster, Layne et al. (2007) describe and contrast five possible trajectories from the moment of impact forward: resistance, resilience, protracted recovery, chronic severe distress, and posttraumatic growth.

Resistance. Some individuals, the rarest of the breed, maintain a near-steady course of highly adaptive functioning throughout the disaster episode. It seems as if they are psychologically impervious to the potentially traumatizing effects of disaster. Using a materials science analogy, it appears as if the disaster is psychologically deflected or merely "bounces off" without making a mark. This response is vanishingly rare.

Resilience. Resilience is the ability to rebound in the face of adversity and represents a more typical response pattern in which function and performance are negatively affected during disaster impact, but the individual rapidly reestablishes psychological equilibrium and regains function in short order. This is the psychological equivalent of what happens to an automobile bumper during a low-velocity collision; on impact the bumper deforms and contorts, but almost immediately springs back to its original contour. Psychological resilience is surprisingly common, and according to Shalev and Errera (2008), actually the "default" outcome for the majority of survivors.

Protracted recovery. A number of survivors will need a notable period of time to recover from the traumatic event. They will ultimately resume full function, but the path back will take time.

Chronic, severe, debilitating distress. These individuals remain at a decreased level of function over time and do not recover without intervention. Many meet criteria for psychopathology such as PTSD, MDD, or a combination.

Posttraumatic growth. Some individuals, despite bludgeoning disaster impact, followed by adversity and obstacles, overcome and strengthen through the experience, emerging healthier and more functional after the disaster than before. Posttraumatic growth describes those who resurface from disaster with an enhanced repertoire of coping skills and an enlivened appreciation for newfound capabilities forged by hardship.

Resilience

One of the trajectories described, resilience, is a topic that is garnering considerable attention in the disaster field. Remarkably, most disaster-exposed individuals are minimally affected by the adversities and are frequently able to adapt to the circumstances. This capacity has been termed resilience, defined as the human ability to maintain stable, healthy levels of psychological and physical functioning following a potentially highly-disruptive event (Bonanno, 2004). Resilient individuals post-disaster manifest only transient mild stress reactions which are not likely to significantly interfere with continued functioning and are typically of short duration.

According to the American Psychological Association (2005), "resilience is the process of adapting well in the face of adversity, trauma, tragedy, threats, or even significant sources of stress." In disasters, resilience is the rule rather than the exception (Shultz et al., 2007b). Resilience is frequently the most common outcome among survivors exposed to natural disasters and among responders who provide support to these survivors. Resilience can be learned and enhanced.

Some defining characteristics of resilience expand its description. Resilience appears to be a common phenomenon that results from positive adaptation to life challenges (Shalev and Errera, 2008). If human abilities to adapt remain intact, individuals are able to adjust healthfully to severe adversity. Many pathways to positive adaptation exist. When faced with traumatizing events, individuals may be resilient in some domains of life but not all; resilience is not a binary – all-or-none – phenomenon (Layne et al., 2007). "Ordinary magic" is a phrase that captures the surprising and optimistic finding that resilience is the expectable outcome for a large proportion of disaster survivors (Masten, 2001).

Summary

The psychological impacts of natural disasters are widespread, expand across a spectrum of severity, extend along a range of duration, and relate to the nature of the disaster event. The psychological consequences of disasters are spawned by, and directly proportional to, the degree of exposure to hazards, loss, and change, the "forces of harm" that characterize natural disasters. High-risk populations in harm's way, those that are particularly vulnerable to the ravages of disaster and the combination of physical and psychological consequences, can only be partially defined before disaster strikes. Disaster impact, compounded by adversities in the aftermath, "reshuffles the deck" by creating new special populations of persons needing medical and psychological support composed of those who have sustained extreme exposure to trauma and harm. While most persons exposed to disaster rebound quickly from transient distress reactions, others progress to psychopathology including PTSD, major depression, anxiety disorders, and substance abuse. Those who lose loved ones in a natural disaster are likely to grapple with complicated grief.

Prevention of psychological consequences of disaster holds great promise but is untried and untested. Early intervention is being redefined as psychological debriefing is supplanted by evidence-informed approaches; psychological first aid is the current contender. A stepped-care approach is advocated for moving survivors through a progression of early to intermediate psychological support and beyond this, for those whose distress is unabated, into psychological and psychiatric treatment.

While focus is understandably drawn to timely, empirically based support and treatment for those who are impacted psychologically, some of the most affirmative guidance to emerge is that resilience, positive adaptation in the face of disaster's adversity, is the most common and expectable outcome. Some survivors even emerge from the disaster experience stronger and more vital psychologically, a recently-recognized phenomenon known as posttraumatic growth. This sets the future agenda for the field; integrating disaster mental and behavioral health with the disciplines of public health, public safety, and emergency response to enhance preparedness for future catastrophic events.

Bibliography

Ahern, J., and Galea, S., 2006. Social context and depression after a disaster: the role of income inequality. *Journal of Epidemiology and Community Health*, **60**, 766–770.

American Psychiatric Association, 2004. *Diagnostic and Statistical Manual of Mental Disorders IV, TR*. Washington, DC: American Psychiatric Association.

American Psychological Association, 2005. The road to resilience. Available at http://www.apa.org/helpcenter/road-resilience.aspx.

Antonovsky, A., 1979. *Health, Stress, and Coping*. San Francisco, CA: Jossey-Bass.

Australian Psychological Society Ltd., 2009. *Guidelines for the Provision of Psychological Services to People Affected by the 2009 Victorian Bushfires*. Victoria.

Basoglu, M., Salcioglu, E., and Livanou, E., 2007. A Randomized controlled study of single-session behavioral treatment of earthquake-related post-traumatic stress disorder using an earthquake simulator. *Psychological Medicine*, **37**, 203–213.

Bonanno, G. A., 2004. Loss, trauma, and human resilience: have we underestimated the human capacity to thrive after extremely aversive events? *The American Psychologist*, **59**, 20–28.

Breslau, N., Peterson, E. L., Poisson, L. M., et al., 2004. Estimating post-traumatic stress disorder in the community: lifetime perspective and the impact of typical traumatic events. *Psychological Medicine*, **34**(5), 889–898.

Bryant, R. A., and Litz, B., 2009. Mental health treatments in the wake of disaster. In Neria, Y., Galea, S., and Norris, F. H. (eds.), *Mental Health and Disasters*. Cambridge, UK: Cambridge University Press, pp. 321–335.

Butler, A. S., Panzer, A. M., and Goldfrank, L. R., 2003. *Preparing for the Psychological Consequences of Terrorism: A Public Health Approach*. Washington, DC: National Academies Press.

Drayer, C. S., Cameron, D. C., Woodward, W. D., and Glass, A. J., 1954. Psychological first aid in community disasters: prepared by the American Psychiatric Association Committee on Civil Defense. *Journal of the American Medical Association*, **156**(1), 36–41.

Everly, G. S., Jr., and Mitchell, J. T., 1999. *Critical Incident Stress Management: A New Era and Standard of Care in Crisis Intervention*, 2nd edn. Ellicott City, MD: Chevron.

Foa, E. B., and Meadows, E. A., 1997. Psychosocial treatments for posttraumatic stress disorder: a critical review. *Annual Review of Psychology*, **48**, 449–480.

Galea, S., and Resnick, H., 2005. Posttraumatic stress disorder in the general population after mass terrorist incidents: considerations about the nature of exposure. *CNS Spectrums*, **10**(2), 107–115.

Galea, S., Vlahov, D., Resnick, H., et al., 2003. Trends of probable post-traumatic stress disorder in New York City after the September 11 terrorist attacks. *American Journal of Epidemiology*, **158**(6), 514–524.

Galea, S., Nandi, A., and Vlahov, D., 2005. The epidemiology of post-traumatic stress disorder after disasters. *Epidemiologic Reviews*, **27**, 78–91.

Hobfoll, S. E., Watson, P., Bell, C. C., et al., 2007. Five essential elements of immediate and mid-term mass trauma intervention: empirical evidence. *Psychiatry*, **70**, 283–315.

Horowitz, M. J., Siegel, B., Holen, A., et al., 1997. Diagnostic criteria for complicated grief disorder. *The American Journal of Psychiatry*, **154**, 904–910.

Kilic, C., Aydin, I., Taskintuna, N., et al., 2006. Predictors of psychological distress in survivors of the 1999 earthquakes in Turkey: effects of relocation after the disaster. *Acta Psychiatrica Scandinavica*, **114**, 194–202.

Layne, C. M., Warren, J. S., Watson, P. J., and Shalev, A. Y., 2007. Risk, vulnerability, resistance, and resilience: towards an integrative conceptualization of posttraumatic adaptation. In Friedman, M. J., et al. (eds.), *Handbook of PTSD: Science and Practice*. New York: Guilford Press.

Lerner, E. B., Cone, D, C., Weinstein, E. S., Schwartz, R. B., Coule, P. L., Cronin, M., Wedmore, I. S., Bulger, E. M., Mulligan, D. A., Swienton, R. E., Sasser, S. M., Shah, U. A., Weireter, L. J. Jr, Sanddal, T. L., Lairet, J., Markenson, D., Romig, L., Lord, G., Salomone, J., O'Connor, R., and Hunt, R. C., 2011. Mass casualty triage: an evaluation of the science and refinement of a national guideline. *Disaster Med Public Health Prep.*, **5**(2), 129–137.

Lichtenthal, W. G., Cruess, D. G., and Prigerson, H. G., 2004. A case for establishing complicated grief as a distinct mental disorder in DSM-V. *Clinical Psychology Review*, **24**, 637–662.

Litz, B. T., 2008. Early intervention for trauma: where are we and where do we need to go? A commentary. *Journal of Traumatic Stress*, **21**(6), 503–506.

Masten, A. S., 2001. Ordinary magic: resilience processes in development. *The American Psychologist*, **56**, 227–238.

McNally, R. J., Bryant, R. A., and Ehlers, A., 2003. Psychological debriefing and its alternatives: a critique of early intervention for trauma survivors. *Psychological Science in the Public Interest*, **4**, 45–79.

Mitchell, J. T., 1983. When disaster strikes...The critical incident stress debriefing process. *Journal of Emergency Medical Services*, **8**, 36–39.

National Biodefense Science Board, 2008. *Disaster Mental Health Recommendations: Report of the Disaster Mental Health Subcommittee of the National Biodefense Science Board*. Washington, DC.

National Child Traumatic Stress Network and National Center for PTSD, 2006. *Psychological First Aid: Field Operations Guide*, 2nd edn. http://www.nctsn.org.

National Institute of Mental Health, 2002. *Mental Health and Mass Violence: Evidence-based Early Psychological Intervention for Victims/Survivors of Mass Violence: A Workshop to Reach Consensus on Best Practices*. Washington, DC: U.S. Government Printing Office. (NIH Publication No. 02-5138).

Neria, Y., and Litz, B., 2004. Bereavement by traumatic means: the complex synergy of trauma and grief. *Journal of Loss and Trauma*, **9**, 73–87.

Neria, Y., Gross, R., Olfson, M., et al., 2006. Posttraumatic stress disorder in primary care one year after the 9/11 attacks. *General Hospital Psychiatry*, **28**(3), 213–222.

Neria, Y., Gross, R., Litz, B., et al., 2007. Prevalence and psychological correlates of complicated grief among bereaved adults 2.5–3.5 years after 9/11 attacks. *Journal of Traumatic Stress*, **20**, 251–262.

Neria, Y., Nandi, A., and Galea, S., 2008. Posttraumatic stress disorder following disasters: a systematic review. *Psychological Medicine*, **38**(4), 467–480.

Norris, F. H., and Wind, L. H., 2009. The experience of disaster: trauma, loss, adversities, and community effects. In Neria, Y., Galea, S., and Norris, F. H. (eds.), *Mental Health and Disasters*. Cambridge, UK: Cambridge University Press, pp. 7–28.

Norris, F. H., Friedman, M., and Watson, P., 2002. 60,000 disaster victims speak. Summary and implications of the disaster mental health research. *Psychiatry*, **65**, 240–260.

Norris, F. H., Byrne, C. M., Diaz, E., and Kaniasty, K., 2007. *Risk Factors for Adverse Outcomes in Natural and Human-Caused Disasters: A Review of the Empirical Literature*. A National Center for PTSD Fact Sheet.

Ozer, E. J., Best, S. R., Lipsey, T. L., and Weiss, D. S., 2003. Predictors of posttraumatic stress disorder and symptoms in adults: a meta-analysis. *Psychological Bulletin*, **129**, 52–73.

Pynoos, R. S., and Nader, K., 1988. Psychological first aid and treatment approach to children exposed to community violence: research implications. *Journal of Traumatic Stress*, **1**(4), 445–473.

Raphael, B., and Maguire, P., 2009. Disaster mental health research, past, present, and future. In Neria, Y., Galea, S., and Norris, F. H. (eds.), *Mental Health and Disasters*. Cambridge, UK: Cambridge University Press, pp. 7–28.

Reissman, D. B., Schreiber, M. D., Shultz, J. M., and Ursano, R. J., 2010. Disaster mental and behavioral health. In Koenig, K. L., and Schultz, C. H. (eds.), *Disaster Medicine*. Cambridge, UK: Cambridge University Press, pp. 103–112.

Roberts, N. P., Kitchiner, N. J., Kenardy, J., and Bisson, J., 2009. Multiple session early psychological interventions for the prevention of post-traumatic stress disorder. Cochrane Database of Systematic Reviews. *The Cochrane Library*, 3:Art. No. CD006869. doi:10.1002/14651858.CD006869.pub3, Oxford, UK.

Rothbaum, B. O., Hodges, L. F., Ready, D., et al., 2001. Virtual reality exposure therapy for Vietnam veterans with posttraumatic stress disorder. *The Journal of Clinical Psychiatry*, **62**, 617–622.

Schlenger, W. E., Caddell, J. M., Ebert, L., et al., 2002. Psychological reactions to terrorist attacks. findings from a national study of Americans' reactions to September 11. *Journal of the American Medical Association*, **288**, 581–588.

Schmitz, S., Bustamante, H., Espinel, Z., Allen, A., and Shultz, J., 2009. *SAFETY FUNCTION ACTION Family Disaster Plan Guidebook*. Miami, FL: DEEP Center, University of Miami School of Medicine.

Shalev, A. Y., and Errera, Y. L. E., 2008. Resilience is the default. How not to miss it. In Blumenfield, M., and Ursano, R. J. (eds.), *Intervention and Resilience after Mass Trauma*. Cambridge, UK: Cambridge University Press, pp. 149–172.

Shultz, J. M., Espinel, Z., Flynn, B. W., et al., 2007a. *DEEP PREP: All-Hazards Disaster Behavioral Health Training*. Tampa, FL: Disaster Life Support Publishing.

Shultz, J. M., Espinel, Z., Galea, S., and Reissman, D. B., 2007b. Disaster ecology: implications for disaster psychiatry. In Ursano, R. J., et al. (eds.), *Textbook of Disaster Psychiatry*. Cambridge, UK: Cambridge University Press, pp. 69–96.

Shultz, J. M., Allen, A., Bustamante, H., and Espinel, Z., 2009. *SAFETY FUNCTION ACTION for Disaster Responders: Training Module Guidebook*. Miami, FL: DEEP Center, University of Miami Miller School of Medicine.

Tak, S., Driscoll, R., Bernard, B., and West, C., 2007. Depressive symptoms among firefighters and related factors after the response to hurricane Katrina. *Journal of Urban Health*, **84**, 153–161.

United States Department of Health and Human Services, 2004. *Mental Health Response to Mass Violence and Terrorism: A Training Manual*. Rockville, MD: Substance Abuse and Mental Health Services Administration. DHHS Publication No. SMA 3959.

Ursano, R. J., Fullerton, C. S., Weisaeth, L., and Raphael, B., 2007. Individual and community responses to disasters. In Ursano, R. J., et al. (eds.), *Textbook of Disaster Psychiatry*. Cambridge, UK: Cambridge University Press, pp. 190–250.

van Griensven, F., Chakkraband, M. L., Thienkrua, W., et al., 2006. Thailand post-tsunami mental health study group: mental health problems among adults in tsunami-affected areas in Southern Thailand. *Journal of the American Medical Association*, **296**, 537–548.

Watson, P. J., and Shalev, A. Y., 2005. Assessment and treatment of adult acute responses to traumatic stress following mass traumatic events. *CNS Spectrums*, **10**(2), 123–131.

Young, B. H., 2006. Adult psychological first aid. In Ritchie, E. C., Watson, P. J., and Friedman, M. J. (eds.), *Interventions Following Mass Violence and Disasters*. New York: Guilford Press.

Zatzick, D., 2007. Interventions for acutely injured survivors of individual and mass trauma. In Ursano, R. J., et al. (eds.), *Textbook of Disaster Psychiatry*. Cambridge, UK: Cambridge University Press, pp. 190–205.

Cross-references

Casualties Following Natural Hazards
Classification of Natural Disasters
Coping Capacity
Critical Incident Stress Syndrome
Exposure to Natural Hazards
Hazard
Mass Media and Natural Disasters
Mortality and Injury in Natural Disasters
Perception of Natural Disasters and Hazards
Post-Traumatic Stress Disorder (PTSD)
Resilience
Risk Perception and Communication
Social–Ecological Systems

PYROCLASTIC FLOW

Robert Buchwaldt
Massachusetts Institute of Technology, Cambridge, MA, USA

Synonyms

Ash flow; Nuées ardentes; Pyroclastic density current

Definition

Pyroclastic density currents are rapidly moving mixtures of hot volcanic particles and gas that flow across the ground under the influence of gravity. These processes are heavier-than-air emulsions that move much like a snow avalanche, except that they are fiercely hot, contain toxic gases, and move at phenomenal, hurricane-force speeds, often over 100 km/h.

Introduction

Pyroclastic flows are some of the most mysterious and dangerous volcanic features. They are also one of the most dramatic and picturesque ways to see the energy and destruction stored in our planet (Figure 1). Since the 1700s these currents have caused the death of several tens of thousands of people. During the eruption of Montagne Pelée in 1902, 28,000 people were killed when a cloud of hot gas and ash destroyed the peaceful village of Saint Pierre, of Martinique in the Lesser Antilles and the sole survivor was found days later in an underground prison (e.g., Anderson and Flett, 1903; Scarth, 2002). Pyroclastic flows generated during the August 25, 79 AD eruption of Mount Vesuvius, Italy, contributed to the devastation of the Roman cities of Pompeii and Herculaneum and the deaths of 10,000–25,000 people (e.g., Sigurdsson et al., 1982; Pliny, 79 AD). During the eruption of Mount St. Helens on May 18, 1980, which was the deadliest and most economically destructive volcanic event in the history of the USA, 57 people were killed; 250 homes, 47 bridges, 15 miles (24 km) of railways, and 185 miles (298 km) of highway were destroyed by a combination of a lateral blast and pyroclastic density currents (Lipman and Mullineaux, 1981). Therefore, a detailed understanding of these natural phenomena is essential for the mitigation of volcanic hazards.

Pyroclastic Flow, Figure 1 Pyroclastic flows descend the southeastern flank of Mayon Volcano, Philippines during the September 23, 1984 eruption (Photograph by C.G. Newhall/USGS).

Terminology

The name Pyroclastic is derived from the Greek "pyro" (πῦρ), meaning fire; and "clastic" (κλαστός), meaning broken and was apparently first used by Francesco Serao (1738) in a short account written on the eruption of Vesuvius between May 14 and June 4, 1736. After this relatively straightforward term the terminology of ground-hugging flows of hot, variably inflated, lava particles, and variable amounts of magmatic gas, steam, and ingested air has been very confusing over the years, predominantly due to the fact that until recently scientists tried to develop genetic terms for something they inferred to have happened. Nevertheless, today the scientific community has largely agreed to use the term pyroclastic flow or better pyroclastic density current for inflated mixtures of hot volcanic particles that flow in variable concentrations and varying velocities along the ground. Based on the textural characteristics of their deposits, there are two end member types distinguished, the products of a dilute suspension called pyroclastic surges, in which particles are carried in turbulent suspension in a thin layer close to the ground, and a generally "thicker," more massive pyroclastic flows, (Fisher and Schmincke, 1984). The ranges of structures which can be found in pyroclastic surge deposits immediately evoke the idea of high energy and a very dynamic depositional environment and therefore have a great destructive potential. The most common depositional characteristics in this group are well-defined bedding, which show erosional unconformities, good to moderate sorting of the particles, the pumice particles are well rounded, and last but not least local lateral grading into massive flow deposits. Spatially, pyroclastic surges are usually not confined to a channeling valley as their path downslope but they can overflow these and can mantle the surrounding landscape. The typical distribution is nevertheless restricted to proximal area around volcanoes (circa 8 km from the vent).

In contrast, the deposits from pyroclastic flows are generally poorly sorted, massive, and are more confined to the valleys used to move downslope. The deposits thicken markedly toward the center and since they fill the channels they tend to flatten the existing morphology. The deposits are created by sedimentation from particle rich flows. Pyroclastic flow deposits containing pumice as a major constituent are called ignimbrites, sometimes ash flow tuffs. Smaller deposits are referred to as block and ash deposits or if the clasts are vesiculated then they are called pumice and ash deposits. From these two end member cases of pyroclastic density currents we can infer that the controlling factor of the deposition style is the amount of particles relative to the dilution medium, the containing gas. The more particles that are present the less the flow is able to travel because the particles try to settle, bouncing against each other and disturb the continuous laminar flow, whereas the less particles that are present the flow is more undisturbed and can travel farther. Therefore, in the first case we preferentially would get pyroclastic flows and in the second we would get pyroclastic surges.

In practice, it is often impossible to conclude with any certainty which type has been deposited due to a process known as "aggradation." After an eruption of a pyroclastic density current, the behavior and character of the flow changes over time. The moving energy decreases, the incorporated particles have difficulty to stay in suspension, start to settle out and change the morphology of the path, and therefore the flow encounters fewer barriers. The temperature of the flow cools along its path, which in turn changes the flow behavior. As a consequence, the deposits and its architecture can be very complex and probably reflects more processes at the base. Branney and Kokelaar (2003) suggested changing the terminology, to reflect this dynamic behavior, to just two end members. The first is a fully dilute pyroclastic flow in which collision between particles play a limited role. The second end member is a granular fluid-based pyroclastic flow in which the particle density is high enough that interaction is important and clasts can be supported by collision and by the flux of the surrounding dusty gas forced upward by the settling particles.

Origin

There are three common mechanisms for the generation of pyroclastic flows: (1) Simple gravitational collapse of a growing lava dome (Lava dome is a mound of volcanic rock extruded from a volcanic vent and the lava piles into a heap rather than flowing away and forms a round plug on top of the volcanic vent) or lava flow on a volcano, known as Merapi type, (2) The Peléean type is the explosion within a growing lava dome which trigger an eruption and collapse of the dome and (3) the Soufrière type is the collapse from eruption columns (The eruption column is a column of gases, ash, and larger rock fragments rising from a crater or other vent. If it is of sufficient volume and velocity, this gaseous column may reach many miles into the stratosphere), the most impressive and hazardous type.

Gravitational collapse of Lava Domes (Meriapi type) generates small pyroclastic flows rich in dense poorly vesicular components. The high temperature rockfalls formed by dome collapse transform rapidly into dusty, relatively small pyroclastic density currents due to the low mechanical strength of hot gas-poor lava (Mellors et al., 1988). Such flows commonly produce block and ash flows. In some cases, dome collapse is preceded by vigorous gas venting (Cole et al., 1998), which empties the shallow magma chamber (Magma chamber is a reservoir of molten rock material beneath the Earth's surface) and allows its roof to collapse (Peléean type). Block and ash flows are emplaced as highly concentrated avalanches overlain by a dilute ash cloud (Cole et al., 1998). The basal avalanche is gravity-driven and follows valleys. Large blocks bounce and roll in the avalanche. The accompanying ash cloud acts as a surge, so is less controlled topographically, allowing it to outrun the avalanche and spill over watersheds into adjacent valleys. The ability of ash cloud surges to decouple from the basal avalanche makes them particularly unpredictable and thus hazardous.

Pyroclastic density currents formed by column collapse (Soufrière type) are rich in gas components and are termed scoria flows if the juvenile material is moderately vesicular, or pumice flows or ash flows if the vesicularity is high. To understand the emplacement of these flows, we have to understand what changes the evolution of the erupting column or jet from a stable environment to an unstable environment. The stability of the erupting column is controlled by two main parameters. Either the erupting magma loses the gas phase and progresses into magma poorer in volatiles (the driving force for the thrust of explosive eruptions), or the size of the vent is increasing due to vent erosion. If either of these parameters change then the erupting column will become unstable and leads to a collapse of the column. When this occurs, part or the entire eruption column does not have sufficient thrust or energy to continue rising. The gas and pyroclastic particles lose buoyancy as they are now denser than the surrounding atmosphere and thus begin to fall under gravity forming pyroclastic flows. Column collapse generates a current in which the particles are suspended by turbulence.

Transport

Small pyroclastic density currents travel at speeds up to a few tens of meters per second (m/s). Particularly detailed measurements have been published by Hoblitt (1986) for the flow of August 7, 1986 Mount St. Helens, which reached a maximum speed of 30 m/s. Pumice flows from the August 1997 explosive eruptions at Monserrat had initial speeds of approximately 60 m/s and some pyroclastic density currents have been reported to be as fast as 100 m/s.

Taylor (1958) in his famous and groundbreaking analysis of the 1951 Mount Lamington eruption emphasized the importance of gravity as an energy source for pyroclastic flows, whether from vertical eruption columns or from collapsing domes growing in craters. He concluded that the thermal energy contained within hot magma may thrust an initial explosion and the ascent of the eruption column, but after the initial loss of energy it does not contribute much to the lateral movement of the pyroclastic flows. A gravity-driven pyroclastic flow may even accelerate and become more hazardous as it continues downslope. This simple and fundamental principle has withstood the test of time, and also it has been refined using more sophisticated computer models, has not changed in modern history and gravity is still considered to be the driving mechanism for pyroclastic density currents.

To understand the dynamics of the pyroclastic density current better, we have to understand first the morphology and architectures of these features (Figure 2). When we observe a single turbulent density current one can see that it consists of a leading edge, with various possible forms, and a long trailing body. The leading edge of the current typically has an overhanging nose and sometimes is thicker than the body. In more detail we can recognize that pyroclastic flows are composed of two parts. A basal "ground hugging avalanche" in which most particles are transported and deposited. This avalanche is usually hidden by an overriding billowing cloud of hot gases and fine ash particles, which can rise tens of kilometers into the atmosphere. This cloud is often termed as Phoenix column or Phoenix plume.

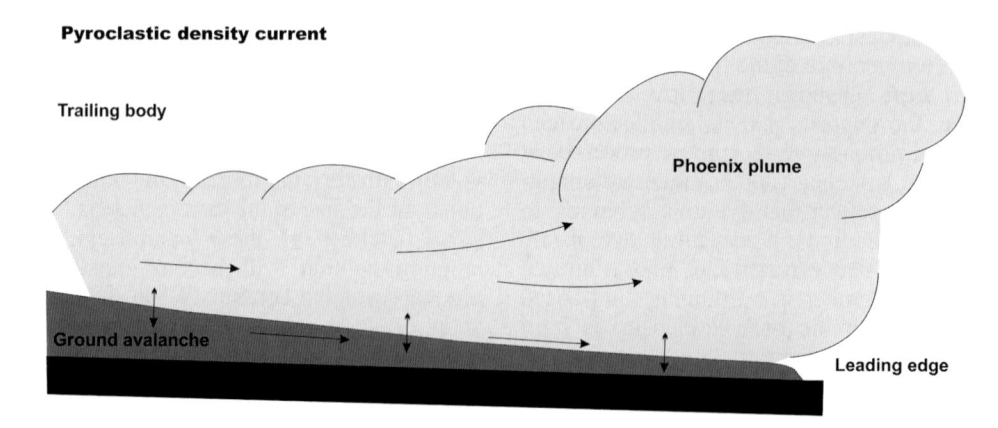

Pyroclastic Flow, Figure 2 Schematic cross-section through a pyroclastic density current. *Arrows* indicate particle movement direction (Adapted from Branney and Kokelaar, 2003).

In general, to see more details in situ and to understand the transport mechanism from pyroclastic density currents is very difficult because most of the dynamic processes within the currents are impossible to observe and are commonly inferred from the associated deposits or derived from numerical models. By examining the deposits of many pyroclastic density flows, we can often observe a thin fine grained basal layer 10–50 cm thick in which both pumice and rock fragments are inversely graded. This layer forms by the interaction of the flow with its substrate. The main division of the flow unit is massive or crudely stratified. In many cases, large pumices are reversely graded forming a concentration zone at the top.

To produce such a deposit we have to envision the following scenario. The gas particle mixture is following gravity and accelerates downhill along the volcanic slopes. If the avalanche would behave like a solid block the friction on the base of the flow should slow the system down, but rather than slowing the whole body the friction creates a small gas layer on which the rest of the flow rides relatively frictionless, similar to a Hovercraft. This mechanism allows for the incredible velocity and their ability to spread laterally. The nature and origin of the gas phase is still debatable. The principle source of the gas is thought to be from degassing of magmatic particles in transit, which is possibly not the major source, gases incorporated during the collapse of the eruption column, air ingested and heated at the front of the moving pyroclastic flows, or vaporized moisture from the ground (possibly only locally important). The top of the flow only slows through contact with the surrounding stagnant air. Thus, the flow travels fastest toward the middle and slower on the outside edges. The small gas layer beneath the flow allows the flow to travel much farther and faster resulting in greater devastation along the volcano's flanks.

Lateral blasts

Lateral Blasts like those at Mount St. Helens, Washington USA, in 1980 and Bezymianny, Kamchatka Russia, in 1955 are caused by the sudden decompression of magma by flank collapse. In these cases, the explosive event distributed the eruption energy horizontally instead of vertically as in a traditional eruption column. Nevertheless, the initial driving force of this glowing cloud is from the blast, at one point gravity takes over as the driving force and pyroclastic density currents form. Lateral blasts are less well understood than fountains, and computer simulations of these scenarios are in their infancy. The Mount St. Helens blast had an internal velocity as high as 235 m/s. This is consistent with simple theoretical models and is comparable to speed of currents formed by column collapse. The Mount St. Helens blast was originally supersonic relative to the internal sound speed of the erupting mixture. One difference between column collapse and lateral blast is that in the latter there is no initial vertical component. Thus, for a given exit velocity the largest block discharged might be smaller than the fountain. Pyroclastic currents formed by lateral blasts are short lived and highly unsteady. Peak discharge of the Mount St. Helens blast lasted no more than a few tens of seconds.

Volcanic hazard mitigation

Volcanic eruptions cannot be influenced by humans. They are governed by processes that start deep in the Earth and often reach the stratosphere. When a volcanic catastrophe occurs, the reasons are often rooted in society itself. There are, however, limited possibilities to reduce some of the impacts of volcanic eruptions. Whether a volcanic eruption leads to major loss in human lives or not depends on the population density in the proximity of a volcano and its state of preparedness and organization. It does not generally depend on the volume of magma erupted or the type of eruptive processes.

All types of mass flows pose a great hazard to communities because people tend to live in valleys that are convenient, support good soil, and have water resources nearby. High velocity, hot, particle-poor blasts are especially dangerous because their pathways are not confined to valleys. The main loss of lives and the largest economic losses in the last few centuries from volcanism resulted from pyroclastic density currents hot blasts. Hot pyroclastic density currents, which consist of gas and particles, at temperatures sometimes exceeding 800°C, speed down the flanks of a volcano and may cover thousands of square kilometers. People caught within the flow path have little chance of surviving. Buildings, crops, and forests are destroyed and warning periods are extremely short. On the other hand the most likely pathways can be clearly delineated in hazard maps. Administrators responsible for land planning can effectively reduce the risk when restricting settlements in valleys that are likely pathways for volcanic mass flows. Each volcano behaves differently, requiring careful analysis of local factors. Effective prediction of pathways and evacuation is currently the main method for mitigating the risk of pyroclastic density currents. Most important in this endeavor is to define hazard zones, which are based on a very detailed analysis of the early history of a volcano, as well as an estimate of the possible scale and nature of eruptive processes at a particular volcano.

The path of some lava flows can be influenced slightly by building barriers. This costly mitigation technique has so far only worked on eruptions that are small and slow but are ineffective when the eruptive volume is very large. These dams are low concrete wall-type structures constructed perpendicular to the proposed flow direction. Such structures break the velocity and therefore the erosional energy of the pyroclastic density currents. In addition, they also act as temporary sediment traps. Dams parallel to valleys have been constructed on several volcanoes to direct the course of the flows.

Overall, the most successful mitigation of pyroclastic density currents is its integration into the volcanic hazard plan as a whole which includes monitoring, prediction, and prevention. A volcanic eruption is the result of a large number of factors and a reliable prediction can only be made in well-instrumented and well-monitored volcanic situations. The areas potentially affected during volcanic eruptions can usually be relatively well predetermined, depending on how well a volcano and its deposits have been analyzed, allowing early mitigation planning and significant minimizing of damage. Finally, communication between the public, scientists, and the administrative body is essential; to understand the risk a community has to know how to prepare for an eventual event. Two of the largest volcanic catastrophes in the last century could have been avoided if the political administrative bodies had taken seriously the fear and determination and of the people, or the warning of scientists.

Summary

In summary, we can define pyroclastic flows as high-density mixtures of hot, dry rock fragments and hot gases that move away from their source vents at high speeds. These flows are driven by gravity on a layer of hot air. During their travel, they damage or destroy structures and vegetation by impact of rock fragments moving at high speeds and may bury the ground surface with a layer of ash and coarser debris tens of centimeters or more thick. Because of their high temperatures, hot pyroclastic flows may start fires and kill or burn people and animals. Consequently, it is essential to comprehend the structure and dynamic of these devastating events. Also active mitigation is difficult for most of the pyroclastic density currents, but casualties and losses can be avoided by a careful integrated approach of the volcano as a whole, using historical data as well as modern computer models to create hazard maps, to understand the pathways of pyroclastic density currents and to create escape routes for the communities. Similarly, the prediction of pyroclastic density currents is complicated; well-monitored volcanoes have in recent years increased the success rate of warnings. Finally, it is an old story that the messenger is the culprit who brings the bad news; recently, the communication between the different parties has improved by bringing the message directly to the endangered populations, involving modern technologies, and thus ensuring that the scientific knowledge has a better chance to be integrated in the decision-making process for the good of society.

Bibliography

Anderson, T., and Flett, J. S., 1903. Report on the eruption of the Soufrière in St. Vincent in 1902 and a visit to Montagne Pelée in Martinique: Part I. *Philosophical Transactions of the Royal Society of London, Series A*, **200**, 353–553.

Branney, M. J., and Kokelaar, P., 2003. Pyroclastic density currents and sedimentation of ignimbrites. *Geological Society of London Memoirs*, **27**, 152.

Cole, P., Calder, E. S., Druitt, T. H., Hoblitt, R., Lejeune, A. M., Robertson, R., Smith, A. L., Stasiuk, M. V., Sparks, R. S. J., Young, S. R., and Team, M. V. O., 1998. Pyroclastic flows generated by gravitational instability of the 1996–97 lava dome of the Soufriere Hills Volcano, Montserrat. *Geophysical Research Letters*, **25**, 3425–3428.

Fisher, R. V., and Schmincke, H.-U., 1984. *Pyroclastic Rocks*. Heidelberg: Springer.

Hoblitt, R. P., 1986. *Observations of the Eruptions of July 22 and August 7, 1980, at Mount St. Helens*, Washington. U.S. Geological Survey Professional Paper 1335, 44p.

Lipman, P. W., and Mullineaux, D. R., 1981. *The 1980 Eruptions of Mount St. Helens*, Washington. USGS Professional Paper 1250, 844p.

Mellors, R. A., Waitt, R. B., and Swanson, D. A., 1988. Generation of pyroclastic flows and surges by hot-rock avalanches from the dome of Mount St. Helens volcano, USA. *Bulletin of volcanology*, **50**, 14–25.

Pliny, 79 AD, 1999. The Destruction of Pompeii, 79 AD. EyeWitness to History. www.eyewitnesstohistory.com.

Scarth, A., 2002. *La Catastrophe: The Eruption of Mount Pelee, the Worst Volcanic Disaster of the 20th Century*. Oxford: Oxford University Press.

Serao, F., 1738, *Reale Accademia delle Scienze Fisiche e Matematiche di Napoli. Istoria dell' incendio del Vesuvio accaduto nel mese di maggio dell'anno MDCCXXXVI*. Naples: Nella stamperia di Novello de Bonis.

Sigurdsson, H., Cashdollar, S., and Sparkes, S. R. J., 1982. The eruption of vesuvius in A. D. 79: reconstruction from historical and volcanological evidence. *American Journal of Archaeology*, **86**, 39–51.

Taylor, G. A. M., 1958. The 1951 eruption of Mount Lamington volcano, Papua. *Australia, Bureau of Mineral Resources, Geology and Geophysics Bulletin*, **38**, 1–117.

Cross-references

Base Surge
Galeras Volcano
Krakatau
Lahar
Montserrat Eruptions
Mt. Pinatubo
Nevado del Ruiz
Nuée Ardente
Stratovolcano
Vesuvius
Volcanic Ash
Volcanic Gas
Volcanoes and Volcanic Eruptions

Q

CASE STUDY

QUEENSLAND FLOODS (2010–2011) AND "TWEETING"

France Cheong, Christopher Cheong
RMIT University, Melbourne, VIC, Australia

Introduction

Australia experienced its worst flooding disaster in 2010 and 2011 with a series of floods occurring in several states between March 2010 and February 2011. Worst of all were the Queensland floods which caused three-quarters of the state to be declared a disaster zone (*Brisbane Times*, 2011).

In times of mass emergencies, a phenomenon known as collective behavior becomes apparent (Dynes and Quarantelli, 1968). It consists of socio-behaviors that include intensified information search and information contagion (Starbird et al., 2010). In these situations, people want to know where exactly their families and friends are, as not being able to reach them or knowing they might not be able to contact you, can be very frightening moments during such situations. Information is critical during emergencies as the availability of immediate information can save lives. People share information about approaching threats, where to evacuate, where to go for help, etc. Not only do they want to know about the destruction that has occurred, but they are also eager to help those affected by giving a helping hand and raise funds from donations. Thus, there is a need to keep abreast of the latest developments; however, this is difficult since information produced under crisis situations is usually scattered and of varying quality.

Social media is media for social interaction enabled by communication technologies such as the web and smartphones. The distributed, decentralized, and real-time nature of these interactions provide the necessary breadth and immediacy of information required in times of emergencies (Palen and Vieweg, 2008). Since social media offer a uniquely rapid and powerful way to disseminate information, accurate and inaccurate, good and bad information spread equally alike as incorrect information can spread like wild fire. However, there is indication that social networks tend to favor valid information over rumors (Castillo et al., 2011).

Twitter is a microblogging service, a form of lightweight chat allowing users to post and exchange short 140-character-long messages known as *tweets*. Although most tweets are conversation and chatter, they are also used to share relevant information and report news (Castillo et al., 2011). Twitter is becoming a valuable tool in disaster and emergency situations as there is increasing evidence that it is not just a social network, it is also a news service (Yates and Paquette, 2011). In emergency situations, tweets provide either first-person observations or bring relevant knowledge from external sources (Vieweg, 2010). Information from official and reputable sources is regarded as valuable and hence is actively sought and propagated. Other users then elaborate and synthesize this pool of information to produce derived interpretations.

We studied the community of Twitter users disseminating information during the crisis caused by the Queensland floods in 2010–2011 in order to reveal interesting patterns and features within this online community. Our aim was to develop an understanding of the online community that was active during that period by answering the following questions: What was the online social behavior during that period? In particular, who were the active players communicating information and how effective were they? What type of information was of importance? How can the information discovered be useful for the management of such situations in the future?

P.T. Bobrowsky (ed.), *Encyclopedia of Natural Hazards*, DOI 10.1007/978-1-4020-4399-4,
© Springer Science+Business Media Dordrecht 2013

Techniques from Social Network Analysis (SNA) were used to analyze the online community. SNA is a sociological approach for analyzing patterns of relationships and interactions between social actors in order to discover underlying social structure such as: central nodes that act as hubs, leaders or gatekeepers; highly connected groups; and patterns of interactions between groups (Wasserman and Faust, 1994). SNA was used to gain an understanding of two types of networks. First, when a Twitter user (or *twitterer*) responds to a tweet, a network of twitterers is created with nodes (or vertices) representing twitterers and edges (or links) representing responses to particular tweets of particular twitterers. Since a response flows from a responder to a recipient, the links in this network are directed links. A second type of network that can be constructed from tweets is an "online resources" network as very often tweets contain links to web pages due to the 140-word limit of tweets preventing more detailed description of "what is happening." Such a network contains two types of nodes, namely, twitterers and resources and these networks are known as bimodal or bipartite networks in the SNA and graph theory literature (Borgatti, 2010; Borgatti and Everett, 1997).

Data collection

Tweets were harvested from Twitter using the *#qldfloods* hashtag by means of an in-house script created for the purpose. For the period February 3–20, 2011, 6,014 tweets were collected. In order to obtain the nodes and links (edgelists) required to generate the "users" and "users-resources" networks, individual tweets were parsed to extract the identity of the owner of the current tweet, the identity of the owner of the tweet that triggered a response, and the URL of the link contained in the tweet. Tweets that were not a response to another tweet or did not contain a URL were ignored for the purpose of the analysis.

Visual network analysis

The "users" network generated is shown in Figure 1. Nodes represent users and are shown as circles whereas directed links represent responses from one user to another for a particular tweet. Although it can be seen from the graph that there are individuals with high number of links (a couple of them is shown circled with thick lines), the diagram is too dense to clearly identify these individuals. Thus, as a next step in the analysis, the main component (the largest subnetwork in which there is a path from a user to any other user) was extracted for further visual analysis. The main component is shown in Figure 2. This network still has too many nodes to be able to identify the major players (again a few popular players are shown by means of thick circles). Thus, the next step was to resort to ego analysis, that is, quantitative analysis of the individual actors or "egos" that make up the nodes of the network, for identification of influential and popular twitterers.

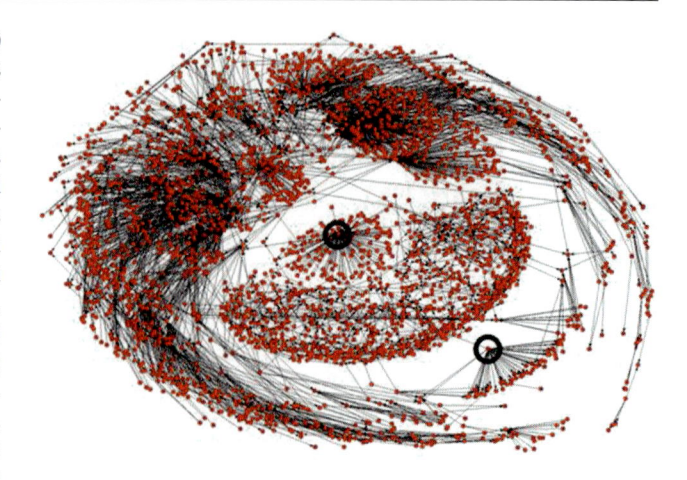

Queensland Floods (2010–2011) and "Tweeting", Figure 1 Users network.

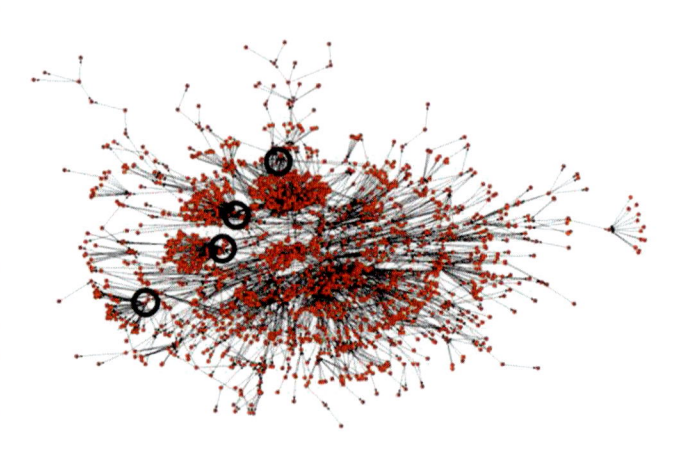

Queensland Floods (2010–2011) and "Tweeting", Figure 2 Main component of users network.

The "users-resources" network is shown in Figure 3. Two types of nodes are involved in this network. User nodes are represented as circles and online resources as squares. Since tweets of users are referring to these resources, the direction of the link is from the user node to the resource node. It can be seen from the figure that some of the resources were quite popular as many users were referring to them and at the same time some users were quite prolific in terms of the number of resources they were suggesting to others. Similarly to the "users" network, ego analysis was required to objectively determine the important nodes as this could not be done visually in a network of this size.

Ego analysis

Twitter users (nodes) were analyzed in terms of their centrality in the online community in order to find popular users or users at the center of attention (Scott, 2007). The centrality of a node can be measured locally in terms

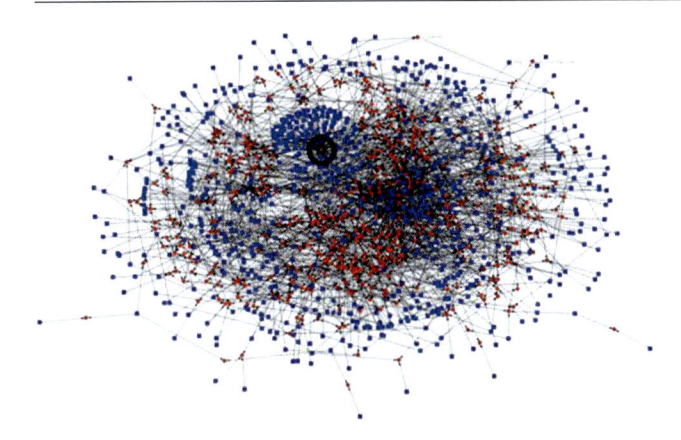

Queensland Floods (2010–2011) and "Tweeting",
Figure 3 Users-resources network.

of the number of direct connections to the node or globally to include indirect connections. Global centrality measures include:

- Closeness (Freeman, 1979), which is based on distances among nodes.
- Betweenness (Freeman, 1979), which is based on the extent to which a particular node lies between other nodes. A node of relatively low degree may play an important intermediary role (e.g., broker, gatekeeper, etc.) and hence be a central node in the network.
- Eigenvector (Bonacich, 1972), which is based on local centrality scores of nodes to which the node is connected, with high-score nodes contributing more than low-score nodes.
- Structural hole (Burt, 1992), which is based on the positional advantage (or disadvantage) of the node in the network.

Table 1 shows the top 24 users in the main component of the online community ranked by order of importance on previously discussed centrality measures, namely: (1) centrality, (2) betweenness, (3) closeness, (4) eigenvector, and (5) structural holes. In order to have a better idea who these users are, we visited their Twitter page and extracted their real names and biographies whenever these details were publicly available.

Since we represented the "users-network" as a directed network, a centrality degree analysis yielded two scores: *out degree* (number of tweets sent out by a particular user when responding to tweets of other users) and *in degree* (number of tweets received by a particular user as response from other users). The first section of Table 1 shows the ranking of the top 24 individuals on their *out degree* score while the second section of the table ranks individuals by their *in degree* score. Top scorers in terms of *out degree* were users having high influence in the network and just to name a few of them, they were: Sean Robertson (Australian Extreme Weather Event & Disaster Updates), Alexandra Worlson (Not-for-profit organization), Isagold Button (virtual name), Wilson Voight (virtual name), Cathy Border (Ten News), George Hall (social media volunteer), etc. Although these users and others from news channels and humanitarian organizations were busy posting tweets, online shopping organizations were also doing the same. The time of tweet collection was during the aftermath of the flood and these organizations were busy promoting their products.

Users with high *in degree* scores are regarded as prestigious or popular individuals and some of them were: Wilson Voight (virtual name), Tony Abbott (Leader of the Opposition), Queensland Police Service (QPS) Media Unit, Andrew Bartlett (ex-Senator), Rove McManus (media personality), Anna Bligh (Queensland Premier), ABC News, Julia Gillard (Prime Minister), Michael Bubble (Canadian singer), etc. In regards to betweenness centrality, the top individuals were: QPS Media Unit, Tony Abbott, Anna Bligh, Andrew Bartlett, Wilson Voight, Rove MacManus, ABC News, Sean Robertson, Animal Welfare League, Julia Gillard, etc. These users can be viewed as leaders in the online network since being on the shortest paths between other users they were able to control the flow of information in the network.

In terms of closeness centrality, leaders were: QPS Media Unit, George Hall (social media volunteer), Liz Baillie (anonymous user), Alexandra Worlson, Wilson Voight, Sean Robertson, Tony Abbott, Operation Angel (Humanitarian Org), Anna Bligh (Queensland Premier), etc. Since closeness centrality measures the distance of a node to all others in the network, the closer a node is to others, the more favored that node is. Nodes with high closeness scores are likely to receive information quicker than others as there are fewer intermediaries between them. It is well known that the Queensland Police Service played a very active role in the network and is thus acknowledged as the leader in terms of closeness centrality. Tony Abbott was the leading user when the eigenvector centrality criterion was used. This means that he was connected to many other users who were well connected and thus was most likely to receive new ideas. This fits well with his role as the leader of the opposition.

Structural holes were measured in terms of Effective size of the network, i.e., the number of connections a user has, minus the average number of connections that each individual has to other users. Tony Abbott, QPS Media Unit, Andrew Bartlett, Rove MacManus, Anna Bligh, etc. led on this criterion suggesting that they had more opportunities to act as brokers or coordinators.

The users-resources network was analyzed as a 2-node network (Borgatti, 2010; Borgatti and Everett, 1997) using centrality measures for online resources as well as users. The top 25 resources as measured by the degree centrality measure are shown in Table 2. At the top of the list, were online resources discussing the proposed flood levy (as this was the post-flood period) followed by web pages related to the following: people with disability, animal rescue, donations, legal help, damaged

Queensland Floods (2010–2011) and "Tweeting", Table 3 Centrality of users in the users-resources network

Users centrality							
1	wilsonvoight	Wilson Voight	9	ljloch	LJ Loch	17	tomtomprince	Tommy Prince
2	seldomsean63	Sean Robertson	10	lyndsayfarlow	Lyndsay Farlow	18	net_hues	Annette
3	babysgotstyle2	Alexandra Worlson	11	greengadflyaus	DianeO'Donovan	19	winecountrydog	
4	askkazza	Karen S	12	annfinster	Online shopping	20	servicecentralz	Service Central
5	visitvineyards	Visit Vineyards	13	ecrameri	Emma Crameri	21	findatoowoomba	Finda Toowoomba
6	qldonline	News and information	14	minxyferret		22	collectiveact	Rosie Williams
7	dmentedpollyana	Kath Cantarella	15	liz_baillie	Liz Baillie	23	autoday	
8	geehall1	George Hall	16	digellabakes	Danielle Crismani	24	alertnetclimate	Alertnet Climate

proved itself to be a valuable platform for disseminating vital information.

Although the tweets were collected after the impact stage, we believe they are still valuable since social network analysis revealed a number of users who were known to be active in that online community (StreetCorner, 2011). Additionally, using several local and global centrality measures, SNA helped to identify the effectiveness of these users who were identified as: local authorities (mainly the Queensland Police Services), political personalities (Queensland Premier, Prime Minister, Opposition Leader, Member of Parliament), social media volunteers, traditional media reporters, and people from not-for-profit, humanitarian, and community associations.

It is well known that Queensland Police took a very active role on Twitter, providing the public with regular updates on the situation every few minutes as well as dealing with the spread of misinformation on Twitter (StreetCorner, 2011). Queensland Police was also very active on Facebook, providing more detailed updates than is possible with 140-character tweets.

Although the analysis of the users-resources network identified a wide range of important resources, they were mostly web pages and blogs providing information of a more general nature rather than vital information and updates on the disaster. Since it is more effective to disseminate critical information on Facebook (because of high penetration) and mining Facebook was not part of the study, we missed such information. If this was the case, it makes sense to conclude that the resources identified supplemented the resources posted on Facebook.

The Federal Emergency Management Agency (FEMA) of the US Department of Homeland Security recognizes the usefulness of Twitter (and other social media) during emergencies and uses Twitter during all stages of a disaster.

Given the positive results obtained by the involvement of the local authorities and government officials in Queensland, and the increasing adoption of Twitter in other parts of the world for emergency situations, it seems reasonable to push for greater adoption of Twitter by local and federal authorities during periods of mass emergencies. This will help to ensure that vital information of an official and reliable nature is quickly propagated throughout the network and false rumors dealt with as they emerge.

Summary

SNA was used to study interactions between Twitter users during the Queensland 2010–2011 floods, one of the worst Australian flooding disasters. Influential members of the online community that emerged during the floods as well as important resources being referred to were identified. The analysis confirmed the active part taken by local authorities, namely, Queensland Police, government officials, and volunteers. Concerning online resources suggested by users, no sensible conclusion can be drawn as important resources identified were more of a general rather than critical nature. This might be comprehensible as it was the post-flood period.

Bibliography

Bonacich, P., 1972. Factoring and weighting approaches to status scores and clique identification. *Journal of Mathematical Sociology*, **2**, 113–120.

Borgatti, S. P., 2010. 2-mode concepts in social network analysis (to appear in *Encyclopedia of Complexity and System Science*). Retrieved May 9, 2011, from http://www.steveborgatti.com/papers/2modeconcepts.pdf.

Borgatti, S. P., and Everett, M. G., 1997. Network analysis of 2-mode data. *Social Networks*, **19**, 243–269.

Brisbane Times, 2011. Three-quarters of Queensland a disaster zone. Retrieved March 8, 2011, from http://www.brisbanetimes.com.au/environment/weather/threequarters-of-queensland-a-disaster-zone-20110111-19mf8.html.

Burt, R. S., 1992. *Structural holes: the social structure of competition*. Cambridge, MA: Harvard University Press.

Castillo, C., Mendoza, M., and Poblete, B., 2011. *Information Credibility on Twitter.* Paper presented at the WWW 2011, Hyderabad.

Dynes, R. R., and Quarantelli, E., 1968. Group behavior under stress: a required convergence of organizational and collective behavior perspectives. *Sociology and Social Research*, **52**, 416–429.

Freeman, L., 1979. Centrality in social networks: conceptual clarification. *Social Networks*, **1**, 215–239.

Mendoza, M., Poblete, B., and Castillo, C., 2010. *Twitter Under Crisis: Can We Trust What We RT?* Paper presented at the SOMA'10, Washington, DC.

Palen, L., and Vieweg, S., 2008. *The Emergence of Online Widescale Interaction in Unexpected Events: Assistance, Alliance & Retreat.* Paper presented at the CSCW'08.

Scott, J., 2007. *Social Network Analysis.* London: Sage Publications.

Starbird, K., Palen, L., Hughes, A. L., and Vieweg, S., (2010). *Chatter on the Red: What Hazards Threat Reveals About the Social Life of Microblogged Information.* Paper presented at the CSCW'10, Savannah.

StreetCorner, 2011. Police and public turn to social media & maps in the Queensland flood crisis. Retrieved March 11, 2011, from http://www.streetcorner.com.au/news/showPost.cfm?bid=20421&mycomm=ES.

Vieweg, S., 2010. *Microblogged Contributions to the Emergency Arena: Discovery, Interpretation and Implications.* Paper presented at the Computer Supported Collaborative Work.

Wasserman, S., and Faust, K., 1994. *Social Network Analysis: Methods and Applications.* New York: Cambridge University Press.

Yates, D., and Paquette, S., 2011. Emergency knowledge management and social media technologies: a case study of the 2010 Haitian earthquake. *International Journal of Information Management*, **31**(1), 6–13.

Cross-references

Communicating Emergency Information
Community Management of Hazards
Disaster Risk Reduction
Education and Training for Emergency Preparedness
Emergency Management
Federal Emergency Management Agency (FEMA)
Global Network of Civil Society Organizations for Disaster Reduction
Information and Communications Technology
Integrated Emergency Management System
Internet, World Wide Web, and Natural Hazards
Mass Media and Natural Disasters
Pacific Tsunami Warning and Mitigation System (PTWS)
Red Cross and Red Crescent, International Federation of
Warning systems

QUICK CLAY

Marten Geertsema
British Columbia Forest Service, Prince George, BC, Canada

Synonyms

Glaciomarine sediment; Leda clay; Sensitive clay

Definition

Quick clay is a special type of clay prone to sudden strength loss upon disturbance. From a relatively stiff material in the undisturbed condition, an imposed stress can turn such clay into a liquid gel.

Discussion

A quick clay is defined as a clay where the undisturbed shear strength of the soil is at least 30 times greater than the remolded (or disturbed) shear strength. The ratio of undisturbed to disturbed strength is termed *sensitivity*. Thus, a quick clay is defined as very sensitive.

Quick clays tend to occur along previously glaciated coastlines, especially in parts of Canada and Scandinavia, but they have also been found in Japan and in Alaska. Coastlines in these areas were once submerged by the weight of glaciers during glaciation. As glaciers retreated, the sea migrated inland in concert with the retreating icefronts. Glacially ground sediments composed of rock flour, silt, and clay minerals were deposited in the saltwater of these ice marginal seas by glacial meltwater. These sediments had a porous flocculated structure (Figure 1).

As the glaciers disappeared the land began to rebound isostatically, rising as much as 300 m above present day sea level in some parts of the world such as the Hudson Bay area of Canada. The uplifted glaciomarine sediments were then exposed to rainfall and groundwater. Salts in the clays were gradually leached out of the sediments by freshwater. Salt contents would decrease from initial concentrations as high as 30 g/l in the sea to less than 1 g/l. With a lower salt content, repulsive forces between particles increased, leaving the saturated, porous sediment prone to collapse. Given the ideal clay mineralogy, some sediments that underwent this process can become quick clays.

In freshwater, clay particles settle more slowly than the larger silt particles. In saltwater, clays and silts flocculate and settle with a random orientation. Negative, repulsive charges on the clay particles are neutralized by cations such as Na^+ and Ca^{2+} in seawater. The resulting sediment has an open structure with high water content. The positive charges of the salts maintain the interparticle bonds.

An imposed load, vibration, or bank erosion can trigger collapse of the sedimentary structure in sensitive glaciomarine soil, often causing liquefaction. During liquefaction, the weight of the soil is transferred from the solids to the porewater.

Quick clays are hazardous because they can lead to sudden rapid landslides on extremely low gradients. An example of a quick clay landslide is shown in Figure 1. Here some 2.5 million m^3 of sediment travelled 1200 m on a 2 degree slope, near terrace, British Columbia in January 1994. Other significant quick clay landslides occurred in Norway and Eastern Canada. On April 29, 1978, a sudden landslide occurred near the town of Rissa, Norway. The largest landslide of the century in Norway, it covered 33 ha and involved 5–6 million m^3 of quick clay. The landslide was triggered by a small external load: earth fill from the excavation of a barn. Seven farms and five houses were destroyed.

Quick Clay, Figure 1 Liquefaction of quick clays may generate large, rapid landslides, such as this one at Mink Creek in British Columbia. This 1994 landslide covered 43 ha and involved 2.5 million cubic meters of soil. Photo Marten Geertsema.

On May 4, 1971, 7 million m³ of quick clay at Saint Jean Vianney, Quebec, Canada, suddenly began to flow at a rate of more than 25 km/hour into the Rivière du Petit-Bras carrying with it some 40 homes. The crater left by the landslide was 32 ha in area and up to 30-m deep.

Bibliography

Geertsema, M., Cruden, D. M., and Schwab, J. W., 2006. A large rapid landslide in sensitive glaciomarine sediments at Mink Creek, nortwestern British Columbia, Canada. *Engineering Geology*, **83**, 36–63.

Norwegian Geotechnical Institute, 1982. The Rissa Landslide. Video tape.

Tavenas, F., Chagnon, J.-Y., and LA Rochelle, P., 1971. The Saint-Jean-Vianney landslide: observations and eyewitnesses accounts. *Canadian Geotechnical Journal*, **8**, 463–478.

Torrance, J. K., 1983. Towards a general model of quick clay development. *Sedimentology*, **30**, 547–555.

Cross-references

Collapsing Soil Hazards
Erosion
Erosivity
Expansive Soils and Clays
Hydrocompaction Subsidence
Land-Use Planning
Landslide
Landslide Types
Lateral Spreading
Liquefaction
Mass Movement
Mudflow
Pore-Water Pressure
Quick Sand

QUICK SAND

János Kovács
University of Pécs, Pécs, Hungary

Synonyms

Running sand

Definition

A mass or bed of fine sand, as at the mouth of a river or along a seacoast, or in the desert, that consists of smooth rounded grains with little tendency to mutual adherence and that is usually thoroughly saturated with water flowing upward through the voids, forming a soft, shifting, semiliquid, highly mobile mass that yields easily to pressure and tends to suck down and readily swallow heavy objects resting on or touching its surface.

Discussion

According to Yamasaki (2003), quicksand is a mixture of sand and water, or sand and air that looks solid, but becomes unstable when disturbed by any additional stress. In normal sand, grains are packed tightly together to form a rigid mass, with about 25–30% of the space (voids) between the grains filled with air or water. Because many sand grains are elongate rather than spherical, loose packing of the grains can produce sand in which voids make up 30–70% of the mass (Yamasaki, 2003). This arrangement is similar to a house of cards in that the space between the cards is significantly greater than the space occupied by the cards. The sand collapses, or becomes "quick," when additional force from loading, vibration, or the upward migration of water overcomes the friction holding the grains together (Yamasaki, 2003).

Most quicksand occurs in settings where there are natural springs, either at the base of alluvial fans (cone-shaped bodies of sand and gravel formed by rivers flowing from mountains), along riverbanks, or on beaches at low tide (Yamasaki, 2003). In such cases, the loose packing is maintained by the upward movement of water. Quicksand does occur in deserts, but only very rarely: where loosely packed sands occur, such as on the down-wind sides of dunes, the amount of sinking is limited to a few centimeters, because once the air in the voids is expelled, the grains are too densely packed to allow further compaction (Yamasaki, 2003).

The nature and danger of quicksand have been disputed for a long time (Freundlich and Juliusburger, 1935; Khaldoun et al., 2005; Matthes, 1953). Despite widespread belief that humans can be swallowed or even sucked in, engineers of soil mechanics have typically asserted that, since the density of sludge is larger than that of water, a person cannot fully submerge (Kadau et al., 2009). The fluidization of a soil due to an increase in ground water pressure, which in fact is often responsible for catastrophic failures at construction sites, is called by engineers the "quick condition" and can theoretically happen to any soil. Another way for fluidization can result from vibrations either from mechanical engines or an earthquake (Kadau et al., 2009).

Bibliography

Freundlich, H., and Juliusburger, F., 1935. Quicksand as a thixotropic system. *Transactions of the Faraday Society*, **31**, 769–773.

Kadau, D., Herrmann, H. J., Andrade, J. S., Jr., Araújo, A. D., Bezerra, L. J. C., and Maia, L. P., 2009. Living quicksand. *Granular Matter*, **11**, 67–71.

Khaldoun, A., Eiser, E., Wegdam, G. H., and Bonn, D., 2005. Rheology: liquefaction of quicksand under stress. *Nature*, **437** (29 Sept), 635.

Matthes, G. H., 1953. Quicksand. *Scientific American*, **188**, 97–102.

Yamasaki, S., 2003. What is quicksand? *Scientific American*, **288**, 95–95.

Cross-references

Collapsing Soil Hazards
Expansive Soils and Clays
Hydrocompaction Subsidence
Liquefaction
Quick Clay

R

RADIATION HAZARDS

Lev I. Dorman
Israel Cosmic Ray and Space Weather Center and Emilio Segre' Observatory, Tel Aviv University, TECHNION, and Israel Space Agency, Qazrin, Israel
IZMIRAN of Russian Academy of Sciences, Troitsk, Moscow region, Russia

Definition

Radiation hazards for people and technology are determined by electromagnetic radiation (from intense radio waves up to UV, X-rays, and γ-rays) and by fluxes of corpuscular radiation (energetic protons, neutrons, nucleous, electrons, pions, muons, and in case of nearby Supernova even by great fluxes of neutrino, see *Supernova*).

Main sources

Radiation hazards for people and technology on the ground and at different altitudes in atmosphere are determined by natural sources (mainly from galactic and solar cosmic rays and from radioactive elements in soil and in air) and by artificial (man-made) sources which generate different types of electromagnetic and corpuscular radiation (from home-used devices such as TV, microwave heaters, mobile phones to atomic and H-bomb explosions). With an increase of altitude, the natural radiation hazard from cosmic rays increases considerably and from radioactive elements – sufficiently decreases. For satellites and astronauts situated in the magnetospheres of the Earth and other planets (especially as Jupiter and Saturn), important natural sources of radiation hazard are cosmic rays and radiation belts; for spaceprobes and astronauts inside the Heliosphere – galactic, solar, planetary, and anomaly cosmic rays, and outside the Heliosphere – mainly galactic cosmic rays.

The basic unit of radiation dose

The basic unit for radiation dose is *rad*, defined in terms of the energy (in ergs or J) deposited by radiation per unit mass (gramm or kg) of exposed matter: 1 *rad* = 100 erg/g = 0.01 J/kg.

Relative biological effectiveness of different types of radiation

Different types of radiation have very different influence on human health. As a result the special factor relative biological effectiveness (RBE), averaged for full body, was introduced. For X-rays, γ-rays, and energetic electrons RBE = 1, for thermal neutrons RBE = 3, for energetic neutrons, protons, and α-particles RBE = 10, and for energetic nucleus heavier than α-particles RBE = 20. Therefore, the unit of radiation, effectively influenced on people is rem (Roentgen Equivalent to Man), defined as *rem* = RBE $\times$ *rad* = 0.01 *seivert*. The unit *seivert* was named after Rolf Sievert (1898–1966), a pioneering Swedish radiation physicist.

Effects of radiation on people

The large radiation dose which can kill outright is about ***5,000 rem*** (so-called Instant Death). The radiation dose near ***900 rem*** leads to death over the course of one day (so-called Overnight Death). The dose approximating ***500 rem*** causes severe radiation sickness (nausea, hair loss, skin lesions, etc.) as the body's short-lived cells fail to provide new generations to replace their normal mortality (cell reproductive death). It is not this trauma which usually kills, however, but the complications that arise from a lack of resistance to infection, in turn due to the lack of new generations of white blood cells. If one can survive the initial radiation sickness and avoid infection, one will probably recover completely in the short term but will very likely develop cancer (especially leukemia) in some 10–20 years with a high probability for genetic mutations. From a whole-body dose of about **100 rem** delivered in less than about a week, it is unlikely one can notice any immediate severe symptoms. However, an individual is likely to develop leukemia in 10–30 years, and will have a significant chance of genetic mutations. The average exposure from natural sources of radiation on the ground is about **0.3 rem** per year. The part of exposure caused by local radioactivity decreases very much with increasing altitude and varies sufficiently from place to place. The part of exposure caused by cosmic rays of galactic and solar origin increases many times with increasing altitude and geomagnetic latitude, and varies considerably with time during the 22-year magnetic solar activity cycle and during great solar flare events.

Different resistance to radiation of different parts of body

It is important to note that different organs or body parts have dramatically different resistance to radiation. The hands, in particular, are able to withstand radiation doses that would kill if the whole body were subjected to this level. The lens of the eye and the gonads are considered to be the most vulnerable and should be protected the most.

Effects of radiation on high level technology systems

Large fluxes of energetic particles (from cosmic rays and radiation belts) very much influence electronics and computers on satellites with different orbits (especially characterized at altitudes $\geq 1,000$ km and inclinations to the equatorial plane $\geq 45°$) and airplanes on different airlines (especially characterized by altitudes ≥ 10 km and those crossing high latitudes). Energetic particles produced in electronic device ionic tracks and along these tracks can discharge with the destruction of some part of electronics. With increasing energetic particle fluxes the probability of electronic systems destruction sufficiently increases and leads to satellite malfunctions and even satellite loss. The destruction of airplane electronic systems may lead to problems in operations and even in the continuation of flight.

Protection from radiation

For protection from dangerous radiation generated by artificial sources there are two main methods: (1) increasing the distance from the source of radiation (e.g., increasing the distance by 10 times leads to a decrease of the level of radiation by 100 times); and (2) shielding people and electronic devices by thin plates of lead (from X-rays and γ-rays) and/or by thick concrete (from neutrons and high-energy charged particles). For protection of people and electronic devices from natural dangerous radiation the first method cannot be used and is only effective with the second method. In this case it is very important to produce ALERT (by using real time 1-min data of continued registration of cosmic rays through a ground network of neutron monitors and on satellites) and other parameters of space environment, (see *Automated Local Evaluation in Real Time (ALERT)*) with information on the expected time and level of natural radiation hazard. According to collected data astronauts should go for several hours inside special structures comprising shielding and to save electronic devices they should switch off some parts from electric power. To save passengers and crew on airplanes dispatchers should decide what to do (e.g., for airplanes at high latitudes to decrease the altitude for some short time) after receiving an ALERT. In case of ALERT for radiation from a nearby Supernova explosion (see *Supernova*) and with an estimation of the expected radiation hazard, Governments would need to decide how to save humanity and the biosphere (e.g., to prepare in several tens of years special buildings to protect them from significant radiation dose and/or to live under few meters below ground for several hundreds of years).

Cross-references

Automated Local Evaluation in Real Time (ALERT)
Dose rate (of risk)
Electomagnetic radiation (EMR)
Solar flare
Space weather
Sunspots
Supernova
Tohoku, Japan, earthquake, tsunami and Fukushima accident (2011)

RADON HAZARDS*

James D. Appleton
British Geological Survey, Nottingham, UK

Definition

Radon (^{222}Rn) is a natural radioactive gas produced by the radioactive decay of radium (^{226}Ra), which in turn is derived from the radioactive decay of uranium. Uranium is found in small quantities in all soils and rocks, although the amount varies from place to place. Radon occurs in the ^{238}U decay series, has a half-life of 3.82 days, and emanates from rocks and soil. Radon provides about 50% of the total ionizing radiation dose to the general world population although the dose received by an individual depends upon where one lives and ones' life style (see *Dose Rate*). Most exposure to radon results from living indoors. Radon decays to form very small, solid, short-lived radioactive decay products (^{218}Po and ^{214}Po) that become attached to natural aerosol and dust particles.

---

*©British Geological Survey

When inhaled, these can irradiate the bronchial epithelial cells of the lung with alpha particles and cause DNA damage. Epidemiological studies confirm that exposure to radon in homes increases the risk of lung cancer in people. It is estimated that between 3% and 14% of all lung cancers are linked to radon. Radon is the second most important cause of lung cancer after smoking and is much more likely to cause lung cancer in people who smoke, or who have smoked, than in lifelong nonsmokers for whom radon is the primary cause of lung cancer. There is no evidence for a threshold radon concentration below which radon exposure is safe. The majority of radon-induced lung cancers are calculated to be caused by low and moderate radon concentrations rather than by high radon concentrations. This is because less people are, in general, exposed to high indoor radon concentrations (WHO, 2009). It is estimated that radon in drinking water causes about 170 cancer deaths per year in the USA, about 90% from lung cancer caused by breathing radon released from water, and about 10% from stomach cancer caused by drinking radon-containing water.

The concentration of radioactivity in the air due to radon is measured in becquerels per cubic meter (Bq m^{-3}) of air. Radon concentrations in outdoor air are generally low (4–8 Bq m^{-3}) whereas radon in indoor air ranges from less than 20 Bq m^{-3} to about 110,000 Bq m^{-3} with a population-weighted world average of 40 Bq m^{-3}. Country arithmetic means range from 20 Bq m^{-3} in the UK, 22 Bq m^{-3} in New Zealand, 44 Bq m^{-3} in China, and 46 Bq m^{-3} in the USA, to 120 Bq m^{-3} in Finland (UNSCEAR, 2009). Radon in the air that occupies the pores in soil commonly varies from 5 to 50 Bq L^{-1} but may be <1 or more than 2,500 Bq L^{-1}. The amount of radon dissolved in ground water ranges from about 3 to nearly 80,000 Bq L^{-1} (Appleton, 2007).

Geological associations

Geology exerts a strong control on indoor radon and radon hazard (Appleton and Miles, 2010). Relatively high concentrations of radon are associated with particular types of bedrock and unconsolidated deposits, for example, some granites, uranium-enriched phosphatic rocks and black shales, limestones, sedimentary ironstones, permeable sandstones, and uraniferous metamorphic rocks. Fault shear zones, glacial deposits derived from uranium-bearing rock, and even clay-rich sediments from which high radon emanation reflects large surface area and high permeability caused by cracking may be radon prone along with uranium and radium-enriched soils derived from limestone (especially karstic terrain), uranium mining residues, and mine tailings. In SW England and Scotland (Scheib et al., 2009), there is a correlation between areas where it is estimated that more than 30% of the house radon levels are above 200 Bq m^{-3} and the major granite areas. In England and Wales, 10% to more than 30% of houses built on Carboniferous limestones have radon concentrations greater than 200 Bq m^{-3}. Similar associations between high radon and Lower Carboniferous limestone, Namurian uraniferous and phosphatic black shales, some granites, and highly permeable fluvioglacial deposits have also been recorded in Ireland (Cliff and Miles, 1997).

In the Czech Republic, the highest indoor and soil gas radon concentrations are associated with the Variscan granites, granodiorites, syenites (12–20 mg kg^{-1} U in bedrock), and phonolites (10–35 mg kg^{-1} U in bedrock) of the Bohemian massif as well as Palaeozoic metamorphic and volcanic rocks and also with uranium mineralization (Barnet et al., 2008). The highest radon in Germany occurs over the granites and Palaeozoic basement rocks (Kemski et al., 2009). High radon potential in Belgium is associated with strongly folded and fractured Cambrian to Lower Devonian bedrocks in which uranium preferentially concentrated in ferric oxyhydroxides in fractures and joints is considered to be the main source of radon (Zhu et al., 2001). In France, some of the highest radon levels occur over peraluminous leucogranites or metagranitoids (8 mg kg^{-1} U in bedrock) in a stable Hercynian basement area located in South Brittany (Ielsch et al., 2001). Granite and alum shale have high radon potential in Sweden, Norway and Belgium.

Permeability and uranium composition of unconsolidated deposits are the main factors influencing their radon potential. Fragments and mineral grains of uranium-rich granites, pegmatites, and black alum shales are dispersed in till and glaciofluvial deposits leading to high radon in soils and dwellings in Sweden, especially when the glaciofluvial deposits are highly permeable sands and gravels (Clavensjö and Åkerblom, 1994). In Norway, high radon levels are associated with highly permeable fluvioglacial and fluvial deposits derived from all rock types and also with moderately permeable unconsolidated sediments (mainly basal till) containing radium-rich rock fragments (Smethurst et al., 2008).

Release and migration of radon gas

The rate of release of radon from rocks and soils is largely controlled by their uranium and radium concentrations, grain size, and by the types of minerals in which the uranium and radium occur (Ball et al., 1991). Most of the radon formed from the decay of radium remains in the mineral grains. Normally 20–40% of the newly generated radon emanates to pore spaces where it is mixed in the soil air or water that fill the pore spaces and from which it can be transported by diffusion or by flow in carrier fluids (soil air or water). After uranium and radium concentration, the porosity, permeability, and moisture content of rocks and soils are probably the next most significant factors influencing the concentration of radon in soil gas and buildings. Soil air containing radon is transported along natural pathways, which include planar discontinuities and openings such as bedding planes, joints, shear zones, and faults as well as potholes and swallow holes in limestone. Artificial migration pathways underground include mine workings, disused tunnels, and shafts as well as near-surface installations for electricity, gas, water, sewage, and telecommunications services.

Radon from soils and rocks is transported into buildings through cracks in solid floors and walls below construction level; through gaps in suspended concrete and timber floors and around service pipes; through crawl spaces, cavities in walls, construction joints and small cracks or pores in hollow-block walls. Warm air rising within a building, called the stack effect, and the suction effects on the roof caused by wind action draws radon in from the ground. Whereas soil air is the dominant source of indoor radon and most houses draw less than 1% of their indoor air from the soil, some houses with low indoor air pressures, poorly sealed foundations, permeable ground underneath, and several entry points for soil air may draw as much as 20% of their indoor air from the soil. Consequently, radon levels inside the house may be very high even in situations where the soil air has only moderate amounts of radon. The design, construction, and ventilation of the home affect indoor radon levels as do both the season and the weather.

Homes that use surface water do not have a radon problem from their water because radon decays in a few days; so water reservoirs usually contain very little radon. In addition, water processing in large municipal systems aerates the water, which allows radon to escape, and also delays the use of water until most of the remaining radon has decayed. However, in some parts of the USA and Sweden with high levels of uranium in the underlying rocks and where ground water is the main water supply for homes and communities, radon from the domestic water released during showering and other household activities could add radon to the indoor air.

Building materials generally contribute only a very small percentage of the indoor air radon concentrations, although in a few areas, concrete, blocks, or wallboard made using radioactive shale or waste products from uranium mining will make a larger contribution to the indoor radon.

Radon hazard mapping

Accurate mapping of radon-prone areas can help to ensure that the health of occupants of new and existing dwellings and workplaces is adequately protected. Radon potential maps can be used to assess whether radon preventive measures may be required in new buildings, for the cost-effective targeting of radon monitoring in existing dwellings and workplaces, and to provide a radon assessment for homebuyers and sellers. National authorities are recommended to set a reference level for radon that represents the maximum accepted radon concentration in a residential dwelling. Remedial actions may be recommended or required for homes in which radon exceeds the reference level. Various factors such as the distribution of radon concentrations, the number of existing homes with high radon concentrations, the arithmetic mean indoor radon level, and the prevalence of smoking are usually considered when setting a national radon reference level. WHO (2009) proposed a reference level of 100 Bq m^{-3} to minimize health hazards due to indoor radon exposure although if this concentration cannot be reached, the reference level should not exceed 300 Bq m^{-3}. This represents approximately 10 mSv per year according to recent calculations and recommendations by the International Commission on Radiation Protection (ICRP, 2007, 2009) which also recommends 1,000 Bq m^{-3} as the entry point for applying occupational radiological protection requirements.

Two main procedures have been used for mapping radon-prone areas. The first uses radon measurements in existing dwellings to map the variation of radon potential between administrative or postal districts, or grid squares, or within geological polygons. Since the main purpose of maps of radon-prone areas is to indicate radon levels in buildings, maps based on actual measurements of radon in buildings are generally preferable to those based on other data. Requirements for mapping radon-prone areas using indoor radon data are similar whether the maps are made on the basis of grid squares or geological units. These requirements include accurate radon measurements made using a reliable and consistent protocol; centralized data holdings; sufficient data evenly spread; and automatic conversion of addresses to geographical coordinates. Great Britain is currently the only country that meets all of these requirements for large areas (Miles and Appleton, 2005). Where lesser quality or quantity of indoor radon data are available, there is greater reliance on proxy data for radon potential mapping (e.g., Czech Republic, Germany, Sweden, and USA). The factors that influence radon concentrations in buildings are largely independent and multiplicative, the distribution of radon concentrations is usually lognormal. Therefore, lognormal modeling was used to produce accurate estimates of the proportion of homes above a reference level in the UK (Miles et al., 2007). Radon maps based on indoor radon data grouped by geological unit have the capacity to accurately estimate the percentage of dwellings affected together with the spatial detail and precision conferred by the geological map data. Combining the grid square and geological mapping methods gives more accurate maps than either method can provide separately (Miles and Appleton, 2005). Radon potential mapping using indoor radon measurements has been carried out in other European countries, including Austria, France, Ireland, and Luxembourg (Dubois, 2005). In Austria, indoor radon data, normalized to a defined "standard room" to reduce variations related to house type, are combined with geological information to produce a national radon potential map (Friedmann and Gröller, 2010).

In other countries, geological radon potential maps predict the average indoor radon concentration (USA; Gundersen et al., 1992) or give a more qualitative indication of radon risk (Germany and the Czech Republic). Radon hazard mapping of the Czech Republic is based upon airborne radiometry, geology, pedology, hydrogeology, ground radiometry, and soil gas radon data (Barnet et al., 2008). Soil gas radon measurements combined with an assessment of ground permeability have been used to map geological radon potential in the absence of sufficient indoor radon measurements. Ten to fifteen soil gas radon measurements are generally required to characterize a site or geological unit. Ground classifications based on geology, permeability, and soil gas radon measurement have been developed in Sweden (Clavensjö and Åkerblom, 1994), Finland, Germany (Kemski et al., 2009), Estonia, and the Czech Republic (Barnet et al., 2008). In some cases, soil gas radon data may be difficult to interpret due to the effects of large diurnal and seasonal variations in soil gas radon close to the ground surface as well as variations in soil gas radon on a scale of a few meters.

Uranium and radium concentrations in surface rocks and soils are a useful indicator of the potential for radon emissions from the ground. Uranium can be estimated by gamma spectrometry either in the laboratory or by ground, vehicle, or airborne surveys. The relatively close correlation between airborne and ground radiometric measurements and indoor radon concentrations has been demonstrated in Virginia and New Jersey in the USA, Nova Scotia in Canada, and also in parts of the UK (Appleton et al., 2008) and Norway (Smethurst et al., 2008). Areas with high permeability tend to have significantly higher indoor radon levels than would be otherwise expected from the equivalent ^{226}Ra concentrations, reflecting an enhanced radon flux from permeable ground.

Summary

Radon is a natural radioactive gas produced by the radioactive decay of radium, which in turn is derived from the radioactive decay of uranium found in all rocks and soils. Radon is the second most important cause of lung cancer after smoking. Radon emanated from rocks and soil tends to concentrate in enclosed spaces such as houses and underground spaces including mines and tunnels. Infiltration of radon in soil gas is the most important source of radon in buildings. Radon levels often vary widely between adjacent buildings due to differences in the radon potential of the underlying ground as well as differences in construction style and use. Although a radon potential map can indicate the relative radon risk for a building in a particular locality, it cannot predict the radon risk for an individual building. Radon from building materials and from water extracted from wells is generally less important.

Bibliography

Appleton, J. D., 2007. Radon: sources, health risks, and hazard mapping. *Ambio*, **36**(1), 85–89.

Appleton, J. D., and Miles, J. C. H., 2010. A statistical evaluation of the geogenic controls on indoor radon concentrations and radon risk. *Journal of Environmental Radioactivity*, **101**(10), 799–803.

Appleton, J. D., Miles, J. C. H., Green, B. M. R., and Larmour, R., 2008. Pilot study of the application of Tellus airborne radiometric and soil geochemical data for radon mapping. *Journal of Environmental Radioactivity*, **99**(10), 1687–1697.

Ball, T. K., Cameron, D. G., Colman, T. B., and Roberts, P. D., 1991. Behaviour of radon in the geological environment – a review. *The Quarterly Journal of Engineering Geology*, **24**(2), 169–182.

Barnet, I., Pacherová, P., Neznal, M., and Neznal, M., 2008. *Radon in Geological Environment – Czech Experience*. Prague: Czech Geological Survey. Czech Geological Survey Special Paper 19.

Clavensjö, B., and Åkerblom, G., 1994. *The Radon Book*. Stockholm: The Swedish Council for Building Research.

Cliff, K. D., and Miles, J. C. H. (eds.), 1997. *Radon Research in the European Union*. EUR 17628. Chilton, UK: National Radiological Protection Board.

Dubois, G. 2005. *An Overview of Radon Surveys in Europe*. EUR 21892 EN. Luxembourg: Office for Official Publications of the European Community.

Friedmann, H., and Gröller, J., 2010. An approach to improve the Austrian Radon Potential Map by Bayesian statistics. *Journal of Environmental Radioactivity*, **101**(10), 804–808.

Gundersen, L. C. S., Schumann, E. R., Otton, J. K, Dubief, R. F., Owen, D. E., and Dickenson, K. E. 1992. Geology of radon in the United States. In Gates, A. E., and Gundersen, L. C. S. (eds.), *Geologic Controls on Radon*. Geological Society America Special Paper 271, pp 1–16.

ICRP, 2007. *The 2007 Recommendations of the International Commission on Radiological Protection*. ICRP Publication 103. Ann. ICRP 37 (2–4).

ICRP, 2009. *International Commission on Radiological Protection Statement on Radon*. ICRP Ref 00/902/09 http://www.icrp.org/docs/ICRP_Statement_on_Radon(November_2009).pdf. Last accessed at 14.34 on January 13, 2010.

Ielsch, G., Thieblemont, D., Labed, V., Richon, P., Tymen, G., Ferry, C., Robe, M. C., Baubron, J. C., and Bechennec, F., 2001. Radon (Rn-222) level variations on a regional scale: influence of the basement trace element (U, Th) geochemistry on radon exhalation rates. *Journal of Environmental Radioactivity*, **53**(1), 75–90.

Kemski, J., Klingel, R., Siehl, A., and Valdivia-Manchego, M., 2009. From radon hazard to risk prediction – based on geological maps, soil gas and indoor measurements in Germany. *Environmental Geology*, **56**, 1269–1279.

Miles, J. C. H., and Appleton, J. D., 2005. Mapping variation in radon potential both between and within geological units. *Journal of Radiological Protection*, **25**, 257–276.

Miles, J. C. H., Appleton, J. D., Rees, D. M., Green, B. M. R., Adlam, K. A. M., and Myers, A. H., 2007. *Indicative Atlas of Radon in England and Wales*. Chilton: HPA.

Scheib, C., Appleton, J. D., Miles, J. C. H., Green, B. M. R., Barlow, T. S., and Jones, D. G., 2009. Geological controls on radon potential in Scotland. *Scottish Journal of Geology*, **45**(2), 147–160.

Smethurst, M. A., Strand, T., Sundal, A. V., and Rudjord, A. L., 2008. Large-scale radon hazard evaluation of the Oslofjord region of Norway utilizing indoor radon concentrations, airborne gamma ray spectrometry and geological mapping. *Science of the Total Environment*, **407**, 379–393.

UNSCEAR, 2009. *United Nations Scientific Committee on the Effects of Atomic Radiation (UNSCEAR)*. UNSCEAR 2006 Report. Annex E. Sources-to-Effects Assessment for Radon in Homes and Workplaces. New York: United Nations.

World Health Organisation (WHO), 2009. *WHO Handbook on Indoor Radon: A Public Health Perspective*. Geneva: WHO Press.

Zhu, H. C., Charlet, J. M., and Poffijn, A., 2001. Radon risk mapping in southern Belgium: an application of geostatistical and GIS techniques. *The Science of the Total Environment*, **272** (1–3), 203–210.

Cross-references

Dose Rate
Hazard and Risk Mapping
Natural Radioactivity

RECOVERY AND RECONSTRUCTION AFTER DISASTER

Michael K. Lindell
Texas A&M University, College Station, TX, USA

Definition

Disaster recovery. Disaster recovery has three distinct but interrelated meanings. First, it is a *goal* that involves the restoration of normal community activities that were disrupted by disaster impacts – in most people's minds, exactly as they were before the disaster struck. Second, it is a *phase* in the emergency management cycle that begins with stabilization of the disaster conditions (the end of the emergency response phase) and ends when the community has returned to its normal routines. Third, it is a *process* by which the community achieves the goal of returning to normal routines. The recovery process involves both activities that were planned before disaster impact and those that were improvised after disaster impact.

Disaster impacts. These are the physical and social disturbances that a hazard agent inflicts when it strikes a community. Physical impacts comprise casualties (deaths, injuries, and illnesses) and damage to agriculture, structures, infrastructure, and the natural environment. Social impacts comprise psychological impacts, demographic impacts, economic impacts, and political impacts.

Incident stabilization. This is the point in time at which the immediate threats to human safety and property resulting from the physical impacts of the hazard agents have been resolved and the community as a whole can focus on disaster recovery.

Disaster impacts

As noted earlier, disaster impacts comprise both physical and social impacts. The physical impacts are casualties (deaths and injuries) and property damage, and both vary substantially across hazard agents. The physical impacts of a disaster are usually the most obvious, easily measured, and first reported by the news media. Social impacts include psychosocial, demographic, economic, and political impacts. A very important aspect of disaster impacts is their *impact ratio* – the amount of damage divided by the amount of community resources. Long-term social impacts tend to be minimal in the USA because most hazard agents have a relatively small scope of impact and tend to strike undeveloped areas more frequently than intensely developed areas simply because there are more of the former than the latter. Thus, the numerator of the impact ratio tends to be low and local resources are sufficient to prevent long-term effects from occurring. Even when a disaster has a large scope of impact and strikes a large developed area (causing a large impact ratio in the short term), state and federal agencies and NGOs (nongovernmental organizations such as the American Red Cross) direct recovery resources to the affected area, thus preventing long-term impacts from occurring. For example, 1992 Hurricane Andrew inflicted $26.5 billion in losses to the Miami Florida USA area, but this was only 0.4% of the US GDP (Charvériat, 2000). However, the fact that communities *as a whole* recover does not mean that specific neighborhoods or households within those neighborhoods recover at the same rate or even at all. Similarly, it does not mean specific economic sectors or individual businesses within those sectors will be able to maintain or even resume operations. Thus, it is important to anticipate which population segments and economic sectors will have the most difficulty in recovering. This will enable community authorities to intervene with technical and financial assistance when it is needed, monitor their recovery, and encourage them to adopt hazard mitigation measures to reduce their hazard vulnerability.

Disaster impacts vary among households and businesses because of preexisting variation in the vulnerability of social units within each of these categories. Specifically, social vulnerability is people's "capacity to anticipate, cope with, resist and recover from the impacts of a natural hazard" (Wisner et al., 2004, p. 11). Whereas people's physical vulnerability refers to their susceptibility to biological changes (i.e., impacts on anatomical structures and physiological functioning), their social vulnerability refers to limitations in their physical assets (buildings, furnishings, vehicles) and psychological (knowledge, skills, and abilities), social (community integration), economic (financial savings), and political (public policy influence) resources. The central point of the social vulnerability perspective is that, just as people's occupancy of hazard prone areas and the physical vulnerability of the structures in which they live and work are not randomly distributed, neither is social vulnerability randomly distributed – either geographically or demographically. Thus, just as variations in structural vulnerability can increase or decrease the effect of hazard exposure on physical impacts (property damage and casualties), so too can variations in social vulnerability (Bolin, 2006; Enarson et al., 2006). In particular, households that are elderly, female-headed, lower income, and ethnic minority are likely to have high vulnerability to disasters.

Physical impacts

Casualties. According to the EM-DAT database (www.emdat.be/database), there were 25 geophysical, hydrological, or meteorological disasters that produced more than 50,000 deaths between 1900 and 2011. Of these, 12 were earthquakes (maximum = 242,000), seven were tropical cyclones (maximum = 300,000), and six were floods (maximum = 3,700,000). There is significant variation by region, with Asia experiencing 54% of the earthquakes but 71% of the casualties from these events, 41% of the floods but 98% of the casualties, and 41% of the storms but 92% of the casualties. By contrast, the Americans experienced 22% of the earthquakes but 17% of the casualties from these events, 24% of the floods but less than 2% of the casualties, and 33% of the storms and 8% of the casualties. Berke (1995) found that developing countries in Asia, Africa, and South America accounted for approximately 3,000 deaths per disaster, whereas the corresponding figure for high-income countries was approximately 500 deaths per disaster. Moreover, these disparities appear to be increasing because the average annual death toll in developed countries declined by at least 75% between 1960 and 1990, but the same time period saw increases of over 400% in developing countries.

Damage. Losses of structures, animals, and crops also are important measures of physical impacts, and the EM-DAT database shows that these have been rising exponentially throughout the world since 1970. Moreover, Berke (1995) reported that the rate of increase is even greater in developing countries such as India and Kenya. Such losses usually result from physical damage or destruction of property, but they also can be caused by losses of land use to chemical or radiological contamination or loss of the land itself to subsidence or erosion. Damage to the built environment can be classified broadly as affecting residential, commercial, industrial, infrastructure (water, waste disposal, electric power, fuel, telecommunications, and transportation), or community services (public safety, health, education) sectors. Moreover, damage within each of these sectors can be divided into damage to structures and damage to contents. It usually is the case that damage to contents results from collapsing structures (e.g., hurricane winds failing the building envelope and allowing rain to destroy the furniture inside the building). Because collapsing buildings are a major cause of casualties as well, this suggests that strengthening the structure will protect the contents and occupants. However, some hazard agents can damage building contents without affecting the structure itself (e.g., earthquakes striking seismically resistant buildings whose contents are not securely fastened). Thus, risk area residents may need to adopt additional hazard adjustments to protect contents and occupants even if they already have structural protection.

Other important physical impacts include damage or contamination to cropland, rangeland, and woodlands. Such impacts may be well understood for some hazard agents but not others. For example, ashfall from the 1980 Mt. St. Helens (USA) eruption was initially expected to devastate crops and livestock in downwind areas, but no significant losses materialized (Warrick et al., 1981). There also is concern about damage or contamination to the natural environment (wild lands) because these areas serve valuable functions such as damping the extremes of river discharge and providing habitat for wildlife. In part, concern arises from the potential for indirect consequences such as increased runoff and silting of downstream river beds, but many people also are concerned about the natural environment simply because they value it for its own sake.

Social impacts

Psychosocial impacts. Research reviews conducted over a period of 25 years have concluded that disasters can cause a wide range of negative psychological responses (Gerrity and Flynn, 1997; Norris et al., 2002a, b). In most cases, the observed effects are mild and transitory – the result of "normal people, responding normally, to a very abnormal situation" (Gerrity and Flynn, 1997, p. 108). The vast majority of disaster victims experience only mild psychological distress. For example, Bolin and Bolton (1986) found negative impacts such as upsets with storms (61%), time pressures (48%), lack of patience (38%), and strained family relationships (31%) after the 1982 Paris Texas USA tornado. However, victims also experienced positive impacts including strengthened family relationships (91%), decreased importance of material possessions (62%), and increased family happiness (23%). The data showed only minor differences between Blacks and Whites in the prevalence of psychosocial impacts.

Researchers have also examined public records in their search for psychological impacts of disasters. For example, Morrow's (1997) examination of vital statistics (births, marriages, deaths, and divorce applications) had no significant long-term trends due to Hurricane Andrew. However, domestic violence rates remained constant for about 6 months after the hurricane but increased about 50% for nearly 2 years after that. Nonetheless, there are especially vulnerable groups that might need extra attention if they show signs of long-standing problems due to the disaster. This includes youth, people with preexisting mental conditions, and victims who have witnessed the death or severe injury of loved ones. Single minority female heads of household who have limited psychosocial resources for coping with severe exposures and secondary stressors have the most adverse outcomes, especially in developing countries (Morrow, 1997; Norris et al., 2002a).

The negative psychological impacts described above, which Lazarus and Folkman (1984) call *emotion focused coping*, generally disrupt the social functioning of only a very small portion of the victim population. Instead, the majority of disaster victims engage in adaptive *problem focused coping* activities to save their own lives and those of their closest associates. Further, there is an increased incidence in prosocial behaviors such as donating material aid and a decreased incidence of antisocial behaviors such as crime (Drabek, 1986; Siegel et al., 1999). In some cases, people even engage in altruistic behaviors that risk their own lives to save the lives of others (Tierney et al., 2001).

There also are psychological impacts with long-term adaptive consequences, such as changes in risk perception (beliefs in the likelihood of the occurrence a disaster and its personal consequences for the individual) and increased hazard intrusiveness (frequency of thought and discussion about a hazard). In turn, these beliefs can affect risk area residents' adoption of household hazard adjustments that reduce their vulnerability to future disasters.

However, these cognitive impacts of disaster experience do not appear to be large in aggregate, resulting in modest effects on household hazard adjustment (see Lindell and Perry, 2000, and Spittal et al., 2008, for literature on seismic hazard adjustment).

Demographic impacts. The demographic impact of a disaster can be assessed by adapting the *demographic balancing equation*, $P_a - P_b = B - D + IM - OM$, where P_a is the population size after the disaster, P_b is the population size before the disaster, B is the number of births, D is the number of deaths, IM is the number of immigrants, and OM is the number of emigrants (Smith et al., 2001). The magnitude of the disaster impact, $P_a - P_b$, is computed at two specific points in time for the population of a specific geographical area. The identification of the "impact area" is especially important in assessing demographic impacts because early US research (Friesma et al., 1979; Wright et al., 1979) suggested disasters have negligible demographic impacts. However, the highly aggregated level of analysis in these studies did not preclude the possibility of significant impacts at lower levels of aggregation (census tracts, block groups, or blocks). For example, casualties and emigration decreased the population of Lampuuk in the Aceh province of Indonesia to approximately 6% of its preimpact population immediately after the 2004 Indian Ocean tsunami (Fanany, 2010). Although there can be a major loss of life in some disasters, the most likely demographic impacts are the emigration of population segments that have lost housing and the (temporary) immigration of construction workers. In many cases, housing-related emigration is also temporary, but census data showed that the city of New Orleans dropped to 44% of its preimpact population in the year after Hurricane Katrina and only returned to 78% of its preimpact population 6 years later. Moreover, there are documented cases in which housing reconstruction has been delayed indefinitely – leading to "ghost towns" (Comerio, 1998). Other potential causes of emigration are psychological impacts (belief that the likelihood of disaster recurrence is unacceptably high), economic impacts (loss of jobs or community services), or political impacts (increased neighborhood or community conflict).

Economic impacts. The property damage caused by disaster impact creates losses in asset values that can be measured by the cost of repair or replacement (CACND, 1999). In most cases, disaster losses are initially borne by the affected households, businesses, and local government agencies whose property is damaged or destroyed. However, some of these losses are redistributed during the disaster recovery process through insurance and disaster relief. In addition to direct economic losses, there are indirect losses that arise from the interdependence of community subunits. Research on the economic impacts of disasters (Alesch et al., 1993; Rose and Limb, 2002; Tierney, 2006) suggests the relationships among the social units within a community can be described as a state of dynamic equilibrium involving a steady flow of resources, especially money. Specifically, a household's linkages with the community are defined by the money it must pay for products, services, and infrastructure support. This money is obtained from the wages that employers pay for the household's labor. Similarly, the linkages that a business has with the community are defined by the money it provides to its employees, suppliers, and infrastructure in exchange for inputs such as labor, materials and services, electric power, fuel, water/wastewater, telecommunications, and transportation. Conversely, it provides products or services to customers in exchange for the money it uses to pay for its inputs.

It also is important to recognize the financial impacts of recovery (in addition to the financial impacts of emergency response) on local government. Costs must be incurred for tasks such as damage assessment, emergency demolition, debris removal, infrastructure restoration, and replanning stricken areas. In addition to these costs, there are decreased revenues due to loss or deferral of sales taxes, business taxes, property taxes, personal income taxes, and user fees.

Political impacts. There is substantial evidence that disaster impacts can cause social activism resulting in political disruption, especially during the seemingly interminable period of disaster recovery. The disaster recovery period is a source of many victim grievances and this creates many opportunities for community conflict (Bates and Peacock, 1993; Bolin, 1982, 1993). Victims usually attempt to recreate preimpact housing patterns, which can thwart government attempts at relocation to less hazardous areas (Dove, 2008). Such attempts also can be problematic for their neighbors if victims attempt to site temporary housing, such as mobile homes, on their own lots while awaiting the reconstruction of permanent housing. Conflicts arise when such housing is considered to be a blight on the neighborhood and neighbors are afraid the "temporary" housing will become permanent. Neighbors also are pitted against each other when developers attempt to buy damaged or destroyed properties and build multifamily units on lots previously zoned for single family dwellings. Such rezoning attempts are a major threat to the market value of owner-occupied homes but tend to have less impact on renters because they have less incentive to remain in the neighborhood. There are exceptions to this generalization because some ethnic groups have very close ties to their neighborhoods, even if they rent rather than own.

Attempts to change prevailing patterns of civil governance can arise when individuals sharing a grievance about the handling of the recovery process seek to redress that grievance through collective action. This is most likely when government agencies or NGOs relate to local communities through manipulation or consultation rather than partnership or ownership (Kenny, 2010). Consistent with Dynes's (1970) typology of organizations, existing community groups with an explicit political agenda can *expand* their membership to increase their strength, whereas community groups without an explicit political agenda can *extend* their domains to include disaster-related grievances. Alternatively, new groups can *emerge* to influence local, state, or federal government agencies and legislators to take actions that they support and to terminate actions that they disapprove. Indeed, such was the case for Latinos in Watsonville, California following the 1989 Loma Prieta earthquake (Tierney et al., 2001). Usually, community action groups pressure government to provide additional resources for recovering from disaster impact, but may oppose candidates' reelections or even seek to recall some politicians from office (Olson and Drury, 1997; Prater and Lindell, 2000; Shefner, 1999). The point here is not that disasters produce political behavior that is different from that encountered in normal life. Rather, disaster impacts might only produce a different set of victims and grievances and, therefore, a minor variation on the prevailing political agenda (Morrow and Peacock, 1997).

Disaster recovery goals

Most people's goal in disaster recovery is to restore household, business, and government activity to the "normal" patterns that existed before the disaster struck. To do this, they typically assume they must restore the buildings and infrastructure as they were before the disaster. However, it is increasingly understood that restoring the community to its previous condition will also reproduce its previous hazard vulnerability. When cities allow too much development in floodplains or allow substandard housing to be built that collapses in an earthquake, "normal" is an unsustainable condition. Consequently, a disaster resilient community learns from experience which areas of the community have excessive levels of hazard exposure. It also identifies the buildings and infrastructure that have inadequate designs, construction methods, and construction materials. In short, communities need to incorporate hazard mitigation into their disaster recovery. That is, they need to adopt hazard source control, community protection works, land-use practices, building construction practices, and building contents protection practices (Federal Emergency Management Agency, 1986; Lindell et al., 2006, Chapter 7). *Hazard source control* involves intervention at the point of hazard generation to reduce the probability or magnitude of an event. By contrast, *community protection works* attenuate disaster impact by altering the hazard transmission process, especially by confining or diverting materials flows to reduce the hazard exposure of target locations and populations. *Land-use practices* limit hazard exposure by minimizing development in areas where the likelihood of hazard impact is high. By contrast, *building construction practices* limit physical vulnerability by building structures whose resistance to hazard impact is high. Finally, *building contents protection* prevents furniture, equipment (e.g., furnaces, air conditioners, washers, dryers), and other building contents from being damaged or destroyed. In many cases, but not all, appropriate building construction practices protect building contents at the same time as they protect the structure itself. For example, preventing wind damage to a structure will also prevent damage to its contents. However, seismic shaking can overturn water heaters and refrigerators without causing any damage to the building structure itself.

Disaster recovery stages and functions

The identification of disaster recovery as an emergency management phase has led some authors to divide it into stages, but there has been little agreement on the number and definitions of recovery stages (Alexander, 1993; Kates and Pijawka, 1977; Sullivan, 2003; UNDRO, 1984; Schwab et al., 1998). It is now generally accepted that disaster recovery encompasses multiple activities, some implemented sequentially and others implemented simultaneously. At any one time, some households and businesses might be engaged in one set of recovery activities whereas others are engaged in other recovery activities. Indeed, some households and businesses might be fully recovered months or years after others and there might be others that never recover at all. Thus, it is more useful to think of disaster recovery in terms of four functions: disaster assessment, short-term recovery, long-term reconstruction, and recovery management (see Table 1). The recovery phase's *disaster assessment* function should be integrated with the emergency response phase's emergency assessment function in identifying the physical impacts of the disaster. *Short-term recovery* focuses on the immediate tasks of securing the impact area, housing victims, and establishing conditions under which households and businesses can begin the process of recovery.

Recovery and Reconstruction After Disaster, Table 1 Disaster recovery functions

Disaster assessment
- Rapid assessment
- Preliminary damage assessment
- Site assessment
- Victims' needs assessments
- "Lessons learned"

Short-term recovery
- Impact area security
- Temporary shelter/housing
- Infrastructure restoration
- Debris management
- Emergency demolition
- Repair permitting
- Donations management
- Disaster assistance

Long-term reconstruction
- Hazard source control and area protection
- Land-use practices
- Building construction practices
- Public health/mental health recovery
- Economic development
- Infrastructure resilience
- Historic preservation
- Environmental recovery
- Disaster memorialization

Recovery management
- Agency notification and mobilization
- Mobilization of recovery facilities and equipment
- Internal direction and control
- External coordination
- Public information
- Recovery legal authority and financing
- Administrative and logistical support
- Documentation

Long-term reconstruction actually implements the reconstruction of the disaster impact area and manages the disaster's psychological, demographic, economic, and political impacts. Finally, *recovery management* monitors the performance of the disaster assessment, short-term recovery, and long-term reconstruction functions. It also ensures they are coordinated and provides the resources needed to accomplish them.

Disaster recovery processes

There are three relatively distinct types of social units that should be considered in disaster recovery: households, businesses, and government agencies. Households and businesses focus primarily on their own recovery but government agencies must address the recovery needs of the entire community.

Household recovery

There are three basic components to household recovery. These are housing recovery, economic recovery, and psychological recovery (Bolin and Trainer, 1978). All three of these components require resources to recover, but households must invest time to obtain these resources. This includes time to find and purchase alternate shelter, clothing, food, furniture, and appliances to support daily living (Yelvington, 1997). Time is also needed to file insurance claims, apply for loans and grants, and search for jobs. The time required for these tasks is increased by multiple trips to obtain required documentation and understaffing of providers (Morrow, 1997). Finally, victims need skill and self confidence to cope with the disaster assistance bureaucracy (Morrow, 1997).

Housing recovery. Households typically use four types of housing recovery following a disaster (Quarantelli, 1982). The first type, *emergency shelter*, consists of unplanned and spontaneously sought locations that are intended only to provide protection from the elements, typically open yards and cars after earthquakes (Bolin and Stanford, 1991, 1998). The second type is *temporary shelter*, which includes food preparation and sleeping facilities that usually are sought from friends and relatives or are found in commercial lodging, although mass care facilities in school gymnasiums or church auditoriums are acceptable as a last resort. The third type is *temporary housing*, which allows victims to reestablish household routines in nonpreferred locations or structures. The last type is *permanent housing*, which reestablishes household routines in preferred locations and structures. There is no single pattern of progression through the stages of housing because households vary in number and sequence of movements and the duration of their stays in each type of housing (Cole, 2003). Indeed, the transition from one stage to another can be delayed unpredictably, as when it took 9 days for shelter occupancy to peak after the 1987 Whittier Narrows California earthquake (Bolin, 1993). Yelvington (1997) reported that temporary shelters experienced increased demand as buildings were condemned by authorities or landlords begin reconstruction on damaged structures.

There are significant variations among households in their housing recovery and these are correlated with households' demographic characteristics (Peacock et al., 2006). Severity of damage and the availability of relatives nearby predict who stays with relatives, whereas income, homeownership, and availability of relatives nearby predict who accepts relatives (Morrow, 1997). Moreover, kin networks are likely to seek temporary shelter together, especially if all relatives became victims because they lived so close together (Yelvington, 1997). Households with higher incomes who lack nearby friends and relatives with undamaged homes seek commercial facilities, whereas lower-income households in such conditions are forced to accept mass care facilities.

Sites for temporary housing include homes of friends and relatives, commercial facilities such as rental houses and apartments, and mass facilities such as trailer parks. Some of these sites are in or near the stricken community, but others are hundreds or even thousands of kilometers away. Lack of alternative housing within an acceptable distance of jobs or peers led some households to leave the Miami area after Hurricane Andrew. The population loss was 18% in South Dade County, 33% in Florida City, and 31% in Homestead (Dash et al., 1997). Other households remained in severely damaged units – or even condemned units – without electric power or telephone service for months (Yelvington, 1997) or doubled up with relatives (Morrow, 1997).

Households encounter many problems during reconstruction, including high prices for repairs, poor quality work, and contract breaches (Bolin, 1993). The rebuilt structures do benefit from improved quality and hazard resistance (Bolin, 1993, indicates 50% of respondents reported this) and this is especially true for public housing (Morrow, 1997). However, few victims think the improvements are worth the inconvenience they experienced. Lower-income households tend to have higher hazard exposure because they often live in more hazard prone locations. They also have higher physical vulnerability because they live in structures that were built according to older, less stringent building codes, used lower quality construction materials and methods, and have been less well maintained (Bolin and Bolton, 1986). Because lower-income households have fewer resources on which to draw for recovery, they also take longer to return to permanent housing, sometimes remaining for extended periods of time in severely damaged homes (Girard and Peacock, 1997). Indeed, they sometimes are forced to accept as permanent what originally was intended as temporary housing (Peacock et al., 1987). Consequently, there might still be low-income households in temporary sheltering and temporary housing even after high-income households all have relocated to permanent housing (Berke et al., 1993; Rubin et al., 1985).

Economic recovery. Some households' economic recovery takes place quickly, but others' takes much longer. For example, the percentage of households reporting complete economic recovery after the Whittier earthquake was 50% at the end of the first year but 21% reported little of no recovery even at the end of 4 years (Bolin, 1993). Economic recovery was positively related to household income and negatively related to structural damage, household size, and the total number of moves (Bolin, 1993). In some cases, this is due to the loss of permanent jobs that are replaced only by temporary jobs in temporary shelter management, debris cleanup, and construction – or are not replaced at all (Yelvington, 1997).

There are systematic differences in the rate of economic recovery among ethnic groups. For example, Bolin and Bolton (1986) found that Black households (30%) lagged behind Whites (51%) in their return to preimpact economic conditions 8 months after the Paris, Texas, tornado. However, the variables affecting economic recovery were relatively similar for Black and White families. In both ethnic groups, economic recovery was negatively related to family size (larger families had lower levels of recovery), but positively related to socioeconomic status (SES – education, profession, and income), use of disaster assistance, insurance adequacy, and aid adequacy. In addition, Black household recovery was negatively related to primary group aid and the number of household moves. The direct effect of family size and SES on economic recovery was compounded by the indirect effects of these variables via their impacts on the use of disaster assistance, insurance adequacy, aid adequacy, and household moves. The variables that had positive direct effects on economic recovery (use of disaster assistance, insurance adequacy, aid adequacy) were negatively related to family size and positively related to SES. That is, larger households were less likely – and higher SES households were more likely – to use disaster assistance, have adequate insurance, or receive adequate aid. Moreover, these variables were positively related to family size and negatively related to SES. That is, larger households made more moves and higher SES households made fewer moves. The overall effect of this complex pattern of relationships is for large poor households to be doubly handicapped in their economic recovery.

Psychological recovery. Few disaster victims require psychiatric diagnosis and most benefit more from a *crisis counseling* orientation than from a *mental health treatment* orientation, especially if their normal social support networks of friends, relatives, neighbors, and coworkers remain largely intact. However, there are population segments requiring special attention and active outreach. These include children, frail elderly, people with preexisting mental illness, racial and ethnic minorities, and families of those who have died in the disaster. Emergency workers also need attention because they often work long hours without rest, have witnessed horrific sights, and are members of organizations in which discussion of emotional issues may be regarded as a sign of weakness (Rubin, 1991). Thus, professionals involved in particularly difficult search operations and medical personnel who handle extraordinary workloads during disaster periods might also benefit from post-disaster counseling. However, there is little evidence of emergency workers needing directive therapies. In particular, there appears to be little scientific evidence that some widely used programs such as *Critical Incident Stress Debriefing* are effective (McNally et al., 2003).

In summary, the majority of victims and responders recover relatively quickly from the stress of disasters without psychological interventions. Those who suffer the greatest losses to their material resources (e.g., the destruction of their homes) and their social networks (e.g., spouses and other family members) are likely to experience the most psychological distress, but not necessarily an amount that is personally unmanageable. Thus, the appropriate strategy for psychological recovery by victims and first responders seems to be one of minimal intervention to provide information about sources of material support (for victims) and to facilitate optional involvement in social and emotional support groups (for victims and first responders).

Business recovery

Several studies of the economic impacts of environmental disasters have examined the ways in which individual businesses prepare for, are disrupted by, and recover from these events (see Tierney, 2006 and Zhang et al., 2009 for reviews). These studies found older, larger (measured by the number of employees), and more financially stable businesses are more likely to adopt hazard adjustments, as are businesses in the manufacturing, professional services, finance, insurance, and real estate sectors. Small businesses are more physically vulnerable because they are more likely than large businesses to be located in nonengineered buildings and are less likely to have the capacity to design and implement hazard management programs to reduce this physical vulnerability. At the same time as they face increased costs to repair structures and replace contents, these businesses also face reduced patronage if they must move far from their previous locations. Ultimately, many small businesses have failed by the time the space is available for reoccupancy at their original locations.

There also is variation among business sectors in their patterns of recovery. Whereas wholesale and retail businesses generally report experiencing significant sales losses, manufacturing and construction companies often show gains following a disaster (Webb et al., 2000). Moreover, businesses that serve a large (e.g., regional or international) market tend to recover more rapidly than those that only serve local markets (Webb et al., 2002). Small businesses, in particular, have been found to experience more obstacles than large firms and chains in their attempts to regain their predisaster levels of operations. Compared to their large counterparts, small firms are more likely to depend primarily on neighborhood customers, lack the financial resources needed for recovery, and lack access to governmental recovery programs (Alesch et al., 2001; Tierney, 2006). Thus, business sector and business size can be seen as indicators of operational vulnerability that are equivalent to the demographic indicators of social vulnerability in households.

Sources of recovery assistance

There are three fundamental patterns of household recovery that are defined by the three corresponding sources of assistance (Bolin and Trainer, 1978). These are autonomous (own resources), kinship (extended family resources), or institutional (governmental) – although few households actually rely on only one source.

Autonomous recovery depends on the household's available human, material, and financial resources. Human resources are available to the extent household members can continue to derive generate income from employment, rental of physical assets, or interest/dividends from financial assets. Moreover, household recovery depends on the degree to which material resources are available. This includes the extent to which its possessions – land, buildings, equipment, furniture, clothes, vehicles, crops, and animals – are undamaged or can be restored at reasonable expense. A household's recovery also depends on the degree to which its financial resources are available. This includes an ability to withdraw savings quickly from banks, to quickly liquidate stocks and bonds at a fair price, and to receive adequate compensation from its insurer. In some cases, household recovery also depends on the degree to which creditors will accept delayed payments on financial liabilities such as loans, mortgages, and credit card debt. Finally, household recovery depends on the degree to which members can reduce consumption such as purchases of shelter, food, clothing, medical care, entertainment, and other goods and services. Kinship recovery depends on the physical proximity of other nuclear families in the kin network, the closeness of the psychological ties within the network, the assets of the other families and, of course, the extent to which those families also suffered losses. Institutional recovery depends on whether victims meet the qualification standards, usually documented residence in the impact area and proof of loss.

Some aspects of household recovery are relatively similar across ethnic groups, but others reveal distinct differences. After Hurricane Andrew, Anglos, Blacks, and Hispanics experienced similar levels of frustration in coping with the challenges of living in damaged homes, job relocation, dealing with agencies, behavioral problems with children, and loss of household members (Morrow, 1997). However, there were significant differences in the experience of other problems. For example, Blacks reported the greatest frequency of frustration about living in temporary quarters, whereas Hispanics experienced the greatest frustrations in dealing with building inspectors.

Hazard insurance

Hazard insurance is important because it decreases government workload and expense after disasters by shifting part of the administrative burden for evaluating damage to insurance companies in the private sector. Moreover, hazard insurance defines the terms of coverage in advance, thus reducing opportunities for politicians to increase benefits after disaster. Unfortunately, the potential contribution of hazard insurance remains to be fully realized. There are many difficulties in developing and maintaining an actuarially sound hazard insurance program, so hazard insurance varies significantly in its availability and cost – flood, hurricane, and earthquake insurance being particularly problematic (Kunreuther and Roth, 1998). In particular, the National Flood Insurance Program has made significant strides over the past 30 years, but it continues to require operational subsidies.

One of the basic problems is that those who are most likely to purchase flood insurance are, in fact, those who are most likely to file claims (Kunreuther, 1998). This problem of *adverse selection* makes it impossible to sustain a market in private flood insurance. In some cases, the homeowners are underinsured or lack any insurance because they cannot afford quality insurance or were denied access to it (Peacock and Girard, 1997). In addition, there are cognitive obstacles to developing a comprehensive hazard insurance program. Building on earlier hazards research (see Burton et al., 1993, for a summary) and psychological research on judgment and decision making (see Slovic et al., 1974, for an early statement and Baron, 2000, for a more recent summary), researchers have identified numerous logical deficiencies in the ways people process information in laboratory studies of risk. One important issue concerns what economists call *moral hazard* and psychologists refer to as a *lack of perceived personal responsibility for protection*. The concept of moral hazard has important policy implications because the Interagency Floodplain Review Committee (1994) report concluded federal disaster relief policy creates this condition by relieving households of the responsibility for providing their own disaster recovery resources.

Hazard insurance can make a significant contribution to household recovery but coverage varies by hazard agent, with Bolin and Bolton (1986) reporting 86% coverage for a tornado and Bolin (1993) reporting 25% for an earthquake. Risk area residents are particularly likely to forego earthquake insurance because they consider premiums to be too high and deductibles too large (Palm et al., 1990). Income, education, and occupational status all correlate with earthquake insurance purchase (Bolin, 1993). Strategies for coping with uninsured losses include obtaining government or commercial loans, obtaining government or nongovernmental organization (NGO) grants, withdrawing savings, and deciding not to replace damaged items (Bolin, 1993). Loans can be problematic because they involve long-term debt that takes many years to repay (Bolin, 1993). Government grants require households to meet specific standards, including proof that they are indeed residents of the disaster impact area. However, there can be problems in registering people who evacuated or were rescued without identification (Yelvington, 1997). Relaxed standards seem humane but can allow the chronically homeless and out of area construction workers to obtain access to services intended only for disaster victims. In turn, resentment toward "freeloaders" can curtail services to victims.

Higher levels of government

Countries across the world vary in the relative roles of local, state (provincial), and federal (national) levels of government. In some cases, the entire burden of government assistance falls on the federal/national government whereas, in others, responsibility is distributed across multiple levels of government. Because of this diversity, this section focuses on the system in the United States because it illustrates types of complexities that arise in a federal system of distributed responsibility.

In the USA, state and federal agencies can play significant roles in disaster recovery, but the burden most frequently falls on local governments because only about 19% of all disasters receive state disaster declarations and 1% qualifies for Presidential Disaster Declarations (PDDs). Thus, local governments should prepare to undertake a variety of functions during a disaster recovery process, understanding that they might not receive any aid from higher levels of government for minor disasters. The main factor affecting the level of involvement of state and federal government is the scope of the event. After a major disaster, a PDD opens a broad range of programs for relief and reconstruction. In such cases, the state plays a coordinating role, working with both federal and local governments.

The lead agency at the federal level is FEMA, but other federal agencies might be called upon when a PDD is granted, including the Small Business Administration, the US Army Corps of Engineers, the Department of Housing and Urban Development, the National Oceanographic and Atmospheric Administration, and the Economic Development Administration, among others. Each of these agencies funds specific disaster recovery programs. According to the National Response Framework, these agencies can be housed in Disaster Field Offices (DFOs) in the vicinity of the disaster. Emergency Response Teams (ERTs) are located in the DFOs. These include an Operations Section that coordinates federal, state, and voluntary efforts, and an Infrastructure Support Branch to facilitate restoration of public utilities and other infrastructure services.

The main types of programs providing recovery assistance are Individual Assistance, Infrastructure Support (formerly Public Assistance), and Hazard Mitigation Grant Program. Individual Assistance is available to households through the Temporary Housing Assistance program, Individual and Family Grants, Disaster Unemployment assistance, legal services, special tax considerations, and crisis counseling programs. Individuals and businesses can receive aid through the Small Business Administration Disaster Loans program, which can provide loans for repairs to housing and businesses, and also for operating expenses. In the past, many loan programs have been inaccessible to low-income households, which tend to rent rather than own their housing. Thus, they failed to qualify for loans because of their low incomes and lack of collateral. The Individual and Family Grant Program was intended to fill the need for a program targeting those whose needs were not being met by the SBA loan program, private insurance, or NGO assistance. However, the amounts awarded tend to be small.

Assistance provided under the Hazard Mitigation Grant Program has increased in importance since the passage of the Disaster Mitigation Act of 2000. This legislation requires local governments to identify potential mitigation measures that could be incorporated into the repair of damaged facilities in order to be eligible for pre- and post-disaster funding. These activities include hazard mapping, mitigation planning, development of building codes, development of training and public education programs, establishing Reconstruction Information Centers, and assisting communities to promote sustainable development. In addition, there may be state programs that also provide assistance to households and local governments for recovery from disasters that do not receive a PDD.

Nongovernmental organizations and community-based organizations

The role of NGOs such as the American Red Cross, Salvation Army, and Mennonite Disaster Service is widely publicized and the role of community-based organizations (CBOs) such as local churches and service organizations is increasingly recognized. Such organizations provide housing, food, clothing, medicine, and financial assistance to disaster victims. In most cases, the *existing* government social service agencies are supplemented by NGOs that *expand* their membership to perform the tasks they are expected to perform during disaster recovery (Dynes, 1970). By contrast, existing CBOs typically *extend* themselves beyond their normal tasks to perform novel activities. In addition, there are situations in which existing, expanding, and extending organizations cannot successfully meet the recovery needs of disaster victims. In such cases, government agencies, NGOs, and CBOs form an *Unmet Needs Committee*, which is an *emergent* organization that is designed to serve those whose needs are not being addressed by existing programs. In some cases, the need for such emergent organizations arises from political organization and activism by population segments that believe they are being neglected (Morrow and Peacock, 1997; Phillips, 1993).

Local government recovery functions

Rubin (1991) accounted for community recovery in terms of six variables – *federal influences and conditions, state influences and conditions, community-based needs and demands for action, personal leadership, ability to act,* and *knowing what to do.* One important commonality among the 14 cases Rubin et al. (1985) studied is that the speed, efficiency, and equity of community recovery depended significantly upon local government's ability to improvise effective recovery strategies. That is, communities recovered more quickly and effectively if they could identify and respond to the specific problems that arose from its unique circumstances. More recently, practitioners and researchers have begun to agree that community disaster recovery is even faster and more effective if it is based on a recovery plan that has been developed prior to disaster impact (Olson et al., 1998; Schwab et al., 1998; Wilson, 1991; Wu and Lindell, 2004). The recovery plan needs to establish clear goals and an implementation strategy (Smith and Wenger, 2006), preferably one that does not reproduce the community's preimpact hazard vulnerability.

After a disaster, local government needs to perform many tasks very quickly, and many of these must be performed simultaneously. It is therefore critical to plan for disaster recovery, as well as for disaster response (Schwab et al., 1998). The line between emergency response and disaster recovery is not clear because some sectors of the community might be in response mode whereas others are moving into recovery, and some organizations will be carrying on both types of activity at the same time. This means that there will be little time to plan for disaster recovery once the emergency response has begun. By planning for recovery before disaster strikes, resources can be allocated more effectively and efficiently, increasing the probability of a rapid and full recovery.

Local government agencies will frequently find during disaster recovery that some households and businesses fail to perform the tasks that are required to recover from the disaster. Whether households and businesses lack the knowledge of how to recover or the resources needed to recover, government can provide assistance. Local government must also perform specific tasks during disaster recovery, some of which involve restoring services it performed before the disaster (e.g., providing functioning roads, street lights and signs, and traffic control devices). In addition, local government must rebuild any critical facilities (e.g., police and fire stations) that were damaged or destroyed. Finally, local government has a heightened need to perform its regulatory functions regarding land-use and building construction. These two functions require rapid action under conditions of a greatly multiplied workload, so special provisions are required to expedite the procedures for reviewing and approving the (re)development of private property.

In approaching the task of preimpact recovery planning, a community must overcome three major misconceptions about disaster recovery. The first misconception is that the entire recovery effort can be improvised after the emergency response is complete. In fact, a timely and effective disaster recovery requires a significant amount of data collection and planning that will delay the recovery if they are postponed until after the emergency response is over. It is important to recognize that the disaster response phase's uncertainty and urgency about human safety has been replaced by households' and businesses' urgency to return to normal patterns of functioning and government agencies' uncertainty about how to organize the community to accomplish this.

The second misconception is that there will be ample time to collect data and plan the recovery during the emergency response. It is true that some recovery relevant data must be collected during the emergency response. However, an assessment of "lessons learned" from the disaster impact should be used to make final adjustments to a recovery process that has been designed before the disaster strikes. Finally, the third misconception is that the objective of disaster recovery should be to restore the community to the conditions that existed before the disaster. As noted earlier, this will simply reproduce the community's existing disaster vulnerability.

In many ways, the process of preparedness for disaster recovery is quite similar to the process of preparedness for emergency response. Thus, the community should establish a Recovery/Mitigation Committee before disaster strikes that will establish a vision of community disaster recovery and articulate the basic strategies that will be implemented before and after disaster impact. In addition, the committee should assign each recovery function to a specific organization, develop a Recovery Operations Plan, and acquire any necessary resources to implement it. Finally, the committee should conduct the training and tabletop exercises needed to ensure the ROP can be implemented effectively.

The recovery/mitigation committee

The Recovery/Mitigation Committee can be an important part of an effective, rapid disaster recovery process. This committee should be established before a disaster during the preimpact recovery planning process. The Recovery/Mitigation Committee should examine the findings from the community hazard/vulnerability analysis to identify the locations having the highest levels of hazard exposure, physical vulnerability, and social vulnerability. The committee should begin to work with the rest of the community, and especially with those at greatest risk, to formulate a vision of the disaster recovery it intends to implement.

The Recovery/Mitigation Committee needs to work with the community before and after a disaster to articulate a vision of community disaster recovery. The recovery process needs to strike a balance between corporate centered and community-based economic development (Blair and Bingham, 2000). According to a *corporate centered economic development*, usually advocated by the local business community, government provides resources such as land and money to the private sector to invest without any restrictions. This market-based strategy tends to produce results that are good in aggregate but produces an inequitable recovery. By contrast, *community-based economic development* involves active participation by government to ensure that the benefits of recovery will also be shared by economically disadvantaged segments of the community.

The short-term recovery following a major disaster can generate an economic boom as state and federal money flows into the community to reconstruct damaged buildings and infrastructure. These funds are used to pay for construction materials and the construction workforce and, to the extent that the materials and labor are acquired locally, they generate local revenues. In addition, the building suppliers hire additional workers and these, along with the construction workers, spend their wages on places to live, food to eat, and entertainment. Unless there are undamaged communities within commuting distance that can compete for this money, it will all be spent within the community.

Communities must also consider the long-term economic consequences of disaster recovery. What will happen after the reconstruction boom is over? They can attract new businesses if they have a skilled labor pool and good schools – especially colleges whose faculty and students can support knowledge-based industries. Other assets include low crime rates, low cost of living, good housing, and environmental amenities such as mountains, rivers, or lakes (Blakely, 2000). A community can also enhance its economic base if it can attract businesses that are compatible with the ones that are already there. Such firms can be identified by asking existing firms to identify their suppliers and distributors. These new firms might be attracted by the newer buildings and enhanced infrastructure that has been produced during disaster reconstruction.

If a disaster stricken community does not already have such assets, they can invest in four fundamental components of economic development – locality development, business development, human resources development, and community development. *Locality development* enhances a community's existing physical assets by improving roads or establishing parks on river and lakefronts. *Business development* involves efforts to retain existing businesses or attract new ones. Although it is not easy, this can be accomplished working with businesses to identify their critical needs. In some cases, this might involve establishing a business incubator that allows startup companies to obtain low cost space and share meetings rooms. *Human resources development* expands the skilled workforce, possibly through customized worker training. Finally, *community development* utilizes NGOs, CBOs, and local firms that will hire current residents of the community whose household incomes are below the poverty level. For example, a comprehensive program for developing small businesses, affordable housing, community health clinics, and inexpensive child care can help to eliminate some of what new businesses might consider to be one of the risks of relocating to the community.

Developing a recovery operations plan

There are six important features of a preimpact recovery operations plan. First, it should define a disaster recovery organization. Second, it should identify the location of temporary housing because resolving this issue can cause conflicts that delay consideration of longer-term issues of permanent housing and distract policymakers altogether from hazard mitigation (Bolin and Trainer, 1978; Bolin, 1982). Third, the plan should indicate how to accomplish essential tasks such as damage assessment, condemnation, debris removal and disposal, rezoning, infrastructure restoration, temporary repair permits, development moratoria, and permit processing because all of these tasks must be addressed before the reconstruction of permanent housing can begin (Schwab et al., 1998).

Fourth, preimpact recovery plans also should address the licensing and monitoring of contractors and retail price controls to ensure victims are not exploited and also should address the jurisdiction's administrative powers and resources, especially the level of staffing that is available. It is almost inevitable that local government will have insufficient staff to perform critical recovery tasks such as damage assessment and building permit processing, so arrangements should be made to borrow staff from other jurisdictions (via preexisting Memoranda of Agreement) and to use trained volunteers such as local engineers, architects, and planners. Fifth, these plans also need to address the ways in which recovery tasks will be implemented at historical sites (Spennemann and Look, 1998). Finally, preimpact recovery plans should recognize the recovery period as a unique time to enact policies for hazard mitigation and make provision for incorporating this objective into the recovery planning process.

Communities that develop recovery plans addressing these functions will be better positioned to move promptly from emergency response into disaster recovery and, in turn, from disaster recovery to normal social and economic functioning. Not only will housing recovery be accelerated, but there is a greater likelihood that hazard mitigation will be integrated into the recovery process. The inclusion of hazard mitigation means that homes, businesses, and critical facilities such as schools and hospitals can be moved out of hazard prone areas. Moreover, any residential, commercial, or industrial structures that remain in hazard prone areas can be retrofitted to higher standards of hazard resistance. Thus, a preimpact plan can not only accelerate recovery but also decrease the community's vulnerability to future disasters.

Summary

Disaster recovery is a *phase* in the emergency management cycle that frequently overlaps with the emergency response. Its *goal* is to restore normal community activities that were disrupted by disaster impacts through a *process* involving both activities that were planned before disaster impact and those that were improvised after disaster impact. Disaster recovery is most rapidly and effectively achieved when communities engage in a preimpact planning process that addresses the major recovery functions and incorporates hazard mitigation and hazard insurance into a recovery operations plan.

Bibliography

Alesch, D. J., Taylor, C., Ghanty, S., and Nagy, R. A., 1993. Earthquake risk reduction and small business. In Committee on Socioeconomic Impacts (ed.), *1993 National Earthquake Conference Monograph 5: Socioeconomic Impacts*. Memphis, TN: Central United States Earthquake Consortium, pp. 133–160.

Alesch, D. J., Holly, J. N., Mittler, E., and Nagy, R., 2001. When small businesses and not-for-profit organizations collide with environmental disasters. Paper presented at The First Annual IIASA-DPRI Meeting Integrated Disaster Risk Management: Reducing Socio-Economic Vulnerability, IIASA, Laxenburg, Austria.

Alexander, D., 1993. *Natural Disasters*. New York: Chapman and Hall.

Baron, J., 2000. *Thinking and Deciding*. New York: Cambridge University Press.

Bates, F. L., and Peacock, W. G., 1993. *Living Conditions, Disasters and Development: An Approach to Cross-Cultural Comparisons*. Athens, GA: University of Georgia Press.

Berke, P. R., 1995. Natural-hazard reduction and sustainable development: a global assessment. *Journal of Planning Literature*, **9**, 370–382.

Berke, P. R., Kartez, J., and Wenger, D. E., 1993. Recovery after disaster: achieving sustainable development, mitigation and equity. *Disasters*, **17**, 93–109.

Blair, J. P., and Bingham, R. D., 2000. Economic analysis. In Hoch, C. J., Dalton, L. C., and So, F. S. (eds.), *The Practice of Local Government Planning*, 3rd edn. Washington, DC: International City/County Management Association, pp. 119–137.

Blakely, E. J., 2000. Economic development. In Hoch, C. J., Dalton, L. C., and So, F. S. (eds.), *The Practice of Local Government Planning*, 3rd edn. Washington, DC: International City/County Management Association, pp. 283–305.

Bolin, B., 2006. Race, Class, Ethnicity, and Disaster Vulnerability. In Rodríguez, H., Quarantelli, E. L., and Dynes, R. R. (eds.), *Handbook of Disaster Research*. New York: Springer, pp. 113–129.

Bolin, R. C., 1982. *Long-Term Family Recovery From Disaster*. Boulder, CO: University of Colorado Institute of Behavioral Science.

Bolin, R. C., 1993. *Household and Community Recovery After Earthquakes*. Boulder, CO: University of Colorado Institute of Behavioral Science.

Bolin, R. C., and Bolton, P., 1986. *Race, Religion, and Ethnicity in Disaster Recovery*. Boulder, CO: University of Colorado Institute of Behavioral Science.

Bolin, R. C., and Stanford, L., 1991. Shelters, housing and recovery: a comparison of U.S. disasters. *Disasters*, **45**, 25–34.

Bolin, R. C., and Stanford, L., 1998. Community-based approaches to unmet recovery needs. *Disasters*, **22**, 21–38.

Bolin, R., and Trainer, P. A., 1978. Modes of family recovery following disaster: a cross-national study. In Quarantelli, E. L. (ed.), *Disasters: Theory and Research*. Beverly Hills, CA: Sage, pp. 233–247.

Bourque, L. B., Siegel, J. M., Kano, M., and Wood, M. M., 2006. Morbidity and, mortality associated with disasters. In Rodríguez, H., Quarantelli, E. L., and Dynes, R. R. (eds.), *Handbook of Disaster Research*. New York: Springer, pp. 97–112.

Burton, I., Kates, R. W., and White, G. F., 1978. *The Environment as Hazard*. New York: Oxford University Press.

Burton, I., Kates, R. and White, G.F., 1993. *The environment as hazard*, 2nd ed. New York: Guildford Press.

CACND—Committee on Assessing the Costs of Natural Disasters, 1999. *The Impacts of Natural Disasters: A Framework for Loss Estimation*. Washington, DC: National Academy Press.

CDRSS—Committee on Disaster Research in the Social Sciences, 2006. *Facing Hazards and Disasters: Understanding Human Dimensions*. Washington, DC: National Academy of Sciences.

Charvériat, C., 2000. *Natural Disasters in Latin America and the Caribbean: An Overview of Risk*. Working paper #434. Washington, DC: Inter-American Development Bank.

Cole, P. M., 2003. *An Empirical Examination of the Housing Recovery Process Following Disaster*. College Station, TX: Texas A and M University.

Comerio, M. C., 1998. *Disaster Hits Home: New Policy for Urban Housing Recovery*. Berkeley, CA: University of California Press.

Dash, N., Peacock, W. G., and Morrow, B. H., 1997. And the poor get poorer: a neglected black community. In Peacock, W. G., Morrow, B. H., and Gladwin, H. (eds.), *Hurricane Andrew: Ethnicity, Gender and the Sociology of Disaster*. London: Routledge, pp. 206–225.

Dove, M. R., 2008. Perception of volcanic eruption as agent of change on Merapi Volcano, Central Java. *Journal of Volcanology and Geothermal Research*, **172**(2008), 329–337.

Drabek, T. E., 1986. *Human System Responses to Disaster: An Inventory of Sociological Findings*. New York: Springer.

Dynes, R., 1970. *Organized Behavior in Disaster*. Lexington, MA: Heath-Lexington Books.

Enarson, E., Fothergill, A., and Peek, L., 2006. Gender and disaster: foundations and directions. In Rodríguez, H., Quarantelli, E. L., and Dynes, R. R. (eds.), *Handbook of Disaster Research*. New York: Springer, pp. 130–146.

Fanany, I., 2010. Towards a model of constructive interaction between aid donors and recipients in a disaster context: the case of Lampuuk. In Clarke, M., Fanany, I., and Kenny, S. (eds.), *Post-Disaster Reconstruction: Lessons from Aceh*. London: Earthscan, pp. 107–125.

Federal Emergency Management Agency, 1986. *Making Mitigation Work: A Handbook for State Officials*. Washington, DC: Author.

Friesma, H. P., Caporaso, J., Goldstein, G., Linberry, R., and McCleary, R., 1979. *Aftermath: Communities after Natural Disasters*. Beverly Hills, CA: Sage.

Gerrity, E. T., and Flynn, B. W., 1997. Mental health consequences of disasters. In Noji, E. K. (ed.), *The Public Health Consequences of Disasters*. New York: Oxford University Press, pp. 101–121.

Girard, C., and Peacock, W. G., 1997. Ethnicity and segregation: post-hurricane relocation. In Peacock, W. G., Morrow, B. H., and Gladwin, H. (eds.), *Hurricane Andrew: Ethnicity, Gender and the Sociology of Disasters*. New York: Routledge, pp. 191–205.

Interagency Floodplain Review Committee, 1994. *Sharing the Challenge: Floodplain Management into the 21st Century*. Washington, DC: U.S. Government Printing Office.

Kates, R. W., and Pijawka, D., 1977. From rubble to monument: the pace of reconstruction. In Haas, J. E., Kates, R. W., and Bowden, M. J. (eds.), *Reconstruction Following Disaster*. Cambridge, MA: MIT Press, pp. 1–23.

Kenny, S., 2010. Reconstruction Through Participatory Practice? In Clarke, M., Fanany, I., and Kenny, S. (eds.), *Post-Disaster Reconstruction: Lessons from Aceh*. London: Earthscan, pp. 79–104.

Kunreuther, H., and Roth, R. J., Sr., 1998. *Paying the Price: The Status and Role of Insurance Against Natural Disasters in the United States*. Washington, DC: Joseph Henry Press.

Kunreuther, H., et al., 1998. Insurability conditions and the supply of coverage. In Kunreuther, H., and Roth, R. J., Sr. (eds.), *Paying the Price: The Status and Role of Insurance Against Natural Disasters in the United States.* Washington, DC: Joseph Henry Press, pp. 17–50.

Lazarus, R. S., and Folkman, S., 1984. *Stress, Appraisal, and Coping.* New York: Springer.

Lindell, M. K., and Perry, R. W., 2000. Household adjustment to earthquake hazard: a review of research. *Environment and Behavior*, **32**, 590–630.

Lindell, M. K., Prater, C. S., and Perry, R. W., 2006. *Fundamentals of Emergency Management.* Emmitsburg, MD: Federal Emergency Management Agency Emergency Management Institute. Available at archone.tamu.edu/hrrc/Publications/books/index.html.

McNally, R. J., Bryant, R. A., and Ehlers, A., 2003. Does early psychological intervention promote recovery from posttraumatic stress? *Psychological Science in the Public Interest*, **4**, 45–79.

Mileti, D. S., 1999. *Disasters by Design: A Reassessment of Natural Hazards in the United States.* Washington, DC: Joseph Henry Press.

Morrow, B. H., 1997. Stretching the bonds: the families of Andrew. In Peacock, W. G., Morrow, B. H., and Gladwin, H. (eds.), *Hurricane Andrew: Ethnicity, Gender and the Sociology of Disaster.* London: Routledge, pp. 141–170.

Morrow, B. H., and Peacock, W. G., 1997. Disasters and social change: hurricane Andrew and the reshaping of Miami. In Peacock, W. G., Morrow, B. H., and Gladwin, H. (eds.), *Hurricane Andrew: Ethnicity, Gender and the Sociology of Disaster.* London: Routledge, pp. 226–242.

Norris, F. H., Friedman, M. J., Watson, P. J., Byrne, C. M., Diaz, E., and Kaniasty, K., 2002a. 60,000 disaster victims speak: Part I. an empirical review of the empirical literature, 1981–2001. *Psychiatry*, **65**, 207–239.

Norris, F. H., Friedman, M. J., and Watson, P. J., 2002b. 60,000 disaster victims speak: Part II. Summary and implications of the disaster mental health research. *Psychiatry*, **65**, 240–260.

Olson, R. S., and Drury, A. C., 1997. Untherapeutic communities: a cross-national analysis of post-disaster political unrest. *International Journal of Mass Emergencies and Disasters*, **15**, 221–238.

Olson, R. S., Olson, R. A., and Gawronski, V. T., 1998. Night and day: mitigation policymaking in Oakland California, before and after the Loma Prieta Earthquake. *International Journal of Mass Emergencies and Disasters*, **16**, 145–179.

Palm, R., Hodgson, M., Blanchard, R. D., and Lyons, D., 1990. *Earthquake Insurance in California.* Boulder, CO: Westview Press.

Peacock, W. G., and Girard, C., 1997. Ethnic and racial inequalities in disaster damage and insurance settlements. In Peacock, W. G., Morrow, B. H., and Gladwin, H. (eds.), *Hurricane Andrew: Ethnicity, Gender and the Sociology of Disaster.* London: Routledge, pp. 171–190.

Peacock, W. G., Killian, L. M., and Bates, F. L., 1987. The effects of disaster damage and housing on household recovery following the 1976 Guatemala earthquake. *International Journal of Mass Emergencies and Disasters*, **5**, 63–88.

Peacock, W. G., Dash, N., and Zhang, Y., 2006. Sheltering and housing recovery following disaster. In Rodríguez, H., Quarantelli, E. L., and Dynes, R. R. (eds.), *Handbook of Disaster Research.* New York: Springer, pp. 258–274.

Phillips, B. D., 1993. Cultural diversity in disasters: sheltering, housing, and long term recovery. *International Journal of Mass Emergencies and Disasters*, **11**, 99–110.

Prater, C. S., and Lindell, M. K., 2000. Politics of hazard mitigation. *Natural Hazards Review*, **1**, 73–82.

Quarantelli, E. L., 1982. *Sheltering and housing after major community disasters.* Columbus, OH: Ohio State University Disaster Research Center.

Rose, A., and Limb, D., 2002. Business interruption losses from natural hazards: conceptual and methodological issues in the case of the Northridge Earthquake. *Environmental Hazards*, **4**, 1–14.

Rubin, C. B., 1991. Recovery from disaster. In Drabek, T. E., and Hoetmer, G. J. (eds.), *Emergency Management: Principles and Practice for Local Government.* Washington, DC: International City Management Association, pp. 224–259.

Rubin, C. B., Saperstein, M. D., and Barbee, D. G., 1985. *Community Recovery From a Major Natural Disaster.* Monograph #41. Boulder, CO: University of Colorado, Institute of Behavioral Science.

Schwab, J., Topping, K. C., Eadie, C. C., Deyle, R. E., and Smith, R. A., 1998. *Planning for Post-disaster Recovery and Reconstruction*, PAS Report 483/484. Chicago, IL: American Planning Association.

Shefner, J., 1999. Pre- and post-disaster political instability and contentious supporters: a case study of political ferment. *International Journal of Mass Emergencies and Disasters*, **17**, 137–160.

Siegel, J. M., Bourque, L. B., and Shoaf, K. I., 1999. Victimization after a natural disaster: social disorganization or community cohesion? *International Journal of Mass Emergencies and Disasters*, **17**, 265–294.

Slovic, P., Kunreuther, H., and White, G. F. W., 1974. Decision processes, rationality, and adjustment to natural hazards. In White, G. F. (ed.), *Natural Hazards: Local, National and Global.* New York: Oxford University Press, pp. 187–204.

Smith, G. P., and Wenger, D. E., 2006. Sustainable disaster recovery: operationalizing an existing agenda. In Rodríguez, H., Quarantelli, E. L., and Dynes, R. R. (eds.), *Handbook of Disaster Research.* New York: Springer, pp. 234–257.

Smith, S. K., Tayman, J., and Swanson, D. A., 2001. *State and Local Population Projections: Methodology and Analysis.* New York: Kluwer.

Spennemann, D. H. R., and Look, D. W., 1998. *Disaster Management Programs for Historic Sites.* San Francisco, CA: Association for Preservation Technology.

Spittal, M. J., McClure, J., Siegert, R. J., and Walkey, F. H., 2008. Predictors of two types of earthquake preparation. *Environment and Behavior*, **40**, 798–817.

Sullivan, M., 2003. Integrated emergency management: a new way of looking at a delicate process. *Australian Journal of Emergency Management*, **18**, 4–27.

Tierney, K. J., 2006. Businesses and disasters: vulnerability, impact, and recovery. In Rodríguez, H., Quarantelli, E. L., and Dynes, R. R. (eds.), *Handbook of Disaster Research.* New York: Springer, pp. 275–296.

Tierney, K., Lindell, M. K., and Perry, R. W., 2001. *Facing the Unexpected: Disaster Preparedness and Response in the United States.* Washington DC: Joseph Henry Press.

UNDRO-United Nations Disaster Relief Organization, 1984. *Disaster Prevention and Mitigation: A Compendium of Current Knowledge. Vol. 11, Preparedness Aspects.* Geneva: Office of the United Nations Disaster Relief Coordinator.

Warrick, R. A., Anderson, J., Downing, T., Lyons, J., Ressler, J., Warrick, M., and Warrick, T., 1981. *Four communities under ash after Mount St. Helens.* Boulder, CO: University of Colorado Institute of Behavioral Science.

Webb, G. R., Tierney, K. J., and Dahlhamer, J. M., 2000. Businesses and disasters: empirical patterns and unanswered questions. *Natural Hazards Review*, **1**, 83–90.

Webb, G. R., Tierney, K. J., and Dahlhamer, J. M., 2002. Predicting long-term business recovery from disasters: a comparison of the Loma Prieta earthquake and hurricane Andrew. *Environmental Hazards*, **4**, 45–58.

Wilson, R. C., 1991. *The Loma Prieta Quake: What One City Learned*. Washington, DC: International City Management Association.

Wisner, B., Blaikie, P., Cannon, T., and Davis, I., 2004. *At Risk: Natural Hazards, People's Vulnerability and Disasters*, 2nd edn. London: Routledge.

Wright, J. D., Rossi, P. H., Wright, S. R., and Weber-Burdin, E., 1979. *After the Clean-Up: Long-Range Effects of Natural Disasters*. Beverly Hills, CA: Sage.

Wu, J. Y., and Lindell, M. K., 2004. Housing reconstruction after two major Earthquakes: the 1994 Northridge Earthquake in the United States and the 1999 Chi-Chi Earthquake in Taiwan. *Disasters*, **28**, 63–81.

Yelvington, K. A., 1997. Coping in a temporary way: the tent cities. In Peacock, W. G., Morrow, B. H., and Gladwin, H. (eds.), *Hurricane Andrew: Ethnicity, Gender and the Sociology of Disaster*. London: Routledge, pp. 92–115.

Zhang, Y., Lindell, M. K., and Prater, C. S., 2009. Vulnerability of community businesses to environmental disasters. *Disasters*, **33**, 38–57.

Cross-references

Civil Protection and Crisis Management
Coastal Zone, Risk Management
Community Management of Hazards
Critical Incident Stress Syndrome
Cultural Heritage and Natural Hazards
Disaster Relief
Disaster Risk Reduction (DRR)
Disease Epidemiology of Natural Disasters
Economics of Disasters
Emergency Management
Emergency Planning
Emergency Shelter
Insurance
Integrated Emergency Management System
Mitigation
Mortality and Injury in Natural Disasters
Natural Hazard in Developing Countries
Post-Traumatic Stress Disorder (PTSD)
Psychological Impacts of Natural Disasters
Risk Assessment
Vulnerability
World-Wide Trends in Disasters Caused by Natural Hazards

RECURRENCE INTERVAL

Glenn Biasi
University of Nevada Reno, Reno, NV, USA

Synonyms

Earthquake period; Recurrence period

Definition

Recurrence interval is the expected length of time between occurrences of a geologic event.

Discussion

Usage of the term recurrence interval varies somewhat among geoscience disciplines. In paleoseismology, recurrence interval refers to the time between ground-rupturing events at a point on a fault. Strong ground shaking is presumed, but in most cases, the magnitude of the earthquake is poorly known. For instrumentally measured earthquakes, the recurrence interval may be more precisely applied, such as with the specification of both the earthquake size and location. When applied to hydrological events, the recurrence interval refers to the average period between floods of a given size or greater. The distinction leads to some difference in statistical definition. Because usage of the term varies, the definition of the recurring event should be provided with the estimate itself.

Estimation: The most common estimator of the recurrence interval is the mean recurrence time based on a record over which detection of the events is considered certain. A useful estimate of the uncertainty of the recurrence interval is given by the standard error of the mean (the standard deviation divided by the square root of the number of intervals). If the number of intervals is small, an adjustment may be made for the distinction between sample and population statistics. The estimate may also be adjusted if it is known that no events occurred for a significant period before the first or after the last event in the sequence.

Application to hazard estimation: The recurrence interval estimate by itself implies nothing about the probability of an event in any particular period of time. Conditional probabilities of occurrence of the event depend on the probability distribution of the intervals themselves. Only when the events are random in time and thus modeled as a Poisson distribution can the reciprocal of the recurrence interval in years be interpreted as the annual probability of an event. If it is known that the process is clustered or quasi-periodic, the distribution model for intervals will affect the conditional probability of a future event. The term "return period" is sometimes used interchangeably with recurrence interval, but this should be done with care. In seismic engineering, return period can refer to the return of some level of ground motion, but when more than one fault contributes to the hazard, the return period may not match the recurrence interval on any contributing faults.

Bibliography

Cornell, C. A., and Winterstein, S. R., 1988. Temporal and magnitude dependence in earthquake recurrence models. *Bulletin of the Seismological Society of America*, **78**, 1522–1537.

McCalpin, J., 1996. *Paleoseismology*. San Diego: Academic Press.

Cross-references

Debris Flow
Earthquake
Earthquake Prediction and Forecasting
Flood Deposits
Hazard and Risk Mapping
Landslide
Land-Use Planning
Neotectonics
Paleoseismology

Probable Maximum Flood (PMF)
Risk assessment
Seismic Gap

RED CROSS AND RED CRESCENT

Donald J. Shropshire
Municipality of Chatham-Kent, Chatham, ON, Canada

Synonyms

International Federation of Red Cross and Red Crescent Societies; IFRC; League of Red Cross Societies

Definition

The International Federation of Red Cross and Red Crescent Societies is the world's largest humanitarian organization.

Discussion

The Federation's work focuses on four core areas: promoting humanitarian values, disaster response, disaster preparedness, and health and community care. The Red Cross was formed in 1864, at the Geneva Convention, to provide humanitarian assistance in times of war. In response to the many conflicts in Europe, the International Committee of the Red Cross (ICRC) grew dramatically. Following the cessation of World War 1, members of the Red Cross recognized the potential to apply the Red Cross's unparalleled capacity to provide humanitarian assistance in response to other tragedies such as disasters. The League of Red Cross Societies was formed in 1919 to carry out this work (Moorehead 1999, p. 261).

Renamed in 1991, the International Federation of Red Cross and Red Crescent Societies, the Federation now comprises 186 Red Cross and Red Crescent national societies that engage almost 100 million volunteers and staff, a Secretariat in Geneva and more than 60 delegations strategically located to support activities around the world. There are more societies in formation. The unique network of National Societies is the Federation's principal strength. The Federation, together with National Red Cross or Red Crescent Societies and the ICRC, makes up the International Red Cross and Red Crescent Movement (IFRC 2009, p. 7). The Movement works in cooperation with UN agencies, governments, donors, and other aid organizations to assist vulnerable people around the world.

National Societies support the public authorities in their own countries as independent auxiliaries to the government in the humanitarian field. Their local knowledge and expertise, access to communities, and infrastructure enable the Movement to rapidly mobilize to get the right kind of help where it is most needed. Its local presence and community-based approach, coupled with the Movement's global outreach, resources, and know-how, give the Red Cross and Red Crescent a distinct advantage when it comes to dealing with today's complex humanitarian challenges. (www.ifrc.org)

Strategy 2020

Strategy 2020 was adopted by the Federation's General Assembly in November 2009 to guide the actions of the Federation during the next ten years in order to achieve its common vision:

"To inspire, encourage, facilitate, and promote at all times all forms of humanitarian activities by National Societies, with a view to preventing and alleviating human suffering, and thereby contributing to the maintenance and promotion of human dignity and peace in the world."

The *strategic aims* of *Strategy 2020* are:

1. Save lives, protect livelihoods, and strengthen recovery from disasters and crises.
2. Enable healthy and safe living.
3. Promote social inclusion and a culture of non-violence and peace.

The *enabling actions* to deliver our Strategic Aims are:

1. Build strong National Red Cross and Red Crescent Societies.
2. Pursue humanitarian diplomacy to prevent and reduce vulnerability in a globalized world.
3. Function effectively as the International Federation.

Guided by the statutes and strategy of the Movement, *Strategy 2010* consolidates previous policies and strategies and provides the basis for doing more, doing better, and reaching further. (IFRC 2009, p. 9)

Bibliography

Moorehead, C., 1999. *Dunant's Dream War, Switzerland and the History of the Red Cross*. New York: Carroll & Graff Publishers. ISBN 0-7867-0609-0.

International Committee of the Red Cross, 1994. *Handbook of the International Red Cross and Red Crescent Movement*, 13th edn. Geneva: ICRC. ISBN 2-88145-074-1.

International Federation of Red Cross and Red Crescent Societies, 2006. *The Federation of the Future, Working Together for a Better Tomorrow*, 1st edn. Geneva.

The International Federation of Red Cross and Red Crescent Societies, 2009. *Strategy 2010, Saving Lives, Changing Minds*, Geneva.

The International Federation of Red Cross and Red Crescent Societies, 2009. *90 Years of Improving the Lives of the Most Vulnerable*, Geneva.

Cross-references

Casualties Following Natural Hazards
Civil Protection and Crisis Management
Community Management of Hazards
Disaster Diplomacy
Disaster Relief
Education and Training for Emergency Preparedness
Emergency Shelter
Federal Emergency Management Agency (FEMA)
Global Network of Civil Society Organizations for Disaster Reduction

Hospitals in Disaster
Insurance
Integrated Emergency Management System
Mortality and Injury in Natural Disasters
Post Disaster Mass Care Needs
Recovery and Reconstruction After Disaster
Rights and Obligations in International Assistance
United Nations Organization and Natural Disasters

RED TIDES

Philip Weinstein
University of South Australia, Adelaide, SA, Australia

Synonyms
Harmful algal bloom

Definition
A "red tide" or "harmful algal bloom (HAB)" is a local overgrowth of particular species of plankton. Plankton are single-celled microscopic organisms that include algae and dinoflagellates, many of which are capable of photosynthesis while floating in the water column. Their colors vary from green to reddish-brown depending on which photosynthetic pigments they contain. Unfortunately, they can also contain toxins, and when large coastal blooms occur, they pose a hazard to recreational water users, local seafood supplies, and sometimes downwind coastal residents.

Causes
The mechanisms that drive red tides are complex, and include both natural and anthropogenic phenomena. Essentially plankton are, like all organisms, nutrient dependent, and blooms occur in response to increased nutrient availability. Coastal upwelling (deep, nutrient-rich waters rising) thus provides a natural driver of blooms, but so does fertilizer-rich runoff from agricultural lands. Different species also depend on specific wind and sea surface temperature conditions for optimal growth. The interactions of these factors are not yet well enough understood to allow public health officials to predict blooms and issue proactive warnings. The hazard can, therefore, only be managed reactively, generally by public warnings not to engage in recreational activities in or on the water, and sometimes by the imposition of a withholding period for affected seafood (filter-feeding shellfish can accumulate the toxin, rendering them unfit for human consumption). Many reasons have been postulated for toxin production by plankton, ranging from incidental by-products of metabolism to creation of a competitive advantage by poisoning other organisms utilizing the same resources. The recently discovered *Pfiesteria* is a controversial organism that may have evolved the ability to capitalize on the sudden increase in food availability from the fish killed by its own toxins – an intriguing story that has been popularized elsewhere (Barker, 1998).

Effects
The toxins produced include extremely potent neurotoxins (e.g., brevetoxin and saxitoxin, commonly from species of *Gymnodinium* and *Alexandrium,* respectively) and various gastrointestinal irritants (e.g., dinophysistoxin and domoic acid, commonly from species of *Dinophysis* and *Pseudo-nitzschia,* respectively) (Heymann, 2004). The effect on humans depends on the dose and route of exposure, commonly involving the consumption of toxic shellfish because the dose is large and intestinal absorption high. "Neurotoxic shellfish poisoning (NSP)" is one classic outcome associated with brevetoxin exposure, giving rise to tingling around the mouth and extremities as well as dizziness. It tends to be a much milder disease than "Paralytic shellfish poisoning (PSP)," which can kill by paralyzing the respiratory muscles, sometimes within 12 hours of saxitoxin ingestion (Heymann, 2004). In windy conditions, toxins can also be aerosolized as sea spray, with eye and respiratory irritation occurring in beachgoers or coastal residents. Because of the complexities of bloom ecology as well as human behavior, both disease and disease intervention can be highly regionally variable. To plan hazard mitigation strategies, it is therefore important to regionalize recommendations based on the relevant scientific literature (e.g., Florida; Kirkpatrick et al., 2004).

Outlook
The hazard posed by red tides and other HABs is likely to increase substantially in the future as a result of: increasing nutrient runoff from expanding agriculture; increasing sea surface temperatures with global *Climate Change*; and increasing human contact with coastal waters as populations expand, and seek more food and recreational opportunities.

Bibliography
Barker, R., 1998. *And the Waters Turned to Blood*. New York: Touchstone.
Heymann, D. L., 2004. *Control of Communicable Diseases Manual*. Washington: American Public Health Association.
Kirkpatrick, B., Fleming, L. E., Squicciarini, D., Backer, L. C., et al., 2004. Literature review of Florida red tide: implications for human health effects. *Harmful Algae*, **3**, 99–115.

Cross-references
Beach Nourishment
Biblical Events
Challenges to Agriculture
Climate Change
Coastal Zone, Risk Management
El Niño/Southern Oscillation
Global Change
Land-Use Planning
Marine Hazards
Monitoring and Prediction of Natural Hazards

REFLECTIONS ON MODELING DISASTER

David A. Etkin
York University, Toronto, ON, Canada

Synonyms

Catastrophe models; Emergency models

Definition of disaster by United Nations

A serious disruption of the functioning of a community or a society causing widespread human, material, economic, or environmental losses which exceed the ability of the affected community or society to cope using its own resources.

Introduction

People and institutions, especially those that employ scientists and analysts, use models to help them understand, predict, and manage aspects of their environment. Within the context of models used to represent disasters, this process has traditionally been rooted in notions of cause and effect, determinism, free will, and power. When combined these notions result in the belief that people can understand, predict, and control the systems in order to create more desirable outcomes. Evidence of this in modern technological society is abundant. However, there is also contrary evidence – of massive failures with catastrophic consequences – and the traditional paradigm has, at least in part, been challenged by the advent of theories that incorporate notions of chaos, complexity, and adaptive systems. As a result, disaster models have become broader and more sophisticated over time. It should also be noted that fatalistic attitudes are not uncommon, reflected at times in the belief that disasters are, for example, sent by God to punish people because of moral transgressions (Taylor, 1999), or that people have no power to control futures predetermined by Fate.

How people, communities, and institutions conceptualize disaster has important consequences for the selection of coping strategies. Different strategies vary, for example, according to culture and professional discipline (Quarantelli, 1989). Engineering and physical sciences tend to emphasize hazard management, whereas social sciences emphasize reducing vulnerability and enhancing resilience. Some strategies emphasize formal plans, whereas others focus on the development of informal networks. Sometimes strategies are formalized as, or result from, explicit conceptual models (particularly by academics and disaster professionals), and at other times they are better described as nebulous attitudes (particularly with respect to the general public). Each strategy is useful to some degree and depends upon the larger social, political, economic, environmental, and geographical context. One could argue that many of the models used in the field of disaster management suffer from the generic problem of narrowly defining the context within which disasters should be understood, with the result of leading users into perilous overconfidence (Hewitt, 2010, personal communication). An example is the use of levees to reduce flood risk; the well-known levee effect, described by Gilbert White, occurs when the presence of flood protection spurs new development within the flood plain. This explains much of the development in New Orleans. Disaster management is best accomplished by utilizing and applying a spectrum of models that are situationally effective.

What is a disaster?

Disasters occur over a wide range of space and time scales and can be of many different types. Historically, they have been categorized as natural, technological, or human-caused (according to the trigger that initiates the event; even so, many disasters have significant components of two or all three of these categories), though the phrase "unnatural disaster" is sometimes used to emphasize the notion that the causes are more rooted in social processes than natural ones (Steinberg, 2000). Even within this structure there is great complexity, and a disaster may or may not cause deaths or economic loss.

Defining disaster is far from simple. For some organizations the answer is direct and quantitative. The Centre for Research on the Epidemiology of Disasters (CRED, 2007), for example, whose data are frequently quoted along with that of reinsurance companies, defines disaster using a threshold approach, as an event that satisfies at least one of the following criteria:

- Ten or more people killed
- One hundred or more people affected
- Declaration of a state of emergency
- Call for international assistance

This and other similar definitions have the advantage of being straight-forward, measurable, and useful from an operational perspective, though attributions of cause of death and the definition of "affected" can be challenging.

Other definitions of disaster are less specific, particularly those in the academic realm. Disciplinary perspectives often frame the meaning of disaster. For example, physical scientists have historically defined disaster in terms of physical processes, whereas sociologists have defined it in terms of the meaning that the communities give to it and how it affects normal community functioning (Quarantelli, 1998). Much of the literature on this topic assumes that a disaster is an externally triggered event that comes and goes, after which people and institutions recover. For some, a disaster is not constrained to a specific damaging event. Hewitt (1983, 1997) emphasizes the "normal" everyday sets of decisions and actions that create vulnerable communities, and views disaster less as a particular damaging event and more in terms of day-to-day community functions. Along these lines Mulvihill and Ali (2007) discuss the importance of "disaster incubation" (the process by which communities create vulnerable conditions) as a precursor to the actual event. Papers and books on this topic (e.g., Quarantelli, 1998; Perry and Quarantelli, 2005) demonstrate a lack of common agreement on a definition. Although there are recurring themes, many definitions emphasize disasters as cultural events that are "*rationalized and interpreted according to the canons and preoccupations of the contemporary period*" (Alexander, 2005).

Some definitions are, at least in part, political in nature. For example, in the Stafford Act, USA "*Major disaster means any natural catastrophe..., or, regardless of cause, any fire, flood, or explosion, in any part of the United States, which in the **determination of the President** causes damage of sufficient severity and magnitude to warrant major disaster assistance under this Act to supplement the efforts and available resources of States, local governments, and disaster relief organizations in alleviating the damage, loss, hardship, or suffering caused thereby.*" (According to the Federal Emergency Management Agency (FEMA), during the period 1953–2012 there were 2060 major disaster declarations in the USA, an average of 34 per year. The state with the greatest number is Texas, with 86 declarations.)

Most discussions about disaster are underlain by the assumption that a disaster is an external, existential event, filtered through the worldview of the observer. One differing perspective is that a disaster is entirely interpretive; Erikson (1994) takes this point of view when he notes that "*instead of classifying a condition as trauma because it was induced by a disaster, we would classify an event as disaster if it had the property of bringing about traumatic reactions.*"

Disasters are complex and often chaotic events (Koehler, 1995). Unlike definitions in geometry they are not amenable to precise, neat, unambiguous descriptions. Quarantelli (1998) argued that disasters are understood or defined intuitively. The notion is that the word has meaning and can be used successfully without reference to a precise definition; though one may not be able to articulate unambiguously what a disaster is, there would be little doubt of its meaning when experiencing it.

Jigyasu (2005) discusses an "eastern" perspective toward disaster, emphasizing a concept of disaster not as an external event that happens to people, but as being caused by an internal chaos that does not allow people to see the true nature of reality, which can only come from achieving enlightenment. In this view, the external world is an illusion that confuses people and takes them away from the right path to God. This is unlike the discussion above, which has a typical western flavor. Along a very different line, Alexander (2008) presents a conceptual model of "disaster as spectacle" (also referred to as the "Hollywood" model), where disaster mythology defines perception and creates a self-fulfilling prophecy.

The UN International Strategy for Disaster Reduction (ISDR) provides one commonly accepted definition that will be used in the remainder of this entry: "*A serious disruption of the functioning of a community or a society causing widespread human, material, economic, or environmental losses, which exceed the ability of the affected community or society to cope using its own resources.*"

The differing views of disaster discussed above have resulted in a variety of models, including ones that consider the environment as cause, others that emphasize social processes, or still others that incorporate various political or ecological systems. The following sections review some of the issues that are important to disaster models.

Disaster models
Model limitations

A model, whether it is conceptual, physical, mathematical, statistical, or visual, is essentially a way of perceiving or thinking about something. As such, it is a representation of a system, entity, phenomenon, or process. It is not the "real" world but rather a construct to aid us in understanding it. Users of disaster models should be cognizant of their limitations (Whitehead, 1997; Hewitt, 2010). Too limited an approach to disaster management can lead to, for example, the "paper plan syndrome," which is the illusion of preparation based only upon planning documents. Along these lines, Clarke (1999) presents an interesting discussion of extreme examples of this, calling them fantasy documents, which he defines as "plans as forms of rhetoric, tools designed to convince audiences that they ought to believe what an organization says."

The discussion that follows describes established relationships, thresholds, routings, sectors, or disciplines of study and flows of information/energy/goods. In this sense, they are all similar, being typical representations of western rationalist thought that tends to view the world as machines of greater or lesser complexity that can be tinkered with and adjusted in order to create desired outcomes. Certainly, models are ubiquitous and can be very useful in classrooms and institutions. But, though patterned after successful approaches used in the physical sciences, given the human nature of this subject area, they should be considered more as mechanical metaphors than reality. When applied to the real world, they invite perilous interventions where society assumes that we have more control over the occurrence or outcome of a disaster than we actually do (Scott, 1998). In this way, disaster models expose us to Type III errors, which relates not to the rejection of a hypothesis when true (Type I) or its acceptance when false (Type II), but rather to a hypothesis construction that is poorly related to the real issue at stake.

The disaster models generally used in academia and institutions, which are discussed in this entry, can be critiqued in that they provide a limited view of disasters that tend to be ahistorical (Hewitt, 2010), and present these events as amoral, impersonal happenings. Mostly absent (though abundantly present in media stories) are the visceral reactions of people to tragedy, notions of good and evil, or the narratives that people and peoples tell to understand and define their place in the world and their relationships to others.

Examples of models

Comprehensive disaster/emergency management

Comprehensive emergency management (CEM) is based upon the four (sometimes five) pillars of mitigation (prevention), preparedness, response, and recovery. Numerous textbooks on the fundamentals of disaster management base their structure on this model, and many government organizations have adopted it (e.g., Coppola, 2006; FEMA, n.d.; ADP, 2007). This model (Figure 1) is normally presented in a cyclic format.

Mitigation refers to long-term actions that reduce the risk of natural disasters such as constructing dams or prohibiting construction in high-risk areas. *Preparedness* involves planning for disasters and putting in place resources needed to cope with them when they happen. Examples include stockpiling water and essential goods and preparing emergency plans to follow in the event of a disaster. *Response* refers to actions taken after a disaster has occurred. The activities of police, firefighters, and medical personnel, during and immediately after a disaster, fall into this category. *Recovery* encompasses longer-term activities to rebuild and restore the community to its pre-disaster state. This period can also provide opportunities to engage in activities that reduce vulnerability and mitigate future disasters, such as strengthening building codes or modifying risky land-use policies.

The development of this model represented the acceptance of a much needed and broader philosophy of emergency management from the traditional one, which had emphasized the preparedness and response pillars because of the historical dominance of civil defense and first responders. CEM incorporates an all-hazards approach and emphasizes proactive actions to risk reduction. It is the model that is probably most commonly used by emergency management organizations. Though it is accepted as best practice within many organizations, its shortcomings include that of creating silos of the different pillars, despite their interdependent functions. This is often reflected institutionally, where different departments or institutions are responsible for each of the pillars.

As well, the CEM model does not explicitly differentiate or address issues of capacity/resilience/vulnerability, informal networks, or formal arrangements. Figure 3 is a modification of the CEM cycle that portrays the platforms upon which the pillars rest. It still suffers, however, by representing the pillars as separate when, in fact, they have fuzzy boundaries.

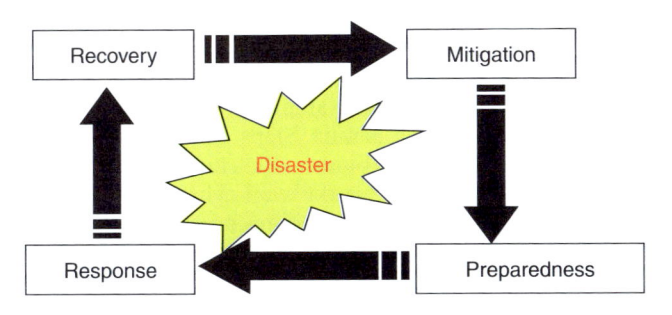

Pressure and release model

This important model was developed largely in response to the hazards' paradigm, which views extreme hazardous events as the cause of disasters (Figure 2). Its greatest contribution was to emphasize the role of society in creating vulnerable communities. It views disasters from a macro-scale perspective of the interaction between social and physical systems.

Within the disaster community, the most common interpretation of "risk" is as a potentially harmful event that is a function of both vulnerability (V) and hazard (H) (i.e., $R = V \times H$) (Sometimes the equation is written as Risk = Hazard $\times$ Consequence). This pseudo equation is discussed at length in Wisner et al. (2004) and is the basis of their Pressure and Release (PAR) Model of disaster. This model portrays disaster as an event that is damaging to society due to the complex interaction between hazard and vulnerability. Vulnerability is portrayed as a multifaceted complex process beginning with root causes, which then creates dynamic pressures that result in unsafe conditions. Hazard is portrayed in a simpler manner, as a trigger that exposes vulnerability. These notions are widely referenced in disaster literature (Rodrigeuz et al., 2006) and have been adopted institutionally, for example by the UN International Strategy on Disaster Reduction (UN, 2006).

Etkin (2009) presents an expanded version of the PAR model (Figure 4), where (a) hazard is portrayed as having a level of dynamic complexity similar to vulnerability, and (b) hazard and vulnerability are linked through box 1 (endogenous root causes). An example of a common root cause is "economic systems and development," which can result in environmental degradation that makes many hazards worse, but that also creates systems of critical infrastructure that lack resilience when short-term profit-making is emphasized. Exogenous causes that are natural in origin are categorized into a separate box at the bottom of the figure. The "Pressure" part of the model is shown by the direction of the arrows, where hazard triggers vulnerability, resulting in a damaging event that may be considered a disaster, depending upon a variety of factors, including cultural filters. The "Release" part of the model comes into play if the arrows related to cause and effect

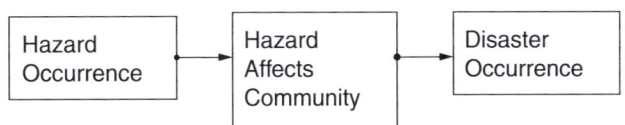

Reflections on Modeling Disaster, Figure 1 The comprehensive disaster/emergency management cycle.

Reflections on Modeling Disaster, Figure 2 Dominant disaster paradigm.

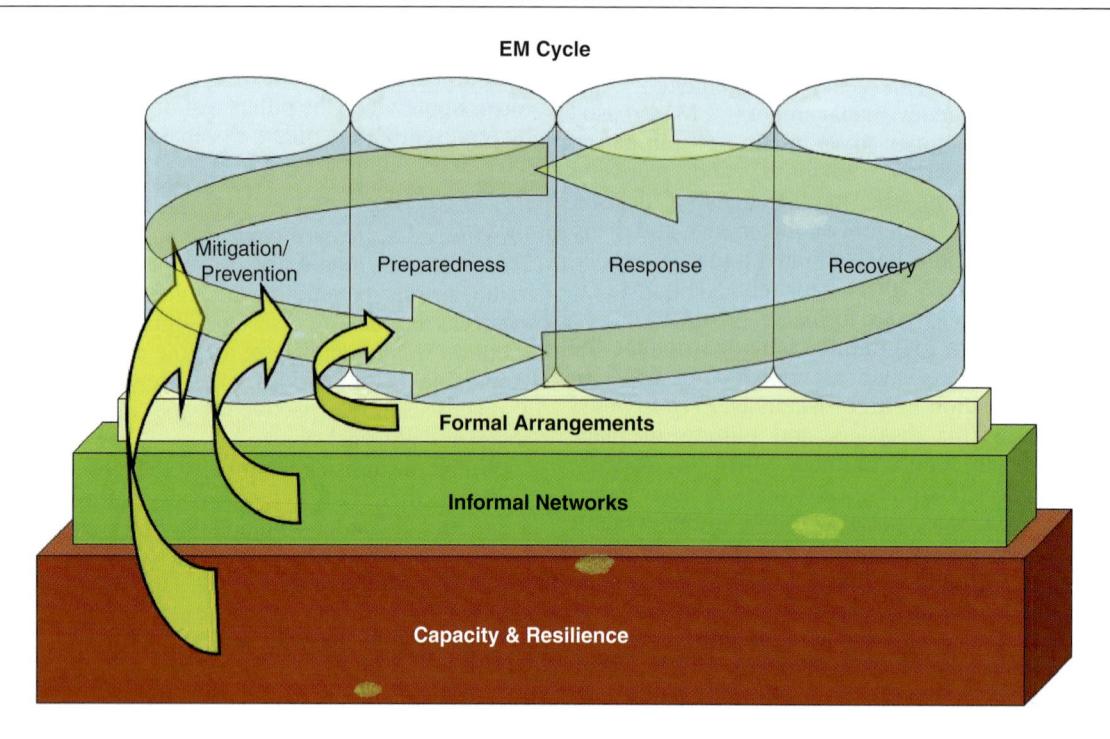

Reflections on Modeling Disaster, Figure 3 Comprehensive disaster/emergency management cycle and the platforms that support it.

reverse direction; by reducing either hazard or vulnerability the "pressure" in the system in lessened.

Though the presentation of Risk as an equation R = V × H might give the impression that it is quantitative and objective, it is best viewed as being socially constructed. This can be interpreted in two ways: The first is that many aspects of vulnerability are created through sets of social decisions (such as how and where we construct our communities and the systems that support them). The second is that risk is, in part, based upon values and subjective assessments as a result of people's and communities' psychological and cultural filters.

Quantifying risk invariably requires subjective judgments. How, for example, can one combine vulnerability due to gender, income, and other factors in an objective way? Inevitably in such assessments there are subjective decisions regarding about what should be included, and how different factors should be measured and weighted (Slovic, 2000).

CARE model
CARE is a nongovernmental organization involved in humanitarian relief and developmental issues. In contrast to the PAR model that emphasizes a macro-/meso- approach to disaster management, the CARE disaster model focuses on the household level (as embedded within larger scales), and views disasters as one of the many external factors (shocks and stresses) within the context box on the left side of the Figure 5. It was developed "to identify constraints to family and community livelihood security and design grassroots programs to overcome them." (Lindenburg, 2002).

Though its focus on the household level differentiates this from the PAR model, it is structurally and topologically similar in that it uses the notion of cause and effect to understand system constraints, and how people use their resources to cope with hazards. It is less explicit in terms of dynamic processes, which must be inferred. For studies focusing on a micro-level where a rapid community-based assessment is required, it is a very useful alternative to the PAR model (Lindenburg, 2002).

Capacity (in the middle under Assets) and vulnerability are explicit in this model, particularly in the livelihood outcomes box that examines such issues as food security. For more information about this model, readers are referred to CARE (2002).

Risk management model
A typical risk management process, such as the Canadian Standards Association (CSA-Q634-91) can be adapted for disaster management. An example of this is the "Emergency Management Australia Steps in the Emergency Management Process" (Figure 6). This type of linear decision-making model works well when uncertainties are small, hazards and vulnerabilities are well understood and subject to known and available controls, and stakeholder buy-in exists in terms of risk management strategies. A typical risk management strategy that might be used within the "Treat Risks" box (Burton et al., 1993) is shown in Figure 7.

Disaster: Chains of Cause & Effect

1. Endogenous Root Causes:

Limited access to:
- power, structures, resources.

Ideologies:
- political systems, economic systems, social systems, technological systems, environmental systems

The Progression of Vulnerability

2. Chains of Cause & Effect:

Lack of: institutions, training, skills, investments, markets, press freedom, ethical standards.

Macro Forces: population growth, urbanization, debt, decline in soil productivity

3. Unsafe Conditions:
- Fragile physical environment
- Fragile local economy
- Vulnerable society
- Public actions

The Progression of Hazard

4. Chains of Cause & Effect:

Lack of: Institutions, land use planning, conservation, environmental controls, resilient technological systems

Macro Forces: Climate warming, deforestation, species extinction, technological system complexity and connectedness, use of chemicals

5. Hazards:
- Earthquakes
- Pollution
- Hurricanes
- Landslides
- Droughts
- Nuclear plants
- Terrorism

Damaging Event → Cultural Filters → Disaster

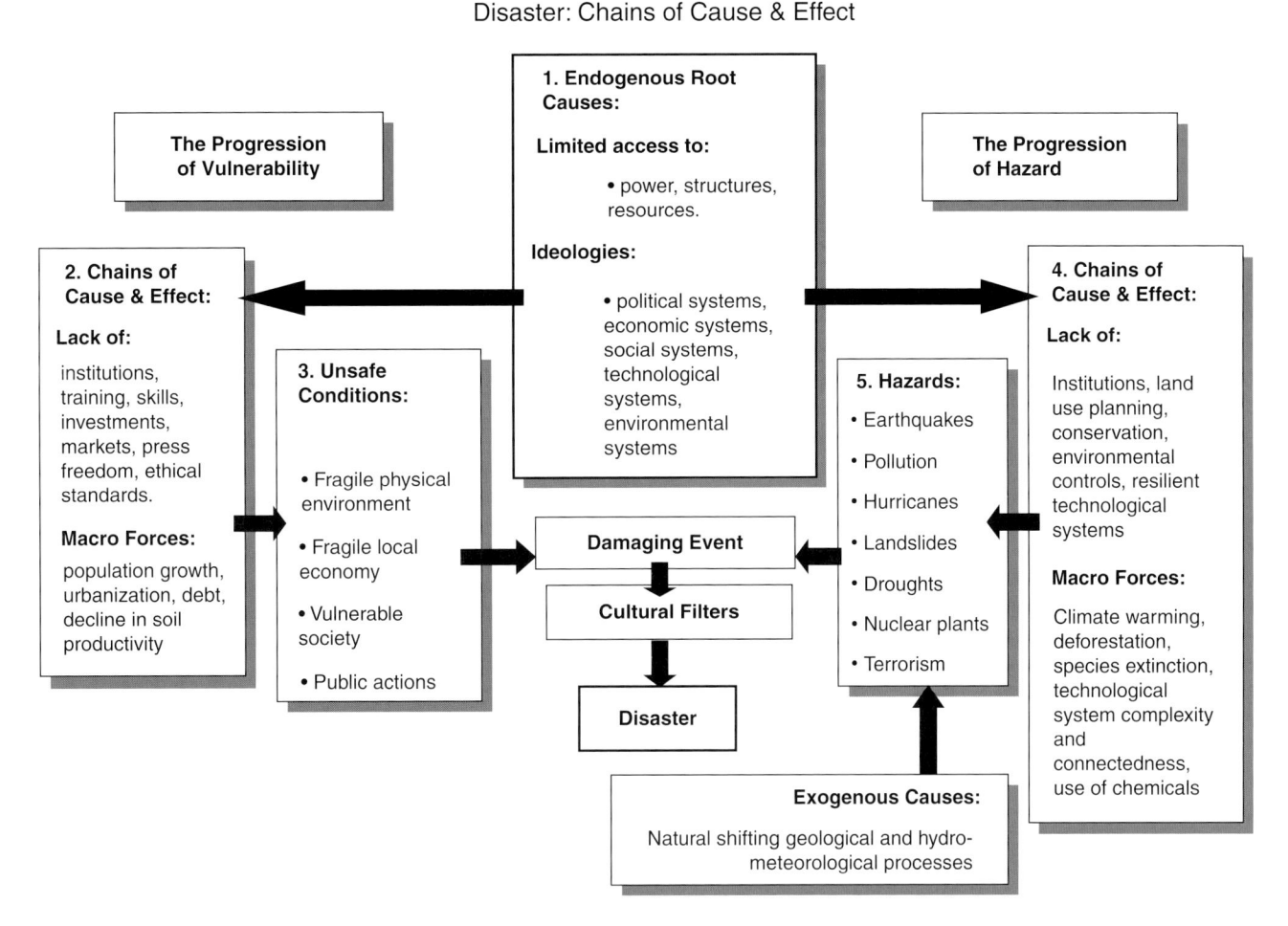

Exogenous Causes: Natural shifting geological and hydro-meteorological processes

Reflections on Modeling Disaster, Figure 4 Modified pressure and release model (Etkin, 2009).

Readers interested in a critique of this type of risk management process may wish to explore the Garbage Can Model (Cohen et al., 1972).

Ecological models
Etkin and Stefanovic (2005) discuss natural disasters from an eco-ethical perspective (Figure 8) emphasizing the interactions and reliance of society on the natural world. This model is, to a large degree, an extension of the work of Burton et al. (1993), which considers the natural environment as both a resource and a hazard, and how mankind's relationships with nature have the potential to increase vulnerability.

Work by Gunderson and Holling (2002) view disaster as part of the evolution of complex systems. Originating from the field of ecology, their work provides a metaphor for any complex system, including society. Figure 9 illustrates their basic model of the pathways that systems move through as they evolve.

An example is the ecology of forest systems. New forests (r) exploit the earth's resources and over time become old forests (K), which are stable. However, as they evolve the potential for fire increases due to increasing amounts of litter on the forest floor and the development of dense canopies. Eventually a trigger (such as lightening or arson) will create a fire that begins the destructive (this is where a disaster may occur) release phase (Ω) after which the system reorganizes itself (α) and the cycle repeats. During the Ω phase, if a critical threshold is passed, a system may flip into a different state, such as a clear lake dominated by fish to a murky one dominated by plankton. In these cases the process can be irreversible.

A complete description of this model is beyond the scope of this entry, and thus far it is not much referred to in the disaster literature, though there is potential for its use at the conceptual if not the applied level. The authors of Panarchy frankly acknowledge that it should be considered as an incomplete metaphor; the greatest advantage of this model may be by broadening perspectives to include issues of instability of complex systems, the significance of cross-scale interactions, and the importance of adaptive change and learning.

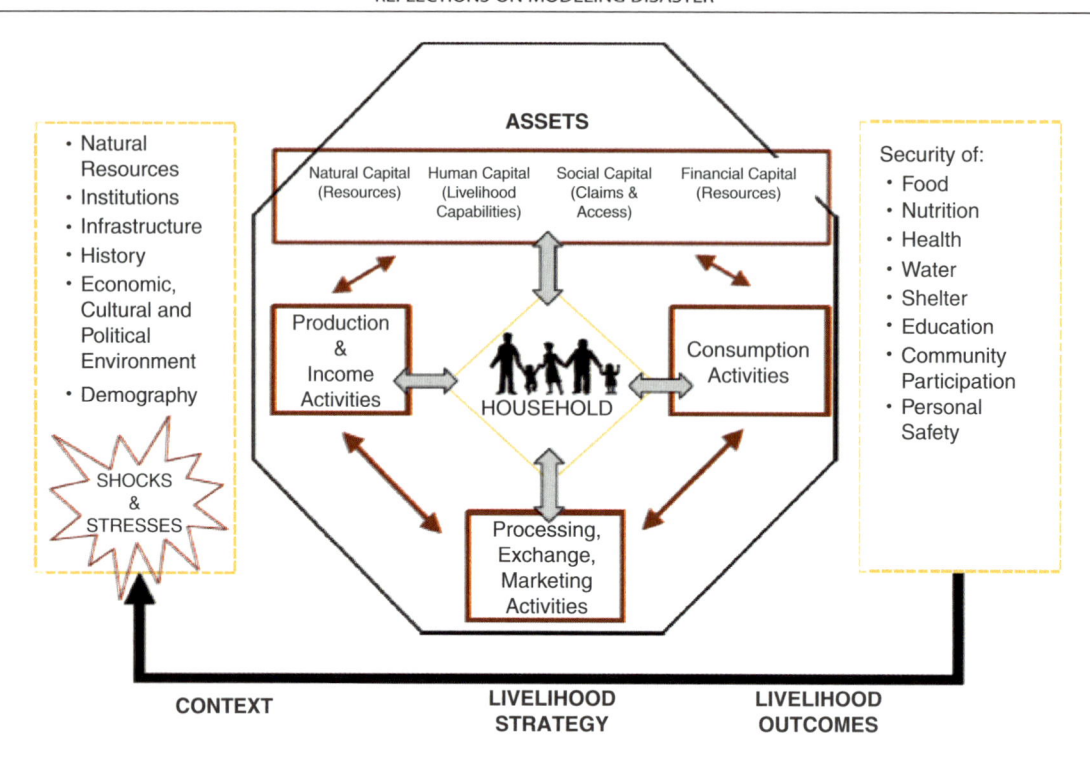

Reflections on Modeling Disaster, Figure 5 CARE household model (CARE, 2002).

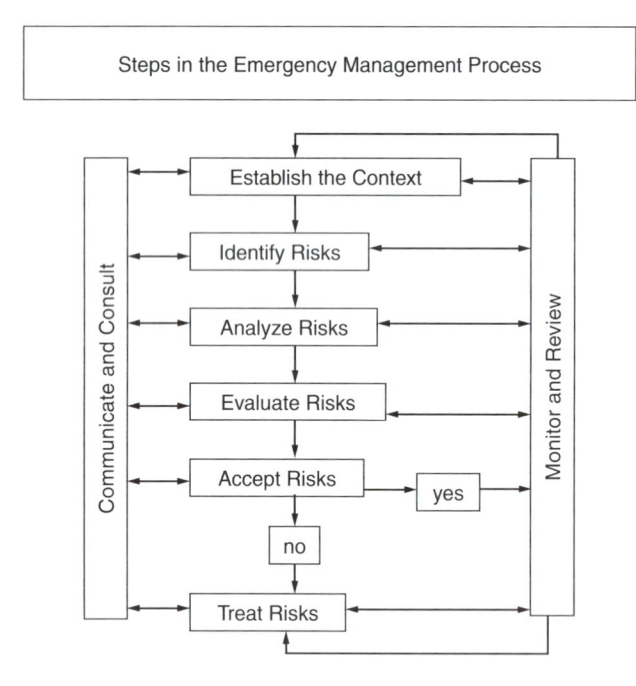

Reflections on Modeling Disaster, Figure 6 Australian emergency management model based upon a risk assessment standard.

Reflections on Modeling Disaster, Figure 7 Choices in "treat risks" Box of Figure 6.

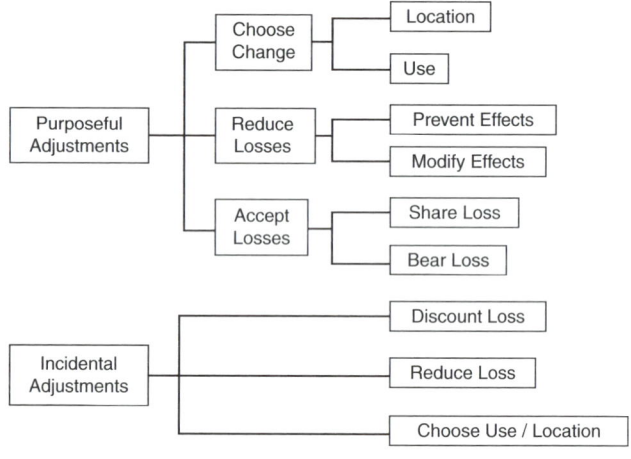

Source: Burton, Kates and White: The Environment as Hazard

Normal accidents

In 1984, Charles Perrow challenged the way in which large accidents or disasters were viewed (Perrow, 1984). The traditional perspective was that accidents were abnormal, unnecessary events that could be prevented if sufficient management was engendered in the form of training, quality control, and standard operating procedures. This view is called "High Reliability Theory" (La Porte and Consolini, 1991).

Perrow's hypothesis is that in certain types of systems, accidents are a fundamental property of the system and therefore, cannot be totally prevented, with such

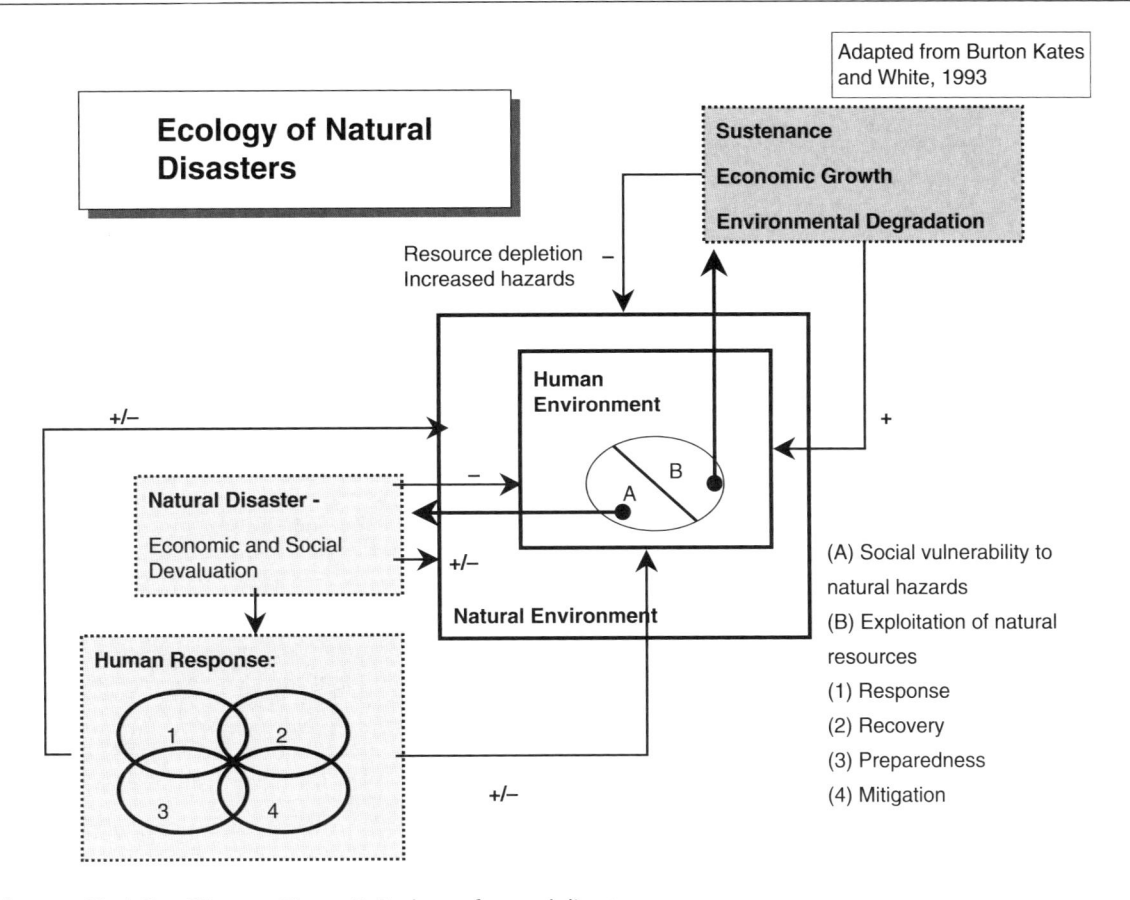

Reflections on Modeling Disaster, **Figure 8** Ecology of natural disasters.

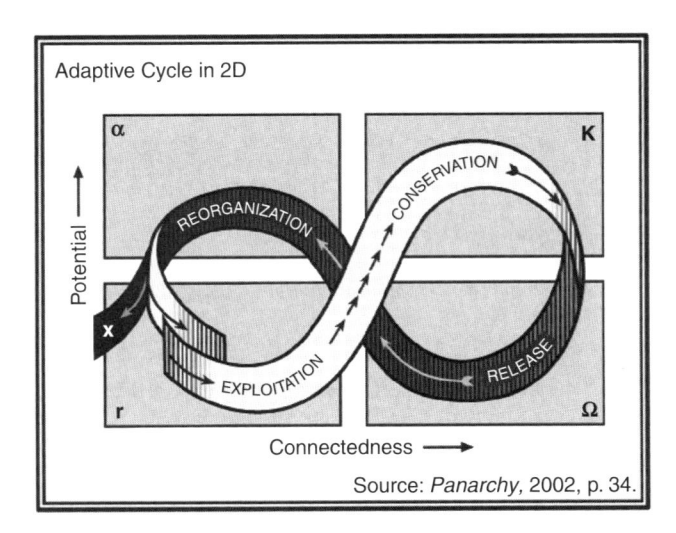

Reflections on Modeling Disaster, **Figure 9** Panarchy model.

events sometimes resulting in disasters. These systems have the characteristics of being complex and tightly coupled. Perrow further argues that some of the methods used to make systems safer (such as redundancy) can actually have the opposite effect, by increasing system complexity. Such systems can reach critical states as they self-organize, where a small perturbation can cause the system to collapse (Per Bak, 1996). Examples of this type of disaster include Three Mile Island and the Challenger explosion. Where the systems are of very high risk, Perrow recommends that those technologies be abandoned. One of the main reasons for this is that complex systems require decentralized management whereas tightly coupled systems require centralized management. When a system is both complex and tightly coupled, a fundamental contradiction exists in terms of how to manage it. In a useful critique of both approaches, Marais et al. (2004) suggest that high-reliability theory "oversimplifies the problems faced by engineers and organizations building safety-critical systems and following some of the recommendations could lead to accidents. Normal accident theory, on the other hand, does recognize the difficulties involved but is unnecessarily pessimistic about the possibility of effectively dealing with them."

Catastrophe models
Partly in response to growing payouts by insurance and reinsurance companies during the 1980s and 1990s,

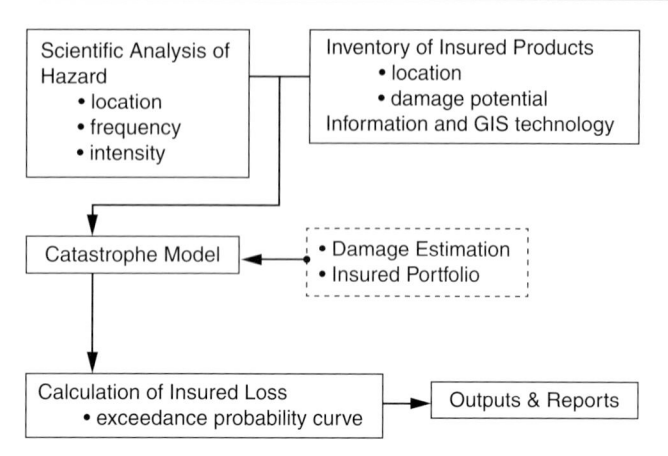

Reflections on Modeling Disaster, Figure 10 Elements of a catastrophe model (Adapted from Grossi and Kunreuther, 2005).

and particularly due to such events as the case with 11 insurance companies going bankrupt after Hurricane Andrew in 1992, various insurance and private sector companies began to finance the development of catastrophe (CAT) models so that they could better assess their risk posed by disasters. The main proprietary models are AIR Worldwide, EQECAT and RMS, Inc. The US government has developed an open source model called Hazards United States (HAZUS) under the auspices of FEMA. There is an International Society of Catastrophe Managers that promotes this profession within the insurance industry.

The main purpose of these models is to estimate the amount of insured damage that would occur due to a catastrophic event, such as a hurricane, flood, or earthquake. HAZUS estimates include not only physical damage but economic loss and social impacts. These models are based upon meteorology, seismology, engineering, and actuarial science.

The basic structure of the CAT models is shown in Figure 10 (adapted from Grossi and Kunreuther, 2005). Final outputs typically include an exceedance probability curve of financial loss. The main sources of uncertainty relate to a lack of empirical data on hazard and vulnerability, and limited scientific knowledge of hazards and engineering. A deficiency of information with respect to business interruption costs and repair costs affect the accuracy of the loss component of the model (Grossi and Kunreuther, 2005).

Conclusions

Disasters are complex human-centric events that are difficult to define. How they are conceptualized has important consequences in terms of strategies used to manage them. There are a number of conceptual models that are used by academics, professionals, and the public in terms of how they are understood – each has its strengths and weaknesses and can be situationally useful. To some extent these models must be considered perilous, as they can exclude other useful ways of viewing disasters. By not adhering rigidly to one particular model, by analyzing model limitations (sometimes serious ones), and by using them according to their different strengths, the management of disasters can be made more flexible and effective.

Bibliography

ADP, 2007. Small Group workshop on preparing for large-scale emergencies. In *Disaster and Emergency Assistance Policy*. Manila: Asian Development Bank.

Alexander, D., 2005. An interpretation of disaster in terms of changes in culture, society, and international relations. In Perry, R. W., and Qarantelli, E. L. (eds.), *What Is A Disaster? New Answers to Old Questions*. Philadelphia: Xlibris Corporation, pp. 25–38.

Alexander, D. (2008). *Symbolic and Practical Interpretations of the Hurricane Katrina Disaster in New Orleans*. http://emergency-planning.blogspot.com/2008/10/symbolic-and-practical-interpretations.html. Accessed April 8, 2009.

Bak, P., 1996. *How Nature Works: The Science of Self-Organized Criticality*. New York: Copernicus. ISBN 0-387-94791-4.

Burton, I., Kates, R., and White, G., 1993. *Environment as Hazard*. New York: The Guilford Press, Vol. 2.

CARE (2002). *Household Livelihood Security Assessments: A Toolkit for Practitioners*. Tucson: TANGO International. http://www.proventionconsortium.org/themes/default/pdfs/CRA/HLSA2002_meth.pdf.

Clarke, L., 1999. *Mission Improbable: Using Fantasy Documents to Tame Disaster*. Chicago: University of Chicago Press.

Cohen, M. D., March, J. G., and Olsen, J. P., 1972. A garbage can model of organizational choice. *Administrative Science Quarterly*, **17**(1), 1–25.

Coppola, D. P., 2006. *Introduction to International Disaster Management*. New York: Elsevier.

CRED (2007). *Centre for Research in the Epidemiology of Disasters*. Emergency Events Database. http://www.emdat.be/.

Erikson, K., 1994. *A New Species of Trouble: Explorations in Disaster, Trauma, and Community*. New York: W.W. Norton.

Etkin, D., 2009. Patterns of risk. NATO book on spatial planning as a strategy for the mitigation of risk from natural hazards. In Fra Paleo, U. (ed.), *Building Safer Communities. Risk Governance, Spatial Planning, and Responses to Natural Hazards*. Amsterdam: IOS Press.

Etkin, D., and Stefanovic, I. L., 2005. Mitigating natural disasters: the role of eco-ethics. *Mitigation and Adaptation Strategies for Global Change*, **10**, 469–490.

FEMA (n.d.). http://www.fema.gov/government/mitigation.shtm. Accessed March 30, 2012.

Grossi, P., and Kunreuther, H., 2005. Catastrophe Modeling: A New Approach to Managing Risk. New York: Springer.

Gunderson, L. H., and Holling, C. S. (eds.), 2002. *Panarchy: Understanding Transformations in Human and Natural Systems*. Washington, DC: Island Press. 507 pp.

Hewitt, K., 1983. *Interpretations of Calamity: From the Viewpoint of Human Ecology*. Boston: Unwin Hyman. The Risks & Hazards Series, Vol. 1.

Hewitt, K., 1997. *Regions of Risk: A Geographical Introduction to Disasters*. London: Longman Harlow.

Hewitt, K., 2010. Personal Communication, April 15, 2010.

Jigyasu, R., 2005. Disaster: a "reality" or "construct"? Perspective from the "East". In Perry, R. W., and Qarantelli, E. L. (eds.), *What Is A Disaster? New Answers to Old Questions*. Philadelphia: Xlibris Corporation, pp. 49–59.

Koehler, G. A., ed. (1995). *What Disaster Response Management Can Learn From Chaos Theory. Conference Proceedings*, May 18–19, 1995. California Research Bureau, http://www.library.ca.gov/crb/96/05/index.html

La Porte, T. R., 1996. High reliability organizations: unlikely, demanding, and at risk. *Journal of Contingencies and Crisis Management*, **63**(4).

Lindenburg, M., 2002. Measuring household livelihood security at the family and community level in the developing world. *World Development*, **30**(2), 301–318.

Marais, K., Dulac, N., and Leveson, N. (2004). Beyond Normal Accidents and High Reliability Organizations: The Need for an Alternative Approach to Safety in Complex Systems. Presented at the Engineering Systems Division Symposium, MIT, Cambridge, MA, March 29–31, 2004.

Mulvihill, P. R., and Ali, S. H., 2007. Disaster incubation, cumulative impacts and the urban/ex-urban/rural dynamic. *Environmental Impact Assessment Review*, **27**(4), 343–358.

Perrow, C., 1984. *Normal Accidents: Living with High-Risk Technologies*. Princeton: Princeton University Press.

Perry, R. W., and Quarantelli, E. L., 2005. *What Is A Disaster? New Answers to Old Questions*. Philadelphia: Xlibris Corporation.

Porte, L., and Consolini, P. M., 1991. Working in practice but not in theory. *Journal of Public Administration Research and Theory*, **1**, 19–47.

Quarantelli, E. L., 1989. Conceptualizing disasters from a sociological perspective. *International Journal of Mass Emergencies and Disasters November*, **7**(3), 243–251.

Quarantelli, E. L., 1998. *What Is A Disaster? Perspectives on the Question*. London: Routledge.

Quarantelli, E. L., 2003. Urban vulnerability to disasters in developing countries: managing risks. In Kreimer, A., Arnold, M., and Carlin, A. (eds.), *Building Safer Cities: The Future of Disaster Risk*. Washington, DC: The International Bank for Reconstruction and Development/The World Bank. Disaster Risk Management Series, pp. 211–232.

Quarantelli, E. L. (2006). Catastrophes are Different from Disasters: Some Implications for Crisis Planning and Managing Drawn from Katrina. Understanding Katrina, Perspectives from the Social Sciences. Social Sciences Research Council. http://understandingkatrina.ssrc.org/Quarantelli/. Accessed March 2009.

Rodrigeuz, H., Quarantelli, E. L., and Dynes, R. R., 2006. *Handbook of Disaster Research*. New York: Springer.

Scott, J. C., 1998. *Seeing Like a State: How Certain Schemes to Improve the Human Condition Have Failed*. New Haven: Yale University Press.

Slovic, P., 2000. *The Perception of Risk*. Earthscan, Virginia.

Slovic, P., and Weber, E. U., 2002. Perception of risk posed by extreme events. In *Risk Management Strategies in an Uncertain World Conference*, Palisades, NY, April 12–13, 2002.

Steinberg, T., 2000. *Acts of God: The Unnatural History of Natural Disasters in America*. New York: Oxford University Press.

Taylor, A. J. W. (1999). Value Conflict Arising from a Disaster. *The Australian Journal of Disaster and Trauma Studies*, **2**. http://www.massey.ac.nz/~trauma/

UN, (2006). *Disaster Assessment Portal*. UN Habitat. http://www.disasterassessment.org/section.asp?id=20 Accessed March 27, 2009.

Whitehead, A. N. 1997/1925. *Science and the Modern World*. New York: N.Y. Free Press (Simon & Schuster).

Wisner, B., Blaikie, P., Cannon, T., and Davis, I., 2004. *At Risk: Natural Hazards, People's Vulnerability and Disasters*, 2nd edn. New York: Routledge.

Wittgenstein, L., 1953/2001. *Philosophical Investigations*. Oxford: Blackwell Publishing.

RELEASE RATES

Pat E. Rasmussen
Health Canada, University of Ottawa, Ottawa,
ON, Canada

Definition

In the context of the natural environment, "release rates" refer to the amount of material that flows through a unit area per unit time.

Discussion

Material is continuously transferred between different components of the Earth System including rocks, soils, plants, waters, and the atmosphere. The rate at which material (such as a metal ion) is transferred from a host component (such as soil) to another component (such as water) depends on how easily it is mobilized, or released, from the host component (Larocque and Rasmussen, 1998). The mobilization of material is controlled partly by the physiochemical conditions existing in the environment, and partly by the characteristics of the material itself.

Knowledge about the release rates of nutrient ions, such as potassium, calcium, and magnesium, as they transfer from bedrock rock to soil, is essential to understanding nutrient cycling in ecosystems (Kolka et al., 1996). Nutrient ions are released by the dissolution of minerals to the overlying soil horizon during bedrock weathering. The rates of release vary depending on the climate, the type of soil, and the type of underlying bedrock.

Release rates are not necessarily constant. For example, the concentration of phosphorus in overland and subsurface water is related to the concentration and release rate of phosphorus in soil. The release of phosphorus from soil occurs simultaneously in two processes: rapid release on contact with solution, and a slower diffusion from inside soil particles (McDowell and Sharpley, 2003).

There are strong seasonal variations in release rates. Measurement of soil carbon dioxide release rates measured in a Chinese fir plantation showed that the rate of carbon dioxide release was higher in the summer than in other seasons. It was found that the soil carbon dioxide release rate increases as air temperature increases within the plantation, and also with increased soil temperature, increased carbon/nitrogen ratios, and increased soil water content (Xi et al., 2005).

Release rates may fluctuate widely over the course of a single day. For example, the release rate of gaseous elemental mercury from mercury-enriched black shale reaches maximum values during the day, depending on sunlight, air and soil temperature, and local meteorological conditions, and decreases to minimum values at night (Rasmussen et al., 2005).

Bibliography

Kolka, R. K., Grigal, D. F., and Nater, E. A., 1996. Forest soil mineral weathering rates: use of multiple approaches. *Geoderma*, **73**, 1–21.

Larocque, A. C. L., and Rasmussen, P. E., 1998. An overview of trace metals in the environment. *Environmental Geology*, **33**, 85–91.

McDowell, R. W., and Sharpley, A. N., 2003. Phosphorus solubility and release kinetics as a function of soil test P concentration. *Geoderma*, **112**, 143–154.

Rasmussen, P. E., Edwards, G. C, Schroeder, W. H., Ausma, S., Steffen, A., Kemp, J., Hubble Fitzgerald, C., El Bilali, E., and Dias, G., 2005. Measurement of gaseous mercury flux in terrestrial environments. In Parsons, M. B., and Percival, J. B. (eds.), *Mercury: Sources, Measurements, Cycles, and Effects*. Mineralogical Association of Canada Short Course Series 34(7), pp. 123–138.

Xi, F., DaLun, T., WenHua, X., WenDe, Y., and WenXing, K., 2005. Soil CO_2 release rate and its effect factors in Chinese fir plantation. *Scientia Silvae Sinicae*, **41**, 1–7.

Cross-references

Climate Change

RELIGION AND HAZARDS

Heather Sangster[1], Angus M. Duncan[2], David K. Chester[1]
[1]University of Liverpool, Liverpool, UK
[2]University of Bedfordshire, Luton, UK

Definition

Religious reactions to disasters are to be found within most traditions of faith, across societies located in different parts of the world, and over the long span of human history. In some cases, it is possible to extend the timescale and interpret archaeological evidence to show how prehistoric societies often used theistic frames of reference to make sense of the suffering caused by catastrophic natural events. Geomythology involves the investigation of oral traditions that preserve information on prehistoric geological events, and there is overwhelming evidence that a wide range of theistic responses to disasters have been preserved (Vitaliano, 2007).

The religious interpretations of disasters are neither confined to archaeological examples, nor are they encountered merely as geomyths within oral traditions of faith, but can be found across all the major religions of the world. Within Judaism and Christianity, there have been two prominent models of theodicy: *free-will* (*Augustinian*) and *best-possible worlds* (*Irenaean*). The notion of free-will considers human beings to have freedom and that suffering is not only a result of the operation of free-will, but may also involve sinfulness because a person has acted against God's purpose, with divine punishment sometimes being stressed. The notion of the best-possible world considers that to design a system in nature that does not have the potential to injure unsuspecting humans would be impossible; therefore, God's purpose is to use disasters to enable a greater good to be achieved. In Buddhism, Jainism, Hinduism, and Shinto, there is a common belief that a person's behavior leads irrevocably to an appropriate reward or punishment (Kogen, 1987) and this has often encouraged fatalistic attitudes toward disasters. Present day suffering is frequently believed to be due to *karma* or fate, which is influenced by conduct in former lives. The word Islam means submission to the will of God, and in the *Qur'an*, earthquakes are interpreted within two different yet related frameworks, being perceived as either signs of a future apocalypse or as a punishment of limited duration for a particular group of people (Akasoy, 2009).

Academic scholarship has been highly critical of the influence of religion on the human understanding of disasters (for example, Sigurdsson, 1999). According to a conventional reading of intellectual history, superstition largely replaced the naturalistic explanations of disaster that were embryonic in the classical era and following the Eighteenth Century Enlightenment, religious explanations became less prominent and were progressively replaced by more scientific and social scientific explanations of disasters and their human impacts. Evidence collected by the present authors from disasters that have occurred over the past hundred years (Chester and Duncan, 2008) shows that this is far from being the case and people frequently employ religious frames of reference to explain both extreme natural events and losses caused by them. Indeed religious explanations were apparent following both the Indian Ocean earthquake and tsunami disaster in 2004 and the Haitian earthquake of 2010.

Although there are some examples of people resisting relief efforts due to their religious beliefs, these are exceptions and for most people belief neither inhibits more practical measures being taken to reduce individual or group exposure to hazards, nor does it hinder people accepting help from the civil authorities. Believing at the same time in two mutually incompatible worldviews, or holding one opinion and acting contrary to it, is known in the literature as *parallel practice*, and this has been discussed in detail with reference to countries with a predominant Christian ethos, where it remains a major feature of faith-based responses (Chester and Duncan, 2009), although more detailed research on *parallel practice* in parts of the world where other religions are in the majority is required.

Bibliography

Akasoy, A., 2009. Interpreting earthquakes in medieval Islamic texts. In Mauch, C., and Pfister, C. (eds.), *Natural Disaster, Cultural Responses: Case Studies Towards a Global Environmental History*. Lanham: Lexington, pp. 183–196.

Chester, D. K., and Duncan, A. M., 2008. Geomythology, theodicy and the continuing relevance of religious worldviews on responses to volcanic eruptions. In Grattan, J., and Torrence, R. (eds.), *Living Under the Shadow*. Walnut Creek: Left Coast, pp. 203–224.

Chester, D. K., and Duncan, A. M., 2009. The Bible, theodicy and Christian responses to historic and contemporary earthquakes and volcanic eruptions. *Environmental Hazards*, **8**, 304–332.

Kogen, M., 1987. Karman. In Eliade, M. (ed.), *The Encyclopedia of Religion*. New York: Macmillan, Vol. 8, pp. 261–268.

Sigurdsson, H., 1999. *Melting the Earth: The History of Ideas on Volcanic Eruptions*. Oxford: Oxford University Press.

Vitaliano, D. B., 2007. Geomythology: Geological origins of myths and legends. In Piccardi, L., and Masse, W. B. (eds.), *Myth and Geology*. Geological Society Special Publication, 273, pp 1–7. London

Cross-references

Biblical Events
Historical Events
Myths and Misconceptions

REMOTE SENSING OF NATURAL HAZARDS AND DISASTERS

Norman Kerle
Faculty of Geo-Information Science and Earth Observation (ITC), University of Twente, Enschede, The Netherlands

Synonyms

Earth observation

Definition

Remote sensing is a summary term for the instrumentation, techniques, and methods to observe the Earth's surface at a distance and to interpret the images or numerical values obtained to acquire meaningful information concerning the nature or state of the observed features. We distinguish between spaceborne, aerial, and terrestrial remote sensing, each of which offers excellent possibilities to observe and study natural hazards and disasters.

Introduction

To be able to mitigate or effectively respond to disasters, a sound understanding of all relevant aspects is needed: the hazards that may lead to losses when events spatially intersect with elements at risk (EaR; i.e., people, infrastructure, assets), their vulnerability toward the specific hazard at different event magnitudes, and the resulting risk (see *Risk*). Such knowledge allows existing hazards to be effectively monitored and early warnings to be given in case of impending hazardous situations. If a disastrous event does occur, its total losses and damages will not only depend on the severity of the event and the vulnerability of the EaR, but also on how effectively the event is responded to. In the long term, future risk – and renewed disaster occurrences and losses – then depend on lessons learnt from the event, that is, how much risk reduction is incorporated in rehabilitation and reconstruction. This means that disaster risk reduction (DRR) – the ultimate objective – is highly contingent on comprehensive, suitable, timely, and accurate information. All of the above aspects are spatial in nature. By that we mean that they have a certain location and extent, thus can be put in relation with one another, and can be associated with attributes that are linked to a geographic place or area. Therefore, for DRR we require spatial information. The most efficient way to obtain this knowledge, such as on where hazards exist, where people live or their assets are located, where immediate assistance is required after a disaster, or where best to resettle a community after a destructive event, is by using remote sensing (RS), also called Earth observation. This entry explains what RS is, why specifically it is suitable as a basis for DRR, and how we can use it to study the various hazard and disaster aspects. The entry finishes with an assessment of currently promising developments that may facilitate DRR in the coming years.

Overview of remote sensing concepts

RS can be described as the process of making measurements or observations without direct contact with the object under investigation. Thus, while in the context of RS satellites often first come to mind, even amateur photography or similar noncontact measurements made on the ground can be considered RS. It usually results in images, but also includes other measurements, such as of temperature or gravity.

Sensors and platforms

For remote sensing, we normally require a sensor (i.e., a camera or scanner), and also something that carries the device. Such platforms can be airplanes or satellites, but also other devices that allow us to place the sensor in a way that the area or object of interest is exposed, such as balloons or kites. The choice of platform directly affects what we can observe and how. Airplanes and helicopters are flexible in their operation, and with relatively low flying heights provide good spatial detail. However, such surveys can be expensive and regular imaging of the same area thus costly. Satellites fly in a fixed orbit, and are thus less flexible, but can provide data at regular intervals (comparable to trains on a track following a fixed schedule). One platform type are the so-called polar orbiters, where the satellites continuously circle the Earth at an altitude of some 500–900 km, passing over or near the poles. Normally, only a relatively narrow strip of Earth underneath the sensor is observed, depending on the sensor type ranging between less than 10 km and several hundred kilometer in width. Adjacent strips are then visited in subsequent orbital passes. Modern satellites can also point the sensor sideways for greater flexibility. The other class of satellites is positioned in geostationary orbit. This means that the satellite is always directly above a designated place on the equator, moving with the rotating Earth at an altitude of approximately 36,000 km. At that height the sensor can usually observe an entire hemisphere (the side of the Earth facing it), and provide data at any desired frequency. Many weather and communication satellites fall in this category, while most Earth observation satellites are polar orbiters.

Collecting information

The data obtained depend primarily on the sensor type, just as one might take color or black/white photographs with a film camera. The principle behind collecting different data lies in the electromagnetic energy (EMR; see *Electromagnetic Radiation (EMR)*), which is what the sensors detect. The most common source of energy is reflected sunlight, which contains not only visible light, but also ultraviolet (UV), infrared (IR), thermal, and other energy (Figure 1). Which part of this continuous energy band, also called electromagnetic spectrum, we capture depends on the sensor? While a consumer camera might only capture visible light, other scanner types record UV, IR, thermal, or other energy ranges.

The data

The data recorded by a sensor typically have the form of a grid or raster (Figure 2a), made up of rows and columns that are populated by cells. These cells, also known as picture elements (pixels), contain the information recorded. A sensor can also have several bands, meaning that different sections of the electromagnetic spectrum are imaged. Thus, for the area observed, an image can contain several bands, and the cell corresponding to a small part on the ground will have one data value for each band. The most important point to understand here is that different materials on the ground reflect or emit energy in a characteristic spectral pattern (see *Electromagnetic Radiation (EMR)*, Figure 2b). For example, vegetation is characterized by high energy in the near infrared (NIR), whereas for water the NIR energy is very low. In Figure 2 this results in high values (digital numbers [DN]) for vegetation and low values for water in the band corresponding to the NIR (Figure 3c).

Other factors influencing the data

RS data come in many forms, often described by sensor type, as well as their spatial, temporal, and spectral resolution. Sensors recording reflected sunlight or energy emitted by the Earth are called passive sensors. Others emit their own energy, which is reflected by the target and recorded, comparable to a camera flash. These are active sensors, well-known examples being radar or laser scanning. The spatial resolution describes the size of the ground area represented in a single pixel. This largely depends on the distance between the sensor and the object. While aerial photos may have a resolution of a few cm, data from polar orbiters range between about 50 cm and 1 km per pixel. Being very far away, sensors on geostationary platforms, such as weather satellites, typically record data at resolutions of several kilometers. The temporal resolution describes the possible frequency of repeat observations. For aerial surveys, this can be years due to the high cost and logistics involved. With more recent and lower cost devices, such as unmanned drones (see below), near-permanent observation is, in principle, possible. This is especially useful to monitor ongoing events, such as wildfires (Ambrosia et al., 2003). Depending on the type of polar orbiter, currently operational sensors have temporal resolutions varying between approximately 1 and 44 days, whereas modern geostationary sensors record data up to every 15 min. The spectral resolution describes how narrow a slice of the electromagnetic spectrum a sensor band records.

Visual information extraction

The acquired data are only of use if they reveal the required information, which can be extracted by visual assessment or digital image processing. The data can be displayed on a computer monitor using three different color channels (blue, green, red) to generate any color (including black and white). At the simplest level, the data can be displayed one band at a time as a grayscale image (Figures 2c, 3a). Alternatively, the three visual bands (blue, green, and red) can be assigned to their respective monitor colors to create a so-called true-color composite

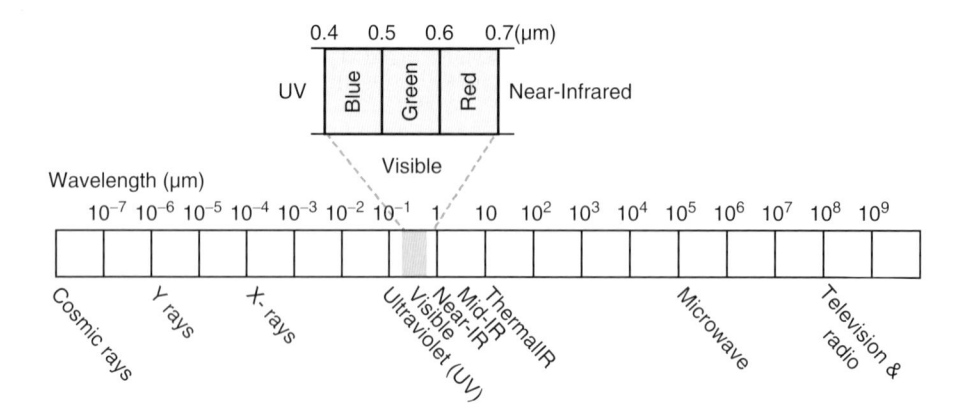

Remote Sensing of Natural Hazards and Disasters, Figure 1 Electromagnetic spectrum (Modified from Lillesand et al., 2004).

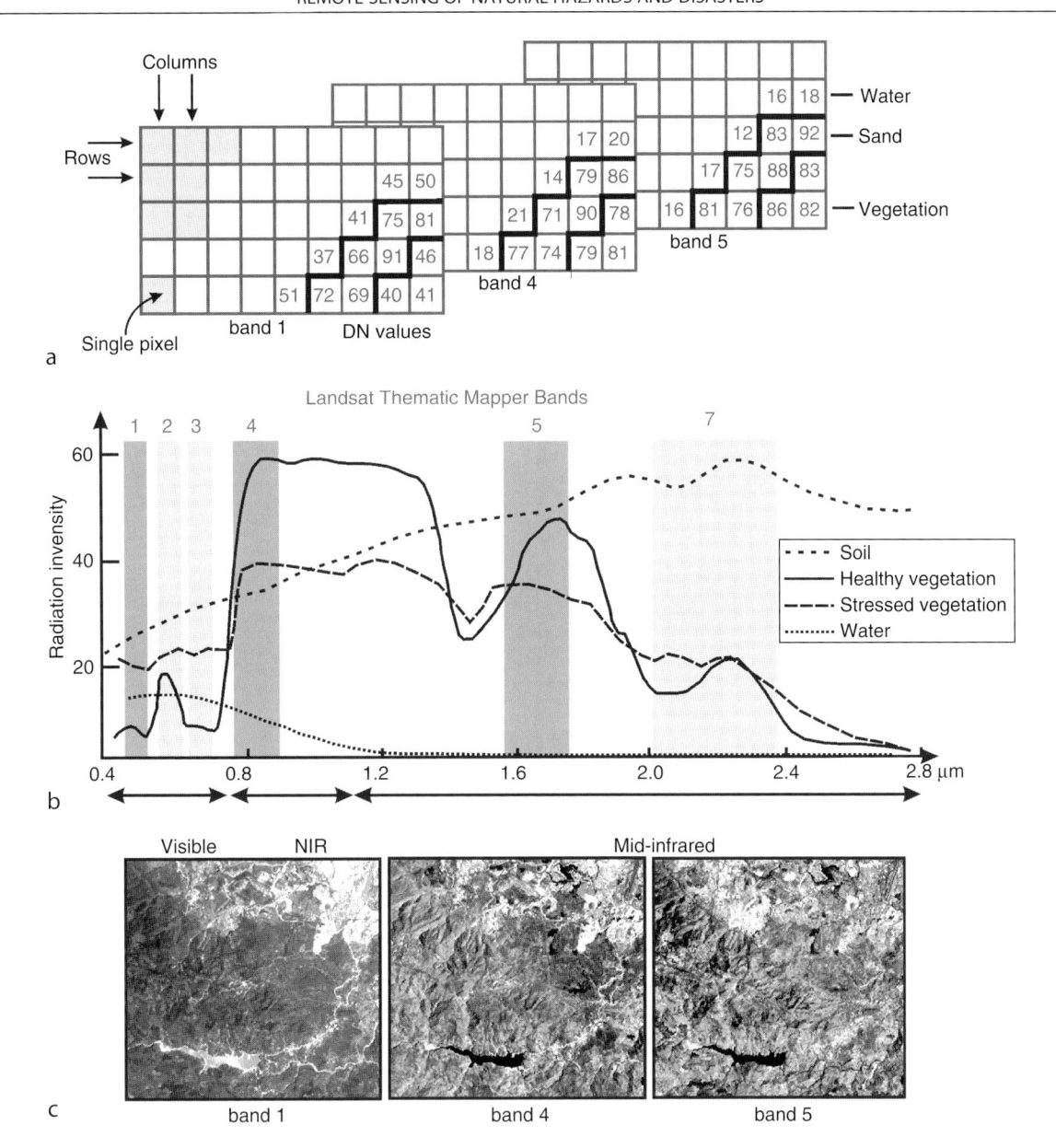

Remote Sensing of Natural Hazards and Disasters, Figure 2 Structure of remote sensing images, where rows and columns form a grid. Images typically have several spectral bands (**a**). Notice how different materials have different spectral values (digital numbers = DN values) that reflect their spectral-signature (**b**). (**c**) shows bands 1, 4, and 5 of a Landsat Thematic Mapper (TM) image, which also illustrate the DN values shown in (**a**). The materials and their DN values shown in (**a**) illustrate what a typical water-sand-vegetation progression, such as at the bottom of (**c**), looks like.

(Figure 3b), similar to what the scene would look like to a human eye from space. However, essentially any band-color combination is possible and can reveal different information about the area observed. A typical combination, called a false-color composite, is shown in Figure 3c, where the information from the NIR band is displayed in red. Since vegetation leads to high DN values in the NIR, a strong vegetation signal leads to a dominant red color wherever there is vigorous vegetation. Figure 3d shows another type of false-color composite.

Remote sensing of natural hazards
Information requirements

The information required to identify or monitor hazards, EaR, or actual disaster damage varies dramatically depending on the hazard type or what are considered features of value that may get affected during an event. Thus, given the diversity of RS data types and observation methods described above, as well as the variable hazard or disaster information needed, an essentially unlimited

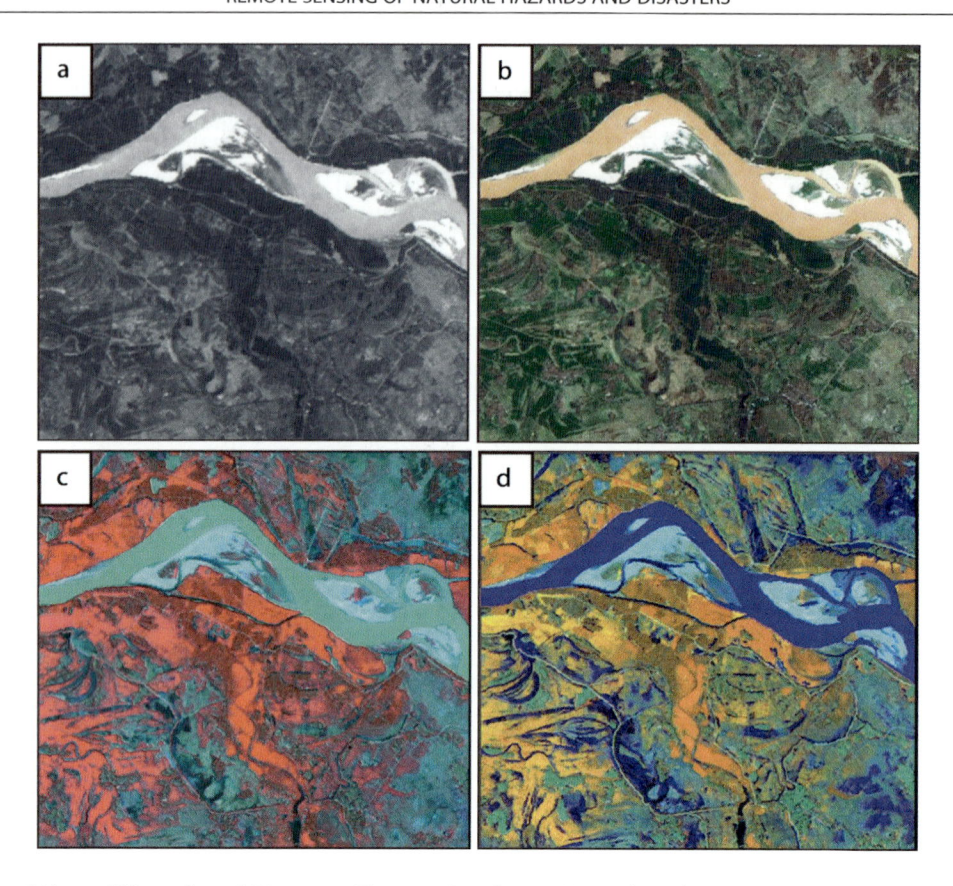

Remote Sensing of Natural Hazards and Disasters, Figure 3 Panchromatic Landsat Thematic Mapper image (**a**), true-color composite (**b**), and different false-color composites (**c** and **d**).

number of ways to deploy RS exists. At times, this is easy, for example, earthquake damage in Google Earth pictures tends to be very obvious, similar to the widely discussed damage caused by the 2004 Asian tsunami. However, sound knowledge of both hazard and disaster characteristics, as well as technical image analysis skills, are usually needed, as the difference between identifying broad damage patterns and a specific damage level classification, which is what disaster responders require, is substantial.

RS-based identification of hazards that may lead to disasters is even more complicated, and requires a clear understanding of what information is specifically required, how frequently and at which accuracy levels. Hazards can typically be characterized by a range of features. For example, RS can be deployed to study wildfire hazards by detecting (1) easily combustible vegetation, (2) its water content (i.e., dryness), (3) wind direction and speed, (4) real-time rainfall, or (5) topography, as fires on hill slopes spread more quickly. A sound approach would consider all of the above aspects (Syphard et al., 2008). For other hazards, it is useful not to consider the main hazard type, but the subtypes actually present. For example, *the* volcanic hazard does not exist. Instead, more than a dozen major subhazards can be identified, such as lava flows, pyroclastic flows, gas emissions, glacier melting, crater lake breach, ash clouds (that constitute an aviation hazard but can also circle the Earth, with consequences in other parts of the world), or more local hazards such as fumaroles. Each of these constitutes a hazard in its own right, and has different characteristics in terms of their overall extent, maximum reach, onset time, and duration. With the exception of seismic activity that often precedes or accompanies volcanic activity, essentially all hazard subtypes can be identified or monitored with RS. Because of their highly variable nature, however, each type needs to be well understood to identify a suitable RS strategy, considering also that some can occur simultaneously. For example, a crater lake that is heating due to rising magma can be identified with a thermal scanner (Oppenheimer, 1993). To study quantity and composition of emitted gases, however, observations in another part of the infrared spectrum are needed (Burton et al., 2000). Magma intruding into a volcanic structure can also lead to flank deformation that can be identified with radar observations, such as so-called interferometry that detects cm-scale vertical changes (Massonnet and Feigl, 1998). Alternatively, ground-based RS methods such as electronic distance meters (EDMs) or GPS measurements (see *Global Positioning Systems (GPS) and Natural Hazards*) can also be used (Sturkell et al., 2006).

For some hazards, it is sufficient simply to identify an actual event or ongoing process, such as magma being issued from a crater. More often, however, the above are mere parts of a more complex hazard evaluation that requires multi-data integration and modeling. For example, to understand better the potential area of spatial intersection of hazards and EaR, or to determine hazard magnitude, models are needed. The role of RS then is to provide input to these models. Kerle and Alkema (2012) described how RS is used to model flood hazard. Knowledge of rainfall in an upstream watershed alone is insufficient; more important is how water subsequently infiltrates, accumulates, and moves toward and through exposed areas, in turn a function of local topography and land cover (the material actually covering the ground, such as trees or buildings). While this is done by sophisticated hydrodynamic models, the contribution of RS is varied. It provides land cover information, real-time rainfall estimates (using ground-based radar RS or satellite observations), and the topographic information needed. The latter is also critical in the modeling of local seismic site effects, as ridges amplify seismic waves, leading to higher ground acceleration values than in depressions. Similarly, the spreading of lava or pyroclastic flows can be modeled using digital topographic information.

Selecting an appropriate strategy

Figure 4 illustrates which considerations determine a suitable RS strategy. As explained above, different hazard types, such as earthquakes or hurricanes, have different (1) spatial, (2) spectral, and (3) temporal characteristics, which are briefly evaluated here. (1) A hazard can be very local and spatially confined (e.g., an unstable hill slope), it can be very extensive (e.g., areas affected by tsunamis, hurricanes, or drought), or there can be a large distance between the actual hazard source and the exposed area in question. Examples of that can be earthquakes, where the responsible fault may be a long distance away from areas that may still experience strong shaking during an event, or the breaking of a dam that may lead to flooding far downstream. The data chosen need to reflect those dimensions and the details that need to be seen. The chosen platform type largely determines how extensive an area can be observed, and what spatial resolution the resulting data have. (2) Remote sensing is very sensitive to the surface characteristics of the object or area under investigation, resulting from the different spectral characteristics of different materials (see Figures 1 and 2), and different sensors have been built that are especially suitable for specific surface materials. It was already described how water, gases, or vegetation can be mapped with appropriate sensors. In situations where clouds, smoke, or night-time conditions prevent a clear view on the surface, active sensors, such as radar, can be used. However, here it is particularly important to understand that radar data primarily reflect the surface physics (structure/roughness, moisture, topography), compared to the surface chemistry that optical data correspond to (mineral type, chlorophyll content in leaves, etc.) (Lillesand et al., 2004). Hence, using radar is only useful if it not only

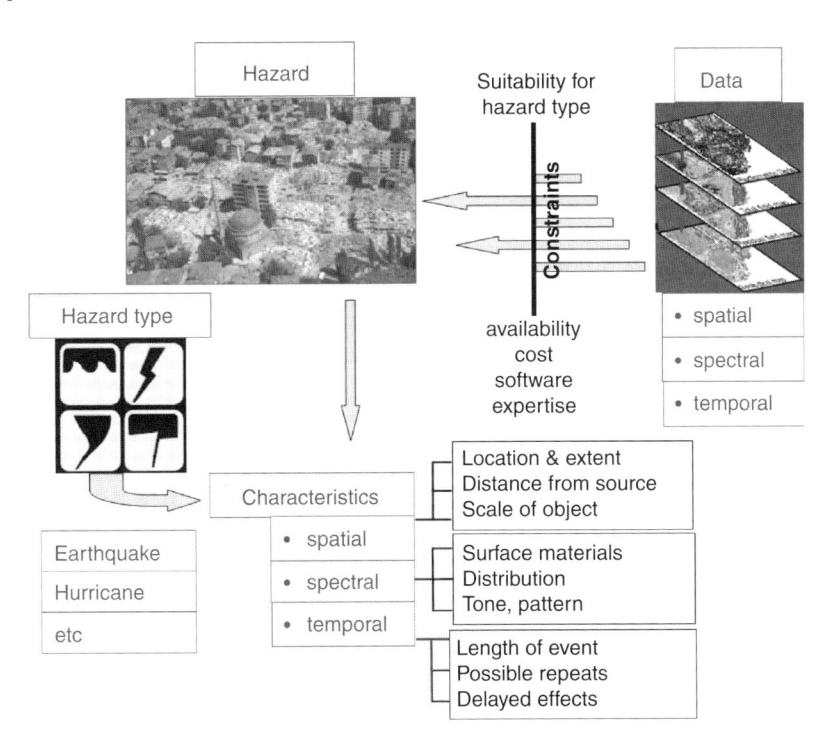

Remote Sensing of Natural Hazards and Disasters, Figure 4 Spatial data suitability depending on hazard or disaster type and their respective spatial, temporal, and spectral characteristics, as well as other constraints that may affect actual availability of the data.

penetrates difficult observing conditions, but also provides the information required. (3) Hazard events can be sudden and brief (e.g., earthquakes or landslides), sudden but of long duration (e.g., a dam break leading to prolonged flooding), but can also show precursory signs (e.g., volcanic activity or hurricanes). Some events, such as earthquakes, may also show a repetitive pattern, where violent aftershocks affect areas already destabilized by the primary event. Some effects may also be delayed, such as disease outbreak after a flood or earthquake. This is also a good example of one hazard type event leading to secondary effects. Others are slope or dam instability caused by earthquakes. A suitable RS strategy reflects the different consequences.

A solid understanding of the hazard(s) in question is thus needed before deciding on a specific analysis type and data that have suitable spatial and temporal characteristics, in case of image data also in the spectral domain. Once suitable data types have been identified, additional constraints have to be considered: data availability and cost, software, and expertise. There can be a large difference between suitable and actually available data, such as historic images taken at different times that may allow hazards to be traced temporally, but that were never acquired. Some sensors are operated by governments, others by private companies, and cost differences for RS data are substantial. As a general trend, costs have been declining, with many current and archive datasets being either free of charge or at a nominal fee (such as Landsat Thematic Mapper [Figures 2 and 3] or ASTER data). However, data from commercial high resolution sensors, such as from IKONOS or Quickbird, cost several thousand US dollars per image. Many software packages exist to process RS data, including a range of powerful open source solutions at zero or low cost. As a general rule, more advanced RS methods, such as radar or laser scanning, require sophisticated processing tools that, like the data, tend to be expensive. In addition, while basic image processing functions, especially applied to optical data, require limited knowledge, the processing of other data types also call for expert knowledge that may not always be available or affordable.

From hazard to risk

To mitigate the consequences of disasters, suitable DRR measures need to be implemented. In addition to knowledge on all hazards present, also information on EaR and their respective vulnerabilities is required. The term EaR includes all elements of value that may suffer damage when exposed to a hazard. As such it not only comprises people, their infrastructure, economic activities and services, but also environmental assets deemed valuable, such as nature reserves or engendered species. RS, in particular high resolution optical imagery, is well suited to identify physical features in direct line of sight of a sensor, such as buildings and other infrastructure (e.g., van Westen et al., 2008). Mapping of infrastructure elements in urban areas has also benefited strongly from increasingly available airborne laser scanning data, another active RS technique. The data collected are characterized by highly accurate and dense measurements that simultaneously allow the actual ground surface (digital terrain model, DTM) and a model that includes all above-ground features such as buildings (digital surface model, DSM) to be generated. This, for example, allows one not only to identify a building (Dash et al., 2004; Rottensteiner, 2003), but via its height also to estimate the number of floors and thus the approximate number of inhabitants. This is an example of how RS-based proxies – building height in this case – can be used to map EaR that are not directly visible. For risk calculation, the monetary value of all EaR should also be considered. For some elements, such as public buildings or infrastructure, such value information may be readily available. In many Western countries similar information also exists for residential buildings. However, there are clear limits to the value information RS can contribute. As no currently operational sensing method can observe the inside of buildings, value information on assets other than the actual building cannot be obtained. Additionally, assigning values to EaR such as human beings or endangered species that can be included in risk equations leads to obvious problems.

While most EaR are physical in nature and thus mappable, assessing vulnerability has always been more difficult. Defined as the susceptibility to suffer loss (see *Vulnerability*), it differs for physical assets, people, and their social structures, and economic and environmental systems, and is strongly dependent on hazard magnitude and event duration. While early risk concepts only considered physical vulnerability, the susceptibility of other critical elements, such as social structures, economic activity, or natural systems, have only gradually become integrated in conceptual risk frameworks, and RS only recently started making a contribution in the assessment of nonphysical vulnerabilities. Using physical proxies such as building type and density, or distance to lifelines, Ebert et al. (2009) showed how social vulnerability can be quantified using high-resolution optical imagery and elevation data derived from laser scanning. A comprehensive conceptualization of how diverse vulnerability types can be included in risk assessment aided by RS was done by Taubenbock et al. (2008). RS data types and processing methods have become increasingly specific to match detailed information needs, and a combination of different methods is typically needed in RS-based risk assessment. For example, while a satellite image may cover large areas, such as the catchment from which a flood might originate, very detailed imagery is needed to map buildings and other structural elements that may be affected by a flood. Since image detail and coverage are usually inversely related (similar to a camera zoom), a multi-data approach may be needed. Assessing vulnerabilities of natural systems or economic processes may also require specific models (Rapicetta and Zanon, 2009; Schmidt-Thome et al., 2006).

Hazard monitoring and early warning

Hazard events can be seen as anomalies in natural processes. The state of such processes can be quasi-binary, that is, have a normal and a non-normal state (e.g., either there is wildfire or not), or the process can follow more or less well-defined patterns. An example for the latter is the annual growing cycle of plants (phenology), whereby a drought condition constitutes an anomaly, one that is plant species-specific. Especially spaceborne RS is ideally suited to provide periodic surveillance to identify such anomalies. However, as in risk assessment a solid understanding of the hazard process in question and its natural variability is essential, as is a suitably matching RS strategy. A number of questions can be asked to guide the development of an appropriate strategy:

1. Which observable features precede and accompany a hazardous events?
2. What is their spatial extent and detailed expression?
3. What is the speed of onset and evolution, as well as duration?

A good example to illustrate these questions is tropical cyclone forecasting. These weather systems develop over warm ocean stretches during certain times of the year, thus it is clear when and on which areas detection has to focus. Storm systems also have a characteristic shape, often with a central, cloud-free eye. Great spatial detail is not needed, but significant developments, especially around the time of landfall can happen rapidly. Therefore, a suitable RS approach is to use geostationary weather satellites that provide data every 15–30 min for an entire hemisphere. A spatial resolution of 3–5 km is sufficient. The instruments on weather satellites also provide information on cloud top temperatures, atmospheric pressure values, wind speed and direction, and rainfall intensity, thus enough information to assess the strength, evolution, and direction of a storm (Katsaros et al., 2002). This can be coupled with other atmospheric models to provide storm track forecasting.

Tropical storms are easily observed during formation, which is also true for other hazards. Volcanoes heating up or bulging were already mentioned, processes that can occur weeks before an eventual eruption. Other hazardous processes proceed on longer timescales. For example, monitoring of desertification may only require yearly or decadal image acquisition. If the process being observed is only one aspect of a hazard, such as deforestation affecting slope stability, detection of significant changes indicates that previously quantified hazards have to be reassessed. Many hazard types have only short buildup periods and short durations, making early warning difficult, yet still possible if integrated with prior information. For example, if coupled with knowledge of watershed size and runoff behavior, RS-based rainfall detection can provide real-time information on impending floods (Asante et al., 2007), potentially allowing several hours to prepare or evacuate. There are currently no proven RS methods to provide early warnings for earthquakes. A comprehensive assessment of seismic hazard is best done through seismic microzonation, where the potential for different hazards (ground shaking, liquefaction, etc.) is quantified. However, due to the detailed data needed, this is typically done at more local to regional scales (see e.g., Fat-Helbary and Tealb, 2002). Space technology, in the form of GPS (see *Global Positioning Systems (GPS) and Natural Hazards*), can help to identify mm-scale tectonic plate movements, useful to identify areas where seismic stress is building up (Sturkell et al., 2006). Tsunami hazard is one type where RS can contribute little. For tsunamis caused by island flanks sliding into the sea, slope stability monitoring is possible (Puglisi et al., 2005). Monitoring of deep-sea fault lines with RS cannot be done, and real-time observation of tsunamis that propagate at speeds of 600–900 km h^{-1} is only a theoretical, but currently not operational option (Fujii and Satake, 2007). Pressure gauges on the ocean floor, together with an effective international warning system, are a better option.

Existing monitoring systems

A number of satellite-based monitoring systems have been established, such as for drought hazards, volcanic activity, wildfires, tropical storms, and other weather phenomena (already described above). The US National Oceanic and Atmospheric Administration (NOAA) operates a system based on multiple Advanced Very High Resolution Radiometers (AVHRR) that observe all land areas at least twice a day at a resolution of 1.1 km. These data are useful for drought monitoring and even yield estimation for specific crops. The critical aspect of drought is less the actual precipitation amount, but rather its effect on vegetation growth, i.e., more when it rains instead of how much it rains. Hence, products such as rainfall anomalies can be used to assess where water shortages may exist. Even more specific is an assessment of actual vegetation health. This can be done via the Normalized Difference Vegetation Index (NDVI), which measures the chlorophyll content of vegetative matter, and thus its health (Lillesand et al., 2004). This knowledge has to be integrated into a more comprehensive framework that considers local agriculture patterns, regional and national drought and famine mitigation strategies, and relief distribution systems. For example, in Africa, the most famine-affected continent, NOAA collaborates with the USAID/Famine Early Warning System (FEWS, www.fews.net) to provide satellite-based precipitation anomaly and food security information.

A simpler system is used to detect magmatic activity at active volcanoes. A global hotspot detection system based on data from the Moderate Resolution Imagining Spectroradiometer (MODIS) is operated by the Hawai'i Institute of Geophysics and Planetology (HIGP, http://modis.higp.hawaii.edu/) at the University of Hawaii. The HIGP MODIS Thermal Alert System provides worldwide near-real-time information on thermal anomalies, using the MODVOLC algorithm (Wright et al., 2004). Based on 1 km data, it is not only sensitive to significant heat emissions such as those occurring during magmatic activity, but also wildfires. As part of the Hyogo Framework for Action, it was agreed to set up a Global Multi-Hazard Early Warning System, which also includes a dedicated Global Early Warning System for Wildland Fire that is currently being set up (http://www.fire.uni-freiburg.de/).

Emergency response

RS images are most frequently seen in the public domain following a disaster. Remote sensing, especially space-based monitoring, has proved to be an excellent information source for emergency response at local, regional, and global scales, due to its (1) synoptic (i.e., large area) coverage, (2) frequent and repetitive collection of data of the Earth's surface, (3) diverse spectral, spatial, and potentially three-dimensional information, and (4) relatively low cost for per unit coverage. Hence, often images have already been acquired before ground-based support reaches a disaster site, and can then be used to guide response operations. Until recently, those data were only accessible to select organizations responsible for post-disaster image processing. Beginning with the 2010 Haiti earthquake, high-resolution post-disaster images (although not the individual spectral image bands) were made available in Google Earth shortly after acquisition, making the information accessible to anyone.

As with hazards, the availability and suitability of RS data depend on the disaster type, as is illustrated in Figure 5. To observe the 1998 volcanic mudflow at Casita volcano (Nicaragua), and even to determine principal structural features, even 50 m resolution imagery is sufficient (Figure 5d). However, to identify individual buildings that may have been affected by smaller flows and landslides (Figure 5a), much higher resolution is needed (Kerle et al., 2003). In addition to the many spatial, temporal, and spectral parameters that determine the value of a given dataset, yet more constraints apply in the aftermath of a disaster event. While rapid data acquisition is desirable, space infrastructure is tied to predefined orbital paths, only partly alleviated by the pointability of more recently launched sensors that can increase the temporal resolution, and a gradually growing number of satellites that provide more frequent observation opportunities. To answer the increasingly detailed information demands from the multitude of stakeholders responding to large disaster events, data from high spatial resolution

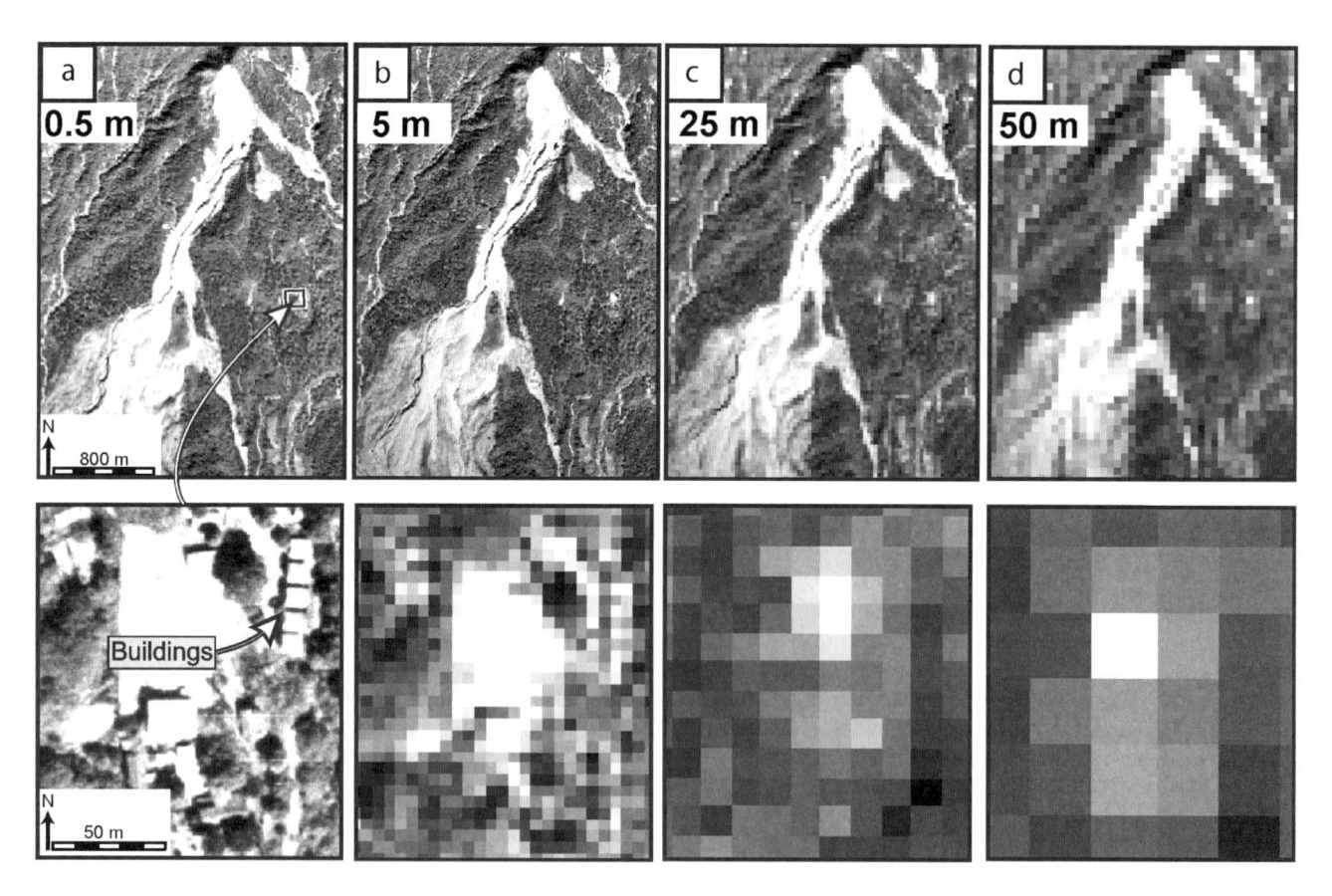

Remote Sensing of Natural Hazards and Disasters, Figure 5 Aerial photo of the 1998 mudflow at Casita volcano, Nicaragua, progressively degraded from 0.5 m (**a**) to 50 m (**d**). While the flow structure remains well identifiable, individual buildings (**a**) are only recognizable at very high resolution.

sensors are needed. These, however, only cover narrow strips of land, thus several days or weeks can still pass before a larger emergency scene, such as after flooding or an earthquake, is completely imaged at high resolution. Radar remote sensing has proved useful for several disaster response scenarios, such as to detect earthquake damage (Arciniegas et al., 2007), map flood extent (Tralli et al., 2005), or following technological disasters such as oil spills (Brekke and Solberg, 2005). In most other areas, optical data are used, yet those require cloud-free conditions, a further serious constraint on rapid data availability.

Disaster constellations and international disaster support

Individual and largely uncoordinated satellites have limited utility in global emergency response efforts. Similar to GPS (see Global Positioning Systems (GPS) and Natural Hazards), which is based on a network of 24 satellites, a constellation comprising a number of coordinated satellites would thus be more appropriate. Such a solution, the Disaster Monitoring Constellation (DMC), was initially proposed in 1996 and led by Surrey Satellite Technology Ltd (SSTL), UK. The objective of the constellation is daily global imaging capability by means of a network of five low-cost microsatellites (da Silva Curiel et al., 2005). They are owned by Algeria, China, Nigeria, Turkey, and the UK, an unlikely alliance for an international effort. In addition to daily revisits, another substantial strength of the DMC satellites is the large ground coverage in tiles up to 600 km wide, as well as its global reach. For example, NigeriaSat-1 was the first civilian satellite to image the Gulf areas affected by Hurricane Katrina in 2005. However, the price for the large coverage and high temporal resolution is the medium spatial resolution (28–32 m). The initial network has since been expanded to include the Spanish Deimos-1 satellite, while the Turkish instrument failed in 2007. Instead of launching dedicated satellites, the Sentinel Asia program focuses on coordinating the use of existing space infrastructure, together with auxiliary spatial data, for disaster risk management. It is coordinated by JAXA, the Japan Aerospace Exploration Agency.

In addition to technical solutions for improved disaster response, a successful organizational approach also exists. The International Charter "Space and Major Disasters" aims at a more efficient use of existing space technology during a disaster response phase. Initiated by the European and French space agencies (ESA and CNES) in 1999, it became operational in 2000. It provides a unified system of space data acquisition and delivery for a rapid response to natural or man-made disasters. So far, the Canadian Space Agency (CSA), the Indian Space Research Organization (ISRO), the Argentine Space Agency (CONAE), JAXA, the United States Geological Survey (USGS), NOAA, the DMC-operator DMCII, and the China National Space Administration have joined the Charter. Each member agency has committed resources to support the data collection of the Charter to improve the efficiency of data provision for emergency responses globally (www.disasterscharter.org). To date, the Charter has been activated nearly 200 times, responding to a variety of natural and technological emergencies (see e.g., Bessis et al., 2004). Once the Charter is activated, the most suitable and most quickly available space resources of the participating partners are used to obtain imagery of the affected areas. These data are then further processed by Charter partners, such as the German Space Agency (DLR), UNOSAT or SERTIT, who provide map products that are made available via the Global Disaster Alert and Coordination System (GDACS, http://www.gdacs.org/), ReliefWeb (http://www.reliefweb.int) or Reuters's AlertNet (http://www.alertnet.org/). Such organizational improvements are arguably of greater value than the mere launching of more satellites.

Further detailed information on the utility of airborne and spaceborne remote sensing in emergency response is given by Kerle et al. (2008) and Zhang and Kerle (2008), respectively. Examples of how satellite remote sensing has been used for emergency responses are also given by Metternicht et al. (2005) on landslides, and by Tralli et al. (2005) on earthquakes, floods, and other disasters.

Unmanned aerial vehicles (UAV)

Given the dynamics involved in disaster response, permanent monitoring during the emergency response phase is desirable. This, however, can only be provided by geostationary (and low spatial resolution) sensors, or by airborne systems. The latter, in particular, the development of unmanned aerial vehicles (UAV), has seen substantial attention in recent years. UAVs range from the very small (weighing only grams) to full-scale aircraft, and offer substantial potential for hazard monitoring and emergency response applications. Most are tropospheric (i.e., flying at low altitudes below approx. 12 km), whereas others are semiautonomous High Altitude Long Endurance (HALE) UAVs, designed to operate on solar power and allowing autonomous missions of up to several months duration in altitudes of 12–30 km (Baldock and Mokhtarzadeh-Dehghan, 2006). Smaller and less expensive platforms can facilitate rapid scene assessments, while large versions provide a stable platform for multi-instrument surveys. UAVs have become indispensable in fighting of large wildfires, particularly in the USA. While the focus initially was on the detection of hotspots, fire perimeters, and burned ground extent, fire intensity and fuel consumption are now also quantified to aid in the understanding of fire spread behavior. This can only be achieved with detailed and frequent data no other platforms can provide (Ambrosia et al., 2003). The RS instruments carried on those platforms are typically integrated into complete sensing systems with direct satellite or radio link for fast data processing.

Definitions

Reservoir – An artificial pond or lake used for the storage and regulation of water.

Dam – A barrier constructed across a waterway to control the flow or raise the level of water.

Seismicity – Earthquake activity experienced in a given region.

"Reservoir-induced seismicity" (RIS) is the seismic activity recorded in the vicinity of an impounded water reservoir. RIS is caused by stress change comparable in magnitude to the ambient shear stress acting on a fault to cause slip (McGarr et al., 2002). McGarr et al. (2002) define the seismicity around a reservoir as "Reservoir-triggered seismicity" (RTS) if the stress change is only a small fraction of the ambient level. In other words, RIS occurs if the stress perturbations cause an earthquake (it would not have occurred without the perturbation) whereas RTS implies that the perturbations barely "advanced the clock" of a seismic event that would have eventually occurred. Most cases of earthquake activity following reservoir impounding would fall in the RTS category. At least 70 cases of recognized RTS are recognized worldwide (Gupta, 1992) but it is possible that much low level seismicity had escaped detection as one goes back in time. Damaging RTS has occurred at Hsinfengkiang (China), Kariba (Zaire), Kremasti (Greece), Koyna (India), Oroville (USA), and Aswan (Egypt) (Gupta, 1992).

Some characteristics of RTS can be explained by the complicated interactions of a number of rock mechanics factors. Earthquakes are dynamic slip instabilities usually occurring on preexisting sliding surfaces. In a highly fractured intraplate environment subject to deviatoric stresses, fractured rocks can slide abruptly, generating earthquakes. To cause slip on a fracture, the shear stress, induced by the existence of a tectonic stress field, has to overcome friction. The frictional law can be written as the Navier-Coulomb criterion (Figure 1):

$$|\tau| = \tau_o + \mu_o(\sigma_N - P_f) = \tau_o + \mu_o \sigma_N'$$

where τ and σ_N are the shear and normal stress acting on the plane of fracture, μ_o denotes the friction coefficient, τ_o is the cohesion, P_f is the pore fluid pressure, and σ_N' is the effective normal stress. Failure occurs when the shear stress τ exceeds the shear strength of the material. Given the fault orientation, failure can be triggered by changes in σ_N' which can be brought about by: (1) an increase in σ'_1 (maximum effective compressive stress), (2) a decrease in σ'_3 (minimum effective compressive stress), or an equal decrease in all effective principal stresses, for example, due to an increase in P_f (Saar and Manga, 2003). The first two mechanisms cause failure by increasing the effective differential stress, $\sigma'_1 - \sigma'_3$, and thus the shear stress, τ. The third mechanism induces failure by decreasing the effective strength of the material caused by a decrease in σ_N', which unlocks the fault and moves the Mohr circle closer to the Mohr-Coulomb failure envelope, while leaving τ unchanged.

Note that RTS involves the complex interaction between all three of the factors τ, σ_N, and P_f. Factors that affect RIS are: ambient stress field conditions, availability of fractures, hydromechanical properties of the underlying rocks, geology of the area, together with reservoir dimensions and the nature of lake-level fluctuations (Talwani, 1997). In real situations, most of these geological factors are poorly determined, notably the permeability of lineaments and fractured zones. Finally, if the mid to upper crust of intraplate areas is in a state of stress close to failure, as is generally assumed, then even small perturbations can trigger seismicity (McGarr et al., 2002).

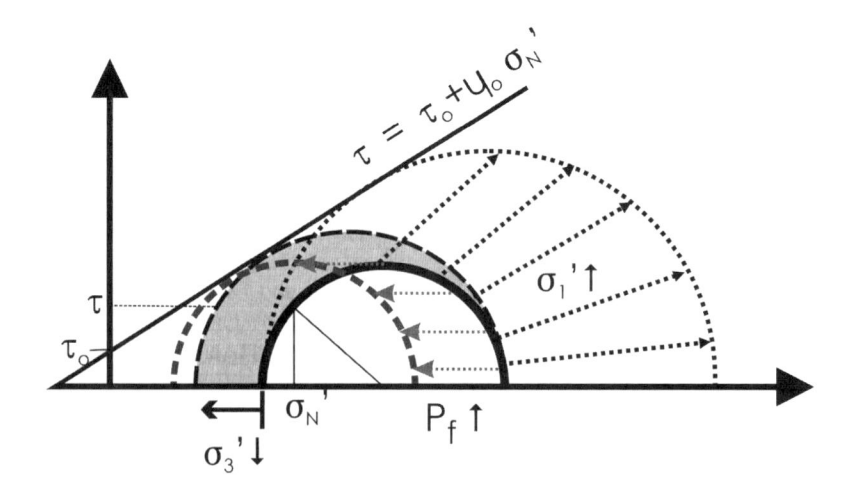

Reservoir, Dams, and Seismicity, Figure 1 Mohr circle with Mohr-Coulomb failure envelope (diagonal line). τ shear stress, τ_o cohesion, μ_o friction coefficient, $\sigma N'$ effective normal stress, Pf pore-fluid pressure. Fault reactivation occurs when the Mohr circle touches the Mohr-Coulomb failure envelope. This can occur by increasing the effective principal stress, decreasing the effective minimum stress or by increasing the pore fluid pressure. For additional description see main text.

Two types of RTS are recognized: "rapid response" and "delayed response" (Simpson et al., 1988; Guha, 2000). Rapid response can be explained by an instantaneous change in the elastic stress following impoundment, leading to elastic compression, which increases the pore-fluid pressure on faults at depth. These induced fluctuations in pore-fluid pressure (P_f), due to an increase in σ_1 (e.g., by the weight of a filling reservoir that contracts the pore space) occur immediately. For this type of RTS, shallow ($\leq$10 km) low-magnitude seismicity in the vicinity of the reservoir follows immediately on first loading. In contrast, delayed response is explained by the diffusion of pore-pressure along fractured zones in a near-failure state caused by reservoir impoundment. This delayed response type is characterized by deep ($\geq$10 km) moderate to large magnitude earthquakes that occur beyond the confines of the reservoir with a significant delay after first filling.

Most cases of RTS are related to the impounding of large reservoirs. Studies have shown that it is not so much the weight of the water that is important as much as the increase of hydrostatic pressure along faults. There is no fixed rule about the timing of this RTS: it can occur at any stage during the impounding or years afterward. Although most RTS is of small magnitude, there have been a few cases of large magnitude RTS, such as the 1967 M 6.3 Koyna reservoir earthquake (Gupta, 1992). The latter example caused damage and casualties. The current accepted number of RTS worldwide is somewhere between 40 and 100 (ICOLD, 2011). Recognizing RTS versus "normal" seismicity is not always evident. To help define the nature of the recorded earthquakes, it is generally recommended that earthquake monitoring with a local network of seismographs starts a few years prior to the beginning of reservoir impounding.

Although no tectonic environment is immune from RTS following reservoir impounding, areas of normal and strike-slip faulting appear most favorable to its occurrence (ICOLD, 2011). RTS in the Canadian Shield (an intraplate environment) is either absent or of small magnitude (maximum magnitude 4.1; Lamontagne et al., 2008). There is no absolute rule to evaluate RTS potential beforehand, it is generally accepted that the potential is greater for reservoirs exceeding 100 m of depth (ICOLD, 2011). The potential for RTS should be included in the design of dams and appurtenants structures for large or deep reservoirs (ICOLD, 2011).

Bibliography

Guha, S. K., 2000. *Induced Earthquakes*. Dordrecht/Boston, MA: Kluwer.

Gupta, H. K., 1992. *Reservoir-Induced Earthquakes*. Amsterdam: Elsevier Scientific Publishing. 364 p.

ICOLD (International Commission on Large Dams) 2011. "Reservoir and Seismicity: State of Knowledge". International Commission on Large Dams Bulletin 137, p. 50.

Lamontagne, M., Hammamji, Y., and Peci, V., 2008. Reservoir-triggered seismicity at the toulnustouc hydroelectric project, Québec North Shore, Canada. *Bulletin of the Seismological Society of America*, **98**, 2543–2552.

McGarr, A., Simpson, D., and Seeber, L., 2002. Case histories of induced and triggered seismicity. In Lee, W., Kanamori, H., Jennings, P., and Kisslinger, C. (eds.), *International Handbook of Earthquake and Engineering Seismology*. New York: Academic, pp. 647–661.

Saar, M. O., and Manga, M. 2003. Seismicity induced by seasonal groundwater recharge at Mt. Hood, Oregon. *Earth and Planetary Science Letters*, **214**, 605–618.

Simpson, D. W., Leith, W. S., and Scholz, C. H., 1988. Two types of reservoir-induced seismicity. *Bulletin of the Seismological Society of America*, **78**, 2025–2040.

Talwani, P., 1997. On the nature of reservoir-induced seismicity. *Pure and Applied Geophysics*, **150**, 473–492.

Cross-references

Earthquake
Induced Seismicity
Pore-Water Pressure
Seismology
Triggered Earthquakes

RESILIENCE

Adriana Galderisi[1], Floriana F. Ferrara[2]
[1]University of Naples "Federico II", Naples, Italy
[2]Environmental Engineering Consultant, Naples, Italy

Synonyms

Adaptability; Flexibility; Robustness; Transformability

Definition

In disaster studies, resilience is defined as "the ability of a system, community or society exposed to hazards to resist, absorb, accommodate to and recover from the effects of a hazard in a timely and efficient manner, including through the preservation and restoration of its essential basic structures and functions" (UNISDR, 2009).

Discussion

The concept originated in the field of Ecology: in 1973, Holling defined resilience as a measure of the ability of a system to absorb changes and still persist. Later, he refined this definition, emphasizing that systems, namely complex ones, can evolve toward different states of equilibrium. Therefore, resilience can be interpreted as a measure of the magnitude of disturbance that can be absorbed before a system changes its structure, by changing variables and processes that control its behavior. This approach, labeled as "ecological resilience," contrasts with "engineering resilience" (Holling, 1996), related to the capacity to withstand the external stresses and to the rapidity of restoring previous equilibrium (Pimm, 1984).

Research on complex adaptive systems drove to a widening of the resilience concept – also thanks to the studies carried out by the Resilience Alliance, a multidisciplinary research group established in

1999 – including relevant properties, such as self-organization, learning, and adaptation (Folke et al., 2002).

Starting from the 1990s, resilience has gained prominence in the disaster field: most international reports devoted to risk reduction have emphasized the need for focusing on resilience rather than *Vulnerability*, due to the positive meaning of the former, relative to the negative implications of the latter. Nevertheless, the relation between resilience and vulnerability is still largely debated and at least two different schools of thought can be distinguished:

- Resilience as the opposite of vulnerability (Fortune and Peters, 1995);
- Resilience and vulnerability as independent, or partially overlapping, processes (Paton and Johnston, 2006).

Currently, the main challenge is represented by turning resilience into operational terms. The main features of a system that may induce resilience have been identified (Godschalk, 2003) and some attempts to quantify some of these features have been carried out (Tierney and Bruneau, 2007). Nevertheless, the term is still so vague that "is in danger of becoming a vacuous buzzword from overuse and ambiguity" (Rose, 2007, 384).

Bibliography

Folke, C. et al., 2002. Resilience and sustainable development: building adaptive capacity in a world of transformations. Scientific Background Paper on Resilience: The World Summit on Sustainable Development.

Fortune, J., and Peters, G., 1995. Learning from failure. The System Approach. England: John Wiley & Sons Ltd.

Godschalk, D. R., 2003. Urban hazard mitigation: creating resilient cities. *Natural Hazards Review*, ASCE, August: 136–143.

Holling, C. S., 1973. Resilience and stability of ecological systems. *Annual Review of Ecology and Systematics*, **4**, 1–23.

Holling, C. S., 1996. Engineering resilience versus ecological resilience. In Schulze, P. (ed.), *Engineering with Ecological Constraints*. Washington, DC: National Academy.

Paton, D., and Johnston, D., 2006. *Disaster Resilience. An Integrated Approach*. USA: Thomas.

Pimm, S. L., 1984. The complexity and stability of ecosystems. *Nature*, **307**, 321–326.

Rose, A., 2007. Economic resilience to natural and man-made disasters: multidisciplinary origins and contextual dimensions. *Environmental Hazards*, **7**, 383–398.

Tierney, K., and Bruneau, M., 2007. Conceptualizing and measuring resilience. A key to disaster loss reduction. *TR News*, **250**, 14–17.

UNISDR, 2009. Terminology on disaster risk reduction. Available at: www.unisdr.org/eng/library/UNISDR-terminology-2009-eng.pdf.

Cross-references

Coping Capacity
Emergency Management
Recovery and Reconstruction After Disaster
Vulnerability

RICHTER, CHARLES FRANCIS (1900–1985)

Susan Hough
Southern California Earthquake Center, Pasadena, CA, USA

The one seismologist with global, iconic name recognition was in fact an accidental seismologist with a complicated life story. Charles Francis Richter was born near Hamilton, Ohio, USA on April 26, 1900 and raised by his mother and maternal grandfather. In 1909, the family, including older sister Margaret, moved to Los Angeles, where Richter had his first and not entirely happy experience with formal education. By his teens, Richter had developed a keen interest in astronomy and an abiding passion for the local mountains. Following a year at the University of Southern California, he transferred to Stanford University, earning a degree in physics in 1920. Socially awkward and painfully introspective, Richter's emotional struggles escalated during his college years, landing him back in Los Angeles under the care of a psychiatrist. By 1923, he had regained enough equilibrium to apply to the Ph.D. program at Caltech. Working under Paul Epstein, in 1928, Richter completed his Ph.D. thesis on the spin of hydrogen atom. Around that time, Harry Oscar Wood, head of the Pasadena Seismological Laboratory, asked Caltech head, Robert Millikan, about potential candidates for a position analyzing the seismograms from newly developed instruments. Millikan suggested Richter. Richter accepted the position, viewing it as a temporary assignment until he could move on to a position in the field of physics. In 1928, he married Lillian Brand, a writer and writing teacher. Brand had a son from a previous marriage; she and Richter had no children. The couple shared eclectic interests, including avante garde theater, hiking, writing poetry, and social nudism. As a young assistant in the Seismo Lab, Richter began to analyze recordings of local earthquakes, within a few years tackling the problem of a first-ever scale to measure *Earthquake* size. He borrowed the term "magnitude" from astronomy, and drew on earlier work by Japanese seismologist K. Wadati to develop distance corrections for seismograms recorded by a network of stations. He incorporated Caltech colleague Beno Gutenberg's suggestion to develop a logarithmic scale. The Richter scale is open-ended: Negative magnitudes are possible, and there is no prescribed upper limit. Richter tuned the scale so that magnitude 0 would be the smallest earthquake that he calculated could be recorded under normal conditions. It is not clear if the scale was tuned at the upper end, but in his seminal 1935 publication, Richter wrote that the 1906 San Francisco earthquake "may have been of magnitude 8." In later years, Richter worked in collaboration with Gutenberg to extend the scale beyond southern California, and to better estimate the size of large earthquakes. Many magnitude scales were developed later by others, all designed to follow Richter's original formulation. Richter's later publications focused on global earthquakes, the magnitude distribution of earthquakes in a given region, and seismotectonics of southern California. The 1933 Long Beach earthquake inspired a lifelong commitment to earthquake hazard and preparedness. Late in his career, Richter was an ardent and vocal skeptic of earthquake prediction research. Charles Richter died of congestive heart failure on September 30, 1985.

Bibliography

Hough, S., 2007. *Richters Scale: Measure of an Earthquake, Measure of a Man*. Princeton: Princeton University Press, 336 p.

Cross-references

Earthquake
Intensity Scales
Magnitude Measures
Modified Mercallit (MM) Scale
Seismography/Seismometer
Seismology

RIGHTS AND OBLIGATIONS IN INTERNATIONAL HUMANITARIAN ASSISTANCE

George Kent
University of Hawai'i, Honolulu, HI, USA

Definition

Rights and obligations of different parties in relation to international humanitarian assistance need to be clarified. There are many different issues, in relation to disaster and non-disaster situations, conflict and nonconflict situations, and many other contingencies. There is confusion regarding the rights of those who provide assistance and the rights of the needy to receive assistance.

This contribution describes the current framework regarding international humanitarian assistance and the troubled status of the concept of humanitarian intervention. That is followed by a brief account of the right to assist and the right to assistance. Then a framework is suggested in which rights and obligations of providers and receivers can be related to one another. The conclusion calls for more systematic formulation of rights and obligations regarding international humanitarian assistance, to be established through agreements among all countries, strong and weak.

Framework

International humanitarian assistance may be needed in many different kinds of circumstances. The United Nations International Strategy for Disaster Reduction (ISDR) is the focal point in the UN System to "promote links and synergies between, and the coordination of, disaster reduction activities in the socioeconomic, humanitarian and development fields, as well as to support policy integration." ISDR's definition of disaster is

> A serious disruption of the functioning of a community or a society causing widespread human, material, economic or environmental losses which exceed the ability of the affected community or society to cope using its own resources (United Nations. International Strategy for Disaster Reduction, 2006).

There is no distinction here between so-called natural and human-made disasters. Assistance is called for when there is a serious disruption of the community's functioning, regardless of the underlying cause of that disruption.

International human rights law does not explicitly address the right to protection and relief from disasters, but this is clearly implied. Article 3 of the *Universal Declaration of Human Rights* states, "Everyone has the right to life, liberty, and security of person." Article 25 states, "Everyone has the right to a standard of living adequate for the health and well-being of himself and of his family, including food, clothing, housing and medical care and necessary social services, and the right to security in the event of unemployment, sickness, disability, widowhood, or old age or other lack of livelihood in circumstances beyond his control." These rights are further elaborated in subsequent human rights agreements, particularly the *International Covenant on Economic, Social, and Cultural Rights*.

At an international colloquium on rights to humanitarian assistance held at UNESCO in Paris, one participant proposed seven rules that summarize international law relating to rights and duties to humanitarian assistance:

1. States have a *duty* to provide humanitarian assistance to victims in their territory or under their control.
2. States, IGO's [intergovernmental organizations such as UN agencies] and NGO's [nongovernmental organizations] have a *right* to *offer* humanitarian assistance to other States.
3. States, IGO's and NGO's have a *right* to provide humanitarian assistance to victims in other States with the *consent* of these States – in case of disintegration of governmental authority and of civil war – with the consent of the relevant local authorities.
4. States have *no duty* to provide humanitarian assistance to victims in other States but they have a *duty* to *facilitate* humanitarian assistance lent by other States, IGO's or NGO's. If measures of coercion are taken against a particular State, supplies for essential humanitarian needs have to be exempted from them.
5. The Security Council, by virtue of Chapter VII of the Charter, may determine that the magnitude of a human tragedy constitutes a threat to international peace and security and *authorize* States or UN forces to take all measures necessary to bring humanitarian assistance to the victims.
6. States have a *duty* to *admit* humanitarian assistance furnished by other States, IGO's or NGO's in accordance with international law. They may not arbitrarily refuse their consent.

7. Individuals have a *right* against the State under whose control they are to receive humanitarian assistance insofar as this State has a duty to provide humanitarian assistance or to permit its distribution according to rules 3, 4, and 6 (Schindler, 1993).

Thus, states have a right but not a duty to provide international humanitarian assistance. The central argument of this entry is that states should have a duty to provide international humanitarian assistance under some circumstances.

Humanitarian intervention

The idea of inviolable national sovereignty, based on the Peace of Westphalia of 1648, served the international system well from the seventeenth century into the twentieth century. That agreement established the basic principles of the modern nation-state system: countries are sovereign, and thus have no formal authority above them, and they are not permitted to involve themselves in the internal affairs of other countries without the latter's permission. Currently, however, we see violations of human rights. Many now feel that the international community should act to protect those whose rights are violated, at least when there are extensive violations.

One response has been the emergence of the doctrine of humanitarian intervention. Under this doctrine, the traditional Westphalian principle of immunity from outside interference remains in place, but with the qualification that under some extreme circumstances and with appropriate legal processes, the international community may forcibly intervene to protect human rights.

In international law, intervention generally refers to forcible intrusion, usually with military force, into the affairs of nations by outsiders (Haass, 1994; Lyons and Mastanduno, 1994). Despite extensive debate, the doctrine of *humanitarian* intervention remains ill-formed. Some individuals equate humanitarian intervention with any sort of humanitarian assistance in armed conflict. Some use the term to refer to military action to free civilians from situations in which there are serious violations of human rights. Perhaps humanitarian intervention is best understood as *humanitarian assistance provided to people within a nation by outsiders without the consent of the national government.*

Humanitarian intervention occurs when there is a claimed right to deliver humanitarian assistance despite the absence of consent from the government of the receiving nation. A distinction should be made between the simple absence of consent to the delivery of assistance, and the clear refusal of such assistance. In the case of the US-led intervention into Somalia in October 1993, for example, there was neither approval nor refusal. Rather, civil order had collapsed to such a degree that there was no Somali national government in place to grant or deny consent to the delivery of assistance. Perhaps intervention in the absence of consent should have different guiding rules than intervention in the face of clear refusal of consent.

While many individuals define humanitarian intervention as armed intervention into states for humanitarian purposes, it has been argued that this historical understanding "has no place in the system established by the UN." In the legal framework based on the UN Charter, unilateral state intervention is allowed solely for protecting national independence (Sandoz, 1992, 217–218).

The International Commission on Intervention and State Sovereignty tried to address the legitimacy issue by advancing a new doctrine on The Responsibility to Protect, clarifying guidelines for humanitarian intervention. The approach was highlighted in a UN report on *A More Secure World* (High-level Panel, 2004), a UN declaration in 2005, and again in the G8 Summit in 2006. On April 28, 2006, in Resolution 1674, the UN Security Council asserted the right of the international community to provide protection to people whose human rights were being violated. It acknowledged that under some circumstances the international community has a responsibility to provide such protection, but this responsibility was not spelled out.

The responsibility to protect has been viewed mainly as the duty of governments of the countries on the receiving end of the intervention. The terminology could suggest that the countries that do the intervening have specific *obligations* to intervene when necessary for humanitarian purposes, but a careful reading of the discussions indicates that the international community really uses this doctrine to assert its *right* to intervene. The *Guardian* recognized this when it referred to "the UN declaration of a right to protect people from their governments (Williams, 2005)." Similarly, the African Union proclaimed, in its Constitutive Act, "the right of the Union to intervene in a Member State... in respect of grave circumstances, namely war crimes, genocide and crimes against human humanity." It did not speak of a duty to intervene in such circumstances.

The assertion of a right to assist under some circumstances, with no counterpart obligation to assist under any circumstances, implicitly invites the politicization of assistance decisions. The view taken here is that those who provide assistance on humanitarian grounds should not be entirely free to choose who and when they help. Assistance in situations like the genocide in Darfur or the widespread malnutrition in North Korea or Zimbabwe should not be optional, at least so long as they can be undertaken with reasonable safety.

In October 2006, Vaclav Havel, Kjell Magne Bondevik, and Elie Wiesel issued a report arguing that because of the widespread malnutrition and other humanitarian problems in North Korea, "the international community now has an obligation to intervene through regional bodies and the United Nations, up to and including the Security Council (Havel et al., 2006a, b)." Their call for recognition of obligations, and not only rights, on the part of those who would intervene was not well received.

It is not surprising that the providers of assistance tend to emphasize their rights rather than their obligations. However, one would think that if the powerful are going to claim a right to assist under some conditions, they also should have an obligation to assist under some conditions (Kent, 2008, 1–31).

Right to assist versus right to assistance

A right to assistance means that people meeting criteria specified in the law are entitled to receive services specified in the law. For example, in some countries women with children whose incomes fall below a particular level are entitled to monthly stipends calculated according to an established formula. Or children in a particular age bracket and meeting certain residency requirements might be entitled to free immunizations. Entitlements of this form are the basis for human services (welfare) programs in many countries.

If there is a right of those in need to receive assistance under specified conditions, then there must also be an obligation for others to render assistance. These obligations are specific responsibilities for action. Typically, in the area of human rights, individuals have specific rights, and the government has specific obligations to fulfill those rights. When rights are described in the law, the counterpart obligations of government, and the specific agencies responsible for their implementation, ought to be described as well. The rights/obligations nexus can be understood as a kind of contract, explicit or implicit, that establishes who is to do what under what conditions.

Whether within nations or internationally, the challenge is to determine the nature of the contract: who should be entitled to what sort of assistance from whom under what conditions at whose expense? Different sorts of answers would be appropriate for different kinds of situations or needs: poverty, armed conflict, refugees, famine, chronic malnutrition, floods, droughts, terrorism, and so on. Some general principles would apply across broad categories of cases.

Many would agree that there are some extreme situations in which needy people should have a clear right to receive help. In any decent social order, if a child falls down a well, there should be a requirement that the child will be rescued. But the idea of the right to assistance has a very checkered history. For example, the law in the United States is characterized by "the missing language of responsibility (Glendon, 1991, 76–108)." Under the no-duty-to-rescue principle, bystanders are not required to come to the assistance of strangers in peril if they did not cause that peril.

This principle of no-duty-to-rescue is peculiarly American. In contrast, "most European countries do impose a legal duty on individuals to come to the aid of an imperiled person where that can be done without risk of harm to the rescuer. And the constitutions of many other liberal democracies do obligate government to protect the health and safety of citizens (Glendon, 1991, 77)."

In the case of ships in distress on the high seas, there is a well-established international duty to come to the assistance of the needy. Captains failing to meet this obligation have been prosecuted. However, there is no general duty of nations to respond to distress in other nations. The major international human rights instruments are concerned primarily with the responsibilities of States Parties to their own people, not to people elsewhere. They do call for international cooperation in working toward the realization of human rights. For example, the *International Covenant on Economic, Social, and Cultural Rights* requires States Parties "individually and through international cooperation to take the measures needed" to implement the agreement. In practice, however, there is no clear hard duty to provide humanitarian assistance internationally. There is no history of case law with respect to international humanitarian assistance.

The needy are sometimes viewed as having a right to assistance in the limited sense that no third party may interfere with its delivery. This is different from a right to assistance as an entitlement, in which the providers of assistance have an obligation to provide it. Table 1 illustrates the relationship between the concepts.

Providers of humanitarian assistance generally emphasize their right to provide assistance, cell A in the table. Where that right is claimed under the right to protect doctrine, it is implied that governments of receiving countries have an obligation to accept assistance, cell D. This relationship is in the A-D diagonal of the matrix. Little attention has been given to the idea that under some conditions providers might be obligated to give assistance (cell B), because under some conditions others might have a right to receive assistance (cell C). This relationship is in the B-C diagonal.

If a party has a right to receive assistance under some conditions, arguably that party also should accept an obligation to receive assistance under some conditions. And if a party has a right to provide assistance under some conditions, perhaps that party also ought to have the obligation to provide assistance under some conditions. These relationships are in the rows of the table, A-B and C-D.

It can be argued that parties receiving assistance should also recognize obligations to monitor and report on the distribution and impact of that assistance, or at least allow such monitoring and reporting. There have been cases in

Rights and Obligations in International Humanitarian Assistance, Table 1 Rights and obligations of providers and receivers of assistance

	Rights	Obligations
Providers	A. right to provide	B. obligation to provide
Receivers	C. right to receive	D. obligation to receive

which countries receiving assistance did not allow any tracking of it after it arrived in national ports, thus raising suspicions about its possible misuse.

While people of poor countries would be the primary beneficiaries of rights-based international assistance, their governments sometimes resist because of their concern with guarding their sovereignty. They fear that humanitarian intervention might be used against them for political purposes. The governments of weak countries do not want powerful countries, which might have ulterior political motives, intervening without their consent under the pretense of providing assistance.

Why should weak countries accept the idea that outsiders can intervene when they see fit, but not be under any obligation to assist when it is not politically convenient for them? There should be more symmetry in the doctrine. If the international community is to have the right to intervene to provide assistance in some circumstances, there also should be some circumstances in which the international community has an obligation to provide assistance.

Rights of the needy to receive international assistance, as distinguished from rights of outsiders to provide assistance, are rarely discussed. Where the rights of the needy are considered it seems it is mainly to clarify the conditions under which intervention may be undertaken (Guiding Principles, 1993). In 1988 France proposed a General Assembly resolution for disaster relief based on explicit recognition of the rights of the needy to receive assistance, but that aspect disappeared by the time Resolution 43/131 of December 8, 1988 was finalized (Beigbeder, 1991, 354, 380–383).

If it were agreed that the international community had an obligation to assist under some circumstances, the obligation would have to be mitigated in the face of extreme danger, as in armed conflict situations. However, in such situations, consent is often given by the combatants, together with assurances of safe passage, to the International Committee for the Red Cross, UNICEF, and other humanitarian agencies. Indeed, ICRC specializes in obtaining such consent. The key problem in providing international humanitarian assistance in conflict situations is not so much the presence of the conflict as the absence of consent.

There is now no hard duty to provide international assistance based on explicit rights of the needy to receive assistance. There should be not only a right but also an obligation to provide international humanitarian assistance under some circumstances. There should be an obligation on the part of the international community to provide assistance at least when such assistance is welcomed by the receiving nation. The international community could recognize an obligation to *offer* assistance even in conflict situations or other situations in which there are doubts about whether it would be accepted. There would then be an obligation to deliver that assistance if the parties controlling the situation consented and safe passage was assured.

There is a need to create a new global regime of clear rights and obligations for those who provide humanitarian assistance and those who receive it. This would be likely to make the system more effective and more just, and thus contribute to the steady strengthening of overall global governance.

Summary

Currently nations have a right but not a duty to provide international humanitarian assistance. It is argued here that nations should have a duty to provide international humanitarian assistance under some circumstances.

The doctrine of the *responsibility to protect* has been used by the international community to assert its *right* to assist, not as the recognition of a *duty* to assist, in certain circumstances. The assertion of such a right with no counterpart obligation implicitly invites the politicization of decisions to assist. As in the case of domestic emergency services, the provision of assistance should not always be optional. There should be an obligation on the part of the international community to provide assistance at least in some circumstances.

A clear system of rights and obligations for those who provide humanitarian assistance and those who receive it would be likely to make international humanitarian assistance more effective and more just, and thus contribute to the steady strengthening of overall global governance.

Bibliography

Beigbeder, Y. 1991. *The Role and Status of International Humanitarian Volunteers and Organizations: The Right and Duty to Humanitarian Assistance.* Dordrecht/Holland: Martinus Nijhoff.

Glendon, M., 1991. *Rights Talk: The Impoverishment of Political Discourse.* New York: Free Press.

Guiding Principles on the Right to Humanitarian Assistance, (1993). *International Review of the Red Cross,* No. 297 (November–December), pp. 519–525.

Haass, R. N., 1994. *Intervention: The Use of American Military Force in the Post-Cold War World.* Washington, DC: Carnegie Endowment.

Havel, V., Bondevik, K. M., and Wiesel, E. 2006a. Turn North Korea into a human rights issue. *New York Times.* Editorial. October 30. http://www.nytimes.com/2006/10/30/opinion/30havel.html?_r=3&oref=slogin&pagewanted=print.

Havel, V., Bondevik, K. M., and Wiesel, E. 2006b. *Failure to Protect: A Call for the UN Security Council to Act in North Korea.* Washington, DC: DLA Piper/U.S. Committee for Human Rights in North Korea. http://www.dlapiper.com/files/upload/North%20Korea%20Report.pdf.

Kent, G. (ed.), 2008. *Global Obligations for the Right to Food.* Lanham, MD: Rowman & Littlefield.

Lyons, G. M., and Mastanduno, M. (eds.), 1994. *Beyond Westphalia? National Sovereignty and International Intervention.* Baltimore: Johns Hopkins University Press.

Sandoz, Y. 1992. 'Droit' or 'Devoir d'Ingerence' and the right to assistance: the issues involved. *International Review of the Red Cross,* 288, 215–227.

Schindler, D., 1993. Humanitarian assistance, humanitarian interference and international law. In Macdonald, R. S. J. (ed.), *Essays in Honour of Wang Tieya.* London: Kluwer, pp. 689–701.

High-level Panel on Threats, Challenges and Change, 2004. *A More Secure World: Our Shared Responsibility.* New York: United Nations. http://www.un.org/secureworld/.

United Nations. International Strategy for Disaster Reduction. 2006. *Terminology: Basic Terms of Disaster Risk Reduction.* Geneva: ISDR. http://www.unisdr.org/eng/library/lib-terminology-eng-p.htm.

Williams, I. 2005. Annan has paid his dues: the UN declaration of a right to protect people from their Governments is a millennial change. *Guardian.* September 20. http://www.guardian.co.uk/news/2005/sep/20/mainsection.commentanddebate2.

Cross-references

Civil Protection and Crisis Management
Disaster Research and Policy Paradigms, History
Disaster Relief
Emergency Management
Emergency Planning
Evacuation
International Strategies for Disaster Reduction
Natural Hazard in Developing Countries
Planning Measures and Political Aspects
Post Disaster Mass Care Needs
Recovery and Reconstruction After Disaster
Red Cross, Red Crescent, International Federation of
United Nations Organization and Natural Disasters
World Economy, Impact of Disasters on
World-Wide Trends in Natural Disasters

RIP CURRENT

Wayne Stephenson
University of Otago, Dunedin, New Zealand

Definition

Rip currents are narrow currents initiated close to the shoreline that flow strongly in a seaward direction through the surf zone and beyond and are an important mechanism for the offshore transport of water and sediment and consequently play a major role in shaping beach morphology.

Discussion

Rip currents are a common occurrence on beaches and are a significant threat to beach users since they are capable of moving bathers offshore into water depths greater than wading depth. Rips should not be confused with the notion of "under tow"; in reality, there are no surf zone processes capable of pulling a person under water. Fatalities from drowning usually result after exhaustion because the panicked swimmer has tried to swim back to shore against the current. If caught in a rip, the best thing to do is relax and go with the flow and wait to be rescued. Stronger swimmers may be able to return to shore once the rip takes them beyond the surf.

Rip currents result from longshore variations in wave heights. Although there are competing theories as to why wave heights vary alongshore. The shoreward transport of water by shoaling waves results in longshore feeder currents flowing from areas of high waves to areas of low waves, at areas of low wave height where two feeder currents converge, water returns seaward as a discrete current. Commonly observed regular spacing of rips is attributed to these longshore variations in wave height. Generally rip spacing increases with wave height and the width of the surf zone. Rip current velocities are commonly 0.5–1.0 m/s but can exceed 2 m/s. There are a number of different types of rips depending largely on wave conditions and surf zone topography. Some beaches exhibit almost fixed rips or persistent rip currents typically controlled by rip channels. The most dangerous rips are "flash rips" that suddenly appear and drain water from the beach face then disappear; such rips are a significant threat to unwary swimmers and present significant challenges to life saving operations. Permanent rips are common on bay head beaches and are usually located at one end of the beach along the embayment wall.

Since 1949, with the advent of record keeping, it is estimated that Australian lifesavers have rescued some 300,000 people from the New South Wales surf, 90% of those rescues were associated with rip currents. Approximately 100 people die from drowning associated with rip currents each year in the USA and 80% of the surf rescues in the USA are as a result of rip currents. Research from the USA also shows that 84% of the fatalities associated with rips were male. In the UK, rip currents were shown to play a role in 71% of incidents occurring within the counties of Devon and Cornwall during the 2005 summer. Rip currents are therefore one of the most significant causes of rescues and drownings on beaches.

Management of rip current hazards involves a two-prong approach, public education about rips, and active patrolling of popular swimming beaches by surf life saving organizations. Public education takes the form of warning signs on beaches and in the media. For example, each summer, Surf Life Saving Australia runs television and print advertisements showing people how to recognize rips, warning never to swim alone and always swim between flags on patrolled beaches. Given the large number of unpatrolled beaches worldwide and ever increasing beach use, rips will remain a significant threat to peoples' safety at the beach.

Bibliography

Gensini, V. A., and Ashley, W. S., 2009. An examination of rip current fatalities in the United States. *Natural Hazards*, doi:10.1007/s11069-009-9458-0.

Scott, T., Russell, P., Masselink, G., Wooler, A., and Short, A., 2007. Beach rescue statistics and their relation to nearshore morphology and hazards: a case study for southwest England. *Journal of Coastal Research*, Special Issue **50**, 1–6.

Short, A. D., 2007. Australian rip systems – friend or foe? *Journal of Coastal Research*, Special Issue **50**, 7–11.

Short, A. D., and Hogan, C. L., 1994. Rip currents and beach hazards: Their impact on public safety and implications for coastal management. *Journal of Coastal Research*, Special Issue **12**, 197–209.

Cross-references

Breakwater
Coastal Erosion
Coastal Zone, Risk Management
Communicating Emergency Information
Hurricane (Cyclone, Typhoon)
Hurricane Katrina
Rogue Wave
Seiche
Storm Surges
Tidal Bores
Tsunami

RISK

Jörn Birkmann
United Nations University, Bonn, Germany

Definition

Risk and risk analysis constitute a multidisciplinary research field. In this context, the term "risk" is defined differently by various disciplines and fields of work. In natural hazard and disaster risk reduction research, *risk* is most commonly defined as the result of the interaction of a hazard (e.g., flood, hurricane, earthquake, etc.) and the vulnerability of the system or element exposed, including the probability of the occurrence of the hazard phenomena (UNDRO, 1980; Cardona, 1990; UN/ISDR, 2004, 2009; Birkmann, 2006b). Risk is estimated by combining the probability of a hazard occurrence, such as the likelihood of a flood (with a specific magnitude or intensity) and the potential scale of consequences (e.g., injury, damage, and loss) that would arise if the event strikes society or exposed elements. The analysis of the potential consequences cannot solely be based on an evaluation of the hazard phenomena. Rather, the vulnerability of the exposed element, e.g., the society at risk, has to be examined. Exposure, susceptibility, or fragility of the system, as well as its coping capacities, are key factors that determine its vulnerability (Birkmann, 2006b). While *risk* is a continuum, a disaster is one of its many "moments" or "materializations" (ICSU-LAC, 2009). However, risk and vulnerability are dynamic; hence, they may differ before, during, and after a disaster (IPCC 2012).

Introduction

Interest in the issue of risk can be traced back to Babylonian times. The development of modern risk analysis was closely linked to the establishment of scientific methodologies for assessing causal links between different types of hazardous events and adverse health effects and the mathematical theories of probability (Covello and Mumpower, 1985, pp. 33–36). Today, risk and its analysis constitute a multidisciplinary research field. In this regard, definitions of risk vary between specific disciplines such as social science, geography, engineering, environmental science, mathematics (risk/loss calculations), as well as between different fields of work, such as disaster management, risk insurance, climate change adaptation, environmental management, and disaster medicine – to name just a few (see, e.g., overview in Thywissen, 2006 and, e.g., different risk definitions in ADRC, 2005; Alexander, 2000; Alwang et al., 2001; Cardona et al., 2003; Crichton, 1999; Disaster Recovery Journal, 2005; Einstein, 1988; Garatwa and Bollin, 2002; Hori et al., 2002; *Journal of Prehospital and Disaster Medicine*, 2004; Natural Disaster Risk Management, 2001; Rashed and Weeks, 2003; Schneiderbauer and Ehrlich, 2004; Shrestha, 2002; Swiss, 2005; Tiedemann, 1992; UNDP, 2004; UNEP, 2002; Burton et al., 1993; Banse, 1996; Greiving, 2002; as well as Stewart and Melchers, 1997; Renn 1992, 2008; Cardona et al., 2003, Wisner et al., 2004, IPCC 2012).

Within the broad field of risk research related to natural or technological hazards and climate change, two major schools of thought can be differentiated that represent larger streams of discussion and discourses. The first school of thought views risk as a product of the interaction of a hazard and vulnerability – which is widely represented within disaster risk reduction research (e.g. UN/ISDR 2004). In contrast, the second school of thought views risk primarily within the broader context of decision and game theory. Consequently, risk is seen as closely linked to decisions (decision options) that could lead to loss or harm, in the light of a hazard event. These two schools of thought will be presented and discussed.

The first school of thought: risk as a product of hazard and vulnerability

The first school of thought, applied by the majority of disaster risk researchers, defines *risk* as a result of the interaction of hazard phenomena and the vulnerability of the exposed elements. This includes the probability of the hazard occurrence and the interaction with the society or elements exposed (see, e.g., UN/ISDR, 2004). Hence, risk is linked to the potential occurrence of negative consequences, in terms of physical, social, economic, and environmental losses, in a given area over a period of time, resulting from interactions between physical events and vulnerable conditions of a society or socio-ecological system.

The report of the United Nations Disaster Relief Organization (UNDRO, 1980) marks an important milestone in this context. The report defines risk as a product of hazard (H) and vulnerability (V) as they affect a series of elements (E) exposed. Examples of such elements could be different social groups and their livelihoods, properties, economic sectors, and public services as well as services of critical infrastructures. Key publications, such as the report "Living with Risk" of the United Nations International Strategy for Disaster Reduction (UN/ISDR), promote this notion of risk. The report presents the formula of *risk* being a function of hazard and vulnerability (Risk = Hazard × Vulnerability) (UN/ISDR, 2004, p. 36). Compared to this formula, some researchers incorporate exposure as an own element and view risk as a function of hazard, exposure, and vulnerability (see, e.g., ADRC, 2005). In addition the IPCC Special Report SREX also underscores that risks are a product of the complex interaction between hazards, such as climate or weather events, on the one hand and the vulnerability and exposure of societies on the other (IPCC 2012, p. 4). Despite differences in terminology, particularly with regard to the definition of exposure, key components of the term "risk" remain the same in nearly all formulas and so-called equations. Hence, the first school of thought and the respective formulas of risk underline, that risk is determined by both, the hazard event (and its magnitude) and the vulnerability of the exposed elements. The Hyogo Framework for Action 2005–2015 – the key document of the World Conference for Disaster Risk Reduction – stresses the starting point for reducing disaster risk and promoting a culture of disaster resilience lies in the knowledge of hazards and of the physical, social, economic, and environmental vulnerabilities most societies face, in addition to the ways in which hazards and vulnerabilities are changing in the short and long term, followed by action taken on the basis of that knowledge (UN, 2005).

Overall, risks related to natural hazards and climate change are not autonomous or externally generated circumstances to which society reacts, adapts, or responds. They are, rather, the result of the interaction of society and the natural or built environment. Consequently, effective risk management essentially requires an improved understanding of this relationship, and the factors influencing it (Susman et al., 1983; Renn, 1992; Vogel and O'Brien, 2004; Birkmann, 2006a, b, c). Understanding the various types of relationships and interactions between nature and society is, therefore, key for better understanding risk. In this context, the "Science Plan for Integrated Research on Disaster Risk" of the International Council for Science (ICSU, 2008) formulates that:

> Risk depends not only on hazards but also on exposure and vulnerability to these hazards, making risk an inherently interdisciplinary issue. In order to reduce risk, there needs to be integrated risk analysis, including consideration of relevant human behaviour, its motivations, constraints and consequences, and decision-making processes in face of risks. This inevitably requires that natural scientists and engineers work together with social or behavioural scientists in promoting relevant decision-making in the risk management area. Moreover, the understanding of risk patterns and risk-management decisions and their promotion require the integration and consideration of scales that go from the local through to the international level. (ICSU, 2008, p. 14)

The second school of thought: risk from a sociological perspective

The second major school of thought evolved in sociological and behavioral science. According to this school of thought, *risk* is defined as an inherent characteristic of decisions, in the light of hazardous events (see Luhmann, 2003; Renn et al., 2007; Renn, 2008). Luhmann, a key authority within the sociological risk community, defined risk related to a decision of an individual or a collective to act in such a way that the outcome of these decisions can be harmful or could, in case of an event (e.g., hazard event), lead to losses and harm (Luhmann, 2003, p. 25). Consequently, *risk* is seen as an inherent element of a decision-making process particularly resulting from having options to decide in one way or another. In other words, if an individual or collective has the opportunity to decide, in the light of adverse consequences, the related decision-making process implies taking a risk. The aforementioned definitions and perspectives link risk and its theoretical and conceptual basis with the broader context of decision and game theory. According to Luhmann (1991), the term "risk" can be solely used, if a decision can be identified that could lead to loss or harm. Furthermore, he differentiated the terms "risk" and "threats." While *risk* exists if loss and harm can be linked to an outcome of a decision, *threats*, in contrast, represent loss and harm that are externally caused, without a decision of an individual or a collective (Luhmann, 1991, p. 31)

Additionally, important impulses for the risk discussion also emerged from Beck, particularly his books "Risk Society" (1986, 1992) and "World Risk Society" (2008). From the perspective of modernization processes in developed countries, Beck underscores that societies are transforming mainly due to their complex and rapid technological developments from industrial societies towards risk societies. Hence, he argues that technology and respective development pathways of societies are increasingly leading to more threats and insecurities induced and introduced by modernization itself (Beck, 1992, p. 21). Many of these new risks can no longer be directly experienced. Rather, they are inherent threats closely embedded within societal development processes. Furthermore, some of these risks involve lifestyle (choices), whereas others encompass risks of unknown statistical probability, such as Severe Acute Respiratory Syndrome (SARS) or terrorist attacks (see, in detail, Beck, 1992). In this regard, Beck calls for a reflexive modernity, an institutionalized activity and state-of-mind, involving constant monitoring and reflections upon and confrontation with these risk phenomena. The idea of a reflexive modernization can be linked to some extend to the broader discussion of root causes of risk within disaster risk reduction research (see, e.g., Wisner et al., 2004, p. 18).

The outline of the two major schools of thought regarding *risk* underlines the different viewpoints of what constitutes and defines risk. Although these schools cannot be merged, it is interesting to note that a certain overlap can be identified, particularly when dealing with new phenomena of climate change. In terms of adaptation strategies, sea level rise and other phenomena, linked to climate change, might imply both, risks, in terms of the potential interaction of a hazard event and a vulnerable society, as well as the inherent risk of different decision options that could lead to harm or losses. Consequently, the dichotomy between the first and second school of thought might melt in the light of natural hazards and decision-making processes, related to climate change and extreme events, as well as irreversible changes.

In the following sections, further facets of risk will be explored, particularly regarding different risk formulas used within the first school of thought and the issue of timescales. Finally, challenges for improving our understanding of risk will be outlined.

Risk formulas

With regard to the first school of thought – represented particularly by the disaster risk reduction research community – different formulas of risk and its factors have been developed. According to UNDRO (1980), the specific risk of an element exposed (e.g., human life, house, infrastructure services, etc.) is seen as a product of the interaction of a hazard (e.g., floods, landslides, etc.) – including its probability and magnitude – and the vulnerability of the element exposed. In this regard, risk represents the potential damages and losses, such as the number of lives likely to be lost, the persons injured, damage to property, and disruption to activities caused by a particular natural phenomenon (UNDRO, 1980).

Overall, risk is expressed – in many publications of the disaster risk research community – by the formula *Risk = Hazard(s) x Vulnerability* (UN/ISDR, 2004). Some authors also include the concept of exposure to refer particularly to the physical aspects of vulnerability or to an own component that is situated in between the nexus of hazards and vulnerability (see e.g., IPCC 2012). Villagrán de Leon (2006) shows different *risk equations* that underline similarities and differences of the respective definitions of risk (see Figure 1).

Important differences in these equations center around the notions of societal response capacities, expressed, e.g., as coping capacities or deficiencies in preparedness.

Additionally, some equations explicitly differentiate exposure from vulnerability, whereas others view exposure as a subcategory of vulnerability itself (see Turner et al., 2003). Despite these differences, the overall understanding that risk and disaster risk are a complex function of the interplay of hazards (e.g., weather and climate events) and vulnerability can be seen as a kind of basic consensus. Regarding different risk equations, UN/ISDR (2004) underlines the importance to also consider the social contexts in which risks occur, as people do not necessarily share the same perceptions of risk, nor their underlying causes. The framing of the societal response capacities remains a difficult task. This cannot be discussed here in detail; however, it might be necessary to further address the potential differences of coping, preparedness, and adaptation linked to risk to natural hazards and climate change within the scientific community. These and related questions are addressed within the IPCC special report on "Managing the Risk of Extreme Events and Disasters to Advance Climate Change Adaptation," published in March 2012.

Timescales

Timescales are an important issue, when dealing with risk and its dynamics. In terms of different hazards, e.g., low-frequency hazards (e.g., earthquakes) occur on a different timescale than high frequency hazards (e.g., floods). Thus, information required to detect and evaluate the different hazard types might need to consider different timescales. Likewise, the identification of vulnerability has to deal with various timescales with regard to different vulnerability patterns and processes. Even if the hazard profile and intensity remains the same, over a period of time, the vulnerability of a society or community – as one of the key components of risk – differs if the analysis focuses on a certain point in time or on a time period, since particularly the social conditions such as susceptibility are likely to change over time (see Kienberger 2010, IPCC 2012).

Overall, it is essential to consider the dynamic nature of risk and vulnerability (IPCC 2012). Considering timescales and aspects in risk definitions and risk analysis requires, first and foremost, addressing these issues explicitly in respective definitions and risk assessments (Figure 3).

Probability

The probability of a hazard event to occur is another important aspect that should be addressed when dealing with risk in the context of natural hazards. The probability of a hazard (hazard event) to occur is often estimated by using the frequency of past similar events or by event-tree methods. However, probabilities for rare events may be difficult to estimate, if an event tree cannot be formulated. The exact identification of the frequency of tsunamis is still a challenge, particularly in regions where they occur very rarely. Likewise, creeping natural hazards and

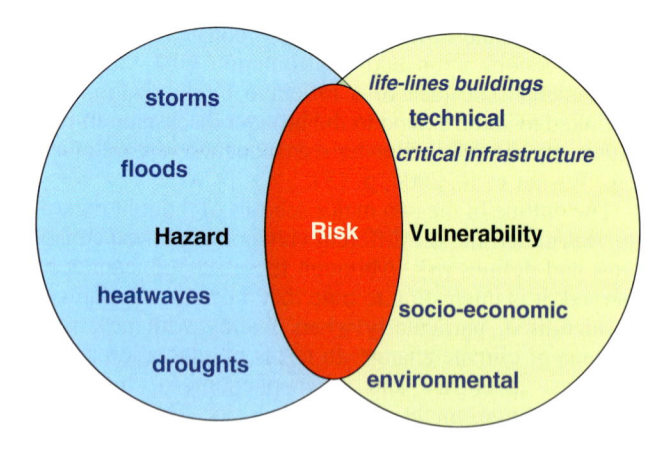

Risk, Figure 1 Risk as a product of the interaction of hazard(s) and vulnerability (Source: Birkmann, 2008, p. 10).

Different Risk Equations
(modified after Villagrán de León (2006)):

Risk = Hazard * Vulnerability
(mathematical function defined in *)

Risk = Hazard x Vulnerability
(UN/ISDR 2004, Wisner et al. 2004)

Total Risk = (∑ elements at risk) x Hazard x Vulnerability
(Alexander 2000)

Risk = (Hazard x Vulnerability)/Coping Capacity
("different agencies")

Risk = Hazard x Vulnerability x Deficiencies in Preparedness
(Villagrán de León 2001)

Risk = Hazard x Exposure x Vulnerability
(Dilley et al. 2005)

Risk = Hazard + Exposure + Vulnerability–Coping Capacities
(Hahn 2003)

Risk = *f*(Hazard, Vulnerability, Value)
(UMESCO 2003)

Risk, Figure 2 Different risk formulas (Source: Villagran de Leon, 2006).

irreversible changes, in the light of climate change, such as sea level rise, as well as man-made hazards, such as terrorist attacks, pose problems to the traditional hazard frequency-probability relation. Also, the cascading hazard in March 2011 in Japan that was triggered by an earthquake followed by a tsunami and an accident in a nuclear power plant (Fukushima) shows that new approaches are needed that go beyond a single hazard probability function. Therefore, irreversible changes might have to be characterized differently. Some probability functions also focus on loss or harm, in other words, the probability of loss of human life. These calculations, however, need to be carefully conducted, when dealing with low-frequency and very specific hazard phenomena (e.g., cascading hazards).

Future challenges: improving our understanding of risk

Understanding risk requires an improved understanding of risk factors, underlying vulnerabilities, and societal coping and response capacities, as well as changes in hazard phenomena and cascades. This includes acknowledging issues such as complexity, nonlinearity, uncertainty, and limits of knowledge, that risk to natural hazards and climate change imply (see, e.g., ICSU-LAC, 2009, p. 15; Birkmann et al., 2009; Renn, 2008, p. 289; Bohle and Glade, 2008; Patt et al., 2005). Additionally, factors that influence and determine risk perceptions, such as beliefs, values, and norms, need to be better understood, in terms of improving knowledge on the social construction of risk (see, e.g., Adger et al., 2009).

Overall, appropriate information and knowledge are essential prerequisites for promoting a culture of risk reduction and resilience. Specific information must first be collected on the dynamic interactions of exposed and vulnerable elements, e.g., persons, their livelihoods and critical infrastructures, and potentially damaging hazard events, such as extreme weather events or potential irreversible changes as sea level rise. In this context, understanding risks related to natural hazards and climate change requires among other aspects:

- Knowledge of the processes and driving forces by which persons, livelihoods, property, infrastructure, goods, and the environment itself are exposed to potentially damaging events, e.g., understanding exposure in its spatial and temporal dimensions.
- Knowledge of the factors and processes which influence and determine the vulnerability of persons and their livelihoods or of socio-ecological systems; Understanding the increases or decreases in susceptibility and societal response capacity, including the role of governance.
- Knowledge on how environmental and climate change impacts are transformed into hazards, particularly regarding processes by which human activities in the natural environment or changes in socio-ecological systems lead to the creation of new hazards (e.g., natural-technical hazards, NaTech), irreversible changes (e.g., undermining supporting ecosystem services), or increasing probabilities of hazard occurrence.
- Knowledge regarding different tools, methodologies, and sources of knowledge (e.g., expert and scientific knowledge, local or indigenous knowledge) that allow capturing new hazards, risk and vulnerability profiles, as well as different risk perceptions. In this context, new tools and methodologies are needed that enable the evaluation, e.g., of current and planned adaptation and risk reduction strategies.
- Knowledge on how risks and vulnerabilities – and, in part, also hazard phenomena – can be modified through forms of governance, particularly risk governance – encompassing formal and informal rule systems and actor-networks at various levels. Furthermore, it is essential to improve knowledge on how to promote adaptive governance within the framework of risk assessment and risk management.

(ICSU-LAC, 2009, p. 15; Birkmann et al., 2008, 2009, Birkmann and Fernando, 2008; Cutter and Finch, 2008; Renn, 2008, p. 289; Bohle and Glade, 2008; Biermann, 2007; Biermann et al., 2009; Füssel, 2007; Renn and Graham, 2006; Patt et al., 2005; Cardona et al., 2005; and Kasperson et al., 2005).

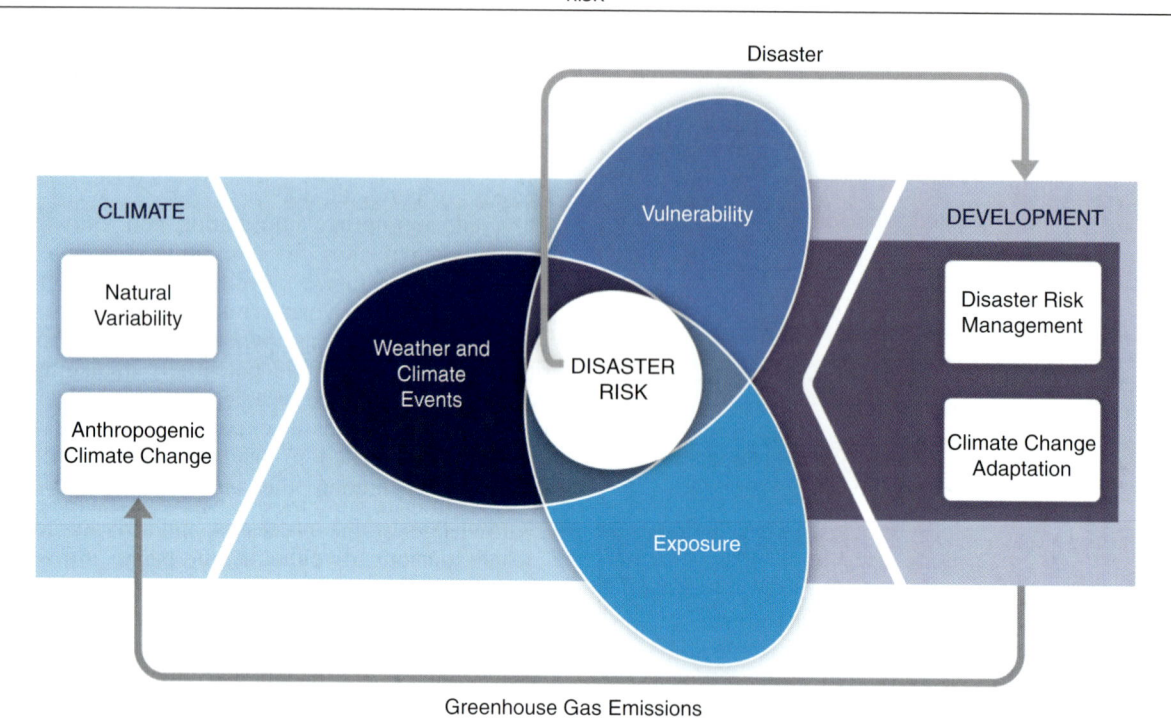

Risk, Figure 3 Systematization of disaster risk, vulnerability and weather and climate events (hazards) according to the new IPCC special report SREX (Source: IPCC 2012, p. 4).

Future research, as well as strategies for improving our understanding of risk, should encompass at least six knowledge demands:

1. Identification of (new) hazards and irreversible changes, as well as changes in known hazards
2. Examination of vulnerability and its core components
3. Identification of risk, as a product of the interaction of a hazardous event (including the probability of occurrence), and vulnerability
4. Risk perception (particularly regarding "unexperienced" hazards such as sea level rise)
5. Evaluation and assessment methodologies and tools, that for example capture actual and (potential) future dynamics of vulnerability and hazard phenomena
6. Risk management and adaptive governance

These knowledge demands also underline that expertise from various disciplines will be needed to address the challenges in a systematic and integrative way in order to improve the knowledge of risk to natural hazards. Additionally, both schools of thought, in terms of risk definitions, can contribute to these challenges and demands within their specific perspectives.

Bibliography

Adger, W. N., Dessai, S., Goulden, M., Hulme, M., Lorenzoni, I., Nelson, D. R., Naess, L. O., Wolf, J., and Wreford, A., 2009. Are there social limits to adaptation to climate change? *Climatic Change*, **93**, 335–354.

ADRC, 2005. *Total Disaster Risk Management – Good Practices*. Available from: http://www.adrc.asia/publications/TDRM2005/TDRM_Good_Practices/PDF/PDF-2005e/Chapter1_1.2.pdf.

Alexander, D. E., 2000. *Confronting Catastrophe*. Harpenden: Terra.

Alwang, J., Siegel, P. B., and Jørgensen, S. L., 2001: *Vulnerability: A View From Different Disciplines*. Social Protection Discussion Paper Series, No. 0115, World Bank. Available from: http://siteresources.worldbank.org/SOCIALPROTECTION/Resources/SP-Discussion-papers/Social-Risk-Management-DP/0115.pdf.

Banse, G., (Hrsg.) 1996. *Risikoforschung zwischen Disziplinarität und Interdisziplinarität – Von der Illusion der Sicherheit zum Umgang mit Unsicherheit*. Berlin: Edition Sigma.

Beck, U., 1986. *Risikogesellschaft. Auf dem Weg in eine andere Moderne*. Frankfurt am Main: Suhrkamp.

Beck, U., 1992. *Risk Society. Towards a New Modernity*. London: Sage.

Beck, U., 2008. *World at Risk*. Cambridge: Polity Press.

Bieri, S., 2006. *Disaster Risk Management and the Systems Approach*. Available from: http://www.drmonline.net/drmlibrary/pdfs/systemsapproach.pdf.

Biermann, F., 2007. "Earth system governance" as a crosscutting theme of global change research. *Global Environmental Change*, **17**(3–4), 326–337.

Biermann, F., Betsill, M., Gupta, J., Kanie, N., Lebel, L., Livermann, D., Schroeder, H., and Siebenhüner, B., 2009. *Earth System Governance – People, Places, and the Planet*. Science and Implementation Plan of the Earth System Governance Project, Report No. 1, Bonn, Available from: http://www.earthsystemgovernance.org/publications/2009/Earth-System-Governance_Science-Plan.pdf.

Birkmann, J., 2006a. *Measuring Vulnerability to Natural Hazards – Towards Disaster Resilient Societies*. Tokyo/New York/Paris: United Nations University Press, p. 450.

Birkmann, J., 2006b. Measuring vulnerability to promote disaster-resilient societies: conceptual frameworks and definitions. In Birkmann, J. (ed.), *Measuring Vulnerability to Natural Hazards: Towards Disaster Resilient Societies*. Tokyo: United Nations University Press, pp. 9–54.

Birkmann, J., 2006c. Indicators and criteria for measuring vulnerability: theoretical bases and requirements. In Birkmann, J. (ed.), *Measuring Vulnerability to Natural Hazards – Towards Disaster Resilient Societies*. Tokyo/New York/Paris: United Nations University Press, pp. 55–77.

Birkmann, J., 2008. Globaler umweltwandel, vulnerabilität und disaster resilienz – erweiterung der raumplanerischen perspektiven. *Raumforschung und Raumordnung*, 66(1), 5–22.

Birkmann, J., and Fernando, N., 2008. Measuring revealed and emergent vulnerabilities of coastal communities to tsunami in Sri Lanka. *Disasters*, 32(1), 82–104.

Birkmann, J., Buckle, P., Jäger, J., Pelling, M., Setiadi, N., Garschagen, M., Fernando, N., and Kropp, J., 2008. Extreme events and disasters: a window of opportunity for change? – analysis of changes, formal and informal responses after mega disasters. *Natural Hazards*, doi:10.1007/s11069-008-9319-2.

Birkmann, J., von Teichman, K., Aldunce, P., Bach, C., Binh, N. T., Garschagen, M., Kanwar, S., Setiadi, N., and Thach, L. N., 2009: Addressing the challenge: recommendations and quality criteria for linking disaster risk reduction and adaptation to climate change. In: Birkmann, J., Tetzlaff, G., and K.-O., Zentel, (eds.), *DKKV Publication Series*, No. 38, Bonn.

Blanchard, W., 2005. *Select Emergency Management-Related Terms and Definitions. Vulnerability Assessment Techniques and Applications (VATA)*. Available from: http://www.csc.noaa.gov/vata/glossary.html.

Bohle, H.-G., and Glade, T., 2008. Vulnerabilitätskonzepte in sozial- und naturwissenschaften. In Felgentreff, C., and Glade, T. (eds.), *Naturrisiken und Sozialkatastrophen*. Berlin/Heidelberg: Spektrum Verlag, pp. 99–119.

Burton, I., Kates, R. W., and White, G. F. (eds.), 1993. *The Environment as Hazard*. New York: Oxford University Press.

Cardona, O. D., 1990. *Terminología de Uso Común en Manejo de Riesgos*, AGID Reporte No. 13, EAFIT, Medellín, actualizado y reimpreso en *Ciudades en Riesgo*, M. A. Fernández (ed.), La RED, USAID.

Cardona, O. D., 2006. A system of indicators for disaster risk management in the Americas. In Birkmann, J. (ed.), *Measuring Vulnerability to Hazards of Natural Origin: Towards Disaster Resilient Societies*. Tokyo: UNU Press, pp. 189–209.

Cardona, O. D., Hurtado, J. E., Duque, G., Moreno, A., Chardon, A. C., Velásquez, L. S., and Prieto, S. D., 2003. *The Notion of Disaster Risk: Conceptual Framework for Integrated Management*. National University of Colombia/Inter-American Development Bank. Available from: http://idea.unalmzl.edu.co.

Cardona, O. D., Hurtado, J. E., Chardon, A. C., Moreno, A. M., Prieto, S. D., Velásquez, L. S., and Duque, G., 2005. *Indicators of Disaster Risk and Risk Management. Program for Latin America and the Caribbean*, National University of Colombia/Inter-American Development Bank.

Comfort, L., Wisner, B., Cutter, S., Pulwarty, R., Hewitt, K., Oliver-Smith, A., Wiener, J., Fordham, M., Peacock, W., and Krimgold, F., 1999. Reframing disaster policy: the global evolution of vulnerable communities. *Environmental Hazards*, 1, 39–44.

Covello, V., and Mumpower, J., 1985. Risk analysis and risk management: an historical perspective. *Risk Analysis*, 5(2), 103–120.

Crichton, D., 1999. The risk triangle. In Ingleton, J. (ed.), *Natural Disaster Management*. London: Tudor Rose, pp. 102–103.

Cutter, S., and Finch, C., 2008. Temporal and spatial changes in social vulnerability to natural hazards. *Proceedings of the National Academy of Sciences*, 105(7), 2301–2306.

Disaster Recovery Journal, 2005. Business continuity glossary. Available from: http://www.drj.com/glossary/glossleft.htm.

Einstein, H. H., 1988. Landslide risk assessment procedure. In *Proceedings of Fifth International Symposium on Landslides*. Lausanne, pp. 1075–1090.

Füssel, H.-M., 2007. Vulnerability: a generally applicable conceptual framework for climate change research. *Global Environmental Change*, 17, 155–167.

Garatwa, W., and Bollin, C., 2002. *Disaster Risk Managment – A Working Concept*. Eschborn: Deutsche Gesellschaft für Technische Zusammenarbeit (GTZ). Available from: http://www2.gtz.de/dokumente/bib/02-5001.pdf.

Greiving, S., 2002. *Räumliche Planung und Risiko*. München: Gerling Akademie Verlag.

Hori, T., Zhang, J., Tatano, H., Okada, N., and Iikebuchi, S., 2002. Micro-zonation based flood risk Assessment in urbanized floodplain. In *Second Annual IIASA-DPRI Meeting Integrated Disaster Risk Management Megacity Vulnerability and Resilience*. Laxenburg, Austria, 29–31 July, 2002. Available from: http://www.iiasa.ac.at/Research/RMS/dpri2002/Agenda/Agenda.pdf.

ICSU (International Council for Science), 2008. *A Science Plan for Integrated Research on Disaster Risk: Addressing the challenge of natural and human-induced environmental hazards*. Report of ICSU Planning Group on Natural and Human-induced Environmental Hazards and Disasters.

ICSU-LAC, 2009. *Science for a better life: Developing regional scientific programs in priority areas for Latin America and the Caribbean*. Vol 2, Understanding and Managing Risk Associated with Natural Hazards: An Integrated Scientific Approach in Latin America and the Caribbean. Cardona, O. D., Bertoni, J. C., Gibbs, A., Hermelin, M., and A., Lavell, Rio de Janeiro and Mexico City, ICSU Regional Office for Latin America and the Caribbean.

IPCC (Intergovernmental Panel on Climate Change) 2012. *Managing the Risk of Extreme Events and Disasters to Advance Climate Change Adaptation*. Cambridge: Cambridge University Press.

Journal of Prehospital and Disaster Medicine, 2004. Glossary of terms. Available from: http://pdm.medicine.wisc.edu/vocab.htm.

Kasperson, J., Kasperson, R., Turner, B. L., Hsieh, W., and Schiller, A., 2005. Vulnerability to global environmental change. In Kasperson, J., and Kasperson, R. (eds.), *The Social Contours of Risk*. London: Corporations & the Globalization of Risk. Earthscan. Risk Analysis, Vol. II, pp. 245–285.

Kienberger, S., 2010. *Spatial Vulnerability Assessment: Methodology for the Community and District Level Applied to Floods in Buzi, Mozambique*. PhD Dissertation, University of Salzburg. Salzburg.

Luhmann, N., 1991. *Soziologie des Risikos*. Berlin: Walter de Gruyter.

Luhmann, N., 2003. *Soziologie des Risikos*. Berlin, Germany/New York: Walter De Gruyterl.

Natural Disaster Risk Management, 2001. *Natural Disaster Risk Management - Guidelines for Reporting*. Brisbane, Queensland: Department of Emergency Services. Available from: http://www.disaster.qld.gov.au/publications/pdf/NDRM_guidelines.pdf, 24 Jan 2006.

Patt, A., Klein, R., and Vega-Leinert, A., 2005. Taking the uncertainty in climate-change vulnerability seriously. *Geoscience*, 337, 411–424.

Rashed, T., and Weeks, J., 2003. Assessing vulnerability to earthquake hazards through spatial multicriteria analysis of urban areas. *International Journal of Geographical Information Science*, 17(6), 547–576.

Renn, O., 1992. Concepts of risk: a classification. In Krimsky, S., and Golding, D. (eds.), *Social Theories of Risk*. Westport: Praeger, pp. 53–79.

Renn, O., 2008. *Risk Governance – Coping with Uncertainty in a Complex World*. London: Earthscan.
Renn, O., and Graham, P., 2006. Risk Governance – Towards an integrative approach. *International Risk Governance Council*. White paper no. 1, Geneva.
Renn, O., Schweizer, P., Dreyer, M., and Klinke, A. (eds.), 2007. *Risiko – Über den gesellschaftlichen Umgang mit Unsicherheit*. München: Ökom verlag.
Schneiderbauer, S., and Ehrlich, D., 2004. *Risk, Hazard and People's Vulnerability to Natural Hazards – A Review of Definitions, Concepts and Data*. Ispra: Office for Official Publication of the European Communities.
Shrestha, B. P., 2002. Uncertainty in risk analysis of water resources systems under climate change. In Bogardi, J. J., and Kundzewicz, Z. W. (eds.), *Risk, Reliability, Uncertainty, and Robustness of Water Resources Systems*. Cambridge: Cambridge University Press, pp. 153–161.
Stewart, M., and Melchers, R. E., 1997. *Probabilistic risk assessment of engineering systems*. London: Chapman and Hall.
Susman, P., O'Keefe, P., and Wisner, B., 1983. Global disasters: a radical interpretation. In Hewitt, K. (ed.), *Interpretations of Calamity*. Winchester: Allen & Unwin, pp. 264–283.
Swiss, R., 2005. Online Glossary. Available from: http://www.swissre.com, 05 March 2010. *The Northridge Earthquake: Vulnerability and Disaster*. London: Routledge, p. 288.
Thywissen, K., 2006. Core terminology of disaster reduction: a comparative glossary. In Birkmann, J. (ed.), *Measuring Vulnerability to Natural Hazards: Towards Disaster Resilient Societies*. Tokyo: United Nations University Press, pp. 448–484.
Tiedemann, H., 1992. *Earthquakes and Volcanic Eruptions: A Handbook on Risk Assessment*. Geneva: SwissRe, p. 951.
Turner, B. L., Kasperson, R. E., Matson, P. A., McCarthy, J. J., Corell, R. W., Christensen, L., Eckley, N., Kasperson, J. X., Luers, A., Martello, M. L., Polsky, C., Pulsipher, A., and Schiller, A., 2003. A framework for vulnerability analysis in sustainability science. *Proceedings of the National Academy of Sciences*, **100**(14), 8074–8079.
UN/ISDR (United Nations/International Strategy for Disaster Reduction), 2004. *Living with Risk: A Global Review of Disaster Reduction Initiatives*. Genf: United Nations.
UN/ISDR (United Nations/International Strategy for Disaster Reduction), 2009. *Global Assessment Report on Disaster Risk Reduction*. Genf: United Nations.
UNDP, 2004. *Reducing Disaster Risk: A Challenge for Development, A Global Report*. New York: United Nations.
UNDRO, 1980. *Natural disasters and vulnerability analysis*. Report of Experts Group Meeting of 9–12 July 1979. Geneva: UNDRO.
UNEP, 2002. *Assessing human vulnerability due to environmental change: concepts, issues, methods and case studies*. UNEP/DEWA/ RS.03-5, United Nations Environment Programme, Nairobi, Kenya.
United Nations (UN), (Hrsg.), 2005. *Hyogo Framework for Action 2005–2015: Building the Resilience of Nations and Communities to Disasters, World Conference on Disaster Reduction*. 18–22 Jan 2005, Kobe.
Villagrán de Leon, J. C., 2006. Vulnerability. A conceptual and methodological review. UNU-EHS, Source-4, UNU-EHS. Germany: Bonn.
Vogel, C., and O'Brien, K., 2004. Vulnerability and global environmental change: rhetoric and reality. In *AVISO 13*. Available from: http://www.gechs.org/aviso/13/.
Wisner, B., Blaikie, P., Cannon, T., and Davis, I., 2004. *At Risk, Natural Hazards, People's Vulnerability and Disasters*. London/New York: Routledge.

Cross-references

Adaptation
Civil Protection and Crisis Management
Climate Change
Complexity Theory
Critical Infrastructure
Disaster Risk Management
Disaster Risk Reduction
Evacuation
Exposure
Hazard
Hazard and Risk Mapping
HYOGO Framework for Action
Natural Hazards
Resilience
Risk Assessment
Risk Governance
Vulnerability

RISK ASSESSMENT

Suzanne Lacasse
Norwegian Geotechnical Institute, Oslo, Norway

Synonyms

Risk analysis; Risk analysis process; Risk qualification; Risk quantification

Definition

Risk assessment is a process of comprehending the nature of risk and determining its level. The risk is assessed by analyzing the potential hazards and evaluating the conditions of vulnerability of the elements potentially at risk from these hazards, including exposed people, property, services and infrastructure, livelihood, culture, quality of life, environment, and even potential loss of reputation.

Risk assessment

To determine the nature and extent of the risk, the risk assessment methodology includes: (1) the analysis of the characteristics of the hazard(s) such as location, intensity, frequency, and probability of occurrence, and (2) the analysis of exposure and vulnerability including the physical, social, health, economical, environmental, and perception dimensions and the assessment of the coping capacities to likely scenarios of occurrence.

Risk is obtained from the probability of failure weighted by the consequence of such failure. The risk assessment can be done in a qualitative or quantitative manner. Risk assessment is part of an integrated risk management process. Figure 1 illustrates such integrated process for the assessment and management of the risk associated with a landslide. The risk management process starts with an inventory of landslides, for example, at the location of interest, and then considers the following:

environmental, cultural, institutional, and political assets to suffer damage from hazardous events. The vulnerable assets use quantities such as loss of life, moneys involved, environmental damage, etc., of the elements at risk.

Risk
Risk is expressed as the product of the probability of a hazard occurring and the potential for adverse consequences, loss, harm, or detriment. Risk can be presented in terms of curves of cumulative frequencies of harmful events causing increasing level of losses, using so-called F-N curves. On these curves, the annual cumulative frequency, F, of events causing N or more losses is shown. The losses can be expressed as number of fatalities, monetary units of damage, level of pollution, time loss, etc. Both axes are usually plotted on logarithmic scales. As part of the risk management, one can use F-N curves to compare the risk level associated with a danger with risk acceptance and tolerability criteria.

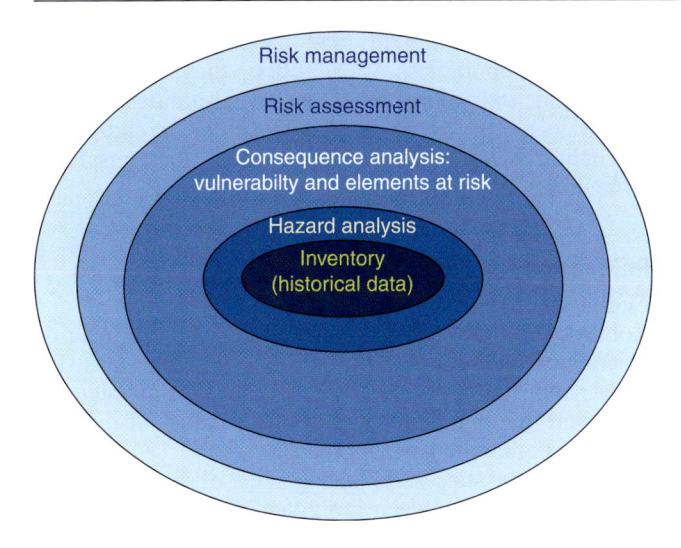

Risk Assessment, Figure 1 Integrated risk management process including risk assessment, starting with inventory of landslides at a location.

1. What can cause harm? (i.e., danger identification)
2. How often can it happen or how likely is the danger to happen? (i.e., estimation of frequency of occurrence)
3. What can go wrong? (i.e., evaluation of loss based on assessment of vulnerability of all elements at risk)
4. How bad are the consequences? (i.e., assessment of severity of consequences)
5. What should be done about the risks, that is, the hazard and vulnerability? (risk management, including comparison with acceptable and/or tolerable levels of risk and risk mitigation)

One typically assesses risk based on a number of plausible scenarios. In the case of a landslide, the following steps would be used: (1) Define scenarios for triggering the landslide and evaluate its probability of occurrence. (2) Compute the run-out distance, volume, and extent of the landslide for each scenario. (3) Estimate the losses for all elements at risk for each scenario. (4) Estimate the risk.

Hazard analysis
A landslide hazard analysis, for example, includes characterizing the danger of a slide in terms of type, size, velocity, location, travel distance, pre-failure deformations and failure mechanism and their frequency. A landslide hazard analysis usually involves an analysis of the likelihood that a slide will occur within a given period of time based on the geology, slope, gradient, elevation, geotechnical properties, vegetation cover, weathering, and drainage pattern.

Vulnerability analysis
The analysis of the vulnerability considers the propensity of the human, social, physical, economical,

Cross-references
Casualties Following Natural Hazards
Civil Protection and Crisis Management
Cognitive Dissonance
Coping Capacity
Critical Infrastructure
Damage and the Built Environment
Disaster Risk Management
Disaster Risk Reduction
Education and Training for Emergency Preparedness
Expert (knowledge-based) Systems for Disaster Management
Exposure
Geohazards
Hazard
Hazard and Risk Mapping
Hazardousness of Place
International Strategies for Disaster Reduction (IDNDR and ISDR)
Natural Hazard
Perception of Natural Hazards and Disasters
Risk
Risk Governance
Risk Perception and Communication
Uncertainty
Vulnerability

RISK GOVERNANCE

Stefan Greiving[1], Thomas Glade[2]
[1]TU Dortmund University, Dortmund, Germany
[2]University of Vienna, Vienna, Austria

Definition
Modern social systems are characterized by complex patterns of interdependencies between actors, institutions, functional activities, and spatial organizations. Controlling, managing, or even steering the complex, fragmented, and often competing societal interests is beyond the capacity of the state as an agent of authority. This is important

for dealing with risks, in particular in cases of high uncertainty and ambiguity (Klinke and Renn, 2002).

The term "governance" in its widest sense can be understood as the following. At the national level, it characterizes the structure and processes for collective decision making involving governmental and non-governmental actors (Nye and Donahue, 2000). At the global level, "governance embodies a horizontally organized structure of functional self-regulation encompassing state and non-state actors bringing about collectively binding decision without superior authority" (see Rosenau, 1992; Wolf, 2002).

This understanding can also be applied in the area of "risk." "Risk governance" aims to enhance the disaster resilience of a society (or a region) and includes "the totality of actors, rules, conventions, processes, and mechanisms concerned with how relevant risk information is collected, analysed and communicated and management decisions are taken" (IRGC, 2005). This definition points at three elements of risk governance: risk assessment and risk management that have to be embedded in a risk communication process among scientists, politicians, and the public (public and private stakeholders) as shown by the following Figure 1.

Aiming at the development of integrative models and concepts that link the different phases of risk governance mentioned before, attention has to be paid to the given differences in characteristics of the several risk types, both on the collective level and the individual perception of risk (see, e.g., IRGC, 2005). These factors might contribute in each single case in a different manner to the perception and estimation of risk. In addition, they are strongly interlinked with more collective social-political factors (WBGU, 2000). Moreover, differences in the individual perception and estimation of risks should be considered (Dubreuil et al., 2002).

Relevance for natural hazards

The limitation of research, which is fragmented and isolated, i.e., between natural sciences and engineering disciplines on the one hand and social sciences on the other hand, the importance and difficulty of maintaining trust, and the complex, socio-political nature of risk call for more comprehensive approaches.

Actually the successful management of risks is limited. The interactions between individual sectors, disciplines, locations, levels of decision-making and cultures are not known or not considered. Similarly, the linkages between the natural processes posing a threat to society are often not well understood, or considered. For example, a forest fire might demolish the trees on a slope. The next rainstorm event will trigger debris flows which will consequently enter the streams. These might be blocked, with consequent danger of breakout floods, or might be additionally loaded with vast amounts of debris with consequent damage farther downstream. To even enhance this problem: such interactions can, but might not, act within a single storm event. These interactions can also been temporally distributed over some days, months, or even years.

Risk Governance, Figure 1 Communication and the disaster cycle (Adapted from Greiving et al. (2006, p. 11).

Therefore, required societal responses to environmental changes are manifold and need careful assessments.

Additionally current risk management research and practice is fragmented by subject and by level of decision making (IRGC, 2005; Greiving et al., 2006). Inadequate information about risks in terms of societal and natural dimensions, incomprehensible procedural steps as well as insufficient involvement of the public in the risk-related decision-making process lead to severe criticism and distrust regarding relevant decisions with regard to a specific risk.

The tremendous increase in annual losses, caused by extreme events, indicates the urgency of inclusiveness. It is noteworthy, however, that this increase is not automatically caused by the increase of occurrences of the hazards. This increase might also be purely attributed to the extension of the societal usage of an area, for example, increase of house values, suburbanization, more expensive – and often vulnerable – infrastructure, etc. The insurability of economic assets becomes more and more questionable, if the so-called probable maximum loss (PML), the insurance industry is able to deal with, would be exceeded due to global change. Global change is herein understood as changes in both the social dimension with all its economic, political, and administrative changes and the natural dimension with the widely discussed climate change, but also vegetation change, changes in the hydrological regimes through water reservoirs, extensive ground water usage, etc.

Often assessed criteria are primarily relevant for climate change and those natural hazards which are triggered by meteorological conditions such as floods, droughts, forest fires, landslides, but also creeping change processes like the loss of biodiversity, soil erosion, or water pollution, to name a few.

Decisions in the area of so-called traditional risks like flooding, taken mainly on the basis of engineering expertise, are normally based on probabilities because they are past-oriented and informed by statistics. However, analyzed data are only available for a specific period – and are thus not representative for longer periods. It is important to recognize that such calculated return periods cannot be easily extended over longer periods. For example, a river changes its flow behavior dramatically, if internal thresholds are exceeded (e.g., changes of bed roughness, different flow dynamics with overtopping riverbanks, etc.). Nearly all natural systems show such nonlinear behavior and consequently, it is most important to determine and identify the limits, within which the analysis is valid, and for which not. This principle problem is also enlarged by observed climate changes. Related effects on temperature and precipitation will certainly lead to new uncertainties, because past events might not be representative anymore. Similarly, changes in the catchments (e.g., deforestation, surface sealing through suburban development, etc.) will lead to high uncertainties. Here, the perspective changes from probabilities to just possibilities. With public decision making not having any precise information at hand, restrictions for private property rights are probably no longer legally justifiable. Hereby, justification of actions and consensus about thresholds and response actions becomes more important, which calls naturally for inclusiveness.

Within the global change debate, the field of climate change in general, but particularly as a triggering factor for many natural hazards, is of special importance for Europe with its dominant existing settlement structures, cultural landscapes, and infrastructures which have been developed over centuries. Prevention actions, carried out by spatial planning, are under these circumstances less effective than in countries which are still growing rapidly in terms of population and the built environment. Here, disaster-prone areas can be kept free from further development whereas most of these areas are in Europe already built up. However, this calls for authorities with risk awareness and means to mitigate this problem. And in the developing countries with high potential to avoid the adverse effects, strong authorities with the required power are needed – and this is often not available in such countries.

Moreover, measures, based on mandatory decisions of public administration as well as measures which are in the responsibility of private owners, need to be accepted widely for their implementation. This is clearly visible when looking at evacuation orders, building protection measures to be taken by private households, risk awareness, etc. Having these facts in mind, the "active involvement," propagated, e.g., by the European Communities Flood Management Directive (European Communities, 2007), has to be seen as crucial for the success of the Directive's main objective: the reduction of flood risks. Within the European Community, it has also been recognized that the risk approach has also to be applied to other natural hazards such as coastal hazards or soil erosion and landslide hazards (e.g., Soil Thematic Strategy, 2006).

This makes clear that risk governance has become increasingly politicized and contentious. The main reasons are controversies concerning risk that are not about science versus misguided public perceptions of science, where the public needs to be educated about "real" risks (Armaş and Avram, 2009). Rather, risk controversies are disputes about *who* will define risk in view of existing ambiguity. Technology policy discourse is not about who is correct about assessment of danger but whose assumptions about political, social, and economic conditions as well as natural or technological forces win in the risk assessment debate. Thus, the hazard is real, but risk is socially constructed.

This understanding of risk used here is different than the definition within engineering and natural sciences. Herein, risk is not understood as socially constructed, but "real" – real in terms of potential monetary losses. All these approaches try to quantify the risk in terms of determination of the expected frequency and magnitude of the relevant hazard (e.g., earthquake, flood, landslide) and the elements at risk (e.g., infrastructure, buildings, etc.). The degree of expected damage as a consequence from a specific event magnitude is herein expressed as vulnerability, and various vulnerability functions exist (e.g., Fuchs et al., 2007). The overall and ultimate aim of natural and engineering science–based approaches is always to provide "objective," reliable, repeatable and comparable information on both hazard and potential consequences.

Scientific literacy and public education are important but not central to risk controversies. Emotional response by stakeholders to issues of risk is truly influenced by distrust in public risk assessment as well as management. Due to this fact, those who manage and communicate risks to the public need to start with an understanding of emotional responses toward risk. It is a matter of the definition of risk how risk policy is carried out. Moreover, defining risk is an expression of power. Slovic (1999) thereby argues that whoever controls the definition of risk controls technology policy.

Within the communication strategies in all approaches, trust can be seen as a central term in this respect. The regressiveness of trust of public in for example, policy makers appears to be related to "a number of factors, including social alienation; a lack of social capital; higher levels of education and greater availability of information resulting in a more sceptical public; increased scientific pluralism leading to confusing messages; cronyism in government; growth of citizen activism in an era of complex and uncertain risks and multiple messengers; regulatory scandals [...] and a hyper-critical media" (Löfstedt, 2005).

Further, public decision making that is based only on the factual "scientific" dimension of risk, not taking into account the "sociocultural" dimension, which includes how a particular risk is viewed when values and emotions are concerned leads to distrust (e.g., whether a risk is judged acceptable, tolerable or intolerable by society is partly influenced by the way it is perceived to intrude upon the value system of society). Regarding this, Löfstedt (2005) summarizes three present distinct problems for risk regulators as following: "(a) it is easier to lose or destroy trust than to gain it; (b) in an 'era of distrust', the public will turn to other sources of information and (in many cases) believe this information more than that provided by the public risk managers, even when the latter may be more accurate; (c) when the public has access to many more sources of information, such as the Internet and 24-h television, they are no longer dependent on policy-makers or risk managers for information. The result is a more knowledgeable but more skeptical public."

Trust has a key role in dealing with risks and should be regarded as fundamental for risk interpretation of the public between "real" and "perceived" risks. (Interpretations of "risk" differ according to individual and social contexts.) Especially "trust is particularly relevant in conditions of ignorance or uncertainty with respect to unknown or unknowable actions of others. [...] In this respect, trust concerns not future actions in general, but all future actions which condition our present decisions" (Gambetta, 2000).

Consequently, ensuring a stakeholder-focused process means consulting and involving stakeholders. In this regard, research on risk governance has to be understood as cooperative research: a form of research process, which involves both researchers and non-researchers in close cooperative engagement. This also meets the demand of the White Paper on European Governance (European Commission, 2001) where a better involvement and more openness as well as better policies, regulation, and delivery have been identified as key objectives of "good governance." Strong stakeholder-oriented elements have also been integrated in recent disaster management strategies like the United Nations International Strategy for Disaster Reduction (ISDR, 2005). The risk governance approach has recently been regarded as important by the new Territorial Agenda of the EC, launched in 2007 in Leipzig, Germany, by the Member States Ministers for Spatial Planning as part of the priority 5 "Promoting Trans-European Risk Management" (European Commission, 2007).

Distrust makes institutional settings vulnerable as it lowers the efficiency and effectiveness of management actions. The whole disaster cycle from mitigation, preparedness, response to recovery is embedded in an institutional system. Institutional vulnerability can in principle be understood as the lack of ability to involve all relevant stakeholders and effectively coordinate them right from the beginning of the decision-making process. It refers to both organizational and functional form as well as guiding legal and cultural rules.

Generally speaking, a good institution design leads to effectiveness, equality, and resilience. Narrow social networks and relations increase the possibility of social learning and adaptation through cooperation and communication. Besides the accurate assessment(s) of the hazard(s), the relevance of informal social networks, local institutions and authorities (traditional and administrative), and local-based organizations is a prerequisite for the reduction of impacts of the disaster.

Existing approaches for assessing good/bad risk governance

The development of integrative models and concepts that link the elements of risk governance (risk assessment, management and communication), from an interdisciplinary perspective can be interpreted as the overall objective for integrative approaches to risk governance. Therefore, it is important to use a methodology that overbears the mentioned tendency to insularity between disciplines, to find a proper way to deal with uncertainty and ambiguity and to create resilient communities as well as reduce the institutional vulnerability.

The elaborated approach is a comprehensive risk governance concept which aims – as already mentioned in literature (Löfstedt, 2005) – at a broad and active involvement of decision makers at the relevant political and administrative levels and/or of stakeholders. In addition it offers a better understanding and acceptance of research by society and vice versa bringing the legitimate interests of society and single stakeholders into research and decision making. The concept is supported by a tool that is able to monitor the performance of a risk governance process and consists of an "indicator system" as a methodological concept that brings together the state-of-the-art concepts in risk governance with methodological and procedural needs, identified by those who are close to the daily practice in risk communication. In addition, an "interest analysis" is useful being a practical method for preparing stakeholder involvement processes.

The concept is based on a comprehensive investigation on the state of the art of knowledge and methodologies applied for risk governance, having analyzed existing risk governance concepts with a special focus on Europe and some input for non-European settings. Accordingly, 14 risk resilience projects and initiatives were analyzed to develop a comprehensive and inclusive framework and indicators for governance of risk resilience, sustainability and performance.

This number and range of the analyzed projects and initiatives was satisfactory because all important elements of risk governance were covered. Furthermore, the multitude of analyzed projects/initiatives was characterized by a diversity of addressed risks which was of great relevance for the project in order to guarantee the transferability to various risk settings. Originally 35 risk governance principles, derived from a first literature screening, had been used for this analysis that was based on two parts (Wanczura et al., 2007):

1. Overall information like, for example, description of the risk governance approach; characterization of the risk governance approach and conclusions.
2. Detailed table where 35 principles/indicators were checked if they were covered in the analyzed approaches. These principles were grouped under five perspectives: Basic/Content, Procedure, Stakeholder, Resources, and Expertise. They allow a better general observation if important aspects were considered appropriately. Additionally, each of the indicators had clarifying questions to reach a better understanding of the indicator and its aim.

The elaboration of the key aspects was enabled by means of a literature research, a set of interviews, and a scientific colloquium. It has to be pointed out that the hazard itself was of minor importance herein. The societal and administrative settings were the focus in this survey. Thus, the aim was to list all significant indicators for a risk governance process in order to enable a detailed description as well as a comparison between them in consideration of existing standards (see Audit Commission, 2000). In addition, it allowed the development of a balanced set of indicators. Nevertheless, it was not feasible to consider all indicators which were available due to the difficulty of handling such a variety and quantity. Especially concerning stakeholder-communication, the number of indicators should be low. Scientific literature proposes only few (approximately 20 indicators) aggregated and comprehensible indicators (Weiland, 1999). Consequently, a core set of the 12 mostly accepted principles/indicators were chosen ("Key Performance Indicators"; KPI), representing the resilience framework (see Table 1).

Indicators are used especially by decision-making stakeholders in order to justify and improve their decisions as well as to improve their impact/implementation of concepts and benchmark the given performance of risk management systems in terms of the attention they paid to the above-listed governance principles. Therefore, such a system is able to assess the progress of management setting over time, but also benchmarking different organizations at a certain point of time.

Old wine in new glasses?

Risk governance can be understood as a holistic way of understanding and dealing with risk. Consequently, it might be just "old wine in new glasses" if a risk assessment and management process is embedded into a comprehensive communication and stakeholder involvement process. However, today's risk management practice is often not inclusive at all, i.e., in the context of traditional risks where natural as well as technological hazards are causing factors. At the same time, the experts of the hazard analysis define what is right and wrong for the affected party involved. The quality of the hazard analysis and assessment is not questioned here. Light is thrown on the way, how the respective results are presented. For example, the experts often require specific management actions in terms of technical measurement, etc., which are sometimes not at all in line with the needs of the affected parties. Here, it belongs to the duties of social sciences to develop proper ways for applying inclusiveness in risk management practices where the role of the affected public is wholly neglected. For that purpose, the concept of risk governance makes inclusiveness more explicit.

It should be kept in mind that different addressees have different attitudes/interests which are relevant when spreading inclusiveness: other scientists (i.e., geoscientists, engineers), politicians, mass media, public, and affected individuals. There is also a question how to communicate inclusiveness to a wider community. Social networking through ICT/WEB technologies and modern visualization tools (i.e., Geo Information Systems, GIS) are quite helpful for target group–oriented communication. Any pedagogy of inclusive risk governance should be designed for different target groups (for instance, pupils) and calls for an adequate language to achieve a mutual understanding. Indeed, further research is necessary to accommodate the different needs and expectations of involved parties.

Risk Governance, Table 1 Definition of risk governance indicators for the integrative approach

Perspective	Keyword	Key question	Objective	Key performance indicator
Basic/content	*Principles*	What are the guiding principles?	Definition of guiding principles and a consistent "target system"	Degree of operationalization of the guiding principles
	Trust	How far is attention paid to relevance of an atmosphere of mutual respect and trust?	Between all relevant stakeholders and decision makers, an atmosphere of mutual respect and trust exists	Reflection of trust concerning people/institutions
	Objectives	What are the concrete protection goals for subjects of the protection?	Definition of a comprehensive and obligatory understanding of the damage-protection relation	Degree of obligation concerning the protection goals for the subjects of the protection
Procedure	*Accountability principle*	How far is accountability defined at each level (process, each risk)?	Each actor knows his responsibilities and acts accordingly	Definition of the responsibility
	Justification	How far is the activity concerning the management of existing risks justified?	Justification of action in the area of risk management	Definition and agreement on a justification concerning the exposure to risk
Stakeholder	*Representation*	How far are all relevant social groups (and their representatives, stakeholder, respectively) and their expectations known?	Identification of all relevant social groups and their expectations	Degree of high profile of all social groups and their expectations
	Access to information	How far is information for all stakeholders accessible?	Access for all stakeholders to the relevant information	Degree of the availability and understandability of the relevant information for stakeholders
	Tolerance process & outcome	How far do the stakeholders tolerate/accept the risk governance process and its outcomes?	All involved stakeholder tolerate/accept the risk governance process and its outcomes	Degree of the tolerance/acceptance on the part of involved stakeholder
	Dialogue	To what extent is a constructive dialogue with the relevant stakeholders available or conducted?	Establishment of custom discourse-processes concerning risk topics	Quality of discourse-processes with relevant stakeholders (i.e., public or private representatives)
Resources	*Financial resources*	To what extent do the available financial resources meet the requirements of the defined Risk Governance Process?	Allocation of sufficient financial resources for a successful risk governance process	Degree of realization of a financial concept
	Staff resources	To what extent do the staff resources (technical qualification and number of people) meet the requirements of the defined Risk Governance Process?	Allocation of adequate staff resources	Realization of a staff assignment concept
Expertise	*Role*	How far has the role of experts been defined?	If experts are involved, their role within the decision-making process has to be defined	Degree of definition and agreement concerning the role of experts

Source: own elaboration

The Strategic Environmental Assessment (SEA) offers a suitable procedural frame for implementing a risk governance approach (European Communities, 2001). Damage potential as a key term in the definition of risk can be understood as a relevant issue, covered by SEA issues of human health, material assets, and cultural heritage (See Annex I, Directive 2001/42/EC). Moreover, EU Directive 2001/42/EC in Annex II points out the characteristics of the effects and the areas likely to be affected. Projects that have to obtain a permit from a certain plan or program might have significant effects on the environment because of specific consequences, for example, increasing damage potential regarding certain hazards that threaten the area where the project is planned.

When looking at the EU directive 2007/60/EC on the assessment and management of flood risks, it becomes obvious, how important risk governance principles are for today's EU policy. Art 10 § 2 states: "Member States shall encourage active involvement of interested parties in the production, review and updating of the flood risk

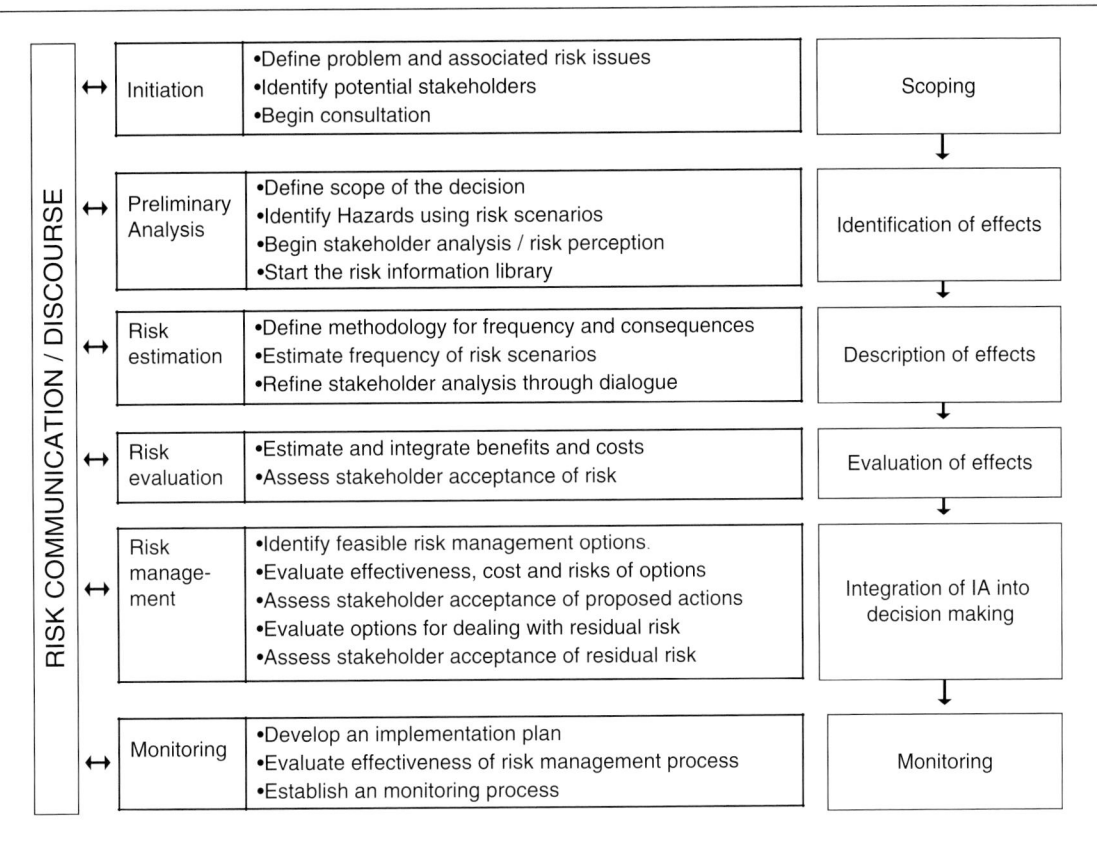

	Stage	Details	Assessment
RISK COMMUNICATION / DISCOURSE	Initiation	• Define problem and associated risk issues • Identify potential stakeholders • Begin consultation	Scoping
	Preliminary Analysis	• Define scope of the decision • Identify Hazards using risk scenarios • Begin stakeholder analysis / risk perception • Start the risk information library	Identification of effects
	Risk estimation	• Define methodology for frequency and consequences • Estimate frequency of risk scenarios • Refine stakeholder analysis through dialogue	Description of effects
	Risk evaluation	• Estimate and integrate benefits and costs • Assess stakeholder acceptance of risk	Evaluation of effects
	Risk management	• Identify feasible risk management options. • Evaluate effectiveness, cost and risks of options • Assess stakeholder acceptance of proposed actions • Evaluate options for dealing with residual risk • Assess stakeholder acceptance of residual risk	Integration of IA into decision making
	Monitoring	• Develop an implementation plan • Evaluate effectiveness of risk management process • Establish an monitoring process	Monitoring

Risk Governance, Figure 2 Integration of stakeholder into the process of assessing and managing risks (Source: own table).

management plans." This underlines the proposed active stakeholder involvement from the early beginning as outlined more in detail by the following Figure 2.

Conclusions

Risk governance becomes more important the more a risk setting is complex, characterized by uncertainty and ambiguity, or when many stakes have to be balanced/weighted up. This includes the proper assessment of hazards, independent of whether we are dealing with natural, technological, or traffic hazards, to name a few. Herein, the major future challenge is the incorporation of the dynamics in all developments, and how to include these dynamics properly into the risk governance cycle. For tackling institutional weaknesses, the procedural approach of risk governance has to be linked with a defined material goal stakeholders are in line with as a process does not end in itself. Here, resilience and therefore proactivity come into play, because resilience is determined by the degree to which the social system is capable of organizing itself to increase its capacity for learning from past events for better future protection and to improve proactively risk reduction measures. The material goal "resilience" can be seen as a widely accepted strategy within the natural disaster community (e.g., Fuchs, 2009; Paton et al., 2001). On the contrary, the more procedural approach "risk governance" has been created and adapted first in the area of new emerging, mostly man-made risks. The usage of both terms in the same context has to be seen as an innovative approach to combining an appropriate path (risk governance – including identification, assessment, management, and communication of risk) toward the material goal of creating resilient communities, able to deal with the whole range of risks, nature-made as well as man-made ones. A fully adopted "risk governance" strategy will indeed cause more work, discussion, and even contradiction during the assessment, but is then accepted by the parties involved and designed for a successful and long-lasting strategy which might support more sustainable communities.

Bibliography

Armaş, J., and Avram, E., 2009. Perception of flood risk in Danube Delta, Romania. *Natural Hazards*, **50**(2), 269–287.

Audit Commission, 2000. *Management Paper on Target the Practice of Performance Indicators*. London: Audit Commission.

Dubreuil, G. H., Bengtsson, G., Bourrelier, P. H., Foster, R., Gadbois, S., and Kelly, G. N., 2002. A report of TRUSTNET on risk governance – lessons learned. *Journal of Risk Research*, **5**(1), 83–95.

European Commission, 2001. *European Governance – A White Paper*. Brussels: European Commission.

European Commission, 2007. *Territorial Agenda of the European Union*. Brussels: European commission.

European Communities, 2001. *Directive 2001/42/EC of the European Parliament and of the Council of 27 June 2001 on*

the assessment of the effects of certain plans and programmes on the environment. OJ L 197, 21.7.2001.
European Communities, 2007. *DIRECTIVE 2007/60/EC OF THE EUROPEAN PARLIAMENT AND OF THE COUNCIL of 23 October 2007 on the assessment and management of flood risks.* Official Journal of the European Union L 288/27.
Fuchs, S., 2009. Susceptibility versus resilience to mountain hazards in Austria – paradigms of vulnerability revisited. *Natural Hazards and Earth System Sciences*, **9**, 337–352.
Fuchs, S., Heiss, K., and Hübl, J., 2007. Towards an empirical vulnerability function for use in debris flow risk assessment. *Natural Hazards and Earth System Sciences*, **7**, 495–506.
Gambetta, D. (ed.), 2000. *Can We Trust Trust?. Trust: Making and Breaking Cooperative Relations.* University of Oxford – Department of Sociology, 213–217. http://www.sociology.ox.ac.uk/papers/gambetta213-237.pdf.
Greiving, S., Fleischhauer, M., and Wanczura, S., 2006. Management of natural hazards in Europe: the role of spatial planning in selected EU Member States. *Journal of Environmental Planning and Management*, **49**(5), 739–757.
IRGC – International Risk Governance Council, 2005. *White Paper on Risk Governance: Towards an Integrative Approach.* Geneva: IRGC.
International Strategy for Disaster Reduction (ISDR), 2005. Hyogo Framework for Action 2005-1015: Building the Resilience of Nations and Communities to Disasters, Tech. Report, World Conference on Disaster Reduction.
Klinke, A., and Renn, O., 2002. A new approach to risk evaluation and management: risk-based, precaution-based and discourse-based strategies. *Risk Analysis*, **22**, 1071–1094.
Löfstedt, R., 2005. *Risk Management in Post-Trust Societies.* Basingstoke/Hampshire/New York: Houndmills.
Nye, J. S. Jr., and Donahue, J. d. (eds.), 2000. Governance in a globalizing world. Cambridge, Mass: Brookings Institution Press.
Paton, D., Millar, M., and Johnston, D., 2001. Community resilience to volcanic hazard consequences. *Natural Hazards*, **24**(2), 157–169.
Rosenau, J. N., 1992. Governance, order, and change in world politics. In Rosenau, J. N., and Czempiel, E. O. (eds.), *Governance Without Government. Order and Change in World Politics.* Cambridge: Cambridge University Press, pp. 1–29.
Slovic, P., 1999. Trust, emotion, sex, politics, and science: surveying the risk-assessment battlefield. *Risk Analysis*, **19**(4), 689–701.
Soil Thematic Strategy, 2006. Communication from the Commission to the Council, the European Parliament, the European Economic and Social Committee and the Committee of the Regions – Thematic Strategy for Soil Protection [SEC (2006)620] [SEC(2006)1165].
Wanczura, S., Fleischhauer, M., Greiving, S., Fourman, M., Sucker, K., d'Andrea, A., 2007. *Analysis of recent EU, international and national research and policy activities in the field of risk governance.* Del. 1.1 MIDIR Project.
WBGU – German Advisory Council on Global Change, 2000. *World in Transition: Strategies for Managing Global Environmental Risks.* Annual Report 1998. Berlin: Springer.
Weiland, U., 1999. Indikatoren einer nachhaltigen Entwicklung – vom Monitoring zur politischen Steuerung? In Weiland, U. (ed.), *Perspektiven der Raum- und Umweltplanung – angesichts Globalisierung, Europäischer Integration und Nachhaltiger Entwicklung.* Berlin: Verlag für Wissenschaft und Forschung, pp. 245–262.
Wolf, K. D., 2002. Contextualizing normative standards for legitimate governance beyond the state. In Grote, J. R., and Gbikpi, B. (eds.), *Participatory Governance. Political and Societal Implications.* Opladen: Leske + Budrich, pp. 35–50.

Cross-references

Disaster Risk Management
Livelihoods and Disasters
Perception of Natural Hazards and Disasters
Risk Assessment
Risk Perception and Communication

RISK PERCEPTION AND COMMUNICATION

Michael K. Lindell
Texas A&M University, College Station, TX, USA

Definitions

Risk perception – the result of an informal personal assessment of adverse consequences that could be caused by extreme environmental events.

Risk assessment – the result of a formal scientific analysis that systematically identifies the causes and consequences of environmental extremes.

Risk – the likelihood that an event of a given magnitude will occur at a given location within a given time period, with that event's associated casualties and damage, as well as its psychological, demographic, economic, and political impacts.

Risk communication – a process by which stakeholders provide (one-way) or exchange (two-way) information about hazards.

Hazard adjustments – mitigation actions such as land use and building construction practices that reduce vulnerability, emergency preparedness actions such as emergency plans that facilitate emergency response, and recovery preparedness actions such as insurance purchase that redistribute losses.

Introduction

Some people, because of their access to data or their specialized expertise in interpreting those data, have scientific risk assessments of environmental hazards and specialized knowledge of ways in which those risks can be managed. These risk analysts engage in risk communication to convey their assessments to decision makers, such as government officials and risk-area residents, whose risk perceptions might or might not be consistent with the analysts' risk assessments. Decision makers often have limited knowledge about what actions to take in response to the environmental risks, so it is important for analysts to communicate information about emergency response actions or long-term hazard adjustments that are both effective and feasible. Finally, people's acceptance of analysts' risk assessments depends on their perceptions of the sources of that information, so it is important for risk analysts to ensure that they establish and maintain credibility with their audiences.

Perception of environmental hazards

The essential dimensions of risk are probability and consequences, but some well-known approaches to perceived risk include qualitative "outrage factors," such as *dread* and *unknown* risks (Slovic et al., 2001). These additional dimensions of perceived personal risk appear to be useful in explaining people's responses to technological hazards, but it is unclear if they are related to the perception of, and response to, natural hazards. Most research on natural hazards has defined perceived risk in terms of people's expectations about the probability of personally experiencing the adverse physical and social impacts that an extreme environmental event causes (Lindell and Perry, 2004). Thus, these studies have typically defined risk perception in terms of specific consequences, such as death, injury, property damage, and disruption to daily activities (work, school, shopping, etc.). Moreover, most research on natural hazards risk perception has examined the ability of risk perception to predict warning responses such as evacuation (Sorensen, 2000) and long-term hazard adjustment such as insurance purchase (Zahran et al., 2009). This research has studied earthquakes (Lindell and Perry, 2000), hurricanes and other storms (Dash and Gladwin 2007), floods (Terpstra and Gutteling, 2008), and volcanic eruptions (Perry and Lindell, 2008).

Perceived personal risk is related to the recency, frequency, and intensity of people's personal experience with hazard events (Weinstein 1989). Such experience can involve casualties or damage experienced by the respondent himself/herself, by members of the immediate or extended family, or by friends, neighbors, or coworkers (Lindell and Prater, 2000). Hazard experience is a function of the annual probability of the occurrence of a major event and the number of years of exposure (past tenure at that location).

Perceived personal risk also is related to proximity to earthquake (Palm et al., 1990), hurricane (Peacock et al., 2005), and flood (Preston et al., 1983) sources. The relation between hazard proximity and personal risk is determined by a perceived risk gradient that relates increasing distance to decreased risk (Lindell and Earle, 1983). However, in addition to abstract beliefs about risk gradients, people who are farther away from hazard sources are also less likely to have direct experience with hazard impacts and, thus, have lower levels of perceived personal risk for this reason. Specifically, those who are closer to a hazard source are likely to observe environmental cues of the hazard (Preston et al., 1983). Finally, risk perception is significantly related to some demographic characteristics. These include lower education and income (Fothergill and Peek, 2004), female gender (Fothergill, 1996), and ethnic minority status (Fothergill et al., 1999).

Risk perception also is influenced by social communication because social groups, such as authorities, news media, and peers, can serve as sources of hazard information (Lindell and Perry, 2004). However, hazard information is not distributed uniformly over the entire population. Authorities, in particular, have limited resources, so they tend to focus on those at greatest risk (Baker, 1991; Lindell et al., 2005). This tendency is compounded by the tendency for those at greater risk to be more likely to recognize the relevance of such information and, thus, pay greater attention to it.

Perceptions of hazard adjustments

A substantial amount of research has focused on people's perceptions of natural *hazards*, but Lindell and Perry (2000) called attention to the need for also studying their perceptions of natural *hazard adjustments*. Such studies are needed because the theory of reasoned action (Fishbein and Ajzen, 1975) posits that one's attitude toward an object (e.g., seismic hazard) is less predictive of behavior than one's attitude toward an act (seismic hazard adjustments) relevant to that object. Thus, to understand the adoption of hazard adjustments, it is just as important to understand the perceived attributes of the hazard adjustments as the perceived attributes of the hazard itself.

The identity of these perceived attributes can be surmised from studies on the adoption of hazard adjustments, which have found support for attributes such as effectiveness (Mulilis and Duval, 1995), cost (Peacock, 2003), required knowledge (Davis, 1989), and utility for other purposes (Russell et al., 1995). These are consistent with the protective action decision model (Lindell and Perry, 2004), which proposes that hazard adjustments can be defined by *hazard-related* and *resource-related* attributes. Hazard-related attributes, such as efficacy in protecting people and property and usefulness for other purposes, have been found to be significantly correlated with adoption intention and actual adjustment (Lindell and Whitney, 2000; Lindell and Prater, 2002). However, resource-related attributes (cost, knowledge and skill requirements, time requirements, effort requirements, and required cooperation with others) generally have the predicted negative correlations with both adoption intention and actual adjustment, but these are small and nonsignificant. The hazard adjustments that have been studied to date have generally had small resource requirements, so it is unclear if the lack of support for the significance of these attributes is due to this factor (Lindell et al., 2009).

Perceptions of information sources

Previous research has characterized stakeholders as authorities (federal, state, and local government), evaluators (scientists, medical professionals, universities) watchdogs (news media, citizens' and environmental groups), industry/employers, and households (Drabek, 1986; Pijawka and Mushkatel, 1991; Lang and Hallman, 2005). The interrelationships among stakeholders can be defined by the power they wield over each other's decisions to adopt protective actions and hazard adjustments. French and Raven (1959, Raven 1965) posited that power relationships can be defined in terms of six bases – reward, coercive, expert, information, referent, and legitimate power. Reward and coercive power are the principal bases of regulatory approaches, but Raven (1993) noted these require continuing surveillance to ensure rewards are received only for compliance and punishment will follow noncompliance. Unfortunately, many state mandates are hampered by a lack of formal reporting or review by state officials, and limited or no penalties for failing to enforce their provisions (Burby et al., 1998). Consequently, there is a need to better understand the ways in which households can be influenced by bases of power other than reward and coercion. French and Raven's conception of expert (i.e., understanding of cause-and-effect relationships in the environment) and information (i.e., knowledge about states of the environment) power suggests assessing perceptions of stakeholders' hazard *expertise*. French and Raven's conception of referent power is defined by a person's sense of shared identity with another (Eagly and Chaiken, 1993), which is related to that person's *trustworthiness*. Although trust has been defined many different ways (see Arlikatti et al., 2007 for a review), fairness, unbiasedness, willingness to tell the whole story, and accuracy are central (Meyer, 1988).

French and Raven (1959) defined legitimate power by the rights and responsibilities associated with each role in a social network, which raises questions about what households consider to be the responsibility of different stakeholders for protecting them from seismic hazard. This is reinforced by research on stakeholders' perceived *protection responsibility*, which dates from Jackson's (1981) research that attributed low rates of seismic adjustment adoption to respondents' beliefs that the federal government was the stakeholder most responsible for coping with earthquakes. Much later, Garcia (1989) found respondents had come to believe earthquake preparedness was an individual's responsibility. Her conclusion that a perception of personal protection responsibility leads to a higher level of seismic adjustment adoption is supported by similar findings on tornado adjustment adoption (Mulilis and Duval, 1997). Recent research has found that these perceived stakeholder characteristics (expertise, trustworthiness, and protection responsibility) have significant positive correlations with hazard adjustment intentions and actual adjustment adoption (Lindell and Whitney, 2000; Arlikatti et al., 2007).

Risk communication

Based on an extensive review of research on risk perception, hazard adjustment perception, stakeholder perception, and risk communication, Lindell and Perry (2004) proposed that risk communication should be a process by which stakeholders *share* information about hazards affecting a community. Communication among stakeholders can be analyzed in terms of who (source) says what (message), via what medium (channel), to whom (receiver), and directed at what kind of change (effect).

As noted earlier, sources are perceived primarily in terms of expertise, trustworthiness, and protection responsibility. Messages vary in their content – especially their information about a hazard, its characteristics (e.g., magnitude, location, and time of impact), alternative protective actions and their attributes (efficacy; cost; safety; requirements for time and effort, knowledge and skill, tools and equipment; and cooperation from others), as well as contacts for further information and assistance.

The primary information channels available for use include print media such as newspapers, magazines, and brochures; electronic media such as television, radio, telephone, and the internet; and face-to-face interaction through personal conversations and public meetings. However, other channels are available for emergency response (e.g., sirens and tone alert radios, see Lindell and Prater, 2010). The distinctions among these information channels are important because they differ in the ways in which they accommodate the information processing activities of the receivers. For example, orally presented information is ephemeral and is easily lost unless otherwise recorded, whereas written information inherently provides a record that can be referred to at a later time. Receivers differ in many respects, but the most important of these are their perceptions of source credibility, access to communication channels, prior beliefs about hazards and protective actions, ability to understand and remember message content, and access to resources needed to implement protective action. Message effects include attention, comprehension, acceptance, retention, and action.

Finally, feedback is an important component of the communication model because some attempts are unidirectional, whereas others are interactive. Unidirectional communications appeal to many hazard managers because they appear to be less time consuming, and sometimes, this actually is the case. Frequently, however, interactive communication is needed for receivers to indicate that they have not comprehended the message that was sent or that the message sent by the source did not satisfy their information needs. It is important to recognize that risk information is transmitted through informal social networks (Lindell et al., 2007), so a warning message is likely to be relayed from the original source (e.g., an authority) through an intermediary (e.g., a risk-area resident) to an ultimate receiver (e.g., a peer such as a friend, relative, neighbor, or coworker) in addition to – or even instead of – being transmitted directly from the original source to the ultimate receiver.

There are six basic functions that should be addressed in a community risk communication program (Lindell and Perry, 2004). These are strategic analysis, operational analysis, resource mobilization, program development, program implementation for the continuing hazard phase, and program implementation for the escalating crisis and emergency response phases. The first function, strategic analysis, lays the foundation for later risk communication activities. Thus, the strategic analysis involves conducting analyses to identify hazards and locations at risk as well as examining the community to identify important characteristics such as its ethnic composition and communication channels. In addition, hazard managers should work with community agencies to identify the community's prevailing perceptions of environmental hazards and hazard adjustments and set appropriate goals for the risk communication program.

Operational analysis seeks to identify appropriate hazard adjustments for community households and businesses and to work with the community agencies to encourage these hazard adjustments. Hazard managers should also work with community agencies to identify available risk communication sources and channels in the community as well as to identify the differences among audience segments in terms of their access to and preferences for different types of media (e.g., radio) and specific channels (e.g., specific radio stations) within each medium.

During resource mobilization, hazard managers should obtain the support of senior management by "selling" the importance of assessing community hazard vulnerability and identifying hazard mitigation, emergency preparedness, emergency response, and disaster recovery as effective solutions. They should also enlist the participation of government agencies, businesses, and nongovernmental organizations in a collaborative strategy to coordinate all of the community's risk communication programs. Hazard managers should also work with these organizations to develop collaborative relationships with the news media and with neighborhood associations and service organizations.

During program development for all phases, hazard managers should staff, train, and exercise a crisis communications team and establish procedures for maintaining an effective communication flow in an escalating crisis and during emergency response. Hazard managers should also work with community organizations to develop a comprehensive risk communication program that presents information about hazards and hazard adjustments in a form that attracts attention and is easily understood and retained. They should also encourage these organizations to use informal communication networks in the community. Hazard managers should also establish procedures for obtaining feedback from the news media and the public.

Program implementation for the continuing hazard phase involves building source credibility by increasing perceptions of expertise and trustworthiness and using a variety of channels to disseminate hazard information. Hazard managers should also describe community-level hazard adjustments being planned or implemented – such as land use regulations or building codes. They should also work with local organizations to describe feasible household hazard adjustments risk-area residents can take to protect themselves. Finally, hazard managers should evaluate risk communication program effectiveness by measuring the degree to which the program has achieved its objectives.

Program implementation for the escalating crisis and emergency response phases involves working with community officials to classify the situation in terms of its severity and activate the crisis communication team promptly so that its members can make all appropriate contacts. Hazard managers should determine the appropriate time to release sensitive information by developing procedures that define when information is to be released and select the communication channels appropriate to the situation. Hazard managers also need to provide timely and accurate news releases for the public that are supplemented by fact sheets containing basic background information that is appropriate to any incident. Hazard managers should maintain source credibility with the news media and the public by being honest about what is and is not known and state that they do not know the answer to a question when this is the case. However, they should make a commitment to answer any unresolved questions at a later time. Last, they should evaluate performance through post-incident critiques involving all members of the crisis communication team.

Summary

Risk perception is the result of people's informal assessment of the adverse personal consequences – such as death, injury, illness, and property damage – that could be caused by extreme environmental events. Perceived personal risk is related to the recency, frequency, and intensity of people's experience with hazard events. It also is influenced by social communication because social groups, such as authorities, news media, and peers, can serve as sources of hazard information. Thus, to understand people's risk perceptions, it is necessary to understand their interpretations of their own hazard experiences and their perceptions of information sources. In addition, to understand people's protective responses to their risk perceptions, it is also important to understand the perceived attributes of those protective actions. These attributes include hazard-related attributes, such as efficacy in protecting people and property and usefulness for other purposes, as well as resource-related attributes (cost, knowledge and skill requirements, time requirements, effort requirements, and required cooperation with others). To be most effective, risk communication programs should engage a process in which stakeholders *share* information about hazards affecting a community. Such programs should address six basic functions – strategic analysis, operational analysis, resource mobilization, program development, program implementation for the continuing hazard phase, and program implementation for the escalating crisis and emergency response phases.

Bibliography

Arlikatti, S., Lindell, M. K., and Prater, C. S., 2007. Perceived stakeholder role relationships and adoption of seismic hazard adjustments. *International Journal of Mass Emergencies and Disasters*, **25**, 218–256.

Baker, E. J., 1991. Hurricane evacuation behavior. *International Journal of Mass Emergencies and Disasters*, **9**, 287–310. Available at www.ijmed.org.

Burby, R. J., French, S. P., and Nelson, A. C., 1998. Plans, code enforcement, and damage reduction: evidence from the Northridge Earthquake. *Earthquake Spectra*, **14**, 59–74.

Dash, N., and Gladwin, H., 2007. Evacuation decision making and behavioral responses: individual and household. *Natural Hazards Review*, **8**, 69–77.

Davis, M. S., 1989. Living along the fault line: an update on earthquake awareness and preparedness in Southern California. *Urban Resources*, **5**, 8–14.

Drabek, T., 1986. *Human System Responses to Disaster: An Inventory of Sociological Findings*. New York: Springer Verlag.

Eagly, A. H., and Chaiken, S., 1993. *The Psychology of Attitudes*. Fort Worth, TX: Harcourt Brace.

Fishbein, M., and Ajzen, I., 1975. *Belief, Attitude, Intention and Behavior*. Reading, MA: Addison-Wesley.

Fothergill, A., 1996. Gender, risk, and disaster. *International Journal of Mass Emergencies and Disasters*, **14**, 33–56. Available at www.ijmed.org.

Fothergill, A., and Peek, L., 2004. Poverty and disasters in the United States: a review of recent sociological findings. *Natural Hazards*, **32**, 89–110.

Fothergill, A., Maestes, E. G. M., and Darlington, J. D., 1999. Race, ethnicity and disasters in the United States: a review of the literature. *Disasters*, **23**, 156–173.

French, J. R. P., and Raven, B. H., 1959. The bases of social power. In Cartwright, D. (ed.), *Studies in Social Power*. Ann Arbor, MI: Institute for Social Research, pp. 150–167.

Garcia, E. M., 1989. Earthquake preparedness in California: a survey of Irvine residents. *Urban Resources*, **5**, 15–19.

Jackson, E. L., 1981. Response to earthquake hazard: the West Coast of North America. *Environment and Behavior*, **13**, 387–416.

Lang, J. T., and Hallman, W. K., 2005. Who does the public trust? The case of genetically modified food in the United States. *Risk Analysis*, **25**, 1241–1252.

Lindell, M. K., and Earle, T. C., 1983. How close is close enough: public perceptions of the risks of industrial facilities. *Risk Analysis*, **3**, 245–253.

Lindell, M. K., and Perry, R. W., 2000. Household adjustment to earthquake hazard: a review of research. *Environment and Behavior*, **32**, 590–630.

Lindell, M. K., and Perry, R. W., 2004. *Communicating Environmental Risk in Multiethnic Communities*. Thousand Oaks, CA: Sage.

Lindell, M. K., and Prater, C. S., 2000. Household adoption of seismic hazard adjustments: a comparison of residents in two states. *International Journal of Mass Emergencies and Disasters*, **18**, 317–338. Available at www.ijmed.org.

Lindell, M. K., and Prater, C. S., 2002. Risk area residents' perceptions and adoption of seismic hazard adjustments. *Journal of Applied Social Psychology*, **32**, 2377–2392.

Lindell, M. K., and Prater, C. S., 2010. Tsunami preparedness on the Oregon and Washington Coast: recommendations for research. *Natural Hazards Review*, **11**, 69–81.

Lindell, M. K., and Whitney, D. J., 2000. Correlates of seismic hazard adjustment adoption. *Risk Analysis*, **20**, 13–25.

Lindell, M. K., Lu, J. C., and Prater, C. S., 2005. Household decision making and evacuation in response to hurricane Lili. *Natural Hazards Review*, **6**, 171–179.

Lindell, M. K., Prater, C. S., and Peacock, W. G., 2007. Organizational communication and decision making in hurricane emergencies. *Natural Hazards Review*, **8**, 50–60.

Lindell, M. K., Arlikatti, S., and Prater, C. S., 2009. Why people do what they do to protect against earthquake risk: perceptions of hazard adjustment attributes. *Risk Analysis*, **29**, 1072–1088.

Meyer, P., 1988. Defining and measuring credibility of newspapers: developing an index. *Journalism Quarterly*, **65**(567–574), 588.

Mulilis, J.-P., and Duval, T. S., 1995. Negative threat appeals and earthquake preparedness: a person-relative-to-event PrE model of coping with threat. *Journal of Applied Social Psychology*, **25**, 1319–1339.

Mulilis, J. P., and Duval, T. S., 1997. The PrE model of coping with threat and tornado preparedness behavior: the moderating effects of felt responsibility. *Journal of Applied Social Psychology*, **27**, 1750–1766.

Palm, R., Hodgson, M., Blanchard, R. D., and Lyons, D., 1990. *Earthquake Insurance in California*. Boulder, CO: Westview Press.

Peacock, W. G., 2003. Hurricane mitigation status and factors influencing mitigation status among Florida's single-family homeowners. *Natural Hazards Review*, **4**, 149–158.

Peacock, W. G., Brody, S. D., and Highfield, W., 2005. Hurricane risk perceptions among Florida's single family homeowners. *Landscape and Urban Planning*, **73**, 120–135.

Perry, R. W., and Lindell, M. K., 2008. Volcanic risk perception and adjustment in a multi-hazard environment. *Journal of Volcanology and Geothermal Research*, **172**, 170–178.

Pijawka, K. D., and Mushkatel, A. H., 1991. Public opposition to the siting of the high-level nuclear waste repository: the importance of trust. *Policy Studies Review*, **10**, 180–194.

Preston, V., Taylor, S. M., and Hedge, D. C., 1983. Homeowner adaptation to flooding: an application of the general hazards coping theory. *Environment and Behavior*, **15**, 143–164.

Raven, B. H., 1965. Social influence and power. In Steiner, I. D., and Fishbein, M. (eds.), *Current Studies in Social Psychology*. New York: Holt, Rinehart and Winston, pp. 371–382.

Raven, B. H., 1993. The bases of power: origins and recent developments. *Journal of Social Issues*, **49**, 227–251.

Russell, L., Goltz, J. D., and Bourque, L. B., 1995. Preparedness and hazard mitigation actions before and after two earthquakes. *Environment and Behavior*, **27**, 744–770.

Slovic, P., Fischhoff, B., and Lichtenstein, S., 2001. Facts and fears: understanding perceived risk. In Slovic, P. (ed.), *The Perception of Risk*. London: Earthscan, pp. 137–153.

Sorensen, J. H., 2000. Hazard warning systems: review of 20 years of progress. *Natural Hazards Review*, **1**, 119–125.

Terpstra, T., and Gutteling, J., 2008. Households' perceived responsibilities in flood risk management in The Netherlands. *International Journal of Water Resources Development*, **24**, 555–565.

Weinstein, N. D., 1989. Effects of personal experience on self-protective behavior. *Psychological Bulletin*, **105**, 31–50.

Zahran, S., Weiler, S., Brody, S. D., Lindell, M. K., and Highfield, W. E., 2009. Modeling national flood insurance policy holding at the county scale in Florida, 1999–2005. *Ecological Economics*, **68**, 2627–2636.

Cross-references

Community Management of Hazards
Disaster Risk Management
Early Warning Systems
Earthquake Prediction and Forecasting
Emergency Management
Emergency Planning
Evacuation
Exposure to Natural Hazards
Frequency and Magnitude of Events
Hazard
Hazard and Risk Mapping
Integrated Emergency Management System
Prediction of Hazards
Recurrence Interval

Risk
Risk Assessment
Sociology of Disasters
Uncertainty
Warning Systems

ROCK AVALANCHE (STURZSTROM)

Reginald L. Hermanns
Geological Survey of Norway, Trondheim, Norway

Synonyms
Rock avalanche; Rock-fall avalanche; Rock-fall generated debris stream; Sturzstrom

Definition
Rock avalanche (sturzstrom) were defined by Hsü (1975) based on Heim's (1932) description of phenomena described with the German terms "Bergsturz," "Trümmerstrom," "Sturzstrom" as a stream of very rapidly moving debris derived from the disintegration of a fallen rock mass of very large size; the speed of a rock avalanche often exceeds 100 km/h, and its volume is commonly greater than 1×10^6 m^3.

Discussion
Rock avalanches are among the most hazardous *landslides* phenomena due to the speed, size, and run-out distance. Rock avalanches have destroyed entire villages and killed thousands. The run-out distance of a rock avalanche often exceeds several kilometers and the mobility becomes visible by the run up on opposite valley slopes, which is related to the volume of the initial rock failure (Scheidegger, 1973) and the morphology of the flow path (Nicoletti and Sorriso Valvo, 1991) but can also be influenced by the entrainment of saturated soil material or ice along the flow path (Hungr and Evans, 2004).

Systematic regional analyses of the temporal spatial distribution of rock avalanches have shown that rock avalanches do not distribute randomly but occur along lithologically, structurally preconditioned mountain fronts during climatic phases of higher run off or following deglaciation (e.g., Hermanns et al., 2006a, Blikra et al., 2006). Due to post failure slope deformation, rock avalanches often occur repeatedly at the same location with short recurrence intervals while neighboring slopes remain stable (Hermanns et al., 2006b). In addition, in tectonically active mountain regions, rock avalanches are often triggered by *earthquake*s and cause the formation of *landslide dams* adding to the *Disasters* (e.g., Petley et al., 2007).

Rock avalanche *Hazard* assessment involves characterizing potentially unstable slopes followed by *Slope Stability* analyses of the slope and run-out modeling of the rock avalanche. Due to the enormous energy released during the event, rock avalanches cannot be mitigated, and unstable rock slopes, which might form rock avalanches, can rarely be prevented from failure (e.g., artificial draining of slope). If a potential failure is imminent, complete *Evacuation* of the potentially affected run-out area is the only *Disaster Risk Management* measure.

Bibliography
Blikra, L. H., Longva, O., Braathen, A., Anda, E., Dehls, J. F., and Stalsberg, K., 2006. Rock slope failures in Norwegian fjord areas: examples, spatial distribution and temporal pattern. In Evans, S. G., Scaraascia Mugnozza, G., Strom, A., and Hermanns, R. L. (eds.), *Landslides from Massive Rock Slope Failures*. Dodrecht: Springer, pp. 475–496.

Dunning, S. A., Mitchell, W. A., Rosser, N. J., and Petley, D. N., 2007. The Hattian rock avalanche associated landslides triggered by the Kashmir earthquake of 8 October 2005. *Engineering Geology*, **93**, 130–144.

Heim, A., 1932. *Bergsturz und Menschenleben*. Zurich: Fretz & Wasmuth Verlag.

Hermanns, R. L., Niedermann, S., Villanueva Garcia, A., and Schellenberger, A., 2006a. Rock avalanching in the NW Argentine Andes as a result of complex interactions of lithologic, stuructural and topographic boundary conditions, climate change and active tectonics. In Evans, S. G., Scaraascia Mugnozza, G., Strom, A., and Hermanns, R. L. (eds.), *Landslides from Massive Rock Slope Failures*. Dodrecht: Springer, pp. 497–520.

Hermanns, R. L., Blikra, L. H., Naumann, M., Nilsen, B., Panthi, K. K., Stromeyer, D., and Longva, O., 2006b. Examples of multiple rock-slope collapses from Köfels (Ötz valley, Austria) and western Norway. *Engineering Geology*, **83**, 94–108.

Hsü, K. J., 1975. Catastrophic Debris Streams (Sturzstroms) generated by rockfalls. *Geological Society of America Bulletin*, **86**, 129–140.

Hungr, O., and Evans, S. G., 2004. Entrainment of debris in rock avalanches: An analysis of long run-out mechanism. *Geological Society of America Bulletin*, **116**, 1240–1252.

Nicoletti, P. G., and Sorriso-Valvo, M., 1991. Geomorphic controls of the shape and mobility of rock avalanches. *Geological Society of America Bulletin*, **103**, 1365–1373.

Scheidegger, A. E., 1973. On the prediction of reach and velocity of catastrophic landslides. *Rock Mechanics*, **5**, 231–236.

Cross-references
Disaster
Disaster Risk Management
Earthquake
Evacuation
Hazard
Landslide
Landslide Dam
Recurrence Interval
Slope Stability

ROCKFALL

Fausto Guzzetti
CNR – IRPI, Perugia, Italy

Synonyms
Rock fall

Definition

A rockfall is a type of fast-moving landslide common in mountain areas worldwide (Cruden and Varnes, 1996).

Discussion

An individual rockfall is a fragment of rock detached from the bedrock along new or preexisting discontinuities (e.g., bedding, joints, fractures, cleavage, foliation) by creeping, toppling, sliding, or falling, that falls along a cliff, proceeds down slope by bouncing, flying, tumbling, or rolling. A rockfall stops when it has lost enough energy through impacts or by friction (Fig. 1) (Dorren, 2003; Guzzetti and Reichenbach, 2010).

Creeping is limited to the pre-failure stage of a rockfall. Toppling consists of the outward or sideway rotation from the slope of a single or multiple adjacent blocks. Sliding occurs in the initial stage of a rockfall or at impact, with significant loss of energy due to high friction. Driven by gravity, free fall occurs along ballistic trajectories at high velocity. In nature, the motion of a boulder is most often by tumbling (a rapid sequence of short, low flying parabolas), and rolling is rarely observed. Irregularities in the shape of the block or the terrain facilitate impact and bouncing, and prevent or interrupt rolling. At impact, energy is lost and the direction of motion of the rockfall changes. Upon impact, a block can break into multiple fragments that proceed along separate trajectories. In forests, collisions against trees can deflect or stop rocks.

For primary rockfalls, the fall immediately follows detachment. For secondary failures, the time and trigger of the fall are different from those of the detachment. A rockfall can involve single or multiple blocks. When multiple blocks are involved in a failure, there is little or no interaction among the individual fragments. Rockfalls travel at speeds ranging from a few to tens of meters per second, and range in size from small cobbles to large boulders hundreds of cubic meters in size. The statistics of rock fall volume obey the behavior of a negative power law.

Natural triggers of rockfalls include earthquakes, freeze–thaw cycles of water, melting of snow or permafrost, temperature changes, intense or prolonged rainfall, stress relief following deglaciation, volcanic activity, and root penetration and wedging. Human-induced causes of rockfalls include undercutting of rock slopes, mining activities, pipe leakage, inefficient drainage, and vibrations caused by excavations, blasting, or traffic. Due to their high mobility, rockfalls are a particularly destructive type of failure, and in several areas, they represent the primary cause of landslide fatalities. A good predictor of rockfall occurrence in an area is the evidence of previous rockfalls.

Bibliography

Cruden, D. M., and Varnes, D. J., 1996. Landslide types and processes. In Turner, A. K., and Schuster, R. L. (eds.), *Landslides, Investigation and Mitigation. Special Report 247*. Washington DC: Transportation Research Board, pp. 36–75.

Dorren, L. K. A., 2003. A review of rockfall mechanics and modelling approaches. *Progress in Physical Geography*, **27**, 69–87.

Guzzetti, F., and Reichenbach, P., 2010. Rockfalls and their hazard. In Stoffel, M., Bollschweiler, M., Butler, D. R., and Luckman, B. H. (eds.), *Tree Rings and Natural Hazards: A Stare-of-the-Art*. Advances in Global Change Research, 41: DOI 10.1007/978-90-481-8736-2_12, Dordrecht: Springer, pp. 129–137.

Rockfall, Figure 1 Rainfall induced, meter-scale rockfall blocks blocking State Road SS3 in Umbria, Italy, in the winter 2006.

Cross-references

Creep
Debris Avalanche
Debris Flow
Deep-Seated Gravitational Slope Deformations
Landslide (Mass Movement)
Landslide Impacts
Landslide Types
Mass Movement
Mudflow
Rock Avalanche
Slide and Slump
Usoi Landslide and Lake Sares
Vaiont Landslide, Italy

ROGUE WAVE

Norm Catto
Memorial University, St. John's, NL, Canada

Synonyms

Anomalous wave; Extreme wave; Freak wave; Killer wave

Definition

A relatively large wave, often not directly associated with immediate weather conditions, which occurs as an isolated, anomalous event.

Discussion

Rogue waves are distinguished both by their height and the relatively unexpected nature of their occurrence. Heights in excess of 30 m have been recorded offshore. At sea, a wave is considered rogue if it has a minimum height of twice the mean significant wave height, defined as the mean of the largest 33% of wave heights recorded for a specific area. In addition, a rogue wave is a single large wave in isolation, not accompanied by a prolonged period of similar waves or wave train. Rogue waves thus can occur during relatively calm conditions, and are not always directly associated with storm activity. They can be generated in both deep, open ocean, or in coastal waters. Rogue waves, locally in excess of 10 m in height, have struck confined coastal embayments directly facing the open ocean, causing damage and loss of life.

Although well known by mariners and in coastal communities for centuries, scientific investigation of rogue waves began relatively recently. Postulated causes for individual events vary, indicating the anomalous nature of the phenomenon. Most explanations involve the focusing and combination of individual smaller waves, generated by normal wind activity or marine swell, into large features. Focusing may be triggered by diffraction of smaller waves from adjacent coastlines, possibly accounting for coastal events; interaction between waves traveling in one direction and an opposing current; or nonlinear addition of smaller clapotis waves in offshore areas, responsible for deep-water events. Rogue waves formed from combinations of storm-induced wind waves may develop some days after the triggering event.

Rogue waves pose hazards for vessels at sea, fixed offshore infrastructure (such as oil-drilling platforms), and coastal communities located in confined embayments. The unpredictable, anomalous nature of the phenomenon, and the lack of detailed explanations linking specific meteorological conditions to genesis, hinders efforts to reduce the hazard posed by rogue waves.

Bibliography

Cartwright, J. E., and Nakamura, H., 2009. What kind of a wave is Hokusai's great wave off Kanagawa? *Notes and Records of the Royal Society*, **63**, 119–135.
Gemmrich, J., and Garrett, C., 2008. Unexpected Waves. *Journal of Physical Oceanography*, **38**, 2330–2336.
Kharif, C., Pelanovsky, E., and Slunayev, A., 2009. *Rogue Waves in the Ocean*. Berlin: Springer, p. 216.

Cross-references

Breakwater
Coastal Erosion
Coastal Zone, Risk Management
Critical Infrastructure
Displacement Wave, Landslide Triggered
Fetch
Impact Tsunami
Levee
Marine Hazards
Pacific Tsunami Warning and Mitigation System (PTWS)
Sea Level Change
Seiche
Storm Surges
Tsunami
Vaiont Landslide, Italy
Waterspout

ROTATIONAL SEISMOLOGY

William H. K. Lee
U.S. Geological Survey, Menlo Park, CA, USA

Definition

Rotational seismology is an emerging study of all aspects of rotational motions induced by earthquakes, explosions, and ambient vibrations. It is of interest to several disciplines, including seismology, earthquake engineering, geodesy, and earth-based detection of Einstein's gravitation waves.

Rotational effects of seismic waves, together with rotations caused by soil–structure interaction, have been observed for centuries (e.g., rotated chimneys, monuments, and tombstones). Figure 1a shows the rotated monument to George Inglis observed after the 1897 Great Shillong earthquake. This monument had the form of an obelisk rising over 19 metres high from a 4 metre base.

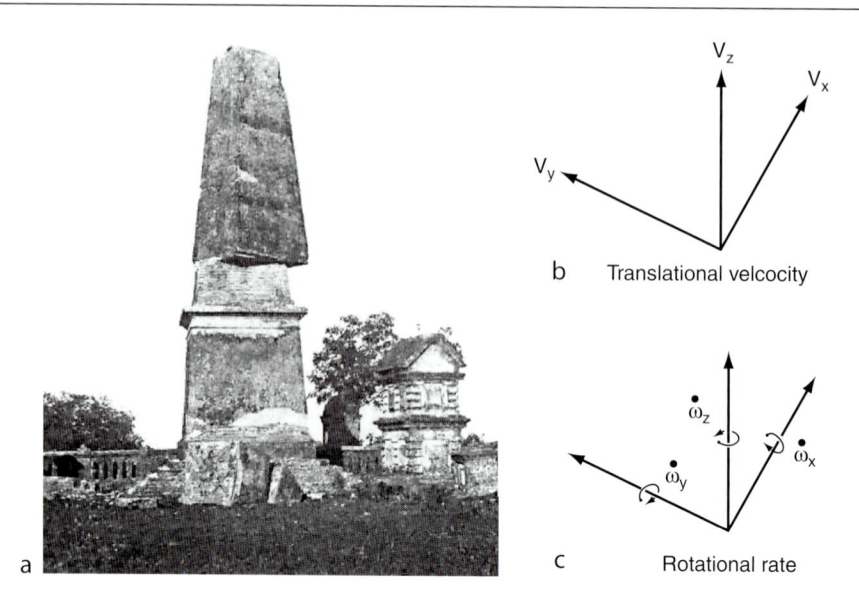

Rotational Seismology, Figure 1 (a) Rotation of the monument to George Inglis (erected in 1850 at Chatak, India) observed after the 1897 Great Shillong earthquake. (b) Coordinate system for translational velocity. (c) Coordinate system for rotational rate.

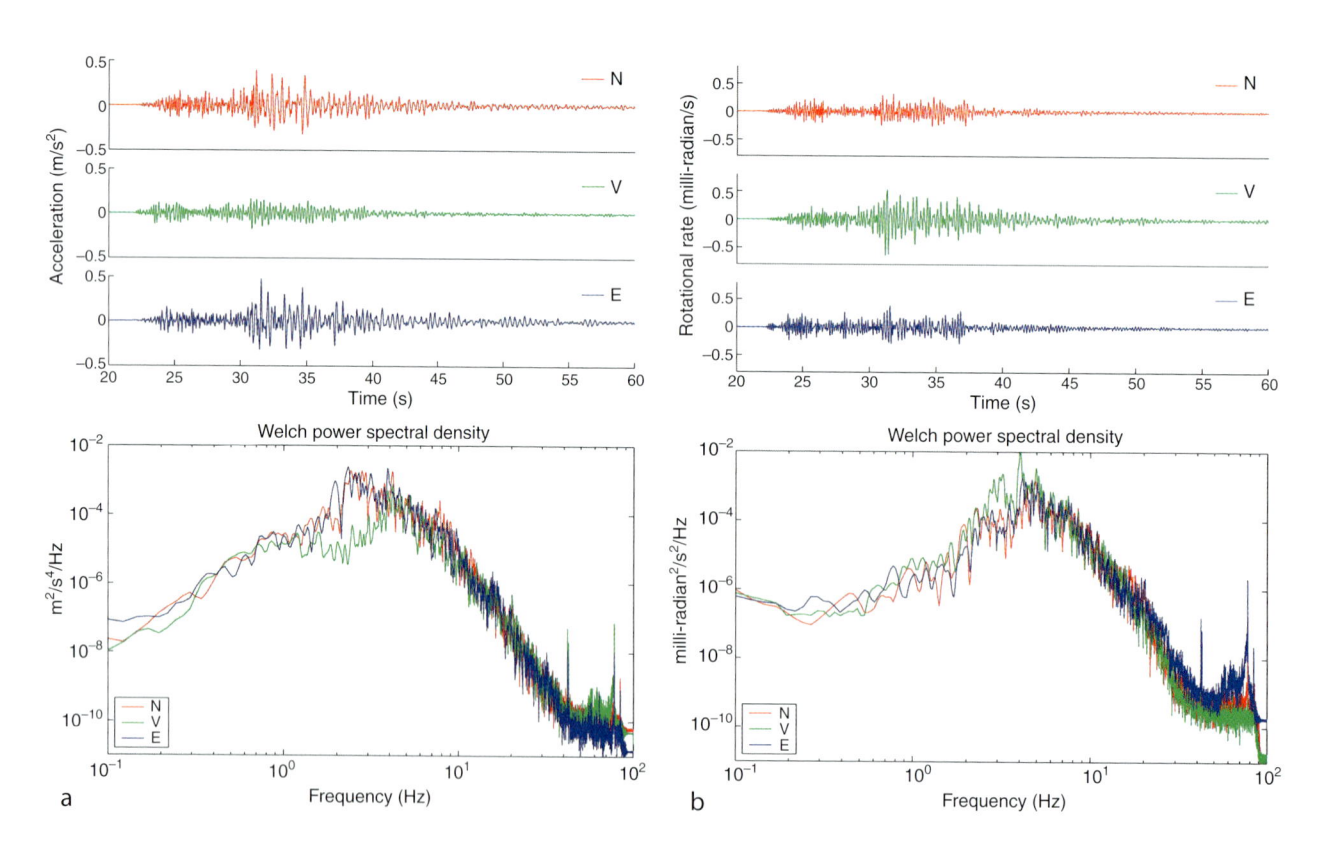

Rotational Seismology, Figure 2 (a) Recorded translational acceleration (*top*) and its spectra (*bottom*) from an M_W 5.1 earthquake in Taiwan on July 23, 2007. (b) Recorded rotational rate (*top*) and the corresponding spectra (*bottom*) from the same earthquake. E, N, and Z refer to orthogonal sensor orientations.

During the earthquake, the top part broke off and the remnant of some 6 metres rotated about 15° relative to the base. The study of rotational seismology began only recently when sensitive rotational sensors became available due to advances in aeronautical and astronomical instrumentations.

Rotational ground motions

The general motion of a particle or a small volume in a solid body can be divided into three parts: translation along and rotation about the orthogonal X-, Y-, and Z-axes, and strain. Figure 1b shows the axes in a Cartesian coordinate system for *translational velocity* measured by the traditional seismometers, and Figure 1c shows the corresponding axes of *rotation rate* measured by rotational sensors.

Rotational ground motions can be measured directly by gyroscopic sensors, or inferred indirectly from an array of translational sensors. In recent years, rotational motions from small local earthquakes to large teleseisms were successfully recorded by sensitive rotational sensors in several countries. Observations in Japan and Taiwan show that the amplitudes of rotations can be one to two orders of magnitude greater than expected from the classical elasticity theory. Figure 2a and b shows, respectively, the amplitudes and spectra of translational acceleration and of rotational rate recorded 51 km away from an M_W 5.1 earthquake in Taiwan.

Relation to traditional seismology

Seismology is traditionally based on classical elasticity theory under infinitesimal strain. Rotation is then defined as the curl of the displacement field, and the rotational components of motion are contained in the S-waves (Lee et al., 2009b). However, modern elasticity theories, for example, that of the Cosserat brothers incorporates local rotation and couple stress (a torque per unit area). Applications of modern elasticity theories to rotational seismology began only recently.

Seismic networks traditionally deploy only translational seismometers. To improve our understanding of damaging earthquakes, rotational sensors must also be deployed, particularly in the *near field* where rotational motions are expected to be the largest. A first step towards such extensive seismic instrumentation has been taken in southwestern Taiwan, where two 32-element arrays of both translational and rotational instruments with station spacing of tens of meters have been deployed.

Bibliography

Lee, W. H. K., Celebi, M., Igel, H., and Todorovska, M. I. (eds.), 2009a. Special issue on Rotational Seismology and Engineering Applications. *Bulletin of the Seismological Society of America*, **99**(2B), 945–1485.

Lee, W. H. K., Igel, H., and Trifunac, M. D., 2009b. Recent advances in rotational seismology. *Seismological Research Letters*, **75**, 486–497.

Teisseyre, R., Takeo, M., and Majewski, E. (eds.), 2006. *Earthquake Source Asymmetry, Structural Media and Rotation Effects*. Heidelberg: Springer.

Cross-references

Body Wave
Secondary Wave (S Wave)
Seismograph/Seismometer
Tiltmeters

SACKUNG

Michael J. Bovis
University of British Columbia, Vancouver, BC, Canada

Synonyms

Deep-seated gravitational deformation; Gravitational spreading; Slope sagging

Definition

Sackung is a German term denoting slow, deep-seated gravitational deformation of slopes.

Discussion

Sackung-type movements have typical surface expression as uphill-facing (antislope) scarps, tension cracks, grabens, and anomalous ridge-top depressions running roughly parallel to the contours in steep mountain topography. Individual scarps, grabens, and cracks (collectively referred to here as "linears") have a typical relief of 1–10 m and may be traced over distances of 100 m to more than 3 km. Linears may be arranged subparallel to one another, or en echelon, and comprise slope-movement complexes covering areas of 1–10 km^2 that are clearly visible on air photographs (Figure 1).

Material properties and movement mechanisms

Sackung features were first recognized in the European Alps, but have since been recognized in many other mountain areas. Hürlimann et al. (2006) provide a comprehensive citation of papers from the past 40 years. Linears occur over a wide range of rock types, ranging from ductile sedimentary rocks to hard jointed intrusive and metamorphic rocks, but often are best displayed and preserved in hard rocks with closely spaced discontinuities dipping steeply into a slope. This structure favors flexural toppling, likely an important mechanism for development of many antislope scarps. The feasibility of flexural toppling may be assessed kinematically for joint poles, α, falling outside a friction circle drawn ϕ degrees flatter than the slope angle β, where ϕ is the friction angle along the through-going discontinuities, i.e., $\alpha < (\beta - \phi)$ (Bovis and Evans, 1996, Nicol et al., 2002).

Numerical modeling of large-scale sackung movements is now fairly common, using both distinctive-element and continuum methods (Nicol et al., 2002; Eberhardt, 2008). Modeled ground-motion vectors agree fairly well with long-term measured ground movements. Such 2-D modeling also shows that sackung movements are sensitive to water pressure fluctuations and to slope steepening by either Neoglacial or fluvial undercutting of slopes (Bovis and Stewart, 1998; Ambrosi and Crosta, 2006).

Hazard implications

Sackung linears have been reported adjacent to many *Rock Avalanche* detachment zones, which has provoked discussion of a probable connection between slow, ductile sackung-type movements and sudden rock avalanche detachments. Some linears probably developed in response to an abrupt change in longitudinal stress as the avalanche detached, but in many cases, the morphology and degree of weathering of linears indicate a far greater age than the rock avalanche. This implies that sackung linears are an expression of premonitory movements in stressed slopes, some of which fail catastrophically, though Hewitt et al. (2008) caution that sackung movements are not a necessary precursor for all catastrophic detachments. Where minimal erosion of linears has occurred since their development, cosmogenic ^{10}Be dating may be used to establish very long-term stability for some sackung slopes, which therefore constitute a minimal hazard at present (Hippolyte et al., 2009).

Sackung, Figure 1 Vertical air photograph of a sackung slope, Affliction Creek, British Columbia, Canada. Note glacial undercutting of the slope. Tension cracks and grabens shown by dashed lines. Antislope scarps shown by dotted lines.

Some sackung linears have been mapped as neotectonic faults, but detailed field investigations supplemented by numerical modeling show that many of these features represent sackung movements controlled by much older tectonic features (Ambrosi and Crosta, 2006).

Bibliography

Ambrosi, C., and Crosta, G. B., 2006. Large sackung along major tectonic features in the Central Italian Alps. *Engineering Geology*, **83**, 183–200.
Bovis, M. J., and Evans, S. G., 1996. Extensive deformations of rock slopes in southern Coast Mountains, southwest British Columbia, Canada. *Engineering Geology*, **44**, 163–182.
Bovis, M. J., and Stewart, T. W., 1998. Long-term deformation of a glacially undercut rock slope, southwest British Columbia. In *Proceedings of the 8th International Congress on International Association of Engineering Geology*, Balkema, Rotterdam, pp. 1267–1276.
Eberhardt, E., 2008. The role of advanced numerical methods and geotechnical field measurements in understanding complex deep-seated rock slope failure mechanisms. *Canadian Geotechnical Journal*, **45**, 484–510.
Hewitt, K., Clague, J. J., and Orwin, J. F., 2008. Legacies of catastrophic rock slope failures in mountain landscapes. *Earth Science Reviews*, **87**, 1–38.
Hippolyte, J.-C., Bourlès, D., Braucher, R., Carcaillet, J., Léanni, L., Arnold, M., and Aumaitre, G., 2009. Cosmogenic ^{10}Be dating of a sackung and its faulted rock glaciers in the Alps of Savoy (France). *Geomorphology*, **108**, 312–320.
Hürlimann, M., Ledesma, A., Corominas, J., and Prat, P. C., 2006. The deep-seated slope deformation at Encampadana, Andorra: representation of morphologic features by numerical modelling. *Engineering Geology*, **83**, 343–357.
Nicol, S. L., Hungr, O., and Evans, S. G., 2002. Large-scale brittle and ductile toppling of rockslopes. *Canadian Geotechnical Journal*, **39**, 773–788.

Cross-references

Deep-Seated Gravitational Slope Deformations
Geohazards
Landslide
Lateral Spreading
Neotectonics
Rock Avalanche
Slope Stability

SAFFIR–SIMPSON HURRICANE INTENSITY SCALE

Ilan Kelman
Center for International Climate and Environmental Research – Oslo (CICERO), Oslo, Norway

Synonyms
Hurricane classification scale

Definition
The Saffir–Simpson Hurricane Scale rates the danger and damage potential from hurricanes.

Discussion
The Saffir–Simpson Hurricane Intensity Scale rates the danger and damage potential from tropical cyclones for the Atlantic, Caribbean, and Northeast Pacific. In those regions, tropical cyclones are called hurricanes. Historically, the scale has reported wind speed, storm surge depth, and barometric pressure. Revisions in 2009 focused the renamed Saffir–Simpson Hurricane Wind Scale on wind speed only, reporting the maximum sustained surface wind speed as the peak one-minute wind at a height of ten meters above ground. Hurricane categories, focusing on infrastructure in the USA, are:

- Category 1: 119–153 km/h. Damage to poorly constructed or poorly anchored buildings and other infrastructure such as signs. Loose items can become lethal projectiles.
- Category 2: 154–177 km/h. Additionally, damage to roofs, doors, and windows.
- Category 3: 178–209 km/h. Additionally, wall failures will occur and trees will be broken or uprooted.
- Category 4: 210–249 km/h. Complete destruction of some buildings along with other infrastructure such as signs and lines for electricity and communications.

Most trees are blown down. Extensive windborne debris contributes to damage, deaths, and injuries.
- Category 5: over 249 km/h. Additionally, many buildings and most other infrastructure suffer complete destruction.

The shorthand "Cat." is frequently used for "Category". Sometimes, the scale includes Tropical Storms defined as having winds of 63–118 km/h. Some media commentators suggest adding Category 6, cutting Category 5 off within the range of 280–290 km/h. This change is not necessarily helpful because Category 5 already suggests widespread destruction, so a further category adds little.

This Saffir–Simpson Scale was invented by an American consulting engineer, Herbert Saffir, in 1969, focusing on wind speed. Over the following years, with the then director of the USA's National Hurricane Center, Bob Simpson, the scale was expanded to include storm surge height and barometric pressure.

Pressure and storm surge have now been phased out of the scale due to concerns about the accuracy and usefulness of the information. In particular, storm surge height on land depends on factors such as near-shore bathymetry and topography. As well, in places with large tidal variations, a storm surge peaking at low tide can be the same as a high tide whereas the same storm surge at high tide can yield devastating flooding. Similarly, the Saffir–Simpson scale does not fully reflect the possibilities for hurricane-related inland flooding. During hurricanes, flooding from rainfall or storm surge has historically caused the majority of fatalities, rather than wind speed, although the statistics have been changing as messages about hurricane-related flooding have become more prominent. The scale should therefore be interpreted as an indicator of wind-related destruction only, not of all dangers from a hurricane.

Cross-references

Dvorak Classification of Hurricanes
Hurricane (Cyclone, Typhoon)
Intensity Scales

SAN ANDREAS FAULT

William A. Bryant
California Geological Survey, Sacramento, CA, USA

Definition

The 1,100-km-long San Andreas Fault is the principal element of the San Andreas Fault System, a network of faults with dextral strike-slip displacement that collectively accommodates the majority of relative northwest-southeast transform motion between the North American and Pacific crustal plates. This major transform *Fault* extends along western California from the Salton Trough near the California-Mexico border northwest to its complex junction with the Mendocino Fault Zone near Punta Gorda.

Discussion

Andrew Lawson first mapped the fault in northern California in 1893. Lawson identified the fault as the San Andreas rift, based on its expression in the San Andreas Valley in San Mateo County. Schuyler (1896–1897) described parts of the fault in southern California and referred to the fault not as the San Andreas but as the "great earthquake crack," referring to surface displacement associated with the Mw ~7.9 1857 Fort Tejon earthquake. The fault first gained international scientific attention following the Mw 7.8 1906 San Francisco earthquake (Lawson, 1908). Lawson's 1908 report summarizing the investigation of the *Earthquake* contained the first integrated description of the San Andreas Fault and formed the underlying basis for modern studies of paleoseismology and earthquake geology (Prentice, 1999).

Noble (1926) was the first to suggest a large amount of dextral slip (38 km) along the San Andreas Fault. Stanley (1987) documented 325–330 km of post-late Oligocene dextral slip. The fastest, generally accepted Holocene dextral slip rate of 33.9 ± 2.9 mm/yr for the San Andreas Fault has been documented along the Cholame-Carrizo Section, which lies in the central portion of the fault zone (Sieh and Jahns, 1984). Average recurrence intervals of large magnitude earthquakes, in general, range from a little more than 100 years to as much as 450 years (Bryant and Lundberg, 2002).

Fault creep as high as 32 mm/yr characterizes the San Andreas Fault along the 132-km-long Creeping Section in central California. The northwestern and southeastern ends of the Creeping Section are transitional to the surface-rupture termination points of the 1906 earthquake to the northwest and 1857 earthquake to the southeast. Creep up to 4 mm/yr also has been measured near the southern end of the fault east of the Salton Sea.

Bibliography

Bryant, W. A., and Lundberg, M. M., (compilers), 2002. San Andreas fault zone (Fault numbers 1a-1j), California. In *Quaternary Fault and Fold Database of the United States*. U.S. Geological Survey Internet database, (http://earthquakes.usgs.gov/regional/qfaults).

Lawson, A. C., (chairman), 1908. The California earthquake of April 18, 1906, Report of the State Earthquake Investigation Commission. Washington, DC: Carnegie Institution of Washington Publication 87.

Noble, L. F., 1926. The San Andreas rift and some other active faults in the desert region of southeastern California. Washington, DC: Carnegie Institution of Washington Year Book 25, pp. 415–428.

Prentice, C. S., 1999. San Andreas fault – 1906 earthquake and subsequent evolution of ideas. In Sloan, D., Moores, E. M., and Stout, D. (eds.), *Classic Cordilleran Concepts: A View From California. Geological Society of America Special Paper 338*, pp. 79–85.

Schuyler, J. D., 1896–1897. Reservoirs for irrigation. *U.S. Geological Survey, 18th Annual Report, Part IV*, pp. 711–712.

Sieh, K., and Jahns, R., 1984. Holocene activity of the San Andreas fault at Wallace Creek, California. *Geological Society of America Bulletin*, **95**, 883–896.

Stanley, R. G., 1987. New estimates of displacement along the San Andreas fault in central California based on paleobathymetry and paleogeography. *Geology*, **15**, 171–174.

Cross-references

Creep
Earthquake
Elastic Rebound Theory
Fault
Geohazard
Modified Mercalli (MM) Scale
Neotectonics
North Anatolian Fault
Paleoseismology
Plate Tectonics
Tectonic and Tectono Seismic Hazard

CASE STUDY

SANTORINI, ERUPTION

Yuri Gorokhovich
Lehman College, City University of New York, Bronx, NY, USA

Introduction

The volcanic island of Santorini is both an ancient and a modern legend. Its history is linked with ancient maritime life of the Mediterranean Sea, the Minoan civilization on Crete and Greek mythology. But the most dramatic page of Santorini's history is directly tied to the volcanic explosion of the ancient island of Thera between 1627 and 1600 B.C.E. (Friedrich et al., 2006) that changed its entire shape after one of the strongest explosions that mankind has ever witnessed. Modern Santorini is a remnant of that explosion, a mighty monument to the grandiose forces of nature.

The formation of the Santorini volcanic complex we see today began after the explosion of Thera which coincided with social and political changes on the nearby island of Crete during the Late Bronze Age. Many historians have long believed that, since Santorini is less than 100 km from Crete, the explosion of Thera triggered the end of the Minoan civilization in the Late Bronze Age (Marinatos, 1939). Thus, the eruption of Thera is also known in human history as the Minoan eruption. Although widely disputed, this point of view still prevails in media documentaries and popular history. Scientific data from models and field observations are ambiguous and do not provide an unequivocal answer.

Many references to the Santorini complex can be found in ancient literature. Its two islands, Thera and Therasia were mentioned in works of Strabo (66 B.C.E.–46 A.D.), Herodotus (c. 484–425 B.C.E.), Pliny (24–79 A.D.), Seneca and many other ancient writers, geographers, historians, and philosophers (Fouquè, 1879). The Pythian Ode of Pindar (522–441 B.C.E.) mentions Thera in verse 10 and Calliste (the ancient name of Santorini) in verse 258 (Friedrich, 2000). During the fourteenth century Venetian period in the Mediterranean region, merchants from Venice built a chapel of Santa Erene (Saint Irene) on the main island of Thera. Since that time, the island complex has come to be known as Santorini.

According to ancient texts, Santorini hosted Argonauts, Phoenicians, Lacedemonians, and Dorians. In Ptolemy's "Dialogs," it was suggested as a potential location of the mysterious Atlantis. It minted its own coins beginning in 635 B.C.E. Due to its central geographic location, Santorini played an important role in Mediterranean maritime trading. It thrived through the Roman period and the early Christianity era, and eventually was owned by the Venetian family Crispo until 1537 and then by the Ottoman Empire until the Greek War of Independence in 1821, when it became a Greek territory.

An abundance of rich volcanic soil and volcanic tuff (which was used as construction material) provided jobs and income for Santorini's inhabitants. Hydraulic (i.e., water resistant) cement, a strong mixture of volcanic ash and lime, was manufactured from Santorini's highly siliceous volcanic ash, called pozzolana (from Pozzuoli, a village near Naples where ancient Romans started mixing volcanic ash and lime). The island's rich volcanic soil helps Santorini's farming sector to produce wines, grapes, vegetables, and fruits for export. The island is also a prime destination for tourists, who are lured by the dramatic geologic features formed by volcanic eruptions in the 1950s, the discovery of the ancient city of Akrotiri and its splendid frescoes in the 1960s, and its mysterious link to the story of Atlantis.

Geographic settings

The Theran (Minoan) eruption in seventeenth century B.C.E. left four separate islands: Thera, Therasia, Aspronisi, and Paleo Kameni (Figure 1). The fifth volcanic island known as Neo Kameni appeared in the eighteenth century CE. Thera, Therasia, and Aspronisi are remnants of the rim of the former volcanic caldera. Paleo and Neo Kameni are small volcanic islands that occupy the center of the caldera. The largest island Thera forms the eastern section of Santorini. Its highest elevation is Mt. Profitis Elias (708 m a.s.l.). The western section of Santorini is the island of Therasia (or *little Thera*, gr.), a volcanic rock complex with elevation of 295 m. a.s.l. Aspronisi (*white island*, gr.) forms the southwestern section of the Santorini complex.

The highest elevations of the caldera walls reach 300 m a.s.l. where the largest towns of the island (Oia and Fira) stretch along the rim and provide picturesque views of

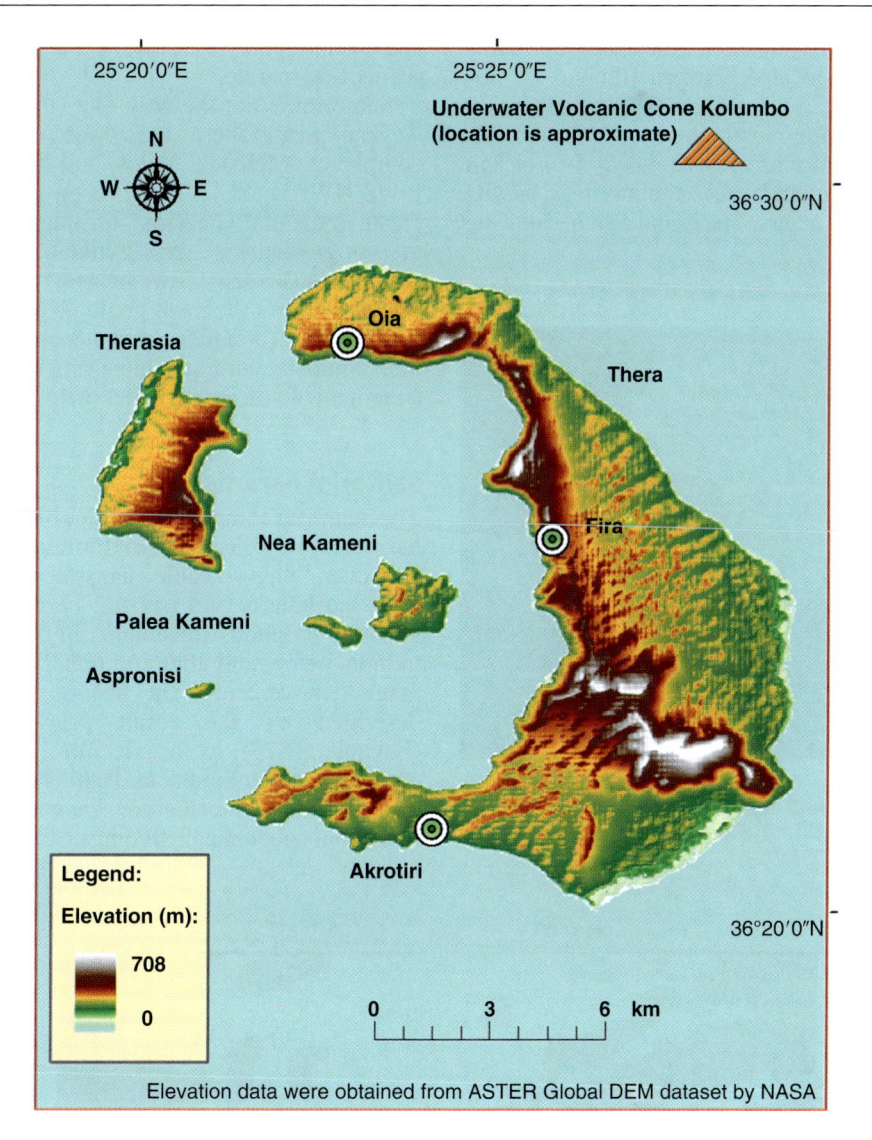

Santorini, Eruption, Figure 1 Santorini volcanic complex.

the surrounding sea and volcanic islands of Nea Kameni and Palea Kameni. Many houses, shops, hotels, restaurants, and even swimming pools are perched on the inner slopes of caldera to provide even more breathtaking views to tourists.

In addition to the main Santorini volcanic complex there is also another volcanic feature, an underwater volcanic cone called Kolumbo, described as a "parasitic cone on a flank of a mass of Santorini" by Fouquè (1879). It is located 7 km northeast from Santorini (Figure 1) at the depth of 500 m. This volcanic complex was studied in 2006 by Robert Ballard and Haraldur Sigurdsson (Gramling, 2007) who used underwater remotely operated vehicles (ROV). They found that this small cone is active with hydrothermal springs and fumaroles (Figure 2). The last explosion of Kolumbo occurred in 1950.

The soils of Santorini were derived from the parental volcanic rocks and according to earlier studies were classified as andisols (Misopolinos et al., 1994). The latest study by Moustakas and Georgoulias (2005) define Santorini soil type as Eentisols due to the very slow rate of weathering, restricted leaching, the longevity of the dry season, and high pH. Rich in nutrients, Santorini soils support approximately 550 taxa of ferns and flowering plants (Biel, 2005).

Geological settings of Santorini

The geological settings of Santorini are tightly coupled with the plate tectonics structure of the Mediterranean region. The northeastern motion of the African plate toward Greece created a convergent ocean-continental margin with corresponding volcanic arc complex, deep-sea trenches and associated with them volcanism and earthquakes with epicenters located between 100 and 170 km below the land surface. This margin where Santorini is located is known as the South Aegean Volcanic Arc (Figure 3). Modern Santorini consists mostly of volcanic rocks such as dacites, andesites, ignimbrites, basalts, tephra, and tuffs among others. The island also has igneous and metamorphic rocks, mostly granites, schists, quartzites, scarns, and marble.

Santorini is not the only one volcanic complex in the Aegean Sea. There are three other large volcanic complexes: Methana, Milos, and Nisyros (Figure 3, see from NW to SE). These complexes are active and create potential hazards for the region. According to Vougioukalakis and Fytikos (2005) all volcanic complexes present low hazard risk except Nisyros. Figure 4 shows the main geologic elements of the Santorini volcanic complex: magma chambers, pre-volcanic rock complex, current caldera rim, and the fractured zone (fragmented volcanics) under the Nea Kameni.

Santorini hazards

The Santorini complex is a source of three main geological hazards: (1) Volcanic; (2) Earthquakes; (3) Tsunamis. However, the volcanic hazard of Santorini induces both, earthquake and tsunami hazards. During previous volcanic eruptions inhabitants of Santorini felt multiple tremors before and after eruptions, most likely due to the magma movements. The historic tsunami effects were described by the Jesuit priest Francois Richard (Fouquè, 1879) in his recollection of the volcanic explosion of Kolumbo in 1680. Historical hazards and disasters on Santorini are known from both various texts and geological findings. Contemporary hazards

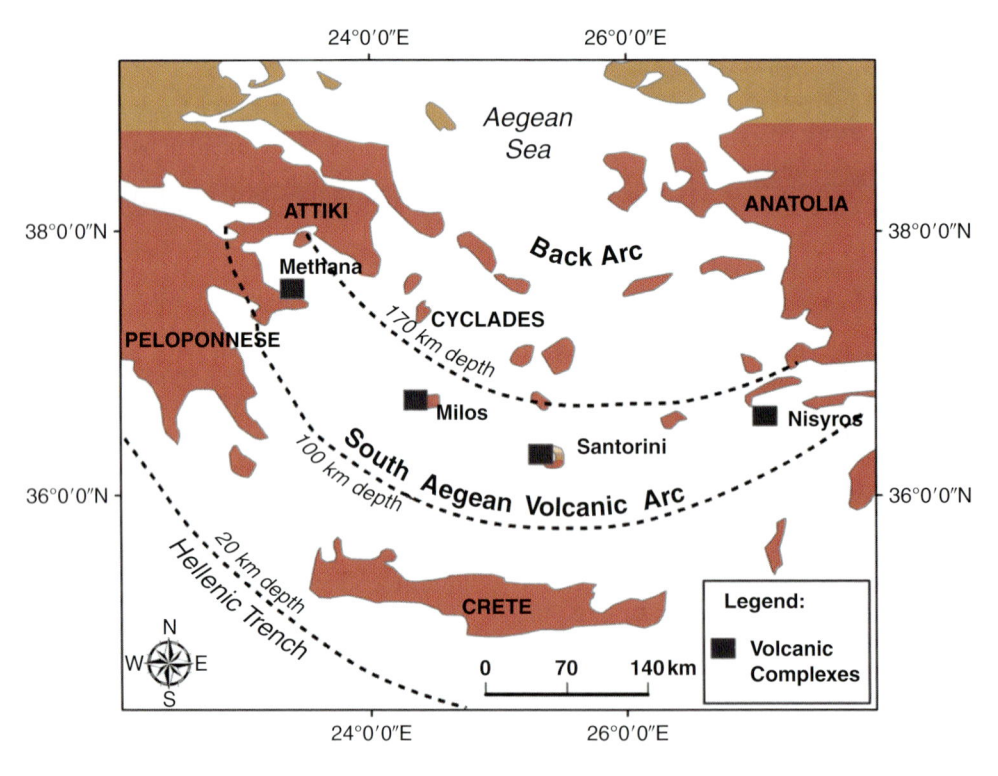

Santorini, Eruption, Figure 2 Hydrothermal activity near Kolumbo (Courtesy of the Institute for the Study and Monitoring of Santorini Volcano (ISMOSAV)).

Santorini, Eruption, Figure 3 Elements of the South Agean Active Arc (Adapted from Pe-Piper and Piper, 2005). *Dashed line* with corresponding numbers show depths of earthquake epicenters.

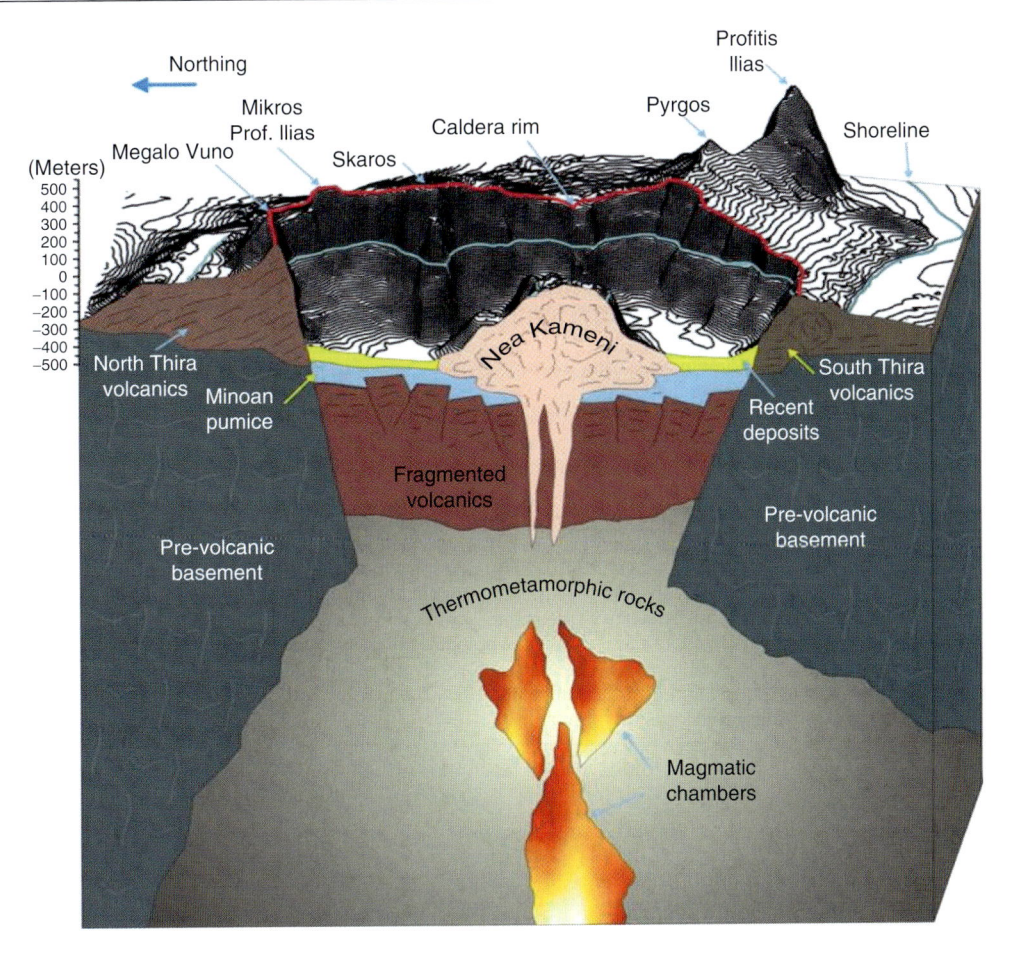

Santorini, Eruption, Figure 4 Schematic diagram showing current structure of Santorini volcanic complex (Courtesy of the ISMOSAV, 2008).

are documented by observations, monitoring, and measurements.

Volcanic hazards caused many disastrous events (Table 1) on Santorini in the past. Combined with recent geologic data these events help scientists to evaluate possible occurrences of similar eruptions. Pyle (1997) suggests periodicity of several such eruptions per thousand years. According to McCoy and Heiken (2000a) a recurrence interval for explosive eruption on Santorini similar to the Minoan is 20,000 years with smaller explosive eruptions every 5,000 years. Eruptions in the twentieth century were recorded on photos (Figure 5) and documented by scientific measurements.

Volcanic eruption of Thera (aka Minoan or Late Bronze Age eruption)

Among many eruptions during the history of Santorini the most outstanding by force and its place in human history was the eruption of Thera during the Late Bronze Age. That period in the Mediterranean was characterized also by intensive earthquake activity (Monaco and Tortorici, 2004; Nur and Cline, 2000). The latest C^{14} analysis of organic material from an olive branch defines the date of the Minoan eruption as 1627–1600 B.C.E. with 95.4% probability (Friedrich et al., 2006). Data from Egyptian medical papyrus collection provides an earlier and more exact date of August 1603 B.C.E. (Trevisanto, 2007).

The multi-hazardous character of Thera's eruption manifested itself in geological and archaeological findings. While recovering remnants of the ancient city Akrotiri that survived Thera's eruption, archaeologists found a damaged staircase with visible effects of the compressive force (Figure 6) from the earthquake (seismic) wave.

Many walls of the dwellings in Akrotiri collapsed due to the earthquake effects and the excessive weight of the volcanic ash on the roofs.

Despite the abundance of discussions and scientific studies, the destructive effect of the volcanic ash from the Thera eruption on the demise of Minoan civilization on nearby Crete is not supported by field data (Vitaliano and Vitaliano, 1974). Earlier, Ninkovitch and Heezen

Santorini, Eruption, Table 1 Historical eruptions on Santorini (Summarized from Friedrich, 2000)

Volcanic event	Description	Source
Approx. 1,640 B.C.E.	Original round volcanic island (Stroghyle) experienced phreatic eruption that created the rim of the modern Santorini	Geologic interpretation
Approx. 197 B.C.E.	New island Hiera was formed as the result of the eruption	Strabo (Geography, 1.3.16)
Approx. 46 A.D.	Island of Theia appeared	Seneca, Cassiodorus (468–562 A.D.)
Approx. 726 A.D.	New island appeared next to Hiera. Ash from this volcanic eruption reached the Straits of Dardanelles, coast of Asia Minor, and Macedonia	Nicephoros (758–828 A.D.) and Theophanes (752–818 A.D.)
Approx. 1457–1458	New island appeared between Thirasia and Thera	Buondelmonte (1465–1466), Athanasius Kircher (1665)
1570–1573	New island Mikra Kameni was formed 4 km northeast of Paleo Kameni	Athanasius Kircher (on account of the Jesuit priest Ricardus)
Sept. 27, 1650	Eruption of Kolumbo, an underwater volcano, northeast from Thera. After 4 months Kolumbo disappeared under the water. "Fifty persons died from the harmful effects of the volcanic gas and more than a thousand animals died that way" (Fouquè, 1879)	Jesuit priests Francois Richard (1657) and Father Goree (1712)
1707–1711	New island (Neo Kameni) appeared west of Mikra Kameni. Sulfurous springs appeared	Father Goree (1712), Barskii (1745)
1866–1870	Neo Kameni and Micro Kameni were joined by newly formed lava that filled out the space between them. The process accompanied by ash eruption and explosions. Gas and fumaroles were formed	Reiss, Stubel, Schmidt, Fouquè, Karl Von Seebach
1925–1928	Increase of sea water temperature, steam, appearance of lava, phreatic eruption, formation of large lava field across Neo Kameni (Figure 5)	Reck (1936)
1939–1941	New vents, two domes, and lava field appeared on Nea Kameni	
1950	Phreatic explosions and extrusion of lava	Georgalas (1959)

Santorini, Eruption, Figure 5 (a) Historical eruption on Nea Kameni in 1926. (b) Eruption on Paleo Kameni, 1950.

(1965) used core analysis of sea-floor sediments that showed that distribution of volcanic material (tephra) from the Thera eruption was not uniform and that its majority was ejected toward the Turkey. Sullivan (1988) reported finding of tephra in Turkish lakes, Francaviglia (1990) reported finding tephra on Turkish coastal beaches, and Guichard et al. (1993) reported findings of Santorini tephra in sediments of the Black Sea.

The analogy between the Thera eruption and the volcanic eruption of Krakatau in Sunda Straight (Indonesia) on August 26, 1883, prompted a hypothesis of the disastrous effect of tsunami waves on Minoan civilization on Crete (Marinatos, 1939). Later, the findings of concave walls in one of the late Minoan houses near Amnissos supported this hypothesis (Marinatos, 1968). Recent studies by Minoura et al. (2000), Koenig (2001), and Dominey-Howes (2004) do not support the disastrous effects of tsunami generated by the Thera eruption on Crete.

At the same time the archaeological study in Gouves (Vallianou, 1996), an analysis of the tsunami on Santorini (McCoy and Heiken, 2000b) and the surrounding area and geological findings of potential tsunami deposits from Thera eruption near Minoan palace Palaikastro (Bruins et al., 2007) and coast of Israel (Goodman-Tchernov et al., 2009) support the conclusion on tsunami distribution and consequent inundation of the coastal areas of Crete and Aegean within several hundred meters (Minoura et al., 2000). Still this evidence does not yet explain the demise of the Minoan civilization on Crete.

It is also hard to assess the impact of the Thera eruption on the local Theran population since only one excavation at Akrotiri (Figure 1) revealed remnants of the ancient town buried under the volcanic ash. Though no skeletons or human remains were found (Doumas, 1990) in the excavated ruins and only one precious object (golden ibex figurine) was discovered (Pantazis et al., 2002). On a global scale Pyle (1997) argues that the Thera eruption was not a catastrophe "unparalleled in all of history" and that its erupted volume and intensity was lower in comparison with the eruption of Tambora (Indonesia) in 1815 CE. The study also argues that the contribution of sulfur emissions and climatic effects from the Thera eruption was even lower than during the Tambora and Mt. Pinatubo (Philippines, 1991) eruptions.

Three important conclusions from the archaeological and geological studies of the Thera eruption help to assess the consequences of this disaster on the ancient population: (1) No human and animal skeletons and no considerable valuables have been found in the excavated ruins of the ancient city of Akrotiri; (2) Distribution of ash and volcanic debris from the Thera eruption was directed toward the Turkish and eastern Mediterranean coast; (3) While the tsunami and inundation of coastal areas most probably occurred, they did not have a disastrous effect on the population and even more, on the collapse of the Minoan civilization on Crete. Apparently the ancient population of Thera was able to react to the early warning from the initial volcanic tremors and leave the island safely.

Santorini, Eruption, Figure 6 Broken staircase in Akrotiri.

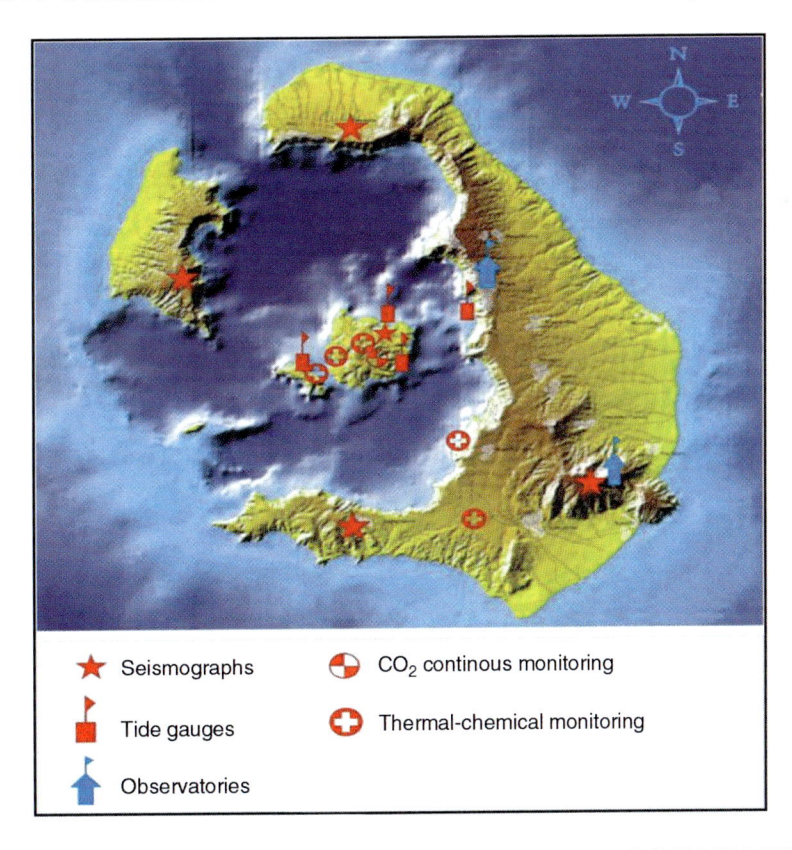

Santorini, Eruption, Figure 7 Monitoring stations and types of observations (Courtesy of ISMOSAV, 2008).

Volcano monitoring

The current geological state of the island is being monitored by the Institute for The Study and Monitoring of Santorini Volcano (ISMOSAV) founded in 1996. Figure 7 shows the locations of various monitoring stations and associated types of measurements.

Ground deformations due to the rising magma in magmatic chambers (Figure 4) are primary indicators of volcanic activity. Ground deformations are usually measured by electronic distance meters (EDM) to detect the slightest inflation of magma with accuracy of up to a few millimeters. EDM devices receive and send electromagnetic signals to established benchmarks on the volcanic surface within the range of 10–50 km. An EDM monitoring system on Santorini was established in 1994. Measurements between 1994 and 2003 revealed gradual inflation (up to 10 cm) of the area between Nea Kameni and Therasia (Stiros et al., 2005). Other areas of the Santorini complex do not show any significant changes. The latest data from ISMOSAV (Figure 8) show the highest vertical (130 mm) and horizontal (56 mm) movements recorded on Therasia.

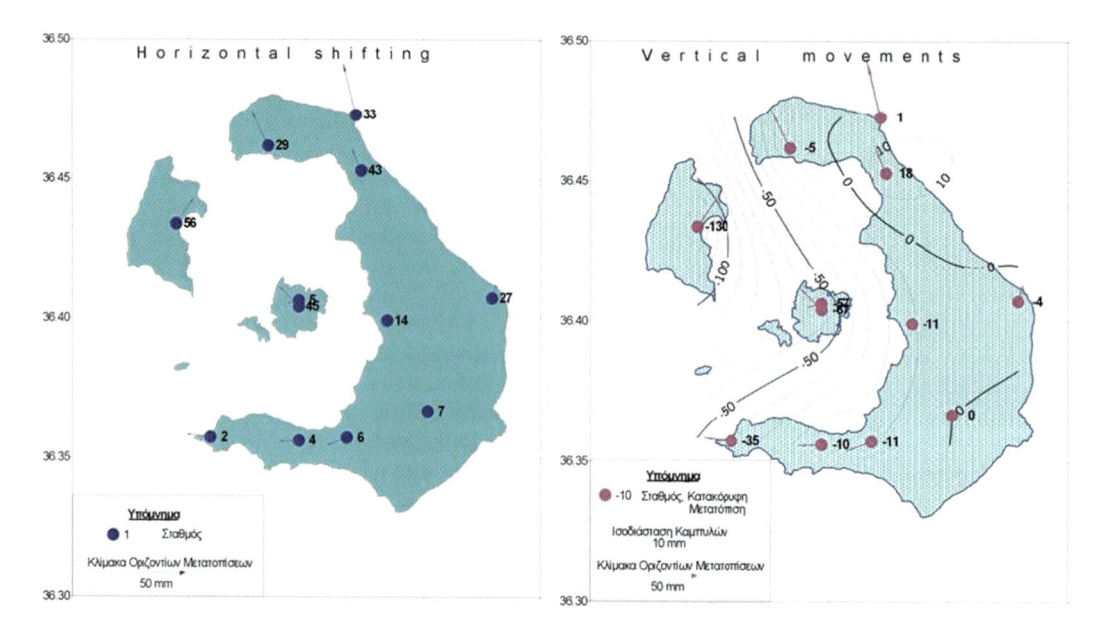

Santorini, Eruption, Figure 8 Map of horizontal and vertical deformations (Courtesy of ISMOSAV, 2008).

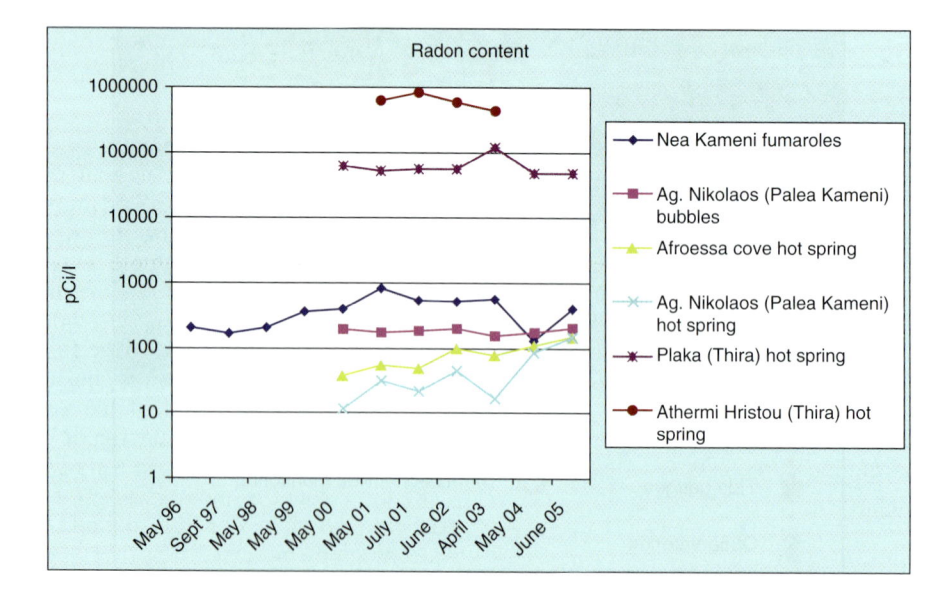

Santorini, Eruption, Figure 9 Changes in radon concentration, 1996–2005 (Courtesy ISMOSAV, 2008).

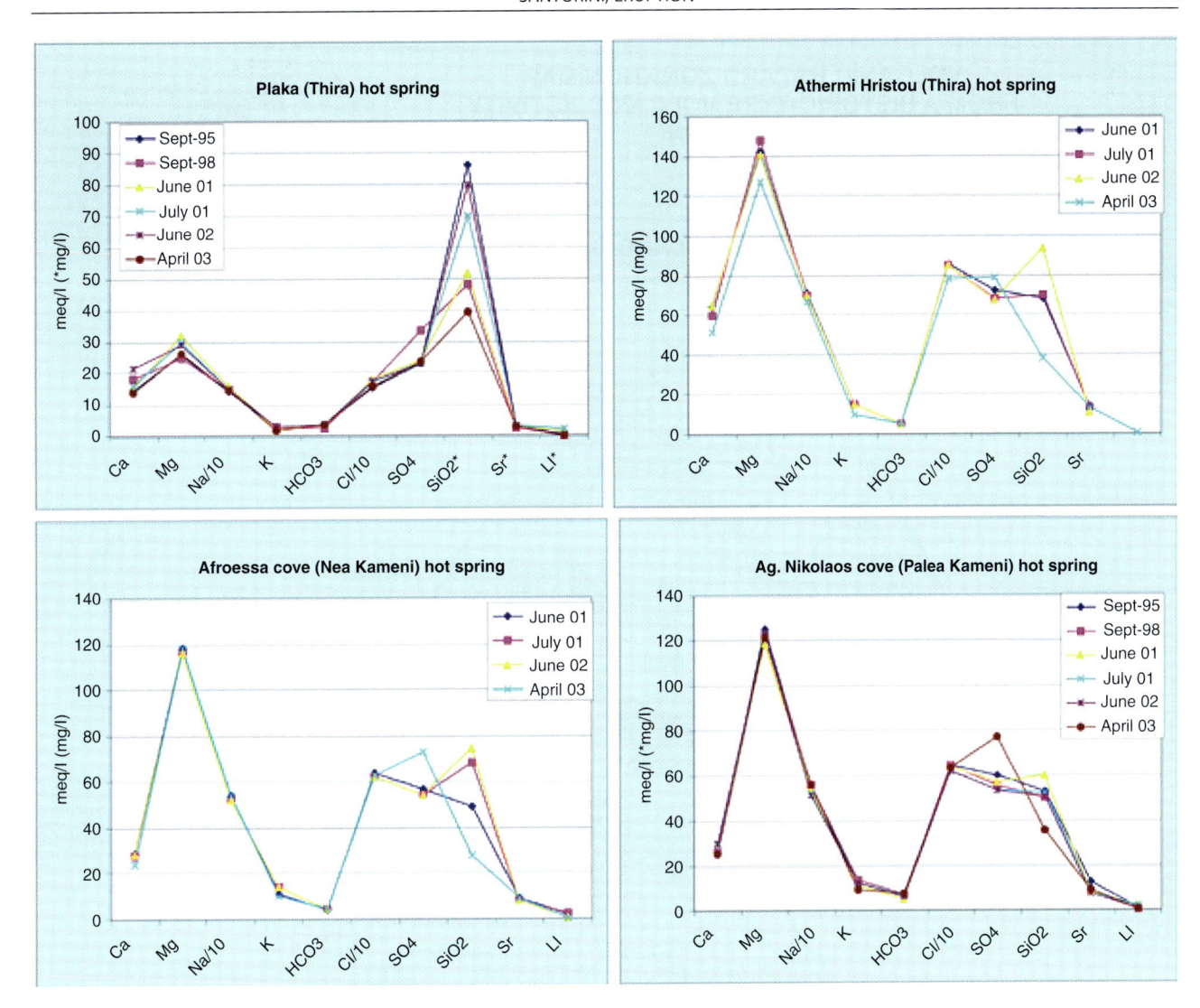

Santorini, Eruption, Figure 10 Chemical composition of hot springs (Courtesy ISMOSAV, 2008).

Seismic activity on the island is monitored by five permanent and four temporary stations (Dimitriadis et al., 2005). For the period 1994–2002 these stations recorded multiple earthquakes with maximum magnitude of 5.0. Most seismic activity occurred at the center of the Santorini complex (i.e., Paleo Kameni) and the northern part near Kolumbo volcanic cone. These locations are associated with two main faults. However, according to authors of the study available data are still limited to explain this activity.

The thermal-chemical monitoring of Santorini volcano consists of the following measurements:

- Hot fluids composition
- Gas composition
- Temperatures
- Radon content in hot gasses and waters
- CO_2 flux

These measurements represent the five main parameters of volcanic activity. The goal of thermal-chemical monitoring is to find fluctuations in measured values and associate them with periods of activity or repose of volcanic processes. For example, in relation to CO_2 flux, ISMOSAV (2008) reported that the main flux values per day range between 5 and 50 ppm/s, whereas the maximum registered values did not exceed the 150 ppm/s. According to ISMOSAV this fluctuation is considered characteristic for the repose time period of the volcano. During the volcano reactivation the magnitude of CO_2 flux will increase and might serve as an early warning sign.

Radon content is another indicator of earthquake and volcanic activity. For the period 1996–2005 radon concentration increased in the Afroessa cove (Nea Kameni) and the Ag. Nikolaos (Palea Kameni) hot springs (Figure 9).

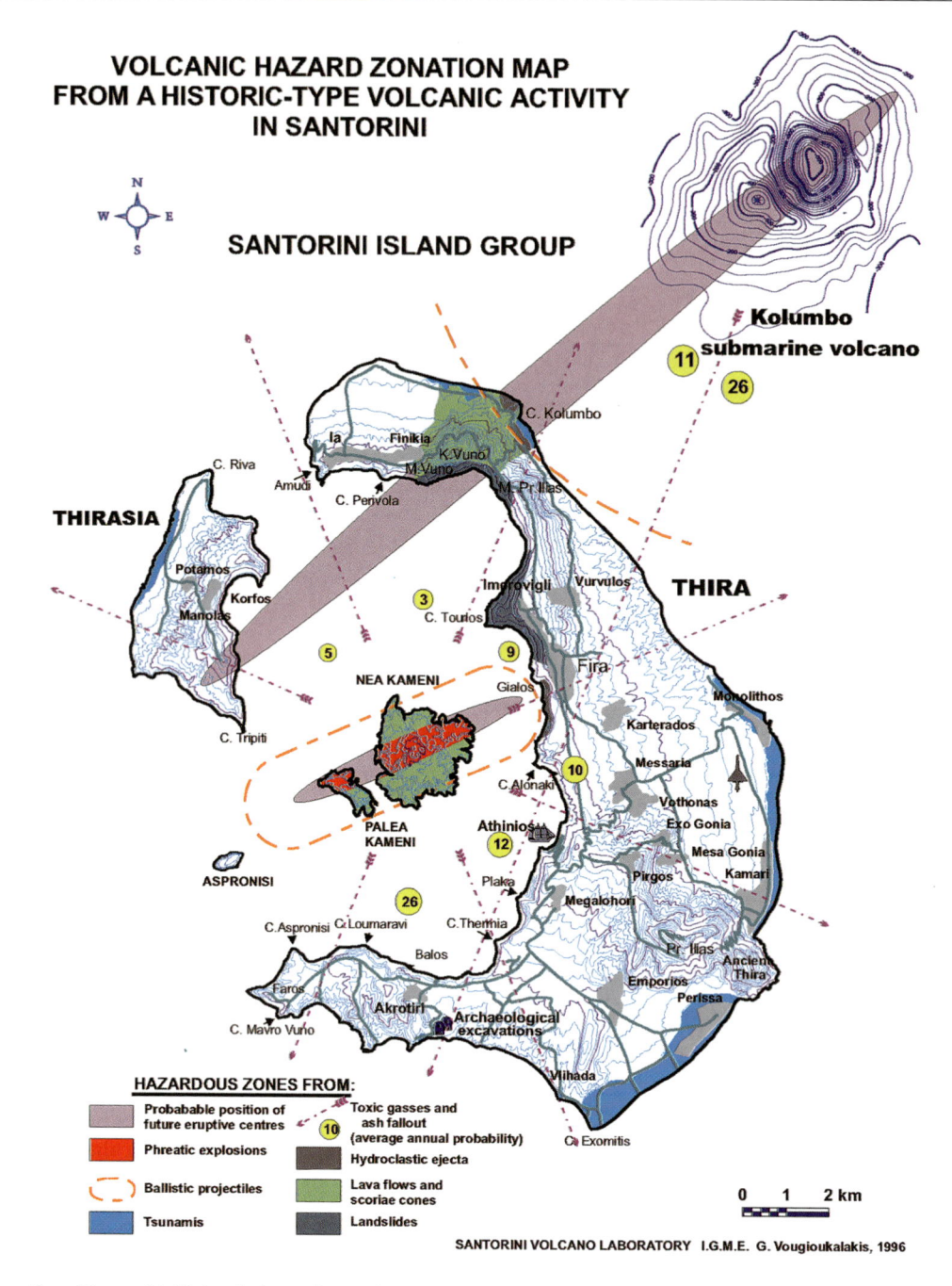

Santorini, Eruption, Figure 11 Volcanic hazard zonation map (Vougioukalakis and Fytikas, 2005).

The maximum registered temperature fluctuation on Nea Kameni fumaroles is about 3°C (94°C ∼ 97°C). The same fluctuation was measured on Palea Kameni. Nea Kameni fumaroles consist mainly of heated atmospheric air and CO_2 with minor concentrations of CH_4, H_2, and CO. Palea Kameni gasses consist mainly of CO_2 (99.9 vol.%) with minor concentrations of N_2, O_2, CH_4, and CO. Their fluctuations since 1995 are not significant (Figure 10).

According to ISMOSAV (2008) findings there are no significant fluctuations in the chemical composition of hot gasses and waters, which could indicate any deep feeding process inside the magma chamber and possible eruption. They also suggest that the decrease of the total gas outflow and the minor gas constituents measured during the last 2 years could indicate ongoing self-sealing processes for the uppermost Kameni area.

While monitoring of geochemical parameters is necessary for the short-term (in geologic sense) analysis and hazard assessment, its use for predicting earthquake and volcanic activity does not appear to be feasible

(Thomas, 1988). Seismic and geodetic components of ongoing monitoring help to acquire additional clues and complement overall conclusions.

Hazard assessment and zonation

Hazard assessment of the Santorini complex is based on historical records, geological studies, and results from current monitoring. Trends in seismic, temperature, gas, and ground deformation data help to make short-term predictions and recognize early warning signs of volcanic activity. Analysis of historical and geological records helps to predict the most probable scenario and map potentially vulnerable areas of ash fallout, zones of inundation and landslides, etc.

The assessment of volcanic hazard is important for Santorini because approximately 70% of tourist establishments are in Oia and Fira (Fytikas et al., 1998). At the same time there is no emergency plan and coordination between departments responsible for emergency measures (Dominey-Howes and Minos-Minopoulos, 2004). With the total number of local population close to 13,600 and an additional 1,000,000 population of tourists during the year (Vougioukalakis and Fytikas, 2005), the absence of the emergency plan reduces the resilience of the Santorini population and makes future disaster management uncertain and potentially ineffective.

According to Vougioukalakis and Fytikas (2005), the most probable volcanic eruption scenario on Santorini will be similar to ones that occurred during the last millennia. The eruption with maximum possible effect will follow the path of the Minoan eruption. In their analysis authors consider a 20,000-year lag between maximum possible events in Santorini history, the number apparently taken from (McCoy and Heiken 2000a). This makes the probability of eruption similar to the one that occurred in Minoan time low. In addition, an eruption of that type will make the island completely obsolete and, therefore, there is no need for a relevant volcanic hazard map.

In case of the most probable eruption scenario, the analysis of historical eruptions and their effects, combined with recently obtained geologic and monitoring data allows drafting a potential hazard zonation map (Figure 11). This zonation includes the following hazard zones:

- Probable position of the future eruptive centers
- Phreatic explosions
- Ballistic ejecta
- Tsunamis
- Toxic gasses and ash fallout
- Hydroclastic ejecta
- Lava flows and scoriae cones
- Landslides

Summary

The Santorini volcanic complex is known for its strong historic and modern volcanic eruptions. While the probability of strong eruption, similar to the Minoan event is low, it is possible to expect less disastrous events, analogous to the post-Minoan eruptions during the CE period with low human and property losses. Current data from the existing monitoring network show relatively stable geochemical trends, clustering of seismic activity along two existing faults and strong ground deformations in the northern part of the Santorini volcanic complex between Nea Kameni and Therasia.

Archaeologic and historic data suggest that during the Minoan eruption the population of Thera was able to leave the island safely and timely. Most likely the early signs of volcanic activity triggered prompt evacuation of the Theran people. The absence of human and animal skeletons and valuable objects in Akrotiri might be a sign of organized and controlled retreat from the island. These are lessons from the ancient past that we need to respect. Although the size of affected population and its distribution on the island is not known because Akrotiri is the only excavated Minoan settlement found on Santorini.

The absence of an emergency plan and lack of coordination between departments responsible for emergency measures weakens the resilience of the modern day Santorini population approaching 14,000 people. Considering the influx of tourists in the amount of 1,000,000 people annually, even the probable volcanic event with low impact can bring the havoc and disorientation of people seeking shelter and safety. More importantly, the absence of a coordinated governmental action plan prevents now any kind of pre-disaster education of potential actors and building strong local capacity of Santorini population to cope with potential disaster.

Acknowledgments

This work would be incomplete without the help of Mr. Ioannis Kotiadis (ISMOSAV) who maintains the monitoring network on Santorini and provided valuable graphs and images for this paper.

Bibliography

Biel, B., 2005. Contributions to the flora of the Aegean Islands of Santorini and Anafi (Kiklades, Greece). Willdenowia, 35(1), 87–96. Botanischer Garten und Botanisches Museum, Berlin-Dahlem. Available at JSTOR, http://www.jstor.org/stable/3997603. Accessed December, 2009.

Bruins, H. J., MacGillivray, J. A., Synolakis, C. E., Benjamini, C., Keller, J., Kisch, H. J., Klügel, A., and van der Plicht, J., 2007. Geoarchaeological tsunami deposits at Palaikastro (Crete) and the Late Minoan IA eruption of Santorini. *Journal of Archaeological Science*, **35**(1), 191–212.

Dimitriadis, I. M., Panagiotopoulos, D. G., Papazachos, C. B., Hatzidimitriou, P. M., Karagianni, E. E., and Kane, I., 2005. Recent seismic activity (1994–2002) of the Santorini volcano using data from local seismological network. In Fytikas, M., and Vougioukakakis, G. E. (eds.), *The South Aegean Active Volcanic Arc: Present Knowledge and Future Perspectives*. Amsterdam: Elsevier, pp. 185–203.

Dominey-Howes, D., 2004. A re-analysis of the Late Bronze Age eruption and tsunami of Santorini, Greece, and the implications for the volcano-tsunami hazard. *Journal of Volcanology and Geothermal Research*, **130**, 107–132.

Dominey-Howes, D., and Minos-Minopoulos, D., 2004. Perceptions of hazard and risk on Santorini. *Journal of Volcanology and Geothermal Research*, **137**, 285–310.

Doumas, C., 1990. Archaeological observations at Akrotiri relating to the volcanic destruction. In Hardy, D., and Renfrew, A. C. (eds.), *Thera and the Aegean World III*, Vol. 3, pp. 48–50.

Fouquè, F. A., 1879. *Santorin et Ses Èruptions* (*Santorini and Its Eruptions*) (trans: A. R. McBirney). Foundations of natural history. The Johns Hopkins University Press, pp. 560

Francaviglia, V., 1990. Sea-borne pumice deposits of archaeological interest on Aegean and Eastern Mediterranean beaches. In Hardy, D. A. (ed.), *Thera and the Aegean World III, 3 Chronology*. London: The Thera Foundation, pp. 127–134.

Friedrich, W. L., 2000. *Fire in the Sea*. Cambridge: Cambridge University Press, p. 258.

Friedrich, W. L., Kromer, B., Friedrich, M., Heinemeier, J., Pfeiffer, T., and Talamo, S., 2006. Santorini eruption radiocarbon dated to 1627–1600 B.C. *Science*, **312**(5773), 548.

Fytikas, M., Vougioukalakis, G., Dalampakis, P., Bardintzeff, J. M., 1998. Volcanic hazard assessment and civil defence planning on Santorini. In Casale, R., Fytikas, M., Sigvaldasson, G., Vougioukalakis, G. (eds.), *Volcanic Risk: the European Laboratory Volcanoes*. The European Commission, Directorate General: Science, Research and Development, Environment and Climate Programme, pp. 339–351.

Goodman-Tchernov, B. N., Dey, H. W., Reinhardt, E. G., McCoy, F., and Mart, Y., 2009. Tsunami waves generated by the Santorini eruption reached Eastern Mediterranean shores. *Geology*, **37**(10), 943–946.

Gramling, C., 2007. Buried beneath the Black Sea: cities and ships submerged. Discovering volcanoes old and new. *Geotimes*. http://www.geotimes.org/jan07/feature_BlackSea.html#sidebar1- Accessed December, 2009.

Guichard, F., Carey, S., Arthur, M. A., Sigurdsson, H., and Arnold, M., 1993. Tephra from the Minoan eruption of Santorini in sediments of the Black Sea. *Nature*, **363**, 610–612.

Institute for the Study and Monitoring of Santorini Volcano, (ISMOSAV). 2008. http://ismosav.santorini.net. Accessed December, 2009.

Koenig, R., 2001. Modeling a 3600-year-old tsunami sheds light on the Minoan past. *Science*, **293**, 1252.

Marinatos, S., 1939. The volcanic destruction of Minoan Crete. *Antiquity*, **13**, 425–439.

Marinatos, S., 1968. The volcano of Thera and the states of the Aegean. In *Acts of the Second Cretological Congress, 1967*, Athens, Vol. 1, pp. 198–216.

McCoy F. W., and Heiken, G., 2000a. The Late Bronze Age explosive eruption of Thera (Santorini), Greece: Regional and Local Effects. In: *Volcanic Hazards and Disasters in Human Antiquity*. Boulder, CO: Geological Society of America. Special Paper 345, pp. 43–70

McCoy, F., and Heiken, G., 2000b. Tsunami generated by the Late Bronze Age eruption of Thera (Santorini), Greece. *Pure and Applied Geophysics*, **157**, 1227–1256.

Minoura, K., Imamura, F., Kuran, U., Nakamura, T., Papadopoulos, G. A., Takahasi, T., and Yalciner, A. C., 2000. Discovery of Minoan tsunami deposits. *Geology*, **28**, 59–62.

Misopolinos, N., Syllaios, N., Prodromou, K., 1994. Physiographic and soil mapping of Santorini Island. Laboratories of Soil Science and Remote Sensing of the School of Agronomy, Aristotle University of Thessaloniki, p. 110 (in Greek).

Monaco, C., and Tortorici, L., 2004. Faulting and effects of earthquakes on Minoan archaeological sites in Crete (Greece). *Tectonophysics*, **382**, 103–116.

Moustakas, N. K., and Georgoulias, F., 2005. Soils developed on volcanic materials in the island of Thera, Greece. *Geoderma*, **129**(3–4), 125–138.

Ninkovitch, D., and Heezen, B. C., 1965. *Santorini Tephra, Proceedings of the 17th Symposium of the Colston Research Society*. London: Butterworths Scientific Publications, pp. 413–453.

Nur, A., and Cline, E. H., 2000. Poseidon's horses: plate tectonics and earthquake storms in the Late Bronze Age Aegean and Eastern Mediterranean. *Journal of Archaeological Science*, **27**(1), 43–63.

Pantazis T., Karydas A. G., Chr. Doumas, A. Vlachopoulos, P. Nomikos, M. Dinsmore. 2002. X-ray fluorescence analysis of a gold ibex and other artifacts from Akrotiri. In *Proceedings of 9th International Aegean Conference – Metron, Measuring the Aegean Bronze Age*, New Haven: Yale University, 2002, pp. 1–7

Pe-Piper, G., and Piper, D. J. W., 2005. The South Aegean active volcanic arc: relationships between magmatism and tectonics. In Fytikas, M., and Vougioukakakis, G. E. (eds.), *The South Aegean Active Volcanic Arc: Present Knowledge and Future Perspectives*. Amsterdam: Elsevier, pp. 113–133.

Pyle, D. M., 1997. The global impact of the Minoan eruption of - Santorini, Greece. *Environmental Geology*, **30**(1–2), 59–61.

Stiros, S., Chasapis, A., and Kontogianni, V., 2005. Geodetic evidence for slow inflation of the Santorini caldera. In Fytikas, M., and Vougioukakakis, G. E. (eds.), *The South Aegean Active Volcanic Arc: Present Knowledge and Future Perspectives*. Amsterdam: Elsevier, pp. 205–210.

Sullivan, D. G., 1988. The discovery of Santorini Minoan tephra in Western Turkey. *Nature*, **333**, 552–554.

Thomas, D., 1988. Geochemical precursors to seismic activity. *Journal of Pure and Applied Geophysics*, **126**(2–4), 241–266.

Trevisanto, S. I., 2007. Medical papyri describe the effects of the Santorini eruption on human health, and date the eruption to August 1603–March 1601 BC. *Medical Hypotheses*, **68**(2), 446–449.

Vallianou, D., 1996., New evidence of earthquake destructions in Late Minoan Crete. In Stiros, S., and Jones, R. E., (eds.), *Archaeroseismology*. Athens: British School of Athens, Fitch Laboratory Occasional Paper 7, pp. 153–167.

Vitaliano, C. J., and Vitaliano, D. B., 1974. Volcanic tephra on Crete. *American Journal of Archaeology*, **1**, 19–24.

Vougioukakakis, G. E., and Fytikas, M., 2005. Volcanic hazards in the Aegean area, relative risk evaluation, monitoring and present state of the active volcanic centers. In Fytikas, M., and Vougioukakakis, G. E. (eds.), *The South Aegean Active Volcanic Arc: Present Knowledge and Future Perspectives*. Amsterdam: Elsevier, pp. 161–183.

Cross-references

Biblical Events
Caldera
Casualties Following Natural Hazards
Cultural Heritage and Natural Hazards
Disaster
Disaster Risk Management
Disaster Risk Reduction (DRR)
Early Warning Systems
Earthquake
Emergency Management
Emergency Mapping
Emergency Planning
Eruption Types (volcanic)
Evacuation
Frequency and Magnitude of Events
Geohazards
Geological/Geophysical Disasters
Hazard
Hazard and Risk Mapping
Historical Events
History of Natural Disasters
Monitoring and Prediction of Natural Hazards

Natural Hazards
Perception of Natural Hazards and Disasters
Plate Tectonics
Risk Assessment
Risk Perception and Communication
Tsunami
Volcanic Ash
Volcanic Gas
Volcanoes and Volcanic Eruptions
Vulnerability
Warning Systems
Zoning

SEA LEVEL CHANGE

Peter J. Hawkes
HR Wallingford Limited, Wallingford, Oxfordshire, UK

Definitions

Mean sea level (MSL). An oceanographic term, referring to sea surface level, possibly averaged over an area or a period of time, but with short-period variations such as waves filtered out.

Monthly mean sea level. Average sea surface level, usually at a single location, preferably taken over an integer number of tidal cycles, averaged over a period of about one calendar month.

Global mean sea level. Average sea level, averaged over all sea areas.

Sea level change. Referring to the long-term trend in mean sea level, which may be calculated from a single location or across a region. One can distinguish between *absolute* sea level change and *relative* (to the local land level) sea level change.

Projection. Extrapolation to what would happen under given starting conditions, for given changes and given physical assumptions. Climate change projections are often expressed relative to a particular time period, presently usually 1961–1990.

Introduction

The sea surface is constantly changing, on several different time and length scales. Each type of sea level change results from a range of forcing factors, and in turn produces its own risks to people and property.

Waves produce the largest and most rapid changes to the sea surface level, on a scale of meters of change in a few seconds, threatening ships, sea defenses, and offshore structures. Tides and surges are also on scale of meters of change, but over a few hours. Tides and surges alone (i.e., without waves) tend not to cause damage, but if flooding does occur, there is an almost unlimited source of water, so flooding may be widespread in low-lying areas. Tidal processes are driven by the interactions of the Sun-Earth-Moon system on large water bodies, whereas surges result from atmospheric low pressure cells. A tsunami is a rare event, on a scale of meters on land in an extreme case, occurring over a period of minutes, but sometimes causing widespread damage because it cannot be forecast and has not been planned for. In many parts of the world there is a seasonal variation in mean sea level, meaning that the mean level at one time of the year may be up to about one quarter of a meter higher than at another time of year.

The issue of most interest here is mean sea level rise, an extremely slow increase in the global mean sea level undetectable by eye but measurable over a period of years. Centimeters or even tens of centimeters rise in mean sea level might appear to be unimportant in the face of tides and waves which cause meters of variation. However, a gradual rise in mean sea level means that, over a long period of time, the same tides and waves act at a slightly higher elevation at the coastline and therefore are more likely to cause damage to people and property.

Causes of mean sea level change

The causes of mean sea level change can be divided into two main types (see, e.g., Sect. 5.5 of IPCC, 2007). One is thermal expansion, under which water volume increases very slightly with increase in water temperature (see, e.g., Ishii et al., 2006). This thermosteric effect can be calculated based on ocean volume and change in temperature, although ocean temperature change is not easy to estimate and is not closely related to air temperature change. The other cause of mean sea level change is the exchange of water between oceans (including floating ice) and land reservoirs (including glaciers, ice caps, ice sheets, and lakes). This can be estimated based on observations of increasing or decreasing ice extent and depth (see, e.g., Dyurgerov and Meier, 2005) but again prediction of what will happen in the future is highly uncertain.

At the end of the last ice age, sea level would have been about 120 m lower than it is today (Gornitz, 2009). If all the remaining ice on land (mainly in Greenland and Antarctica) were to melt and join the oceans, then mean sea level would be about 65 m higher than today (Permanent Service for Mean Sea Level website: Frequently Asked Questions and Answers).

Tide gauge and satellite measurements of mean sea level change

The Permanent Service for Mean Sea Level (PSMSL) has monthly mean sea level data (http://www.psmsl.org/data/obtaining) from hundreds of tide gauges around the world, some going back over a century. Individual tide gauge records should be treated with some caution where, for example, there is a break in the record, or the tide gauge has been moved during the record, or the land level is known to be moving. The PSMSL website gives access to best estimates of annual global mean sea levels by Church and White (2011). These are plotted in Figure 1, with linear trend lines for 1910–2009

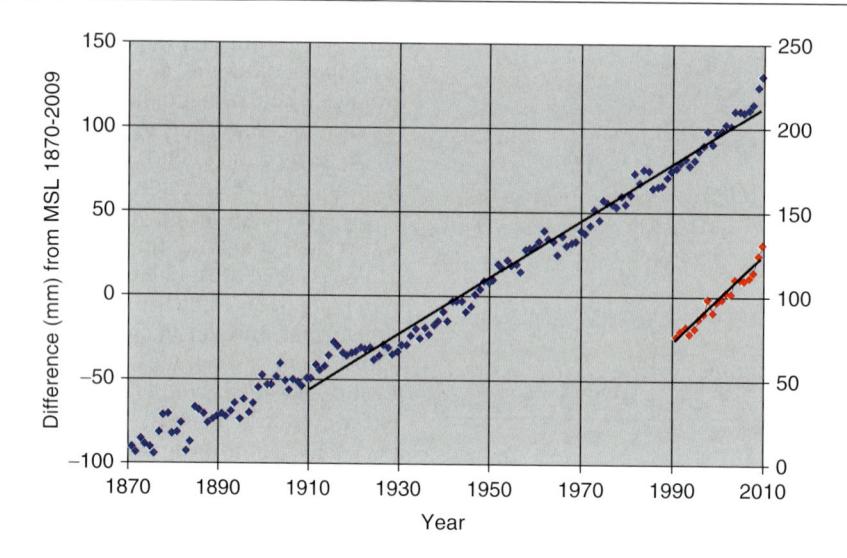

Sea Level Change, Figure 1 Global annual mean sea level 1870–2009 (From tide gauge data available via the PSMSL website).

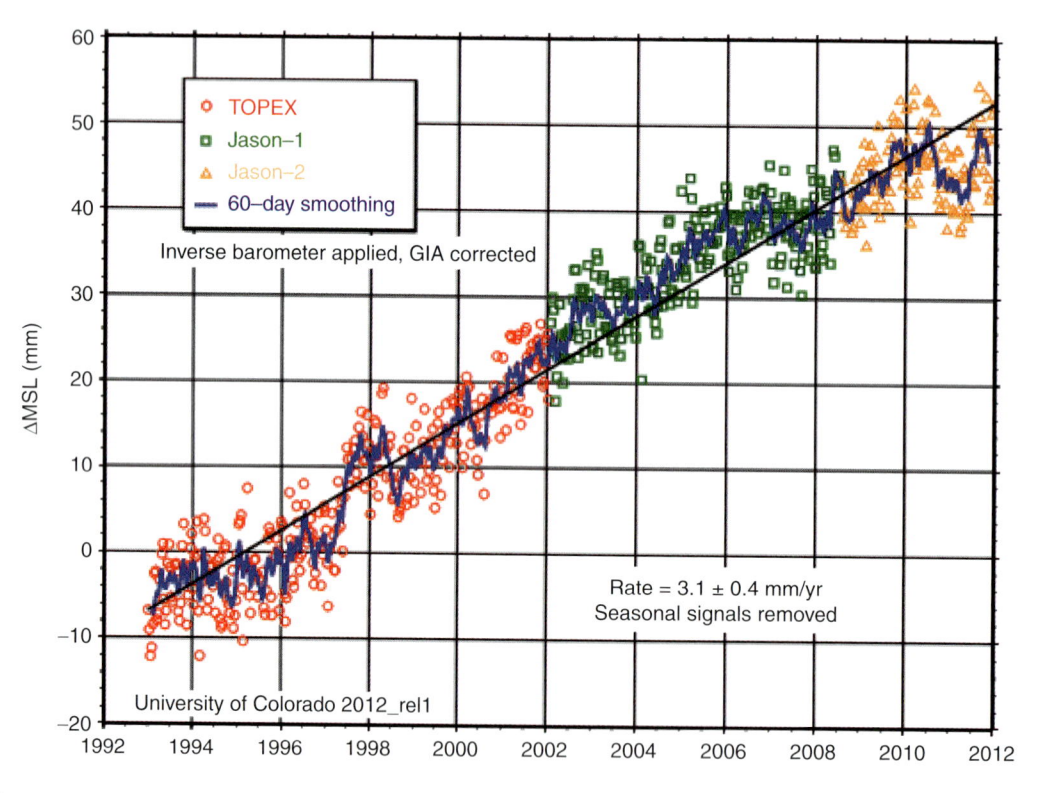

Sea Level Change, Figure 2 Global mean sea level (in millimeters relative to the average level 1993–1999) from satellite measurements (From the University of Colorado Sea Level Research Group website).

(blue dots and left-hand scale, 1.7 mm/year) and for 1990–2009 (red dots and right-hand scale, 2.6 mm/year).

Figure 2, from the University of Colorado Sea Level Research Group website, shows global mean sea level 1993–2011 based on satellite measurements, indicating an upward trend of 3.1 mm/year.

The intergovernmental panel on climate change

The Intergovernmental Panel on Climate Change (IPCC) is the largest international group looking at evidence of climate change and projections of potential future climate change and impacts. The last major revision of its position was published in 2007, and the next update is expected in 2013 and 2014. The reports are available at http://www.ipcc.ch/.

Of all the metocean variables, including temperature, mean sea level probably shows the best evidence of past change, because a number of different geological, biological, and instrumental records are available, and continued sea level rise in the future might be considered a prediction rather than a projection. The report of Working Group 1, the Physical Science Basis (IPCC, 2007) is the most relevant of the IPCC reports for the study of mean sea level change. Its Sect. 5.5 discusses the evidence for and causes of mean sea level change, including regional variations.

Section 11.3.1 of IPCC (2001) notes that "for the past 3,000–5,000 years oscillations in global sea level on time-scales of 100–1,000 years are unlikely to have exceeded 0.3–0.5 m" and "the rate over the past 3,000 years was only about 0.1–0.2 mm/year". Section 5.5.2.1 of IPCC (2007) notes 15–20 cm rise in global mean sea level during the twentieth century.

IPCC projection of future mean sea level change

IPCC (2007) concludes that global mean sea level rose at an average rate of about 1.7 ± 0.5 mm/year during the twentieth century and that the rate has been slightly higher over the period 1961–2003. Climate model projections in IPCC (2007) suggest that the global average rate of rise over the twenty-first century will be 2–5 mm/year, meaning that mean sea level will be 0.2–0.5 m higher in 2,100 than in 2000 (see Figure 3). These figures primarily relate to the thermal expansion of the oceans, excluding rapid dynamic changes such as any contribution from the Antarctic and Greenland ice sheets.

Estimates of future sea level rise by 2100, including an uncertain allowance for ice melt based on present-day melting rates, are given in Figure 4, and are in the range 0.2–0.6 m. The diagram shows thermal expansion of existing sea water (red bar) as the largest contribution to sea level rise, and the Antarctic (dark blue bar) as a negative contribution, meaning that it is projected to increase its ice content.

The projected sea level rise in Figure 4 assumes that the part of the present-day ice sheet mass imbalance that is due to recent ice flow acceleration will persist unchanged. This is probably the greatest uncertainty and several more recent papers have projected higher rates of sea level rise in the twenty-first century caused by higher assumed rates of ice melt. Some researchers argue that uncertainty in the ice melt effect means that there could be over 1 m increase in sea level by 2100: for example, Rahmstorf (2007) projects 0.5–1.4 m rise, Horton et al. (2008) up to 1 m rise, Jevrejeva et al. (2010) 0.6–1.6 m rise, and Pfeffer et al. (2008) 0.8–2.0 m rise.

In combination with other factors, such as subsidence and glacial rebound, sea level rise relative to the land could be slightly different from place to place. Modeling reported in IPCC (2007) indicates some global variation in the projected future rate of mean sea level rise, suggesting (see Figure 5) a lower than global average rate of rise around the Antarctic and a higher than average rise in the Arctic. At mid-latitudes, the mean sea level rise will be generally higher than in the equatorial area due to changes in ocean density distribution (steric sea level rise).

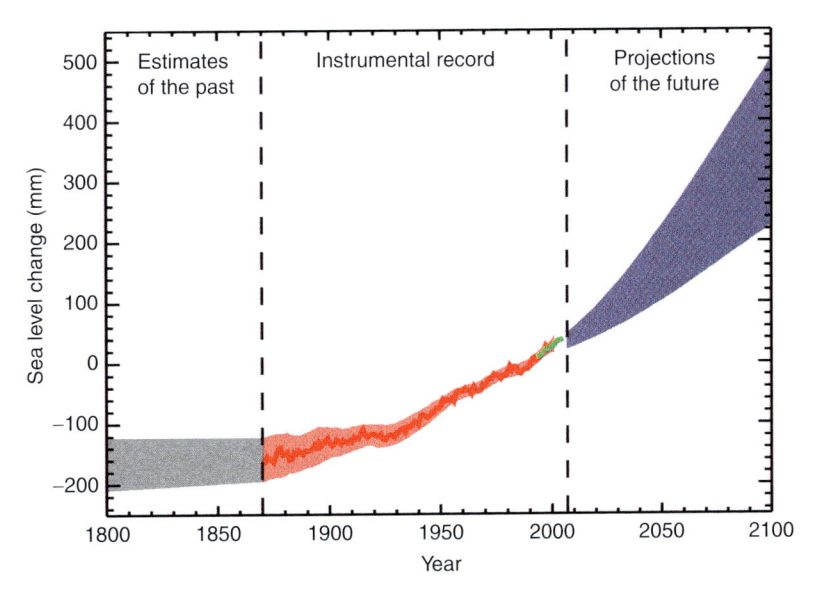

Sea Level Change, Figure 3 Projections of future mean sea level rise due mainly to thermal expansion (Reproduced from Fig. FAQ 5.1 of IPCC, 2007).

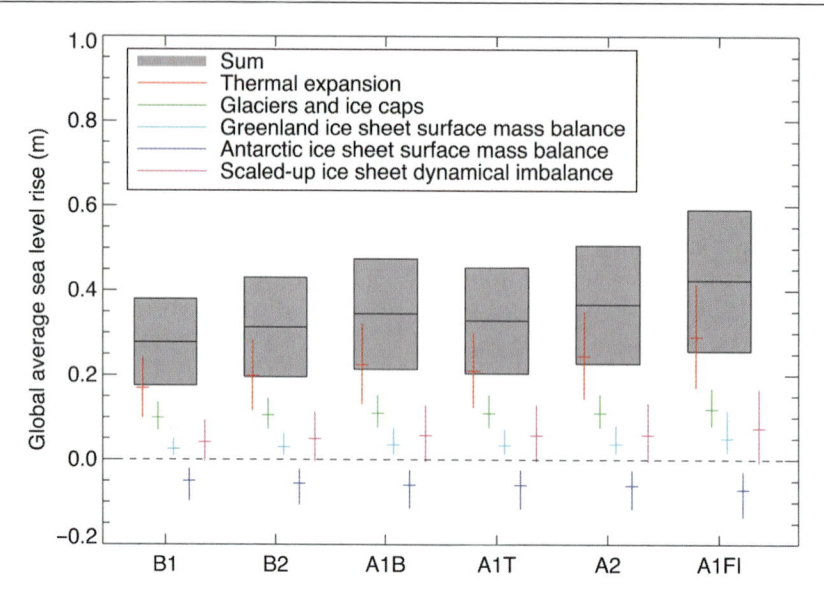

Sea Level Change, Figure 4 Projections and uncertainties (5–95% ranges) of global average sea level rise and its components in 2090–2099, relative to 1980–1999, based on six different emissions scenarios (Reproduced from Fig. 10.33 of IPCC, 2007).

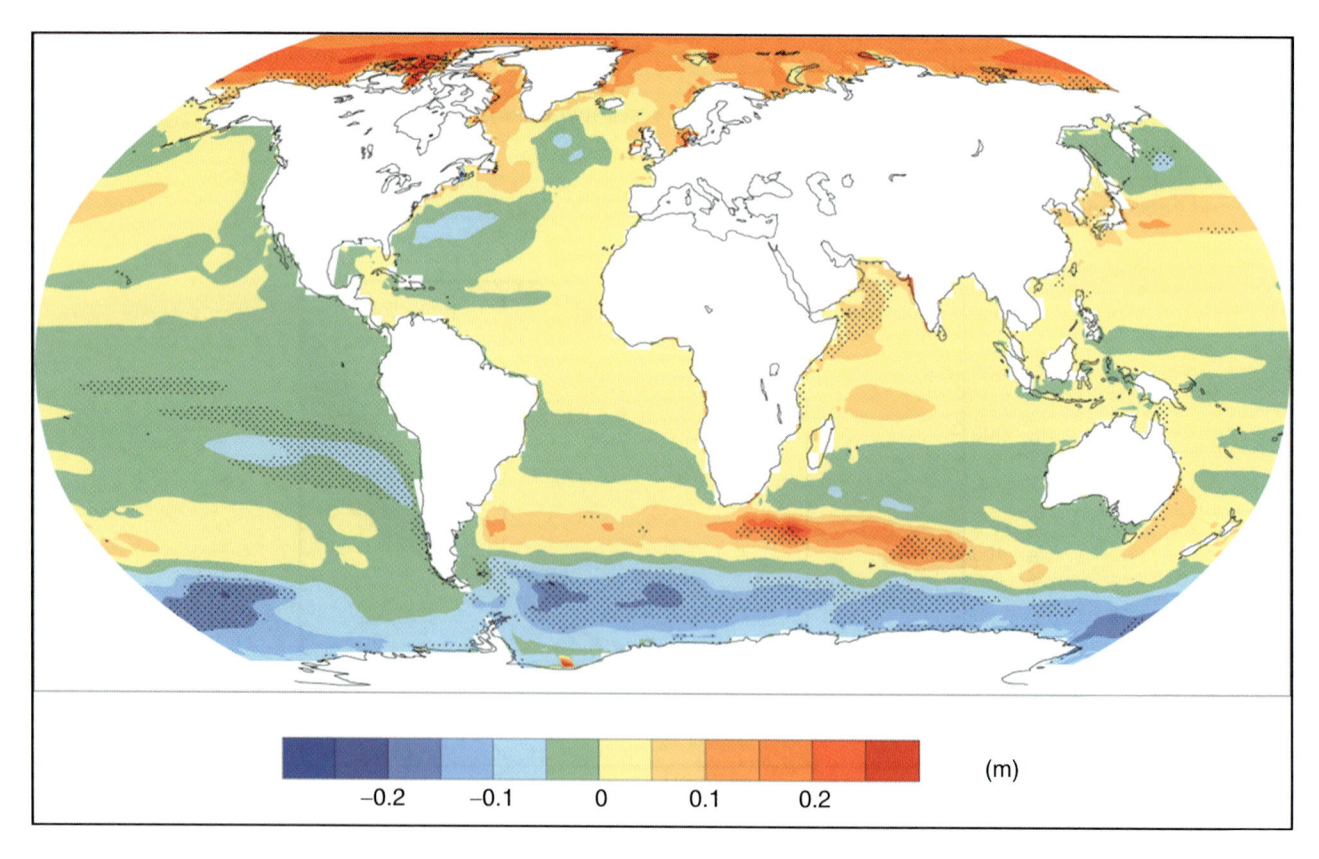

Sea Level Change, Figure 5 Local mean sea level change relative to the global average change projected to 2100 (Reproduced from Fig. 10.32 of IPCC, 2007).

Relative (to the local land level) mean sea level change

Locally, the concern is likely to be with relative, rather than absolute, mean sea level change, that is to say the change in mean sea level relative to the local land level. In combination with other factors, like subsidence and glacial isostatic adjustment, sea level rise relative to the land will vary regionally. Venice, for example, would be particularly vulnerable to mean sea level rise, because the local land level is dropping slowly, creating a larger than average relative mean sea level rise. However, in Upper Canada, glacial rebound is up to about a meter a century (see Figure 6) meaning that the land is rising faster than the sea level.

Impacts

Sea level change is of obvious interest to those living near the coast or on low-lying islands. Without investment in improved sea defenses, the probability of flooding would increase with sea level rise, and the damage done when flooding occurs is also likely to increase.

An increase in mean sea level would mean a reduction in the distance between the sea level and the level at which flooding and/or damage may begin to occur, but this alone would be unlikely to cause flooding. Significant impacts of mean sea level rise, for example, involving loss of life or property, would tend to occur on extreme (very high sea level) events rather than under average conditions. As an illustration of this (see Figure 7) consider extreme sea levels for a location on the south coast of England. At the present-day mean sea level, an extreme level of 3.7 m above mean sea level is expected to occur once, on average, per year; 4.1 m once in 10 years; and 4.5 m once in 100 years. If there is a significant flood risk when sea level reaches 4.5 m above present-day mean level, then this risk will occur once, on average, in 100 years. If mean sea level rises by, say, half a meter over the next 100 years, and the extreme levels are assumed to rise by the same amount, then the present-day flood risk level would occur about once, on average, in 5 years. This is a fairly typical result that, if no change is made to sea defenses, the probability of the coastal flood threshold being reached is around twenty times higher following a century of projected future mean sea level rise.

Mitigation

The obvious methods of mitigation are to build higher and stronger sea defenses, and/or move people and assets to higher ground, possibly abandoning very low-lying areas where it is not cost effective to protect them. For the magnitude and very low rate of mean sea level rise that has occurred in recent centuries, these approaches have been adequate and there has been time enough to apply them. If sea level rise continues for several more centuries, possibly at an increasing rate, the meters of rise that may occur may mean abandonment of large areas of low-lying land.

Stabilization of global mean air temperature by limiting carbon emissions would offer only a partial solution, as sea water would continue warming and expanding for some time even if air temperature did not increase further. If and when sea level rise begins to become a more serious and widespread problem, technology may find a way of putting water from the sea back into the land-ice reservoirs of Antarctica and Greenland, for example, through weather controls.

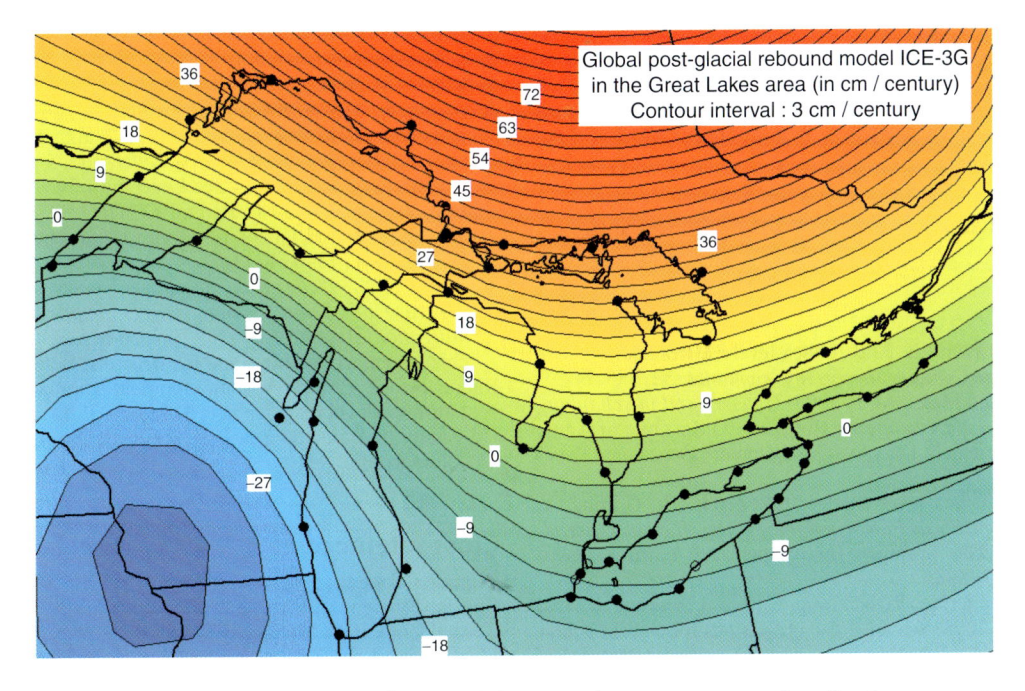

Sea Level Change, Figure 6 Postglacial rebound in the Great Lakes area, shown as contours of centimeters per century (Reproduced from Fig. 3.2 of PIANC, 2008).

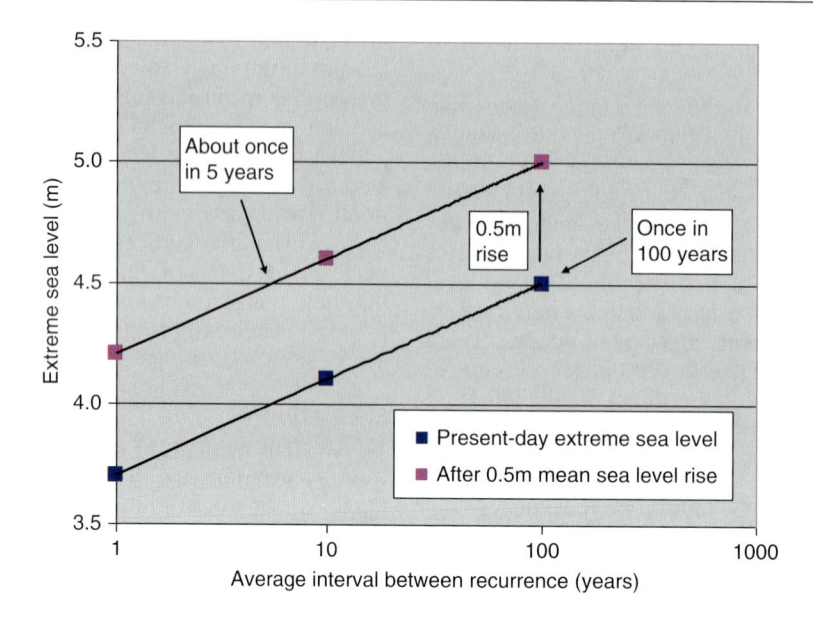

Sea Level Change, Figure 7 Illustration of the effect of mean sea level rise on the frequency at which a flood threshold may be exceeded.

Summary

Sea level change is a natural process that has occurred throughout history and prehistory. Over the last century, the 15 cm increase in global mean sea level was small enough to have had little impact on humans, but steady enough to make it probably the most reliably documented of any climate-related trend.

Future climate projections suggest an increased rate of rise in mean sea level, likely to cause about 0.5 m rise in total over the coming century. This would require considerable coastal defense expenditure to prevent the present frequency of coastal flooding from increasing, possibly leading to abandonment of some areas of very low-lying land. Although the rate of sea level rise is very low, it is cumulative, perhaps amounting to a few meters over several centuries – potentially a huge problem for our distant descendants.

Bibliography

Church, J. A., and White, N. J., 2011. Sea-level rise from the late 19th to the early 21st Century. *Surveys in Geophysics*, **32**, 585–602.

Dyurgerov, M., and Meier, M. F. 2005. *Glaciers and the changing earth system: a 2004 snapshot*. Boulder, CO: Institute of Arctic and Alpine Research, University of Colorado. Occasional Paper 58, 118 pp.

Gornitz, V., 2009. Sea level change, post-glacial. In Gornitz, V. (ed.), *Encyclopedia of Paleoclimatology and Ancient Environments*. Dordrecht: Springer. Encyclopedia of Earth Sciences Series, pp. 887–893.

Horton, R., Herweijer, C., Rosenzweig, C., Liu, J., Gornitz, V., and Ruane, A. C., 2008. Sea level rise projections for current generation CGCMs based on the semi-empirical method. *Geophysical Research Letters*, **35**, L02715, doi:10.1029/2007/GL032486.

Intergovernmental Panel on Climate Change (IPCC), 2001. *Climate change 2001: IPCC third assessment report: working group I. The scientific basis*. Web published by GRID Arendal in 2003. http://www.grida.no/publications/other/ipcc_tar/.

Intergovernmental Panel on Climate Change (IPCC), 2007. *Climate Change 2007: The Physical Science Basis: Contribution of Working Group I to the Fourth Assessment Report of the IPCC*. Solomon, S., Qin, D., Manning, M., Marquis, M., Averyt, K., Tignor, M. M. B., Miller, H. L., and Chen, Z. (eds.) Cambridge, UK/New York: Cambridge University Press, 996 pp. http://www.ipcc.ch/ipccreports/ar4-wg1.htm.

Ishii, M., Kimoto, M., Sakamoto, K., and Iwasaki, S. I., 2006. Steric sea level changes estimated from historical ocean subsurface temperature and salinity analyses. *Journal of Oceanography*, **62**(2), 155–170.

Jevrejeva, S., Moore, J. C., and Grinsted, A., 2010. How will sea level respond to changes in natural and anthropogenic forcings by 2100. *Geophysical Research Letters*, **37**, 5 pp.

Permanent Service for Mean Sea Level. http://www.psmsl.org.

Pfeffer, W. T., Harper, J. T., and O'Neel, S., 2008. Kinematic constraints on glacier contributions to 21st century sea level rise. *Science*, **321**, 1340–1343.

PIANC, 2008. *Waterborne transport, ports and waterways: a review of climate change drivers, impacts, responses and mitigation*. Report of PIANC EnviCom Task Group 3, Climate change and navigation, PIANC, Brussels, Belgium.

Rahmstorf, S., 2007. A semi-empirical approach to projecting future sea level rise. *Science*, **315**, 368–370.

University of Colorado Sea level Research Group. http://sealevel.colorado.edu/.

Cross-references

Climate Change
Coastal Erosion
Coastal Zone Risk Management
Flood Hazard and Disaster
Hydrometeorological Hazards
Marine Hazards
Storm Surges

SECONDARY WAVE (S-WAVE)

Allison Bent
Natural Resources Canada, Ottawa, ON, Canada

Synonyms
S wave; Shear wave; Transverse wave

Definition
Secondary waves are elastic shear waves that travel through the Earth.

Discussion
S waves are seismic body waves meaning they travel through the Earth's interior. Their velocity is slower than that of P waves, and they are normally the second major phase to be observed on a seismogram, and are therefore also referred to as secondary waves. In the Earth's crust, S wave velocities are typically 3–4 km/s. S waves are usually larger in amplitude than P waves and may cause strong shaking and/or damage. The particle motion associated with S waves is perpendicular to the direction of propagation. S waves can be subdivided into two groups: SV waves, which are recorded by seismographs on the vertical and radial components; and SH waves, which appear on the tangential component. S waves cannot propagate through liquids or gases, the knowledge of which helped lead to the discovery that the outer core was liquid.

Bibliography
Aki, K., and Richards, P. G., 1980. *Quantitative Seismology Theory and Methods*. San Francisco: W. H. Freeman.
Bolt, B. A., 1993. *Earthquakes*. New York: W. H. Freeman.

Cross-references
Body Wave
Primary Wave
Seismograph/Seismometer

SEDIMENTATION OF RESERVOIRS

Anton J. Schleiss
Ecole polytechnique fédérale de Lausanne (EPFL), Lausanne, Switzerland

Synonyms
Filling up of reservoirs by sediments; Reservoir sedimentation; Siltation of reservoir

Definition
Process of filling up of reservoirs by sediments, which are transported by rivers or overland flows.

Introduction
Sedimentation is known as the process which fills up natural lakes and man-made reservoirs by sediments to become the end land again. The main reason for this process is the sediment yield transported by the rivers as suspended or bed load into the reservoirs. Bed and suspended sediment load originate from soil and rock erosion in the catchment area of the reservoir. Suspended fine sediments are also the result of surface erosion as well as of crashing and abrasion of coarser sediments transported by rivers. When entering lakes and reservoirs, the coarser sediments such as sand and gravel settle down and form a delta. The finer suspended sediments are deposited over the whole reservoir. During floods, they are periodically transported as turbidity currents like an underwater avalanche directly from the delta along the reservoir to the deepest point in front of the dam.

Today, the worldwide yearly loss of storage capacity due to sedimentation is already higher than the increase of capacity by the construction of new reservoirs for irrigation, drinking water, and hydropower. In Asia, for example, 80% of the useful storage capacity for hydropower production will be lost by 2035. Thus, the sustainable use of the reservoirs is not guaranteed in long term. In the case of deep and long reservoirs, the sedimentation rate is much below the world mean value. Nevertheless, the sedimentation threatens also these reservoirs since the mentioned turbidity currents are sporadically transporting large volumes of sediments down to the dam. There, the concentrated deposits are hindering the safe operation of the outlet structures as intakes and bottom outlets. Thus, after only 30–40 years of operation, sedimentation has become a serious problem in many reservoirs located even in catchment areas with moderate surface erosion as in the Alps.

The problem of reservoir sedimentation
Reservoir sedimentation is a problem that will keep those responsible for water resources management occupied more than usual during the decades to come. All sorts of impounding structures are concerned: large dams, in the form of fill or concrete dams, as well as river barrages comprising weirs, power plants, locks, impounding dams, and dykes (Figure 1). Impounding facilities are always costly, but this is justified by their various potential uses.

Although the aim behind the efforts to create reservoirs is storing water, other substances are carried along by the water and are usually deposited there as well. This is a result of dam construction, dramatically altering the flow behavior and leading to transformations in the fluvial process with deposition of solid particles transported by the flow. Each reservoir created on natural rivers, independent of its use (water supply, irrigation, energy, or flood control), can have its capacity decreased due to deposition over the years. In an extreme case, this may result in the reservoir becoming filled up with sediments, and the river flows over land again.

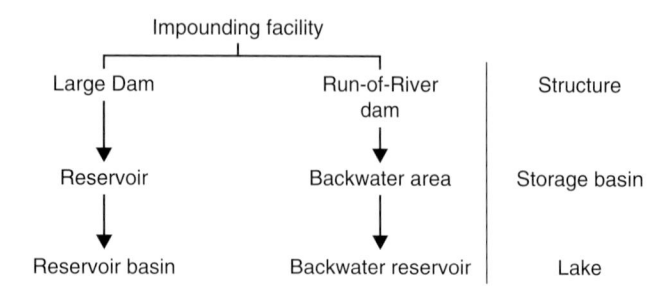

```
Impounding facility
    |
Large Dam    Run-of-River dam         Structure
    |             |
Reservoir    Backwater area           Storage basin
    |             |
Reservoir basin  Backwater reservoir  Lake
```

Sedimentation of Reservoirs, Figure 1 Classification of impounding facilities.

Sedimentation of Reservoirs, Table 1 Average sedimentation rates in different regions (Basson, 2009)

Region	Average sedimentation rate (%/year)
Africa	0.85
Asia	0.79
Australia and Oceania	0.94
Central America	0.74
Europe	0.73
Middle East	1.02
North America	0.68
South America	0.75

A reservoir, like a natural lake, silts up more or less rapidly. In actual fact, reservoirs may completely fill with sediments even within just a few years, whereas natural lakes, for example, in the Alpine foreland, may remain as stable features of the landscape for as much as 10,000–20,000 years after they were formed during the last Ice Age.

Reservoir sedimentation reduces the value of or even nullifies the dam construction investment. The use for which a reservoir was built can be sustainable or represent a renewable source of energy only where sedimentation is controlled by adequate management and suitable mitigation measures. Lasting use of reservoirs in terms of water resources management involves the need for reduction of sediment yield and sediment removal.

The planning and design of a reservoir require the accurate prediction of erosion, sediment transport, and deposition in the reservoir. For existing reservoirs, more and wider knowledge is still needed to better understand and solve the sedimentation problem and hence improve reservoir operation.

Consequences of reservoir sedimentation

The accumulating sediments successively reduce the water storage capacity (Fan and Morris, 1992). Consequently, at long term, the reservoir operates only at reduced functional efficiency. Declining storage volume reduces and eventually eliminates the capacity for flow regulation and with it all water supply, energy and flood control benefits (Graf, 1984; International Committee on Large Dams (ICOLD), 1989). Reservoir sedimentation can even lead to a perturbation of the operating intake and to sediment entrainment in waterway systems and hydropower schemes. Depending on the degree of sediment accumulation, the outlet works may be clogged by the sediments. Blockage of intake and bottom outlet structures or damage to gates that are not designed for sediment passage is also a severe security problem. Other consequences are sediments reaching intakes and greatly accelerating abrasion of hydraulic machinery, decreasing their efficiency and increasing maintenance costs.

Sedimentation rate

The worldwide installed capacity of all reservoirs is about 7,000 km^3, whereas 4,000 km^3 can be used for energy

Sedimentation of Reservoirs, Table 2 Dates when 80% of the useful reservoir volumes for hydropower production are lost by sedimentation and 70% of the reservoirs for other uses (Basson, 2009), respectively

Region	Hydropower dams: Date when 80% filled with sediment	Non-hydropower dams: Date when 70% filled with sediment
Africa	2,100	2,090
Asia	2,035	2,025
Australasia	2,070	2,080
Central America	2,060	2,040
Europe and Russia	2,080	2,060
Middle East	2,060	2,030
North America	2,060	2,070
South America	2,080	2,060

production, irrigation, and water supply. (Basson, 2009). The average age of existing reservoirs is about 30–40 years. It is estimated that about 0.8% of the worldwide storage capacity is lost annually by sedimentation. The highest average sedimentation rate is found in arid regions as in the Middle East, Australia, and Oceania as well as Africa (Table 1). A detailed collection of sedimentation rates in regions all over the world can be found in Batuca and Jordaan (2000).

In Table 2, the date prediction is given when 80% of the useful reservoir volumes for hydropower production are lost by sedimentation and 70% of the reservoirs for other uses (Basson, 2009), respectively. It may be seen that in Asia, for example, 80% of the useful storage capacity for hydropower production will be lost in 2035. In view of the increasing energy demand, this is a serious problem; 70% of the storage volumes, used for irrigation, will be filled up by sediments already in 2025. This will be the case in the Middle East in 2030 and in Central America in 2040. This underlines that reservoir sedimentations endanger the sustainable energy and food production in many regions in the world.

The sedimentation rate of each particular reservoir is very variable. It depends more particularly on the climatic situation, the geomorphology, and the conception of the reservoir including its outlet works.

Reservoir sedimentation by turbidity currents

In long, deep, narrow reservoirs, turbidity currents are often the governing process in reservoir sedimentation by transporting fine materials in high concentrations (Jenzer Althaus et al., 2009). The erosion of the soil within a catchment area is at the origin of the material transported by a river. The erosion process starts in the high mountainous regions and continues in the highlands and plains and ends in the lakes or in the sea respectively where it comes – due to the decreasing flow velocity – to sedimentation. Depending upon the sediment supply from the watershed and flow intensity in terms of velocity and turbulence, rivers usually carry sediment particles within a wide range of sizes. During flood events, the fraction of sediments smaller than sand reaches 80–90 % of the total sediment carried by the river (Alam, 1999), and the total sediment discharge is usually significant. If the sediment concentration is high enough, it may come to turbidity current.

The turbidity currents belong to the family of sediment gravity currents. These are flows of water laden with sediment that move downslope in otherwise still waters like oceans, lakes, and reservoirs. Their driving force is gained from the suspended matter (fine solid material), which renders the flowing turbid water heavier than the clear water above. When a sediment-laden river flows into a large reservoir, the coarser particles deposit gradually and form a delta in the headwater area of the reservoir that extends farther into the reservoir as deposition continues (Figure 2). Finer particles, being suspended, flow through the delta stream and pass the lip point of the delta. If after the lip point of the delta, the difference in density between the lake water and inflowing water is high enough, it may cause the flow to plunge, and turbidity current can be induced. During the passage of the reservoir, the turbidity current may unload or even resuspend granular material. Subsequently, the sediments are deposited along the path due to a decrease in flow velocity caused by the increased cross-sectional area. Fine sediments (clay and silt sizes) are usually the only sediments that remain in suspension long enough, following over long distances the reservoir bottom along the thalweg through the impoundment down to the deepest point in the lake normally near the dam, to reach the outlets. At the dam, the sediments settle out.

Measures against reservoir sedimentation

Over the years, several measures against reservoir sedimentation have been proposed (Schleiss and Oehy, 2002). But not all of them are sustainable, efficient, and affordable. For example, raising the height dams and outlet works does not provide a long-term and sustainable solution.

There is a strong need to limit sediment accumulation in reservoirs in order to ensure their sustainable use. Management of sedimentation in Alpine reservoirs cannot be apprehended by a standard generalized rule or procedure. Furthermore, sediment management is not limited to the reservoir itself, it begins in the catchment areas and extends to the downstream river. Every situation has to be analyzed for itself in order to determine the best combination of solutions to be applied. The possible measures are summarized in Figure 3 and grouped according to the areas where they can be applied.

An integrated approach to sediment management that includes all feasible strategies is required to balance the sediment budget across reservoirs (Morris, 1995). Integrated sediment management includes analysis of the complete sediment problem and application of the range of sediment strategies as appropriate to the site. It implies that the dam and the impoundment are operated in a manner consistent with the preservation of sustainable long-term benefits, rather than the present strategy of developing and operating a reservoir as a nonsustainable

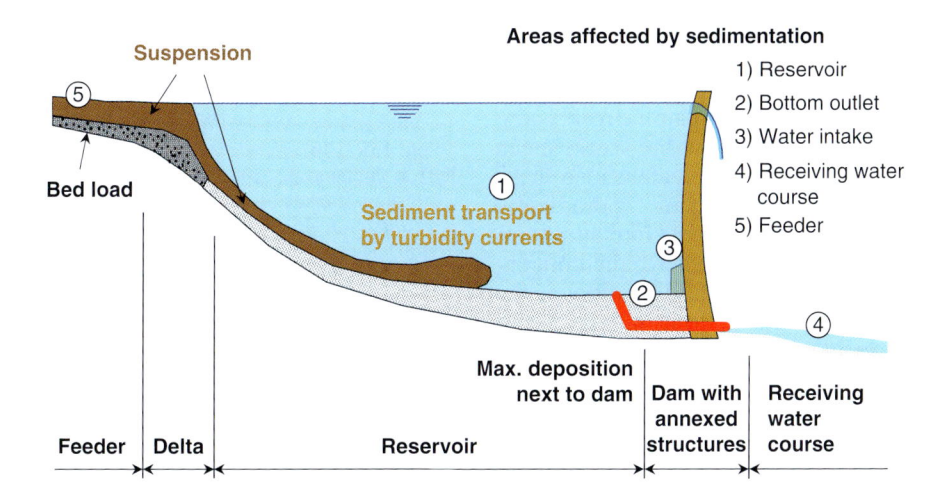

Sedimentation of Reservoirs, Figure 2 Areas affected by sedimentation in the surroundings of a reservoir (De Cesare et al., 2001).

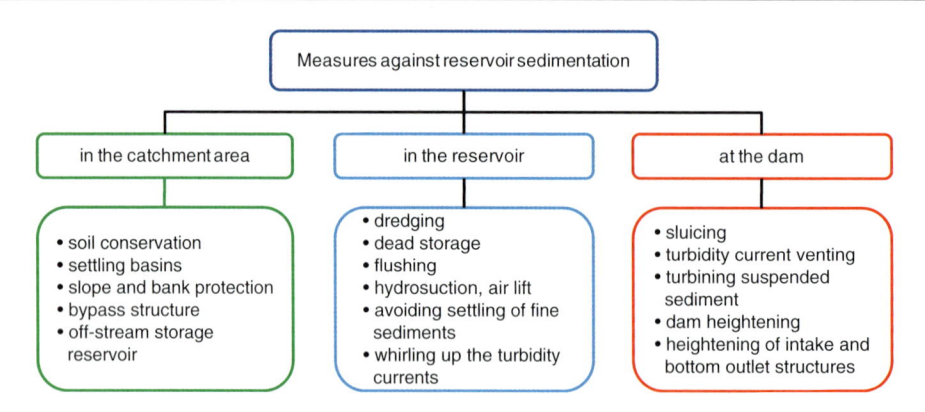

Sedimentation of Reservoirs, Figure 3 Inventory of possible measures for sediment management (Schleiss and Oehy, 2002).

source of water supply (Morris, 1996). A sustainable sediment strategy should also include the downstream reaches; therefore, monitoring data should also include downstream impacts as well as sedimentation processes in the reservoir (Morris and Fan, 1997).

The known remedial measures can be subdivided in those taken in the catchment area, in the reservoir, and at the dam itself as shown in Figure 3. Oehy (2003) and Oehy and Schleiss (2007) proposed and studied several technical measures against turbidity currents. Turbidity currents may be stopped and forced to settle by obstacles situated in the upper part of the reservoir in order to keep the outlet structures free of sediments. They can also be whirled up near the dam and intakes and kept in suspension, which allows a continuous evacuation through the turbines (Jenzer Althaus, 2011). In certain cases, fully venting of turbidity currents through outlet structures is possible.

Summary

Today, the worldwide yearly mean loss of storage capacity due to sedimentation is already higher than the increase of capacity by the construction of new reservoirs for irrigation, drinking water, and hydropower. Thus, the sustainable use of the reservoirs is not guaranteed in long term. In the case of alpine reservoirs, the sedimentation rate is much below the world mean value. Nevertheless, sedimentation threatens also these reservoirs since turbidity currents are sporadically transporting large volumes of sediments like an underwater avalanche down to the dam. There, the concentrated deposits are hindering the safe operation of the outlet structures as intakes and bottom outlets. Many possible measures against sedimentation are known from practice, but they are strongly depending on the local conditions. For reservoirs in rather narrow valleys, technical measures, which can govern turbidity currents, are of special interest. The problematic of sedimentation and sediment management should be considered in the early stage of the design of the reservoir in order to obtain sustainable solutions. Although methods for erosion volume estimation and empirical relationships for trap efficiency estimation are available, this is still not the case for many reservoirs built recently all over the world.

Bibliography

Alam, S., 1999. The influence and management of sediment at hydro projects. *Hydropower & Dams*, **3**, 54–57.

Basson, G. R., 2009. Management of siltation in existing and new reservoirs. General Report Q. 89. In *Proceedings (on CD) of the 23rd Congress of the International Commission on Large Dams CIGB-ICOLD*, May 25–29, 2009, Brasilia, Vol. II.

Batuca, D. G., and Jordaan, J. M., 2000. *Silting and Desilting of Reservoir*. Rotterdam: A.A. Balkema.

De Cesare, G., Schleiss, A., and Hermann, F., 2001. Impact of turbidity currents on reservoir sedimentation. *Journal of Hydraulic Engineering*, **127**(1), 6–16.

Fan, J., and Morris, G. L., 1992. Reservoir sedimentation. II: reservoir desiltation and long-term storage capacity. *ASCE, Journal of Hydraulic Engineering*, **118**(3), 370–384.

Graf, W. H., 1984. Storage losses in reservoirs. *International Water Power & Dam Construction*, **36**(4), 37–40.

International Committee on Large Dams (ICOLD), 1989. *Sedimentation Control of Reservoirs – Guidelines*. Paris: ICOLD, Bulletin Vol. 67.

Jacobsen, T., 1998. *New Sediment Removal Techniques and Their Applications*. Aix-en-Provence, France: Hydropower & Dams, pp. 135–146.

Jenzer Althaus, J., 2011. Sediment evacuation from reservoirs through intakes by jet induced flow. In Schleiss, A. (ed), *Communication 45 du Laboratoire de constructions hydrauliques*, PhD Thesis No. 4927. Lausanne, Switzerland, EPFL.

Jenzer Althaus, J. M. I., De Cesare, G., Boillat, J.-L., Schleiss, A. J., 2009. Turbidity currents at the origin of reservoir sedimentation, case studies. In *Proceedings (on CD) of the 23rd Congress of the International Commission on Large Dams CIGB-ICOLD*, May 25–29, 2009, Brasilia, Vol. II, Q.89-R.24.

Morris, G. L., 1995. Reservoir sedimentation and sustainable development in India: problem scope and remedial strategies. In *Sixth International Symposium on River Sedimentation, Management of Sediment: Philosophy, Aims, and Techniques*, New Delhi, November 7–11, 1995. Rotterdam, The Netherlands: Balkema, pp. 53–61.

Morris, G. L., 1996. Reservoirs and integrated management. In Bruk S., and Zebidi, H. (eds.), *Reservoir Sedimentation, Proceedings of the St. Petersburg Workshop*, May 1994. IHP-V, Technical Documents in Hydrology No. 2. Paris: UNESCO, pp. 135–148.

Morris, G. L., and Fan, J., 1997. *Reservoir Sedimentation Handbook: Design and Management of Dams, Reservoir and Watersheds for Sustainable Use*. New York: McGraw-Hill, xxiv+805 pp.

Oehy, C., 2003. Effects of obstacles and jets on reservoir sedimentation due to turbidity currents. In Schleiss, A. (ed.), *Communication 15 du Laboratoire de Constructions Hydrauliques*. PhD Thesis No. 2684. Lausanne, Switzerland, EPFL.

Oehy, Ch, and Schleiss, A., 2007. Control of turbidity currents in reservoirs by solid and permeable obstacles. *Journal of Hydraulic Engineering ASCE*, **133**(6), 637–648.

Schleiss, A., and Oehy, C., 2002. Verlandung von Stauseen und Nachhaltigkeit. *Wasser, Energie, Luft-Eau, Énergie, Air*, **94**(7/8), 227–234.

Cross-references

Climate Change
Erosion
Erosivity
Global Change and Its Implications for Natural Disasters
Reservoir Dams and Natural Hazards

SEICHE

Giovanni Cuomo
Hydraulics Applied Research & Engineering Consulting (HAREC) s.r.l, Rome, Italy

Synonyms

Harbor wave; Long period waves; Tidal wave

Definition

A seiche is a standing wave in an enclosed or partially enclosed body of water. Seiches have been observed in lakes, harbors, reservoirs, swimming pools, bays and seas.

Discussion

The word originates in a Swiss French dialect word that means "to sway back and forth" and was first used by the Swiss hydrologist François-Alphonse Forel in 1890, to make scientific observation of the seiches occurring in Lake Geneva, Switzerland.

Seiches can be caused by meteorological forcing (wind and atmospheric pressure variations), wind waves and swells, seismic activity, landslides, and tsunamis.

Once originated, the wave is reflected from the ends and the water surface oscillates around its rest (horizontal) position. Repeated reflections produce standing waves whose shape corresponds to one of the natural modes of oscillation of the water body, as a function of its geometry.

For a rectangular basin of length L, width W, and depth h, the period of each mode can be evaluated using the generalized Merian's formula:

$$T_{mn} = \frac{2L}{\sqrt{gh}}(\alpha^2 m^2 + n^2)$$

where $\sqrt{gh}$ is the speed of a shallow water wave, $\alpha = L/W$ and m and n define the harmonic nodes in the transverse and longitudinal directions.

Generally, the motion of the water body develops as a combination of the natural modes of the basin. If the forcing is impulsive (i.e., tsunamis or pressure front), the disturbed body of water tends to return at rest after dissipating its energy within a number of free oscillations; if the forcing is periodic, with a frequency close to that of one or more of the natural frequencies of the basin, the body of water can resonate locally amplifying the amplitude of the seiche.

Seiches are often imperceptible to the naked eye due to their usually small amplitude and long wavelength, resulting in a small steepness of the water surface. Small amplitude seiches are almost always present on large lakes; harbors, bays, and estuaries are often prone to seiches with amplitudes of a few centimeters and periods of a few minutes. Seiches can also form in semi-enclosed seas. For instance, in the Adriatic Sea and the Baltic Sea, they contribute to flooding of Venice and St. Petersburg, respectively.

Tsunami can also excite seiches in bays and harbors, and these can in turn be amplified due to the interaction of consecutive tsunami waves with local bathymetric and topographic features, originating very large waves. For instance, the tsunami that struck Hawaii in 1946 had a 15-min interval between wave fronts; since the natural period of oscillation of Hilo Bay is about 30 min, consecutive waves were in phase with the motion of Hilo Bay, and excited a resonant seiche in the bay, which reached a height of up to 26 ft (8 m circa), killing 96 people in the city alone.

Harbors are usually prone to wave-induced seiching (Los Angeles and Long Beach, Rotterdam, Barbers Point and Marina di Carrara) with standing wave patterns resulting in strong localized currents at the nodes (points that experience no vertical motion) reducing the operability of berths and eventually inducing breakage of mooring lines.

Bibliography

Kim, Y. C., 2009. Seiches and harbor oscillations. In *Handbook of Coastal and Ocean Engineering* (1192 p). Singapore: World Scientific Publishing Company, pp. 193–236.

Okihiro, M., Guza, R. T., and Seymour, R. J., 1993. Excitation of Seiche observed in a Small Harbor. *Journal of Geophysical Research*, **98**(C10), 18201–18211.

Wiegel, R. L., 1964. Tsunamis, storm surges, and harbor oscillations. In *Oceanographical Engineering*. Englewood Cliffs, NJ: Prentice-Hall, Chap. 5, pp. 95–127.

Cross-references

Breakwater
Coastal Erosion
Coastal Zone, Risk Management
Displacement Wave
Fetch
Hydrograph, Flood
Marine Hazards
Reservoir Dams and Natural Hazards
Rogue Wave
Storm Surges
Tidal Bores
Tsunami
Vaiont Landslide, Italy

SEISMIC GAP

John F. Cassidy
Natural Resources Canada, Sidney, BC, Canada

Synonyms

Earthquake forecasting; Seismic potential

Definition

Seismic Gap. A segment of an active plate boundary that, relative to rest of the boundary, has not recently ruptured and is considered to be more likely to produce an *Earthquake* in the future.

The seismic gap theory (for a summary of early work in this field see McCann et al., 1979) states that a segment of a plate boundary that has not ruptured recently has the greatest chance of rupturing in the future (relative to other segments that have experienced large earthquakes). This is based on the recognition that tectonic plates move relative to one another at an approximately constant speed and the assumption that the slip of plate boundary faults occurs primarily during major (magnitude 7 or greater) earthquakes. There are clear applications of this method to the estimation of earthquake hazards and to "earthquake forecasting" (e.g., McCann et al., 1979). However, there are questions about the applicability of this method (e.g., see Kagan and Jackson, 1991, 1995). Some variables that affect this hypothesis include the possibility of aseismic slip, an incomplete earthquake record, distribution of slip along several faults on a plate boundary (e.g., the San Andreas/Hayward fault systems in the San Francisco area), earthquake triggering and clustering, and the uncertainty in the recurrence interval for large earthquakes along plate boundaries. In recent years, Global Positioning Satellite (GPS) data and *Paleoseismology* (that provides a better record of the earthquake history) have been used to better understand plate boundaries, plate motions, fault segmentation, and the possibility of aseismic slip. Some recent examples of the applications of earthquake forecasting include:

1. Using the earthquake catalog of Chile and recent GPS measurements, Ruegg et al. (2009) identified the Concepción–Constitución region of central Chile as a mature seismic gap. In their article published in June 2009, they concluded, "a worst-case scenario of a magnitude 8–8.5 earthquake should it happen in the near future." Eight months after that article was published, a magnitude 8.8 subduction earthquake occurred in this seismic gap and ruptured nearly 700 km of the plate boundary. Madariaga et al. (2010) summarize research on this seismic gap and the subsequent earthquake.
2. Using GPS data, detailed geological information from faults, paleoseismology, and *Seismology*, a 59% probability in the next 30 years for one of more magnitude 6.7 or greater earthquakes on the San Andreas Fault in southern California has been calculated by the 2007 Working Group on California Earthquake Probabilities (2008).

With improved seismic, GPS, and paleoseismological data, and better earthquake catalogs and understanding of fault interaction, earthquake triggering, and rupture, we can expect to see improvements in the area of understanding "seismic gaps" and forecasting the location of future large earthquakes. This, in turn, should lead to improved earthquake hazard estimations and may help focus earthquake preparedness and mitigation efforts.

Bibliography

2007 Working Group on California Earthquake Probabilities, 2008. The Uniform California Earthquake Rupture Forecast, Version 2 (UCERF 2): *U.S. Geological Survey Open-File Report 2007-1437 and California Geological Survey Special Report 203*, 104 pp. [http://pubs.usgs.gov/of/2007/1437/].

Kagan, Y. Y., and Jackson, D. D., 1991. Seismic gap hypothesis: ten years after. *Journal of Geophysical Research*, **96**, 21419–21431.

Kagan, Y. Y., and Jackson, D. D., 1995. New seismic gap hypothesis: five years after. *Journal of Geophysical Research*, **100**, 3943–3959.

Madariaga, R., Metois, M., Vigny, C., and Campos, J., 2010. Central Chile finally breaks. *Science*, **328**, 181–182.

McCann, W. R., Nishenko, S. P., Sykes, L. R., and Krause, J., 1979. Seismic gaps and plate tectonics: seismic potential for major boundaries. *Pageoph*, **117**, 1082–1147.

Ruegg, J. C., Rudloff, A., Vigny, C., Madariaga, R., de Chabalier, J. B., Campos, J., Kausel, E., Barrientos, S., and Dimitrov, D., 2009. Interseismic strain accumulation measured by GPS in the seismic gap between Constitución and Concepción in Chile. *Physics of the Earth and Planetary Interiors*, **175**, 78–85, doi:10.1016/j/pepj.2008.02.015.

Cross-references

Earthquake
Earthquake Prediction and Forecasting
Geological/Geophysical Hazards
Global Positioning System (GPS) and Natural Hazards
Paleoseismology
Seismology

SEISMOGRAPH/SEISMOMETER

Allison Bent
Natural Resources Canada, Ottawa, ON, Canada

Definition

A seismograph is an instrument used to detect and record ground motion caused by seismic waves.

Discussion

A seismograph typically consists of three components – a sensor (seismometer), an amplifier or digitizer, and a recording device. Although seismographs have evolved over the years, all operate on the same basic principles by measuring ground motion relative to something that remains fixed or unaffected by the shaking. Early seismographs usually consisted of a mass suspended by a spring within a case. The mass would remain still while the case shook. The relative movement was usually recorded on paper by a pen or light source. In most modern seismographs, the mass has been replaced by a permanent magnet. Ground shaking causes the magnet to generate electric signals in the wires that are wound tightly in the surrounding case. The ground motions are digitally recorded and either stored on site or transmitted to another site for archiving. The record showing the recorded waveform is referred to as a seismogram.

Some seismographs are designed for use over a narrow range of frequencies but broadband instruments, which detect and record signals over the full range of frequencies of interest to seismologists are becoming more commonly used. Typically a seismograph station will host three seismographs to detect ground motion in three orthogonal directions – vertical and two horizontal directions, most commonly north–south and east–west – although there are some stations that have only a vertical seismograph. Seismographs can detect ground motions much smaller than that can be felt by a human. Although designed principally for recording ground motions caused by earthquakes, seismographs record seismic waves triggered by other sources, such as underground explosions.

Bibliography

Bolt, B. A., 1993. *Earthquakes*. New York: W. H. Freeman.
Incorporated Research Institutions for Seismology, 2010. GSN Instrumentation, http://www.iris.edu/hq/programs/gsn/instrumentation.
Natural Resources Canada, 2010. How we record earthquakes, http://earthquakescanada.nrcan.gc.ca/info-gen/smters-smetres/seismograph-eng.php.
Richter, C. F., 1958. *Elementary Seismology*. San Francisco: W. H. Freeman.

Cross-references

Accelerometer
Body Wave
Earthquake
Epicenter
Harmonic Tremor
Intensity Scales
Isoseismal
Magnitude Measures
Mercalli, Giuseppe (1850–1914)
Modified Mercalli (MM) Scale
Primary Wave (P Wave)
Richter, Charles F.
Secondary Wave (S Wave)
Seismology

SEISMOLOGY

Alik T. Ismail-Zadeh
Karlsruher Institut für Technologie, Karlsruhe, Germany
Institut de Physique du Globe de Paris, Paris, France
Institute of Earthquake Prediction Theory and Mathematical Geophysics, Russian Academy of Sciences, Moscow, Russia

Definition

Seismology is a branch of geophysics dealing with earthquakes and other Earth's vibrations, with the sources that generate the vibrations, with seismic wave propagation, and with the structures through which they propagate.

Discussion

Seismology can be considered as a tool for exploring the Earth's interior, and it is deeply rooted in continuum and fracture mechanics, theoretical physics, applied mathematics, and statistics. Seismology has a remarkable diversity of applications in natural hazard research including seismic hazard assessment, *Earthquake* forecasting, volcano and *Tsunami* warning systems, and detections of sudden movements of glaciers and ice sheets.

There are several geophysical disciplines to explore the Earth's interior, including studies of gravity, magnetic and electric properties of the planet, but seismology is perhaps the most powerful and informative, especially in studies of the deep mantle and core. Seismologists study the internal structure of the Earth by recording and analyzing the signals arising from natural (e.g., earthquakes) and human-made (e.g., explosions) energy sources. Such seismic signals contain a wealth of information that enables scientists to quantify active wave sources and to reveal structures in the planetary interior associated with dynamic processes that occurred in the geological past and are now active. Additionally, if an earthquake is of sufficient magnitude, the Earth as a whole may be set into oscillations of various modes, and the periods of these free oscillations can yield useful information.

Earthquakes occur as a sudden release of stresses. When an earthquake occurs, part of the released energy generates elastic waves propagating through the Earth. These waves can be detected by a seismograph that measures and records the ground motion. There are two principal types of elastic waves: body waves, which travel through the body of the

Earth, and surface waves, which are guided along and near the surface of the Earth (e.g., Stein and Wysession, 2003).

Seismology can be divided into several sub-disciplines including earthquake seismology, *Paleoseismology* (studies of past large earthquakes by analyzing near-surface sediments and rocks), statistical seismology (statistical analysis and modeling of earthquake occurrence in time and space), and computational seismology (e.g., quantitative modeling of seismic wave propagation, earthquake occurrences, and other seismological phenomena).

Although the origin of seismology dates back to the East Han Dynasty in China, it is misleading to pretend that state-of-the-art seismology is a well-developed branch of science. To the contrary, most of the seismologists clearly understand the grand challenges for seismology. Several important questions still remain pressing, including: How do faults slip? How do magmas ascend? How does the near-surface environment affect natural hazards? (Forsyth et al., 2009). The fundamental difficulty in answering these questions comes from the fact that no earthquake or magma ascent from its chamber has ever been observed directly and just a few of them were subject to an in situ verification of their physical parameters.

Bibliography

Forsyth, D. W., Lay, T., Aster, R. C., and Romanowicz, B., 2009. Grand challenges for seismology. *EOS, AGU Transactions,* **90**(41): doi:10.1029/2009EO410001.

Stein, S., and Wysession, M., 2003. *An Introduction to Seismology, Earthquakes, and Earth Structure*. Oxford: Blackwell.

Cross-references

Accelerometer
Body Wave
Earthquake
Earthquake Prediction and Forecasting
Epicenter
Geological/Geophysical Disasters
Harmonic Tremor
Intensity Scales
Isoseismal
Macroseismic Survey
Magnitude Measures
Paleoseismology
Plate Tectonics
Primary Wave (P Wave)
Rotational Seismology
Seismic Gap
Seismograph/Seismometer
Tectonic Tremor
Tsunami

SHEAR

Murat Saatcioglu
University of Ottawa, Ottawa, ON, Canada

Definition

Shear is force applied parallel to, or in the plane of a cross section of a structural member. Shear stresses are typically computed across a section that is perpendicular to the member length. Therefore, shear forces used in structural design are often perpendicular (transverse) to member length.

Discussion

Shear forces generated by a natural hazard can occur in horizontal or vertical directions depending on the type of hazard. Earthquake-induced inertia forces are generated from accelerations that occur during seismic excitations. Although seismic shear forces on structures can be developed in any direction, it is generally accepted that they are more critical in the horizontal direction.

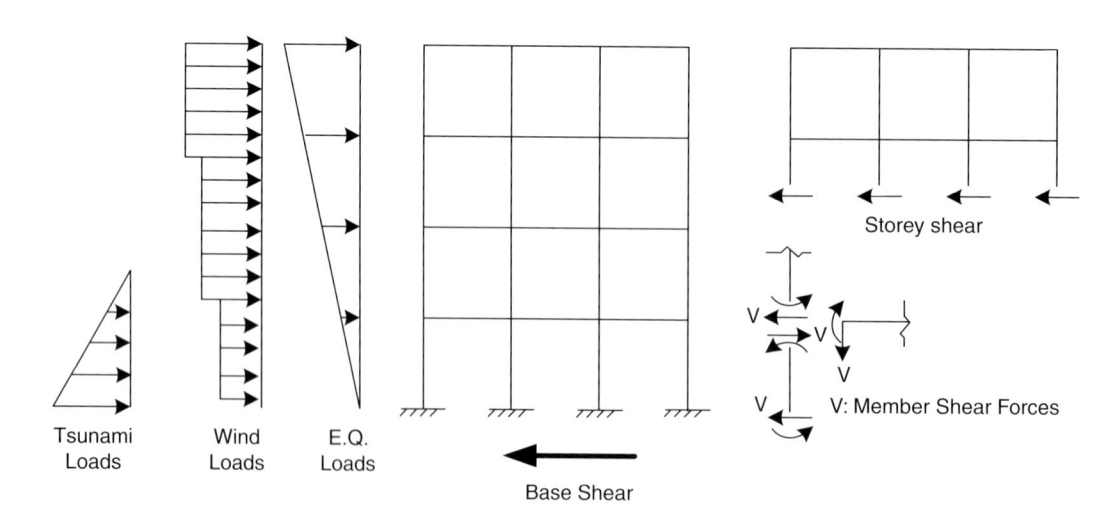

Shear, Figure 1 Examples of base shear, story shear, and member shear forces.

Therefore, structures are routinely analyzed using horizontal (lateral) forces in two orthogonal directions of structural plan. These forces are obtained either from dynamic analysis or equivalent static load analysis. Equivalent static loads are computed by distributing "base shear force" computed from the code of practice. "Story shear" is computed by accumulating lateral seismic forces above the story level. Vertical seismic excitations also contribute to member shear. Shear forces associated with vertical earthquake accelerations are considered to be a fraction of those generated by horizontal accelerations, though very large vertical accelerations have been recorded during previous earthquakes. Shear usually occurs in combination with flexure (bending), and sometimes in combination with axial load. Shear forces either produce flexural (bending), or they are produced by flexure. Typical shear forces caused by natural hazards are illustrated in Figure 1.

Tsunami-induced hydraulic bores, storm surges, and floods generate shear forces primarily in the horizontal direction. These shear forces may be caused by hydrostatic and hydrodynamic water pressures, impulsive hydraulic bores, and/or impacting debris. Tsunamis and floods can also result in vertical uplift (buoyant) forces, generating vertical shear forces in horizontal elements. Shear created by these forces are computed in much the same manner as those shown in Figure 1. Similarly, shear forces generated by wind pressures associated with windstorms and other extreme wind hazards are primarily applied in the horizontal direction. These forces are calculated by multiplying external wind pressures and suctions on windward and leeward surfaces by applicable surface areas over which they act, generating story shears as shown in Figure 1.

Shear forces generate shear stresses and shear distortions, as illustrated in Figure 2. Shear stress distribution across a section is parabolic within the elastic range of materials. Shear capacity of a member is dictated either by its direct shear capacity parallel to the direction of shear force, as in the case of bolts connecting two plates, or by principal tensile or compressive stresses that develop from shear along an inclined plane, as in the case of reinforced concrete elements. Structural members often fail in shear due to diagonal tension or compression. This is illustrated in Figure 3.

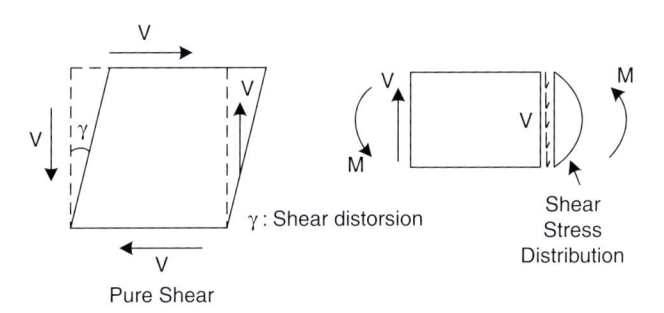

Shear, Figure 2 Shear distortions and shear stress distribution.

a Bolts in direct shear

b Reinforced concrete beam in shear
t: Diagonal tension
c: Diagonal compression

c Shear failure of bolt (Aleszka 2010)

d Diagonal tension crack in reinforced concrete column (Saatcioglu et al. 2001)

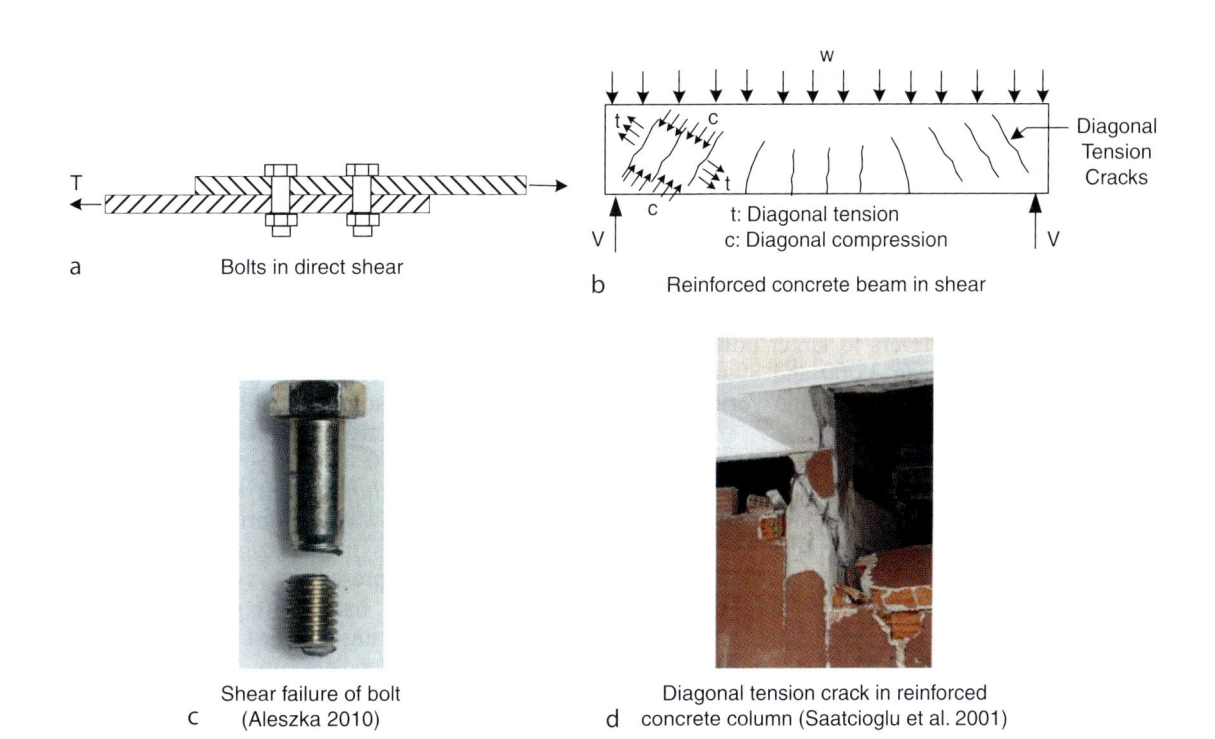

Shear, Figure 3 Direct shear versus diagonal tension and compression caused by shear.

Sinkhole, Figure 1 Several solution sinkholes developed on limestones in the Sinkhole Plain (Mammoth area) in Kentucky (USA). Some of them contain water.

Sinkhole, Figure 2 Collapse sinkhole located in Chichén Itzá (Mexico) Locally named *cenotes*.

2. Collapse of the roof of a cavern or conduit is propagated upward until reaching the surface. Consolidated rocks (Figure 2) and unconsolidated cover (Figure 3) can be affected. The slopes of the collapse sinkhole are vertical, but its later evolution generates funnel and even basin morphologies.
3. Suffosion is the evacuation of unconsolidated sediments through solution conduits. Such sinkholes have varying slopes.

There are genetic classifications of sinkholes (Williams, 2003; Waltham et al., 2005) that basically consider the type of mechanism involved and the type of material affected. As an example, caprock collapse sinkhole indicates that the upward propagation of collapse affects non-karst rocks.

The occurrence of sinkholes is worldwide: examples can be found in the USA, Canada, Mexico, Cuba, Italy, Great Britain, Spain, Israel, Turkey, the Russian

Sinkhole, Figure 3 Cover collapse sinkhole in the Ebro Basin (Spain) caused by dissolution of evaporite rocks.

Federation, Iran, and elsewhere. They can cause severe economic problems in populated areas.

Specific terms include cenote (collapse sinkhole flooded with water) and blue holes (large collapse flooded with seawater).

Bibliography

Ford, D., and Williams, P., 2007. *Karst Hydrogeology and Geomorphology*. Chichester: Wiley. 562 pp.

Waltham, T., Bell, F., and Culshaw, M., 2005. *Sinkholes and Subsidence. Karst and Cavernous Rocks in Engineering and Construction*. Berlin: Springer. 382 pp.

Williams, P., 2003. Dolines. In Gunn, J. (ed.), *Encyclopedia of Caves and Karst Science*. New York and London: Taylor & Francis, pp. 304–310.

Cross-references

Collapsing Soil Hazards
Creep
Deep-Seated Gravitational Slope Deformations
Dispersive Soil Hazards
Expansive Soils and Clays
Land Subsidence
Mining Subsidence Induced Fault Reactivation
Subsidence Induced by Underground Extraction
Urban Environments and Natural Hazards

SLIDE AND SLUMP

Lionel E. Jackson, Jr.
Geological Survey of Canada, Vancouver, BC, Canada

Definitions

Slide A slide is the displacement of a rigid or semi-rigid mass of soil or rock so that it descends along a distinct underlying failure surface under the influence of gravity. If the failure surface is planar, it is referred to as a *translational slide*.

Slump. A slump is the name given to a *slide* when the underlying failure surface is curved rather than planar. The motion of a slump is rotational, so that portions of the slide drop and rise with respect to the adjacent stable slopes. It is more properly referred to as a *rotational slide*.

Introduction

Slides and rotational slides (slumps) include downward and horizontal displacement of rigid or semi-rigid masses of soil (earth and debris), or rock under the influence of gravity. This includes movements of land and the sea bottom. It differs from a *collapse* where displacement is entirely vertical. Landslides are classified according to the type of material that is moving or has moved, the type of movement involved, and the velocity of movement (Cruden and Varnes, 1996; Multinational Andean Project, 2009). In slides, rigid or semi-rigid masses are displaced vertically and horizontally along well-defined underlying failure surfaces. This is in contrast to flow-type failures where earth material behaves like a liquid (Hungr et al., 2001) or failures called *topples* or *falls* that involve transport in part by free fall (Cruden and Varnes, 1996). Movements of slides are further subdivided based upon the geometry of the underlying failure surface (Table 1). If it approximates a plane (Figure 1, diagram 1, Figure 2), it is classified as a *translational slide*. If it approximates a curved surface (Figure 1, diagrams 4, 5; Figure 3), it is classified as a *rotational slide*. The latter is commonly referred to as a *slump*. Although the failure plain of a rotational slide is seldom directly observable, its curved geometry is inferred in the field on the basis of rotational movement in the landslide. The upper portion of the landslide drops below the level of the adjacent land surface whereas the middle or lower part of the landslide rises above its pre-failure level (Figure 3). Velocities of translational and rotational landslides can range from very slow (millimeters/year) to very rapid (meters/second). Translational and rotational slides are further subdivided based upon the material involved (Tables 1, 2).

In order for earth or debris to move as a semirigid mass, water content must be below the plastic limit so that the mass behaves as a solid. With increasing water content beyond the plastic limit or remolding of earth and debris (which contain fine particles) by landslide movement, translational and rotational slides become increasingly

Slide and Slump, Table 1 Classification of translational and rotational slides (After Cruden and Varnes, 1996)

Movement type	Subtype based upon material
Slide (translational)	Rock slide, debris slide, earth slide, rock wedge failure
Slide (rotational)	Rotational rock slide, rotational debris slide, rotational earth slide

plastic in behavior. They can eventually grade into flow-type landslides such as *earth flows* and *debris avalanches*. Translational or rotational failures of steep rock slopes can transition into *rock avalanches*.

Mechanics of translational slides and environments conducive to them

In rock, failure planes underlying translational slides are typically preexisting natural planes of weakness. They

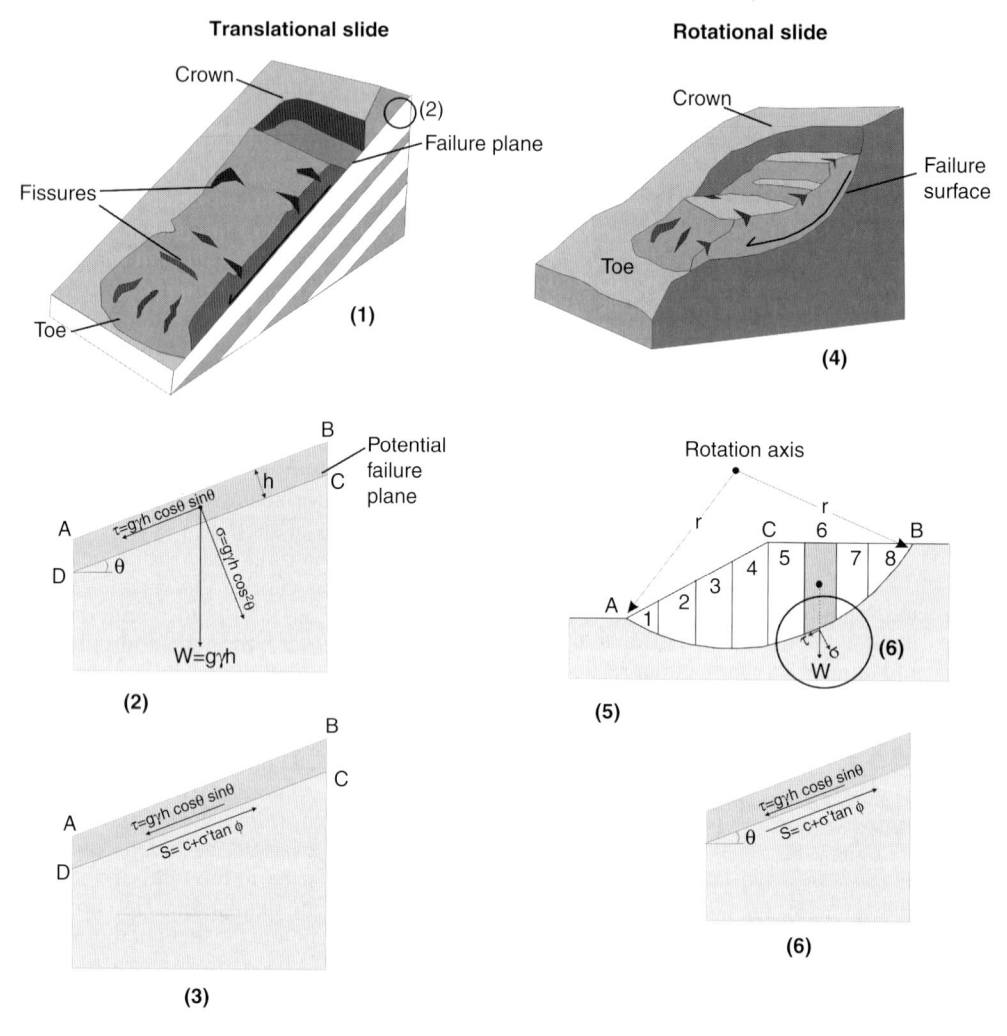

Slide and Slump, Figure 1 Diagrams illustrating translational and rotational slides and forces determining stability within the geologic and geomorphologic environments in which they occur. *Diagram 1* depicts a translational slide in bedded sedimentary rock. The slide has occurred along a bedding plane dipping at an angle, θ. *Diagram 2* summarizes gravitational forces acting on polygon A-B-C-D. The *dot* indicates its center of mass. W is the gravitational force acting on the polygon as a whole; shear stress (τ) is the component acting parallel to potential failure plane C-D; and σ is the normal stress acting normal to C-D. *Diagram 3* shows the opposed forces of shear strength (S) and shear stress (τ). The ratio of the two is the *factor of safety* which must be >1 if it is to remain stable. *Diagram 4* depicts a rotation slide. Arc A-B in *diagram 5* is a potential failure surface in a homogenous earth material such as thick marine clay or mudstone. For the sake of explanation, A-B is the potential failure surface where shear stress is a maximum for slope A-C. The stability of the slope can be determined by dividing it into elements (*Diagrams 5 and 6*) and totaling shear strength and shear stress.

include bedding planes in sedimentary rocks (Figure 2), foliation (parallel orientation of the long axes of minerals) and cleavage planes in metamorphic rocks, and joints and faults in rocks of all types. Where two or more planes of weakness intersect in steep rock slopes, failure of part or all of the mass defined by their intersection is a *wedge failure* (Figure 4). Unconsolidated sediments, such as colluvium, till (boulder clay), and tephra, commonly fail as translational slides where they overlie rock. Forces involved in translational slides in rock or overburden on rock for an idealized slope are summarized in Figure 1, diagrams 1–3 (modified from Ritter, 1986).

In Figure 1, diagram 2, polygon A-B-C-D (light gray) represents a mass of rock or sediment overlying rock (dark gray) along potential failure plane C-D (sedimentary contact, bedding plane, joint, etc.). The black dot represents its center of mass. It has a thickness h (m), a unit mass γ (kg/m^3). C-D is inclined at an angle θ. Gravity (g) pulls the slope vertically toward the center of the earth with a force, W (in Newtons (N)):

$$W = g\gamma h \quad (1)$$

For the purpose of calculating forces that drive and resist translational slides, W (Equation 1) can be resolved into the force acting parallel to the potential failure plane in a descending direction (shear stress (τ)) and the force normal to it (normal stress (σ)) so that:

$$\tau = g\gamma h \, \cos\theta \, \sin\theta \quad (2)$$

and

$$\sigma = g\gamma h \, \cos^2\theta \quad (3)$$

Normal stress across C-D (σ) in Equation 3 assumes absence of water in the slope. However, water is

Slide and Slump, Figure 2 Failure surface of the Qianjiangping landslide, Three Gorges Reservoir, China, December, 2009. The failure surface is a bedding plane (see Wang et al., 2004; Photo by Lionel Jackson, Geological Survey of Canada).

Slide and Slump, Figure 3 Large rotational slides, Yocarhuaya, Bolivia, March, 2009 (defined by *dashed lines*). The slides occur in massive landslide colluvium at the toe of a 5-km-long earthflow complex. *Arrows* show relative vertical movement. Movement of the upper slide caused abandonment of part of the Yocarhuaya community (between the *upper set of arrows*). The church steeple (S) is about 8 m in height. (Photo by Lionel Jackson, Geological Survey of Canada.)

Slide and Slump, Table 2 Definition of Landslide Materials (After Varnes, 1978 and Couture, 2011)

Rock	Material made up of strongly bonded minerals or cemented aggregations of rock fragments and mineral material. In soil engineering practice, rock is commonly differentiated from debris and earth in that it cannot be ripped or ripped with difficulty by earth-moving machinery in its unfailed state and usually requires blasting if it is to be excavated
Soil	Aggregate of solid, typically inorganic particles that was either transported or formed *in situ* by weathering of rock; subdivided into *earth and debris* *Earth*: Generally unconsolidated material dominated by sand and finer sediments. It usually exhibits low plasticity. Earth is usually the breakdown product of silt-and-clay-rich sedimentary rocks such as shale and mudstone and igneous or metamorphic rocks prone to chemical breakdown *Debris*: Disordered mixtures of fragments of rock and sediment (sand and finer sized particles). Colluvium (product of rock breakdown), glacial sediments, volcanic ejecta, and former landslide material are types of debris. It can also include a significant content of organic detritus such as logs and organic-rich soil

Slide and Slump, Figure 4 Scar created by a wedge failure in a bedrock slope, Vancouver Island, British Columbia, Canada. The failure occurred at the intersection of two joint systems A and B in plutonic rock. Instrument in foreground is approximately 30 cm across. (Photo courtesy of Marc-Andre Brideau, Simon Fraser University.)

universally present in slopes from trace amounts to total saturation. It is referred to as *pore pressure* (μ). It may be positive when the slope is saturated or negative when water is held in pores by capillary attraction. Pore pressure is an important factor in many slides. Thus, σ must be adjusted for it because μ can reduce or increase it. The term *effective normal stress* is used and its symbol is σ':

$$\sigma' = \sigma - \mu \tag{4}$$

A component of the effective normal stress is not the only force acting to resist τ. There are frictional and cohesive forces that act along the failure plane. Cohesion (c) can be thought of as molecular attractive forces between soil and rock particles along the potential failure plane. These have to be overcome in order for translational movement to begin. Movement must also overcome frictional forces between grains in rock or sediment. This is referred to as *internal friction* which is expressed as an angle φ which is determined by testing in soil or rock mechanics laboratories. The total resistance to shear stress (τ) is called shear strength (S). Shear strength is related to cohesion, effective stress, and internal friction:

$$S = c + \sigma' \tan \varphi \tag{5}$$

Equation 5 is known as the *Coulomb equation*. Shear strength acts in opposition to shear stress and is depicted in Figure 1, diagram 3. The ratio of shear stress and shear strength along a failure plane is called the *factor of safety* (F). F must be >1 or the slide begins to move:

$$F = \frac{\text{Shear strength}}{\text{Shear stress}} \tag{6}$$

Translational slides are among the largest in mountain regions. Once in motion, these may continue to travel slowly as a broken slab like the Downie slide in British Columbia (Piteau et al., 1978), or transition into rapidly moving, fluid-like rock avalanches like the 1903 Frank Slide in Alberta, Canada (Cruden and Krahn, 1978). Over time, various processes work to reduce the factors that act together to produce shear strength (Equation 5). This is why mountain and hill slopes can remain stable for thousands of years and then fail as translational slides or other types of landslides. For example, weathering processes may slowly alter the minerals that cement or provide cohesion along bedding planes, faults, and joints. Such alteration can reduce cohesion or the angle of internal friction, reducing shear strength. Hydrothermal alteration can also be a factor in reducing shear strength in volcanic (Siebert, 2002) and nonvolcanic settings (Piteau et al., 1978). With time, the combination of long-term reduction in these factors and transient reduction in effective stress due to high pore pressure (e.g., prolonged or heavy rainfall or snow melt) can reduce F to 1 causing failure. Similar processes occur with time in unconsolidated sediments on bedrock. Shear stress τ can also increase naturally with time. Tectonic movements or erosion at the base of slopes can steepen slopes (increase θ). These act to increase τ and decrease effective stress σ', thus reducing shear strength S (Equations 2, 3, 4). In regions prone to earthquakes, acceleration of slopes due to seismic shaking can trigger slides by increasing τ or decreasing σ' so that the factor of safety F drops to 1 on normally stable slopes (Keefer, 1984). Lastly, slopes can be loaded from above by landslides up-slope. Such loading increases τ. Poorly sited artificial fills on slopes can have similar effects.

Mechanics of rotational slides (slumps) and environments conducive to them

Rotational slides (Figure 1, diagram 4) typically occur in thick, relatively homogenous sediments such as clay, silt, and artificial fills. They also occur in massive and weak sedimentary rocks such as mudstone or claystone. In such slopes, a semicircular or curved surface defines a potential surface of failure where shear stress is at a maximum relative to shear strength. The successive rotational slides in Figure 3 occur in landslide colluvium of a 5-km-long landslide complex. The landslide colluvium is many tens of meters in thickness.

In reality, there is a continuum between translational and rotational slides. Rotational slides can have relatively planar segments interspersed with curved segments and still display rotational movement at the surface. Rotational slide colluvium is commonly remolded as a result of slide movement. This remolding can weaken and raise pore pressure within the slide colluvium so that it begins to flow. Transformation of rotational slides into *earthflows* or *debris flows* is common.

Rotational slides are triggered when the ratio of shear strength to shear stress along a critical curved surface approaches 1 (Figure 1, diagrams 4–6). Because the failure occurs along a curved surface (A-B), movement approximates rotation around an axis with a radius r. Sophisticated numerical approaches are used to locate and evaluate potential failure surfaces (e.g., U.S. Army Corps of Engineers, 2003). These subdivide slopes into elements and compute forces acting on them. In Figure 1, diagrams 5 and 6 illustrate the general principle (modified from Terzaghi and Peck (1967) and Krynine and Judd (1957)). The potential slide in diagram 5 is subdivided into elements (1–8 in the example). In element 6, the dot represents the center of mass of this element and W, τ, and σ vectors are shown at the base of the element similar to diagram 2 since it is assimilated to a planar failure surface. To compute the factor of safety for the slope, shear strength and shear stress are evaluated for each element. Because of the curved surface, the shear stresses of some elements are opposed to other elements. Elements 1–3 would tend to rotate the mass to the right (up-slope) whereas elements 5–8 would rotate it to the left (down-slope). Element 4 straddles the lowest part of the curve: Shear stresses in it cancel themselves. The factor of safety F is the sum of the shear strength for all the elements over the sum of the shear stress for all the elements.

Natural processes discussed with respect to translational slides can alter shear strength and shear stress in slopes subject to rotational slides.

Human activities and translational and rotational slides

Human activities can trigger new translational and rotational slides, or reactivate dormant or inactive ones. Timber harvesting and conversion of forest to pasture or other vegetation types are probably the most widespread human activities responsible for creating new or aggravating existing translational and rotational slides. The roots of trees anchor unconsolidated overburden to underlying rock, thus adding to shear strength S at the overburden rock contact. Within a few years after timber harvesting, roots decomposed and lose their strength. A significant increase in translational landslide activity follows (Swanson and Dryness, 1975; Sidle, 2005; Figure 5). This is particularly significant in steep areas where translational sliding quickly accelerates to form debris avalanches (Rollerson et al., 2005). The transpiration

Slide and Slump, Figure 5 Soil slides following logging and haul road construction on a mountainside on Vancouver Island, British Columbia, Canada. Glacial till and colluvium overlie bedrock and failure occurred along the overburden/bedrock contact (Photo courtesy of Brent Ward, Simon Fraser University).

of trees also acts to lower the water table in slopes thus increasing effective stress (σ'). Rising water tables is a factor in reactivating inactive or dormant landslides where their tree cover is removed (Sidle, 2005). Logging also requires the construction of a road network. In hilly or mountainous areas, road cuts must be made across slopes as well as the emplacement of fills across valleys. Road cuts reduce resisting forces on natural hillslopes: This can be visualized by removing elements 1–3 in Figure 1, diagram 5. Poorly engineered fills commonly fail in rotational slides or load underlying slopes causing them to fail. The combined effect of poor road construction, logging, and vegetation conversions in upland areas is a significant factor in increasing landslide and related disasters in the developing world (Fisher and Vasseur, 2000; Sidle, 2005; Mora, 2009).

Increasing pore water pressure through the creation of artificial reservoirs is another prominent way in which human activities can trigger translational and rotational slides, sometimes with disastrous results. The raising and lowering the reservoir behind the Vaiont Dam, Italian Alps, in 1963, destabilized adjacent slopes and triggered a rapidly moving translational slide with a volume of 2.7×10^8 m^3. It crossed the reservoir creating a wave that overtopped the dam and killed nearly 2,000 people downstream (Kiersch, 1964). At present, translational and rotational slides are a major problem associated with the flooding of the Three Gorges Reservoir, China. The Qianjiangping landslide (Figure 2) was triggered during the initial flooding of the reservoir to the 134 m level on July 13, 2003. The landslide was a translational slide that failed along a bedding plane in sandstone of Jurassic age. It had an estimated volume of 2.4×10^7 m^3 (Wang et al., 2004). Direct impact by the landslide and the displacement wave that it generated in the reservoir killed 24 people and left 1,100 people homeless. The level of the Three Gorges Reservoir will fluctuate 30 m seasonally in order to control flood flows of the Yangtze River. The fluctuation of the reservoir has proven to accelerate or reactivate large, creeping translational or rotational slides. New cities such as Wushan with populations of hundreds of thousands are situated on landslide complexes that experience movement with reservoir level fluctuations and require constant monitoring (Wang et al., 2009).

Summary

In a slide, a rigid or semi-rigid mass of soil or rock is displaced so that it descends vertically and horizontally along a distinct underlying failure surface under the influence of gravity. If the failure surface is planar, it is referred to as a translational slide; if it approximates a curved surface, it is classified as a rotational slide that is commonly referred to as a *slump*. Translational and rotational slides are further subdivided based upon the material involved. The failure surfaces underlying translational slides usually follow preexisting planes of weakness such as bedding planes, faults, joints, and foliation or contacts between overburden and bedrock. Rotational slides typically occur in massive rock or sediments where a curved surface defines maximum values of shear stress. Shear strength that resists shear stress is a function of cohesion, internal friction, and effective stress. Effective stress is the component of force operating normally to a potential failure surface minus pore water pressure. The ratio of shear strength to shear stress, called the *factor of safety*, must be greater than 1 for a slide to be stable. Natural geomorphic processes may act to reduce shear strength and increase shear stress with time. Human activities can induce translational and rotational slides by reducing shear strength through steepening slopes by excavation (road cuts, excavations, logging, and vegetation conversion), raising pore pressure (impounding reservoirs, removing trees), or increasing shear stress (placing fills on slopes).

Bibliography

Couture, R., 2011. Landslide terminology–national technical guidelines and best practices on landslides. Geological Survey of Canada, Open File 6824, 12 p.

Cruden, D. M., and Krahn, J., 1978. Frank Rockslide, Alberta, Canada. In Voight, B. (ed.), *Rockslides and Avalanches*. Amsterdam: Elsevier Scientific. Natural Phenomena, Vol. 1, pp. 97–112.

Cruden, D. M., and Varnes, D. J., 1996. Landslide types and processes. In Turner, K., and Schuster, R. L. (eds.), *Landslides Investigation and Mitigation*. Washington, DC: National Academy Press. Transportation Research Board Special Report 247, pp. 36–75.

Fisher, A., and Vasseur, L., 2000. The crisis in shifting cultivation practices and the promise of agroforestry: a review of the Panamanian experience. *Biodiversity and Conservation*, **9**, 739–756.

Hungr, O., Evans, S. G., Bovis, M. J., and Hutchinson, J. N., 2001. A review of the classification of landslides of the flow type. *Environmental and Engineering Geoscience*, **7**, 221–238.

Keefer, D., 1984. Landslides caused by earthquakes. *Geological Society of America Bulletin*, **95**, 406–421.

Kiersch, G. A., 1964. Vaiont Reservoir disaster. *Civil Engineering*, **34**, 32–39.

Krynine, D. P., and Judd, W. R., 1957. *Engineering Geology and Geotechnics*. New York: McGraw-Hill, p. 730.

Mora, S., 2009. Disasters are not natural: risk management, a tool for development. In Culshaw, M. G., Reeves, H. J., Jefferson, I., and Spink, T. W. (eds.), *Engineering Geology for Tomorrow's Cities*. London: Geological Society. Geological Society Engineering Geology Special Publication No. 22, pp. 101–112.

Multinational Andean Project 2009. *Field description of a landslide and its impact*. Geological survey of Canada Open File 5991. CD-ROM.

Piteau, D. R., Mylrea, F. H., and Blown, I. G., 1978. Downie slide, Columbia River, British Columbia, Canada. In Voight, B. (ed.), *Rockslides and Avalanches*. Amsterdam: Elsevier Scientific, Vol. 1, pp. 365–392.

Ritter, D. F., 1986. *Process Geomorphology*. Dubuque: Wm. C Brown, p. 579.

Rollerson, T. P., Millard, T. H., and Collins, D., 2005. Debris flows and debris avalanches in Clayoquot Sound. In Jakob, M., and Hungr, O. (eds.), *Debris Flow Hazards and Related Phenomena*. Chichester: Springer-Praxis, pp. 595–613.

Sidle, R. C., 2005. Influence of forest harvesting activities on debris avalanches and flows. In Jakob, M., and Hungr, O. (eds.), *Debris Flow Hazards and Related Phenomena*. Chichester: Springer-Praxis, pp. 387–409.

Siebert, L., 2002. Landslides resulting from structural failure of volcanoes. In Evans, S. G., and DeGraff, J. D. (eds.), *Catastrophic Landslides: Effects, Occurrence, and Mechanisms*. Boulder: Geological Society of America. Reviews in Engineering Geology, Vol. XV, pp. 209–236.

Swanson, F. J., and Dryness, C. T., 1975. Impact of clearcutting and road construction on soil erosion by landsliding in the Western Cascade Range, Oregon. *Geology*, **7**, 393–396.

Terzaghi, K., and Peck, R. B., 1967. *Soil Mechanics in Engineering Practice*. New York: Wiley, p. 729.

U.S Army Corps of Engineers, 2003. Slope stability. US Army Corps of Engineers, Engineering and Design EM 1110-2-1902; http://140.194.76.129/publications/eng-manuals/EM_1110-2-1902/basdoc.pdf.

Varnes, D. J., 1978. Slope movements. Types and processes, en Schuster. In Krizker, R. J. (ed.), *Landslides: Analysis and Control*. Washington, DC: National Research Council, pp. 9–33.

Wang, F. W., Zhang, Y. M., Huo, Z. T., Matsumoto, T., and Huang, B. L., 2004. The July 14, 2003 Qianjiangping landslide, Three Gorges Reservoir, China. *Landslides*, **1**, 157–162.

Wang, F., Huo, Z., and Zang, Y., 2009. Recent five-year displacement monitoring of Shuping landslide in the Three Gorges Dam Reservoir, China. In Sassa, K. (ed.), *Early Warning of Landslides. Proceedings of the International Workshop on Early Warning for Landslide Disaster Risk Reduction in the Eastern Asian Region, Kunming-Xinping, Yunnan China, 2009*. International Consortium on Landslides, pp. 106–116.

Cross-references

Debris Avalanche (sturzstrom)
Earthquake
Humanity as an Agent of Geological Disaster
Landslide (Mass Movement)
Landslide Types
Mass Movement
Pore-Water Pressure
Rock Avalanche (Sturzstrom)
Rockfall
Shear

SLOPE STABILITY

Kaare Høeg
Norwegian Geotechnical Institute, Oslo, Norway

Definition

In sloping terrain *landslides* (downhill mass movements) may be triggered by a number of different factors, natural as well as man-made. To prevent landslides from occurring or to mitigate potential consequences, *slope stability* analyses are required.

Slope failure mechanisms

The general term "landslide" is used to describe any type of downslope movement of soil and rock under the effects of gravity. There are many different types of mass movements (*landslide; mass movement; avalanche; debris avalanche*) in clay, silt, sand, gravel, rock, and snow slopes. However, internationally, it has been agreed to classify the different mass movements into five main categories as illustrated in Figure 1 (e.g., Cruden and Varnes, 1996; Turner and Schuster, 1996; Highland and Bobrowsky, 2008): (a) slide; (b) spread; (c) flow; (d) toppling; (e) fall.

Sliding may occur as a translational movement along a fairly plane sliding surface or as a rotational movement along a spoon-shaped failure surface (*slump and slide*). Combined mass movements also occur with both translation and rotation. The mass moving above the sliding surface is commonly referred to as the sliding body (Figure 1a).

Spread (*lateral spreading*) is a special form of sliding and has therefore been given a specific term. The sliding body slides on a plane of weakness most often caused by high pore-water pressures (*pore-water pressure*). Earthquakes commonly cause landslides of this type. Often, the sliding body will split into separate smaller blocks and spread out during the mass movement (Figure 1b).

A flow will often start as a slide or spread with a limited mass movement, but if the soil or rock mass is broken down (degrades/disintegrates) during the initial motion and deformation, the sliding mass will gradually be transformed into a flowing mass that can attain a very high velocity and travel a long distance (run-out). As an example, after an initial slide has been triggered in a sensitive and brittle clay (*quick clay*), the clay will lose its strength, turn into a thick fluid, and flow even in rather flat terrain (Figure 1c). There are many other examples of dramatic flows (*debris flow; debris avalanche; lahar; liquefaction; mudflow; quick sand*.) Flows also occur in dry masses where the air pressure between soil and rock particles and collisions among particles cause dynamic motions and very long run-out distances.

Toppling, rather than sliding, may develop in masses of rock or cemented soils if subvertical cracks or joints have developed in the ground (Figure 1d). The depth of the cracks and the distance between them will govern whether failure will occur by sliding or toppling.

Fall often starts as the sliding or toppling of a block of rock in a very steep slope, followed by free fall and bouncing down the valley side until it stops on flatter ground (Figure 1e). A fall of several rocks at the same time may cause such a large impact and induced vibrations and pore-water pressures that a more extensive landslide is triggered (*impact triggered landslides*).

The most dangerous and damaging snow avalanches are denoted as sheet failures and consist of an extensive snow cover which is released almost all at once along a weak layer in the snow pack or along the ground surface. This type of mass movement belongs to the category of slide or spread and is transformed to a flow with high velocity and high air pressures inside and in front of the avalanche (*avalanches*).

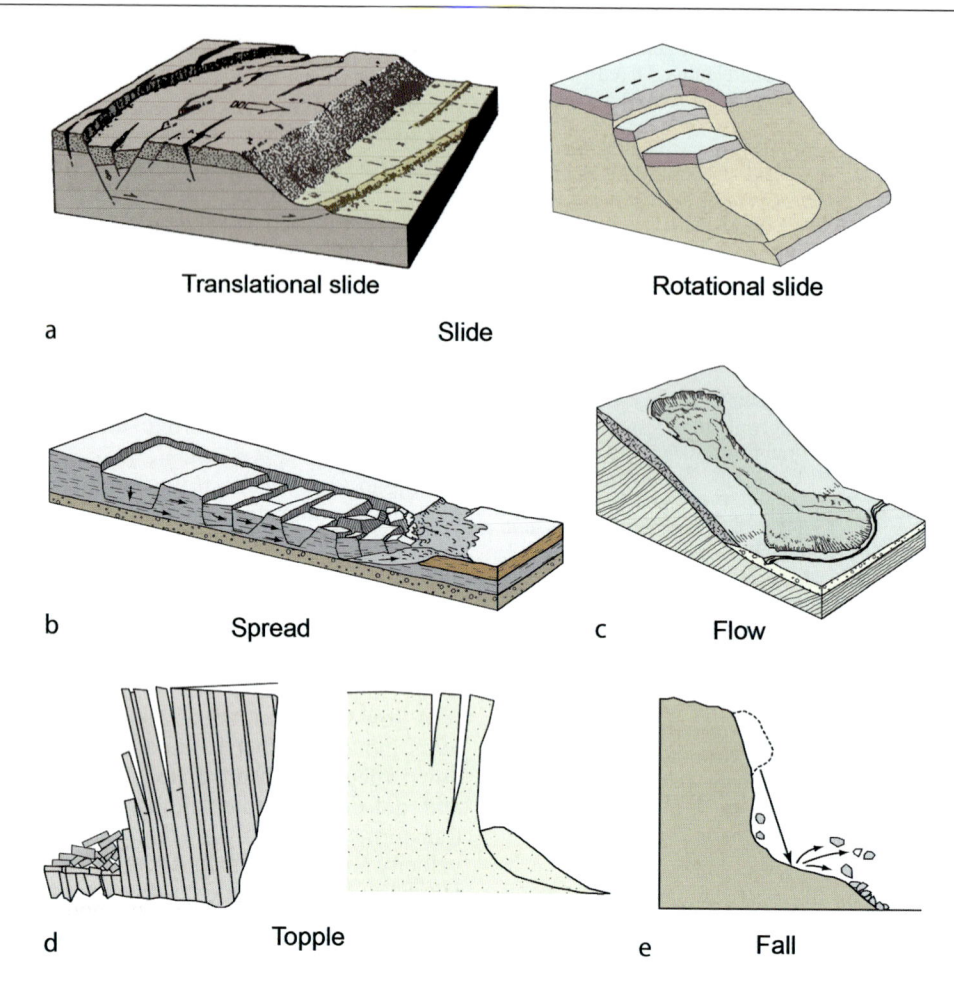

Slope Stability, Figure 1 Different types of mass movements classified in five main categories (Modified after Cruden and Varnes, 1996).

In general, a mass movement is better and more fully described by combining terms from the five main landslide categories shown in Figure 1. For instance, a mass movement should be described by combining the terms "slide" and "flow" into "soil slide-debris flow," when the failure starts as a slide of an intact sliding soil body, but is broken down during the initial shear displacement and internal deformations and is transformed into a disaggregated flowing mass.

In some geologic materials which exhibit brittle deformation properties, e.g., quick clay or stiff, fissured clay, initial small slides may develop into extensive landslides due to a gradual retrogressive failure mechanism which causes the back scarp to gradually move upslope until firm ground is reached.

Evaluating slope stability

The stability of a slope is evaluated by considering the force equilibrium of the assumed sliding, spreading, or toppling body (Figure 1). The following description refers to the likelihood of initiating/triggering slope instability, not the subsequent development of mass movements. Usually, there is considerable uncertainty with respect to the prediction of the critical failure mechanism and volume of mass that may initially fail. Several analyses must be performed to make sure that one has considered the most critical situation. The slope will find the path of least resistance to failure, and the stability computations must include that situation, otherwise the analyses will be misleading. Figure 2 shows a simplified sketch of a translational slide. The vertical gravitational force, W, of the sliding body has a component along the sliding plane, i.e., a destabilizing force. In addition, there may be other destabilizing forces, as, for instance, water pressure in a crack in the back of the slope or additional loads from buildings constructed on the slope. The resistance against sliding, i.e., the stabilizing forces, consists of the shear resistance between the sliding block plus the resistance of any man-made supporting structure or anchorage that may have been provided to prevent sliding. In Figure 2 it is assumed that there is a resulting destabilizing force, P, along the sliding plane. Thus, the shear force between the sliding plane and the block required to prevent sliding (T_{req}) is:

$$T_{req} = W\sin\beta + P \qquad (1)$$

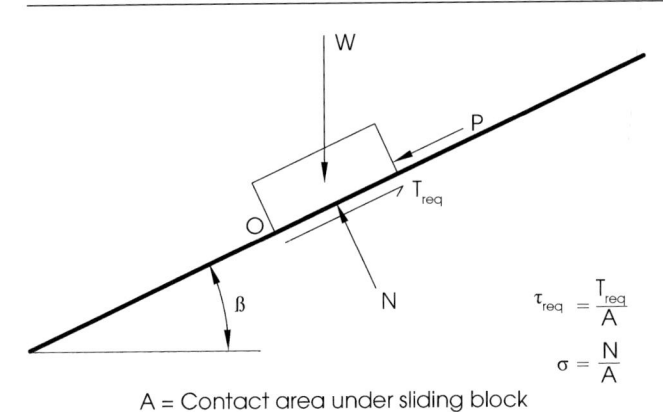

Slope Stability, Figure 2 Principle sketch for the situation of translational sliding.

If T_{req} is larger than the shear resistance that may be mobilized (T_f), instability and sliding will take place. The subscript "f" indicates failure. In engineering computations one tends to use shear stress (τ) rather than shear force, where $\tau = T/A$, and A is the area over which the shear force acts. In Figure 2 that is the contact area between the sliding block and the sliding plane.

The shear stress at failure, τ_f, is called the shear strength. In slope stability evaluations the most important and difficult task is reliable determination or estimate of the shear strength of the rock, soil, or snow involved in the slope stability analysis. This aspect is discussed under a subsequent subheading. In addition to material properties, the magnitude of the normal stress ($\sigma = N/A$) minus the water pressure between the block and the sliding plane determines the value of the shear strength, τ_f.

As shown in Figure 1a, for many slopes a rotational slide may be more critical than a translational slide. In a cross section through the sliding body in Figure 1a, the sliding surface may be approximated by a circular arc. The slope stability evaluation consists of comparing the destabilizing moment with the stabilizing moment around an axis of rotation through the center of the circle. The destabilizing moment is caused by the gravitational force, W, of the sliding mass and any applied loads toward the top of the slope. The stabilizing moment is created by the shear forces along the sliding surface opposing any rotational motion as well as any man-made retaining structure at the toe of the slide or anchorage into firm ground behind the sliding surface. The forces normal to a circular sliding surface have no moment arm and therefore cause no moment around the axis of rotation.

Thus, as for the translational slide in Figure 2, the stability evaluation for the rotational slide consists of comparing the average shear stress required for equilibrium, τ_{req}, with the average τ_f along the sliding surface. In advance, one does not know the location of the center of the circle or the radius of the critical circle, and all conceivable combinations must be considered to find the critical sliding surface.

In general, the sliding surface does not consist of a plane or a circular arc but is composed of a more complex surface combining translation and rotation accounting for any existing planes of weakness in the ground. The stability computations then become more complicated and laborious, but the limit equilibrium principle outlined above still holds by comparing τ_{req} to the maximum available average τ_f along the sliding surface.

If the sliding block on the plane in Figure 2 is high compared to the width, the critical failure mechanism is toppling (Figure 1d) rather than sliding. Then one has to compare the stabilizing moment with the destabilizing moment, for instance, around an axis through the point 0.

Definition of safety factor and probability of failure

A commonly used measure for evaluating the safety of a slope is the term "safety factor," $F = \tau_f/\tau_{req}$ which must be larger than unity to have a stable situation. This equation may also be written as $\tau_{req} = \tau_f/F$, which says that the shear strength may be reduced by the factor F before instability occurs and a slide is triggered. It emphasizes the importance of being able to determine shear strength. The safety margin (SM) is expressed as: $SM = F - 1$.

If the safety factor could be computed with great accuracy and reliability, a value of, for instance, 1.05 may be considered satisfactory. However, there are uncertainties associated with all factors that enter into the stability analyses; the properties of the soil or rock, pore-water pressures in the ground, intensity, and duration of future rainfall and snow melts, location of any planes of weakness and cracks in the ground, magnitude of destabilizing loads, and other factors described under a separate subheading below. The probability (likelihood) of failure may be estimated if the uncertainties in the different parameters may be quantified. By using the computational model for the stability evaluation and by introducing the uncertainties in the different parameters, one may find the combined effect of all the uncertainties and a probability of failure. Reliability index (RI), rather than safety factor, is a term that gradually is being more used to reflect the uncertainties in slope stability evaluations. It is defined as: $RI = (F_{avg} - 1)/SD_F$, where F_{avg} = most probable value of F; SD_F = standard deviation of F. The reliability index may be directly related to the probability of failure.

Determination of shear strength

The process to determine shear strength of soils, rock, and snow, respectively, has many similarities and common features. The basic linear Mohr-Coulomb equation is used as a first approximation:

$$\tau_f = c' + \sigma'_f \tan\phi' \qquad (2)$$

where

τ_f = shear strength (on the failure plane)
c' = cohesion (shear strength parameter)

$\emptyset'$ = angle of shearing resistance (friction angle) (shear strength parameter)
tan$\emptyset'$ = friction coefficient
σ'_f = effective normal stress on failure plane; $\sigma'_f = \sigma_f - u_f$
σ_f = total stress on failure plane to satisfy force equilibrium
u_f = pore-water pressure on failure plane

Soils (sediments) that have some cementation between particles have a c'-parameter larger than zero. The magnitude depends on chemical reactions that may have occurred in the ground and on earlier preloading by sediments which later have been eroded away (*erosion*) or by overlying glaciers that later have melted. Sediment that has been preloaded in this way is termed *overconsolidated*, because earlier in its geologic history it has carried higher vertical effective stresses than those existing at present. Sediment that has not been preloaded is called *normally consolidated*. It is in general much more compressible and has lower shear strength than an *overconsolidated* sediment of the same mineral composition at the same effective stress level.

The product of effective normal stress and the friction coefficient tan$\emptyset'$ provides the frictional resistance along the sliding surface (failure surface). It is essential to note that it is not the total stress but the effective normal stress, defined as $\sigma'_f = \sigma_f - u_f$, that contributes to shear strength which points out the importance of pore-water pressures on slope stability. It is difficult to accurately estimate the value of u_f, especially in relatively impervious soils like clays, where the pore pressure may consist of a stationary as well as a nonstationary component caused by a change in loads acting on the slope (*pore-water pressure*). Field measurements are required to determine reliable values.

Capillary pressures in the pore water above the ground water level are lower than atmospheric pressure (i.e., under-pressure). The capillary pressures therefore cause an increase in effective stresses and thereby an increase in strength and soil stability. They give the impression that the sediment has significant cohesion. However, this is termed apparent cohesion as it disappears when the capillary under-pressures disappear during precipitation. Thus, the shear strength is reduced, and slope instability may occur.

The magnitude of the angle of shearing resistance ($\emptyset'$) depends on the type of soil (clay, silt, sand, gravel), how densely packed the soil is, degree of overconsolidation, grain size distribution, mineralogy, chemical weathering, and many other factors. Laboratory and/or field testing is required to determine this parameter.

To be able to mobilize the maximum shear resistance (shear strength τ_f), some shear deformation along the sliding surface must take place. Many types of soils and joints in rock masses exhibit reduced shear resistance after the peak in a shear stress vs. shear displacement diagram is passed, and the shear resistance is reduced from the value τ_f to a value τ_r. The residual strength (τ_r) is expressed in the same way as peak strength:

$$\tau_r = c'_r + s'_r \tan\emptyset'_r \qquad (3)$$

However, the strength parameter c'_r is usually zero because any cementation between particles is broken during the previous shear deformations beyond the peak level, and $\emptyset_r'$ may be significantly smaller than $\emptyset'$ at peak resistance. In many slopes there are layers or seams of soils that already have been subjected to large shear deformations due to previous movements of over-riding glaciers, due to slope deformations during a previous period when the slope geometry was different, or movements caused by creep or freezing/thawing cycles (*creep; solifluction*). The shear strength along these potential sliding surfaces may therefore be closer to τ_r than τ_f. The stability of many slopes has been incorrectly analyzed by failing to understand the geologic history, geomorphology, and engineering geology of the site under consideration, and unexpected landslides have occurred.

Most rock slopes contain planes of weakness (e.g., joints, cracks, shears, faults) (*shear; fault*), and it is the shear strength along such planes that governs the slope stability. The shear strength is determined by a nonlinear form of the Mohr-Coulomb, because the friction angle is a function of the magnitude of effective stress on the plane (e.g., Barton and Choubey, 1977). The friction angle is primarily governed by the roughness of the joint, and as shear deformations take place, the roughness is worn down and reduced, and so is the friction angle. There is a significant difference between the friction angle at peak strength and at residual strength. The determination of joint roughness, the degree of weathering in the joint, and the type and thickness of infill material (clay and silt) is essential to the estimation of the shear strength of the plane of weakness (sliding plane).

Causes of slope instability and main triggering factors

Slope instability occurs when the stabilizing forces are reduced, the destabilizing forces are increased, or a combination thereof. After a landslide has taken place it is usually very difficult to pinpoint just one triggering factor. Water, for instance, affects the stability of slopes in many different ways.

Reduction of stabilizing forces by a reduction in shear strength

– Increase in pore-water pressure (or in water pressure in a rock joint) and thus reduction of effective normal stress. This may be caused by heavy precipitation, snowmelt, blockage of drainage paths (for instance, due to frost), or leakage from a water main. Earthquake vibrations (*earthquake*) lead to a temporary increase in pore-water pressure. This is particularly significant for loosely deposited silts and sands which may liquefy (*liquefaction*) and lose most of their strength. Even thin layers (seams) of saturated silts and sands in clay slopes

may create planes of weakness and cause slope instability during or shortly after an earthquake.
- Opening of cracks in the extension zone at the top of the slope allows ingress of water. When water enters into the fissures of stiff, overconsolidated clays, swelling occurs, and there may be a significant reduction in shear strength. Smectitic clays in a rock joint may swell considerably and reduce the shear strength and stability of the joint.
- Chemical changes in the pore water can lead to weakening of chemical and electrostatic forces between soil particles. A classic example is that the ductile behavior of a marine clay deposited in salt water may turn into a brittle quick clay (*quick clay*) when fresh water flow leaches out the salt ions.
- Reduction of shear strength from peak value down to residual strength due to large shear deformations. This may imply reduction of friction angle from, for instance, 30° to 10° and a complete loss of cohesion. Gradual downslope creep deformations (*creep; solifluction*) caused by shear stresses, freeze/thaw cycles, and swell/shrink cycles (*expansive soils and clays*) may cause deformations large enough to reduce the peak strength down to residual strength.

Reduction of stabilizing forces by erosion and excavation at toe of slope

- Erosion at the toe of the slope due to flowing water leads to undermining, loss of toe support, and instability. For many submarine slopes, such erosion is the main cause of instability (*marine hazards*). In a brittle material like quick clay, river or brook erosion at the toe may lead to a small initial slide that can start a retrogressive development in the back of the initial slope failure, and a large quick clay flow may develop. Similarly, excavation at the toe of a slope for the construction of a road or railway without providing other means of slope support may trigger a landslide.

Increase in destabilizing forces

- Loading the top part of the slope by placing landfills, buildings, or other structures. Submarine slopes, in areas of rapid deposition of sediments near river outlets, often become unstable due to such sediment loading. The situation is worsened by the fact that fine-grained sediments have very low permeability and excess pore-water pressures build up in the slope (*marine hazards*).
- Buildup of water pressure in cracks developed in the extension zone in the top part of the slope. The water pressure tends to push the slope downhill, and if the cracks remain filled with water for a long time, destabilizing seepage forces develop in the slope.
- When it rains the ground becomes saturated, and there may be a significant increase in density of the slope material and the destabilizing forces. This factor comes in addition to the increase in pore-water pressures caused by the precipitation.
- Forces developed due to expansion when water freezes in cracks in the extension zone. A similar phenomenon occurs when tree roots grow into joints and cracks of rock masses. This has been the cause of many rock slides, topples, and falls.
- Lowering of external water level against a slope extending into a lake or reservoir can trigger instability. The external water pressure against the lower part of the slope constitutes a stabilizing force, and if the water level is lowered, the safety factor against instability is reduced. Furthermore, if the lowering takes place rapidly such that the pore-water pressures inside the slope along the potential failure surface have no time to drain (dissipate), the destabilizing effect of the external water level lowering is even more detrimental.
- Earthquake vibrations cause dynamic destabilizing forces which come in addition to the already existing static forces. Even if the dynamic loads may not cause slope instability, they can cause an accumulation of downhill movements during the earthquake, the opening of cracks, and a reduction of the peak strength toward residual strength. In addition, as described above, the earthquake vibrations may cause a buildup of pore-water pressures.

Landslides triggered by volcanic activity result in some of the most devastating types of failure (Highland and Bobrowsky, 2008). Volcanic lava melts snow quickly, which can form a deluge of ash, soil, rock, and water that accelerates rapidly on the steep slopes of the volcano. These volcanic debris flows are known as lahars (*lahar*). Volcanic edifices are young and geologically weak structures that in many cases can collapse and cause rock and soil slides which may develop into debris avalanches (*debris avalanche*) (Devoli et al., 2009).

Soil slope instability after forest fires is a fairly common occurrence. When the vegetation disappears, the tree roots no longer prevent shallow slides from being triggered, water penetrates into the ground, and debris flows may develop more easily (Turner, 1996).

Summary

Various types of landslides and slope failure mechanisms are described. The evaluation of slope stability is commonly done by performing limit equilibrium analyses. The result is presented in the form of a computed factor of safety against instability or a probability of failure considering all the uncertainties in the evaluation. The main uncertainty usually lies in the determination of shear strength of the material in the slope, whether it consists of soil/sediment, jointed rock, or snow. Instability occurs when the stabilizing forces are reduced, the destabilizing forces are increased, or a combination thereof. Many examples of potential triggering factors are described. After a landslide has taken place, it is usually very difficult to pinpoint just one triggering factor. Water, for instance, affects slope stability in many different ways.

Bibliography

Barton, N., and Choubey, V., 1977. The shear strength of rock joints in theory and practice. *Rock Mechanics*, **1/2**, 1–54.

Cruden, D. M., and Varnes, D. J., 1996. Landslide types and processes. In Turner, A. K., and Schuster, R. L. (eds.), *Landslides – Investigation and Mitigation*. Washington, DC: National Academy Press. Special Report (Transportation Research Board, National Research Council), Vol. 247, pp. 36–75.

Devoli, G., Cepeda, J., and Kerle, N., 2009. The 1998 Casita volcano flank failure revisited – new insights into geological setting and failure mechanisms. *Engineering Geology*, **105**, 65–83.

Highland, L. M., and Bobrowsky, P., 2008. *The Landslide Handbook – A Guide to Understanding Landslides*. Reston: U.S. Geological Survey. U.S. Geological Survey Circular, Vol. 1325. 129p.

Turner, A. K., 1996. Colluvium and tallus. In Turner, A. K., and Schuster, R. L. (eds.), *Landslides - Investigation and Mitigation*. Washington, DC: National Academy Press. Special Report (Transportation Research Board, National Research Council).

Turner, A. K., and Schuster, R. L., 1996. *Landslides – Investigation and Mitigation*. Washington, DC: National Academy Press. Special Report (Transportation Research Board, National Research Council), Vol. 247.

Cross-references

Avalanches
Creep
Debris Avalanche
Debris Flow
Earthquake
Erosion
Expansive Soils and Clay
Fault
Lahar
Landslide
Lateral Spreading
Liquefaction
Marine Hazards
Mass Movement
Mudflow
Quick Clay
Quick Sand
Rock Avalanche
Rockfall
Shear
Slide and Slump
Solifluction

SNOWSTORM AND BLIZZARD

Thomas W. Schmidlin
Kent State University, Kent, OH, USA

Synonyms

White-out

Definition

Snowstorm. An atmospheric disturbance that produces snowfall of sufficient intensity or depth to cause disruption to society. An accumulation of 15 cm of snow in 24 h is generally sufficient to be considered a snowstorm, although lesser thresholds may apply early and late in the snow season or in locations where snowfall is rare.

Blizzard. An extreme snowstorm with strong winds and blowing or drifting snow that obscure visibility, obstruct travel, and create a hazardous environment to people and animals. The definition of a blizzard varies. In the USA, the National Weather Service defines a blizzard as winds over 35 miles h^{-1}, falling or blowing snow reducing visibility to ¼ mile or less, for a duration of at least 3 h. A blizzard in Canada is defined as wind speed over 40 km h^{-1}, a wind chill greater than 1,600 W m^{-2}, and visibility less than 1 km in snow, lasting for at least 4 h. The U.K. Met Office defines a blizzard as wind speed over 56 km h^{-1} causing blowing and drifting snow reducing visibility to 200 m (Wild, 1997). The term "blizzard" was first used by Henry Ellis along Hudson Bay in 1746, although Ellis's meaning at the time was ambiguous (Wild, 1997). It was used in South Dakota, USA, by 1867 and the term was in general use by the 1880s (Wild, 1997).

Discussion

Snow is frozen precipitation that originates as an ice crystal and grows as a snowflake through deposition, accretion, or aggregation. Snow is a winter feature of most of the Earth's middle and high latitudes. Snow occurs year around in portions of the Arctic and Antarctic and at high elevations. Snow covers about 78 million square kilometers of the Earth's land area each year, including about one half of the land in the northern hemisphere. Snow is an important component of the energy and water balances, in plant and animal ecology, and in human recreation. A large accumulation of snow within a short time causes disruption to human activities and is called a snowstorm. A severe snowstorm with strong winds and reduced visibility is a blizzard.

Snowstorms and blizzards originate as an area of low atmospheric pressure, typically a mid-latitude cyclone with associated fronts and other mesoscale features. Heavy snowfall rates result from uplift and a supply of moist air that produce condensation and snowfall. Uplift may result from orographic lift over terrain, warm air advection, an upper-level jet streak, diabatic processes, and cyclogenetic lift (Mote et al., 1997). Local areas of intense snowfall (5–15 cm h^{-1}) within the snowstorm may result from mesoscale processes in the atmosphere. These mesoscale bands of intense snowfall are associated with sharp temperature gradients, frontogenesis at the 700 hPa pressure level (∼3 km height), and a deep layer of negative equivalent potential vorticity (Nicosia and Grumm, 1999). The resulting upward motion and convection may even include lightning and thunder along with the snowfall. Most snowstorms are large and affect a long swath across tens of thousands of square kilometers. Other snowstorms may be quite localized and affect just a few hundred square kilometers. These may occur as convective bands of snowfall downwind of large lakes or bays (e.g., Laird et al., 2009).

Blizzards with their extreme winds and reduced visibility often occur along a band within the larger snowstorm. The strong wind that accompanies a blizzard requires a sharp pressure gradient and extreme low atmospheric pressure in a mid-latitude cyclone. An ordinary winter storm intensifies into a blizzard when the favorable conditions exist in the atmosphere. These include interaction of the surface cyclone with upper-level short-wave troughs and jet streaks and rapid amplification of upper-level ($\sim$5.5 km height) troughs (Kocin et al., 1995).

The average area affected by blizzards in the USA is 150,000 km^2 although the 24–27 January 1978 Midwest Blizzard covered over 1 million square kilometers (Schwartz and Schmidlin, 2002). The USA (excluding Alaska) averages about 11 blizzards annually with an average population of 2.5 million in the area affected by each blizzard. The March 1993 Superstorm had blizzard conditions across a large region of the eastern United States with a population of 72 million people.

Snowstorm records

The deepest snowfalls occur in the moist mountainous regions of the middle and high latitudes. The greatest 24-h snowfall in North America was 193 cm (76 in.) at Silver Lake, Colorado, at 3,115 m elevation on 14–15 April 1921 (Krause and Flood, 1997). Most other North American snowstorm records were set at Thompson Pass at 855 m elevation in the Chugach Range of southern Alaska. A snowstorm there in December 1955 gave 306 cm in 2 days, 373 cm in 3 days, and 414 cm in 4 days. A 7-day snowstorm in February 1959 produced 480 cm of snow at Mt. Shasta Ski Bowl, California, at 1,084 m elevation. Bessans, France, at 1,710 m elevation in the Alps recorded 172 cm of snowfall in just 19 h on 5 April 1959 (Krause and Flood, 1997). Canada's deepest 1-day snowstorm was 118 cm at Lakelse Lake, British Columbia, on 17 January 1974. Similar extreme snowfalls occur in the mountains of Norway and northwestern Japan.

Societal impacts

Snowstorms create slippery surfaces and loss of roadway friction along with low visibility to limit surface transportation by automobile, truck, bus, and train (Changnon et al., 2006). Vehicle accidents occur on the snowy and icy surfaces. Air travel is also disrupted by poor visibility and traction at airports. Deep snow simply restricts movement. Deep snowfalls can lead to medical emergencies when people in need of assistance cannot travel to hospitals. Social functions, such as school, churches, athletic events, concerts, and other businesses, may close or restrict functions during and after snowstorms. Electricity may be disrupted during snowstorms by collapsed poles or wires or by fallen trees due to high winds or the weight of snow. Removal of snow from transportation routes is a costly endeavor for individuals, businesses, and governments. The weight of snow can cause the collapse of structures and trees. Deep snowfalls on sloping surfaces can result in avalanches of snow onto buildings or transportation routes. Rapid snowmelt after large snowstorms may result in flooding. Persons are injured during and after snowstorms in vehicle accidents and by falling on snow or ice, they suffer heart attacks from exertion in snow removal, and are injured by collapsed structures or fallen trees. Delays in transportation lead to losses of perishable products. Most retail sales, business, and tourism are diminished during and after snowstorms. Among positive impacts, snowstorms may bring needed moisture in watersheds and agricultural landscapes. Some businesses, such as winter recreation, snow removal, and the sales of winter-related household, auto, or recreational goods, may benefit from snowstorms (Schmidlin, 1993).

The impacts of snowstorms have been quantified by combining the snowfall depth with the area and population affected by the snowstorm with an emphasis on the urban northeastern United States (Kocin and Uccellini, 2004). By this system, the 12–14 March 1993 blizzard was the most extreme snowstorm on record.

Summary

Snowstorms with 15 cm or more of snow in 1 day cause societal disruptions and may create hazardous situations. They are also an important part of the winter environment for much of the northern middle and high latitudes of earth. Blizzards are extreme forms of snowstorms that result from very intense mid-latitude cyclones. They paralyze transportation and pose extreme risks to persons outdoors.

Bibliography

Changnon, S. A., Changnon, D., and Karl, T. R., 2006. Temporal and spatial characteristics of snowstorms in the contiguous United States. *Journal of Applied Meteorology and Climatology*, **45**, 1141–1155.

Kocin, P. J., Schumacher, P. N., Morales, R. F., Jr., and Uccellini, L. W., 1995. Overview of the 12–14 March 1993 Superstorm. *Bulletin of the American Meteorological Society*, **76**, 165–182.

Kocin, P. J., and Uccellini, L. W., 2004. A snowfall impact scale derived from northeast storm snowfall distributions. *Bulletin of the American Meteorological Society*, **85**, 177–194.

Krause, P. F., and Flood, K. L., 1997. Weather and Climate Extremes. Alexandria: U.S. Army Corps of Engineers Topographic Engineering Center, Report TEC-0099.

Laird, N., Sobash, R., and Hodas, N., 2009. The frequency and characteristics of lake-effect precipitation events associated with the New York State finger lakes. *Journal of Applied Meteorology and Climatology*, **48**, 873–886.

Mote, T. L., Gamble, D. G., Underwood, S. J., and Bentley, M. L., 1997. Synoptic-scale features common to heavy snowstorms in the Southeast United States. *Weather and Forecasting*, **12**, 5–23.

Nicosia, D. J., and Grumm, R. H., 1999. Mesoscale band formation in three major northeastern United States snowstorms. *Weather and Forecasting*, **14**, 346–368.

Schmidlin, T. W., 1993. Impacts of severe winter weather during December 1989 in the Lake Erie snowbelt. *Journal of Climate*, **6**, 759–767.

Schwartz, R. M., and Schmidlin, T. W., 2002. A climatology of blizzards in the conterminous United States, 1959–2000. *Journal of Climate*, **15**, 1765–1772.

Wild, R., 1997. Historical review on the origin and definition of the word blizzard. *Journal of Meteorology*, **22**, 331–340.

Cross-references

Avalanches
Ice Storm
Impact Winter

SOCIAL–ECOLOGICAL SYSTEMS

Fabrice G. Renaud
United Nations University, UN Campus, Bonn, Germany

Synonyms

Human–environment systems; Social-environmental systems; Socio-ecological systems

Definition

Social–ecological systems are complex, integrated systems where human activities and environmental processes are inter-dependent, co-evolving, and linked through various feedback loops.

Discussion

Social–ecological systems are composed of a diversity of agents such as microbes, plants and humans, a set of actions related to their physical or behavioral characteristics, and a physical substrate (Anderies et al., 2006). These systems are highly dynamic and are characterized by the mutual interactions between the societal and ecological subsystems (Gallopín, 2006). The interactions between elements of the system are typically stochastic and nonlinear.

The coupling of the two systems emanates from the fact that the delineation between social and biophysical systems is artificial and subjective since human action and ecological structures are closely linked and dependent on each other. It is therefore considered more relevant to consider the subsystems within a common analytical framework where the interactions, regulations, and feedback mechanisms are captured at various spatial and temporal scales (e.g., Turner et al., 2003a). The systems can be defined at various spatial scales ranging from the local, such as communities and their immediate environment, to global. Because the systems are highly dynamic and processes within them operate at various speeds, several temporal scales have to be considered to characterize them.

Social–ecological systems are an element of analysis in resilience theory and in sustainability science and are increasingly considered in the context of vulnerability and risk assessment linked to natural hazards and climate change (Turner et al., 2003b; Walker and Meyers, 2004; Renaud et al., 2010).

Bibliography

Anderies, J. M., Walker, B. H., and Kinzig, A. P., 2006. Fifteen weddings and a funeral: case studies and resilience-based management. *Ecology and Society*, **11**(1), 21.

Gallopín, G. C., 2006. Linkages between vulnerability, resilience, and adaptive capacity. *Global Environmental Change*, **16**, 293–303.

Renaud, F. R., Birkmann, J., Damm, M., and Gallopín, G. C., 2010. Understanding multiple thresholds of coupled social–ecological systems exposed to natural hazards as external shocks. *Natural Hazards*, doi:10.1007/s11069-010-9505-x.

Turner, B. L., II, Kasperson, R. E., Matson, P. A., McCarthy, J. J., Corell, R. W., Christensen, L., Eckley, N., Kasperson, J. X., Luerse, A., Martello, M. L., Polsky, C., Pulsipher, A., and Schiller, A., 2003a. A framework for vulnerability analysis in sustainability science. *Proceedings of the National Academy of Sciences*, **100**, 8074–8079.

Turner, B. L., II, Matson, P. A., McCarthy, J. J., Corell, R. W., Christensen, L., Eckley, N., Hovelsrud-Broda, G. K., Kasperson, J. X., Kasperson, R. E., Luers, A., Martello, M. L., Mathiesen, S., Naylor, R., Polsky, C., Pulsipher, A., Schiller, A., Selin, H., and Tyler, N., 2003b. Illustrating the coupled human–environment system for vulnerability analysis: three case studies. *Proceedings of the National Academy of Sciences*, **100**, 8080–8085.

Walker, B. H., and Meyers, J. A., 2004. Thresholds in ecological and social–ecological systems: a developing database. *Ecology and Society*, **9**(2), 3.

Cross-references

Resilience
Risk Assessment

SOCIOLOGY OF DISASTER

Alison Herring
University of North Texas, Denton, TX, USA

Introduction

Disasters are viewed as an ever-present threat in our lives and appear to be on the rise; the past few decades have seen a growth in the number of disasters and the devastation caused by disasters (Mileti, 1999; Quarantelli, 1998; Waugh, 2007). The past decade alone has seen multiple disasters around the world: the 2004 South Asian tsunami which killed an estimated 230,000 people (Stern, 2007); Hurricane Katrina which hit the Gulf Coast in August, 2005 killed an estimated 1,325 people (Bullard and Wright, 2009); the 2005 northern Pakistan earthquake killed an estimated 80,000 people, and the more recent Japanese earthquake of 2011, killed an estimated 15,835 people (Saito and Kunimitsu, 2011). While these disasters are seen as having had a natural physical trigger, there are a variety of man-made disasters with comparable fatality rates, such as Chernobyl which occurred in the Ukraine in 1986 where the release of radiation reportedly affected

50,000 people, and the World Trade Center Bombings in New York in 2001 that killed an estimated 2,996 people (Ai et al., 2005).

Disasters are not only measured in terms of death rates but also in monetary costs which have been escalating in the USA (Mileti, 1999). Between 1950 and 1959, the USA experienced 20 disasters costing roughly $38 billion (1998 value), whereas between 1990 and 1999 the USA experienced 82 disasters costing an estimated $535 billion (Perrow, 2007). The cost of Hurricane Katrina is estimated at $200 billion (Burby, 2006), and is the first disaster in which the government has paid out more $100 billion for damages (Sylves, 2006). There are several reasons which may contribute to this increase in disasters; a population explosion has resulted in greater concentrations of urbanization in places where hazards are prevalent (Dash, 2010; Phillips and Fordham, 2010), both in the Western World and the developing world; and a reliance on technology to solve problems in our environment may increase the likelihood of failure down the road, as evidenced by the catastrophic failure of the levees during Hurricane Katrina (Bullard and Wright, 2009; Burby, 2006; Campanella, 2008; White, 1975).

While the agent of the disaster may be appealing to study, it is not the only aspect of disasters which is important. Disaster research has become a growing field resulting in interdisciplinary study with researchers from other fields such as geography, political science, engineering, psychology, international affairs, and others all seeking to add their knowledge to the overall picture of disasters.

Sociology of disaster investigates the social aspects of disaster, specifically issues related to how people perceive disasters, how they are able or unable to prepare for, respond, and recover from disasters. Topics which sociologists study range from how race, class, gender, and age may influence our risks to disaster, our survival chances, and our response to disaster; whether we evacuate in the event of an impending disaster, how we evacuate, and why we may not be able or willing to evacuate; how popular culture influences our perceptions of disaster, and the myths that prevail about disasters in our culture. The importance of sociology to the study of disasters is significant not only in terms of understanding human behavior in situations where no social norms exist, but also in an applied perspective where emergency management professionals may use the information to make better planning decisions to prepare for the disaster and its anticipated outcomes with the limited resources they have at their disposal.

Definitions and paradigms

The first acknowledged disaster research study was conducted by Samuel Henry Prince in 1917, following an explosion in the Halifax Harbor when a French munitions ship collided with a Belgium ship in the harbor (Perry, 2007; Scanlon, 1996). An estimated 1,963 people were killed and 9,000 people injured in the vicinity. Prince's observations of people at the scene, which included people trying to get to family members, rumors beginning shortly after the hazard, the fear of looting, the lack of preparation by the city, and resources converging upon the scene, still hold true today and form the basis for many sociological studies (Drabek and Key, 1976; Fitzpatrick and Mileti, 1991; Kendra and Wachtendorf, 2003; Neal, 1994).

It was not until the 1950s and the Cold War era that disaster research began in earnest when the military approached researchers to study possible civilian reactions to a nuclear attack. The prevailing belief of the military held that civilians would be traumatized by disasters that would inevitably make them prone to psychological breakdowns and lead to social control issues following the event. This assumption was the cornerstone of the military's command and control model, which was prevalent in emergency management offices (Quarantelli, 1987). As researchers were unable to conduct experiments in bombing the population, it was determined that natural disasters would provide a means of studying people's reactions and behaviors to adverse conditions which would provide suitable knowledge that could be used in the event of war (Quarantelli, 1987). Since many of the disaster researchers at the time came from the field of collective behavior, analysis of people's behavior in disaster situations was more often than not interpreted through symbolic interactionist theories (Quarantelli, 1987, 2002). The military's concentration on fast-onset, single event disasters coupled with the researchers focus on collective, or group dynamics, within a disaster concentration can be clearly seen in the definition of disaster produced by Fritz (1961, 655) at the time

> actual or threatened or uncontrollable events that are concentrated in time and space, in which a society, or a relatively self-sufficient subdivision of a society undergoes severe danger, and incurs such losses to its members and physical appurtenances that the social structure is disrupted and the fulfillment of all or some of the essential functions of the society, or its subdivision, is prevented.

The criticism now leveled against this definition is that it assumes disaster events as being limited in terms of time and space, thus disaster research focused on hazards such as hurricanes, tornadoes, and earthquakes, while not applying to other types of hazards such as famine or drought (Kroll-Smith and Gunter, 1998). Albeit more recent definitions of disaster do still focus on the "event" perspective defined by Fritz, qualifiers have been introduced to emphasize social processes (Perry, 2007). For example McEntire (2007, 2) states that "[d]isasters are the deadly, destructive, and/or disruptive events that occur whenever a hazard (or multiple hazards) interact with vulnerability." A disaster may be viewed then, as the actual impact of an external hazard upon the social environment and the consequences that result (Hewitt, 1998; Oliver-Smith, 2002). The key is the resulting disruption to the social structure that a hazard causes (Bullard, 2008; Wisner et al., 2004).

Sociology of Disaster, Figure 2 A young man walks through chest deep water after looting a grocery store in New Orleans on Tuesday, August 30, 2005. Flood waters continue to rise in New Orleans after Hurricane Katrina did extensive damage when it made landfall on Monday (AP Photo/Dave Martin).

Coupled with exposure to more environmental hazards and related health issues, educational attainment is often lower than in wealthier areas and job prospects limited by knowledge and skill levels. Fewer resources are available for education in these areas than is the case in more wealthy areas. Traditionally, the poor are more likely to accept low-paid unskilled labor which is hourly paid and more often than not without benefits (Johnson, 2008). This means that many people will not evacuate at all, or in time, because of fear of losing not only their pay but also their jobs in a market that can easily replace unskilled labor. When a disaster hits, their jobs are frequently the first to be lost and the last to be replaced. Five years after Hurricane Katrina hit New Orleans it is estimated that only 85% of the pre-Katrina jobs have been recreated (Louisiana Workforce Commission, 2010). In comparison, the wealthier members of society are more likely to have salaried positions, with benefits, are less likely to lose that income in the event of a disaster and thus have more resources available to use.

The poorer populations are more constrained by their environment and typically stay close to their roots (Dash et al., 2010). This has an adverse effect on evacuation for this population. For the wealthier populations, moving away from family in search of jobs means that there are family members outside of the region of the disaster who can provide support and aid in recovery; a resource that is not as available to poorer members of society who live around their family (Freudenburg et al., 2009). The social networks that are available to the lower socioeconomic classes during normal daily activity cannot be accessed during a disaster situation when everyone is in the same boat (Bourdieu, 2005; Bourque et al., 2006; Elliot et al., 2010).

After a disaster, housing is often in short supply and the demand is high, not only from those displaced by the disaster and having nowhere else to stay, but by the influx of construction workers who can afford to pay higher rents. Rents inevitably increase to such a point that the poor are forced out of the area. It is estimated that, after hurricane Katrina, an estimated 58% of renters were paying more than 35% of their pre-tax income on rent and utilities, an increase of 43% since 2004. Paradoxically, the number of homes classified as vacant rose to 25% of all residential units. This is, in part, due to the classification of a housing unit as having a roof, doors and windows, which many of the severely damaged homes in New Orleans did, but they were still uninhabitable from water damage and boarded up by residents unable to afford to repair them (Louisiana Workforce Commission, 2010).

In general, the lower socioeconomic classes are less likely to have the resources prior to the disaster to be able to mitigate the effects of the disaster and subsequently have fewer resources after the disaster to be able to recover (Blaikie et al., 1994). The majority of the costs of a disaster are borne by the victims of the disaster and disproportionately by low income groups (Beatley, 1989; Cutter et al., 2003; Dash, 2010; Mileti, 1999).

Gender
Sociological research in the field of disasters has shown that men and women experience disasters differently due to the differing roles and responsibilities expected of them. Women are more likely to live longer, but have less secure incomes throughout their lives, whereas men are more likely to be homeless and experience higher levels of violence (Enarson, 2010). These types of inequalities have large impacts on the ways in which gender experiences disasters, especially as women are less likely to have access to much needed resources. During any phase of the disaster, women are more likely to experience a dramatic expansion of their caregiver roles regardless of whether they are married or single. For example, women are more likely to heed any warnings that they hear and take preventive action, especially if they have children (Enarson, 2010; Gladwin and Peacock, 1997). Men on the other hand, are more likely to be risk tolerant and therefore less likely to take action to protect themselves (Enarson, 2010).

Women are more likely to live in poverty, live in single female head of households, and so lack access to resources when a disaster strikes. In New Orleans, at the time Hurricane Katrina struck in 2005, one in four women lived in poverty (Comfort, 2006; Drier, 2006; Enarson, 2010; Morrow and Enarson, 1996). This translates to women being less able to take mitigative action and more likely to experience higher death and injury rates than men regardless of the fact that they are more likely to heed the warnings (Enarson, 2010).

Recovery from a disaster is also experienced differently by gender. Divorce rates and spousal abuse tend to increase dramatically after a disaster often leaving women in charge of the children (Enarson, 2010). At the time of Hurricane Katrina, more than a half (56%) of families with children were headed by women and two fifths of these lived in poverty. This population not only found it difficult to recover from the effects of Katrina but was disadvantaged in trying to negotiate through a bureaucratic system which did not recognize their head of household status when multiple families lived in one housing unit (Enarson, 2010; Morrow, 1997).

Gender is not alone in how it shapes an individual's experience, nor does it guarantee that all women, or all men, will experience the same thing. As stated earlier, individuals are a conglomerate of several characteristics of which gender and race/ethnicity are master statuses. That is, they often define the individual regardless of the achievements of that individual.

Race/ethnicity
Racial stereotypes allow us to categorize people into groups, often based on false assumptions, and allow an "us" and "them" attitude, which limits the power given to "them." This trend often shows up dramatically in disaster situations. New Orleans, a predominantly African American city, had an African-American population of 67% prior to the storm (Johnson, 2008) and was one of the poorest cities in the USA ranking third in the nation (Drier, 2006) with a poverty rate twice the national average (Bourque et al., 2006). Over a third of African-American women in New Orleans were officially poor prior to the storm (Enarson, 2010).

As a consequence of racial stereotyping and discrimination, minorities tend to live in more hazardous areas either from natural hazard, or man-made hazardous sites, where they trade safety for employment (Dash, 2010). Their perception of risk due to the presence of the nearby hazard, or working within the hazard site may actually raise their tolerance for risk, and thus, they are less likely to evacuate in the event of an impending disaster (Bullard, 2008). In terms of New Orleans, although African-Americans represented over two thirds of the population they only comprised a fifth of the urban population being situated in the surrounding high flood risk, low-lying areas (Drier, 2006).

The map below shows the percentage of African-Americans living in census blocks in New Orleans Parish: The higher the percentage of African-Americans in the block the darker the shade of blue. From this map it can clearly be seen that African-Americans are not concentrated on the high land near the banks of the Mississippi, but are in fact positioned farther back from the river in the lower lying swamp areas. As this was the area that sustained most damage from the flooding, it can be appreciated that in the aftermath of a disaster, the damage patterns are often directly related to ethnic/racial boundaries within the environment, which highlights the social vulnerability of certain groups (Bolin and Bolton, 1986; Sharkey, 2007) (Figure 3).

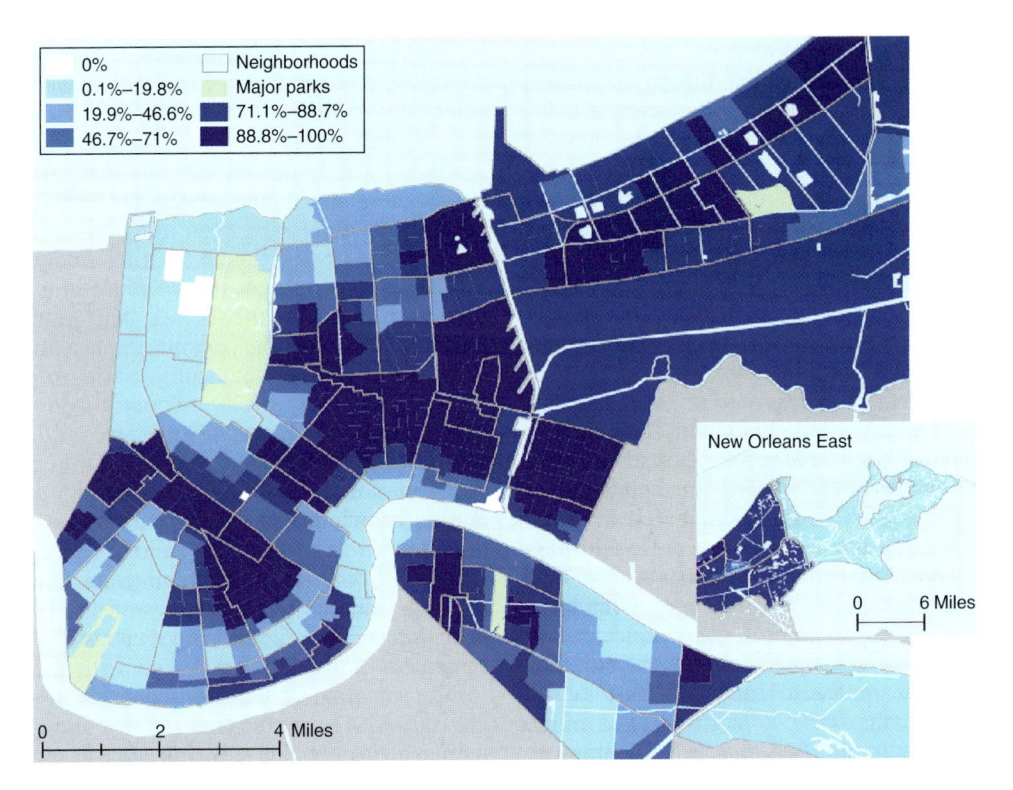

Sociology of Disaster, Figure 3 Percent African-American by Census block group in Orleans Parish (Adapted from Census 2000 and compiled by Greater New Orleans Community Data Center (GNOCDC, 2011)).

Although there are obvious disadvantages to belonging to certain groups, there are some less obvious aspects of group membership which also have significant consequences in disaster situations. For example, minorities do not always "hear" the warnings given by emergency managers, partly because of language barriers, but also because minorities tend to view "official warnings" differently (Dash, 2010). Hispanics, for example, are more likely to seek confirmation from other familial groups rather than from official channels and more likely to decide what they should do as a group. Many minorities reside within the same area as their family, and so are more likely to need the use of shelters in the event of an evacuation should they evacuate especially given that people will evacuate as families rather than as individuals (Dash, 2010; Drabek and Key, 1976; Fitzpatrick and Mileti, 1991).

Cultural patterns may also result in a higher degree of resiliency. For example, research conducted by Klinenberg (2003) on the 1995 Chicago heat wave showed that the elderly Latino population was protected from the effects of the heat wave by cultural reverence that holds elders as valuable resources enmeshed within the family system. This is in sharp contrast to African-American and white cultures where the elderly lived alone and isolated from family and neighbors, and consequently resulted in the large number of deaths in this group.

Sociopolitical ecology perspective

This perspective goes beyond just examining the impact of the disaster and the social structures in society. Sociopolitical ecology seeks to take into account the cultural and organizational components that make up communities and the social networks within them (Peacock and Ragsdale, 1997). Communities are an amalgamation of preexisting social structures and complex relationship networks. In a disaster event, those structures and networks can aid or hinder an individual's response and recovery to a disaster. Those same networks extend outside of the community into the larger political sphere and, subsequently, become a resource or deterrent outside of their local effect. Individuals and communities are seen as competing for scarce resources such as food, temporary shelter, and recovery funds. Those with power will be able to garner these resources and recover faster and more thoroughly to the detriment of those with less power who will have more difficulty in recovering. Thus, recovery is not a homogeneous experience within the community.

This brings stratification systems within the community into play. For example, in Hurricane Andrew the Anglos and the Cubans were able to recover much quicker than any other population in the community because they had more power, resources, and networks. It was found in the aftermath of Hurricane Andrew that Florida City, a mainly poor minority neighborhood, which had a weak local government, was ultimately unable to recover from the effects of the hurricane as it had so few resources and no political voice. However, an adjoining neighborhood, Homestead, which was comprised of more middle-class families had more resources and a stronger local government. Because of their political power, the Homestead community was able to acquire funds from the federal government in order to help recovery of the neighborhood (Dash et al., 1997).

By analyzing social structures prior to a disaster, researchers using the sociopolitical ecology or social vulnerability perspective are able to identify the groups of people within an area who will be more likely to be severely impacted by a disaster, and are thus able to advise local governments and emergency managers.

As sociologists we are interested in all aspects of social behavior related to disasters, thus, we also focus on aspects of popular culture such as the media and how the media has influenced the popular perception of disasters and created the myths that abound about disasters.

Media and popular culture in disasters

Researchers have studied popular culture within disasters (Bahk and Neuwirth, 2000; Couch, 2000; Webb et al., 2000). In fact, an entire issue of the *International Journal of Mass Emergencies and Disasters*, March 2000 Volume 18(1), was devoted to popular culture in disaster. These researchers look at how a culture represents the disasters that they have been a part of. After a disaster has struck we often see considerable graffiti, much of which "mocks" the hazard that passed through. For example "Remodeled by Andrew" was found on a nearly destroyed house after Hurricane Andrew had passed through the area. Popular culture can also be seen in merchandized items; for example, after Hurricane Katrina many t-shirts were produced with logos that represented resentment toward FEMA or the belief that looting had been widespread after the event. American culture also tends to remember the anniversaries of the disasters. These dates are usually memorialized for a number of years after the event by people revisiting the site on the day and reconnecting with others and producing some way of remembering that date. It is not known how long such dates will be remembered; after the first 5 years, the tenth anniversary is remembered. Hurricane Andrew did celebrate its 15th anniversary. However, there are usually fewer and fewer people at each succeeding anniversary as the event becomes a memory and people move on with their lives.

The sociological focus on behavior in disasters rather than attitudes has led to considerable understanding about human behavior, even showing that a hazard does not necessarily need to be present in reality for people to perceive its danger. One case in point was the reaction to the Three Mile Island incident in 1979. Response to the incident at the nuclear power plant was more intense and fearful because of the release of *The China Syndrome* a few weeks prior. The premise of the movie concerned the meltdown of a nuclear power plant including people's reactions to the meltdown and the efforts of authorities to avert the crisis (Wills, 2006). The influence of the media has inevitably become a significant topic of research amongst sociologists concerned with how the public perceive and react to disasters (Scanlon, 2007; Tierney et al., 2006).

Disaster movies have been analyzed by numerous researchers (Wenger et al., 1975; Jones, 1993; Fischer, 1998; Mitchell et al., 2000) for the myths that, invariably, are proven to be false by researchers. For instance, disaster movies tend to show that people will panic when they realize the danger they are in; research shows that panic is very rare. In fact, even though the authority figures in the movies tend to stress the need to wait to inform the public so as not to cause panic, in reality it is very hard to get people to evacuate even when they are asked to do so.

Death tolls are used by the movie industry to provide the extent of the disaster by focusing on high numbers of casualties. However, in reality, since the Galveston Hurricane of 1900 which killed 8,000, death tolls in America have been very low in most disaster situations (Mitchell et al., 2000), until hurricane Katrina in 2005 (Brunkard et al., 2008; Sharkey, 2007). However, even in real news programs, the death statistics reported at the beginning of the event will be drastically overstated. As the days pass, the statistics get substantially lower as more and more people are accounted for. Disaster movies focus on the pre-impact and impact stages of the disaster, and the lives of various characters throughout these stages, showing high energy small events which happen to them so as to keep audience's attention. Little attention is paid to the post-impact and recovery stages of disaster, thus ignoring the lingering consequences of disasters that continue well after the encounter with a hazard. The need for the audience to connect with characters in the movie also places emphasis on certain characteristics within the scene which are unrealistic. For example, the Director of FEMA, or the head of the local emergency management office, single-handedly saving the region from catastrophe as in the movie *Volcano*, or *Category 6*, when in reality he would be directing and coordinating the response and recovery efforts from a secure location.

Some researchers have focused on the news media and their disaster coverage showing that reporters tend to believe the disaster myths too (Scanlon, 2006). Many reporters believe that looting will occur in a disaster and will report either that it is occurring as did happen in hurricane Katrina, as evidenced by footage, or they will report the lack of looting. Interestingly, the perception of "looting" changes with race/ethnicity emphasizing the disparities and prejudices of the population. The two photos shown here were found in the Associated Press a day or two after Hurricane Katrina.

With the advancement in technology, the news media are able to provide coverage from the affected site almost immediately. During hurricanes, reporters actively enter the area where it is most likely to hit and report on the situation. Although this act is an attempt to show how bad the situation is their televised presence often has the reverse effect. People will not evacuate presuming that since the media is there then they can ride it out as well while shaking their heads at the weatherman holding onto to the palm tree with both hands to prevent himself being blown away. The convergence of the media into a disaster zone also puts pressure on local resources which are already stretched thin by the disaster and subsequently increases the risk to emergency workers (Kendra and Wachtendorf, 2003). The rate at which new technology and social media is advancing also means that live action video is now making its way immediately out of the disaster-stricken area. This will have significant impact not only on being able to respond to the populations affected but, hopefully, will also begin to allay some of the disaster myths.

Bibliography

Ai, A. L., Cascio, T., Santangelo, L. K., and Evans-Campbell, T., 2005. Hope, meaning, and growth following the September 11, 2001, terrorist attacks. *Journal of Interpersonal Violence*, **20**, 523–548, doi:10.1177/0886260504272896.

Bahk, C. M., and Neuwirth, K., 2000. Impact of movie depictions of volcanic disaster on risk perception and judgments. *International Journal of Mass Emergencies and Disasters*, **18**(1), 63–84.

Bankoff, G., 2006. *The Tale of the Three Pigs: Taking Another look at Vulnerability in the Light of the Indian Ocean Tsunami and Hurricane Katrina*. http://understandingkatrina.ssrc.org/Bankoff/. Accessed April 14, 2010.

Beatley, T., 1989. Toward a moral philosophy of natural disaster mitigation. *International Journal of Mass Emergencies and Disasters*, **7**(1), 5–32.

Blaikie, P., Cannon, T., Davis, I., and Wisner, B., 1994. *At Risk: Natural Hazards, People's Vulnerability, and Disasters*. London: Routledge.

Blinn-Pike, L., 2010. Households and families. In Phillips, B. D., Thomas, D. S. K., Fothergill, A., and Blinn-Pike, L. (eds.), *Social Vulnerability to Disaster*. Boca Raton: CRC, pp. 257–278.

Bolin, R., and Bolton, P. 1986. *Race, Religion, and Ethnicity in Disaster Recovery*. Boulde: University of Colorado/Institute of Behavioral Science, Program on Environment and Behavior. Monograph #42.

Bourdieu, P. 2005. *Outline of a Theory of Practice* (R. Nice, Trans.). Cambridge, UK: Cambridge Open University Press (Original work published 1972).

Bourque, L. B., Siegel, J. M., Kano, M., and Wood, M. M., 2006. Weathering the storm: the IMpact of hurricanes on physical and mental health. *ANNALS, AAPSS*, **604**, 129–151, doi:10.1177/0002716205284920.

Boyce, J. K., 2000. Let them eat risk? Wealth, rights and disaster vulnerability. *Disasters*, **24**(3), 254–261.

Branshaw, J., and Trainer, J., 2007. Race, class, and capital amidst the hurricane Katrina Diaspora. In Brusma, D. L., Overfelt, D., and Picou, J. S. (eds.), *The Sociology of Katrina*. Lanham: Rowan and Littlefield, pp. 91–105.

Brezina, T., and Kaufman, J. M., 2008. What really happened in New Orleans? Estimating the threat of violence during the Hurricane Katrina disaster. *Justice Quarterly*, **25**(4), 701–722, doi:10.1080/07418820802290504.

Brunkard, J., Namulanda, G., and Ratard, R., 2008. Hurricane Katrina deaths, Louisiana, 2005. *Disaster Medicine and Public Health Preparedness*, **1**, 1–9.

Bullard, R. D., 2008. Differential vulnerabilities: environmental and economic inequality and government response to unnatural disasters. *Social Research*, **75**(3), 753–784.

Bullard, R. D., and Wright, B., 2009. The color of toxic debris: the racial injustice in the flow of poison that followed the flood. *The American Prospect*, **20**(2), 9–11.

Burby, R. J., 2006. Hurricane Katrina and the paradoxes of government policy: bringing about wise governmental decisions for hazardous waste. *ANNALS, AAPSS*, **604**, 171–191, doi:10.1177/0002716205284676.

Burkle, F. M., Jr., 2006. Globalization and disaster: issues of public health, state capacity and political action. *Journal of International Affairs*, **59**(2), 241–265.

Campanella, R., 2008. *Bienville's Dilemma: A Historical Geography of New Orleans*. Lafayette, LA: University of Louisiana at Lafayette.

Comfort, L. K., 2006. Cities at risk: Hurricane Katrina and the drowning of New Orleans. *Urban Affairs Review*, **41**(4), 501–516, doi:10.1177/1078087405284881.

Couch, S., 2000. The cultural scene of disasters: conceptualizing the field of disasters and popular culture. *International Journal of Mass Emergencies and Disasters*, **18**(1), 21–37.

Cutter, S. L., Boruff, B. J., and Shirley, W. L., 2003. Social vulnerability to environmental hazards. *Social Science Quarterly*, **84**(2), 242–261.

Dash, N., 2010. Race and ethnicity. In Phillips, B. D., Thomas, D. S. K., Fothergill, A., and Blinn-Pike, L. (eds.), *Social Vulnerability to Disaster*. Boca Raton: CRC, pp. 75–100.

Dash, N., Peacock, W. G., and Morrow, B. H., 1997. And the poor get poorer: a neglected black community. In Peacock, W. G., Morrow, B. H., and Gladwin, H. (eds.), *Hurricane Andrew: Ethnicity, Gender, and the Sociology of Disasters*. Miami, FL: Florida International University, pp. 206–225.

Dash, N., McCoy, B. G., and Herring, A., 2010. Class. In Phillips, B. D., Thomas, D. S. K., Fothergill, A., and Blinn-Pike, L. (eds.), *Social Vulnerability to Disaster*. Boca Raton: CRC, pp. 75–100.

Drabek, T. E., 2002. Following some dreams: recognizing opportunities, posing interesting questions, and implementing alternative methods. In Stallings, R. A. (ed.), *Methods of Disaster Research*. Philadelphia: Xlibris, pp. 127–153.

Drabek, T. E., and Key, W. H., 1976. The impact of disaster on primary group linkages. *Mass Emergencies*, **1**, 89–105.

Drier, P., 2006. Katrina and power in America. *Urban Affairs Review*, **41**(4), 528–549, doi:10.1177/1078087405284886.

Elliot, J. R., Haney, T. J., and Sams-Abiodum, P., 2010. Limits to social capital: comparing network assistance in two New Orleans neighborhoods Devastated by Hurricane Katrina. *The Sociological Quarterly*, **51**, 624–648.

Enarson, E., 2010. Gender. In Phillips, B. D., Thomas, D. S. K., Fothergill, A., and Blinn-Pike, L. (eds.), *Social Vulnerability to Disaster*. Boca Raton: CRC, pp. 123–154.

Fischer, H. W., 1998. *Response to Disaster Fact Versus Fiction and Its Perpetuation: The Sociology of Disaster*, 2nd edn. New York: University Press of America.

Fitzpatrick, C., and Mileti, D. S., 1991. Motivating public evacuation. *International Journal of Mass Emergencies and Disasters*, **9**(2), 137–152.

Freudenburg, W. R., Gramling, R., Laska, S., and Erikson, K. T., 2009. Disproportionality and disaster: Hurricane Katrina and the Mississippi River-Gulf outlet. *Social Science Quarterly*, **90**(3), 497–515.

Fritz, C. E., 1961. Disasters. In Merton, R. K., and Nisbet, R. A. (eds.), *Contemporary Social Problems*. New York: Harcourt.

Gladwin, H., and Peacock, W. G., 1997. Warning and evacuation: as night for hard horses. In Peacock, W. G., Morrow, B. H., and Gladwin, H. (eds.), *Hurricane Andrew: Ethnicity, Gender, and the Sociology of Disasters*. Miami, FL: Florida International University, pp. 52–74.

GNOCDC (Greater New Orleans Community Data Center). 2011. Information retrieved 3 Feb 2011 from http://www.gnocdc.org.

Grenier, G. J., and Morrow, B. H., 1997. Before the storm: the socio-political ecology of Miami. In Peacock, W. G., Morrow, B. H., and Gladwin, H. (eds.), *Hurricane Andrew: Ethnicity, Gender and the Sociology of Disasters*. New York: Routledge, pp. 36–51.

Hewitt, K., 1997. *Regions of Risk: A Geographical Introduction to Disasters*. Essex, UK: Longman.

Hewitt, K., 1998. Excluded perspectives in the social construction of disaster. In Quarantelli, E. L. (ed.), *What is a Disaster? Perspectives on the Question*. London: Routledge, pp. 75–91.

Hoffman, S., and Oliver-Smith, A. (eds.), 1999. *The Angry Earth: Disaster in Anthropological Perspective*. New York: Routledge.

Johnson, G. S., 2008. Environmental justice and Katrina: a senseless environmental disaster. *The Western Journal of Black Studies*, **32**(1), 42–52.

Jones, D., 1993. Environmental hazards in the 1990s: problems, paradigms, and prospects. *Geography*, **78**, 161–165.

Kendra, J. M. 2007. Geography's contributions to understanding hazards and disasters. In McEntire, D.A., (ed.) *Disciplines, Disasters and Emergency Management*. Illinois: Charles C. Thomas. Retrieved from http://training.fema.gov?EMIWeb/edu/ddemtextbook.asp.

Kendra, J., and Wachtendorf, T., 2003. Reconsidering convergence and converger legitimacy in response to the world trade center disaster. *Research in Social Problems and Public Policy*, **11**, 97–122.

Klinenberg, E., 2003. *Heatwave: A Social Autopsy of Disaster in Chicago*. Chicago: The University of Chicago Press.

Kroll-Smith, S., and Gunter, V., 1998. Legislators, interpreters, and disasters. In Quarantelli, E. L. (ed.), *What is a Disaster: Perspectives on the Question*. London: Routledge, pp. 160–176.

Louisiana Workforce Commission, 2010. *Louisiana Labor Force Diversity Data*. Retrieved 4 February 2011: http://www.laworks.net/Downloads/Downloads_LMI.asp#EmployWageAnnual.

McEntire, D. A., 2007. The importance of multi- and inter-disciplinary research on disasters and for emergency management. In McEntire, D. A. (ed.), *Disciplines, Disasters and Emergency Management*. Illinois: Charles C. Thomas.

Mileti, D., 1999. *Disasters By Design: A Reassessment of Natural Hazards in the United States*. Washington, DC: Joseph Henry Press.

Miller, D. S., and Rivera, J. D., 2006. Landscapes of disaster and place orientation in the aftermath of Hurricane Katrina. In Brusma, D. L., Overfelt, D., and Picou, J. S. (eds.), *The Sociology of Katrina*. Lanham: Rowan and Littlefield, pp. 141–154.

Mitchell, J. T., Thomas, D. S. K., Hill, A. A., and Cutter, S. L., 2000. Catastrophe in reel life versus real life: perpetuating disaster myth through hollywood films. *International Journal of Mass Emergencies and Disasters*, **18**(3), 383–402.

Morrow, B. H., 1997. Stretching the bonds: the families of Andrew. In Peacock, W. G., Morrow, B. H., and Gladwin, H. (eds.), *Hurricane Andrew: Ethnicity, Gender, and the Sociology of Disasters.* Miami, FL: Florida International University, pp. 141–170.

Morrow, B. H., and Enarson, E., 1996. Hurricane Andrew through women's eyes: issues and recommendations. *International Journal of Mass Emergencies and Disasters,* **14**(1), 1–22.

Neal, D. M., 1994. The consequences of excessive unrequested donations: the case of Hurricane Andrew. *Disaster Management,* **6**(1), 23–28.

Oliver-Smith, A., 1998. Global changes and the definition of disaster. In Quarantelli, E. L. (ed.), *What is a Disaster? Perspectives on the Question.* New York: Routledge, pp. 177–194.

Oliver-Smith, A., 2002. Theorizing disasters: nature, power, and culture. In Oliver-Smith, A., and Hoffman, S. (eds.), *Catastrophe and Culture: The Anthropology of Disaster.* Santa Fe: School of American Research Press, pp. 23–47.

Peacock, W. G., and Girard, C., 1997. Ethnic and racial inequalities in hurricane damage and insurance settlements. In Peacock, W. G., Morrow, B. H., and Gladwin, H. (eds.), *Hurricane Andrew: Ethnicity, Gender, and the Sociology of Disasters.* Miami, FL: Florida International University, pp. 171–190.

Peacock, W. G., and Ragsdale, A. K., 1997. Social systems, ecological networks and disasters: toward a socio-political ecology of disasters. In Peacock, W. G., Morrow, B. H., and Gladwin, H. (eds.), *Hurricane Andrew: Ethnicity, Gender, and the Sociology of Disasters.* Miami, FL: Florida International University, pp. 20–35.

Perrow, C., 2007. *The Next Catastrophe: Reducing Our Vulnerabilities to Natural, Industrial, and Terrorist Disasters.* Princeton, NJ: Princeton University Press.

Perry, R. W., 2007. What is a disaster? In Rodríguez, H., Quarantelli, E. L., and Dynes, R. R. (eds.), *Handbook of Disaster Research.* New York: Springer, pp. 1–15.

Phillips, B. D., and Fordham, M., 2010. Introduction. In Phillips, B. D., Thomas, D. S. K., Fothergill, A., and Blinn-Pike, L. (eds.), *Social Vulnerability to Disaster.* Boca Raton: CRC, pp. 1–23.

Quarantelli, E. L., 1981. An Agent Specific or an All Disaster Spectrum Approach to Socio-Behavioral Aspects of Earthquakes? Paper presented at the Third International Conference: Social and Economic Aspects of Earthquakes and Planning to Mitigate their Impacts. Held at Lake Bled, Yugoslavia, June 29–July 2, 1981.

Quarantelli, E. L., 1987. Disaster studies: an analysis of the social and historical factors affecting the development of research in the area. *International Journal of Mass Emergencies and Disasters,* **5**, 285–310.

Quarantelli, E. L., 1994. Disaster studies: the consequences of the historical use of a sociological approach in the development of research. *International Journal of Mass Emergencies and Disasters,* **12**(1), 25–49.

Quarantelli, E. L., 1998. *What is A Disaster? Perspectives on the Question.* London: Routledge.

Quarantelli, E. L., 2002. The Disaster Research Center (DRC) field studies of organized behavior in the crisis time period of disasters. In Stallings, R. A. (ed.), *Methods of Disaster Research.* New york: Xlibris, pp. 94–126.

Ritzer, G., 1979. Toward an integrated sociological paradigm. In Snizek, W., Fuhrman, E., and Miller, N. (eds.), *Contemporary Issues in Theory and Research.* Westport, CT: Greenwood Press, pp. 25–46.

Saito, T., and Kunimitsu, A., 2011. Public health response to the combined great East Japan earthquake, tsunami and nuclear power plant accident: perspective from the ministry of health, labour and welfare of Japan. *Western Pacific Surveillance and Response Journal,* **2**(4), doi:10.5365/wpsar.2011.2.4.008.

Scanlon, J., 1996. Not on the record: disasters, records and disaster research. *International Journal of Mass Emergencies and Disasters,* **14**, 265–280.

Scanlon, J., 2006. Unwelcome irritant or useful ally? the mass media in emergencies. In Rodríguez, H., Quarantelli, E. L., and Dynes, R. R. (eds.), *Handbook of Disaster Research.* New York: Springer, pp. 413–429.

Scanlon, J., 2007. Research about the mass media and disaster: never (well hardly ever) the twain shall meet. In McEntire, D. A. (ed.), *Disciplines, Disasters and Emergency Management.* Illinois: Charles C. Thomas.

Sharkey, P., 2007. Survival and death in New Orleans: an empirical look at the human impact of Katrina. *Journal of Black Studies,* **37**, 482–501.

Stallings, R. A., 2002. Weberian political sociology and sociological disaster studies. *Sociological Forum,* **17**(2), 281–305.

Stern, G., 2007. *Can God Intervene? How Religion Explains Natural Disasters.* Westport: Praeger.

Sylves, R. T., 2006. President Bush and Hurricane Katrina: a presidential leadership study. *ANNALS, AAPSS,* **604**, 26–56.

Tierney, K., Bevc, C., and Kuligowski, E., 2006. Metaphors matter: disaster myths, media frames, and their consequences in Hurricane Katrina. *ANNALS, AAPSS,* **604**, 57–81.

Tobin, G., and Montz, B., 1997. *Natural Hazards.* New York: Guildford.

Waugh, W. L., Jr., 2007. Public administration, emergency management, and disaster policy. In McEntire, D. A. (ed.), *Disciplines, Disasters and Emergency Management.* Illinois: Charles C. Thomas.

Webb, R. G., Wachtendorf, T., and Eyre, A., 2000. Bringing culture back in: exploring the cultural dimensions of disaster. *International Journal of Mass Emergencies and Disasters,* **18**(1), 5–19.

Wenger, D. E., Dykes, J. D., Sebok, T. D., and Neff, J. L., 1975. It's a matter of myths: an empirical examination of individual insight into disaster response. *Mass Emergencies,* **1**, 33–46.

White, G. F., 1975. *Flood Hazard in the United States: A Research Assessment.* Boulder, CO: Institute of Behavioral Science, University of Colorado.

Wills, J., 2006. Celluloid chain reactions: *the China syndrome* and three mile Island. *European Journal of American Culture,* **25**(2), 109–122.

Wisner, B., Blaikie, P., Cannon, T., and Davis, I., 2004. *At Risk: Natural Hazards, People's Vulnerability and Disasters,* Secondth edn. New york: Routledge.

Wolff, E. N. 2007. Recent Trends in Household Wealth in the U.S.: Rising Debt and the Middle-Class Squeeze. New York: The Levy Economics Institute of Bard College. Working Paper No. 502.

Cross-references

Causalities Following Natural Hazards
Civil Protection and Crisis Management
Communicating Emergency Information
Community Management of Hazards
Disaster
Disaster Research and Policy, History
Emergency Management
Emergency Shelter
Evacuation
Federal Emergency Management Agency (FEMA)
Hazard
Heat Wave
Historical Events
History of Natural Disasters
Human Impact of Hazards
Humanity as Agent of Natural Disaster
Hurricane Katrina
Indian Ocean Tsunami, 2004
Insurance

SOLAR FLARES

David H. Boteler
Earth Science Sector, Natural Resources Canada, Ottawa, ON, Canada

Definition

Solar flares are a burst of electromagnetic radiation from the Sun.

Discussion

The first flare was observed in 1859 when Richard Carrington, while making sunspot drawings, saw a sudden brightening. Now, with modern instruments, we know that the burst of radiation from a solar flare lasts from a few minutes to several hours and spans the entire electromagnetic spectrum. The intensity of a flare is classified according to the X-ray flux, F, in the 0.1–0.8-nm [1–8 Å] wavelength range measured on the GOES satellite (see Table 1). The size of a flare is given by the flux value, using the classification letter as a multiplier: For example, a X4.5 flare has a peak flux of 4.5×10^{-4} W m^{-2}.

Solar flares are often thought to be the cause of a wide range of space weather effects. The radiation from solar flares is directly responsible for increased ionization on the dayside of the Earth that interferes with radio communications. However, other effects, such as magnetic storms, are due to associated phenomena like coronal mass ejections. Nevertheless, observations of solar flares continue to be a valuable precursor of such phenomena for space weather forecasting.

Solar Flares, Table 1 Solar flare classification

Class	X-ray flux (W.m^{-2})
X	$F \geq 10^{-4}$
M	$10^{-5} \geq F > 10^{-4}$
C	$10^{-6} \geq F > 10^{-5}$

Bibliography

Carrington, R. C., 1859. Description of a singular appearance seen on the Sun on September 1, 1859. *Monthly Notices of the Royal Astronomical Society*, **20**, 13–15.
Tandberg-Hanssen, E., and Emslie, A. G., 2009. *The Physics of Solar Flares*. Cambridge: Cambridge University Press, p. 288.

Cross-references

Critical Infrastructure
Electromagnetic Radiation (EMR)
Space Weather
Sunspots

SOLIFLUCTION

Piotr Migoń
University of Wrocław, Wrocław, Poland

Synonyms

Gelifluction; Soil flow

Definition

Solifluction, literally "soil flow," is a category of shallow mass movement, which affects saturated unconsolidated deposits and results from reduction of internal friction and cohesion due to excess water.

Discussion

Precise definition of solifluction has not been agreed upon, and three views can be found in the literature: (1) solifluction is a laminar flow of soil in any environment, soil saturation being an essential prerequisite; (2) solifluction is a specific cold-climate phenomenon and involves soil flow of water-saturated active layer above the permafrost table. Others call this type of movement "gelifluction" and consider it a variant of solifluction; (3) solifluction is again a cold-climate phenomenon and involves two mechanisms: gelifluction (understood as in (2) above) and slow soil creep due to volume changes imposed by alternating freezing and thawing. In addition, sliding of soil on a frozen substrate is included in solifluction. The final definition (3) is most widely accepted today (Matsuoka, 2001; French, 2007).

Solifluction is usually associated with periglacial environments of high latitudes and high altitudes, particularly with permafrost terrain, and concerns downslope movement of seasonally unfrozen ground of the active layer. The rates of solifluction vary with slope gradient, water content in soil, and climatic conditions. In frigid, polar regions, it occurs at a rate of <5 cm year^{-1}, whereas in alpine environments, yearly displacements may exceed 1 m. Solifluction is present on slopes as gentle as ~2°. The thickness of deposit affected by solifluction is usually <1 m, but

Land-Use Planning
Mass Media and Natural Disasters
Mitigation
Myths and Misconceptions
Natural Hazard
Psychological Impacts of Natural Disaster
Risk Perception and Communication
Social-Ecological Systems
Vulnerability

lower hill slopes may carry composite solifluction sheets a few meters thick. Lobes and terraces, ending with distinct risers 0.2–1.5 m high, are landforms indicative of solifluction. However, under more severe climatic conditions rather uniform sheets develop.

Hazards associated with solifluction need to be considered in respect to active and relict solifluction. Active solifluction poses little hazard in the short term, and its negative impacts are easy to mitigate. Insertion of piles into perennially frozen ground to support building foundations or other constructions makes near-surface solifluction practically harmless. More hazardous are relict solifluction sheets and lobes in the former periglacial terrains of Europe or North America. They are often degraded, vegetated, and subdued in appearance, hence not easy to recognize and map. Nevertheless, relict shear planes may be present in solifluction slope deposits, especially if these are clayey, and are capable of reactivation. Slope undercutting or extra load disturbs the state of equilibrium and may lead to renewed sliding or flow, particularly if water content in the soil is high. Examples are known where reactivation of Late Pleistocene solifluction lobes caused considerable engineering problems during road and dam constructions, including the necessity of relocation of structures (Hutchinson, 1991).

Bibliography

French, H. M., 2007. *The Periglacial Environment*. Chichester: Wiley.

Hutchinson, J. N., 1991. Theme lecture: periglacial and slope processes. In Forster, A., Culshaw, M. G., Cripps, J. C., Little, J. A., Moon, C. F. (eds.), *Quaternary Engineering Geology*. Geological Society Engineering Geology Special Publication, 7, pp 283–331.

Matsuoka, N., 2001. Solifluction rates, processes and landforms: a global review. *Earth-Science Reviews*, **55**, 107–134.

Cross-references

Creep
Cryological engineering
Permafrost

SPACE WEATHER

David H. Boteler
Earth Science Sector Natural Resources Canada, Ottawa, ON, Canada

Definition

Space weather is the chain of processes from eruptions on the Sun, their passage through interplanetary space, and the interaction with the Earth's magnetic field that leads to disturbances in the Earth's magnetosphere, ionosphere, and on the ground that represent a hazard to man-made technology and human life.

Introduction

Space weather starts at the Sun. As well as the heat and light that the Sun provides to sustain life on Earth, the Sun occasionally emits bursts of electromagnetic radiation and high-energy particles that impact the Earth. The particles can damage satellite electronics and present a hazard for manned space flight. Disturbances produced in the ionosphere disrupt radio communication and satellite navigation systems, and magnetic disturbances on the ground interfere with aeromagnetic surveys and directional drilling and are a threat to power systems and pipelines.

Space weather processes

A typical space weather event would start with a sunspot group crossing the Sun's disk. Energy buildup associated with the sunspots is released in a solar flare and associated solar energetic particles, and often a coronal mass ejection. The radiation travels at the speed of light and reaches the Earth 8 min later, solar energetic particles travel at a significant fraction of the speed of light, and start arriving an hour later, whereas the particles in coronal mass ejections and high speed streams travel at speeds from 500 to 2,000 km/s and take 1–4 days to travel the 149 million kilometers to Earth. At other times coronal holes, sometimes lasting for months, produce streams of high-speed particles that sweep past the Earth with every solar rotation (Schwenn, 2006).

At the Earth the solar radiation and particles produce effects extending from the space environment down through the ionosphere to the ground. The electromagnetic radiation from solar flares produces extra ionization on the dayside of the Earth. Solar energetic particles are shielded by the Earth's magnetic field (Smart et al., 2006) from entry to the upper atmosphere except at high latitudes where the magnetic field lines are nearly vertical. Particles from coronal mass ejections and high speed streams interact with the Earth's magnetic field causing a worldwide depression of the field strength (a magnetic storm) lasting 1–2 days and short-lived (15–60 min) magnetic excursions (magnetic substorms) at high latitudes. Substorm particles also enter the upper atmosphere creating extra ionization and causing the atoms to emit light (the aurora) (Pulkkinen, 2007).

Magnetic effects

Magnetic disturbances have long been identified by the deflections they cause to compass needles. Modern electronic compasses are still used for navigation and applications like directional drilling where magnetic disturbances can cause errors. Magnetic disturbances also create magnetic field variations that interfere with aeromagnetic surveys.

Magnetic field variations also induce electric fields in the Earth and in electrical conductors at the Earth's surface. The telegraph was the first system using long conductors and magnetically induced voltage created false signals that prevented messages being sent. Following the solar flare observed by Carrington in 1859, the magnetic storm caused worldwide disruption to the operation of the telegraph system (Boteler, 2006). As technology evolved and new cable communication systems were introduced, magnetically induced voltages were found to affect both land cables and undersea cables (Lanzerotti, 2001). Pipelines also experience magnetically induced voltage swings that interfere with corrosion protection of the pipeline.

Power transmission networks are another system using long conductors that are exposed to magnetic disturbances. Geomagnetically induced currents (GIC) in transmission lines flow to ground at substations where they interfere with operation of the power transformers (Kappenman, 1996, Pirjola, 2007). Extra harmonics are generated that can cause misoperation of protective relays, increased reactive power demand causes voltage dips, and overheating can damage transformers. In March 1989, a magnetic storm caused widespread power system problems including a blackout in Quebec and transformers burnt out in the USA and UK. Modern power systems are more heavily loaded than in 1989; so, there is concern that a similar size magnetic disturbance could produce widespread power system problems.

Ionospheric effects

The ionosphere affects the propagation of radio waves: both those reflected by the ionosphere and those passing through it. Radio signals travel in straight lines and the curvature of the Earth's surface would restrict radio communication to short distances if it was not for the signals being reflected by the ionosphere. It was the presence of this reflecting layer at the top of the atmosphere that made possible Marconi's transatlantic radio transmissions and all subsequent long-distance high-frequency (HF) radio communication.

Solar illumination of the atmosphere creates the normal ionosphere. This consists of the E and F regions that reflect radio waves and the D region that absorbs radio waves. The amount of reflection or absorption depends on the electron density of the region and the frequency of the radio wave. At high enough frequencies the radio signal passes right through the ionosphere and these frequencies are used for satellite communication and navigation systems.

All phases of a space weather event create ionospheric disturbances that impact radio communications (Goodman, 2005). X rays emitted during solar flares create extra ionization in the dayside D region causing increased absorption and a shortwave fade-out (SWF) of HF radio communications. Solar energetic particles penetrate into the ionosphere at high latitudes, creating polar cap absorption (PCA) that can block HF radio communications for days. During geomagnetic storms, precipitation of auroral particles creates absorption in the auroral zone that sporadically blocks HF radio communication. Disturbances in the E and F regions can also change the reflecting properties of the ionosphere making it difficult for radio operators to know which frequencies to use or making HF communications impossible.

At frequencies above 500 MHz which are used for satellite communications and positioning systems, ionospheric absorption is small, so the signal is not lost; however, signals are slowed going through the ionosphere, so for global navigation satellite systems (GNSS) that rely on signal timing the extra delays can lead to positioning errors during space weather events. Patches of ionization mean that signals may travel different paths to reach a receiver. Destructive and constructive interference between these signals can cause scintillation of the signal at the receiver causing loss of lock of GNSS receivers and loss or degradation of satellite communication links. Because of local characteristics of the ionosphere, scintillation is a particular problem in equatorial and auroral regions.

Space environment effects

The energetic particles in the space environment around the Earth and in interplanetary space can be a hazard to spacecraft and manned spaceflight. In interplanetary space, spacecrafts are directly exposed to the "solar wind" particles streaming from the Sun and the bursts of particles associated with coronal mass ejections. Close to the Earth the geomagnetic field acts as a shield against these particles (Smart et al., 2006), creating a cavity in the solar wind called the magnetosphere. However, inside the magnetosphere, particles magnetically trapped in the radiation belts can be energized to levels where they present a hazard. Satellites in high altitude orbits are exposed to these trapped particles. Satellites in low altitude orbits are below the radiation belts except when they cross the South Atlantic Anomaly, a region of low magnetic field strength that allows the radiation belt particles to come down to lower altitudes.

Satellites are affected by the space environment in a variety of ways (Tribble, 2003). Energetic particles from solar events and in the radiation belts can penetrate the walls of a spacecraft damaging components inside. Sometimes the deposited charge does not cause damage but is enough to change the state of memory levels, leading to false signals known as Single Event Upsets. Energetic electrons can penetrate into the dielectrics of electrical circuits inside the spacecraft. This "deep dielectric charging" can build up to a point where a sudden electrical discharge occurs that damages the satellite electronic systems. These energetic particles can also damage human tissue, so presenting a danger for manned spaceflight, particularly during activities like spacewalks that are outside the shielding of the spacecraft.

Satellites in low earth orbit also experience drag from the upper reaches of the atmosphere. This is something that is routinely taken into account in managing the orbit. However, during geomagnetic storms the particles

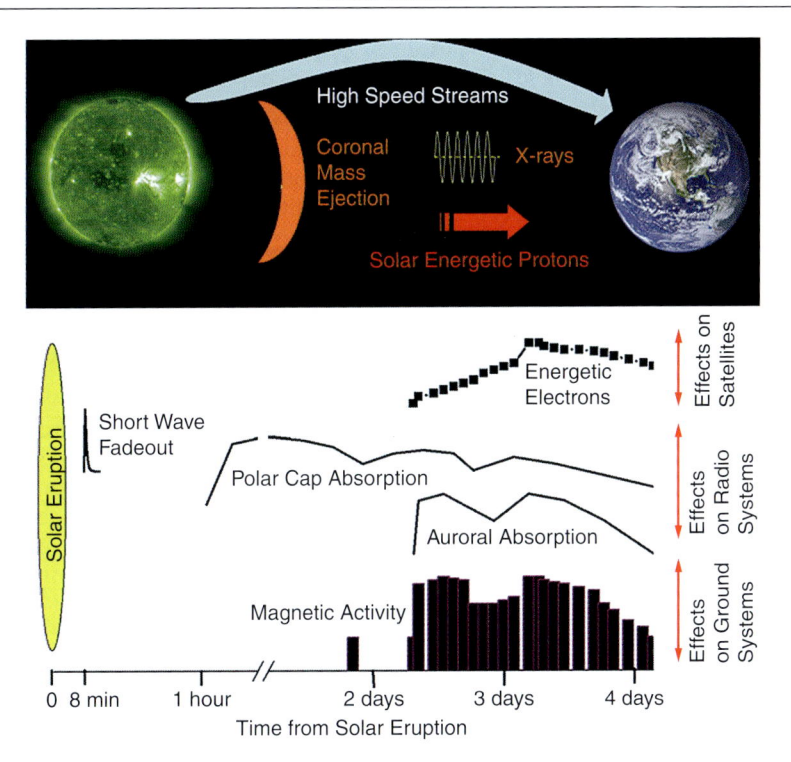

Space Weather, Figure 1 Solar eruptions produce electromagnetic and particle radiation which cause disturbances at different times on the Earth that affect a variety of technologies.

deposited into the ionosphere produce extra heating that causes the atmosphere to expand. This causes extra drag on the low altitude satellites. In some cases, this can cause premature de-orbiting of spacecraft: a notable example being when Skylab descended earlier than planned. The connection with space weather was recognized and led to headlines such as "Skylab coming down, sunspots to blame." Thus, space weather can affect satellite operations from interplanetary space all the way down to the top of the atmosphere. Energetic particles penetrating into the atmosphere in polar regions are also being recognized as a possible hazard for aircraft electronics and frequent fliers on high altitude flights.

Mitigation and forecasting

Modern systems, because of greater sophistication or reduced operating margins, are becoming more vulnerable to space weather. Engineering solutions to make systems immune to the problems are difficult or not cost effective. For example, satellite electronics could be protected from energetic particles by adding more shielding to the satellite but the added weight increases the launch costs. With space weather phenomena, like many natural hazards, the more extreme events occur less frequently, so the optimum solution is a careful assessment of the probability of occurrence and a decision made to protect against an event that could reasonably be expected to occur within the lifetime of the system. Where systems cannot be protected special operating procedures are used to put the system into "safe" mode when major space weather events are forecast. Forecasting space weather is a major scientific challenge and is the focus of considerable research work around the world. Operational space weather services are coordinated by the International Space Environment Service. Consult their web site (www.ises-spaceweather.org) for links to the ISES Regional Warning Centres providing space weather services around the world.

Conclusions

Space weather is a natural hazard of the technological age. The hazards originate with disturbances on the Sun that travel through space to impact the Earth. A range of systems are affected: power systems, pipelines, radio communications and satellite positioning systems, and satellites and manned space missions. As society places greater reliance on these systems, there is an increasing requirement for forecasts of space weather so that action can be taken to protect affected systems.

Bibliography

Boteler, D. H., 2006. The super storms of August/September 1859 and their effects on the telegraph system. *Advances in Space Research*, **38**, 159–172.

Goodman, J. M., 2005. *Space Weather & Telecommunications*. New York: Springer.

Kappenman, J. G., 1996. Geomagnetic storms and their impact on power systems. *IEEE Power Engineering Review*, **16**, 5–8.

Lanzerotti, L. J., 2001. Space weather effects on communications. In Daglis, I. A. (ed.), *Space Storms and Space Weather Hazards*. Dordrecht/Boston: Kluwer, pp. 313–334.

Pirjola, R., 2007. Space weather effects on power grids. In Bothmer, V., and Daglis, I. A. (eds.), *Space Weather – Physics and Effects*. Berlin: Springer, pp. 269–288.

Pulkkinen, T., 2007. Space weather: terrestrial perspective, *Living Reviews in Solar Physics*, **4**, (Online Article: cited March 10, 2010, http://www.livingreviews.org/lrsp-2007-1).

Schwenn, R., 2006. Space weather: the solar perspective. *Living Reviews in Solar Physics*, **3**, (Online Article: cited March 10, 2010, http://www.livingreviews.org/lrsp-2006-2).

Smart, D. F., Shea, M. A., and Flückiger, E. O., 2006. Magnetospheric models and trajectory computations. *Space Science Reviews*, **93** (1-2), 305–333. Revised version published online, 2006.

Tribble, A. C., 2003. *The Space Environment: Implications for Spacecraft Design*. Princeton: Princeton University Press. Revised and expanded edition.

STORM SURGES

Gonéri Le Cozannet[1], Hormoz Modaressi[2]
Rodrigo Pedreros[1], Manuel Garcin[1], Yann Krien[1], Nicolas Desramaut[1]
[1]BRGM-French Geological Survey – Natural Risks and CO2 Storage Division, Orléans, France
[2]BRGM-French Geological Survey – RNSC/D, Orléans, France

Definition

A storm surge is a rise of water above (1) the predicted astronomical sea tide or (2) a typical lake level. Storm surges occur because of severe hydrometeorological conditions associated with storms and tropical cyclones.

Discussion

Synonym

"Storm tide" refers to a rise of seawater above a reference altitude, thus including the astronomical tide.

Processes involved

Storm surges are primarily caused by:

(a) Reduced atmospheric pressure: This process is known as the "inverse barometric effect," by which a reduction of 1 hPa of pressure causes 1-cm water level elevation in stationary conditions.

(b) Strong winds that may push water coastward, by creating currents that accumulate waters in shallow areas.

(c) Strong waves that create a mean sea level rise as they are breaking (waves setup).

Typically, atmospheric forcing due to pressure and wind dominates for the large continental shelf, whereas wave setup becomes more important for steep submarine slopes.

Storm surges are gravity waves, and are thus affected by the rotation of the Earth:

(d) The Coriolis effect creates an acceleration of water currents to the right in the Northern Hemisphere, thus contributing to sea level rise when the coast is to the right of a storm surge track.

As they hit the coast, storm surge intensity also interacts with:

(e) Boundary water level, including water level in estuaries and coastal swamps and lowlands, often affected by storm precipitations.

While these processes and coastal patterns control the elevation of a timely averaged sea level above astronomical tide, coastal engineers also need to take into account the instantaneous sea level change caused by waves (swash), in order to estimate, e.g., how much water runs on (run-up), or over (overtopping) a natural or man-made protection.

Related hazards

Storm surges can cause: (1) coastal flooding of low-lying areas by simple submersion of the shoreline, dunes overwash, or dikes overflowing; (2) wave impacts on infrastructures, by the collection of debris between them, by scouring (i.e., submersion waters carrying debris behind the coastal protections and destroy their foundations), or, in high latitudes, by the involvement of ice; (3) rapid erosion of the shoreline and damages or destruction of assets or defenses located on eroded areas; and (4) salinization of coastal lowland.

Storm surge impact

Storm surges effects are most damaging when they occur at high tide. Storm surges are considered by NOAA as the most deadly coastal hazard associated to hurricanes: Hurricane Katrina, which affected the New-Orleans in 2005 (see section Cross-references) caused more than 1,800 deaths. More than 300,000 people lost their lives during the Bhola cyclone in 1970 primarily because of the storm surges it caused. With respect to extra-tropical storms, the 1953 North Sea storm caused more than 2,000 deaths in the Netherlands, England, and Belgium, also causing damages to infrastructures in France.

Cross-references

Coastal Erosion
Coastal Zone, Risk Management
El Niño/Southern Oscillation
Extreme Value Theory
Flash Flood
Flood Deposits
Flood Hazard and Disaster
Flood Protection
Global Change and Its Implications for Natural Disaster
Hurricane (Cyclone, Typhoon)
Hurricane Katrina
Hydrometeorological Hazards

Monsoon
Sea Level Change
Storms
Surge

STORMS

Norm Catto
Memorial University of Newfoundland,
St. John's, NL, Canada

Synonyms

Abrupt; Severe weather events

Definition

Types of storms include tropical cyclones (hurricanes, typhoons); mid-latitude cyclones; extratropical storms; thunderstorms (including lightning and hail storms), tornadoes, dust storms, blizzards (snowstorms), and ice storms (verglas).

Tropical cyclones include hurricanes in the Atlantic and eastern Pacific Oceans, typhoons in the western Pacific Ocean and along the coasts of Southeast Asia, and cyclones in the Indian Ocean and along the coasts of Australia. They are responsible for large death tolls in heavily populated, less economically developed countries, particularly in Southeast Asia, Eastern Africa, Central America, and the Caribbean. They are also costly in financial terms. Including indirect and ongoing recovery costs, Hurricane Katrina in 2005 cost the USA in excess of $500 billion.

Mid-latitude cyclones can affect any mid-latitude area between 30° and 70° latitude. Mid-latitude cyclones develop by frontal movement, and can occur at any time of the year. Although many follow the trajectory of the Jet Stream, others do not, moving along the varying frontal systems that exist at the time. They thus differ from tropical cyclones in their location, track direction, and the method of formation. Mid-latitude cyclones can generate winds as strong as tropical cyclones. In coastal areas, they can create storm surges capable of causing damage. Their effects are widespread in interior areas. Because they do not depend on warm water as a source of energy, they do not dissipate as they cross land. In Atlantic Canada and New England (USA), strong autumn and winter mid-latitude cyclones are common. Extratropical storms, which form when weakening hurricanes interact with mid-latitude cyclones and other weather systems, also can cause severe flooding.

Thunderstorms form as unstable, warm, and moist air rapidly rises into colder air and condenses. They are generated by several mechanisms, including convection, frontal activity, and uplift due to orographic effects. In coastal areas, thunderstorms can form when moist air is heated, either over warmer land or over warm ocean waters.

The hazard from lightning strikes largely depends on the number of thunderstorms. In the USA, between 70 and 80 people are killed each year. Lightning strikes cost the insurance industry more than $5 billion annually. In the past 10 years, the value of personal computers destroyed by electrical surges from lightning strikes has exceeded $200 million annually in the USA. Throughout North America, hail causes more than $3 billion in annual damages to cars, roofs, crops, and livestock.

Although tornadoes can occur in several countries, more than 90% are developed in the USA and adjacent southern Canada. Tornadoes are the most significant natural hazard in much of the midwestern United States. However, other areas are not immune. The most deadly tornado in the last 100 years, causing 922 deaths, occurred in Bangladesh in 1969.

Blizzards, ice storms, and other hazards of winter can affect more than one billion people each year. Although winter hazards do not always receive the attention devoted to volcanic eruptions or earthquakes, humans die from exposure and hypothermia every year. For people unaccustomed to winter hazards, a snowfall may come as a profound shock and pose difficulties for citizens and public services.

Hazards posed by storms have been greatly reduced, although not eliminated, by improvements in meteorological forecasting, particularly involving remote sensing. However, storms continue to kill people and cause significant property damage throughout the world.

Bibliography

Hyndman, D., Hyndman, D., and Catto, N. R., 2008. *Natural Hazards and Disasters*. Toronto: Nelson.

Cross-references

Dust Storm
Hurricane (Cyclone, Typhoon)
Hurricane Katrina
Ice Storm
Lightning
Snowstorm and Blizzard
Thunderstorms
Tornado

STRATOVOLCANOES

Shane J. Cronin
Massey University, Palmerston North, New Zealand

Synonyms

Composite volcano; Stratocone

Stratovolcanoes, Figure 1 Mt. Taranaki, New Zealand, 2,518 m above sea level, is a classic stratovolcano with a construction of lava flows and pyroclastic fragmental deposits. Its steep upper slopes are only a few thousand years old, whereas the low outermost slopes contain mass flow and volcanic ash deposits dating back over 200,000 years.

Stratovolcanoes, Figure 2 Mt. St. Helens, Washington State, USA, site of a massive eruption in May 1980 generated when a huge part of the volcano's flank collapsed outward in a debris avalanche, unleashing a supersonic volcanic blast that devastated the surrounding landscape and took over 60 lives. The scar left by the landslide collapse is slowly healing with the growth of a lava dome.

Definition

A volcanic construct, built up into the form of a tall mountain over thousands to hundreds of thousands of years by accumulation of successive layers (=strata) of lava, pyroclastic deposits (from explosive eruptions), and related mass-flow deposits (Davidson and DeSilva, 2000; Schmincke, 2004). In their simplest form, these are conical-shaped mountains with steep upper slopes curving gradually down into low-sloping outer flanks (Figure 1). Older examples may show complex morphologies with irregular edifices being scarred by several generations of partial collapses (Figure 2). Their wide bases pass outward into overlapping fans of volcano-sedimentary deposits that may join to form a radiating ring-plain around the edifice (Cotton, 1944). Stratovolcanoes may vary greatly in volume between 10 and 500 km^3, with surrounding ringplains containing up to 80% of the erupted volume of long-lived examples (Zernack et al., 2009).

Introduction

Stratovolcanoes or composite volcanoes are the most recognizable of volcanic landforms. These distinctive and very high mountains are often picturesque and snow capped. Despite their fairy-tale appearances, stratovolcanoes have a destructive and unpredictable side to their character, matched in violence by few other volcano types. This dark side is well demonstrated by disasters resulting from widely different types and scales of eruptions from these volcanoes around the world. Stratovolcanoes have collectively killed >200,000 people since AD 1783 (Tanguy et al., 1998) and make up eight out of the top-ten killer volcanoes over the twentieth century (Witham, 2005).

Occurrence

Stratovolcanoes are one of the most common volcano types on Earth, with at least 700 recognized as being active (Simkin and Siebert, 1994). Long chains of stratovolcanoes, separated by 50–100 km, occur along volcanic arcs and continental margins. One of the most extensive chains wraps around the entire Pacific Ocean from the Aleutians through the Americas, and from Kamchatka through Southeast Asia, the Southwest Pacific Islands, and into New Zealand. They are common to all zones of tectonic plate subduction and are also formed occasionally in other settings, including intraplate localities (e.g., Nyiragongo, Congo).

Construction and magmatic systems

As their name indicates, stratovolcanoes are composite features, built up gradually from repetitively stacked beds of eruption products that accumulate around a single vent system (Figure 1). These volcanic mountains contain a range of volcanic deposits, including explosively produced pyroclastic fall units (tephra), pyroclastic density current deposits (pyroclastic surge or flow units), along with lava flows, lava domes, and a large range of reworked volcanic deposits (volcaniclastic sediments). The diversity of these eruptive products is characteristic for stratovolcanoes, engendering both variation in the mountain shapes and a bewildering array of volcanic hazard processes. They may also undergo several phases of growth, weathering, collapse, and rebirth, with active vents migrating and occasionally multiple vents developing (Davidson and DeSilva, 2000).

Stratovolcanoes are most commonly produced by andesitic magmas derived from the process of subduction (Gill, 1981). Cold and water-saturated subducted lithospheric crust descends into the Earth's mantle beneath convergent margins, releasing volatiles (mainly water) as temperatures increase. The volatiles rise into the mantle wedge above the subducting slab, promoting its partial melting, and magmas form, accumulate, and rise due to the effects of buoyancy (McKenzie, 1985). Complex assimilation, fractionation, and mixing processes undergo these typically basaltic melts as they rise from the mantle through the crust (e.g., Grove et al., 2002). In general terms, melts often pond at the base of the lithospheric crust, anywhere between 15 and 35 km below the volcano (Annen et al., 2006). Upon further evolution to lower densities, melts from this zone rise and may pond higher within the crust, where they again reach neutral buoyancy, often between 9 and 14 km below the volcano (Carmichael, 2002; Devine et al., 2003). Here magmas continue to fractionate (forming crystals that separate out due to density differences) and assimilate (melt and incorporate) surrounding rock, possibly also mixing with other magma batches (Price et al., 1999). Eruption-triggering processes may involve tectonic motion, injection of further heat/magma from below, and crystallization of rising magma (Blundy et al., 2006). These complex magmatic roots produce a range of erupted compositions, dominantly in the andesitic range. If a volcano mainly erupts low-viscosity magmas, a shield-like form will result from stacked lava flows, whereas high-viscosity magmas generally produce caldera volcanoes or lava dome complexes. The volumes and compositions of andesitic magmas favor explosive eruptions; but these are generally not powerful enough to distribute the bulk of their products very far; hence, they build up around the vent and a composite mountain forms.

Stratovolcanoes, Figure 3 Gunung Merapi, Central Java, Indonesia, in the aftermath of the 2006 eruption where the summit lava dome partially collapsed down the southern volcano flanks, generating a dense pyroclastic density current that filled the Kali Gendol valley, spilling out to devastate the village in this area, 5 km from the summit.

Why are stratovolcanoes so hazardous?

There are a number of reasons why stratovolcanoes have repeatedly proven deadly to populations that surround them. First, they typically spend the majority of their life span in periods of dormancy, lasting years, decades, or several centuries. An old adage is that a volcano is ready to strike, once the memory of the last event has faded. Volcanoes with long periods of dormancy slip from public attention, and especially when written records in many parts of the world are <200 years old, many volcanoes are never recognized as being a source of threat (e.g., Mt Pinatubo prior to its 1991 eruption; Newhall and Punongbayan, 1996).

A second cause for stratovolcano-induced disasters is that human populations are naturally drawn to living on and near such structures. Especially during dormant periods, pyroclastic deposits from stratovolcanoes engender extremely fertile, friable, and well-drained soils. The edifices also host high-quality and plentiful water supplies and may have other economic value in terms of building aggregate, hot springs, mineral deposits, and forest/wildlife resources (Schmincke, 2004).

The third major risk factor of stratovolcanoes is their unpredictability. In some cases, where magma flux is high, this variability is limited and eruption magnitudes and types are relatively uniform for long periods (e.g., historic activity at Merapi Volcano, Java, Indonesia; Voight et al., 2000; Figure 3). However, typical variations in the complex magma generation mechanisms throughout a volcano's lifespan, along with environmental factors, including the height, saturation, and stability of the volcanic edifice, combine to produce a huge array in magnitude, composition, and style of eruptions from any single stratovolcano (e.g., Vesuvius; Andronico and Cioni, 2002). This means that populations preparing for the next eruption based only on the last event could be in for an unfortunate surprise.

The final major basis for society's vulnerability to stratovolcanoes is their height and inherent instability. Even small eruptions of these volcanoes are capable of generating destructive mass flows, because they have so much gravitational potential energy when generated 2–4 km above communities at risk. In addition, because stratovolcanoes are formed from a variety of solid (lava) and loose strata, the latter layers are easily saturated and weakened. With increasing height and hydrothermal alteration, mountain volcanoes become very unstable. In addition, snow and glacier ice caps, along with crater lakes, provide ready sources of water to rapidly magnify mass flows (e.g., Cronin et al., 1996; Figure 4). Small magmatic intrusions, earthquakes, and major rainstorms may readily trigger massive collapses, generating lahars and debris avalanches that obliterate everything in their paths with little warning (e.g., Lowe et al., 1986).

Stratovolcanoes, Figure 4 Ruapehu Volcano, New Zealand, a typical snow-capped stratovolcano showing also a crater lake over its active vent. Explosions through this lake frequently generate devastating lahars that build up with snow and sediment to flow over 200 km to the sea. The black material surrounding the lake and forming the stripe down the glacier is the remains of pyroclastic density currents and lahars from the September 2007 eruption.

Volcanic processes, hazard impacts, and hazard evaluation at stratovolcanoes

Hazard maps for stratovolcanoes are often complex and contain a range of diverse information due to the large range in volcanic processes associated with such volcanoes. The main threats posed by stratovolcanoes include:

Pyroclastic fall (ashfall or tephra fall): Stratovolcanoes typically produce eruptions that are intermediate in magnitude between caldera volcanoes and scoria cones, involving <1 km^3 of magma and typically only a tenth to a hundredth of this. Despite their moderate volume, stratovolcanoes may produce sustained and unstable eruption plumes of up to 35-km high. Being such high mountains, their eruption columns may efficiently reach high into the atmosphere and wind typically disperses ash over several hundred square kilometers (Carey and Sparks, 1986). Ashfall impacts are typically defined via mapping the dispersal, thickness, and grain size of products of past eruptions and determining the prevailing wind directions and strength conditions that may affect future eruptions (e.g., Pyle, 1989). Hazard zones may be based on this mapping along with numerical simulations that combine particle settling and wind/climate models (Bonadonna, 2006). These zones are normally ellipsoidal, stretched along the main axis of ash dispersal. Ashfalls can be heavy enough to cause building collapse and smother crops. Even light falls may disrupt air and road transport, electricity lines and cause human and animal health problems (Cronin et al., 2003). Under ashfall, visibility is extremely low and the ash is a hard, abrasive, and an extreme irritant to breathing. Adsorbed volcanic gases commonly adhere to volcanic ash, including some poisonous compounds, such as sulfuric, hydrochloric, and hydrofluoric acid (Hansell et al., 2006).

Pyroclastic density currents: They are rapidly moving hot mixtures of volcanic gas and particles (Branney and Kokelaar, 2002), generated by a number of mechanisms on andesitic stratovolcanoes. These are by far the greatest cause of casualties from volcanic eruptions (Witham, 2005). Due to erupting magmas of lower-viscosity and higher density than rhyolitic eruptions, andesitic eruption columns are often less stable, with frequent partial collapses generating localized pyroclastic density currents. Even small eruption columns that collapse on the upper volcano flanks may efficiently travel long distances as they descend the steep volcanic slopes (Lacroix, 1904). Further cascading hazards of huge potential consequence are generated when pyroclastic flows travel across snow- and glacier-capped areas. Andesitic pyroclastic flows are also often generated from the collapse or explosion of lava domes (Abdurachman et al., 2000). These viscous bodies of magma ooze slowly out at the summit or upper flanks of many andesitic volcanoes, becoming very unstable as they grow and build up internal magmatic gas pressure (Voight and Elsworth, 2000). These types of pyroclastic flow hug the ground and remain concentrated in the valleys over a runout of 5–15 km from source. Pyroclastic density currents sweep away, bury, or burn most objects in their paths, with the famous AD79 eruption of Vesuvius claiming most of its victims with a combination of repeated pyroclastic flows and heavy ashfalls (Sigurdsson et al., 1982). Mont Peleé in Martinique extinguished 29,000 lives in a matter of hours in May 1902 with a series of pyroclastic flows and directed blasts that destroyed the town of St Pierre, eerily leaving most of the buildings standing (Lacroix, 1904). Typically, pyroclastic flow hazard zones are drawn concentrically around a central volcanic edifice, although recognizing that these flows are strongly topographically controlled on stratovolcanoes, higher-level hazard zones may be concentrated along major valley systems.

Directed blasts: These are a special type of pyroclastic density current that are characterized by extremely highly energetic explosions that expel country rock and fresh pyroclasts radially at supersonic velocities. They are usually generated when there is a major collapse of a volcano's flank or large lava dome (Siebert et al., 1987). At Mt. St. Helens in 1980, the flank collapse and subsequent blast were among the most violent natural events that modern volcanologists have ever witnessed. These swept over forests in an area of 600 km^2, killing over 60 people and tossing heavy machinery around like children's toys (Lipman and Mullineaux, 1981).

Lahars: These fast-flowing mixtures of water and rock debris (also called mudflows, debris flows, and hyperconcentrated flows) are ubiquitous to stratovolcanoes and are a natural consequence of eruptions at high elevations where there is often snow cover, crater lakes, or simply just unstable and saturated lithologies (Vallance, 2000).

Appearing like flowing concrete with huge boulders, lahars extend the deadly range of stratovolcanoes well beyond that typically affected by any other process except ashfall. They can flow for so long in cases (>300 km) that people in the river valleys where lahars are focused may be taken by surprise without knowing even if the volcano has erupted (Neall, 1996). This tragic circumstance occurred in the town of Amero in Colombia, where 23,000 people were swept away and buried in their houses overnight by huge lahars (Voight, 1990). The lahars were produced 73 km away by small pyroclastic flows traveling across a glacier on the summit of Nevado del Ruiz Volcano. The eruption was so small, and the alarms too late for the residents to be warned of these huge flows of mud and rock. Lahars also occur after major eruptions as rainstorms trigger sudden remobilization of loose and unstable pyroclastic deposits high on the flanks of mountain volcanoes or if crater lakes break out from their impoundments (Manville and Cronin, 2007). Lahars have killed thousands of people from eruptions of Mts. Merapi, Kelut, and Semeru in Indonesia, along with devastating large areas of the Philippines following the 1991 eruption of Mt. Pinatubo (Newhall and Punongbayan, 1996).

Debris avalanches: These are major "dry" collapses of volcanic slopes, usually resulting from the failure of unstable flank areas of the edifice. These sweep aside and bury everything in their path and may change the landscape irrevocably, by leaving behind major amphitheater like holes and scarps on the edifice and filling valleys and lowland areas with deposits tens of meters thick (Siebert, 1996). In extreme cases, over half the volume of a stratovolcano may collapse and cover hundreds of square kilometers with several cubic kilometers of deposit. The fact that stratovolcanoes are built so high with both strong (lava) and weak (volcaniclastic and pyroclastic breccias) strata making up their flanks makes them vulnerable to collapse. In most areas of the world, these edifices also act like large sponges, soaking up rainwater that at times can be heated by geothermal activity below to gradually soften and weaken the volcano's core. Once an edifice becomes too high, intrusion of new magma into it, changes in the hydrothermal system, major earthquakes, or even increasing water pressure within it may trigger catastrophic failure and debris avalanche (McGuire, 2003). The propensity to collapse depends on the dominant lithology of the strata, with some pyroclastic-dominated volcanoes, such as Mt. Taranaki growing and collapsing more than 12 times over a lifespan of >200,000 years (Zernack et al., 2009). The 1980 catastrophic eruption of Mt. St. Helens began with a "small" debris avalanche of only 1 km^3 (Lipman and Mullineaux, 1981).

Tsunami: While these great ocean waves are most commonly generated by large submarine earthquakes, several have been generated by pyroclastic flows, debris avalanches, and major collapses of stratovolcanoes into the sea. Collapse of the Unzen-Dake volcano in Japan in 1792 killed over 6,000 people when the avalanche entered the ocean to generate huge tsunami (Siebert et al., 1987). Similarly, during the massive 1883 Krakatau eruption, part of the stratovolcano collapsed into the sea and a large caldera formed below it, generating huge pyroclastic flows and extremely deadly tsunami that killed over 36,000 people in the surrounding region (Simkin and Fiske, 1983).

Monitoring at stratovolcanoes

A large range of geophysical, remote sensing, and geochemical techniques are employed at stratovolcanoes to detect the signs of unrest preceding eruptions (e.g., Scarpa and Tilling, 1996; Carn and Oppenheimer, 2000; Bonforte et al., 2007; Scott and Travers, 2009). Unfortunately, they are also among the most challenging volcanoes to monitor due to their height, snow cover, and the extreme climates of their upper slopes. The rise of magma below and into stratovolcanoes generates earthquake, gravity, electromagnetic, geodetic, and thermal changes. Hence, the slopes of these edifices are monitored by tilt meters, stationary satellite global-positioning systems, and other surveying techniques to detect swelling of the flanks, such as that preceding the catastrophic Mt. St. Helens eruption. If crater lakes or hot springs are present, temperature and chemical records are carried out. Once unrest is determined, the detection and measurement of gas released from the volcano (especially SO_2) via satellite or ground-based measurements is normally the most reliable means of warning for an imminent eruption (e.g., Lipman and Mullineaux, 1981; Newhall and Punongbayan, 1996).

Current investigations in stratovolcano hazards

Hazards research on stratovolcanoes has historically started with the documentation of past events in order to develop the most rigorous forecast of potential hazard scenarios for the future, such as what was carried out before the major 1980 eruption of Mt. St. Helens (e.g., Crandell and Mullineaux, 1978). The key danger factors for stratovolcanoes can be lessened if they are treated as real hazard sources during dormant periods. This depends on accurate determinations of the length of dormancy periods and developing probabilistic forecasts of scale and types of future eruptions (e.g., Turner et al., 2008). To this end research is ongoing to integrate the latest geochemically based models of magma generation, storage, and ascent processes with eruption outcomes. By combining these theoretical models with geophysical methods for determining rise of magma prior to eruption at stratovolcanoes (e.g., Curilem et al., 2009), more precise forecasts on the likely nature of volcanic eruptions can be made. Several research groups are also working on ways of treating missing data in geological records and developing methods to apply the histories of well-known volcanoes to improving forecasts at those with few known historic or well-dated prehistoric event records. Other areas of ongoing research include the development of numerical models and forecasts of the potential areas of eruption impact (e.g., Procter et al., 2010). By probabilistically determining the specific sectors of a stratovolcano that may be affected by mass flows and the likelihood and severity of ashfalls, warning and community hazard preparations can be made more effective.

Summary

Stratovolcanoes are the typical snow-capped mountains of fairy tales. They are also the unpredictable agents of destruction that would not be out of place in nightmares. Many historical eruptions from such volcanoes have led to significant casualties and societal disruption from a large range of volcanic processes. Their danger lies not only in their unpredictable eruption processes, but also because they attract population growth nearby due to associated fertile soils and excellent water supplies. All the while, these volcanoes lull surrounding communities into a false sense of security with long periods of dormancy. Research into the volcanic histories, underlying magmatic processes, and statistics is helping to develop more effective hazard mapping and management strategies for these, the most common volcano type on Earth.

Bibliography

Abdurachman, E. K., Bourdier, J. L., and Voight, B., 2000. Nuées ardentes at Merapi Volcano, Indonesia: distribution, deposits and damage. *Journal of Volcanology and Geothermal Research*, **100**, 345–361.

Andronico, D., and Cioni, R., 2002. Contrasting styles of Mount Vesuvius activity in the period between the Avellino and Pompeii Plinian eruptions and some implications for the assessment of future hazards. *Bulletin of Volcanology*, **64**, 372–391.

Annen, C., Blundy, J. D., and Sparks, R. J. S., 2006. The genesis of intermediate and silicic magma in deep crustal hot zones. *Journal of Petrology*, **47**, 505–539.

Blundy, J., Cashman, K., and Humpreys, M., 2006. Magma heating by decompression-driven crystallization beneath andesite volcanoes. *Nature*, **443**, 76–80.

Bonadonna, C., 2006. Probabilistic modelling of tephra dispersal. In Mader, H., Cole, S., and Connor, C. B. (eds.), *Statistics in Volcanology*. London: Geological Society. IAVCEI Series, Vol. 1, pp. 243–259.

Bonforte, A., Gambino, S., Guglielmino, F., Obrizzo, F., and Palano, M., 2007. Ground deformation modelling of flank dynamics prior to the 2002 eruption of Mt. Etna. *Bulletin of Volcanology*, **69**, 757–768.

Branney, M. J., and Kokelaar, B. P., 2002. *Pyroclastic Density Currents and the Sedimentation of Ignimbrites*. London: Geological Society, Memoir 27, 143 pp.

Carey, S., and Sparks, R. S. J., 1986. Quantitative models of the fallout and dispersal of tephra from volcanic eruption columns. *Bulletin of Volcanology*, **48**, 109–125.

Carmichael, I. S. E., 2002. The andesite aqueduct: perspectives on the evolution of intermediate magmatism in west–central (105–99°W) Mexico. *Contributions to Mineralogy and Petrology*, **143**, 641–663.

Carn, S. A., and Oppenheimer, C., 2000. Remote monitoring of Indonesian volcanoes using satellite data from the Internet. *International Journal of Remote Sensing*, **21**, 873–910.

Cotton, C. A., 1944. *Volcanoes as Landscape Forms*. Christchurch, New Zealand: Whitcombe and Tombs, 415 pp.

Crandell, D. R., and Mullineaux, D. R., 1978. *Potential Hazards from Future Eruptions of Mt. St. Helens Volcano, Washington*. Washington, DC: U.S. Government Printing Office. U.S. Geological Survey Bulletin 1383-C, 26 pp.

Cronin, S. J., Neall, V. E., Lecointre, J. A., and Palmer, A. S., 1996. Unusual "snow slurry" lahars from Ruapehu volcano, New Zealand, September 1995. *Geology*, **24**, 1107–1110.

Cronin, S. J., Neall, V. E., Lecointre, J. A., Hedley, M. J., and Loganathan, P., 2003. Environmental hazards of fluoride in volcanic ash: a case study from Ruapehu volcano, New Zealand. *Journal of Volcanology and Geothermal Research*, **121**, 271–291.

Curilem, G., Vergara, J., Fuentealba, G., Acuna, G., and Chacon, M., 2009. Classification of seismic signals at Villarrica volcano (Chile) using neural networks and genetic algorithms. *Journal of Volcanology and Geothermal Research*, **180**, 1–8.

Davidson, J., and DeSilva, S., 2000. Composite volcanoes. In Sigurdsson, H., Houghton, B. F., McNutt, S. R., Rymer, H., and Stix, J. (eds.), *Encyclopedia of Volcanoes*. San Diego, CA: Academic, pp. 663–681.

Devine, J. D., Rutherford, M. J., Norton, G. E., and Young, S. R., 2003. Magma storage region processes inferred from geochemistry of Fe-Ti oxides in andesitic magma, Soufriere Hills Volcano, Montserrat, WI. *Journal of Petrology*, **44**, 1375–1400.

Gill, J. B., 1981. *Orogenic Andesites and Plate Tectonics*. Heidelberg, Germany: Springer, 390 pp.

Grove, T. L., Parman, S. W., Bowring, S. A., Price, R. C., and Baker, M. B., 2002. The role of an H_2O-rich fluid component in the generation of primitive basaltic andesites and andesites from Mt. Shasta region, N California. *Contributions to Mineralogy and Petrology*, **142**, 375–396.

Hansell, A., Horwell, C. J., and Oppenheimer, C., 2006. The health hazards of volcanoes and geothermal areas. *Occupational and Environmental Medicine*, **63**, 149–156.

Lacroix, A., 1904. *La Montagne Pelée et ses Eruptions*. Paris: Masson, 622 pp.

Lipman, P. W., and Mullineaux, D. R. (eds.), 1981. *The 1980 Eruptions of Mt. St. Helens, Washington*. Washington, DC: U.S. Government Printing Office. US Geological Survey Professional Paper 1250, 843 pp.

Lowe, D. R., Williams, S. N., Leigh, H., Connor, C. B., Gemmell, J. B., and Stoiber, R. E., 1986. Lahars initiated by the 13 November 1985 eruption of Nevado del Ruiz, Colombia. *Nature*, **324**, 51–53.

Manville, V., and Cronin, S. J., 2007. Breakout lahar from New Zealand's Crater Lake. *Eos Transactions, American Geophysical Union*, **88**(43), 441–442.

McGuire, W. J., 2003. Volcanic instability and lateral collapse. *Revista*, **1**, 33–45.

McKenzie, D., 1985. The extraction of magma from the crust and mantle. *Earth and Planetary Science Letters*, **74**, 81–91.

Neall, V. E., 1996. Hydrological disasters associated with volcanoes. In Singh, V. P. (ed.), *Hydrology of Disasters*. Dordrecht, The Netherlands: Kluwer, pp. 395–425, 442p.

Newhall, C. G., and Punongbayan, R. (eds.), 1996. *Fire and Mud: Eruptions and Lahars of Mount Pinatubo, Philippines. Philippine Institute of Volcanology and Seismology, Quezon City*. Seattle, WA: University of Washington Press, 1126 pp.

Price, R. C., Stewart, R. B., Woodhead, J. D., and Smith, I. E. M., 1999. Petrogenesis of high-K arc magmas: evidence from Egmont Volcano, North Island, New Zealand. *Journal of Petrology*, **40**, 167–197.

Procter, J. N., Cronin, S. J., Platz, T., Patra, A., Dalbey, K., Sheridan, M., and Neall, V., 2010. Mapping block-and-ash flow hazards based on Titan2D simulations; a case study from Mt. Taranaki, NZ. *Natural Hazards*, **53**, 483–501.

Pyle, D. M., 1989. The thickness, volume and grainsize of tephra fall deposits. *Bulletin of Volcanology*, **51**, 1–15.
Scarpa, R., and Tilling, R. I. (eds.), 1996. *Monitoring and Mitigation of Volcano Hazards*. Berlin/Heidelberg: Springer, pp. 541–572.
Schmincke, H. -U., 2004. *Volcanism*. Berlin: Springer, 324 pp.
Scott, B. J., and Travers, J., 2009. Volcano monitoring in NZ and links to SW Pacific via the Wellington VAAC. *Natural Hazards*, **51**, 263–273.
Siebert, L., 1996. Hazards of large volcanic debris avalanches and associated eruptive phenomena. In Scarpa, R., and Tilling, R. I. (eds.), *Monitoring and Mitigation of Volcano Hazards*. Berlin/Heidelberg: Springer, pp. 541–572.
Siebert, L., Glicken, H., and Ui, T., 1987. Volcanic hazards from Bezymmianny- and Bandai-type eruptions. *Bulletin of Volcanology*, **49**, 435–459.
Sigurdsson, H., Cashdollar, S., and Sparks, R. S. J., 1982. The eruption of Vesuvius in AD79: reconstruction from historical and volcanological evidence. *American Journal of Archeology*, **86**, 39–51.
Simkin, T., and Fiske, R. S., 1983. *Krakatau 1883: The Volcanic Eruption and Its Effects*. Washington, DC: Smithsonian Institute, 464 pp.
Simkin, T., and Siebert, L., 1994. *Volcanoes of the World*. Tuscon, AZ: Geoscience Press, 368 pp.
Tanguy, J. -C., Ribière, Ch, Scarth, A., and Tjetjep, W. S., 1998. Victims from volcanic eruptions, a revised database. *Bulletin of Volcanology*, **60**, 137–144.
Turner, M. B., Cronin, S. J., Bebbington, M. S., and Platz, T., 2008. Developing a probabilistic eruption forecast for dormant volcanoes; a case study from Mt Taranaki, New Zealand. *Bulletin of Volcanology*, **70**, 507–515.
Vallance, J. W., 2000. Lahars. In Sigurdsson, H., Houghton, B., McNutt, S., Rymer, H., and Stix, J. (eds.), *Encyclopedia of Volcanoes*. San Diego, CA: Academic, pp. 601–616.
Voight, B., 1990. The 1985 Nevado del Ruiz Volcano catastrophe: anatomy and retrospection. *Journal of Volcanology and Geothermal Research*, **44**, 349–386.
Voight, B., and Elsworth, D., 2000. Instability and collapse of hazardous gas-pressurised lava domes. *Geophysical Research Letters*, **27**, 1–4.
Voight, B., Constantine, E. K., Siswowidjoyo, S., and Torley, R., 2000. Historical eruptions of Merapi Volcano, Indonesia, 1768–1998. *Journal of Volcanology and Geothermal Research*, **100**, 69–138.
Witham, C. S., 2005. Volcanic disasters and incidents: a new database. *Journal of Volcanology and Geothermal Research*, **148**, 191–233.
Zernack, A., Procter, J., and Cronin, S. J., 2009. Sedimentary signatures of cyclic growth and destruction of stratovolcanoes: a case study from Mt. Taranaki, New Zealand. *Sedimentary Geology*, **220**, 288–305.

Cross-references

Base Surge
Debris Avalanche
Galeras Volcano, Colombia
Krakatoa
Lahar
Lava
Magma
Mt. Pinatubo
Nevado del Ruiz Volcano, Colombia
Nuee Ardente
Pyroclastic Flow
Santorini
Shield Volcano
Vesuvius
Volcanic Ash
Volcanic Gas
Volcanoes and Volcanic Eruptions

STRUCTURAL DAMAGE CAUSED BY EARTHQUAKES

Murat Saatcioglu
University of Ottawa, Ottawa, ON, Canada

Definitions

Drift (*Drift Ratio*). Horizontal displacement of a building, story level, or vertical element divided by respective height.
Inelastic deformability. Ability to deform in the inelastic range of deformations without significant strength decay (analogous to ductility).
Spalling. Crushing or separation of exterior segments of concrete in compression, usually in cover concrete of structural elements.
Yield. Deformation level beyond which the rate of deformations increase rapidly under very little increase in force.

Introduction

Earthquakes may cause structural and nonstructural damage during seismic excitations. Structural damage consists of distress induced in structural components of lateral and gravity-load-resisting systems, such as beams, columns, load-bearing walls, and shear walls, as well as horizontal diaphragms, such as slabs and roofs.

Seismic damage in structures is caused either by lack of sufficient strength or lack of inelastic deformability. Lack of strength is associated with insufficient member size, material strength, and/or reinforcement against seismic forces. It often results in damage to the critical regions of structures under different types of stresses. The lack of strength may also be attributed to the instability of members, which may be triggered by local or global buckling, or lack of adequate anchorage, connection, and continuity between the adjoining members. Seismic force demands in such structures are higher than the structural capacities provided in design and/or construction, imposing high inelastic deformation demands. Once the force capacity is exceeded, those structures that do not have the ability to deform in the inelastic range of deformations may suffer severe seismic damage. The ability to develop inelastic deformations without significant strength decay is referred to as "ductility" or "inelastic deformability," and it is often expressed in the form of a ductility ratio (maximum inelastic deformation divided by deformation at yield). Brittle construction materials, such as adobe, unreinforced masonry, or concrete without seismic detailing, do not have sufficient ductility against seismic forces. Well-designed and detailed reinforced concrete and steel structures tend to perform in a ductile manner and dissipate seismic-induced energy, reducing vulnerability against earthquake damage. Lack of strength and/or deformability creates seismically deficient structures that often suffer significant damage during strong earthquakes.

Seismic demands on structures are associated with seismic hazards (magnitude, distance, and characteristics of earthquake), soil conditions, and structural layout (including structural mass and stiffness and associated dynamic characteristics of the structure). Irregularities in structures tend to increase seismic force and deformation demands. Vertical and horizontal discontinuities in buildings, asymmetry in floor plans, and especially the eccentricity between the centers of rigidity and mass generating torsional effects, increase seismic demands significantly, resulting in seismic damage. Seismic demands are also dictated by the intimate relationship that exists between the dynamic characteristics of earthquake excitation and those of the structure. High-frequency ground motions tend to excite low-rise buildings, which tend to have short natural periods (high frequencies), increasing structural demands. Inelasticity in a structure reduces stiffness and elongates structural period, which may reduce seismic demands, as the period of the structure moves away from the period of excitation. Therefore, ductility on a structural system also helps reduce seismic demands and associated damage.

Typical structural damage in buildings caused by earthquakes

Earthquake reconnaissance investigations after past earthquakes show a pattern of seismic damage in different types of structures (Saatcioglu and Bruneau, 1993; Mitchell et al., 1995a, b; Saatcioglu et al., 2001, 2006; Gillies et al., 2001). Seismic-force-resisting systems that form the majority of building inventory worldwide consists of; (1) adobe and masonry buildings, (2) reinforced concrete buildings, (3) structural steel systems, and (4) timber construction.

Adobe and masonry buildings

Structures built using brittle construction materials, such as unreinforced masonry and adobe units, experience essentially elastic behavior until they fail in a brittle manner. These types of structures do not possess the required ductility and energy dissipation capacity. Figures 1 and 2 illustrate typical failures of adobe and masonry buildings. A common use of unreinforced masonry in building construction includes infill panels as nonstructural elements in other lateral-force-resisting systems. Though used as nonstructural elements, during structural response, these walls may interact with the enclosing frame. In spite of their brittle characteristics, unreinforced masonry infill walls may provide lateral bracing to the enclosing frame, controlling lateral drift and reducing structural and nonstructural damage until their elastic limit is exceeded, depending on how well they are integrated with the

Structural Damage Caused by Earthquakes, Figure 1 Collapse of adobe buildings in Talca, Chile, after the 2010 Earthquake.

structure, their strength, and the duration of earthquake. For older buildings with sufficiently high wall-to-floor area ratios, especially in regions of the world where masonry walls are used as partition walls, these walls provide significant lateral bracing and damage control to otherwise seismically deficient structural systems. Figure 2b illustrates an apartment building located in an area that suffered significant damage during the 1999 Kocaeli Earthquake in Turkey. The building performed reasonably well due to the high wall-to-floor area ratio, whereas many other nearby buildings suffered partial or full collapses. Another building during the same earthquake, shown in Figure 2c, with a smaller amount of masonry infill wall only around the perimeter, suffered complete loss of masonry, though the nonductile framing system survived without structural damage. This could be explained by the duration of strong ground shakings that was not long enough to damage the frame beyond the failure of masonry and the elongation of building period which may have reduced seismic forces. In cases where the masonry walls are not able to sustain seismic forces during the full duration of earthquake and the structural system does not have sufficient inelastic deformability, structural collapse may occur as shown in Figure 2d. In-plane damage in masonry walls is typically caused by excessive diagonal tension, resulting in associated diagonal cracking. This is illustrated in Figure 2e, f. The same shear forces also produce diagonal compressive stresses, sometimes crushing masonry units along wall diagonals. When the ground shaking is significant in a direction perpendicular to the plane of the wall, out-of plane failures may develop. This type of failure occurs at different levels of severity, which ranges between the dislocations of a few masonry units and partial or complete collapse of the wall panel, posing significant threat to life safety. Figure 2c shows out-of plane failure of the entire exterior masonry enclosure of a frame building.

Structural Damage Caused by Earthquakes, Figure 2 Masonry damage observed during previous earthquakes. (**a**) Collapse of an unreinforced masonry church in Curico, Chile during the 2010 Chile Earthquake. (**b**) Reinforced concrete frame building with minor damage to masonry infill walls, during the 1999 Kocaeli Earthquake in Turkey. (**c**) Out-of-plane failure of exterior masonry walls during the 1999 Kocaeli Earthquake in Turkey. (**d**) Collapse of a non-ductile reinforced concrete frame building after the loss of masonry infills during the 1999 Earthquake in Turkey. (**e**) Typical diagonal tension cracking in a masonry wall, Curico, Chile, after the 2010. Chile Earthquake. (**f**) Diagonal tension crack in masonry infill wall after the 1999 Earthquake in Turkey.

Reinforced concrete buildings

Damage to reinforced concrete frames occurs in critical regions that are often located at the ends of flexure-dominant members where plastic hinges may form, in beam-column connections, and shear deficient regions of elements. An important parameter that affects damage-control in potential plastic hinge regions of reinforced-concrete elements is the confinement of compression concrete by closely placed and properly designed transverse reinforcement. The confinement reinforcement overcomes tendency of concrete to expand laterally when subjected to axial compression. This in turn prevents the crushing of concrete, while improving ductility. Compression elements such as columns and walls, which are often responsible for overall strength and stability of the entire structural system, benefit tremendously from concrete confinement. Confinement reinforcement has to be placed with due considerations given to the spacing and detailing requirements of current building codes. Figure 3 shows examples of column damage caused by lack of concrete confinement. Such column damage is especially common in soft stories where deformation demands become very high, as illustrated in Figure 3.

Lack of sufficient transverse reinforcement in reinforced concrete structural elements also reduces shear capacity substantially. In the absence of sufficient shear reinforcement, the degradation of shear transfer mechanism in concrete under reversed cyclic loading leads to

Structural Damage Caused by Earthquakes, Figure 3 Lack of sufficient confinement reinforcement in columns. (**a**) Column hinging at the base of a multi-storey building during the 1999 Earthquake in Turkey. (**b**) Column hinging underneath column capital during the 2008 Wenchuan Earthquake in China. (**c**) Column hinging of the Imperial Valley Services Building during the 1979 Imperial Valley Earthquake (Photo by V. Bertero). (**d**) Close-up view of column hinging of the Imperial Valley Services Building (Photo by V. Bertero).

a brittle shear failure that takes place along an inclined shear plane. This is illustrated in Figure 4. Sometimes, shear failures may be caused by diagonal compression, resulting in the crushing of concrete. Shear stresses may be critical in plastic hinge regions of flexure dominant members due to the deterioration of concrete in these heavily stressed regions. Furthermore, some elements, though designed as predominantly flexural members, may experience shear distress because of the amplification of shear stresses associated with reduced shear spans resulting from the presence of unintentional lateral supports. A nonstructural wall placed adjacent to a column around a window opening is a good example of an unintentional reduction in shear span, resulting in a "short" or "captive" column. Such interference of a nonstructural element is often not considered in

Structural Damage Caused by Earthquakes, Figure 4 (Continued)

Structural Damage Caused by Earthquakes, Figure 4 Column shear failures. (**a**) Column shear failure during the 2010 Earthquake in Chile. (**b**) Column shear failure during the 1999 Kocaeli Earthquake in Turkey. (**c**) Short column effect during the 1972 Managua Earthquake in Guatamala (Photo by V. Bertero). (**d**) Short column effect during the 1999 Kocaeli Earthquake in Turkey. (**e**) Lateral force exerted by a stairway landing slab on a column, causing shear failure during the 1980 El Asnam Earthquake in Algeria (Photo by V. Bertero). (**f**) Stairway landing slabs causing column shear failure during the 1999 Kocaeli Earthquake in Turkey.

structural design. These columns are notorious for suffering shear damage during seismic response. Examples of shear damage in captive columns are included in Figure 4c through Figure 4f.

Beam-column joints of reinforced concrete frames are another region where shear stress reversals take place during an earthquake. Joint shear is induced by the flexural action in attached beams, coupled with shear forces transferred from the columns. Joints without adequate shear reinforcement may be damaged, undermining the integrity of the entire framing system. Figure 5 depicts joint damage observed in previous earthquakes. Furthermore, a good seismic design practice calls for strong columns and weak beams at each joint. This promotes flexural hinging in beams for the dissipation of seismic energy while protecting the columns from damage. Strong beams and weak columns force the plastic hinges to occur in the columns, and may result in the collapse of the entire structural system in a manner that is often referred to as "pan caking." Figure 6 shows examples of structural damage and building collapse resulting from the use of strong beams and weak columns.

It is always a good practice to use a regular structural system, without vertical and plan irregularities. Any sudden change in strength and stiffness of a building along its height, as well as asymmetrical floor layout with potentials for torsional effects result in increased deformation demands on elements. Figure 7 shows damage induced in reinforced concrete buildings due to vertical and horizontal irregularities.

One of the best structural systems used in buildings against earthquakes is a shear wall system. Shear walls provide strength, stiffness, and ductility to the building, which are the qualities required during seismic response. Because shear walls tend to be very rigid as compared to the accompanying frames, they tend to attract a major portion of the total seismic force. Though shear walls generally perform well under seismic forces, they may develop diagonal tension and/or diagonal compression failures if not designed properly. Figure 8 illustrates seismic damage observed in reinforced concrete shear walls. These walls typically possess high flexural capacities because of the high internal lever arm associated with their geometry. However, most older shear walls lack boundary elements containing properly tied concentration of longitudinal reinforcement. These walls may suffer from the buckling of compression bars, as well as the crushing of unconfined concrete.

Structural steel systems

Steel structures generally perform well during earthquakes. This is attributed to the relatively low structural mass that exists in steel systems, compared with concrete and masonry structures, as well as the inherent ductility of steel as a material. Therefore, moment-resisting steel frames and braced frames are used successfully in

Structural Damage Caused by Earthquakes, Figure 5 Damage in beam-column joints. (**a**) Joint damage in a frame building in Banda Aceh, Indonesia during the 2004 South East Asian Earthquake. (**b**) Joint damage in a frame building in Banda Aceh, Indonesia during the 2004 South East Asian Earthquake. (**c**) Joint failure during the 1999 Kocaeli Earthquake in Turkey. (**d**) Collapse of a reinforced concrete frame building triggered by joint failures during the 1999 Kocaeli Earthquake in Turkey.

earthquake-prone areas. Structural damage in steel structures tend to be limited to the buckling of light steel braces and inadequate connections. The 1994 Northridge earthquake exposed one of the deficiencies in welded connections of steel structures. Over 200 buildings during this earthquake alone suffered weld ruptures, mostly at the beam-column connections where full penetration welds were used, resulting in a change in the practice.

Timber construction

Timber structures generally have superior performance during earthquakes. Low structural mass inherent to these structures limit seismic-induced inertia forces, while the nailed connections promote deformability. Single-family timber homes and multistory timber buildings have had a successful history of seismic performance, except for their brittle nonstructural components. A commonly observed seismic damage in such structures results from lack of proper anchorage to adjoining stiff and rigid elements, as in the case of single-family houses built on concrete foundations.

Seismic damage in bridges

Previous earthquakes indicate that bridges, as essential components of the transportation infrastructure, can be

Structural Damage Caused by Earthquakes, Figure 6 Strong beam-weak column joints resulting in damage during previous earthquakes. (**a**) Strong beam-weak column resulting in column hinging. (**b**) Total collapse of a frame building due to the use of strong beams and weak columns during the 1999 Kocaeli Earthquake. (**c**) Use of strong beams and weak columns, resulting in partial collapse of a shopping centre in Banda Aceh Indonesia during the 2004 Earthquake.

vulnerable to seismic damage. Bridges are simple structures that are comprised of a support system that consists of abutments and piers, and a bridge superstructure that consists of either simply supported single or multiple-span girders, or continuous girders supporting a bridge deck. Two types of seismic damage are common in bridges: (1) support failures, and (2) unseating and instability of bridge superstructure.

Bridge supports consist of bridge piers and abutments. Bridge piers with older concrete columns often do not have sufficient transverse reinforcement. This results in lack of shear capacity and flexural deformability. Figure 9 illustrates various bridge column failures observed during past earthquakes. Short and stubby shear dominant columns attract high shear force reversals and suffer from diagonal tension failures in the absence of adequate transverse reinforcement. Long and flexure-dominant columns develop plastic hinges at member ends, and fail through the crushing of concrete if the core concrete is not confined by sufficient confinement reinforcement.

Structural Damage Caused by Earthquakes, Figure 7 Effects of vertical and horizontal irregularities on seismic damage. (**a**) Vertical irregularity in a multi-storey building in Concepcion, Chile. (**b**) Front view of building in (**a**) that suffered damage at locations of sudden strength and stiffness change during the 2010 Earthquake. (**c**) Floor irregularity resulting from the core wall at the far end, triggering torsional failure during the 1976 Guatemala Earthquake (Photo by V. Bertero). (**d**) Close-up view of column damage in building in (**c**) Photo by V. Bertero.

Bridge abutments are vulnerable against soil instability. Vertical, horizontal, and rotational movements and settlements of abutments during seismic excitations result in the separation of superstructure from the road surface, leading to the unseating of bridge girders. Similarly, excessive deformations and movements in bridge piers and columns may result in the unseating of bridge girders, causing the collapse of bridge superstructure. Figure 10 illustrates examples of seismic damage associated with support movements and unseating of bridge girders.

Building performance as affected by seismic damage

Post-earthquake use and occupancy of a structure is dependent on the level of damage sustained during the earthquake. Some buildings can be occupied immediately after the earthquake whereas others need to be repaired and strengthened before they can be used, and yet some others may have to be condemned for demolition. Different types of structures have different requirements for their functionality after a seismic event, that is, bridges must be passable with some minimum load, liquid containers must not leak and pose safety and environmental threats, transmission towers should maintain safe transmission capabilities, etc. While different structures have different performance requirements and may have to be assessed differently for post-earthquake use, building structures may be assessed on the basis of a commonly accepted performance criteria developed by the Federal Emergency Agency (FEMA) of the USA (FEMA 273, 1997; FEMA 356,

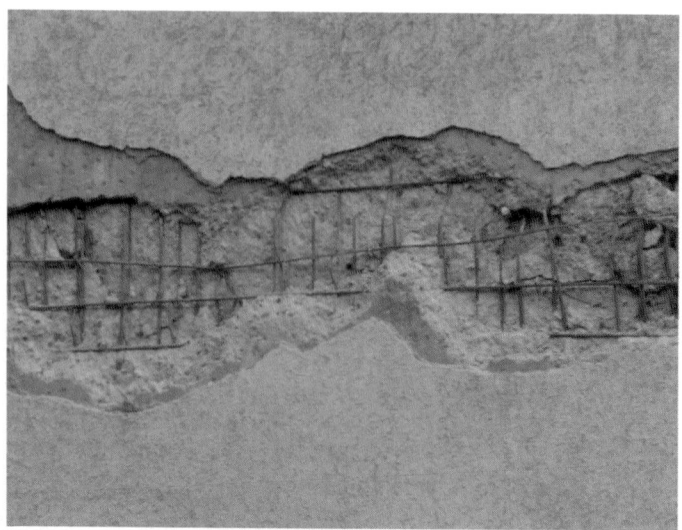

Structural Damage Caused by Earthquakes, Figure 8 Damage to shear walls. (**a**) Sliding shear damage in a structural wall in Santiago, Chile after the 2010 Earthquake. (**b**) Close-up view of structural wall in (**a**). (**c**) Shear wall damage during the 1999 Kocaeli Earthquake.

2000; ASCE 41-06, 2007) for seismic retrofit of buildings. These criteria may be adapted for post-earthquake assessment of buildings as presented below.

Structural damage associated with operational level of performance

This level of performance reflects very light overall damage. Structural elements essentially remain elastic with some minor hairline cracks. They retain their original strength and stiffness and do not experience permanent drift. Buildings demonstrating these characteristics can continue operation during and after an earthquake as expected from post-disaster buildings such as hospitals, and police and fire stations.

Structural damage associated with immediate occupancy performance level

This level of performance describes post-earthquake damage where the building remains safe to be reoccupied. The primary lateral-force- and gravity-load-resisting systems

Structural Damage Caused by Earthquakes, Figure 9 Bridge column damage observed during previous earthquakes. (**a**) Lack of confinement reinforcement and resulting column damage during the 1971 San Fernando Earthquake (Photo by V. Bertero). (**b**) Lack of confinement reinforcement in a bridge column that suffered damage during the 1999 Northridge Earthquake. (**c**) Bridge column shear failure during the 1995 Kobe Earthquake. (**d**) Shear failure in a bridge column, 1999 Chi-Chi Earthquake in Taiwan (Photo by D. Lau).

retain almost their entire pre-earthquake design strength with light damage. Some minor structural repairs may be appropriate, though these can be done while the building is occupied.

Structural damage associated with life safety performance level

Life Safety level indicates a post-earthquake damage state in which significant damage to the structure has occurred, even though some margin of safety remains against partial or total structural collapse. Some structural elements and components may be severely damaged, resulting in injuries. However, the overall risk of life-threatening injury due to such damage is expected to be low. Some residual strength and stiffness is left in all stories. Gravity-load-carrying elements continue fulfilling their functions. Extensive damage is anticipated in beams, with hinge formation in ductile elements. Unreinforced masonry infill walls experience extensive cracking and some crushing of masonry units is expected. The structure requires a thorough structural evaluation before deciding whether it can be repaired and reused.

Structural Damage Caused by Earthquakes, Figure 10 Examples of bridge superstructure failure. (**a**) Unseating of Gaoyuan Bridge during the 2008 Wenchuan Earthquake in China. (**b**) Close-up view of the seat width for the bridge in (**a**). (**c**) Unseating of Xiao Yu Dong Bridge during the 2008 Wenchuan Earthquake in China. (**d**) Unseating of a bridge in Santiago, Chile during the 2010 Earthquake. (**e**) Unseating of bridge segments during the 1999 Northridge Earthquake. (**f**) Unseating of a bridge during the 2010 Chile Earthquake.

Structural damage associated with collapse prevention performance level

This level of performance indicates that the building is on the verge of partial or total collapse as a result of earthquake damage. Substantial damage to structural elements is expected, potentially accompanied by significant strength and stiffness degradation of the lateral-load-resisting system. Large permanent lateral deformations of the structure and limited degradation of the vertical-load-carrying capacity are expected. However, the gravity-load-resisting system must continue carrying its gravity load demands. The structure is not repairable and not safe to reoccupy.

Bibliography

ASCE/SEI 41-06, 2007. *Seismic rehabilitation of existing buildings*. American Society of Civil Engineers.
FEMA 273, 1997. *NEHRP Guidelines for the seismic rehabilitation of buildings*. Federal Emergency Management Agency, Washington, D.C.
FEMA 356, 2000. *Prestandard and commentary for the seismic rehabilitation of buildings*. Federal Emergency Management Agency, Washington, D.C.
Gillies, A., Anderson, D. A., Saatcioglu, M., Tinawi, R., and Mitchell, D., 2001. The August 17, 1999 Kocaeli (Turkey) earthquake – Lifelines and earthquake preparedness. *Canadian Journal of Civil Engineering*, **28**(6), 881–890.
Mitchell, D., De Vall, R., Saatcioglu, M., Simpson, R., Tinawi, R., and Tremblay, R., 1995a. Damage to concrete structures due to the 1994 Northridge earthquake. *Canadian Journal of Civil Engineering*, **22**(2), 361–377.
Mitchell, D., Bruneau, M., Williams, M., Anderson, D. L., and Sexsmith, R. G., 1995b. Performance of bridges in the 1994 Northridge earthquake. *Canadian Journal of Civil Engineering*, **22**, 2.
Saatcioglu, M., and Bruneau, M., 1993. Performance of structures during the 1992 Erzincan earthquake. *Canadian Journal of Civil Engineering*, **20**(2), 305–325.
Saatcioglu, M., Mitchell, D., Tinawi, R., Gardner, N. J., Gillies, A. G., Ghobarah, A., Anderson, D. L., and Lau, D., 2001. The August 17, 1999 Kocaeli (Turkey) earthquake – damage to structures. *Canadian Journal of Civil Engineering*, **28**(4), 715–737.
Saatcioglu, M., Ghobarah, A., Nistor, I., 2006. Performance of structures in Indonesia during the December 2004 Great Sumatra earthquake and tsunami. *Earthquake Spectra, the Journal of the Earthquake Engineering Research Institute*, **22**.

Cross-references

Accelerometer
Concrete structures
Critical Infrastructure
Damage and the Built Environment
Earthquake
Earthquake Damage
Earthquake Resistant Design
Federal Emergency Management Agency (FEMA)
Haiti Earthquake 2010 Psychosocial Impacts
Harmonic Tremor
High Rise Buildings in Natural Disaster
Isoseismal
Macroseismic Survey
Magnitude Measures
Mercalli, Giuseppe (1850–1914)
Modified Mercallit (MM) Scale
Plate Tectonics
Primary wave (P wave)
Richter, Charles Francis (1900–1985)
Rotational Seismology
Secondary Wave (S wave)
Seismograph/Seismometer
Seismology
Shear
Structural Mitigation
Tangshan, China (1976 Earthquake)
Tectonic Tremor
Tohoku, Japan, Earthquake, Tsunami and Fukushima Accident (2011)
Unreinforced Masonry Building
Wenchuan, China (2008 Earthquake)

STRUCTURAL MITIGATION

Murat Saatcioglu
University of Ottawa, Ottawa, ON, Canada

Definitions

Base isolation (device). A device used to isolate structures from earthquake motions.
Beam. Horizontal structural element that usually supports the floor or roof slab system.
Bracing element. Structural element that provides lateral support to increase stiffness and reduce deformations, while strengthening the structure.
Column. Vertical structural element that supports the attached beams while transferring loads to foundation.
Confinement. Containment of compression concrete in a properly designed steel cage of a structural element for improved strength and inelastic deformability.
Damping. Any effect that gradually reduces vibration amplitude or motion.
Damper. A device that introduces damping.
Drift. Horizontal displacement of a building, storey level, or vertical element divided by the respective height.
Ductility. Ability to deform in the inelastic range of deformations without a significant loss of strength.
Energy dissipation (device or mechanism). A device or mechanism that dissipates seismic-induced energy by transforming it to another form of energy, usually to inelastic strain energy.
FRP. Fiber reinforced polymer composite material that consists of carbon, glass, or aramid fibers in a polymer-based (usually epoxy) matrix used to strengthen/retrofit structural or nonstructural elements.
Inelastic deformability. Ability to deform in the inelastic range of deformations without significant strength decay (analogous to ductility).
Jacket. Enclosure around a member (usually column) that has sufficient strength and stiffness to enhance the performance of the member.
Retrofit. Improvement/rehabilitation introduced to enhance the performance of a structure, structural element, or nonstructural element.
Shear wall. Vertical element that is intended to resist lateral forces in the plane of the wall.
Slab. Horizontal structural element to serve as building floor or ceiling, or bridge deck.

Structure. An assembly of elements put together to carry or resist externally applied or internally generated loads and forces.
Structural element. A component of a structure that resists stresses induced by applied loads/forces.
Nonstructural element. An element designed to fulfill certain functional requirement, but not intended to carry or resist stresses induced by loads/forces.
Subassembly. A combination of structural components that are usually connected (e.g., beam-column subassembly).
Unseating of bridge girder. Sliding off of a bridge girder from its support due to seismic movement.
Wall. Vertical structural element that is designed primarily for vertical load (load-bearing wall), or primarily for lateral load (shear wall).
Yield. Deformation level beyond which the rate of deformations increase rapidly under very little increase in force.

Introduction

A large proportion of existing infrastructure worldwide consists of seismically deficient structural systems. These structures often perform poorly during strong earthquakes. The majority of buildings that were built prior to the enactment of modern seismic codes of 1970s, as well as those built more recently but located in areas where code enforcement is not ensured, fall in this category. Seismic design practices for bridges have a more recent history, leaving the majority of bridge infrastructure vulnerable to seismic damage. These structures constitute significant seismic risk, especially in large metropolitan centers. Because it is economically not feasible to replace seismically vulnerable infrastructure with those conforming to modern codes and standards, seismic retrofitting remains to be a viable structural mitigation strategy against earthquake hazards.

Seismic deficiency in a structural system occurs either due to lack of adequate strength or inelastic deformability (ductility). Individual elements, their connections, or assemblies may not have sufficient strength to resist seismic forces. These structures may exceed their elastic design capacities in critical regions, and may experience local or global structural damage, sometimes leading to partial or total structural collapse. Structures forced to experience inelastic deformations can withstand their inelastic deformation demands only if they are designed and detailed to have adequate seismic deformation capacity. Seismic-induced energy in such structures may be dissipated through the formation of plastic hinges. Most seismically deficient structures lack both strength and inelastic deformability. Non-ductile structures are especially prone to sudden and brittle failures during strong earthquakes. Lateral drift control protects brittle elements and reduces seismic damage. Structures with adequate lateral bracing perform well while ensuring structural stability and survivability of brittle elements.

Retrofitting of seismically deficient structures is implemented either at the system or component level, depending on the nature of seismic deficiency. If the entire structural system consists of brittle elements with strength deficiency, it may be more feasible to implement seismic retrofitting at the system level, and implement global structural strengthening and bracing strategy. This type of retrofit ensures sufficient strength against seismic forces, while ensuring elastic or near-elastic response, preventing damage associated with lack of inelastic deformability. If the structure suffers seismic deficiency in an element or group of elements, it may be possible to strengthen these elements locally while also potentially improving their inelastic deformability for ductile response.

Structural mitigation at the system level

Structures that have a large number of seismically deficient non-ductile elements are good candidates for retrofits at the system level. This retrofit strategy includes lateral bracing and strengthening, base isolation, passive and semi-active supplementary damping devices, and active energy dissipation devices and force control methods, all of which aim to reduce seismic demands on the entire structural system. Interventions at the system level can be categorized into three different groups: (a) passive retrofit techniques, (b) semi-active techniques, and (c) active control systems. Adding shear-walls, braces, or installing base-isolation systems or passive dampers form examples of passive retrofit methods. Semi-active control methods involve the installation of a central processing unit that can adjust the properties of the structure according to the nature of the earthquake excitation. Different types of controllable dampers fall under this category. Active control systems involve energy dissipation devices that are revised and repositioned in real time based on measured seismic response, or apply corrective control forces directly on the structure to oppose the seismic action, following a control algorithm. This method brings in an external source of energy into the system to derive the control mechanism (often in the form of hydraulic or electromechanical actuators). It is highly sensitive to time lags, but if utilized properly, offers a great potential, once fully developed.

The system level seismic retrofit strategy can be explained by using the principles of the conservation of energy (Uang and Bertero, 1988). Accordingly,

$$E = E_k + E_s + E_h + E_d$$

where;

E : Total input energy by earthquake;
E_k: Absolute kinetic energy;
E_s: Recoverable elastic strain energy;
E_h: Irrecoverable energy dissipated by the structural system through inelastic deformation or other intrinsic forms of damping;
E_d: Energy dissipated by structural protective systems.

Seismic retrofit techniques discussed in the following sections are aimed at enhancing one or more of the energy terms on the right-hand side of the above equation.

Structural Mitigation, Figure 1 Examples of adding reinforced concrete structural walls (Thermou and Elnashai, 2005).

Lateral bracing and strengthening: One of the conventional seismic retrofit strategies involves strengthening the structure through lateral bracing to promote elastic or near elastic response while controlling lateral drift and limiting associated seismic damage. Hence, this approach aims to increase the recoverable elastic strain energy in the system. A large number of existing buildings have non-ductile reinforced concrete frames with or without unreinforced masonry (URM) infill walls as their lateral load resisting systems. These buildings are primarily designed and detailed to resist gravity loads. A common approach to providing lateral strength and stiffness is to provide properly designed and detailed reinforced concrete shear walls that are well integrated with the existing structural system. This approach has been applied successfully over the years when it was possible to fill in existing bays (sometimes replacing existing masonry walls) with properly designed reinforced concrete structural walls. Three considerations become important for successful performance of these walls; (a) suitable locations within the plan of the building so that the required strength and stiffness (drift control) can be provided without creating significant torsional effects, (b) continuity in elevation without much strength and stiffness taper, and (c) integration with existing vertical and horizontal structural elements to ensure proper force transfer through existing floor diaphragms. Figure 1 illustrates examples of adding structural walls to existing buildings. The walls are designed as new structural walls following the seismic provisions of current building codes, with due consideration given to wall boundary elements, and the confinement of these elements in potential plastic hinge regions. A major challenge in this retrofit scheme is to ensure appropriate strength in the foundation. Often, the addition of a new structural wall may require strengthening of the existing foundation, which may be a significant undertaking.

Another effective approach to lateral bracing involves the use of steel braces. Diagonal bracing provided in one or more bays in critical direction(s) of buildings provide substantial increase in lateral force capacity and drift control. Steel braces provide diagonal tension ties, as well as diagonal compression struts, resisting seismic-induced inertia forces at floor levels. These members are often sufficiently strong to remain elastic during seismic response. However, if seismic force capacity is exceed, they may yield in tension and buckle in compression, while continuing to provide lateral force resistance under reversed deformation cycles. Figure 2 shows examples of multi-storey buildings retrofitted with structural steel bracing elements. Research is currently underway for developing buckling-restrained bracing elements (Berman and Bruneau, 2009; Xi, 2009; Di Sarno and Manfredi, 2010), as well as developing compression-only braces to avoid the implementation of expensive and difficult connection details for substandard old structural elements (Caron, 2010). Figure 3 shows recent research on steel braces as seismic retrofit elements.

A significant proportion of non-ductile frame buildings have existing unreinforced masonry (URM) infill walls. These walls are often not intended to resist lateral forces, and therefore are not designed for seismic resistance. However, they are often not separated sufficiently from the structural framing system, and interact with the rest of the structure. When their elastic threshold is exceeded under in-plane seismic forces, or their stability is compromised under out-of-place excitations, they fail in a brittle manner, not only providing safety hazard, but also leaving the structural framing system without much lateral bracing. A convenient approach to enhance the lateral bracing capacity of these URM infill walls is the use of surface-bonded fiber-reinforced polymer (FRP) sheets. The use of FRP, either covering the entire wall surface area or by means of diagonally placed strips, well anchored to the enclosing frame, may provide sufficient diagonal tension and compression capacity to brace the entire frame (Saatcioglu et al., 2005). Figure 4 shows two test specimens, with carbon FRP sheets having carbon fibers aligned with wall diagonals, providing significant strength enhancement to the structure. This type of retrofitting can only provide strength and stiffness increase, without any ductility enhancement, as both the URM and the FRP are non-ductile materials. Hence, this retrofit scheme is most suited for structures that are expected to remain

Structural Mitigation, Figure 2 Steel bracing as seismic retrofit strategy; (**a**) University of California dormitory, Berkeley, Ca., USA, (**b**) Parking garage in Berkeley, CA., USA, (**c**) Multi-storey building in San Francisco, Ca., U.S.A.

elastic during seismic response after the implementation of FRP strengthening. The major challenge in such application is the debonding of FRP during reversed cyclic loading. While epoxy-bonded FRP may have limited bond strength, the use of FRP anchors proves to be an effective means of providing additional physical anchorage (Saatcioglu et al., 2005; Ozbakkaloglu, 2009). Figure 5 shows the FRP anchors developed at the University of Ottawa, which ensure the development of full FRP strength when used to anchor surface-bonded FRP sheets on the enclosing concrete frame elements.

Another approach for lateral bracing involves the use of diagonal prestressing strands or cables, providing tension ties (Miranda and Bertero, 1989; Pincheira and Jirsa, 1992; Shalouf and Saatcioglu, 2006). High-strength steel cables secured at beam-column joints enable the development of diagonal tension ties in single or multiple bays, sometimes covering the entire height of the building

Structural Mitigation, Figure 3 Research on compression only steel bracing (Caron, 2010).

Structural Mitigation, Figure 4 Tests of R/C frames with URM infill walls, retrofitted with FRP (Saatcioglu et al., 2005).

as illustrated in Figure 6 (Pincheira and Jirsa, 1995). The cables are provided either snug-tight, allowing the development of passive resistance during seismic response, or prestressed to resist higher forces as one diagonal develops increased tensile forces and the opposite diagonal loses its prestressing force, allowing redistribution of forces among the cables, thereby increasing their effectiveness. Figure 7 shows a test frame with brick URM infill wall, retrofitted with prestressed diagonal cables, increasing the lateral force resistance of a non-ductile system significantly (Shalouf and Saatcioglu, 2006).

Perhaps one of the most vulnerable structural systems against earthquake effects is the URM load-bearing structures. URM buildings form a significant proportion of building inventory worldwide, especially in the developing world. They also form the majority of heritage building inventory in the form of stone masonry structures. The URM buildings do not have sufficient inelastic deformability, and fail in a brittle manner under in-plane and out-of plane seismic excitations. Though these buildings may benefit from their inherent lateral rigidity in controlling lateral drift, especially in buildings with high wall-to-floor area ratios, they exhibit extremely brittle behavior upon reaching their elastic strength threshold. These structural systems need to be retrofitted for increased strength and integrity, and if possible, for improved deformability, though replacement of the URM may be a strong possibility as a viable seismic risk mitigation strategy. Structural retrofit measures include

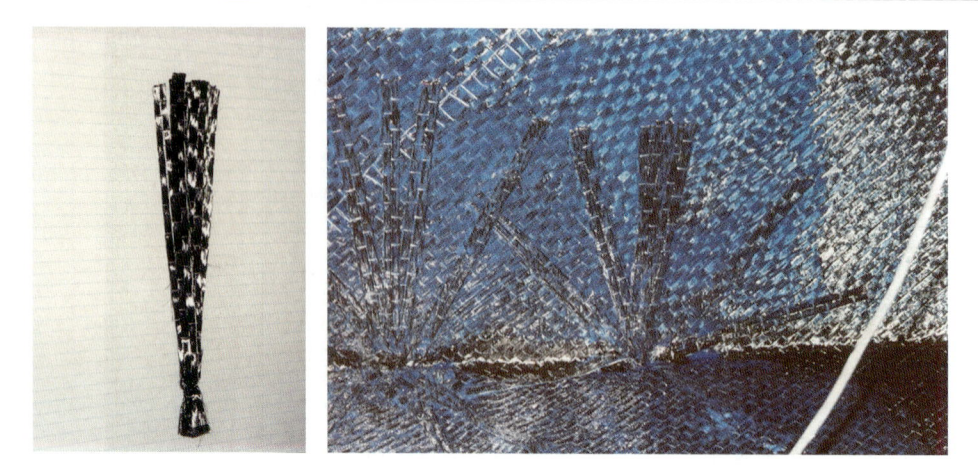

Structural Mitigation, Figure 5 FRP anchors developed at the University of Ottawa (Saatcioglu et al., 2005; Ozbakkaloglu et al., 2009).

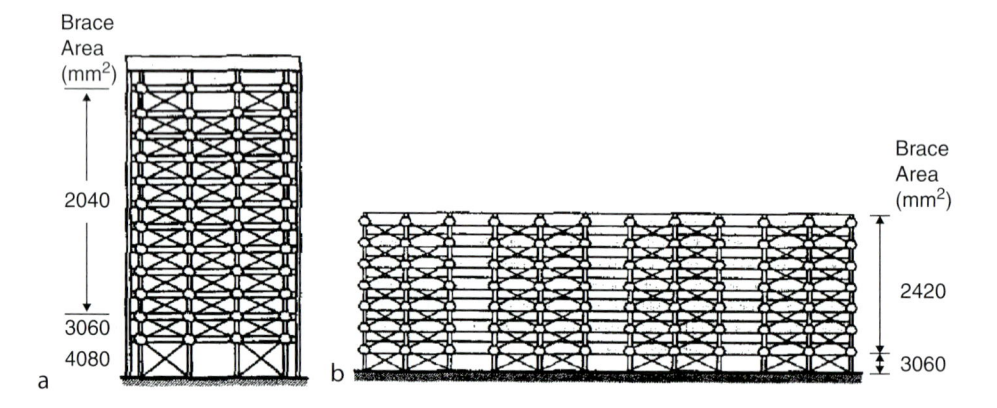

Structural Mitigation, Figure 6 Buildings with diagonal post-tensioning braces (Pincheira and Jirsa, 1995); (**a**) 12-storey building, (**b**) 7-storey building.

improving the integrity of the structure as a whole by attaching slabs and floor beams securely to the load-bearing walls. Buildings may have to be reinforced with additional vertical structural elements in the form of steel or reinforced concrete columns and walls to ensure vertical strength and stability of the entire lateral load resisting system. An increasingly accepted recent seismic retrofit technique involves the use of glass or carbon FRP systems. FRP sheets may be surface bonded on vertical URM elements for increased flexural and shear resistance. FRP sheets maintain out-of-plane stability while providing sufficient diagonal tension reinforcement against seismic forces, as the masonry continues providing diagonal compression struts. Figure 8 shows the application of FRP sheets on URM walls. Figure 9 shows URM test specimens retrofitted with carbon and glass FRP sheets/strips (Vandergrift et al., 2010; ElGawady et al., 2006). The primary shortcoming of this system is the potential for premature debonding of FRP from the masonry substrate, as well as the invasive nature of the application creating challenges in preserving the heritage value of historical buildings. An effective alternative to using FRP sheets is the use of steel strips as vertical and diagonal elements, providing both flexural and shears strength enhancements (Taghdi et al., 2000a, b). Figure 10 illustrates the use of steel plates on both sides of a masonry wall, connected by drilled-through bolts to resist flexural tension and compression, as well as shear-induced diagonal tension and compression as the bolts prevent global buckling. The figure also illustrates the improvement attained in the hysteretic force-force deformation relationship.

Base isolation: The traditional approach to seismic retrofitting involves improving resistance to higher seismic forces and/or enhancing the energy dissipation capacity through ductile response. These improvements enhance both the recoverable elastic strain energy, and the dissipated inelastic energy. Seismic strengthening often involves lateral bracing, which inevitably results in reduced structural period and increased seismic forces. An alternative approach is to reduce seismic force and deformation demands through the use of a base isolation

Structural Mitigation, Figure 7 Reinforced concrete frame with infill masonry retrofitted with diagonal prestressing (Shalouf and Saatcioglu, 2006).

Structural Mitigation, Figure 8 Application of QuakeWrap FRP strips on URM walls in a school building in La Jolla, CA., USA (QuakeWrap, 2010).

system. In this approach the building is placed on a base isolation system, which is designed to prevent or control the transmission of ground motion from the building foundation to the superstructure. This essentially isolates the structure from the ground motion, reducing seismic demands imparted on the building superstructure significantly. The majority of seismic deformations occur in the isolator, while the building remains almost entirely elastic as a rigid body, experiencing only a limited motion, while ensuring the survival of often brittle nonstructural elements and building contents. This is why base isolation is most suited to rigid and non-ductile structures, including historical and heritage buildings, as well as buildings with motion-sensitive contents. Figure 11 illustrates schematically the function of a base isolation system. Figure 12 illustrates the application in the field.

Base isolation techniques have evolved quickly over the last decade, and many different types of isolators have become available for use in practice. Commonly used base isolators consist of laminated rubber bearings that provide sufficient strength against vertical loads while performing in a flexible manner under lateral seismic forces. Typical base-isolation systems in this category consist of elastometric bearings, lead rubber bearings, and high damping rubber bearings. These systems are often equipped with passive dampers or energy dissipation

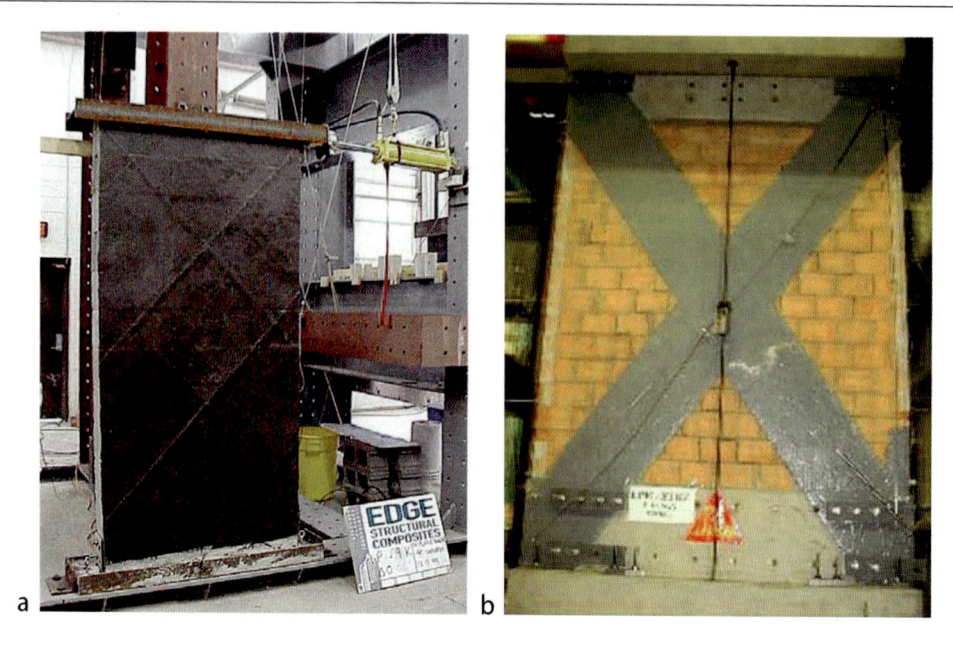

Structural Mitigation, Figure 9 Experimental research on FRP retrofitted URM walls; (**a**) Carbon FRP sheets on concrete block masory (Vandergrifit et al., 2010) (**b**) Glass FRP strips on brick masonry (ElGawady et al., 2006).

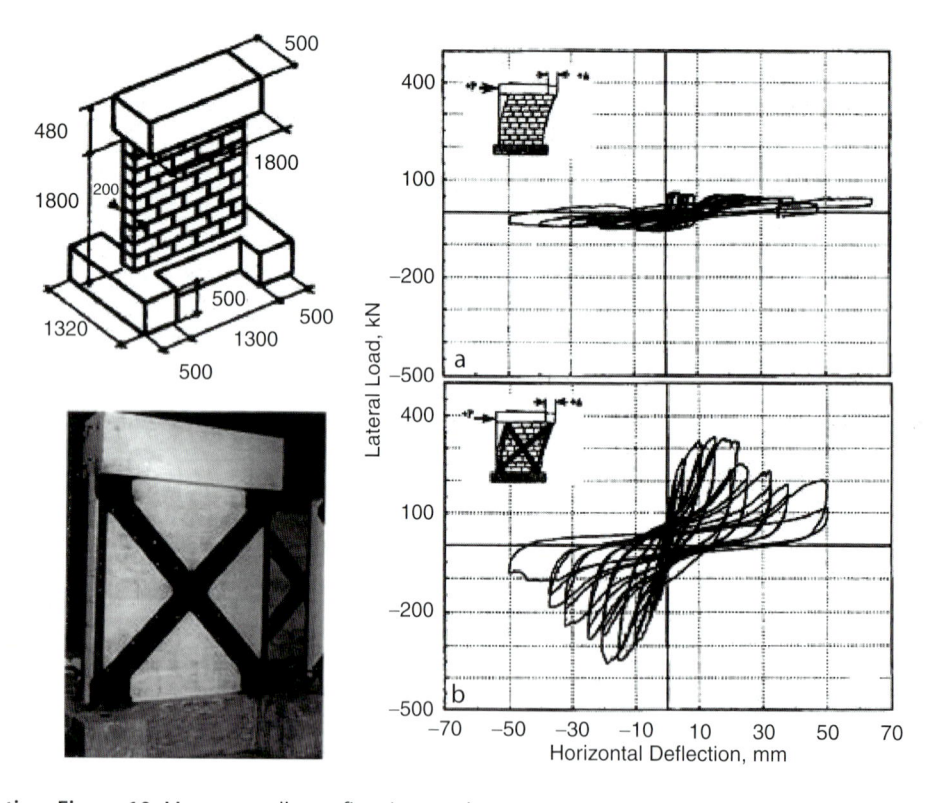

Structural Mitigation, Figure 10 Masonry wall retrofit using steel strips (Taghdi et al., 2000a).

capabilities to control excessive deformations. Other commonly used types of isolation systems involve roller, sliding, or rocking mechanisms. An example of the latter category is a sliding friction pendulum bearing. Some applications include high-tension springs; rubber bearings; or concave plates to control excessive lateral displacements and to re-center the isolators. During the past 2 decades, numerous structures have been equipped

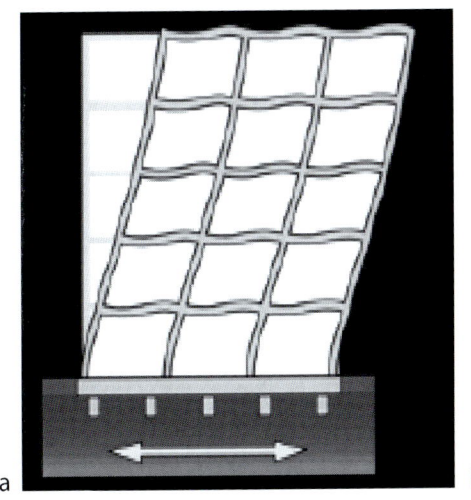

Structural Mitigation, Figure 11 Schematic view of base isolation.

Structural Mitigation, Figure 12 Application of base isolation system in practice; (**a**) A hospital building (DIS, 2010) (**b**) An airport hangar (EMKE, 2010).

with base-isolation systems for seismic protection, especially in Japan and the USA (Bertero, 2010).

Supplementary damping devices: Seismic energy in structures can be dissipated either through inelastic action in structural members (hysteretic energy), or by enhancing structural damping through supplemental damping devices. Supplemental damping devices have been developed since the early 1980s for the purpose of retrofitting seismically deficient structures. These devices are often built into a bracing element, enhancing damping, while also contributing to the strength and stiffness of the structure, controlling deformations, and limiting structural and nonstructural damage during earthquakes. They are generally used as viscous dampers, friction dampers, yielding elements or viscoelastic dampers. Viscous dampers consist of a piston and a cylinder filled with viscous fluid. The energy is dissipated through the piston moving in the cylinder. Friction dampers consist of steel plates tightened against each other to generate sufficient friction to dissipate energy. Yielding dampers are equipped with metallic components that yield and dissipate hysteretic energy. Viscoelastic dampers consist of bonded layers of solids that dissipate energy through the shearing of elements. Figure 13 illustrates different types of supplementary damping devices schematically. Figure 14 shows the application of friction dampers in buildings.

Another form of a passive damping device is the use of sloshing fluid or a column of fluid to dissipate the motion. This method is used as a tuned sloshing damper to absorb energy through the sloshing of waves in a partially filled tank of liquid placed in a vibrating structure. The liquid tank in such a system is tuned to have approximately the same period as the fundamental period of the building. The period of the liquid is adjusted by the density, length, width, and depth of fluid.

Sometimes, dynamic vibration absorbers are employed to transfer energy from one mode of vibration to another, with a view of distributing them approximately equally among the modes of vibration. Examples include tuned mass dampers and tuned liquid dampers (Housner et al., 1997). Supplementary damping devices are more effective in flexible structures where deformations are large enough to activate the energy dissipation device.

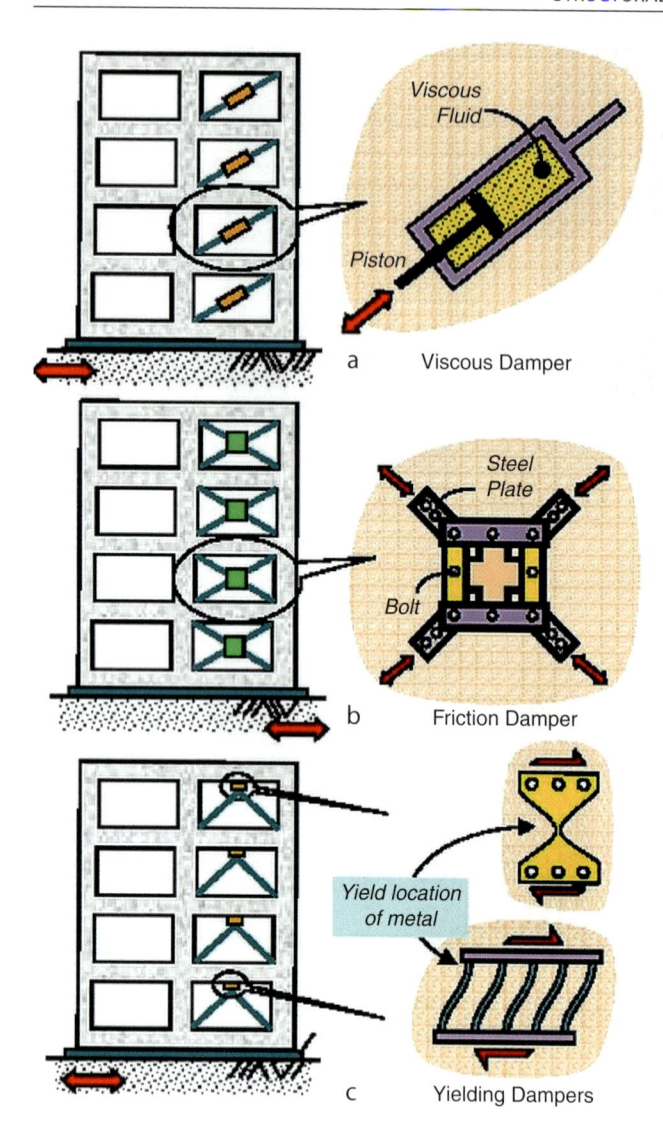

Structural Mitigation, Figure 13 Supplemental damping devices (Murthy, 2010).

Comprehensive reviews of passive dampers are presented by Soong and Dargush (1997), Housner et al. (1997), Symans and Constantinou (1999), and Soong and Spencer (2002). The theoretical concepts and behavior patterns of supplementary damping devices are well understood and their favorable performance in mitigating seismic damage is widely recognized.

Semi-active control devices: A semi-active control device may be defined as a system that can optimally reduce the vibration amplitude by adjusting structural properties during seismic response without injecting any mechanical energy into the structure. Sensors are used to measure the excitation and/or the response of the structure. The readings from sensors are used to monitor building response and adjust the mechanical properties of the energy dissipation device to achieve the required building response. Such devices are often referred to as controllable passive devices (Housner et al., 1997; Symans and Constantinou, 1999; Spencer and Nagarajaiah, 2003). Examples of these devices include variable-orifice fluid dampers, variable friction dampers, smart tuned mass dampers, tuned liquid dampers, variable stiffness devices, and controllable fluid dampers.

In 1990, Feng and Shinozuka proposed a controllable, electromechanical, variable-orifice valve to adjust hydraulic fluid flow resistance in the damper. They suggested that the implementation of such a device on bridges could control the bridge movement during an earthquake. A modified version of the same semi-active damper was developed by Kobori et al. (1993) for buildings. Full-scale dampers of this type were installed in a five-storey steel structure in Shizuoka, Japan as illustrated in Figure 15 (Kurata et al., 1999, 2000). Each of the installed dampers in the building can produce up to a 1,000 kN damping force, and requires only about 70 W of power to operate. These systems usually require low power, and can be operated through the use of a battery.

A semi-active control system was proposed by Akbay and Aktan (1991) and Kannan et al. (1995) based on surface friction to dissipate structural vibrations. A friction shaft is rigidly connected to a structural bracing system and the magnitude of the friction force is adjusted through changing the magnitude of the preload on the friction interface. A similar device was developed at the University of British Colombia (Cherry, 1994; Dowdell and Cherry, 1994). The researchers developed a control device that used an "OFF-ON" algorithm. In this algorithm, the device maintains a constant pressure over the friction interface (ON) until the inter-storey velocity comes close to zero; then, the device releases the pressure (OFF) to let the brace position itself, and after that it returns to its original condition again (ON).

Research during the past 2 decades has indicated that semi-active control systems, when appropriately installed, can show significantly better performance than passive control systems and have the potential to achieve the performance level of a fully active system (Dyke et al., 1996; Spencer and Sain, 1997; Spencer et al., 2000; Yi and Dyke, 2000; Jansen and Dyke, 2000).

Active control devices: An active control system may be defined as a system in which a control actuator applies control forces to the structure on the basis of real-time seismic response and/or earthquake motion to either improve the energy dissipation characteristics during response or to resist applied seismic forces. Control forces are computed in real time on the basis of measured structural responses and/or ground motion, depending on the algorithm employed. The control process is illustrated in Figure 16. Sensors that are mounted on the structure and/or ground record response and send the required data to the controller as feedback. The control forces are then computed on the basis of a prescribed control algorithm and are applied on the structure for the required corrective active. The actuator(s) are normally supported by an

Structural Mitigation, Figure 14 Application of friction dampers in buildings.

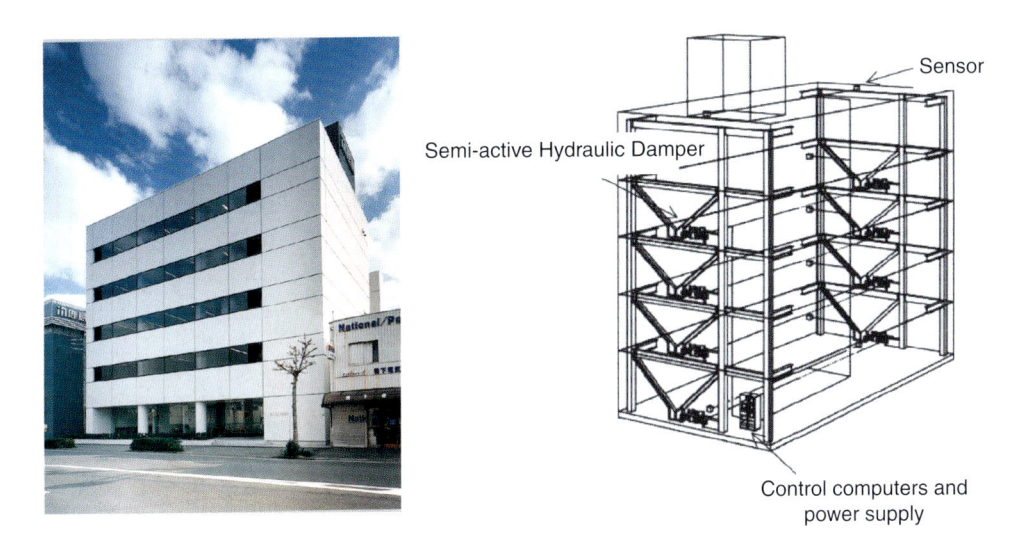

Structural Mitigation, Figure 15 Semi-active dampers installed in a five-storey building in Japan.

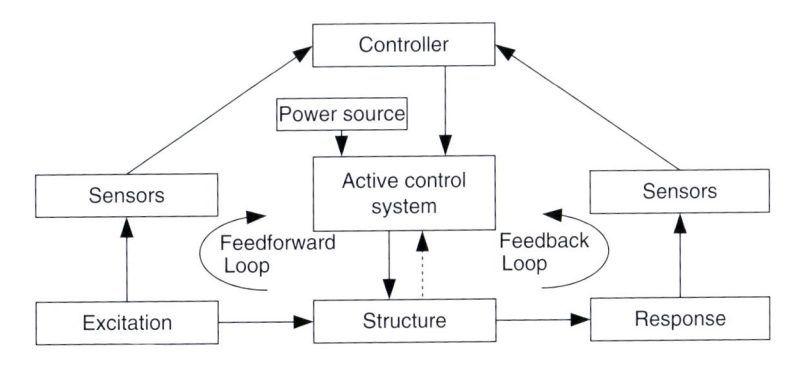

Structural Mitigation, Figure 16 Active structural control (Housner et al., 1997).

external power source due to the high operational power demand of electrohydraulic or electromechanical systems (Housner et al., 1997; Symans and Constantinou, 1999), and hence are susceptible to power failure during earthquakes. Different control algorithms are used to optimize the control action. A variety of active control systems were designed, developed, and modified during the past few decades. The most common active control systems

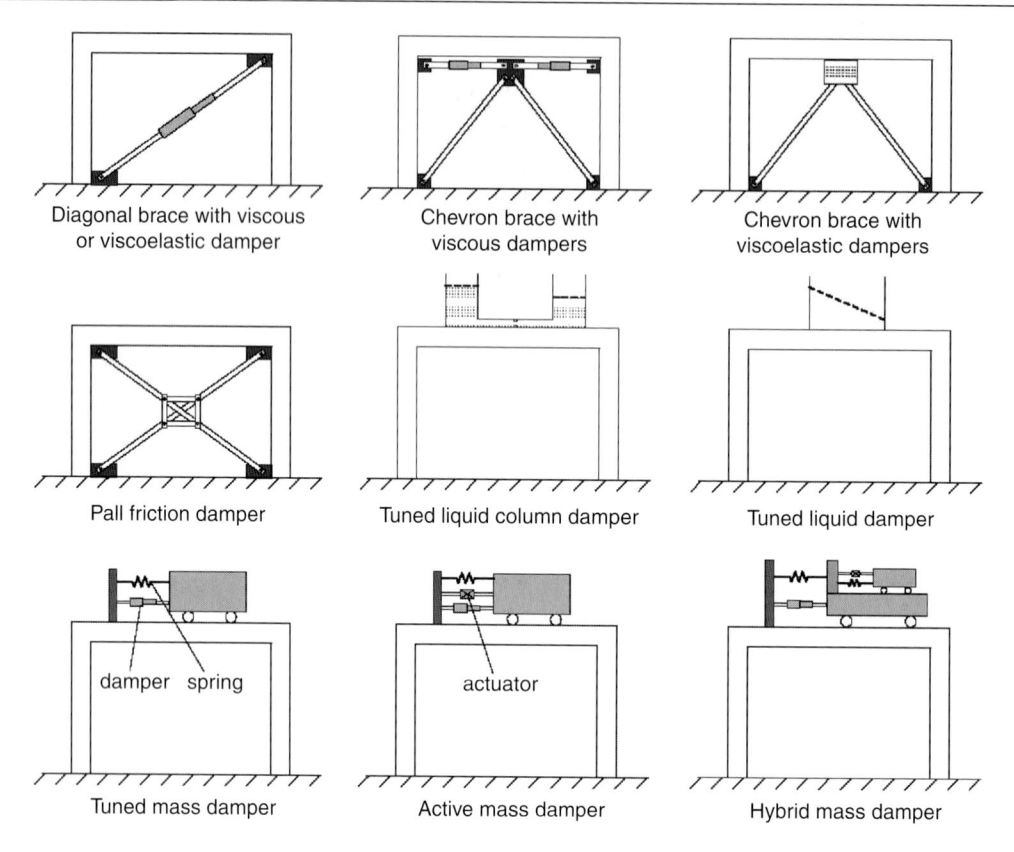

Structural Mitigation, Figure 17 Summary of commonly used supplemental energy dissipation devices (IST Group, 2010).

include, active mass dampers and active mass drivers, active tuned mass dampers, active variable stiffness or damping systems, active bracing systems, pulse generators, and active tendon controls. Problems associated with time lag, reliability of the computed response, and the possibility of power outage during earthquakes provide challenges for the widespread acceptance of active control techniques, though a number of buildings currently have active control systems installed for seismic risk mitigation, mostly in the form of active mass dampers.

Figure 17 provides a schematic summary of supplemental energy dissipation systems.

Structural mitigation at the element or subassembly level

Structures with seismically deficient members can be retrofitted at the element or subassembly level to address local deficiencies that may otherwise trigger significant damage during seismic response, potentially leading to partial or total collapse. The most vulnerable members are vertical structural elements, which are responsible for overall strength and stability of the entire structure. The following section provides an overview of commonly employed seismic retrofit techniques for different types of building and bridge components.

Column retrofits

Reinforced concrete columns may suffer seismic damage primarily due to: (a) insufficient shear strength, (b) lack of concrete confinement, and (c) improper splicing of longitudinal reinforcement within the potential plastic hinge region. Jacketing is used to address these deficiencies as an effective seismic retrofit strategy for both bridge and building columns. Column jacketing is often implemented in the form of steel, reinforced concrete, or FRP jackets. Steel jacketing was advanced through a comprehensive research program (Priestley et al., 1994a, b; Chai et al., 1991; Chai, 1996; Daudey and Filiatrault, 2000). The technique was used in the bridge retrofit program of the California Transportation Department in the USA, especially after the 1994 Northridge Earthquake. The approach is fully effective in circular columns due to hoop tension. Preformed circular steel jackets are built in two half shells, rolled into a radius that is usually 20–30 mm larger than the column diameter. These two sections are then welded at the seams in the field. A gap of about 50 mm is left in the vertical direction between the jacket and the support to prevent bearing of the jacket on the supporting element. The most common steel casings are manufactured from 8 to 25 mm thick plates. The gap between the jacket and the column is filled with grout. The grout is pumped from the bottom in order to remove air gaps. Steel elliptical jackets are used to retrofit rectangular columns, and round

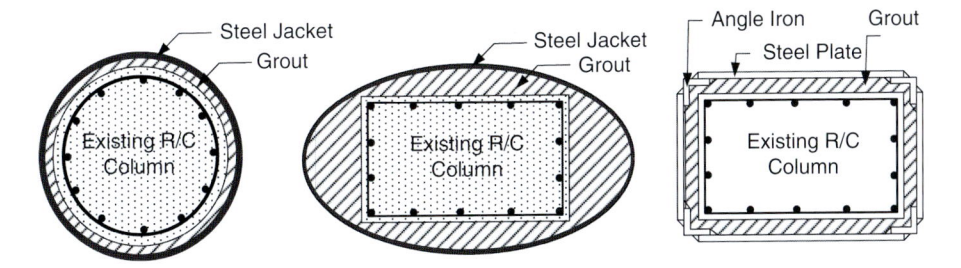

Structural Mitigation, Figure 18 Steel jacketing of columns.

Structural Mitigation, Figure 19 Steel jackets on bridge columns; (**a**) Steel jacket under construction in California, (**b**) Steel jacked bridge column survived the Kobe Earthquake.

jackets are used for square columns to achieve hoop tension. These applications are similar to that used for circular columns. However, the larger gaps that exist between the column and the steel jacket require more effort in the placement of the grout or in-filling concrete. Sometimes steel jackets of rectangular geometry are used, consisting of steel plates or angle irons that are field welded. The jackets can be partially prefabricated into two L sections with the final assembly completed in the field. They may be secured along the sides of columns by means of anchor bolts to improve the confining action. As in the previous application, a gap of about 25 mm is left between the steel jacket and the concrete column, which is filled with nonshrink grout. Figure 18 illustrates various steel jacketing configurations for circular and rectangular columns. Figure 19 illustrates steel jackets in the field.

Concrete jacketing involves the addition of a thick layer of reinforced concrete around the column (Rodriguez and Park, 1994; Priestley et al., 1996). New longitudinal bars must be dowelled into the footing or the storey below with sufficient anchorage depth to achieve full enhancement of flexural strength. In most cases, the footing must also be retrofitted to enhance the footing strength so that the plastic hinging develops in the column. The construction of a reinforced concrete jacket is similar to the construction of a new column. The longitudinal bars are installed first, followed by the placement of transverse reinforcement and the formwork prior to casting the concrete. Figure 20 illustrates the application of a reinforced concrete column jacket in the field.

The third type of column jacketing involves the use of FRP, and is a relatively new technology. The use of advanced composite materials for retrofitting reinforced concrete columns has been the focus of significant research during the last decade (Mirmiran and Shahawy, 1997; Saafi et al., 1999; Demers and Neale, 1999; Saatcioglu and Elnabelsy, 2001; Iacobucci et al., 2002; Elnabelsy and Saatcioglu, 2004; Saatcioglu et al., 2008). FRP systems can be assembled and constructed in a number of ways depending on the materials used and the prevailing site conditions. They can be classified as "wet lay-up systems" and "pre-cured systems." The installation begins with repairing the substrate and preparing the surface prior to the application of the jacket. In the wet lay-up system, wet fibers are hand applied on the column while making sure that entrapped air is removed. Subsequent layers are placed before the previous layer is completely cured. In machine-applied systems, the wrapping machine is placed around the column, automatically wrapping the tow material around the perimeter of the column while moving up and down. Pre-cured jackets are placed around a column that has been covered with an adhesive and clamped until cured. Figure 21 shows the application of FRP jackets in practice.

The column jacketing applications discussed above result in shear strength enhancements. This is especially important for short and stubby columns, as well as the segments of columns that develop short column behavior due to unintentional lateral supports provided by adjacent nonstructural elements (captive columns). These columns are especially vulnerable to shear failure during earthquakes. The jackets also improve concrete confinement for increased inelastic deformability and energy dissipation. However, some of the jacketing techniques employed may not be suitable for strengthening reinforcement splice deficient regions, especially in rectangular columns. These jacketing techniques rely on passive lateral resistance during seismic response that develops due to the expansion of compression concrete. A method of column retrofitting has been developed recently based on transverse prestressing of columns,

Structural Mitigation, Figure 20 Reinforced concrete jacket during construction (Thermou and Elnashai, 2006).

Structural Mitigation, Figure 21 Application of FRP jackets in the field.

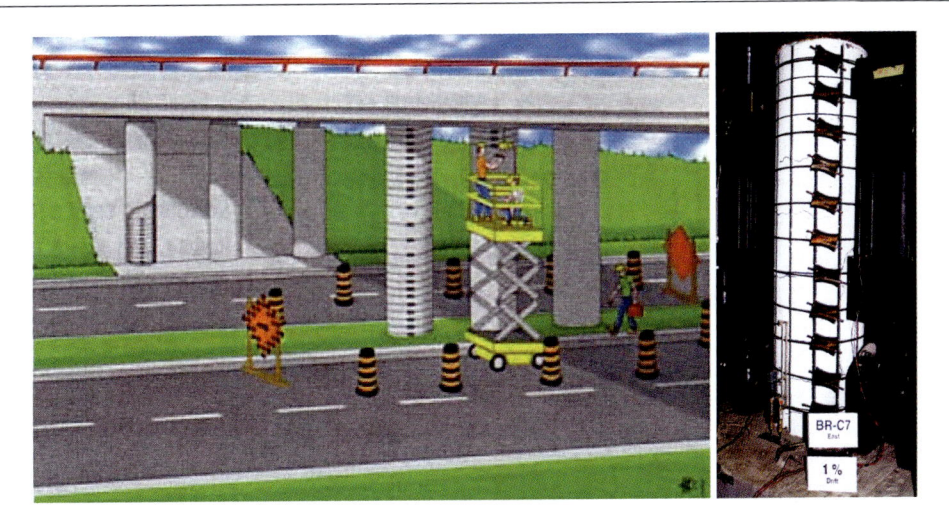

Structural Mitigation, Figure 22 Application of RetrfoBelt (transverse prestressing) on columns.

providing active lateral pressure (Saatcioglu et al., 2002; Saatcioglu and Yalcin, 2003; Yarandi et al., 2004). The technique is known as "Retro Belt System," and provides active and passive lateral confinement pressures, improving concrete confinement while also improving concrete shear capacity and reinforcement bond in splice deficient regions. Transverse prestressing is applied by placing 7-wire strands around columns in the form of individual hoops. The strand ends are tied using specially developed anchors. While circular columns can be prestressed directly on the concrete, producing uniform active lateral pressure, square and rectangular columns require the use of steel spreader elements to distribute the transverse prestressing force on concrete reasonably uniformly. Figure 22 illustrates the application of the RetroBelt retrofit system on concrete columns.

Structural Mitigation, Figure 23 Installation of FRP sheets on a reinforced concrete structural wall.

Wall retrofits

Shear walls are primary lateral load resisting elements. They often possess the required strength and ductility. However, older walls without adequate web reinforcement, or properly confined boundary elements may be critical during strong earthquakes. A number of retrofit techniques have been developed for such walls. These include surface-bonded FRP and externally applied steel plates or bars. Figure 23 shows a concrete wall specimen retrofitted with carbon FRP sheets for shear and flexural strengthening. In such applications, the anchorage of FRP sheets on concrete pose challenges. Steel brackets or FRP anchors like the one shown in Figure 5 may be used to anchor the FRP to the concrete substrate. Another retrofit technique for concrete flexural walls, developed by Elnashai and Salama (1992), involves the use of external unbonded re-bars. The researchers also used externally unbonded steel plates as tension elements. Figure 24 illustrates the technique on test specimens. Steel strips were used by Taghdi et al. (2000a, b) to retrofit seismically deficient reinforced concrete shear walls for strength and ductility enhancements. The strips were placed vertically on opposite sides of walls at the ends by means of closely spaced drilled-through bolts to create boundary elements. The researchers extended the same technique to retrofitting unreinforced and partially reinforced masonry load-bearing walls successfully. This is illustrated in Figure 10. In all cases, the wall strength increased substantially, with significant improvements in inelastic deformability.

Beam retrofits

Though not as critical as vertical elements, seismically deficient beams may experience sudden and brittle shear failures when they lack adequate transverse reinforcement, resulting in local failures. Flexure dominant beams may also suffer from the crushing of compression concrete and buckling of compression bars near the face of the

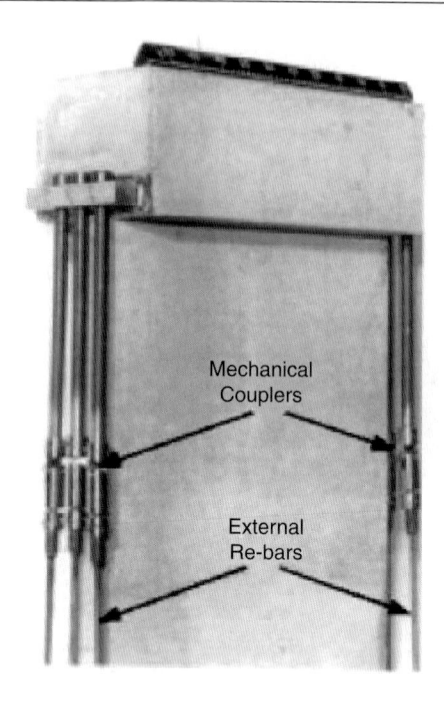

Structural Mitigation, Figure 24 Reinforced concrete structural wall retrofit with external unbonded tension elements.

Structural Mitigation, Figure 25 Retrofitting beam end region with FRP sheets (Thermou and Elnashai 2006).

column where plastic hinges form, if not confined by properly designed transverse confinement reinforcement. Seismic retrofitting concrete beams can be performed by adding transverse reinforcement externally in the form of steel brackets clamped on the beams, surface-bonded FRP strips or as surface mounted steel plates. When appropriate, adding a reinforced concrete jacket improves beam capacity to desired levels. However, care must be exercised to ensure that the beams do not become stronger than the framing columns so that the strong-column weak-beam design principle is not violated. Figure 25 illustrates the application of FRP sheets on a concrete beam.

Beam-column joint retrofits

One of the critical areas of reinforced concrete frame buildings is the beam-column joint region. Under lateral seismic forces, these connection regions are subjected to shear stress reversals, resulting in excessive diagonal cracking and concrete deterioration, compromising the overall rigidity of the structure. Joint retrofits often involves steel or reinforced concrete jackets built around existing joints, including passing longitudinal reinforcement through the slabs (Alcocer and Jirsa, 1993; Beres et al., 1992; Corazao, 1989). Because of the difficulties associated with working around a connection area where the columns, the beams, and the slabs join, the use of surface-bonded FRP has gained popularity in recent years as a seismic retrofit technique (Pantelides and Gergely, 2008; Ghobarah, et al., 1996; Ghobarah and Said, 2001; Mosalam, 2008). Figure 26 shows the application of FRP strips on a bridge joint.

The 1994 Northridge Earthquake exposed one of the deficiencies of welded connections in steel frame buildings. After the earthquake, widespread weld failures were observed in steel buildings. A number of solutions were developed for retrofitting such buildings, including weld strengthening, the addition of a welded haunch at the bottom flange of beams, and reducing beam sections near the connection by cutting the flange (dog-bone shaped flange) to promote beam yielding prior to joint weld failure (Engelhardt and Husain, 1993; Engelhardt

Structural Mitigation, Figure 26 Retrofitting beam-column joint with FRP in a bridge structure (Pandelides and Gergely, 2008).

Structural Mitigation, Figure 27 Retrofitting steel beam-column connections with weld deficiencies (Civjan et al., 2000); (**a**) Specimen with bottom flange weakened by cutting to form a dog bone shape, (**b**) Specimen with hunched bottom.

et al., 1996; Tremblay et al., 1997; Civjan et al., 2000). Sample test specimens involved in such development are shown in Figure 27.

Slab retrofits

Reinforced concrete slabs may be vulnerable against punching shear failures when subjected to shear reversals caused by moment transfer from the columns. They may also be subjected to direct shear due to vertical earthquake excitations. Therefore, column-slab connections may require seismic retrofitting. A number of techniques have been developed to increase punching shear perimeter by adding concrete or steel drop panels (Masri, 1996; Lou, 1994; Martines, 1994). Adding shear reinforcement to concrete slabs by means of through-bolts placed between steel plates was also suggested (Martines, 1994). The use of surface-bonded FRP on concrete slabs has gained popularity in recent years due to ease in application (Mosalam, 2002; Binici and Bayrak, 2008; Stark et al., 2005).

Bridge superstructure retrofits

Two types of seismic damage dominate bridge performance: (a) column damage due to lack of adequate transverse reinforcement and longitudinal reinforcement splice length, and (b) unseating of bridge superstructure. The column retrofit methodologies discussed earlier provide effective solutions for bridge columns. In fact, some of these techniques were originally developed for bridge columns. The separation and unseating of bridge superstructure from its supports (columns and abutments) are frequently observed after major earthquakes. The retrofit techniques for this type of deficiency involve increasing

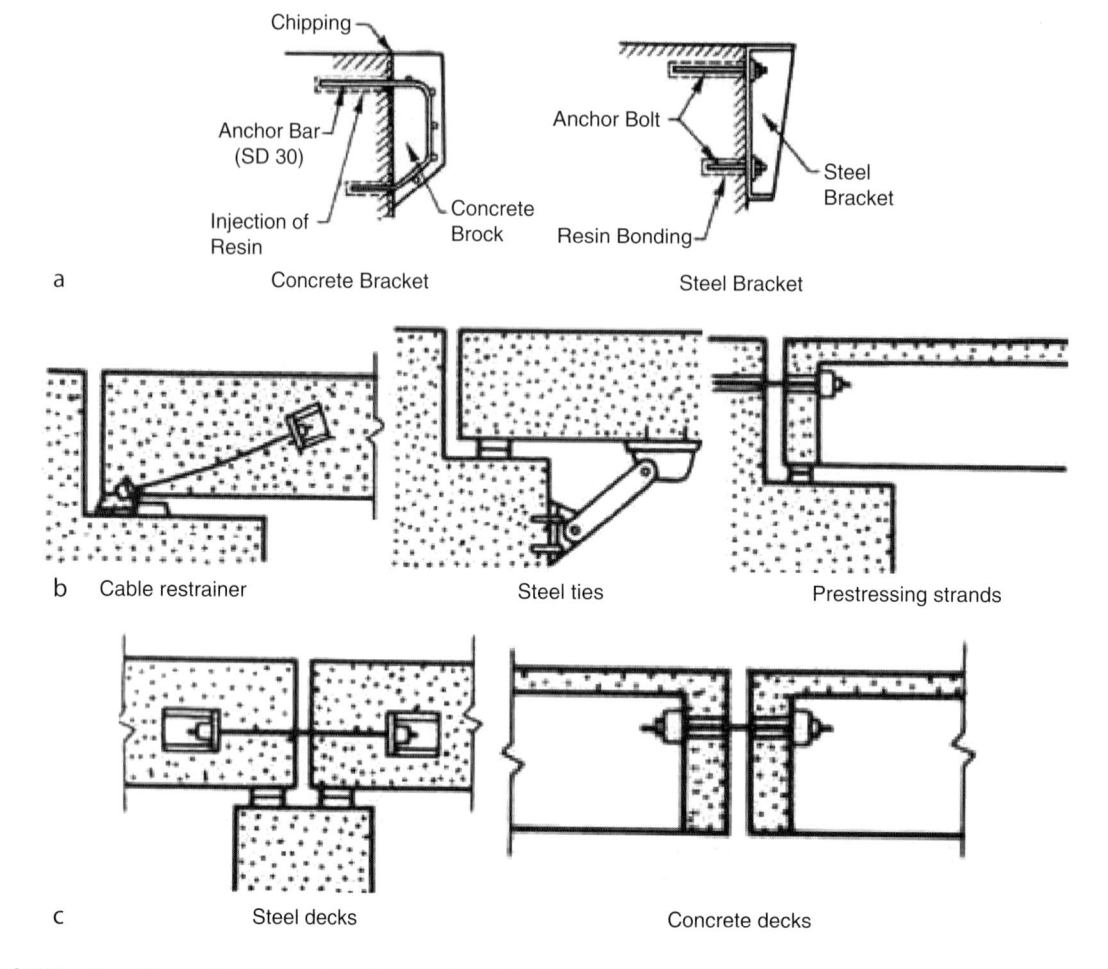

Structural Mitigation, Figure 28 Prevention of unseating of bridge superstructure; (**a**) Extension of seat width, (**b**) Connection of superstructure to substructure, (**c**) Connection between decks.

the support seat width, retrofitting the abutments, including stabilizing the soil around the abutments, and use of restrainers to ensure the integrity of superstructure with its support. Figure 28 illustrates the retrofit methods available for bridge superstructure.

Single and multifamily residential house retrofits

Single family and multifamily residential houses are often built using timber construction. Timber buildings offer superior performance during earthquakes because of their relatively light mass and flexible nailed connections. However, this does not guarantee damage-free performance. Primary seismic deficiency in such buildings is the lack of continuity between the timber framing system and the concrete foundation. Racking action caused by seismic motion can free up the building frame from its foundation (basement or crawl space) wall, or its footings, causing horizontal sliding, and, under extreme cases, overturning of the building. Seismic retrofit against these actions involves securing the floor beams to load-bearing walls and the walls to foundations.

Sometimes, the addition of timber shear walls or timber bracings may be appropriate for areas where parking garages may create soft stories.

A large proportion of residential buildings, including houses, is enclosed with stone or brick masonry veneers. These masonry veneers do not have a structural function and provide additional mass to the building, increasing seismic force demands. While they need to be tied back and secured to vertical structural elements by means of metal straps, the connections may not be sufficiently strong during seismic action. The metal straps may be completely missing in older buildings. This may cause safety hazard for the occupants and they need to be secured along the height. Another nonstructural element that may be a seismic safety hazard is the brittle masonry chimneys, which may have to be retrofitted by means of steel straps and additional supports.

Conclusions

Earthquake risk reduction policy in any municipality should include retrofitting seismically deficient structures.

Structural retrofits prevent or minimize damage in buildings, bridges, and other infrastructure, mitigating seismic risk on society. Research over the last four decades has produced seismic retrofit methodologies that can be applied either at the system level, or at the element or subassembly level, enhancing structural performance during earthquakes. Conventional retrofit techniques involve modifications to the structural system or individual elements for improved strength and ductility. While these techniques have improved over the years, new approaches have also evolved in the form of base isolation, supplementary energy dissipation devices, and active control. The suitability of a retrofit approach depends on the characteristics of the structure and the prevailing conditions. The decision for seismic retrofitting, including the selection of the technology, requires the consideration of multiple factors; including structural safety, sustainability, and economic feasibility. The infrastructure owner, engineer, and decision maker should weigh the costs and benefits of seismic safety and decide what measures to take.

Bibliography

Akbay, Z., and Aktan, H. M., 1991. Actively regulated friction slip devices. In *Proceedings of 6th Canadian Conference of Earthquake Engineering*, pp. 367–374.

Alcocer, S., and Jirsa, J., 1993. Strength of reinforced concrete frame connections rehabilitated by Jacketing. *ACI Structural Journal*, **90**(3), 249–261.

Beres, A., El-Borgi, S., White, R., and Gergely, P., 1992. *Experimental Results of Repaired and Retrofitted Beam-Column Joint Tests in Lightly Reinforced Concrete Frame Buildings*. NCEER Report No. NCEER-92-0025.

Berman, J. W., and Bruneau, M., 2009. Cyclic testing of a buckling restrained braced frame with unconstrained gusset connections. *Journal of Structural Engineering*, **135**(12), 1499–1510.

Bertero, V. V., 2010. *Seismic Upgrading of Existing Buildings*. www.funvisis.gob.ve/archivos/www/terremoto/Papers/Doc028/doc028.htm.

Binici, B., and Bayrak, O., 2008. FRP retrofitting of two-way slabs. FIP Bulletin 35: retrofitting of concrete structures by externally bonded FRPs, pp. 89–108.

Caron, F., 2010. *Repair and Retrofit of Non-Ductile Reinforced Concrete Frames with Diagonal Steel Compression Struts*. M.A.Sc. Thesis, Ottawa, Canada, Department of Civil Engineering, University of Ottawa.

Chai, Y. H., 1996. An analysis of the seismic characteristics of steel jacketed circular bridge columns. *Earthquake Engineering and Structural Dynamics*, **25**, 149–161.

Chai, Y. H., Priestley, M. J. N., and Seible, F., 1991. Seismic retrofit of circular bridge columns for enhanced flexural performance. *ACI Structural Journal*, **88**, 572–584.

Cherry, S., 1994. Research on friction damping at the University of British Columbia. In *Proceedings of International Workshop on Structural Control*, pp. 84–91.

Civjan, S. A., Engelhardt, M. D., and Gross, J. L., 2000. Experimental program and proposed design method for the retrofit of steel moment connections. In *Proceedings of 12WCEE*, Paper 257.

Corazao, M., and Durrani, A., 1989. *Repair and Strengthening of Beam-to-Column Connections Subjected to Earthquake Loading*. NCEER Report No. NCEER-89-13.

Daudey, X., and Filiatrault, A., 2000. Seismic evaluation and retrofit with steel jackets of reinforced concrete bridge piers detailed with lap splices. *Canadian Journal of Civil Engineering*, **27**, 1–16.

Demers, M., and Neale, K. W., 1999. Confinement of reinforced concrete columns with fibre-reinforced composite sheets-an experimental study. *Canadian Journal of Civil Engineering*, **26**, 226–241.

Di Sarno, L., and Manfredi, G., 2009. Seismic retrofitting with buckling restrained braces: application to an existing non-ductile RC framed building. *Soil Dynamics and Earthquake Engineering*, **30**(11), 1279–1297.

DIS (Dynamic Isolation Systems), 2010. www.dis-inc.com.

Dowdell, D. J., and Cherry, S., 1994. Structural control using semiactive friction dampers. In *Proceedings of First World Conference on Structural Control*, Los Angeles, CA, pp. 59–68.

Dyke, S. J., Spencer, B. F., Jr., Sain, M. K., and Carlson, J. D., 1996. Modeling and control of magnetorheological dampers for seismic response reduction. *Smart Materials and Structures*, **5**(5), 565–575.

ElGawady, M. A., Lestuzzi, P., and Badoux, M. A., 2006. Seismic retrofitting of unreinforced masonry walls using FRP. *Composites: Part B*, **37**, 148–162. Elsevier.

Elnabelsy, G., and Saatcioglu, M., 2004. Design of FRP Jackets for seismic retrofit of circular concrete columns. *Emirates Journal for Engineering Research*, **9**(2), 65–69. Published by United Arab Emirates University.

Elnashai, A. S., and Salama, A. I., 1992. *Selective Repair and Retrofitting Techniques for RC Structures in Seismic Regions*. Research Report ESEE/92-2, Engineering Seismology and Earthquake Engineering Section, Imperial College, London.

EMKE, 2010. www.emke.com.tr/products.php.

Engelhardt, M. D., and Husain, A. S., 1993. Cyclic-loading performance of welded flange-bolted web connections. *Journal of Structural Engineering*, **119**(12), 3537–3550.

Engelhardt, M. D., Winneberger, T., Zekany, A. J., and Potyraj, T. J., 1996. The dogbone connection: part II. *Modern Steel Construction*, **36**, 46–55.

Feng, Q., and Shinozuka, M., 1990. Use of a variable damper for hybrid control of bridge response under earthquake. In *Proceedings of U.S. National Workshop on Structural Control Research*. USC Publication No. CE-9013, pp. 107–112.

Ghobarah, A., and Said, A., 2001. Seismic rehabilitation of beam–column joints using FRP laminates. *Journal of Earthquake Engineering*, **15**(1), 113–129.

Ghobarah, A., Tarek, S. A., and Biddah, A., 1996. Seismic rehabilitation of reinforced concrete beam–column connections. *Earthquake Spectra*, **12**(4), 761–780.

Housner, G. W., Bergman, L. A., Caughey, T. K., Chassiakos, A. G., Claus, R. O., Masri, S. F., Skelton, R. E., Soong, T. T., Spencer, B. F., and Yao, J. T. P., 1997. Structural control: past, present and future. *Journal of Engineering Mechanics*, **123**(9), 897–971.

Iacobucci, R. D., Sheikh, S. A., and Bayrak, O., 2002. Retrofit of square concrete columns with carbon fibre-reinforced polymer for seismic resistance. *ACI Structural Journal*, **100**(6), 785–794.

IST Group, 2010. www.web.mit.edu/istgroup/ist/documents/earthquake/Part5.pdf.

Kannan, S., Uras, H. M., and Aktan, H. M., 1995. Active control of building seismic response by energy dissipation. *Earthquake Engineering and Structural Dynamics*, **24**(5), 747–759.

Kobori, T., Takahashi, M., Nasu, T., Niwa, N., and Ogasawara, K., 1993. Seismic response controlled structure with active variable stiffness system. *Earthquake Engineering and Structural Dynamics*, **22**(11), 925–941.

Kurata, N., Kobori, T., Takahashi, M., Niwa, N., and Midorikawa, H., 1999. Actual seismic response controlled building with semi-active damper system. *Earthquake Engineering and Structural Dynamics*, **28**(11), 1427–1447.

Kurata, N., Kobori, T., Takahashi, M., Ishibashi, T., Niwa, N., Tagami, J., and Midorikawa, H., 2000. Forced vibration test of a building with semi-active damper system. *Earthquake Engineering and Structural Dynamics*, **29**(5), 629–645.

Miranda, E., and Bertero, V. V., 1989. Performance of low-rise buildings in Mexico City. *Earthquake Spectra*, **5**(1), 121–143.

Mirmiran, A., and Shahawy, M., 1997. Behaviour of concrete columns confined by fibre composites. *Journal of Structural Engineering, ASCE*, **123**(5), 583–590.

Mosalam, A., 2002. *Structural Evaluation of SCCI-FRP Composite Systems (Tuflam) for Flexure Repair and Upgrade of Reinforced Concrete Floor Slabs, Slabs-on-Grade, and Out-of-Plane RC Walls*. Fullerton, California State University. Test Report No. SRRS-SCCI, 79 p.

Mosalam, K. M., 2008. Seismic retrofitting of RC beam-column joints using FRP. FIP Bulletin 35: retrofitting of concrete structures by externally bonded FRPs, pp. 143–166.

Murthy, C. V. R., 2010. *How to Reduce Earthquake Effects on Buildings*. New Delhi, India: Indian Institute of Technology Kanpur, BMTPC.

Ozbakkaloglu, T., and Saatcioglu, M., 2009. Tensile behavior of FRP anchors in concrete. *ASCE Journal of Composites*, **13**(2), 82–92.

Pantelides, C. P., and Gergely, J., 2008. Seismic retrofit of reinforced concrete beam-column T-joints in bridge piers with FRP composite jackets. *SP258-1, ACI Special Publication on Seismic Strengthening of Concrete Buildings using FRP Composites*.

Pincheira, J. A., and Jirsa, J. O., 1992. Post-tensioned bracing for seismic retrofit of RC frames. In *Proceedings of the Tenth World Conference on Earthquake Engineering*, Madrid, Spain, Vol. 9, pp. 5199–5204.

Pincheira, J. A., and Jirsa, J. O., 1995. Seismic response of RC frames retrofitted with steel braces or walls. *Journal of Structural Engineering*, **121**(8), 1225–1235.

Priestley, M. J. N., Seible, F., Xiao, Y., and Venna, R., 1994a. Steel jacket retrofitting of reinforced concrete bridge columns for enhanced shear strength – part 1: theoretical considerations and test design. *ACI Structural Journal*, **91**(4), 394–405.

Priestley, M. J. N., Seible, F., Xiao, Y., and Venna, R., 1994b. Steel jacket retrofitting of reinforced concrete bridge columns for enhanced shear strength – part 2. Test results and comparison with theory. *ACI Structural Journal*, **91**(5), 537–551.

Priestley, M. J. N., Seible, F., and Calvi, G. M., 1996. *Seismic Design and Retrofit of Bridges*. New York: Wiley.

QuakeWrap, 2010. www.quakewrap.com.

Rodriguez, M., and Park, R., 1994. Seismic load tests on reinforced concrete columns strengthened by jacketing. *ACI Structural Journal*, **91**, 150–159.

Saafi, M., Toutanji, H. A., and Li, Z., 1999. Behaviour of concrete columns confined with fibre reinforced polymer tubes. *ACI Structural Journal*, **96**(4), 500–509.

Saatcioglu, M., and Elnabelsy, G., 2001. *Seismic Retrofit of Bridge Columns with CFRP Jacket. FRP Composites in Civil Engineering*. Elsevier Science, pp. 833–838.

Saatcioglu, M., and Yalcin, C., 2003. External prestressing concrete columns for improved seismic shear resistance. *ASCE Journal of Structural Engineering*, **129**, 8.

Saatcioglu, M., Chakrabarti, S., Selbuy, R., and Mes, D., 2002. Improving ductility and shear capacity of reinforced concrete columns with the retro-belt retrofitting system. In *Proceedings of the 7th U.S. National Conference on Earthquake Engineering*, Earthquake Engineering Research Institute.

Saatcioglu, M., Serrato, F., and Foo, S., 2005. Seismic performance of masonry infill walls retrofitted with CFRP sheets. *ACI Special Publication SP-230*, **230**, 341–354.

Saatcioglu, M., Ozbakkaloglu, T., and Elnabelsy, G., 2008. Seismic behavior and design of reinforced concrete columns confined with FRP stay-in-place formwork. *ACI-SP 257-09, American Concrete Institute*, **230**, 149–170.

Shalouf, F., and Saatcioglu, M., 2006. Seismic retrofit of nonductile reinforced concrete frames with diagonal prestressing. In *Proceedings of the 8th U.S. National Conference on Earthquake Engineering, San Francisco*.

Soong, T. T., and Dargush, G. F., 1997. *Passive Energy Dissipation Systems in Structural Engineering*. Chichester, England: Wiley.

Soong, T. T., and Spencer, B. F., 2002. Supplemental energy dissipation: State-of-the-art and state-of-the-practice. *Engineering Structures*, **24**(3), 243–259.

Spencer, B. F., Jr., and Nagarajaiah, S., 2003. State of the art of structural control. *Journal of Structural Engineering*, **129**(7), 845–856.

Spencer, B. F., Jr., and Sain, M. K., 1997. Controlling buildings: a new frontier in feedback. *IEEE Control Systems Magazine*, **17**(6), 19–35.

Spencer, B. F., Jr., Johnson, E. A., and Ramallo, J. C., 2000. Smart isolation for seismic control. *JSME International Journal, Series C: Mechanical Systems, Machine Elements and Manufacturing*, **43**(3), 704–711.

Stark, A., Binici, B., and Bayrak, O., 2005. Seismic upgrade of reinforced concrete slab-column connections using carbon fibre-reinforced polymers. *ACI Structural Journal*, **102**(2), 324–333.

Symans, M. D., and Constantinou, M. C., 1999. Semi-active control systems for seismic protection of structures: a state-of-the-art review. *Engineering Structures*, **21**(6), 469–487.

Taghdi, M., Bruneau, M., and Saatcioglu, M., 2000a. Seismic retrofitting of low-rise masonry and concrete walls using steel strips. *Journal of Structural Engineering*, **126**(9), 1017–1025.

Taghdi, M., Bruneau, M., and Saatcioglu, M., 2000b. Analysis and design of low-rise masonry and concrete walls retrofitted using steel strips. *Journal of Structural Engineering*, **126**(9), 1026–1032.

Tremblay, R., Tchebotarev, N., and Filiatrault, A., 1997. Seismic performance of RBS connections for steel moment resisting frames: influence of loading rate and floor slab. In *Proceedings – STESSA 1997*, Kyoto, Japan, pp. 4–7.

Uang, C-M., and Bertero, V. V., 1988. Use of energy as a design criterion in earthquake-resistant design. *Report No. UCB/EERC-88/18*, Berkeley, University of California.

Vandergrifit, J., Gergely, J., and Young, D. T., 2010. *CFRP Retrofit of Masonry Walls*. www.quakewrap.com/technical-papers-on-Fiber-Reinforced-Polymer.php.

Xie, Q., 2005. Sate of the art of buckling-restrained braces in Asia. *Journal of Constructional Steel Research*, **61**(6), 727–748.

Yarandi, M. S., Saatcioglu, M., and Foo, S., 2004. Rectangular concrete columns retrofitted by external prestressing for seismic shear resistance. In *Proceedings of the 13th World Conference on Earthquake Engineering*, Vancouver, Canada.

Cross-references

Accelerometer
Concrete structures
Critical Infrastructure
Damage and the Built Environment
Earthquake
Earthquake Damage
Earthquake Resistant Design
Federal Emergency Management Agency (FEMA)
Haiti Earthquake 2010 Psychosocial Impacts
Harmonic Tremor
High Rise Buildings in Natural Disaster

Isoseismal
Macroseismic Survey
Magnitude Measures
Mercalli, Giuseppe (1850–1914)
Modified Mercallit (MM) Scale
Plate Tectonics
Primary wave (P wave)
Richter, Charles Francis (1900–1985)
Rotational Seismology
Secondary Wave (S wave)
Seismograph/Seismometer
Seismology
Shear
Structural Damage caused by Earthquakes
Tangshan, China (1976 Earthquake)
Tectonic Tremor
Tohoku, Japan, Earthquake, Tsunami and Fukushima Accident (2011)
Unreinforced Masonry Building
Wenchuan, China (2008 Earthquake)

SUBDUCTION

Alik T. Ismail-Zadeh
Karlsruhe Institute of Technology, Karlsruhe, Germany
International Institute of Earthquake Prediction Theory and Mathematical Geophysics, Russian Academy of Sciences, Moscow, Russia
Institut de Physique du Globe de Paris, Paris, France

Synonyms

Descending of the lithosphere; Lithosphere subduction

Definition

The word "subduction" came from the Latin *subductus*, which means "carry off" or "transfer." In plate tectonics, the process in which an oceanic lithosphere (plate) descends into the Earth's interior is referred to as *subduction*. The descending plate generates seismicity and volcanism.

Discussion

Observations of large gravity anomalies and high seismic activity, associated with oceanic trenches, resulted in the hypothesis that trenches are the sites of crustal convergence and consumption (e.g., Hess, 1962). The loss of the surface area at trenches (compared to surface augmentation at seafloor spreading zones) became later known as *subduction*, one of the basic elements of plate tectonics. As the lithosphere moves away from an oceanic ridge, it cools, densifies, and thickens. Once the lithosphere becomes sufficiently dense compared to the underlying mantle rocks, it bends, founders, and begins sinking into the hot mantle due to gravitational instability. As results of subduction, an oceanic trench forms in the front of the overriding lithosphere, and an accretionary prism evolves from the sediments that scraped off from the descending oceanic crust.

Because the lithosphere behaves elastically at short time scales, it can transmit stresses. The downward buoyancy forces (generated due to the excess density of the rocks of the descending lithosphere) promote the sinking of the lithosphere, but elastic, viscous, and frictional forces resist the descent. The combination of these forces produces shear stresses high enough to cause earthquakes. Other processes contributing to stress generation in the descending lithosphere and its release in earthquakes can be plastic instability at high temperature, faulting due to metamorphic phase transitions, and dehydration-induced embrittlement. Oceanic trenches are the sites of the world largest earthquakes. The earthquakes at oceanic trench zones can occur along the descending lithosphere to depths of about 660 km depending on the thermal state in the mantle. This seismogenic region called the Wadati–Benioff zone (Wadati, 1928; Benioff, 1949) delineates the shape of the descending lithosphere.

Volcanoes in the overriding lithosphere associated with subduction lie normally parallel to the oceanic trench. These volcanoes may form an island arc or may generate a volcanic chain on a continent. The volcanoes exist due to melting of the subducting lithosphere and/or of mantle rocks above the descending lithosphere and subsequent upward magmatic migration.

Bibliography

Benioff, H., 1949. Seismic evidence for the fault origin of oceanic deeps. *Bulletin of the Geological Society of America*, **60**, 1837–1856.
Hess, H., 1962. History of ocean basins. In Engeln, A., James, H. L., and Leonard, B. F. (eds.), *Petrologic Studies*. New York: Geological Society of America, pp. 599–620.
Wadati, K., 1928. Shallow and deep earthquakes. *Geophysical Magazine*, **1**, 161–202.

Cross-references

Earthquake
Plate Tectonics
Volcanoes and Volcanic Eruptions

SUBSIDENCE INDUCED BY UNDERGROUND EXTRACTION

Devin L. Galloway
US Geological Survey, Sacramento, CA, USA

Synonyms

Anthropogenic subsidence; Land subsidence; Mining subsidence; Subsidence

Definition

Subsidence. Sinking or settlement of the land surface caused by natural or human-induced processes. As commonly used, the term principally relates to the vertical downward movement of natural surfaces although some generally small-scale horizontal movement may be present.

Underground extraction. Purposeful withdrawal of mass (solids, liquids, or gases) and (or) heat from the subsurface.

Introduction

Subsidence induced by underground extraction is a class of human-induced (anthropogenic) land subsidence that principally is caused by the withdrawal of subsurface fluids (groundwater, oil, and gas) or by the underground mining of coal and other minerals. Subsidence in general (see *Land Subsidence*) is a surface manifestation of the mass-wasting or rearrangement of subsurface materials that reduces their bulk volume, and is caused by a variety of anthropogenic and natural processes (National Research Council, 1991) including:

1. Subsurface fluid and heat withdrawal
2. Underground mining
3. Drainage of organic soils
4. Sinkholes (see Sinkhole)
5. Hydrocompaction (see Hydrocompaction Subsidence)
6. Thawing permafrost (see Permafrost)
7. Natural consolidation
8. Underground erosion (see Piping Hazard)
9. Volcanism and tectonism

In many developed areas, most anthropogenic subsidence, as measured by the area affected, is caused by the extraction of subsurface fluids (principally groundwater from porous granular media) and by the underground mining of coal. Subsidence accompanies a variety of natural processes (e.g., items 3–9 above) that constitute the geologic and climatic history of many areas. These natural processes often are accelerated or amplified by anthropogenic factors such as excavation (mining), surface mechanical loading, drainage of wetlands, redirected or focused surface-water drainage and groundwater infiltration, and the extraction of subsurface fluids.

Globally, many subsidence areas have been identified, mapped, and documented (see, e.g., Poland (1984) and Galloway et al. (1999, 2008)). Though regional subsidence, such as that caused by the extraction of groundwater in susceptible aquifer systems, may be subtle and difficult to identify, various factors aid identifying unrecognized subsiding regions, such as increased incidences of damaged or protruding wells, a history of adjustments to local geodetic controls, increasing incidences of coastal or riverine flooding, local conveyance and drainage problems, and ground failures – surface faulting and earth fissuring. Measuring, mapping, and monitoring subsidence are necessary to assess subsidence hazards. Analysis and simulation of subsidence processes, constrained by the available data, often are used to assess present and potential future hazards.

Subsidence contributes to permanent inundation of coastal lands as they settle with respect to sea level (see "*Sea Level Change*," "*Coastal Zone Risk Management*," and "*Coastal Erosion*"), aggravates riverine flooding, alters topographic gradients, ruptures the land surface, disrupts cultural infrastructure (see "*Damage and the Built Environment*"), and causes other hazards related to deterioration of land and water resources. The assessment of costs related to subsidence is complicated by difficulties in identifying and mapping the affected areas, establishing cause-and-effect relations, assigning economic value to environmental resources, and by inherent conflicts in the legal system regarding the recovery of damages under established land and water rights.

Subsidence related to subsurface fluid withdrawal and underground mining

Subsurface fluid withdrawal

Extraction of fluids from subsurface formations perhaps is the best understood of all causes of subsidence. Many areas of subsidence caused by withdrawal of groundwater, hydrocarbons, and geothermal fluids have been identified, surface and subsurface changes have been monitored, and corrective measures have been devised. Decades ago, the topic of "Land Subsidence Due to Fluid Withdrawal" was reviewed by Poland and Davis (1969). A brief overview of groundwater pumping–induced subsidence is given by Phillips and Galloway (2008).

Experience in identifying and coping with subsidence caused by withdrawing groundwater in many parts of the world is reported in Poland (1984). Countries where documented subsidence is attributed to groundwater withdrawal include Australia, Great Britain, Hungary, India, Indonesia, Iran, Italy, Japan, Mexico, Philippines, the Peoples Republic of China (PRC), Spain, Taiwan (Republic of China), Thailand, the USA, and Vietnam. Subsidence exceeding 9 m attributed to groundwater withdrawal has occurred in Mexico City and in the San Joaquin Valley in California (Figure 1). By areas affected, the USA and PRC are most impacted by subsidence attributed to groundwater withdrawal. Extraction of subsurface fluids, principally groundwater, from clastic sediments has permanently lowered the elevation of about 26,000 km² of land in the conterminous USA – an area of similar extent to the State of Massachusetts (Holzer and Galloway, 2005). In recent decades, the occurrence, recognition, and assessment of subsidence in the PRC have increased dramatically. In 2003, the total subsiding area was more than 90,000 km² affecting more than 50 cities and other areas primarily in the Yangtze River Delta, the North China plain, and Fen-Wei Faulted Basin (He et al., 2005; Yang et al., 2005b). The maximum subsidence reported was more than 3 m in Tianjin.

Subsidence Induced by Underground Extraction, Figure 1 Approximate location of maximum measured subsidence in the San Joaquin Valley, California, USA. Land surface subsided approximately 9 m from 1925 to 1977 due to aquifer-system compaction from groundwater withdrawal. Signs on the pole are positioned at the approximate former elevations of the land surface in 1925 and 1955 (Photograph by R.L. Ireland, USGS, ca 1977).

Subsidence attributed to the extraction of oil, gas, and associated water and brines from petroleum reservoirs has been documented in many basins of the world (Yerkes and Castle, 1969; Martin and Serdengecti, 1984; Chilingarian et al., 1995; Nagel, 2001). Subsidence at the Wilmington (California) and Ekofisk (North Sea, Norwegian sector) oil fields are well-known examples due both to the magnitude of subsidence as well as the cost of remediation. Parts of the city and port of Long Beach suffered major problems owing to rapid (as much as 750 mm/year) subsidence during 1937–1962 related to extraction of oil, gas, and associated water from the underlying Wilmington oil field (Gilluly and Grant, 1949; Mayuga and Allen, 1969). Total subsidence in Long Beach reached as much as 9 m before the land surface was stabilized by an integrated program of fluid injection to balance the extraction. Elsewhere in California, the Lost Hills and Belridge oil fields in the San Joaquin Valley subsided at a rate of about 400 mm/year during 1995–1996 (Fielding et al., 1998). In the area of Daqing, the largest oil-producing region in the PRC, as much as 1.5 m of subsidence has occurred since 1959 owing to oil and natural gas production. The subsidence in Daqing is further complicated by groundwater withdrawal associated with waterflooding applied to maintain peak rates of oil production (Zhou et al., 2005). In the Netherlands, the large Groningen gas field began production in the 1960s. The maximum measured subsidence by 2003 was about 0.25 m and it was subsiding at an average rate of about 7.5 mm/year (Ketelaar et al., 2005). Even small amounts of subsidence pose significant challenges in the Netherlands because large portions of the country are below sea level and protected by dikes.

Subsidence due to subsurface fluid extraction in coastal regions such as those near Bangkok, Houston, Jakarta, Long Beach, Manila, New Orleans, Santa Clara Valley in California, Shanghai, and Venice plays a role in the rise of Local Mean Seal Level (LMSL). LMSL is affected by land movements attributed to isostasy, and sea level changes attributed to eustasy (see "*Sea Level Change*"). The observed rate of eustatic rise is 1–2 mm/year (IPCC, 2001), of which global climate change likely is a significant contributor (see "*Climate Change*"). Estimated average subsidence rates vary widely and are confounded by estimates of the relative (subsidence and uplift) contributions from isostasy (González and Törnqvist, 2006): natural consolidation, tectonic uplift and downwarping, and continental rebound from glacial retreats and from local and regional land subsidence caused by anthropogenic factors. Locally, during the latter part of the twentieth century, much of the isostatic decline can be attributed to subsidence caused by the compaction of sediments owing to the withdrawal of subsurface fluids; sediment compaction rates greater than 100 mm/year have been measured in coastal basins while compaction rates attributed to other processes contributing to relative sea level rise were nearly one to two orders of magnitude less. Some of the hazards and environmental consequences associated with coastal subsidence include damage to engineered structures, enhanced coastal and riverine flooding, loss of saltwater and freshwater marsh ecosystems, formation of earth fissures, and reactivation of surface faults.

Several areas have experienced subsidence caused by the extractions of fluids and heat from geothermal fields, including Wairakei, New Zealand, Cerro Prieto, Mexico, and the East Mesa, Casa Diablo and the Geysers in California (Massonnet et al., 1997; Carnec and Fabriol, 1999; Singhal and Gupta, 1999; Howle et al., 2003). In Wairakei, ground surface has subsided by as much as 5–10 m (Elder, 1981). In geothermal fields, the cooler reinjected fluids can exacerbate thermally induced subsidence.

Underground mining

Underground mining of coal and other minerals has caused subsidence of an estimated 8,000 km^2 of land in the USA; most of this fraction is associated with underground mining for coal (National Research Council, 1991). Coal mining in the PRC, India, Germany, Great Britain, and the Netherlands has resulted in widespread subsidence (Saxena, 1978; Yang et al., 2005a). In the PRC, the largest coal producer in the world, subsidence occurs in many coal mining fields. Subsidence caused by coal extraction at Fushun in the northeastern PRC is as much as 16.4 m with a subsiding area of 18.4 km^2 (Yang et al., 2005a). The chief cause of documented subsidence in India is due to the underground mining of coal deposits from Lower Gondwana coalfields in the Jharia and Raniganj coalfields (Saxena and Singh, 1991; Anon, 1997). In these two coalfields, an area of about 100 km^2 has subsided 1–5 m. Such subsidence has led to severe alteration of surface topography, development of mine fires [see "*Coal Fire (Underground)*"], damage to buildings, roads, and other structures, and serious damage to surface water and groundwater systems.

Subsidence related to subsurface fluid withdrawal and underground mining: mechanisms

Subsurface fluid withdrawal

Permanent subsidence can occur when fluids stored beneath the Earth's surface are removed by wells or drains. Because most of the identified subsidence attributed to subsurface fluid withdrawal has been caused by groundwater extraction, the following discussions focus on the development of groundwater resources in aquifer systems. Withdrawing groundwater from the aquifer system decreases pore-fluid pressures which reduce pore volume and the volume of the aquifer system causing subsidence. Support for the overlying material – the overburden or equivalent geostatic stress – is provided by pore-fluid pressure and the compliant granular structure, the so-called skeleton of the aquifer system. A shift in the balance of support provided by fluid pressure to support provided by the skeleton – the intergranular or "effective stress" – causes the skeleton to deform slightly. Both the aquifers (coarse-grained sediments) and aquitards (fine-grained sediments) that constitute the aquifer systems undergo deformation, but aquitards deform much more than aquifers. This phenomenon, known to hydrogeologists as compaction and to soil engineers as consolidation, is based on the theory of primary one-dimensional consolidation of clays (Terzaghi, 1925; Terzaghi and Peck, 1967). This theory commonly is used to estimate the magnitude and rate of settlement or compaction that will occur in fine-grained clay deposits under a given change in stress. According to the theory, compaction in low-permeability, fine-grained deposits results from the transfer of stress from the pore water to the skeleton as pore water gradually is released from the stressed deposits. When the effective stress exceeds the previous maximum stress (preconsolidation stress) on the skeleton, the compaction is essentially permanent and irreversible. Almost all the permanent subsidence in aquifer systems is attributable to the compaction of aquitards during the typically slow process of aquitard drainage (Tolman and Poland, 1940). This concept, known as the *aquitard drainage model*, has formed the theoretical basis of many successful subsidence investigations (Holzer, 1998), and is the approach used to simulate subsidence (Hoffman et al., 2003; Leake and Galloway, 2007) using the MODFLOW groundwater flow model (Harbaugh, 2005).

In compacting aquifer systems, the deformation is generally spread over a large area so that regional-scale lateral (sub-horizontal) strains are small. Locally, however, lateral strains may be large, such as near pumping wells where hydraulic gradients are large, where the aquifer system thins abruptly above inflections in the basement topography of the aquifer system, and near the boundaries of hydrogeologic units with contrasting material properties. The conventional groundwater theory approach to evaluating aquifer-system compaction and subsidence may be suitable for evaluating regional groundwater resources, but it has limitations in evaluating the hazards associated with ground displacements on local scales (Burbey, 2001, 2002). With respect to evaluating hazards, local factors should guide the evaluation and where relevant, the application of multidimensional poroelastic models based on Biot (1941) consolidation theory should be considered.

Underground mining

Subsidence attributed to underground coal mining, generally classified as pit subsidence or sag/trough subsidence (Figure 2), is time dependent with vertical (generally largest) and horizontal components of movement, depending on the type and extent of mining. Pit subsidence is a circular hole in the ground with essentially vertical to belled-outward sidewalls. The diameter of subsidence pits ranges from about 1 to 12 m and generally occurs over shallow mines (depths less than about 50 m) with incompetent bedrock overburden. Sag subsidence is a rectangular depression with gently sloping sides and is typically developed over room-and-pillar mines at greater depths (20–100 m or more) and with more competent overburden. Trough subsidence is similar in surface geometry to sag subsidence.

Pit and sag subsidence occurs more or less randomly and unexpectedly when pillars of coal collapse under overburden loading, typically long after an underground mine has been abandoned (Figure 3). Pit subsidence may involve collapse of only a few pillars, while sag subsidence may involve progressive failure of many pillars. Trough subsidence typically occurs in conjunction with longwall mining, which involves use of moveable hydraulic roof supports that are advanced behind the excavation so the mine roof and overlying rock fracture

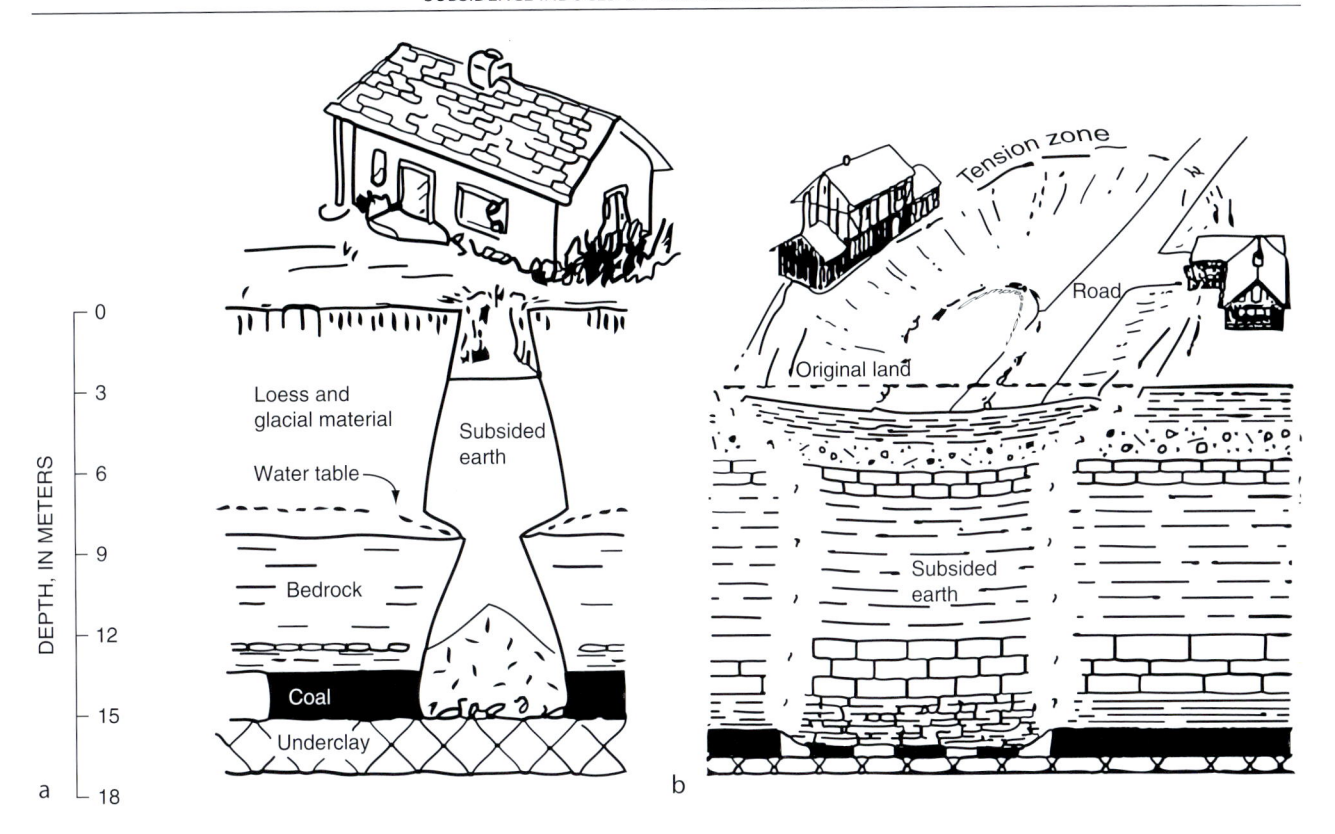

Subsidence Induced by Underground Extraction, Figure 2 Typical types of subsidence associated with underground coal mining: (**a**) Pit subsidence, and (**b**) Sag subsidence (Modified from Bauer and Hunt, 1982).

Subsidence Induced by Underground Extraction, Figure 3 Subsidence depressions and pits above the abandoned Old Monarch Mine (coal) in operation from 1904 to 1921 near Sheridan, Wyoming, USA. Rectangular depressions, some of which are bounded by pits, are evident on the *right* (Photograph by C.R. Dunrud, USGS, May 1978).

and collapse into the void behind the supports. Caving and fracturing propagate up through the overlying rock mass and bulking occurs until the collapsed rock supports the overlying strata. The overlying rock mass subsides and the ground surface ultimately deforms into a trough.

Prediction of mining subsidence involves the use of mathematical models to compute the subsidence profile. Models based on the influence function approach assume that extracting a tiny element of an underground seam, for example, a coal seam, will cause the ground surface to subside into a predefined shape such as a normal probability distribution. The ground surface directly above the extracted element receives the greatest amount of influence. Coal seam elements offset from this location have less influence on that surface point. The final subsidence at this point is the summation of the influence of each mined-out element. Explicit techniques that employ finite-element or distinct-element models are capable of accounting for the engineering characteristics of various rock strata and the orientation of bedding planes and preexisting faults. The more explicit modeling techniques require much more information than the influence models, and the numerous input parameters can make it much more difficult to calibrate to case histories.

Early underground mining was less efficient than more recent underground mining. Subsidence over early mines can occur tens to hundreds of years after mining has ceased, whereas subsidence over more recent mines where virtually total extraction is practiced tends to occur contemporaneously with mining. Subsidence over underground coal workings develops as a gradual downwarping of the overburden into mine voids and is generally unrelated to subsurface water conditions.

Abandoned tunnels and underground mining for metallic ores, limestone, gypsum, and salt contribute a small percentage of the subsidence attributed to underground mining. These mined areas are subject to downwarping of the overburden, but limestone, salt, and gypsum whether undermined or not are susceptible to extensive dissolution by water, frequently leading to sinkholes (see "*Sinkhole*") and in some cases catastrophic collapses (Galloway et al., 1999).

Summary

Subsidence induced by underground extraction is a class of anthropogenic land subsidence that principally is caused by the withdrawal of subsurface fluids (groundwater, oil, and gas) and the underground mining of coal and other minerals. Permanent subsidence can occur when fluids stored beneath the Earth's surface are removed by wells or drains. Most of the identified subsidence attributed to subsurface fluid withdrawal has been caused by groundwater extraction. Underground mining of coal and other minerals has caused widespread subsidence; most of this fraction is associated with underground mining for coal.

Globally, many subsidence areas have been identified, mapped, and documented. Measuring, mapping, and monitoring subsidence are necessary to assess subsidence hazards. Analysis and simulation of subsidence processes, constrained by the available data, often are used to assess present and potential future hazards.

Bibliography

Anon, 1997. *State of the art on land subsidence, Govt. of India*. New Delhi: Indian National Committee on Rock Mechanics and Tunnelling Technology, Govt. of India.

Bauer, R. A., and Hunt, S. R., 1982. Profile, strain and time characteristics of subsidence from coal mining in Illinois. In Peng, S. S., and Harthill, M. (eds.), *Proceedings Workshop on Surface Subsidence due to Underground Mining, November 1981*. Morgantown, WV: West Virginia University, pp. 207–219.

Biot, M. A., 1941. General theory of three-dimensional consolidation. *Journal of Applied Physics*, **12**, 155–164.

Burbey, T. J., 2001. Storage coefficient revisited – is purely vertical strain a good approximation? *Ground Water*, **39**(3), 458–464.

Burbey, T. J., 2002. The influence of faults in basin-fill deposits on land subsidence, Las Vegas, Nevada, USA. *Hydrogeology Journal*, **10**(5), 525–538.

Carnec, C., and Fabriol, H., 1999. Monitoring and modeling land subsidence at the Cerro Prieto Geothermal Field, Baja California, Mexico, using SAR interferometry. *Geophysical Research Letters*, **26**(9), 1211–1214.

Chilingarian, G. V., Donaldson, E. C., and Yen, T. F. (eds.), 1995. *Subsidence Due to Fluid Withdrawal*. Amsterdam: Elsevier Science. Developments in Petroleum Science, Vol. 41, 498 p.

Elder, J. (1981). Geothermal systems. London, Academic Press, 508 p.

Fielding, E. J., Blom, R. G., and Goldstein, R. M., 1998. Rapid subsidence over oil fields measured by SAR interferometry. *Geophysical Research Letters*, **27**, 3,215–3,218.

Galloway, D. L., Jones, D. R., and Ingebritsen, S. E., 1999. *Land Subsidence in the United States*. Reston, VA: U.S. Geological Survey. U.S. Geological Survey Circular, 1182, 175 p. http://pubs.usgs.gov/circ/circ1182/. Cited November 25 2009.

Galloway, D. L., Bawden, G. W., Leake, S. A., and Honegger, D. G., 2008. Land subsidence hazards. In Baum, R. L., Galloway, D. L., and Harp, E. L. (eds.), *Landslide and Land Subsidence Hazards to Pipelines*. Reston, VA: U.S. Geological Survey. U.S. Geological Survey Open-File Report, 2008–1164, pp. 40–106. http://pubs.usgs.gov/of/2008/1164/. Cited November 28, 2009.

Gilluly, J., and Grant, U. S., 1949. Subsidence in the Long Beach Harbor area, California. *Geological Society of America Bulletin*, **60**, 461–530.

González, J. L., and Törnqvist, T. E., 2006. Coastal Louisiana in crisis – subsidence or sea level rise? *EOS Transactions of the American Geophysical Union*, **87**(45), 493–498.

Harbaugh, A. W., 2005. *MODFLOW-2005, the US Geological Survey Modular Ground-Water Model – The Ground-Water Flow Process*. Reston, VA: U.S. Geological Survey. Techniques and Methods, 6-A16, variously paged. http://pubs.usgs.gov/tm/2005/tm6A16/. Cited December 1, 2009.

He, Q., Ye, X., Liu, W., Li, Z., and Li, C., 2005. Land subsidence in the Northern China Plain (NCP). In Zhang, A., Gong, S., Carbognin, L., and Johnson, A. I. (eds.), *Land Subsidence: Proceedings of the Seventh International Symposium on Land Subsidence*. Shanghai: Shanghai Scientific and Technical Publishers, Vol. 1, pp. 18–29.

Hoffman, J., Leake, S. A., Galloway, D. L., and Wilson, A. M., 2003. *MODFLOW-2000 Ground-Water Model – User Guide to the Subsidence and Aquifer-System Compaction (SUB) Package.* Tucson, AZ: U.S. Dept. of the Interior, U.S. Geological Survey. U.S. Geological Survey Open-File Report, 03-233, 46 p. http://pubs.usgs.gov/of/2003/ofr03-233/. Cited January 15, 2010.

Holzer, T. L. (ed.), 1984. *Man-Induced Land Subsidence.* Boulder, CO: Geological Society of America. Reviews in Engineering Geology, Vol. 6, 221 p.

Holzer, T. L., 1998. History of the aquitard-drainage model in land subsidence case studies and current research. In Borchers, J. W. (ed.), *Land Subsidence Case Studies and Current Research. Proceedings of the Dr. Joseph F. Poland Symposium on Land Subsidence.* Belmont, CA: Star Publications. Association of Engineering Geologists Special publication no, 8, pp. 7–12.

Holzer, T. L., and Galloway, D. L., 2005. Impacts of land subsidence caused by withdrawal of underground fluids in the United States. In Ehlen, J., Haneberg, W. C., and Larson, R. A. (eds.), *Humans as Geologic Agents.* Boulder, CO: Geological Society of America. Reviews in Engineering Geology, Vol. 16, pp. 87–99. doi: 10.1130/2005.4016(08).

Holzer, T. L., and Serdengecti, S., 1984. *Man-Induced Land Subsidence.* Boulder, CO: Geological Society of America. Reviews in Engineering Geology, Vol. 6, pp. 23–34.

Howle, J. F., Langbein, J. O., Farrar, C. D., and Wilkinson, S. K., 2003. Deformation near the Casa Diablo geothermal well field and related processes Long Valley caldera, Eastern California, 1993–2000. *Journal of Volcanology and Geothermal Research,* **127**(3–4), 365–390.

IPCC [Houghton, J. T., Ding, Y., Griggs, D. J., Noguer, M., van der Linden, P. J., Dai, X., Maskell, K., and Johnson, C. A. (eds.)], 2001. *Climate Change 2001 – The Scientific Basis. Contribution of Working Group I to the Third Assessment Report of the Intergovernmental Panel on Climate Change.* Cambridge/New York: Cambridge University Press, 881 p.

Ketelaar, G., van Leijen, F., Marinkovic, P., and Hanssen, R., 2005. On the use of point target characteristics in the estimation of low subsidence rates due to gas extraction in Groningen, the Netherlands. In *Proceedings, Advances in SAR Interferometry from ENVISAT and ERS missions, ESA ESRIN,* Frascati, Italy, November 28–December 2, 2005. http://earth.esa.int/fringe2005/proceedings/papers/472_ketelaar.pdf. Cited January 12, 2010.

Leake, S. A., and Galloway, D. L., 2007. *MODFLOW Ground-Water Model – User Guide to the Subsidence and Aquifer-System Compaction Package (SUB-WT) for Water-Table Aquifers.* Reston, VA: U.S. Department of the Interior, U.S. Geological Survey Techniques and Methods, 6–A23, 42 p. http://pubs.usgs.gov/tm/2007/06A23/. Cited January 12, 2010.

Massonnet, D., Holzer, T., and Vadon, H., 1997. Land subsidence caused by the East Mesa geothermal field, California, observed using SAR interferometry. *Geophysical Research Letters,* **24**(8), 901–904.

Martin, J. C., and Serdengecti, S., 1984. Subsidence over oil and gas fields. In Holzer, T.L., (ed.), Man-induced land subsidence. *Reviews in Engineering Geology,* Vol. 6, pp. 23–34.

Mayuga, M. N., and Allen, D. R., 1969. Subsidence in the Wilmington oil field, Long Beach, California, U.S.A. In Tison, L. J. (ed.), *Land Subsidence: Proceedings of the Tokyo Symposium.* Paris: IASH. IAHS 88(1), pp. 66–79. http://iahs.info/redbooks/a088/088013.pdf. Cited January 12, 2010.

Nagel, N. B., 2001. Compaction and subsidence issues within the petroleum industry – from Wilmington to Ekofisk and beyond. *Physics and Chemistry of the Earth,* **26**, 3–14.

National Research Council, 1991. *Mitigating Losses from Land Subsidence in the United States.* Washington, DC: National Academy Press, 58 p.

Phillips, S. P., and Galloway, D., 2008. Groundwater: pumping and land subsidence. In Trimble, S. W., Stewart, B. A., and Howell, T. A. (eds.), *Encyclopedia of Water Science,* 2nd edn. Boca Raton, FL: Taylor and Francis, pp. 466–470. doi: 10.1081/E-EWS2-120010084. http://www.informaworld.com/10.1081/E-EWS2-120010084. Cited February 13, 2009.

Poland, J. F. (ed.), 1984. *Guidebook to Studies of Land Subsidence due to Ground-Water Withdrawal.* Paris: UNESCO. UNESCO Studies and Reports in Hydrology, 40, 305 p. http://wwwrcamnl.wr.usgs.gov/rgws/Unesco/. Cited November 24, 2009.

Poland, J. F., and Davis, G. H., 1969. Land subsidence due to withdrawal of fluids. In Varnes, D. J., and Kiersch, G. (eds.), *Reviews in Engineering Geology.* New York: Geological Society of America, Vol. 2, pp. 187–269.

Saxena, S. K. (ed.), 1978. *Evaluation and Prediction of Subsidence. Proceedings of International Conference on Evaluation and Prediction of Subsidence, Florida.* New York: ASCE.

Saxena, N. C., and Singh, B., 1991. Subsidence research in India. In Singh, B., and Saxena, N. C. (eds.), *Land Subsidence – International Symposium, 11–15 December, 1989. Central Mining Research Station (Dhanbad, India), Council of Scientific & Industrial Research.* Rotterdam: A.A. Balkema, pp. 178–187.

Singhal, B. B. S., and Gupta, R. P., 1999. *Applied Hydrogeology of Fractured Rocks.* Dordrecht: Kluwer, pp. 325–346.

Terzaghi, K., 1925. *Erdbaumechanik auf Bodenphysikalisher Grundlage [Earthworks Mechanics Based on Soil Physics].* Vienna, Austria: Deuticke, 399 p.

Terzaghi, K., and Peck, R. B., 1967. *Soil Mechanics in Engineering Practice,* 2nd edn. New York: Wiley, 729 p.

Tolman, C. F., and Poland, J. F., 1940. Ground-water infiltration, and ground-surface recession in Santa Clara Valley, Santa Clara County, California. *American Geophysical Union Transactions,* **21**(1), 23–34.

Yang, F., Li, X., Diao, C., Lin, L., and Zhang, Y., 2005a. The recent state and cause analysis of land subsidence disaster in Fushun, Liaoning. In Zhang, A., Gong, S., Carbognin, L., and Johnson, A. I. (eds.), *Land Subsidence: Proceedings of the Seventh International Symposium on Land Subsidence.* Shanghai: Shanghai Scientific and Technical Publishers, Vol. 1, pp. 188–195.

Yang, G., Wei, Z., and Huang, C., 2005b. Advances and challenge in research on land subsidence in China. In Zhang, A., Gong, S., Carbognin, L., and Johnson, A. I. (eds.), *Land Subsidence: Proceedings of the Seventh International Symposium on Land Subsidence.* Shanghai: Shanghai Scientific and Technical Publishers, Vol. 1, pp. 3–9.

Yerkes, R. F., and Castle, R. O., 1969. Surface deformation associated with oil and gas field operation in the United States. In Tison, L. J. (ed.), *Land Subsidence: Proceedings of the Tokyo Symposium.* Paris: IASH. IAHS 88(1), pp. 55–64. http://iahs.info/redbooks/a088/088012.pdf. Cited December 16, 2009.

Zhou, Y., Li, B., Chen, J., Yang, J., Hu, F., Du, D., and Don, Q., 2005. Genesis analysis on land subsidence in Daqing Oil Field. In Zhang, A., Gong, S., Carbognin, L., and Johnson, A. I. (eds.), *Land Subsidence: Proceedings of the Seventh International Symposium on Land Subsidence.* Shanghai: Shanghai Scientific and Technical Publishers, Vol. 1, pp. 196–206.

Cross-references

Climate Change
Coal Fire (Underground)
Coastal Erosion
Coastal Zone, Risk Management
Damage and the Built Environment
Hydrocompaction Subsidence
Land Subsidence
Permafrost

Piping Hazards
Sea Level Change
Sinkhole

SUNSPOTS

David H. Boteler
Earth Science Sector Natural Resources Canada, Ottawa, ON, Canada

Definition

Sunspots are dark regions on the Sun's surface. The number of sunspots varies with an approximately 11-year cycle and is a used as an indicator of the level of solar activity that drives space weather.

Discussion

When Galileo pointed his telescope at the Sun and observed dark regions instead of a uniform disk, it triggered an upheaval in thinking about the Sun. We now know that sunspots are regions of strong magnetic fields that inhibit the heat convection from the Sun's interior resulting in regions (4,500 K) that are cooler than the surrounding areas (6,000 K) and so appear darker. The strong magnetic fields within sunspot groups lead to solar flares and eruptions on the Sun producing "space weather," which is a hazard to modern technology. The number of sunspots provides a good indicator of the level of solar activity and the space weather hazard to a variety of systems. Regular drawings of sunspots have been made for many years and used to produce a daily sunspot number calculated from the number of sunspot groups multiplied by ten added to the number of individual spots. The International Sunspot Number is now produced by the Solar Influences Data Centre in Belgium (www.sidc.be). A plot of the sunspot number shows the cycles of solar activity which last between 10 and 14 years with an average length of 11 years. During peak years of the solar cycle, there are increased hazards from magnetic storms to power systems and ionospheric disturbances affecting radio communication and positioning systems.

Bibliography

Clette, F., Berghmans, D., Vanlommel, P., Van der Linden, R., Koeckelenbergh, A., and Wauters, L., 2007. From the Wolf number to the International Sunspot Index: 25 years of SIDC. *Advances in Space Research*, **40**, 919–928.

Van der Linden, R., and Vanlommel, P., (eds.), *Sunspot Bulletin*, Solar Influenes Data Analysis Center, http://sidc.oma.be.

Cross-references

Solar Flare
Space Weather

SUPERNOVA

Lev I. Dorman
Tel Aviv University, Qazrin, Israel
Russian Academy of Sciences, Troitsk, Moscow, Russia

Synonyms

Stellar explosion

Definition

A supernova is a stellar explosion that is extremely luminous and may cause a burst of radiation that often briefly outshines as a whole galaxy, before fading from view over several weeks or months. The explosion ejects the main part of star's material at a velocity of up to 30,000 km/s driving a shock wave into the surrounding interstellar medium. This shock wave sweeps up an expanding shell of gas and dust called a supernova remnant and generates huge fluxes of cosmic rays (CR) with energy up to about several millions GeV (1 GeV = 10^9 eV).

Supernovae as main source of cosmic rays in galaxy and estimation of their average frequency

From the energetic balance of CR in the Galaxy (taking into account the energy density of CR at about 1 eV/cm^3, and an average life of CR in the Galaxy about 30 million years), it follows that the full power for CR production is about 3×10^{27} MW. Now, it is commonly accepted that only supernovae explosions may supply such significant power to be the main source of galactic CR. At each explosion, the average energy transferred to CR is about 10^{43}–10^{44} J. From this, it follows that the average frequency of supernovae explosions must be about 10^{-1}–10^{-2} per year.

Types of supernovae

Most supernovae are of types Ia, Ib, Ic (e.g., type Ia is formatted from double stars with a white dwarf as an accreting star and different masses slightly more than mass of the Sun). Supernovae of type II are formatted from an evolving star of a mass more than about nine times the Sun's mass. In a typical type II supernova, the newly formed neutron core has an initial temperature of about 100 GK, that is, 6,000 times the temperature of the Sun's core. The core temperature of a star of over about 140 solar masses can become so high that photons convert spontaneously to electron–positron pairs, reducing the photon pressure supporting the star's outer layers and triggering a collapse that vaporizes the star. In this case, about 10^{46} J of energy is converted into a ten-second burst of neutrinos, which is the main output of this type of event.

Interstellar impact of supernovae and role in formation of the sun and solar system

Supernovae are a key source of elements heavier than oxygen in the Universe and, particularly, during formation of the Sun and solar system about 4.6 billion years ago. These elements are produced by nuclear fusion (for iron-56 and lighter elements), and by nucleo-synthesis during the supernova explosion for elements heavier than iron.

Nearby supernovae and their possible effects

Supernova between 30 and 1,000 pc, depending upon type and energy output, can be dangerous for the Earth's biosphere, human life, and high-level technology. Gamma rays from a supernova could induce chemical reactions in the upper atmosphere converting molecular nitrogen into nitrogen oxides, depleting the ozone layer enough to expose the surface to harmful solar and cosmic radiation. It was proposed that a nearby Supernova was the cause of the terminal Ordovician extinction, which resulted in the death of nearly 60% of the oceanic life on Earth. It has also been suggested that traces of past supernovae might be detectable on Earth in the form of metal isotope signatures in rock strata. Indeed, iron-60 enrichment was recently found in deep-sea rock of the Pacific Ocean. Type Ia supernovae are thought to be potentially the most dangerous if they occur close enough to the Earth. Because these supernovae arise from dim, common white dwarf stars, it is likely that a supernova that can affect the Earth will occur unpredictably and in a star system that is not well studied. One theory suggests that a Type Ia supernova would have to be closer than a 1,000 pc to affect the Earth. Several large stars within the Galaxy have been suggested as possible supernovae within the next 100 million years (Rho Cassiopeiae, Eta Carinae, RS Ophiuchi, U Scorpii, VY Canis Majoris, Betelgeuse, Antares, Spica, and many Wolf-Rayet stars). The nearest known supernova candidate is IK Pegasi (HR 8210), located at a distance of 150 light-years. In case a nearby supernova should occur, the information obtained from optical, x-ray, γ-ray, and cosmic ray observations will help to produce ALERT on expected different types of natural hazards (including radiation hazard) for people, biosphere, technology (especially for satellites and space-ships with astronauts), so that experts could decide what to do next (see also *Automated Local Evaluation in Real Time (ALERT)* and *Radiation Hazards*).

Cross-references

Automated Local Evaluation in Real Time (ALERT)
Radiation Hazards
Solar Flare
Space Weather

SURGE

Giovanni Cuomo
Hydraulics Applied Research & Engineering Consulting (HAREC) s.r.l., Rome, Italy

Synonyms

Sea level rise; Wave setup

Definition

A surge is a relatively long lasting increase in water level. Surges result from the combination of a series of factors including: sea level rise, astronomical tide, wind setup, atmospheric pressure gradient, wave setup, tsunami, and seiches. These phenomena become more evident in shallow water and coastal regions where their effects are enhanced by local bathymetric and topographic features.

Discussion

The (eustatic) sea level rise results from melting of the polar ice caps and thermal expansion of the water mass with temperature change. Observation suggests that it has occurred at a mean rate of 1.8 mm/year for the past century with more recent estimates (1993–2003) indicating an actual rise of 3 mm/year.

Astronomical tides are periodic oscillations of the sea level caused by the gravitational attraction of the Moon, the Sun, and the other astronomical bodies acting on the rotating Earth. Tides are the largest sources of short-term sea-level fluctuations and generate level variations that can exceed 10 m; their prediction is therefore essential to ensure safety of coastal navigation and the correct design of coastal structures. The world's largest tides have been recorded in Minas Basin, Nova Scotia, with a maximum tidal range recorded at 16.8 m (54.6 ft).

Wind setup is an increase of the water level due to the action (shear stress) exerted by inshore wind over the water surface and can build up the water level by up to 3 m in height. An initial estimate of the setup Δh of the water level induced by the wind over a distance Δx is given by:

$$\Delta h = K \frac{U^2}{gD} \Delta x$$

where $K = 3.2 \times 10^{-6}$ is a constant, U is the wind speed, g is the gravitational acceleration, and D is the water depth.

Similarly, wave setup is the increase in water level forced by the onshore mass transport of water by wave action. In the surf zone, wave setup can reach up to 30% of the incident (deepwater) wave height.

A barometric surge (Δh) is induced by the passage of a low atmospheric pressure front to balance the reduction of hydrostatic pressure (Δp) exerted by the air column on the water surface:

$$\Delta h = \frac{\Delta p}{\rho g}$$

where ρg is the specific weight of the water body.

Wind-induced setups are usually accompanied by a fast movements of a pressure front (for example during tropical cyclones) and major wave storms, and have caused some of the most disastrous flooding and coastal catastrophes in the world's history including: the 1990 flood in Bangladesh, the 1953 flood in UK and the Netherlands, and the 2005 flood of New Orleans during passage of Hurricane Katrina. The highest storm surge noted in historical accounts was 13 m (43 ft) during the passage of the 1899 Cyclone Mahina over Bathurst Bay, Australia.

Tsunami, seiches, and other long period waves can also cause a sudden rise of the water surface. The largest ever observed tsunami was triggered on July 9, 1958 by an earthquake generated landslide in Lituya Bay, originating a giant wave measuring 524 m (1,719 ft) in height.

Bibliography

Harris, D. L., 1963. *Characteristics of the Hurricane Storm Surge*. Technical Paper No. 48. U.S. Dept. of Commerce, Weather Bureau, Washington, DC, 139 pp. Available online at: http://www.csc.noaa.gov/hes/images/pdf/CHARACTERISTICS_STORM_SURGE.pdf

Wiegel, R. L., 1964. Tsunamis, storm surges, and harbor oscillations. In *Oceanographical Engineering*. Englewood Cliffs, NJ: Prentice-Hall, Chap. 5, pp. 95–127.

Cross-references

Breakwater
Coastal Erosion
Coastal Zone, Risk Management
Displacement Wave
Fetch
Hurricane Katrina
Hydrograph, Flood
Marine Hazards
Rogue Wave
Seiche
Storm Surges
Tidal Bores
Tsunami

SUSCEPTIBILITY

María José Domínguez-Cuesta
Universidad de Oviedo, Oviedo, Spain

Synonyms

Lack of resistance; Hazardousness of place; Spatial hazard

Definition

Susceptibility means "the state of being susceptible" or "easily affected." In natural hazards terms, susceptibility is related to spatial aspects of the hazard. It refers to the tendency of an area to undergo the effects of a certain hazardous process (e.g., floods, earthquakes, tsunamis, subsidence, etc.) without taking into account either the moment of occurrence or potential victims and economic losses.

Discussion

Susceptibility linked to slope instabilities, for instance, indicates the tendency of an area to breakdown. According to Brabb (1984), susceptibility is the probability of an event happening in a specific zone, depending on the correlation of the instability-determining factors with the distribution of the past movements.

By applying different methods, including heuristic, probabilistic, nearest neighbor distance, matrix, or density methods among others, it is possible to draw up susceptibility maps that classify a territory into classes depending on the probability of undergoing future hazardous events. Good examples of sinkhole susceptibility maps can be seen in Galve et al. (2009).

Bibliography

Brabb, E. E., 1984. Innovative approaches to landslide hazard and risk mapping. In *Proceedings of the Fourth International Symposium on Landslides*, Toronto, Vol. 1, pp 307–324.

Galve, J. P., Gutiérrez, F., Remondo, J., Bonachea, J., Lucha, P., and Cendrero, A., 2009. Evaluating and comparing methods of sinkhole susceptibility mapping in the Ebro Valley evaporite karst (NE Spain). *Geomorphology*, **111**, 160–172.

Cross-references

Coping Capacity
Emergency Management
Hazardousness of Place
Historical Events
Models of Hazard and Disaster

T

CASE STUDY

TANGSHAN, CHINA (1976 EARTHQUAKE)

Zhengwen Zeng[1], Chenghu Wang[1,2]
[1]University of North Dakota, Grand Forks, ND, USA
[2]China Earthquake Administration, Beijing, China

Introduction
Seismic characteristics

The 1976 Tangshan Earthquake occurred at 03:42:56 a.m. on July 28, 1976 (local time). The epicenter was on Jixiang Road, Lunan District, Tangshan City, Hebei Province, China (Figure 1); geographic coordinates is 39.63°N 118.18°E. The focal depth was about 11 km. The sequence of the 1976 Tangshan Earthquake included five major events: two of magnitude >Ms 7, and three between Ms 6 and Ms 7. The Main seismic parameters of Tangshan Earthquake are as shown in Table 1, and the statistics of the earthquake sequence is in Table 2.

Tectonically, the Tangshan Earthquake occurred at the intersection of the east–west active faulting belt of Yinshan-Yanshan Mt. and the NNE active tectonic belt of North China Plain. Tangshan Earthquake ruptured 100 km (62 mile) along the Tangshan Fault, a right-lateral strike-slip fault in the north-northeastern direction. The epicenter area experienced 10 km (6 miles) extensive faulting. The fault ran through the city centre of Tangshan with a horizontal displacement up to 1.5 m (5 ft). During the main shock, much of northern China felt the tremors. The disaster-affected areas including Tangshan City and Tangshan Prefecture of Hebei Province, Tianjin Municipality, Beijing (Peking) Municipality, and adjacent areas. The most severely affected areas were Tangshan City, Tangshan Prefecture, and Tianjin Municipality. These three areas account for 99% of all casualties and property losses. Figure 2 is the intensity contour map of the Tangshan Earthquake (Guo, 2008).

Overview of the affected areas before the earthquake
Tangshan

Tangshan included two independent administrative units: Tangshan City and Tangshan Prefecture; both were under the direct administration of Hebei Province.

Tangshan City is located in the northeast of the province. In 1976, it was composed of two downtown districts (Lunan and Lubei), one industrial district (Kailuan Mine), and one suburban district (Kaiping). It was an important industrial city in the province; its industrial production output value was the second largest in the province. By 1976, Tangshan City had also become a very important part of Beijing-Tianjin-Tangshan economic zone. As of 1975, the gross industrial production of Tangshan City reached 2.2 billion RMB (1.2 billion USD), which accounted for nearly 1% of the national gross industrial production of China at that time.

Tangshan Prefecture included one city (Qinghuangdao) and 12 counties (Changli, Fengnan, Fengrun, Funing, Leting, Luanxian, Luan'nan, Lulong, Qian'an, Qianxi, Yutian, and Zunhua).

Tianjin

In 1976, Tianjin was one of the three municipalities directly under the jurisdiction of the central government in China. It is located in the northeastern part of North China Plain, close to Yanshan Mt. in the north, and adjacent to the Bohai Bay in the east. The entire city was composed of six downtown districts (Hebei, Hedong, Heping, Hexi, Hong Qiao, and Nakai), two industrial districts (Hangu and Tanggu), four suburban districts (Beijiao, Dongjiao, Nanjiao, and Xijiao), and five rural counties (Baodi, Jinghai, Jixian, Ninghe, and Wuqing). By population, Tianjin was the third

Tangshan, China (1976 Earthquake), Figure 1 Location of the 1976 Tangshan Earthquake.

Tangshan, China (1976 Earthquake), Table 1 Seismic parameters of the five major quakes

Date of occurrence	Time (local)	Latitude	Longitude	Depth (km)	Magnitude (Ms)	Epicenter	Intensity at epicenter
July 28, 1976	03:42:56	39.63°N	118.18°E	11	7.8	Downtown, Tangshan	XI
July 28, 1976	07:17:32	39.45°N	117.00°E	19	6.2	No data	No data
July 28, 1976	18:45:37	39.83°N	118.65°E	10	7.1	Shangjialin, Luanxian	IX
November 15, 1976	21:53:01	39.40°N	117.70°E	17	6.9	Ninghe, Tianjin	VIII–IX
May 12, 1977	19:17:54	39.40°N	117.80°E	18	6.2	No data	No data

Tangshan, China (1976 Earthquake), Table 2 Statistics of Tangshan Earthquake sequence (July 28, 1976–December 31, 1979)

Magnitude, Ms	1.0 ~ 1.9	2.0 ~ 2.9	3.0 ~ 3.9	4.0 ~ 4.9	5.0 ~ 5.9	6.0 ~ 6.9	7.0 ~ 7.9
Frequency	12,290	5,387	933	279	25	3	2

largest city in China, following Shanghai and Beijing. The gross industrial output value in 1975 was up to 14.4 billion RMB (7.7 billion USD). At that time, Tianjin had become one of the largest industrial, trading, finance, and communication centers in North China.

Casualties

The accuracy of casualties have been investigated and discussed by many specialists all over the world since the earthquake, the most recent research findings show

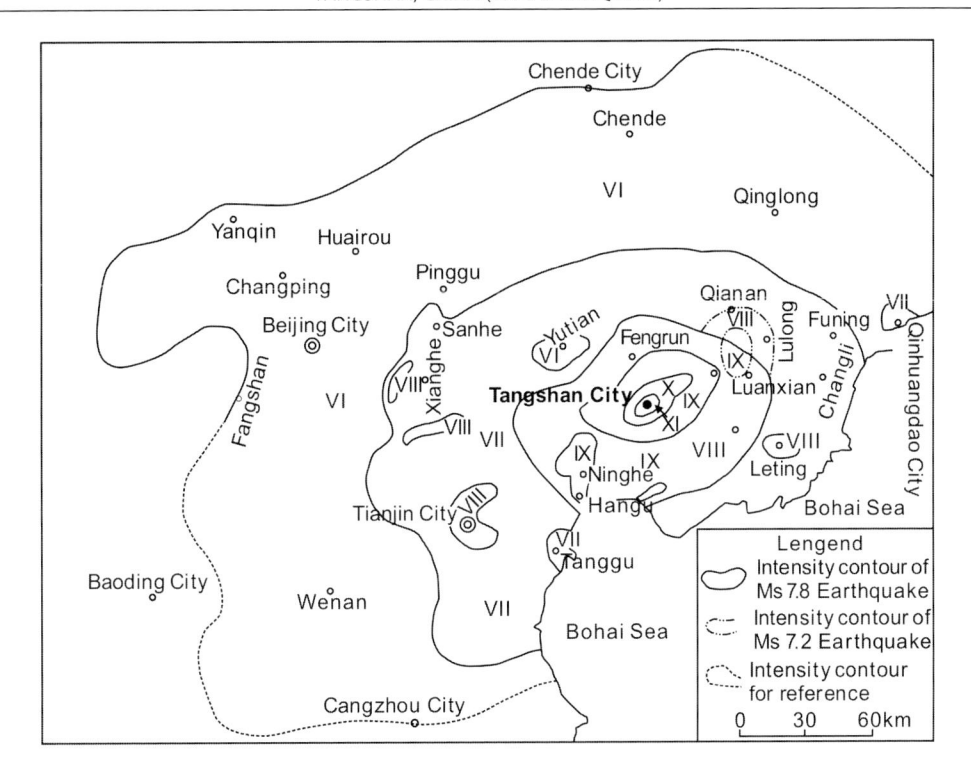

Tangshan, China (1976 Earthquake), Figure 2 Intensity contour map of Ms 7.8 Tangshan Earthquake.

Tangshan, China (1976 Earthquake), Table 3 Casualties of Tangshan Earthquake

Area	Dead	Severe injury	Light injury	Subtotal
Downtown Tangshan City	135,919	81,630	257,384	474,933
Tangshan Prefecture	69,065	63,620	284,079	416,764
Visitors in Tangshan	12,248	–	–	12,248
Tianjin	24,398	21,874	–	46,272
Beijing and other adjacent areas	839	–	–	839
Total	242,469	167,124	541,463	951,056

that the Tangshan Earthquake had caused the death of 242,469 and 730,876 injuries across China, as detailed in Table 3 (Guo, 2008; Zou et al., 1990).

Statistics of casualties in Tangshan city

Pre-earthquake population of Tangshan City was 1,062 million. Due to the earthquake, 135,919 died, accounting for 12.8% of the total population; 81,630 people were severely injured, accounting for 7.7% of the total population; and the ratio of death to injury was 1.7. Lunan District of Tangshan City was the meizoseismal area; 34,089 people died, at a death rate of 26.7%. In Lubei District, the death rate was 16.4%; in suburban area, the death rate dropped to 10.2% elsewhere. Tables 4–7 show more detailed statistics on earthquake casualties. Table 5 shows that the death ratio between male and female was 1:1.21; the number of dead females was much greater than that of males. One of the reasons for that was there were more

Tangshan, China (1976 Earthquake), Table 4 Casualties in Tangshan City

Casualty	Number	Percentage (%)
Original population	1,061,926	
Dead	135,919	12.8
Severe injury	81,630	7.7
Slight injury	257,384	24.2
Original household	294,247	
Destroyed household	7,210	2.5

Tangshan, China (1976 Earthquake), Table 5 Casualties by gender in Tangshan City

Gender	Dead	Percentage (%)	Severe injury	Percentage (%)
Male	61,423	45.19	41,770	51.17
Female	74,496	54.81	39,860	48.83

Tangshan, China (1976 Earthquake), Table 6 Fatalities by age in Tangshan City

Age	All age	<7	8–15	16–25	26–40	41–55	>56
Number	135,919	20,044	26,950	22,403	22,197	22,326	21,999
Percentage (%)	100	14.8	19.8	16.5	16.3	16.4	16.2

Tangshan, China (1976 Earthquake), Table 7 Fatalities by occupation in Tangshan City

Occupation	All	Industrial worker	Official	Peasant	Student	Others
Number	135,919	33,684	5,193	19,749	32,261	45,032
Percentage (%)	100	24.8	3.8	14.5	23.8	33.1

Tangshan, China (1976 Earthquake), Table 8 Casualties in Tianjin

Items		Original number (ON)	Loss in Earthquake (LE) Total	Dead	Severe injury	Ratio of LE to ON (%) Total	Dead	Severe injury
Household	Total	1,589,626	31,545	–	–	1.98	–	–
	Agricultural	749,042	24,647	–	–	3.29	–	–
	Nonagricultural	840,584	6,898	–	–	0.82	–	–
Overall population	Total	7,012,756	56,272	34,398	21,874	0.80	0.49	0.31
	Agricultural	3,548,465	35,500	18,172	17,336	1.0	0.51	0.49
	Nonagricultural	3,464,291	20,772	16,226	4,538	0.59	0.46	0.13

Data source: Statistic Bureau of Tianjin, China

male than female workers in the night shift when the earthquake occurred.

Statistics of casualties in Tianjin

In total 24,398 people died from the main earthquake of Ms 7.8 on July 28, 1976, and the strongest aftershock of Ms 7.2 in the afternoon on the same day, accounting for 0.49% of the entire population in Tianjin at that time (Guo, 2008; Zou et al., 1990). In addition, 21,874 people were severely injured, accounting for 0.31% of the population in the municipality. The ratio between death and severe injury was 1.12. There were 1,589,626 households in Tianjin; 31,545 households experienced dead or injured family members, accounting for 1.98% of the entire households (Table 8).

Economic loss

The high-intensity area of Tangshan Earthquake included Tangshan City and suburban, and Tianjin and suburban. The economic loss was concentrated in these areas.

Economic loss in Tangshan City

According to detailed investigation into Tangshan City downtown districts (Lubei and Lunan), industrial district (Kailuan Coal Mine), suburban district (Kaiping), and the City's orchards, the total economic loss caused by the earthquake was about 2.8 billion RMB (1.4 billion USD), as shown in Table 9 (Guo, 2008; Zou et al., 1990).

Before the earthquake, buildings and infrastructures in Tangshan City were built according to totally different codes and standards to meet diversified needs. Most were below poor construction standards. The entire city was virtually non-earthquake-resistant. The downtown districts were severely damaged. The damage rates of urban and rural buildings were 96% and 91%, respectively, as shown in Table 10 (Guo, 2008; Zou et al., 1990).

Economic loss in Tianjin

Based on investigation of the six urban districts (Hebei, Hedong, Heping, Hexi, Hongqiao, and Nankai) and the two industrial districts (Hangu and Tanggu), the total direct economic loss of Tianjin was 6.2 billion RMB

Tangshan, China (1976 Earthquake), Table 9 Economic loss of Tangshan City

Items	Fixed assets	Current assets	Houses	Personal valuables	Production interruption	Incomplete projects	Total
Value (billion RMB)	0.4	0.1	0.2	0.1	1.9	0.1	2.8

Data source: Statistic Bureau of Tianjin, China

Tangshan, China (1976 Earthquake), Table 10 Loss of urban and rural houses in Tangshan

Location	Original houses m²	Units	%	Destroyed houses m²	Units	%	Damaged houses m²	Units	%
All	10,932,272	683,267	100	10,501,056	656,316	96	431,216	26,951	4
Urban	5,448,960	340,560	100	5,419,568	338,723	99	29,392	1,837	1
Rural	5,483,312	342,707	100	5,081,488	317,593	93	401,824	25,114	7

Tangshan, China (1976 Earthquake), Table 11 Economic loss of Tianjin

Item	Total	Fixed assets	Current assets	Houses	Personal valuables	Industrial production
Value (billion RMB)	6.2	2.3	0.2	0.6	0.2	2.9

(3.2 billion USD), as shown in Table 11 (Guo, 2008; Zou et al., 1990).

According to statistical data, more houses in rural areas were destroyed or severely damaged than that in urban areas. In rural areas of Tianjin City, over 50% of the houses collapsed, and more than 30% of the houses were severely damaged.

Earthquake prediction and preparation
Seismic zoning and earthquake-resistant design

Before the 1976 earthquake, Tangshan was a city with considerable low-earthquake-resistance capabilities. According to the 1957 version of "China Earthquake Intensity Table" and "China Seismic Zoning Map," Tangshan was classified as seismic intensity VI area, because no big earthquakes had occurred in that region in documented history. By China's building code at that time, earthquake-resistant design was not compulsory in areas of seismic intensity VI and below. Therefore, before 1976 most buildings, old and new, in Tangshan did not have earthquake-resistant capacity.

By 1976, Tangshan was a middle-sized city growing from the integration and expansion of small towns and rural villages. Before the 1976 earthquake, buildings and infrastructures were built following different construction standards in different periods of time. Until 1949, when the People's Republic of China was founded, most downtown buildings were single-story houses owned by individual inhabitants; and coal miners were living in extremely simple work sheds. During the 1950s, Kailuan Coal Mine and other major state-owned enterprises built many single-story brick-and-stone residential houses for to be rented by their employees. Since the end of 1950s, local government and enterprises started building multi-level brick-and-concrete office buildings, most were 3 to 4 stories, some up to 7 to 8 stories; industrial buildings erected in that period of time were mainly single-story factory workshops, a few were frame structured. Except for some of the multi-story buildings constructed after 1960s which had been designed to resist seismic intensity of VI, the rest of Tangshan buildings were not designed for earthquake resistance. Therefore, Tangshan was virtually a city without any earthquake-resistance measures designed for such a strong earthquake.

Earthquake prediction

Scientific study of earthquake prediction in Tangshan started in 1968, 2 years after the destructive Xintai earthquake(s). In 1975, China predicted the Ms 7.3 Haicheng earthquake in which evacuation was ordered 1 day in advance. After this successful earthquake prediction, the first in human history, monitoring and prediction work for possible earthquakes in Tangshan and the nearby area was greatly intensified due to its closeness to Haicheng. At the beginning of 1976, the Tangshan Earthquake Administration Office issued its annual prediction stating that "between July and August or later in the year, strong earthquakes of Ms 5–7 may occur within 50 km around Tangshan City." Due to this prediction, many local governments and organizations established earthquake

earthquake deformation cycle (Figure 1). After the earthquake, crustal deformation continues at a slow rate during the interseismic part of the earthquake deformation cycle, until finally stress again exceeds fault strength and another earthquake occurs. Interseismic deformation is described later under Tectonic Hazards.

Vibratory ground shaking (strong ground motion)
When earthquake elastic waves reach the Earth's surface they produce vertical and horizontal accelerations of the surface and of all objects that lie on or beneath the surface (buildings, roads, bridges, pipelines, etc.). These waves are strongest nearest the earthquake epicenter and weaken with distance. Motions can be further amplified by certain geological site conditions (e.g., soft sediments overlying bedrock) or topographic site conditions ("whipping" of hilltops and ridge crests).

Fault surface rupture
In large earthquakes (Moment Magnitude [Mw] greater than 6.5–7; see *Magnitude Measures*), displacement on the fault plane is so great that it propagates upward to the ground surface and displaces the ground surface along the mapped fault trace (Figure 2). Fault surface rupture increases in height and length with increasing earthquake magnitude. Vertical displacement along a normal or reverse fault can create a surface fault scarp as high as 10 m along the fault trace (e.g., 1897 Assam, India earthquake). On strike-slip faults, the largest historic coseismic displacement was 18.7 m during the 1855 Wairarapa earthquake, New Zealand (pure horizontal, right-lateral movement).

Coseismic geodetic changes (uplift, subsidence)
During an earthquake the large crustal blocks on either side of the coseismic fault slip rapidly past each other, and also bend (rebound) to try to regain their original, pre-strain shape (see *Elastic Rebound Theory*). This rapid coseismic rebound causes vertical movements of the crust far away from the fault during very large earthquakes. For example, in the 1964 Mw 9.2 Alaska earthquake, an area of 150,000 km^2 landward of the coast subsided up to 6 m (Figure 3) and a corresponding area of 150,000 km^2 offshore rose up to 30 m (see *Subduction*).

Seismic hazards, how processes create hazards
Vibratory ground shaking (strong ground motion)
Every earthquake produces some ground motion, but for the vast majority of earthquakes on the planet, this motion is so small that it can only be measured by sensitive seismographs. Humans seldom report earthquakes as "felt" if the magnitude is less than 2.0, and the ground acceleration less than about 0.02 g. In general, earthquakes smaller than about M5.0 rarely generate strong enough ground motions to cause damage, even near the epicenter. Lee et al. (2003) suggest that an acceleration of 0.1 g is the approximate threshold for damage to old structures and non-earthquake-resistant structures.

Tectonic and Tectono-Seismic Hazards, Figure 2 The magnitude 6.9 earthquake of October 28, 1983, near Borah Peak, Idaho, produced this normal-oblique fault scarp. The scarp and fracture zone are located on Rock Creek, near Dickey, Idaho, and Double Springs Pass Road. The wooden pole is 1.9 m high. This normal fault shows characteristics of oblique slip, where there is both vertical and horizontal displacement. The horizontal movement is left-lateral. The fault scarp extended for more than 35 km, with vertical displacements up to 2.5 m observed between MacKay and Challis, Idaho (Photo credit: G. Reagor, U.S. Geological Survey. Source: NGDC, 1997).

Strong earthquake ground motion can exceed the design tolerances of the structures and can cause damage of various sorts (see *Earthquake Damage*). This exceedance often occurs because the dominant period of the seismic waves matches either the resonant period of the structure, causing amplified resonant motion in the structure, or the resonant frequency of the structure's geological foundation. Most of the deaths, injuries, and damage in historic earthquakes have been caused by collapse of buildings due to strong ground acceleration (Figure 4). This is particularly common for unreinforced masonry buildings (see *Unreinforced Masonry Building*). Vibratory ground motion also creates a suite of "secondary hazards" (described later).

Fault surface rupture
Fault surface displacement will rupture any structure located on or across the fault trace, such as a building, road, bridge, dam, buried utility lines, etc. For buildings,

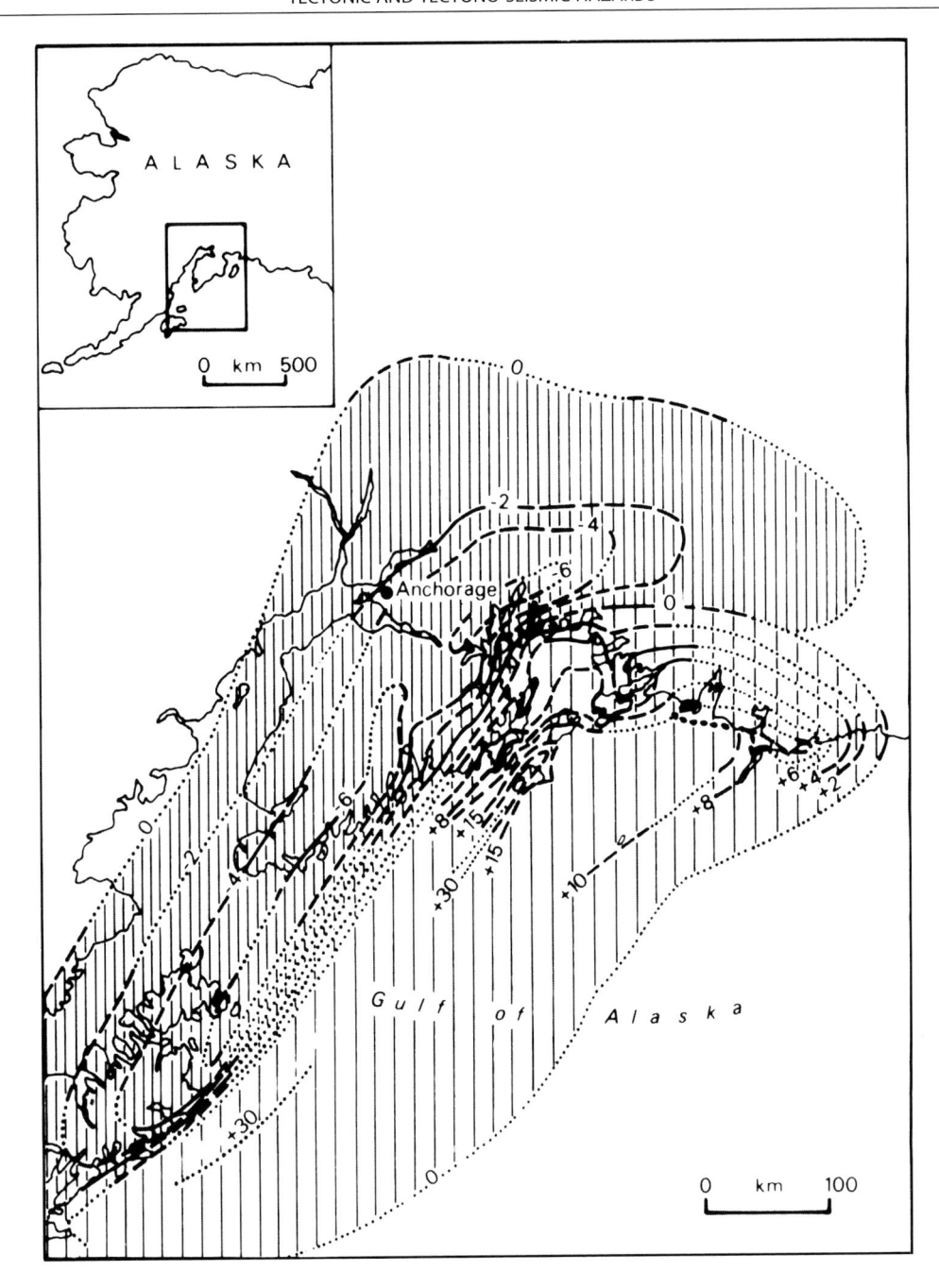

Tectonic and Tectono-Seismic Hazards, Figure 3 Regional coseismic deformation produced by the Mw 9.2 1964 Alaskan earthquake. Regional subsidence (*closely spaced lines*) and uplift (widely *spaced lines*) encompassed most of the forearc along the length of the rupture. The axis of maximum regional subsidence and maximum regional uplift were located near the zero isobase (no land-level change). Measurements are in feet (From Vita-Finzi (1986), after data in Plafker (1969)).

vertical displacement of more than 30 cm typically causes collapse; smaller displacements can often be repaired (Youd, 1980). Buried utility lines can break by decimeter to meter-scale displacements, either vertical or horizontal. Roads have the greatest resilience to fault surface rupture, because unpaved roads can simply be rebuilt.

Coseismic geodetic changes (uplift, subsidence)

Coseismic geodetic changes cause the most damage along shorelines of oceans or lakes because they cause a shift in the area of inundation. Coastal regions that permanently subside are invaded by water, which can submerge coastal buildings, roads, and docks. In coastal regions that

Tectonic and Tectono-Seismic Hazards, Figure 4 Collapse of diaphragms (floors) leaving walls standing in building in Leninakan, Armenia, during the 1988 Spitak earthquake (M 6.9). This three story building is one of the older buildings. Most of the newer ones and two story dwellings and two and three story commercial buildings in Leninakan received little damage. The debris of a nine story building is to the right. Such collapses left no spaces for occupant survival. One hundred and thirty-two nine story precast-concrete-frame buildings collapsed or were heavily damaged in this city (Photograph Credit: U.S. Geological Survey (C.J. Langer). Source: NGDC, 1997).

permanently rise (uplift), docks and harbor facilities become stranded above the water line and become unusable (Figure 5).

Secondary hazards

Vibratory ground motion during the earthquake can cause ground failures such as landslides (see *Landslide*; *Mass Movement*); lateral spreads (a type of low-gradient landslide caused by liquefaction) (see *Lateral Spreading*); snow avalanches (see *Avalanches*); and liquefaction (see *Liquefaction*). Because these ground failures require high ground accelerations, they occur mainly in the epicentral region, but in very large earthquakes the epicentral region can be significant in size. For example, in the Alaska earthquake of 1964, landslides (Figure 6) were triggered in an area of 300,000 km^2. Likewise, liquefaction can cause sudden loss of foundation strength and foundation bearing failures in the epicentral area (Figure 7). In such cases, even though the building withstands the strong ground shaking during the earthquake and does not collapse, it subsides or tilts to such a degree it becomes unusable.

In many earthquakes, relief efforts are obstructed by secondary ground failures (landslides, rockfalls, lateral spreading), which block roads and railroads into the epicentral area. Added to this is the breakage of critical utility lines (water and gas pipelines, electrical lines), also due to secondary ground failures. The lack of water supply, electricity, and heat further exacerbates the so-called "follow-on" hazards such as fire and disease.

Coseismic geodetic changes can cause tsunamis (see *Tsunami*; *Indian Ocean Tsunami, 2004*) and seiches (see *Seiche*). A estimated 228,000 persons were killed or missing (and presumed dead) after the 2004 Sumatra earthquake (Mw 9.1), mostly due to a tsunami that reached a height near 35 m along the Sumatran coast, and more than 5 m on the east coast of India, 2,000 km from the epicentral region.

Numerous "Follow-on" Secondary Hazards typically occur during the aftermath of a large earthquake. These include fire (see *Fire and Firestorms*); disease (see *Epidemiology of Disease in Natural Disasters*); looting, etc. Although these hazards are not strictly tectonic hazards, they are ultimately caused by a tectonic hazard.

Seismic hazards, the spatial distribution

The worldwide distribution of earthquakes and volcanoes shows a strong spatial association with plate boundaries (Figure 8; see Plate Tectonics). Specifically, divergent plate boundaries (mid-ocean spreading centers and their on-land extensions) are characterized by extensional tectonic faulting, shallow earthquakes, and basaltic volcanism (e.g., Iceland on the Mid-Atlantic Ridge). In contrast, convergent plate boundaries are typified by reverse and thrust faulting, subduction, folding, and

Tectonic and Tectono-Seismic Hazards, Figure 5 Uplifted dock, Prince William Sound, Alaska. The dock on Hinchinbrook Island, Prince William Sound, shows uplift. Land in this area rose about 8 ft (2.4 m) during the earthquake, and the dock could then be used only at extremely high tides. Tsunami damage also occurred in this area (Photo credit: U.S. Geological Survey, Menlo Park, CA. Source: NGDC, 1997).

Tectonic and Tectono-Seismic Hazards, Figure 6 The Government Hill Elementary School in Anchorage, Alaska was torn apart by subsidence of the graben at the head of the Government Hill landslide, during the 1964 M 9.2 earthquake. The south wing of the school dropped about 30 ft (9 m); the east wing split lengthwise and collapsed. The playground became a chaotic mass of blocks and fissures. Part of this slide became an earth flow that spread 150 ft (45.5 m) across the flats into the Alaska Railroad yards. During the earthquake, the shaking loosened clay beneath Government Hill and the clay began to move toward the flats. On the hill, 400 ft (121.2 m) back from the rim of the bluff, the earth cracked on a front 1,180 ft (357.6 m) wide (Photo credit: U.S. Geological Survey, Menlo Park, CA. Source: NGDC, 1997).

Tectonic and Tectono-Seismic Hazards, Figure 7 Aerial view of leaning apartment houses in Niigata, Japan, produced by soil liquefaction during the June 14, 1964 Niigata earthquake (M 7.4). Most of the damage was caused by cracking and unequal settlement of the ground such as is shown here. About one third of the city subsided by as much as 2 m as a result of sand compaction. The magnitude 7.4 earthquake killed 26 people and destroyed 3,018 houses and moderately or severely damaged 9,750 in Niigata prefecture (Source: NGDC, 1997).

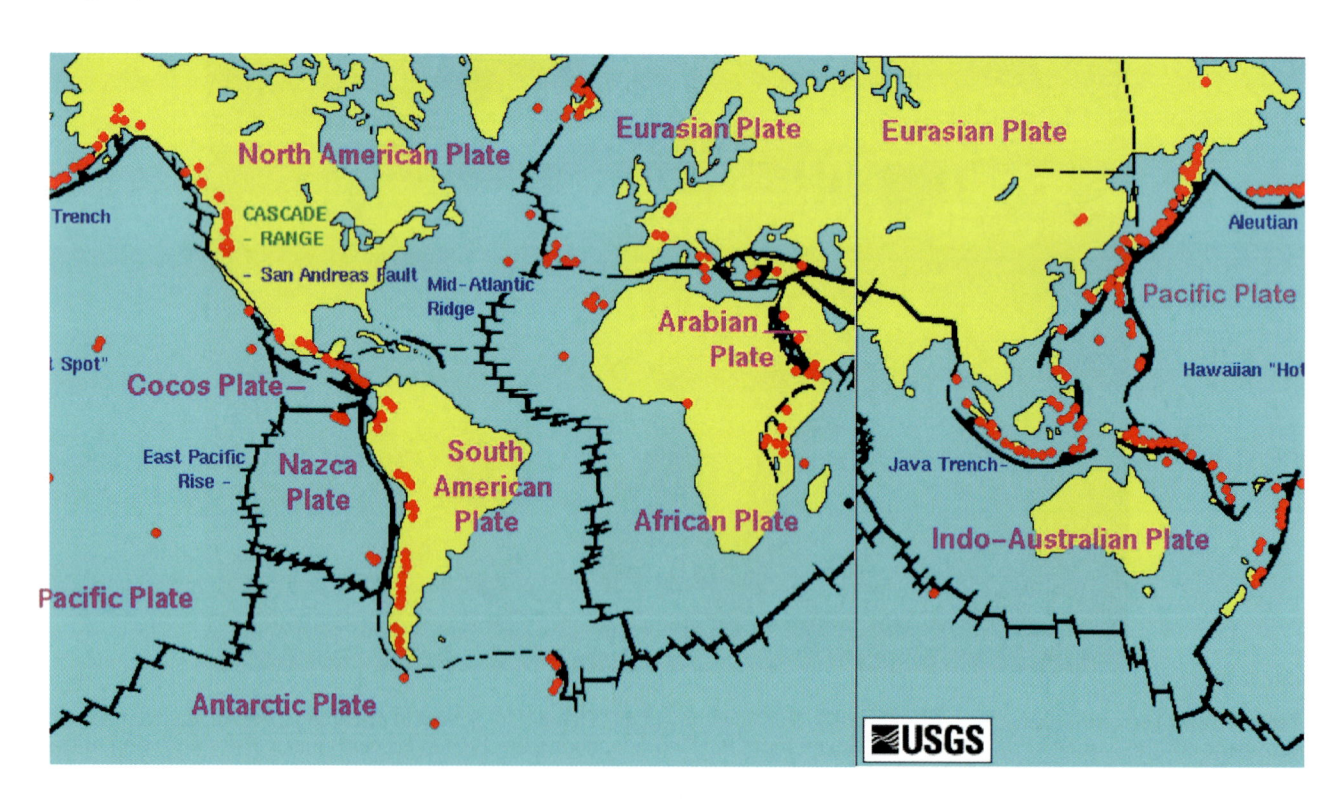

Tectonic and Tectono-Seismic Hazards, Figure 8 World map of major tectonic plates (names in *purple*), divergent plate boundaries (*black lines* segmented by numerous cross-faults), convergent plate boundaries (*black lines* with *triangles*), and active volcanoes (*red circles*) (Source: U.S. Geological Survey, Cascades Volcano Observatory).

explosive island-arc volcanism (e.g., west coast of South America and the Andes volcanic chain).

Within the interiors of the major tectonic plates, both earthquakes and volcanoes are less abundant. This is particularly true of the so-called "stable continental regions," the core areas of continents composed of Precambrian basement rocks, such as the Canadian and Fennoscandian Shields. However, many major tectonic plates also contain sub-plates (microplates) in which contemporary deformation is concentrated. An example is the Basin and Range extensional province of the North American plate.

Figure 9 shows the uneven spatial distribution of worldwide seismic hazards. Large parts of the continental interiors of North America, South America, Africa, and northern Asia have very low seismic hazard (white areas in Figure 9). Surrounding these areas are bands of low seismic hazard (green). The high hazard areas (pink, red, brown) comprise narrow bands parallel to major active faults at plate boundaries. Thus, high hazard zones on the western margins of North and South America coincide with the ocean-continent subduction and transform faults of the eastern margin of the Pacific plate. The high hazard zones extending from southern Europe to southeast Asia coincide with the great zone of ongoing continent-continent collision between African and Indo-Australian plates to the south, and the Eurasian plate to the north. High hazard zones of the western Pacific rim (from Kamchatka, south to New Zealand) coincide with ocean-ocean subduction and transform faults. The moderate seismic hazard in eastern Africa coincides with the East African rift zone, a zone of intra-continental spreading (incipient divergent plate boundary). Other anomalous areas include Australia, which displays moderate seismic hazard despite being composed of older plate interior rocks.

Seismic hazards, the temporal distribution

In those regions where most of the seismic hazard emanates from a single fault, the true seismic hazard in the near future may depart significantly from the hazard shown in Figure 9. This arises due to the temporal character of the earthquake deformation cycle, in which stress on a fault builds up over time, eventually culminating in a large earthquake, after which stress falls to low levels and must rebuild until the next earthquake. From this periodic model of large earthquake occurrence, it follows that each fault has a "memory" of when it last ruptured. The seismic hazard on such a fault varies through time, being the lowest just after a major earthquake (near-total stress release), and progressively higher until the next earthquake.

The seismic hazard values shown in Figure 9 do not reflect the fact that faults have a "memory." Thus, the accelerations shown are derived from long-term average slip rates on the faults, with no knowledge of when the latest earthquake occurred on them. Such a seismic hazard model is "memoryless," and assumes that the probability of a large earthquake occurring on a fault is the same in every year within the earthquake deformation cycle, from the year after a large earthquake, to hundreds or thousands of years later. The latest generation of seismic hazard models do incorporate memory for selected faults, and

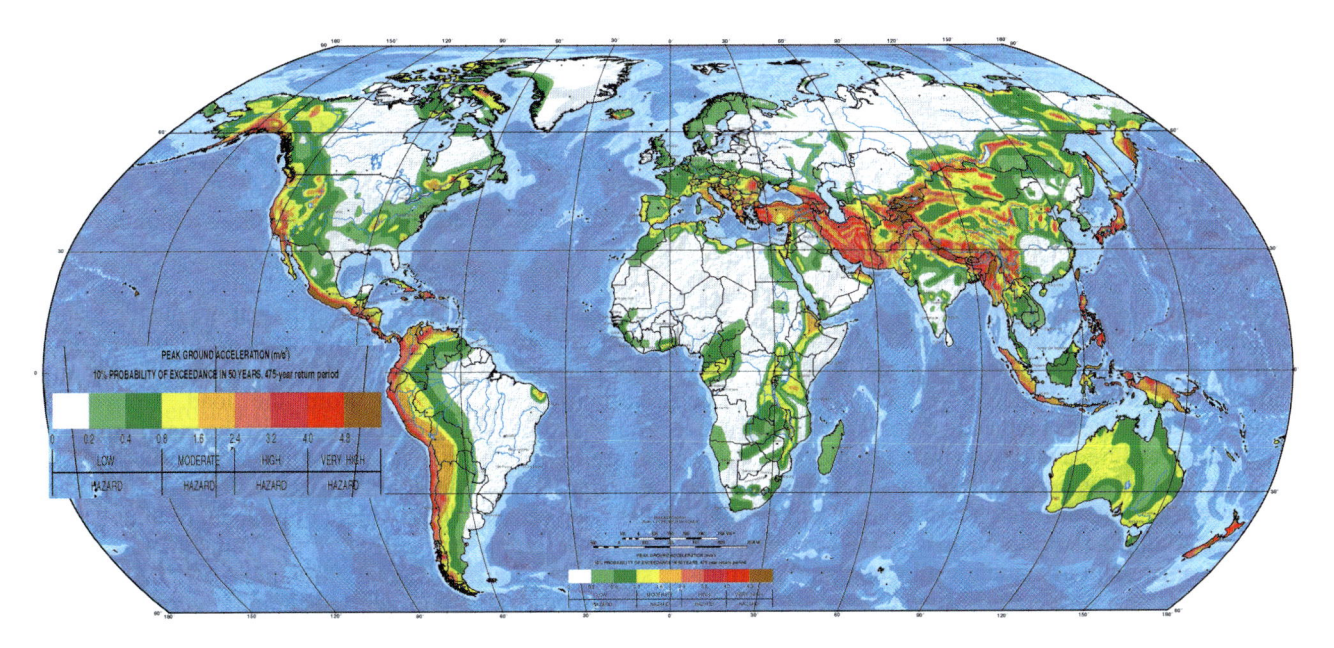

Tectonic and Tectono-Seismic Hazards, Figure 9 Seismic hazard map of the world. Color shading represents classes of predicted peak ground acceleration (m/s^2), which have only a 10% probability of being exceeded in any 50-year time span. This is equivalent to the largest ground motion expected in a 475-year return period. (Note: to convert these accelerations to g, divide by 9.8. Source: Global Seismic Hazard Assessment Program (www.seismo.ethz.ch/GSHAP)).

Crustal uplift

Crustal uplift can be caused by a variety of mechanisms, from interseismic deformation near an active fault, to slow movement on an upwarping fold (anticline), to rebound from crustal unloading.

Slow crustal uplift does not generally pose a hazard except:

1. Along the margins of water bodies such as lakes or oceans. Coastal uplift will eventually cause the shoreline to retreat seaward, leaving shore facilities separated from the shoreline, and decreasing the water depth at docks and slips. For example, docks and other loading facilities may no longer have sufficient water depth to be used, requiring dredging.
2. Where the lateral gradient of uplift is high, such that the ground surface tilts. Many engineered structures are sensitive to small tilts, such as water tanks, high-rise buildings, etc.

Tectonic hazards, the spatial distribution

Those tectonic hazards associated with active faults and folds have a similar geographic distribution to tectono-seismic hazards, which were described previously. However, due to their slow movement rate, tectonic processes create their most significant hazards in coastal regions, where land elevation changes cause shifts in shorelines. Crustal loading and unloading is greatest in areas where glacial ice thickness is rapidly changing today, or has changed in the past. The latter includes the large continental ice sheets such as the Laurentide in North America and the Fennoscandian in Europe.

Tectonic hazards, mitigation

Slow tectonic hazards are seldom mitigated, because they pose little threat to life safety or to the structural soundness of buildings. Accordingly, they are dealt with more as nuisance problems by routine maintenance.

Summary

Tectonic processes result in movements of the Earth's crust, and these movements can cause hazards to life via earthquakes when the movement is very rapid (tectono-seismic hazards), and to certain structures when movement is slower. Tectonic hazards in general result from three types of physical processes:

1. Rapid fault rupture and its associated earthquake (see *Fault*; *Earthquake*).
2. Physical displacement or warping of the Earth's crust (uplift, depression, folding, tilting, and horizontal motion) that occurs either rapidly or slowly.
3. Movement of molten rock material (magma) within the Earth's crust (intrusive igneous activity) and at the Earth's surface (extrusive volcanic activity). These are "volcanic hazards." (see *Volcanoes and Volcanic Eruptions*).

Most injuries and damages in historic time have been caused by tectono-seismic hazards (related to earthquakes), rather than to the slower tectonic processes such as fault creep, warping, tilting, uplift, and subsidence.

Bibliography

Gilbert, G.K., 1890. *Lake Bonneville*. Washington, DC: U.S. Geological Survey, Monograph 1, p. 438.
Lee, W. H. K., Kanamori, H., Jennings, P. C., and Kisslinger, K. (eds.), 2003. *The International Handbook of Earthquake and Engineering Seismology, Part B*. Orlando, FL: Academic. p. 1942.
Morner, N.-A., 2003. *Paleoseismicity of Sweden; A Novel Paradigm*. Sweden: Paleogeophysics & Geodynamics, Stockholm University. p. 320.
NGDC, 1997. *Geologic Hazard Photos*. Washington, DC: National Geophysical Data Center, U.S. National Oceanographic and Atmospheric Administration. 3 CD-ROMs.
Plafker, G., 1969. *Tectonics of the March 27, 1964 Alaska Earthquake*. U.S. Geological Survey, Professional Paper 543-I, Washington, DC, pp. I1–I74.
Vita-Finzi, C., 1986. *Recent Earth Movements-An Introduction to Neotectonics*. Orlando, FL: Academic. 226 p.
Youd, T.L., 1980, Ground failure displacement and earthquake damage to buildings. In *Proceedings of 2nd ASCE Conference On Civil Engineering and Nuclear Power*, Knoxville, Tennessee, September 15–17, 1980, p. 26.

Cross-references

Avalanches
Earthquake
Earthquake Damage
Elastic Rebound, Theory of
Fault
Geological/Geophysical Disasters
Indian Ocean Tsunami
Landslide (Mass Movement)
Lateral Spreading
Liquefaction
Magnitude Measures
Paleoseismology
Plate Tectonics
Seiche
Seismology
Structural Mitigation
Subduction
Tsunami
Unreinforced Masonry Building
Volcanoes and Volcanic Eruptions

TECTONIC TREMOR

David Shelly
U.S. Geological Survey, Menlo Park, California, CA, USA

Synonyms

Deep low-frequency tremor; Non-volcanic tremor

Definition

Tectonic, non-volcanic tremor is a weak vibration of ground, which cannot be felt by humans but can be detected by sensitive seismometers. It is defined empirically as a low-amplitude, extended duration seismic signal associated with the deep portion (~20–40 km depth) of some major faults. It is typically observed most clearly in the frequency range of 2–8 Hz and is depleted in energy at higher frequencies relative to regular earthquakes.

Discussion

Origin

Tectonic "non-volcanic" tremor was first reported in 2002, when it was identified in the Nankai Trough subduction zone of southwest Japan (Obara, 2002). Since then, tremor has been identified in other subduction zones (Rogers and Dragert, 2003) and the strike-slip San Andreas fault (Nadeau and Dolenc, 2005), yet it remains incompletely understood (Schwartz and Rokosky, 2007; Rubinstein et al., 2010). Tremor appears to be composed of numerous small overlapping earthquakes, sometimes called low-frequency earthquakes (LFEs) (Katsumata and Kamaya, 2003). Growing evidence suggests that tremor is generated by shear slip on these deep faults (Ide et al., 2007; Shelly et al., 2007a; Wech and Creager, 2007), often as part of larger slow slip events (Rogers and Dragert, 2003; Obara et al., 2004). Slip from larger multi-day events can often be observed geodetically with Global Positioning System (GPS) (Rogers and Dragert, 2003) or borehole tilt instruments (Obara et al., 2004). In Cascadia and southwest Japan, tremor and slip events occur semi-regularly with periods ranging from 3 to 20 months, depending on the location (Rogers and Dragert, 2003; Obara et al., 2004; Brudzinski and Allen, 2007). The coupled phenomenon is sometimes called "Episodic tremor and slip" (ETS) (Rogers and Dragert, 2003).

Triggering

Unlike regular earthquakes, tremor is commonly triggered by minute stresses imparted by teleseismic waves (especially surface waves) (e.g., [Miyazawa and Mori, 2005; Rubinstein et al., 2007; Peng et al., 2009]) and tidal stresses (e.g., [Nakata et al., 2008; Thomas et al., 2009]). These observations suggest that the affected faults are critically stressed and they support the notion that high pore fluid pressures exist within the fault zone, as suggested by some seismic imaging studies (Shelly et al., 2006).

Unanswered questions

Despite recent study, many aspects of the underlying physics of tremor remain unexplained. In particular, why are such events slow? Observations of tremor migration in subduction zones at ~10 km/day along strike (Obara, 2002; Kao et al., 2007) and 30–150 km/h (Shelly et al., 2007b) in the slip direction suggest that fault geometry may play a role, but the physics behind these migration velocities is still poorly understood.

The relationship between tremor (and slow slip) and shallower earthquakes is of particular interest to earth scientists, yet also remains poorly constrained. Future observations will help understand whether tremor behavior might evolve during a seismic cycle and indicate time of increase hazard or whether tremor can be used to delineate the downward extent of rupture in a major earthquake.

Future potential

Tremor provides exciting information regarding the behavior of deep faults and represents a new tool for measuring subtle deformations that have previously remained hidden. In addition, because tremor occurs on portions of the fault without regular earthquakes, it gives new constraints on physical properties in its source region and could eventually be used to improve knowledge of neighboring earth structure.

Bibliography

Brudzinski, M. R., and Allen, R. M., 2007. Segmentation in episodic tremor and slip all along Cascadia. *Geology*, **35**, 907–910.

Ide, S., Shelly, D. R., and Beroza, G. C., 2007. Mechanism of deep low frequency earthquakes: Further evidence that deep non-volcanic tremor is generated by shear slip on the plate interface. *Geophysical Research Letters*, **34**, L03308, doi:10.1029/2006GL028890.

Kao, H., Shan, S.-J., Rogers, G., and Dragert, H., 2007. Migration characteristics of seismic tremors in the northern Cascadia margin. *Geophysical Research Letters*, **34**, L03304, doi:10.1029/2006GL028430.

Katsumata, A., and Kamaya, N., 2003. Low-frequency continuous tremor around the Moho discontinuity away from volcanoes in the southwest Japan, *Geophysical Research Letters*, **30**, doi:10.1029/2002GL015981.

Miyazawa, M., and Mori, J., 2005. Detection of triggered deep low-frequency events from the 2003, 2005 Tokachi-oki earthquake, *Geophysical Research Letters*, **32**, doi:10.1029/2005GL022539.

Nadeau, R. M., and Dolenc, D., 2005. Nonvolcanic tremors deep beneath the San Andreas fault. *Science*, **307**, 389; published online 9 December 2004 (10.1126/science.1107142).

Nakata, R., Suda, N., and Tsuruoka, H., 2008. Non-volcanic tremor resulting from the combined effect of Earth tides and slow slip events. *Nature Geoscience*, **1**, 676–678, doi:10.1038/ngeo288.

Obara, K., 2002. Nonvolcanic deep tremor associated with subduction in southwest Japan. *Science*, **296**, 1679–1681.

Obara, K., Hirose, H., Yamamizu, F., and Kasahara, K., 2004. Episodic slow slip events accompanied by non-volcanic tremors in southwest Japan subduction zone. *Geophysical Research Letters*, **31**, doi:10.1029/2004GL020848.

Peng, Z., Vidale, J. E., Wech, A., Nadeau, R. M., and Creager, K. C., 2009. Remote triggering of tremor around the Parkfield section of the San Andreas fault. *Journal of Geophysical Research*, **114**, B00A06, doi:10.1029/2008JB006049.

Rogers, G., and Dragert, H., 2003. Episodic tremor and slip on the Cascadia subduction zone: The chatter of silent slip. *Science*, **300**, 1942–1943.

Rubinstein, J. L., Vidale, J. E., Gomberg, J., Bodin, P., Creager, K. C., and Malone, S. D., 2007. Non-volcanic tremor driven by large transient shear stresses. *Nature*, **448**, 579–582, doi:10.1038/nature06017.

Rubinstein, J. L., Shelly, D. R., and Ellsworth, W. L., 2010. Non-volcanic tremor: A window into the roots of fault zones. In Cloetingh, S., and Negendank, J., (eds.), *New Frontiers in*

Integrated Solid Earth Sciences, International Year of Planet Earth. Netherlands: Springer, pp. 287–314.

Schwartz, S. Y., and Rokosky, J. M., 2007. Slow slip events and seismic tremor at circum-pacific subduction zones. *Reviews of Geophysics,* **45**, RG3004, doi:10.1029/2006RG000208.

Shelly, D. R., Beroza, G. C., Ide, S., and Nakamula, S., 2006. Low-frequency earthquakes in Shikoku, Japan and their relationship to episodic tremor and slip. *Nature,* **442**, 188–191.

Shelly, D. R., Beroza, G. C., and Ide, S., 2007a. Non-volcanic tremor and low frequency earthquake swarms. *Nature,* **446**, 305–307.

Shelly, D. R., Beroza, G. C., and Ide, S., 2007b. Complex evolution of transient slip derived from precise tremor locations in western Shikoku, Japan. *Geochemistry Geophysics Geosystems,* **8**, Q10014, doi:10.1029/2007GC001640.

Thomas, A. M., Nadeau, R. M., and Bürgmann, R., 2009. Tremor-tide correlations and near-lithostatic pore pressure on the deep San Andreas fault. *Nature,* **462**, 1048–1051.

Wech, A. G., and Creager, K. C., 2007. Cascadia tremor polarization evidence for plate interface slip. *Geophysical Research Letters,* **34**, L22306, doi:10.1029/2007GL031167.

Cross-references

Early Warning Systems
Earthquake
Earthquake Prediction and Forecasting
Fault
Global Positioning System (GPS)
Harmonic Tremor
Hypocentre
Macroseismic Survey
Plate Tectonics
Primary Wave (P wave)
Sand Andreas Fault
Secondary Wave (S wave)

THUNDERSTORMS

Colin Price
Tel Aviv University, Ramat Aviv, Israel

Synonyms

Convective storms; Deep convection; Electrical storms; MCS; Squall lines; Supercells

Definition

Thunderstorms are defined as weather storms that are associated with lightning discharges, which result in the production of acoustic waves called thunder.

Discussion

There are approximately 2,000 thunderstorms active at any time around the globe, with all these thunderstorms producing between 50 and 100 lightning discharges per second. Thunderstorms are the result of strong vertical air currents in the lower atmosphere, produced when the atmosphere becomes unstable. More than 90% of all thunderstorms occur within the tropical regions, where the air becomes unstable due to the daily solar heating of the surface, and the resulting rising thermals. The air condenses as it rises and cools, with additional heating of the air occuring due to the release of latent heat within the clouds, enhancing the development of the thunderstorms. The larger the instability, the larger the vertical winds (updrafts) and the greater the hazards associated with the thunderstorm. These hazards include heavy rain, lightning, hail, strong winds, and even tornados.

Lightning is responsible for thousands of deaths every year, and many more injuries. Lightning is also a major cause of wildfires in temperate and high latitudes during summer months, while being a major hazard to commercial aviation. Hailstorms produce significant agricultural damage every year across the globe, with additional damage to property, cars, aircraft, etc. Tornados can result in severe damage to property and loss of life, and are always associated with thunderstorms. Whereas lightning is most prominent in tropical regions, tornados are more prominent in mid-latitude regions. Heavy rainfall from thunderstorms can often result in flash floods that appear with very short warnings. Finally, downbursts and straight-line winds can be extremely hazardous around airports, causing extensive property damage as a result of the intense winds.

Whereas most thunderstorms occur in the tropics, the majority of these storms occur over the continental landmasses, and not over the oceans. Furthermore, thunderstorms occur primarily in the summer months, although some regions experience winter thunderstorms over warm ocean waters (e.g., Gulf Stream, Mediterranean Sea, Koroshiu Current). By definition, thunderstorms are associated with lightning, and lightning activity can be detected and tracked by monitoring the electromagnetic pulses emitted by lightning channels. These radio waves can travel thousands of kilometers before decaying into the background noise, and hence we can use ground-based lightning networks to track in real time the lightning activity over large regions of the globe, allowing us to monitor such hazardous storms in close to real time.

Bibliography

Betz, H. D., Schumann, U., and Laroche, P., 2009. *Lightning: Principles, Instruments and Applications.* Amsterdam: Springer.
MacGorman, D. R., and Rust, W. D., 1998. *The electrical nature of storms.* Oxford: Oxford University Press.

Cross-references

Beaufort Wind Scale
Cloud Seeding
Dust Storm
Dvorak Classification of Hurricanes
Fog Hazards
Fujita Tornado Scale
Hurricane (Cyclone, Typhoon)
Lightning
Monsoon
Storm Surges
Storms
Tornadoes
Waterspout

TIDAL BORES

Hubert Chanson
The University of Queensland, Brisbane, QLD, Australia

Synonyms

Benak (Malaysia); Bono (Indonesia); Burro (Mexico); Mascaret (France); Pororoca (Brazil)

Definition

A tidal bore is a positive surge of tidal origin that may occur in an estuary when the tidal flow turns to rising; the existence of a tidal bore is linked with a large tidal range, an estuarine bathymetry that amplifies the tidal wave and a low freshwater level.

Discussion

A tidal bore is an unsteady flow motion generated by the rapid water level rise at a river mouth during the early flood tide when the flood tide waters rush into a funnel-shaped river mouth that amplifies the tidal range. A bore is a sudden increase of the water depth as illustrated in Figures 1 and 2. Figure 1 shows a tidal bore in the Bay of Mont Saint Michel (France). The tidal bore advances in the river channel and on the surrounding sand flats. Figure 2 presents the tidal bore of the Dordogne River (France). The surfers give the scale of the bore front. Worldwide, it is estimated that over 400 estuaries are affected by a tidal bore process, on all continents but Antarctica. Some famous tidal bores include the "pororoca" of the Amazon River in Brazil, the bore of the Qiantang River in China, and the 'mascaret' of the Seine River in France (Malandain, 1988).

A tidal bore is almost a mythical phenomenon because it is rare to observe. It occurs only during the flood tide under spring tidal conditions and low freshwater levels. Its passage is very rapid, that is, a few minutes at most and it is easily missed. The bore is a sharp front that propagates upstream into the river mouth and may travel several dozens of kilometers inland before vanishing. The presence of a tidal bore indicates some macro-tidal conditions (tidal range > 4.5–6 m) associated with an asymmetrical tide. The flood tide is typically shorter than the ebb tide period and the flood flow is much faster. A feature of the tidal bore is its rumble noise that can be heard from far away. Some field measurements show that the generated sounds have a low-pitch comparable to the sounds generated by bass drums and locomotive trains (Chanson, 2009).

Tidal Bores, Figure 1 Tidal bore in the Mont Saint Michel Bay in France on October 19, 2008 morning – Bore propagation from right to left.

Tidal Bores, Figure 2 Tidal bore of the Dordogne River (France) at Port de Saint Pardon on September 2, 2008 evening – Looking downstream at the incoming tidal bore.

Theoretical considerations

A tidal bore may occur when the tidal range exceeds 4.5–6 m and the bathymetry of the river mouth amplifies the tidal wave. The driving process is the large tidal amplitude. The tides are forced oscillations generated by the attractions of the Moon and Sun, and have the same periods as the motion of the Sun and Moon relative to the Earth. At full moon or new moon, the attraction forces of the Sun and Moon reinforce one another, and these conditions give the spring tide conditions. The tidal range may be locally amplified further by a number of factors, such as when the natural resonance of the bay and estuary is close

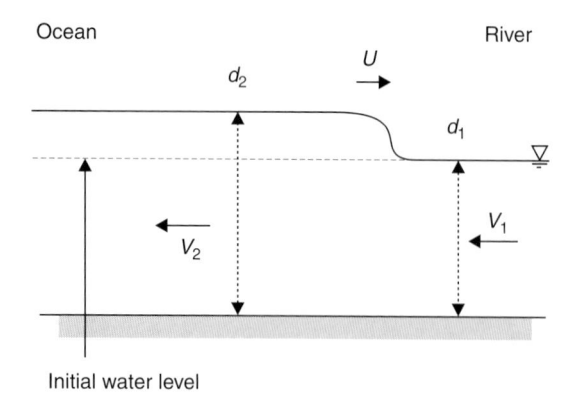

Tidal Bores, Figure 3 Definition sketch of a tidal bore propagating upstream.

to the tidal period. This coincidence implies that the general sloshing of the waters around the inlet or bay becomes synchronized with the lunar tides and amplifies their effect, yielding often the best tidal bores a couple of days after the date of the maximum tidal range.

When the sea level rises with time during the flood tide, the tidal wave becomes steeper and steeper, until it forms an abrupt front: the tidal bore. The inception and development of a tidal bore may be predicted using the Saint-Venant equations and the method of characteristics (Peregrine, 1966; Chanson, 2004). After the formation of the bore, the flow properties directly upstream and downstream of the tidal bore front must satisfy the equations of conservation of mass and momentum. The integral form of the continuity and momentum principles gives a well-known relationship between the flow depth in front of and behind the tidal bore front:

$$\frac{d_2}{d_1} = \frac{1}{2}\left(\sqrt{1 + 8 \times \text{Fr}_1^2} - 1\right) \quad (1)$$

where Fr_1 is the tidal bore Froude number defined as: $\text{Fr}_1 = (V_1 + U)/\sqrt{g \times d_1}$ with g the gravity acceleration, V the flow velocity positive downstream towards the river mouth, U the bore speed for an observer standing on the bank, d the water depth, and the subscript 1 refers to the initial flow conditions whereas the subscript 2 refers to the new flow conditions (Figure 3).

The Froude number of the tidal bore is always greater than unity and the quantity (Fr_1-1) is a measure of the strength of the bore. If the Froude number is less than unity, then the tidal wave cannot become a tidal bore. For a tidal bore Froude number between unity and 1.5–1.8, the bore front is followed by a train of well-formed, quasi-periodic free-surface undulations, also called whelps. For larger Froude numbers, the tidal bore is characterized by a breaking front as seen in Figure 1 (Koch and Chanson, 2009). Some simple energy considerations show that a tidal bore can occur only with a net flux of mass from downstream to upstream. This characteristic sets apart the tidal bore from a wave or soliton.

Impacts of tidal bores

The tidal bores can be dangerous and some have had a sinister reputation. For example, in the Seine River estuary (France), more than 220 ships were lost between 1789 and 1840 in the Quilleboeuf–Villequier section. Similarly, the bores of the Petitcodiac River (Bay of Fundy, Canada) and Colorado River (Mexico) are feared by some of the populace. In China, some tidal bore warning signs are erected along the Qiantang River banks and yet a number of tragic accidents happen every year. The tidal bores affect the shipping and navigation in the estuarine zone as in Papua New Guinea (Fly and Bamu Rivers), Malaysia (Benak at Batang Lupar), and India (Hoogly bore).

However, the tidal-bore affected estuaries are the feeding zone and breeding grounds of several forms of wildlife. For example, some large predators feed behind the bore: sharks in Australia, whales in Alaska, seals in France, crocodiles in Australia and Malaysia. The estuarine zones are the spawning and breeding grounds of several fish species, while the turbulent mixing and aeration induced by the tidal bore contribute to the abundant growth of many species of fish and shrimps.

Related processes

A number of geophysical, as well as man-made, processes are related to the tidal bore. In the Bay of Bengal, the development of a *storm surge* during the early flood tide with spring tidal conditions may yield a rapid rise in water levels generating a bore front. The wind shear amplifies the tidal range and the phenomenon has been observed in Bangladesh where the storm events are called locally "tidal bores." Another related process is the *tsunami-induced bore*. After breaking, a tsunami wave propagating in shallow-water regions is led by a positive surge. In shallow rivers, the process is somehow similar to a tidal bore and the tsunami-induced bore may propagate far upstream in a river mouth as observed in Hawaii, in Japan, and more recently during the December 26, 2004 Indian Ocean tsunami catastrophe in Malaysia, Thailand, and Sri Lanka. At a smaller scale, some *swash*-induced bores may be observed on beaches when the wave run-up enters into a small creek or channel.

Positive surges and bores may be observed in irrigation channels and water power canals during gate operation. Some bores are also observed at the leading edge of violent *flash floods* propagating downstream narrow canyons. Lastly, some water theme parks include large *artificial beaches* in which man-made waves somehow similar to a bore are generated for the agreement of the visitors.

Summary

A tidal bore is a series of waves propagating upstream in the river mouth as the tide turns to rising. It forms during the spring tide conditions with a tidal range in excess of 4.5–6 m in a narrow funneled estuary with low freshwater levels. The presence of a tidal bore indicates some macrotidal conditions associated with an asymmetrical tide. Two key features of a tidal bore are (a) its rumble noise that can be heard from far away, and (b) the turbulent mixing induced by the bore propagation that stirs the sediments and matters.

Bibliography

Chanson, H., 2004. *Environmental Hydraulics of Open Channel Flows*. Oxford: Elsevier Butterworth-Heinemann, p. 483.

Chanson, H., 2009. The rumble sound generated by a tidal bore event in the Baie du Mont Saint Michel. *Journal of Acoustical Society of America*, **125**(6), 3561–3568, doi:10.1121/1.3124781.

Koch, C., and Chanson, H., 2009. Turbulence measurements in positive surges and bores. *Journal of Hydraulic Research*, **47**(1), 29–40. doi:10.3826/jhr.2009.2954.

Malandain, J.J., 1988. La Seine au Temps du Mascaret. ('The Seine River at the Time of the Mascaret.') Le Chasse-Marée, No. 34, pp. 30–45 (in French).

Peregrine, D. H., 1966. Calculations of the development of an Undular Bore. *Journal of Fluid Mechanics*, **25**, 321–330.

Cross-references

Flash Floods
Storm Surge
Surge
Tides
Tsunami

TILTMETERS

Erik Eberhardt
University of British Columbia, Vancouver, BC, Canada

Synonyms

Clinometers; Tilt sensors

Definition

Tiltmeters are devices used to monitor the change in inclination of a ground surface point; see Dunnicliff (1993) for a detailed description. The device consists of a gravity sensing transducer (e.g., servo-accelerometer, electrolytic tilt sensor, pendulum-actuated vibrating wire, etc.) capable of measuring changes in inclination as small as one arc second (0.00028 degrees). They are used to monitor slope movements where the landslide failure mode is expected to contain a rotational component. Advantages of using tiltmeters are their light weight, simple operation, and relatively low cost; tiltmeters can be read manually or automated by connecting to a data logger.

Bibliography

Dunnicliff, L., 1993. *Geotechnical Instrumentation for Monitoring Field Performance*. New York: Wiley.

Cross-references

Extensometer
Landslides
Mass Movement

TIME AND SPACE IN DISASTER

Thomas Glade[1], Michael James Crozier[2], Nick Preston[2]
[1]University of Vienna, Vienna, Austria
[2]Victoria University of Wellington, Wellington, New Zealand

Disasters in time and space

Early attempts to define disasters were based on the exceedence of certain loss thresholds. For instance, Sheehan and Hewitt (1996) classified as disasters all those events that killed or injured at least 100 people or caused at least US $1 million damage. This definition was further developed in more qualitative terms, e.g., by UNDRO (1984) "... an event, concentrated in time and space, in which a community undergoes severe danger and incurs such losses to its members and physical appurtenances that the social structure is disrupted and the fulfillment of all or some of the essential functions of the society is prevented." Other definitions reduce the term disaster to those events where ".. large numbers of people exposed to hazard are killed, injured or damaged in some way ..." (Smith, 2004, p. 5). In this context, Smith also states, that "there is no universally agreed definition of the scale on which loss has to occur in order to qualify as a disaster." Further, Smith (2004, p. 22) writes that "... a disaster generally results from the interaction, in time and space, between the physical exposure to a hazardous process and a vulnerable human population." For statistical purposes some authorities require the impact of a natural event to exceed certain thresholds of areal extent, as well as lives lost, or economic costs before they are classified as disasters. In this contribution, disasters are defined as those damaging events that exceed the coping capacity of affected individuals, groups, or institutions and, in some cases, even nations. This definition avoids the use of absolute quantitative measures, which can vary dramatically between different countries, or in more general terms, between different social groups.

Thus, irrespective of the magnitude of the natural event, disasters are defined in terms of human impact and related consequences. In the contextual framework of natural hazards, disasters can be localized. They occur at a specific location or in a region as a sudden onset or as slow creeping, often unstoppable processes. Sources and affected areas can be very distinct with easy to delineate boundaries (e.g., a debris flow with source area, travel path, and deposition) or difficult to assess (e.g., pollution of ground water). Whereas the boundaries of source and impact areas may be identifiable after an event, it is not always possible to predict where a disaster may occur. Some hazards that give rise to disasters tend to recur in the same locality; these are described as *location-specific*, e.g., lahars, debris flows, snow avalanches, and in some cases earthquakes and volcanic eruptions. *Non–location-specific* hazards which are more or less random in terms of place of occurrence include events such as drought, epidemic, and many weather related phenomena. However, because vulnerability and resilience of human communities have a large influence on the magnitude of consequences resulting from a hazard event, most disasters occur in the poorer less-developed countries of the world (Table 1). Nevertheless industrialized regions can also suffer from major disasters, for example, when design thresholds of mitigation structures are exceeded (refer to Hurricane Katrina in USA). Although economic losses can be large in industrialized regions, in contrast, in the transient states, loss of life and other direct effects on the population are generally much higher (Table 1). These differences do not simply represent a decadal trend, but can be observed over much longer periods (e.g., OFDA/CRED International Disaster Database).

Because of the human element implicit in the notion of disasters an understanding of their causes and behavior requires information not only on the properties and patterns of the natural event, but also on the socioeconomic conditions of the affected area. In numerous regions of the world, people are unable to divert resources toward counter measures against natural hazards. They have to face much more dramatic problems such as unemployment, famine, crime, and so on. These problems become much more severe with constantly growing cities and urban agglomerations and thus, these social groups become increasingly vulnerable toward natural events. Some socioeconomic factors that turn an event into a disaster relate to:

- Demographic characteristics
- GDP
- Urbanization
- Emergency preparedness
- Insurance coverage
- Community perception and awareness

These factors alone are all subject to constant, often rapid change, producing dramatic transformations of the human condition within time and space. Consequently, risk is changing as well – and as a result the magnitude and areal extent of disasters have tended to increase with time. Thus, not only do the characteristics of the physical process change (e.g., more intense rainstorms, stronger winds, higher waves), but also the elements at risk undergo a continuous change (Hufschmidt et al., 2005; Keiler, 2004).

Another important issue is the time lag between the triggering input, the occurrence of the process and the resulting disaster. In the case of a debris flow, it is straightforward. Heavy rain accumulates in the flow lines and starts to move erodible material until there is sufficient sediment that the debris flow is formed, travels down a channel and affects the downstream people or infrastructure. Other processes such as soil erosion caused by human activity are much more difficult to assess. The time lag between deforestation, start of soil erosion and erosion cycles that are based on the timing of the precipitation event and the agricultural usage is often very large. Also, the onset of the associated disaster is gradual rather than sudden. In such cases agricultural productivity slowly decreases and although the affected social groups might be able to cope with these changes in the beginning, the continuous increase of pressure and then the sudden drop of productivity can lead also to a disaster. Therefore, it is important to consider the chain of cause – consequence for disasters (Figure 1).

As indicated earlier, both slow and fast-onset natural hazards can cause disasters. The consequences of the fast-onset processes are mostly clearly visible and these disasters are often quantifiable in terms of their impact. In contrast, slow onset disasters continue over long periods. Besides desertification and soil erosion, other examples include water pollution or subsidence through extensive ground water removal. These "creeping" or gradual processes still cause disasters in the above defined sense – at some stage, there may be no soil left for agricultural use and the farmers have to move, or the ground water has been extensively extracted to an extent, where there is no readily available water. The now nearly dry Aral lake (Waltham and Sholji, 2001) is a dramatic example of excessive water usage in the upper catchment for irrigation purposes to the extent, that in certain years virtually no water reaches the lake (Cai et al., 2003). The lake now has more or less disappeared causing a dramatic disaster for the affected population – not only in terms of water shortage and depressed economy, but also in terms of an increase in the impact of pesticide polluted dust storms (O'Hara et al., 2000). Therefore, the time lag between input and consequences can be several years, and in some cases, even decades.

Another issue in this context is difference between the source area and the potential effects. Although snow avalanches, rock falls, and hurricanes have distinct and localized occurrences and consequential damage potential, a debris flow or a flash flood might be initiated high up in the catchment area but will cause destructive damages far away from the source. Similarly tsunami with travel distances of thousands of kilometers or ash clouds from volcanic eruptions with consequent and long-lasting flight interruptions are other examples (e.g., eruption of Icelandic volcano Eyjafjallajökull in March and April 2010).

Different perspectives

Assessing the temporal and spatial distribution of disasters is often very difficult particularly for events that have taken place in the past when instrumental and other records are limited. The human memory and associated observations can be useful sources of information. However, the larger the time lags between event occurrence and the recording of the event, the vaguer the information. In addition, smaller events are more often forgotten in

Time and Space in Disaster, Table 1 Selected entries of natural disasters for the period 1999–2009, ordered by largest numbers of (A) fatalities, (B) injuries, and (C) economic damages. (Note: *Gray shaded boxes* are not relevant for the respective entry. Data extracted from the EM-DAT: The OFDA/CRED International Disaster Database – www.emdat.be, maintained by CRED (Centre for Research on the Epidemiology of Disasters), Université Catholique de Louvain, Brussels (Belgium), and accessed 05.05.2010)

Date		Location		Type of event			Consequences		
Start	End	Country	Region	Main	Subtype	Name	Killed	Totally affected ($\times 10^6$)	Est. damage (Mio US$)

(A) Disasters with largest fatalities

Start	End	Country	Region	Main	Subtype	Name	Killed
26.12.2004	26.12.2004	Indonesia	Aceh province (Sumatra)	Earthquake (seismic activity)	Tsunami		165,708
02.05.2008	03.05.2008	Myanmar	Ngapadudaw, Labutta	Storm	Tropical cyclone	Cyclone Nargis	138,366
12.05.2008	12.05.2008	China P Rep	Wenchuan country	Earthquake (seismic activity)	Earthquake (ground shaking)		87,476
08.10.2005	08.10.2005	Pakistan	Bagh, Muzzafarabad	Earthquake (seismic activity)	Earthquake (ground shaking)		73,338
26.12.2004	26.12.2004	Sri Lanka		Earthquake (seismic activity)	Tsunami		35,399
26.12.2003	26.12.2003	Iran Islam Rep	Bam (Kerran province)	Earthquake (seismic activity)	Earthquake (ground shaking)		26,796
16.07.2003	15.08.2003	Italy	Milan, Turin (Piemont)	Extreme temperature	Heat wave		20,089
26.01.2001	26.01.2001	India	Kachch-Bhuj, Ahmedabad	Earthquake (seismic activity)	Earthquake (ground shaking)		20,005
01.08.2003	20.08.2003	France	Paris region – all countries	Extreme temperature	Heat wave		19,490
26.12.2004	26.12.2004	India	Tamil Nadu state, Andaman	Earthquake (seismic activity)	Tsunami		16,389

(B) Disasters with most affected people

Start	End	Country	Region	Main	Subtype	Totally affected ($\times 10^6$)
00.07.2009	00.08.2009	India	Bongaigaon, Cachar	Drought	Drought	300
23.06.2003	28.07.2003	China P Rep	Zhejiang, Jiangsu	Flood	General flood	150
15.06.2007	00.07.2007	China P Rep	Sichuan, Anhui, Hubei	Flood	General flood	105
14.03.2002	31.03.2002	China P Rep	North	Storm	Local storm	100
08.06.2002	18.06.2002	China P Rep	Shanxi, Sichuan, Hubei	Flood	Flash flood	80
10.01.2008	05.02.2008	China P Rep	Zhejiang, Sichuan	Extreme temperature	Extreme winter conditions	77
00.04.2002	00.00.2002	China P Rep	Guangdong, Fujian	Drought	Drought	60
00.10.2009	00.03.2010	China P Rep	Yunnan, Guizhou, Sichuan	Drought	Drought	51
00.04.2000	00.00.2001	India	Gujarat, Rajasthan	Drought	Drought	50
00.01.2003	00.01.2003	China P Rep	Inner Mongolia Autonomous	Drought	Drought	48

Time and Space in Disaster, Table 1 (Continued)

Date		Location		Type of event			Consequences	
Start	End	Country	Region	Main	Subtype	Name	Killed	Totally affected ($\times 10^6$)	Est. damage (Mio US$)

(C) Disasters with largest economic damage

Start	End	Country	Region	Main	Subtype	Name	Est. damage (Mio US$)
29.08.2005	19.09.2005	United States	Mobile, Bayou La Batre	Storm	Tropical cyclone		125,000
12.05.2008	12.05.2008	China P Rep	Wenchuan country, Wencgua	Earthquake (seismic activity)	Earthquake (ground shaking)		85,000
12.09.2008	16.09.2008	United States	Galvestin, Brazoria	Storm	Tropical cyclone		30,000
23.10.2004	25.10.2004	Japan	Niigata	Earthquake (seismic activity)	Earthquake (ground shaking)		28,000
10.01.2008	05.02.2008	China P Rep	Zhejiang, Sichuan	Extreme temperature	Heat wave		21,100
15.09.2004	16.09.2004	United States	Alabama, Louisiana	Storm	Tropical cyclone	Ivan	18,000
23.09.2005	01.10.2005	United States	Louisiana, Texas	Storm	Tropical cyclone	Rita	16,000
13.08.2004	13.08.2004	United States	Florida	Storm	Tropical cyclone	Charley	16,000
24.10.2005	24.10.2005	United States	Florida Keys, Naples	Storm	Tropical cyclone	Hurricane "Wilmna"	14,300
16.07.2007	16.07.2007	Japan	Niiagata prefecture	Earthquake (seismic activity)	Earthquake (ground shaking)		12,500

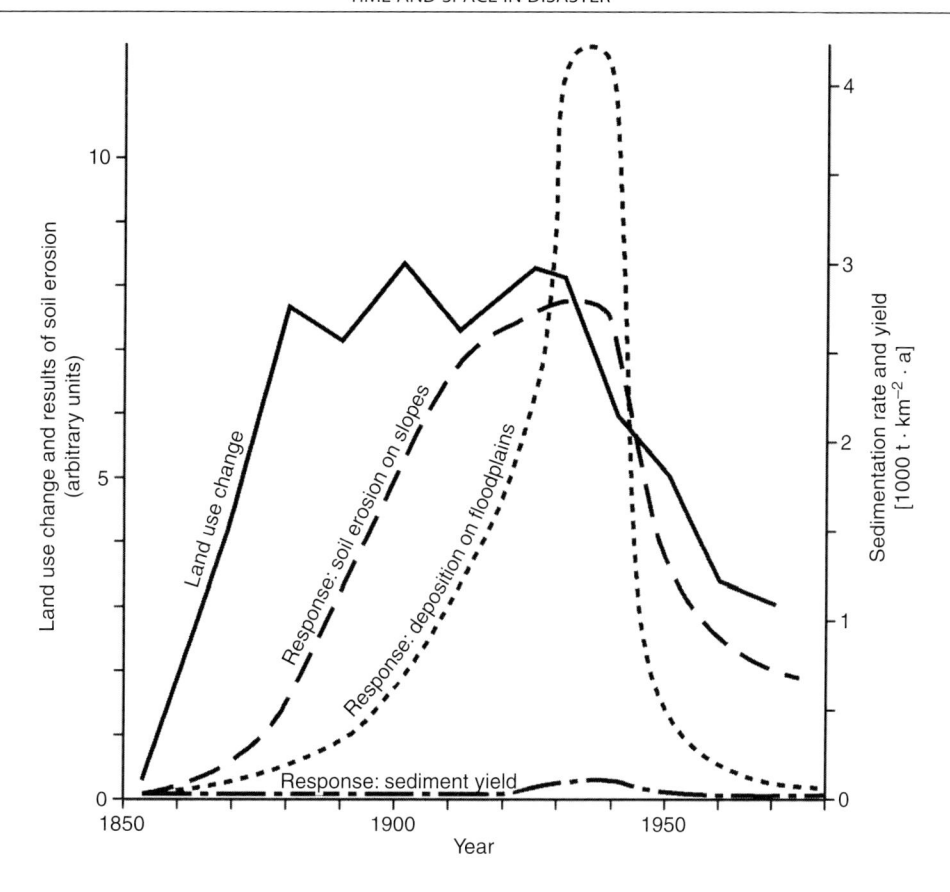

Time and Space in Disaster, Figure 1 Potential time lag between cause and different responses (Dearing et al., 2006), for the example, of soil erosion. Please note that such time lags operate as well in the social system.

time. Within historical research on former disastrous events, this is often a major problem (refer to entry "*Disaster Research and Policy, History*"). Therefore, graphs showing the development of disasters over time have to be treated with care (e.g., Figure 2). Such trends might reflect a number of factors unrelated to actual occurrence, such as increased awareness and thus enhanced reporting, better data availability, higher exposure of elements at risk, and so on. It is therefore important to carefully analyze temporal records to ensure any apparent trends are indeed real.

In recent years, media coverage has changed the public perception of disasters. For example, in some parts of the world, very small and localized events receive prominent media attention and provide a false impression of the magnitude of the event (e.g., snow avalanches in Galtür, Austria on the 22.02.1999). On the other hand, significant disasters such as desertification in certain regions often do not receive equivalent reporting representation and are thus not perceived by the public as large disasters.

Media, of course, play an important role in emergency management and disaster communication as well as being an important educational source about the causes and consequences of disasters. For instance, in Germany two large floods occurred in the Rhine valley within the 2 years (1995 and 1996). The result of comprehensive media coverage on the first flood meant that the public were well informed and were better prepared for the second flood and as a result the damages of the second flood were much lower (Engel, 1997). This again demonstrates the need to examine media reports carefully before using these in any form of magnitude frequency record, particularly noting the effect of reporting on events closely associated in time.

Future trends

There is a need for a better understanding of the causative factors of disasters, not only in terms of increased knowledge within natural sciences issues, but also within the social sciences. In this respect, of critical importance is the need to investigate the relationships between these two systems, the interconnections, the dependencies, the different reaction and response times, and the spatial implications associated with each system.

Therefore, studies of disasters should not confine themselves solely to post-event analysis and single-case studies. In order to understand better the root-cause-consequence principle in all its dimensions long-term investigations are necessary. Monitoring is a crucial part

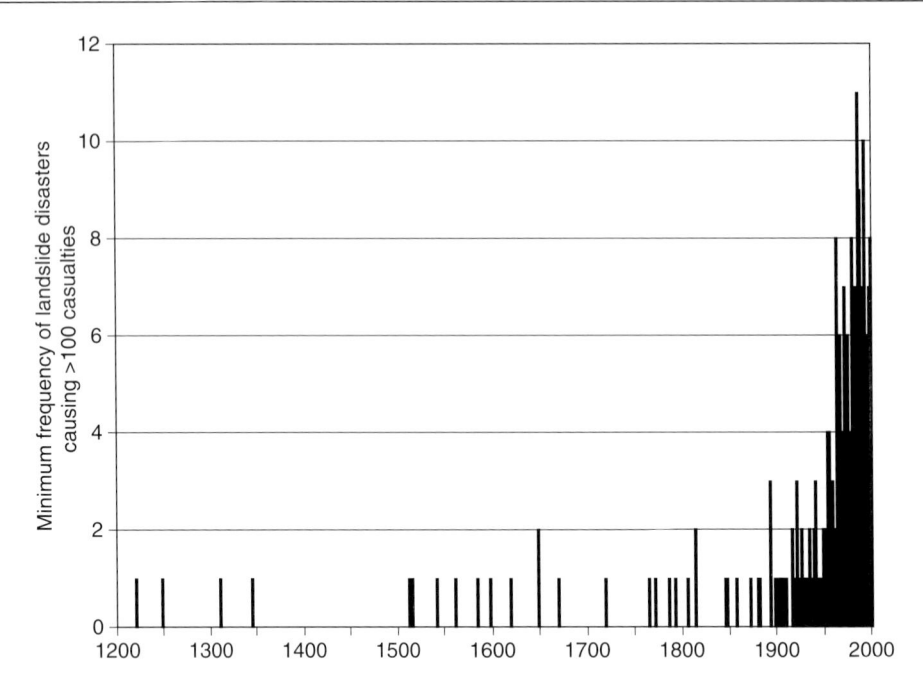

Time and Space in Disaster, Figure 2 Historical data of landslide disasters causing >100 causalities (Glade and Dikau, 2001). Note: This graph does not necessarily express a real increase of landslide disasters, but is purely reflecting the available reports and the better reporting within the last decades.

of this process, in particular, monitoring the natural system, the social system and – most importantly – the linkages between these elements. The resultant understanding of the basic underlying causes, the factors enforcing or reducing adverse affects, and – in principle – how disasters happen can support decision and policy makers in evaluating potential developments and promoting sustainable development for potentially disaster prone regions.

Summary

It has been stressed, that for a detailed and useful understanding of time and space in disasters, all factors have to be taken into consideration, the natural science, the social science, and the inherent interrelationships. It is evident, that disasters do not stop at any pre-subscribed boundaries, whether ethical, governmental, or topographic. Physical hazards can change their behavior, onset time, processes, and intensity in time and space. The human condition and state of development is also changing with implications for vulnerability and resilience. Associated risks and disaster occurrence can consequently change dramatically in time and space. The changing dynamic of disaster occurrence represents one of the most important and concerning elements of global change facing mankind.

Bibliography

Cai, X., McKinney, D. C., and Rosegranta, M. W., 2003. Sustainability analysis for irrigation water management in the Aral Sea region. *Agricultural Systems*, **76**(3), 1043–1066.

Dearing, J. A., Battarbee, R. W., Dikau, R., Larocque, I., and Oldfield, F., 2006. Human-environment interactions: towards synthesis and simulation. *Regional Environmental Change*, **6**, 115–123.

Engel, H., 1997. The flood events of 1993/1994 and 1995 in the Rhine River basin. In *Destructive Water: Water-Caused Natural Disasters, their Abatement and Control (Proceedings of the Conference held at Anaheim, California, June 1996)*. IAHS Publ. No. 239, pp. 21–32.

Glade, T., and Dikau, R., 2001. Gravitative massenbewegungen von naturereignis zur naturkatastrophe. *Petermanns Geographische Mitteilungen*, **145**, 42–55.

Hufschmidt, G., Crozier, M., and Glade, T., 2005. Evolution of natural risk: research framework and perspectives. *Natural Hazards and Earth System Sciences*, **5**, 375–387.

Keiler, M., 2004. Development of the damage potential resulting from avalanche risk in the period 1950-2000, case study Galtur. *Natural Hazards and Earth System Sciences*, **4**, 249–256.

O'Hara, S. L., Wiggs, G. F. S., Mamedov, B., Davidson, G., and Hubbard, R. B., 2000. Exposure to airborne dust contaminated with pesticide in the Aral Sea region. *The Lancet*, **355**(9204), 627–628.

Smith, K., 2004. *Environmental Hazards: Assessing Risk and Reducing Disaster*. London/New York: Routledge.

Sheehan, L., and Hewitt, K. 1996. A pilot study of global natural disasters of the past twenty years. Working Paper No. 11, Boulder, CO: Institute of Behavioural Science, University of Colorado.

UNDRO, 1984. *Disaster Prevention and Mitigation*. New York: Office of the Disaster relief Coordinator, United Nations. Preparedness Aspects, Vol. 11.

Waltham, T., and Sholji, I., 2001. The demise of the Aral Sea an environmental disaster. *Geology Today*, **17**, 218–228.

Cross-references

Antecedent Conditions
Civil Protection and Crisis Management
Classification of Natural Disasters
Communicating Emergency Information
Community Management of Hazards
Coping Capacity
Disaster
Economics of Disasters
Exposure to Natural Hazards
History of Natural Disasters
Mass Media and Natural Disasters
Natural Hazards in Developing Countries
Perception of Natural Hazards and Disasters
Risk Perception and Communication
Vulnerability

CASE STUDY

TOHOKU, JAPAN (2011 EARTHQUAKE AND TSUNAMI)

Kenji Satake
Earthquake Research Institute, University of Tokyo, Bunkyo-ku, Tokyo, Japan

Definition

- The giant earthquake (Magnitude 9.0) off Tohoku, Japan, was the largest earthquake in Japan's history.
- It caused nearly 20,000 casualties, mostly from devastating tsunamis.
- The earthquake and tsunami also caused serious damage to the Fukushima nuclear power station, causing meltdown of the reactor, hydrogen explosion, and release of radioactive materials.
- Similar tsunami hazards occurred in the past, but the experience was not utilized to reduce the disaster.
- The lessons learned include the reexamination for preparedness for such infrequent hazards.

Introduction

A giant earthquake (official name: off the Pacific coast of Tohoku earthquake, abbreviated as the Tohoku earthquake) occurred near northern Honshu, Japan, on March 11, 2011. This earthquake, with magnitude M 9.0, was the largest in Japan's history, and produced a devastating tsunami disaster, as well as serious damage to the nearby Fukushima Dai-ichi nuclear power station.

The earthquake source parameters provided by the Japan Meteorological Agency (JMA) are: the epicenter was at 38° 06.2′ N, 142° 51.6′ E, depth 24 km, the origin time 14:46:18.1 JST. The earthquake was felt in more than a half of the Japanese islands, with the largest seismic intensity of 7 (the highest) on JMA's scale, with nearly 3,000 gals of peak ground acceleration.

The earthquake and tsunami caused about 15,900 deaths, 3,100 missing, and 6,000 injured. The damaged houses consisted of 129,000 totally collapsed, 255,000 partially collapsed, and about 697,000 partially damaged. Among them, some 7,600 houses were destroyed by ground shaking, 19,000 were damaged by liquefaction, and the rest affected by the tsunami. The total economic loss is estimated as 16,900 billion yen (about 200 billion U.S. dollars) according to the Central Disaster Management Council.

The 2011 Tohoku earthquake occurred on the boundary between the Pacific plate and overlying plate (Figure 1). The earthquake focal mechanism solution shows a thrust-type fault movement on a shallowly dipping plane. The Pacific plate subducts beneath northern Honshu at the Japan trench at a rate of about 8 cm per year. This movement usually causes westward movement of northern Honshu, as observed before 2011 by the land-based GPS network (Ozawa et al., 2011). At the time of the 2011 Tohoku earthquake, the GPS network recorded large movements in the direction opposite to the previous observations, with a maximum of 5.3 m eastward and 1.2 m downward motions (Ozawa et al., 2011). The marine geodetic measurements showed that the seafloor near the epicenter moved as much as 24 m in a horizontal direction and 3 m in a vertical direction (Sato et al., 2011). Repeated multi-beam sonar soundings also indicated that seafloor horizontally moved nearly 50 m near the trench axis (Fujiwara et al., 2011).

Unforeseen earthquake

After the 1995 Kobe earthquake, the Japanese government made long-term forecasts of large earthquakes in and around Japan based on past earthquake records. On the basis of the long-term forecast, national seismic hazard maps were published. Off Miyagi prefecture, near the epicenter of the 2011 Tohoku earthquake, large ($M{\sim}7.5$) earthquakes have repeatedly occurred since 1793 with an average interval of 37 years. On the basis of this recurrence, the probability of a great ($M{\sim}8$) earthquake in the next 30 years was estimated as 99 % (Earthquake Research Committee, 2009). Similar forecasts were also made in the neighboring regions, assuming that characteristic earthquakes repeat at each region. The forecast also estimated that a tsunami earthquake, an unusual earthquake that produces much larger tsunamis than expected from seismic waves, may occur anywhere along the Japan trench with 20% probability in the next 30 years.

The 2011 Tohoku earthquake, however, was much larger than the forecast, both in magnitude and the source area. The rupture started off Miyagi but propagated into neighboring regions. The source area was about 500-km long and 200-km wide, including the region along the Japan trench. The long-term forecast thus failed to predict the occurrence of the 2011 Tohoku earthquake, because it was based on past earthquake records and the occurrence of giant ($M{\sim}9$) earthquakes was not evident in Japan.

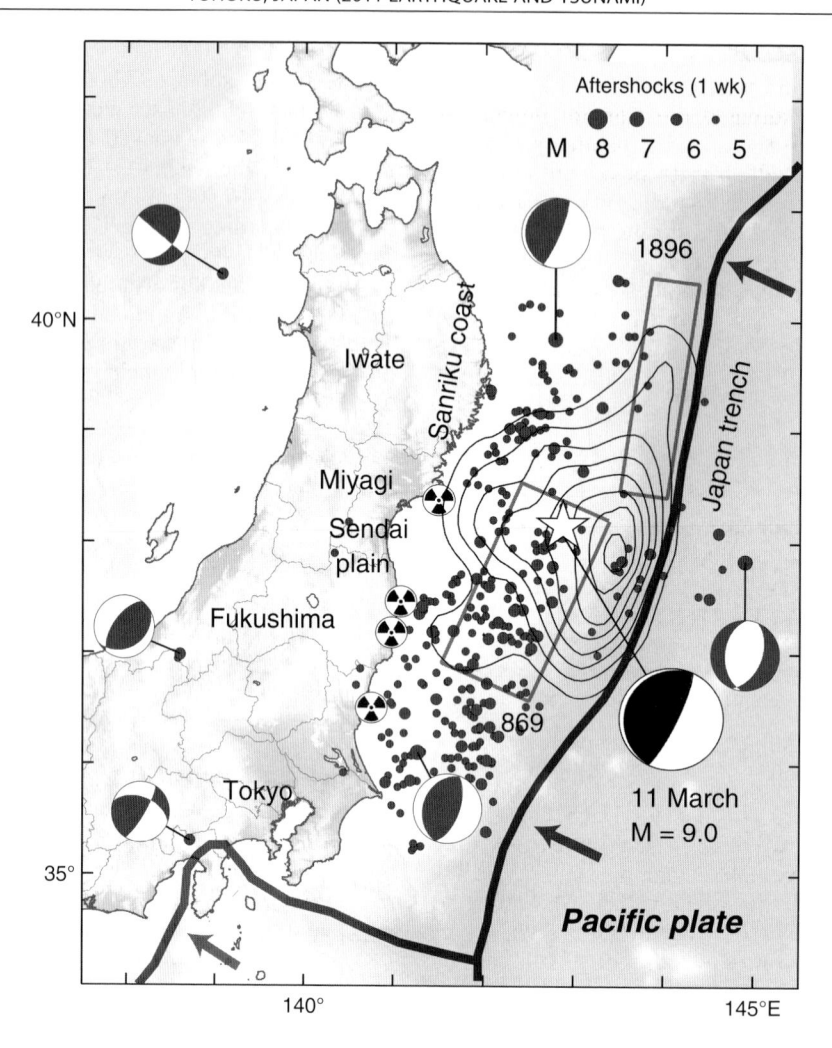

Tohoku, Japan (2011 Earthquake and Tsunami), Figure 1 The source region of the 2011 Tohoku earthquake. The mainshock (*white star*) and earthquakes with $M \geq 5.0$ occurred within a week (*black circles*) are shown with focal mechanism solutions for large ($M \geq 6.0$) earthquakes, according to Japan Meteorological Agency. Thick curves are plate boundaries and the arrows are the relative motions. The contours show the slip distribution with 4-m interval estimated from the tsunami waveforms (Satake et al., 2012). *Gray rectangles* are the fault models of the 1896 Sanriku and 869 Jogan earthquakes. Locations of four nuclear power stations are also shown.

Forgotten past tsunamis

The Sanriku coast of Tohoku had been devastated by previous tsunamis. The 1896 Sanriku earthquake caused a large tsunami, with a maximum height of 38 m, despite its weak ground shaking. It was a typical tsunami earthquake. The 2011 tsunami heights along the Sanriku coast were as high as nearly 40 m, roughly similar to the 1896 tsunami heights. The 1896 tsunami caused about 22,000 casualties, somewhat more than the 2011 tsunami. Study of tsunami waveforms indicates that the 1896 earthquake was generated from a fault motion near the trench axis (Tanioka and Satake, 1996). The 1933 Sanriku earthquake also caused tsunami, up to 24 m, and caused about 3,000 casualties.

To the south, in Sendai plain in Miyagi prefecture, the 2011 tsunami inundated about 5 km from the coast, whereas the past Sanriku tsunamis produced a few meters of coastal heights and did not flood the plain. A similarly large earthquake and tsunami occurred there in the past. A national history book depicts strong shakings, collapse of houses, kilometers of tsunami flooding with 1,000 drowned people in Sendai plain in AD 869 in Jogan era. Tsunami deposits from the Jogan earthquake had been found as far as 3 km from the coast lines in Sendai plain (Minoura and Nakaya, 1991; Sawai et al., 2008). Older tsunami deposits were also found, indicating the recurrence interval of 500–800 years if they are from the same type of earthquakes. Based on the distribution of tsunami deposits, the 869 Jogan earthquake was modeled as an interplate earthquake with $M = 8.4$ (Satake et al., 2008).

Tsunami warning: success and failure

The JMA issued a tsunami warning at 14:49, 3 min after the earthquake. They estimated the tsunami heights as 6 m on Miyagi coast, and 3 m on Fukushima and Iwate coasts, based on the initial estimate of magnitude ($M = 7.9$) and tsunami numerical simulation results stored in a database (Ozaki, 2011). Very strong ground shaking and the tsunami warning urged many coastal residents to evacuate to high ground and thus saved their lives.

The 2011 tsunami was first recorded on ocean bottom pressure and GPS wave gauges. The bottom pressure gauge about 76 km off Sanriku coast at a 1,600-m water depth recorded ∼2 m water rise in about 6 min starting immediately after the earthquake, followed by an impulsive wave with additional 3 m rise within 2 min (Fujii et al., 2011). Similar two-step tsunami waveforms were also recorded on a GPS wave gauge near the coast 12 min later, just before tsunami arrival on the coast.

The JMA upgraded the tsunami warning messages at 15:14 (28 min after the earthquake), after detecting the large offshore tsunami data. However, this updated information did not reach all the coastal communities, because a power failure occurred and the residents had already started evacuation.

The tsunami arrived at the Sanriku coast about 30 min after the earthquake, with the maximum heights of nearly 40 m. The tsunami reached the Sendai plain about 1 h after the earthquake. Despite the considerable time delay between the earthquake and tsunami arrival, nearly 20,000 people lost their lives. More than two thirds of the casualties were elderly (60 years or older), who experienced difficulties in prompt evacuation. The total area of the 2011 tsunami inundation was estimated as 561 km^2 by Geospatial Information Authority of Japan, and the population in the inundation area was about 600,000.

The JMA, based on lessons from the 2011 Tohoku earthquake, attempts to improve the tsunami warning system, including technical developments to estimate the earthquake size in a few minutes by using various and redundant information, to deploy and utilize the offshore tsunami observations, to issue a warning based on the worst case scenario if a possibility of giant earthquake exists, and to announce the expected tsunami heights in simpler expressions considering the uncertainties.

Tsunami hazard maps were made and distributed to the coastal residents. On the Sanriku coast, the 2011 tsunami inundation areas were roughly similar to the estimated flood areas. In the Sendai plain, the tsunami hazard maps assumed the M∼8 earthquake with 99% probability, and estimated much smaller inundation areas than the 2011 tsunami.

The 2011 earthquake model: deep and shallow slips

The 2011 Tohoku earthquake source has been modeled by various geophysical data, including seismic waves (e.g., Ide et al., 2011), land-based GPS data (Ozawa et al., 2011), tsunami data (Fujii et al., 2011), or combination of these (e.g., Simons et al., 2011). A common feature of the source models is that huge (30–50 m) slip occurred on a shallow plate interface near the trench axis. This was a surprise to many geophysicists, because the shallow plate interface has been considered to be weakly coupled and unable to accumulate strain.

The tsunami modeling indicates that the 2011 earthquake was a combination of a tsunami earthquake and a deeper interplate earthquake (Fujii et al., 2011). A huge slip near the trench axis, similar to the 1896 Sanriku tsunami earthquake, caused the impulsive tsunami waves to be recorded on the pressure and GPS gauges, and was responsible for the high tsunamis along the Sanriku coast. The fault motion along the deeper plate interface, similar to the previous model of the 869 Jogan earthquake, produced long-wavelength seafloor deformation and caused the first gradual rise at the gauges, as well as the large tsunami inundation in Sendai plain. The modeling also confirmed that the deeper interplate slip was essential for the 869 Jogan earthquake, but it is currently unknown whether or not the shallow huge slip also occurred in 869.

Activated seismicity in Japan

Many aftershocks followed the mainshock of March 11. Within 2 months (until May 10), five $M \geq 7$ earthquakes, 76 $M \geq 6$ earthquakes, and 449 $M \geq 5.0$ earthquakes occurred in the source region, about 500-km long and 200-km wide.

Although many aftershocks had similar mechanism to the mainshock, or a thrust faulting, some had different mechanisms. Among three $M > 7$ aftershocks within 1 h of the mainshock, two in the northern and southern ends of aftershock area had similar mechanisms to the mainshock, but one occurred in the east that had a normal fault mechanism, indicating an opposite causative stress.

The 2011 Tohoku earthquake also triggered seismicity outside the source region. Three inland earthquakes with $M > 6$ at large distances (more than a few hundred kilometers away) occurred within 7 days of the mainshock. In Fukushima prefecture, a normal fault type earthquake with M 7.0 occurred on April 11 and was followed by many aftershocks, due to east-west extensive stress, opposite to the stress condition before March 11. The number of large $M \geq 6$ earthquakes in Japan within 1 year since March 2011 was 116, including the Tohoku aftershocks, which is much larger than the annual average of 14.

Fukushima nuclear plant crisis

The 2011 tsunami also impacted four nuclear power stations (NPS) located near the source area. At these stations, the strong ground shaking automatically shut down the reactors, and the diesel generators started to cool down the reactors. The strong ground shaking also damaged the external power supply system at the Fukushima Dai-ichi station, making the station blacked out. The 15-m high tsunamis flooded and damaged the diesel generators at Fukushima Dai-ichi and Dai-ni stations. Without external power and diesel generator, the Fukushima Dai-ichi station failed to cool down and caused a meltdown of the reactors, then hydrogen explosions of three reactors, and release of radioactive materials. Areas about 20 km from the station became off limit, due to high radioactivity, and the residents were forced to evacuate. At the Fukushima Dai-ni station, the external power was used to cool down the system. At the Onagawa station, the tsunami was about 14-m high, but did not reach the level of major facilities. At the Tokai station, the tsunami height was about 6 m, but the construction of a breakwater was almost completed, which prevented major flooding.

Earthquake and tsunami hazard assessment

After the Tohoku earthquake and tsunami, the Japanese government established a general policy for future tsunami hazards. The possible future tsunamis are classified into two levels: L1 and L2. The L2 tsunamis are the largest possible tsunamis with low frequency of occurrence, but cause devastating disaster once they occur. For such events, saving people's lives is the first priority and soft measures such as tsunami hazard maps, evacuation facilities, or disaster education need to be prepared. The L1 tsunamis are expected to occur more frequently, typically once in a few decades, for which hard countermeasures such as breakwater must be prepared to protect lives and properties of residents as well as economic and industrial activities.

Conclusions

The 2011 Tohoku earthquake was the largest earthquake in Japan's history, and such a giant ($M = 9.0$) earthquake was not foreseen. However, similar tsunamis occurred in the past on the Sanriku coast from the 1896 Sanriku tsunami earthquake and in Sendai plain from the 869 Jogan earthquake. The tsunami analysis indicated that the 2011 earthquake was a combination of a tsunami earthquake with a huge shallow slip near the trench axis and a deeper interplate earthquake similar to the previously proposed model of the Jogan earthquake. The Tohoku earthquake not only caused many aftershocks but also activated the seismic activity in Japan.

Most of nearly 20,000 casualties were due to the tsunami. Although the tsunami warning saved many lives, the tsunami warning system will be further improved to help residents take more immediate actions. Education to coastal residents is important: instructing each individual to run to higher ground when they feel strong ground motion. The experience from the past tsunamis was not adequately utilized to reduce the tsunami damage.

The long-term forecast of large earthquake should incorporate paleoseismological studies such as tsunami deposits for infrequent giant earthquakes. Preparations for such infrequent hazard include hazard maps and education to save lives, but hard countermeasures are also needed for critical facilities including nuclear power stations.

Bibliography

Earthquake Research Committee, 2009. *Long-Term Forecast of Earthquakes from Sanriku-oki to Boso-oki (revised) (in Japanese)*. Headquarters for Earthquake Research Promotion, Tokyo, 63 pp.

Fujii, Y., Satake, K., Sakai, S., Shinohara, M., and Kanazawa, T., 2011. Tsunami source of the 2011 off the Pacific coast of Tohoku, Japan, earthquake. *Earth Planets Space*, **63**, 815–820.

Fujiwara, T., Kodaira, S., No, T., Kaiho, Y., Takahashi, N., and Kaneda, Y., 2011. The 2011 Tohoku-Oki earthquake: displacement reaching the trench axis. *Science*, **334**, 1240.

Ide, S., Baltay, A., and Beroza, G. C., 2011. Shallow dynamic overshoot and energetic deep rupture in the 2011 Mw 9.0 Tohoku-oki earthquake. *Science*, **332**, 1426–1429.

Minoura, K., and Nakaya, S., 1991. Traces of tsunami preserved in intertidal lacustrine and marsh deposits. *Journal of Geology*, **99**, 265–287.

Ozaki, T., 2011. Outline of the 2011 off the Pacific coast of Tohoku earthquake (Mw 9.0) -Tsunami warnings/advisories and observations. *Earth Planets Space*, **63**, 827–830.

Ozawa, S., Nishimura, T., Suito, H., Kobayashi, T., Tobita, M., and Imakiire, T., 2011. Coseismic and postseismic slip of the 2011 magnitude-9 Tohoku-Oki earthquake. *Nature*, **474**, 373–376.

Satake, K., Namegaya, Y., and Yamaki, S., 2008. Numerical simulation of the AD 869 Jogan tsunami in Ishinomki and Sendai plains (in Japanese with English abstract). *Annual Report Active Fault Paleoearthquake Research*, **8**, 71–89.

Satake, K., Fujii, Y., Harada, T. and Namegaya, Y., 2012, Time and space distribution of coseismic slip of the 2011 Tohoku earthquake as inferred from tsunami waveform data. Bulletin of the Seismological Society of America, submitted.

Sato, M., Ishikawa, T., Ujihara, N., Yoshida, S., Fujita, M., Mochizuki, M., and Asada, A., 2011. Displacement above the hypocenter of the 2011 Tohoku-oki earthquake. *Science*, **332**, 1395.

Sawai, Y., Fujii, Y., Fujiwara, O., Kamataki, T., Komatsubara, J., Okamura, Y., Satake, K., and Shishikura, M., 2008. Marine incursions of the past 1500 years and evidence of tsunamis at Suijin-numa, a coastal lake facing the Japan Trench. *The Holocene*, **18**, 517–528.

Simons, M., Minson, S. E., Sladen, A., Ortega, F., Jiang, J. L., Owen, S. E., Meng, L. S., Ampuero, J. P., Wei, S. J., Chu, R. S., Helmberger, D. V., Kanamori, H., Hetland, E., Moore, A. W., and Webb, F. H., 2011. The 2011 magnitude 9.0 Tohoku-Oki earthquake: mosaicking the megathrust from seconds to centuries. *Science*, **332**, 1421–1425.

Tanioka, Y., and Satake, K., 1996. Fault parameters of the 1896 Sanriku tsunami earthquake estimated from tsunami numerical modeling. *Geophysical Research Letters*, **123**, 1549–1552.

Cross-references

Early Warning Systems
Earthquake
Earthquake Prediction and Forecasting
Tsunami

TORINO SCALE

Norm Catto
Memorial University of Newfoundland, St. John's, NL, Canada

Torino Scale

A numerical scale measuring the statistical chance that a specific Near-Earth Object will impact Earth.

The Torino Scale (http://neo.jpl.nasa.gov/risk) was developed by researchers with NASA (USA) following a conference in Torino, Italy. It assesses the likelihood that any particular Near-Earth Object (NEO), including asteroids in Earth orbit or near the Earth, and meteorites, and comets, will impact Earth. The scale includes both the statistical chance of an impact and the consequences, based on the size of the NEO.

The trajectory for each known, individual NEO can be plotted, and the statistical probability of an impact with Earth assessed. Identification of the trajectories of NEO is one component of a response strategy to the impact hazard of NEO. Groups and agencies working to identify NEO and calculate their orbits include the Jet Propulsion Laboratory (JPL), part of NASA; the United States Air Force; the Meteorite and Impact Advisory Committee of the Canadian Space Agency; the Anglo–Australian Near-Earth Asteroid Survey; the European Asteroid Research Agency (EARA); and the Spaceguard Foundation (based in Europe).

Although the Torino Scale includes all levels of potential impacts, no rating higher than 0 has yet been determined for any NEO.

Bibliography

Bobrowsky, P., and Rickman, H. 2007. Comet/Asteroid Impacts and Human Society: an interdisciplinary approach. Berlin: Springer Verlag Publishing.

Cross-references

Asteroids
Comets

TORNADOES

Matthew R. Clark[1,2], R. Paul Knightley[2]
[1]Exeter, Devon, UK
[2]TORRO, Thelwall, Warrington, UK

Definition

Tornado: A violently rotating column of air, in contact with the ground, either pendant from a cumuliform cloud or underneath a cumuliform cloud, and often (but not always) visible as a funnel cloud.

Severe thunderstorm: A thunderstorm producing one or more of the following; hail of diameter 25.4 mm (1 in.) or greater; wind gusts of 25 ms^{-1} or greater; one or more tornadoes.
Buoyant instability: A state of the atmosphere such that an air parcel, when displaced vertically, would be accelerated in the direction of the displacement.
Temperature lapse rate: The rate of decrease of temperature with height in the atmosphere.
Vorticity: A measure of the local rotation in a fluid.
Wind shear: The local variation of the wind vector or any of its components in a given direction.

Introduction

Tornadoes are arguably the most violent of all meteorological phenomena on Earth. Indeed, wind speeds estimated from observed damage, or as measured by radar, may be higher than those generated by any other type of weather system, reaching 110–135 ms^{-1} (250–300 mph) in extreme cases (Bluestein et al., 1993; Wurman et al., 2007). Such violent tornadoes result in the wholesale destruction of natural and man-made obstacles along their narrow but well-defined damage tracks, and on occasion, very considerable loss of life. Fortunately, most tornadoes do not attain such violent intensity, though, even in the case of relatively weak tornadoes, the threat to life and property is not negligible. Owing to their small scale, however, the destruction resulting from tornadoes is extremely localized and only affects small areas. Even the widest tornadoes rarely exceed a kilometer in diameter. This may be compared, for example, with the typical diameter of a hurricane, which is on the order of hundreds of kilometers. Commensurate with their small scale, tornadoes are also generally short-lived phenomena. The longest-lived tornadoes may have lifetimes on the order of an hour or two and path lengths exceeding 100 km; for example, the infamous "Tri-State" tornado of March 18, 1925, affecting the US states of Missouri, Illinois, and Indiana, traveled over 320 km during its three and a half hour lifetime (though there is still debate as to whether this was a single tornado or in fact a series of several tornadoes along an extended track). During the so-called "Super-outbreak" of April 3–4, 1974, also in the USA, tornadoes with path lengths of up to 145 km were recorded (Corfidi et al., 2010). Such examples are exceptions to the rule however. More typically, damage may occur for periods of a few minutes, along tracks ranging from several hundred meters to several kilometers long, and lifetimes of individual tornadoes very seldom exceed an hour. Due to their generally short lifetimes and small scale, the chances of any one location being hit by a tornado are remote, even in the most tornado-prone areas on Earth. Despite this, on regional and national scales, the tornado hazard may be considerable. For example, in the USA, tornadoes are responsible for an average of around 60 deaths per annum, with most deaths attributable to injury from flying or falling debris (SPC, 2009). Annual death tolls and even insurance losses may be heavily dominated by a few, large events. So whilst the mean frequency, per unit area, of events may be low, the effect of occasional, large events can be devastating.

Tornado intensity scales

Tornadoes are most commonly classified according to their strength. In the 1970s two tornado damage scales were devised; the Fujita (F) scale (Fujita, 1971) (see *Fujita Tornado Scale*), and the International Tornado Intensity Scale (subsequently referred to as the "T scale") (Meaden, 1983; http://www.torro.org.uk/site/tscale.php). Both scales assign a rating based on the maximum intensity of observed damage, which equates, theoretically, to a given wind speed range. Since wind speed is estimated from a subjective judgment of damage intensity as opposed to being directly measured, the wind speed estimates contain much uncertainty. The nature of the object subject to damage must be considered. For example, factors such as building type, size, material, construction methods, and condition may have a significant effect on the level of damage sustained at any given wind speed. This makes standardization of the relationship between damage and wind speed essentially impossible. Other factors such as the availability or absence of objects to damage, and the translational speed of the tornado may also impact substantially on the apparent level of damage. To help address some of these problems, the Enhanced Fujita (EF) Scale was adopted as the official tornado intensity scale in the USA from 2007 (Doswell et al., 2009). The reader is referred to Doswell and Burgess (1988) for a more detailed discussion of issues related to tornado intensity rating.

The Fujita scale, comprising six categories ranging from F0 (weakest) to F5 (strongest), has gained the most widespread acceptance globally. The T scale, which comprises 11 categories ranging from T0 (weakest) to T10 (strongest), is the official tornado intensity scale in the UK and is also frequently used, for example, in parts of Europe. Tornadoes rated as F0–F1 (T0–T3) are classified as "weak" and are associated with winds of up to approximately $51~\text{ms}^{-1}$ (115 mph). Tornadoes rated at F2–F3 (T4–T7) are considered "strong," whereas those rated at F4–F5 (T8–T10), equivalent to estimated wind speeds higher than approximately $93~\text{ms}^{-1}$ (200 mph), are described as "violent." On the evidence of engineering studies of building damage, estimates of wind speed in the more intense tornadoes were revised down on introduction of the EF scale; for example, the EF3-4 threshold was revised down to $74~\text{ms}^{-1}$ (166 mph). The vast majority of tornadoes globally are rated as weak. Violent tornadoes are very rare. For example, even in the USA, which has the highest known frequency of violent tornadoes of any country in the world, only about ten such tornadoes occur across the whole country each year on average, equating to only 1% of all recorded tornadoes. F5-rated tornadoes are even rarer, accounting for around 0.1% of the total.

In addition to the damage intensity rating, tornado statistics typically recorded by research bodies and organizations such as the Tornado and Storm Research Organization (TORRO), the European Severe Weather Database (ESWD), and the US National Climatic Data Center (NCDC), include location, damage track length, maximum track width, and path direction. In some cases, more detailed surveys may be conducted which allow intensity ratings to be assigned at various points along the tornado's damage track. The overall rating given to a tornado is always that corresponding to the maximum observed damage along the track. The documentation of such information is an essential requirement for severe weather research, and for construction of accurate tornado climatologies, which permit quantitative assessments of the tornado hazard and its variation in time and space.

Meteorological conditions supportive of tornadoes

In order to understand the observed distribution of tornadoes across the globe, it is useful to consider the basic meteorological environments supportive of tornado development. A key concept is that tornadoes can occur anywhere that meteorological conditions become suitable; it naturally follows that higher tornado frequencies will tend to be found at locations experiencing those suitable conditions more frequently.

For convenience, tornadoes may be split into two main types; those associated with supercell thunderstorms, and those not associated with supercell thunderstorms (hereafter "supercell" and "non-supercell" tornadoes, respectively). The defining characteristic of a supercell thunderstorm is a deep, persistent, rotating updraft. This rotation is associated with a dynamically-induced area of low pressure, known as a mesocyclone. Rotation typically develops initially at mid-levels within the storm (i.e., 3–7 km above ground level) and the comparatively low pressure at these levels enhances the updraft at lower levels within the storm. This enhanced updraft helps to explain why supercell thunderstorms produce a disproportionate amount of other severe convective weather, including very large hail and damaging non-tornadic ("straight line") wind gusts (Moller et al., 1994; Thompson et al., 2003). In some supercells, rotation remains at mid-levels and fails to develop closer to the ground. However, in other cases, rotation does subsequently develop at ground level (a low-level mesocyclone). The development of a low-level mesocyclone involves processes related to the downdraft within the supercell, unless significant preexisting rotation about a vertical axis (vertical vorticity) is present in the storm environment. This low-level rotation can sometimes lead to tornadogenesis, via mechanisms that are still not fully understood. For a detailed review of current understanding concerning tornadogenesis within supercells, the reader is referred to Markowski and Richardson (2009).

Supercells require a rather specific combination of meteorological conditions in order to form. As is the case for all thunderstorms, three basic ingredients are required (Doswell et al., 1996): moisture, instability, and a source of lift (ascent). In addition, supercells also require strong vertical wind shear in the lowest few kilometers above ground level, that is, gradients in wind speed and direction with height above the ground. Rotation about a vertical axis develops via tilting, by the storm updraft, of initially horizontal vorticity, which is present in these strong shear environments. Observational and modeling studies in the USA have shown that supercell storms generally occur only when the "deep-layer" wind shear (0–6 km above ground level) exceeds around 15–20 ms^{-1} (30–40 knots).

Although there is an association between violent tornadoes and supercells, it should be noted that most supercells do not in fact produce tornadoes. This is because the required combination of conditions for *tornadic* supercells is even more specific than that required for the development of supercells per se; studies suggest tornadoes are more likely in environments which additionally exhibit very strong wind shear in the lowest layers of the atmosphere (in particular, combinations of strong directional and speed shear), significant instability at low levels, and low cloud base heights (indicative of high relative humidity at ground level) (e.g., Markowski et al., 2002). Hence, many conditions must be satisfied in order for supercell tornadoes to occur, which perhaps helps explain their relative rarity.

While weak to strong tornadoes occur in many different meteorological environments, it is generally accepted that the vast majority of violent, long-lived tornadoes are associated with supercells. Indeed, the relatively high frequency of violent tornadoes over the Great Plains of the USA can be linked to the relatively high frequency of occurrence of supercell thunderstorms in this region. Accordingly, the vast majority of what is currently known about supercell storms comes from research conducted in the USA. However, tornadic supercell thunderstorms have also been observed in many other parts of the World, including Canada, China, Australia, and most European countries, although in some of these places such storms are rare and supercell tornadoes may account for only a very small fraction of all reported occurrences. In northeast India and more especially Bangladesh, although scientific accounts of tornado producing storms have historically been lacking, the occurrence of strong and even violent tornadoes in environments characterized by strong instability and substantial wind shear (Yamane and Hayashi, 2006), suggests that tornadic supercells almost certainly occur from time to time. Further study is required in order to reveal how frequently such storms occur in this region.

Although supercell storms have been the focus of much research owing to their association with strong and violent tornadoes, and indeed with other types of significant severe weather, many tornadoes are not associated with supercell thunderstorms. Unfortunately, the ratio of supercell to non-supercell tornadoes, globally, is not currently known. Although non-supercell tornadoes are typically weak, they may occasionally be strong. The ingredients required for non-supercell tornadoes may arise in a number of different meteorological environments. Given that all tornadoes occur in association with moist convection, the basic ingredients of moisture, instability, and a source of lift are of course required, as is the case for supercell tornadoes. A further requirement for non-supercell tornadoes is strong preexisting vertical vorticity around and below cloud base height, and some mechanism by which this vorticity can be stretched. Stretching is required in order to amplify the preexisting vertical vorticity, which is usually rather weak, to tornadic speeds and strengths. This is generally achieved by strong updrafts, and so non-supercell tornadoes also tend to be associated with vigorous convection. Steep temperature lapse rates in the sub-cloud layer are particularly favorable, since this promotes strong stretching at low levels, where the preexisting vertical vorticity resides. Therefore, non-supercell tornadoes are possible when preexisting vertical vorticity and a convective updraft become collocated. Strong vertical wind shear is not essential. In fact, weak shear (which usually occurs in a situation of light winds throughout the depth of the troposphere) may increase the chances of non-supercell tornadoes in some circumstances, by increasing the length of time that a given updraft can reside over the area of preexisting vertical vorticity (owing to small storm motions associated with the generally weak winds).

Substantial vertical vorticity is often found along mesoscale or synoptic-scale boundaries that exhibit substantial and abrupt changes in wind speed and/or direction within a short distance. Examples occurring frequently include synoptic-scale fronts (air-mass boundaries), sea breeze fronts, outflow boundaries from preexisting convection, and terrain-induced convergence lines. Of course, boundaries of one type or another occur at almost all locations over the globe. The relative frequency with which each type occurs varies substantially from region to region however, depending on such factors as topography, homogeneity of the land surface, frequency of deep convection, and proximity to coastlines.

A further important tornado environment is the landfalling hurricane or tropical storm. Tornadoes occurring in such situations comprise a reasonable portion of the total number of observed tornadoes in Japan, eastern China, and parts of the USA adjacent to the Gulf Coast, and to a lesser extent along the Eastern Seaboard. Tornadoes are most likely in the right, front quadrant of the land-falling storm, owing to favorable configurations of vertical wind shear (McCaul, 1991). Instability in such situations is typically weak, but tornadoes tend to be associated with environments exhibiting some instability at low levels, coincident with the layer of strongest wind shear (e.g., McCaul and Weisman, 1996). Tornadoes in such environments generally occur in association with small supercell storms. Although hurricane-associated tornadoes are generally weaker than those associated with supercell storms in continental, midlatitude regions, reasonably large outbreaks of tornadoes may occur on occasion (e.g., McCaul, 1987).

Global distribution of tornado frequency

Existing databases and their limitations

One way to assess the magnitude of the tornado hazard across the globe is to consider the distribution of tornado frequency by country. Before discussing this, the limitations of the available data should be considered. A primary limitation is the absence, in many countries, of a systematic method of reporting and recording of events, which prevents reliable tornado climatologies from being constructed. Encouragingly, this situation has recently improved in many countries, with the formation of national and international bodies responsible for the collection of tornado event data. Examples include the recently established European Severe Weather Database (Dotzek et al., 2009) and TORRO (Rowe, 1985). The increased ownership of mobile phones and digital cameras and the advent of the internet has allowed for more detailed documentation and greater ease of reporting of tornadoes in recent years. Whereas in the long term this should lead to more accurate estimates of true tornado frequency, in the short term it introduces temporal trends in the apparent tornado frequency, making the construction of reliable climatologies challenging.

Since the frequency of tornadoes varies substantially on annual and even decadal timescales, climatologies must be constructed from datasets spanning at least several decades in order to be considered representative. At present, such long-period data series are only available in a few countries around the world. Even in these countries, many issues surround the quality of the data. One important consideration is underreporting. Reported frequency is always lower than the true frequency, because a certain proportion of events inevitably go unreported. This is more likely in the case of weak or short-lived tornadoes. The extent of underreporting may vary substantially from region to region, which makes meaningful comparison difficult. In sparsely populated regions, a higher proportion of events are likely to go unreported, since events are less likely to directly affect people or infrastructure. Conversely, in densely populated areas where established reporting mechanisms exist, it is likely that the large majority of events will be recorded. This bias can explain the clusters of apparently high tornado frequency occurring in and around major cities, as has been documented in the tornado climatologies of many countries around the world (e.g., Niino et al., 1997; Hanstrum et al., 2002; Meaden and Chatfield, 2009). Other issues include, but are not limited to, the misclassification of non-tornadic wind damage as tornadic and vice versa, issues relating to public education and awareness, the availability and coverage of meteorological observations such as Doppler radar, trends in population density, and changes in warning and reporting procedures. The reader is referred to Verbout et al. (2006) and Dotzek (2003) for more detailed discussions of these and other related issues.

Table 1 shows the reported tornado frequency, expressed as the number of tornadoes per 10,000 km^2 per year, for a selection of countries which possess climatologies sufficiently well developed to allow reasonably meaningful quantitative estimates to be made. When considering strong and violent tornadoes only, by far the highest reported frequencies are found in the USA, east of the Rocky Mountains (e.g., note the high frequency of F2 and stronger tornadoes in Oklahoma). When considering tornadoes of all intensities, there are some interesting differences. Some of the highest frequencies are found over parts of northwestern Europe. Indeed, reported frequencies per unit area

Tornadoes, Table 1 Reported mean annual tornado totals and frequencies (per 10^5 km^2) for a selection of countries and regions worldwide.

Location	Mean annual total (all)	Mean annual total (F2+)	Per 10,000 km^2 per year	F2+ per 10,000 km^2 per year	Period	Source
Oklahoma	57	17	3.15	0.938	1950–1995	NCDC, 2000
England	33.3	1.9	2.55	0.146	1980–2004	Kirk, 2007
UK	51	2.9	2.08	0.118	1980–2004	Kirk, 2007
Ireland	10.3	0.2	1.47	0.028	1950–2001	Tyrrell, 2003
US (contiguous 48 States)	934	215	1.16	0.266	1950–1995	NCDC, 2000
Japan	25	N/A	0.67	N/A	1961–1993	Niino et al., 1997
New Zealand	15.7	N/A	0.59	N/A	1961–1975	Tomlinson and Nicol, 1976
Taiwan	1.5	0.54	0.42	0.150	1951–1978	Wang, 1979
Austria	2.7	1.03	0.33	0.125	1946–1971	Holzer, 2001
Germany	7	2.5	0.20	0.070		Dotzek, 2001
Argentina	7.4	1.1	0.03	0.004	1930–1979	Schwarzkopf and Rosso, 1982
Australia (south and west)	7.7	N/A	0.02	N/A	1987–1996	Hanstrum et al., 2002
France	N/A	2	N/A	0.037	1960–1988	Dessens and Snow, 1989

Tornadoes, Figure 1 World map showing known areas of relatively high tornado frequency (*shaded*). *Darker shaded* areas denote regions of highest tornado frequency, according to currently available climatologies.

in the UK and the Netherlands are comparable to those in the most tornado-prone parts of the USA, including Oklahoma. Fortunately, strong and violent tornadoes are comparatively rare in these countries, as shown by the relatively low frequencies of F2 and stronger tornadoes. A further issue with derivation of quantitative estimates of tornado frequency over specified areas is illustrated by the tornado frequency for Argentina, as quoted in Table 1. Although the figure is very low for the country as a whole, there exists a much smaller area in which tornado frequency is far higher than the national average (Schwarzkopf and Rosso, 1982), outside of which tornadoes apparently seldom occur. Such regional variations occur within all countries. Further examples can be seen in Table 1 by comparison of figures for Oklahoma with those for the whole conterminous USA, and those of England with those of the whole UK. This illustrates that the obtained frequency can be highly dependant on the exact location and size of the area which is chosen for inclusion in the analysis, a further limitation of quantification and comparison of tornado frequencies by country, and indeed, regions of any arbitrary size.

Although these limitations preclude the possibility of obtaining quantitative estimates of tornado frequency in most countries, it is possible to qualitatively identify regions of enhanced tornado frequency across the globe. Fujita (1973) was the first to produce a global map showing areas of enhanced tornado frequency. Figure 1 shows a modified version of this map, as can best be constructed from current climatologies. No attempt has been made to quantify tornado frequencies within the "high frequency" areas highlighted here, owing to the uncertainties previously discussed. It can be seen that tornadoes occur in many regions of the world. In fact, tornadoes of strong and violent intensity (i.e., F2 (T4) and greater) have been documented on every continent, with the exception of Antarctica. At the broadest level, tornadoes occur most frequently in the midlatitudes, that is, between approximately 25° and 55° north and south of the equator. Regions of enhanced tornado frequency within these midlatitude belts include large parts of North America, northern and central parts of Europe, Japan, South Africa, parts of the Indian subcontinent, parts of Southern and Western Australia, New Zealand and parts of Argentina. Some explanations for the observed distribution of the more tornado-prone areas are offered in the following sections.

Areas of low tornado frequency

Bearing in mind the range of tornado-producing environments described previously, and the principal conditions required for tornado formation, it is possible to explain some of the broader aspects of the observed global distribution of tornado frequency. First consider areas with lowest observed frequencies. Many of the areas of apparently low tornado frequency are also areas of very low population density. This is no coincidence. In fact, it is almost certain that population bias has a strong impact on the apparent global extent and distribution of areas of low tornado frequency, as shown in Figure 1. However, the population density itself is influenced strongly by climate, and there are good meteorological reasons to expect low tornado frequencies in some of the sparsely populated parts of the world. Figure 1 shows that tornadoes tend to be absent from areas with desert or semiarid climates. Low tornado frequencies would be expected in these areas simply because deep, moist convection occurs rarely, if at all, owing to the lack of moisture. Occasional storms which do form are usually high based owing to low relative humidity, and such storms are not conducive to tornado development. This helps to explain the minima in tornado activity over large parts of North Africa, the west coast of South America, much of inland western North America, and parts of the interiors of Russia and Australia.

Figure 1 also shows that tornadoes are rare or absent at latitudes higher than about 60° in both hemispheres. Antarctica is the only continent on which tornadoes have not been observed. This is because incursions of warm, moist (and originally potentially unstable) air are strongly modified by the time they reach these latitudes. The surface layers are cooled as they travel over relatively cold ocean or land surfaces, and so surface-based convection is suppressed. Cold air also holds less moisture than warm air; consequently, particularly cold climates are also usually arid climates. As in other arid regions, population density tends to be low. In these respects, the polar regions represent the extreme case; intensely cold, dry conditions, and weak solar heating entirely prohibit surface-based convection, while vast tracks of land are entirely uninhabited.

Inspection of Figure 1 also shows a general minimum in tornado activity over tropical regions. This is perhaps surprising, given the fact that thunderstorm activity is frequent and widespread there. A major limiting factor for tornadoes in tropical regions is the lack of significant wind shear. The tropics are well removed from the upper level jet streams which are found at higher latitudes, and the associated baroclinic disturbances which act to periodically increase vertical wind shear. This means that environments conducive to supercell thunderstorms do not generally occur. This likely explains an apparent lack of strong and violent tornadoes in the tropics. A further result of the distance from baroclinic disturbances is that well-marked air mass boundaries associated with frontal systems, one important environment for the development of non-supercell tornadoes, also seldom occur in the tropics. On the other hand, the presence of widespread instability and weak wind shear suggests that, despite the lack of recorded events, non-supercell tornadoes do occur from time to time. Again, population bias likely plays a role; underreporting could contribute significantly to the apparently low tornado frequencies in some tropical regions. For example, the lack of violent supercell tornadoes may have resulted in a general lack of awareness of the phenomenon, reducing the chances that events will be reported. It remains to be seen whether tornado reports increase in the future as awareness and documentation of tornado events improves.

Areas of high tornado frequency

Northwest Europe

Some of the highest frequencies of tornadoes are to be found over northwestern Europe, including countries such as the UK and the Netherlands. In this area, the prevailing wind is off the Atlantic Ocean which results in a reliable and abundant supply of moisture throughout the year. Midlatitude depressions (low pressure systems) also affect this region with a high frequency. These baroclinic systems, deriving additional energy from warmth and moisture provided by the Atlantic Ocean, are often well developed and vigorous as they reach northwest Europe. They are frequently associated with strong vertical wind shear, sufficient for supercell storms. Instability in the maritime air masses is generally small, however, which often limits the potential for supercells. Nevertheless, such storms do occasionally occur. With the abundance of low pressure systems also comes an abundance of well-defined air mass boundaries, such as cold fronts. Strong cold fronts, along which lines of convection develop, are characterized by strong low-level convergence and vertical vorticity. Tornadoes are frequently observed in association with such cold fronts and, less commonly, post-frontal squall lines. In some cases, the development of small-scale (diameter ≤ 4 km) vortices and tornadoes may be associated with horizontal shearing instability along the vertical vortex sheet associated with the narrow zone of wind veer at the cold front (e.g., Carbone, 1982; Smart and Browning, 2009). However, in general, these cold frontal and post-frontal squall lines are not well understood from a dynamical perspective, and much remains to be discovered about tornadogenesis in this type of environment.

A substantial number of the UK's tornadoes occur in association with cold fronts and post-frontal squall lines (Bolton et al., 2003). Such tornadoes tend to be rather weak and short-lived, which largely explains why deaths resulting from tornadoes are very rare despite the high overall tornado frequency. For example, only two tornado fatalities were documented in the UK between 2000 and 2009, though approximately 30 injuries were recorded during the same period (G. T. Meaden, personal communication 2011). Strong tornadoes have, however, been documented in this type of situation. An example is the T5 (F2) tornado which affected northwest London on December 7, 2006 (Clark, 2011). Tornadoes associated with cold fronts also occur rather frequently in other midlatitude west coastal localities, including parts of western Canada, California, and the west coast of Australia. Island nations of the midlatitudes may also be susceptible to such events; examples include Japan and New Zealand.

Another potentially relevant factor in parts of northwest Europe is the coastline. Relative to the size of each country, the coastlines of the UK and Netherlands, for example, are very long, which means that sea breeze boundaries commonly affect substantial areas, especially in the summer months. Such boundaries frequently act as focal regions for the development of deep, moist convection. It is likely that tornadoes are also favored in such situations, owing to the strong vertical vorticity often present, associated with strong shear of the horizontal wind across the sea breeze front. However, no clear link has yet been demonstrated between areas of frequent sea breeze formation and elevated tornado frequencies. In the winter, cold air outbreaks moving south over the relatively warm waters (the Gulf Stream ensuring that waters close to northwest Europe are unusually warm for their latitude) result in strong instability at low levels. Deep convection is often observed in such situations, and occasionally tornadoes have been documented. Again, it could be speculated that coastline-generated convergence lines provide environments favorable for tornado development. However, such cold air outbreaks are also frequently associated with rather strong vertical wind shear; consequently, small supercell storms may also be responsible for some of these tornadoes. Therefore, further research is required in order to elucidate the relevant physical processes in these situations. Strong instability generated by cold air flowing over relatively warm waters also occurs quite regularly over places such as the Mediterranean Sea and the waters surrounding Japan, which can help explain more frequent cool-season tornado activity in areas adjacent to these water bodies.

US Great Plains and Midwest

Parts of the central third of the USA, including the Great Plains and much of the Midwest, have the dubious distinction of experiencing the highest documented frequency of strong and violent tornadoes in the world. Within this region tornadoes constitute a substantial hazard to life and property. As a result, tornadoes and tornadic storms in this area have been the subject of a tremendous amount of scientific study in recent decades. This has led to a much greater understanding of the types of situation in which tornadoes occur. Public awareness of the tornado hazard in this region is high, and this, together with an established and comprehensive forecasting and warning infrastructure significantly reduces the number of deaths and injuries sustained as a result of tornadoes. Within this region, tornado activity reaches a well-defined maximum in the spring and early summer months. In basic terms, this is because increasing solar elevation raises instability, while the upper level midlatitude jet stream, though on average retreating gradually northward at this time of year, is still far enough south to permit episodes of sufficiently strong vertical shear over the region. Therefore, the required combination of instability and strong wind shear occurs most frequently at this time of year. As the mean position of the jet stream moves northward through spring and into early summer, so the location of maximum tornado activity tends to shift northward across the region.

Peak activity in Oklahoma and northern Texas, for example, typically occurs in May, while the peak occurs in June and July over the Dakotas and other northern States (Brooks et al., 2003).

The combination of instability and substantial vertical wind shear favorable for supercell thunderstorms occurs from time to time across many continental regions of the midlatitudes, especially during the warm season. Accordingly, tornadic supercell storms occasionally occur in these regions too; for example, over central and eastern Europe (e.g., Caspar et al., 2009) and the Canadian Prairies (Kumjian and Ryzhkov, 2008). However, there are a number of additional factors which set the central US region apart from other midlatitude continental regions. One is the presence of the Rocky Mountains to the west. A consequence of westerly flow across the Rockies in middle and upper levels of the troposphere is the development, at the surface, of a lee trough over the high Plains region. This results in southerly flow over the Plains to the east, which transports warm and very moist air originating over the Gulf of Mexico hundreds or even thousands of miles northward (Figure 2). Instability is further increased by the frequent presence of an elevated mixed layer over the same region. This warm, dry layer of air originates over the high Plateaux of the southwestern USA and parts of Mexico and is advected northeastward over the Plains above the surface layer of moist air (Lanicci and Warner, 1991). This initially caps convection, thus allowing the instability associated with the surface moist layer to build further over the Plains, until it is released explosively as convection is triggered by the approach of atmospheric disturbances from the west.

Because the land slopes gently upward toward the west over the whole region, the depth of the moist surface layer decreases to the west. The western boundary of the moist air is marked by a feature called the "dryline." This boundary frequently acts as a focus for the initiation of deep convection, which then typically moves northeast over the Plains. This kind of setup sometimes favors concentrated "outbreaks" of severe thunderstorms and tornadoes over parts of the Plains, especially when stronger middle and upper level disturbances arrive from the west. These outbreaks occasionally result in the occurrence of several tens or even in excess of 100 tornadoes, within the space of 24–48 h. Such outbreaks have the potential to cause significant loss of life, but fortunately they can now usually be forecast in general terms several days in advance. To give an example of a significant, though unexceptional, tornado outbreak over the US Plains, Figure 2 also shows the locations of tornado reports on June 11, 2008. Forty-six separate tornadoes occurred on this day, resulting in 7 deaths, 52 injuries, and over $24 million of damages in total (NCDC, 2009).

Bangladesh and Northeast India

It is interesting to note that some of the features which make parts of the USA favorable for violent tornado development may also be found in one or two other regions around the world. One example is Bangladesh and the northeast region of India. In this area, the Bay of

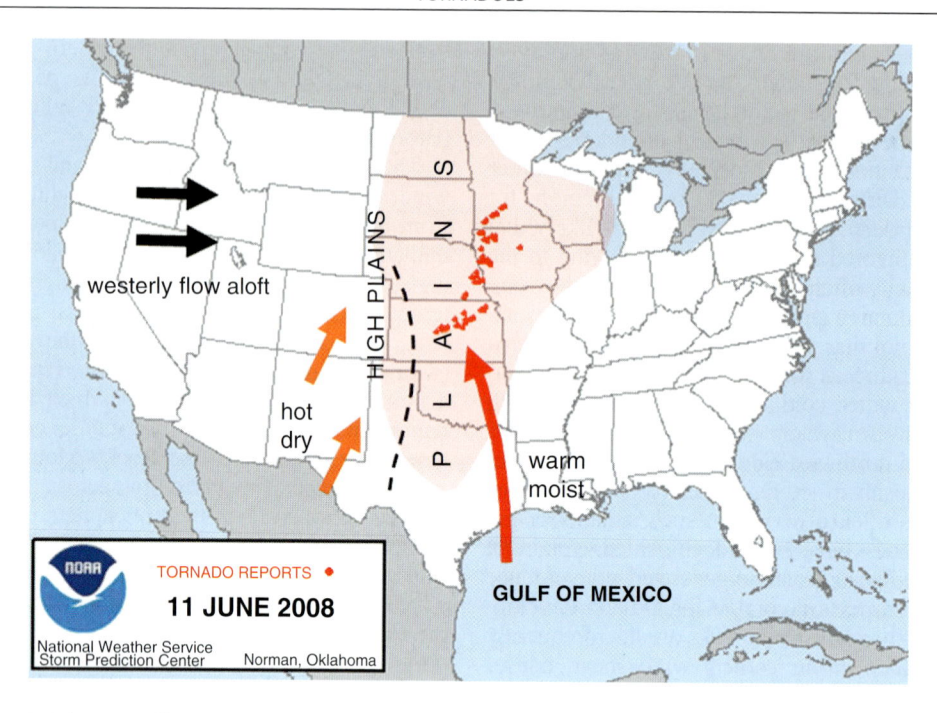

Tornadoes, Figure 2 Schematic illustration of a typical synoptic setup for severe thunderstorm and tornado outbreaks over the US Great Plains region. *Red circles* show the locations of tornado reports received by the Storm Prediction Center (SPC) during the outbreak of June 11, 2008.

Bengal provides a source of warmth and abundant moisture at low levels. Elevated terrain to the west of the region allows for the development of an elevated mixed layer which, when advected into the region from the west, helps to cap instability associated with low level moisture allowing it to build until it may be released explosively, in the same way that an elevated mixed layer originating in the southwestern US Plateaux helps to cap instability over the US Plains. A lee trough also frequently occurs east of this elevated terrain inducing surface flow off the Bay of Bengal and transporting high moisture further inland. Locations at such low latitudes as Bangladesh and northeast India are normally well removed from the midlatitude westerly jet streams. However, in this region, the Himalayas and Tibetan Plateau to the north act as a huge barrier, diverting a portion of the jet stream well to the south of its typical latitudes. This not only increases wind shear to levels which may support supercell thunderstorm development, but larger-scale ascent associated with disturbances within the jet stream, also occasionally diverted over the region, act to trigger severe thunderstorm systems periodically. A brief but well-marked period of tornado activity occurs over the region, peaking in April as the pre-monsoon heat and humidity begin to build, but before the upper level jet stream retreats well to the north of the Himalayas. Singh (1981) also notes that tornadoes have very occasionally been associated with landfalling tropical cyclones (hurricanes) over Bangladesh.

As mentioned previously, violent tornadoes have been documented in this region (Singh, 1981; Mandal and Saha, 1983). Although the total number of such tornadoes in a typical year is not as great as that in the US Plains and Midwest, a combination of high population density in parts of the region, low public awareness of the hazard, and an absence of warning systems means that such events occasionally result in extremely high numbers of fatalities. The worst recent example occurred in Daultipur, Bangladesh, on April 26, 1989. This event is estimated to have resulted in around 1,300 deaths. In the period 1978–1998, nine tornadoes have been documented which resulted in over 100 deaths each (Finch, 2009). Tornadoes resulting in this many deaths are comparatively rare in the USA. Unfortunately however, such events still occur on occasion, as demonstrated most recently on May 22, 2011 when a violent tornado struck the city of Joplin, Missouri, resulting in 158 deaths.

Forecasting and warning
History
Given the hazards associated with tornadoes, it is desirable for weather forecasting agencies to attempt to predict them. Tornado forecasting could be said to have had its roots in the nineteenth century, largely out of the work undertaken by J.P. Finley (Galway, 1985). European settlers advancing into the frontier land of the US Great Plains would have started to experience thunderstorms and tornadoes of ferocity far greater, in general, than they had ever witnessed beforehand. However, it was not until the 1950s that tornado forecasting really began to be taken seriously, and

scientifically (Doswell et al., 1993). Prior to this, the word "tornado" was banned from public forecasts. Perhaps the most famous early attempt at a tornado forecast by meteorologists was undertaken by Major Ernest Fawbush and Captain Robert Miller (of the US Air Force), who went on to produce pioneering work on the subject.

Their initial forecast was issued on March 25, 1948, for Tinker Air Force Base, Oklahoma. Five days prior to this, a tornado had torn through the base, causing more than $10 million damage (in 1948 dollars), injuring several men, and destroying aircraft. It was estimated by Miller that the odds of another tornado hitting the base within a few days of this were around 20 million to 1. On the day of this initial tornado, the forecast had been for gusty winds, and an Air Force inquiry the next day ruled that a tornado "was not forecastable given the present state of the art." On March 22, and for the next 3 days, Fawbush and Miller made a highly concentrated effort to observe and document every piece of available meteorological information which had preceded the March 20 tornado, including as much data as they could find about the upper air. In addition, they scrutinized data surrounding previous tornado outbreaks to attempt to ascertain which parameters were present, and whether there was any set "pattern" to the weather beforehand. The weather maps for March 25 showed remarkably similar patterns to those which had turned up in their research. Based on this information, at 2.50 pm on that afternoon, they issued a prediction for a tornado. This triggered emergency procedures, which included rolling as many aeroplanes as possible into the hangers, and locations such as the control tower were evacuated. Incredibly, at 6 pm another tornado did hit Tinker Air Force Base. It did $6 million damage, but thanks to the warning, no one was injured. From that point on, the Air Force began to use this so-called Fawbush and Miller technique to issue tornado forecasts, and this attracted attention from the media. Several subsequent forecasts were deemed successful.

To put this into perspective, it must be appreciated that it is extremely unusual to have successfully predicted a tornado *for a specific site*. Modern forecasters would not attempt to predict a tornado strike for a specific site, 3 h hence; this is despite a much greater understanding of the atmosphere and the processes involved in severe storms and tornadogenesis. Ironically, it is for this very reason that tornadoes are not predicted for specific sites – meteorologists understand that this is simply not possible beyond guesswork and luck. So, although pioneering in many respects, the successful outcome of this initial tornado forecast also relied very heavily on chance. However, an important outcome of this, and subsequent successful Air Force forecasts was in making many people's minds up: tornado forecasts should be made publicly available so that people and businesses can take evasive and protective action. In 1952, the US Weather Bureau was forced to establish a tornado forecasting center. This was the forerunner of today's Storm Prediction Center (SPC; www.spc.noaa.gov).

Modern forecasting

Formal forecasting of severe weather for civilian use started in the 1950s. A specialized unit was set up in March 1952, and in early 1953 this became known as the Severe Local Storms (SELS) Center, and moved to Kansas City, Missouri in 1954. This is what subsequently became the Storm Prediction Center (SPC).

Early forecasts by SELS was a largely empirical process. It had been observed by forecasters and researchers that certain meteorological elements in the larger scale were apparent in several tornado outbreaks. Examples of these include the presence of significant areas of low pressure (extratropical cyclones), abundant low-level moisture, and the proximity jet streams. However, it soon became apparent, even in these early days, that no single set of features was present in each case. Certain features were present in certain groups of cases, and this led to pattern recognition in forecasting severe weather.

While, undoubtedly, some major outbreaks of severe weather across the Great Plains region have very similar features present, pattern recognition is not a desirable method of severe weather forecasting, since it will not reliably identify all outbreaks of severe weather.

With the advent of numerical weather prediction and hugely powerful computers, day-to-day weather forecasting has become much more accurate than was the case even 15–20 years ago. A trend away from pattern recognition and empirical techniques, toward more physical principles, has taken place. However, tornadoes are very small-scale and transient phenomena and so cannot be resolved by current computer modeling, and certainly not from a forecasting point of view. This is unlikely to improve soon, not least because the processes which cause tornadoes are still not fully understood. For these reasons, the method of severe thunderstorm and tornado forecasting typically employed today uses an "ingredients-based" approach for identifying areas at risk of severe storm or tornado development. The forecaster will use a number of weather forecast models, along with current weather observations at both the surface, and in the upper atmosphere. In simple terms, the forecast is based on identification of the four key ingredients required for severe thunderstorm formation: moisture, instability, lift, and wind shear. Where these are coincident in space and time, severe thunderstorms and tornadoes *might* occur. However, there are no thresholds over which the values of these parameters guarantee the development of severe weather; the forecaster is charged with quantifying the *risk* of severe weather for a particular area, based on all the available information.

As the USA suffers from severe thunderstorms and tornadoes on a regular basis, it is no surprise to find that this is the country with the most advanced severe weather forecast dissemination system. The Storm Prediction Center (SPC – a division of the National Weather Service), in Norman, Oklahoma, is responsible for the forecasting of organized severe thunderstorms and tornadoes across the contiguous 48 states. It is also responsible for monitoring heavy rain, heavy snow, and fire weather events, and warning the public as necessary.

The SPC issues a range of forecasts which cover varying timescales in order to give timely notice of possible severe weather. "Outlooks" cover the 1–8 day period. There is one forecast product covering the 4–8 day period. This typically highlights possible risks in the medium term, but at that range, detail on the exact locations, timing, and nature of the threat is usually not possible. There are also individual outlooks for days one, two, and three. The "Day Three" forecast is updated once per day; "Day Two," twice per day; and "Day One," five times daily. The "Day One" Convective Outlook provides guidance in technical meteorological language along with maps to highlight areas at risk from organized severe thunderstorms. The text provides information regarding the specific nature of the threats, timing of the threads, and their expected severity. Graphics include a categorical forecast of the severe risk, broken down into regions of slight, moderate, and high risk, along with an area depicting 10% or greater chance of thunderstorms. There are separate maps depicting the probabilities for severe hail, wind, and tornadoes.

During a typical severe weather day, as conditions become more favorable for severe weather, the SPC may issue a *mesoscale discussion (MCD)*. Once conditions become favorable for the development of organized severe thunderstorms and tornadoes, the SPC issues a severe thunderstorm or tornado watch. Although tornadoes can develop in both types of watch, tornado watches are usually issued when conditions become favorable for multiple tornadoes or strong tornadoes. These watches are designed to alert the public to the risk of rapidly changing weather conditions, and to review severe weather safety rules, as well as listening out for possible warnings. For storm spotters, broadcast media, and emergency managers, watches can allow time to bring in extra staff and gear up operations. A watch does not guarantee severe weather, and is certainly not a warning of imminent danger. The watch area covers specified counties within one or more states and typically covers an area of 50,000–100,000 square kilometers (20,000–40,000 square miles), although this can vary rather widely. Even in a successful case, only a tiny fraction of the area covered by the tornado watch would be expected to experience a direct tornado "hit." Typically, between 800 and 1,000 severe thunderstorm and tornado watches will be issued each year. They are numbered sequentially starting afresh on January 1 each year.

When severe weather is either imminent or actually underway, weather warnings will be issued to alert communities in the projected path of the severe weather-producing storm or storms. One of the most useful tools for deciding when and where to issue warnings is Doppler weather radar. Doppler radar is capable of the detection of rotation within a storm, something which may be a precursor to tornado development. To this end, tornado warnings are often issued when strong rotation has been detected in the storm, even if no tornado has yet been reported. However, such remote sensing techniques are not fail-safe. Radar-based warning relies on detectable rotation, or other known severe weather "signatures," developing before tornado occurrence, something which does not occur in every event. On the other hand, false alarm rates may be high since the development of storm rotation does not guarantee the development of a tornado. Also consider that it typically takes an operational weather radar unit around 5 min to complete a full volume scan of the sky around it, while tornadoes can develop within seconds. Despite these limitations however, Doppler radars are probably the most useful tool available to forecasters to aid in the warning making process, in real time. For this reason, and because of the existence of other applications for radar data, Doppler radar networks are now being set up in many countries worldwide. Figure 3 shows an example of Doppler radar data from the UK Met Office, showing an area of rotation associated with the tornado of December 7, 2006 which affected parts of London.

Warnings may also be issued based on actual reports of severe weather. In the USA, *Storm spotters* will be activated when severe weather is expected. These are trained individuals who voluntarily observe conditions during periods of heightened risk. They are charged with calling or radioing-in severe weather reports, and are coordinated by a meteorologist at the local NWS office. Law enforcement officials also act as storm spotters, as do some storm chasers. In a rapidly developing situation, such reports are essential to maintain real-time monitoring of the storms.

Warnings are issued via NOAA Weather Radio, as well as to broadcast media, and to government officials. Emergency managers in towns and cities will usually sound civil defense sirens when a tornado warning has been issued. In addition, many TV stations and radios will switch to continuous severe weather coverage, with their own storm chasers/spotters in the field, and sometimes helicopters. These can all bring live footage of tornadoes to viewers' TV screens. Indeed, in many of the more tornado prone parts of the USA, severe weather coverage is deemed a ratings-winner, and the stations make a big point of advertising their coverage as "the best."

There are a number of other organizations around the world who attempt to forecast severe thunderstorms and tornadoes in their own countries. In the UK and Eire, the Tornado and Storm Research Organization (TORRO) is one such organization. Severe thunderstorm and tornado watches have been issued since 1992, and since 2006 they have been publicly available of the TORRO website (www.torro.org.uk). Being a voluntary organization means that coverage is not provided 24 h per day, but forecasts are issued as often as is necessary, and possible, by professional meteorologists. Across Europe, a group of meteorologists and meteorology students have set up the European Storm Forecast Experiment (Estofex). The group issues daily forecasts of severe thunderstorms

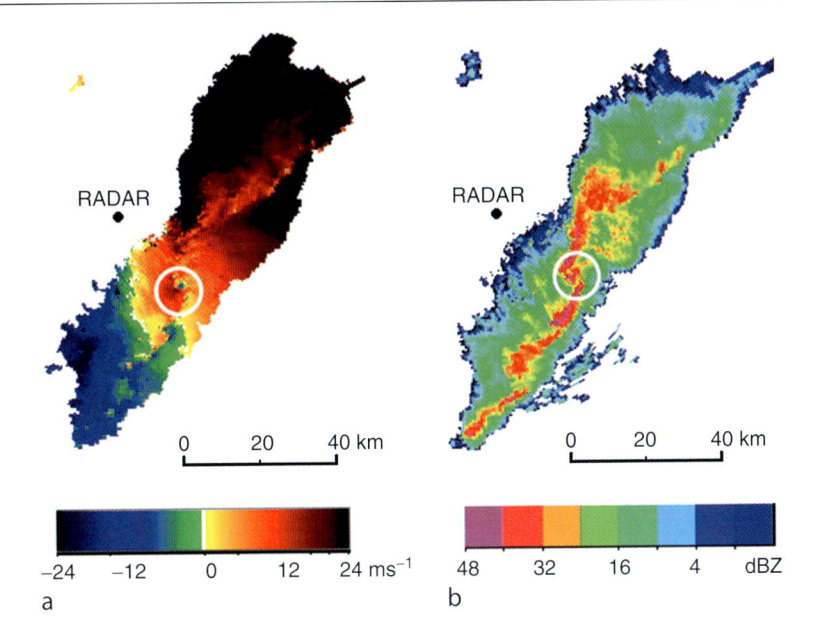

Tornadoes, Figure 3 Doppler radar observations of a tornadic squall line in the UK. (**a**) radial winds: positive (negative) values indicate a component of wind directed away from (toward) the radar, and (**b**) radar reflectivity (precipitation intensity) at 1056 UTC December 7, 2006. At this time, a T5 tornado was on the ground in west London. The parent circulation associated with this tornado is shown by the region of strong inbound and outbound winds located in close proximity to each other (*circled*). Radar location is shown by the *black dot* in each panel (**a**) (Crown Copyright Met Office 2009).

Tornadoes, Figure 4 The Baca County, Colorado, tornado of May 31, 2010. This tornado was rated EF2 following a damage survey by the US National Weather Service. Fortunately, since the tornado moved over open ground, the damage was mostly to power poles and no injuries were reported (Copyright Matthew Clark 2010).

across Europe, in a similar manner to the SPC's "Day One" outlooks. Much like TORRO, this is a voluntary group.

The future of tornado forecasting will likely focus on attempting to determine which thunderstorms have the best potential for producing tornadoes, with the aim of reducing warning false alarm rates. Currently, many supercells which do not produce tornadoes become tornado warned due to the detection of rotation. Higher resolution radar and a denser network of radars may help to discriminate between tornadic and non-tornadic supercells in future. Future forecasting will also likely focus on trying to better anticipate those areas most likely to see thunderstorm development in the short term (i.e., the next few hours) as this will allow more targeted watches to be issued. However, given that we are far from understanding the full complexity of the tornado, storm spotters in the field, issuing real-time reports, will likely be required for many years to come and possibly indefinitely.

For more information about tornado forecasting in the USA, the reader is referred to Doswell et al. (1993) (Figure 4).

Summary

Although even the largest outbreaks of tornadoes do not generally result in widespread destruction and large death tolls on the scale sometimes associated with other natural hazards, such as earthquakes and volcanic eruptions, the relatively high frequency of tornadoes, the locally intense nature of their damage, and their possible occurrence in many of the inhabited regions on the planet ensures that they are justifiably considered as one of the Earth's primary natural hazards. Owing to their small scale and short lifetimes, observing and forecasting tornadoes remains a challenging task. In spite of some remarkable advances in these areas in recent decades, much remains to be discovered. Continued improvements in our understanding of the exact mechanisms responsible for tornadogenesis, the range of environments in which tornadoes can occur, and more accurate and complete observation and documentation of events, amongst other things, are vital if further advances are to be made.

To this end, tornado research is still very much an active and developing area of meteorological science. In the USA, where much of the tornado research conducted thus far has been based, dedicated field campaigns continue (e.g., VORTEX2; www.vortex2.org/home/), with the aim of observing tornadoes and their immediate environments at higher temporal and spatial resolution than has yet been possible. For example, the use of phased array, polarimetric radars, capable of providing data at very high temporal resolution, has provided additional insight into sub-storm scale processes whose evolution is typically very rapid (e.g., Heinselman et al., 2008; Heinselman and Torres, 2011). High-resolution numerical models, used in conjunction with polarimetric radar observations, provide insight into microphysical aspects of severe storms (i.e., processes taking place on the scale of individual aerosol or precipitation particles), which also likely influence the potential for severe weather, including tornadoes (e.g., Dawson et al., 2010; Bryan and Morrison, 2012). Collectively, these efforts will help to address some of the remaining questions regarding tornadogenesis, in particular. Outside of the USA, interest in tornadoes and severe convective weather has also increased in recent years. Advances in observing capability and development of warning procedures are now occurring in some countries; the benefits of these improvements in terms of reduction of injury and death tolls should soon be realized. In other countries however, little, if anything has yet been done to address the hazard posed by tornadoes.

Bibliography

Bluestein, H. B., Ladue, J. G., Stein, H., Speheger, D., and Unruh, W. F., 1993. Doppler radar wind spectra of supercell tornadoes. *Monthly Weather Review*, **121**, 2200–2222.

Bolton, N., Elsom, D. M., and Meaden, G. T., 2003. Forecasting tornadoes in the United Kingdom. *Atmospheric Research*, **67–68**, 53–72.

Brooks, H. E., Doswell, C. A., III, and Kay, M. P., 2003. Climatological estimates of local daily tornado probability for the United States. *Weather and Forecasting*, **18**, 626–640.

Bryan, G. H., and Morrison, H., 2012. Sensitivity of a simulated squall line to horizontal resolution and parameterization of microphysics. *Monthly Weather Review*, **140**, 202–225.

Carbone, R. E., 1982. A severe frontal rainband. Part I. Stormwide hydrodynamic structure. *Journal of the Atmospheric Sciences*, **39**, 258–279.

Caspar, R., Labbe, L., and Jakob, E., 2009. Les tornades en France: generalites et analyse de l'evenement du 3 aout 2008 en val de Sambre. *La Meteorologie*, **8**, 31–42.

Clark, M. R., 2011. Doppler radar observations of mesovortices within a cool-season tornadic squall line over the UK. *Atmospheric Research*, **100**, 749–764.

Corfidi, S. F., Weiss, S. J., Kain, J. S., Corfidi, S. J., Rabin, R. M., and Levit, J. J., 2010. Revisiting the 3–4 April 1974 super outbreak of tornadoes. *Weather and Forecasting*, **25**, 465–510.

Dawson, D. T., II, Xue, M., Milbrandt, J. A., and Yau, M. K., 2010. Comparison of evaporation and cold pool development between single-moment and multimoment bulk microphysics schemes in idealized simulations of tornadic thunderstorms. *Monthly Weather Review*, **138**, 1152–1171.

Dessens, J., and Snow, J. T., 1989. Tornadoes in France. *Weather and Forecasting*, **4**, 110–132.

Doswell, C. A., III, and Burgess, D. W., 1988. On some issues of United States tornado climatology. *Monthly Weather Review*, **116**, 495–501.

Doswell, C. A., III, Weiss, S. J., and Johns, R. H., 1993. Tornado forecasting: a review. In *The Tornado: Its Structure, Dynamics, Prediction, and Hazards*. Washington, DC: American Geophysical Union. Geophysical Monograph, Vol. 79, pp. 557–571.

Doswell, C. A., III, Brooks, H. E., and Maddox, R. A., 1996. Flash flood forecasting: an ingredients-based methodology. *Weather and Forecasting*, **11**, 560–581.

Doswell, C. A., III, Brooks, H. E., and Dotzek, N., 2009. On the implementation of the enhanced Fujita scale in the USA. *Atmospheric Research*, **93**, 554–563.

Dotzek, N., 2001. Tornadoes in Germany. *Atmospheric Research*, **56**, 233–251.

Dotzek, N., 2003. An updated estimate of tornado occurrence in Europe. *Atmospheric Research*, **67–68**, 153–161.

Dotzek, N., Groenemeijer, P., Feuerstein, B., and Holzer, A. M., 2009. Overview of ESSL's severe convective storms research using the European Severe Weather Database ESWD. *Atmospheric Research*, **93**, 575–586.

Finch, J. D., 2009. Bangladesh and East India tornadoes. Available online, http://bangladeshtornadoes.org/bengaltornadoes.html. Accessed December 15, 2009.

Fujita, T. T., 1971. Proposed characterisation of tornadoes and hurricanes by area and intensity. *Satellite and Mesometeorology Research paper* No. 91.

Fujita, T. T., 1973. Tornadoes around the world. *Weatherwise*, **26**, 56–62.

Galway, J. G., 1985. J.P. Finley: the first severe storms forecaster. *Bulletin of the American Meteorological Society*, **66**, 1389–1395.

Hanstrum, B. N., Mills, G. A., Watson, A., Monteverdi, J. P., and Doswell, C. A., III, 2002. The cool-season tornadoes of California and Southern Australia. *Weather and Forecasting*, **17**, 705–722.

Heinselman, P. L., and Torres, S. M., 2011. High-temporal-resolution capabilities of the national weather radar testbed phased-array radar. *Journal of Applied Meteorology and Climatology*, **50**, 579–593.

Heinselman, P. L., Priegnitz, D. L., Manross, K. L., Smith, T. M., and Adams, R. W., 2008. Rapid sampling of severe storms by the national weather radar testbed phased array radar. *Weather and Forecasting*, **23**, 808–824.

Holzer, A. M., 2001. Tornado climatology of Austria. *Atmospheric Research*, **56**, 203–211.

Kirk, P., 2007. UK tornado climatology 1980–2004. *The International Journal of Meteorology*, **32**, 158–172.

Kumjian, M. R., and Ryzhkov, A. V., 2008. Polarimetric signatures in supercell thunderstorms. *Journal of Applied Meteorology and Climatology*, **47**, 1940–1961.

Lanicci, J. M., and Warner, T. T., 1991. A synoptic climatology of the elevated mixed-layer inversion over the southern great plains in spring. Part III: relationship to severe-storms climatology. *Weather and Forecasting*, **6**, 214–226.

Mandal, G. S., and Saha, S. K., 1983. Characteristics of some recent north Indian tornadoes. *Vayu Mandal*, **13**, 74–80.

Markowski, P. M., and Richardson, Y. P., 2009. Tornadogenesis: our current understanding, forecasting considerations, and questions to guide future research. *Atmospheric Research*, **93**, 3–10.

Markowski, P. M., Straka, J. M., and Rasmussen, E. N., 2002. Direct surface thermodynamic observations within the rear-flank downdrafts of nontornadic and tornadic supercells. *Monthly Weather Review*, **130**, 1692–1721.

McCaul, E. W., 1987. Observations of the hurricane "Danny" tornado outbreak of 16 August 1985. *Monthly Weather Review*, **115**, 1206–1223.

McCaul, E. W., 1991. Buoyancy and shear characteristics of hurricane-tornado environments. *Monthly Weather Review*, **119**, 1954–1978.

McCaul, E. W., and Weisman, M. L., 1996. Simulations of shallow supercell storms in landfalling hurricane environments. *Monthly Weather Review*, **124**, 408–429.

Meaden, G. T., 1983. The TORRO tornado intensity scale. *Journal of Meteorology*, **8**, 151–153.

Meaden, G. T., and Chatfield, C. R., 2009. Tornadoes in Birmingham, England, 1931 and 1946 to 2005. *The International Journal of Meteorology*, **34**, 155–162.

Moller, A. R., Doswell, C. A., III, Foster, M. P., and Woodall, G. R., 1994. The operational recognition of supercell thunderstorm environments and storm structures. *Weather and Forecasting*, **9**, 327–347.

NCDC, 2000. U.S. tornado climatology. Available online, http://www.ncdc.noaa.gov/oa/climate/severeweather/tornadoes.html. Accessed December 30, 2009.

NCDC, 2009. NCDC storm events database. Available online, http://www4.ncdc.noaa.gov/cgi-win/wwcgi.dll?wwEvent~Storms. Accessed January 20, 2010.

Niino, H., Fujitani, T., and Watanabe, N., 1997. A statistical study of tornadoes and waterspouts in Japan from 1961 to 1993. *Journal of Climate*, **10**, 1730–1752.

Rowe, M. W., 1985. TORRO, the tornado and storm research organization. The main objectives and scope of the network. Part B. The work of the tornado division of TORRO. *Journal of Meteorology*, **10**, 186–187.

Schwarzkopf, M. L., and Rosso, L. C., 1982. Severe storms and tornadoes in Argentina. In *Preprints 12th Conference on Severe Local Storms*, San Antonio, pp. 59–62.

Singh, R., 1981. On the occurrence of tornadoes and their distribution in India. *Mausam*, **32**, 307–314.

Smart, D. J., and Browning, K. A., 2009. Morphology and evolution of cold-frontal misocyclones. *Quarterly Journal of the Royal Meteorological Society*, **135**, 381–393.

SPC, 2009. Frequently asked questions about tornadoes. Available online, http://www.spc.noaa.gov/faq/tornado/. Accessed November 10, 2009.

Thompson, R. L., Edwards, R., Hart, J. A., Elmore, K. L., and Markowski, P., 2003. Close proximity soundings within supercell environments obtained from the rapid update cycle. *Weather and Forecasting*, **18**, 1243–1261.

Tomlinson, A. I., and Nicol, B., 1976. Tornado reports in New Zealand 1961–1975. *New Zealand Meteorological Service Technical Note 229*.

Tyrrell, J., 2003. A tornado climatology for Ireland. *Atmospheric Research*, **67–68**, 671–684.

Verbout, S. M., Brooks, H. E., Leslie, L. M., and Schultz, D. M., 2006. Evolution of the U.S. tornado database: 1954–2003. *Weather and Forecasting*, **21**, 86–93.

Wang, G. C. Y., 1979. Tornadoes in Taiwan. In *Preprints Eleventh Conference on Severe Local Storms*, Kansas City, pp. 216–221.

Wurman, J., Alexander, C., Robinson, P., and Richardson, Y., 2007. Low-level winds in tornadoes and potential catastrophic tornado impacts in urban areas. *Bulletin of the American Meteorological Society*, **88**, 31–46.

Yamane, Y., and Hayashi, T., 2006. Evaluation of environmental conditions for the formation of severe local storms across the Indian subcontinent. *Geophysical Research Letters*, **33**, L17806.

Cross-references

Aviation, Hazards to
Building, Structures and Public Safety
Doppler Weather Radar
Fujita Tornado Scale
Thunderstorms
Waterspout

TRIGGERED EARTHQUAKES

Harsh K. Gupta

Government of India, NDMA Bhawan, New Delhi, India
National Geophysical Research Institute, Hyderabad, India

Introduction

Under certain suitable geological conditions, anthropogenic activity can trigger or induce earthquakes. The triggered/induced earthquakes are known to have occurred due to gold and coal mining, petroleum production, filling of artificial water reservoirs, high-pressure liquid injection into ground, and natural gas production. The largest triggered, scientifically accepted earthquake of magnitude 6.3 occurred on December 10, 1967 in the vicinity of Koyna Dam near the west coast of India. It is debated whether the M 7 Gazli earthquakes of May 1976 and

March 19, 1984 were induced due to the production of large quantities of gas at the Gazli Oil Field in Uzbekistan. A good account of the above and the mechanism of triggered/induced seismicity can be found in a review by McGarr et al. (2002).

Triggered vis-a-vis induced earthquakes

For a long time, the adjectives "induced" and "triggered" were used interchangeably whenever one talked of artificially simulated earthquakes. McGarr and Simpson (1997) have addressed this question and suggested that it would be important to draw a distinction between the two. They proposed that the adjective "triggered seismicity" should be used only when a small fraction of stress change or energy associated with earthquakes is accounted for by the causative activity. The term "induced seismicity" should be used where the causative activity is responsible for a substantial part of the stress change. In case of triggered seismicity, tectonic loading plays an important role. The stress level changes associated with filling of some of the deepest artificial water reservoirs are only of the order of 1 M Pa or so, whereas the stress drop associated with the earthquakes is much larger. Therefore, all cases of earthquakes occurring subsequent to filling of the artificial water reservoirs fall in the category of "triggered earthquakes," and hence it is appropriate to call it "reservoir triggered seismicity" (RTS).

In the present entry, we concentrate on triggered earthquakes caused by the filling of the artificial water reservoirs.

Reservoir triggered seismicity (RTS)

Generation of hydroelectric power, irrigation, and flood control has necessitated creation of artificial water reservoirs globally. Triggering of earthquakes was for the first time pointed out by (Carder, 1945) at Lake Mead in the USA. Figure 1 depicts Lake Mead water levels and local seismicity. The rise in water levels and the corresponding bursts in seismicity are numbered. The correspondence is indeed remarkable.

Over 100 reservoir sites are now known globally, where triggered earthquakes occurred after filling of the artificial water reservoirs (Gupta, 2002, 2008). Earthquakes exceeding magnitude 6 have occurred at Kariba, Zambia-Zimbabwe border; Hsinfengkiang, China; Kremasta, Greece and Koyna, India. Koyna earthquake of M 6.3 that occurred on December 10, 1967 is so far the largest triggered earthquake. It claimed over 200 human lives, injured about 1,500 and rendered thousands homeless. The occurrence and potential of triggered earthquakes has caused major modification of civil works and engineering projects. Anticipating a large triggered earthquake, the Hsinfengkiang Dam was strengthened twice before the occurrence of M 6.1 earthquake on March 20, 1962 (Shen et al., 1974). The disposal of waste fluid through injection into the ground at Rocky Mountain Arsenal had to be discontinued due to triggered earthquakes (Evans, 1966). The possibility of high magnitude triggered seismicity was responsible for terminating the Auburn Dam project in California (Allen, 1978). A number of researchers believe that the M 7.9 Wenchuan earthquake of May 12, 2008 in China, which claimed around 90,000 lives, may have been triggered by the near by Zipingu reservoir (Kerr and Stone, 2009).

In Table 1, the reservoir sites where triggered earthquakes exceeding M 4 occurred are listed. Many studies examined the correspondence among several possible correlates such as the size of the reservoir, time interval for the largest RTS event to occur from the first filling of the reservoir, height of the water column in the reservoir, rate of loading, and the RTS. The most important correlate is the depth of the water column in the reservoir (Baecher and Keeny, 1982). Figure 2 shows that when the water column depth exceeds 150 m, a quarter of reservoirs experienced RTS. A review of recent global review supports this finding. Artificial water reservoirs with water column exceeding 100 m and/or volume exceeding 1 m^3 are called large reservoirs (ICOLD, 2008). There are over 1,000 large reservoirs in the world and only a small percentage of these reservoirs have evidenced RTS. One should remember, however, that many pre-1990s hydroelectric

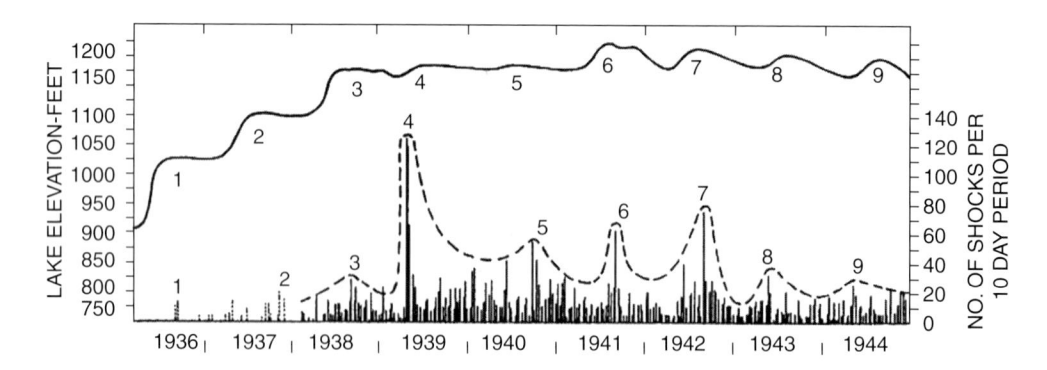

Triggered Earthquakes, Figure 1 Lake Mead water levels and the local seismicity. For 1936 and 1937, only the felt shocks are plotted. The rises in water levels and the corresponding bursts of seismic activity are numbered. General trend of tremor-frequency variation is shown by *dotted lines* (After Carder, 1945).

Triggered Earthquakes, Table 1 Reported cases of reservoir-triggered seismicity (RTS) where M $\geq$ 4 earthquake occurs.

Name of the dam/Reservoir	Country	Height of dam (m)	Reservoir volume (10^6 m^3)	Year of impounding	Year of the largest earthquake	Magnitude/Intensity	References
Sites where earthquakes having magnitude $\geq$ 6.0 were triggered							
Hsinfengkiang	China (PRC)	105	13,896	1959	1962	6.1	Gupta and Rastogi (1976); Packer et al. (1979); Shen et al. (1974)
Kariba	Zambia-Zimbabwe	128	175,000	1958	1963	6.2	Gupta and Rastogi (1976); Packer et al. (1979); Gough and Gough (1970b)
Koyna	India	103	2,780	1962	1967	6.3	Gupta and Rastogi (1976); Packer et al. (1979); Rothe (1970, 1973); Bozovic (1974)
Kremasta	Greece	160	4,750	1965	1966	6.2	Gupta and Rastogi (1976); Packer et al., (1979); Rothe (1970, 1973); Bozovic (1974)
Sites where earthquakes having magnitude between 5.0 and 5.9 were triggered							
Aswan	Egypt	111	1,64,000	1964	1981	5.6	Packer et al. (1979); Toppozada (1982)
Benmore	New Zealand	110	2,040	1964	1966	5.0	Gupta and Rastogi (1976); Packer et al. (1979); Adams (1974)
Charvak	Uzbekistan	148	2,000	1971	1977	5.3	Plotnikova et al. (1992)
Eucumbene	Australia	116	4,761	1957	1959	5.0	Packer et al. (1979)
Geheyan	China	151	3,400	1993	1997	VI	Chen et al. (1996)
Hoover	USA	221	36,703	1935	1939	5.0	Gupta and Rastogi (1976); Packer et al. (1979); Carder (1945)
Marathon	Greece	67	41	1929	1938	5.7	Gupta and Rastogi (1976); Packer et al. (1979); Rothe (1970, 1973); Bozovic (1974)
Oroville	USA	236	4,400	1967	1975	5.7	Packer et al. (1979); Bufe et al. (1976)
Srinagarind	Thailand	140	11,750	1977	1983	5.9	Chung and Liu (1992)
Warna	India	80	1,260	1985	1993	5.0	Rastogi et al. (1997)
Sites where earthquakes having magnitude between 4.0 and 4.9 were triggered							
Akosombo Main	Ghana	134	148,000	1964	1964	V	Packer et al. (1979); Simpson (1976)
Bajina Basta	Yugoslavia	90	340	1966	1967	4.5–5.0	Packer et al. (1979); Bozovic (1974)
Bhatsa	India	88	947	1981	1983	4.9	Rastogi et al. (1986)
Bratsk	Russia	100	169		1996	4.2	Pavlenov and Sherman (1996)
Camarillas	Spain	49	37	1960	1964	4.1	Packer et al. (1979); Rothe (1970, 1973); Bozovic (1974)
Canelles	Spain	150	678	1960	1962	4.7	Packer et al. (1979); Rothe (1970, 1973); Bozovic (1974)
Capivari–Cachoeira	Brazil	58	180	1970	1971	VI	Berrocal (personal communication, 1989)
Clark Hill	USA	60	3,517	1952	1974	4.3	Packer et al. (1979); Talwani (1976)
Dahua	China (PRC)	74.5	420	1982	1993	4.5	Guang (1995)
Danjiangkou	China (PRC)	97	16,000	1967	1973	4.7	Oike and Ishikawa (1983)
Foziling	China (PRC)	74	470	1954	1973	4.5	Oike and Ishikawa (1983)
Grandval	France	88	292	1959	1963	V	Gupta and Rastogi (1976); Packer et al. (1979); Rothe (1970, 1973); Bozovic (1974)
Hoa Binh	Vietnam	125		1988	1989	4.9	Tung (1996)
Kastraki	Greece	96	1,000	1968	1969	4.6	Packer et al. (1979)
Kerr	USA	60	1,505	1958	1971	4.9	Gupta and Rastogi (1976); Packer et al. (1979); Simpson (1976)
Komani	Albania	130	1,600	1985	1986	4.2	Muco (1991)
Kurobe	Japan	186	149	1960	1961	4.9	Packer et al. (1979); Hagiwara and Ohtake (1972)
Lake Baikal	Russia					4–4.8a	Djadkov (1997)

Triggered Earthquakes, Table 1 (Continued)

Name of the dam/ Reservoir	Country	Height of dam (m)	Reservoir volume (10^6 m^3)	Year of impounding	Year of the largest earthquake	Magnitude/ Intensity	References
Lake Pukaki	New Zealand	106	9,000	1976	1978	4.6	Reyners (1988)
Manicouagan 3	Canada	108	10,423	1975	1975	4.1	Packer et al. (1979)
Marimbondo	Brazil	94	6,150	1975	1975	IV	Veloso et al. (1987)
Monteynard	France	155	275	1962	1963	4.9	Gupta and Rastogi (1976); Packer et al. (1979); Rothe (1970, 1973); Bozovic (1974)
Nurek	Tajikistan	317	1,000	1972	1972	4.6	Gupta and Rastogi (1976); Packer et al.(1979); Soboleva and Mamadaliev 1976)
P. Colombia/ V.Grande	Brazil	40/56	1,500/2,300	1973–1974	1974	4.2	Berrocal et al. (1984)
Piastra	Italy	93	13	1965	1966	4.4	Packer et al. (1979); Rothe (1970, 1973); Bozovic (1974)
Pieve de Cadore	Italy	116	69	1949	1950	V	Packer et al. (1979); Caloi (1970)
Shenwo	China (PRC)	50	540	1972	1974	4.8	Oike and Ishikawa (1983)
Vouglans	France	130	605	1968	1971	4.4	Packer et al. (1979); Rothe (1970, 1973); Bozovic (1974)
Karun-III	Iran	185	2,970	2005	2005	4.3	Kangi and Heidari (2008)

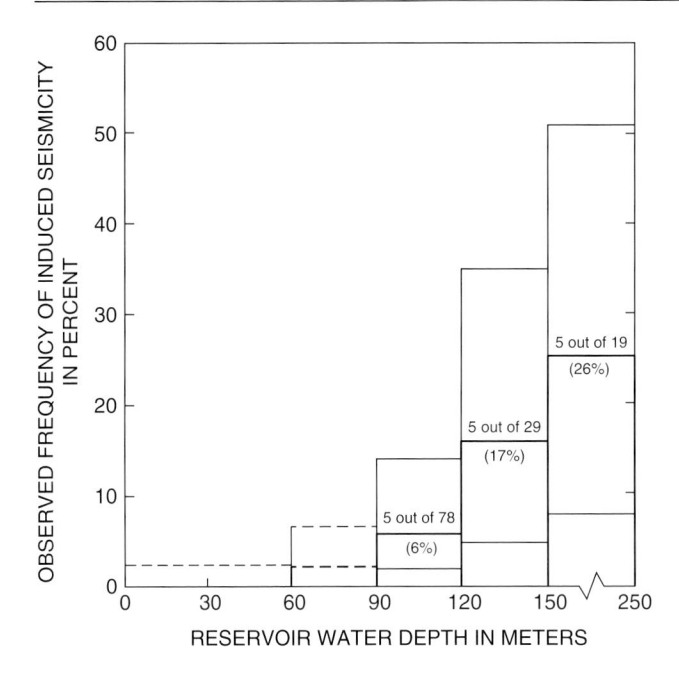

Triggered Earthquakes, Figure 2 Height of water column is the most important correlate (After Alexander and Mark, 1976).

developments did not have seismographic capacity to detect small induced earthquakes. Hence, one cannot discard the possibility that small earthquakes below the detection threshold of some areas could have been triggered.

How to assess the RTS potential and what would be the largest RTS event at a given reservoir site, is frequently debated. Although the phenomenon of RTS is not yet fully understood, it is clear that the stress changes caused by the reservoir are small and that the region must be stressed close to critical for triggered earthquakes to occur. It is possible to estimate the likelihood of RTS by measuring the in situ stresses at a site at depth and see how close to failure it is. Such experiments were conducted by (Zoback and Hickman, 1982) at Monticello Reservoir at South Carolina, USA, and are still used. The magnitude of an RTS event would not exceed the maximum credible earthquake of the concerned region. The possibility of mitigating triggered earthquakes through the manipulation of the water levels in the reservoirs was first demonstrated by (Simpson and Negmatullaev, 1981) at the Nurek dam in Tajikistan and (Gupta, 1983) at Koyna, India. Similar suitable approaches can be made at other sites taking into consideration the locale specific situations and parameters.

Koyna, India, continues to be the most significant RTS site where since the impoundment of the reservoir in 1962, 22 M $\geq$ 5 earthquakes (including the December 10, 1967 M 6.3 earthquake), over 200 M $\geq$ 4 and several thousand smaller earthquakes have occurred. All this RTS is confined to a small area of 30 × 20 km, and there is no other seismically active area within 50 km of the dam site (Gupta, 2008).

Summary

Anthropogenic activities can trigger/induce earthquakes. Earthquakes triggered by filling of the artificial water reservoirs have been damaging at several locations. Proper geoscientific evaluation of site conditions and in situ stress measurements can help in finding safer sites.

Bibliography

Adams, R. D., 1974. The effect of Lake Benmore on local earthquakes. *Engineering Geology*, **8**, 155–169.

Allen, C. R., 1978. *Evaluation of Seismic Hazard at the Auburn Damsite*. California, U.S. Bureau of Reclamation Report, Denver, CO, 10 pp.

Baecher, B. G., and Keeney, R. L., 1982. Statistical examination of reservoir induced seismicity. *Bulletin Seismology Society America*, **72**, 553–569.

Berrocal, J., Assumpcao, M., Antezana, R., Dias Neto, C. M., Ortega, R., Franca, H., and Velose, J., 1984. *Sismicidade do Brasil Instituto Astronomico e Geofisico*. Sao Paulo: Universidade de Sao Paulo, p. 320.

Berrocal, J., Fernandes, C., Antezana, R., Shukowsky, R., Barbosa, J. R., Shayani, S., and Pereira, E. S., 1989. *Induced Seismicity by the Sobradinho Reservoir*. Bahia (personal communication).

Bozovic, A., 1974. Review and appraisal of case histories related to seismic effects of reservoir impounding. *Engineering Geology*, **8**, 9–27.

Bufe, C. G., Lester, F. W., Lahr, K. M., Lahr, J. C., Seekins, L. C., and Hanks, T. C., 1976. Oroville earthquakes: normal faulting in Sierra Nevada foothills. *Science*, **192**, 72–74.

Caloi, P., 1970. How nature reacts on human intervention—responsibilities of those who cause and who interpret such reaction. *Anna Geofisica*, **23**, 283–305.

Carder, D. S., 1945. Seismic investigations in the Boulder Dam area, 1940–1944, and the influence of reservoir loading on earthquake activity. *Bulletin of the Seismological Society of America*, **35**, 175–192.

Chen, B., Li, S., and Yin, Z., 1996. *On the Characteristics and Prediction of Induced Earthquakes of the Geheyan Reservoir*. Abstract Vol. IASPEI Regional Assembly in Asia, August 1–3, Tangshan.

Chung, W. Y., and Liu, C., 1992. The reservoir-associated earthquakes of April 1983 in Western Thailand: source modeling and implications for induced seismicity. *PAGEOPH*, **138**(1), 17–41.

Djadkov, P. G., 1997. *Induced Seismicity at the Lake Baikal; Principal Role of Load Rate*. Abstract Vol. IASPEI General Assembly, August 18–28, Thessaloniki.

Evans, M. D., 1966. Man made earthquakes in Denver. *Geotimes*, **10**, 11–17.

Gough, D. I., and Gough, W. I., 1970. Load induced earthquakes at Kariba. *Geophysical Journal of the Royal Astronomical Society*, **21**, 79–101.

Guang, Y. H., 1995. Seismicity induced by cascade reservoirs in Dahan, Yantan Hydroelectric Power Stations. In *Proceedings International Symposium on Reservoir-Induced Seismicity*. State Seismological Bureau, Beijing, pp. 157–163.

Gupta, H. K., 1983. Induced seismicity hazard mitigation through water level manipulation at Koyna, India: a suggestion. *Bulletin of the Seismological Society of America*, **73**, 679–682.

Gupta, H. K., 2002. A review of recent studies of triggered earthquakes by artificial water reservoirs with special emphasis on earthquakes in Koyna, India. *Earth-Science Reviews*, **58**(3–4), 279–310.

Gupta, H. K., (2008), *Artificial Water Reservoir Triggered Earthquakes, with Special Emphasis on Koyna Earthquakes, India*. Memoir 66, Golden Jubilee Memoir of the Geological Society of India, pp. 395–422.

Gupta, H. K., and Rastogi, B. K., 1976. *Dams and Earthquakes*. Amsterdam: Elsevier, p. 229.

Hagiwara, T., and Ohtake, M., 1972. Seismic activity associated with the failing of the reservoir behind Kurobe Dam, Japan 1963–1970. *Tectonophysics*, **15**, 241–254.

ICOLD (International Commission on Large Dams), 2008. "Reservoir and Seismicity: State of Knowledge" Bulletin (International Commission on Large Dams); Rough version of Bulletin 137, 50 p.

Kangi, A., and Heidari, N., 2008. Reservoir-induced seismicity in Karun III dam (Southwestern Iran). *Journal of Seismology*, **12**, 519–527.

Kerr, R. A., and Stone, R., 2009. A human trigger for the great quake of Sichuan? *Science*, **323**(5912), 322, doi:10.1126/science.323.5912.322.

McGarr, A., and Simpson, D., 1997. Keynote lecture: a broad look at induced and triggered seismicity, Rockbursts and seismicity in mines. In Gibowicz, S. J., and Lasocki, S. (eds.), *Proceedings of 4th International Symposium On Rockbursts and Seismicity in Mines*, Poland, August 11–14, A.A. Balkema, Rotterdam, pp. 385–396.

McGarr, A., Simpson, D., and Seeber, L., 2002. Case histories of induced and triggered seismicity. *International Handbook of Earthquake and Engineering Seismology*, Amsterdam: Elsevier, Vol. 81 A, pp. 647–661.

Muco, B., 1991. The swarm of Nikaj-Merturi, Albania. *Bulletin of the Seismological Society of America*, **81**, 1015–1021.

Oike, K., and Ishikawa, Y., 1983. Induced earthquakes associated with large reservoirs in China. *Chinese Geophysics*, **II**(2), 383–403.

Packer, D. R., Cluff, L. S., Knuepfer, P. L., and Withers, R. J., 1979. *A study of reservoir induced seismicity*. Woodward-Clyde Consultants USA. U.S. Geological Survey Contract 14-08-0001-16809 (unpublished report).

Pavlenov, V. A., and Sherman, S. I., 1996. *Premises of Induced Seismicity on the Reservoirs of the Angare River*. Abstract Vol. IASPEI Regional Assembly in Asia, August 1–3, Tangshan.

Plotnikova, L. M., Makhmudova, V. I., and Sigalova, O. B., 1992. Seismicity associated with the Charvak reservoir. *Pure and Applied Geophysics*, **139**, 607–608.

Rastogi, B. K., Chadha, R. K., and Raju, I. P., 1986. Seismicity near Bhatsa Reservoir, Maharashtra, India. *Physics Earth Planet International*, **44**, 179–199.

Rastogi, B. K., Chadha, R. K., Sarma, C. S. P., Mandal, P., Satyanarayana, H. V. S., Raju, I. P., Kumar, N., Satyamurthy, C., and Nageswara Rao, A., 1997. Seismicity at Warna reservoir (near Koyna) through 1995. *Bulletin of the Seismological Society of America*, **87**(6), 1484–1494.

Reyners, M., 1988. Reservoir induced seismicity at Lake Pukaki. *Geophysical Journal*, **93**, 127–135.

Rothe, J. P., 1970. The seismic artificials (man-made earthquakes). *Tectonophysics*, **9**, 215–238.

Rothe, J. P., 1973. A geophysics report. In Ackermann, W. C., White, G. F., and Worthington, E. B. (eds.), *Man-Made Lakes: Their Problems and Environmental Effects*. Washington, DC: American Geophysical Union. Geophysical Monograph, Vol. 17, pp. 441–454.

Shen, C., Chang, C., Chen, H., Li, T., Hueng, L., Wang, T., Yang, C., and Lo, H., 1974. Earthquakes induced by reservoir impounding and their effect on the Hsinfengkiang Dam. *Scientia Sinica*, **17**(2), 239–272.

Simpson, D. W., 1976. Seismicity changes associated with reservoir loading. *Engineering Geology*, **10**, 123–150.

Simpson, D. W., and Negmatullaev, S. K., 1981. Induced seismicity at Nurek Reservoir, Tadjikistan, USSR. *Bulletin of the Seismological Society of America*, **71**(5), 1561–1586.

Soboleva, O. V., and Mamadaliev, U. A., 1976. The influence of the Nurek Reservoir on local earthquake activity. *Engineering Geology*, **10**, 293–305.

Stuart-Alexander, D. E., and Mark, R. K., 1976. *Impoundment-Induced Seismicity Associated with Large Reservoirs*. U.S. Geological Survey, Open file Report, pp. 76–770.

Talwani, P., 1976. Earthquakes associated with Clark Hill Reservoir, South Carolina—a case of induced seismicity. *Paper Presented at the 1st International Symposium on Induced Seismicity*. Engineering Geology, Vol. 10, pp. 239–253.

Toppozada, T. R., 1982. *UNDP/Token Report on Aswan Earthquakes*.

Tung, N. T., 1996. *The Induced Seismicity at Hoa Binh Reservoir Region*. Abstract Vol. IASPEI Regional Assembly in Asia, August 1–3, Tangshan.

Veloso, J. A. V., Assumpcao, M., Concalves, E. S., Reis, J. C., Duarte, V. M., and Motta da, C. B. G., 1987. *Registro de sismicidade induzida em reservatorios da CEMIG e FURNAS*. An 50 Congr. Bras. Geol. Eng. Vol. 1, pp. 135–146.

Zoback, M. D., and Hickman, S., 1982. Physical mechanisms controlling induced seismicity at Monticello Reservoir, South Carolina. *Journal of Geophysical Research*, **87**, 6959–6974.

Cross-references

Body Wave
Earthquake
Earthquake Damage
Earthquake Prediction and Forecasting
Elastic Rebound Theory
Harmonic Tremor
Hazardousness of Place
Induced Seismicity
Isoseismal
Mercalli, Giuseppe (1850–1914)
Primary Wave (P Wave)
Reservoir Dams and Seismicity
Secondary Waves (S Waves)
Seismography/Seismometer
Seismology

TSUNAMI

William Power, Graham S. Leonard
GNS Science, Lower Hutt, New Zealand

Synonyms

Seismic sea-wave; Tidal wave (obsolete)

Definition

Tsunami. From Japanese *tsu*, harbor, and *nami*, wave. A wave, or series of waves, generated when a large volume of water is vertically/horizontally displaced by an impulsive disturbance such as an earthquake, landslide, or volcanic eruption. Tsunami are distinguished from regular sea waves by their long wavelength and period. "Tsunami" and "tsunamis" are both used for the plural in

English. There is no pluralizing suffix "s" used in the Japanese language.
Tsunami run-up height. The elevation above sea level at a point along the maximum inundation extent of a tsunami. The sea level datum should be specified; often the ambient sea level at the time of the tsunami is used.
Tsunami run-up distance. The horizontal distance from the coast line to a point along the maximum inundation extent of a tsunami.
Tsunami wave height. The height of a tsunami wave, measured either relative to the ambient sea level or from the peak to the trough of the wave. These are referred to as the *zero to peak wave height* or the *peak to trough wave height*, respectively.

Introduction

Tsunami are a natural process recorded every year on tide gauges around the world. They can occur in any ocean, sea, or lake provided there is some way the water body (or a connected water body) can be suddenly displaced vertically/horizontally, usually by an *earthquake*, and less often by a *volcanic eruption* or *landslide*. Tsunami large enough to cause injury to people and damage to property recur globally on the order of years to decades. Tsunami are often mistakenly called "tidal waves" and although tides play no part in the source of tsunami, the arrival of a tsunami at high tide can be significantly more hazardous than if it arrives at low tide – especially on coasts with a large tidal range. The word "tsunami" is adopted from Japanese – tsunami disasters have punctuated Japan's history due to that country's proximity to large faults associated with plate boundaries and the presence of many volcanoes. For example, in 1792 a volcanic earthquake on Mt Unzen triggered a landslide that descended into the nearby bay causing a tsunami with run-up heights of 35–55 m in Shimbara; the tsunami and landslide killed approximately 15,000 people in total (Lockridge, 1990). In recent years, the Indian Ocean tsunami of December 26th, 2004, and the Tohoku, Japan, tsunami of March 11, 2011, have raised public awareness of the hazard from very large subduction zone earthquake-generated tsunami.

Tsunami can propagate across the deep ocean at high speed, sometimes exceeding 700 km/h, and with relatively little dissipation of energy. They can consequently be dangerous far from the source, for example, the tsunami caused by the giant Chilean earthquake of 1960 claimed 61 lives in Hawaii and 138 in Japan (Atwater et al., 2005). Despite the high propagation speed there is still the potential to provide warning ahead of a long-range tsunami, as it takes, e.g., 12–24 h for a tsunami to travel across the Pacific basin. A Pacific-wide tsunami warning system has been in operation since the late 1960s.

The primary geological causes of tsunami are earthquakes, landslides, and volcanic eruptions. Earthquakes initiate tsunami principally through coseismic uplift or subsidence of the seafloor, which displaces the water above. Submarine landslides cause tsunami by displacing water to make way for the descending landslide material, and leaving a volume to be filled by water in the landslide's wake. Volcanic events displace water by a wide variety of mechanisms, some of which involve gravitational collapse similar to landslides, whereas others are associated with explosive events. Weather systems are also capable of creating tsunami-like waves, known as storm-surges or "meteorological tsunami" (Monserrat et al., 2006). The flooding of New Orleans associated with Hurricane Katrina in 2005 was one such event (Fritz et al., 2007). Although frequent and sometimes devastating events, they are generally regarded as a different category of event to tsunami, because they have no geological cause.

Physical properties of tsunami

When a body of water is suddenly displaced vertically, gravity acts to try and restore equilibrium. The dynamics of this process cause waves to propagate away from the initial disturbance. Tsunami waves are thus an example of gravity waves, for which the restoring force is gravity. In a tsunami these waves are able to propagate long distances with little dissipation, and this property of tsunami means that, for example, a tsunami caused by an earthquake on one side of the Pacific can be destructive to people and property on the other side of the Ocean.

A distinguishing feature of a tsunami is the long period of its waves; these can last from several minutes to an hour or more, in contrast to wind waves which have periods ranging from a few seconds to about a minute. One consequence of this is that tsunami inundation can be much more dangerous than inundation caused by wind waves of the same amplitude. Another distinguishing feature of tsunami waves is that they involve the motion of water all the way from the seabed to the surface, whereas shorter period waves typically only involve the first few tens of meters. This contributes to the large amount of energy that can be transferred within a tsunami.

The propagation speed of a tsunami is controlled by the water depth. Provided the wavelength is long compared to the water depth, as it is for most earthquake-generated tsunami, the speed, c, is given by:

$$c = \sqrt{gh}$$

where g is the gravitational acceleration and h is the water depth. In deep water very high propagation speeds are possible, e.g., in 4,000 m of water the speed is approximately 200 m/s. It should be noted that this is the speed with which the wave propagates, rather than the speed of individual particles within the wave (which is generally much slower, except in shallow water where it may be comparable).

When a tsunami approaches land its speed reduces causing the wave to "bunch-up," increasing in amplitude as it does so. This is known as shoaling, and is the reason that a tsunami that goes unnoticed by ships at sea can rise to be tens of meters high at the shore. Overland speeds for tsunami inundation flows have been measured in the 10–75 km/h range (Matusutomi et al., 2006; Choowong et al., 2008).

Because the tsunami speed is controlled by water depth, the tsunami waves are subject to processes familiar from optics such as reflection, refraction, and wave-guiding, but for tsunami it is the bathymetric profile that controls the direction of propagation rather than the refractive index as in the case of light waves. The effects of bathymetry on tsunami propagation have been studied extensively by Mofjeld et al. (2000).

Sources of tsunami

Since tsunami involve the motion of large volumes of water they require similarly large sources of initial displacement. The most frequent and well-known source of tsunami are earthquakes, but landslides and volcanic eruptions can also cause them, as well as more exotic phenomena such as asteroid impacts.

Earthquakes

The majority of destructive tsunami are caused by earthquakes; in a typical year there may be one or two damaging earthquake-caused tsunami, and many more nondestructive smaller events. The process by which an earthquake initiates a tsunami is principally one in which the earthquake dislocation causes sudden and persistent uplift or subsidence to the seabed over a wide area. This lifts (or lowers) the water column above the seabed, putting the water out of equilibrium and causing a tsunami to propagate away (Figure 1). A variety of factors determine the area, timing, and degree of coseismic vertical displacement; these include depth, dip-angle, slip-distribution, rupture velocity, and the rigidity of surrounding rock. The influence of factors such as these on the tsunamigenic potential has been studied in depth by Geist (1999).

The most well-known examples of an earthquake-caused tsunami, and indeed of any tsunami, are the 2004 Sumatra-Andaman Islands tsunami and the 2011 Tohoku, Japan tsunami. The former was caused by an earthquake on the subduction zone boundary between the India plate and Burma microplate that lies in the northeast of the Indian Ocean. In this earthquake, which had an estimated moment magnitude of 9.3 (Stein and Okal, 2007), the average slip was over 9 m along a 1,300–1,600 km segment of plate boundary, with a maximum slip of 25–30 m near northern Sumatra. The subsequent tsunami produced run-ups in excess of 30 m along part of the Sumatra coast (Borrero, 2005), devastated the city of Banda Aceh, and caused huge loss of life (>220,000 deaths) and damage to property around the Indian Ocean.

One particular type of earthquake that causes tsunami is worthy of special mention is known as a "tsunami-earthquake" (Kanamori, 1972; Bilek and Lay, 2002) as it causes large tsunami relative to the felt shaking. This is particularly problematic in situations where the earthquake shaking is relied upon to provide a natural early warning of tsunami, as tsunami-earthquakes are liable to go unnoticed or ignored by much of the population. Seismic characteristics of tsunami-earthquakes include long-rupture durations, shallow locations close to the trench, low rupture velocities, and disproportionate energy release at low frequencies. A recent example of

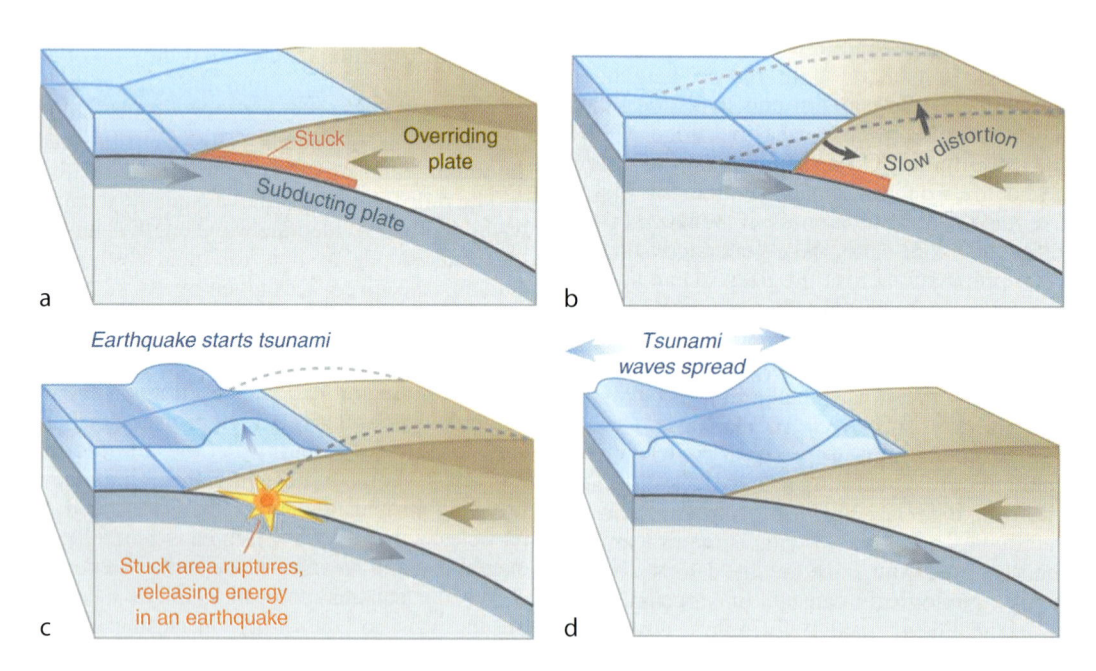

Tsunami, Figure 1 Stages in tsunami generation by a subduction zone earthquake: (**a**) coupling between plates, (**b**) strain accumulation, (**c**) strain release during earthquake, and (**d**) tsunami propagation (Credit: USGS; not subject to copyright).

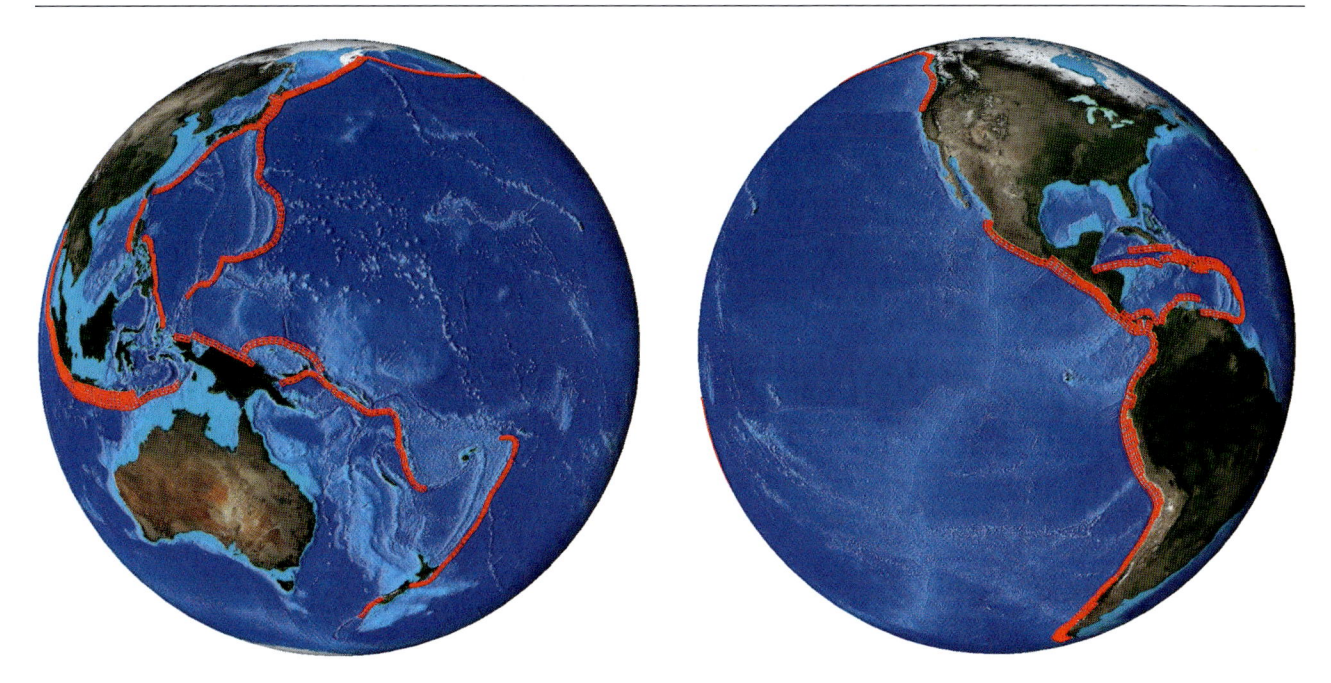

Tsunami, Figure 2 International tsunami source map. Identified subduction zone source locations are shown in *red* (Credit: NOAA; not subject to copyright).

a tsunami caused by a tsunami-earthquake is the 2006 M7.7 south Java tsunami (Reese et al., 2007).

Most major tsunami-generating earthquakes occur on subduction zones (Figure 2). In general, earthquakes with moment magnitudes of less than 6.5 are unlikely to cause tsunami directly. However such smaller events may still cause ground shaking sufficient to act as a trigger for landslides, especially in areas where large earthquakes are rare.

Landslides

Landslides are generally believed to be less frequent causes of tsunami than earthquakes, but as most large landslides are likely to be triggered by earthquakes it is not always possible to determine the ultimate source. A well-known instance of a landslide-caused tsunami is the Grand Banks tsunami of 1929; in this case although there was a triggering earthquake, we know that a substantial submarine landslide took place because of the sequential severing of a sequence of telephone cables (Heezen and Ewing, 1952).

Tsunami-causing landslides may be entirely submarine, or they may be subaerial, i.e., starting above the water surface but descending into it. In both cases it is the displacement of water by the descending body of dense material that initiates the tsunami. The classification of landslides is quite complicated which reflects the variety of different failure mechanisms and the physical properties of the landslide material (Figure 3).

The scale of earthquake-generated tsunami is limited by the size of the largest earthquakes which is itself constrained by the length of subduction zones (McCaffrey, 2008). Landslides, on the other hand, appear to have fewer constraints on the maximum size, and consequently may become the dominant source of tsunami at scales beyond the reach of tectonic sources. While it is simplistic to reduce the scale of a landslide source to just the volume of material in the landslide, it does provide a first approximation. The 1929 Grand Banks tsunami, which produced waves 3–8 m high and claimed 29 lives, was estimated to be caused by a turbidity current of volume 200 km^3 (Fine et al., 2005). This is dwarfed by comparison to the volumes of material implied by submarine paleo-landslides: the Ruatoria debris flow in New Zealand is estimated to have a volume of 3,000 km^3 (Collot et al., 2001) and the Storegga debris flow in the North Sea is thought to be similar (Haflidason et al., 2004). Fortunately, the timescales between such events is estimated to range from tens to hundreds of thousands of years. The Nuuanu slide near Hawaii is estimated to have contained more than 5,000 km^3 of material (Moore et al., 1989), but is thought to have occurred more than 1.5 Ma years ago (Moore et al., 1989).

The largest recorded tsunami run-up heights are due to a subaerial landslide source, namely, the Lituya Bay (Alaska) landslide in 1958 (Miller, 1960). The landslide was triggered by an earthquake, and although the volume of material was relatively small at about .04 km^3 the sudden, fast, descent into the confined body of the Bay caused water to rush to over 500 m elevation on the opposite banks.

Volcanoes

Volcanic eruptions are an infrequent, but occasionally catastrophic source of tsunami. There are thought to be many

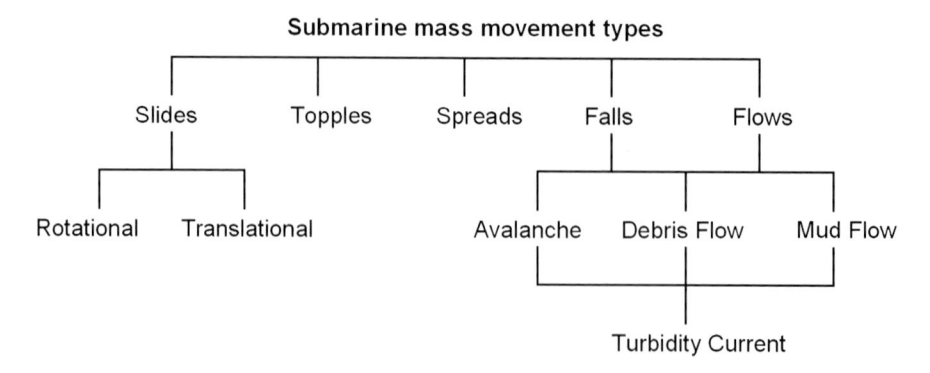

Tsunami, Figure 3 Classification of submarine mass movements (Locat and Lee 2000).

different possible modes by which a volcanic event can trigger a tsunami (Latter, 1981), and it is often difficult to establish the mechanism for historical events due largely to the self-destruction of the volcanoes involved and the limited prospects for eyewitnesses. The most well-known tsunami following a volcanic event was that which followed the eruption of Krakatau in 1883 (Verbeek, 1884). The tsunami was the principal cause of loss of life in this eruption, waves which ran-up to about 40 m in the Sunda Strait caused a similar level of devastation to that of the 2004 Sumatra-Andaman Islands tsunami. Other notable tsunami-causing eruptions are that of Santorini in the Aegean Sea in about 1470 BC, and the eruption of Mt Unzen in southern Japan in 1792.

Proposed mechanisms by which a volcano can create a tsunami are varied, Latter (1981) lists ten: earthquakes accompanying eruptions, displacement of water by submarine explosions, displacement of water by pyroclastic flows, caldera collapse and subsequent in-filling with water, landslides on volcano flanks, shock waves caused by base-surges, hot-rock avalanches, lahars entering the water, air waves following explosions, and lava avalanches.

Others

In addition to the three main source types of earthquakes, landslides, and volcanoes, there are some additional sources of tsunami. The impact of extraterrestrial objects such as asteroids and comets (collectively referred to as bolides) into the ocean has the potential to create tsunami. These tsunami could conceivably be larger than those due to any terrestrial source, though such large events are believed to be extremely infrequent. The extinction of the dinosaurs is linked to the impact of a large comet near Chicxulub in Mexico approximately 65 million years ago (Bryant, 2001), but tsunami are just one of the many devastating global consequences of such an event.

Impacts of tsunami

Tsunami damage and casualties are usually caused by five main factors:

The impact of a swiftly flowing torrent (up to 70 km/h), or traveling bores, on vessels in navigable waterways, canal estates and marinas, and on buildings, infrastructure and people where rivers break their banks or coastal margins are inundated. Torrents (inundating and receding) and bores can also cause substantial erosion both of the coast and the seafloor. They can scour roads and railways, land, and associated vegetation. The receding flows, or "out-rush," when a large tsunami wave recedes, are often the main cause of drowning, as people are swept out to sea.

Debris impacts – many casualties and much building damage arise from the high impulsive impacts of floating debris picked up and carried by the in-rush (inundating) and out-rush (receding) flows. Building materials, boats, cars, and other vehicles are some of the most common and destructive debris in urban areas.

Fire and contamination – fire may occur when fuel installations are floated or breached by debris, or when home heaters are overturned. Breached fuel tanks, and broken or flooded sewerage pipes or works can cause contamination. Homes and many businesses contain many harmful chemicals that can be spilled.

Inundation and saltwater-contamination – by the ponding of potentially large volumes of seawater will cause medium to long-term damage to buildings, electronics, fittings, and to farmland.

Sedimentation and erosion – advancing and retreating tsunami pulses can be highly erosive, removing large amounts of sand, soil, and even loose rock. This material is then deposited as sediments in other locations. Material, especially sand, eroded from the sea floor can be deposited on land, and vice versa. Changes in the coastal seabed due to erosion and deposition can change shipping channels and affect the operability of ports.

Buildings

The response of buildings to tsunami varies widely depending on building construction and wave dynamics. In general, reinforced concrete buildings fare far better than unreinforced buildings. For example, in Samoa in 2009 flow heights of 3–4 m caused the complete destruction of many unreinforced timber buildings, whereas those built nearby with cement and reinforcing iron, often schools and churches, often remained standing. Water depth and velocity are important; there is less structural damage from slow moving deep water, or faster moving but shallow (a fraction of wall height) water. Unreinforced buildings are often destroyed by the force of water alone, whereas the majority of structural damage to reinforced buildings comes from debris impact (especially vehicles and boats). Buildings with reinforced structural members and unreinforced in-fill walls (often brick) usually have partial or complete loss of the walls with relatively little damage to the structure, even with water depth greater than one story. For multistory buildings, this means that people often survive unscathed on the roof or upper stories (Figure 4).

People

Even shallow (depth less than knee-height) tsunami inundation can be very dangerous or deadly. Water velocity and turbulence means that it is hard to remain standing; drowning is a significant risk. Additionally, the water usually contains sharp or heavy debris, such as roofing iron or cars, and fine sand and silt. This leads to a high incidence of cut, crush, and abrasion injuries, all of which can be lethal through internal or external bleeding and shock (e.g., Prasartritha et al., 2008). Disease and infection are a high risk in the hours and days following a tsunami, compounded by the entrainment of sewage and the damage to health and sanitation facilities.

Infrastructure

Tsunami can damage anything they touch, and infrastructure and lifeline utilities are often significantly impacted. Even buried utilities are vulnerable due to the erosion generated by advancing and retreating tsunami. Roads, bridge approaches, foundations, and airport runways are often eroded; utility poles, bridge structures, wharves and piers are damaged by erosion and debris impacts; water and sewage pipes are broken, and electrical and communications lines severed, especially where they cross bridges; water supply, storage, and treatment facilities are damaged or filled with material; lifeline service buildings are damaged or contaminated (e.g., hospitals) and emergency responders and their vehicles may be directly impacted making them unavailable for relief efforts. Dependant services are also affected, for example, an ambulance may be undamaged but the roads are not drivable, or a hospital intact but without power or water.

Boats and shipping

Boats in shallow water and marinas are often damaged by collision with other boats, debris, wharves, or buildings. They are also often rafted inland and stranded, lost out to the open water, capsized or otherwise swamped and sunk. Even undamaged boats, and especially larger ships, may have difficulty usefully operating due to damage to wharves and piers. Loss of the use of boats and ships can have flow on consequences such as hindrance of relief efforts and aid supply, and loss of trade (from commodity shipping to local fishing).

Historical tsunami and paleotsunami

Historical tsunami databases are an important source of information on past events. The National Geophysical Data Center (NGDC) maintains an online database of historically recorded tsunami from 2000 BC to 2008 (NGDC, 2008). The database contains over 2,300 events, of these 1,126 are considered to be of high validity

Tsunami, Figure 4 Damaged house in Leone, American Samoa 2009. The ground floor has been destroyed by the tsunami apart from the reinforced pillars. The occupants survived the tsunami on the upper floor (Credit: GNS Science).

(probable or definite tsunami), and 902 have taken place since 1800 (Power and Downes, 2009). Approximately 73% of source events were in the Pacific Ocean, 14% in the Mediterranean Sea, 6% in the Caribbean Sea and Atlantic, and 5% in the Indian Ocean (NGDC, 2008). NGDC also maintains a database of tsunami run-up measurements covering approximately 7,000 locations. The Novosibirsk Tsunami Laboratory (NTL) also maintains a comprehensive historical tsunami database (NTL, 2010).

Prehistoric tsunami may be studied via their geological impacts, for example, by the inland deposition of sand and other materials from the coast. It is however difficult to distinguish between inundation events with geological sources from those arising from weather events. Many smaller tsunami do not leave paleotsunami traces, and those that do may be hard to detect due to the shifting coastal environment. Consequently the known record of paleotsunami is much less complete than the recent historical record of tsunami. A well-studied paleotsunami event is the Cascadia tsunami of 1700, which was also observed in Japan and recorded in the myths of Native Americans (Atwater, 2005).

Tsunami modeling

Tsunami modeling allows us to reconstruct, and better understand, past tsunami; and to look into the consequences of possible future events. Creating a model of a tsunami can be achieved both through scaled-down physical models and more commonly via numerical calculations of the processes involved using a computer.

Tsunami modeling can be performed for all stages in the generation, spreading and impact of a tsunami. Numerical modeling of earthquake sources is a fairly mature field; typically Okada's (1985) formulae are used to estimate the seabed deformation following slip on portions of a fault plane. A large area of uncertainty here is in estimating the initial slip-distribution, as this is often poorly constrained by real data and can have a big influence on the resulting tsunami, especially close to the source (Geist, 1999). Landslide and volcanic source modeling does not follow a single well-established procedure and to some extent this reflects the variety of mechanisms by which they can initiate tsunami. In the case of landslide sources a sliding-block approach is sometimes used to represent the failure of large blocks, at the other extreme the landslide may be described as a dense, viscous fluid flow. Real landslides form a continuum between these extremes, with, for instance, debris avalanches, having some properties of both solid and fluid.

Numerical modeling of tsunami propagation is usually achieved by solving the shallow-water wave equations. These algorithms may include nonlinear terms in situations where the tsunami propagates over shallow water, and additional corrections may be included to account for Coriolis effects and to approximate frequency-dispersion effects. Inundation modeling is often carried out in conjunction with propagation modeling, as once an area becomes wet the evolution is described by similar equations. Key differences include the optional incorporation of bottom-friction, and equations to describe the wetting and drying processes at the front of a tsunami. Examples of software packages that combine propagation and inundation modeling are MOST (Titov and Gonzalez, 1997), COMCOT (Liu et al., 1995; Wang and Liu, 2006) and ANUGA (Roberts et al., 2008).

Tsunami mitigation

Hazard assessment

Important precursors for efficient tsunami mitigation are *risk assessment* and *hazard and risk mapping*, to establish the likelihood and probable consequences for a tsunami to impact a particular area. This helps to ensure that mitigation measures are applied where they can have the greatest benefit. The approach used to achieve this is combination of vulnerability cataloging (where people and assets are located and how susceptible to tsunami they are), and tsunami hazard assessment (Power and Downes, 2009).

Tsunami hazard assessment can take two main forms. It can be based on a scenario approach, where models are made to represent one or more likely situations; or it can be based on a probabilistic approach, in which a spectrum of possible events are analyzed and weighted according to their likelihood. This latter approach is in its infancy for tsunami modeling, but allows for a more systematic comparison of hazards between different locations and across different types of phenomena.

Whether scenario or probabilistic, it is usually important for a hazard assessment to also estimate the likely consequences of the tsunami impact. This is then usually referred to as a risk assessment, and takes into consideration the fragility of buildings and people; that is how likely they are to be damaged by the impact of a tsunami wave of a particular height or velocity. Tsunami *risk management* is often best conducted through a combination of *land-use planning* and *building codes*, possibly *structural mitigation* (with some caveats) and *warning systems* (e.g., Jonientz-Trisler et al., 2005; NTHMP, 2001). Deciding on the exact tsunami hazard at a location is very difficult, because future tsunami inundations are models, and there is usually only a short historical record, and incomplete geological record, for calibration.

Land-Use planning and building codes

Land-use planning for tsunami can greatly reduce the risk, if people do not live near the coast. However, coastal land is some of the most desirable and expensive because of access to the coast, attractive views, and lifestyle factors. Some areas are of high tsunami risk to most people (e.g., beach-front facing a subduction zone fault line), but deciding on the line inland/above which the risk is acceptable is very difficult due to uncertainties in modeling tsunami hazard. As a result virtually no place has instituted tsunami land-use planning rules. Buildings can in theory be built to codes that will protect people inside them from tsunami, or be built to be elevated above the tsunami risk height (FEMA, 2008). Such buildings would need to be water-proof and very strong. Tall reinforced buildings can be used as evacuation structures in tsunami warnings, where evacuation routes to higher ground require travel too far. Specific tsunami evacuation structures and dual use buildings with evacuation access have been implemented in Japan and recently Indonesia, for example.

Structural mitigation

Barrier *structural mitigation*, such as sand dunes and sea walls may be effective against tsunami of a similar or lower height. Because of the large volume of sand dunes they can also act as a sacrificial barrier, being eroded by a tsunami and reducing velocity and distance of inundation. Sea walls tend to create a high level of perceived landward safety, leading to increased development. Walls can still fail or be exceeded and the additional development has in the meantime added vulnerability, increasing the consequences in these larger events. Walls are typically very expensive and also have major cultural (community division), life quality (view and access) and safety (barrier to evacuation) problems.

Warning

In most cases *warning systems* are the default option relied upon to mitigate tsunami risk, but do not mitigate against other damage – they only aim to ensure life safety. Tsunami can reach the shore from sources only minutes away (e.g., a local fault or landslide) or take 12 h or more to arrive as they travel across an ocean. Warnings can come from natural, informal, or official sources (Gregg et al., 2006). Natural and informal warnings may be the only warnings for sources close to the impact location, as there may not be enough time for official warning systems to activate and communicate, or if the official system fails to notify people (due, for example, to equipment fault, or patchy coverage). Natural warnings for tsunami are usually:

Earthquakes – either strong earthquakes that it is difficult to stand up in or gentler rolling earthquakes that last for a minute or more (indicative of large earthquakes farther away, and of "tsunami-earthquakes")
Unusual ocean behavior – water rushing in or out
Unusual noises from the ocean – often described as "roaring like a jet engine," or "like an animal roaring"

Informal warnings may propagate from one person to another after natural signs have been observed, or an official warning message is received.

Forecasting

Official warning systems are typically technological and rely on geophysical monitoring and scientific evaluation to generate tsunami forecasts. The precursor to the Pacific Tsunami Warning Center (PTWC) in Hawaii was established in 1949 in response to the 1946 Aleutians tsunami, and its remit extended to cover the Pacific Basin in 1968. The West Coast and Alaska Tsunami Warning Center (WC/ATWC) was established in 1967 in response to the 1964 Alaskan tsunami. The WC/ATWC also acts as a backup for PTWC within the wider Pacific Basin. Since the 2004 Indian Ocean tsunami PTWC has also taken responsibility for the Indian Ocean, South China Sea, and parts of the Caribbean. The Japan Meteorological Agency has operated a tsunami warning system since 1952.

The initial basis for tsunami warnings comes from seismic data alone, since seismic waves propagate much faster than tsunami waves. This provides information on the location and magnitude of the source earthquake, from which an initial evaluation of tsunami threat can be made. More accurate modeling requires time and detailed earthquake parameters.

Subsequent information comes in the form of DART buoy (Gonzalez et al., 1998; Figure 5) and tide gauge measurements, which can be used to confirm the presence of a tsunami and then to refine estimates of the tsunami size. Software such as SIFT (Gica et al., 2008) is under development to provide more accurate tsunami forecasting using DART buoy data.

Communication and response

For warnings to be effective they must reach an aware and prepared public with enough time to take protective actions. Protective actions for tsunami are usually to evacuate, either to higher ground and inland or up a tsunami evacuation structure. Warning systems only activate a decision-making process, so for an appropriate decision to be made the public must be aware of the hazard, believe the potential consequences, know the appropriate protective actions and know that they will be effective (Gregg et al., 2007). They must believe that their actions will mitigate the risk and that they (the public) are capable of completing these actions in time.

The official warning process always includes a lead time comprised of the time taken to detect tsunami data (e.g., earthquake location and magnitude, wave arrivals), to decide on the message to deliver, to prepare the message, and to communicate the message to the public. To keep this lead time as short as possible detailed planning of the system, decision-making thresholds, and evacuation are needed. The public need to know where the official message is coming from, be able to get that message at any time of the day or night, regardless of what they are doing, know if the message is trustworthy, and know what to do in response. Evacuation zones, routes, and safe areas need to be preplanned and widely known.

Given that evacuation is the usual protective action for tsunami, and that destructive tsunami are infrequent (often decades or more apart at a given location) evacuation exercise drills are essential to (1) test the system for reliability and effectiveness and (2) maintain a high level of awareness as to the correct actions. For successful evacuation

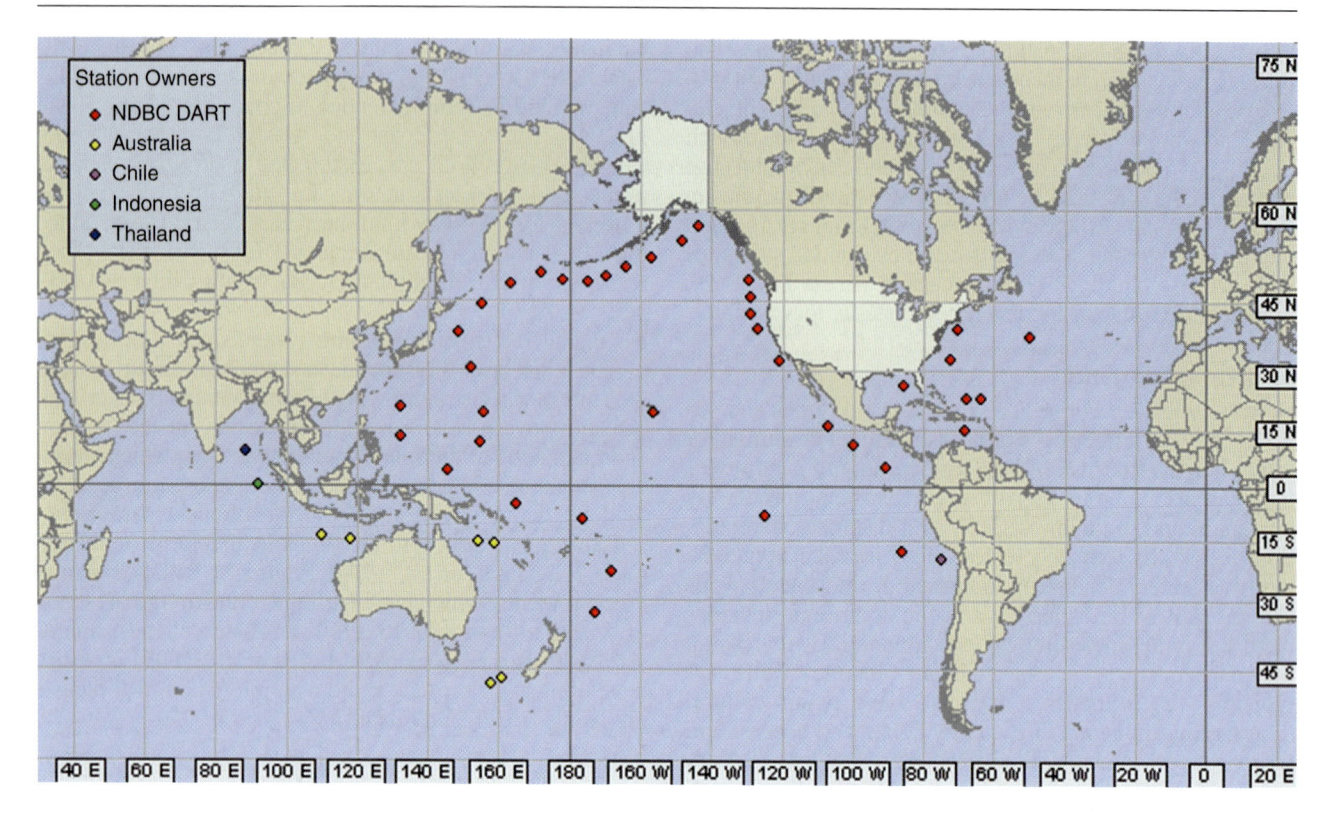

Tsunami, Figure 5 Worldwide distribution of DART buoys (Credit: NOAA; not subject to copyright).

from tsunami the mode of transport is an important criterion. Cars may be effective with many hours between warning and impact, but quickly jam roads in short time-frame events. For sources less than a few hours away it is probably best to aim for evacuation on foot and bicycle wherever possible, to avoid congestion.

Summary

Tsunami are naturally occurring long period waves in oceans, seas, and lakes generated by vertical displacement of water. They are most-often driven by earthquakes, and less often by volcanic eruptions and landslides. Earthquake sources tend to displace water over a larger area and so generate tsunami that are hazardous at a much larger distance from source. Tsunami can be highly destructive, reaching up to tens of meters above sea level on steep coastal land, and up to several kilometers inland over flat land. They can travel up to 200 m/s in the open ocean and go virtually unnoticed, shoaling and slowing down as they reach shallow coastal waters. Damage is caused by the force of water, entrained debris (cars and building materials) and fine particles (sand, etc.), erosion and sedimentation by withdrawing water, and fouling, contamination or salination from the water and impurities in the water. Land-use planning and structural mitigation (such as dune restoration rather than sea walls) can be effective, and warning systems are often implemented. It is hard to achieve high warning effectiveness and reliance on warning hardware often reduces the focus on community planning and options such as land-use planning, natural and informal warnings, and evacuation exercises, to the detriment of risk reduction.

Bibliography

Atwater, B. F., Musumi-Rokkaku, S., Satake, K., Tsuji, Y., Ueda, K., and Yamaguchi, D. K., 2005. *The Orphan Tsunami of 1700*. Seattle, WA: University of Washington Press.

Bilek, S. L., and Lay, T., 2002. Tsunami earthquakes possibly widespread manifestations of frictional conditional stability. *Geophysical Research Letters*, **29**, 1673.

Borrero, J. C., 2005. Field survey of northern Sumatra and Banda Aceh, Indonesia after the tsunami and earthquake of 26 December 2004. *Seismological Research Letters*, **76**, 312.

Bryant, E., 2001. *Tsunami; the Underrated Hazard*. New York: Cambridge University Press.

Choowong, M., Murakoshi, N., Hisada, K. I., Charusiri, P., Charoentitirat, T., Chutakositkanon, V., Jankaew, K., Kanjanapayont, P., and Phantuwongraj, S., 2008. 2004 Indian Ocean tsunami inflow and outflow at Phuket, Thailand. *Marine Geology*, **248**(3–4), 179–192.

Collot, J.-Y., Lewis, K., Lamarche, G., and Lallemand, S., 2001. The giant Ruatoria debris avalanche on the northern Hikurangi margin, New Zealand; results of oblique seamount subduction. *Journal of Geophysical Research*, **106**, 19.

Federal Emergency Managemeny Agency (FEMA), 2008. Guidelines for design of structures for vertical evacuation from tsunamis. *FEMA P646*. 158p.

Fine, I. V., Rabinovich, A. B., Bornhold, B. D., Thomson, R. E., and Kulikov, E. A., 2005. The Grand Banks landslide-generated tsunami of November 18, 1929: preliminary analysis and numerical modeling. *Marine Geology*, **215**, 45.

Fritz, H. M., Blount, C., Sokoloski, R., Singleton, J., Fuggle, A., McAdoo, B. G., Moore, A., Grass, C., and Tate, B., 2007. Hurricane Katrina storm surge distribution and field observations on the Mississippi Barrier Islands. *Estuarine, Coastal and Shelf Science*, **74**, 12–20.

Geist, E. L., 1999. Local tsunamis and earthquake source parameters. *Advances in Geophysics*, **39**, 117.

Gica, E., Spillane, M., Titov, V. V., Chamberlin, C., and Newman, J. C., 2008. Development of the forecast propagation database for NOAA's short-term inundation forecast for tsunamis (SIFT). *NOAA Technical Memorandum OAR PMEL-139*, 89 pp.

Gonzalez, F. I., Milburn, H. M., Bernard, E. N., and Newman, J. C., 1998. Deep-ocean assessment and reporting of tsunamis (DART®): brief overview and status report. In *Proceedings of the International Workshop on Tsunami Disaster Mitigation* January 19–22 1998, Tokyo.

Gregg, C. E., Houghton, B. F., Paton, D., Lachman, R., Lachman, J., Johnston, D. M., and Wongbusarakum, S., 2006. Natural warning signs of tsunamis: human sensory experience and response to the 2004 Great Sumatra earthquake and tsunami in Thailand. *Earthquake Spectra*, **22**, 671–691.

Gregg, C. E., Houghton, B. F., Paton, D., Johnston, D. M., Swanson, D. A., and Yanagi, B. S., 2007. Tsunami warnings: understanding in Hawai'i. *Natural Hazards*, **40**, 71–87.

Haflidason, H., Sejrup, H. P., Nygård, A., Mienert, J., Bryn, P., Lien, R., Forsberg, C. F., Berg, K., and Masson, D., 2004. The storegga slide: architecture, geometry and slide development. *Marine Geology*, **213**, 201.

Heezen, B. C., and Ewing, W. M., 1952. Turbidity currents and submarine slumps, and the 1929 Grand Banks [Newfoundland] earthquake. *American Journal of Science*, **250**, 849.

Johnston, D., Paton, D., Crawford, G. L., Ronan, K., Houghton, B., and Burgelt, P., 2005. Measuring tsunami preparedness in coastal Washington, United States. *Natural Hazards*, **35**, 173–184.

Jonientz-Trisler, C., Simmons, R. S., Yanagi, B. S., Crawford, G. L., Darienzo, M., Eisner, R. K., Petty, E., and Priest, G. R., 2005. Planning for tsunami-resilient communities. *Natural Hazards*, **35**, 121–139.

Kanamori, H., 1972. Mechanism of tsunami earthquakes. *Physics of the Earth and Planetary Interiors*, **6**, 346–359.

Latter, J. H., 1981. Tsunamis of volcanic origin: summary of causes, with particular reference to Krakatoa, 1883. *Bulletin Volcanologique*, **44**, 467.

Liu, P. L. F., Cho, Y.-S., Briggs, M. J., Synolakis, C. E., and Kanoglu, U., 1995. Run-up of solitary waves on circular island. *Journal of Fluid Mechanics*, **302**, 259–285.

Locat, J., and Lee, H. J., 2000. Submarine landslides: advances and challenges. iN *Proceedings of the 8th International Symposium on Landslides*, June 2000, Cardiff.

Lockridge, P. A., 1990. Nonseismic phenomena in the generation and augmentation of tsunamis. *Natural Hazards*, **3**, 403.

Matusutomi, H., Sakakiyama, T., Nugroho, S., and Matsuyama, M., 2006. Aspects of inundated flow due to the 2004 Indian Ocean tsunami. *Coastal Engineering Journal*, **48**, 167–195.

McCaffrey, R., 2008. Global frequency of magnitude 9 earthquakes. *Geology*, **36**, 263.

Miller, D. J., 1960. Giant waves in Lituya Bay Alaska. *Geological Survey Professional Paper* 354-C.

Mofjeld, H. O., Titov, V. V., Gonzalez, F. I., and Newman, J. C., 2000. Analytical theory of tsunami wave scattering in the open ocean with application to the North Pacific. *NOAA Technical Memorandum OAR PMEL-116*.

Monserrat, S., Vilibic, I., and Rabinovich, A. B., 2006. Meteotsunamis: atmospherically induced destructive ocean waves in the tsunami frequency band. *Natural Hazards and Earth System Sciences*, **6**, 1035–1051.

Moore, J. G., Clague, D. A., Holcomb, R. T., Lipman, P. W., Normark, W. R., and Torresan, M. E., 1989. Prodigious submarine landslides on the Hawaiian Ridge. *Journal of Geophysical Research*, **94**, 17.

National Tsunami Hazard Mitigation Program (NTHMP), 2001. Designing for tsunamis: seven principles for planning and designing for tsunami hazard. *NTHMP*. 60p.

NGDC, 2008. national geophysical data center, tsunami data and information. http://www.ngdc.noaa.gov/hazard/tsu.shtml.

NOAA, USGS, FEMA, NSF, Alaska, California, Hawaii, Oregon, and Washington.

NTL, 2010. Novosibirsk Tsunami Laboratory, Historical Tsunami Database for the World Ocean. http://tsun.sscc.ru/nh/tsunami.php.

Okada, Y., 1985. Surface deformation due to shear and tensile faults in a half-space. *Bulletin of the Seismological Society of America*, **75**, 1135.

Oregon Emergency Management and the Oregon Department of Geology and Mineral Industries (OEM&ODGAMI), 2001. *Tsunami Warning Systems and Procedures Guidance for Local Officials*. Oregon Department of Geology and Mineral Industries ODGAMI Special Paper 35. 41p.

Paton, D., Houghton, B. F., Gregg, C. E., Gill, D. A., Ritchie, L. A., McIvor, D., Larin, P., Meinhold, S., Horan, J., and Johnston, D. M., 2008. Managing tsunami risk in coastal communities: Identifying predictors of preparedness. *Australian Journal of Emergency Management*, **23**, 4–9.

Power, W. L., and Downes, G. L., 2009. Tsunami hazard assessment. In Connor, C. B., Chapman, N. A., and Connor, L. J. (eds.), *Volcanic and Tectonic Hazard Assessment for Nuclear Facilities*. Cambridge: Cambridge University Press, pp. 276–306.

Prasartritha, T., Tungsiripat, R., and Warachit, P., 2008. The revisit of 2004 tsunami in Thailand: characteristics of wounds. *International Wound Journal*, **5**, 8–19.

Reese, S., Cousins, W. J., Power, W. L., Palmer, N. G., Tejakusuma, I. G., and Nugrahadi, S., 2007. Tsunami vulnerability of buildings and people in South Java: field observations after the July 2006 Java tsunami. *Natural Hazards and Earth System Sciences*, **7**(5), 573–589.

Roberts, S. G., Nielsen, O. M., and Jakeman, J., 2008. Simulation of tsunami and flash floods. In Bock, H. G., Kostina, E., Phu, H. X., and Rannacher, R. (eds.), *Modeling, Simulation and Optimization of Complex Processes*. Berlin/Heidelberg: Springer, pp. 489–498.

Stein, S., and Okal, E. A., 2007. Ultralong period seismic study of the December 2004 indian ocean earthquake and implications for regional tectonics and the subduction process. *Bulletin of the Seismological Society of America*, **97**, S279.

Tang, Z., Lindell, M. K., Prater, C. S., and Brody, S. D., 2008. Measuring tsunami planning capacity on U.S. pacific coast. *Natural Hazards Review*, **9**, 91–100.

Verbeek, R. D. M., 1884. The Krakatoa eruption. *Nature*, **30**, 10.

Wang, X., and Liu, P. L.-F., 2006. An analysis of 2004 Sumatra earthquake fault plane mechanisms and Indian Ocean tsunami. *Journal of Hydraulic Research*, **44**(2), 147–154.

Cross-references

Breakwater
Civil Protection and Crisis Management
Coastal Erosion
Coastal Zone, Risk Management

Communicating Emergency Information
Critical Infrastructure
Damage and the Built Environment
Disaster Risk Management
Displacement Wave
Early Warning Systems
Earthquake
Education and Training for Emergency Management
Emergency Shelter
Federal Emergency Management System (FEMA)
Flood Protection
Impact Tsunami
Land-Use Planning
Pacific Tsunami Warning and Mitigation System (PTWS)
Risk Assessment
Rogue wave
Seiche
Storm Surge
Structural Mitigations
Tidal Surge
Tohoku, Japan, Earthquake, Tsunami and Fukushima Accident (2011)
Tsunami Loads on Infrastructure
Warning Systems

TSUNAMI LOADS ON INFRASTRUCTURE

Dan Palermo, Ioan Nistor, Murat Saatcioglu
University of Ottawa, Ottawa, ON, Canada

Definition

Tsunami. The Japanese word for "harbor wave."
Coastal bathymetry. The study and mapping of the submarine ocean floor in near-shore areas.
Inundation. The overflowing of water onto normally dry land.
Loading combinations. The summation of individual force components occurring simultaneously.

Introduction

Tsunami, meaning "harbor wave" in Japanese, is the outcome of a vertical displacement of a large body of water. It can be triggered by various geological or astronomical phenomena, including: underwater earthquakes occurring along tectonic boundaries, volcanic eruptions, submerged or aerial landslides, and impact from asteroids or comets. In deep, open waters, tsunamis have small amplitudes (wave height), but very long wavelengths. However, as tsunami waves advance toward shorelines they transform. First, the amplitude of the tsunami wave increases due to shoaling, which occurs as the wave is "squeezed" by the up-sloping seabed. Second, the celerity and the wavelength decrease. However, the wave period remains constant. Depending on coastal bathymetry, tsunami waves can break offshore and advance in the form of a hydraulic bore, which is a turbulent, foamy wall of water, or surge in the form of a sudden increase in water level. Both bores and surges cause inundation of low-lying coastal areas. This in turn can significantly impact infrastructure located in the path of the advancing tsunami. The risks associated with tsunami hazard have increased in recent years due to the rapid development of coastal regions. The risk is more severe in low-lying coastal areas in developing countries, as shown in Figure 1, where structures, specifically residential, are often nonengineered and inadequately designed and constructed, thus prone to extensive damage when subjected to extreme events such as earthquakes, wind storms, and tsunamis. Even in developed countries, however, where structures are typically designed for gravity loads, wind-induced lateral loads, and earthquake excitations, they are not generally designed for tsunami-induced loading.

Tsunami forces on infrastructure

The impact of tsunami-induced forces on coastal protection structures, such as breakwaters, seawalls, reefs, etc., has been previously analyzed by researchers and engineers, particularly in Japan. However, understanding of the adverse effects of the impact of tsunami-induced flooding on near-shoreline infrastructure, such as bridges and buildings, is significantly less developed. Building codes do not explicitly consider tsunami loading, as it is understood that inland structures can be protected by proper site planning and site selection. Therefore, forces generated by tsunami are often neglected in structural design practice. Furthermore, code developers consider tsunami to be a rare event with a long return period. However, depending on the geographical location and tectonic characteristics of the underlying fault lines, major tsunamis can have a recurrence in the order of tens to hundreds of years; therefore, they should be given more attention in building codes. Recent catastrophic events (2004 Indian Ocean Tsunami; 2007 and 2010 Solomon Islands Tsunamis; 2010 Chile Tsunami; 2011 Tohoku, Japan Tsunami) have brought to light the destructive power of tsunami-induced flooding on near-shoreline structures. These events caused major structural damage to infrastructure, devastating coastal communities and resulting in widespread fatalities. Figure 2 illustrates the damage sustained by reinforced concrete structures during the 2004 Indian Ocean Tsunami. The research community has been responding with significant efforts to better understand the phenomenon of tsunami-induced forces and the interaction with structures to provide guidelines for engineers to design or assess infrastructure against such actions. Recent research indicates that forces imposed on structures due to impact of tsunami-induced flooding can be significantly higher than those associated with wind and comparable to or in excess of forces due to earthquake ground shaking (Nouri et al., 2007; Palermo and Nistor, 2009; Saatcioglu, 2009).

Tsunami Loads on Infrastructure, Figure 1 Overall damage after the 2004 Indian Ocean tsunami (Saatcioglu et al., 2006a).

Existing design guidelines

While several design codes explicitly provide guidelines for flood-induced loads (UBC, 1997; ASCE, 2006; IBC, 2006), a survey of current design codes, design standards, and design guidelines indicates that limited attention has been given to tsunami-induced forces. Four pioneering design documents specifically account for tsunami-induced forces, namely: the Federal Emergency Management Agency Coastal Construction Manual, FEMA 55 (FEMA, 2003), which provides recommendations for tsunami-induced flood and wind wave loads; the City and County of Honolulu Building Code (CCH, 2000), which contains regulations that apply to districts located in flood and tsunami-risk areas; the Structural Design Method of Buildings for Tsunami Resistance (SMBTR) proposed by the Building Center of Japan (Okada et al., 2005), outlining structural design for tsunami refuge buildings; and Guidelines for Structures that Serve as Tsunami Vertical Evacuation Sites, prepared by Yeh et al. (2005) for the Washington State Department of Natural Resources to estimate tsunami-induced forces on structures. Recently, the Federal Emergency Management Agency published Guidelines for Design of Structures for Vertical Evacuation from Tsunamis, FEMA P646, (FEMA, 2008). This document focuses on high-risk tsunami-prone areas, and provides design guidance for vertical evacuation structures. Conservative assumptions have been incorporated in FEMA P646 to ensure safety and security for the public requiring shelter from tsunami flood waters.

Tsunami-induced force components

A tsunami wave imposes significant loading on structures. The parameters defining the magnitude and application of these forces include inundation depth, flow velocity, and flow direction. These parameters mainly depend on tsunami wave height and wave period, near-shore bathymetry, coastal topography, and roughness of the coastal inland. The inundation depth at a specific location can be estimated using various tsunami scenarios (magnitude and direction) and by numerically modeling coastal inundation. The estimation of flow velocity and direction, however, is much more difficult to quantify. Flow velocities can vary in magnitude, whereas flow directions can vary due to onshore local topographic features, as well as soil cover and obstacles. The force components associated with tsunami-induced flows consist of: (1) hydrostatic force, (2) hydrodynamic force, (3) buoyant and uplift forces, (4) impulsive force, (5) debris impact and damming forces, and (6) gravity forces. The reader is referred to Nistor et al. (2009) and FEMA P646 for a comprehensive review of the individual force components.

Hydrostatic force

The hydrostatic force, generated by still or slow-moving water, acts perpendicular to the surface of the structural element of interest. The hydrostatic force, F_{HS}, can be calculated using the expression in Equation 1, where ρ is the seawater density, g is the gravitational acceleration, h is the maximum water depth or flood level, and b is the width of the structure or structural element. The force arises from a difference in water levels on opposite sides of the structural element. Equation 1 is based on water being present on one side of a structural element; however, it can be applied for cases where there is a difference in water elevation on two sides of an element.

Tsunami Loads on Infrastructure, Figure 2 Damage to reinforced concrete buildings after the 2004 Indian Ocean tsunami (Saatcioglu et al., 2006b).

$$F_{HS} = \frac{1}{2}\rho g h^2 b \quad (1)$$

Equation 1 is based on a triangular pressure distribution, as shown in Figure 3, with height of h and maximum pressure of pgh at the base. The point of application of the resultant hydrostatic force is located at one third from the base of the pressure distribution. In the case of a hydraulic bore, the hydrostatic force has a smaller magnitude compared to the hydrodynamic and impulsive forces. However, for surge-type tsunamis, the hydrostatic force may be substantial.

Hydrodynamic (drag) force

As tsunami-induced flow encounters a building or structural element, hydrodynamic forces, F_D, are applied to the building. The force includes the effect of the flow velocity on all sides of the building or structural element. The general expression for this force is given in Equation 2. Existing codes suggest different drag coefficient, C_D, values.

$$F_D = \frac{\rho C_D h u^2 b}{2} \quad (2)$$

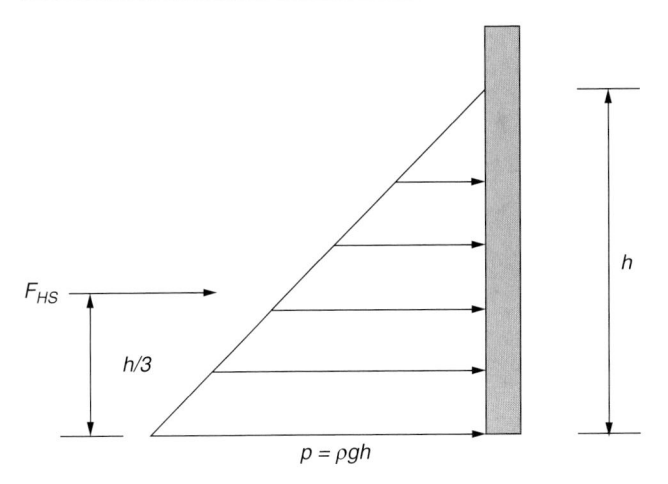

Tsunami Loads on Infrastructure, Figure 3 Hydrostatic force.

where u is the tsunami-induced flow velocity (see section "Tsunami Flow Velocity" below). The flow is assumed to be uniform, and therefore, the pressure is constant through the depth of the flow. The resultant force is applied at the centroid of the projected area. The FEMA 55 document permits the hydrodynamic force to be converted to an equivalent hydrostatic force for flow velocities not exceeding approximately 3.0 m/s. Figure 4 illustrates the hydrodynamic force on a structural element.

Buoyant and uplift forces

The buoyant force, F_B, is a vertical force acting through the center of mass of a submerged or partially submerged structure. Its magnitude is equal to the weight of the volume of water displaced by the structure. Buoyant forces can induce stability problems by reducing the resistance of a structure to sliding and overturning. The buoyant force is calculated as follows:

$$F_B = \rho g V \qquad (3)$$

where V is the volume of water displaced by the submerged or partially submerged structure. The effect of buoyancy in combination with hydrodynamic forces result in uplift forces on horizontal structural elements that have been submerged by tsunami inundation. The contribution of the hydrodynamic force occurs from the rapidly rising water level. It can be estimated using Equation 2 by replacing the flow velocity with the vertical component of the flow velocity, and applying an appropriate hydrodynamic coefficient. Figure 5 demonstrates the effects of uplift forces on concrete slab panels after the 2004 Indian Ocean Tsunami due to buoyant and hydrodynamic forces.

Impulsive force

The impulsive force, F_S, is a short duration load generated by the initial impact of the leading edge of a tsunami bore on a structure. Due to a lack of detailed experiments specifically applicable to tsunami bores running up the shoreline, the calculation of the impulsive force exerted on a structure is subject to substantial uncertainty and has not been universally validated. Dames and Moore (1980) suggested an impulsive force, known as surge force, as follows:

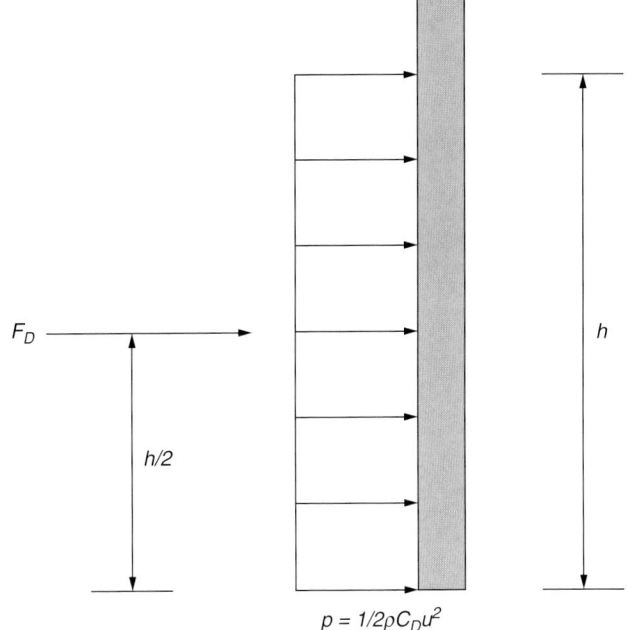

$$F_S = 4.5 \rho g h^2 b \qquad (4)$$

Tsunami Loads on Infrastructure, Figure 4 Hydrodynamic force.

where h is the surge height, usually assumed equal to the inundation depth or flood level. This expression is based on a triangular pressure distribution, as illustrated in Figure 6, extending $3h$ in height, with a corresponding maximum pressure of $3\rho g h$ at the base. Thus, the point of application of the resultant surge force is located at a distance h from the base of the pressure distribution. The surge force as given in Equation 4 results in excessively large forces. Conversely, FEMA P646 proposes an impulsive force equal to 1.5 times the hydrodynamic force, based on experimental results reported by Ramsden (1996) and Arnason (2005), as provided in Equation 5:

$$F_S = 1.5 F_D \qquad (5)$$

Debris impact and damming forces

Tsunami-induced flooding traveling inland carries debris such as floating automobiles (as illustrated in Figure 7), floating pieces of buildings, drift wood, boats, and ships. The impact of floating debris can induce significant forces on a building, leading to structural damage or collapse (Saatcioglu et al., 2006a). The debris impact forces, F_i,

Tsunami Loads on Infrastructure, Figure 5 Displaced slab panels due to uplift forces (Saatcioglu et al., 2006b).

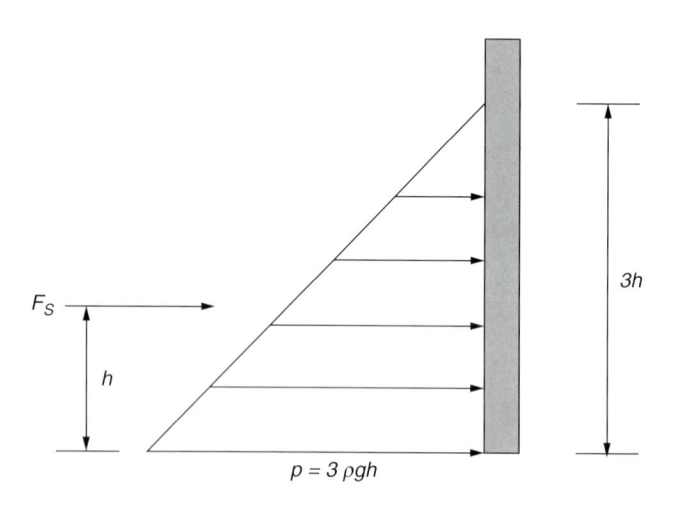

Tsunami Loads on Infrastructure, Figure 6 Surge force.

in its simplest form, can be estimated from the following momentum expression:

$$F_i = m \frac{u}{\Delta t} \quad (6)$$

where m is the mass of the body impacting the structure, u is the approach velocity of the impacting body (assumed equal to the flow velocity), and Δt is the impact duration taken equal to the time between the initial contact of the floating body with the building and the time the floating body comes to rest. FEMA P646 provides additional methods for calculating the debris impact force. The impact force acts horizontally at the flow surface or at any point below it. The impact force is to be applied to the structural element at its most critical location. Depending on the assumed debris mass, this force may not represent a significant contribution to the total lateral tsunami load relative to the other force components. However, it is significant in the design of the structural member that is subjected to the impact.

Debris impacting a structure can cause accumulation of debris, as depicted in Figure 7, leading to a damming effect. The forces generated due to damming can be estimated from the hydrodynamic force (Equation 2) by replacing b with the width of the debris dam.

Gravity forces

Drawdown of the tsunami-induced flooding can result in retention of water on structural flooring systems. This phenomenon imposes additional gravity loading on the structure, which must be considered in design.

Wave-breaking

Classic wave-breaking formulas are applicable for the case of wave breaking directly onto coastal structures, such as breakwaters, piers, and docks. Tsunami waves, however, depending on the near-shore bathymetry, tend to break offshore and approach the shoreline in the form of a rapidly moving hydraulic bore. Furthermore, inland infrastructure is generally not affected by the action of wave breaking occurring at the shoreline.

Tsunami flow velocity

The hydrodynamic force is proportional to the square of the flow velocity. Thus, uncertainties in estimating velocities result in large differences in the magnitude of

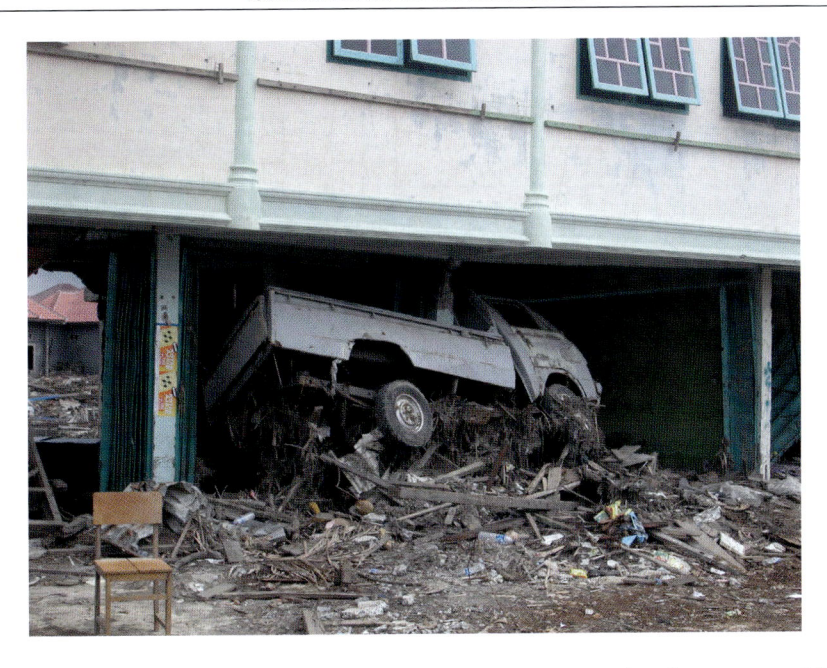

Tsunami Loads on Infrastructure, Figure 7 Impact of a floating vehicle during 2004 Indian Ocean tsunami (Saatcioglu et al., 2006a).

the resulting hydrodynamic force. Tsunami inundation velocity magnitude and direction can vary significantly during a major tsunami. Current estimates of the velocity are crude; a conservatively high flow velocity impacting the structure at a normal angle is usually assumed. Also, the effects of run-up, backwash, and direction of velocity are not addressed in current design documents. A number of guidelines and researchers have proposed estimates of velocity for given tsunami inundation levels, such as Murty (1977), Camfield (1980), FEMA 55 (Dames and Moore, 1980), Kirkoz (1983), CCH (2000), Iizuka and Matsutomi (2000), Bryant (2001), and FEMA P646 (2008).

Tsunami-induced loading combinations

The design documents previously discussed do not explicitly provide loading combinations to estimate the maximum tsunami load for design. In the case of SMBTR, the tsunami load is determined from a single force component that is equivalent to the surge force. FEMA 55 provides load combinations for flood loads, which include wave breaking. However, modifications are necessary to derive loading combinations that are directly applicable to tsunamis. Yeh et al. (2005) recommended that tsunami shelters located in the inundation zone, but inland, be designed for hydrodynamic (drag) and debris impact. The surge force that is generated due to the formation of a turbulent bore is neglected, since Yeh et al. (2005) consider dry-bed test conditions only where the initial impulsive force does not exceed the drag force. Dias et al. (2005) proposed two loading combinations: point of impact and post-submergence. The point of impact considers the initial impact of the tsunami wave and is estimated as the sum of hydrodynamic (drag) and hydrostatic force components on the upstream face of the structure. The post-submergence includes hydrodynamic (drag) on the upstream face, hydrostatic forces on the upstream and downstream faces, and buoyancy. The impact of debris is not explicitly included in either of the load combinations. Pacheco and Robertson (2005) analyzed structures to various inundation levels. In the estimation of the tsunami load, FEMA 55 was followed and wave-breaking forces were omitted. For columns directly exposed to the tsunami wave, the load was estimated as a combination of hydrodynamic and debris impact forces. The tsunami load for structural walls placed parallel to the shoreline, (and perpendicular to the flow of the tsunami), was considered as the maximum of two combinations: (1) The combined effect of hydrodynamic and debris impact forces, and (2) the surge and debris impact forces. Nouri et al. (2007) proposed loading combinations specifically for turbulent bores generated by tsunamis, as shown in Figure 8. Two combinations were developed, which were based on modifications of those recommended by Dias et al. (2005). The first combination (Initial Impact) considers the first arrival of the tsunami bore on a structure, and includes the combined effect of surge and debris impact forces. The second combination (Post Impact) considers the flow of the tsunami bore around the structure. Hydrodynamic, debris impact, and hydrostatic forces are combined to determine the lateral loading. Consideration is also given to buoyancy, which can cause sliding and overturning instability. The more recent FEMA P646 document provides separate tsunami force combinations for a structure and the individual structural

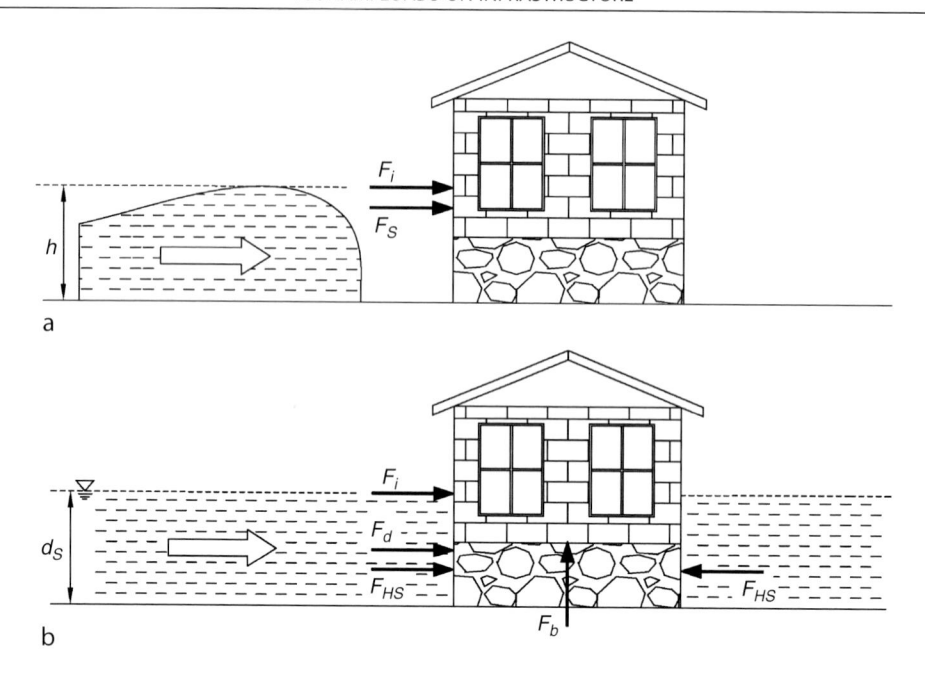

Tsunami Loads on Infrastructure, Figure 8 Tsunami loading combinations: (a) Initial impact; (b) Post impact (Nistor et al., 2009).

elements. For the structure as a whole, three loading combinations are described. The first is a combination of the impulsive forces on structural members located at the leading edge of the bore and drag forces on all previously submerged members behind the leading edge. The second combines a single impact force with drag forces on all structural members. Finally, the third considers the effect of debris damming with drag forces on all structural members. In, addition, the buoyant and hydrodynamic uplift forces should be considered in all load combinations. The design of tsunami load can be readily incorporated in building codes and combined with other loads. Given that a tsunami is considered to be an extreme event, load cases adopting the philosophy of seismic loading have been suggested (Palermo et al., 2009). FEMA P646 has also provided load combinations consistent with ASCE (2006).

Design considerations

Appropriate construction and layout design of a structure located in a tsunami-prone area can reduce the risk of damage during a tsunami event. Tsunami forces increase proportionally with exposed area and nonstructural elements that remain intact during the impact of the tsunami-induced flooding. Therefore, it is prudent to orient buildings with the shorter side parallel to the shoreline. Further, structural walls should also be oriented, if possible, to minimize the exposed area. Exterior nonstructural elements located at lower levels should be designed with a controlled failure mechanism that is triggered by the initial impact of the tsunami. This concept, known as *breakaway walls*, reduces the amount of lateral load that is transferred to the lateral force resisting system of the structure. Conversely, however, breakaway walls may result in an increase in debris loading. The use of rigid nonstructural exterior components, while providing protection to buildings from flooding, increases the lateral loading.

Summary

Recent catastrophic tsunamis (2004 Indian Ocean Tsunami, 2007 and 2010 Solomon Islands Tsunamis, 2010 Chile Tsunami, 2011 Tohoku, Japan Tsunami) have emphasized the destructive power of tsunami-induced flooding as it propagates overland and impacts near-shoreline infrastructure. As a result, research has evolved to improve our understanding of the forces associated with tsunamis and the interaction between tsunami-induced flow and infrastructure. Currently, force components and loading combinations have been proposed to assess and design structures against tsunami forces. The force components include hydrostatic, hydrodynamic, buoyant and uplift, impulsive, debris impact and damming, and gravity. There is, however, uncertainty in both the estimation of the component forces, as well as the total tsunami load that should be considered. Future efforts, including experimental and analytical studies, are being directed toward a better understanding of the forces that should be considered in design of infrastructure located in tsunami-prone areas.

Bibliography

Arnason, H., 2005. *Interactions Between an Incident Bore and a Free-Standing Coastal Structure*. Ph.D. thesis, Seattle, WA, University of Washington.

ASCE, 2006. *Standard, minimum design loads for buildings and other structures.* SEI/ASCE 7-05.

Bryant, E., 2001. Tsunami: The Underrated Hazard. Cambridge University Press, London, UK.

Camfield, F., 1980. *Tsunami engineering.* Coastal Engineering Research Center, US Army Corps of Engineers, Special Report SR-6.

CCH, 2000. *City and County of Honolulu Building Code (CCH).* Honolulu, HI: Department of Planning and Permitting of Honolulu Hawaii, Chap. 16, Article 11.

Dames and Moore, 1980. *Design and construction standards for residential construction in tsunami prone areas in Hawaii.* Prepared for the Federal Emergency Management Agency.

Dias, P., Fernando, L., Wathurapatha, S., and De Silva, Y., 2005. Structural resistance against sliding, overturning and scouring caused by tsunamis. In *Proceedings of the International Conference of Disaster Reduction on Coasts*, Melbourne, Australia.

FEMA, 2003. *Coastal Construction Manual (3 vols, FEMA 55)*, 3rd edn. Jessup, MD: Federal Emergency Management Agency.

FEMA, 2008. *Guidelines for Design of Structures for Vertical Evacuation from Tsunamis, (FEMA P646)*, Jessup, MD., US: Federal Emergency Management Agency.

Ghobarah, A., Saatcioglu, M., and Nistor, I., 2006. The impact of the 26 December earthquake and tsunami on structures and infrastructure. *Engineering Structures*, **28**, 312–326.

IBC, 2006. *International Building Code (IBC).* Country Club Hills, IL: International Code Council.

Iizuka, H., and Matsutomi, H., 2000. Damage due to flood flow of tsunami. *Proceedings of the Coastal Engineering of JSCE*, **47**, 381–385 (in Japanese).

Kirkoz, M. S. 1983. Breaking and run-up of long waves, tsunamis: their science and engineering. In *Proceedings of the 10th IUGG International Tsunami Symposium*, Sendai-shi/Miyagi-ken, Japan. Tokyo, Japan: Terra Scientific Publishing.

Murty, T. S., 1977. Seismic sea waves: tsunamis. Bulletin of the Fisheries Research Board of Canada No. 198, Department of Fisheries and the Environment, Fisheries and Marine Service. Ottawa, Canada: Scientific Information and Publishing Branch.

NBCC, 2005. *National Building Code of Canada (NBCC).* Ottawa: National Research Council of Canada.

Nistor, I., Palermo, D., Nouri, Y., Murty, T., and Saatcioglu, M., 2009. Tsunami-induced forces on structures. In Kim, Y. C. (ed.), *Handbook of Coastal and Ocean Engineering.* Singapore: World Scientific, pp. 261–286.

Nouri, Y., Nistor, I., Palermo, D., and Saatcioglu, M., 2007. Tsunami-induced hydrodynamic and debris flow forces on structural elements. In *Proceedings of 9th Canadian Conference of Earthquake Engineering*, Ottawa, Canada, pp. 2267–2276.

Okada, T., Sugano, T., Ishikawa, T., Ohgi, T., Takai, S., and Hamabe, C., 2005. *Structural Design Methods of Buildings for Tsunami Resistance (SMBTR).* Japan: The Building Centre of Japan.

Pacheco, K. H., and Robertson, I. N., 2005. *Evaluation of tsunami loads and their effect on reinforced concrete buildings.* University of Hawaii Research Report, HI.

Palermo, D., and Nistor, I., 2009. Quantifying tsunami loads for design and assessment of infrastructure. In *Proceedings of WCCE-ECCE-TCCE Earthquake & Tsunami*, Istanbul, Turkey.

Palermo, D., Nistor, I., Nouri, Y., and Cornett, A., 2009. Tsunami loading of near-shoreline structures: a primer. *Canadian Journal of Civil Engineering*, **36**(11), 1804–1815.

Ramsden, J. D., 1996. Forces on a vertical wall Due to long waves, bores, and dry bed surges. *Journal of Waterway, Port, Coastal, and Ocean Engineering*, **122**(3), 134–141.

Saatcioglu, M., 2009. Performance of structures during the 2004 Indian Ocean tsunami and tsunami induced forces for structural design. In *Proceedings of WCCE-ECCE-TCCE Earthquake & Tsunami*, Istanbul, Turkey.

Saatcioglu, M., Ghobarah, A., and Nistor, I., 2006a. Performance of structures in Indonesia during the December 2004 Great Sumatra earthquake and Indian Ocean tsunami. *Earthquake Spectra*, **22**(S3), S295–S319.

Saatcioglu, M., Ghobarah, A., and Nistor, I., 2006b. Performance of structures in Thailand during the December 2004 Great Sumatra earthquake and Indian Ocean tsunami. *Earthquake Spectra*, **22**(S3), S355–S375.

UBC, 1997. *Uniform Building Code (UBC).* California: International Conference of Building Officials.

Yeh, H., Robertson, I., and Preuss, J., 2005. *Development of design guidelines for structures that serve as tsunami vertical evacuation sites.* Report No 2005-4. Olympia, WA: Washington Department of Natural Resources.

Cross-references

Asteroid Impact
Building Codes
Casualties Following Natural Hazards
Coastal Zone, Risk Management
Debris Flow
Disaster
Early Warning Systems
Earthquake
Federal Emergency Management Agency (FEMA)
Flood Hazard and Disaster
Impact Tsunami
Indian Ocean Tsunami
Natural Hazard
Pacific Tsunami Warning and Mitigation System (PTWS)
Tsunami

U

UNCERTAINTY

Philipp Schmidt-Thomé
Geological Survey of Finland (GTK), Espoo, Finland

Synonyms
Incertitude; Insecurity

Definition
Uncertainty encompasses all factors of the lack of knowledge towards the exact probability, the timing, magnitude and potential frequency of return of a natural hazard event.

Discussion
Uncertainty extends to the imprecise knowledge of the risk, that is, the precise knowledge of vulnerabilities at any given time of a hazardous event. Uncertainty comprises all unknown inaccuracies. The term is not unanimously defined but it certainly comprises a larger concept than error, the statistical expression for known inaccuracies. Natural hazards are complex phenomena that cannot be forecasted precisely. Allegedly one of the most descriptive manners to describe the risk types attached to natural hazards, including uncertainty aspects, was developed by the German Advisory Council on Global Change – WBGU (2000). WBGU risk types are based on prominent figures from the Greek mythology, and most natural hazards fall into the "cyclope" type risks. Cyclopes are mighty giants with only one eye, meaning that the extent of damage is well known but that the perspective is lost, that is, the probability (or timing) of occurrence. Beyond this, uncertainty is a concept that includes imperfect knowledge, inaccuracy, lack of reliability and inconsistency, and so on of the data (Pang, 2008). Uncertainty is mainly grouped into two types: (1) Aleatory (external) uncertainty is the unpredictability and randomness of the precise moment of an event or process (rock fall, climate change), and (2) Epistemic (internal) uncertainty is the inaccuracy of data and the shortcomings in the understanding of complex processes (models). According to this distinction epistemic uncertainty can be encountered by improving data sets and models. Aleatory uncertainty is subject to probability analysis (e.g., return periods) and epistemic uncertainty is encountered by expert knowledge.

The complexity of uncertainty plays a vital role in the design and estimation of mitigation and adaptation efforts. The cost-benefit analysis of measures to minimize risks related to natural hazards, or the potential impacts of climate change, is greatly dependent on data accuracy. The higher the uncertainty the higher is the potential to invest in inappropriate measures or to take unsustainable decisions. There are several approaches to visualize uncertainty in hazards to better inform about the complexity of the problem and ultimately to support decision making (Pang, 2008). The integration of uncertainty in hazard maps has direct effects on, and may be used to support, the delineation of hazard zones (zoning (further reading: Bostrom et al., 2008), Hoffmann & Hammonds, 1994). Traditionally hazard maps have sharp borders between for instance, "high" and "medium" hazard areas. Since natural events seldom follow strict borders introduced by human concepts, the introduction of uncertainty concepts into hazard maps assists the perception of the potential spatial extent and impacts of hazards such as floods (see also MacEachren et al., 2005).

Bibliography
Bostrom, A., French, S., and Gottlieb, S. (eds.), 2008. *Risk Assessment, Modeling and Decision Support*. Heidelberg: Berlin.
German Advisory Council on Global Change (WBGU), 2000. World in Transition: Strategies for managing global environmental risks.

Hoffman, F. O., and Hammonds, J. S., 1994. Propagation of uncertainty in risk assessments: the need to distinguish between uncertainty due to lack of knowledge and uncertainty due to variability. *Risk Analysis*, **14**(5), 707–712.

MacEachren, A. M., Robinson, A., Hopper, S., Gardner, S., Murray, R., Gahegan, M., and Hetzler, E., 2005. Visualizing geospatial information uncertainty: what we know and what we need to know. *Cartography and Geographic Information Science*, **32**, 3.

Pang, A., 2008. Visualizing uncertainty in natural hazards. *Risk Assessment, Modeling and Decision Support*, Berlin: Heidelberg, pp 261–294.

Cross-references

Land-Use Planning
Zoning

UNITED NATIONS ORGANIZATIONS AND NATURAL DISASTERS

Badaoui Rouhban
UNESCO, Paris, France

Introduction and historical background

The content of this entry does not necessarily reflect the views of the United Nations Educational, Scientific, and Cultural Organization (UNESCO). The United Nations (UN) and its system of organizations is an international ensemble that addresses various areas of concern to Member States and member communities. The UN system includes specialized agencies, autonomous bodies, and programs having specific mandates and carrying out varied missions. This entry provides an overview of different United Nations entities' involvement in activities related to natural disasters. One of the UN's key roles is managing international partnerships and promoting multilateral cooperation with respect to disaster risk attenuation and post-disaster situations. The UN plays a leading role in global collaboration for the understanding and assessment of natural hazards, the mitigation of their consequences, as well as the provision and coordination of disaster relief and emergency response. Through their expertise in social, health, economic, environmental, and technical sectors, including disaster management and emergency response, UN organizations have responsibilities for creating an enabling environment to promote the study of natural hazards such as earthquakes, windstorms (cyclones, hurricanes, tornadoes, typhoons), tsunamis, floods, landslides, volcanic eruptions, droughts, and wildfires and in the development of techniques and measures to mitigate risks arising therefrom. They engage in supporting and implementing activities aimed at assisting disaster-prone countries, notably developing ones, in disaster risk reduction and advancing integrated approaches to building disaster resilient communities. They help countries to set up national strategies and plans of action and programs for disaster risk reduction including prevention, prediction, early warning, and preparedness and to develop their institutional and technical capacities in this field. They encourage stronger linkages, coherence, and integration of disaster risk reduction elements into the humanitarian and sustainable development fields. The responsibilities of the UN system of agencies encompasses coordination of disaster relief. Disaster response is an integral part of the humanitarian mandate of the UN which enjoys representatives and resources around the world ready to get mobilized rapidly in the aftermath of disasters. The system provides logistical coordination of relief efforts; assists with needs assessments to ensure that help is directed where it is needed; and uses its own resources to deliver food, medicine, and other emergency supplies. The UN actively plans the rehabilitation of areas affected by disaster. This part of the mandate includes support for risk reduction activities in post-disaster recovery and rehabilitation processes and sharing of good practices, knowledge, and technical support. Natural hazards do not recognize geographical or political boundaries. When disaster affects people on a multinational scale, it is the mission of the UN to coordinate response activity and perform disaster-response planning across borders.

The engagement of the UN in activities related to natural disasters may be traced back to the early 1970s with the launch of initiatives related to both hazard surveys and disaster relief coordination. In December 1987, the General Assembly of the UN decided to "designate the 1990s as a decade in which the international community, under the auspices of the United Nations, will pay special attention to fostering international co-operation in the field of natural disaster reduction." This decision was behind the International Decade for Natural Disaster Reduction (IDNDR), 1990–2000, (A/RES/42/169 of 11 December 1987), which had mobilized the UN system around the objective of reducing losses from natural hazards. The strategic focus at the start of the Decade was on disaster reduction through the scientific understanding of natural disasters; the assessment of their damage potential; and the mitigation and reduction of damage through technical assistance and technology transfer, education, and training. The IDNDR did serve as a global partnership for the development of disaster reduction activities among different stakeholders inside and outside the United Nations. Several actors concerned with natural hazards have enhanced disaster reduction efforts that had already been in place prior to the Decade. At the same time, and in the early 1990s, the General Assembly of the United Nations underlined the need to strengthen the coordination of emergency humanitarian assistance of the United Nations system. It therefore designated a high-level official as emergency relief coordinator. Subsequently, the UN Department of Humanitarian Affairs was established to "mobilize and coordinate the collective efforts of the international community, in particular those of the UN system, to meet in a coherent and timely manner the needs of those exposed to human suffering and material destruction in disasters and emergencies. This involves reducing vulnerability, promoting solutions to root causes, and facilitating the smooth transition from relief to rehabilitation and development."

At the conclusion of the IDNDR, it was generally accepted that the momentum generated by it must be maintained. The global mobilization that occurred during the Decade confirmed that the United Nations is well placed to offer "a suited framework for bringing the various interests together" in the mutual interest of all concerned with the reduction of effects of natural hazards. Furthermore, the rise in the occurrence of natural disasters and the evolution in their intensity and complexity compel the UN system to adapt and strengthen its mandate and intervention with respect to natural hazards. This mandate is progressively evolving. In 1999, the UN General assembly initiated the International Strategy for Disaster Reduction (ISDR) to succeed the program of the IDNDR. The Strategy aims at "building disaster resilient communities by promoting increased awareness of the importance of disaster reduction as an integral component of sustainable development, with the goal of reducing human, social, economic and environmental losses due to natural hazards and related technological and environmental disasters." It is thus crucially important that a platform such as the ISDR be maintained and in place for cooperation through coordination, and to promote synergy among stakeholders, rather than compartmentalize them.

Over the past decade, the involvement of the United Nations system in activities related to natural disasters has increased markedly. A number of UN system entities carry out active programs in support of disaster reduction and many of them have strengthened their disaster reduction capacity in their respective areas of competency during recent years. All work with regional, national, or local authorities and in many cases with civil society organizations and groups. Coordination and cooperation among these entities are progressing well. Still improved coherence and cooperation is needed among United Nations entities in disaster risk reduction and disaster response.

Overview on role of United Nations Organizations

An overview is given of the role of the main UN players in relation to natural disasters. The essence of this text is derived from a recent reference publication produced by the Inter-Agency Secretariat of the International Strategy for Disaster Reduction, ISDR, 2009: *Disaster Risk Reduction in the United Nations- Roles, mandates and areas of work of key United Nations entities* (Geneva, Switzerland, UNISDR). The description does not provide an exhaustive inventory of all relevant activities of the UN. The purpose is to illustrate the work rather than be comprehensive. Relevant website citations are given as sources of further information.

United Nations General Assembly

http://www.un.org/ga/

The General Assembly is the chief deliberative, policymaking, and representative organ of the United Nations and is composed of representatives of all Member States. It determines the policies and the main lines of work of the UN. It provides a unique forum for multilateral discussion of the full spectrum of international issues covered by the Charter of the UN. The Assembly recognized in 1971 the need to reduce the impact of disasters and created the Office of the United Nations Disaster Relief Coordinator (UNDRO) for the improved coordination of "assistance in cases of natural disaster and other disaster situations," including disaster mitigation. The Assembly recognized in 1987 the need to focus on disaster reduction as an activity in itself and launched the IDNDR in 1989, followed by ISDR in 1999.

Deliberations concerning natural disasters take place under diverse items of the Assembly's agenda. Disaster risk reduction is generally discussed under the segment *Environment and Sustainable Development* whereas the Humanitarian segment of the Assembly covers discussions related to coordination of humanitarian and disaster relief assistance.

Economic and Social Council (ECOSOC)

http://www.un.org/en/ecosoc/

ECOSOC was established under the United Nations Charter as the principal organ to coordinate economic, social, and related work of the 14 UN specialized agencies, functional commissions, and five regional commissions. ECOSOC serves as the central forum for discussing international economic and social issues, and for formulating policy recommendations addressed to Member States and the United Nations system.

ECOSOC promoted and approved the establishment of an Office of the Emergency Relief Co-ordinator, the subsequent UN Department of Humanitarian Affairs (DHA) and the current Office for the Coordination of Humanitarian Affairs (OCHA). It endorsed the founding documents of the IDNDR and the ISDR. Post-disaster responses and disaster risk reduction are issues which are regularly covered by deliberations and decisions of ECOSOC. Furthermore, the ministerial declarations of the high-level segment (Annual Ministerial Review) of ECOSOC deal with these issues. ECOSOC Commissions work actively in humanitarian assistance aspects and disaster risk reduction.

United Nations System Chief Executives Board for Coordination (CEB)

http://www.ceb.unsystem.org

The Chief Executives Board (CEB) brings together the executive heads of all United Nations organizations to further coordination and cooperation on the whole range of substantive and management issues facing the United Nations system. In addition to its regular reviews of contemporary political issues and major concerns facing the UN system, CEB approves policy statements on behalf of the UN system as a whole.

In its work, CEB periodically considers disaster reduction, humanitarian assistance, and related development and environmental issues. It issues statements and takes decisions regarding the coordination among agencies and organizations active in these areas. It ensures that disaster risk reduction is mainstreamed into the UN system's policies and practices.

United Nations Development Group (UNDG)

http://www.undg.org

UNDG is a committee composed of the 33 UN funds, programs, agencies, departments, and offices. It supports the work of CEB, providing a framework for greater coherence and cooperation in United Nations development operations, notably at the country level. As most of the UNDG members carry out activities toward disaster reduction, this group offers a mechanism to integrate disaster reduction into other areas of concern, in particular sustainable development.

Inter-Agency Standing Committee (IASC)

http://www.humanitarianinfo.org/iasc/

The Inter-Agency Standing Committee (IASC) is a forum for coordination, policy development, and decision making involving key UN and non-UN agencies and entities concerned with humanitarian assistance. It develops system-wide humanitarian policies and ensures effective response to post-disaster needs at the onset of a crisis.

Office for the Coordination of Humanitarian Affairs (OCHA)

http://ochaonline.un.org/

OCHA is a department of the United Nations whose mission is to mobilize and coordinate humanitarian action in partnership with national and international actors in order to alleviate human suffering in disasters and emergencies, advocate for the rights of people in need, promote preparedness and prevention, and facilitate sustainable solutions. OCHA is led by the Under-Secretary-General for Humanitarian Affairs and Emergency Relief Coordinator, who serves also as the Chair of the ISDR system.

One of OCHA's objectives is greater incorporation of disaster risk reduction approaches and strengthened preparedness in humanitarian response.

International Strategy for Disaster Reduction (ISDR)

http://www.unisdr.org

The ISDR is a global framework in which countries, institutions, and individuals can cooperate in order to promote and achieve disaster risk reduction. It is a global platform for the United Nations and other organizations to coordinate and guide disaster risk reduction and its integration into development planning and action and to ensure synergy among the activities of partners concerned with disaster risk reduction. The guiding document behind the ISDR is the Hyogo Framework for Action 2005–2015: Building the Resilience of Nations and Communities to Disasters which was adopted in January 2005 by 168 countries.

The ISDR is coordinated within the United Nations by an Inter-Agency Secretariat located in Geneva, Switzerland.

Department of Economic and Social Affairs (UN/DESA)

http://www.un.org/esa/desa/

The DESA multidimensional program promotes broad-based and sustainable development through an integrated approach to economic, social, environmental, population, and gender-related aspects of development. The Department serves as the secretariat for the UN Commission on Sustainable Development. Many of its divisions are concerned with disaster reduction, particularly the Division for Sustainable Development (DSD).

Food and Agriculture Organization (FAO)

http://www.fao.org

The FAO's general mandate is to raise the levels of nutrition, to improve agricultural productivity and the condition of rural populations. Based in Rome, Italy, FAO develops programs to strengthen the capacity of communities and local institutions in preparing for and addressing natural disasters, notably in order to reduce the vulnerability of agricultural production systems to disasters. It helps alleviate the impact of emergencies that affect food security and therefore strengthens programs for agricultural relief and rehabilitation in the aftermath of disasters.

International Telecommunication Union (ITU)

http://www.itu.int/en/pages/default.aspx

ITU, with its headquarters in Geneva, Switzerland, is an international organization where governments and the private sector coordinate global telecommunication networks and services. ITU work emphasizes the importance of telecommunications for disaster mitigation and disaster relief operations. It addresses the impact of disasters on communication facilities and information flows and works toward rapid availability and access to telecommunication resources.

United Nations Educational, Scientific and Cultural Organization (UNESCO)

http://www.unesco.org

The main objective of UNESCO is to promote collaboration among States through education, science, culture, and communication. UNESCO is based in Paris, France. The Organization provides intergovernmental coordination, advice to governments, and policy support for the establishment and operation of monitoring networks and early warning and risk mitigation systems for natural hazards, with particular emphasis on earthquakes, tsunamis, floods, volcanoes, and landslides. It promotes activities to develop a better scientific understanding of natural hazards and the mitigation of their effects. It promotes joint multi-stakeholder strategies for enhancing disaster education. Integrated approaches and synergy between natural sciences, social sciences, culture, education, and information systems lay the basis for interdisciplinary platforms to manage disaster risks. UNESCO provides technical advice on the construction of hazard-resistant schools and for the protection of cultural heritage. Through its Intergovernmental Oceanographic Commission, UNESCO coordinates tsunami early warning systems.

In the aftermath of disasters, UNESCO contributes to the rehabilitation of the educational establishment and the restoration of cultural heritage and to long-term reconstruction processes in its fields of competence.

The World Bank Group

http://www.worldbank.org

The mandate of the World Bank Group, based in Washington, D.C., USA, is to alleviate poverty and improve quality of life. It furthers strategies and procedures to promote proactive ways to integrate disaster prevention and mitigation into its development work. The Group promotes disaster risk management as a priority for poverty reduction, linked to environmental management. It concentrates on reconstruction measures that strengthen resilience to future disaster and identify innovations in risk transfer and financing.

The World Bank's work in integrating disaster risk reduction in development is carried out primarily through its Global Facility for Disaster Reduction and Recovery (GFDRR). It offers a model for advancing disaster risk reduction based on ex ante support to high-risk countries and ex post assistance for accelerated recovery and risk reduction after a disaster. Interventions of the World Bank catalyze greater cooperation between the humanitarian and development actors for accelerated recovery and risk reduction.

World Health Organization (WHO)

http://www.who.int/en

The World Health Organization is the directing and coordinating authority for health within the United Nations system. Based in Geneva, Switzerland, WHO deals with disaster preparedness connected with health. Its purpose is to reduce avoidable loss of life and the burden of disease and disability in disaster-affected countries. It works for emergency preparedness and response, as well as for advocacy for health and humanitarian action. WHO is concerned with the safety of hospitals in hazard-prone areas.

World Meteorological Organization (WMO)

http://www.wmo.ch

Based in Geneva, Switzerland, WMO is an intergovernmental organization mandated to facilitate international collaboration in meteorology, including coordinated observations and standardized instruments. It provides world leadership in expertise and international cooperation in weather, climate, hydrology, and water resources and related environmental issues. The Organization deals with hazards related to weather, climate, and water and coordinates global scientific activity to provide the advance warnings.

United Nations Children's Fund (UNICEF)

http://www.unicef.org

UNICEF is mandated by the United Nations General Assembly to advocate for the protection of children's rights, to help meet their basic needs. UNICEF generally works on warning, prevention, preparedness, relief, and recovery activities for the care of children and women in disaster-prone areas. The Fund sustains and enhances its support of national risk reduction efforts, specifically, those most directly related to threats to children, both in noncrisis and emergency situations. Policy and procedural guidelines for UNICEF staff in emergencies include both emergency response and preparedness/prevention activities.

United Nations Development Programme (UNDP)

http://www.undp.org

The United Nations Development Programme (UNDP) is the main development branch of the United Nations. It plays a coordinating role at country level. UNDP considers crisis prevention and disaster mitigation as integral parts of sustainable human development strategies. It therefore helps countries prevent and recover from natural disasters through advocacy, capacity building, conflict sensitive development, development of tools and methodologies, gender equality, knowledge networking, strategic planning and programming, and policy and standard setting. The UNDP has operational responsibilities at national level for natural disaster mitigation, prevention and preparedness, as well as disaster response. It works to ensure that disaster risk considerations are factored into national and regional development programs, and that countries take advantage of disaster recovery to mitigate future risks and vulnerabilities. UNDP focuses on the national and sub-national levels, where it assists the national and local governments in implementing disaster risk reduction activities.

United Nations Environment Programme (UNEP)

http://www.unep.org

Based in Kenya, Nairobi, the United Nations Environment Programme (UNEP) is the environmental agency of the United Nations. UNEP provides worldwide environmental information about the necessity of a viable sustainable ecosystem and early warning about the hazards to ecosystem health and environmental hazards. Objectives of its action are to minimize environmental threats to human well-being from the environmental causes and consequences of existing and potential natural and man-made disasters. It deals with emergency prevention, preparedness, assessment, mitigation, and response and implements programs on disaster reduction at all levels.

UN-Habitat

http://www.unhabitat.org

With its headquarters in Kenya, Nairobi, the United Nations Human Settlements Programme, UN-HABITAT promotes socially and environmentally sustainable towns and cities with the goal of providing adequate shelter for all. As such the Programme contributes to reducing the vulnerabilities of human settlements and to strengthening their capacities for managing disasters at all levels as well as to responding to immediate needs in the aftermath of crises that are linked to Agency mandated interventions supporting sustainable human settlements (they help rebuild a settlement after a disaster).

World Food Programme (WFP)

http://www.wfp.org/

WFP is mandated by the United Nations to combat global hunger. Based in Rome, Italy, the Programme meets emergency needs regarding food security, and provides the necessary logistics to deliver food. The consolidated framework of the WFP policies emphasizes the mitigation of the effects of recurring disasters in vulnerable areas. WFP assesses measures to prevent and mitigate disasters that pose threats to food production or livelihoods as part of country programming in areas subject to recurring disasters. WFP develops standard procedures jointly with government counterparts for borrowing from and replenishing national food stocks.

United Nations Population Fund (UNFPA)

http://www.unfpa.org/

UNFPA is an international development agency "that promotes the right of every woman, man and child to enjoy a life of health and equal opportunity." UNFPA has a main goal to ensure adequate emergency preparedness and contingency planning at country level and to improve environmental protection. Its strategy for emergency preparedness, humanitarian response, transition, and recovery includes various measures to strengthen technical and institutional capacities to incorporate population, reproductive health, and gender concerns into overall emergency preparedness, response, transition, and recovery.

United Nations Institute for Training and Research (UNITAR)

http://www.unitar.org

UNITAR is an autonomous body within the United Nations with a mandate to enhance the effectiveness of the United Nations through training and research activities. Its headquarters is located in Geneva, Switzerland. Training and capacity-building programs are organized for policy and institutional development. Its environmental program addresses risk issues in the areas of chemical and waste management, climate change, biodiversity, land degradation.

United Nations University (UNU)

http://www.unu.edu/

United Nations University (UNU) is an autonomous organ whose overall mission is to contribute, through research and capacity building, to efforts to resolve the pressing global problems that are the concern of the United Nations Member States. It mobilizes an international community of scholars, engaged in research, postgraduate training, and dissemination of knowledge to provide alternative perspective on sustainable development challenges. UNU, with headquarters in Tokyo, Japan, provides useful knowledge about, and effective training on, important issues related to human-environmental-climate security, peace, and development including crises and disasters.

United Nations Centre for Regional Development (UNCRD)

http://www.uncrd.or.jp/

The UNCRD head office, located in Nagoya, Japan, encourages training and research in regional development as well as information dissemination. The Centre's work includes basic research programs for the design of community-based projects for disaster management planning, and disaster management capacity building introducing best practices case studies in developing countries.

United Nations Office for Outer Space Affairs (UNOOSA)

http://www.oosa.unvienna.org/

The United Nations Office for Outer Space Affairs (UNOOSA) is responsible for promoting international cooperation in the peaceful uses of outer space, and assisting developing countries in using space science and technology. Space applications and space-based services are used for disaster mitigation, relief, and prevention and space technologies can play important roles in the reduction of disasters. The use of such technologies can be particularly useful in the risk assessment, mitigation, and preparedness phases of disaster management. Space technologies are also vital to the early warning and management of the effects of disasters.

International Labor Organization (ILO)

http://www.ilo.org

ILO is the United Nations agency specialized in matters related to labor. ILO established a special In-Focus Programme on Crisis Response and Reconstruction that concentrates on various types of crises including natural disasters. The Programme promotes employment in post-disaster situations.

United Nations Volunteers (UNV)

http://www.unv.org/

The United Nations Volunteers program contributes to peace and development through volunteerism worldwide. UNV mobilize volunteers and strengthen volunteerism initiatives in support of disaster risk reduction and management, especially to strengthen community capacity to respond to and prevent disasters. In emergency and post-disaster situations, it also mobilizes and places national and international UNV volunteers in response to requests from UN entities and government partners. UNV also works with partners to integrate volunteerism into programming for disaster risk reduction and management. UNV works in disaster response, risk reduction, crisis prevention, and community-based adaptation to climate change.

International Atomic Energy Agency (IAEA)

http://www.iaea.org

The IAEA is the world's center of cooperation in the nuclear field. The IAEA Secretariat is headquartered in Vienna, Austria. The agency is concerned with the zoning of nuclear power plants in areas prone to seismic activity, and it has been actively concerned with the design of reactors that can withstand the most severe natural disasters. A core element of the IAEA's work is to help countries to upgrade nuclear safety and to prepare for and respond to emergencies.

Conclusions

The UN system is widely acknowledged as the central multilateral framework through which the international community can address global challenges, including by providing a coherent approach to global climate change, preparing for, and responding to natural disasters. Action in these fields increasingly depends on the active involvement and support by all major stakeholders.

Bibliography

Brown, B. J., 1979. *Disaster Preparedness and the United Nations: Advance Planning for Disaster Relief.* New York: Paragon Press.
http://www.un.org/apps/news/story.asp?NewsID=33031&Cr=climate+change&Cr1=disaster.
http://www.un.org/apps/news/test/story.asp?NewsID=28506&Cr=DISASTER&Cr1=.
http://www.un.org/en/events/tenstories/08/climatechange.shtml.
http://www.un.org/en/globalissues/humanitarian/index.shtml.
http://www.un.org/esa/desa/desaNews/v12n06/global.html.
http://www.un.org/ga/president/62/news/news.asp?NewsID=33480.
http://www.un.org/News/Press/docs/2003/sgsm8909.doc.htm.
http://www.un.org/News/Press/docs/2008/sgsm11841.doc.htm.
ISDR, 2003. *United Nations Documents Related to Disaster Reduction.* Geneva: United Nations, Vol. 1, 2.
ISDR, 2009a. *Disaster Risk Reduction in the United Nations- Roles, Mandates and Areas of Work of Key United Nations Entities.* Geneva: UNISDR.
ISDR, 2009. Biennial work programme for 2010–2011. *Invest Today for a Safer Tomorrow.* Geneva: UNISDR.
Katoch, A., 2003. International disaster response and the United Nations. *International Disaster Response Laws, Principles and Practice: Reflections, Prospects, and Challenges*, International Federation of the Red Cross and Red Crescent Societies, pp. 47–56.
Living with Risk: A Global Review of Disaster Reduction Initiatives, Vols. I and II, Inter-Agency secretariat of the International Strategy for Disaster Reduction, 2004 (A/54/136-E/1999/89, June 18, 1999).
Strengthening the coordination of emergency humanitarian assistance of the United Nations. Report of the Secretary-General (A/59/93-E/2004/74 of June 11, 2004).
UN (United Nations), 1994. Yokohama strategy and plan of action for a safer world, guidelines for natural disaster prevention, preparedness and mitigation. In *World Conference on Natural Disaster Reduction*, Yokohama, May 23–27, 1994, United Nations, New York.
UNISDR (United Nations International Strategy for Disaster Reduction Secretariat), 2007. *Disaster Risk Reduction: Global Review 2007.* Geneva: United Nations.
UNISDR (United Nations International Strategy for Disaster Reduction Secretariat), 2009. *Terminology on Disaster Risk Reduction.* http://www.unisdr.org/eng/library/lib-terminology-eng.htm.
UNISDR (United Nations Secretariat of the International Strategy for Disaster Reduction), 2008. *Indicators of Progress: Guidance on Measuring the Reduction of Disaster Risks and the Implementation of the Hyogo Framework for Action.* Geneva: United Nations.
UN-OCHA, 2009. Compilation of United Nations Resolutions on Humanitarian Assistance, Office for the Coordination of Humanitarian Affairs, Policy Development and Studies Branch, Policy and Studies Series, Selected resolutions of the General Assembly, Economic and Social, Council and Security Council Resolutions and Decisions.
UN-OCHA, 2009. Reference Guide, Office for the Coordination of Humanitarian Affairs, Policy Development and Studies Branch, Policy and Studies Series, Vol. I N°2, Normative Developments on the coordination of humanitarian assistance in the General Assembly, the Economic and Social Council, and the Security Council since the adoption of General Assembly resolution 46/182.

Cross-references

Civil protection and Crisis Management
Communicating Emergency Information
Disaster Diplomacy
Disaster Risk Reduction
Education and Training for Emergency Preparedness
Emergency Management
Federal Emergency Management Agency (FEMA)
Hyogo framework for action (2005–2015)
International strategies for disaster reduction (IDNDR and ISDR)
Red Cross and Red Crescent, International Federation of

UNIVERSAL SOIL LOSS EQUATION (USLE)

Armand LaRocque
University of New Brunswick, Fredericton, NB, Canada

Synonyms

Revised Universal Soil Loss Equation (RUSLE); RUSLE2; USLE

Definition

The Universal Soil Loss Equation (USLE) is a mathematical model developed to predict the soil erosion by rainfall and surface runoff on a field. The empirical result of the USLE corresponds to a long-term average annual rate of soil losses under a variety of climatic conditions, soil types, topographic characteristics, crop systems, and conservation practices. However, USLE only predicts the amount of soil loss resulting from sheet or rill erosion on a single slope and does not account for additional soil losses that might occur.

Discussion

The USLE is based on soil erosion data collected from experiments in erosion plots and rainfall simulators since the 1930s by the USDA (United States Department of Agriculture) Soil Conservation Service (now the "USDA Natural Resources Conservation Service"). The main purpose of this model was to preserve cropland from erosion and the diminution of the agricultural productivity by devastating drought, wind erosion, and dust storms, such as the famous "Dust Bowl," affecting the American and Canadian prairie lands between 1930 and 1940. The Revised Universal Soil Loss Equation (RUSLE and now RUSLE2), which is a computerized version of USLE, includes improvements in many of the factor estimates.

The USLE is composed of six major factors used to predict the long-term average annual soil loss (A), in tons per acre per year for a specific site. The equation takes the simple product form as follows:

$$A = R \times K \times L \times S \times C \times P.$$

The potential soil loss (A) is calculated using the rainfall erosivity factor determined for each geographic location (R), the soil erodibility factor (K), which is mainly based on the soil texture, the slope length factor (L), the slope steepness factor (S), the cover type and management factor (C), and the conservation practices factor (P). Each of these factors is a numerical estimate of specific conditions affecting the severity of soil erosion at the studied site. In the case of factors K, L, and S, the estimates are based on measures realized on a "standard" slope (steepness of 9% and slope length of 22.1 m or 72.6 ft).

The calculated soil loss (A) can then be compared with a tolerable soil loss that is the maximum annual amount of soil that can be removed before the natural productivity in the long term of soil may be affected. This comparison allows determining the adequacy of conservation measures in farm planning. USLE is also used to guide conservation planning of land use for nonagricultural conditions such as construction sites.

Bibliography

Foster, G. R., 2008. *Draft Science Documentation: Revised Universal Soil Loss Equation Version 2*. Washington (DC): USDA-Agriculture Research Service. 349 p. http://www.ars.usda.gov/SP2UserFiles/Place/64080530/RUSLE/RUSLE2_Science_Doc.pdf.

Wischmeier, W. H., Smith, D. D., 1978. Predicting rainfall erosion losses: a guide to conservation planning. U.S. Department of Agriculture, Agriculture Handbook number 537, U.S. Government Printing Office, Washington (DC), 58 p. http://www.ars.usda.gov/SP2UserFiles/ad_hoc/36021500USLEDatabase/AH_537.pdf

Cross-references

Dust Bowl
Erosion
Erosivity

UNREINFORCED MASONRY BUILDINGS

Fabio Taucer
European Commission – Joint Research Centre, Ispra (VA), Italy

Synonyms

Unreinforced concrete block masonry (UCB) (Jaiswal and Wald, 2008); Unreinforced fired brick masonry (UFB); URM

Definition

A structure whose load-bearing system consists of an assemblage of masonry units generally made of stone, brick, or concrete blocks laid in a specific pattern and joined together with mortar.

Discussion

Unreinforced masonry (URM) buildings are found all over the world and have been used since ancient times for providing shelter with a number of advantageous properties, such as thermal and acoustic insulation, fire resistance, and weather protection. URM buildings are commonly used for the construction of low-rise residential and office buildings with the floor area subdivided into a large number of spaces that repeat along the height from the foundation to the roof level (Hendry et al., 1997). This allows for a flexible arrangement of the load-bearing walls, which form the backbone of the structural system of resisting forces.

The structural performance of the masonry fabric depends on the performance of the masonry units, the mortar, and their composite behavior. Modern building codes provide guidelines for the optimum combination of these, namely, the quality of bond between bricks and mortar, the connection among the width of brick walls, the connection among the walls at the corners and junctions, and the connection between the walls and the roof and floor structures (D'Ayala, 2010).

In spite of their good performance in resisting vertical loads, URM buildings are particularly vulnerable to seismic loading, due to their intrinsic low capacity in resisting tensile stresses that develop during an earthquake. The collapse of URM buildings was responsible for most of the 242,000 people killed during the 1976 Tangshan earthquake in the People's Republic of China (Grossi et al., 2010). In Italy, the 115,619 deaths registered between 1900 and 2010 (EM-DAT, 2010) were due in large part to the collapse of URM buildings that constitute a large part of the building stock constructed in Italy before the 1950s.

The damage patterns commonly observed in URM buildings range from the collapse of chimneys and plaster cracks, to the development of cracks in the walls starting at window openings and corners, to the partial or total collapse of the building due to the out-of-plane collapse of walls. The lack of effective connections between the walls and the horizontal structures of roofs and floors that act as diaphragms that keep together the masonry fabric has been associated as the main factor in leading to the collapse of URM buildings. The use of heavy reinforced concrete roofs and floors to replace older horizontal structures made of light wood construction has exacerbated the amount of damage suffered by these structures.

Possible ways to strengthen and retrofit URM buildings to enhance their seismic performance include ensuring adequate connections of the masonry fabric, efficient connections between the walls and the horizontal structures, stitching and grouting of cracks, installation of steel ties for anchoring the walls, reinforced concrete or composite polymer coating, and the addition of new members to prevent the out-of-plane collapse of walls.

Bibliography

D'Ayala, D., 2010. Unreinforced Brick Masonry Construction. World Housing Encyclopedia. www.world-housing.net.
EM-DAT, 2008. The International Disaster Database. CRED – Centre for Research on the Epidemiology of Disasters. Brussels: Université Catholique de Louvain, www.emdat.be.
Grossi, P., Del Re, D., and Wang, Z., 2006. *The 1976 Great Tangshan Earthquake 30-Year Retrospective*. Newark: Risk Management Solutions, Inc. www.rms.com.
Hendry, A. W., Sinha, B. P., and Davies, S. R., 1997. *Design of Masonry Structures*. London: Chapman and Hall.
Jaiswal, K., and Wald, D. J., 2008. Creating a global building inventory for earthquake loss assessment and risk management. Open-file report 2009-1160. Reston, Virginia: U.S. Geological Survey.

Cross-references

Accelerometer
Building Codes
Building Failure
Buildings, Structures, and Public Safety
Collapsing Soil Hazards
Concrete Structures
Damage and the Built Environment
Dispersive Soil Hazards
Earthquake Damage
Earthquake Resistant Design
High-rise Buildings in Natural Disaster
Land Subsidence
Liquefaction
Seismology
Structural Damage Caused by Earthquakes
Tangshan, China (1976 Earthquake)

URBAN ENVIRONMENTS AND NATURAL HAZARDS

Pat E. Rasmussen
University of Ottawa and Health Canada, Ottawa, ON, Canada

Definitions

Chronic obstructive pulmonary disease (COPD). A respiratory disease characterized by a gradual loss of lung function.
Elephantiasis. Symptom of an infectious tropical disease filariasis (philariasis) caused by parasitic nematodes and spread by mosquitoes.
Natural hazard. Any naturally occurring event or agent that has the potential to negatively impact human and environmental health.
Slums. A socioeconomic term describing intensely populated parts of cities characterized by poverty, inadequate housing, water, and sanitation.

Introduction

Natural hazards have disastrous consequences when they impact cities. Recent examples are the tens of thousands of Americans left homeless when Hurricane Katrina flooded 80% of the city of New Orleans, Louisiana, on August 29, 2005; the widespread destruction of schools, hospitals, and factories caused by the May 12, 2008, earthquake in Sichuan Province, China; and the death of some quarter million Haitians in the city of Port-au-Prince caused by the earthquake of January 12, 2010.

Natural hazards in the urban environment may be physical, climatic, biological, or chemical in origin, and often occur in combination with anthropogenic hazards. For example, extreme weather conditions such as *heat waves* further aggravate the negative health effects of urban air pollution. Degradation of the urban environment by *flooding* is another example, where the natural hazard triggers increased pollution of surface water and groundwater from sewage and surface runoff. Therefore, when we consider natural hazards in urban environments, we must consider the additive effects of interactions between human activities and natural forces.

Large-scale urbanization

In his book "Crucibles of Hazard: Mega-Cities and Disasters in Transition" Mitchell (1999) warned that the world's large cities were rapidly becoming more exposed and more vulnerable to natural hazards. By the dawn of the millennium, the world had changed from predominantly rural to predominantly urban: at the time, more than 300 cities counted a million or more inhabitants. In Europe more than 70% of the population had moved to urban areas, whereas in the USA, an estimated 90% of the population resided in cities (Santamouris, 2001). Mitchell (1999) warned that rapid large-scale urbanization would be a major contributor to the rising global toll of disaster losses.

By 2010 the United Nations recognized the increasing urban population at risk, and undertook to engage city mayors and other local authorities in designing and building disaster-resilient cities, schools, and hospitals as a priority under the 2010–2011 World Disaster Reduction Campaign (United Nations, 2009). Preventative measures, emergency preparedness, and sound environmental management practices, which are put into place prior to the occurrence of a natural disaster, help to reduce the probability of damage, minimize deaths, illnesses and economic loss, and assist the city in its recovery (see *Disaster Risk Reduction*, and *Recovery and Reconstruction After Disaster*).

Natural physical hazards

The construction of human settlements on hazardous lands represents an interaction between the natural and urban environment which creates an increased risk to humans. Examples include practices such as locating schools on slopes with high risks of *landslides*; building residential developments on ocean cliffs or *floodplains*; and situating high-rise office buildings in *earthquake* zones. Earthquakes strike violently and suddenly, causing extensive property damage and many deaths and injuries when they occur in or near cities (FEMA, 2010a).

Extreme weather conditions, such as *heat waves* and *hailstorms*, are important natural physical hazards affecting urban environments. Inadequate protection from the elements is a common cause of death in cities all over the world, whether it is the urban homeless facing the extreme cold of a winter night in northern cities; homes collapsing during monsoon rains in tropical cities; or the combination of stifling heat and air pollution in under-ventilated slums of southern and temperate climates.

Degradation of air quality

The impact of air temperature on urban environments is an example of the interaction between natural and anthropogenic hazards. Many cities are located in valleys where thermal inversions cause air pollutants to become trapped, increasing both the concentration of pollutants and the duration of exposure. The combination of high temperatures and air pollution affects the health of vulnerable groups living in cities, especially the elderly, small infants, and people with respiratory disorders. In general, densely built urban environments experience elevated temperatures compared to the rural surroundings, a phenomenon known as the "urban heat island" (Santamouris, 2001). The "urban heat island" effect is caused by the radiation of heat from surfaces of building and pavement surfaces.

According to a World Health Organization (WHO, 2006) assessment of the burden of disease due to air pollution, more than two million premature deaths each year can be attributed to the effects of urban outdoor air pollution and indoor air pollution (caused by the burning of solid fuels). Globally, more than 40% of all pulmonary disease (COPD) cases, and 41% of all cases of lower respiratory infections, are attributed to air pollution, both indoors and outdoors (WHO, 2006).

Where cities are located in the vicinity of *volcanic eruptions*, increased respiratory problems occur in vulnerable populations who inhale ash particles and acidic gas vapors. Even distant cities may experience problems, as wind can carry ash particles from volcanic eruptions that occur hundreds of kilometers away (FEMA, 2010b). *Volcanic ash* also damages engines and electrical equipment, and ash accumulations mixed with water become heavy and can collapse roofs (FEMA, 2010b).

Biological hazards in the urban environment

The World Health Organization attributes 94% of the global burden of disease to poor urban environmental management practices, mainly unsafe water, sanitation, and hygiene (WHO, 2006). Natural physical disasters such as *flooding*, *hurricanes*, and *earthquakes* interrupt urban services, leading to further degradation of environmental conditions in cities. The result is increased transmission of diseases from food and water, including diarrhea, gastroenteritis, cholera, dysentery, hepatitis-A, hepatitis-E, and typhoid fever. Overcrowded and poorly ventilated slums suffer inordinately from these diseases (United Nations, 2003).

A large proportion of deaths caused by malaria (42% of all cases globally) are attributed to poor management of water resources, housing, and land use, which result in a failure to control *insect* populations effectively (WHO, 2006). For example the malaria-spreading *Anopheles* mosquito breeds in standing water and the elephantiasis-spreading *Culex* mosquitoes breed in blocked drains, latrines, and septic tanks.

Urban geochemical hazards

The United Nations General Assembly declared 2008 as the International Year of Planet Earth. One of the main themes "Earth and Health – Building a Safer Environment" served to inspire earth scientists and medical researchers to work together toward understanding and mitigating geochemical hazards (Rasmussen and Gardner, 2008).

There is a long history of health effects related to urban exposures to metals and minerals, the most famous being lead plumbing used by the ancient Romans (Skinner and Berger, 2003; Selinus et al., 2005). Mielke et al. (2003) describe the neurological effects of childhood exposures to lead-contaminated soil in modern cities, and make the case for urban geochemical mapping efforts. Concentrations of lead and other metals are commonly higher in urban house dust than in soil, partly caused by the tendency for house dust to accumulate metals that are tracked indoors from outside sources (Rasmussen, 2004). Radon from underlying bedrock is a natural household source of radiation in air and tap water (Appleton, 2005). Lung disease caused by inhalation of asbestos fibers is mainly related to occupational exposures, but urban environmental exposures have been documented, notably during the demolition of buildings following the 1995 earthquake in Kobe, Japan (Nolan et al., 2001).

Summary

- Natural hazards are most likely to become catastrophic when they hit populated urban environments.
- In urban environments, natural hazards are interactive processes that involve both people and natural systems.
- Natural hazards disproportionately hurt the urban poor.
- Natural hazards have the greatest impact in cities that are vulnerable due to location and/or inadequate attention to prevention and mitigation.
- When natural hazards strike a city, they cause further degradation of the urban environment and exacerbate preexisting biological and chemical hazards.
- Interrupted urban services and destruction of urban infrastructure increase the transmission of disease, contamination of surface and groundwater, and human exposures to toxic wastes and chemical spills.
- Urban geochemical hazards include radon, soil and dust contaminated with lead and other metals, and environmental pollution caused by destruction of buildings and factories following earthquakes and other physical disasters.
- The rapid growth of large cities has contributed to the increased rate of death and destruction when natural hazards interact with urban settlements.

Bibliography

Appleton, J. D., 2005. Radon in air and water. In Selinus, O., Alloway, B., Centeno, J. A., Finkelman, R. B., Fuge, R., Lindh, U., and Smedley, P. (eds.), *Essentials of Medical Geology. Impact of the Natural Environment on Public Health*: Academic, pp. 227–262.
FEMA, 2010a. Website information on earthquakes. Federal Emergency Management Agency, an agency of the United States Department of Homeland Security. http://www.fema.gov/hazard/earthquake/index.shtm. Accessed Jan 21 2010.
FEMA, 2010b. Website information on volcanoes. Federal Emergency Management Agency, an agency of the United States Department of Homeland Security. http://www.fema.gov/hazard/volcano/index.shtm. Accessed Jan 21 2010.
Mielke, H. W., Gonzales, C., Powell, E., Coty, S., and Shah, A., 2003. Anthropogenic distribution of lead. In Catherine, H., Skinner, W., and Berger, A. R. (eds.), *Geology and Health: Closing the Gap*. Cary, NC: Oxford University Press. 192 pp.
Mitchell, J. K. (ed.), 1999. *Crucibles of Hazard: Mega-Cities and Disasters in Transition*. Tokyo: United Nations University Press. 544 pp.
Nolan, R. P., Langer, A. M., Ross, M., Wicks, F. J., and Martin, R. F. (eds.), 2001 *Health Effects of Chrysotile Asbestos: Contribution of Science to Risk-Management Decisions*. The Canadian Mineralogist, Mineralogical Association of Canada, Special Publication 5, 304 pp.
Rasmussen, P. E., 2004. Elements and their compounds in indoor environments. In Merian, E., Anke, M., Ihnat, M., and Stoeppler, M. (eds.), *Elements and their Compounds in the Environment*. Weinheim: Wiley, Vol. 1(11), pp. 215–234.
Rasmussen, P. E., and Gardner, H. D., 2008. Earth and health – building a safer Canadian environment. *Geoscience Canada*, **35**(2/3), 61–72.
Santamouris, M. (ed.), 2001. *Energy and Climate in the Urban Built Environment*. London: James & James, 402 pp.
Selinus, O., Alloway, B., Centeno, J. A., Finkelman, R. B., Fuge, R., Lindh, U., and Smedley, P., 2005. *Essentials of Medical Geology: Impacts of the Natural Environment on Public Health*. Amsterdam: Academic. 832 pp.
Skinner, H. W. C., and Berger, A. R., 2003. *Geology and Health: Closing the Gap*. Cary: Oxford University Press. 192 pp.
United Nations, 2003. *The Challenge of Slums: Global Report on Human Settlements*. Nairobi: UN Habitat.
United Nations, 2009. Invest Today for A Safer Tomorrow. International Strategy for Disaster Reduction: 2010–2011 Biennial Work Programme UNISDR secretariat. http://www.unisdr.org/news/v.php?id=11801. Accessed Jan 20 2010.
WHO, 2006. *Air Quality Guidelines for Particulate Matter, Ozone, Nitrogen Dioxide and Sulfur Dioxide – Global Update 2005*. Geneva: World Health Organization, 2006.

Cross-references

Disaster Risk Reduction (DRR)
Earthquake
Flood
Floodplains
Heat Wave
Insect
Landslide
Recovery and Reconstruction After Disaster
Tsunami
Volcanic Ash
Volcanoes and Volcanic Eruptions

CASE STUDY

USOI LANDSLIDE AND LAKE SAREZ

Alexander Strom
Geodynamic Research Center – Branch of JSC "Hydroproject Institute", Moscow, Russia

Introduction

On February 18, 1911, a strong *earthquake* in Pamirs (Tajikistan) caused a catastrophic failure of about 2.2 km^3 (six billion tons) of rock (Figure 1) and resulted

Usoi Landslide and Lake Sarez, Figure 1 Headscarp of the Usoi landslide.

in the formation of the Usoi landslide dam named after the small village buried by the *landslide*. Fifty eight inhabitants lost their lives in this event. The Usoi landslide located at 38°16.5′N, 72°36′E is the world's largest non-volcanic landslide ever recorded in historical times. Despite the remoteness and inaccessibility of the site, Russian researchers performed their first studies of this feature soon after the event (Bukinich, 1913; Shpilko, 1915; Preobrajensky, 1920). Regular studies started in the 1960s (Sheko and Lekhatinov, 1970; Agakhanjanz, 1989).

Main parameters of the dam and the lake

The Usoi *landslide* occurred in bedrock and formed a 567-m-high, 5-km-long (across the valley), and 3.75-km-wide dam in the Murgab River valley. The upper part of the dam's body is composed of rocky blocks about 1.5 × 1.5 km in size, intensively fractured but, at the same time, retaining original layering. Distal and proximal parts of the dam are composed of heavily crushed material. No direct data on the structure and grain size composition of the dam's interiors are available (Ischuk, 2011). The downstream slope of the dam is cut by the deep canyon eroded by the filtering water and by debris flows that originate from the glaciers remaining above the headscarp (in 1947, these flows were diverted toward the lake by rockfall from the headscarp wall). The lake that has been impounded was named after the Sarez village submerged by rising water. Three years after the dam formation, in 1914, water was detected to seep through the blockage. Annual mean discharge passing through the dam is 47 m^3/s at present. During the flood periods when water rises to about 5 m above the mean annual level, it increases up to 85 m^3/s (Alford and Schuster, 2000; Ischuk, 2011). According to regular observations, lake level increases gradually (up to 20 cm/year) and is now 3261 m.a.s.l., only 38 m below the lowermost part of the blockage at maximal water level (Ischuk, 2011). The 500-m-deep and 60-km-long Lake Sarez contains almost 17 km^3 of water and is the world's deepest existing landslide-dammed lake (Figure 2).

Usoi Landslide and Lake Sarez, Figure 2 Google Earth image of Lake Sarez.

Right-bank landslide

In the 1960s A.I. Sheko (Sheko, 1968; Sheko and Lekhatinov, 1970) hypothesized that a large-scale slope failure may occur on Lake Sarez' right bank, 4–5 km from the blockage, which could cause a huge *displacement wave* that could spill over the dam at its lowermost section, resulting in the dam's partial or complete breach and a downstream *flood* as happened, for example, in the Las Conchas valley (Argentina) in prehistoric times (Hermanns et al., 2004). Many researchers estimated the volume of this "right-bank landslide" to range from 0.3 to 2.0 km^3. However, the possibility of such a large-scale failure, its volume, and velocity are still controversial and require additional studies (Alford and Schuster, 2000; Ischuk, 2011). Additional uncertainty of this slope stability assessment is due to the high seismic activity of the Central Pamirs.

Safety measures

Though studies performed since 1960s revealed that the dam in its present state should be considered as a stable feature that could not be destroyed by water pressure or by gradual overtopping (Ischuk, 2011), an *early warning system* has been installed recently aiming to detect various indicators of the dam's instability (seismic strong motion, rapid increase of water level, etc.) and to allow people living downstream to escape to shelters constructed above the flood level calculated for outburst scenarios (Zaninetti, 2000).

Nevertheless, in a long-term perspective, Lake Sarez endangers sustainable development of the communities living in the Bartang, Pianj, and Amu-Daria River valleys in Tajikistan, Afghanistan, Uzbekistan, and Turkmenistan with a population over 5.5 million people (Gaziev, 1984; Alford and Schuster, 2000; Schuster, 2002). Thus, since the potential risk of the landslide dam breach and of a devastating outburst flood exists, special measures should be undertaken to ensure long-term safety of Lake Sarez regardless of any dynamic effects such as strong *earthquake*s or the impact of a landslide-triggered *displacement wave* caused by large-scale slope failure.

Possible long-term solutions, which are under discussion, envisage not only risk reduction measures but also use of the Lake Sarez water for irrigation and power production. Construction of a spillway tunnel system through the left-bank bedrock massif seems to be the most reliable variant that allows not only lowering of the lake up to the safe level but also the integration of a powerhouse.

Bibliography

Agakhanjanz, O. E., 1989. Sarez. Leningrad: Leningrad Press, pp. 110 (in Russian).

Alford, D., and Schuster, R. L., 2000. Introduction and summary. In Alford, D., and Schuster, R. L. (eds.), *Usoi Landslide Dam and Lake Sarez. An Assessment of Hazard and Risk in the Pamir Mountains, Tajikistan*. ISDR Prevention Series, No 1. New York/Geneva: UN, pp. 1–18.

Bukinich, D. D., 1913. Usoi earthquake and its consequences. *Russian Gazette*, No. 187 (in Russian).

Gaziev, E., 1984. Study of the Usoi landslide in Pamir. In *Proceedings of 4th International Symposium on Landslides*, Toronto, Vol. 1, pp. 511–515.

Hermanns, R. L., Niedermann, S., Ivy-Ochs, S., and Kubik, P. W., 2004. Rock avalanching into a landslide-dammed lake causing multiple dam failure in Las Conchas valley (NW Argentina) – evidence from surface exposure dating and stratigraphic analyses. *Landslides*, **1**, 113–122.

Ischuk, A. R., 2011. Usoi rockslide dam and Lake Sarez, Pamir mountains, Tajikistan. In Evans, S. G., Hermanns, R., Scarascia-Mugnozza, G., and Strom, A. L. (eds.), *Natural and Artificial Rockslide Dams*. New York/London: Springer. Lecture Notes in Earth Sciences, Vol. 133, pp. 423–440.

Preobrajensky, J., 1920. The Usoi landslide. *Geological Communications, Papers on Applied Geology*, **14**, 21 (in Russian).

Schuster, R. L., 2002. Usoi landslide dam, southeastern Tajikistan. In *Proceedings of International Symposium on Landslide Risk Mitigation and Protection of Cultural and Natural Heritage*, Kyoto, pp. 489–505.

Sheko, A. I., 1968. The Usoi blockage stability and the Lake Sarez breach assessment. *Bulletin of Moscow Nature Investigation's Society, Geological Section*, **4**, 151–152 (in Russian).

Sheko, A. I., and Lekhatinov, A. M., 1970. Current state of the Usoi blockage and tasks of future studies. In *Materials of Scientific-Technical Meeting on the Problems of Study and Forecast of the Mudflows, Rockfalls and Landslides*, Dushanbe, pp. 219–223 (in Russian).

Shpilko, G. A., 1915. New data on the Usoi blockage and the Sarez Lake. *Proceedings of the Turkestan Department of Russian Geographical Society*, **11**, 11–17 (in Russian).

Zaninetti, A., 2000. Monitoring and early warning system. In Alford, D., and Schuster, R. L. (eds.), *Usoi Landslide Dam and Lake Sarez. An Assessment of Hazard and Risk in the Pamir Mountains, Tajikistan*. ISDR Prevention Series, No 1. New York/Geneva: UN, pp. 63–72.

V

CASE STUDY

VAIONT LANDSLIDE, ITALY

Monica Ghirotti[1], Doug Stead[2]
[1]Alma Mater-University of Bologna, Bologna, Italy
[2]Simon Fraser University, Burnaby, BC, Canada

Synonyms
Vajont landslide

Definition
The Vaiont landslide (northern Italy) is one of the best known and most tragic examples of a natural disaster induced by human activity. On October 9, 1963, a catastrophic landslide occurred on the northern slope of the Mount Toc; a rock mass of approximately 270 million m^3 collapsed into the reservoir at velocities up to 30 m/s generating a wave that overtopped the dam and swept into the Piave valley below, with the loss of about 2,000 lives.

Introduction
Landslides, as major natural hazards, account for extremely significant property damage/losses in terms of direct/indirect costs, especially in hilly or mountainous areas. Triggering factors can be intensive rainfall, earthquake shaking, groundwater changes, or rapid stream erosion promoting a sudden decrease in the shear strength and stability of slope-forming materials. Landslides induced by reservoir impounding can also damage dams and result in considerable loss of life. The Vaiont reservoir landslide is one of the best-known examples of a natural disaster induced by human activity.

Many questions have been posed and remain concerning the legal, economic, social, and scientific issues associated with the history of the dam and in particular emergency management of the instability in the Vaiont reservoir slope up to the time of catastrophic failure. The global impact of the event has been to stimulate a large body of research on the stability of natural rock slopes and in particular the development of geotechnical risk protocols for the construction of hydroelectric projects in mountainous topography. The 1963 Vaiont rock slide represents a dramatic example of the consequences of limitations in available data (parameter uncertainty) and our understanding of slope failure processes (model uncertainty). It provides a clear example of the importance of fully understanding the complex mechanics and dynamics of large rock slope instabilities.

Chronology of events
The Vaiont Dam, constructed between 1957 and 1960, is located on the Vaiont River in northern Italy, about 100 km north of Venice. The double-curved arch dam, at 265.5 m above the valley floor, was in 1963 the highest thin arch dam in the world. Its abutments were founded on the steep flanks of a deep canyon cut into dolomitic limestones of Malm and Dogger age. The planned full reservoir capacity was to reach a volume of 169 million m^3.

The slopes of Mount Toc underwent nearly 3 years of intermittent, slow slope movements, beginning at the time of the first filling of the reservoir. On October 9, 1963, at 22.39 local time, during the third reservoir emptying event, a catastrophic landslide (Figure 1) suddenly occurred on the southern slope of the Vaiont dam reservoir (northern slope of Mount Toc) and the whole mass collapsed into the reservoir in less than 45 s. The failed mass drove the water of the reservoir forward, giving rise to

Vaiont Landslide, Italy, Figure 1 (a) Mount Toc before October 9, 1963 (Semenza and Ghirotti 2000); (b) the failure scar of the 1963 Vaiont landslide (Mount Toc behind)

a wave, which overtopped the dam at a height of more than 100 m above the crest and hurtled down the Vaiont Gorge to the bottom of the Piave River. The flood destroyed the villages of Pirago, Villanova, Rivalta, and Faé and most of the town of Longarone; almost 2,000 people lost their lives (Figure 2). The event produced seismic shocks which were recorded throughout Europe. Remarkably the dam remained relatively intact, with only minor damage at the dam crest. The landslide was characterized by a long-term phase of creep deformation lasting 2–3 years, clearly related to the reservoir water levels and followed by the catastrophic failure. Three years prior to the catastrophic failure, the presence of an M-shaped tension crack on Mount Toc, 1 m wide and 2.5 km long, delineated the eventual failure.

However, even if Mount Toc provided clear important evidence to suspect the instability of its northern slope, technicians and experts of the time incorrectly hypothesized a very large and slow-moving landslide that could be controlled by reservoir operations (Müller, 1964, 1968, 1987).

The landslide

The landslide involved Jurassic and Cretaceous rocks (limestones and marls) showing varying degrees of fracturing. Movement occurred along a chair-shaped failure surface (Figure 3) in part corresponding to a preexisting slip surface at or close to residual strength, as indicated by the geological evidence recognized before 1963 (Semenza and Ghirotti 2000). The failure surface was largely confined within 0.5–18-cm-thick clay-rich layers (Hendron and Patton, 1985) which were observed to be continuous over large areas of the failure surface.

Geological and tectonic evidence suggests that parts of both the 1963 landslide perimeter and the prehistoric slide closely correspond to one or more faults (Hendron and Patton, (1985).

During the third reservoir emptying operation, the southern rock slope of Mount Toc failed suddenly over a length of 2 km and a surface area of 2 km^2. The slide moved a 250-m-thick mass of rock some 300–400 m horizontally with an estimated velocity of 20–30 m/s, before running up and stopping against the opposite side of the Vaiont Valley. The majority of the slide moved as a whole and reached the opposite side of the valley without any change in shape apart from a general rotation evident from both the surface morphology and the stratigraphic sequence that remained essentially unchanged after the movement. Kinematic release of the rock mass is suggested to have required internal yielding and fracturing with surface faults and a graben forming within the rock slope (Mencl, 1966; Hutchinson, 1987).

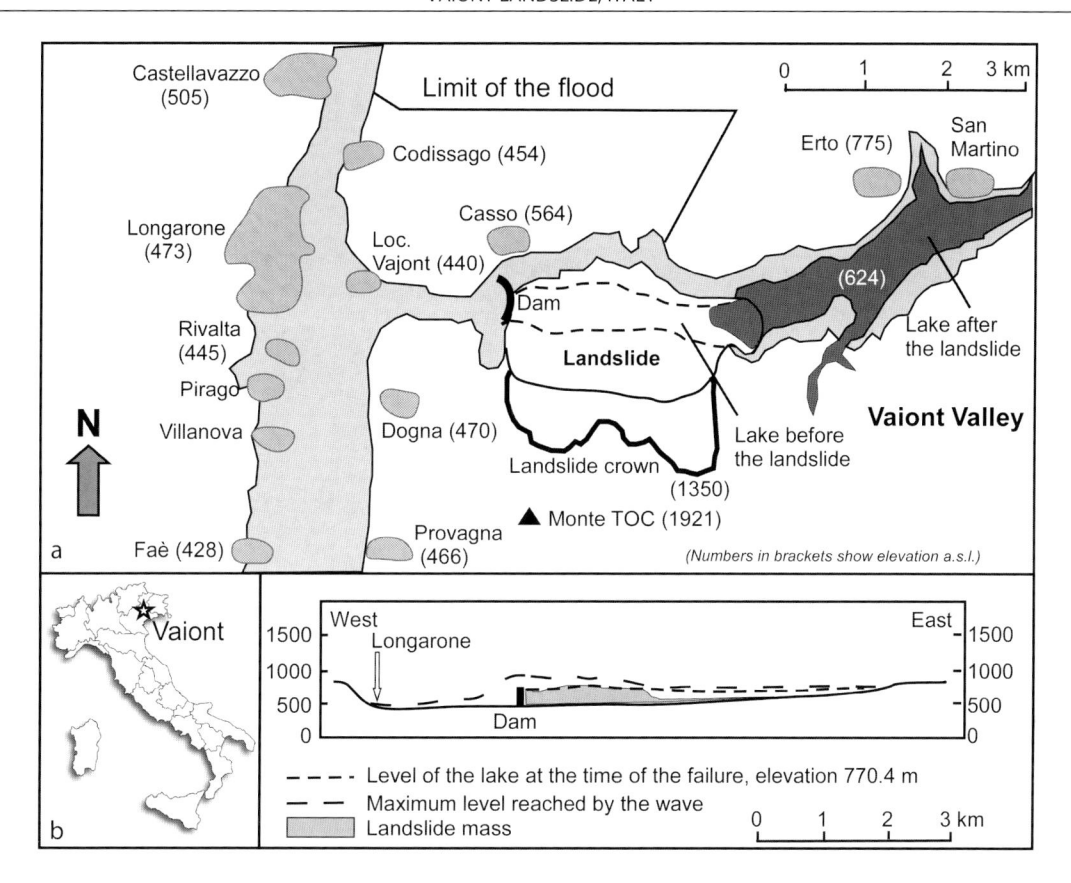

Vaiont Landslide, Italy, Figure 2 General plan showing the Vaiont landslide and the limit of the flood (**a**); schematic longitudinal section showing the original lake level and the elevation of the flood wave (**b**) (After Selli et al. 1964).

Summary

Considerable engineering geological research has been undertaken as a consequence of both the dramatic scientific and human impact of the Vaiont landslide disaster and the complex dynamics involved. In spite of the volume and number of research investigations undertaken to date, the Vaiont landslide continues to provide an engineering case study of great scientific interest and technological challenge. As emphasized by Mencl in 1966: "Anyone working in the field of slope stability may meet a similar problem and no pains should be spared to discover an explanation of the Mount Toc landslide."

Papers on the Vaiont landslide, published in the international literature after 1963, can be schematically subdivided into the following groups: papers based on geological and geomorphological data collected or providing detailed engineering geological and rock engineering descriptions of the Vaiont landslide (Müller, 1964; Selli et al., 1964; Semenza, 1965; Broili, 1967; Müller, 1968, 1987; Semenza and Ghirotti, 2000); papers mainly dealing with specific aspects including the geotechnical properties of the failure material, the physical and rheological behavior of the failure mass, and the use of varied methods of stability analysis (limit equilibrium and numerical modeling) as a means to understanding the complex role of the many factors involved in the triggering and development of the landslide (Mencl, 1966; Skempton 1966; Voight and Faust, 1982; Hendron and Patton, 1985; Hutchinson, 1987; Tika and Hutchinson, 1999; Vardoulakis, 2002; Crosta and Agliardi, 2003; Helmstetter et al., 2003; Sitar et al., 2005; Veveakis et al., 2007); papers on microseismic and other instrumentation data (Belloni and Stefani, 1987; Kilburn and Petley, 2003); and papers on the landslide-generated impulsive wave (Tinti et al., 2002; Panizzo et al., 2005).

Several interpretations of the event have been attempted during the last 45 years, but a comprehensive and convincing explanation of both the triggering and dynamics of the phenomenon remains elusive, the most comprehensive work to date being by Hendron and Patton (1985) who notwithstanding noted the need for further research. A review of all research on the Vaiont landslide has been published (Genevois and Ghirotti, 2005). Readers are referred to Superchi et al. (2010) who provide a comprehensive electronic bibliographic database on the significant number of publications to date on the Vaiont landslide.

At present, the area affected by the 1963 landslide is subject to environmental and town planning restraints.

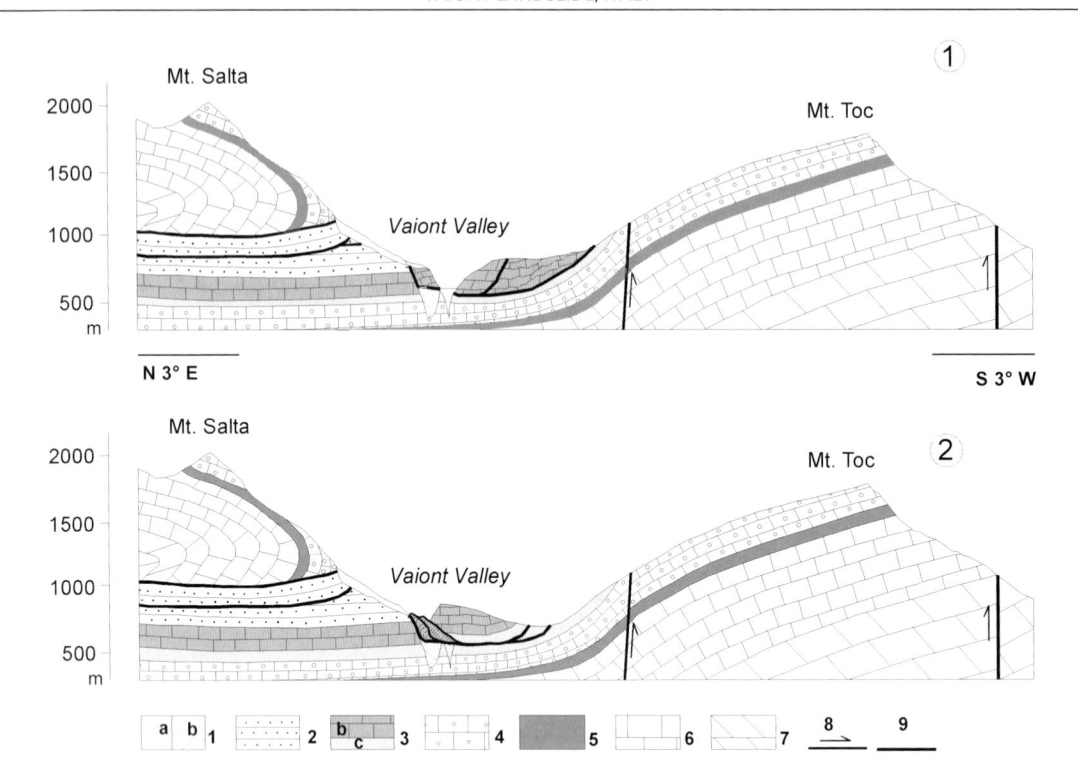

Vaiont Landslide, Italy, Figure 3 North-South geological sections of the Vaiont landslide: (**1**) before October 9, 1963; (**2**) after October 9, 1963 (Semenza and Ghirotti 2000). Legend: 1. *a* Quaternary, *b* stratified alluvial gravels; 2. Scaglia Rossa Fm. (Upper Cretaceous – Lower Paleocene); 3. *a* Cretaceous-Jurassic Fms. (Socchér Formation sensu lato and coeval), *b* Socchér Fm. sensu stricto, *c* Ammonitico Rosso and Fonzaso Fms.; 4. Calcare del Vaiont Fm. (Dogger); 5. Igne Fm. (Upper Liassic); 6. Soverzene Fm. (Lower and Middle Liassic); 7. Dolomia Principale (Upper Triassic); 8. faults and overthrusts; 9. failure surfaces of landslide

Numerous initiatives have been carried out in order to assess this territory and contribute to keeping the memory of the catastrophic disaster alive. Among these are included the creation of a permanent laboratory for the study of hydrogeological hazards, the dissemination of information concerning these kinds of risk, and the creation of a "multicenter museum," consisting of historical-natural science paths, permanent (e.g., memorial chapel) exhibitions, and other educational and/or popular multimedia material concerning the sites where the catastrophe occurred.

Bibliography

Belloni, L. G., and Stefani, R., 1987. The Vaiont slide: instrumentation past experience and the modern approach. In Leonards, G. A. (ed.), *Dam Failures. Engineering Geology*, **24**(1–4), 445–474.

Broili, L., 1967. New knowledge on the geomorphology of the Vaiont slide slip surface. *Rock Mechanics & Engineering Geology*, **5**(1), 38–88.

Crosta, G. B., and Agliardi, F., 2003. Failure forecast for large rock slides by surface displacement measurements. *Canadian Geotechnical Journal*, **40**, 176–191.

Genevois, R., and Ghirotti, M., 2005. The 1963 Vaiont landslide. *Giornale di Geologia Applicata*, **1**, 41–52, doi:10.1474/GGA.2005-01.0-05.0005.

Helmstetter, A., Sornette, D., Grasso, J. R., Andersen, J. V., Gluzman, S., and Pisarenko, V., 2003. A slider block model for landslides: application to Vaiont and La Clapierre landslides. *Journal of Geophysical Research*, **109**, B02409, doi:10.1029/2002JB002160.

Hendron, A. J., and Patton, F. D., 1985. The Vaiont slide, a geotechnical analysis based on new geologic observations of the failure surface. *Technical Report GL-85-5, U.S. Army Corps of Engineers, Waterways Experiment Station*, I, II, Vicksburg, MS.

Hutchinson, J. N., 1987. Mechanisms producing large displacements in landslides on pre-existing shears. In *1st Sino-British Geological Conference*, Memoir of the Geological Survey of China, Tapei, Vol. 9, pp. 175–200.

Kilburn, C. R. J., and Petley, D. N., 2003. Forecasting giant, catastrophic slope collapse: lessons from Vajont, Northern Italy. *Geomorphology*, **54**(1–2), 21–32.

Mencl, V., 1966. Mechanics of landslides with non-circular slip surfaces with special reference to the Vaiont slide. *Geotechnique*, **XVI**(4), 329–337.

Müller, L., 1964. The rock slide in the Vaiont valley. *Rock Mechanics & Engineering Geology*, **2**, 148–212.

Müller, L., 1968. New considerations on the Vaiont Slide. *Rock Mechanics & Engineering Geology*, **6**, 1–91.

Müller, L., 1987. The Vaiont catastrophe – a personal review. *Engineering Geology*, **24**, 423–444.

Panizzo, A., De Girolamo, P., Di Risio, M., Maistri, A., and Petaccia, A., 2005. Great landslide events in Italian artificial reservoirs. *Natural Hazards and Earth System Sciences*, **5**, 733–740.

Selli, R., Trevisan, L., Carloni, C. G., Mazzanti, R., and Ciabatti, M., 1964. La Frana del Vajont. *Giornale di Geologia*, **XXXII**(I), 1–154.

Semenza, E., 1965. Sintesi degli studi geologici sulla frana del Vaiont dal 1959 al 1964. *Museo Tridentino di Scienze Naturali*, **16**, 1–52.

Semenza, E., and Ghirotti, M., 2000. History of 1963 Vaiont slide. The importance of the geological factors to recognise the ancient landslide. *Bulletin of Engineering Geology*, **59**, 87–97.

Sitar, N., MacLaughlin, M. M., and Dolin, D. M., 2005. Influence of kinematics on landslide mobility and failure mode. *Journal of Geotechnical and Geoenvironmental Engineering*, **131**(6), 716–728.

Skempton, A. W., 1966. Bedding-plane slip, residual strength and the Vaiont landslide. *Geotechnique*, **16**, 82–84.

Superchi, L., Floris, M., Ghirotti, M., Genevois, R., Jaboyedoff, M., and Stead, D., 2010. Technical note: implementation of a geodatabase of published and unpublished data on the catastrophic Vaiont landslide. *Natural Hazards and Earth System Sciences*, **10**, 865–873.

Tika, Th. E., and Hutchinson, J. N., 1999. Ring shear tests on soil from the Vaiont landslide slip surface. *Geotechnique*, **49**(1), 59–74.

Tinti, S., Zaniboni, F., Manucci, A., and Bortolucci, E., 2002. A 2D block model for landslide simulation: an application to the 1963 Vajont case. *Abstracts: 27th EGS General Assembly*, Nice, France, *Geophysical Research Abstracts*, Vol. 4.

Vardoulakis, I., 2002. Dynamic thermo-poro-mechanical analysis of catastrophic landslides. *Geotechnique*, **52**(3), 157–171.

Veveakis, E., Vardoulakis, I., and Di Toro, G., 2007. Thermoporomechanics of creeping landslides: the 1963 Vaiont slide, northern Italy. *Journal of Geophysical Research*, **112**, F03026, doi:10.1029/2006JF000702.

Voight, B., and Faust, C., 1982. Frictional heat and strength loss in same rapid slides. *Geotechnique*, **32**(1), 43–54.

Cross-references

Casualties Following Natural Hazards
Disaster
Displacement Wave, Landslide Triggered Tsunami
History of Natural Hazards
Human Impact of Hazards
Landslide (Mass Movement)
Landslide Dam
Landslide Impacts
Reservoir Dams and Natural Hazards
Slope Stability

CASE STUDY

VESUVIUS

Bill McGuire
Aon Benfield UCL Hazard Centre, University College London, London, UK

Synonyms

Somma-Vesuvius

Introduction

Vesuvius is the only volcano on the European mainland to have erupted within the last 100 years, although it has currently been inactive for more than half a century. The volcano is best known for its catastrophic 79 AD eruption, during which a number of Roman settlements and buildings, including Pompeii and Herculaneum, were buried in ash and pumice. Since its discovery in 1748, excavations at Pompeii have provided a unique view of a first century Roman town frozen in time.

Europe's most dangerous volcano

Vesuvius (40.821°N 14.426°E) is located on the Bay of Naples, directly to the southeast of the city of Naples in the Italian province of Campania. It is a 1,281-m high composite or stratovolcano, constructed from the products of both explosive (ash and pumice) and effusive (lava flows) eruptions. Vesuvius is also the type example of a *Somma* volcano, in which the active center is built up within a caldera resulting from the collapse of an earlier center of activity. Vesuvius is sometimes allocated to a so-called "Campanian volcanic arc," which includes Campi Flegrei to the west of Naples and the island of Ischia in the Bay of Naples itself. Vesuvius and neighboring volcanic centers sit close to the northern, steeply dipping edge of a subducting slab reflecting the northwestward subduction of the Ionian Sea plate beneath southern Italy. In the Bay of Naples region, the subducting slab has become detached, contributing to the somewhat anomalous chemistry and mineralogy of the volcanic products. These range from mafic basalts to trachytes and phonolites and are particularly rich in potassium, resulting in the occurrence of distinctive minerals such as leucite, hauyne (nosean), and nepheline.

Volcanic activity in the Vesuvius area stretches back to 400,000 years, although the Somma-Vesuvius central edifice has been constructed over just the last 25,000 years. The ancestral Somma volcano was built largely from effusive eruptions interspersed with mildly explosive eruptions of ash and other pyroclastic materials. At 18,300 y BP, the style of eruption changed dramatically with the first of several *Plinian* eruptions – major, explosive events named after Pliny the Younger, who described such an eruption in 79 AD. The 18,300 y BP eruption was probably the largest eruption in the history of the volcano, and it deposited the widespread marker horizon known as the Pomici di Base (basal pumice). Collapse of the Somma edifice was also initiated during this eruption, forming a caldera within which the current Vesuvius cone has developed. Between 18,300 y BP and 79 AD, there occurred a further three major Plinian eruptions, together with six sub-Plinian events. One of the most notable and violent was the Pomici di Avellino event at 3,800 y BP, which deposited pyroclastic flow and surge deposits up to 3 m deep in the Naples area, and left half a meter of pumice in the vicinity of what is now the city of Avellino, 35 km away. Archeological excavations reveal that the eruption had a considerable and detrimental impact on a flourishing Bronze Age society in the region.

The 79 AD "Pompeii" eruption is the first to be historically documented in any detail, thanks to descriptions of the event contained in two letters sent by observer Pliny the Younger to the Roman senator and historian Tacitus. Little is known about precursory signs, but earthquake swarms seem to have begun on August 20 and increased over the next 4 days. After 800 years of quiescence, the eruption started on August 24, first with steam blasts as rising magma came into contact with groundwater and then with the development of an eruption column that reached a height of 15 km. Following a pause, a second phase ensued involving a major Plinian eruption. Gravitational collapse of the eruption column generated pyroclastic flows and surges that devastated the towns of Herculaneum (Ercolano), Pompeii, and Stabia to the south and west. Pyroclastic flows and heavy pumice and ash fall continued until the eruption ended during late morning of the 25th, further burying the inundated communities. During the 19 h of the eruption, the volcano ejected an estimated 4 km^3 of ash and pumice, registering the event at 5 on the volcanic explosivity index (VEI). The numbers of people killed in the eruption are not known, although estimates vary from 10,000 to 25,000. To date, the remains of some 1,500 bodies have been recovered from Pompeii and Herculaneum, most showing evidence of having died due to the intense heat of pyroclastic surges.

Following the 79 AD event, less violent explosive eruptions continued sporadically, to be replaced in the eleventh century by predominantly effusive events. A period of quiescence, which started in the thirteenth century, was ended by a large (VEI 4 or 5) sub-Plinian eruption in 1631, which took more than 3,000 lives. The volcano was in almost continuous eruption over the next 300 years, culminating in lava-dominated eruptions in 1906, which led to more than 100 deaths, and in 1944, which destroyed several villages. Vesuvius has since been quiet for 66 years – the longest period of quiescence in more than half a millennium.

Looking ahead, Vesuvius presents one of the greatest volcanic threats to a major urban center on Earth. This is becoming increasingly recognized, and an emergency plan is in place, based upon a future eruption on the scale of 1631. The plan assumes that monitoring will provide an early warning of between 14 and 20 days, during which time 600,000 people will require evacuation from a *zona rossa* (red zone) that is most at risk from pyroclastic flows and surges. In order to reduce the numbers requiring evacuation, financial incentives are being offered to inhabitants of the red zone, and new building is banned. Whether this can have any significant impact before the next eruption is, however, questionable.

Summary

Because of the resulting inundation, and subsequent exhumation, of Pompeii, the 79 AD eruption has arguably helped to make Vesuvius the most famous of all the world's volcanoes. At present, the volcano has the dubious distinction of presenting the greatest potential volcanic threat to a major urban center in the developed world. The current, long period of quiescence will ultimately be terminated by an eruption that is liable to be moderately violent and that will test the emergency plans of the civil authorities to the limit.

Bibliography

De Natale, G., Troise, C., Pingue, F., Mastrolorenzo, G., and Pappalardo, L., 2006. The Somma-Vesuvius volcano (southern Italy): structure, dynamics and hazard evaluation. *Earth-Science Reviews*, **74**, 73–111.

Guest, J., Cole, P., Duncan, A., and Chester, D., 2003. *Volcanoes of Southern Italy*. London: The Geological Society. 282 pp.

Kilburn, C. R. J., and McGuire, W. J., 2001. *Italian Volcanoes*. Harpenden: Terra Publishing. 166 pp.

Mastrolorenzo, G., Petrone, P., Pappalardo, L., and Sheridanm, M. F., 2006. The Avellino 3780 y BO catastrophe as a worst-case scenario for a future eruption at Vesuvius. *Proceedings of the National Academy of Sciences*, **103**, 4366–4370.

Pesce, A., and Rolandi, G., 1994. *Vesuvio 1944 L'ultima eruzione*, San Sebastiano al Vesuvio, Naples.

Scandone, R., Giacomelli, L., and Gasparini, P., 2000. Mount Vesuvius: 2000 years of volcanological observations. *Journal of Volcanology and Geothermal Research*, **58**, 5–25.

Scarth, A., 2009. *Vesuvius a Biography*. London: Terra Publishing. 342 pp.

Cross-references

Casualties Following Natural Hazards
Cultural Heritage and Natural Hazards
Early Warning Systems
Eruption Types (Volcanic)
Galeras Volcano, Colombia
Hazardousness of Place
Krakatoa (Krakatau)
Montserrat Eruptions
Nevado del Ruiz, Colombia (1985)
Pyroclastic Flow
Santorini
Stratovolcano
Volcanic Ash
Volcanoes and Volcani Eruptions

VOLCANIC ASH

Thomas Wilson[1], Carol Stewart[2]
[1]University of Canterbury, Christchurch, New Zealand
[2]GNS Science/Massey University, New Zealand

Synonyms

Tephra

Definition

Volcanic ash is the material produced by explosive volcanic eruptions that is <2 mm in diameter. Fine ash is <0.063 mm; coarse ash is from 0.063 to 2 mm.

Generation

Volcanic ash is formed during explosive volcanic eruptions. Explosive eruptions occur when magma decompresses as it rises, allowing dissolved volatiles (dominantly H_2O and CO_2) to exsolve into gas bubbles.

These expand the fluid magma into foam and accelerate its ascent to the surface. A *magmatic eruption* occurs when the foamy magma fragments at the volcanic vent, with violently expanding gas bubbles tearing the silicate lattice apart to erupt into the atmosphere, where it then solidifies into fragments of volcanic rock and glass (Figure 1). Ash can also be produced when rising magma contacts water (e.g., surface water, groundwater, snow, and ice). The water explosively flashes into steam, causing shattering of the magma and resulting in a *phreatomagmatic eruption*. A single eruption may be driven by both the release of magmatic gases and interaction with water. The violence of the eruption is determined by the viscosity of the magma (a function of silica content, crystallinity, temperature, and the dissolved volatile content) and the exsolved volatile content (Heiken and Wohletz, 1985; Sigurdsson et al., 2000).

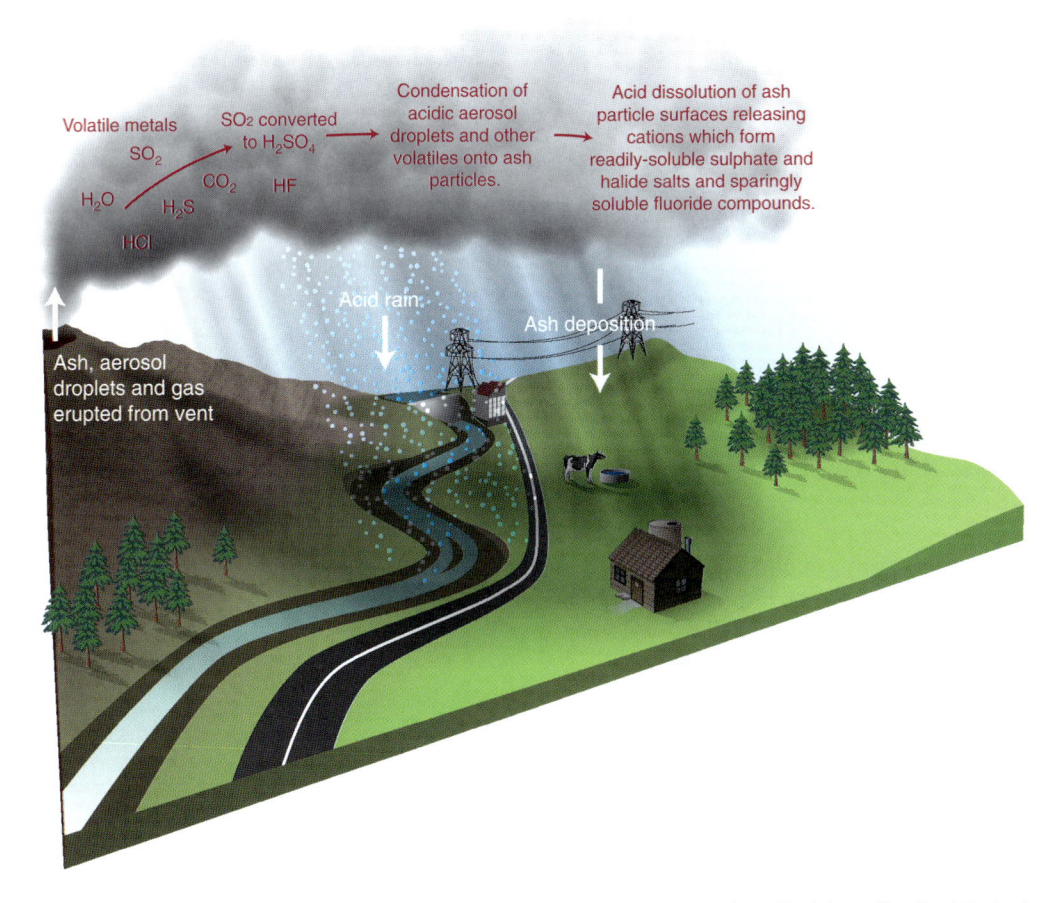

Volcanic Ash, Figure 1 A volcanic ash particle (vesiculated glass).

Volcanic Ash, Figure 2 The principal gases released by volcanic eruptions are water, carbon dioxide, sulfur dioxide, hydrogen sulfide, carbon monoxide, hydrochloric acid, and hydrofluoric acid. Volcanic ash particles are thought to have a very thin surface coating comprised of readily soluble sulfate and chloride salts, and more sparingly-soluble fluoride compounds. These are thought to be formed following the condensation of acidic aerosols (H_2SO_4, HCl and HF) onto ash particles, followed by the rapid acid dissolution of Mg, Ca, Na, and K from the ash surface, and then precipitation of the salts at the ash/liquid interface (Witham et al., 2005; Delmelle et al., 2007). Recent work suggests gas/aerosol–ash interactions scavenge around 30–40% of the emitted sulfur and 10–20% of the emitted chlorine in the volcanic plume (Delmelle et al., 2007).

Dispersal and deposition

Volcanic ash is the most widely distributed product of explosive volcanic eruptions. Ash particles ejected from an eruption vent are typically incorporated into an eruption column that may buoyantly rise tens of kilometers (and up to 50 km) into the atmosphere. Eruption plumes are dispersed by prevailing winds, and the ash can be deposited hundreds to thousands of kilometers from the volcano, depending on wind strength, ash grain size, and eruption magnitude.

Ash deposit thicknesses and median grain size generally decrease exponentially with distance from a volcano (Pyle, 1989). Ash deposit thicknesses can be measured, plotted on a map, and contoured. Lines of equal ash thickness are termed *isopachs*. Unconsolidated ash deposits may be subject to remobilization by fluvial or wind processes.

Characteristics

Due to their violent and rapid formation, volcanic ash particles are made up of various proportions of vitric (glassy, noncrystalline), crystalline or lithic (non-magmatic) particles. They are typically very hard and angular, making them abrasive. Vitric particles typically contain small voids formed by expansion of magmatic gas before the enclosing magma solidified, known as vesicles (Figure 1). Ash particles can have varying degrees of vesicularity. Vesicular particles can have extremely high surface area-to-volume ratios. Exsolved magmatic gases condense onto ash particle surfaces while they are in the conduit and ash plume (Figure 2). Surface coatings on fresh ash can be highly acidic, due to the presence of strong mineral acids H_2SO_4 (formed from the oxidation of SO_2), HCl, and HF. Over 55 soluble components have been measured in volcanic ash leachates. Freshly deposited volcanic ash is, therefore, potentially corrosive and electrically conductive. Some components, such as fluoride, may be toxic (Witham et al., 2005; Stewart et al., 2006; Cronin et al., 2003).

Impacts

Volcanic ash can affect many people because of the large areas that can be covered by ashfall. Although ashfalls rarely endanger human life directly, respiratory health hazards of volcanic ash, disruption of critical infrastructure such as electricity and water supplies, transport routes, wastewater treatment and drainage systems and communications networks, disruptions to aviation, building damage and impacts on crops and livestock can all lead to significant societal impacts (Horwell and Baxter, 2006; Johnston et al., 2000; Stewart et al., 2006; Wilson et al., 2010). For more information on impacts, refer to http://volcanoes.usgs.gov/ash/ and http://www.ivhhn.org/.

Bibliography

Cronin, S. J., Neall, V. E., Lecointre, J. A., Hedley, M. J., and Loganathan, P., 2003. Environmental hazards of fluoride in volcanic ash: a case study from Ruapehu volcano, New Zealand. *Journal of Volcanology and Geothermal Research*, **121**, 271–291.

Delmelle, P., Lambert, M., Dufrêne, Y., Gerin, P. A., and Óskarsson, O., 2007. Gas/aerosol-ash interaction in volcanic plumes: new insights from surface analysis of fine ash particles. *Earth and Planetary Science Letters*, **259**, 159–170.

Heiken, G., and Wohletz, K. H., 1985. *Volcanic Ash*. Berkeley: University of California Press.

Horwell, C. J., and Baxter, P. J., 2006. The respiratory health hazards of volcanic ash: a review for volcanic risk mitigation. *Bulletin of Volcanology*, **69**, 1–24.

Johnston, D. M., Houghton, B. F., Neall, V. E., Ronan, K. R., and Paton, D., 2000. Impacts of the 1945 and 1995–1996 Ruapehu eruptions, New Zealand: An example of increasing societal vulnerability. *Geological Society of America Bulletin*, **112**(5), 720–726.

Pyle, D. M., 1989. The thickness, volume and grain size of tephra fall deposits. *Bulletin of Volcanology*, **51**, 1–15.

Sigurdsson, H., Houghton, B., Rymer, H., and Stix, J., 2000. *Encyclopedia of Volcanoes*. San Diego: Academic.

Stewart, C., Johnston, D. M., Leonard, G. S., Horwell, C. J., Thordarson, T., and Cronin, S. J., 2006. Contamination of water supplies by volcanic ash fall: A literature review and simple impact modelling. *Journal of Volcanology and Geothermal Research*, **158**, 296–306.

Wilson, T. M., Cole, J. W., Cronin, S. J., Stewart, C., and Johnston, D. M., 2010. Impacts on agriculture following the 1991 eruption of Vulcan Hudson, Patagonia: Lessons for Recovery. *Natural Hazards*, doi 10.1007/s11069-010-9604-8.

Witham, C. S., Oppenheimer, C., and Horwell, C. J., 2005. Volcanic ash leachates: a review and recommendations for sampling methods. *Journal of Volcanology and Geothermal Research*, **141**, 299–326.

Cross-references

Aviation, Hazards to
Base Surge
Critical Infrastructure
Eruption Types (Volcanic)
Galeras Volcano, Colombia
Krakatoa (Krakatau)
Lahar
Montserrat Eruptions
Mt. Pinatubo
Nevado del Ruiz Volcano, Colombia (1985)
Nuee Ardente
Pyroclastic Flow
Santorini
Stratovolcano
Volcanic Gas
Volcanoes and Volcanic Eruptions

VOLCANIC GAS

Travis W. Heggie

University of North Dakota, Grand Forks, ND, USA

Synonyms

Volcanic emissions; Volcanic fumes

Definition

Volcanic gases refer to a range of substances emitted by volcanoes during and between eruptive phases.

Discussion

Origins: Volcanic gas is contained within magma. The gases are originally dissolved in magma under high-pressure conditions deep beneath the surface of the Earth. As magma rises to lower pressure conditions near the surface of the Earth, gases held in the magma form tiny bubbles that increase in volume. The increasing volume taken up by gas bubbles makes the magma less dense than the surrounding rock which in turn allows magma to continue rising to the Earth's surface.

Composition: The most common volcanic gases released into the atmosphere are water vapor (H_2O), carbon dioxide (CO_2), sulfur dioxide (SO_2), hydrogen chloride (HCl), hydrogen sulfide (H_2S), hydrogen fluoride (HF), carbon monoxide (CO), nitrogen (N_2), hydrogen (H_2), helium (He), methane (CH_4), radon (Rn), and heavy metals such lead and mercury (Hansell et al., 2006; Heggie, 2009).

Hazards: Volcanic gases can impact the climate at local, regional, and global scales. For example, sulfur aerosols in the atmosphere can lead to lower surface temperatures and ozone depletion. Volcanic gases are also harmful to humans, animals, and agricultural crops. SO_2 can produce acid rain and contribute to air and water pollution downwind of volcanoes, CO_2 is heavier than air and may flow into low-lying areas and collect in soil, and a collection of volcanic gases can instantly corrode aircraft engines (Heggie, 2005).

Direct Health Hazards: Human death resulting from volcanic gases is usually the result of asphyxiation and/or acidic corrosion (Baxter, 1990). Volcanic gases are estimated to account for <1–4% of recorded volcano-related fatalities (Hansell and Oppenheimer, 2004). However, these figures are likely underestimated as reported studies tend to focus more on fatalities during volcanic eruptions and neglect degassing events between eruptive phases. CO_2 is particularly dangerous because it is odorless and poses the risk of asphyxiation, vomiting, dizziness, visual disturbances, headaches, sweating, mental depression, and tremors (Heggie, 2009). H_2S is a colorless gas with a sewer or rotten egg smell than can cause upper respiratory irritation and pulmonary edema at higher concentrations and irritate the eyes and act as a depressant at lower concentrations (Heggie, 2009). Common gases such as SO_2 and HCl are irritating to the eyes, throat, mucous membranes, and respiratory tract (Heggie et al., 2009). HCl can also be irritating to the skin and those with asthma can suffer severe consequences to SO_2 (Heggie et al., 2009).

Bibliography

Baxter, P. J., 1990. Medical effects of volcanic eruptions. *Bulletin of Volcanology*, **52**, 532–544.

Hansell, A. L., and Oppenheimer, C., 2004. Health hazards from volcanic gases: a systematic literature review. *Archives of Environmental Health*, **59**, 628–639.

Hansell, A. L., Horwell, C. J., and Oppenheimer, C., 2006. Health hazards of volcanoes and geothermal areas. *Occupational and Environmental Medicine*, **63**, 149–156.

Heggie, T. W., 2005. Reported Fatal and non-fatal incidents involving tourists in Hawaii Volcanoes National Park. *Travel Medicine and Infectious Disease*, **3**, 123–131.

Heggie, T. W., 2009. Geotourism and volcanoes: health hazards facing tourists at volcanic and geothermal destinations. *Travel Medicine and Infectious Disease*, **7**, 257–261.

Heggie, T. W., Heggie, T. M., and Heggie, T. J., 2009. Death by volcanic laze. *Wilderness and Environmental Medicine*, **20**, 101–103.

Cross-references

Acid Rain
Climate Change
Gas-Hydrates
Global Dust
Methane Release from Hydrate
Mt. Pinatubo
Ozone
Ozone Loss
Volcanic Ash
Volcanoes and Volcanic Eruptions

VOLCANOES AND VOLCANIC ERUPTIONS

Sue C. Loughlin
British Geological Survey, Edinburgh, UK

Definitions

Volcano: There are volcanoes wherever magma rising from the Earth's interior reaches the surface. In terms of landforms, a single dome, cinder cone, tuff cone, or maar may be described as a volcano, but these are *monogenetic volcanoes* arising from a single eruption. They may form on the summit and flanks of larger, long-lived volcanic edifices (*polygenetic volcanoes*) or as part of extensive volcanic fields that result from several eruptive periods. Here, in order to report numbers of active volcanoes (i.e., those that have erupted in the last 10,000 years and could erupt again), we follow the same broad definition as the Global Volcanism Program (Simkin and Siebert, 2002–2009) where all products of a single magmatic system are considered part of one "volcano" or "volcanic system." So, a volcanic field tens to hundreds of kilometers across, comprising hundreds or thousands of monogenetic volcanoes, is here considered to be a single volcanic system.

Eruption: An eruption is usually described as the explosive expulsion of fragmental material (tephra) and/or lava from a volcano (Simkin and Siebert, 2002–2009). An eruption may last for hours, days, years, decades, or, in the case of Stromboli, Italy, thousands of years. A pause in an eruption longer than 3 months would constitute the end of the eruption (Simkin and Siebert, 2002–2009). Periods of dormancy between eruptions may extend to thousands of years.

Introduction

Volcanoes are awe inspiring, beautiful, and deadly. They may provide benefits such as good quality agricultural soils, geothermal energy, stunning landscapes, and tourist attractions, but equally they may cause devastating destruction and numerous fatalities. In the twentieth century, 91,724 people were killed by volcanic phenomena (see *Casualties Following Natural Hazards*), and the best estimate for people affected is 5.6 million (Witham, 2005). Almost one in ten people live within possible range of an active volcano, and most of these people are in urban centers in the developing world (Chester et al., 2001). As the world's population increases and urban populations increase, the number and proportion of vulnerable people will also increase.

Some communities have learned to live alongside and benefit from frequently erupting volcanoes (e.g., Mauna Loa in Hawaii and Etna in Italy). Typically, such volcanoes are well monitored, and emergency plans and communication networks are well established (see *Emergency Planning*). However, more commonly, volcanoes are dormant for hundreds or thousands of years and then awaken suddenly to the surprise of local populations. All too often, emergency plans and monitoring networks are established in response to the onset of volcanic activity, and this lack of preparedness can have deadly consequences.

Distribution of volcanoes

The distribution of volcanoes is strongly controlled by plate tectonics (see entry *Plate Tectonics*) which is driven largely by convection of the Earth's mantle (Figure 1). Volcanoes occur mainly at plate margins: above subduction (see entry *Subduction*) zones (destructive plate margins), or at mid-ocean ridges and continental rifts (constructive plate margins). They also occur within plates and/or associated with "hot spots." Each tectonic setting results in unique processes of magma (see entry *Magma*) generation resulting in differing types and styles of volcanic eruption (see entry *Eruption Types (Volcanic Eruptions)*). The most dangerous volcanoes tend to be associated with subduction zones (island arcs and continental margins) where two tectonic plates collide and one is forced underneath the other.

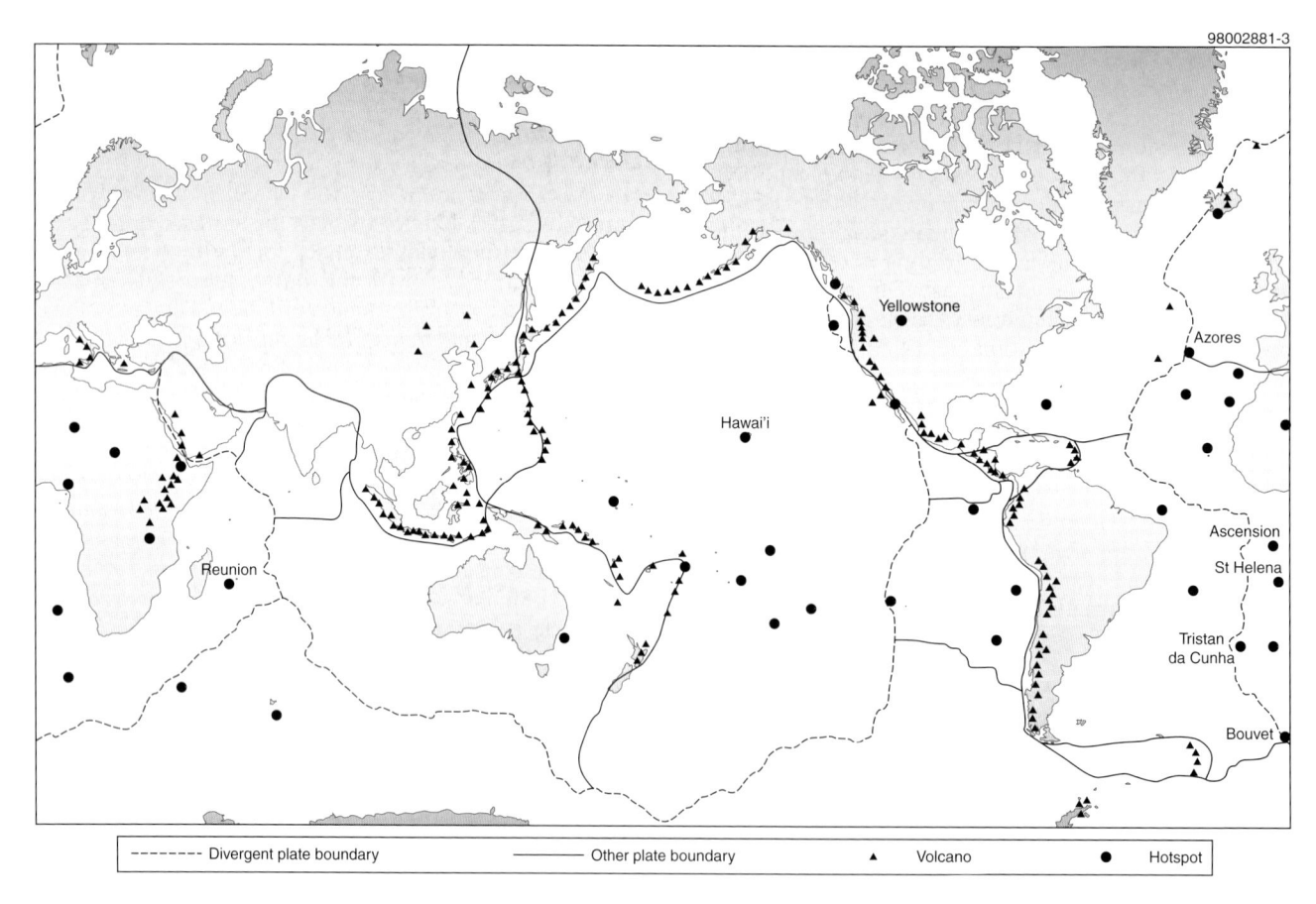

Volcanoes and Volcanic Eruptions, Figure 1 Global distribution of historically active volcanoes (Simplified from Simkin and Siebert, 2000).

Magma generation, storage and transport

Basaltic magma is generated in the Earth's mantle when rocks undergo partial melting by decompression, hydration, or heating. The exact composition of primitive basaltic magma varies depending on the tectonic environment and consequent depth and degree of partial melting although it always has low silica content ($<52\%$ SiO_2) and low viscosity. As magma moves toward the surface, it is modified by a variety of processes (see papers in Sigurdsson, 2000) and can evolve into magma of very different composition. At mid-ocean ridges, where minimal magma modification occurs, large volumes of basaltic lava are erupted at the surface, whereas at subduction zones, high silica magmas are more commonly erupted. In magma chambers, basaltic magma cools and crystallizes various minerals, some of which are dense and sink (fractional crystallization), leaving magmas of more silicic composition. Additional processes may contribute to magma evolution such as the partial melting of crustal rocks (heat supplied by the hot primitive magmas) and mixing of different magma compositions (e.g., Blake, 1981; Marsh, 2000). The evolution to silicic compositions can only take place if the magma is stored for some time; so long repose periods between eruptions facilitate the process. Silicic magmas such as dacite (e.g., Mount St. Helens, USA), trachyte (e.g., Fogo, Azores), or rhyolite (e.g., Chaitén, Chile) have higher potential for cataclysmic volcanic explosions.

Eruptions are commonly triggered when basaltic magma enters the base of a long-lived cooling magma chamber. As well as a space problem, the fresh magma supplies heat and volatiles that trigger convection and mixing (e.g., Murphy et al., 1998). Magma makes its way toward the surface along zones of high deformation (at depth) and fracturing (near the surface). Some volcanoes develop on large rift systems where lava (see entry Lava) is erupted along dyke systems; other volcanoes have a near-surface cylindrical conduit, often where two fracture zones intersect.

Number of active volcanoes

Based on the records at the Smithsonian Institution, 50–70 volcanoes are in eruption each year and about 160 each decade. There are at least 1,300 (and possibly more than 1,500) volcanoes known to have erupted in the Holocene (last 10,000 years) and 550 of these have had historically documented eruptions (Simkin and Siebert, 2002–2009). Based on the definition of a "volcano" given above, Simkin and Siebert (2002–2009) suggest that there are at least 1,000 known magma systems on land that might erupt in the future. It is entirely possible that there are active volcanoes (volcanoes with the potential to erupt) that have not yet been identified due to ice or vegetation cover. If a volcano has not erupted for hundreds or thousands of years, our understanding of it is entirely dependent on geological mapping and the dating of eruption products. There are volcanoes that have not been mapped in detail (and some not at all), and the products of many past eruptions remain to be identified let alone dated.

The number of volcanoes on the seafloor and ocean rift system is unknown, but it far exceeds the number on land. A seamount is defined as an edifice at least 50–100 m above the seafloor of which there are more than one million; highly productive seamounts form ocean islands of which there are a few thousand (Schmidt and Schmincke, 2000). It is estimated that three-fourths of the magma reaching the Earth's surface is erupted along submarine ocean ridges (Crisp, 1984).

Morphology and volcano type

Volcanoes have a significant variety of forms from ice-capped symmetrical cones like Mount Fuji (see *Stratovolcanoes*) and broad shield volcanoes (see entry *Shield Volcano*) like Mauna Loa to potentially devastating caldera (see entry *Calderas*) volcanoes (e.g., Toba, Sumatra) which have barely any topography; so despite their great size, some were only identified from space (e.g., Francis, 1993).

The shape, size, and type of volcano depend on tectonic setting, magma supply rates, and eruption frequency. The features and deposits generated by a single eruption depend on several factors including lava chemistry, gas content, and eruption style which may change through the lifespan of a volcano or even during a single eruption. Sector collapse and eruption into, or erosion by, water or ice also significantly modify volcano shape, giving an enormous variety of volcanic forms (see papers in Sigurdsson, 2000; Smellie and Chapman, 2002; Francis and Oppenheimer, 2003).

The simplest forms are monogenetic volcanoes created during a single eruption. They include scoria (cinder) cones (often associated with lava flows), lava domes, maars, and tuff cones. Monogenetic landforms may form on the summit or flanks of a large polygenetic volcano (e.g., cinder cones on Etna, Italy), as part of a volcanic field (e.g., Michoacan-Guanajuato filed, Mexico), or independently (e.g., Inyo craters, California).

Polygenetic volcanoes have undergone repeated eruptive episodes. Large complex, constructional features that are long-lived and comprise both lava and fragmentary eruption products (tephra) are termed composite volcanoes (Davidson and Da Silva, 2000). This broad category includes stratovolcanoes (see *Stratovolcanoes*), central volcanoes, shield volcanoes (see *Shield Volcano*), and compound volcanoes. Their eruption products are diverse in type and composition, and many have experienced sector or caldera collapse. They are widely distributed, developing above subduction zones (e.g., Pinatubo, Philippines; Mount St. Helens, Washington, USA) but also in intraplate environments (e.g., Mount Kilimanjaro, Tanzania) and associated with "hot spots" (e.g., Hekla, Iceland). Intrusive dykes and sills build up the edifice from the inside. Klyuchevskoy in Kamchatka, Russia, is one of the tallest composite volcanoes, standing 4,725 m above sea level.

If a magmatic system supplying a composite volcano shifts a limited distance during the volcano's lifetime, a compound volcano is formed. It is asymmetric and generally has the appearance of one or several cones superimposed upon one another. The result may be an elongated ridge (e.g., Lascar, N. Chile), or a more complex massif with nonsystematic migration of the eruption center (e.g., Ojos del Salado, Chile). In some cases, two distinct composite volcanoes are formed much closer than average volcano spacing (<10 km) and one cone is commonly older than the other (e.g., Nevado de Colima and Volcán de Colima, Mexico).

A "volcanic field" may be active for up to 5 Ma and contain thousands of individual monogenetic volcanoes. Clusters or extensive fields of monogenetic volcanoes may extend for tens or hundreds of kilometers.

Small summit calderas on basaltic volcanoes (e.g., Piton de la Fournaise, Réunion Island) form when magma from a shallow summit chamber is ejected laterally into dykes, leaving the summit to collapse, often incrementally rather than catastrophically. A *caldera* (*cf*) like Crater Lake, Oregon, formed when a Plinian explosion rapidly evacuated a large, shallow magma chamber and resulted in collapse of the roof of the magma chamber along ring fractures. Such eruptions are usually associated with a preexisting edifice; in the case of Crater Lake, it was originally Mount Mazama. In contrast, resurgent calderas are the largest volcanic structures on Earth formed by enormous volcanic eruptions. They are up to 100 km across, dwarfing any preexisting structure, and are characterized by post-eruption resurgence of the caldera floor, sometimes by up to 1 km (e.g., Cerro Galan (Argentina), Long Valley (California, USA), and Yellowstone (Wyoming, USA)).

Eruption style

The style of eruption depends mainly on the gas pressure in bubbles growing in the magma as it moves toward the surface, the rheology of the magma, and magma ascent rates (see *Eruption types (Volcanic Eruptions)* and papers in Sigurdsson, 2000). Magma ascent rate and decompression in the last few km below the surface has a significant effect on the properties of magma and dynamics of conduit flow (e.g., Massol and Jaupart, 1999; Melnik and Sparks, 1999, 2002, 2005, 2007). It is in the shallow subsurface that gas exsolution and degassing occurs, controlling whether a volcanic eruption is effusive (lava flows) or explosive. If gases are unable to separate effectively during ascent, the pressure in the bubbles increases and then may blow the magma apart during explosive eruptions (e.g., see papers in Gilbert and Sparks, 1998; Freundt and Rosi, 1998).

A volcano may switch between effusive and explosive styles during a single eruption (e.g., Vesuvius 1944; Soufrière Hills 1995-ongoing). Melnik and Sparks (2005, 2007) have shown that nonlinear dynamic processes in volcanic conduits at andesitic volcanoes can lead to substantial changes in style simply as a result of a change in a single parameter such as volatile content. The same nonlinear dynamics may make volcanoes inherently unpredictable although many volcanoes exhibit repetitive cyclic patterns of activity on various timescales that can assist with effective forecasting (e.g., Voight et al., 1998, 1999; Druitt et al., 2002a; Sparks and Young, 2002).

Scale of eruptions

It is useful to compare eruptions, even though they cover an enormous range of types and impacts, and several scales have been developed for this purpose (see *Intensity Scales* and *Magnitude Measures*).

The Volcanic Explosivity Index (VEI) is a commonly used logarithmic scale devised by Newhall and Self (1982) that is useful to compare magnitudes of explosive eruptions (Table 1). Most volcanic eruptions are explosive to some degree, and although the eruptive products are different, the dynamics are the same. The VEI scale includes eruption column height, erupted volume of tephra (which may be controlled by eruption duration), and qualitative descriptions to define eruptions. On the VEI scale, fire fountaining might score 0–1, strombolian explosions 1–2, vulcanian explosions 2–4, and plinian explosions 4 and above (see *Eruption Types (Volcanic Eruptions)*). The VEI scale is often used to compare recent explosive eruptions (defined by erupted tephra volume and/or column height) and historic or prehistoric explosive eruptions (defined by erupted tephra volume only).

Volcanoes and Volcanic Eruptions, Table 1 The VEI based on Newhall and Self, 1982, showing approximate frequencies of eruptions based on the Smithsonian Institution Holocene database (Simkin and Siebert, 2000)

VEI	Maximum volume of erupted tephra m^3	Eruption column height	Description	Approximate frequency
0	10^4	<0.1	Nonexplosive	
1	10^6	0.1–1	Small	
2	10^7	1–5	Moderate	
3	10^8	3–15	Moderate-large	10/year
4	10^9	10–25	Large	1/year
5	10^{10}	>25	Very large	1/10 years
6	10^{11}	>25		1/100 years
7	10^{12}	>25		1/1,000 years
8	10^{13}	>25		1/100,000 years

There are also logarithmic scales to define magnitude and intensity (e.g., Pyle, 2000) which are useful to compare effusive and explosive eruptions:

Magnitude = $\log_{10}$(erupted mass, kg) − 7
Intensity = $\log_{10}$(mass eruption rate, kg/s) + 3

For example, the Rosa lava flow which is a single lava flow from the Columbia River basalts (USA) had a magnitude of 8.5, and the explosive Toba caldera-forming eruption of 75 ka had a magnitude of 8.8.

Another way to compare otherwise dissimilar events is a destructiveness index: = $\log_{10}$(total area affected by lava, lahar, pyroclastic flows, debris avalanche, or tephra accumulations exceeding 100 kg/m^2).

The Taupo eruption of 180 AD affected 45,000 km^2 of land, whereas the Mount St. Helens eruption of 1980 affected 710 km^2 (Pyle, 2000).

Magnitude, frequency, duration and interval

Based on the Smithsonian Institution Holocene database, small to moderate size eruptions are frequent (in geological time) but large eruptions are not, in other words, there is a link between eruption magnitude and eruption frequency (Figure 2, Simkin and Siebert, 2000). Nevado del Ruiz (1985) was a small eruption with devastating consequences, but about 10 eruptions of this size could be expected every year. Human civilization has not yet experienced the largest possible explosive eruptions (VEI 8+: super-eruptions, see below) which occur on average every 100,000 years (Simkin and Siebert, 2000).

Long repose periods (hundreds to thousands of years) precede large eruptions (e.g., Mason et al., 2004). The repose period is often longer than documented history at that volcano; so communities may be unprepared. Simkin and Siebert (2000) pointed out that of the 16 largest explosive eruptions of the nineteenth and twentieth centuries, 11 were the first historical eruptions known from that volcano (e.g., Santa Maria, Guatemala, and El Chichon, Mexico).

Some volcanic eruptions have a "paroxysmal" explosive phase, and these tend to cause fatalities. In recent history, such eruptions include Mount St. Helens (1980; Lipman and Mullineaux, 1981) and Pinatubo (1990; Newhall and Punongbayan, 1996), both of which reached their cataclysmic stage after months of precursory activity. However, Simkin and Siebert (2000) showed that more than 40% of well-documented explosive volcanic eruptions reached their peak in the first day of the eruption, and where detailed data exist, half reached the climax within 1 h of the start of the eruption (e.g., Tarawera, 1886; Hekla, 1947; Shiveluch, 1964). Such rapid development was demonstrated at Chaitén (Chile) in 2008 when only 1 day of seismic unrest was felt before a large Plinian eruption. The apparent rapid ascent rate of silicic magmas had been a dilemma to geologists due to the assumption that they are highly viscous – we now know that near-liquidus hydrous rhyolite is in fact very fluid (Castro and Dingwell, 2009). There had been no historical eruptions at Chaitén, and so the eruption was not expected, and this scenario will be repeated at other volcanoes in the future.

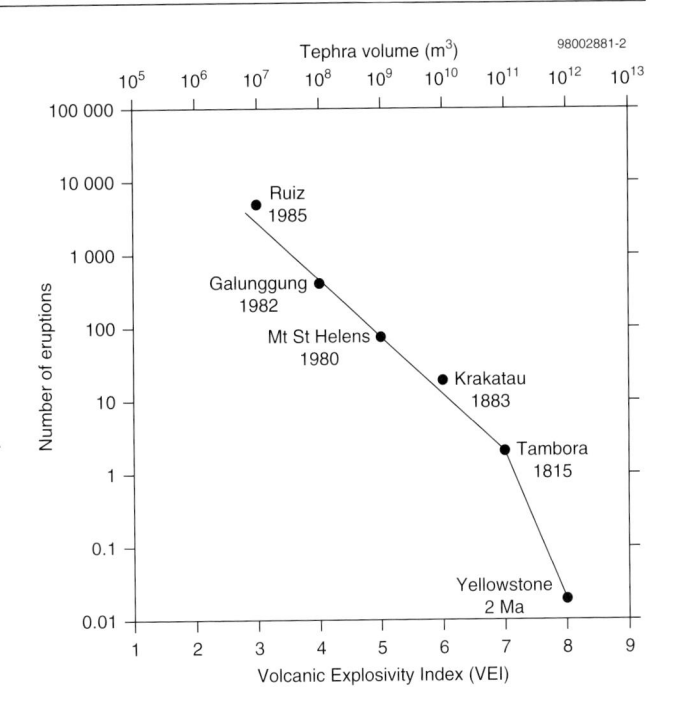

Volcanoes and Volcanic Eruptions, Figure 2 Magnitude and frequency of Holocene eruptions normalized to show eruptions per 1,000 years. Data for VEI 8 eruptions is based on geological not historical data.

For those historical eruptions in the Smithsonian Institution database with start and end dates, there is a median eruption duration of about 7 weeks, most last for less than 3 months and a few last for longer than 3 years (Simkin and Siebert, 2000–2009). Some ongoing long-lived eruptions include Santa Maria (Guatemala) and Arenal (Costa Rica).

Super-eruptions

A super-eruption is defined as an eruption of at least 100 km^3 of magma (equivalent to about 250 km^3 of volcanic ash) that has global consequences (Sparks et al., 2005). Super-eruptions are up to hundreds of times larger than the biggest eruptions of the last 200 years (Tambora, 1815; Krakatoa, 1883) that both caused major global climatic anomalies for several years after the eruptions (see *Global Dust*). The most recent super-eruption was at Toba (Sumatra, Indonesia) 73,500 years ago which may have released up to 6,000 km^3 of tephra and up to five million tons of sulfuric acid (Rampino and Self, 1994). The largest flood basalt province (Ontong Java plateau) reached emplacement rates of up to 12 km^3 per year, and the global impacts of such large-scale magmatism and volcanism must be severe (White and Saunders, 2005; Thordarson et al., 2009).

It is tempting to link these catastrophic events to biotic extinctions (e.g., Rampino and Self, 2000), but conclusive evidence is lacking (see *Extinction*). The impacts of a super-eruption would include the devastation of world agriculture, severe disruption of food supplies, social unrest, mass starvation, disease, and financial collapse.

Volcanic hazards and impacts

Volcanic ash (*cf*) fall has significant impact on agriculture (see *Challenges to Agriculture*), human health, and environment (e.g., Hansell et al., 2006; Martin et al., 2009) and has led to the greatest number of volcano-related fatalities (by famine and disease) in the last 400 years (Table 2; e.g., Baxter, 2000). Volcanic ash (see entry *Volcanic Ash*) was also responsible for the highest numbers of people made homeless/evacuated or otherwise affected by volcanic eruption in the twentieth century (Witham, 2005). It can disrupt critical infrastructure, such as power, communication, and aviation (see *Aviation (Hazards to)*) (e.g., Miller and Casadevall, 2000), and contaminate water. Volcanic ash is likely to have cross-border impacts (e.g., Watt et al., 2009), making countries which may not have a volcano vulnerable if *emergency planning* (*cf*) is not in place. As an example of unexpected impacts, the total losses to global GDP of the small-moderate 2010 eruption of Eyjafjallajökull in Iceland were US$5 billion (Oxford Economics, 2010).

When considering numbers of immediate fatalities, pyroclastic flows/surges (commonly referred to as "pyroclastic density currents," Druitt, 1998; Branney and Kokelaar, 2002) are the most devastating hazards (Figure 3; Table 2). Recent eruptions have provided many opportunities to observe pyroclastic flows (see *Pyroclastic Flow* and *Nuée Ardente*) and study their behavior and deposits (e.g., Fujii and Nakada, 1999; Ui et al., 1999; Druitt and Kokelaar, 2002; Saucedo et al., 2002). Pyroclastic surges (see "*Surge*") are highly mobile; they do not necessarily follow valleys and may travel uphill or across the surface of the ocean catching observers by surprise (e.g., Fisher, 1995; Loughlin et al., 2002; Trofimovs et al., 2006). An additional hazard known as surge-derived pyroclastic flows was identified at both Mount St. Helens and Soufrière Hills (e.g., Druitt et al., 2002b). The best risk reduction strategy for *pyroclastic flows* (*cf*) and *surges* (*cf*) is to identify areas at risk and then avoid them (e.g., Crandell et al., 1984). However, identification of potentially affected areas is nontrivial, and various statistical methods and dynamic models have

Volcanoes and Volcanic Eruptions, Table 2 Fatalities greater than 1,000 in the last 400 years (Data from Fisher et al., 1997; Blong, 1984)

Year	Volcano	Pyroclastic flows/surges	Lava flows	Debris flows/lahars	Tsunami	Famine
1586	Kelut, Indonesia			10,000		
1600	Huaynaputina, Peru	>1,000				
1631	Vesuvius, Italy		<4,000			
1669	Etna, Italy		?			
1711	Awu, Indonesia			3,200		
1741	Oshima, Japan				1,480	
1741	Cotopaxi, Ecuador			1,000		
1772	Papadian, Indonesia	2,960				
1783	Laki, Iceland					9,340
1783	Asama, Japan	1,150				
1792	Unzen, Japan				15,190	
1814	Mayon, Philippines	1,200				
1815	Tambora, Indonesia	12,000				80,000
1822	Galunggung, Indonesia			<4,000		
1825	Mayon, Philippines			1,500		
1826	Awu, Indonesia			3,000		
1845	Nevado del Ruiz, Columbia			1,000		
1856	Awu, Indonesia			1,530		
1877	Cotopaxi, Ecuador			1,000		
1883	Krakatau, Indonesia				36,417	
1892	Awu, Indonesia			1,532		
1902	Soufriere, St. Vincent	1,680				
1902	Pelee, Martinique	29,025				
1902	Santa Maria, Guatemala	6,000				
1911	Taal, Philippines	1,335				
1919	Kelut, Indonesia			5,110		
1930	Merapi, Indonesia	1,300				
1951	Lamington, New Guinea	2,942				
1963	Agung, Indonesia	1,900				
1982	El Chichon, Mexico	1,700				
1985	Nevado del Ruiz, Columbia			25,000		

Volcanoes and Volcanic Eruptions, Figure 3 Block-and-ash flows entering the sea off the coast of Montserrat 20 May 2006. The surges traveled across the water surface.

so far been employed (e.g., Calder et al., 1999; Denlinger and Iverson, 2001; Patra et al., 2005; Esposito Ongaro et al., 2005; Charbonnier and Gertisser, 2009; Wadge, 2009). The essential parameters needed for effective dynamic modeling of flows, especially the surge (see entry *Surge*) component, remain elusive.

Pyroclastic density currents (see *Pyroclastic Flow* and *Nuée Ardente*) are also formed during lateral blasts when part of a volcanic edifice collapses to (a) expose pressurized magma (e.g., Mount St. Helens, 1980, Lipman and Mullineaux, 1981) or (b) remove support for a pressurized lava dome (e.g., Soufrière Hills, Montserrat, 1997, see papers in Druitt and Kokelaar, 2002). The enormous impact of such events may not be immediately recognizable in the geological record when only localized hummocky debris avalanche deposits may remain. The Mount St. Helens collapse was preceded by an obvious bulge (cryptodome) on the southern flank. At Soufrière Hills, open fractures and frequent rock falls (see Rockfall) were evident signs of instability 14 months before the collapse. It was also obvious that the southern part of the growing lava dome was resting on a deeply hydrothermally altered flank. The longest known debris avalanche occurred associated with an eruption at Volcán de Colima 4,000 years ago; even at a distance of 90 km, it was estimated to have had a velocity of 44 m/s (158 km/h) (Stoopes and Sheridan, 1992). Debris avalanches can cause tsunami (see entry *Tsunami*) if they enter the sea.

Lahars (see *Lahar*) have also been devastating in terms of fatalities. There were 25,000 fatalities at Nevado del Ruiz (*cf*) volcano, Columbia, in 1985 (Voight, 1988; Table 1). A *lahar* (*cf*) is controlled by topography; it may be hot or cold and can precede, accompany, or follow volcanic activity. They are particularly voluminous and hazardous at volcanoes where pyroclastic deposits are generated. Lahars typically travel much farther than pyroclastic flows; 30 km is not unusual at larger volcanoes, and so towns and cities developed on a coastal plain or river delta tens of kilometers from a volcano may be at risk. This is especially the case if the volcano is snow or ice-covered (e.g., Mount Rainier, USA, Fisher et al., 1997). GIS has been used effectively to deal with potential lahar inundation zones (e.g., Schilling, 1998; Iverson et al., 1998). A *jökulhlaup* (*cf*) is a flood of meltwater generated at a subglacial volcano; it is similar to a lahar but may be more dilute and often carries large blocks of ice.

Basaltic lavas (see *AA-Lava* and *Pahoehoe-lava*) typically travel at several kilometers per hour; so an observer can walk away (but there have been fatalities when flows advance over wet ground causing small explosions) (Manga and Ventura, 2005). Slowly advancing lava nevertheless destroys everything in its path. Alkaline basaltic lavas erupted in continental environments have very different properties. In 1977, at Nyiragongo in Congo, lava flows generated by draining of the lava lake traveled at 60 km/h (and possibly up to 100 km/h) and hundreds of people were killed in Goma. In 2002, another eruption killed over a 100 (see papers in Capaccioni and Vaselli, 2004). Walker (1973) determined that effusion rate is the single most important factor in determining the morphology and length of lava flows. The largest basaltic eruption of the last 1,000 years was from the Laki fissure in 1783 which extruded at a rate of 5,000 m^3/s, produced lava flows that spread over 565 km^3, and flowed for distances up to 65 km (e.g., Thordarson and Self, 2003). Quaternary basalt lavas have been reported with lengths of up to 181 km (Pasquare, 2008). Mauna Loa's axial rift has been the source of numerous lavas in the recent past that have extended for over 30 km to Hilo.

Andesitic to rhyolitic lava tends to be viscous and forms either short lava flows or lava domes. They are often associated with explosive eruptions and pyroclastic flows. Extrusion rates are highly variable and lava domes at different volcanoes show differing morphologies (e.g., Fink and Anderson, 2000; Watts et al., 2002). At Soufrière Hills Volcano, distinct cycles relating to magma depressurization, degassing and crystallization were recognized (Voight et al., 1999). Each pulse generated a recognizable shear lobe in the dome, facilitating assessment of the most likely direction of dome-collapse pyroclastic flows (Calder et al., 2002). At Soufrière Hills Volcano (*Montserrat eruptions*), several lava dome collapses have taken place during heavy rainfall and have lacked other precursors such as seismic activity. In these cases, rainfall has been implicated in the triggering of the collapse. Several mechanisms have been proposed, and it was apparent that most collapses took place during the first heavy rains after a long dry season (e.g., Matthews et al., 2002; Elsworth et al., 2004).

Volcanogenic tsunami (see entry *Tsunami*) may be caused by voluminous pyroclastic flows and/or debris avalanches entering the sea, underwater explosions, or submarine caldera collapse. Pyroclastic flows entering the sea may be accompanied by hydrovolcanic explosions (e.g., Edmonds and Herd, 2005). During the eruption of Krakatau in 1883, many were killed on Sumatra and Java by tsunami up to 40 m high. The caldera-forming eruption at Santorini in about 1620 BC generated tsunami in the eastern Mediterranean up to 50 m high and probably contributed to the downfall of the Minoan civilization (see entry *Santorini, Eruption*).

Potentially hazardous volcanic gas (see entry *Volcanic Gas*) is released at elevated concentrations almost continuously from active volcanoes and at lower levels from fumaroles (see *Fumarole*) when there is no eruption. Volcanic gases include sulfur dioxide, hydrogen sulfide, carbon dioxide, and hydrogen fluoride, all of which can have significant environmental and health impacts. Human fatalities have been mainly caused by carbon dioxide and hydrogen sulfide both of which are denser than air and so collect in dangerous concentrations in depressions (e.g., Baxter, 2000). At Lake Nyos in Cameroon, carbon dioxide seeping up into the lake bed was dissolved in lake waters in huge quantities. The gas suddenly exsolved causing the lake waters to overturn on 21 August 1986, and killed at least 1,700 people. The conversion of volcanic gases to acid aerosol in the atmosphere can result in acid rain (see entry *Acid Rain*) and damage to agriculture, environment, and water supplies.

Climate and volcanoes

There is evidence that major volcanic activity can cause a change in the Earth's climate for several years after an eruption (e.g., Rampino and Self, 1994; Rampino and Self, 2000; Robock and Oppenheimer, 2003; Thordarson et al., 2009). In 1783, a dry fog across much of Europe and a severe winter has been attributed to gases and the resulting sulfate aerosols from the Laki eruption of 1783 (Thordarson and Self, 2003; Oman et al., 2006). There were severe environmental and health impacts (e.g., de Boer and Sanders, 2002). Similar effects such as a dry fog and reduced sunlight were reported in 1816 after the plinian (VEI 7) eruption of Tambora in 1815. The average global temperature fell by 0.4–0.7°C, and there is a clear acidity spike in the Greenland and Antarctic ice cores. The summer of 1816 in the northern hemisphere was known as the "year without a summer." Crop failure was widespread and famine resulted in epidemics of cholera (India) and typhus (Europe) and mass migrations. More recently, the 1982 eruption of El Chichon and the 1991 Pinatubo eruption (McCormick et al., 1995) both added considerable masses of sulfate aerosol to the stratosphere.

It is the release of sulfur dioxide and hydrogen sulfide and production of sulfuric acid aerosols that have significant impact on the atmosphere, and in general, these gases are more abundant in basaltic eruptions. The impact of a given eruption on the climate depends on the height of the eruption column, the erupted mass of gas (which relates to erupted volume of magma), the geographical location, composition, and eruption duration (e.g., Parfitt and Wilson, 2008). Flood basalt eruptions could repeatedly inject large masses sulfur dioxide into the stratosphere over years or even decades giving little opportunity for recovery, whereas the aerosol injection of a cataclysmic silicic eruption might just last several days (Thordarson et al., 2009).

The impact of changes in sea level and varying degrees of glacial loading on crustal dynamics is a growing area of research.

Volcano observatories and international collaboration

The first volcano observatory was established at Vesuvius in 1841 (Krafft, 1993). Since then, significant advances in volcanology have tended to follow major volcanic eruptions and also technological advances. There are now more than 70 institutes signed up to the World Organization of Volcano Observatories (a commission of the International Association of Volcanology and Chemistry of the Earth's Interior), some with extremely sophisticated monitoring equipment enabling scientists to forecast volcanic activity and thus enabling populations at risk to take action (e.g., Scarpa and Tilling, 1996). Nevertheless, most potentially active volcanoes are not monitored.

The monitoring of inactive volcanoes is essential in order to establish long-term baseline measurements for each volcano so that precursory signs can be identified (see Monitoring Natural Hazards). Technological advances have increased the reliability and sensitivity of monitoring equipment and allow real-time collection, interrogation (e.g., Mader et al., 2006), and communication of vast datasets. Nevertheless, equipment and monitoring comes at a cost; at many volcanoes, access to effective, durable, low-cost monitoring techniques is more practical. International, interdisciplinary collaboration has been successful and, combined with more focused use of remote sensing capability (see *Remote Sensing of Natural Hazards and Disasters*), may in time facilitate improved monitoring of more volcanoes.

Data gathered by volcano observatories is being collated in a database of volcanic unrest (WOVOdat) which is facilitating international collaboration in identifying eruption precursors. Other international, web-based initiatives include "VHub" focusing on the development of computational models of volcanic processes and

"VOGRIPA" which will focus on collating and building volcanic hazards databases. Another valuable international initiative includes the International Volcanic Health Hazard Network (IVHHN). An effort to provide a platform for global hazard and risk assessments is underway called the Global Volcano Model.

Hazards, risk and uncertainty

Effective hazards assessments require a combination of detailed geological mapping and dating of historic and ancient deposits, recognition of all possible hazards, good dynamic modeling, and effective expert elicitation. A thorough hazard analysis must also be combined with a consideration of vulnerability in order to constitute a *risk assessment* (*cf*) (e.g., Blong, 2000). Effective volcanic *risk management* (*cf*) involves strong partnerships between scientists, communities, and various authorities. There are many opportunities for the reduction of risk posed by volcanic hazards (e.g., Blong, 2000), but many volcanoes still require basic monitoring and mapping (e.g., Scarpa and Tilling, 1996). In the twentieth century, there were notable "successes" for the volcanological community (e.g., Pinatubo, 1991) and failures (e.g., Nevado del Ruiz, 1985). Long-lived eruptions like Soufrière Hills (see *Montserrat Eruptions*) have brought different challenges and offer the opportunity to develop and test new methodologies in probabilistic hazard and risk assessment (e.g., Aspinall et al., 2002; Aspinall, 2006). Probabilistic methods are also being developed elsewhere (e.g., Newhall and Hoblitt, 2002; Marzocchi et al., 2007; Neri et al., 2008). Nevertheless, the communication of the findings in a way that is useful and meaningful to the community stakeholder remains a challenge (e.g., Johnston and Ronan, 2000) (see *Risk Perception and Communication*). The concept of uncertainty (see entry *Uncertainty*) remains a major difficulty to overcome.

Summary

The most significant volcanic eruptions of the last 50 years have led to great advances in understanding how volcanoes work. Nevertheless, there is still a great deal to be done in terms of short-term forecasting of eruption onset and dynamic changes in an eruption, understanding how eruptions are triggered and sustained, and how magma is supplied to the surface, and improving multiparameter monitoring to track magma movement and to identify when unrest will become an eruption. Recognizing all potential hazards and delineating the likely impact areas of pyroclastic flows, surges, and lahars in particular remains an elusive goal with potential for more sophisticated modeling.

In order to reduce volcanic risk, baseline monitoring at the very least should be established at all volcanoes with potentially at-risk populations. During periods of dormancy, scientists, policy and decision makers, emergency managers, and communities potentially at risk can build relationships, knowledge, and preparedness planning. The importance of effective communication between these groups cannot be understated (see *Risk Perception and Communication*).

The concept of "multi-hazards" requires a more holistic view of volcanic hazards. For example, heavy rainfall is known to trigger collapse of lava domes, and hence planning for a hurricane should include preparation for a possible concurrent volcanic collapse, lahars, landslides, and tsunami.

The planet is undergoing significant rapid environmental change at the same time as great population growth; so it is urgent that we understand the short- and long-term effects and consequences of volcanic activity in order to mitigate the risk (see entry *Risk*) as far as we can (Sparks and Aspinall, 2004). We need to appreciate our vulnerability (see entry *Vulnerability*) in the face of volcanic hazards and reduce that vulnerability wherever possible.

Bibliography

Aspinall, W. P., 2006. Structured elicitation of expert judgment for probabilistic hazard and risk assessment in volcanic eruptions. In Mader, H. M., Coles, S. G., Connor, C. B., and Connor, L. J. (eds.), *Statistics in Volcanology*. London: Geological Society of London on behalf of IAVCEI, pp. 15–30.

Aspinall, W. P., Loughlin, S. C., Michael, F. V., Miller, A. D., Norton, G. E., Rowley, K. C., Sparks, R. S. J., and Young, S. R., 2002. The Montserrat Volcano Observatory: its evolution, organisation, role and activities. In Druitt, T. H., and Kokelaar, B. P. (eds.), *The Eruption of Soufrière Hills Volcano, Montserrat, from 1995 to 1999*. London: Geological Society. Memoir number, 21, pp. 71–92.

Baxter, P. J., 2000. Impacts of volcanism on human health. In Sigurdsson, H. (ed.), *Encyclopedia of Volcanoes*. San Diego: Academic, pp. 1035–1044.

Blake, S., 1981. Volcanism and the dynamics of open magma chambers. *Nature*, **289**, 783–785.

Blong, R., 1984. *Volcanic Hazards: A Sourcebook on the Effects of Eruptions*. Sydney, Australia: Academic.

Blong, R., 2000. Volcanic hazards and risk management. In Sigurdsson, H. (ed.), *Encyclopedia of Volcanoes*. San Diego: Academic, pp. 1215–1228.

Branney, M. J., and Kokelaar, P., 2002. *Pyroclastic Density Currents and the Sedimentation of Ignimbrites*. Bath: Geological Society. Memoir number, 27.

Calder, E. S., Cole, P. D., Dade, W. B., et al., 1999. Mobility of pyroclastic flows and surges at the Soufriere Hills Volcano, Montserrat. *Geophysical Research Letters*, **26**, 537–540.

Capaccioni, B., and Vaselli, O. (eds.), 2004. *The January 2002 Eruption of Nyiragongo Volcano and the Socio-Economical Impact*. Pisa: Istituti editoriali E Poligrafici Internazionali. *Acta Vulcanologica*, **14**(1–2), **15**(1–2).

Castro, J. M., and Dingwell, D. B., 2009. Rapid ascent of rhyolitic magma at Chaiten volcano, Chile. *Nature*, **461**, 780–783, doi:1038/nature08458.

Charbonnier, S. J., and Gertisser, R., 2009. Numerical simulations of block-and-ash flows using the Titan2D flow model: examples from the 2006 eruption of Merapi volcano, Java, Indonesia. *Bulletin of Volcanology*, **71**(8), 953–959, doi:10.1007/s00445-009-0299-1.

Chester, D. K., Degg, M., Duncan, A. M., and Guest, J. E., 2001. The increasing exposure of cities to the effects of volcanic eruptions: a global survey. *Environmental Hazards*, **2**, 89–103.

Crandell, D. R., Booth, B., Kusumadinata, K., Shimozuru, D., Walker, G. P. L., and Westercamp, D., 1984. *Source-Book for Volcanic-Hazards Zonation*. Paris: UNESCO, p. 97.

Crisp, J., 1984. Rates of magma emplacement and volcanic output. *Journal of Volcanology and Geothermal Research*, **20**, 177–211.

Davidson, J., and Da Silva, S., 2000. Composite volcanoes. In Sigurdsson, H. (ed.), *Encyclopedia of Volcanoes*. San Diego: Academic, pp. 663–682.

De Boer, J. Z., and Sanders, D. T., 2002. *Volcanoes in Human History*. Princeton: University Press.

Denlinger, R. P., and Iverson, R. M., 2001. Flow of variably fluidized granular masses across three-dimensional terrain 2 numerical predictions and experimental tests. *Journal of Geophysical Research*, **106**(1B), 553–566.

Druitt, T. H., 1998. Pyroclastic density currents. In Gilbert, J. S., and Sparks, R. S. J. (eds.), *The Physics of Explosive Volcanic Eruptions*. London: Geological Society. Special publication number, 145, pp. 145–182.

Druitt, T. H., and Kokelaar, B. P., 2002. *The Eruption of Soufriere Hills Volcano, Montserrat from 1995 to 1999*. London: Geological Society. Memoir number, 21.

Druitt, T. H., et al., 2002a. Episodes of cyclic Vulcanian explosive activity with fountain collapse at Soufriere Hills Volcano, Montserrat. In Druitt, T. H., and Kokelaar, P. (eds.), *The eruption of Soufriere Hills Volcano, Montserrat from 1995 to 1999*. London: Geological Society. Memoir number, 21, pp. 281–306.

Druitt, T. H., et al., 2002b. Small-volume, mobile pyroclastic flows formed by rapid sedimentation from pyroclastic surges at Soufriere Hills Volcano, Montserrat: an important volcanic hazard. In Druitt, T. H., and Kokelaar, R. (eds.), *The eruption of Soufriere Hills Volcano, Montserrat from 1995 to 1999*. London: Geological Society. Memoir number, 21, pp. 263–279.

Edmonds, M., and Herd, R. A., 2005. Inland-directed base surge generated by the explosive interaction of pyroclastic flows and seawater at Soufriere Hills volcano, Montserrat. *Geology*, **33**, 245–248, doi:10.1130/G21166.1.

Elsworth, D., Voight, B., Thompson, G., and Young, S. R., 2004. Thermal-hydrologic mechanism for rainfall-triggered collapse of lava domes. *Geology*, **32**, 969–972, doi:10.1130/G20730.1.

Esposito Ongaro, T., Clarke, A. B., Neri, A., Voight, B., and Widiwijayanti, C., 2008. Fluid dynamics of the 1997 Boxing Day volcanic blast on Montserrat, West Indies. *Journal of Geophysical Research*, **113**, B03211, doi:10.1029/2006JB004898.

Fink, J., and Anderson, S. W., 2000. Lava Domes and Coulees. In Sigurdsson, H. (ed.), *Encyclopedia of Volcanoes*. San Diego: Academic, pp. 307–320.

Fisher, R. V., 1995. Decoupling of pyroclastic currents: hazards assessment. *Journal of Volcanology and Geothermal Research*, **66**, 257–263.

Fisher, R. V., Heiken, G., and Hulen, J. B., 1997. *Volcanoes, Crucibles of Change*. Princeton, NJ: Princeton University Press.

Francis, P., 1993. *Volcanoes: A Planetary Perspective*. Oxford: Oxford University Press.

Francis, P., and Oppenheimer, C., 2003. *Volcanoes*. Oxford: Oxford University Press.

Francis, P. W., Hammill, M., Kretzschmar, G. A., and Thorpe, R. S., 1978. The Cerro Galan Caldera, northwest Argentina. *Nature*, **274**, 749–751.

Freundt, A., and Rosi, M. (eds.), 1998. *From Magma to Tephra: Modelling Physical Processes of Explosive Volcanic Eruptions*. Amsterdam: Elsevier.

Fujii, T., and Nakada, S., 1999. The 15 September 1991 pyroclastic flows at Unzen Volcano (Japan): a flow model for associated ash-cloud surges. *Journal of Volcanology and Geothermal Research*, **89**, 159–172.

Gilbert, J. S., and Sparks, R. S. J. (eds.), 1998. *The Physics of Explosive Volcanic Eruptions*. London: Geological Society. Special publication number, 145.

Hansell, A. L., Horwell, C. J., and Oppenheimer, C., 2006. Health risks from volcanoes. *Occupational and Environmental Medicine*, **63**, 149–156, doi:10.1136/oem.2005.022459.

Iverson, R. M., Schilling, S. P., and Vallance, J. W., 1998. Objective delineation of lahar inundation hazard zones. *Geological Society of America, Bulletin*, **110**, 972–984.

Johnston, D., and Ronan, K., 2000. Risk education and intervention. In Sigurdsson, H. (ed.), *Encyclopedia of Volcanoes*. San Diego: Academic, pp. 1229–1242.

Krafft, M., and Bahn, P. G., 1993. *Volcanoes: Fire from the Earth*. New Horizons.

Lipman, P. W., and Mullineaux, D. R. (eds.), 1981. *The 1980 Eruptions of Mount St. Helens*. Washington, DC: United States Geological Survey. Professional Paper 1250.

Loughlin, S. C., Baxter, P. J., Aspinall, W. A., Darroux, B., Harford, C. L., and Miller, A. D., 2002. Eyewitness accounts of the 25 June 1997 pyroclastic flows and surges at Soufrière Hills Volcano, Montserrat, and implications for disaster mitigation. In Druitt, T. H., and Kokelaar, P. (eds.), *The Eruption of Soufriere Hills Volcano, Montserrat from 1995 to 1999*. London: Geological Society. Memoir number, 21, pp. 211–230.

Mader, H., Coles, S., Connor, C., and Connor, L. (eds.), 2006. *Statistics in Volcanology*. London: Geological Society. IAVCEI Publications, Vol. 1.

Manga, M., and Ventura, G. (eds.), 2005. *Kinematics and Dynamics of Lava Flows*. Boulder, CO: Geological Society of America. Special Paper 396.

Marsh, B., 2000. Magma chambers. In Sigurdsson, H. (ed.), *Encyclopedia of Volcanoes*. San Diego: Academic, pp. 191–206.

Martin, R. S., Watt, S. F. L., Pyle, D. P., Mather, T. A., Matthews, N. E., Georg, R. B., Day, J. A., Fairhead, T., Witt, M. L. I., and Quayle, B. M., 2009. Environmental effects of ashfall in Argentina from the 2008 Chaiten volcanic eruption. *Journal of Volcanology and Geothermal Research*, **184**, 462–472.

Marzocchi, W., Sandri, L., and Selva, J., 2007. BET_EF: a probabilistic tool for long- and short-term eruption forecasting. *Bulletin of Volcanology*, **70**(5), 623–632.

Mason, B. G., Pyle, D. M., and Oppenheimer, C., 2004. The size and frequency of the largest explosive eruptions on Earth. *Bulletin of Volcanology*, **66**, 735–748.

Massol, H., and Jaupart, C., 1999. The generation of gas overpressure in volcanic eruptions. *Earth and Planetary Science Letters*, **166**, 57–70.

Matthews, A. J., Barclay, J., Carn, S., Thompson, G., Alexander, J., Herd, R., and Williams, C., 2002. Rainfall-induced volcanic activity on Montserrat. *Geophysical Research Letters*, **29**(13), doi:10.1029/2002GL014863.

McCormick, M. P., Thomason, L. W., and Trepte, C. R., 1995. Atmospheric effects of the Mt Pinatubo eruption. *Nature*, **373**, 399–404.

Melnik, O., and Sparks, R. S. J., 1999. Nonlinear dynamics of lava dome discharge. *Nature*, **402**(6757), 37–41.

Melnik, O., and Sparks, R. S. J., 2002. Dynamics of magma ascent and lava extrusion at Soufriere Hills Volcano, Montserrat. In Druitt, T. H., and Kokelaar, P. (eds.), *The eruption of Soufriere Hills Volcano, Montserrat from 1995 to 1999*. London: Geological Society. Memoir number, 21, pp. 153–171.

Melnik, O. E., and Sparks, R. S. J., 2005. Controls on conduit magma flow dynamics during lava dome building. *Journal of Geophysical Research: Solid Earth*, **110**(2), B02209, 1–21. ISSN: 0148-0227.

Melnik, O. E., and Sparks, R. S. J., 2007. Transient models of conduit flows during volcanic eruptions. In Mader, H. M., Coles, S. G., Connor, C. B., and Connor, L. J. (eds.), *Statistics in Volcanology, 1*. London: Geological Society on behalf of IAVCEI, pp. 201–214. ISBN 1862392080.

Miller, T. P., and Casadevall, T. J., 2000. Volcanic ash hazards to aviation. In Sigurdsson, H. (ed.), *Encyclopedia of Volcanoes*. San Diego: Academic, pp. 915–930.

Murphy, M. D., Sparks, R. S. J., Barclay, J., Carroll, M. R., Lejeune, A.-M., Brewer, T. S., Macdonald, R., Black, S., and Young, S., 1998. The role of magma mixing in triggering the current eruption at the Soufriere Hills volcano, Montserrat, West Indies. *Geophysical Research Letters*, **25**(18 & 19), 3433–3436.

Neri, A., Aspinall, W. P., Cioni, R., Bertagnini, A., Baxter, P. J., Zuccaro, G., Andronico, D., Barsotti, S., Cole, P. D., Esposti Ongaro, Y., Hincks, T. K., Macedonio, G., Papale, P., Rosi, M., Santacroce, R. A., and Woo, G., 2008. Developing an event tree for probabilistic hazard and risk assessment at Vesuvius. *Journal of Volcanology and Geothermal Research*, **178**(3), 397–415, doi:10.1016/j.jvolgeores.2008.05.014.

Newhall, C. G., and Hoblitt, R. P., 2002. Constructing event trees for volcanic crises. *Bulletin of Volcanology*, **64**, 3–20.

Newhall, C. G., and Punongbayan, R. S. (eds.), 1996. *Fire and mud: eruptions and lahars of Mount Pinatubo, Philippines*. Quezon City/Seattle: Philippine Institute of Volcanology and Seismology/University of Washington Press.

Newhall, C. G., and Self, S., 1982. The volcanic explosivity index (VEI): an estimate of the explosive magnitude for historical volcanism. *Journal of Geophysical Research*, **87**, 1231–1238.

Oman, L., Robock, A., Stenchikov, G. L., Thordarson, T., Koch, D., Shindell, D. T., and Gao, C., 2006. Modeling the distribution of the volcanic aerosol cloud from the 1783–1784 Laki eruption. *Journal of Geophysical Research*, **111**, D12209, doi:10.1029/2005JD006899,2006.

Oppenheimer, C., Pyle, D. M., and Barclay, J. (eds.), 2003. *Volcanic Degassing*. London: Geological Society. Special publication number, 213.

Oxford Economics, 2010. The economic impacts of air travel disruptions due to volcanic ash. A report prepared for airbus.

Parfitt, E. A., and Wilson, L., 2008. *Fundamentals of Physical Volcanology*. Oxford: Blackwell.

Pasquare, G., 2008. Very long pahoehoe inflated basaltic lava flows in the Payenia volcanic province (Mendoza and La Pampa, Argentina). *Revista de la Asociación Geológica Argentina*, **63**(1), 131–149. ISSN: 0004-4822.

Patra, A. K., Bauer, A. C., Nichita, C. C., Pitman, E. B., Sheridan, M. F., Bursik, M., Rupp, B., Webber, A., Stinton, A. J., Namikawa, L. M., and Renschler, C. S., 2005. Parallel adaptive simulation of dry avalanches over natural terrain. *Journal of Volcanology and Geothermal Research*, **139**, 1–22.

Pyle, D. M., 2000. Sizes of volcanic eruptions. In Sigurdsson, H. (ed.), *Encyclopedia of Volcanoes*. San Diego: Academic, pp. 263–270.

Rampino, M. R., and Self, S., 1994. Climate-volcanic feedback and the Toba eruption of 74,000 years ago. *Quaternary Research*, **40**, 69–80.

Rampino, M. R., and Self, S., 2000. Volcanism and biotic extinctions. In Sigurdsson, H. (ed.), *Encyclopedia of Volcanoes*. San Diego: Academic, pp. 1083–1094.

Robock, A., and Oppenheimer, C. (eds.), 2003. *Volcanism and the Earth's atmosphere*. San Francisco: American Geophysical Union. Geophysical Monograph, Vol. 139.

Rogers, N., and Hawkesworth, C., 2000. Composition of magmas. In Sigurdsson, H. (ed.), *Encyclopedia of Volcanoes*. San Diego: Academic, pp. 115–132.

Saucedo, R., Macías, J. L., Bursik, M. I., Mora, J. C., Gavilanes, J. C., and Cortes, A., 2002. Emplacement of pyroclastic flows during the 1998–1999 eruption of Volcán de Colima, México. *Journal of Volcanology and Geothermal Research*, **117**, 129–153.

Saucedo, R., Macías, J. L., Sheridan, M. F., Bursik, M. I., and Komorowski, J. C., 2005. Modeling of pyroclastic flows of Colima Volcano, Mexico: implications for hazard assessment. *Journal of Volcanology and Geothermal Research*, **139**, 103–115.

Scarpa, R., and Tilling, R. I. (eds.), 1996. *Monitoring and Mitigation of Volcano Hazards*. Berlin: Springer.

Schilling, S. P., 1998. LAHARZ: GIS programs for automated mapping of lahar inundation zones. United States Geological Survey, Open-file Report, 98.

Schmidt, R., and Schmincke, H. U., 2000. Seamounts and island building. In Sigurdsson, H. (ed.), *Encyclopedia of Volcanoes*. San Diego: Academic, pp. 383–402.

Sigurdsson, H. (ed.), 2000a. *Encyclopedia of Volcanoes*. San Diego: Academic.

Sigurdsson, H., 2000b. Volcanic episodes and rates of volcanism. In *Encyclopedia of Volcanoes*. San Diego: Academic, pp. 271–282.

Simkin, T., and Siebert, L., 2000. Earth's volcanoes and eruptions: an overview. In Sigurdsson, H. (ed.), *Encyclopedia of Volcanoes*. San Diego: Academic, pp. 249–262.

Simkin, T., and Siebert, L., 2002–2009. *Volcanoes of the World: An Illustrated Catalog of Holocene Volcanoes and their Eruptions*. Smithsonian Institution, Global Volcanism Program Digital Information Series, GVP-3. http://www.volcano.si.edu/world/.

Smellie, J. L., and Chapman, M. G. (eds.), 2002. *Volcano-Ice Interaction on Earth and Mars*. London: Geological Society of London. Special publication number, 202.

Sparks, R. S. J., and Aspinall, W. P., 2004. Volcanic activity: frontiers and challenges in forecasting prediction and risk assessment. In Sparks, R. S. J., and Hawkesworth, C. J. (eds.), *The State of the Planet: Frontiers and Challenges in Geophysics*. Washington, DC: American Geophysical Union. Geophysical Monograph, Vol. 150. IUGG, Vol. 19, pp. 359–373.

Sparks, R. S. J., and Young, S. R., 2002. The eruption of Soufriere Hills Volcano, Montserrat (1995–1999): overview of scientific results. In Druitt, T. H., and Kokelaar, B. P. (eds.), *The Eruption of Soufriere Hills Volcano, Montserrat, from 1995 to 1999*. London: Geological Society. Memoirs number, 21, pp. 115–152.

Sparks, S., Self, S., Pyle, D., Oppenheimer, C., Rymer, H., and Grattan, J., 2005. *Super-eruptions: global effects and future threats*. Report of a Geological Society of London Working Group. http://www.geolsoc.org.uk/gsl/education/page2965.html.

Stoopes, G. R., and Sheridan, M. F., 1992. Giant debris avalanches from the Colima Volcanic Complex, Mexico: implications for long–runout landslides (>100 km) and hazard assessment. *Geology*, **20**, 299–302.

Thordarson, T. S., and Self, S., 2003. Atmospheric and environmental effects of the 1783–1784 Laki eruption: a review and reassessment. *Journal of Geophysical Research*, **108**(D1), 4011, doi:10.1029/2001JD002042.

Thordarson, T., Rampino, M., Keszthelyi, L. P., and Self, S., 2009. *Effects of Megascale Eruptions on Earth and Mars*. San Francisco: The Geological Society of America. Special Paper 453.

Trofimovs, J., et al., 2006. Submarine pyroclastic deposits formed at the Soufriere Hills volcano, Montserrat (1995–2003): what happens when pyroclastic flows enter the ocean? *Geology*, **34**, 549–552.

Ui, T., Matsuwo, N., Sumita, M., and Fujinawa, A., 1999. Generation of pyroclastic flows during the 1990–1995 eruption of Unzen Volcano, Japan. *Journal of Volcanology and Geothermal Research*, **89**, 123–137.

Voight, B., 1988. Countdown to catastrophe. *Earth and Mineral Sciences*, **57**(2), 17–30.

Voight, B., Hoblitt, R. P., Clarke, A. B., Lockhart, A. B., Miller, A. D., Lynch, L., and McMahon, J., 1998. Remarkable cyclic ground deformation monitored in real-time on Montserrat, and its use in eruption forecasting. *Geophysical Research Letters*, **25**(18), doi:10.1029/98GL01160.

Voight, B., Sparks, R. S. J., Miller, A. D., et al., 1999. Magma flow instability and cyclic activity at Soufriere Hills volcano, Montserrat, British West Indies. *Science*, **283**, 1138–1142.

Wadge, G., 2009. Assessing the pyroclastic flow hazards from dome collapse at Soufriere Hills Volcano, Montserrat. In Thordarson, T., Self, S., Larsen, G., Rowlands, S. K., and Hoskuldsson, A. (eds.), *Studies in Volcanology: The Legacy of George Walker*. London: Geological Society and IAVCEI. Special publication number, 2, pp. 211–224.

Walker, G. P. L., 1973. Lengths of lava flows. *Philosophical Transactions of the Royal Society, London*, **274**, 107–118.

Watt, S. F. L., Pyle, D. M., Mather, T. A., Martiin, R. S., and Matthews, N. E., 2009. Fallout and distribution of volcanic ash over Argentina following the May 2008 explosive eruption of Chaitén, Chile. *Journal of Geophysical Research*, **114**, B04207, doi:10.1029/2008JB006219.

Watts, R. B., Herd, R. A., Sparks, R. S. J., and Young, S. R., 2002. Growth patterns and emplacement of the andesite lava dome at the Soufrière Hills Volcano, Montserrat. In Druitt, T. H., and Kokelaar, B. P. (eds.), *The Eruption of Soufrière Hills Volcano, Montserrat, from 1995 to 1999*. London: Geological Society. Memoir number 21, pp. 115–152.

White, R. V., and Saunders, A. D., 2005. Volcanism, impact and mass extinctions: incredible or credible coincidences? *Lithos*, **79**, 299–316, doi:10.1016/j.lithos.2004.09.016.

Witham, C. S., 2005. Volcanic disasters and incidents: a new database. *Journal of Volcanology and Geothermal Research*, **148**, 191–233.

Yamamoto, T., Takarada, S., and Suto, S., 1993. Pyroclastic flows from the 1991 eruption of Unzen volcano, Japan. *Bulletin of Volcanology*, **55**, 166–175.

Cross-references

Aa Lava
Acid Rain
Airphoto and Satellite Imagery
Aviation, Hazards to
Base Surge
Caldera
Casualties Following Natural Hazards
Challenges to Agriculture
Civil Protection and Crisis Management
Climate Change
Communicating Emergency Information
Community Management of Hazards
Cost-Benefit Analysis of Natural Hazard Mitigation
Costs (Economic) of Natural Hazards and Disasters
Damage and the Built Environment
Debris Flow
Disaster Risk Management
Education and Training for Emergency Preparedness
Emergency Management
Emergency Planning
Epidemiology of Disease in Disasters
Eruption Types (Volcanic)
Evacuation
Expert (Knowledge-Based) Systems for Disaster Management
Exposure to Natural Hazards
Extinction
Frequency and Magnitude of Events
Fumarole
Galeras Volcano
Geographic Information Systems (GIS) and Natural Hazards
Geological/Geophysical Disasters
Global Change and Its Implications for Natural Disaster
Global Dust/Aerosol Effects
Hazard
Hazard and Risk Mapping
Historical Events
History of Natural Disasters
Human Impact of Hazards
Intensity Scales
International Strategies for Disaster Reduction (IDNDR and ISDR)
Internet, World Wide Web, and Natural Hazards
Jökulhlaup (Débâcle)
Krakatoa (Krakatau)
Lahar
Lava
Magma
Magnitude Measures
Megacities and Natural Disasters
Mitigation
Monitoring and Prediction of Natural Hazards
Montserrat Eruptions
Mt. Pinatubo
Mud Volcano
Nevado del Ruiz
Nuee Ardente (Glowing Avalanche)
Pahoehoe Lava
Perception of Natural Hazards and Disasters
Plate Tectonics
Psychological Impacts of Natural Disasters
Pyroclastic Flow
Remote Sensing of Natural Hazards and Disasters
Risk Assessment
Risk Perception and Communication
Santorini, Eruption of
Seismic Gap
Seismograph/Seismometer
Seismology
Shield Volcano
Slope Stability
Stratovolcano
Subduction
Surge
Tiltmeters
Tsunami
Uncertainty
Vesuvius
Volcanic Ash
Volcanic Gas
Vulnerability
Warning Systems
World Economy, Impact of Disasters on
Worldwide Trend in Natural Disasters

VULNERABILITY

Susan L. Cutter
University of South Carolina, Columbia, SC, USA

Synonyms

Exposure; Susceptibility

Definition

Vulnerability is potential for harm or loss.

> The characteristics and circumstances of a community, system, or asset that make it susceptible to the damaging effects of a hazard (ISDR 2009:30).

Introduction

Vulnerability is a key concept in natural hazards and disasters research. It is used to describe the potential for harm or loss in a physical system (e.g., erosion of beaches, glacial melting); the fragility of the built environment and infrastructure; or the susceptibility to harm in social and economic systems. Vulnerability can be studied at the individual level, for example, an individual person, a house, a bridge. It can also be aggregated to the group level to study the vulnerability of low-income populations, commercial establishments, and infrastructure systems such as the transportation networks. Vulnerability can also be examined spatially with a focus on particular places such as communities, regions, or nations. From the spatial vantage point, vulnerability examines the potential for harm at the place as a function of the intersection between the built environment and infrastructure, social and economic systems, and physical systems (Cutter, 2010).

Moving from risk-hazards to vulnerability

There is a rich multidisciplinary tradition that focuses on disasters, risk, and hazards. As noted elsewhere in this volume, natural hazards arise from the interaction between society and natural systems (earthquakes and tsunamis, for example). Risk is another widely used term, and here we use it to define the likelihood of incurring harm from a particular hazard event. This risk–hazard concept permeated the early work in the field and has now given way to vulnerability, which recognizes the degree to which an individual or system is likely to experience harm to due to some exposure from natural, technological, or human-induced sources (National Research Council 2007).

While there are numerous definitions, frameworks, conceptual models, and assessment techniques used for understanding vulnerability (Turner et al., 2003b; Wisner et al., 2004; Kasperson et al., 2005; Adger, 2006; Eakin and Luers, 2006) several common principles or understandings have emerged. First, vulnerability is examined from a social-ecological perspective. In this viewpoint, natural processes by themselves do not create hazards, rather it is the intersection with human system that produces the potential for harm, and ultimately the vulnerability to the natural hazard. Second, there is a clear recognition of the importance of place-based studies in examining vulnerability. Finally, vulnerability is seen as an equity concern, who is vulnerable is equally important as what is vulnerable. There are three important concepts emerging in the vulnerability literature: exposure, sensitivity, and resilience. Exposure primarily relates to the proximity to the source of the threat and environmental characteristics. Sensitivity focuses on the characteristics of individuals, social groups, or places and examines the differential susceptibility to harm. Resilience pertains to both the physical system and the social system and is used to describe the coping capacity of individuals, communities, or places, and their ability to respond and recover after a hazard event (Wood et al., 2010).

There are differences in the literature on the underlying causes of vulnerability. For some, the causal structure is exposure to the hazard, which is often measured in terms of proximity (Alexander, 1993). A different perspective suggests that vulnerability is caused by the underlying social conditions that give rise to unsafe conditions (Wisner et al., 2004). The third perspective incorporates elements from both. This view suggests that vulnerability includes both the exposure from the physical system, and the social response and how these are produced locally to create hazardousness of places (Hewitt and Burton, 1971; Cutter et al., 2000; Turner et al., 2003b).

Vulnerability assessments

A solid body of evidence exists on how vulnerabilities in physical systems interact with social conditions to produce hazard vulnerability (Birkmann, 2006). Many of these examine single stressors such as drought (Polsky, 2004) or earthquakes and tsunamis (Rashed and Weeks, 2003; Wood et al., 2010), but multi-stressor vulnerability assessments, while more difficult, are slowly emerging (O'Brien et al., 2004; Boruff and Cutter, 2007). In each case, the assessments begin with an understanding of the interactions between physical systems, human systems, and the built environment, and how these vary locally – the differential exposure of people and places to natural hazards.

Practical applications

The value of vulnerability assessments for emergency preparedness is increasing as local, regional, and national governments are more focused on risk reduction in preparedness and planning. The delineation of hazard zones (exposure) has a long tradition within the research community and as the science of hazards improves as well as our computational power, we are increasingly adept at mapping such high-risk zones from the perspective of the threat. However, the science of vulnerability vis-à-vis social vulnerability is less advanced. Social vulnerability frames hazards and their impacts within broader social contexts (Tierney, 2006; Laska and Morrow, 2007) and is equally important in understanding the differential impacts of natural hazards on people and places (Yarnal, 2007; Zahran et al., 2008). Yet, the development of social vulnerability metrics and models is still in its infancy. There are a variety of social vulnerability indices available, among the most significant are the Prevalent Vulnerability Index (PVI) developed by Cardona and colleagues (http://www.iadb.org/exr/disaster/pvi.cfm?language=EN&parid=4); the Environmental Vulnerability Index (EVI) developed by South Pacific Applied Geoscience Commission (SOPAC) (http://www.vulnerabilityindex.net/); and the Social Vulnerability Index (SoVI) developed by Cutter and colleagues (http://webra.cas.sc.edu/hvri/products/sovi.aspx).

Conclusion

The study of vulnerability to natural hazards is one of the emerging fields within the hazards and disasters community. New advancements in our conceptual understanding and measurement are rapidly occurring. The need for such information is critical as localities and nations cannot develop disaster risk-reduction strategies in the absence of knowing who and what is most vulnerable.

Bibliography

Adger, W. N., 2006. Vulnerability. *Global Environmental Change*, **16**, 268–281.

Alexander, D., 1993. *Natural Disasters*. New York: Chapman and Hall.

Birkmann, J., 2006. *Measuring Vulnerability to Natural Hazards: Towards Disaster Resilient Societies*. Tokyo: United Nations University Press.

Boruff, B. J., and Cutter, S. L., 2007. The environmental vulnerability of Caribbean island nations. *Geographical Review*, **97**(1), 24–45.

Cutter, S. L., 2010. Social science perspectives on hazards and vulnerability science. In Beer, T. (ed.), *Geophysical Hazards*. New York: Springer, pp. 17–30.

Cutter, S. L., Mitchell, J. T., et al., 2000. Revealing the vulnerability of people and places: a case study of Georgetown County, South Carolina. *Annals of the Association of American Geographers*, **90**(4), 713–737.

Eakin, H., and Luers, A. L., 2006. Assessing the vulnerability of social-environmental systems. *Annual Review of Environment and Resources*, **31**, 365–394.

Hewitt, K., and Burton, I., 1971. *The Hazardousness of a Place: A Regional Ecology of Damaging Events*. Toronto: University of Toronto.

ISDR International Strategy for Disaster Reduction, 2009. *2009 UNISDR Terminology on Disaster Risk Reduction*. Geneva: United Nations.

Kasperson, J. X., Kasperson, R. E., et al., 2005. Vulnerability to global environmental change. In Kasperson, J. X., and Kasperson, R. E. (eds.), *Social Contours of Risk, Vol. II*. London: Earthscan, pp. 245–285.

Laska, S., and Morrow, B. H., 2007. Social vulnerabilities and Hurricane Katrina: an unnatural disaster in New Orleans. *Marine Technology Society Journal*, **40**(4), 16–26.

National Research Council, 2007. *Tools and Methods for Estimating Populations at Risk from Natural Disasters and Complex Humanitarian Crises*. Washington DC: National Academies Press.

O'Brien, K. L., Leichenko, R., et al., 2004. Mapping vulnerability to multiple stressors: climate change and globalization in India. *Global Environmental Change*, **14**, 303–313.

Polsky, C., 2004. Putting space and time in Ricardian climate change impact studies: the case of agriculture in the U.S. Great Plains. *Annals of the Association of American Geographers*, **94**(3), 549–564.

Rashed, T., and Weeks, J., 2003. Assessing vulnerability to earthquake hazards through spatial multicriteria analysis of urban areas. *International Journal of Geographical Information Science*, **17**(6), 547–576.

Tierney, K., 2006. Social inequality, hazards, and disasters. In Daniels, R. J., Kettl, D. F., and Kunreuther, H. (eds.), *On Risk and Disaster: Lessons from Hurricane Katrina*. Philadelphia: University of Pennsylvania Press.

Turner, B. L., II, Kasperson, R. E., et al., 2003. Framework for vulnerability analysis in sustainability science. *Proceedings of the National Academy of Sciences of the United States of America*, **100**, 8074–8079.

Wisner, B., Blaikie, P., et al., 2004. *At Risk: Natural Hazards, People's Vulnerability and Disasters*. New York: Routledge.

Wood, N. J., Burton, C. G., et al., 2010. Community variations in social vulnerability to Cascadia-related tsunamis in the U.S. Pacific Northwest. *Natural Hazards*, **52**, 369–389.

Yarnal, B., 2007. Vulnerability and all that jazz: addressing vulnerability in New Orleans after Hurricane Katrina. *Technology in Society*, **29**, 249–255.

Zahran, S., Brody, S. D., et al., 2008. Social vulnerability and the natural and built environment: a model of flood casualties in Texas. *Disasters*, **32**(4), 537–560.

Cross-references

Adaptation
Cognitive Dissonance
Community Management of Hazards
Complexity Theory
Coping Capacity
Disaster Risk Reduction (DRR)
Global Network of Civil Society Organizations for Disaster Reduction
Hazard
Hazardousness of Place
International Strategies for Disaster Reduction (IDNDR and ISDR)
Land-Use Planning
Marginality
Myths and Misconceptions
Risk
Risk Perception and Communication
Sociology of Disasters
Susceptibility
Uncertainty

W

WARNING SYSTEMS

Graham S. Leonard[1], David M. Johnston[2], Chris E. Gregg[3]
[1]Massey University, Lower Hutt, New Zealand
[2]GNS Science, Massey University, Lower Hutt, New Zealand
[3]East Tennessee State University, Johnson City, TN, USA

Synonyms

Alerting system; Early warning systems; End-to-end warning system; Immediate warning system; Short fuse warning

Definition

A *warning system* is a network of interrelated sensors and processes that detect signals of a possible or imminent dangerous event and provide *information* that people can use to make protective action decisions before the moment of impact.

A *natural hazard warning system* is usually technology based, monitors signs of a natural hazard, evaluates the signs against rules and notifies people, triggering a human response.

End-to-end warning systems consider, or are owned by, responding communities.

Early warning systems are often synonymous with warning systems.

Public notification systems are the mechanisms by which the public are notified within a warning system.

Natural warnings are provided by nature. They are often synonymous with *environmental* cues, but are distinguished from them in that some geological or meteorological phenomena, for example, serve as a warning of an event, whereas other phenomena serve only as a cue that an event may occur.

Informal warnings are provided by people not acting in an official warning capacity (e.g., a friend calling another friend to alert them of a dangerous event).

Official warnings are provided by people acting in an official warning capacity.

Social warnings or social cues are provided via observations of other people's behavior.

Introduction

Warnings are a natural part of everyday life for humans. When we detect fear or pain our built-in biological warning system is letting us know something unusual is happening and that it may be detrimental to our health. *Natural hazard warning systems* are usually technological, designed to detect hazardous events and inform us of those events in a time frame that permits us to take protective action, perhaps through evacuation, sheltering, or protecting property (see Lindell and Perry, 2004). The use of a canary to detect gas in a coal mine is one early and well-known warning system. In this case, a miner's canary encapsulates two core components of most warning systems: monitoring (canary biology) and notification (the canary stops singing). Warning system research has increasingly recognized the need for "people centered" warning systems (summarized by Basher, 2006), which are not considered effective unless they trigger appropriate human behavior. This requires that the hazard detection and monitoring aspects of a warning system be integrated with emergency management and the public. Such an integrated warning system focused on reaching at-risk communities, ideally managed and owned by those communities, is often referred to as an "end-to-end" warning system.

A warning system is only as effective as its weakest link. That is, a failure of one component may render the whole system ineffective. An effective warning system requires the integration of several features of warnings in the design of the system. These include (a) Standard Operating Procedures (SOPs) of scientists and emergency managers, (b) channels of message dissemination, and (c) public response (Sorensen, 2000). Redundancy, especially in communication and notification technologies and SOPs, reduces the chance of a broken link. Warning systems that are solely "hardware based" with no focus on community response or participation have proven ineffective.

Warning systems and risk management

Warning systems are one *risk management* option available to reduce the *risk* posed by a *hazard*. It is therefore advisable to complete a *risk assessment* before embarking on developing a warning system. A 100% reliable warning system does not exist for any hazard, which leaves some level of residual risk (Sorensen, 2000). This is especially true in the case of infrequent, difficult to detect, or short lead-time hazards. Development of a warning system may also increase risk by perpetuating the occupation of marginal land (Sorensen, 2000). A flood warning system, for example, may encourage development in the floodplain where it would otherwise not be desirable. *Land use planning* and *structural mitigation* are additional risk reduction options, which for some hazards have the potential to remove the risk entirely. For example, in the case of land use planning, if a hazardous location is not developed or otherwise occupied then risk is reduced or eliminated. In such a situation, no warning system is needed. Some combination of land use planning, structural mitigation, and warning systems is often recommended to maximize risk reduction and use of land. Increasing development and population growth in coastal areas and increasing diversity in global travelers, especially visitors to coastal areas, demands that warnings are issued for multiple languages, a point underscored by the loss of life of visitors from some two dozen countries in the 2004 Indian Ocean tsunami. This requires additional complexity in warnings and other risk communications.

Monitoring, forecasting, and prediction

A warning system is dependent upon reliable *detection and monitoring* of an event or hazardous processes. In order to provide maximum response time, a warning system is designed to detect a change in a monitored parameter and provide a clear indication (forecast or prediction) when a hazardous event is expected to occur. Warning for natural hazards relies on lead-time between detection of a precursor sign(s), or the event or hazards themselves, and the impact of the hazard. Longer lead-times correspond to improved decision making, notification, and human response. The "lead-time" of a hazard varies considerably across hazards. For example, hurricanes have a particularly long lead-time measured in days, whereas earthquakes have little lead-time, often measured in a few seconds at best. Onset of the hazard event must be able to be predicted with a degree of certainty, and within a relatively small time window, to be of use in decision making concerning response. Some hazards, such as earthquakes, can only be predicted to within decades or centuries. Although this is not considered to be a traditional warning system, it can be very useful to inform other risk management strategies with enough time for changes, such as land use planning, to be implemented (Sorensen, 2000).

Hazards with warnings

A given warning system is often proposed and developed for a single natural hazard, despite the fact that most communities are exposed to multiple hazards. Warning systems and other risk management strategies should be considered in an all-hazard context wherever practical to achieve maximum community resilience (Paton, 2003). Only a few hazards regularly have no currently detectable precursors: earthquakes; dry landslides; landslide-induced tsunamis; and mud volcanoes, for example. Other hazards usually have detectable precursors, but in some cases such as small volcanic eruptions, the precursors are below detectable levels. Warnings based upon the onset of the hazard event itself, as opposed to a precursor, are theoretically possible for all hazards. These post-onset warning systems have typically short lead-times with the extreme case being earthquakes. Several countries (e.g., Mexico City, Japan, USA, and Taiwan) have implemented or are testing an earthquake warning system that gives seconds of lead-time once an earthquake is detected by issuing warnings after non-damaging seismic Primary waves are detected but in advance of the damaging Secondary and Surface waves (Cyranoski, 2004). This may provide sufficient time to shut down critical infrastructure such as trains or gas lines, or start alternate generators in hospitals and for people to drop, cover and hold. Such warnings are more likely to be useful with increasing distance from the epicenter, but only up to a distance where the intensity of ground shaking is potentially damaging. It is not very useful for near source areas because the time between the arrival of the Primary waves and subsequent damaging waves is too short.

Decision to warn

For a warning system to notify the public, a specific warning message needs to be generated and a notification disseminated. Issuing a warning notification requires that a criteria or threshold within monitored parameters must usually be reached. Such criteria often include geographic location and magnitude, or the expected intensity. Determination as to whether or not the criteria have been met must be rapid, but some decision time is nearly always needed and this should not be underestimated. The most effective warning system will minimize the frequency of warnings for events that either do not happen or are much smaller in magnitude or intensity than expected. Emergency managers are often overcautious of the potential to "cry wolf." However, research has shown that only repetitive false alarms will begin to reduce human behavioral response rates, as long as the reason for false alarms is disclosed by officials and understood by recipients (Dow and Cutter, 1998).

Interpretation of initial data for the purpose of evaluating whether an event will actually occur may have high uncertainty. As time progresses more data often become available; but the time until the impact of the event is also reduced during this time. There is a minimum time period before an event within which a warning must be given if it is to achieve the expected response actions, such as evacuation of a population. Setting criteria for the decision to warn involves a trade-off over time between decreasing uncertainty or error and decreasing the time remaining to respond once a warning is issued. These criteria influence critical response actions such as evacuation. Emergency managers often find the criteria particularly difficult to set. This can be compounded in cases where a decision is either "yes" (evacuate) or "no" (do not evacuate), because scientists usually often provide information to emergency managers in the form of probabilities. Decision support systems, with preplanned probability thresholds and critical timeframes, reduce the time required to make decisions and issue warnings or evacuations.

Dissemination of warning messages

The process of detecting and monitoring a hazardous event and disseminating a notification requires time, which must be allowed for when planning response times. Notifications are usually technological and sent through telephone, radio, television, etc., but warnings may also be perceived directly from natural processes in one's environment. With the correct prior knowledge, a wide range of natural phenomena can be interpreted by members of the public either as a warning of an imminent event or as a cue to simply be alert. These notifications are termed "natural warnings" or "environmental cues" and include earthquake-generated ground shaking preceding the arrival of a tsunami and heavy rainfall or a high rate of *river stage* rise preceding a flood. If the time between onset of precursor signs and a hazard is too short for official warnings to be issued or if a warning system does not exist or fails to disseminate a warning message, natural warnings or environmental cues may be the only warning (Gregg et al., 2007).

Warnings from one person to another can be classed as either "official" or "informal." Official warnings are those communicated by people acting in an official capacity to do so. They are preplanned and the most likely to be accurate. Informal warnings are warnings communicated by individuals not acting in an official capacity. They may increase the reach of, or occur in place of, official warnings (which may have failed or been too slow to be useful). Informal warnings are often diffused through a population in advance of official warnings (Sorensen and Mileti, 1989; Gregg et al., 2007). Because natural warnings or environmental cues sometimes provide the first and only warning of an event and informal warnings are often received before official warnings, it is significant that those developing official warning systems understand how receipt of different warnings and cues in time and space influences response.

Technological notification systems can be grouped into either "third party" systems that are usually used for a non-warning-related purpose or "dedicated" systems that exist for the specific purpose of warning notification, such as those used in industrial settings where hazardous material is handled. There are wide ranges of warning notification technologies available and new possibilities continuously emerge.

Third party systems include:

- Radio and television station broadcasts
- Break-in broadcast
- Landline telephone
- Mobile telephone/device (Short Message Service (SMS) text message, broadcast, multicast)
- Pagers
- Websites
- Inserted website banners
- E-mail
- Emergency responders and their hardware (e.g., route alert with car/appliance loud-hailers)
- GPS receiver messaging
- Power-line messaging
- Loudspeakers (e.g., those separately used for public address or prayer calls)

Dedicated systems include:

- Tone-activated alert radio
- Private radio
- Sirens
- Flares and explosives.

The advantage of third party systems is that they are often self-maintaining and self-testing through regular use (e.g., telephones). Dedicated notification systems generally have a higher rate of failure, unless very regularly tested in realistic conditions and designed with redundancy (Gruntfest and Huber, 1989). Notifications that catch a person's attention even though they are focused on something else (e.g., telephone or siren) are more effective than those where the target person will only get the message if they are actively using the notification pathway (e.g., watching television, viewing a website). A person's location and activity changes throughout the day, so one notification technology that might reach them at work might not reach them while they are recreating or asleep at home. Warnings at night are also more slowly disseminated and acted upon (Sorensen, 2000). Transient populations, especially tourists, may need notification via different means than local residents. This can be accentuated from season to season as tourists visit coastal areas in summer months or snowy mountains in winter months.

Weather conditions can alter the effectiveness of a system. Loud-hailers are harder to hear in wind, and cold weather may mean more people are inside insulated from outdoor notification systems. Mobile technologies

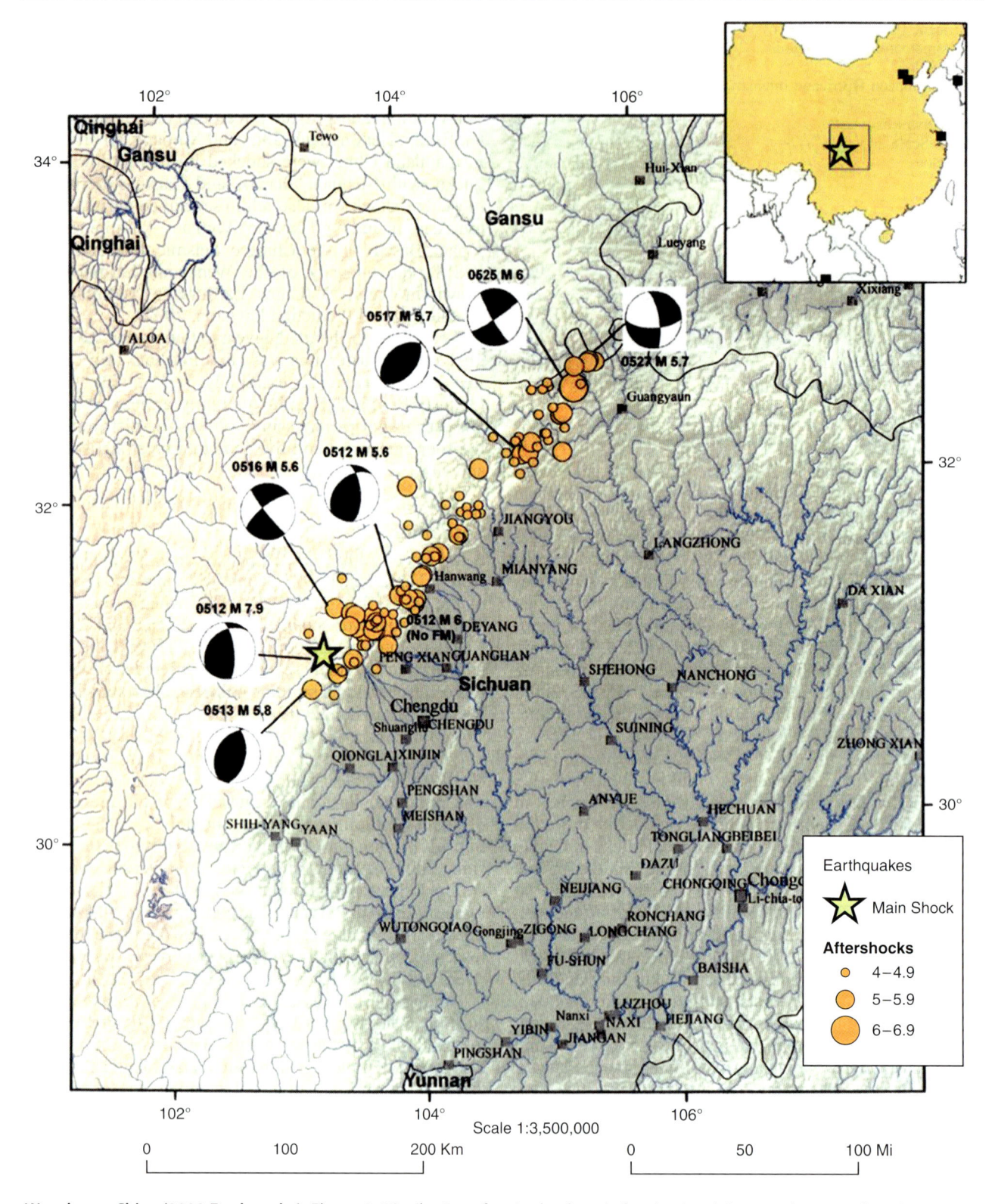

Wenchuan, China (2008 Earthquake), Figure 1 Distribution of main shock and aftershocks of the Wenchuan Earthquake, M - moment magnitude, preferably being used for M8+ shocks (After USGS, 2008; China Earthquake Administration Monitoring and Prediction Division, 2009).

Wenchuan, China (2008 Earthquake), Figure 2 Epicenter of the Wenchuan Earthquake in China.

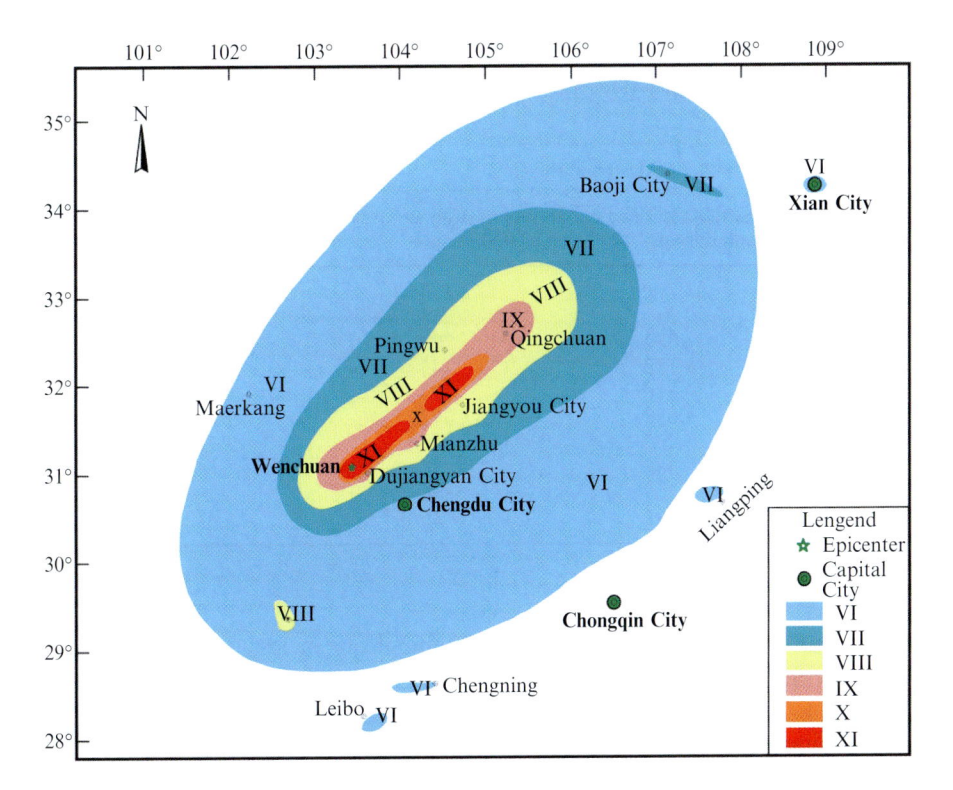

Wenchuan, China (2008 Earthquake), Figure 3 CSIS-based intensity map of the Ms 8.0 Wenchuan Earthquake (China Earthquake Administration Monitoring and Prediction Division, 2009).

Wenchuan, China (2008 Earthquake), Table 1 Demographics of each category of the Wenchuan Earthquake severity (National Disaster Relief Commission, 2008)

Category	Number of districts and counties affected	Area (km^2)	Population ($\times 10,000$)
Extremely severe	10 in Sichuan	26,000	365
Severe	26 in Sichuan	61,473	1,332
	7 in Gansu	20,293	190
	3 in Shaanxi	8,480	97
Heavily affected	103 in Sichuan	383,615	About 8,600
	37 in Shaanxi,		
	33 in Gansu,		
	10 in Chongqing,		
	5 in Ningxia, and		
	3 in Yunnan		
Slightly affected	180	10,410	About 8,100
Total	417	510,271	About 18,684

The population in the heavily and slightly affected areas was estimated

Wenchuan, China (2008 Earthquake), Table 2 Casualty statistics by provinces (National Disaster Relief Commission, 2008)

Province	Sichuan	Gansu	Shaanxi	Chongqing	Henan	Yunnan	Guizhou	Hubei	Hunan
Dead	68,712	365	121	16	2	1	1	1	1
Missing	17,921	2							
Injured	360,341	10,158	2,948	637	7	51	15	14	

Wenchuan, China (2008 Earthquake), Table 3 Casualties and building collapses in the category of "extremely severe" (National Disaster Relief Commission, 2008)

		Number of dead and missing		Number of collapsed buildings	
County or City	Total population ($\times 10,000$)	Number of dead and missing	Number of dead and missing per 10,000 people	Number of building collapses	Number of collapses per 10,000 people
Wenchuan	11	23,871	2,170	608,198	55,291
Beichuan	16	20,047	1,253	347,856	21,741
Mianzhu	51	11,380	223	1,397,925	27,410
Shifang	43	6,132	143	1,006,921	23,417
Qinchuan	25	4,819	193	714,804	28,592
Maoxian	11	4,088	372	300,229	27,294
Anxian	50	3,295	66	774,896	15,498
Dujiangyan	61	3,388	56	655,265	10,742
Pingwu	19	6,565	346	299,557	15,766
Pengzhou	78	1,131	15	622,066	7,975

Earthquake prediction and preparation

Seismic zoning and earthquake preparation

China's Code for Earthquake Resistant Design of Buildings and the Seismic Zoning Map (State Seismological Bureau, 1992) have been put into force to set standards for the buildings and infrastructures all over the country, but adherence to these codes and regulations were not compulsory in less-developed areas until 2006 because of economic factors. Before this earthquake, these codes and regulations set the maximum seismic intensity at CSIS VI for all the extremely severe disaster areas and at CSIS VII for a few of the more potentially severe disaster areas (Shi and Li, 2001), because there had been no large earthquakes in this area before these codes was established.

According to these earthquake resistant design codes, infrastructure in areas of CSIS VI and below would be built with no earthquake-resistant design. Earthquake-resistant design was only mandatory in areas of CSIS VII and above. Therefore, seismic resistant capabilities of buildings designed by such standards would not have met the required strength to withstand such a shallow focused Ms 8.0 earthquake.

Earthquake prediction

Whether the Wenchuan Earthquake was predicted or not aroused many controversies and rumors in both international and Chinese societies. Unlike the 1976 Tangshan earthquake in China, the middle section of the Longmenshan faulting belt – the meizoseismal area of the Wenchuan Earthquake – had been considered inactive before this earthquake according to analyses of Chinese mainstream earthquake experts. Compared with the middle section of the Longmenshan faulting belt, the Xianshuihe faulting belt, the Anninghe-Zemuhe-Xiaojiang faulting belt, and the Songpan-Ganzi faulting zone in the southwest China had more tectonic activities, and attracted much more research attention; thus, many seismic monitoring measures and research efforts were applied to those areas. However, prior to the Wenchuan Earthquake, a few Chinese specialists and American experts did give a fairly accurate middle-term prediction for the possible earthquake, and some advice for future research. Unfortunately, such predictions and research findings were not accepted by the mainstream earthquake experts in China. Obviously, earthquake prediction is still not an exact science and remains an unsolved world-class challenge. A large gap still exists between the actual problem-solving ability of scientists and the need for community disaster preparation.

Emergency responses and relief measures

After the occurrence of the Wenchuan Earthquake, the Chinese government immediately began an emergency rescue and relief program (China Earthquake Administration, et al., 2010). Headquarters for victim rescue and disaster relief were established all around the affected areas.

According to statistics issued by the Headquarters at the State Council, all relevant ministries and organizations of the Chinese central government had allocated a large amount of resources and exerted many efforts during the disaster relief campaign. As of May 24, 2008, the Headquarters had mobilized 111,278 soldiers, 22,970 armed policemen, 24,401 special policemen, 4,000 specialized earthquake rescue workers, 88,341 medical workers, and 6,013 volunteers to help in the disaster areas. As of June 14, 2008, the central government and the local governments had allocated 1.31 million tents, 4.82 million quilts, 14.09 million articles of clothing, 1.04 million tons of liquid fuel, and 2.23 million tons of coal to the disaster area. As of August 4, 2008, the central government and the local governments had devoted 64.16 billion RMB to the disaster areas, and about 59.41 billion RMB was received from public and private donations for the disaster relief.

After the occurrence of the Wenchuan Earthquake, the international society expressed its sincere sympathy to the Chinese people through the central government, together with various kinds of support. As of July 18, 2008, the Chinese Ministry of Foreign Affairs and other Chinese embassies and diplomatic missions all over the world had received about 1.71 billion RMB in donations, including 0.77 billion RMB from foreign governments, international, and regional organizations; 1.99 million RMB from foreign embassies, diplomatic missions, and individuals in China; and 0.94 billion RMB from foreign civilian organizations, foreign enterprises and individuals, overseas Chinese, Chinese scholars and students living overseas, overseas branches of China-based companies and organizations, and other individuals.

Recovery and reconstruction after the earthquake

One month after the Wenchuan Earthquake, the Chinese central government began to implement recovery and reconstruction work by establishing basic guidelines (Wen, 2008). At the same time, the relevant ministries jointly issued specific instructions to direct the recovery and reconstruction work. In September 2008, the State Overall Planning for Post-Wenchuan Earthquake Restoration and Reconstruction (SOP) was issued (State Planning Group, 2008).

According to the statistics issued by the China Ministry of Finance, by May 4, 2009, the Chinese central government had supplied almost 85 billion RMB to the disaster area for the recovery and reconstruction. By the end of 2009, 160 billion RMB from the central government and 17 billion RMB from the provincial government were allocated to the affected areas in Sichuan province. In addition, a total of 250 billion RMB was available from other provincial governments through the "Twin Assistance" program, social donations, contributions from Hong Kong, Macao, and Taiwan, and rehabilitation funds from the county and city governments of the affected areas.

Following the SOP, 29,704 recovery and reconstruction projects were planned for Sichuan Province, the worst-struck province, including 1,500,000 rural dwelling houses, 259,000 urban dwelling houses, 3,002 schools, 1,362 medical and health institutes (including hospitals), and 38 towns and cities.

As of March 2010, 95% of these projects were being implemented, and 74% were completed; special efforts were made to care for the most vulnerable groups (China Earthquake Administration, et al., 2010).

Parallel to the Chinese central government's efforts, recovery projects assisted by international loans were also being implemented. The World Bank loaned a total of US $710 million to the Chinese central government for recovery and reconstruction projects in Sichuan and Gansu Provinces for a period of 5 years from March 2009 to March 2014 (World Bank, 2009). As of June 2010,

US$116 million had been disbursed for infrastructure, health, and educational projects (World Bank, 2010). One year later, the cumulative disbursement reached 35% of the total loan, and the progress on the project was moderately satisfactory (World Bank, 2011).

In March 2012, the Chinese central government announced that the SOP recovery and reconstruction tasks were completed; the basic working and living conditions and the level of economic and social development in the disaster areas were significantly better than that before the earthquake (NDRC, 2012).

Bibliography

China Earthquake Administration Monitoring and Prediction Division, 2009. *Scientific Research Report of Ms 8.0 Wenchuan Earthquake (in Chinese)*. Beijing: Earthquake Press.

China Earthquake Administration, Department of Civil Affairs of Sichuan Province, International Labor Organization, United Nations International Strategy for Disaster Reduction and the International Recovery Platform, 2010. *Wenchuan Earthquake 2008: Recovery and Reconstruction in Sichuan Province, Recovery Status Report (#04)*. International Recovery Platform Secretariat, Kobe.

National Disaster Relief Commission, 2008. *Comprehensive Analysis and Evaluation of the Wenchuan Earthquake Disaster (in Chinese)*. Beijing: Science Press.

NDRC, 2012. *Report on China's Economic and Social Development Plan*. National Development and Reform Commission, Beijing.

Shi, Z., and Li, Y., 2001. Seismic Zonation in China. *Engineering Sciences in China*, **3**(6), 65–68.

State Planning Group, 2008. *The State Overall Planning for Post-Wenchuan Earthquake Restoration and Reconstruction*. National Development and Reform Commission, State Council, Beijing.

State Seismological Bureau, 1992. *China Seismic Intensity Map (1:4,000,000) and Instructions (in Chinese)*. Beijing: Earthquake Press.

USGS, 2008. Earthquake Summary-Magnitude 7.9 Eastern Sichuan, China.

Wen, J., 2008. *Decree on Post-Wenchuan Earthquake Recovery and Reconstruction*. State Council, Beijing.

World Bank, 2009. *Loan Agreement (Wenchuan Earthquake Recovery Project) between People's Republic of China and International Bank for Reconstruction and Development*. Washington, DC.

World Bank, 2010. *Implementation Status & Results of China Wenchuan Earthquake Recovery Project (P114107)*. Washington, DC.

World Bank, 2011. *Implementation Status & Results of China Wenchuan Earthquake Recovery Project (P114107)*. Washington, DC.

Yuan, Y., 2008. Loss assessment of Wenchuan earthquake. *Journal of Earthquake Engineering and Engineering Vibration*, **28**(5), 10–19.

Cross-references

Casualties Following Natural Hazards
Earthquake
Earthquake Damage
Earthquake Prediction and Forecasting
Earthquake Resistant Design
Tangshan, China (1976 Earthquake)

WILDFIRE

Brigitte Leblon, Laura Bourgeau-Chavez
University of New Brunswick, Fredericton, NB, Canada

Synonyms

Burn; Bush fire; Fire; Forest fire; Natural fire; Wildland fire

Definition

A wildfire is an unplanned or unwanted natural or person-caused fire occurring in a natural setting or wilderness (Merrill and Alexander 1987: 44). A fire is a simultaneous release of heat, light, and flame generated by the combustion of flammable material (Merrill and Alexander, 1987: 12). It requires the so-called fire triangle elements: fuel, oxygen, and a heat source. Naturally occurring wildfires are less frequent than man-made wildfires, but usually burn large areas.

Types of wildfires

Several types of wildfires can be defined as a function of the fuel layer involved:

Type	Fuel layer	Remark
Ground or *subsurface*	Below the litter layer of the forest floor (duff, roots, buried punky wood, peat)	Difficult to detect
Crawling or *surface*	Low-lying vegetation (leaf and timber litter, debris, herbaceous vegetation, low-lying shrubbery)	Control depends on the type of fuels involved
Ladder	Between low-level vegetation and tree canopies (medium shrubs, tree seedlings, stumps, small trees, downed-dead logs)	Control depends on the type of fuels involved
Crown, canopy, or *aerial*	Suspended combustible material at the canopy level (tree crowns, vines, mosses)	Difficult to control because it depends on wind; Three types can be defined depending on whether it is dependent or independent of a surface fire
Spotting	Firebrands that are thrown ahead of the main fire by the wind	Difficult to control because it depends on wind

Fire regime

Fire regime refers to the general type of fire activity and pattern of burns that characteristically occur in a given region. Some important elements of fire regimes include fire frequency, fire intensity, type of fire, and burn severity (Merrill and Alexander, 1987). Fire regimes can range from low-intensity, high-frequency surface fires to high-intensity, low-frequency stand-replacing fires. The fire regime is determined by fire climate, lightning incidence, physiography, ecosystem type, and a balance between dry matter production and decay.

Fire risk, hazard, and danger

Ignition and spread of wildfires depends on four main types of factors (Leblon, 2005):

1. The state and nature of the fuel: proportion of live and dead vegetation, compactness, morphology, species, density, stratification, fuel arrangement, continuity of fuel, and moisture content;
2. The physical environment: weather conditions and topography
3. Causing factors: human or nature (lightning)
4. Fire prevention and suppression means

The first two factors define *fire hazard*, whereas the last two factors define *fire risk*. *Fire danger* includes both hazard and risk.

Fire danger rating systems

Wildfire is the one of the dominating disturbances in the world's forests and grasslands. Wildfire is part of the natural cycle. For some species, it is necessary to their reproductive cycle. For example, jack, pitch, red, yellow (ponderosa), and lodgepole pines, as well as sequoia and black spruce store for years live seed in their crown cones that are glued by resin. The resin will melt by heat from a fire, allowing the cone opening. Seed germination for red and white pines, white spruce and Douglas fir require ground that has been exposed by fire creating a good seed bed. Exposure to fire smoke promotes seed germination in some plants, like *Eragrostis tef (Zucc.) Trotter*, by inducing production of butenolide (3-methyl-2H-furo[2,3-c]pyran-2-one) (Ghebrehiwot et al., 2008). Despite their ecological benefits, wildfires can also become a threat to property, human life, and economy. For this reason, fire danger predicting systems have been developed for use in fire management, among others for fire suppression. Among others, there are the National Fire Danger Rating System (NFDRS) in USA (see the review of Hardy and Hardy, 2007) and the Canadian Forest Fire Danger Rating System (CFFDRS) in Canada (Van Wagner, 1987; Stocks et al., 1989; Taylor and Alexander, 2006). The CFFDRS is also used in Alaska and in some other parts of the world (see the review of Taylor and Alexander, 2006). It is a semiempirical modular system that has the structure shown in Figure 1.

The relative mid-afternoon fire potential is rated using the Fire Weather Index (FWI). It represents the intensity of a spreading fire. It is computed daily from noon standard time weather records (dry-bulb temperature, relative humidity, 10 m high open wind speed, 24-h precipitation), by one of the *CFFDRS* subsystems, the Fire Weather Index (FWI) system. It considers, as forest type, a generalized pine forest similar to the jack pine and lodgepole pine forest, but three fuel layers (Van Wagner, 1987; Stocks et al., 1989), each of them being characterized by its moisture content represented by a code: (1) *Fuel Moisture Code (FFMC)* for the 1–2 cm depth fine surface litter and other cured fine fuels (mosses, needles, small twigs) having a time lag of 2/3 day (16 h); (2) *Duff Moisture Code (DMC)* for the loosely compact duff of moderate depth (5–10 cm) having a time lag of 15 days; and (3) *Drought Code (DC)* for the deep (15–20 cm) compacted duff layer (organic matter of the soil, large logs) having a time lag of 52 days. FFMC determines the potential for fire ignition, DMC, the resistance to control of fire and the ability of the fire to spread, and DC, the seasonal drought, the smoldering in deep duff or large logs and the resistance to extinguishment and ability to mop-up. These fuel moisture codes are combined into two fire behavior indices: (1) *Initial Spread Index (ISI)* representing the relative fire spread expected immediately after ignition and (2) *Build-Up Index (BUI)* indicating the total amount of fuel available for combustion by a moving flame front. It is used in pre-suppression planning, indicates fire's resistance to control. *ISI* and *BUI* are combined together to produce *FWI* (see Figure 2).

Another *CFFDRS* subsystem, the *Fire Behavior Prediction (FBP) system*, predicts the fire behavior in 17 specific fuel types (Stocks et al., 1989; Forestry Canada Fire Danger Group, 1992; Wotton et al., 2009).

Use of remote sensing in wildfire management

The availability of remote sensing technology coupled with the development of geostatistics and spatial analyses using geographic information technology allows moving fire danger rating from point-based estimates based on weather stations to spatially explicit estimates. Indeed, remote sensing has the advantages of larger sampling areas, lack of destruction of the studied resource, gathering data on less accessible areas and is representing, in essence, the integrated response of vegetation to environmental influences. Such a technology can be first used for fire danger monitoring. The first fire danger variable that can be derived from remote sensing is the fuel type, which can be mapped from high spatial resolution optical or radar images, as in classical land use (see the review of Leblon, 2005). Such maps can then be linked within a wildfire threat system, to other fire danger parameters, such as topography, proximity to roads and to urban areas (see the review of Leblon 2005). Fuel moisture is another fire danger parameter, which can be estimated by remote

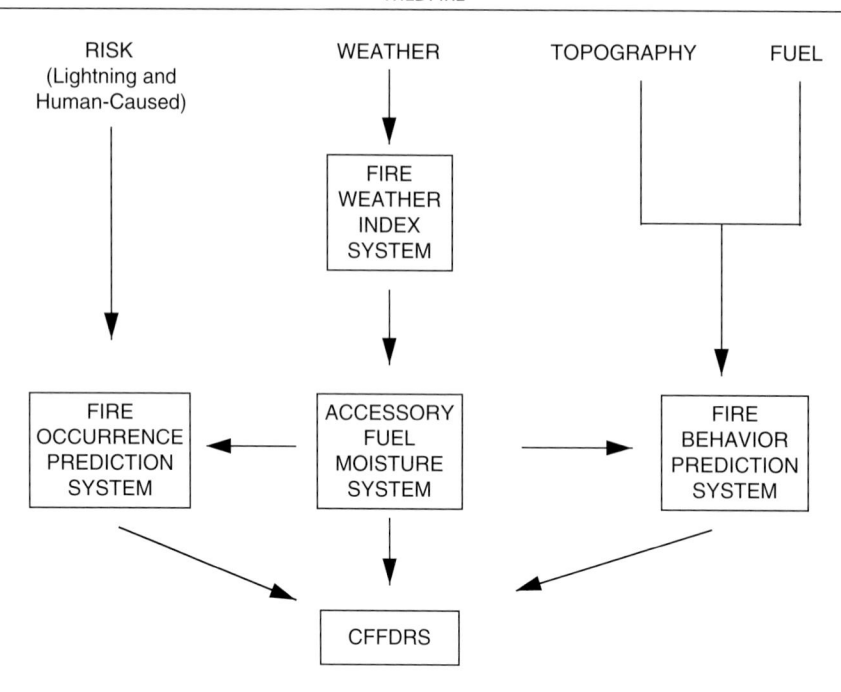

Wildfire, Figure 1 The Canadian Forest Fire Rating System (CFFDRS) (After Stocks et al., 1989).

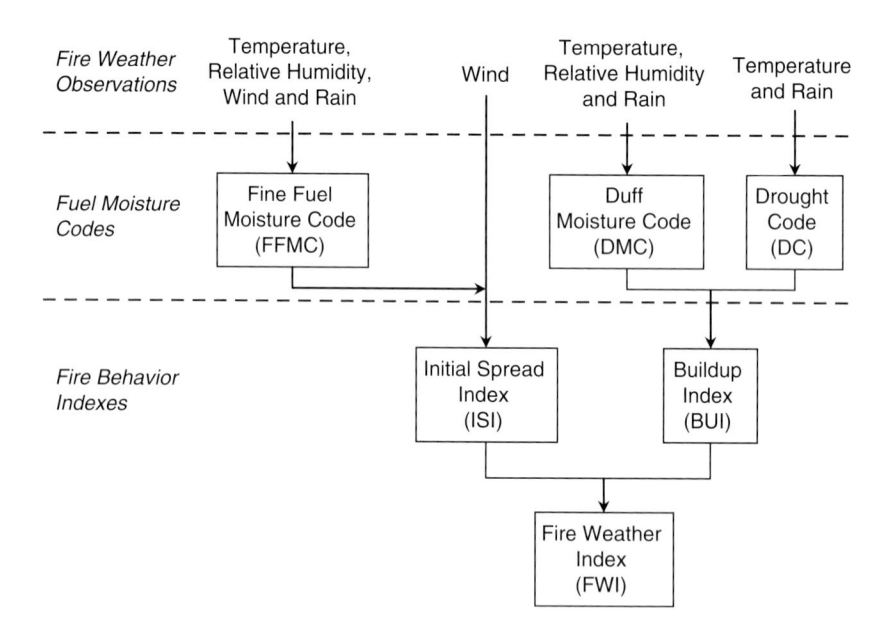

Wildfire, Figure 2 The Fire Weather Index of the CFFDRS (Adapted from Van Wagner, 1987).

sensing. Remote sensing–based fuel moisture monitoring has been done primarily using indices derived from NOAA-AVHRR normalized difference vegetation index (NDVI) images, like the *relative greenness* (Burgan et al., 1998) or the *Vegetation Condition Index* (Kogan, 1990). They were also used to estimate the degree of curing, which is an important fuel behavior variable for herbaceous fuel type (e.g., Paltridge and Mitchell, 1990). Thermal infrared data were better correlated than NDVI to fire danger codes (see the review of Leblon, 2005; Oldford et al., 2003; Oldford et al., 2006; Leblon et al., 2007) and to foliar moisture content (see the review of Leblon, 2005). There were also studies using indices that combine NDVI and thermal infrared NOAA-AVHRR images, like the *Vegetation and Temperature Condition Index* of Kogan (2001) and the index of Chuvieco et al. (2001). However, both *NDVI* and thermal infrared images have limited image availability during cloudy days. By contrast, active microwave (or radar or Synthetic Aperture Radar (*SAR*)) images can be acquired in any weather conditions. Good correlations were obtained between *CFFDRS* fuel moisture codes and backscatter images from *ERS-1* (Bourgeau-Chavez et al., 1999; Leblon et al., 2002) or *RADARSAT-1* (Abbott et al., 2007) single-polarized SAR images acquired over boreal forests. The operational use of *SAR* images in fuel moisture monitoring is limited by the influence of interfering factors, including the type, species, structure, and biomass of the vegetation, as well as the topography and surface roughness (see the review of Leblon, 2005). It has been shown that polarimetric *SAR* images are better for estimating the surface roughness independently from the soil moisture and vice versa in the case of agricultural bare soils or crop canopies (Mattia et al., 1997; Hajnsek et al., 2003; Hajnsek et al., 2009). Polarimetric *SAR* images are provided by several existing satellites, for instance, *RADARSAT-2* and *ALOS-PALSAR*. These images are currently being tested for fuel moisture mapping in boreal forests of the interior Alaska (Bourgeau-Chavez et al., 2009). Fuel moisture mapping is not only needed for assessing fire danger probabilities, but also for post-fire regeneration assessment, because vegetation regrowth highly depends on soil moisture availability.

Remote sensing can also be used for detecting active fires. Such detection is based on the detection of either a smoke plume produced by fire emissions, hot surface temperatures above normal environmental temperatures, or hot surface temperatures with respect to the surroundings. Detection of smoke plumes is achieved by optical sensors that are sensitive to atmospheric interferences. The detection of hot surface temperatures due to wildfires is done first using satellite sensors (*NOAA-AVHRR*, *MODIS*) operating in the mid-infrared (3–5 μm) spectral window. Indeed, as shown in Figure 3, wildfires have surface temperatures around 1,000 K leading to a peak of the spectral radiant exitance around 3 μm. By comparison, the Earth has a surface temperature of 300 K (leading to a maximum of exitance around 9.7 μm) and the Sun has a surface temperature of 6,000 K (leading to a maximum of exitance around 0.5 μm). The thermal infrared (8–12 μm) spectral window can also be used, but confusion with the Earth exitance peak can occur. A detailed description of the methods used for detecting active fires on satellite images can be found in the review by San-Miguel-Ayanz et al. (2005).

Finally, remote sensing technology can also be used to map burn scars, either from optical images, such as LANDSAT-TM or SPOT-HRV or from *SAR* images. Burned areas are mapped on the image either as spectrally homogeneous areas that are distinct from the surroundings or by comparison of prefire and postfire satellite imagery. In the optical and mid-infrared spectral bands, burned areas have usually a significant low spectral reflectance. Based on these observations, Fraser et al. (2000) designed the *Hotspot And NDVI Differencing Synergy (HANDS)* algorithm that uses *NOAA-AVHRR NDVI* and mid-infrared images to map burn scars. In their algorithm, the burned area mapping is helped by identifying potential active fires based on satellite-detected hot spots. However, similarly for other remote sensing applications, the use of optical and thermal infrared images is limited, when fire smokes or clouds mask the observed areas. For this reason, the use of single-polarized *SAR* imagery has been successfully tested both in the case of Mediterranean forests (Gimeno and San-Miguel-Ayanz, 2004a, b) and boreal forests (Bourgeau-Chavez et al., 1997). Both the optical and *SAR* images are also used for postfire regeneration assessment.

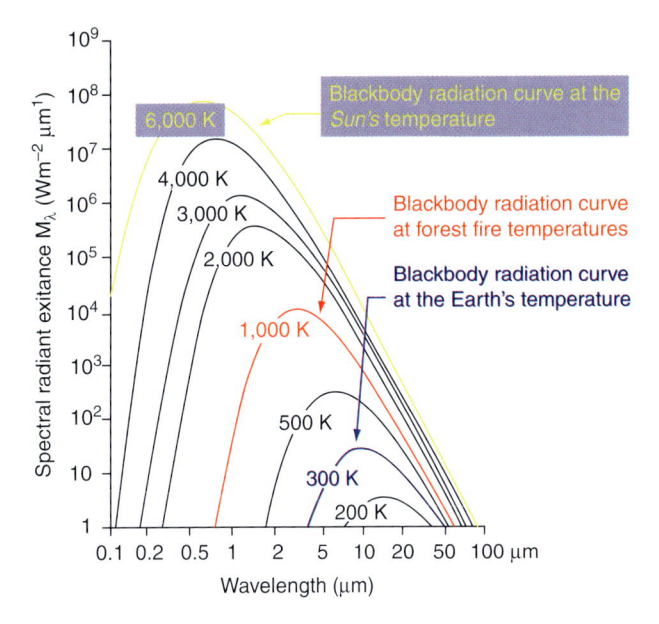

Wildfire, Figure 3 Spectral distribution of the emitted radiation of black bodies computed by Planck's law.

Summary

A fire is a simultaneous release of heat, light, and flame generated by the combustion of flammable material. The fire is a wildfire, when it corresponds to an unplanned or unwanted natural or person-caused fire, occurring in a natural setting or wilderness. Types of wildfires include ground, surface, ladder, crown, and spotting fires. Although wildfires have an ecological benefit for some species, they can also become a threat to property, human life, and economy. For this reason, fire danger predicting systems, including the National Fire Danger Rating System (NFDRS) in USA and the Canadian Forest Fire Danger Rating System (CFFDRS) in Canada, have been developed for use in fire management, among others in fire suppression. Fire danger is determined as a function of the state and nature of the fuel, of the physical environment, of the causing factors, and of the fire prevention and suppression means. With the availability of remote sensing technology coupled with the development of geostatistics and spatial analyses using geographic information technology, fire danger rating systems have been moved from point-based estimates from weather stations to spatially explicit estimates. Remote sensing can be used to map fuel types and fuel moisture, as well as to detect active fires and to map fire scars and post-fire regeneration.

Bibliography

Abbott, K., Leblon, B., Staples, G., Alexander, M., and MacLean, D., 2007. Fire danger monitoring in a northern boreal forest region using RADARSAT-1 imagery. *International Journal of Remote Sensing*, **28**(5–6), 1317–1338.

Bourgeau-Chavez, L. L., Harrell, P. A., Kasischke, E. S., and French, N. H. F., 1997. The detection and mapping of Alaskan wildfires using a spaceborne imaging radar system. *International Journal of Remote Sensing*, **18**(2), 355–373.

Bourgeau-Chavez, L. L., Kasischke, E. S., and Rutherford, M. D., 1999. Evaluation of ERS SAR data for prediction of fire danger in a boreal region. *International Journal of Wildland Fire*, **9**(3), 183–194.

Bourgeau-Chavez, L. L., Leblon, B., Charbonneau, F., Buckley, J., 2009. *Use of RADARSAT-2 polarimetric SAR images for fuel moisture mapping over boreal forests in interior Alaska*. Interim Report to the Canadian Space Agency, June 2009, 50 p.

Burgan, R. E., Klaver, R. W., and Klaver, J. M., 1998. Fuel models and fire potential from satellite and surface observations. *International Journal of Wildland Fire*, **8**(3), 159–170.

Chuvieco, E., Aguado, I., Cocero, D. Riaño, D., 2001. Design of an empirical index to estimate fuel moisture content from NOAA-AVHRR analysis in forest fire danger studies. In Proceedings 3rd International Workshop of the European Association of Remote Sensing Laboratories (EARSel) on Remote Sensing and GIS applications to Forest Fire Management, Ghent, Belgium, pp. 36–39.

Forestry Canada Fire Danger Group, 1992. Development and structure of the Canadian Forest Fire Behavior Prediction System. Forest Canada, Ottawa, Ont., Inf. Rep. ST-X-3, 63 p.

Fraser, R., Li, Z., and Cihlar, J., 2000. Hotspot and NDVI differencing synergy: a new method for mapping fire scars. *Remote Sensing of Environment*, **74**, 362–375.

Ghebrehiwot, M., Kulkarni, M. G., Kirkman, K. P., and Van Staden, J., 2008. Smoke-water and a smoke-isolated butenolide improve germination and seedling vigour of Eragrostis Tef (Zucc.) Trotter under high temperature and low osmotic potential. *Journal of Agronomy and Crop Science*, **194**(4), 270–277.

Gimeno, M., and San-Miguel-Ayanz, J., 2004a. Evaluation of RADARSAT-1 data for identification of burnt areas in Southern Europe. *Remote Sensing of Environment*, **92**, 370–375.

Gimeno, M., and San-Miguel-Ayanz, J., 2004b. Identification of burnt areas in Mediterranean forest environments from ERS-2 SAR time series. *International Journal of Remote Sensing*, **25**(22), 4873–4888.

Hajnsek, I., Pottier, E., and Cloude, S. R., 2003. Inversion of surface parameters from polarimetric SAR. *IEEE Transactions on Geoscience and Remote Sensing*, **41**(4), 727–744.

Hajnsek, I., Jaghuber, T., Schon, H., and Papathanassiou, K., 2009. Potential of estimating soil moisture under vegetation cover by means of PolSAR. *IEEE Transactions on Geoscience and Remote Sensing*, **47**(2), 442–454.

Hardy, C. C., and Hardy, C. E., 2007. Fire danger rating in the United States of America: an evolution since 1916. *International Journal of Wildland Fire*, **16**, 217–231.

Kogan, F. N., 1990. Remote sensing of weather impacts on vegetation in non homogeneous areas. *International Journal of Remote Sensing*, **11**, 1405–1419.

Kogan, F. N., 2001. Operational space technology for global vegetation assessment. *Bulletin American Meteorology Society*, **82**(9), 1949–1964.

Leblon, B., Kashike, E. S., Alexander, M. E., Doyle, M, and Abbott M., 2002. Fire danger monitoring using ERS_1SAR images over northern boreal forests. *Natural Hazards*, **27**, 231–255.

Leblon, B., 2005. Using remote sensing for fire danger monitoring. *Natural Hazards*, **35**(3), 343–359.

Leblon, B., Fernandez-Garcia, P. A., Oldford, S., MacLean, D., and Flannigan, M., 2007. Using NOAA-AVHRR cumulative indices for estimating fire danger codes in northern boreal forests. *International Journal of Applied Earth Observations and Geoinformation*, **9**, 335–342.

Mattia, F., Le Toan, T., Souyris, J.-C., De Carolis, C., Floury, N., Posa, F., and Pasquariello, N. G., 1997. The effect of surface roughness on multifrequency polarimetric SAR data. *IEEE Transactions on Geoscience and Remote Sensing*, **35**(4), 954–966.

Merrill, D. F., and Alexander, M. E., 1987. Glossary of forest fire management terms, 4th edition, National Research Council of Canada, Canadian Committee on Forest Fire Management, Ottawa, Ontario, Publication NRCC n°26516, 91 p.

Oldford, S., Leblon, B., Gallant, L., and Alexander, M., 2003. Mapping pre-fire conditions in the Northwest Territories, Canada, using NOAA-AVHRR images. *Geocarto International*, **18**(4), 21–32.

Oldford, S., Leblon, B., MacLean, D., and Flannigan, M., 2006. Predicting slow drying fuel moisture codes using NOAA-AVHRR images. *International Journal of Remote Sensing*, **27**(18), 3881–3902.

Paltridge, G. W., and Mitchell, R. M., 1990. Atmospheric and viewing angle correction of vegetation indices and grassland fuel moisture content derived from NOAA-AVHRR. *Remote Sensing of Environment*, **31**, 121–135.

San-Miguel-Ayanz, J., Ravail, N., Kelha, V., and Ollero, A., 2005. Active fire detection for fire emergency management: potential and limitations for the operational use of remote sensing. *Natural Hazards*, **35**(3), 361–376.

Stocks, B. J., Lawson, B. D., Alexander, M. E., Van Wagner, C. E., McAlpine, R. S., Lynham, T. J., and Dubé, D. E., 1989. The Canadian Forest Fire Danger Rating System: an overview. *Forestry Chronicle*, **65**(6), 450–457.

Taylor, S. W., and Alexander, M. E., 2006. Science, technology, and human factors in fire danger rating: the Canadian experience. *International Journal of Wildland Fire*, **15**, 121–135.

Van Wagner, C. E., 1987. Development and structure of the Canadian Forest Fire Weather Index System. Canadian Forest Service, Ottawa, Ont., For. Tech. Rep. 35, 37 p.

Wotton, B. M., Alexander, M. E., and Taylor, S. W., 2009. Updates and revisions to the 1992 Canadian Forest Fire Behavior Prediction System. Natural Resources Canada, Great Lakes Forestry Centre, Sault Ste. Marie, Ontario, Info. Rep. GLC-X-10, 52 p.

Cross-references

Airphoto and Satellite Imagery
Drought
Electromagnetic Radiation (EMR)
Fire and Firestorms
Forest and Range Fires (Wildfires)

Geographic Information Systems (GIS) and Natural Hazards
Geographic Information Technology
Hazard and Risk Mapping
Landsat (Satellite)
Lightning
Monitoring and Prediction of Natural Hazards
Remote Sensing of Natural Hazards and Disasters

WORLD ECONOMY, IMPACT OF DISASTERS

Ilan Noy
University of Hawaii, Manoa, Honolulu, HI, USA
Victoria Business School, Wellington, New Zealand

Definition

Human and economic catastrophes that are associated with natural hazards are by no means new. Recent very larger events, such as the Indian Ocean tsunami of 2004 or the 2005 Kashmir earthquake on the de facto Pakistan-India border, have also been experienced across national boundaries. Most of our current and rapidly evolving understanding regarding their relevance to economic dynamics is focused on the domestic economic and socioeconomic impacts (Cavallo and Noy, 2009, survey this literature). Here, the focus is on the international dimension of such economic and socioeconomic impacts, specifically the channels through which a disaster in one region/country can have an impact elsewhere.

Introduction

Barro (2006) has shown that the occurrence of infrequent economic disasters has a much larger welfare cost than continuous economic fluctuations of lower amplitude. For less developed countries, which suffer from a larger propensity to disasters of all types, and of disasters of larger magnitude than advanced economies, such events have an even greater effect on the welfare of the average citizen. These very large events are also the ones that are bound to have repercussions beyond the national borders of the affected countries.

In analyzing the economics of disasters, one typically distinguishes between direct damages, indirect damages, and secondary effects. Direct damage is the damage to fixed assets and capital, to raw materials and extractable natural resources and, of course, the mortality and morbidity that are a direct consequence of the natural hazards. Indirect impacts may be caused by the direct damage to physical infrastructure and productive capacity, or be a consequence of the fact that reconstruction pulls resources away from normal production. Such indirect damages also include the additional costs that are incurred because of the need to use alternative and less efficient means of production and/or spend dire resources for the provision of basic goods and services. At the household level, indirect costs include the loss of income resulting from the non-provision of goods and services or from the destruction of previously used means of production.

The "secondary effects" are in essence the aggregate measures of the secondary impact of the indirect effects, that is, the effects on other geographical regions, or at a later date, which originate from the indirect impact on the economy. In these secondary effects, one can identify the potential impact of disasters on the world economy, and the latter's effect on the secondary impacts of the disaster. These can, potentially, be even more severe than the initial negative disaster shock.

The secondary consequences of disasters can affect or be affected by other economies through two channels: (1) trade in both goods and services, and (2) flow of capital as aid, debt, or investment. Each of these channels is discussed in turn. An additional potential channel is migration. Disasters, however, have rarely generated large-scale migration; it is likely that only slow climate-related environmental changes, such as extended drought or desertification, will result in future migrations, but rapid-onset disasters most likely will not have such effects. Disasters, however, do frequently lead to internal displacement.

The only publicly available dataset on the economic impact of disasters worldwide is the Emergency Events Database (EM-DAT), maintained by the Center for Research on the Epidemiology of Disasters (http://www.emdat.be). The EM-DAT data reports on the number of people killed, the number of people affected, and the dollar amount of direct damages for each recorded disaster. The amount of damage reported in the database, however, consists only of direct damages (e.g., immediate damage to infrastructure, crops, and housing) and does not include either indirect or secondary damages. Thus, much of the following discussion is speculative, and the few research projects that have attempted to shed a more rigorous light on this issue rely on statistical analysis that is not always robust to permutations and changes in underlying assumptions. Much work remains to be done.

The often-cited data from EM-DAT appears to suggest that the frequency of disasters has been increasing rapidly in the last couple of decades (see, e.g., Figure 1 in Strömberg, 2007). However, much of the apparent increased incidence of disasters is due to an increase in the number of reported "small events." Figure 2 shows the frequency of large events with no apparent trend over time. Whether the reason for the increased frequency in the number of reported disasters is because small events are happening more frequently or because they are reported more often remains an open question.

Disaster fatalities are not distributed evenly across geographical regions, with a large majority occurring in the Asia-Pacific region. The Asia-Pacific region is an important region for the global economy: It includes more than 60% of the world population and more than 60% of world fatalities from disasters; it trades heavily in both goods and services (trade is 64% of the region's GDP, whereas the comparable figure for the United States is 25%); and it is

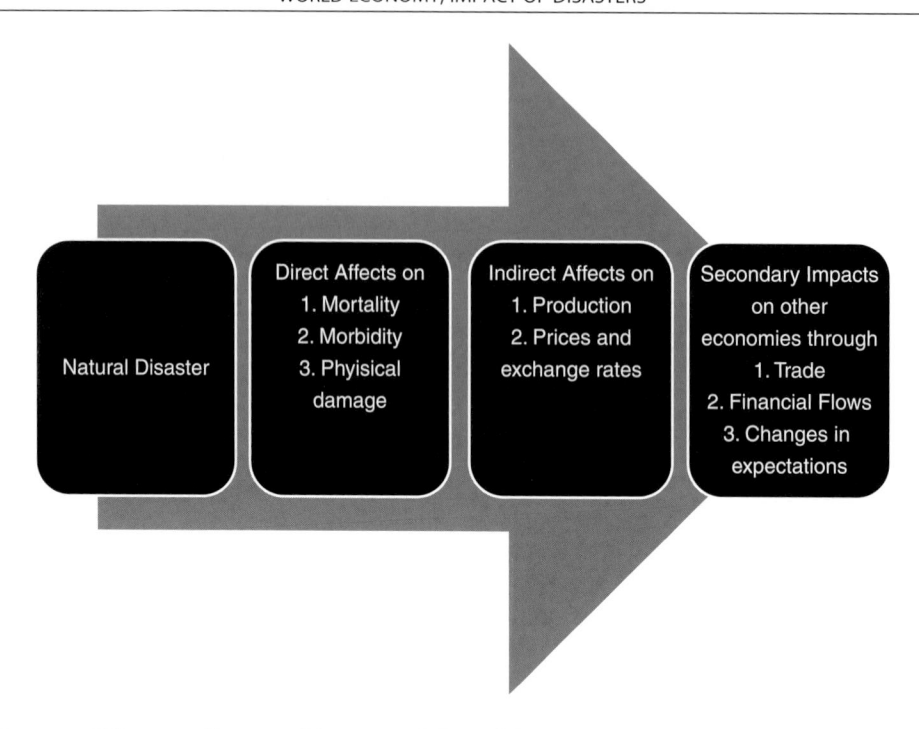

World Economy, Impact of Disasters, Figure 1 Disasters and their global impacts.

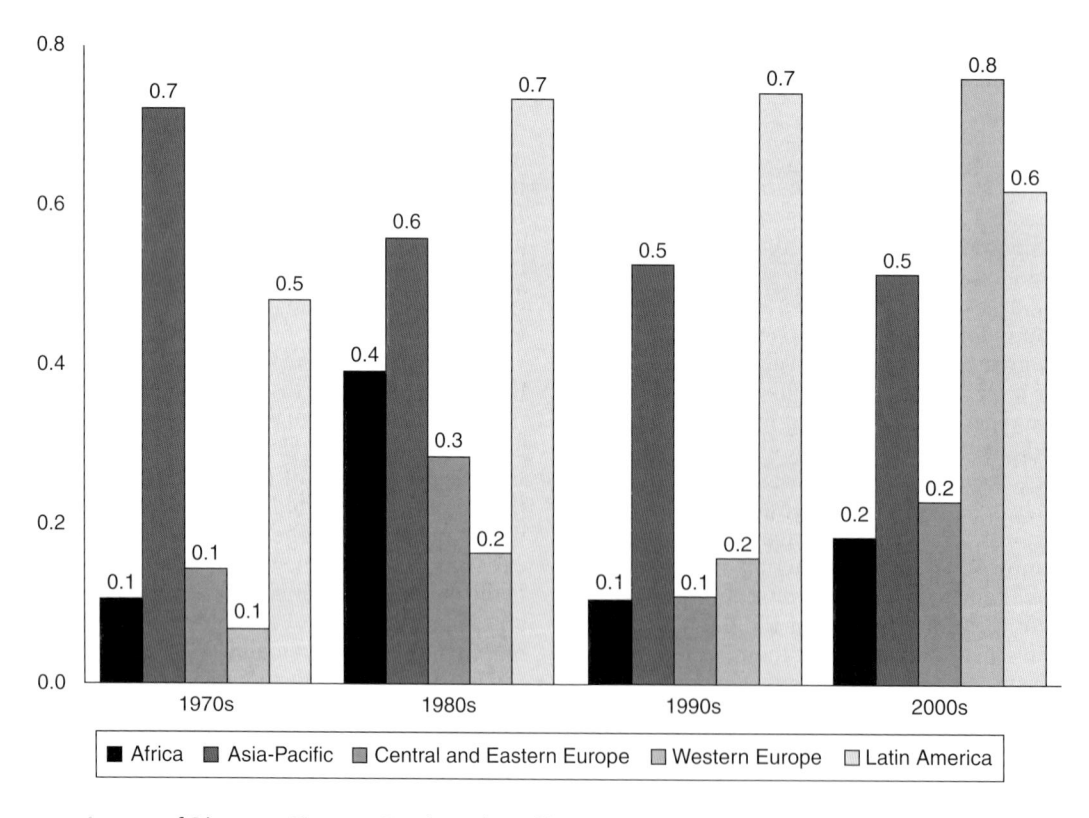

World Economy, Impact of Disasters, Figure 2 Total number of large disasters per country by region. Note for figure: 2000s measures include data up to 2008 and were adjusted for comparability. For more details see Cavallo and Noy (2009) (Data was taken from the EM-DAT database).

the largest recipient of capital flow outside the richest countries, and increasingly also a source of capital outflow to other regions (especially Africa).

Trade

The initial disaster impact leads to mortality, morbidity, and loss of physical infrastructure (housing, roads, telecommunication and electricity networks, and other infrastructure). The main issue here is how these initial direct effects lead to changes in trade patterns. There is some evidence that the trade deficit increases in the aftermath of a disaster. A disaster is unlikely to generate increases in output; it is also likely that a decrease in output will result in decrease in exports. Service exports, especially tourism, are uniquely vulnerable and will likely decrease in a post-disaster period. However, there may be a boost to exports if the disaster causes the local currency to depreciate. Currency depreciation decreases the price of exports and can therefore lead to an increase in their volume. Because of these contradictory effects, empirical work has not yet reached any conclusions regarding the "export channel."

There is some evidence that post-disaster reconstruction leads to an increase in import. Imports, especially of machinery and reconstruction materials, but also potentially of non-aid foodstuffs, are all likely to increase in the post-disaster reconstruction period. In that case, a disaster may have potential benefits for trading partners since demand for their products will increase. This is especially true if there is no discernible impact on the exchange rate that negates some of the increase in demand.

The possibility of major disasters in oil-exporting regions, such as the North Sea or the Arabian Gulf, is another concern that can have a dramatic impact on the world economy. Previous crises that led to a decrease in oil supplies for export have led to dramatic increases in oil prices, and to contractions in economic activity (e.g., 1973 or 1979–1980). A major disaster event in a large oil-exporting region is likely to have a similar detrimental impact on the world economy.

Noy (2009) concludes that countries that are more open to trade (measured by the amount of trade they conduct, relative to the size of their economies), but that have less open capital accounts, are better able to withstand the initial disaster shock and prevent further spillovers. This may be because countries that have been more open to trade historically may find it easier to absorb and finance the additional imports that are normally necessary for reconstruction. The reason why countries that are less open to capital flows may be less vulnerable is less obvious.

Capital flows

As for trade flows, one can identify several ways in which disasters may have an impact on the flow of financial capital and vice versa. One obvious impact is that a *large* disaster generates inflows of official and private aid. In some cases, these aid flows can be quite sizable, especially if the disaster has been widely reported in the international media (see Eisensee and Strömberg, 2007). For example, the 2004 South-East Asian tsunami led to an initial inflow of $US 6 billion in foreign aid to the affected countries (Economist Intelligence Unit, 2005). There is some evidence that, as happened in this case, aid inflows are useful in reducing the indirect damage to the national economy that may result from a disaster. This is unlike the very mixed evidence as to the effect of aid flows at "normal" times.

Beyond an increase in post-disaster aid, it is less clear what will be the effects of a disaster on other types of capital flows. Official lending may also increase as a form of aid to finance reconstruction; however, private lending may decrease if the disaster has destroyed infrastructure, so that the productive capacity of the country/region decreases. Other types of investment such as direct foreign investment – both in the form of purchases of existing productive assets and in new "green-field" investment creating new production facilities – may increase or decrease, depending on the actual range of damages and their implications for economic opportunities in the affected regions. The same can be expected for capital outflows. Whereas domestic residents may see new opportunities as a result of the destruction, the disaster may also generate capital flight if it changes perceptions regarding the likelihood of future similar events.

Yang (2008) and Bluedorn (2005) have both attempted to investigate the evolution of capital flows following disasters empirically, and both conclude that disasters generate some inflow. Bluedorn finds that, in the case of Caribbean hurricanes, the post-disaster inflow is quite large (5% of GDP). Yang, using a broader dataset, argues that the magnitude of these inflows is relatively small. He observes that international aid and remittances to poorer countries increase, whereas other types of flows are rarely affected in empirically observable magnitudes. He also finds that there is some increase in private capital outflows from the affected countries. This reported increase in capital outflows may be the reason why a closed capital account, and specifically limitations on capital outflows, may enhance an economy's resiliency in the face of a natural hazard.

Given the beneficial impact of foreign capital in financing post-disaster reconstruction and assistance, it is not surprising that governments may be interested in insuring against disaster onset, in order to guarantee this flow of funds. The simplest way to guarantee the availability of post-disaster reconstruction funding is with precautionary savings. However, Borensztein et al. (2008) argue that, in the case of less developed countries exposed to large natural disasters, insurance – or debt contracts with insurance-like features – provides an attractive alternative to a fiscal rainy-day funds. To illustrate this, they examine the vulnerability of Belize's public finance to the occurrence of hurricanes and the potential impact of insurance instruments on reducing that vulnerability. Through numerical simulations they show that the provision of catastrophic risk insurance has the potential to significantly improve Belize's debt sustainability.

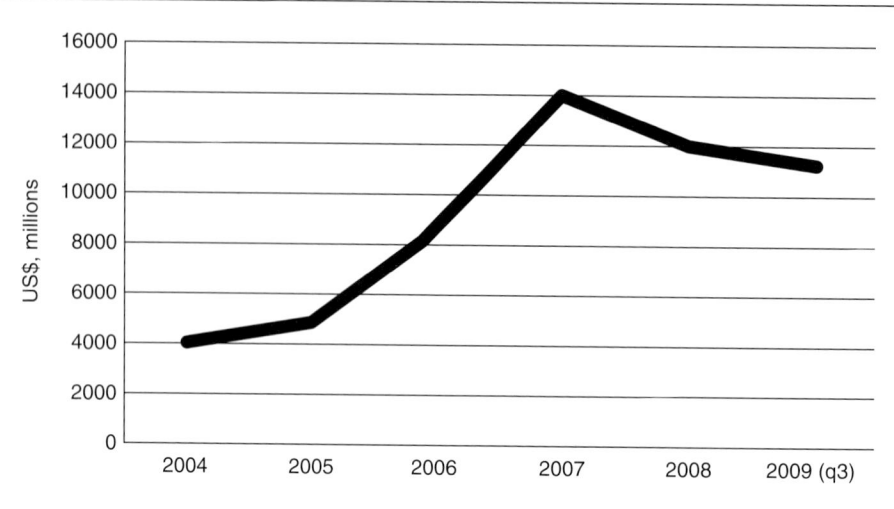

World Economy, Impact of Disasters, Figure 3 Cat bonds outstanding.

Implementing disaster insurance faces two types of obstacles: paucity of markets and political resistance (Kunreuther and Pauly, 2009). For a number of reasons, certain markets have traditionally been insufficiently developed, whereas elsewhere they are simply nonexistent. Recently, however, advances such as the development of parametric insurance policies, in which payments are based on externally verifiable disaster magnitude indices (such as the Richter measure for earthquakes), have the potential of increasing the availability of insurance. Political reluctance to engage in insurance purchase derives from the fact that there is little benefit for a political leadership with short-term interests from entering into insurance contracts. Insurance involves immediate costs and a possible payoff in an undetermined future when the government may already have changed hands (Healy and Malhotra, 2009). Since disasters are natural phenomena, and politicians cannot be blamed for their occurrence, the incentives to take relatively complex measures, such as insurance, to offset some of the costs are indeed weak. Political incentives to invest in *ex ante* mitigation are even weaker if the perceived likelihood of the disaster is relatively small.

In less developed countries, an additional obstacle to any *ex ante* preparation that includes the provision of insurance is an inadequate institutional framework – this relates to low government policymaking capabilities, opaque and potentially corrupt management practices of public assets, and systems that may be unable to provide efficient post-disaster disbursing of funds.

Catastrophe debt instruments (cat bonds) are another recently introduced means to transfer risks to the international capital markets. The main difference between a cat bond and a plain-vanilla bond is that if a prespecified disaster strikes, the creditor's claim is extinguished and the issuer's debt is essentially erased. The use of cat bonds is increasing rapidly, as can be seen in Figure 3, using data from GC Securities. However, it is not clear whether the increased use of these assets is indeed motivated by a demand for insurance or is merely speculative.

Conclusion

It seems clear that a large disaster, such as the Indian Ocean tsunami of 2004, will also cause people to change their evaluation of the probability of a future disaster occurring. In Hawaii, for example, the fear of a tsunami has increased markedly since 2004, but whether that has led to behavioral changes is still an open question. Disasters, thus, may have an impact on the global economy, not only because of what they do, but also because of the expectations they modify.

Bibliography

Barro, R., 2006. Rare disasters and asset markets in the twentieth century. *Quarterly Journal of Economics*, **121**, 823–866.

Bluedorn, J. C., 2005. Hurricanes: Intertemporal Trade and Capital Shocks. *Nuffield College Economics Paper* 2005–22.

Borensztein, E., Cavallo E., and Valenzuela P., 2008. *Debt Sustainability Under Catastrophic Risk: The Case for Government Budget Insurance*. IMF Working Paper WP/08/44.

Cavallo, E., and Noy, I., 2009. *The Economics of Natural Disasters – A Survey*. Inter-American Development Bank Working Paper.

Economist Intelligence Unit, 2005. *Asia's Tsunami: The Impact*. EIU Special Report.

Eisensee, T., and Strömberg, D., 2007. News floods, news droughts, and U.S. disaster relief. *Quarterly Journal of Economics*, **122**(2), 693–728.

Healy, A., and Malhotra, N., 2009. Myopic voters and natural disaster policy. *American Political Science Review*, **103**, 387–406.

Kunreuther, H., and Pauly, M., 2009. Insuring against catastrophes. In Diebold, F. X., Doherty, N. J., and Herring, R. J. (eds.), *The Known, the Unknown and the Unknowable in Financial Risk Management*. Princeton: Princeton University Press.

Noy, I., 2009. The macroeconomic consequences of disasters. *Journal of Development Economics*, **88**(2), 221–231.

Sen, A., 1981. *Poverty and Famines: An Essay on Entitlement and Deprivation*. New York: Oxford University Press.

Strömberg, D., 2007. Natural disasters, economic development, and humanitarian aid. *Journal of Economic Perspectives*, **21**(3), 199–222.

Yang, D., 2008. Coping with disaster: the impact of hurricanes on international financial flows. *B.E. Journal of Economic Analysis and Policy*, **8**(1), 13.

Cross-references

Economics of Disasters
World-Wide Trends in Natural Disasters

WORLDWIDE TRENDS IN NATURAL DISASTERS

Margreth Keiler
Duke University, Durham, NC, USA

Definition

Multiple definitions of the term "disaster" exist, which is rooted in different conceptualizations (authorities, scientists, journalists) and the context in which these definitions are used (Perry and Quarantelli, 2005; Perry, 2007). In the context of worldwide trends in natural disasters, the United Nations define disaster within the "International Strategy for Disaster Reduction" (UNISDR, 2010a) as "a serious disruption of the functioning of a community or a society involving widespread human, material, economic, or environmental losses and impacts, which exceeds the ability of the affected community or society to cope using its own resources." This UNISDR definition provides the base for different worldwide databases on natural disasters, although it does not provide the basis for NatCat and Sigma, whereas the Centre for Research on the Epidemiology of Disasters (CRED) declares more precisely when the local capacity is exceeded by "necessitating a request to a national or international level for external assistance" (CRED, 2010a).

Background: Worldwide database

Worldwide databases of disasters and losses are maintained by, e.g., the Centre of Research on Epidemiology of Disasters in Brussels (CRED) with the sponsorship of the USAID's Office of Foreign Disaster Assistance (OFDA), the Emergency Events Database (EM-DAT), as well as by reinsurance companies, such as Munich Reinsurance Group (NatCatSERVICE) and Swiss Reinsurance (Sigma). These databases contain essential core data on the occurrence and effects (casualties, people affected, economic losses) of disasters over the globe. While the focus of EM-DAT is to serve the purposes of humanitarian action (disaster preparedness, vulnerability assessment, priority setting) at national and international levels as well as scientific research (CRED, 2010b), the reinsurance companies give more emphasis to economic loss and especially to insured losses. Therefore, due to different foci and, above all, due to a lack of clear internationally accepted standards, definition, and criteria for the disaster data compilation, these global databases are limited by inconsistent reliability and poor interoperability (Below et al., 2009). The following table (Table 1) gives a short overview on the different information contained and criteria applied by taking the EM-DAT and NatCatSERVICE databases as examples.

Recent efforts were undertaken between the provider of global and regional databases by comparing the datasets in the different databases to identify inconsistencies and gaps, to improve data quality, and to allow for a comparison and exchange of data on a detailed level (Guha-Sapir and Below, 2002). The result of a comparative study between data of EM-DAT, NatCatSERVICE, and Sigma, carried out on four countries, showed that the data collected vary significantly (up to 37% difference for casualties, 66% for people affected, and 35% for economic damage according to Guha-Sapir and Below, 2002; Peduzzi et al., 2005). One result to overcome the above-mentioned limitations is an agreement (between CRED and Munich Re) on a common hierarchy and terminology on natural disasters and on a definition of disaster groups. It should be

Worldwide Trends in Natural Disasters, Table 1 Information and criteria applied in worldwide databases EM-DAT (CRED) and NatCatSERVICE (Munich Re), after Guha-Sapir and Below (2002) and Below et al. (2009).

	EM-DAT	NatCatSERVICE
Type of disasters	Natural and technological disasters	Natural disasters
Entry criteria	≥ 10 deaths and/or ≥ 100 affected and/or declaration of a state of emergency/call for international assistance	Entry if any property damage and/or any person sincerely affected (injured, dead) It is distinguished between six categories (two loss events, four catastrophes) Only major events prior to 1970
Methodology	Country entry	Country and event entry, all disasters geo-coded for GIS evaluation
Main sources	UN agencies, US Government agencies, official governmental sources, IFRC, research centers, Lloyd's, reinsurance sources, press, private	Munich Re branch offices; insurance associations, insurance press, scientific sources, governmental and nongovernmental organizations
Priority source	UN agencies	For monetary losses, priority is given to Munich Re branch offices and insurance associations
Period covered	1900–present	0079–present
Number of entries	17,000	26,000
Access	Public	Partially accessible

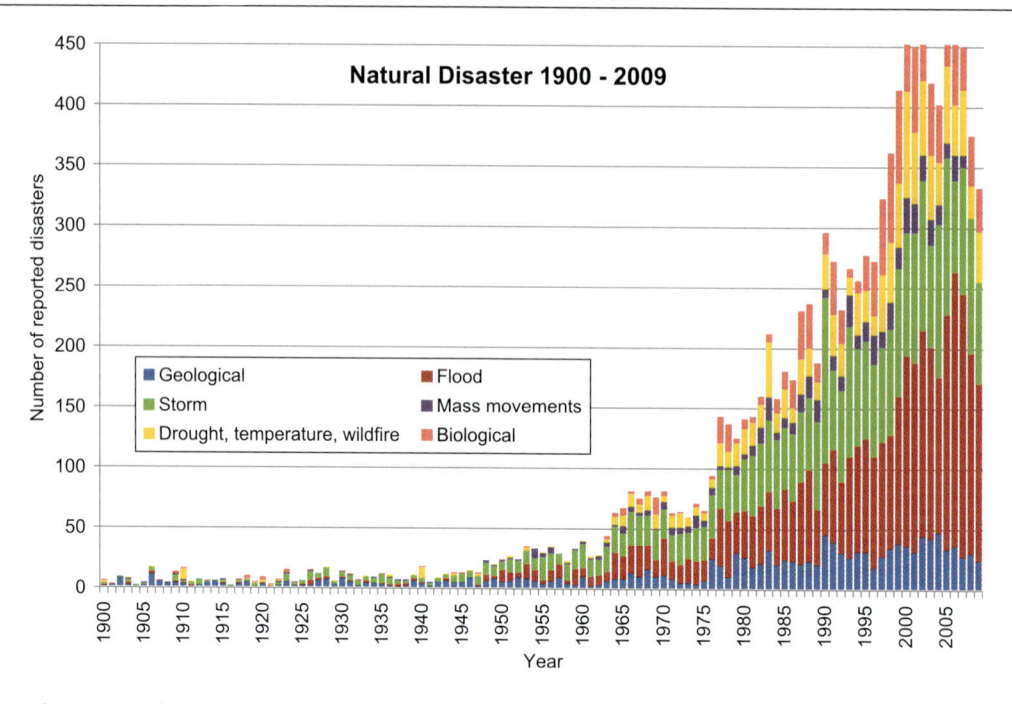

Worldwide Trends in Natural Disasters, Figure 1 Number of natural disasters from 1900 to 2009 classified in geological (earthquakes and volcanic eruptions), flood, storm, mass movement, drought (including extreme temperature and wild fire), and biological (epidemics, insect infestations) triggers (Source: CRED, 2010b).

mentioned that a common understanding about the regional scope and the time scale (start and end) of a disaster would also be required to harmonize data. The natural disaster categories are divided into six groups: biological, geophysical, meteorological, hydrological, climatological, and extraterrestrial, and each group includes several disaster main types and subtypes (Below et al., 2009). Still, after redistribution of datasets, differences between the databases exist as a result of the different entry criteria.

Worldwide trends

Disaster events

The analysis of the data recorded in the EM-DAT database indicates the following worldwide trend (CRED, 2010b). The number of reported disasters has increased, from about 10 per year between 1900 and 1949, 60 per year in the 1960s, 180 per year in the 1980s, almost 300 per year in the 1990s to about 440 per year for the decade 2000–2009. A similar exponentially rising trend is presented by Munich Re (Munich Re, 2009). In the last two decades, 64% of the total disaster numbers were recorded (Figure 1). The high increase in the number of hazardous events reported is probably mainly due to significant improvements in information access and reporting, especially after the establishment of EM-DAT in 1988. The most frequent triggers for disasters are floods (32%) and storms (28%, including hurricanes, typhoons, tornadoes, mid-latitude winter storms); earthquakes and volcanic eruptions (geological) caused about 12% of the disasters.

Disaster mortality and people affected

The related losses due to these natural disasters are concentrated in a small number of infrequently occurring events (UNISDR, 2009). The number of people reported killed by disasters since 1900 has been decreasing since the 1940s, whereas the reported number of people affected, defined as those requiring survival needs such as medical care, food, and shelter during and in the wake of disasters, increased strongly since the 1960s. The average number of people affected rose from about 25 million per year in the 1960s to 300 million in the 2000s due to floods, storms, and droughts (CRED, 2010b; UNISDR, 2010b). Regarding the time period between 1975 and October 2008 (with a more comparable global reporting coverage), 8,866 natural disaster events which were recorded by EM-DAT (excluding epidemics) caused more than two million killed people (UNISDR, 2009). However, only 23 M-disasters (0.26% of the events) accounted for 78.2% of the mortality, mostly in developing countries. Half of the 10 disasters with the highest death tolls since 1975 have occurred between 2003 and 2008 (UNISDR, 2009). For the period between 1975 and 2009, earthquakes (36%) are largest cause of natural disaster mortality, followed by storms (27%), droughts (23%), and floods (9%) (CRED, 2010b). The average number of people reported killed per million inhabitants is presented in Figure 2, indicating the highest number in Asia caused by "geological disasters" (earthquakes and volcanic eruptions) and "hydrometeorological disasters," including floods, storms, droughts and related disasters (extreme temperatures and wild fires), and mass movements

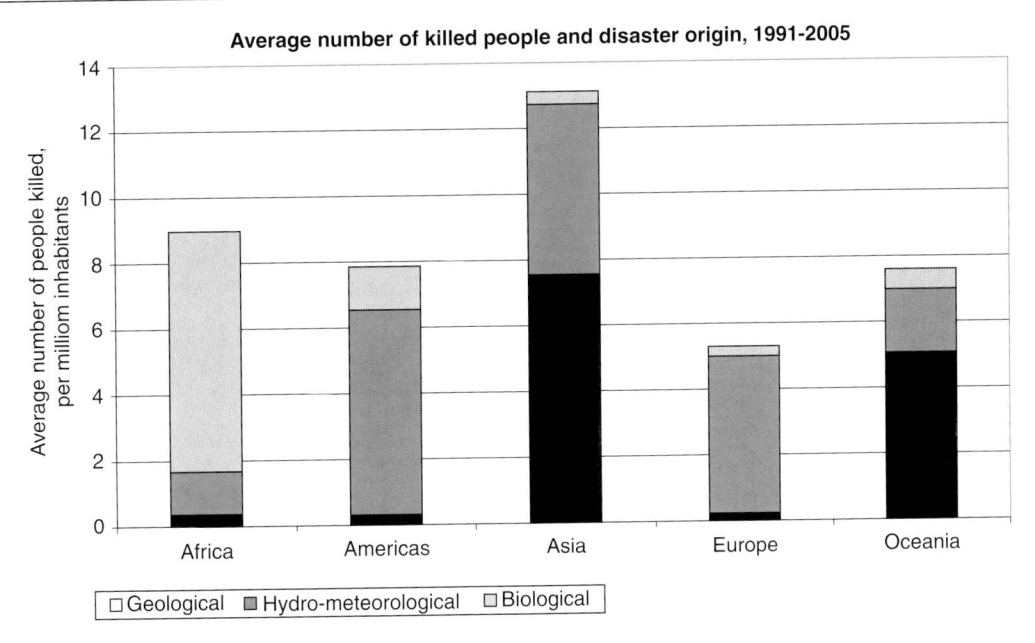

Worldwide Trends in Natural Disasters, Figure 2 Average number of people killed per million inhabitants by continent and disaster origin (Source: UNISDR, 2010b).

(UNISDR, 2010b). Africa shows the highest contribution to biological disasters covering epidemics and insect infestations. The greatest portion of individuals killed in Europe and the Americas are related to hydrometeorological disasters.

Economic losses

Economic losses due to natural disasters highlighted an exponentially rising trend since the mid-1970s regarding the period 1900–2009 (CRED, 2010b). The recorded economic losses in EM-DAT added up more than US$ 1.5 trillion between 1975 and 2008, with high peaks about US$ 160 billion in 1995 (Kobe earthquake), US$ 220 billion in 2005 (hurricane Katrina), and US$ 190 billion in 2008 (Sichuan earthquake and hurricane Ike). Only 25 M-disasters representing 0.28% of the events accounted for 40% of that loss which were located mainly in developed countries (UNISDR, 2009). The annual average of economic damage for the years 2000–2008 was about US$ 94 billion (CRED, 2010b). According to inflation-adjusted data from Munich Re, the average annual losses from the period 1977–1986 to the period 1997–2006 increased at a decadal rate of about 125% (Bouwer et al., 2007). Furthermore, Bouwer et al. (2007) highlighted that global economic losses caused by weather-related disasters have increased from an annual average of US$ 8.9 billion (1977–1986) to US$ 45.1 billion (1997–2006) based on the Munich Re database. The highest economic losses occurred since the 1990s in Asia by hydrometeorological (about 60%) and geological disasters (about 40%), followed by the Americas whose losses were dominated by hydrometeorological disasters (UNISDR, 2010b).

Discussion and conclusion

Trends of the disaster numbers, mortality, and people affected by disaster as well as economic losses show different developments or rate of increase especially if the records are analyzed by different "hazards" and regions. UNDP/BCPR (2004) highlighted the disproportionate impact on developing countries, that while only 11% of those exposed to hazards live in low human development countries, 53% of disaster mortality is concentrated in those countries. Mortality and direct economic losses appear to be highly clustered geographically; these are areas with a major concentration of people and economic assets and associated with a very small number of disaster events (UNISDR, 2009). More frequently occurring lower impact events are often not covered by the global databases, but in contrast, these widespread low-intensity losses are coupled to a large number of affected people and damage to housing and local infrastructure (McBean and Ajibade, 2009; UNISDR, 2009). The following example will illustrate the geographical disparity of disasters due to exposure and vulnerability. A comparable hazard event of a similar severity will generally result in lower mortality and smaller losses in countries with higher income and higher human development levels than in a less developed country when measured against the country's total wealth, despite the higher economic losses in absolute terms (UNISDR, 2009).

Beside the fact that a part of worldwide trends in natural disasters can be explained by improved disaster reporting (UNISDR, 2009), drivers for rising trends of affected people and economic losses are population growth (Strömberg, 2007) and increase in population and housing units in vulnerable areas (Pielke et al., 2008), particularly in urban and coastal areas. The Intergovernmental Panel on Climate Change concluded that the rising number of disasters triggered by hydrometeorological events is likely to have been partly driven by anthropogenic global climate change altering the hazard pattern. Furthermore, it is very likely that hot extremes, heat waves, and heavy precipitation events will continue to become more frequent and future tropical cyclones will become more intense (Solomon et al., 2007). Nevertheless, the increasing trend of weather-related disaster losses such as stated by Bouwer et al. (2007) should be carefully considered, as already indicated earlier by Downton and Pielke (2005). Even if flood damages continued to increase despite extensive flood management efforts since 1900, particularly when measured in constant currency units, the trend is not as obvious once normalized. If flood data related to the United States of America are presented in terms of damage per unit wealth, a slight and statistically insignificant downward trend is observed (Loucks and Stedinger, 2007), which suggests that floods might have a lessening or neutral impact on the overall personal wealth of citizens in the United States of America over the course of the past decades. Similarly, no significant loss trends were found for flood and wind storm in Europe (Barredo, 2009; Barredo, 2010), tropical cyclones in India (Raghavan and Rajesh, 2003), and weather-driven disasters in Australia (Crompton and McAneney, 2008) if they were normalized by eliminating the socioeconomic influence of growing exposure in areas affected.

In summary, multiple, interacting drivers and records of disaster that are of poor quality, inhomogeneous, and collected using a wide range of methods for different purposes make the analysis of worldwide trends in natural disasters extremely challenging.

Bibliography

Barredo, J., 2009. Normalised flood losses in Europe: 1970–2006. *Natural Hazards and Earth System Sciences*, **9**, 91–104.

Barredo, J., 2010. No upward trend in normalised windstorm losses in Europe: 1970–2008. *Natural Hazards and Earth System Sciences*, **10**, 97–104.

Below, R., Wirtz, A., and Guha-Sapir, D., 2009. *Disaster Category, Classification and peril – Terminology for Operational Purposes*. Working Paper 264. Brussels/Munich: CRED/Munich Re.

Bouwer, L. M., Crompton, R. P., Faust, E., Höppe, P., and Pielke, R. A., Jr., 2007. Confronting disaster losses. *Science*, **318**, 753.

CRED 2010a. Glossary. *Emergency Events Database, Centre for Research on the Epidemiology of Disasters*. Université Catholique de Louvain, Belgium. http://www.emdat.be/glossary/9.

CRED 2010b. *EM-DAT - The OFDA/CRED International Disaster Database, Database Advanced Search, Centre for Research on the Epidemiology of Disasters*. Université Catholique de Louvain, Brussels, Belgium. http://www.emdat.be/Database/AdvanceSearch/advsearch.php.

Crompton, R. P., and McAneney, K. J., 2008. Normalised Australian insured losses from meteorological hazards: 1967–2006. *Environmental Science Policy*, **11**, 371–378.

Downton, M., and Pielke, R. A., Jr., 2005. How accurate are disaster loss data? The case of U.S. flood damage. *Natural Hazards*, **35**, 211–228.

Guha-Sapir, D., and Below, R., 2002. Quality and accuracy of disaster data. A comparative analysis of 3 global data sets. *Working Document prepared for the Disaster Management Facility of The World Bank*. Brussels: The World Bank.

Loucks, D., and Stedinger, J., 2007. Thoughts on the economics of floodplain development in the U.S. In Vasiliev, O., van Gelder, P., Plate, E., and Bolgov, M. (eds.), *Extreme Hydrological Events: New Concepts for Security*. Dordrecht: Springer. NATO Science Series IV: Earth and Environmental Sciences, pp. 3–19.

McBean, G., and Ajibade, I., 2009. Climate change, related hazards and human settlements. *Current Opinion in Environmental Sustainability*, **1**, 179–186.

Munich Re, 2009. *Topics Geo, Natural Catastrophes 2008 – Analyses, Assessments, Positions*. Munich: Munich Re.

Peduzzi, P., Dao, H., and Herold, C., 2005. Mapping disastrous natural hazards using global datasets. *Natural Hazards*, **35**, 265–289.

Perry, R. W., 2007. What is a disaster? In Rodríguez, H., Quarantelli, E. L., and Dynes, R. R. (eds.), *Handbook of Disaster Research*. New York: Springer, pp. 1–15.

Perry, R. W., and Quarantelli, E. L., 2005. *What is a Disaster? New Answers to Old Questions*. Philadelphia: Xlibris.

Pielke, R. A., Jr., Gratz, J., Landsea, C. W., Saunders, M. A., and Musulin, R., 2008. Normalized hurricane damage in the United States: 1900–2005. *Natural Hazard Review*, **9**(1), 29–42.

Raghavan, S., and Rajesh, S., 2003. Trends in tropical cyclone impact: a study in Andhra Pradesh, India. *Bulletin of the American Meteorological Society*, **84**, 635–644.

Solomon, S., Qin, D., Manning, M., Chen, Z., Marquis, M., Averyt, K. B., Tignor, M., and Miller, H. L. (eds.), 2007. *Climate Change 2007: The Physical Science Basis. Contribution of Working Group I to the Fourth Assessment Report of the Intergovernmental Panel on Climate Change*. Cambridge: Cambridge University Press.

Strömberg, D., 2007. Natural disaster, economic development, and humanitarian aid. *Journal of Economic Perspectives*, **21**(3), 199–222.

UNDP/BCPR (United Nations Development Programme, Bureau for Crisis Prevention and Recovery), 2004. *Reducing Disaster Risk: A Challenge for Development*. New York: UNDP/BCPR.

UNISDR, 2009. *Global Assessment Report on Disaster Risk Reduction*. Geneva: United Nations.

UNISDR, 2010a. *Terminology on Disaster Risk Reduction, 2009*. Geneva: United Nations. http://www.unisdr.org/eng/terminology/terminology-2009-eng.html.

UNISDR, 2010b. *Disaster Statistics 1991-2005*. Geneva: United Nations. http://www.unisdr.org/.

Cross-references

Classification of Natural Disasters
Cultural Heritage and Natural Hazards
Disaster Diplomacy
Disaster Relief
Disaster Research and Policy, History
Disaster Risk Reduction (DRR)
Global Change and Its Implications for Natural Disaster
Natural Hazard in Developing Countries

Z

ZONING

Philipp Schmidt-Thomé[1], Stefan Greiving[2]
[1]Geological Survey of Finland (GTK), Espoo, Finland
[2]TU Dortmund University, Dortmund, Germany

Synonyms
Outlining; Planning; Regulation

Definition
Natural hazard zoning is the division of any determined space into areas which could be affected by hazardous phenomena to variable degrees.

Discussion
Zoning is used as an instrument in land use planning to regulate land use in areas affected by specified hazard patterns. Usually, the determination of hazard zones vary gradually from imminent hazard zones, with very restrictive land use or determined building regulations, toward zones that are considered outside the statistically relevant reoccurrence or magnitude of a given hazard, which do not require land use regulations that pay special attention to a defined hazard. There are different national approaches on the definition and the use of hazard zones. The legal binding of hazard zones depends not only on the potential occurrence of hazards, but also on land planning regulations and respective responsibilities in the case of the occurrence of a hazard. In some countries, a municipality issuing land use plans can be held responsible in the case of a natural hazard, whereas in other countries the land developer has to ensure and guarantee safety.

Hazard zoning aims mainly at influencing or avoiding future developments in affected areas, but is less effective for existing structures due to private property rights. For that purpose, there are different options in the insurance sector. In some areas, official hazard zones and respective risk patterns of insurance companies do not match, whereas in other countries legal institutions tightly cooperate with the insurance sector. Since the second half of the twentieth century, hazard maps have increasingly been incorporated into legal acts that regulate land use, for instance, in Austria and Switzerland, hazard zones related to avalanches have been mapped since the 1960s and following legal regulations were installed officially as a practice in the 1970s. The rapid rise of costs related to damages caused by natural hazards since the 1990s have led to an increase in national activities to regulate land use in hazard-prone areas. This development is in line with a continued population pressure and land take toward hazard-prone areas, a general understanding that natural hazards cannot be fully mitigated by technical means, as well as growing demands from the insurance industry to lower vulnerabilities related to hazards.

The determination of hazard zones depends on the nature of a hazard. In the case of avalanches, usually return periods, maximum run out extensions, and magnitudes are used to determine the borderlines. Mass movements (landslides, rock falls, soil creep, etc.) are influenced by many factors such as the steepness of slopes, the underlying geology, potential run out areas, and shear zones. Flood and storm surge hazard zones are also related to reoccurrence, e.g., 100 year floods, the potential depth of inundation and the runoff velocity (the latter is especially important in mountainous areas). Tsunami hazard zones follow elevation above ground for potential inundation areas. Volcanic hazard zoning focuses on potential pathways of lava, pyroclastic flows, and/or lahars. Depending on the type of volcano, main wind directions influencing the directions of ashfall may be taken into account. Earthquake zones are related to the potential magnitudes, tectonics (location of fault lines), and the underlying geology (liquefaction and ground acceleration potential). Storm hazard zones depend on peak wind speeds and forest fire hazard zones potentials of ignition and spreading.

Bibliography

AGS, 2007. Guideline for landslide susceptibility, hazard and risk zoning for land use management. Australian geomechanics society landslide taskforce landslide zoning working group. *Australian Geomechanics*, **42**(1), 13–36.

Arnalds, P., Jónasson, K., and Sigurdsson, S., 2004. Avalanche hazard zoning in Iceland based on individual risk. *Annals of Glaciology*, **38**, 285–290.

Fell, R., Corominas, J., Bonnard, C., Cascini, L., Leroi, E., and Savage, W., 2008. Guidelines for landslide susceptibility, hazard and risk zoning for land-use planning. *Engineering Geology*, **102**, 99–111.

IcelandMinistry for the Environment, 2000. Regulation on hazard zoning due to snow- and landslides, classification and utilization of hazard zones, and preparation of provisional hazard zoning.

Musson, R. M. W., Sargeant, S. L., 2007. Eurocode 8 seismic hazard zoning maps for the UK. British Geological Survey Technical Report, CR/07/125, 70 pp.

Schanze, J., Zeman, E., and Marsalek, J. (eds.), 2006. *Flood Risk Management: Hazards, Vulnerability and Mitigation Measures*. Dordrecht: Springer, 330 p.

Cross-references

Coastal Zone, Risk Management
Community Management of Hazards
Damage and the Built Environment
Emergency Mapping
Emergency Planning
Evacuation
Floodplain
Hazard and Risk Mapping
Hazardousness of Place
Land-Use Planning
Land Use, Urbanization and Natural Hazards
Megacities and Natural Hazards
Mining Subsidence Induced Fault Reactivation
Uncertainty
Urban Environments and Natural Hazards

Author Index

A
Adhikari, Pradeep, 324, 326
Alexander, David E., 78
Allen, Andrea, 419, 779
Ammann, Walter J., 170
Ancey, Christophe, 706
Anderson, Peter S., 536
Andreychouk, Viacheslav, 571
Appleton, James D., 808
Arenson, Lukas U., 132
Artemieva, Natalia, 522

B
Basabe, Pedro, 508
Basu, Arindam, 13
Beer, Tom, 42, 741
Benito, Gerardo, 748
Bent, Allison, 417, 777, 901, 907
Biasi, Glenn, 824
Biernacki, Wojciech, 655
Birkmann, Jörn, 305, 856
Blikra, Lars H., 611
Boano, Camillo, 280
Bokwa, Anita, 346, 711
Boteler, David H., 936, 937, 986
Bourgeau-Chavez, Laura, 1102
Bovis, Michael J., 881
Brown, Peter, 524
Brown, Rodger A., 188
Bryant, William A., 317, 883
Buchwaldt, Robert, 1, 623, 748, 791
Burkett, Virginia R., 119
Burnham, Gilbert M., 285

C
Calvache, Marta L., 369, 732
Cancio, Leopoldo C., 625
Carresi, Alejandro López, 706
Cassidy, John F., 208, 906
Catto, Norm, 97, 322, 519, 741, 877, 941, 1019
Cavalli, Marco, 378
Ceudech, Andrea, 141
Chanson, Hubert, 1007
Chapman, Clark R., 28
Cheong, Christopher, 797
Cheong, France, 797
Chester, David K., 836
Clague, John J., 400, 594
Clark, Matthew R., 1019

Cole, James W., 57
Comerci, Valerio, 284, 565, 671, 683
Cook, Benjamin I., 197
Croft, Paul J., 342
Cronin, Shane J., 941
Crozier, Michael James, 10, 606, 764, 1009
Cruden, David, 615
Cuomo, Giovanni, 48, 905, 987
Curt, Corinne, 300
Cutter, Susan L., 1088

D
Dasog, Ghulappa S., 297
Davies, Timothy R.H., 678
Derbyshire, Edward, 409
Derron, Marc-Henri, 686
Desramaut, Nicolas, 223, 405, 940
Dix, Andreas, 478
Dodson, John, 189
Domínguez-Cuesta, María José, 988
Donnelly, Laurance, 673
Donovan, Katherine, 697, 703
Doocy, Shannon, 699
Dore, Mohammed H.I., 240
Dorman, Lev I., 31, 807, 986
Duffour, Philippe, 231
Duncan, Angus M., 836
Dzialek, Jaroslaw, 98, 542, 756

E
Eberhardt, Erik, 306, 529, 1009
Eftychidis, George, 346, 664
Espinel, Zelde, 419, 779
Etkin, David A., 827

F
Fa, Lin, 2, 47
Ferrara, Floriana F., 849
Fiedrich, Frank, 272, 544, 773
Finkl, Charles W., 42
Fuchs, Sven, 121

G
Gaillard, J.C., 160, 529, 633
Galderisi, Adriana, 141, 849
Galloway, Devin L., 979
Garcin, Manuel, 940
Gayà, Miquel, 1096
Geertsema, Marten, 567, 803

Gerz, Thomas, 34
Ghil, Michael, 250
Ghirotti, Monica, 1069
Giardino, John Rick, 435
Giardino, Marco, 145
Gibson, Terry, 416
Giles, David, 544, 640
Gisler, Galen, 525
Glade, Thomas, 10, 78, 478, 552, 606, 764, 863, 1009
Goldin, Tamara, 525
Gorokhovich, Yuri, 884
Gracheva, Raisa, 452
Gregg, Chris E., 207, 1091
Greiving, Stefan, 618, 863, 1115
Gupta, Harsh K., 377, 1031
Guthrie, Richard, 387
Guzzetti, Fausto, 875

H
Haldorsen, Sylvi, 764
Hall, Minard L., 732
Harrald, John R., 773
Harris, Alan W., 9, 14
Hartwell, William T., 112
Hawkes, Peter J., 895
Heggie, Travis W., 367, 1076
Hermanns, Reginald L., 602, 611, 875
Herring, Alison, 926
Hervás, Javier, 610
Hickson, Catherine J., 41, 290, 639, 740
Higgins, Wayne, 696
Høeg, Kaare, 919
Hollins, Suzanne, 189
Hong, Yang, 324, 326
Horton, Pascal, 686
Hough, Susan, 850
Hungr, Oldrich, 149
Hunter, William, 280

I
Ingram, Jane Carter, 159
Ismail-Zadeh, Alik T., 225, 907, 979

J
Jaboyedoff, Michel, 686
Jackson, Jr., Lionel E., 359, 913
Jefferson, Theresa, 773

Jigyasu, Rohit, 49
Johal, Sarb, 474
Johnston, David M., 207, 474, 1091
Juhola, Sirkku, 3

K
Kameg, Kirstyn, 130
Keiler, Margreth, 1111
Kelman, Ilan, 118, 158, 160, 882
Kelman, Melanie, 425
Kent, George, 851
Kerle, Norman, 250, 416, 660, 837
Knight, Jasper, 82, 750
Knightley, R. Paul, 1019
Koeberl, Christian, 18, 523
Komac, Blaž, 288, 289, 387, 580
Kong, Laura S.L., 747
Korty, Robert, 481
Kovács, János, 325, 637, 804
Kramer, Steven L., 623, 629
Krien, Yann, 940
Krings, Susanne, 132
Kunreuther, Howard, 125
Kvalstad, Tore Jan, 652

L
Lacasse, Suzanne, 862
LaDochy, Steve, 338
Lamontagne, Maurice, 516, 535, 847
Lancaster, Nicholas, 155
LaRocque, Armand, 338, 777, 1062
Lavigne, Franck, 529
Le Cozannet, Gonéri, 223, 405, 940
Leblon, Brigitte, 385, 1102
Lee, William H.K., 117, 877
Leonard, Graham S., 207, 1036, 1091
Leone, Frédéric, 529
Leroy, Suzanne A.G., 452
L'Heureux, Jean-Sébastien, 611
Lindell, Michael K., 263, 812, 870
Logan, John M., 51
Longchamp, Céline, 686
Loughlin, Sue C., 1077

M
MacPhee, Ross D.E., 307
Madanes, Sharon B., 419
Marcelin, Louis Herns, 419
March, Julie A., 65
Marker, Brian R., 426, 583, 590, 765
Marsden, Brian G., 29
McBean, Gordon, 497
McCalpin, James P., 730, 994
McGuire, Bill, 576, 1073
Melosh, Jay, 672
Menoni, Scira, 69, 276
Mermut, Ahmet R., 297
Michel-Kerjan, Erwann, 125
Michoud, Clément, 686
Migoń, Piotr, 129, 135, 936
Mitchell, Ann M., 130
Mitchell, Jerry T., 45
Modaressi, Hormoz, 223, 405, 940

Mollard, J.D., 5
Monsalve, Maria Luisa, 732
Montz, Burrell E., 293
Morin, Julie, 529
Mossa, Joann, 40, 494, 624
Müller, Annemarie, 660
Murton, Julian B., 759
Musson, Roger M.W., 639

N
Nadim, Farrokh, 425, 682
Nardi, Fernando, 336, 497
Nelson, Alan R., 749
Neria, Yuval, 419, 779
Nirupama, N., 164
Nistor, Ioan, 1046
Norris, Fran H., 776
Noy, Ilan, 1107

P
Palermo, Dan, 1046
Panahi, Behruz M., 705
Paris, Raphaël, 529, 910
Parisi, Vincent R., 321
Paron, Paolo, 718
Paton, Douglas, 474
Pedreros, Rodrigo, 940
Power, William, 1036
Preston, Nick, 10, 606, 764, 1009
Price, Colin, 1006
Pugliano, Antonio, 69

R
Radke, John, 323
Rasmussen, Pat E., 835, 1063
Regmi, Netra Raj, 435
Reid, Mark E., 772
Renaud, Fabrice G., 926
Rennó, Nilton O., 201, 202
Ristau, John, 249, 769
Rockwell, Thomas, 738
Ronan, Kevin R., 247
Rossetto, Tiziana, 50, 231
Rouhban, Badaoui, 1056
Rubin, Jeffrey N., 471

S
Saatcioglu, Murat, 118, 451, 908, 947, 959, 1046
Sain, Kalachand, 377
Salvadori, Gianfausto, 310
Sangster, Heather, 836
Satake, Kenji, 1015
Scheib, Cathy, 188, 726
Schieb, Pierre-Alain, 242
Schleiss, Anton J., 901
Schmidlin, Thomas W., 366, 924
Schmidt-Thomé, Philipp, 3, 618, 1055, 1115
Schumann, Ulrich, 34
Seager, Richard, 197
Shelly, David, 1004
Shropshire, Donald J., 825
Shultz, James M., 419, 779

Sidle, Roy C., 657
Siems, Steven T., 92
Sigmundsson, Freysteinn, 311
Silva, Viviane, 696
Singh, Rajiv G., 240
Snyder, Richard L., 363
Soldati, Mauro, 151
Sorensen, John H., 110
Soriano, María Asunción, 594, 911
Springman, Sarah M., 132
Spurgeon, T.C., 290, 639, 740
Stanbrough, Lucy, 563
Stead, Doug, 1069
Stephenson, Wayne, 94, 855
Stethem, Chris, 31
Stewart, Carol, 1074
Stewart, Ian, 175
Stewart, Ronald E., 520
Stracher, Glenn B., 92
Strom, Alexander, 1065
Stumpf, Andrew J., 99, 186, 496

T
Tacnet, Jean-Marc, 300
Tarolli, Paolo, 378
Taucer, Fabio, 1062
Tetzlaff, Gerd, 447, 778
Thornbush, Mary J., 2, 743
Tilling, R.I., 290, 639, 740
Tobin, Graham A., 293
Tockner, Klement, 337
Toon, Owen Brian, 528
Tyc, Andrzej, 571

V
Varga, György, 637
Vitek, John D., 435
Voight, Barry, 369, 732

W
Waitt, Richard B., 579
Wang, Chenghu, 989, 1097
Watt, Kerrianne, 59
Webb, Leanne, 363
Weinstein, Philip, 59, 540, 826
Weissman, Paul R., 105
Westbrook, Graham, 672
Whiteford, Linda M., 293
Wilson, Thomas, 1074
Wisner, Ben, 651
Witiw, Michael R., 338

Y
Yang, Song, 696

Z
Zaliapin, Ilya, 250
Zehr, Raymond, 203
Zeng, Zhengwen, 2, 47, 989, 1097
Zentel, Karl-Otto, 552
Zimmer, Janek, 778
Zlatanova, Sisi, 272
Zorn, Matija, 288, 289, 387, 580

Subject Index

A
Aa-lava, 1
Absorbed dose, 188–189
Accelerometer, 2
Access to Resources (AR) model, 165
Accretionary lava balls, 1
Aceh province, Sumatra Indonesian Island, 530, 532, 534, 814
Acid rain, 2–3, 8, 54, 438, 1077
Active and inactive landslides, 616
Active control devices, 968–970
Active fault, 216–218, 235, 319, 1002–1004
Adaptation, 3–5
  adaptive capacity determinants, 4
  anticipatory, 4
  climate change, 3–4
  disasters, 4
  flood-prone areas, 3
  living environment, 3
  natural hazards, 3
  protective measurements, 3
  reactive, 4
Adriatic sea, 905
Advective frosts, 363, 365
African Centre of Meteorological Application for Development (ACMAD), 513
Agglomerations, 661
AGRHYMET Regional Centre (ARC), 513
Agricultural drought, 191–192, 195
Agriculture systems, 65–69
  diversification spreads risk, 68
  farmers, 65
  floods, 66
  plant pests and diseases, 67–68
  rainfall irregularity and drought, 65–66
  storm surges, 66, 67
  subsistence and commercial, 65
  tsunamis, 66
  volcanic eruptions, 66–67
Air and water pollution, 443
Air fluidization phenomena, 147
Airphoto and satellite imagery, 5–9
  earthquake hazards, 7
  global natural hazards, 5–7
  ground collapse, 8
  ground subsidence, 8–9
  ground swelling, 8
  natural hazards, 9
  slope failure (landslide) hazards, 8
  terrain analysis, 5
  terrain interpretation, 5
  volcanic hazards, 7–8
Air quality degradation, 1064
Albedo, 9–10, 15, 16, 83, 411
Aleatory uncertainty, 1055
Alquist-Priolo Earthquake Fault Zoning Act, 320
American Heart Association (AHA), 628
Anatolian Plate, 738
Andesitic magmas, 943
Antecedent conditions, 10–12, 85, 499
  antecedent soil water status index, 12
  decay factor, 12
  factors influencing, 12
  hazard history, 11
  rainfall-triggered landslides, 11
Antecedent soil water status index, 11–12
Anthropogenic global warming, 82, 83, 88
Anticipatory adaptation, 4
Aphelion, 14, 15, 29
Apophis, 29, 30
Aquitard drainage model, 982
Arias intensity, 551
Armenian earthquake, 62
Arsenic in groundwater, 13
Artificial levees, 494, 624
Ash fall, 66–67, 372, 474, 735, 736, 944, 1076, 1115
Ashgabat earthquake, 467
Asteroid, 14–17
  classification, 14
  future aspects, 16–17
  near-Earth asteroids, 14–16
  physical properties, 14–16
  planet embryos, 14, 16
  taxonomic classification and mineralogy, 14
Asteroid impact, 18–27
  formation of, 19–21
  hazards of, 25–26
  history of, 18–19
  microtektites, 24–25
  mitigation, 28
  predictions, 29–30
  recognition criteria for, 21–22
  shock metamorphic effects, 22–24
  tektites, 24–25
Astronomical unit, 14, 29
Atchafalaya River in south Louisiana, 40
Atmosphere-ocean global climate model simulations, 192, 193
Atmospheric state parameters, 35, 39
Australian classification of tropical cyclone severity, 481, 482
Automated Local Evaluation in Real Time (ALERT), 31, 808, 987
Automatic intensity assessment algorithm, 639
Avalanches, 31–33
  human interaction with, 32
  phenomena of, 32
  prediction, 33
  terrain, 32–33
Avalanche size, 32, 33
Aviation (hazards to), 34–39
  aircrafts, 34
  atmospheric state parameters, 35
  downburst, 36
  dust storms and volcanic ash, 38
  flight envelope and hazardous weather, 35
  hail and heavy rain, 36–37
  in-flight icing, 37
  lightning, 36
  low visibility, 35–36
  microburst, 36
  severe storms, 35
  thunderstorms and tornadoes, 36
  turbulence, 37–38
  weather-related accidents, 35
  wind shear, 36
Avulsion, 40, 494, 608, 624
Aztec Wash alluvium in southwestern Colorado, USA, 100

B
Baltic sea, 905
Banda Aceh in Northern Sumatra, Indonesia, 530–532, 953, 1038
Bangladesh cyclone, 441, 701
Bangladesh floods, 180
Barometric surge, 987–988
Basaltic lavas, 624, 1083
Base isolation device, 238–239
Base surges, 41, 1040
Beach nourishment (replenishment), 42, 96
Beam-column joint retrofits, 974–975
Beam retrofits, 973–974
Beaufort wind scale, 42–45
Berm/reshaping breakwaters, 49

Bhola cyclone in East Pakistan, 463, 467, 940
Biblical events, 45–47
  natural hazards, 45
  power, 46
  prophecy and prediction, 46–47
  punishment, 46
Biblical flood, 59
Bingham plastic flow models, 147
Bioenergy-harvesting technology, 348
Bjerknes hypothesis, 253
Black frost, 363
Black Sea Basin, 3
Blind thrust faults, 318–320
Blizzard, 498, 924–925, 941
Body wave, 47–48, 644, 646, 777
Bolometric Bond albedo, 10
Bracing element, 959, 961, 967
Breakwaters, 48–49, 96
Bridge abutments, 955
Bridge superstructure retrofits, 975–976
Bromus tectorum (cheat grass), 156
Buffer layer, 762
Building codes, 49–50, 218, 219, 224, 231, 233, 235, 243, 245
Building failure, 50–51, 235
Buildings, structures, and public safety, 51–55
  earthquakes and tsunamis, 51–53
  exteriors, environmental fatigue, 53–55
Build-Up Index (BUI), 1103
Bushfires in Victoria, Australia, 293, 354, 711, 712, 716, 788

C
Caldera collapse, 910
Calderas, 57–59
California bearing ratio (CBR), 298
Campanian volcanic arc, 1073
Camp Management Toolkit, 282
Canadian Forest Fire Danger Rating System (CFFDRS), 1103, 1104
Canadian Roundtable on Crisis Management, 279
Capelinhos volcano eruption, Azores, 41
Cardiopulmonary resuscitation, 627–628
CARE household model, 830, 832
Carnot cycle, 490
Carnot heat engine, 490
Cassava mosaic virus/cassava mosaic disease (CMD), 68
Casualty, 59–63
  Biblical flood, 59
  Disability Adjusted Life Years (DALYS), 59
  disaster types, 60–62
  earthquakes, 60–61
  effective disaster management and preparedness, 62
  Emergency Disasters Data Base (EM-DAT), 60
  long-term sequelae, 62
  meteorological, 61–62
  public health approach, 63
  reported natural disasters, 59
  volcanoes, 61
Catastrophic subsidence sinkhole, 573
Centre for Research on the Epidemiology of Disaster (CRED), 60, 327, 423, 498, 1111
Chaco-Pampean plain, Argentina, 13
Chicxulub impactor, 25
Chihuahuan Desert, USA, 156
Chilean earthquake in Chile, 390, 420, 1037
Chiloe island, 3
China Seismic Intensity Scale, 549
China's Sichuan Province, 184

Chlorofluorocarbons, 743
Cholera outbreak in London, 272
Chondrites, 672
Civil Defence Unit of Chatham Islands Council of New Zealand, 167
Civil protection, 69–77
  coping capacity, 70
  emergency management (EMA), 70
  nutshell, 74–75
Classification of natural disasters, 78–81
  atmospheric phenomena, 80
  disaster cycle, 81
  disaster risk reduction (DRR), 82
  drought, 79
  earthquake main shocks, 79
  emergencies and contingencies, 78, 79
  extreme natural phenomena, 79
  fat-tailed distribution problem, 80
  magnitude of impact, 78
  river flooding, 80
  sudden-impact disaster, 79
  volcanic eruptions, 79
Clausius-Clapeyron relation, 87
Clean Air Act, 743–744
Clear-air turbulence (CAT), 37, 38
Climate change, 82–90
  adaptation, 171, 174
  anthropogenic global warming, 82, 83
  Clausius-Clapeyron relation, 87
  climate variability, 83–84
  Ferrel cell, 87
  hazards and normal distribution, 88–89
  high-latitude areas, 86
  human systems impacts, 87–90
  hydrological cycle, 87
  low-latitude areas, 85–86
  macroscale atmospheric circulation patterns, 87
  mean annual air temperature (MAAT), 83
  mid-latitude areas, 86
  palaeoclimatology, 83
  projections, 895, 897–898
  proxy biological/geological evidence, 83
  resources and hazards, 84–85
  surface and subsurface circulation patterns, 87
  United Nations Framework Convention on Climate Change (UNFCCC), 82–83
Clinkers, 1
Cloud seeding, 92
Coal, 676
Coal fire (underground), 92–93
Coastal erosion, 94–97
  causes, 94–96
  erosion mitigation, 97–98
  impacts, 95, 96
  measuring coastal erosion, 96
  natural geomorphic process, 94
  water spring/mapped feature, 94, 95
Coastal megacity, 662
Coastal zone risk management, 97–98
Coefficient of Linear Extensibility (COLE), 298, 300
Cognitive dissonance, 98–99
Cold fog, 340
Collapsed buildings in Port-au-Prince in Haiti, 390, 393
Collapsing soil hazards, 99–104
  Aztec Wash alluvium, 100
  cemented particles/dispersive clays, 101–103
  dispersive soil hazard, 103
  distribution, 103
  flood deposits, 103
  geologic and man-made materials, 104
  hydrocompaction subsidence, 102

internal structure, 99, 100
karst hazard, 101
microscopic scale, 101
slumping, 103
Column retrofits, 970–973
Comet, 14, 17, 105–109
  cometary atmospheres, 107–108
  dynamics, 108
  impact hazard, 109
  solar nebula, 105
Cometary comae, 105, 107
Cometary nuclei, 105–107
Comet Hale-Bopp, 106, 109
Comet impact hazard, 109
Comet tails, 105, 106, 108
Common Alerting Protocol (CAP), 564
Communicating emergency information, 110–111
Community-Based Disaster Risk Reduction (CBDRR), 162
Community-based organizations, 269, 819
Community Environmental Monitoring Program, 115
Community management of natural hazards, 112–115
  earthquakes, 114–115
  fire and police departments, 114
  hazard identification and mitigation, 113
  hurricanes, coastal erosion, and coastal flooding, 115
  internet technology role, 114
  landslides, 115
  natural radioactivity, radon hazard, 115
  preparedness, 113
  recovery and reconstruction, 114
Community preparedness analysis, 265
Community Rating System (CRS), 543
Complex Humanitarian Emergencies (CHE), 285
Complexity theory, 117
Composite breakwaters, 48
Computer-based information system, 378
Concrete jacketing, 971
Concrete structures, 118
Conflict-related hazards, 443
Constructed beach, 42
Convective turbulence, 38
Conventional weather radar, 188
Convergence, 118–119
Convergent boundaries, 770–771
Coping capacity, 65, 70, 73, 119–120, 779
Coriolis force, 251, 491–492
Cosmic radiation, 726
Cosmic rays, 986
Cosmogenic radionuclides, 726
Cost-benefit analysis (CBA) of natural hazard mitigation, 121–124
  determination of benefit, 122–123
  determination of costs, 122
  human capital approach, 123
  mitigation approaches, 122
  mitigation measures, 122, 123
  private goods, 121
  sensitivity analysis, 122
  societal and political decisions, 122
Costs (economic) of natural hazards and disasters, 125–128
  catastrophes, 125–126
  catastrophic losses, 125
  Cyclone Nargis, Myanmar in Burma, 125
  hurricanes, 127–128
  storms, 127
Coupled human-natural environments, 143
Creep, 129, 152, 435, 617, 753, 1003
Creeping phenomenon, 712

Cretaceous marine clay shales, 8
Crisis management, 69–77
　adjective multisite, 72
　ash crises, 72
　conditions, 76–77
　inter/intraorganizational challenges, 76
　National Incident Management System (NIMS), 73–74
　nutshell, 74–75
　plain floods, 73
　regional events, 71
　spatial and time factors, 71
　vulnerability and resilience, 73
Critical incident stress debriefing (CISD), 786, 787
Critical incident stress syndrome, 130–131
Critical infrastructure, 132
Critical water content (CWC), 11
Crop Moisture Index (CMI), 192
Cryological engineering, 132–135
　cryosphere, 133
　material properties, 133, 134
　mechanical engineers, 133
　road engineers, 133
　solutions, 133–135
Cryptic survivorship, 308
Cryptosporidium, 285
Cultural heritage and natural hazards, 135–140
　ancient natural catastrophes, 137–139
　catastrophic events, 136
　cliff-top sites, 139
　flood hazard, 139
　humankind, 136
　natural processes and loss, 136–137
　seismic zones, 139
　surface processes, 139
　tsunamis, 139
Cyclone Nargis, Myanmar in Burma, 61, 125, 176, 328, 498, 500, 701

D

Damage and the built environment, 141–144
　chains of damages, 142–143
　coupled human-natural environments, 143
　direct damages, 141
　hazards and exposed elements and systems, 141
　indirect damages, 141, 143
　man-made structures, 141, 144
　measuring damages, 143–144
　nonmarket effects, 141
　Northridge earthquake, 143
　physical features, 143
　typologies of damages, 142
Dam failure, 326, 444, 765, 778
Damping, 232, 959
DART buoys, 1043, 1044
Dead Sea Fault Zone, 459–460
Debris avalanche, 79, 145–149, 612, 914, 917, 945, 1083
　air fluidization phenomena, 147
　basal lubricant, 148
　catastrophic landslides, 145
　destructive mass movement, 145
　dynamic emplacement models, 147
　geo-environmental contexts and typology, 145
　geomorphological and sedimentological characteristics, 145–147
　hazards and risks, 148–149
　mechanical fluidization, 147
　mountain debris avalanches, 147
　natural hazard phenomena, 145
　pseudo-stratification, 148
　sturzstroms, 145, 146
Debris fans, 150, 151
Debris flow, 149–151, 596–597
　boulder deposit, 150, 151
　debris fans, 150, 151
　debris flood, 150
　debris slides, 149
　flow-like mass movement, 149
　levee deposit, 150
　multiple debris flow initiation points, 149, 150
　rapid undrained loading mechanism, 149
　soil creep, 149
　surge, 150
Decay factor, 12
Deep-seated gravitational slope deformation, 151–154
　causes, 152
　gravity-induced process, 151
　hazard implications, 153–154
　lateral spreading, 153
　mass movements/favor collateral landslide processes, 152
　sackung, 152, 154
Deforestation and construction, 443–444
Deltaic plains of Bangladesh, 179
Department of Economic and Social Affairs (UN/DESA), 1058
Desertification, 155–158, 443, 580, 741
　causes, 155
　climate change, 157–158
　ecological effect, 155–156, 158
　land degradation in drylands, 155
　occurrence, 155, 157
　physical effects, 156–157
　social dimensions, 157
　United Nations Convention to Combat Desertification (UNCCD), 155
Destructive earthquakes, 212, 228, 460, 652
Destructive mass movement, 145
Devastating displacement waves, 611
Devil's staircase, 254, 256, 259
Digital Terrain Models (DTMs), 379
Digital Worldwide Standardized Seismograph Network (DWWSSN), 417
Dinaric Karst in Mediterranean Europe, 573
Dip-slip faults, 317, 318
Direct damages, 141–144
Disability Adjusted Life Years (DALYS), 59
Disaster cycle, 81
Disaster diplomacy, 158–159, 184
Disaster Ecology Model, 779, 780
Disaster incubation, 827–828
Disaster models, 828
Disaster preparedness, 263, 264, 266–267, 286
Disaster recovery, 263, 264, 266–268
Disaster relief, 159–160
Disaster research and policy, history, 160–163
　climate change and conceptual regressing, 162–163
　Community-Based Disaster Risk Reduction (CBDRR), 162
　hazard paradigm, 161
　Lisbon earthquake and tsunami, 160
　vulnerability paradigm, 161–162
Disaster risk management, 164–169
　Access to Resources (AR) model, 165
　Civil Defence Unit of Chatham Islands Council of New Zealand, 167
　comprehensive disaster risk management system, 165, 169
　economic, political, and institutional support considerations, 168
　Federal Emergency Management Agency (FEMA) model, 166–167
　Geographic Information System (GIS), 167
　Hazard, Risk, and Vulnerability Analysis (HRVA) method, 166
　Hyogo Framework for Action, 165
　International Strategy for Disaster Reduction (ISDR), 164
　National Oceanic and Atmospheric Administration (NOAA), 167
　perceptions of hazard/disaster risk, 165, 166
　Pressure and Release (PR) model, 169
　qualitative and quantitative frameworks and method, 165
　resilience building and community participation, 168
　response, recovery, reconstruction, and rehabilitation, 168
　risk and vulnerability identification, 165
　Risk Management Index (RMI), 168
　Seriousness, Manageability, Urgency, and Growth (SMUG) model, 167
　standard risk formula, 165, 166
　structural, nonstructural, cost/benefit analysis, 167
Disaster risk reduction (DRR), 81, 170–174, 837
　climate change adaptation, 171
　climate change and land degradation, 170
　disaster management, 170, 171
　Hyogo Framework for Action (HFA), 170
　integrative risk management (IRM), 171–172
　planning of measures, 172, 173
　risk analysis, 173
　risk assessment, 173–174
　risk culture, 172
　risk dialogue and strategic controlling, 174
　risk governance, 172
　safety measures, 174
　UNISDR, 170
　UN Millennium Development Goals (UN MDGs), 170
　World Conference on Disaster Risk Reduction in Kobe, Japan, 170
Disaster risk theory, 663
Disasters, 175–184
　anthropogenic climate change, 177
　Bangladesh floods, 180
　coastalization, 178
　cultures of catastrophe, 181–183
　deltaic plains of Bangladesh, 179
　earthquake-prone shores, California and Washington, 178
　hot spots, 179
　human disasters, 179
　Hurricane Katrina, 181
　hydrometeorological hazards, 177
　International Decade of Natural Disaster Reduction. (IDND), 176
　Izmit earthquake, 181
　landscapes of chronic vulnerability, 181
　line of defense, 180
　Lisbon earthquake, 176
　natural hazard, 176
　Payatas rubbish dump in central Manila, Philippines, 179
　politics of disaster, 183–184
　progression of vulnerability, 180
　self protection from hazards, 181
　social protection, 181
　socioeconomic pressures and physical exposures, 180
　statistics, 176–177
　storm surges and tsunamis, 178, 179

Disasters (*Continued*)
    Tohoku earthquake, Japan, 179
    US flood hazard, 176
    Western technocratic methods, 176
Dispersive soil hazard, 101, 103, 187–188
Divergent boundaries, 770
Diversification spreads risk, 68
Doppler weather radar, 188
Dose rate, 188–189
Downburst, 36
Drift ratio, 947
Drought, 65, 66, 68, 72, 79, 84, 87, 89, 136, 189–196, 699–700
    agricultural drought, 191, 192
    drought indices, 191–192
    El Niño-Southern Oscillation, 194
    historical impact of drought, 192–194
    hydrological drought, 190–191
    hydrometeorological drought, 194
    impacts of drought, 192
    intensity, duration, and spatial extent, 190
    meteorological drought, 190, 191
    meteorological records, 194
    mitigating agricultural drought, 195
    mitigating hydrological drought, 195
    natural hazard, 189
    Palmer Drought Severity Index (PDSI), 194
    socioeconomic drought, 191
Drought Code (DC), 1103
Drought hazard, 382
Dry snow avalanches, 32
Ductility, 959
Duff Moisture Code (DMC), 1103
Duration magnitude scale, M_D, 647–648
Dust Bowl, USA, 197–200, 463, 1062
    agricultural economy and farming, 199
    climate model simulations, 199
    dust storms, 199
    Great Plains, 197, 198
    meteorological origins, 197–199
    rain-fed agriculture, 157
    sea surface temperatures (SSTs), 199
    soil erosion control, 199
Dust devil, 201–202
Dust pneumonia, 199
Dust storm, 38, 202–203
Dvorak classification of hurricanes, 203–205
Dynamic emplacement models, 147

E
Early warning systems, 207–208
Earth impacts, 526
Earthquake, 51–53, 114–115, 208–219, 653, 700–701
    active fault zones, 139, 217
    ancient earthquakes, 138
    building codes and standards, 218–219
    catastrophes, 117, 125
    causes, 208–210
    community management of hazards, 114–115
    concrete structures, 118
    continuous GPS monitoring, 215–216
    conventional earthquake resistant design, 236–237
    cultural legacy, 139
    earthquake damage, 223–225
    earthquake early warning (EEW), 219
    earthquake ground shaking, 629
    earthquake hazards, 7
    earthquake monitoring, 690
    earthquake prediction, 218
    episodic tremor and slip, 216
    fault displacement, 214
    fires, 215
    fractals, 117
    geophysical, 60–61
    ground shaking, 217, 219
    Gutenberg-Richter relation, 117
    history of earthquakes, 210–212
    human-made disasters, 79
    injuries, 60–61
    intensity scales, 212–213
    landslides, 214
    LIDAR, 216–217
    liquefaction, 215
    magnitude scales, 213
    Nankai drilling project in Japan, 217
    paleoseismology, 216
    psychological effects, 215
    real-time seismology, 217
    real-time warning systems, 219
    recording earthquakes, 212
    seismology, 215
    soil conditions on structure, 234
    strong shaking, 213–214
    subduction zone, 215
    sudden-impact disasters, 79
    triggering and stress changes, 217
    tsunamis, 214
    wave propagation and ground motion, 216
Earthquake prediction and forecasting, 225–229
    accuracy and testing, 228–229
    earthquake precursors, 228
    earthquake-prone region, 218, 227
    Earth's lithosphere, 225
    Haicheng earthquake, 226
    M8 algorithm, 227
    Reid's model, 227
    San Andreas fault system, 227
    Short-Term Earthquake Probability (STEP) method, 228
    stress-induced variations, 228
Earthquake-prone shores, California and Washington, 178
Earthquake resistant design, 231–240
    base-isolation devices, 238–239
    displacement-based design frameworks, 237
    dissipative devices, 239
    effect of earthquakes on structures, 231–232
    force-based earthquake resistant design, 235–236
    Hyogo-ken-Nanbu earthquake, Japan, 237
    Messina Earthquake, Italy, 231
    performance-based design approach, 237
    San Francisco earthquake, USA, 231
    seismic codes of practice, 237–238
    seismic loading, 232–235
Earth Resources Technology Satellite (ERTS), 594
Earth's geosystem, 711
East African Community (EAC), 512
East African Rift zone, 211
Eastern Canada Acid Rain Control Program, 3
East Han Dynasty in China, 226
Economic and Social Council (ECOSOC), 1057
Economic Community of Central African States (ECCAS), 512
Economic Community of West African States (ECOWAS), 512
Economic laws, 245
Economics of disasters, 242–246
    catastrophe, 242
    conventional economic methodologies, 245
    economic laws, 245
    econophysics, 246
    OECD, 243
Economic valuation of life, 240–242
Econophysics, 246
Education and training for emergency preparedness, 247–248
Effective dose, 189
Ego analysis, 798–801
Elastic rebound theory, 227, 249
Elastic strain energy, 232
Electromagnetic radiation (EMR), 5, 250
Electronic distance meters (EDM), 890
Electrosphere, 625
Elm rockfall in Canton Glarus, Switzerland, 479
El Niño/Southern Oscillation (ENSO), 194, 250–261
    atmospheric and oceanic processes, 251
    Bjerknes hypothesis, 253
    climate models, 254
    coupled oscillator, 255–257
    delayed oceanic wave adjustments, 253
    deterministically chaotic, nonlinear paradigm, 255
    Devil's staircase, 259
    global climatic and socioeconomic impacts, 252–253
    Kelvin wave, 253
    quasi-periodic behavior, 257–259
    Rossby wave, 253
    seasonal forcing, 253
    sea surface temperature (SST), 252
    Southern Oscillation Index (SOI), 254
    stochastic, linear paradigm, 255, 261
    thermocline, 251
    thermohaline circulation, 251
    threshold-crossing episodes, 252
    Tropical-Ocean-Global-Atmosphere (TOGA) Program, 253
    walker circulation, 251
Emergency/contingency planning, 276–280
    communication, 279
    crisis management, plans, 277
    development and structure, 277–278
    multi-hazard plans, 277
    preparation and production processes, 279
    spatial dimension, 277
    structure and main ingredients, 278–279
    turbulent environment, 276
Emergency Disasters Data Base (EM-DAT), 327
Emergency Events Database (EM-DAT), 1107, 1111, 1112
Emergency management, 263–271
    community preparedness analysis, 265
    disaster preparedness, 266–267
    disaster recovery, 267–268
    Emergency Management Accreditation Program (EMAP), 268–269
    emergency response, 267
    hazard mitigation, 266
    LEMA/LEMC member, 269–270
    losses, 263
    National Fire Protection Association (NFPA), 268
    performance evaluations, 268
    physical environments, 263
    policy development, 269
    preimpact disaster planning, 268
    principles of community emergency planning, 265
    technological fixes, 264
Emergency Management (EMA), 70

Emergency Management Accreditation Program (EMAP), 268–269
Emergency mapping, 272–275
  cholera outbreak in London, 272
  Geographical Information Systems (GIS), 272
  geospatial data, 273
  Haiti earthquake, 272
  hazard and risk mapping, 272
  Hurricane Katrina, 272
  Indian Ocean tsunami, 272
  innovative systems, 274–275
  interaction with emergency maps and visual analytics, 274
  post-event emergency maps, 272–273
  pre-event emergency maps, 272
  remote sensing, 273–274
  symbols, 273
  World Trade Center, 272
Emergency operations plan (EOP), 472
Emergency planning, 768
Emergency response, 263, 267
Emergency shelter, 280–283
  adequate housing, 281
  Camp Management Toolkit, 282
  crisis recovery development, 280
  host families, 282
  Humanitarian community, 282, 283
  IASC Emergency Shelter Cluster, 282
  permanent housing, 281
  refugee camp, 282
  rural self-settlement and collective centers, 282
  self-settled and planned camps, 282
  temporary housing, 282
  transitional shelter/settlements, 282
  urban self-settlement, 282
End-to-end warning systems, 1091
Energy dissipation, 959
Energy magnitude scale, M_e, 646–647
Enhanced Fujita Scale (EF-scale), 367
Enriquillo-Plantain Garden Fault Zone (EPGFZ), 420
Environmental effects of airborne dust, 410–414
  atmospheric effects, 410–412
  impacts of mineral particles on human health, 413
  mineral particle toxicity, 413
  particle size, mineral dust composition, and potential health impact, 413–414
Environmental Protection Agency (EPA), 2
Environmental Seismic Intensity Scale (ESI–2007), 546, 549–551
Epicenter, 284
Epidemiology of disease in natural disasters, 285–287
  communicable disease, 285
  Complex Humanitarian Emergencies (CHE), 285
  endemic and epidemic diseases, 285, 287
  existing endemic disease patterns, 285–286
  health of population, 286
  population characteristics, 286
  primary prevention, 286
  response capacity, 286
  secondary prevention, 286–287
  surveillance system, 287
  tertiary prevention, 287
Episodic tremor and slip (ETS), 1005
Epistemic uncertainty, 1055
Equilibrium-disturbance models, 751
Equivalent dose, 189

Equivalent Lateral Load Method, 233
Erosion, 288–289
Erosivity, 289
Eruption types (volcanic eruptions), 290–292
Etna eruption, 457
European Commission Humanitarian Office (ECHO), 553
European Macroseismic Scale (EMS–98), 213, 224, 547
European-Mediterranean earthquakes, 211
European Union Solidarity Fund, 143
Evacuation, 293–296
  bushfires in Victoria, Australia, 293
  emergency management, 293
  hazardous environments, 293
  perception of risk, 295
  resources and social factors, 295–296
  spatial concerns, 294–295
  temporal concerns, 294
  Tungurahua volcano, Ecuador, 293
Event water, 11
Excess rainfall, 12
Expansion index (EI), 298
Expansive soils and clays, 297–300
  active zone, 298–299
  characterization of expansive soils, 297–298
  shrinkage and swelling, 300
  solutions, 299–300
  swelling clay minerals, 297
  vertisols, 297
Expert (knowledge-based) systems for disaster management, 300–304
  decision support systems (DSS), 301, 302
  expertise and expert systems, 301–302
  geological hazards, 301
  knowledge-based systems, 302–304
  spatial and temporal, 301
Explosive eruptions, 910
Exposure to natural hazards, 305–306
Extensometers, 306–307
Extinction, 307–310
  biological extinction, 308
  causes, 309
  IUCN Red List, 307
  mass extinctions, 308
  rates of extinction, 308–309
Extreme Value Theory, 310–311
Eyjafjallajökull eruptions 2010, 311–315
  basaltic eruptions, 311
  deformation and elevated seismicity, 312
  divergent plate boundary, 311
  effusive basaltic eruption, 312
  effusive eruption, 311
  eruption response and impact, 315
  explosive summit eruption, 311
  Gígjökull outlet glacier, 314
  ice-capped summit, 313
  ice-capped volcano in South Iceland, 311
  seismic signals, 312
  tectonic setting and volcanic unrest, 311–312

F
Famine Early Warning System Network (FEWS NET), 65
Fault, 51, 317–320
  active fault, 319
  blind thrust faults, 318–319
  dip-slip faults, 317, 318
  fault creep, 319
  high-angle faults dip, 317
  normal faults, 317–319
  oblique-slip faults, 317
  reverse faults, 318

  seismic hazard assessment, 319–320
  strike-slip faults, 319
Fault scarp, 675
Federal Emergency Management Agency (FEMA) model, 166–167, 266, 321–322, 334, 802, 955
Federated States of Micronesia (FSM), 66
FEMA's HAZUS damage scale, 144
Fen-Wei faulted basin, 979
Ferrel cell, 86, 87
Fetch, 322
Fiber-reinforced polymer (FRP) sheets, 959, 961–964
Fimmvörduháls eruption, 311
Fire and firestorms, 323
Fire Behavior Prediction (FBP) system, 1103
Fire danger rating systems, 1103
Fire Weather Index (FWI), 1103
Flash flood monitoring and prediction algorithms (FFMPA), 324–325
Flood deposits, 325
Flood frequency analysis (FFA), 360–361, 749
Flood hazard, 380–381
Flood hazard and disaster, 326–334
  dams, 332–333
  flood forecasting, 333–334
  flood generation, 326, 327
  flood measurement, 326–327
  flood monitoring, 333, 690
  floodplain, 326, 334
  floodwaters, 326
  levees/dykes, 333
  river channel improvement, 333
  socioeconomic impact of floods, 328–330
  spatiotemporal distribution of flood, 327–328
  wetland management, 334
Floodplain, 326, 337
Flood-prone areas, 3
Flood protection, 336
Flood stage, 336–337
Flood waves, 619
Floodway, 338
Flow slide in Lake Merced, San Francisco, 629, 630
Fluvial outburst flood deposits, 325
Fluvial sediment yield changes, 750
Fog hazard mitigation, 338–341
  aviation, 340
  dense/heavy fog, 339
  droplets/ice crystals, 338
  forecasting, 340
  health, 341
  Laser telecommunications, 340–341
  marine transport, 339–340
  remote sensing, 340
  road transport, 339
Fog hazards, 342–345
  fog impact, 342–343
  fog investigation, 343–344
  fog mitigation and applications, 344–345
  operational support, 345
Föhn, 346
Food and Agriculture Organization (FAO), 582, 1058
Food security, 65–67
Force-based earthquake resistant design, 235–236
Forecast plume, 260
Forest and range fires, 346–358
  bioenergy-harvesting technology, 348
  crawling/surface fires, 348
  crown, canopy/aerial fires, 348
  fire effects, 357–358

Forest and range fires (*Continued*)
  fire intensity, 354
  Fire Manager, 348
  fire modeling, 355
  fire paradox, 351–352
  fire season, 347
  fire suppression, 355, 357
  fire tetrahedron, 353
  forest fire, 353
  forest fuels and fuel management, 352–353
  grass fire, 347
  ground fires, 348
  Integrated Forest Fire Management system, 348
  mobile fire detection station, 349, 350
  post-fire management decisions, 349
  prevention, 349–351
  satellite-mounted sensors, 349
  unmanned aerial vehicles (UAV), 349, 350
  vegetation/bushfire, 347
  wildfires, 346, 353
  wireless sensors, 348
Forest fires, 694
Fractals, 117
Frank Slide in Alberta, Canada, 598, 916
Free space optical (FSO) communications, 340
Frequency and magnitude of events, 359–362
  flood frequency analysis, 360–361
  global earthquake, 359–361
  landslide-type hazards, 361–362
  log-normal frequency, 359, 360
  moment magnitude (M_W) scale, 359, 360
Friction melt, 22
Friction-pendulum bearings, 239
Frost hazard, 363–366
  advective frosts, 363
  agricultural products, 363
  biological impact, 364
  climate, 366
  ice crystals formation, 363
  plant damage, 363
  probability and risk, 365
  protection from frost hazard, 364–365
  radiative frosts, 363
Froude number of tidal bore, 1008
Fuel Moisture Code (FFMC), 1103
Fujita scale, 45, 367, 1020
Fujita Tornado Scale, 366–367
Fukushima nuclear plant crisis, 1017
Fumarole, 367–368
Functional damages, 142
Fuvial sediment yield changes, 750
Fuzzy boundary, 142

G
Gacial-interglacial timescales, 750
Galeras volcano, Colombia, 369–376, 732
  amphitheater-like crater, 370, 371
  crisis management issues, 376
  hazards assessments, 374–375
  horseshoe-shaped amphitheater, 369
  INGEOMINAS, 371, 373
  institutional structure and political reactions, 371–374
  lava dome, 374
  long-period (LP) events, 370–371
  narino province, 369, 370
  Nevado del Ruiz eruption, 371
  ONAD, 371, 373
  orange warning, 373
  seismic swarms and damage, 374
  social and economic impacts and feedback, 375–376
  tornillo-type events, 371, 373
  volcano-tectonic (VT) events, 370
  voluntary preventive evacuations, 372
Ganges delta, 13
Gas emissions, 910
Gas hydrates, 377–378, 654
Gas hydrate stability zone (GHSZ), 673
Gasodynamic karst hazards, 575
Gazli earthquakes, Uzbekistan, 1031–1032
Gazli Oil Field in Uzbekistan, 1031–1032
Geochemical techniques, 945
Geographic Information Systems (GIS), 167, 272, 538
Geographic Information Systems (GIS) and natural hazards, 378–382
Geographic information technology, 385–386
Geohazards, 387
Geological/geophysical disasters, 387–399
  asteroids and near-earth objects, 399
  climatological and hydrometeorological events, 388
  destructive events, 388
  earthquakes, 389–390
  floods, 397–399
  International Decade for Natural Disaster Reduction (IDNDR), 388
  landslides, 395–397
  Shaanxi earthquake in China, 388
  snow avalanches, 399
  subsidence, 399
  supervolcanoes, 388
  tsunami, 390–392
  volcanic hazard, 381–382
  volcanoes, 393–395
Geological hazards, 711
Geomagnetically induced currents (GIC), 938
Geometric albedo, 9
Geophone, 2
Geophysical Fluid Dynamics Laboratory (GFDL), 258
Gígjökull outlet glacier, 314
Glacial-interglacial timescales, 750
Glacial lake outburst floods (GLOFs), 398–399, 462
Glacier-dammed lakes, 567
Glacier hazards, 400–404
  alpine glaciers, 400
  climate change, and sea-level rise, 401
  glacial outburst floods, 401–404
  glacier ice, 400
  landslides and mass movements, 401
  natural Earth processes, 400
  sediment yield, 404
  stream discharge, 404
Global Assessment of Human Induced Soil Degradation (GLASOD), 580
Global change and its implications for natural disasters, 405–407
Global Digital Seismograph Network (GDSN), 417
Global dust, 409–414
  airborne dust, 409–410
  blood rain, 409
  dust pathways and deposition, 410
  dust sources, 410
  environmental effects of airborne dust, 410–414
  erosion, airborne transport and deposition, 409
  loess, 409
  volcanic dust plumes, 409
Global Flood Inventory (GFI), 327
Global mean sea level, 895, 896
Global natural hazards, 5–7
Global navigation satellite systems (GNSS), 938
Global Network of Civil Society Organizations for Disaster Reduction (GNDR), 162, 416
Global Positioning Systems (GPS) and natural hazards, 416–417, 906
Global Seismic Hazard Assessment Program (GSHAP), 218–219
Global Seismic Hazard Map, 219
Global Seismograph Network (GSN), 212, 417–418
Global Telemetered Seismograph Network (GTSN), 417–418
Gonave microplate, 419
Gondwana coalfields, 282
Grand Banks tsunami, 1039
Granite, 51, 54
Granulometric heterogeneities, 147
Gravidynamic karst hazards, 572–576
Gravimeter, 2
Gravitational slope-forming processes, 657
Grazing, 156
Groningen gas field, 981
Gros Ventre landslide, 399
Grote Mandränken, 479
Ground movement measurements, 686
Ground subsidence, 585
Gutenberg-Richter relation, 117

H
Haicheng earthquake, 226
Hail storms, 694
Haiti 2010 earthquake, psychosocial impacts, 419–424
  disaster severity, 420–422
  disaster type, 419–420
  hazard profile, 419, 421
  mass mortality, 422–423
  mental health consequences, 419
  psychological stressors, 419, 421, 422
  scope and social context, 423–424
Halley-type comets, 108
Harbor wave, 390, 905
Harmful algal bloom (HAB), 826
Harmonic tremor, 425
Hattian Bala landslide dam, 605
Hawaiian eruption, 290
Hayabusa spacecraft, 15–16
Hazard and risk mapping, 426–432
  Geographic Information Systems (GIS), 427
  land use and development, 431–432
  preparation of hazard maps, 427–431
  preparation of vulnerability and risk maps, 431
  primary damage and injuries, 427
  seismic hazard maps, 427
  traffic light approach, 432
Hazard identification and mitigation, 113
Hazard insurance, 818–819
Hazard mitigation, 263, 264, 266
Hazardousness of a place, 435–445
  anthropogenic activities, 435, 443–444
  assessment of hazards and risks, 444
  atmospheric events, 437, 441–445
    climate change, 439–440
    drought/desertification, 442–443
    freeze, 442
    geomagnetic storms, 440
    hail, 442
    hurricane, 440
    lightning, 442
    rainfall, 441
    storm surge, 440–441
    tornadoes, 441–442

winds, 440
geophysical events, 435–439
    coastal erosion, 439
    earthquakes, 436
    fire, 439
    flooding, 438
    landslides, 435–436
    land subsidence, 436
    tsunamis, 438–439
    volcanoes, 436–438
mitigation of hazards, 444–445
Hazard paradigm, 161–163
Hazard vulnerability analysis (HVA), 471–472
Hazard, vulnerability, and risk analysis (HVRA), 166, 536
HAZUS's earthquake casualty model, 775
Heat waves, 447–450
    blocking action, 449
    energy balance and transport, 447
    energy budget, 450
    heat comfort conditions, 447
    heat stress, 448, 449
    heat stroke, 448
    temperature control mechanisms, 447
    thermal-regulatory activities, 447
    thermoregulatory system, 450
    weather and environmental conditions, 447
    wet-bulb global temperature (WBGT), 448
High-angle faults dip, 317
High-rise buildings in natural disaster, 451–452
Hilo bay, 905
Historical events, 452–467
    archaeological evidence, 453, 456, 457
    Ashgabat earthquake, 467
    Bhola cyclone in East Pakistan, 467
    bolides and near-earth objects, 464–465
    drought, 462–463
    earthquakes and tsunamis, 459–460
    earth sciences, 456
    floods, 461–462
    geomorphological-environmental impact, 452
    global climatic events, 463
    historical documents and phenological indicators, 452–453, 455
    historical natural disasters, 452
    human environment and human society, 466
    landslides, 460–461
    limnic eruptions and lake bursts, 465
    meteorological events, 463
    proxy records, 456, 457
    Randa rockslide, 467
    sinkholes, 463–464
    volcanic eruptions and lahars, 457–459
    wildfires, 465–466
Hoar frost, 363
Hospitals in disaster, 471–473
Hot spots, 179
Huang He River (Yellow River) flood, 461
Huascaran debris avalanches, Perù, 147–149
Human capital approach (HK), 240
Human impacts of hazards, 474–477
    hazard characteristics, 474–475
    hazard-scape, 474
    human populations, 475
    preparation/readiness strategies, 475
    reducing the impact of hazards on communities, 476–477
    reducing the impact of hazards on infrastructure, 475
    reducing the impact of hazards on people, 475–476
    resources and amenities, 474
    volcanic processes, 474
Humanitarian community, 282, 283

Humanity as an agent of natural disasters, 478–480
    Elm rockfall in Canton Glarus, Switzerland, 479
    Grote Mandränken, 479
    historical events, 478–479
    humans and natural disasters, 478
    intensified land use, 479
    Lisbon earthquake, 479
    non-geophysical hazards, 479
Hummocky topography, 145, 147, 148
Hunza dam, 605
Hurricane (typhoon, cyclone), 481–493
    Australian classification, 482
    genesis and climatology, 487–489
    Hurricane Mitch, 481, 482
    intensity, 489–492
    locations of tropical cyclone genesis, 481, 483
    paleotempestology, 492
    radar image in Hurricane Floyd, 485, 488, 489
    radial-height cross sections, 485–487
    radial-height plane, 482, 484
    Saffir-Simpson Hurricane Wind Scale, 481
    storm formation, 482, 483
    structure and size, 482, 485, 487
    tracks and motion, 489
    tropical cyclone, 481
Hurricane Andrew, 125, 126, 184, 812, 834, 932
Hurricane Ike, 125, 177, 441, 543
Hurricane Katrina, 143, 160, 164, 170, 181, 494–495
Hurricane Mitch, 114, 183, 481, 482
Hurricane Rita, 294
Hydrocompaction subsidence, 102, 496
Hydrodynamic karst hazards, 573–575
Hydrofluorocarbons (HFCs), 744
Hydrograph, flood, 497
Hydrological drought, 190–192, 194, 195
Hydrological hazards, 711
Hydrometeorological hazards, 177, 497–505
    building capacity, 503
    Center for Research on the Epidemiology of Disasters (CRED), 498
    climate, 498
    Cyclone Nargis, 498
    disaster risk reduction, 502
    drought, 499
    global sea level rise, 500
    heat waves, 498, 499
    human influence assessment, 498, 499
    impacts of hydrometeorological hazards, 500–501
    Integrated Research on Disaster Risk (IRDR), 504–505
    Intergovernmental Panel on Climate Change (IPCC), 498, 499
    International Decade for Natural Disaster Reduction, 503
    ocean surface temperature, 499
    scientific predictions, 502–503
    storms, 498
    United Nations General Assembly, 503
    World Climate Research Program, 503
Hydrostatic force, 1047–1048
Hymenoptera, 540
Hyogo Framework for Action (HFA), 165, 170, 508–516, 558–559
    Africa ministerial declaration on DRR, 2010, 515–516
    Africa strategy and program of action for disaster risk reduction (DRR), 510–515
    framework for disaster risk reduction, 508–510

Hyogo-ken-Nanbu earthquake, Japan, 237
Hypocenter, 284, 516–517
Hypothermia, 62

I
IASC Emergency Shelter Cluster, 282
Ice and icebergs, 519–520
Ice-capped volcano in South Iceland, 311
Icelandic Meteorological Office, 315
Ice storms, 520–521
Impact airblast, 522
Impact crater
    formation of, 19–21
    history of, 18–19
    recognition criteria for, 21–22
Impact ejecta, 523–524
Impact fireball, 524
Impact firestorms, 525
Impact tsunamis, 525–528
    asteroid/comet impacts, 526
    classical tsunamis, 526–528
    crown splash, 526
    dangers from oceanic impacts, 528
    Earth impacts, 526
    impact craters, 526
    physics of water impacts, 526
    wave production in impacts, 526
Impact winter, 528
Impedance, 234
Inclinometers, 529
Incorporated Research Institutions for Seismology (IRIS), 212
Indian Ocean Commission (IOC), 512
Indian ocean tsunami, 2004, 529–534
    Aceh province, Sumatra Indonesian Island, 530
    disaster, 532–534
    geomorphological effects, 531–532
    precursory signs, 530–531
    tsunami-induced damages, 532
    tsunami sources and offshore tsunami propagation, 530
Indirect damages, 141, 143
Induced seismicity, 535
Inelastic deformability, 947, 959
In-flight icing, 37
Informal/unofficial convergers, 119
Information and communication technology (ICT), 536–539
    decision support and incident management systems, 538–539
    disaster management, 536
    disaster risk reduction approach, 537
    early warning systems, 539
    electronic information systems, 536
    emergency management, 536
    Geographic Information Systems (GIS), 538
    hazard, vulnerability, and risk analysis (HVRA), 536
    internet and new social media, 539
    prevention and mitigation, 536
    remote and in situ sensing and data acquisition, 538
    space and terrestrial communication methods, 537
Initial Spread Index (ISI), 1103
INQUA Environmental Seismic Intensity Scale, 549
Insect hazards, 540–542
    bites and disease, 541
    imagined hazards, 542
    mosquito-borne disease, 541
    physical impacts, 541–542
    stings and allergies, 540

Instrument meteorological conditions (IMC), 34
Insurance, 542–543
Integrated Emergency Management System (IEMS), 544
Integrated Forest Fire Management system, 348
Integrated Research on Disaster Risk (IRDR), 504–505
Integrative risk management (IRM), 171–172, 174
Intensity scales, 544–551
　Arias intensity, 551
　earthquake intensity, 545
　elastic energy, 545
　Environmental seismic intensity scale (ESI–2007), 549–551
　intensity and isoseismal maps, 551
　macroseismic scales, 545–547, 550
　meizoseismal zone, 545
Inter-Agency Standing Committee (IASC), 1058
Intergovernmental Authority on Development (IGAD), 512
Intergovernmental Panel on Climate Change (IPCC), 498, 896–898
Intermediate Technology Development Group (ITDG), 50
International Agency for Research on Cancer (IARC), 13
International Atomic Energy Agency (IAEA), 1061
International Civil Aviation Organization (ICAO), 315
International Decade for Natural Disaster Reduction (IDNDR), 176, 388, 1056–1057
International Labor Organization (ILO), 1061
International Monetary Fund (IMF), 162
International Potato Center (CIP), 65
International Society for Soil Mechanics and Geotechnical Engineering (ISSMGE), 426
International strategies for disaster reduction (IDNDR and ISDR), 552–562
　Brundtland Commission, 553
　Decade on natural disasters, 555–556
　disaster risk reduction, 552–553
　Early Warning, 553, 559–561
　El Nino Southern Oscillation (ENSO) phenomenon, 553, 561
　European Commission Humanitarian Office (ECHO), 553
　Geneva Mandate, 555, 556
　Hyogo Framework for Action (HFA), 558–559
　Munich Reinsurance Company, 555
　National Platform for Disaster Risk Reduction, 555
　natural disasters, 553
　Scientific and Technical Committee (STC), 554
　Secretary General Kofi Annan, 555
　Tokyo Declaration, 553
　United Nations Development Programme (UNDP), 554
　United Nations Education and Scientific Organisation (UNESCO), 554
　United Nations proclaimed IDNDR, 554
　United Nations system, 554
　World Commission on Environment and Development, 553
　World Conference on Disaster Reduction (WCDR), Kobe 2005, 557–558
　World Conference on Natural Disaster Reduction, Yokohama, Japan, 555
　World Meteorological Organisation (WMO), 554
　World Summit on Sustainable Development, South Africa, 2002, 556–557
International Strategy for Disaster Reduction (ISDR), 164, 851, 1058
International Telecommunication Union (ITU), 1058
International Union for the Conservation of Nature (IUCN) Red List, 210, 308
International Volcanic Health Hazard Network (IVHHN), 1085
Internet, world wide web and natural hazards, 563–565
Intraplate earthquakes, 209, 210
Ionosphere effects, 938
Isopachs, 1076
Isoseismal, 565–566
Izmit earthquake, 181

J
Jacopo Gastaldi, 545
Japanese Meteorological Agency (JMA), 549, 649
Joaquin valley, 979, 981
Jökulhlaups, 567–569
Jurassic Neothethyan suture zone, 738

K
Karst hazards, 101, 571–577
　gasodynamic karst hazards, 575
　gravidynamic, 572–576
　hydrodynamic, 573–575
　natural and human-induced hazards, 571
　risk assessment and mitigation, 572
　soluble rocks, 571
Katrina hurricane, 143, 160, 164, 170, 181
Kelvin wave, 251, 253, 255, 256
Khait earthquake, Tajikistan, 397
Kilauea volcano, 910
Killer bees, 540
Killer Smog (smoke and fog) of London, 341
Kinetic energy, 289
Kolumbo volcanic cone, 885, 891
Koyna Dam, India, 1031
Koyna reservoir earthquake, 849
Krakatoa (Krakatau), 576

L
Lahar, 579
Lake Sarez, Tajikistan, 1065–1067
Lake Tong Le Sap, 3
Laminated rubber bearings, 239
Land degradation, 155–157, 170, 192, 195, 580–583
　anthropogenic causes, 580–582
　Food and Agriculture Organization (FAO), 582
　mechanism, 579
　natural ecosystem and human system, 580
　natural/human-induced process, 579, 580
　time and regional overview, 582
　United Nations Development Programme (UNDP), 582
　United Nations Environmental Programme (UNEP), 582
　World Meteorological Organisation (WMO), 582
Landsat satellite, 594

Landslide, 594–601, 657
　causes and triggers, 598–600
　corrective and defensive works, 600
　debris flows, 596–597
　falls, 595
　Frank Slide, 598
　Landslide hazard assessment, 153
　landslide-prone unconsolidated sediments, 595
　land-use restrictions, 600
　monitoring and early warning, 600
　mudflows, 596–597
　rock avalanches, 597–598
　sackung and lateral spreads, 598
　sediment flows, 597
　Sichuan (Wenchuan) earthquake in southwest China, 595
　slides, 595–596
　slumps, 596, 598
　topples, 595
Landslide dam, 602–605
　failure and outburst floods, 603
　Hattian Bala landslide dam, 605
　hazard assessment, 603–604
　Hunza dam, 605
　mountain environments, 602
　mountain terrains, 605
　moving landslides, 602
　spillways, 604
　three-dimensional distribution, rockslide debris, 603, 604
　Usoi dam, 602
　valley-damming landslides, 602
Landslide hazard, 379–380
Landslide impacts, 606–609
　assessing and managing impacts, 607
　physical and human context, 607
　physical and human controls of impact severity, 607–608
　process-based landslide, 609
Landslide inventory, 610
Landslide Management by Community-Based Approach in the Republic of Armenia, 115
Landslide-prone unconsolidated sediments, 595
Landslides monitoring, 691–693
Landslides reactivation, 674, 677
Landslide triggered tsunami, displacement wave, 611–613
　devastating displacement waves, 611
　glacier, 611–612
　hazard assessment of non-seismic tsunamis, 612–613
　non-seismic tsunami, 611
　ocean and lakes surrounding volcanoes, 612
　seiche, 611
　submarine landslides, 612
　submarine mass movement, 611
Landslide types, 615–618
　active and inactive landslides, 616
　displaced material, 617
　distribution of activity, 616
　inactive landslides, 616
　IUGS Working Group, 617
　liquefaction/flow, 617
　Multilingual Landslide Glossary, 615
　slide, 618
　soil/rock fall, 617
　style of the landslide activity, 616
　topple, 617
Land subsidence, 583–589
　building of surface structures and tunneling, 584, 586
　ground subsidence, 585

man-made, 584, 585
remediation and treatment, 586
responsibility and liability, 586, 589
unconsolidated soils, 584, 587–588
underground cavities, 584
vertical downward displacement, 583
voids, 584
Land-use planning, 618–622
flood waves, 619
intra-generational conflict, 619
natural hazards and relevance, 618–619
natural variability, 620
risk management, 620–622
Theory of Justice, 619
Veil of Ignorance, 619
Land use, urbanization and natural hazards, 590–593
Laser telecommunications, 340–341
Lateral spreading, 153, 623
Lava, 623–624
Lena River at Lensk City, 462
Levee, 624
Light detecting and ranging (LiDAR), 7, 96, 154, 215–217, 379, 380, 600
Lightning, 36, 625–629
electrical activity, 625
electroporation and flashover, 626
electrosphere, 625
epidemiology, 626–627
global electric circuit, 625
injury and treatment, 627–629
prevention of injury, 627
Linear shrinkage, 297, 300
Liquefaction, 629–633
flow slide, Lake Merced, San Francisco, 629, 630
ground shaking, 632
initiation, 631–632
instability, 632
liquefaction-induced foundation failure, 630
mitigation, 632–633
post-earthquake settlement, 629, 631
settlement-related building damage, 632
soil susceptibility, 629–631
Liquid core pellets, 521
Lisbon earthquake, 46, 160, 176, 211, 215, 459, 479
Lithosphere subduction, 979
Lituya Bay, Alaska, 392, 611, 988, 1039
Livelihoods and disasters, 633–636
Local mean seal level (LMSL), 981
Local wave magnitude scale, M_L, 643–644
Loess, 637–638
Long Beach earthquake, 851
Long-period earthquakes (harmonic tremor), 210
Loose snow avalanches, 32
Los Angeles International Airport (LAX), 341
Los Angeles metropolitan area, 662
Low-frequency earthquakes, 1005

M
Macroscale atmospheric circulation patterns, 87
Macroseismic scales, 545–551
China Seismic Intensity Scale (CSIS), 549
European Macroseismic Scale (EMS-98), 547
Japanese Meteorological Agency, 549
Mercalli-Cancani-Sieberg Scale, 547
12-point scale, 547
Rossi-Forel Scale, 545
significance, 549, 550
Macroseismic survey, 639
Magma, 639–640, 910

Magmatic eruption, 1075
Magnitude measures, 640–650
body wave magnitude scale, m_B and M_B, 646
duration magnitude scale, M_D, 647–648
earthquake magnitude, 640
energy magnitude scale, M_e, 646–647
local wave magnitude scale, M_L, 643–644
magnitude measurements scales, 643
magnitude of Japanese earthquakes, M_{JMA}, 649
moment magnitude scale, M_w, 643, 645
Nuttli magnitude scale, M_N, 648–649
quantification of earthquake size, 641
seismic energy, 642–643
seismic event characterises, 641
seismic moment, 641–642
seismic waves, 641
surface wave magnitude scale, M_S, 644
Magnitude of Japanese earthquakes, M_{JMA}, 649
Major depressive disorder (MDD), 780, 785
M8 algorithm, 227
Managua earthquake, Nicaragua, 188
Man-made land subsidence, 584, 585
Mapping radon prone areas, 810–811
Marble, 51
Marginality, 651
Marine hazards, 652–655
downslope mass transport, 652
earthquakes, 653
gas hydrates, 654
geohazard identification and assessment, 654
geological processes, 652–653
mud diapirs and volcanoes, 653–654
offshore petroleum industry, 652
salt diapirs, 654
seabed displacements, 652
shallow gas and shallow water flow, 654
slope stability, 653
Mass care events, 773–774
European Union (EU), 774
post-disaster needs assessment, 774
scenario-based models, 774–775
United Nations (UN), 773–774
Massive flank failures, 910–911
Mass media and natural disasters, 655–656
Mass movement, 657–660
geomorphic agents, 657
gravitational slope-forming processes, 657
landslides, 657
mountainous landscapes, 657
rapid, deep-seated slides and flows, 658–659
role of climate, 658
shallow, rapid mass movements, 658
slower, deep-seated landslides, 659
slow flows and deformations, 659–660
surficial mass movements, 660
trigger mechanisms, 657
Varnes classification system, 658
Mayon Volcano, Philippines, 791, 793
Mean sea level (MSL) change, 895–900
coastal flood threshold probability, 899, 900
Intergovernmental Panel on Climate Change (IPCC), 897–898
local land level, 899
mitigation, 899
satellite measurements, 896
thermal expansion, 895
tide gauge, 895–896
water exchange, 895
Mechanical fluidization, 147
Medvedev-Sponheuer-Karnik Scale (MSK–64), 224

Megacities and natural hazards, 660–664
agglomerations, 661
assets, 663–664
coastal megacity, 662
disaster risk theory, 663, 664
global climate change, 661
hydrometeorological hazards, 662–663
Los Angeles metropolitan area, 662
sea-level rises, 663
substantial seismic hazard, 662
urbanization growth rates, 661
urbanization process, 661
Mega-fires in Greece (2007), 664–671
Meizoseismal zone, 545
Mercalli Scale, 144, 547
Mercalli-Cancani-Sieberg Scale, 547
Mercalli, Giuseppe (1850–1914), 671–672
Messina earthquake, Italy, 231, 460
Meteorite, 672
Meteorological drought, 190, 191
Meteorological monitoring, 687–690
Methane release from hydrate, 672–673
Microblogging service, 797
Microburst, 36
Microtektites, 24–25
Mineral particle toxicity, 413
Minimum orbit intersection distance (MOID), 29, 30
Mining and gas flaring, 443
Mining subsidence induced fault reactivation, 673–678
coal, 676
fault scarp, 675
landslides reactivation, 674, 677
mechanisms, 676
physical damage, 674, 676, 677
scarp, graben, fissure/zone of compression, 674, 675
topographic expression, 674
Minor planet center, 29
Mirounga angustirostris, 308
Misconceptions about natural disasters, 678–682
causes of, 678
cost-benefit analysis, 680
Earth's natural systems, 679
meteorological disasters, climate change, 680
mudslide, 680
natural process, 679
non-constructive reaction, 681
people resist hazard mitigation, 680–681
reliable modification, 679
river correction, 679
sustainable, 679
worst-case scenario, 681
Mississippi River, 40
Mitigation, 682–683
Mixed motion avalanche, 32
Moderate Resolution Imagining Spectroradiometer (MODIS), 843
Modified Mercalli Intensity (MMI) scale, 53, 213, 219, 421–422, 683–685
MODVOLC algorithm, 844
Monitoring natural hazards, 686–694
early warning system, 686
earthquake monitoring, 690
floods monitoring, 690
forest fires, 694
ground movement measurements, 686
hail storms, 694
hurricane statistics, 686
instruments and measured variables, 686–687
landslides monitoring, 691–693

Monitoring natural hazards (*Continued*)
  meteorological monitoring, 687–690
  snow avalanche monitoring, 693–694
  tsunamis monitoring, 690–691
  volcanoes, 691
Monsoons, 696–697
Monthly mean sea level, 895
Montreal Protocol, 742–744
Montserrat eruptions, 697–699
Mortality and injury in natural disasters, 699–702
  drought, 699–700
  earthquakes, 700–701
  extreme temperature events, 701–702
  floods, 701
  human populations impacts, 699
  rapid-onset natural disasters, 700
  storms, 701
Moscow-Prague Formula, 644
Most Likely Landslide Initiation Points (MLIPs), 379
Mountain glacier, 755
Mountainous landscapes, 657
Mountain permafrost, 760
Mount Merapi, 183
Mount Pinatubo, 703–704
Mount St. Helens, USA, 21, 394, 396
  cataclysmic eruption, 148
  composite volcanoes, 8
  debris avalanche, 147
  devastating displacement waves, 612
  high internal velocity, 794
  lateral blasts, 794
  loose sand, and flow downvalley, 579
  paroxysmal explosive phase, 1081
  pyroclastic density currents, 793
  surge-derived pyroclastic flows, 1082
  volcanic-tectonic earthquake activity, 210
Mount Vesuvius (Italy) eruption, 457
Mud diapirs, 653–654
Mudflow, 596–597, 706
Mudslide, 680
Mud volcanoes, 653–654, 705
Multilingual Landslide Glossary, 615
Munich Reinsurance Company, 555
Muong Nongtype tektites, 24, 25
Murgab River valley, 1066
Myths and misconceptions in disasters, 706–709
  dead bodies, epidemics and disease, 707–708
  displacement and disaster-stricken populations, 708–709
  donations, 709
  looting and social unrest, 708
  Pan-American Health Organization (PAHO), 707
  social unrest, 709

N

Nankai drilling project in Japan, 217
Nankai Trough subduction zone of southwest Japan, 1005
Narino Province, 369, 371
National Adaptation Programmes of Action (NAPAs), 515
National Building Code of Canada (NBCC), 218
National Centers for Environmental Prediction (NCEP), 259
National Fire Protection Association (NFPA), 268
National Flood Insurance Program (NFIP), 543
National Geophysical Data Center (NGDC), 1041

National Incident Management System (NIMS), 73–74
National Oceanic and Atmospheric Administration (NOAA), 127, 167, 843
National Platform for Disaster Risk Reduction, 552–553
National Science Foundation (NSF), 417
Natural hazard, 711–718
  coastal movement, 713–714
  communicating natural hazards, 716–717
  creeping phenomenon, 712
  droughts, 712
  Earth's geosystem, 711
  economic framework, 713
  geological and atmospheric hazards, 711
  geological hazards, 711
  geophysical processes, 711
  human-induced reasons, 713
  hydrological hazards, 711
  impact and predictability, 712
  meteorological hazards, 711
  natural disaster, 712–713
  oceanographical hazards, 711
  physical effects, 712
  primary and secondary phenomena, 712
  public perception, 715–716
  risk, vulnerability, and mitigation, 714–715
  SHIELD Project, 711
  spatial distribution, 712
  time of occurrence, 712
Natural hazards, 378
Natural hazards in developing countries, 718–725
Natural hazard warning system, 1091
Natural levees, 624
Natural radioactivity, 726–729
  areas of high natural radioactivity, 729
  cosmic radiation, 726
  cosmogenic radionuclides, 726
  Oklo natural nuclear reactor, 729
  primordial radionuclides, 726
  radioactive decay, 726
  secondary radionuclides, 726
  Technologically enhanced naturally occurring radioactive materials (TENORM), 729
  terrestrial natural radioactivity, 726–729
Navier-Coulomb criterion, 848
Navier-Stokes equations, 117
Near-Earth Asteroids (NEAs), 10, 14, 15, 28, 29
Near-Earth Object (NEO), 1019
Near-earth object search programs, 16
Neotectonics, 730–731
  active seismicity and active faulting, 730
  late Cainozoic tectonics, 730
  measuring neotectonic motions, 730–731
  modeling neotectonic motions, 731
  seismogenic and non-seismogenic neotectonic deformation, 731
  tectonic geomorphology, 730
  tectonic movements, 730
Neurotoxic shellfish poisoning (NSP), 826
Nevado del Ruiz Volcano, Colombia 1985, 732–738
  analysis of a catastrophe, 735–736
  Armero died, 735
  emergency management system, 732
  fumarolic activity, 733
  Galeras and Puracé volcanoes, 732
  geology-mines bureau, 733
  INGEOMINAS, 733
  lahars, 732
  Lessons from Armero, 737
  map of, 732, 733
  phreatic eruption, 735

re-awakening of Nevado del Ruiz, 732–734
summit crater, 732
volcanic unrest, 735
Newton's law, 2, 232
Nongovernmental organizations (NGOs), 269
Nonmarket effects, 141
Nonprofit organizations (NPOs), 263, 269
Normal faults, 209, 317319–
Normalized Difference Vegetation Index (NDVI), 843
North Anatolian Fault, Turkey, 738–739
North China Plain, 989
Northridge earthquake, California, 126, 143, 184, 953, 970, 974
Nourished beach, 42
Novel Strong-Motion Seismic Network for Community Participation in Earthquake Monitoring, 114–115
Nuclear winter, 528
Nuée ardente, 740
Nuttli magnitude scale, M_N, 648–649

O

Oblique-slip faults, 317
Oceanic trenches, 979
Oceanographical hazards, 711
Office for the Coordination of Humanitarian Affairs (OCHA), 1058
Oklo natural nuclear reactor, 729
Open-source software systems, 379
Osceola Mudflow, Mount Rainier, USA, 579
Overgrazing, 741
Ozone, 741–743
Ozone depleting substances (ODS), 742
Ozone loss, 743–745
Ozone repair, 744

P

Pacific Tsunami Warning and Mitigation System (PTWS), 747–748
Pacific Tsunami Warning Center (PTWC) in Hawaii, 1043
Pahoehoe lava, 748
Paleoflood hydrology, 748–749
Paleomagnetism, 770
Paleoseismology, 749–750, 906
Palermo scale, 29, 30
Palmer Drought Severity Index (PDSI), 191, 192
Palmer Hydrological Drought Severity Index (PHDI), 191
Panarchy model, 831, 833
Paper plan syndrome, 828
Paraglacial, 750–755
  climate change, 754–755
  equilibrium-disturbance models, 751
  fluvial sediment yield changes, 750
  glacial-interglacial timescales, 750
  glacially conditioned sediment availability, 750
  mass movements, 752
  mountain glacier, 755
  natural hazards, 754
  proglacial processes, 750
  rainfall intensity, 753
  response of paraglacial landscapes, 754
  river responses to ice retreat, 751–752
  rockslides, rock avalanches and rock falls, 752–753
  sediment yield changes, 750
  solifluction, 753, 754
Paralytic shellfish poisoning (PSP), 826
Parameterless Scale of Intensity (PSI), 144

Paricutín volcano, Mexico, 138, 139
Paroxysmal explosive phase, 1081
Peléan eruptions, 290–291
Perception of natural hazards and disasters, 756–759
  cognitive factors, 758
  decision making process, 758
  experts and non-experts assess, 757
  hazard perceivers, 758
  human perception, 757
  media reports/school education, 758
  psychological phenomenon, 757–758
  risk perception analysis, 756
  types of communities, 758
  types of perception and behavior, 758
Perihelion, 29
Permafrost, 759–762
  buffer layer, 762
  creep, 129
  degradation, 145
  ground ice and carbon, 760
  mountain permafrost, 760
  plateau/montane permafrost, 760
  polar permafrost, 760
  subsea permafrost, 760
  thaw and C/N release, 760–761
  thermokarst activity and global warming, 761–762
Permanent Service for Mean Sea Level (PSMSL), 895–896
Perrow's hypothesis, 832–833
Person-Relative-to-Event-Model, 476
Philippine Institute of Volcanology and Seismology (PHILVOLCS), 703
Phreatic eruption, 290
Phreatomagmatic eruption, 290, 1075
Phreatoplinean, 290
Phytophthora infestans, 69
Piezometer, 764
Piping hazard, 764–765
Piuro landslide, Italy, 146, 461
Planar deformation features (PDFs), 23–24
Planar fractures (PFs), 23
Planning measures and political aspects, 765–769
  communication, 767
  emergency planning, 765, 768
  environmental impact assessment, 765–766
  government and governance, 766–767
  policy options and development, 767–768
  political systems, 766
  potential hazards, 766
  restoration and rebuilding, 768
  risk reduction, 769
  spatial and land-use planning, 765, 767
  strategic environmental assessment, 766
  sustainability appraisal, 766
Plant pathogens, 68
Plant pests and diseases, 67–68
Plate tectonics, 769–771
  convergent boundaries, 770–771
  divergent boundaries, 770
  major and minor tectonic plates, 771
  paleomagnetism, 770
  subduction zones, 770
  supercontinent, 771
  transform boundaries, 771
Plinian eruptions, 1073
Pneumatic piezometers, 764
Polar permafrost, 760
Policy process model, 269, 270
Polygenetic volcanoes, 1077
Pompeii eruption, 1073

Pore fluid vaporization, 147
Pore-water pressure, 772
Post disaster mass care needs, 773–775
Post-disaster recovery, 637
Post-earthquake settlement, 629, 631
Posttraumatic stress disorder (PTSD), 134, 423, 475, 629, 776–777, 780, 783–785
Potentially hazardous asteroids (PHAs), 16
Potential volume change (PVC), 298
Pre-disaster vulnerability, 635
Pre-event damage assessment methods, 143
Preimpact disaster planning, 268
Presidential Disaster Declarations (PDDs), 819
Pressure and Release (PR) model, 165
Primary wave (P-Wave), 777
Primordial radionuclides, 726, 727
Probable maximum flood (PMF), 749, 777–778
Probable maximum precipitation (PMP), 777–778
Process-based landslide, 609
Proximal ejecta, 21
Pseudotachylitic breccia, 22
Psychological impacts of natural disasters, 779–790
  chronic, severe, debilitating distress, 789
  critical incident stress debriefing (CISD), 786–787
  disaster-induced stress and distress reactions, 782, 784
  empirically informed early intervention, 787
  forces of harm, 781–782
  harm's way, 782–783
  intermediate-term intervention, 788
  long-term intervention, 788
  major depressive disorder (MDD), 785
  mental and behavioral health consequences, 779–781
  post-disaster mental and behavioral health interventions, 786
  posttraumatic growth, 789
  posttraumatic stress disorder (PTSD), 784–785
  prevention, 786
  protracted recovery, 789
  psychiatric disease, 783
  psychological first aid, 787
  resilience, 789
  resistance, 789
  spectrum of severity, 780
  triage, screening, and referral, 785–786
Public notification systems, 1091
Purgatorio ratio (PR), 29, 30
Pyramid of damages, 142
Pyroclastic density current (PDC), 41, 381, 740, 791–795, 944, 1082, 1083
Pyroclastic flow, 740, 791–795
  lateral blasts, 794
  Mayon Volcano, Philippines, 791, 793
  origin, 792–793
  rock fragments and hot gases, 795
  terminology, 792
  transport, 793–794
  volcanic hazard mitigation, 794–795
Pyroclastic surges, 41

Q

Qianjiangping landslides, 914–915
Quasi-periodic behavior, 257–259
Queensland floods (2010–2011) and tweeting
  data collection, 798
  ego analysis, 798–801
  microblogging service, 797
  social media, 797
  social network analysis (SNA), 798
  visual network analysis, 798, 799

Quick clay, 803–804
Quick sand, 804–805

R

Radar systems, 6, 7
Radiation dose, 807
Radiation hazards, 807–808
Radiative frosts, 363
Radon hazards, 808–811
  geological associations, 809
  lung cancer, 809
  mapping radon prone areas, 810–811
  natural radioactive gas, 808
  radioactive decay of uranium, 808
  radon gas, 809–810
Radon prone areas, 810–811
Rainfall intensity, 753
Rainfall irregularity and drought, 65–66
Rain gauge data, 197
Randa rockslide, Switzerland, 467
Rapid undrained loading mechanism, 149
Reactive adaptation, 4
Recovery and reconstruction after disaster, 812–822
  business recovery, 817–818
  disaster recovery goals, 815
  household recovery, 816–817
  impact ratio, 812
  physical impacts, 812–813
  social impacts, 813–815
  sources of recovery assistance, 818–821
    autonomous recovery, 818
    community-based organizations (CBO), 819
    hazard insurance, 818–819
    higher levels of government, 819
    local government recovery functions, 819–820
    nongovernmental organizations (NGO), 819
    preimpact recovery operations plan, 821
    recovery/mitigation committee, 820–821
  stages and functions, 815–816
Recovery/mitigation committee, 820–821
Recurrence interval, 824
Red Crescent national societies, 825
Red cross and red crescent, 825
Red River Floodway, Canada, 338
Red tides, 826
Reflections on modeling disaster, 827–834
  CARE household model, 830, 832
  catastrophe models, 833–834
  comprehensive disaster/emergency management (CEM) model, 829
  disaster as spectacle, 828
  disaster incubation, 827–828
  disaster management, 827
  disaster models, 827
  ecological models, 831, 833
  explicit conceptual models, 827
  normal accidents, 832–833
  paper plan syndrome, 828
  pressure and release model, 829–831
  risk management model, 830, 832
  unnatural disaster, 827
Reflectivity measurement, albedo, 10
Reid's model, 227
Relative biological effectiveness (RBE), 807
Release rates, 835
Religion and hazards, 836
Remote sensing, 5
Remote sensing of natural hazards and disasters, 837–846
  data record, 838, 839
  disaster information requirements, 839–841

Remote sensing of natural hazards and disasters (*Continued*)
   disaster risk reduction (DRR), 837
   electromagnetic spectrum, 838
   hazard monitoring and early warning, 843–845
      disaster constellations and international disaster support, 845
      emergency response, 844–845
      satellite-based monitoring systems, 843–844
      tropical cyclone forecasting, 843
      tropical storms, 843
      unmanned aerial vehicles (UAV), 845
   hazard to risk, 842
   hazard types, 841
   natural/industrial hazards, 846
   sensors and platforms, 837–838
   temperature/gravity, 837
   visual information extraction, 838–840
Reservoir, dams and seismicity, 847–849
Reservoir triggered seismicity (RTS), 848–849, 1032
Resilience, 849–850
Resonant return, 29, 30
Response Modification Factor, 233
Retro belt system, 973
Reverse faults, 318
Revised Universal Soil Loss Equation (RUSLE), 1062
Richter, Charles Francis (1900–1985), 850–851
Rights and obligations in international humanitarian assistance, 851–854
   humanitarian intervention, 852–853
   International Strategy for Disaster Reduction (ISDR), 851
   natural and human-made disasters, 851
   right to assist *vs.* right to assistance, 853–854
   UNESCO in Paris, 851–852
Ring of Fire, 8
Rip current, 855
Risk, 856–861
   formulas, 858
   hazard and vulnerability, 856–857
   hazard occurrence, 856
   knowledge demands, 860
   natural hazards and climate change, 859
   probability, 858–859
   sociological perspective, 857–858
   timescales, 858, 860
Risk assessment, 862–863
Risk culture, 171–172
Risk governance, 172, 863–869
   analyzed projects/initiatives, 867
   communication and disaster cycle, 864
   comprehensive risk governance concept, 866
   decision-making stakeholders, 867
   indicator system, 867
   integration of stakeholder, 869
   integrative models, 864
   relevance for natural hazards, 864–866
   resilience framework, 867, 868
   strategic environmental assessment (SEA), 868
Risk management cycle, 174
Risk Management Index (RMI), 168
Risk perception and communication, 870–873
   decision makers, 871
   environmental hazards perception, 871
   hazard adjustments perceptions, 871
   information sources, 871–872
   risk communication, 872–873
River flooding, 80
Rock avalanches, 597–598, 875
   landslides, 597–598

   sackung linears, 881
Rockbursts, 535
Rock creep, 129
Rockfall, 875–876
Rockfall shadow, 595
Rockslide-debris avalanche, 397
Rogue wave, 877
Rosa lava flow, 1081
Rossby wave, 251, 256
Rossi-Forel scale, 213, 224, 545
Rotational seismology, 877–879
Rubble mound breakwaters, 48
Running sand, 804

S
Sackung, 152, 154, 881–882
Saffir-Simpson Hurricane Intensity Scale, 882–883
Saffir-Simpson Hurricane Wind Scale, 481
Saffir-Simpson scale, 127, 481
Saharan Air Layer, 410
Saint-Venant equations, 1008
Salinization, 157
Saltation, 201–203
Salt diapirs, 654
San Andreas fault, California, 7, 80, 216–218, 227, 249, 319, 771, 883, 906, 928, 1002, 1005
San Andreas Fault Observatory at Depth (SAFOD) project, 217
San Francisco earthquake, USA, 231, 249, 437, 460
Santorini, eruption, 884–893
   chemical composition, 892
   electronic distance meters, 890
   geochemical parameters monitoring, 892–893
   geographic settings, 884–885
   geological setting, 885–886
   hazard assessment and zonation, 893
   temperature fluctuation, 892
   Thera's eruption, 887–889
   thermal-chemical monitoring, 891
   tsunami effects, 886
   volcanic hazards, 886–888
Satellite imagery, 5–9
Sea and lake erosion, 288
Seabed displacements, 652
Sea ice, 519
Sea level change, 895–900
   Intergovernmental Panel on Climate Change (IPCC), 896–897
   Mean sea level change (*see* Mean sea level change)
   mitigation, 896
   tides and surges, 895
   tsunami, 895
Sea surface temperature (SST), 194, 252, 254, 255, 257, 259–261
Second Africa Ministerial Conference, Nairobi, Kenya, 515
Secondary radionuclides, 726, 727
Secondary wave (S-wave), 901
Sedimentation of reservoirs, 901–904
   impounding facilities, 901, 902
   sedimentation rate, 902–903
   sediment deposition, 901
   sediment management, 903–904
   suspended sediments, 901
   turbidity currents, 903
   water storage capacity reduction, 901, 902
Seiche, 611, 905
Seine River estuary, France, 1008
Seismic gap, 906

Seismic hazard, 381
Seismic hazard assessment, 319–320
Seismic zoning, 993, 1100
Seismic waves, 47
Seismographs, 217, 907
Seismology, 907–908
Semi-active control devices, 968
Seriousness, Manageability, Urgency, and Growth (SMUG) model, 167
Severe storms, 36
Shaanxi earthquake in China, 388
Shear, 908–909
SHIELD Project, 711
Shield volcano, 911–912
Shock metamorphic effects, 22–24
Short-Term Earthquake Probability (STEP) method, 228
Sichuan earthquake, China, 176, 177, 595
Sichuan province of China, 396
Sinkhole, 911
Sinkhole in Berezniki, Russia, 399
Slab retrofits, 977
Slide and slump, 913–918
   flow-type failures, 913
   rotational slides, 913, 914, 917–918
   topples and falls, 913
   translational slides, 914–916
Slope failure (landslide) hazards, 8
Slope stability, 919–923
   fall, 919, 920
   flow, 919, 920
   increased destabilizing forces, 323
   lateral spreading, 919, 920
   reliability index, 921
   rotational slide, 919–921
   safety factor, 921
   shear strength, 921–922
   stabilizing force reduction, 922–323
   topple, 919, 920
   translational slide, 919–921
Smithsonian Institution Holocene database, 1081
Snow avalanche monitoring, 693–694
Snowstorm, 924–925
Social-ecological systems, 926
Social Network Analysis (SNA), 798
Socioeconomic drought, 191
Sociology of disaster, 926–933
   death rates and monetary costs, 927
   hazards paradigm, 928
   media and popular culture in, 932–933
   military's command and control model, 927–928
   natural physical trigger, 926–927
   social vulnerability paradigm, 928–932
   sociopolitical ecology perspective, 932
Soil Conservation Service, 199, 200
Soil creep, 129, 149
Soil erosion, 288
Soil susceptibility, 629–631
Solar dimming process, 412
Solar flares, 936
Solifluction, 753, 754, 936–937
Somma volcano, 1073–1074
South Agean Active Arc, 886
Southern African Development Community (SADC), 512
Southern Oscillation Index (SOI), 251, 254
South Pole-Aitken basin, 25
Space and terrestrial communication methods, 537
Spaceguard, 26
Space weather, 937–939
   earth solar energetic particles, 937
   hazard, 986

ionosphere effects, 938
magnetic disturbances, 937–938
mitigation and forecasting, 939
space environment effects, 938–939
Spalling, 947
Spectral modal analysis method, 233
Spillways, 604
Standardized precipitation index (SPI), 191
Standpipe piezometers, 764
Steeply dipping interbedded sedimentary rock strata, 8
Stellar explosion, 986–987
Stir cross-border tensions, 184
Storms, 701, 941
Storm surges, 940
  agricultural production, 66
  concrete structures, 118
  Federated States of Micronesia (FSM), 66
  injuries, 62
  sea level rise, 66
Strategic environmental assessment (SEA), 622, 766, 767, 868
Stratospheric ozone, 742–743
Stratovolcanoes, 941–946
  andesitic magmas, 943
  chains wraps, 942
  debris avalanches, 945
  destructive and unpredictable side, 942
  directed blasts, 944
  dormancy periods, 943
  geo-chemical techniques, 945
  hazards research, 945–946
  height and inherent instability, 943–944
  lahars, 944–945
  lava flows and pyroclastic fragmental deposits, 942
  pyroclastic density currents (PDC), 944
  pyroclastic fall, 944
  stacked beds of eruption products, 942
  tsunami, 945
  unpredictability, 943
Strike-slip earthquakes, 209
Strike-slip faults, 319
Strombolian eruptions, 290
Structural damage caused by earthquakes, 947–959
  adobe and masonry buildings, 948–949
  building performance
    collapse prevention performance level, 958–959
    Federal Emergency Agency (FEMA), 955
    immediate occupancy performance level, 956–957
    life safety performance level, 957
    operational level of performance, 956
  high-frequency ground motions, 948
  inelastic deformability, 947–948
  lack of strength, 947
  reinforced concrete buildings, 949–952
    beam-column joint damage, 952, 953
    lack of confinement reinforcement, 950–951
    plastic hinges, 949–950
    shear failures, 951–952
    shear wall damage, 952, 956
    strong beam-weak column joints, 952, 954
    vertical and horizontal irregularities, 955
  seismic damage in bridges, 953–955
    bridge abutments, 955
    bridge superstructure, 954
    bridge superstructure failure, 954, 958
    unseating of bridge girders, 955
  seismic hazards, 948
  structural steel systems, 952–953

timber construction, 953
vertical and horizontal discontinuities, 948
Structural Design Method of Buildings for Tsunami Resistance (SMBTR), 1047
Structural mitigation, 959–977
  beam-column joint retrofits, 974–975
  beam retrofits, 973–974
  bridge superstructure retrofits, 975–976
  column retrofits, 970–973
    concrete jacketing, 971
    fiber reinforced polymer jackets, 971–972
    retro belt system, 973
    seismic damage, 970
    steel jacketing, 970–971
  non-ductile structures, 960
  seismically deficient structural systems, 960
  single and multifamily residential house retrofits, 976
  slab retrofits, 975
  system level seismic retrofit strategy (see System level seismic retrofit strategy)
  wall retrofits, 973, 974
Structural shrinkage, 297
Structure failure, 50
Sturzstroms, 145, 146, 875
Subduction, 979
Subduction earthquakes, 209
Submarine landslides, 612, 1037
Subsea permafrost, 760
Subsidence induced by underground extraction, 979–984
  anthropogenic and natural processes, 980
  inundation of coastal lands, 980
  regional subsidence, 980
  subsurface fluid withdrawal
    aquifer system, 982
    aquitard drainage model, 982
    conventional groundwater theory approach, 982
    geothermal fields, 981
    Groningen gas field, 981
    groundwater pumping, 980
    Joaquin valley, 981
    local mean seal level, 981
    Wilmington oil field, 981
  underground coal mining, 981–282
    coal seam, 284
    distinct-element models, 984
    Gondwana coalfields, 282
    mathematical models, 984
    pit and sag subsidence, 982–984
    sinkholes, 984
Sudbury lakes, 2
Sudden-impact disaster, 79
Sumatra-Andaman Islands tsunami, 1038
Sumatra-Andaman megathrust (Mw 9.2) earthquake, 392
Sunspots, 986
Super-eruption, 1081–1082
Superficial gravidynamic hazards, 571–572
Supernova, 986–987
  average frequency, 986
  cosmic rays, 986
  interstellar impact, 987
  nearby supernovae and effects, 987
  types II, 986
Supplementary damping devices, 967–968
Surface Water Supply Index (SWSI), 191
Surface wave magnitude scale, M_S, 644
Surge, 987–988
Surge force, 1049, 1050
Surtseyan eruptions, 41
Survival Commission (SSC), 307
Susceptibility, 988

Sustainable Coastal Communities and Ecosystems (SUC-CESS), 115
Sustainable sediment strategy, 904
Systemic damages, 142
System level seismic retrofit strategy
  active control devices, 968–970
  active control systems, 968–970
  base isolation, 964–967
    ground motion, 965
    isolators, 965
    lateral bracing, 964
    passive dampers, 965
    rigid and non-ductile structures, 965, 967
    seismic deformations, 965
    sliding friction pendulum bearing, 966
  conservation of energy, 960
  lateral bracing and strengthening, 961–964
    diagonal prestressing strands and cables, 962–963
    recoverable elastic strain energy, 960
    reinforced concrete structural walls, 961
    steel braces, 961–963
    surface-bonded fiber-reinforced polymer sheets, 961–962, 964, 966
    unreinforced masonry (URM) infill walls, 961, 963–965
  semi-active control devices, 968
  supplementary damping devices, 967–969

T

Tangshan earthquake, China, 212, 218
  casualties, 990–992
  earthquake emergency response, 994
  earthquake prediction, 993–994
  economic loss, 990–992
  intensity contour map, 989, 990
  location, 989, 990
  seismic characteristics, 989
  statistics of casualties, 991–992
  Tangshan City, 989
  Tangshan Prefecture, 989
Technological fixes, 264
Technologically enhanced naturally occurring radioactive materials (TENORM), 729
Technological notification systems, 1093–1094
Tectonic and tectono-seismic hazards, 994–1004
Tectonic tremor, 1004–1005
Tektites, 24–25
Teleconnections, 252, 253
Tenerife airport disaster, 343
Terrain analysis and interpretation, 5
Terrestrial natural radioactivity, 726–729
Theory of Justice, 619
Thera's eruption, 887–889
Thermal infrared images, 6
Thermokarst activity and global warming, 761–762
Thunderstorms, 36, 1006
Tianjin Municipality
  economic loss, 992–993
  North China Plain, 989
  statistics of casualties, 992
Tidal bores in Mont Saint Michel Bay, France, 1007–1009
Tide gauge, 895–896
Tides, 987
Tiltmeters, 1009
Timber construction, 953
Timber harvesting and forest conversion, 917–918
Time and space in disaster, 1009–1014
Tohoku earthquake and tsunami, Japan, 179, 1015–1018

SUBJECT INDEX

Tokyo Declaration, 553
Tomnod Disaster Mapper, 565
Tong Le Sap lake, 3
Topography, morphological factors, 379
Torino scale, 29, 30, 1019
Tornadoes, 36, 115, 1019–1030
Total Ozone Mapping Spectrometer (TOMS), 744
Toxic elements, 61
Transform boundaries, 771
Transient cavity, 21
Treat risks box, 830, 832
Triggered earthquakes, 1031–1035
Tropical cyclone forecasting, 843
Tropical-Ocean-Global-Atmosphere (TOGA) Program, 253
Tropical Rainfall Measuring Mission (TRMM), 334
Tropical storms, 843
Tsunami, 51–53, 1036–1044
  boats and shipping, 1041
  buildings, 1040–1041
  communication and response, 1043–1044
  earthquakes, 1038–1039
  extraterrestrial objects, 1040
  forecasting, 1043
  Grand Banks tsunami, 1039
  hazard assessment, 1042
  historical tsunami and paleotsunami, 1041–1042
  impacts of tsunami, 1040
  impulsive disturbance, 1036
  Indian Ocean Tsunami, 66
  infrastructure, 1041
  landslides, 1039
  land-use planning, 1042–1043
  monitoring, 690–691
  physical properties, 1037–1038
  seiches, 905
  storm-surges/meteorological tsunami, 1037
  structural mitigation, 1043
  submarine landslides, 1037
  sudden displacement, ocean water, 79
  tsunami modeling, 1042
  tsunami run-up distance, 1037
  tsunami run-up height, 1037
  tsunami wave height, 1037
  volcanoes, 1039–1040
  warning systems, 1043
Tsunami loads on infrastructure, 1046–1052
  breakaway walls, 1052
  buoyant and uplift forces, 1049
  catastrophic events, 1046
  coastal bathymetry, 1046
  coastal protection structures, 1046
  debris impact and damming forces, 1050–1051
  flood-induced loads, 1047
  flow velocity, 1047, 1050–1051
  gravity forces, 1051
  high-risk tsunami-prone areas, 1047
  hydrodynamic (drag) force, 1048–1049
  hydrostatic force, 1047–1048
  impulsive force, 1049, 1050
  initial impact and post impact, 1051–1052
  inundation depth, 1047
  low-lying coastal areas, 1046, 1047
  reinforced concrete structures, 1046, 1048
  wave-breaking, 1051
Tune-mass dampers (TMD), 239
Tungurahua volcano, Ecuador, 293
Turbidity currents, 903
Turbulence, 37–38

U

Ultra-Plinian eruption, 290
Uncertainty, 1055
Underground gravidynamic hazards, 571–572
UN-Habitat, 1060
United Nations Centre for Regional Development (UNCRD), 1060
United Nations Children's Fund (UNICEF), 1059
United Nations Convention to Combat Desertification (UNCCD), 155
United Nations Development Group (UNDG), 1058
United Nations Development Programme (UNDP), 554, 582, 1059
United Nations Disaster Relief Organization (UNDRO), 856
United Nations Educational, Scientific and Cultural Organization (UNESCO), 554, 1058
United Nations Environment Programme (UNEP), 582, 1059–1060
United Nations General Assembly, 503, 1057
United Nations Institute for Training and Research (UNITAR), 1060
United Nations International Strategy for Disaster Reduction (UN ISDR), 828, 1111–1113
United Nations Millennium Development Goals (UN MDGs), 170
United Nations Office for Outer Space Affairs (UNOOSA), 1060
United Nations Organizations and natural disasters, 1056–1061
United Nations Population Fund (UNFPA), 1060
United Nations Strategy for Disaster Reduction, 181
United Nations University (UNU), 1060
United Nations Volunteers (UNV), 1061
Universal Soil Loss Equation (USLE), 289, 1062
Unmanned aerial vehicles (UAV), 845
Unreinforced masonry (URM) buildings, 961, 963, 964, 1062–1063
Urban air quality, 742
Urban environments and natural hazards, 1063–1065
  air quality degradation, 1064
  biological hazards, 1064
  flooding, 1063
  heat waves, 1063
  large-scale urbanization, 1064
  natural physical hazards, 1064
  urban geochemical hazards, 1064–1065
Urban heat island, 1064
USA Clean Air Act Amendments, 3
Users-resources network, 798, 799, 801, 802
US flood hazard, 176
US Homeland Security Mapping Standard, 273
Usoi dam, 602
Usoi landslide and Lake Sarez, 1065–1067

V

Vaiont landslide, Italy, 1069–1072
Väisälä orbit, 29
Vajont dam, Italy, 611
Valley-damming landslides, 602
Value of a statistical life (VSL), 241
Varnes classification system, 658
Vegetation and Temperature Condition Index, 1105
Vegetation Condition Index, 1104
Vertical breakwaters, 48
Vertisols, 297
Vesuvius volcano, 458, 1073–1074
Vibrating wire piezometers, 764
Views from the Frontline (VFL), 416
Viscous magmas, 1
Visual network analysis, 798, 799
Volcanic amphitheaters, 145
Volcanic ash, 38, 1074–1076
  ashfalls, 1076
  deposit thickness, 1076
  dispersion, 1076
  fresh ash, 1076
  gases, 1075, 1076
  magmatic eruption, 1074–1075, 1079–1080
  vesicularity, 1076
  vitric particles, 1076
Volcanic Ash Advisory Centre in London, 311
Volcanic eruptions, 290–292, 1064
  ash fall, 65, 66
  lava flows, 67
  predictability and warning potential, 79
  pyroclastic flows, 139
Volcanic explosivity index (VEI), 290, 393, 1073
Volcanic gas, 1076–1077
Volcanic hazard, 381–382
Volcanic hazards, 7–8
Volcanic-tectonic (VT) earthquakes, 210
Volcanoes, 61, 115
Volcanoes and volcanic eruptions, 1077–1085
  active volcanoes, 1079
  andesitic to rhyolitic lava, 1083–1084
  basaltic lavas, 1083
  climate changes, 1084
  destructiveness index, 1081
  effusive and explosive styles, 1080
  eruption magnitude and frequency, 1081
  geological mapping and dating, 1085
  inactive volcanoes monitoring, 1084
  lahars, 1083
  logarithmic scales, 1081
  long repose periods, 1081
  magma, 1079
  monogenetic volcanoes, 1077, 1079
  paroxysmal explosive phase, 1081
  plate tectonics and margins, 1078
  polygenetic volcanoes, 1077, 1079
  pyroclastic density current, 1082–1083
  small summit calderas, 1080
  super-eruption, 1081–1082
  tsunami, 1084
  volcanic ash, 1082
  volcanic explosivity index (VEI), 1080
  volcanic gas, 1084
  volcanic hazards databases, 1084–1085
World Organization of Volcano Observatories, 1084
Vulcanian eruption, 290
Vulnerability, 1088–1090

W

Wadati-Benioff zone, 979
Walker circulation, 251
Wall retrofits, 973
Warning systems, 1091–1095
  community-based planning, 1094
  decision support systems, 1092–1093
  detectable precursors, 1092
  flood warning system, 1092
  informal warnings, 1093
  land use planning and structural mitigation, 1092
  lead-times of hazard, 1092
  natural hazard warning system, 1091
  natural warnings, 1093

official warnings, 1093
people centered warning system, 1091
primary waves, 1092
standard operating procedures, 1092
technological notification systems, 1093–1094
Water erosion, 288
Water exchange, 895
Water reservoir impounding, 535
Waterspout, 1096
Weather-related accidents, 35
Wenchuan, China (2008 earthquake), 1097–1102
  casualty statistics and building collapses, 1097, 1100
  CSIS-based intensity map, 1097, 1099
  demographics and severity, 1097, 1100
  earthquake prediction, 1101
  emergency responses and relief measures, 1101
  epicenter, 1099
  recovery and reconstruction, 1101–1102
  seismic zoning and earthquake preparation, 1100–1101
  shock and aftershocks, 1097, 1098
West Coast and Alaska Tsunami Warning Center (WC/ATWC), 1043
Western technocratic methods, 176
Wet-bulb global temperature (WBGT), 448
Wet snow avalanches, 32, 33

Wildfire, 1102–1106
  fire danger rating systems, 1103
  fire regime, 1103
  fuel layer, 1102
  ignition and spread, 1103
  remote sensing technology, 1103–1105
  simulator, 355, 356
  triangle elements, 1102
Wildland-urban interface (WUI), 323
Willingness-to-pay method (WTP), 241
Willi Willi/Dust Devil, 442
Wind erosion, 288
Wind shear, 36
Windstorm disasters, 61
World Bank Group, 1059
World Climate Research Program, 503
World Commission on Environment and Development, 553
World Conference on Disaster Reduction (WCDR), Kobe 2005, 170, 557–558
World Conference on Natural Disaster Reduction, Yokohama, Japan, 555
World economy, impact of disasters, 1107–1110
  Asia-Pacific region, 1107, 1109
  direct damages, 1107
  Emergency Events Database (EM-DAT), 1107, 1108
  financial capital flows, 1109–1110
  indirect damages, 1107

  secondary consequences, 1107
  trade patterns, 1109
World Food Programme (WFP), 1060
World Health Organization (WHO), 1059
World Meteorological Organisation (WMO), 554, 582, 714, 1059
World Organization of Volcano Observatories, 1084
World Summit on Sustainable Development, South Africa, 2002, 556–557
World Trade Center, 272, 274
World Trade Organization (WTO), 162
World-Wide Standard Seismograph Network (WWSSN), 212, 417
Worldwide trends in natural disasters, 1111–1114
  disaster events, 1112
  disaster mortality and people affected, 1112–1113
  economic losses, 1113
  Emergency Events Database (EM-DAT), 1111
  NatCatSERVICE databases, 1111

Y
Yangtze river delta, China, 979
Yungay province of Peru, 397

Z
Zoning, 1115–1116

Printing and Binding: Stürtz GmbH, Würzburg